S. CHAND'S SUCCESS GUIDES
(Questions and Answers)

ORGANIC CHEMISTRY

For the Students of B.Sc. I, II and III of All Indian Universities

S. CHAND'S SUCCESS GUIDES
(Questions and Answers)

ORGANIC CHEMISTRY

For the Students of B.Sc. I, II and III of All Indian Universities

Also Covering UGC Model Curriculum

[Including Solved Long Answer Type, Short Answer Type Questions and Numerical Problems]

THOROUGHLY UPDATED EDITION

Dr. R.L. MADAN
M.Sc., Ph.D. H.E.S.(I)
Former Principal, Govt. College Panchkula
and Former Head of Chemistry Department
Govt. Post-Graduate College
Faridabad

S Chand And Company Limited
(ISO 9001 Certified Company)

S Chand And Company Limited

(ISO 9001 Certified Company)

Head Office: Block B-1, House No. D-1, Ground Floor, Mohan Co-operative Industrial Estate, New Delhi – 110 044 | Phone: 011-66672000

Registered Office: A-27, 2nd Floor, Mohan Co-operative Industrial Estate, New Delhi – 110 044 Phone: 011-49731800

www.**schandpublishing.com**; e-mail: **info@schandpublishing.com**

Branches

Ahmedabad	:	Ph: 27542369, 27541965; ahmedabad@schandpublishing.com
Bengaluru	:	Ph: 22354008, 22268048; bangalore@schandpublishing.com
Bhopal	:	Ph: 4274723, 4209587; bhopal@schandpublishing.com
Bhubaneshwar	:	Ph: 2951580; bhubaneshwar@schandpublishing.com
Chennai	:	Ph: 23632120; chennai@schandpublishing.com
Guwahati	:	Ph: 2738811, 2735640; guwahati@schandpublishing.com
Hyderabad	:	Ph: 40186018; hyderabad@schandpublishing.com
Jaipur	:	Ph: 2291317, 2291318; jaipur@schandpublishing.com
Jalandhar	:	Ph: 4645630; jalandhar@schandpublishing.com
Kochi	:	Ph: 2576207, 2576208; cochin@schandpublishing.com
Kolkata	:	Ph: 23357458, 23353914; kolkata@schandpublishing.com
Lucknow	:	Ph: 4003633; lucknow@schandpublishing.com
Mumbai	:	Ph: 25000297; mumbai@schandpublishing.com
Nagpur	:	Ph: 2250230; nagpur@schandpublishing.com
Patna	:	Ph: 2260011; patna@schandpublishing.com
Ranchi	:	Ph: 2361178; ranchi@schandpublishing.com
Sahibabad	:	Ph: 2771238; info@schandpublishing.com

First Edition 2001
Subsequent Editions and Reprints 2003, 2005, 2012, 2013, 2014, 2015, 2016 (Twice), 2017, 2018
Reprint 2020

ISBN : 978-81-219-2050-6 **Product Code** : H6OCQ68CHEM10ENZX0XO

PRINTED IN INDIA

By Vikas Publishing House Private Limited, Plot 20/4, Site-IV, Industrial Area Sahibabad, Ghaziabad – 201 010 and Published by S Chand And Company Limited, A-27, 2nd Floor, Mohan Co-operative Industrial Estate, New Delhi – 110 044.

PREFACE TO THE REVISED EDITION

Overwhelming response to previous editions and reprints has encouraged me to prepare this third revised edition of the book. I hope the book in the present form will be found even more useful to the students. A book in the question-answer form has its own advantages. It guides the students about the type of questions that are expected and how to answer them appropriately. It instils a sense of confidence in the students and helps them get maximum score.

Some noteworthy additions to various chapters are :

Ch 4: Assigning formal charge on intermediates and ionic species.

Ch 6: Threo and erythro diastereomers, geometric isomerism in oximes and alicyclic compound. Fischer and flying wedge formula.

Ch 11: Role of σ-and π-complexes, Birch reduction, Methods of formation and chemical reactions of alkylbenzenes, alkenyl benzenes and alkynylbenzenes.

Ch 12: Dichloroethenes, chloroform, iodoform and carbon tetrachloride.

Ch 14: Organozinc compounds

Ch 15: Synthesis of epoxides. Reactions with epoxides

Ch 16: Hauben-Hoesch and Lederer-Mannase reactions

Ch 17: Epoxidation, Ziesel method

Ch 19: Ketenes

Ch 21: Urea

Ch 24: Detergents

Ch 28: Preparation and synthetic uses of diazomethane

Ch 30: Methods of formation and reactions of cycloalkenes.

Ch 34: Bischler-Napieralski synthesis, Mechanism of electrophilic substitution reactions of indole, quinoline and isoquinoline.

Questions from latest papers of different universities have been added. Suggestions for further improvement of the book are welcome.

Dr. R. L. MADAN

Mobile : 09971775666

PREFACE

It gives me pleasure to place, at the hands of students and teachers, Simplified Course in Organic Chemistry for B.Sc. I, II, III (Pass and Honours) Students of Indian Universities. In my long teaching career, I have noticed that students are scared of this branch of chemistry because of tough nomenclature of organic compounds and unique mechanistic steps involved in the reactions. I have tried to present the matter in a lucid style using the simplest possible language.

A large number of well-labelled diagrams and sketches have been provided to enable the students to follow the subject easily. My aim has been to create interest in the study of organic chemistry. One chapter has been devoted to nomenclature of organic compounds, so that the student feels comfortable in dealing with them. Mechanisms of reactions, which form a sizeable part of a University question paper, have been explained in an easy-to-understand manner. Structures of compounds in natural products have been discussed stepwise in a lucid style.

A large number of questions set in University examinations have been solved in order to apprise the students of the type of questions generally asked and the appropriate answers to them.

I hope that the book will satisfy the long-felt need of the students for a friendly book on Organic Chemistry. Any suggestions for the improvement of the book are welcome and will be gratefully acknowledged.

I express my thanks to the editorial staff of S. Chand & Company Ltd., New Delhi for the cooperation extended to me in bringing out the book.

Dr. R. L. MADAN

CONTENTS

1

OCCURRENCE AND CHARACTERISTICS OF ORGANIC COMPOUNDS

Q. 1. What is vital force theory of organic compounds? How was this theory discounted?

Ans. Berzelius, a Swedish Chemist postulated in 1815 that there exists a mysterious force, called *vital force*, in living organism, which is responsible for the formation and properties of organic compounds in plants and animals. At that time, Berzelius ruled that organic compounds can only be obtained from plants and animals, they cannot be synthesised in the laboratory. This is called *vital force theory*.

This belief was, however, shattered by a German Chemist, Wohler, who was a student of Berzelius. He prepared urea, an organic compound by heating ammonium cyanate, an inorganic compound.

$$\underset{\text{Ammonium cyanate}}{NH_4CNO} \xrightarrow{\Delta} \underset{\text{Urea}}{NH_2CONH_2}$$

Vital force theory of Berzelius was thus disproved.

Q. 2. List some important characteristics of organic compounds.

Ans. Some important characteristics of organic compounds are given below:

1. Composition. Organic compounds are made up of a few elements viz., C, H, O, N, S, P and halogens. This differentiates them from inorganic compounds which are made up out of 109 elements approximately.

2. Large number. The number of organic compounds exceeds the number of inorganic compounds despite the fact that organic compounds are constituted of a few elements.

3. Type of linkage. Most of the organic compounds contain covalent bonds in contrast to inorganic compounds which are generally electrovalent.

4. Complex nature. Organic compounds are highly complex and possess higher molecular weights. For example, molecular formula for starch is $(C_6H_{10}O_5)_{200}$.

5. Catenation (self-linkage). Carbon has a unique property of combining with other carbon atoms to form long chains and rings of different sizes. *This property of carbon to combine with other carbon atoms to form long chains or large rings is called* **catenation** and is illustrated below:

–C–C–C–C–

Straight chain compound

–C–C–C–C–
    |
  –C–

Branched chain compound

Closed chain compound

6. Melting points and boiling points. Organic compounds are usually volatile having low melting and low boiling points. This is because various atoms in the molecule are held together by covalent bonds.

7. Solubility. These compounds have generally *low solubility in water and high solubility in organic solvents.* Some of the organic compounds such as lower alcohols, sugar, etc., dissolve freely in water.

8. Conductance. Aqueous solutions of organic compounds have *lower conductances*, which is explained in terms of the bonding.

9. Rates of reactions. Reactions involving organic compounds are generally *slower* than those involving inorganic compounds. This is because of the formation of intermediate compounds.

10. Combustibility. Organic compounds are combustible and they generally leave no residue on burning.

11. Isomerism. Organic compounds exhibit the phenomenon of *isomerism i.e.*, compounds with the same molecular formula possess widely different physical and chemical properties. This is due to the difference in the arrangement of their atoms.

Q. 3. What is meant by homologous series?

Ans. A homologous series *is a group of related compounds in which the formula of each member differs from that of its preceding or succeeding member by one CH_2 group.* The individual members of a homologous series are called homologues. The phenomenon is known as homology. Consider, simple hydrocarbons of *alkane series* viz., CH_4 (methane), C_2H_6 (ethane), C_3H_8 (propane), C_4H_{10} (butane), C_5H_{12} (pentane). Each member differs from the preceding or following member in composition by —CH_2. It is also evident that all members of the same series have the same general formula. Thus, the general formula of the alkane series is C_nH_{2n+2} where *n* is the number of carbon atoms.

Similarly, there are other homologous series such as alcohols, aldehydes, ketones, fatty acids, amines, etc. Study of exceedingly large number of compounds becomes easier by grouping them into homologous series.

Q. 4. List some characteristics of homologous series.

Ans. (*i*) Various members of the series conform to the same general formula.

(*ii*) Different members of the series contain the same functional group and, therefore, show similar chemical reactions.

(*iii*) Physical properties such as melting point, boiling point, density, etc., change gradually as we move from lower to higher members or vice-versa.

(*iv*) The members of a series can be prepared by common methods.

Consider the homologous series of alcohols:

CH_3OH	*Methyl alcohol*
C_2H_5OH	*Ethyl alcohol*
C_3H_7OH	*Propyl alcohol*
C_4H_9OH	*Butyl alcohol*

Any two adjacent members of the series differ by — CH_2.

The members can be represented by the general formula $C_nH_{2n+1}OH_3$ have the same functional group, *i.e.*, hydroxy group (OH), and can be prepared by similar methods of preparation.

Q. 5. Describe the occurrence of organic compounds.

Ans. Principal natural sources of organic compounds are listed below:

1. Plants. These form the richest sources of organic compounds. Substances like starch, sugar, cellulose, oils, etc., are isolated from various plants, their leaves, fruits and bark. Distillation of wood also yields organic compounds like methanol, acetic acid and acetone.

2. Animals. Many organic compounds like milk, fats, proteins, urea, uric acid, etc. are obtained from animals.

3. Petroleum. Petroleum is a mixture of liquid hydrocarbons, which can be separated into various fractions, such as petroleum ether, gasoline, kerosene oil, lubricating oil, etc. by distillation of petroleum at different temperatures. These fractions are used for various purposes.

4. Coal. The destructive distillation of coal gives a very large number of organic compounds such as benzene, toluene, naphthalene, pyridine, and so on. These products are further used in the manufacture of dyes, drugs, perfumes, explosives and plastics.

5. Fermentation. Some organic reactions take place with the help of bacteria and enzymes. Such a reaction is called **fermentation**. Fermentation forms an important source of many organic compounds. Thus, ethyl alcohol is prepared by fermentation of sugar and acetic acid is made by fermentation of ethyl alcohol. *Antibiotics* such as penicillin, streptomycin, aureomycin, tetramycin, etc., are the products of fermentation processes.

6. Natural gas. It is obtained from petroleum wells. It mainly contains methane, CH_4 which is used as household fuel. In addition to this, natural gas *i.e.*, methane is employed for the preparation of other organic compounds like methyl alcohol CH_3OH, methyl chloride CH_3Cl, chloroform $CHCl_3$, etc.

Q. 6. Name different techniques for isolation and purification of organic compounds.

Ans. The techniques used to isolate and purify the organic compounds are:

1. Crystallization. This method is used for the purification of solid organic compounds. For example, sugar is obtained from its saturated solution by crystallization.

2. Fractional Crystallization. This method is used for the separation of a mixture of solids having different solubilities.

3. Sublimation. This method is used for the purification of volatile solids from non-volatile solids. Organic compounds which can be purified by this method are naphthalene, benzoic acid, camphor, etc.

4. Simple Distillation. This method is used for the purification of those liquids which contain non-volatile impurities. For example, organic liquids like benzene, acetone, ether can be purified by this method.

5. Fractional Distillation. This method is used for the separation of a mixture of organic liquids which possess different boiling points.

6. Steam Distillation. This method is used for the purification of organic solids or liquids which are insoluble in water, volatile in steam and contain non-volatile impurities. For example, purification of aniline and nitro benzene can be done by this method.

7. Distillation under the reduced pressure. Those liquids which decompose at or below their boiling points under atmospheric pressure are purified by this method. For example, recovery of glycerol from spent lye is done by this process.

8. Extraction with solvent. Extraction of oils and fats from seeds and essences from flowers is done by using suitable organic solvents.

9. Chromatography. This is a modern technique used to separate and to purify mixtures of organic compounds.

Q. 7. What is the criteria of purity of an organic compound? (*Agra 2004*)

Ans. Criteria of purity of an organic compound

A pure substance possesses fixed physical constants and shows a variation with the impurities. *A pure solid organic substance possesses a sharp melting point, i.e., it liquefies at one fixed temperature. Similarly a pure organic liquid will boil at a fixed temperature, i.e., it will possess a sharp boiling point.*

Generally it is a common practice to check up the purity of organic solids and liquids by determining their melting and boiling points respectively.

Q. 8. How can you use melting point property to check the purity of a substance?

Ans. Purity of a substance can be checked by using the melting point property. Every pure organic substance has a fixed melting point. These melting points of pure substances are available in literature. Two techniques may be used to check the purity of a substance:

1. Repeated crystallisation. Melting point of the substance to be checked is determined with the help of apparatus shown in Fig. 1.1. Then the substance is recrystallised using a suitable solvent. Melting point of the dry, recrystallised sample is determined again. This is repeated three times. If the melting point remains the same and does not change on recrystallisation, the substance is a pure substance.

2. Mixed melting point. Let us say we want to know whether the given sample of acetanilide is pure or not. To establish this, we shall take a pure sample of a acetanilide and mix to it, the given sample of acetanilide thoroughly. Determine the melting point of this mixture. If the melting point is the same as that of acetanilide (from the literature value), the given sample of acetanilide is pure. If it is different, the sample is impure. Impurities always change the melting point of a substance.

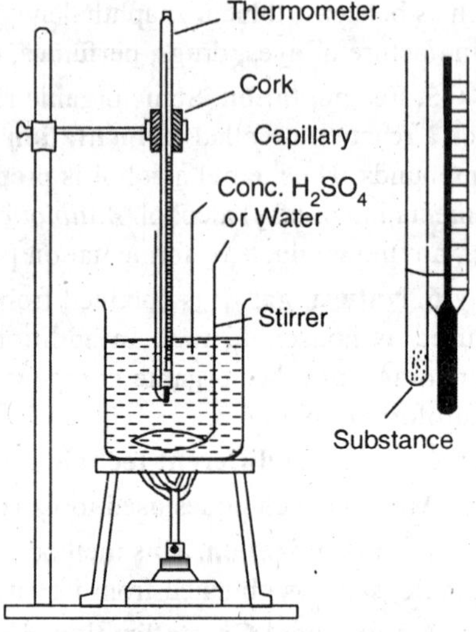

Fig. 1.1. Determination of melting point.

Q. 9. What is the criteria of purity of an organic liquid?

Ans. The purity of an organic liquid can be checked by determining its boiling point. Every liquid has a fixed characteristic boiling point. A pure liquid boils sharply at this temperature. If a liquid does not boil sharply but boils over a range, it is an impure liquid.

Boiling point is the temperature at which the vapour pressure of a liquid is equal to the atmospheric pressure.

Boiling point of a liquid is determined with the help of the apparatus shown in Fig. 1.2.

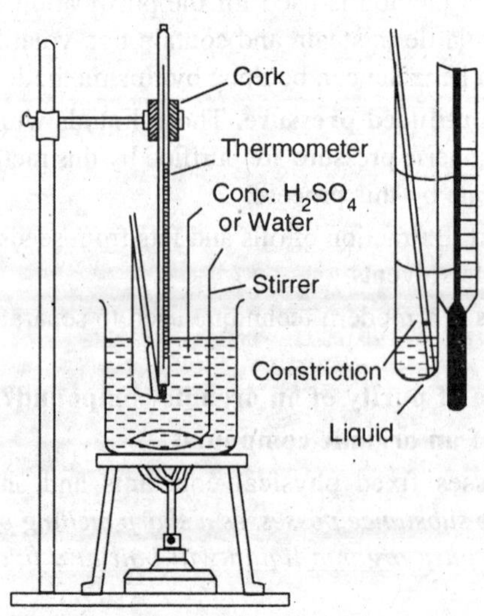

Fig. 1.2. Determination of boiling point.

Q. 10. How do you establish the tetrahedral nature of carbon?

Ans. Carbon has the electronic configuration $1s^2\ 2s^1\ 2p_x^{\ 1}\ 2p_y^{\ 1}$ With this configuration, carbon appears to be divalent. But we come across compounds like methane and carbon tetrachloride which indicate that carbon is tetravalent. This can be explained by assuming that one electron from 2*s* orbital is excited to $2p_z$ orbital giving the electronic configuration as

$$1s^2\ 2s^1\ 2p_x^{\ 1}\ 2p_y^{\ 1}\ 2p_z^{\ 1}$$

This is called configuration in the *excited state*. With the help of this configuration, we can say that carbon is tetravalent as there are four orbitals with single electrons.

Although a certain amount of energy is required to excite an electron from 2*s* to 2*p* orbital, but this is compensated by the formation of two more bonds (from 2 to 4). It is well known that energy is liberated when bond formation takes place.

The next question is how the four valencies are oriented with respect to one another. There are three possibilities of orientation as shown below:

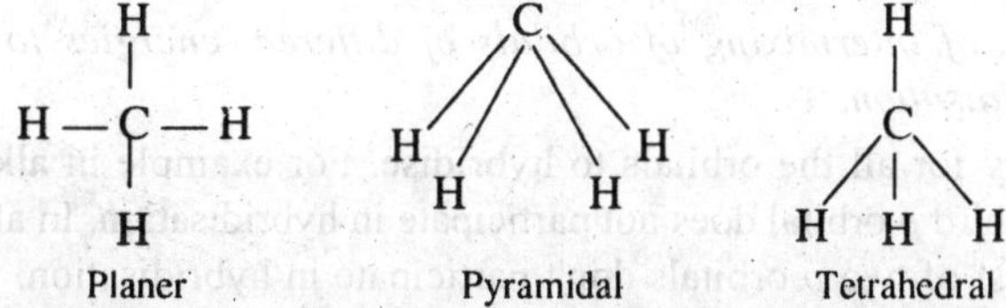

Taking the case of methane one possibility is that the four valencies are in the same plane, angle between neighbouring bonds being 90°.

Second possibility is that the four valencies make a pyramid, carbon being situated at the top and hydrogen at the corners of a square. Third possibility is that carbon is in the centre of a tetrahedron and the hydrogens are occupying the corners of the tetrahedron.

Out of these possibilities, the last one is most probable because it is most symmetrical. In this orientation, any two bonds are inclined at an angle of 109° 28′, which is not so if we consider the planer or pyramidal arrangement. This is further supported by the following observations. We find that we obtain only one kind of disubstitution product CH_2X_2 when we substitute two hydrogens of methane by two –X groups. This can be explained only on the tetrahedral orientation, because planer and pyramidal orientations would provide more than one disubstitution products as shown below.

Thus carbon compounds have tetrahedral bonding in them and are thus three-dimensional in structure.

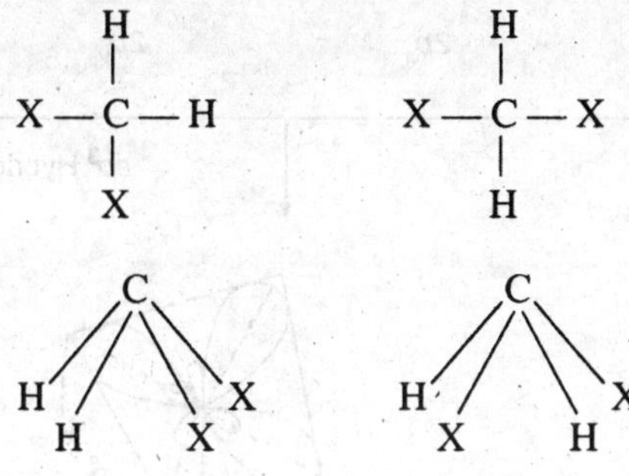

HYBRIDISATION

Q. 11. What is meant by hybridisation of atomic orbitals? (*Delhi 2003, Nagpur, 2005*)

Ans. Tetravalency of carbon can be explained by assuming that one electron from orbital is excited to $2p$ orbital, thus making available four orbitals having single electrons, so that four bonds could be formed.

Consider the case of methane. Here the four C — H bonds have been formed out of one *s* and three *p* orbitals($2s^1$, $2p_x^{\ 1}$, $2p_y^{\ 1}$ and $2p_z^{\ 1}$). Therefore the four bonds that are formed should be different from one another as regards bond length and bond energy. But we observe that all the C — H bonds in methane, or for that purpose in any organic compound, are equivalent in all respects.

How to explain that? This is explained in terms of **hybridisation.** We assume that the four orbitals ($2s$, $2p_x$, $2p_y$ and $2p_z$) mix with one another to produce four equivalent types of orbitals called sp^3 hybrid orbitals. These equivalent orbitals then overlap with the orbitals of hydrogen to form four equivalent C — H bonds.

The phenomenon of intermixing of orbitals of different energies to produce equivalent orbitals is called hybridisation.

It is not necessary for all the orbitals to hybridise. For example in alkenes, one *s* and two orbitals hybridise, the third *p* orbital does not participate in hybridisation. In alkynes, one *s* and one *p* orbitals hybridise, rest of two *p* orbitals don't participate in hybridisation.

Q. 12. Explain clearly sp^3 or tetrahedral hybridisation.

(*Cochin 2003, Kalyani 2003, Udaipur, 2004, Nagpur, 2005*)

Ans. Tetrahedral or sp^3 hybridisation. In this type of hybridisation all the three $2p$ orbitals mix with the $2s$ orbital to produce four identical orbitals. The four newly formed orbitals are directed towards the four corners of a regular tetrahedron and make an angle of 109° 28′ with one another. The new orbitals are termed as *tetrahedral or sp^3 orbitals.* **This type of hybridisation takes place in the formation of compounds containing only single bonds.**

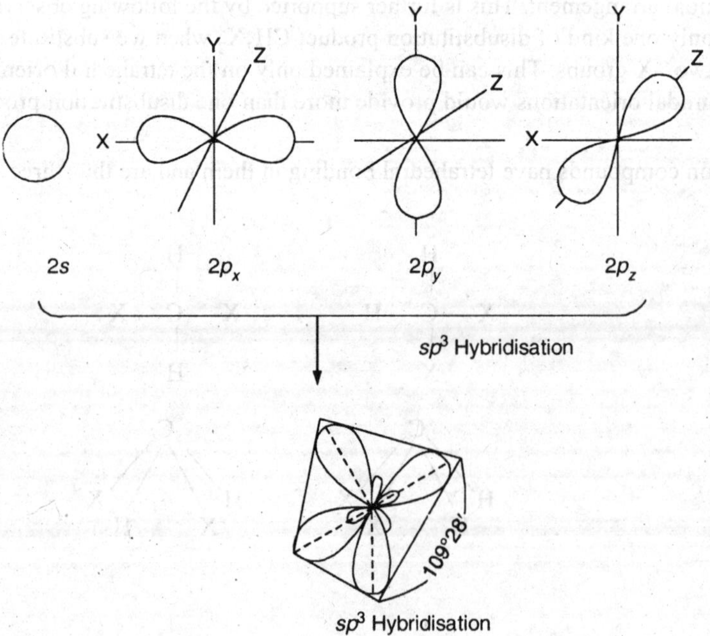

sp^3 Hybridisation

In the formation of a molecule of methane, the four sp^3 hybridised orbitals of a carbon atom overlap the four $1s$ orbitals of the four hydrogen atoms located at the corners of the tetrahedron giving rise to four σ bonds. See figure below:

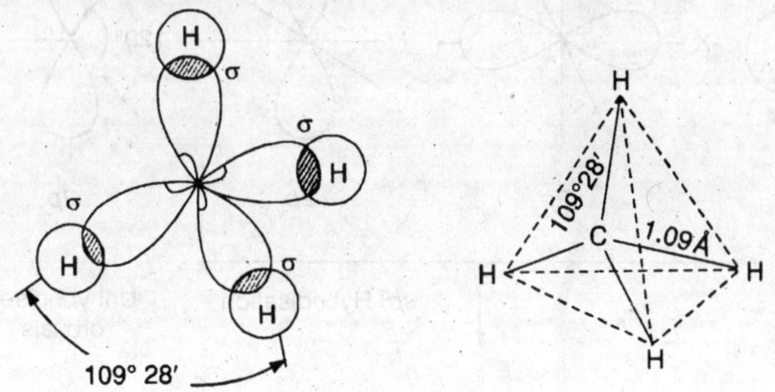

A molecule of methane

Each carbon hydrogen bond has a length of 1.09 Å. The bond angle between any two carbon-hydrogen bonds is 109°28'.

When a single bond is set up between any two carbon atoms as in the case of ethane, one of the sp^3 orbitals of one carbon atom overlaps one of the sp^3 orbitals of the other to establish a σ bond. The bond length for carbon-carbon bond is 1.54 Å.

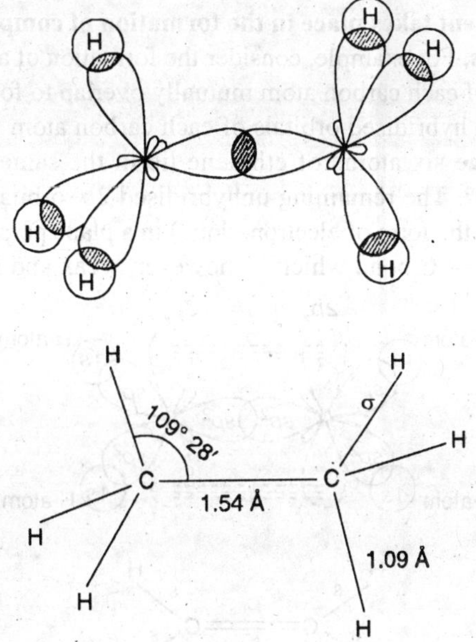

A molecule of ethane

Q. 13. Describe trigonal or hybridisation.

Ans. Trigonal or sp^2 hybridisation. In this type of hybridisation, the $2s$ and two of the three $2p$ orbitals (say $2p_x$ and $2p_y$) of the carbon atom are hybridised leading to the formation of three equivalent sp^2 orbitals which lie in the same plane and make angle of 120° with one another. The third $2p_z$ orbital remains in its unhybridised state and is perpendicular to the plane containing three sp^2 hybrid orbitals as shown on next page.

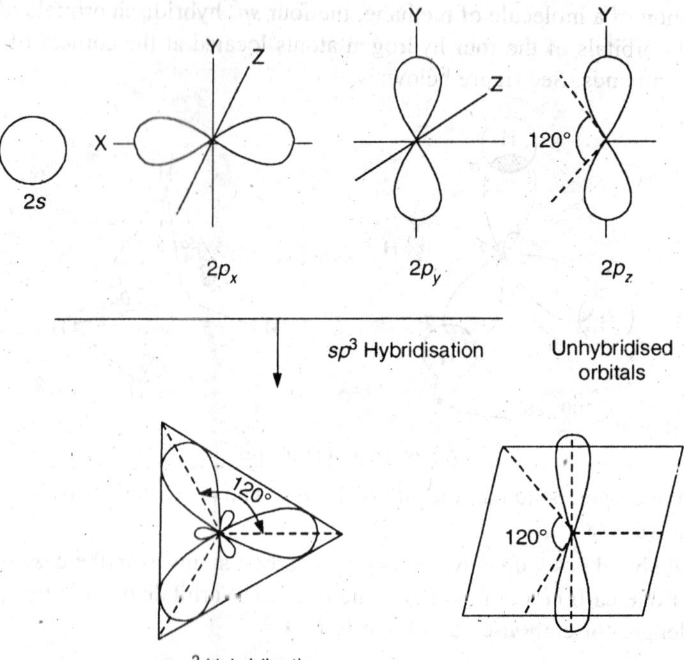

sp^2 Hybridisation

The **trigonal arrangement takes place in the formation of compounds containing a double bond between carbon atoms.** For example, consider the formation of a molecule of ethylene. Here one of the three sp^2 orbitals of each carbon atom mutually overlap to form σ-bond between the two carbon atoms. The other two hybridised orbitals of each carbon atom form σ-bonds with the four hydrogen atoms. Thus all the six atoms of ethylene lie in the same plane and the bond angle between two σ-bonds is 120°. The remaining unhybridised $2p_z$ orbitals of the two carbon atoms **overlap to a small extent** (in the form of electron-cloud) in a plane perpendicular to the plane of the six atoms to form another C — C bond which is, however, weak and is termed π-bond.

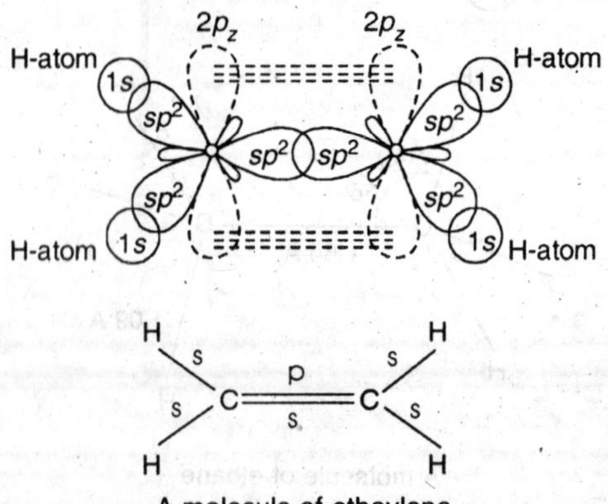

A molecule of etheylene

It is clear that the two bonds comprising the double bond between carbon atoms are not equally strong. The σ-bond between the two carbon atoms has a bond strength of 250 kJ/mole while the π-bond has a strength of 160 kJ/mole and is responsible for the reactivity of the double bond. The C — C distance in the double bond is 1.33 Å as the two carbon atoms are held more tightly together than in a single bond.

Q. 14. Describe diagonal or *sp* hybridisation. (*Nagpur 2003*)

Ans. Diagonal or *sp* hybridisation. In this type of hybridisation only 2*s* and $2p_x$ orbitals hybridise forming two equivalent *sp* orbitals which are collinear. The $2p_y$ and $2p_z$ orbitals remain in their natural state. This gives rise to two equivalent bonds pointing in opposite directions along a straight line. Of the other two bonding orbitals one is oriented along the *y*-axis and the other along *z*-axis. The diagonal hybridisation occurs in compounds containing a triple bond between two carbon atoms.

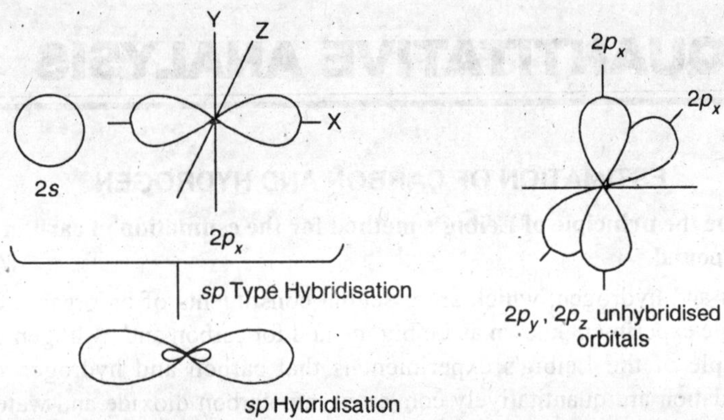

sp Hybridisation

In the formation of a molecule of acetylene, one of the *sp* orbitals of each carbon atom overlaps to form a σ-bond between the two carbon atoms while the other hybridised orbitals of each carbon atom form σ-bonds with two hydrogen atoms. Since all the four atoms lie along the same straight line, we get a linear molecule. The $2p_y$ and $2p_z$ orbitals of each of the two carbon atoms which are oriented along the *y*-axis and *z*-axis, at right angles to each other, overlap slightly along their sides resulting in the formation of two π-bonds in planes at right angles to each other. This produces a cylinder of negative charge around C — C -bond.

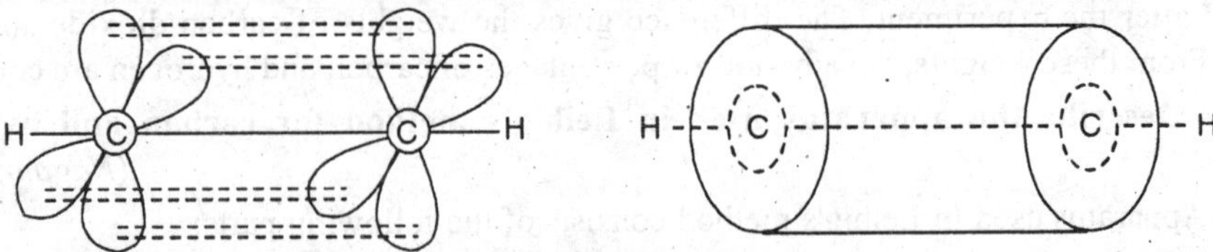

Molecule of acetylene

As is obvious from the above discussion, the carbon-carbon triple bond is made up of one strong σ-bond and two weak π-bonds. Its total strength is 510 kJ/mole. The carbon-carbon distance in the triple bond is 1.20 Å.

Q. 15. What is the hybridisation of different carbon atoms in isoprene ? (*Delhi 2002*)

$$\overset{1}{C}H_2 = \overset{2}{C}(\overset{5}{C}H_3) - \overset{3}{C}H = \overset{4}{C}H_2$$

Ans. (structure above)

Carbon atoms 1, 2, 3 and 4 have sp^2 hybridisation while carbon-5 has sp^3 hybridisation.

2

QUANTITATIVE ANALYSIS

ESTIMATION OF CARBON AND HYDROGEN

Q. 1. Describe the principle of Leibig's method for the estimation of carbon and hydrogen in an organic compound. (*Nagpur, 2005*)

Ans. Carbon and hydrogen, which are essential constituents of an organic compound, are estimated in a single experiment known as Leibig method for carbon and hydrogen estimation. The underlying principle of the Leibig's experiment is that carbon and hydrogen of the organic compound on oxidation are quantitatively converted into carbon dioxide and water respectively. For complete oxidation of carbon and hydrogen, a known quantity of organic compound is heated in a furnace and oxygen is passed through it. The following reactions take place

$$C + O_2 \xrightarrow{CuO} CO_2$$

$$2H + O \xrightarrow{CuO} H_2O.$$

The weights of carbon dioxide and water are obtained by passing them through calcium chloride and potassium hydroxide respectively. Calcium chloride absorbs water and potassium hydroxide absorbs carbon dioxide. Weights of calcium chloride tube and potash bulbs are recorded before and after the experiment. The difference gives the weights of carbon dioxide and water produced. From these weights, the amounts or percentages of carbon and hydrogen are computed.

Q. 2. Describe the apparatus used in Leibig's method for carbon and hydrogen estimation. (*Nagpur, 2002*)

Ans. Apparatus used in Leibig's method consists of the following parts:

1. Combustion Tube: It is a hard glass tube approx. 1 m long and 1-2 cm in diameter (see Fig. 2.1). From left to right, it is filled with materials as described below.

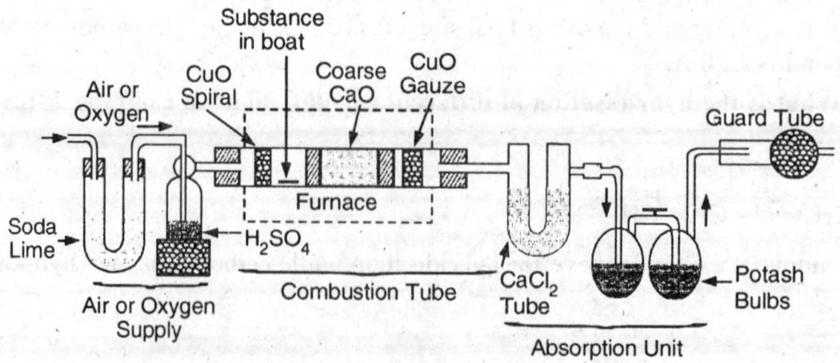

Fig. 2.1. Apparatus for the estimation of C and H.

(*i*) A coil of oxidised copper spiral to oxidise any vapour of the substance diffusing backwards.

(*ii*) A porcelain boat containing a known weight of organic substance.

(*iii*) Coarse copper oxide packed in major portion of the tube on the right and kept in position by loose asbestos pads.

(*iv*) A roll of oxidised copper gauze to oxidise any vapour left unoxidised. The combustion tube is enclosed in a furnace which is heated electrically or by gas burners.

On the left and right ends of the combustion tube *oxygen supply apparatus and absorption apparatus* are attached respectively.

2. Oxygen supply apparatus: Oxygen or air to be sent in is first passed through a *concentrated sulphuric acid bubbler*, which dries it and then passed through a *U-tube containing soda-lime*, which detains carbon dioxide. Thus oxygen or air completely free from moisture and carbon dioxide is passed through the combustion tube.

3. Absorption apparatus: It consists of the following parts:

(*a*) *Anhydrous calcium chloride tube* for the absorption of water.

(*b*) *Potash bulb* containing KOH solution for the absorption of carbon dioxide.

(*c*) *Guard tube* containing soda-lime and anhydrous calcium chloride to prevent carbon dioxide and moisture from entering the absorption tubes from outside.

Working. A known weight of the organic substance is placed in the porcelain boat, absorption apparatus is weighed and attached, oxygen is sent in and the combustion tube is again heated by turning on the furnace. Organic substance vaporises and is oxidised by CuO to water and carbon dioxide, which are absorbed by anhydrous calcium chloride tube and potash bulbs respectively. The whole operation takes about three hours. The absorption apparatus is detached and weighed again.

Increase in weight of $CaCl_2$ tube = Weight of H_2O produced.

Increase in weight of potash bulb = Weight of CO_2 produced.

Calculations:

Percentage of Carbon:

Let the weight of organic substance $= w$ g

Let the weight of CO_2 produced $= x$ g

1 mole of CO_2 contains 1 g atom of C

or 44 g of CO_2 contain = 12 g carbon

$$x \text{ g of } CO_2 \text{ contain} = \frac{12}{44} \times x \text{ g carbon}$$

$$\text{Weight of C in } w \text{ g of organic substance} = \frac{12}{44} \times x \text{ g}$$

$$\text{Weight of C in 100 g of organic substance} = \frac{12}{44} \times x \times \frac{100}{w} \text{ g}$$

or

$$\boxed{\text{Percentage of Carbon} = \frac{12}{44} \times \frac{\text{weight of } CO_2}{\text{weight of organic substance}} \times 100}$$

Percentage of Hydrogen:

Let the weight of organic substance $= w$ g

Let the weight of H_2O produced $= y$ g

1 mole of H_2O contains 2 g atoms of H

or 18 g of H_2O contain $=$ 2 g hydrogen

$$y \text{ g of } H_2O \text{ contain} = \frac{2}{18} \times y \text{ g}$$

$$\text{Weight of H in 100 g of organic substance} = \frac{2}{18} \times y \times \frac{100}{w} \text{ g}$$

or

$$\text{Percentage of Hydrogen} = \frac{2}{18} \times \frac{\text{weight of } H_2O}{\text{weight of organic substance}} \times 100$$

Q. 3. What modifications are made in the Liebig's apparatus when the compound contains N, S and halogens also?

Ans. Modifications :

(*i*) **For compounds containing nitrogen, in addition to C and H.** On combustion, oxides of nitrogen are also formed which are absorbed by caustic potash solution

$$2KOH + N_2O_3 \longrightarrow 2KNO_2 + H_2O$$

The difference in the weights of potash bulbs before and after the experiment, will give the sum of weights of CO_2 and N_2O_3.

This difficulty is overcome by placing at right end, a reduced copper spiral which will reduce oxides of nitrogen into free nitrogen

$$N_2O_3 + 6Cu \longrightarrow N_2 + 3Cu_2O$$

$$2NO + 2Cu \longrightarrow N_2 + 2CuO$$

Free nitrogen is not absorbed by caustic potash.

(*ii*) **For compounds containing sulphur, in addition to C and H.** Sulphur is oxidized to sulphur dioxide. It is also absorbed by caustic potash solution.

$$SO_2 + 2KOH \longrightarrow K_2SO_3 + H_2O$$

To overcome this difficulty, a mixture of cupric oxide and lead chromate is employed. Lead chromate will absorb SO_2 to form lead sulphate and thus prevents its absorption by KOH.

$$2PbCrO_4 + 2SO_2 \longrightarrow 2PbSO_4 + Cr_2O_3 + \frac{1}{2}O_2$$

(*iii*) **For compounds containing halogen, in addition to C and H.** Halogens present in the compound are converted into free halogens which are absorbed by hot caustic potash solution.

$$3Cl_2 + 6KOH \longrightarrow 5KCl + KClO_3 + 3H_2O$$

This difficulty is overcome by placing a silver gauze at the extreme right end of combustion tube. Silver absorbs halogens to form silver halide.

$$2Ag + X_2 \longrightarrow 2AgX$$

NUMERICAL PROBLEMS ON ESTIMATION OF C AND H

Q. 4. 0.26 g of an organic compound gave on combustion 0.039 g of water and 0.245 g of carbon dioxide. Calculate the percentage of carbon and hydrogen in it.

Ans. Weight of compound = 0.26 g

Weight of water = 0.039 g

Weight of carbon dioxide = 0.245 g

18 g of water contain hydrogen = 2 g

$\therefore$ 0.039 g of water contain hydrogen = $\frac{2}{18} \times 0.039$ g

Hence percentage of hydrogen = $\frac{2}{8} \times 0.039 \times \frac{100}{0.26}$ = **1.66**

44 g of CO_2 contain C = 12 g

$\therefore$ 0.245 g of CO_2 contain C = $\frac{12}{44} \times 0.245$ g

Hence percentage of carbon = $\frac{12}{44} \times 0.245 \times \frac{100}{0.26}$ = **25.69**

Q. 5. 0.25 g of an organic compound containing C, H and oxygen was analysed by the combustion method. The increase in weights of calcium chloride tube and the potash bulbs at the end of the operation was found to be 0.15 g and 0.1873 g respectively. Calculate the percentage composition of the compound.

Ans. Weight of CO_2 (increase in wt. of potash bulbs) = 0.1873 g

Weight of H_2O (increase in wt. of calcium chloride tubes) = 0.15 g

Mol. wt. of CO_2 = 12 + 32 = 44

Mol. wt. of H_2O = 2 + 16 = 18

44 parts of CO_2 contain carbon = 12 parts

or 44 gram of CO_2 contain carbon = 12 g

0.1873 g of CO_2 will contain carbon = $\frac{0.1873}{44} \times 12$ g

Hence percentage of carbon = $\frac{0.1873 \times 12 \times 100}{44 \times 0.25}$ = **20.43%**

Again, 18 parts of H_2O contain hydrogen = 2 parts

or 18 g of H_2O contain hydrogen = 2 g

0.15 g of H_2O contain hydrogen = $\frac{2 \times 0.15}{18}$ g

Hence percentage of hydrogen = $\frac{2 \times 0.15 \times 100}{18 \times 0.25}$ = 6.66%

Percentage of oxygen (by difference) = 100 – (20.43 + 6.66)

= **72.91**

PROBLEMS FOR PRACTICE

1. *0.2346 g of an organic compound containing carbon, hydrogen and oxygen was put to combustion by Liebig's method and the increase in weight of the U-tube and potash*

bulbs at the end of the operation was found to be 0.2754 g and 0.4488 g respectively. Calculate the percentage composition of the compound.

[**Ans.** *C = 52.17%, Hydrogen = 13.04%, Oxygen = 34.79%*]

2. *0.465 g of an organic substance gave on combustion 1.32 g of* CO_2 *and 0.315 g of* H_2O. *Calculate the percentage of carbon and hydrogen in the compound.*

[**Ans.** *C = 77.42%, H = 7.53%*]

3. *A compound which is believed to have the molecular formula* C_4H_8O *is submitted for elemental analysis. What percentage of carbon and hydrogen would you expect to be reported if* C_4H_8O *is the correct formula ?* (*Panjab, 2005*)

[**Ans.** *C = 66.7%, H = 11% and O = 22.2%*]

ESTIMATION OF NITROGEN

Q. 6. Describe Duma's method for the estimation of nitrogen in the organic compound.

(*Rajasthan, 2004, Agra, 2004, Kalyani, 2004*)

Ans. Principle. A known weight of the organic compound is heated with cupric oxide in an atmosphere of carbon dioxide (free from nitrogen) when carbon, hydrogen and sulphur are respectively oxidised to carbon dioxide, water and sulphur dioxide. A small amount of nitrogen may be oxidised to oxides of nitrogen but these are reduced back to nitrogen with the help of a reduced copper gauze. All other gases except nitrogen are absorbed in an aqueous solution of potassium hydroxide while nitrogen being insoluble in potassium hydroxide gets collected over it. From the volume of nitrogen thus obtained, the percentage of nitrogen can be easily determined.

Apparatus. It consists of the following parts:

1. Combustion tube: It consists of a hard glass tube about 1 metre long and 1-2 cm in diameter. It is filled with the following materials from left to right:

(*a*) Oxidised copper spiral, which oxidises any vapour of the substance diffusing backwards.

(*b*) A mixture of a known weight of the organic substance under analysis and coarse cupric oxide.

(*c*) A packing of coarse cupric oxide held in the major portion of the tube by loose asbestos pads.

(*d*) A reduced copper spiral to reduce any oxides of nitrogen passing over.

The combustion tube is enclosed in a furnace which is heated electrically or by gas burners.

On the left and right sides, it is connected to the *carbon dioxide supply apparatus and nitrometer* respectively.

2. Carbon dioxide supply apparatus: It consists of *carbon dioxide generator i.e.*, a big flask in which sodium bicarbonate or magnesite is heated producing CO_2.

$$2NaHCO_3 \longrightarrow Na_2CO_3 + H_2O + CO_2$$

$$MgCO_3 \longrightarrow MgO + CO_2$$

CO_2 produced is passed through a concentrated sulphuric acid bubbler which dries the gas. The gas is then passed to the combustion tube.

3. Schiff's nitrometer: It is a long wide graduated tube provided with a funnel and a tap at the top and two side-tubes at the bottom. The upper side tube is connected to a reservoir containing 30-40% KOH solution and the lower side tube is connected to the combustion tube. A little mercury is placed at the bottom of the Schiff's nitrometer which acts as a valve *i.e.*, it allows the gases from the combustion tube to enter into the nitrometer but does not allow the KOH solution to flow back into the combustion tube.

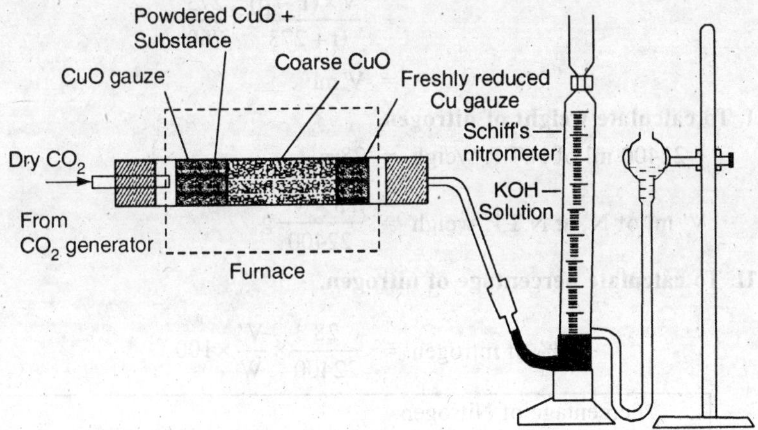

Fig. 2.2. Duma's method for nitrogen estimation

Procedure. The various steps involved are:

1. A known weight of organic substance is mixed thoroughly with excess of fine copper oxide and the mixture is placed in the position, as shown in Fig. 2.2.

2. The nitrometer is filled with 30-40% KOH solution and the tap of the nitrometer is opened. A slow current of dry CO_2 is passed into the combustion tube so as to replace the air inside by the CO_2 gas. When no more air bubbles are produced in the nitrometer this indicates that all the air in the tube has been replaced by CO_2 gas.

3. The reservoir is now raised so as to fill the nitrometer completely with the KOH solution and the tap is then closed.

4. The combustion tube is then heated gently but continuously. A slow current of CO_2 is maintained throughout to prevent the flow in the backward direction. As a result C, H, S etc., get oxidised to CO_2, H_2O, SO_2 etc., which are absorbed by KOH while nitrogen gets collected over KOH solution.

5. Towards the end, a strong current of CO_2 gas is passed so as to sweep away any traces of nitrogen left in the combustion tube.

6. The nitrometer is now disconnected and allowed to cool to room temperature. The volume of the nitrogen produced is measured by making the levels of the KOH solution in the nitrometer and reservoir equal by lowering or raising the latter.

7. The room temperature and the atmospheric pressure (from the barometer) are noted. The aqueous tension at the room temperature is also noted from the tables.

Calculations:

Let the weight of the organic compound be = W g

Let the volume of N_2 evolved be V ml at t°C and P mm pressure.

Let the Aq. tension at t°C be p mm.

Step I. To convert the volume of N_2 to N.T.P. conditions.

By applying the formula $$\frac{P_1V_1}{T_1} = \frac{P_2V_2}{T_2}$$

or $$V_2 = \frac{P_1V_1}{T_1} \times \frac{T_2}{P_2}$$

∴ The volume of dry nitrogen at N.T.P. is calculated.

The volume of dry nitrogen at N.T.P.

$$= \frac{V \times (P-p)}{t+273} \times \frac{273}{760}$$

$$= V' \text{ ml}$$

Step II. To calculate weight of nitrogen.

$\because$ 22400 ml of N.T.P. weigh = 28 g

$$V' \text{ ml of } N_2 \text{ at N.T.P. weigh} = \frac{28 \times V'}{22400} \text{ g}$$

Step III. To calculate percentage of nitrogen.

$$\% \text{ of nitrogen} = \frac{28}{22400} \times \frac{V'}{W} \times 100$$

or

Percentage of Nitrogen

$$= \text{Volume of Nitrogen at N.T.P.} \times \frac{28}{22400} \times \frac{100}{\text{wt. of organic compound}}$$

NUMERICAL PROBLEMS ON DUMA'S METHOD

Q. 7. 0.2313 g of an organic substance gave 27.4 ml of nitrogen at 0°C and 760 mm pressure. Calculate percentage of nitrogen in the organic compound.

Ans. Weight of the organic substance = 0.2313 g

The volume of given here is under N.T.P. conditions.

Percentage of nitrogen

22400 ml of nitrogen at N.T.P. weigh = 28 g

$$27.4 \text{ ml of nitrogen of N.T.P. weigh} = \frac{28}{22400} \times 27.4 \text{ g}$$

$$\% \text{ of N} = \text{weight of N} \times \frac{100}{\text{weight of compound}}$$

Hence percentage of nitrogen in the compound

$$= \frac{28}{22400} \times 27.4 \times \frac{100}{0.2313} = \mathbf{14.80}$$

Q. 8. 0.25 g of organic substance gave 38 ml of moist nitrogen measured at 27°C and 746.5 mm pressure. Calculate the percentage of nitrogen in the substance.

(Aq. tension at 27°C = 26.5 mm)

Ans. Weight of the organic compound = 0.25 g

(*i*) Volume of N_2 at N.T.P.

(Given conditions)	(At N.T.P.)
$V_1 = 38.0$ ml	$V_2 = ?$ ml
$P_1 = 746.5 - 26.5 = 720$ mm	$P_2 = 760$ mm
$T_1 = 27 + 273 = 300$ K	$T_2 = 273$ K

Applying the general gas equation:

$$\frac{P_1V_1}{T_1} = \frac{P_2V_2}{T_2}$$

Substituting the values, we have

$$\frac{38 \times 720}{300} = \frac{V_2 \times 760}{273}$$

or $$V_2 = \frac{38 \times 720}{300} \times \frac{273}{760}$$

$$= \mathbf{32.76\ ml}$$

(*ii*) Percentage of nitrogen

22400 ml of nitrogen at N.T.P. weigh = 28 g

32.76 ml of nitrogen at N.T.P. weigh $= \frac{28}{22400} \times 32.76$ g

$$\%\text{ of nitrogen} = \text{weight of N} \times \frac{100}{\text{wt. of compound}}$$

$$= \frac{28}{22400} \times 32.76 \times \frac{100}{0.25}$$

$$= \mathbf{16.38}$$

PROBLEMS FOR PRACTICE

1. *0.1124 g of an organic compound gave 19 ml of N_2 at 16°C and 753.5 mm pressure. Aqueous tension at 16°C = 13.5 mm. Find out the % of nitrogen.* **[Ans.** *19.43%***]**
2. *0.300 g of an organic substance gave 37.9 ml of dry N_2 at 290 K and 770 mm pressure. Calculate the percentage of nitrogen in the compound.* **[Ans.** *15.06%***]**
3. *0.3 g of an organic compound gave 50.0 cm³ of nitrogen collected at 300 K and 715 mm pressure in Duma's method. Calculate the percentage of nitrogen in the compound. (Vapour pressure of water at 300 K is 15 mm)* **[Ans.** *17.46%***]**
4. *0.2046 g of an organic compound gave 30.4 cm³ of moist nitrogen measured at 288 K and 732.7 mm pressure. Calculate the percentage of nitrogen in the substance. Aqueous tension at 288 K is 12.7 mm.* **[Ans.** *16.68%***]**

Q. 9. Describe Kjeldahl's method for the estimation of nitrogen in an organic compound.

(Kerala, 2001; Kumaon, 2000; Delhi, 2003; Madurai, 2003; Osmania, 2004; Nagpur 2008)

Ans. Principle. This method is based upon the fact that when an organic compound containing nitrogen is heated with concentrated sulphuric acid, the nitrogen present in it is quantitatively converted into ammonium sulphate. Ammonium sulphate when distilled with excess of sodium hydroxide solution yields free ammonia. The ammonia evolved is passed through known excess of standard sulphuric acid and the quantity of unreacted acid is determined by back titration with some standard alkali.

$$\text{N} \xrightarrow{\text{Conc. } H_2SO_4} (NH_4)_2SO_4$$

$$(NH_4)_2SO_4 + 2NaOH \longrightarrow Na_2SO_4 + 2H_2O + 2NH_3\uparrow$$

Procedure. The following steps are involved in the estimation of N.

(*a*) **Formation of ammonium sulphate:** A mixture of .3–.5 g of the accurately weighed substance with about 10 g of potassium sulphate, 25 ml of concentrated sulphuric acid and a drop of mercury (or $CuSO_4$) is placed in a long-necked Kjeldahl's flask and heated (Fig. 2.3). Potassium sulphate is added to raise the boiling point of sulphuric acid whereas mercury or ($CuSO_4$) acts as a catalyst. The heating is continued till the brown colour first produced disappears and the contents become clear again. At this stage, the conversion of nitrogen into ammonium sulphate is complete.

(*b*) Liberation of ammonia and its absorption in acid: The Kjeldahl's flask is then cooled. Its contents are diluted and transferred along with its washings into one litre round bottom flask. To this is added 40% caustic soda solution and the flask is fitted with Kjeldahl's trap and a water condenser as shown in Fig. 2.3. The lower end of the condenser is made to dip in known excess of standard sulphuric acid taken in a titration flask. The function of the Kjeldahl's trap is to prevent the mixture in the flask to pass into the absorption acid as a result of bumping. The flask is heated and the ammonia evolved is passed through standard acid solution.

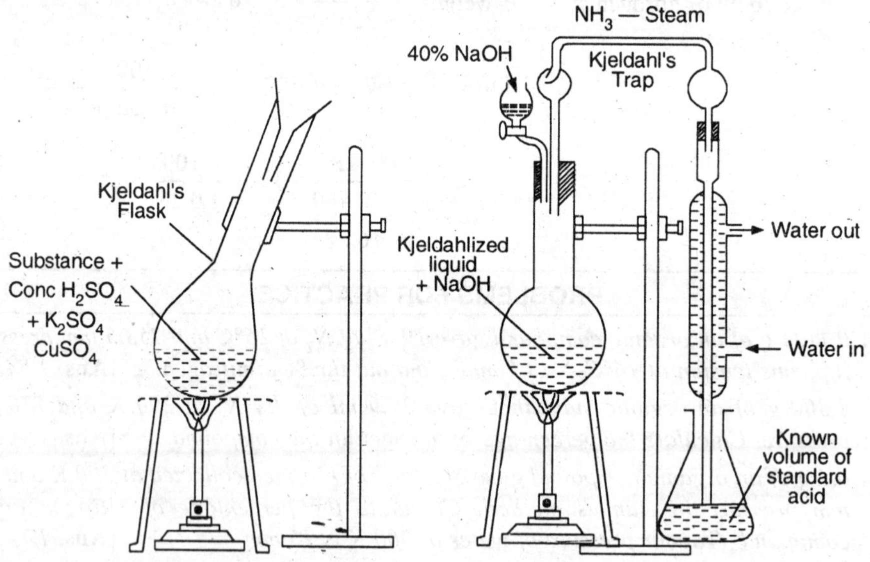

Fig. 2.3. Estimation of nitrogen by Kjeldahl's method.

(*c*) Titration of the excess acid: The amount of the acid left unused in the conical flask can be estimated either by titrating it directly against standard alkali (NaOH) solution or by first diluting it to a known volume (say 250 ml) and then titrating against standard alkali solution

Observations

Let the weight of the organic substance taken = W g

Volume of the standard acid solution of normality N_1 taken = V ml.

Volume of the standard alkali solution of normality say, N_1, required for titration of the excess acid (unused acid) = v ml.

Calculations. (Because acids and alkalies neutralise in equivalent amounts)

Therefore, the acid solution of normality N_1 left unused = v ml

∴ Acid solution of normality N_1 used for neutralising ammonia = $(V - v)$ ml = x (say)

But x ml of N_1 acid solution = x ml of N_1 NH_3 solution.

Now, according to the definition of **normal solution.**

$$1000 \text{ ml of 1N } NH_3 \text{ solution} \equiv 17 \text{ g of } NH_3$$
$$\equiv 14 \text{ g of nitrogen}$$

$\therefore \quad x \text{ ml of } N_1\ NH_3 \equiv \frac{14}{1000} \times x \times N_1$ g of nitrogen

Hence **percentage of nitrogen** $= \frac{14}{1000} \times x \times N_1 \times \frac{100}{W} = \frac{1.4 \times N_1 \times x}{W}$

Thus

Percentage of nitrogen
$= \frac{1.4 \times \text{Normality of acid} \times \text{Vol. of acid used}}{\text{Weight of organic substance}}$

Note: In solving problems on Kjeldahl's method, it is convenient to convert the normality of residual acid to the normality of the original acid by using normality equation.

NUMERICAL PROBLEMS ON KJELDAHL'S METHOD

Type I. ***When the ammonia evolved completely neutralises the standard acid taken.***

Q. 10. 0.3 g of organic compound on Kjeldahl's analysis gave enough ammonia to just neutralise 30 ml of 0.1 N H_2SO_4. Calculate the percentage of nitrogen in the compound.

Ans. Weight of the substance = 0.3 g

Volume of 0.1 N H_2SO_4 required for neutralisation of ammonia = 30 ml

30 ml of 0.1 N H_2SO_4 ≡ 30 ml of 0.1 N NH_3

Now 1000 ml of 1 N NH_3 ≡ 17 g of ammonia

≡ 14 g of nitrogen

$\therefore \quad$ 30 ml of 0.1 N NH_3 $\equiv \frac{14}{1000} \times 30 \times 0.1$ g of nitrogen

$\therefore \quad$ Percentage of nitrogen $= \frac{14}{1000} \times \frac{30 \times 0.1 \times 100}{0.3} = \mathbf{14.0}$

Q. 11. 1.0 g of ethanamine is Kjeldahlised. How much N/2 sulphuric acid is neutralised by the ammonia evolved?

Ans. $CH_3CH_2NH_2 \equiv NH_3$

45 g 17 g

45 g compound evolve = 17 g NH_3

1 g compound evolves $= \frac{17}{45}$ g NH_3

17 g NH_3 are contained in = 1000 ml 1 N NH_3

17/45 g NH_3 is contained in $= \frac{1000}{17} \times \frac{17}{45}$ ml 1 N NH_3

$= \frac{1000}{45}$ ml 1 N NH_3

= 22.2 ml 1 N NH_3

22.2 ml 1N NH_3 ≡ 44.4 ml N/2 NH_3

To neutralise 44.4 ml N/2 NH_3, an equivalent amount *i.e.*, 44.4 ml of N/2 H_2SO_4 will be required.

Type II. ***When the ammonia evolved neutralises part of the standard acid and the acid left unused is titrated against a standard NaOH.***

Q. 12. Ammonia obtained from 0.4 g of an organic substance by Kjeldahl's method was absorbed in 30 ml of seminormal H_2SO_4. The excess of the acid was neutralised by the addition of 30 ml of N/5 NaOH. Calculate the percentage of nitrogen in the substance.

Ans. Weight of the organic substance = 0.4 g

Volume of N/2 H_2SO_4 taken = 30 ml

Let us say x ml of N/2 NaOH = 30 ml of N/5 NaOH

or $$x \times \frac{1}{2} = 30 \times \frac{1}{5} \text{ or } x = 12 \text{ ml}$$

Hence 12 ml of N/2 NaOH was used to neutralise excess acid

or Vol. of N/2 acid left unused = 12 ml

∴ Volume of N/2 acid used in

neutralising ammonia = (30 – 12)

= 18 ml

18 ml of N/2 H_2SO_4 = 18 ml of N/2 NH_3

But 1000 ml of 1 NH_3 = 17 g of NH_3

= 14 g of nitrogen

$$\% \text{ of N} = \frac{14 \times \text{Normality} \times \text{Vol. of acid used}}{\text{Weight organic substance}}$$

$$= \frac{1.4 \times 0.5 \times 18}{0.4} = \mathbf{31.5}$$

Q. 13. 1.0 g of an organic compound was Kjeldahlised and ammonia evolved was absorbed in 25.00 ml of N/5 HCl. The excess of acid required 12.40 ml of N/10 sodium hydroxide solution for neutralisation. Determine the amount of nitrogen present in 1 g of the compound and also the percentage of nitrogen.

Ans. Weight of organic compound = 1.0 g

Volume of standard acid taken = 25.0 ml of N/5 HCl

= 5.0 ml of 1 N/HCl

Volume of NaOH required for the excess acid

$$= 12.40 \text{ ml of } \frac{N}{10} \text{ sol.}$$

But 1.24 ml of 1 N – NaOH ≡ 1.24 ml of 1 N HCl

∴ Acid left unused = 1.24 ml of 1 N HCl

Volume of acid used for neutralising ammonia

= 5 – 1.24 = 3.76 ml of 1 N HCl

Now. 3.76 ml of 1 N HCl solution

= 3.76 ml of 1 N NH_3 solution

(*i*) Weight of nitrogen present in 1 g of compound = $\frac{14 \times 3.76}{1000}$

= 0.0526 g

(*ii*) Percentage of nitrogen $= \frac{14 \times 3.76 \times 100}{1000 \times 1} =$ **5.26.**

Type III. ***When the residual acid in the conical flask is diluted to a known volume with water and then titrated against standard alkali solution.***

Q. 14. 0.6 g of an organic compound was Kjeldahlised and the ammonia evolved was passed into 100 ml of semi-normal H_2SO_4. The residual acid was diluted to 500 ml with distilled water. 25 ml of diluted acid required 20.4 ml of decinormal caustic soda solution for complete neutralisation. Calculate the percentage of nitrogen in the compound.

Ans. Weight of substance taken = 0.6 g

Vol. of N/2 H_2SO_4 taken = 100 ml

Let the vol. of N/2 H_2SO_4 acid unused = x ml

Now 25 ml of the diluted acid solution = 20.4 ml of N/10 NaOH

$\therefore$ Normality of the diluted acid $= 20.4 \times \frac{1}{10} \times \frac{1}{25} = \frac{51}{625}$ N

500 ml of $\frac{51}{625}$ N of diluted acid = x ml of N/2 H_2SO_4 (unused)

$\therefore$ x (*i.e.*, vol. of N/2 H_2SO_4 left unused) $= 500 \times \frac{51}{625} \times \frac{2}{1}$

= 81.6 ml

$\therefore$ Vol. of N/2 acid used for ammonia = 100 – 81.6

= 18.4 ml

Hence the percentage of nitrogen

$$= 1.4 \times \text{Normality of acid} \times \frac{\text{vol. of acid used}}{\text{wt. of substance}}$$

$$= \frac{1.4 \times \frac{1}{2} \times 18.4}{0.6}$$

= **21.46**

Q. 15. 0.42 g of a substance was Kjeldahlised and the ammonia evolved was passed into 60 ml of decinormal sulphuric acid. The excess of the acid was diluted and the volume made upto 250 ml. 25 ml of this solution required 10 ml of N/40 NaOH for neutralization. Calculate percentage of nitrogen in the compound.

Ans. Weight of substance = 0.42 g

Vol. of N/10 sulphuric acid taken = 60 ml

Let the volume of N/10 H_2SO_4 left unused = x ml

This volume has been diluted to 250 ml

25 ml of the diluted acid ≡ 10 ml of N/40 NaOH

$\therefore$ Normality of diluted acid $= 10 \times \frac{1}{40} \times \frac{1}{25} = \frac{1}{100}$ N

250 ml of $\frac{1}{100}$ N of diluted acid ≡ x ml of N/10 H_2SO_4 (unused)

$\therefore$ x (*i.e.*, vol. of N/10 H_2SO_4 left unused) $= 250 \times \frac{1}{100} \times \frac{10}{1}$

$= 25.0$ ml.

Now, volume of N/10 H_2SO_4 neutralised by ammonia $= 60 - 25.0$

$= 35.0$ ml

Hence, the percentage of nitrogen

$$= 1.4 \times \text{Normality of acid} \times \frac{\text{vol. of acid used}}{\text{Wt. of substance}}$$

$$= \frac{1.4 \times \frac{1}{10} \times 35.0}{0.42}$$

$$= \mathbf{11.66}$$

Q. 16. 0.4 g of an organic compound was Kjeldahlised and ammonia evolved was absorbed into 50 ml of semi-normal solution of sulphuric acid. The residual acid was diluted with distilled water and volume was made up to 150 ml. 20 ml of this acid solution required 31 ml of N/20 NaOH solution for complete neutralisation. Find out the percentage of nitrogen in the compound.

Ans. Weight of substance taken = 0.4 g

Vol. of N/2 acid taken = 50 ml

Now 20 ml of the diluted acid solution ≡ 31 ml of N/20 NaOH

$\therefore$ Normality of diluted acid $= 31 \times \frac{1}{20} \times \frac{1}{20}$

$= \frac{31}{400}$ N

150 ml of 31/400 N of diluted acid ≡ V ml of N/2 H_2SO_4 (unused)

$\therefore$ V (*i.e.*, Vol of N/2 H_2SO_4 left unused) $= 150 \times \frac{31}{400} \times \frac{2}{1}$

$= \mathbf{23.25}$ **ml**

Now Vol. of N/2 acid used for ammonia $= 50 - 23.25$

$= 26.75$ ml

Hence percentage nitrogen

$$= \frac{1.4 \times \text{Normality of acid} \times \text{Vol. of acid used}}{\text{Weight of substance}}$$

$$= \frac{1.4 \times \frac{1}{2} \times 26.75}{0.4}$$

$$= \mathbf{46.81}$$

PROBLEMS FOR PRACTICE

1. *0.56 g of a nitrogenous organic compound when distilled by Kjeldahl's method produced 0.34 g of ammonia. What is the percentage of nitrogen in the compound?* [**Ans.** 50%]

2. *0.3 g of an organic compound on Kjeldahl's analysis gave enough ammonia to just neutralise 30 ml of 0.1 N H_2SO_4 Calculate the percentage of nitrogen in the compound.* [**Ans.** 14%]

3. *0.24 g of an organic compound containing nitrogen was Kjeldahlised and ammonia formed was absorbed in 50 ml of N/4 H_2SO_4 The excess of the acid required 77.0 ml of N/10 NaOH for complete neutralisation. Calculate the percentage of nitrogen in the organic compound.* [**Ans.** 28%]

4. *0.59 g of an organic compound was Kjeldahlised. The ammonia evolved was passed into 50 ml of 1 N sulphuric acid. The unreacted acid was diluted to 100 ml. 10 ml of this solution required 16 ml of N/4 sodium hydroxide. Find out the percentage of nitrogen.* [**Ans.** 23.7%]

ESTIMATION OF HALOGENS

Q. 17. How do you estimate halogen in an organic compound?

(Meerut, 2003; Utkal, 2004; Sambalpur, 2004)

Ans. Halogens like chlorine, bromine or iodine are estimated by **Carius** method described below:

Principle. A known weight of the organic substance is heated with fuming nitric acid along with a few crystals of silver nitrate. Carbon, hydrogen and sulphur are respectively oxidised to carbon dioxide, water and sulphur dioxide while halogen combines with silver nitrate to form a precipitate of silver halide.

$$\text{C [From organic compound]} \xrightarrow{\text{Fuming HNO}_3} CO_2$$

$$\text{2H [From organic compound]} \xrightarrow{\text{Fuming HNO}_3} H_2O$$

$$\text{S [From organic compound]} \xrightarrow{\text{Fuming HNO}_3} H_2SO_4$$

$$\text{X [From organic compound]} \xrightarrow[\text{AgNO}_3]{\text{Fuming HNO}_3} \underset{\text{Silver halide}}{\text{Ag X}}$$

The precipitate of silver halide is filtered, washed, dried and is finally weighed. Knowing the weight of the organic compound taken and the weight of the silver halide formed, the percentage of halogen can be easily calculated.

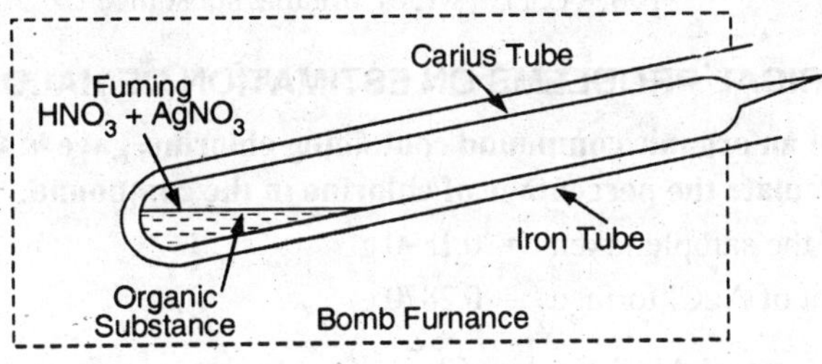

Fig. 2.4. Carius method for estimation of halogens.

Procedure: (*i*) About 0.5 g of silver nitrate crystals and 5 ml of fuming nitric acid are taken in a Carius tube which is a double walled hard glass tube about 40 cm long and 2 cm in diameter.

(*ii*) A known weight (0.2 to 0.3 g) of the organic substance is taken in a small tube which is carefully introduced into the Carius tube.

(*iii*) The open end of Carius tube is then sealed by heating.

(*iv*) The Carius tube is then placed in a iron jacket and gradually heated in a bomb furnace at 530 – 540 K for about 5 hours.

(*v*) Thereafter, the tube is allowed to cool for 6 hours. Then a small hole is made in the sealed end of the tube to allow the gases to escape slowly. The precipitate is filtered, washed, dried and weighed.

Calculations. Let the weight of organic substance = W g

Let the weight of silver halide formed = w g

Now $\underset{(108+X)\text{ parts}}{\text{Ag X}} \equiv \underset{\text{X parts}}{\text{X}}$

Here X stands for the halogen

(108 + X) g of silver halide contain halogen = X g

*w*g of silver halide will contain halogen $= \dfrac{X}{(108+X)} \times w\text{ g}$

Percentage of halogen

$$= \frac{\text{wt. of halogen}}{\text{wt. of the organic substance}} \times 100 = \frac{X \times w \times 100}{(108+X) \times W}$$

$$= \frac{\text{At. wt. of halogen}}{(108 + \text{At. wt. of halogen})} \times \frac{\text{wt. of silver halide}}{\text{wt. of organic substance}} \times 100$$

Percentage of chlorine

$$= \frac{35.5}{(108+35.5)} \times \frac{\text{wt. of silver chloride}}{\text{wt. of organic substance}} \times 100$$

Percentage of bromine

$$= \frac{80}{(108+80)} \times \frac{\text{wt. of silver bromide}}{\text{wt. of organic substance}} \times 100$$

Percentage of iodine

$$= \frac{127}{(108+127)} \times \frac{\text{wt. of silver iodide}}{\text{wt. of organic substance}} \times 100$$

NUMERICAL PROBLEMS ON ESTIMATION OF HALOGENS

Q. 18. 0.1890 g of an organic compound containing chlorine, gave in Carius determination 0.2870 g of AgCl. Calculate the percentage of chlorine in the compound.

Ans. Weight of the sample taken = 0.1890 g

Weight of AgCl formed = 0.2870 g

$$\underset{108+35.5=143.5}{\text{AgCl}} = \underset{35.5}{\text{Cl}}$$

∴ Weight of Cl in 0.2870 g of AgCl

$$= \frac{35.5}{143.5} \times 0.2870$$

$$\text{Hence the percentage of Cl} = \frac{35.5 \times 0.2870}{143.5 \times 0.1890} \times 100$$

$$= \mathbf{37.56\%}$$

Q. 19. 0.185 g of an organic compound when treated with conc. nitric acid and silver nitrate gave 0.320 g of silver bromide. Calculate percentage of bromine in the compound.

Ans. Weight of substance = 0.185 g

Weight of silver bromide = 0.320 g

Now $\underset{108+80=188}{AgBr} = \underset{80}{Br}$

188 g of AgBr contain = 80 g Br

$$0.32 \text{ g of AgBr contain} = \frac{80}{188} \times 0.32 \text{ g Br}$$

$$\text{Hence the percentage of bromine} = \frac{80}{188} \times 0.32 \times \frac{100}{0.185}$$

$$= \mathbf{73.60}$$

Q. 20 0.40 g of an iodo substituted organic compound by Carius method gave 0.235 g of silver iodide. Calculate the percentage of iodine in the compound.

Ans. Weight of substance = 0.40 g

Weight of silver iodide = 0.235 g

Now $\underset{108+127=235}{AgI} \equiv \underset{127}{I}$

235 g of AgI contain = 127 g I

$$0.235 \text{ g of AgI contain} = \frac{127}{235} \times 0.235 \text{ g I}$$

$$\text{Hence the percentage of iodine} = \frac{127}{235} \times 0.235 \times \frac{100}{0.40}$$

$$= \mathbf{31.75}$$

PROBLEMS FOR PRACTICE

1. *0.284 g of an organic substance gave 0.287 g of silver chloride in a Carius estimation of halogen. Find out the percentage of chlorine in the compound (Ag = 108, Cl = 35.5)*

   [**Ans.** *25%*]

2. *0.394 g of an organic compound containing iodine on treatment by Carius method gave 0.705 g of silver iodide. Find the percentage of halogen in the compound.*

   [**Ans.** *96.7%*]

ESTIMATION OF SULPHUR

Q. 21. Describe Carius method for the estimation of sulphur in a compound.

(*Agra, 2004*)

Ans. Princple. A known weight of the organic compound is heated strongly with fuming nitric acid. Carbon and hydrogen are oxidised to carbon dioxide and water vapours respectively while sulphur is oxidised to sulphuric acid.

$$\text{C} \quad \text{[From the organic compound]} \xrightarrow{\text{Fuming } HNO_3} CO_2$$

$$\text{2H} \quad \text{[From the organic compound]} \xrightarrow{\text{Fuming } HNO_3} H_2O$$

$$\text{S} \quad \text{[From the organic compound]} \xrightarrow{\text{Fuming } HNO_3} SO_2 \longrightarrow H_2SO_4$$

The sulphuric acid is then treated with excess of $BaCl_2$ to give a precipitate of barium sulphate. From the weight of barium sulphate formed, the percentage of sulphur is calculated.

Procedure. (*i*) A known weight of the organic compound (about 0.2 g) is taken in a Carius tube.

(*ii*) About 5 ml of fuming nitric acid is added into it.

(*iii*) The Carius tube is then sealed, placed in the iron tube and heated gradually for about 5 hours.

(*iv*) Thereafter, the tube is cooled and a small hole is made to release the gases trapped in it.

(*v*) Then the Carius tube is broken and its contents are transferred to a beaker.

(*vi*) Excess of barium chloride solution is added to precipitate sulphuric acid as barium sulphate.

(*vii*) The precipitate of barium sulphate is filtered, washed, dried and weighed.

(*viii*) From the weight barium sulphate, the percentage of sulphur is calculated as described below:

Calculations:

Let the weight of organic compound = W g

Let the weight of barium sulphate = w g

$$\underset{\substack{(137+32+64)\\233}}{BaSO_4} \equiv \underset{32}{S}$$

233 g of barium sulphate contain sulphur ≡ 32 g

$$\therefore \; w \text{ g of barium sulphate will contain sulphur} = \frac{32}{233} \times \text{w g}$$

$$\therefore \qquad \text{Percentage of sulphur} = \frac{\text{weight of sulphur}}{\text{weight of compound}} \times 100$$

$$= \frac{32}{233} \times \frac{w}{W} \times 100$$

$$= \frac{32}{233} \times \frac{\text{weight of barium sulphate}}{\text{weight of compound}} \times 100$$

NUMERICAL PROBLEMS ON ESTIMATION OF SULPHUR

Q. 22. 0.2595 g of an organic compound by Carius method gave 0.350 g of barium sulphate. Calculate the percentage of sulphur in the compound.

Ans. Weight of the organic compound taken = 0.2595 g

Weight of formed = 0.350 g

Now $233 \text{ g of } BaSO_4 \equiv 32 \text{ g of S}$

$\therefore \quad 0.350 \text{ g of } BaSO_4 \equiv \frac{32 \times 0.350}{233} \text{ g of S}$

$$\text{Percentage of sulphur in the compound} = \frac{32}{233} \times 0.350 \times \frac{100}{0.2595}$$

$$= \mathbf{18.52}$$

Q. 23. In an estimation of sulphur by Carius method 0.2175 g of the substance gave 0.5825 g of barium sulphate. Calculate the percentage of sulphur in the compound.

Ans. Weight of organic compound = 0.2175 g

Weight of barium sulphate = 0.5825 g

$$\underset{\substack{(137 + 32 + 64) \\ 233}}{BaSO_4} = \underset{32}{S}$$

233 g of barium sulphate contain sulphur = 32 g

$$0.5825 \text{ g of barium sulphate contain sulphur} = \frac{32}{233} \times 0.5825 \text{ g}$$

$$\text{Percentage of sulphur} = \frac{32}{233} \times \frac{0.5825}{0.2175} \times 100$$

$$= \mathbf{36.78}$$

PROBLEMS FOR PRACTICE

1. *0.16 g of an organic substance was heated in a Carius tube and sulphuric acid formed was precipitated as $BaSO_4$ with $BaCl_2$ The weight of dry $BaSO_4$ was 0.35 g. Find the percentage of sulphur.* **[Ans.** 30.04%**]**
2. *0.4037 g of an organic compound containing S was treated with conc. HNO_3 in a Carius tube. On precipitation with $BaCl_2$ 0.1936 g of $BaSO_4$ was produced. Determine the percentage of sulphur in the compound.* **[Ans.** *6.58%***]**

DETERMINATION OF MOLECULAR WEIGHT

Q. 24. Describe silver salt method for the determination of molecular weight of an organic acid. *(Nagpur, 2003 ; Delhi, 2004)*

Ans. Principle. Organic acids react with silver nitrate to produce silver salt of the acid. When the silver salt is ignited, it quantitatively gets decomposed to metallic silver. Knowing the weight of silver metal and basicity of the acid, molecular weight can be calculated.

$$RCOOH + AgNO_3 \longrightarrow RCOOAg\downarrow + HNO_3$$

Procedure. A small amount of the organic acid is dissolved in NH_4OH. The solution is boiled to remove excess of ammonia. Then excess of silver nitrate solution is added and the solution is heated on a water bath after wrapping the beaker with black paper. A white precipitate of silver salt of the organic acid is obtained. The precipitate is filtered, washed and dried.

A known weight of the dried precipitate is then taken in a weighed crucible. The precipitate is then ignited in the furnace at 700 K to a constant weight. Silver salt is quantitatively decomposed to silver metal.

$$\text{RCOOAg} \xrightarrow[700\text{ K}]{\Delta} \underset{\text{[Silver metal]}}{\text{Ag}} \downarrow + [CO_2 + H_2O]$$

From the weight of silver metal obtained the molecular weight of the acid can be calculated as follows.

Calculations

Let the weight of silver salt taken = W g

Weight of metallic silver left (residue) = w g

Now, $$\frac{\text{Equivalent weight of silver salt}}{\text{Equivalent weight of silver (108)}} = \frac{\text{Weight of silver salt taken (W)}}{\text{Weight of silver left (w)}}$$

$\therefore$ $$\text{Equivalent weight of silver salt} = \frac{W}{w} \times 108 = E \text{ (say)}$$

But equivalent weight of the acid = Equivalent weight of the silver salt – Equivalent weight of silver + Equivalent weight of hydrogen

$$= E - 108 + 1$$

$$= E - 107$$

$\therefore$ Molecular weight of the acid = Equivalent Weight × Basicity

$$= (E - 107) \times n$$

$$= \left[\left(\frac{W}{w} \times 108\right) - 107\right] \times n$$

where '*n*' is the basicity of the acid.

NUMERICAL PROBLEMS ON MOLECULAR WEIGHT

Q. 25. When 0.76 g of the silver salt of a dibasic acid was ignited, it gave 0.54 g of pure silver. Determine the molecular weight of the acid.

Ans. Weight of the silver salt = 0.76 g

Weight of metallic silver = 0.54 g

$$\frac{\text{Equivalent weight of silver salt}}{\text{Equivalent weight of silver}} = \frac{\text{weight of silver salt}}{\text{weight of silver}}$$

$\therefore$ $$\text{Equivalent weight of silver salt} = \frac{0.76 \times 108}{0.54} = 152$$

$\therefore$ Equivalent weight of the acid = 152 – 107 = 45

$\therefore$ Molecular weight of the acid = Eq. weight × Basicity

= **45 × 2 = 90.**

Q. 26. 0.607 g of the silver salt of a tribasic acid on combustion deposited 0.370 g of pure silver. Calculate the molecular weight of the acid.

Ans. Weight of silver salt = 0.607 g

Weight of silver left = 0.370 g

$$\frac{\text{Equivalent weight of silver salt}}{\text{Equivalent weight of silver}} = \frac{\text{Weight of silver salt}}{\text{Weight of silver}}$$

$$\therefore \quad \text{Equivalent weight of silver salt} = \frac{0.607 \times 108}{0.370} = 177.1$$

$$\therefore \quad \text{Equivalent weight of the acid} = 177.1 - 108 + 1 = 70.1$$

$$\therefore \quad \text{Molecular weight of the acid} = 70.1 \times 3 = \mathbf{210.3}.$$

PROBLEMS FOR PRACTICE

1. *The silver salt of a dibasic organic acid contains 71.05% silver. Calculate the molecular mass of the acid.* [**Ans.** *90*]
2. *On ignition 0.2299 g of the silver salt of a monobasic acid yields 0.1188 g of silver. Calculate molecular mass of the acid.* [**Ans.** *102*]

Q. 27. Describe platinichloride method for determination of molecular weight of organic bases. (*Karnataka, 2004*)

Ans. Principle. The principle underlying the process is that the platinichloride salt of organic base when heated leaves metallic platinum as residue. Knowing the weight of the salt and of the metallic residue, the molecular weight of the base can be determined.

Procedure. A small amount of the organic base is dissolved in excess of concentrated hydrochloric acid. To this solution, platinic chloride ($PtCl_4$) solution is added when the platinichloride salt of the base gets precipitated.

$$\underset{\text{(Base)}}{B} + HCl \longrightarrow \underset{\text{Base hydrochloride}}{B.HCl}$$

$$2B.HCl + \underset{\text{Platinic chloride}}{PtCl_4} \longrightarrow \underset{\text{Base chloroplatinate}}{B_2H_2PtCl_6\uparrow}$$

The precipitate is washed and dried. A known weight of the chloroplatinate is ignited and the weight of the metallic platinum left is determined.

('*B*' stands for one equivalent of the base.)

$$B_2H_2PtCl_6 \xrightarrow{\text{Ignite}} \text{Pt (metal)}$$

From the weight of platinum metal the molecular weight of the base is calculated as follows.

Chloroplatinic acid is a dibasic acid, therefore it combines with two equivalents of the base.

Calculations :

Let the weight of the chloroplatinate taken = W g

And the weight of platinum (residue) left after ignition = W g

Now, 1 molecule of $B_2H_2\,PtCl_2$ ≡ 1 atom of Pt

i.e., one gm molecule of platinichloride salt contains one gram atom of platinum.

$$\frac{\text{Mol. wt. of platinichloride}}{\text{At. wt. of platinum}} = \frac{\text{Wt. of platinichloride}}{\text{Wt. of platinum left}}$$

But molecular weight of platinichloride ($B_2H_2PtCl_2$)

$$= 2B + 2 + 195 + 6 \times 35.5$$
$$= 2B + 410$$

(where B represent one equivalent of the base)

$$\therefore \quad \frac{2B + 410}{195} = \frac{W}{w}$$

or
$$B = \frac{1}{2}\left(\frac{W}{w} \times 195 - 410\right)$$

If n is the acidity of base, then molecular weight of the base = B × n.

NUMERICAL PROBLEMS

Q. 28. 0.984 g of the chloroplatinate of a diacid base gave 0.39 g of platinum. Calculate the molecular weight of the base.

Ans. Weight of chloroplatinate = 0.984 g

Weight of platinum left behind = 0.39 g

Let the chloroplatinate be represented by $B_2H_2PtCl_6$, where B stands for one equivalent of the base.

Since, $$\frac{\text{Mol. wt. of the chloroplatinate}}{\text{At. wt. of platinum (195)}} = \frac{\text{Wt. of the chloroplatinate } (B_2H_2PtCl_6)}{\text{Wt. of platinum left}}$$

∴ Molecular wt. of the chloroplatinate (2B + 410)

$$= \frac{0.984}{0.39} \times 195 = 492$$

$$\therefore \quad \text{Eq. weight (B) of the base} = \frac{1}{2}(492 - 410) = 41$$

Acidity of the base = 2 (given)

∴ Molecular weight = 41 × 2 = **82.**

Q. 29. A chloroplatinate of a diacid base contains 39% of platinum. What is the molecular weight of the base?

Ans. Let the weight of chloroplatinate taken = 100 g

The % of Pt. in the chloroplatinate = 39

$$\therefore \quad \text{The weight of Pt. left as a residue} = \frac{100 \times 39}{100} = 39 \text{ g}$$

Now $$\frac{\text{Mol. weight of chloroplatinate}}{\text{At. weight of platinum}} = \frac{\text{Weight of chloroplatinate taken}}{\text{Weight of platinum left}}$$

or
$$\frac{2B + 410}{195} = \frac{109}{39}$$

$$2B + 410 = \frac{100}{39} \times 195 = 500$$

or $$2B = 500 - 410 = 90$$

$\therefore$ $$= 45$$

$$\text{Acidity of the base} = 2 \text{ (given)}$$

$\therefore$ $$\text{Molecular weight of the acid} = 45 \times 2 = \mathbf{90}.$$

PROBLEMS FOR PRACTICE

1. *0.595 g of platinichloride of a monoacid organic base left on ignition 0.195 g of platinum. Calculate the molecular weight of the base and also give the name of aromatic amine.* **(Ans.** *92.5, aniline)*
2. *0.44 g of the chloroplatinate of an organic base when heated to a constant mass left 0.1375 g of metallic platinum. Find the mol. mass of the base, if the acidity of base is one.* **(Ans.** *107)*
3. *0.75g of platinichloride of a mono acidic base on ignition gave 0.245g of platinum. Find the molecular weight of the base. At weights : Pt = 195, Cl =35.5 (Jiwaji, 2003)* **(Ans.** *93.46)*

Q. 30. Explain the principle of volumetric method of determination of molecular weights of organic acids and bases.

Ans. Principle. Volumetric method is based upon the principle that acids and bases neutralise each other in equivalent amounts.

(*i*) **Molecular weight of organic acids**

Procedure. A known weight of the acid is dissolved in water or alcohol, and titrated against a standard alkali (*caustic soda solution*), using phenolphthalin as indicator. From the amount of the alkali required to neutralise the acid, the equivalent weight and, then the molecular weight of the acid are calculated.

Suppose V ml of N/10 alkali neutralise = *w* g of the acid.

$$\therefore \text{ 1000 ml of 1 N-alkali neutralise} = \frac{1000}{V} \times w \times \frac{10}{1} \text{ g of acid}$$

Now 1000 ml of the normal alkali contain one gram equivalent of the alkali and, therefore, it must neutralise one gram equivalent of the acid.

$$\therefore \quad \text{Gram equivalent weight of the acid} = \frac{1000}{V} \times w \times \frac{10}{1} \text{ g}$$

and $$\text{Equivalent weight of the acid} = \frac{1000}{V} \times w \times 10$$

Hence, $$\textbf{mol. weight of the acid} = \text{Eq. weight} \times \text{basicity}$$

$$= \left(\frac{1000}{V} \times w \times 10\right) \times n$$

(*ii*) **Molecular weight of organic bases**

The molecular weight of the base is also determined in the same way by titrating a known weight of the base against a standard acid (say HCl). The weight of the base that would neutralise one gram equivalent weight of the acid is the equivalent weight of the base.

Molecular weight of base = Equivalent weight × acidity

NUMERICAL PROBLEMS

Q. 31. 0.115 g of a mono-basic organic acid required 25 ml of decinormal caustic soda solution for complete neutralisation. Calculate the molecular weight of organic acid.

Ans. 25 ml of $\frac{N}{10}$ NaOH $\equiv$ 0.115 g of the acid

$$1000 \text{ ml of 1 N-NaOH} = 0.115 \times \frac{1000}{25} \times \frac{10}{1}$$

$$= 46 \text{ g acid}$$

But 1000 ml of normal NaOH contains 1 gm equivalent of NaOH and, therefore, it must neutralise 1 gm equivalent of the acid.

$\therefore$ Equivalent. wt. of acid = 46

Molecular weight of the acid = Equivalent wt. × Basicity = 46 × 1

= **46**

Q. 32. 0.25 g of a dibasic organic acid was dissolved in water and the volume made to 100 ml. 10 ml of this solution required 12.3 ml of N/30 NaOH for complete neutralisation. Find the molecular weight of the acid.

Ans. 10 ml of the acid $\equiv$ 12.3 ml of N/30 NaOH

Let N_1 be the normality of the acid solution.

Applying normality equation $N_1V_1 \equiv N_2V_2$

$$10 \times N_1 = 12.3 \times \frac{1}{30}$$

$$\therefore \quad N_1 = \frac{12.3}{30 \times 10} = 0.041$$

We have the relation: Strength/litre = Normality × Eq. weight

$\therefore$ Strength of the acid = 0.041 × Equivalent weight

$$\text{But strength of the acid} = \frac{0.25}{100} \times 1000 = 2.5 \text{ g. per litre}$$

$\therefore$ 0.041 × Equivalent weight of the acid = 2.5

$$\text{or} \quad \text{Equivalent weight of the acid} = \frac{2.5}{0.041} = 60.98$$

Basicity = 2 (given)

$\therefore$ **Molecular weight of the acid** = 60.98 × 2 = **121.96**

PROBLEMS FOR PRACTICE

1. *0.2709 g of a mono-basic acid required 25.20 ml of N/12 NaOH for complete neutralisation. What is its molecular weight?* **(Ans.** *129***)**
2. *0.225 g of a dibasic acid required 20 ml of 0.25 N-NaOH solution for complete neutralisation. What is its molecular weight?* **(Ans.** *90***)**
3. *0.225 g of a dibasic acid required 10.84 ml of 0.25 N-NaOH for complete neutralisation. What is the molecular mass?* **(Ans.** *166.05***)**

4. *0.637 g of a tri-acid base required for exact neutralisation 21.6 ml of N/2-HCl. Calculate the mol mass of the base.* (**Ans.** 176.9)

5. *0.25 g of a dibasic organic acid was dissolved in water and the volume was made to 100 ml 10 ml of the solution required for neutralisation 24.6 ml of N/60 NaOH. Calculate the molecular weight of the acid.* (**Ans.** 122)

6. *0.16 g of a dibasic organic acid required 25 ml of decinormal NaOH for complete neutralisation. Find out the molecular weight of the acid.* (**Ans.** 128)

EMPIRICAL AND MOLECULAR FORMULAE

Q. 33. What is meant by empirical and molecular formulae? How do you determine them from elemental analysis data?

Ans. Empirical Formula. *It is the simplest whole number ratio between the atoms of various elements present in one molecule of the compound.* Empirical formula is not the actual formula of the substance, it only tells the relative number of atoms present in the substance. For example, empirical formula of glucose is CH_2O. It means the relative ratio of carbon, hydrogen and oxygen atoms in glucose is 1 : 2 : 1 whereas a molecule of glucose contains actually six carbon, twelve hydrogen and six oxygen atoms.

Determination of Empirical Formula:

Following steps are employed to calculate the empirical formula:

(*i*) The percentage of each element is divided by the atomic weight of the element to obtain the relative number of elements.

(*ii*) The figures obtained from above are divided by the lowest figure to get the simplest ratio of elements.

(*iii*) If whole numbers are not obtained and fractions are obtained then multiply the figures by a suitable number to obtain the whole number simplest ratio.

(*iv*) The empirical formula is written by writing the symbol of elements along with the number of atoms of elements in the subscript.

Molecular Formula: *It is the actual number of atoms of various elements present in one molecule of the substance.* Molecular formula is either the same as empirical formula or its integral multiple

$$n = \frac{\text{Molecular weight}}{\text{Empirical formula wt.}}$$

where n is an integer 1, 2, 3 etc.

Take for example the case of glucose. It has empirical formula CH_2O and molecular formula $C_6H_{12}O_6$.

$$\text{Empirical formula weight} = 12 + (2 \times 1) + 10 = 30$$

$$\text{Molecular weight} = (6 \times 12) + (12 \times 1) + (6 \times 16)$$
$$= 180$$

In this case

$$n = \frac{180}{30} = 6$$

Determination of molecular formula

1. First calculate the empirical formula with the help of percentage values of elements.

2. Add up to get empirical formula weight.
3. Divide the molecular weight by the empirical formula weight to obtain the value of *n*.

$$\text{Molecular formula} = (\text{Empirical formula})_n$$

Multiply the relative number of atoms of elements in the empirical formula by *n*.

NUMERICAL PROBLEMS ON EMPIRICAL AND MOLECULAR FORMULAE

Q. 34. An inorganic compound on analysis gave the following percentage composition. Potassium = 57.24%, C = 8.69% and Oxygen = 34.78%. Calculate the Empirical formula of the substance.

Ans. *Calculation of simplest whole number ratio of the atoms.*

Elements	*Symbol*	*Percentage of element*	*At. mass*	*Relative no. of atoms* $= \frac{percentage}{At.\ mass}$	*Simplest atomic ratio*
Potassium	K	57.24	39	$\frac{57.24}{39} = 1.468$	$\frac{1.468}{0.724} = 2$
Carbon	C	8.69	12	$\frac{8.69}{12} = 0.724$	$\frac{0.724}{0.724} = 1$
Oxygen	O	37.78	16	$\frac{34.78}{16} = 2.174$	$\frac{2.174}{0.724} = 3$

The simplest whole number ratio of the atoms is

$$2 : 1 : 3$$

Hence the formula of the compound can be written as

$$K_2C_1O_3$$

Omitting the unit subscript, the empirical formula is K_2CO_3.

Q. 35. An organic compound contains C = 12.76%, H = 2.13% and Br = 85.11%. Its V.D. = 94. Find its molecular formula.

Ans. Calculation of Empirical formula:

Elements	*Percentage*	*Atomic weight*	*Relative number of atoms*	*Simplest whole number ratio*
C	12.76	12	$\frac{12.76}{12} = 1.06$	$\frac{1.06}{1.06} = 1$
H	2.13	1	$\frac{2.13}{1} = 2.13$	$\frac{2.13}{1.06} = 2$
Br	85.11	80	$\frac{85.11}{80} = 1.06$	$\frac{1.06}{1.06} = 1$

Therefore empirical formula is CH_2Br

$$\text{Empirical formula weight} = 1 \times 12 + 2 \times 1 + 1 \times 80$$
$$= 12 + 2 + 80 = 94$$

$$\text{Molecular weight} = 2 \times \text{V.D.}$$
$$= 2 \times 94 = 188$$
$$n = \frac{\text{Molecular weight}}{\text{Empirical formula weight}} = \frac{188}{94} = 2$$

Hence, molecular formula = $(\text{Empirical formula})_n$

$$= [CH_2Br]_2 = C_2H_4Br_2$$

MORE SOLVED PROBLEMS

Q. 36. 0.45 g of an organic compound gave on combustion 0.792 g CO_2 and 0.324 g of water. 0.24 g of the same substance was Kjeldahlised and ammonia formed was absorbed in 50.0 cm³ of N/4 H_2SO_4. The excess acid required 77 cm³ of N/10 NaOH for complete neutralisation. Calculate the empirical formula of the compound.

Ans. *Percentage of carbon*

$$= \frac{12}{44} \times \frac{\text{Wt. of } CO_2}{\text{Wt. of compound}} \times 100$$

$$= \frac{12}{44} \times \frac{0.792}{0.45} \times 100 = 48\%$$

Percentage of hydrogen

$$= \frac{2}{18} \times \frac{\text{Weight of } H_2O}{\text{Wt. of compound}} \times 100$$

$$= \frac{2}{18} \times \frac{0.324}{0.45} \times 100 = 8\%$$

Determination of percentage of nitrogen

$$77 \text{ cm}^3 \text{ of N/10 NaOH} = \text{V ml of N/4 NaOH}$$

or $$77 \times \frac{1}{10} = V \times \frac{1}{4} \quad \text{or} \quad V = \frac{77 \times 4}{10} = 30.8 \text{ ml}$$

$$30.8 \text{ ml N/4 NaOH} = 30.8 \text{ ml of N/4 } H_2SO_4$$
$$\text{Excess acid} = 30.8 \text{ ml of N/4 } H_2SO_4$$

Acid neutralised by ammonia = 50 – 30.8 = 19.2 cm³ of N/4 acid

$$19.2 \text{ cm}^3 \text{ of N/4 acid} \equiv 19.2 \text{ cm}^3 \text{ of N/4 } NH_3$$

Now 1000 cm³ of 1 N NH_3 contain = 14 g N

$$19.2 \text{ cm}^3 \text{ of N/4 } NH_3 \text{ contain} = \frac{14}{1000} \times 19.2 \times \frac{1}{4} \text{ g N}$$
$$= 0.0672 \text{ g}$$

$$\% \text{ of nitrogen} = \frac{0.0672}{0.24} \times 100 = 28$$

$$\% \text{ of oxygen (by difference)} = 100 - (48 + 8 + 28) = 16$$

Calculation of empirical formula

Element	%	*At. wt.*	*Relative no. of atoms*	*Simplest ratio*
C	48	12	48/12 = 4	4
H	8	1	8/1 = 8	8
N	28	14	28/14 = 2	2
O	16	16	16/16 = 1	1

The empirical formula is $C_4H_8N_2O$.

Q. 37. 0.45 g of an organic compound containing only C, H and N on combustion gave 1.1 g of CO_2 and 0.3 g of H_2O. What is the empirical formula of the compound?

Ans. *Percentage of carbon*

$$= \frac{12}{44} \times \frac{\text{Weight of } CO_2}{\text{Wt. of compound}} \times 100$$

$$= \frac{12}{44} \times \frac{1.1}{0.45} \times 100 = 66.66\%$$

Percentage of hydrogen

$$= \frac{2}{18} \times \frac{\text{Weight of } H_2O}{\text{Wt. of compound}} \times 100$$

$$= \frac{2}{18} \times \frac{0.3}{0.45} \times 100 = 7.40\%$$

Percentage of nitrogen (by difference)

$$= 100 - (66.66 + 7.40)$$

$$= 25.94$$

Determination of empirical formula

Element	*Percentage*	*At. mass*	*Relative no, of atoms*	*Simplest ratio*
C	66.66	12	66.66/12 = 5.55	5.55/1.85 = 3
H	7.40	1	7.40/1 = 7.40	7.40/1.85 = 4
N	25.94	14	25.94/14 = 1.85	1.85/1.85 = 1

The empirical formula is C_3H_4N.

Q. 38. 1.01 g of an organic compound containing 41.37% of C, 5.75% of H on Kjeldahlising required 11.6 cm³ of 1 N HCl. In Carius determination, 0.2066 g of the substance gave 0.5544 g of barium sulphate. Find the formula of the compound.

Ans. *Percentage of N*

$$11.6 \text{ cm}^3 \text{ of } 1 \text{ N HCl} \equiv 11.6 \text{ cm}^3 \text{ of } 1 \text{ N } NH_3$$

$$1000 \text{ cm}^3 \text{ of } 1 \text{ N } NH_3 \text{ contain} = 14 \text{ g N}$$

$$11.6 \text{ cm}^3 \ 1 \text{ N } NH_3 \text{ contain} = \frac{14}{1000} \times 11.6$$

$$= 0.1624 \text{ g N}$$

$$\% \text{ of nitrogen} = \frac{0.1624}{1.01} \times 100$$

$$= 16$$

Percentage of sulphur $= \frac{32}{233} \times \frac{\text{weight of } BaSO_4}{\text{wt. of compound}} \times 100$

$$= \frac{32}{233} \times \frac{0.5544}{0.2066} \times 100$$

$$= 36.85$$

Calculation of empirical formula

Element	*Percentage*	*At. mass*	*Relative no. of atoms*	*Simplest ratio*
C	41.37	12	41.37/12 = 3.45	3.45/1.14 = 3
H	5.75	1	5.75/1 = 5.75	5.75/1.14 = 5
N	16.00	14	16/14 = 1.14	1.14/1.14 = 1
S	36.85	32	37.95/32 = 1.15	1.15/1.14 = 1

The empirical formula of the compound is C_3H_5NS.

Q. 39. 0.246 g of an organic compound containing 58.53% C, 4.06% H gave 22.4 cm³ of N_2 at S.T.P. What is the empirical formula of the compound?

Ans. *Calculation of percentage of N*

22400 ml of N_2 at S.T.P. weigh $= 28$ g

22.4 ml of N_2 at S.T.P. weigh $= \frac{28}{22400} \times 22.4$

or weight of N $= 0.028$ g

$$\% \text{ of nitrogen} = \frac{0.028}{0.246} \times 100$$

$$= 11.38\%$$

Percentage of oxygen (by difference)

$$= 100 - (58.53 + 4.06 + 11.38)$$

$$= 26.03$$

Calculation of empirical formula

Element	*Percentage*	*At. mass*	*Relative no. of atoms*	*Simplest ratio*
C	58.53	12	58.53/12 = 4.87	4.87/0.81 = 6
H	4.06	1	4.06/1 = 4.06	4.06/0.81 = 5
N	11.38	14	11.38/14 = 0.81	.81/0.81 = 1
O	26.03	16	26.03/16 = 1.62	1.62/0.81 = 2

Hence the empirical formula is $C_6H_5NO_2$.

Q. 40. A monoacid organic base contains 78.49% C, 8.41% H and rest is nitrogen. 0.369 g of its chloroplatinate gave 0.1156 g platinum. Calculate molecular formula of the base.

Ans. Determination of empirical formula

Element	*Percentage*	*At. mass*	*Relative no. of atoms*	*Simplest ratio*
C	78.49	12	78.49/12 = 6.54	7
H	8.41	1	8.41/1 = 8.41	9
N	13.10	14	13.10/14 = 0.93	1

The empirical formula is C_7H_9N.

Empirical formula weight = 84 + 9 + 14 = 107

Determination of molecular mass

$$2B + 410 = \frac{0.369}{0.1156} \; 02150 \times 195 = 622.4$$

or $$2B = 212.4$$

or $$B = 106.2$$

Since the base is monoacidic, the molecular mass = 106.2

$$\frac{\text{Molecular mass}}{\text{Empirical formula mass}} = \frac{106.2}{107} \simeq 1$$

Hence molecular formula is the same as empirical formula *viz.*, C_7H_9N.

Q. 41. 0.246 g of an organic compound gave 0.198 g of CO_2 and 0.1014 g of H_2O on complete combustion. 0.37 g of this compound gave 0.638 g of AgBr. What is the molecular formula of the compound if its V.D. is 54.4?

Ans. *Percentage of carbon*

$$= \frac{12}{44} \times \frac{\text{weight of } CO_2}{\text{wt. of compound}} \times 100$$

$$= \frac{12}{44} \times \frac{0.198}{0.246} \times 100 = 22\%$$

Percentage of hydrogen

$$= \frac{2}{18} \times \frac{\text{weight of } H_2O}{\text{wt. of compound}} \times 100$$

$$= \frac{2}{18} \times \frac{0.1014}{0.246} \times 100 = 4.57\%$$

Percentage of bromine

$$= \frac{80}{188} \times \frac{\text{weight of AgBr}}{\text{wt. of compound}} \times 100$$

$$= \frac{80}{188} \times \frac{0.638}{0.37} \times 100 = 73.43\%$$

Calculation of empirical formula

Element	*Percentage*	*At. mass*	*Relative no. of atoms*	*Simplest ratio*
C	22.00	12	22/12 = 1.83	1.83/.91 = 2
H	4.57	1	4.57/1 = 4.57	4.57/.91 = 5
Br	73.43	80	73.43/80 = 0.91	0.91/.91 = 1

Hence the empirical formula is C_2H_5Br.

Q. 42. 0.1 g of an organic monobasic acid gave 0.2545 g of CO_2 and 0.04428 g H_2O on complete combustion. 0.122 g of the acid required for complete neutralisation 10 cm^3 of N/10 alkali. Determine the molecular formula of the acid.

Ans. *Percentage of carbon*

$$= \frac{12}{44} \times \frac{\text{weight of } CO_2}{\text{wt. of compound}} \times 100$$

$$= \frac{12}{44} \times \frac{0.2545}{0.1} \times 100 = 69.40\%$$

Percentage of hydrogen

$$= \frac{2}{18} \times \frac{\text{weight of } H_2O}{\text{wt. of compound}} \times 100$$

$$= \frac{2}{18} \times \frac{0.04428}{0.1} \times 100 = 4.92\%$$

Percentage of O (by difference)

$$= 100 - (69.40 + 4.92)$$

$$= 25.68\%$$

Calculation of empirical formula

Element	*Percentage*	*At. mass*	*Relative no. of atoms*	*Simplest ratio*	*Whole no. ratio*
C	69.4	12	69.4/12 = 5.78	5.78/1.61 = 3.5	7
H	4.92	1	4.92/1 = 4.92	4.92/1.61 = 3	6
O	25.68	16	25.68/16 = 1.61	1.61/1.61 = 1	2

Hence the empirical formula is $C_7H_6O_2$.

$$\text{Empirical formula weight} = 7 \times 12 + 6 \times 1 + 2 \times 16$$

$$= 84 + 6 + 32 = 122$$

Calculation of mol. weight

$$10 \text{ cm}^3 \text{ of N/10 alkali} \equiv 0.122 \text{ g acid}$$

$$1000 \text{ cm}^3 \text{ of 1 N alkali} \equiv \frac{0.122}{10} \times 1000 \times 10 \text{ g acid}$$

$$= 122 \text{ g acid}$$

or $$\text{Mol. weight} = 122$$

$$n = \frac{\text{Mol. weight}}{\text{Empirical formula weight}} = \frac{122}{122} = 1$$

Hence $$\text{Molecular Formula} = (\text{Empirical Formula})_1$$

$$= C_7H_6O_2$$

Q. 43. An acid of molecular mass 104 contains 34.6% C and 3.85% H. 3.812 mg of this acid required 7.33 cm^3 of 0.01 N NaOH for neutralisation. Suggest a structure for this acid.

Ans. Percentage of C = 34.6

Percentage of H = 3.85

Percentage of O (By difference)

$$= 100 - (34.6 + 3.85)$$
$$= 61.55$$

Calculation of empirical formula

Element	*Percentage*	*At. mass*	*Relative no. of atoms*	*Simplest ratio*	*Whole no. ratio*
C	34.60	12	34.60/12 = 2.88	2.88/2.88 = 1	3
H	3.85	1	3.85/1 = 3.85	3.85/2.88 = 1.33	4
O	61.55	16	61.55/16 = 3.84	3.85/2.88 = 1.33	4

Hence empirical formula of the compound is $C_3H_4O_4$.

Empirical formula weight = 3 × 12 + 4 × 1 + 4 × 16 = 104

Calculation of mol. weight

7.33 cm^3 of 0.01 N NaOH≡.003812 g acid

$$1000 \text{ cm}^3 \text{ of 1 N NaOH} = \frac{.003812}{7.33} \times 1000 \times 100 \text{ g acid}$$

$$= 52 \text{ g acid}$$

Hence Eq. wt. of the acid = 52

Mol. weight (empirical formula weight) = 104

$$\text{Basicity} = \frac{\text{Mol. wt.}}{\text{Eq. wt.}} = \frac{104}{52} = 2$$

Hence the acid is dibasic.

The possible structure is: $CH_2(COOH)_2$ (malonic acid)

PROBLEMS FOR PRACTICE

1. *A chemical compound is found to have the following composition: C = 65.3%, H = 3.40%, N = 9.52% and O = 21.77%.*

   0.882 g of the substance occupy 134.4 ml at N.T.P. Calculate the molecular formula of the compound. **(Ans. $C_8H_5NO_2$)**

2. *A compound contains 57.8% C, 3.6% H and the rest oxygen. Its vapour density is 83 Find its empirical and molecular formula.* **(Ans. $C_4H_3O_2$, $C_8H_6O_4$)**

3. *The percentage analysis of a certain compound is as follows: C = 54.5%, O = 36.4% and H = 9.1%. If its V.D. be 44, find the molecular formula of the compound.* **(Ans. $C_4H_8O_2$)**

4. *Determine the molecular formula of a compound which contains H = 2.1%, C = 12.8%, Br = 85.2% and 1.0 g of the compound occupies 119 ml of N.T.P.* **(Ans. $C_2H_4Br_2$)**

5. *A compound whose mol. wt. is 126, on analysis gave the following results: Na = 36.5%, S = 25.4% and O = 38.1%. Calculate the empirical and molecular formula of the compounds.* **(Ans. Na_2SO_3)**

6. *A compound has the following percentage composition: Na = 19.3%, S = 26.9% and N = 53.8%. Its molecular weight is found to be 238. Derive the molecular formula.*

(**Ans.** $Na_2S_2O_3$)

7. *A compound was found to contain 40% C, 6.67% H and 53.33% oxygen. Its vapour density is 30. Calculate the molecular formula of the compound.* (**Ans.** $C_2H_4O_2$)

8. *On analysis, a substance was found to have the following percentage composition: Na = 29.11, S = 40.5 and 0 = 30.38. Calculate the empirical formula. If the mol. wt. be 158, find out its mol. formula. Name this compound.* (**Ans.** $Na_2S_2O_3$)

9. *0.21 g of an organic substance containing C, H, O and N gave on combustion 0.462 g of CO_2 and 0.1215 g H_2O 0.104 g of it when distilled with caustic soda evolved ammonia which was neutralised by 30 ml of N/20 H_2SO_4. Calculate its empirical formula.* (**Ans.** $C_7H_9NO_2$)

10. *An organic compound on analysis yielded the following results:*
    (a) *On combustion, C and H were found to be 41.37% and 5.75% respectively.*
    (b) *On Kjeldahlising the ammonia evolved from 1.01 g of the substance was neutralised by 11.6 ml of 1 N HCl.*
    (c) *In the Carius estimation of sulphur, 0.2066 g of the substance resulted in the precipitation of 0.5544 g of $BaSO_4$ Determine the empirical formula.*

(**Ans.** C_3H_5NS)

11. *The elementary analysis of organic compound yielded the following results: C = 39.98%, H = 6.72%, O = 53.30%, 0.151 g of the compound on vaporisation displaced 33.8 ml of air measured at 298 K over water and at barometric pressure of 745.0 mm pressure. Determine the empirical formula and molecular formula.*

(**Ans.** CH_2O, $C_4H_8O_4$)

12. *On combustion 0.1579 g of organic compound gave 0.2254 g of CO_2 and 0.0769 g of H_2O. The same amount of the original substance in a Carius estimation on treatment with nitric acid and silver nitrate yielded 0.245 g of silver chloride. Determine the empirical formula of the compound.* (**Ans.** C_3H_5ClO)

3

SYSTEMATIC NOMENCLATURE

Q. 1. Describe the classification of organic compounds.

Ans. The two main categories of these compounds are:

1. Open chain compounds: These are aliphatic compounds and derive their name from Greek word meaning **Fatty**, because the earlier compounds of this category were obtained from fats. In such compounds, carbon atoms are joined to form open chains, straight or branched. A large **number of compounds of everyday use belong to this category. A few examples of this category are:**

CH_3-CH_2OH (Ethyl alcohol) $\quad$ CH_3-COOH (Acetic acid) $\quad$ (Straight chain)

$CH_3-CH(CH_3)-CH_3$ (Isobutane) $\quad$ $CH_3-C(CH_3)(OH)-CH_3$ (Tert.butyl alcohol) $\quad$ (Branched chain)

2. Closed chain or Cyclic compounds or Ring compounds: Organic compounds with closed chains of atoms are termed as **closed chains or cyclic compounds:** When a molecule contains two or more rings, it is called polycyclic. The cyclic compounds may be further sub-divided into two types Viz., Homocyclic and Heterocyclic Compounds.

Homocyclic Compounds

The homocyclic ring compounds are made up of carbon atoms only. These include two types of compounds:

(*i*) **Aromatic compounds:** Compounds having benzene rings of six carbon atoms (*i.e.*, ring with single and double bonds at alternate positions) are called **aromatic compounds.** Aromatic compounds occur chiefly in plants and many of them are sweet smelling. Hence the name (aroma in Greek = sweet smell). Some examples of aromatic compounds are:

Benzene (C_6H_6) $\quad$ Toluene ($C_6H_5CH_3$) $\quad$ Benzoic acid (C_6H_5COOH)

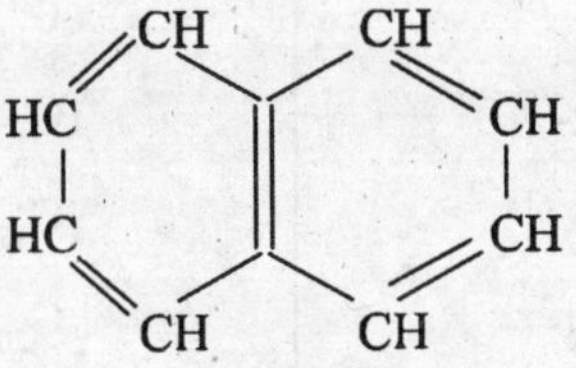

Naphthalene
($C_{10}H_8$)

(*ii*) **Alicyclic compounds:** These are homocyclic ring compounds which behave more like aliphatic than aromatic compounds. These have, therefore, been named as **aliphatic cyclic or alicyclic compounds.** These include polymethylenes such as cyclopropane, cyclobutane, etc. Such compounds contain no double bonds in their rings. Examples are:

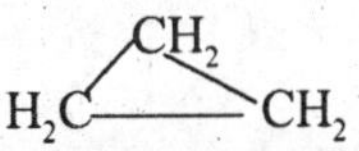

Cyclopropane (C_3H_6) Cyclobutane (C_4H_8) Cyclopentane (C_5H_{10}) Cyclohexane (C_6H_{12})

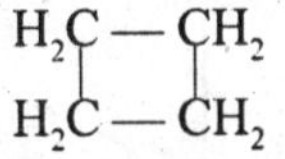

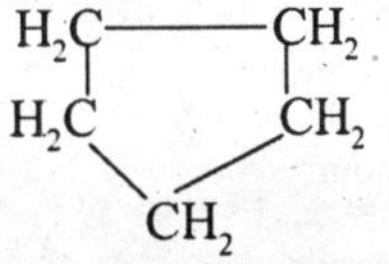

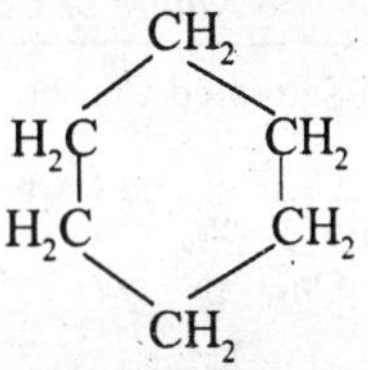

Heterocyclic Compounds

Compounds having elements like O, N and S besides carbon at ring positions are called **heterocyclic compounds.** Examples are:

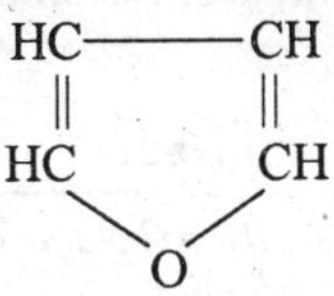

Furan (C_4H_4O) Pyrrole (C_4H_4NH) Thiophene (C_4H_4S) Pyridine (C_5H_5N)

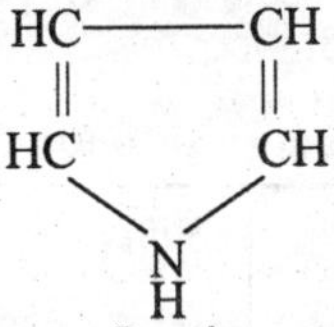

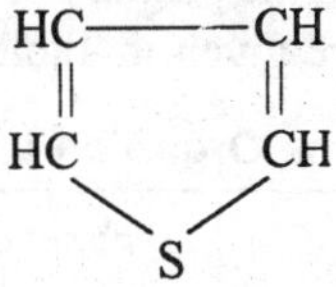

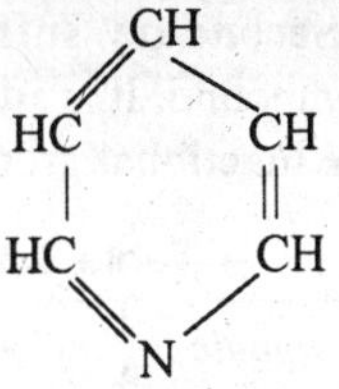

Q. 2. Explain the following terms: (*i*) Word root (*ii*) Primary suffix (*iii*) Secondary suffix (*iv*) Prefix

Ans. (*i*) Word root. The word root represents the number of carbon atoms in the *parent chain.* The parent chain in the compound is selected by following certain rules as discussed later. For the chains upto four carbon atoms special word roots are used, but for the chains containing more than four carbon atoms, Greek numerals are used. The general word root for different aliphatic compounds is **ALK.** The word roots for carbon chains of different lengths are given below in Table 3.1.

Table 3.1. Word Roots for Carbon Chains of Different Lengths

Chain Length	*Word root*	*Chain Length*	*Word root*
C_1	Meth-	C_9	Non-
C_2	Eth-	C_{10}	Dec-
C_3	Prop-	C_{11}	Unidec-

C_4	But-	C_{12}	Dodec-
C_5	Pent-	C_{13}	Tridec-
C_6	Hex-	C_{18}	Octadec-
C_7	Hept-	C_{20}	Icos-
C_8	Oct-	C_{30}	Tricont-

***(ii)* Primary suffix.** Primary suffix is used to represent saturation or unsaturation in the carbon chain. While writing the name, primary suffix is added to the word root. Some of the primary suffixes are given below in Table 3.2.

Table 3.2. Some Primary Suffixes.

Nature of Carbon Chain		*Primary Suffix*
Saturated Carbon Chain		*ane*
Unsaturated Carbon Chains:		
One	C = C bond	*ene*
two	C = C bonds	*adiene*
three	C = C bonds	*atriene*
one	C ≡ C bonds	*yne*
two	C ≡ C bonds	*adiyne*

***(iii)* Secondary suffix.** Secondary suffix is used to indicate the functional group in the organic compound. It is added to the primary suffix by dropping its terminal *e*. Secondary suffixes for various functional groups are given in Table 3.3.

Table 3.3. Some Organic Families and Secondary Suffixes.

Class of organic compound	*General formula*	*Functional Group*	*Suffix*	*IUPAC Name of the family* (*Word root* + *prim. suffix* + *Sec. suffix*)
Alcohols	R–OH	–OH	-ol	Alkanol
Thioalcohols	R–SH	–SH	-thiol	Alkanethiol
Amines	R–NH_2	–NH_2	-amine	Alkanamine
Aldehydes	R–CHO	–CHO	-al	Alkanal
Ketones	R–COR	>CO	-one	Alkanone
Carboxylic acids	R–COOH	–COOH	-oic acid	Alkanoic acid
Amides	R–$CONH_2$	–$CONH_2$	-amide	Alkanamide
Acid chlorides	R–COCl	–COCl	-oyl chloride	Alkanoyl chloride
Esters	R–COOR	–COOR	-oate	Alkyl alknoate
Nitriles	R–C ≡ N	–C ≡ N	-nitrile	Alkane nitrile

If the name of the secondary suffix begins with a consonent, then the terminal 'e' of the primary suffix is not dropped while adding secondary suffix to it.

The terminal 'e' of primary suffix is retained if some numerical prefix like di, tri, etc., is used before the secondary suffix.

***(iv)* Prefix.** Prefix is a part of the name which appears before the word root. Prefixes are used to represent the names of alkyl groups or some functional groups as discussed below:

***(a)* Alkyl groups.** These are formed by the removal of H atom from the alkanes. These are represented by the general formula C_nH_{2n+1} – or R –. Some alkyl groups along with their prefixes are given in Table 3.4.

Table 3.4. Some Alkyl Groups along with their Prefixes

Alkane	*Alkyl Group*	*Prefix*
CH_4	$CH_3–$	Methyl
C_2H_6	$CH_3CH_2–$	Ethyl
C_3H_8	$CH_3CH_2CH_2–$	*n*-Propyl
C_3H_8	$CH_3 - \underset{\displaystyle CH_3}{\underset{\vert}{CH}} -$	isopropyl or (1-Methyl ethyl)

***(b)* Some functional groups** are always indicated by the prefixes instead of secondary suffixes. These functional groups along with their prefixes are listed below in Table 3.5.

Table 3.5. Functional Groups which are always Represented by Prefixes

Functional Group	*Prefix*	*Family*	*IUPAC Name*
$–NO_2$	Nitro	$R–NO_2$	Nitroalkane
–OR	Alkoxy	R–OR	Alkoxyalkane
–Cl	Chloro	R–Cl	Chloroalkane
–Br	Bromo	R–Br	Bromoalkane
–I	Iodo	R–I	Iodoalkane
–F	Fluoro	R–F	Fluoroalkane
–NO	Nitroso	R–NO	Nitrosoalkane

***(c)* In poly functional compound,** i.e., compound with more than one functional groups, one of the functional groups is treated as principal functional group and is indicated by the secondary suffix whereas other functional groups are treated as substituents and are indicated by the prefixes. The prefixes for various functional groups are given in Table 3.6.

Table 3.6. Prefixes for Functional Groups in Poly Functional Compound

Functional Group	*Prefix*	*Functional Group*	*Prefix*
–OH	Hydroxy	–COOR	Carbalkoxy
–CN	Cyano	–COCl	Chloroformyl
–NC	Isocyano	$–CONH_2$	Carbamoyl
–CHO	Formyl	$–NH_2$	Amino
–SH	Mercapto	= NH	Imino
–SR	Alkylthio	>CO	Keto or Oxo
–COOH	Carboxy		

Q. 3. How are prefixes, word root and suffixes arranged in naming an organic compound?

Ans. The prefixes, word root and suffixes are arranged as follows while writing the name

Prefix(es) + Word root + prim. suffix + sec. suffix

The following examples will illustrate

$CH_3-CH_2-CH(CH_3)-CH_2-COOH$ — **Methyl......pent.........an.........oicacid** (Prefix, word Root, primary suffix, secondary suffix)

(3 -Methylpentanoic acid)

$CH_3-C(CH_3)=CH-CHO$ — **Chloro..........but............en...........al** (Prefix, word Root, primary suffix, secondary suffix)

(3 -Chlorobut-2-enal)

Notes. 1. *The name of organic compound may or may not contain prefix or secondary suffix but it always contains word root and primary suffix. For example, the names of straight chain alkanes, alkenes and alkynes do not contain any prefix and secondary suffix.*

$CH_3CH_2CH_2CH_3$ (Butane) — **But..........** (Word root) **.....ane.........** (primary suffix)

$CH_3-CH_2CH=CHCH_3$ (Pent - 2- ene) — **Pent..........** (Word root) **.....ane..........** (primary suffix)

2. *Numerical prefixes such as* ***di, tri,*** *etc., are used before the prefixes or secondary suffixes, if the compound contains more than one similar substituents or similar functional groups.*

$CH_3-C(CH_3)=C(CH_3)COOH$ — **Di.......** (Numerical Prefix) **Methyl...** (prefix) **but.........** (word root) **An............** (prim. suffix) **oic acid........** (Sec.suffix)

(2, 3-dimethylbutanoic acid)

While adding numerical prefix before the secondary suffix, the terminal 'e' of the primary suffix is not removed.

$CH_3-CH(OH)-CH_2(OH)$ — **Prop...** (Word Root) **ane...** (prim. suffix) **di.....** (Numerical prefix) **ol** (sec.suffix)

(Propane 1, 2-diol)

$COOH-COOH$ — **Eth.....** (word root) **ane......** (prim. suffix) **di........** (numerical prefix) **oic acid** (sec.suffix)

(Ethane dioic acid)

3. *In case of* ***alicyclic compounds,*** *a separate secondary prefix cyclo is used immediately before the word root. Word root, here depends upon the number of carbon atoms in the ring.*

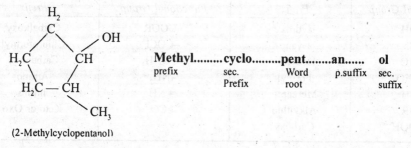

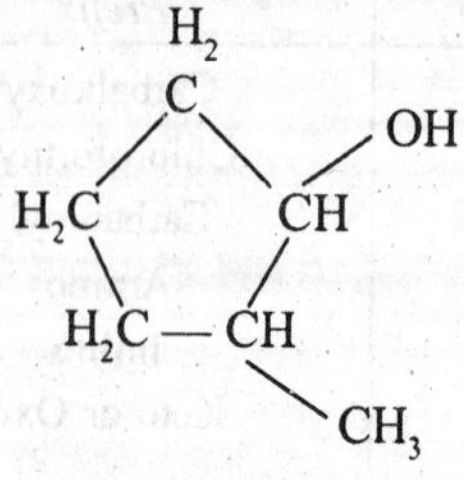

Ring: CH_2 (top), H_2C, $CH(OH)$, H_2C, $CH(CH_3)$ — **Methyl.........cyclo.........pent.........an......** **ol** (prefix, sec. Prefix, Word root, p.suffix, sec. suffix)

(2-Methylcyclopentanol)

Q. 4. Discuss IUPAC nomenclature of organic compounds.

Ans. We describe below the rules for IUPAC nomenclature.

1. Longest possible chain rule. The longest possible continuous chain of carbon atoms containing the main functional group and also as many of the carbon-carbon multiple bond(s) is selected. This chain is taken as the parent chain. The name of the compound is derived from the alkane having the same number of carbon atoms as the parent chain.

For example,

Structure	*Parent chain*	*Alkane from which name of the compound is derived*
$\overset{1}{C}H_3-\overset{2}{C}H_2-\overset{3}{C}H(-\overset{4}{C}H_2-\overset{5}{C}H_2-\overset{6}{C}H_3)-CH_3$	Contains six carbon atoms	Hexane
$\overset{1}{C}H_2=\overset{2}{C}(CH_3)-\overset{3}{C}H_2=\overset{4}{C}H_2$	Contains four carbon atoms	Butane

2. Lowest possible number rule for longest chain containing only substituents/side chains:

(*a*) The longest chain selected is numbered from that end which gives lowest number to the carbon bearing the substituents or side chains.

(*b*) The location of the substituent is indicated by a numeral which precedes the suffix. The numeral is the number of the carbon atom to which the substituents/side chain is attached. For example,

$$\underset{1}{\overset{7}{H_3C}}-\underset{2}{\overset{6}{H_2C}}-\underset{3}{\overset{5}{H_2C}}-\underset{4}{\overset{4}{H_2C}}-\underset{5}{\overset{3}{HC}}(CH_3)-\underset{6}{\overset{2}{H_2C}}-\underset{7}{\overset{1}{CH_3}} \qquad \text{(I)}$$

3-Methyl heptane

The numbering is done from the right side as it gives lowest number (number 3) to the carbon bearing methyl group.

(*c*) If the same substituents/side chains occur more than once, prefixes di, tri, tetra, etc., are used to indicate their number. They are preceded by a number (one number for each group present) to indicate the carbon atom bearing them. For example,

$$\overset{6}{H_3C}-\overset{5}{H_2C}-\overset{4}{H_2C}-\overset{3}{H_2C}-\overset{2}{C}(CH_3)_2-\overset{1}{CH_3} \qquad \text{(II)}$$

2, 2-Dimethylhexane

(*d*) If there are more than one side chains/substituents, they should be arranged in alphabetical order. However, prefixes (di, tri, etc.) or any other prefix that is hyphenated (n-, sec-, tert- etc.) should be ignored while arranging the side chains/substituents alphabetically. But the prefix like cyclo, iso, neo should not be ignored.

$$\overset{6}{H_3C}-\overset{5}{CH_2}-\overset{4}{CH_2}-\overset{3}{CH}(Br)-\overset{2}{CH}(CH_3)-\overset{1}{CH_3} \qquad \text{(III)}$$

3-bromo-2-methylhexane

The methyl is attached to carbon number 2 and bromine atom to carbon number 3. Since bromine comes first in alphabetical order, the exact name is 3-Bromo-2-methyl hexane and not 2-methyl-3-bromo hexane. Thus, the substituents are written in alphabetical order.

(*e*) If identical alkyl groups (side chains), are at equal distances from both the ends, the chain is numbered from the end where there are more side chains/substituents. For example,

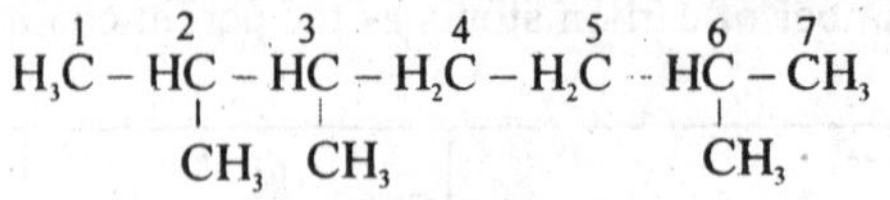

$$\overset{1}{H_3C} - \underset{\displaystyle CH_3}{\overset{2}{HC}} - \underset{\displaystyle CH_3}{\overset{3}{HC}} - \overset{4}{H_2C} - \overset{5}{H_2C} - \underset{\displaystyle CH_3}{\overset{6}{HC}} - \overset{7}{CH_3} \qquad \text{(IV)}$$

2, 3, 6-Trimethylheptane
and not 2, 5, 6-Trimethylheptane

(*f*) If different alkyl groups are in equal position w.r.t. the ends of the chain, the chain is numbered from the end which gives the smaller number to the smaller alkyl substituents. For example,

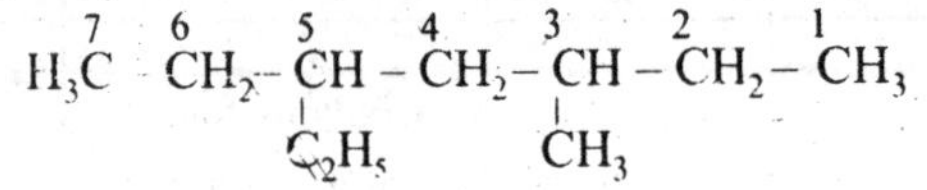

$$\overset{7}{H_3C} - \overset{6}{CH_2} - \underset{\displaystyle C_2H_5}{\overset{5}{CH}} - \overset{4}{CH_2} - \underset{\displaystyle CH_3}{\overset{3}{CH}} - \overset{2}{CH_2} - \overset{1}{CH_3} \qquad \text{(V)}$$

5-Ethyl-3-methylheptane
and not 3-Ethyl-5-methylheptane

Note: A hyphen (-) is always put between the numeral and the side chain/substituent.

3. Lowest sum rule. The longest chain selected is numbered from that end which keeps the sum of numbers used to indicate the position of the side chains/substituents and functional group(s) as small as possible.

In example IV, the lowest sum = 2 + 3 + 6 = 11.

The name 2, 5, 6-Trimethyl heptane is incorrect as the sum of the number = 2 + 5 + 6 = 13.

Similarly in following example VI, the correct name is 2, 4, 4-Trimethyl hexane and not 3, 3, 5-Trimethyl hexane as the sum of the first set of numbers is 10 (2 + 4 + 4) and that of second set of numbers is 11 (3 + 3 + 5).

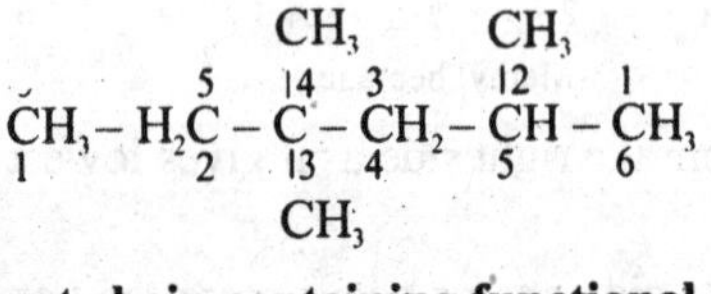

$$\underset{1}{\overset{6}{CH_3}} - \underset{2}{\overset{5}{H_2C}} - \underset{\displaystyle \underset{3}{|}\; CH_3}{\overset{\displaystyle CH_3}{\overset{4\,|\,3}{C}}} - \underset{4}{\overset{3}{CH_2}} - \underset{5}{\overset{\displaystyle CH_3}{\overset{2}{CH}}} - \underset{6}{\overset{1}{CH_3}} \qquad \text{(VI)}$$

4. Lowest number for largest chain containing functional groups. When a compound has a functional group including a multiple bond and one or more side chains/substituents, the lowest number should be given to the functional group even if it violates the lowest sum rule. For example,

(*i*)

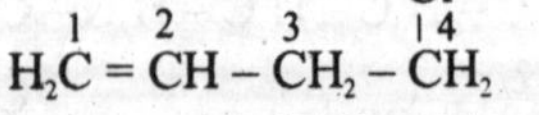

$$\overset{1}{H_2C} = \overset{2}{CH} - \overset{3}{CH_2} - \overset{\displaystyle Cl}{\overset{4}{CH_2}} \qquad \text{(VII)}$$

4-Chlorobutene-1 or 4-Chloro-1-butene

The chain is numbered from left side so as to give lowest number to the double bond.

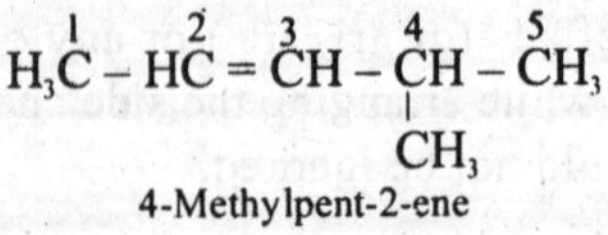

$$\overset{1}{H_3C} - \overset{2}{HC} = \overset{3}{CH} - \underset{\displaystyle CH_3}{\overset{4}{CH}} - \overset{5}{CH_3} \qquad \text{(VIII)}$$

4-Methylpent-2-ene

(*ii*)

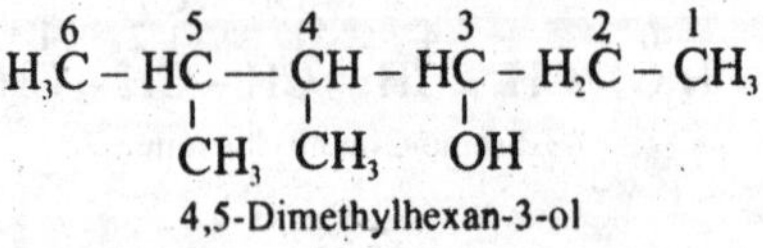

$$\overset{6}{H_3C} - \underset{\displaystyle CH_3}{\overset{5}{HC}} - \underset{\displaystyle CH_3}{\overset{4}{CH}} - \underset{\displaystyle OH}{\overset{3}{HC}} - \overset{2}{H_2C} - \overset{1}{CH_3} \qquad \text{(IX)}$$

4,5-Dimethylhexan-3-ol

(*iii*)

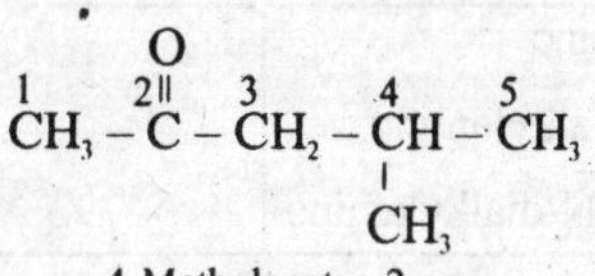

$$\overset{1}{CH_3}-\overset{2}{\underset{}{\overset{O}{\overset{\|}{C}}}}-\overset{3}{CH_2}-\overset{4}{\underset{CH_3}{\underset{|}{CH}}}-\overset{5}{CH_3} \quad (X)$$

4-Methylpentan-2-one

If funtional groups such as $-\overset{H}{\overset{|}{C}}=O$, $-\overset{O}{\overset{\|}{C}}-O-H$

$-\overset{O}{\overset{\|}{C}}-NH_2$, $-\overset{O}{\overset{\|}{C}}-O-R$, $-C\equiv N$, $-\overset{O}{\overset{\|}{C}}-Cl$ etc.

are present in the molecule, the numbering of the parent chain in such cases must start from the carbon atom of the functional group and the number 1 given to the carbon atom of the above functional groups. For example,

(*a*)

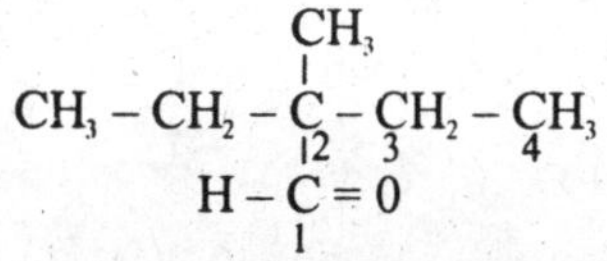

$$CH_3-CH_2-\overset{CH_3}{\overset{|}{\underset{H-\underset{1}{C}=O}{\underset{|2}{C}}}}-\overset{}{\underset{3}{CH_2}}-\underset{4}{CH_3} \quad (XI)$$

2-Ethyl-2-methylbutanal

(*b*)

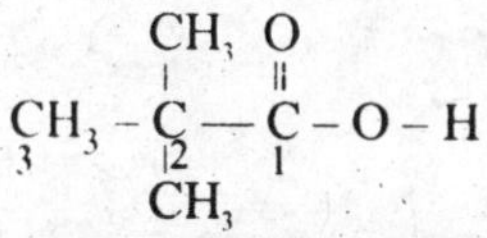

$$\underset{3}{CH_3}-\overset{CH_3}{\overset{|}{\underset{CH_3}{\underset{|2}{C}}}}-\overset{O}{\overset{\|}{\underset{1}{C}}}-O-H \quad (XII)$$

2, 2-Dimethylpropanoic acid

(*c*)

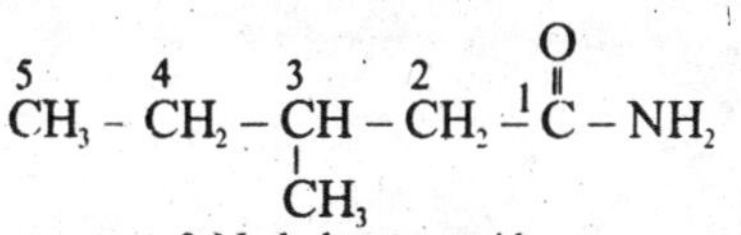

$$\overset{5}{CH_3}-\overset{4}{CH_2}-\overset{3}{\underset{CH_3}{\underset{|}{CH}}}-\overset{2}{CH_2}-\overset{O}{\overset{\|}{\overset{1}{C}}}-NH_2 \quad (XIII)$$

3-Methylpentanamide

5. For longest chain containing two or more different functional groups (Poly functional compounds). When a compound contains two or more different functional groups, one of the functional groups is chosen as the **Principal functional group** and the remaining functional groups (secondary functional groups) are considered as the substituents. The following rules should be noted down:

When two or more functional groups are present in a compound, the principal functional group is chosen according to the following order of priority.

Sulphonic acid > carboxylic acid > acid anhydride > ester > acid chloride > amide > nitrile > isocyanide > aldehyde > hetone > alcohol > amine > alkene > alkyne > halo, nitro, alkoxy, alkyl

The prefixes for secondary functional groups are:

–OH	Hydroxy	–COOH	Carboxy
–OR	Alkoxy	–CN	Cyno
>C = O	Keto or oxo	–COOR	Carbalkoxy
–CHO	Formyl	$-CONH_2$	Carboxamide
$-NH_2$	Amino	COCl	Chloroformyl

–NH	Imino		
–NHR	N-alkylamino		
$–NR_2$	N,N-dialkylamino		

Examples:

(*i*) $H_3CO\overset{2}{C}H_2\overset{1}{C}H_2-OH$

Prefix : Methoxy
Word Root : eth
Pri. suffix : an
Sec. suffix : ol

2-Methoxyethanol

(*ii*)
$$\begin{array}{ccccc} & CH_3 & & CH_3 & \\ & | & & | & \\ CH_3- & C & — & C & -CH_3 \\ & | & & | & \\ & OH & & OH & \end{array}$$

Numerical prefix : di
Prefix : methyl
Word root : but
Pri.suffix : ane
Numerical suffix : di
Suffix : ol

2, 3-Dimethylbutane-2, 3-diol

(*iii*)
$$\overset{6}{C}H_3-\overset{5}{C}H=\overset{4}{C}H-\overset{3}{C}H_2-\overset{\overset{O}{\|}}{\underset{2}{C}}-\overset{1}{C}H_3$$

Word root : Hex
Pri. suffix : ene
See. suffix : one

(*iv*) $HO\overset{3}{C}H_2\overset{2}{C}H_2-\overset{1}{C}OOH$

Prefix : Hydroxy
Word root : prop
Pri.suffix : an
Sec. suffix : oic acid

3-Hydroxypropanoic acid

(*v*)
$$\begin{array}{l} \overset{8}{CH_3}\diagdown \\ \qquad\quad \underset{7}{C}=\underset{6}{C}H-\underset{5}{C}\equiv\underset{4}{C}-\overset{OH}{\underset{3}{C}H}-\overset{CH_3}{\underset{\underset{CH_3}{|}}{\underset{2}{C}}}-\underset{1}{C}H_3 \\ CH_3\diagup \end{array}$$

Prefix : Methyl
Word root : oct
Pri.suffixes : ene, yne
sec. suffix : ol

2, 2, 7-Trimethyloct-6-en-4yn-3-ol

(*vi*)
$$H\overset{4}{C}\equiv\overset{3}{C}-\overset{OH}{\underset{2}{C}H}-\overset{1}{\underset{\underset{O}{\|}}{C}}-O-H$$

2-Hydroxybut-3-ynoic acid

(*vii*) $H_2\overset{4}{C} = \overset{3}{C}H - \underset{2}{\overset{NH_2}{C}}H - \overset{1}{C}H_2OH$

2-Amino but-3-en-l-ol

(*viii*) $\overset{3}{C}H_2 = \overset{2}{C}H - \overset{1}{C}H_2OH$

Prop-2-en-l-ol

6. If the chain of the carbon atom selected as the branched chain also contains multiple bonds or functional groups, the branched chain is separately numbered. This is done in such a way that the carbon of the branched chain which is attached to the parent chain is assigned no. 1. Also the names of such branched chains are written in brackets. For example:

$$\overset{5}{C}H_3 - \overset{4}{C}HOH - \overset{3}{C}H(\overset{1}{C}H_2\overset{2}{C}H_2OH) - \overset{2}{C}H_2 - \overset{1}{C}OOH$$

4-Hydroxy-3 (2-hydroxy ethyl)-pentanoic acid

Select that chain of carbon atoms that includes the maximum number of functional groups. It is numbered from that end which gives lowest number to the principal functional group. Examples:

(*a*) $\overset{3}{C}H_2(OH) - \overset{2}{C}H(OH) - \overset{1}{C}H_2(OH)$

1, 2, 3-Propanetriol or propane-1, 2 3-triol

(*b*) $\overset{5}{C}H_2 = \overset{4}{C}H - \overset{3}{C}H_2 - \overset{2}{C}(=O) - \overset{1}{C}H_3$

Pent-4-en-2-one

(*c*) $\overset{4}{C}H_3 - \overset{3}{C}H = \overset{2}{C}H - \overset{1}{C}OOH$

But-2-enoic acid or 2-Butenoic acid

(*d*) $\overset{4}{C}H_3 - \overset{3}{C}H(OH) - \overset{2}{C}H_2 - \overset{1}{C}OOH$

3-Hydroxybutanoic acid

(*e*) $\overset{4}{C}H_3 - \overset{3}{C}H(OH) - \overset{2}{C}H(CH_3) - \overset{1}{C}HO$

3-Hydroxy-2-methylbutan-1-al

(*f*) $\overset{5}{C}H_3 - \overset{4}{C}H(CH_3) - \overset{3}{C}H = \overset{2}{C}H - \overset{1}{C}H_2OH$

4-Methylpent-2-en-l-ol

(*g*) $\overset{4}{C}H_3 - \overset{3}{C}H(Cl) - \overset{2}{C}H(OH) - \overset{1}{C}H_3$

3-Chlorobutan-2-ol

(*h*) $\overset{5}{C}H_3 - \overset{4}{C}H_2 - \overset{3}{C}H_2 - \overset{2}{C}H(NH_2) - \overset{1}{C}OOH$

2-Aminopentanoic acid

Q. 5. How are alicyclic organic compounds named?

Ans. Names of alicyclic compounds are derived by putting another prefix '*cyclo*' before the word root which depends upon the number of carbon atoms in the ring. The suffixes *ane, ene* or *yne* are written depending upon saturation or unsaturation in the ring, as usual.

If some substituent or functional group is present, it is indicated by some appropriate prefix

Cyclohexane Cyclopentene

or suffix and its position is indicated by numbering the carbon atoms of the ring. *The numbering is done in such a way so as to assign least possible number to the functional group or substituent in accordance with the rules already discussed. Some examples are:*

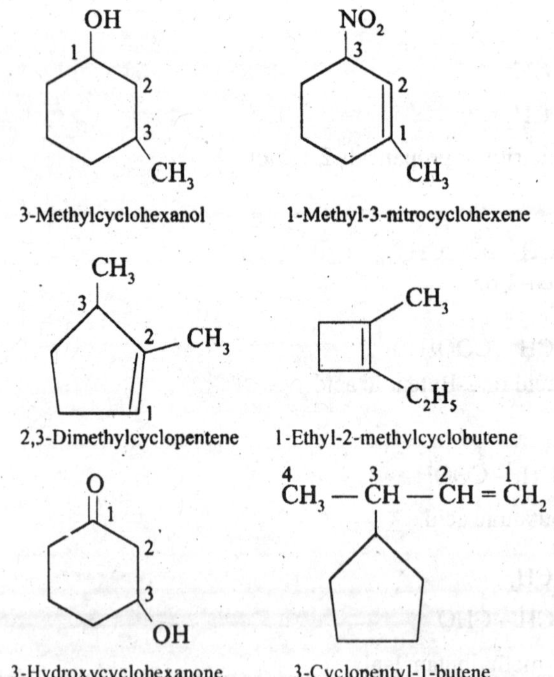

3-Methylcyclohexanol 1-Methyl-3-nitrocyclohexene

2,3-Dimethylcyclopentene 1-Ethyl-2-methylcyclobutene

3-Hydroxycyclohexanone 3-Cyclopentyl-1-butene

Q. 6. Explain bond line notation of organic compounds.

Ans. It is a simple, brief and convenient method of representing organic molecules. In these notations, the bonds between the carbon atoms are represented by lines. A single line (–) represents single bond, two parallel lines (=) represent a double bond and three parallel lines (≡) represent a triple bond. The intersection of lines represents carbon atoms carrying appropriate number of H atoms. For example, 1, 3-butadiene ($CH_2 = CH - CH_2 = CH_2$) can be represented as follows:

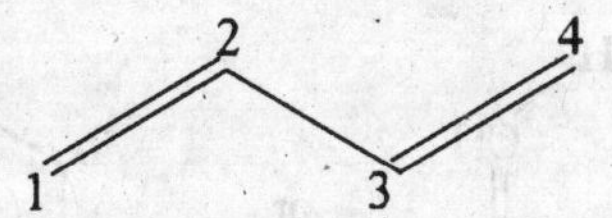

2 4
1 3

Some bond line structures along with their IUPAC names are given below:

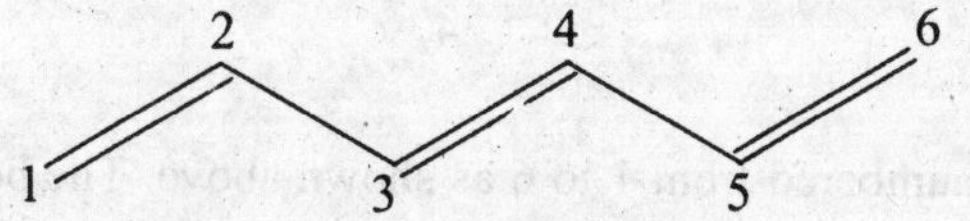

1,3,5-Hexatriene

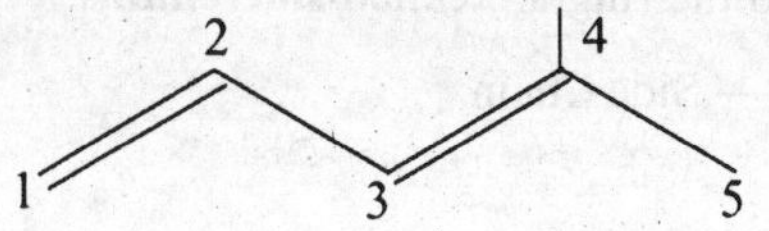

4-Methyl-1,3-pentadiene

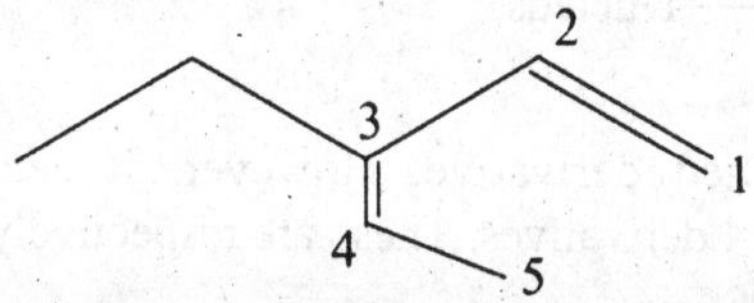

3-Ethyl-1,3-pentadiene

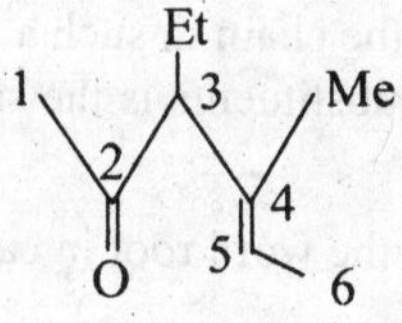

3-Ethyl-4 methylhex-4-en-2-one

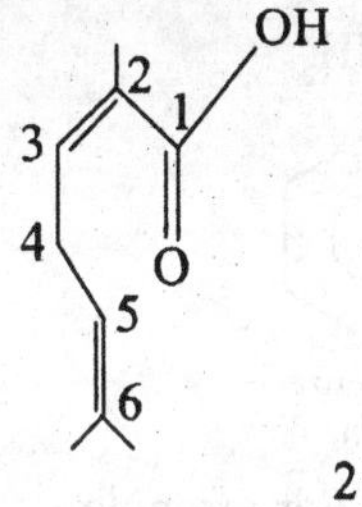

2,6-Dimethyl-2, 5-heptadienoic acid

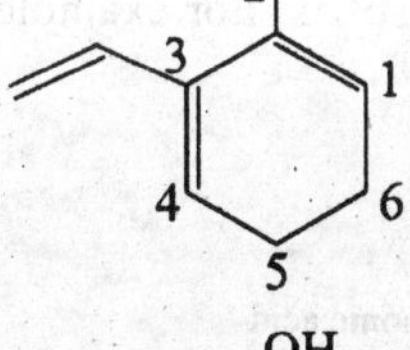

3-Ethenyl-2-methyl-1, 3-hexadiene

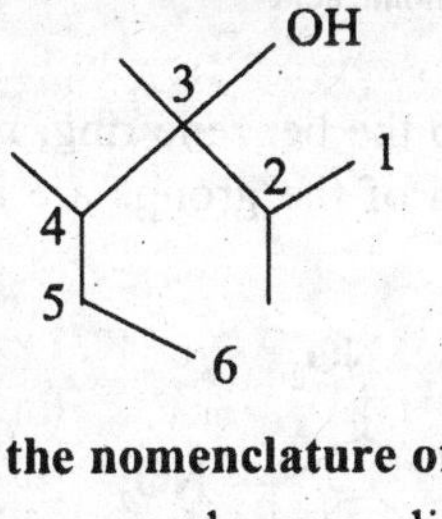

2,3,4-Trimethylhex-1-en-3-ol

Q. 7. Describe the nomenclature of aromatic compounds.

Ans. Aromatic compounds are cyclic compounds which contain one or more benzene type rings. Benzene is a simplest hydrocarbon of aromatic series which has a planar cyclic ring of six carbon atoms having three double bonds in alternate positions as shown below:

Benzene

The carbon atoms of benzene are numbered from 1 to 6 as shown above. The benzene ring is called the nucleus and alkyl groups attached to the ring are called **side chains**.

CH_3 —— Side Chain

—— Nucleus

Benzene forms only one mono substituted derivative. However, it can form three disubstituted derivatives; namely 1, 2; 1, 3 and 1, 4 derivatives. These are respectively called *ortho* (or *o*-), *meta* (or *m*-) and *para* (or *p*-) derivatives.

Tri and poly substituted derivatives are named by numbering the chain in such a way that the parent group gets the lowest number and sum of the positions of substituents is the smallest.

Rule 1. The word root for benzene derivatives is benzene.

Rule 2. The name of substituent group is added as a prefix to the word root in case of mono-substituted benzenes. For example,

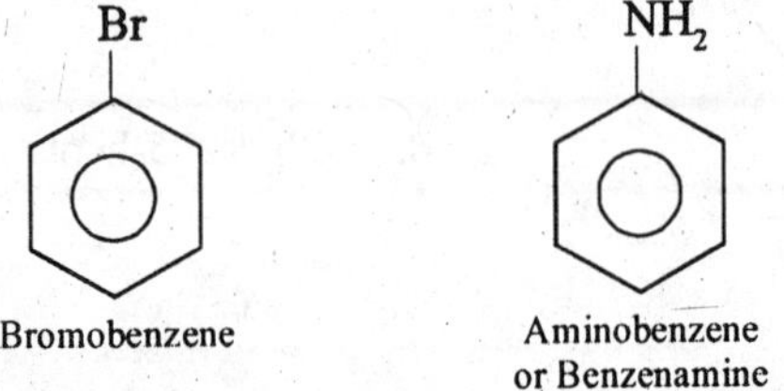

Bromobenzene   Aminobenzene or Benzenamine

In some cases the name of the group is written as suffix. For example,

SO_3H

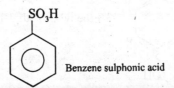

Benzene sulphonic acid

Rule 3. When two similar groups are attached to the benzene ring, numerical prefix *di* is placed before the name of the group, relative portion of the groups are indicated by suitable numbers or by the symbols *o*, *m* or *p*. For example,

1, 3-Dibromobenzene or *m*-Dibromobenzene

1, 2-Dinitrobenzene or *o*-Dinitrobenzene

Rule 4. When two different groups are attached to the ring, the names of both groups are added as prefixes, in alphabetical order, to the word root and their relative positions are indicated. For example,

2-Bromoiodobenzene
or *o*-Bromoiodobenzene

If one of the groups a special name to the compound, than the name of the other group only is written as prefix. For example,

3-Chlorotoluene

4-Nitrophenol

Rule 5. In case of tri-substituted or higher-substituted derivatives, the positions of groups are indicated by numbers.

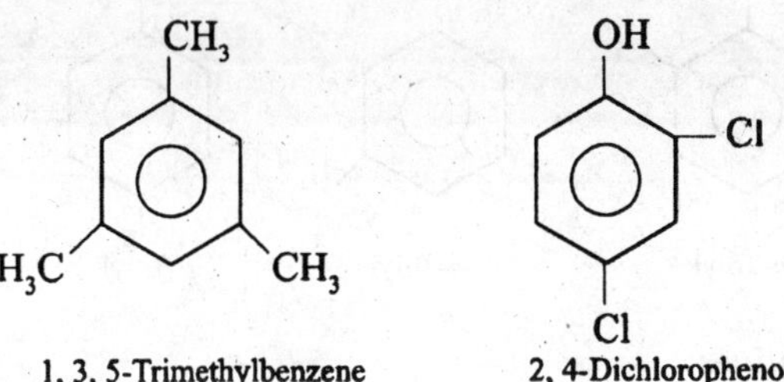

1, 3, 5-Trimethylbenzene

2, 4-Dichlorophenol

Names of Some Aromatic Compounds

1. Hydrocarbons (Arenes)

(a) Hydrocarbons containing condensed rings

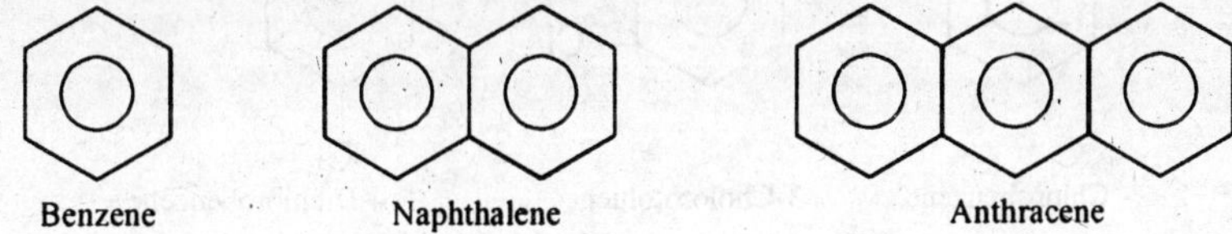

Benzene

Naphthalene

Anthracene

(b) Hydrocarbons containing one ring only

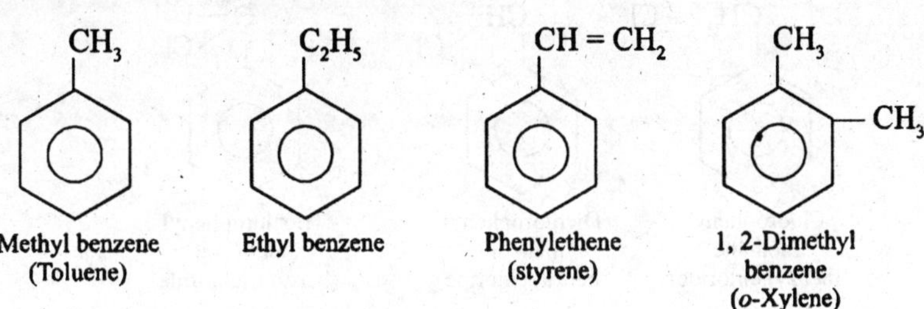

Methyl benzene (Toluene)

Ethyl benzene

Phenylethene (styrene)

1, 2-Dimethyl benzene (*o*-Xylene)

CH_3 CH_3

CH_3 H_3C CH_3

1, 3-Dimethyl benzene (*m*-Xylene)

1, 3, 5-Trimethyl benzene (Mesitylene)

2. Aromatic or Aryl radicals:

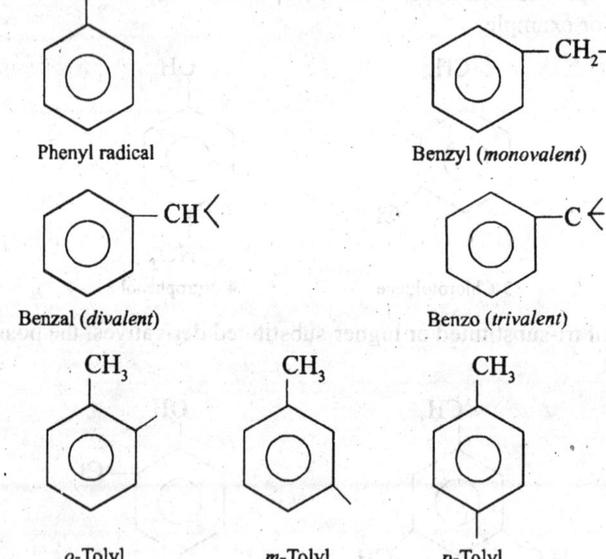

Phenyl radical

Benzyl (*monovalent*)

Benzal (*divalent*)

Benzo (*trivalent*)

o-Tolyl

m-Tolyl

p-Tolyl

3. Halogen Derivatives:

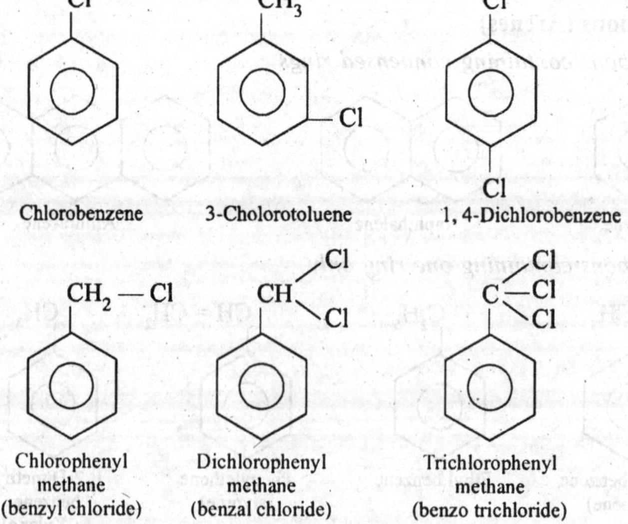

Chlorobenzene

3-Cholorotoluene

1, 4-Dichlorobenzene

Chlorophenyl methane (benzyl chloride)

Dichlorophenyl methane (benzal chloride)

Trichlorophenyl methane (benzo trichloride)

4. Phenols:

Hydroxy benzene (Phenol)   3-Hydroxy toluene (*m*-Cresol)   Phenylmethanol (Benzyl alcohol)

Here benzyl alcohol is not phenol but an aromatic alcohol.

5. Amino Derivatives:

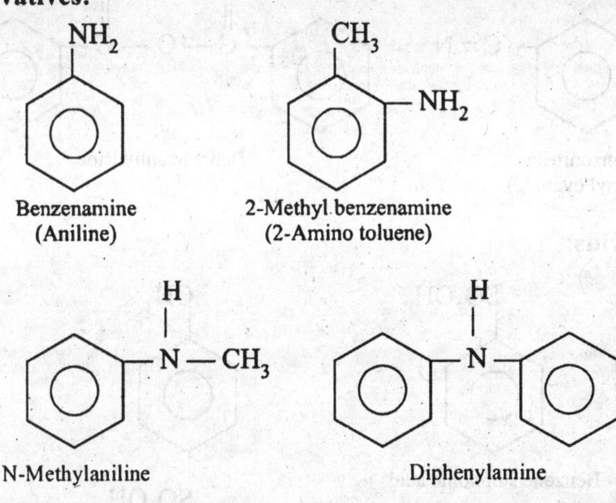

Benzenamine (Aniline)   2-Methyl benzenamine (2-Amino toluene)

N-Methylaniline   Diphenylamine

6. Ketones:

$COCH_3$

Methyl phenyl ketone or Acetophenone   Diphenyl ketone or Benzophenone

7. Aldeydes:

Benzaldehyde (Benzenecarbaldehyde)   2-Hydroxy benzaldehyde (Salicyladehyde)

8. Carboxylic acids:

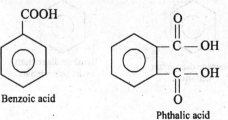

Benzoic acid   Phthalic acid (1, 2-Benzene dicarboxylic acid)

9. Acid derivatives :

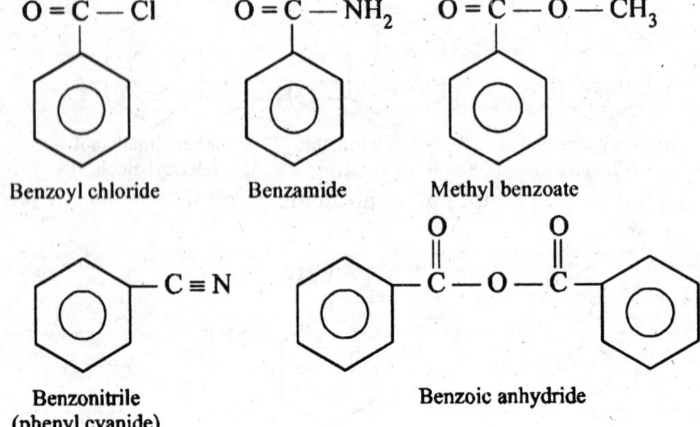

Benzoyl chloride Benzamide Methyl benzoate

Benzonitrile (phenyl cyanide) Benzoic anhydride

10. Sulphonic acids:

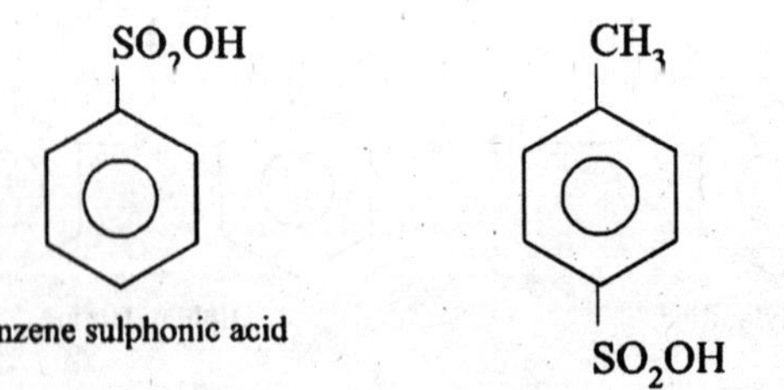

Benzene sulphonic acid

p-Toluene sulphonic acid (PTS)

11. Nitro derivatives:

NO_2, NO_2 — 1, 3-Dinitro benzene

CH_3, O_2N, NO_2, NO_2 — 2, 4, 6-Trinitro toluene (T.N.T.)

OH, O_2N, NO_2, NO_2 — 2, 4, 6-Trinitro phenol (Picric acid)

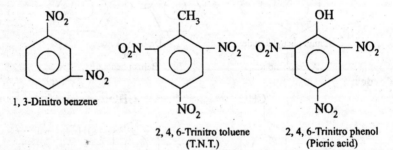

12. Diazonium salts:

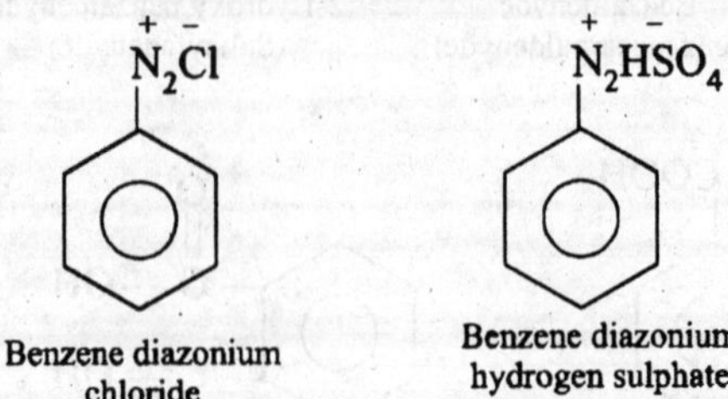

Benzene diazonium chloride

Benzene diazonium hydrogen sulphate

13. Grignard reagents:

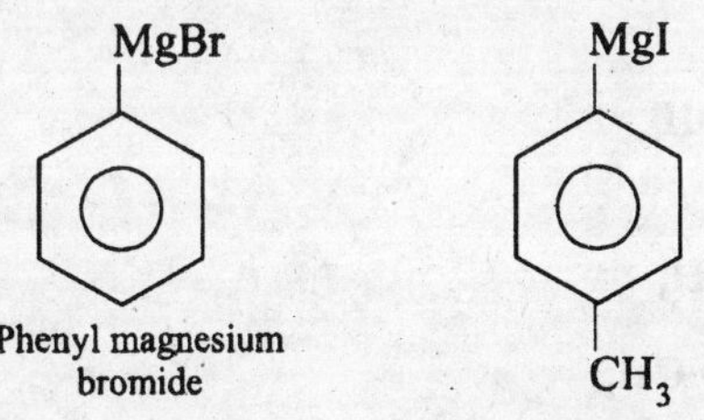

Phenyl magnesium bromide

p-Tolyl magnesium iodide

14. Isocyanides:

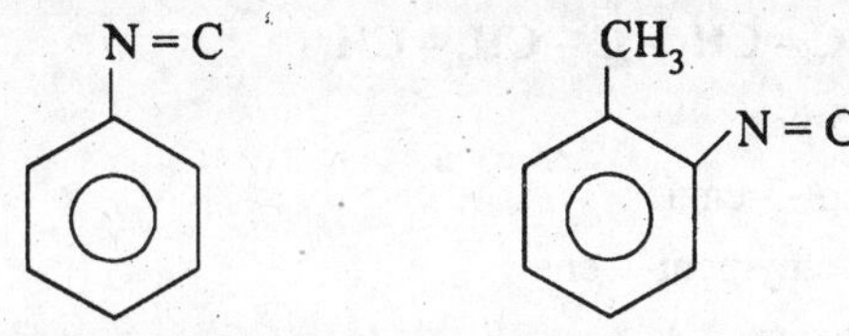

Benzene carbylamine

o-Toluene carbylamine

Q. 8. Give the IUPAC name of the following compounds:

(*i*) $(CH_3CH_2CH_2)_4C$

(*ii*) $CH_2 = CH - CH_2 - CH = CH_2$

(*iii*) $CH_3 - \underset{CH_3}{\underset{|}{CH}} - \underset{CH_3}{\underset{|}{CH}} - \underset{CH_3}{\underset{|}{CH}} - CH_2CH_3$

(*iv*) $OHCH_2 - C \equiv C - CH_2OH$

(*v*) $(CH_3)_2\,CH - CH_2 - COOH$

(*vi*) $CH_2 = CH - CH_2 - \underset{OH}{\underset{|}{CH}} - CH_2 - CH_3$

Ans. (*i*) 4, 4-di *n*-propylheptane

(*ii*) 1, 4-pentadiene

(*iii*) 2, 3, 4-trimethylhexane

(*iv*) But-2-yne-1, 4-diol

(*v*) 3-methylbutanoic acid

(*vi*) Hex-5-en-3-ol

Q. 9. Give IUPAC names of the following :

(*i*) $CH_3 - CH_2 - CH = CH - C \equiv CH$

(*ii*) $CH_3 - (CH_2)_2 - CH = CH - \underset{CH_3}{\underset{|}{CH}} - CHO$

(*iii*) $C_2H_5 - \underset{CH_2}{\underset{\|}{C}} - CH_2 - \underset{CH_3}{\underset{|}{CH}} - NH_2$

(*iv*) $CH_3 - CH = CH - \underset{O}{\underset{\|}{C}} - CH_3$

(*v*) $CH_3 - CO - CH_2 - CH_2 - COOH$

(*vi*) $CH_3-\underset{OH}{\underset{|}{CH}}-CH=\underset{CH_3}{\underset{|}{C}}-CH_3$

(*vii*) $CH_3-CH_2-CH\begin{matrix}\diagup OH\\ \diagdown CH_3\end{matrix}$

(*viii*) $CH_3-\underset{Br}{\underset{|}{CH}}-\underset{NO_2}{\underset{|}{CH}}-CH=CH_2$

(*ix*) $CH_3-CH_2-C\equiv C-CH_2-\overset{O}{\overset{||}{C}}-CH_2-CH_3$

Ans. (*i*) Hex-3-ene-5-yne

(*ii*) 2-Methylhept-3-enal

(*iii*) 4-Amino-2-ethylpent-1-ene

(*iv*) Pent-3-en-2-one

(*v*) 4-Ketopentanoic acid

(*vi*) 4-Methylpent-3-en-2-ol

(*vii*) 1-Methylpropan-1-ol

(*viii*) 4-Bromo-3-nitro pent-1-ene

(*ix*) Oct-5-yn-3-one

Q. 10. Write IUPAC names of the following:

(*i*) $HOOC-CH_2-\underset{Br}{\underset{|}{CH}}-COOH$

(*ii*) $HC\equiv C-CH_2-CH=CH_2$

Ans. (*i*) 2-Bromobutane-1, 4-dioic acid

(*ii*) Pent-1-en-4-yne

Q. 11. Write IUPAC names of the following:

(*i*) $(CH_3)_2\,CH-N(CH_3)_2$

(*ii*) $O_2N-CH_2-\underset{OCH_3}{\underset{|}{CH}}-CH_2-COOCH_2-CH_3$

(*iii*) $H-\underset{O}{\underset{||}{C}}-\underset{CH_3}{\underset{|}{CH}}-CH_2-\underset{O}{\underset{||}{C}}-OH$

(*iv*) $H_3C-\underset{CH_2-CH_2-CH_3}{\underset{|}{N}}-CH_2-CH_3$

Ans. (*i*) Isopropyl dimethylamine

(*ii*) Ethyl-3-methoxy-4-nitrobutanoate

(*iii*) 3-Formyl-3-methylpropanoic acid

(*iv*) N-Ethyl-N-methyl-1-aminopropane

Q. 12. Wrtie IUPAC names of the following:

(*i*) $CH_2\,(COOH)\,CH\,(COOH)\,CH_2COOH$

(*ii*) [benzene ring with two adjacent substituents] —COOH, —COOH

Ans. (*i*) 3-Carboxypentane-1, 5-dicarboxylic acid

(*ii*) Benzene-1, 2-dicarboxylic acid

Q. 13. Write IUPAC names of the following:

(*i*) $Cl-CH_2-CH=CH-\underset{\underset{OCH_3}{|}}{CH}-CHO$

(*ii*) $HO-\overset{\overset{CH_3}{|}}{CH}-\underset{\underset{Br}{|}}{CH}-\overset{\overset{NH_2}{|}}{CH}-COOCH_3$

(*iii*) $CH_3-\underset{\underset{CH_3-CH-CH_3}{|}}{CH}-CH_2-\overset{\overset{Cl}{|}}{\underset{\underset{OCH_3}{|}}{C}}-CH_3$

(*iv*) [cyclobutane ring] 1 COOH; 2 OC_2H_5; 3 NH_2

Ans. (*i*) 5-Chloro-2-methoxypent-3-enal

(*ii*) Methyl 2-amino-3-bromo-4-hydroxypentanoate

(*iii*) The compound can be written as

$$CH_3-\overset{\overset{CH_3}{|}}{CH}-\overset{\overset{CH_3}{|}}{CH}-CH_2-\overset{\overset{Cl}{|}}{\underset{\underset{OCH_3}{|}}{C}}-CH_3$$

2-Chloro-2-methoxy-4,5 dimethylhexane

(*iv*) 3-Amino-2-ethoxycyclobutanoic acid

Q. 14. Write the structural formulae of the following compounds:

(*i*) **2-Methyl-3-pentynoic acid**

(*ii*) **3-Ethyl-5-hydroxy-3-hexenal**

(*iii*) **2-Methyl-1, 3-butadiene**

(*iv*) **Hept-1-en-4-yne**

(*v*) **Pent-3-enoic acid**

(*vi*) **1-Amino-4-methylpentan-2-one**

Ans. (*i*) $\overset{5}{C}H_3-\overset{4}{C}\equiv\overset{3}{C}-\overset{2}{\underset{\underset{CH_3}{|}}{C}}H-\overset{1}{C}OOH$

(*ii*) $\overset{6}{C}H_3-\overset{5}{\underset{\underset{OH}{|}}{C}}H-\overset{4}{C}H=\overset{3}{\underset{\underset{C_2H_5}{|}}{C}}-\overset{2}{C}H_2-\overset{1}{C}HO$

(*iii*) $\overset{4}{C}H_2 = \overset{3}{C} - \overset{2}{C}H = \overset{1}{C}H_2$ (with CH_3 on C-3)

(*iv*) $\overset{7}{C}H_3 - \overset{6}{C}H_2 - \overset{5}{C} = \overset{4}{C} - \overset{3}{C}H_2 - \overset{2}{C}H = \overset{1}{C}H_2$

(*v*) $\overset{5}{C}H_3 - \overset{4}{C}H = \overset{3}{C}H - \overset{2}{C}H_2 - \overset{1}{C}OOH$

(*vi*) $\overset{5}{C}H_3 - \overset{4}{C}H - \overset{3}{C}H_2 - \overset{2}{C}(=O) - \overset{1}{C}H_2NH_2$ (with CH_3 on C-4)

Q. 15. Give the IUPAC names of:

(*i*) $CH_3 - CH_2 - CH(CH_3) - CH_2 - CH(CH_2 - CH_2 - CH_3) - CH(CH_3) - CH_3$

(*ii*) $CH_3 - C(CH_3)_2 - CH(CH(CH_3) - CH_3) - C(CH_3) = CH_2$

(*iii*) $CH_3 - (CH_2)_3 - C \equiv C - (CH_2)_3 - CH_3$

(*iv*) $CH_2 = CH - C(CH_3)_2 - CH = CH - CH(CH_3)(CH_3)$

(*v*) $CH_3 - C \equiv C - CH(CH_3) - CH_3$

Ans. (*i*) $\overset{1}{C}H_3 - \overset{2}{C}H_2 - \overset{3}{C}H(CH_3) - \overset{4}{C}H_2 - \overset{5}{C}H(-\overset{1}{C}H(CH_3) - \overset{2}{C}H_3) - \overset{6}{C}H_2 - \overset{7}{C}H_2 - \overset{8}{C}H_3$ ⟵ Side chain

5-(1-methyl ethyl)-3-methyloctane

(*ii*) $\overset{5}{C}H_3 - \overset{4}{C}(CH_3)_2 - \overset{3}{C}H - \overset{2}{C}(CH_3) = \overset{1}{C}H_2$, with side chain $CH(CH_3) - CH_3$ on C-3

Side chain

3-(1-methyl ethyl)-2, 4, 4 Trimethylpent-1-ene

(*iii*) 5-Decyne

(*iv*) 3, 3, 6-Trimethyl-1, 4-heptadiene

(*v*) 4-Methylpent-2-yne

Q. 16. Write the structural formulae of the following compounds:

(*i*) **2-Methyl pent-2-ene-1-ol**

(*ii*) **Hex-1-en-4-yne**

(*iii*) **4-Cyano-3-methoxybutanoic acid**

(*iv*) **4-Amino-2-ethyl-2-pentenal**

(*v*) **3, 5-Octadiene**

(*vi*) **4-Chloro-2-isopropyl-3-methylcyclopentanone**

Ans. (*i*)
$$\overset{5}{C}H_3-\overset{4}{C}H_2-\overset{3}{C}H=\underset{\displaystyle CH_3}{\overset{2}{C}}-\overset{1}{C}H_2OH$$

(*ii*)
$$\overset{6}{C}H_3-\overset{5}{C}\equiv\overset{4}{C}-\overset{3}{C}H_2-\overset{2}{C}H=\overset{1}{C}H_2$$

(*iii*)
$$NC-\overset{4}{C}H_2-\underset{\displaystyle OCH_3}{\overset{3}{C}H}-\overset{2}{C}H_2\overset{1}{C}OOH$$

(*iv*)
$$\overset{5}{C}H_3-\underset{\displaystyle NH_2}{\overset{4}{C}H}-\overset{3}{C}H=\underset{\displaystyle C_2H_5}{\overset{2}{C}}-\overset{1}{C}HO$$

(*v*)
$$\overset{1}{C}H_3-\overset{2}{C}H_2-\overset{3}{C}H=\overset{4}{C}H-\overset{5}{C}H=\overset{6}{C}H-\overset{7}{C}H_2-\overset{8}{C}H_3$$

(*vi*) Cyclopentanone ring: C=O at C-1; at C-2 a $CH(CH_3)_2$ group (CH bearing two CH_3); Cl on each of the two lower ring carbons.

O, CH, CH_3, CH_3, Cl, Cl

Q. 17. Write IUPAC names of the following:

(*i*)
$$CH_3-\underset{\displaystyle \overset{\|}{O}}{C}-CH_2-\underset{\substack{|\\ \displaystyle C=O\\ |\\ \displaystyle H}}{CH}-CH_3$$

(*ii*)
$$CH_3-\underset{\displaystyle C_2H_5}{CH}-CH=CH-\underset{\displaystyle OH}{CH}-CH_3$$

Ans. (*i*)
$$\overset{5}{C}H_3-\underset{\displaystyle O}{\overset{4}{C}}-\overset{3}{C}H_2-\underset{\substack{|\\ \displaystyle {}^{2}C=O\\ |\\ \displaystyle H}}{\overset{2}{C}H}-CH_3$$

2-methyl-4-ketopentanal

(*ii*)
$$CH_3-\underset{\substack{|\\ \displaystyle {}^{6}CH_2\\ |\\ \displaystyle {}^{7}CH_3}}{CH}-CH=CH-\underset{\displaystyle OH}{CH}-CH_3$$

5-methylhept-3-en-2-ol

Q. 18. Write IUPAC names of the following:

(*i*)
$$CH_3-\underset{\displaystyle H-C=O}{CH}-CH_2-C\begin{matrix}{}^{\nearrow}O\\ \searrow O-H\end{matrix}$$

(*ii*) $CH_3-C(=O)-CH_2-C(=O)-CH_2-CH_3$

Ans. (*i*) $\overset{4}{C}H_3-CH(CHO)-\overset{2}{C}H_2-\overset{1}{C}(=O)-O-H$ (side chain: $H-C=O$)

3-Formylbutanoic acid

(*ii*) $\overset{1}{C}H_3-\overset{2}{C}(=O)-\overset{3}{C}H_2-\overset{4}{C}(=O)-\overset{5}{C}H_2-\overset{6}{C}H_3$

Hexane-2, 4 dione

Q. 19. Write IUPAC names of the following:

(*i*) Benzene ring bearing $CH(CH_3)_2$ (top), Cl and OH (bottom)

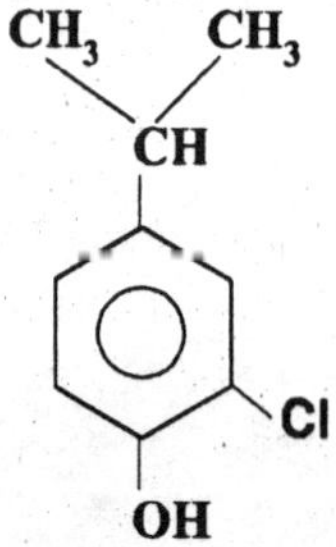

(*ii*) $CH_3-CH_2-CH(CH_2CH_3)-CH(OH)-C(=O)-CH_3$

Ans. (*i*) Benzene ring: C-1 OH, C-2 Cl, C-4 $CH(CH_3)_2$

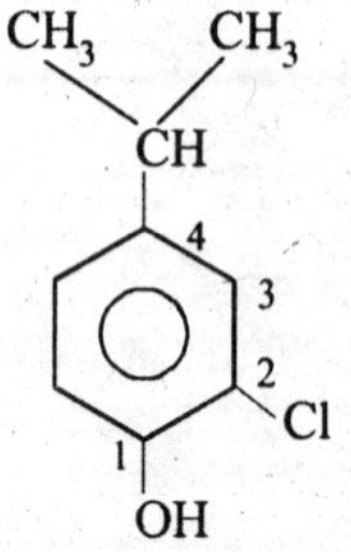

2-Chloro-4-isopropyl phenol

(*ii*) $CH_3-CH_2-\overset{4}{C}H(\overset{5}{C}H_2\overset{6}{C}H_3)-\overset{3}{C}H(OH)-\overset{2}{C}(=O)-\overset{1}{C}H_3$

4-Ethyl-3-hy[illegible]oxy-hexan-2-one

Q. 20. Write IUPAC names of the following:

(*i*) $CH_3-CH(CH_3)-C(CH_2-CH_3)=CH_2$ (*ii*) $CH_3-CH(H-C=O)-CH_2-COOH$

Ans. (*i*) $\overset{4}{C}H_3-\overset{3}{C}H(CH_3)-\overset{2}{C}(CH_2-CH_3)=\overset{1}{C}H_2$ (*ii*) See Q. 18.

2 Ethyl-3-methylbut-2-ene

Q. 21. Write IUPAC names of the following:

(*i*) $CH_2(COOH)-CH(COOH)-CH_2(COOH)$ (*ii*) $CH_3-C(CH_3)=C(CH_3)-COOC_2H_5$

Ans. (*i*)

$$\overset{4}{CH_2}-\overset{3}{CH}-\overset{2}{CH_2}$$
$$\overset{5}{COOH}\quad COOH\quad COOH$$

3-Carboxypentane-1, 5-dioic acid

(*ii*)

$$CH_3-\underset{}{\overset{CH_3}{C}}=\overset{CH_3}{C}-COOC_2H_5$$

Ethyl-2-dimethylbut-2-enoate

Q. 22. Write systematic names for

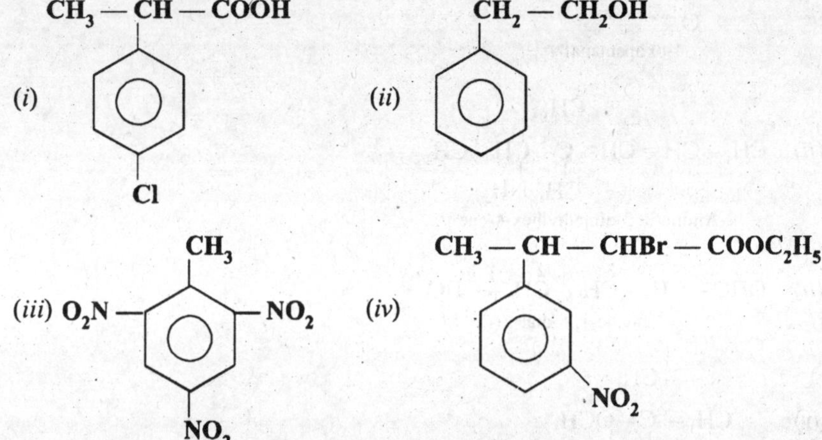

(*i*) $CH_3-CH-COOH$ with 4-chlorophenyl on CH (*ii*) CH_2-CH_2OH with phenyl on CH_2

(*iii*) O_2N, CH_3, NO_2, NO_2 substituted benzene (*iv*) $CH_3-CH-CHBr-COOC_2H_5$ with 3-nitrophenyl on CH

Ans. (*i*) 2-(4-Chorophenyl) propanoic acid. (*ii*) 2-phenylethanol

(*iii*) 2, 4, 6-trinitrotoluene (*iv*) Ethyl-2-bromo-3-(3-nitrophenyl) butanoate

Q. 23. (*a*) Write IUPAC names of the following :

(*i*) $\mathbf{Br-CH_2-CH=CH-\overset{O}{\overset{\|}{C}}-CH_3}$ (*ii*) $\mathbf{CH_2=CH-C\equiv CH}$

(*b*) Draw Structural formulate of the following compounds :

(*i*) 4-Tert. butyl-5-isopropyl decane

(*ii*) 2, 4-Dimethyl-3-ethylhexan-2-ol

Ans. (*a*) (*i*) $Br-\underset{5}{CH_2}-\underset{4}{CH}=\underset{3}{CH}-\underset{2}{\overset{O}{\overset{\|}{C}}}-\underset{1}{CH_3}$

5-Bromopent-3-en-2-one

(*ii*) $CH_2=CH-C\equiv CH$

But-1-en-3-yne

(*b*) (*i*)

$$\overset{10}{CH_3}-\overset{9}{CH_2}-\overset{8}{CH_2}-\overset{7}{CH_2}-\overset{6}{CH_2}-\overset{5}{\overset{CH_3-CH-CH_3}{CH}}-\overset{4}{\underset{CH_3-\underset{CH_3}{C}-CH_3}{CH}}-\overset{3}{CH_2}-\overset{2}{CH_2}-\overset{1}{CH_3}$$

(*ii*)

$$\overset{6}{CH_3}-\overset{5}{CH_2}-\overset{CH_3}{\overset{4}{CH}}-\underset{C_2H_5}{\overset{3}{CH}}-\underset{OH}{\overset{CH_3}{\overset{2}{C}}}-\overset{1}{CH_3}$$

Q. 24. Give IUPAC names for the following :

(*i*) $\mathbf{CH_3-\underset{O}{\underset{\|}{C}}-CH_2-CH_2-CHO}$

(*ii*) $\mathbf{CH_3-CH=CH-\underset{CH_3}{\overset{CH_3}{C}}-\underset{NH_2}{CH}-CH_3}$

(*iii*) $OHC - CH_2 - CH_2 - CH_2 - CHO$ (*iv*) $CH_3 - \underset{CH_3}{\overset{CH_3}{|}}\!\!\!C - O - CH_3$

Ans. (*i*) $\overset{5}{C}H_3 - \overset{4}{\underset{\underset{O}{\|}}{C}} - \overset{3}{C}H_2 - \overset{2}{C}H_2 - \overset{1}{C}HO$

4-oxopentanal-1

(*ii*) $CH_3 - CH = CH - \underset{CH_3}{\overset{CH_3}{C}} - \underset{NH_2}{CH} - CH_3$

2-Amino-3, 3-dimethylhex-4-ene

(*ii*) $O\overset{5}{H}C - \overset{4}{C}H_2 - \overset{3}{C}H_2 - \overset{2}{C}H_2 - \overset{1}{C}HO$

Pentane-1, 5-dial

(*iv*) $CH_3 - {}^{2}\underset{{}^{3}CH_3}{\overset{{}^{1}CH_3}{C}} - OCH_3$

2-methoxy-2-methylpropane

Q. 25. Write structural formulae of the following compounds:

(*i*) **Hexanedioic acid** (*ii*) **2-N, N-Dimethylaminobutane**

(*iii*) **Pent-3-ynal** (*iv*) **2-Ethylbut-1-ene**

Ans. (*i*) $HOOC - CH_2 - CH_2 - CH_2 - CH_2 - COOH$

(*ii*) $CH_3 - CH_2 - \underset{N(CH_3)_2}{CH} - CH_3$

(*iii*) $CH_3 \equiv C - CH_2CHO$

(*iv*) $CH_3 - CH_2 - \underset{C_2H_5}{C} = CH_2$

Q. 26. Write IUPAC names of the following :

(*a*) $CH_2 = CH - CH_2 - \overset{\overset{O}{\|}}{C} - CH_3$

(*b*) $CH_3 - C \equiv C - \overset{CH_3}{CH} - CH_3$

(*c*) $CH_3 - \overset{OH}{CH} - CH_2 - \overset{\overset{O}{\|}}{C} - OH$

Ans. (*a*) 4-Penten-2-one

(*b*) 4-Methylpent-2-yne

(*c*) 3-Hydroxybutanoic acid

Q. 27. Write down IUPAC names of the following :

(*i*) **OHC – COOH**

(*ii*) $\triangleright$ $- O - CH_3$ (*ii*) $H_3C - \underset{\large C_2H_5}{\overset{|}{C}} = CH - CH_3$

Ans. (*i*) Formylmethanoic acid (*ii*) Methoxycyclopropane (*iii*) 3-Ethylbut-2-ene

Q. 28. (*a*) Write IUPAC names of the following formulae

(*i*)

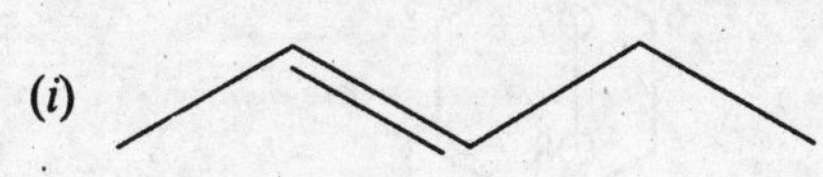

(*ii*) CH_3

(*b*) Write structural formulae of the following:

(*i*) Pent-1, 4-diene (*ii*) 3-hydroxypentanal (*Guwahati 2006*)

Ans. (*a*)

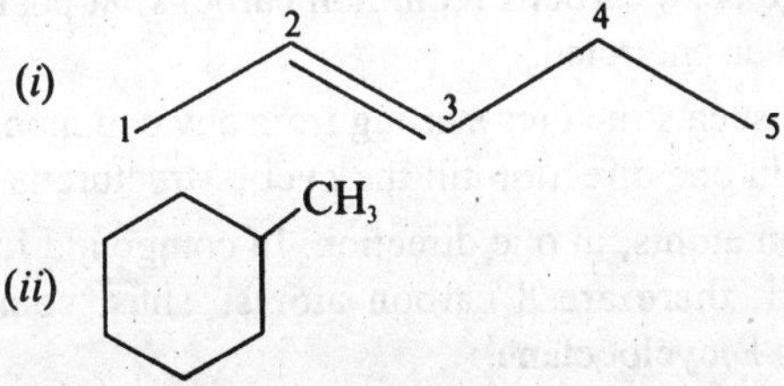

(*i*) Pent-2-ene

(*ii*) Methylcyclohexane

(*b* (*i*) $\overset{1}{H_2C} = \overset{2}{CH} - \overset{3}{CH_2} - \overset{4}{CH} = \overset{5}{CH_2}$

Pent-1, 4-diene

(*ii*) $\overset{5}{CH_3} - \overset{4}{CH_2} - \underset{\large OH}{\overset{3}{CH}} - \overset{2}{CH_2} - \overset{1}{CHO}$

Q. 29. (*a*) Write IUPAC name of the following :

(*i*)

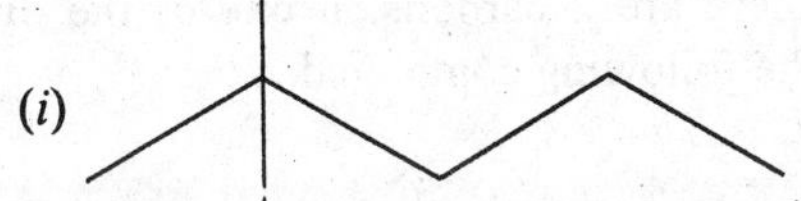

(*ii*) $\triangleright$ $- CH_2 - CH_2 - CH_3$

(*b*) Write structural formulae of the following

(*i*) Pent-1-en-4-yne

(*ii*) But-2-en-1-ol (*Guwahati 2007*)

Ans. (*a*)

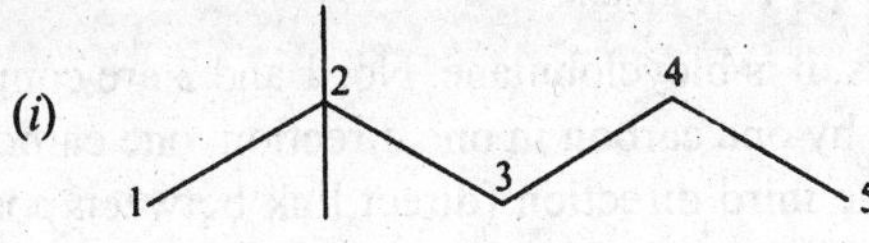

(*i*) 2, 2-dimethylpentane

(*ii*) $\triangleright$ $- CH_2 - CH_2 - CH_3$ *n*-propylcyclopropane

(*b*) (*i*) $\overset{5}{CH} \equiv \overset{4}{C} - \overset{3}{CH_2} - \overset{2}{CH} = \overset{1}{CH_2}$

Pent-1-en-4-yne

(*ii*) $\overset{4}{CH_3} - \overset{3}{CH} = \overset{2}{CH} - \overset{1}{CH_2OH}$

Q. 30. Explain the nomenclature of polycyclic hydrocarbons.

Ans. We come across compounds, containing two or more rings fused together or having two or more carbon atoms in common. These common carbon atoms are called **bridgehead** carbons. Consider, for example, the following bicyclic compounds (containing two rings).

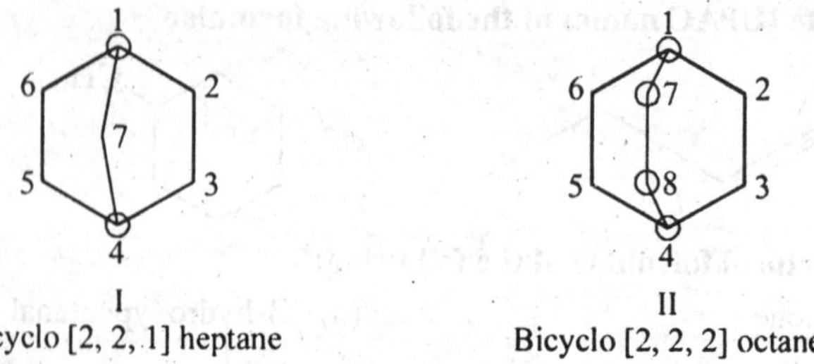

I
Bicyclo [2, 2, 1] heptane

II
Bicyclo [2, 2, 2] octane

Both these compounds have two bridge head carbons (common carbons) at positions 1 and 4. These common carbons have been shown as encircled.

1. We number various carbon atoms in such structures starting from any common carbon and go on numbering all carbon atoms in one direction till the cyclic structure is complete.
2. Then, we number the left out carbon atoms, in one direction. In compound I, there are 7 carbon atoms and in compound II, there are 8 carbon atoms. Thus, compound I is Bicycloheptane and compound II is Bicyclooctane.
3. To name these compounds, we find out, what are the numbers of carbon atoms separating the common carbons, when seen through different routes. Thus, in compound I, there are two carbons through one route, two carbons through the second route and one carbon through the third route, separating the common carbons. This compound will be named as Bicyclo [2, 2, 1] heptane. Here, the numbers 2, 2, 1 represent the number of carbon atoms separating the common carbons in different directions

These numbers in decreasing order are written in *square brackets* after the word bycyclo.

4. Coming to compound II, we find that there are 2 carbons in one of the directions, separating the common points. Consider the following compound.

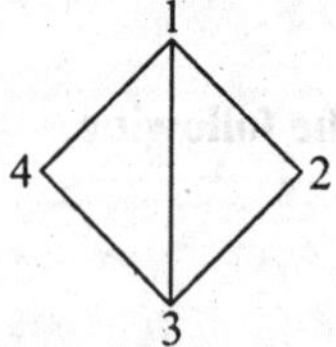

Bicyclo [1, 1, 0] butane

There are 4 carbon atoms in all. Thus, it is bicyclobutane. No. 1 and 3 are common points. These common points are separated by one carbon in one direction, one carbon in second direction and no (zero) carbon in the third direction (direct link between position 1 and 2). Thus, the name of the compound is Bicyclo [1, 1, 0] butane.

Q. 31. Write names of the following bicyclic compounds.

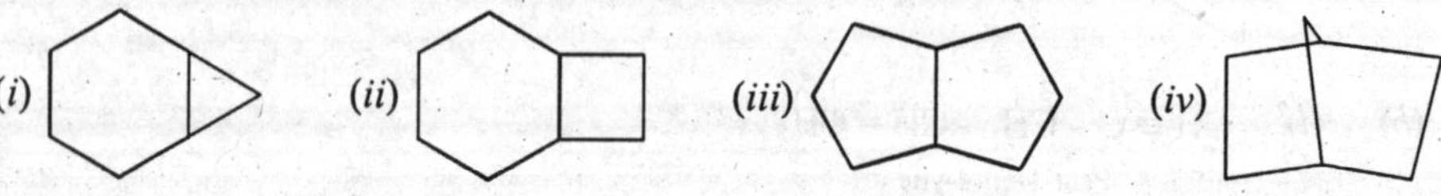

Ans. (*i*) There are two common points. These points are separated by 4, 1 and zero carbon atoms in different directions. Total no. of carbon atoms in the compound is 7. Thus, the name will be Bicyclo [4, 1, 0] heptane.

(*ii*) There are two common carbons in the compound, which are separated by 4, 2 and 0 carbon atoms in different directions. There are a total of 8 carbon atoms. Therefore, it is named as Bicyclo [4, 2, 0] octane.

(*iii*) Following the same procedure, the name of compound (*iii*) is Bicyclo [3, 3, 0] octane.

(*iv*) This compound is the same as compound I discussed earlier.

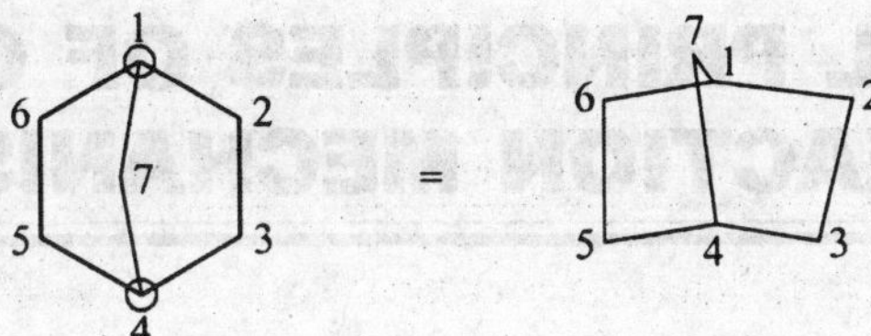

Positions 1 and 4 are common points. These are separated by 2, 2 and 1 carbon atoms in different directions. Therefore, the compound is named as Bicyclo [2, 2, 1] heptane.

Q. 32. Name the following compounds.

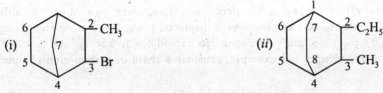

Ans. (*i*) There is a methyl group at position 2 and bromo group at position 3. Therefore, the compound can be named as 3-Bromo-2-methylbicyclo [2, 2, 1] heptane.

(*ii*) There is a methyl group at position 3 and ethyl group at position 2. Therefore, the name of the compound is : 2-Ethyl-3-methyl bicyclo [2, 2, 2] octane.

Q. 33. Write the names of the following compounds.

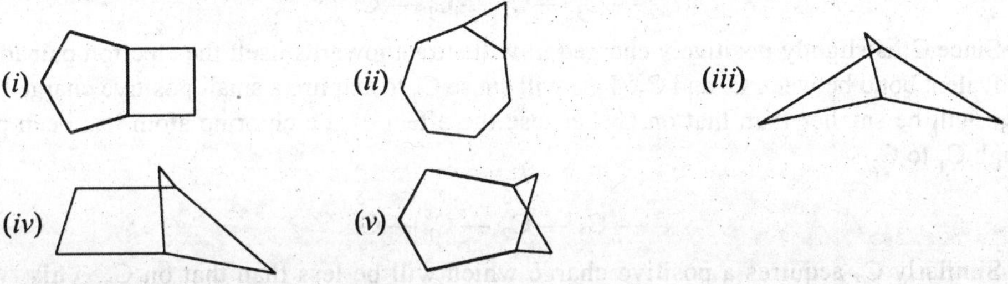

Ans. (*i*) Bicyclo [3, 2, 0] heptane

(*ii*) Bicyclo [5, 1, 0] octane

(*iii*) Bicyclo [1, 1, 1] pentane

(*iv*) Bicyclo [2, 1, 1] heptane

(*v*) Bicyclo [3, 1, 1] heptane

4

GENERAL PRINCIPLES OF ORGANIC REACTION MECHANISM

INDUCTIVE EFFECT

Q. 1. What is inductive effect? *(Marathwada, 2003; Madurai, 2004; Guwahati, 2005; Nagpur, 2008; Patna, 2010)*

Ans. Inductive effect: When a covalent bond ininvolves two atoms of different electronegativities, the shared pair of electrons is displaced towards the more electronegative element. As a result, the more electronegative atom gets a small negative charge and the other atom gets equal small positive charge. For example, consider a chain of carbon atoms joined to a chlorine atom

$$—C_3—C_2—C_1—Cl$$

Chlorine has a greater electronegativity than carbon, therefore the electron pair forming the covalent bond between the chlorine atom and C_1 will be displaced towards the chlorine atom. This causes the chlorine atom to acquire a small negative charge, $-\delta$ and C_1 a small positive charge, $+\delta$ as shown below :

$$—C_3—C_2—\overset{+\delta}{C_1}—\overset{-\delta}{Cl}$$

Since C_1 is slightly positively charged, it will attract towards itself the electron pair forming the covalent bond between C_1 and C_2. This will cause C_2 to acquire a small positive charge but the charge will be smaller than that on C_1 because the effect of the chlorine atom has been passed through C_1 to C_2.

$$—C_3—\overset{+\delta}{C_2}—C_1—\overset{-\delta}{Cl}$$

Similarly C_3 acquires a positive charge which will be less than that on C_2. **This type of electron shift or displacement along a carbon chain is known as inductive effect.** Inductive effect is permanent and decreases rapidly as the distance from the source (Cl) increases. Practically it is almost negligible beyond two carbon atoms from the active atom or group. The important thing to note is that the electron pairs although permanently displaced, remains in the same valency shell. The inductive effect is shown by a line marked with arrow head (→) pointing towards the electron attracting atoms or groups.

Types of Inductive Effect

1. Negative Inductive Effect: If the atom or group of atoms attached to a carbon atom is such that it attracts the shared pair of electrons towards itself it is said to exert negative or –I effect. The decreasing order of the –I effect of some atoms or groups is:

$$NO_2 > CN > COOH > F > Cl > Br > I > OCH_3 > C_6H_5$$

2. Positive Inductive Effect: If the atom or group of atoms attached to the carbon atom is such that it pushes the electrons away from it, it is said to exert positive or +I effect. The decreasing order of + I effect of some groups is:

$$(CH_3)_3C \ > \ (CH_3)_2CH \ > \ CH_3CH_2 \ > \ CH_3$$

For comparison of inductive effect hydrogen (H) is chosen as the standard. Atoms or groups with –I effect are more electron withdrawing than H while with +I effect are less electron withdrawing than H or in other words they more electron releasing than H.

Q. 2. Define inductive effect. Explain with the help of this effect that chloroacetic acid is stronger than acetic acid. *(Kerala 2000)*

Ans. (*i*) For inductive effect see Q. 1.

(*ii*) Chloroacetic acid is stronger than acetic acid because of the inductive effect of chlorine.

$$Cl \longleftarrow \underset{H}{\overset{H}{\underset{|}{\overset{|}{C}}}} \longleftarrow \overset{O}{\overset{||}{C}} \longleftarrow O \longleftarrow H \longrightarrow Cl-\underset{H}{\overset{H}{\underset{|}{\overset{|}{C}}}}-\overset{O}{\overset{||}{C}}-O^- + H^+$$

Chlorine exerts negative inductive effect, as a result of which, the electron displacement takes place away from carboxy hydrogen. This facilitates the removal of hydrogen as proton.

ELECTROMERIC EFFECT

Q. 3. Explain clearly electromeric effect.

(Meerut, 2000; G. Nanakdev, 2000; Awadh 2000)

Ans. It is temporary effect which involves a complete and instantaneous transfer of a shared pair of electrons to one or the other atom joined by multiple bond at the requirement of an attacking reagent.

The electromeric effect is represented by a curved arrow beginning at the original position of the electron pair and ending where the pair has migrated. Let us consider a molecule AB having a double bond between atoms A and B. In the presence of attacking reagent, one of the shared pair of electrons is completely transferred to one of atoms (say B) as shown below :

$$A = B \ \underset{\text{Reagent withdrawn}}{\overset{\text{Reagent added}}{\rightleftharpoons}} \ A^+ — B^-$$

Thus A becomes positively charged while B becomes negativily charged,. The removal of the attacking reagent causes the charged molecule to revert to its original condition.

Electronegativity plays an important role in knowing the direction of the transfer of the shared pair of electrons to one of the atoms.

For example,

(*i*) In a carbonyl group > C = O, present in aldehydes or ketones, the displacement is towards the oxygen atom. This is due to greater electronegativity of oxygen than carbon.

$$>C = O \longrightarrow \ > C^+ - O^- \qquad (-\text{ E effect})$$

(*ii*) In propylene, the displacement of shared electron pair is towards the carbon atom which is away from methyl group. This is due to the inductive effect of methyl group which is electron repelling.

$$\underset{\text{Propylene}}{CH_3 - CH = CH_2} \longrightarrow CH_3 - \overset{+}{C}H - \overset{-}{C}H_2 \qquad (+\text{ E effect})$$

Thus the electromeric effect is of two types. When the electron displacement is away from the group, it is denoted by +E and when the displacement is towards the group, it is denoted by – E.

RESONANCE OR MESOMERIC EFFECT

Q. 4. Describe resonance or mesomeric effect. Give conditions necessary for resonance

(Shivaji, 2004; West Bengal, 2004; Punjab, 2005; Banaras, 2004; Purvanchal 2007; Patna. 2010; Pune, 2010

Ans. Resonance and Resonance effect

Sometimes it is found that a single structural formula cannot satisfactorily explain all the properties of a given compound. In such a case the compound is represented by two or more structural formulae which differ only in the arrangement of electrons. None of the structural formula alone can explain all the observed properties of the compound. The compound is then said to show **resonance.** The various structures are called **resonating structures or canonical structures.** The true structure of the molecule is not represented by any of the resonating structures but is considered to be a resonance hybrid (intermediate) of the various resonating structures.

For example, the most frequently used Lewis structure for carbon dioxide molecule is

$$:\ddot{O}::C::\ddot{O}: \quad \text{or} \quad \underset{\text{I}}{O=C=O}$$

On the basis of this structure, the heat of formation of carbon dioxide from its elements should have been 1463 kJ/mole and C = O bond length should have been 1.21 Å. But the experimental value for the heat of formation of CO_2 is 1588 kJ/mole and C = O bond length is 1.15Å. Thus, structure I for CO_2 cannot explain its properties completely. Structures II and III contain triple bonds which have shorter bond lengths,. But they have shorter bond lengths than 1.15Å. Thus structures II and III also fail to account for all the observed properties of carbon dioxide.

$$:\overset{-}{\ddot{O}}:C:::\overset{+}{O}: \quad \text{or} \quad \underset{\text{II}}{\overset{-}{O}-C\equiv\overset{+}{O}}$$

and

$$:\overset{+}{O}:::C:\ddot{\underset{..}{O}}^{-}: \quad \text{or} \quad \underset{\text{III}}{\overset{+}{O}\equiv C-\overset{-}{O}}$$

In the light of the above concept, CO_2 is regarded as a resonance hybrid of the structures I, II and III, *i.e.*,

$$\underset{\text{I}}{O=C=O} \longleftrightarrow \underset{\text{II}}{\overset{-}{O}-C\equiv\overset{+}{O}} \longleftrightarrow \underset{\text{III}}{\overset{+}{O}\equiv C-O^{-}}$$

Resonance is also called mesomerism.

Conditions for resonance

Main conditions for resonance are:

(*i*) The positions of the nuclei (of each atom) in each structure must be the same.

(*ii*) The number of unpaired electrons in each structure must be the same.

(*iii*) Each structure must have about the same internal energy.

(*iv*) The contributing structures, which involve separation of positive and negative charges are of high energy and hence contribute slightly towards the resonance hybrid.

(*v*) The greater the number of contributing structures, the greater will be the stability of the molecule.

(*vi*) The larger the number of bonds in the contributing structure, the greater is its stability.

Consequences of resonance

(*i*) **Bond length.** The bond length in actual molecule is different from that of any of the contributing structures.

Let us consider the example of benzene molecule. Had I or II been the structure of benzene then it should have shown two bond lengths, *i.e.*, three C = C double bonds should have been 1.34 Å long and three C — C single bonds should have been 1.54 Å long. But the X-ray analysis of benzene shows that all the carbon-carbon bond lengths are identical and equal to 1.39 Å. This can be explained if we consider benzene to be a resonance hybrid of two Kekule's structures I and II.

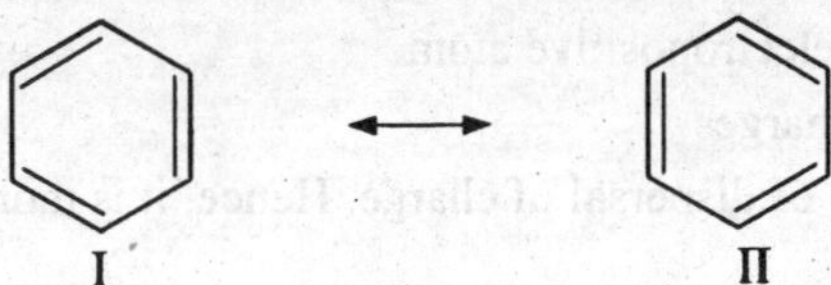

In other words, the real structure of benzene is neither represented by Kekule structure I nor by II but is somewhat in between these two structures. This means that any two adjacent carbon atoms of the benzene molecule are joined neither by a pure single bond nor by a pure double bond. As a result, all the carbon-carbon bond lengths should not only be equal but should also lie in between C = C bond length of 1.34 Å and C — C bond length of 1.54 Å. This is in excellent agreement with the experimental value of 1.39 Å.

(*ii*) **Stability.** A resonance hybrid is always more stable than any of its canonical structures. For example, the observed heat of formation of carbon dioxide is 125 kJ/mole than the value calculated on the basis of structure O = C = O. This implies that the internal energy of CO_2 is less than the calculated value by 125 kJ/mole. In other words the real molecule of CO_2 is 125 kJ/mole more stable than the Lewis structure O = C = O.

The magnitude of stability conferred on a molecule as a result of resonance is expressed in terms of **resonance energy or delocalization energy.** It is defined as the difference between the actual energy of the molecule capable of exhibiting resonance and the energy calculated for the resonating structures.

Q. 5. By what electronic effect can you explain the low reactivity of halogen atom in vinyl bromide.

Or

On the basis of resonance, how will you explain low reactivity of vinyl bromide as compared to ethyl bromide?

Ans. Low reactivity of halogen in vinyl bromide can be explained on account of the phenomenon or resonance or mesomerism.

$$CH_2 = CH - \ddot{\underset{\cdot\cdot}{Br}}: \longleftrightarrow \bar{C}H_2 - CH = \overset{+}{\underset{\cdot\cdot}{Br}}:$$

Halogen compounds generally give nucleophilic substitution reactions, in which the halogen is removed as halide ion.

But in the case of vinyl bromide, resonance takes place as illustrated above. This creates a double bond between carbon and bromine. Removal of bromine thus becomes difficult. Moreover, bromine, acquires a positive charge and hence cannot be substituted by a nucleophile. That is why vinyl bromide shows low reactivity.

Q. 6. Which of the following canonical forms would contribute most towards resonance? Explain

$$\underset{(a)}{CH_2 = CH - CH = CH_2} \longleftrightarrow \underset{(b)}{\overset{-}{C}H_2 - CH = CH - \overset{+}{C}H_2} \longleftrightarrow \underset{(c)}{\overset{-}{C}H_2 - \overset{+}{C}H_2 - CH = CH_2}$$

(*Delhi 2006*)

Ans. The resonance structures fulfilling the following conditions are more stable :

1. Greater number of covalent bonds
2. Less separation of charges
3. Negative charge on electronegative atom
4. Position charge on electropositive atom
5. More dispersal of charges

Structure (*b*) takes care of dispersal of charge. Hence, it is more stable and contributes more towards resonance.

HYPERCONJUGATION

Q. 7. What is meant by hyperconjugation? Why is it also termed no-bond resonance?

(*Kerala, 2001; Manipal, 2002; Punjab, 2003; Nagpur, 2008; Patna, 2010; Andhra, 2010*)

Ans. It is a special type of resonance which involves the overlapping of σ-orbital with a π-orbital or *p*-orbital. Hyperconjugation takes place in the following carbonium ion as depicted

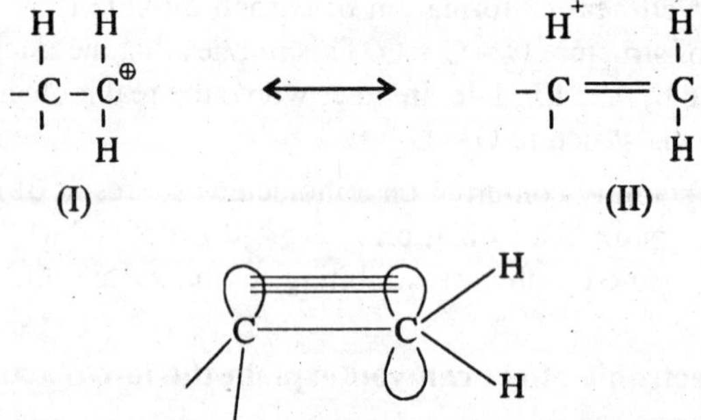

Since in structure (II), there is no bond between H^+ and C, hyperconjugation is also referred to as no-bond resonance.

Q. 8. In what ways a covalent bond can be fissioned and what are its results?

Or

Explain :

(*i*) Homolytic fission

(*ii*) Heterolytic fission

(*Kerala, 2001; Delhi, 2003; Nagpur, 2003; Goa, 2004; Garhwal, 2010*)

Ans. Consider a covalent bond between atoms A and B.

A : B or A — B

The cleavage (or breaking) of this bond can take place in three possible ways depending upon the relative electronegativities of A and B.

(*i*) A : B $\longrightarrow$ A• + B• (A and B of equal electronegativity)

(*ii*) A : B $\longrightarrow$ $\overset{-}{A}: + \overset{+}{B}$ (A more electronegative than B)

(*iii*) A : B $\longrightarrow$ $\overset{+}{A} + :\overset{-}{B}$ (B more electronegative than A)

The first type of cleavage is called *homolytic fission* or homolysis and leads to the formation of very reactive species called '*free radicals*' (atoms or groups of atoms containing odd or unpaired electrons).

In homolytic fission the covalent bond breaks in such a way that each fragment carries one unpaired electron.

The second and third types of cleavage is called *heterolytic fission* and leads to the formation of ionic species. These ionic species are also very reactive and carry charges on carbon. Cationic species carrying positive charge on a carbon atom are called *carbonium ions* or *carbocations*. Anionic species carrying negative charge on carbon atom are called *carbanions*.

In heterolytic fission the covalent bond breaks in such a way that the pair of electrons stays on the more electronegative atom.

Q. 9. What is meant by reaction intermediates? List all of them.

(*Kerala 2000; Nagpur 2003*)

Ans. In organic reactions, reactants do not change into products in one step. The change normally takes place via an intermediate product, which is short-lived. From the intermediate product, the reaction passes on to the products. Various reaction intermediates that we come across in the study of organic reactions are:

(*i*) Carbonium ion (or carbocation) (*ii*) Carbanion (iii) Free radial

(*iv*) Carbene (*v*) Nitrene

FREE RADICALS

Q. 10. What are free radicals? Discuss their characteristics, structure and stability.

(*G. Nanakdev, 2000; Garhwal, 2000; Meerut, 2000; Kanpur 2001, Magadh, 2002; Nagpur 2008*)

Ans. Free radical is an atom or group of atoms containing odd or unpaired electron. Methyl free radical ($CH_3\bullet$), ethyl free radical $CH_3—CH_2\bullet$), chlorine free radical ($Cl\bullet$) are some of the examples.

Characteristics of Free Radicals

(1) Free radicals ae electrically neutral and highly reactive species formed by homolytic fission of a covalent bond.

(2) They are paramagnetic in nature due to the presence of odd electrons.

(3) They are extremely reactive since they have tendency to complete the octet of the atom carrying the unpaired electron.

(4) They have short life period.

(5) They are usually formed in reactions which are carried out at very high temperatures or under the influence of ultraviolet light or in the presence of radical initiators such as peroxides.

For example,

$$:\ddot{\underset{..}{Cl}}:\ddot{\underset{..}{Cl}}: \xrightarrow[\text{or } \Delta]{h\nu} :\ddot{\underset{..}{Cl}}\cdot + \cdot\ddot{\underset{..}{Cl}}:$$

Structure of an Alkyl Free Radical

The carbon atom of an alkyl free radical is sp^2 hybridised. Its three hybridised orbitals are bonded to three atoms or groups of atoms. Thus free radicals have planer structures and have bond angles of 120°. The unpaired electron is present in unhybridised *p*-orbitals as shown in the figure.

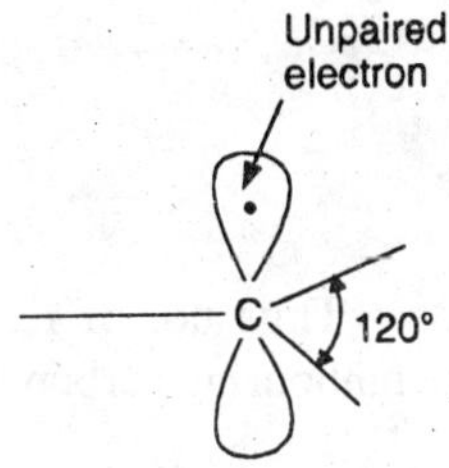

Structure of free radical

Stability of Free Radicals

The bond dissociation energies for various alkanes are given below:

$$CH_4 \longrightarrow \bullet CH_3 + H^\bullet \qquad \Delta H = +426 \text{ kJ}$$

$$C_2H_6 \longrightarrow \underset{\text{A primary free radical}}{\bullet CH_2 - CH_3} + H^\bullet \qquad \Delta H = 406 \text{ kJ}$$

$$CH_3 - CH_2 - CH_3 \longrightarrow \underset{\text{A primary free radical}}{CH_3 - CH_2CH_2^\bullet} + H^\bullet \qquad \Delta H = 406 \text{ kJ}$$

$$CH_3 - CH_2 - CH_3 \longrightarrow \underset{\text{A secondary free radical}}{CH_3 - \dot{C}H - CH_3} + H^\bullet \qquad \Delta H = 393 \text{ kJ}$$

$$CH_3 - \underset{CH_3}{\overset{H}{C}} - CH_3 \longrightarrow \underset{\text{Tert. free radical}}{CH_3 - \underset{CH_3}{\dot{C}} - CH_3} + H^\bullet \qquad \Delta H = 380 \text{ kJ}$$

It is obvious from the above that tertiary alkyl radical is more stable than secondary alkyl radical which in turn is more stable than primary alkyl radical.

The order of stability is:

$$\text{Tert. alkyl radical} > \text{Sec. alkyl radical} > \text{Prim. alkyl radical} > \dot{C}H_3$$

Q. 11. Arrange the following free-radicals in increasing order of stability and explain your answer.

$$\dot{C}_2H_5,\ (C_2H_5)_2\dot{C}(C_2H_5),\ (C_2H_5)_2\dot{C}H,\ \dot{C}H_3$$

Ans. The order of stability can be given considering the hyperconjugation phenomenon.

(*i*) Ethyl radical is a hybrid of four resonating structures.

$$\underset{\text{(I)}}{H - \underset{H}{\overset{H}{C}} - \underset{H}{\dot{C}} - H} \longleftrightarrow \underset{\text{(II)}}{H - \underset{H}{\overset{\dot{H}}{C}} = \underset{H}{C} - H} \longleftrightarrow \underset{\text{(III)}}{\dot{H}\ \underset{H}{\overset{H}{C}} = \underset{H}{C} - H}$$

$$\updownarrow$$

$$\underset{\text{(IV)}}{H - \underset{\dot{H}}{\overset{H}{C}} = \underset{H}{C} - H}$$

Thus no. of resonating structures is one more than the no. of hydrogen atoms on the neighbouring carbon atoms.

(*ii*) $(C_2H_5)_2\ \dot{C}\ (C_2H_5)$ is a hybrid of seven resonating structures, as there are six hydrogens marked * which can participate in hyperconjugation.

$$CH_3-\overset{*}{C}H_2-\dot{C}-\overset{*}{C}H_2-CH_3$$
$$\quad\ |$$
$$\quad CH_2^*$$
$$\quad\ |$$
$$\quad CH_3$$

(*iii*) $(C_2H_5)_2\ \dot{C}H$ is a hybrid of five resonating structures.

(*iv*) There is one and only one structure of $\dot{C}H_3$.

Greater the no. of resonating structures of a species, greater is the stability. Hence the order of increasing stability is

$$\dot{C}H_3 < CH_3\dot{C}H_2 < (C_2H_5)\dot{C}H < (C_2H_5)_2\ \dot{C}(C_2H_5)$$

Q. 12. Arrange the following free-radicals in order of increasing stability.

$CH_3-\dot{C}H_2$, $(CH_3)_2\ \dot{C}H$, $\dot{C}H_3$ and $(CH_3)_3\ \dot{C}$. Explain your answer.

(*Kurukshetra, 2001*)

Ans. Proceed as explained in Q. 11.

CARBOCATIONS (CARBONIUM IONS)

Q. 13. What are carbocations? Discuss their characteristics, stability and structure.

(*Delhi, 2005; Purvanchal, 2007; Nagpur, 2008; Andhra, 2010; Garhwal, 2010*

Ans. A species which has a carbon atom bearing only six electrons and having a positive charge is called *carbonium ion* or *carbocation*.

For example,

$$\underset{\text{3° Alkyl bromide}}{R_3-C-Br} \longrightarrow \underset{\text{3° Carbocation}}{R-\overset{R}{\underset{R}{C^+}}} + :\bar{Br}$$

When a proton adds to a carbon-carbon double bond, a carbonium ion gets formed.

$$H_2C=CH_2 + H^+ \longrightarrow \underset{\text{Ethyl carbocation}}{H_2C^+-\underset{H}{CH_2}}$$

Characteristics of Carbocation

1. Carbocations have a very strong tendency to complete the octet and consequently they are exceedingly reactive. Either they combine with nucleophiles such as OH^-, CN^- ions, or may lose a proton to form an alkene. The loss of proton takes place from the adjacent atom.

$$\underset{\text{Nucleophile}}{\bar{Br}:} + \underset{\text{Ethyl carbocation}}{H-\overset{+}{\underset{H}{\overset{H}{C}}}-\underset{H}{\overset{H}{C}}-H} \longrightarrow \underset{\text{Ethyl bromide}}{Br-\underset{H}{\overset{H}{C}}-\underset{H}{\overset{H}{C}}-H}$$

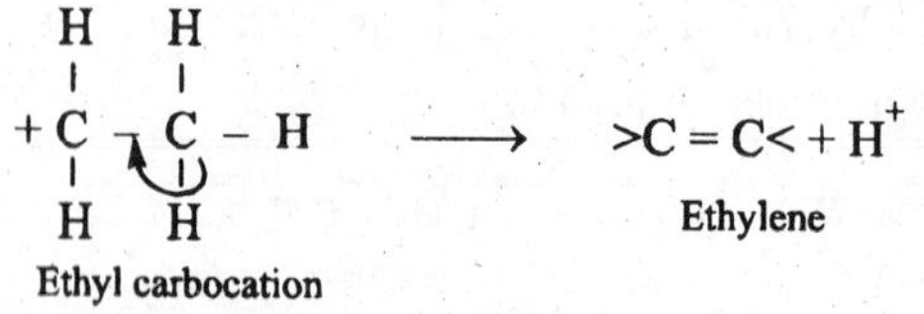

$$\overset{+}{C}H_2 - \underset{H}{\overset{H}{C}} - H \longrightarrow \;>C=C< + H^+$$

Ethyl carbocation → Ethylene

2. They are short lived.

Stability of Carbocations

Like free radicals, the relative stabilities of carbocations is of the following order.

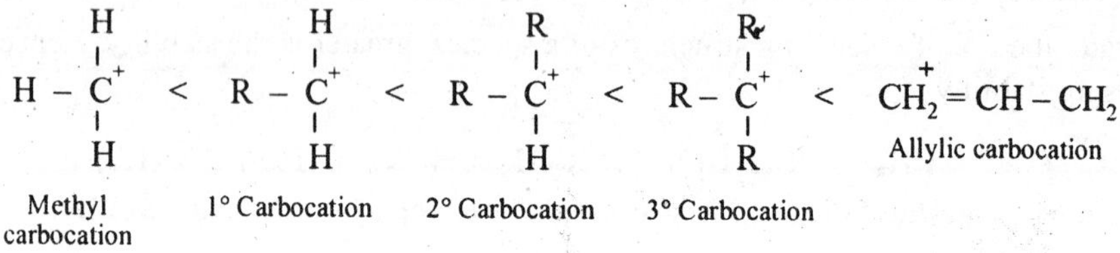

$$H-\underset{H}{\overset{H}{C^+}} < R-\underset{H}{\overset{H}{C^+}} < R-\underset{H}{\overset{R}{C^+}} < R-\underset{R}{\overset{R}{C^+}} < \overset{+}{C}H_2=CH-CH_2$$

Methyl carbocation, 1° Carbocation, 2° Carbocation, 3° Carbocation, Allylic carbocation

Any factor (Inductive effect, Resonance or both) which helps in the dispersal of the charge increases the stability of the carbocation.

The above mentioned order can be easily explained in terms of the electron releasing (+I) inductive effect of the alkyl groups. An alkyl group attached to the positively charged carbon atom of the carbocation tends to release electrons to the positively charged carbon. As a result, the +ve charge on the carbon atom decreases. In other words, positive charge on the carbon atom gets dispersed. According to the laws of electrostatics, dispersal of the charge leads to the stability of the ionic species. *Thus, greater the number of alkyl groups, greater in the dispersal of the +ve charge and hence greater is the stability of the carbocation.*

This explains why 3° carbocation with three alkyl groups are more stable than 2° carbocation with two alkyl groups which in turn are more stable than 1° carbocation with one such group. A methyl carbocation which does not contain any alkyl group is, the least stable.

The resonance stabilisation of allylic carbocation enhances its stability (delocalisation of π electrons).

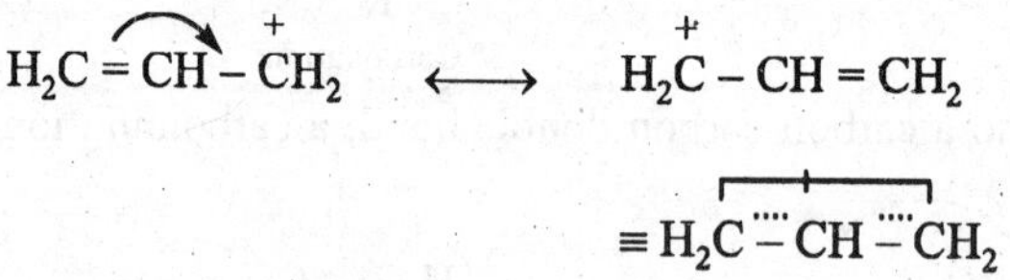

$$H_2C=CH-\overset{+}{C}H_2 \longleftrightarrow H_2\overset{+}{C}-CH=CH_2$$

$$\equiv H_2C \overset{+}{\text{---}} CH \text{---} CH_2$$

Structure of Carbocations

The carbon atom of a carbocation is sp^2 hybridized. The three sp^2 orbitals form three σ-bonds with the three substituents. The unhybridized p-orbital which is perpendicular to the plane of the three σ-bonds, however, remains empty. *Thus, carbocations are planer chemical species having all the three σ-bonds in one plane with a angle of 120° between them* as shown in the figure.

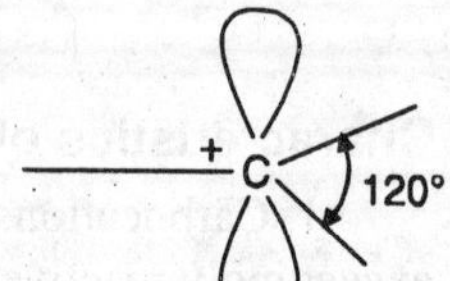

120°

Structure of carbocation ion.

Q. 14. Explain why ethyl carbocation ($CH_3 - \overset{+}{C}H_2$) is more stable than n-propyl carbocation ($CH_3 - CH_2 - \overset{+}{C}H_2$)? *(Delhi 2006)*

Ans. Greater the number of hyperconjugation structures of a carbocation, greater is the stability

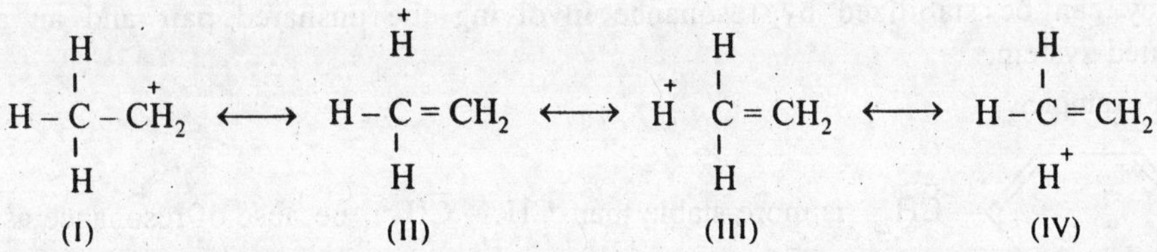

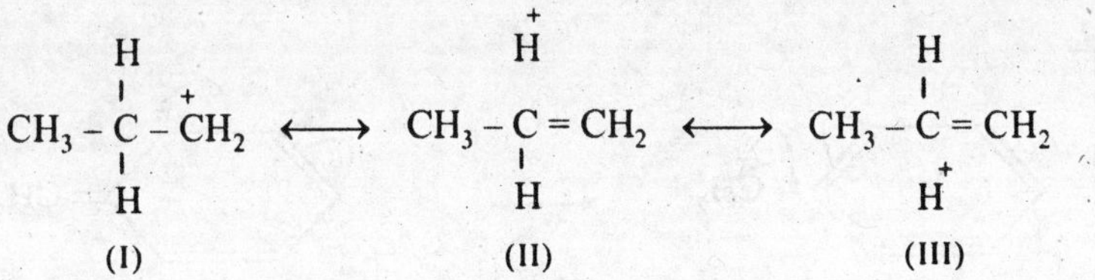

$$\underset{\text{(I)}}{H-\overset{H}{\underset{H}{C}}-\overset{+}{C}H_2} \longleftrightarrow \underset{\text{(II)}}{H-\overset{\overset{+}{H}}{\underset{H}{C}}=CH_2} \longleftrightarrow \underset{\text{(III)}}{\overset{+}{H}\ \overset{H}{\underset{H}{C}}=CH_2} \longleftrightarrow \underset{\text{(IV)}}{H-\overset{H}{\underset{\overset{+}{H}}{C}}=CH_2}$$

$$\underset{\text{(I)}}{CH_3-\overset{H}{\underset{H}{C}}-\overset{+}{C}H_2} \longleftrightarrow \underset{\text{(II)}}{CH_3-\overset{\overset{+}{H}}{\underset{H}{C}}=CH_2} \longleftrightarrow \underset{\text{(III)}}{CH_3-\overset{H}{\underset{\overset{+}{H}}{C}}=CH_2}$$

We find that ethyl carbocation has four hyperconjugative structures while *n*-propyl carbocation has three. Thus, ethyl carbocation is more stable .

CARBANIONS

Q.15. What are carbanions? Discuss their formation, stability and structure.

(G. Nanakdev 2000; M. Dayanand 2000; Kerala, 2003; Osmania, 2003, Sri Venkateswara, 2003; Guwahati, 2005, Nagpur, 2005)

Ans. An organic anion with the negative charge on carbon is known as *carbanion.* It is generated by the removal of an atom or group of atoms from a molecule without the bonding electrons.

Formation of Carbanion

$$\underset{\text{Acetylene}}{H-C\equiv C-H} \xrightarrow{:\overline{N}H_2} \underset{\substack{\text{Acetylide ion}\\ \text{(Carbanion)}}}{HC\equiv \overline{C}:} + NH_3$$

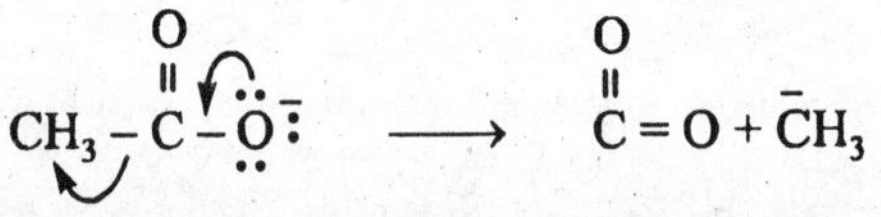

$$CH_3-\overset{O}{\overset{\|}{C}}-\ddot{\underset{..}{O}}:^{-} \longrightarrow \overset{O}{\overset{\|}{C}}=O+\overline{C}H_3$$

Stability of carbanion. The relative stability of carbanions is just the opposite of carbonium ions. Thus $\overline{C}H_3$ (Methyl carbanion) is the most stable among 1°, 2° and 3° carbanions.

$$\underset{3^\circ\text{ Carbanion}}{R_3\overline{C}:} < \underset{2^\circ\text{ Carbanion}}{R_2\overline{C}H} < \underset{1^\circ\text{ Carbanion}}{R\overline{C}H_2} < \underset{\substack{\text{Methyl}\\ \text{carbanion}}}{\overline{C}H_3}$$

This is due to the fact that electron releasing alkyl groups tend to intensify the negative charge and hence destabilize the carbonium ions while electron withdrawing substituents such as > C = O and –C ≡ N groups enhance the stability of carbonium ion due to the dispersal of –ve charge (delocalisation).

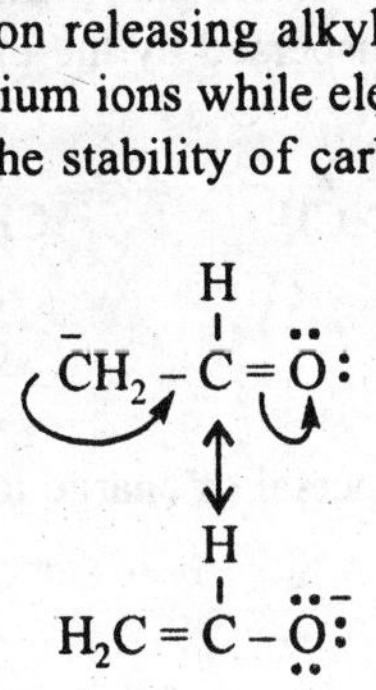

$$\overline{C}H_2-\overset{H}{C}=\ddot{O}:$$

$$\updownarrow$$

$$H_2C=\overset{H}{C}-\ddot{\underset{..}{O}}:^{-}$$

They can be stabilized by resonance involving the unshared pair and an adjacent unsaturated system.

For example,

$C_6H_5-\bar{\ddot{C}}H_2$ is more stable than $CH_3-\bar{C}H_2:$ because of resonance as shown:

Structure of Carbanions

The negatively charged carbon atom of simple carbanions is sp^3-hybridized. Three of the sp^3 orbitals form three σ-bonds with the three substituents whereas the fourth contains the non-bonding pair of electrons. Thus, *carbanions are isoelectronic with ammonia.* Like NH_3, *they appear to have pyramidal structure.*

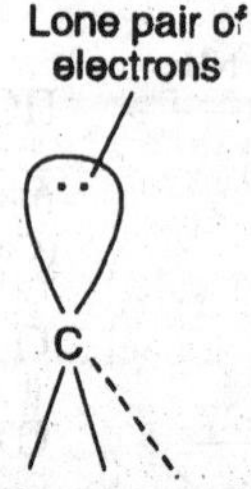

Structure of carbanion

Q. 16. Define carbonium ion and carbanion and explain that a secondary carbonium ion is more stable than primary but a secondary carbanion is less stable than primary.

(*Dayanand, 2000*)

Ans. For the definition of carbonium ion and carbanion, see Q. No. 13 and 15.

A secondary carbonium ion is more stable than primary carbonium ion because in secondary carbonium ion, the positive charge is dispersed by the inductive effect of two methyl groups.

$(CH_3)_2\overset{+}{C}H$ — Secondary Carbonium ion; $CH_3 \rightarrow \overset{+}{C}H_2$ — Primary Carbonium ion

In primary carbonium ion, the dispersal of charge takes place to a smaller extent as there is only one alkyl group.

However, when we go to carbanions, the situation is just opposite.

$$\begin{matrix} CH_3 \searrow \\ \quad\quad \bar{C}H \\ CH_3 \nearrow \end{matrix} \qquad CH_3 \rightarrow \bar{C}H_2$$

Here the alkyl groups tend to destabilize the carbanion. Instead of dispersal of charge, there is rather concentration of charge on the carbanion. Two alkyl groups will destabilize carbanion more than one alkyl group. Hence secondary carbanion is less stable than primary carbanion.

CARBENES

Q. 17. What are carbenes? Discuss their formation and structure.

(Meerut, 2000; Kurukshetra, 2001; Kerala 2001; Punjab 2003; Delhi 2006, Nagpur 2008)

Ans. Carbenes are neutral divalent chemical species which contain a carbon atom having six electrons in their valence shell out of which two are unshared. The simplest carbene is methylene $:CH_2$, other examples are $:CCl_2$ dichlorocarbene, $:CBr_2$ dibromocarbene, $:C(C_6H_5)_2$ diphenyl carbene etc.

Formation of Carbenes. Carbenes are obtained by the following methods:

(*i*) Methylene $:CH_2$ is obtained by the action of UV light on diazomethane

$$\underset{\text{Diazomethane}}{CH_2N_2} \xrightarrow[\text{light}]{\text{UV}} :CH_2 + N_2$$

It is also obtained by the action of UV light on ketene

$$CH_2 = C = O \xrightarrow[\text{light}]{\text{UV}} :CH_2 + CO$$

(*ii*) Dichloro carbene is obtained by the action of chloroform on sodium ethoxide

$$HCCl_3 + C_2H_5ONa \longrightarrow \underset{\text{Dichloro carbene}}{:CCl_2} + C_2H_5OH + NaCl$$

Types of carbenes. Depending upon whether the two electrons on carbene are paired or not, they can be classified into two types.

Singlet carbenes are those in which the unshared electrons are paired. Singlet carbene may be written as

$$\uparrow\downarrow CH_2$$

Triplet carbenes are those in which the unshared electrons are unpaired, it may be represented as

$$\uparrow\uparrow CH_2$$

Triplet carbenes are lower in energy and hence more stable than singlet carbenes.

Structure of Carbene

The structures of two types of carbenes are different : Singlet carbenes have their carbon in sp^2 hybridised state. Of the three sp^2 hybrid orbitals, two are used in forming two single bonds with monovalent atoms or groups attached to carbon. The unshared pair of electrons is present in the third sp^2 hybrid orbital and the unhybridised *p*-orbital is emtpy.

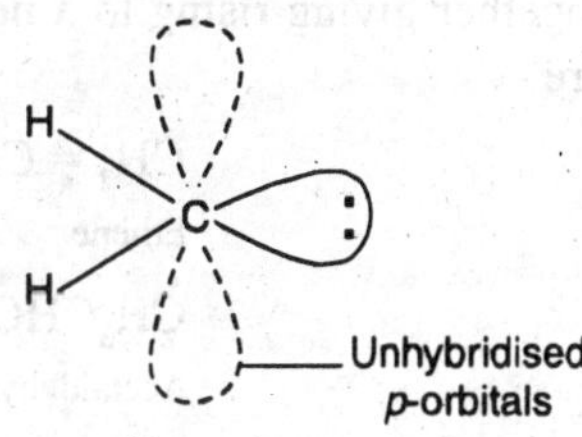

Singlet methylene

Triplet carbene has its carbon in *sp* hybridised state. The two *sp* hybrid orbitals form two bonds with monovalent atoms or groups attached to carbon. The two unhybridised *p*-orbitals contain one electron each.

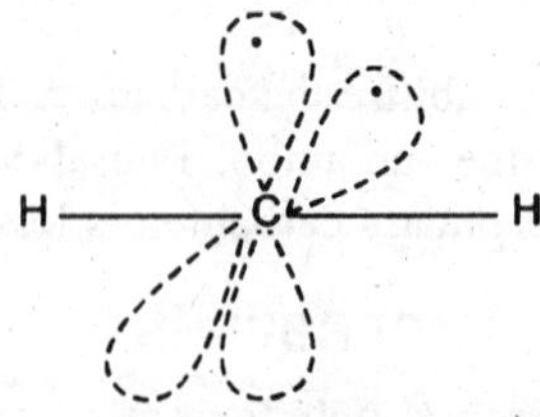

Triplet methylene

Q. 18. Describe different types of reaction that we come across.

(*Kerala 2001, Delhi 2003, Nagpur 2008*)

Ans. Organic reactions may be classified into four main types:

(*a*) Substitution reactions

(*b*) Addition reactions

(*c*) Elimination reactions

(*d*) Rearrangements.

These are separately described as under.

(*a*) Substitution reactions: A substitution reaction is one in which a part of one molecule is replaced by other atom or group without causing a change in the rest of the molecule. Following are some examples of substitution reactions.

(*i*) $CH_4 + Cl_2 \longrightarrow CH_3Cl + HCl$

Methane → Methyl chloride

(*ii*) $C_2H_5Br + KOH \longrightarrow C_2H_5OH + KBr$

(*iii*) Benzene $+ Cl_2 \xrightarrow{FeCl_3}$ Chlorobenzene $+ HCl$

The substitution reactions may be brought about by free-radicals, nucleophilic or electrophilic reagents.

(*b*) Addition reactions: When two molecules of same or different substances combine together giving rising to a new product, it is a addition reaction. Examples of addition reactions are:

$CH_2 = CH_2 + Br_2 \longrightarrow CH_2BrCH_2Br$

Ethene → Ethylene dibromide

$CH_3CHO + HCN \longrightarrow CH_3CH(OH)CN$

Acetaldehyde → Acetaldehyde cyanohydrin

Addition reactions could be brought about by free-radical, electrophilic or nucleophilic reagents.

(c) Elimination reactions: These reactions involve the removal of atoms or groups from a molecule to form a new compounds containing multiple bonds. Dehydrohalogenation of alkyl halides is a common example of this type of reaction.

$$\underset{\text{n-Propyl chloride}}{CH_3 - CH_2CH_2Cl} + KOH \xrightarrow{\text{Alc.}} \underset{\text{Propene}}{CH_3CH = CH_2} + KCl + H_2O$$

(d) Rearrangement reactions: Rearrangement reactions involve the migration of an atom or a group from one atom to the other within the same molecule.

It is interesting to note that the first organic compound, *i.e.*, urea synthesized in the laboratory by Wohler actually involved a rearrangement reaction.

$$\underset{\text{Amm. cyanate}}{NH_4CNO} \xrightarrow{\Delta} \underset{\text{Urea}}{NH_2 - CO - NH_2}$$

Another important example of such reactions is *Hoffmann bromamide reaction.* This reaction involves the migration of an alkyl groups from the carbon to the nitrogen atom of an amide with the simultaneous elimination of CO as carbonate ion under the influence of Br_2/KOH.

$$\underset{\text{Acetamide}}{CH_3 - CONH_2} + Br_2 + 4KOH \longrightarrow \underset{\text{Methylamine}}{CH_3 - NH_2} + K_2CO_3 + 2KBr + 2H_2O$$

Similarly maleic acid, when heated in a sealed tube, is converted into fumaric acid.

$$\underset{\text{Maleic acid}}{\begin{array}{c} H - C - COOH \\ \| \\ H - C - COOH \end{array}} \xrightarrow{\Delta} \underset{\text{Fumaric acid}}{\begin{array}{c} H - C - COOH \\ \| \\ HOOC - C - H \end{array}}$$

Q. 19. Write notes on:

(a) Nucleophilic reagents

(b) Electrophilic reagents. (*Kerala, 2000; Nagpur, 2005; Patna, 2010*)

Ans. (a) Nucleophilic reagents or nucleophiles

A nucleophilic reagent is a reagent with an atom having an unshared or lone pair of electrons. Such a reagent is in search of a point where it can share these electrons to form a bond. Nucleophiles are of two types:

(i) Neutral nucleophiles. These are the nucleophiles which are neutral in charge. But they carry some unshared electrons which they like to share with some positive centre or electron deficient centre. Ammonia $\ddot{N}H_3$, water $H_2\ddot{O}$ and alcohols $R - \ddot{O} - H$ are examples of neutral nucleophiles.

(ii) Negative nucleophiles. These are the nucleophiles which carry negative charge. Examples of this type of nucleophiles are hydroxyl ions (OH^-), halide ion (X^-), alkoxide ion (RO^-) and cyanide ion (CN^-). Carbanions also come in the category of negative nucleophiles.

(b) Electrophilic Reagents or Electrophiles

An electrophile is a reagent containing electron deficient atoms. Such species have a tendency to attach themselves to centres of high electron density. There are two types of electrophiles:

(i) Neutral electrophiles. These electrophile don't carry any net charge. Lewis acids like $AlCl_3$, $FeCl_3$ and BF_3 belong to this category of electrophiles. Sulphonium ion (SO_3) carries no net charge, but it acts as an electrophile for sulphonation in benzene rings. This is because of its structure.

$$\overset{-}{O} - \overset{++}{S} - \overset{-}{O}$$
$$\| $$
$$O$$

As the positive charge is concentrated and the negative charge is scattered, it acts as an electrophile. Substances like $SnCl_4$ which have vacant *d*-orbitals would like to accommodate electrons in them. Thus such substances also act as electrophiles.

***(ii)* Positive electrophiles.** These electrophiles carry a net positive charge. Examples of this category of electrophiles are hydrogen ion (H^+), hydronium ion (H_3O^+), nitronium ion (NO_2^+), and chloronium ion (Cl^+). In the halogenation and nitration of aromatic systems, these electrophiles are involved.

STRENGTH OF AN ORGANIC ACID

Q. 20. What are the factors that determine the strength of an organic acid?

(*Kerala 2000*)

Ans. *(i)* Nature of the atom holding the hydrogen atom. According to Arrhenius and Bronstead-Lowry concepts an acid is a substance which gives out protons. The strength of an acid is determined by the concentration of hydrogen ions. Strength of the acid actually depends upon the capacity of the atom holding the protons, to accommodate the electrons after, hydrogen has been released Greater the electronegativity and size of the atom holding hydrogen, greater will be the strength of the acid.

$$A - H \longrightarrow \overset{-}{A} + \overset{+}{H}$$

Thus A must be highly electronegative in order that A – H exhibits strong acidic properties.

***(ii)* Inductive and resonance effects in the molecules.** Presence of groups which give rise to the above effects influence the strength of the acid. We observe that electron withdrawing groups like $-NO_2$, –CN etc. increase the strength of the acid electron donating groups like alkyl groups decrease, the strength.

Q. 21. What is meant by K_a and pK_a values of an acid?

Ans. Consider the dissociation of an acid HA as given below

$$HA + H_2O \rightleftharpoons H_3\overset{+}{O} + A^-$$

For this reaction in equilibrium

$$K = \frac{[H_3O^+][A^-]}{[HA][H_2O]}$$

Water is taken in large excess, so its concentration $[H_2O]$ may be taken as constant

$$K\,[H_2O] = \frac{[H_3O^+][A^-]}{[HA]}$$

or
$$K_a = \frac{[H_3O^+][A^-]}{[HA]}$$

where K_a is dissociation constant of the acid. Higher the value of K_a, stronger is the acid.

These days, the strength of an acid is represented as pK_a where

$$pK_a = -\log K_a$$

consider the case of acetic acid. Its K_a value is 1.75×10^{-5}. pK_a value of acetic acid will be given by

$$pK_a = -\log K_a$$
$$= -\log (1.75 \times 10^{-5})$$
$$= 4.75$$

It may be noted that a higher value of K_a signifies a stronger acid, but a higher value of pK_a signifies a weaker acid.

STRENGTH OF A BASE

Q. 22. What are the factors that determine the strength of a base? *(Kerala, 2000)*

Ans. The following factors determine the strength of a base:

(*i*) Nature of the atom holding the lone pair of electrons. If the atom holding the lone pair of electrons has smaller electronegativity, smaller size and a negative charge, the electron pair will be more easily available for reaction. Consequently, the base will be a strong base.

(*ii*) Magnitude of negative charge or no. of lone pairs of electrons. Greater the magnitude of negative charge or greater the no. of lone pairs of electrons, greater will be the basic strength.

(*iii*) Inductive or resonance effects. Presence of groups exhibiting inductive or resonance effects influence the basic strength. Groups having electron donating effect increase the basic strength whereas groups which show electron withdrawing effect decrease the basic strength.

Q. 23. What is meant by K_b and pK_b value of a base?

Ans. The dissociation of a base can be represented as

$$B + H_2O \rightleftharpoons \overset{+}{B}H + \overset{-}{O}H$$

The equilibrium constant of this reaction is given by

$$K = \frac{[B^+H][OH^-]}{[B][H_2O]}$$

or
$$K[H_2O] = \frac{[B^+H][OH^-]}{[B]}$$

Water is taken in large excess, so its concentration may be taken as constant, therefore, the product of K and $[H_2O]$, which are both constants may be taken as K_b, another constant K_b is called the dissociation constant of the base, higher the K_b value of a base, greater is the basic strength.

These days the basic strength is represented by pK_b.

where
$$pK_b = -\log K_b$$

For example for methylamine $K_b = 4.4 \times 10^{-4}$

$$pK_b = -\log K_b$$
$$= -\log (4.4 \times 10^{-4})$$
$$= 3.36$$

It is noteworthy that a greater value of K_b signifies greater basic strength whereas a greater value of pK_b signifies smaller basic strength.

Q. 24. Classify the following as nucleophiles and electrophiles: H_3O^+, NH_3, $AlCl_3$, ROH, BF_3, CN^-, SO_3.

Ans.	**Nucleophiles**	**Electrophiles**
	NH_3	H_3O^+
	ROH	$AlCl_3$
	CN^-	BF_3
		SO_3 (Sulphenium ion)

Q. 25. Pick up from the following, the electrophiles and nucleophiles:

$AlCl_3$, PH_3, HNO_3, R_3N

Ans. Electrophiles: $AlCl_3$, HNO_3.

Nucleophiles: PH_3, R_3N.

Q. 26. Pick up from the following, the electrophiles and nucleophiles:

BF_3, NH_3, ROH, $SnCl_4$.

Ans. Electrophiles : BF_3, $SnCl_4$.

Nucleophiles : NH_3, ROH.

ASSIGNING FORMAL CHARGE

Q. 27. Discuss the method for assigning formal charges on intermediates and ionic species.

Ans. Formal charge on a intermediate or an ionic species is calculated from Lewis structure. The main features of this method are given below :

1. The average electrons around an atom in a given species is equal to half the number of electrons present in bond pairs plus the number of non-bonding electrons.

$$\text{Average electrons} = \left(\frac{1}{2} \times \text{No. of electrons in bond pairs}\right) + \text{No. of non-bonding electrons.}$$

2. If the no. of average electrons present around an atom is equal to the no. of valence electrons in the neutral atom, there is no charge on the species.
3. If the average electrons present around an atom is less than the no. of valence electrons in the neutral atom, the species carries a positive charge.
4. If the average electrons on an atom in a species is more than the valence electrons on that atom, the species carries a negative charge.

The examples that following will illustrate the point.

Average Charge on Methane

The Lewis structure of methane is :

```
      C
      ..
  H : C : H
      ..
      C
```

There are four C – H bonds. Each bond means one pair of electrons shared between carbon and hydrogen. In each one electron is possessed by carbon and one electron is possessed by hydrogen.

Thus electrons possessed by carbon = 1 × 4 = 4

The no. of valence electrons in carbon atom = 4. Since the two values are the same, there is no **net charge on the carbon.**

Average Charge on Ammonia

The Lewis structure of ammonia is :

```
    H
    ..
  H:N:
    ..
    H
```

There are three bond pairs and one lone pair of electrons in ammonia. Bond pair electrons are shared between N and H whereas lone pair electrons are solely possessed by N.

Average electrons on N = $\left(\frac{1}{2}\times 6\right)+2=5$

This is equal to the number of valence electrons in neutral N. Hence there is no net charge on ammonia.

Average Charge on $\dot{C}H_3$ Radical

The structure of methyl radical is

```
     .
  H :N: H
     ..
     H
```

There are three bond pairs (or 6 electrons) shared between C and H. Additionally one electron is solely possessed by C.

Therefore, Average electrons present on C = 3 + 1 = 4

Again there is no net charge on methyl radical.

Average Charge on Carbocation

Let us consider the charge on methyl carbocation, which has the structure

```
     +
  H :C: H
     ..
     H
```

Here average charge on carbon = 3.

This is one less than the valence electrons in neutral carbon.

Therefore charge on CH_3 = 4 – 3 = 1

Average Charge on Carbanion

Let us consider the charge on methyl carbanion having structure

```
  [   ..    ]-
  [ H :C: H ]
  [   ..    ]
  [   H     ]
```

It carries extra electron which is responsible for its negative charge. No. of average electrons on C = 4 + 1 = 5.

This is one more than the no. of valence electrons.

Hence charge on methyl carbanion = 5 – 4 = 1.

Q. 28. Classify the following species as nucleophile, electrophile or free radical : (*i*) Iodine anion (*ii*) Nitronium ion (*iii*) Ammonia (*iv*) Atomic Chlorine. (*Kerala 2001*)

Ans. (*i*) Iodide anion — Nucleophile

(*ii*) Nitronium ion — Electrophile

(*iii*) Ammonia — Nucleophile

(*iv*) Atomic chlorine — Free radical

Q. 29. Define "mechanism" of a chemical reaction. (*Kerala 2001*)

Ans. The various steps through which an organic reaction takes place is called the mechanism of the reaction.

Q. 30. Differentiate between inductive effect and electromeric effect. (*Nagpur 2003*)

Ans. See Q. 1 and Q. 3

5

CHROMATOGRAPHY

Q. 1. What is chromatography? On what principle does it work? What are its applications? (*Delhi 2005*)

Ans. Chromatography is a modern separation technique. It was first discovered by a Russian scientist Tswett. In this technique, separation of a mixture of substances is brought about by differential movement of individual components through a porous medium under the influence of a solvent. Since the discovery of this technique, it has undergone many changes and is now a very sophisticated technique.

Principle. Chromatography is based upon the principle of selective adsorption of various components of a mixture between two phase, **one fixed phase**, and the other **moving phase**. The mixture to be separated is taken in the moving phase and passed over the fixed phase. Different components are adsorbed to different extents by the fixed phase. This results in separation of the mixture.

Applications. This technique finds applications in qualitative and quantitative chemical analysis.

Q. 2. Write notes on: (*i*) Adsorption chromatography (*ii*) Partition chromatography. (*Delhi 2002*)

Ans. (*i*) Adsorption chromatography. When the chromatographic technique involves a solid as the stationary (fixed) phase and a liquid or gas as moving phase, it is a case of adsorption chromatography. Column chromatography, in which a solid adsorbent is taken in a column is an example of adsorption chromatography. T.L.C. (Thin Layer Chromatography) is another example of adsorption chromatography in which the adsorbent is supported on a glass plate.

(*ii*) Partition chromatography. When the chromatographic technique involves a liquid supported on an inert solid as stationary phase and a liquid or gas as moving phase, it is a case of partition chromatography. Paper chromatography is an example of partition chromatography. Here paper acts as inert support.

Q. 3. Which chromatographic technique is based upon liquid-liquid partition? Explain its principle.

Ans. Paper chromatography is an example of chromatographic technique involving liquid partition. In this technique, the stationary phase is a liquid supported on paper. The moving phase is also a liquid. The components off the mixture are retained on the paper to different extents and these are obtained at different levels on the paper.

Q. 4. What is meant by R_f value in a chromatographic process?

Ans. It is observed during the development of a chromatogram that the solvent rises at a higher rate than the components of the mixture. R_f value for a particular components depends upon the following two factors:

(*i*) Nature of the eluting liquid.

(*ii*) Nature of the adsorbent

R_f stands for **Retention Factor** and can be mathematically expressed as:

$$R_f = \frac{\text{Distance travelled by the component}}{\text{Distance travelled by the solvent in the same time}}$$

The eluting liquid and adsorbents are chosen in such a way so as to give a wide difference in the R_f values of the components. This facilities the separation and identification of the components.

Q. 5. What is paper chromatography? How it is carried out? *(Delhi, 2003)*

Ans. It is an important separation technique based upon partition chromatography. The principle used is that separation takes place as a result of difference in partition coefficients between two solvents.

Procedure. (*i*) The solution of the sample is prepared in a suitable solvent.

(*ii*) The sample is spotted on a strip of filter paper using pipette. A fine capillary tube can also be used, and chromatogram is "developed" by keeping the bottom of the paper in a suitable solvent. See Fig. 5.1.

(*iii*) The solvent rises up by the capillary action and the components of the solvent rise up the paper at different rates depending totally on their solubility and their degree of retention by the paper. The paper is then properly developed and the individual spots are carefully noted.

The spots can be made visible by treating with a reagent which gives a coloured derivative.

(*iv*) It will be observed that the spots generally move at a certain fraction of the rate at which the solvent rises and are characterised by the R_f value. The R_f value stands for the relative rate of the movement of solute and solvent.

$$R_f = \frac{\text{Distance by which solute moves}}{\text{Distance by which solvent moves}}$$

(*v*) The distances are measured from the centre of the spot where the sample was spotted just near the bottom of the paper. The distance travelled by the solute is measured upto the centre of the solute spot, or where its maximum density occurs, in case tailing appears. *The R_f value is very much characteristic of a solute, paper and solvent combination.*

This type of chromatography is liquid-liquid partition chromatography. As the cellulosic filter paper is hydrophilic, it absorbs the stationary thin coating of water from the air. The sample, thus, distributes itself between the stationary water phase and the developing solvent. Mostly a mixture of an organic solvent with water (buffered at a definite pH) is employed as a developing solvent. Sometimes a water immiscible solvent is also employed to develop the chromatograms.

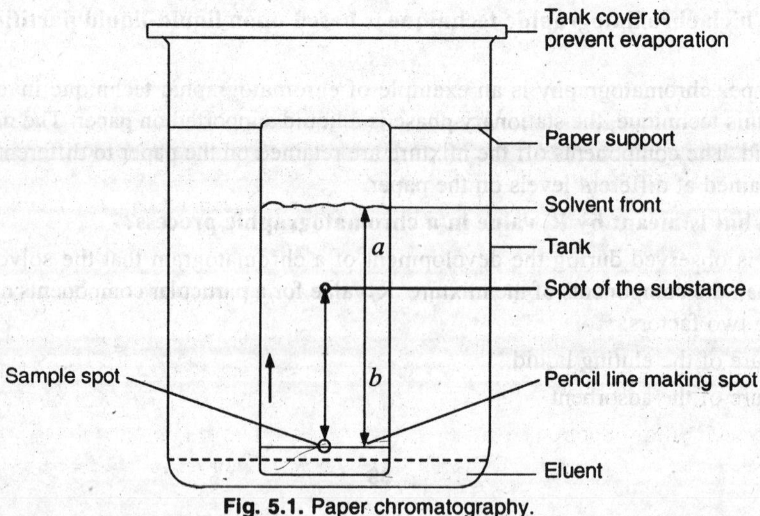

Fig. 5.1. Paper chromatography.

A labelled diagram for paper chromatography is shown in Fig. 5.1. For R_f measurement, first of all a pencil line is drawn across the paper a few centimeters from the bottom and the sample is spotted on this line. The spot must be made as small as possible for getting maximum separation and minimum tailing. The paper is kept in a chamber having its end dipping in the developing solvent. A closed chamber should be employed to saturate the atmosphere with the solvent and prevent it from evaporating from the surface of the paper as it rises up. It may take one hour or so for developing. In case wider paper is used, several samples and standards can be easily spotted along the bottom and developed simultaneously.

When the spots have been identified, they may be cut to get the solutes eluted and identified quantitatively.

Q. 6. What are the advantages and limitations of paper chromatography?

Ans. Advantages

(*i*) It is a rapid and reliable separation and identification technique when the quantity of sample available is very small.

(*ii*) It is possible to have perfect separation by using two dimensional paper chromatography.

(*iii*) This technique can perfectly work when compounds are slightly soluble in organic solvents but moderately soluble in water. Thin layer chromatography fails in such circumstances.

(*iv*) This technique although meant primarily for qualitative work can also be extended for quantitative studies using special micro methods.

Limitations

(*i*) Large scale separation of components is not possible.

(*ii*) Spots sometimes spread out and tailing is produced, which acts as a hindrance in separation of components.

Q. 7. Explain the terms (*i*) one-dimensional (*ii*) two-dimensional chromatographs.

Ans. The appearance that we get, after the chromatographic technique has been subjected to elution is called *chromatograph*. It shows the separation of a mixture into different components, in the form of bands in the case of column chromatography, in the form of spots in the case of TLC and paper chromatography.

One-dimensional chromatograph. Normally, the techniques that are employed viz., column, thin-layer and paper chromatograph, give us one-dimensional chromatographs. The movement of the components takes place in one direction only. That is why it is known as one-dimensional chromatograph. The disadvantage of one-dimensional chromatograph is that it does not lead to perfect separation.

Two-dimensional chromatograph. The chromatographic technique which is performed in two mutually perpendicular directions leads to two-dimensional chromatographs. This technique can give us complete separation of mixture. In this technique, paper is spotted with a drop of mixture solution near the linear left hand corner and the paper is developed with a solvent. The paper is removed from the development chamber and dried. It is then turned at right angle to the original direction of flow of solvent. The development is done again with a different solvent. This displaces the components of the mixture in two different directions from the original spot and, thus, facilitates their separation into different components.

Fig. 5.2 shows how valine and histidine have been separated and identified using two-dimensional technique.

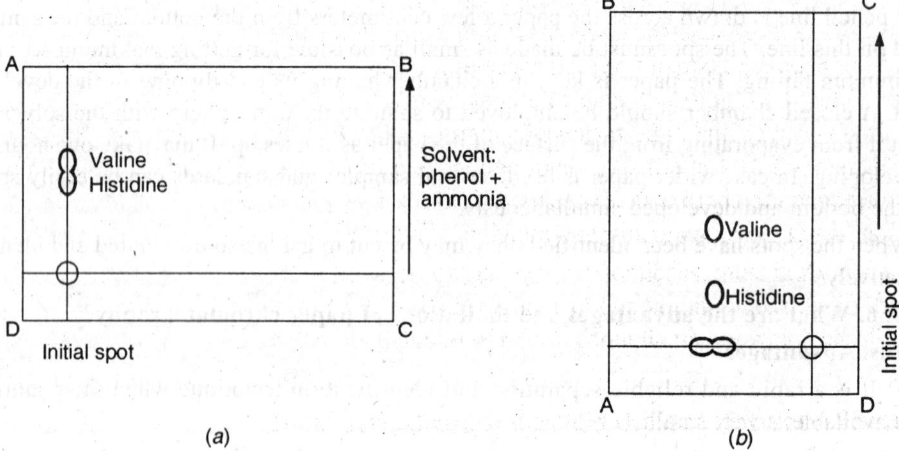

Fig. 5.2. Two dimensional chromatography.

The first elution is done with phenol + ammonia as the solvent. The two spots overlap as shown in Fig. 5.2 (*a*) . The paper is then turned at 90° and eluted with a second solvent consisting of butanol + ethanoic acid. We achieve a complete separation of two components in the form of two spots as shown in Fig. 5.2 (*b*).

Q. 8. What is thin layer chromatography? Explain. (*Madurai, 2003*)

Ans. It is a chromatographic technique in which the adsorbent layer is obtained on glass plates. Elution is carried out as in the case of paper chromatography. It involves the following steps:

Preparation of Chromatoplates

Chromatoplates are prepared by applying a uniform layer (0.25–0.3 mm) of an appropriate adsorbent material in the form of a thin aqueous paste to clean glass plates. The size of a chromatoplate may vary according to requirements of an experiment. For all ordinary work, microscope slides (2.5 × 10 cm) are used.

The glass slides are first washed with chromic acid and then freely flushed with water. They are then rinsed with alcohol and dried by means of hot air. A uniform paste is made by treating alumina or silica gel with a little binder like calcium sulphate and water in a mortar. The uniform thin paste or slurry thus obtained is applied on the glass slides by means of a special mechanical *applicator*. But this may not be available in most laboratories. The coating may, alternatively, be done by dipping as explained below.

Various steps involved in TLC experiment are as follows:

(*i*) Preparation of plates (By dipping). In this method, the adsorbent paste is taken in a beaker or deep vessel. Two slides are held together back to back and dipped into the paste for a little while and taken out. The paste is wiped from the edges with thumb and forefinger. Then they are separated and kept in to the oven for drying.

(*ii*) Application of sample. The sample to be analysed is dissolved in a *suitable* solvent. The solution is then applied by means of a fine capillary on a pencilled line near one edge of the chromatoplate, as was done previously in paper chromatography. The spot is allowed to dry and to make it bold, more solution is delivered to it in a number of instalments, followed by drying by means of a drier every time.

(*iii*) Developing. In TLC, invariably the ascending technique (rising upwards) is used for developing. After application of the sample chromatoplate is made to stand in a layer of the pure

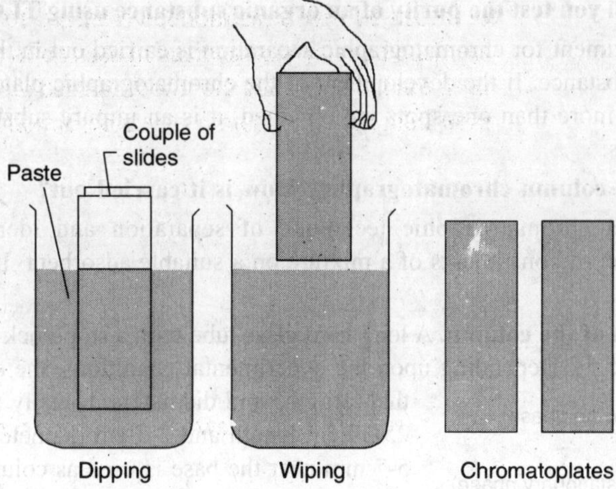

Fig. 5.3. Coating the slides by dipping.

and selected solvent taken in a jar (Fig. 5.4). Better results are obtained if the plate is kept vertically erect.

The jar is then covered with the lid or glass plate. The solvent (or element) rises at the rate of one cm in 3-4 minutes. When the solvent has nearly reached the top of the absorbent layer, the chromatoplate is taken out and dried.

***(iv)* Location.** After developing, the spot is located at its new position. If it is a coloured one it can be easily seen with the naked eye. For colourless spots, the developed chromatoplate is placed in a covered container having iodine crystals.

The iodine vapour is adsorbed into the areas of the plate containing organic compounds; brown spots due to iodine interaction appear on a background. These spots are known as *chromatograms*. Almost all organic compounds can be detected with this technique. The method is usually non-destructive. The spots may be marked and the plate gently warmed to allow the iodine to sublime out of the layer, leaving the compound unchanged. R_f is then calculated in the same way as explained in paper chromatography.

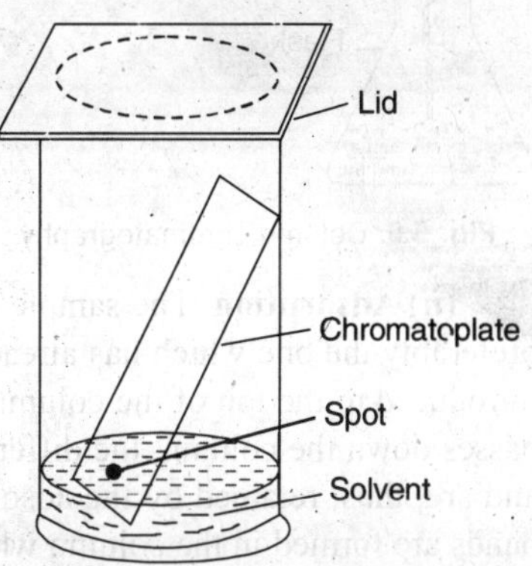

Fig.5.4. Developing of spot on a chromatoplate

Q. 9. List the advantages of thin-layer chromatography.

Ans. 1. Very sharp spots are obtained in this technique whereas in paper chromatography the spots are usually diffused.

2. It is far more rapid than paper chromatography and gives quick and reliable results.

3. Acidic or alkaline solutions can be used for location of spots on the chromatoplates, while this is not possible when filter paper is used.

4. For further analysis, the spot can be scrapped out and subjected to desired treatment.

5. If need be the TLC chromatoplates can be heated. No heating is possible in case of filter-paper chromatogram which may be burnt.

Q. 10. How will you test the purity of an organic substance using TLC?

Ans. The experiment for chromatographic separation is carried out in the usual manner with the given organic substance. If the development of the chromatographic plate gives one spot, it is a pure substance. If more than one spots are obtained, it is an impure substance i.e., it contains additional substances.

Q. 11. What is column chromatography? How is it carried out? *(Delhi, 2004)*

Ans. This is a chromatographic technique of separation and identification based on differential adsorption of constituents of a mixture on a suitable adsorbent. It is carried out in the following steps:

***(i)* Preparation of the column.** A long hard glass tube with a stop-cock at the bottom is used as the column (Fig. 5.5). Depending upon the experimental conditions, the columns may vary in their lengths and diameters. Usually a glass tube of about 25-30 cm length and 2-3 cm diameter narrowing down to 5-7 mm near the base is used as column. A burette can be used for this purpose. The tube used as column is thoroughly washed and dried. It is then plugged with glass or cotton wool placed at the narrow end of the tube. It is then clamped in a vertical position for filling with an adsorbent such as alumina, silica gel etc. A slurry of the adsorbent is prepared in a suitable solvent (preferably non-polar) such as hexane or petroleum ether and poured into the tube. The solvent is allowed to pass down slowly to let the adsorbent settle in the tube. The solvent which flows is collected in a flask and can be used again. **To prevent the drying and the channeling of column, the upper surface should always remain covered with the solvent right from the beginning of experiment.**

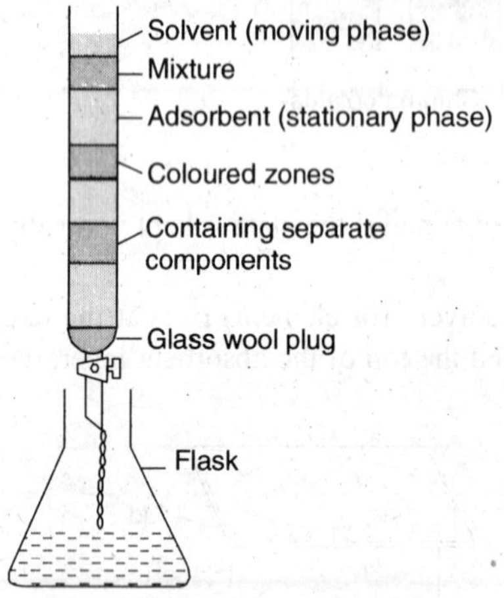

Fig. 5.5. Column chromatography

***(ii)* Adsorption.** The sample of the mixture to be separated is dissolved in a suitable solvent preferably the one which has already been used in the packing of the column. The solution is then introduced to the top of the column and is allowed to be adsorbed slowly over the adsorbent. As it passes down the column, the different components of the mixture are adsorbed to different extents and are, thus, retained by the adsorbent at different levels of the column. Thus, different zones or bands are formed in the column which contain different components of the mixture. As soon as the last portion of the solution containing the mixture enters the column, the selected solvent called *eluent* is added to the column. This acts as the moving phase.

***(iii)* Elution.** The adsorbed component of the mixture from the adsorbent is recovered with the help of suitable solvent of increasing polarity.

As the eluent passes through the column, it dissolves different components from the various bands. The least adsorbed component is eluted first by the least polar solvent while more strongly adsorbed ones are eluted later by the solvent of higher polarity. Thus, the various components of the mixture are eluted and their solutions are collected separately. The distillation of the solvent from the different fractions gives different components of the mixture in pure form.

There is one more method of affecting the separation. The various zones are cut at their separating boundaries and the components are then recovered by extraction with suitable solvents. The process of recovering the components is called *elution*.

Q. 12. Explain the importance of choice of solvent and adsorbent in chromatographic separation.

Ans. Choice of the solvents. In chromatographic separation, the solvents used play a multiple role performing three important functions:

(*i*) They are used in making a slurry of adsorbent to be taken in the column.

(*ii*) They are used to introduce the mixture into the column by dissolving its components. Generally, non-polar solvents such as petroleum ether and benzene are used for this purpose since adsorption occurs more quickly from such solvents.

(*iii*) They are used in the development of chromatogram by separating the various zones. So these are also known as developers. The solubility of the components in the developer should not be very high. Usually, the solvents with low molecular weights are used for the purpose. The same solvent is used for introducing the mixture and as a developer, as far as possible.

(*iv*) They are used to remove various components from the various zones after the development of chromatograms. As such they are known as *eluents*. The eluent should be capable of dissolving the components readily and should get itself adsorbed only to a limited extent. More than one eluents can be used for different components if required. For this purpose generally a low boiling liquid is selected, because it helps the quick evaporation from the fractions collected.

The commonly used solvents are petroleum ether, benzene, cyclohexane, carbon tetrachloride, chloroform, ether, acetone, ethyl alcohol and ethyl acetate. Two or more solvents mixed in different proportions can also be used.

Choice of adsorbents. An adsorbent should have the following characteristics for giving best results:

(*i*) It should not react with the components to be separated and solvent selected.

(*ii*) It should be immiscible with the solvents selected.

(*iii*) It should preferably be colourless, porous solid, so that the zones containing coloured components can be visualized locally.

Usually the inorganic substances, before being used as adsorbents, are activated (water content reduced) by heating to 473 K to 533 K. Activated alumina is the most commonly used adsorbent. The other adsorbents are activated charcoal, Fuller's earth, calcium carbonate, calcium phosphate, magnesia, starch, sucrose etc.

Q. 13. What is gas chromatography?

Ans. This is the latest technique for separation and identification of components of a mixture. In this technique the mobile phase is a gas (carrier gas) like nitrogen, hydrogen, helium, argon etc. It could also be mixture of two such gases. The gas is passed over a solid which acts as a support for the stationary liquid phase. For this purpose an inert solid is taken to support the stationary liquid phase. It is packed into a steel tube of the shape of U or W or a coil kept in a thermostated oven to maintain a constant temperature, as shown in Fig. 5.6.

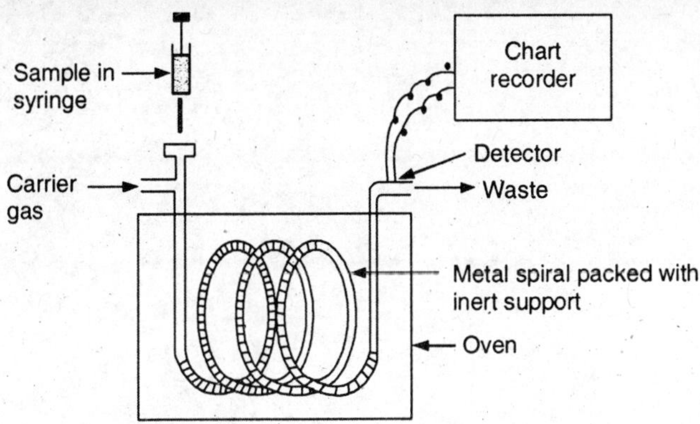

Fig. 5.6. Key parts of a gas chromatography.

The sample is injected into the coil with the help of a syringe through a rubber septum. The oven is maintained at a temperature high enough to vaporise the sample, which is pushed through the coil by the carrier gas. After leaving the coil, the sample and carrier gas mixture passes through a detector. The detector works usually on the principle of measuring electrical conductivity of the gaseous mixture. The signals are amplified and sent to a chart recorder. A separate peak is given for each component in the mixture. The area under the peak is proportional to the amount of the substance present. Thus by measuring area under the peak, we can also make a quantitative estimation of the components present in the mixture.

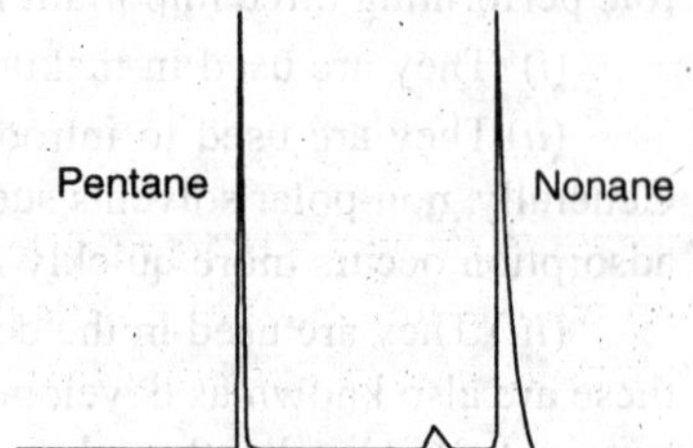

Fig.5.7. Peaks of pentane and nonane in a gas chromatogram

Fig. 5.7 shows the peaks of pentane and nonane obtained from a gas chromatograph.

6

STEREOCHEMISTRY

Q. 1. What is structural isomerism? (*Devi Ahilya, 2001; Kanpur, 2001; Kurukshetra, 2001; A.N. Bahuguna, 2000; Kumaon, 2000; Garhwal, 2000; Madras, 2004*)

Ans. If two or more compounds differ in the relative arrangement of atoms in the molecule, they are said to be structural isomers and this phenomenon is known as **structural isomerism.** There are different kinds of structural isomerisms as under:

(*i*) Chain isomerism or Nuclear isomerism.

(*ii*) Position isomerism.

(*iii*) Functional isomerism.

(*iv*) Metamerism.

(*v*) Tautomerism.

Each one of these is discussed as under.

***(i)* Chain Isomerism or Nuclear Isomerism.** If different compounds of the same class of organic compounds, having the same molecular formula, differ in the structure of carbon chain, they are called chain isomers. Examples of this type of isomerism are:

(*a*) *n-Butane and Isobutane (Mol Formula =* (C_4H_{10})

```
    H   H   H   H                    H     H       H
    |   |   |   |                    |     |       |
H - C - C - C - C - H      and   H - C --- C ----- C - H
    |   |   |   |                    |     |       |
    H   H   H   H                    H  H- C - H   H
                                           |
                                           H
       n-Butane                        Isobutane
```

(*b*) *n-pentane, isopentane and neo-pentane (Mol. Formula =* C_5H_{12}*)*

```
    H   H   H   H   H                 H     H       H   H
    |   |   |   |   |                 |     |       |   |
H - C - C - C - C - C - H  ;      H - C --- C ----- C - C - H
    |   |   |   |   |                 |     |       |   |
    H   H   H   H   H                 H  H- C - H   H   H
                                            |
                                            H
         n-Pentane                      Isopentane

                        H
                        |
                H   H - C - H   H
                |       |       |
            H - C ----- C ----- C - H
                |       |       |
                H   H - C - H   H
                        |
                        H
                   Neo-pentane
```

***(ii)* Position isomerism.** If different compounds, belonging to same homologous series, with same molecular formula have same carbon skeleton but differ in the position of substituent or functional group; these are known as *position isomers.* Examples of this type of isomers are:

(*a*) 1-Propanol and 2-propanol (Mol. Formula = C_3H_8O)

$$\underset{\text{1-Propanol}}{CH_3-CH_2-CH_2OH} \quad \text{and} \quad \underset{\text{2-Propanol}}{CH_3-\underset{\displaystyle OH}{\underset{|}{C}H}-CH_3}$$

(*b*) 1-Butene and 2-Butene (Mol. Formula = C_4H_8)

$$\underset{\text{1-Butene}}{CH_2=CH_2-CH_2-CH_3} \quad \text{and} \quad \underset{\text{2-Butene}}{CH_3-CH=CH-CH_3}$$

***(iii)* Functional isomerism.** Different compounds, with same molecular formula but different functional groups are known as *functional isomers.* For example:

(*a*) Ethyl alcohol and Dimethyl ether (Mol. Formula = C_2H_6O)

$$\underset{\text{Ethyl alcohol}}{CH_3CH_2OH} \quad \text{and} \quad \underset{\text{Dimethyl ether}}{CH_3-O-CH_3}$$

(*b*) Propionaldehyde and acetone (Mol. Formula = C_3H_6O)

$$\underset{\text{Propionaldehyde}}{CH_3CH_2CHO} \quad \text{and} \quad \underset{\text{Acetone}}{CH_3-CO-CH_3}$$

***(iv)* Metamerism.** This is a special kind of structural isomerism in which different compounds, with same molecular formula, belong to same homologous series but differ in the distribution of alkyl groups around a central atom. Examples are:

(*a*) Diethyl ether and methyl propyl ether (Mol. Formula = $C_4H_{10}O$)

$$\underset{\text{Diethyl ether}}{C_2H_5-O-C_2H_5} \quad \text{and} \quad \underset{\text{Methyl propyl ether}}{CH_3-O-C_3H_7}$$

(*b*) Diethyl ketone and methyl propyl ketone (Mol. Formula = $C_5H_{10}O$)

$$\underset{\text{Diethyl ketone}}{C_2H_5-CO-C_2H_5} \quad \text{and} \quad \underset{\text{Methyl propyl ketone}}{CH_3-CO-C_3H_7}$$

This type of isomers are known as *metamers* and the phenomenon is known as *metamerism.*

***(v)* Tautomerism.** Compounds whose structures differ in the arrangement of atoms but which exist simultaneously in dynamic equilibrium with each other, are called *tautomers.*

In most of the cases tautomerism is due to shifting of a hydrogen atoms from one carbon (or oxygen or nitrogen) to another with the rearrangement of single or double bonds. For example,

$$\underset{\text{Ethyl acetoacetate (Keto form)}}{CH_3-\overset{\overset{\displaystyle O}{\|}}{C}-CH_2-COOC_2H_5} \rightleftharpoons \underset{\substack{\text{Ethyl acetoacetate}\\\text{(Enolic form)}}}{CH_3-\overset{\overset{\displaystyle O}{|}}{C}-CH-COOC_2H_5}$$

$$\underset{\text{Nitroethane}}{CH_3-CH_2-N\begin{matrix}\nearrow O\\ \searrow O\end{matrix}} \rightleftharpoons \underset{\text{Isonitroethane}}{CH_3-CH=N\begin{matrix}\diagup OH\\ \searrow O\end{matrix}}$$

This phenomenon is called *tautomerism.*

Q. 2. Define structural isomers which exist in equilibrium with each other. Give one example. (*Bangalore 2004*)

Ans. See Q. 1 under tautomerism.

Q. 3. Write the structures of all possible isomers of C_3H_8O giving their IUPAC names.

Ans. Position isomers

$CH_3CH_2\,CH_2OH$ (Propanol–1) $CH_3-CHOHCH_3$ (Propanol–2)

Functional isomers

$CH_3CH_2CH_2OH$ (Propanol–1) $CH_3OCH_2CH_3$ (Methoxy ethane)

Q. 4. What is stereoisomerism? What are different kinds of stereoisomerism? *(Kerala, 2000)*

Ans. Compounds having different three-dimensional relative arrangement of atoms in space are called stereoisomers. This phenomenon is called *stereoisomerism.* These compounds are said to have different configurations. Stereoisomerism is of the following different kinds:

(*i*) Optical isomerism (*ii*) Geometrical isomerism
(*iii*) Conformational isomerism.

Q. 5. What is meant by plane polarised light? *(Delhi 2002)*

Ans. Ordinary light has vibrations taking place at right angles to the direction of propagation of light spread in all the possible planes. If we pass ordinary light through **nicol** prism, vibrations in all planes except one are cut off. Thus light coming out of nicol prism has vibrations only in one plane. Such a light is called *plane polarised light.*

Q. 6. What is optical activity? *(Garhwal, 2010)*

Ans. Behaviour of certain substances is strange. When a plane polarised light is passed through the solution of such substances, the light coming out of the solution is found to be in a different plane. The plane of polarised light is rotated. *Such substances, which rotate the plane of plane polarised light when placed in its path are known as optically active substances and the phenomenon* is known as *optical activity.* The angle of rotation (α) of plane polarised light is known as *Optical rotation*. The substances which rotate the plane of polarised light to the clockwise or right direction are known as *dextrorotatory* or having positive (+) rotation and those which rotate the plane polarised light to the anticlockwise or left direction are known as *laevorotatory* or having negative (–) rotation. Substances which do not rotate the plane of polarised light are said to be optically inactive.

The instrument used for measuring optical rotation is called polarimeter. It consists of a light source, two nicol prisms and in between a tube to hold the solution of organic substance. The schematic representation of a polarimeter is given in Fig. 6.1.

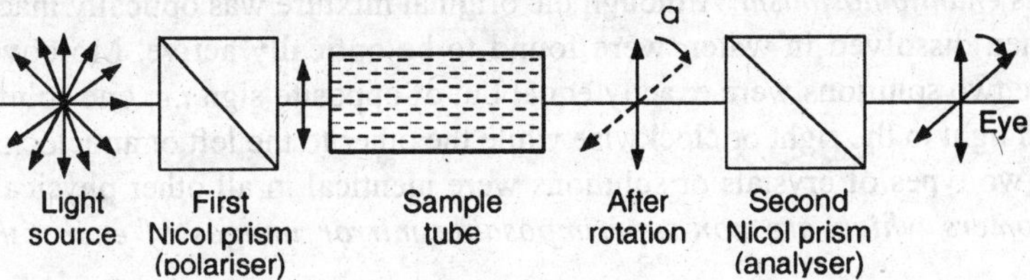

Q. 7. What is specific rotation of an optically active substance? *(Madras, 2011)*

Ans. Specific rotation. The angle of rotation of plane-polarised light or optical rotation (α) of an organic substance depends not only on the kind of molecules but also varies considerably with the number of molecules that light encounters in its path which in turn depends on the concentration of the solution used and the length of polarimeter tube containing it. Besides this, it depends on temperature, wavelength of light and nature of solvent used.

The optical activity of a substance is expressed in terms of *specific rotation.* $[\alpha]_\lambda^t$ which is a constant quantity, characteristics of a particular substance,

$$[\alpha]_\lambda^t = \frac{\alpha}{l \times c}$$

where, α = observed rotation in degrees

l = length of polarimeter tube in decimeter

c = concentration of substance in gm per ml of solution

t and λ signify the temperature and wavelength of light used.

When $l = 1$ and $c = 1$; $[\alpha]_\lambda^t = \alpha$

Specific rotation is thus defined as the optical rotation produced by a compound when plane polarised light passes through one decimeter length of the solution having concentration one gram per millilitre. Usually the monochromatic light used is D line of sodium (λ = 589 nm). Thus specific rotation of cane sugar can be expressed as

$$[\alpha]_D^{20°C} = +66.5° \text{ (water)}$$

In this expression D stands for D line of sodium, 20°C is temperature of measurement, + sign shows the dextrorotation and water is the solvent used.

Q. 8. Write a note on enantiomerism. (*Coimbatore, 2000; Kanpur, 2001; Kerala, 2001, Awadh, 2000; Madurai 2004; Jiwaji, 2004; Nagpur, 2008*

Or

Define optical isomerism. Mention its conditions and selecting lactic acid as example, how many isomers are possible in it.

(*Baroda, 2003; Marathwada, 2004; Mumbai, 2004; Delhi, 2004; Bharathiar 2011*)

or Write the structure of any three different kinds of molecules whose optical activity is due to a chiral axis. (*Kerala, 2000*)

Ans. Louis Pasteur, while studying the crystallography of salts of tartaric acid made a peculiar observation. He observed that optically inactive sodium ammonium tartarate existed as a mixture of two different types of crystals which were mirror images of each other. With the help of a hand lens and a pair of forceps, he carefully separated the mixture into two different types of crystals. These crystals were mirror images of each other and were called *enantiomorphs* and the phenomenon as *enantiomorphism*. Although the original mixture was optically inactive; each type of crystals when dissolved in water, were found to be optically active. Moreover the specific rotations of the two solutions were exactly equal but of opposite sign *i.e.* one solution rotated the plane polarised light to the right or clockwise while the other to the left or anticlockwise and to the same extent. Two types of crystals or solutions were identical in all other physical and chemical properties. *Isomers which are non-superimposable mirror images of each other are called enantiomers.*

According to La Bell and Van't Hoff the four valencies of a carbon atom are directed towards the four corners of a regular tetrahedron at the centre of which lies the carbon atom. Consider a compound of formula C_{LMNO} having four different groups L, M, N and O attached to a carbon atom. This compound can be represented by two models which look like mirror images of each other.

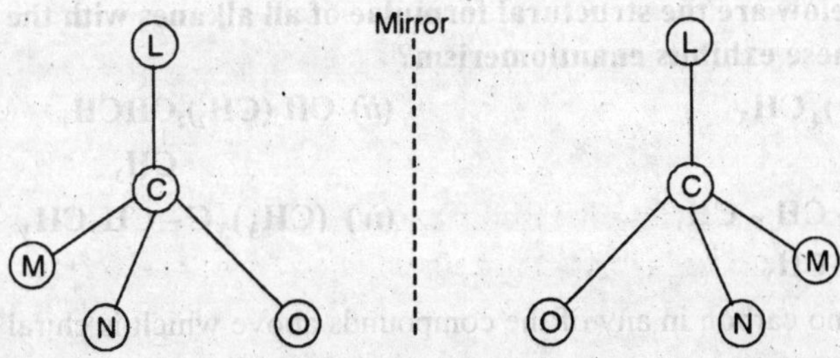

It is important to note here that these two molecules cannot be superimposed on each other *i.e.*, they will not coincide in all their parts. We may turn them in as many ways as we like but we find that though two groups of each may coincide, the other two do not. Hence these must represent two isomers of formula C_{LMNO}. Lactic acid $CH_3CHOHCOOH$ and sec-Butyl chloride

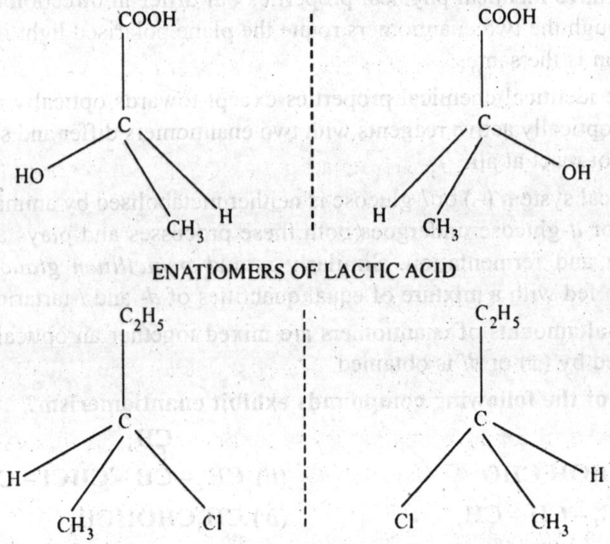

ENATIOMERS OF LACTIC ACID

ENANTIOMERS OF SEC. BUTYL CHLORIDE

$C_2H_5CHClCH_3$, exist as two optically active isomers which are enantiomers *i.e.*, mirror images of each other. Mirror images of the two compounds are represented as above.

The carbon atom to which four different groups are attached, is known as *asymmetric* or *chiral carbon atom.*

If two of the groups attached to carbon are same, we shall observe that it is possible to superimpose the mirror images on each other. Such a compound will not show optical isomerism or *enantiomerism*.

Hence non-superimposability of the mirror images is responsible and essential for the type of stereoisomerism known as enantiomerism.

The term optical isomerism is used for the existence of stereoisomers which differ in their behaviour towards the plane polarised light. Thus enantiomeric molecules are always non-superimposable mirror images of each other. The non-superimposability of mirror images arises due to **chiral** or **asymmetric** nature of molecule. A molecule is said to be chiral if it has no plane of symmetry and is therefore non-superimposable on its mirror image.

It may be concluded with the remarks that chirality is the fundamental condition of enantiomerism or optical isomerism.

Q. 9. Given below are the structural formulae of all alkanes with the molecular formula C_6H_{14} Which of these exhibits enantiomerism?

(*i*) $CH_3(CH_2)_4CH_3$ **(*ii*) $CH_3(CH_2)_2\underset{\substack{|\\CH_3}}{C}HCH_3$**

(*iii*) $CH_3\underset{\substack{|\\CH_3}}{C}H - \underset{\substack{|\\CH_3}}{C}H - CH_3$ **(*iv*) $(CH_3)_3C - CH_2CH_3$**

Ans. There is no carbon in any of the compounds above which is chiral *i.e.* attached to four different groups.

Chirality is the necessary condition for a molecule to exhibit enantiomerism. Hence none of the compounds above shows enantiomerism.

Q. 10. Describe some properties of enantiomers. (*Himachal, 2000; Kerala, 2001*)

Ans. (*i*) They have identical physical properties but differ in direction of rotation of plane-polarised light. Though the two enantiomers rotate the plane polarised light in opposite direction, the extent of rotation is the same.

(*ii*) They have identical chemical properties except towards optically active reagents. The rates of reaction of optically active reagents with two enantiomers differ and sometimes one of the enantiomers does not react at all.

(*iii*) In biological system (–) or *l*-glucose is neither metabolised by animals nor fermented by yeast whereas (+) or *d*-glucose undergoes both these processes and plays an important role in animal metabolism and fermentation. Similarly mould *penicillium glaucum* consumes only *d*-tartaric acid when fed with a mixture of equal quantities of *d*- and *l*-tartaric acid.

(*iv*) When equal amounts of enantiomers are mixed together an optically inactive racemic modification denoted by (±) or *dl* is obtained.

Q. 11. Which of the following compounds exhibit enantiomerism?

(*i*) $CH_2OH\ CHOH\ CHO$ **(*ii*) $CH_3 - \overset{\substack{CH_3\\|}}{C}H - CHCl - CH_3$**

(*iii*) $CH_3 - CH_2 - \underset{\substack{|\\CH_3}}{C}H - CH_3$ **(*iv*) $CH_3CHOHCH_3$**

(*v*) CH_2NH_2COOH

Ans. Compounds (*i*), (*ii*) and have chiral carbons marked with asterisks and hence these two compounds show enantiomerism.

$$CH_2OH\overset{*}{C}HOHCHO$$

$$CH_3 - \overset{\substack{CH_3\\|}}{C}H - \overset{*}{C}HCl - CH_3$$

Compounds (*iii*), (*iv*) and (*v*) have no chiral carbon atom in the molecule. Hence they do not show enantiomerism.

Q. 12. Point out the optically active compounds out of the following:

(*i*) $CH_2OH - CHOH - CHO$

(*ii*) $CH_3 - CHOH - CH_2OH$

(*iii*)

$$\begin{array}{c} CHOH \\ | \\ H - C - OH \\ | \\ HO - C - H \\ | \\ CHOH \end{array}$$

Ans. Compounds (*i*) and (*ii*) have one chiral carbon (the middle one) and hence both these compounds are optically active compounds. (*iii*) has two chiral carbons (the middle ones) and therefore this compound is also optically active.

Q. 13. Describe in detail the Fischer's projection formula for planar representation of three dimensional molecules.

Ans. Emil Fischer in 1891 introduced a simple method for representing three dimensional molecules in one plane. It is known as *Fischer projection formula.* Following points are to be observed for this purpose:

(*i*) The chiral molecule is imagined in such a way that two groups point towards the observer and two away from the observer. The groups pointing towards the observer are written along the horizontal line (shown as thick wedge-like bonds) and those pointing away are written along the vertical line. The central carbon is present at the crossing of the horizontal and vertical lines.

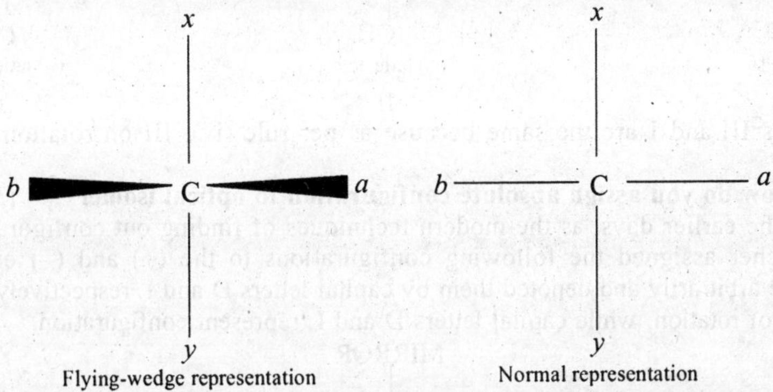

Flying-wedge representation Normal representation

Thus if *a, b, x* are *y* four groups attached to a carbon, the molecule will be represented by the projection formula as above. Here *a* and *b* groups point towards the observer (or above the plane) and groups *x* and *y* are away from the observer (or below the plane).

(*ii*) The longest chain of carbon atoms in the molecule should be represented along the vertical line. Lactic acid, therefore, according to the above conventions will be represented as

$$\begin{array}{c} COOH \\ | \\ H-C-OH \\ | \\ CH_3 \end{array} \quad \text{or} \quad \begin{array}{c} COOH \\ | \\ H-\!\!\!+\!\!\!-OH \\ | \\ CH_3 \end{array}$$

(*iii*) We can avoid writing carbon at the crossing of the vertical and horizontal lines. A crossing automatically means the presence of a carbon.

(*iv*) If necessary, planar formula may be imagined to be rotated from end to end without lifting it from the plane of the paper. Rotation by 180° in the plane of the paper does not create any change in the configuration of the molecule.

$$\begin{array}{c} COOH \\ | \\ H-\!\!\!+\!\!\!-OH \\ | \\ CH_3 \end{array} \xrightarrow[\text{through } 180^\circ]{\text{Rotation}} \begin{array}{c} CH_3 \\ | \\ HO-\!\!\!+\!\!\!-H \\ | \\ COOH \end{array}$$

I (–) lactic acid II (–) lactic acid

In the above rotation by 180°, II has been obtained from I. There has been no change in configuration of the molecule. I and II are infact the same thing.

(*v*) Rotation by 90° or 270° brings about a change in configuration of the molecule. Consider the following rotation.

```
        COOH                              H
         |                                |
  H -----+----- OH   --Rotation by-->  CH3 ---+--- COOH
         |               90°                  |
        CH3                                   OH
  (–) lactic acid                      (+) lactic acid
```

(*vi*) **If the positions of two groups across the chiral atom are interchanged, it leads to inversion of configuration. Two consecutive such changes neutralise the effect.**

```
       COOH                          COOH                          CH3
        |                             |                             |
  H ----+---- OH  --First-->   HO ----+---- H  --Second-->  HO -----+----- H
        |       Interchange           |       Interchange           |
       CH3                           CH3                           COOH
  (–) latic acid               (+) latic acid               (–) latic acid
        I                             II                           III
```

Structures III and I are the same because as per rule (iv), III on rotation through 180° gives I.

Q. 14. How do you assign absolute configuration to optical isomers? (*Nagpur 2003*)

Ans. In the earlier days, as the modern techniques of finding out configuration were not available, Fischer assigned the following configurations to the (+) and (–) enantiomers of glyceraldehyde arbitrarily and denoted them by capital letters D and L respectively. Small letters represent sign of rotation, while capital letters D and L represent configuration.

```
                        MIRROR
            CHO            |            CHO
             |             |             |
      H -----+----- OH     |     HO -----+----- H
             |             |             |
           CH2OH           |           CH2OH
    D (+) glyceraldehyde       L (–) glyceraldehyde
```

The relative configurations of a number of other optically active compounds have been established by correlating them with D(+) or L(–) glyceraldehyde. All those optically active compounds, which are obtained from D(+) glyceraldehyde through a sequence of reactions *without breaking the bonds of asymmetric carbon atom*, are designated as D configuration irrespective of their sign of rotation and the other enantiomer as L configuration.

For example,

```
          CHO                                  COOH
           |                                    |
    H -----+----- OH   --Br2/HO2-->      H -----+----- OH
           |                                    |
         CH2OH                                CH2OH
   D(+) Glyceraldehyde                  D(–) Glyceric acid
                                                |
                                                | PBr3
                                                v
          COOH                                 COOH
           |                                    |
    H -----+----- OH   <--Zn, H+--       H -----+----- OH
           |                                    |
          CH3                                 CH2Br
   D(–) Lactic acid                      D(–) 3-Bromo-
                                      2-Hydroxy Propanoic acid
```

In all the D configurations –OH attached to asymmetric carbon atom is written on the right hand side of Fischer projection formula.

Q. 15. Describe R and S specification for the configuration of an optically active compound. *(Kerala, 2001; Punjab, 2002; Nagpur, 2005; Andhra, 2010)*

Ans. Cahn, Ingold and Prelog developed a method which can be used to designate the configuration of all the molecules containing asymmetric carbon atom (chiral centre). This system is known as *Cahn-Ingold-Prelog system* or R and S system and involves two steps.

Step I. The four different atoms or groups of atoms attached to chiral carbon atom are assigned a sequence of priority according to the following set of sequence rules.

Sequence Rule 1. *If the four atoms, directly attached to asymmetric carbon atom, are all different, the priority depends on their atomic number. The atom of higher atomic number gets higher priority.* For example, in chloroiodomethane sulphonic acid the priority sequence is I, Cl, SO_3H, H.

```
         SO3H (3)
           |
       I - C - Cl
      (1)  |  (2)
           H
          (4)
```

We consider the atom of the group which is directly linked to the central carbon.

Sequence Rule 2. *If Rule 1 fails to decide the relative priority of two groups it is determined by similar comparison of next atoms in the group and so on.* In other words, if two atoms directly attached to chiral centre are same, the next atoms attached to each of these atoms, are compared. For example in 2-butanol two of the atoms directly attached to chiral centre are carbon themselves. To decide the priority between the two groups $-CH_3$ and $-CH_2CH_3$, we proceed like this. Methyl carbon is further linked to H, H and H. The sum of atomic numbers of three H is 3. The methylene carbon of the ethyl group is linked to two hydrogens and one carbon directly. The sum of at. no. of two H and one C is 8. Thus ethyl group gets the priority over methyl. Hence the priority sequence is OH, C_2H_5, CH_3, H.

```
                 OH (1)
                  |
   CH3 - CH2  -  C  - CH3
          (2)     |    (3)
                 H (4)
```

2-Butanol

In 2-methyl 3-pentanol, the C, C, H of isopropyl gets priority over the C, H, H of ethyl, so the priority sequence is OH, isopropyl, ethyl, H.

```
         CH3  H (4)
          |   |
   CH3 - CH - C - CH2 - CH3
         (2)  |   (3)
              OH
             (1)
```

2-Methyl-3-pentanol

In 1, 2-dichloro-3-methyl butane the Cl, H, H of CH_2Cl of gets priority over the C, C, H of isopropyl due to atomic number of Cl being higher than that of C. So the priority sequence is Cl, CH_2Cl, isopropyl, H.

```
         CH3  H (4)
          |   |
   CH3 - CH - C - CH2Cl
         (3)  |   (2)
              Cl
             (1)
```

1, 2-dichloro-3-methylbutane

Sequence Rule 3. *A doubly or triply bonded atom is considered equivalent to two or three such atoms; but two or three atoms, if attached actually, get priority over doubly or triply bonded atom.* In glyceraldehyde, O, O, H of –CHO gets priority over the O, H, H of $-CH_2OH$; so the priority sequence is –OH, – CHO, $-CH_2OH$, –H.

$$\begin{array}{c} (2)\ CHO \\ | \\ H - C - OH\ (1) \\ (4) \quad | \quad\quad \\ CH_2OH \\ (3) \end{array}$$

Glyceraldehyde

Step II. After deciding the sequence of priority for four atoms or groups attached to asymmetric carbon atom; the molecule is visualised in such a way that the atom or group of lowest or last (*i.e.*, fourth) priority is directed away from us, while the remaining three atoms or groups are pointing towards us. Now if on looking at these three groups (pointing towards us) in the order of their decreasing priority, our eye moves in clockwise direction, the configuration is specified as R (from Latin word *rectus* meaning right) and on the other hand if our eye moves in anticlockwise direction the configuration is specified as S (from Latin word *sinister* meaning left).

The following examples illustrate the above method for specification of configuration as R and S to molecules of compounds containing an asymmetric or chiral carbon atom.

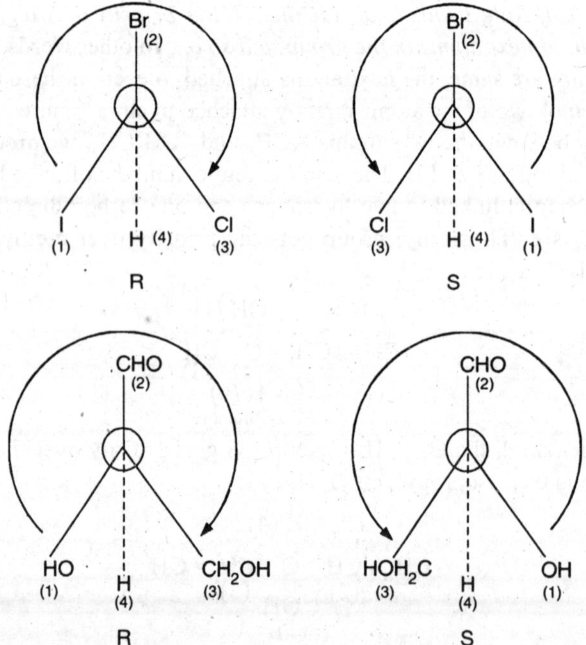

Configuration on the Basis of Projection Formula

When a compound is represented by the Fischer projection formula, the configuration can be easily determined without constructing the model. To determine whether the eye travels clockwise or anticlockwise, we have to place the group or atom of the lowest priority at the bottom of the Fischer projection formula. The following four situations arise:

(*i*) *The atom/group of lowest priority is at the bottom.* In such case, simply rotate the eye in the order of decreasing priorities. The configuration is R if the eye travels in clockwise direction and S if the eye travels in *anticlockwise* direction.

For example, Glyceraldehyde, represented by the following projection formula has R configuration:

(1) OH
(3) | (2)
$HOH_2C - C - CHO$
|
H

R-Glyceraldehyde

(*ii*) *The atom/group of lowest priority is at the top.* In such case, rotate the molecule by 180° so as to bring atom/group of lowest priority at the bottom. This can be done by reversing the position of all the atoms or groups. Then find the direction in the order of decreasing priorities. For example, the compound CHBrClI, represented by following projection formula, has S configuration:

H
|
I – C – Cl ≡ (2) Br, (3) Cl – C – I (1), H

S-Bromochloroiodomethane

(*iii*) *The atom/group of lowest priority is at the right hand side of the horizontal line.* In such case, change the position of atoms or groups in clockwise direction so that atom/group of the lowest priority comes at the bottom but do not change the position of the atom/group at the top of the vertical end. Then find the direction in the order of decreasing priorities. For example, CHBrClI, represented by following projection formula, has S configuration.

Br
|
I – C – H
|
Cl

≡ (2) Br, (3) Cl – C – I (1), H

S-Bromochloroiodo methane

(*iv*) *The atom/group of lowest priority is at the left hand side of the horizontal line.* In such case, without changing the position of atom/group at the top of the vertical end, change the position of other atoms/groups in the anticlockwise direction so that atom/group of lowest priority comes at the bottom. Then find the direction in the order of decreasing priorities. For example, CHBrClI, represented by following projection formula, has R configuration.

Br
|
H – C – I
|
Cl

≡ (2) Br, (1) I – C – Cl (3), H

R-Bromochloroiodo methane

Q. 16. Assign R and S configuration to the following: (*Kerala, 2000*)

(*i*)
Cl CH$_3$
| |
H – C – C – H
| |
CH$_3$ H

(*ii*) $HO-\overset{\overset{NH_2}{|}}{\underset{\underset{Cl}{|}}{C}}-CH_3$

(*iii*) $HO-\overset{\overset{H}{|}}{\underset{\underset{COOH}{|}}{C}}-CH_3$

(*iv*) $CH_3-\overset{\overset{C_2H_5}{|}}{\underset{\underset{CHO}{|}}{C}}-COOH$

Ans. (*i*) $H-\overset{\overset{Cl}{|}}{\underset{\underset{CH_3}{|}}{C}}-\overset{\overset{CH_3}{|}}{\underset{\underset{H}{|}}{C}}-H \longrightarrow H-\overset{\overset{CH_3}{|}}{\underset{\underset{H}{|}}{C}}-\overset{\overset{Cl}{|}}{\underset{\underset{H}{|}}{C}}-CH_3$ [S]

(*ii*) $HO-\overset{\overset{NH_2}{|}}{\underset{\underset{Cl}{|}}{C}}-CH_3 \longrightarrow Cl-\overset{\overset{NH_2}{|}}{\underset{\underset{CH_3}{|}}{C}}-OH$ [S]

(*iii*) $OH-\overset{\overset{H}{|}}{\underset{\underset{COOH}{|}}{C}}-CH_3 \xrightarrow[\text{Through } 180°]{\text{Rotate}} CH_3-\overset{\overset{COOH}{|}}{\underset{\underset{H}{|}}{C}}-OH$ [S]

(*iv*) $CH_3-\overset{\overset{C_2H_5}{|}}{\underset{\underset{CHO}{|}}{C}}-COOH \longrightarrow HOOC-\overset{\overset{C_2H_5}{|}}{\underset{\underset{CH_3}{|}}{C}}-CHO$ [S]

Q. 17. Assign R and S configuration to the following: (*M. Dayanand, 2000*)

(*i*) $C_2H_5-\overset{\overset{Br}{|}}{\underset{\underset{OH}{|}}{C}}-H$ (*ii*) $H-\overset{\overset{CH_3}{|}}{\underset{\underset{C_2H_5}{|}}{C}}-NH_2$ (*iii*) $C_2H_5-\overset{\overset{CHO}{|}}{\underset{\underset{NH_2}{|}}{C}}-H$

Ans. (*i*) $C_2H_5-\overset{\overset{Br}{|}}{\underset{\underset{OH}{|}}{C}}-H \longrightarrow HO-\overset{\overset{Br}{|}}{\underset{\underset{H}{|}}{C}}-C_2H_5$ [S]

(*ii*) $H-\overset{\overset{CH_3}{|}}{\underset{\underset{C_2H_5}{|}}{C}}-NH_2 \longrightarrow H_2N-\overset{\overset{CH_3}{|}}{\underset{\underset{H}{|}}{C}}-C_2H_5$ [S]

$$(iii)\quad C_2H_5 - \overset{\displaystyle CHO}{\underset{\displaystyle NH_2}{C}} - H \longrightarrow H_2N - \overset{\displaystyle CHO}{\underset{\displaystyle H}{C}} - C_2H_5 \quad [S]$$

Q. 18. Assign R and S configuration to the following compounds.

$$(i)\quad CH_3 - \overset{\displaystyle C_2H_5}{\underset{\displaystyle CH_2OH}{C}} - H \qquad (ii)\quad HO - \overset{\displaystyle H}{\underset{\displaystyle COOH}{C}} - CH_3$$

Ans. (*i*)

$$CH_3 - \overset{\displaystyle C_2H_5}{\underset{\displaystyle CH_2OH}{C}} - H \longrightarrow HOH_2C - \overset{\displaystyle C_2H_5}{\underset{\displaystyle H}{C}} - CH_3 \quad [S]$$

(*ii*)

$$OH - \overset{\displaystyle H}{\underset{\displaystyle COOH}{C}} - CH_3 \xrightarrow[180^\circ]{\text{Rotate by}} H_3C - \overset{\displaystyle COOH}{\underset{\displaystyle H}{C}} - OH \quad [S]$$

Q. 19. Assign R and S configuration to the following:

$$(i)\quad H - \overset{\displaystyle COOH}{\underset{\displaystyle CH_3}{C}} - NH_2 \qquad (ii)\quad OH - \overset{\displaystyle CHO}{\underset{\displaystyle CH_2OH}{C}} - H$$

Ans. (*i*)

$$H - \overset{\displaystyle COOH}{\underset{\displaystyle CH_3}{C}} - NH_2 \longrightarrow H_2N - \overset{\displaystyle COOH}{\underset{\displaystyle H}{C}} - CH_3 \quad [R]$$

(*ii*)

$$HO - \overset{\displaystyle CHO}{\underset{\displaystyle CH_2OH}{C}} - H \longrightarrow HOH_2C - \overset{\displaystyle CHO}{\underset{\displaystyle H}{C}} - OH \quad [S]$$

Q. 20. A carboxy acid of the formula $C_3H_5O_2Br$ is optically active. What is its structure?

Ans. Two structures are possible for a carboxy acid with the above molecular formula

$$CH_3 - \overset{\displaystyle H}{\underset{\displaystyle Br}{C}} - COOH \qquad BrCH_2 - \overset{\displaystyle H}{\underset{\displaystyle H}{C}} - COOH$$

$$\text{I} \qquad\qquad\qquad \text{II}$$

Of the two structures, structure I has a chiral (or asymmetric) carbon atom whereas structure II has none. Therefore the structure showing optical activity is I.

Q. 21. How will you establish the configuration of compounds containing more than one chiral carbon? (*Calcutta 2000*)

Ans. In such a case, firstly the configuration about each of the chiral carbon is specified and then with the help of numbers, the specification pertaining to the carbon atom of that number is written. Thus the configurations of isomers of 2, 3, 4-Trihydroxybutanal are:

(A)	(B)
^{1}CHO	^{1}CHO
$H-^{2}C-OH$	$HO-^{2}C-OH$
$H-^{3}C-OH$	$HO-^{3}C-OH$
$^{4}CH_2OH$	$^{4}CH_2OH$
(2R, 3R) -2, 3, 4-Trihydroxy butanal (A)	(2S, 3S)-2, 3 4-Trihydroxy butanal (B)

The configurations are explained as follows:

In the compound A above, sequence of groups attached to C_2 is OH, CHO, CHOH – CH_2OH and H.

Now in order to fix specification, first we consider C_2 and ignore C_3

$$\begin{array}{c} CHO \\ | \\ H-C_2-OH \\ | \\ CHOHCH_2OH \end{array} \underset{\text{Interchanges}}{\equiv} \begin{array}{c} (2) \\ CHO \\ (1)\quad | \quad (3) \\ HO-C_2-CHOHCH_2OH \\ | \\ H \end{array}$$

Ignoring (C_3) (2R)

Similarly we consider C_3 and we ignore C_2. Thus

$$\begin{array}{c} CHOHCHO \\ | \\ H-C_3-OH \\ | \\ CH_2OH \end{array} \underset{\text{Interchanges}}{\equiv} \begin{array}{c} (2) \\ CHOHCHO \\ (1)\quad | \quad (3) \\ HO-C_3-CH_2OH \\ | \\ H \end{array}$$

Ignoring (C_2) (3R)

(Here the sequence of groups attached to C_3 is OH, CHOHCHO, CH_2OH and H)

Hence the configuration of compound A is (2R, 3R)–2, 3, 4-Trihydroxy butanal.

Similarly the configuration of compound B can be derived as explained below:

For C_2 :

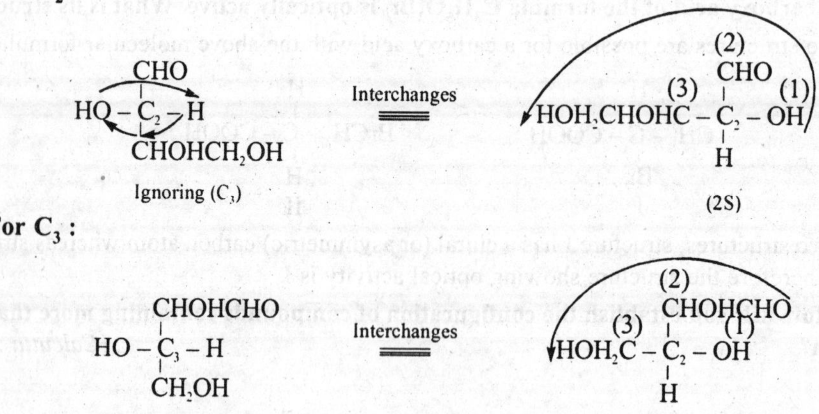

$$\begin{array}{c} CHO \\ | \\ HO-C_2-H \\ | \\ CHOHCH_2OH \end{array} \underset{\text{Interchanges}}{\equiv} \begin{array}{c} (2) \\ CHO \\ (3)\quad | \quad (1) \\ HOH_2CHOHC-C_2-OH \\ | \\ H \end{array}$$

Ignoring (C_3) (2S)

For C_3 :

$$\begin{array}{c} CHOHCHO \\ | \\ HO-C_3-H \\ | \\ CH_2OH \end{array} \underset{\text{Interchanges}}{\equiv} \begin{array}{c} (2) \\ CHOHCHO \\ (3)\quad | \quad (1) \\ HOH_2C-C_2-OH \\ | \\ H \end{array}$$

Ignoring (C_2) (3S)

Hence configuration of compound B is (2S, 3S)Å2, 3, 4-Trihydroxybutanal.

Q. 22. Prove that the presence or absence of chiral carbon atom in a molecule is not the necessary and sufficient criterion for existence of optical activity. (*Bangalore 2002*)

Ans. Optical activity is a property which is related to dissymmetry in the molecule. Dissymmetry occurs normally in compounds with chiral carbon atoms. But sometimes there are deviations. For example,

(*i*) Consider the case of meso tartaric acid

```
     COOH
      |
  H - C* - OH
      |
  H - C* - OH
      |
     COOH
```

Although there are two chiral carbons marked with asterisks, still the compound does not show optical activity. This is because the molecule has a plane of symmetry and is thus non-dissymmetric.

(*ii*) There are molecules which do not contain a chiral carbon but still show optical activity. Consider the optically active substance

$$RCH = C = CHR$$

Optical activity of this molecule is explained on the basis of dissymmetry in the molecule. Thus it can be remarked that presence or absence of chiral C-atom is not the necessary and sufficient criterion for the existence of optical activity.

In fact it is the dissymmetry criterion which is responsible for the same.

A dissymmetric molecule is that which has no plane of symmetry.

Q. 23. Assign R and S configuration to the following: (*Kerala, 2001*)

```
(i)       COOH              (ii)       NH2              (iii)          Cl
           |                            |                               |
      HO - C2 - H                  H - C - COOH              ClCH2 - C - CH(CH3)2
           |                            |                               |
       H - C3 - OH                     COOC2H5                         CH3
           |
          COOH
```

Ans. (*i*) **First consider C_2.** The groups in order of priority attached to C_2 are – OH, –COOH, –CH(OH) COOH, H

```
       COOH                               COOH
        |                                  |
  HO - C2 - H        ≡     HOOC(OH)CH - C - OH
        |                                  |
    CH(OH)COOH                             H
```

Thus the configuration around C_2 is S

Now consider C_3. The groups in order of priority attached to C_3 are –OH, –COOH, –CH(OH)COOH, H

```
   CH(OH)COOH                 CH(OH)COOH
        |                          |
   H - C3 - OH       ≡      HO - C3 - COOH
        |                          |
      COOH                         H
```

Thus the configuration around C_3 is S

Thus this structure has the configuration 2S, 3S

(*ii*)

```
        NH2
         |
    H - C - COOH
         |
       COOC2H5
```

The groups attached to central carbon, in order of priority are $–NH_2$, $–COOC_2H_5$, –COOH and H

$$\begin{array}{c} NH_2 \\ | \\ H-C^3-COOH \\ | \\ COOC_2H_5 \end{array} \equiv \begin{array}{c} NH_2 \\ | \\ HOOC-C-COOC_2H_5 \\ | \\ H \end{array}$$ It has R configuration

(*iii*)
$$\begin{array}{c} Cl \\ | \\ ClCH_2-C-CH(CH_3)_2 \\ | \\ CH_3 \end{array}$$

The groups attached, in order of priority, to the central carbon are —Cl, —CH_2Cl, —$CH(CH_3)_2$ and CH_3.

$$\begin{array}{c} Cl \\ | \\ ClCH_2-C-CH(CH_3)_2 \\ | \\ CH_3 \end{array}$$ The compound has thus S configuration

Q. 24. Assign R and S configuration to the following Fischer projection.

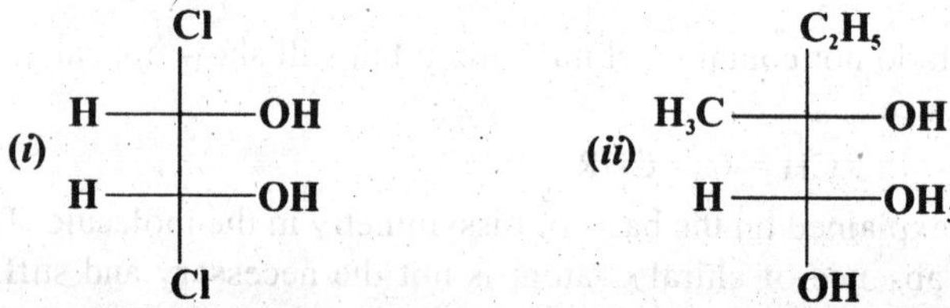

Ans. (*i*) The groups attached to C-2, in order of priority are –Cl, –OH, –CH(OH) Cl and –H

$$\begin{array}{c} Cl \\ | \\ H-C^2-OH \\ | \\ CH(OH)Cl \end{array} \equiv \begin{array}{c} Cl \\ | \\ HO-C^2-CH(OH)Cl \\ | \\ H \end{array}$$ Configuration around C-2 is S

Now consider configuration around C-3.

The groups around C-3 in order of priority are –Cl, –OH, –CH(OH)Cl and H

$$\begin{array}{c} CH(OH)Cl \\ | \\ H-C^3-OH \\ | \\ Cl \end{array} \equiv \begin{array}{c} CH(OH)Cl \\ | \\ HO-C^3-Cl \\ | \\ H \end{array}$$ The configuration around C-3 is *R*.

Thus complete configuration of the compound is 2S, 3R

(*ii*)
$$\begin{array}{c} C_2H_5 \\ H_3C - + - OH \\ H - + - CH_3 \\ OH \end{array}$$

The groups around C_2, in order of priority are –OH, –$CH(OH)CH_3$, C_2H_5 and CH_3

$$\begin{array}{c} C_2H_5 \\ | \\ H_3C-C^2-OH \\ | \\ CH(OH)CH_3 \end{array} \equiv \begin{array}{c} C_2H_5 \\ | \\ HO-C-CH(OH)CH_3 \\ | \\ CH_3 \end{array}$$ The configuration around C-2 is S

Configuration around C-3

$$\begin{array}{c} C(OH)(CH_3)C_2H_5 \\ | \\ H-C^3-CH_3 \\ | \\ OH \end{array}$$ The groups around C-3, in order of priority are –OH, –$C(OH)(CH_3)C_2H_5$, –CH_3 and –H

$$\begin{array}{c} C(OH)(CH_3)C_2H_5 \\ | \\ H-C^3-OH \\ | \\ OH \end{array} \equiv \begin{array}{c} C(OH)(CH_3)C_2H_5 \\ | \\ H_3C-C^3-OH \\ | \\ H \end{array}$$

The configuration around C-3 is S

Thus complete configuration of the compound is 2S, 3S.

Q. 25. A compound $C_4H_{10}O$ shows optical activity. Identify the compound and write the possible stereoisomers.

Ans. With the molecular formula $C_4H_{10}O$ a no. of alcohols and ethers are possible. But we are interested in a compound with a chiral carbon atom so as to give optical activity. Such a compound with the formula $C_4H_{10}O$ is 2-Butanol. It exists in two enantiomeric forms.

$$\begin{array}{c} CH_3 \\ | \\ HO-C-C_2H_5 \\ | \\ H \end{array} \qquad \begin{array}{c} CH_3 \\ | \\ C_2H_5-C-OH \\ | \\ H \end{array}$$

S-2-Butanol R-2-Butanol

Q. 26. What is meant by racemic modification? *(Kerala, 2001)*

Ans. Racemic modification is the term used for a mixture of equal amounts of enantiomers. A racemic mixture is optically inactive because of external compensation. The optical activity caused by one enantiomer is neutralised by the activity of the other enantiomer. The notation for a racemic modification or mixture is ± or *dl*. A racemic mixture may also be denoted by the letters R and S. For example RS-sec. butyl chloride.

When a chiral compound is synthesised from an achiral reactant, a racemic variety of products is obtained. For example, when propionic acid is brominated, α-bromo propionic acid (a chiral product) is obtained. The two enantiomers (+) and (–) α-bromopropionic acids are formed in equal quantities and the product is a racemic mixture. It is optically inactive.

Q. 27. Discuss optical isomerism (enantiomerism) in compounds having two dissimilar hiral carbon atoms. *(Kerala, 2001; Punjab, 2002; Nagpur, 2008; Patna, 2010)*

Ans. Two asymmetric carbon atoms are said to be dissimilar when atoms or groups attached to one asymmetric carbon atom are different from those attached to the other. Compounds of this type exist in 2^2 *i.e.*, 4 stereoisomers. For example 3-chloro-2-butanol is a compound with two dissimilar chiral carbons marked 1 and 2 and exists in the following four forms:

$$\begin{array}{c} CH_3 \\ | \\ H-{}^1C-OH \\ | \\ Cl-{}^2C-H \\ | \\ CH_3 \\ (I) \end{array} \quad \begin{array}{c} CH_3 \\ | \\ HO-C-OH \\ | \\ H-C-Cl \\ | \\ CH_3 \\ (II) \end{array} \quad \begin{array}{c} CH_3 \\ | \\ H-C-OH \\ | \\ H-C-Cl \\ | \\ CH_3 \\ (III) \end{array} \quad \begin{array}{c} CH_3 \\ | \\ HO-C-H \\ | \\ Cl-C-H \\ | \\ CH_3 \\ (IV) \end{array}$$

The structures I and II are non-superimposable mirror images, so these are a pair of enantiomers. Similarly structures III and IV are another pair of enantiomers. These two pairs of enantiomers give rise to two racemic modifications.

The structures I and III are neither enantiomers nor superimposable. Such type of stereoisomers are called diastereomers. Similarly I and IV, II and III, and II and IV, are the other pairs of diastereomers. *Diastereomers can be defined as those stereoisomers which are not mirror images of each other.*

Unlike enantiomers, diastereomers have different physical properties and may rotate the plane of polarised light in the same or different directions and to different extent.

Another example of this type is 2, 3-dichloropentane.

```
    C2H5            C2H5            C2H5            C2H5
    |               |               |               |
H - C - Cl     Cl - C - H      H - C - Cl     Cl - C - H
    |               |               |               |
Cl - C - H      H - C - Cl     H - C - Cl     Cl - C - H
    |               |               |               |
    CH3             CH3             CH3             CH3
    (I)             (II)            (III)           (IV)
```

A pair of enantiomer (I, II)   Second pair of enantiomers (III, IV)

Properties of Diastereomers

(1) They show similar, but not identical, chemical properties (as they contain the same functional groups). Rates of reactions of *diastereomers* with a given reagent are generally different.

(2) They have different physical properties like m.p. and b.p., densities, refractive indices, specific rotations, solubilities etc. in a given solvent.

(3) They can be separated by techniques like fractional crystallisation, fractional distillation and chromatography.

Q. 28. Write four configurations of tartaric acid and select the pairs forming Enantiomers, Diastereomers. Which of them are optically active and which of them are not. Why? (*Meerut, 2000; Kanpur, 2001; Awadh, 2000; Bangalore, 2002; Nagpur 2003 Karnataka 2004 ; Vidyasagar 2007 ; Delhi 2007*)

Ans. Tartaric acid is an example of compounds with two similar chiral carbon atoms. Here the four groups attached to both carbons (C-2 and C-3) are the same. Various configurations of tartaric acid are:

```
    COOH            COOH            COOH            COOH
    |               |               |               |
H - C - OH     HO - C - H      H - C - OH     HO - C - H
    |               |               |               |
OH - C - H      H - C - OH     H - C - OH     HO - C - H
    |               |               |               |
    COOH            COOH            COOH            COOH
    (I)             (II)            (III)           (IV)
```

If we rotate configuration IV by an angle of 180°, we get configuration III. Hence structures III and IV represent the same configuration. Thus there are only three configurations of tartaric acid viz., I, II & III.

(*i*) Structures I and II are enantiomers, they are mirror images of each other. There is no plane of symmetry in either of them. Therefore both I and II are optically active but display the optical activity in opposite directions.

(*ii*) Structures I and III are not mirror images of each other. Hence they are not enantiomers. Such pairs of compounds having the identical molecular formula and identical groups are called distereomers.

Similarly structures II and III also form a pair of diastereomers.

Compound III has two chiral carbons, but still it optically inactive. This is because there is a plane of symmetry as indicated by dotted line. There being no dissymmetry, the compound is not optically active. Such a compound is caused meso comp. Thus III and IV represent meso tartaric acid.

Q. 29. Explain, giving example the phenomenon of racemisation.

(*Coimbatore, 2000; A.N. Bhauguna, 2000; Garhwal, 2000; Bhartiar, 2000; Calcutta, 2000;Panjab 2002*)

Ans. Under suitable conditions, most of the optically active compounds can lose their optical activity without undergoing any change in their structure *i.e.*, the two enantiomeric forms are convertible into each other so that the final result is racemic modification. **The transformation of an optically active enantiomer into the optically inactive racemic modification under the influence of heat, light or chemical reagents is known as racemisation.** Thus if the starting material is the (+) form, then after treatment, half will be converted into (–) form. If the starting material is (–) form, half will be converted into (+) form. For example, (+) or (–) lactic acid on warming with sodium hydroxide gets converted into lactic acid.

Though different mechanisms have been presented for the racemisation of different types of compounds in most of the cases it occurs via the formation of an intermediate which is no longer dissymmetric or chiral. For example, a ketone in which chiral carbon atom is joined to a hydrogen atom can undergo racemisation via the formation of enolic form by toutomeric change.

$$\begin{array}{c} C_2H_5 \\ | \\ CH_3-C-H \\ | \\ C=O \\ | \\ C_6H_5 \\ \text{(+) form} \end{array} \rightleftharpoons \begin{array}{c} C_2H_5 \\ | \\ CH_3-C \\ \| \\ C-OH \\ | \\ C_6H_5 \\ \text{loss of chirality} \end{array} \rightleftharpoons \begin{array}{c} C_2H_5 \\ | \\ H-C-CH_3 \\ | \\ C=O \\ | \\ C_6H_5 \\ \text{(–) form} \end{array}$$

The intermediate enolic form, which is no longer chiral when reverts to the stable keto form, it is equally likely to produce (+) or (–) forms and thus racemisation takes place.

Q. 30. Giving examples distinguish between the following: Meso and Racemic forms.

(*Kurukshetra 2001; Panjab, 2003*)

Ans. Meso form of a compound is optically inactive form, in spite of the presence of asymmetric carbon atoms in it. This is because there is a plane of symmetry in the molecule. The activity of one part is neutralised by the activity of the other part. An example is meso tartaric acid.

$$\begin{array}{c} COOH \\ | \\ H-C-OH \\ | \\ H-C-OH \\ | \\ COOH \end{array} \quad \text{meso tartaric acid}$$

Racemic form means a mixture of equal amounts of *d* forms *l* of a compound, like mixture of equal amounts of *d* and *l* lactic acid or ± tartaric acid.

Such a mixture does not show any optical activity.

It may be mentioned that optical inactivity of the meso form is due to internal compensation as the activity of one part of the molecule is neutralised by that of the other part.

Optical inactivity of racemic form is due to external compensation as the two forms *d*- and *l*-neutralise the optical activity of each other.

Q. 31. Predict whether 3-chlorohexane will be optically active or not? Give reasons for your answer.

Ans.

$$\begin{array}{c} \quad\quad\quad\quad\quad\quad H \\ \quad\quad\quad\quad\quad\quad | \\ CH_3CH_2CH_2-C^*-CH_2-CH_3 \\ \quad\quad\quad\quad\quad\quad | \\ \quad\quad\quad\quad\quad\quad Cl \end{array}$$

3-chlorohexane

The compound contains one asymmetric carbon atom (marked with *). The four different groups attached are

1. –H
2. –Cl
3. $-CH_2-CH_3$
4. $-CH_2-CH_2-CH_3$

Hence the compound shows optical activity.

Q. 32. Describe various methods for the resolution of racemic mixtures.

(Punjab, 2002; Ranchi, 2003; Purvanchal, 2007, Nagpur, 2008; Andhra, 2010)

Ans. *The separation of racemic modification into enantiomers is called resolution.* Since the two enantiomers in a racemic mixture have identical physical and chemical properties, these cannot be separated by usual methods of fractional distillation or fractional crystallisation. Special methods are adopted for their separation as given below:

***(i)* Mechanical separation.** This method was first adopted by Pasteur for separating the enantiomers of ammonium tartarate. When racemic modification is crystallized from a solution, two types of crystals are obtained. These are mirror images of each other consisting of (+) and (–) forms which can be separated by hand picking with the help of a pair of tweezers and a powerful lens. This is a very laborious method and can be applied only to those compounds which give well defined distinguishable crystals of enantiomers.

***(ii)* Biochemical method.** Certain micro-organisms grow in a racemic mixture, consuming only one of the enantiomers while leaving the other unaffected. Thus *Penicillium glaucum* when placed in (±) tartaric acid, consumes only (+) tartaric acid and leaves (–) tartaric acid unused. The major disadvantage of this method is that one of the enantiomers get destroyed.

***(iii)* Chemical method.** This method is mostly used for the resolution of racemic modification. In this method the racemic modification is treated with an optically active reagent to get a pair of *diastereomers.* Since diastereomers differ in their physical properties, it is possible to separate them by physical methods such as fractional crystallisation, fractional distillation etc. The pure diastereomers are then decomposed, into a mixture of optically active reagent and corresponding enantiomer, which can be separated.

Suppose the racemic modification is an (±) acid. When it is treated with an optically active, say (–) base, it gives a mixture of two salts, one of (+) acid (–) base, the other of (–) acid (–) base.

These salts are neither superimposable nor mirror images; so these are diastereomers having different physical properties and can be separated by fractional crystallisation. After separation the optically active acids can be recovered in pure forms by adding mineral acid.

$$\left.\begin{matrix}(+)\text{ acid}\\ \\ (-)\text{ acid}\end{matrix}\right] + (-)\text{ base} \longrightarrow \begin{bmatrix}\text{Salt of}\\ (+)\text{ acid }(-)\text{ base}\\ \text{Salt of}\\ (-)\text{ acid }(-)\text{ base}\end{bmatrix} \begin{matrix}\xrightarrow{H^+} (+)\text{ acid}\\ \\ \xrightarrow{H^+} (-)\text{ acid}\end{matrix}$$

Racemic modification — Diastereomers easily separable — Enantiomers in pure form

The commonly used optically active bases for the purpose are naturally occurring alkaloids such as (–) brucine, (–) quinine, (–) strychnine and (+) cinchonine.

Similarly, the resolution of racemic bases can be carried out using a naturally occurring optically active acid such as (–) malic acid. Alcohols can be resolved in a similar way by ester formation using an optically active acid.

Q. 33. What is asymmetric synthesis? *(Madras, 2004; Nagpur, 2008; Andhra, 2010)*

Or

How will you prepare (–) lactic acid from pyruvic acid? *(M. Dayanand, 2000)*

Ans. *The preparation of an optically active dissymmetric (chiral) compound from non-dissymmetric molecules under the influence of an optically active substance is known as asymmetric synthesis.*

Whenever a dissymmetric product is synthesised from a non-dissymmetric reactants, it is always an optically inactive racemic modification (as described earlier). However, by the use of an optically active reagent, one of the enantiomers can be obtained in excess so that the resulting product is optically active.

For example pyruvic acid on direct reduction yields the optically inactive racemic lactic acid, but pyruvic acid, pre-esterified with an optically active alcohol say (–) menthol, on reduction and subsequent hydrolysis yields predominantly (–) lactic acid.

$$\underset{\text{Pyruvic acid}}{CH_3-\overset{\overset{O}{\|}}{C}-COOH} \xrightarrow[\text{reduction}]{2H} \underset{(\pm)\text{ Lactic acid}}{CH_3-CHOH-COOH}$$

$$\underset{\text{Pyruvic acid}}{CH_3-\overset{\overset{O}{\|}}{C}-COOH} + \underset{(-)\text{ Menthol}}{C_{10}H_{19}OH} \longrightarrow \underset{(-)\text{ Methyl pryuvate}}{CH_3-\overset{\overset{C}{\|}}{C}-COOC_{10}H_{19}}$$

$$\downarrow \text{reduction, } 2H$$

$$\underset{(-)\text{ Menthol}}{C_{10}H_{19}OH} + \underset{\substack{(-)\text{ Latic acid}\\ \text{(predominantly)}}}{CH_3-CHOH-COOH} \xleftarrow{H_2O,\ H} \underset{(-)\text{ Menthyl lactate}}{CH_3-CHOH-COOC_{10}H_{19}}$$

Q. 34. What happens when pyruvic acid is first treated with 1-menthol and the product reduced with H_2/Ni this is followed by hydrolysis.

Ans. Refer to question above.

Q. 35. Explain Walden inversion with examples.

(Kerala, 2001; M. Dayanand, 2000; Madurai, 2003 ; Nagpur, 2003)

Ans. In 1893, Walden was able to convert an optically active compound into its enantiomer by a series of replacement reactions. For example,

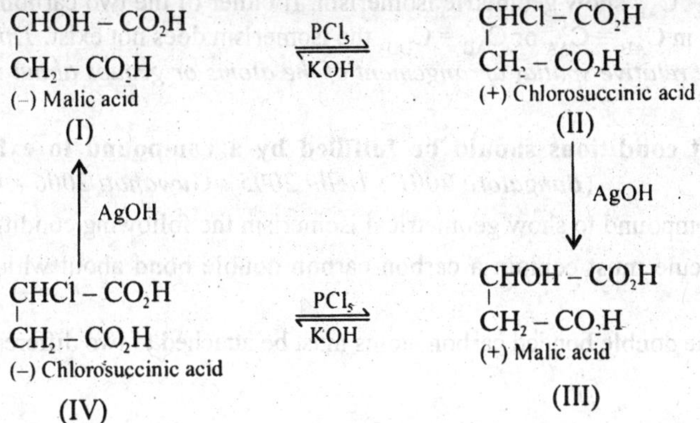

During the change of (–) malic acid (I) to (+) malic acid (III) there must be a change in configuration in one of the two steps. If the configuration of II and III is same, the change must have taken place between I and II. Any single reaction in which change of configuration takes place, is termed as Walden Inversion.

In other words if an atom or group of atoms directly attached to chiral carbon atom is replaced by another atom or group of atoms the configuration of product is sometimes found to be different from that of starting compound. Such phenomenon which involves inversion of configuration during a reaction is called *Walden Inversion or Optical Inversion.*

For example, in SN^2 hydrolysis of 2-bromo octane inversion of configuration takes place.

$$\begin{array}{c} C_6H_{13} \\ | \\ H-C-Br \\ | \\ CH_3 \end{array} \xrightarrow{OH^-} \begin{array}{c} C_6H_{13} \\ | \\ HO-C-H \\ | \\ CH_3 \end{array}$$

(–) -Bromoctane (+) 2-Octanol

It must be kept in mind here that inversion of configuration may or may not lead to a change in direction of rotation. The change, if any, is only a matter of chance as in the example given above.

GEOMETRICAL ISOMERISM

Q. 36. Show that hindered rotation around a carbon-carbon double bond gives rise to geometrical isomerism. *(Delhi, 2004 ; Punjab, 2005; Nagpur, 2005; Andhra, 2010)*

Ans. Two carbon atoms joined by a single bond (σ bond) are capable of free rotation around each other, but this rotation is hindered in case of compounds containing carbon-carbon double bond. According to molecular orbital theory, carbon atoms involved in double bond formation are sp^2 hybridised so that each carbon atom has three planar sp^2 hybridised orbitals and fourth p orbital having its lobes at right angles to the plane of sp^2 orbitals. The formation of π bond involves the overlapping of p orbitals. With the formation of a π bond between C—C along with a bond which is already existing, there remains no possibility of rotation along C—C axis. Neither of the two doubly bonded carbon atoms can be rotated about double bond without destroying π orbital which requires large amount of energy. Thus at ordinary temperature, the rotation about a carbon-carbon double bond is restricted or hindered and gives rise to a kind of stereoisomerism known as *Geometrical Isomerism.*

Geometrical Isomerism, also known as *cis-trans* isomerism takes place in compounds containing carbon-carbon double bond in which each of the two doubly bonded carbon atoms is attached to two different atoms or groups. All the compounds with general formula of the type $C_{AB} = C_{DE}$ or $C_{AB} = C_{AB}$ show geometric isomerism. If either of the two carbon atoms carries two identical groups as in $C_{AB} = C_{AA}$ or $C_{AB} = C_{DD}$, the isomerism does not exist. *This isomerism is due to difference in the relative spatial arrangement of the atoms or groups about the doubly bonded carbon atoms.*

Q. 37. What conditions should be fulfilled by a compound to exhibit geometrical isomerism? *(Bangalore 2001 ; Delhi 2005 ; Guwahati 2006 ; Purvanchal 2007)*

Ans. For a compound to show geometrical isomerism the following conditions are necessary:

(*i*) The molecule must contain a carbon-carbon double bond about which there is no free rotation.

(*ii*) Each of the double bonded carbon atoms must be attached to two different atoms or groups.

In case of compounds with formula of the type $C_{AB} = C_{AB}$; if two similar groups are on the same side of double bond, the isomer is known a cis- and if two similar groups are on the opposite sides of the double bond the isomer is known as *trans-* such as:

A(A)C=C(B)B	and	B(A)C=C(B)A
cis-isomer		*trans*-isomer

For example

2-butene exists in two isomeric forms.

H(CH_3)C=C(H)CH_3	and	CH_3(H)C=C(H)CH_3
cis-2-butene		*trans*-2-butene

Similarly butene-dioic acid exists in two isomeric forms; cis- form is called maleic acid and trans- form is called fumaric acid

H(CO_2H)C=C(H)CO_2H	and	H(CO_2H)C=C(HO_2C)H
Maleic acid (*cis*-isomer)		Fumaric acid (*trans*-isomer)

Q. 38. How will you determine the configuration of a geometrical isomer?

Or

Describe two methods of distinguishing between isomers.

Ans. Different methods available for determination of configuration of a geometrical isomers are described below:

***(i)* From Dipole moments.** Generally cis-isomer has greater dipole moment as compared to trans-isomer. In case of cis- the similar groups being on the same side, the electronic effects are additive; while in case of trans-isomer, the similar groups being on opposite side, the electronic effects cancel each other.

H(CH_3)C=C(H)CH_3	H_3C(H)C=C(H)CH_3
cis-2-butene	*trans*-2-butene
$\mu = 0.4$	$\mu = 0$

H Cl H Cl
C=C C=C
H Cl Cl H

cis-1, 2-Dichloroethene
$\mu = 1.85$

trans-1, 2-Dichloroethene
$\mu = 0$

***(ii)* From boiling point.** Generally speaking, a cis isomer has a higher boiling point compared to the trans isomer. This is because of higher dipole moment and higher polarity in the molecule which acts as the binding force and is responsible for higher b.p. of the cis-isomer. Boiling point of cis-2-butene is 277 K while the trans-2-butene boils at 274 K.

CH_3 H CH_3 H
C=C C=C
CH_3 H H CH_3

cis-2-butene
b.p. 277 K

trans-2-butene
b.p. 274 K

***(iii)* From melting point.** The isomers show a reverse trend here. A cis-isomer has a lower melting point compared to the trans isomer. Here the factor that is important is the size of the molecule. It can be realised that a cis isomer will occupy a smaller volume compared to the trans isomer as illustrated by dotted lines. Thus maleic acid melts at 403 K whereas fumaric acid melts at 575 K.

H COOH H COOH
C=C C=C
H COOH HOOC H

Maleic acid (*cis*)
m.p. 403 K

Fumaric acid (*trans*)
m.p. 575 K

***(iv)* From the formation of cyclic compounds.** Two geometric isomers (*cis* and *trans*) can be distinguished through reactions that lead to formation of ring. cis-isomer undergoes ring closure more readily than the trans-isomer. For example maleic acid readily loses water when heated to about 423 K, to give an anhydride; while fumaric acid does not give anhydride at this temperature. Rather it must be heated to 573 K to get the same anhydride. Hydrolysis of anhydride yields only maleic acid.

H, CO_2H / C=C / H, CO_2H	$\xrightarrow{423\ K}$	H, CO / C=C / H, CO (–O–)	$\xleftarrow{573\ K}$	H, CO_2H / C=C / HO_2C, H
Maleic acid (*cis*-2-isomer)		Maleic anhydride		Fumaric acid (*trans*-2-isomer)

Maleic anhydride $\xrightarrow{\text{hydrolysis}}$ H, CO_2H / C=C / H, CO_2H

only maleic acid and not fumaric acid

(*v*) From the formation of the type of optical isomer. Maleic acid and fumaric acid, both on treatment with $KMnO_4$ or OsO_4 yield optically inactive variety of tartaric acid. Maleic acid yields meso tartaric acid while fumaric acid yields racemic (± or *dl*) tartaric acid.

Maleic acid (*cis*-2-isomer): H, CO_2H / C=C / H, CO_2H $\xrightarrow{KMnO_4 \text{ or } OsO_4}$ Meso-tartaric acid: OH | H—C—CO_2H | H—C—CO_2H | OH

Fumaric acid (*trans*-isomer): H, CO_2H / C=C / HO_2C, H $\xrightarrow{KMnO_4 \text{ or } OsO_4}$ OH | H—C—CO_2H | HO_2C—C—H | OH and OH | HO_2C—C—H | H—C—CO_2H | OH

± or *dl*-tartaric acid

(*vi*) From the method of preparation. Method of preparation of a compound sometimes leads to its configuration. The isomer obtained by the rapture of a ring must be the cis-isomer, *eg.*, maleic acid can be prepared by the oxidation of benzene or quinone, so it must be a cis-isomer.

Benzene or Quinone $\xrightarrow{[O]}$ H, CO_2H / C=C / H, CO_2H

Benzene | Quinone | Maleic acid (*cis*-2-isomer)

Q. 39. Explain clearly E and Z designations of geometrical isomers.

(*Punjab, 2003; Nagpur, 2008; Garhwal, 2010*)

Ans. The *cis* and *trans-* designation can be used only for the compounds in which two doubly bonded carbon atoms are having similar atoms or groups *e.g.*, of the type $C_{AB} = C_{AB}$. But, when the two doubly bonded carbon atoms are having different atoms or groups attached to them *e.g.*, of the type $C_{AB} = C_{DE}$; it is not possible to assign them cis or trans configurations. To overcome this difficulty, a more general system for designating the configuration of geometric isomers has been adopted. This system developed by Cahn, Ingold and Prelog originally for the absolute configuration of optical isomers is known as *E* and *Z system* and is based on priority of attached groups. The atoms or groups attached to each carbon of the double bond, are assigned first and second priority. If the atoms or groups having higher priority attached to two carbons are on the same side of double bond the configuration is designated as Z (derived from German word *Zusammen* meaning together) and if the atoms or groups of higher priority are on the opposite side of the double bond, the configuration is designated as E (derived from German word -*entgegen* meaning across or opposite).

Priorities of atoms or groups are determined in the same way as for R & S configurations of optical isomers. At. weights or atomic numbers of atoms directly linked with ethylenic carbon atoms are taken into consideration.

Let us consider an example in which two doubly bonded atoms are attached to four different halogens such as $C_{BrF} = C_{ICl}$. Since Br is having higher priority over F and I is having priority over Cl (due to their higher atomic numbers). The isomer in which Br and I are on the same side of double bond will be called Z and the isomer in which Br and I are on the opposite sides of double bond will be called E.

F, Br / C=C / Cl, I — *Z*-isomer; F, Br / C=C / I, Cl — *E*-isomer

In the same way cis and trans-isomers of 2-butene can be called Z and E-2-butenes respectively.

H, CH_3 / C=C / H, CH_3 — *Z*-2-Butene; H, CH_3 / C=C / H_3C, H — *E*-2-Butene

Similarly maleic acid can be specified as Z-isomer and fumaric acid as E-isomer.

H, CO_2H / C=C / H, CO_2H — Maleic acid (Z-isomer); H, CO_2H / C=C / HO_2C, H — Fumaric acid (*E*-isomer)

In determining the configuration, we have to select the group of higher priority on one carbon. Similarly we select the group of higher priority on the other carbon atom. If these two groups are on the same side of double bond, the configuration is Z, otherwise it is E.

Some more examples are:

CH_3CH_2 and $CH_2CH_2CH_3$ on C; C=C; H and CH_3 on C
Z-3-Ethyl-2-hexene

and

$CH_3CH_2CH_2$ and CH_2CH_3 on C; C=C; H and CH_3 on C
E-3-Ethyl-2-hexene

Cl and F on C; C=C; I and Br on C
Z-1-Bromo-2-chloro 2-fluoro-1-iodoethene

Cl and F on C; C=C; Br and I on C
E-1-Bromo-2-chloro 2-fluoro-1-iodoethene

Q. 40. Assign E and Z configurations to the following compounds.

(*Kurukshetra, 2001; M.Dayanand, 2001*)

(*i*) $CH_3(H)C=C(Br)(Cl)$ — CH_3 and H on one carbon; Br and Cl on the other

(*ii*) $CH_3(H)C=C(CH_2CH_3)(CH(CH_3)_2)$ — CH_3 and H on one carbon; CH_2CH_3 and $CH(CH_3)_2$ on the other

(*iii*) $H((CH_3)_2CH)C=C(C_2H_5)(C_6H_5)$ — H and $(CH_3)_2CH$ on one carbon; C_2H_5 and C_6H_5 on the other

(*iv*) $CH_3(H)C=C(H)(CHO)$ — CH_3 and H on one carbon; H and CHO on the other

(*v*) $CH_3(H)C=C(H)(COOH)$ — CH_3 and H on one carbon; H and COOH on the other

(*vi*) $I(Br)C=C(C_6H_5)(NHCH_3)$ — I and Br on one carbon; C_6H_5 and $NHCH_3$ on the other

Ans. (*i*) $CH_3(H)C=C(Br)(Cl)$ — CH_3 and Br above; H and Cl below

Out of $-CH_3$ and $-H$, $-CH_3$, has higher priority, out of $-Br$ and $-Cl$, $-Br$ has higher priority. Groups of higher priority lie on the same side. Therefore the configuration is Z.

(*ii*) $CH_3(H)C{=}C(CH_2CH_3)CH(CH_3)_2$

Out of $-CH_3$ and $-H$, $-CH_3$ has higher priority. Out of two groups on the second carbon, $-CH\ (CH_3)_2$ has higher priority. The groups of higher priority lie on opposite sides of the double bond. Therefore the configuration is E.

Following the same analogy, the students can see the configurations of compounds (*iii*), (*iv*), (*v*) and (*vi*) are Z, E, E and E respectively.

Q. 41. Assign E and Z configurations to the following:

I: $CH_3(H)C{=}C(Cl)C_2H_5$ — CH_3 and Cl on top, H and C_2H_5 below.

II: $CH_3(H)C{=}C(H)D$ — CH_3 and H on top, H and D below.

Ans. In compound I, the groups of higher priority on the two carbons linking double bond are CH_3- and Cl. Since these are on the same side of the double bond it has Z configurations.

In compound II, the groups of higher priority on the two carbons linking the double bond are CH_3- and D-(Deuterium) since these are on opposite sides of the double bond, it has E configuration.

Q. 42. Assign E and Z configuration to the following:

(*i*) Maleic acid **(*ii*) *trans*-But-2-enoic acid.**

Maleic acid: upper C bears H and COOH; lower C bears H and COOH (H on the same side, COOH on the same side).

trans-But-2-enoic acid: upper C bears CH_3 and H; lower C bears H and COOH.

Ans. In maleic acid, groups of higher priority on the two carbon atoms are $-COOH$ and $-COOH$. These two groups are on the same side of the double bond. Hence configuration is Z.

In *trans*-but-2-enioc ancd, the groups of higher priority on the two carbons are $-CH_3$ and $-COOH$, since these are on the opposite sides of the double bond, it has E configuration.

Q. 43. Assign E or Z configuration to the following:

(***a***) $CH_3(CH_3-CH_2)C{=}C(H)CH_2-\overset{O}{\overset{\|}{C}}-H$ — CH_3 and H on top, CH_3-CH_2 and CH_2-CHO below.

(***b***) $HC{\equiv}C(H)C{=}C(D)CH{=}CH_2$ — $HC\equiv C$ and D on top, H and $CH=CH_2$ below.

(*Delhi 2006*)

Ans. (*a*) Groups of higher priority on the two doubly bonded carbon atoms are on the same side of the double bond.

Hence, its configuration is Z.

(*b*) Groups of higher priority on the two doubly bonded carbon atoms are on the opposite of the double bond (D stands for heavy hydrogen). Hence, its configuration is E.

Q. 44. Assign R or S configuration to the followings :

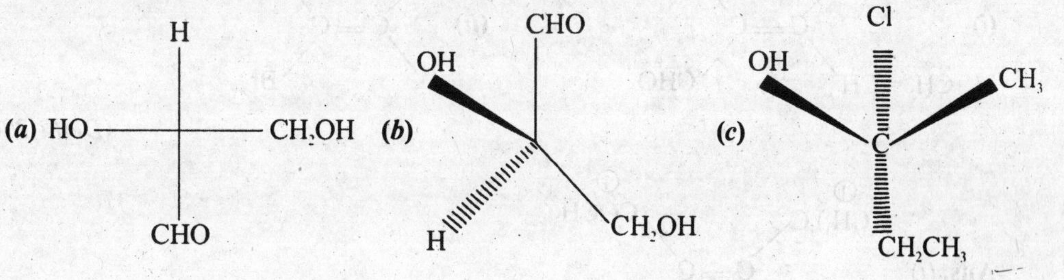

(*Delhi 2006*)

Ans. We need to bring the group of lowest priority *i.e.*, H at the bottom (Follow the rules given in the chapter)

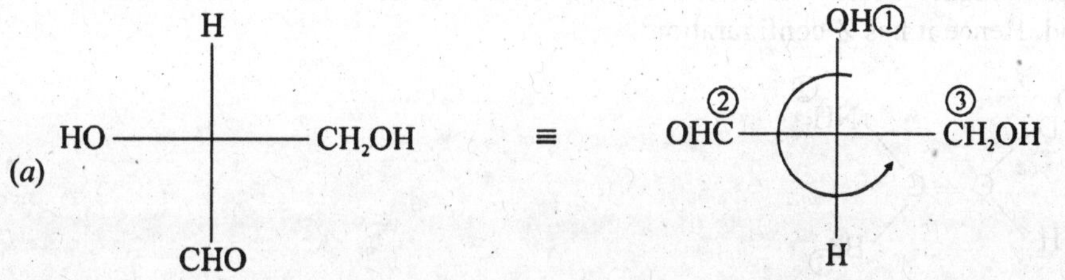

We move in the anticlockwise direction in moving from the group of highest priority to the group of lowest priority (1 → 2 → 3). Hence, it has S configuration.

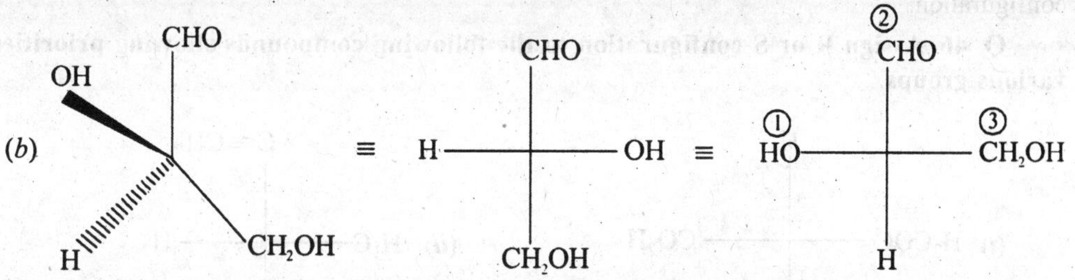

We move in the clockwise direction in moving from group of highest priority to the groups of lowest priority (leaving H) - Hence, it has R configuration.

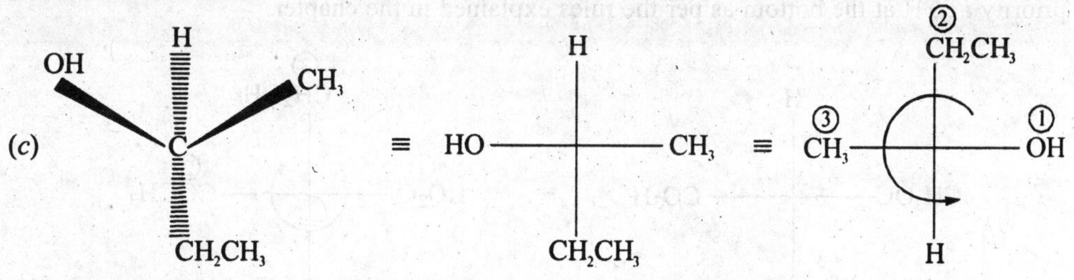

We follow anticlockwise direction in moving from group of highest priority to the group of lowest priority. Hence, it has S configuration.

Q. 45. Assign E or Z notations to the following compounds showing priorities for various groups:

(*i*)

$$\begin{array}{ccc} (CH_3)_3C & & COCH_3 \\ & C=C & \\ CH_2=CH & & CHO \end{array}$$

(*ii*)

$$\begin{array}{ccc} D & & NH_2 \\ & C=C & \\ H & & Br \end{array}$$

(*Delhi 2007*)

Ans. (*i*)

$$\begin{array}{ccc} \overset{①}{(CH_3)_3C} & & \overset{①}{COCH_3} \\ & C=C & \\ \underset{②}{CH_2=CH} & & \underset{②}{CHO} \end{array}$$

Groups of higher priority on the two doubly bonded carbon atoms are on the same side of the double bond. Hence it has Z configuration.

(*ii*)

$$\begin{array}{ccc} \overset{①}{D} & & \overset{②}{NH_2} \\ & C=C & \\ \underset{②}{H} & & Br_{①} \end{array}$$

Here, D stands for *deuterium i.e.,* heavy bydrogen. Groups of higher priority on the two doubly bounded carbon atoms are on the opposite sides of the double bond. Hence, it is E configuration.

Q. 46. Assign R or S configuration to the following compounds showing priorities for various groups.

(*i*)

$$\begin{array}{ccc} & H & \\ H_2COC & — & CO_2H \\ & CH_2CH_3 & \end{array}$$

(*ii*)

$$\begin{array}{ccc} & C\equiv CH & \\ H_3C & — & H \\ & CH=CH_2 & \end{array}$$

(*Delhi 2007*)

Ans. To assign the configuration to the compounds, we need to bring the group of lowest priority *i.e.,* H at the bottom as per the rules explained in the chapter.

(*i*)

$$\begin{array}{ccc} & H & \\ CH_3OC & — & CO_3H \\ & CH_2CH_3 & \end{array} \equiv \begin{array}{ccc} & \overset{③}{CH_2OH} & \\ \overset{①}{HO_2C} & — & \overset{②}{COCH_3} \\ & \underset{④}{H} & \end{array}$$

We have to trace anti-clockwise direction in moving from the group of lowest priority to the group of highest priority. Hence, it has S configuration.

(ii) H_3C—[C: top C≡CH, bottom CH=CH₂]—H ≡ $CH_2=CH$ ②—[C: ① C≡CH top, ④ H bottom]—③ CH_3

$$C \equiv CH$$

We have to trace anti-clockwise direction in moving from group of highest priority to group of lowest priority (leaving H). Hence, it has S configuration.

Q. 47. Define a " Plane of Symmetry". *(Kerala 2000 ; Nagpur 2003 ; Purvanchal 2007)*

Ans. Plane of symmetry is an imaginary plane passing through the molecule in such a way that it divides the molecule into two equal parts and one part appears to be the mirror image of the other part. Mesotartaric acid and 1, 2-dichloro-1, 2-dihydroxy ethane are the two examples of the compounds having the plane of symmetry.

```
   COOH                    H
    |                      |
H - C - OH           HO - C - Cl
    |                      |
H - C - OH           HO - C - Cl
    |                      |
   COOH                    H
```

meso-tartaric acid   1, 2-dichloro-1, 2-dihydroxyethane

Q. 48. Assign E or Z configuration to each of the following compounds:

(A) $(Me)(Et)C=C(H)(Cl)$ — Me and H on top, Et and Cl on bottom

(B) $(Br)(I)C=C(Cl)(H)$ — Br and Cl on top, I and H on bottom

(Calcutta 2007)

Ans. In compound (A), groups of higher priority on the doubly bonded carbon atoms are Et and Cl, which lie on the same side of the double bond. Hence, it is Z-isomer.

In compound (B), groups of higher priority on the doubly bonded carbon atoms are I and Cl, which lie on the opposite sides of the double bond. Hence, it is E-isomer.

Q. 49. Indicate which of the following compounds has E or Z designation.

(i) $(CH_3)(H)C=C(Cl)(Br)$ — CH_3 and Cl on top, H and Br on bottom

(ii) $(CH_3)(H)C=C(C_3H_7)(C_2H_5)$ — CH_3 and C_3H_7 on top, H and C_2H_5 on bottom

(iii) [skeletal formula of an alkene]

(iv) $(H)(HOOC)C=C(COOH)(H)$ — H and COOH on top, HOOC and H on bottom

(Guwahati 2006)

Ans. (i) $(CH_3)(H)C=C(Cl)(Br)$ — E-isomer

(ii) $(CH_3)(H)C=C(C_3H_7)(C_2H_5)$ — Z-isomer

(iii) E-isomer (iv) $\underset{\text{HOOC}}{\overset{\text{H}}{}}\!\!>C=C<\!\!\underset{\text{H}}{\overset{\text{COOH}}{}}$ E-isomer

Q. 50. Write geometrical isomers of the following and assign E and Z designation.

(i) $\underset{\text{Br}}{\overset{\text{HO}}{}}\!\!>C=C<\!\!\underset{C_2H_5}{\overset{\text{Cl}}{}}$ (ii) $\underset{\text{H}}{\overset{\text{Cl}}{}}\!\!>C=C<\!\!\underset{CH_3}{\overset{\text{I}}{}}$

(iii) $CH_3-CH_2-CH=CH-CH_3$ (iv) $CH_3-CH=\underset{\underset{Br}{|}}{C}-Cl$

(Guwahati 2007)

Ans. (i) $\underset{\text{Br}}{\overset{\text{HO}}{}}\!\!>C=C<\!\!\underset{C_2H_5}{\overset{\text{Cl}}{}}$ E-isomer; $\underset{\text{Br}}{\overset{\text{HO}}{}}\!\!>C=C<\!\!\underset{\text{Cl}}{\overset{C_2H_5}{}}$ Z-isomer

(ii) $\underset{\text{H}}{\overset{\text{Cl}}{}}\!\!>C=C<\!\!\underset{CH_3}{\overset{\text{I}}{}}$ Z-isomer; $\underset{\text{H}}{\overset{\text{Cl}}{}}\!\!>C=C<\!\!\underset{\text{I}}{\overset{CH_3}{}}$ E-isomer

(iii) $\underset{CH_3-CH_2}{\overset{\text{H}}{}}\!\!>C=C<\!\!\underset{CH_3}{\overset{\text{H}}{}}$ Z-isomer; $\underset{CH_3-CH_2}{\overset{\text{H}}{}}\!\!>C=C<\!\!\underset{\text{H}}{\overset{CH_3}{}}$ E-isomer

(iv) $\underset{CH_3}{\overset{\text{H}}{}}\!\!>C=C<\!\!\underset{\text{Br}}{\overset{\text{Cl}}{}}$ Z-isomer; $\underset{CH_3}{\overset{\text{H}}{}}\!\!>C=C<\!\!\underset{\text{Cl}}{\overset{\text{Br}}{}}$ E-isomer

Q. 51. How many stereoisomers are possible for a compound which has three chiral carbon atoms? *(Delhi 2006)*

Ans. The number of stereoisomers for a compound containing *n* chiral carbon atoms is given by 2^n.

∴ No. of stereoisomers for a compound containing 4 chiral carbon atoms = $2^3 = 8$.

Q. 52. Consider the pair of structures and identify the relationship between them, whether enantiomers or two orientations of the same molecule.

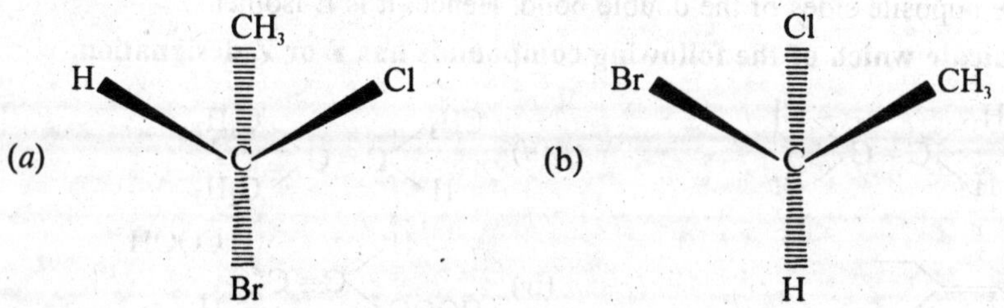

(Delhi 2006)

Ans. Enantiomers are non-superimposable mirror images of each other. In the present case, it is not so. We get the projection formulae of the two compounds as under:

CH_3
H — — Cl
Br

Cl
Br — — CH_3
H

Clearly, these are not the enantiomers but two orientations of the same molecule.

Q. 53. Discuss geometrical isomerism in oximes.

Ans. Oximes of aldehydes having the general formulae R–CH = N–OH are capable of exhibiting geometrical isomerism as –H and –OH groups may be present on the same side or opposite sides of the double bond. The two stereoisomers thus obtained are names as *syn* (equivalent of *cis*) and *anti* (equivalent of *trans*). Two geometrical isomers of an aldoxime may be represented as :

R(H)C=N–OH (OH on the same side as H)
***syn* - aldoxime**

R(H)C=N–OH (OH on the side of R)
***anti* - aldoxime**

Geometrical isomers of benzaldoxime have been actually isolated. *syn*-benzaldoxime has m.p 35°C while *anti* - benzaldoxime has m.p. 130°C. The two compounds may be shown as under:

C_6H_5(H)C=N–OH
***syn* - benzaldoxime**

C_6H_5(H)C=N–HO
***anti* - benzaldoxime**

Oximes of ketones (RR′C = NOH) can also show geometrical isomerism provide R and R′ are not the same. Thus benzophenone oxime does not exhibit geometrical isomerism because the two groups attached to carbonyl carbon are the same. Phenyl tolyl ketoxime is known to exist in two geometrical forms.

$CH_3C_6H_4$(C_6H_5)C=N–OH
***syn* -phenyl tolyl oxime**

$CH_3C_6H_4$(C_6H_5)C=N–OH
***anti* - phenyl tolyl oxime**

Q. 54. Discuss stereochemistry (geometrical and optical isomerism) of alicyclic compounds.

Ans. The discussion can be made under two headings

Even-membered rings (containing 4, 6 carbon atoms)

If the substituents are across the ring from each other, the molecule possesses a plane of symmetry, Therefore, no optical forms exist. However geometrical isomerism is possible. *cis* and *trans* isomers of 3-methylyclobutane carboxylic acid and cyclobutane-1, 2-dicarboxylic acid are shown below:

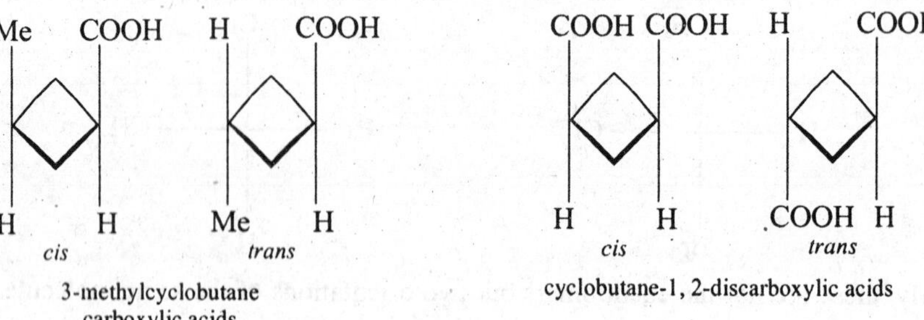

Odd membered rings (Containing 3, 5 carbon atoms)

A disubstituted (on different carbons) odd membered rings has two asymmetric carbon atoms. Therefore two diastereomeric pairs of enantiomers are possible. Take the case of 2-methylcyclopropane carboxylic acid. Geometrical and optical isomers have been represented as under :

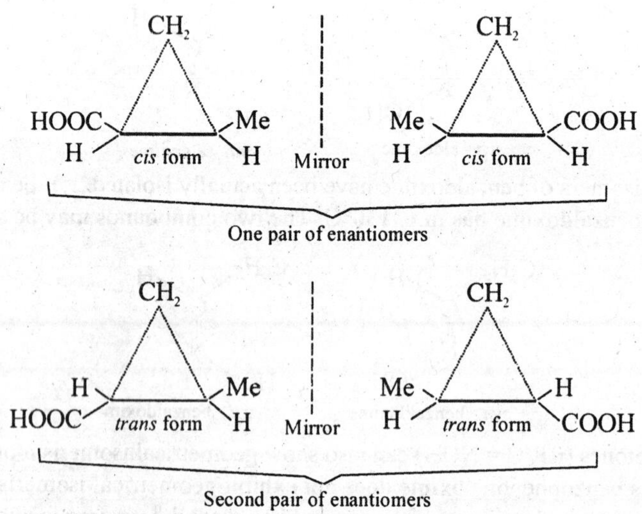

If hower, the two substituents are the same, there will be one pair of enantiomers and a meso form. Take the case of cyclopropane-1, 2-dicarboxylic acid. *trans* form exists as a pair of enantiomers while the *cis* form as meso compound.

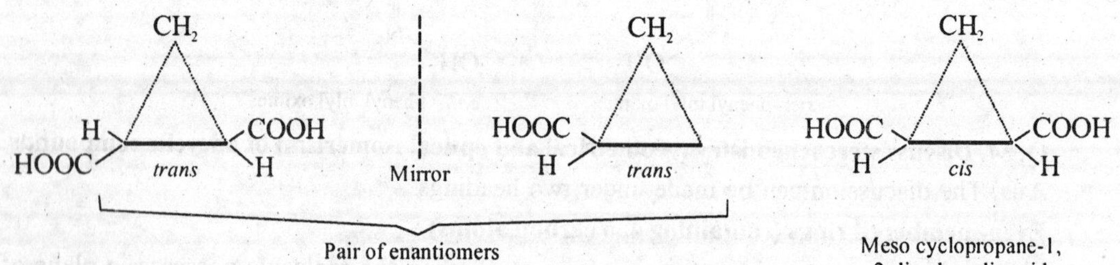

Q.55. Discuss optical activity in organic compounds without asymmetric carbons

Or

Give an account of optical activity in (*i*) allene derivatives (*ii*) biphenyl derivatives.

(*Mumbai, 2010*)

Ans. Compounds containing an asymmetric carbon can exist in optically active forms. However, compounds which do not possess an asymmetric carbon atom can also exist in optically active forms provided that the molecule is dissymmetric. Example are :

(1) **Allene Derivatives**

Some derivatives of allene ($CH_2 = C = CH_2$) exhibit optical isomerism. Example is 1,3-diphenylpropadiene. In allenes, the central carbon is *sp* hybridized, and the terminal carbons are sp^2 hybridized. The central carbon forms two sp-sp^2 σ bonds. The central carbon also has two *p* orbitals which are mutually perpendicular. They form π bonds with the *p* orbitals on the other carbon atoms. As a result, the substituents at one end of the molecule are in a plane which is perpendicular to that of the substituents at the other end, so that the compound exists in two forms which are non-superimposable mirror images and are optically active.

Mirror

C_6H_5, H, C = C = C, H, C_6H_5 | H, C_6H_5, C = C = C, C_6H_5, H

1,3-Diphenylpropadiene

(2) **Biphenyl Derivatives**

Substituted bihenyls show optical isomerism when substituents in the 2-positions are large enough to prevent rotation about the bond joining the two benzene rings. For example, biphenyl 2,2′ -disuplhonic acid exists in two forms.

Mirror

SO_3H, SO_3H | SO_3H, HO_3S

Biphenyl-2-2′ -disulphonic acid
(a biphenyl)

These two forms are non-superimposable mirror images. They do not interconvert at room temperature because the energy required to twist one ring through 180° relative to the other is too high. This in turn is because, during the twisting process, the two-SO_3H groups must come into very close proximity when the two benzene rings become coplanar and strong repulsive forces are introduced.

Q. 56. What do the notations threo and erythro stand for?

Explain with reference to the stereochemistry of diastereomers.

Ans. Diastereomers are optical isomers of a compound which are not mirror images of each other and hence are not enantiomers. For example, *d*-tartaric acid is a diastereomer of mesotartaric acid. But the two are not enantiomers.

As far as the absolute configuration of a compound is concerned we derive it from the configuration of glyceraldahyde which has been taken as arbitrary standard.

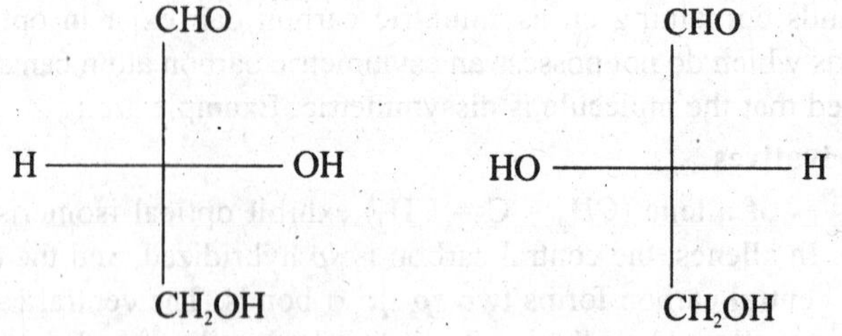

D (+) glyceraldehyde L (–) glyceraldehyde

D and L stand for the configuration while (+) and (–) signs denote the actual direction of rotation of plane polarised light. If H and OH are on the LHS and RHS of central carbon, it denotes D configuration. If OH and H are on the LHS and RHS of central carbon it denotes L configuration.

Any compound that can be obtained from or converted into D(+) glyceraldelydc has D configuration. Similarly any compound that can be obtained from or converted into L(–) glyceraldehyde has L configuration.

Thus, when we have to decide the configuration of compounds (particularly sugars) containing more than three carbon, we shall check the configuration of lower two carbons. as shown below:

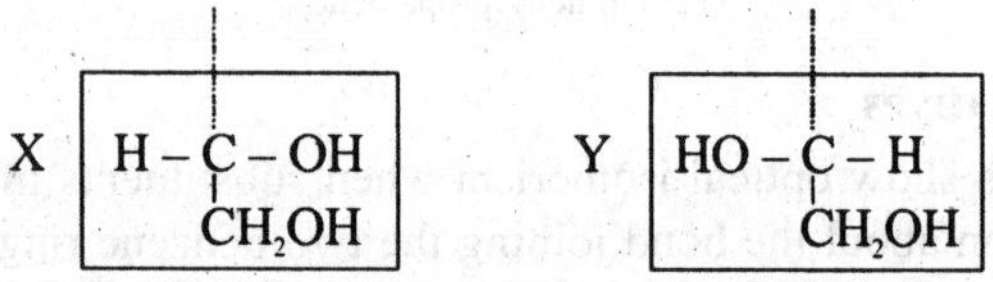

D-series L-series

If the arrangement corresponds to X, the compound has D configuration.

If the arrangement corresponds to Y, the compound belongs to L configuration.

Let us take the example of $CHO\overset{*}{C}HOH\overset{*}{C}HOH\ CH_2\ OH$. This contains two asymmetric carbon atoms marked with asterisks. There are four optical isomers possible, all of which are known.

If the H on the third carbon atom (from the bottom) is on the left hand side, the compound is *erythro*, while if the H on the third carbon atom is on the right hand side, the compound is *threo.* The structures of D-erythrose and D-threose which may be assumed to be obtained from D-glyceraldelyde are given below :

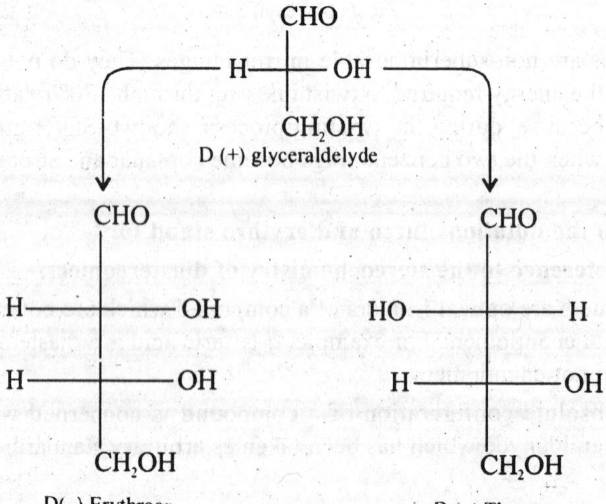

D(–) Erythrose (I) D (–) Threose (II)

Compounds I and II are diastereomers of each other from the D-series. Similarly there will be another pair of diastereomers from the L-series. It must be noted that erythro compound (I) on oxidation gives mesotartaric acid.

```
          CHO                                  COOH
           |                                    |
   H ------+------ OH        [O]        H ------+------ OH
           |              ------>               |
   H ------+------ OH                   H ------+------ OH
           |                                    |
         CH2OH                                 COOH
          (I)                            mesotartaric acid
```

Similarly the erythro compound from L-series on oxidation will also produce meso tartaric acid.

Q. 57. Draw the Fischer projection formulas for all possible stereoisomers of 2, 3, 4-trihydroxyglutaric acid. Comment on the stereogenicity of C-3 centre in the active and the meso isomers. *(Calcutta, 2000)*

Ans. Following are the stereoisomers of 2, 3, 4-tridhydroxyglutaric acid :

```
     COOH                 COOH                 COOH
      |                    |                    |
 HO - C - OH           H - C - OH           H - C - OH
      |                    |                    |
 HO - C - OH          HO - C - H            H - C - OH
      |                    |                    |
  CH (OH) COOH         CH (OH) COOH         CH (OH) COOH

      I                    II                   III
```

The Fischer projection formula of compound I is :

```
            COOH
             |
   HO -------+------- H
             |
   CH (OH)-CH (OH) COOH
```

The Fischer projection formula of compound II is :

```
            COOH
             |
   H --------+------- OH
             |
   CH (OH)-CH (OH) COOH
```

Following points are to be observed :

(*i*) The central carbon is present at the crossing of the horizontal and vertical lines.

(*ii*) The longest chain of carbon atoms, *i.e.*, CH(OH) – CH (OH) – COOH in the molecule is present along the vertical line.

MISCELLANEOUS QUESTIONS

Q. 58. Write the structures of isomeric compounds with molecular formula C_3H_8O. What type of isomers are these?

Ans. The Compound has the following isomers

(*i*) $CH_3\,CH_2\,CH_2\,OH$ (*ii*) $CH_3\,CHOH\,CH_3$

(*iii*) $CH_3OCH_2\,CH_3$

(*i*) and (*ii*) are position isomers while (*ii*) and (*iii*) are functional isomers.

Q. 59. Discuss geometrical isomerism in maleic acid and fumaric acid. (*Nagpur 2003*)

Ans. See Q 37

Q. 60. Explain the optical isomerism of lactic acid. (*Nagpur 2002; Vidyasagar 2007*)

Ans.

```
     COOH                HOOC
      |                    |
  H – C – OH          OH – C – H
      |                    |
     CH3                  CH3
```

d-lactic acid *l*-lactic acid

Q. 61. How will you distinguish between maleic and fumaric acid. (*Delhi 2003*)

Ans. The two compounds are geometrical isomers of each other. Maleic acid is *cis* while fumaric acid is *trans*. Maleic acid easily forms an anhydride on heating while fumaric acid does not. It reacts only at a higher temperature.

For Conformational Isomerism See Chapter 31

7

ALKANES

Q. 1. Give IUPAC names of the following compounds:

(*i*)
$$\begin{array}{l} CH_3 - \underset{\underset{\displaystyle CH_3}{|}}{\underset{|}{CH_2}}\!\!\!\!\!\!\!\!\overset{}{} \end{array}$$

$$\begin{array}{cccccccc} CH_3 & - & CH & - & CH & - & CH_3 \\ & & | & & | & & \\ & & CH_2 & & CH_2 & & \\ & & | & & | & & \\ & & CH_3 & & CH_3 & & \end{array}$$

(*ii*)
$$\begin{array}{ccccccccccc} & & CH_3 & & CH_2 - CH_3 & & & & & & \\ & & | & & | & & & & & & \\ CH_3 & - & CH & - & CH & - & CH_2 & - & CH_2 & - & CH_3 \end{array}$$

(*iii*)
$$\begin{array}{ccccccccc} & & CH_3 & & & & & & \\ & & | & & & & & & \\ CH_3 & - & C & - & CH_2 & - & CH & - & CH_3 \\ & & | & & & & | & & \\ & & CH_3 & & & & CH_3 & & \end{array}$$

(*iv*)
$$\begin{array}{ccccccccccccc} & & & & & & & & & & CH_3 & & \\ & & & & & & & & & & | & & \\ CH_3 & - & CH & - & CH_2 & - & CH & - & CH_2 & - & CH & - & CH_2CH_3 \\ & & | & & & & | & & & & & & \\ & & CH_3 & & & & CH\,(CH_3)_2 & & & & & & \end{array}$$

Ans. (*i*) 3, 4 Dimethylhexane.

(*ii*) 3-Ethyl-2 methylhexane.

(*iii*) 2, 2, 4-Trimethylpentane.

(*iv*) 2, 6- Dimethyl-4 isopropyl octane.

Q. 2. Write the structural formulae and IUPAC names of various compounds possible with the molecular formula C_6H_{14}. (*Nagpur, 2005*)

Ans. Various structures possible are given below:

(*i*) $CH_3CH_2CH_2CH_2CH_2CH_3$ — *n*-hexane

(*ii*) 2-methylpentane
$$\begin{array}{ccccccccc} CH_3 & - & CH & - & CH_2 & - & CH_2 & - & CH_3 \\ & & | & & & & & & \\ & & CH_3 & & & & & & \end{array}$$

(*iii*) 2, 3-dimethylbutane
$$\begin{array}{ccccccc} CH_3 & - & CH & - & CH & - & CH_3 \\ & & | & & | & & \\ & & CH_3 & & CH_3 & & \end{array}$$

(*iv*) 2, 2-dimethylbutane
$$\begin{array}{ccccccc} & & CH_3 & & & & \\ & & | & & & & \\ CH_3 & - & C & - & CH_2 & - & CH_3 \\ & & | & & & & \\ & & CH_3 & & & & \end{array}$$

Q. 3. Give the general methods of preparation of alkanes. (*Nagpur, 2005*)

Ans. Different methods of preparation of alkanes are given below :

(1) Hydrogenation of unsaturated hydrocarbons (Sabatier-Senderen's reaction). A mixture of unsaturated hydrocarbons and hydrogen is passed our finely divided platinum, palladium or nickel at 523-573 K. Alkenes and alkynes are reduced to alkanes.

$$\begin{array}{c} CH \\ \| \\ CH_2 \end{array} + H_2 \xrightarrow[523-573\,K]{Pt, Pd \text{ or } Ni} \begin{array}{c} CH_3 \\ | \\ CH_3 \end{array}$$

$$\underset{\text{Propene}}{CH_3-CH=CH_2} + H_2 \xrightarrow[523-573\,K]{Pt} \underset{\text{Propane}}{CH_3-CH_2-CH_3}$$

$$\underset{\text{Acetylene}}{CH\equiv CH} + 2H_2 \xrightarrow[523-573\,K]{Pt} \underset{\text{Ethane}}{CH_3-CH_3}$$

(2) Reduction of alkyl halides. This may be done in different ways:

(*i*) Chemical reducing agents like zinc and acetic acid or hydrochloric acid, zinc-copper couple and alcohol; magnesium-amalgam and water etc., convert alkyl halides into alkanes.

$$\underset{\text{Alkyl halide}}{R-X} \xrightarrow[\text{Zn / Acetic acid}]{\text{Zn - Cu / alchol}} \underset{\text{Alkane}}{R-H} + H-X$$

$$C_2H_5-Br \xrightarrow[\text{Zn / Acetic acid}]{\text{Zn - Cu / alchol}} \underset{\text{Ethane}}{C_2H_6} + HBr$$

(*ii*) Alkyl iodides are easily reduced by heating with concentrated HI at about 423 K in a sealed tube.

$$R-I + HI \xrightarrow[423\,K]{P} R-H + I_2$$

$$C_2H_5I + HI \xrightarrow[423\,K]{P} C_2H_6 + I_2$$

(3) Wurtz reaction. This reaction involves the interaction of two melecules of alkyl halides and sodium metal in dry ether.

$$R-\boxed{X + 2Na + X}-R \xrightarrow[\text{ether}]{\text{Dry}} R-R + 2NaX$$

$$CH_3-\boxed{Br + 2Na + Br}-CH_3 \xrightarrow[\text{ether}]{\text{Dry}} CH_3-CH_3 + 2NaBr$$

$$C_2H_5-\boxed{Br + 2Na + Br}-C_2H_5 \xrightarrow[\text{ether}]{\text{Dry}} CH_3-CH_2-CH_2-CH_3 + 2NaBr$$

When different alkyl halides are used, a mixture of three alkenes is obtained as shown below:

$$\underset{\text{Methyl bromide}}{CH_3-Br} + 2Na + \underset{\text{Ethyl bromide}}{Br-C_2H_5} \longrightarrow \underset{\text{Propane}}{CH_3-CH_2-CH_3} + 2NaBr$$

$$CH_3-Br + 2Na + Br-CH_3 \longrightarrow \underset{\text{Ethane}}{CH_3-CH_3} + 2NaBr$$

$$C_2H_5-Br + 2Na + Br-C_2H_5 \longrightarrow CH_3-CH_2-CH_2-CH_3 + 2NaBr$$

Mechanism of Reaction

Two mechanisms have been suggested for the reaction.

Mechanism I

The reaction takes place in two steps as shown below:

$$RX + 2Na \longrightarrow R-Na + NaX$$

$$RNa + XR \longrightarrow R-R + NaX$$

Mechanism II

Again the reaction takes place in two steps as under:

$$RX + Na \longrightarrow \dot{R} + NaX$$

$$\dot{R} + \dot{R} \longrightarrow \underset{\text{Hydrocarbon}}{R-R}$$

(4) Corey-House Synthesis

It is a better method than Wurtz reaction. An alkyl halide and a lithium dialkyl copper are reacted to give a higher hydrocarbon.

$$\underset{\substack{\text{Alkyl}\\\text{halide}}}{R'-X} + \underset{\substack{\text{Lithium}\\\text{dialkyl}\\\text{copper}}}{R_2\,CuLi} \longrightarrow \underset{\text{Alkane}}{R-R'} + R-Cu + LiX$$

(R and R′ may be same or different)

For example,

$$\underset{n\text{-Propyl bromide}}{CH_3CH_2CH_2Br} + \underset{\substack{\text{Lithium dimethyl}\\\text{copper}}}{(CH_3)_2\,CuLi} \longrightarrow \underset{n\text{-Butane}}{CH_3CH_2CH_2CH_3} + \underset{\text{Methyl copper}}{CH_3Cu} + LiBr$$

A notable feature of this reaction is that it can be used for preparing symmetrical, unsymmetrical, straight-chain or branched-chain alkanes.

For a better yield of the product, the alkyl halide used should be primary whereas lithium dialkyl copper may be primary, secondary or tertiary.

(5) Indirect reduction of alkyl halides. Alkyl halides react with magnesium in dry ether to form alkyl magnesium halides (Grignard reagent). Decomposition of the Grignard reagent with water yield alkanes.

$$\underset{\text{Alkyl halide}}{R-X + Mg} \xrightarrow{\text{Dry ether}} \underset{\text{Grignard reagent}}{R-MgX}$$

$$R-MgX + H-OH \longrightarrow \underset{\text{Alkane}}{R-H} + Mg\begin{matrix}\diagup OH\\ \diagdown X\end{matrix}$$

$$\underset{\substack{\text{Methyl magnesium}\\\text{bromide}}}{CH_3-MgBr} + H-OH \longrightarrow CH_4 + Mg\begin{matrix}\diagup OH\\ \diagdown X\end{matrix}$$

(6) Decarboxylation of Carboxylic Acid Salts. Sodium salts of a fatty acid are heated with soda lime (NaOH + CaO) to form alkanes.

$$R-\boxed{COONa + NaO}-H \longrightarrow R-H + Na_2CO_3$$

$$CH_3-COONa + NaOH \longrightarrow CH_4 + Na_2CO_3$$

(7) Kolbe's Electrolytic Process. By the electrolysis of conc. solution of sodium or potassium salts of fatty acids, alkanes are obtained.

$$\underset{(aq)}{2R-COONa} \longrightarrow 2RCOO^- + 2Na^-$$

$$2R-COO^- \longrightarrow RCOO + 2e^- \quad \text{[At anode]}$$

$$\underset{\text{(Unstable)}}{RCOO} \downarrow R-R+2CO_2$$

$$2Na^+ + 2e^- \longrightarrow 2Na \quad \text{[At cathode]}$$

$$2Na \xrightarrow{2H_2O} 2NaOH + H_2$$

$$2CH_3COONa \xrightarrow{\text{Electrolysis}} \underbrace{CH_3-CH_2 + 2CO_2}_{\text{[At anode]}} + \underbrace{2Na + 2H_2O}_{\text{[At cathode]}}$$

(8) By Reduction of Alcohols, Aldehydes, Ketones and Acids. The reduction is carried out with the help of HI and P.

(*i*)
$$\underset{\text{Alcohol}}{R-OH + 2HI} \xrightarrow[423\text{ K}]{P} R-H+H_2O+I_2$$

$$\underset{\text{Ethyl alcohol}}{C_2H_5-OH+2HI} \xrightarrow[423\text{ K}]{P} \underset{\text{Ethane}}{C_2H_6} + H_2O + I_2$$

(*ii*)
$$\underset{\text{Aldehyde}}{R-\overset{\overset{\displaystyle O}{\|}}{C}-H+4HI} \xrightarrow[423\text{ K}]{P} \underset{\text{Alkane}}{R-CH_3} + H_2O + 2I_2$$

$$\underset{\text{Acetaldehyde}}{CH_3-\overset{\overset{\displaystyle C}{\|}}{C}-H+4HI} \xrightarrow[423\text{ K}]{P} \underset{\text{Ethane}}{CH_3-CH_3} + H_2O + 2I_2$$

(*iii*)
$$\underset{\text{Ketone}}{R-\overset{\overset{\displaystyle O}{\|}}{C}-R+4HI} \xrightarrow[423\text{ K}]{P} \underset{\text{Alkane}}{R-CH_2-R} + H_2O + 2I_2$$

$$\underset{\text{Acetone}}{CH_3-\overset{\overset{\displaystyle O}{\|}}{C}-CH_3+4HI} \xrightarrow[423\text{ K}]{P} \underset{\text{Propane}}{CH_3-CH_2-CH_3} + H_2O + 2I_2$$

(*iv*)
$$\underset{\text{Acid}}{R-\overset{\overset{\displaystyle O}{\|}}{C}-OH+6HI} \xrightarrow[423\text{ K}]{P} \underset{\text{Alkane}}{R-CH_3} + 2H_2O + 3I_2$$

$$\underset{\text{Formic acid}}{H-COOH} + 6HI \xrightarrow[423\,K]{P} CH_4 + 2H_2O + 3I_2$$

$$CH_3-COOH + 6HI \xrightarrow[423\,K]{P} CH_3-CH_3 + 2H_2O + 3I_2$$

Q. 4. Write the structural formula of *n*-pentane, isopentane and neopentane. Which of these has highest boiling point and why? (*Guwahati 2006*)

Ans. Structural formula of pentanes

$$\underset{n\text{-Pentane}}{CH_3-CH_2-CH_2-CH_2-CH_3} \qquad \underset{\text{Isopentane}}{CH_3-CH_2-\underset{\displaystyle CH_3}{\underset{|}{CH}}-CH_3} \qquad \underset{\text{Neopentane}}{CH_3-\overset{\displaystyle CH_3}{\overset{|}{\underset{\displaystyle CH_3}{\underset{|}{C}}}}-CH_3}$$

Highest boiling point. Of the three pentanes, *n*-pentane has the highest boiling point.

Explanation. All the three pentanes are non-polar compounds having only weak intermolecular forces of attraction (van der Waals' forces). The molecular weights of the three isomers are naturally the same but they have different surface areas. As the surface area of *n*-pentane is larger than those of the other two, it has strongest intermolecular forces of attraction. Therefore it has highest boiling point followed by iso- and neopentanes.

Q. 5. As we move up from *n*-pentane to *n*-hexane, there is an increase of about 35° in the melting point but on moving from *n*-hexane to *n*-heptane the increase is only about 4°. How can you account for these observations?

Or

Discuss the variation of m.p. in alkanes. (*Punjab 2003*)

Ans. Disproportionate variation in melting points of the three alkanes can be explained on the basis of their molecular weights and their ability to fit in the crystal structures as follows.

In alkanes, the carbon atoms form zig-zag chains as depicted below for *n*-pentane, *n*-hexane and *n*-heptane.

$$\underset{n\text{-Pentane}}{CH_3 \diagdown CH_2 \diagup CH_2 \diagdown CH_2 \diagup CH_3} \qquad \underset{n\text{-Hexane}}{CH_3 \diagdown CH_2 \diagup CH_2 \diagdown CH_2 \diagup CH_2 \diagdown CH_3}$$

$$\underset{n\text{-Heptane}}{CH_3 \diagdown CH_2 \diagup CH_2 \diagdown CH_2 \diagup CH_2 \diagdown CH_2 \diagup CH_3}$$

It is apparent that in *n*-hexane, the two terminal methyl groups lie on opposite sides of the zig-zag chain which enables this chain to fit in closely in the crystal structure. But in *n*-pentane and *n*-heptane, the two terminal groups lie on the same side so that the crystal structure is not so closely packed.

As we move from *n*-pentane to *n*-hexane, the molecular weight increases as well as the packing arrangement becomes more closed. Both these factors require greater energy to melt the solid. As such there is a large increase in melting point.

On the other hand, as we move from *n*-hexane to *n*-heptane, the molecular weight increases but the packing arrangement does not remain so close. While the melting point would tend to increase due to increase in molecular weight, the decrease in close packing would tend to have the opposite effect. The net result is that there is only a small increase in melting point in this case.

Q. 6. Describe chemical properties of alkanes.

Ans. Chemical properties of alkanes are described as under:

1. Halogenation. Displacement or substitution of hydrogen atoms by halogens is known as *halogenation*. Methane on treatment with chlorine in the presence of diffused sunlight or by heating the reaction mixture to 523-673 K gives chloromethane

$$CH_4 + Cl_2 \xrightarrow[\text{or } 523-673\text{ K}]{\text{Sunlight}} \underset{\text{Chloromethane}}{CH_3Cl} + HCl$$

The reaction does not stop here. Chloromethane is further chlorinated to dichloromethane. Dichloromethane is chlorinated to trichloromethane and trichloromethane is finally chlorinated to tetrachloromethane. This sequence of reactions is given below:

$$CH_3Cl + Cl_2 \longrightarrow \underset{\text{Dichloromethane}}{CH_2Cl_2} + HCl$$

$$CH_2Cl_2 + Cl_2 \longrightarrow \underset{\text{Trichloromethane}}{CHCl_3} + HCl$$

$$CHCl_3 + Cl_2 \longrightarrow \underset{\text{Tetrachloromethane}}{CCl_4} + HCl$$

Bromination of alkanes takes place in similar manner but less readily. Direct iodination of alkanes is not possible as the reaction is reversible.

$$R-H + I_2 \rightleftharpoons R-I + HI$$

However, iodination can be carried out in the presence of oxidising agent such as iodic acic (HIO$_3$), which destroys the hydrogen iodide as soon as it is formed and so drives the reaction to the right.

$$5HI + 4HIO_3 \longrightarrow 3I_2 + 3H_2O$$

Direct fluorination is usually explosive and brings about rupture of C – C and C – H bonds leading to a mixture of products. However, fluorination of alkanes can be carried out by diluting fluorine with an inert gas such as nitrogen.

2. Nitration. It is a process in which hydrogen atom of alkane is replaced by nitro group (– NO$_2$). Alkanes undergo nitration when treated with fuming HNO$_3$ in the vapour phase between 423–748 K. For example,

$$CH_3 - \boxed{H + HO} - NO_2 \longrightarrow \underset{\text{Nitromethane}}{CH_3 - NO_2} + H_2O$$

Higher alkanes on nitration give a mixture of all possible mono nitro derivatives. For example,

$$CH_3-CH_2-CH_3 + HNO_3 \xrightarrow{673\text{ K}} \underset{\text{1-Nitropropane}}{CH_3-CH_2-CH_2-NO_2} +$$

$$\underset{\text{2-Nitropropane}}{CN_3-\overset{\overset{NO_2}{|}}{CH}-CH_3} + \underset{\text{Nitroethane}}{CH_3-CH_2-NO_2} + \underset{\text{Nitromethane}}{CH_3-NO_2}$$

3. Sulphonation. It is a process in which hydrogen atom of alkane is replaced by sulphonic acid group ($-SO_3H$). Sulphonation of hexane and higher members may be carried out by treating the alkane with oleum (fuming sulphuric acid). For example,

$$\underset{n\text{-Hexane}}{C_6H_{13}-H + HO-SO_3H} \xrightarrow{\Delta} \underset{\text{Hexane sulphonic acid}}{C_6H_{13}-SO_3H + H_2O}$$

The ease of replacement of hydrogen atom is

Tertiary hydrogen > Secondary hydrogen > Primary hydrogen

4. Oxidation. (*i*) *Combustion.* Alkanes burn readily in excess of air or oxygen to form CO_2 and H_2O. This process is called combustion and is accompanied by the evolution of a large amount of heat. For example,

$$C_nH_{2n+2} + O_2 \xrightarrow{\Delta} nCO_2 + (n+1)H_2O + \text{Heat}$$

$$CH_4 + 2O_2 \xrightarrow{\Delta} CO_2 + 2H_2O + 890 \text{ kJ}$$

Alkanes are, therefore, used as fuels in the form of kerosene oil, gasoline, LPG.

(*ii*) *Catalytic oxidation.* When alkanes are heated in a limited supply of air or oxygen at high pressure and in the presence of suitable catalyst, they are oxidised to alcohols, aldehydes or fatty acids. For example,

$$2CH_4 + O_2 \xrightarrow[\text{Cu Tube}]{523\text{ K, 100 Atm}} 2CH_3OH$$

$$CH_4 + O_2 \xrightarrow[Mo_2O_3]{673\text{ K, 200 Atm}} HCHO + H_2O$$

(*iii*) Alkanes containing tertiary hydrogens can be oxidised by such reagents to form alcohols. For example,

$$\underset{Iso\text{-butane}}{CH_3-\underset{CH_3}{\overset{CH_3}{C}}-H} + O \xrightarrow{KMnO_4} \underset{tert\text{-Butyl alcohol}}{CH_3-\underset{CH_3}{\overset{CH_3}{C}}-OH}$$

5. Aromatisation. When alkanes containing six or more carbon atoms are heated to high temperature (773 K) under high pressure and in the presence of a catalyst (oxides of chromium, vanadium, molybdenum supported on alumina), they are converted into aromatic hydrocarbons. The process involves cyclisation, isomerisation and dehydrogenation and is known as *aromatisation*. For example,

$$\underset{n\text{-hexane}}{CH_3-CH_2-CH_2-CH_2-CH_2-CH_3} \xrightarrow[\text{Cyclisation}]{-H_2} \underset{\text{Cyclohexane}}{C_6H_{12}} \xrightarrow[-3H_2]{\text{Dehydrogenation}} \underset{\text{Benzene}}{C_6H_6}$$

Similarly,

$$\underset{n\text{-heptane}}{\begin{matrix} & CH_2-CH_3 & \\ CH_2 & & CH_3 \\ | & & | \\ CH_2 & & CH_2 \\ & CH_2 & \end{matrix}} \xrightarrow[\text{2-Dehydrogenation}]{\text{1-Cyclisation}} \underset{\text{Toluene}}{C_6H_5CH_3} + 4H_2$$

6. Thermal decomposition or pyrolysis or cracking. The process of decomposition of bigger molecules into a number of simpler molecules by the action of heat is called *pyrolysis* or *cracking*. This process involves the breaking of C – C and C – H bonds leading to the formation of lower alkanes and alkenes. For example,

$$CH_3-CH_2-CH_3 \xrightarrow{733\text{ K}} \underset{\text{Ethylene}}{CH_2=CH_2} + \underset{\text{Methane}}{CH_4} \text{ or } \underset{\text{Propene}}{CH_3-CH=CH_2} + H_2$$

Q. 7. Explain the mechanism of halogenation of alkane.

(Punjab 2003; Nagpur, 2003; Delhi, 2004 . Guwahati, 2006; Andhra, 2010)

Ans. The halogenation of alkanes occurs by a free radical mechanism. It involves three steps: (*a*) Chain Initiation, (*b*) Chain propagation, (*c*) Chain termination. Mechanism of halogenation is explained by considering the chlorination of methane.

(*a*) **Chain Initiation.** When a mixture of CH_4 and Cl_2 is heated or subjected to diffused sunlight, Cl_2 absorbs energy and undergoes homolytic fission producing chlorine free radicals. One molecule gives rise to two radicals.

Step (1)

$$:\ddot{\underset{..}{Cl}}\cdot\cdot\ddot{\underset{..}{Cl}}: \xrightarrow[\text{or }\Delta]{h\nu} \underset{\text{Chlorine free radical}}{:\ddot{\underset{..}{Cl}}\cdot + \cdot\ddot{\underset{..}{Cl}}:}$$

(*b*) **Chain propagation.** The chlorine free radical produced above collides with a molecule of methane forming hydrogen chloride and a methyl free-radical. The methyl free-radical in turn reacts with a molecule of chlorine forming methyl chloride and chlorine free-radical. The newly formed chlorine radical can react with another molecule of methane as in step (2) generating methyl free-radical and hydrogen chloride. The methyl free-radical can again repeat step (3) and so on. Thus, the sequence of reactions in steps (2) and (3) is repeated over and over again and thus the chain is propagated. In other words, a single photon of light initially absorbed by chlorine can bring about the conversion of a large number of molecules of methane into methyl chloride.

Step (2)

$$H-\overset{H}{\underset{H}{\overset{|}{\underset{|}{C}}}}-H + Cl\cdot \longrightarrow \underset{\text{Methyl free-radical}}{H-\overset{H}{\underset{H}{\overset{|}{\underset{|}{C}}}}\cdot} + HCl$$

Step (3)

$$H-\overset{H}{\underset{H}{\overset{|}{\underset{|}{C}}}}\cdot + Cl-Cl \longrightarrow H-\overset{H}{\underset{H}{\overset{|}{\underset{|}{C}}}}-Cl + Cl\cdot$$

However, the above reaction does not stop at methyl chloride stage but proceeds further till all the H–atoms of methane are replaced by chlorine atoms giving a mixture of mono-, di-, tri- and tetra-chloromethane

Step (4) $H-\underset{H}{\overset{H}{C}}-Cl + \cdot Cl \longrightarrow \cdot\underset{H}{\overset{H}{C}}-Cl + HCl$

Step (5) $\cdot\underset{H}{\overset{H}{C}}-Cl + Cl-Cl \longrightarrow Cl-\underset{H}{\overset{H}{C}}-Cl + Cl\cdot$

Methyledine chloride

Step (6) $C-\underset{H}{\overset{H}{Cl}}-Cl + Cl\cdot \longrightarrow Cl-\underset{\bullet}{\overset{H}{C}}-Cl + HCl$

Step (7) $Cl-\underset{\bullet}{\overset{CH}{C}}-Cl + Cl-Cl \longrightarrow Cl-\underset{Cl}{\overset{H}{C}}-Cl + Cl\cdot$

Chloroform

Step (8) $Cl-\underset{Cl}{\overset{H}{C}}-Cl + \cdot Cl \longrightarrow Cl-\underset{Cl}{\overset{\bullet}{C}}-Cl + HCl$

Step (9) $Cl-\underset{Cl}{\overset{\bullet}{C}}-Cl + Cl-Cl \longrightarrow Cl\underset{Cl}{\overset{Cl}{C}}Cl + \cdot Cl$

Carbon tetrachloride

Chain Termination

The chain reactions mentioned above, however, come to an end if the free radicals combine amongst themselves to form neutral molecules. Some of the chain terminating steps are:

$$\cdot Cl + \cdot Cl \longrightarrow Cl-Cl$$

$$\cdot CH_3 + \cdot Cl \longrightarrow CH_3-Cl$$

$$\cdot CH_3 + \cdot CH_3 \longrightarrow CH_3-CH_3$$

Q. 8. Give evidence in support of free radical mechanism of halogenation of alkanes.

(*Delhi 2005*)

Ans. The following points support the free radical mechanism:

(*i*) Reaction does not take place in dark at room temperature but requires energy in the form of heat or light. This is due to the fact that the chain initiation step (1) is endothermic and hence needs a large amount of energy to break the Cl – Cl bond into radicals.

(*ii*) The reaction has a high quantum yield *i.e.*, many thousand molecules of methyl chloride (or alkyl halide in general) are formed for each photon of light absorbed. This fact can be explained on the basis of chain propagation step (2) and (3).

(*iii*) Oxygen acts as an inhibitor. This is due to the fact oxygen combines with the alkyl free radical to form peroxyalkyl radical, (R – O – O •). This radical is much less reactive than alkyl free radical (R •) to continue the chain. As a result, the halogenation of alkyl in the presence of oxygen is slowed down or stopped. Thus the role of inhibitors like oxygen in this reaction gives support to the above mechanism.

(*iv*) If the above mechanism actually involves free radicals as reactive intermediate, then the addition of substances which are source of free radicals should initiate the reaction even in the dark at room temperature or much below 523 K. This has actually been found to be so. Thus chlorination of methane can be carried out in the dark at room temperature in the presence of small amount of benzoyl peroxide.

$$\underset{\text{Benzoyl peroxide}}{C_6H_5CO-O-O-COC_6H_5} \xrightarrow[298\text{ K}]{\text{dark}} \underset{\text{Benzoyl free radical}}{C_6H_5-CO-\ddot{\underset{..}{O}}\cdot}$$

$$C_6H_5-CO-\ddot{\underset{..}{O}}\cdot \longrightarrow \underset{\text{Phenyl free radical}}{C_6H_5\cdot} + CO_2$$

$$C_6H_5^{\cdot} + Cl-Cl \longrightarrow C_6H_5-Cl + Cl\cdot$$

Once, the chlorine free radicals are produced, the reaction can take place in the manner explained above.

(*v*) Chlorination of methane takes place at 423 K in dark in the presence of a little $(C_2H_5)_4$ Pb (Tetraethyl lead). Tetraethyl lead yields free radicals, when heated to 423 K.

$$(C_2H_5)_4\,Pb \xrightarrow[\text{dark}]{423\text{ K}} 4C_2H_5\cdot + Pb$$

The ethyl free radical then reacts with chlorine molecule forming ethyl chloride and chlorine free radical.

$$\cdot C_2H_5 + Cl-Cl \longrightarrow C_2H_5-Cl + Cl\cdot$$

The chlorine free radical thus formed brings about chlorination of methane as explained above.

Q. 9. What different products and in what proportion are obtained on chlorination and bromination of propane, butane and isobutane?

Ans. Halogenation of an alkane containing more than one type (primary, secondary, tertiary) of hydrogens, gives a mixture of isomeric products. For example, chlorination of propane, butane and isobutane give the following products.

(*i*) $$CH_3-CH_2-CH_3 \xrightarrow[\text{light, 298 K}]{Cl_2} \underset{\substack{n\text{-Propyl chloride}\\(45\%)}}{CH_3-CH_2-CH_2-Cl} + \underset{\substack{\text{Isopropyl chloride}\\(45\%)}}{CH_3-\overset{\overset{Cl}{|}}{C}H-CH_3}$$

(ii) $CH_3-CH_2-CH_2-CH_3 \xrightarrow[\text{light, 298 K}]{Cl_2} \underset{\substack{n\text{-Butyl chloride} \\ (28\%)}}{CH_3-CH_2-CH_2-CH_2-Cl} + \underset{\substack{\text{sec-Butyl chloride} \\ (72\%)}}{CH_3-CH_2-\overset{\overset{Cl}{|}}{CH}-CH_3}$

(iii) $\underset{\text{Isobutane}}{CH_3-\overset{\overset{CH_3}{|}}{CH}-CH_3} \xrightarrow{Cl_2} \underset{\substack{\text{Isobutyl chloride} \\ (64\%)}}{CH_3-\overset{\overset{CH_3}{|}}{CH}-CH_2-Cl} + \underset{\substack{\text{tert-Butyl chloride} \\ (36\%)}}{CH_3-\overset{\overset{CH_3}{|}}{\underset{\underset{CH_3}{|}}{C}}-Cl}$

Bromination gives the corresponding bromides but in different proportions:

(iv) $\underset{\text{Propane}}{CH_3-CH_2-CH_3} \xrightarrow[\text{light, 400 K}]{Br_2} \underset{\substack{n\text{-Propyl bromide} \\ (3\%)}}{CH_3-CH_2-CH_2-Br} + \underset{\substack{\text{Isopropyl bromide} \\ (97\%)}}{CH_3-\overset{\overset{Cl}{|}}{CH}-CH_3}$

(v) $\underset{n\text{-Butane}}{CH_3-CH_2-CH_2-CH_3} \xrightarrow[\text{light, 400 K}]{Br_2} \underset{\substack{n\text{-Butyl bromide} \\ (2\%)}}{CH_3-CH_2-CH_2-CH_2-Br} + \underset{\substack{\text{sec-Butyl chloride} \\ (98\%)}}{CH_3-CH_2-\overset{\overset{Br}{|}}{CH}-CH_3}$

(vi) $\underset{\text{Iso-Butane}}{CH_3-\overset{\overset{CH_3}{|}}{CH}-CH_3} \xrightarrow[\text{light, 400 K}]{Br_2} \underset{\substack{\text{Isobutyl bromide} \\ (1\%)}}{CH_3-\overset{\overset{CH_3}{|}}{CH}-CH_2-Br} + \underset{\substack{\text{Tert-Butyl bromide} \\ (99\%)}}{CH_3-\overset{\overset{CH_3}{|}}{\underset{\underset{CH_3}{|}}{C}}-Br}$

The result given above shows that the relative amounts of the different isomeric products differ largely depending upon the halogen used. It is also important to note that the bromination, in contrast to chlorination, leads to the formation of only one of the possible isomeric products. This is reflected in the percentages like 97%, 98% and 99% for one of the products in each reaction. Thus bromine atom is more selective in the site of attack than chlorine.

Q. 10. What are the factors that influence the orientation of halogenation of alkanes?

Ans. Factors that influence the halogenation process are:

(i) Collision frequency

(ii) Probability factor

(iii) Energy factor

Let us take the example of chlorination of propane.

$$\begin{array}{c} H\ \ H\ \ H \\ |\ \ \ |\ \ \ | \\ H-C-C-C-H \\ |\ \ \ |\ \ \ | \\ H\ \ H\ \ H \end{array}$$

$$\xrightarrow{Cl\cdot}\begin{cases} \xrightarrow{\text{Removal of 1° H}} H-\underset{H}{\overset{H}{C}}-\underset{H}{\overset{H}{C}}-\underset{H}{\overset{H}{C}}\cdot \ (n\text{-propyl radical}) \xrightarrow{Cl_2} H-\underset{H}{\overset{H}{C}}-\underset{H}{\overset{H}{C}}-\underset{H}{\overset{H}{C}}-Cl\ (n\text{-propyl chloride}) \\ \xrightarrow{\text{Removal of 2° H}} H-\underset{H}{\overset{H}{C}}-\overset{H}{\underset{|}{C}}-\underset{H}{\overset{H}{C}}-H\ (\text{Isopropyl radical}) \xrightarrow{Cl_2} H-\underset{H}{\overset{H}{C}}-\underset{Cl}{\overset{H}{C}}-\underset{H}{\overset{H}{C}}-H\ (\text{Isopropyl chloride}) \end{cases}$$

The relative amounts of *n*-propyl chloride and isopropyl chloride depend upon the relative rates at which the intermediate *n*-propyl and isopropyl radicals are formed as the rate-determining step is the formation of *n*-propyl and isopropyl radicals by the attack of chlorine radical on propane at the proper site.

The rates at which the two propyl radicals are formed depend upon the above factors.

(*i*) **Collision frequency.** The collision frequency is same for the two reactions because both involve the collisions of the same particles viz., propane and chlorine atom.

(*ii*) **Probability factor.** Since propane has six primary hydrogens and two secondary hydrogens, the probability of removal of primary hydrogens as compared to the secondary hydrogens should be in the ratio of 6 : 2 or 3 : 1.

If we consider the probability factor and collision frequency only, the chlorination of propane would be expected to yield *n*-propyl chloride and isopropyl chloride in the ratio of 3 : 1. But in actual practice, the two chloride are formed roughly in equal amounts *i.e.*, the ratio is 1 : 1 or 3 : 3 (see the experimental data).

(*iii*) **Energy factor.** The probability of formation of isopropyl chloride is about three times as large as that of formation of *n*-propyl chloride. This means that collisions with secondary hydrogens are about three times more effective than the collisions with primary hydrogens. This implies that isopropyl radicals are formed more easily than *n*-propyl radicals in the above reactions. We know that isopropyl radical is secondary radical while *n*-propyl radical is a primary radical and the decreasing order of stability is:

Tert-radical > sec-radical > primary radical

It is, therefore, clear that isopropyl chloride would be more stable (and hence easily formed) than *n*-propyl chloride. Now the stability or the ease of formation of an alkyl radical is related to the E_{act} (energy of activation) of the reaction leading to the formation of that radical. Hence E_{act} of the reaction leading to the formation of isopropyl radical is smaller than E_{act} of the reaction leading to the formation of *n*-propyl radical. In other words, Eact for the abstraction of a secondary hydrogen is less than E_{act} for the abstraction of a primary hydrogen. Hence the orientation in chlorination of propane is determined by the energy factor, *i.e.*, the E_{act} for the abstraction of secondary and primary hydrogens in propane.

Chlorination of isobutane presents a similar problem. Here there are nine primary hydrogens and one tertiary hydrogen. The abstraction of one of the nine primary hydrogens leads to the formation of isobutyl chloride while the abstraction of single tertiary hydrogen leads to the formation of tertbutyl chloride. We would, therefore, expect that the probability factor favours the formation of *iso*-butyl chloride (where primary hydrogen is replaced by chlorine) by the ratio of 9 : 1. But isobutyl chloride and tert-butyl chloride are formed roughly in the ratio of 2 : 1, or 9 : 4.5. Naturally, about 4.5 times as many collisions with tertiary hydrogens are effective as with primary hydrogens.

This means that E_{act} for the abstraction of a tertiary hydrogn (3° H) is less than E_{act} for the abstraction of a primary hydrogen (1° H) and, in fact, even less than for the abstraction of secondary hydrogen (2° H).

$$CH_3-\underset{|}{\overset{CH_3}{CH}}-CH_3 \xrightarrow{Cl\bullet} \begin{cases} \xrightarrow[\text{1° H}]{\text{Removal of}} CH_3-\overset{CH_3}{\overset{|}{CH}}-\dot{C}H_2 \ (\text{Isobutyl radical}) \xrightarrow{Cl_2} CH_3-\overset{CH_3}{\overset{|}{CH}}-CH_2Cl \ (\text{Isobutyl chloride}) \\ \xrightarrow[\text{3° H}]{\text{Removal of}} CH_3-\overset{CH_3}{\overset{|}{\dot{C}H}}-CH_3 \ (\text{tert-Butyl radical}) \xrightarrow{Cl_2} CH_3-\underset{Cl}{\underset{|}{\overset{CH_3}{\overset{|}{C}}}}-CH_3 \ (\text{tert-Butyl chloride}) \end{cases}$$

Study of chlorination of many alkanes have shown that at room temperature, the relative rates of abstraction of tertiary, secondary and primary hydrogens are in the ratio 5.0 : 3.8 : 1.0. With the help of these values, we can predict fairly well the ratio of isomeric chlorination products obtained from a given alkane. The expected ratio of *n*-propyl chloride and isopropyl chloride in the chlorination of propane can be calculated as follows:

$$\frac{\text{yield of } n\text{-propyl chloride}}{\text{yield of isopropyl chloride}} = \frac{\text{No. of 1° H}}{\text{No. of 2° H}} \times \frac{\text{Reactivity of 1° H}}{\text{Reactivity of 2° H}}$$

$$= \frac{6}{2} \times \frac{1}{3.8} = \frac{6}{7.6}$$

or
$$\frac{\text{yield of isopropyl chloride}}{\text{yield of } n\text{-propyl chloride}} = \frac{7.6}{6} = 1.26$$

Thus yield is isopropyl chloride = 1.26 yield of *n*-propyl chloride.

$$\frac{\text{Actual yield of isopropyl chloride}}{\text{Actual yield of } n\text{-propyl chloride}} = \frac{55}{45} = 1.22$$

Similarly, in case of isobutance, we get the following result:

$$\frac{\text{yield of isobutyl chloride}}{\text{yield of tert-butyl chloride}} = \frac{\text{No. of 1° H}}{\text{No. of 3° H}} \times \frac{\text{Reactivity of 1° H}}{\text{Reactivity of 3° H}}$$

$$= \frac{9}{1} \times \frac{1}{5} = \frac{9}{5}$$

or
$$\frac{\text{yield of isobutyl chloride}}{\text{yield of tert-butyl chloride}} = \frac{9}{5} = 1.8$$

Thus yield of isobutyl chloride = 1.80 × yield of tert-butyl chloride

$$\frac{\text{Actual yield of isobutyl chloride}}{\text{Actual yield of tert-butyl chloride}} = \frac{16}{9} = 1.78$$

In bromination of alkane, the same sequence of reactivity *i.e.*, 3° > 2° > 1° is found, but the relative reactivity rates are much larger. At 400 K, the relative rates for abstraction of tertiary. Secondary and primary hydrogens are 1600 : 82 : 1. The observed ratio of *n*-propyl bromide and iso-propyl bromide can be calculated as follows:

$$\frac{\text{yield of } n\text{-propyl bromide}}{\text{yield of isopropyl bromide}} = \frac{\text{No. of } 1°\text{ H}}{\text{No. of } 2°\text{ H}} \times \frac{\text{Reactivity of } 1°\text{ H}}{\text{Reactivity of } 2°\text{ H}}$$

$$= \frac{6}{2} \times \frac{1}{82} = \frac{3}{82}$$

or

$$\frac{\text{yield of } n\text{-propyl bromide}}{\text{yield of iso-propyl bromide}} = \frac{3}{82}$$

Similarly,

$$\frac{\text{yield of iso-butyl bromide}}{\text{yield tert-butyl bromide}} = \frac{\text{No. of } 1°\text{ H}}{\text{No. of } 3°\text{ H}} \times \frac{\text{Reactivity of } 1°\text{ H}}{\text{Reactivity of } 3°\text{ H}}$$

$$= \frac{9}{1} \times \frac{1}{1600} = \frac{9}{1600}$$

Q. 11. An alkane with molecular wieght 72 formed only one monochloro substitution product. Suggest a structure for the alkane. (*Kurukshetra, 2001; Madurai, 2004*)

Ans. Molecular weight 72 suggests that the alkane has the molecular formula C_5H_{12}. Out of three isomeric pentanes viz., *n*-pentane, isopentane and neopentane the last one is the answer of the question. This is because only neopentane has one type of hydrogen (primary) and hence would give one monochloro substitution product.

$$CH_3-\underset{\underset{CH_3}{|}}{\overset{\overset{CH_3}{|}}{C}}-CH_3 \xrightarrow[-HCl]{Cl_2} CH_3-\underset{\underset{CH_3}{|}}{\overset{\overset{CH_3}{|}}{C}}-CH_2Cl$$

Neopentane (2,2-Dimethylpropane) — Neopentyl chloride

Q. 12. Give the structure of pentane that would be expected to produce the largest number of isomeric monochloro derivatives.

Ans. There are three isomers of pentanes.

$$CH_3-CH_2-CH_2-CH_2-CH_3$$

n-pentane

$$CH_3-\underset{\underset{CH_3}{|}}{CH}-CH_2-CH_3$$

Isopentane

$$H_3C-\underset{\underset{CH_3}{|}}{\overset{\overset{CH_3}{|}}{C}}-CH_3$$

Neopentane

Out of the three isomers, isopentane would give largest number of isomeric derivatives as it contains three types of replaceable hydrogen (primary, secondary and tertiary).

Q. 13. Calculate the percentage of expected isomers during monobromination of *n*-butane. The relative rates of substitution per 3°, 2° and 1° hydrogen are 1600 : 82 : 1.

Ans.

$$\underset{n\text{-butane}}{CH_3 - CH_2 - CH_2 - CH_3} \xrightarrow{Br_2} \underset{n\text{-butyl bromide}}{CH_3 - CH_2 - CH_2 - CH_2Br} + \underset{\text{sec-butyl bromide}}{CH_3CH_2CHBrCH_3}$$

$$\frac{\text{yield of } n\text{-butyl bromide}}{\text{yield of sec-butyl bromide}} = \frac{\text{No. of 1° H} \times \text{Reactivity of 1° H}}{\text{No. of 2° H} \times \text{Reactivity of 2° H}}$$

$$= \frac{6 \times 1}{4 \times 82} = \frac{6}{328}$$

$$\% \text{ of } n\text{-butyl bromide} = \frac{6}{334} \times 100 = 1.79\%$$

$$\% \text{ of sec-butyl bromide} = 100 - 1.79 = 98.21\%$$

Q. 14. Predict the proportions of isomeric products from chlorination at room temperature of (*i*) *n*-Butane (*ii*) 2, 3-dimethyl butane.

Ans. (*i*) *n*-Butane

$$CH_3 - CH_2 - CH_2 - CH_3 \xrightarrow{Cl_2} \underset{n\text{-Butyl chloride}}{CH_3 - CH_2 - CH_2 - CH_2Cl} + \underset{\text{sec-Butyl chloride}}{CH_3 - CH_2 - \overset{\overset{Cl}{|}}{CH} - CH_3}$$

$$\frac{\text{yield of } n\text{-Butyl chloride}}{\text{yield of sec-Butyl chloride}} = \frac{\text{No. of 1° H}}{\text{No. of 2° H}} \times \frac{\text{Reactivity of 1° H}}{\text{Reactivity of 2° H}}$$

$$= \frac{6}{4} \times \frac{1}{3.8} = \frac{3}{7.6} = \frac{28\%}{72\%}$$

(*ii*) 2, 3-dimethyl butane

$$H_3C - \overset{\overset{CH_3}{|}}{HC} - \overset{\overset{CH_3}{|}}{HC} - CH_3 + Cl_2 \longrightarrow \underset{\text{1-Chloro-2,3-dimethyl butane}}{CH_3 - \overset{\overset{CH_3}{|}}{CH} - \overset{\overset{CH_3}{|}}{CH} - CH_2Cl \text{ (I)}}$$

$$+ \underset{\text{2-Chloro-3-methyl butane}}{CH_3 - \overset{\overset{CH_3}{|}}{CH} - \overset{\overset{Cl}{|}}{CH} - CH_3 \text{ (II)}}$$

$$\frac{\text{yield of I (Primary)}}{\text{yield of II (Sec.)}} = \frac{\text{No. of 1° H}}{\text{No. of 3° H}} \times \frac{\text{Reactivity of 1° H}}{\text{Reactivity of 3° H}}$$

$$= \frac{12}{2} \times \frac{1}{5} = \frac{6}{5} = \frac{55\%}{45\%}$$

Q. 15. Chlorination reaction of certain higher alkanes can be used for laboratory preparations. For example, we can prepare neopentyl chloride from neopentane and cyclopentyl chloride from cyclopentane. How do you account for this fact?

Ans. Chlorination of neopentane $\left[CH_3-\underset{CH_3}{\overset{CH_3}{\underset{|}{\overset{|}{C}}}}-CH_3\right]$ and cyclopentane [cyclopentane ring] can be carried out because all the hydrogens of each compound are equivalent and replacement of any one yields the same product, *i.e.* neopentyl chloride $\left[CH_3-\underset{CH_3}{\overset{CH_3}{\underset{|}{\overset{|}{C}}}}-CH_2Cl\right]$ and cyclopentyl chloride [cyclopentane ring bearing Cl] respectively.

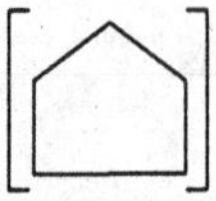

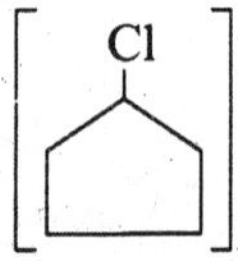

Q. 16. Monochlorination of an equimolar mixture of methane and ethane does not yield a mixture of methyl chloride and ethyl chloride in equimolar proportions. How do you account for this observation?

Ans. Monochlorination of an equimolar mixture of methane and ethane yields a mixture of methyl chloride and ethyl chloride in which ethyl chloride is formed about 400 times as much as methyl chloride. This can be attributed to two factors as discussed below:

(1) Probability factor. Ethane contains six hydrogen atoms each of which can be replaced by chlorine while methane contains only four hydrogen atoms. Therefore the probabilities of formation of ethyl chloride and methyl chloride will be in the ratio 6 : 4 or 3 : 2.

(2) Relative reactivities of different classes of hydrogen atoms. This is the most important factor which determines the relative amount of products in the halogenation of methane and ethane (or other alkanes). This in turn depends upon the relative stabilities of alkyl radicals formed in the step which controls the overall process of halogenation. In case of methane and ethane, this step may be written as:

$$CH_4 + \dot{C}l \longrightarrow \dot{C}H_3 + HCl$$

$$CH_3-CH_3 + Cl\cdot \longrightarrow CH_3-\dot{C}H_2 + HCl$$

Now the stability of ethyl radical (relative to ethane) is more than the stability of methyl radical (relative to methane). Therefore the transition state involved in the formation of ethyl radical from ethane would be more stable than the transition state involved in the formation of methyl radical from methane. As such transition state of ethyl radical requires lesser Eact than transition state of methyl radical and therefore the former is formed more readily than the latter (Fig. 7.1). This means that carbon hydrogen bond in ethane breaks more readily than carbon hydrogen bond in methane or the primary hydrogen atoms in ethane are more reactive than hydrogen atoms in methane. As a consequence ethane undergoes chlorination much more readily than methane so that the mixture formed contains a very high proportion of ethyl chloride as compared to methyl chloride.

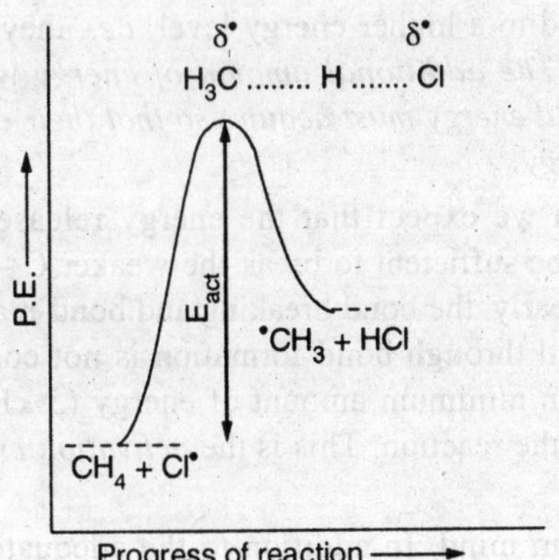

Greater E_{act} of transition state, lesser ease of formation of free radical, lesser reactivity of hydrogens.

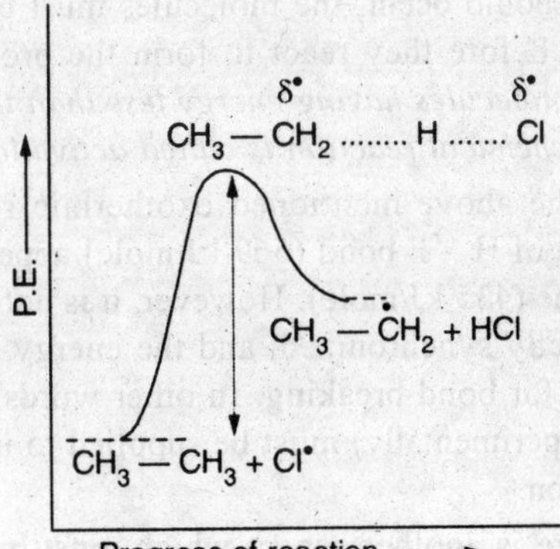

Lesser E_{act} of transition state, greater ease of formation of free radical, greater reactivity of hydrogens.

Fig. 7.1. Illustration of relative reactivities of hydrogens of methane and primary hydrogens of ethane

Q. 17. On chlorination, an equimolar mixture of neopentane and ethane yields neopentyl chloride and ethyl chloride in the ratio of 2.3 : 1. How does the reactivity of a primary hydrogen in neopentane compare with that of primary hydrogen in ethane?

Ans. No. of primary hydrogen atoms in ethane = 6

No. of primary hydrogen atoms in neopentane

$$\left[\begin{array}{c} CH_3 \\ | \\ CH_3-C-CH_3 \\ | \\ CH_3 \end{array}\right] = 12$$

$$\frac{\text{yield of neopentyl chloride}}{\text{yield of ethyl chloride}} = \frac{\text{No. of 1° H} \times \text{Reactivity of 1° H in neopentane}}{\text{No. of 1° H} \times \text{Reactivity of 1° H in ethane}}$$

$$\frac{2.3}{1} = \frac{12 \times \text{Reactivity in neopentane}}{6 \times \text{Reactivity in ethane}}$$

or $$\frac{\text{Reactivity in neopentane}}{\text{Reactivity in ethane}} = \frac{2.3 \times 6}{12} = \frac{1.15}{1}$$

or $$1.15 : 1.$$

Q. 18. Write a note on activation energy.

Ans. Consider the reaction

$$F° + CH_3-H \longrightarrow \cdot CH_3 + H-F \qquad \Delta H = -134 \text{ kJ}$$

$$\Delta H° = 435 \text{ kJ} \qquad 569 \text{ kJ} \qquad E_{act} = 5 \text{ kJ}$$

For a reaction to occur, collision between reactant molecules is a must. Thus the above mentioned reaction takes place only if the collision between a chlorine atom and a methane molecule takes place. But all the collisions are not effective. For a collision to be effective, the reactant molecules must be associated with a certain minimum amount of energy called *threshold energy*. The no. of such molecules constitute only a small fraction of the total no. of molecules or most of the molecules have kinetic energy lower than the threshold energy. Thus, in order that the

reaction should occur, the molecules must be raised to a higher energy level, *i.e.*, they must be activated before they react to form the products. *The additional amount of energy which the reactant molecules having energy less than threshold energy must acquire so that their collisions result in chemical reaction is called activation energy.*

In the above mentioned exothermic reaction we expect that the energy released in the formation of H – F bond (569 kJ/mole) appears to be sufficient to break the weaker C – H bond of methane (435 kJ/mole). However, it is not so. Clearly, the bond breaking and bond-making are not perfectly synchronized, and the energy released through bond formation is not completely available for bond breaking. In other words, certain minimum amount of energy (5 kJ/mole as shown experimentally) must be supplied to initiate the reaction. This is the *activation energy* for the reaction.

There is another aspect which must be kept in mind. In addition to the adequate energy (kinetic energy) which the colliding molecules possess at the time of collision, they must be properly oriented before the collision can be really effective in causing the reaction. The number of properly oriented collisions in a reaction is quite small as compared to the total number of collisions.

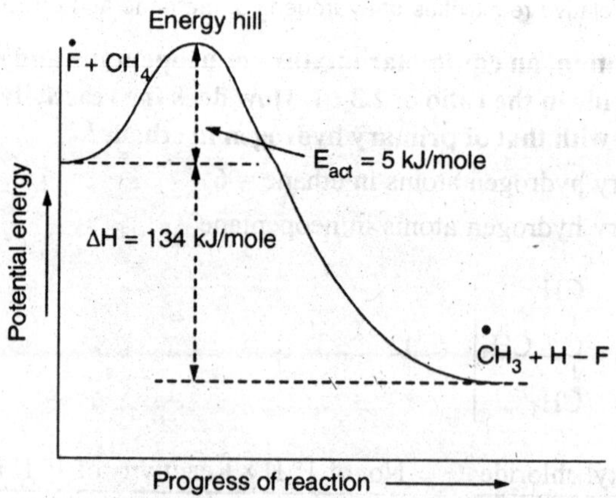

Fig. 7.2. Concept of activation energy

Thus in the above mentioned example, the methane molecule must be oriented in such a way that it presents a hydrogen atom to the full force of the impact. In this case, only about one collision in eight is properly oriented.

In general, *in collisions, sufficient energy (E_{act}) and proper orientation are essential for a reaction to take place.*

As the reaction proceeds, changes in the potential energy takes place and these energy changes are shown in the figure above.

The potential energy of methane molecule and fluorine atom is shown on the left and that of the methyl free radical (CH_3) and hydrogen fluoride on the right (Fig. 7.2). As the reaction begins, the kinetic energy of the reactants (due to their motion) is changed into potential energy.

With the increase in potential energy, we move up the energy hill and move down the other side. During the descent, the potential is changed back into kinetic energy till we reach the level of the products (CH_3 • and HF). As the potential energy of the product is less than that of the reactants; there must be a corresponding increase in kinetic energy. In other words, the molecules of the products (• CH_3 and HF) move faster than the molecules of the reactants (CH_4 and • F) so there would be a rise in temperature or heat will be evolved. The height of the hill top or energy hill (in kJ/mole) above the level of the reactant is the energy of activation (5 kJ/mole).

Q. 19. Discuss briefly the relative reactivities of halogens in the halogenation of alkanes.

(G. Nanakdev, 2000; Kumaon 2000)

Ans. The magnitude of E_{act} determines the rate of a reaction. Reactions with low E_{act} proceed at faster rate than the one with a high E_{act} at the same temperature. This relationship between the rate of a reaction an the magnitude of the energy of activation can be explained with the help of halogenation of methane as follows:

$$CH_4 + X_2 \xrightarrow{(X_2 = F_2, Cl_2, Br_2, I_2)} CH_3 - X + HX$$

The rate determining step here is the abstraction of hydrogen by halogen atom *i.e.*, how fast methyl radical is formed.

$$CH_4 + X\cdot \longrightarrow CH_3^{\bullet} + HX$$

Formation of radical in chain propagation step is difficult but once formed it gets readily converted into the alkyl halide.

Thus, the overall rate of halogenation of methane depends upon now fast the methyl radical is formed which, in turn, depends on the E_{act} for the formation of methyl radical. Thus,

$$\underset{\phantom{435\text{ kJ}}}{F\cdot} + \underset{435\text{ kJ}}{CH_3 - H} \longrightarrow \underset{569\text{ kJ}}{H - F} + \cdot CH_3 \qquad \Delta H = -134\text{ kJ},\ E_{act} = 5\text{ kJ}$$

$$Cl\cdot + \underset{435\text{ kJ}}{CH_3 - H} \longrightarrow \underset{431\text{ kJ}}{H - Cl} + \cdot CH_3 \qquad \Delta H = +4\text{ kJ},\ E_{act} = 17\text{ kJ}$$

$$Br\cdot + \underset{435\text{ kJ}}{CH_3 - H} \longrightarrow \underset{368\text{ kJ}}{H - Br} + \cdot CH_3 \qquad \Delta H = +67\text{ kJ},\ E_{act} = 78\text{ kJ}$$

$$I\cdot + \underset{435\text{ kJ}}{CH_3 - H} \longrightarrow \underset{297\text{ kJ}}{H - I} + \cdot CH_3 \qquad \Delta H = +138\text{ kJ},\ E_{act} = 138\text{ kJ}$$

Potential energy diagrams for the above mentioned reactions are given in Fig. 7.3.

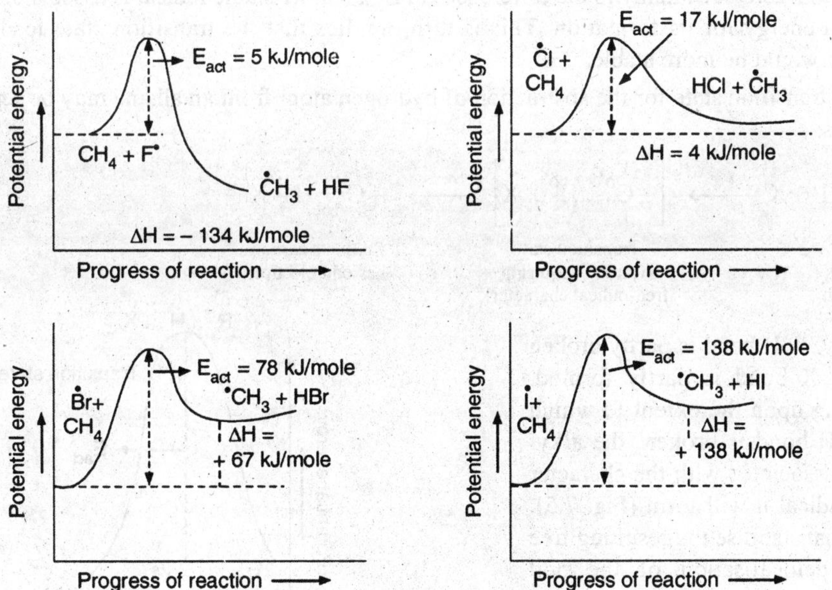

Fig. 7.3. Potential energy diagrams

It is clear the decreasing order of E_{act} for chain propagation step is:

Iodination > Bromination > Chlorination > Fluorination

Hence order of halogenation should be:

Fluorination > Chlorination > Bromination > Iodination

This order is in conformity with the observed order of reactivity of F_2, Cl_2, Br_2 and I_2 in halogenation of methane in particular and alkanes in general.

Q. 20. What is the effect of activation energy on the structure of transition state in halogenation of alkanes?

Ans. For a collision to be effective collision, the reactant molecules must be associated with a certain amount of energy called threshold energy. Thus, they also have to cross an energy barrier (hill top) between the reactants and products before reaction can be accomplished. The top of the energy hill (energy barrier) represents an intermediate structure called the *transition state*. The transition state corresponds to a particular arrangement of atoms of reacting species as they are converted from reactants into product. It represents a configuration in which old bonds are partially broken and new bonds are partially formed. *The activation energy (E_{act}) is the difference in energy content between the transition state and reactants*, as shown in Fig. 7.4.

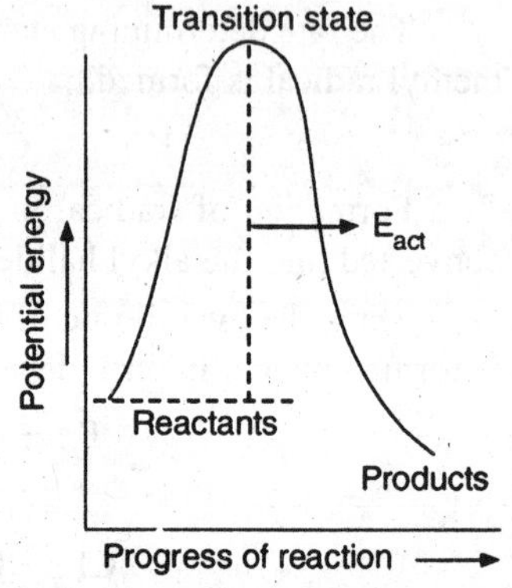

Fig. 7.4. Transition state

Hence the transition state of a reaction is the state of highest potential energy acquired by the reactants during their change into the products. The transition state is unstable due to its high energy content and cannot be isolated.

Let us see how the concept of transition state helps to explain the difference in the rates of halogenation of alkanes. We have already seen that the difference in reactivity of alkanes towards halogen atoms are due mainly to the differences in E_{act}; a more stable radical is associated with low activation energy for its formation. This is turn, implies that the transition state leading to its formation would be more stable.

The transition state for the abstraction of hydrogen atom from an alkane may be represented as follows:

$$-\overset{|}{\underset{|}{C}}-H + \cdot X \longrightarrow \left[-\overset{|}{\underset{|}{C}}\overset{\delta^\bullet}{\ldots\ldots}H\overset{\delta^\bullet}{\ldots\ldots}X\right] \longrightarrow -\overset{|}{C}\cdot + H-X$$

Reactants halogen has free electron — Transition state (Carbon acquiring free-radical character) — Products carbon has odd electron

The C – H bond is partly broken and H – X bond is partly formed. Depending upon the extent to which the C – H bond is broken, the alkyl group is associated with the character of free radical it will form (Fig. 7.5). Factors that stabilise the resulting free radical (delocalisation of the odd electron) tend to stabilize the nascent free radical in the transition state.

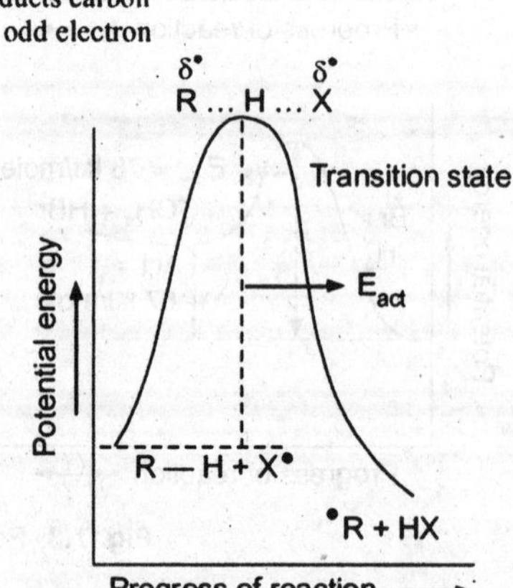

Fig. 7.5. Transition state in halogenation of alkane

Q. 21. Explain: Bromine is less reactive but more selective whereas chlorine is more reactive and less selective in its attack on alkanes.

Ans. Bromine is less reactive towards alkanes than chlorine but bromine is more selective than chlorine. The relative rates of substitution by bromine per hydrogen atom (3°, 2°, 1° hydrogen) are 1600 : 82 : 1 as compared to 5.0 : 3.8 : 1 in case of chlorine.

Thus the reaction of isobutane and bromine, for example, gives mainly tert-butyl bromide.

$$(CH_3)_2\underset{CH_3}{C}-H + Br_2 \xrightarrow[400\text{ K}]{\text{light}} \underset{(99\%)}{CH_3-\overset{CH_3}{\underset{CH_3}{C}}-Br} + \underset{(1\%)}{CH_3-\overset{CH_3}{\underset{H}{C}}-CH_2Br}$$

Further, bromine atom is less reactive than chlorine atom. Thus, selectivity is related to reactivity and can be generalised that the less reactive the reagent the more selective it is in its attack under a set of similar reaction. The greater selectivity of bromine can be explained in terms of transition state theory. According to a postulate made by Professor G.S. Hammond the structure of transition state of endothermic step (with high value of E_{act}) of a reaction resembles the products of that step more than it does the reactants. The structure of transition state of exothermic state with low value of E_{act} of a reaction resembles the reactant of that step more than it does the products.

This principle can be explained with the help of potential energy diagrams for the chlorination and bromination of isobutane.

The abstraction of hydrogen by the highly reactive chlorine atom is exothermic and has low E_{act}. According to the above postulate, the transition state resembles the reactants more than it does the products. In other words, the transition state is reached early and the carbon-hydrogen bond is slightly stretched. Atoms and electrons are distributed almost in the same way as in the reactants; carbon is still tetrahedral. The alkyl radical has developed very little free radical character. The transition state for the hydrogen (1° and 3°) abstraction steps may be shown in the following way:

$$CH_3-\overset{CH_3}{CH}-CH_3 + Cl\cdot \longrightarrow CH_3-\overset{CH_3}{CH}-\overset{\delta\cdot}{CH_2}\ldots\ldots H\ldots\ldots\overset{\delta\cdot}{Cl}$$

Reactant like structure

$$\downarrow$$

$$Cl\cdot + CH_3-\overset{CH_3}{CH}-CH_2Cl \xleftarrow{Cl_2} CH_3-\overset{CH_3}{CH}-\dot{C}H_2 + HCl$$

$$CH_3-\overset{CH_3}{\underset{CH_3}{C}}-H + \cdot Cl \longrightarrow CH_3-\overset{CH_3}{\underset{CH_3}{\overset{\delta\cdot}{C}}}\ldots\ldots H\ldots\ldots\overset{\delta\cdot}{Cl}$$

Reactant like structure

$$\downarrow$$

$$Cl\cdot + CH_3-\overset{CH_3}{\underset{CH_3}{C}}-Cl \xleftarrow{Cl_2} CH_3-\overset{CH_3}{\underset{CH_3}{C}}\cdot + HCl$$

The transition state in both cases resembles the reactants. Since the reactants in both the cases are the same and the same type of C – H bonds are broken (primary or tertiary), it has a relatively small influence on the relative rates of the reactions. The two reactions proceed with similar (but not identical) rates because their respective activation energies are similar (See Fig. 7.6).

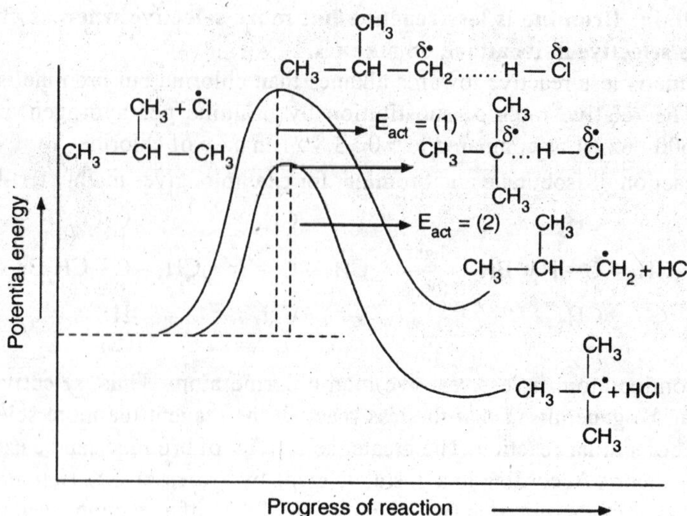

Fig. 7.6. Activation energies in exothermic reactions

In the above diagram it is clear that are similar, but because tertiary C – H bond is broken more easily than primary C – H bond, the reaction (2) has low and proceeds comparatively at faster rate.

On the other hand, the abstraction of hydrogen by the less reactive bromine atom is endothermic and has a very high E_{act}. The transition state resembles the products more than it does the reactants. In other words, the transition state is reached late and the carbon-hydrogen bond is broken to a considerable extent. Atoms and electrons are distributed almost in the same way as in the products; the carbon is almost trigonal. The alkyl radical has developed considerable free radical character.

The transition states may be shown in the following ways:

$$CH_3-\underset{}{\overset{CH_3}{\overset{|}{C}H}}-CH_3 + Br\cdot \longrightarrow CH_3-\overset{CH_3}{\overset{|}{C}H}-\overset{\delta\cdot}{C}H_2 \ldots\ldots H \ldots\ldots \overset{\delta\cdot}{Br}$$

Product like structure

$$Br\cdot + CH_3-\overset{CH_3}{\overset{|}{C}H}-CH_2Br \xleftarrow{Br_2} CH_3-\overset{CH_3}{\overset{|}{C}H}-\dot{C}H_2 + HBr$$

$$CH_3-\underset{CH_3}{\underset{|}{\overset{CH_3}{\overset{|}{C}H}}} + \cdot Br \longrightarrow CH_3-\underset{CH_3}{\underset{|}{\overset{CH_3}{\overset{|\,\delta\cdot}{C}}}} \ldots\ldots H \ldots\ldots \overset{\delta\cdot}{Cl}$$

Product like structure

$$Br\cdot + CH_3-\underset{CH_3}{\underset{|}{\overset{CH_3}{\overset{|}{C}}}}-Br \xleftarrow{Br_2} CH_3-\underset{CH_3}{\underset{|}{\overset{CH_3}{\overset{|}{C}}}}\cdot \quad + HBr$$

The transition states in both steps resemble the products in energy and structure. Since the products in both the cases are different, the type of C – H bond being broken has a lot of influence on the relative rates of reactions. They proceed with different rates. Abstraction of tertiary hydrogen takes place much faster. Bromine is, therefore, more selective in its attack. See Fig. 7.7 ahead.

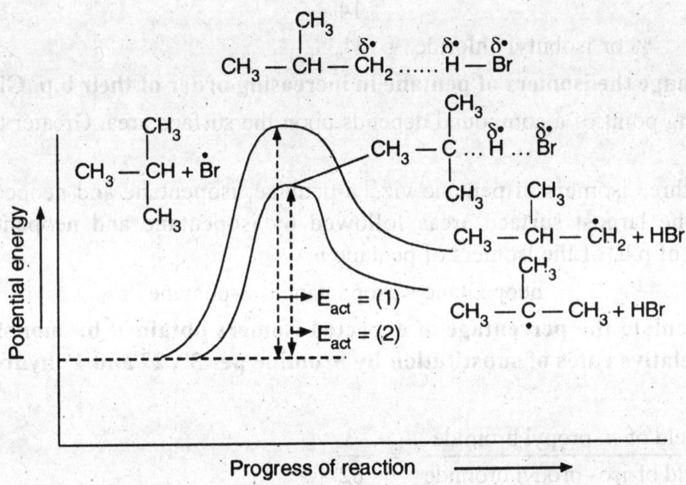

Fig. 7.7. Activation energies in endothermic reactions.

$$CH_3-\underset{|}{\overset{CH_3}{CH}}-\overset{\delta\bullet}{C}H_2 ...... H ...... \overset{\delta\bullet}{Br}$$

It is clear from the diagram that transition state for reaction (1) resembles a less stable primary radical while the transition state for reaction (2) resembles a more stable radical. The E_{act} for reaction (2) is much lower that for reaction (1). The product contains mostly tert-butyl bromide

Q. 22. Write short notes on the following related to halogenation of alkanes:

(i) Reactivity and selectively

(ii) Relative rates of halogenation in terms of bond dissociation energy.

Ans. See Q. 21.

Q. 23. Calculate the percentage of the expected isomers obtained by monochlorination of isobutane. The relative rates of substitution by chlorine per 3°, 2° and 1° hydrogen atoms are 5 : 3.8 : 1.

Ans.
$$\frac{\text{yield of isobutyl chloride}}{\text{yield of tert. butyl chloride}} = \frac{\text{No. of } 1^\circ \text{ H}}{\text{No. of } 3^\circ \text{ H}} \times \frac{\text{Reactivity of } 1^\circ \text{ H}}{\text{Reactivity of } 3^\circ \text{ H}}$$

$$= \frac{9}{1} \times \frac{1}{5} = \frac{9}{5}$$

$$1 + \frac{\text{yield of isobutyl chloride}}{\text{yield of tert. butyl chloride}} = 1 + \frac{9}{5}$$

$$\frac{\text{yield of tert. butyl chloride and isobutyl chloride}}{\text{yield of tert. butyl chloride}} = \frac{14}{5}$$

Take the reciprocal of above

$$\frac{\text{yield of tert. butyl chloride}}{\text{Total yield}} = \frac{5}{14}$$

$$\text{\% of tert. butyl chloride} = \frac{5}{14} \times 100 = 35.7$$

$$\text{\% or isobutyl chloride} = 64.3.$$

Q. 24. Arrange the isomers of pentane in increasing order of their b.p. Give reasons.

Ans. Boiling point of a compound depends upon the surface area. Greater the surface area, greater the b.p.

There are three isomers of pentane viz., *n*-pentane, isopentane and neopentane. Of these, *n*-pentane has the largest surface areas followed by isopentane and neopentane. Therefore increasing order of p.b. of the isomers of pentane is

neopentane < isopentane < *n*-pentane

Q. 25. Calculate the percentage of expected isomers obtained by monobromination of propane. The relative rates of substitution by bromine per 3°, 2° and 1° hydrogen atoms are 1600 : 82 : 1.

Ans. $$\frac{\text{yield of } n\text{-propyl bromide}}{\text{yield of iso-propyl bromide}} = \frac{3}{82}$$

Add 1 to both sides

$$1 + \frac{\text{yield of } n\text{-propyl bromide}}{\text{yield of iso-propyl bromide}} = 1 + \frac{3}{82}$$

$$\frac{\text{Total yield}}{\text{yield of iso-propyl bromide}} = \frac{85}{82}$$

or $$\frac{\text{yield of iso-propyl bromide}}{\text{Total yield}} = \frac{82}{85}$$

$$\text{\% of iso-propyl bromide} = 196.5$$

$$\text{\% of } n\text{-propyl bromide} = 3.5.$$

PETROLEUM

Q. 26. What is the composition of petroleum?

Ans. Petroleum is a complex mixture of aliphatic hydrocarbons including alkanes ($C_1 - C_{40}$), cycloalkanes and aromatic hydrocarbons. Small quantities of nitrogen, sulphur and oxygen compounds are also present. However the exact composition varies with the place of origin. It is usually covered with a gaseous mixture known as natural gas which is a mixture of low boiling alkanes like methane, ethane, propane, butane and pentane but major constituent is methane. It is the source for preparing a large number of chemical products such as medicines, rubber, dyes, perfumes, plastics, explosives etc.

Q. 27. What is modern theory of origin of petroleum?

Ans. According to this theory, oil is formed by the bacterial decomposition of plant remains and animals (which got buried under the surface of the earth as a result of earthquakes, upheavals etc., millions of years ago) under high pressure and temperature. This theory is supported by the following facts:

(*i*) Association of brine and salt with it.

(*ii*) Presence of coal mines near the oil fields.

(*iii*) Chlorophyll, haemin and optically active compounds of nitrogen and sulphur are present in petroleum.

Q. 28. How is petroleum mined?

Ans. The oil usually floats over a layer of salt water and has a layer of natural gas upon it. It is present deep below the impervious rocks (See Fig. 7.8). Mining is done by drilling holes in the earth's crust and sinking pipes up to the oil bearing porous rocks. The oil rushes up itself through the pipes due to the high pressure exerted by the natural gas inside. When the pressure subsides, the oil is brought to the surface by means of lift pumps.

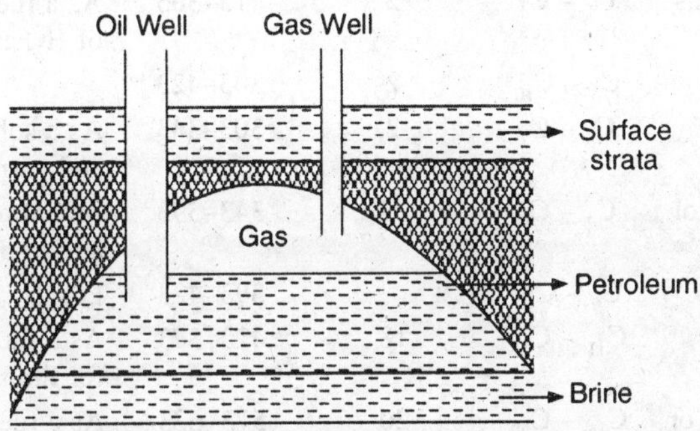

Fig. 7.8. Mining of petroleum

The crude oil from these wells is sent to refineries through underground pipelines for separation into its constituents.

Q. 29. What is refining of petroleum? What different fractions are obtained on refining? (*Assam*, *2003*)

Ans. The crude petroleum obtained from the oil fields is a dark viscous liquid having unpleasant smell due to the presence of sulphur containing compounds. It is also associated with impurities of sand and brine. It is subjected to fractional distillation to remove these impurities and to get various useful fractions. *The process of separating petroleum into different useful fractions and the removal of undesirable impurities is called refining or processing of petroleum.*

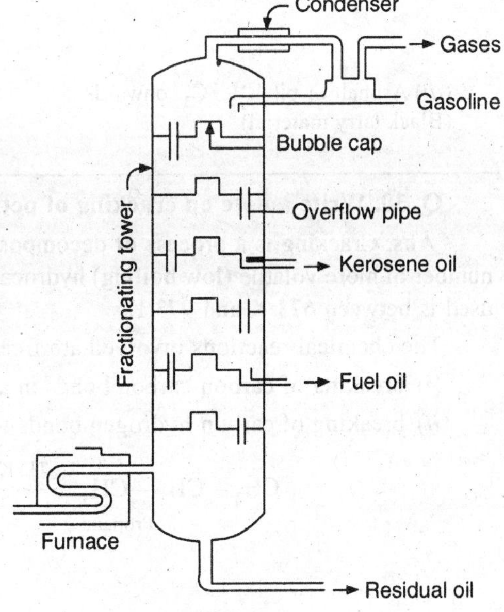

Fig. 7.9. Refining of petroleum

The crude petroleum is first washed with acid and then with alkali to remove the basic and acidic compounds respectively. It is then heated to 675 K in coiled pipes in a gas heated furnace and the vapours so obtained are admitted into the fractionating column made of steel (See Fig.

7.9). The tower is divided into a number of compartments by means of shelves having openings which are covered by caps known as bubble caps. Each shelf is provided with overflow pipe in order to keep the liquid at a constant level.

Temperature at the bottom of the fractionating column is much higher than at the top. As the vapours of the oil rise up the fractionating column, they cool down and condense into different fractions. These fractions are withdrawn continuously through the outlets provided in the tower at various heights as shown in the figure.

The different fractions so obtained are refractionated and purified to get various products. Some of the important fractions along with their approximate composition, boiling range and uses are given in the following table.

Fraction	*Approximate*		*Boiling range*	*Uses*
	Composition	*Percentage*	*(K)*	
1. Gaseous hydrocarbons	$C_1 - C_4$	2	113–303	As a fuel gas, in the production of H_2, carbon black.
2. Crude naphtha	$C_5 - C_{10}$	2	303–423	
(*i*) Petroleum ether	$C_5 - C_7$	2	303–343	As a solvent for rubber, oils, fats, and varnishes.
(*ii*) Gasoline or petrol	$C_7 - C_9$	32	343–393	As a motor oil, for dry cleaning and in preparation of petrol gas.
(*iii*) Benzine	$C_9 - C_{10}$		393–423	For dry cleaning.
3. Kerosene oil	$C_{11} - C_{15}$	18	423–573	As a fuel, for making oil gas and for illumination purposes.
4. Diesel oil or Fuel oil or Heavy oil	$C_{16} - C_{20}$	20	573–623	As a fuel for diesel engines, for making gasoline by cracking.
5. Residual oil				
(*i*) Lubricating oil	$C_{20} - C_{24}$		623–673	As a lubricant.
(*ii*) Paraffin wax	$C_{24} - C_{29}$		above 673 (melts between 325–330)	In the manufacture of candles, ointments, toilet soaps, polishes, varnishes and waxed papers.
(*iii*) Asphalt or pitch (Black tarry material)	C_{30} onwards		Residue	Used for water proofing, road metalling and in the manufacture of paints.

Q. 30. Write a note on cracking of petroleum.

Ans. Cracking is a process of decomposing less volatile (high boiling) hydrocarbons into a number of more volatile (low boiling) hydrocarbons by the action of heat. The temperature usually used is between 673 K and 973 K.

The chemical reactions involved are free radical reactions associated with

(*i*) breaking of carbon-carbon bonds in alkanes to form simple alkanes and alkenes

(*ii*) breaking of carbon-hydrogen bonds to form hydrogen and alkenes. For example,

$$\underset{\text{Propane}}{CH_3 - CH_2 - CH_3} \xrightarrow{733\text{ K}} \underset{\text{Ethyl radical}}{CH_3 - \dot{C}H_2} + \underset{\text{Methyl radical}}{\dot{C}H_3}$$

$$\downarrow \text{Disproportionation}$$

$$\underset{\text{Ethylene}}{CH_2 = CH_2} + \underset{\text{Methane}}{CH_4}$$

$$CH_3-CH_2-CH_2-CH_2-H \xrightarrow{873\text{ K}}$$
n-Butane

$$\longrightarrow CH_3-\dot{C}H_2+H_2\dot{C}-CH_3$$
↓ Disproportionation
$$CH_2=CH_2+CH_3-CH_3$$
$$CH_3-CH_2-\dot{C}H_2+\dot{C}H_3$$
n-Propyl radical
↓ Disproportionation
$$CH_3-CH_2=CH_2+CH_4$$
$$CH_3-CH_2-\dot{C}H_2+CH_2^{\bullet}+H^{\cdot}$$
n-Butyl radical
↓ Disproportionation
$$\longrightarrow CH_3-CH_3-CH=CH_2+H_2$$
1-Butene

Cracking may be carried out in two different ways:

(*a*) Liquid-phase cracking

(*b*) Vapour-phase cracking

(*a*) Liquid-phase cracking. Heavy oil or fuel oil is cracked by heating to a temperature of 748–803 K under a pressure of about 7–10 atmosphere. The heavy oil is converted into gasoline to the extent of 60–65% of the oil. The cracking material is kept in liquid state under these conditions due to high pressure.

(b) Vapour-phase cracking. Kerosene oil or other oils of similar boiling range are heated to a temperature of 873 K under a pressure of about 3–4 atmospheres. The cracking material is kept in vapour state under these conditions.

Q. 31. Write a note on reforming or aromatization.

Ans. The thermal treatment of gasoline to increase the octane number (a term associated with quality) is called reforming. This process is carried out by heating the gasoline for a short time at temperature of 723–823 K and under a pressure of 10–20 atmospheres in the presence of a suitable catalysts. Catalyst usually used are the oxides of chromium, vanadium and molybdenum supported on alumina.

It involves the conversion of *n*-alkanes to cycloalkanes by cyclisation, cycloalkanes to aromatic hydrocarbons by dehydrogenation or aromatization and straight chain hydrocarbons to branched chain hydrocarbons by isomerisation.

$$CH_3-CH_2-CH_2-CH_2-CH_2-CH_3 \xrightarrow[\substack{723\text{—}823\text{ K}, -H_2\\ 10\text{-}20\text{ Atom}\\ \text{(Cyclization)}}]{Cr_2O_3/V_2O_5/MoO_3} \text{Cyclohexane } (C_6H_{12})$$

Cyclohexane

↓ Dehydrogenation or aromatization

Benzene $+ 3H_2$

$$\underset{n\text{-heptane}}{CH_3CH_2CH_2CH_2CH_2CH_2CH_3} \xrightarrow{\text{Isomerization}} \underset{\text{2-Methylhexane}}{CH_3-CH_2-CH_2-CH_2-\underset{|}{\overset{CH_3}{CH}}-CH_3}$$

$$n\text{-heptane} \xrightarrow[-H_2]{\text{Cyclisation}} \text{Methylcyclohexane} \xrightarrow{\text{Aromatization}} 3H_2 + C_6H_5CH_3$$

This method is not only used for making gasoline with high octane number but is also used to prepare aromatic hydrocarbons such as benzene, toluene, xylenes etc.

Q. 32. What is meant by octane number? (*Madrass, 2003*)

Ans. The gasoline or petrol is mostly used in internal combustion engines in automobiles. The mixture of the fuel vapours and air is compressed and burnt within the cylinder of the engine by spark obtained from spark plug to produce CO_2 and water. Small amounts of C and CO are also formed due to incomplete combustion. As a result of combustion, a large expansion takes place which causes the movement of the piston and consequently of the automobile.

The efficiency of the engine depends upon the compression to which the fuel-air mixture is subjected at the time of combustion; greater the compression, greater the efficiency. But when the compression increases beyond a certain limit, the mixture burns suddenly and very quickly giving a violent jerk against the piston. This is known as **'knocking'**. This lowers the efficiency of the fuel and causes damage to the cylinder and the piston.

It has been found that branched chain alkanes have higher antiknocking property than straight chain alkanes. Two pure hydrocarbons, namely, *n*-heptane and iso-octane are taken as standards. *n*-Heptane, which has anti-knocking property lower than any other hydrocarbons, is arbitrarily given the value of zero. On the other hand, 2, 2, 4-trimethyl pentane or iso-octane, which has anti-knocking property higher than any other hydrocarbon, is arbitrarily given the value of 100.

$$CH_3-CH_2-CH_2-CH_2-CH_2-CH_2-CH_3$$

n-heptane (octane number = zero)

$$CH_3-\underset{\underset{CH_3}{|}}{\overset{\overset{CH_3}{|}}{C}}-CH_2-\underset{\underset{CH_3}{|}}{CH}-CH_3$$

2,2,4-trimethyl pentane or iso-octane
(Octane number = 100)

The octane number may be defined as the percentage of iso-octane by volume in the mixture of n-heptane and iso-octane which has the same anti-knocking property as the fuel under consideration.

For example, a fuel is assigned an octane number 80 if it has the same anti-knocking properties as a mixture of 80 per cent iso-octane and 20 per cent *n*-heptane.

The study of octane numbers of several hydrocarbons has revealed the following facts:

(*i*) Straight chain hydrocarbons have rather low octane number. The octane number decreases with the length of the chain.

(*ii*) Branched chain hydrocarbons have higher octane number than the isomeric straight chain hydrocarbons.

(*iii*) Alkenes, cycloalkanes and aromatic hydrocarbons have higher octane number than alkanes.

The gasoline obtained by the process of cracking has higher octane number than gasoline obtained by direct distillation *due to the fact that the cracked gasoline contains higher percentage of branched chain hydrocarbons, alkenes and aromatic hydrocarbons.*

It has been found that the octane number of a fuel can be increased by the addition of certain compounds called anti-knocking compounds. The most important compound is tetraethyl lead (TEL), $Pb(C_2H_5)_4$. *Such gasoline is called ethyl gasoline or leaded gasoline.*

TEL decomposes to produce ethyl radical which combines with the radicals produced due to irregular combustion. As a result, reaction chains are broken and smooth burning of the fuel takes place. Thus, knocking is prevented.

$$Pb(C_2H_5)_4 \xrightarrow{\Delta} Pb + \underset{\text{Ethyl radical}}{4CH_3CH_2^{\bullet}}$$

Q. 33. What is meant by cetane number? *(Andhra, 2002)*

Ans. The principle of working of diesel engines is different from that of internal combustion engine.

In diesel engines, the mixture of diesel and air is ignited by high temperature produced by the compression inside the engine rather using a spark from spark plug. It has been found that straight chain hydrocarbons are better than branched chain hydrocarbons. Thus, a separate scale is required for grading diesel fuels. The scale used for grading diesel fuels is called *cetane number.*

Cetane or *n*-hexadecane ($C_{16}H_{34}$) ignites rapidly and is assigned a cetane number of 100 while α-methylnaphthalene ignites slowly and is assigned a cetane number of zero. Thus, *cetane number is defined as the percentage of cetane by volume in a mixture of cetane and α-methyl naphthalene which has the same ignition qualities as the fuel under test.* For example, a diesel oil is assigned a cetane number of 80, if it ignites as rapidly as a mixture of 80% cetane and 20% α-methyl naphthalene.

Q. 34. Describe Bergius process for the synthesis of petrol. *(Delhi, 2003)*

Ans. In this method, coal is finely powdered and converted into paste with 50 per cent heavy oil. The paste is heated to 673-773 K in hydrogen under a pressure of 200-250 atmosphere in the presence of a catalyst (usually an organic compound of tin). Under these conditions, hydrogenation of coal takes place and a mixture of higher hydrocarbons (synthetic crude oil) is obtained. This is then put to fractional distillation to get gasoline or petrol as one of the fractions.

The line sketch of the plants is shown in figure 7.10.

The hydrogenation of coal takes place in the strong reacting chamber (converter) which can withstand a pressure of 200-250 atmospheres. It is maintained at a temperature 673-773 K. The resulting crude is put to fractional distillation in fractionating column to get various fractions as shown in the figure. The heavy oil obtained by this method is again used to make paste with coal dust and the process is repeated.

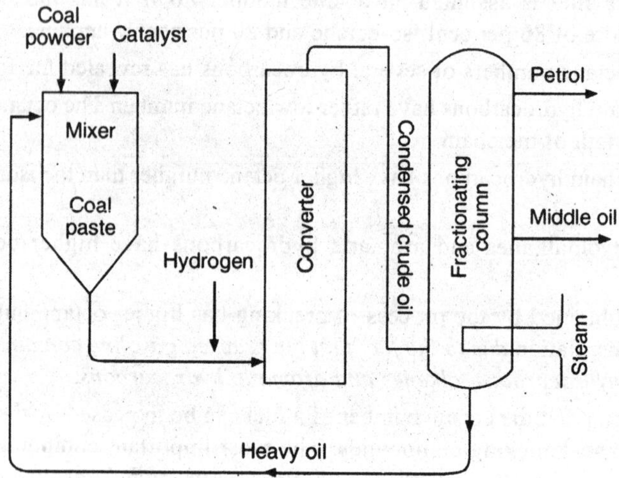

Fig. 7.10. Synthetic petroleum from coal by Bergius process.

Q. 35. Explain the principle of Fischer Tropsch process for synthetic petrol.

Ans. In this method, water gas is produced by passing steam over red hot coke.

$$\underset{\text{Red hot coke}}{C} + \underset{\text{Steam}}{H_2O} \longrightarrow \underbrace{CO + H_2}_{\text{Water gas}}$$

The water gas so produced in mixed with half its volume of hydrogen so that the gaseous mixture consist of carbon monoxide and hydrogen in the ratio of 1 : 2. This mixture is then heated to 473-523 K under a pressure of 1-10 atmosphere in the presence of a catalyst consisting of cobalt oxide, thoria, magnesia and keiselguhr, when petroleum is obtained. This is then put to fractional distillation to get gasoline as one of the fractions. The reaction may be represented by the following equations:

$$nCO + 2nH_2 \longrightarrow C_nH_{2n} + nH_2O$$

$$nCO + (2n + 1)H_2 \longrightarrow C_nH_{2n+2} + nH_2O$$

The octane number of this oil is low but can be raised by cracking.

MISCELLANEOUS QUESTIONS

Q. 36. The halogination of alkanes in the presence of tetraethyl lead proceeds at a lower temperature than when it is carried in its absence. Why ? *(Kerala, 2000)*

Ans. Tetraethyl lead decomposes to produce ethyl radical as under

$$Pb(C_2H_5)_4 \xrightarrow{\Delta} Pb + 4\dot{C}_2H_5$$

As the halogenation of alkanes follows a free radical mechanism the reaction can take place at a lower temperature.

Q. 37. Photoiodination of methane is not feasible. Why ? *(Kumaon, 2000)*

Ans. See Q. 6.

Q. 38. What happens when n-hexane is treated with Cr_2O_3 supported over alumina at 600°C. *(Punjab 2005)*

Ans. Benzene is formed. For reaction see Q.6.

Q. 39. Fluorocarbon (C_5F_{12}) has lower boiling point than pentane (C_5H_{12}) even though it has a far higher molecular weight. Explain. *(Delhi 2006)*

Ans. Molecular weight is not the only criterion that affects the boiling point of a compound. There is a wide gap between the electronegativities of C and F. This results in separation of charges and creation of dipoles on the molecules of C_5F_{12}. This causes repulsions between the molecules of C_5F_{12}.

C_5H_{12} on the other hand does not exhibit this type of repulsive interactions between the molecules because of comparable electronegativities of C and H.

Q. 40. Which would have a higher boiling point and why?

$$CH_3CH_2CH_2CH_2CH_3 \quad \text{or} \quad \begin{array}{c} CH_3 \\ | \\ H_3C-C-CH_3 \\ | \\ CH_3 \end{array}$$

(Delhi 2007)

Ans. Straight chain compounds arrange themselves in zig-zag chains. Different molecules of such compounds overlap on each other due to van der Waals forces as shown below.

one molecule of *n*-pentane

second molecule of *n*-pentane

Due to association between the molecules on account of van der Waals forces, the boiling point rises.

This type of association does not take place in branched chain compounds like neopentane. These compounds consequently boil at lower temperatures.

Thus, *n*-pentane has a higher boiling point than neopentane.

8

ALKENES

Q. 1. What are alkenes? Discuss their nomenclature.

Ans. Alkenes are unsaturated hydrocarbons having the general formula C_nH_{2n}. They contain two hydrogen atoms less than required to form alkanes with the same number of carbon atoms. This is made possible by introducing a double bond between two carbon atoms. This symbolises unsaturation in the compound. Example of alkenes are ethylene and propylene.

Nomenclature

Nomenclature of alkenes is explained in the form of table below:

S.No.	*Structure*	*Common name*	*IUPAC name*
1.	$CH_2 = CH_2$	Ethylene	Ethene
2.	$CH_3CH = CH_2$	Propylene	Propene
3.	$CH_3CH_2CH = CH_2$	β-butylene	Butene-1
4.	$CH_3CH = CH - CH_3$	α-butylene	Butene-2
5.	$CH_3 - \underset{\underset{CH_3}{\mid}}{C} = CH_2$	Isobutylene	2-Methylpropene

Q. 2. Write the IUPAC names of isomers having the molecule formula C_5H_{10}.

Ans. Names and structure of five isomers having the molecules formula are given below.

1. $CH_3 - CH_2 - CH_2 - CH = CH_2$ — 1-Pentene

2. $\begin{matrix} CH_3-CH_2\diagdown & & \diagup CH_3 \\ & C = C & \\ H\diagup & & \diagdown H \end{matrix}$ — *cis*-2-Pentene

3. $\begin{matrix} CH_3-CH_2\diagdown & & \diagup H \\ & C = C & \\ H\diagup & & \diagdown CH_3 \end{matrix}$ — *trans*-2-Pentene

4. $\overset{4}{C}H_3 - \overset{3}{C}H = \underset{\underset{CH_3}{\mid}}{\overset{2}{C}} - \overset{1}{C}H_3$ — 2-Methylbutene-2

Q. 3. What is meant by hindered rotation around carbon-carbon double bond? What is its explanation?

Ans. Hindered rotation. The lack of freedom of rotation of the bonded atoms around a carbon-carbon double bond is known as hindered rotation.

Explanation. A double bond is constituted of a σ bond and a π bond between two carbon atoms.

$$\diagdown\!\!\!\!\diagup C \overset{\pi}{\underset{\sigma}{=\!=}} C \diagup\!\!\!\!\diagdown$$

It is thus associated with greater bond dissociation energy compared to a single bond. Bond length is also shorter compared to a single bond. It is therefore much more difficult to break a double bond.

Rotation round a double bond would involve breaking the double bond which requires a large amount of energy. Hence rotation around the double bonds is difficult.

Q. 4. Discuss the orbital structure of ethylene giving the various bond angles and bond lengths. *(Kumaon, 2000; Devi Ahilya, 2001, Awadh, 2000; Magadh, 2003 ; Sri Venkateswara, 2003 ; Calicut, 2004 ; Patna, 2004)*

Ans. *Orbital structure of ethylene.* In the molecule of ethylene (C_2H_4), the two carbon atoms are linked to each other through a double bond while each carbon atom is separately linked to two hydrogen atoms through single bonds.

Since hybridised orbitals lie in the same plane all the carbon and hydrogen atoms of ethylene also lie in the same plane.

The unhybridised 2*p* orbitals of the two carbon atoms overlap in a plane perpendicular to the plane of carbon and hydrogen atoms, to form carbons-carbon π bonds as shown below.

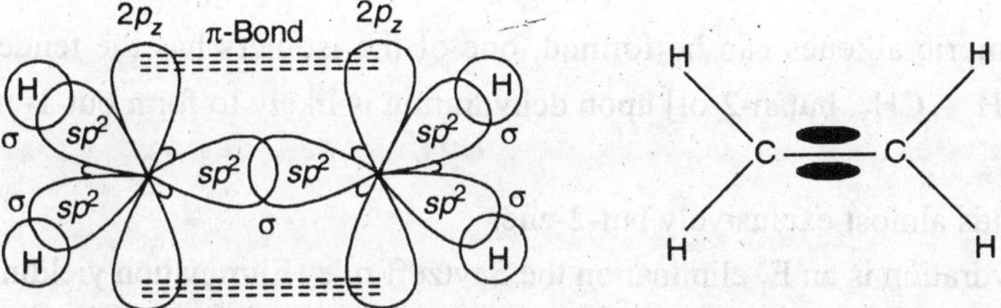

Q. 5. How do you account for the following observation? The carbon-carbon double bond length in ethylene is smaller than the carbon-carbon single bond length in ethane.

Ans. The carbon-carbon double bond length in ethylene is somewhat smaller than the carbon-carbon single bond length in ethane because of the following reasons:

(*i*) In ethylene, the two carbon atoms are sp^2 hybridised while in ethane the carbon atoms are sp^3 hybridised. The size of sp^2 hybrid orbitals is slightly smaller than that of sp^3 orbitals. Therefore, the length of the bond formed by the mutual overlapping of sp^2 orbitals would be a smaller than bond formed by the mutual overlapping of sp^3 obitals.

(*ii*) Carbon-carbon double bond also involves the additional sidewise overlap of unhybridised 2*p*. orbitals. As a result, the carbon atoms are brought still closer and the bond length decreases.

Q. 6. Describe the preparation of alkene by dehydration of alcohol, giving the mechanism of the reaction. *(Kurukshetra, 2001; Garhwal, 2010)*

Ans. An alcohol is converted into an alkene by dehydration. Dehydration is brought about by the use of an acid say conc. H_2SO_4 and application of heat. It involves (1) heating the alcohol with H_2SO_4 or (2) by passing the vapours of alcohol over alumina Al_2O_3 at 623-673 K.

$$\underset{\boxed{\text{H} \quad \text{O}-\text{H}}}{-\overset{|}{\underset{|}{C}}-\overset{|}{\underset{|}{C}}-} \xrightarrow[\Delta]{\text{Acid}} -\overset{|}{C}=\overset{|}{C}- + H_2O$$

1. $CH_3CH_2OH \xrightarrow[433\text{–}443\ K]{H_2SO_4} \underset{\text{Ethylene}}{CH_2 = CH_2} + H_2O$

2. $\underset{\text{Butan-1-ol}}{CH_3 - CH_2 - CH_2 - CH_2OH} \xrightarrow[413\ K]{H_2SO_4} \underset{\text{But-2-ene (major product)}}{CH_3 - CH = CH - CH_3}$

3. $$CH_3-\overset{\overset{1}{CH_3}}{\underset{\underset{3}{CH_3}}{\overset{2}{C}}}-OH \xrightarrow[363\,K]{20\%\ H_2SO_4} \begin{matrix}CH_3\\CH_3\end{matrix}\!>C=CH_2 - H_2O$$

2-Methyl propan-2-ol — 2-Methyl prop-1-ene

4. $$\begin{matrix}CH_3\\CH_3\end{matrix}\!>CHOH \xrightarrow[623\,K]{Al_2O_3} CH_3-CH=CH_3$$

Isopropyl alcohol — Propylene

Reactivity and Orientation

From the above examples it is observed that the ease of dehydration is usually the greatest in 3° alcohols, that is why the 3° alcohols require milder conditions for dehydration while simple 1° alcohols are dehydrated only with difficulty. Therefore, the order of reactivity and the ease of dehydration of alcohols is as follows:

$$3^\circ > 2^\circ > 1^\circ$$

When isomeric alkenes can be formed, one of the isomers has the tendency to dominate [$CH_3-\underset{\underset{OH}{|}}{CH}-CH_2-CH_3$, butan-2-ol] upon dehydration is likely to form but-1-ene and but-2-ene, but actually yields almost exclusively but-2-ene.

Since dehydration is an E_1 elimination the Saytzeff rule [Elimination yielding the most highly alkylated olefins] applies to those cases where more than one olefins can be formed.

$$CH_3-\overset{\overset{O-H}{|}}{\underset{\underset{CH_3}{|}}{C}}-CH_2-CH_3 \xrightarrow{\text{Conc. } H_2SO_4} CH_3-\underset{\underset{CH_3}{|}}{C}=CH-CH_3 + H_2C=\underset{\underset{CH_3}{|}}{C}-CH_2-CH_3$$

2-Methylbutan-2-ol — 2-Methylbut-2-ene (Major product) — 2-Methylbut-1-ene (Minor)

Mechanism of dehydration of alcohols. The mechanism of dehydration of alcohols is given schematically below:

1. Attachment of the proton to alcoholic oxygen

$$-\overset{|}{\underset{\underset{H}{|}}{C}}-\overset{|}{\underset{|}{C}}-\ddot{O}-H+\overset{+}{H} \rightleftharpoons -\overset{|}{\underset{\underset{H}{|}}{C}}-\overset{|}{\underset{|}{C}}-\overset{+}{\underset{\underset{H}{|}}{O}}-H$$

Alcohol — Protonated alcohol

2. Remove of water molecule

$$-\overset{|}{\underset{\underset{H}{|}}{C}}-\overset{|}{\underset{|}{C}}-\overset{+}{\underset{\underset{H}{|}}{O}}-H \rightleftharpoons -\overset{|}{\underset{\underset{H}{|}}{C}}-\overset{|+}{\underset{|}{C}}+H_2O$$

Carbonium ion

3. Removal of proton

$$-\overset{|}{\underset{\underset{H}{|}}{C}}-\overset{|+}{\underset{|}{C}} \rightleftharpoons \;>C=C<\; + H^+H$$

Alkene

*For those alcohols that yield volatile olifins, the reaction can always be driven to completion by distilling off the olefin as it is formed.

Q. 7. With suitable example, explain alkyl shift.

or

Explain the rules that govern the orientation of double bonds during the elimination reaction of alkyl halides leading to isomeric olefins. (*Kerala, 2000*)

Ans. Sometimes it is found that the alkene obtained by dehydration of alcohols does not fit in the mechanism.

For example,

$$\underset{\text{3,3-Dimethyl butan-2-ol}}{CH_3-\overset{\overset{CH_3}{|}}{\underset{\underset{CH_3}{|}}{C}}-\underset{\underset{OH}{|}}{CH}-CH_3} \xrightarrow{\text{Conc. } H_2SO_4} \underset{\substack{\text{2,3-Dimethyl but-2-ene} \\ \text{(Major product)}}}{CH_3-\underset{\underset{CH_3}{|}}{C}=\underset{\underset{CH_3}{|}}{C}-CH_3}$$

This can be explained by considering a rearrangement of carbonium ions. A carbonium ion can arrange to form a more stable carbonium ion. The mechanism of rearrangement is given below:

$$CH_3-\overset{\overset{CH_3}{|}}{\underset{\underset{CH_3}{|}}{C}}-\underset{\underset{\ddot{O}-H}{|}}{CH}-CH_3+\overset{+}{H} \rightleftharpoons CH_3-\overset{\overset{CH_3}{|}}{\underset{\underset{CH_3}{|}}{C}}-\underset{\underset{\underset{\underset{H}{|}}{\overset{+}{O}-H}}{|}}{CH}-CH_3$$

$$\Downarrow\!\Uparrow$$

$$\underset{\substack{3^\circ \text{ Carbonium ion} \\ \text{(more stable)}}}{CH_3-\overset{+}{\underset{\underset{CH_3}{|}}{C}}-\overset{\overset{CH_3}{|}}{\underset{\underset{H}{|}}{CH}}-CH_3} \underset{\text{Rearrangement}}{\overset{\text{1,2-Alkyl shift}}{\rightleftharpoons}} \underset{\substack{2^\circ \text{ Carbonium ion} \\ \text{(less stable)}}}{CH_3-\overset{\overset{CH_3}{|}}{\underset{\underset{CH_3}{|}}{C}}-\overset{+}{CH}-CH_3} + H_2O$$

$$\Downarrow\!\Uparrow \; -H^+$$

$$\underset{\substack{\text{2,3-Dimethyl but-2-ene} \\ \text{(major product)}}}{CH_3-\underset{\underset{CH_3}{|}}{C}=\overset{\overset{CH_3}{|}}{C}-CH_3}$$

Thus, we see that the carbonium ion which is the intermediate product in the dehydration of alcohol changes to a more stable carbonium ion by the shifting of alkyl group from one position to another. If the alkyl group shifts to the neighbouring carbon atom, it is called *1, 2 alkyl shift*.

Somethings, a hydride ion shifts from the neighbouring carbon to the carbonium ion, thereby producing a more stable carbonium ion. This is called *1, 2 hydride shift*.

Q. 8. Predict the major product obtained on dehydration of the following alcohols:

(i) $(CH_3)_2C(OH)CH_2-CH_3$

(ii) $CH_3-\underset{CH_3}{\overset{OH}{\underset{|}{\overset{|}{C}}}}-\underset{CH_3}{\underset{|}{CH}}-CH_3$

(iii) $CH_3-\underset{CH_3}{\underset{|}{CH}}-CH_2-CH_2OH$

Ans. (i)

$$(CH_3)_2\underset{OH}{\underset{|}{C}}-CH_2-CH_3 \xrightarrow{H^+} (CH_3)_2\underset{H^+\,OH}{\underset{|}{C}}-CH_2-CH_3$$

$$\downarrow -H_2O$$

$$(CH_3)_2C=CH-CH_3 \xleftarrow{-H^+} (CH_3)_2C^{\oplus}-\underset{H}{\underset{|}{CH}}-CH_3$$

2-Methyl butene-2 — Tert. carbonium ion

(ii)

$$CH_3-\underset{CH_3}{\overset{OH}{\underset{|}{\overset{|}{C}}}}-\underset{CH_3}{\underset{|}{CH}}-CH_3 \xrightarrow{H^+} CH_3-\underset{CH_3}{\overset{H\overset{\oplus}{O}H}{\underset{|}{\overset{|}{C}}}}-\underset{CH_3}{\underset{|}{CH}}-CH_3$$

$$\downarrow -H_2O$$

$$CH_3-\underset{CH_3}{\underset{|}{C}}=\underset{CH_3}{\underset{|}{C}}-CH_3 \xleftarrow{-H^+} CH_3-\underset{CH_3}{\underset{|}{\overset{+}{C}}}-\underset{CH_3}{\overset{H}{\underset{|}{\overset{|}{C}}}}-CH_3$$

2,3-Dimethylbutene-2

(iii)

$$CH_3-\underset{CH_3}{\underset{|}{CH}}-CH_2-CH_2-OH \xrightarrow{H^+} CH_3-\underset{CH_3}{\underset{|}{CH}}-CH_2-CH_2-\overset{\oplus}{O}H_2$$

$$\downarrow -H_2O$$

$$CH_3-\overset{H}{\underset{CH_3}{\underset{|}{\overset{|}{C}}}}-\overset{\oplus}{C}H-CH_3 \xleftarrow[\text{shift}]{\text{1,2-Hydride}} CH_3-\underset{CH_3}{\underset{|}{CH}}-\underset{H}{\underset{|}{CH}}-\overset{\oplus}{C}H_2$$

Sec. carbonium ion — Primary carbonium ion

1,2-Hydride shift ↓

$$CH_3-\underset{CH_3}{\underset{|}{\overset{\oplus}{C}}}-\underset{H}{\underset{|}{CH}}-CH_3 \longrightarrow CH_3-\underset{CH_3}{\underset{|}{C}}=CH-CH_3$$

Tert. carbonium ion — 2-Methylbutene-2

Q. 9. What is dehydrohalogenation of alkyl halides? How can we obtain alkenes by this method? Explain the mechanism of dehydrohalogenation. (*Kumaon, 2000; Kerala, 2000*)

Ans. Alkyl halides can form alkenes by the loss of molecule of hydrogen halide under the influence of a base catalyst (alcoholic potash). The reaction is known as *dehydrohalogenation of alkyl halides*. It involves the removal of the halogen atom together with the hydrogen atom from the adjacent carbon atom.

$$CH_3 - CH_2X \xrightarrow{OH^-} CH_2 = CH_2 + H_2O + X^-$$

Mechanism of reaction (E_2 mechanism)

$$\underset{\text{Alkyl halide}}{-\underset{H}{\overset{|}{C}}-\underset{X}{\overset{|}{C}}-} \xrightarrow{OH^-} \underset{\text{Alkene}}{-\underset{|}{\overset{|}{C}}=\underset{|}{\overset{|}{C}}-}$$

Orientation in dehydrohalogenation. In some cases, this reaction yields a single product (alkene) and in other cases yields a mixture of alkenes. For example, out of 1-chlorobutane and 2-chlorobutane the former yields only but-1-ene while the latter gives but-1-ene and but-2-ene. Thus, the preferred product is the alkene that has the greater number of alkyl groups attached to the doubly bonded carbon atoms. In other words, the ease of formation of alkene has the following order:

$$R_2C = CR_2 > R_2C = CHR > R_2C = CH_2 > RCH = CHR > RCH = CH_2 > CH_2 = CH_2$$

This order is based on heat of hydrogenation values.

Hence in dehydrohalogenation, the more stable the alkene more easily it is formed.

This is also clear from the transition state formed in the dehydronaıogenation of alkyl halides as discussed below. The double bond is partly formed and the transition state acquires alkene character. Factors that stabilise an alkene also stabilise an *incipient* alkene in the transition state.

$$-\underset{|}{\overset{\overset{H}{|}}{C}}-\underset{X}{\overset{|}{\underset{|}{C}}}- \xrightarrow{OH^-} \left[\begin{array}{c} \overset{\delta-}{HO}\cdots H \\ \vdots \\ -\underset{|}{\overset{|}{C}}-\underset{\vdots}{\overset{|}{C}}- \\ X^{\delta-} \end{array} \right] \rightarrow \rangle C = C \langle \; + X^- + H_2O$$

Reactivity of alkyl halides in dehydrohalogenation. The decreasing order of reactivity of alkyl halides in this reaction is :

Tert-Alkyl halides	>	Sec-Alkyl halides	>	Primary Alkyl halides
(3°)		(2°)		(1°)

As we move from primary to secondary and from secondary to tertiary halides, the structure becomes more branched at carbon atom bearing the halogen. This increased branching has two results (*i*) the number of hydrogen atoms available for attack by a base is more and thus there is greater probability towards elimination, (*ii*) it leads to the formation of more highly branched (hence more stable) alkene hence more stable transition state and low E_{act}.

As a result of these two factors, the decreasing order of reactivity of alkyl halides for dehydrohalogenation is:

$$3° > 2° > 1°$$

The mechanism discussed above is called E_2 type of mechanism of dehydrohalogenation.

Mechanism of Dehydrohalogenation (E_1 Mechanism)

E_1 mechanism. Some secondary and tertiary alkyl halides undergo dehydrogenation in a solution of low base concentration by a different mechanism known as E_1 mechanism (E for elimination; 1 for unimolecular). *This mechanism operates through a two-step process in which the rate determining step involves only one molecule.* It is believed that in the first step, which is a slow rate-determining step, the alkyl halide dissociates into halide ion and carbocation. In the second step which is a fast one, the carbocation loses a proton to the OH^- ion to from the alkene.

The complete mechanism is as shown below:

$$-\overset{H}{\underset{|}{\overset{|}{C}}}-\underset{X}{\overset{|}{C}}- \xrightleftharpoons{\text{Slow}} -\overset{H}{\underset{|}{\overset{|}{C}}}-\overset{|}{\underset{+}{C}}- + X^-$$

Carbocation

$$HO^- + -\overset{H}{\underset{|}{\overset{|}{C}}}-\underset{|}{\overset{|}{C}}- \xrightarrow{\text{Fast}} >C=C< + H_2O$$

In certain alkyl halides (say 3°), there can be slight variation in the mechanistic approach as the carbon cation initially formed undergoes rearrangement to form more stable carbocation and thus yields highly branched alkene as the chief product of the reaction.

For example, action of alcoholic potash on 2-chloro-3, 3-dimethyl butane yields chiefly 2, 3-dimethyl but-2-ene as the carbocation initially formed undergoes rearrangement as shown below:

$$CH_3-\underset{CH_3}{\overset{CH_2Cl}{C}}-CH-CH_3 \xrightarrow{-Cl^-} CH_3-\underset{CH_3}{\overset{CH_3}{C}}-\overset{\oplus}{C}H-CH_3$$

2° Carbocation

$$\Downarrow$$

$$\begin{matrix}CH_3\\CH_3\end{matrix}>C=C<\begin{matrix}CH_3\\CH_3\end{matrix} \xleftarrow{-H^+} CH_3-\overset{\oplus}{\underset{CH_2H}{C}}-\overset{CH_3}{C}-CH_3$$

2,3-Dimethylbut-2-ene (Chief product) | 3° Carbocation (More stable)

This is an example of type of mechanism of dehydrohalogenation.

Q. 10. Discuss E_1 mechanism of dehydrohalogenation of alkyl halides.

(*Kerala, 2000; Guwahati 2006*)

Ans. See Q. 9.

Q. 11. For a given halogen, what is the order of reactivity of various alkyl halides?

Ans. For a given halogen, the order of reactivity of alkyl halides is:

tertiary > secondary > primary

This is because in a tertiary alkyl halide, there is maximum number of hydrogens on the neighbouring carbon atom. So the chances of attack by hydroxyl ion (E_2 mechanism) are maximum leading to maximum reactivity. This is followed by secondary and primary alkyl halides.

Even if the reaction takes place by E_1 mechanism, tertiary alkyl halide has greatest reactivity because the carbonium ion formed in the first step has the maximum stability leading to completion of reaction. A tertiary carbonium ion (carbocation) has maximum stability followed by sec. and followed by primary carbonium ion.

Q. 12. For a given alkyl group, what is the order of reactivity with different halogens?

Ans. Order of reactivity with different halogens, with a given alkyl group is

$$I > Br > Cl > F$$

Alkyl iodide has the maximum reactivity because C – I bond can be broken most easily leading to the formation of products, C – I bond has the minimum bond dissociation energy. Reactivity of alkyl halide decreases as we go to bromide, chloride and fluoride.

Q. 13. What is Saytzeff's Rule? *(Kerala, 2001; Garhwal, 2010; Patna, 2010 Pune 2010*

or

In the dehydrohalogenation of alkyl halides with alk. KOH more stable alkene is obtained. It is according to which rule?

Ans. When an alkyl halide is subjected to elimination reaction to form alkene, the preferred product is the alkene which is more highly alkylated at the doubly bound carbon atoms.

This is the statement of Saytzeff's Rule.

For example of Saytzeff's Rule, see Q.9.

Q. 14. How can you obtain alkenes by the dehalogenation of dihalides? Explain with mechanism.

Ans. Dehalogenation of 1, 2 dihaloalkanes (vic-dihalides) involves the treatment to the vic-dihalide with reactive metals like Zn in acetic acid and it yields the corresponding alkenes.

$$\underset{\text{1,2-dihaloalkane}}{-\underset{X}{\overset{|}{\underset{|}{C}}}-\underset{X}{\overset{|}{\underset{|}{C}}}} + Zn \longrightarrow ZnX_2 + \underset{\text{Alkene}}{>C=C<}$$

For example:

$$\underset{\text{1,2-Dibromoethane}}{\begin{matrix} CH_2-Br \\ | \\ CH_2-Br \end{matrix}} + Zn \longrightarrow ZnBr_2 + \underset{\text{Ethylene}}{\begin{matrix} CH_2 \\ \| \\ CH_2 \end{matrix}}$$

$$\underset{\text{1,2-Dibromo propane}}{\begin{matrix} CH_3 \\ | \\ CH-Br \\ | \\ CH_2-Br \end{matrix}} + Zn \longrightarrow ZnBr_2 + \underset{\text{Propylene}}{\begin{matrix} CH_3 \\ | \\ CH \\ \| \\ CH_2 \end{matrix}}$$

Mechanism. The divalent Zn metal possesses a pair of electrons in its outermost shell and therefore it acts as a nucleophile, Br^- are formed in the reaction solution and the reaction can be represented by a concerted process.

$$\begin{matrix} Zn : Br \\ | \\ -C - C \\ \quad | \\ \quad Br \end{matrix} \longrightarrow >C=C< + :Br^- + [ZnBr]^+$$

(Trans elimination process)

$$[Zn\,Br]^+ \longrightarrow Zn^{2+} + Br^-$$

Q. 15. How can you obtain alkenes by

(*i*) Reduction of alkynes (*G. Nanak Dev, 2000*)

(*ii*) Dehydrogenation of alkanes. Explain.

Ans. (*i*) Alkenes can be prepared by the partial reduction of alkynes using metallic *sodium* or *lithium* in liquid ammonia, or by using a calculated quantity of hydrogen over palladium catalyst.

Reduction with sodium in liquid ammonia yields trans alkene, whereas cis alkene results from H_2 in the presence of Pd.

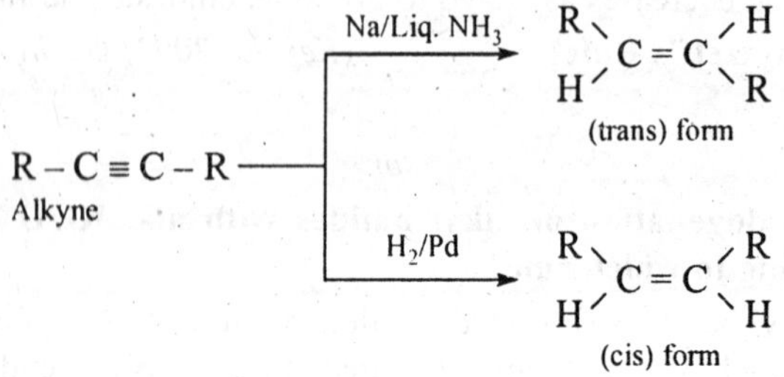

$$\underset{\text{Alkyne}}{R-C\equiv C-R} \xrightarrow{\text{Na/Liq. } NH_3} \underset{\text{(trans) form}}{\begin{matrix} R & & H \\ & C=C & \\ H & & R \end{matrix}}$$

$$\underset{\text{Alkyne}}{R-C\equiv C-R} \xrightarrow{H_2/Pd} \underset{\text{(cis) form}}{\begin{matrix} R & & R \\ & C=C & \\ H & & H \end{matrix}}$$

(*ii*) **Dehydrogenation of alkanes.** It involves the passing of the alkane in the vapour phase over a bed of catalysts (oxides of chromium and aluminium).

$$\underset{\text{Iso butane}}{CH_3-\underset{\substack{|\\CH_3}}{CH}-CH_3} \xrightarrow[773\text{ K}]{Cr_2O_3,\,Al_2O_3} \underset{\text{Iso butylene}}{\begin{matrix} CH_3 & \\ & C=CH_2 \\ CH_3 & \end{matrix}} + H_2$$

Q. 16. Which alcohol of each pair would you expect to be more easily dehydrated? Also name the main products of dehydration.

(1) $CH_3-CH_2-CH_2-CH_2OH$, $CH_3CH_2\underset{\substack{|\\OH}}{CH}-CH_3$

(2) $(CH_3)_2CH-\overset{\substack{CH_3\\|}}{\underset{\substack{|\\OH}}{C}}-CH_3$, $(CH_3)_2CH-CH(CH_3)CH_2OH$

Ans.

	Alcohol	*Main product of dehydration*
(1)	Butan-2-ol	$CH_3-CH=CH-CH_3$ But-2-ene
(2)	$CH_3-\underset{\substack{|\\CH_3}}{CH}-\overset{\substack{CH_3\\|}}{\underset{\substack{|\\O-H}}{C}}-CH_3$ 2,3-Dimethylbutan-2-ol	$CH_3-\underset{\substack{|\\CH_3}}{CH}=\overset{\substack{CH_3\\|}}{C}-CH_3$ 2,3-Dimethylbut-2-ene

Q. 17. When neopentyl alcohol $(CH_3)_3\cdot C\cdot CH_2OH$ is heated with acid, it is slowly converted into a mixture of two isomeric alkenes of the formula C_5H_{10}. Name these alkenes according to IUPAC system of nomenclature and also show how they are formed. Which one of them is the major product and why?

Ans.

$$\underset{\text{Neo-pentyl alcohol}}{CH_3-\overset{\overset{CH_3}{|}}{\underset{\underset{CH_3}{|}}{C}}-CH_2OH} \xrightleftharpoons{+H^+} CH_3-\overset{\overset{CH_3}{|}}{\underset{\underset{CH_3}{|}}{C}}-CH_2-\overset{+}{\underset{\underset{H}{|}}{O}}-H$$

$$\Big\Updownarrow -H_2O$$

$$\underset{\substack{\text{3° Carbonium ion}\\ \text{(More stable)}}}{CH_3-\overset{\overset{CH_3}{|}}{\underset{\oplus}{C}}-CH_2-CH_3} \rightleftharpoons \underset{\text{1° Carbonium ion}}{CH_3-\overset{\overset{CH_3}{|}}{\underset{\underset{CH_3}{|}}{C}}-\overset{+}{C}H_2}$$

$$\downarrow -H^+$$

$$\underset{\substack{\text{2-Methylbut-2-ene}\\ \text{(Major product)}}}{CH_3-\overset{\overset{CH_3}{|}}{C}=CH-CH_3} \qquad \underset{\substack{\text{2-Methylbut-1-ene}\\ \text{(Minor)}}}{CH_2=\overset{\overset{CH_3}{|}}{C}-CH_2-CH_3}$$

Q. 18. State and explain briefly the main distinguishing properties of geometrical isomers.

Ans. Main distinguishing properties of geometrical isomers. The geometrical isomers generally differ from each other in the following poperties:

(*i*) **Dipole moment.** In case of compounds having polar bonds in their molecules, the trans isomer has generally smaller dipole moment than cis isomer. For example:

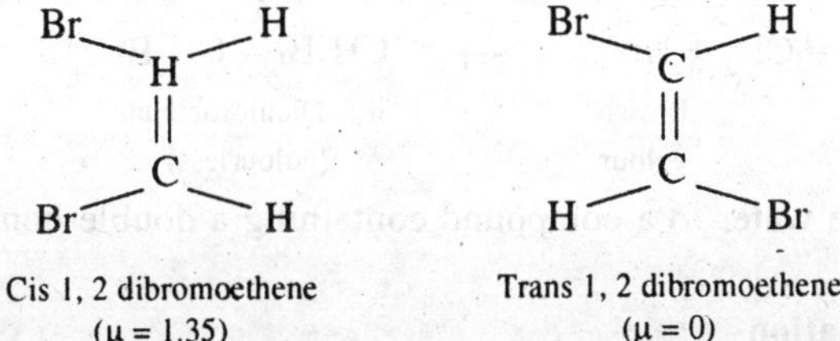

Cis 1, 2 dibromoethene (μ = 1.35) Trans 1, 2 dibromoethene (μ = 0)

This is due to the fact that in trans isomer, similar groups are on opposite sides of the molecule and their bond polarities tend to cancel each other. But in cis isomers the bond polarities are not cancelled out.

(*ii*) **Boiling points.** Generally speaking cis isomers have higher boiling points than trans isomers. This is due to the fact that cis isomers have greater intermolecular forces of attraction on account of higher dipole moments of their molecules. For example, boiling point of cis-2-butene is 277 K while that of trans-2-butene is 274 K.

(*iii*) **Melting points.** Out of cis and trans isomers, the trans compound has higher melting point. This is because trans compounds are more symmetrical than their cis isomers and are, therefore, more closely packed in the crystal lattice. For example, melting point of maleic acid (cis isomer) is 403 K while fumaric acid (trans isomer) is 503 K.

$$\begin{array}{c} H \qquad COOH \\ \diagdown \quad \diagup \\ C \\ \| \\ C \\ \diagup \quad \diagdown \\ H \qquad COOH \end{array} \qquad\qquad \begin{array}{c} H \qquad COOH \\ \diagdown \quad \diagup \\ C \\ \| \\ C \\ \diagup \quad \diagdown \\ HOOC \qquad H \end{array}$$

Maleic acid (cis isomer, less symmetrical) — Fumaric acid (trans isomer, more symmetrical)

(*iv*) **Formation of ring compounds.** Whenever ring formation is possible, cis isomer changes into ring compound much more easily than the trans isomer. For example, maleic acid (*cis* isomer) forms maleic anhydride (ring compound) on heating to 423 K while fumaric acid (*trans* isomer) does not form ring compound at this temperature.

$$\begin{array}{c} H \qquad COOH \\ \diagdown \quad \diagup \\ C \\ \| \\ C \\ \diagup \quad \diagdown \\ H \qquad COOH \end{array} \xrightarrow[-H_2O]{423\ K} \begin{array}{l} CHCO \diagdown \\ \| \qquad\quad O \\ CHCO \diagup \end{array}$$

Maleic acid — Maleic anhydride

Q. 19. Describe the mechanism of halogenation of alkenes.

or

Explain that addition of bromine to ethene is a two-step reaction.

(*Punjab, 2004; Nagpur, 2008*)

Ans. Alkenes add a molecule of halogen to form alkyl halides.

$$\underset{}{CH_2 = CH_2} + \underset{\text{Brown colour}}{Br_2} \longrightarrow \underset{\text{1,2-Dichloroethane (colourless)}}{CH_2Br - CH_2Br}$$

On adding bromine water to a compound containing a double bond, brown colour due to bromine disappears.

Mechanism of Halogenation

Halogenation of an alkene takes place by electrophilic addition mechanism. A halogen molecule (Cl_2, Br_2 or I_2) is non-polar in nature. But when this molecule approaches the alkene molecule, the bonds have the effect of polarising the halogen molecule. The positive end of the polarised halogen molecule is attached to one ethylenic carbon atom with the simultaneous attachment of the negative end to the second ethylenic carbon atom forming a halonium ion, with the relase of halide ion. The halide ion then attacks the halonium ion from the back side to avoid steric hindrance, thus giving rise to a dihalide. The steps of the mechanism are shown as under, taking the example of bromination.

$$Br—Br \xrightarrow[\text{cloud}]{\pi\ \text{Electron}} \overset{\delta+}{Br}—\overset{\delta-}{Br}$$

Polarised bromine molecule

Step 1. $C{=}C + \overset{\delta+}{Br}—\overset{\delta-}{Br} \rightleftharpoons$ π-Complex $\longrightarrow$ Bromonium ion $Br^+ + Br^-$

Alkene π-Complex Bromonium ion

Step 2. $Br^- +$ Bromonium ion $\longrightarrow$ Dibromide

Backside attack Dibromide

Q. 20. Give evidence to prove that addition of halogen to an alkene proceeds through a halonium ion and not through carbonium ion (carbocation). (*Kurukshetra, 2001*)

Ans. Addition of halogen to an alkene proceeds through the formation of a halonium ion and not through a carbonium ion. This can be proved like this:

It is observed that halogenation of alkenes gives rise to products that are optically active. This phenomenon can be explained by halonium ion mechanism and not by carbonium ion mechanism.

In halonium ion mechanism, the halide (in the second step) attacks from the back of halonium ion, giving rise to trans dihalide product.

In carbonium ion mechanism, the halide can attack from both sides of the carbonium ion (as it is flat in shape), giving rise to both cis and trans isomers.

Trans product obtained from halonium ion mechanism explains optical activity of the actual product obtained.

If we assume the carbonium ion mechanism, the mixture of *cis* and *trans* products will give us a racemic modification which will have no net optical activity. This is against the observations. Practically, we get an optically active compound.

This is schematically shown as under taking the example of bromination of cyclohexene.

Carbonium Ion-mechanism

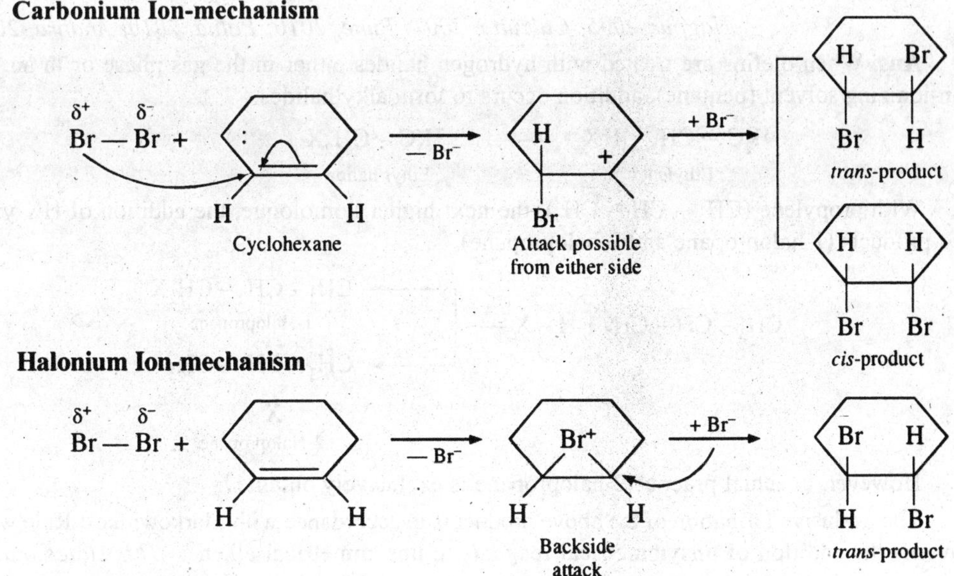

Hence halonium ion mechanism is confirmed.

Q. 21. How do alkenes react with halogens in the presence of water? Give the mechanism of the reaction.

Ans. Reaction of alkenes with halogens in the presence of water. Alkenes react with halogens in the presence of water to form halo-hydrins.

For example:

$$>C=C< + Cl_2 + H_2O \longrightarrow -\underset{\underset{Cl}{|}}{\overset{|}{C}}-\underset{\underset{OH}{|}}{\overset{|}{C}}- + HCl$$

Chlorohydrins

Mechanism of the Reactions

The reaction takes place through the intermediate formation of a halonium ion as shown below :

$$>C=C< + \overset{\delta+}{Cl}-\overset{\delta-}{Cl} \rightleftharpoons \text{[C=C}\cdots\overset{\delta+}{Cl}-\overset{\delta-}{Cl}] \longrightarrow \text{(cyclic)} Cl^+ + Cl^-$$

$$H_2\ddot{O}: + \text{(cyclic chloronium ion)} Cl^+ \longrightarrow H-\overset{+}{\underset{\underset{H}{|}}{O}}-\overset{|}{C}-\,,\ -\overset{|}{C}-Cl \longrightarrow HO-\overset{|}{C}-\,,\ -\overset{|}{C}-Cl + H^+$$

Q. 22. Name and explain the mechanism of addition of an unsymmetrical molecule over an unsaturated double bond. *(Kerala, 2001; Panjab 2003, Mysore 2003)*

or

Describe the mechanism of Markownikoff rule. *(Kalyani, 2004; Guwahati, 2005; Nagpur, 2005; Calcutta, 2007; Pune, 2010; Patna, 2010; Andhra, 2010)*

Ans. When olefins are treated with hydrogen halides either in the gas phase or in an inert non-ionizing solvent (pentane), addition occurs to form alkylhalides.

$$\underset{\text{Ethylene}}{H_2C=CH_2} + HX \longrightarrow \underset{\text{Ethyl halide}}{H_3C-CH_2X}$$

With propylene ($CH_3-CH=CH_2$), the next higher homologue, the addition of HX yields two products (1-halopropane and 2-halopropane).

$$CH_3-CH=CH_2 + H-X \longrightarrow \underset{\text{1-Halopropane}}{CH_3-CH_2-CH_2X} \quad \underset{\text{2-Halopropane}}{CH_3-\underset{\underset{X}{|}}{CH}-CH_3}$$

However, in actual practice 2-halopropane is exclusively obtained.

The exclusive formation of the above product is in accordance with Markownikoff Rule which governs the addition of unsymmetrical reagents to unsymmetrical alkenes. *This states that the ionic addition of unsymmetrical reagents to unsymmetrical olefins proceeds in such a way that the more positive part of the reagent, becomes attached to the carbon atom with larger number of hydrogen atoms.*

When the hydrogen halides or water is added, hydrogen atom constitutes the more positive part or in other words, the hydrogen atom becomes attached to the olefinic carbon atom which carries the larger number of hydrogen atoms, for example:

$$(CH_3)_2C=CH-CH_3 + HBr \longrightarrow CH_3-\underset{Br}{\overset{CH_3}{C}}-CH_2-CH_3$$

2-Methyl-1-but-1-ene 2-Bromo-2-methylbutane

Theoretical explanation of the orientation of addition to olefins is related to the relative stability of carbonium ions. We know that relative stability of carbonium ions is: tertiary > secondary > primary > CH_3^+. If there is the probability of more then one carbonium ions being formed, the addition of electrophile yields the more stable one.

That is why 2 bromopropane is exclusively formed by the addition of HBr to propylene (an unsymmetrical alkene) and not *1-bromo-propane* because secondary isopropyl carbonium ion is more stable than primary *n*-propyl carbonium ion.

$$CH_3-CH=CH_2+\overset{+}{H} \longrightarrow \left(CH_3-\overset{+}{C}H-CH_3\right)$$

Isopropyl Carbonium ion (27°) (More stable)

$$\xrightarrow{Br^-} CH_3-\underset{Br}{CH}-CH_3$$

2-Bromopropane

$$\left(CH_3-CH_2-\overset{+}{C}H_2\right)$$

n-propyl Carbonium ion (1°) (Less stable)

$$\xrightarrow{Br^-} CH_3-CH_2-CH_2Br$$

1-Bromopropane

Considering the electronic effect of methyl group in propylene the polarization of the double bond due to the (+I) electron-repelling inductive effect of the methyl group, is depicted below.

$$CH_3 \rightarrow \overset{\delta^+}{C}H=\overset{\delta^-}{C}H_2$$

Thus, the +ve and –ve ends of the dipole (H – X) will add to the –ve and +ve ends of the double bonds, respectively yielding 2-halopropane.

Mechanism

$$CH_3-\overset{\delta^+}{C}H=\overset{\delta^-}{C}H_2 \quad \overset{\delta^+}{H}-\overset{\delta^-}{H}$$

Electrophilic attack

$$CH_3-\overset{+}{C}H-CH_3$$

Intermediate 2° carbonium ion

(Nucleophilic attack by $:X^-$)

$$CH_3-\underset{}{\overset{X}{CH}}-CH_3$$

2-Halopropane

Thus the modern statement of Markownikoff's rule is:

In the ionic addition of an unsymmetrical reagent to a double bond, the positive portion of the adding reagent becomes attached to the carbon atom of the double bond so as to yield the more stable carbonium ion.

Q. 23. What is meant by anti-Markownikoff's addition? Give its mechanism.

(*Kerala, 2001; Garhwal, 2000; A.N. Bahuguna, 2000; Kanpur, 2001; G. Nanakdev, 2000*)

or

What is peroxide effect? Explain giving mechanism.

(*Punjab, 2003; Guwahati, 2007; Nagpur, 2008*)

Ans. If has been observed that the addition of H – Br to propylene in the presence of peroxides, yields predominantly 1-bromo-propane, (that is the reagent adds onto the olefins under these conditions in a manner contrary to Markownikoff's rule thus suggesting a change in the mechanism.

This change in the orientation of addition due to the presence of peroxides is known as the *peroxide effect.*

Here the change in the mode of addition of the reagent is due to a change from an ionic *mechanism to a free-radical mechanism.* Markownikoff's addition requires the initiation by H^+. Anti-Markownikoff's addition requires initiation by $\dot{B}r$. Each species attacks the olefin molecule at the centre of highest electron density to yield the most stable intermediate carbonium ion or free radical.

Mechanism of addition of HBr to propylene in the presence of peroxides.

It involves the following steps:

(*i*) **Chain initiation step.** The reaction is initiated by the alkoxy radical produced by the homolytic fission of peroxides, which abstracts an atom of hydrogen from HBr generating bromine free radicals ($\dot{B}r$).

$$\underset{\text{Peroxides}}{R-O-O-R} \longrightarrow \underset{\text{Alkoxy radical}}{2R\dot{O}}$$

$$R-\dot{\underset{\cdot}{O}} + H-\dot{B}r \longrightarrow \underset{\text{(Chain initiation)}}{R-OH+\dot{B}r}$$

(*ii*) **Chain propagation step.** The then attacks the propylene molecule to give a more stable secondary radical addition species:

$$CH_3-CH=CH_2+\dot{B}r \longrightarrow \underset{\substack{\text{2° free radical}\\\text{(More stable)}}}{CH_3-CH=\dot{C}H_2Br} + \underset{\substack{\text{2° free radical}\\\text{(Less stable)}}}{CH_3-\underset{\substack{|\\Br}}{CH}-CH_2^{\cdot}}$$

The secondary radical then reacts with another H – Br molecule to yield the product, and another $\dot{B}r$ which can further propagate the reaction:

$$\underset{\text{(Chain propagation)}}{CH_3-\dot{C}H-CH_2Br+H-Br} \longrightarrow \underset{\text{1-Bromopropane}}{CH_3-CH_2-CH_2Br} + \dot{B}r$$

This mechanism is supported by the fact that small amount of peroxide can influence addition to a large number of molecules of an alkene and a small amount of an inhibitor such as hydroquinone or diphenyl amine can prevent this change.

Q. 24. Explain why peroxide oxide effect or anti-Markownikoff's rule (or Kharasch effect) is observed in the addition of H – Br and not H – Cl or H – I?

(*Kurukshetra 2001; Himachal, 2000; Panjab, 2003*)

Ans. In case of HCl, it is probably due to the fact that H – Cl bond (430 kJ/mole) is stronger than H – Br bond (368 kJ/mole) and is not broken homolytically by the free radicals generated by peroxides. As such free radical addition of HCl to alkenes is not possible. In case of HI, H – I bond (297 kJ/mole) is no doubt weaker than H – Br bond and can be broken easily. But the iodine atoms (I ·) thus formed readily combine amongst themselves to form iodine molecules rather than add to the olefins.

Q. 25. Give the machanism of addition of

(*i*) sulphuric acid (*ii*) water to an alkene

Ans. (*i*) **Mechanism of addition of sulphuric acid to alkene.** It is a two step electrophilic addition reaction which takes place as follows:

$$\underset{\text{Alkene}}{>C=C<} + \overset{\delta^+}{H}-\overset{\delta^-}{OSO_3H} \longrightarrow \underset{\text{Carbocation}}{-\overset{|}{C}H-\overset{+}{C}<} + \overline{O}SO_3H$$

$$-\overset{|}{C}-OH\;\; -\overset{+}{C}< + \overline{O}SO_3H \longrightarrow \underset{\text{Alkyl hydrogen sulphate}}{-\overset{|}{C}-H \;\; -\overset{|}{C}-OSO_3H}$$

(*ii*) **Mechanism of addition of water to an alkene.** Alkenes react on water in the presence of an acid to form alcohols.

$$\underset{\text{Alkene}}{>C=C<} + H_2O \xrightarrow{H_3O^+} \underset{\text{Alcohol}}{-\underset{H}{\overset{|}{C}}-\underset{OH}{\overset{|}{C}}-}$$

$$\underset{\text{Alkene}}{>C=C<} + H_3O^+ \rightleftharpoons \underset{\text{Carbocation}}{-\overset{|}{C}-H \;\; -\overset{+}{C}<} + H_2O$$

$$-\overset{|}{C}-H \;\; -\overset{+}{C}< + :\ddot{O}-H\,(H) \rightleftharpoons \underset{\text{Protonated alcohol}}{-\overset{|}{C}-H \;\; -\overset{|}{C}-\overset{+}{O}(H)-H}$$

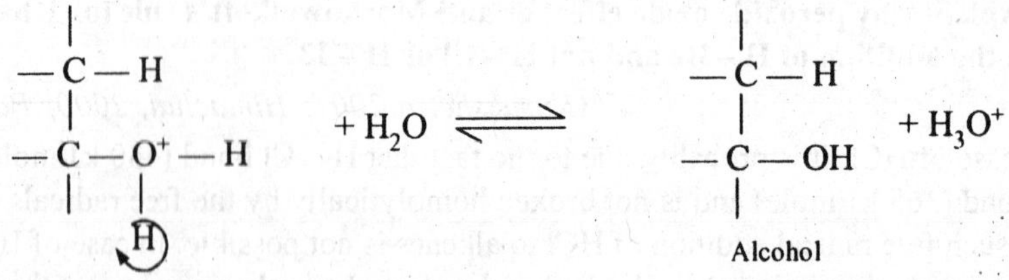

$$\begin{matrix} | \\ -C-H \\ | \\ -C-O^{+}-H \\ | \quad | \\ \quad H \end{matrix} + H_2O \rightleftharpoons \begin{matrix} | \\ -C-H \\ | \\ -C-OH \\ | \end{matrix} + H_3O^+$$

Alcohol

The reaction takes place by a electrophilic mechanism as depicted below:

Q. 26. Write a detailed note on ozonolysis of alkenes.

(Gorakhpur, 2000; Lucknow, 2000; A.N. Bhauguna, 2001; Guwahati, 2005; Nagpur, 2008)

Ans. Ozone undergoes a reaction with olefins in an inert solvent at low temperatures to yield unstable addition compounds called 'ozonides'. These ozonides are not easily isolated and yield carbonyl compounds on further treatment either with boiling water and Zn or zinc and acetic acid.

This overall reaction is called as *Ozonolysis*.

Ozonolysis is a very useful reaction for illucidating the structures of olefinic compounds.

Ozone has a dipolar resonance structure with a bond angle of about 120°.

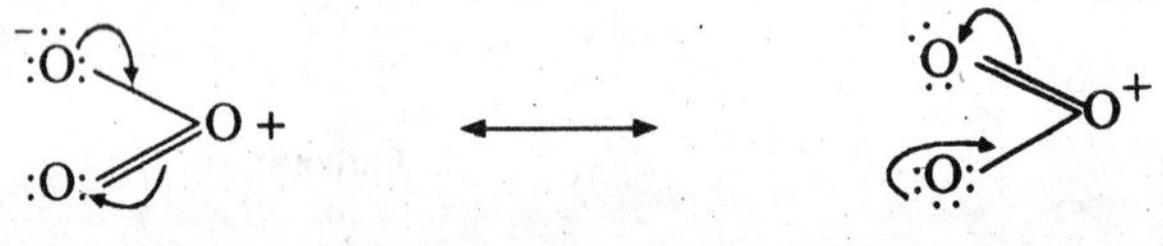

$$:\ddot{O}:^{-} - O^{+} = \ddot{O}: \longleftrightarrow \ddot{O} = O^{+} - :\ddot{O}:^{-}$$

The central atom (oxygen) is relatively electron deficient and thus can act as an electrophile for attack on the π electrons of the olefins. The final structure of the ozonide has been shown to have a structure resulting from the complete rupture of the carbon-carbon double bond.

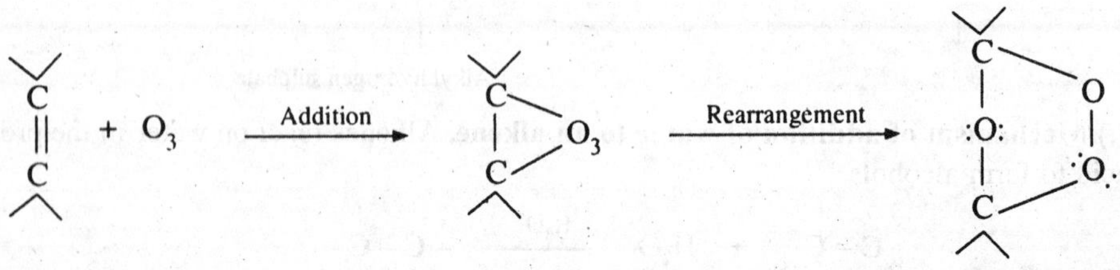

$$>C=C< + O_3 \xrightarrow{\text{Addition}} \begin{matrix} >C \\ | \\ >C \end{matrix} O_3 \xrightarrow{\text{Rearrangement}} \text{Ozonoide}$$

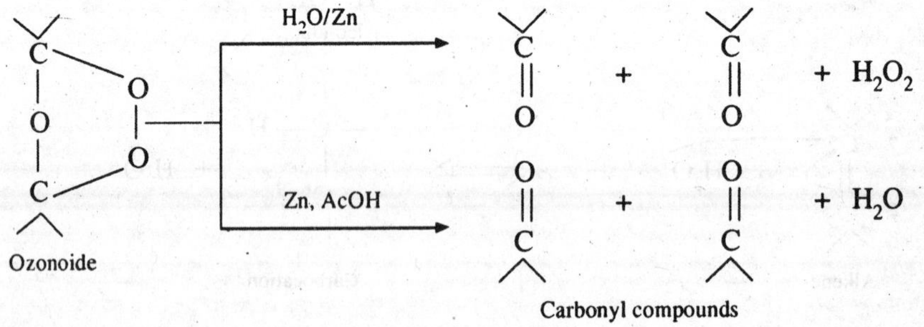

Ozonoide

$$\text{Ozonoide} \xrightarrow{H_2O/Zn} >C=O + >C=O + H_2O_2$$

$$\text{Ozonoide} \xrightarrow{Zn,\ AcOH} O=C< + O=C< + H_2O$$

Carbonyl compounds

The important thing to note about ozonides is that the *carbon-carbon bond has been broken*. For example:

$$CH_3-CH=C\langle^{CH_3}_{CH_3} + O_3 \longrightarrow \underset{\text{Ozonide}}{CH_3-CH(-O-O-)C\langle^{CH_3}_{CH_3}} \text{ (with bridging O)}$$

2-Methyl but-2-ene

$$\xrightarrow{H_2/Pt} \underset{\text{Acetaldehyde}}{\overset{CH_3}{\underset{H}{>}}C=O} + \underset{\text{Acetone}}{O=C\langle^{CH_3}_{CH_3}}$$

By identifying the products of ozonlysis, it is possible to locate the double bond in alkene.

Q. 27.On reductive ozonolysis, an unsaturated hydrocarbon gave the following compounds:

(*i*) Ethanediol (*ii*) Propanone (*iii*) Ethanol

Write the structural formula of the hydrocarbon and write its IUPAC name.

(*Kurukshetra, 2001*)

Ans. The hydrocarbon is

$$\overset{1}{C}H_3-\underset{\displaystyle CH_3}{\overset{2}{C}}=\overset{3}{C}H-\overset{4}{C}H=\overset{5}{C}H-\overset{6}{C}H_3$$

2-methylhex-2,4-diene

$$\Big\downarrow O_3, H_2O$$

$$\underset{\text{Propanone}}{CH_3-\underset{\displaystyle CH_3}{C}=O} + \underset{\text{Glyoxal}}{\underset{\displaystyle CHO}{CHO}} + \underset{\text{Acetaldehyde}}{CH_3CHO}$$

Glyoxal and acetaldehyde on reduction change to ethanediol and ethyl alcohol respectively.

Q. 28. Complete the following reaction

$$\mathbf{CH_3CH=CHCH_3+O_3 \longrightarrow ? \xrightarrow{Zn/H_2O}}$$

Ans. $$CH_3CH=CHCH_3 \xrightarrow{O_3} CH_3CH(-O-O-)CH-CH_3 \text{ (with bridging O)}$$

$$\Big\downarrow Zn/H_2O$$

$$2CH_3CHO+H_2O_2$$

Q. 29. What is meant by the term glycollisation or hydroxylation? Give the mechanism of this reaction. (*M. Dayanand, 2000; Nagpur, 2008*)

Ans. Conversion of alkene into 1, 2-diols is known as 1, 2 glycollisation reaction; an – OH group is added on to each carbon atom of the double bond. The most commonly used reagents are potassium permanganate and osmium tetroxide (OsO_4).

$$>C=C< \xrightarrow[\text{or } OsO_4]{KMnO_4} -\underset{OH}{\overset{|}{C}}-\underset{OH}{\overset{|}{C}}- \qquad \text{(1, 2 diol)}$$

The two reagents ($KMnO_4$ and OsO_4) are known to react by similar mechanism which involves cyclic intermediates of the following type which are subsequently hydrolyzed in aqueous solutions to glycol and the reduced form of the reagent.

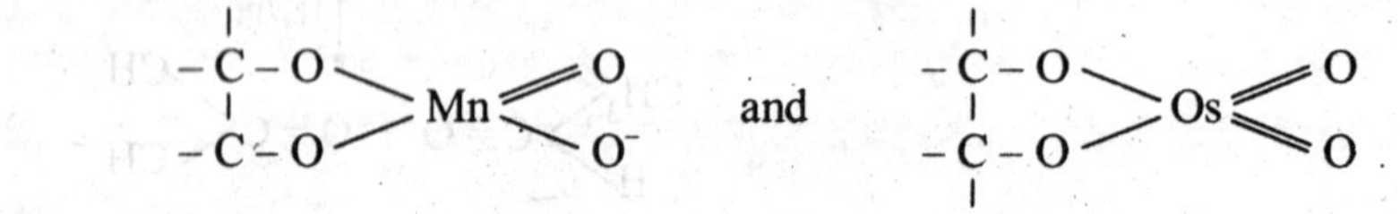

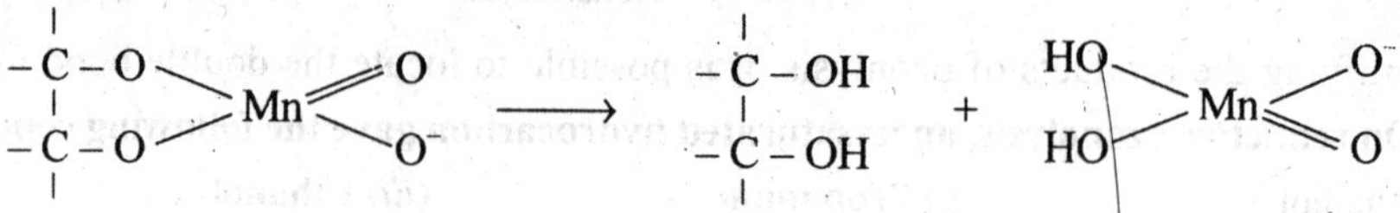

$$\begin{matrix} -C-O \\ -C-O \end{matrix} > Mn \begin{matrix} =O \\ -O^- \end{matrix} \quad \text{and} \quad \begin{matrix} -C-O \\ -C-O \end{matrix} > Os \begin{matrix} =O \\ =O \end{matrix}$$

$$\begin{matrix} -C-O \\ -C-O \end{matrix} > Mn \begin{matrix} =O \\ -O^- \end{matrix} \longrightarrow \begin{matrix} -C-OH \\ -C-OH \end{matrix} + \begin{matrix} HO \\ HO \end{matrix} > Mn \begin{matrix} -O^- \\ =O \end{matrix}$$

Cyclic manganese ester

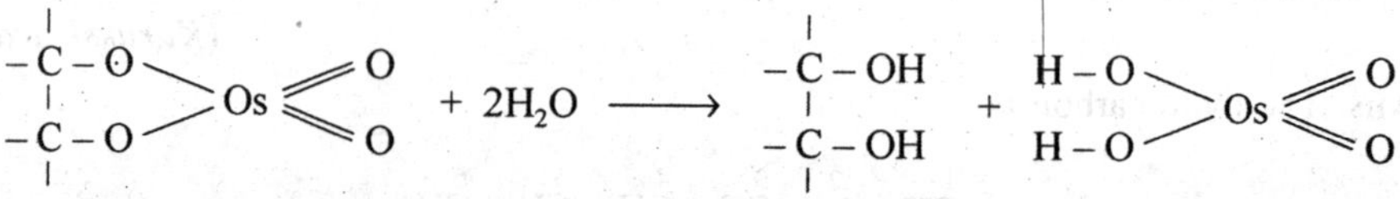

$$\begin{matrix} -C-O \\ -C-O \end{matrix} > Os \begin{matrix} =O \\ =O \end{matrix} + 2H_2O \longrightarrow \begin{matrix} -C-OH \\ -C-OH \end{matrix} + \begin{matrix} H-O \\ H-O \end{matrix} > Os \begin{matrix} =O \\ =O \end{matrix}$$

Cyclic osmium ester — *Cis* Glycol — Osmic acid

The permanganate reaction (Baeyer's test) is usually carried out in cold, aqueous alkaline solution.

$$\begin{matrix} CH_2 \\ \| \\ CH_2 \end{matrix} + M_2O + O \text{ (from } KMnO_4) \longrightarrow \begin{matrix} CH_2OH \\ | \\ CH_2OH \end{matrix}$$

Glycol

$$\underset{\text{Propylene}}{CH_3-CH=CH_2} + H_2O + O \text{ (from } KMnO_4) \longrightarrow \underset{\text{(Propylene glycol)}}{CH_3-\underset{OH}{CH}-\underset{OH}{CH_2}}$$

Q. 30. What is hydroboration of alkenes? Explain with mechanism. (*Panjab, 2002*)

Ans. On treatment with diborane $(BH_3)_2$, alkenes form alkyl boranes which are further converted into di and tri alkylboranes. For example:

$$\underset{\text{Propylene}}{2CH_3-CH=CH_2} + (BH_3)_2 \longrightarrow \underset{n\text{-Propylborane}}{2CH_3-CH_2-CH_2BH_2}$$

$$CH_3-CH_2-CH_2BH_2 + CH_3-CH=CH_2 \longrightarrow \underset{\text{Di-}n\text{-propylborane}}{(CH_3-CH_2-CH_2)_2BH}$$

$$(CH_3-CH_2-CH_2)_2BH + CH_3-CH=CH_2 \longrightarrow \underset{\text{Tri-}n\text{-propylborane}}{(CH_3-CH_2-CH_2)_3B}$$

Alkylboranes in turn are readily oxidised by alkaline solution of to form alcohols. For example:

$$(CH_3CH_2CH_2)_3B + 3H_2O \xrightarrow{H_2O_2/OH^-} 3CH_3-CH_2-CH_2OH + H_3BO_3$$

This reactions of preparation of alcohol from alkene is called **hydroboration.**

Mechanism. Diborane participates in the reaction in its monomeric form *i.e.*, BH_3. It behaves as an electrophile since the boron atom in the molecule is electron deficient as it has only six valence electrons.

Since BH_3 is an electrophile, hydroboration of alkanes also involves an electrophilic addition mechanism. But unlike other addition reactions of alkenes, hydroboration takes place in a single step through a transition state as shown below by considering the hydroboration of propylene.

$$CH_3 \rightarrow CH=CH_2 + BH_3 \longrightarrow CH_3 \rightarrow \overset{\delta-}{CH} \cdots\cdots CH_2 \quad (\text{with } H \ldots \overset{\delta-}{B}(H)-H) \longrightarrow$$

$$\overset{3}{CH_3} \rightarrow \overset{2}{CH}(\delta^+) \cdots\cdots \overset{1}{CH_2} \quad (H \cdots\cdots \overset{\delta-}{B}(H)-H) \longrightarrow \underset{n\text{-propyl borane}}{CH_3-CH(H)-CH(BH_2)}$$

Q. 31. Write a note on alkylation of alkenes taking the case of dimerisation of isobutylene. *(Lucknow, 2000)*

Ans. Isobutylene may be converted into iso-octane by the addition of isobutane in the presence of H_2SO_4 as a catalyst. This process is known as '*alkylation*' and is of industrial importance as it is extensively used in oil refineries in the production of high grade motor fuels.

The probable route adopted may be shown taking place as:

Initiation

$$(CH_3)_2C=CH_2 + H^+ \xrightarrow[\text{From } H_2SO_4]{} \underset{3^\circ \text{ Butyl carbonium ion}}{CH_3-\overset{CH_3}{\underset{CH_3}{C^+}}}$$

Propagation. There are two propagation steps. In the first step; addition of 3° butyl carbonium ion with isobutylene to form a new carbonium ion ($^+C_8H_{17}$) takes palce.

$$CH_3-\overset{CH_3}{\underset{CH_3}{C^+}} + H_2C=C(CH_3)_2 \longrightarrow \underset{[C_8H_{17}]^+}{CH_3-\overset{CH_3}{\underset{CH_3}{C}}-CH_2-\overset{CH_3}{\underset{CH_3}{C^+}}}$$

In the second step this new carbocation then abstracts a hydride ion from iso-butane to form isooctane and a new 3° butyl carbonium ion.

$$CH_3-\overset{\overset{CH_3}{|}}{\underset{\underset{CH_3}{|}}{C}}-CH_2-\overset{\overset{CH_3}{|}}{\underset{\underset{CH_3}{|}}{C^+}}+\underset{\text{Isobutane}}{CH_3-\overset{\overset{H}{|}}{\underset{\underset{CH_3}{|}}{C}}-CH_3}$$

$$\longrightarrow \underset{\substack{\text{Iso-octane}\\\text{(2,2,4-tri-methyl pentane)}}}{CH_3-\overset{\overset{CH_3}{|}}{\underset{\underset{CH_3}{|}}{C}}-CH_2-\overset{\overset{CH_3}{|}}{\underset{\underset{CH_3}{|}}{C^+}}}+CH_2-\overset{\overset{H}{|}}{\underset{\underset{CH_3}{|}}{C}}-H$$

Termination. It involves the loss of a proton from the carbonium ions to form an alkene.

$$CH_3-\overset{\overset{CH_3H}{|\;\;|}}{\underset{\underset{CH_3}{|}}{C}}-CH-\overset{\overset{CH_3}{|}}{\underset{\underset{CH_3}{|}}{C^+}} \xrightarrow{-H^+} \overset{5}{C}H_3-\overset{\overset{CH_3}{|}}{\underset{\underset{CH_3}{|}}{\overset{4}{C}}}-\overset{3}{C}H=\overset{2}{C}\begin{matrix}\overset{1}{C}H_3\\CH_3\end{matrix}$$

2,4,4-Tri methylpent-2-ene
(Higher %)
(A)

Or

$$CH_3-\overset{\overset{CH_3}{|}}{\underset{\underset{CH_3}{|}}{C}}-CH_2-\overset{\overset{CH_3}{|}}{\underset{\underset{H-CH_2}{|}}{C^+}} \xrightarrow{-H^+} CH_3\overset{\overset{CH_3}{|}}{\underset{\underset{CH_3}{|}}{C}}-CH_2-\overset{\overset{CH_3}{|}}{C}=CH_2$$

2,4,4-Tri methylpent-1-ene
(B)

Q. 32. What is mustard gas? How is it prepared?

Ans. Ethylene and sulphur monochloride produce the powerful blistering agent (Vesicant) known as mustard gas used during World War I. Though it is a liquid (b.pt. 490 K), small quantities of its vapours in air are highly toxic.

$$\begin{matrix}CH_2=CH_2\\+\\H_2C=CH_2\\\text{(Ethylene)}\end{matrix} + \underset{\substack{\text{Sulphur}\\\text{monochloride}}}{S_2Cl_2} \longrightarrow \underset{\substack{\text{Mustard gas}\\[\beta,\beta'\text{-Dichloro-}\\\text{diethyl sulphide]}}}{\begin{matrix}CH_2Cl & & CH_2Cl\\| & & |\\CH_2 & -S- & CH_2\end{matrix}} + S$$

Q. 33. Explain the action of carbenes on alkenes? *(G. Nanakdev, 2000)*

Ans. When treated with diazomethane (CH_2N_2) or ketene ($CH_2 = C = O$) in the presence of light, alkenes add on carbene, a methylene, to form substituted cyclopropane.

$$\underset{\text{Alkene}}{>C=C<} + \underset{\text{Diazomethane}}{CH_2N_2} \xrightarrow{\text{Light}} \underset{\substack{\text{Substituted}\\\text{cyclopropane}}}{>C\underset{\underset{CH_2}{\vee}}{—}C<} + N_2$$

$$\underset{\text{But-2-ene}}{CH_3-CH=CH-CH_3} + \underset{\text{Diazomethane}}{CH_2N_2} \xrightarrow{\text{Light}} \underset{\substack{\text{1,2-Dimethyl}\\\text{cyclopropane}}}{CH_3-CH\underset{\underset{CH_2}{\vee}}{—}CH-CH_3} + N_2$$

It is of theoretical interest that this reaction is stereospecific with singlet methylene occurring exclusively in one spatial direction that is; addition of carbene to cis-but-2-ene yields cis 1, 2-dimethyl cyclopropane whereas the reaction with trans-but-2-ene yields trans 1, 2-dimethyl cyclopropane.

Q. 34. Describe with mechanism the addition of carbenes on alkenes.

Ans. Alkenes undergo addition reactions with carbenes, both in the singlet and triplet forms*, to form cyclopropanes. Since a ring is generated as a result of such an addition reaction, the reaction is known as *cyclcoaddition.* For example, when treated with diazomethane (CH_2N_2) or ketene ($CH_2 = C = O$) which produce carbene in the presence of light, alkenes add on methylene to from cyclopropanes.

$$>C=C< \; + \; :CH_2 \longrightarrow \text{cyclopropane}$$

Carbene (Singlet or triplet) → Cyclopropane ($>C-C<$ bridged by CH_2)

Addition of **singlet methylene** takes place in a *sterospecific* manner to yield product having the same stereochemistry as the alkene. For example, singlet methylene reacts with *cis*-2 butene to form *cis*-1, 2-dimethyl cyclopropane. Similarly, it reacts with *trans*-2 butene to form *trans*-1, 2-dimethylcyclopropane.

$$\underset{cis\text{-2-Butene}}{CH_3(H)C=C(H)CH_3} + \underset{\text{Singlet methylene (obtained from } CH_2N_2 \text{ in liquid } cis\text{-2-butene)}}{:CH_2} \longrightarrow \underset{cis\text{-1,2-Dimethylcyclopropane}}{CH_3(H)C-C(H)CH_3 \ (CH_2)}$$

$$\underset{trans\text{-2-Butene}}{CH_3(H)C=C(H)CH_3} + \underset{\text{Singlet methylene (obtained from } CH_2N_2 \text{ in liquid } trans\text{-2-butene)}}{:CH_2} \longrightarrow \underset{trans\text{-1,2-Dimethylcyclopropane}}{CH_3(H)C-C(CH_3)H \ (CH_2)}$$

Addition of **triplet methylene** is *non-stereospecific* and produces a mixture of products. For example, addition of triplet methylene to *either* cis-2-butene or *trans*-2 butene produces a *mixture of cis and trans*-1, 2-dimethylcyclopropanes.

$$\underset{\text{2-Butene } (cis \text{ or } trans)}{CH_3-CH=CH-CH_3} + \underset{\text{Triplet methylene (obtained from } CH_2N_2 \text{ in gaseous butene)}}{\cdot\dot{C}H_2} \longrightarrow cis\text{-}CH_3(H)C-C(H)CH_3\,(CH_2) + trans\text{-}CH_3(H)C-C(CH_3)H\,(CH_2)$$

Mixture of *cis* and *trans* 1,2-methyl cyclopropanes

---

* For difference between singlet and triplet forms of carbenes, see chapter 2, under carbenes.

Different mechanistic routes in the two cases are responsible for obtaining different results.

Mechanism of Addition of Singlet Methylene

Addition of singlet methylene involves single step *electrophilic addition*. Being electron deficient in nature, singlet methylene seeks electrons from carbon-carbon double bond. As a result singlet methylene gets simultaneously attached to both the doubly bound carbon atoms. Therefore, stereochemistry of the reactant remains preserved in the products, *cis*-alkenes yield *cis*-dialkyl cyclopropanes while *trans*-alkenes yield *trans*-dialkyl cyclopropanes. The complete mechanism may be depicted as:

$$>C=C< \; + \; :CH_2 \longrightarrow \left[>C \overset{\cdots}{=} C< \; (CH_2) \right] \longrightarrow >C-C< \; (CH_2)$$

Mechanism of Addition of Triplet Methylene

Triplet methylene reacts with alkenes in a two step free radical mechanism. Triplet methylene is a diradical and it attacks the alkene to form an intermediate diradical having the conformation A. This diradical has a life time long enough for rotation of groups to take place around central carbon-carbon bond. This gives rise to conformation B. Ring closures of A and B lead to the formation of both *cis* and *trans* products as shown below:

$$\underset{cis\text{-alkene}}{RRC=CRR} + \underset{\text{Triplet Methylene}}{\cdot\dot{C}H_2} \longrightarrow \underset{A}{R_2\dot{C}-CR_2(\cdot CH_2)} \xrightarrow{\text{Rotation}} \underset{B}{R_2\dot{C}-CR_2(\cdot CH_2)}$$

A → *cis*-cyclopropane (R, H on one carbon; R, H on the other, bridged by CH_2)

B → *trans*-cyclopropane (H, R on one carbon; R, H on the other, bridged by CH_2)

Q. 35. What is meant by allylic substitution? Illustrate your answer by a suitable example and also discuss the mechanism of the reaction.

Ans. Allylic substitution. It is a substitution reaction which takes place at the carbon atom next to the double bond. Higher homologues of ethylene undergo allylic substitution under specific conditions. For example, when propene is treated with chlorine at 773 K, it undergoes allylic substitution to yield 3-chloropropene.

$$CH_3-CH=CH_2 + Cl_2 \xrightarrow{773\,K} \underset{\text{3-Chloropropene}}{CH_2Cl-CH=CH_2} + HCl$$

Mechanism of the reaction. Allylic substitution takes place by free radical chain mechanism as given below:

(1) Chain initiating step

$$Cl-Cl \xrightarrow{773\,K} 2Cl\cdot$$

(2) Chain initiating step

$$\cdot Cl + CH_3-CH=CH_2 \longrightarrow HCl + \underset{\text{Allyl radical}}{[\cdot CH_2-CH=CH_2 \leftrightarrow CH_2=CH-CH_2\cdot]}$$

$$Cl-Cl + \cdot CH_2-CH=CH_2 \longrightarrow CH_2Cl-CH=CH_2 + Cl\cdot$$

(3) Possible chain terminating steps

$$Cl\cdot + Cl\cdot \longrightarrow Cl_2$$

$$Cl\cdot + \cdot CH_2-CH=CH_2 \longrightarrow CH_2Cl-CH=CH_2$$

More conveniently, allylic bromination is done with the help of N-Bromosuccinimide (abbreviated as NBS). The reaction is carried in the presence of light

$$\underset{\text{Propene}}{CH_3CH=CH_2} + \underset{\text{N.B.S.}}{\begin{matrix}CH_2-CO\\ |\\ CH_2-CO\end{matrix}\!\!>NBr} \xrightarrow{h\nu} \underset{\text{Allyl bromide}}{BrCH_2CH=CH_2} + \underset{\text{Succinimide}}{\begin{matrix}CH_2CO\\ |\\ CH_2CO\end{matrix}\!\!>NH}$$

Q. 36. How do you account for the fact that at high temperature alkenes undergo free radical substitution and not free radical addition of halogens?

Ans. When an alkene is treated with a halogen at high temperature, what happens is that the halogen molecule readily decomposes due to the high temperature to form halogen free radicals or atoms. The halogen atom in turn attacks the alkene molecule at the allylic carbon to form an allyl free radical. The allyl free radical then abstracts a halogen from another halogen molecule to form the substitution product (as shown in Q. No. 35).

The question is why the halogen atom does not attack the alkene at the double bond to form the free radical and then add another halogen atom to form an addition product. The reason is that halogen atom does attack the alkene even at the double bond to form a free radical. But due to very high temperature, this free radical readily eliminates the halogen atom before another halogen atom could be added to form the addition product.

That is:

$$Cl\cdot + CH_3-CH=CH_2 \underset{\text{Elimination of Cl}}{\overset{\text{Addition of Cl}\cdot}{\rightleftharpoons}} CH_3-\dot{C}H-CH_2Cl$$

Thus the formation of addition product is prevented.

Q. 37. What happens when:

(*i*) an alkene is treated with peroxybenzoic acid.

(*ii*) 2, 3-Dimethyl-2-pentene is subjected to ozonolysis.

(*iii*) when propylene is treated with sodium borohydride and boron trifluoride followed by the treatment of product thus formed with alkaline hydrogen peroxide.

(*Lucknow, 2000; M. Dayanand, 2000; Awadh, 2000; G. Nanakdev, 2000; Kerala, 2001; Kurukshetra, 2001*)

Write equations for the reactions involved.

Ans. (*i*) When an alkene is treated with peroxybenzoic acid, an epoxide is formed. This reaction is called **epoxidation** reaction.

$$\underset{\text{Alkene}}{>C=C<} + \underset{\text{Peroxybenzoic acid}}{C_6H_5-\overset{O}{\overset{\|}{C}}-O-OH} \longrightarrow \underset{\text{Epoxide}}{-\overset{|}{C}-\overset{|}{C}-\text{ (bridged by O)}} + \underset{\text{Benzoic acid}}{C_6H_5-\overset{O}{\overset{\|}{C}}-OH}$$

Mechanism of the reaction

$$C_6H_5\overset{O}{\overset{\|}{C}}-O-O-H \overset{\text{Slow}}{\rightleftharpoons} R-\overset{O}{\overset{\|}{C}}-O^- + \underset{\text{Electrophile}}{\overset{+}{O}H}$$

$$>C=C< + OH^+ \longrightarrow >\overset{+}{C}-\underset{}{C}(\ddot{O}-H)- \longrightarrow >C-C<(\overset{+}{\ddot{O}}-H) \xrightarrow{-H^+} >C-C<(O) \text{ (Epoxide)}$$

(*ii*) When 2, 3-dimethyl-2 pentene is subjected to ozonolysis, a mixture of acetone and ethyl methyl ketone is obtained.

$$\underset{\text{2,3 Dimethyl-2-pentene}}{CH_3-CH_2-\overset{CH_3}{\overset{|}{C}}=\overset{CH_3}{\overset{|}{C}}-CH_3} + O_3 \longrightarrow CH_3-CH_2-\overset{H_3C}{\overset{|}{C}}(-O-)\overset{CH_3}{\overset{|}{C}}-CH_3 \; (O-O)$$

$$\xrightarrow{Zn/H_2O} \underset{\text{Ethyl methyl ketone}}{CH_3-CH_2-\overset{CH_3}{\overset{|}{C}}=O} + \underset{\text{Acetone}}{CH_3-\overset{CH_3}{\overset{|}{C}}=O}$$

(*iii*) On treatment with sodium borohyride and boron tri-fluoride propylene yields tripropyl borane which on treatment with alkaline hydrogen peroxide yields *n*-propyl alcohol.

$$3NaBH_4 + 4BF_3 \xrightarrow{273\ K} 2(BH_3)_2 + 3NaBF_4$$

$$2CH_3CH=CH_2 + (BH_3)_2 \longrightarrow 2CH_3-CH_2-CH_2BH_2$$

$$CH_3-CH_2-CH_2BH_2 \xrightarrow{CH_3CH=CH_2} (CH_3-CH_2-CH_2)_2BH$$

$$\xrightarrow{CH_3-CH=CH_2} \underset{\text{Tri } n\text{-propyl borane}}{(CH_3-CH_2-CH_2)_3B}$$

$$(CH_3-CH_2-CH_2)_3B + 3H_2O_2 \xrightarrow{NaOH} \underset{n\text{-propyl alcohol}}{3CH_3CH_2OH} + H_3BO_3$$

Q. 38. Write equations for the reactions and name the product formed when isobutylene reacts with: ***(G Nanakdev, 2000)***

(*i*) Cl_2 and water **(*ii*) HBr**

(*iii*) HBr in the presence of peroxide **(*iv*) H_2O in the presence of H^+**

(*v*) HI in the presence of peroxide **(*vi*) H_2SO_4**

Ans. (*i*)

$$CH_3-\overset{CH_3}{\overset{|}{C}}=CH_2 + Cl_2 + H_2O \longrightarrow \underset{\text{1-Chloro-2-methyl-2-propanol}}{CH_3-\overset{CH_3}{\overset{|}{\underset{OH}{\underset{|}{C}}}}-CH_2Cl} + HCl$$

(*ii*)

$$CH_3-\overset{CH_3}{\overset{|}{C}}=CH_2 + HBr \longrightarrow \underset{\text{2-Bromo-2-methyl propane}}{CH_3-\overset{CH_3}{\overset{|}{\underset{Br}{\underset{|}{C}}}}-CH_3}$$

(*iii*)

$$CH_3-\overset{CH_3}{\overset{|}{C}}=CH_2 + HBr \xrightarrow{\text{Peroxide}} \underset{\text{1-Bromo-2-methyl propane}}{CH_3-\overset{CH_3}{\overset{|}{CH}}-CH_2Br}$$

$$(iv) \quad CH_3-\underset{\underset{CH_3}{|}}{\overset{\overset{CH_3}{|}}{C}}=CH_2+H_2O \xrightarrow{H^+} CH_3-\underset{\underset{OH}{|}}{\overset{\overset{CH_3}{|}}{C}}-CH_3$$

2-methyl-2-propanol

$$(v) \quad CH_3-\overset{\overset{CH_3}{|}}{C}=CH_2+HI \xrightarrow{\text{Peroxide}} CH_3-\underset{\underset{I}{|}}{\overset{\overset{CH_3}{|}}{C}}-CH_3$$

2-Iodo-2-methylpropane

$$(vi) \quad CH_3-\overset{\overset{CH_3}{|}}{C}=CH_2+H_2SO_4 \longrightarrow CH_3-\underset{\underset{OSO_3H}{|}}{\overset{\overset{CH_3}{|}}{C}}-CH_3$$

Tert. butyl hydrogen sulphate

Q. 39. Giving at least one example in each case, define the terms: Polymer, Polymerisation and Monomer.

Ans. Polymer. It is a substance having a very high molecular weight which is formed by the joining together of a very large number of simple molecules in a regular fashion. For example, combination of a very large number of ethylene molecules gives rise to the polymer known as *polythene.*

Polymerisation. It is the process of formation of a large molecule from small molecules. For example, when heated at a high temperature and pressure in the presence of traces of oxygen, ethylene undergoes polymerisation to form polythene.

$$nCH_2=CH_2 \xrightarrow[\text{Traces of } O_2]{\text{High temp. pressure}} (-CH_2-CH_2-)_n$$

Monomer. The simple molecules which are used as starting substance in the process of polymerisation are known as *monomers.* For example ethylene constitutes the monomer used in the formation of polythene.

Q. 40. Describe the free radical chain mechanism of addition polymerisation with help of a suitable example. *(Delhi, 2003)*

Ans. Free radical polymerisation. Many alkenes and substituted alkenes undergo polymerisation by free radical chain mechanism. Such reactions usually occur at high temperature under pressure or in the presence of catalysts such as peroxides and salts of peracids. The catalyst provides the free radical to initiate the chain process which is carried on by chain propagating steps. The mechanism is illustrated below with the help of polymerisation of ethylene in the presence of peroxide.

(1) *Chain initiating state*

$$\underset{\text{Peroxide}}{R-O-O-R} \xrightarrow{h\nu} \underset{\text{Free radical}}{2R\dot{O}}$$

(2) *Chain propagating stage*

$$R\dot{O}+CH_2=CH_2 \longrightarrow RO-CH_2=\dot{C}H_2$$

$$RO-CH_2-\dot{C}H_2+CH_2=CH_2 \longrightarrow RO-CH_2-CH_2-CH_2-\dot{C}H_2$$

The free radical formed in each step adds to a fresh molecule of alkene and so the chain progresses till a large molecule of the polymer is obtained and the chain is terminated.

(3) *Possible chain terminating stage*

(*i*) **Coupling** of free radicals to form a deactivated molecule.

For example:

$$RO-(CH_2-CH_2)_n-CH_2-\dot{C}H_2+\dot{C}H_2-(CH_2-CH_2)_n-OR$$
$$\longrightarrow RO-(CH_2-CH_2)_n-CH_2-CH_2+CH_2-(CH_2-CH_2)_n-OR$$

(*ii*) **Disproportionation** of free radicals in which one free radical acquires a hydrogen from another and both get deactivated.

$$RO-(CH_2-CH_2)_n-CH_2-\dot{C}H_2+\dot{C}H_2-CH_2-(CH_2\,CH_2)_n-OR$$
$$\longrightarrow RO-(CH_2-CH_2)_n-CH_2=CH_2+CH_3-CH_2-(CH_2-CH_2)_n-OR$$

Q. 41. Discuss the following:

(a) Cationic mechanism of polymerisation (*G.Nankdev, 2000; Panjab, 2002*)

(*b*) Anionic mechanism of polymerisation.

Ans. (*a*) **Cationic mechanism.** Alkene molecules having electron releasing groups (propylene, isobutylene etc.) in the acidic medium may polymerise by cationic mechanism. In fact, the electrophile attack of proton of acid on the alkene molecule leads to a carbocation. Such cations may further take part in the chemical combination leading to the formation of polyalkene. The cationic polymerisation is illustrated with the example of isobutyne.

$$\underset{\text{Isobutylene}}{CH_3-\overset{\overset{CH_3}{|}}{C}=CH_2}+\underset{\text{Acid}}{H^{\oplus}}\longrightarrow \underset{\text{Carbocation}}{CH_3-\overset{\overset{CH_3}{|}}{\underset{\oplus}{C}}=CH_2+H^{\oplus}}$$

$$CH_3-\overset{\overset{CH_3}{|}}{\underset{\underset{CH_3}{|}}{C}}{\oplus}+CH_2=\overset{\overset{CH_3}{|}}{C}-CH_3\longrightarrow CH_3-\overset{\overset{CH_3}{|}}{\underset{\underset{CH_3}{|}}{C}}-CH_2-\overset{\overset{CH_3}{|}}{\underset{\oplus}{C}}-CH_3$$

$$\Big\downarrow \text{Repetition}$$

$$CH_3-\overset{\overset{CH_3}{|}}{\underset{\underset{CH_3}{|}}{C}}-(-CH_2-\overset{\overset{CH_3}{|}}{\underset{\underset{CH_3}{|}}{C}}-)_n-CH_2-\overset{\overset{CH_3}{|}}{\underset{\oplus}{C}}-CH_3$$

$$\Big\downarrow -H^{\oplus}$$

$$\underset{\text{Poly isobutylene}}{CH_3-\overset{\overset{CH_3}{|}}{\underset{\underset{CH_3}{|}}{C}}-(-CH_2-\overset{\overset{CH_3}{|}}{\underset{\underset{CH_3}{|}}{C}}-)_n-CH=\overset{\overset{CH_3}{|}}{C}-CH_3}$$

(*b*) **Anionic mechanism.** The anionic mechanism is noticed in alkene molecules having some electron withdrawing groups present in them *e.g.* vinyl chloride, etc. The anionic polymerisation is carried in the presence of a suitable base such as sodamide ($NaNH_2$), *n*-butyl lithium (*n*- C_4H_9Li) etc.

$$B: + CH_2 = CH(Cl) \longrightarrow B - CH_2 - \overset{\oplus}{C}H(Cl)$$

Base — Carbanion

$$B - CH_2 - \overset{\oplus}{C}H(Cl) + CH_2 = CH(Cl) \longrightarrow B - CH_2 - CH(Cl) - CH_2 - \overset{\oplus}{C}H(Cl)$$

$\downarrow$ Repetition

Polyvinyl chloride (P.V.C.)

Polymerisation in alkenes may take place by both free radical or ionic mechanism. The free radical polymerisation takes place in the presence of an organic peroxide. In the ionic polymerisation, the cationic polymerisation takes place in the presence of an acid while the anionic polymerisation proceeds in the presence of base acting as the catalyst.

Q. 42. How is it possible to know whether the given hexene is hexene-1, hexene-2 or hexene-3?

Ans. It is possible to distinguish between the three compounds by carrying out ozonolysis of the substance. An alkene forms an ozonide with ozone. The ozonide is subjected to hydrolysis when we obtain a mixture of carbonyl compounds. By analysing the carbonyl compounds, we can tell about the position of double bond in the molecule. The products obtained on ozonolysis of hexene-1 are as given below:

$$CH_3CH_2CH_2CH_2CH = CH_2 \xrightarrow{O_3} \text{ozonide } CH_3CH_2CH_2CH_2CH\overset{O}{\diagup\diagdown}CH_2 \text{ (with O—O bridge)}$$

$\downarrow$ Zn, Acetic acid

$$HCHO + CH_3CH_2CH_2CH_2CHO$$

Methanal — Pentanal

Products obtained on ozonolysis of hexene-2 are as follows:

$$\underset{\text{Hexene-2}}{CH_3CH_2CH_2CH = CHCH_3} \xrightarrow{O_3} CH_3CH_2CH_2CH\overset{O}{\diagup\diagdown}CHCH_3 \text{ (with O—O bridge)}$$

$\downarrow$ Zn, Acetic acid

$$CH_3CH_2CH_2CHO + CH_3CHO$$

Butanal — Ethanal

Products obtained on ozonolysis of hexene-3 are as follows:

$$CH_3CH_2CH = CHCH_2CH_3 \xrightarrow{O_3} CH_3CH_2CH\overset{O}{\diagup\diagdown}CHCH_2CH_3 \text{ (with O—O bridge)}$$

$\downarrow$ Zn, Acetic acid

$$CH_3CH_2CHO + CH_3CH_2CHO$$

Propanal — Propanal

Thus by analysing the carbonyl compounds produced it is possible to tell whether the given hexene is hexene-1, hexene-2 or hexene-3.

MISCELLANEOUS QUESTIONS

Q. 43. What happens when:

(*i*) 2,2,2-trimethyl-1-bromoethane reacts with alcoholic potassium hydroxide. (*Kumaon, 2000*)

(*ii*) A mixture of ethylene (1 mole) and acetylene (1 mole) is treated with bromine (1 mole) in carbon tetrachloride solvent. (*Calcutta, 2000*)

(*iii*) But-1-ene is treated with $CBrCl_3$ in presence of catalytic amount of benzoyl peroxide. (*Calcutta, 2000*)

(*iv*) Propene-1 is heated with bromine followed by the treatment of product thus formed with sodium and methyl bromide. (*Kumaon, 2000*)

Ans. (*i*) When 2,2,2-trimethyl-1-bromoethane reacts with alc. potassium hydroxide, 1,2,2-trimethyl ethene is obtained.

$$CH_3-\underset{CH_3H}{\overset{CH_3H}{C-C}}-Br \xrightarrow[-Br^-]{\text{alc. KOH}} CH_3-\underset{CH_3}{\overset{CH_3}{C}}-\overset{\oplus}{C}H_2 \longrightarrow CH_3-\overset{CH_3H}{\underset{\oplus}{C}-C}-CH_3 \text{ (with H below)} \longrightarrow CH_3-\overset{CH_3H}{C=C}-CH_3$$

2,2,2-trimethyl-1-bromoethane ... 1,1,2-trimethyl ethene

(*ii*) When a mixture of ethylene (1 mole) and acetylene (1 mole) is treated with bromine (1 mole) in carbon tetrachloride solvent, 1,2-dibromoethene is obtained.

$$\underset{\text{ethylene (1 mole)}}{CH_2=CH_2} + CH\equiv CH \xrightarrow[\text{acetylene (1 mole)}]{Br_2\text{ (1 mole)}} \underset{\text{1,2-dibromoethene}}{CHBr=CHBr} + CH_2=CH_2$$

(*iii*) When But-1-ene is treated with $CBrCl_3$ in presence of benzoyl peroxide, 3-bromo-1,1,1-trichloropentane is formed.

$$\underset{\text{but-1-ene}}{CH_3CH_2CH=CH_2} + CBrCl_3 \xrightarrow{\text{benzoyl peroxide}} \underset{\text{3-bromo-1,1,1-trichloropentane}}{CH_3CH_2\underset{Br}{CH}-CH_2-CCl_3}$$

(*iv*) When propene-1 is treated with bromine, it yields 3-bromopropene which on treatment with sodium and methyl bromide yields butene-1.

$$CH_3CH=CH_2 \xrightarrow{Br_2} BrCH_2CH=CH_2 \xrightarrow[CH_3Br]{2\,Na} \underset{\text{butene-1}}{CH_3CH_2CH_2=CH_2}$$

Q. 44. How will you differentiate between 1-pentene and 2-pentene? (*Lucknow, 2000; Awadh, 2000*)

Ans. The products obtained on ozonolysis of pentene-1 are as given below:

$$CH_3CH_2CH_2CH=CH_2 \xrightarrow{O_3} CH_3CH_2CH_2\overset{O}{CH}\quad CH_2 \text{ (ozonide, O—O bridge)} \xrightarrow{\text{Zn / Acetic acid}} \underset{\text{methanal}}{HCHO} + \underset{\text{butanal}}{CH_3CH_2CH_2CHO}$$

Products obtained on ozonalysis of pentene-2 are as follows:

$$CH_3CH_2CH=CHCH_3 \xrightarrow{O_3} CH_3CH_2CH\overset{O}{\diagup\diagdown}CHCH_3 \text{ (ozonide, with O—O bridge)} \xrightarrow[\text{Acetic acid}]{Zn} \underset{\text{propanal}}{CH_3CH_2CHO} + \underset{\text{ethanal}}{CH_3CHO}$$

By identifying the reaction products, 1-pentene and 2-pentene can be differentiated.

Q. 45. Compare with proper justification the stabilities of isobutene, Z-2-butene and E-2-butene. *(Calcutta, 2000)*

Ans. In isobutene, two methyl groups are attached to same carbon atom due to which it experiences steric hinderance. Hence, it is least stable. In Z-2 butene, two methyl groups are attached to two different carbon atoms but on the same side of the double bond due to which it experiences steric hinderance but less than in isobutene. Hence, it is more stable than isobutene.

In E-2-butene, the two methyl groups are attached to two different carbon atoms on the opposite side of the double bond. Hence, it experiences least steric hinderance. So, it is stable to the maximum extent.

$$\underset{\text{Z-2-butene}}{(H)(CH_3)C=C(CH_3)(H)} > \underset{\text{E-2-butene}}{(CH_3)(H)C=C(CH_3)(H)} > \underset{\text{isobutene}}{(CH_3)_2C=CH_2}$$

Q. 46. Write the major product (products) in the following reaction: ***(Kanpur 2000)***

$$CH_3-CH_2-\underset{CH_3}{\overset{Br}{C}}-CH_3 \xrightarrow{\text{alc. KOH}/\Delta}$$

Ans. 2-methyl-butene-2 is the major product which is formed in the reaction as under :

$$CH_3-CH_2-\underset{CH_3}{\overset{Br}{C}}-CH_3 \xrightarrow[\Delta]{\text{alc. KOH}} CH_3CH_2-\underset{CH_3}{\overset{\oplus}{C}}-CH_3 \longrightarrow CH_3-\underset{H}{\overset{H}{C}}-\underset{CH_3}{\overset{\oplus}{C}}-CH_3$$

$$\longrightarrow \underset{\substack{\text{2-methylbutene-2} \\ \text{(major product)}}}{CH_3-CH=\underset{CH_3}{C}-CH_3}$$

Q. 47. How will you bring about the following conversion? *(Kerala, 2000)*

$$(CH_3)_3C\cdot CH_2Br \xrightarrow{\text{aq. NaOH}} (CH_3)_2-\underset{OH}{C}-CH_2-CH_3$$

Ans. $CH_3-\underset{CH_3}{\overset{CH_3}{C}}-\underset{H}{\overset{H}{C}}-Br \xrightarrow{\text{aq. NaOH}} CH_3-\underset{CH_3}{\overset{CH_3}{C}}-\overset{\oplus}{C}H_2 \longrightarrow CH_3-\underset{\oplus}{\overset{CH_3}{C}}-CH_2-CH_3 \xrightarrow{OH^-} (CH_3)_2-\underset{OH}{C}-CH_2-CH_3$

Q. 48. Identify A and B in the following reaction :

$$CH_2=CH-CH_2-Cl+HBr \xrightarrow{\text{Peroxide}} A \xrightarrow[\text{Et}]{\text{Zn}} B+ZnBrCl$$

(Lucknow, 2000)

Ans. The complete reaction is as under:

$$CH_3=CH-CH_2-Cl+HBr \xrightarrow{\text{Peroxide}} \underset{(A)}{BrCH_2-CH_2-\dot{C}H_2-Cl}$$

$$\xrightarrow{\text{Zn / EtOH}} ZnBrCl + \underset{(B)}{\text{cyclopropane } (CH_2-CH_2-CH_2 \text{ ring})}$$

Q. 49. Draw geometrical isomers of the following: *(Lucknow, 2000)*

(i) Monobromopropenes

(ii) Hept-2-en-5-yne

(iii) Pentene-2

Ans. (*i*) *Geometrical isomers of monobromopropene:*

cis-form *trans*-form

(*ii*) *Geometrical isomers of hept-2-ene-5-yne:*

$$\underset{\textit{cis}\text{-form}}{\begin{matrix} CH_3-C\equiv C-CH_2 & & H \\ & C & \\ & \Vert & \\ & C & \\ CH_3 & & H \end{matrix}} \qquad \underset{\textit{trans}\text{-form}}{\begin{matrix} CH_3-C\equiv C-CH_2 & & H \\ & C & \\ & \Vert & \\ & C & \\ H & & CH_3 \end{matrix}}$$

(*iii*) *Geometrical isomers of pentene-2:*

$$\underset{\textit{cis}\text{-form}}{\begin{matrix} CH_3-CH_2 & & H \\ & C & \\ & \Vert & \\ & C & \\ CH_3 & & H \end{matrix}} \qquad \underset{\textit{trans}\text{-form}}{\begin{matrix} CH_3-CH_2 & & H \\ & C & \\ & \Vert & \\ & C & \\ H & & CH_3 \end{matrix}}$$

Q. 50. How will you distinguish between 1-hexene and *n*-hexane? *(Lucknow, 2000)*

Ans. This can be done with the help of following tests.

1. **Baeyer's test.** (reaction with cold dil. $KMnO_4$)

n-hexane does not react. 1-hexene reacts with Baeyer' reagent. Purple colour of $KMnO_4$ is discharged.

$$CH_3CH_2CH_2CH_2CH = CH_2 \xrightarrow[H_2O + O]{\text{dil. } KMnO_4} CH_3CH_2CH_2CH_2\underset{\displaystyle OH}{\underset{|}{CH}} - \underset{\displaystyle OH}{\underset{|}{CH_2}}$$

2. **Reaction with Br_2 in CCl_4.**

1-hexene will discharge the yellow colour of bromine

$$CH_3CH_2CH_2CH_2CH = CH_2 \xrightarrow[CCl_4]{Br_2 \text{ in}} CH_3CH_2CH_2CH_2CHBrCH_2Br$$

n-hexane will react only on heating to a high temperature.

9

ALKYNES

Q. 1. What are alkynes? Give their nomenclature.

Ans. Alkynes are unsaturated hydrocarbons having the general formula C_nH_{2n-2}. Such compounds are more unsaturated than alkenes. This is because alkynes contain a triple bond in a molecule. One triple bond is equivalent to two double bonds. First member of this series is acetylene C_2H_2 with two carbon atoms.

Nomenclature of alkynes, according to common system and IUPAC system is given below:

Compound	*Common Name*	*IUPAC Name*
$CH \equiv CH$	Acetylene	Ethyne
$CH_3C \equiv CH$	Methyl acetylene	Propyne
$CH_3CH_2C \equiv CH$	Ethyl acetylene	Butyne-1
$CH_3C \equiv CCH_3$	Dimethyl acetylene	Butyne-2
$CH_3CH_2C \equiv CCH_3$	Ethyl methyl acetylene	Pentyne-2

Q. 2. Draw the structures of all isomeric alkynes having the formula C_6H_{10} and give their IUPAC names.

Ans. $CH_3 - CH_2 - CH_2 - CH_2 - C \equiv CH$

Hex-1-yne

$CH_3 - CH_2 - CH_2 - C \equiv C - CH_3$

Hex-2-yne

$CH_3 - CH_2 - C \equiv C - CH_2 - CH_3$

Hex-3-yne

$$CH_3 - \underset{}{\overset{CH_3}{\overset{|}{C}H}} - C \equiv C - CH_3$$

4-Methylpent-2-yne

$$CH_3 - CH_2 - \overset{CH_3}{\overset{|}{C}H} - C \equiv CH$$

3-Methylpent-1-yne

$$CH_3 - \overset{CH_3}{\overset{|}{C}H} - CH_2 - C \equiv CH$$

4-Methylpent-1-yne

Q. 3. How will you prepare alkynes by

(*i*) Dehydrohalogenation of dihalides

(*ii*) Dehalogenation of tetrahalides

(*iii*) Action of alkyl halides on acetylides

(*iv*) Action of water on calcium carbide

Ans. (*i*) Dehydrohalogenation of vicinal dihalides. When 1, 2-dihaloalkane is heated with alcoholic potash, it undergoes dehydrohalogenation, yielding an alkyne. The reaction takes place in two stages involving the intermediate formation of vinyl halide (haloalkene).

The second stage of reaction generally requires a stronger base (sodium amide).

$$R-\underset{X}{\overset{H}{C}}-\underset{X}{CH}-R' \xrightarrow{\text{KOH (alcoholic)}} R-CH=\overset{X}{C}-R'$$

Haloalkene (Vinyl halide)

$$R-CH=\overset{X}{C}-R' \xrightarrow{NaNH_2} RC\equiv CR'$$

Alkyne

X = Cl, B or I R, R′ may be H or alkyl group.

For example,

$$\overset{3}{C}H_3-\underset{Br}{\overset{H}{\overset{2}{C}}}-\underset{Br}{\overset{1}{C}H}-H \xrightarrow[-HBr]{\text{KOH (Alc.)}} \overset{3}{C}H_3-\overset{2}{C}H-\overset{1}{C}HBr$$

1,2-dibromopropane (Propylene bromide) → 1,2-dibromopropane (Propylene bromide)

$$\downarrow -HBr \quad NaNH_2$$

$$CH_3-C\equiv CH$$

Methyl acetylene

From Gem dihalide. When 1, 1-dihaloalkane is heated with alcoholic potash or sodium amide, it undergoes dehydrohalogenation yielding an alkyne. The reaction proceeds in two stages involving the intermediate formation of haloalkene.

The sequence of the reactions is similar to that in dehydrohalogenation of vic. dihalides.

$$R-CH_2-\underset{X}{\overset{R'}{C}}-X \xrightarrow{\text{KOH (Alc.)}} RCH=\overset{X}{C}-R'$$

1,1-Dihaloalkane (gem dihalide) → 1-Haloalkene (vinyl halide)

$$\downarrow NaNH_2,\ -NH_3,\ -NaX$$

$$RC\equiv C-R'$$

Alkyne

For example,

$$CH_3-CH\begin{matrix}\diagup Br\\ \diagdown Br\end{matrix} \xrightarrow[-HBr]{\text{alc. KOH}} CH_2=CH-Br$$

1,1-Dibromo ethane or Ethylidene bromide → Vinyl bromide

$$\downarrow NaNH_2,\ NH_3,-NaBr$$

$$HC\equiv CH$$

Acetylene

(*b*) $$CH_3-CH_2-\underset{Br}{\overset{H}{C}}-Br \xrightarrow[-HBr]{\text{alc. KOH}} CH_3-CH=CH-Br$$

1,1-Dibromopropane or propylidene bromide → 1-Bromoprop-1-ene

$$\downarrow NaNH_2,\ -NH_3,\ -NaBr$$

$$CH_3-C\equiv CH$$

Methyl acetylene

(*ii*) **Dehalogenation of tetra halides.** Dehalogenation of tetra halides (1, 1, 2, 2-tetra haloalkanes) is carried out by passing their vapours over heated zinc and it results in the formation of alkyne.

$$R-\underset{X}{\overset{X}{C}}-\underset{X}{\overset{X}{C}}-R' \xrightarrow[-ZnX_2]{Zn} \underset{X}{\overset{R}{}}\!\!>C=C<\!\!\underset{R'}{\overset{X}{}} \xrightarrow[-ZnX_2]{Zn} RC\equiv C-R'$$

$$\underset{\text{1,1,2,2-Tetrabromoethane}}{H-\underset{Br}{\overset{Br}{C}}-\underset{Br}{\overset{Br}{C}}-H} + 2Zn \longrightarrow 2ZnBr_2 + \underset{\text{Acetylene}}{HC\equiv CH}$$

$$\underset{\text{1,1,2,2-Tetrabromopropane}}{CH_3-\underset{Br}{\overset{Br}{C}}-\underset{Br}{\overset{Br}{C}}-H} + 2Zn \longrightarrow 2ZnBr_2 + \underset{\text{Prop-1-yne, Methyl acetylene}}{CH_3-C\equiv CH}$$

(*iii*) **Action of acetylides on alkyl halides.** The metallic acetylides yield higher alkynes by reacting with alkyl halides. It is a very good method of converting lower alkynes into higher alkynes.

For example,

$$\underset{\text{Sodium acetylide}}{HC\equiv\overset{-}{C}\overset{+}{Na}} + CH_3I \longrightarrow NaI + \underset{\text{Methyl acetylene}}{HC\equiv C-CH_3}$$

Acetylides used in the reaction are obtained from alkynes with terminal triple bond ($-C\equiv C-H$) by the action of sodium or sodium amide.

$$-C\equiv C-H + Na \longrightarrow -C\equiv\overset{-}{C}\overset{+}{Na} + 1/2\,H_2$$

(*iv*) **Hydrolysis of calcium carbides.** Acetylene is prepared by the hydrolysis of calcium carbide. The latter is obtained by heating lime stone with coke at 2000°C in an electric furnace.

$$CaCO_3 + 3C \xrightarrow{2000^\circ C} CaC_2 + CO_2 + CO$$

$$CaC_2 + 2H_2O \longrightarrow HC\equiv CH + Ca(OH)_2$$

Q. 4. Alkynes do not exhibit geometrical isomerism. Explain.

Ans. The carbon atoms involved in the formation of triple covalent bond in alkynes are *sp* hybridised. As a result, alkynes have linear structure. Each triply bonded carbon atom is further linked to only one atom or group of atoms. Hence inspite of hindered rotations, the possibility of different relative spatial arrangement of the groups does not arise. Thus, alkynes do not exhibit geometrical isomerism. This type of isomerism is possible when there are at least two groups attached to one carbon.

$$\underset{\text{Geometrical isomerism possible}}{\underset{Y}{\overset{X}{}}\!\!>C=C<\!\!\underset{Y}{\overset{X}{}}} \qquad \underset{\text{Geometrical isomerism not possible}}{X-C\equiv C-X \;(\text{or } Y)}$$

Q. 5. Describe the orbital structure of acetylene.

(Madras, 2003; Delhi, 2004; Nagpur, 2008; Garhwal, 2010)

Ans. In acetylene molecule both the carbon atoms are *sp* hybridised. One of the *sp* hybridised orbitals of carbon overlaps along internuclear axis with similar orbital of the other carbon atom to form C – C, σ-bond. The second *sp* orbital of each carbon atom overlaps along internuclear axis with half filled 1*s* orbital of hydrogen atom to form C – H, σ-bond each. Since *sp* orbitals of each carbon atom lie along a straight line and overlapping of these orbitals takes place along their internuclear axis, all the four atoms of acetylene lie along the single straight line. Thus, acetylene is a linear molecule.

Each carbon atom is still left with two unhybridised *p*-orbitals which are perpendicular to each other as well as to the plane of carbon and hydrogen atoms. Each of these unhybridised orbitals of one carbon atom overlaps sideways with similar orbitals of the other carbon atom to form two π-bonds. There is overlapping between these two π electron clouds. As a result, the four lobes of two π-bonds merge to form a single electron cloud which is cylindrically symmetrical about the internuclear axis.

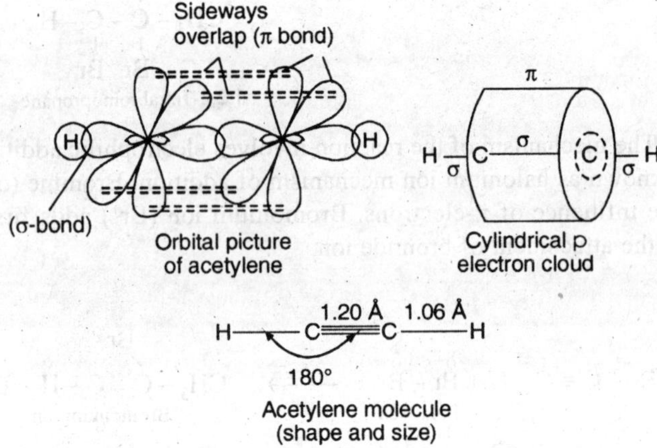

The carbon-carbon triple bond is thus made up of one strong σ-bond and two weak π-bonds.

Bond parameters. The various bond lengths and energies for the molecule of acetylene are given below:

(I) C ≡ C bond length = 1.20 Å

(II) C – H bond length = 1.06 Å

(III) H – C – C bond angle = 180°

(IV) C ≡ C bond dissociation energy = 830 kJ mol^{-1}

(V) C – H bond dissociation energy = 522 kJ mol^{-1}

Q. 6. Explain why are alkynes less reactive than alkenes towards electrophilic addition reactions.

Ans. Alkynes are less reactive than alkenes towards electrophilic addition reactions. Following factors are responsible for this:

(*i*) **Cylindrical -electron cloud.** In alkynes, the four lobes of two π-bonds merge to form a single electron cloud which is cylindrically symmetrical about the internuclear axis and occupies a big volume. Thus, electron density per unit volume becomes low. Due to decrease in electron density, π-electrons are not easily available to an electrophile. Hence alkynes are less reactive than alkenes towards electrophilic addition reactions.

(*ii*) ***sp*-Hybridisation of carbon.** The carbon atoms in alkynes are *sp*-hybridised while in alkenes are sp^2-hybridised. Greater the *s* character of an orbital, the more closely the electrons in that orbital are held by the nucleus. Thus π-electrons in alkynes are more strongly held by the carbon atoms than in case of alkenes and are less easily available for reactions with electrophiles. Thus makes alkynes less reactive than alkenes in electrophilic reactions.

Q. 7. Describe the mechanism of addition of Br_2 to propyne.

Ans. Bromine readily adds to propyne first forming trans-1, 2-dibromopropane and then 1, 1, 2, 2-tetrabromopropane.

$$CH_3-C\equiv C-H \xrightarrow{Br_2/CCl_4} \underset{\text{Trans-1,2-Dibromopropane}}{(Br)(CH_3)C=C(H)(Br)} \xrightarrow{Br_2/CCl_4} \underset{\text{1,2,2-Tetrabromopropane}}{CH_3-CBr_2-CBr_2-H}$$

Mechanism. The mechanism of the reaction involves electrophilic addition. It takes place in two steps. This is known as halonium ion mechanism of addition. Bromine (or any halogen) gets polarised under the influence of π-electrons. Bromonium ion (Br^+) adds first forming a bridge bond, followed by the attachment of bromide ion.

First step

$$CH_3-C\equiv C-H+\underset{\text{Polarised}}{\overset{\delta+}{Br}-\overset{\delta+}{Br}} \longrightarrow \underset{\text{Bromonium ion}}{CH_3-C\equiv C-H \text{ (bridged by } Br^+)}+Br^-$$

$$CH_3-C\equiv C-H \text{ (bridged by } Br^+)+Br^- \longrightarrow \underset{\text{trans-1,2-Dibromopropane}}{(Br)(CH_3)C=C(H)(Br)}$$

This sequence is repeated in the addition of another bromine molecule.

Second step

$$(Br)(CH_3)C=C(H)(Br)+\underset{\text{Polarised}}{\overset{\delta+}{Br}\longrightarrow\overset{\delta+}{Br}} \longrightarrow CH_3-\underset{Br}{C}-\underset{Br}{C}-H \text{ (bridged by } Br^+)+Br^-$$

$$CH_3-\underset{Br}{C}-\underset{Br}{C}-H \text{ (bridged by } Br^+)+Br^- \longrightarrow \underset{\text{1,1,2,2-Tetrabromopropane}}{CH_3-CBr_2-CBr_2-H}$$

Q. 8. Explain the acidic nature of acetylenic protons.

(*Hyderabad 2003; Delhi 2005 Nagpur, 2005*)

Ans. In acetylene and terminal alkynes, the hydrogen atom is attached to a *sp* hybridised carbon which is more electronegative because of increase in *s*-character. Due to greater electronegativity of the *sp* hybridised carbon, the electron pair of C – H bond gets displaced more towards carbon and this helps in the release of proton by strong bases. Consequently acetylene and terminal alkynes behave as acids.

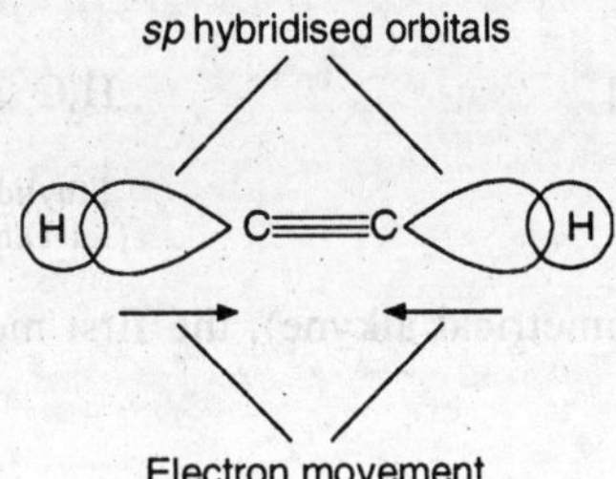

Following reactions illustrate the acidic nature of 1-alkynes:

$$2HC \equiv C-H + 2Na \longrightarrow 2HC \equiv \bar{C}\overset{+}{N}a + H_2$$

Acetylene → Sodium acetylide

$$CH_3-C \equiv C-H + NaNH_2 \longrightarrow CH_3-C \equiv \bar{C}\overset{+}{N}a + NH_3$$

Propyne → Sodium methyl acetylide

Q. 9. Compare the acidic strength of acetylene, ethylene and ethane. (*Lucknow, 2000*)

Ans. Strength of an acid depends upon the ease with which it can lose a proton. This will depend upon the type of bonding between carbon and hydrogen.

$$\underset{\text{Ethane}}{H_3C-CH_3}\ (sp^3-s\ \text{overlapping}) \qquad \underset{\text{Ethylene}}{H_2C=CH_2}\ (sp^2-s\ \text{overlapping}) \qquad \underset{\text{Acetylene}}{H-C\equiv C-H}\ (sp-s\ \text{overlapping})$$

(*i*) In the case of ethane, C – H bond involves overlapping of sp^3 hybrid orbital of carbon and *s*-orbital of H.

(*ii*) In case of ethylene, overlapping is between sp^2 hybrid orbital of carbon and *s*-orbital of H.

(*iii*) In case of acetylene, overlapping in C – H bond is between *sp* hybrid orbital of carbon and *s*-orbital of hydrogen, *s*-orbital has greater electron density compared to a *p*-orbital, which can be explained in terms of their shape. Out of sp^3, sp^2 and *sp* hybrid orbitals, the *sp* orbital has the greatest *s*-character followed by sp^2 and sp^3. Thus carbon atom possessing *sp* hybrid orbital will have maximum electronegativity followed by carbon having sp^2 hybrid orbitals and then followed by carbon having sp^3 hybrid orbital.

In view of this, the hydrogen atom in acetylene will be replaceable with maximum ease followed by that in ethylene and ethane. Hence acidic strength decreases in the order.

acetylene > ethylene > ethane

Q. 10. Describe with mechanism addition of halogen acids on alkynes.

Ans. Treatment of alkynes with halogen acid initially yields vinyl halides and finally *alkylidene halides*. Combination with second molecule takes place in accordance with Markownikoff's rule.

$$H-C\equiv C-H+HX \longrightarrow \underset{\text{Vinyl halide}}{H_2C=CHX}$$

$$\downarrow HX \text{ (2nd molecule)}$$

H = Cl, B or I

$$\underset{\substack{\text{Ethylidene halide} \\ \text{(1,1-Dihalo ethane)}}}{H_3C-CHX_2}$$

In case of propyne (unsymmetrical alkyne), the first molecule of HX adds according to Markownikoff's rule.

For example,

$$\underset{\text{1-Propyne}}{CH_3 \rightarrow\!\!- C\equiv CH + H-X} \longrightarrow \underset{\text{2-Halo propene-1}}{CH_3-\overset{X}{\overset{|}{C}}=CH_2}$$

$$\downarrow HX$$

$$\underset{\text{(2,2-Dihalo propane)}}{CH_3-\overset{CH_3}{\overset{|}{\underset{X}{\underset{|}{C}}}}-CH_2}$$

Mechanism of the reaction is depicted below:

$$H-C\equiv C-H+\overset{\delta+}{H}-\overset{\delta-}{X} \longrightarrow H-\overset{H}{C{=}C}^{+}-H+:\bar{X}$$

$$H-\overset{H^{+}}{C{=}C}-H \;\; :\bar{X} \longrightarrow \underset{\text{Vinyl halide}}{H-\overset{C}{\overset{|}{C}}=\overset{C}{\overset{|}{C}}-H}$$

$$\downarrow H-X$$

$$-\overset{H}{\overset{|}{\underset{H}{\underset{|}{C}}}}-\overset{X}{\overset{|}{\underset{X}{\underset{|}{C}}}}-H \longleftarrow -\overset{H\;H^{+}X}{C{=}C}-H \;\; :X$$

Q. 11. Write a note on hydrogenation of alkynes.

Ans. Catalytic hydrogenation of alkynes yields alkanes through the intermediate formation of an *alkene*.

The reaction can be stopped at the alkene stage with the proper choice of reagent. Predominantly, *trans alkene* is obtained if the reduction of alkyne is carried out in the presence of sodium or lithium in liquid ammonia. Almost entirely *cis alkene* is obtained if the hydrogenation of alkyne is carried out with Lindlar's catalyst (Palladium supported over $BaSO_4$ partially poisoned with quinoline) or Nickel Boride catalyst. Thus, the reduction of alkynes is a stereoselective reaction.

For example,

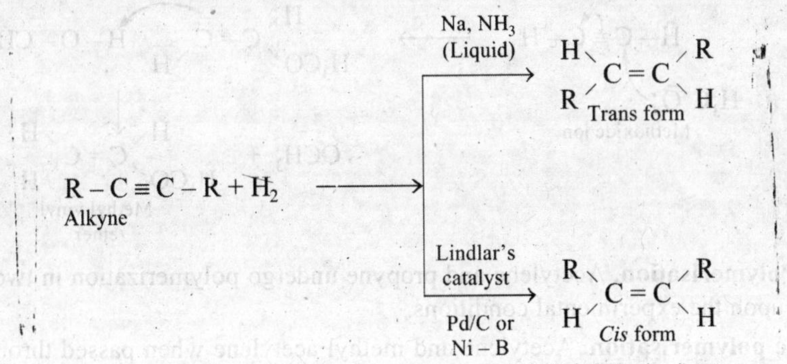

Q. 12. What happens when ethyne reacts with hydrogen in the presence of Lindlar's catalyst. (*Kurukshetra, 2001; M. Dayanand, 2000*)

Ans.

$$\underset{\text{Ethyne}}{CH \equiv CH} + H_2 \xrightarrow[\text{catalyst}]{\text{Lindlar's}} \underset{\text{Ethene}}{CH_2 = CH_2}$$

Q. 13. Describe with mechanism the following reactions of acetylene:

(*i*) Addition of HCN **(*ii*) Addition of methanol**

(*iii*) Polymerisation **(*iv*) Ozonolysis**

Ans. (*i*) **Addition of alkynes.** Acetylene reacts with hydrogen cyanide in the presence of barium cyanide yielding vinyl cyanide (acrylonitrile).

$$H - C \equiv C - H + HCN \xrightarrow{Ba^{++}} \underset{\substack{\text{Vinyl cyanide} \\ \text{(acrylonitrile)}}}{H_2C = CH - CN}$$

Acrylonitrile is used in the manufacture of Buna N.

The probable mechanism of the reaction is depicted below :

$$H - C \equiv C - H \;\; + \;\; :\bar{C} \equiv N \longrightarrow \overset{H}{\underset{NC}{}} C = \bar{\ddot{C}} - H$$

$$\overset{H}{\underset{NC}{}} C = \bar{\ddot{C}} \diagdown R \xrightarrow{H - CN} \underset{\substack{\text{Vinyl cyanide} \\ \text{(Acrylonitraile}}}{\overset{H}{\underset{NC}{}} C = C \overset{H}{\underset{H}{}}} + CN^-$$

(*ii*) **Addition of Methanol.** Acetylene adds methanol in the presence of sodium methoxide yielding vinyl ether.

$$H - C \equiv C - H + CH_3OH \xrightarrow{CH_3ONa} \underset{\text{Methyl vinyl ether}}{H_2C = CH \cdot OCH_3}$$

Methyl vinyl ether is used in the manufacture of polyvinyl ether plastics.

The probable mechanism of the reaction is shown below:

$$H-C\equiv C-H \xrightarrow{H_3C\ddot{O}:^-\ \text{(Methoxide ion)}} \begin{matrix} H \\ H_3CO \end{matrix}\!\!>C=\ddot{C}^-<\!\!\begin{matrix} \\ H \end{matrix} \quad H-O-CH_3$$

$$\downarrow$$

$$^-\!:OCH_3 + \begin{matrix} H \\ H_3CO \end{matrix}\!\!>C=C<\!\!\begin{matrix} H \\ H \end{matrix}$$

Methyl vinyl ether

(*iii*) **Polymerisation.** Acetylene and propyne undergo polymerization in two different ways depending upon the experimental conditions.

Cyclic polymerisation. Acetylene and methyl acetylene when passed through red hot iron tube, polymerize to form benzene and mesitylene respectively.

$$3\,CH\equiv CH \xrightarrow{500°C} C_6H_6$$

Benzene

$$3\,CH_3-C\equiv CH \xrightarrow{500°C} C_6H_3(CH_3)_3$$

Mesitylene

Acetylene when passed through under high pressure and in the presence of catalyst $Ni(CN)_2$, tetramerises to form cyclo octatetraene.

$$4\,HC\equiv CH \xrightarrow[\text{Under Presure, T.H.F.}]{Ni(CN)_2} C_8H_8$$

Cyclooctatetraene

Linear polymerization. Acetylene when passed through a solution of cuprous chloride and ammonium chloride at 343 K, dimerises to form *vinyl acetylene* and finally trimerises to form divinyl acetylene.

$$HC\equiv CH + HC\equiv CH \xrightarrow{Cu_2Cl_2,\ NH_4Cl} \underset{\text{Vinyl acetylene}}{H_2C=CH-C\equiv CH}$$

$$HC\equiv CH + HC\equiv CH + HC\equiv CH \xrightarrow{Cu_2Cl_2,\ NH_4Cl} \underset{\text{Divinyl acetylene}}{H_2C=CH-C\equiv C-CH=CH_2}$$

(*iv*) **Ozonolysis.** Alkynes add on ozone to form ozonides. The ozonides are hydrolysed by water to form dicarbonyl compounds (1, 2-diketones) which undergo oxidative cleavage by H_2O_2 to form *acids*. The identification of the acids formed helps to locate the position to triple bond in the original alkyne. Thus, ozonolysis followed by oxidative cleavage can be used as an unambiguous method for locating the position of triple bond in the original alkyne.

$$\underset{\text{Alkyne}}{-C\equiv C-} + O_3 \longrightarrow \text{Ozonide} \xrightarrow{H_2O} \underset{\text{Diketone}}{-\overset{O}{\overset{\|}{C}}-\overset{O}{\overset{\|}{C}}-} + H_2O_2 \xrightarrow{\text{Oxidative cleavage}} -\overset{O}{\overset{\|}{C}}-O-H + -\overset{O}{\overset{\|}{C}}-O-H$$

Q. 14. Explain why alkynes undergo nucleophilic addition reactions but alkenes do not.

(*M. Dayanand, 2000*)

Ans. Alkynes undergo nucleophilic addition reactions because of the greater stability of the resulting carbanion formed by the attack of nucleophile. Thus, consider the possible formation of carbanions formed by the attack of nucleophile, $\ddot{N}\bar{u}$.

$$\underset{\text{Alkyne}}{-C\equiv C} + \ddot{N}\bar{u} \longrightarrow \underset{\text{Vinylic carbanion}}{-\underset{Nu}{\underset{|}{C}}=\ddot{\bar{C}}-}$$

$$\underset{\text{Alkene}}{-\underset{|}{C}=\underset{|}{C}-} + \ddot{N}\bar{u} \longrightarrow \underset{\text{Alkyl carbanion}}{-\underset{Nu}{\underset{|}{\overset{|}{C}}}-\underset{|}{\ddot{\bar{C}}}-}$$

In vinylic carbanion, the negative charge can be dissipated more easily by the electronegative hybridised carbon. In other words, vinylic carbanion is more stable than the alkyl carbanion. Since the rate determining step involves the formation of carbanion, it is clear that alkynes are more susceptible to nucleophilic attack than alkenes.

Q. 15. Explain the mechanism of nucleophilic addition to alkynes.

Ans. It is observed that such reactions are generally catalysed by the presence of heavy metal salts such as those of mercury and barium. This suggests that metal ions form some sort of complex with alkyne by co-ordinating with the electrons of the pie (π) bonds as shown below:

$$-\underset{|}{C}\equiv\underset{|}{C}- + Hg^{2+} \longrightarrow \underset{\downarrow Hg^{2+}}{-C\equiv C-}$$

This complex formation decreases the electron density around the triply bonded carbon atoms and thus helps in the attack by the nucleophilic reagent.

Like electrophilic addition reactions, nucleophilic addition also takes place in two steps.

In the first step, the nucleophile attacks the alkyne to form a vinyl carbanion. In the second step, the carbonion takes up a proton to form the final product. The complete mechanism may be depicted as follows:

$$\bar{N}u: + R - C \equiv C - R' \xrightarrow{\text{Slow}} \underset{Nu}{\overset{R}{>}}C = \ddot{C}\langle^{R}$$

Nucleophile Alkyne

$$\xrightarrow[\text{Fast}]{H - Nu} \underset{Nu}{\overset{R}{>}}C = C\langle^{R'}_{H} + \bar{N}u:$$

Q. 16. What happens when acetylene is passed through dil. sulphuric acid containing mercuric sulphate at 343 K? Also discuss the mechanism of this reaction.

Or

Explain the mechanism of the following reaction:

$$HC \equiv CH \xrightarrow[HgSO_4]{\text{dil. } H_2SO_4}$$

(*Kanpur, 2004; Guwahati 2006*)

Ans. When treated with dil. in the presence of mercuric sulphate, acetylene gets hydrated to form first an enol which readily tautomerises to form acetaldehyde.

$$CH \equiv CH + HOH \xrightarrow{HgSO_4/H_2SO_4} \underset{\text{Vinyl alcohol}}{[CH_2 = CH - OH]} \xrightarrow{\text{Tautomerises}} \underset{\text{Acetaldehyde}}{CH_3 - \overset{O}{\overset{||}{C}} - H}$$

Mechanism. The mechanism involves the nucleophilic attack of water molecule on acetylene to form a carbanion which changes into vinyl alcohol. Since vinyl alcohol is unstable, it tautomerises to form stable aldehyde.

$$H - C \equiv C - H + H - \underset{H}{\ddot{O}}: \xrightarrow{Hg^{2+}} H - \ddot{C} = \underset{H - O^+}{C} - H \longrightarrow \underset{\text{(Carbanion)}}{}$$

$$H - \bar{C} = C(OH) - H \;\text{(Carbanion)} \longrightarrow \underset{\text{Vinyl alcohol}}{H - \overset{H}{C} = \underset{OH}{C} - H} \xrightarrow{\text{Tautomerises}} \underset{\text{Acetaldehyde}}{H - \underset{H}{\overset{H}{C}} - \overset{O}{\overset{||}{C}} - H}$$

Q. 17. How will you synthesise the following compounds from acetylene?

(I) Propyne (II) Dichloroacetaldehyde (III) Benzene (IV) Acrylonitrile (V) Ethylidene chloride. (*Guwahati, 2005*)

Ans. Synthesis of the given compounds from acetylene:

(I) Propyne

$$CH \equiv CH + NaNH_2 \longrightarrow \underset{\text{Sodium acetylide}}{CH \equiv \overset{-}{C}\overset{+}{Na}} + NH_3$$

$$CH_3-Cl+\overset{+}{Na}\overset{-}{C}\equiv CH \longrightarrow \underset{\text{Propyne}}{CH_3-C\equiv CH}+NaCl$$

(II) Dichloroacetaldehyde

$$Cl_2+H_2O \longrightarrow HCl+HOCl$$

$$CH\equiv CH+\overset{\delta-\ \delta+}{HOCl} \longrightarrow \underset{\text{Acetylene chlorohydrin}}{\underset{OH\quad Cl}{CH=CH}}$$

$$\underset{\text{Acetylene chlorohydrin}}{CH(OH)=CH(Cl)} \xrightarrow{\overset{\delta-\ \ \delta+}{HO-Cl}} \underset{\text{Unstable}}{\left[\underset{OH}{HO-CH-CHCl_2}\right]} \xrightarrow{-H_2O} \underset{\text{Dichloroacetaldehyde}}{\underset{H}{O=C-CHCl_2}}$$

(III) Benzene

$$3\,CH\equiv CH \xrightarrow[773\text{ K}]{\text{Red hot iron tube}} \underset{\text{Benzene}}{C_6H_6}$$

(IV) Acrylonitrile

$$CH\equiv CH+HCN \xrightarrow{Ba(CN)_2} \underset{\text{Acrylonitrile (Vinyl cyanide)}}{CH_2=CH-CN}$$

(V) Ethylidene chloride

$$CH\equiv CH \xrightarrow{HCl} \underset{\text{Vinyl chloride (Chloroethene)}}{CH_2=CHCl} \xrightarrow{HCl} \underset{\text{Ethylidene chloride (1,1-Dichloroethane)}}{CH_3-CHCl_2}$$

Markownikoff rule is followed in the second step.

Q. 18. Explain Hydroboration-oxidation of alkynes.

Ans. When treated with diborane, *terminal alkynes* (*i.e.*, alkynes which contain a $-C\equiv C-H$ group) undergo addition in two stages, forming firstly a *vinylborane* and finally a *gem-dibora derivative. It is not possible to* **stop** *the reaction at the stage of vinyl borane.*

$$\underset{\text{Terminal alkyne}}{2R-C\equiv C-H}+(BH_3)_2 \longrightarrow \underset{\text{Vinyl borane}}{\left[\underset{BH_2}{2RCH=CH}\right]} \xrightarrow{(BH_3)_2} \underset{\text{gem-diborane derivative}}{\underset{BH_2}{2RCH_2-CH-BH_2}}$$

In case of *internal alkynes* (*i.e.*, alkynes in which triply bonded carbons do not carry any hydrogen), the reaction can be stopped at the vinylic stage by using suitable amounts of the reagent.

If in place of diborane, a bulky sterically hindered borane such as bis (1, 2-dimethylpropyl) borane $\left[\begin{matrix} & CH_3 & CH_3 \\ & | & | \\ CH_3 - & CH - & CH - \end{matrix}\right]_2$ BH commonly known as *disiamylborane* (abbreviated as Sia_2 BH) is used, reaction can be stopped at the vinyl borane stage for all alkynes whether terminal or internal. That is:

$$\underset{\text{Terminal alkyne}}{R - C \equiv C - H} + Sia_2\,BH \longrightarrow R - CH = CH - BSia_2$$

$$\underset{\text{Internal alkynes}}{R - C \equiv C - R} + Sia_2\,BH \longrightarrow R - CH = \underset{\underset{BSia_2}{|}}{C} - R$$

The vinylic boranes can then be oxidised with H_2O_2 in basic solution to form aldehydes or ketones via enolic intermediates.

$$- CH = \overset{|}{C} - B\,SiA_2 \xrightarrow{H_2O_2,\ OH^-} \underset{\text{Enol}}{\left[- CH = \overset{|}{C} - OH\right]} \longrightarrow \underset{\text{Aldehyde or ketone}}{- CH_2 - \overset{|}{C} = O}$$

For example, hydroboration of 1-hexyne with Sia_2 BH followed by oxidation with H_2O_2/OH^- to yield hexanal is shown below.

$$\underset{\text{1-Hexyne}}{CH_3(CH_2)_3C \equiv CH} \xrightarrow{Sia_2BH} \underset{\text{Vinylic borane}}{CH_3(CH_2)_3CH = CHBSia_2} \xrightarrow{H_2O_2/OH^-}$$

$$\underset{\text{Enol}}{[CH_3(CH_2)_3CH = CH - OH]} \rightleftharpoons \underset{\text{Hexanal}}{CH_3(CH_2)_3CH_2 - \overset{\overset{O}{||}}{C} - H}$$

Similarly hydroboration-oxidation of 4-octyne gives rise to 4-octanone.

$$\underset{\text{4-Octyne}}{C_3H_7 - C \equiv C - C_3H_7} \xrightarrow{Sia_2BH} \underset{\text{Vinylic borane}}{C_3H_7 - CH = \underset{\underset{BSia_2}{|}}{C} - C_3H_7} \xrightarrow{H_2O_2/OH^-}$$

$$\underset{\text{Enol}}{\left[C_3H_7 - CH = \underset{\underset{OH}{|}}{C} - C_3H_7\right]} \rightleftharpoons \underset{\text{4-Octanone}}{C_3H_7 - CH_2 = \underset{\underset{O}{||}}{C} - C_3H_7}$$

It may be noted that hydroboration oxidation leads to **anti-Markownikoff** *addition of H – and – OH. It may also be observed that hydroboration oxidation of* **terminal alkynes** *furnishes aldehydes while that of internal alkynes gives rise to ketones.*

Addition of boranes to carbon-carbon triple bond can also produce cis-alkenes

$$R - C \equiv C - R \xrightarrow{B_2H_6} \begin{matrix} R - C - H \\ || \\ R' - C)_3 - B \end{matrix} \xrightarrow{CH_3COOH} \underset{\textit{cis}\text{-alkene}}{\begin{matrix} R \diagdown & & \diagup R' \\ & C = C & \\ H \diagup & & \diagdown H \end{matrix}}$$

Q. 19. What happens when:

(I) 2-Butyne is treated with Pd/C (Lindlar calalyst and hydrogen)

(II) Propyne is treated with chlorine water

(III) Acetylene is treated with Br_2 dissolved in CCl_4 *(Delhi. 2002*

(IV) Acetylene is treated with ammoniacal cuprous chloride solution *Nagpur 2008)*

(V) Acetylene is treated with ammoniacal $AgNO_3$ *(Burdwan 2003 Nagpur 2008)*

(VI) 2-Butyne is treated with sodium dissolved in liquid ammonia *(M. Dayanand, 2000)*

(VII) Acetylene is subjected to ozonolysis.

Ans. (I) $$\underset{\text{2-Butyne}}{H_3C-C\equiv C-CH_3} + H_2 \xrightarrow{Pd/C} \underset{\text{cis-2-butene}}{\overset{CH_3 \qquad CH_3}{\underset{H \qquad\quad H}{C=C}}}$$

(II) $$Cl_2 + H_2O \longrightarrow HCl + HOCl$$

$$CH_3-C\equiv CH + \overset{\delta-}{HO}-\overset{\delta+}{Cl} \longrightarrow CH_3-\underset{OH}{C}=\underset{Cl}{CH}$$

$$\xrightarrow{\overset{\delta-}{HO}-\overset{\delta+}{Cl}} \underset{\text{Unstable}}{\left[CH_3-\underset{OH}{\overset{OH}{C}}-CHCl_2\right]} \xrightarrow{-H_2O} \underset{\text{Dichloroacetone}}{CH_3-\overset{O}{\overset{\|}{C}}-CHCl_2}$$

(III) $$CH\equiv CH \xrightarrow{Br_2/CCl_4} \underset{\substack{\text{1,2-Dibromoethene}\\\text{(Acetylene dibromide)}}}{\underset{Br\quad Br}{CH=CH}} \xrightarrow{Br_2/CCl_4} \underset{\substack{\text{1,1,2,2-Tetrabromoethane}\\\text{(Acetylene tetrabromide)}}}{\overset{Br\quad Br}{\underset{Br\quad Br}{CH-CH}}}$$

(IV) $$CH\equiv CH + Cu_2Cl_2 + 2NH_4OH \longrightarrow \underset{\substack{\text{Copper acetylide}\\\text{(Red ppt.)}}}{CuC\equiv CCu} + 2NH_4Cl + 2H_2O$$

(V) $$CH\equiv CH + 2NH_4OH + 2AgNO_3 \longrightarrow \underset{\text{Silver acetylide}}{AgC\equiv CAg} + 2NH_4NO_3 + 2H_2O$$

(VI) $$CH_3-C\equiv C-CH_3 + H_2 \xrightarrow{Na/liq.\,NH_3} \underset{\textit{trans}\text{-2-butene}}{\overset{CH_3 \qquad H}{\underset{H \qquad CH_3}{C=C}}}$$

(VII) $$CH\equiv CH + O_3 \longrightarrow \underset{\substack{\text{Acetylene}\\\text{ozonide}}}{\overset{O}{\underset{O-O}{CH-CH}}} \xrightarrow[-H_2O_2]{H_2O/Zn} \underset{\text{Glyoxal}}{\begin{matrix}CHO\\|\\CHO\end{matrix}}$$

Q. 20. How will you bring about the following conversions?

(I) Acetylene into oxalic acid

(II) Propyne into 2, 2-dibromopropane

(III) 2-Butyne into acetic acid
(IV) Propyne into propanone (*Kanpur, 2001*)
(V) Acetylene into chloroprene
(VI) Acetylene into vinyl acetate.

Ans. (I) Acetylene into oxalic acid

$$\underset{\text{Acetylene}}{\begin{matrix} CH \\ ||| \\ CH \end{matrix}} + 4[O] \xrightarrow{\text{Alk. } KMnO_4} \underset{\text{Oxalic acid}}{\begin{matrix} COOH \\ | \\ COOH \end{matrix}}$$

(II) Propyne into 2, 2-dibromopropane

$$\underset{\text{Propyne}}{CH_3-C\equiv CH} \xrightarrow{HBr} \underset{\text{2-Bromopropene}}{CH_3-\overset{\overset{Br}{|}}{C}=CH_2} \xrightarrow{HBr} \underset{\text{2,2-dibromopropane}}{CH_3-\overset{\overset{Br}{|}}{\underset{\underset{Br}{|}}{C}}-CH_3}$$

(III) 2-Butyne into acetic acid

$$\underset{\text{2-Butyne}}{CH_3-C\equiv C-CH_3} + H_2O + 3[O] \xrightarrow{\text{Alk. } KMnO_4} \underset{\text{Acetic acid}}{CH_3COOH} + \underset{\text{Acetic acid}}{HOOCCH_3}$$

(IV) Propyne into propanone

$$CH_3-C\equiv CH + HOH \xrightarrow[333\text{ K}]{HgSO_4/H_2SO_4} \underset{\text{Unstable}}{\left[CH_3-\overset{\overset{OH}{|}}{C}=CH_2\right]} \rightleftharpoons \underset{\text{Propanone}}{CH_3-\overset{\overset{O}{||}}{C}-CH_3}$$

(V) Acetylene into chloroprene

$$CH\equiv CH + CH\equiv CH \xrightarrow[NH_4Cl]{CuCl} \underset{\text{Vinyl acetylene}}{CH_2=CH-C\equiv CH}$$

$$CH_2=CH-C\equiv CH + HCl \longrightarrow \underset{\text{Chloroprene}}{CH_2=CH-\underset{\underset{Cl}{|}}{C}=CH_2}$$

(VI) $$CH\equiv CH + CH_3COOH \xrightarrow{Hg^{2+}} \underset{\text{Vinyl acetate}}{CH_2=CHOCOCH_3}$$

Q. 21. Complete the following reactions

(*i*) $HC\equiv CH + H_2O \xrightarrow{H_2SO_4}$

(*ii*) $HC\equiv CH + HBr \xrightarrow{HgSO_4}$

Ans. (*i*)

$$HC \equiv CH + H_2O \xrightarrow{HgSO_4} CH_2 = CHOH$$

Acetylene — Vinyl alcohol

$$\downarrow \text{Rearranges}$$

$$CH_3CHO$$

Acetaldehyde

(*ii*)

$$HC \equiv CH + HBr \xrightarrow{HgSO_4} CH_2 = CHBr$$

$$\downarrow HBr$$

$$CH_3CHBr_2$$

Ethylidene bromide

Q. 22. Justify with example the following statement: Acetylene undergoes electrophilic as well as nucleophilic addition reactions. (*Kurukshetra, 1995; G. Nanakdev 2000*)

Ans. Acetylene undergoes nucleophilic reaction like addition of water in the presence of dil. H_2SO_4 and $HgSO_4$ as follows:

$$CH \equiv CH + H_2O \xrightarrow[HgSO_4]{HgSO_4} CH_3CHO$$

Nucleophilic reaction with acetylene takes place because Hg^{2+} forms a complex with the π electrons of carbon-carbon triple bond thereby reducing election density. This consequently encourages attack by nucleophiles. Moreover the intermediate allylic carbanion in the nucleophilic attack gets stabilized.

Acetylene undergoes electrophilic reaction as given by alkenes because of π electrons of the carbon-carbon triple bond.

$$CH \equiv CH + Br_2 \longrightarrow CHBr = CHBr$$

$$\downarrow Br_2$$

$$CHBr_2 - CHBr_2$$

Q. 23. Ethyne forms metallic salt but 'ethene' does not. Why? (*Awadh, 2000*)

Ans. See Q. 8.

Q. 24. How will you bring about the following conversion?

$$\mathbf{CH_3C \equiv CH \longrightarrow CH_3C \equiv CCH_3}$$ (*Kumaon, 2000*)

Ans.

$$CH_3C \equiv CH + Na \longrightarrow CH_3C \equiv CNa$$

$$\downarrow CH_3Br$$

$$CH_3C \equiv C.CH_3$$

Q. 25. How will you differentiate propyne-1 from butyne-2? (*Kumaon, 2000*)

Ans. The compounds will be treated with ammoniacal cuprous chloride and ammoniacal silver nitrate separately. Propyne-1 gives a red precipitate with ammoniacal cuprous chloride and gives a white precipitate with ammoniacal silver nitrate. While butyne-2 does not give any precipitate with these reagents.

$$2CH_3CH_2 \equiv CH + Cu_2Cl_2 + 2NH_4OH \longrightarrow \underset{\text{(red ppt.)}}{2CH_3CH_2\ C.Cu} + 2NH_4Cl + 2H_2O$$

$$CH_3CH_2 \equiv CH + 2NH_4OH + 2AgNO_3 \longrightarrow \underset{\text{(white ppt.)}}{2CH_3CH_2 \equiv C.Ag} + 2NH_4NO_3 + 2H_2O$$

$$CH_3.C \equiv C.CH_3 + Cu_2Cl_2 + 2NH_4OH \longrightarrow \text{No reaction}$$

$$CH_3.C \equiv C.CH_3 + 2NH_4OH + 2AgNO_3 \longrightarrow \text{No reaction}$$

Q. 26. How will you synthesise acetone from propene? (*Aligarh, 2004*)

Ans. Following steps are involved

$$\underset{\text{Propene}}{CH_3CH = CH_2} \xrightarrow{Br_2} CH_3\underset{}{\overset{Br}{\overset{|}{C}H}} - \overset{Br}{\overset{|}{C}H_2} \xrightarrow{NaNH_2} \underset{\text{Propyne}}{CH_3C \equiv CH} \xrightarrow[H_2SO_4/HgSO_4]{H_2O} \underset{\text{Acetone}}{CH_3 - \overset{O}{\overset{||}{C}} - CH_3}$$

Q. 27. How will you differentiate between but-1-yne and but-2-yne acetone from propene? (*Guwahati, 2005*)

Ans. The compounds will be treated with ammoniacal cuprous chloride and ammoniacal silver nitrate. Butyne-1-gives red ppt and silver mirror respectively. Butyne-2 reacts with neither. For reactions, see Q.25.

10

DIENES

Q. 1. Write notes on

(*a*) Isolated dienes **(*b*) Comulated dienes**

(c) Conjugated dienes (*Kerala, 2001; Sri Venkateswara 2005; Nagpur 2008*)

Ans. (*a*) Isolated Dienes: The dienes in which the double bonds are separated by more than one single bond are isolated dienes.

For example:

$$\underset{5}{H_2C} = \underset{4}{CH} - \underset{3}{CH_2} - \underset{2}{CH} = \underset{1}{CH_2} \qquad \text{1,4-pentadiene}$$

$$\underset{6}{H_2C} = \underset{5}{CH} - \underset{4}{CH_2} - \underset{3}{CH_2} - \underset{2}{CH} = \underset{1}{CH_2} \qquad \text{1,5-hexadiene}$$

(*b*) Comulated dienes. These dienes contain the grouping >C = C = C< and are characterized by the presence of at least one carbon atom joined to both the neighbouring C-atoms by double bonds.

For example,

$$H_2C = C = CH_2 \qquad \text{allene} \quad \text{or} \quad \text{1, 2-propadiene}$$

$$CH_3 - CH = C = CH_2 \qquad \text{1, 2-butadiene}$$

(*c*) Conjugated dienes. In these dienes, a single bond intervenes two double bonds or in other words, they have an arrangement of alternate single and double bonds.

For example,

$$CH_2 = CH - CH = CH_2 \qquad \text{1, 3-butadiene}$$

'β' carotene which is found in carrots and green leaves, contains a system of eleven conjugate double bonds.

Q. 2. Describe methods of preparation of 1, 3-butadiene.

Methods of preparation of 1, 3-butadiene are given as under.

(*Kerala, 2001; Sri Venkateswara 2005; Nagpur 2008*)

Ans. (*i*) Dehydrogenation of alkenes. 1, 3-butadiene is prepared on an industrial scale by passing the vapours of *n*-butane over heated Cr_2O_3 and Al_2O_3 catalysts.

$$\underset{n\text{-butane}}{CH_3 - CH_2 - CH_2 - CH_3} \xrightarrow[Cr_2O_3,\ Al_2O_3]{\text{Heat}} \underset{\text{1,3-butadiene}}{CH_2 = CH - CH = CH_2} + 2H_2O$$

(*ii*) Dehydration of diols. 1, 3-butadiene can be prepared by the dehydration of butane 1, 4-diol (or butane-1, 3-diol) in the presence of H_2SO_4 or phosphoric acid.

$$\underset{\text{OH}}{\underset{|}{CH_2}} - CH_2 - CH_2 - \underset{\text{OH}}{\underset{|}{CH_2}} \xrightarrow[\Delta]{\text{Conc. } H_2SO_4} \underset{\text{1,3-butadiene}}{CH_2 = CH - CH = CH_2} + 2H_2O$$

(*iii*) **Pyrolysis of Cycloalkenes.** When the vapours of cyclohoxene are passed over heated Nichrome (Ni – Cr – Fe) alloy, 1, 3-butadiene results.

$$\text{Cyclohexene} \xrightarrow[\text{Nichrome}]{\Delta} \underset{\text{1, 3-Butadiene}}{CH_2=CH-CH=CH_2} + \underset{\text{Ethylene}}{CH_2=CH_2}$$

Q. 3. Give the mechanism of addition of Br_2 to 1, 3-Butadiene.

(*Panjab, 2002; Baroda 2004; Gulberga 2004*)

Or

Explain regioselectivity in the bromination of 1, 3-butadiene

(*Kurukshetra, 2001; Mumbai, 2010*)

Ans. When one mole of the addendum such as Br_2 is added to a conjugated diene, the resulting product obtained is a mixture of 1, 2 and 1, 4-addition products. Such an addition is known as 1, 2- and 1, 4-addition to conjugated dienes. Consider the addition of 1 mole of Br_2 to 1, 3-butadiene.

$$\underset{\text{1,3-Butadiene}}{CH_2=CH-CH=CH_2} \xrightarrow{+Br_2} \underset{\substack{\text{3,4-Dibromo-1-butene}\\ \text{(1,2-addition product)}}}{CH_2=CH-\underset{Br}{\underset{|}{CH}}-\underset{Br}{\underset{|}{CH_2}}} + \underset{\substack{\text{1,4-Dibromo-2-Butene}\\ \text{(1,4-addition product)}}}{\underset{Br}{\underset{|}{CH_2}}-CH=CH-\underset{Br}{\underset{|}{CH_2}}}$$

In order to understand the occurrence of 1, 2- and 1, 4-addition reactions, the mechanism of the above reaction is discussed.

Mechanism. It involves two steps:

(*i*) Formation of carbocation.

$$CH_2=CH-CH=CH_2 + \underset{\text{(Polarised)}}{\overset{\delta^+}{Br}-\overset{\delta^-}{Br}} \longrightarrow \underset{\text{(Carbocation)}}{CH_2=CH-\overset{+}{CH}=\underset{Br}{\underset{|}{CH_2}}} + Br^-$$

The above intermediate carbocation is allylic in nature and hence can be considered to be a resonance hybrid of two equivalent resonating structures I and II.

$$\left[\underset{\text{I}}{CH_2=CH-\overset{\oplus}{CH}-\underset{Br}{\underset{|}{CH_2}}} \longleftrightarrow \underset{\text{II}}{\overset{\oplus}{CH_2}-CH=CH-\underset{Br}{\underset{|}{CH_2}}}\right] \equiv \underset{\text{Resonance stabilised carbocation}}{\overset{+}{CH_2\cdots CH\cdots CH}-\underset{Br}{\underset{|}{CH_2}}}$$

(*ii*) **Combination of bromide with carbocation.**

Bromide ion combines with carbocation I and II to form 1, 2 and 1, 4 addition products respectively.

$$\underset{\text{I}}{CH_2 = CH - \overset{\oplus}{C}H - CH_2 + Br^-} \longrightarrow \underset{(1,2\text{-addition})}{CH_2 = CH - \underset{Br}{\underset{|}{C}H} - \underset{Br}{\underset{|}{C}H_2}}$$

$$\underset{\text{II}}{\overset{\oplus}{C}H_2 - CH = CH - CH_2 + Br^-} \longrightarrow \underset{(1,4\text{-addition})}{\underset{Br}{\underset{|}{C}H_2} - CH = CH - \underset{Br}{\underset{|}{C}H_2}}$$

Q. 4. Discuss the mechanism of the addition of HBr to 1, 3-butadiene.

(*G. Nanakdev 2000; Tamilnadu 2004*)

Ans. Addition of HBr to 1, 3-butadiene gives a mixture of 1, 2 and 1, 4 addition products as given below:

$$\underset{\text{1,3-Butadiene}}{CH_2 = CH - CH = CH_2} \xrightarrow{HBr} \underset{\substack{\text{3-Bromo-1-butene} \\ (1,2\text{-addition})}}{CH_2 = CH - \underset{Br}{\underset{|}{C}H} - \underset{Br}{\underset{|}{C}H_3}}$$

$$\xrightarrow{HBr} \underset{\substack{\text{1-Bromo-2-butene} \\ (1,4\text{-addition})}}{\underset{Br}{\underset{|}{C}H_2} - CH = CH - CH_3}$$

This can be explained in terms of the mechanism which assumes the formation of carbocations I and II.

Mechanism.

(*i*) $$\underset{\text{1,3-Butadiene}}{CH_2 = CH - CH = CH_2} \; \overset{\delta^+}{H} - \overset{\delta^-}{Br} \longrightarrow CH_2 = CH - \overset{+}{C}H - CH_3 + Br^-$$

The above carbocation is a resonance hybrid of the following two equivalent structure (I and II).

$$\underset{\text{I}}{CH_2 = CH - \overset{\oplus}{C}H - CH_3} \longleftrightarrow \underset{\text{II}}{\overset{\oplus}{C}H_2 - CH = CH - CH_3} \text{ or } \underset{\substack{\text{Resonance stabilised} \\ \text{carbocation}}}{\overset{\delta^+}{C}H_2 \cdots CH \cdots \overset{\delta^-}{C}H - CH_3}$$

(*ii*) Carbocations I and II react with bromide ion to give 1, 2 and 1, 4 products respectively.

$$\underset{\text{I}}{CH_2 = CH - \overset{+}{C}H - CH_3 + Br^-} \longrightarrow \underset{(1,2\text{-addition})}{CH_2 = CH - \underset{Br}{\underset{|}{C}H} - CH_3}$$

$$\underset{\text{II}}{\overset{+}{C}H_2 - CH = CH - CH_3 + Br^-} \longrightarrow \underset{(1,4\text{-addition})}{\underset{Br}{\underset{|}{C}H_2} - CH = CH - CH_3}$$

Q. 5. Discuss the effect of temperature on 1, 2- and 1, 4-addition to conjugated dienes.

(*Panjab, 2002*)

Ans. It has been observed that temperature has a marked effect on the proportions in which 1, 2- and 1, 4-products are obtained as a result of electrophilic additions to conjugated dienes. For example:

$$CH_2 = CH - CH = CH_2$$

1,3-Butadiene

+ HBr

193 K →

$CH_3 - CH(Br) - CH = CH_2$ 1,2-Adduct (80%)

$CH_3 - \overset{+}{C}H = CH - CH_2 - Br$ 1,4-Adduct (20%)

313 K →

$CH_3 - CH(Br) - CH = CH_2$ 1,2-Adduct (20%)

$CH_3 - \overset{+}{C}H = CH - CH_2 - Br$ 1,4-Adduct (80%)

(The 193 K mixture converts to the 313 K mixture on heating at 313 K.)

The observations can be summed up as follows:

(*a*) At 193 K, a mixture containing 80% of the 1, 2-product and 20% of the 1, 4-product is obtained.

(*b*) Reaction at 313 K yields a mixture containing 80% of the 1, 4-product and 20% of the 1, 2-product.

(*c*) At intermediate temperatures, mixtures of intermediate compositions are obtained.

(*d*) Although each isomer is quite stable at low temperatures, prolonged heating of either 1, 2-product or 1, 4-product yields the same mixture.

Interpretation of the above Observations

The fact that 1, 2-product dominates at 193 K, indicates that 1, 2-product is formed faster than 1, 4-product. Thus E_{act} leading to 1, 2-product is less than that leading to 1, 4-product.

At 313 K, 1, 4-product dominates over 1, 2-product. This shows that 1, 4-product is more stable.

Both the isomers undergo ionisation forming a common carbocation (I). Thus, an equilibrium between the isomers and the carbocation (I) is established as indicated below:

$$\underset{(1,2\text{-product})}{HC_3 - CH(Br) - CH = CH_2} \rightleftharpoons \underset{(I)}{CH_3 - \overset{\delta^+}{C}H - CH - \overset{\delta^+}{C}H_2} + Br^- \rightleftharpoons \underset{(1,4\text{-product})}{CH_3 - CH = CH - CH_2Br}$$

Potential energy changes during the progress of the reaction have been shown in Fig. 10.1.

It is clear that E_{act} for the ionisation of 1, 2-product to carbocation (I) is less than that for 1, 4-product. As a result, 1, 2-product will ionise much more readily than 1, 4-product. Equilibrium is reached when the rates of these two opposing reactions become equal.

When temperature is low, ionisation of the product is quite low and thus 1, 2-product which is formed rapidly is the main product. At high temperature, less stable 1, 2-product readily ionises while the more stable, 1, 4-product ionises very slowly. Thus, at high temperature, 1, 4-product tends to stay and hence constitutes the major product of the reaction.

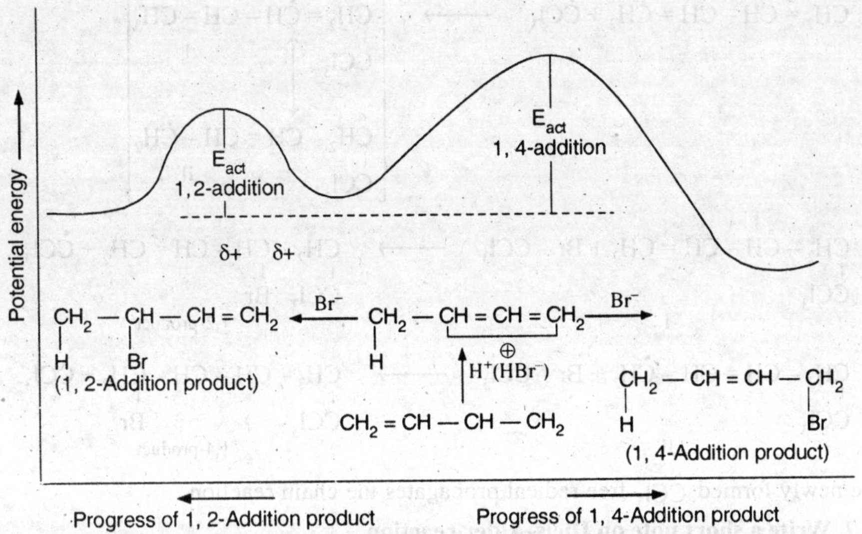

Fig. 10.1. Potential Energy Changes during the progress of 1, 2 and 1, 4-addition reactions

Q. 6. What happens when 1, 3-butadiene is treated with $BrCCl_3$ in the presence of some organic peroxide? Give the suitable mechanism.

Or

Explain the mechanism of free-radical addition in butadiene.

Ans. When $BrCCl_3$ is added to 1, 3-butadiene in the presence of some organic peroxide, a mixture containing 1, 2- and 1, 4-addition products is obtained.

$$\underset{\text{1,3-Butadiene}}{CH_2=CH-CH=CH_2} + \underset{\text{Bromotri-chloromethane}}{BrCCl_3} \xrightarrow{\text{Peroxide}} \begin{cases} \underset{\substack{\text{(1,2-addition)} \\ \text{3-Bromo-4-trichloromethyl-1-butene}}}{CH_2=CH-\underset{Br}{CH}-\underset{CCl_3}{CH_3}} \\ \underset{\substack{\text{(1,4-addition)} \\ \text{1-Bromo-4-trichloromethyl-2-butene}}}{\underset{CCl_3}{CH_2}-CH=CH-\underset{Br}{CH_2}} \end{cases}$$

Mechanism. The above reaction is assumed to occur by a free radical mechanism as illustrated below:

$$\text{(I)}\quad \underset{\text{Benzoyl peroxide}}{C_6H_5-\overset{O}{\overset{\|}{C}}-O-O-\overset{O}{\overset{\|}{C}}-C_6H_5} \xrightarrow{\Delta} 2C_6H_5-\overset{O}{\overset{\|}{C}}-\dot{O} \longrightarrow \underset{\text{Phenyl free radical}}{2C_6\dot{H}_5} + 2CO_2$$

$$\text{(II)}\quad C_6\dot{H}_5 + Br-CCl_3 \longrightarrow C_6H_5Br + \underset{\text{Trichloromethyl free radical}}{\dot{C}Cl_3}$$

(III) 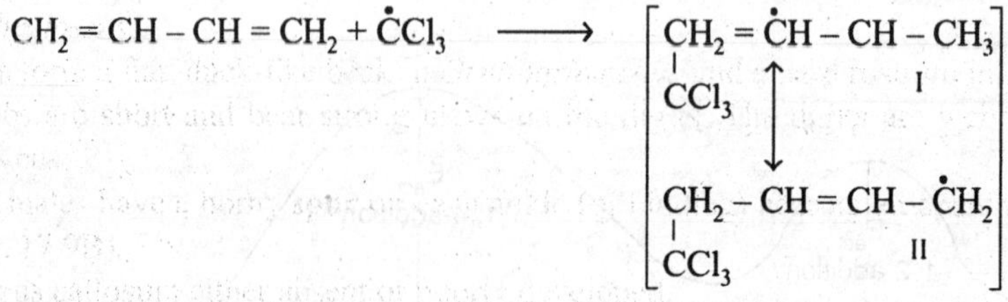

(IV) 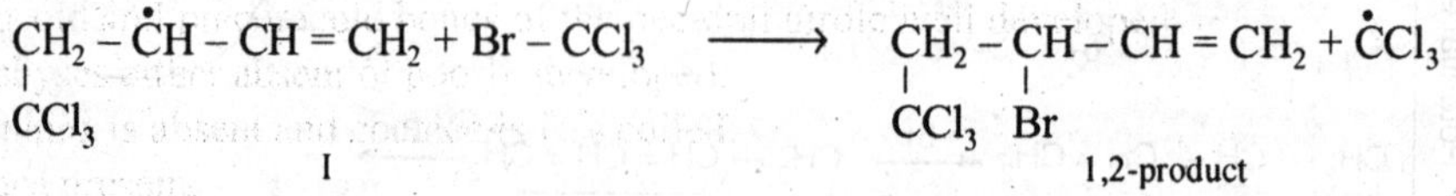

(V) 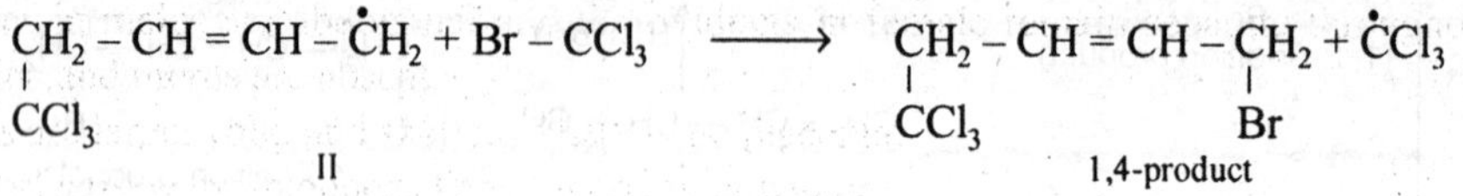

The newly formed $\dot{C}Cl_3$ free radical propagates the chain reaction.

Q. 7. Write a short note on Diels-Alder reaction.

(*Kerala, 2001; Devi Ahilya 2001; Panjab, 2003; Kalyani, 2004; Nagpur, 2005; Guwahati, 2007; Purvanchal, 2007*)

Ans. Diels-Alder Reaction. This reaction involves the formation of a six membered ring by the 1, 4-addition of an alkene to a conjugated diene. Alkene used in this reaction is generally referred to as **dienophile** and the product formed is called *Diels-Alder adduct*. For example:

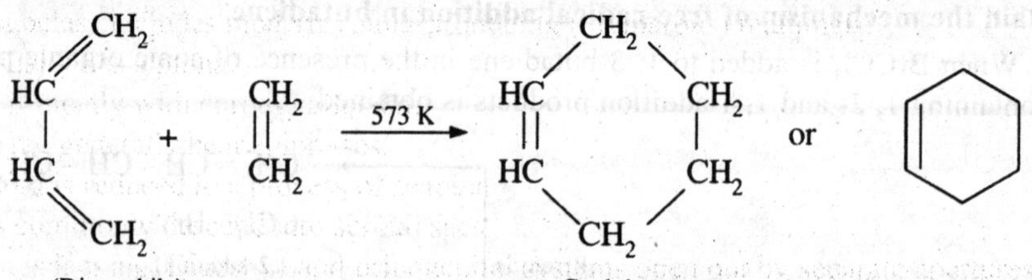

(Dienophile) Cyclohexene

This reaction proceeds faster when the dienophile has electron withdrawing substituents such as – COOH, – CHO – NO_2 and – CN. Thus following reactions take place at low temperatures.

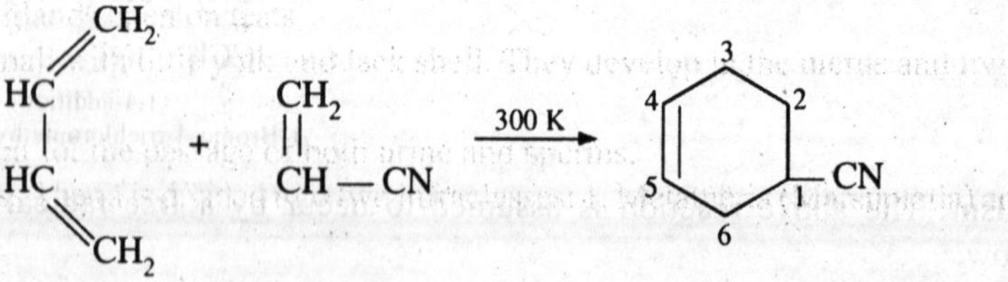

1, 3-Butadiene Acrylonitrile 1, 2, 3, 6-Tetrahydrobenzonitrile

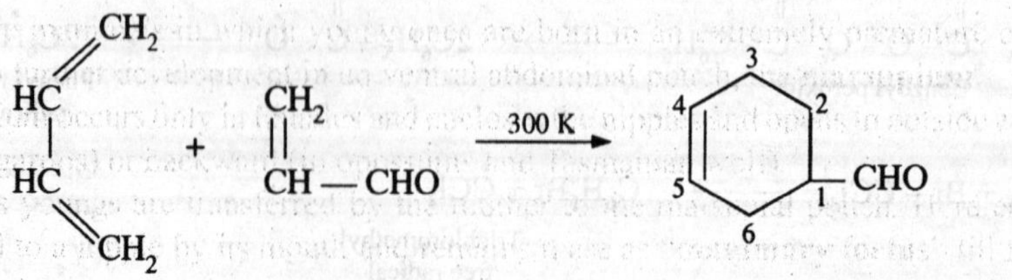

Acrolein 2, 3, 6-Tetrahydrobenzaldehyde

$CH_2=CH-CH=CH_2$ + Maleic anyhydride $\xrightarrow{293\ K}$ *Cis*-1, 2, 3, 6-Tetrahydrophthalic anhydride

Maleic anyhydride

Cis-1, 2, 3, 6-Tetrahydrophthalic anhydride

Q. 8. Discuss the stereochemistry of Diels-Alder reaction. *(Kanpur, 2001)*

Ans. 1. 1, 3-butadiene exists in two planar conformations called *s*-cis and *s*-trans as shown below:

s-cis-1, 3-Butadiene *s*-trans-1, 3-Butadiene

2. The reaction takes place by syn-addition of the diene to the dienophile and the configuration of the dienophile is retained in the product.

1, 3-Butadiene + Dimethylmaleate (*cis*-configuration) $\longrightarrow$ Dimethyl *cis*-4-cyclohexane-1, 2-dicaryboxylate

1, 3-Butadiene + Dimethyl fumarate (*trans*-configuration) $\longrightarrow$ Dimethyl *trans*-4-cyclohexane-1, 2-dicaryboxylate

3. Diels-Alder reaction takes place in an endo manner. This implies that any other unsaturated group in the dienophile prefers to be oriented near the developing double bond in the product.

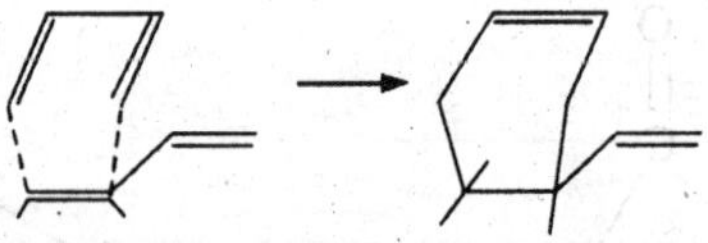

Endo: Preferred for Diels-Alder reaction

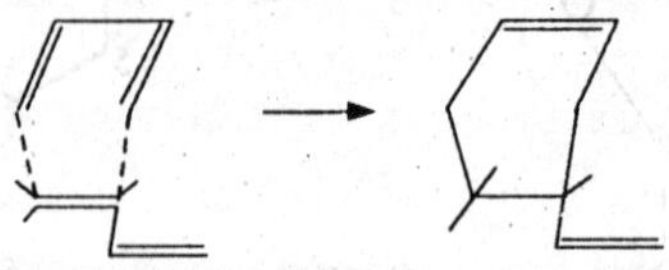

Exo: Not favoured for Diels-Alder reaction

Q. 9. Discuss the mechanism for free radical polymerisation of 1, 3-butadiene.

OR

Complete the rection:  $\xrightarrow[\text{sensitiser}]{h\nu}$ (*Kerala, 2001*)

Ans. 1, 3-Butadiene polymerises to form polybutadiene.

$$\underset{\text{1,3-Butadiene}}{nCH_2 = CH - CH - CH_2} \longrightarrow \underset{\text{Polybutadiene}}{(- CH_2 - CH = CH - CH_2 -)_n}$$

The formation of polybutadiene involves mainly 1, 4-addition of butadiene units to each other so that the polymer obtained still contains a double bond for each butadiene unit.

Mechanism. In the presence of some organic peroxides, the polymerisation of 1, 3-butadiene proceeds through the formation of free radicals as illustrated below:

(*i*) Organic peroxide $\longrightarrow$ Rad $\cdot$

(*ii*) $CH_2 = CH - CH = CH_2 + \text{Rad}\cdot \longrightarrow \left[\begin{array}{c} CH_2 = CH - \dot{C}H - CH_2\ \text{Rad} \\ \updownarrow \\ \dot{C}H_2 - CH = CH - CH_2\ \text{Rad} \end{array}\right]$

(*iii*) $CH_2 = CH - CH = CH_2 + \dot{C}H_2 - CH = CH - CH_2\ \text{Rad}$

$\longrightarrow \dot{C}H_2 - CH = CH - CH_2 - CH_2 - CH = CH - CH_2$ Rad and so on.

Q. 10. Write the structures for the synthetic and natural rubbers. What is the difference between the two? (*Garhwal, 2000; Delhi, 2003*)

Ans. Synthetic rubber. It is obtained by free radical polymerisation of isoprene. It has trans-configuration at almost every double bond.

$$\underset{\substack{\text{2-Methyl-1,3-butadiene} \\ \text{(Isoprene)}}}{CH_2 = \overset{\overset{\large CH_3}{|}}{C} - CH = CH_2} + CH_2 = \overset{\overset{\large CH_3}{|}}{C} - CH = CH_2 \xrightarrow[\text{peroxide}]{\text{Organic}}$$

$$\begin{array}{ccccc} CH_3 \diagdown & & \diagup CH_2 - CH_2 \diagdown & & \diagup H \\ & C = C & & C = C & \\ -CH_2 \diagup & & \diagdown H \quad\quad CH_3 \diagup & & \diagdown CH_2 - \end{array}$$

Trans-Polyisoprene or Gutta-percha
(Synthetic rubber)

Natural rubber. It is also a polymer of isoprene in which all the double bonds have mainly cis-stereochemistry as indicated below:

$$\begin{array}{ccccc} CH_3 \diagdown & & \diagup H \quad CH_3 \diagdown & & \diagup H \\ & C = C & & C = C & \\ -CH_2 \diagup & & \diagdown CH_2 - CH_2 \diagup & & \diagdown CH_2 - \end{array}$$

Cis-Polyisoprene
(Natural rubber)

It is obtained from *latex* which is a milky white liquid.

Latex contains 33-40% rubber which is coagulated by adding salt and acetic acid to it.

Q. 11. Write a short note on vulcanization of rubber.

Ans. The products obtained from raw natural and synthetic rubbers become sticky in hot weather and stiff in cold weather. This drawback can be removed by heating rubber with sulphur. The process of heating rubber with 5–10% of sulphur is called *Vulcanization*. During vulcanization, cross links are produced between polymer chains through reactive allylic positions as indicated below:

$$\begin{array}{c} \quad CH_3 \qquad\qquad\qquad\qquad CH_3 \\ -CH_2-\overset{|}{C}=CH-\underset{\underset{S}{|}}{\underset{S}{|}}{CH}-CH_2-\underset{}{\overset{|}{CH}}-\underset{\underset{S}{|}}{\underset{S}{|}}{CH}-CH_2- \\ \qquad\qquad\qquad | \qquad\qquad\qquad\qquad | \\ -CH_2-\underset{|}{C}=CH-CH-CH_2-\underset{|}{CH}-CH-CH_2- \\ \quad CH_3 \qquad\qquad\qquad\qquad CH_3 \end{array}$$

Vulcanized Rubber

Vulcanisation removes the drawbacks of natural rubber and makes it tougher and stronger.

Q. 12. Write notes on

***(i)* Buna** *(Kerala, 2001; Punjab 2003)*

***(ii)* Buna S** *(Awadh, 2000)*

***(iii)* Buna N** ***(iv)* Polychloroprene**

Ans. (*i*) 1, 4 addition polymerisation of 1, 3 butadiene yields two products as given below.

$$nCH_2=CH-CH=CH_2 \longrightarrow \begin{cases} (-CH_2\diagdown \quad \diagup CH_2-)_n \\ \qquad C=C \\ \quad H\diagup \quad \diagdown H \\ \textit{cis}\text{-addition} \\[1em] (-CH_2\diagdown \quad \diagup H \\ \qquad C=C \\ \quad H\diagup \quad \diagdown CH_2-)_n \\ \textit{Trans}\text{-addition} \\ \text{(Poly butadiene)} \end{cases}$$

$$\underset{\text{Butadiene}}{nCH_2=CH-CH=CH_2} \xrightarrow[\Delta \text{ or Peroxide}]{Na} \underset{\text{Buna}}{(-CH_2-CH=CH-CH_2-)_n}$$

Buna (Bu-Butadiene, Na-Natrium) is a rubber like polymer.

(*ii*) Synthetic rubber (Buna S) is formed by the copolymerisation (two or more different monomers for polymerisation) of butadiene and styrene (phenyl ethylene)

$$\underset{\text{Butadiene (75\%)}}{CH_2=CH-CH=CH_2} + \underset{\underset{\text{Styrene (25\%)}}{C_6H_5}}{CH_2=\overset{}{\underset{|}{CH}}}$$

$$\Big\downarrow Na$$

$$(-CH_2-CH=CH-CH_2-\underset{\underset{C_6H_5}{|}}{CH}-CH_2-\underset{\underset{CH=CH_2}{|}}{CH}-)_n$$

Buna S (S for styrene)

(*iii*) **Buna N.** [Copolymer of butadiene and acrylonitrile or vinyl cyanide)

$$CH_2 = CH - CH = CH_2 + CH_2 = \underset{\underset{CN}{|}}{CH}$$

Butadiene (75%) Vinyl cyanide (acrylonitrile) (25%)

$$\updownarrow$$

$$(- CH_2 - CH = CH - CH_2 - \underset{\underset{CN}{|}}{CH} - CH_2 - \underset{\underset{CH = CH_2}{|}}{CH} -)_n$$

Buna N is a useful polymer as it is resistant to organic solvents.

(*iv*) Polymerisation of 2-chloro 1, 3-butadiene or 2-chloroprene yields polychloroprene, a polymer superior to rubber as it resists the action of oils and other organic solvents.

$$nCH_2 = \overset{\overset{Cl}{|}}{C} - CH = CH_2 \longrightarrow (- CH_2 = \overset{\overset{Cl}{|}}{C} - CH = CH_2 -)_n$$

2-Chloro-1,3-butadiene or Chloroprene → polychloroprene

The double bonds in natural rubber are known to have the *cis* configuration whereas various methods of polymerisation of isoprene give the more stable *trans* arrangement.

Q. 13. What is conjugation? How do you explain greater stability of conjugated dienes? (*Himachal, 2000*)

Or

Which is more stable, 1 4-pentadiene or 1, 3-pentadiene?

Ans. Conjugated dienes are more stable than the non-conjugated dienes. The greater stability of conjugated dienes may be explained in terms of (1) **Resonance** or **delocalisation of electrons** and (2) **heat of hydrogenation**.

Unlike isolated dienes, a conjugated diene exists as a resonance hybrid of various canonical structures (a $\longrightarrow$ d etc.)

$$\underset{(a)}{CH_2 = C - CH = CH_2} \longleftrightarrow \underset{(b)}{\bar{C}H_2 = C - CH = \overset{+}{C}H_2}$$

$$\underset{(c)}{\overset{+}{C}H_2 - CH = CH - \bar{C}H_2} \longleftrightarrow \underset{(d)}{\dot{C}H_2 - CH = CH - \dot{C}H_2} \longrightarrow \text{etc.}$$

In canonical structures (*b, c* and *d*) $C_2 - C_3$ bonds has some double bond character and $C_1 - C_2$ bond and $C_3 - C_4$ bonds some single bond character as confirmed by bond length measurements; $C_2 - C_3$ (1.46 Å) as against 1.54 Å for C – C single bond; $C_1 - C_2$ and $C_3 - C_4$ have bond length 1.35 Å. In structure *d* it has got only a formal bond between C_1 and C_4 due to the tendency of the odd electron on each of these carbons to get paired up.

$$[\dot{C}H_2 - CH = CH - \dot{C}H_2]$$

The formation of an effective covalent bond between is not possible due to the large distance between them.

The canonical structures *b* and *c* involve charge separation and structure *a* which does not involve any charge separation and has more number of effective covalent bonds (π-bonds) is most stable and therefore, the most important of all the contributing structures.

Thus, the actual structure of 1, 3-butadiene is a resonance hybrid which lies somewhere in between the canonical structures ($a \rightarrow d$).

Greater the number of resonating structures of a compound, greater is the stability. Since non-conjugated dienes do not exist in more than one form, they are less stable than conjugated dienes.

Heat of hydrogenation

Heats of hydrogenation of conjugated dienes when compared with the heats of hydrogenation of non-conjugated dienes, reveal that in case of conjugated dienes, the observed values are less than the anticipated values.

Heats of hydorgenation of monosubstituted alkenes ($RCH = CH_2$), disubstituted alkenes ($R_2C = CH_2$ or $RCH = CHR$) and trisubstituted alkenes ($R_2C = CHR$) have been found to be 125.5, 117 and 113 kJ mol^{-1} respectively. For a diene, the total heat of hydrogenation is expected to be sum of heats of hydrogenation of individual double bonds. This is found to be nearly so for non-conjugated bonds. For example, heat of hydrogenation of 1, 4-pentadiene ($CH_2 = CH - CH_2 - CH = CH_2$) is 254.4 kJ mol^{-1} which is almost equal to the calculated value ($2 \times 125.5 = 251$).

But the situation is different in case of conjugated dienes. Consider the case of 1, 3-butadiene. It has observed that heat of hydrogenation is equal to 238.0 kJ mol^{-1}. This is 12.0 kJ less than the calculated value. Similarly 1, 3-pentadiene has heat of hydrogenation 16.7 kJ less than expected value.

$$\underset{\text{1,4-Pentadiene}}{CH_2 = CH - CH_2 - CH = CH_2} \xrightarrow{H_2} \underset{n\text{-Pentane}}{CH_3CH_2CH_2CH_2CH_3}; \quad \Delta H = -254.4 \text{ kJ}$$

$$\underset{\text{1,3-Pentadiene}}{CH_3 - CH = CH - CH = CH_2} \xrightarrow{H_2} \underset{n\text{-Pentane}}{CH_3CH_2CH_2CH_2CH_3}; \quad \Delta H = -234.3 \text{ kJ}$$

Smaller heat of hydrogenation means greater stability of the diene. Hence 1, 3-pentadiene is more stable than 1, 4-pentadiene. In general, a conjugated diene is more stable than non-conjugated diene.

Q. 14. Explain the stability of conjugated dienes with the help of molecular orbital theory. (*Madras, 2004*)

Or

Discuss molecular orbital structure of 1, 3-butadiene. (*Calcutta, 2000 ; Panjab 2005*)

Ans. Each carbon atom in 1, 3-butadiene involve sp^2 hybridization thus leaving on each carbon atom an unhybridized *p*-orbital which overlaps sideways, with the *p*-orbitals of the two adjacent carbon atoms forming π bond(s).

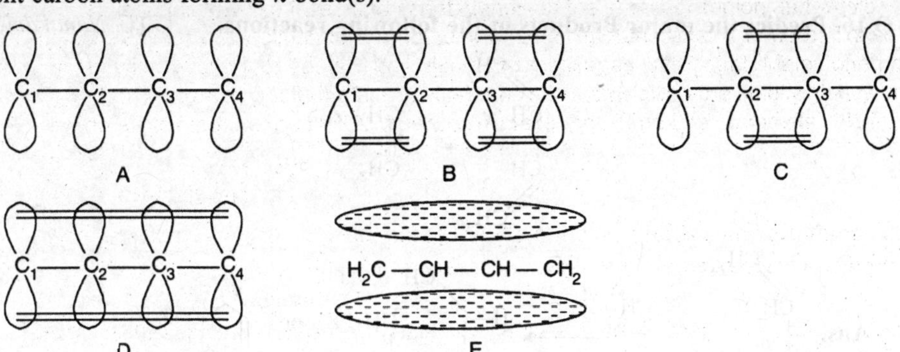

Since the four carbons and the hydrogens attached to them lie in the same plane, the *p*-orbital of C_2 and C_3 can also overlap.

Thus, considering the various methods of overlapping of *p*-orbitals, a π molecular orbital will be formed which covers all the four carbon atoms (Fig. D/E) and the formation of delocalized molecular orbitals (M.O. spreading over the nuclei of more than two C-atoms) results in *delocalized electrons* and delocalized electrons account for the extra stability of conjugated dienes. As a result of delocalisation, all the bonds acquire intermediate character between C – C and C = C.

Thus, the extra stability imparted to the molecule (1, 3-butadiene) is due to the presence of delocalized electrons as each pair of delocalized electrons attracts and is attracted by four, and not two nuclei.

Recently, it has been pointed out that resonance stabilization in case of 1, 3-butadiene is negligible and stability to the extent of 15 kJ mol^{-1} is due to the hybridization of $C_2 - C_3$ bond from $sp^3 - sp^3$ to $sp^2 - sp^2$ hybridization. This also accounts for the shortening of $C_2 - C_3$ bond length.

Q. 15. Complete the following reactions;

(*i*) $\underset{\text{OH}}{CH_2} - CH_2 - CH_2 - \underset{\text{OH}}{CH_2} \xrightarrow{H_2SO_4}$

(*ii*) $CH_2 = CH - CH = CH - CH_3 \xrightarrow{O_3,\ H_2O,\ Zn}$

Ans. (*i*) $\underset{\text{OH}}{CH_2} - CH_2 - CH_2 - \underset{\text{OH}}{CH_2} \xrightarrow[-2H_2O]{H_2SO_4} CH_2 = CH - CH = CH_2$

Butane-1,4-diol 1,3-Butadiene

(*ii*) $CH_2 = CH - CH = CH - CH_3$

1,3-pentadiene

$\downarrow O_3$

Ozonides: $CH_2(-O-O-)(O)CH - CH(-O-O-)(O)CH - CH_3$

$\downarrow$ Zn, H_2O

$HCHO + \underset{\text{CHO}}{CHO} + CH_3CHO$

Formaldehyde Glyoxal Acetaldehyde

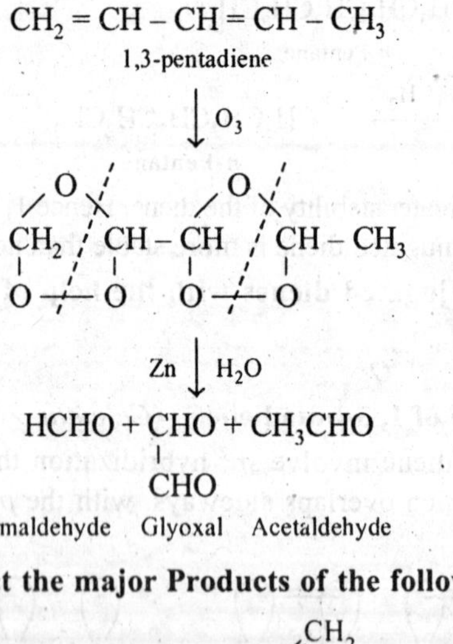

Q.16. Predict the major Products of the following reactions. (*G. Nanakdev, 2000*)

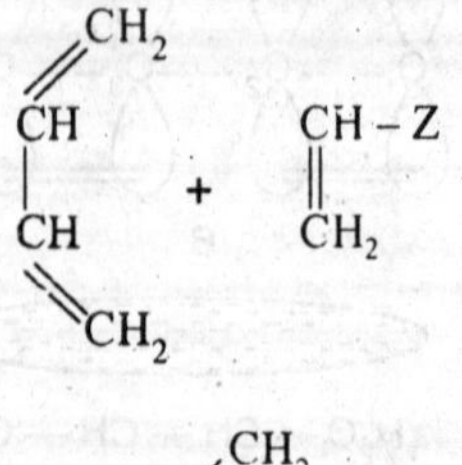

$CH_2{=}CH{-}CH{=}CH_2 + CH_2{=}CH{-}Z$

Ans. $CH_2{=}CH{-}CH{=}CH_2 + CH_2{=}CH{-}Z \longrightarrow$ cyclohexene ring: CH_2, H–C=C–H, CH_2, CH–Z, CH_2 or 4-Z-cyclohexene

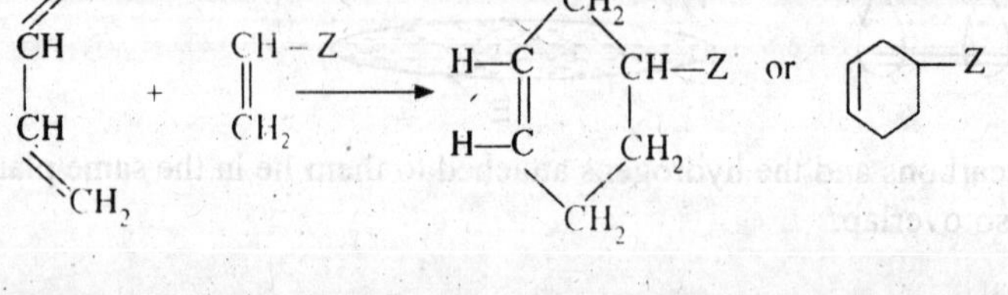

11

ARENES

Q.1. What are arenes?

Ans. Arenes are mixed aromatic aliphatic compounds. An alkyl, alkenyl or alkynyl group attached to a benzene ring constitutes an arene. Thus, arenes can be classified into the following types:

Alkyl benzenes

$C_6H_5-CH_3$ — Toluene

$C_6H_5-CH_2CH_3$ — Ethyl benzene

$C_6H_5-C(CH_3)_3$ — *t*-Butyl benzene

Alkenyl benzenes

$C_6H_5-CH=CH_2$ — Styrene

$C_6H_5-CH_2-CH=CH_2$ — 3-Phenylpropene-1

Alkynyl benzenes

$C_6H_5-C\equiv CH$ — Phenylacetylene

Q.2. Describe methods of obtaining benzene on industrial scale.

(Nagpur 2003, Magadh 2003 ; Gorakhpur 2003, Assam 2004)

Ans. (*i*) From Coal Tar (Industrial method). The light oil fraction (boiling range 353 K to 443 K) of coal tar which contains mainly benzene, toluene and xylenes, is treated with conc. H_2SO_4 to remove impurities of pyridine and thiophene. To remove acidic impurities such as phenol, the organic layer is washed successively with water and aq. KOH and finally after washing with water it is subjected to fractional distillation to obtain three main fractions as given below:

(*a*) *90% Benzol* (boiling range 353 K – 383 K) consists of 70% benzene, 24% toluene and rest xylenes. Its careful fractionation gives benzene, toluene and xylene.

(*b*) *90% Toluol* (boiling range 383 K– 413 K) consists mainly of xylenes and small amounts of benzene and toluene.

(*c*) *Solvent naphtha or Benzine* (boiling range 413 K – 443 K) mainly consists of xylenes.

***(ii)* From Petroleum (Industrial method).** Aromatic hydrocarbons are synthesized by catalytic reforming of alkanes obtained from petroleum which involves cyclization and aromatization. For example, hexane can be converted to cyclohexane and then to benzene by passing its vapours over Cr_2O_3, V_2O_5 or MoO_3 at 723 K – 823 K. Similarly, *n*-heptane gives toluene.

$$\underset{n\text{-Hexane}}{C_6H_{14}} \xrightarrow[\text{723-823 K, 10-20 atm.}]{Cr_2O_3/V_2O_5/MoO_3} \underset{\text{Benzene}}{C_6H_6} + 4H_2$$

***(iii)* From acetylene.** Acetylene on passing through red hot tube trimerizes to benzene

$$\underset{\text{Acetylene}}{3CH \equiv CH} \xrightarrow[\text{723 K}]{\text{Red hot tube}} \underset{\text{Benzene}}{C_6H_6}$$

Q.3. Describe laboratory methods of preparing benzene.

Ans. Laboratory methods of preparing benzene are given as under:

***(a)* Decarboxylation of benzoic acid/phathalic acid with soda lime.**

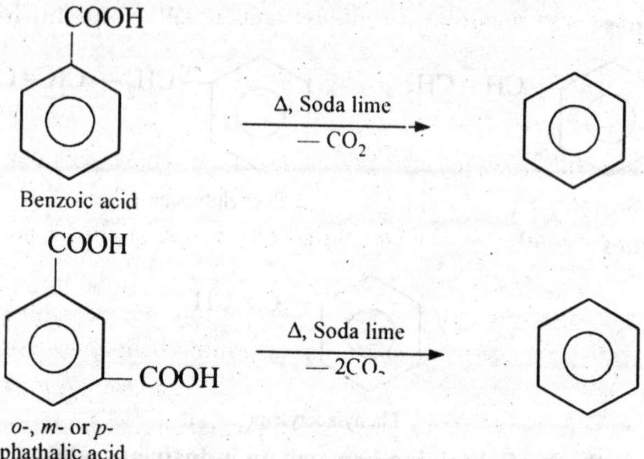

***(b)* Distillation of phenol with zinc dust (reduction).**

$$C_6H_5OH + Zn \xrightarrow{\Delta} C_6H_6 + ZnO$$

***(c)* By hydrolysis of benzene sulphonic acid with super-heated steam or boiling with HCl under pressure.**

$$C_6H_5SO_3H + HOH \xrightarrow{\text{Steam}} C_6H_6 + H_2SO_4$$

(*d*) By reduction of benzene diazonium chloride with alkaline sodium stannate.

$$C_6H_5N_2Cl + Na_2SnO_4 + NaOH \longrightarrow C_6H_6 + Na_2SnO_3 + N_2 + HCl$$

Benzene

Q.4 Describe substitution reactions of benzene.

Ans. Substitution reactions can be discussed under the following headings:

(*a*) **Halogenation.** Benzene reacts with chlorine or bromine in presence of a catalyst (iron, iron halide or aluminium halide) and in absence of sunlight to form chlorobenzene or bromo-benzene. As a hydrogen atom is displaced by a halogen, the reaction is called *halogenation.*

$$C_6H_6 + Cl_2 \xrightarrow{\text{Fe, in absence of sunlight}} C_6H_5Cl + HCl$$

Chlorobenzene

$$C_6H_6 + Br_2 \xrightarrow{\text{Fe, in absence of sunlight}} C_6H_5Br + HBr$$

Bromobenzene

(*b*) **Nitration.** Benzene on warming with a mixture of nitric acid and sulphuric acid forms nitrobenzene. As a hydrogen atom is replaced by a nitro group ($-NO_2$), the reaction is called nitration.

$$C_6H_6 + HNO_3 \xrightarrow{H_2SO_4,\ 323\ K} C_6H_5NO_2 + H_2O$$

Nitrobenzene

(*c*) **Friedel-Crafts alkylation.** This reaction is used for introducing an alkyl group in benzene ring.

$$C_6H_6 + RX \xrightarrow{\text{Anhydrous } AlCl_3} C_6H_5R + HX$$

Alkyl benzene

(*d*) **Friedel-Crafts acylation.** This reaction is used for the preparation of aromatic aldehydes and ketones. Benzene is treated with acid chloride in the presence of anhydrous aluminium chloride.

$$C_6H_6 + RCOCl \xrightarrow{\text{Anhydrous } AlCl_3} C_6H_5COR + HCl$$

Acid chloride — Aryl alkyl ketone

$$C_6H_6 + HCOCl \xrightarrow{\text{Anhy. } AlCl_3} C_6H_5CHO + HCl$$

Benzaldehyde

(*e*) **Sulphonation.** Benzene on warming with conc. sulphuric acid forms benzene sulphonic acid. As a hydrogen atom is replaced by a sulphonic acid group (–SO_3H), the reaction is called *sulphonation*.

$$C_6H_6 + H_2SO_4 \xrightarrow{\Delta} C_6H_5SO_3H + H_2O$$

Benzene sulphonic acid

Q.5. Describe addition reactions of benzene.

Ans. (*a*) Addition of hydrogen. When benzene vapours mixed with hydrogen are passed over nickel catalyst at 473 K, cyclohexane is formed.

$$C_6H_6 + 3H_2 \xrightarrow{\text{Ni, 473 K}} C_6H_{12}$$

Cyclohexane

On reduction with sodium and ethanol in presence of liquid ammonia, benzene gives 1, 4-cyclohexadiene. This reaction is called ***Birch reduction.***

$$C_6H_6 + 2\,[H] \xrightarrow[\text{Na, } C_2H_5OH]{\text{Liquid } NH_3} C_6H_8$$

1, 4-cyclohexadiene

At higher temperature, even in the absence of ethanol, Birch reduction takes place and cyclohexene is formed.

$$C_6H_6 + 4[H] \xrightarrow[\text{Na, NH}_3]{433\text{-}443\text{ K}} \text{Cyclohexene}$$

Cyclohexene

(*b*) Addition of halogens. Benzene on treatment with halogens in the presence of sunlight gives benzene hexahalides.

$$C_6H_6 + 3Cl_2 \xrightarrow{\text{In presence of sunlight}} C_6H_6Cl_6$$

Benzene hexachloride (BHC)
or Hexachlorocyclohexane

(*c*) Addition of ozone. Benzene reacts with ozone to form a triozonide, which is decomposed by water to give three molecules of glyoxal.

$$\text{Benzene} + O_3 \longrightarrow \text{Ozonide} \xrightarrow[\text{Zn}]{3H_2O} 3\,\begin{matrix}CHO\\|\\CHO\end{matrix} + 3H_2O_2$$

Benzene — Ozonide — Glyoxal

Q.6. Describe oxidation reactions of benzene.

Ans. Oxidation. When benzene vapours mixed with air are passed over heated vanadium pentoxide catalyst, maleic anhydride is obtained.

$$C_6H_6 + O_2 \xrightarrow{V_2O_5,\ \Delta} \begin{matrix}CH-CO\\ \| \quad\quad\ \ O\\ CH-CO\end{matrix}$$

Maleic anhydride

Combustion. When burnt in air or oxygen, carbon dioxide and water are formed

$$2C_6H_6 + 15O_2 \longrightarrow 12\,CO_2 + 6H_2O$$

Q.7. What is Kekule structure of benzene? What are its drawbacks?

(*Bangalore, 2006; Nagpur, 2008*)

OR

How will you show that benzene molecule contains three double bonds?

(*Calcutta 2007*)

Ans. Kekule structure. According to Kekule, six carbon atoms of benzene are linked to each other by alternate single and double covalent bonds to form a hexagonal ring as shown in the following figure:

Kekule structure of benzene

Each carbon atom is linked to one hydrogen atom thus conforming to its molecular formula.

Drawbacks of Kekule structure. Kekule structure does not explain the following observations:

***(i)* Chemical reactions.** Benzene does not give addition reactions and fails to decolorise Baeyer's reagent. It readily undergoes electrophilic substitution reactions in which the benzene ring is retained.

***(ii)* Heat of combustion.** On the basis of Kekule structure, the heat of combustion of benzene is expected to be 3449.0 kJ mol^{-1} . But the experimental value is 3298.5 $kJmol^{-1}$. Thus, benzene has 150.5 $kJmol^{-1}$ of energy less than Kekule structure. Benzene is, therefore, more stable than the structure proposed by Kekule.

***(iii)* Heat of hydrogenation.** On the basis of Kekule structure, the heat of hydrogenation of benzene is expected to be 358.0 $kJmol^{-1}$. But the experimental value is 208.5 $kJmol^{-1}$. This again shows that benzene is more stable than expected from Kekule structure.

***(iv)* Disubstituted product.** Benzene forms only one *ortho* disubstituted product whereas Kekule structure predicts two *o*-disubstituted products as shown below:

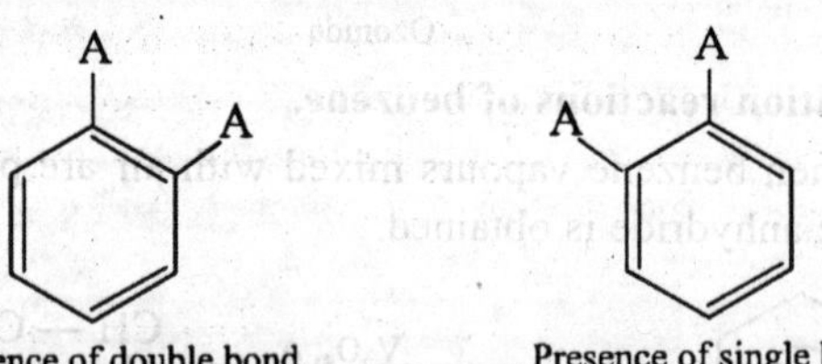

Presence of double bond between the substituents

Presence of single bond between the substituents

***(v)* Carbon-carbon bonds lengths.** X-ray diffraction studies show that all the carbon-carbon bond lengths are identical, *viz.* 1.397 Å and lie in between that of single and double bonds. This is not in accordance with Kekule structure which contains two kinds of carbon-carbon bonds.

Q.8. Discuss molecular orbital structure of benzene. (*Kerala, 2001; Kanpur, 2001; Kathmandu 2004; Kalyani, 2004; Purvanchal, 2007; Nagpur, 2008*)

Ans. Orbital structure. Structure of benzene can be best described by using the orbital concept. Each carbon atom in benzene is sp^2 hybridised and thus forms three bonds, two with adjacent carbon atoms and one with hydrogen. Thus, all the six carbons and six hydrogen atoms lie in the same plane and the angle between two adjacent σ bonds is 120° [Fig. (a)].

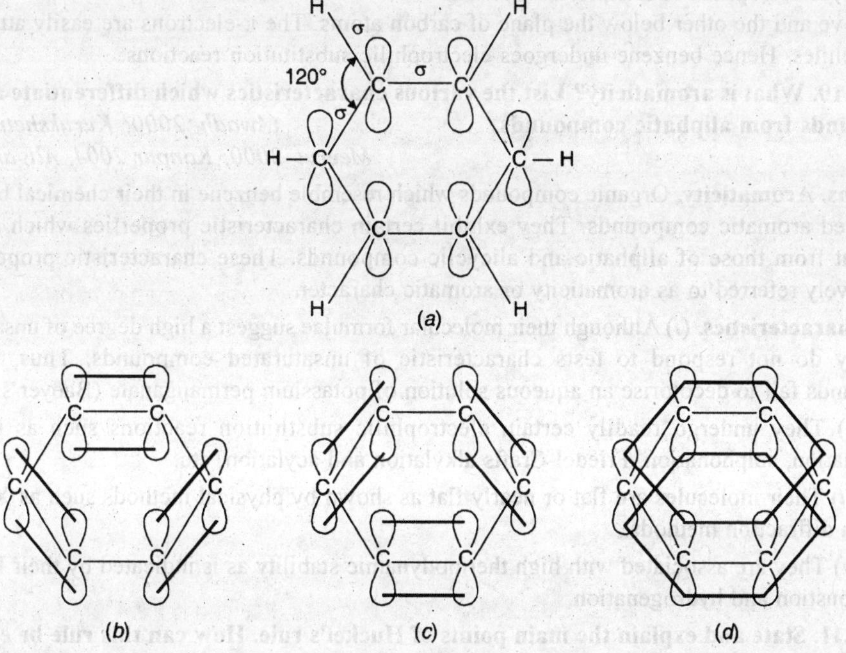

Each carbon is still left with an unhybridised *p*-orbital lying above and below the plane of benzene ring. Each one of these *p*-orbitals overlaps sidewise on either sides to form two sets of π -electron clouds. [Figs. (*b*) and (*c*)]. π electrons are delocalised as these can move over all the six carbon atoms [Fig. (*d*)]. As a result of this delocalisation two continuous ring like electron clouds one above and the other below the plane of carbon atoms [Fig. (*e*)] are formed. Bond angles and bond lengths in the molecule of benzene are depicted in [Fig. (*f*)].

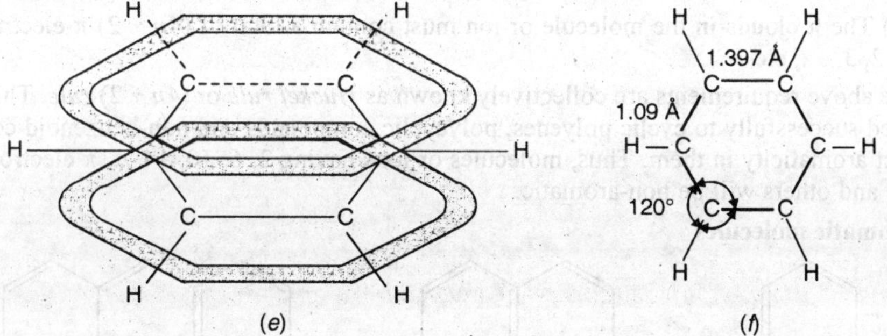

Q.9. What evidence can you produce in support of orbital structure of benzene?

Ans. Evidence in support of orbital structure.

(*i*) Unusual stability. Benzene molecule exhibits unusual stability and resists the formation of addition products. This can easily be understood in terms of delocalisation of π-electrons which is responsible for aromaticity.

(*ii*) Isomer number. According to orbital concept all the six carbons in benzene are completely equivalent. Similarly, all the six hydrogen atoms also occupy identical positions. Thus, benzene should form only one monosubstituted and three disubstituted products. This has been found to be so in actual practice.

(*iii*) Electrophilic substitution reactions. There are two continuous ring like π-electron clouds one above and the other below the plane of carbon atoms. The π-electrons are easily attacked by electrophiles. Hence benzene undergoes electrophilic substitution reactions.

Q.10. What is aromaticity? List the various characteristics which differentiate aromatic compounds from aliphatic compounds. *(Awadh, 2000; Kurukshetra, 2000; Meerut, 2000; Kanpur 2004, Aligarh 2004)*

Ans. Aromaticity. Organic compounds which resemble benzene in their chemical behaviour are called aromatic compounds. They exhibit certain characteristic properties which are quite different from those of aliphatic and alicyclic compounds. These characteristic properties are collectively referred to as aromaticity or aromatic character.

Characteristics. (*i*) Although their molecular formulae suggest a high degree of unsaturation, yet they do not respond to tests characteristic of unsaturated compounds. Thus, aromatic compounds fail to decolorise an aqueous solution of potassium permanganate (Baeyer's test).

(*ii*) They undergo readily certain electrophilic substitution reactions such as nitration, halogenation, sulphonation, Friedel-Crafts alkylation and acylation, etc.

(*iii*) Their molecules are flat or nearly flat as shown by physical methods such as X-ray and electron diffraction methods.

(*iv*) They are associated with high thermodynamic stability as is indicated by their low heats of combustion and hydrogenation.

Q.11. State and explain the main points of Huckel's rule. How can this rule be employed to explain the aromaticity of organic compounds? *(Punjab, 2002; Nagpur, 2005; Garhwal, 2010; Bharathiar, 2011)*

Ans. Huckel's rule. This rule is based upon molecular orbital treatment and is employed for predicting aromaticity in organic compounds. The main theoretical requirements for a substance to possess aromaticity are:

(*i*) The molecule or ion must be flat or nearly flat.

(*ii*) It must have cyclic clouds of delocalised π electrons above and below the plane of the molecule.

(*iii*) The π clouds in the molecule or ion must contain a total of $(4n + 2)$ π electrons where $n = 0, 1, 2, 3, ...$, etc.

The above requirements are collectively known as *Huckel rule* or *(4n + 2) rule*. This rule can be applied successfully to cyclic polyenes, polycyclic compounds and non-benzenoid compounds to predict aromaticity in them. Thus, molecules or ions having 2, 6, 10, 14 .... π electrons will be aromatic and others will be non-aromatic.

Aromatic molecules

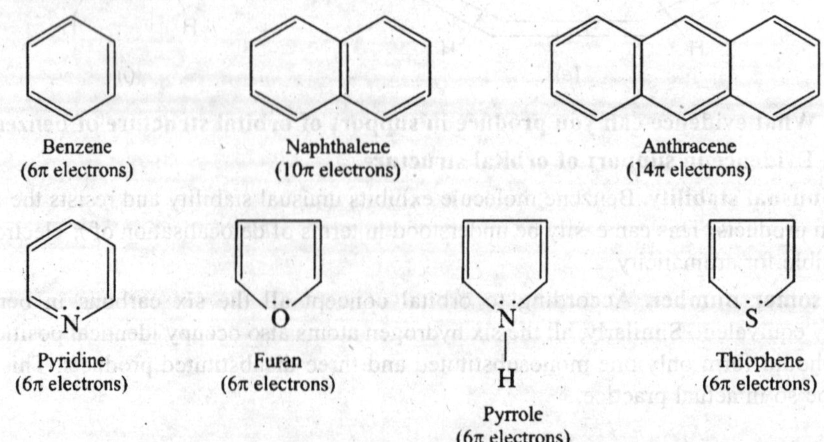

Benzene (6π electrons) | Naphthalene (10π electrons) | Anthracene (14π electrons)

Pyridine (6π electrons) | Furan (6π electrons) | Pyrrole (6π electrons) | Thiophene (6π electrons)

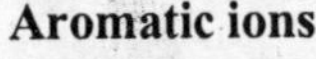

Aromatic ions

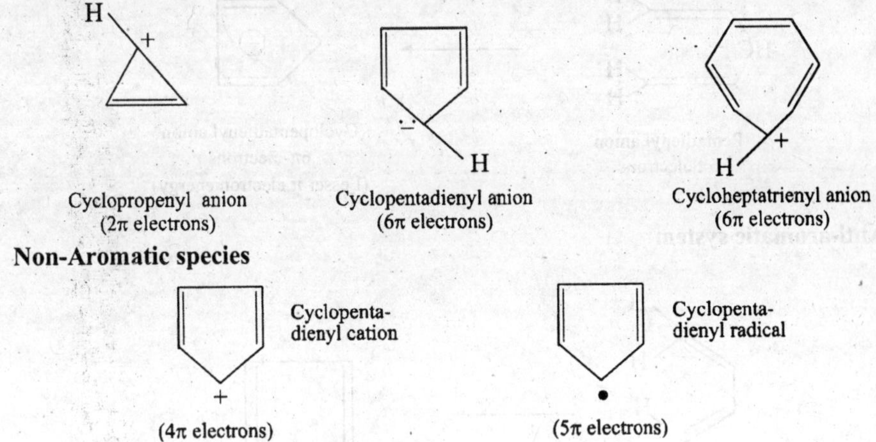

Cyclopropenyl anion (2π electrons)   Cyclopentadienyl anion (6π electrons)   Cycloheptatrienyl anion (6π electrons)

Non-Aromatic species

Cyclopentadienyl cation (4π electrons)   Cyclopentadienyl radical (5π electrons)

Q.12. What are annulenes? (*G.Nanakdev, 2000; Himachal, 2000; Kurukshetra, 2001*)

Ans. Annulenes are monocyclic compounds having a system of single and double bonds in alternate positions. They may be called cyclic polyenes and can be represented by a general formula $(\text{–CH = CH–})_n$ where is n a positive integer. They are named by writing the ring size followed by the word "annulene". Some examples of annulenes are given below:

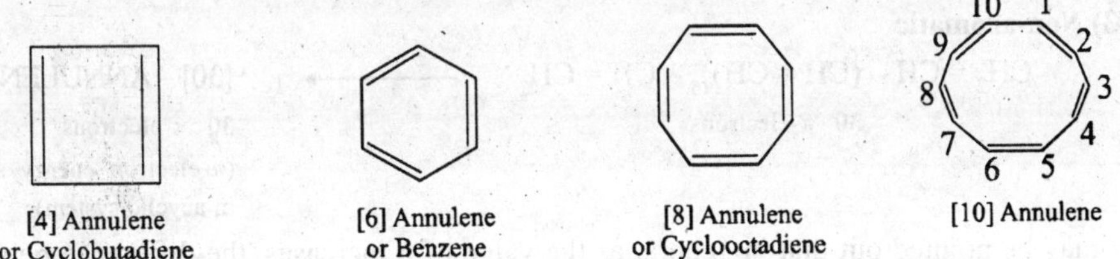

[4] Annulene or Cyclobutadiene   [6] Annulene or Benzene   [8] Annulene or Cyclooctadiene   [10] Annulene

According to Huckel's rule, Annulenes containing $(4n + 2)$ π electrons and having a coplaner ring should be aromatic in nature. The synthesis of a no. of annulenes has confirmed this. Thus [6] annulene or benzene is aromatic. [4] annulene and [8] annulene are not aromatic because they don't conform to Huckel's rule. They are rather anti-aromatic. [10] annulene although expected to be aromatic as per Huckel's rule is not so because the ring is not coplaner.

(*G. Nanakdev, 2000; Kerala, 2001*)

Q.13. What are aromatic, anti-aromatic and non-aromatic compounds?

Ans. (*i*) If on removing hydrogen from each end of the chain and joining the ends to form a ring, the ring formed has **lesser π-electron energy** than the open chain, the ring is *aromatic*.

(*ii*) If the ring has the **same amount of π-electron energy** as the open chain, the ring is *non-aromatic.*

(*iii*) If the ring has **greater π-electron energy** than the acyclic chain, the ring is anti-aromatic. The example given below illustrate the point.

(1) Aromatic system

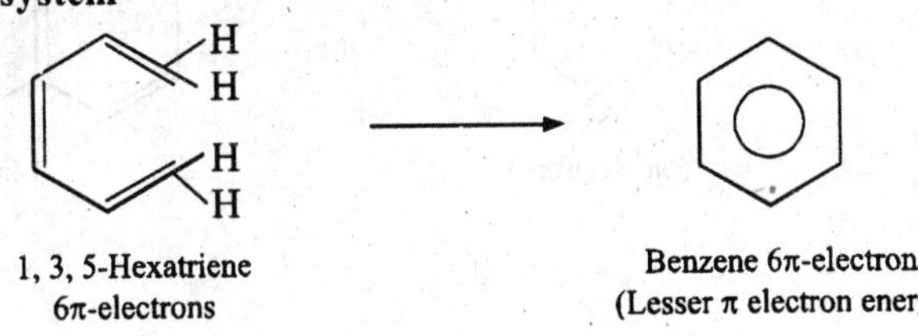

1, 3, 5-Hexatriene 6π-electrons   Benzene 6π-electrons (Lesser π electron energy)

Pentadienyl anion
6π electrons

Cyclopentadienyl anion
6π-electrons
(Lesser π electron energy)

(2) Anti-aromatic system

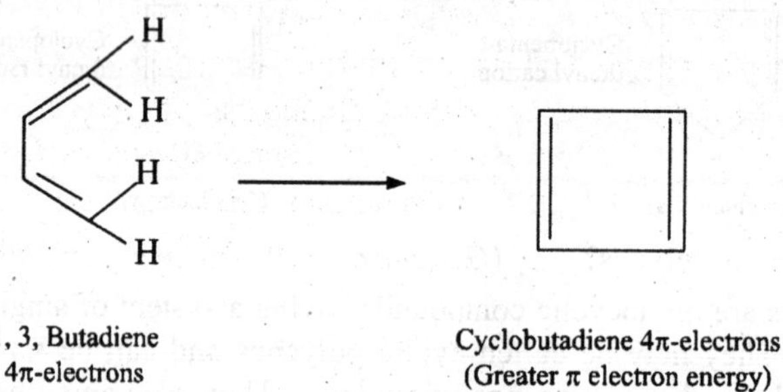

1, 3, Butadiene
4π-electrons

Cyclobutadiene 4π-electrons
(Greater π electron energy)

Similarly other systems having $4\,n\,\pi$ electrons are anti-aromatic.

(3) Non-aromatic

$$CH_2 = CH - (CH = CH)_{15} - CH = CH_2 \longrightarrow [30] - \text{ANNULENE}$$

30 π-electrons

30 π-electrons
(π-electron energy same as in acyclic system)

It may be pointed out that in general as the value of n increases, the difference between the cyclic and corresponding acyclic system decreases both in system having $(4n)$ π-electrons and $(4n + 2)$ π-electrons. In other words, such cyclic systems show non-aromatic behaviour.

Q.14. Which of the following species will exhibit aromatic character? Write their structural formulae.

Cyclooctatetraene, Tropylium ion, Cycloheptatrienyl anion, Furan, Cyclopropeny anion, Cyclopentadienyl cation, Cyclobutadiene, Anthracene, Phenanthrene, Pyridine.

(*Devi Ahilya, 2001; Bharathiar, 2011*)

Ans. (*a*) The compounds mentioned below are aromatic in nature in terms of Huckel's rule. No. of π electrons in these compounds is $(4n + 2)$ where $n = 0, 1, 2$ etc.

I Tropylium ion
(cycloheptatrienyl cation)

II Furan

III Anthracene

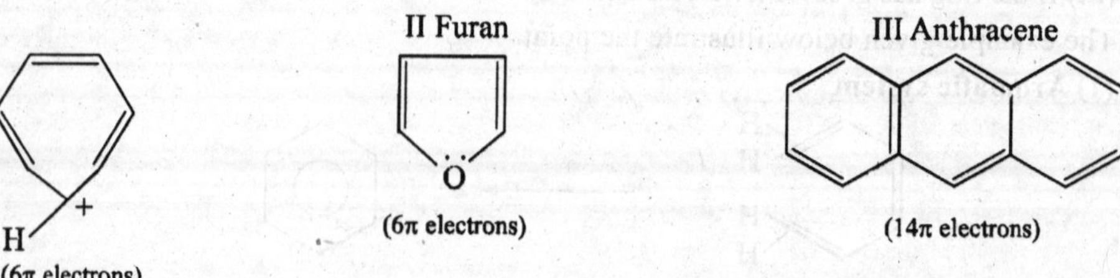

(6π electrons)

(6π electrons)

(14π electrons)

IV Phenanthrene

(14π electrons)

V Pyridine

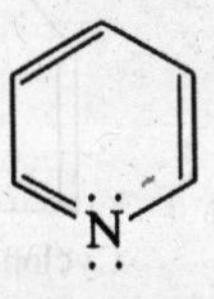

(10π electrons)

(*b*) Following compounds are **not aromatic** in nature as they do not comply with the $(4n + 2)$ rule.

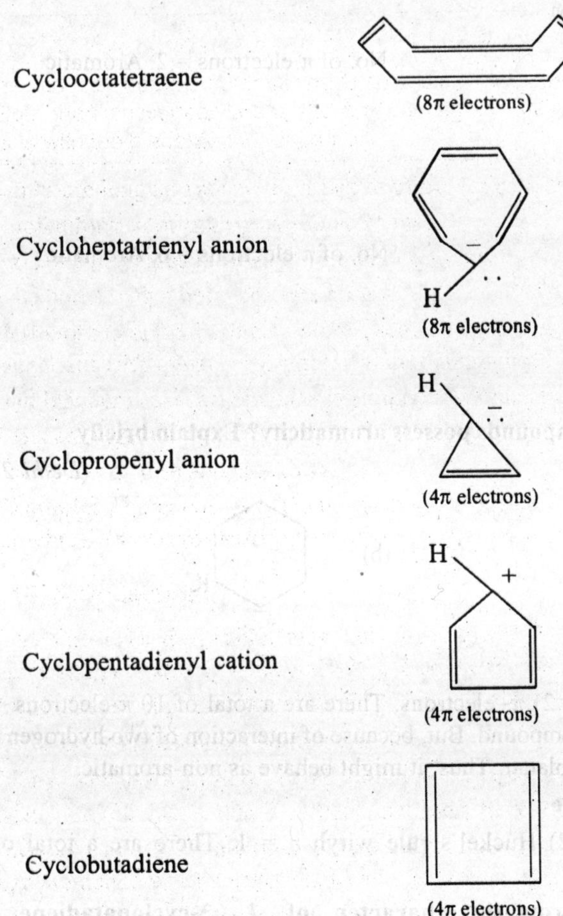

Q.15. Which of the following have aromatic character and why?

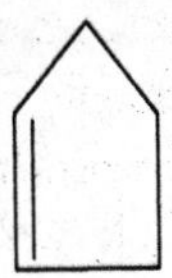   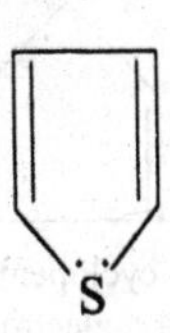

Ans. Apply ($4n + 2$) Huckel's Rule

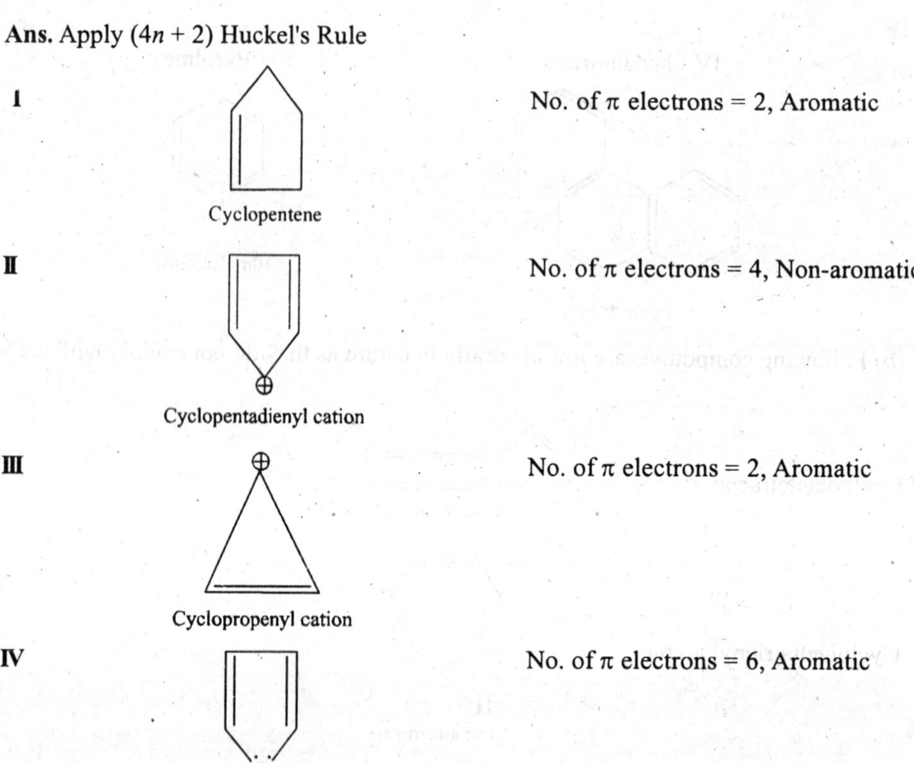

Q.16. Which of the following compounds possess aromaticity? Explain briefly

(*Delhi 2006*)

Ans. (a) [10} - Annulene

It satisfies the Huckl's rule ($4n + 2$) π–electrons. There are a total of 10 π-electrons. Thus $n = 2$. Theoretically, it is an aromatic compound. But, because of interaction of two hydrogen atom inside the planar ring, it becomes non-planar. Thus, it might behave as non-aromatic.

(b) Potassium Cyclopentadienide

This compound follows ($4n + 2$) Huckel's rule wityh $n = 1$. There are a total of six π- electrons.

Q.17. Comment on the aromatic character of 1, 3-cyclopatadiene and 1, 3-cyclopentadienyl anion. (*Delhi 2007*)

Ans.

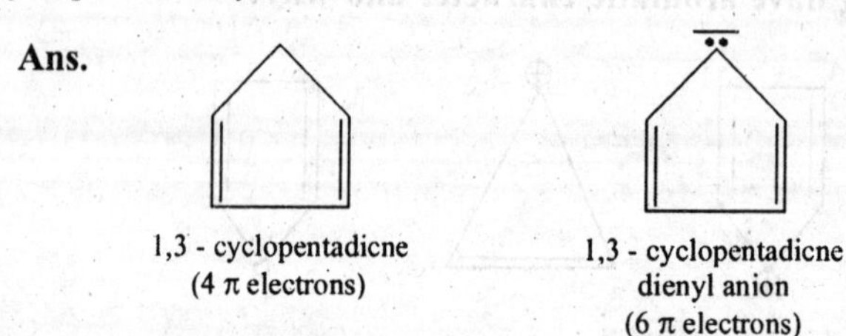

1,3-cyclopentadiene does not conform to $(4n + 2)$ Huckel's rule. Thus it is not aromatic.

1,3-cyclopentadienyl anion conforms to $(4n + 2)$ Huckel's rule with $n = 1$. Thus, it is an aromatic species.

Q.18. Describe general mechanism of electrophilic substitution in benzene. Give evidence in support of this mechanism. *(Kerala, 2001; Devi Ahilya, 2001; Garhwal, 2000)*

Ans. General mechanism. Electrophilic substitution reactions are initiated by substances which are either electrophilic themselves or which generate some electrophilic species. General mechanism involves the following sequence of steps:

(I) Generation of electrophile. To start with, there is preliminary reaction which generates an electrophile.

$$A - B \longrightarrow \underset{\text{Electrophile}}{A^{\oplus}} + B^-$$

(II) Formation of intermediate carbocation (Carbonium ion). The electrophile attacks the electron cloud of the benzene ring and thus brings about an electronic displacement. This results in the formation of **intermediate carbocation** which is resonance stabilized as shown below:

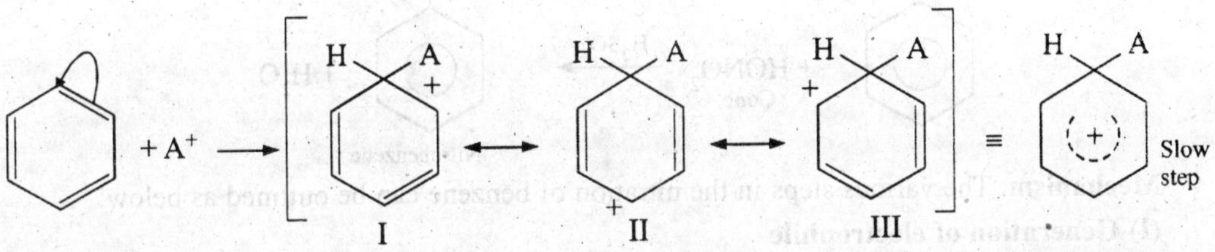

+ A⁺ → [I ↔ II ↔ III] ≡ Slow step

Resonating structures

Resonance stabilised carbocation

During the formation of the carbocation, the aromaticity of the benzene ring is destroyed. Consequently, the formation of carbocation is slow and hence is the **rate determining step.**

(III) Abstraction of a proton from the carbocation. The base present in the reaction mixture abstracts the proton from the carbocation to form the final product. Since the aromatic character of the benzene ring is restored, this step is **fast** and hence is not the rate determining step.

+ :B⁻ → + BH Fast step

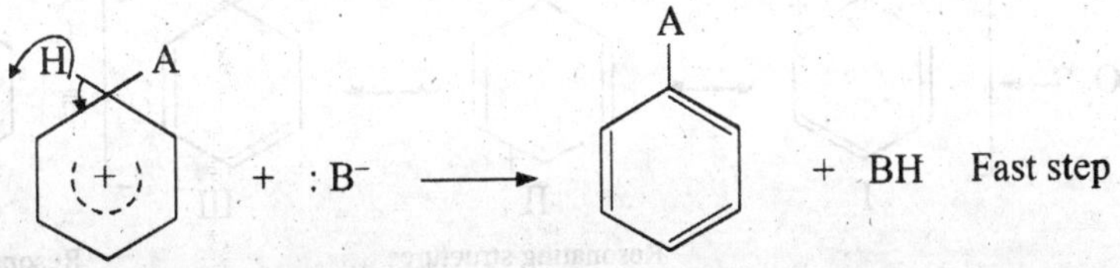

Evidence in support of the mechanism. Electrophilic aromatic substitution involves two steps viz. formation of carbocation and abstraction of proton from carbocation, after the generation of electrophile. The first step involving the formation of intermediate carbocation is slow and hence is the rate determining step of the reaction. The second step which involves the abstraction of a proton from the carbocation is fast.

The above mechanism is supported by **isotope tracer technique** as described below.

A carbon-deuterium bond is broken more slowly than a bond between carbon and hydrogen. Thus, if a hydrogen is lost in the rate determining step of a reaction, it will show an isotope effect on replacing hydrogen by deuterium. In other words, a deuterated compound should undergo substitution more slowly than a non-deuterated compound. But this isotope effect has not been observed in aromatic electrophilic substitution reactions. For example, the rate of nitration of deutero-benzene is the same as the rate for benzene. Thus, proton elimination is not the rate determining step of the reaction. The step leading to the formation of carbocation does not involve the cleavage of carbon-hydrogen bond and hence it is the rate determining step of the reaction.

Q.19. Define electrophilic substitution and explain the mechanism taking nitration as example. (*Punjab, 2004; Tripura, 2004; Delhi, 2004; Purvanchal, 2007; Calcutta, 2007; Pune, 2010*)

Or

With the help of an example, give the mechanism of electrophilic substitution in benzene. (*Nagpur 2005, Venkateswara 2005*)

Ans. Benzene has a tendency to give electrophilic substitution reactions because of the π electron cloud above and below benzene ring. In these reactions, a hydrogen atom of benzene is substituted by $-NO_2$, $-X$ (halogen), $-SO_3H$, $-R$ (alkyl) or $-COR$ group, under different conditions and by taking a suitable electrophile. Such reactions are called electrophilic substitution reactions.

Nitration. This reaction involves the treatment of an aromatic compound with a mixture of nitric acid and sulphuric acid. For example:

$$C_6H_6 + \underset{\text{Conc.}}{HONO_2} \xrightarrow{H_2SO_4} \underset{\text{Nitrobenzene}}{C_6H_5NO_2} + H_2O$$

Mechanism. The various steps in the nitration of benzene can be outlined as below:

(I) Generation of electrophile

$$HONO_2 + 2H_2SO_4 \rightleftharpoons \underset{\text{Nitronium ion}}{NO_2^+} + H_3O^+ + 2HSO_4^-$$

(II) Formation of carbocation

$$C_6H_6 + NO_2^+ \longrightarrow \left[\text{I} \longleftrightarrow \text{II} \longleftrightarrow \text{III} \right] \equiv C_6H_6NO_2^+ \quad \text{Slow step}$$

Resonating structures (I, II, III: H and NO_2 on the same ring carbon, positive charge on the ring)

Resonance stabilised carbocation

(III) Abstraction of proton from the carbocation

$$C_6H_6NO_2^+ + HSO_4^- \longrightarrow \underset{\text{Nitrobenzene}}{C_6H_5NO_2} + H_2SO_4$$

Q.20. What is sulphonation? Give its mechanism. (*Guwahati, 2002; Nagpur, 2008; Pune, 2010*)

Ans. Sulphonation. The sulphonation of an aromatic compound can be brought about by the action of conc. H_2SO_4 or oleum. For example:

$$C_6H_6 + HOSO_3H \longrightarrow C_6H_5SO_3H + H_2O$$

Benzene sulphonic acid

Mechanism. The attacking electrophilic reagent in this reaction is believed to be sulphur trioxide which is present as such in oleum or may be formed by the dissociation of sulphuric acid. The electrophilic nature of molecule is due to the presence of electron deficient sulphur atom.

$$O = \overset{++}{S}(-O^-)(-O^-)$$

As the positive charge is concentrated on sulphur and negative charges are scattered on two oxygens, SO_3 molecule acts as electrophile.

The complete mechanism for this reaction may be outlined as follows:

(I) Generation of electrophile

$$2H_2SO_4 \rightleftharpoons \underset{\text{Electrophile}}{SO_3} + HSO_4^- + H_3O^+$$

(II) Formation of carbocation

$$C_6H_6 + SO_3 \rightleftharpoons \left[\text{I} \longleftrightarrow \text{II} \longleftrightarrow \text{III} \right] \equiv \text{Resonance stabilised carbocation}$$

Resonating structures

Resonance stabilised carbocation

(III) Abstraction of proton from the carbocation

$$C_6H_6(H)(SO_3^-)^+ + HSO_4^- \rightleftharpoons C_6H_5SO_3^- + H_2SO_4$$

(IV) Formation of the final product

$$C_6H_5SO_3^- + H_3O^+ \rightleftharpoons C_6H_5SO_3H + H_2O$$

Q.21. Give the mechanism of Friedel-Crafts acylation.

(Garhwal, 200: M. Dayanad, 2000; Kerala 2001; Vikram 2004; Bangalore 2006)

Or

Give the mechanism of the following reaction:

$$C_6H_6 + CH_3COCl \xrightarrow[AlCl_3]{Anhyd.} C_6H_5COCH_3 + HCl$$

(*North Eastern Hill, 2004*)

Ans. When benzene is treated with acetyl chloride in the presence of anhydrous aluminium chloride, the formation of acetophenone takes place. This reaction is known as *Friedel-Craft's acylation.*

$$C_6H_6 + CH_3-\overset{O}{\overset{\|}{C}}-Cl \xrightarrow{\text{Anhydrous } AlCl_3} C_6H_5-\overset{}{C}(=O)-CH_3 + HCl$$

Acetyl chloride — Acetophenone

Mechanism. The most probable mechanism of the above reaction is discussed as under:

(I) Generation of electrophile

$$Ch_3-\overset{O}{\overset{\|}{C}}-Cl + AlCl_3 \rightleftharpoons CH_3-\overset{O}{\overset{\|}{C}}{}^{\oplus} + AlCl_4^-$$

Acetyl cation

(II) Formation of carbocation

$$C_6H_6 + CH_3-\overset{O}{\overset{\|}{C}}{}^{+} \rightarrow [\text{I} \leftrightarrow \text{II} \leftrightarrow \text{III}] \equiv \text{Resonance stabilised carbocation}$$

I, II, III: Resonating structures (cyclohexadienyl cation bearing H and $-\overset{O}{\overset{\|}{C}}-CH_3$ on the same carbon, with + at positions ortho, para, ortho respectively)

Resonance stabilised carbocation

(III) Abstraction of proton from the carbocation

$$\text{Carbocation (H, } -\overset{O}{\overset{\|}{C}}-CH_3) + AlCl_4^- \longrightarrow C_6H_5-C(=O)-CH_3 + AlCl_3 + HCl$$

Acetophenone

Q.22. Discuss the mechanism of Friedel-Craft's alkylation of benzene.

(*Osmania, 2004; Punjab, 2002; Nagpur, 2003; Guwahati, 2006; Andhra, 2010; Pune, 2010*)

Ans. Friedel-Craft's alkylation. The reaction consists in treating benzene or substituted benzenes with an alkyl halide in the presence of small amounts of Lewis acids ($AlCl_3$, BF_3, $FeCl_3$. etc.). It leads to the direct introduction of an alkyl group into the benzene ring.

$$C_6H_6 + \underset{\text{Alkyl chloride}}{RCl} \xrightarrow{\text{Anhy. } AlCl_3} \underset{\text{Alkyl benzene}}{C_6H_5R} + HCl$$

Mechanism. The reaction is believed to take place through the following steps:

(I) Generation of electrophile

$$R-Cl + AlCl_3 \rightleftharpoons \underset{\text{Electrophile}}{\overset{+}{R}} + Al\overline{Cl}_4$$

(II) Formation of carbocation

I, II, III — Resonating structures; Resonance stabilised carbocation

(III) Abstraction of proton from the carbocation

$$+ AlCl_4^- \longrightarrow \underset{\text{Alkyl benzene}}{C_6H_5R} + AlCl_3 + HCl$$

Q.23. Describe the mechanism of halogenation. (*Kerala, 2001; Purvanchal 2003*)

Ans. Halogenation of benzene. Halogenation of benzene and other aromatic hydrocarbons can be brought about by treating with halogens in the presence of Lewis acids such as ferric halides, anhydrous aluminium chloride. Chlorination of benzene is considered below:

$$C_6H_6 + Cl_2 \xrightarrow{FeCl_3} \underset{\text{Chlorobenzene}}{C_6H_5Cl} + HCl$$

Mechanism. Various steps involved in the reaction are as follows:

(I) Generation of electrophile

$$Cl-Cl + FeCl_3 \rightleftharpoons \overset{+}{Cl} + Fe\overline{Cl}_4$$

$+ Cl^+ \longrightarrow$ I, II, III — Resonating structures; $\equiv$ Resonance stabilised carbocation

Cl

H Cl

(+) + $FeCl_4^-$ ⟶ + $FeCl_3$ + HCl

Chlorobenzene

Q.24. Discuss effect of substituents on orientation and reactivity of benzene ring.

(*G. Nanakdev, 2000; Awadh, 2000; Meerut, 2000; Devi Ahilya, 2001; Kurukshetra, 2001*)

Ans. It has been observed that substituents already attached to the benzene ring not only govern the orientation of further substitution but also affect the reactivity of the benzene ring. It is discussed in brief as below:

(I) Effect of substituents on orientation. As stated above, the nature of the group already attached to benzene ring determines the position of the incoming group. In general, groups have been classified into two categories:

(a) Ortho-para directing groups. The groups which direct the incoming group towards ortho and para positions are called ortho-para directing groups. Groups such as –R, $-C_6H_5$,–OH, –SH, –OR, $-NH_2$, – NHR,– NR_2, –Cl, –Br, –I, etc. are all ortho-para directing groups.

(b) Meta directing groups. The groups which direct the incoming groups towards meta position are called meta directing groups. Groups such as –COOH, –CHO, –CN, ,$-NO_2$, COR, – SO_3H, etc., are all meta directing groups.

It may be mentioned that groups which contain double or triple bond are usually meta directing while those which do not contain multiple bonds are ortho-para directing. However, there are certain exceptions to this rule.

(II) Effect of substituents on reactivity. Reactivity of the benzene ring in electrophilic substitution reactions depends upon the tendency of the substituent group already present in the benzene ring to release or withdraw electrons. A group that releases electrons activates benzene ring while the one which draws electrons deactivates the benzene ring. It is found that except halogens all ortho-para directing groups activate the ring and all meta directing groups deactivate the ring towards further electrophilic substitution. Thus, nitration of toluene can be carried out at room temperature, while that of nitrobenzene requires more drastic conditions. This is illustrated as below:

CH_3 (toluene) $\xrightarrow[300\ K\ (-H_2O)]{Dil.\ HNO_3/H_2SO_4}$ CH_3, NO_2 (*o*-Nitrotoluene) + CH_3, NO_2 (*p*-Nitrotoluene)

o-Nitrotoluene

p-Nitrotoluene

NO_2 (nitrobenzene) $\xrightarrow[300\ K\ (-H_2O)]{Conc.\ HNO_3/H_2SO_4}$ NO_2, NO_2 (*m*-Dinitrobenzene)

m-Dinitrobenzene

Q.25. Discuss theory of reactivity in aromatic compounds on the basis of inductive and resonance effects.

Ans. Theory of reactivity. It has been observed that the rate determining step in electrophilic aromatic substitution is the formation of intermediate resonance stabilised carbocation.

Benzene + A^+ (Electrophile) —Slow→ Resonance stabilised carbocation (H, A)

It is thus clear that any factor which stabilises the intermediate carbocation will also stabilise the transition state leading to its formation. Consequently, the carbocation will be formed more quickly and the rate of the overall reaction increases. On the other hand, factors which destabilise the carbocation will decrease the ease of its formation and hence decrease the rate of the reaction.

It is a familiar fact that dispersion of charge leads to the stability of the system. Now electron releasing groups tend to decrease the positive charge on the carbocation and thus stabilise the ion. As a result, the rate of the reaction increases. On the other hand, electron withdrawing groups destabilise the carbocation by intensifying the positive charge on it. Consequently the rate of further electrophilic substitution reaction decreases. It may be mentioned that release or withdrawal of electrons may occur due to inductive effect alone or through the net result of inductive and resonance effect.

The effect of substituents on reactivity can be illustrated by considering electrophilic substitution in benzene, toluene and nitrobenzene. The relative rates in these three substitution eactions will depend upon the relative stabilities of their corresponding intermediate carbocations ormed as shown below:

Toluene (CH_3) + A^+ ⟶ I (CH_3, +, H, A)

Benzene + A^+ ⟶ II (+, H, A)

Nitrobenzene (N, O, O) + A^+ ⟶ III (N, O, O, +, H, A)

Due to its electron releasing inductive effect, methyl group tends to disperse the positive charge and thus stabilise the carbocation (I). As a result, the carbocation (I) is formed more quickly than the carbocation (II) from benzene. Consequently toluene undergoes electrophilic substitution at a faster rate than does benzene.

On the other hand, the nitro group tends to intensify the positive charge due to its electron withdrawing inductive and resonance effects. This destabilises the carbocation (III) which is, therefore, formed slowly than the carbocation (II) from benzene. Consequently nitrobenzene undergoes electrophilic substitution at a slower rate as compared to benzene.

Q. 26. Give electronic interpretation of the ortho-para directing influence of amino groups.
(Kerala, 2001; Devi Ahilya, 2001; Nagpur, 2003; Ranchi, 2003)

Ans. Amino group exerts electron withdrawing (–I) and electron releasing (+M) effects. Of the two opposing effects, the resonance effect dominates and thus overall behaviour of $-NH_2$ group is electron releasing, it acts as an activator.

The ortho and para directing influence of $-NH_2$ group can be explained by assuming that nitrogen can share more than a pair of electrons with the benzene ring and can accommodate a positive charge. Thus, consider the case of further electrophilic substitution in aniline. The various resonating structures of the carbocations formed by ortho, para and meta attack are given below:

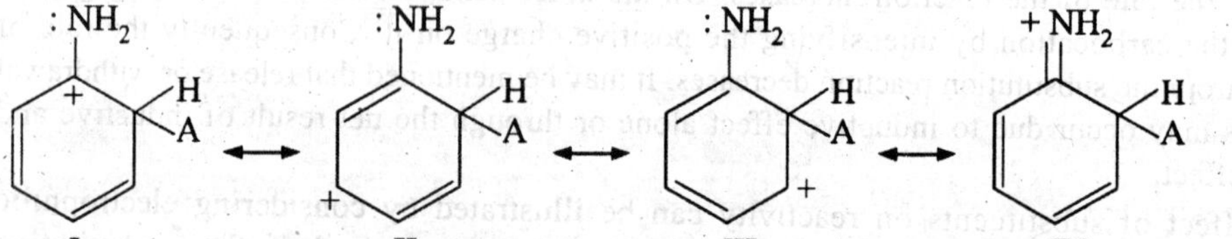

I ⟷ II ⟷ III ⟷ IV Ortho attack

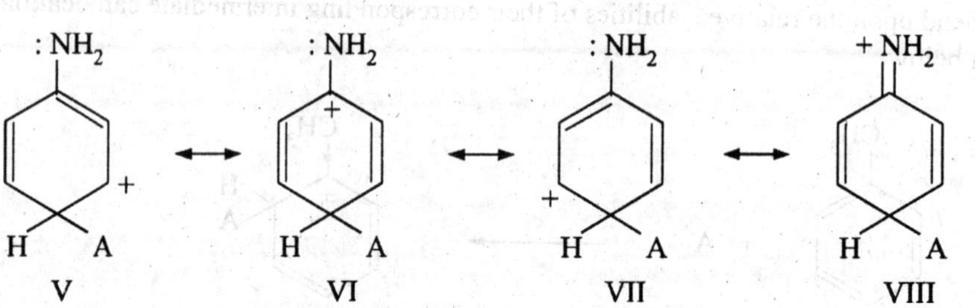

V ⟷ VI ⟷ VII ⟷ VIII Para attack

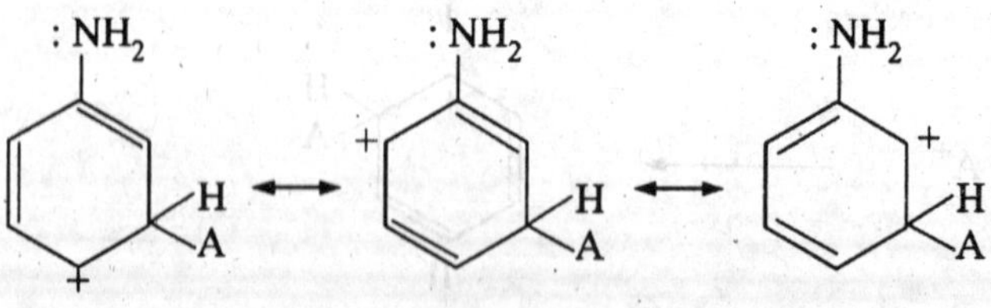

IX ⟷ X ⟷ XI Meta attack

The intermediate carbocation resulting from ortho as well as para attack is a resonance hybrid of four structures while the one formed by meta attack is a resonance hybrid of three structures. Further, in structures IV and VIII, the positive charge is carried by nitrogen. These structures are more stable since in every atom (except hydrogen) has a complete octet of electrons in them. No such structure is, however, possible in case of meta attack. It is, therefore, clear that the resonance hybrid carbocations resulting from ortho and para attack are more stable than the carbocation formed by attack at the meta position. Consequently further electrophilic substitution in aniline occurs faster at the ortho and para positions than at the meta position. In other words $-NH_2$ group is an ortho and para directing group.

Q.27. Halogens are electron withdrawing and yet they direct the incoming group to ortho and para positions. Explain. (*Delhi 2002; Panjab 2005; Guwahati 2005*)

Or

Explain why chlorine in chlorobenzene is ortho and para directing although it has –I effect during electrophilic substitution in benzene ring. (*Patna, 2010*)

Ans. Halogens exert electron withdrawing (–I) and electron releasing (+M) effects. Due to high electronegativities of halogens, the inductive effect predominates over the mesomeric effect and thus the overall behaviour of halogens is electron withdrawing. In other words, halogens act as deactivators for further substitution.

In order to account for their ortho and para directing nature, it has been assumed that halogens can share more than one pair of electrons with the benzene ring and can accommodate positive charge. Thus, consider the case of further electrophilic substitution in chlorobenzene. The various resonating structures of the carbocation formed by ortho, para and meta attack are given below:

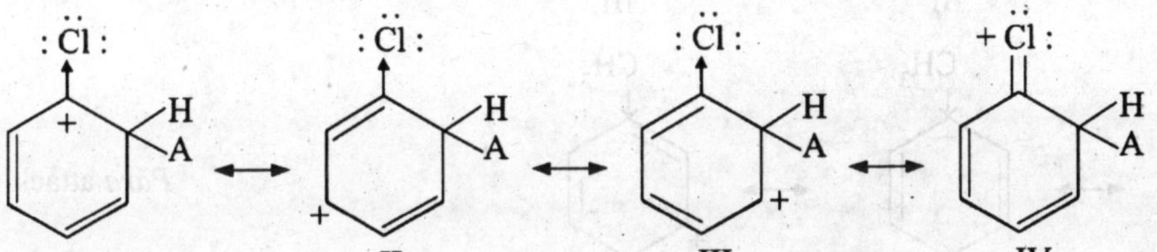

I ⟷ II ⟷ III ⟷ IV *Ortho* attack

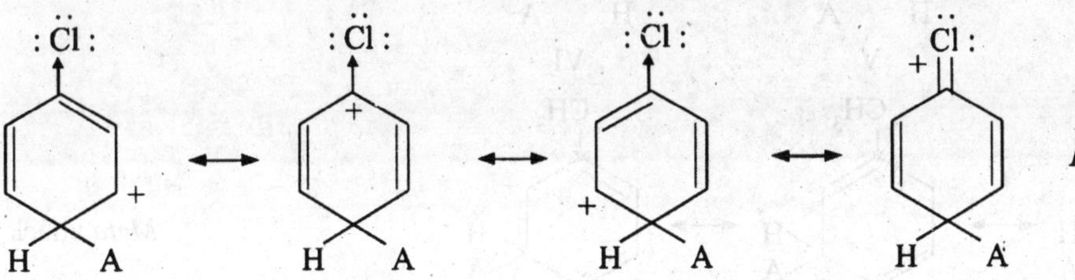

V ⟷ VI ⟷ VII ⟷ VIII *Para* attack

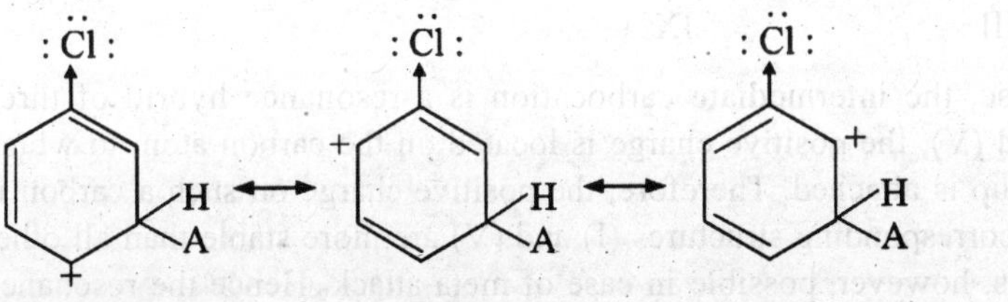

IX ⟷ X ⟷ XI *Meta* attack

It is clear that the intermediate carbocation resulting from ortho as well as para attack is a resonance hybrid of four structures while the one formed by meta attack is a resonance hybrid of three structures. Structures (I) and (VI) are highly unstable since in these the positive charge is carried by that carbon which is linked to electron withdrawing chlorine atom. However in structures (IV) and (VIII), the positive charge is carried by chlorine. These structures are extra stable since in every atom (except hydrogen) has a complete octet of electrons in them. No such structure is, however, possible in case of meta attack. Thus, the resonance hybrid carbocations resulting from ortho and para attack are more stable than that formed by attack at the meta position. As a result, chlorine is ortho and para directing. The same is true for other halogens. It may thus be concluded that in case of halogens, the reactivity is controlled by the stronger inductive effect and the orientation is determined by mesomeric effect.

Q.26. Give electronic interpretation of ortho and para directing influence of alkyl groups.
(Meerut, 2000; Lucknow, 2003)

Or

Why methyl group in toluene is o- and p-directing towards electrophilic substitution?

Ans. Alkyl groups exert electron releasing inductive effect (+I effect). Consider the case of further electrophilic substitution in toluene which contains an electron releasing methyl group. The various resonating structures of the carbocations formed by ortho, para and meta attack are given below:

I ⟷ II ⟷ III — *Ortho* attack

IV ⟷ V ⟷ VI — *Para* attack

VII ⟷ VIII ⟷ IX — *Meta* attack

We find that in each case, the intermediate carbocation is a resonance hybrid of three structures. In structures (I) and (V), the positive charge is located on the carbon atom to which electrons releasing methyl group is attached. Therefore, the positive charge on such a carbon is highly dispersed and thus the corresponding structures (I) and (V) are more stable than all other structures. No such structure is, however, possible in case of meta attack. Hence the resonance hybrid carbocations resulting from ortho and para attack are more stable than the one formed by attack at the meta position. Therefore further electrophilic substitution in toluene occurs faster at the ortho and para position than at the meta position. In other words, methyl group is an ortho and para directing group. Other alkyl groups behave similarly.

Q.29. Explain deactivating and meta directing nature of nitro group towards electrophilic aromatic substitution.
(Kerala, 2001; Delhi, 2004; Bangalore, 2003; Nagpur, 2008; Andhra, 2010)

Ans. Nitro group is electron withdrawing in nature. In this case, the electron withdrawal occurs through electron withdrawing effect (–I effect) as well as electron withdrawing resonance effect (–M effect). Due to its electron withdrawing character, nitro group deactivates the benzene ring towards further electrophilic substitution.

Let us examine the directing influence of $-NO_2$ group by considering electrophilic substitution in nitrobenzene. The various resonating structures of the carbocations formed by ortho, para and meta attack are given below:

I (Specially unstable) ⟷ II ⟷ III — *Ortho* attack

IV ⟷ V (Specially unstable) ⟷ VI — *Para* attack

VII ⟷ VIII ⟷ IX — *Meta* attack

In the contributing structures (I) and (V), the positive charge is located on that carbon atom which is directly linked to electron withdrawing nitro group. Although $-NO_2$ group withdraws electrons from all positions, it does so most from the carbon directly attached to it. Hence this carbon atom, already made positive by nitro group has little tendency to accommodate the positive charge of the carbocation. Consequently structures (I) and (V) are unstable and their contribution towards stabilization of the carbocation is almost negligible. Thus carbocations formed by ortho and para attack are virtually resonance hybrids of only two structures while the one formed by meta attack is a resonance hybrid of three structures. Therefore the resonance hybrid carbocation resulting from meta attack is more stable than the carbocations resulting from ortho and para attack. Consequently, further electrophilic substitution takes place at the meta position. Thus, nitro group is meta directing.

Q.30. Alkoxy group is *o-p* directing even though oxygen is more electronegative than carbon. Explain.

Ans. Alkoxy group is *o-p* directing because the lone-pair of electrons on oxygen takes part in resonance as illustrated below:

$\ddot{O}CH_3$ ⟷ $\overset{+}{O}CH_3$ ⟷ $\overset{+}{O}CH_3$ ⟷ $\overset{+}{O}CH_3$

Ortho and para positions relative to methoxy group are negative. Hence in further substitution the electrophile will preferentially attach at these negative points.

Q.31. Describe the methods of preparation of toluene. (*Nagpur, 2005*)

Ans. (*i*) Friedel-Crafts methylation of benzene. Benzene on treatment with methyl chloride in the presence of a Lewis acid catalyst such as anhydrous aluminium chloride yields toluene.

$$\underset{\text{Benzene}}{C_6H_6} + CH_3Cl \xrightarrow{\text{Anhydrous } AlCl_3} \underset{\text{Toluene}}{C_6H_5CH_3}$$

(*ii*) Wurtz-Fittig reaction. When an arylhalide and alkyl halide are warmed with metallic sodium in ethereal solution, an alkyl benzene is formed.

$$\underset{\text{Bromobenzene}}{C_6H_5Br} + 2Na + CH_3Br \xrightarrow[\Delta]{\text{Ether}} \underset{\text{Toluene}}{C_6H_5CH_3} + 2NaBr$$

(*iii*) Reduction of benzaldehyde. Benzaldehyde on refluxing with zinc amalgam and conc. hydrochloric acid (*Clemmensen reduction*) yields toluene.

$$\underset{\text{Benzaldehyde}}{C_6H_5CHO} \xrightarrow{\text{Zn-Hg and HCl}} \underset{\text{Toluene}}{C_6H_5CH_3}$$

The reduction can also be carried out by heating benzaldehyde with hydrazine and a strong base like potassium hydroxide or potassium *tert- butoxide* (*Wolff-Kishner reduction*).

$$C_6H_5CHO \xrightarrow[KOC(CH_3)_3 \text{ or } KOH]{NH_2NH_2 \text{ and}} C_6H_5CH_3$$

(*iv*) Grignard reaction. Phenyl magnesium bromide on treatment with methyl halide yields toluene.

$$C_6H_5MgBr + CH_3-X \longrightarrow C_6H_5CH_3 + Mg\begin{matrix} X \\ Br \end{matrix}$$

(*v*) Decarboxylation of toluic acids. Toluene can also be prepared by heating *ortho-*, *meta-* or *para*-toluic acid with soda lime.

CH_3 / COOH — o-toluic acid; CH_3 / COOH — m-toluic acid; or CH_3 / COOH — p-toluic acid $\xrightarrow{\text{Soda lime, }\Delta}$ CH_3 (toluene)

(*vi*) **Reduction of cresols with zinc dust.** *ortho, meta or para-cresol* on heating with zinc dust yields toluene.

CH_3 / OH — o-cresol or CH_3 / OH — m-cresol; CH_3 / OH — p-cresol $\xrightarrow[\text{D}]{\text{Zn dust}}$ CH_3 (toluene)

Q.32. Describe chemical properties of toluene. (*Nagpur, 2005*)

Ans. (*i*) Electrophilic aromatic substitution reactions. Methyl group in toluene activates the aromatic ring and directs the incoming group to *ortho* and *para-* positions. Some important reactions of this type are as given below:

(*a*) **Halogenation.** *In the absence of sunlight and in the presence of catalyst*, toluene reacts with halogens to give *ortho-* and *para-* substituted toluene such as

$$C_6H_5CH_3 + Cl_2 \xrightarrow[\text{of sunlight}]{\text{Fe, in absence}} \underset{o\text{-chlorotoluene}}{o\text{-}ClC_6H_4CH_3} + \underset{p\text{-chlorotoluene}}{p\text{-}ClC_6H_4CH_3}$$

(*b*) **Nitration.** Toluene reacts with nitrating mixture to yields *ortho-* and *para*-nitrotoluenes. The reaction takes place at a lower temperature as compared to benzene.

$$C_6H_5CH_3 + HNO_3 \xrightarrow{H_2SO_3,\ 303\text{ K}} \underset{o\text{-nitrotoluene}}{o\text{-}NO_2C_6H_4CH_3} + \underset{p\text{-nitrotoluene}}{p\text{-}NO_2C_6H_4CH_3}$$

(*c*) **Friedel-Crafts alkylation.** In this reaction nature of product depends on reaction temperature. At 273 K toluene reacts with methyl chloride to yield a mixture of *ortho*-and *para-* xylenes while at 353 K, it gives *meta*-xylene.

CH_3 (toluene) $+ CH_3Cl \xrightarrow{AlCl_3,\ 273\ K}$ o-xylene + p-xylene

o-xylene

p-xylene

CH_3 (toluene) $+ CH_3Cl \xrightarrow{AlCl_3,\ 353\ K}$ m-xylene

m-xylene

(*d*) Friedel-Crafts acylation

CH_3 (toluene) $+ CH_3COCl \xrightarrow{AlCl_3}$ o-methyl acetophenone ($COCH_3$) + p-methyl acetophenone ($COCH_3$)

o-methyl acetophenone

p-methyl acetophenone

(*e*) Sulphonation

CH_3 (toluene) $+ H_2SO_4 \xrightarrow{\Delta}$ o-toluene sulphonic acid (SO_3H) + p-toluene sulphonic acid (SO_3H)

o-toluene sulphonic acid

p-toluene sulphonic acid

(*ii*) Substitution at the side chain (methyl group). *Halogenation.* In presence of sunlight and in absence of a catalyst or at high temperature in presence of free radical initiators, halogens displace the hydrogen atoms of the side chain *i.e.*, methyl group.

$C_6H_5CH_3 \xrightarrow[hv]{Cl_2} C_6H_5CH_2Cl \xrightarrow[hv]{Cl_2} C_6H_5CHCl_2 \xrightarrow[hv]{Cl_2} C_6H_5CCl_3$

Benzyl chloride — Benzal chloride — Benzo trichloride

This reaction takes place via free radical chain mechanism as in the case of halogenation of alkanes.

(*iii*) Oxidation of the side chain. Though benzene and alkanes are quite stable towards oxidation, toluene or any monoalkyl substituted benzene on oxidation with potassium permanganate, potassium dichromate or even with dilute nitric acid, yields benzoic acid.

$$C_6H_5CH_3 + 3\,[O] \xrightarrow[\text{or dil. } HNO_3\ \Delta]{KMnO_4,\ \text{or } K_2Cr_2O_7} C_6H_5COOH + H_2O$$

Benzoic acid

$$C_6H_5CH_2CH_2CH_3 + 9\,[O] \xrightarrow{KMnO_4} C_6H_5COOH + 2CO_2 + 3H_2O$$

On treatment with a milder oxidising agent such as chromyl chloride in carbon tetrachloride solution, toluene yields benzaldehyde.

$$C_6H_5CH_3 + 2\,[O] \xrightarrow[\text{in } CCl_4]{CrO_2Cl_2} C_6H_5CHO + H_2O$$

Benzaldehyde

(*iv*) Addition of hydrogen. Like benzene, toluene also add on hydrogen in the presence of catalyst to form methyl cyclohexane.

$$\underset{\text{Toluene}}{C_6H_5CH_3} + 3H_2 \xrightarrow[\Delta]{Ni} \underset{\text{Methylcyclohexane}}{C_6H_{11}CH_3}$$

Q.33. How can you obtain xylenes from toluene?

Ans. By alkylation of toluene

$$C_6H_5CH_3 + CH_3Cl \xrightarrow[AlCl_3]{\text{Anhy.}} \underset{o\text{-xylene}}{C_6H_4(CH_3)_2} + \underset{p\text{-xylene}}{C_6H_4(CH_3)_2}$$

$$C_6H_5CH_3 + CH_3Cl \xrightarrow[AlCl_3\ 353\ K]{\text{Anhyd.}} \underset{m\text{-xylene}}{C_6H_4(CH_3)_2}$$

Q.34. Describe chemical properties of xylenes.

Ans. (*i*) Electrophilic substitution reactions. Xylenes show all the characteristic aromatic reactions. The incoming group attacks preferentially the positions *para* to any of the two methyl groups in case of *o*-and *m*-isomers.

In case of *p*-isomer attack can take place any of the four available positions and we get the same compound because all the four positions are equivalent.

The various positions susceptible for electrophilic attack are as shown below:

CH_3 CH_3 CH_3 CH_3 CH_3 CH_3

o-xylene (same product is obtained in both cases) *m*-xylene (same product is obtained in both cases) *p*-xylene (same product is obtained in four cases)

(*ii*) Substitution at the side chains. In the presence of sunlight and at high temperature, hydrogen atoms of methyl groups can be displaced by halogens. The substitution takes place at both the methyl groups.

$$C_6H_4(CH_3)_2 \xrightarrow[\text{Sunlight}]{Cl_2} C_6H_4(CH_2Cl)(CH_3) \xrightarrow[\text{Sunlight}]{Cl_2} C_6H_4(CH_2Cl)_2$$

o, m or *p*-xylene — Methyl benzyl chloride — Xylene chloride

(*iii*) Isomerisation. Any one of the three isomeric xylenes on heating with HF – BF_3 yields a mixture of *o*, *m* and *p*-xylenes in fixed ratio *i.e.*, 18%, 60% and 22% respectively, *m*-xylene, being thermodynamically more stable, is formed in larger amounts.

$$C_6H_4(CH_3)_2 \xrightarrow[\Delta]{HF\text{-}BF_3} o\text{-xylene} + m\text{-xylene} + p\text{-xylene}$$

o, m or *p*-xylene — *o*-xylene 18% — *m*-xylene 60% — *p*-xylene 22%

(*iv*) Oxidation. Xylenes on oxidation with potassium permanganate or dil. nitric acid are oxidised to corresponding phthalic acids. The oxidation takes place via the formation of toluic acid.

$$C_6H_4(CH_3)_2 \xrightarrow{[O]} C_6H_4(CH_3)(COOH) \xrightarrow{[O]} C_6H_4(COOH)_2$$

o-xylene — *o*-toluic acid — Phthalic acid m.p. = 504 K

m-xylene $\xrightarrow{[O]}$ m-toluic acid $\xrightarrow{[O]}$ Iso-phthalic acid (m.p. = 619 K)

p-xylene $\xrightarrow{[O]}$ p-toluic acid $\xrightarrow{[O]}$ Terephthalic acid

Q.35. Benzyl chloride is more reactive towards nucleophilic substitution reactions than methyl chloride. Explain.

Or

Both benzyl and methyl carbocations are primary but benzyl cation is much more stable than methyl cation.

Ans. The greater reactivity of benzyl chloride is mainly due to the fact that it readily ionises to form a resonance stabilised benzyl carbocation. On the other hand, methyl chloride does not ionise to form methyl carbocation because there is no resonance.

$C_6H_5CH_2Cl \rightleftharpoons Cl^- +$ [I $\leftrightarrow$ II $\leftrightarrow$ III $\leftrightarrow$ IV]

Resonating structures of benzyl carbocation

Q.36. What is side chain halogenation? Discuss the mechanism of side chain halogenation of toluene. *(Lucknow, 2000)*

Ans. Side chain halogenation. The reaction in which one or more hydrogen atoms of the side chain are replaced by the halogen atoms is called *side chain halogenation.* It can be brought about by treating an arene with a suitable halogen in the presence of light or heat and the absence of halogen carriers.

$C_6H_5CH_3$ (Toluene) $\xrightarrow[\text{(heat or light)}]{Cl_2}$ $C_6H_5CH_2Cl$ (Benzyl chloride) $\xrightarrow[\text{(heat or light)}]{Cl_2}$ $C_6H_5CHCl_2$ (Benzal dichloride) $\xrightarrow[\text{(heat or light)}]{Cl_2}$ $C_6H_5CCl_3$ (Benzotrichloride)

Mechanism of halogenation. Side chain halogenation of toluene occurs by a free radical mechanism as given below:

(*a*) Chain initiation step

$$Cl-Cl \xrightarrow[\text{(Homolytic fission)}]{\text{Heat or light}} 2\,Cl.$$

(*b*) Chain propagating steps

(I)

$$C_6H_5-CH_3 + Cl^{\bullet} \longrightarrow C_6H_5-\dot{C}H_2 + HCl$$

Benzyl free radical

(II)

$$C_6H_5-\dot{C}H_2 + Cl-Cl \longrightarrow C_6H_5-CH_2-Cl + Cl^{\bullet}$$

Benzyl chloride

The above steps (I) and (II) are repeated again and again.

(*c*) Chain terminating steps. When the free radicals combine with each other, the reaction comes to a stop and thus the chain of reaction gets erminated.

$$Cl^{\bullet} + Cl^{\bullet} \longrightarrow Cl-Cl$$

When excess of chlorine is used, other hydrogen atoms are also replaced in the same manner. Thus,

$$\underset{\text{Benzyl chloride}}{C_6H_5-CH_2Cl} + Cl^{\bullet} \longrightarrow C_6H_5-\dot{C}HCl + HCl$$

$$C_6H_5-\dot{C}HCl + Cl-Cl \longrightarrow \underset{\text{Benzal chloride}}{C_6H_5-CHCl_2} + Cl^{\bullet}$$

$$C_6H_5-CHCl_2 + Cl^{\bullet} \longrightarrow C_6H_5-\dot{C}Cl_2 + HCl$$

$$C_6H_5-\dot{C}Cl_2 + Cl-Cl \longrightarrow \underset{\text{Benzotrichloride}}{C_6H_5-CCl_3} + \dot{C}l$$

Q.37. Explain why bromination of ethylbenzene gives only 1-bromo-1-phenylethane without a trace of 2-bromo-1-phenylethane.

Ans. Side chain halogenation of alkyl benzene occurs only at the carbon atom directly attached to the benzene ring. This is because the intermediate benzylic free radicals are more stable and hence easily formed. For example, benzylic free radical ($C_6H_5-\dot{C}HCH_3$) obtained from ethylbenzene is a resonance hybrid of the following structures:

H—Ċ—CH_3 ⟷ H—C—CH_3 ⟷ H—C—CH_3 ⟷ H—C—CH_3

On the other hand, the second free radical ($C_6H_5 - CH_2 - \dot{C}H_2$) that can be obtained from ethylbenzene is a resonance hybrid of only two structures.

$CH_2 - \dot{C}H_2$ ⟷ $CH_2 - \dot{C}H_2$

Due to its greater number of contributing structures, benzylic free radical is more resonance stabilised than non-benzylic free radical. Thus, bromination of ethylbenzene occurs exclusively at the **benzylic position.**

$$C_6H_5CH_2-CH_3 \xrightarrow[\text{heat or light}]{Br_2} C_6H_5CH(Br)-CH_3 + HBr$$

1-Bromo-1-phenylethane

Q.36. What are (*i*) σ complexes (*ii*) π complexes. (ferrocene) (*Nagpur 2993*)

Ans. (*i*)- σ-complexes. σ-*complexes* are obtained by the *formation of a σ-bond between the donor and acceptor atoms.* An important example of a σ-complex is the **non-aromatic carbocation** or **arenium ion** formed as an intermediate during electrophilic aromatic substitution. In the first step of electrophilic aromatic substitution, the electrophile attacks the π electron cloud of benzene ring and takes away two electrons of the π system to form a σ bond between itself and one carbon of the benzene ring. This carbon becomes sp^3 hybridised and is not left with any unhybridised π-orbital. As a results, the cyclic π electron cloud is interrupted and the remaining four π electrons are delocalised over the remaining five sp^2 hybrid carbon atoms.

Y^+ + C_6H_6 ⟶ (H, Y, +) ← sp^3 hybridised

Aremium ion or σ-complex

This is an example of commonly formed σ-complex intermediate in organic reaction mechanism.

Wilkinson's catalyst which is a complex of rhodium metal with the ligand triphenyl phosphine (PPh_3) having the formula RhCl (PPh_3) is an example of transition metal σ complex. It is used as a catalyst in the hydrogenation of alkenes by forming a π-complex with the alkene by means of π electron cloud of alkene.

(*ii*) **π-complexes.** *These are complexes formed by the overlap of a filled π-orbital of compounds like alkenes with the orbital of the acceptor atom.* The addition of halogens to alkenes involves the formation of a π-complex precursor of the cyclic halonium ion as given below.

As the halogen molecule (say bromine) approaches the alkene, it gets polarised. The positively polarised bromine end forms a π-complex with the π-electrons of the alkene. This π-complex is reversibly changed into the bromonium ion as shown below.

It may be noted that in the formation of above π-complex, bromine gets bonded, not only to one of the alkene carbons, but both by π-electrons of alkene.

$$>C=C< + \overset{\delta+}{Br}-\overset{\delta-}{Br} \longrightarrow >C{=}C< \cdots \overset{\delta+}{Br}-\overset{\delta-}{Br} \rightleftharpoons >C-C< \,(\text{bridged } Br^+) + :Br^-$$

π-complex Bromonium ion

Dicyclopentadienyl iron $[(C_5H_2^-)]_2$ Fe, commonly known as ferrocene is an example of transition metal π complex. It is a stable molecule in which iron is "sandwiched" between two flat cyclodienyl rings. The carbon iron bond results by overlap between π-cloud of cyclopentadienyl anions and 3*d* orbitals of iron.

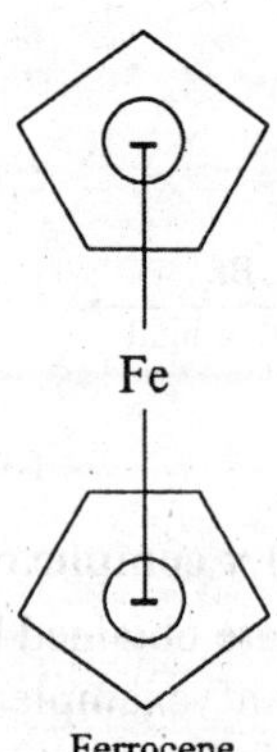

Ferrocene

Q.39. What happens when:

(i) Benzyl chloride is subjected to oxidation with lead nitrate.

(ii) Toluene is treated with methyl chloride in the presence of anhydrous aluminium chloride.

(iii) Benzyl chloride is treated with silver acetate.

(iv) Chlorine is passed through boiling toluene. (*Kalyani 2004, Rajasthan 2004*)

(v) *o*-Xylene is treated with hot aqueous $KMnO_4$ solution.

Ans.

(*i*)

$$C_6H_5CH_2Cl \xrightarrow[(O)]{Pb(NO_3)_2} C_6H_5CHO$$

Benzyl chloride Benzaldehyde

(*ii*)

$$C_6H_5CH_3 \text{ (Toluene)} + CH_3Cl \xrightarrow[-\,HCl]{\text{Anhydrous } AlCl_3} o\text{-xylene} + p\text{-xylene}$$

Toluene — *o*-xylene — *p*-xylene

(*iii*)

$$C_6H_5CH_2Cl + CH_3COOAg \longrightarrow AgCl + CH_3COOCH_2C_6H_5$$

Benzyl chloride — Silver accetate — Benzyl acetate

(*iv*)

$$C_6H_5CH_3 \xrightarrow[-\,HCl]{Cl_2\text{ (heat)}} C_6H_5CH_2Cl \xrightarrow[-\,HCl]{Cl_2\text{ (heat)}} C_6H_5CHCl_2 \xrightarrow[-\,HCl]{Cl_2\text{ (heat)}} C_6H_5CCl_3$$

Toluene — Benzylchloride — Benzaldichloride — Benzotrichloride

(*v*)

$$C_6H_4(CH_3)_2 + 6\,[O] \xrightarrow[\text{(heat)}]{KMnO_4/H_2O} C_6H_4(COOH)_2 + H_2O$$

o-Xylene — Phthalic acid

Q.40. Explain the following:

(*i*) *n*-Butylbenzene on oxidation gives benzoic acid but *t*-butylbenzene does not.

(*ii*) Benzene is stable towards oxidising agents but toluene can be oxidised easily to benzoic acid.

Ans. (*i*) It is believed that oxidation involves the attack on benzylic hydrogen. *t*-Butylbenzene does not possess benzylic hydrogen and hence it can't undergo oxidation easily. On the other hand, *n*-butylbenzene is easily oxidised due to the presence of benzylic hydrogens. Thus,

$$C_6H_5-CH_2-CH_2-CH_2-CH_3 + 12\,[O] \xrightarrow[H_2O]{\text{Hot } KMnO_4/} C_6H_5COOH + 3CO_2 + 4H_2O$$

n-Butylbenzene

$$C_6H_5-C(CH_3)_2-CH_3 \xrightarrow{\text{Hot } KMnO_4/H_2O} \text{No reaction}$$

t-Butylbenzene

(*ii*) Benzene ring exhibits unusual stability and resists oxidation even by strong oxidising agents This can be explained in terms of delocalisation of π electrons and its high value of resonance energy.

Toluene can easily be oxidised due to the presence of benzylic hydrogens *i.e.*, hydrogens attached to carbon linked directly to the benzene ring which are highly reactive. Thus

$$C_6H_5CH_3 + 3[O] \xrightarrow[370\text{ K}]{KMnO_4/H_2O} C_6H_5COOH + H_2O$$

Toluene → Benzoic acid

Q.41. Fill in the blanks, giving names of products.

(*i*) $C_6H_5CH_3 + H_2SO_4 \longrightarrow ?$

(*ii*) $C_6H_5CH_3 + Cl_2 \xrightarrow{h\nu} ?$

Ans.

(*i*) $C_6H_5CH_3 + H_2SO_4 \longrightarrow o\text{-}CH_3C_6H_4SO_3H + p\text{-}CH_3C_6H_4SO_3H$

Toluene → *o*-toluene sulphonic acid + *p*-toluene sulphonic acid

(*ii*) $C_6H_5CH_3 + Cl_2 \xrightarrow{h\nu} C_6H_5CH_2Cl + HCl$

Toluene → Benzyl chloride

Q. 42. Identify the products A and B.

$$C_6H_5Br \xrightarrow[\text{Dry ether}]{CH_3Br/Na} A \xrightarrow{KMnO_4} B$$

Ans.

Bromobenzene $\xrightarrow[\text{Dry ether}]{CH_3Br/Na}$ Toluene $\xrightarrow[[O]]{KMnO_4}$ Benzoic acid

Q. 43. Give the mechanism of nitration of toluene? *(Guwahati 2007)*

Ans.

Toluene ($C_6H_5CH_3$) $\xrightarrow[H_2SO_4]{HNO_3}$ o-nitrotoluene + p-nitrotoluene

Mechanism. (*i*) Generation of an electrophile

$$HONO_2 + 2H_2SO_4 \longrightarrow \overset{+}{N}O_2 + H_3\overset{+}{O} + 2HSO_4^-$$

Formation of carbocation

Alkyl groups exert electron releasing inductive effect (+I effect). Consider the case of further electrophilic substitution in toluene which contains an electron releasing methyl group. The various resonating structures of carbocations formed by ortho, para, meta attacks given below:

I ⟷ II ⟷ III Ortho attack

IV ⟷ V ⟷ VI Para attack

VII ⟷ VIII ⟷ IX Meta attack

We find that in each case, the intermediate carbocation is a resonance hybrid of three structures. In structures (I) and (V), the positive charge is located on the carbon atom to which electron releasing methyl group is attached. Therefore, the positive charge on such a carbon is highly dispersed and thus the corresponding structures (I) and (V) are more stable than all other structures. No such structure is, however, possible in case of meta attack. Hence the resonance hybrid carbocations resulting from ortho and para attack are more stable than the ones formed by attack at the meta position. Therefore, further electrophilic substitution in toluene occurs faster at ortho and para position than at meta position.

Q.44. Give the methods of preparation and properties of styrene (alkenyl benzene) and phenylacetylene (alkynyl benzene)

Ans. Styrene (Phenylethylene, vinyl benzene), $C_6H_5 - CH = CH_2$

Methods of preparation

By heating phenylmethylmethanol with sulphuric acid

$$C_6H_5MgBr + CH_3CHO \longrightarrow C_6H_5\underset{CH_3}{CHOMgBr} \longrightarrow C_6H_5CHOHCH_3$$

Phenylmethylmethanol

$$\xrightarrow[-H_2O,\ H_2SO_4]{\Delta} C_6H_5CH = CH_2$$

Styrene

2. *By heating 2-phenylethanol with alkali*

$$C_6H_5CH_2OH \xrightarrow{NaOH} C_6H_5CH = CH_2 + H_2O$$

2-Phenylethanol

3. *By heating cinnamic acid in the presence of a small quantity of quinol*

$$C_6H_5CH = CHCOOH \xrightarrow{\Delta} C_6H_5CH = CH_2 + CO_2 \uparrow$$

Cinnamic acid

Quinol prevents the polymeristion of styrene

4. *Styrene is obtained on industrial scale by dehydrogenating ethylbenzane in the presence of ZnO*

$$C_6H_5CH_2CH_3 \xrightarrow[600°C]{ZnO} C_6H_5CH{=}CH_2 + H_2$$

Ethyl benzene

Properties

1. It adds molecule of bromine to form dibromide

$$C_6H_5CH = CH_2 + Br_2 \longrightarrow C_6H_5CHBrCH_2Br$$

(Styrene dibromide)

2. It polymerises when exposed to sunlight in the presence of sodium

$$nC_6H_5CH = CH_2 \xrightarrow{sunlight} (C_8H_8)_n$$

3. Substitution in the side chain gives two possibilities

$$C_6H_5CH = CH_2 \xrightarrow{Cl_2} C_6H_5CH = CHCl \quad \text{(β-chlorostyrene)}$$

$$C_6H_5CH = CH_2 \xrightarrow{Cl_2} C_6H_5CCl = CH_2 \quad \text{(α-chlorostyrene)}$$

Phenylacetylene, $C_6H_5C \equiv CH$

Methods of preparation

1. **By decarboxylation of phenylpropiolic acid**

$$C_6H_5C \equiv CCOOH \xrightarrow{\Delta} C_6H_5C \equiv CH + CO_2$$

Propiolic acid   Phenyl acetylene

2. By heating ω–bromostyrene with ethanolic potassium hydroxide

$$C_6H_5CH = CHBr \xrightarrow{\text{Alc.KOH}} C_6H_5\,C{\equiv}CH + HBr$$

ω-bromostyrene

Properties

1. It forms metallic compounds with alkali metals

$$C_6H_5\,C \equiv CH + Na \longrightarrow C_6H_5\,C \equiv C\,Na + \frac{1}{2}H_2$$

Sodium phenylacetylide

2. It is reduced to styrene in the presence of zinc and acetic acid

$$C_6H_5C \equiv CH + 2[H] \xrightarrow{\text{Zn/Acetic acid}} C_6H_5CH = CH_2$$

Styrene

3. It adds a water molecule in the presence of sulphuric acid to form acetophenone

$$C_6H_5C \equiv CH + H_2O \xrightarrow{H_2SO_4} C_6H_5\,\underset{\text{unstable}}{\overset{OH}{\overset{|}{C}}} = CH_2 \longrightarrow \underset{\text{acatophenone}}{C_6H_5COCH_3}$$

Q.45. Identify the compounds A, B and C.

$$\textbf{Toluene} \xrightarrow{KMnO_4} A \xrightarrow[H^+]{MeOH} B \xrightarrow[H_2SO_4]{HNO_3} C$$

Ans. Toluene ($C_6H_5CH_3$) $\xrightarrow{KMnO_4}$ C_6H_5COOH (A) $\xrightarrow[H^+]{MeOH}$ $C_6H_5COOCH_3$ (B) $\xrightarrow[H_2SO_4]{HNO_3}$ m-$NO_2C_6H_4COOCH_3$ (C)

Q.46. Starting from benzene how will you prepare each of the following?

(i) Tribromobenzene *(Garhwal 2000; Awadh 2000)*

(ii) Insecticide *(Awadh, 2000; Garhwal, 2000; Meerut 2000)*

Ans. (*i*) Benzene $+ HOSO_3H \longrightarrow C_6H_5SO_3H \xrightarrow[\text{Fuse}]{NaOH} C_6H_5OH \xrightarrow{3Br_2}$ 2,4,6-tribromophenol (OH with Br, Br, Br) $\xrightarrow{Zn}$ $ZnO +$ 1,3,5-tribromobenzene (Br, Br, Br)

Tribromobenzene

(ii) Benzene + $3\,Cl_2$ $\xrightarrow{\text{In Presence of sunlight}}$ Benzene hexachloride (insecticide)

Q.47. Describe mechanism involved in the following reaction: (*Kumaon, 2000*)

$$C_6H_6 + CH_3CH_2CH_2Br \xrightarrow[\Delta]{AlCl_3} C_6H_5-CH(CH_3)-CH_3$$

Ans. $CH_3CH_2CH_2Br \xrightarrow[\Delta,\ Br]{AlCl_3} \overset{\oplus}{C}H_2-CH(H)-CH_3 \longrightarrow CH_3-\overset{\oplus}{C}H-CH_3$

$$C_6H_6 + CH_3-\overset{\oplus}{C}H-CH_3 \longrightarrow C_6H_5-CH(CH_3)-CH_3$$

Q.48. When chlorine is passed into boiling toluene three different chlorinated products are obtained. What products will be obtained if each of these chlorinated products is separately treated with aqueous sodium bydroxide solution? (*Kumaon, 2000*)

Ans. When chlorine is passed into boiling toluene chlorine displaces the hydrogen atoms of the side chain *i.e.,* methyl group, one by one.

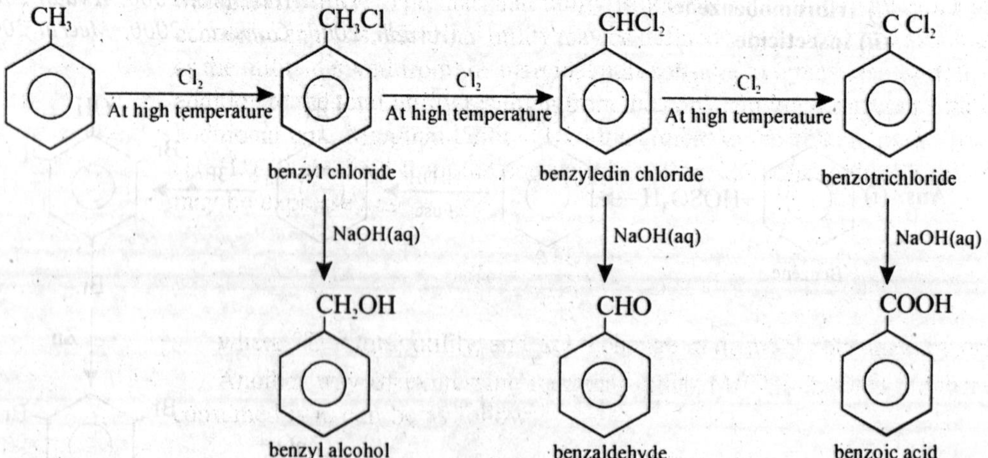

Q.49. Indicate the factors that determine the proportion of ortho and para isomers when a second substituent enters into a monosubstituted benzene. Substantiate with one example each (*Kerala, 2001*)

Ans. Since there are two ortho positions and one para position, ortho product is obtained in greater amount.

Q.50. Give the preparation of toluene from (*i*) benzene (*ii*) chlorobenzene. How does toluene react with (*i*) Cl_2 in presence of Fe Cl_3 (*ii*) Cl_2 in sunlight? (*Nagpur 2003*)

Ans.

$$C_6H_6 + CH_3Cl \xrightarrow[AlCl_3]{Anhy} C_6H_5CH_3$$

$$C_6H_5Cl + 2\,Na + CH_3Cl \xrightarrow{Ether} C_6H_5CH_3$$

Reactions

$$C_6H_5CCl_3 \xleftarrow[hv]{Cl_2} C_6H_5CH_3 \xrightarrow[FeCl_3]{Cl_2} o\text{-}ClC_6H_4CH_3 + p\text{-}ClC_6H_4CH_3$$

Q.51. How will you synthesise benzoic acid from benzene? (*Panjabi, 2005*)

Ans. Follwing steps are involved

$$\underset{\text{Benzene}}{C_6H_6} \xrightarrow[AlCl_3]{CH_3Cl} \underset{\text{Toluene}}{C_6H_5CH_3} \xrightarrow[[O]]{KMnO_4} \underset{\text{Benzoic acid}}{C_6H_5COOH}$$

12

ALIPHATIC HALOGEN COMPOUNDS

Q.1. Give the classification of aliphatic halogen compounds.

Ans. Aliphatic halogen compounds can be classified as follows:

(*a*) Monohalogen compounds. Compounds such as methyl chloride, which contain one halogen in the molecule are called monohalogen compounds.

(*b*) Dihalogen compounds. Compounds such as ethylene dichloride which contain two halogen atoms in the molecule are called dihalogen compounds. These can be further classified into two types:

(*i*) Vicinal dihalides. Compounds containing two halogens on neighbouring carbon atoms are called vicinal dihalides.

(*ii*) Geminal dihalides. Compounds containing two halogens on the same carbon in the molecule are called geminal dihalides. Vicinal (or vic.) and geminal (or gem.) dihalides are discussed in detail later in this chapter.

(*c*) Trihalogen compounds. Compounds such as chloroform or iodoform which contain three halogens in the molecule are called trihalogen compounds.

(*d*) Tetrahalogen compounds. Compounds such as carbon tetrachloride, which contain four halogen atoms in the molecule are called tetrahalogen compounds.

Monohalogen compounds are further classified into primary, secondary and tertiary alkyl halides.

Primary alkyl halides are those compounds in which the halogen is linked to a carbon which is further linked to one carbon atom. For example, $CH_3 - CH_2 - Cl$.

Secondary alkyl halides are those compounds in which the halogen is linked to a carbon which is further linked to two carbon atoms. For example.

$$\begin{array}{c} CH_3-CH-CH_3 \\ | \\ Cl \end{array}$$

Tertiary alkyl halides are those compounds in which the halogen is linked to a carbon which is further linked to three carbon atoms. For example,

$$\begin{array}{c} CH_3 \\ | \\ CH_3-C-CH_3 \\ | \\ Cl \end{array}$$

Q.2. Discuss the nomenclature of monohalogen aliphatic compounds.

Ans. Nomenclature of monohalogen compounds is explained as under by taking examples of a few compounds.

Structural formula	**Common name**	**IUPAC name**
CH_3Cl	Methyl chloride	Chloromethene
CH_3-CH_2Br	Ethyl bromide	Bromoethane
$CH_3-CH_2-CH_2I$	*n*-Propyl iodide	1-Iodopropane
$CH_3-\overset{\overset{\displaystyle Cl}{\vert}}{CH}-CH_3$	*Iso*-propyl chloride	2-Chloropropane
$CH_3-CH_2-CH_2-CH_2-Br$	*n*-Butyl bromide	1-Bromobutane
$CH_3-CH_2-\overset{\overset{\displaystyle Br}{\vert}}{CH}-CH_3$	*Sec*-butyl bromide	2-Bromobutane
$CH_3-\underset{\underset{\displaystyle CH_3}{\vert}}{CH}-CH_2-Br$	*Iso*-butyl bromide	2-Methyl-1-bromopropane

Q. 3. Write all the possible structures for a compounds having molecular formula $C_5H_{11}Cl$. Name these according to IUPAC system. How many are optically active?

Ans. Structural formula	**IUPAC Name**
(*a*) $CH_3-CH_2-CH_2-CH_2-CH_2-Cl$	1-Chloropentane
(*b*) $CH_3-CH_2-CH_2-\underset{\underset{\displaystyle Cl}{\vert}}{CH}-CH_3$	2-Chloropentane
(*c*) $CH_3-CH_2-\underset{\underset{\displaystyle Cl}{\vert}}{CH}-CH_2-CH_3$	3-Chloropentane
(*d*) $CH_3-CH_2-\overset{\overset{\displaystyle CH_3}{\vert}}{CH}-CH_2-Cl$	1-Chloro-2 methylbutane
(*e*) $CH_3-CH_2-\overset{\overset{\displaystyle CH_3}{\vert}}{\underset{\underset{\displaystyle Cl}{\vert}}{C}}-CH_3$	2-Chloro-2-methylbutane
(*f*) $CH_3-\underset{\underset{\displaystyle Cl}{\vert}}{CH}-\underset{\underset{\displaystyle CH_3}{\vert}}{CH}-CH_3$	2-Chloro-3-methylbutane
(*g*) $CH_3-\overset{\overset{\displaystyle CH_3}{\vert}}{CH}-CH_2-CH_2-Cl$	1-Chloro-3-methylbutane
(*h*) $CH_3-\overset{\overset{\displaystyle CH_3}{\vert}}{\underset{\underset{\displaystyle CH_3}{\vert}}{C}}-CH_2-Cl$	1-Chloro-2, 2-dimethylpropane

Optically active forms. The isomers (*b*), (*d*) and (*f*) are optically active due to the presence of **chirality** in their structures.

Q.4. How does the boiling point of an alkyl halide vary

(i) with the halogen

(ii) with the alkyl group.

Ans. (*i*) For a given alkyl group, the boiling point decreases in the order.

$$RI > RBr > RCl$$

(*ii*) For a given halogen, the boiling point of alkyl halide increases with increase in the size of alkyl group.

(*iii*) In case of isomeric halides, the boiling point decreases with branching. This is because surface area decreases with branching. And the inter-molecular forces of attraction decrease with surface area.

Q.5. What are nucleophilic substitution reactions? Describe some typical nucleophilic reactions of alkyl halides.

Ans. Nucleophilic substitution reaction. The reaction involving the replacement of an atom or a group of atoms by a suitable nucleophile is called nucleophilic substitution reaction. The carbon-halogen bond in alkyl halides is polar due to greater electronegativity of the halogen as compared to carbon. It is depicted as below:

$$-\overset{|}{\underset{|}{C}}^{\delta^+}\!\!-\!\!-\!\!-\overset{\delta^-}{X}$$

The presence of partial positive charge on carbon atom makes it susceptible to nucleophilic attack. Thus, when a stronger nucleophile approaches the positively charged carbon of the alkyl halide, the halogen is readily displaced as halide ion as shown below:

$$-\overset{|}{\underset{|}{C}}^{\delta^+}\!\!-\!\!-\!\!-\overset{\delta^-}{X} + \underset{\text{Attacking nucleophile}}{:Nu^-} \longrightarrow -\overset{|}{\underset{|}{C}}-Nu + \underset{\text{Halide ion}}{X^-:}$$

Nucleophilic Substitutions of Alkyl halides

A few important nucleophilic substitution reactions are discussed below:

(*i*) Replacement by hydroxyl group (Formation of alcohols). Alkyl halides react with aqueous KOH or moist silver oxide, *i.e.*, AgOH to form alcohols.

$$RX + :OH^- \longrightarrow ROH + :X^-$$

$$\underset{\text{Ethyl bromide}}{C_2H_5Br} + KOH\ (aq) \longrightarrow \underset{\text{Ethyl alcohol}}{C_2H_5OH} + KBr$$

(*ii*) Replacement by cyano group (Formation of cyanides). Alkyl halides react with alcoholic potassium cyanide to form cyanides or nitriles.

$$RX + CN^-: \longrightarrow RCN + X^-:$$

$$C_2H_5Br + KCN \longrightarrow \underset{\substack{\text{Ethyl cyanide} \\ \text{(Propanenitrile)}}}{C_2H_5CN} + KBr$$

(*iii*) Replacement by alkoxy group (Formation of ethers). When treated with sodium alkoxide, alkyl halides yield ethers.

$$RX + :OR^- \longrightarrow R-O-R + :X^-$$

$$C_2H_5Br + \underset{\text{Sodium ethoxide}}{C_2H_5ONa} \longrightarrow \underset{\text{Diethyl ether}}{C_2H_5-O-C_2H_5} + NaBr$$

(iv) Replacement by alkynyl group (Formation of alkynes). When an alkyl halide is treated with sodium salt of an alkyne, higher alkynes are formed.

$$RX + :\bar{C} \equiv OR' \longrightarrow R - C \equiv C - R' + X^-:$$

$$\underset{\text{Sodium methyl acetylide}}{CH_3Br + NaC \equiv C - CH_3} \longrightarrow \underset{\text{2-Butyne}}{CH_3 - C \equiv C - CH_3} + NaBr$$

$$\underset{\text{Sodium acetylide}}{CH_3Br + NaC \equiv CH} \longrightarrow \underset{\text{Propyne}}{CH_3 - C \equiv CH} + NaBr$$

(v) Replacement by carboxylate group (Formation of esters). When heated with silver salts of carboxylic acids, alkyl halides give esters.

$$RX + R'CO\bar{O}: \longrightarrow R'COOR + \bar{X}:$$

$$C_2H_5Br + \underset{\text{Silver acetate}}{CH_3COOAg} \longrightarrow \underset{\text{Ethyl acetate (Ethylethanoate)}}{CH_3COOC_2H_5} + AgBr$$

Q.6. Alkyl halides contain a C – X polar bond, yet they are insoluble in water. Explain.

Ans. There is no possibility of formation of hydrogen bond between the molecule of alkyl halide and water. This is due to insufficient electronegativity difference between the halogen and hydrogen. This is in contrast to the case of alcohols which are water soluble due to the existence of hydrogen bonds as shown below:

$$....H - \underset{\underset{\text{Alcohol}}{R}}{O} .... H - \underset{\underset{\text{Water}}{H}}{O} ....H - \underset{\underset{\text{Alcohol}}{R}}{O} .... H - \underset{\underset{\text{Water}}{H}}{O} ....$$

Q.7. Halogenation of alkanes in the presence of tetraethyl lead proceeds at low temp. than in its absence. *(Himachal, 2000)*

Ans. Halogenation of alkanes is a free-radical reaction in which the reaction is started by halogen free-radical (say Cl•). Tetraethyl lead decomposes into ethyl radicals on slight heating as follows:

$$(C_2H_5)_4Pb \xrightarrow{\Delta} 4C_2H_5\bullet + Pb$$

Ethyl radicals combine with chlorine molecule to produce chlorine radicals

$$C_2H_5\bullet + Cl_2 \longrightarrow C_2H_5Cl + Cl\bullet$$

Chlorine radicals so produced, carry out halogenation of alkanes.

Q.8. What are SN[1] and SN[2] reactions of alkyl halides?

(Devi Ahilya, 2001; Dibrugarh, 2004)

Ans. reactions means nucleophilic substitution reaction of first order.

In such reactions, there is only one species in the rate determining step. An example of reaction is:

$$\underset{\text{Tert. butyl chloride}}{(CH_3)_3C - Cl} \xrightarrow{\bar{OH}} \underset{\text{Tert. butyl alcohol}}{(CH_3)_3C - OH} + Cl^-$$

Here the rate-determining step of the reaction is:

$$(CH_3)_3C - Cl \longrightarrow (CH_3)_3\overset{+}{C} + Cl^-$$

It involves only one substance on the reactant side.

SN[2] reaction means nucleophilic substitution reaction of second order. In such reactions there are two species in the rate-determining step.

An example of reaction is:

$$CH_3Br + O\bar{H} \longrightarrow CH_3OH + Br^-$$

In this reaction the rate determining step of the reaction is:

$$CH_3Br + O\bar{H} \longrightarrow [HO - - - CH_3 - - - Br]^-$$

As it involves two species on the reactant's side, it is a bimolecular or SN^2 reaction.

Q.9. Give the mechanism of the following reaction: (*Purvanchal 2007; Nagpur 2008*)

$$\mathbf{CH_3Br + KOH\ (aq) \longrightarrow CH_3OH + KBr}$$

Or

Discuss the SN^2 reaction mechanism of substitution in alkyl halides. Why does SN^2 reaction take place with stereochemical inversion?

(*Bangalore, 2002; Burdwan, 2004; Guwahati, 2006; Garhwal, 2010; Bharathiar, 2011*)

Ans. Hydrolysis of methyl bromide proceeds by SN^2, (meaning substitution nucleophilic reaction of the second order), mechanism. SN^2 reaction may also be termed as bimolecular substitution reaction. The term bimolecular implies that there are two reacting species in the rate determining step of the reaction.

A simple and typical example of SN^2 substitution is the hydrolysis of methyl bromide with aqueous sodium hydroxide.

$$CH_3Br + OH^- \longrightarrow CH_3OH + Br^-$$

The kinetic data reveal that the rate of the reaction is dependent upon the concentration of both these reactants

$$\text{Rate} = k\,[CH_3Br]\,[OH]^-$$

The reaction is of second order and thus it proceeds by a direct displacement mechanism in which both the reactants are present in the rate determining step. It is assumed that the nucleophile attacks the side of the carbon atom opposite to that of bromine. This is referred to as **backside attack.** As a result of this attack, a transition state is formed in which carbon atom is partially bonded to both –OH and –Br groups. In the transition state, the central carbon is *sp*² hybridised and the three hydrogens attached to it lie in the same plane with mutual bond angles of 120°. The reaction may be depicted as:

$$OH^- + H_3C\text{—}Br \longrightarrow \left[\overset{\delta-}{HO}\cdots CH_3 \cdots \overset{\delta-}{Br}\right] \longrightarrow HO\text{—}CH_3 + Br^-$$

Transition State

It may be noted that in the transition state, the hydroxide ion has diminished negative charge since it has started sharing its electrons with carbon. Similarly bromine develops a partial negative charge as it is tending to depart with the bonding electrons. When carbon-oxygen bond is completely formed, the carbon-bromine bond is altogether broken. The energy required for breaking a bond is compensated by the formation of a new bond. The overall reaction is thus a concerted process occurring in one step through the intervention of a single transition state. In a single step, C–Br bond is broken and C–OH bond is formed resulting in the formation of alcohol molecule.

Stereochemistry of SN^2 reactions. In SN^2 reactions, it is assumed that the nucleophile (OH^- in the present case) attacks the side of the carbon opposite to that of the leaving group. As a result, the configuration of the resulting product is inverted. In other words reaction proceeds with **stereochemical inversion.** This has been found to be so in actual practice. Thus when (–)-2 -bromooctane is hydrolysed under SN^2 conditions, it gives exclusively (+)-2-octanol. The reaction may thus be represented as:

$$H\bar{O} + \underset{(-)\text{-2-Bromooctane}}{(H)(C_6H_{13})(CH_3)C\text{—}Br} \longrightarrow \underset{\substack{\text{Transition State}\\ \text{(Inversion in } SN^2 \text{ reaction)}}}{\left[\overset{\delta^-}{HO}\text{----}C(C_6H_{13})(H)(CH_3)\text{----}\overset{\delta^-}{Br}\right]} \longrightarrow \underset{(+)\text{-2-Octanol}}{HO\text{—}C(C_6H_{13})(H)(CH_3)}$$

We have cited the example of (–)-2-bromooctane because it contains a chiral carbon atom (a carbon linked to four different groups). The configurations of the reactant and the products are different in such SN^2 reaction. We normally check the configuration by measuring the direction of rotation of plane polarised light although it is not necessary always that inversion of configuration should be accompanied by change in direction of rotation of plane-polarised light.

Q.10. Discuss relative reactivities of alkyl halides in substitution.

Ans. In SN^2 reactions, the nucleophile attacks the side of the carbon opposite to that of the leaving group because both groups are to be accommodated in the transition state. Thus consider the alkaline hydrolysis of methyl bromide which follows SN^2 mechanism. It involves the backside attack of the nucleophile (OH^-) on the carbon bearing the bromine atom. Evidently, when the hydrogen atoms of methyl group are successively replaced by bulkier alkyl groups such as methyl, ethyl, etc., there is an increased crowding around the central carbon both in the reactant and transition state. Due to steric hindrance of the alkyl groups, the nucleophile finds it more and more difficult to attack the carbon carrying the halogen. Thus, in the reaction of hydrolysis of an alkyl bromide, simultaneous presence of –OH and –Br in the transition state becomes difficult because of overcrowding due to bulky groups. Greater the size of the alkyl group attached to halogen bearing carbon atom, greater is the difficulty in the transition state formation and thus slower will be the rate of such SN^2 reaction. The overcrowding of alkyl groups also leads to strong non-bonded interactions in the transition state in which carbon is bonded to five atoms.

In general, the order of relative reactivities of alkyl halides in SN^2 reactions follows the sequence:

Methyl halides > Primary halides > Secondary halides > Tertiary halides

Q.11. Reactions of alkyl halides with acetylide ion to form higher alkynes is successful only for primary halides. Explain.

Ans. Reactions of alkyl halide with acetylide ion is nucleophilic substitution reaction (SN^2). It is successful only with primary halides. With secondary and tertiary halides, the halogen carrying carbon is heavily crowded. The nucleophile (acetylide ion) finds it difficult to attach to this carbon because of steric hindrance.

Acetylide ion --------C(R)(R)(R)-------- X

Hence, reaction with secondary and tertiary halides does not take place easily. With primary halides, there is comparatively less crowding around the halogen carrying carbon. Here R-groups are replaced by H-groups. Hence, the reaction is comparatively easier.

Q. 12. Arrange the following compounds in increasing order of reactivity towards SN^2 reaction.

***t*-butyl chloride, *n*-propyl chloride.**

Ans.

$$CH_3-\overset{\displaystyle CH_3}{\underset{\displaystyle CH_3}{\overset{|}{\underset{|}{C^*}}}}-Cl \qquad CH_3\,CH_2\,\overset{*}{C}H_2\,Cl$$

t-Butyl chloride (I) *n*-Propyl chloride (II)

In SN^2 reaction, the nucleophile attacks the halide from the side opposite to that where halogen is attached. Attachment of the nuclophile becomes easier if the carbon carrying the halogen is not heavily crowded *i.e.*, if minimum number of alkyl groups are attached to this carbon. There is no crowding with hydrogen atoms. In the two compounds above, the carbon atom marked with an asterisk, where the nucleophile will attach itself, contains three and one alkyl groups respectively in compounds I and II. Hence SN^2 reaction proceeds faster in compounds II.

Q.13. SN^2 reaction is accompanied by Walden inversion. Explain.

Ans. See Q. No. 9.

Q.14. Why are SN^2 displacements more difficult with 2-chloro-2-methylbutane than with 1-chloro-2-methylbutane?

Ans. 2-chloro-2-methylbutane and 1-chloro-2-methylbutane are tertiary and primary halides respectively as indicated below:

$$CH_3-CH_2-\overset{\displaystyle CH_3}{\underset{\displaystyle Cl}{\overset{|}{\underset{|}{C}}}}-CH_3 \qquad CH_3-CH_2-\overset{\displaystyle CH_3}{\overset{|}{C}H}-CH_2-Cl$$

2-Chloro-2-methylbutane (Tertiary alkyl halide) 1-Chloro-2-methylbutane (Primary alkyl halide)

In SN^2 displacements, the nucleophile attacks the side of the carbon opposite to that of the leaving group. In tertiary halides, there is steric hindrance because of the presence of bulky (two methyl and one ethyl) groups and hence the nucleophile finds it difficult to attack the carbon carrying chlorine. On the other hand, there is less steric hindrance because of the presence of two hydrogen and one sec. butyl groups in the primary halide. Consequently the backside attack of the nucleophile on the carbon bearing halogen in 1-chloro-2-methylbutane is easier. Hence 2-chloro-2-methylbutane is less reactive as compared to 1-chloro-2-methylbutane in SN^2 displacement reactions.

Q.15. Give the kinetics and stereochemistry of unimolecular nucleophilic substitution in alkyl halides. (*Panjab, 2002; Nagpur, 2005*)

Or

SN^1 reaction leads to recemisation. Explain.

Or

Give mechanism of SN^1 reaction of R–X.

(*Madras 2003; Bangalore 2006; Guwahati 2007; Purvanchal 2007, Pune 201(*

Ans. An example of unimolecular substitution reaction (SN^1) is the hydrolysis of *tert.* butyl bromide

$$(CH_3)_3\,C-Br+OH^- \longrightarrow (CH_3)_3\,C-OH + \bar{B}r$$

Chemical kinetic studies reveal that the rate of this reaction depends only on the concentration of tert-butyl bromide. Thus,

$$\text{Rate} = k\,[(CH_3)_3\,C-Br]$$

The reaction is of first order and thus it is believed to occur in two steps.

$$CH_3-\underset{CH_3}{\overset{CH_3}{\underset{|}{\overset{|}{C}}}}-Br \xrightarrow{\text{Slow}} CH_3-\underset{CH_3}{\overset{CH_3}{\underset{|}{\overset{|}{C^+}}}} + Br^-$$

$$CH_3-\underset{CH_3}{\overset{CH_3}{\underset{|}{\overset{|}{C^+}}}} + O\bar{H} \xrightarrow{\text{Fast}} CH_3-\underset{CH_3}{\overset{CH_3}{\underset{|}{\overset{|}{C}}}}-OH$$

It is thus clear that the rate determining step of the reaction is the slow ionisation of *tert*-butyl bromide to form tert-butyl carbocation. The energy required for ionisation is supplied by the formation of many ion-dipole bonds between the ions produced and the polar solvent molecules. The second step involving the combination of carbocation with hydroxide ion to form alcohol is fast.

Stereochemistry of SN^1 reactions. In unimolecular nucleophilic substitution, the rate determinings step involves the formation of a carbocation. The carbon atom in the carbocation is in the sp^2 hybridised state. Thus the carbocation has a flat structure in which all the three substituents attached to carbon lie in a plane, making angles of 120° between them. The empty *p*-orbital lies perpendicular to the plane.

The attachment of the nucleophile to the flat carbonium ion can occur probably from both sides of carbonium ion. Nucleophile will attach itself with 50% probability from the front and 50% probability from the back of carbonium ion. This may not make a difference in a simple case, but the difference is clearly made when the halogen carrying carbon is chiral, *i.e.*, linked to four different groups. Hydrolysis of an alkyl halide in which the halogen bearing carbon is chiral will produce a product which is racemic, *i.e.*, it contains 50% each of the *d* and *l* forms. Attachment of the nucleophile from front produces one configuration and that from the back produces the other configuration. As a result, the final product obtained is expected to be racemic, containing equal number of molecules with retention and inversion of configuration. However, in actual practice, the product as a whole is not racemic. Usually there is a larger proportion of molecules with **inverted configuration** than those of the same configuration. This can be explained on the basis that initial ionisation of alkyl halide does not form a free carbocation. In fact ionisation of alkyl halide leads first to the formation of an **ion pair** in which the departing halide ion is still in close proximity to the carbocation. Consequently, the attack on the front side of the carbocation which leads to a product with retention of configuration is slightly hindered. On the other hand, the attack on the backside leading to a product with inversion of configuration is somewhat preferred. Thus, the actual product formed is **partially recemised** and the enantiomer with inverted configuration predominates. For example, when (–)-2-bromooctane is hydrolysed under SN^1 conditions (low concentration of OH^- ions), a partially racemised product is formed. This is further illustrated as follows:

C_6H_{13}, H, Br, CH_3 (–)-2-Bromooctane → [C_6H_{13}, H, Br, CH_3, +]

OH^- Back Side Attack (a) — OH^- Front Side Attack (b)

(+) –2-Octanol (a) Inversion of confi Predominates — HO, C_6H_{13}, H, CH_3

(–) –2-Octanol (b) Retention of confi — H, C_6H_{13}, OH, CH_3

Enantiomers

(Partial racemisation in SN^1 reaction)

Q.16. Give evidence for the formation of carbocation intermediates in SN^1 reactions?

Ans. It is well known that carbocations undergo rearrangement to form more stable carbocations wherever possible. Since the carbocations are the intermediates in SN^1 reactions, it is expected that these intermediates will undergo rearrangement if the structure permits. As a result, the rearranged substitution product will be obtained. This is actually found to be so. For example, when 2-bromo-3- methylbutane is hydrolysed under SN^1 conditions, the formation of rearranged product, 2-methyl-2-butanol and not 3-methyl butanol-2 takes place as shown below:

$$\underset{\text{2-Bromo-3-Methylbutane}}{CH_3-\underset{H}{\overset{CH_3}{C}}-\underset{Br}{CH}-CH_3} \xrightarrow[\text{(Polar Solvent)}]{SN^1\ \text{Conditions}} \underset{\textit{Sec-}\text{Carbocation}}{\left[CH_3-\underset{H}{\overset{CH_3}{C}}-\overset{+}{C}H-CH_3\right]} + Br^-$$

Rearrangement (Hydride shift)

$$\underset{\textit{Tert-}\text{Carbocation}}{CH_3-\underset{+}{\overset{CH_3}{C}}-CH_2-CH_3} \xrightarrow{OH^-} \underset{\text{2-Methyl-2-Butanol}}{CH_3-\underset{OH}{\overset{CH_3}{C}}-CH_2-CH_3}$$

Q.17. Discuss the relative reactivities of alkyl halides in unimolecular nucleophilic substitution (SN^1) reactions.

Ans. Relative reactivities of alkyl halides. Formation of the carbocation is the rate determining step in unimolecular nucleophilic substitution (SN^1). More the stability, more rapidly is the carbocation formed and hence faster is the reaction. Now the relative stability of the carbocations follows the sequence:

Benzyl, allyl > Tertiary > Secondary > Primary > Methyl carbocations. Therefore, the reactivity of alkyl halides is SN^1 reactions follows the same order *i.e.*,

Benzyl, allyl > Tertiary > Secondary > Primary > Methyl halides

Q.18. Discuss the effect of polarity of solvent on the rate of SN^1 and SN^2 reactions.

(Bharathiar, 2011)

Ans. Effect of solvent in SN^1 reactions. The rate determining step in these reactions involves the ionisation of alkyl halide to form carbocation and halide ion. The energy required for the ionisation is provided by the solvation of these ions. Now polar solvent such as water, alcohols, etc., have a greater capacity as compared to non-polar solvents to solvate the ions and thus liberate considerable amount of energy. The energy released facilitates the ionisation and hence increases the rate of the reaction. It is thus obvious that reactions are favoured by polar solvents. For example,

$$R - X \xrightarrow{\text{Ionization}} R^+ + X^- \quad \text{slow}$$

$$R^+ + \bar{O}H \xrightarrow{\text{Polar solvent}} ROX \quad \text{fast}$$

Effect of solvent in SN^2 reactions. SN^2 reaction is a concerted process occurring in one step through the intervention of a transition state.

$$\bar{N}u + R - X \longrightarrow \underset{\text{Transition state}}{\left[\overset{\delta^-}{Nu} \ldots\ldots R \ldots\ldots \overset{\delta^-}{X}\right]} \longrightarrow Nu - R + \underset{\text{Halide ion}}{\bar{X}}$$

It is clear that in the transition state, the negative charge is dispersed over nucleophile and halogen while in the reactants, it is concentrated only on the nucleophile. Reactants are thus more polar than the transition state. Evidently polar solvent will solvate the reactants more strongly than the transition state and hence slow down the rate of SN^2 reaction. In other words SN^2 reactions are favoured by solvents of low polarity.

Q.19. Enlist in a tabular form the distinguishing points of SN^1 and SN^2 mechanism.

(Lucknow, 2000; Kurukshetra, 2001; Andhra, 2010)

Or

Draw a comparison between SN^1 and SN^2 reactions.

Ans. Comparison between SN^1 and SN^2 reactions is given in the form of table below:

S.No.	*SN^1 reactions*	*SN^2 reactions*
1.	SN^1 reactions follow first order kinetics.	SN^2 reactions follows second order kinetics.
2.	Reaction rate is determined by electronic factors mainly.	Reaction rate is determined by steric factors mainly.
3.	The nucleophile attacks the carbocation from both sides although backside attack dominates.	The attack of nucleophile takes place from the backside only.
4.	Partial racemisation of optically active halides takes place.	Inversion of configuration takes place.
5.	It is favoured by solvents of high polarity.	It is favoured by solvents of low polarity.
6.	Rearrangement of products takes place.	There is no possibility of rearrangement.
7.	It is favoured by mild nucleophiles.	It is favoured by strong nucleophiles.
8.	The reactivity follows the order: Tert. > Sec. > Primary > Methyl halide.	The reactivity follows the order: Methyl > Primary > Sec. > Ter. halides.

Q.20. What are the factors that affect SN^1 and SN^2 reactions? (*Kerala, 2001.*)

Ans. The factors that affect SN^1 and SN^2 reactions are described hereunder.

(*i*) **Nature of alkyl groups.** *SN^1 reactions.* The relative reactivity of alkyl halides is the same as the ease of formation and stability of carbonium ion. The stability of carbonium ion is in the order.

Tert. > Sec. > Primary > Methyl carbonium ion

Therefore, the reactivity of alkyl halides is also in the same order.

SN^2 reactions. Here the steric factors are to be taken into consideration. Lesser the crowding around halogen carrying carbon, greater the reactivity. For such reactions, the reactivity follows the order:

Methyl carbonium ion > Primary > Sec. > tert

(*ii*) **Nature of halogen atom.** *SN^1 reactions.* The ease of elimination of the leaving group is $I^- > Br^- > Cl^- > F^-$, therefore, the order of reactivity of alkyl haldies is RI > RBr > RCl > RF.

SN^2 reactions. Same as in SN^1.

(*iii*) **Nature of the nucleophile.** *SN^1 reaction.* As the nucleophile is not involved in the rate determining step on reaction, its nature does not affect the reaction.

SN^2 reaction. Stronger the nucleophile, greater will be the rate of reaction. Increasing strength of a few nucleophiles is given below:

$$H_2O < OH^- < RO^- < CN^-$$

(*iv*) **Nature of the solvent.** *SN^1 reaction.* Polarity of the solvent is important in the ease of formation of carbocation. Greater the polarity of the solvent, greater is the rate of reaction.

SN^2 reaction. Polarity of the solvent has a role to play in this reaction also but in a reverse manner. Higher polarity of the solvent is likely to destroy less polar intermediate product. Hence, less polar the solvent, greater is the reactivity.

Q.21. Explain giving reason why electronic factors have no significant effect on SN^2 reaction.

Ans. SN^2 reaction is a reaction which depends upon the steric environment around halogen carrying carbon. More the crowding of alkyl groups around this carbon, smaller the reactivity.

By electronic effects, we mean the electron withdrawing or repelling effect of various substituents. Such electron withdrawing or repelling groups have no effect on the stability of the intermediate complex, therefore, they don't affect SN^2 reactions.

Q.22. Hydrolysis of ethyl chloride is SN^2 but hydrolysis of tert. butyl chloride is SN^1. Explain with mechanism.

Ans. Hydrolysis of Ethyl Chloride

Hydrolysis of ethyl chloride is a bimolecular reaction (SN^2). Here there is no steric hindrance to the attachment of hydroxide nucleophile to form the transition state. For complete mechanism, see question no. 7.

Hydrolysis of Tert. Butyl Chloride

The nucleophile OH^- faces steric hindrance in attaching itself to the carbon bearing halogen. This is because of the steric hindrance of three bulkyl methyl groups. Hence it follows SN^1 mechanism. For complete mechanism, see question no.11.

Q.23. Which alkyl halide to do you expect to react more rapidly by SN^2 mechanism?

(*a*) $CH_3—CH_2—CH_2—CH_2Cl$ or $CH_3—CH(Cl)—CH_2—CH_2—CH_3$

(*b*) $CH_3—CH_2—CH(Cl)—CH_3$ or $CH_3—C(CH_3)(Cl)—CH_3$

Ans. In SN^2 reaction, the nucleophile and the halogen are both attached to the halogen bearing carbon in the transition state as given below:

$$\overset{\delta^-}{Nu} \cdots\cdots \text{C}(R')(R'')(R''') \cdots\cdots X^-$$

where R′, R″, R‴ are alkyl or hydrogen groups. Obviously, greater the size of three alkyl groups, greater will be the hindrance faced by the nucleophile in forming the transition state.

Therefore out of 1-chloro and 2-chlorobutane, the former will offer less hindrance to the nucleophile and therefore will react more rapidly.

Similarly, out of 2-chlorobutane and 2-chloro-2-methyl propane, the former will offer less hindreance to the nuclophile in forming transition state and therefore will react more rapidly.

Q.24. Predict the products in the following reactions:

(*i*) $CH_3—C(CH_3)_2—CH_2—Br \xrightarrow[SN^1]{\bar{O}H}$ (*Kumaon, 2000*)

(*ii*) $CH_3—C(CH_3)_2—CH_2—Br \xrightarrow[SN^2]{\bar{O}CH_3}$

Ans. (*i*)

$$\underset{\text{Neopentyl bromide}}{CH_3—C(CH_3)_2—CH_2Br} \longrightarrow CH_3—C(CH_3)_2—\overset{+}{C}H_2 + \bar{B}r$$

$$\downarrow \text{Alkyl Shift}$$

$$\underset{\text{2-Methylbutanol-2}}{CH_3—C(CH_3)(OH)—CH_2CH_3} \xleftarrow{OH^-} CH_3—\underset{\oplus}{C}(CH_3)—CH_2CH_3$$

(*ii*)

$$CH_3-\underset{CH_3}{\overset{CH_3}{\underset{|}{\overset{|}{C}}}}-CH_2Br \xrightarrow{\bar{O}CH_3} \overset{\delta^-}{O}CH_3\ldots\overset{H\ \ H}{C}\ldots\overset{\delta^-}{Br} \text{ (with } C(CH_3)_3 \text{ on C)}$$

Neopentyl bromide

$$\downarrow$$

$$CH_3O-CH_2-\underset{CH_3}{\overset{CH_3}{\underset{|}{\overset{|}{C}}}}-CH_3-Br^-$$

Methyl neopentyl ether

Q.25. Neopentyl alcohol on reaction with HCl forms tert. pentyl chloride and not neopentyl chloride.

Ans.

$$CH_3-\underset{CH_3}{\overset{CH_3}{\underset{|}{\overset{|}{C}}}}-CH_2OH \xrightarrow{HCl} CH_3-\underset{Cl}{\overset{CH_3}{\underset{|}{\overset{|}{C}}}}-CH_2-CH_3$$

Neopentyl alcohol Tert.pentyl chloride

It does not form neopentyl chloride having the formula

$$CH_3-\underset{CH_3}{\overset{CH_3}{\underset{|}{\overset{|}{C}}}}-CH_2Cl$$

This reaction is SN[1] reaction in which, the carbocation obtained in the first step rearranges to more stable carbocation.

$$CH_3-\underset{(CH_3)}{\overset{CH_3}{\underset{|}{\overset{|}{C}}}}-\overset{\oplus}{C}H_2 \xrightarrow{\text{Rearranges}} CH_3-\underset{\oplus}{\overset{CH_3}{\overset{|}{C}}}-CH_2-CH_3$$

$$\downarrow Cl^-$$

$$CH_3-\underset{Cl}{\overset{CH_3}{\underset{|}{\overset{|}{C}}}}-CH_2-CH_3$$

Tert.pentyl chloride

Q.26. *n*-butyl chloride and *tert*-butyl chloride behave differently when treated with aqueous sodium hydroxide. Discuss reaction path in both cases. (*Kurukshetra, 2001*)

Ans. $CH_3-CH_2-CH_2-CH_2Cl$ (*n*-Butyl chloride)

(I)

$$CH_3-\underset{CH_3}{\overset{CH_3}{\underset{|}{\overset{|}{C}}}}-Cl$$ (*tert*-Butyl chloride)

(II)

Compound (I) when treated with aqueous NaOH gives *n*-butyl alcohol by SN^2 mechanism because the nucleophile attaches itself to the carbon carrying chlorine, as there is not much crowding on this carbon.

Compound (II) when treated with aqueous NaOH gives *tert*-butyl alcohol by SN^1 mechanism. Three alkyl groups attached to central carbon exert their inductive effect and polarise the molecule as:

$$CH_3 \rightarrow \underset{\underset{CH_3}{\uparrow}}{\overset{\overset{CH_3}{\downarrow}}{C}} \ggg Cl \longrightarrow CH_3-\underset{\underset{CH_3}{|}}{\overset{\overset{CH_3}{|}}{C^+}} + Cl^-$$

Tert-Butyl carbocation

For reaction path, see questions 8 and 13.

Q.27. What is meant by nucleophilicity or nucleophilic strength?

Ans. Nucleophilicity of reagent represents its ability to donate a pair of electrons to the electron deficient carbon. Greater the nucleophilicity, greater will be its tendency to participate in SN^2 reaction.

Comparison of nucleophilic strength of reagents can be made with the help of following general rules.

(*i*) A negatively charged nucleophile is stronger than its conjugate acid. For example OH^- is a stronger nucleophile than its conjugate acid H_2O and RO^- is a stronger nucleophile than its conjugate acid ROH.

(*ii*) In case of nucleophiles containing the same nucleophilic atom, the order of basicity determines the order of nucleophilicity. For example, oxygen containing nucleophiles have the order of nucleophilicty as follows:

$$RO^- > OH^- > RCOO^- > ROH > H_2O$$

(*iii*) In case of nucleophiles having different nucleophilic atoms, nucleophilicity does not necessarily, have the same order as basicity. It is observed that larger the nucleophilic atom, stronger the nucleophile. As an example RSH is a stronger nucleophilic than ROH. On the same analogy, halide ions have the nucleophilic strength in the following order

$$I^- > Br^- > Cl^- > F^-$$

Q.28. Give the mechanism of (Elimination reaction of second order) reactions of alkyl halides. *(Andhra, 2004, Mumbai 2010)*

Ans. An alkyl halide on treatment with alcoholic KOH gives an alkene. This is an elimination reaction in which a molecule of hydrogen halide is eliminated.

$$H-\underset{\underset{H}{|}}{\overset{\overset{H}{|}}{C}}-\underset{\underset{H}{|}}{\overset{\overset{H}{|}}{C}}-Cl \xrightarrow{\text{Alc.KOH}} CH_2=CH_2 + HCl$$

This reaction could take place either by E_2 or E_1 mechanism. When it takes place by E_2, it follows the following steps.

Mechanism of E_2 Reaction

$$B: \quad -\underset{|}{\overset{\overset{H}{|}}{C}}-\underset{\underset{X}{|}}{\overset{|}{C}}-Cl \longrightarrow \;\rangle C=C\langle \; + H:B + X^-$$

This reaction is found to follow second order kinetics *i.e.*, the rate of reaction is dependent upon the concentration of both the alkyl halide and the base.

Factors Influencing E_2 Reaction

(*i*) It is found that for a given alkyl group, the rate of the E_2 reaction varies in the order.

$$RI > RBr > RCl > RF$$

(*ii*) For a given halogen, the rate of reaction follows the order.

tert. halide > sec. halide > prim. halide

(*iii*) Stronger the base, greater will be the rate of reaction. A notable feature of such reactions is that the alkene which has maximum possible branching (Saytzeff's Rule) is obtained.

Q.29. Give the mechanism of E_1 (Elimination reaction of first order) reactions of alky halides. (*Kerala, 2000; Andhra, 2004; Guwahati, 2006; Patna, 2010*)

Ans. Some alkyl halides particularly secondary and tertiary alkyl halides lose a molecule of hydrogen halide by E_1 mechanism.

In such reactions, it is believed that in the first step, which is rate determining step, the molecule of alkyl halide undergoes ionisation to give a halide ion and a corbonium ion.

In the second step, which is a fast step, the carbonium ion (or carbocation) loses a proton from the neighbouring carbon to yield the alkene.

The mechanism of the reaction is illustrated as under:

$$-\overset{\overset{\large H}{|}}{\underset{|}{C}}-\overset{|}{\underset{\underset{\large X}{|}}{C}}-Cl \xrightarrow{\text{Slow}} -\overset{\overset{\large H}{|}}{\underset{|}{C}}-\overset{|}{C^{+}}- + X^{-}$$

Alkyl halide

$$B: \curvearrowright \; -\overset{\overset{\large H}{|}}{\underset{|}{C}}-\overset{|}{C}- \xrightarrow{\text{Fast}} \; >C=C< \; + H:B$$

Factors Influencing E_1 Mechanism

(*i*) For a given alkyl group, it is found that rate of reaction varies in the order.

$$RI > RBr > RCl > R$$

(*ii*) For a given halogen, the order of reactivity of alkyl halides for E_2 reaction follows the order.

Tert. halides > sec. halides > prim. halides

It is noteworthy that unlike in SN^1 and SN^2 reactions, the order of reactivity in E_2 and E_1 is the same.

Q.30. Discuss SN^2 versus E_2 reactions of alkyl halides. (*Pune, 2010*)

Ans. It is observed that when we are carrying out an SN^2 reaction, some amount of E_2 products are also obtained. Similarly when it is intended to carry out E_2 reaction, some amounts of SN^2 products are also obtained. Thus the two reactions proceed concurrently although with different speeds.

It is possible to direct the reaction in one or the other way by regulating some parameters as discussed hereunder.

(*i*) The order of reactivity for SN^2 reaction is

1° halide > 2° halide > 3° halide

The order of reactivity for E_2 reaction is

3° halide > 2° halide > 1° halide

Thus E_2 reaction is increasingly favoured as we move from primary to secondary to tertiary alkyl halide.

(*ii*) The size of the alkyl group and branching at β-carbon influences the relative rates of SN^2 and E_2 reactions due to steric reasons. Formation of intermediate product in SN^2 reactions becomes difficult with bulky groups. Thus rate of reaction decreases when we move from CH_3CH_2Br to $CH_3CH_2CH_2Br$ to $(CH_3)_2$ $CHCH_2Br$.

It is observed that nature of halogen does not affect the relative ratio of SN^2 products.

It is observed that by deceasing the polarity of the solvent, we can shift more towards elimination (E_2).

By increasing the temperature also, it is possible to shift towards E_2 reaction.

As far as the solvent is concerned, SN^2 reactions are favoured by strong nucleophiles and E_2 reactions are favoured by strong bases. And in general, strong nucleophiles are strong bases as well.

Q.31. Discuss SN^1 versus E_1 reactions of alkyl halides. (*Kerala, 2001*)

Ans. Whether it is SN^1 reaction or E_1 reaction, the first step is the slow rate-determining step of dissociation of alkyl halide into carbocation and the halide ion.

It is the second step of the two reactions that is different. In the elimination reaction, the proton is removed from the neighbouring carbon of the carbocation. In the substitution reactions, the nucleophile attaches itself to the carbocation.

In the elimination reaction, the stability of the alkene to be formed will determine the rate of reaction. Keeping in view the order of stability of alkenes, it is found that the rate of elimination follows the order.

3 ° > 2° > 1°

In the substitution reactions, the rate of reaction will be governed by the stability of the carbocation. More unstable the carbocation, greater will be the rate of reaction. Accordingly the rate of substitution follows the order.

1 °> 2° > 3°

Coming to the temperature, it is found that a higher temperature favours E_1 reaction over SN^1 reaction.

Q.32. Why allyl chloride is more reactive than alkyl halide?

Ans. Allyl chloride (in fact any halide) is more reactive than alkyl halide because it can give more stable intermediate carbocation as illustrated below.

$$\underset{\text{Allyl chloride}}{CH_2=CH-CH_2-Cl} \longrightarrow CH_2=CH-\overset{+}{C}H_2 + Cl^- \quad \rightleftharpoons \quad {}^{+}CH_2-CH=CH_2$$

The two resonating structures are of equal energies thereby making it most stable intermediate carbocation. Here stabilization is by means of resonance which is more effective than in case of alkyl halides. Take the case of *t*-butyl chloride. Here the carbocation is stabilized by a weaker phenomenon of inductive effect.

$$H_3C-\underset{CH_3}{\overset{CH_3}{C}}-Cl \longrightarrow H_3C\rightarrow\underset{CH_3\uparrow}{\overset{CH_3\downarrow}{C^+}} + Cl^-$$

Tert-butyl carbocation

A substance that can furnish more stable carbocation is naturally more reactive. Therefore allyl chloride is more reactive than alkyl halide.

Even if the reaction takes place by SN^2 mechanism, allyl chloride will show greater reactivity because being linear in shape, will not hinder the attachment of a nucleophile to form the intermediate product.

Q.33. Why vinyl chloride is less reactive than alkyl halide? *(Devi Ahilya, 2001; Gulberga 2003; West Bengal 2004; Purvanchal 2007)*

Ans. Vinyl chloride is less reactive than alkyl halide because of the following resonance plenomenon.

$$\overset{\frown}{CH_2} = CH - \ddot{Cl}: \rightleftharpoons \bar{C}H_2 - CH = \overset{+}{Cl}$$

This will result in creating a double bond between carbon and chloride. A double bond is more difficult to break than a single bond in the case of alkyl halides. Hence the chances of a reaction with a nucleophile become much less.

Whether the reaction takes place by SN^2 or SN^1 mechanism, vinyl chloride will show lower reactivity towards nucleophilic substitution reactions. It is difficult to attach a nucleophile to carbon because there is no +ve charge on the carbon linked to the halogen.

DIHALOGEN COMPOUNDS (VICINAL AND GEMINAL)

Q.34. Give two methods of preparation of vicinal and geminal dihalogenated alkenes.

Ans. Preparation of Vicinal Dihalides

1. From alkenes Vicinal dehalides can be prepared by the addition of halogen to alkenes

$$R - CH = CH_2 + X_2 \longrightarrow R - \underset{X}{\underset{|}{CH}} - \underset{X}{\underset{|}{CH_2}}$$

$$CH_2 = CH_2 + Cl_2 \longrightarrow \underset{Cl}{\underset{|}{CH_2}} - \underset{Cl}{\underset{|}{CH_2}}$$

Ethylene chloride

$$CH_2 = CH_2 + Br_2 \longrightarrow CH_3 - \underset{Br}{\underset{|}{CH}} - \underset{Br}{\underset{|}{CH_2}}$$

Proplylene bromide

2. From glycols. (*i*) By the action of halogen acids.

$$R - \underset{OH}{\underset{|}{CH}} - \underset{OH}{\underset{|}{CH_2}} + 2HX \xrightarrow{\Delta} R - \underset{X}{\underset{|}{CH}} - \underset{X}{\underset{|}{CH_2}} + 2H_2O$$

R = alkyl group)

$$\begin{matrix} CH_2OH \\ | \\ CH_2OH \end{matrix} + 2HCl \xrightarrow{473\ K} \begin{matrix} CH_2Cl \\ | \\ CH_2Cl \end{matrix} + 2H_2O$$

Ethylene glycol

(*ii*) By action of PCl_5

$$\begin{matrix} CH_2OH \\ | \\ CH_2OH \end{matrix} + 2PCl_5 \longrightarrow \begin{matrix} CH_2Cl \\ | \\ CH_2Cl \end{matrix} + 2POCl_3 + 2HCl$$

Ethylene glycol — Phosphorus oxytrichloride

$$\underset{\text{Propylene glycol}}{CH_3-\overset{OH}{\overset{|}{C}H}-\overset{OH}{\overset{|}{C}H_2}} + 2PCl_5 \longrightarrow \underset{\text{Propylene chloride}}{CH_3-\overset{Cl}{\overset{|}{C}H}-\overset{Cl}{\overset{|}{C}H_2}} + 2POCl_3 + 2HCl$$

Preparation of Geminal Dihalides

1. By the action of PCl_5 on aldehydes and ketones

$$\underset{\text{Aldehyde}}{\begin{matrix} R \\ H \end{matrix}\!\!>C=O} + POCl_3 \longrightarrow \begin{matrix} R \\ H \end{matrix}\!\!>C<\!\!\begin{matrix} Cl \\ Cl \end{matrix} + \underset{\text{Phosphorus oxytrichloride}}{POCl_3}$$

$$CH_3CHO + PCl_5 \longrightarrow \underset{\text{Ethylidene chloride or 1, 1-Dichloroethane}}{CH_3\,CHCl_2} + POCl_3$$

$$\underset{\text{Ketone}}{\begin{matrix} R \\ R \end{matrix}\!\!>C=O} + POCl_3 \longrightarrow \begin{matrix} R \\ R \end{matrix}\!\!>C<\!\!\begin{matrix} Cl \\ Cl \end{matrix} + POCl_3$$

$$\underset{\text{Acetone}}{CH_3COCH_3} + PCl_5 \longrightarrow \underset{\text{2, 2-Dichloropropane}}{CH_3CCl_2CH_3} + POCl_3$$

2. From alkynes

$$\underset{\text{Alkyne}}{R-C\equiv CH} + H-X \longrightarrow R-\underset{X}{\underset{|}{C}}=CH_2 \xrightarrow{H-X} R-\overset{X}{\overset{|}{\underset{X}{\underset{|}{C}}}}-CH_3$$

$$CH\equiv CH + HCl \longrightarrow \underset{\text{Vinyl chloride}}{CH_2=CHCl} \xrightarrow{HCl} \underset{\text{Ethylidene chloride}}{CH_3-CHCl_2}$$

Q.35. Describe the action of the following on ethylene chloride

(*i*) Alcoholic KOH

(*ii*) Zn dust

(*iii*) Aqueous KOH

(*iv*) KCN and hydrolysis

Ans. (1) Action with alcoholic KOH. With alcoholic KOH, it first forms an unsaturated halogen compounds and finally gives an ethyne (dehydro-halogenation).

$$\begin{matrix} CH_2Cl \\ | \\ CH_2Cl \end{matrix} + \underset{\text{(Alc)}}{KOH} \longrightarrow \underset{\text{Vinyl chloride}}{\begin{matrix} CH_2 \\ \| \\ CH-Cl \end{matrix}} + KCl + H_2O \xrightarrow[\text{Alc.}]{KOH} \begin{matrix} CH \\ ||| \\ CH \end{matrix} + KCl + H_2O$$

(2) Action of Zn dust. When heated with Zn dust, it gives ethylene (Dehalogenation).

$$\begin{matrix} CH_2-Cl \\ | \\ CH_2-Cl \end{matrix} + Zn \xrightarrow{\Delta} \begin{matrix} CH_2 \\ \| \\ CH_2 \end{matrix} + ZnCl_2$$

(3) Action of aqueous alkali. Glycol is obtained.

$$\begin{array}{l} CH_2-Cl \\ | \\ CH_2-Cl \end{array} + \underset{(aq)}{2NaOH} \longrightarrow \underset{\text{Ethylene glycol}}{\begin{array}{l} CH_2OH \\ | \\ CH_2OH \end{array}} + 2NaCl$$

(4) Action of alkali cyanide. Ethylene cyanide is formed. On hydrolysis, it gives dibasic acid which does not evolve carbon dioxide on heating.

$$\begin{array}{l} CH_2-Cl \\ | \\ CH_2-Cl \end{array} + KCN \longrightarrow \underset{\text{Ethylene cyanide}}{\begin{array}{l} CH_2-CN \\ | \\ CH_2-CN \end{array}} + 2KCl$$

$$\xrightarrow{H_2O \mid H^+} \underset{\text{Succinic acid}}{\begin{array}{l} CH_2-COOH \\ | \\ CH_2-COOH \end{array}}$$

Q.36. How does ethylidene chloride react with the following:

(*i*) Alcohalic KOH

(*ii*) Zn dust

(*iii*) Aqueous KOH

(*iv*) KCN

Ans. (*i*) Action of alcoholic KOH.

$$CH_3-CHCl_2 + KOH\ (alc.) \longrightarrow \underset{\text{Vinyl chloride}}{CH_2=CHCl} + KCl + H_2O$$

$$\downarrow \text{Alc. KOH}$$

$$\underset{\text{Acetylene}}{CH=CH} + KCl + H_2O$$

(*ii*) Action of Zn dust

$$CH_3\ CHBr_2 + Zn \xrightarrow[\Delta]{\Delta} \underset{\text{Acetylene}}{CH \equiv CH} + ZnBr_2$$

(*iii*) Action of aqueous alkali

$$CH_3-CHCl_2 \xrightarrow{2KOH} \left[CH_3-CH\begin{matrix} OH \\ OH \end{matrix} \right] + 2KCl$$

$$\downarrow$$

$$\underset{\text{Acetaldehyde}}{CH_3-CHO} + H_2O$$

(*iv*) Action of alkali cyanide

$$CH_3-CHCl_2 + 2KCN \longrightarrow \underset{\text{Ethylene cyanide}}{CH_3-CH\begin{matrix} CN \\ CN \end{matrix}} \xrightarrow{H_2O \mid H^+} CH_3-CH\begin{matrix} COOH \\ COOH \end{matrix}$$

$$\downarrow -CO_2 \mid \Delta$$

$$\underset{\text{Propionic acid}}{CH_3-CH_2-COOH}$$

Q.37. Give the preparation and uses of 1, 1- and 1, 2- Dichloroethene

Ans. These are substitution products of ethene in which two hydrogen atoms of ethene are replaced by chlorine atoms. Dichloroethene exists in two different structural forms.

$$\underset{\substack{\text{1, 2- Dichlorethene} \\ (\textit{Acetylene dichloride})}}{ClCH=CHCl} \quad \text{and} \quad \underset{\substack{\text{1, 1-Dichloroethene} \\ (\textit{Vinylidene dichloride})}}{CH_2=CCl_2}$$

1,2-Dichloroethene can exhibit geometric isomerism but 1, 1-Dechloroethene cannot.

$$\underset{\textit{Trans}\text{-1, 2-Dichloroethene}}{\begin{matrix} Cl \\ H \end{matrix} \!\!> C=C <\!\! \begin{matrix} H \\ Cl \end{matrix}} \qquad \underset{\textit{Cis}\text{-1, 2-Dichloroethene}}{\begin{matrix} H \\ Cl \end{matrix} \!\!> C=C <\!\! \begin{matrix} H \\ Cl \end{matrix}}$$

Preparation of 1, 2-Dichloroethene

(*i*) It can be obtained by the addition of chlorine in CCl_4 to acetylene.

$$HC\equiv CH+Cl_2 \xrightarrow{CCl_4} ClCH=CHCl$$

(*ii*) From Westron. By the action of zinc on westron (1, 1, 2, 2-tetrachloroethane).

$$\underset{\text{1, 1, 2, 2-Tetrachloroethane}}{Cl_2CH-CHCl_2} + Zn \longrightarrow \underset{\text{1, 2-Dichloroethene}}{ClCH=CHCl} + ZnCl_2$$

Preparation of 1, 1-Dichloroethene

$$\underset{\text{Vinyl chloride}}{CH=CH-Cl} \xrightarrow[CCl_4]{Cl_2} ClCH_2-CHCl_2 \xrightarrow[363K]{Ca(OH)_2} \underset{\text{1, 1-Dichloroethene}}{CH_2=C<\begin{matrix} Cl \\ Cl \end{matrix}} + HCl$$

Uses of Dichloroethenes :

1. As solvent for rubber.

2. Vinylidene dichloride and vinyl chloride polymerize to give a copolymer, *saran,* which is used for food packaging.

Q.38. Give the methods of preparation, properties and uses of trichloromethane (chloroform), $CHCl_3$

Ans. Chloroform is prepared by the following methods :

(*i*) From Chloral hydrate. ***Pure Chloroform*** can be obtained by distilling chloral or chloral hydrate with concentrated aqueous NaOH solution.

$$NaOH+\underset{\text{Chloral}}{CCl_3CHO} \longrightarrow \underset{\text{Sod. formate}}{HCOONa}+CHCl_3$$

$$NaOH+\underset{\text{Chloral hydrate}}{CCl_3CH(OH)_2} \longrightarrow HCOONa + \underset{\text{Chloroform}}{CHCl_3} + H_2O$$

(*ii*) From Methane. Chloroform is ***manufactured*** by chlorination of methane in the presence of light or catalysts when a mixture of methyl chloride, methylene dichloride, chloroform and carbon tetrachloride is obtained.

$$CH_4+Cl_2 \longrightarrow CH_3Cl+HCl$$
$$CH_3Cl+Cl_2 \longrightarrow CH_2Cl_2+HCl$$
$$CH_2Cl_2+Cl_2 \longrightarrow CHCl_3+HCl$$
$$CHCl_3+Cl_2 \longrightarrow CCl_4+HCl$$

The mixture of CH_3Cl, CH_2Cl_2, $CHCl_3$ and CCl_4 can be separated by fractional distillation

(*iii*) Laboratory Method. ***Ethanol or acetone is treated with a paste of bleaching powder and water.***

In this reaction, bleaching powder serves as a source of chlorine which first oxides ethanol to acetaldehyde, which is then further chlorinated to chloral. Chloral is hydrolysed with $Ca(OH)_2$ given by $CaOCl_2$ to give chloroform and calcium formate.

(a) In the case of ethanol, the following reactions are involved :

$$CaOCl_2 + H_2O \longrightarrow Ca(OH)_2 + Cl_2$$

$$\underset{\text{Ethanol}}{CH_3CH_2OH} + Cl_2 \xrightarrow{\text{Oxidation}} \underset{\text{Ethanol}}{CH_3CHO} + 2HCl$$

$$CH_3CHO + 3Cl_2 \xrightarrow{\text{Chlorination}} CCl_3CHO + 3HCl$$

$$Ca(OH)_2 + 2CCl_3CHO \xrightarrow{\text{Hydrolysis}} \underset{\text{Chloroform}}{2CHCl_3} + \underset{\text{Calcium formate}}{(HCOO)_2Ca}$$

(b) In the case of acetone, the following reactions are involved :

$$CaOCl_2 + H_2O \longrightarrow Ca(OH)_2 + Cl_2$$

$$\underset{\text{Acetone}}{CH_3COCH_3} + 3Cl_2 \xrightarrow{\text{Chlorination}} \underset{\text{Trichloroacetone}}{CCl_3COCH_3}$$

$$Ca(OH)_2 + CCl_3COCH_3 \xrightarrow{\text{Hydrolysis}} \underset{\text{Chloroform}}{CHCl_3} + \underset{\text{Calcium acetate}}{(CH_3COO)_2Ca}$$

(iv) From carbon tetrachloride. *By partial reduction of carbon tetrachloride with iron filings and water.*

$$CCl_4 + 2[H] \xrightarrow[\text{heat}]{Fe/H_2O} CHCl_3 + HCl$$

Physical Properties

1. Chloroform is a colourless, heavy liquid which has sweetish sickly odour and burning taste
2. It is heavier than water.
3. It boils at 334 K and is slightly soluble in water.
4. Inhaling of the vapours of chloroform induces unconsciousness.

Chemical Properties

1. Action of sunlight and air. In the presence of sunlight, chloroform gets oxidized by air to produce highly poisonous compound phosgene, $COCl_2$.

$$2CHCl_3 + O_2 \xrightarrow{\text{Sunlight}} 2COCl_2 + 2HCl$$

As chloroform is used for anaesthetic purposes, therefore, high level of purity is desirable. To prevent oxidation of chloroform it is stored in dark bottles, completely filled upto brim. *The use of dark bottles cuts radiations and filling upto brim keeps out air.* A small amount of enthanol (1%) is usually added to bottle of chloroform. *Addition of a little ethanol converts the toxic $COCl_2$ to non-poisonous diethyl carbonate.*

$$COCl_2 + 2C_2H_5OH \longrightarrow \underset{\text{Diethyl carbonate}}{O{=}C\begin{cases}OC_2H_5\\OC_2H_5\end{cases}} + 2HCl$$

2. Reduction. Chloroform can be reduced to methylene chloride with zinc and HCl.

$$Zn + 2HCl \longrightarrow ZnCl_2 + 2[H]$$

$$CHCl_3 + 2[H] \longrightarrow \underset{\substack{\text{Dichloromethane}\\(\textit{Methylene chloride})}}{CH_2Cl_2} + HCl$$

Reduction of chloroform with zinc and water (neutral medium) produces methane.

$$CHCl_3 \xrightarrow{Zn + H_2O} CH_4 + 3HCl$$

3. Hydrolysis. When boiled with aqueous KOH, chloroform is hydrolysed to potassium formate.

$$H-C\begin{matrix}Cl\\Cl\\Cl\end{matrix} + \begin{matrix}K\ OH\\K\ OH\\K\ OH\end{matrix} \xrightarrow{-3KCl} H-C\begin{matrix}OH\\OH\\OH\end{matrix} \xrightarrow{-H_2O} H-\overset{\overset{O}{\|}}{C}-OH \xrightarrow[-H_2O]{+KOH} HCOOK$$

Unstable; Potassium formate

4. Reaction with acetone. Chloroform reacts with acetone in the presence of a base such as KOH and forms the addition product, *chloretone.*

$$\begin{matrix}CH_3\\CH_3\end{matrix}\!\!>C=O + CHCl_3 \xrightarrow{KOH} CH_3-\underset{\underset{CCl_3}{|}}{\overset{\overset{CH_3}{|}}{C}}-OH$$

Acetone; Chloretone

Chloretone is used as sleep inducting drug.

5. Reaction with primary amines (*Carbylamine reaction*). On warming chloroform with a primary amine (aliphatic or aromatic) in the presence of alc. KOH, an offensive smell of *isocyanide* or *carbylamine* is obtained.

$$C_6H_5-NH_2 + CHCl_3 + 3KOH \longrightarrow C_6H_5-NC + 3KCl + 3H_2O$$

6. Reaction with nitric acid. Chloroform reacts with conc. HNO_3 on heating.

$$CHCl_3 + HONO_2 \longrightarrow CCl_3{\cdot}NO_2 + H_2O$$

Chloropicrin

Chloropicrin is used as an insecticide and war-gas.

7. Chlorination. Further chlorination of chloroform gives CCl_4.

$$CHCl_3 + Cl_2 \xrightarrow{hv} CCl_4 + HCl$$

8. Reimer-Tiemann reaction. Chloroform reacts with phenol and alc. KOH to form salicylaldehyde which is used to prepare a number of drugs.

$$C_6H_5OH + CHCl_3 + 3NaOH \xrightarrow{323-343\ K} C_6H_4(OH)CHO + NaCl + 2H_2O$$

Phenol; Salicylaldehyde (2-Hydroxybenzaldehyde)

9. Reaction with silver powder. On heating with silver powder, chloroform forms acetylene.

$$2CHCl_3 + 6Ag \xrightarrow{heat} CH \equiv CH + 6AgCl$$

Acetylene

Use of Chloroform

1. As laboratory reagent.
2. As solvent in oils and varnishes.
3. As an anaesthetic.
4. As preservative for anatomical specimens.
5. In medicines such as cough syrups.

Q. 39. Give the methods of preparation, properties and uses of triiodomethane (iodoform).

Ans. Preparation: Iodoform is prepared by the action of iodine and alkali on ethyl alcohol or acetone.

Following reactions are involved in the preparation of CHI_3 from ethyl alcohol :

$$2NaOH + I_2 \longrightarrow NaI + \underset{\text{Sod. hypoiodite}}{NaOI} + H_2O$$

$$\underset{\text{Ethanol}}{C_2H_5OH} + NaOI \xrightarrow{\text{Oxidation}} CH_3CHO + NaI + H_2O$$

$$\underset{\text{Ethanal}}{CH_3CHO} + 3NaOI \xrightarrow{\text{Iodination}} CI_3CHO + 3NaOH$$

$$\underset{\text{Triodoethanal}}{CI_3CHO} + NaOH \xrightarrow{\text{Hydrolysis}} \underset{\text{Iodoform}}{CHI_3} + \underset{\text{Sod.formate}}{HCOONa}$$

The complete reaction may be written as :

$$CH_3CH_2OH + 4I_2 + 6NaOH \longrightarrow CHI_3 + 5NaI + HCOONa + 2H_2O$$

With sodium carbonate in place of sodium hydroxide, the complete reaction may be written as:

$$CH_3CH_2OH + 4I_2 + 3Na_2CO_3 \longrightarrow CHI_3 + 5NaI + HCOONa + 3CO_2 + 2H_2O$$

Following reactions are involved in the preparation of CHI_3 *from acetone.*

$$CH_3 - \underset{\underset{O}{\|}}{C} - CH_3 + 3NaOI \longrightarrow \underset{\text{Triiodoacetone}}{CI_3 - \underset{\underset{O}{\|}}{C} - CH_3} + 3NaOH$$

$$CI_3 - \underset{\underset{O}{\|}}{C} - CH_3 + 3NaOH \longrightarrow \underset{\text{Iodoform}}{CHI_3} - \underset{\text{Sod. Acetate}}{CH_3\ \underset{\underset{O}{\|}}{C} - ONa}$$

Physical Properties

It is a pale yellow solid having melting point 392 K. The compound has a characteristic odour. It is insoluble in water but readily soluble in alcohol and ether.

Chemical Properties

1. Reaction with alc. KOH. On boiling with potassium hydroxide solution, potassium formate is formed as shown below :

$$\underset{\text{Iodoform}}{CHI_3} + 3KOH \xrightarrow[-3KI]{} \left[CH \begin{matrix} OH \\ OH \\ OH \end{matrix} \right] \xrightarrow[-H_2O]{} \underset{\text{Formic acid}}{HCOOH} \xrightarrow{KOH} \underset{\text{Pot.formate}}{HCOOK} + H_2O$$

2. Reaction with amines (carbylamine reaction). An offensive smell of isocyanide (carbylamine) is obtained when iodoform is heated with primary amine and potassium hydroxide.

$$CHI_3 + \underset{\text{Methylamine}}{CH_3NH_2} + 3KOH \xrightarrow{\text{heat}} \underset{\text{Methyl isocyanide}}{CH_3NC} + 3KI + 3H_2O$$

3. Reduction. On heating with HI/P, chhloroform is reduced to methylene chloride

$$CHI_3 + HI \xrightarrow{P} CH_2I_2 + I_2$$

4. Oxidation. On heating in air in the presence of moisture, carbon dioxide, idodine and water are formed.

$$4CHI_3 + 5O_2 \longrightarrow 4CO_2 + 2H_2O + 6I_2$$

5. Reaction with Ag powder. On heating iodoform with silver powder, acetylene gas is produced.

$$CHI_3 + 6Ag + I_3HC \xrightarrow{heat} \underset{\text{Acetylene}}{CH \equiv CH} + 6AgI$$

Use of Iodoform

1. It is uded as an antiseptic for dressing wounds.
2. It is used to manufacture certain pharmaceutical formulations.

Q.40. What is iodoform test? What is its utility?

Ans. This is an important test in organic chemistry and is used for the identification of functional groups like.

$CH_3\overset{O}{\overset{\|}{C}}-$ and $CH_3\overset{OH}{\overset{|}{C}}H-$ in organic compounds. *It is noticed that when the compounds containing above functional groups are heated with an aqueous solution of Na_2CO_3 or NaOH containing iodine, a yellow precipitate of iodoform is obtained.*

Thus, the test helps us to identify such type of groupings in the organic compounds. The reaction involved can be written as :

$$2NaOH + I_2 \longrightarrow NaI + NaOI + H_2O$$

$$CH_3 - \underset{OH}{\underset{|}{\overset{H}{\overset{|}{C}}}} - R + NaOI \longrightarrow CH_3 - \underset{O}{\underset{\|}{C}} - R + NaI + H_2O$$

$$CH_3 - \underset{O}{\underset{\|}{C}} - R + 3NaOI \longrightarrow CI_3 - \underset{O}{\underset{\|}{C}} - R + 3NaOH$$

$$CI_3 - \underset{O}{\underset{\|}{C}} - R + NaOH \longrightarrow \underset{\text{Iodoform}}{CHI_3} + \underset{\text{Sodium salt of acid}}{R - \underset{O}{\underset{\|}{C}} - ONa}$$

Here, R can be H atom or some alkyl group.

The following pairs of compounds can be distinguished by iodoform test :

***(i)* Ethanal and propanal :** Ethanal gives the iodoform test but propanal does not.

***(ii)* Ethyl alcohol and methyl alcohol :** Ethyl alcohol gives the iodoform test but methyl alcohol does not

***(iii)* 2-Pentanone and 3-pentanone :** Here, 2-pentanone gives iodoform test but 3-pentanone does not

Q.41. Give the methods of preparation, properties and uses of tetrachloromethane (carbon tetrachloride)

Ans. Carbon tetrachloride is prepared by the following methods:

***(i)* From Methane.** *By chlorination of methane* in the presence of sunlight. Chlorine gas is taken in excess in the reaction.

$$CH_4 + 4Cl_2 \xrightarrow{Hv} CCl_4 + 4HCl$$

***(ii)* From Carbon disulphide.** By reaction of chlorine with carbon disuphide in the presence of aluminium chloride as catalyst.

$$CS_2 + 3Cl_2 \xrightarrow{AlCl_3} CCl_4 + \underset{\text{Sulphur monochloride}}{S_2Cl_2}$$

CCl_4 separated from S_2Cl_2 by fractional distillation.

Physical Properties

1. Carbon tetrachloride is a colorless, heavy liquid with characteristic sickly smell.
2. It is heavier than water.
3. It is insoluble in water but is soluble in organic solvents such as ether.

Chemical Properties

***(i)* Stability.** CCl_4 is stable to red heat but when the vapours come in contact with water, phosgene is liberated.

$$CCl_4 + H_2O \longrightarrow COCl_2 + 2HCl$$

It is because of thermal stability that CCl_4 is used as a fire extinguisher under the name pyrene.

***(ii)* Hydrolysis.** On heating with alc. KOH, it undergoes hydrolysis and gives CO_2 which dissolves in KOH to yield potassium carbonate.

$$CCl_4 + 4KOH \xrightarrow[-4KCl]{\Delta} \underset{\text{unstable}}{[C(OH)_4]} \longrightarrow CO_2 + 2H_2O$$

$$2KOH + CO_2 \longrightarrow K_2CO_3 + H_2O$$

***(iii)* Reduction.** On reduction with moist iron fillings, CCl_4 gives chloroform.

$$CCl_4 + 2H \xrightarrow{Fe/H_2O} CHCl_3 + HCl$$

***(iv)* Reimer-Tiemann reaction.** On heating with phenol and sodium hydroxide, salicyclic acid is obtained.

$$\underset{\text{Phenol}}{C_6H_5OH} + \underset{\text{Carbon tetrachloride}}{CCl_4} \xrightarrow[340\ K]{NaOH} C_6H_4(ONa)CCl_3 \xrightarrow{3NaOH} C_6H_4(ONa)C(OH)_3$$

$$\xrightarrow[-H_2O]{} C_6H_4(ONa)COOH \xrightarrow{NaOH} C_6H_4(ONa)COONa \xrightarrow{Dil.HCl} \underset{\text{2-Hydroxybenzoic acid (salicylic acid)}}{C_6H_4(OH)COOH}$$

Uses of CCl_4

1. As a laboratory reagent.
2. As a fire extinguisher.
3. As an industrial solvent.
4. In dry cleaning.
5. For the manufacture of chloroform.

Q.42. Write notes on the following :

(1) Dichlorodifluromethane (Freon)

(2) Westrosol

Ans. (1) Freon. It is prepared by the action of antimony fluoride on carbon tetrachloride in the presence of antimony pentachloride. It is used as refrigerant. It is a non toxic and non inflammable liquid.

$$3CCl_4 + 2SbF_3 \xrightarrow{SbCl_5} 2SbCl_3 + 3CCl_2F_2$$

Since freon has been found to be one of the factors responsible for the depletion of ozone layer, it is being replaced by alternative refrigerants in many countries.

(2) Westrosol or Trichloroethylene. It is prepared by passing 1, 1, 2, 2-tetrachloroethane (Westron) over heated barium chloride.

$$\underset{\text{Westron}}{CHCl_2 - CHCl_2} \xrightarrow[\Delta]{BaCl_2} \underset{\text{Westrosol}}{CHCl = CCl_2} + HCl$$

Uses. It is used commercially as a solvent for oils, paints, warnishes and rubber.

MISCELLANEOUS QUESTIONS

Q.43. What is the action of alc. alkali on secondary butyl bromide ? Which is the predominant product and explain the reason. (*Kerala 2001*)

Ans. See Q. 15.

Q.44. What are vicinal and germinal dihalides ? Give one important reaction of each. (*Himachal, 2000*)

Ans. See Q.1, Q.35 and Q. 36.

Q.45. Identify compounds X and Y in the following reactions : (*Lucknow, 2000*)

(*i*) $CH_3 - CH - Cl_2 \xrightarrow{alc.KOH} X \xrightarrow{Hg^{++}/H_2SO_4} Y$

(*ii*) $\underset{\quad\quad\quad\;|\quad\;\;|}{CH_3 - CH - CH_2}$ (with I on CH and I on CH_2) $\xrightarrow{\Delta} X \xrightarrow{HBr} Y$

Ans. (*i*) $CH_3\,CH\,Cl_2 \xrightarrow{alc.KOH} \underset{\text{Acetylene}}{CH \equiv CH} \xrightarrow{Hg^{++}/H_2SO_4} CH_3CHO$

(*ii*) $CH_3 - \underset{I}{\underset{|}{CH}} - \underset{I}{\underset{|}{CH_2}} \xrightarrow{\Delta} \underset{\text{Propyne}}{CH_3 - C \equiv CH} \xrightarrow{HBr} \underset{\text{2-bromopropene}}{CH_3 - \underset{Br}{\underset{|}{C}} = CH_2}$

Q.46. How will you distinguish between ethylene dichloride and ethylidene dichloride ? (*Lucknow, 2000*)

Ans. See Q. 35 and Q. 36.

Q.47. How will you convert ethyl iodide to propionic acid ? *(Kerala, 2001)*

Ans. $\underset{\text{Ethyl Iodide}}{C_2H_5I} \xrightarrow{KCN} CH_3CH_2CN \xrightarrow{H_2O} \underset{\text{propionic acid}}{CH_3CH_2COOH}$

Q.48. Discuss the different mechanisms associated with aliphatic nucleophilic substitution reactions. Explain the effect of structure on the reactivity of alkyl halide towards nucleophilic substitution. *(Kerala, 2001)*

Ans. See Q.9 and Q. 15.

Q.49. How will you bring about the following conversions :

(i) n–propyl bromide → n–propane *(Kerala, 2001)*

(ii) $CH_3\ CH\ Br\ CH_3 \rightarrow CH_3\ CH_2\ CH_2\ OH$ *(Kumaon, 2000)*

Ans. (i) $\underset{n\text{-propyl bromide}}{CH_3 - CH_2 - CH_2Br} \xrightarrow[\text{Zn/Acetic acid}]{\text{Zn-Cu/alcohol}} \underset{n\text{-propane}}{CH_3\ CH_2 - CH_3} + Br$

(ii) $CH_3 - CH_2 - BrCH_3 \xrightarrow{\text{alc.KOH}} CH_3\ CH = CH_2 \xrightarrow[\text{HBr}]{\text{Peroxide}} CH_3CH - CH_2\ Br \xrightarrow{H_2O}$
$CH_3 - CH_2 - CH_2OH$

Q.50. Identify the compounds P, Q and R in the following reaction : *(Merrut, 2000)*

$$CH_3 - CH_2 - CH_2I \xrightarrow{\text{alc.KOH}} P \xrightarrow{H_2O/H^+} Q \xrightarrow{SOCl_2} R$$

Ans. $CH_3 - CH_2 - CH_2I \xrightarrow{\text{alc.KOH}}$

$$\underset{\substack{(P)\\ \text{Propene-2}}}{CH_3 - CH = CH_2} \xrightarrow{H_2O/H^+} \underset{\substack{(Q)\\ \text{Propan 2-ol}}}{CH_3 - \underset{OH}{CH} - CH_3} \xrightarrow{SOCl_2} \underset{\substack{(R)\\ \text{2-chloropropane}}}{CH_3 - \underset{Cl}{CH} - CH_3}$$

Q.51. Complete the following reaction. *(Kanpur, 2001)*

$$CH_3 - CH_2 - \overset{Br}{\underset{CH_3}{C}} - CH_3 \xrightarrow{\text{alc.KOH}}$$

Ans. $CCH_3 - CH_2 - \overset{Br}{\underset{CH_3}{C}} - CH_3 \xrightarrow{\text{alc.KOH}} \underset{\text{2-methyl-butene-2}}{CH_3 - CH_2 = \underset{CH_3}{C} - CH_3}$

Q.52. Compare the reactivity of allyl chloride, vinyl chloride and ethyl chloride towards nucleophiles. *(Purvanchal 2003)*

Ans. $CH_2 = CH - \ddot{C}l \longleftrightarrow \overset{-}{C}H_2 - CH = \overset{+}{C}l$

Because of the above transition vinyl chloride will be least reactive

$$CH_3 \rightarrow CH_2 \rightarrow Cl \longleftrightarrow CH_2 - \overset{\delta+}{C}H_2 - \overset{\delta-}{C}l$$

Allyl chloride will be most reactive because of the stabilisation of allyl carbacation producec by the release of Cl^-

$$CH_2 = CH - CH_2 - Cl \longrightarrow CH_2 = CH - \overset{+}{C}H_2 + Cl^-$$

$$CH_2 = CH - \overset{+}{C}H_2 \longleftrightarrow \overset{+}{C}H_2 - CH = CH_2$$

Thus, the order of reactivity is : allyl chloride > ethyl chloride > vinyl chloride

13

ARYL HALIDES

Q. 1. Give the classification and nomenclature of aromatic halogen compounds.

Ans. Aromatic halogen compounds are those compounds which contain at least one benzene ring and one halogen atom in the molecule. They are classified as under:

Nuclear Halogen Compounds

These are the compounds in which the halogen is directly attached to the benzene ring. For example,

Cl Br CH_3

Chlorobenzene *o*-Bromotoluene

Side-chain Halogen Compound

These are the compounds in which the halogen is attached to side chain of benzene ring. For example,

CH_2Cl CH_2Br Cl

Benzyl chloride *o*-Chloro benzyl bromide

More than one halogen could be present in the same side-chain in a compound.

Nomenclature

Following rules are generally followed in the nomenclature of aromatic halogen compounds.

(*i*) In monohalogen compounds, the name of the compound is written by mentioning the halogen followed by the word "benzene". For example,

Bromobenzene Chlorobenzene

All the six positions of benzene are equivalent. Hence, monohalogen compounds could be represented by attaching the halogen to any of the six positions.

(*ii*) In nuclear dihalogen compounds, there are two halogen groups attached. Now two groups can be attached in different relative positions. If the two groups are located in neighbouring positions, we say that the two groups are *ortho* to each other. If there is a gap of one position, we say that the two groups are *meta* to each other. If there is a gap of two positions we say that the two groups are *para* to each other. The names of a few compounds are given below. For *ortho, meta* and *para*, we use abbreviations *o*–, *m*- and *p*-respectively.

Cl, Cl — *o*-Dichlorobenzene or 1, 2 Dichlorobenzene

Cl, Br — *m*-Bromo chlorobenzene

Br, I — *p*-Iodo bromobenzene

Cl, NO_2 — *o*-Nitro chlorobenzene

(*iii*) For naming trihalogen compounds, we use the number system. Number 1 is given to the position of one of the groups and then the other groups are numbered accordingly. Clockwise direction is followed for this purpose.

1, 2, 3 Trichlorobenzene   1, 2, 4 Trichlorobenzene   1, 3, 5 Trichlorobenzene

(*iv*) Side-chain halogen compounds are named as follows:

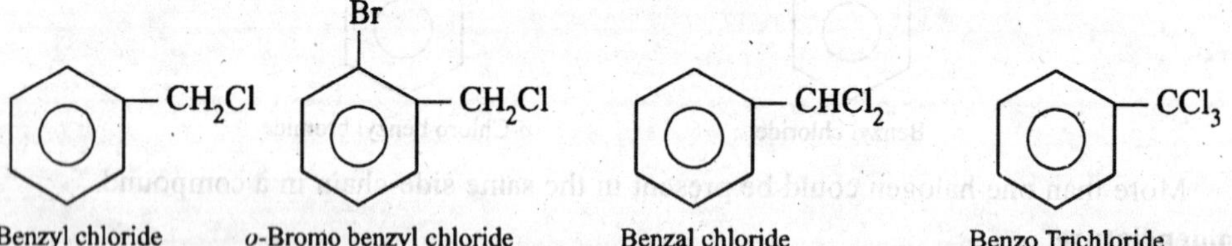

Benzyl chloride   *o*-Bromo benzyl chloride   Benzal chloride   Benzo Trichloride

Q. 2. Write the structures and names of all possible isomers of an aromatic compound with molecular formula C_7H_7Cl. (*M. Dayanand, 1993*)

Ans. The structures and names of possible isomers of the compound with molecular formula C_7H_7Cl are given below:

(*i*) CH_2Cl — Benzyl chloride

(*ii*) CH_3, Cl — *o*-Chlorotoluene (or *o*-Methyl chlorobenzene)

(iii) m-Chlorotoluene (structure: CH_3, Cl on benzene ring)

(iv) p-Chlorotoluene (structure: CH_3, Cl on benzene ring)

Q. 3. Describe three methods of preparation of aryl halides.

Ans. Aryl halides are prepared by the following methods:

(1) Direct halogenation. It consists in treating the aromatic compound with halogen (chlorine or bromine only) at low temperature and in the absence of sunlight and in the presence of halogen carriers like ferric halides or aluminium halides. Iron is mostly used which is converted into the corresponding ferric halide. For example:

$$C_6H_6 + Cl_2 \xrightarrow{Fe} C_6H_5Cl + HCl$$

Chlorobenzene

The reaction stops at the mono substituted stage due to the deactivating influence of the halogen atom on the aromatic ring. However, if chlorine is used in excess, further substitution takes place and a mixture of *o*- and *p*- dichlorobenzene is obtained.

$$C_6H_5Cl + Cl_2 \xrightarrow{Fe} o\text{-}C_6H_4Cl_2 + p\text{-}C_6H_4Cl_2$$

o– *p*–

Dichlorobenzene

Similarly, when toluene is chlorinated or brominated (using 1 molecule of chlorine or bromine), a mixture *o*-, *p*- chlorotoluene or bromotoluene is obtained.

$$C_6H_5CH_3 + Cl_2 \xrightarrow{Fe} o\text{-}ClC_6H_4CH_3 + p\text{-}ClC_6H_4CH_3$$

o– *p*–

Chlorotoluene

$$C_6H_5CH_3 + Br_2 \xrightarrow{Fe} \text{o-Bromotoluene} + \text{p-Bromotoluene}$$

o– p–

Bromotoluene

When aromatic compounds contain strongly deactivating groups such as $-NO_2$, $-COOH$, etc., high temperature is required and the halogen enters the meta position.

$$C_6H_5NO_2 + Br_2 \xrightarrow[413\ K]{Fe} m\text{-}BrC_6H_4NO_2$$

Nitrobenzene → *m*-Bromo nitrobenzene

(2) From diazonium salts (Sandmeyer's reaction). This is an important method for the preparation of aryl halides, specially those which cannot be prepared by the direct halogenation. Aryl chlorides and bromides are obtained by treating the diazonium salt solution with cuprous chloride or bromide in the corresponding halogen acid.

$$C_6H_5\overset{+}{N_2}\overset{-}{Cl} + HCl \xrightarrow{CuCl} C_6H_5Cl + N_2 + HCl$$

Benzene diazonium chloride → Chlorobenzene

$$C_6H_5\overset{+}{N_2}\overset{-}{Cl} + HBr \xrightarrow{CuBr} C_6H_5Br + N_2 + HCl$$

Bromobenzene

Modified form of above mentioned reaction, called **Gattermann** reaction, involves the use of copper powder and halogen acid instead of cuprous chloride or bromide.

$$C_6H_5\overset{+}{N_2}\overset{-}{Cl} + HCl \xrightarrow{Cu} C_6H_5Cl + N_2 + HCl$$

$$C_6H_5\overset{+}{N_2}\overset{-}{Cl} + HBr \xrightarrow{Cu} C_6H_5Br + N_2 + HCl$$

Aryl iodides are obtained by warming the diazonium salt solution with potassium iodide solution.

When aromatic compounds contain highly activating groups such as –OH, $-NH_2$ etc. they undergo halogenation in the absence of halogen carries and halogen enters the ortho and para positions.

Aniline ($C_6H_5NH_2$) + Br_2 (*aq.*) ⟶ 2, 4, 6-Tribromoaniline

Phenol (C_6H_5OH) + Br_2 (*aq.*) ⟶ 2, 4, 6-Tribromophenol

(3) From Phenols. Phenols react with PCl_5 to form aryl chlorides.

Phenol (C_6H_5OH) + PCl_5 ⟶ C_6H_5Cl + $POCl_3$ + HCl

o-Cresol + PCl_5 ⟶ *o*-Chlorotoluene + $POCl_3$ + HCl

Q. 4. Aromatic nucleophilic substitution in aryl halides is difficult. Explain.

Or

Give with mechanism the cause of low reactivity of chlorobenzene. (*Panjab, 2002*)

Or

Why is chlorobenzene less reactive than ethyl chloride during nucleophilic substitution?
(*Madras 2003, Panjab 2004*)

Ans. In chlorobenzene, C–Cl bond is not easily broken because chlorobenzene is a resonance hybrid of various contributing structures such as I–V shown below:

Thus, the molecule of chlorobenzene is resonance stabilised. Moreover, the contribution of structures like III to V imparts a partial double bond character to the carbon halogen bond in the resonance hybrid of the chlorobenzene. Carbon and chlorine are thus held together by little more than a single bond pair of electrons. On the other hand, carbon and chlorine are attached by a single bond in alkyl halides (say CH_3Cl). As a result, the carbon-chlorine bond in chlorobenzene is stronger than if it were a pure single bond and hence cannot be easily broken. The low reactivity of aryl halides is also partly due to resonance stabilization of halide which increases the energy of activation (E_{act}) for displacement and thus slows down the reaction.

I ⟷ II ⟷ III ⟷ IV ⟷ V

There is an alternative explanation for the low reactivity of aryl halides. The carbon atom attached to the halogen in aryl halides is sp^2 hybridised while that in alkyl halides is sp^3 hybridised. Since a sp^2 hybridised orbital is smaller in size than sp^3 hybridised orbital, therefore, the C–Cl bond in chlorobenzene is shorter and hence stronger than in methyl chloride.

This has been confirmed by the X-ray analysis which shows that the C–Cl bond in chlorobenzene is 1.69 Å while in CH_3Cl it is 1.77 Å.

Q. 5. Give important properties of aryl halides.

Ans. A few important reactions of aryl halides are discussed below:

(1) Nucleophilic substitution reactions.

(*i*) *Reaction with NaOH.* When heated with aqueous solution of NaOH at 573 K and under a pressure of 200 atmosphere, aryl halides form phenol.

$$C_6H_5Cl + NaOH \xrightarrow[200\ atom]{573\ K} C_6H_5ONa \xrightarrow{HCl} C_6H_5OH$$

Sod. phenate   Phenol

(*ii*) *Reaction with NH_3.* When aryl halides are heated with aqueous NH_3 at 473 K in the presence of Cu_2O as catalyst under a pressure of 60 atmosphere, amino compounds are obtained.

$$2\,C_6H_5Cl + 2NH_3 + Cu_2O \xrightarrow[200\ atom]{473\ K} 2\,C_6H_5NH_2 + Cu_2Cl_2 + H_2O$$

(*iii*) *Reaction with cuprous cyanide.* When aryl halides are heated with cuprous cyanide at 473 K in the presence of pyridine, the halogen atom is replaced by –CN group.

$$C_6H_5Br + CuCN \xrightarrow[Pyridine]{473\ K} C_6H_5CN + CuBr$$

Cynobenzene (Benzonitrile)

(2) Wurtz-Fittig reaction. When aryl halide is treated with ethereal solution of alkyl halide, in the presence of sodium, an alkyl benzene is formed.

$$C_6H_5\text{-}Cl + 2Na + Cl\text{-}CH_3 \xrightarrow{\text{Dry ether}} C_6H_5\text{-}CH_3 + 2NaCl$$

Chlorobenzene   Methyl chloride   Toluene

But diaryls are produced when only aryl halide treated with sodium. This reaction is called **Fittig's reaction.**

$$C_6H_5\text{-}Cl + 2Na + Cl\text{-}C_6H_5 \xrightarrow{\text{Ether}} C_6H_5\text{-}C_6H_5$$

Chlorobenzene   Diphenyl

(3) Formation of Grignard reagent. Aryl bromides and iodides form Grignard reagent when they are treated with magnesium turnings in dry ether. Aryl chlorides form Grignard reagents only when the reaction is carried out in dry tetrahydrofuran (THF) as solvent

$$C_6H_5Br + Mg \xrightarrow{\text{Dry ether}} C_6H_5MgBr$$

Bromobenzene   Phenyl magnesium bromide

$$C_6H_5Cl + Mg \xrightarrow{\text{Dry THF}} C_6H_5MgCl$$

Chlorobenzene   Phenyl magnesium chloride

THF stands for tetrahydrofuran.

(4) Reduction. Aryl halides are reduced with difficulty by nickel-aluminium alloy in alkali or sodium amalgam and aqueous alcohol in the presence of alkali.

$$C_6H_5Cl + 2H \xrightarrow[\text{or Na(Hg)/Alcohol}]{\text{Ni-Al/NaOH}} C_6H_6 + HCl$$

Benzene

(5) Electrophilic aromatic substitution. Aryl halides undergo the typical electrophilic aromatic substitution reactions though the halogens have a deactivating influence on the aromatic ring. It may be mentioned that inspite of being deactivators, the halogens are *o*- and *p*-directing groups.

(*i*)
$$C_6H_5Cl + Cl_2 \xrightarrow{FeCl_3} o\text{-}C_6H_4Cl_2 + p\text{-}C_6H_4Cl_2 + HCl$$

Chlorobenzene   *o*–   *p*–   Dichlorobenzene

(ii) Chlorobenzene + HNO_3, H_2SO_4 →(Δ) o-Chloronitrobenzene + p-Chloronitrobenzene + H_2O

Chlorobenzene

o– p–

Chloronitrobenzene

(iii) Chlorobenzene + H_2SO_4 →(Δ) o-Chlorobenzene sulphonic acid + p-Chlorobenzene sulphonic acid + H_2O

Chlorobenzene

o– p–

Chlorobenzene sulphonic acid

(iv) Chlorobenzene + CH_3Cl →(Anhy. $AlCl_3$) o-Chlorotoluene + p-Chlorotoluene + HCl

Chlorobenzene

o-Chlorotoluene

p-Chlorotoluene

This particular reaction is Friedel Crafts reaction.

(6) Ullmann reaction

C_6H_5–I + 2Cu + I–C_6H_5 →(Cu Powder) C_6H_5–C_6H_5 + 2CuI

Diphenyl

Q. 6. Discuss nucleophilic substitution in aryl halides in the presence of activating and deactivating groups. (*M. Dayanand, 2000*)

Ans. Aryl halides undergo nucleophilic substitution under drastic conditions. On the other hand, when powerful electron-withdrawing groups such as $-NO_2$, > C=O, –CHO, –COOH and $-SO_3H$ etc. are present in o- and/or p-position with respect to the halogen atom, replacement of the latter by nucleophilic reagents take place under moderate or even ordinary conditions. For example.

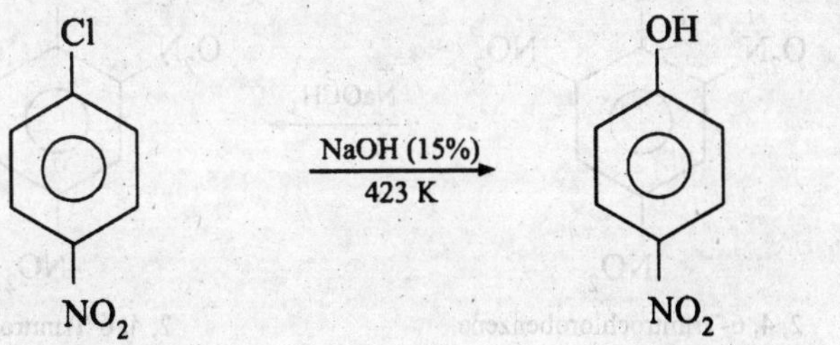

$$\underset{p\text{-Chloronitrobenzene}}{\text{1-Cl-4-}NO_2C_6H_4} \xrightarrow[423\ K]{NaOH\ (15\%)} \underset{p\text{-Nitrophenol}}{\text{1-OH-4-}NO_2C_6H_4}$$

p-Chloronitrobenzene → (NaOH (15%), 423 K) → **p-Nitrophenol**

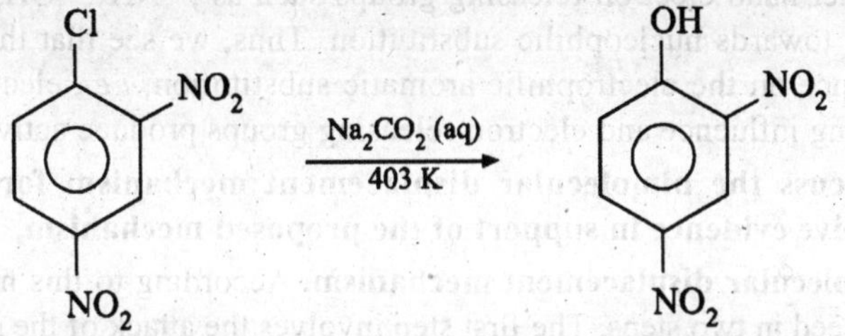

2, 4-Dinitrochlorobenzene → (Na_2CO_3 (aq), 403 K) → **2, 4- Dinitrophenol**

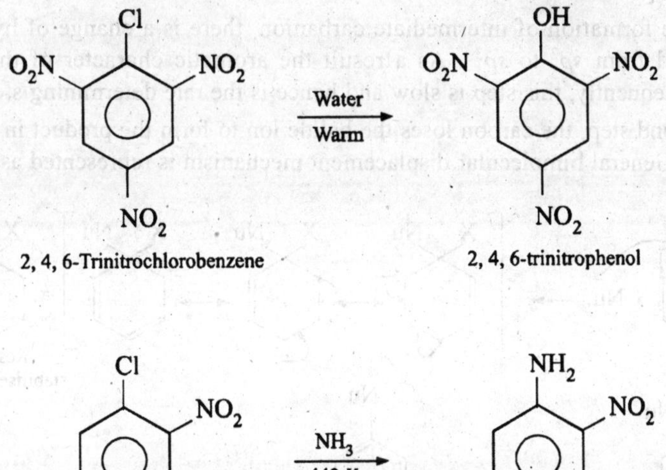

2, 4, 6-Trinitrochlorobenzene → (Water, Warm) → **2, 4, 6-trinitrophenol**

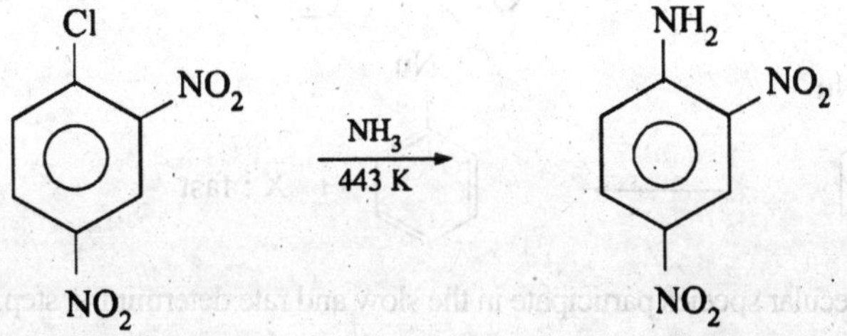

2, 4-Dinitrochlorobenzene → (NH_3, 443 K) → **2, 4-Dinitroaniline**

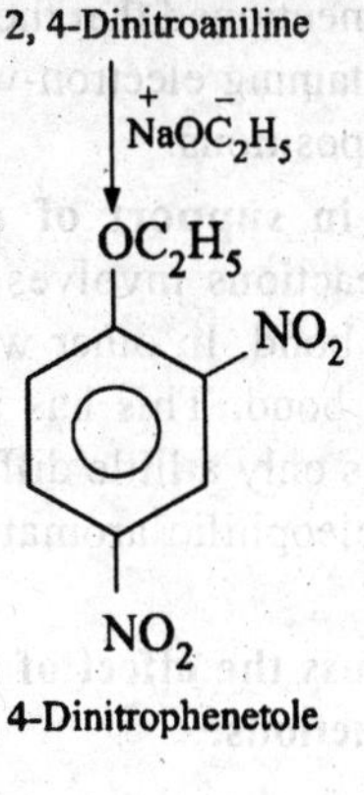

2, 4-Dinitroaniline → ($Na^{+}\,{}^{-}OC_2H_5$) → **2, 4-Dinitrophenetole**

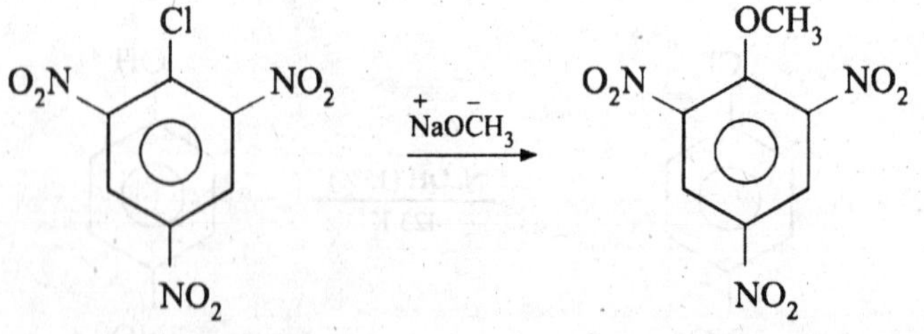

2, 4, 6-Trinitrochlorobenzene $\xrightarrow{\overset{+}{Na}\overset{-}{O}CH_3}$ 2, 4, 6-Trinitro anisole

On the other hand electron releasing groups such as , $-NH_2$, $-OH$, $-OR$, $-R$ etc. deactivate the aryl halides towards nucleophilic substitution. Thus, we see that these groups show just the opposite influence on the electrophilic aromatic substitution, *i.e.*, electron-withdrawing groups show deactivating influence and electron-releasing groups produce activating influence.

Q. 7. Discuss the bimolecular displacement mechanism for nucleophilic aromatic substitution. Give evidence in support of the proposed mechanism. (*G. Nanakdev, 2000*)

Ans. Bimolecular displacement mechanism. According to this mechanism the reaction is believed to proceed in two steps. The first step involves the attack of the nucleophile on the carbon carrying the halogen. This results in the formation of carbanion which is stabilised by resonance.

During the formation of intermediate carbanion, there is a change of hybridisation of the carbon involved from sp^2 to sp^3 . As a result the aromatic character of the benzene ring is destroyed. Consequently, this step is slow and hence is the rate determining step of the reaction.

In the second step, the carbon loses the halide ion to form the product in which aromaticity is regenerated. General bimolecular displacement mechanism is represented as:

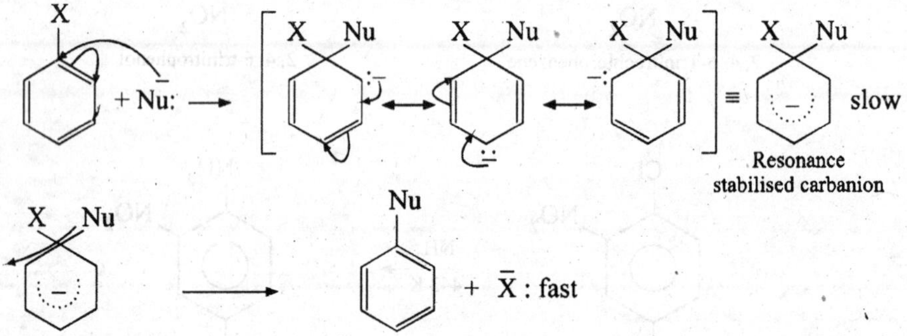

$C_6H_5X + \overline{Nu}: \longrightarrow$ [resonance structures] $\equiv$ slow

Resonance stabilised carbanion

$\longrightarrow C_6H_5Nu + \overline{X}$: fast

As two molecular species participate in the slow and rate determining step, it is a bimolecular reaction.

It may be mentioned that the above bimolecular mechanism is applicable mainly to activated aryl halides containing electron-withdrawign groups such as $-NO_2$, $-C \equiv N$, $-COOH$, etc. at the *ortho* and *para* positions.

Evidence in support of this mechanism. The rate determining step in bimolecular displacement reactions involves the formation of carbanion in which there is no cleavage of carbon-halogen bond. In other words, the rate of the reaction is independent of the strength of carbon-halogen bond. This has actually been verified by the absence of **element effect.** For example, there is only a little difference in the reactivity of aryl iodides, bromides, chlorides and fluorides in nucleophilic aromatic substitution reactions. This lends supports to the proposed mechanism.

Q. 8. Discuss the effect of substituents on the reactivity of aryl halides in nucleophilic substitution reactions.

Ans. In terms of bimolecular displacement mechanism, the nucleophile forms a carbanion with the aryl halide.

$$\bar{N}u : + Ar - X \longrightarrow \underset{\text{Carbanion}}{Nu - \bar{A}r - X}$$

Greater the stability of the carbanion, more rapidly the carbanion is formed and hence faster is the reaction. Thus, factors which stabilize the carbanion will increase the rate of the reaction, while factors which destabilise the carbanion decrease the rate of the reaction.

Consider the nucleophilic substitution reaction of *p*-nitrochlorobenzene, chlorobenzene and *p*-chloroanisol.

The relative rates of these substitution reactions depend upon the relative stabilities of the intermediate carbanions.

$$\underset{p\text{-Nitrochlorobenzene}}{O_2N - C_6H_4Cl} + \bar{N}u : \longrightarrow \underset{(1)}{O_2N - C_6\bar{H}_4 \langle^{Cl}_{Nu}}$$

$$\underset{\text{Chlorobenzene}}{C_6H_5Cl} + N\bar{u} : \longrightarrow \underset{(2)}{C_6\bar{H}_5 \langle^{Cl}_{Nu}}$$

$$\underset{p\text{-Chloroanisol}}{CH_3O - C_6H_4Cl} + N\bar{u} : \longrightarrow \underset{(3)}{CH_3O - \bar{C}_6H_4 \langle^{Cl}_{Nu}}$$

Due to its electron-withdrawing inductive and resonance effects, the $-NO_2$ group stabilises the carbanion (1) by dispersing its negative charge. Hence carbanion (1) is more stable than carbanion (2). On the other hand, the electron-releasing $- CH_3O$ group destabilises the carbanion (3) by intensifying its negative charge. As a result, carbanion (3) is less stable than the carbanion (2). Thus, relative stabilities of these carbanions follow the sequence:

$$(1) > (2) > (3)$$

Therefore the order of reactivity of the above aryl halides in nucleophilic aromatic substitution reaction is:

p-Nitrochlorobenzene > Chlorobenzene > *p*-Chloroanisol.

It may, thus, be concluded that electron-withdrawing groups ($-NO_2$, $-CN$, $-COOH$ etc.) increase the reactivity while electron-releasing groups , ($-CH_3$, $-OH$, $-OCH_3$ etc.) decrease the reactivity of aryl halides towards nucleophilic substitution reactions.

Q. 9. Explain, why electron withdrawing substituents affect the reactivity of aryl halides towards nucleophilic substitution reactions only when present at the ortho and para positions with respect to halogen?

Ans. Influence of substituents on orientation in nucleophilic substitution. In order to explain the effect of substituents on orientation, a comparison of the carbanions formed during the attack of *ortho*, *para* and *meta*-nitrochlorobenzenes by a nucleophilic is made.

It is evident from the structures given below that the carbanions, resulting from *ortho* and *para*-nitrochlorobenzenes are resonance hybrid of four structures and that resulting from *meta*-nitrobenzene is a resonance hybrid of three structures. It is also clear that contributing structures (III) and (VI) are particularly stable because the negative charge in these is located on the carbon directly attached to the electron-withdrawing $- NO_2$ group. It is further noted that in the structures (IV) and (VIII), the negative charge is carried by highly electronegative oxygen and hence such structures are especially stable. No such structures are possible for the carbanion resulting from *m*-nitrochlorobenzene.

Consequently, *o*-nitrochlorobenzene and *p*-nitrochlorobenzene undergo nucleophilic substitution much faster than m nitrochlorobenzene.

o-Nitrochlorobenzene I II III IV

p-Nitrochlorobenzene V VI VII VIII

o-Nitrochlorobenzene IX X XI

It can be said, therefore, that electron-withdrawing groups present at *ortho* and *para* positions with respect to the halogen activate the aryl halide towards nucleophilic substitution reaction.

Q. 10. 2, 4, 6-Trinitrochlorobenzene is easily hydrolysed with water but chlorobenzene is not hydrolysed. Explain

Ans. It has been observed that the presence of electron-withdrawing groups at the *o*- and *p*- positions with respect to the halogen, always activates the aryl halide towards nucleophilic aromatic substitution reactions. Greater the number of such groups at *o*- and *p*- positions, more reactive is the corresponding aryl halide. Since there are three electron-withdrawing nitro groups present at the *ortho*- and *para*-positions, trinitochlorobenzene is highly active and hence can undergo nucleophilic substitution reaction easily. There is no such group present at the *o*- and *p*- positions in chlorobenzene. Hence it requires drastic conditions to change into phenol.

$\xrightarrow{H_2O \text{ (warm)}}$

2, 4, 6-Trinitrochlorobenzene (Picryl chloride) → **2, 4, 6-Trinitrophenol (Picric acid)**

$\xrightarrow[347 \text{ atm}]{7\text{-}8\% \text{ NaOH, 623 K}}$

Chlorobenzene → **Phenol**

Q. 11. What is benzyne?

Ans. Benzyne. It is a highly reactive chemical species which contains an additional bond between two carbon atoms of the benzene ring. This new bond of benzyne is formed by the sideways overlap of sp^2 orbitals belonging to two neighbouring carbon atoms. It is depicted as:

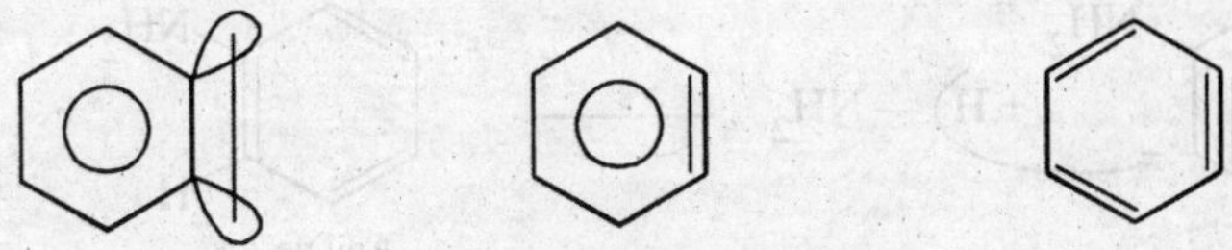

Representation of benzyne

It is evident that the new extra bond orbital lies along the side of the benzene ring and it has little interaction with the π-electron cloud of the benzene ring. Since the sideways overlapping is not very effective, the new bond is a weak one and hence benzyne is highly reactive molecule.

Q. 12. Describe the benzyne mechanism for nucleophilic aromatic substitution. Give evidence in support of the mechanism.

(*Kurukshetra, 2001; M. Dayanand, 2000; Panjab 2003*)

Ans. Benzyne mechanism. It has been found that electron-withdrawing groups activate the aryl halides towards bimolecular nucleophilic substitution reactions. In the absence of such activation, aryl halides can also be made to undergo nucleophilic substitution in the presence of strong nucleophiles. Thus, when chlorobenzene is treated with very strong nucleophile, *viz.* amide ion, it is converted into aniline.

$$C_6H_5Cl \text{ (Chlorobenzene)} + \overset{+}{Na}\overset{-}{NH_2} \xrightarrow{\text{Liquid } NH_3} C_6H_5NH_2 \text{ (Aniline)} + \overset{+}{Na}\overset{-}{Cl}$$

The above type of nucleophilic substitution reaction occurs by benzyne mechanism. It involves both **elimination** and **addition** as represented below:

(I) Elimination. In the elimination step, the amide ion abstracts a proton from one of the *ortho*- positions with respect to the halogen. The resulting carbanion loses the halide ion to form the benyzne, as illustrated below:

$$C_6H_5Cl + :\overset{-}{N}H_2 \longrightarrow \text{Carbanion} + NH_3$$

$$\text{Carbanion} \longrightarrow \text{Benzyne} + :\overset{-}{Cl}:$$

(II) Addition. In the addition step, the amide ion attacks the benzyne molecule to form the carbanion which abstracts a proton from solvent ammonia to yield the final substituted product. These reactions are given below:

$$\text{Benzyne} + :\bar{N}H_2 \longrightarrow \text{Carbanion (NH}_2\text{ substituted)}$$

Carbanion

$$\text{Carbanion} + H-NH_2 \longrightarrow \text{Aniline} + :\bar{N}H_2$$

Aniline

Evidence in support of benzyne mechanism

There is ample experimental evidence to prove the truth of benzyne mechanism of aromatic substitution.

(*i*) **Isotope effect.** The amination of *o*-deuteriobromobenzene is slower than that of bromobenzene. This observation indicates that the cleavage of *ortho* hydrogen is involved in the rate determining step during the elimination stage. This can further be illustrated as follows:

o-Deuteriobromobenzene $\xrightarrow[\text{Relatively slow}]{\bar{N}H_2 / NH_3}$ Benzyne $\xleftarrow[\text{Relatively fast}]{\bar{N}H_2 / NH_3}$ Bromobenzene

Benzyne $\xrightarrow{\bar{N}H_2 / NH_3}$ Aniline (NH_2)

(*ii*) **Absence of ortho hydrogen.** Aryl halides containing two groups ortho to hydrogen, like 2-bromo-3-methylanisol, do not react at all. This is because benzyne intermediate cannot be formed due to the absence of ortho hydrogens with respect to the halogen.

2-Bromo-3-methylanisol (Br, H_3C, OCH_3) $\xrightarrow{NH_2 / NH_3}$ No reaction

(*iii*) **Tracer studies.** When labelled chlorobenzene in which chlorine is linked to C^{14} isotope (C^*) is allowed to react with amide ion, two types of aniline are formed. In one type, the $-NH_2$ group is bonded to C^{14} while in the other it is attached to carbon *ortho* to the labelled carbon. This can be explained as follows:

Chlorobenzene (labelled) $\xrightarrow{\bar{N}H_2 / NH_3}$ Benzyne $\xrightarrow{\bar{N}H_2 / NH_3}$ aniline with NH_2 on C^* + aniline with NH_2 ortho to C^*

Fig. st13-40

(*iv*) **Identical product from different aryl halides.** When *o*-bromo-anisol and *m*-bromo-anisol are allowed to react separately with $\bar{N}H_2$ ion in liquid NH_3, the same product, *i.e.*, *m*-anisidine is formed. The formation of the same product is due to the formation of the same intermediate benzyne as shown below. The amide ion attacks the benzyne species at *meta* position and not *ortho* position probably because of steric factor.

$$o\text{-Bromoanisole} \xrightarrow{\bar{N}H_2/NH_3} \text{Benzyne} \xrightarrow{\bar{N}H_2/NH_3} m\text{-Anisidine}$$

$$m\text{-Bromoanisole} \xrightarrow{\bar{N}H_2/NH_3} \text{Benzyne}$$

Q. 13. How will you distinguish between benzyl chloride and chlorobenzene?

(*Awadh 2000; Nagpur, 2003; Udaipur 2003; Delhi, 2005*)

Ans. Benzyl chloride is easily hydrolysed with aqueous KOH to form water soluble KCl which gives white precipitate of AgCl on treating with an aqueous solution of silver nitrate. Thus,

$$C_6H_5-CH_2-Cl + \underset{(Aq.)}{KOH} \longrightarrow C_6H_5-CH_2Cl + \underset{(Aq.)}{KCl}$$

$$\underset{(Aq.)}{KCl} + \underset{(Aq.)}{AgNO_3} \longrightarrow \underset{(Aq.)}{KNO_3} + \underset{(White\ ppt.)}{AgCl}$$

Chlorobenzene is less reactive towards nucleophilic substitution reactions. Hence it is not easily hydrolysed by alkalies. Therefore, it does not respond to the above test.

Q. 14. Starting from bromobenzene how will you prepare each of the following:

(*i*) Iodobenzene (*ii*) Aniline (*iii*) Benzyl alcohol (*iv*) Benzoic acid (*v*) Fluorobenzene (*vi*) Benzene (*Nehru 2004; Shivaji 2004*)

Ans. (*i*) Iodobenzene

$$C_6H_5Br + \overset{+}{Na}\overset{-}{N}H_2 \xrightarrow[-\ NaBr]{\text{Liquid } NH_3} C_6H_5NH_2 \xrightarrow[273\ K]{\text{Diazotisation} \atop (NaNO_2 + HCl)} \underset{\text{Benzene diazonium chloride}}{C_6H_5\overset{+}{N_2}\overset{-}{Cl}}$$

$$C_6H_5\overset{+}{N_2}\overset{-}{Cl} \xrightarrow{KI} \underset{\text{Iodobenzene}}{C_6H_5I} + KCl + N_2$$

(ii) Aniline

$$C_6H_5Br + \overset{+}{Na}\overset{-}{NH_2} \xrightarrow{\text{Liquid } NH_3} C_6H_5NH_2 + NaBr$$

Aniline

(iii) Benzoic acid

$$C_6H_5Br + 2Na + CH_3Br \xrightarrow[\text{— 2NaBr}]{\text{Dry ether}} C_6H_5CH_3$$

$$C_6H_5CH_3 \xrightarrow{KMnO_4 \quad [O]} C_6H_5COOH$$

Benzoic acid

(iv) Benzyl alcohol

$$C_6H_5Br + 2Na + CH_3Br \xrightarrow[\text{(Dry ether)}]{\text{Wurtz-Fittig reaction}} C_6H_5CH_3 + 2NaBr$$

$$C_6H_5CH_3 \xrightarrow[\text{— HCl}]{Cl_2 \text{ (1 mole) sunlight}} C_6H_5CH_2-Cl$$

$$C_6H_5CH_2-Cl \xrightarrow[\text{— KCl}]{\text{Aq. KOH}} C_6H_5CH_2OH$$

Benzyl alcohol

(v) Fluorobenzene

$$C_6H_5Br \xrightarrow[\text{— NaBr}]{NaNH_2 / NH_3} C_6H_5NH_2 \xrightarrow[273\text{ K}]{(NaNO_2 + HCl)} C_6H_5\overset{+}{N_2}\overset{-}{Cl} \xrightarrow[\text{— HCl}]{HBF_4} C_6H_5\overset{+}{N_2}\overset{-}{BF_4} \xrightarrow{\text{Heat}} C_6H_5F + BF_3 + N_2$$

Benzene diazonium fluoborate — Fluorobenzene

(*vi*) **Benzene**

$$C_6H_5Br + 2[H] \xrightarrow[NaOH]{Ni/Al} C_6H_6 + HBr$$

Q.15. Give two methods of obtaining side-chain aromatic halogen compounds.

Ans. (1) Direct halogenation. When chlorine or bromine is passed through toluene or higher homologues at boiling point in the presence of sunlight but in the absence of a halogen carrier, halogenation takes place in the side chain. Thus,

$$\underset{\text{Toluene}}{C_6H_5CH_3} \xrightarrow[\text{Boiling}]{Cl_2} \underset{\text{Benzyl chloride}}{C_6H_5CH_2Cl} + HCl$$

On continued chlorination, benzal chloride and benzotrichloride are formed.

(2) From aromatic alcohols and aldehydes. By the action of phosphorus pentachloride. Thus,

$$\underset{\text{Benzyl alcohol}}{C_6H_5CH_2OH} + PCl_5 \longrightarrow \underset{\text{Benzyl chloride}}{C_6H_5CH_2Cl} + POCl_3 + HCl$$

$$\underset{\text{Benzaldehyde}}{C_6H_5CHO} + PCl_5 \longrightarrow \underset{\text{Benzal chloride}}{C_6H_5CHCl_2} + POCl_3$$

Q. 16. Give some important reactions of side-chain aromatic halogen compounds.

Ans. 1. Nucleophillic reactions

(*i*) Replacement by hydroxyl group

$$\underset{\text{Benzyl chloride}}{C_6H_5CH_2Cl} + \underset{\text{(aq.)}}{KOH} \longrightarrow \underset{\text{Benzyl alcohol}}{C_6H_5CH_2OH} + KCl$$

$$\underset{\text{Benzal chloride}}{C_6H_5CHCl_2} + \underset{\text{(aq.)}}{2KOH} \longrightarrow \underset{\text{Benzaldehyde}}{C_6H_5CHO} + H_2O + 2KCl$$

$$C_6H_5CCl_3 + 3KOH\ (aq.) \longrightarrow C_6H_5COOH + H_2O + 2KCl$$

Benzo trichloride → Benzoic acid

(*ii*) Replacement by amino group

$$C_6H_5CH_2Cl + 2NH_3\ (alc.) \longrightarrow C_6H_5CH_2NH_2 + NH_4Cl$$

Benzyl chloride → Benzyl amine

(*iii*) Replacement by cyano group

$$C_6H_5CH_2Cl + KCN \longrightarrow C_6H_5CH_2CN + KCl$$

Benzyl chloride → Benzyl cyanide

(*iv*) Replacement by hydrogen (Reduction)

$$C_6H_5CH_2Cl + 2[H] \xrightarrow{\text{Zn-Cu /alcohol}} C_6H_5CH_3 + HCl$$

Benzyl chloride → Toluene

(*v*) Replacement by an alkyl or an aryl radical. On treating side-chain halogen derivative with alkyl halides or aryl halides in the presence of sodium in dry ether, the halogen atom is replaced by an alkyl or aryl groups. (Wurtz-Fittig reaction).

$$C_6H_5CH_2Cl + 2Na + ClCH_3 \xrightarrow[\text{reaction}]{\text{Wurtz Fittig's}} C_6H_5CH_2CH_3 + 2NaCl$$

Benzyl chloride, Methyl chloride → Ethyl benzene

$$C_6H_5CH_2Cl + 2Na + ClC_6H_5 \xrightarrow[\text{reaction}]{\text{Fittig's}} C_6H_5CH_2C_6H_5 + 2NaCl$$

Benzyl chloride, Chlorobenzene → Diphenyl methane

2. Reactions of the benzene nucleus. The side-chain halogen derivatives give the usual substitution reactions of benzene nucleus, *viz.*, they can be nitrated, sulphonated and halogenated, the new entrant occupying the *ortho* and *para* positions.

$$C_6H_5CH_2Cl + Cl_2 \longrightarrow o\text{-}ClC_6H_4CH_2Cl + p\text{-}ClC_6H_4CH_2Cl + H_2O$$

Benzyl chloride → *o*-Chlorobenzyl chloride + *p*-Chlorobenzyl chlorides

$$C_6H_5CH_2Cl + HNO_3 \longrightarrow o\text{-}NO_2C_6H_4CH_2Cl \text{ and } p\text{-}NO_2C_6H_4CH_2Cl + H_2O$$

o- and *p*- Nitrobenzyl chlorides

$$C_6H_5CH_2Cl + H_2SO_4 \longrightarrow o\text{-}HOSO_2C_6H_4CH_2Cl \text{ and } p\text{-}HOSO_2C_6H_4CH_2Cl + H_2O$$

o- and *p*- Benzyl chloride sulphonic acids

$$C_6H_5CH_2Cl + H{-}C_6H_5 \xrightarrow[\text{AlCl}_3]{\text{Anhy.}} C_6H_5{-}CH_2{-}C_6H_5$$

Diphenyl methane

This is Friedel Craft type of reaction.

(3) Other reactions. (*i*) Formation of Grignard Reagents. Like nuclear halogen compounds, side chain derivatives form Grignard reagents on treatment with dry magnesium powder in dry ether.

$$C_6H_5CH_2Br + Mg \xrightarrow{\text{Dry ether}} C_6H_5CH_2MgBr$$

Benzyl bromide → Benzyl magnesium bromide

(ii) **Oxidation.** Action of oxidising agents results in the oxidation of the side chain including the halogen atom present in the chain. Mild oxidising agents like lead nitrate convert the side chain into –CHO group while reagents like nitric acid change it into –COOH group.

$$C_6H_5COOH \xleftarrow[HNO_3]{O} C_6H_5CH_2Cl \xrightarrow[Pb(NO_3)_2]{O} C_6H_5CHO$$

Benzoic acid — Benzaldehyde

(iii) **Halogenation of side chain.** On treatment with halogens at higher temperature, the presence of sunlight and absence of a halogen carrier, the side chain gets further halogenated.

$$C_6H_5CH_2Cl + Cl_2 \xrightarrow[\text{High temp.}]{\text{Sunlight}} C_6H_5CHCl_2 \xrightarrow{Cl_2} C_6H_5CCl_3$$

Benzyl chloride — Benzal chloride — Benzo trichloride

Q. 17. Starting with a suitable alkyl or aryl halide, how will you obtain (*i*) Diphenyl (*ii*) Butyne (*iii*) Benzaldehyde. (*Nagpur, 2005*)

Ans. (*i*)

$$C_6H_5-Cl + 2Na + Cl-C_6H_5 \quad \text{(Fittig's Reaction)}$$

Chlorobenzene

$$\downarrow \text{Dry ether}$$

$$C_6H_5-C_6H_5$$

Diphenyl

(*ii*) Butyne –1 can be obtained as follows:

$$CH_3\,CH_2\,Cl + NaC \equiv CH \longrightarrow CH_3CH_2C \equiv CH + NaCl$$

Chloroethane — Sod.acetylide — Butyne-1

Butyne –2 can be obtained as follows:

$$CH_3 - Cl + NaC \equiv CNa + Cl - CH_3$$

Chloro methane — Disodium acetylide

$$\downarrow$$

$$CH_3 - C \equiv C - CH_3 + 2NaCl$$

Butyne-2

(*iii*) Benzaldehyde can be obtained by the hydrolysis of benzal chloride, which can be obtained by the chlorination of benzyl chloride

$$C_6H_5CH_2Cl \xrightarrow[\text{Sunlight}]{Cl_2} C_6H_5CHCl_2 \xrightarrow{H_2O} C_6H_5CH(OH)_2 \longrightarrow C_6H_5CHO + H_2O$$

Benzyl chloride — Benzal chloride — (Unstable) — Benzaldehyde

Q. 18. Complete the following reactions, giving the names of products.

(*i*) $C_6H_5Cl + CH_3Cl + Na \longrightarrow$

(*ii*) $CH_3\ CHO + PCl_5 \longrightarrow$

Ans. (*i*) $C_6H_5Cl + 2Na + Cl—CH_3 \xrightarrow[\text{ether}]{\text{Dry}} C_6H_5CH_3 + 2NaCl$

Chlorobenzene — Toluene

(*ii*) $CH_3CHO + PCl_5 \longrightarrow CH_3CHCl_2 + POCl_3$

Acetaldehyde — Benzal chloride — Phosphorus oxy trichloride

Q. 19. How would you distinguish between *o*-chlorobenzyl bromide and o-bromobenzyl chloride?

Ans.

o-Chlorobenzyl bromide (CH_2Br with Cl at ortho position) — *o*-Bromobenzyl chloride (CH_2Cl with Br at ortho position)

The two compounds can be distinguished by performing the hydrolysis reaction of the two compounds with aqueous NaOH under ordinary conditions. Under ordinary reaction conditions *i.e.*

at room temp. and atmospheric pressure, nuclear chloro and bromo groups are not replaced. Only the halogen in side chain is substituted as given below:

$$\underset{\text{I}}{o\text{-}ClC_6H_4CH_2Br} + NaOH \longrightarrow \underset{o\text{-Chlorobenzyl alcohol}}{o\text{-}ClC_6H_4CH_2OH} + NaBr$$

$$\underset{\text{II}}{o\text{-}BrC_6H_4CH_2Cl} + NaOH \longrightarrow \underset{\text{o-Bromobenzyl alcohol}}{o\text{-}BrC_6H_4CH_2OH} + NaCl$$

Thus compounds I and II give sodium bromide and sodium chloride respectively in the solution. These are ionic compounds and furnish bromide and chloride ions in solutions. These ions can be analysed by reaction with silver nitrate solution as given below:

$$Br^- + AgNO_3 \longrightarrow \underset{\text{Pale yellow ppt.}}{AgBr} + NO_3^-$$

$$Cl^- + AgNO_3 \longrightarrow \underset{\text{White ppt.}}{AgCl} + NO_3^-$$

Thus, *o*-chlorobenzyl bromide on hydrolysis with aqueous NaOH will produce bromide ions which give a *pale*-yellow ppt. with $AgNO_3$.

o-bromobenzyl chloride on hydrolysis with aqueous NaOH will produce chloride ions which give a white ppt. with $AgNO_3$. This is how the two compounds can be distinguished.

Q. 20. Describe the structure of aryl and vinyl halides and discuss their low reactivity as compared to alkyl halides.

(*Devi Ahilya 2001; Awadh, 2000; Kurukshetra, 2001; G Nanakdev, 2000*)

Or

Explain how resonance theory explains low reactivity of vinyl and aryl halides.

Ans. (*i*) Aryl halides are less reactive than alkyl halides towards nucleophilic substitution reactions because of the resonance stabilization of aryl halide.

$$C_6H_5\ddot{X}: \longleftrightarrow \overset{+}{X}{=}C_6H_5^{-}\ (ortho) \longleftrightarrow \overset{+}{X}{=}C_6H_5^{-}\ (para) \longleftrightarrow \overset{+}{X}{=}C_6H_5^{-}\ (ortho')$$

Since C – X double bond is not easily broken, SN reaction does not take place easily.

(*ii*) Vinyl halides are also less reactive than alkyl halides because of a similar reason.

$$CH_2{=}CH{-}\ddot{X}: \longrightarrow \bar{C}H_2{-}CH{=}\overset{+}{X}$$

Double bond between C and X and presence of a positive charge on X prevent the SN reaction from taking place.

Q. 21. Vinyl chloride is an aliphatic halide but it resembles chlorobenzene in its properties. Explain.

Ans. Because of low reactivity of both towards nucleophilic substitution reactions, they resemble each other.

In both compounds, because of resonance phenomenon, a double bond is created between carbon and halogen and a positive charge is created on the halogen and a negative charge is created on the rest of the molecule. These two factors go against the nucleophilic substitution reactions. For the resonating structures of aryl and vinyl halide, see Q. 20.

Q. 22. Aryl halides are less reactive than alkyl halides. Explain.

Ans. See Q. 20.

Q. 23. Describe with the help of molecular orbital structure that an allyl radical is more stable as compared to primary radical.

Or

How do you account for unusual high stability of allyl free radical.

Ans. The orbital structure of allyl radical is illustrated as under.

$$CH_2 = CH - \dot{C}H_2 \longrightarrow \dot{C}H_2 - CH = CH_2$$

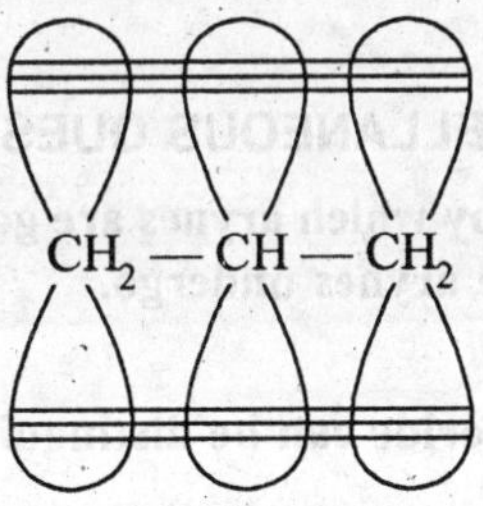

$$CH_2 - CH - CH_2$$

The delocalisation of π electron cloud lends stability to the allyl radical. It is believed that *p* orbitals of the doubly bonded carbon atoms and the single electron of the third carbon are in the same plane and overlap. This type of delocalisation is not possible in alkyl radicals

$$CH_2 - CH - \dot{C}H_2$$

Q. 24. Allyl chloride is more reactive than vinyl chloride. Explain.

Ans. The intermediate carbocation formed during nucleophilic substitution reaction is more stable in the case of allyl chloride than in case of vinyl chloride.

$$\underset{\text{Allyl carbocation}}{CH_2 = CH - \overset{+}{C}H_2} \longleftrightarrow \overset{+}{C}H_2 - CH = CH_2$$

$$CH_2 = \overset{+}{C}H \qquad \text{No resonance}$$

With vinyl carbocation, there is no possibility of resonance and hence no resonance stabilization. As the allyl carbocation is resonance stabilized, nucleophilic reaction takes place more readily.

Q. 25. Bromobenzene does not give precipitate even on prolonged heating with alcoholic silver nitrate. Explain.

Ans. 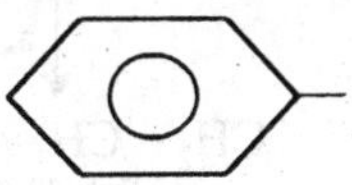 $C_6H_5-Br \xrightarrow{\text{Alc. } AgNO_3} \text{No precipitate}$

Aromatic halides, do not give nucleophilic substitution reactions easily.

As a result, bromine is not displaced as bromide ion in the solution consequently a precipitate is not obtained.

Q. 26. An alcoholic solution of butyl bromide when heated with silver nitrate gives a white precipitate but pentyl bromide does not give precipitate of silver bromide even on prolonged heating with $AgNO_3$. Explain why.

Ans. Butyl bromide and pentyl bromide are expected to give butyl and pentyl alcohol respectively on treatment with alcoholic silver nitrate SN^2 by mechanism. The bromide ions produced in the reaction react with $AgNO_3$ to give a precipitate of silver bromide.

The success of an SN^2 reaction depends upon the ease with which a nucleophile, can attach itself to the halogen carrying carbon.

In the case of *n*-butyl-chloride, the groups attached to this carbon are H, H and C_3H_7 whereas

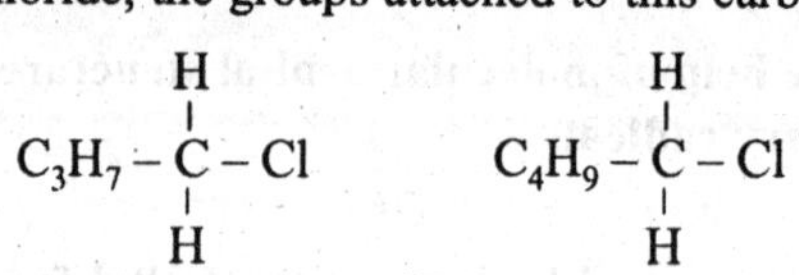

$$C_3H_7-\overset{\displaystyle H}{\underset{\displaystyle H}{\overset{|}{\underset{|}{C}}}}-Cl \qquad\qquad C_4H_9-\overset{\displaystyle H}{\underset{\displaystyle H}{\overset{|}{\underset{|}{C}}}}-Cl$$

Butyl chlorde Pentyl chloride

in case of pentyl chloride, the groups attached are H, H and C_4H_9. Thus there is greater crowding and smaller chance of SN^2 reaction to take place in the case of pentyl chloride. Therefore, no precipitate is obtained.

MISCELLANEOUS QUESTIONS

Q.27. State any two methods by which arynes are generated, comment on their stability and mention any two reactions the arynes undergo. (*Kerala 2000*)

Ans. See Q. 11 and Q. 12

Q.28. Explain how benzyl chloride can be distinguished from *p*-chlorotulene? (*Kerala 2001*)

Ans. See Q. 3.

Q.29. Give the equation for a reaction involving benzyne intermediate. (*Kerala 2001*)

Ans. See Q. 12.

Q.30. How will you obtained vinyl benzene from bromobenzene ? (*Lucknow, 2000*)

Ans.

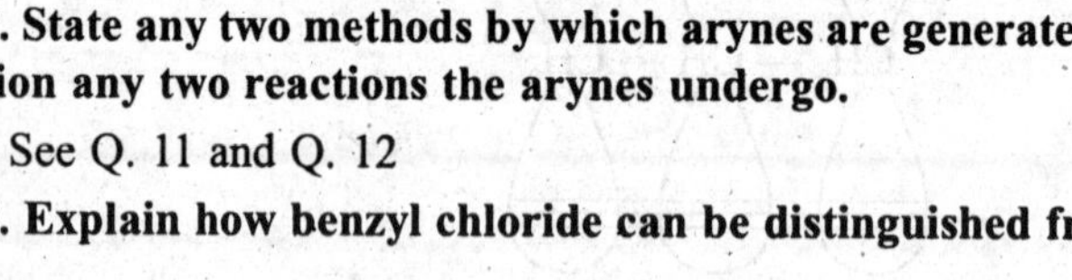

$$C_6H_5-Br \xrightarrow[\text{2 Na in dry ether}]{Br-CH=CH_2} C_6H_5-CH=CH_2$$

Bromobenzene Vinyl benzene

Q.31. Write a note on Sandmeyer's reaction. (*Himachal, 2000*)

Ans. See Q. 3(2).

Q.32. How will you convert propane to allyl chloride? (*Meerut, 2000*)

Ans. $CH_3-CH_2-CH_3 \xrightarrow{Cl_2} CH_3\,CH_2\,CH_2\,Cl \xrightarrow[KOH]{alc} CH_3-CH=CH_2$

Propane Propane

$$CH_3-CH=CH_2 \xrightarrow{Cl_2 \text{ at high temp}} CH_2=CH-CH_2\,Cl$$

Allyl chloride

Q.33. Benzyl chloride is more reactive than chlorobenzene. Explain.

(*Himachal, 2000, Negpur, 2005*)

Ans. The electron pair on chlorine in chlorobenzene takes part in rasonance with the benzene ring. This creates a double bond between the chlorine and ring carbon. This makes nucleophilic substitution difficult.

Q.34. How would you synthesize DDT from chlorobenzene? Give Chemical equation.

(*Panjab, 2002; Jammu, 2003; Agra, 2004; Nagpur 2008*)

Ans.

$$2\,Cl-C_6H_4-H + O=CH-CCl_3 \xrightarrow[-H_2O]{\text{Conc } H_2SO_4} H-C(C_6H_4Cl)_2-CCl_3$$

DDT (p, p- Dichlorodiphenyl trichloromethane)

Q.35. Compare the reactivity of benzye chloride with ethyl chloride with ethyl chloride.

(*Nagpur 2008*)

Ans. Benzyl chloride as well as ethyl chloride react via SN^2 mechanism with different reagents to give different products. This involves the formation of an intermediate product in which he –Cl group and the reacting nucleophiles are attached. Benzyl chloride ($C_6H_5CH_2Cl$) is a bigger molecule, therefore, there will be steric hindrance in the formation of the intermediate. Thus. benzyl chloride will show smaller reactivity than ethvl chloride.

14

ORGANOMETALLIC COMPOUNDS

Q. 1. What do you mean by organometallic compounds? Give some examples.

(*Kalyani, 2002; Jammu, 2003; Mumbai, 2010*)

Ans. Organometallic compounds are those organic compounds in which there is a bond between carbon and metal. There is a wide variation in the nature of this bond in different organometallic compounds. Main factor that is responsible for difference in the nature of carbon-metal bond is the nature of metal itself. Highly electropositive metals like sodium and potassium tend to make this bond ionic, with the metal carrying positive charge and carbon carrying negative charge. Organometallic compounds of magnesium and lithium have this bond with partial ionic character.

We come across organometallic compounds with metal constituents such as sodium, potassium, lithium, magnesium, lead and zinc.

Some examples of these compounds are:

CH_3MgBr	C_6H_5MgI	C_2H_5MgCl
Methyl magnesium bromide	Phenyl magnesium iodide	Ethyl magnesium chloride
CH_3Li	C_2H_5Li	$(CH_3)_2Zn$
Methyl lithium	Ethyl lithium	Dimethyl zinc
$(C_2H_5)_2Cd$	$(C_2H_5)_4Pb$	
Diethyl cadmium	Tetraethyl lead	

The following sequence indicates the variation in ionic character in carbon-metal bond.

$$C-K > C-Na > C-Li > C-Mg > C-Zn > C-Cd$$

Q. 2. What are Grignard reagents? Give the method of preparation of Grignard reagent.

(*Kerala, 2001; Panjab, 2002; Assam, 2003; Agra, 2004; Baroda, 2004; Calcutta, 2007; Guwahati, 2007*)

Ans. Organic compounds having the general formula RMgX are called Grignard reagents. Here R stands for some alkyl or aryl group and X denotes some halogen.

Preparation of Grignard Reagents. Grignard reagents are prepared by reacting alkyl or aryl halides and Mg metal in dry ether.

$$R-X + Mg \xrightarrow[\text{ether}]{\text{Dry}} RMgX$$

$$Ar-X + Mg \longrightarrow ArMgX$$

$$\underset{\text{Ethyl bromide}}{C_2H_5-Br} + Mg \xrightarrow{\text{Dry ether}} \underset{\text{Ethyl magnesium bromide}}{C_2H_5MgBr}$$

Ether is used as a solvent because it can solvate the reagent by acting as a base towards the acidic magnesium. For obtaining Grignard reagents from inactive halogen compounds, tetrahydrofuran (THF) is used as the solvent in place of dry ether.

$$C_6H_5Cl + Mg \xrightarrow{THF/\Delta} C_6H_5MgCl$$

The magnesium metal disappears and Grignard reagents is formed with evolution of heat.

For a given alkyl group, the order of reactivity of halogen is

$$I > Br > Cl$$

Similarly, for a given halogen atom, the order of reactivity of alkyl group is:

$$CH_3 > C_2H_5 > C_3H_7$$

Grignard reagents are seldom isolated in solid state as they ignite spontaneously in air. They are usually used in solution.

Q. 3. Give the mechanism of formation of Grignard reagents. What precautions are necessary in preparing Grignard reagents?

Ans. Mechanism. The mechanism by which Grignard reagents are formed is still not fully understood. The most likely mechanism appears to be free radical mechanism as shown below:

$$R - X + Mg \longrightarrow R\cdot + \cdot MgX$$

$$R\cdot + \cdot MgX \longrightarrow RMgX$$

Precautions. (*i*) The reagents used should be pure and dry as moisture tends to stop the reactions.

(*ii*) Ether should be washed with water to free it from alcohol and dried over anhydrous $CaCl_2$ for 2-3 days to remove alcohol and moisture. It should be finally distilled over sodium and P_4O_{10} to remove last traces of water.

(*iii*) Magnesium turnings should be treated with ether to remove grease and then dilute HCl to remove oxide film. They should be finally dried in an oven at 383-393 K.

(*iv*) Alkyl halide should be dried by distilling over P_4O_{10}

Q. 4. How is Grignard reagent prepared and what precautions must be taken during its preparation?

Ans. See Qs. 1 and 2.

Q. 5. Discuss the structure of Grignard reagent.

Ans. Structure of Grignard reagent.

During the preparation of Grignard reagents, ether is used as a solvent. Accordingly it has been assumed that ether molecules are present as ether of crystallisation. Thus, two most probable structures have been proposed for the Grignard reagent.

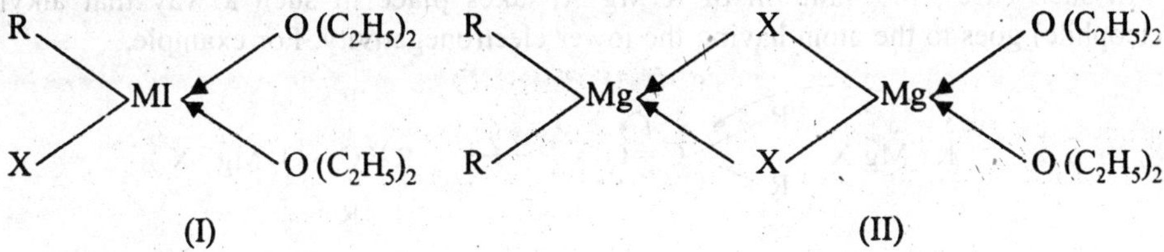

It has also been proposed that the following equilibria exist for Grignard reagent in ether.

$$R_2Mg \cdot MgX_2 \rightleftharpoons R_2Mg + MgX_2 \rightleftharpoons 2\,RMgX$$

However, the general formula of Grignard reagents is represented as RMgX for the sake of convenience.

Q. 6. What kind of reactions are given by Grignard reagents?

Ans. The reactions of Grignard reagents can be studied under two headings.

***(i)* Double decomposition with compounds containing an active hydrogen atom.**

An active hydrogen atom is one which is joined to oxygen, nitrogen or sulphur. When such compounds are treated with Grignard reagent, the alkyl group is converted into alkane

$$H-O-\boxed{H+R}-Mg-X \longrightarrow R-H+Mg\begin{matrix} X \\ OH \end{matrix}$$

$$R-O-\boxed{H+R}-Mg-X \longrightarrow R-H+Mg\begin{matrix} X \\ OR \end{matrix}$$

$$NH_2-\boxed{H+R}-Mg-X \longrightarrow R-H+Mg\begin{matrix} X \\ NH_2 \end{matrix}$$

$$HCl + RMg-X \longrightarrow R-H+Mg\begin{matrix} X \\ Cl \end{matrix}$$

Carbon-magnesium bond in RMgX has considerable ionic character. Because of this, the carbon atom bonded to the metal is a strong base or we can say that Grignard reagents are strong bases. They react with those compounds which have a hydrogen more acidic than the hydrogen of the hydrocarbon from which the Grignard reagent is derived. The reactions of Grignard reagent with above mentioned compounds are acid-base reactions and lead to the formation of conjugate acid and conjugate base:

$$\underset{\text{(Stronger base)}}{\overset{\delta-\ \ \delta+}{R:Mg}}\ X + \underset{\text{(Stronger acid)}}{H-\ddot{O}-R} \longrightarrow \underset{\text{(Weaker acid)}}{R-H} + \underset{\text{(Weaker base)}}{R-\ddot{O}\!:^-} + Mg^{+2} + X$$

$$\underset{\text{(Stronger base)}}{\overset{\delta-\ \ \delta+}{R:Mg}}\ X + \underset{\text{(Stronger acid)}}{H-\ddot{O}-H} \longrightarrow \underset{\text{(Weaker acid)}}{R-H} + \underset{\text{(Weaker base)}}{H-\ddot{O}\!:^-} + Mg^{+2} + X^-$$

***(ii)* Addition to compounds containing multiple bonds.**

In such cases, the addition of R Mg X, takes place in such a way that alkyl group (nucleophile) goes to the atom having the lower electronegativity. For example,

$$\overset{\delta-\ \ \delta+}{R:Mg\,X}\quad \begin{matrix} R \\ R \end{matrix}\!\!>C=\ddot{O}: \longrightarrow R-\underset{R}{\overset{R}{C}}-\ddot{O}\!:Mg^{+2}X^-$$

When water or dilute acid is added to the reaction mixture, an acid-base reaction takes place to produce an alcohol.

$$R-\underset{R}{\overset{R}{\mid}}{C}-\ddot{O}\!:\,Mg^{+2}X^{-}+H-\ddot{O}-H \longrightarrow R-\underset{R}{\overset{R}{\mid}}{C}-\ddot{O}-H+MgX_2+H_2O$$

Q. 7. Describe synthetic applications of Grignard reagents.

(*Kalyani, 2003; Bangalore, 2004; Burdwan, 2004; Guwahati, 2007; Calcutta, 2007; Andhra, 2010; Bharathiar, 2011*)

Ans. We can prepare almost every type of organic compound from them by selecting a suitable reagent.

(1) **Alkanes.** Grignard reagents are decomposed by water, alcohol, ammonia to form alkanes. Thus:

$$\underset{\text{Methylmag. iodide}}{CH_3\,Mg-I} + \underset{\text{Water}}{H-OH} \longrightarrow \underset{\text{Methane}}{CH_4} + \underset{\text{Magnesium hydroxy iodide}}{Mg\begin{matrix} I \\ OH\end{matrix}}$$

$$\underset{\text{Methylmag. Iodide}}{C_2H_5-Mg-I} + \underset{\text{Ethanol}}{HOC_2H_5} \longrightarrow \underset{\text{Ethane}}{C_2H_6} + \underset{\text{Magnesium ethoxy iodide}}{Mg\begin{matrix} I \\ OC_2H_5\end{matrix}}$$

$$CH_3-Mg-Br + \underset{\text{Ammonia}}{H-NH_2} \longrightarrow CH_4 + \underset{\text{Magnesium amino bromide}}{Mg\begin{matrix} Br \\ NH_2\end{matrix}}$$

(2) **Alkenes.** These are obtained by reaction with unsaturated halogen derivatives. Thus:

$$CH_3\,Mg\,I + ICH_2-CH=CH_2 \longrightarrow \underset{\text{Butene-1}}{CH_3-CH_2-CH=CH_2} + MgI_2$$

(3) **Alkynes.** Higher alkynes can be obtained by treating lower alkynes with a Grignard reagent and then alkyl halide. Thus:

$$\underset{\text{Propyne}}{CH_3-C\equiv CH} + CH_3\,MgI \longrightarrow CH_3-C\equiv C-Mg\,I + CH_4$$

$$CH_3-C\equiv C-Mg\,I \xrightarrow{CH_3-CH_2-I} \underset{\text{2-Pentyne}}{CH_3-C\equiv C-CH_2-CH_3} + Mg\,I_2$$

(4) **Alcohols.** (*a*) *Primary alcohols*. Primary alcohols can be obtained by one of the following methods:

(*i*) By treating formaldehyde with Grignard reagent followed by decomposition with dilute acid.

$$\overset{\delta-}{R}:\overset{\delta+}{Mg}X + H-\underset{H}{\underset{\mid}{C}}=O \longrightarrow H-\underset{H}{\underset{\mid}{\overset{R}{\overset{\mid}{C}}}}-OMg\,X \xrightarrow{H_2O} \underset{1^\circ\text{ Alcohol}}{H-\underset{H}{\underset{\mid}{\overset{R}{\overset{\mid}{C}}}}-OH}$$

$$C_2H_5-Mg-Br + H-\underset{\underset{H}{|}}{C}=O \longrightarrow H-\overset{\overset{C_2H_5}{|}}{\underset{\underset{H}{|}}{C}}-OMgBr$$

Ethyl magnesium bromide; Formaldehyde; Addition product

$$\xrightarrow{H_2O} CH_3CH_2CH_2OH + Mg(OH)Br$$

Propanol-1; Mag. hydroxy bromide

(*ii*) By treating ethylene oxide with Grignard reagent and subjecting the product to hydrolysis.

$$\overset{\delta-\ \ \delta+}{CH_3MgBr} + \underset{\diagdown O \diagup}{CH_2-CH_2} \longrightarrow CH_3CH_2CH_2OMgBr$$

Methyl mag. bromide; Ethylene oxide

$$\downarrow H_2O$$

$$CH_3CH_2CH_2OH + Mg(OH)Br$$

n-Propyl alcohol; Mag. hydroxy bromide

(*b*) **Secondary alcohols.** Secondary alcohols can be obtained by one of the following methods.

(*i*) By the action of an aldehyde other than formaldehyde followed by hydrolysis.

$$R\,Mg\,X + R-\underset{\underset{H}{|}}{C}=O \longrightarrow R-\overset{\overset{R}{|}}{\underset{\underset{H}{|}}{C}}-OMgX$$

Addition product

$$\downarrow H_2O$$

$$\begin{matrix}R\\R\end{matrix}\!\!>CHOH + Mg\,(OH)\,X$$

Sec. alcohol

$$CH_3MgBr + CH_3-\underset{\underset{H}{|}}{C}=O \longrightarrow CH_3-\overset{\overset{CH_3}{|}}{\underset{\underset{H}{|}}{C}}-OMgBr$$

Methyl mag. bromide; Acetaldehyde; Addition product

$$\xrightarrow{H_2O} CH_3-\overset{\overset{CH_3}{|}}{\underset{\underset{H}{|}}{C}}-OH + Mg\,(OH)\,Br$$

Isopropyl alcohol

(*ii*) **By the action of ethyl formate on two molecules of Grignard reagent.** Ethyl formate with one molecule of Grignard reagent gives a molecule of aldehyde which then reacts further with the second molecule of Grignard reagent to produce secondary alcohol.

$$\underset{\text{Methyl mag. bromide}}{C_2H_5MgBr} + \underset{\text{Ethyl formate}}{H-\overset{\overset{O}{\|}}{C}-OC_2H_5} \longrightarrow \underset{\text{Addition product}}{H-\overset{\overset{OMgBr}{|}}{\underset{\underset{C_2H_5}{|}}{C}}-OC_2H_5}$$

$$\xrightarrow{H_2O} \underset{\text{Propanal}}{C_2H_5CHO} + \underset{\text{Magnesium ethoxy bromide}}{Mg\,(OC_2O_5)\,Br}$$

$$\underset{\text{Ethyl mag. bromide}}{C_2H_5MgBr} + \underset{\text{Propanal}}{C_2H_5-\underset{\underset{H}{|}}{C}=O} \longrightarrow \underset{\text{Addition product}}{C_2H_5-\overset{\overset{C_2H_5}{|}}{\underset{\underset{H}{|}}{C}}-OMgBr}$$

$$\xrightarrow{H_2O} \underset{\text{Pentanol-3}}{C_2H_5-\overset{\overset{C_2H_5}{|}}{\underset{\underset{H}{|}}{C}}-OH} + Mg(OH)Br$$

(*c*) **Tertiary alcohols.** Tertiary alcohols can be obtained from Grignard reagents by either of the two methods as given below:

(*i*) *By action of a ketone with Grignard reagent*

$$\underset{\text{Gringard reagent}}{R'MgX} + \underset{\text{Ketone}}{R-\underset{\underset{R}{|}}{C}=O} \longrightarrow \underset{\text{Addition product}}{R-\overset{\overset{R'}{|}}{\underset{\underset{R}{|}}{C}}-OMgX}$$

$$\downarrow H_2O$$

$$\underset{\text{Tert. alcohol}}{R-\overset{\overset{R'}{|}}{\underset{\underset{R}{|}}{C}}-OH} + Mg\,(OH)\,X$$

R and R′ could be same or different

$$CH_3MgBr + CH_3-\underset{\underset{CH_3}{|}}{C}=O \longrightarrow \underset{\text{Addition product}}{CH_3-\overset{\overset{CH_3}{|}}{\underset{\underset{CH_3}{|}}{C}}-OMgBr}$$

$$\xrightarrow{H_2O} \underset{\text{Tert. butyl alcohol}}{CH_3-\overset{\overset{CH_3}{|}}{\underset{\underset{CH_3}{|}}{C}}-OH} + Mg\,(OH)\,Br$$

(*ii*) Tertiary alcohols may be prepared by the action of a molecule of ester (other than formic ester) with two molecules of Grignard reagent. First one molecule each of the ester and Grignard reagent react to form a ketone, which subsequently reacts with another molecule of Grignard reagent to produce a tertiary alcohol.

$$\underset{\text{Methyl mag. bromide}}{CH_3MgBr} + \underset{\text{Ethyl acetate}}{CH_3-\overset{\overset{O}{||}}{C}-OC_2H_5} \longrightarrow \underset{\text{Addition product}}{CH_3-\overset{\overset{OMgBr}{|}}{\underset{\underset{CH_3}{|}}{C}}-OC_2H_5}$$

$$\xrightarrow{H_2O} \underset{\text{Acetone}}{CH_3-\overset{\overset{O}{||}}{\underset{\underset{CH_3}{|}}{C}}} + \underset{\text{Magnesium ethoxy bromide}}{Mg(OC_2H_5)Br}$$

$$C_2H_5MgBr + \underset{\text{Acetone}}{CH_3-\underset{\underset{CH_3}{|}}{C}=O} \longrightarrow \underset{\text{Addition product}}{CH_3-\overset{\overset{CH_3}{|}}{\underset{\underset{CH_3}{|}}{C}}-OMgBr} \xrightarrow[-\,Mg(OH)Br]{H_2O} \underset{\text{Tert. butyl alcohol}}{CH_3-\overset{\overset{CH_3}{|}}{\underset{\underset{CH_3}{|}}{C}}-OH}$$

(5) Aldehydes. (*i*) By the action of ethyl formate with Grignard reagent (in equimolar amount)

$$\underset{\text{Gringard reagent}}{R\,MgX} + \underset{\text{Ethyl formate}}{H-\overset{\overset{O}{||}}{C}-OC_2H_5} \longrightarrow \underset{\text{Addition product}}{H-\overset{\overset{OMgBr}{|}}{\underset{\underset{R}{|}}{C}}-OC_2H_5}$$

$$\xrightarrow{H_2O} \underset{\text{Aldehyde}}{H-\overset{\overset{O}{||}}{\underset{\underset{R}{|}}{C}}} + \underset{\text{Mag. ethyoxy bromide}}{Mg(OC_2H_5)Br}$$

(*ii*) A better yield of aldehyde is obtained if ethyl ortho formate is used instead of ethyl formate

$$R-MgX + \underset{\text{Ethyl ortho formate}}{H-C(OC_2H_5)_3} \longrightarrow RCH(OC_2H_5)_2 + Mg\begin{matrix} X \\ OC_2H_5 \end{matrix}$$

$$\downarrow H_2O$$

$$RCHO + 2C_2H_5OH$$

(6) **Ketones.** Ketones may be obtained by either of the two methods given below:

(*i*) *By the action of acid chloride on Grignard reagent*

$$CH_3MgCl + CH_3-\underset{\substack{|\\Cl}}{C}=O \longrightarrow \underset{\text{Addition product (unstable)}}{CH_3-\overset{\substack{CH_3\\|}}{\underset{\substack{|\\Cl}}{C}}-OMgCl}$$

$$\xrightarrow{\text{Changes to}} \underset{\text{Acetone}}{CH_3-\overset{\substack{CH_3\\|}}{C}=O} + MgCl_2$$

(*ii*) By the action of alkyl cyanide on Grignard reagent.

$$CH_3MgBr + CH_3-C\equiv N \longrightarrow \underset{\text{Addition product}}{CH_3-\underset{\substack{|\\CH_3}}{C}=NMgBr}$$

$$\xrightarrow{H_2O} \underset{\text{Acetone}}{CH_3-\underset{\substack{|\\CH_3}}{C}=O} + \underset{\text{Magnesium amino bromide}}{Mg\begin{matrix}\diagup Br\\ \diagdown NH_2\end{matrix}}$$

(7) **Carboxylic acids.** By the action of CO_2 on Grignard reagent followed by hydrolysis.

$$RMg\,X + O=C=O \longrightarrow \underset{\text{[Addition product]}}{O=\overset{\substack{R\\|}}{C}-OMgX}$$

$$\xrightarrow{H_2O} \underset{\text{Acid}}{RCOOH} + Mg(OH)\,X$$

(8) Primary amines. Primary amines may be prepared by the action of chloramine on Grignard reagent.

$$R\,MgX + \underset{\text{Chloramine}}{Cl-NH_2} \longrightarrow \underset{\text{Primary amine}}{R-NH_2} + Mg\begin{matrix}\diagup X\\ \diagdown Cl\end{matrix}$$

(9) Esters. On treatment with chloroformic ester, Grignard reagents produce esters.

$$Cl-\overset{\substack{O\\||}}{C}-OC_2H_5 + CH_3MgBr \longrightarrow Cl-\overset{\substack{OMgBr\\|}}{\underset{\substack{|\\CH_3}}{C}}-OC_2H_5$$

$$\xrightarrow{-Mg(Br)Cl} \underset{\text{Ethyl acetate}}{CH_3\overset{\substack{O\\||}}{C}-OC_2H_5}$$

Q. 8. How can you obtain the following from Grignard reagent?

(*i*) Dithionic acid

(*ii*) Tert. butyl alcohol (*Kurukshetra, 2001, Assam 2003, Agra 2004*)

Ans. (*i*) Dithionic acid may be prepared by the action of carbon disulphide on a Grignard reagent, followed by hydrolysis.

$$R-MgX + \overset{S}{\overset{\|}{C}}=S \longrightarrow R-\overset{S}{\overset{\|}{C}}-SMgX \quad \text{(Addition product)}$$

$$\xrightarrow{H_2O} R-\overset{S}{\overset{\|}{C}}-SH + Mg\begin{matrix}OH\\X\end{matrix} \quad \text{(Dithionic acid)}$$

$$CH_3\,MgBr + \overset{S}{\overset{\|}{C}}=S \longrightarrow CH_3-\overset{S}{\overset{|}{C}}-SMgX$$

$$\xrightarrow{H_2O} CH_3-\overset{S}{\overset{\|}{C}}-SH + Mg\,(OH)\,Br \quad \text{(Dithionic acid)}$$

(*ii*) Tert. butyl alcohol

See question 7 under serial No. 4 (tertiary alcohols)

Q. 9. What happens when methyl magnesium bromide is reacted with (*i*) CO_2 (*ii*) $HCOOC_2H_5$ (*iii*) CH_3CHO.

Ans. (*i*) $CH_3\,MgBr + \overset{O}{\overset{\|}{C}}=O \longrightarrow CH_3-\overset{O}{\overset{\|}{C}}-OMgBr$

$$\xrightarrow{H_2O} CH_3-\overset{O}{\overset{\|}{C}}-OH + Mg\begin{matrix}OH\\Br\end{matrix} \quad \text{(Acetic acid)}$$

(*ii*) $CH_3\,MgBr + H-\overset{O}{\overset{\|}{C}}-OC_2H_5 \longrightarrow H-\underset{CH_3}{\underset{|}{\overset{OMgBr}{\overset{|}{C}}}}-OC_2H_5$

$$\longrightarrow \underset{\text{Acetaldehyde}}{CH_3CHO} + \underset{\text{Mag. ethoxy bromide}}{Mg(OC_2H_5)Br}$$

(*iii*) $CH_3\,MgBr + CH_3-\underset{H}{\underset{|}{C}}=O \longrightarrow CH_3-\underset{H}{\underset{|}{\overset{CH_3}{\overset{|}{C}}}}-OMgBr$

$$\xrightarrow{H_2O} CH_3-\underset{H}{\underset{|}{\overset{CH_3}{\overset{|}{C}}}}-OH + Mg\,(OH)Br \quad \text{(Isopropyl alcohol)}$$

Q. 10. Selecting a suitable organometallic compound, how will you prepare the following?

(*i*) Tert. butyl alcohol (*Devi Ahilya 2001, Kurukshetra 2001*)

(*ii*) Isobutyric acid (*iii*) Acetone. (*M. Dayanand, 2000; Meerut, 2003*)

Ans. (*i*) Tert. butyl alcohol will be obtained by reacting acetone with methyl magnesium bromide (See Q. 7).

(*ii*) Isobutyric acid will be obtained by treating isopropyl magnesium bromide with CO_2, followed by hydrolysis.

$$(CH_3)_2CHMgBr + O=C=O \longrightarrow (CH_3)_2CH-\overset{O}{\overset{\|}{C}}-OMgBr$$

$$\xrightarrow{H_2O} \underset{\text{Isobutyric acid}}{(CH_3)_2CH-\overset{O}{\overset{\|}{C}}-OH} + Mg(OH)Br$$

(*iii*) Acetone can be prepared by treating methyl magnesium bromide with either acetyl chloride or methyl cyanide followed by hydrolysis (*See Q. No. 7 under serial no. 6*).

Q. 11. Write equations to show the reaction of CH_3MgBr with each of the following:

(*i*) $SiCl_4$ **(*ii*) C_2H_5OH**

(iii) CH_2-CH_2 (bridged by O) **(iv) $CH_3COOC_2H_5$** (*Indore 2005*)

Ans. (*i*)
$$4CH_3MgBr + SiCl_4 \longrightarrow \underset{\text{Tetra methyl silane}}{(CH_3)_4Si} + \underset{\text{Magnesium chloro bromide}}{4Mg(Cl)Br}$$

(*ii*)
$$CH_3MgBr + C_2H_5OH \longrightarrow \underset{\text{Methane}}{CH_4} + Mg\begin{cases}OC_2H_5\\Br\end{cases}$$

(*iii*)
$$CH_3MgBr + \underset{\text{Ethylene oxide}}{CH_2-CH_2\ (\text{–O–})} \longrightarrow CH_3CH_2CH_2OMgBr$$

$$\xrightarrow{H_2O} \underset{\text{Propanol-1}}{CH_3CH_2CH_2OH} + Mg(OH)Br$$

(*iv*)
$$CH_3MgBr + CH_3COOC_2H_5 \longrightarrow CH_3\overset{OMgBr}{\overset{|}{C}}-OC_2H_5 \ (\text{with } \underset{|}{}CH_3)$$

$$\xrightarrow[\text{to}]{\text{changes}} \underset{\text{Acetone}}{CH_3-\overset{O}{\overset{\|}{C}}-CH_3} + Mg(OC_2H_5)Br$$

Q. 12. Using suitable RMg X prepare the following:

(*i*) $CH_3CH_2CHOHCH_3$

(*ii*) $(CH_3)_3COH$ (*Viswa Bharti 2003; Ranchi 2004*)

(*iii*) $CH_3CH_2CH_2COOH$

Ans. (*i*) $CH_3CH_2CHOHCH_3$
Butanol-2

Secondary alcohol can be obtained by treating Grignard reagent with an aldehyde other than formaldehyde.

$$CH_3MgBr + CH_3CH_2CHO \longrightarrow CH_3CH_2-\overset{\displaystyle H}{\underset{\displaystyle CH_3}{\overset{|}{\underset{|}{C}}}}-OMgBr$$

$$\xrightarrow{H_2O} \underset{\text{Butanol-1}}{CH_3CH_2-\overset{\displaystyle H}{\underset{\displaystyle CH_3}{\overset{|}{\underset{|}{C}}}}-OH} + Mg(OH)Br$$

(*ii*) Tert. butyl alcohol

See questions no. 7.

(*iii*) $CH_3CH_2CH_2COOH$ (Butyric acid)

$$\underset{\text{Propyl mag. bromide}}{CH_3CH_2CH_2MgBr} + \overset{O}{\overset{||}{C}}=O \longrightarrow CH_3CH_2CH_2-\overset{O}{\overset{||}{C}}-OMgBr$$

$$\xrightarrow{H_2O} \underset{\text{Butyric acid}}{CH_3CH_2CH_2-\overset{O}{\overset{||}{C}}-OH} + Mg(OH)Br$$

Q. 13. How can you obtain thioalcohols and sulphanilic acid using Grignard reagents?

Ans. (*i*) Preparation of thioalcohols.

$$R\,MgX + S \longrightarrow RSMgX \xrightarrow{H_2O} \underset{\text{Thio alcohol}}{RSH} + Mg(OH)X$$

$$CH_3MgBr + S \longrightarrow CH_3SMgBr \xrightarrow{H_2O} \underset{\text{Methyl thioalcohol}}{CH_3SH} + Mg(OH)Br$$

(*ii*) Preparation of sulphanilic acid

$$CH_3MgBr + \overset{O}{\overset{||}{S}}=O \longrightarrow CH_3-\overset{O}{\overset{||}{S}}-OMgBr$$

$$\xrightarrow{H_2O} \underset{\text{Sulphanilic acid}}{CH_3-\overset{O}{\overset{||}{S}}-OH} + Mg\genfrac{}{}{0pt}{}{Br}{OH}$$

Q. 14. Starting from ethyl magnesium iodide, how will you obtain each of the following:

(*i*) 2-Pentyne **(*ii*) Propanoic acid** (*Kerala, 2001; Awadh, 2000*)
(*iii*) 3-Pentanone **(*iv*) 2-Methyl-2-butanol** (*Baroda 2003, Sambalpur, 2004*)
(*v*) Ethane **(*vi*) 1-Pentene**

Ans. (*i*) **2-Pentyne**

$$\underset{\text{Propyne}}{CH_3-C\equiv CH} + CH_3-CH_2-MgI \longrightarrow \underset{\text{Propynyl magnesium iodide}}{CH_3-C\equiv C-MgI} + CH_3-CH_3$$

$$CH_3-C\equiv C-MgI + I-CH_2-CH_3 \longrightarrow CH_3-C\equiv C-CH_2-CH_3 + MgI_2$$

Ethyl iodide 2-Pentyne

(*ii*) Propanoic acid

$$O=C=O + CH_2-CH_2-MgI \longrightarrow \left[CH-CH_2-\overset{O}{\overset{\|}{C}}-OMgI\right] \xrightarrow[-Mg(OH)]{H_2O,H^+} CH_3-CH_2-\overset{O}{\overset{\|}{C}}-OH$$

Ethyl magnesium iodide; Propanoic acid

(*iii*) 3-Pentanone

$$CH_3-CH_2-C\equiv N + CH_3-CH_2-MgI \longrightarrow \left[CH_3-CH_2-\underset{\underset{CH_3}{|}}{\underset{CH_2}{|}}{C}=NMgI\right]$$

Propanenitrile; Addition product

$$\xrightarrow[Mg(OH)]{HOH,\ H^+} \left[CH_3-CH_2-\underset{\underset{CH_3}{|}}{\underset{CH_2}{|}}{C}=NH\right] \xrightarrow[-NH_3]{HOH,\ H} CH_3-CH_2-\underset{\underset{CH_3}{|}}{\underset{CH_2}{|}}{C}=O$$

3-Pentanone

(*iv*) 2-Methyl-2-butanol

$$CH_3-\overset{CH_3}{\overset{|}{C}}=O + CH_3-CH_2-MgI \longrightarrow CH_3-\overset{CH_3}{\overset{|}{\underset{\underset{CH_3}{|}}{\underset{CH_2}{|}}{C}}}-OMgI$$

Acetone; Addition product

$$\xrightarrow[-Mg(OH)]{HOH,\ H^+} CH_3-\overset{CH_3}{\overset{|}{\underset{\underset{CH_3}{|}}{\underset{CH_2}{|}}{C}}}-OH$$

2-Methyl-2-butanol

(*v*) Ethane

$$HOH + CH_3-CH_2-MgI \longrightarrow CH_3-CH_3 + Mg(OH)I$$

Ethane

(*vi*) **1-Pentene**

$$CH_2=CH-CH_2-I+CH_3-CH_2-MgI \longrightarrow \underset{\text{1-Pentene}}{CH_2=CH-CH_2-CH_2-CH_3}+MgI_2$$

Q. 15. Explain, giving equations, what happens when:

(*i*) Phenyl magnesium bromide is treated with ethylene oxide and the product formed is hydrolysed. (*Kerala, 2001*)

(*ii*) Ethyl magnesium bromide is treated with ethyl formate and the product is hydrolysed.

(*iii*) Ethyl magnesium iodide is treated with chloromethyl ether.

(*iv*) Ethyl magnesium bromide is treated with cadmium chloride.

(*v*) Methyl magnesium iodide is treated with ethyl amine.

(*vi*) Methyl magnesium iodide is treated with acetyl chloride and the product formed is hydrolysed.

Ans. (*i*)

$$\underset{\text{Phenyl magnesium bromide}}{C_6H_5-MgBr} + \underset{\text{Ethylene oxide}}{\begin{matrix} CH_2 \\ | \\ CH_2 \end{matrix}\!\!>O}$$

$$\longrightarrow C_6H_5-CH_2-CH_2-OMgBr$$

$$\downarrow \; -Mg(OH) \quad HOH, H^+$$

$$\underset{\text{Phenylethanol}}{C_6H_5-CH_2-CH_2-OH}$$

(*ii*)

$$\underset{\text{Ethyl formate}}{H-\overset{\overset{O}{\|}}{C}-OC_2H_5} + CH_3-CH_2-MgBr \longrightarrow \left[\begin{matrix} OMg\,Br \\ | \\ H-C-OC_2H_5 \\ | \\ CH_2 \\ | \\ CH_3 \end{matrix} \right]$$

$$\downarrow \; -Mg(OC_2H_5)\,Br$$

$$\begin{matrix} OMgBr \\ | \\ H-C-CH_2-CH_3 \\ | \\ CH_2 \\ | \\ CH_3 \end{matrix} \xleftarrow{CH_3\,CH_2\,MgBr} \underset{\text{Propanal}}{H-\overset{\overset{O}{\|}}{C}-CH_2-CH_3}$$

$$\downarrow \; -Mg(OH) \quad HOH, H^+$$

$$\underset{\text{3-Pentanol}}{CH_3-CH_2-\overset{\overset{OH}{|}}{CH}-CH_2-CH_3}$$

(*iii*) $CH_3-CH_2-MgI + Cl-CH_2-O-CH_3 \longrightarrow CH_3-CH_2-CH_2-O-CH_3 + Mg(I)Cl$

Ethyl magnesium iodide; Chloromethyl ether; 1-Methoxypropane (Methyl n-propyl ether)

(*iv*) $2C_2H_5MgBr + CdCl_2 \longrightarrow (C_2H_5)_2Cd + 2Mg(Br)Cl$

Ethyl magnesium bromide; Diethyl Cadmium

(*v*) $C_2H_5NH_2 + CH_3MgI \longrightarrow CH_4 + Mg(NHC_2H_5)I$

Ethylamine

(*vi*)

$$\underset{\text{Acetyl chloride}}{CH_3-\overset{O}{\overset{\|}{C}}-Cl} + CH_3MgI \longrightarrow \left[CH_3-\underset{CH_3}{\underset{|}{\overset{OMgI}{\overset{|}{C}}}}-Cl\right] \xrightarrow{-Mg(I)Cl} \underset{\text{Acetone}}{CH_3-\underset{CH_3}{\underset{|}{\overset{O}{\overset{\|}{C}}}}}$$

$$\underset{\text{Acetone}}{CH_3-\underset{CH_3}{\underset{|}{\overset{O}{\overset{\|}{C}}}}} \xrightarrow{CH_3MgI} CH_3-\underset{CH_3}{\underset{|}{\overset{OMgI}{\overset{|}{C}}}}-CH_3 \xrightarrow[-Mg(I)Cl]{HOH,\ H^+} \underset{\substack{\text{2-Methyl-2-propanol} \\ \text{(tert-Butyl alcohol)}}}{CH_3-\underset{CH_3}{\underset{|}{\overset{OH}{\overset{|}{C}}}}-CH_3}$$

ORGANOLITHIUM COMPOUNDS

Q. 16. What are organolithium compounds?

Ans. Organolithium compounds. Organic compounds in which lithium is bonded directly to a carbon atom are called organolithium compounds. These are represented by the general formula R–Li where R is the organic radical. For example

CH_3Li (Methyl lithium) $\quad CH_3-CH_2-Li$ (Ethyl lithium) $\quad CH_3-CH_3-CH_2-CH_2-Li$ (*n*-Butyl lithium)

C_6H_5-Li (Phenyl lithium)

Due to greater ionic character of C–Li bond, organolithium compounds are more reactive than Grignard reagent and thus undergo a few additional useful reactions.

Q. 17. How are organolithium compounds prepared? (*Kerala, 2000*)

Ans. Preparation. These compounds are generally prepared by the following methods:

(1) By reaction between lithium and an alkyl halide. Alkyl lithium can be prepared by treating a suitable alkyl halide with lithium in the presence of dry ether or benzene at low temperature and under an inert atmosphere of nitrogen or argon. For example,

$$R-Br + 2Li \xrightarrow[263\ K]{\text{Dry ether, } N_2} \underset{\text{Alkyl lithium}}{R-Li} + LiBr$$

$$CH_3CH_2CH_2CH_2Cl + 2Li \xrightarrow[263\ K]{Benzene,\ N_2} \underset{n\text{-Butyllithium}}{CH_3CH_2CH_2CH_2Li} + LiCl$$

(2) By exchange method. Aryl lithium are usually prepared by treating an aryl halide with an alkyl lithium. For example,

$$C_6H_5Br + \underset{n\text{-Butyl lithium}}{CH_3-CH_2-CH_2-CH_2-Li} \longrightarrow \underset{\text{Phenyl lithium}}{C_6H_5Li} + CH_3-CH_2-CH_2-CH_2-Br$$

Q. 18. Give the synthetic applications of organolithium compounds.

Ans. Organolithium compounds are more reactive than Grignard reagents as the former have greater ionic character than the latter.

Important reactions of organolithium compounds are given below.

1. Reactions with compounds containing active hydrogen.

$$R-Li + H_2O \longrightarrow R-H + LiOH$$

$$\underset{n\text{-Butyl lithium}}{CH_3CH_2CH_2CH_2Li} + H_2O \longrightarrow \underset{n\text{-Butane}}{C_4H_{10}} + LiOH$$

2. Formation of alcohols.

$$R-Li + \underset{\text{Formaldehyde}}{H-\overset{H}{\overset{|}{C}}=O} \longrightarrow \underset{\text{Addition compounds}}{H-\overset{H}{\overset{|}{\underset{R}{\underset{|}{C}}}}-OLi} \xrightarrow{H_2O} \underset{\text{Primary alcohol}}{H-\overset{H}{\overset{|}{\underset{R}{\underset{|}{C}}}}-OH} + LiOH$$

$$\underset{n\text{-Butyl lithium}}{C_4H_9Li} + H-\overset{H}{\overset{|}{C}}=O \longrightarrow H-\overset{H}{\overset{|}{\underset{C_4H_9}{\underset{|}{C}}}}-OLi \xrightarrow[-\ LiOH]{H_2O} \underset{\text{Pentanol-1 (Primary alcohol)}}{H-\overset{H}{\overset{|}{\underset{C_4H_9}{\underset{|}{C}}}}-OH} + LiOH$$

$$\underset{\text{Phenyl lithium}}{C_6H_5Li} + H-\overset{H}{\overset{|}{C}}=O \longrightarrow H-\overset{H}{\overset{|}{\underset{C_6H_5}{\underset{|}{C}}}}-OLi \xrightarrow[-\ LiOH]{H_2O} \underset{\text{Benzyl alcohol}}{C_6H_5CH_2OH}$$

$$R-Li + \underset{\text{Aldehyde}}{R-\overset{H}{\overset{|}{C}}=O} \longrightarrow \underset{\text{Addition compounds}}{R-\overset{H}{\overset{|}{\underset{R}{\underset{|}{C}}}}-OLi} \xrightarrow{H_2O} \underset{\text{Secondary alcohol}}{R-\overset{H}{\overset{|}{\underset{R}{\underset{|}{C}}}}-OH} + LiOH$$

$$R-Li + \underset{\text{Ketone}}{R-\overset{R}{\overset{|}{C}}=O} \longrightarrow R-\overset{H}{\overset{|}{\underset{R}{\underset{|}{C}}}}-OLi \xrightarrow{H_2O} \underset{\text{Tertiary alcohol}}{R-\overset{R}{\overset{|}{\underset{R}{\underset{|}{C}}}}-OH} + LiOH$$

3. Formation of carboxylic acid.

$$R-Li + O=C=O \longrightarrow O=\underset{}{\overset{R}{\overset{|}{C}}}-OLi \xrightarrow{H_2O} O=\overset{R}{\overset{|}{C}}-OH + LiOH$$

Carboxy acid

If excess of alkyl lithium is used, another molecule of R–Li may add to the addition product leading to the formation of ketone.

$$R-Li + R-\overset{O}{\overset{||}{C}}-OLi \longrightarrow R-\underset{R}{\underset{|}{\overset{OLi}{\overset{|}{C}}}}-OLi \xrightarrow{2H_2O} R-\underset{R}{\underset{|}{C}}=O + 2LiOH$$

Ketone

4. Reaction with epoxides.

$$R-Li + \underset{\diagdown O \diagup}{CH_2-CH_2} \longrightarrow R-CH_2-CH_2OLi \xrightarrow{H_2O} RCH_2CH_2OH + LiOH$$

Primary alcohol

5. Reaction with alkenes.

$$R-Li + CH_2=CH_2 \longrightarrow R-CH_2-CH_2Li \xrightarrow{CH_2=CH_2} R-CH_2-CH_2-CH_2-CH_2-Li$$

and so on.

This reaction can be used for preparing long-chain alkanes.

6. Formation of aldehydes.

Organolithium compounds react with dimethyl formamide to form aldehydes.

$$C_6H_5Li + H-\overset{O}{\overset{||}{C}}-N\begin{matrix} CH_3 \\ CH_3 \end{matrix} \longrightarrow C_6H_5-\underset{H}{\underset{|}{\overset{OLi}{\overset{|}{C}}}}-\underset{CH_3}{\underset{|}{N}}-CH_3$$

$$\xrightarrow{H_2O} C_6H_5-\overset{O}{\overset{||}{C}}-H + LiOH + (CH_3)_2NH$$

Benzaldehyde Dimethyl amine

7. Formation of ketones.

$$\underset{\text{Phenyl lithium}}{C_6H_5-Li} + \underset{\text{Methyl cyanide}}{CH_3-C\equiv N} \longrightarrow CH_3-\overset{C_6H_5}{\overset{|}{C}}=NLi \xrightarrow{H_2O} \underset{\text{Methyl phenyl Ketone}}{CH_3-\overset{C_6H_5}{\overset{|}{C}}=O} + \underset{\text{Lithium amide}}{LiNH_2}$$

8. Reaction with pyridine.

$$\underset{\text{Pyridine}}{C_5H_5N} + C_6H_5Li \longrightarrow \underset{\text{2-Phenyl pyridine}}{C_5H_4N-C_6H_5} + LiH$$

Q. 19. Compare the addition reactions of Grignard reagent and organolithium compounds, on α, β-unsaturated carbonyl compounds.

Ans. Organolithium compounds give **1, 2-addition** reactions as follows:

$$\underset{\text{Benzal acetophenone}}{C_6H_5-CH=CH-\overset{\overset{1}{O}\,\|}{^{2}C}-C_6H_5} \xrightarrow{C_6H_5Li} \underset{\text{Addition product}}{C_6H_5-CH=CH-\overset{OLi}{\overset{|}{\underset{\underset{C_6H_5}{|}}{C}}}-C_6H_5}$$

$$\xrightarrow{H_2O} C_6H_5-CH=CH-\overset{OH}{\overset{|}{\underset{\underset{C_6H_5}{|}}{C}}}-C_6H_5$$

1, 2-addition means that the two parts of alkyl lithium add to oxygen and carbon of the carbonyl group *i.e.*, on neighbouring sites.

Grignard reagents give 1, 4-addition reactions. For example:

1. 4-addition

$$C_6H_5-\overset{4}{C}H=\overset{3}{C}H-\overset{\overset{1}{O}\,\|}{^{2}C}-C_6H_5+C_6H_5MgBr$$

$$\longrightarrow C_6H_5-\underset{\underset{C_6H_5}{|}}{CH}-CH=\overset{OMgBr}{\overset{|}{C}}-C_6H_5 \xrightarrow{H_2O} C_6H_5-\underset{\underset{C_6H_5}{|}}{CH}-CH=\overset{OH}{\overset{|}{C}}-C_6H_5$$

$$\Updownarrow$$

$$C_6H_5-\underset{\underset{C_6H_5}{|}}{CH}-CH_2=\overset{O}{\overset{\|}{C}}-C_6H_5$$

Different behaviour of Grignard reagents and organo lithium compounds is due to greater nucleophilic character of lithium compounds.

Due to weaker nucleophilic strength, Grignard reagent attaches itself to C–4 instead of C–2.

Q. 20. How will you prepare C_6H_5Li in the laboratory? Give the synthesis of *m*-phenyl anisole from it.

Ans. Preparation of phenyl lithium

(*i*) This compound may be prepared by mixing phenyl bromide and metallic lithium in the molar ratio of 1 : 2 respectively in dry ether as solvent at –10°C in an inert atmosphere of N_2.

$$C_6H_5-Br+2Li \xrightarrow[-10°C]{\text{Dry ether}} C_6H_5-Li+LiBr$$

(*ii*) A more convenient and efficient method would be to treat bromobenzene with *n*-butyl lithium.

$$C_6H_5-Br+C_4H_9Li \longrightarrow \underset{\text{Phenyl Lithium}}{C_6H_5-Li}+C_4H_9Br$$

Synthesis of *m*-phenyl anisole

This reaction involves benzyne mechanism

$$\text{o-Fluoro anisole } (C_6H_4(OCH_3)F) + C_6H_5Li \longrightarrow C_6H_3(OCH_3)(F)(Li) \longrightarrow \text{Benzyne } (C_6H_3OCH_3) + LiF$$

$$\text{Benzyne} \xrightarrow{C_6H_5Li} C_6H_3(OCH_3)(Li)(C_6H_5) \xrightarrow[-\,LiOH]{H_2O} \text{o-Phenyl anisole } (C_6H_4(OCH_3)C_6H_5)$$

Q. 21. With the help of methyllithium how can you prepare the following compounds:

(*i*) **Ethanol** (*ii*) **Iso-propyl alcohol**

(*iii*) **tert-Butyl alcohol** (*iv*) **Acetaldehyde**

(*v*) **Methane** (*vi*) **Acetophenone**

Ans. (*i*) **Ethanol**

$$\underset{\text{Formaldehyde}}{H-\overset{H}{\overset{|}{C}}=O} + CH_3-Li \longrightarrow H-\underset{CH_3}{\underset{|}{\overset{H}{\overset{|}{C}}}}-OLi \xrightarrow[-\,LiOH]{HOH,\ H^+} \underset{\text{Ethanol}}{CH_3-CH_2-OH}$$

(*ii*) **Iso-propyl alcohol**

$$\underset{\text{Acetaldehyde}}{CH_3-\overset{H}{\overset{|}{C}}=O} + CH_3-Li \longrightarrow CH_3-\underset{CH_3}{\underset{|}{\overset{H}{\overset{|}{C}}}}-OLi \xrightarrow[-\,LiOH]{HOH,\ H^+} \underset{\text{Iso-propyl alcohol}}{CH_3-\underset{CH_3}{\underset{|}{\overset{H}{\overset{|}{CH_2}}}}-OH}$$

(*iii*) **Tert-butyl alcohol**

$$\underset{\text{Acetone}}{CH_3-\overset{CH_3}{\overset{|}{C}}=O} + CH_3-Li \longrightarrow CH_3-\underset{CH_3}{\underset{|}{\overset{CH_3}{\overset{|}{C}}}}-OLi \xrightarrow[-\,LiOH]{H_2O,\ H^+} \underset{\text{Tert. Butyl alcohol}}{CH_3-\underset{CH_3}{\underset{|}{\overset{CH_3}{\overset{|}{CH_2}}}}-OH}$$

(*iv*) **Acetaldehyde**

$$\underset{\text{Hydrogen cyanide}}{H-C\equiv N} + CH_3-Li \longrightarrow H-\underset{CH_3}{\underset{|}{C}}=N-Li \xrightarrow[-\,LiOH]{H_2O,\ H^+} \underset{\text{Acetaldehyde}}{H-\underset{CH_3}{\underset{|}{C}}=O} + NH_3 + LiOH$$

(*v*) **Methane**

$$HOH + CH_3-Li \longrightarrow \underset{\text{Methane}}{CH_4} + LiOH$$

(*vi*) **Acetophenone**

$$\underset{\text{Benzoic acid}}{C_6H_5-\overset{O}{\overset{\|}{C}}-OH} \xrightarrow[-CH_4]{CH_3-Li} C_6H_5-\overset{O}{\overset{\|}{C}}-OLi \xrightarrow{CH_3-Li} C_6H_5-\underset{CH_3}{\underset{|}{\overset{OLi}{\overset{|}{C}}}}-OLi$$

$$\xrightarrow[\text{HOH, } H^+]{-2LiOH} \left[C_6H_5-\underset{CH_3}{\underset{|}{\overset{OH}{\overset{|}{C}}}}-OH \right]_{\text{Unstable}} \xrightarrow{-H_2O} \underset{\text{Acetophenone}}{C_6H_5-\overset{O}{\overset{\|}{C}}-CH_3}$$

Q. 22. Alkyl lithiums add to sterically hindered ketones while Grignard reagents do not. Explain.

Ans. Organolithiums are stronger nucleophiles and smaller in size than the corresponding Grignard reagents. Hence organolithium compounds react with sterically hindered ketones to form tertiary alcohols.

$$\underset{\text{Di-isopropyl ketone}}{CH_3-\overset{CH_3}{\overset{|}{CH}}-\underset{\underset{CH_3}{\underset{|}{CH_3-C}}}{\underset{|}{C}}=O} + \underset{\text{Iso-Propylithium}}{CH_3-\overset{CH_3}{\overset{|}{CH}}-Li} \xrightarrow{\text{Dry ether}} CH_3-\overset{CH_3}{\overset{|}{CH}}-\underset{\underset{CH_3}{\underset{|}{CH_3-C}}}{\underset{|}{\overset{\overset{CH_3}{\overset{|}{CH-CH_3}}}{\overset{|}{C}}}}-OLi$$

$$\xrightarrow[\text{HOH, } H^+]{-LiOH} \underset{\substack{\text{Tri-isopropyl carbinol} \\ (3^\circ \text{ alcohol})}}{CH_3-\overset{CH_3}{\overset{|}{CH}}-\underset{\underset{CH_3}{\underset{|}{CH_3-CH}}}{\underset{|}{\overset{\overset{CH_3}{\overset{|}{CH-CH_3}}}{\overset{|}{C}}}}-OH}$$

We cannot expect such a reaction, involving bulky groups, from Grignard reagents.

Q. 23. How will you achieve the following conversions? (*Garhwal, 2000; Awadh, 2000*)

(*i*) Ethyl magnesium bromide into ethanethiol (Ethyl mercaptan).

(*ii*) Methyl magnesium bromide into methyl amine. (*Kerala, 2001*)

(*iii*) Methyl lithium into ethane nitrile.

(*iv*) Ethyllithium into butanone.

(*v*) Ethyl magnesium iodide into ethanesulphinic acid.

Ans. (*i*) Ethanethiol

$$CH_3-CH_2-Mg\,Br + S \longrightarrow CH_3-CH_2-S-MgBr \xrightarrow[-\,Mg\,(OH)\,Br]{HOH,\ H^+} \underset{\text{Ethanethiol}}{CH_3-CH_2-SH}$$

(*ii*) Methylamine

$$CH_3-Mg\,Br + \underset{\text{Chloramine}}{Cl-NH_2} \longrightarrow \underset{\text{Methylamine}}{CH_3-NH_2} + Mg\,(Br)\,Cl$$

(*iii*) Ethanenitrile

$$CH_3-Li + \underset{\text{Cyanogen chloride}}{Cl\,CN} \longrightarrow \underset{\text{Ethanenitrile}}{LiCl + CH_3\,CN}$$

(*iv*) Butanone

$$\underset{\substack{\text{Ethanenitrile}\\\text{(Acetonitrile)}}}{CH_3-C\equiv N} + \underset{\text{Ethyl lithium}}{CH_3-CH_2-Li} \longrightarrow CH_3-\underset{\substack{|\\CH_2\\|\\CH_3}}{C}=N-Li$$

$$\xrightarrow{HOH,\ H^+} \underset{\text{Butanone}}{CH_3-\overset{\overset{O}{||}}{C}-CH_2-CH_3} + NH_3 + LiOH$$

(*v*) Ethanesulphinic acid

$$\overset{\overset{O}{||}}{S}=O + CH_3-CH_2-MgI \longrightarrow CH_3-CH_2-\overset{\overset{O}{||}}{S}-OMgI$$

$$\xrightarrow[-\,Mg\,(OH)I]{HOH,\ H^+} \underset{\text{Ethanesulphinic acid}}{CH_3-CH_2-\overset{\overset{O}{||}}{S}-OH}$$

Q. 24. How will you prepare the following using Grignard reagent? Give chemical equations.

(*i*) a tertiary alcohol (*ii*) an alkane (*iii*) a ketone

Ans. *See Q. 7.*

Q. 25. How will you prepare the following compounds with the help of Grignard's reagent?

(*i*) An alkanamine (*ii*) An alkyne (*iii*) A thioalcohol

Ans. *See Q. 7 and Q. 13.*

Q. 26. How will you prepare the following starting from phenyllithium

(*i*) 2-Phenyl pyridine (ii) Benzoic acid

Ans. Preparation of 2-Phenylpyridine

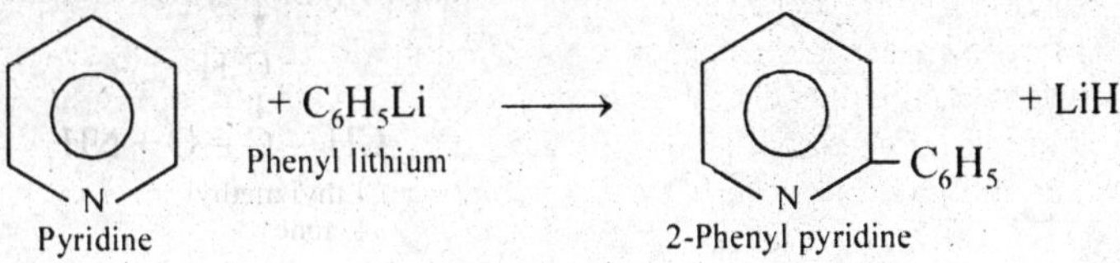

Preparation of benzoic acid

$$\underset{\text{Phenyl Lithium}}{C_6H_5Li} + O=C=O \longrightarrow O=\underset{}{\overset{C}{\overset{|}{C}}}-OLi \xrightarrow{H_2O} \underset{\text{Benzoic acid}}{C_6H_5COOH} + LiOH$$

Q. 27. Starting from a suitable Grignard reagent, how will you prepare the following:
(*i*) Ethanol (*ii*) 1-Butene (*iii*) Ethane (*iv*) Ethane sulphinic acid (*Kurukshetra, 1996*)

Ans. (*i*) Ethanol

$$CH_3MgBr + \underset{\underset{H}{|}}{\overset{\overset{H}{|}}{C}}=O \longrightarrow CH_3-\underset{\underset{H}{|}}{\overset{\overset{C_6H_5}{|}}{C}}-O\,MgBr \xrightarrow{H_2O} \underset{\text{Ethanol}}{CH_3CH_2OH} + Mg(OH)Br$$

(*ii*) 1-Butene

$$\underset{\text{Allyl chloride}}{CH_2=CH-CH_2Cl} + CH_3\,MgCl \longrightarrow \underset{\text{1-Butene}}{CH_2=CH-CH_2-CH_3} + MgCl_2$$

(*iii*) Ethane

$$CH_3CH_2Mg\,Br + H_2O \longrightarrow CH_3CH_3 + Mg\,(OH)\,Br$$

(*iv*) Ethane sulphinic acid

$$\overset{\overset{O}{||}}{S}=O + CH_3CH_2-MgBr \longrightarrow CH_3-CH_2-\overset{\overset{O}{||}}{S}-OMg\,Br \xrightarrow{H_2O} \underset{\text{Ethanesulphinic acid}}{CH_3CH_2-\overset{\overset{O}{||}}{S}-OH}$$

Q. 28. How will you prepare the following compounds using appropriate organometallic compounds:

(*i*) Ethyl methyl ketone (*ii*) Acetic acid
(*iii*) Ethyl propionate

Ans. (*i*) Ethyl methyl ketone

$$CH_3CH_2MgBr + CH_3\,C\equiv N \longrightarrow CH_3-\overset{\overset{C_2H_5}{|}}{C}=N\,Mg\,Br \xrightarrow[-\,Mg\,(OH)\,Br]{2H_2O} \underset{\text{Ethyl methyl ketone}}{CH_3-\overset{\overset{C_2H_5}{|}}{C}=O} + NH_3$$

***(ii)* Acetic acid**

$$CH_3MgBr + \overset{O}{\overset{||}{C}} = O \longrightarrow CH_3 - \overset{O}{\overset{||}{C}} - OMgBr \xrightarrow[-Mg(OH)Br]{2H_2O} \underset{\text{Acetic acid}}{CH_3 - \overset{O}{\overset{||}{C}} - OH}$$

***(iii)* Ethyl propionate**

$$Cl - \overset{O}{\overset{||}{C}} - OC_2H_5 + CH_3CH_2MgBr \longrightarrow Cl - \underset{CH_2CH_3}{\underset{|}{\overset{OMgBr}{\overset{|}{C}}}} - OC_2H_5 \xrightarrow{Mg(Br)Cl} \underset{\text{Ethyl propionate}}{CH_3CH_2\overset{O}{\overset{||}{C}} - OC_2H_5}$$

Q. 29. How are organozinc compounds prepared ? Give their physical properties and synthetic applications.

Ans. Preparation. Dialkyl Zinc Compounds were prepared by Frankland in the year 1849 in an attempt to prepare the ethyl radical by removing iodine from ethyl iodide by means of zinc. However, it ended with the formation of dialkyl zinc.

$$RI + Zn \longrightarrow \underset{\text{Alkyl Zinc iodide}}{R - Zn - I}$$

$$2RZnI \longrightarrow \underset{\text{Dialkyl Zinc}}{R_2Zn} + ZnI_2$$

The yield of dialkyl zinc may be increased by carrying out distallation in vacuum

Physical Properties

1. Alkylzinc compounds are volatile liquids, spontaneously inflammable in air.
2. They burn the skin and posses an unpleasant smell

Synthetic applications

1. Preparation of hydrocarbons containing a quaternary carbon atom. Neopentane may be prepared by the action of dimethylzinc on *Tert*.-butyl Chloride.

$$\underset{\textit{Tert.}\text{-butyl chloride}}{(CH_3)_3CCl} + \underset{\text{Dimethyl zin}}{(CH_3)_2Zn} \longrightarrow \underset{\text{Neopentane}}{(CH_3)_4C} + \underset{\text{Methylzinc chloride}}{CH_3ZnCl}$$

2. Preparation of Ketones

Alkyl zinc compounds react with acyl chlorides to form ketones

$$\underset{\text{Acetyl chloride}}{CH_3COCl} + \underset{\text{Dimethyl zinc}}{(CH_3)_2Zn} \longrightarrow \underset{\text{Acetone}}{CH_3COCH_3} + CH_3ZnCl$$

$$CH_3COCl + \underset{\text{Diethyl zinc}}{(C_2H_5)_2Zn} \longrightarrow \underset{\text{Methylethyl ketone}}{CH_3COC_2H_5} + C_2H_5ZnCl$$

3. Preparation of long-chain fatty acids (or their esters)

First a keto-ester is prepared by reaction between as alkyl zinc chloride and the acid chloride ester derivative. The keto-ester is then reduced by Clemmenson reduction

$$\underset{\text{Alkyl zinc chloride}}{CH_3(CH_2)_x\,ZnCl} + \underset{\text{acid chloride-ester derivative}}{ClCO(CH_2)_y\,COOC_2H_5} \longrightarrow$$

$$CH_3(CH_2)_x\,CO(CH_2)_y\,COOC_2H_5 \xrightarrow[HCl]{Zn/Hg} CH_3(CH_2)_x\,CH_2(CH_2)_y\,COOC_2H_5$$

$$\xrightarrow[H^+]{Hydrolysis} CH_3(CH_2)_{x+y+1}\,COOH + C_2H_5\,OH$$

MISCELLANLEOUS QUESTIONS

Q. 30. Using CH_3 MgBr, How will you obtain the following?
(*i*) CH_3CH_2OH (*ii*) $CH_3CH_2CH_2OH$ (*iii*) $CH_3CHOHCH_3$

(*Kanpur, 2001; Kumaon, 2000; G. Nanakdev, 2000*)

Ans. See Q.7

Q. 31. Using organomentallic compound, how will you obtain the following?

(*Garhwal, 2000*)

(*i*) Propyl alcohol
(*ii*) Tetraethyl lead
(*iii*) Ethane
(*iv*) Ethane thiol (mercaptan)
(*v*) Ethyl methyl ketone

Ans. (*i*) Q.7.

(*ii*) $4C_2H_5MgBr + 2PbCl_2 \longrightarrow (C_2H_5)_4\,Pb + Pb + 4Mg\,(Br)\,Cl$

(*iii*) See Q. 7

(*iv*) See Q.23 (*i*)

(*v*) See Q. 28 (*i*)

Q. 32. How can you obtain following compounds using Grignard reagents?

(*Kerala, 2001*)

(*i*) A secondary alcohol
(*ii*) a paraffin
(*iii*) a ketone
(*iv*) mono carboxylic acid

Ans. (*i*) See Q. 7, (4) (*b*), (*i*)

(*ii*) See Q. 7 (1)

(*iii*) See Q. 7, (4), C (*ii*)

(*iv*) See Q. 14 (*ii*)

Q. 33. What happens when methyl magnesium bromide is reacted with
(*i*) Lead Chloride (*ii*) Propyne (*iii*) Ethyl formate (*iv*) Allylbromide.

Ans. (*i*) $4CH_3MgBr + 2PbCl_2 \longrightarrow (CH_3)_4\,Pb + Pb + 4Mg\,(Br)Cl.$ (*Kerala, 2000*)

(*ii*) See Q. 7 (3)

(*iii*) See Q. 7 (4) (*b*), (*ii*) and Q. 7 (5) (*i*)

(*iv*) $CH_3MgBr + BrCH_2-CH=CH_2 \longrightarrow \underset{\text{Butene-1}}{CH_3-CH_2-CH=CH_2} + MgBr_2$

Q. 34. Using suitable R-Mg-X, prepare the following compounds: (*Lucknow, 2000*)
(*i*) $C_6H_5CH_2-OH$ (*ii*) $CH_3-CO-CH_2-CH_2-CH_3$

(iii) $CH_3-\overset{\overset{CH_3}{|}}{\underset{\underset{OH}{|}}{C}}-CH_2-CH_3$ **(iv)** $(CH_3)_3-C-COOH$

Ans.

(i) $C_6H_5-MgBr + H-\overset{\overset{H}{|}}{C}=O \longrightarrow H-\overset{\overset{H}{|}}{\underset{\underset{C_6H_5}{|}}{C}}-OMgBr \xrightarrow{HOH} C_6H_5CH_2OH + Mg\begin{matrix} Br \\ OH \end{matrix}$

Phenyl magnesium bromide

(ii) $CH_3CH_2CH_2Cl + CH_3\underset{\underset{Cl}{|}}{C}=O \longrightarrow CH_3CH_2CH_2-\overset{\overset{CH_3}{|}}{\underset{\underset{Cl}{|}}{C}}-OMgCl \longrightarrow C_3H_7-\overset{\overset{CH_3}{|}}{C}=O + MgCl_2$

n-Propyl chloride

(iii) $CH_3CH_2MgBr + \begin{matrix} CH_3 \\ CH_3 \end{matrix}\!\!>C=O \longrightarrow CH_3-\overset{\overset{CH_3}{|}}{\underset{\underset{CH_3}{|}}{C}}-OMgBr \xrightarrow{H_2O} CH_3-\overset{\overset{CH_3}{|}}{\underset{\underset{OH}{|}}{C}}-C_2H_5 + Mg(OH)Br$

(iv) $(CH_3)_3CMgBr + O=C=O \longrightarrow O=\underset{\underset{C(CH_3)_3}{|}}{C}-OMgBr \xrightarrow{HOH} (CH_3)_3C\ COOH$

Q. 35. Using organometallic compound, how will you obtain the following?

(i) Dimethyl mercury *(Awadh, 2000)*
(ii) Ethyl mercaptan *(Awadh, 2000)*
(iii) 2–Butyne *(Awadh, 2000)*
(iv) Acetone *(Awadh, 2000, Himachal, 2000)*
(v) Methyl n-propyl ether *(Himachal, 2000)*
(vi) Acetaldehyde *(Himachal, 2000)*
(vii) Diethylamine *(Himachal, 2000)*
(viii) Trimethyl carbinol *(Himachal, 2000)*

Ans. (i) $2\ C_2H_5MgI + HgCl_2 \longrightarrow (C_2H_5)_2\ Hg + 2Mg(I)\ Cl$

Mercuric chloride — Mercury dimethyl

(ii) $CH_3-CH_2-MgBr + S \longrightarrow CH_3-CH_2-S-MgBr \xrightarrow[-\ Mg(OH)Br]{HOH,\ H^+} CH_3-CH_2-SH$

(iii) $CH\equiv CH + 2CH_3MgI \longrightarrow IMgC\equiv CMgI$

$\downarrow + 2CH_3I$

$CH_3C\equiv C-CH_3 + 2MgI_2$

(iv) $CH_3MgBr + CH_3-\overset{\overset{O}{||}}{C}-OC_2H_5 \longrightarrow CH_3-\overset{\overset{OMgBr}{|}}{\underset{\underset{CH_3}{|}}{C}}-OC_2H_5 \xrightarrow{HOH} CH_3\ \overset{\overset{O}{||}}{\underset{\underset{CH_3}{|}}{C}} + Mg(OC_2H_5)Br$

ethyl acetate — acetone

(*v*) $CH_3-CH_2-MgI + Cl-CH_2-O-CH_3 \longrightarrow CH_3-CH_2-CH_2-O-CH_3 + MgI(Cl)$

Chloromethyl ether — methyl n-propyl ether

(*vi*) $CH_3-MgBr + H-\overset{O}{\overset{\|}{C}}-OC_2H_5 \longrightarrow H-\underset{CH_3}{\underset{|}{\overset{OMgBr}{\overset{|}{C}}}}-OC_2H_5 \longrightarrow CH_3CHO + Mg(OC_2H_5)Br$

ethyl formate — acetaldehyde

(*vii*) Trimethyl carbinal is the same compound as t-butyl alcohol. See Q. 7, (4), (*C*) (*i*)

Q. 36. Starting from ethyl magnesium bromide, how will you prepare the following compounds?

(*i*) Butanol-1 **(*ii*) 3-methylbutanol-3** (*Delhi, 2003*)
(*iii*) Pentanone-3 **(*iv*) Propanoic acid**

Ans. See Q.7.

Q. 37. Give the use of Grignard reagents in preparing:

(*a*) an alkane **(*b*) a primary alcohol**
(*c*) a secondary alcohol **(*d*) a tertiary alcohol**
(*e*) a carboxylic acid (*North Eastern Hill, 2004*)

Ans. See Q. 7.

15

ALCOHOLS AND EPOXIDES

Q. 1. What are alcohols? How are they classified?

Ans. Alcohols are organic compounds having one or more –OH groups attached to carbon in the molecule.

Classification. Alcohols may be classified as under:

(*a*) Aliphatic alcohols.

(*i*) **Monohydric alcohols.** Alcohols having one alcoholic group (–OH) in the molecule are called monohydric alcohols. For example,

$$\underset{\text{Methyl alcohol}}{CH_3 - OH} \qquad \underset{\text{Ethyl alcohol}}{C_2H_5OH}$$

(*ii*) **Dihydric alcohols.** Alcohols having two alcoholic groups in the molecule are called dihydric alcohols. For example,

$$\begin{array}{l} CH_2OH \\ | \\ CH_2OH \end{array} \quad \text{Ethylene glycol}$$

(*iii*) **Trihydric alcohols.** Alcohols having three alcoholic groups in the molecule are called trihydric alcohols. For example,

$$\begin{array}{l} CH_2OH \\ | \\ CHOH \\ | \\ CH_2OH \end{array} \quad \text{Glycerol}$$

It may be mentioned that more than one alcoholic groups attached to one carbon make the molecule unstable with liberation of a water molecule.

(*b*) Aromatic alcohols. These are aromatic compounds in which the alcohol group is attached to the side-chain of the benzene nucleus. For example,

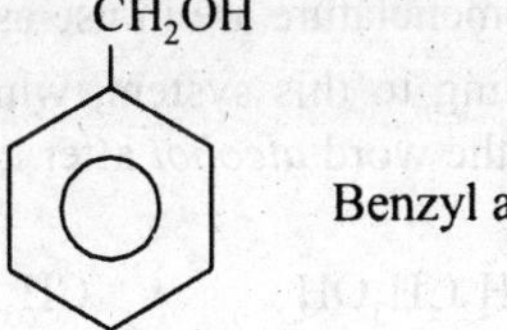

Benzyl alcohol

It may be noted that aromatic compounds containing –OH group directly attached to nucleus don't come in the category of alcohols. Such compounds are phenols and are different from alcohols in many respects.

Q. 2. What are primary, secondary and tertiary alcohols?

Or

How are monohydric alcohols classified?

Ans. Monohydric alcohols are classified as under:

(*a*) **Primary alcohols** *are those alcohols which have just one (or none) alkyl group linked to the carbon attached directly to hydroxyl group.*

They have the general formula R – CH_2 – OH

For example,

$$H-\overset{\displaystyle H}{\underset{\displaystyle H}{C}}-OH \qquad CH_3-\overset{\displaystyle H}{\underset{\displaystyle H}{C}}-OH$$

Methyl alcohol Ethyl alcohol

Therefore, primary alcoholic group is – CH_2OH.

(*b*) **Secondary alcohols** *are those alcohols which have two alkyl groups linked to the carbon attached directly to hydroxyl group*. They have the general formula $\begin{matrix}R\\R\end{matrix}\!\!>CHOH$

For example, $\begin{matrix}CH_3\\CH_3\end{matrix}\!\!>CHOH$ *Iso*-propyl alcohol

Therefore, secondary alcoholic group is > CHOH

(*c*) **Tertiary alcohols** *are those alcohols which have three alkyl groups linked to the carbon attached directly to hydroxyl group*. They have the general formula

$$\begin{matrix}R\\R\end{matrix}\!\!>\underset{\displaystyle R}{C}-OH$$

For example, $CH_3-\overset{\displaystyle CH_3}{\underset{\displaystyle CH_3}{C}}-OH$ *tert*-Butyl alcohol

Therefore, tertiary alcoholic group is C – OH

Q. 3. Discuss nomenclature of alcohols.

Ans. Three different systems of nomenclature are in use as described hereunder.

(*i*) **The common system.** According to this system, which is used only for the simpler alcohols, alcohols are named by adding the word *alcohol* after the name of alkyl group present in the molecule. For example:

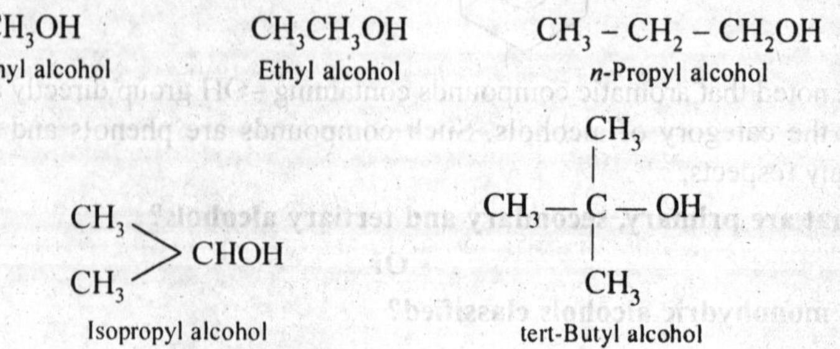

CH_3OH Methyl alcohol

CH_3CH_2OH Ethyl alcohol

$CH_3-CH_2-CH_2OH$ *n*-Propyl alcohol

$\begin{matrix}CH_3\\CH_3\end{matrix}\!\!>CHOH$ Isopropyl alcohol

$CH_3-\overset{\displaystyle CH_3}{\underset{\displaystyle CH_3}{C}}-OH$ tert-Butyl alcohol

(*ii*) **The carbinol system.** According to this system, all alcohols are named as derivatives of methyl alcohol, which itself is known as carbinol.

For example,

$CH_3 — CH_2OH$
Methyl carbinol

$(CH_3)_2CHOH$
Dimethyl carbinol

$(C_6H_5)_3C—O—H$
Triphenyl carbinol

However, this system is used only rarely and has been replaced by IUPAC system.

(*iii*) **The IUPAC system.** IUPAC names of alcohols end with the suffix *ol*. Of course, in case of higher alcohols, the longest chain of carbon atoms bearing the –OH groups is selected as the parent chain and the positions of –OH groups are indicated by suitable numbers.

CH_3OH
Methanol

$CH_3CHOHCH_3$
2-Propanol

$CH_3—C(CH_3)_2—OH$
2-Methylpropan-2-ol

$C_6H_5CH_2CH_2OH$
2-Phenylethanol

The common and IUPAC names of a few alcohols are given in the table ahead.

Formula	*Common Name*	*IUPAC Name*
CH_3OH	Methyl alcohol	Methanol
CH_3CH_2OH	Ethyl alcohol	Ethanol
$CH_3CH_2CH_2OH$	n-Propyl alcoho	Propan-1-ol
$CH_3CHOHCH_3$	Iso-Propyl alcohoı	Propan-2-ol
$CH_3 - CH_2 - CH_2 - CH_2OH$	n-Butyl alcohol	Butan-1-ol
$CH_3 - CH_2 - CHOH - CH_3$	Sec-Butyl alcohol	Butan-2-ol
$CH_3—CH(CH_3)—CH_2OH$	Iso-butyl alcohol	2-Methylpropan-1-ol
$CH_3—C(CH_3)_2—OH$	Tert-butyl alcohol	2-Methylpropan-2-ol
$(CH_3)_2CHCH_2CH_2OH$	Isoamyl alcohol	3-Methylbutan-1-ol

Q. 4. Describe general methods of preparation of alcohols.

Ans. Some important methods of preparation of alcohols are described as follows:

1. From Grignard reagent. The most important method for the preparation of primary, secondary and tertiary alcohols is by the reaction between a Grignard reagent and a carbonyl compound such as an aldehyde or a ketone. The addition product initially formed is decomposed with water, or preferably dilute mineral acid such as HCl or H_2SO_4 to get the alcohols. The general reaction may be depicted as follows:

$$\underset{\text{Carbonyl compound}}{>C=O} + \underset{\text{Grignard reagent}}{RMgX} \xrightarrow{\text{Dry ether}} \underset{\text{Addition product}}{-\overset{|}{\underset{R}{\underset{|}{C}}}-OMgX} \xrightarrow[-Mg(OH)X]{H_2O} \underset{\text{Alcohol}}{-\overset{|}{\underset{R}{\underset{|}{C}}}-OH}$$

(*i*) To obtain primary alcohol, formaldehyde is taken as the carbonyl compound.

$$\underset{\text{Formaldehyde}}{H-\overset{H}{\overset{|}{C}}=O} \xrightarrow[\text{(Dry ether)}]{RMgX} H-\overset{H}{\overset{|}{\underset{CH_3}{\underset{|}{C}}}}-OMgX \xrightarrow[-Mg(OH)X]{H_2O} \underset{\text{Primary alcohol}}{H-\overset{H}{\overset{|}{\underset{R}{\underset{|}{C}}}}-OH}$$

For example,

$$H-\overset{H}{\overset{|}{C}}=O \xrightarrow[\text{(Dry ether)}]{CH_3MgI} H-\overset{H}{\overset{|}{\underset{CH_3}{\underset{|}{C}}}}-OMgI \xrightarrow[-Mg(OH)X]{H_2O} \underset{\text{Ethyl alcohol}}{H-\overset{H}{\overset{|}{\underset{CH_3}{\underset{|}{C}}}}-OH}$$

(*ii*) To obtain a secondary alcohol, the Grignard reagent is treated with an aldehyde other than formaldehyde.

$$\underset{\text{Some aldehyde other than Formaldehyde}}{R'-\overset{H}{\overset{|}{C}}=O} \xrightarrow[\text{(Dry ether)}]{RMgX} R'-\overset{H}{\overset{|}{\underset{R}{\underset{|}{C}}}}-OMgX \xrightarrow[-Mg(OH)X]{H_2O} \underset{\text{Secondary alcohol}}{R'-\overset{H}{\overset{|}{\underset{R}{\underset{|}{C}}}}-OH}$$

R and R′ could be same or different.

For example,

$$\underset{\text{Acetaldehyde}}{CH_3-\overset{H}{\overset{|}{C}}=O} \xrightarrow[\text{(Dry ether)}]{CH_3MgI} CH_3-\overset{H}{\overset{|}{\underset{CH_3}{\underset{|}{C}}}}-OMgI \xrightarrow[-Mg(OH)I]{H_2O} \underset{\text{Isopropyl alcohol}}{CH_3-\overset{H}{\overset{|}{\underset{CH_3}{\underset{|}{C}}}}-OH}$$

(*iii*) Use of a ketone to react with Grignard reagent leads to the ultimate formation of a **tertiary alcohol**. For obtaining tertiary alcohol, a suitable Grignard reagent and a ketone are taken together.

$$\underset{\text{Ketone}}{R''-\overset{R'}{\overset{|}{C}}=O} \xrightarrow[\text{(Dry ether)}]{R-MgX} R'-\overset{R''}{\overset{|}{\underset{R}{\underset{|}{C}}}}-OMgX \xrightarrow[-Mg(OH)X]{H_2O} \underset{\text{Tertiary alcohol}}{R'-\overset{R''}{\overset{|}{\underset{R}{\underset{|}{C}}}}-OH}$$

For example,

$$\underset{\text{Acetone}}{CH_3-\overset{\overset{\large CH_3}{|}}{C}=O} \xrightarrow[\text{(Dry ether)}]{CH_3MgI} CH_3-\overset{\overset{\large CH_3}{|}}{\underset{\underset{\large CH_3}{|}}{C}}-OMgI \xrightarrow[-\,Mg(OH)I]{H_2O} \underset{\textit{Tert}\text{-Butyl alcohol}}{CH_3-\overset{\overset{\large CH_3}{|}}{\underset{\underset{\large CH_3}{|}}{C}}-OH}$$

It may be noted that in all these reactions, the number of carbon atoms in the alcohol obtained is equal to the sum of the carbon atoms in the aldehyde or ketone and Grignard reagent.

Mechanism of reaction

The carbon-magnesium bond of the Grignard reagent is a highly polar bond; carbon being negatively charged and magnesium being positively charged. The negatively charged carbon of the alkyl group becomes attached to the positively polarised carbon of the carbonyl group. Simultaneously, the positively charged magnesium end of the remaining part of Grignard reagent gets attached to negatively polarised oxygen. Magnesium salt of the feebly acidic alcohol is formed as shown below.

$$>C=O + R-MgX \longrightarrow\ >C<^{OMgX}_{R}$$

The addition complex thus formed is decomposed by water in an acidic solution to form alcohols.

$$>C<^{OMgX}_{R} \xrightarrow{H_2O^+}\ >C<^{OH}_{R} + Mg(OH)X$$

(*iv*) If a higher homologue of alcohol is required to be prepared, use is made of ethylene oxide.

$$RMgX + \underset{\text{Ethylene oxide}}{\underset{\diagdown\, O\, \diagup}{CH_2-CH_2}} \xrightarrow{\text{Dry ether}} RCH_2\,CH_2\,OMgX \xrightarrow[-\,Mg(OH)X]{H_2O} RCH_2CH_2\,OH$$

For example,

$$CH_3MgBr + \underset{\diagdown\, O\, \diagup}{CH_2-CH_2} \xrightarrow{\text{Dry ether}} CH_3CH_2CH_2\,OMgI \xrightarrow[-\,Mg(OH)X]{H_2O} \underset{n\text{-Propyl alcohol}}{CH_3CH_2CH_2OH}$$

It may be noted that the alcohol obtained contains two carbon atoms more than the alkyl group of Grignard reagent.

Mechanism of reaction

The nucleophilic alkyl group of the Grignard reagent attaches itself to positively polarised carbon while the electrophilic magnesium attaches itself to negatively polarised oxygen of the C–O bond of ethylene oxide; the three-membered ring getting ruptured during the process. The resulting addition product is then decomposed as usual.

$$\overset{-\delta\quad +\delta}{R:MgX} + \underset{\diagdown\, O\, \diagup}{CH_2-CH_2} \longrightarrow R-CH_2-CH_2-OMgX \xrightarrow{H_2O} RCH_2CH_2OH + Mg\,(OH)X$$

(2) By the hydration of alkenes. An alkene adds a molecule of sulphuric acid to give alkyl hydrogen sulphate which is hydrolysed with water to obtain an alcohol

$$\underset{\text{Alkene}}{>C=C<} \xrightarrow{H_2SO_4} \underset{\text{Alkyl hydrogen sulphate}}{-\overset{|}{\underset{\underset{H}{|}}{C}}-\overset{|}{\underset{|}{C}}-HSO_4} \xrightarrow{H_2O} \underset{\text{Alcohol}}{-\overset{|}{\underset{\underset{H}{|}}{C}}-\overset{|}{\underset{|}{C}}-OH} + H_2SO_4$$

For example,

$$\underset{\text{Propylene}}{CH_3-CH=CH_2} \xrightarrow{H_2SO_4} \underset{\text{Isopropyl hydrogen sulphate}}{CH_3CH(HSO_4)CH_3} \xrightarrow[\text{Heat}]{H_2O} \underset{\text{Isopropyl alcohol}}{CH_3CHOHCH_3} + H_2SO_4$$

In case of unsymmetrical alkenes, addition of suiphuric acid takes place according to Markownikoff's rule.

Ethanol can be obtained by the hydration of ethene.

Mechanism of Hydration

$$\underset{\text{Alkene}}{>C=C<} + H_3O^+ \rightleftharpoons \underset{\text{Carbocation}}{-\overset{|}{C}-H \atop -\underset{|}{C^+}} + H_2O$$

$$-\overset{|}{C}-H \atop -\underset{|}{C^+} \quad + :\ddot{O}-H \atop \quad\ \ \underset{H}{|} \rightleftharpoons \underset{\text{Protonated Alcohol}}{-\overset{|}{C}-H \atop -\underset{|}{C}-\overset{+}{\underset{\underset{H}{|}}{O}}-H}$$

$$-\overset{|}{C}-H \atop -\underset{|}{C}-\overset{+}{\underset{\underset{H}{|}}{O}}-H \quad + H_2O \rightleftharpoons \underset{\text{Alcohol}}{-\overset{|}{C}-H \atop -\underset{|}{C}-O-H} + H_3O^+$$

The first step in the hydration of alkenes involves the formation of a carbocation. The carbocation thus formed may sometimes undergo rearrangement to a more stable carbocation. *This leads the formation of* **unexpected products** *in the reaction.* For example, hydration of 3, 3-dimethyl-1-butene gives rise to 2, 3-dimethyl-2-butanol as the major product while the product expected is 3, 3-dimethyl-2-butanol.

$$\underset{\text{3, 3-Dimethyl-1-butene}}{CH_3-\overset{CH_3}{\overset{|}{\underset{\underset{CH_3}{|}}{C}}}-CH=CH_2} \longrightarrow \underset{\text{(Major product obtained)}}{CH_3-\overset{OH}{\overset{|}{\underset{\underset{CH_3}{|}}{C}}}-\underset{\underset{CH_3}{|}}{CH}-CH_3}$$

$$\longrightarrow \underset{\text{(Expected product)}}{CH_3-\overset{CH_3}{\overset{|}{\underset{\underset{CH_3}{|}}{C}}}-\overset{OH}{\overset{|}{CH}}-CH_3}$$

This is due to the fact that the carbocation formed in the first step of the reaction is a secondary carbocation which rearranges to a more stable tertiary carbocation by alkyl shift as shown below.

$$CH_3-\underset{CH_3}{\overset{CH_3}{\underset{|}{\overset{|}{C}}}}-CH=CH_2+H_3O^+ \rightleftharpoons CH_3-\underset{CH_3}{\overset{CH_3}{\underset{|}{\overset{|}{C}}}}-\overset{+}{C}H-CH_3 \longrightarrow CH_3-\underset{+}{\overset{CH_3}{\overset{|}{C}}}-\underset{CH_3}{\underset{|}{CH}}-CH_3$$

3, 3-Dimethyl-2-Butyl ion   2, 3-Dimethyl-2-Butyl ion (More Stable)

The tertiary ion formed by rearrangement then gives rise to 2, 3-dimethyl-2-butanol as the major product.

(3) By the hydroboration-oxidation of alkenes. When treated with diborane $(BH_3)_2$, alkenes are converted into alkyl boranes which on oxidation with alkaline hydrogen peroxide yield alcohols. For example:

$$6CH_2=CH_2+(BH_3)_2 \longrightarrow \underset{\text{Triethylborane}}{2(CH_3-CH_2)_3B}$$

$$(CH_3-CH_2)_3B+3H_2O_2 \xrightarrow{OH} 3CH_3CH_2OH+H_3BO_3$$

Diborane required in the reaction is prepared *in situ* by reaction between sodium borohydride $(NaBH_4)$ and boron trifluoride (BF_3). It is not necessary to isolate the alkyl boranes, rather they are treated in the same reaction vessel with alkaline hydrogen peroxide.

A significant feature of hydroboration-oxidation is that it always leads to **anti-Markownikoff** addition of water to alkenes. As such this method can be used to prepare alcohols which cannot be prepared by other methods of hydration of alkenes. Thus, we have:

$$6CH_3-CH=CH_2 \xrightarrow{(BH_3)_2} \underset{\text{Tri-n-propylborane}}{2(CH_3-CH_2-CH_2)_3B} \xrightarrow{H_2O_2\,/\,OH^-} \underset{n\text{-Propyl alcohol}}{6CH_3CH_2CH_2OH}+2H_3BO_3$$

Due to anit-Markownikoff orientation, only primary alcohols can be obtained by this mehtod.

Mechanism of the reaction is illustrated as under: In a normal HX addition, H-would have attached itself to C_1 and X– to C_2. But in this case, electron-deficient boron of attaches itself to C_1 giving rise to an intermediate product as illustrated below:

$$\underset{3}{CH_3}\rightarrow\underset{2}{CH}=\underset{1}{CH_2}+BH_3 \longrightarrow \underset{3}{CH_3}\rightarrow\underset{2}{\overset{\delta^+}{CH}}-\underset{1}{CH_2}\cdots\overset{\delta^-}{B}H_3$$

$$\underset{3}{CH_3}\rightarrow\underset{2}{\overset{\delta^+}{CH}}\cdots\cdots\underset{1}{CH_2}$$
$$\vdots \qquad\qquad \vdots$$
$$H\cdots\cdots\overset{\delta^-}{B}-H$$
$$|$$
$$H$$

Intermediate product

The transition state ultimately generates the alkyl borane in which B–H bond has been cleaved and the C_2 – H bond has been completely formed.

$$\underset{\text{H}}{\overset{}{CH_3 - CH}} - \underset{\underset{\text{H}}{|}}{\overset{}{CH_2}}\text{B}-H \quad \text{or} \quad CH_3 - CH_2 - CH_2BH_2$$

4. Oxymercuration-demercuration of alkenes. It is a convenient two-step method for preparing alcohols from alkenes.

When treated with mercuric acetate in a water-tetrahydrofuran solution, alkenes form hydroxy mercurial compounds. These mercury compounds are reduced by sodium borohydride ($NaBH_4$) to yield alcohols.

$$(i)\ \underset{\text{Alkene}}{>C=C<} + H_2O + \underset{\text{Mercuric acetate}}{Hg(OOCCH_3)_2} \xrightarrow[\text{(Oxymercuration)}]{\text{THF}} -\underset{OH}{\overset{|}{C}}-\underset{HgOOCCH_3}{\overset{|}{C}}- + CH_3COOH$$

$$(ii)\ -\underset{OH}{\overset{|}{C}}-\underset{HgOOCCH_3}{\overset{|}{C}}- \xrightarrow[\text{(Demercuration)}]{NaBH_4} \underset{\text{Alcohol}}{-\underset{OH}{\overset{|}{C}}-\underset{H}{\overset{|}{C}}-} + Hg + CH_3COO^-$$

In the first step water and mercuric acetate add to be double bond. This is known as **oxymercuration.** In the second step, sodium borohydride reduces the mercurial compound and the mercuri-acetate group gets replaced by hydrogen. This is known **demercuration.** Both the steps take place very rapidly and conveniently at room temperature. The organo-mercurial compound formed in the first step is not isolated but is reduced *in situ* by sodium borohydride to form alcohol.

The alcohol formed is in *accordance with Markownikoff addition* of water (*i.e.* H– and –OH) to the carbon-carbon double bond.

Thus the overall reaction may be written as follows:

$$R - CH = CH_2 \xrightarrow[(ii)\ NaBH_4]{(i)\ Hg(OAc)_2/THF - H_2O} R - \underset{OH}{\underset{|}{CH}} - CH_3$$

The examples given below illustrate the reaction

$$\underset{\text{1-Pentene}}{CH_3 - CH_2 - CH_2 - CH = CH_2} \xrightarrow[THF - H_2O]{Hg(OAc)_2} CH_3 - CH_2 - CH_2 - \underset{OH}{\underset{|}{CH}} - \underset{Hg(OAc)}{\underset{|}{CH_2}}$$

$$\xrightarrow{NaBH_4} \underset{\text{2-Pentanol}}{Ch_3 - CH_2 - CH_2\underset{OH}{\underset{|}{C}}HCH_3}$$

$$\underset{\text{2-Methyl-1-butene}}{CH_3 - CH_2 - \overset{CH_3}{\overset{|}{C}} = CH_2} \xrightarrow[THF - H_2O]{Hg(OAc)_2} CH_3 - CH_2 - \underset{OH}{\underset{|}{\overset{CH_3}{\overset{|}{C}}}} - \underset{Hg(OAc)}{\underset{|}{CH_2}}$$

$$\xrightarrow{NaBH_4} \underset{\text{2-Methyl-2-butanol}}{Ch_3 - CH_2 - \underset{OH}{\underset{|}{\overset{CH_3}{\overset{|}{C}}}} - CH_3}$$

Mechanism of Oxymercuration

Electrophilic attack by the mercury species, $\overset{+}{H}gOAc$, at the less substituted carbon of the double bond (*i.e.* the carbon which carries greater number of hydrogens) takes place to form a mercury-bridged carbocation. This cation is immediately attacked by nucleophilic solvent *i.e.* water to yield the hydroxymercurial compound as shown below.

$$\underset{\text{Alkene}}{R-CH=CH_2} + \overset{+}{H}gOAc \longrightarrow \underset{\text{Mercury bridged carbocation}}{R-\overset{\delta+}{CH}-CH_2 \cdots \overset{\delta+}{HgOAc}}$$

The mechanism of demercuration step is still not clearly understood.

(5) By the hydrolysis of alkyl halides. An alcohol may be obtained by the hydrolysis of an alkyl halide with sod. or pot. hydroxide solution or moist silver oxide

$$\underset{\text{Alkyl halide}}{R-X} + \underset{aq}{KOH} \longrightarrow \underset{\text{Alcohol}}{R-OH} + KX$$

For example,

$$\underset{\text{Ethyl bromide}}{C_2H_5Br} + \underset{aq}{KOH} \longrightarrow \underset{\text{Ethyl alcohol}}{C_2H_5OH} + KBr$$

(6) By the reduction of aldehydes, ketones, acids and esters. This is another useful method for the preparation of alcohols. The reduction may be brought about with hydrogen in the presence of a catalyst, with nascent hydrogen from some suitable source such as reaction between sodium and ethanol or with complex metal hydrides such as sodium borohydride ($NaBH_4$) or lithium aluminium hydride ($LiAlH_4$).

For example,

$$\underset{\text{Acetaldehyde}}{CH_3-\overset{H}{\overset{|}{C}}=O} + H_2 \xrightarrow{\text{Pt or Ni}} \underset{\text{Ethyl alcohol}}{CH_3-\overset{H}{\overset{|}{\underset{H}{\underset{|}{C}}}}-OH}$$

$$\underset{\text{Acetone}}{CH_3-\overset{O}{\overset{||}{C}}-CH_3} + 2H \xrightarrow{Na\,/\,C_2H_5OH} \underset{\text{Isopropyl alcohol}}{CH_3-\overset{OH}{\overset{|}{\underset{H}{\underset{|}{C}}}}-CH_3}$$

$$\underset{\text{Butyric acid}}{CH_3-CH_2-CH_2-\overset{O}{\overset{||}{C}}-OH} \xrightarrow{LiAlH_4} \underset{n\text{-Butyl alcohol}}{CH_3-CH_2-CH_2-\overset{H}{\overset{|}{\underset{H}{\underset{|}{C}}}}-OH}$$

$$\underset{\text{Ethyl acetate}}{CH_3-\overset{O}{\overset{||}{C}}-OC_2H_5} \xrightarrow{LiAlH_4} \underset{\text{Ethyl alcohol}}{2C_2H_5OH}$$

$NaBH_4$ is particularly suited for the preparation of unsaturated alcohols by the reduction of unsaturated carbonyl compounds as it does not affect the carbon-carbon double bond.

For example,

$$\underset{\text{Crotonaldehyde}}{CH_3-CH=CH-CHO} \xrightarrow{NaBH_4} \underset{\text{But-2-ene-1-ol}}{CH_3-CH=CH-CH_2OH}$$

(7) By the hydrolysis of esters. The hydrolysis of esters with alkali solution is generally employed for the preparation of alcohols from naturally occurring esters.

$$\underset{\text{Ester}}{RCOOR'} + NaOH \longrightarrow \underset{\text{Sodium salt of acid}}{RCOONa} + \underset{\text{Alcohol}}{R'OH}$$

For example,

$$\underset{\text{Cetyl palmitate}}{C_{15}H_{31}-COOC_{16}H_{33}} + \underset{aq}{NaOH} \longrightarrow \underset{\text{Sodium palmitate}}{C_{15}H_{31}COONa} + \underset{\text{Cetyl alcohol}}{C_{16}H_{33}OH}$$

(8) By the fermentation of carbohydrates. This is one of the oldest methods of obtaining alcohols particularly ethyl alcohol. Glucose and fructose are obtained by the enzymatic hydrolysis of canesugar. Then there is further conversion to ethyl alcohol. Similarly starch can be fermented into ethyl alcohol.

The reactions taking place are described below.

From sugar:

$$C_{12}H_{22}O_{11} + H_2O \xrightarrow{\textit{Invertase}} \underset{\text{Glucose}}{C_6H_{12}O_6} + \underset{\text{Fructose}}{C_6H_{12}O_6}$$

$$C_6H_{12}O_6 \xrightarrow{\textit{Zymase}} \underset{\text{Ethyl alcohol}}{2C_2H_5OH} + 2CO_2$$

From starch:

$$2(C_6H_{10}O_5)_n + nH_2O \xrightarrow{\textit{Diastase}} \underset{\text{Maltose}}{nC_{12}H_{22}O_{11}}$$

$$C_{12}H_{22}O_{11} + H_2O \xrightarrow{\textit{Maltase}} 2C_6H_{12}O_6$$

$$C_6H_{12}O_6 \xrightarrow{\textit{Zymase}} \underset{\text{Ethyl alcohol}}{2C_2H_5OH} + 2CO_2$$

Q. 5. Name the following alcohols by Carbinol and by IUPAC methods:

(a) $CH_3-\underset{\underset{OH}{|}}{CH}-CH=CH_2$

(b) $C_6H_5-\overset{\overset{CH_3}{|}}{\underset{\underset{CH_2CH_3}{|}}{C}}-OH$

(c) $CH_3-CH_2-\overset{\overset{CH_3}{|}}{\underset{\underset{CH_2CH_2CH_3}{|}}{C}}-OH$

Ans. Carbinol System	**IUPAC System**
(*a*) Methyl vinyl carbinol	But-3-en-2-ol
(*b*) Ethyl methyl phenyl carbinol	2-Phenylbutan-2-ol
(*c*) Methyl-ethyl-n-propyl carbinol	3-Methylhexan-3-ol.

Q. 6. (*a*) Alcohol has higher boiling point than alkyl halide or alkane of comparable mol. wt. Why? (*Andhra 2003*)

(*b*) Why alcohols have higher boiling points than ethers of comparable molecular weights (*Kerala, 2001; Panjab 2004*)

Ans. (*a*) Since in alcohol (ROH), strongly electronegative oxygen atom is directly linked to hydrogen atom, its molecules get associated through intermolecular hydrogen bonding as shown below. This increases the molecular mass and hence the boiling point of alcohol rises.

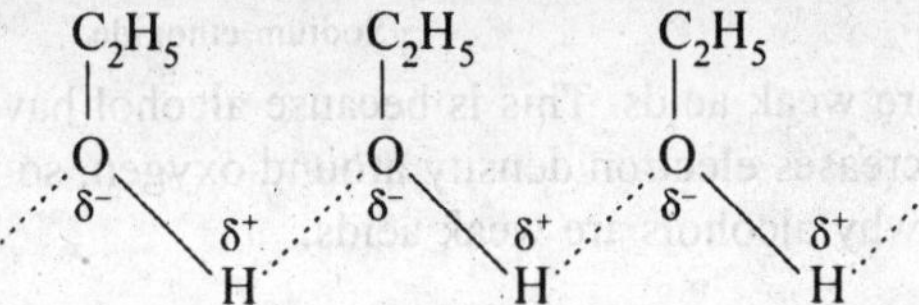

(*b*) Alcohols form intermolecular hydrogen bonding whereas ethers do not. That is why alcohols have higher b.p.

Q. 7. Unlike propane or butane, propanol is soluble in water, why?

Ans. Propane molecules cannot join with water molecules through hydrogen bonding as propane has no oxygen-hydrogen bond in its molecule. Consequently, propane is not soluble in water.

On the other hand, propanol is soluble in water as it molecule contains –OH bond, which causes propanol molecules to form hydrogen bonds with water molecules as shown below:

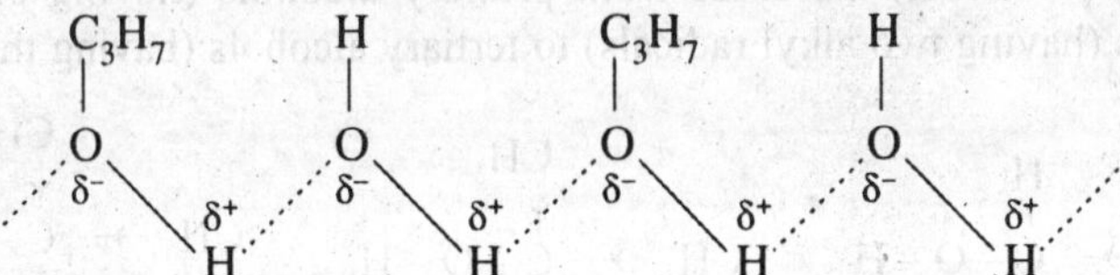

Conseqnelty propanol is soluble in water.

Q. 8. Why is *n*-hexanol insoluble in water?

Ans. Alcohols have the general formula ROH. As the R group becomes larger, ROH resembles more closely with the hydrocarbon.

In *n*-hexanol molecule, the carbon content is quite high which causes it to resemble closely with hexane (which is not soluble in water). Moreover, the large hexyl (C_6H_{13}–) group obstructs the formation of hydrogen bonds between water and *n*-hexanol molecules. As a result, *n*-hexanol is not soluble in water.

Q. 9. Why are lower members of alcohols soluble in water while higher members are not?

Ans. Lower members of alcohols are soluble in water because they form hydrogen bonds with water. Higher members behave more like hydrocarbons which are insoluble in water. Formation of hydrogen bonds in the case of higher members is hindered because of steric reasons as the alkyl groups in such cases are quite bulky. As the hydrogen bond is not formed, they are insoluble in water.

Q. 10. The boiling point of 1-propanol (molecular mass 60) is 370 K while the boiling point of 1, 2-ethanediol (molecular mass 62) is 470 K. Explain the reason for high boiling point of 1, 2-ethanediol. (*Delhi, 2006*)

Ans. Since 1, 2-ethanediol molecule contains two –OH groups, it is capable of forming more intermolecular hydrogen bonds and hence is very highly associated substance. Therefore, its boiling point is very high.

On the other hand, 1-propanol molecule has only one –OH group, its molecules are comparatively less associated through hydrogen bonding. Consequently, its boiling point is lower than that of 1, 2-ethanediol.

Q. 11. "Alcohols are weak acids." Explain.

Ans. Alcohols have acidic character as they react with active metals like sodium or potassium liberating hydrogen. For example,

$$C_2H_5OH + Na \longrightarrow \underset{\text{Sodium ethoxide}}{C_2H_5ONa} + 1/2\ H_2$$

However, alcohols, are weak acids. This is because alcohol have a electron releasing alkyl group (+I effect) which increases electron density around oxygen, so that the release of proton is rendered difficult. That is why alcohols are weak acids.

$$C_2H_5 \rightarrow \ddot{\underset{\cdot\cdot}{O}} - H$$

Electron releasing group (+ I effect)

Q. 12. Compare the acidity of primary, secondary and tertiary alcohols.

Or

Why reactivity of alcohols with sod. metal is as under:

Primary alcohol > Secondary alcohol > Tertiary alcohol (*M. Dayanand, 2000*)

Ans. The acidic character of alcohols depends upon the release of proton from O–H bond. Electron releasing (+I effect) increases from primary alcohols (having one alkyl radical) to secondary alcohols (having two alkyl radicals) to tertiary alcohols (having three alkyl radicals).

$$CH_3 \rightarrow \underset{H}{\overset{H}{\underset{|}{\overset{|}{C}}}} - O - H \qquad CH_3 \rightarrow \underset{H}{\overset{CH_3}{\underset{|}{\overset{\downarrow}{C}}}} - O - H \qquad CH_3 \rightarrow \underset{CH_3}{\overset{CH_3}{\underset{\uparrow}{\overset{\downarrow}{C}}}} - O - H$$

Primary alcohol (Acidic)   Secondary alcohol (Less acidic)   Tertiary alcohol (Least acidic)

As a result, in tertiary alcohols, the release of proton is hindered making them weakest acids of the three types of alcohols. Therefore tertiary alcohols are the weakest acids.

Q. 13. Giving reasons arrange the following in decreasing order of acidity.

Tert. butyl alcohol, ethyl alcohol, methyl alcohol and isopropyl alcohol.

Ans. In view of the discussion in the last question, the acidity decreases in the following order:

$$CH_3OH > C_2H_5OH > CH_3 - \underset{H}{\overset{CH_3}{\underset{|}{\overset{|}{C}}}} - OH > CH_3 - \underset{CH_3}{\overset{CH_3}{\underset{|}{\overset{|}{C}}}} - OH$$

Methyl alcohol (*Most acidic*)   Ethyl alcohol (*Less acidic*)   Isopropyl alcohol (*Still less acidic*)   *tert.* butyl alcohol (*Least acidic*)

It may be noted that ethyl group has greater + I effect compared to methyl group.

Q. 14. How does an alcohol react with the following:

(*i*) Grignard reagents (*Nagpur, 2008*)

(*ii*) Organic acids

(*iii*) Acid halides and anhydrides

Ans. (*i*) Reaction with Grignard reagents. Alcohols react with Grignard reagents to produce hydrocarbons as shown below.

$$\underset{\text{Alcohol}}{R'OH} + \underset{\text{Grignard reagent}}{RMgX} \longrightarrow \underset{\text{Hydrocarbon}}{RH} + R'OMgX$$

For example,

$$\underset{\text{Methyl alcohol}}{CH_3OH} + \underset{\text{Ethyl magnesium iodide}}{C_2H_5MgI} \longrightarrow \underset{\text{Ethane}}{C_2H_6} + CH_3OMgI$$

(*ii*) **Reaction with organic acids.** Alcohols react with organic acids to form esters. This reaction is termed as esterification. It is generally carried out in the presence of concentrated sulphuric acid or dry hydrogen chloride.

$$\underset{\text{Acid}}{RCOOH} + \underset{\text{Alcohol}}{HOR'} \xrightleftharpoons{\text{Conc. } H_2SO_4} \underset{\text{Ester}}{RCOOR'} + H_2O$$

For example,

$$\underset{\text{Acetic acid}}{CH_3COOH} + \underset{\text{Ethyl alcohol}}{HOC_2H_5} \xrightleftharpoons{H^+} \underset{\text{Ethyl acetate}}{CH_3COOC_2H_5} + H_2O$$

Mechanism. The reaction is believed to proceed through the following mechanism:

$$R-C(=O)-OH \xrightleftharpoons{H^+} R-\overset{+}{C}(-O-H)(-OH) \xrightleftharpoons{:\ddot{O}R'-H} R-C(-O-H)(-\overset{+}{O}R'-H)(-:OH) \rightleftharpoons$$

$$R-C(-OH)(-OR')(-H:\overset{+}{O}H) \xrightleftharpoons{-H_2O} R-\overset{+}{C}(-O-H)-OR' \rightleftharpoons R-C(=O)-OR' + H^+$$

It has been proved beyond doubt by isotopic tracer technique that the reaction involves the cleavage of O–H bond of alcohols and the C–OH bond of organic acids.

(*iii*) **Reaction with acid halides or acid anhydrides.** When alcohols are heated with an acid halide or acid anhydride, the hydrogen of –OH group is replaced by acyl group (RCO–) and, an ester is formed. The reaction takes place more rapidly than esterification with carboxylic acids. Thus, we have:

$$\underset{\text{Acetyl chloride}}{CH_3COCl} + HOC_2H_5 \longrightarrow \underset{\text{Ethyl acetate}}{CH_3COOC_2H_5} + HCl$$

$$\underset{\text{Acetic anhydride}}{(CH_3CO)_2O} + HOC_2H_5 \longrightarrow \underset{\text{Ethyl acetate}}{CH_3COOC_2H_5} + \underset{\text{Acetic acid}}{CH_3COOH}$$

Mechanism. The reaction mechanism may be depicted as follows:

$$R-\overset{\overset{\ddot{O}:}{\|}}{C}-Cl + :\ddot{O}R' (H) \longrightarrow R-\overset{:\ddot{O}:^-}{\underset{:\overset{+}{O}R(H)}{C}}-Cl \xrightarrow[-Cl^-]{} R-\overset{\overset{O}{\|}}{C}-\overset{+}{\ddot{O}}R'(H) \xrightarrow[-H^+]{} R-\overset{\overset{O}{\|}}{C}-OR'$$

Q. 15. How do alcohols react with the following:
(*i*) Halogen acids
(*ii*) Phosphorus halides
(*iii*) Thionyl chloride *(Nagpur, 2008*

Ans. (*i*) Reaction with hydrogen halides. Alcohols react with hydrogen halides to form alkyl halides and water.

$$\underset{\text{Alcohol}}{ROH} + HX \longrightarrow \underset{\text{Alkyl halide}}{RX} + H_2O$$

For example,

$$\underset{\text{Ethyl alcohol}}{C_2H_5OH} + HCl \xrightarrow{ZnCl_2} \underset{\text{Ethyl chloride}}{C_2H_5Cl} + H_2O$$

Hydrogen bromide and hydrogen iodide react in the same way but without the presence of a catalyst. However, if an alcohol is heated with hydrogen iodide in the presence of red phosphorus, it gets reduced to a hydrocarbons. For example,

$$\underset{\text{Methyl alcohol}}{CH_3OH} + 2HI \xrightarrow{P} \underset{\text{Methane}}{CH_4} + H_2O + I_2$$

The order of reactivity of halogen acids is HI > HBr > HCl. This is because I^- is a stronger nucleophile than Br^- which in turn is stronger than Cl^-. The order of reactivity of alcohols is tertiary > secondary > primary.

Mechanism. Separate mechanisms are proposed for primary alcohols and for secondary and tertiary alcohols. For secondary and tert. alcohols, SN^1 mechanism, as given below is proposed.

$$R-OH \xrightarrow{\text{Slow}} \underset{\text{Carbonium ion}}{R^+} + OH^-$$

$$R^+ + \underset{\text{Halide ion}}{X^-} \xrightarrow{\text{Fast}} R-X$$

This mechanism is supported by the fact that the observed order of relative reactivities of alcohols is the same as the order of stabilities of carbonium ions.

For primary alcohols, SN^2 mechanism involving a transition state is proposed.

$$R-OH+X^- \xrightarrow{\text{Slow}} \underset{\text{Transition state}}{[X......R.....OH]^-}$$

$$\downarrow \text{Fast}$$

$$R-X+OH^-$$

(*ii*) Reaction with phosphorus halides. Alcohols react with phosphorus halides, *i.e.* PCl_5, PCl_3, PBr_3 or PI_3, to yield corresponding alkyl halides.

$$\underset{\text{Alcohol}}{ROH} + PCl_5 \longrightarrow \underset{\text{Alkyl chloride}}{RCl} + POCl_3 + HCl$$

$$\underset{\text{Alcohol}}{3\,ROH + PX_3} \longrightarrow \underset{\text{Alkyl halide}}{3\,RX} + \underset{\text{Phosphorus acid}}{H_3PO_3}$$

$$\underset{\text{Ethyl alcohol}}{3\,C_2H_5OH + PI_3} \longrightarrow \underset{\text{Ethyl iodide}}{3C_2H_5I} + H_3PO_3$$

$$\underset{\text{Benzyl alcohol}}{C_6H_5-CH_2OH} + PCl_5 \longrightarrow \underset{\text{Benzyl chloride}}{C_6H_5-CH_2Cl} + POCl_3 + HCl$$

Mechanism. The mechanism of reaction with $\mathbf{PX_3}$ may be illustrated by considering the reaction with PCl_3 as given below:

$$R-\ddot{O}(H): + PCl_3 \longrightarrow \underset{\text{Protonated Alkyl Dichloro phosphite}}{R-\overset{+}{O}(H)-PCl_2} + :Cl^-$$

$$\bar{C}l: + R-\overset{+}{O}(H)-PCl_2 \longrightarrow R-Cl + HOPCl_2$$

In the first step, nucleophilic attack of alcohol on PCl_3 takes place to form a protonated alkyl dichlorophosphite. In the second step, chloride ion acts as a nucleophile and displaces $HOPCl_2$ from the phosphite initially formed.

The mechanism for reaction between alcohol and is as follows:

$$2\,PCl_5 \longrightarrow \overset{+}{P}Cl_4 + \overset{-}{P}Cl_6$$

$$R-\ddot{O}(H): + \overset{+}{P}Cl_4 \longrightarrow R-\overset{+}{O}(H)-PCl_4$$

$$\underset{(\text{From } \overline{P}Cl_5)}{\bar{C}l:} + R-\overset{+}{O}(H)-PCl_4 \longrightarrow R-Cl + H-O-PCl_4$$

$$H-O-PCl_4 \longrightarrow HCl + O=PCl_3$$

***(iii)* Reaction with thionyl chloride.** Alcohols react with thionyl chloride to form alkyl chlorides.

$$\underset{\text{Alcohol}}{ROH} + \underset{\text{Thionyl chloride}}{SOCl_2} \longrightarrow \underset{\text{Alkyl chloride}}{RCl} + SO_2 + HCl$$

$$\underset{\text{Ethyl alcohol}}{C_2H_5OH} + SOCl_2 \longrightarrow \underset{\text{Ethyl chloride}}{C_2H_5Cl} + SO_2 + HCl$$

Mechanism. The reaction is supposed to have the following mechanism.

$$R-\ddot{O}(H): + Cl-S(=O)-Cl \longrightarrow R-O-S(=O)-Cl + HCl$$

Alkyl Chlorosulphite

$$Cl:^- + R-O-S(=O)-Cl \longrightarrow R-Cl + \bar{O}-S(=O)-Cl \longrightarrow SO_2 + Cl^-$$

Q. 16. Describe the following reaction of alcohols:

(Purvanchal, 2003; Kalyani, 2004; Nagpur, 2005; Garhwal, 2010

(i) Dehydration

(ii) Oxidation

(iii) Dehydrogenation.

Ans. (i) Dehydration. When alcohols are heated with sulphuric acid, they get dehydrated to alkenes. For example,

$$\underset{\text{Ethyl alcohol}}{CH_3CH_2OH} \xrightarrow{H_2SO_4\ (433-443\ K)} \underset{\text{Ethylene}}{CH_2=CH_2} + H_2O$$

Mechanism of dehydration

The mechanism of dehydration of alcohols involves the formation of a carbocation as an intermediate. The reaction is found to take place in three steps.

1. In the first step alcohol combines with a hydrogen ion (from the acid) to form an oxonium ion.

2. The second step involves the dissociation of oxonium ion and formation of the carbocation.

3. Finally, the carbocation loses a hydrogen ion to yield the alkene as illustrated below.

(1) $$\underset{\text{Ethyl Alcohol}}{H-CH_2-CH_2-\ddot{O}-H} + \underset{\text{From Acid}}{H^+} \rightleftharpoons \underset{\text{Oxonium ion}}{H-CH_2-CH_2-\ddot{O}^+(H)-H}$$

(2) $$H-CH_2-CH_2-\ddot{O}^+(H)-H \xrightleftharpoons{\text{Rate Determining Step}} \underset{\text{Carbocation}}{H-CH_2-\overset{+}{C}H_2} + H_2O$$

(3) $$H-CH(H)-\overset{+}{C}H_2 \longrightarrow \underset{\text{Ethylene}}{H_2C=CH_2} + \underset{\text{Taken up by A new molecule of alcohol}}{H^+}$$

Relative ease of dehydration of alchols. It has been observed that the ease of dehydration of different classes of alcohols follows the order: tertiary > secondary > primary.

Rate determining step of the reaction may be depicted as under.

$$-\overset{|}{\underset{\underset{H}{|}}{C}}-\overset{|}{\underset{\underset{^{+}OH_2}{|}}{C}}- \longrightarrow \overset{|}{\underset{\underset{H}{|}}{C}}-\overset{|}{\underset{+}{C}}- + H_2O$$

The transition state leading to the formation of a tertiary cation (which is the most stable ion) would be most stable and, therefore, have least E_{act}. In other words, such a transition state and the carbocation generated by it would be formed most readily. Consequently, the dehydration of a tertiary alcohol would take place most readily. The dehydration of a secondary alcohol which involves the formation of less stable secondary cation would naturally take place less readily. Similarly the dehydration of a primary alcohol involving the least stable primary cation would be most difficult to take place. Thus the relative ease of dehydration of alcohols follows the same order as that of ease of formation of carbocations.

Orientation of dehydration. Many alcohols give rise to mixtures of isomeric alkenes on dehydration. The ratio of the alkenes formed in a particular reaction depends upon the reaction conditions. However in general it may be stated that dehydration of an alcohol favours the formation of more stable alkene. For example:

$$\underset{\text{Sec- Butyl alcohol}}{CH_3-CH_2-CHOH-CH_3} \xrightarrow[-H_2O]{H_3PO_4-Al_2O_3,\ 298\ K} \begin{cases} \underset{\substack{\text{1-Butene (24.1\%)} \\ \text{(less stable)}}}{CH_3-CH_2-CH=CH_2} \\ \underset{\substack{\text{2-Butene (75.9\%)} \\ \text{(more stable)}}}{CH_3-CH=CH-CH_3} \end{cases}$$

(*ii*) **Oxidation.** Oxidation of alcohols can be carried out with aqueous, acidified or alkaline potassium permanganate, acidified potassium dichromate or dilute nitric acid. The ease of oxidation and the nature of the products obtained depend on the type of alcohol employed.

(*a*) *Primary alcohols* are easily oxidised to aldehydes and then to acids containing the same number of carbon atom as the original alcohol. For example,

$$\underset{\text{Ethyl alcohol}}{CH_3-\overset{\overset{C}{|}}{\underset{\underset{C}{|}}{C}}-OH} \xrightarrow[-H_2O]{[O]} \underset{\text{Acetaldehyde}}{CH_3-\overset{\overset{H}{|}}{C}=O} \xrightarrow{[O]} \underset{\text{Acetic acid}}{CH_3-\overset{\overset{OH}{|}}{C}=O}$$

(*b*) *Secondary alcohols* on oxidation first give ketones with the same number of carbon atoms. Ketones are further oxidised only on drastic treatment to give acids containing *lesser* number of carbon atoms. For example:

$$\underset{\text{Isopropyl alcohol}}{\begin{matrix} CH_3 \\ CH_3 \end{matrix}\!\!>C<\!\!\begin{matrix} H \\ OH \end{matrix}} \xrightarrow[-H_2O]{[O]} \underset{\text{Acetone}}{\begin{matrix} CH_3 \\ CH_3 \end{matrix}\!\!>C=O} \xrightarrow[-CO_2,\ -H_2O]{4[O]} \underset{\text{Acetic acid}}{CH_3COOH}$$

(*c*) *Tertiary alcohols* fail to undergo oxidation in neutral or alkaline solution. However, with acid oxidising agents they are oxidised to give ketones and acids containing smaller number of carbon atoms than the original alcohols. For example:

$$\begin{matrix} CH_3 \\ CH_3 \\ CH_3 \end{matrix} \!\!> C-OH \xrightarrow{4[O]} \begin{matrix} CH_3 \\ CH_3 \end{matrix}\!\!> C=O + CO_2 + 2H_2O$$

Tert-Butyl alcohol → Acetone

$$\begin{matrix} CH_3 \\ CH_3 \end{matrix}\!\!> C=O \xrightarrow{4[O]} CH_3COOH + CO_2 + H_2O$$

Acetic acid

(*iii*) **Dehydrogenation.** Different classes of alcohols furnish different products when their vapours are passed over copper at high temperature 473–573 K.

(*a*) *Primary alcohols* are dehydrogenated to form aldehydes.

For example:

$$CH_3-\underset{H}{\overset{H}{|}}{C}-OH \xrightarrow{Cu,\ 473-573\ K} CH_3-\overset{H}{\overset{|}{C}}=O + H_2$$

Acetaldehyde

(*b*) *Secondary alcohols* are also dehydrogenated to give ketones. For example:

$$\begin{matrix} CH_3 \\ CH_3 \end{matrix}\!\!> C <\!\!\begin{matrix} H \\ OH \end{matrix} \longrightarrow \begin{matrix} CH_3 \\ CH_3 \end{matrix}\!\!> C=O + H_2$$

Isopropyl alcohol → Acetone

(*c*) *Tertiary alcohols* are dehydrated to give alkenes. For example:

$$CH_3-\underset{CH_3}{\overset{CH_3}{\overset{|}{\underset{|}{C}}}}-OH \longrightarrow CH_3-\overset{CH_3}{\overset{|}{C}}=CH_2 + H_2O$$

Tert-Butyl alcohol → *Iso*-butylene

Q. 17. The ease of dehydration of alcohols is as under; tert. alcohol > sec. alcohol > primary alcohol. Explain.

Ans. As per the mechanism of dehydration of alcohols, carbonium ion is produced as intermediate product. The reactivity of the alcohol will depend upon the stability of the carbonium ion. It is found that the stabilities of carbonium ions follows the order :

tert. carbonium ion > sec. carbonium ion > primary carbonium ion

Therefore, the ease of dehydration of alcohols also follows the same order

tert. alcohol > sec. alcohol > primary alcohol

Q. 18. 2-Butene is the major product when *n*-butyl alcohol is heated with conc. H_2SO_4. Explain.

Ans.

$$CH_3-CH_2-\underset{H}{\overset{H}{\overset{|}{\underset{|}{C}}}}-\underset{H}{\overset{H}{\overset{|}{\underset{|}{C}}}}-OH + H^+$$

$$\downarrow$$

$$CH_3-CH_2-\underset{H}{\overset{H}{\overset{|}{\underset{|}{C}}}}-\underset{H}{\overset{H}{\overset{|}{\underset{|}{C}}}}-\overset{+}{O}-H \longrightarrow CH_3-CH_2-\underset{H}{\overset{H}{\overset{|}{\underset{|}{C}}}}-\underset{H}{\overset{H}{C^+}}$$

Protonated alcohol → *n*-Butyl carbonium ion

n-Butyl carbonium ion is converted into more stable sec. butyl carbonium ion by hydride shift from the neighbouring carbon atom as follows:

$$CH_3-\overset{H}{\underset{H}{C}}-\overset{H}{\underset{H}{C}}-\overset{H}{\underset{H}{C^+}} \longrightarrow CH_3-\overset{H}{\underset{H}{C}}-\overset{H}{\underset{\oplus}{C}}-\overset{H}{\underset{H}{C}}-H$$

$$\longrightarrow CH_3-\overset{H}{C}=\overset{H}{C}-CH_3 \quad \text{(2-Butene)}$$

This is how 2-butene and not 1-butene is obtained.

Q. 19. How will you distinguish between primary, secondary and tertiary alcohols?

(*Kerala, 2001, Garhwal, 2000; A.N. Bhauguna, 2000; Bangalore 2004; Anna 2004; Delhi 2005;Purvanchal 2007*)

Ans. The following methods enable us to distinguish between primary, secondary and tertiary alcohols.

(1) Oxidation. Primary, secondary and tertiary alcohols yield different products on oxidation. A primary alcohol gives an aldehyde and then an acid containing the same number of carbon atoms as the original alcohol. Secondary alcohols form ketones containing the same number of carbon atoms as the original alcohol but further oxidation of ketones occurs only with strong oxidising agents to form acids containing lesser number of carbon atoms. Tertiary alcohols are very difficult to oxidise but if oxidised they yield ketones and then acids, both containing lesser number of carbon atoms than the alcohol. Thus, the identification of oxidation products can reveal the nature of alcohols.

$$R-CH_2OH \xrightarrow{[O]} R-CHO \qquad \begin{matrix}R\\R\end{matrix}\!\!>CHOH \xrightarrow{[O]} \begin{matrix}R\\R\end{matrix}\!\!>C=O$$

$$\begin{matrix}CH_3\\CH_3\\CH_3\end{matrix}\!\!>C-OH \xrightarrow{4[O]} CH_3COCH_3+CO_2+2H_2O$$

(2) Reaction with hot reduced copper. When the vapours of alcohols are passed over hot reduced copper at 473–573 K, primary and secondary alcohols lose hydrogen to form aldehydes and ketones respectively while tertiary alcohols get dehydrated to form alkenes as already explained.

(3) Lucas test. This test consists in treating the alcohol with Lucas reagent which is an equimolar mixture of concentrated hydrochloric acid and anhydrous zinc chloride. Appearance of cloudiness in the reaction mixture indicates the conversion of alcohol into alkyl chloride. It has been observed that a tertiary alcohol reacts immediately, a secondary alcohol reacts within five minutes while a primary alcohol does not react appreciably at room temperature.

(4) Victor Meyer's method. This reaction is carried out as follows:

The given alcohol is treated with phosphorous and iodine when an alkyl iodide is obtained. It is distilled with silver nitrite to yield the corresponding nitroalkane. The nitroalkane is finally treated with nitrous acid (*i.e.*, sodium nitrite + dil. sulphuric acid) and the solution made alkaline.

The formation of *blood red* colouration shows the original alcohol to be primary, a *blue* colouration indicates a secondary alcohol while the solution remains *colourless* in case of tertiary alcohol. The reactions are explained in tabular form as under:

Primary alcohol	*Secondary alochol*	*Tertiary alcohol*
RCH_2OH	R_2CHOH	R_3C-OH
$\downarrow$ P + I_2	$\downarrow$ P + I_2	$\downarrow$ P + I_2
RCH_2I	R_2CHI	R_3CI
$\downarrow$ $AgNO_2$	$\downarrow$ $AgNO_2$	$\downarrow$ $AgNO_2$
RCH_2NO_2	R_2CHNO_2	R_3CNO_2
$\downarrow$ HONO Nitrous acid	$\downarrow$ HONO Nitrous acid	$\downarrow$ HNO_2
$R-C(=NOH)-NO_2$	$R_2C(NO)-NO_2$	No action
Nitrolic acid	Pseudo nitrol	
Produces blood red colour with alkali	*Blue colour*	*Colourless solution*

Q. 20. How will you achieve the following conversion:

(*i*) Ethyl alcohol ⟶ Sec. butyl alcohol

(*ii*) *n*-propyl alcohol ⟶ Isopropyl alcohol (*Calcutta 2007*)

(*iii*) Isopropyl alcohol ⟶ Tert. butyl alcohol

(iv) Isopropyl alcohol ⟶ *n*-Propyl alcohol (*Calcutta 2007*)

(*v*) 2-Methyl-2-butanol ⟶ 3-Methyl-2-butanol

(*vi*) Isobutyl alcohol ⟶ Tert. butyl alcohol.

Ans. (*i*) Ethyl alcohol into sec. butyl alcohol

$$\underset{\substack{\text{Ethyl alcohol}\\(\textit{Primary})}}{CH_3CH_2OH} \xrightarrow{K_2Cr_2O_7} \underset{\text{Acetaldehyde}}{CH_3CHO}$$

$$CH_3CHO + CH_3CH_2MgBr \longrightarrow CH_3-\overset{H}{\underset{CH_2CH_3}{C}}-OMgBr \xrightarrow{H_3O^+}$$

$$\underset{\substack{\textit{sec.}\text{ Butyl alcohol}\\(\textit{Secondary})}}{CH_3-CHOH-CH_2-CH_3} + Mg(OH)Br$$

(*ii*) *n*-Propyl alcohol into isopropyl alcohol

$$\underset{\substack{n\text{-Propyl alcohol}\\(\textit{Primary})}}{CH_3CH_2CH_2OH} \xrightarrow[Al_2O_3,\ 625\ K]{\text{Dehydration}} \underset{\text{Propylene}}{CH_3-CH=CH_2} \xrightarrow{HI}$$

$$\underset{\text{Isopropyl iodide}}{CH_3-CHI-CH_3} \xrightarrow[\text{(Moist silver oxide)}]{AgOH} \underset{\substack{\text{Isopropyl alcohol}\\(\textit{Secondary})}}{CH_3-CHOH-CH_3} + AgI$$

(*iii*) Isopropyl alcohol into tert. butyl alcohol

$$\underset{\substack{\text{Isopropyl alcohol}\\(\textit{Secondary})}}{CH_3-CHOH-CH_3} \xrightarrow{K_2Cr_2O_7} \underset{\text{Acetone}}{CH_3COCH_3} \xrightarrow{CH_3MgI} CH_3-\overset{\overset{CH_3}{|}}{\underset{\underset{CH_3}{|}}{C}}-OMgI$$

$$\xrightarrow{H_2O^+} \underset{\substack{\textit{tert}\text{-Butyl alcohol}\\(\textit{Tertiary})}}{CH_3-\overset{\overset{CH_3}{|}}{\underset{\underset{CH_3}{|}}{C}}-OH} + Mg(OH)I$$

(*iv*) Isopropyl alcohol into n-propyl alcohol

$$\underset{\substack{\text{Isopropyl alcohol}\\(\textit{Secondary})}}{CH_3-CHOH-CH_3} \xrightarrow[Al_2O_3,\ \text{Heat}]{\text{Dehydration}} CH_3-CH=CH_2 \xrightarrow{(BH_3)_2} (CH_3-CH_2-CH_2)_3B \longrightarrow \underset{\substack{n\text{-Propyl alcohol}\\(\textit{Primary})}}{CH_3-CH_2-CH_2OH}$$

(*v*) 2-Methyl-2-butanol into 3-methyl-2-butanol

$$\underset{\substack{\text{2-Methyl-2-butanol}\\(\textit{Tertiary})}}{CH_3-\overset{\overset{CH_3}{|}}{\underset{\underset{OH}{|}}{C}}-CH_2-CH_3} \xrightarrow[Al_2O_3,\ 625\ K]{\text{Dehydration}} \underset{\text{2-methyl-2-butene}}{CH_3-\overset{\overset{CH_3}{|}}{C}=CH-CH_3} \xrightarrow{(BH_3)_2}$$

$$\left[CH_3-\overset{\overset{CH_3}{|}}{CH}-\underset{\underset{CH_3}{|}}{CH}-\right]_3 B \xrightarrow{H_2O_2\ /\ OH^-} \underset{\substack{\text{3-methyl-2-butanol}\\(\textit{Secondary})}}{CH_3-\overset{\overset{CH_3}{|}}{CH}-CHOH-CH_3}$$

(*vi*) Isobutyl alcohol into tert. butyl alcohol

$$\underset{\substack{\text{Isobutyl alcohol}\\(\textit{Primary})}}{CH_3-\overset{\overset{CH_3}{|}}{C}-CH_2OH} \xrightarrow[Al_2O_3,\ 625\ K]{\text{Dehydration}} \underset{\text{Isobutylene}}{CH_3-\overset{\overset{CH_3}{|}}{C}=CH_2} \xrightarrow{HI}$$

$$\underset{\textit{Tert}\text{-Butyl iodide}}{CH_3-\overset{\overset{CH_3}{|}}{\underset{\underset{I}{|}}{C}}-CH_3} \xrightarrow{AgOH} \underset{\substack{\textit{Tert}\text{-Butyl alcohol}\\(\textit{Tertiary})}}{CH_3-\overset{\overset{CH_3}{|}}{\underset{\underset{OH}{|}}{C}}-CH_3} + AgI$$

Q. 21. Why does dehydration of 1-phenyl-2-propanol in acid form 1-phenyl-1-propene rather than 1-phenyl-2-propene?

Ans. The structure of 1-phenyl-2-propanol is

$$\begin{array}{c} \text{H}\ \ \text{H}\ \ \text{H} \\ | \quad | \quad | \\ \text{H}-\text{C}-\text{C}-\text{C}-\text{C}_6\text{H}_5 \\ | \quad | \quad | \\ \text{H}\ \ \text{OH}\ \text{H} \end{array}$$

I

On reaction with a protonic acid, it first forms an oxonium ion (II) which then loses a water molecule to form a carbocation (III).

$$\begin{array}{c} \text{H}\ \ \text{H}\ \ \text{H} \\ | \quad | \quad | \\ \text{H}-\text{C}-\text{C}-\text{C}-\text{C}_6\text{H}_5 \\ | \quad | \quad | \\ \text{H}\ \ \overset{+}{\text{O}}\ \ \text{H} \\ / \ \backslash \\ \text{H} \quad \text{H} \end{array}$$

II

$$\begin{array}{c} \text{H}\ \ \text{H}\ \ \text{H} \\ | \quad | \quad | \\ \text{H}-\overset{3}{\text{C}}-\overset{2}{\underset{+}{\text{C}}}-\overset{1}{\text{C}}-\text{C}_6\text{H}_5 \\ | \qquad \quad | \\ \text{H} \qquad \quad \text{H} \end{array}$$

III

The carbocation having structure III loses a proton to stabilize it. This proton can be ejected either from carbon atom No. 1 or carbon atom No. 3. Accordingly, two different alkenes are possible as:

$$\text{H}-\overset{\text{H}}{\underset{\text{H}}{\text{C}^3}}-\overset{\text{H}}{\underset{+}{\text{C}^2}}-\overset{\text{H}}{\underset{\text{H}}{\text{C}^1}}-\text{C}_6\text{H}_5 \xrightarrow[\text{from C}-1]{\text{Loss of H}^+} \text{H}-\overset{\text{H}}{\underset{\text{H}}{\text{C}}}-\overset{\text{H}}{\text{C}}=\overset{\text{H}}{\text{C}}-\text{C}_6\text{H}_5$$

1-Phenyl-1-propene
IV

Or

$$\text{H}-\overset{\text{H}}{\underset{\text{H}}{\text{C}^3}}-\overset{\text{H}}{\underset{+}{\text{C}^2}}-\overset{\text{H}}{\underset{\text{H}}{\text{C}^1}}-\text{C}_6\text{H}_5 \xrightarrow[\text{from C}-3]{\text{Loss of H}^+} \text{H}-\overset{\text{H}}{\underset{\text{H}}{\text{C}}}-\overset{\text{H}}{\text{C}}=\overset{\text{H}}{\underset{\text{H}}{\text{C}}}-\text{C}_6\text{H}_5$$

1-Phenyl-2-propene
V

Alkene IV is more substituted and therefore more stable than alkene V. Moreover, alkene IV has additional stability as its double bond is conjugated with the benzene ring. Consequently, only alkene IV will be formed.

Q. 22. Outline all the steps in the synthesis of following compounds from *n*-butyl alcohol using necessary inorganic reagents.

(*a*) *n*-Butyl bromide **(*b*) 1-Butene**
(*c*) *n*-Butyl hydrogen sulphate **(*d*) *n*-Butyraldehyde**
(*e*) *n*-Butyric acid **(*f*) 1-Butyne.**

Ans. (*a*) *n*-Butyl alcohol into *n*-Bromide.

$$\underset{n\text{-Butyl alcohol}}{CH_3CH_2CH_2CH_2OH} \xrightarrow{P + Br_2} \underset{n\text{-Butyl bromide}}{CH_3CH_2CH_2CH_2Br}$$

(*b*) *n*-Butyl alcohol into 1-butene.

$$\underset{n\text{-Butyl alcohol}}{CH_3CH_2CH_2CH_2OH} \xrightarrow{SOCl_2} \underset{n\text{-Butyl chloride}}{CH_3CH_2CH_2CH_2Cl} \xrightarrow[KOH]{\text{Alcoholic}} \underset{\text{1-Butene}}{CH_3CH_2CHCH=CH_2}$$

(*c*) *n*-Butyl alcohol into *n*-butyl hydrogen sulphate.

$$\underset{n\text{-Butyl alcohol}}{CH_3CH_2CH_2CH_2OH} \xrightarrow[383\text{ K}]{H_2SO_4\text{ at}} \underset{n\text{-Butyl hydrogen sulphate}}{CH_3CH_2CH_2CH_2HSO_4} + H_2O$$

(*d*) *n*-Butyl alcohol into *n*-butyraldehyde.

$$\underset{n\text{-Butyl alcohol}}{CH_3CH_2CH_2CH_2OH} \xrightarrow[573\text{ K}]{\text{Hot reduced}} \underset{n\text{-Butylaldehyde}}{CH_3CH_2CH_2CHO} + H_2$$

(*e*) *n*-Butyl alcohol into *n*-butyric acid.

$$\underset{n\text{-Butyl alcohol}}{CH_3CH_2CH_2CH_2OH} \xrightarrow[\text{Cu, 573 K}]{\text{Hot reduced}} \underset{n\text{-Butyraldehyde}}{CH_3CH_2CH_2CHO} \xrightarrow[K_2Cr_2O_7 + H_2SO_4]{\text{Oxidation}} \underset{n\text{-Butyric acid}}{CH_3CH_2CH_2COOH}$$

(*f*) *n*-Butyl alcohol into 1-butyne.

$$\underset{n\text{-Butyl alcohol}}{CH_3CH_2CH_2CH_2OH} \xrightarrow{SOCl_2} \underset{n\text{-Butyl chloride}}{CH_3CH_2CH_2CH_2Cl} \xrightarrow[KOH]{\text{Alcoholic}} \underset{\text{1-Butene}}{CH_3CH_2CHCH=CH_2}$$

$$\xrightarrow{Br_2} \underset{\text{1, 2-Dibromobutane}}{CH_3CH_2\underset{Br}{\underset{|}{CH}}-\underset{Br}{\underset{|}{CH_2}}} \xrightarrow[\text{liq. }NH_3]{NaNH_2\text{ in}} \underset{\text{Butyne-1}}{CH_3CH_2C\equiv CH}$$

Q. 23. How will you convert:

(*a*) Isopropyl alcohol into 2-methyl-2-hexanol.

(*b*) Secondary butyl alcohol into 3-methyl-3-pentanol.

Ans. (*a*) Isopropyl alcohol into 2-methyl-2-hexanol

$$\underset{\text{Isopropyl alcohol}}{CH_3-\underset{OH}{\underset{|}{CH}}-CH_3} \xrightarrow[H_2SO_4]{K_2Cr_2O_7} \underset{\text{Acetone}}{CH_3-\underset{O}{\underset{||}{C}}-CH_3} \xrightarrow{CH_3CH_2CH_2CH_2MgI}$$

$$CH_3-\overset{CH_3}{\overset{|}{\underset{OMgI}{\underset{|}{C}}}}-CH_2CH_2CH_2CH_3 \xrightarrow{H_3O^+} \underset{\text{2-Methyl-2-hexanol}}{CH_3-\overset{CH_3}{\overset{|}{\underset{OH}{\underset{|}{C}}}}-CH_2CH_2CH_2CH_3}$$

(*b*) Secondary butyl alcohol into 2-methyl-3-pentanol.

$$\underset{\text{Secondary butyl alcohol}}{CH_3CH_2\underset{OH}{\underset{|}{CH}}CH_3} \xrightarrow[H_2SO_4]{K_2Cr_2O_7} CH_2CH_2\underset{O}{\underset{||}{C}}CH_3 \xrightarrow{CH_3CH_2MgI}$$

$$CH_3CH_2\overset{CH_2CH_3}{\overset{|}{\underset{OMgI}{\underset{|}{C}}}}-CH_3 \xrightarrow{H_3O^+} \underset{\text{3-Methyl-3-Pentanol}}{CH_3CH_2\overset{CH_2CH_3}{\overset{|}{\underset{OH}{\underset{|}{C}}}}-CH_3}$$

Q. 24. Predict the products of the following reactions. Write the mechanism involved:

$$\text{(a)}\quad CH_3CH_2\overset{\overset{CH_3}{|}}{C}H.CH_2OH \xrightarrow{H^+}$$

$$\text{(b)}\quad CH_3\overset{\overset{CH_3}{|}}{C}H-\underset{\underset{OH}{|}}{\overset{\overset{CH_3}{|}}{C}}-CH_3 \xrightarrow{H^+}$$

Ans. (*a*)

$$CH_3CH_2\overset{\overset{CH_3}{|}}{C}H.CH_2OH + H^+ \longrightarrow CH_3CH_2\overset{\overset{CH_3}{|}}{C}H-\underset{\underset{H}{|}}{\overset{\overset{H}{|}}{C}}-\overset{\overset{H}{|}}{\overset{+}{O}}H \xrightarrow{-H_2O}$$

$$CH_3-CH_2\overset{\overset{CH_3}{|}}{C}H-\underset{\underset{H}{|}}{\overset{\overset{H}{|}}{C^+}} \xrightarrow[\text{shift}]{1,2\text{-Methyl}} CH_3CH_2-\underset{\underset{H}{|}}{\overset{+}{C}}-\underset{\underset{H}{|}}{\overset{\overset{CH_3}{|}}{C}}-H \xrightarrow{-H^+} \underset{\text{2-Pentene}}{CH_3CH=CHCH_2CH_3}$$

A more substituted alkene is more stable.

(*b*)

$$CH_3\overset{\overset{CH_3}{|}}{C}H-\underset{\underset{OH}{|}}{\overset{\overset{CH_3}{|}}{C}}-CH_3 \xrightarrow{H^+} CH_3-\overset{\overset{CH_3}{|}}{C}H-\underset{\underset{\underset{H}{|}}{^+O-H}}{\overset{\overset{CH_3}{|}}{C}}-CH_3 \xrightarrow{-H_2O}$$

$$CH_3-\overset{\overset{CH_3}{|}}{C}H-\underset{+}{\overset{\overset{CH_3}{|}}{C}}-CH_3 \xrightarrow{-H^+} \underset{\text{2, 3-Dimethyl-2 butene}}{CH_3\overset{\overset{CH_3}{|}}{C}=\overset{\overset{CH_3}{|}}{C}-CH_3}$$

Q. 25. Identify the products (A, B, C and D) in the following sequence of reactions.

$$\textbf{Isopropyl alcohol} \xrightarrow[H_2SO_4]{K_2Cr_2O_7} A \xrightarrow{CH_3MgBr} B \xrightarrow[H^+]{H_2O} C \xrightarrow{\text{Hot reduced copper}} D$$

Ans.
$$\underset{\text{Isopropyl alcohol}}{CH_3-\underset{\underset{OH}{|}}{\overset{\overset{H}{|}}{C}}-CH_3} \xrightarrow[H_2SO_4]{K_2Cr_2O_7} \underset{\text{Acetone}}{CH_3-\underset{\underset{O}{||}}{C}-CH_3} \xrightarrow{CH_3MgBr}$$

$$CH_3-\underset{\underset{OMgBr}{|}}{\overset{\overset{CH_3}{|}}{C}}-CH_3 \xrightarrow{H_2O/H^+} \underset{\text{2-Methyl-2-propanol}}{CH_3-\underset{\underset{OH}{|}}{\overset{\overset{CH_3}{|}}{C}}-CH_3} \xrightarrow[\text{Copper}]{\text{Hot reduced}} \underset{\text{2-Methyl-1-propene}}{CH_3\overset{\overset{CH_3}{|}}{C}=CH_2}$$

Therefore,

A is $CH_3-\underset{\underset{O}{||}}{C}-CH_3$ Acetone or propanone-2

$$\text{B is} \quad CH_3-\underset{\underset{OMgBr}{|}}{\overset{\overset{CH_3}{|}}{C}}-CH_3 \qquad \textit{Tert}\text{-butoxy magnesium bromide}$$

$$\text{C is} \quad CH_3-\underset{\underset{OH}{|}}{\overset{\overset{CH_3}{|}}{C}}-CH_3 \qquad \textit{Tert}\text{-butyl alcohol or 2-Methyl-2-propanol}$$

$$\text{D is} \quad CH_3-\overset{\overset{CH_3}{|}}{C}=CH_2 \qquad \text{2-Methyl-1-propene}$$

Q. 26. Give the mechanism of dehydration of alcohols. (*Guwahati, 2006*)

Ans. The mechanism of dehydration of alcohol is as under:

$$\underset{\text{Ethyl alcohol}}{H-\underset{\underset{H}{|}}{\overset{\overset{H}{|}}{C}}-\underset{\underset{H}{|}}{\overset{\overset{H}{|}}{C}}-\ddot{\underset{\cdot\cdot}{O}}-H} + H^{\oplus} \longrightarrow H-\underset{\underset{H}{|}}{\overset{\overset{H}{|}}{C}}-\underset{\underset{H}{|}}{\overset{\overset{H}{|}}{C}}-\overset{\oplus}{\underset{\underset{H}{|}}{O}}-H$$

$$\overset{\oplus}{H}+\underset{\text{Ethane}}{H-\underset{\underset{H}{|}}{C}=\underset{\underset{H}{|}}{C}-H} \longleftarrow H-\underset{\underset{H}{|}}{\overset{\overset{H}{|}}{C}}-\underset{\underset{H}{|}}{\overset{\overset{H}{|}}{C^{\oplus}}}$$

DIHYDRIC ALCOHOLS

Q. 27. What are glycols? Describe their nomenclature.

Ans. Glycols are organic compounds containing two alcoholic groups. For example, ethylene glycol.

Nomenclature. Common and IUPAC names of a few compounds are given in the following table:

Structure	*Common name*	*IUPAC name*
$\underset{OH}{\underset{\vert}{CH_2}}-\underset{OH}{\underset{\vert}{CH_2}}$	Ethylene glycol	Ethane-1,2-diol
$CH_3-\underset{OH}{\underset{\vert}{CH}}-\underset{OH}{\underset{\vert}{CH_2}}$	Propylene glycol	Propane-1,2-diol
$\underset{OH}{\underset{\vert}{CH_2}}-CH_2-\underset{OH}{\underset{\vert}{CH_2}}$	Trimethylene glycol	Propane-1,3-diol

Q. 28. Describe methods of preparation of ethylene glycol.

(*Shivaji 2003; Magadh 2004, Nagpur 2008*)

Ans. Ethylene glycol may be prepared by one of the following methods:

(1) Hydroxylation of ethylene. When ethylene is treated with a cold alkaline solution of potassium permanganate (known as Baeyer's reagent), a hydroxyl group is added to each of the doubly bonded carbon atoms resulting in the formation of ethylene glycol.

$$\underset{\text{Ethylene}}{CH_2=CH_2} + \underset{\text{From alk. } KMnO_4}{H_2O + O} \longrightarrow \underset{\text{Ethylene glycol}}{2\ \underset{OH}{\underset{|}{CH_2}} - \underset{OH}{\underset{|}{CH_2}}}$$

(2) Hydrolysis of ethylene oxide. In this method, ethylene is first oxidised by heating in air at 525 K in the presence of silver catalyst to form ethylene oxide or ethylene epoxide. The epoxide, thus, formed is hydrolysed in the presence of an acid or base to form ethylene glycol.

$$CH_2=CH_2 + \frac{1}{2}O_2 \xrightarrow{525\text{ K, Ag}} \underset{\text{Ethylene epoxide}}{CH_2 - CH_2 \text{ (bridged by O)}}$$

$$\underset{\text{(bridged by O)}}{CH_2 - CH_2} + H_2O \xrightarrow{\text{Acid or base}} \underset{\text{Ethylene glycol}}{\underset{OH}{\underset{|}{CH_2}} - \underset{OH}{\underset{|}{CH_2}}}$$

Ethylene can also be converted into ethylene epoxide by treatment with peroxyformic acid (HCO_2 OH)

$$CH_2=CH_2 + HCO_2OH \longrightarrow \underset{\text{(bridged by O)}}{CH_2 - CH_2} + HCOOH$$

$$\xrightarrow{\text{Acid or base}} \underset{OH}{\underset{|}{CH_2}} - \overset{OH}{\overset{|}{CH_2}}$$

It may be noted that hydroxylation by $KMnO_4$ gives cis compound whereas peroxyformic acid produces *trans* compound.

(3) By the hydrolysis of ethylene dichloride or dibromide or chlorohydrin.

$$\begin{matrix} CH_2Br \\ | \\ CH_2Br \end{matrix} + Na_2CO_3 + H_2O \longrightarrow \begin{matrix} CH_2OH \\ | \\ CH_2OH \end{matrix} + 2NaBr + CO_2$$

The yield of ethylene glycol is, however, not good as the hydrolysis is also accompanied by the formation of vinyl bromide through an elimination reaction.

$$\begin{matrix} CH_2Br \\ | \\ CH_2Br \end{matrix} + Na_2CO_3 \longrightarrow \begin{matrix} CH_2 \\ \| \\ CHBr \end{matrix} + NaBr + NaHCO_3$$

(4) Bimolecular reduction of carbonyl compound. Symmetrical glycols, other than ethylene glycol, can be obtained by the bimolecular reduction of aldehydes or ketones, with magnesium in the presence of benzene.

e.g.

$$2\ \begin{matrix} \diagdown\diagup \\ C \\ \| \\ O \end{matrix} \xrightarrow[\text{Benzene (Bimolecular reduction)}]{\text{Mg.}} \underset{\text{Pinacol}}{-\underset{OH}{\underset{|}{\overset{|}{C}}} - \underset{OH}{\underset{|}{\overset{|}{C}}}-}$$

For example,

$$\begin{matrix} CH_3 \\ | \\ CH_3-C=O \\ \\ CH_3-C=O \\ | \\ CH_3 \end{matrix} + Mg \xrightarrow{\text{Benzene}} \begin{matrix} CH_3CH_3 \\ |\quad | \\ H_3C-C-C-CH_3 \\ |\quad | \\ O\quad O \\ \diagdown\diagup \\ Mg \end{matrix}$$

Acetone (2-molecules)

$$\downarrow 2H_2O$$

$$\begin{matrix} CH_3CH_3 \\ |\quad | \\ CH_3-C-C-CH_3 \\ |\quad | \\ OH\ OH \end{matrix}$$

Pinacol (2, 3-Dimethylbutane-2, 3-diol)

Physical properties of ethylene glycol. It is a colourless syrupy liquid (b.p. 470 K) having a sweet taste. Its high boiling point is due to the presence of two sites for hydrogen bonding per molecule (*i.e.*, two –OH groups). For the same reason it is also highly soluble in water and ethyl alcohol.

Q. 29. Describe chemical properties of ethylene glycol.

Ans. Chemical properties of ethylene glycol. Like monohydric alcohols, ethylene glycol exhibits the usual reactions of hydroxyl group. However, it is not necessary that both the alcoholic groups must react with a particular reagent, or in other words, it is possible to carry out reactions in which only one of the two alcoholic groups is involved.

(1) Reaction with metals. Ethylene glycol reacts with metals like sodium to form mono and disodium derivatives.

$$\begin{matrix} CH_2OH \\ | \\ CH_2OH \end{matrix} \xrightarrow[-\frac{1}{2}H_2]{Na,\ 333\ K} \begin{matrix} CH_2ONa \\ | \\ CH_2OH \end{matrix} \xrightarrow[-\frac{1}{2}H_2]{Na,\ 433\ K} \begin{matrix} CH_2ONa \\ | \\ CH_2ONa \end{matrix}$$

Ethylene glycol — Monosodium glycolate — Disodium glycolate

(2) Reaction with halogen acids. It reacts with halogen acids to form two different products as shown below:

$$\begin{matrix} CH_2OH \\ | \\ CH_2OH \end{matrix} + HCl \xrightarrow[-H_2O]{433\ K} \begin{matrix} CH_2Cl \\ | \\ CH_2OH \end{matrix} \xrightarrow[-H_2O]{HCl,\ 473\ K} \begin{matrix} CH_2Cl \\ | \\ CH_2Cl \end{matrix}$$

Ethylene glycol — Ethylene chlorohydrin — Ethylene dichloride

With HI, ethylene diodide first formed decomposes to give ethylene.

$$\begin{matrix} CH_2OH \\ | \\ CH_2OH \end{matrix} \xrightarrow[-2\,H_2O]{2HI} \begin{matrix} CH_2I \\ | \\ CH_2I \end{matrix} \longrightarrow \begin{matrix} CH_2 \\ \| \\ CH_2 \end{matrix} + I_2$$

Unstable

(3) Reaction with phosphorus trihalides. When treated with PCl_3 or PBr_3, ethylene glycol forms ethylene dichloride or dibromide.

$$3\begin{matrix} CH_2OH \\ | \\ CH_2OH \end{matrix} + 2PCl_3 \longrightarrow 3\begin{matrix} CH_2Cl \\ | \\ CH_2Cl \end{matrix} + 2H_3PO_3$$

(4) Reaction with PCl_5 or thionyl chloride. It reacts with phosphorus pentachloride or thionyl chloride to form ethylene dichloride:

$$\begin{array}{l} CH_2OH \\ | \\ CH_2OH \end{array} + 2PCl_5 \longrightarrow \begin{array}{l} CH_2Cl \\ | \\ CH_2Cl \end{array} + POCl_3 + 2HCl$$

$$\begin{array}{l} CH_2OH \\ | \\ CH_2OH \end{array} + 2SOCl_2 \longrightarrow \begin{array}{l} CH_2Cl \\ | \\ CH_2Cl \end{array} + 2SO_2 + 2HCl$$

(5) Reaction with organic acids. It reacts with organic acids to form mono- or di-esters depending upon whether one or two moles of the acid are used per mole of glycol.

$$\begin{array}{l} CH_2OH \\ | \\ CH_2OH \end{array} \xrightarrow[-H_2O]{HOOCCH_3} \underset{\text{Glycol monoacetate}}{\begin{array}{l} CH_2O.OCCH_3 \\ | \\ CH_2.OH \end{array}} \xrightarrow[H_2O]{HOOCCH_3} \underset{\text{Glycol diacetate}}{\begin{array}{l} CH_2O.OCCH_3 \\ | \\ CH_2O.OCCH_3 \end{array}}$$

(6) Oxidation. Ethylene glycol gives different products on oxidation under different conditions.

(*i*) *When oxidised with nitric acid*, ethylene glycol ultimately gives rise to oxalic acid through the intermediate formation of many products which may also be isolated.

$$\underset{\text{Ethylene glycol}}{\begin{array}{l} CH_2OH \\ | \\ CH_2OH \end{array}} \xrightarrow[(-H_2O)]{[O]} \underset{\text{Glycollic aldehyde}}{\begin{array}{l} CHO \\ | \\ CH_2OH \end{array}} \xrightarrow{[O]} \underset{\text{Glycollic acid}}{\begin{array}{l} COOH \\ | \\ CH_2OH \end{array}}$$

$$\text{Glycollic aldehyde} \xrightarrow{[O]\ (-H_2O)} \underset{\text{Glyoxal}}{\begin{array}{l} CHO \\ | \\ CHO \end{array}} \xrightarrow{[O]} \underset{\text{Glyoxalic acid}}{\begin{array}{l} COOH \\ | \\ CHO \end{array}} \xrightarrow{[O]} \underset{\text{Oxalic acid}}{\begin{array}{l} COOH \\ | \\ COOH \end{array}}$$

$$\text{Glycollic acid} \xrightarrow{[O]\ (-H_2O)} \text{Glyoxalic acid}$$

(*ii*) *When treated with periodic acid* (HIO_4) *or lead tetra acetate* [$Pb(OOCH_3)_4$], compounds containing two (or more) –OH groups attached to adjacent carbon atoms undergo oxidative cleavage. The reaction breaks carbon-carbon bonds and yields carbonyl compounds (aldehydes, ketones, etc) as oxidation products.

$$\begin{array}{c} | \\ -C-OH \\ | \\ -C-OH \\ | \end{array} \xrightarrow[\text{or } Pb(OOCCH_3)_4]{HIO_4} 2\ \begin{array}{c} | \\ -C=O \end{array}$$

For example,

$$\underset{\text{Propane-1, 2-diol}}{\begin{array}{c} H \\ | \\ H-C-OH \\ | \\ H-C-OH \\ | \\ CH_3 \end{array}} + HIO_4 \longrightarrow \begin{array}{c} H \\ | \\ H-C-OH \\ + \\ H-C-OH \\ | \\ CH_3 \end{array} + HIO_3 + H_2O$$

$$\begin{array}{c} R \\ | \\ R-C-OH \\ | \\ H-C-OH \\ | \\ R' \end{array} + Pb(OOCCH_3)_4 \longrightarrow \begin{array}{c} R \\ | \\ R-C=O \\ + \\ H-C=O \\ | \\ R' \end{array} \begin{array}{l} + 2CH_3COOH \\ + Pb(OOCCH_3)_2 \end{array}$$

It may be noted that for *every C–C bond broken, a C–O bond is formed at each carbon.*

It may be added that in these reactions *periodic acid is found to work well in* **aqueous solutions** *while lead tetraacetate gives good results in* **organic solvents.**

Mechanism

(*a*) **Oxidation with periodic acid.** Oxidative cleavage with periodic acid is believed to take place through a cyclic intermediate formed by the attack of periodic ion (IO_4^-) on glycol as shown below:

$$\begin{array}{c} | \\ -C-OH \\ | \\ -C-OH \\ | \end{array} + IO_4^- \longrightarrow \begin{array}{c} | \\ -C-O \\ | \\ -C-O \\ | \end{array}\!\!>\!I(O^-)(=O)_2 \longrightarrow \begin{array}{c} | \\ -C=O \\ \\ -C=O \\ | \end{array} + IO_3^-$$

(*b*) **Oxidation with lead tetraacetate.** This reaciton is also believed to occur through a cyclic intermediate which undergoes an oxidation-reduction process leading to the cleavage of carbon-carbon bond.

$$\begin{array}{c} | \\ -C-OH \\ | \\ -C-OH \\ | \end{array} + \begin{array}{c} CH_3-\overset{O}{\overset{\|}{C}}-O \\ CH_3-\underset{O}{\underset{\|}{C}}-O \end{array}\!\!>Pb(OOCCH_3)_2 \xrightarrow{-2CH_3COOH} \begin{array}{c} | \\ -C-O \\ | \\ -C-O \\ | \end{array}\!\!>Pb(OOCCH_3) \longrightarrow \begin{array}{c} -C=O \\ + \\ -C=O \end{array} + Pb(OOCCH_3)$$

(7) Dehydration. When treated with dilute sulphuric acid or anhydrous zinc chloride under pressure in a sealed tube, ethylene glycol gets dehydrated to form acetaldehyde (formed through arrangement of vinyl alcohol).

$$\underset{\text{Ethylene glycol}}{\begin{array}{c} C \\ | \\ H-C-OH \\ | \\ HO-C-H \\ | \\ H \end{array}} \xrightarrow[-H_2O]{H^+} \underset{\substack{\text{Vinyl alcohol} \\ (\textit{Unstable})}}{\begin{array}{c} CH_2 \\ \| \\ CHOH \end{array}} \xrightarrow{\text{Rearrangement}} \underset{\text{Acetaldehyde}}{\begin{array}{c} CH_3 \\ | \\ CHO \end{array}}$$

But when distilled with conc. H_2SO_4, glycol forms dioxane as shown below.

$$\begin{array}{c} \boxed{H}O-CH_2-CH_2-\boxed{OH} \\ \boxed{HO}-CH_2-CH_2-O\boxed{H} \end{array} \xrightarrow[-2H_2O]{\text{Conc. } H_2SO_4} \underset{\text{Dioxane}}{O\begin{array}{c} CH_2-CH_2 \\ \\ CH_2-CH_2 \end{array}O}$$

(8) Reaction with aldehydes and ketones. Glycol undergoes condensation with aldehydes and ketones in the presence of *p*-toluene sulphonic acid (commonly called PTS) to form cyclic acetals and ketals respectively. For example:

$$\underset{\text{Aldehyde}}{R-\overset{H}{\overset{|}{C}}=O} + \begin{matrix} HO-CH_2 \\ | \\ HO-CH_2 \end{matrix} \longrightarrow \underset{\text{Cyclic acetal}}{R-CH\begin{matrix} OCH_2 \\ | \\ OCH_2 \end{matrix}}$$

$$\underset{\text{Ketone}}{\begin{matrix} R \\ R \end{matrix}\!\!>C=O} + \begin{matrix} HO-CH_2 \\ | \\ HO-CH_2 \end{matrix} \longrightarrow \underset{\text{Cyclic ketal}}{\begin{matrix} R \\ R \end{matrix}\!\!>C\begin{matrix} OCH_2 \\ | \\ OCH_2 \end{matrix}}$$

Q. 30 What is pinacol-pinacolone rearrangement? Give its mechanism.

(*Sardar Patel 2003; Sukhadia 2003; Delhi 2003; Madras, 2004; Kalyani, 2004; Nagpur 2005; Purvanchal 2007*)

Ans. It is found that on treatment with sulphuric acid, pinacol undergoes dehydration with molecular rearrangement to form methyl *tert*-butyl ketone also known as **pinacolone.**

$$\underset{\text{Pinacol}}{CH_3-\overset{CH_3}{\underset{OH}{C}}-\overset{CH_3}{\underset{OH}{C}}-CH_3} \xrightarrow[-H_2O]{H^+} \underset{\text{Methyl tert butyl ketone or Pinacolone}}{CH_3-\underset{O}{\overset{\|}{C}}-\overset{CH_3}{\underset{CH_3}{C}}-CH_3}$$

This reaction is known as pinacol-pinacolone rearrangement.

Many other vicinal glycols are now known to undergo similar acid-catalysed reactions. All these reactions are, therefore, collectively known as pinacol rearrangements.

Mechanism. The pinacol rearrangement takes place through a mechanism as depicted below.

Step 1. Protonation of pinacol takes place.

Step 2. Protonated pinacol loses a water molecule to form carbonium ion.

Step 3. 1, 2 alkyl shift takes place forming a more stable carbonium ion.

Step 4. The carbonium ion loses a hydrogen ion to form the final product. These steps are illustrated as follows:

$$CH_3-\overset{CH_3}{\underset{:\ddot{O}H}{C}}-\overset{CH_3}{\underset{:\ddot{O}H}{C}}-CH_3 \underset{\text{Step 1}}{\overset{H^+}{\rightleftharpoons}} CH_3-\overset{CH_3}{\underset{:\ddot{O}H}{C}}-\overset{CH_3}{\underset{\overset{+}{:O}H_2}{C}}-CH_3 \xrightarrow[\text{Step 2}]{-H_2O} CH_3-\overset{CH_3}{\underset{:\ddot{O}H}{C}}-\overset{CH_3}{\overset{+}{C}}-CH_3$$

$$\xrightarrow{\text{Step 3}} CH_3-\underset{^+\ddot{O}-H}{\overset{\|}{C}}-\overset{CH_3}{\underset{CH_3}{C}}-CH_3 \underset{\text{Step 4}}{\overset{-H^+}{\rightleftharpoons}} CH_3-\underset{O}{\overset{\|}{C}}-\overset{CH_3}{\underset{CH_3}{C}}-CH_3$$

It may be emphasised that the migrating group (*i.e.* methyl group in the above example) is not completely detached from the carbon it is leaving till it has attached itself to the electron-deficient carbon as shown below.

$$-\underset{:\ddot{O}H}{\overset{R}{C}}-\overset{+}{C}- \longrightarrow \left[-\underset{:\ddot{O}H}{\overset{\delta^+}{C}} \overset{\overset{\delta^+}{R}}{\frown} \overset{\delta^+}{C}- \right] \longrightarrow -\underset{^+\ddot{O}H}{\overset{\|}{C}}-\overset{R}{C}-$$

This would naturally mean that the migrating group in pinacol rearrangement will retain its configuration i.e. the new bond formed will have the same relative position as that held previously be the old bond.

Q. 31. List some uses of ethylene glycol.

Ans. (*i*) It is used as an antifreeze for automobile radiators and as a coolant for aeroplane motors under the trade name Prestone.

(*ii*) It is used in the manufacture of Dacron, a well-known synthetic fibre.

(*iii*) It finds use as solvent for stamp inks.

(*iv*) It is used as starting material for the preparation of nitroglycol (an explosive) and solvents such as cellosolve and diglyme.

TRIHYDRIC ALCOHOLS

Q. 32. Describe the preparation of glycerol from fats. (*Garhwal, 2000; Vidyasagar, 2007*)

Ans. Glycerol occurs in many natural fats and oils in the form of glyceryl esters of long chain fatty acids. Hydrolysis of these fats or oils with sodium hydroxide gives sodium salts of fatty acids and glycerol.

$$\underset{\text{Fat or oil}}{\begin{array}{l} CH_2O.OCR \\ | \\ CHO.OCR' + 3NaOH \\ | \\ CH_2O.OCR'' \end{array}} \longrightarrow \underset{\text{Glycerol}}{\begin{array}{l} CH_2OH \\ | \\ CHOH \\ | \\ CH_2OH \end{array}} + \underset{\text{Sodium salts of fatty acids or soap}}{\begin{array}{l} RCOO^-Na^+ \\ R'COO^-Na^+ \\ R''COO^-Na^+ \end{array}}$$

The process of hydrolysis of fats and oils is also known as *saponification*.

Glycerol can also be obtained from fats or oils by transesterification with methyl alcohol in the presence of an acid or a base. The reaction leads to the formation of methyl esters of acids and glycerol.

$$\underset{\text{Fat or oil}}{\begin{array}{l} CH_2O.OCR \\ | \\ CHO.OCR' + 3CH_3OH \\ | \\ CH_2O.OCR'' \end{array}} \longrightarrow \underset{\text{Glycerol}}{\begin{array}{l} CH_2OH \\ | \\ CHOH \\ | \\ CH_2OH \end{array}} + \underset{\text{Methyl esters of fatty acids}}{\begin{array}{l} RCOOCH_3 \\ R'COO^+CH_3 \\ R''COO^+CH_3 \end{array}}$$

Q. 33. How is glycerol obtained from petroleum products?

Or

Work out the following conversion:

Propene ⟶ Glycerol

(*Delhi, 2002; Kashmir, 2003, North Eastern Hill, 2004, Bangalore, 2004; Nagpur, 2008*)

Ans. Glycerol is prepared synthetically from propylene, which is obtained by the cracking of petroleum. The various steps taking place in the process are:

$$\underset{\text{Propylene or propene}}{CH_2 = CH - CH_3} + Cl_2 \xrightarrow{773\ K} \underset{\text{Allyl chloride (3-chloro-propene-1)}}{CH_2 = CH - CH_2Cl} + HCl$$

The allyl chloride is converted into allyl alcohol by hydrolysing with aqueous sodium carbonate at 423 K and at a pressure of 12 atmospheres.

$$\underset{\text{Allyl chloride}}{CH_2 = CH - CH_2Cl} \xrightarrow[\text{423 K 12 atm}]{\text{aq. } Na_2CO_3} \underset{\text{Allyl alcohol}}{CH_2 = CH - CH_2OH}$$

Allyl alcohol is then treated with hypochlorous acid (HOCl) to give rise to chlorohydrin from which glycerol is obtained by treatment with sodium hydroxide.

$$CH_2 = CH - CH_2OH \xrightarrow{HOCl} \underset{\text{Glycerol } \beta\text{-mono chlorohydrin}}{\underset{OH}{CH_2} - \underset{Cl}{CH} - \underset{OH}{CH_2}}$$

$$\xrightarrow{\text{Aq. NaOH}} \underset{\text{Glycerol}}{\underset{OH}{CH_2} - \underset{OH}{CH} - \underset{OH}{CH_2}} + NaCl$$

Q. 34. Describe complete synthesis of glycerol.

(*Baroda 2003; Patna 2003, Hyderabad, 2004*)

Ans. Complete synthesis of glycerol starting from the elements is described as under.

$$2C + H_2 \xrightarrow{\text{Electric arc}} \underset{\text{Acetylene}}{C_2H_2} \xrightarrow[H_2SO_4/HgSO_4,\ 333\ K]{H_2O} \underset{\text{Acetaldehyde}}{CH_3CHO}$$

$$\xrightarrow{\text{Oxidation}} \underset{\text{Acetic acid}}{CH_3COOH} \xrightarrow[\text{of calcium salt}]{\text{Distillation}} \underset{\text{Acetone}}{CH_3COCH_3} \xrightarrow{\text{Reduction}}$$

$$\longrightarrow \underset{\text{Isopropyl alcohol}}{CH_3CHOH.CH_3} \xrightarrow[\text{dehydration}]{H_2SO_4} \underset{\text{Propylene}}{CH_3.CH = CH_2}$$

$$\xrightarrow[773\ K]{Cl_2} \underset{\text{Allyl chloride}}{Cl.CH_2.CH = CH_2} \xrightarrow{\text{dil. NaOH}} \underset{\text{Allyl alcohol}}{HOCH_2 - CH = CH_2}$$

$$\xrightarrow{Cl_2 / H_2O} \underset{\text{Glycerol } \beta\text{-mono chlorohydrin}}{\underset{OH}{CH_2} - \underset{Cl}{CH} - \underset{OH}{CH_2}} \xrightarrow{\text{Aq. NaOH}} \underset{\text{Glycerol}}{\underset{OH}{CH_2} - \underset{OH}{CH} - \underset{OH}{CH_2}}$$

Q. 35. Describe the chemical properties of glycerol.

(Delhi, 2002; Bangalore, 2004; Nagpur, 2008)

Ans. Important reactions of glycerol are described as under:

(1) Reaction with sodium. When treated with metals like sodium at room temperature, one of the two primary alcoholic groups reacts forming monosodium glycerolate. At higher temperature, the second primary alcoholic group also reacts giving disodium glycerolate but the secondary alcoholic group is not affected.

$$\underset{\text{Glycerol}}{\begin{matrix} CH_2OH \\ | \\ CHOH \\ | \\ CH_2OH \end{matrix}} \xrightarrow[H_2]{\text{Na, room temp.}} \underset{\text{Mono sodium glycerolate}}{\begin{matrix} CH_2ONa \\ | \\ CHOH \\ | \\ CH_2OH \end{matrix}} \xrightarrow[H_2]{\text{Na, high temp.}} \underset{\text{Disodium glycerolate}}{\begin{matrix} CH_2ONa \\ | \\ CHOH \\ | \\ CH_2ONa \end{matrix}}$$

(2) Reaction with HCl. When treated with equimolar quantities of hydrogen chloride at 383K, glycerol gives a mixture of α- and β-monochlorohydrins. If excess of hydrogen chloride is used, however, then a mixture of dichlorohydrins is formed.

$$\begin{matrix} CH_2OH \\ | \\ CHOH \\ | \\ CH_2OH \end{matrix} \xrightarrow[-H_2O]{\text{HCl, 383 K}} \underset{\alpha\text{-Glycerol monochlorohydrin}}{\begin{matrix} CH_2Cl \\ | \\ CHOH \\ | \\ CH_2OH \end{matrix}} + \underset{\beta\text{-Glycerol monochlorohydrin}}{\begin{matrix} CH_2OH \\ | \\ CHCl \\ | \\ CH_2OH \end{matrix}}$$

$$\begin{matrix} CH_2OH \\ | \\ CHOH \\ | \\ CH_2OH \end{matrix} \xrightarrow[-2H_2O]{\text{Excess of HCl}} \underset{\alpha,\beta\text{-Glycerol dichlorohydrin}}{\begin{matrix} CH_2Cl \\ | \\ CHCl \\ | \\ CH_2OH \end{matrix}} + \underset{\alpha,\alpha'\text{-Glycerol dichlorohydrin}}{\begin{matrix} CH_2Cl \\ | \\ CHOH \\ | \\ CH_2Cl \end{matrix}}$$

(3) Reaction with HI. Different products are formed under different conditions as depicted below.

(*i*) When heated with a *small* amount of hydrogen iodide, glycerol produces mainly allyl iodide.

$$\underset{\text{Glycerol}}{\begin{matrix} CH_2OH \\ | \\ CHOH \\ | \\ CH_2OH \end{matrix}} \xrightarrow[-3H_2O]{+3HI} \underset{\textit{Unstable}}{\begin{matrix} CH_2I \\ | \\ CHI \\ | \\ CH_2I \end{matrix}} \xrightarrow{-I_2} \underset{\text{Allyl iodide}}{\begin{matrix} CH_2 \\ \| \\ CH \\ | \\ CH_2I \end{matrix}}$$

(*ii*) When a large amount of hydrogen iodide is used, the main product is isopropyl iodide, which, in fact, results by the subsequent action of hydrogen iodide on the allyl iodide first formed.

$$\underset{\text{Allyl iodide}}{\begin{matrix} CH_2 \\ \| \\ CH \\ | \\ CH_2I \end{matrix}} \xrightarrow{HI} \begin{matrix} CH_3 \\ | \\ CHI \\ | \\ CH_2I \end{matrix} \longrightarrow \begin{matrix} CH_3 \\ | \\ CH \\ \| \\ CH_2 \end{matrix} \xrightarrow{HI} \underset{\text{Isopropyl iodide}}{\begin{matrix} CH_3 \\ | \\ CHI \\ | \\ CH_3 \end{matrix}}$$

(4) Reaction with monocarboxylic organic acids. When heated with organic acids like acetic acid, glycerol forms mono-, di- and tri-esters depending upon the amount of acid and reaction conditons employed.

$$\begin{array}{ccc} CH_2O.OCCH_3 & CH_2O.OCCH_3 & CH_2O.OCCH_3 \\ | & | & | \\ CHOH & CHO.OCCH_3 & CHO.OCCH_3 \\ | & | & | \\ CH_2OH & CH_2OH & CH_2O.OCCH_3 \\ \text{Glycerol monoacetate} & \text{Glycerol diacetate} & \text{Glycerol triacetate} \end{array}$$

(5) Reaction with oxalic acid.

(*i*) When heated with oxalic acid at 383 K, it forms glycerol mono-oxalate. This loses CO_2 to give glycerol monoformate which undergoes hydrolysis to yield formic acid.

$$\begin{array}{l} CH_2OH \\ | \\ CHOH + \underset{\text{Oxalic acid}}{HOOC-COOH} \\ | \\ CH_2OH \\ \text{Glycerol} \end{array} \xrightarrow[-H_2O]{383\text{ K}} \begin{array}{l} CH_2OOC-COOH \\ | \\ CHOH \\ | \\ CH_2OH \\ \text{Glycerol monooxalate} \end{array} \xrightarrow[-CO_2]{\text{Heat}}$$

$$\begin{array}{l} CH_2OOCH \\ | \\ CHOH \\ | \\ CH_2OH \\ \text{Glycerol monoformate} \end{array} \xrightarrow{H_2O} \begin{array}{l} CH_2OH \\ | \\ CHOH \\ | \\ CH_2OH \end{array} + \underset{\text{Formic acid}}{HCOOH}$$

(*ii*) At 503 K, oxalic acid and glycerol form glyceryl dioxalate which decomposes to give allyl alcohol and carbon dioxide.

$$\begin{array}{l} CH_2OH \\ | \\ CHOH \\ | \\ CH_2OH \\ \text{Glycerol} \end{array} + \begin{array}{l} HOOC \\ | \\ HOOC \\ \text{Oxalic acid} \end{array} \xrightarrow[-2H_2O]{503\text{ K}} \begin{array}{l} CH_2OOC \\ | \qquad | \\ CH\ OOC \\ | \\ CH_2OH \\ \text{Glyceryl dioxalate} \end{array} \xrightarrow[-2CO_2]{\text{Heat}} \begin{array}{l} CH_2 \\ \| \\ CH \\ | \\ CH_2OH \\ \text{Allyl alcohol} \end{array}$$

(6) Oxidation. Glycerol gives a variety of products on oxidation with different oxidising agents. With a mild oxidising agents such as hydrogen peroxide in the presence of ferrous sulphate (Fenton's reagent) it gives a mixture of glyceraldehyde and dihydroxy acetone. With stronger oxidising agents it is ultimately converted into carbon dioxide and water as shown below.

$$\begin{array}{l} CH_2OH \\ | \\ CHOH \\ | \\ CH_2OH \end{array} \xrightarrow{[O]} \begin{cases} \begin{array}{l} CH_2OH \\ | \\ CHOH \\ | \\ CHO \\ \text{Glyceraldehyde} \end{array} \xrightarrow{[O]} \begin{array}{l} CH_2OH \\ | \\ CHOH \\ | \\ COOH \\ \text{Glyceric acid} \end{array} \xrightarrow{[O]} \begin{array}{l} COOH \\ | \\ CHOH \\ | \\ COOH \\ \text{Tartronic acid} \end{array} \xrightarrow{[O]} \text{(to Mesoxalic acetone)} \\ \begin{array}{l} CH_2OH \\ | \\ C=O \\ | \\ CH_2OH \\ \text{Dihydroxy-acetone} \end{array} \xrightarrow{[O]} \begin{array}{l} COOH \\ | \\ C=O \\ | \\ COOH \\ \text{Mesoxalic acetone} \end{array} \xrightarrow{[O]} \begin{array}{l} COOH \\ | \\ COOH \\ \text{Oxalic acid} \end{array} \xrightarrow{[O]} CO_2 + H_2O \end{cases}$$

(7) Reaction with nitric acid. When added slowly to a mixture of concentrated nitric acid and concentrated sulphuric acid maintained at 283–298 K, glycerol forms glyceryl trinitrate which is commonly known as nitroglycerine.

$$\underset{\text{Glycerol}}{\begin{array}{l} CH_2OH \\ | \\ CHOH \\ | \\ CH_2OH \end{array}} + 3HONO_2 \longrightarrow \underset{\substack{\text{Glycerol trinitrate} \\ \text{(Nitroglycerine)}}}{\begin{array}{l} CH_2O.NO_2 \\ | \\ CHO.NO_2 \\ | \\ CH_2O.NO_2 \end{array}} + 3H_2O$$

It is a powerful explosive discovered by Alfred Nobel.

(8) Dehydration. When heated with potassium hydrogen sulphate, glycerol undergoes dehydration to yield acrolein.

$$\underset{\text{Glycerol}}{\begin{array}{c} H \\ | \\ H-C-OH \\ | \\ HO-C-H \\ | \\ H-C-OH \\ | \\ H \end{array}} \xrightarrow[2H_2O]{KHSO_4,\ Heat} \underset{\substack{\textit{Unstable} \\ \text{enol}}}{\begin{array}{c} CH_2 \\ \| \\ C \\ \| \\ CHOH \end{array}} \longrightarrow \underset{\text{Acrolein}}{\begin{array}{c} CH_2 \\ \| \\ CH \\ | \\ CHO \end{array}}$$

Q. 36. List important uses of glycerol.

Ans. (*i*) As an antifreeze in automobile radiators.

(*ii*) As a sweetening agent in confectionary and beverages.

(*iii*) In the manufacture of glyptal–a polyester of glycerol and pthalic acid–used as an alkyl resin.

(*iv*) In making medicines like cough syrup, lotions, etc.

(*v*) As a preservative for fruits and other eatables which require to be kept moist. This is because glycerol is sufficiently hygroscopic to absorb moisture from air.

(*vi*) In making non-drying printing inks, stamp colours, etc.

(*vii*) In the preparation of good quality toilet soaps and cosmetics.

(*viii*) In the manufacture of nitroglycerine–one of the best known explosives.

Q. 37. How can you obtain Nobel's oil and 2-propenol from glycerol?

Ans. (*i*) Nobel's oil

$$\underset{\text{Glycerol}}{\begin{array}{l} CH_2OH \\ | \\ CHOH \\ | \\ CH_2OH \end{array}} + \underset{\text{Nitric acid}}{3HONO_2} \longrightarrow \underset{\substack{\text{Glycerol trinitrate} \\ \text{(Nobel's oil)}}}{\begin{array}{l} CH_2ONO_2 \\ | \\ CHONO_2 \\ | \\ CH_2ONO_2 \end{array}} + 3H_2O$$

(*ii*) **2-Propenol**

$$\begin{array}{l} CH_2OH \\ | \\ CHOH \\ | \\ CH_2OH \end{array} + \begin{array}{l} HOOC \\ \quad | \\ HOOC \end{array} \xrightarrow[-2H_2O]{503\ K} \underset{\substack{\text{Glycerol} \\ \text{oxalate}}}{\begin{array}{l} CH_2OOC \\ | \qquad\ | \\ CH\ OOC \\ | \\ CH_2OH \end{array}} \xrightarrow[Heat]{-2CO_2} \underset{\substack{\text{2-Propenol} \\ \text{(Allyl alcohol)}}}{\begin{array}{l} CH_2 \\ \| \\ CH \\ | \\ CH_2OH \end{array}}$$

Q. 38. What different structures are represented by the molecular formula $C_4H_{10}O$? Which of these structures can give a positive iodoform test? Is any of these structures capable of showing stereoisomerism?

Ans. Different structures are:

(*i*) $CH_3CH_2CH_2CH_2OH$ (*v*) $CH_3CH_2OCH_2CH_3$

(*ii*) $CH_3CH_2CHOHCH_3$ (*vi*) $CH_3OCH_2CH_2CH_3$

(*iii*) $CH_3-\underset{\underset{CH_3}{|}}{CH}-CH_2OH$ (*vii*) $CH_3O-\underset{\underset{CH_3}{|}}{CH}-CH_3$

(*iv*) $CH_3-\overset{\overset{CH_3}{|}}{\underset{\underset{CH_3}{|}}{C}}-CH_3$

Structure (*ii*) gives positive iodoform test

Structure (*ii*) exhibits optical isomerism.

EPOXIDES

Q. 39. What are epoxides? Give the preparation and properties of ethylene oxide

(*Purvanchal, 2007; Nagpur, 2008*)

Ans. Epoxides are cyclic ethers with three-membered rings. They are called oxiranes according to IUPAC nomenclature. Some examples of epoxides are given below

IUPAC name : oxirane
Common name : ethylene oxide

IUPAC name : 2-methyl oxirane
Common name : Propylene oxide

Preparation of ethylene oxide

1 By distilling ethylene chlorohydrin with potassium hydroxide solution

$$\underset{\text{Etylene chlorohydrin}}{CH_2OH-CH_2Cl+KOH} \longrightarrow \underset{CH_2-CH_2}{\overset{O}{\wedge}} + H_2O + KCl$$

2 It is manufactured by passing ethylene and oxygen under pressure over a silver catalyst at 250°C

$$CH_2=CH_2+\tfrac{1}{2}O_2 \xrightarrow[250°C]{Ag} \underset{CH_2-CH_2}{\overset{O}{\wedge}}$$

Properties of ethylene oxide

1 Ethylene oxide on heating undergoes rearrangement and forms acetaldehyde

$$\underset{CH_2-CH_2}{\overset{O}{\wedge}} \xrightarrow{\Delta} CH_3CHO$$

2 Ring opening reactions in acidic medium. Ethylene oxide reacts with H_2O, HCl, HCN etc. in acidic medium to form addition products by cleavage of the oxide ring.

$$\underset{\text{(O bridging)}}{CH_2 - CH_2} + HOH \xrightarrow{H^+} \underset{\text{Ethylene glycol}}{CH_2OH - CH_2OH}$$

$$\underset{\text{(O bridging)}}{CH_2 - CH_2} + HCl \xrightarrow{H^+} \underset{\text{Ethylene chlorohydrin}}{CH_2OHCH_2Cl}$$

$$\underset{\text{(O bridging)}}{CH_2 - CH_2} + HCN \xrightarrow{H^+} \underset{\text{Ethylene cyanohydrin}}{CH_2OHCH_2CN}$$

3 Ring opening in basic mediun. Ethylene glycol reacts with ethyl alcohol and ammonia in basic medium or in the presence of basic catalysts to form addition products

$$\underset{\text{(O bridging)}}{CH_2 - CH_2} + C_2H_5OH \longrightarrow \underset{\text{Glycol monoethyl ether}}{CH_2OHCH_2OC_2H_5}$$

$$\underset{\text{(O bridging)}}{CH_2 - CH_2} + NH_3 \longrightarrow \underset{\text{Ethanolamine}}{CH_2NH_2CH_2OH}$$

4 Reaction with Grignard reagents. It forms addition products with Grignard reagents, which upon hydrolysis give primary alcohols.

$$\underset{\text{(O bridging)}}{CH_2 - CH_2} + CH_3MgI \longrightarrow CH_3CH_2CH_2OMgI \xrightarrow{H_2O} \underset{\text{Propanol}}{CH_3CH_2CH_2OH} + Mg(OH)I$$

5 Reaction with $LiAlH_4$.

$$\underset{\text{(O bridging)}}{CH_2 - CH_2} + 2[H] \xrightarrow{LiAlH_4} \underset{\text{Ethanol}}{CH_3CH_2OH}$$

MISCELLANLOUS QUESTIONS

Q. 40. How will you prepare allyl alcohol from propene? *(Kerala, 2000)*

Ans. See Q.33

Q. 41. Complete the following reaction *(Himachal, 2000)*

$$C_2H_5OH + I_2 \xrightarrow{NaOH}$$

Ans. $CH_3CH_2OH + 4I_2 + 6NaOH \longrightarrow \underset{\text{iodoform}}{CHI_3} + HCOONa + 5NaI$

Q. 42. How is allyl alcohol obtained from glycerrol? How does it reacts with HBr? *(Kerala, 2001)*

Ans. Preparation : See Q. 35, (5) (*ii*)

Reaction with HBr :

$$\begin{array}{c} CH_2 \\ \| \\ CH \\ | \\ CH_2OH \end{array} + HBr \longrightarrow \begin{array}{c} CH_3 \\ | \\ CHBr \\ | \\ CH_2OH \end{array}$$

Q. 43. How will you prepare *t*-butyl alcohol using Grignard's reagent? *(Kurukshetra, 2001)*

Ans. See Q. 4 (*iii*)

Q. 44. How would you prepare glycerol in the laboratory? Explain reactions of glycerol with : (*a*) PI_3 (excess) (*b*) $HONO_2$ (*c*) $KHSO_4$ (*d*) Ac_2O

Ans. (*a*) See Q. 35, (3) (*ii*) (*b*) See Q. 35 (7) (*c*) See Q.35 (8) (*d*) See Q.35 (4)

Q. 45. How glycerol is recovered from spentlye (soap industry)? How does glycerol reacts with following? *(A.N. Bhauguna, 2000)*

(*i*) oxalic acid (*ii*) acetic acid (*iii*) hydrogen iodide. *(Garhwal, 2000)*

Ans. See Q. 32 and Q. 35.

Q. 46. Discuss hydroboration and oxymercuration methods for the synthesis of alcohols. *(Garhwal, 2000, A.N. Bhauguna, 2000)*

Ans. See Q. 4, (4)

Q. 47. Which alcohol will form alkene on heating with Cu at 300°C?

(*i*) CH_3CH_2OH (*ii*) $(CH_3)_3COH$ (*iii*) $(CH_3)_2CHOH$ (*iv*) $C_6H_5-\underset{\underset{OH}{|}}{CH}-CH_3$

(Kanpur, 2001)

Ans. (*ii*) $CH_3-\underset{\underset{CH_3}{|}}{\overset{\overset{CH_3}{|}}{C}}-OH \xrightarrow[300°C]{Cu} \underset{\text{2-methylbutene}}{CH_3-\overset{\overset{CH_3}{|}}{C}=CH_2} + H_2O$

Q. 48. How will you distinguish between following pairs of the compounds:

(*i*) CH_3OH and C_2H_5OH

(ii) $CH_3CH_2CH_2OH$ and $CH_3CHOHCH_3$

Ans. (*i*) Ethanol gives the iodoform test while methanol does not.

$$CH_3CH_2OH + 4I_2 + 6NaOH \longrightarrow \underset{\text{Iodoform}}{CHI_3} + HCOONa + 5NaI$$

(*ii*) $CH_3CHOHCH_3$ gives the iodoform test while $CH_3CH_2CH_2OH$ does not.

Q. 49. How will you do the following conversions: (*M. Dayanand, 2000*)

(*i*) Propylene $\longrightarrow$ 2 Propanal

(ii) Ethanol $\longrightarrow$ Ethyl bromide

Ans. (*i*) $CH_3CH=CH_2 \xrightarrow[\text{dil } H_2SO_4]{H_2O} CH_3CH(OH)CH_3$

Propylene — 2-propanol

(*ii*) $CH_3CH_2OH \xrightarrow{PBr_3} CH_3CH_2Br$

ethanol — ethyl bromide

Q. 50. Why 3°-alcohols are easily dehydrated as compared to 2° or 1° alcohols ? (*Delhi 2002*)

Ans. The intermediate carbocation formed in case of 3° alcohol is more stable than that formed from 2° or 1° alcohol.

Q. 51. How will you obtain the following from glycerol? Give conditions and equations. (*Delhi 2002*)

(*i*) α, α-dichloroacetone (*ii*) HCOOH (*iii*) Nitroglycerine

Ans. See Q. 35

16

PHENOLS

Q. 1. What are phenols? Give their classification and nomenclature.

Ans. Phenols are organic compounds containing at least one –OH group attached directly to benzene ring.

Classification and Nomenclature

Depending upon the number of hydroxyl groups attached to benzene ring, phenols can be classified as monohydric, dihydric and trihydric phenols.

Monohydric phenols. The simplest member of the series is hydroxy benzene commonly known as phenol, while others are named as substituted phenols. The three isomeric hydroxy toluenes are known as cresols.

OH

Phenol or Hydroxybenzene or Carbolic acid

OH, CH_3 (*o*–) OH, CH_3 (*m*–) OH, CH_3 (*p*–)

Cresols or Hydroxy toluenes

OH, NO_2 OH, NH_2 OH, Cl etc.

o-Nitrophenol *p*-Aminophenol *o*-Chlorophenol

Dihydric phenols. The three isomeric dihydroxy benzenes are better known by their common names as given ahead :

OH, OH OH, OH OH, OH

Catechol or *o*-Dihydroxybenzene

Resorcinol or *m*-Dihydroxybenzene

Quinol or Hydroquinone or *p*-Dihydroxybenzene

Trihydric phenols. Trihydroxy benzenes are known by their common names as given below :

1, 2, 3-Trihydroxybenzene or Pyrogallol

1, 2, 4-Trihydroxybenzene or Hydoxy quinol

1, 3, 5-Trihydroxybenzene or Phloroglucinol

Q. 2. Describe common methods of preparation of monohydric phenols.

Ans. The commonly employed methods of preparation of phenols are described as under :

1. Hydrolysis of diazonium salts. In this method, the reaction is usually carried out by slowly adding diazonium salt solution to excess of boiling dilute sulphuric acid.

$$ArN_2{}^+X^- + H_2O \xrightarrow{H^+,\ heat} Ar - OH + HX + N_2$$

For example :

$$C_6H_5-N_2{}^+Cl^- + H_2O \xrightarrow{H^+,\ heat} C_6H_5-OH + HCl + H_2$$

Benzene diazonium chloride

Since hydrolysis is carried out under very mild conditions, other groups in the molecule of diazonium salt are not affected.

$$m\text{-}NO_2C_6H_4N_2{}^+Cl^- + H_2O \xrightarrow{H^+,\ heat} m\text{-}NO_2C_6H_4OH + HCl + N_2$$

m-Nitrobenzene diazonium chloride → *m*-Nitrophenol

2. Fusion of sulphonates with alkali. In this method, phenols are obtained by fusion of sodium salts of aromatic sulphonic acids with sodium hydroxide, followed by acidification.

$$\underset{\text{Sodium sulphonate}}{ArSO_3Na} + 2NaOH \xrightarrow[-Na_2SO_3,\ -H_2O]{Fuse} AR - ONa \xrightarrow{HCl} \underset{\text{Phenol}}{Ar - OH} + NaCl$$

For example :

$$\underset{\text{Sodium } p\text{-toluene sulphonate}}{H_3C-C_6H_4-SO_3Na} + 2NaOH \xrightarrow[-H_2O]{-Na_2SO_3,} H_3C-C_6H_4-ONa$$

$$H_3C-C_6H_4-ONa \xrightarrow{+HCl} NaCl + \underset{p\text{-cresol}}{H_3C-C_6H_4-OH}$$

3. From cumene hydroperoxide. A recently developed process involves the preparation of phenol from cumene, *i.e.*, isopropyl benzene. Cumene itself is obtained by a Friedel-Craft type reaction between benzene and propylene in the presence of ferric chloride.

$$C_6H_6 + \underset{\text{Propylene}}{CH_3-CH=CH_2} \xrightarrow{FeCl_3} \underset{\text{Cumene}}{C_6H_5-CH(CH_3)_2}$$

Cumene is readily oxidised by air to form cumene hydroperoxide which is converted into phenol and acetone by treatment with an aqueous acid.

$$\underset{\text{Cumene}}{C_6H_5-C(CH_3)_2-H} \xrightarrow[333\text{–}353\ K]{Air} \underset{\text{Cumene hydroperoxide}}{C_6H_5-C(CH_3)_2-OOH} \xrightarrow{H^+,\ H_2O} \underset{\text{Phenol}}{C_6H_5OH} + \underset{\text{Acetone}}{CH_3COCH_3}$$

The mechanism of conversion of cumene hydroperoxide into phenol involves a sequence of five steps as shown below :

$$H_3C-C(C_6H_5)(CH_3)-O-\ddot{O}-H + H^+ \rightleftharpoons H_3C-C(C_6H_5)(CH_3)-O-\overset{+}{\ddot{O}}H_2 \xrightarrow{-H_2O} H_3C-C(C_6H_5)(CH_3)-O^+ \longrightarrow$$

$$\left[H_3C-\overset{+}{C}(H_3C)-O-C_6H_5 \longleftrightarrow H_3C-C(H_3C)=\overset{+}{O}-C_6H_5\right] \xrightarrow{H-\ddot{O}-H} H_3C-C(\overset{+}{O}H_2)(CH_3)-O-C_6H_5 \underset{}{\overset{-H^+}{\rightleftharpoons}} \underset{\text{Hemiacetal}}{H_3C-C(OH)(CH_3)-O-C_6H_5}$$

$$\xrightarrow{H^+} H_3C-\overset{O}{\overset{\|}{C}}-CH_3 + HO-C_6H_5$$

4. From aryl halides. Substituted phenols having electron withdrawing groups like $-NO_2$ group in *ortho-para* positions to the –OH group can be easily prepared from the corresponding aryl halides by treatment with aqueous sodium hydroxide. For example :

$$\underset{p\text{-Nitrochlorobenzene}}{ClC_6H_4NO_2} + 2NaOH \longrightarrow ONaC_6H_4NO_2 + NaCl + H_2O$$

$$ONaC_6H_4NO_2 \xrightarrow{+HCl} \underset{p\text{-Nitrophenol}}{OHC_6H_4NO_2} + NaCl$$

5. From Grignard reagents. Phenols may be prepared by treating aryl magnesium bromides with oxygen followed by hydrolysis of the product.

$$2ArMgBr \xrightarrow{O_2} 2ArOMgBr \xrightarrow{H_2O / H^+} 2ArOH + Mg\begin{matrix}OH\\Br\end{matrix}$$

For example :

$$\underset{\text{Phenyl magnesium bromide}}{C_6H_5MgBr} \xrightarrow{[O]} C_6H_5OMgBr \xrightarrow{H_2O/H^+} \underset{\text{Phenol}}{C_6H_5OH} + Mg\begin{matrix}OH\\Br\end{matrix}$$

6. From phenolic acids. Distillation of sodium salts of phenolic acids with soda lime yields phenols. For example :

$$\underset{\text{Sodium salicylate}}{C_6H_4(OH)COONa} + NaOH \xrightarrow{CaO} C_6H_5OH + Na_2CO_3$$

Q. 3. Why has phenol higher boiling point than toluene?

Ans. Phenol (molecular mass 94) boils at 455 K while toluene (molecular mass 92) boils at 383 K only. The reason for higher boiling point of phenol as compared to toluene is that phenol forms intermolecular hydrogen bonds leading to association of its molecule. Consequently, its molecular mass increases many fold and hence its boiling point is high. Additional energy is required to break the hydrogen bonds which raises its boiling point. Hydrogen bonding in phenol is depicted below :

···O—H ···O—H ···O—H··· O—H···

Hydrogen bonding

On the other hand, toluene does not contain any strong electronegative atom attached to hydrogen. Therefore, it does not form intermolecular hydrogen bonding. Hence, its boiling point is lower than that of phenol.

Q. 4. Why is phenol more soluble in water than toluene?

Ans. Phenol is somewhat soluble in water as phenol molecules can form hydrogen bonds with water molecules as shown below :

···H—O···H—O··· H—O···H—O ···H—O···

H H

Intermolecular hydrogen bonding

On the other hand, toluene is not capable of forming hydrogen bonds with water molecules, hence toluene is not soluble in water. It can be said, in general, that a compound is soluble in water if it can form hydrogen bond with it.

Q. 5. Account for the lower boiling point and decreased water solubility of *o*-nitrophenol as compared with their *m*- and *p*-isomers. (*Meerut, 2000*)

Ans. *o*-Nitrophenol forms *intramolecular* hydrogen bonding resulting in the formation of a six membered ring.

O—H

N = O

↓

O

o-Nitrophenol
Intramolecular hydrogen bonding

Due to intramolecular hydrogen bonding, *o*-nitrophenol is incapable of forming hydrogen bonding with other *o*-nitrophenol molecules, *i.e.*, intermolecular hydrogen bonding. Hence, it does not exist as associated molecule. Moreover, the molecular size shrinks as a consequence of intramolecular bonding. Therefore, it boils at a lower temperature.

Similarly, due to intramolecular hydrogen bonding, *o*-nitrophenol does not form hydrogen bonds with water molecules also. Therefore, it is not very soluble in water.

m-nitrophenol as well as *p*-nitrophenol form intermolecular hydrogen bonding resulting in association of molecules. Consequently, they have high boiling points.

Intermolecular hydrogen bonding of *p*-nitrophenol

Moreover, *m*- and *p*-nitrophenols can also form hydrogen bonds with water molecules. Consequently, they are more soluble in water than *o*-nitrophenol.

Q. 6. Comment on the statement "Phenol is acidic while ethyl alcohol is not".

(Delhi, 2003; Manipur, 2003; Ranchi, 2004; West Bengal, 2004; Karnataka, 2004; Andhra, 2010

Ans. Phenol is slightly acidic as its dissociation constant (K_a) is 1.3×10^{-10} . The acidic character of phenol is indicated by the following reactions :

(*i*) Phenol turns blue litmus red.

(*ii*) Phenol produces hydrogen with sodium.

(*iii*) Phenol neutralises sodium hydroxide to form sodium phenolate and water as:

$$C_6H_5OH + NaOH \longrightarrow C_6H_5\overset{-}{O}\overset{+}{Na} + H_2O$$

Phenol (*acid*) + NaOH (*Base*) → Sodium phenolate (*Salt*) + H_2O (*Water*)

On the other hand, ethyl alcohol ($K_a \sim 10^{-16}$) is almost neutral as it does not turn blue litmus red nor does it neutralise sodium hydroxide.

Reason for acidic character of phenol.

In water, phenol ionises as :

$$C_6H_5\ddot{O}-H + H_2O \rightleftharpoons C_6H_5\ddot{O}:^{-} + H_3O^+$$

Phenol → Phenoxide ion

Now phenol as well as phenoxide ion cannot be represented by a single valence bond formula, rather both are resonance hybrids of various contributing structures as given below :

I ⟷ II ⟷ III ⟷ IV ⟷ V

Resonating structures of phenol

It is evident from the contributing structures III, IV and V that oxygen atom acquires partial positive charge due to resonance. As a result, bond between O–H becomes weak which helps the release of proton and makes phenol acidic.

After the release of proton, the phenoxide ion is formed which is also resonance hybrid of various contributing structures (VI–X).

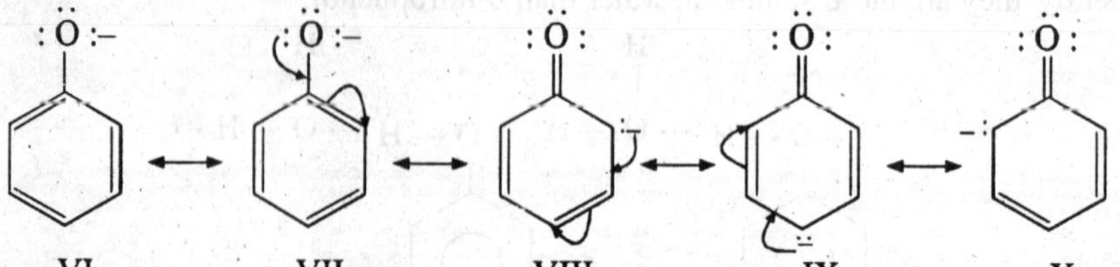

VI ⟷ VII ⟷ VIII ⟷ IX ⟷ X

Thus, phenol as well as phenoxide ion are stabilised by resonance.

However, phenoxide ion is more stabilized by resonance than phenol. This is because in phenol contributing structures III to V involve separation of charge and are therefore less stable than structures I and II. On the other hand, in phenoxide ions none of the contributing structures VI to X involve charge separation and hence all the contributing structures in phenoxide ion are almost equally stable. In short,

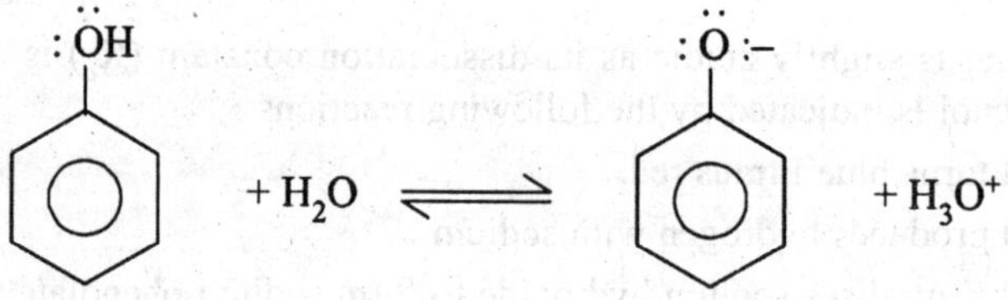

$$\text{Phenol} + H_2O \rightleftharpoons \text{Phenoxide ion} + H_3O^+$$

Phenol (*Less resonance stabilised*)   Phenoxide ion (*More resonance stabilised*)

The greater resonance stabilisation of phenoxide ion as compared to phenol causes the above reaction to take place in the forward direction, *i.e.*, increases acidic character of phenol.

On the other hand, ethyl alcohol is neutral as neither ethyl alcohol molecule nor the ethoxide ion produced from it has resonating structures. In other words, neither ethyl alcohol nor ethoxide ions is resonance stabilised.

$$C_2H_5-\ddot{\underset{\cdot\cdot}{O}}-H + H_2O \rightleftharpoons C_2H_5-\ddot{\underset{\cdot\cdot}{O}}:^- + H_3O^+$$

Ethyl alcohol (*No resonance stabilization*)   Ethoxide ion (*No resonance stabilization*)

Q. 7. Discuss the effect of substituents on the acidity of phenolic group.

Ans. It is observed that the acidic strength of a phenol shows a marked change when another group is introduced in the benzene ring containing phenolic group. The following factors determine the change in the acidic strength of phenol.

1. Nature of the substituent group.
2. Position of the substituent group.

These two factors are discussed separately as under :

1. Nature of substituent in the benzene ring. Electron-withdrawing substituents such as halogens, nitro group, carboxylic group increase the acidic character of phenol. This is because electron-withdrawing groups help in delocalization of the negative charge of phenoxide ion and thus help to stabilize the phenoxide ion and thereby increase the acidic character of phenol. For example, nitrophenols, are more acidic than phenol.

$$G\text{-}C_6H_4\text{-}OH + H_2O \rightleftharpoons G\text{-}C_6H_4\text{-}O^- + H_3O^+$$

G = *Electron-withdrawing group such as* $-NO_2$, $-CN$, $-COOH$, $-Cl$, $-CHO$

Electron-withdrawing group helps delocalisation of charge and stabilises the ion.

On the other hand, electron releasing groups such as $-CH_3$, $-OH$, $-OR$ tend to destabilise the phenoxide ion by adding to its negative charge and thereby reduce the acidic character of phenol. For example, cresols are less acidic than phenol.

$$G\text{-}C_6H_4\text{-}OH + H_2O \rightleftharpoons G\text{-}C_6H_4\text{-}\bar{O} + H_3O^+$$

G = *Electron-releasing groups such as* $-CH_3$, $-OH$, $-OR$, $-NR_2$

Electron-releasing group intensifies the negative charge and destabilizes the ion.

2. Position of substituent relative to hydroxyl group. Not only the nature of substituent, but also the position of the substituent relative to hydroxyl group has influence on the acidic character of phenol. The effect of electron-withdrawing or electron-releasing substituent is felt more at *ortho-* or *para-*positions than *meta-*position. Moreover, the effect at *para-*position is less than that at *ortho-*position due to greater distance from phenolic group in *para* isomer.

Q. 8. Introduction of nitro group in aromatic nucleus increases the acidity of phenols but the introduction of $-CH_3$ group into the nucleus decreases the acidity of phenols. Explain.

Ans. Nitro group increases the acidity of phenol. For example *o*-nitrophenol is a stronger acid than phenol. This is because of the reason that nitro group is able to delocalise the negative charge on phenoxide ion and thus able to stabilize it as illustrated below :

$$o\text{-}O_2N\text{-}C_6H_4\text{-}O\text{-}H \xrightarrow{-H^+} \left[o\text{-}O_2N\text{-}C_6H_4\text{-}\bar{O} \longrightarrow O{=}C_6H_4{=}N(O)\bar{O} \right]$$

For this reason, nitro group increase the acidity of phenol. On the other hand a methyl group, which is electron repelling further intensifies negative charge on the phenoxide ion, making it unstable.

$$o\text{-}CH_3\text{-}C_6H_4\text{-}O\text{-}H \xrightarrow{-H^+} o\text{-}CH_3\text{-}C_6H_4\text{-}\bar{O}$$

Thus *o*-methyl phenol is a weaker acid than phenol.

Q. 9. 2, 4, 6-Trinitrophenol gives effervescence with sodium bicarbonate but phenol does not. Explain.

Or

2, 4, 6-Trinitrophenol is a very strong acid. Explain.

Ans. Nitro group is an electron withdrawing group. Such groups delocalise the negative charge on the phenoxide ion and thus stabilize it. This increases the acidity of the compound. In 2, 4, 6-trinitrophenol, there are three such electron withdrawing groups. Their combined effect is so strong that hydrogen of the phenol is knocked out as proton in high concentration. Thus, 2, 4, 6-trinitrophenol is a very strong acid and gives effervescence with $NaHCO_3$. Only strong acids like mineral acids give effervescence with sodium bicarbonate. In phenol (C_6H_5OH), the concentration of H^+ is not sufficient to react with sod. bicarbonate.

O—H, O_2N, NO_2, NO_2 $\xrightarrow{-H^+}$ O^-, O_2N, NO_2, NO_2

(Highly stabilized by resonance)

Q. 10. Arrange the following in decreasing order of acidity : *p*-nitrophenol, *p*-cresol, *m*-nitrophenol, 2, 4-dinitrophenol.

Ans. Electron-withdrawing groups increase the acidity of phenols and the effect is more pronounced in *ortho* and *para* positions compared to the *meta* positions.

On the other hand, electron-donating groups tend to decrease the acidity of the phenol.

OH, NO_2 — *p*-Nitrophenol; OH, CH_3 — *p*-Cresol; OH, NO_2 — *m*-Nitrophenol; OH, NO_2, NO_2 — 2, 4-Dinitrophenol

In the light of above generalisation, 2, 4-dinitrophenol, which has two electron-withdrawing groups in *ortho* and *para* positions is expected to have maximum acidic character.

Out of *p*- and *m*-nitrophenol, the former is expected to be more acidic because the electron-withdrawing group is in *para* position, *p*-cresol which has an electron donating effect will be a weaker acid than phenol. Thus, the decreasing order of acidity is

2, 4-dinitrophenol > *p*-nitrophenol > *m*-nitrophenol > *p*-cresol.

Q. 11. Explain the order of acidic strength of various nitrophenols (pK_a values are given) : *o*-nitrophenol (2.17), *m*-nitrophenol (3.45), *p*-nitrophenol (3.43). *(Delhi 2007)*

Ans. For an acid HX,

$$HX \rightleftharpoons H^+ + X^-$$

$$K_a = \frac{[H^+][X^-]}{[HX]}$$

$$pK_a = -\log K_a$$

Thus, lower the value of pK_a, greater is the value of K_a and thus greater the [H^+] concentration or greater acidity. The order of acidic strength in the light of pK_a values can be written as.

m-Nitrophenol (OH, NO_2) < p-Nitrophenol (OH, NO_2) < o-Nitrophenol (OH, NO_2)

Nitro group has the effect of withdrawing electrons towards itself through resoance. This facilitates, removal of H^+ and thus increases the acidity. It is found that electron-withdrawing effect works more in *p*-position than in *m*-position. Thus *p*-nitrophenol is more acidic than *m*-nitrophenol. Maximum acidity in the case of *o*-nitrophenol is explained by *ortho-effect.*

Q. 12. Describe the following reactions of phenols :

(*i*) Williamson synthesis

(*ii*) Ester formation

Ans. (i) **Williamson synthesis. Ether formation.** On reaction with alkyl halides or dialkyl sulphates in alkaline medium, phenols are converted into ethers. The alkyl halide or sulphate undergoes nucleophilic substitution by the phenoxide ion present in alkaline solution of phenol. Thus we have :

$$ArOH + NaOH \longrightarrow ArO^- Na^+ + H_2O$$

$$ArONa + RX \longrightarrow Ar-O-R + NaX$$

For example :

$$C_6H_5-ONa + C_2H_5I \longrightarrow C_6H_5-O-C_2H_5 + NaI$$

Sod. phenoxide

Ethyl phenyl ether or Phenetole

This reaction constitutes an important method of synthesising ethers, under the name **Williamson's synthesis,** from phenols (or alcohols).

(ii) **Ester formation.** Like alcohols, phenols are converted into esters by reaction with carboxylic acids, acid chlorides or anhydrides. Acid chlorides or anhydrides give a better yield than the acids.

$$ArOH + R-\overset{O}{\overset{\|}{C}}-Cl \longrightarrow Ar-O-\overset{O}{\overset{\|}{C}}-R + HCl$$

For example :

$$C_6H_5OH + C_2H_5-\overset{O}{\overset{\|}{C}}-Cl \longrightarrow C_6H_5-O-\overset{O}{\overset{\|}{C}}-C_2H_5 + HCl$$

Phenol

Phenyl propionate

Phenol + $(CH_3CO)_2O$ $\xrightarrow{\text{Pyridine}}$ Phenyl acetate ($C_6H_5O{-}CO{-}CH_3$) + CH_3COOH

Q. 13. Give the mechanism of Fries rearrangement.

(*Kerala, 2000; Himachal, 2000; Panjab, 2003; Nagpur 2005; Purvanchal 2007*)

Ans. If an ester of a phenol is heated with aluminium chloride, the acyl group migrates from the phenolic oxygen to the *ortho* or *para* position on the ring. This is called **Fries rearrangement.** For example :

Phenyl propionate ($C_6H_5O{-}CO{-}CH_2CH_3$) $\xrightarrow[413-423\text{ K}]{AlCl_3}$ *p*-Hydroxyphenyl ethyl ketone (About 50%) ($HO{-}C_6H_4{-}CO{-}CH_2CH_3$) + *o*-Hydroxyphenyl ethyl ketone (About 35%) ($HO{-}C_6H_4{-}CO{-}CH_2{-}CH_3$)

Al low temperature, *para* product predominates while at high temperature *ortho* isomer is the main product.

Mechanism. Fries rearrangement probably involves the formation of an acylium ion (RCO^+) which then attacks the aromatic ring as in Friedel-Crafts acylation. The complete mechanism may be summed up as follows :

$C_6H_5{-}\ddot{O}COR \xrightleftharpoons{AlCl_3} C_6H_5{-}\overset{+}{O}(\overset{-}{A}lCl_3)COR \rightleftharpoons C_6H_5{-}\ddot{O}AlCl_3 + RCO^+ \rightleftharpoons$ (cyclohexadienone intermediate with $\overset{+}{O}AlCl_3$, H and COR at ortho carbon)

$\xrightarrow{-H^+Cl^-}$ *o*-$(COR)C_6H_4{-}\ddot{O}AlCl_3 \xrightarrow{H_2O}$ *o*-$(COR)C_6H_4{-}OH$

Q. 14. How does phenol react with the following:

(*i*) Zinc dust **(*ii*) Ammonia**

Ans. (*i*) **Reaction with zinc dust.** When distilled with zinc dust, phenols form corresponding hydrocarbons. For example:

$$C_6H_5OH + Zn \xrightarrow{\text{Heat}} C_6H_6 + ZnO$$

Phenol Benzene

(*ii*) **Reaction with ammonia.** When heated with ammonia at high temperature and pressure in the presence of catalysts like anhydrous zinc chloride, –OH group of phenols is replaced by $-NH_2$ group. For example :

$$C_6H_5OH + NH_3 \longrightarrow C_6H_5NH_2 + H_2O$$

Q. 15. Discuss the electrophilic substitution reactions of phenols. (*Nagpur, 2005*)

Ans. The important ring substitutions are described below :

1. Halogenation. On treatment with halogens, phenols are very rapidly halogenated forming polyhalogen derivatives. For example, reaction of phenol with an aqueous solution of bromine immediately produces tribromophenol.

$$C_6H_5OH + \underset{aq.}{3Br_2} \longrightarrow C_6H_2Br_3OH + HBr$$

Phenol 2, 4, 6 Tribromophenol

Halogenation can, however, be limited to monohalogenation stage by performing it at low temperature and in non-polar solvents such as carbon tetrachloride or carbon disulphide. For example :

$$C_6H_5OH + Br_2 \xrightarrow[273\ K]{CS_2} p\text{-}BrC_6H_4OH + o\text{-}BrC_6H_4OH + HBr$$

p-Bromophenol *o*-Bromophenol

2. Nitration. Nitration of phenols gives poor yields of the expected nitro compounds. This is because phenols are easily oxidised and nitric acid can bring about not only nitration but also oxidation of the starting phenol.

$$C_6H_5OH + HNO_3 \text{ (dil.)} \xrightarrow{290\ K} o\text{-}O_2NC_6H_4OH + p\text{-}O_2NC_6H_4OH + H_2O$$

o-Nitrophenol (40%) *p*-Nitrophenol (13%)

Use of conc. nitric acid in the above reaction leads to the formation of 2, 4, 6-trinitrophenol or picric acid in low yields.

$$C_6H_5OH + 3HNO_3 \text{ (Conc.)} \longrightarrow 2,4,6\text{-}(O_2N)_3C_6H_2OH + 3H_2O$$

Picric acid

3. Sulphonation. Sulphonation of phenols gives either the *ortho* or the *para* isomer as the main product depending upon the temperature of the reaction. In general, low temperature favours the formation of *ortho* isomer while high temperature favours the *para* isomer. For example :

$$p\text{-}HO_3SC_6H_4OH \xleftarrow[-H_2O]{H_2SO_4,\ 373\ K} C_6H_5OH \xrightarrow[-H_2O]{H_2SO_4,\ 288\ K} o\text{-}HO_3SC_6H_4OH$$

p-Phenol sulphonic acid (*Main product*) Phenol *o*-Phenol sulphonic acid (*Main product*)

4. Friedel-Craft's alkylation and acylation. Phenols undergo Friedel-Craft's alkylation as well as acylation forming predominantly *para* isomers.

$$C_6H_5OH + (CH_3)_3C\text{—}Cl \xrightarrow{HF} p\text{-}(CH_3)_3C\text{—}C_6H_4OH + HCl$$

Tert-butyl chloride *p*-Tert. butyl phenol

5. Coupling with diazonium salts. When treated with diazonium salts in weakly alkaline solution, phenols undergo coupling to form hydroxyazo compounds. For example :

$C_6H_5-N_2^+Cl^- + C_6H_5-OH \xrightarrow{\text{Weakly alkaline}} C_6H_5-N=N-C_6H_4-OH + HCl$

Benzenediazonium chloride

p-Hydroxyazobenzene

Q. 16. Give the mechanism of the following reaction:
Phenol ⟶ carboxylic acid

(Panjab 2003; Lucknow 2004; Nagpur 2008)

Or

Give the mechanism of Kolbe's reaction. *(Calcutta 2007)*

Ans. Kolbe's reaction. When heated with carbon dioxide under pressure, sodium salts of phenols lead to the formation of salts of phenolic acids; the ortho isomers being the main product. For example, when sodium phenoxide is heated with carbon dioxide under pressure, sodium salicylate is obtained as the major product.

$C_6H_5OH + CO_2 \xrightarrow{\text{398 K, 4-7 atom.}} o\text{-}HOC_6H_4COONa \xrightarrow{H^+} o\text{-}HOC_6H_4COOH$

Sodium phenoxide — Sodium salicylate (*Main product*) — Salicylic acid

Mechanism. The mechanism of Kolbe's reaction is believed to involve the electrophilic attack by carbon dioxide on the phenoxide ion as shown below :

$C_6H_5O^- + O=C=O \longrightarrow$ (cyclohexadienone intermediate with H and COO^- at the ortho carbon) $\longrightarrow o\text{-}HOC_6H_4COO^-$

Q. 17. What is Reimer-Tlemann reaction? Give its mechanism.

(Punjab, 2003, Delhi, 2003; Purvanchal, 2007 Nagpur, 2008; Andhra, 2010)

Or

Explain the reaction

$C_6H_5-OH + CHCl_3 \xrightarrow{NaOH}$ *(Nagpur, 2005)*

Ans. Reimer-Tiemann reaction. On heating with chloroform and an alkali, phenols are converted into phenolic aldehydes; an aldehyde group is introduced in the aromatic ring generally in *ortho*-position to the –OH group. For example, when phenol is heated with chloroform and aqueous sodium hydroxide, salicylaldehyde is obtained as the main product along with a small amount of the *para* isomer.

$$C_6H_5OH + CHCl_3 + 3NaOH \xrightarrow{333-343\ K} o\text{-}HOC_6H_4CHO + 3NaCl + 2H_2O$$

Salicylaldehyde
(*Main product*)

Mechanism. The mechanism of reaction consists of the electrophilic substitution of phenoxide ion formed in solution by the action of alkali on phenol. The attacking electrophilic reagent is the electron deficient dichloromethylene; : CCl_2 a carbene generated by the action of alkali on chloroform. Dichloromethylene is electrically neutral but it acts as an electrophilic reagent as the carbon atom present in it is surrounded by only six valence electrons. The complete mechanism may be summed up as follows :

$$CHCl_3 + \bar{O}H \rightleftharpoons \bar{C}Cl_3 + H_2O$$
$$\bar{C}Cl_3 \longrightarrow :CCl_2 + Cl^-$$

$$C_6H_5O^- + :CCl_2 \longrightarrow [\text{cyclohexadienone intermediate with H and } \overset{\ominus}{C}Cl_2] \longrightarrow o\text{-}(CHCl_2)C_6H_4O^- \xrightarrow{OH^-} o\text{-}(CHO)C_6H_4O^- \xrightarrow{H^+} o\text{-}HOC_6H_4CHO$$

Salicylaldehyde

If we take CCl_4 in place of $CHCl_3$, salicylic acid is formed.

$$C_6H_5OH + CCl_4 + 4NaOH \longrightarrow o\text{-}HOC_6H_4COOH + 4NaCl + 2H_2O$$

Salicylic acid

Q. 18. What happens when phenol is treated with CCl_4 and alkali?

Ans. This reaction is a type of Reimer-Tiemann reaction with the difference that in place of usual chloroform, carbon tetrachloride has been taken in this case.

It may be seen that the product contains salicylic acid in place of salicylaldehyde. It follows the same mechanism as in Q. 16.

Q. 19. How does phenol react with the following :

(*i*) Formaldehyde (with mechanism) (Lederer - Manasse reaction)

(*ii*) Phthalic anhydride.

Ans. (*i*) Reaction with formaldehyde. When phenol is treated with formaldehyde in the presence of alkali or acid, a resinous polymer, called bakelite, is obtained. The process involves the interlinking of many phenol rings by – CH_2 – groups in *ortho* and *para* positions with respect to the phenolic group as depicted below :

Phenol $\xrightarrow[H^+ \text{ or } OH^-]{HCHO}$ *o*-Hydroxymethyl phenol $\xrightarrow{C_6H_5OH}$ $HOC_6H_4-CH_2-C_6H_4OH$ $\xrightarrow{HCHO,\ C_6H_5OH}$ (cross-linked polymer: $-CH_2-$ bridges at *ortho* and *para* positions of the phenol rings)

o-Hydroxymethyl phenol

Mechanism. The mechanism of the first stage of reaction, *i.e.* formation of *o*-hydroxymethylphenol, involves the electrophilic substitution on the aromatic ring. The reaction may take place between phenol and protonated formaldehyde (*acid catalysis*) or between phenoxide ion and formaldehyde (*base catalysis*) as shown below :

$$H_2C{=}O + H^+ \rightleftharpoons \left[H_2C{=}\overset{+}{O}H \longleftrightarrow H_2\overset{+}{C}-OH \right]$$

Protonated formaldehyde

$$C_6H_5\ddot{O}H + \overset{+}{C}H_2OH \longrightarrow [\text{intermediate: } \overset{+}{O}H{=}C_6H_5(H)(CH_2OH)] \longrightarrow o\text{-}HOC_6H_4CH_2OH$$

Acid catalysis

$$C_6H_5\ddot{O}:^- + H_2C{=}O \longrightarrow [\text{intermediate: } O{=}C_6H_5(H)(CH_2O^-)] \longrightarrow o\text{-}{}^-OC_6H_4CH_2OH$$

Base catalysis

(*ii*) **Phthalein reaction.** On heating with phthalic anhydride in the presence of concentrated sulphuric acid or anhydrous zinc chloride, phenols are converted into **phthaleins.** For example, phenol forms phenolphthalein as shown below :

$$2\,C_6H_5OH + C_6H_4(CO)_2O \xrightarrow{H_2SO_4} \text{Phenolphthalein} + H_2O$$

Phthalic anhydride Phenolphthalein

Q. 20. Why does phenol turn pink on exposure to air and sunlight?

Ans. Phenol turns pink on exposure to air and light due to slow oxidation. It is believed that first of all phenol gets oxidised by air to quinone as :

$$C_6H_5-OH + O_2 \longrightarrow O{=}C_6H_4{=}O + H_2O$$

Phenol Quinone

Then quinone combines with phenol through hydrogen bonding to form a brilliant red addition product known as **phenoquinone.**

$$C_6H_5-OH + O{=}C_6H_4{=}O + HO-C_6H_5$$

$$\downarrow$$

$$C_6H_5-O-H\cdots O{=}C_6H_4{=}O\cdots H-O-C_6H_5$$

Phenoquinone
(Brilliant red compound)

Q. 21. Explain Claisen rearrangement giving its mechanism.

(*Coimbatore, 2000; Kerala, 2001; Purvanchal 2007*)

Ans. Claisen rearrangement. This rearrangement reaction involves aryl allyl ethers. *The allyl group of the ether migrates from ether oxygen to the ring carbon at* **ortho position** at 475 K. However if both the ortho positions are already occupied, the allyl group migrates to *para* position.

$$C_6H_5-O-CH_2-CH=CH_2 \xrightarrow{475\text{ K}} o\text{-}HO-C_6H_4-CH_2-CH=CH_2$$

Allylphenyl ether *o*-Allylphenol

$O-CH_2-CH=CH_2$ (Allyl 2, 6-dimethylphenyl ether) $\xrightarrow{475\ K}$ 4-Allyl-2, 6-dimethylphenol

Allyl 2, 6-dimethylphenyl ether

4-Allyl-2, 6-dimethylphenol

Mechanism of ortho isomerisation

Ortho isomerisation involves a concerted mechanism in which the cleavage of allyl-oxygen bond and formation of allyl-carbon bond at the *ortho* position take place simultaneously. The cyclohexadienone intermediate thus formed undergoes a tautomeric change to give the o-allylphenol in which the ring regains the aromatic character as shown below:

Slow → (cyclohexadienone with H and $CH_2-CH=CH_2$) ⇌ (Fast, Tautomerisation) o-allylphenol ($CH_2-CH=CH_2$)

It may be noted that it is the γ-carbon of the allyl group (with respect to oxygen) which attaches itself to the ring carbon. This is supported by the fact that *ortho migration* involves an **inversion** in the position of substituents in the allyl group with respect to that of the starting compound. The following reaction illustrates this.

$OCH_2-CH=CH-CH_3$ $\xrightarrow{Heat}$ OH, $-CH(CH_3)-CH=CH_2$

Mechanism of para isomerisation

The *para isomerisation* involves two stages. In the first stage, *ortho migration* takes place and the cyclohexadienone intermediate is formed. But this intermediate cannot undergo tautomerisation since there is no hydrogen at *ortho* position. As such this intermediate undergoes another migration of allyl group to yield the final product.

(2,6-disubstituted allyl phenyl ether, R, R) → (cyclohexadienone, R, $CH_2-CH=CH_2$) → (R, R, O, H, $CH_2-CH=CH_2$)

⇅ Tautomerisation

$$\text{(2,6-dialkyl-4-allylphenol: } C_6H_2(OH)(R)_2 - CH_2 - CH = CH_2\text{)}$$

It may be noted that due to two migrations in para-isomerisation there is **no net inversion** in the position of the substituents with respect to the starting compound. For example :

$$R_2C_6H_3 - O - CH_2 - CH = CH - CH_3 \longrightarrow R_2C_6H_2(OH) - CH_2 - CH = CH - CH_3$$

Q. 22. Describe industrial methods for the preparation of phenol (carbolic acid).

Ans. 1. Dow's process. When chlorobenzene is heated with 10 per cent sodium hydroxide solution at 600–625 K and under a pressure of 200 atmospheres in the presence of a copper salt (catalyst), phenol is produced.

$$\underset{\text{Chlorobenzene}}{C_6H_5Cl} + NaOH \xrightarrow[\text{200 atm.}]{600-625\text{ K}} \underset{\text{Phenol}}{C_6H_5OH} + NaCl$$

2. Raschig's process. A recent method consists in preparing chlorobenzene by passing a mixture of benzene vapours, hydrogen chloride and air over copper chloride (catalyst) heated to 500K.

$$2\underset{\text{Benzene}}{C_6H_6} + 2HCl + \underset{\text{(From air)}}{O_2} \xrightarrow[CuCl_2]{500\text{ K}} 2\underset{\text{Chlorobenzene}}{C_6H_5Cl} + 2H_2O$$

Chlorobenzene formed is hydrolysed with steam at 700 K in the presence of silica catalyst.

$$\underset{\text{Chlorobenzene}}{C_6H_5Cl} + H.OH \xrightarrow[SiO_2]{700\text{ K}} \underset{\text{Phenol}}{C_6H_5OH} + HCl$$

Q. 23. Write a note on Houben-Hoesch reaction. *(Nagpur 2008)*

Ans. Highly reactive polyhydric phenols in which –OH groups are located at *meta* to each other, may be acylated when treated with alkyl cyanide in the presence of zinc chloride and hydrogen chloride. Phenolic ketone is obtained.

$$\text{HO}-C_6H_4-\text{OH} + RCN + H_2O \xrightarrow[\text{Ether}]{ZnCl_2/HCl} \text{HO}-C_6H_3(\text{OH})-COR + NH_3$$

Resorcinol → Phenolic ketone

For example

$$\text{HO}-C_6H_4-\text{OH} + CH_3CN \longrightarrow \text{HO}-C_6H_3(\text{OH})-C(=NH)CH_3 \xrightarrow{H_2O} \text{HO}-C_6H_3(\text{OH})-COCH_3 + NH_3$$

ketimine (intermediate) → Resorinol acatophenone

Phenol does not respond to this reaction

Q. 24. How can you prepare *o*-, *m*- or *p*-cresols?

Ans. 1. Cresols are generally prepared by the following methods :

$$C_6H_4\begin{matrix} CH_3 \\ NH_2 \end{matrix} \xrightarrow[273-278\ K]{NaNO_2,\ HCl} C_6H_4\begin{matrix} CH_3 \\ N_2^+Cl^- \end{matrix} \xrightarrow{H_2O,\ Heat} C_6H_4\begin{matrix} CH_3 \\ OH^+ \end{matrix}$$

o-, *m*- or *p*-toluidine → *o*-, *m*- or *p*- Toluene diazonium chloride → *o*-, *m*- or *p*-Cresol

2. From toluenesulphonic acid. Cresols are obtained by the fusion of sodium salts of corresponding toluenesulphonic acids with sodium hydroxide :

$$C_6H_4\begin{matrix} CH_3 \\ SO_3Na \end{matrix} \xrightarrow[\text{Fuse}]{NaOH} C_6H_4\begin{matrix} CH_3 \\ ONa \end{matrix} \xrightarrow{H^+} C_6H_4\begin{matrix} CH_3 \\ OH \end{matrix}$$

o-, *m*- or *p*- Sodium toluenesulphonate → *o*-, *m*- or *p*- Sodium salts of cresol → *o*-, *m*- or *p*-Cresol

3. From chlorotoluenes. When chlorotoluenes are heated with sodium hydroxide under pressure at 575 K, they give the corresponding cresols.

$$C_6H_4\begin{matrix} CH_3 \\ Cl \end{matrix} \xrightarrow[575\ K]{NaOH} C_6H_4\begin{matrix} CH_3 \\ ONa \end{matrix} \xrightarrow{H^+} C_6H_4\begin{matrix} CH_3 \\ OH \end{matrix}$$

o-, *m*- or *p*- Chlorotoluene → *o*-, *m*- or *p*- Sodium salts of cresol → *o*-, *m*- or *p*-Cresol

It may be mentioned that cresols give the usual reactions of phenols in addition to the reactions due to methyl group.

Q. 25. Describe methods of preparation of catechol. *(Kerala 2001)*

Ans. 1. From *o*-chlorophenol. When heated with 20% aqueous sodium hydroxide at 463 K under pressure in the presence of traces of $CuSO_4$, *o*-chlorophenol forms catechol.

$$\text{o-Chlorophenol} \xrightarrow[463\ K]{NaOH} \text{Catechol} + NaCl$$

o-Chlorophenol Catechol

2. From salicylaldehyde. Salicylaldehyde on treatment with alkaline H_2O_2 yields catechol **(Dakin's reaction).**

$$\text{Salicylaldehyde} + H_2O_2 + NaOH \xrightarrow[-HCOONa,\ -H_2O]{} \text{Catechol}$$

Salicylaldehyde Catechol

3. From *o*-phenolsulphonic acid. On fusion with sodium hydroxide, sodium salt of *o*-phenolsulphonic acid gives catechol.

$$\text{Sodium salt of o-phenolsulphonic acid } (SO_3Na) + NaOH \xrightarrow{\text{Fuse}} \text{Catechol} + Na_2SO_3$$

Sodium salt of o-phenolsulphonic acid Catechol

Q. 26. Give two important properties of catechol.

Ans. (*i*) Being readily oxidisable, it acts as a good reducing agent. It reduces silver nitrate solution in the cold and Fehling solution on warming. With silver oxide, it gets oxidised to *o*-benzoquinone.

$$\text{Catechol} \xrightarrow{Ag_2O} \text{o-Benzoquinone}$$

o-Benzoquinone

(*ii*) It condenses with phthalic anhydride in the presence of concentrated sulphuric acid to form alizarin.

$$\text{Phthalic anhydride} + \text{Catechol} \xrightarrow[H_2O]{H_2SO_4} \text{Alizarin}$$

Alizarin

Q. 27. Give the preparation and properties of resorcinol.

Ans. Preparation. It is generally prepared by fusing sodium *m*-benzene disulphonate with sodium hydroxide.

SO_3Na, SO_3Na —NaOH, Fuse→ ONa, ONa —HCl→ OH, OH

Sodium *m*-benzene disulphonate — Sodium salt or Resorcinol — Resorcinol

(*i*) Unlike catechol and quinol, it is only a weak reducing, agent. It reduces ammoniacal silver nitrate solution and Fehling solution on warming.

(*ii*) On treatment with bromine water, it gives a precipitate of 2, 4, 6-tribromoresorcinol.

OH, OH —Br_2 water→ OH, Br, Br, OH, Br

2, 4, 6-Tribromoresorcinol

(*iii*) With nitrous acid, it forms dinitrosoresorcinol which is used as a dye under the name *Fast Green O.*

OH, OH —HNO_2→ OH, NO, OH, NO ⇌ O, NOH, O, NOH

Fast green O

(*iv*) It reacts with nitric acid in the presence of sulphuric acid to form 2, 4, 6-trinitroresorcinol commonly called *styphnic acid*

OH, OH —HNO_3/H_2SO_4→ OH, O_2N, NO_2, OH, NO_2

2, 4, 6-Trinitroresorcinol

(*v*) It shows keto-enol tautomerism which is clear from the fact that it forms a dioxime and a bisulphite compound.

Dienol form ⇌ Diketo form

(*vi*) It condenses with phthalic anhydride in the presence of conc. H_2SO_4 or anhydrous $ZnCl_2$ to form fluorescein.

Conc. H_2SO_4

Fluorescein

Q. 28. Describe methods of preparation of quinol.

Ans. Preparation

1. From aniline. Aniline on oxidation with MnO_2 and conc. H_2SO_4 forms *p*-benzoquinone which on reduction with iron and steam forms quinol.

MnO_2/H_2SO_4 ; Fe/H_2O

p-Benzoquinone ; Quinol

2. From *p*-benzoquinone. Reaction of *p*-benzoquinone with Fe and steam or sulphurous acid gives quinol.

p-Benzoquinone $+ H_2SO_4 + H_2O \longrightarrow$ Quinol $+ H_2SO_4$

3. From *p*-aminophenol. *p*-Aminophenol on diazotisation followed by steam distillation yields quinol.

$$\text{p-}HOC_6H_4NH_2 \xrightarrow[273-298\ K]{NaNO_2/HCl} \text{p-}HOC_6H_4N_2Cl \xrightarrow[H_2O]{Warm} \text{p-}HOC_6H_4OH$$

Q. 29. Give the chemical properties of quinol.

Ans. Chemical Properties. (1) It gives *p*-benzoquinone with $FeCl_3$.

$$\text{Quinol} \xrightarrow{FeCl_3} \text{p-Benzoquinone}$$

(2) It is easily oxidised by Tollen's reagent and Fehling solution. Therefore, it acts as a strong reducing agent.

$$\text{Quinol} \xrightarrow[Ag_2O \text{ or } CuO]{[O]} \text{p-Benzoquinone}$$

(3) It shows keto-enol tautomerism like resorcinal.

$$\text{Dienol form} \rightleftharpoons \text{Diketo form}$$

Dienol form Diketo form

Q. 30. *o*-cresol is less acidic than phenol. Explain.

Ans.

o-Cresol Phenol

Presence of electron repelling group in *o*-cresol destabilizes the phenoxide ion. Consequently *o*-cresol is a weaker acid than phenol.

Q. 31. *o*-nitrophenol is more acidic than *o*-aminophenol. Explain.

Ans. (structures: *o*-nitrophenol, benzene ring with OH and NO_2; *o*-aminophenol, benzene ring with OH and NH_2)

Nitro group is an electron withdrawing group. It helps in delocalising the negative charge on phenoxide ion. The phenoxide ion is stabilized and can exist independently without combining back with H^+ to give undissociated acid. Thus nitrophenol is a stronger acid.

Amino group is electron donating group. It intensifies the negative charge on the phenoxide ion. The phenoxide ion is thus destabilized. It combines back with H^+ to give undissociated acid. Thus aminophenol is a weaker acid.

MISCELLANEOUS QUESTIONS

Q. 32. Give equations for the formation of fluorescein from resorcinol. (*Kerala 2001*)

Ans. See Q. 27 (*vi*)

Q. 33. Give one test to distinguish between phenol and benzyl alcohol. (*Meerut, 2000*)

Ans. Benzyl alcohol reacts with PCl_5 to form benzyl chloride. It is easily hydrolysed with aqueous KOH to form water soluble KCl which gives white precipitate of AgCl on treating with an aqueous solution of silver nitrate thus.

$$C_6H_5\text{-}CH_2\text{-}OH + PCl_5 \longrightarrow C_6H_5\text{-}CH_2\text{-}Cl \xrightarrow{+\ KOH\ (aq.)} C_6H_5\text{-}CH_2\text{-}OH + \underset{(aq.)}{KCl}$$

$$\underset{(aq.)}{KCl} + \underset{(aq.)}{AgNO_3} \longrightarrow \underset{(aq.)}{KNO_3} + \underset{(White\ ppt)}{AgCl}$$

Phenol does not give the above test

Q. 34. Identify A, B and C in the following reaction. (*Lucknow, 2000*)

$$\text{Phenol} \xrightarrow{Ac_2O} A \xrightarrow[AlCl_3]{anhy} B + C$$

Ans.

$$C_6H_5OH \xrightarrow{(CH_3CO)_2O} \underset{A}{C_6H_5OCOCH_3} \xrightarrow[AlCl_3]{anhy} \underset{B\ (o\text{-hydroxyacetophenone; OH, }COCH_3)}{C_6H_4(OH)COCH_3} + \underset{C\ (p\text{-hydroxyacetophenone; OH, }COCH_3)}{C_6H_4(OH)COCH_3}$$

Q. 35. Complete the following reactions :

(*i*) $C_6H_5OH \xrightarrow[303\ K]{H_2SO_4}$ (*M. Dayanand, 2000*)

(*ii*) $C_6H_5OH + Br_2 \xrightarrow{H_2O}$ (*Himachal, 2000*)

Ans. (*i*) See Q. 15 (3)

(*ii*) See Q. 15 (1)

Q. 36. Draw all the possible resonating structures of phenoxide. (*Kerala, 2001*)

Ans. See Q. 6.

Q. 37. *p*-nitrophenol is more acidic than phenol. Explain (*Delhi 2002*)

Ans. This is because phenoxide ion from *p*-nitrophenol is able to stabilize itself more with the help of delocalisation of negative charge.

17

ETHERS

Q. 1. What are ethers? Classify them into different types.

Ans. Ethers *may be regarded as derivatives of water in which both the hydrogen atoms have been replaced by alkyl (or aryl) groups.* Ethers, have the general formula R–O–R′ where R and R′ may be similar or dissimilar alkyl or aryl radicals. For example,

$CH_3 — O — CH_3$ Dimethyl ether

$C_6H_5 — O — C_2H_5$ Ethyl phenyl ether

Ethers may be classified into two types :

(*a*) Simple (or Symmetrical) Ethers *are those ethers in which groups R and R′ are the same.* For example,

$CH_3 — O — CH_3$ Dimethyl ether

$C_2H_5 — O — C_2H_5$ Diethyl ether

(*b*) Mixed (or Unsymmetrical) Ethers *are those ethers in which group R and R′* are different. For example,

$CH_3 - O - C_2H_5$ Ethyl methyl ether

$CH_3 - O - C_6H_5$ Methyl phenyl ether

Q. 2. Discuss the structure of diethyl ether.

Ans. Ethers are expected to have shape and structure similar to that of water as shown below :

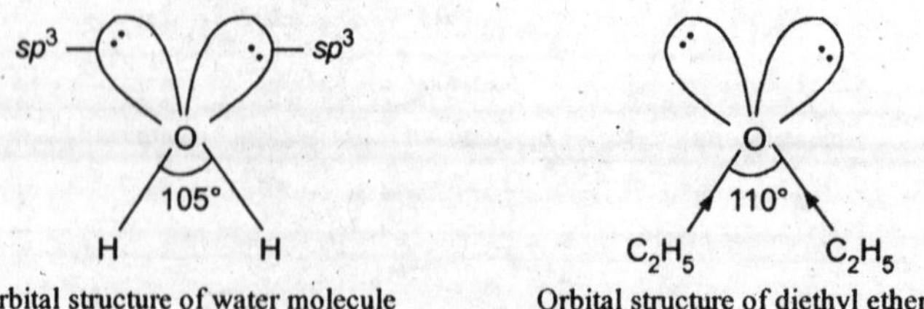

Orbital structure of water molecule

Orbital structure of diethyl ether

The $C_2H_5 - O - C_2H_5$ bond angle has been found to be 110° which is close to the tetrahedral angle of 109°, 28′. However, the $C_2H_5 - O - C_2H_5$ bond angle of 110° is bigger than H – O – H bond angle in water of 105°. This is due to the fact that the steric hindrance between two bulkier alkyl groups tends to increase the bond angle while the lone pair-lone pair repulsions tend to decrease the bond angle. The net effect is that the tetrahedral bond angle is almost preserved.

Diethyl ether has a dipole moment of 1.18 D. The dipole moment value of diethyl ether reveals that ether is not a linear molecule otherwise there would have been no resultant dipole moment. Hence the ether molecule is planar and angular.

Q. 3. Describe briefly nomenclature of ethers.

Ans. There are two systems of naming ethers.

(i) **Common system**

(ii) **IUPAC system**

These are described separately as under :

(*i*) **Common system.** According to this system, the alkyl or aryl groups on either side of oxygen are identified. The compound is named by writing these groups in alphabetical order followed by the word 'ether'.

(*ii*) **IUPAC system.** According to this system, ethers are considered as alkoxy derivatives of hydrocarbons. In the case of unsymmetrical ethers, the group having greater no. of carbon atoms is taken as the parent alkane and the other group along with the oxygen is taken as the alkoxy group. Common and IUPAC names of some common compounds are tabulated below.

Compound	*Common name*	*IUPAC name*
CH_3OCH_3	Dimethyl ether	Methoxymethane
$C_2H_5OC_2H_5$	Diethyl ether	Ethoxyethane
$CH_3OC_2H_5$	Ethyl methyl ether	Methoxyethane
$C_6H_5-O-C_6H_5$	Diphenyl ether	Phenoxybenzene
$C_6H_5-OCH_3$	Methyl phenyl ether	Methoxybenzene
$C_6H_5-OC_2H_5$	Ethyl phenyl ether	Ethoxybenzene

Sometimes special names are also used for some of the ethers. For example,

$C_6H_5-OCH_3$ is popularly known as anisole while $C_6H_5-OC_2H_5$ is called phenetole.

Q. 4. Describe with mechanism the preparation of ethers by dehydration of alcohols.

(*Madras, 2003; Purvanchal 2003*)

Ans. Dehydration of alcohols. When an excess of alcohol is heated with concentrated sulphuric acid (or glacial phosphoric acid) at suitable temperature, two moles of the alcohol lose one mole of water to form a symmetrical ether. Thus, we have :

$$2\ ROH \xrightarrow[413\ K]{H_2SO_4} R-O-R + H_2O$$

For example,

$$\underset{\text{Ethyl alcohol}}{2C_2H_5OH} \xrightarrow{H_2SO_4,\ 413\ K} \underset{\text{Diethyl ether}}{C_2H_5-O-C_2H_5} + H_2O$$

The formation of alkenes, which is a competing reaction, is suppressed by the use of excess of alcohol and regulation of temperature.

Mechanism. The mechanism of formation of ethers involves nucleophilic substitution with the protonated alcohol as the substrate and a second molecule of alcohol as nucleophile. The reaction can follow either first order kinetics (SN^1) or second-order kinetics (SN^2) depending upon the nature of alcohol used. In general, secondary and tertiary alcohols react by SN^1 mechanism. In both the mechanisms, the first step involves the formation of protonated alcohol. This is followed by other steps which differ in the two mechanisms as illustrated below :

SN^1 mechanisms :

(*i*) $R:\ddot{O}:H + H^+$ (From acid) $\rightleftharpoons$ $R:\overset{H}{\ddot{O}^+}:H$ Protonated alcohol

(*ii*) $R:\overset{H}{\ddot{O}^+}:H \xrightarrow{\text{Slow}} R^+ + H_2O$ (Carbonium ion)

(*iii*) $R:\overset{H}{\ddot{O}}: + R^+ \overset{\text{Fast}}{\rightleftharpoons} R:\overset{H}{\ddot{O}^+}:R$ Protonated ether

(*iv*) $R:\overset{H}{\ddot{O}}:R^+ \overset{\text{Fast}}{\rightleftharpoons} R:\ddot{O}:R + H^+$ Ether

SN^2 mechanism

(*i*) $R:\ddot{O}:H + H^+$ (From acid) $\rightleftharpoons$ $R:\overset{H}{\ddot{O}^+}:H$ Protonated alcohol

(*ii*) $R-\overset{H}{\ddot{O}}: + R\overset{+}{O}H_2 \xrightarrow{\text{Slow}} \left[R-\overset{H}{O}\ldots R\ldots OH_2\right]^+$ (Transition state) $\xrightarrow{-H_2O} R-\overset{H}{O^+}-R$

(*iii*) $R-\overset{H}{O^+}-R$ (Protonated ether) $\rightleftharpoons$ $R-O-R + H^+$ (Ether)

Q. 5. Describe preparation of methyl ethers using diazomethane.

Ans. Preparation of methyl ethers by the use of diazomethane. In the presence of fluoroboric acid, diazomethane reacts with alcohols or phenols to produce ethers in excellent yields. For example, *n*-hexyl alcohol, $C_6H_{13}OH$, is easily converted into hexyl methyl ether by this method.

Mechanism

(*a*) $:\bar{C}H_2-\overset{+}{N}\equiv N$ (Diazomethane) $+ \overset{+}{H}BF_4^-$ (Fluoro boric acid) $\longrightarrow \left[CH_3-\overset{+}{N}\equiv N:\bar{B}F_4\right] \longrightarrow \overset{+}{C}H_3 + N_2 + BF_4^-$

(*b*) $CH_3CH_2-\ddot{O}H + \overset{\oplus}{C}H_3 \longrightarrow CH_3CH_2-\underset{+}{\overset{CH_3}{O}}-H \xrightarrow{-H^+} CH_3CH_2OCH_3$

Q. 6. Explain Williamson's synthesis of ethers.

(*Punjab 2003; Viswa Bharti 2003; Madurai 2004; Calcutta 2007*)

Ans. Williamson synthesis. This is the most important and widely used method for the preparation of both symmetrical and unsymmetrical ethers. It is carried out by treating an alkyl or substituted alkyl halide with sodium alkoxide or sodium phenoxide.

$$\underset{\text{Sod.alkoxide}}{R'ONa} + RX \longrightarrow R'-O-R + NaX$$

$$\underset{\text{Sod. phenoxide}}{C_6H_5-ONa} + RX \longrightarrow C_6H_5-O-R + NaX$$

For example,

$$\underset{\text{Sod.ethoxide}}{C_2H_5ONa} + \underset{\text{Ethyl iodide}}{C_2H_5I} \longrightarrow \underset{\text{Diethyl ether}}{C_2H_5-O-C_2H_5} + NaI$$

$$\underset{\text{Sod.ethoxide}}{C_2H_5ONa} + \underset{\text{Methyl bromide}}{CH_3Br} \longrightarrow \underset{\text{Ethyl methyl ether}}{C_2H_5-O-CH_3} + NaBr$$

$$\underset{\text{Sod. phenoxide}}{C_6H_5-ONa} + \underset{\text{Ethyl iodide}}{C_2H_5I} \longrightarrow \underset{\text{Phenetole}}{C_6H_5-OC_2H_5} + NaI$$

The reaction, of course, involves the **nucleophilic substitution** of alkoxide or phenoxide ion for the halide ion of alkyl halide.

$$\overset{\delta+}{R}-\overset{\delta-}{X} + \underset{\text{Alkoxide ion}}{\bar{O}R'} \longrightarrow [R'O \,....R\, ...X]^- \longrightarrow R'-O-R + X^-$$

For the preparation of methyl aryl ethers, methyl sulphate, $(CH_3)_2SO_4$ is generally used in place of methyl halides.

$$2\underset{\text{Sod. phenoxide}}{C_6H_5-ONa} + \underset{\text{Dimethyl sulphate}}{(CH_3)_2SO_4} \longrightarrow 2\underset{\text{Anisole}}{C_6H_5-OCH_3} + Na_2SO_4$$

Q. 7. What are the limitations on the use of alkyl or aryl halide and the alkoxide/phenoxide in Williamson's synthesis?

Ans. Limitations in Williamson's synthesis are explained as under :

(*i*) *Since a halogen attached to an aromatic ring cannot be readily replaced by a nucleophilic reagent, aryl halides are not generally used in the Williamson synthesis.* Therefore in the preparation of alkyl-aryl ethers, the reaction is carried out between an alkyl halide and sodium phenoxide and not between aryl halide and sodium alkoxide. For example, as shown above, phenetole is prepared by reaction between ethyl iodide and sodium phenoxide and not between phenyl iodide and sodium ethoxide.

Diaryl ethers are not generally prepared by this reaction.

(*ii*) As the Williamson synthesis involves the attack of strongly nucleophilic alkoxide or phenoxide ion at the alkyl halide, *elimination can also take place as a competing reaction to substitution.* Now, tertiary alkyl halides have a greater tendency to undergo elimination than secondary or primary alkyl halides. Therefore, in the preparation of unsymmetrical ethers where there is choice of alkoxides and alkyl halides, one should select the alkyl halide that has a greater tendency to take part in substitution and a smaller tendency to undergo elimination. Therefore, primary alkyl halides are used in preference to secondary alkyl halides which in turn are preferred to tertiary halides. On the other hand, tertiary alkoxides are preferred to secondary or primary alkoxides.

For example, in the preparation of ethyl *tert*-butyl ether, we employ ethyl halide and sodium tert-butoxide and not *tert*-butyl halide and sodium ethoxide.

$$\underset{\text{Sod.}\textit{tert}\text{-butoxide}}{CH_3-\underset{CH_3}{\overset{CH_3}{\overset{|}{\underset{|}{C}}}}-ONa} + C_2H_5I \longrightarrow \underset{\text{Ethyl }\textit{tert}\text{-butyl ether}}{CH_3-\underset{CH_3}{\overset{CH_3}{\overset{|}{\underset{|}{C}}}}-OC_2H_5} + NaI$$

Q. 8. Select the preferred combination of halide and oxide in the preparation of following ethers by Williamson synthesis.

(*a*) Ethyl phenyl ether

(*b*) Sec. Isopropyl methyl ether

Ans. (*a*) Preparation of ethyl phenyl ether. There are two combinations of halide and oxide as given below :

(*i*) $C_6H_5-Cl + NaOC_2H_5$

Chlorobenzene Sod. ethoxide

(*ii*) $C_6H_5-ONa + ClC_2H_5$

Sod. phenate Chloroethane

Williamson synthesis involves nucleophilic reaction. In the combination (*i*), the nucleophile $C_2H_5O^-$ attacks the chlorobenzene molecule. But it is difficult to break the C–Cl bond in aromatic compounds. Hence, the reaction does not take place easily.

$$C_6H_5-Cl + \bar{O}C_2H_5 \longrightarrow \left[C_2H_5O \cdots C_6H_5-Cl\right]^-$$

C–Cl bond is difficult to break and hence no reaction.

In combination (*ii*), the nucleophile $C_6H_5O^-$ attacks the chloroethane molecule, knocking out Cl^- easily. The reactions can be depicted as under :

$$C_6H_5O^- + C_2H_5-Cl \longrightarrow [C_6H_5O \ldots C_2H_5 \ldots Cl]^-$$

$$\downarrow$$

$$\underset{\text{Ethyl phenyl ether}}{C_6H_5OC_2H_5} + Cl^-$$

(*b*) Isopropyl methyl ether. There are two combinations of halide and oxide.

(*i*) $\underset{\text{Sod.isopropoxide}}{(CH_3)_2CHONa} + \underset{\text{Methyl bromide}}{BrCH_3}$

(*ii*) $\underset{\text{Isopropyl bromide}}{(CH_3)_2CHBr} + \underset{\text{Sod. methoxide}}{NaOCH_3}$

Out of the two combinations, the first is preferred

$$(CH_3)_2\,CHONa + BrCH_3 \longrightarrow \underset{\text{Methyl isopropyl ether}}{(CH_3)_2\,CHO\,CH_3}$$

With the second combination, elimination instead of substitution will take place resulting in the formation of propene.

$$CH_3-\underset{\displaystyle CH_3}{\underset{|}{CH}}-Br + NaOCH_3 \longrightarrow \underset{\text{Propene}}{Ch_3-\underset{\displaystyle CH_2}{\underset{||}{CH}}} + NaBr + CH_3OH$$

Q. 9. Why cannot diphenyl ether be prepared by dehydration of phenol with conc. H_2SO_4?

Ans. Diphenyl ether cannot be prepared by dehydration of phenol with conc. H_2SO_4 due to the following reasons :

(*i*) Phenol is attacked by sulphuric acid to form hydroxy benzene sulphonic acids.

(*ii*) In case of phenols, initial protonation leading to oxonium ion does not take place readily because the oxygen atom in phenol has some positive charge (*i.e.*, less electron density) as the electrons of the oxygen atom tend to enter the benzene ring.

Q. 10. Describe preparation of ethers by epoxidation of alkenes.

Ans. Preparation of epoxides : Epoxidation of alenes

Epoxides are prepared by reaction between an alkene and an organic *peroxy acid* (also known as **peracids**). The reaction is known as **epoxidation.**

$$\underset{\text{Alkene}}{RCH=CHR} + \underset{\text{Per acid}}{R'-\overset{\displaystyle O}{\overset{||}{C}}-O-OH} \xrightarrow{\text{Epoxidation}} \underset{\text{Epxoxide}}{RCH\underset{O}{-}CHR} + R'-\overset{\displaystyle O}{\overset{||}{C}}-OH$$

Commonly used peroxy acids in this reaction are **peroxyformic** acid $(H-\overset{\displaystyle O}{\overset{||}{C}}-O-OH)$, **peroxyacetic acid** $(CH_3-\overset{\displaystyle O}{\overset{||}{C}}-O-OH)$ and peroxybenzoic acid $(C_6H_5-\overset{\displaystyle O}{\overset{||}{C}}-O-OH)$. When a solution of peroxy acid and alkene in ether or chloroform is allowed to react, epoxide is formed.

$$\underset{\text{2-Butene}}{CH_3-CH=CH-CH_3} + \underset{\text{Peroxybenzoic acid}}{C_6H_5-\overset{\displaystyle O}{\overset{||}{C}}-O-OH} \longrightarrow \underset{\text{2, 3-Dimethyl-ethylene oxide}}{CH_3-CH\underset{O}{-}CH-CH_3}$$

$$\underset{\text{Cyclohexene}}{C_6H_{10}} + \underset{\text{Peroxybenzoic acid}}{C_6H_5-\overset{\displaystyle O}{\overset{||}{C}}-O-OH} \longrightarrow \underset{\text{Cyclohexene oxide}}{C_6H_{10}O} + C_6H_5-\overset{\displaystyle O}{\overset{||}{C}}-OH$$

Mechanism. The mechanism of epoxidation involves the transfer of an oxygen atom from peroxy acid to alkene as shown below :

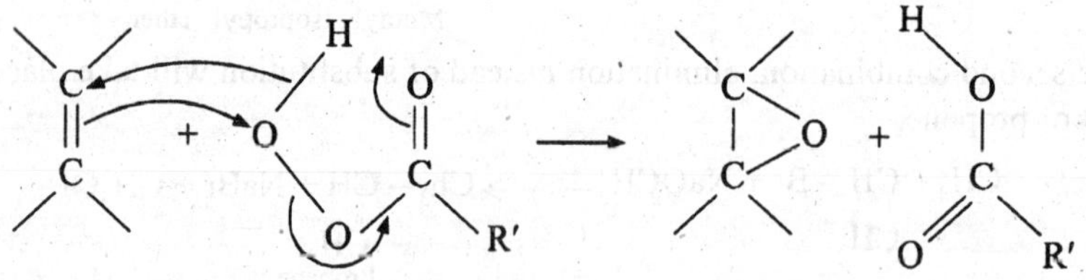

$$>C=C< + H-O-O-C(=O)-R' \longrightarrow \text{epoxide} + H-O-C(=O)-R'$$

Q. 11. Why do ethers have lower boiling points than isomeric alcohols?

(Osmonia, 2004)

Ans. Ethers have lower boiling points than isomeric alcohols. For example, ethyl alcohol boils at 351 K while diethyl ether boils at 308 K. This is because in ether the oxygen atom is not directly linked with hydrogen atom. Therefore, hydrogen bonding is not possible between the ether molecules. Hence, their boiling points are not high.

On the other hand, alcohols such as ethyl alcohol exhibit hydrogen bonding as they contain oxygen atom which is directly linked with hydrogen atom. As a result, the molecules of alcohol get associated due to hydrogen bonding. Due to strong intermolecular interaction between the molecules of alcohol, they have high boiling points.

Q.12. Which compound of the following pair would have the higher boiling point and why?

(Delhi 2006)

$$CH_3-CH_2-CH_2-OH \quad \text{or} \quad CH_3-CH_2-O-CH_3$$

Ans. If the molecules of a compound get associated due to hydrogen bonding, it results in rise of boiling point. Greater the extent of hydrogen bonding, greater the boiling point of the liquid.

n-butyl alcohol has a higher boiling point than methyl ethyl ether. This is because of occurrence of hydrogen bonding in the former. Representing $CH_3-CH_2-CH_2-$ by R–, the hydrogen bonding can be depicted as under :

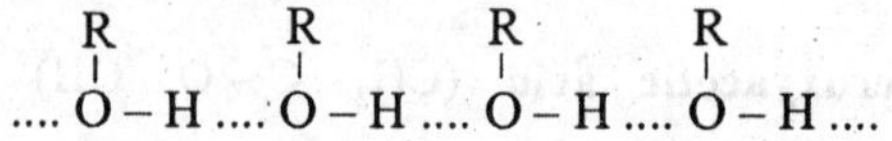

$$\ldots \underset{\text{R}}{O}-H \ldots \underset{\text{R}}{O}-H \ldots \underset{\text{R}}{O}-H \ldots \underset{\text{R}}{O}-H \ldots$$

There is no possibility of hydrogen bond in methyl ethyl ether.

Q. 13. Why do alcohols and isomeric ether have comparable water solubilities?

Ans. Like alcohols, ethers can also form hydrogen bonds with water. Consequently, ethers have some solubility in water.

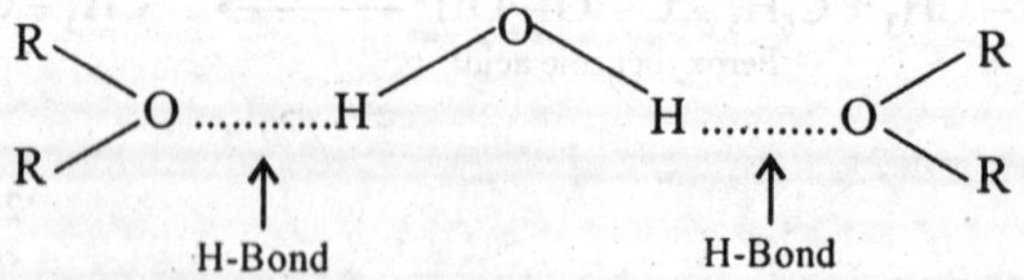

$$R_2O \cdots\cdots H-O-H \cdots\cdots OR_2$$

H-Bond H-Bond

Moreover, ethers are weakly polar molecules. This is also the reason for their solubility in water. However, the solubility of ethers in water decreases as the size of alkyl group increases. Ethers are more soluble in alcohol due to hydrogen bonding and similarity of structure.

Q. 14. Describe important chemical properties of ethers.

Ans. 1. Formation of oxonium salts. Ethers dissolve in concentrated solutions of strong mineral acids to form oxonium salts, *i.e.*, ethers behave as weak bases. For example :

$$CH_3-\ddot{\underset{..}{O}}-CH_3+HCl \longrightarrow \left[CH_3-\overset{H}{\underset{..}{\overset{|+}{O}}}-CH_3\right] Cl^-$$

Dimethyl oxonium chloride

$$C_2H_5-\ddot{\underset{..}{O}}-C_2H_5+H_2SO_4 \longrightarrow \left[C_2H_5-\overset{H}{\underset{..}{\overset{|+}{\ddot{O}}}}-C_2H_5\right] HSO_4^-$$

Diethyl oxonium hydrogen sulphate

The salts are stable at low temperature in strongly acidic medium. However, they get decomposed by water to give back ether and the acid.

2. Formation of peroxides. On standing in air or oxygen, particularly in the presence of light, ethers form peroxides. The structure and composition of ether peroxides are not known with certainty. One of the proposed structures is as shown below.

$$CH_3-CH_2-O-CH_2-CH_3+O_2 \longrightarrow CH_3-CH_2-O-\underset{\underset{O-O-H}{|}}{CH}-CH_3$$

Diethyl ether peroxide

Another structure that has been proposed for diethyl ether peroxide is given as under :

$$C_2H_5-O-C_2H_5+O \xrightarrow[\text{air}]{\text{From}} C_2H_5-\overset{\overset{O}{\uparrow}}{O}-C_2H_5$$

Diethyl peroxide

The ether peroxides are unstable compounds and decompose explosively when heated. Their presence in ethers can be detected by the formation of a red colour on shaking with an aqueous solution of ferrous ammonium sulphate and potassium thiocyanate. Ethers must be made free of peroxides before being distilled.

3. Cleavage by acids. On heating with concentrated acids (particularly hydroiodic acid or hydrobromic acid), the ether linkage breaks to form an alcohol or phenol and an alkyl halide.

$$R-O-R'+HX \longrightarrow RX+R'OH$$

For example,

$$\underset{\text{Ethyl ether}}{C_2H_5-O-C_2H_5}+HI \longrightarrow \underset{\text{Ethyl iodide}}{C_2H_5I}+\underset{\text{Ethyl alcohol}}{C_2H_5OH}$$

The order of reactivity of halogen acids in this reaction is :

$$HI > HBr > HCl$$

In the presence of excess acid, the alcohol formed reacts with halogen acid to form alkyl halide. Thus in that case the overall reaction may be depicted as :

$$R-O-R'+2HX \longrightarrow RX+R'X+H_2O$$

The mechanism of the reaction is believed to be as follows :

(*a*) Initially the ether reacts with the acid to form a protonated ether as shown below.

$$R-O-R'+HX \longrightarrow \underset{\text{Protonated ether}}{\left[R-\overset{H}{\underset{..}{\overset{|}{O^+}}}-R'\right]}+X^-$$

(*b*) The protonated ether is then attacked by the nucleophilic halide ion by SN[1] or SN[2] mechanism depending upon the nature of R and R′ groups and the reaction conditions. Presence of primary alkyl groups favours SN[2] mechanism whereas secondary or tertiary alkyl groups tend to react by SN[1] mechanism. Thus, we have :

SN[1] mechanism

$$\left[R - \overset{\displaystyle H}{\overset{|}{\ddot{O}^+}} - R' \right] \xrightarrow{\text{Slow}} R^+ + R'OH$$

$$R^+ + X^- \longrightarrow RX$$

SN[2] mechanism

$$\left[R - \overset{\displaystyle H}{\overset{|}{O^+}} - R' \right] + X^- \longrightarrow \left[\overset{\delta^-}{X} \ldots\ R \ldots \overset{\displaystyle H}{\overset{|\ \delta^+}{O}} - R' \right] \longrightarrow RX + R'OH$$

Point of cleavage. In case of mixed ethers, the alkyl halide produced depends upon the nature of groups attached to oxygen.

(*i*) If one group is methyl and the other a *primary* or *sec*-alkyl group, it is methyl halide which is produced. For example,

$$CH_3 - O - C_2H_5 + HI \longrightarrow CH_3I + C_2H_5OH$$

It is assumed that the reaction proceeds by SN[2] mechanism in this case and the halide ion attaches itself to a smaller alkyl group on steric consideration.

(*ii*) When the ether is a methyl-*tert*-alkyl ether, we get a *tert*-alkyl halide and methyl alcohol. For example,

$$\underset{\text{Methyl } tert\text{-butyl ether}}{CH_3 - \underset{CH_3}{\overset{CH_3}{\overset{|}{\underset{|}{C}}}} - O - CH_3} + HI \longrightarrow \underset{tert\text{-butyl iodide}}{CH_3 - \underset{CH_3}{\overset{CH_3}{\overset{|}{\underset{|}{C}}}} - I} + CH_3OH$$

This can be explained by SN[1] mechanism, the carbonium ion produced being *tert*-alkyl because it is more stable than primary carbonium ion.

Cleavage of alkyl-aryl and diaryl ethers. In case of an alkyl-aryl ether, the cleavage of *aromatic* carbon-oxygen bond is much more difficult than the cleavage of *alkyl* carbon-oxygen bond. Thus, the cleavage of such ethers by halogen acids yields a phenol and an alkyl halide. For example,

$$\underset{\text{Anisole}}{C_6H_5-OCH_3} + HI \longrightarrow \underset{\text{Phenol}}{C_6H_5-OH} + CH_3I$$

Diaryl ethers are very difficult to cleave as both carbons attached to oxygen are part of aromatic rings. For example, diphenyl ether is not cleaved by HI even at 525 K.

4. Halogenation. On treatment with halogens in the presence of sunlight, ethers are readily halogenated to form polyhalogenated ethers. For example :

$$\underset{\text{Ethyl ether}}{C_2H_5 - O - C_2H_5} \xrightarrow{Cl_2\,/\,h\nu} \underset{\text{Perchlorodiethyl ether}}{C_2Cl_5 - O - C_2Cl_5}$$

If the reaction is carried out in the absence of sunlight, halogenation takes place at the α-position only. For example :

$$CH_3 - CH_2 - O - CH_2 - CH_3 \xrightarrow{Cl_2} \underset{\alpha\text{-}\alpha'\text{-Dichlorodiethyl ether}}{CH_3 - CHCl - O - CHCl - CH_3}$$

5. Reaction with PCl_5

$$\underset{\text{Diethyl ether}}{C_2H_5OC_2H_5} + PCl_5 \longrightarrow \underset{\text{Ethyl chloride}}{2C_2H_5Cl} + POCl_3$$

6. Electrophilic substitution. Aromatic ethers such as anisole and phenetole undergo aromatic electrophilic substitution like nitration, halogenation, etc. It may be recalled that the alkoxy group is moderately activating and *ortho-para* directing.

For example,

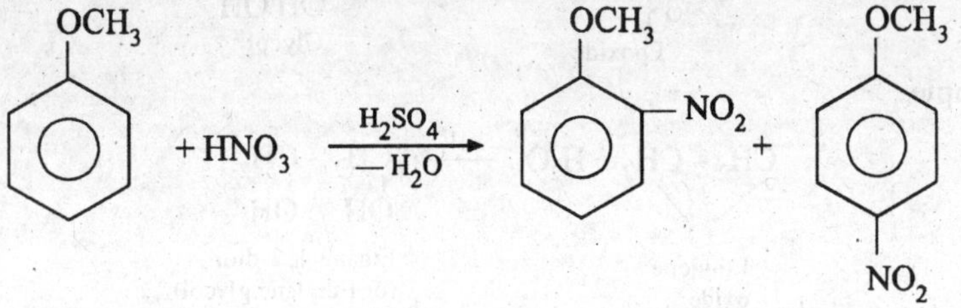

$$C_6H_5OCH_3 + HNO_3 \xrightarrow[-H_2O]{H_2SO_4} o\text{-}NO_2C_6H_4OCH_3 + p\text{-}NO_2C_6H_4OCH_3$$

o- and *p*-Nitroanisols

Alkoxy group has – I inductive effect and + M resonance effect. The latter effect being stronger, there is greater electron density on the ring. This leads to a faster electrophilic substitution.

The *ortho-para* directing effect may be explained in terms of the contributing structures of the carbonium ions formed by *ortho*, *para* and *meta* attacks as shown below, taking Y^+ as the electrophile.

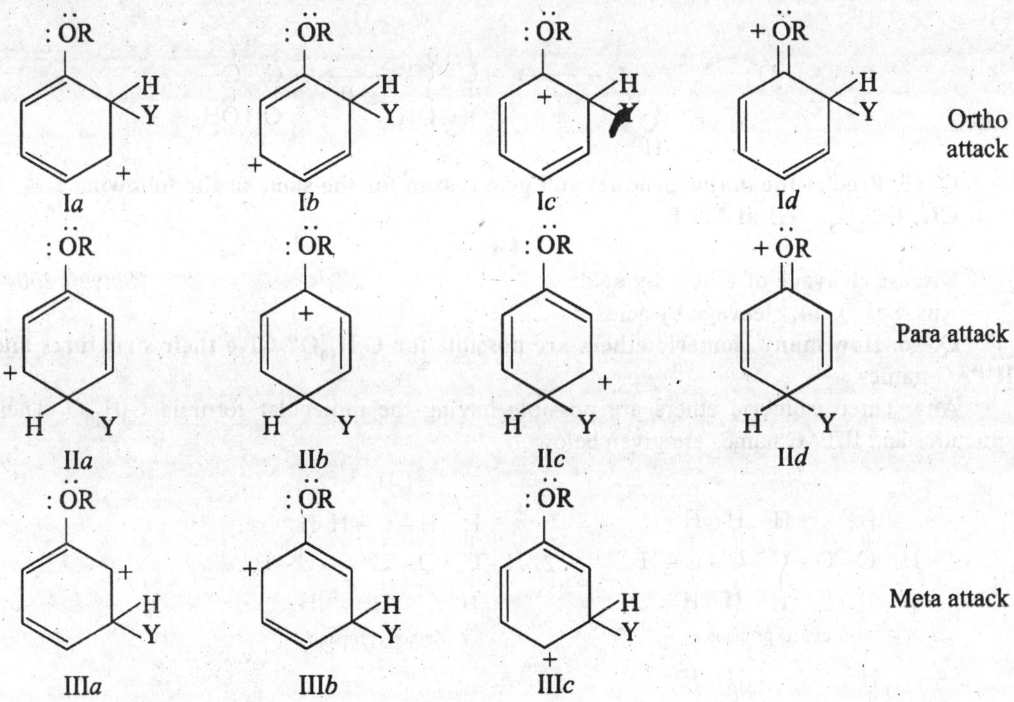

Ortho attack: I*a*, I*b*, I*c*, I*d*

Para attack: II*a*, II*b*, II*c*, II*d*

Meta attack: III*a*, III*b*, III*c*

Carbonium ions formed by *ortho* and *para* attacks are more stable than the meta carbonium ion, due to the additional contributions by particularly stable contributing structures I*d* and II*d* respectively. As a result, *ortho* and *para* carbonium ions are formed in preference to meta ion so that substitution occurs mostly at *ortho* and *para* positions.

7. Hydrolysis of epoxides to glycols. Ethers are relatively unreactive compounds which react with only a few reagents. However, the three-membered ring present in epoxides is highly strained and undergoes ring-opening quite easily. These reactions take place through the cleavage of one of the carbon-oxygen bonds and may be catalysed by either acids or bases.

One of the most important cleavage reactions of epoxides is **acid-catalysed hydrolysis** to form glycols.

$$-\overset{|}{C}-\overset{|}{C}- \text{ (bridged by O) } + H_2O \xrightarrow{H^+} -\underset{OH}{\overset{|}{C}}-\underset{OH}{\overset{|}{C}}-$$

Epoxide Glycol

For example :

$$CH_2-CH_2 \text{ (bridged by O) } + H_2O \xrightarrow{H^+} \underset{OH}{CH_2}-\underset{OH}{CH_2}$$

Ethylene oxide — Ethane-1, 2-diol (or Ethylene glycol)

Mechanism

In the acid-catalysed hydrolysis of epoxides, the epoxide is initially protonated by the acid. The protonated epoxide then undergoes nucleophilic attack by water as shown below :

$$-\overset{|}{C}-\overset{|}{C}- \text{ (bridged by } \ddot{O}) + H^+ \rightleftharpoons -\overset{|}{C}-\overset{|}{C}- \text{ (bridged by } \ddot{O}^+\text{–H)}$$

$$H_2\ddot{O}: \; -\overset{|}{C}-\overset{|}{C}- \text{ (bridged by } O\text{–}H^+) \longrightarrow -\underset{^+OH_2}{\overset{|}{C}}-\underset{OH}{\overset{|}{C}}- \xrightarrow{-H^+} -\underset{OH}{\overset{|}{C}}-\underset{OH}{\overset{|}{C}}-$$

Q. 15. Predict the major product and give reason for the same in the following :
$CH_3OC_2H_5$ + HI at 373 K

Or

Discuss cleavage of ethers by acids. (*Shivaji 2004*)

Ans. See Q. 14, cleavage by acids.

Q. 16. How many isomeric ethers are possible for $C_4H_{10}O$? Give their structures and IUPAC names.

Ans. Three isomeric ethers are possible having the molecular formula $C_4H_{10}O$. Their structures and IUPAC names are given below :

1. $H-\underset{H}{\overset{H}{C}}-O-\underset{H}{\overset{H}{C}}-\underset{H}{\overset{H}{C}}-\underset{H}{\overset{H}{C}}-H$

1-Methoxypropane

2. $H-\underset{H}{\overset{H}{C}}-O-\underset{H}{\overset{H-\overset{H}{C}-H}{C}}-\underset{H}{\overset{H}{C}}-H$

2-Methoxypropane

3. $H-\underset{H}{\overset{H}{C}}-\underset{H}{\overset{H}{C}}-O-\underset{H}{\overset{H}{C}}-\underset{C}{\overset{H}{C}}-H$

Ethoxyethane

Q. 17. Predict the products in the following reactions :

(*a*) $C_6H_5-O-CH_2CH_3 \xrightarrow{HI}$

(*b*) $(CH_3)_2\ CHO\ C_3H_7 \xrightarrow{HBr}$ (*c*) $CH_3\ O\ C(CH_3)_3 \xrightarrow{HI}$

(*d*) $C_6H_5-O-C_6H_5 \xrightarrow{HBr}$

Ans. (*a*) $\underset{\text{Phenetole}}{C_6H_5-O-CH_2CH_3} \xrightarrow{HI} \underset{\text{Phenol}}{C_6H_5-OH} + CH_3CH_2I$

This is because aromatic C–O bond is difficult to break. Instead of this, alkyl carbon-oxygen bond breaks forming ethyl carbonium ion which combines with iodide ion to form ethyl iodide.

(*b*) $\underset{\substack{\text{2-Propoxy propane}\\\text{(Propyl-isopropyl ether)}}}{(CH_3)_2\ CHO\ C_3H_7} \xrightarrow{HBr} \underset{\text{Propanol-2}}{(CH_3)_2\ CHOH} + \underset{\text{1-Bromopropane}}{C_3H_7Br}$

Here we assume the reaction to take place by SN² mechanism. Bromide ion will attach itself to less hindered *n*-propyl ion, so that *n*-propyl bromide will be one of the products.

(*c*) $\underset{\text{Methyl tert. butyl ether}}{CH_3\ OC\ (CH_3)_3} \longrightarrow \underset{\text{Methyl alcohol}}{CH_3OH} + \underset{\text{tert. Butyl iodide}}{(CH_3)_3\ CI}$

In this reaction, SN¹ reaction is assumed. The protonated ether will decompose giving more stable *tert.* butyl carbonium ion which will combine with iodide ion to give the above product.

(*d*) We don't expect the reaction to take place because of the difficulty in breaking aromatic carbon and oxygen bonds.

$C_6H_5-O-C_6H_5 \xrightarrow{HBr}$ No reaction.

Q. 18. How would you convert ethanol into ethoxy ethane and vice-versa?

Ans. (*i*) *Ethanol into ethoxyethane.*

$$\underset{\text{Ethanol}}{2C_2H_5OH} \xrightarrow[413\ K]{Conc.\ H_2SO_4} \underset{\text{Ethoxyethane}}{C_2H_5-O-C_2H_5} + H_2O$$

(*ii*) *Ethyoxyethane into ethanol*

$$\underset{\text{Ethoxyethane}}{C_2H_5-O-C_2H_5} + HI \longrightarrow \underset{\text{Ethanol}}{C_2H_5OH} + C_2H_5I$$

Q. 19. How do you say that ethers have weak basic character?

Ans. Ethers have the structure $R-\ddot{\underset{\cdot\cdot}{O}}-R'$

There are two lone pairs of electrons on oxygen in the molecule. Ethers can donate these electrons and thus they have basic character.

However, they have weak basic nature as they can dissolve only in strong mineral acids like HCl or H_2SO_4 forming salts.

$$C_2H_5-\ddot{\underset{\cdot\cdot}{O}}-C_2H_5 + HCl \longrightarrow \underset{\text{Diethyl oxonium hydrogen chloride}}{[C_2H_5-\overset{H}{\underset{\cdot\cdot}{O}}-C_2H_5]^+\ Cl^-}$$

$$C_2H_5-\ddot{\underset{..}{O}}-C_2H_5+H_2SO_4 \longrightarrow [C_2H_5-\overset{H}{\underset{..}{O}}-C_2H_5]^+ HSO_4^-$$

Diethyl oxonium hydrogen sulphate

Q. 20. Give a brief account of Ziesel's method.

Ans. This method is used for the estimation of methoxyl and ethoxyl groups in a compound.

The compound under examination is heated with concentrated hydroiodic acid when the ether is decomposed as shown below :

$$\underset{\text{Anisole}}{C_6H_5OCH_3} \xrightarrow{HI} C_6H_5OH + CH_3\,I$$

$$\underset{\text{Phenetole}}{C_6H_5OC_2H_5} \xrightarrow{HI} C_6H_5OH + C_2H_5\,I$$

The alkyl halide (CH_3I or C_2H_5I) formed is absorbed by an ethanolic solution of silver nitrate. Silver iodide formed is filtered and weighed.

A general reaction with a compound containing *n* -methoxy groups can be written as :

$$R(OCH_3)_n + n\,HI \longrightarrow R(OH)_n + n\,CH_3I$$

$$n\,CH_3\,I + n\,AgNO_3 \longrightarrow n\,AgI + n\,CH_3\,NO_3$$

From the weight of silver iodide, we can estimate the number of $-OCH_3$ or $-OC_2H_5$ groups present in the molecule. Ziesel method helps us to arrive at the structure and molecular weight of an ethereal compound.

MISCELLANEOUS QUESTIONS

Q. 21. How is ethylene oxide prepared? What are its important reactions? What happens when ethylene oxide is heated? *(Kerala, 2001)*

Ans. See Q. 10 and Q. 14.

Q. 22. Explain the hydrolysis of an epoxide and its mechanism. *(Kurukshetra, 2001)*

Ans. See Q. 14(7).

Q. 23. Complete the following reactions:

(*i*) $C_6H_5OCH_3 + HI$ *(Himachal, 2000)*

(*ii*) $CH_3-O-\underset{CH_3}{\overset{CH_3}{\underset{|}{\overset{|}{C}}}}-CH_3+HI$ *(M. Dayanand, 2000)*

(*iii*) $C_2H_5-O-C_2H_5 \xrightarrow{Cl_2\,/\,hr}$ *(M. Dayanand, 2000)*

Ans. (*i*) See Q. 14(3)

(*ii*) See Q. 14(3)

(*iii*) See Q. 14(4)

Q. 24. The dipole moment of diethylene ether (1.18 D) is lower than that of water (1.84 D). Explain, why? *(Himachal, 2000)*

Ans. Due to the inductive effect of the ethyl groups, the separation of charges becomes weaker and, therefore, dipole moment of diethylene ether becomes smaller.

18

THIOLS AND THIOETHERS

Q. 1. What are thiols? Describe their nomenclature.

Ans. (*i*) Thiols are sulphur analogs of alcohols in which the oxygen has been replaced by sulphur atom.

Thus, the functional group in thiol is —SH. It is also called *mercapto* group. Like H_2S, thiols are weakly acidic They react with mercuric ions to form insoluble salts. Therefore, they were given the name mercaptans (meaning mercury catching). They may be conceived to be derived from hydrogen sulphide as follows

$$\underset{}{H-S-H} \xrightarrow[+R]{-H} \underset{\text{Thiol}}{R-S-H}$$

Nomenclature

Thiols are named as alkyl mercaptans according to the common system. The IUPAC names are obtained by adding the suffix *thiol* to the name of the corresponding alkane.

Compound	Common Name	IUPAC Name
CH_3SH	Methyl mercaptan	Methanethiol
CH_3CH_2SH	Ethyl mercaptan	Ethanethiol
$CH_3CH_2CH_2SH$	*n*-Propyl mercaptan	1-Propanethiol

Q. 2. Give the methods of preparation of thiols.

Ans. Thiols are obtained by the following methods.

1. By heating alkyl halides with potassium hydrosulphide (KSH) solution.

$$\underset{\text{Ethyl iodide}}{CH_3CH_2I} + KSH \xrightarrow{\text{Heat}} \underset{\text{Ethanethiol}}{CH_3CH_2SH} + KI$$

2. By the reaction of Grignard reagent with sulphur followed by hydrolysis using an acid.

$$\underset{\text{Ethyl magnesium bromide}}{CH_3CH_2MgBr} + S \longrightarrow CH_3CH_2SMgBr \xrightarrow[H^+]{H_2O} \underset{\text{Ethanethiol}}{CH_3CH_2SH} + Mg(OH)Br$$

3. By the addition of hydrogen sulphide to alkene in the presence of sulphuric acid as catalyst

$$CH_3-CH=CH_2+H-SH \xrightarrow{H_2SO_4} \underset{\text{2–Propanethiol}}{CH_3-\overset{\overset{SH}{|}}{C}H-CH_3}$$

The addition takes place according to Markovnikov rule.

4. By heating alkyl halides with thiourea followed by alkaline hydrolysis.

$$\underset{\text{Ethyl bromide}}{CH_3CH_2\,Br} + S=\underset{\underset{NH_2}{|}}{C}-NH_2 \longrightarrow \underset{\text{Ethyl isothiourea salt}}{\left[CH_3CH_2-S^+=\overset{\overset{NH_2}{|}}{C}-NH_2\right]Br^-}$$

$$\xrightarrow{NaOH} \underset{\text{Ethylthiol}}{CH_3CH_2SH} + \underset{\text{Urea}}{O = \underset{\underset{NH_2}{|}}{C} - NH_2} + NaBr$$

5. By passing a mixture of alcohol vapours and hydrogen sulphide over thoria (ThO_2) as a catalyst at 673 K

$$\underset{\text{Ethyl alcohol}}{CH_3CH_2OH} + H_2S \xrightarrow[673\,K]{ThO_2} \underset{\text{Ethanethiol}}{CH_3CH_2SH} + H_2O$$

Q. 3. Describe physical properties of thiols.

Ans. 1. Thiols possess a strong odour. The smell coming out of the leaking LPG cylinders is because of methanethiol or ethanethiol that has been added in small quantities (less than 1%) to butane gas in the cylinders. This is done for quick detection of leak in the cylinders.

2. Boiling points of thiols are much lower than those of corresponding alcohols.

Compound	Boiling point (°C)
CH_3SH	6
CH_3OH	56
CH_3CH_2SH	35
CH_3CH_2OH	78
$CH_3CH_2CH_2SH$	68
$CH_3CH_2CH_2OH$	98

This is explained in terms of absence of hydrogen bondings in thiols. Normal alcohols have higher boiling points due to hydrogen bonding.

Q. 4. Explain why *n*-propyl alcohol is more soluble in water than 1-propanethiol.

(Annamalai 2003)

Ans. *n*-propyl alcohol is more soluble compared to 1-propane thiol because of stronger hydrogen bonds with water. There is greater polarisation in the O–H bond compared to S–H bond. This is because of greater electronegativity difference between oxygen and hydrogen than between sulphur and hydrogen. Hence the alcohol molecules form stronger hydrogen bonds with water molecules and therefore are more soluble in water.

Q. 5. Explain why the boiling point of 1-butanol (bp 117°C) is higher than that of 1-butanethiol (bp 98.5°C).

(Annamalai 2004)

Ans. Alcohol molecules are associated because of hydrogen bonding

$$\cdots\underset{\underset{R}{|}}{O} - \overset{\delta+}{H} \cdots \overset{\delta-}{\underset{\underset{R}{|}}{O}} - \overset{\delta+}{H} \cdots \overset{\delta-}{\underset{\underset{R}{|}}{O}} - H$$

The extent of hydrogen bonding depends upon the separation of charges which will in turn depend upon the electronegativity difference between the two elements.

Hydrogen bonding is weaker in case of thiols because of smaller separation of charges. And this is because of smaller difference in electronegativities of S and H than between those of O and H

$$\cdots\underset{\underset{R}{|}}{S} - H \cdots \underset{\underset{R}{|}}{S} - H \cdots \underset{\underset{R}{|}}{S} - H$$

↑ Weaker bond

As the hydrogen bonding is stronger is alcohols, they associate to a greater extent resulting in higher b.p. for alcohols.

Q. 6. Describe chemical properties of thiols.

Ans. There is some similarity between the chemical properties of alcohols and thiols.

1. Reaction with active metals like Na, K and Ca

Thiols react with active metals liberating hydrogen gas

$$2CH_3CH_2SH + Na \longrightarrow 2CH_3CH_2\overset{-}{S}\overset{+}{Na} + H_2\uparrow$$

2. Reaction with alkalis

Thiols are more acidic than alcohols. Reaction takes place with alkalis forming salts

$$CH_3CH_2SH + NaOH \longrightarrow CH_3CH_2\overset{-}{S}\overset{+}{Na} + H_2O$$

3. Reaction with metallic salts and oxides

Thiols react with metallic salts and oxides to form salts

$$2CH_3CH_2SH + HgO \longrightarrow \underset{\text{Mercurydiethyl mercaptide}}{(CH_3CH_2S)_2Hg} + H_2O$$

$$2CH_3CH_2SH + (CH_3COO)_2Pb \longrightarrow \underset{\text{Lead diethyl mercaptide}}{(CH_3CH_2S)_2Pb} + 2CH_3COOH$$

4. Reaction with aldehydes and ketones

Thiols react with aldehydes and ketones in the presence of hydrochloric acid to form mercaptals and mercaptols respectively.

$$2C_2H_5SH + \underset{\text{Acetaldehyde}}{CH_3CHO} \xrightarrow{HCl} \underset{\text{Diethyl methyl mercaptal}}{CH_3-\overset{\overset{SC_2H_5}{|}}{\underset{\underset{SC_2H_5}{|}}{C}}-H} + H_2O$$

$$2C_2H_5SH + \underset{\text{Acetone}}{CH_3-\overset{\overset{O}{\|}}{C}-CH_3} \xrightarrow{HCl} \underset{\text{Diethyl dimethyl mercaptol}}{CH_3-\overset{\overset{SC_2H_5}{|}}{\underset{\underset{SC_2H_5}{|}}{C}}-CH_3} + H_2O$$

Diethyl dimethyl mercaptol on oxidation with an oxidising agent like potassium permanganate gives *sulphonal* which is an important hypnotic

$$CH_3-\overset{\overset{SC_2H_5}{|}}{\underset{\underset{SC_2H_5}{|}}{C}}-CH_3 + 4[O] \xrightarrow{KMnO_4} \underset{\text{Sulphonal}}{CH_3-\overset{\overset{SO_2C_2H_5}{|}}{\underset{\underset{SO_2C_2H_5}{|}}{C}}-CH_3}$$

5. Reaction with acids and acid chlorides

Thiols react with acids and acid chlorides to form thioesters.

$$CH_3CO\boxed{OH + H}SC_2H_5 \longrightarrow \underset{\text{Ethyl thioacetate}}{CH_3COSC_2H_5} + H_2O$$

$$CH_3CO\boxed{Cl + H}SC_2H_5 \longrightarrow \underset{\text{Ethyl thioacetate}}{CH_3COSC_2H_5} + HCl$$

6. Oxidation

(*a*) Thiols get oxidised easily even with mild oxidising agents like halogens and hydrogen peroxide.

$$2C_2H_5SH + I_2 \longrightarrow \underset{\text{Diethyl disulphide}}{C_2H_5 - S - S - C_2H_5} + I_2$$

$$2C_2H_5SH + H_2O_2 \longrightarrow \underset{\text{Diethyl disulphide}}{C_2H_5 - S - S - C_2H_5} + 2H_2O$$

(*b*) Sulphonic acids are formed when oxidation is brought about by a strong oxidising agent like potassium permanganate or conc. nitric acid.

$$C_2H_5SH + 3[O] \xrightarrow{KMnO_4} \underset{\text{Ethanesulphonic acid}}{C_2H_5SO_3H}$$

Q. 7. How will you distinguish between ethanethiol and ethyl alcohol? (*Jamia Millia Islamia 2004*)

Ans. 1. Sodium hydroxide test

Ethanathiol (ethyl mercaptan) dissolves in sodium hydroxide forming a mercaptide

$$C_2H_5SH \xrightarrow{NaOH} \underset{\substack{\text{Sodium ethyl}\\ \text{mercaptide}\\ \text{(clear solution)}}}{C_2H_5\overset{-}{S}\overset{+}{Na}} + H_2O$$

Ethyl alcohol does not react with sodium hydroxide solution and hence does not dissolve in it.

2. Mercuric chloride test

Ethanethiol reacts with mercuric chloride to form a precipitate of mercury diethyl mercaptide.

$$\underset{\text{Ethanethiol}}{2C_2H_5SH} + HgCl_2 \longrightarrow \underset{\substack{\text{Mercury diethyl}\\ \text{mercaptide}}}{(C_2H_5S)_2Hg\downarrow} + 2HCl$$

Ethyl alcohol does not give this reaction.

Thus, by performing these two tests, the two compounds can be distinguished.

Q. 8. How will you prepare 1-propanethiol from propene? (*Marathwada 2003*)

Ans. Method 1

$$\underset{\text{Propene}}{CH_3CH = CH_2} \xrightarrow[\text{Peroxide}]{HBr} CH_3CH_2CH_2Br \xrightarrow[\text{Sod. hydrogen sulphide}]{NaSH} \underset{\text{1-Propanethiol}}{CH_3CH_2CH_2SH}$$

Method 2

$$CH_3CH = CH_2 \xrightarrow[\text{Peroxide}]{HBr} CH_3CH_2CH_2Br \xrightarrow[\text{Dry ether}]{Mg} \underset{\text{Propyl magnesium bromide}}{CH_3CH_2CH_2MgBr}$$

$$\xrightarrow{S} CH_3CH_2CH_2SMgBr \xrightarrow[H]{H_2O} \underset{\text{1-Propanethiol}}{CH_3CH_2CH_2SH} + Mg(OH)Br$$

Q. 9. How will you synthesis sulphonal from propene? (*Rohilkhand 2004*)

Ans. The following steps are involved in the synthesis :

$$\underset{\text{Propene}}{CH_3CH = CH_2} \xrightarrow[H]{H_2O} CH_3\overset{\overset{OH}{|}}{C}H - CH_3 \xrightarrow[K_2Cr_2O_7/H^+]{[O]} CH_3 - \overset{\overset{O}{||}}{C} - CH_3$$

$$\xrightarrow[HCl]{2C_2H_5SH} CH_3 - \underset{\underset{SC_2H_5}{|}}{\overset{\overset{SC_2H_5}{|}}{C}} - CH_3 \xrightarrow[KMnO_4]{[O]} \underset{\text{sulphonal}}{CH_3 - \underset{\underset{SO_2C_2H_5}{|}}{\overset{\overset{SO_2C_2H_5}{|}}{C}} - CH_3}$$

THIOETHERS

Q. 10. What are thioethers? Discuss their nomenclature.(*Nagpur, 2008; Garhwal, 2010*)

Ans. Thioethers are sulphur analogs of ethers. They have general formula R – S – R′ where R and R′ are alkyl groups. These alkyl groups may be same or different. Thioethers may be considered to be derived from H_2S as follows :

$$H - S - H \xrightarrow[+2R]{-2H} R - S - R$$

Functional group in thioethers is – S –.

Nomenclature

Names of some compounds according to common and IUPAC systems are given in the table below :

Compound	Common name	IUPAC name
$CH_3 - S - CH_3$	Dimethyl sulphide	Methylthiomethane
$CH_3 - S - CH_2CH_3$	Ethylmethyl sulphide	Methylthioethane
$CH_3CH_2 - S - CH_2CH_3$	Diethyl sulphide	Ethylthioethane
$CH_3 - S - CH_2CH_2CH_3$	Methyl-*n*-propyl sulphide	1-Methylthiopropane
$CH_3 - \underset{\vert}{\overset{SCH_3}{}}CH - CH_2CH_3$	Methyl-sec.butyl sulphide	2-Methylthiobutane

Smaller alkyl group linked to sulphur atom is regarded as the substituent.

Q. 11. Give the methods of preparation and properties of thioethers.

Ans. Methods of preparation

The following methods are generally employed for the preparation of thioethers

1. **Reaction of alkyl halides with sodium or potassium merceptide**

$$C_2H_5\,SNa + CH_3Br \longrightarrow \underset{\text{Ethylmetyl thioether}}{C_2H_5 - S - CH_3} + Na\,Br$$

The reaction takes place by SN^2 mechanism as shown below

$$RS^- + R' - X \longrightarrow R - S - R' + X^-$$

Primary or secondary alkyl halides and aryl halides give this reaction which is similar to Williamson synthesis of ethers. Tertiary alkyl halides are not used because they have a tendency to undergo elimination rather than substituation reaction.

2. **From ethers**

Ethers on heating with P_2S_5 form thioethers

$$\underset{\text{Diethyl ether}}{5C_2H_5 - O - C_2H_5} + P_2S_5 \longrightarrow \underset{\text{Diethyl thioether}}{5C_2H_5 - S - C_2H_5} + P_2O_5$$

3. **From alkenes**

Addition of thiols to alkenes in the prense of peroxides (anit-Markownikoff's addition) gives thioethers

$$R - CH = CH_2 + R'\,SH \xrightarrow{\text{peroxide}} R - CH_2 - CH_2 - S - R'$$

In the absence of peroxides, the reactions does not take place.

4. **Reaction of alkyl halides or potassium alkyl sulphate with sodium or potassium sulphide**

$$2\,C_2H_5\,I + K_2S \longrightarrow \underset{\text{Diethyl thioether}}{C_2H_5 - S - C_2\,H_5} + 2KI$$

$$2\,C_2H_5\,SO_4\,K + K_2\,S \longrightarrow \underset{\text{Diethyl thioether}}{C_2H_5 - S - C_2\,H_5} + 2\,K_2SO_4$$

5. From thiols

When vapours of a thiol are passed over a mixture of Al_2O_3 and ZnS heated to 575 K, thioethers are obtained.

$$2\,C_2H_5SH \xrightarrow[575\ K]{Al_2O_3/ZnS} \underset{\text{Diethyl thioether}}{C_2H_5-S-C_2H_5} + H_2S$$

Physical properties

1. Thioethers are colourless oily liquids.
2. They have unpleasant smell.
3. Thioethers have higher boiling points compared to those of corresponding ethers.

Compound	Boiling point °C
CH_3SCH_3	37
CH_3OCH_3	24
$C_2H_5SC_2H_5$	92
$C_2H_5OC_2H_5$	35

4. Thioethers are insoluble in water but soluble in ether and alcohol.
5. They show a stretching band at 600–800 cm^{-1}

Q. 12. Describe chemical properties of thioethers.

Ans. Chemical properties of thioethers are given below :

1. Reaction with halogens

Thioethers react with halogens to give dihalides.

$$\underset{\text{(Diethyl sulphide)}}{C_2H_5-S-C_2H_5} + Br_2 \longrightarrow \underset{\text{(Diethyl sulphide dibromide)}}{C_2H_5-\overset{\overset{Br}{|}}{\underset{\underset{Br}{|}}{S}}-C_2H_5}$$

2. Reaction with alkyl halides

Thioethers react with alkyl halides to form sulphonium salts

$$\underset{\text{(Diethyl sulphide)}}{C_2H_5-S-C_2H_5} + C_2H_5I \longrightarrow \underset{\text{(Triethyl sulphonium iodide)}}{C_2H_5-\overset{\overset{C_2H_5}{|}}{S^+}-C_2H_5I^-}$$

Triethyl sulphonium iodide reacts with moist silver oxide (AgOH) to form strongly basic triethylsulphonium hydroxide.

$$(C_2H_5)_3S^+I^- + AgOH \longrightarrow \underset{\text{(Triethyl sulphonium hydroxide)}}{(C_2H_5)_3\overset{+\,-}{S}OH} + AgI$$

3. Hydrolysis

Thioethers get hydrolysed with aqueous NaOH to form H_2S and alcohols

$$\underset{\text{Diethyl sulphide}}{C_2H_5-S-C_2H_5} + 2H_2O \xrightarrow{NaOH} \underset{\text{Ethyl alcohol}}{2C_2H_5OH} + H_2S\uparrow$$

4. Oxidation

(*a*) Mild oxidising agents like dilute nitric acid or hydrogen peroxide convert thioethers into sulphoxides at room temperature. Dimethyl sulphoxide is an excellent solvent for polar or non polar organic compounds. It must be handled carefully as it penetrates the skin readily.

$$CH_3-S-CH_3+H_2O_2 \longrightarrow CH_3-\overset{\overset{O}{\|}}{S}-CH_3+H_2O$$

Dimethyl sulphide → Dimethyl sulphoxide (DMSO)

(*b*) Strong oxidising agent potassium permanganate converts thioethers into sulphones. The same reaction is obtained with H_2O_2 at 100°C.

$$CH_3-S-CH_3+H_2O_2 \xrightarrow{100^\circ C} CH_3-\underset{\underset{O}{\|}}{\overset{\overset{O}{\|}}{S}}-CH_3$$

Dimethyl sulphide → Dimethyl sulphone

Liquid thioethers can be identified by converting them into sulphones.

Q. 13. Explain why dimethyl sulphoxide (bp 189°C) has a much higher boiling point than dimethyl sulphide (bp 37°C). (*Delhi 2005*)

Ans. The bond between S and O in dimethyl sulphoxide is polar because of electronegativity difference.

$$CH_3-\overset{\overset{O}{\|}}{S}-CH_3 \longleftrightarrow CH_3-\overset{\overset{\bar{O}}{|}}{\overset{+}{S}}-CH_3$$

This makes the molecule to associate resulting in increase in its boiling point

$$.....\underset{\underset{CH_3}{|}}{\overset{\overset{CH_3}{|}}{S^+}}-O^-.....\underset{\underset{CH_3}{|}}{\overset{\overset{CH_3}{|}}{S^+}}-O^-.....\underset{\underset{CH_3}{|}}{\overset{\overset{CH_3}{|}}{S^+}}-O^-.....$$

No such polarisation of bond and association between molecules takes place in dimethyl sulphide. Hence, dimethyl sulphoxide has a much higher boiling than dimethyl sulphide.

Q. 14. Write a note on mustard gas.

Ans. Mustard gas is a deadly poisonous gas. It was used as a poison gas in world war (1914-1919). It has the structure :

$$\underset{CH_2-S-CH_2}{\overset{CH_2Cl \quad\quad CH_2Cl}{| \quad\quad\quad\quad |}}$$

Preparation

Mustard gas may be prepared by the following methods :

$$CH_2=CH_2 \xrightarrow[\text{Addition}]{HOCl} CH_2OH-CH_2Cl$$

$$2CH_2OH-CH_2Cl \xrightarrow[-2\,NaCl]{Na_2S} \underset{CH_2-S-CH_2}{\overset{CH_2OH \quad CH_2OH}{| \quad\quad\quad |}} \xrightarrow[-2\,H_2O]{2\,HCl} \underset{CH_2-S-CH_2}{\overset{CH_2Cl \quad CH_2Cl}{| \quad\quad\quad |}}$$

Mustard gas

2. **From ethylene (Addition of Sulphur monochloride)**

$$CH_2=CH_2+\overset{\overset{Cl}{|}}{S}-\overset{\overset{Cl}{|}}{S}+CH_2=CH_2 \longrightarrow \underset{CH_2-S-CH_2}{\overset{CH_2Cl \quad CH_2Cl}{| \quad\quad\quad |}}+S$$

Sulphur monochloride → Mustard gas

Properties

1. It is an oily liquid with b.p. 215°C (Mustard gas is a misnomer)
2. It is insoluble in water, but soluble in ethyl alcohol and ether.
3. It produces painful blisters on the skin and damages lungs.
4. It has a prolonged action and causes death to the exposed persons after four days.

19

ALDEHYDES AND KETONES

Q. 1. What are aldehydes and ketones? What is the cause of similarity between them?

Ans. (*i*) Aldehydes and ketones are organic compounds, containing the carbonyl group *i.e.*, >C=O. Aldehydes can be represented as :

$$\begin{matrix} R \\ H \end{matrix} \!\!>\! C = O$$

where R stands for an alkyl or aryl group.

Ketones can be represented as :

$$\begin{matrix} R \\ R' \end{matrix} \!\!>\! C = O$$

where R and R′ are alkyl or aryl groups. R and R′ could be same or different groups. If R and R′ are same, the ketones are *simple ketones* and if they are different, they are called *mixed ketones.*

Thus, we find that in case of aldehydes, one of the groups that is attached to the carbonyl group is an alkyl or aryl group and the second group that is attached is hydrogen (H). In case of ketones, the two groups that are attached to carbonyl group are alkyl or aryl groups.

(*ii*) The presence of the common carbonyl group in the two classes of compounds makes them display similar chemical properties. However aldehydes are more reactive than ketones because of the presence of free hydrogen atom.

Q. 2. Discuss the nomenclature of aldehydes.

Ans. There are two systems of naming aldehydes.

(*i*) **Common system.** According to this system, an aldehyde is named after the acid which it produces on oxidation. Thus the aldehyde which produces acetic acid on oxidation is called acetaldehyde, CH_3CHO. The aldehyde which produces propionic acid on oxidation is called propionaldehyde CH_3CH_2CHO. The aldehyde which produces benzoic acid on oxidation is called benzaldehyde C_6H_5CHO and so on.

(*ii*) **IUPAC system.** Rules governing IUPAC system of nomenclature are employed in naming aldehydes in this system. The longest chain containing the —CHO group is identified. Total number of carbon atoms in this chain is counted. The aldehyde is named as the derivative of the corresponding hydrocarbon. For example, if there are 4 carbon atoms, it is a derivative of butane. If there are 5 carbon atoms in the chain, it is a derivative of pentane and so on. The ending *-e* of the hydrocarbon is changed into *-al*. The position of other substituents, if any, is indicated by suitable numbers. Aldehydic carbon is given number 1 for this purpose. For example the compound

$$\overset{4}{CH_3} - \overset{3}{CH_2} - \overset{2}{\underset{\underset{CH_3}{|}}{CH}} - \overset{1}{CHO}$$

is named as 2-methylbutanal.

Aromatic aldehydes are named by writing carbaldehyde after the aromatic part.

Table below compares common and IUPAC names of some common aldehydes.

Common and IUPAC names of some aldehydes

Formula	Common Name	IUPAC Name
$HCHO$	Formaldehyde	Methanal
$CH_3 - CHO$	Acetaldehyde	Ethanal
$CH_3 - \underset{\underset{CH_3}{\vert}}{CH} - CHO$	Isobutyraldehyde	2-Methyl propanal
$CH_3 - CH = CH - CHO$	Crotonaldehyde	2-Butenal
C_6H_5-CHO	Benzaldehyde	Benzene carbaldehyde

Q. 3. Discuss the nomenclature of ketones.

Ans. Like in the case of aldehydes, there are two systems of naming ketones.

Common system. Here the two alkyl or aryl groups constituting the compound are written followed by the word 'ketone'. For example $CH_3COC_2H_5$ in named as ethyl methyl ketone, $CH_3COC_6H_5$ is named as methyl phenyl ketone.

IUPAC system. The longest chain containing the —CO— group is identified and total number of carbon atoms in that chain is counted. The ketone is named as the derivative of the corresponding hydrocarbon. The suffix *-e* of the hydrocarbon is converted into —**one.** The positions of other substituents, if any, is indicated by suitable numbers. The ketonic carbon is given the lowest possible number as per the rules. The compound

$$\overset{1}{CH_3} - \overset{2}{CO} - \overset{3}{\underset{\underset{CH_3}{|}}{CH}} - \overset{4}{CH_2} - \overset{5}{CH_3}$$

is named as 3-methylpentanone-2.

Common and IUPAC names of some common ketones are listed in the table below:

Common and IUPAC names of some ketones

Formula	Common Name	IUPAC Name
CH_3COCH_3	Acetone or dimethyl ketone	Propanone
$CH_3COCH_2CH_3$	Methyl ethyl ketone	Butanone
$CH_3CH_2COCH_2CH_3$	Diethyl ketone	Pentan-3-one
$CH_3COCH_2CH_2CH_3$	Methyl propyl ketone	Pentan-2-one
$CH_3 - CO - \underset{\underset{CH_3}{\vert}}{CH} - CH_3$	Methyl isopropyl ketone	3-Methylbutan-2-one
$C_6H_5 - CO - CH_3$	Methyl phenyl ketone	—
$C_6H_5 - CO\ C_6H_5$	Diphenyl ketone	—

Q. 4. (*a*) Discuss the structure of the carbonyl group. (*Nagpur, 2005*)

(*b*) Account for the reactivity of the carbonyl group. (*Nagpur, 2002*)

Ans. (*a*) The bond between carbon and oxygen is a double bond. It comprises a σ and a π bond. Carbonyl carbon atom uses sp^2 hybridisation of orbitals around it, the fourth one remains unhybridised. Three sp^2 hybrid orbitals form three σ bonds, one with oxygen and two with the other groups attached to it.

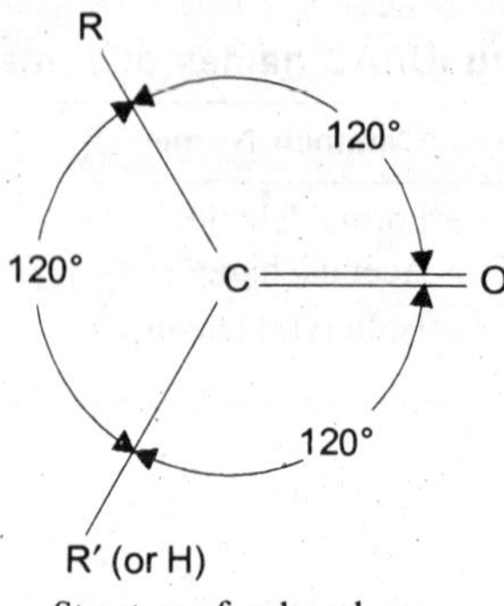

Structure of carbonyl group

The fourth orbital (unhybridised) of carbon forms a π bond with oxygen. As is well known, three sp^2 hybridised orbitals are inclined at an angle of 120° in the same plane. Hence R, R′ (or H), C and O lie in the same plane.

Another observation that we make is that there is a difference of electronegativity between carbon and oxygen, oxygen being more electronegative than carbon. That makes the shared pair of electrons between C and O to shift towards oxygen, thereby making carbon slightly positive and oxygen slightly negative.

$$>\overset{\delta+}{C}=\overset{\delta-}{O}$$

(*b*) There are two reasons for the reactivity of the carbonyl group.

(*i*) As the carbonyl group is planar in shape, there is a possibility for the attacking species to attach themselves to the carbonyl group from above the plane or below the plane of the carbonyl group. That doubles the probability of the attachment of reacting species to the carbonyl group making it a reactive group.

(*ii*) The polar nature of the carbonyl group supplements the rate of reaction. Positive charge on carbon invites the negatively charged groups or ions (nucleophiles) to attack this centre thereby making it highly reactive.

Q. 5. Compare and contrast the properties of the carbonyl group and ethylenic double bond.

Ans. Similarities

1. Both the carbonyl group and the ethylenic double bond consist of one σ bond and one π bond.

2. sp^2 hybridisation takes place on the carbon atom in both cases.

3. Both of them are planar in shape giving an angle of 120°.

4. Carbonyl group as well as the ethylenic double bond give addition reactions.

Dissimilarities

1. Ethylenic double bond is longer and weaker than the carbonyl group bond as shown below :

	Bond Length	*Bond Energy*
$>C=O$	1.23 Å	745 kJ mol^{-1}
$>C=C<$	1.44 Å	610 kJ mol^{-1}

2. Ethylenic double bond is non-polar in nature whereas carbonyl group is polar in nature.

3. Carbonyl group has a higher dipole moment as compared to ethylenic double bond.

4. Addition reactions on ethylenic compounds are initiated by electrophiles whereas in carbonyl compounds, they are initiated by nucleophiles.

Q. 6. Describe the general methods of preparation of aldehydes.

Ans. 1. Oxidation of Primary Alcohols. Primary alcohols on oxidation in the presence of acidified pot. dichromate yield aldehydes.

$$RCH_2OH \xrightarrow[H_2SO_4]{K_2Cr_2O_7} RCHO$$

where R stands for an alkyl or aryl group. For example,

$$\underset{n\text{-Propyl alcohol}}{CH_3CH_2CH_2OH} \xrightarrow[H_2SO_4]{K_2Cr_2O_7} \underset{\text{Propionaldehyde}}{CH_3CH_2CHO}$$

$$\underset{\text{Benzyl alcohol}}{C_6H_5CH_2OH} \xrightarrow[H_2SO_4]{K_2Cr_2O_7} \underset{\text{Benzaldehyde}}{C_6H_5CHO}$$

A demerit of this method of preparation is that the reaction does not completely stop on the production of aldehyde. Further oxidation of the aldehyde to carboxy acid could take place. Further oxidation can be averted by using pyridinium chloro-chromate as the oxidising agent.

2. Dehydrogenation of Primary Alcohols. Alcohol vapours when passed over heated copper at 475–575 K get dehydrogenated and yield aldehydes.

$$RCH_2OH \xrightarrow[-2[H]]{Cu,\ 475-575\,K} RCHO$$

$$C_6H_5CH_2OH \xrightarrow[-2[H]]{Cu,\ 475-575\,K} C_6H_5CHO$$

There is no risk of the aldehyde getting oxidised to carboxy acid in this method of preparation.

3. Oxidation of Methylbenzenes. Aromatic aldehydes like benzaldehyde can be obtained by the oxidation of methylbenzene or toluence using chromium trioxide in acetic anhydride.

$$\underset{\text{Toluene}}{C_6H_5-CH_3} \xrightarrow[(CH_3CO)_2O]{Cr_2O_3} \underset{\text{Benzylidene acetate}}{C_6H_5-CH(OOCCH_3)_2} \xrightarrow{H_2O} \underset{\text{Benzaldehyde}}{C_6H_5-CHO} + 2\,CH_3COOH$$

4. Etard's Reaction. This method of preparation of aldehydes involves treatment of methylbenzene with chromyl chloride (CrO_2Cl_2) which acts as an oxidising agent.

$$C_6H_5CH_3 \text{ (Toluene)} \xrightarrow{CrO_2Cl_2} [C_6H_5CH_3 . 2CrO_2Cl_2] \text{ (Addition complex)}$$

$$\xrightarrow{H_2O} C_6H_5CHO \text{ (Benzaldehyde)}$$

5. Reduction of Acid Chloride (Rosenmund's Reduction). Hydrogen gas is passed through a hot solution of acid chloride in xylene in the presence of palladium as catalyst based on barium sulphate, when the partial reduction of acid chloride to aldehyde takes place.

$$RCOCl + H_2 \xrightarrow[\text{Xylene solution}]{Pd/BaSO_4} RCHO + HCl$$

$$\underset{\text{Acetyl chloride}}{CH_3COCl} + H_2 \xrightarrow{Pd/BaSO_4} \underset{\text{Acetaldehyde}}{CH_3CHO} + HCl$$

$$\underset{\text{Benzoyl chloride}}{C_6H_5COCl} + H_2 \xrightarrow{Pd/BaSO_4} \underset{\text{Benzaldehyde}}{C_6H_5CHO} + HCl$$

6. By the hydrolysis of Gem dihalides. Gem dihalides on hydrolysis with an alkali solution produce aldehydes.

$$\underset{\text{Gem dihalide}}{RCHCl_2} \xrightarrow{NaOH} \underset{\text{Unstable}}{RCH(OH)_2} \xrightarrow{-H_2O} \underset{\text{Aldehyde}}{RCHO}$$

$$\underset{\text{Ethylidene dichloride}}{CH_3CHCl_2} \xrightarrow{NaOH} \underset{\text{Unstable}}{CH_3CH(OH)_2} \longrightarrow \underset{\text{Acetaldehyde}}{CH_3CHO} + H_2O$$

7. Ozonolysis of Alkenes. An alkene on treatment with ozone forms an ozonide which gets decomposed into aldehydes on hydrolysis in the presence of some reducing agents like zinc and acetic acid or H_2 and Pd.

$$CH_3-CH_2-CH=CH_2+O_3 \longrightarrow CH_3-CH_2-\underset{\text{Ozonide}}{CH(-O-)CH_2(O-O)}$$

Butene-1

$$\xrightarrow{H_2O} \underset{\text{Propionaldehyde}}{CH_3-CH_2CHO} + \underset{\text{Formaldehyde}}{HCHO}$$

8. Stephen's Reduction. Alkyl cyanides on reduction with stannous chloride and HCl produce aldehydes. The reaction takes place via formation of aldimine intermediate.

$$CH_3-C\equiv N+2[H]+HCl \xrightarrow{SnCl_2/HCl} \underset{\text{Acetaldimine hydrochloride}}{CH_3CH=NHHCl}$$

$$\downarrow H_2O$$

$$\underset{\text{Acetaldehyde}}{CH_3CHO} + NH_4Cl$$

9. Gattermann-Koch Reaction. Aromatic aldehydes can be prepared by treating benzene with carbon monoxide in the presence of $AlCl_3$ and hydrochloric acid.

$$\underset{\text{Benzene}}{C_6H_6} + CO \xrightarrow[HCl]{AlCl_3} \underset{\text{Benzaldehyde}}{C_6H_5CHO}$$

10. Reimer-Tiemann Reaction. Phenolic aldehydes can be prepared conveniently by this reaction. A phenol is treated with chloroform and aqueous sodium hydroxide solution.

$$\underset{\text{Phenol}}{C_6H_5OH} + CHCl_3 + 3NaOH \xrightarrow{333\ K} \underset{\text{Salicylaldehyde}}{C_6H_4(OH)CHO} + 3NaCl + 2H_2O$$

Q. 7. Describe general methods for the preparation of ketones.

Ans. 1. Oxidation of Secondary Alcohols. Secondary alcohols on oxidation in the presence of acidified potassium dichromate or potassium permanganate or chromium trioxide yield ketones.

$$\underset{\text{Secondary alcohol}}{\begin{matrix} R \\ H \end{matrix} > CHOH} \xrightarrow[\text{[O]}]{K_2Cr_2O_7} \underset{\text{Ketone}}{\begin{matrix} R \\ H \end{matrix} > C = O}$$

$$\underset{\text{Isopropyl alcohol}}{\begin{matrix} CH_3 \\ CH_3 \end{matrix} > CHOH} \xrightarrow[\text{[O]}]{K_2Cr_2O_7} \underset{\text{Acetone}}{\begin{matrix} CH_3 \\ CH_3 \end{matrix} > C = O}$$

2. Oppenauer Oxidation. This method is suitable for the oxidation of alcohols containing a double bond. The double bond is unaffected. Oxidation is carried out by heating the secondary alcohol in presence of acetone as solvent and aluminium isopropoxide $[(CH_3)_2CHO]_3$ Al or Aluminium tert-butoxide $[(CH_3)_3CO]_3$ Al.

$$\underset{\text{pent-3-ene ol-2}}{CH_3CH = CHCHOHCH_3} \xrightarrow[\text{isopropoxide}]{\text{Aluminium}} \underset{\text{Pent-3-ene-one-2}}{CH_3CH = CHCOCH_3}$$

3. Dehydrogenation of Secondary Alcohols. Ketones can be prepared by dehydrogenation of secondary alcohols. Dehydrogenation can be achieved by passing vapours of the alcohols over heated copper at 475-575 K.

$$\underset{\text{Isopropyl alcohol}}{\begin{matrix} CH_3 \\ CH_3 \end{matrix} > CHOH} \xrightarrow[\text{475 - 575 K}]{\text{Cu}} \underset{\text{Acetone}}{\begin{matrix} CH_3 \\ CH_3 \end{matrix} > C = O} + H_2$$

4. Friedel-Crafts Reaction. Aromatic ketones can be obtained by treating an aromatic hydrocarbon with a compound containing —COCl group in the presence of anhyd. $AlCl_3$.

$$\underset{\text{Benzene}}{C_6H_6} + \underset{\text{Acid chloride}}{RCOCl} \xrightarrow[AlCl_3]{\text{Anhyd.}} \underset{\text{Ketone}}{C_6H_5COR} + HCl$$

$$\underset{\text{Benzene}}{C_6H_6} + \underset{\text{Benzoyl chloride}}{ClCOC_6H_5} \xrightarrow[AlCl_3]{\text{Anhyd.}} \underset{\text{Benzophenone}}{C_6H_5—CO—C_6H_5} + HCl$$

5. Reaction of Acid Chlorides with Organocadmium Compounds.

$$\underset{\text{Acid chloride}}{2RCOCl} + \underset{\text{Dialkyl or Diaryl cadmium}}{R'_2Cd} \longrightarrow \underset{\text{Ketone}}{2RCOR'} + CdCl_2$$

where R and R′ stand for alkyl or aryl groups.

$$\underset{\text{Acetyl chloride}}{2CH_3COCl} + \underset{\text{Diethyl cadmium}}{(C_2H_5)_2Cd} \longrightarrow \underset{\text{Methyl ethyl ketone}}{2CH_3COC_2H_5} + CdCl_2$$

6. Hydration of Alkynes. Hydration of alkynes containing three or more carbons produces ketones. The alkyne is passed through a solution of dil. H_2SO_4 containing $HgSO_4$. It may be mentioned that acetylene on hydration produces acetaldehyde.

$$\underset{\text{Propyne}}{CH_3-C\equiv CH} + H_2O \xrightarrow[HgSO_4]{H_2SO_4} \underset{\text{Unstable}}{CH_3-\underset{\underset{OH}{|}}{C}=CH_2} \longrightarrow \underset{\text{Acetone}}{CH_3-\underset{\underset{O}{\|}}{C}-CH_3}$$

$$\underset{\text{2-Butyne}}{CH_3-C\equiv C-CH_3} + H_2O \xrightarrow[HgSO_4]{H_2SO_4} \underset{\text{Unstable}}{CH_3-CH=\underset{\underset{OH}{|}}{C}-CH_3} \longrightarrow \underset{\text{Butanone}}{CH_3-CH_2-\underset{\underset{O}{\|}}{C}-CH_3}$$

7. Ozonolysis of Alkenes. Ozonolysis of alkenes followed by reductive hydrolysis produces ketones.

$$\underset{\text{Alkene}}{R_1-\overset{\overset{R_2}{|}}{C}=\overset{\overset{R_3}{|}}{C}-R_4} + O_3 \longrightarrow \underset{\text{Ozonide}}{R_1-\overset{R_2}{C}(-O-)(O-O)\overset{R_3}{C}-R_4}$$

$$\xrightarrow{Zn/H_2O} \underset{\text{Ketone}}{R_1-\overset{\overset{R_2}{|}}{C}=O} + \underset{\text{Ketone}}{R_3-\overset{\overset{R_4}{|}}{C}=O} + H_2O_2$$

Here R_1, R_2, R_3 and R_4 stand for alkyl groups. It may be mentioned that ozonolysis of alkenes with branching at double bond produces ketones and ozonolysis of alkenes with no branching at the double bond produces aldehydes.

8. From Acetoacetic Ester. Hydrolysis of alkyl derivatives of ethyl acetoacetates and subsequent heating yields methyl ketones.

$$\underset{\text{Alkyl acetoacetic ester}}{CH_3CO\underset{\underset{R}{|}}{CH}COOC_2H_5} \xrightarrow{H_2O} CH_3CO-\underset{\underset{R}{|}}{CH}COOH \xrightarrow{\text{Heat}} \underset{\text{Ketone}}{CH_3-CO-CH_2R} + CO_2$$

Q. 8. Write a note on the physical properties of aldehydes and ketones.

Ans. The boiling points of aldehydes and ketones are higher as compared to hydrocarbons of comparable molecular weights. Due to polar nature of carbonyl compounds, there are intermolecular attraction in them, which results in higher boiling points.

However aldehydes and ketones have lower boiling points as compared to alcohols and carboxy acids of comparable molecular weights. This is because unlike alcohols and carboxy acids, carbonyl compounds have no hydrogen bonding in them. Aldehydes and ketones having upto five carbon atoms are soluble in water whereas higher members are insoluble in water. However all aldehydes and ketones are soluble in organic solvents.

Q. 9. The boiling point of acetone is 329 K while that of propanal is 322 K although both have the same molecular mass.

Ans. Boiling point is a property which is related to the polarity in the molecule. Greater the polarity in the molecule, greater will be the intermolecular attraction and hence higher will be the boiling point. Acetone has a higher boiling point compared to propanal because of double inductive effect of two alkyl groups.

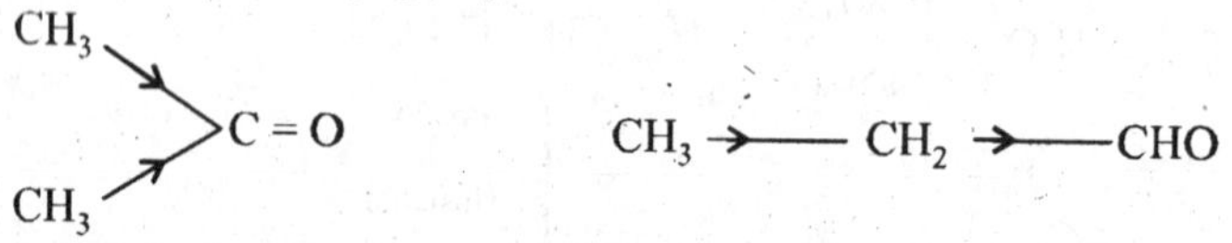

$$\begin{matrix} CH_3 \searrow \\ \quad\; C=O \\ CH_3 \nearrow \end{matrix} \qquad\qquad CH_3 \rightarrow\!\!- CH_2 \rightarrow\!\!- CHO$$

In propanal, there is inductive effect of only one alkyl group. Hence greater polarity will be induced in acetone.

Q. 10. *n*-butyl alcohol boils at 391 K while *n*-butyraldehyde boils at 349 K though their molecular masses are 74 and 72 respectively. Account for the higher b.p. of the alcohol.

Ans. There is hydrogen bonding in the case of *n*-butyl alcohol which results in association of *n*-butyl alcohol molecules.

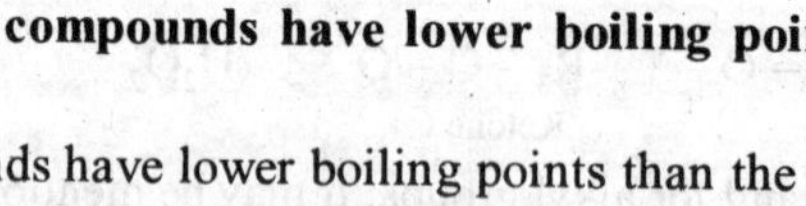

$$\begin{matrix} \cdots\cdots H-O\cdots\cdots & H-O\cdots\cdots & H-O\cdots\cdots \\ \quad | & \quad | & \quad | \\ \quad C_4H_9 & \quad C_4H_9 & \quad C_4H_9 \end{matrix}$$

No such hydrogen bonding takes place in *n*-butyraldehyde. Hence these molecules are not associated. As such, *n*-butyl alcohol boils at a higher temperature compared to *n*-butyraldehyde. Boiling point is a property which is dependent upon the size of the molecule. Associated molecules which become bigger and heavier require a higher temperature to boil.

Q. 11. Why carbonyl compounds have lower boiling points than the alcohols from which they are derived?

Ans. Carbonyl compounds have lower boiling points than the alcohols containing an equal no. of carbon atoms. For example, acetaldehyde has lower boiling point than ethyl alcohol. Acetøne and propionaldehyde have lower boiling points than propyl alcohol etc.

This is explained in terms of hydrogen bonding. In aldehydes and ketones, there is no hydrogen bonding as there is not enough polarity in the molecules. But alcohols exhibit hydrogen bonding between hydrogen of one and oxygen of second molecule.

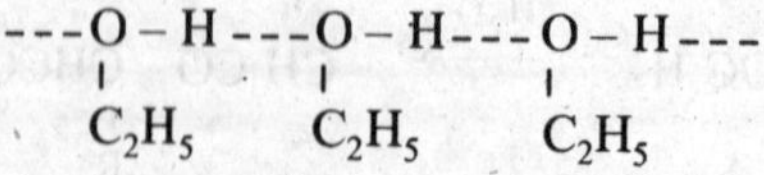

$$\begin{matrix} ---O-H--- & O-H--- & O-H--- \\ \;| & \;| & \;| \\ C_2H_5 & C_2H_5 & C_2H_5 \end{matrix}$$

Associated molecule of ethyl alcohol

As a consequence of hydrogen bonding, the molecular weight of alcohol increases. Hence, it boils at a higher temperature.

As the carbonyl compounds have no such bonding, they boil at temperatures below the boiling point of alcohols.

Q. 12. *o*-salicylaldehyde has a lower boiling point compared to *m*- or *p*- isomer. Explain.

Ans.

o-salicylaldehyde *m*-salicylaldehyde *p*-salicylaldehyde

Because of the proximity of – OH and – CHO groups in *o*-salicylaldehyde, there is possibility of hydrogen bonding which reduces the size of the molecule. Consequently, the boiling point is lowered

Q. 13. What is a nucleophilic addition reaction? Describe the mechanism of nucleophilic addition to a carbonyl compound. (*Nagpur 2008*)

Ans. A nucleophilic addition reaction is a reaction in which the addition of a molecule to the carbonyl compound takes place with the initiation of the negative part of the adding molecule. The negative part of the adding molecule (e.g. CN in HCN) adds first to the positive carbon of the carbonyl group, followed by the attachment of the positive part of the adding molecule to the negative oxygen of the carbonyl group. The addition of a nucleophile (: Z) to a carbonyl compound may be represented as shown in figure below :

Carbonyl compound (R, R′ (or H)) → Transition state → Product

Mechanism of nucleophilic addition

Carbonyl compound has a trigonal configuration. In the transition state, it starts acquiring sp^3 configuration which is accomplished in the final product. Ability of oxygen to acquire negative charge is the reason for nucleophilic reactions in carbonyl compounds.

Q. 14. Explain why acetaldehyde is more reactive than acetone towards nucleophilic addition reactions? (*Punjab 2003; Calcutta 2007*)

Ans. Aldehydes are observed to be more reactive than ketones particularly in giving nucleophilic reactions for the following reasons:

Steric Factors. As has been explained earlier, the nucleophile attaches itself to the positive carbon of the carbonyl group. With the ketone, the transition state (See figure above) will experience *steric hindrance* because of bulky alkyl groups. Consequently the transition state will not be stable. Hence the reaction will not take place easily. With an aldehyde, steric hindrance will be less as there is only one alkyl group in the transition state. This accounts for greater reactivity of aldehyde group compared to the ketonic group.

Electronic Factors. Alkyl groups exert electron releasing *inductive effect* which increases the negative charge density on carbon making it difficult for the nucleophile to attach to carbon. This would destabilise the transition state (See figure below).

Only one alkyl group : lesser destabilisation

Two alkyl groups : greater destabilisation

Destabilising effect of alkyl groups

With ketones, this destabilising effect will be doubled because of two alkyl (or aryl) groups. This is the reason why ketones are less reactive than aldehydes.

Q. 15. Why are aromatic aldehydes and ketones less reactive than aliphatic aldehydes and ketones?

Ans. Phenyl group has electron withdrawing inductive effect. Thus we would expect the transition state to get stabilized in the nucleophilic addition reaction of an aromatic aldehyde or ketone. But this is not actually so. As a matter of fact, aromatic aldehydes and ketones are less reactive than their aliphatic counterparts. This is so because aromatic aldehydes and ketones get stabilised due to resonance as shown below :

This creates a positive charge on the ring instead of carbonyl carbon thus decreasing chances of nucleophilic attack. Moreover, there are steric factors due to bulky size of phenyl group.

Hence they show less tendency for nucleophilic reactions.

Q. 16. Addition of certain nucleophilic reagents to aldehydes and ketones is catalysed by the presence of an acid. Explain.

Ans. It has been observed that a strong nucleophile gets attached to the carbonyl group easily but some difficulty is felt by the weak nucleophile in attaching itself to the carbonyl group. In such cases, the presence of a protonic acid containing (H^+) proves helpful as illustrated below :

Transition state

Product

Role of the acid catalyst in nucleophilic reaction

The proton from the acid attaches itself to the negative oxygen. This paves the way for the attachment of even the weakest nucleophile. The energy of activation for the nucleophilic attack is decreased thereby facilitating the reaction.

Q. 17. Why is α-hydrogen in aldehydes and ketones acidic in nature? (*Nagpur 2002*)

Ans. Electronegative nature of oxygen atom in the carbonyl group directs the flow of electrons towards oxygen. This helps in the removal of hydrogen as in the presence of a base (: B) as shown below :

$$-\overset{\overset{\displaystyle H}{|}}{\underset{|}{C}}_{\alpha}-\overset{\overset{\displaystyle O}{\|}}{C}- + :B \;(\text{Base}) \rightleftharpoons B-H + \left[-\overset{-}{\underset{|}{C}}-\overset{\overset{\displaystyle O}{\|}}{C}- \longleftrightarrow -\underset{|}{C}=\overset{\overset{\displaystyle O^-}{|}}{C}-\right]$$

The carbanion obtained after the removal of hydrogen gets resonance stabilized. This is all due to the ability of oxygen to accommodate the negative charge.

Q. 18. Describe the addition of sodium bisulphite to the carbonyl compound. What is the utility of this reaction?

Ans. All aldehydes and ketones react with a saturated solution of sodium bisulphite to form the solid bisulphate compound.

$$\underset{\text{Aldehyde or ketone}}{>C=O} + NaHSO_3 \rightleftharpoons \underset{\text{Bisulphite compound}}{>C<^{OH}_{SO_3Na}}$$

$$\underset{\text{Acetone}}{(CH_3)_2C=O} + NaHSO_3 \rightleftharpoons \underset{\text{Acetone bisulphite}}{(CH_3)_2C<^{OH}_{SO_3Na}}$$

$$C_6H_5-CHO + NaHSO_3 \rightleftharpoons \underset{\text{Benzaldehyde bisulphite}}{C_6H_5-CH<^{OH}_{SO_3Na}}$$

Ketones containing bulky groups fail to give the reaction due to steric hindrance.

Mechanism of the Reaction

The reaction is initiated by the attack of the bisulphite ion on the carbonyl carbon followed by the shifting of H to carbonyl oxygen. The complete mechanism is illustrated hereunder.

$$>C=O + :\overset{-}{S}O_3H \rightleftharpoons >C<^{O^-}_{SO_3H} \rightleftharpoons >C<^{OH}_{SO_3^-} \overset{Na^+}{\rightleftharpoons} >C<^{OH}_{SO_3Na}$$

Utility of the Reaction. The bisulphite compound as obtained above gets decomposed on treatment with dil. HCl giving back the aldehyde or ketone. Thus the reaction can be used to purify or separate the carbonyl compound from impurities.

$$>C<^{OH}_{SO_3Na} + HCl \longrightarrow >C=O + NaCl + H_2O + SO_2$$

Q. 19. Describe with mechanism the addition of alcohol to carbonyl compounds. (*Purvanchal 2003*)

Ans. Alcohols add to aldehydic compounds forming acetals. Generally the reaction takes place in two steps. In the first step, one molecule of alcohol reacts with the carbonyl compound to form hemi-acetal which is generally unstable. Therefore, hemi-acetal reacts with the second molecule of alcohol to form acetal. The reaction takes place in the presence of HCl gas. Ketones do not participate in such a reaction.

$$\underset{\text{Aldehyde}}{\overset{R}{\underset{H}{>}}C=O} + \underset{\text{Alcohol}}{R'OH} \xrightarrow[\text{gas}]{\text{HCl}} \underset{\text{Hemiacetal}}{\overset{R}{\underset{H}{>}}C\overset{OH}{\underset{OR'}{<}}} \xrightarrow[-H_2O]{R'OH} \underset{\text{Acetal}}{\overset{R}{\underset{H}{>}}C\overset{OR'}{\underset{OR'}{<}}}$$

$$\underset{\text{Acetaldehyde}}{\overset{CH_3}{\underset{H}{>}}C=O} + C_2H_5OH \xrightarrow[\text{gas}]{\text{HCl}} \underset{\text{Hemiacetal}}{\overset{CH_3}{\underset{H}{>}}C\overset{OH}{\underset{OC_2H_5}{<}}} \xrightarrow[-H_2O]{C_2H_5OH} \underset{\text{Acetal}}{\overset{CH_3}{\underset{H}{>}}C\overset{OC_2H_5}{\underset{OC_2H_5}{<}}}$$

$$\overset{C_6H_5}{\underset{H}{>}}C=O + C_2H_5OH \xrightarrow[\text{gas}]{\text{HCl}} \overset{C_6H_5}{\underset{H}{>}}C\overset{OH}{\underset{OC_2H_5}{<}} \xrightarrow[-H_2O]{C_2H_5OH} \underset{\text{Diethyl acetal of benzaldehyde}}{\overset{C_6H_5}{\underset{H}{>}}C\overset{OC_2H_5}{\underset{OC_2H_5}{<}}}$$

R & R′ in the above reaction stand for alkyl groups.

Mechanism of the Reaction

Step 1. A molecule of alcohol adds to the carbonyl group to form a hemiacetal. This is made possible by the presence of H^+ as it produces a positive centre on the carbonyl carbon. Alcohol molecule approaches this positive centre. H^+ is regenerated in this step.

Step 2. The proton now removes the –OH group from the hemi-acetal, thus reproducing positive centre. This enables the second alcohol molecule to attach itself to the hemi-acetal. Finally, the proton is again given out. These steps are illustrated below :

$$\overset{R}{\underset{H}{>}}C=O \overset{H^+}{\rightleftharpoons} \left[\overset{R}{\underset{H}{>}}C=\overset{+}{O}H \longleftrightarrow \overset{R}{\underset{H}{>}}\overset{+}{C}-OH\right]$$

$$\Big\Updownarrow R'OH$$

$$\underset{\text{Hemi-acetal}}{\overset{R}{\underset{H}{>}}C\overset{OH}{\underset{OR'}{<}}} \overset{-H^+}{\rightleftharpoons} \overset{R}{\underset{H}{>}}C\overset{OH}{\underset{\overset{+}{O}R'(H)}{<}}$$

$$\underset{H}{\overset{R}{>}}C\underset{OR'}{\overset{OH}{<}} \underset{}{\overset{H^+}{\rightleftharpoons}} \underset{H}{\overset{R}{>}}C\underset{OR'}{\overset{\overset{H}{|}\overset{+}{O}H}{<}} \xrightarrow{-H_2O} \left[\underset{H}{\overset{R}{>}}\overset{+}{C}-OR' \longleftrightarrow \underset{H}{\overset{R}{>}}C=\overset{+}{O}R'\right]$$

$$\Big\Downarrow R'OH$$

$$\underset{\text{Acetal}}{\underset{H}{\overset{R}{>}}C\underset{OR'}{\overset{OR'}{<}}} \underset{}{\overset{-H^+}{\rightleftharpoons}} \underset{H}{\overset{R}{>}}C\underset{\overset{+}{O}R\atop H}{\overset{OR'}{<}}$$

Q. 20. Illustrate with mechanism the addition of hydrogen cyanide on the carbonyl compounds. (*Delhi 2002*)

Ans. Hydrogen cyanide adds to aldehydes and ketones to form compounds known as *cyanohydrins*. Cyanohydrin contains a cyano group and a hydroxy group.

$$\underset{\text{Aldehyde or ketone}}{\underset{R \text{ (or H)}}{\overset{R}{>}}C=O} + HCN \longrightarrow \underset{\text{Cyanohydrin}}{\underset{R}{\overset{R}{>}}C\underset{CN}{\overset{OH}{<}}}$$

$$\underset{\text{Acetaldehyde}}{\underset{H}{\overset{CH_3}{>}}C=O} + HCN \longrightarrow \underset{\text{Acetaldehyde cyanohydrin}}{\underset{H}{\overset{CH_3}{>}}C\underset{CN}{\overset{OH}{<}}}$$

$$\underset{\text{Acetone}}{\underset{CH_3}{\overset{CH_3}{>}}C=O} + HCN \longrightarrow \underset{\text{Acetone cyanohydrin}}{\underset{CH_3}{\overset{CH_3}{>}}C\underset{CN}{\overset{OH}{<}}}$$

Mechanism of the Reaction

Step 1. Cyanide ion attaches itself to the carbonyl carbon by nucleophilic addition process.

Step 2. H^+ then attaches itself to the oxygen, which carries a negative charge, to produce the final product.

$$>C=O + :CN^- \longrightarrow >C\underset{CN}{\overset{O^-}{<}} \xrightarrow{H^+} >C\underset{CN}{\overset{OH}{<}}$$

Cyanohydrins are important compounds as they yield α-hydroxy carboxylic compounds on hydrolysis.

$$\underset{\text{Acetaldehyde cyanohydrin}}{CH_3CH(OH)CN} \xrightarrow{\text{Hydrolysis}} \underset{\text{Lactic acid}}{CH_3CH(OH)COOH}$$

Q. 21. Describe with mechanism the reaction ammonia derivatives with the carbony compounds. *(Himachal 2000, Mumbai 2010*

Ans. Derivatives of ammonia *viz.* hydroxylamine, hydrazine, phenyl hydrazine and semicarbazide react with the carbonyl compounds to give oximes, hydrazones, phenyl hydrazones and semi-carbazones respectively. as shown below :

(*i*) $>C = O + H_2NOH \longrightarrow >C = NOH + H_2O$

Hydroxyl amine — Oxime

(*ii*) $>C = O + H_2NNH_2 \longrightarrow >C - NNH_2 + H_2O$

Hydrazine — Hydrazone

(*iii*) $>C = O + H_2NNHC_6H_5 \longrightarrow >C = N\,NHC_6H_5 + H_2O$

Phenylhydrazine — Phenylhydrazone

(*iv*) $>C = O + H_2N\,NHCONH_2 \longrightarrow >C = N\,NHCONH_2 + H_2O$

Semicarbazide — Semicarbazone

(*v*) $>C = O + H_2NNH-C_6H_3(NO_2)_2$

2 : 4 Dinitrophenyl hydrazine

$\xrightarrow{H^+} >C = NNH-C_6H_3(NO_2)_2$

2, 4, Dinitrophenylhydrazone

Reaction (*v*) is particularly used to identify the carbonyl group as it gives yellow derivative viz. 2, 4 dinitrophenylhydrazone. 2, 4 dinitrophenylhydrazine is abbreviated as DNPH. Hence this reaction is also known as DNPH Test.

Mechanism of the Reaction. Ammonia derivatives may be represented by H_2NG. Various steps involved in the reaction may be written as under.

Step 1. Attack of Proton. The proton from the acid which is used to catalyse the reaction attaches itself to the electron rich carbonyl oxygen atom as shown below :

$$>C=O \underset{}{\overset{H^+}{\rightleftharpoons}} >\overset{+}{C}-OH \xrightarrow{H_2\ddot{N}-G} \left[>C\begin{matrix}OH \\ \overset{+}{H_2N}-G\end{matrix}\right] \xrightarrow{-H^+} \left[>C\begin{matrix}OH \\ HN-G\end{matrix}\right] \xrightarrow{-H_2O} >C=N-G$$

Step 2. Attack of Nucleophile. The ammonia derivative in which nitrogen has a lone pair of electron acts as the nucleophile, attaches itself to the carbonyl carbon atom. This results in a positive charge on nitrogen atom.

Step 3. Loss of Proton. The intermediate formed is step 2 is unstable and loses a proton.

Step 4. Loss of Water Molecule. The product formed in step 3 loses a water molecule to form a crystalline compound having a double bond between carbonyl carbon and nitrogen ammonia derivative.

Q. 22. Give the mechanism of the reaction

$$CH_3CHO + NH_2NH_2 \longrightarrow$$

Ans.

$$\underset{\text{Acetaldehyde}}{CH_3CHO} + \underset{\text{Hydrazine}}{NH_2NH_2} \longrightarrow \underset{\text{Acetaldehyde hydrazone}}{CH_3CH=N-NH_2} + H_2O$$

The mechanism of the reaction is as follows :

$$>C=O \underset{}{\overset{H^+}{\rightleftharpoons}} >C=\overset{+}{O}H \quad \text{(Attack of proton)}$$

$$>C=\overset{+}{O}H \xrightarrow{H_2\ddot{N}-NH_2} >C<^{OH}_{H_2\overset{+}{N}-NH_2} \quad \text{(Attack of nucleophile)}$$

$$>C<^{OH}_{H_2\overset{+}{N}-NH_2} \xrightarrow{-H^+} >C<^{OH}_{HN-NH_2} \quad \text{(Removal of } H^+\text{)}$$

$$>C<^{OH}_{HNNH_2} \xrightarrow{-H_2O} >C=N-NH_2 \quad \text{(Removal of } H_2O\text{)}$$

Q. 23. Give the reaction of ammonia on

(i) Formaldehyde (*Himachal, 2000; Awadh, 2000; Kanpur, 2001; Nagpur 2008*)

(ii) Acetaldehyde

(iii) Benzaldehyde.

Ans. (*i*) Ammonia reacts with formaldehyde to give a white crystalline solid known as hexamethylene tetramine.

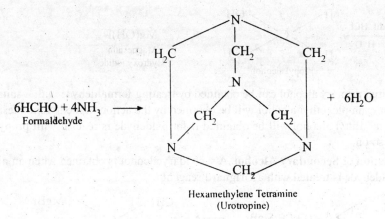

$$\underset{\text{Formaldehyde}}{6HCHO} + 4NH_3 \longrightarrow (CH_2)_6N_4 + 6H_2O$$

Hexamethylene Tetramine
(Urotropine)

Hexamethylene tetramine is used as a urinary antiseptic medicine.

(*ii*) Ammonia reacts with acetaldehyde to form a crystalline product called acetaldehyde ammonia. This product is unstabie and loses a water molecule to yield aldimine.

$$\underset{H}{\overset{CH_3}{>}}C=O + NH_3 \longrightarrow \underset{\substack{\text{Acetaldehyde}\\ \text{ammonia}}}{CH_3(H)C(OH)(NH_2)} \xrightarrow{-H_2O} \underset{\text{Aldimine}}{CH_3(H)C=NH}$$

Aldimine polymerises as under :

$$nCH_3CHNH \longrightarrow (CH_3CHNH)_n$$

(*iii*) Ammonia reacts with benzaldehyde to form hydrobenzamide.

$$3\,C_6H_5CHO + 2NH_3 \longrightarrow \underset{\text{Hydrobenzamide}}{C_6H_5CH=N-CH(C_6H_5)-N=CHC_6H_5} + 3H_2O$$

Q. 24. Discuss the reaction of carbonyl compounds with Grignard's reagents. Give the mechanism and importance of this reaction.

Ans. Carbonyl compounds (aldehydes and ketones) react with Grignard reagents forming addition compounds which get decomposed by dil. HCl to yield different alcohols.

Formation of Primary Alcohols. Primary alcohols are obtained by the reaction of formaldehyde with the Grignard reagent.

$$\underset{H}{\overset{H}{>}}C=O + \underset{\substack{\text{Ethyl mag.}\\ \text{iodide}}}{C_2H_5MgI} \longrightarrow \underset{\substack{\text{Addition}\\ \text{comppound}}}{H_2C(OMgI)(C_2H_5)}$$

$$\xrightarrow[H_2O]{\text{dil. HCl}} \underset{\text{Propyl alcohol}}{H_2C(OH)(C_2H_5)} + \underset{\substack{\text{Magnesium}\\ \text{Hydroxy iodide}}}{Mg(OH)I}$$

A desired primary alcohol can be obtained by treating formaldehyde with a suitable Grignard reagent. For example, ethyl alcohol will be obtained by the action of methyl magnesium iodide on formaldehyde. Butyl alcohol will be obtained if formaldehyde is reacted with propyl magnesium iodide and so on.

Formation of Secondary Alcohol. A secondary alcohol is obtained when an aldehyde other than formaldehyde is treated with a Grignard reagent.

$$\underset{\text{Acetaldehyde}}{\underset{H}{\overset{CH_3}{>}}C=O} + CH_3MgBr \longrightarrow CH_3(H)C(OMgBr)(CH_3)$$

$$\xrightarrow[H_2O]{dil.\,HCl} \underset{\text{Isopropyl alcohol or Propanol-2}}{(CH_3)(H)C(OH)(CH_3)} + \underset{\text{Magnesium Hydroxybromide}}{Mg(OH)Br}$$

Again a desired secondary alcohol can be produced by combining a suitable aldehyde with Grignard reagent.

Formation of Tertiary Alcohol. A tertiary alcohol is formed by the reaction of ketone with a Grignard reagent followed by hydrolysis.

$$(CH_3)_2C=O + CH_3MgBr \longrightarrow \underset{\text{Addition product}}{(CH_3)_2C(OMgBr)(CH_3)}$$

$$\xrightarrow[H_2O]{dil.\,HCl} \underset{\text{2-methylpropanol-2 (Tertiary alcohol)}}{(CH_3)_2C(OH)(CH_3)} + Mg(OH)(Br)$$

Mechanism of the Reaction

The reaction proceeds as a nucleophilic addition reaction. Grignard reagent (R Mg X) consists of two parts R^- and MgX^+. The nucleophile R^- attacks the carbonyl group at carbon followed by the attachment of positive group MgX^+ at carbonyl oxygen, forming the addition product. The product is decomposed by water in the presence of HCl as described above.

$$>C=O + R-MgX \longrightarrow >C(OMgX)(R)$$

Q. 25. Give an account of Roformatsky reaction along with its mechanism.

(Nagpur, 2002; Madras, 2003; Delhi, 2003; Patna, 2005; Lucknow, 201(

Ans. Reformatsky reaction is performed in order to obtain β-hydroxy esters from aldehydes and ketones using α-bromo ester and zinc in the presence of dry ether. The reactions with an aldehyde and a ketone are depicted below :

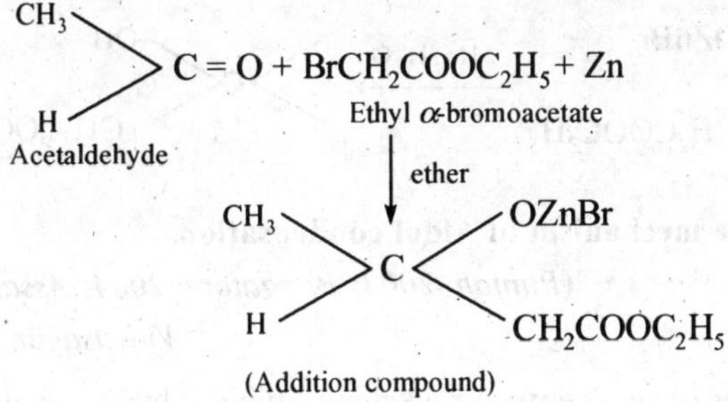

$$\xrightarrow{H^+, H_2O} \begin{matrix} CH_3 \\ H \end{matrix}\!\!>\!C\!<\!\!\begin{matrix} OH \\ CH_2COOC_2H_5 \end{matrix}$$

Ethyl β-hydroxyl butyrate

$$\underset{\text{Acetone}}{\begin{matrix} CH_3 \\ CH_3 \end{matrix}\!\!>\!C = O} + Br\underset{|}{\overset{CH_3}{\overset{|}{C}}}HCOOC_2H_5 + Zn$$

$$\downarrow \text{Ether}$$

$$\begin{matrix} CH_3 \\ CH_3 \end{matrix}\!\!>\!C\!<\!\!\begin{matrix} OZnBr \\ \underset{\underset{CH_3}{|}}{CHCOOC_2H_5} \end{matrix} \quad \text{(Addition product)}$$

$$\downarrow H^+, H_2O$$

$$\begin{matrix} CH_3 \\ CH_3 \end{matrix}\!\!>\!C\!<\!\!\begin{matrix} OH \\ \underset{\underset{CH_3}{|}}{CHCOOC_2H_5} \end{matrix}$$

Ethyl 2, 3-dimethyl 3-hydroxy butyrate.

Mechanism of the Reaction

The following mechanism has been proposed for the Reformatsky reaction.

Step 1. Zinc reacts with the ester to form an organometallic compound.

$$Zn + BrCH_2COOC_2H_5 \xrightarrow{\text{Ether}} BrZnCH_2COOC_2H_5$$

Step 2. Organometallic compound reacts with the carbonyl compound by nucleophilic addition.

$$>C = O + BrZnCH_2COOC_2H_5 \longrightarrow \underset{\text{Addition Product}}{>C\!<\!\!\begin{matrix} OZnBr \\ CH_2COOC_2H_5 \end{matrix}}$$

Step 3. The addition product formed above gets hydrolysed to give the final product.

$$>C\!<\!\!\begin{matrix} OZnBr \\ CH_2COOC_2H_5 \end{matrix} \xrightarrow{H^+, H_2O} >C\!<\!\!\begin{matrix} OH \\ CH_2COOC_2H_5 \end{matrix} + Zn\!<\!\!\begin{matrix} OH \\ Br \end{matrix}$$

Q. 26. Describe the mechanism of Aldol condensation.

(*Punjab, 2003; Bangalore, 2004; Assam, 2004; Patna, 2005; Vidyasagar, 2007; Lucknow, 2010*)

Ans. Two molecules of an aldehyde or a ketone having α hydrogen atom react in the presence of dil. alkali to form β-hydroxyaldehyde or β-hydroxy ketone.

$$CH_3-\overset{\overset{O}{\|}}{\underset{\underset{H}{|}}{C}} + \underset{\text{Acetaldehyde}}{HCH_2CHO} \xrightarrow{OH^-} CH_3-\overset{\overset{OH}{|}}{\underset{\underset{H}{|}}{C}}-CH_2CHO$$

Acetaldehyde 3-hydroxybutanol (Aldol)

$$CH_3-\overset{\overset{O}{\|}}{\underset{\underset{CH_3}{|}}{C}} + \underset{\text{Acetone}}{HCH_2COCH_3} \xrightarrow{OH^-} CH_3-\overset{\overset{OH}{|}}{\underset{\underset{CH_3}{|}}{C}}-CH_2COCH_3$$

Acetone Diacetone alcohol

Mechanism of the Reaction

Step 1. Formation of carbanion. The hydroxide ion from a base removes the proton from the α-carbon atom to give a carbanion. This carbanion gets resonance stabilised.

Step 2. Attack by carbanion. The carbanion thus formed attacks the second molecule by nucleophilic addition mechanism to form an alkoxide ion.

Step 3. Attachment of a hydrogen. The alkoxide ion attaches a hydrogen atom removed from water molecule.

These steps as mentioned above are sketched as under :

$$H\bar{O} + H-CH_2-\underset{\underset{H}{|}}{C}=O \rightleftharpoons \left[\bar{C}H_2-\underset{\underset{H}{|}}{C}=O \longleftrightarrow CH_2=\underset{\underset{H}{|}}{C}-\bar{O}\right] + H_2O$$

$$CH_3-\overset{\overset{H}{|}}{\underset{\underset{O}{\|}}{C}} \rightleftharpoons CH_3-\overset{\overset{H}{|}}{\underset{\underset{O^-}{|}}{C}}-CH_2-CHO \underset{}{\overset{H_2O}{\rightleftharpoons}} CH_3-\overset{\overset{H}{|}}{\underset{\underset{OH}{|}}{C}}-CH_2-CHO + OH^-$$

It may be noted that aldol condensation is a property of aldehydes and ketones having hydrogen atoms attached to α-carbon atoms. Hence this reaction is not observed with compounds like formaldehyde, benzaldehyde which don't process α-hydrogen atoms.

Another point which may be noted is, that β-hydroxy-aldehydes and β-hydroxy ketones which are obtained in the base-catalysed reaction easily get dehydrated on warming with dilute acids to form α - β unsaturated carbonyl compounds.

$$\underset{\text{Aldol}}{CH_3\underset{\underset{OH}{|}}{C}HCH_2CHO} \xrightarrow[-H_2O]{H^+} \underset{\text{Crotonaldehyde}}{CH_3CH=CHCHO}$$

This is the reason we avoid obtaining aldol by acid-catalysed reaction, although, in principle, it is possible to carry out aldol condensation in the presence of an acid.

Aldol reaction provides an important tool in the hands of a chemist to synthesize α, β unsaturated compounds.

Q. 27. Write a note on Crossed Aldol Condensation.

Ans. We may come across cases involving reaction between two different aldehydes or ketones or an aldehyde and a ketone. Such reactions are known as crossed aldol condensation. If both the reacting carbonyl compounds have α-hydrogen atoms, we may get four different products which are difficult to separate. But by adjusting experimental condition it is possible to obtain one product in abundance. Out of the two carbonyl compounds, the one having an α-hydrogen atom forms the carbanion and the other having no α-hydrogen atom acts as the site for attack by the nucleophile carbanion.

$$\underset{\text{Formaldehyde}}{\overset{\overset{\displaystyle H}{|}}{H-C=O}} + \underset{\text{Acetaldehye}}{H-CH_2CHO} \xrightarrow{OH^-} \underset{\text{3-Hydroxypropionaldehyde}}{\overset{\overset{\displaystyle H}{|}}{\underset{\underset{\displaystyle CH_2CHO}{|}}{H-C-OH}}}$$

$$\underset{\text{Formaldehyde}}{\overset{\overset{\displaystyle H}{|}}{H-C=O}} + \underset{\text{Acetone}}{HCH_2COCH_3} \xrightarrow{OH^-} \underset{\text{4-Hydroxy-2-butanone}}{\overset{\overset{\displaystyle H}{|}}{\underset{\underset{\displaystyle CH_2COCH_3}{|}}{H-C-OH}}}$$

Mechanism of crossed aldol condensation in the same as that of aldol condensation.

Q. 28. Briefly discuss Claisen-Schmidt reaction. (*Patna 2005*)

Or

Suggest a mechanism for the following conversion. Name the product

$$\mathbf{C_6H_5CHO + CH_3CHO \longrightarrow C_6H_5CH = CHCHO}$$ (*Kerala, 2000*)

Ans. This reaction may be considered as a special case of crossed aldol condensation. It involves condensation of aromatic carbonyl compound with an aliphatic carbonyl compound in the presence of a base. Hydroxy derivative first obtained loses a water molecule to yield α β unsaturated aldehyde or ketone. For example

$$\underset{\text{Benzaldehyde}}{C_6H_5-CHO} + \underset{\text{Acetaldehyde}}{CH_3CHO} \xrightarrow{OH^-} \underset{\text{Addition compound}}{C_6H_5-\overset{\overset{\displaystyle OH}{|}}{CH}-CH_2CHO} \xrightarrow{-H_2O} \underset{\text{Cinnamaldehyde}}{C_6H_5-CH=CHCHO}$$

$$C_6H_5-CHO + CH_3COCH_3 \xrightarrow{OH^-} \underset{\text{Addition compound}}{C_6H_5-\overset{\overset{\displaystyle OH}{|}}{CH}-CH_2COCH_3} \xrightarrow{-H_2O} \underset{\text{Benzalacetone}}{C_6H_5-CH=CHCOCH_3}$$

Mechanism. It involves the following steps :

Step 1. Formation of carbanion. α-hydrogen of the aliphatic carbonyl compound is removed by the hydroxide ions, thus creating carbanions.

Step 2. Attack by carbanion on the aromatic carbonyl compound. The carbanion obtained in step 1 attacks the aromatic carbonyl group by nucleophilic addition mechanism to form the alkoxide ion.

Step 3. Attachment of a Proton. The alkoxide ion removes a proton from water and attaches to itself to give β-hydroxy compound.

Step 4. α-hydrogen atom in the hydroxy compound, is removed by the base followed by the removal of —OH group, thus removing a water molecule finally. The sequence of the above steps is shown below :

$$HO^- + H-CH_2-CHO \rightleftharpoons \bar{C}H_2-CHO + H_2O$$

$$C_6H_5-\overset{H}{\underset{O}{C}} + \bar{C}H_2CHO \rightleftharpoons C_6H_5-\overset{H}{\underset{O^-}{C}}-CH_2CHO \xrightleftharpoons{H_2O} C_6H_5-\overset{H}{\underset{OH}{C}}-\underset{H}{CH}-CHO + OH^-$$

$$C_6H_5-\overset{H}{\underset{OH}{C}}-\underset{H}{CH}-CHO \xrightleftharpoons{OH^-} C_6H_5-\underset{OH}{CH}-CH-CHO + H_2O$$

$$\downarrow -OH^-$$

$$C_6H_5-CH=CHO$$

Q. 29. Write a note on Benzoin Condensation.

(Calcutta, 2007; Nagpur, 2008; Andhra, 2010; Lucknow, 2010

Ans. Aldol condensation is a property of carbonyl compounds containing α-hydrogen atoms. Thus benzaldehyde which does not contain α-hydrogen atom is not capable of giving this reaction. Such aromatic aldehydes condense in the presence of KCN (which acts as a base) to yield *benzoins*. This reaction is known as *Benzoin condensation*.

$$C_6H_5-CHO + OHC-C_6H_5 \xrightarrow{KCN} C_6H_5-\underset{}{\overset{OH}{CH}}-\overset{O}{\overset{\|}{C}}-C_6H_5$$

Benzaldehyde Benzaldehyde Benzoin

Mechanism. It involves the following steps :

Step 1. Formation of a carbanion. Direct formation of a carbanion is not possible, as there are no α-hydrogen atoms. Hence this carbanion formation takes place after some rearrangement. It involves attachment of CN^- to the carbonyl carbon with simultaneous shifting of electrons to oxygen. Hydrogen is then shifted from carbon to oxygen. This results in the formation of carbanion.

Step 2. Attack of the carbanion (nucleophile) on the second molecule. Carbanion obtained in step 1 above attacks the carbonyl group of the second molecule forming an addition product, an alkoxide.

Step 3. Loss of cyanide ion. The alkoxide loses cyanide ion with simultaneous loss of hydrogen from the same carbon atom. Hydrogen combines with the oxygen on the neighbouring carbon atom giving the final products. The sequence of steps is illustrated below:

$$C_6H_5-\overset{\overset{O}{\|}}{C}-H + CN^- \rightleftharpoons C_6H_5-\underset{\underset{CN}{|}}{\overset{\overset{O^-}{|}}{C}}-H \rightleftharpoons C_6H_5-\underset{\underset{CN}{|}}{\overset{\overset{OH}{|}}{C^-}}$$

$$C_6H_5-\underset{\underset{CN}{|}}{\overset{\overset{OH}{|}}{C^-}} + \underset{\underset{H}{|}}{\overset{\overset{O}{\|}}{C}}-C_6H_5 \rightleftharpoons C_6H_5-\underset{\underset{CN}{|}}{\overset{\overset{O-}{|}}{C}}-\underset{\underset{H}{|}}{\overset{\overset{HO^-}{|}}{C}}-C_6H_5$$

$$\Updownarrow$$

$$C_6H_5-\overset{\overset{O}{\|}}{C}-\underset{\underset{H}{|}}{\overset{\overset{OH}{|}}{C}}-C_6H_5 + CN^-$$

Q. 30. Give the mechanism of Perkin's reaction.

(Purvanchal, 2003; Barathidasan, 2003; Vidyasagar, 2007; Lucknow, 2010

Ans. It is a kind of aldol condensation in which an aromatic aldehyde condenses with acetic anhydride in the presence of sodium acetate to yield unsaturated carboxylic acid. Thus benzaldehyde on treatment with acetic anhydride and sod. acetate yields cinnamic acid.

$$\underset{\text{Benzaldehyde}}{C_6H_5-CHO} + \underset{\text{Acetic anhydride}}{(CH_3CO)_2O} \xrightarrow[180^\circ C]{\text{Sod. acetate}} C_6H_5-\underset{\underset{OH}{|}}{CH}-CH_2CO-O-COCH_3$$

$$\xrightarrow{-CH_3COOH} \underset{\text{Cinnamic acid}}{C_6H_5-CH=CHCOOH}$$

Mechanism. It involves the following steps :

Step 1. Formation of a carbanion. Acetic anhydride in the presence of acetate ion loses a hydrogen forming a carbanion.

Step 2. Attack of the carbanion on the carbonyl group. Carbanion obtained in the above step attacks the carbonyl carbon atom (nucleophilic attack) forming on alkoxide.

Step 3. Protonation of the alkoxide ion. Alkoxide gets protonated to form an aldol type compound.

Step 4. Dehydration. The hydroxy group and neighbouring hydrogen are removed as water forming an unsaturated product.

Step 5. Hydration. The alkenic compound obtained above gets hydrolysed giving the final product.

The above steps are illustrated as given below :

$$C_6H_5-\overset{\displaystyle H}{\underset{}{C}}=O + \overset{\ominus}{C}H_2-\overset{O}{\overset{\|}{C}}\!\!\diagdown\!O \diagup CH_3-\underset{\|}{\overset{}{C}}=O \longrightarrow C_6H_5-\overset{H}{\overset{|}{C}}-O^- \text{ with } CH_2-C(=O)-O-C(=O)-CH_3$$

Carbanion

$\downarrow H_2O$

$$C_6H_5-CH(OH)-CH(H)-C(=O)-O-C(=O)-CH_3 + OH^-$$

$\xleftarrow{-H_2O}$

$$C_6H_5-CH=CH-C(=O)-O-C(=O)-CH_3$$

$\downarrow H_2O$

$$C_6H_5-CH=CH-\overset{O}{\overset{\|}{C}}-O-H + CH_3COO^{\ominus}$$

Cinnamic acid

Q. 31. Discuss the mechanism of Knoevenagal Reaction.

(*Meerut, 2000; Kerala, 2001; Purvanchal 2007*)

Ans. An aldehyde or a ketone on condensation with a compound having an active methylene group (such as malonic ester or acetoacetic ester) in the presence of a base such as pyridine gives an unsaturated compound.

$$\underset{\text{Acetaldehyde}}{CH_3-\overset{H}{\overset{|}{C}}=O} + \underset{\text{Malonic ester}}{H_2C\begin{cases}COOC_2H_5\\COOC_2H_5\end{cases}} \xrightarrow[-H_2O]{\text{Pyridine}} CH_3-\overset{H}{\overset{|}{C}}=C\begin{cases}COOC_2H_5\\COOC_2H_5\end{cases}$$

$\downarrow H_2O, H^+$ and heat

$$\underset{\text{Crontonic acid}}{CH_3CH=CHCOOH}$$

$$CH_3-\overset{CH_3}{\overset{|}{C}}=O + \underset{\text{Acetoacetic ester}}{H_2C\begin{cases}COCH_3\\COOC_2H_5\end{cases}} \xrightarrow[-H_2O]{\text{Pyridine}} CH_3-\overset{CH_3}{\overset{|}{C}}=C\begin{cases}COCH_3\\COOC_2H_5\end{cases}$$

$\downarrow H_2O, H^+$

$$CH_3-\overset{CH_3}{\overset{|}{C}}=\underset{\underset{COOC_2H_5}{|}}{CHCOCH_3}$$

Mechanism. Mechanism of the reaction is explained by taking the example of reaction between an aldehyde and aceto acetic ester.

Step 1. Formation of a carbanion. The base B : attracts a hydrogen from the acetoacetic ester forming a carbanion.

Step 2. Attack by the carbanion. The carbanion as obtained above attacks the carbonyl carbon by nucleophilic chains forming an alkoxide ion.

Step 3. Protonation. The alkoxide draws a proton to convert itself into a hydroxy compound.

Step 4. Dehydration. Hydroxy and hydrogen from the neighbouring carbon atoms are removed as water, and an unsaturated compound is obtained. the above steps are illustrated as follows :

$$B: + H_2C\begin{cases} COCH_3 \\ COOC_2H_5 \end{cases} \longrightarrow H\bar{C}\begin{cases} COCH_3 \\ COOC_2H_5 \end{cases} + BH^+$$

$$R-\overset{\overset{H}{|}}{\underset{\underset{O}{||}}{C}} + H\bar{C}\begin{cases} COCH_3 \\ COOC_2H_5 \end{cases} \longrightarrow R-\overset{\overset{H}{|}}{\underset{\underset{O^-}{|}}{C}}-\underset{\underset{COCH_3}{|}}{CH}-COOC_2H_5$$

$$\xrightarrow{H^+} R-\overset{\overset{H}{|}}{\underset{\underset{OH}{|}}{C}}-\underset{\underset{COCH_3}{|}}{CH}COOC_2H_5 \xrightarrow{-H_2O} R-CH=\underset{\underset{COCH_3}{|}}{C}-COOC_2H_5$$

Q. 32. Write a note on Mannich Reaction.

(*Awadh 2000;; Kumaon 2000; Shivaji 2000; Kurukshetra, 2001; Nagpur, 2005; Purvanchal 2007*)

Ans. Formaldehyde on condensation with an amine and an active methylene compound gives a product which is called *Mannich Base*. This reaction is called Mannich reaction.

$$C_6H_5-\overset{\overset{O}{||}}{C}-CH_3 + HCHO + (CH_3)_2NH \xrightarrow{-H_2O} \underset{\text{Mannich base}}{C_6H_5-\overset{\overset{O}{||}}{C}-CH_2-CH_2-N(CH_3)_2}$$

Mechanism. It involves the following steps :

Step 1. Formation of Imminium ion. Formaldehyde reacts with the amine to form an imminium salt which gives rise to the imminium ion.

Step 2. Formation of the carbanion. The active methylene compound in the presence of a base gives the carbanion.

Step 3. Attack by the carbanion. Carbanion formed in the above step attacks the imminium ion to give the final product. The above steps are illustrated ahead.

$$(CH_3)_2N-H + \underset{H}{\overset{H}{C}}=O \rightleftharpoons (CH_3)_2N-\underset{H}{\overset{H}{C}}-OH \rightleftharpoons (CH_3)_2\overset{+}{N}=CH_2 + OH$$

Imminium salt   Imminium ion

$$R-\overset{O}{\overset{\|}{C}}-CH_3 \overset{Base}{\rightleftharpoons} \left[R-\overset{O}{\overset{\|}{C}}-\overline{C}H_2 \longleftrightarrow R-\overset{O^-}{\overset{|}{C}}=CH_2\right]$$

$$H_2C=\overset{+}{N}(CH_3)_2 \longrightarrow R-\overset{O}{\overset{\|}{C}}-CH_2-CH_2-N(CH_3)_2$$

Q. 33. Discuss the mechanism of Wittig's Reaction.

(Kurukshetra, 2001; M. Dayanand, 2000; Punjab, 2003; Madras, 2004; Nagpur 2008)

Ans. This reaction takes place between an aldehyde or ketone and phosphorus ylides to form substituted alkenes. The phosphorus ylides required in the reaction are obtained by the action of a base on suitable alkyl triphenyl phosphonium halides (prepared from alkyl halide and triphenyl phosphine). For example

$$\underset{\text{Triphenyl phosphorus}}{(C_6H_5)_3P} + \underset{\text{Methyl iodide}}{CH_3I} \longrightarrow [(C_6H_5)_3\overset{+}{P}CH_3]\overset{-}{I} \xrightarrow{C_6H_5Li\,(Base)} \underset{\text{Methylene triphenyl phosphorane (ylide)}}{(C_6H_5)_3P=CH_2} + C_6H_6 + LiI$$

The ylides obtained as shown as above are not separated but are made to react as such with the aldehydes or ketones in solution.

Some examples of Wittig reaction are given below :

$$\rangle C=\boxed{O+(C_6H_5)_3P=}\,CRR' \longrightarrow \underset{\text{Alkene}}{\rangle C=C\langle}^{R}_{R'} + \underset{\text{Triphenyl phosphine oxide}}{(C_6H_5)_3PO}$$

$$C_6H_5-CHO + \underset{\text{Methylene triphenyl Phosphorane}}{(C_6H_5)_3P=CH_2} \longrightarrow \underset{\text{Styrene}}{C_6H_5-CH=CH_2} + (C_6H_5)_3PO$$

$$\frac{CH_3}{CH_3}\rangle C=O + \underset{\text{Ethylidene triphenyl phosphorane}}{(C_6H_5)_3P=CH-CH_3}$$

$$\longrightarrow \underset{\text{2-Methylbutene-2}}{\frac{CH_3}{CH_3}\rangle C=CH-CH_3} + (C_6H_5)_3PO$$

Mechanism. The following steps are involved in the reaction :

Step 1. Attack by ylides. The phosphorus ylides exists as a resonance hybrid of two structures.

$$\underset{[A]}{(C_6H_5)_3\,P = CRR'} \quad \longleftrightarrow \quad \underset{[B]}{(C_6H_5)_3\,\overset{+}{P} - \overset{-}{C}RR'}$$

Structure [B] has a negative charge on carbon, thus it acts as a carbanion and initiates the nucleophilic attack on the carbonyl carbon. The resulting addition product is known as *betaine*.

Step 2. Elimination of Triphenyl phosphine oxide. Betaine obtained above undergoes elimination of triphenyl phosphine oxide to yield the alkene. The above steps are illustrated hereunder.

$$>C=O + \overset{-}{C}RR'\text{—}{}^{+}P(C_6H_5)_3 \longrightarrow \underset{\text{Betaine}}{-\overset{|}{\underset{O^-}{C}}-\underset{{}^{+}P(C_6H_5)_3}{CRR'}} \longrightarrow >C = CRR' + (C_6H_5)_3\,PO$$

Q. 34. Suggest a mechanism for the following conversion. Name the product.

$$\text{Cyclohexanone}{=}O \longrightarrow \text{Cyclohexane}{=}CH_2$$

Ans.

$$\underset{\text{Cyclohexanone}}{C_6H_{10}{=}O} + \underset{\text{Methylene triphenyl phosphorane}}{(C_6H_5)_3P = CH_2} \longrightarrow \underset{\text{Methylene cyclohexane}}{C_6H_{10}{=}CH_2} + \underset{\text{Triphenyl phosphine oxide}}{(C_6H_5)_3PO}$$

Mechanism of the Reaction

$$(C_6H_5)_3P = CH_2 \longrightarrow (C_6H_5)_3\overset{+}{P} - \overset{-}{C}H_2$$

$$C_6H_{10}{=}O \xrightarrow{(C_6H_5)_3\overset{+}{P} - \overset{-}{C}H_2} C_6H_{10}(O^-)\text{—}CH_2\text{—}\overset{+}{P}\text{—}(C_6H_5)_3$$

$$\longrightarrow \underset{\text{Methylene cyclohexane}}{C_6H_{10}{=}CH_2} + (C_6H_5)_3PO$$

Q. 35. Which ylides and carbonyl compounds would you use to prepare

(*a*) $C_6H_5 - CH = \underset{CH_3}{\underset{|}{C}} - CH_3$ **(*b*)** $CH_3 - \underset{CH_3}{\underset{|}{C}} = \underset{C_6H_5}{\underset{|}{C}} - CH_3$

(*c*) $C_2H_5 - \underset{CH_3}{\underset{|}{C}} = \underset{C_6H_5}{\underset{|}{C}} - C_6H_5$

Ans. The following pairs of carbonyl compounds and ylides are used to prepare the above compounds.

	Carbonyl compound		Ylide
(*a*)	C_6H_5-CHO (Benzaldehyde)	+	$(C_6H_5)_3P=C(CH_3)-CH_3$
or	$CH_3-C(CH_3)=O$ (Acetone)	+	$(C_6H_5)_3P=CHC_6H_5$
(*b*)	$CH_3-C(CH_3)=O$ (Acetone)	+	$(C_6H_5)_3P=C(C_6H_5)-CH_3$
or	$CH_3-C(C_6H_5)=O$ (Methyl phenyl ketone)	+	$(C_6H_5)_3P=C(CH_3)-CH_3$
(*c*)	$C_2H_5-C(CH_3)=O$ (Methyl ethyl ketone)	+	$(C_6H_5)_3P=C(C_6H_5)-C_6H_5$
or	$C_6H_5-C(C_6H_5)=O$ (Diphenyl ketone)	+	$(C_6H_5)_3P=C(CH_3)-C_2H_5$

Q. 36. Give the mechanism of Cannizzaro reaction.

(Purvanchal, 2003; Punjab, 2003; Pondicherry, 2003, Karnataka, 2004; Nagpur, 2005; Vidyasagar, 2007; Lucknow, 2010

Ans. Aldehydes having no α-hydrogen like formaldehyde and benzaidehyde undergo selt oxidation and reduction. This reaction is given when an aldehyde is treated with a conc. solution of sodium or potassium hydroxide. An alcohol and a salt of carboxylic acid are produced. For example:

$$2\,C_6H_5-CHO + NaOH \longrightarrow C_6H_5-CH_2OH + C_6H_5-COONa$$

Benzaldehyde — Benzyl alcohol — Sodium bezoate

$$2HCHO + NaOH \longrightarrow CH_3OH + HCOONa$$

Formaldehyde — Methanol — Sod. formate

It is evident that one molecule of the aldehyde is oxidised to acid and the second molecule is reduced to alcohol.

Mechanism. The following steps are involved:

Step 1. Attack by OH^-. Hydroxide ion attacks the aldehyde molecule to form an anion.

Step 2. Hydride ion transfer. Oxidation-reduction takes place between the anion formed in step 1 and the second molecule of aldehyde through transfer of a hydride ion. These two steps are illustrated hereunder.

$$C_6H_5-\overset{O}{\overset{\|}{C}}-H + \bar{O}H \rightleftharpoons C_6H_5-\underset{OH}{\underset{|}{\overset{O^-}{\overset{|}{C}}}}-H$$

$$C_6H_5-\underset{OH}{\underset{|}{\overset{O^-}{\overset{|}{C}}}}-H \quad C_6H_5-\overset{O}{\overset{\|}{C}}-H \longrightarrow C_6H_5-\underset{OH}{\underset{|}{\overset{O}{\overset{\|}{C}}}} + C_6H_5-\underset{H}{\underset{|}{\overset{O^-}{\overset{|}{C}}}}-H$$

$$\xrightarrow{-H^+} C_6H_5COO^- \qquad \xrightarrow{+H^+} C_6H_5CH_2OH$$

Q. 37. Describe with mechanism halogenation of carbonyl compounds.

Or

Suggest a mechanism for the following conversion. Name the product.

$$\mathbf{CH_3-\overset{O}{\overset{\|}{C}}-CH_3 \longrightarrow Br-CH_2-\overset{O}{\overset{\|}{C}}-CH_3}$$

Ans. The α hydrogen atoms in the carbonyl compounds can be replaced by halogens in the presence of acids and bases. These are discussed below separately as base-promoted halogenation and acid-promoted halogenation.

Base-Promoted halogenation. One or more α hydrogen atoms in the carbonyl compound can be replaced by halogens in the presence of a base like OH^- or CH_3COO^- ions.

$$\underset{\text{Acetone}}{CH_3COCH_3} + Br_2 + OH^- \longrightarrow \underset{\text{Bromo acetone}}{CH_3COCH_2Br} + H_2O + Br^-$$

Mechanism

Kinetics studies reveal that the rate of above reaction is proportional to the concentration of the carbonyl compound and the base. It is independent of the concentration of bromine.

Rate of halogenation ∝ [Carbonyl compound] × [Base].

This is explained in terms of formation of a carbanion intermediate as the slow and rate determining step followed by fast bromination process as illustrated hereunder.

$$CH_3-\overset{O}{\overset{\|}{C}}-CH_3 + :B \underset{\text{Slow}}{\overset{-BH}{\rightleftharpoons}} \left[CH_3-\overset{O}{\overset{\|}{C}}-\bar{C}H_2 \longleftrightarrow CH_3-\overset{O^-}{\overset{|}{C}}=CH_2 \right] \underset{\text{Fast}}{\xrightarrow{Br_2}} CH_3-\overset{O}{\overset{\|}{C}}-CH_2Br + Br^-$$

As the base is actually consumed in the rate determining step, we call it base promoted reaction rather than base-catalysed reaction.

Certain observations have been made in respect of base promoted halogenation reaction.

(1) In base promoted halogenation of unsymmetrical ketones (RCOR′), halogenation occurs preferably at the methyl group rather than the methylene group.

$$CH_3CH_2COCH_3 + Br_2 \xrightarrow{\text{Base}} CH_3CH_2COCH_2Br + HBr$$

This is because the electron releasing effect of R alkyl group ($CH_3 - CH_2 -$ in the above case) decreases the acidity of the methylene hydrogen atom and prevents the formation of the carbanion. This results in the preferential formation of carbanion (I) which is more stable than the carbanion (II).

$$CH_3 \rightarrow CH_2 - \overset{\overset{O}{\|}}{C} - \bar{C}H_2 \qquad CH_3 - \bar{C}H - \overset{\overset{O}{\|}}{C} - CH_3$$

(I) (II)

(2) Base promoted halogenation of a monohalogenated ketone takes place preferentially at the carbon already halogenated as long as it has a hydrogen atom. This is because the hydrogen atom attached to carbon containing halogen atoms are more acidic and likely to be replaced by halogen as per the above mechanism. The two carbanions resulting from the halogenated ketone are written below. Obviously carbanion III is more easily formed than IV. Hence further halogenation occurs at carbon already halogenated.

$$CH_3 - \overset{\overset{O}{\|}}{C} - \bar{C}H \rightarrow Br \qquad \bar{C}H_2 - \overset{\overset{O}{\|}}{C} - CH_2Br$$

(III) (IV)

Acid-promoted Halogenation. Hence halogenation of the carbonyl compound takes place in the presence of H^+ ions.

$$CH_3 - CO - CH_3 + Br_2 \xrightarrow{H^+} CH_3 - \overset{\overset{O}{\|}}{C} - CH_2Br + HBr$$

Mechanism. Like in case of base-promoted reaction, here, too, the rate of the reaction is proportional to the concentration of the carbonyl compound and the acid. It is independent of the concentration of the halogen.

Rate of halogenation $\propto$ [Carbonyl compound.] [H^+].

This shows the possibility of formation of intermediate in the rate-determining slow step of the reaction followed by fast attachment of the halogen. The intermediate in acid-promoted halogenation is an *enol*. The steps involved in the reaction are reproduced as follows:

(I) $$CH_3 - \overset{\overset{O}{\|}}{C} - CH_3 + H - Z \xrightleftharpoons{\text{Fast}} CH_3 - \overset{\overset{O}{\|}}{C} - CH_2Br + HBr$$

(II) $$CH_3 - \overset{\overset{\overset{+}{O}H}{\|}}{C} - CH_2 - H + \bar{Z} \xrightarrow{\text{Slow}} \underset{\text{Enol}}{CH_3 - \overset{\overset{:\ddot{O}H}{|}}{C} = CH_2} + H - Z$$

(III) $$CH_3 - \overset{\overset{:\ddot{O}H}{|}}{C} = CH_2 + X - X \xrightarrow{\text{Fast}} \underset{\text{Protonated Haloketone}}{CH_3 - \overset{\overset{OH}{\|}}{C} - CH_2X} + X^-$$

$$\text{(IV)}\quad CH_3-\overset{\overset{+}{O}-H}{\overset{\|}{C}}-CH_2X+Z^- \xrightarrow{\text{Fast}} CH_3-\overset{O}{\overset{\|}{C}}-CH_2X+H-Z$$

Following observations have been made in respect of acid-promoted reactions.

(1) In the acid-catalysed halogenation of ketones, halogenation takes place preferably at the methylene group rather than the methyl group. Thus in the compound $CH_3CH_2COCH_3$, halogenation would yield $CH_3CHBr\ COCH_3$ and not $CH_3CH_2COCH_2Br$. This is explained by the fact that enols which are more branched are more stable. The enols which would be formed with halogenation at the methylene and methyl groups respectively are given below :

$$CH_3CH_2-\overset{OH}{\overset{|}{C}}=CH_2 \qquad\qquad CH_3CH=\overset{OH}{\overset{|}{C}}-CH_3$$

(I) Less branched less stable (II) More branched more stable

Less branched less stableMore branched more stable

(2) Further halogenation of the monohalogen ketones does not take place readily because the halogen already present tends to withdraw the electrons towards itself. This prevents the attachment of hydrogen ion on the carbonyl oxygen.

Aromatic aldehydes and ketones undergo electrophilic substitution reaction with the halogen in the presence of halogen carriers like FeX_3. The halogen enters at the meta position w.r.t. the carbonyl group.

$$C_6H_5CHO + Cl_2 \xrightarrow{FeCl_3} m\text{-}ClC_6H_4CHO + HCl$$

m-chloro-benzaldehyde

Q. 38. Write notes on

(*i*) Clemmensen Reduction

(*ii*) Wolff-Kishner Reduction. (*Punjab, 2003; Vidyasagar, 2007; Purvanchal, 2007; Mumbai, 2010*)

Ans. (*i*) Clemmensen Reduction. Aldehydes and ketones give hydrocarbons on reduction with Zn amalgam and conc. hydrochloric acid. Thus > C = O group is reduced to $-CH_2-$ group.

$$CH_3COCH_3 + 4[H] \xrightarrow[HCl]{Zn-Hg} CH_3CH_2CH_3 + H_2O$$

This reaction is known as Clemmensen Reduction.

(*ii*) Wolff-Kishner Reduction. Aldehydes and ketones on being heated with hydrazine and a strong base like KOH at 450–470 K form hydrocarbon.

$$C_6H_5COCH_3 + 4[H] \xrightarrow[KOH]{NH_2NH_2} C_6H_5CH_2CH_3 + H_2O$$

Methyl phenyl ketone Ethylbenzene

Q. 39. Describe the reduction of carbonyl compounds into alcohols.

Ans. Reduction of aldehydes and ketones into alcohols can be achieved by the following different methods.

1. Hydrogenation. Aldehydes and ketones on hydrogenation with molecular hydrogen in the presence of nickel produce primary and secondary alcohols respectively.

$$\underset{\text{Propanal}}{CH_3CH_2CHO} + H_2 \xrightarrow{Ni} \underset{\text{Propanol}}{CH_3CH_2CH_2OH}$$

$$\underset{\text{Propanone}}{CH_3COCH_3} + H_2 \xrightarrow{Ni} \underset{\text{Propanol-2}}{CH_3CHOHCH_3}$$

However, this method has the disadvantage of reducing some other reducible group such as a double bond in the compound also.

$$\underset{\text{Crotonaldehyde}}{CH_3CH = CHCHO} + H_2 \xrightarrow{Ni} \underset{\text{Butanol-1}}{CH_3CH_2CH_2CH_2OH}$$

2. Reduction with $LiAlH_4$. This reagent reduces the carbonyl group into alcoholic group without touching the double bond.

$$\underset{\text{Crotonaldehyde}}{CH_3CH = CHCHO} \xrightarrow[{[H]}]{LiAlH_4} \underset{\text{Crotonyl alcohol}}{CH_3CH = CHCH_2OH}$$

3. Reduction with $NaBH_4$. Sodium borohydride has the same function as $LiAlH_4$ although the former is much milder than the latter.

$$\underset{\text{Cinnamaldehyde}}{C_6H_5-CH = CHCHO} \xrightarrow{NaBH_2} \underset{\text{Cinnamyl alcohol}}{C_6H_5-CH = CHCH_2OH}$$

Sodium borohydride is uneffective on ethylenic double bond, nitro group and nitrile group.

Q. 40. Describe the mechanism of reduction of acetone with lithium aluminium hydride.

Ans. It involves the following steps :

Step 1. Formation of alkoxide. Hydride ion from $LiAlH_4$ combines with carbonyl carbon as shown ahead.

$$LiAlH_4 \longrightarrow Li^+ + Al^{3+} + 4H^-$$

$$CH_3-\overset{\overset{\displaystyle O}{\|}}{C}-CH_3 + :\bar{H} \longrightarrow CH_3-\overset{\overset{\displaystyle O^-}{|}}{\underset{\underset{\displaystyle H}{|}}{C}}-CH_3$$

Step 2. Lithium ion combines with the alkoxide ion.

$$CH_3-\overset{\overset{\displaystyle O^-}{|}}{\underset{\underset{\displaystyle H}{|}}{C}}-CH_3 + Li^+ \longrightarrow CH_3-\overset{\overset{\displaystyle OLi}{|}}{\underset{\underset{\displaystyle H}{|}}{C}}-CH_3$$

Step 3. Lithium alkoxide is changed into alcohol by adding dilute acid.

$$CH_3-\underset{H}{\overset{OLi}{\underset{|}{\overset{|}{C}}}}-CH_3+\overset{+}{H} \longrightarrow CH_3-\underset{H}{\overset{OH}{\underset{|}{\overset{|}{C}}}}-CH_3+\overset{+}{Li}$$

Isopropyl alcohol

Q. 41. Write notes on (i) Biomolecular reduction (ii) Reductive amination of carbonyl compounds.

Ans. (*i*) **Bimolecular Reduction.** Aldehydes and ketones undergo bimolecular reduction in the presence of Mg amalgam to form symmetrical glycols known as **pinacols.**

$$2\ (CH_3)_2C=O \xrightarrow[\text{Benzene}]{\text{Mg Amalgam}} \begin{matrix} (CH_3)_2C-O \\ | \\ (CH_3)_2C-O \end{matrix}\!\!>Mg \xrightarrow{H_2O} \begin{matrix} (CH_3)_2C-OH \\ | \\ (CH_3)_2C-OH \end{matrix}$$

Acetone — Pinacol

Mechanism. It involves formation of a bond between two carbonyl carbon atoms as shown below:

$$\begin{matrix} R_2C=O \\ R_2C=O \end{matrix} + Mg \longrightarrow \begin{matrix} R_2C-O \\ | \\ R_2C-O \end{matrix}\!\!>Mg \xrightarrow{H_2O} \begin{matrix} R_2C-OH \\ | \\ R_2C-OH \end{matrix}$$

(*ii*) **Reductive Amination.** Aldehydes and ketones on treatment with hydrogen and ammonia in the presence of nickel are converted into amines. This reaction is known as reductive amination.

$$C_6H_5-CHO \xrightarrow[\text{Ni}]{H_2,\ NH_3} C_6H_5-CH_2NH_2$$

Benzaldehyde — Benzylamine

$$CH_3CH_2COCH_3 \xrightarrow[\text{Ni}]{H_2,\ NH_3} CH_3CH_2\underset{}{\overset{NH_2}{\overset{|}{CH}}}-CH_3$$

Methyl ethyl ketone — 2-amino butane

Q. 42. Discuss oxidation of (*a*) aldehydes (*b*) ketones with :

1. Tollen's reagent

2. Fehling solution

Ans. (*a*) Oxidation of Aldehydes. 1. Tollen's reagent is a mixture of silver nitrate and ammonium hydroxide. Aldehydes reduce the Tollens reaent on heating, silver is deposited on the inner walls of the test tube. This is used as a test for the aldehydic group and is known as **silver mirror test.**

$$RCHO + 2[Ag(NH_3)_2]^+[OH]^- \longrightarrow RCOONH_4 + \underset{\text{Silver mirror}}{2Ag} + 3NH_3 + H_2O$$

2. Fehling solution is a mixture of copper sulphate, sodium hydroxide and sod. pot. tartarate. The resulting solution liberates very small amounts of copper ions. Aldehydes on warming with Fehling solution produces a red precipitate of cuprous oxide.

$$RCHO + \underset{\text{Cupric ion}}{2Cu^{2+}} + Na^+ + 5[OH]^- \longrightarrow RCOONa + \underset{\text{Cuprous oxide (Red ppt.)}}{Cu_2O} + 3H_2O$$

(*b*) Oxidation of Ketones. Ketones don't contain an easily oxidisable group. As such they don't react with Tollen's reagent or Fehling solution. However, in the presence of strong oxidising agents and on prolonged heating, these are oxidised to carboxy acids containing fewer carbon atoms than the original ketone.

$$CH_3CH_2CH_2COCH_2CH_3 \xrightarrow[K_2Cr_2O_7/H^+]{[O]} \underset{\text{Butanoic acid}}{CH_3CH_2CH_2COOH} + \underset{\text{Ethanoic acid}}{CH_3COOH}$$

Q. 43. Write a note on Haloform Reaction giving its importance.

Or (*Kerala 2001*)

Complete the reaction

$$\mathbf{CH_3CHOHR + 3\ NaOI \longrightarrow}$$

Ans. It is a reaction given by a compound which contains CH_3CO- group or a group oxidisable to this group. Thus acetone CH_3COCH_3, Acetaldehyde CH_3CHO, acetophenone $CH_3COC_6H_5$ and similar other compounds give this reaction.

In this reaction, the compound containing CH_3CO- group is treated with a halogen (chlorine, bromine or iodine) and a dilute solution of sodium hydroxide to form haloform CHX_3 (chloroform, bromoform or iodoform).

$$NaOH + X_2 \longrightarrow NaOX + HX$$

$$CH_3COCH_3 + 3NaOX \longrightarrow \underset{\text{Haloform}}{CHX_3} + CH_3COONa + 2NaOH$$

Compounds like ethanol and ethanal also give this reaction.

$$CH_3CH_2OH + X_2 \longrightarrow CH_3CHO + 2HX$$

$$CH_3CHO + 3X_2 \longrightarrow \underset{\text{Trihalo acetaldehyde}}{CX_3CHO} + 3HX$$

$$CX_3CHO + NaOH \longrightarrow \underset{\text{Haloform}}{CHX_3} + \underset{\text{Sod. formate}}{HCOONa}$$

Thus any compound which contains a CH_3CO- group or compounds like CH_3CH_2OH or CH_3CHO or $CH_3CHOHCH_3$ which on oxidation would produce CH_3CO- group, give this test.

Generally we use I_2 / NaOH mixture in which case iodoform is produced as result of the reaction. Iodoform being a yellow solid is easily identified.

The mechanism of the reaction is the same as that of base promoted halogenation.

Importance of the Reaction. The reaction assumes analytical importance as it can be used to detect a CH_3CO- group in a compound. If the substance responds to this test, it is certain that there is CH_3CO- group or a group oxidisable to this group, in the compound. As has been explained above, the substance is treated with a mixture of iodine and dil. NaOH when a yellow precipitate of iodoform is formed.

$$CH_3COCH_3 + 3NaOI \longrightarrow CH_3COCI_3 + 3\,NaOH$$

$$CH_3COCI_3 + NaOH \longrightarrow CH_3COONa + \underset{\substack{\text{Iodoform} \\ \text{(yellow ppt.)}}}{CHI_3}$$

Q. 44. Identify the compounds which respond to haloform test : CH_3CO_2H, CH_3CH_2OH, CH_3CONH_2, CH_3COCl, $CH_3CHOHCH_3$, CD_3CHO, $C_6H_5COCD_3$ and $CH_3CH_2COCH_2CH_3$ *(Calcutta 2007)*

Ans. Compounds which contain CH_3CO- group or a group which can be oxidised to this group give haloform reaction.

In the list given above, the compounds that respond to haloform test are :

CH_3CH_2OH, $CH_3CHOHCH_3$, CD_3CHO, $C_6H_5COCD_3$

D stands for deuterium *i.e.* heavy hydrogen. The compounds containing heavy hydrogen give the reaction but at a slower speed.

Q. 45. Complete the following reactions indicating against each the name of the reaction.

(*i*) $C_6H_5OH + CHCl_3 + NaOH \longrightarrow$

(*ii*) $CH_3CHO + HCN \xrightarrow{H_3O^+}$

(*iii*) $C_6H_5CHO + (CH_3CO)_2O \xrightarrow{CH_3COONa}$

(*iv*) ⬡$= O + (C_6H_5)_3P = CH_2 \longrightarrow$

Ans. (*i*) $C_6H_5OH + CHCl_3 + 3NaOH$

$$\longrightarrow \underset{\text{Salicyladehyde}}{C_6H_4(OH)CHO} + 3\,NaCl + 2H_2O$$

This reaction is Reimer-Tiemann Reaction

(*ii*)

$$CH_3CHO + HCN \longrightarrow CH_3\underset{}{\overset{OH}{CH}} - CN \xrightarrow{H_3O^+} CH_3\overset{OH}{CH}COOH$$

α-hydroxy propionic acid

This is nucleophilic addition of HCN to acetaldehyde followed by hydrolysis.

(*iii*)

$$C_6H_5CHO + (CH_3CO)_2O \xrightarrow{CH_3COONa} C_6H_5CH{=}CHCOOH$$

Cinnamic acid

This is a case of Perkin's Reaction.

(*iv*)

$$\text{Cyclohexanone}{=}O + (C_6H_5)_3P{=}CH_2 \longrightarrow \text{Cyclohexane}{=}CH_2 + (C_6H_5)_3PO$$

Cyclohexanone; Methylene cyclohexane

This is Wittig's reaction.

Q. 46. Complete the following reactions and identify the products:

(*i*) Acetic anhydride + CH_3COO^- ⟶ A + ?

(*ii*) A + C_6H_5CHO ⟶ B

(*iii*) B is hydrolysed and heated ⟶ C + ?

Ans.

(*i*)

$$\begin{matrix}CH_3CO\\ CH_3CO\end{matrix}\!\!>O + CH_3COO^- \longrightarrow \begin{matrix}\bar{C}H_2CO\\ CH_3CO\end{matrix}\!\!>O + CH_3COOH$$

Carbanion (A)

(*ii*)

$$\begin{matrix}\bar{C}H_2CO\\ CH_3CO\end{matrix}\!\!>O + C_6H_5CHO \longrightarrow C_6H_5\overset{O^-}{\underset{H}{C}} - \underset{OCOCH_3}{CH_2CO}$$

Addition Product (B)

(*iii*)
$$C_6H_5\underset{\underset{H}{|}}{\overset{\overset{O}{\|}}{C}}-\underset{\underset{OCOCH_3}{|}}{CH_2CO} \xrightarrow[\text{Heat}]{H_3\overset{+}{O}} C_6H_5\,CH=CHCOOH + CH_3COOH$$
(B) Cinnamic acid (C)

Q. 47. What happens when

(*i*) Butanone is treated with $LiAlH_4$?

(*ii*) $CH_3\,CH = CHCHO$ is reduced with $NaBH_4$?

(*iii*) Butanone is reduced with amalgamated Zn and HCl?

(*iv*) Acetophenone is treated with hydrazine and KOH?

Ans.

(*i*)
$$\underset{\text{Butanone}}{CH_3CH_2COCH_3} \xrightarrow[\text{[H]}]{LiAlH_4} \underset{\text{Butanol-2}}{CH_3CH_2CHOHCH_3}$$

(*ii*)
$$CH_3CH=CHCHO \xrightarrow[\text{[H]}]{NaBH_4} \underset{\text{But-2-en-1-ol}}{CH_3CH=CH-CH_2OH}$$

(*iii*)
$$\underset{\text{Butanone}}{CH_3CH_2COCH_3} \xrightarrow[\text{[H]}]{Zn/HCl} \underset{\text{Butane}}{CH_3CH_2CH_2CH_3}$$

(*iv*)
$$C_6H_5COCH_3 \xrightarrow[KOH]{NH_2NH_2} \underset{\text{Ethylbenzene}}{C_6H_5CH_2CH_3}$$

Q. 48. How would you distinguish easily and by chemical means between pure samples of the following pairs of compounds?

(*a*) CH_3CHO and CH_3COCH_3

(*b*) CH_3CHO and CH_3CH_2CHO (*Kerala, 2001*)

(*c*) $CH_3COCH_2CH_2CH_3$ and $CH_3CH_2COCH_2CH_3$

(*d*) C_6H_5 CHO and CH_3CHO (*Annamalai, 2003*)

Ans. (*a*) CH_3CHO and CH_3COCH_3. The above two compounds namely acetaldehyde and acetone can be distinguished from each other by performing Tollen's reagent and Fehling solution tests.

Acetaldehyde gives a silver mirror with Tollen's reagent.

$$CH_3CHO + 2\,[Ag(NH_3)_2]^+\,[OH]^- \longrightarrow CH_3COONH_4 + 3NH_3 + H_2O + \underset{\text{Silver mirror}}{2\,Ag}$$

Also it gives a red ppt. of Cu_2O with Fehling solution.

$$CH_3CHO + 2Cu^{2+} + Na^+ + 5[OH]^- \longrightarrow CH_3COONa + 3H_2O + \underset{\text{Red ppt.}}{Cu_2O}$$

Acetone does not respond to these tests.

(*b*) Acetaldehyde and propionaldehyde. The two compounds can be distinguished from each other by performing the haloform test. Acetaldehyde (CH_3CHO) contains CH_3CO- group. Hence it would give a yellow precipitate of CHI_3 in the haloform test. Propionaldehyde doesn't contain CH_3CO- group. Hence it does not respond to this test.

$$CH_3CHO + 3\,NaOI \longrightarrow CI_3\,CHO + 3\,NaOH$$

$$CI_3CHO + NaOH \longrightarrow \underset{\text{Iodoform}}{CHI_3} + HCOONa$$

(c) Pentanone-2 ($CH_3COC_3H_7$) and pentanone-3 ($C_2H_5COC_2H_5$). Pentanone-2 contains $CH_3CO–$ group, hence it would give the iodoform test. Pentanone-3, does not contain $CH_3CO–$ group. Hence it would not respond to this test. This is how the two compounds can be distinguished from each other.

$$CH_3COC_3H_7 + 3NaOI \longrightarrow CI_3\,COC_3H_7$$

$$CI_3COC_3H_7 + NaOH \longrightarrow \underset{\text{Iodoform}}{CHI_3} + \underset{\text{Sod. butyrate}}{C_3H_7COONa}$$

***(d)* Benzaldehyde and acetaldehyde.** Benzaldehyde on oxidation with alkaline $KMnO_4$ would produce benzoic acid which is a white solid. Acetaldehyde on oxidation under the same condition produces acetic acid which is not a solid and has the typical vineger smell.

$$C_6H_5CHO + [O] \longrightarrow \underset{\text{Benzoic acid}}{C_6H_5COOH}$$

$$CH_3CHO + [O] \longrightarrow \underset{\text{Acetic acid}}{CH_3COOH}$$

Q. 49. Give equations for the reactions, if any, of propionaldehyde with the following reagents:

***(i)* H_2 in the presence of nickel**
***(ii)* Ethanol and HCl gas**
***(iii)* Saturated aqueous $NaHSO_3$ solution**
***(iv)* $K_2Cr_2O_7 + H_2SO_4$**
***(v)* Dil. NaOH solution**
***(vi)* NaOH + Br_2 (or sod. hypobromite)**
***(vii)* Benzaldehyde with dil. NaOH solution**
***(viii)* Phenyl hydrazine**

Ans. *(i)* H_2 in the presence of Ni

$$\underset{\text{Propionaldehyde}}{CH_3CH_2CHO + H_2} \xrightarrow{Ni} \underset{\text{Propyl alcohol}}{CH_3CH_2CH_2OH}$$

***(ii)* Ethanol and HCl gas**

$$CH_3CH_2CHO + 2C_2H_5OH \xrightarrow[-H_2O]{HCl} \underset{\text{Acetal}}{CH_3CH_2CH\begin{cases}OC_2H_5\\OC_2H_5\end{cases}}$$

***(iii)* Saturated aqueous $NaHSO_3$ solution**

$$CH_3CH_2CHO + NaHSO_3 \longrightarrow \underset{\text{Bisulphite compound}}{CH_3CH_2CH\begin{cases}OH\\SO_3Na\end{cases}}$$

***(iv)* $K_2\,Cr_2O_7 + H_2SO_4$**

$$CH_3CH_2CHO + O \xrightarrow[H_2SO_4]{K_2Cr_2O_7} \underset{\text{Propanoic acid}}{CH_3CH_2COOH}$$

(*v*) **Dil. NaOH solution**

$$CH_3CH_2CHO + CH_3CH_2CHO \xrightarrow{NaOH} CH_3CH_2CH(OH)\underset{|}{\overset{CH_3}{CH}}CHO$$

3-hydroxy-2-methyl pentanal

(*vi*) **NaOH + Br_2**

There will be no reaction as there is no $CH_3CO–$ group in propionaldehyde.

(*vii*) $C_6H_5CHO + CH_3CH_2CHO \xrightarrow{NaOH} C_6H_5\overset{OH}{\overset{|}{CH}}—\underset{CH_3}{\underset{|}{CH}}CHO$

3-hydroxy-3-phenyl 2-methylpropanal

$$\downarrow -H_2O$$

$$C_6H_5CH=\underset{CH_3}{\underset{|}{C}}—CHO$$

α-Methyl Cinnamaldehyde

(*viii*) **Phenylhydrazine**

$$CH_3CH_2CHO + C_6H_5NHNH_2 \longrightarrow CH_3CH_2CHNNHC_6H_5$$

Acetaldehyde phenyl hydrazone

Q. 50. Identify A, B and C in the following :

(*i*) **Benzaldehyde** $\xrightarrow{NaHSO_3}$ **A**

(*ii*) **Benzaldehyde** $\xrightarrow[2.\ H_3O^+]{1.\ CH_3MgI}$ **A** $\xrightarrow[Oxid.]{Mild}$ **B** $\xrightarrow[OH^-]{Br_2}$ **C**

Ans. (*i*) $C_6H_5CHO \xrightarrow{NaHSO_3} C_6H_5CH<\begin{matrix} OH \\ SO_3Na \end{matrix}$

Benzaldehyde; Bisulphite Compound [A]

(*ii*) $C_6H_5CHO \xrightarrow[H_3O^+]{CH_3MgI} C_6H_5CH<\begin{matrix} OMgI \\ CH_3 \end{matrix} \xrightarrow[-Mg(OH)I]{H_3O^+} C_6H_5CH(OH)CH_3$

Phenyl Methyl Carbinol [A]

$$C_6H_5CH(OH)CH_3 \xrightarrow[Oxid.]{Mild} \begin{matrix} C_6H_5 \\ CH_3 \end{matrix}>CO \xrightarrow[KOH]{Br_2} C_6H_5COONa + CHBr_3$$

Acetophenone [B]; Sod. benzoate [C]; Bromoform

Q. 51. Complete the following equations and give the names of the reactions involved in each case

(i) $C_6H_5COCH_3 + HCHO + (CH_3)_2NH \longrightarrow$

(ii) $C_6H_5CHO + CH_3COCH_3 \xrightarrow{Base}$

(iii) $2HCHO + NaOH \longrightarrow$ *(Lucknow 2000)*

Ans. (i) $C_6H_5COCH_3 + HCHO + (CH_3)_2NH \xrightarrow{-H_2O} \underset{\text{Mannich base}}{C_6H_5COCH_2CH_2-N(CH_3)_2}$

This reaction is called **Mannich** reaction.

(ii) $C_6H_5CHO + CH_3COCH_3 \xrightarrow{NaOH} C_6H_5CH(OH)CH_2COCH_3$

This reaction is called **Claisen** condensation.

(iii) $HCHO + HCHO + NaOH \longrightarrow \underset{\text{Methanol}}{CH_3OH} + \underset{\text{Sol. formate}}{HCOONa}$

This reaction is called **Cannizzaro** reaction.

Q. 52. Arrange the following in increasing order of reactivity: CH_3CHO, HCHO, CH_3CH_2CHO:

Ans. The reactivity in the case of the carbonyl compounds depends upon the size of the alkyl groups attached to carbonyl groups.

$$\begin{matrix} R \\ \text{(or H) } R' \end{matrix} \Big\rangle C=O$$

Smaller the size of the alkyl groups, greater is the reactivity. Hence with this criteria, the reactivity in the decreasing order will be as follows :

$$CH_3CH_2CHO < CH_3CHO < HCHO$$

Q. 53. An organic compound A, $C_5H_{10}O$ forms chain isomers B and C. B undergoes Cannizzaro reaction and C undergoes aldol condensation and forms 3-hydroxy-2-propyl heptanal. Identify A, B and C and give their IUPAC names. Also give the outlines of the mechanism of Cannizzaro reaction as applicable to B.

Ans. It is obvious from the above that A, B and C are carbonyl compounds. Also B is an aldehyde with no α-hydrogen as only such compounds give Cannizzaro reaction. C is an aldehyde containing α-hydrogen in order to be able to give aldol condensation. Based on the above observations the probable structures of A, B and C are as follows :

[A] $\underset{\text{2-Methylbutanal}}{CH_3CH_2\underset{|}{\overset{}{C}}H\text{—}CHO}$ with CH_3 on the CH; or $\underset{\text{3-Methylbutanal}}{CH_3\underset{|}{C}HCH_2CHO}$ with CH_3 on the CH

$$\text{[A]}\quad \underset{\text{2-Methylbutanal}}{\begin{matrix} CH_3CH_2CH\text{—}CHO \\ | \\ CH_3 \end{matrix}} \quad \text{or} \quad \underset{\text{3-Methylbutanal}}{\begin{matrix} CH_3CHCH_2CHO \\ | \\ CH_3 \end{matrix}}$$

$$\text{[B]}\quad \underset{\text{2, 2 Dimethypropanal}}{\begin{matrix} CH_3 \\ | \\ CH_3\text{—}C\text{—}CHO \\ | \\ CH_3 \end{matrix}} \qquad \text{[C]}\quad \underset{\text{Pentanal}}{CH_3CH_2CH_2CH_2CHO}$$

The compounds as suggested above give the products which tally with those given in the problem

$$2CH_3-\underset{\underset{[B]}{CH_3}}{\overset{CH_3}{\underset{|}{\overset{|}{C}}}}-CHO + NaOH \longrightarrow CH_3-\underset{CH_3}{\overset{CH_3}{\underset{|}{\overset{|}{C}}}}-CH_2OH + CH_3-\underset{CH_3}{\overset{CH_3}{\underset{|}{\overset{|}{C}}}}-COONa$$

Also $\underset{[C]}{CH_3CH_2CH_2CH_2CHO} + \underset{[C]}{CH_3CH_2CH_2CH_2CHO}$

$$\xrightarrow{NaOH} CH_3CH_2CH_2CH_2CH(OH)\underset{\underset{CH_2\,CH_2\,CH_3}{|}}{CH}CHO$$

3-hydroxy-2-propylheptanal

Q. 54. Complete the following reactions:

1. $(CH_3CH_2CH_2CH_2)_2\,Cd + 2\ \begin{matrix} CH_3 \\ CH_3 \end{matrix}\!\!>CHCOCl \longrightarrow$

2. $C_6H_5-CH_2COCl \xrightarrow{H_2,\ Pd-BaSO_4}$

3. $NO_2-C_6H_4-COCl \xrightarrow{[C_4H_9(t)O]_3Al}$

Ans. 1. $\underset{\text{Dibutyl cadmium}}{(CH_3CH_2CH_2CH_2)_2\,Cd} + 2\ \underset{\text{Isobutyl chloride}}{\begin{matrix} CH_3 \\ CH_3 \end{matrix}\!\!>CHCOCl}$

$$\longrightarrow \underset{\text{2-Methylheptan-3-one}}{CH_3CH_2CH_2CH_2COCH\!\!<\!\begin{matrix} CH_3 \\ CH_3 \end{matrix}} + 2\,CdCl_2$$

2. $\underset{\text{Phenyl acetyl chloride}}{C_6H_5-CH_2COCl} \xrightarrow[\text{Rosenmund's Reduction}]{H_2,\ Pd\text{-}BaSO_4} \underset{\text{Phenyl acetaldehyde}}{C_6H_5-CH_2CHO}$

3. $\underset{p\text{-nitro benzoyl chloride}}{O_2N-C_6H_4-COCl} \xrightarrow[\text{tert. butoxide}]{\text{Aluminium}} \underset{p\text{-nitrobenzaldehyde}}{O_2N-C_6H_4-CHO}$

Q. 55. Two isomeric compounds A and B each with molecular formula C_3H_6O react with phenyl hydrazine to form phenyl hydrazones. A reduces Fehling solution and B does not. Upon oxidation A gives an acid of formula $C_3H_6O_2$ and B on similar treatment gives an acid of formula $C_2H_4O_2$. Identify compounds A and B and explain the reactions.

Ans. Reaction with phenyl hydrazine is a property characteristic of carbonyl compounds. Two possible compounds with the formula C_3H_6O are

CH_3CH_2CHO [A]
Propionaldehyde

$(CH_3)_2C=O$ [B]
Acetone

A reduces Fehling solution and B does not, as we know that aldehydes reduce the Fehling solution and the ketones does not give this reaction.

$$\underset{[A]}{CH_3CH_2CHO} + \underset{\text{Cupric ions}}{2Cu^{2+}} + Na^+ + 5[OH]^- \longrightarrow CH_3CH_2COONa + \underset{\text{Cuprous oxide}}{Cu_2O} + 3H_2O$$

$$\underset{[B]}{(CH_3)_2C=O} + 2\,Cu^{2+} + Na^+ + 5\,[OH]^- \longrightarrow \text{No reaction}$$

$$\underset{[A]}{CH_3CH_2CHO} \xrightarrow{[O]} \underset{\text{Propionic acid}}{CH_3CH_2COOH} \quad (\text{Mol. Formula } C_3H_6O_2)$$

$$\underset{[B]}{(CH_3)_2C=O} \xrightarrow{[O]} \underset{\substack{\text{Acetic acid} \\ (\text{Mol. Formula} \\ C_2H_4O_2)}}{CH_3COOH} + CO_2 + H_2O$$

Hence compound A is CH_3CH_2CHO *i.e.* propionaldehyde and compound B is CH_3COCH_3, *i.e.*, acetone

Q. 56. Complete and name the following reactions:

(*i*) $(CH_3)_3\,CCHO + (CH_3)_3\,CCHO \xrightarrow{50\%\ NaOH}$

(*ii*) $(C_2H_5)_2CO + (C_6H_5)_3\,P=CH-CH_3 \longrightarrow$

(*iii*) $C_6H_5CHO + (C_6H_5)_3\,P=CH_2 \longrightarrow$

(*iv*) $C_6H_5CHO + CH_2\,(COOC_2H_5)_2 \xrightarrow{\text{Base}}$

(*v*) $C_6H_5CHO + BrCH_2-COOC_2H_5 \xrightarrow[\text{Ether}]{\text{Zn}}$

(*vi*) $O_2N-C_6H_4-CHO + (CH_3CO)_2O \xrightarrow[453\ K]{CH_3COONa}$

Ans. (*i*) $(CH_3)_3CCHO + (CH_3)_3\,CHO$ Cannizzaro Reaction

$$\xrightarrow{50\%\ NaOH} \underset{\text{Sod. Trimethylacetate}}{(CH_3)_3CCOONa} + \underset{\text{Neopentyl alcohol}}{(CH_3)_3\,CCH_2OH}$$

(*ii*) $\begin{matrix} C_2H_5 \\ C_2H_5 \end{matrix}\!\!>\!C = O + (C_6H_5)_3\,P = CH - CH_3$ Wittig Reaction

Diethyl ether

$\longrightarrow \begin{matrix} C_2H_5 \\ C_2H_5 \end{matrix}\!\!>\!C = CH - CH_3 + (C_6H_5)_3\,PO$

3-Ethylpent-2-ene Triphenyl phosphine oxide

(*iii*) $C_6H_5CHO + (C_6H_5)_3\,P = CH_2$ Wittig Reaction

$\longrightarrow C_6H_5CH = CH_2 + (C_6H_5)_3\,PO$

Styrene Triphenyl phosphine oxide

(iv) $C_6H_5CHO + CH_2\,(COOC_2H_5)_2$ Knoevenagal Reaction

Malonic ester

$\xrightarrow{\text{Base}} C_6H_5CH = C\!\!<\!\begin{matrix} COOC_2H_5 \\ COOC_2H_5 \end{matrix} + H_2O$

Benzal malonic ester

(v) $C_6H_5CHO + BrCH_2COOC_2H_5 + Zn$ Reformatsky Reaction

$\xrightarrow{\text{Ether}} C_6H_5CH(OH)\,CH_2COOC_2H_5 + Zn\,(OH)\,Br$

Ethyl 3-hydroxy-3-phenyl propionate

(*vi*) $O_2N-C_6H_4-CHO + (CH_3CO)_2\,O$ Perkin's Reaction

$\xrightarrow[453]{CH_3COONa} O_2N-C_6H_4-CH = CH - COOH + CH_3COOH$

p-nitro cinnamic acid

Q. 57. What do you know about Meerwin-Ponndorf-Verley reduction?

Ans. An aldehyde or ketone on heating with aluminium isopropoxide in the presence of excess isopropyl alcohol is reduced to corresponding alcohol while isopropyl alcohol is oxidised to acetone.

$$\begin{matrix} R \\ R' \end{matrix}\!\!>\!C = O + CH_3\,\underset{|}{\overset{CH_3}{C}}HOH \longrightarrow R - \overset{R}{\overset{|}{C}}HOH + CH_3\,\overset{CH_3}{\overset{|}{C}} = O$$

Ketone; Secondary alcohol; Acetone (Removed by distillation)

This is a suitable method for reducing unsaturated carbonyl compounds as the double bond as not affected.

Q. 58. How do you distinguish a ketone and an aldehyde by chemical method?

(*Purvanchal 2003*)

Ans. Both aldehydes and ketones give a yellow precipitate of 2, 4-dinitrophenyl hydrazone on treatment with 2, 4-dinitrophenyl hydrazine.

Aldehydes on boiling with ammoniacal silver nitrate solution give a shining layer of silver metal on the inner wall of the test tube (Tollen's reagent test). Similarly aldehydes on boiling with Fehling solution give a red ppt. of cuprous oxide.

Ketones give neither the Tollen's reagent test nor Fehling solution test.

Q. 59. Alkenes undergo electrophilic addition reactions while carbonyl compounds undergo nucleophilic addition reactions. Explain.

Ans. This is due to the electronegative nature of carbonyl oxygen that a positive charge is created on the carbonyl carbon. This attracts nucleophilies towards it giving nucleophilic reactions.

$$>C=O \longrightarrow >\overset{+}{C}-\overset{-}{O}$$

In alkenes, such positive centre is not created. Instead, the π electrons on C = C attract the electrophiles towards it.

Q. 60. Why does not HCN add across carbonyl double bond in benzophenone.

Ans. This is because the positive charge created on carbonyl carbon is removed by the resonance in two benzene rings in the compound.

$$C_6H_5-\overset{O^-}{\overset{|}{\overset{+}{C}}}-C_6H_5 \longleftrightarrow \overset{+}{C_6H_5}=\overset{O^-}{\overset{|}{C}}-C_6H_5$$

As the positive centre on carbonyl carbon is destroyed, nucleophilic addition of HCN becomes difficult.

Q. 61. Nucleophilic addition of ammonia derivatives to carbonyl group is catalysed by acid but in presence of large excess of the acid, the rate of reaction is lowered considerably.

Ans. Presence of a mineral acid promotes the addition of ammonia derivatives to carbonyl group, as this helps to stabilize positive charge on carbonyl carbon.

$$>C=O+\overset{+}{H} \longrightarrow >\overset{+}{C}-OH$$

But in the presence of excess of the acid, the nucleophile also gets protonated and thus loses tendency to attach to the positive centre

$$R-\ddot{N}H_2+H^+ \longrightarrow R-\overset{+}{N}H_3$$

Q. 62 Acid catalyses the addition of semi carbozide to acetone. Yet, too much acidity of the medium is harmful in this reaction.

Ans. In the presence of excess acid, the nucleophile gets protonated and loses ability to approach the positive centre

$$\ddot{N}H_2NHCONH_2+\overset{+}{H} \longrightarrow \overset{+}{N}H_3NHCONH_2$$

Q. 63. Formaldehyde does not give aldol condensation. Explain.

Ans. Aldol condensation is given by compounds containing α-hydrogen atoms. Since formaldehyde does not contain α-hydrogen atoms, it does not give aldol condensation.

Q. 64. *o*-hydroxy benzaldehyde is steam volatile while *p*-hydroxy benzaldehyde is not. Why?

Ans.

o-hydroxy benzaldehyde (structure: benzene ring with O—H and —C=O(H), dashed H-bond between H and O)

p-hydroxy benzaldehyde (structure: benzene ring with O—H at top and C=O(H) at bottom)

Intramolecular hydrogen bonding between hydrogen of phenolic group and oxygen of aldehydic group takes place in *o*-hydroxy benzaldehyde because of proximity of the two groups, thus lowering the boiling point. Hence *o*-hydroxybenzaldehyde is steam volatile. This type of hydrogen bonding does not take place in the para isomer because the two groups are apart.

Q. 65. Write all the possible isomers of the compound with meolcular formula C_4H_8O and give their IUPAC names.

Ans. Possible isomers are:

1. $CH_3CH_2CH_2CHO$	Butanal
2. $CH_3-\underset{\displaystyle CH_3}{\underset{\vert}{CH}}-CHO$	2-Methylpropanal
3. $CH_3COC_2H_5$	Butanone-2

Q. 66. Acetaldehyde undergoes aldol condensation but trimethyl acetaldehyde does not. Explain.

Ans. Refer to Q. 63

Q. 67. How will you distinguish between aldehydes and ketones?

Ans. Both aldehydes and ketones give a yellow precipitate with 2, 4, Dinitrophenylhydrize. Aldehydes respond to Fehling solution test and silver mirror test whereas ketones don't respond to these tests.

KETENE

Q. 68. Give the methods of preparation and synthetic uses of ketene ($CH_2 = C = O$).

(Guwahati, 2006)

Ans. Methods of preparation :

1. By refluxing acetyl chloride with pyridine

$$CH_3COCl + C_5H_5N \longrightarrow \underset{\text{Ketene}}{CH_2 = C = O} + C_5H_5NHCl$$

2. By distillation of α - bromo acetyl bromide with zinc in non-aqueous and non-alcoholic medium.

$$Br - CH_2COBr + Zn \longrightarrow CH_2 = C = O + ZnBr_2$$

3. By pyrolysis of acetone

$$CH_3COCH_3 \xrightarrow{>700°C} CH_2 = C = O + CH_4$$

Properties of ketenes :

1. Ketene oxidises in air to form unstable peroxide

$$CH_2 = C = O + O_2 \longrightarrow \begin{matrix} CH_2 - C = O \\ \vert \quad\quad \vert \\ O - O \end{matrix}$$

2. It rapidly polymerises to diketene

$$2CH_2=C=O \longrightarrow \begin{array}{l} CH_2=C=O \\ | \quad\quad | \\ O - C \end{array}$$

3. Ketenes undergoes photochemical decomposition to give mainly the triplet methylene and carbon monoxide. Methylene is used for insertion to C—H and addition across C = C

$$CH_2=C=O \xrightarrow{U.V} \dot{C}H_2 + CO$$

4. Ketene acts as an acetylating agent. Ketene can acetylate alcohols, phenols, ammonia, primary and secondary amines. Primary amino group is far more readily acetylated compared to hydroxyl group.

(*i*) $\underset{\text{Ketene}}{CH_2=C=O} + H_2O \longrightarrow CH_3COOH$

(*ii*) $CH_2=C=O + CH_3COOH \longrightarrow (CH_3CO)_2O$

(*iii*) $CH_2=C=O + ROH \longrightarrow CH_3COOR$

(*iv*) $\underset{\text{Ketene}}{CH_2=C=O} + \underset{\text{Primary amine}}{RNH_2} \longrightarrow \underset{\text{Anilide}}{RNHCOCH_3}$

(*v*) $CH_2=C=O + \begin{array}{l} R \\ R \end{array}\!\!>NH \longrightarrow \begin{array}{l} R \\ R \end{array}\!\!>NCOCH_3$

(*vi*) $CH_2=C=O + \underset{\text{Acetamide}}{CH_3CONH_2} \xrightarrow{H_2SO_4} (CH_3CO)_2NH$

(*vii*) $CH_2=C=O + CH_3CONH_2 \xrightarrow[H_2SO_4]{\text{Absence of}} CH_3CN + CH_3COOH$

(*viii*) $CH_2=C=O + Br_2 \longrightarrow \begin{array}{l} CH_2 - C=O \\ | \quad\quad\; | \\ Br \quad\; Br \end{array}$

1,2-Dibromoketene

(*ix*) $CH_2=C=O + PCl_5 \longrightarrow \begin{array}{l} CH_2 - C=O \\ | \quad\quad\; | \\ Cl \quad\; Cl \end{array} + PCl_3$

1, 2-Dichloroketene

(*x*) $CH_2=C=O + HX \longrightarrow \begin{array}{l} CH_3 - C=O \\ \quad\quad\;\; | \\ \quad\quad\;\; X \end{array}$

xi) $CH_2=C=O + RMgX \longrightarrow \begin{array}{l} CH_2 - C=O \\ | \quad\quad\; | \\ MgX \;\; R \end{array} \xrightarrow{\text{Hydrolysis}} \begin{array}{l} CH_3 - C=O \\ \quad\quad\;\; | \\ \quad\quad\;\; R \end{array}$

MISCELLANEOUS QUESTIONS

Q. 69. How will you differentiate between acetophenone and benzophenone? (*Sagar, 2002*)

Ans. Acetophenone reacts with hydrogen cyanide, while benzophenone does not.

$$\begin{array}{l} C_6H_5 \\ CH_3 \end{array}\!\!>C=O + HCN \longrightarrow \begin{array}{l} C_6H_5 \\ CH_3 \end{array}\!\!>C<\!\!\begin{array}{l} OH \\ OH \end{array}$$

Q. 70. Mention a reagent to which acetaldehyde and benzaldehyde react similarly and another reagent to which these compounds react differently. (*Kerala, 2001*)

Ans. (*i*) HCN (*ii*) I_2/NaOH

Q. 71. Discuss the methods of preparing benzaldehyde. Explain the reactions of benzaldehyde with.

(*a*) Amm. silver nitrate (*b*) Phenyl hydrazine (*c*) Phosphorus pentachloride (*d*) Aniline (*e*) Potassium cynaide in aq. alcohol and (*f*) a mixture of concentrated sulphuric acid and conc. nitric acid. (*Kerala, 2001*)

Ans. See. Q. 6

(*a*) See Q. 42 (*b*) See Q. 21.

Q. 72. Give the methods of preparations and important uses of benzophenone. (*Meerut, 2000*)

Ans. See Q. 7 (4).

Q. 73. How is butanone prepared from aceto acetic acid? (*Kerala, 2001*)

Ans. See Q. 7 (8).

Q. 74. Formaldehyde gives Cannizzaro reaction but acetaldehyde does not, why? (*Himachal, 2000*)

Ans. See Q. 36.

Q. 75. Complete the following reactions:

(*i*) $CH_3COCH_3 \xrightarrow{NH_2OH}$ (*Himachal, 2000*)

(*ii*) 3 pentanone $\xrightarrow{\text{Mg - Hg. dry benzene}}$ (*Calcutta, 2000*)

Ans. (*i*) $CH_3COCH_3 \xrightarrow{NH_2OH.\Delta} (CH_3)_2C{=}NOH + H_2O$ (Oxime)

(*ii*) See Q. 41.

Q. 76. Predict product (s) formed in the following reactions with proper reaction mechanism. (*Calcutta, 2000*)

(*i*) Salicyldehyde $\xrightarrow[\text{reflux}]{Ac_2O/NaOAc}$

(*ii*) $CHO-C_6H_4-CHO \xrightarrow[\text{solution}]{50\%\ NaOH}$

Ans. (*i*) See Q. 29. In place of benzaldehyde take salicyldehyde.

(*ii*) $CHO-C_6H_4-CHO \xrightarrow[\text{solution}]{50\%\ NaOH}$

$HOH_2C-C_6H_4-CH_2OH + NaOOC-C_6H_4-COONa$

For the mechanism see Q. 36.

Q. 77. *p*-dimethylaminobenzaldehyde fails to undergo benzoin condensation but when mixed with benzaldehyde, the condensation does occurs, explain why ? *(Calcutta, 2000)*

Ans. There is steric hinderance in the attachement of the carbanions formed in the first step to the second molecule.

Q. 78. Complete the following reactions.

(*i*) $CH_3-CH=CH-CHO \xrightarrow{NaBH_4}$ *(Kurukshetra, 2001)*

(*ii*) $\begin{matrix} C_6H_5 \\ CH_3 \end{matrix} \!\!> CO \xrightarrow{Zn/HCl}$ *(M. Dayanand, 2000)*

Ans. (*i*) See Q. 39

(*ii*) $\begin{matrix} C_6H_5 \\ CH_3 \end{matrix} \!\!> CO \xrightarrow{Zn/HCl} C_6H_5CH_2CH_3$

Q. 79. Give two methods of preparation of $CH_3COCH_2CH_3$. *(M. Dayanand, 2000)*

Ans. See Q. 7.

Q. 80. Give suitable reagent (s) for the following conversions :

(*i*) $CH_3-CH=CH-CHO \longrightarrow CH_3-CH=CH-CH_2-OH$ *(Kanpur, 2001)*

(*ii*) $CH_3CHO \longrightarrow CH_3-CH(OH)COOH$ *(Kanpur, 2001)*

Ans. (*i*) See Q. 39

(*ii*) $CH_3CHO \xrightarrow{HCN} CH_3-CH(OH)-CN \xrightarrow{H_2O} CH_3-CH(OH)-COOH$

Q. 81. Explain the application of following reagents in the identification of organic compound (give only reactions) :

(*i*) Phenyl hydrazine

(*ii*) $NH_2OH.HCl$

(*iii*) Semicarbazide.HCl *(Lucknow, 2000)*

Ans. (*i*) See Q. 21 (*iii*)

(*ii*) See Q. 21 (*i*)

(*iii*) See Q. 21 (*iv*)

Q. 82. A ketone (A) gives iodoform test. (A) on hydrogenation gives (B) which on heating with H_2SO_4 gives (C). Action of O_3 on. (C) gives (D) which when treated with water in the presence of Zn dust gives only acetaldehyde. Identify (A), (B), (C), (D) and explain the reactions involved. *(Delhi 2002)*

Ans. $A = CH_3COCH_2CH_3$ $\quad B = CH_3CHOHCH_2CH_3$

$C = CH_3CH=CH-CH_3$

$D = CH_3-CH\overset{O}{\frown}CH-CH_3$ (ozonide: the two CH carbons joined through an upper O and through a lower O—O bridge)

$E = CH_3CHO$

$$CH_3COCH_2CH_3 \xrightarrow{H_2} CH_3CHOHCH_2CH_3 \xrightarrow[-H_2O]{H_2SO_4} CH_3CH=CH-CH_3 \xrightarrow{O_3,\ Zn} 2\,CH_3CHO$$

20

CARBOXYLIC ACIDS

Q. 1. What are carboxylic acids? (*Purvanchal 2003*)

Ans. Carboxylic acids are organic compounds containing –COOH or carboxy group. Such compounds can be represented by the general formula R–COOH, where R stands for some aliphatic or aromatic group. Carboxylic compounds can be classified as follows:

Monocarboxylic compounds	:	Those containing one carboxy group in the molecule such as acetic acid, butyric acid.
Dicarboxylic compounds	:	Those containing two carboxy groups in the molecule such as oxalic acid, malonic acid.
Tricarboxylic compounds	:	Those containing three carboxy groups in the molecule such as citric acid.

The carboxy group $-\overset{\overset{\large O}{\|}}{C}-OH$ consists of two groups viz. $-\overset{\overset{\large O}{\|}}{C}-$ (carbonyl group) and –OH (hydroxyl group). The word carboxy is derived from the two groups – carbonyl and hydroxy.

Q. 2. Describe the nomenclature of aliphatic and aromatic carboxylic compounds.

1. Common system **2. IUPAC system**

Ans. 1. Common System. In this system of naming, the carboxylic acid is named after the source from which it is obtained. For example, HCOOH, formic acid is so named because it is obtained from *formica* i.e. ants. Acetic acid gets its name from *acetum* i.e. vinegar which forms the source of acetic acid. Butyric acid C_3H_7COOH gets its name from *butyrum* i.e. butter.

Substituted carboxylic acids are named by indicating the group and the position where such group is attached. The positions on the carbon chain in a carboxylic acid are determined as under:

$$\overset{\delta}{C}-\overset{\gamma}{C}-\overset{\beta}{C}-\overset{\alpha}{C}-COOH$$

Thus the position next to the carboxy group is α position. Next to that is β position, still next are γ and δ positions.

The compound $CH_3-\overset{\overset{\large Cl}{|}}{CH}-COOH$ is named as α -chloropropionic acid. All the carbons in the longest chain containing carboxylic acid are taken into consideration for determination of the parent carboxylic acid. Some examples are given below:

$$\overset{\gamma}{CH_3}-\underset{\underset{\large CH_3}{|}}{\overset{\beta}{CH}}-\overset{\alpha}{CH_2}-COOH \qquad \beta\text{-methyl butyric acid}$$

$$\overset{\delta}{CH_3} - \underset{\displaystyle CH_3}{\overset{\gamma}{CH}} - \underset{\displaystyle CH_3}{\overset{\beta}{CH}} - \overset{\alpha}{CH_2} - COOH$$ β, γ-dimethyl valeric acid

2. IUPAC System. In this system, carboxylic acids are known as *alkanoic acids.* Total carbon atoms in the longest chain containing the carboxylic group and including the carboxylic carbon atom are counted. The carboxylic compound is the derivative of the alkane with as many carbons as the acid. Thus if there are 4 carbons in the acid, it is a derivative of butane, if there are 5 carbons, it is a derivative of pentane and so on.

The **e** of the alkane is changed into **oic acid.** Some examples are given below:

CH_3COOH Ethanoic acid

CH_3CH_2COOH Propanoic acid (Common name propionic acid)

For naming substituted acids, according to IUPAC system, again we identify groups which are attached and positions where these are attached. But the numbering of the positions is done in different manner. The carboxy group carbon is given no. 1 position, next one no. 2 position and so on.

$$\overset{5}{C} - \overset{4}{C} - \overset{3}{C} - \overset{2}{C} - \overset{1}{COOH}$$

Thus the following compounds will be named, according to IUPAC system, as shown below against the compounds.

$$\overset{3}{CH_3} - \underset{\displaystyle Cl}{\overset{2}{CH}}\overset{1}{COOH}$$

2-chloropropanoic acid

$$H_2N\overset{4}{C}H_2\overset{3}{C}H_2\overset{2}{C}H_2\overset{1}{C}OOH$$

4-aminobutanoic acid

The table below gives the common and IUPAC names of some common carboxylic compounds.

Molecular Formula	*Common name*	*IUPAC name*
$HCOOH$	Formic acid	Methanoic acid
CH_3COOH	Acetic acid	Ethanoic acid
CH_3CH_2COOH	Propionic acid	Propanoic acid
$CH_3CH_2CH_2COOH$	Butyric acid	Butanoic acid
$(CH_3)_2CHCOOH$	Isobutyric acid	2-methylpropanoic acid
$CH_3CH_2CH_2CH_2COOH$	Valeric acid	Pentanoic acid

Nomnclature of Aromatic Carboxylic Acids

The simplest aromatic carboxy acid is benzoic acid.

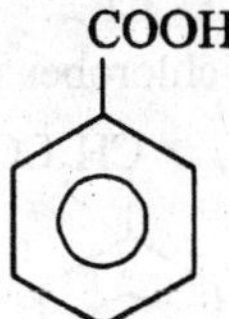

There may be other groups attached to the ring. These are indicated by mentioning the positions where these are attached, the position on the ring next to the position carrying –COOH group is ortho (*o*), next to that is meta (*m*) and still next is para (*p*). Thus

COOH

ortho ortho

meta meta

para

In IUPAC system, carboxy group position is taken as no. 1 and then the positions are counted clockwise.

COOH

6 2

5 3

4

The name benzoic acid is accepted in IUPAC nomenclature. It is also named as benzene carboxylic acid in IUPAC nomenclature.

Name of some carboxylic compounds are given below.

(*i*) COOH, Cl

Ortho chlorobenzoic acid

(*ii*) COOH, CH_3

Para methylbenzoic acid

(*iii*) COOH, NO_2, Cl

2-nitro-4-chlorobenzoic acid

(*iv*) CH_2COOH

Phenyl acetic acid

Q. 3. Describe the general methods of preparation of carboxylic acids giving the mechanism wherever applicable. *(Nagpur 2008)*

Ans. Carboxylic acids (aliphatic and aromatic) can be prepared by the following methods:

1. Oxidation of alcohols and aldehydes. Oxidation can be carried out in the presence of acidified $KMnO_4$ or $K_2Cr_2O_7$ or dilute nitric acid.

$$R-CH_2OH \xrightarrow[K_2Cr_2O_7\,/\,H_2SO_4]{[O]} R-CHO \xrightarrow{[O]} R-COOH$$

Examples

$$\underset{\text{Ethanol}}{CH_3CH_2OH} \xrightarrow[K_2Cr_2O_7\,/\,H_2SO_4]{[O]} \underset{\text{Ethanol}}{CH_3CHO} \xrightarrow[K_2Cr_2O_7\,/\,H_2SO_4]{[O]} \underset{\text{Ethanoic acid}}{CH_3COOH}$$

$$C_6H_5CH_2OH \xrightarrow[K_2Cr_2O_7\,/\,H_2SO_4]{[O]} \underset{\text{Benzaldehyde}}{C_6H_5CHO} \xrightarrow[H_2SO_4]{K_2Cr_2O_7} \underset{\text{Benzoic acid}}{C_6H_5COOH}$$

2. By oxidation of alkyl benzenes. Alkyl benzenes on oxidation with potassium permanganate yield carboxylic acid. Whatever the length of the attached alkyl group, it is converted simply into –COOH.

$$\underset{\text{Toluene}}{C_6H_5CH_3} \xrightarrow[KMnO_4]{[O]} \underset{\text{Benzoic acid}}{C_6H_5COOH}$$

$$\underset{\text{Ethylbenzene}}{C_6H_5C_2H_5} + 6\,[O] \xrightarrow{KMnO_4} \underset{\text{Benzoic acid}}{C_6H_5COOH} + CO_2 + 2H_2O$$

3. By hydrolysis of nitriles or cyanides

$$R-CN + 2H_2O \xrightarrow[\text{Alkali}]{\text{Acid or}} R-COOH + NH_3$$

$$\underset{\text{Methyl cyanide}}{CH_3-CN} + 2H_2O \xrightarrow[\text{Alkali}]{\text{Acid or}} \underset{\text{Acetic acid}}{CH_3COOH} + NH_3$$

$$\underset{\text{Phenyl cyanide}}{C_6H_5CN} + 2H_2O \xrightarrow[\text{Alkali}]{\text{Acid or}} \underset{\text{Benzoic acid}}{C_6H_5COOH} + NH_3$$

Mechanism

Acidic hydrolysis. (*i*) Hydrogen ion attaches itself to the cyano nitrogen atom with the shifting of charge to nitrogen and creation of positive charge on carbon.

(*ii*) Water molecule is attached to the carbonium ion.

The sequence of changes that take place subsequently is shown hereunder.

$$R-C\equiv N \xrightarrow{H^+} R-\overset{+}{C}=NH \xrightarrow{H_2O} R-\underset{}{\overset{\overset{+}{O}H_2}{\overset{|}{C}}}=NH$$

$$\downarrow$$

$$R-\overset{\overset{\overset{+}{O}H}{||}}{C}-NH_2 \longleftrightarrow R-\overset{\overset{:\ddot{O}H}{|}}{C}=\overset{+}{N}H_2$$

$$R-\overset{\overset{\overset{+}{O}H}{//}}{C}\overset{OH_2}{\leftarrow}-NH_2 \rightleftharpoons R-\overset{\overset{OH}{|}}{\underset{\underset{NH_2}{|}}{C}}-\underset{\underset{H}{|}}{\overset{+}{O}}-H \longrightarrow R-\overset{\overset{O}{//}}{C}-OH + NH_4^+$$

Alkaline hydrolysis. (*i*) Here the electromeric change takes place in the presence of the hydroxyl group. Hydroxyl group is attached to cyano carbon atom with a consequent negative charge on nitrogen.

(*ii*) A H^+ is attached to negative nitrogen to produce the imine group C = NH.

Sequence of changes is shown below:

$$R-C\equiv N \xrightarrow{OH^-} R-\overset{\overset{OH}{|}}{C}=N^- \xrightarrow[-OH^-]{H_2O} R-\overset{\overset{O-H}{|}}{C}=NH$$

$$\rightleftharpoons$$

$$R-\overset{\overset{O}{||}}{C}-NH_2$$

$$R-\overset{\overset{\ddot{O}:}{//}}{C}\overset{\bar{O}H}{\leftarrow}-NH_2 \longrightarrow R-\overset{\overset{:\ddot{O}^-}{|}}{\underset{\underset{NH_2}{|}}{C}}-OH \longrightarrow R-\overset{\overset{O}{||}}{C}-\bar{O}+NH_3$$

4. By hydrolysis of acid derivatives viz. acid chlorides, esters, amides and anhydrides. The above acid derivatives on hydrolysis in the presence of an acid or alkali produce carboxylic acids.

$$\underset{\text{Acid chloride}}{RCOCl + H_2O} \longrightarrow RCOOH + HCl$$

$$\underset{\text{Ester}}{RCOOR' + H_2O} \longrightarrow \underset{\text{Acid}}{RCOOH} + R'OH$$

$$RCONH_2 + H_2O \longrightarrow RCOOH + NH_3$$

$$\begin{matrix} RCO \\ RCO \end{matrix}\!\!>\! O + H_2O \longrightarrow \underset{\text{Acid}}{2\,RCOOH}$$
Acid Anhydride

Mechanism. An acid derivative could be represented as $R-\overset{O}{\overset{\|}{C}}-Y$ where Y stands for –Cl, –OR′, $-NH_2$ or –OOCR in case of acid chloride, ester, amide or anhydride respectively.

Acidic hydrolysis

$$R-C(=O)-Y \xrightarrow{H^+} R-C(=\overset{+}{O}H)-Y \xrightarrow{OH_2} R-\underset{Y}{\overset{OH}{C}}-\overset{+}{O}H_2 \longrightarrow R-C(=O)-OH + H-Y + H^+$$

Alkaline hydrolysis

$$R-C(=\ddot{O}:)-Y \xrightarrow{^-OH} R-\underset{Y}{\overset{:\ddot{O}:^-}{C}}-OH \longrightarrow R-\overset{O}{\overset{\|}{C}}-OH + Y^- \xrightarrow{OH^-} R-\overset{O}{\overset{\|}{C}}-O^- + H_2O$$

5. By the action of carbon dioxide on alkyl or phenyl magnesium bromide. Grignard reagent reacts with carbon dioxide forming an addition product which yields carboxylic acid on hydrolysis.

$$\overset{O}{\overset{\|}{C}}=O + \underset{\text{Methyl magnesium bromide}}{CH_3\,Mg\,Br} \longrightarrow \underset{\text{Addition Product}}{CH_3-\overset{O}{\overset{\|}{C}}-OMgBr} \xrightarrow{H_2O} \underset{\text{Acetic acid}}{CH_3-\overset{O}{\overset{\|}{C}}-OH} + Mg\begin{matrix} OH \\ Br \end{matrix}$$

$$\overset{O}{\overset{\|}{C}}=O + C_6H_5Mg\,Br \longrightarrow C_6H_5-\overset{O}{\overset{\|}{C}}-OMgBr \xrightarrow{H_2O} \underset{\text{Benzoic acid}}{C_6H_5-\overset{O}{\overset{\|}{C}}-OH} + Mg\begin{matrix} OH \\ Br \end{matrix}$$

Mechanism. The nucleophile alkyl group is attached to carbon of the carbon dioxide molecule whereas –MgBr group is attached to one of the two oxygen atoms. This is shown below schematically as under:

$$R-MgX + O=C=O \longrightarrow R-\overset{O}{\overset{\|}{C}}-OMgX$$

6. By decarboxylation of dicarboxy acids. Two carboxy groups attached to the same carbon atom are unstable to heat and a molecule of water is lost.

$$RCH\begin{cases}COOH\\COOH\end{cases} \xrightarrow{\Delta} RCH_2COOH + CO_2\uparrow$$

$$\underset{\text{Malonic acid}}{CH_2\begin{cases}COOH\\COOH\end{cases}} \longrightarrow CH_3\,COOH + CO_2$$

$$\underset{\text{Oxalic acid}}{\begin{matrix}COOH\\|\\COOH\end{matrix}} \xrightarrow{\Delta} \underset{\text{Formic acid}}{HCOOH} + CO_2$$

7. By hydrolysis of trihalogen compounds.

$$RCCl_3 + 3H_2O \xrightarrow[-3HCl]{KOH} RC\begin{cases}OH\\OH\\OH\end{cases} \xrightarrow{\Delta} RCOOH + H_2O$$

$$\underset{\text{Trichloro-ethane}}{CH_3CCl_3} + 3H_2O \xrightarrow[-3HCl]{KOH} CH_3\,C\begin{cases}OH\\OH\\OH\end{cases} \xrightarrow{\Delta} \underset{\text{Acetic acid}}{CH_3COOH} + H_2O$$

$$\underset{\text{Benzo-trichloride}}{C_6H_5CCl_3} + 3H_2O \xrightarrow[-3HCl]{KOH} C_6H_5\begin{cases}OH\\OH\\OH\end{cases} \xrightarrow{\Delta} \underset{\text{Benzoic acid}}{C_6H_5COOH} + H_2O$$

8. From Malonic ester and Acetoacetic ester. Malonic ester synthesis. A desired acid can be prepared from malonic ester having the formula

$$CH_2\begin{cases}COOC_2H_5\\COOC_2H_5\end{cases} \qquad \text{Diethyl malonate}$$

It involves the following steps:

(*i*) Diethyl molecule is treated with sod. ethoxide to produce sodium malonic ester.

$$CH_2(COOC_2H_5)_2 \xrightarrow[-C_2H_5OH]{C_2H_5ONa} \overset{+}{Na}\overset{-}{C}H\,(COOC_2H_5)_2$$

(*ii*) Sodiomalonic ester is then treated with an appropriate alkyl halide to produce alkyl malonic ester.

$$\overset{+}{N}a\overset{-}{C}H\,(COOC_2H_5)_2 + RX \longrightarrow RCH\,(COOC_2H_5)_2$$

(*iii*) Alkyl malonic ester is hydrolysed and heated to produce carboxylic acid.

$$RCH\,(COOC_2H_5)_2 \xrightarrow{\text{Hydrolysis}} RCH\,(COOH)_2 \xrightarrow{\Delta} RCH_2COOH$$

By a proper choice of the alkyl halide, any desired carboxylic acid can be produced.

Acetoacetic ester synthesis. Like malonic ester, acetoacetic ester also possesses an α-hydrogen atom. This α-hydrogen can be replaced by sodium to produce sodium acetoacetic ester, which is then treated with an alkyl halide. The addition product so obtained is subjected to alkaline hydrolysis to produce acids.

$$\underset{\text{Acetoacetic ester}}{CH_3COCH_2COOC_2H_5} \xrightarrow{\text{Na / Acohol}} \underset{\text{Sod. salt of the ester}}{CH_3CO\overset{\overset{+}{Na}}{\overset{|}{C}}HCOOC_2H_5}$$

$$\xrightarrow{RBr} \underset{\text{Alkyl acetoacetic ester}}{CH_3\,CO\overset{R}{\overset{|}{C}}HCOOC_2H_5} \xrightarrow{\text{Hydrolysis}} CH_3COOH + \underset{\text{Desired acide}}{RCH_2COOH} + C_2H_5OH$$

A proper choice of the alkyl halide is required in required in order to obtain a particular acid. For example, to obtain propionic acid, the alkyl halide needed is methyl bromide, to obtain valeric acid, the alkyl halide to be used is *n*-butyl bromide and so on.

Q. 4. Give probable mechanism of alkaline hydrolysis of an ester. (*Garhwal 2000*)

Or

Give mechanism in support of the fact that during alkaline hydrolysis of an ester, it is the acyl oxygen bond that cleaves.

Ans. Hydrolysis of an ester can be represented by the following equation:

$$\underset{\text{Ethyl acetate}}{CH_3-\overset{\overset{O}{\|}}{C}-OC_2H_5} + NaOH \longrightarrow CH_3-\overset{\overset{O}{\|}}{C}-ONa + C_2H_5OH$$

The mechanism involves the following steps:

(*i*) Nucleophilic attack of hydroxide ion

$$CH_3-\overset{\overset{O}{\|}}{C}-OC_2H_5 \xrightarrow{\overset{-}{O}H} \underset{\text{(II)}}{CH_3-\overset{\overset{\overset{-}{O}}{|}}{\underset{\underset{OH}{|}}{C}}-OC_2H_5}$$

(*ii*) Elimination of ethoxide ion from (II)

$$CH_3-\overset{\overset{\overset{-}{O}}{|}}{\underset{\underset{OH}{|}}{C}}-OC_2H_5 \longrightarrow \underset{\text{(III)}}{CH_3-\overset{\overset{O}{\|}}{\underset{\underset{OH}{|}}{C}}} + C_2H_5O^-$$

***(iii)* Removal of proton from the acid (III)**

$$CH_3-\overset{\overset{\displaystyle O}{||}}{C}-O-H + C_2H_5O^- \longrightarrow CH_3-\overset{\overset{\displaystyle O}{||}}{C}-O^- + C_2H_5OH$$

Q. 5. Sketch the mechanism of hydrolysis of ethyl acetate in the presence of dil. HCl.

(Panjab 2003; Mumbai, 2004)

Ans. Mechanism of hydrolysis of ethyl acetate in the presence of dil. HCl is an given below.

$$\underset{\text{Ethyl acetate}}{CH_3-C\begin{cases}=O\\ -OC_2H_5\end{cases}} \xrightarrow{H^+} R-\overset{+}{C}\begin{cases}-OH\\ -OC_2H_5\end{cases} \xrightarrow{H_2O} R-\underset{\underset{\displaystyle OC_2H_5}{|}}{\overset{\overset{\displaystyle OH}{|}}{C}}-\overset{+}{O}H_2$$

$$\longrightarrow R-\overset{\overset{\displaystyle O}{||}}{C}-OH + C_2H_5OH + H^+$$

Q. 6. Discuss the structure of carboxylic group. Why are carboxy compounds stronger acids than phenols and alcohols? *(Purvanchal, 2003; Nagpur, 2005)*

Ans. Structure of carboxylic group. Carboxy acid is represented as:

$$R-\overset{\overset{\displaystyle O}{||}}{C}-O-H$$

where R is an alkyl or aryl group. The carboxy group actually exhibits resonance. The two resonating structures of the carboxy group are:

$$\underset{\text{I}}{R-\overset{\overset{\displaystyle O}{||}}{C}-\ddot{O}H} \longleftrightarrow \underset{\text{II}}{R-\overset{\overset{\displaystyle O^-}{|}}{C}=\overset{+}{\ddot{O}}H}$$

Structure II is not so stable energetically because of the separation of charges. It does not contribute much to the actual structure of the carboxylic group. Thus the two structure do not contribute equally to the actual structure. Consequently, resonance energy, which estimates the stability of a compound is less. On the other hand, the carboxylate ion obtained after removing one hydrogen atom is much more stable.

$$R-\overset{\overset{\displaystyle O}{||}}{C}-O-H \longrightarrow \underset{\text{Carboxylate ion}}{R-\overset{\overset{\displaystyle O}{||}}{C}-\bar{O}} + H^+$$

$$\underset{\text{III}}{R-\overset{\overset{\displaystyle O}{||}}{C}-O^-} \longleftrightarrow \underset{\text{IV}}{R-\overset{\overset{\displaystyle O^-}{|}}{C}=O}$$

Structures III and IV which are resonances hydrids of carboxylate ion are similar and contribute equally to the actual structure. In such a case as per the rules of resonance phenomenon, the value of resonance energy is quite high which accounts for the stability of carboxylate ion. This would imply that carboxylate ion and the hydrogen ion can exist separately. This explains the highly acidic nature of the carboxylate ion.

Resonance stabilization does not take place to this extent in the case of phenol or alcohol. Hence carboxylic compounds are stronger in acidic nature than phenols and alcohols.

Since the double bond is changing its position, the carboxylate ion could be represented as

$$\left[R-C \overset{O}{\underset{O}{\lessgtr}} \right]^{-}$$

Q. 7. Explain why the first few members of carboxylic acids are soluble and higher members insoluble in water? *(Purvanchal 2003)*

Ans. Lower members from formic to butyric acid are soluble in water because of the formation of hydrogen bond between water and the acid. Because of electronegativity difference between oxygen and hydrogen, hydrogen bond is formed as shown below:

(i) R – C(=O....H–O(H)....H–O)(–O–H....O(H)–H....O=) C – R

(ii) R–C=O with O—H ...... O(H)—H ..... R–C=O with O—H ...... O(H)—H....

As the size of the alkyl group increases, the solubility in water decreases because the strength of hydrogen bond decreases. This is because a long fatty chain hinders the formation of hydrogen bond. Hence higher fatty acids are insoluble in water. They are, however, soluble in organic solvents like benzene and ether.

Q. 8. Explain why carboxylic acids show much higher boiling point compared to alcohols of comparable molecular masses. *(Delhi 2004)*

Or

Most of the carboxy acids exist as dimers. Explain.

Or

Mol. wt. of acetic acid is double than the calculated value. Explain.

Ans. Carboxylic compounds exhibits hydrogen bonding between two acid molecules as shown below:

R—C(=O ...... H—O)(—O—H .....O=)C—R

Thus the acid molecule exists as a **dimer** in which two molecules have been associated into one molecule. Thus, the size of the molecule is increased which accounts for the higher boiling point of carboxylic compounds. No such dimerisation takes place in between alcohol molecules. Hence carboxy compounds have higher boiling points compared with alcohols of comparable molecular weights.

Q. 9. Write the properties of the carboxylic acids due to ionisable hydrogen.

Ans. The hydroxyl group, which is a part of the carboxylic group, has ionisable hydrogen. This is because of the electronic change that takes place as shown below:

$$-\overset{\overset{\displaystyle O}{\|}}{C}\leftarrow O\leftarrow H$$

This facilitates in the release of protons (H^+).

Various reactions of the carboxylic group due to ionisable hydrogen are given below:

(*i*) **Acidic Nature.** Carboxy compounds turn blue litmus red. This is indicative of acidic nature of the group.

$$RCOOH + H_2O \rightleftharpoons RCOO^- + H_3O^+$$

(*ii*) **Liberation of Hydrogen.** Electropositive elements like Na, K and Ca react with carboxy acids to produce hydrogen gas with the formation of the salts.

$$2CH_3\,COOH + 2Na \longrightarrow 2CH_3COONa + H_2\uparrow$$

$$2CH_3COOH + Ca \longrightarrow (CH_3COO)_2\,Ca + H_2\uparrow$$

(*iii*) **Salt Formation.** Carboxy acids react with alkalies and metallic oxides and decompose bicarbonates with the formation of salts. Reaction takes place also with ammonia.

$$CH_3\,COOH + KOH \longrightarrow \underset{\text{Pot. acetate}}{CH_3COOK} + H_2O$$

$$2CH_3COOH + HgO \longrightarrow \underset{\text{Mercuric acetate}}{(CH_3COO)_2\,Hg} + H_2O$$

$$CH_3COOH + NaHCO_3 \longrightarrow CH_3COONa + CO_2 + H_2O$$

$$CH_3COOH + NH_3 \longrightarrow \underset{\text{Ammonium acetate}}{CH_3COONH_4}$$

Q. 10. Describe reactions giving mechanism, of the carboxy compounds involving replacement of hydroxyl group. (*Andhra, 2010; Mumbai, 2010*)

Ans. The –OH group of the acid can be replaced by –OR′, –Cl, $-NH_2$ and OOCR′ groups to form esters, acid chlorides, amides and anhydrides respectively. These reactions are given separately as under:

Ester. Carboxylic acids on heating with an alcohol in the presence of a strong mineral acid (such as H_2SO_4) form **esters.** This reaction is known as **esterification** and is reversible in nature.

$$\underset{\text{Acid}}{RCOOH} + \underset{\text{Alcohol}}{R'OH} \overset{H^+}{\rightleftharpoons} RCOOR' + H_2O$$

$$\underset{\text{Acetic acid}}{CH_3COOH} + \underset{\text{Ethanol}}{C_2H_5OH} \overset{H^+}{\rightleftharpoons} \underset{\text{Ethyl acetate}}{CH_3COOC_2H_5} + H_2O$$

Mechanism. H^+ from the mineral acid protonates carbonyl oxygen. As a result, carbon atom becomes positive and undergoes nucleophilic attack by the alcohol molecule. Finally ester is obtained as depicted below:

$$R-\overset{O}{\overset{\|}{C}}-O-H \xrightarrow{H^+} R-\overset{OH}{\overset{|}{\underset{+}{C}}}-O-H \xrightarrow{\overset{OR'}{H}} R-\overset{OH}{\overset{|}{\underset{\underset{H}{+OR'}}{C}}}-O-H$$

$$\underset{\text{Ester}}{R-\overset{O}{\overset{\|}{\underset{OR'}{C}}}} \xleftarrow{-H_2O} R-\overset{O-H}{\overset{|}{\underset{OR'}{C}}}-\overset{+}{\underset{H}{O}}-H$$

To summerise esterification involves the cleavage of O–H bond of alcohol and C–OH bond of the carboxy acid. Thus

$$R-\overset{\overset{O}{\|}}{C}-\boxed{OH + H}\ OR' \longrightarrow \underset{\text{Ester}}{R-\overset{\overset{O}{\|}}{C}-OR'}$$

This has been confirmed by readiocative tracer studies.

Acid chloride. An acid chloride is obtained when an acid is treated with PCl_5 or $SOCl_2$ (Thionyl chloride).

$$RCOOH + PCl_5 \longrightarrow \underset{\text{Acid chloride}}{RCOCl} + POCl_3 + HCl$$

$$3RCOH + \underset{\text{Phosphorus trichloride}}{PCl_3} \longrightarrow RCOCl + \underset{\text{Phosphorus acid}}{H_3PO_3}$$

$$CH_3COOH + \underset{\text{Thionyl chloride}}{SOCl_2} \longrightarrow \underset{\text{Acetyl chloride}}{CH_3COCl} + SO_2\uparrow + HCl\uparrow$$

Acid amides. Ammonia reacts with acids to form ammonium salts which on heating liberate a water molecule giving amides.

$$\underset{\text{Acid}}{RCOOH} + NH_3 \longrightarrow \underset{\text{Amm.salt}}{RCOONH_4} \xrightarrow[-H_2O]{\Delta} \underset{\text{Amide}}{RCONH_2}$$

$$\underset{\text{Benzoic acid}}{C_6H_5-COOH} + NH_3 \longrightarrow \underset{\text{Ammonium benzoate}}{C_6H_5-COONH_4} \xrightarrow[H_2O]{\Delta} \underset{\text{Benzamide}}{C_6H_5-CONH_2}$$

Anhydrides. When the vapours of a carboxy acid are passed over P_2O_5, one molecule of water is eliminated from two molecules of the acid resulting in the formation of acid anhydride.

$$\begin{matrix} RCOOH \\ RCOOH \end{matrix} \xrightarrow[-H_2O]{P_2O_5} \underset{\text{Acid anhydride}}{\begin{matrix} RCO \\ RCO \end{matrix}\!\!>O}$$

An anhydride can also be obtained by treating an acid with an acid chloride in the presence of pyridine which removes HCl molecules from the mixture.

$$\underset{\text{Acetic acid}}{CH_3COOH} + \underset{\text{Acetyl chloride}}{CH_3COCl} + \text{Pyridine}$$

$$\longrightarrow \underset{\text{Acetic anhydride}}{\begin{matrix} CH_3CO \\ CH_3CO \end{matrix}\!\!>O} + \text{Pyrdine hydrochloride}$$

Q. 11. (*a*) Acetic acid reacts with the following alcohols separately. Arrange the alcohols in decreasing order of reactivity of reaction:

(*i*) *n*-butyl alcohol (*ii*) isobutyl alcohol (*iii*) tert. butyl alcohol

(*b*) Ethanol reacts with the following acids separately. Arrange the acids in order of reactivity of reaction.

(*i*) *n*-valeric acid (*ii*) isovaleric acid (*iii*) tert. valeric acid.

Ans. (*a*) As per the mechanism of esterification the alcohol molecule attaches itself to the carboxy acid by nucleophilic addition mechanism. This nucleophilic addition will be easy if the alkyl group of the alcohol is small. Large alkyl groups and branched alkyl groups will pose a problem in the nucleophile addition of the alcohol molecule to the carboxy acid. Hence the order of reactivity will be:

n-butyl alcohol > isobutyl alcohol > tert. butyl alcohol.

or $$CH_3CH_2CH_2CH_2OH > \begin{matrix} CH_3 \\ CH_3 \end{matrix}\!\!>\! CHCH_2OH > CH_3-\overset{\displaystyle CH_3}{\underset{\displaystyle CH_3}{\overset{|}{\underset{|}{C}}}}-OH$$

(*b*) By the same analogy as given above, the order of reactivity of the acid with ethanol will be:

n-valeric acid > isovaleric acid > tert. valeric acid

or $$CH_3CH_2CH_2CH_2COOH > \begin{matrix} CH_3 \\ CH_3 \end{matrix}\!\!>\! CHCH_2COOH > CH_3-\overset{\displaystyle CH_3}{\underset{\displaystyle CH_3}{\overset{|}{\underset{|}{C}}}}-OH$$

Q. 12. Describe the reduction and decarboxylation reactions of carboxylic acids.

Ans. Reduction. Carboxylic acids undergo reduction in the presence of lithium aluminium hydride to give primary alchols.

$$RCOOH \xrightarrow[4\,[H]]{LiAlH_4} RCH_2OH + H_2O$$

The reduction can also be brought about by first converting the carboxy compound into ester followed by reduction. This is found to be more convenient. Reduction in this manner is known as **Beauveault Blanc** reaction.

$$RCOOH + R'OH \xrightarrow{H^+} RCOOR' \xrightarrow{Na/Alcohol} RCH_2OH + R'OH$$

Decarboxylation. Carboxylic acid gets decomposed if heated in the presence of soda-lime. The reaction takes place in two steps:

$$CH_3COOH + NaOH \longrightarrow CH_3COONa + H_2O$$

$$CH_3COONa + NaOH \xrightarrow{Heat} \underset{Methane}{CH_4} + Na_2CO_3$$

$$C_6H_5COOH + NaOH \longrightarrow C_6H_5COONa + H_2O$$

$$C_6H_5COONa + NaOH \xrightarrow[Heat]{} \underset{Benzene}{C_6H_6} + Na_2CO_3$$

Carboxylic acid containing electron withdrawing groups such as $-NO_2$ or CH_3CO- get decarboxylated by heating alone. Thus

$$CH_3-\overset{O}{\overset{\|}{C}}-CH_2COOH \xrightarrow{\Delta} CH_3-\overset{O}{\overset{\|}{C}}-CH_3 + CO_2\uparrow$$

Kolbe's electrolytic reaction. If an electric current is passed through an aqueous solution of alkali salt of a carboxy acid, wc obtain hydrocarbons.

$$2RCOONa \longrightarrow 2RCOO^- + 2Na^+$$

At the anode: $2RCOO^- \xrightarrow{-2e} \underset{Unstable}{[2\,RCOO]} \longrightarrow \underset{Hydrocarbon}{R-R} + CO_2$

At the cathode: $2Na^+ \xrightarrow{+2e} 2Na \xrightarrow{2H_2O} 2NaOH + H_2$

Thus, if we take sodium acetate as the salt, the hydrocarbon produced will be ethane.

Distillation of calcium salt. Calcium salts of carboxylic acid on distillation form aldehydes and ketones.

$$\underset{\text{Cal.formate}}{\begin{matrix}HCOO\\HCOO\end{matrix}\!\!>Ca} \xrightarrow{\text{Distil}} \underset{\text{Formaldehyde}}{HCHO} + CaCO_3$$

$$\underset{\text{Cal.acetate}}{\begin{matrix}CH_3COO\\CH_3COO\end{matrix}\!\!>Ca} + \underset{\text{Cal. formate}}{Ca<\!\!\begin{matrix}OOCH\\OOCH\end{matrix}} \xrightarrow{\text{Distil}} \underset{\text{Acetaldehyde}}{2\,CH_3\,CHO} + 2CaCO_3$$

$$\underset{\text{Cal.acetate}}{\begin{matrix}CH_3COO\\CH_3COO\end{matrix}\!\!>Ca} \xrightarrow{\text{Distil}} \underset{\text{Acetone}}{CH_3COCH_3} + CaCO_3$$

Q. 13. Write a note on halogenation of carboxylic acids. (*Punjab, 2003 ;Himachal, 2000*)

Ans. In the presence of a small amount of red phosphorus, α-hydrogen atoms of the alkyl group are replaced by chlorine or bromine to form α-halogen acids. This reaction is known as **Hell-Volhard-Zelinsky** (Abbreviated as HVZ) reaction.

$$\underset{n\text{-Butyric acid}}{CH_3CH_2CH_2COOH} \xrightarrow[\text{Heat}]{Br_2/P} \underset{\alpha\text{-bromo butyric acid}}{CH_3CH_2CHBr\,COOH}$$

$$\xrightarrow{Br_2/P} \underset{\alpha,\,\alpha\text{-dibromobutyric acid}}{CH_3CH_2CBr_2COOH}$$

Halogen substitution will take place as long as α-hydrogen atoms are available. It is worth mentioning that bromination cannot be carried out beyond α-position whereas chlorination beyond α position is possible provided all α-hydrogen atoms have been replaced by chlorine.

Mechanism of halogenation. The reaction is believed to take place bv the following steps:

(*i*) $2P + \underset{\text{Halogen}}{3X_2} \longrightarrow \underset{\text{Phosphorus trihalide}}{2PX_3}$

(*ii*) $3RCH_2COOH + PX_3 \longrightarrow \underset{\text{Acid halide}}{3RCH_2COX} + H_3PO_3$

(*iii*) $RCH_2\,COX + X_2 \longrightarrow RCHXCOX + HX$

(*iv*) $RCHXCOX + RCH_2COOH \longrightarrow \underset{\text{Halogen substituted carboxy acid}}{RCHXCOOH} + RCH_2COX$

Q. 14. A carboxylic acid does not form phenyl hydrazone when treated with phenyl hydrazine. Explain.

Ans. In a carboxylic acid, there is a pair of electron on the oxygen in conjugation with carbonyl oxygen.

$$R-\overset{\overset{O}{\|}}{C}-\ddot{O}-H \longrightarrow R-\overset{\overset{^-O}{|}}{C}=\overset{+}{\ddot{O}}-H$$

This prevents the creation of a positive centre at carboxyl carbon. Therefore nucleophilic addition of phenyl hydrazine to give phenyl hydrazone is not possible.

Q. 15. A carboxy acid does not form an oxime. Explain. (*Awadh, 2000*)

Ans. An oxime is obtained in a reaction between a carboxyl compound and hydroxylamine

$$>C=O + NH_2OH \longrightarrow \underset{\text{Oxime}}{>C=N-OH}$$

This is a nucleophilic reaction in which the hydroxylamine molecule first attaches itself to the carbonyl carbon which is positive due to electronegative oxygen attached to it.

In the case of carboxylic compounds, this positive centre is destroyed because of lone pair of electrons present in conjugation with the carbonyl group

$$R-\overset{\overset{\displaystyle O}{\|}}{C}-\ddot{O}-H$$

Hence the reaction does not take place.

Q. 16. Give a detailed account of acid strength of the carboxy acids. (*Nagpur 2008*)

Or

How is the acidity affected by substituents?

(*Osmonia, 2002; Panjab, 2003; Shivaji, 2003; North Gujarat ,2004*)

Ans. As explained earlier, carboxylic compounds are acidic in nature because they liberate carboxylate and hydrogen ions. Hydrogen ions do not exist as such but they are associated with water molecules as H_3O^+ ions. Carboxylate ion is able to stabilize itself by resonance energy and this is the cause of strongly acidic nature of carboxylic compounds. The acid is dissociated as:

$$RCOOH + H_2O \longrightarrow RCOO^- + H_3O^+$$

This is a reaction in equilibrium and according to the law of mass action, its equilibrium constant, K_a, which may also be referred to as acidity constant, is given by the relation

$$K_a = \frac{[RCOO^-][H_3O^+]}{[RCOOH][H_2O]}$$

The quantities within square brackets represent the molar concentration of these species. Since water exists in large quantities, its concentration almost remains constant and hence may be omitted. The equilibrium constant for acetic acid may be rewritten as:

$$K_a = \frac{[CH_3COO^-][H_3O^+]}{[CH_3COOH]}$$

The above equation shows that greater the value of $[H_3O^+]$ greater will be the value of K_a. It implies that a greater value of K_a signifies a stronger acid. Hence greater the value of K_a, greater is the acid strength of the carboxy compound.

Lately, there has been a practice to represent the strength of an acid in term of pK_a value which is related to K_a by the following equation:

$$pK_a = -\log K_a$$

Thus an acid having K_a value equal to 1.75×10^{-5} will have its pK_a value as 4.75. It may be osbserved that pK_a value changes inversely as the K_a value. Thus a strong acid will have a higher value of K_a but a smaller value of pK_a.

Effect of Substituents on the Acid Strength

The strength of a carboxylic compound changes when certain groups are introduced in it at certain positions. As a matter of fact, we expect an increase in the acid strength if some electron-withdrawing group is introduced in the molecule. A reverse effect is observed when the group introduced is electron repelling. If the extra group attached is represented as –G then the situation in the two cases can be represented as below:

$$G \leftarrow CH_2 \leftarrow C \begin{matrix} \nearrow O \\ \searrow O^- \end{matrix} \qquad \text{Acid strength increases}$$

$$G \rightarrow CH_2 \rightarrow C \begin{matrix} \nearrow O \\ \searrow O^- \end{matrix} \qquad \text{Acid strength decreases}$$

When an electron-attracting group is attached, it will stabilise the carboxylate ion to a greater extent by dissipating the negative charge, it will help in the release of more protons. In the case of an electron-repelling group, there will be a greater concentration of negative charge on the carboxylate group which will ultimately reduce the stability of the carboxylate ion. Hence the liberation of protons will be suppressed. The effect of substituents on the acid strength can be studied under two headings.

Aliphatic acids. The following observations are made here:

(*i*) Consider the case of acetic acid, it has a pK_a value of 4.75. If we replace one of the hydrogens of the methyl group by a halogens group, an electron-attracting group, the halogenated acid is found to have a lower value of pK_a *i.e.* it is found to be stronger than acetic acid. Further, the acid strength changes with the halogen. As fluorine is more electronegative than chlorine and chlorine is more electronegative than bromine, we find that fluoroacetic acid is more acidic than chloroacetic acid which in turn is more acidic than bromoacetic acid.

$$F-CH_2COOH > Cl-CH_2COOH > Br-CH_2COOH$$

(*ii*) The increase in the acid strength of an acid depends upon the no. of such electron withdrawing groups. Greater the no. of electron withdrawing groups, greater is the strength of the acid. Thus dichloroacetic acid is stronger than monochloro acetic acid and trichloroacetic acid is stronger than dichloroacetic acid.

$$Cl \leftarrow \underset{\underset{Cl}{\downarrow}}{\overset{\overset{Cl}{\uparrow}}{C}} \lll COOH > \begin{matrix} Cl \searrow \\ Cl \nearrow \end{matrix} CH \twoheadleftarrow COOH > Cl \leftarrow CH_2 \leftarrow COOH > CH_3 - COOH$$

Trichloroacetic acid; Dichloroacetic acid; Monochloroacetic acid; Acetic acid

(*iii*) The strength of the acid depends upon the position of the electron withdrawing group relative to the carboxy group. Nearer the electron withdrawing group to the carboxy group, greater is the acid strengths. Thus α-chloropropionic acid is stronger than β-chloropropionic acid.

$$CH_3-\overset{\overset{Cl}{|}}{CH}-COOH \quad > \quad ClCH_2-CH_2-COOH$$

α-chloropropionic acid; β-chloropropionic acid

(*iv*) The substitution of an electron repelling group in the carboxy acid decreases the acid strength. Greater the number of such constituents, greater will be the effect.

$$CH_3 \rightarrow \underset{\underset{CH_3}{\uparrow}}{\overset{\overset{CH_3}{\downarrow}}{C}} \ggg COOH < \begin{matrix} CH_3 \searrow \\ CH_3 \nearrow \end{matrix} CH-CH_2 \twoheadrightarrow COOH < CH_3 \rightarrow CH_2 \rightarrow CH_2 \rightarrow CH_2 \rightarrow COOH$$

I; II; III

Thus consider the above three compounds, compound I will experience the maximum electron repelling effect as the three methyl groups are directly attached to carbon next to the carboxy group. In compounds II and III, the number of methyl (or alkyl) groups attached to carbon next to the carboxy group is 2 and 1 respectively. As a consequence compound III will be the strongest acid and compound I the weakest in the series.

It may be mentioned that a propyl group will have a greater electron repelling effect than ethyl group which in turn will have greater repelling effect than methyl group. Thus, the first four homologues of carboxylic compounds have the strength in the following order:

Formic acid > Acetic acid > Propionic acid > Butyric acid

Aromatic acids. The following observations are made here:

(*i*) Electron withdrawing groups like $-NO_2$, $> C = O$, $-CN$, $-Cl$ when substituted into the benzene ring raise the strength of the acid by inductive effect and resonance effect. Resonance effect is known to be stronger than inductive effect. But the position where such groups are linked is more important. The above groups produce a positive charge at the ortho and para positions but not at the meta position as shown below:

$O \leftarrow N = O$ (I) ⟷ $O \leftarrow N - O^{\ominus}$ (II, $\oplus$) ⟷ $O \leftarrow N - O^{\ominus}$ (III, $\oplus$)

I II III

The positive charge at ortho and para positions helps in the removal of proton from the carboxy group. Consequently *o*- and *p*-nitrobenzoic acids are stronger than *m*-nitrobenzoic acid. But *m*-nitrobenzoic acid is definitely stronger than benzoic acid as nitro group draws the electron, although not strongly, by inductive effect at the meta position.

(*ii*) Reverse is the case when electron donating group is linked to the benzene ring. Hydroxy group is an electron donating group in the resonance phenomenon as shown below:

$: OH$ (I) ⟷ $\overset{\oplus}{O}H$ ($\ominus$) (II) ⟷ $\overset{\oplus}{O}H$ ($\ominus$) (III)

I II III

Thus para position, where the carboxy group is attached, become negative. This will decrease the strength of *p*-hydroxy benzoic acids as compared to benzoic acid. But hydroxy group will produce a slight positive charge at the meta position due to inductive effect. Thus *m*-hydroxy benzoic acid is stronger than benzoic acid. Ortho hydroxy benzoic acid, although against expectation, is a stronger acid than benzoic acid. As a general observation, a group in the ortho position always increases the acid strength. This is called *ortho effect*. Thus out of *o*-, *m*- and *p*-substituted benzoic acids, *o*-compound will be strongest acid. However, there is no satisfactory explanation for this effect.

Q. 17. Explain why monochloroacetic acid is stronger than acetic acid?

(*Tejpur 2003, Puvanchal 2003; Bangalore 2004; Mysore, 2004 ; Nagpur, 2005*)

Ans. See Q. 16.

Q. 18. Arrange the following in the order of their increasing acidity. Give reasons for your answer.

COOH COOH COOH

CH_3 NO_2

Ans. The order of increasing acidity is :

COOH < COOH < COOH

CH_3 NO_2

Methyl group is electron donating group. It creates a negative centre on the carbon linking the carboxy group. This creates difficulty in the removal of proton.

Nitro group is electron withdrawing group. This creates a positive charge on the carbon linking the carboxy group. This helps in the release of proton from the carboxy group.

Q. 19. Arrange the following in order of increasing acidity: benzoic acid, *p*-hydroxy benzoic acid, *p*-nitrobenzoic acid.

Ans. The order of increasing acidity is:

p-hydroxy benzoic acid < benzoic acid < *p*-nitrobenzoic acid

An electron donating group like –OH creates negative charge on the carbon linking the carboxy group. This hinders the removal of proton from the carboxy group.

An electron attracting group like --NO_2 creates a positive centre on the carbon carrying the carboxy group. This helps in the release of protons.

Q. 20. Arrange the following compounds in the increasing order of acidity:

(*a*) (*i*) Benzoic acid (*ii*) *p*-nitrobenzoic acid (*iii*) phenol.

(*b*) (*i*) *p*-toluic acid (*ii*) Benzoic acid (*iii*) *p*-chlorobenzoic acid (*iv*) 2, 4, 6-Trichlorobenzoic acid.

(*c*) (*i*) *p*-Toluic acid (*ii*) *p*-chlorobenzoic acid (*iii*) *m*-chlorobenzoic acid (*iv*) Benzoic acid.

Ans. (*a*) Phenol < benzoic acid < *p*-nitrobenzoic acid

(*b*) *p*-Toluic acid < benzoic acid < *p*-chlorobenzoic acid < 2, 4, 6-Trichlorobenzoic acid.

(c) *p*-Toluic acid < benzoic acid < *p*-clorobenzoic acid < *m*-chlorobenzoic acid.

Q. 21. Arrange the following compounds in the increasing order of their acid strength and account for it.

$ClCH_2COOH, BrCH_2COOH, CNCH_2COOH, CH_3COOH, CH_3CH_2CH_2COOH$

Ans. The acid strength of an acid is affected by the substituents. An electron attracting group in the chain increases the acid strength and an electron releasing group decreases the acid strength due to inductive affect of the group. Out of –Cl, –Br and –CN which are all electron-withdrawing group, –CN has the strongest inductive effect followed by –Cl and –Br groups. In butyric acid $CH_3CH_2CH_2COOH$, it is the CH_3-CH_2- group which is present in the chain. It has electron releasing effect. Hence the compounds are arranged in increasing order of acid strength as under:

$CH_3CH_2-CH_2COOH$
$H-CH_2COOH$
$Br-CH_2COOH$ — Acid strength increases
$Cl-CH_2COOH$
$CN-CH_2COOH$ ↓

Q. 22. Complete the following:

(*i*) $CH_2\langle^{COOH}_{COOH} \xrightarrow{\Delta}$

(*ii*) $CH_3CONH_2 + X \longrightarrow CH_3COOH + Y$

(*iii*) $(CH_3COO)_2Ca \xrightarrow{\Delta}$

(*iv*) $CH_3COOH + Cl_2 \longrightarrow X + Y$

(*v*) $CH_3COOH + P_2O_5 \longrightarrow X$

(*vi*) $C_6H_5COOH + SOCl_2 \longrightarrow X + Y + Z$

Ans. (*i*) $CH_2\langle^{COOH}_{COOH} \xrightarrow{\Delta} CH_3COOH + CO_2\uparrow$
Malonic acid

(*ii*) $CH_3CONH_2 + H_2O \longrightarrow CH_3COOH + NH_3$
Acetamide — Acetic acid

(*iii*) $(CH_3COO)_2Ca \xrightarrow{\Delta} CH_3COCH_3 + CaCO_3$
Cal.acetate — Acetone

(*iv*) $CH_3COOH + Cl_2 \longrightarrow CH_2ClCOOH + HCl$
Chloroacetic acid

(*v*) $2CH_3COOH \xrightarrow{P_2O_5} (CH_3CO)_2O + H_2O$
Acetic anhydride

(*vi*) $C_6H_5COOH + SOCl_2 \longrightarrow C_6H_5COCl + SO_2 + HCl$

Q. 23. An amide (A) having molecular formula C_3H_7ON on hydrolysis gives an acid $C_3H_6O_2$ which upon chlorination in the presence of red phosphorus (HVZ reaction) produces chloro acid. The latter on boiling with aqueous NaOH and subsequent acidification forms lactic acid. Trace the reactions and deduce the structural formula of A.

Ans. $\underset{(A)}{RCH_2CONH_2} \xrightarrow{H_2O} \underset{(B)}{RCH_2COOH} \xrightarrow{Cl_2/P} \underset{(C)}{RCHClCOOH}$

$\downarrow H_2O$

$\underset{(D)}{RCH(OH)COOH}$

Lactic acid has the formula $CH_3CH(OH)COOH$.

It means R in the above compounds A, B and C is methyl group (CH_3—).

Hence the compounds A, B and C are

(A) $CH_3CH_2CONH_2$
Propanamide

(B) CH_3CH_2COOH
Propanoic acid

(C) $CH_3CHClCOOH$
2-Chloropropanoic acid

Q. 24. Although *p*-hydroxybenzoic acid is less acidic than benzoic acid, *o*-hydroxybenzoic acid (salicylic acid) is 15 times more acidic than benzoic acid. Explain.

Ans. Carboxylate ions of *o*-hydroxybenzoic acid is stabilized by means of effective hydrogen bonding leading to the formation of a ring.

O—H, C, O, O—H → O=C—O⁻, H, O, + H^+

Salicylic acid

Salicylate ion stabilized by intramolecular hydrogen bonding

Q. 25. Arrange the following in decreasing order of their acid strength:

(*i*) C_6H_5COOH, *p*-$OH.C_6H_4COOH$, *p*-$CH_3.C_6H_4COOH$, *p*-$Cl.C_6H_4COOH$, *p*-$Br.C_6H_4COOH$, *p*-$NO_2.C_6H_4.COOH$

(*ii*) HCOOH, C_6H_5COOH, C_6H_5OH, HCl

Ans. (*i*) *p*-$NO_2.C_6H_4COOH$ (–I and –R effect of –NO_2) > *p*-ClC_6H_4COOH (–I effect of –Cl) > *p*-BrC_6H_4COOH) (–I effect of –Br) > C_6H_5COOH > *p*-$CH_3.C_6H_4.COOH$ (+ I effect of –CH_3) > *p*-$OHC_6H_4.COOH$ (+ R effect of –OH)

Note: – I is electron withdrawing inductive effect

+ I is electron donating inductive effect

– R is electron withdrawing resonance effect

+ R is electron donating resonance effect

(*ii*) HCl > HCOOH > C_6H_5 COOH > C_6H_5OH

Benzoic acid is weaker than formic acid due to electron releasing resonance effect of –C_6H_5.

Q. 26. Explain

(*a*) 2, 4, 6-Trimethylbenzoic acid fails to esterify under normal conditions of esterification, while 2, 4, 6-trimethylphenylacetic acid is esterified readily under normal conditions.

(*b*) 2, 4, 6-Trihydroxybenzoic acid decarboxylates readily on heating, but benzoic acid does not.

Ans. (*a*) 2, 4, 6-trimethylbenzoic acid fails to esterify under normal conditions of esterification, because of steric hindrance of methyl groups.

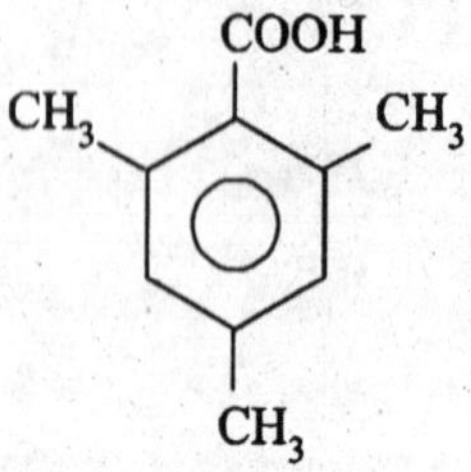

COOH

CH_3 CH_3

CH_3

2, 4, 6-Trimethylbenzoic acid

There is no such hindrance due to methyl groups in trimethylphenylacetic acid as methyl groups are at a distance from the carboxylic group.

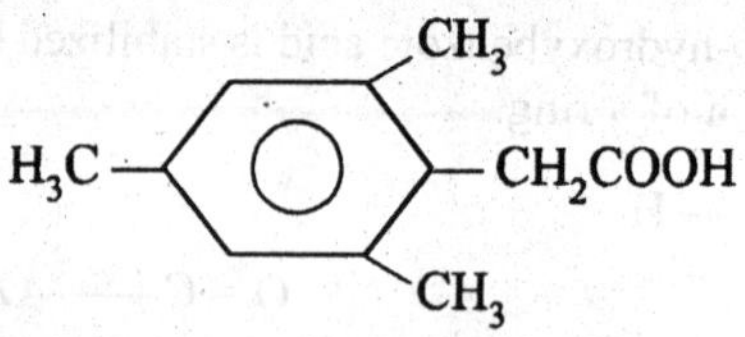

CH_3

H_3C — CH_2COOH

CH_3

(*b*) 2, 4, 6-trihydroxybenzoic acid decarboxylates easily due to electron withdrawing inductive effect of the –OH group.

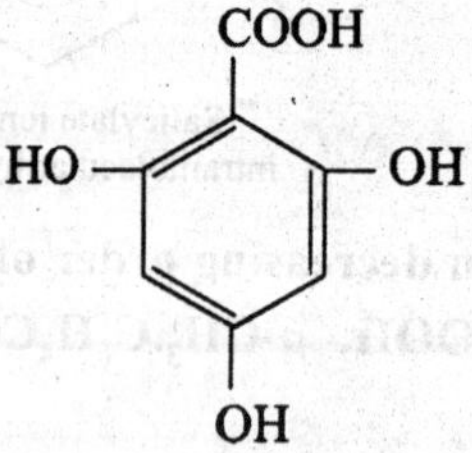

COOH

HO OH

OH

It can also be explained in terms of ortho effect.

Q. 27. Arrange the following in descending order of strength and justify your answer: *p*-toluic acid, phenylacetic acid, *m*-nitrobenzoic acid, benzoic acid.

Ans. *p*-Toluic acid H_3C — COOH

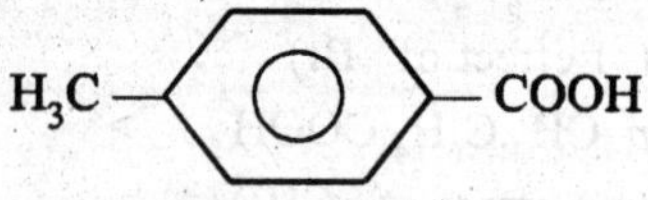

Phenylacetic acid — CH_2COOH

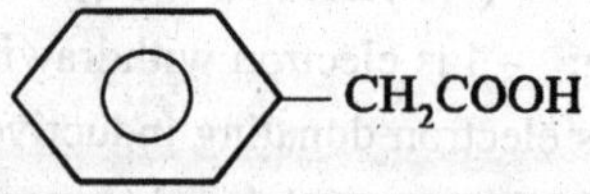

m-Nitrobenzoic acid NO_2 — COOH

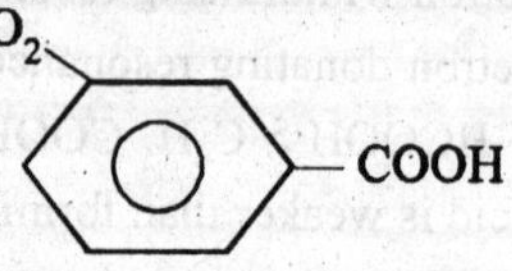

Benzoic acid C_6H_5-COOH

The decreasing order of strength is as follows:

p-$NO_2C_6H_4-COOH$ > C_6H_5-COOH > $C_6H_5-CH_2COOH$ > $H_3C-C_6H_4-COOH$

– I Effect (on NO_2); + I Effect (on H_3C)

Q. 28. Arrange the following in order of decreasing acid strength and account for your answer.

(*i*) *p*-Nitrobenzoic acid, *m*-nitrobenzoic acid

(*ii*) *p*-Hydroxybenzoic acid, *m*-hydroxybenzoic acid.

$O_2N-C_6H_4-COOH$ (para) > m-$NO_2C_6H_4-COOH$

This is because the nitro group brings about a positive charge at para position by inductive and resonance effects, whereas it brings about a positive charge at meta position by inductive effect only. Resonance effect is stronger than inductive effect. Hence *p*-isomer is a stronger acid than *m*-isomer.

m-HOC_6H_4-COOH > $HO-C_6H_4-COOH$ (para)

Hydroxy group brings about a negative charge at para position by resonance effect, thereby decreasing the acid strength. Hence *p*-hydroxybenzoic acid is a weaker acid. On the other hand, hydroxy group at meta position has –I inductive effect, which results in increased acid strength.

Q. 29. Giving reasons, arrange the following in increasing order of their acidity:

Propionic acid, 2-chloropropionic acid, 2-fluoropropionic acid.

Ans. Fluorine has –I inductive effect, therefore, it tends to draw the electrons towards itself. It helps in the release of protons from the acid. Fluorine being more electronegative than chlorine, has greater inductive effect than chlorine.

The increasing order of acid strength is as follows:

$$CH_3CH_2COOH < Cl-CH_2CH_2COOH < F-CH_2CH_2COOH$$

Q. 30. Arrange the following in order of increasing acidity. Give reasons.

Butanoic acid, 2-fluorobutanoic acid, 2-chlorobutanoic acid.

Ans. $CH_3CH_2CH_2COOH < CH_3CH_2CH(Cl)COOH < CH_3-CH_2-CH_2-CH(F)-COOH$

The reason is the same as in the question 29.

Q. 31. Arrange the following in order of increasing acidity. Give reason.

$F-CH_2COOH, Cl-CH_2COOH, CH_3-COOH, H-COOH$

Ans. (*i*) Out of Cl and F, the latter has greater inductive effect (–I). Hence, fluoroacetic acid is more acidic than chloroacetic acid which, in turn, is more acidic than acetic acid.

(*ii*) Out of acetic and formic acids, the former has a weaker acid strength because it has +I inductive effect.

The increasing order of acid strength is as follows:

$$CH_3COOH < Cl-CH_2COOH < F-CH_2COOH < H-COOH$$

Q. 32. Giving reasons, justify that trimethyl acetic acid is a weaker acid whereas trichloroacetic acid is a much stronger acid than acetic acid.

Ans.

$$CH_3 \rightarrow C(CH_3)_2-COOH \qquad Cl \leftarrow C(Cl)_2-COOH$$

Trimethylacetic acid   Trichloroacetic acid

As shown above, three methyl groups together exert +I inductive effect, as a result of which, the release of protons from the acid is suppressed. On the other hand, three chloro groups exert –I inductive effect, which helps in the release of protons from the acid. Therefore trichloroacetic acid is much stronger than acetic acid and trimethyl acetic acid is a much weaker acid than acetic acid.

Q. 33. Give the mechanism of esterification. (*Burdwan 2004; Delhi, 2005; Nagpur 2008*)

Ans.

$$R-C(=O)-OH \xrightarrow{H^+} R-\overset{+}{C}(OH)-O-H \xrightarrow{H-\ddot{O}-R'} R-C(OH)(OH)-\overset{+}{O}(H)-R'$$

$$\downarrow$$

$$R-C(=O)-OR' \xleftarrow{-H_2O} R-C(OR')(O-H)-\overset{+}{O}H_2$$

Q. 34. Give the mechanism of esterification of propionic acid and methanol in the presence of conc. H_2SO_4.

Ans. See Q. 33. Replace R by CH_3CH_2- and R′ by CH_3-. Rest of the scheme is the same.

Q. 35. Chloroacetic acid is stronger acid than acetic acid while acetic acid is itself weaker than formic acid. Why?

Ans. Chloro group exerts inductive effect and pulls the electrons towards itself, thus helping in the release of protons

$$Cl \leftarrow CH_2 \leftarrow \overset{O}{\overset{\|}{C}} \leftarrow O - H$$

CH_3 – group in acetic acid pushes the electrons away from itself thus hindering removal of protons

$$CH_3 \rightarrow \overset{O}{\overset{\|}{C}} \rightarrow O - H$$

In formic acid, there is no electron pushing alkyl group therefore chloroacetic acid is stronger then acetic acid which in turn is weaker than formic acid.

Q. 36. When K_a of an acid is large, pK_a is small, why?

Ans. This is because $pK_a = -\log K_a$.

Hence increase in the value of K_a leads to decrease in the value of pK_a.

Q. 37. Formic acid acts both as an acid as well as an aldehyde. Explain.

Ans. Besides acting as an acid, formic acid also gives the properties of an aldehyde. This is because, like the aldehydes, formic acid also gets oxidised easily.

$$HCOOH \xrightarrow{[O]} CO_2 + H_2O$$

Q. 38. Formic acid is stronger than acetic acid. Why?

(Panjab, 2003; Mysore 2004; Nagpur 2005)

Ans. $CH_3 \rightarrow \underset{\underset{O}{\|}}{C} \rightarrow O \rightarrow H$ (Acetic acid) $\quad H - \underset{\underset{O}{\|}}{C} - O - H$ (Formic acid)

Methyl group in acetic acid displaces the electrons towards hydrogen in acetic acid as shown by arrow mark. This diminishes the chances of removal of hydrogen as protons. There is no such situation in formic acid. Hence formic acid is stronger than acetic acid.

Q. 39. Benzoic acid is more acidic than phenol. Explain, why ? *(Awadh, 2000)*

Ans. See Q. 6.

Q. 40. Esterification of benzoic acid is easier than that of orthodimethyl benzoic acid why ? *(Devi Ahilya, 2001)*

Ans. Esterification of benzoic acid is easier than orthodimethyl benzoic acid because orthodimethyl benzoic acid experiences steric hinderance due to the methyl groups.

Q. 41. Among the following pairs, which is more acidic and why ? *(Kerala, 2000)*

COOH (with ortho **OCH_3**) **and** **COOH** (with para **OCH_3**)

COOH (with ortho **NO_2**) **and** **COOH** (with para **NO_2**)

Ans. See Q. 16.

Q. 42. Explain : trichloroacetic acid is a stronger acid than acetic acid *(Nagpur 2002)*

Ans. See Q. 16.

Q. 43. Compound (A), C_3H_7Cl reacts with alcoholic KOH to from (B), C_3H_6. Compound (B) decolourises Br_2/CCl_4 solution. Reaction of (A) with Mg in ether and subsequent treatment with CO_2 and dil.acid gives a compound (C) whose molecular formula is $C_4H_8O_2$; When we add compound (C) to aqueous $NaHCO_3$ solution, bubbles are evolved. Give the structural formulas of (A), (B) and (C), write equations for all the reactions involved.

(*Gulberga 2004*)

Hint.

$$(A) = CH_3-\underset{}{\overset{Cl}{\overset{|}{C}H}}-CH_3 \quad \text{(2-Chloropropane)}$$

$$(B) = CH_3-CH=CH_2... \quad \text{(Propene)}$$

$$(C) = CH_3-\overset{CH_3}{\overset{|}{C}H}-COOH \quad \text{(2-Methylpropanoic acid)}$$

Q. 44. How will you distinguish between benzoic acid and cinnamic acid ?

(*Baroda 2004*)

Ans. As cinnamic acid has a double bond, it will discharge the colour of Br_2 / CCl_4;

$$C_6H_5CH=CHCOOH + \underset{\text{red}}{Br_2} \xrightarrow{CCl_4} \underset{\text{(Colourless)}}{C_6H_5CHBrCHBrCOOH}$$

Benzoic acid will not give this test.

Q. 45. How will distinguish between phenol and salicyclic acid ? (*Delhi, 2005*)

Ans. Phenol on heating with phthalic anhydride and conc. H_2SO_4 produces phenolphthalein which gives pink colour when NaOH solution is added to the reaction mixture. Salicyclic acid does not give this test.

21

DERIVATIVES OF CARBOXYLIC ACIDS

CONTENTS

Q. 1. What is meant by carboxylic acid derivative? What are the different types of them? How are they obtained?

Ans. A carboxylic acid derivative is compound which is derived or obtained from a carboxylic acid by performing a reaction.

There are four different types of derivatives which we commonly come across. These are:

1. Esters
2. Acid Anhydrides
3. Acid Chlorides
4. Acid Amides.

The four types of acid derivatives are obtained replacing the –OH group of the carboxylic acid by – OR′, –OCOR, –Cl and – NH_2 groups, respectively

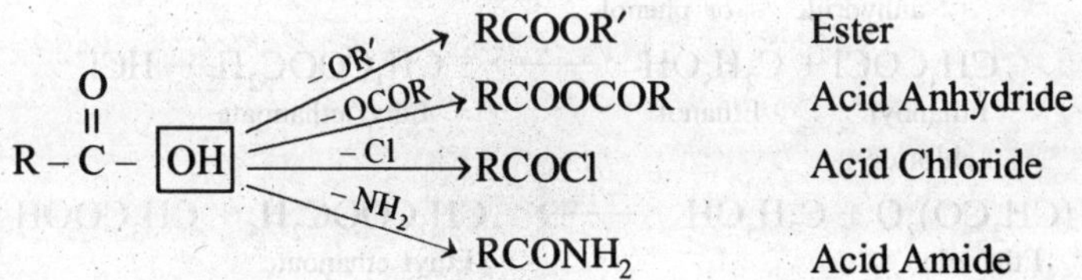

Q. 2. Describe the nomenclature of Esters.

Ans. Esters are named as follows by two different systems of nomenclature:

Common System. As an ester is obtained by the combination of a carboxylic acid and an alcohol (or phenol), it is named accordingly. The alkyl part of the alcohol is written first followed by the acid part (with the ending – *ate*).

IUPAC system. Naming an ester according to this system is not much different from the common system, except that the IUPAC name of the relevant carboxylic acid is used.

Names of a few compounds according to both systems of nomenclature are given below.

Compound	*Common Name*	*IUPAC Name*
$CH_3COOC_2H_5$	Ethyl acetate	Ethyl Ethanoate
$HCOOC_2H_5$	Ethyl formate	Ethyl methanoate
$CH_3COOC_6H_5$	Phenyl acetate	Phenyl ethanoate
$C_6H_5COOC_6H_5$	Phenyl benzoate	Phenyl benzoate
$C_6H_5COOCH_3$	Methyl benzoate	Methyl benzoate

Q. 3. Describe the general methods of preparation of esters giving mechanism wherever necessary. *(Nagpur 2008)*

Ans. Esters can be prepared by the following general methods:

1. By the combination of a carboxylic acid and an alcohol (or phenol)

$$\underset{\text{Carboxylic acid}}{RCOOH} + \underset{\text{Alcohol}}{R'OH} \xrightarrow{H^+} RCOOR' + H_2O$$

$$CH_3COOH + C_2H_5OH \xrightarrow{H^+} \underset{\text{Ethyl ethanoate}}{CH_3COOC_2H_5} + H_2O$$

$$\underset{\text{Benzoic acid}}{C_6H_5COOH} + CH_3OH \xrightarrow{H^+} \underset{\text{Methyl benzoate}}{C_6H_5COOCH_3} + H_2O$$

$$CH_3COOH + \underset{\text{Phenol}}{C_6H_5OH} \xrightarrow{H^+} \underset{\text{Phenyl acetate}}{CH_3COOC_6H_5} + H_2O$$

This reaction is known as esterification. Its mechanism is described in the chapter on carboxylic acids.

2. By the combination of acid chlorides or anhydrides with alcohols or phenols

$$\underset{\text{Acid chloride}}{RCOCl} + \underset{\text{Alcohol or phenol}}{R'OH} \longrightarrow \underset{\text{Ester}}{RCOOR'} + HCl$$

$$\underset{\text{Acid anhydride}}{(RCO)_2O} + \underset{\text{Alcohol or phenol}}{R'OH} \longrightarrow \underset{\text{Ester}}{RCOOR'} + \underset{\text{Acid}}{RCOOH}$$

$$\underset{\text{Ethanoyl chloride}}{CH_3COCl} + \underset{\text{Ethanol}}{C_2H_5OH} \longrightarrow \underset{\text{Ethyl ethanoate}}{CH_3COOC_2H_5} + HCl$$

$$\underset{\text{Ethanoic anhydride}}{(CH_3CO)_2O} + C_2H_5OH \longrightarrow \underset{\text{Ethyl ethanoate}}{CH_3COOC_2H_5} + CH_3COOH$$

$$\underset{\text{Ethanoyl chloride}}{CH_3COCl} + \underset{\text{Phenol}}{C_6H_5OH} \longrightarrow \underset{\text{Phenyl ethanoate}}{CH_3COOC_6H_5} + HCl$$

Unlike esterification involving a carboxylic acid and an alcohol, these reactions do not involve equilibrium between the reactants and products. Hence we can expect better yields in such reactions.

3. Trans-esterification. Treatment of an ester with an alcohol different from the constituent alcohol part of the ester results in the displacement reaction where the two alcohols are interchanged.

$$\underset{\text{Ethyl acetate}}{CH_3COOC_2H_5} + \underset{\text{Butyl alcohol}}{C_4H_9OH} \overset{H^+}{\rightleftharpoons} \underset{\text{Butyl acetate}}{CH_3COOC_4H_9} + C_2H_5OH$$

4. By heating silver salt of an acid with an alkyl halide

$$\underset{\text{Silver salt of acid}}{RCOOAg} + \underset{\text{Alkyl halide}}{R'X} \longrightarrow \underset{\text{Ester}}{RCOOR'} + \underset{\text{Silver halide}}{AgX}$$

$$\underset{\text{Silver acetate}}{CH_3COOAg} + \underset{\text{Ethyl bromide}}{C_2H_5Br} \longrightarrow \underset{\text{Ethyl acetate}}{CH_3COOC_2H_5} + AgBr$$

bromide

5. Reaction of a carboxylic acid with diazomethane

$$\underset{\text{Aceticacid}}{CH_3COOH} + \underset{\text{Diazo-methane}}{CH_2N_2} \xrightarrow{\text{Ether}} \underset{\text{Methyl acetate}}{CH_3COOCH_3} + N_2 \uparrow$$

$$\underset{\text{Benzoic acid}}{C_6H_5COOH} + CH_2N_2 \xrightarrow{\text{Ether}} \underset{\text{Methyl benzoate}}{C_6H_5COOCH_3} + N_2 \uparrow$$

Q. 4. Write a note on the physical properties of esters.

Ans. 1. Sweet smell and fragrance in fruits and flowers is due to the presence of esters. Esters present in various substances are as follows:

Fruit	**Ester**
Orange	Octyl acetate
Banana	Isoamyl acetate
Apple	Isoamyl valerate
Pineapple	Methyl butyrate
Apricot	Amyl butyrate

2. Esters are insoluble in water and show lower boiling points than the corresponding acids. This is due to absence of hydrogen bonding in esters.

Q. 5. Describe the chemical properties of esters.

Ans. 1. Nucleophilic substitution. Esters undergo nucleophilic substitution that is typical of carboxylic acid derivatives. Attack occurs at electron-deficient acyl carbon and results in the replacement of – OR′ group by – OH, – OR″ or – NH_2.

$$R-\overset{\overset{O}{\|}}{C}-OR' + :Z^- \longrightarrow R-\underset{\underset{OR'}{|}}{\overset{\overset{\bar{O}}{|}}{C}}-Z \longrightarrow R-\overset{\overset{O}{\|}}{C}-Z + :\bar{O}-R'$$

The following reactions are examples of nucleophilic substitution:

Hydrolysis. Esters on hydrolysis in the presence of an alkali give the alcohol (or phenol) and the sodium salt of the carboxylic acid.

$$\underset{\text{Ester}}{RCOOR'} + NaOH \longrightarrow RCOONa + \underset{\text{Alcohol}}{R'OH}$$

$$\underset{\text{Ethyl acetate}}{CH_3COOC_2H_5} + NaOH \longrightarrow \underset{\text{Sod. acetate}}{CH_3COONa} + C_2H_5OH$$

In the presence of an inorganic acid, the hydrolysis takes place as follows:

$$\underset{\text{Ethyl acetate}}{CH_3COOC_2H_5} + H_2O \xrightarrow{H^+} \underset{\text{Acetic acid}}{CH_3COOH} + C_2H_5OH$$

The mechanism of hydrolysis of an ester in the presence of an acid and an alkali is explained in the chapter on carboxylic acids.

Action of PCl_5 or $SOCl_2$. Esters, on treatment with phosphorus pentachloride or thionyl chloride, are converted into acid chloride.

$$\underset{\text{Ester}}{RCOOR'} + PCl_5 \longrightarrow \underset{\text{Acid chloride}}{RCOCl} + POCl_3 + R'Cl$$

$$\underset{\text{Ester}}{R\,COOR'} + \underset{\text{Thionyl chloride}}{SOCl_2} \longrightarrow \underset{\text{Acid chloride}}{RCOCl} + R'Cl + SO_2$$

$$\underset{\text{Ethyl acetate}}{CH_3COOC_2H_5} + PCl_5 \longrightarrow \underset{\text{Acetyl chloride}}{CH_3COCl} + \underset{\text{Ethyl chloride}}{C_2H_5Cl} + POCl_3$$

$$\underset{\text{Ethyl acetate}}{CH_3COOC_2H_5} + SOCl_2 \longrightarrow \underset{\text{Acetyl chloride}}{CH_3COCl} + \underset{\text{Ethyl chloride}}{C_2H_5Cl} + SO_2$$

Action of Ammonia

$$\underset{\text{Ester}}{RCOOR'} + NH_3 \longrightarrow \underset{\text{Amide}}{RCONH_2} + \underset{\text{Alcohol}}{R'OH}$$

$$\underset{\text{Ethyl acetate}}{CH_3COOC_2H_5} + NH_3 \longrightarrow \underset{\text{Acetamide}}{CH_3CONH_2} + \underset{\text{Ethanol}}{C_2H_5OH}$$

Mechanism

$$R-\overset{\overset{\ddot{O}:}{||}}{\underset{\underset{:\ddot{O}R'}{|}}{C}} + \ddot{N}H_3 \longrightarrow \left[R-\overset{\overset{\ominus\ddot{O}:}{|}}{\underset{\underset{:\ddot{O}R'}{|}}{C}}-\overset{\oplus}{N}H_3 \right] \longrightarrow R-\overset{\overset{O}{||}}{C}-NH_2 + R'OH$$

2. Reduction of Esters. Esters get reduced to alcohols in the presence of Na and alcohol or $LiAlH_4$.

$$RCOOR' \xrightarrow[\text{or Na/Alcohol}]{LiAlH_4} RCH_2OH + R'OH$$

$$\underset{\text{Ethylacetate}}{CH_3COOC_2H_5} \xrightarrow[\text{or Na/Alcohol}]{LiAlH_4} \underset{\text{Ethylalchol}}{2CH_3CH_2OH}$$

3. Reaction with Grignard reagents. Esters of formic acid give secondary alcohols and those of other acids give tertiary alcohols on reaction with Grignard reagents.

Reaction between ethyl formate and methyl magnesium bromide

$$(i)\ H-\overset{\overset{O}{||}}{C}-OC_2H_5 + \overset{-}{C}H_3\overset{+}{Mg}Br \longrightarrow H-\overset{\overset{\bar{O}}{|}}{\underset{\underset{CH_3}{|}}{C}}-OC_2H_5 \xrightarrow{\overset{+}{Mg}Br}$$

$$H-\overset{\overset{OMgBr}{|}}{\underset{\underset{CH_3}{|}}{C}}-OC_2H_5 \xrightarrow{-Mg(Br)OC_2H_5} \underset{\text{Acetaldehyde}}{H-\overset{\overset{O}{||}}{\underset{\underset{CH_3}{|}}{C}}}$$

$$(ii)\ H-\overset{\overset{O}{\|}}{\underset{CH_3}{C}} + \overset{-}{C}H_3\overset{+}{Mg}B \longrightarrow H-\overset{\overset{\bar{O}}{|}}{\underset{CH_3}{\underset{|}{C}}}-CH_3 \xrightarrow{\overset{+}{Mg}Br} H-\overset{\overset{OMgBr}{|}}{\underset{CH_3}{\underset{|}{C}}}-CH_3$$

$$\xrightarrow{H_2O} H-\overset{\overset{OH}{|}}{\underset{CH_3}{\underset{|}{C}}}-CH_3 + Mg\langle^{OH}_{CH_3}$$

Isopropyl alcohol

Reaction between ethyl acetate and methyl magnesium bromide

$$(i)\ CH_3-\overset{\overset{O}{\|}}{C}-OC_2H_5 + \overset{-}{C}H_3\overset{+}{Mg}Br \longrightarrow CH_3-\overset{\overset{\bar{O}}{|}}{\underset{CH_3}{\underset{|}{C}}}-OC_2H_5$$

$$\xrightarrow{\overset{+}{Mg}Br} CH_3-\overset{\overset{OMgBr}{|}}{\underset{CH_3}{\underset{|}{C}}}-OC_2H_5 \longrightarrow CH_3-\overset{\overset{O}{\|}}{\underset{CH_3}{\underset{|}{C}}} + Mg\langle^{Br}_{OC_2H_5}$$

Acetone — Magnesium ethyoxy bromide

$$(ii)\ CH_3-\overset{\overset{O}{\|}}{\underset{CH_3}{C}} + \overset{-}{C}H_3\overset{+}{Mg}Br \longrightarrow CH_3-\overset{\overset{\bar{O}}{|}}{\underset{CH_3}{\underset{|}{C}}}-CH_3 \xrightarrow{\overset{+}{Mg}Br} CH_3-\overset{\overset{OMgBr}{|}}{\underset{CH_3}{\underset{|}{C}}}-CH_3$$

$$\xrightarrow{H_2O} CH_3-\overset{\overset{OH}{|}}{\underset{CH_3}{\underset{|}{C}}}-CH_3 + Mg\langle^{OH}_{Br}$$

Tert. butyl alcohol

4. Trans-esterification. As mentioned in the methods of preparation of esters, trans-esterification is a reaction involving an ester and an alcohol which is different from the alcohol part of the ester.

$$RCOOR' + R''OH \xrightarrow[\text{or alkali}]{\text{Acid}} RCOOR'' + R'OH$$

The reaction takes place in the presence of an acid or alkali.

Mechanism (acid catalysed)

$$R-\overset{\overset{O}{\|}}{C}-OR' \underset{}{\overset{H^+}{\rightleftharpoons}} R-\overset{\overset{\overset{+}{O}H}{\|}}{C}-OR' \xrightarrow{R''OH} R-\overset{\overset{OH}{|}}{\underset{\underset{R''}{\oplus OH}}{\underset{|}{C}}}-OR'$$

$$\xrightarrow{-H^+} R-\overset{\overset{O-H}{|}}{\underset{\underset{R''}{O}}{\underset{|}{C}}}-OR' \longrightarrow R-\overset{\overset{O}{\|}}{\underset{\underset{R''}{O}}{\underset{|}{C}}} + R'OH$$

Mechanism (Base catalysed)

$$R''OH + NaOH \longrightarrow R''ONa + H_2O$$

$$R''ONa \longrightarrow R''O^- + Na^+$$

$$R-\overset{O}{\overset{||}{C}}-OR' + R''O^- \longrightarrow R-\underset{OR'}{\underset{|}{\overset{O^{\ominus}}{\overset{|}{C}}}}-OR' \xrightarrow{-OR'} \underset{\text{Ester}}{R-\underset{OR'}{\underset{|}{\overset{O}{\overset{||}{C}}}}}$$

Q. 6. Describe the mechanism of acid-catalysed and base-catalysed hydrolysis of an ester.
(Himachal, 2000; Lucknow, 2000; Kerala, 2000; Delhi, 2002; Bundelkhand 2004, Nagpur 2008)

Ans. Both these reactions are acyl nucleophilic substitution reactions.

Acid-catalysed

$$R-\overset{O:}{\overset{||}{C}}-OR' + H^{\oplus} \longrightarrow R-\underset{OR'}{\underset{|}{\overset{\overset{+}{O}H}{\overset{||}{C}}}} \longrightarrow R-\underset{OR'}{\underset{|}{\overset{OH}{\overset{|}{C^{\oplus}}}}}$$

$$\downarrow H_2O:$$

$$R-\overset{O}{\overset{||}{C}}-OH + R'OH \xleftarrow{-H^+} R-\underset{OR'}{\underset{|}{\overset{OH}{\overset{|}{C}}}}-\overset{+}{O}H_2$$

Base-catalysed

$$R-\overset{O}{\overset{||}{C}}-OR' \xrightarrow{OH^-} R-\underset{OR'}{\underset{|}{\overset{O^-}{\overset{|}{C}}}}-OH$$

$$\downarrow$$

$$R-\overset{O}{\overset{||}{C}}-OH + RO^-$$

Q. 7. What are acid anhydrides? How are they obtained? Describe their nomenclature.

Ans. Acid anhydrides are organic substances obtained by the dehydration of carboxylic acids. Generally one water molecule is removed from the two molecules of monocarboxylic acid. In the case of dicarboxylic acids, one water molecule is taken out from one molecule of the acid. Thus

$$\begin{matrix} RCOOH \\ RCOOH \end{matrix} \xrightarrow{-H_2O} \begin{matrix} RCO \\ RCO \end{matrix}\!\!>O$$

2 molecules of the acid → Acid anhydride

$$\begin{matrix} CH_2COOH \\ | \\ CH_2COOH \end{matrix} \xrightarrow{-H_2O} \begin{matrix} CH_2CO \\ | \\ CH_2CO \end{matrix}\!\!>O$$

Succinic acid → Succinic anhydride

Nomenclature. Such compounds are named by adding the word "anhydride" after the name of the acid as indicated below:

Compound	*Common Name*	*IUPAC Name*
CH_3CO, CH_3CO >O or $(CH_3CO)_2O$	Acetic Anhydride	Ethanoic Anhydride
CH_2CO, CH_2CO >O	Succinic Anhydride	Butanedioic Anhydride
C_6H_4(CO)(CO)>O	Phthalic Anhydride	*o*-Benzenedioic Anhydride

Q. 8. Describe general methods of preparation of acid anhydrides.

Ans. Following methods are used for the preparation of acid anhydrides:

1. Dehydration. Vapours of the acid are passed over phosphorus pentaoxide.

$$2\,RCOOH \xrightarrow[-H_2O]{P_2O_5} \begin{matrix} RCO \\ RCO \end{matrix}\!\!>O$$

Acid anhydride

$$2\,CH_3COOH \xrightarrow[-H_2O]{P_2O_5} \begin{matrix} CH_3CO \\ CH_3CO \end{matrix}\!\!>O$$

Acetic anhydride

2. By the action of acid chloride on the sod. salt of acid:

$$\underset{\text{Acid chloride}}{RCOCl} + \underset{\text{Sod. salt of acid}}{NaOOCR} \longrightarrow \underset{\text{Acid anhydride}}{\begin{matrix} RCO \\ RCO \end{matrix}\!\!>O} + NaCl$$

$$CH_3COCl + NaOOCCH_3 \longrightarrow \begin{matrix} CH_3CO \\ CH_3CO \end{matrix}\!\!>O + NaCl$$

3. By the combination of acid and ketene. This reaction takes place in two steps:

(*i*) Dehydration of acid at high temp. gives ketene.

$$\underset{\text{Acetic acid}}{CH_3COOH} \xrightarrow[973\,K]{AlPO_4} \underset{\text{Ketene}}{CH_2 = C = O} + H_2O$$

(*ii*) Ketene reacts with a molecule of acetic acid to produce acetic anhydride.

$$\underset{\text{Ketene}}{CH_2 = C = O} + CH_3COOH \longrightarrow \underset{\text{Acetic anhydride}}{(CH_3CO)_2O}$$

Q. 9. Describe the general properties of acid anhydrides.

Ans. 1. Hydrolysis. Acid anhydrides get hydrolysed by water to produce acids.

$$\underset{\text{Acetic anhydride}}{(CH_3CO)_2O} + H_2O \longrightarrow \underset{\text{Acetic acid}}{2\,CH_3COOH}$$

2. Alcoholysis. With ethyl alcohol, acetic anhydride reacts to form ethyl acetate.

$$\underset{\text{Acetic anhydride}}{(CH_3CO)_2\,O} + \underset{\text{Ethyl alcohol}}{C_2H_5OH} \longrightarrow \underset{\text{Ethyl acetate}}{CH_3COOC_2H_5} + CH_3COOH$$

Better yield of the ester is obtained by this method.

3. Ammonolysis. Ammonia reacts with acetic anhydride to produce acetamide.

$$\underset{\text{Acetic anhydride}}{(CH_3CO)_2\,O} + NH_3 \longrightarrow \underset{\text{Acetamide}}{CH_3CONH_2} + CH_3COOH$$

4. Action of primary amine

$$(CH_3CO)_2O + \underset{\text{Ethyl amine}}{C_2H_5NH_2} \longrightarrow \underset{\text{N-Ethyl acetamide}}{CH_3CONHC_2H_5} + CH_3COOH$$

5. Action of PCl_5 or HCl. With either of these, acid chlorides are formed.

$$\underset{\text{Acetic anhydride}}{(CH_3CO)_2\,O} + PCl_5 \longrightarrow \underset{\text{Acetyl chloride}}{2CH_3COCl} + \underset{\text{Phosphorus oxytrichloride}}{POCl_3}$$

6. Reduction. Acid anhydrides on reduction produce alcohols. Reducing agents used are sodium-alcohol or $LiAlH_4$.

$$(CH_3CO)_2\,O + 8[H] \xrightarrow[\text{or } LiAlH_4]{\text{Na/Alcohol}} 2\,C_2H_5OH + H_2O$$

7. Friedel-Crafts acylation. Acetic anhydride can be used for acylating aromatic compounds in the presence of anhyd. $AlCl_3$

$$\underset{\text{Benzene}}{C_6H_6} + (CH_3CO)_2O \xrightarrow[AlCl_3]{\text{Anhyd.}} \underset{\text{Acetophenone}}{C_6H_5COCH_3} + CH_3COOH$$

Q. 10. What are acid amides? How are they obtained? Give their nomenclature.

Ans. Acid amides are organic compounds having the general formula $RCONH_2$ where R stands for some alkyl or aryl group. These are acid derivatives as these compounds can be obtained from acid.

These compounds are obtained by replacing the – OH group of the carboxy acids RCOOH by NH_2 group. Nomenclature of amides according to common system and IUPAC system are given below:

Compound	*Corresponding acid*	*Common name*	*IUPAC name*
$HCONH_2$	HCOOH	Formamide	Methanamide
CH_3CONH_2	CH_3COOH	Acetamide	Ethanamide
$CH_3CH_2CONH_2$	CH_3CH_2COOH	Propionamide	Propanamide
$C_6H_5CONH_2$	C_6H_5COOH	Benzamide	Benzenamide

Q. 11. Describe the general methods of preparation of amides giving mechanism wherever necessary.

Ans. 1. By the action of NH_3 on acid derivatives

(*i*) $(CH_3CO)_2O + NH_3 \longrightarrow CH_3CONH_2 + CH_3COOH$
Acetic anhydride — Acetamide

(*ii*) $CH_3COCl + NH_3 \longrightarrow CH_3CONH_2 + HCl$
Acetyl chloride — Acetamide

(*iii*) $CH_3COOC_2H_5 + NH_3 \longrightarrow CH_3CONH_2 + C_2H_5OH$
Ethyl acetate — Acetamide

Mechanism

These are all nucleophilic substitution reactions. Mechanism of the reaction is given below:

$$H_3N\colon + R-\overset{O}{\overset{\|}{C}}-X \longrightarrow R-\underset{\oplus NH_3}{\underset{|}{\overset{O^{\ominus}}{\overset{|}{C}}}}-X \longrightarrow R-\overset{O}{\overset{\|}{C}}-\overset{\oplus}{N}H_3 \xrightarrow{-H^{\oplus}} R-\overset{O}{\overset{\|}{C}}-NH_2$$

Here X stands for $-OCOCH_3$, $-Cl$ or $-OC_2H_5$ group.

2. From ammonium salts. Ammonium salts of carboxylic acids on heating give amides.

$$CH_3COOH + NH_3 \longrightarrow CH_3COONH_4$$

$$CH_3COONH_4 \xrightarrow{\Delta} CH_3CONH_2 + H_2O$$

3. Partial hydrolysis of cyanides

$$CH_3CN + H_2O \xrightarrow[H_2O_2]{\text{Alkaline}} CH_3CONH_2$$

Q. 12. Describe the general properties of amides.

Ans. 1. Hydrogen bonding. There is intermolecular hydrogen bonding between amide molecules. This leads to higher boiling points of amides. Hydrogen bonding involves H–O bonds as shown below:

$$\ldots H-NH-\underset{R}{\underset{|}{CO}}\ldots H-NH-\underset{R}{\underset{|}{CO}}\ldots H-NH-\underset{R}{\underset{|}{CO}}\ldots$$

2. Reactivity. Acid amides are quite reactive towards nucleophilic substitution. This is because of the positive charge on acyl carbon atom, which is so necessary for nucleophilic substitution and the intermediate compound is stabilised as shown below:

$$R-\overset{:O}{\overset{\|}{C}}-\ddot{N}H_2 \longleftrightarrow R-\overset{:O^{-}}{\overset{+|}{C}}-\ddot{N}H_2 \longleftrightarrow R-\overset{:O^{-}}{\overset{|}{C}}=\overset{+}{N}H_2$$

3. Amphoteric nature. Amides are neutral to litmus. They show weakly acidic as well as basic properties.

$CH_3CONH_2 + HCl \longrightarrow CH_3CONH_2HCl$ — Basic Property
Acetamide hydrochloride

$CH_3CONH_2 + Na \longrightarrow CH_3CONHNa + 1/2\ H_2$ — Acidic Property
Sod. acetamide

The low basicity is because of the following transition:

$$R-\overset{\overset{O}{\|}}{C}-\underset{\underset{H}{|}}{\ddot{N}}-H \longrightarrow R-\overset{\overset{O^{\ominus}}{|}}{C}=\overset{\oplus}{N}H_2$$

4. Hydrolysis. They are hydrolysed to parent acids in the presence of a mineral acid or alkali.

$$CH_3CONH_2 + H_2O \xrightarrow[\text{or } OH^-]{H^+} CH_3COOH + NH_3$$

Mechanism (In the presence of an acid)

$$R-\overset{\overset{O}{\|}}{C}-NH_2 + \overset{\oplus}{H} \longrightarrow R-\underset{\underset{NH_2}{|}}{\overset{\overset{OH}{|}}{C^{+}}} + H\ddot{O}-H \longrightarrow R-\underset{\underset{NH_2}{|}}{\overset{\overset{OH}{|}}{C}}-\overset{\oplus}{O}H_2$$

$$\xrightarrow{-H^{\oplus}} R-\overset{\overset{O}{\|}}{C}-OH + NH_3 \longrightarrow RCOONH_4$$

Mechanism (In the presence of a base)

$$R-\overset{\overset{O}{\|}}{C}-NH_2 \xrightarrow{OH^{\ominus}} R-\underset{\underset{NH_2}{|}}{\overset{\overset{O^{\ominus}}{|}}{C}}-O-H \longrightarrow RCOO^- + NH_3$$

5. Dehydration. On heating with some dehydrating agent such as phosphorus pentaoxide, amides get converted into cyanides or nitriles.

$$\underset{\text{Acetamide}}{CH_3CONH_2} \xrightarrow{P_2O_5} \underset{\text{Methyl cyanide}}{CH_3CN} + H_2O$$

6. Reduction. On reduction in the presence of sodium and alcohol, or lithium aluminium hydride, amides get converted into primary amines.

$$\underset{\text{Acetamide}}{CH_3CONH_2} \xrightarrow[\text{Na/Alcohol}]{LiAlH_4} \underset{\text{Ethyl amine}}{CH_3CH_2NH_2}$$

7. Action of nitrous acid

$$CH_3CONH_2 + HNO_2 \longrightarrow CH_3COOH + N_2 + H_2O$$

8. Hoffmann Bromamide or Hoffmann Degradation reaction. Amides on treatment with bromine and potassium hydroxide yield amines with one carbon atom less than the original amides. This reaction is employed to step-down the series.

$$\underset{\text{Acetamide}}{CH_3CONH_2} + Br_2 + KOH\text{ (alc.)} \longrightarrow \underset{\text{Methyl amine}}{CH_3NH_2} + 2KBr + K_2CO_3 + 2H_2O$$

For mechanism, see the chapter on amines.

Q. 13. What are acid chlorides? How are they formed? Describe the nomenclature of acid chlorides.

Ans. Acid chlorides are organic compounds having the general formula RCOCl where R stands for some alkyl or aryl group. These are acid derivatives and can be obtained by substituting

– OH group of the carboxy acids by –Cl group. These can also be obtained by replacing – NH_2, – OCOR or – OR group by – Cl group.

Nomenclature

Nomenclature of acid chlorides according to common and IUPAC systems is given below:

Acid chloride	*Corresponding acid*	*Common name*	*IUPAC name*
HCOCl	HCOOH	Formyl chloride	Methanoylchloride
CH_3COCl	CH_3COOH	Acetyl chloride	Ethanoylchloride
CH_3CH_2COCl	CH_3CH_2COOH	Propionyl chloride	Propanoylchloride
C_6H_5COCl	C_6H_5COOH	Benzoyl chloride	–

Q. 14. Describe the general methods of preparation of acid chlorides.

Ans. 1. Using PCl_5 or PCl_3. Acid chlorides can be prepared by the action of PCl_5 or PCl_3 on a carboxylic acid.

$$\underset{}{CH_3COOH} + PCl_5 \longrightarrow \underset{\text{Acetyl chloride}}{CH_3COCl} + \underset{\text{Phosphorus oxytrichloride}}{POCl_3} + HCl$$

$$3CH_3COOH + PCl_3 \longrightarrow 3CH_3COCl + \underset{\text{Phosphorus acid}}{H_3PO_3}$$

2. By the action of thionyl chloride on carboxylic acid

$$CH_3COOH + \underset{\text{Thionyl chloride}}{SOCl_2} \longrightarrow CH_3COCl + SO_2\uparrow + HCl$$

3. By the action of PCl_3 or thionyl chloride on sodium or calcium salts of fatty acids

$$3CH_3COONa + PCl_3 \longrightarrow \underset{\text{Acetyl chloride}}{CH_3COCl} + Na_3PO_3$$

$$\underset{\text{Cal.acetate}}{(CH_3COO)_2\,Ca} + \underset{\text{Thionyl chloride}}{SOCl_2} \longrightarrow \underset{\text{Acetyl chloride}}{2CH_3COCl} + CaSO_3$$

Q. 15. Describe the chemical properties of acid chlorides giving mechanism wherever necessary.

Ans. Acid chlorides are the most reactive of carboxylic acid derivatives. They undergo nucleophilic substitution reactions. Important reactions of acid chlorides are given below:

1. Hydrolysis. Acid chlorides are hydrolysed by water to form the parent carboxylic acids.

$$\underset{\text{Acetyl chloride}}{CH_3COCl} + H_2O \longrightarrow \underset{\text{Acetic acid}}{CH_3COOH} + HCl$$

$$\underset{\text{Benzoyl chloride}}{C_6H_5COCl} + H_2O \longrightarrow \underset{\text{Benzoic acid}}{C_6H_5COOH} + HCl$$

Mechanism of reaction is as follows:

$$CH_3-\overset{\overset{O}{\parallel}}{C}-Cl + H-\ddot{O}-H \longrightarrow CH_3-\underset{\underset{\oplus OH_2}{|}}{\overset{\overset{\ominus}{O}}{\overset{|}{C}}}-Cl \xrightarrow{-Cl^-} CH_3-\overset{\overset{O}{\parallel}}{C}-\overset{\oplus}{O}H_2 \xrightarrow{-H^{\oplus}} CH_3-\overset{\overset{O}{\parallel}}{C}-O-H$$

2. Reaction with alcohol. An acid chloride reacts with alcohol to form an ester.

$$CH_3COCl + C_2H_5OH \longrightarrow CH_3COOC_2H_5 + HCl$$

The reaction takes place as a nucleophilic substitution with the following mechanism:

$$CH_3-\overset{O}{\overset{\|}{\underset{Cl}{C}}}+C_2H_5\ddot{O}-H \longrightarrow CH_3-\overset{O^{\ominus}}{\overset{|}{\underset{Cl}{C}}}-\overset{\oplus}{\underset{C_2H_5}{OH}} \xrightarrow{-Cl^-} CH_3-\overset{O}{\overset{\|}{C}}-\overset{\oplus}{\underset{C_2H_5}{OH}} \xrightarrow{-H^{\oplus}} CH_3-\overset{O}{\overset{\|}{C}}-OC_2H_5$$

3. Ammonolysis. An acid chloride reacts with ammonia to form an amide.

$$\underset{\text{Acetyl chloride}}{CH_3COCl} + NH_3 \longrightarrow \underset{\text{Acetamide}}{CH_3CONH_2} + HCl$$

$$\underset{\text{Benzoyl chloride}}{C_6H_5COCl} + NH_3 \longrightarrow \underset{\text{Benzamide}}{C_6H_5CONH_2} + HCl$$

Mechanism of the reaction is given below. It is again a nucleophilic substitution reaction.

$$CH_3-\overset{O}{\overset{\|}{\underset{Cl}{C}}}+\ddot{N}H_3 \longrightarrow CH_3-\overset{O^{\ominus}}{\overset{|}{\underset{Cl}{C}}}-\overset{\oplus}{N}H_3 \xrightarrow{-Cl^{\ominus}} CH_3-\overset{O}{\overset{\|}{C}}-\overset{\oplus}{N}H_3 \xrightarrow{-H^{\oplus}} CH_3-\overset{O}{\overset{\|}{C}}-NH_2$$

4. Reaction with amines. An acid chloride on reaction with an amine gives substituted amide.

$$CH_3COCl + C_2H_5NH_2 \longrightarrow \underset{\text{N-Ethyl acetamide}}{CH_3CONHC_2H_5} + HCl$$

$$CH_3COCl + (C_2H_5)_2NH \longrightarrow \underset{\text{N,N-Diethyl acetamide}}{CH_3CON\,(C_2H_5)_2} + HCl$$

5. Formation of acid anhydride. There is a reaction with sodium salt of fatty acid and acid anhydride is obtained.

$$\underset{\text{Acetyl chloride}}{CH_3COCl} + \underset{\text{Sod. acetate}}{NaOOCCH_3} \longrightarrow \underset{\text{Acetic anhydride}}{(CH_3CO)_2\,O} + NaCl$$

6. Reaction with Grignard reagent

$$\underset{\text{Acetyl chloride}}{CH_3COCl} + \underset{\text{Methyl mag. chloride}}{ClMgCH_3} \longrightarrow \underset{\text{Acetone}}{CH_3COCH_3} + MgCl_2$$

7. Reaction with organocadmium compound

$$\underset{\text{Diethyl cadmium}}{(C_2H_5)_2\,Cd} + \underset{\text{Acetyl chloride}}{2CH_3COCl} \longrightarrow \underset{\text{Methyl ethyl ketone}}{2CH_3COC_2H_5} + CdCl_2$$

8. Rosenmund's reduction. Hydrogenation of an acid chloride in the presence of palladium based on $BaSO_4$ using xylene as a solvent gives aldehyde.

$$\underset{\text{Acetyl chloride}}{CH_3COCl} + H_2 \xrightarrow[\text{Xylene soln.}]{Pd\,/\,BaSO_4} \underset{\text{Acetaldehyde}}{CH_3CHO} + HCl$$

9. Friedel-Crafts reaction

$$C_6H_6 + CH_3COCl \xrightarrow[AlCl_3]{Anhy.} C_6H_5COCH_3 + HCl$$

Acetophenone

10. Action with halogens. Treatment with a halogen brings about substitution at the alkyl group, α-hydrogen atoms undergo substitution with halogen as shown below:

$$\underset{\text{Propionyl chloride}}{CH_3CH_2COCl} + Cl_2 \longrightarrow \underset{\text{Chloropropionyl chloride}}{CH_3CHCl\ COCl} + HCl$$

11. Reaction with carboxy acid. An acid chloride reacts with a carboxylic acid in the presence of pyridine to produce acid anhydride.

$$\underset{\text{Acetyl chloride}}{CH_3COCl} + CH_3COOH \xrightarrow{\text{Pyridine}} \underset{\text{Acetic anhydride}}{(CH_3CO)_2\ O} + HCl$$

Q. 16. The following compounds are subjected to hydrolysis:

(*a*) Acetic anhydride (*b*) Acetamide (*c*) Ethyl acetate (*d*) Acetyl chloride

Arrange the reactants in decreasing order of reactivity. (*Kurukshetra, 2001*)

Or

Arrange the acid chlorides, amides, anhydrides and esters in order of increasing reactivity towards nucleophilic attack. Give suitable reasons.

Ans. The order of reactivity of the above reactions in the decreasing order is as follows:

Acetyl chloride > Acetic anhydride > Ethyl acetate > Acetamide

1. All the above reactions are nucleophilic substitution reactions where – OCOR group of the anhydride, – NH_2 of the amide, – OC_2H_5 of the ester and – Cl of the acid chloride are replaced by – OH group.

2. We expect greater reactivity in a nucleophilic substitution when the acyl carbon has the maximum positive charge so that it could attract the nucleophile (water in this case).

3. All the groups which are attached to the acyl group (CH_3CO–) donate the electrons to acyl carbon thereby rendering it less positive. However, it is observed that –Cl supplies the minimum electrons to the acyl carbon atom. Thus carbon remains positive to the maximum extent in the case of acid chloride. Hence reactivity is maximum with an acid chloride. This is illustrated as below:

$$R-\overset{\delta^+}{C}(=\overset{\delta^-}{O})-\ddot{Cl}: \;>\; R-\overset{\delta^+}{C}(=\overset{\delta^-}{O})-\ddot{O}-\overset{\delta^+}{C}(=\overset{\delta^-}{O})-R \;>\; R-\overset{\delta^+}{C}(=\overset{\delta^-}{O})-\ddot{O}R' \;>\; R-\overset{\delta^+}{C}(=\overset{\delta^-}{O})-\ddot{N}H_2$$

This can also be explained on the basis of ease of removal of the bases Cl^-, $RCOO^-$, $R'\ O^-$ and NH_2^- respectively.

Q. 17. How acetyl chloride can be converted into (*i*) acetamide (*ii*) Acetic anhydride (*iii*) Ethylacetate?

Ans. See Q. 15.

Q. 18. Complete the following equations:

(*i*) $CH_3CH_2CONH_2 + Br_2 + KOH \longrightarrow$

(*ii*) $CH_3COCl + H_2 \xrightarrow{Pd / BaSO_4}$

(*iii*) $CH_3COCl + CH_3COONa \longrightarrow$

Ans. (*i*) $\underset{\text{Propionamide}}{CH_3CH_2CONH_2} + Br_2 + KOH \longrightarrow \underset{\text{Ethyl amine}}{CH_3CH_2NH_2} + 2KBr + K_2CO_3 + 2H_2O$

(*ii*) $\underset{\text{Acetyl chloride}}{CH_3COCl} + H_2 \xrightarrow{Pd / BaSO_4} \underset{\text{Acetaldehyde}}{CH_3CHO} + HCl$

(*iii*) $CH_3COCl + CH_3COONa \xrightarrow{\text{Pyridine}} \underset{\text{Acetic anhydride}}{(CH_3CO)_2\,O} + NaCl$

Q. 19. Complete the following reactions:

(*i*) **Propionic acid + $SOCl_2$**

(*ii*) **Benzoyl chloride + NH $(CH_2\,CH_3)_2$**

(*iii*) **Butanoyl chloride + 1-butanol**

(*iv*) **Toluene + Acetyl chloride + Anhy. $AlCl_3$**

(*v*) **Propionamide + $NaNO_2$ + HCl**

(*vi*) **$(CH_3)_2CHCOOH + CH_2N_2$**

Ans. (*i*) $\underset{\text{Propionic acid}}{CH_3CH_2COOH} + \underset{\text{Thionyl chloride}}{SOCl_2} \longrightarrow \underset{\text{Propionoyl chloride}}{CH_3CH_2COCl} + SO_2 + HCl$

(*ii*) $\underset{\text{Benzyol chloride}}{C_6H_5COCl} + \underset{\text{Diethyl amine}}{NH\,(CH_2CH_3)_2} \longrightarrow \underset{\text{N, N Diethyl benzamide}}{C_6H_5CON\,(CH_2CH_3)_2} + HCl$

(*iii*) $\underset{\text{Butanoyl chloride}}{CH_3CH_2CH_2COCl} + \underset{\text{1-Butanol}}{C_4H_9OH} \longrightarrow \underset{\text{Butyl butanoate}}{CH_3CH_2CH_2COOC_4H_9} + HCl$

(*iv*) Toluene ($C_6H_5CH_3$) $+ \underset{\text{Acetyl chloride}}{CH_3COCl} \xrightarrow[AlCl_3]{\text{Anhyd.}}$ α-methyl acetophenone (CH_3 and $COCH_3$ on adjacent ring positions) $+ HCl$

Toluene — Acetyl chloride — α-methyl acetophenone

(*v*) $\underset{\text{Propionamide}}{CH_3CH_2CONH_2} + HNO_2 \xrightarrow[HCl]{NaNO_2} \underset{\text{Propionic acid}}{CH_3CH_2COOH} + N_2 + H_2O$

(*vi*) $\underset{\text{Isobutyric acid}}{(CH_3)_2\,CHCOOH} + \underset{\text{Diazo methane}}{CH_2N_2} \longrightarrow \underset{\text{Methyl iso-butyrate}}{(CH_3)_2\,CHCOOCH_3} + N_2\uparrow$

Q. 20. Give the mechanism of alkaline hydrolysis of amides.

Ans. $CH_3CONH_2 + NaOH \longrightarrow CH_3COONa + NH_3 \uparrow$

The mechanism of the reaction is as follows:

$$(i)\ CH_3-\overset{O}{\overset{\|}{C}}-NH_2 \xrightarrow{\bar{O}H} CH_3-\underset{OH}{\overset{-O}{C}}-NH_2$$

$$(ii)\ CH_3-\underset{OH}{\overset{\bar{O}}{C}}-NH_2 \longrightarrow CH_3-\overset{O}{\overset{\|}{C}}-OH + NH_2^-$$

$$(iii)\ CH_3-\overset{O}{\overset{\|}{C}}-OH + NH_2^- \longrightarrow CH_3-\overset{O}{\overset{\|}{C}}O^- + NH_3$$

Q. 21. Discuss the probable mechanism of ester formation from carboxylic acid and alcohol.

Ans. $RCOOH + R'OH \overset{H^+}{\rightleftharpoons} \underset{\text{Ester}}{RCOOR'} + H_2O$

Mechanism of the reaction is as follows:

(*i*) In the first step, proton from the mineral acid is attached to carbonyl oxygen forming protonated carboxy acid with a positive centre on carbon.

$$R-\overset{O}{\overset{\|}{C}}-\ddot{O}-H \xrightarrow{H^+} R-\overset{OH}{\overset{|}{\overset{+}{C}}}-\ddot{O}-H \xrightarrow{:\overset{H}{O}R'} R-\underset{+\overset{H}{O}R}{\overset{OH}{C}}-\ddot{O}-H$$

$$\longrightarrow R-\underset{OR'H}{\overset{O-H}{C}}-\overset{+}{O}-H \xrightarrow{-H_2O} \underset{\text{Ester}}{R-\underset{OR'}{\overset{O}{\overset{\|}{C}}}}$$

(*ii*) Alcohol molecule is attached to the protonated carboxy acid with linking of oxygen of the alcohol to the positive carbon centre.

(*iii*) Hydrogen from the alcohol is shifted to oxygen of hydroxyl group.

(*iv*) Hydrogen ion and water molecule are finally removed.

Q. 22. What is nucleophilic acyl substitution? Give its mechanism.

Ans. These are the substitution reactions of acid derivatives of carboxylic acids in which $-Cl$, $-OOCR$, $-NH_2$ or $-OR'$ of the acid chloride, anhydride, amide and ester is replaced by other nucleophiles. The reaction can be represented in general as:

$$R-C\begin{matrix}\nearrow O \\ \searrow Y\end{matrix} + :Z \longrightarrow R-C\begin{matrix}\nearrow O \\ \searrow Z\end{matrix} + :Y$$

where Y stands for $-Cl$, $-OOCR$, $-NH_2$ or $-OR'$ and $:Z$ stands for the nucleophile like OH^-.

Since the nucleophile attaches itself to the carbon of the acyl group, such reactions are called nucleophilic substitution reactions.

The mechanism of the reaction is as follows:

$$(i)\ R-C(=\ddot{O}:)-Y + :Z \longrightarrow R-\overset{:\ddot{O}:^-}{\underset{Y}{C}}-Z$$

$$(ii)\ R-\overset{:\ddot{O}:^-}{\underset{Y}{C}}-Z \longrightarrow R-C(=\ddot{O}:)-Z + :Y$$

Q. 23. Giving suitable reasons, arrange acetamide, acetyl chloride, ethyl acetate and acetic anhydride in order of increasing reactivity towards nucleophilic attack.

Ans. Refer to Q. 22, Y stands for – Cl of acid chloride or – OOCR of anhydride or – NH_2 of amide or – OR of the ester.

The reactivity of the acid derivative to give the product will depend upon the ease with which Y can be removed from the intermediate product in step (*ii*) of the reaction as shown in Q. 22

The ease of removal of Y is governed by two factors:

1. The ease with which Y can be removed depends upon its basicity. Weaker the base, easier it is removed from the intermediate compound. The basicity of Y increases in the order.

$$Cl < RCOO^- < NH_2 < OR$$

Thus – Cl can be removed most easily followed by RCOO followed by NH_2 and finally followed by OR. This can also be interpreted in terms of inductive effects of Cl, O and N. Hence reactivity follows the order:

Ester < Acid amide < Acid anhydride < Acid chloride

Q. 24. Compare and contrast acyl nucleophilic substitution reactions with alkyl nucleophilic substitution reactions.

Ans. The two types of reactions can be expressed in a general form as follows:

$$R-Y + :Z \longrightarrow R-Z + :Y \qquad \text{(Alkyl nucleophilic substitution reaction)}$$

$$R-C(=O)-Y + :Z \longrightarrow R-C(=O)-Z + :Y \qquad \text{(Acyl nucleophilic substitution reaction)}$$

The main points of difference between the two are as under:

1. In alkyl nucleophilic reaction, the nucleophile attaches itself to the alkyl carbon whereas in acyl nucleophilic reaction it attaches itself to the acyl carbon.

$$R-Y + Z \longrightarrow Z \ldots\ldots R \ldots\ldots Y$$

$$R-C(=O)-Y + Z \longrightarrow R-\overset{O}{\overset{\|}{\underset{\vdots}{\underset{Y}{C}}}} \ldots\ldots\ldots Z$$

2. Acyl nucleophilic reaction takes place faster as compared to alkyl nucleophilic reaction. The intermediate product in the former is less stable because the pentavalent carbon becomes more

crowded. In contrast in acyl reaction, the intermediate product is less crowded with tetravalent carbon and hence more stable.

Alkyl nucleophilic substitution reaction

$$Z:+ \;\; \text{(tetrahedral C–Y)} \xrightarrow{SN^2} Z\cdots\cdots(\text{C})\cdots\cdots Y \longrightarrow Z-(\text{C}) + :Y$$

Acyl nucleophilic substitution reaction

$$Z:+ \;\; R(\text{C}{=}O)Y \longrightarrow \;Z,R(\text{C}(O))-Y \longrightarrow R(\text{C}{=}O)Z \;\; +:Y$$

Further the attachment of nucleophile to acyl carbon requires the breaking of a π-bond whereas attachment of a nucleophile to alkyl carbon requires partial breaking of a σ-bond which requires greater energy.

Q. 25. Discuss the mechanism of acyl nucleophilic substitution reaction. Compare it with alkyl nucleophilic substitution reaction.

Ans. See Qs. 21, 23.

Q. 26. Acetyl chloride is more reactive than ethyl chloride towards, nucleophilic substitution. Discuss why. *(Himachal, 2000)*

Or

SN^2 reaction takes place faster at an acyl carbon atom than at an alkyl carbon atom. Explain.

Ans. See Q. 23.

Q. 27. Explain why acetylating activities of an acid anhydride is lower than that of acid chloride.

Or

Which of the following is more reactive towards nucleophilic attack and why?

Acetic anhydride, acetyl chloride

Ans. Acetic anhydride and acetyl chloride are used to acetylate alcohols, phenols and amines.

$$(i)\quad CH_3-\overset{\overset{O}{\|}}{C}-Cl + \overset{-}{O}R \longrightarrow CH_3-\underset{\underset{OR}{|}}{\overset{\overset{O}{\|}}{C}}-Cl^-$$

$$(ii)\quad \begin{matrix} CH_3-\overset{\overset{O}{\|}}{C} \\ \quad\quad\quad >O \\ CH_3-\underset{\underset{O}{\|}}{C} \end{matrix} + \overset{-}{O}R \longrightarrow CH_3-\underset{\underset{OR}{|}}{\overset{\overset{O}{\|}}{C}} + CH_3CO\overset{-}{O}$$

Reaction (*i*) takes place faster as compared to reaction (*ii*) because $\bar{Cl}$ is a weaker base and can be removed easily compared to $CH_3CO\bar{O}$ which is a strong base.

Conjugate bases of strong acids are weak bases and conjugate bases of weak acids are strong bases. Accordingly $\bar{Cl}$ which is a conjugate base of strong acid HCl is weak and $CH_3CO\bar{O}$ which is a conjugate base of weak acid acetic acid, is strong.

Q. 28. Out of the following, which is more reactive towards nucleophilic attack and why?

(i) Acetic anhydride

(ii) Acetamide

Ans. Acetic anhydride is more reactive than acetamide because after the nucleophile has been linked to carbonyl carbon, it is easier to reomove CH_3COO^- than NH_2^-

$$(CH_3-\overset{O}{\overset{\|}{C}})_2O + \bar{O}R \longrightarrow CH-\overset{O}{\overset{\|}{C}}-OR + CH_3CO\bar{O}$$

$$CH_3-\overset{O}{\overset{\|}{C}}-NH_2 + \bar{O}R \longrightarrow CH_3-\overset{O}{\overset{\|}{C}}-OR + NH_2^-$$

Q. 29. Sketch out the mechanism of acidic hydrolysis of acetamide.

Ans. See Q. 12.

Q. 30. Reason out why acetamide is slowly hydrolysed with water but rapidly in the presence of dil H_2SO_4.

Ans. Presence of dil H_2SO_4 helps in creating positive centre at the carbonyl carbon which further enables the nucleophilic attack of water molecule on this carbon.

$$R-\overset{O}{\overset{\|}{C}}-NH_2 \xrightarrow{H^+} R-\overset{OH}{\overset{|}{\overset{+}{C}}}-NH_2$$

Q. 31. Explain why acetyl chloride reacts vigorously with water at room temperature but benzoyl chloride reacts slowly.

Ans. This is because positive charge developed on carbonyl carbon in benzoyl chloride is destroyed due to resonance in the benzene ring

So the attack of nucleophile H_2O becomes mild.

$$C_6H_5-\overset{O^-}{\overset{|}{\overset{+}{C}}}-Cl \longleftrightarrow {}^{+}C_6H_5=\overset{O^-}{\overset{|}{C}}-Cl \longleftrightarrow {}^{+}C_6H_5=\overset{O^-}{\overset{|}{C}}-Cl$$

Q. 32. Explain why does acetyl chloride boil at lower temperature than acetic acid?

Ans. Acetic acid exists as a dimer because of hydrogen bonding as under

$$CH_3-C\begin{matrix}\nearrow O \cdots\cdots H-O \searrow \\ \searrow O-H \cdots\cdots O \nearrow\end{matrix}C-CH_3$$

This type of hydrogen bonding does not take place in acetyl chloride. Hydrogen bonding results in bigger molecules, boiling at higher temperature.

Q. 33. Amides are much weaker bases than ammonia. Explain.

Ans. This is because the lone pair of electrons on nitrogen which is responsible for basic nature of a compound is drawn towards carbon as given below.

$$R-\overset{O}{\overset{\|}{C}}-\ddot{N}H_2 \longrightarrow R-\overset{O^-}{\overset{|}{C}}=\overset{+}{N}H_2$$

Q. 34. The b.p. of acid anhydrides are higher than those of acids from which they are derived. Explain.

Ans. This is because an anhydride is obtained by the combinations of two molecules of the acid. This is supposed to have a bigger molecular size and thus higher b.p

Q. 35. Give suitable reagent for the following conversion : (*Kanpur, 2001*)

$$C_6H_5\,CONH_2 \longrightarrow C_6H_5\,NH_2$$

Ans. $C_6H_5\,CONH_2 \xrightarrow{Br_2/KOH(alc)} C_6H_5NH_2$

Q. 36. Complete the following reactions

1. $C_6H_5-CH_2-\overset{O}{\overset{\|}{C}}-NH_2 \xrightarrow[NaOH]{Br_2}$ (*Awadh, 2000*)

2. $C_6H_4(OH)(COOH) \xrightarrow{CH_3COCl}$ (*Lucknow, 2000*)

3. $PhCOOH \xrightarrow{PCl_5} A \xrightarrow{NH_3} B \xrightarrow[NaOH]{NaOBr}$ (*Lucknow, 2000*)

Ans. 1. $C_6H_5-CH_2-\overset{O}{\overset{\|}{C}}-NH_2 \xrightarrow[4NaOH]{Br_2}$

$$C_6H_5-CH_2\,NH_2 + 2NaBr + Na_2CO_3 + 2H_2O$$

2. $C_6H_4(OH)(COOH) \xrightarrow{CH_3COCl} C_6H_4(OCOCH_3)(COOH) + HCl$

3. $PhCOOH \xrightarrow{PCl_5} PhCOCl \xrightarrow{NH_3} PhCONH_2 \xrightarrow[NaOH]{NaOBr} PhNH_2$

Q. 37. What is the historical importance of urea ? Give methods of preparation , properties and uses of urea (Carbamide)

Ans. It is the end product of human metabolism of nitrogen containing foods (proteins). In 1773, Roulle isolated it from urine and named it urea. It has historical significance as it was the first organic compound synthesized in the laboratory (Wohler 1828).

Methods of Preparation

1. From urine. Urine is evaporated to a small bulk and nitric acid is added to it to precipitate sparingly soluble urea nitrate ($NH_2CONH_2.HNO_3$). Urea nitrate is treated with barium carbonate to temove the acid and then extracted with alcohol, barium nitrate being insoluble in alcohol.

$$\underset{\text{Urea nitrate}}{2NH_2CONH_2.HNO_3} + BaCO_3 \longrightarrow \underset{\text{Urea}}{2NH_2.CO.NH_2} + Ba(NO_3)_2 + H_2O + CO_2$$

2. By heating a solution containing a mixture of potassium cyanate and ammonium sulphate to dryness (Wohler synthesis). When this mixture is heated tc dryness, ammonium cyanate is formed, which undergoes molecular rearrangement (isomeric change) to give urea. Urea is extracted from this mixture by dissolving in hot absolute alcohol. Ammonium sulphate being insoluble may be filtered off and the filtrate on cooling deposits urea.

$$\underset{\text{Pot. cyanate}}{2KCNO} + (NH_4)_2SO_4 \xrightarrow{\Delta} K_2SO_4 + \underset{\text{Ammonium cyanate}}{2NH_4CNO}$$

$$NH_4CNO \xrightarrow{\text{Rearrangement}} \underset{\text{Urea}}{NH_2CONH_2}$$

3. By partial hydrolysis of cyanamide (Manufacture). Cyanamide for this purpose is obtained from calcium cyanamide which is manufactured by passing nitrogen through heated calcium carbide at about 1073 K.

$$\underset{\text{Calcium carbide}}{CaC_2} \xrightarrow[1073\text{ K}]{N_2} \underset{\text{Calcium cyanamide}}{CaN.C\equiv N} \xrightarrow[-CaSO_4]{H_2SO_4} \underset{\text{Cyanamide}}{H_2N.C\equiv N} \xrightarrow{H_2O} \underset{\text{Urea}}{NH_2CONH_2}$$

4. By the action of ammonia on carbonyl chloride (Laboratory preparation).

$$\underset{\text{Carbonyl chloride}}{O=C\begin{cases}Cl \\ Cl\end{cases}} \; \begin{matrix}H \; NH_2 \\ + \\ H \; NH_2\end{matrix} \; (\text{Ammonia}) \longrightarrow \underset{\text{Urea}}{O=C\begin{cases}NH_2 \\ NH_2\end{cases}} + 2HCl$$

5. By the action of liquid carbon dioxide on liquid ammonia at 423 K and 35 atmospheric pressure (Manufacture).

$$O=C=O + \begin{matrix}H\,NH_2 \\ H\,NH_2\end{matrix} \xrightarrow[423\text{ K}]{35\text{ atm}} \underset{\text{Urea}}{O=C\begin{cases}NH_2 \\ NH_2\end{cases}} + H_2O$$

Physical Properties of Urea

Urea is a white crystalline solid. It melts at 405 K. It is soluble in water and ethyl alcohol but insoluble in ether. It is odourless and has a cooling taste.

Chemical Properties of Urea

Urea contains an amide group attached to an amino group and gives reactions of both these groups.

1. Basic nature (Salt formation). An aqueous solution of urea is neutral to litmus but behaves as a weak monoacid base and forms salt with strong acids. Thus, when nitric acid is added to a strong solution of urea, a crystalline precipitate of urea nitrate is formed.

$$\underset{\text{Urea}}{NH_2.CO.NH_2} + HNO_3 \longrightarrow \underset{\text{Urea nitrate}}{[NH_2.CO.NH_3]^+ NO_3^-}$$

2. Hydrolysis. It is rapidly hydrolysed into ammonia and carbon dioxide when boiled with dilute acids or alkalis.

$$\underset{\text{Urea}}{O=C\begin{cases}NH_2 \\ NH_2\end{cases}} + 2H_2O \xrightarrow{-2NH_3} \underset{\text{Unstable}}{O=C\begin{cases}OH \\ OH\end{cases}} \longrightarrow CO_2 + H_2O$$

Thus, upon hydrolysis with dil. hydrochloric acid urea forms ammonium chloride and with caustic soda it forms sodium carbonate.

$$NH_2.CO.NH_2 + H_2O + 2HCl \longrightarrow CO_2 + 2NH_4Cl$$

$$NH_2.CO.NH_2 + 2NaOH \longrightarrow Na_2CO_3 + 2NH_3$$

Enzyme, urease also brings about the hydrolysis of urea.

$$NH_2.CO.NH_2 \xrightarrow[\text{Urease}]{H_2O} 2NH_3 + CO_2$$

3. Reaction with nitrous acid. Urea reacts with nitrous acid to produce carbon dioxide, nitrogen and water.

$$\begin{matrix} HO\ \ NO \\ H_2N-CO-NH_2 \\ ON-HO \end{matrix} \longrightarrow 2N_2 + 2H_2O + O=C\begin{matrix} OH \\ OH \end{matrix} \longrightarrow H_2O + CO_2$$

Unstable

4. Acetylation. Urea undergoes acetylation with acetyl chloride.

$$\underset{\text{Acetyl chloride}}{CH_3CO}\boxed{Cl + H}\underset{\text{Urea}}{NHCONH_2} \longrightarrow \underset{\text{Acetyl urea}}{CH_3CONHCONH_2} + HCl$$

5. Reaction with sodium hypohalites. With excess of alkaline sodium hypobromite, urea is converted to nitrogen and sodium carbonate.

$$NH_2.CO.NH_2 + 3NaBrO \longrightarrow N_2 + CO_2 + 2NaBr + 2H_2O$$

$$CO_2 + \underset{\text{excess}}{2NaOH} \longrightarrow Na_2CO_3 + H_2O$$

6. Action of heat.

(*a*) On heating gently at about 405 K urea melts and after evolution of ammonia, a solid compound called **biuret** is produced.

Biuret contains – CO – NH – group and hence, gives a violet colouration with alkaline dilute copper sulphate solution (1%).

$$\underset{\text{Urea}}{NH_2-\overset{O}{\overset{\|}{C}}-}\boxed{NH_2 + H}\underset{\text{Urea}}{NH-\overset{O}{\overset{\|}{C}}-NH_2} \xrightarrow[405\text{ K}]{\text{heat}} \underset{\text{Biuret}}{NH_2-\overset{O}{\overset{\|}{C}}-NH-\overset{O}{\overset{\|}{C}}-NH_2} + NH_3$$

(*b*) When heated rapidly, ammonia is liberated and isocyanic acid is produced which trimerizes to cyanuric acid.

$$\underset{\text{Urea}}{NH_2.CO.NH_2} \xrightarrow[-NH_3]{\text{heat}} \underset{\text{Isocyanic acid}}{HNCO} \xrightarrow{\text{Trimerizes}} \underset{\text{Cyanuric acid}}{(HNCO)_3}$$

7. Reaction with formaldehyde. Urea reacts with formaldehyde to form urea formaldehyde resins or plastics.

$$H-\overset{H}{\overset{|}{C}}=O + H_2N-CO-NH_2 \longrightarrow \underset{\text{Methyl urea}}{HO-\overset{H}{\overset{|}{\underset{H}{\underset{|}{C}}}}-NH-CO-NH_2}$$

$$HO-\overset{H}{\overset{|}{\underset{H}{\underset{|}{C}}}}-NH-CO-NH_2 + O=\overset{H}{\overset{|}{C}}-H \longrightarrow$$

$$\underset{\text{Dimethyl urea}}{HO-\overset{H}{\overset{|}{\underset{H}{\underset{|}{C}}}}-NH-CO-NH-\overset{H}{\overset{|}{\underset{H}{\underset{|}{C}}}}-OH}$$

As the end – CHOH groups further react with – NH_2, the chain grows in length. Again the – NH – groups of different chains react with formaldehyde which gives rise to cross linkages forming valuable polymers.

Uses of Urea: It is used:

1. in the manufacture of formaldehyde-urea plastics or resins.
2. as a stablizer for explosives.
3. for making barbiturates used as hypnotics and sedatives.
4. as a fertilizer.

Q. 38. Which dicarboxylic acid gives anhydride on dehydration? (*Lucknow, 2000*)

Ans. See Q.8 (1)

Q.39. What happens when acetamide is treated with Br_2 and KOH (Conc.). (*Kerala, 2000*)

Ans. See Q. 12 (8)

Q. 40. Starting from acetic acid, how will you prepare : (*Kerala, 2000*)

(*i*) acetyl chloride

(*ii*) acetic anhydride

(*iii*) acetamide

(*iv*) ethyl acetate

Ans. (*i*) See Q. 14 (1)

(*ii*) See Q. 8 (1)

(*iii*) See Q. 11 (2)

(*iv*) See Q. 3 (1)

Q. 41 An unknown ester $C_5H_{12}O_2$ was hydrolysed with water and acid to give carboxylic acid (A) and alcohol (B). Treatment of (B) with PBr_3 gave an alkyl bromide (C). When (C) was treated with KCN, a product (D) was formed which on hydrolysis with water and acid gave a carboxylic acid (A). Give the structure and name of the original ester. Identify (A) through (D) and write equations for the reactions involved. (*Delhi 2003*)

Ans. Unknown ester is $C_2H_5COOC_2H_5$: On hydrolysis it gives C_2H_5COOH (A) and C_2H_5OH(B)

$$\underset{(B)}{C_2H_5OH} \xrightarrow{PBr_3} \underset{(C)}{C_2H_5Br} \xrightarrow{KCN} \underset{(D)}{C_2H_5CN} \xrightarrow{H_2O} \underset{(A)}{C_2H_5COOH}$$

Q. 42. What is Hofmann's degradation of amides ? Discuss its mechanism. (*Delhi, 2003; Burdwan, 2004*)

Ans. See Q. 12.

22

SUBSTITUTED ACIDS AND THEIR DERIVATIVES

CONTENTS

Q. 1. What are halo acids? Write a note on nomenclature of aliphatic and aromatic halogen acids.

Ans. Halogen substituted or halo acids are compounds in which one or more hydrogen atoms of the hydrocarbon chain of carboxylic acid are replaced by corresponding number of halogen atoms.

Nomenclature (Common system)

Common names of halogen substituted acids (or halo acids) are derived from the common names of parent carboxylic acids. The position of halogen atoms are indicated by Greek letters α, β, γ, δ etc., where α is the position of carbon adjacent to the carboxylic group, next position being β, still next positions are λ and δ

$$\overset{\delta}{C}-\overset{\gamma}{C}-\overset{\beta}{C}-\overset{\alpha}{C}-COOH$$

IUPAC System

In IUPAC system of nomenclature, the position of halogen is indicated by numerals 1, 2, 3, 4...., number 1 being the position of carboxylic carbon atom

$$\overset{4}{C}-\overset{3}{C}-\overset{2}{C}-\overset{1}{C}OOH$$

The common & IUPAC names of some halogen substituted acids are given in the table below.

Halo Acid	*Common Name*	*IUPAC Name*
$ClCH_2COOH$	Chloroacetic acid	Chloroethanoic acid
$CH_3CHCOOH$ (Cl on C-2)	α-chloropropionic acid	2-chloropropanoic acid
CH_3CHCH_2COOH (Br on C-3)	β-bromobutyric acid	3-bromobutanoic acid

$CH_2CH_2CH_2COOH$ with I on the first CH_2	γ-iodobutyric acid	4-iodobutanoic acid
$CH_3CHCHCOOH$ with Br, Br	α, β-dibromobutric acid	2, 3-dibromobutanoic acid
Benzene ring with Cl and COOH	α-Chlorobenzoic acid	2-chlorobenzene carboxylic acid

Q. 2. Give the general methods of preparation of halo acids.

Ans. General methods of preparation of halo acids are given below

1. Hell-Volhard-Zelinsky Reaction (HVZ Reaction) α-hydrogen atoms of a carboxylic acid can be replaced by halogen on treatment with the halogen in the presence of red phosphorus

$$\underset{\text{Propanoic acid}}{CH_3CH_2COOH} + Br_2 \xrightarrow[-\,HBr]{\text{Red P}} \underset{\text{2-Bromopropanoic acid}}{CH_3CHBrCOOH}$$

$$\xrightarrow{Br_2/P} \underset{\text{2,2-Dibromopropanoic acid}}{CH_3CBr_2COOH}$$

The substitution takes place at the α-position. And the reaction can be stopped at a particular stage by using appropriate quantities of the halogen and phosphorus. It is observed that chlorination can take place beyond positions as well.

In the reaction between acetic acid and Cl_2, all the three hydrogen atoms can be successively replaced by chlorine.

$$CH_3COOH + Cl_2 \xrightarrow[-HCl]{} \underset{\text{Monochloro-acetic acid}}{CH_2ClCOOH} \xrightarrow[-\,HCl]{Cl_2} \underset{\text{Dichloroacetic acid}}{CHCl_2COOH}$$

$$\xrightarrow[-\,HCl]{Cl_2} \underset{\text{Trichloroacetic acid}}{CCl_3COOH}$$

2. Action of sulphuryl chloride on carboxylic acid. Carboxylic acid on treatment with sulphuryl chloride in the presence of iodine yields halo acid.

$$\underset{\text{Propanoic acid}}{CH_3CH_2COOH} + \underset{\text{Sulphuryl chloride}}{2SO_2Cl_2} \xrightarrow{I_2} \underset{\text{2-chloropropanoyl chloride}}{CH_3CHClCOCl} + 2SO_2 + 2HCl$$

$$\xrightarrow[\text{Boil}]{H_2O} \underset{\text{2-Chloropropanoic acid}}{CH_3CHClCOOH + HCl}$$

3. Halogenation of malonic acid and alkyl malonic acids

$$CH_2\begin{matrix}COOH\\COOH\end{matrix} + Br_2 \longrightarrow CHBr\begin{matrix}COOH\\COOH\end{matrix} \xrightarrow[-CO_2]{\Delta} CH_2BrCOOH$$

Malonic acid — Bromomalonic acid — Bromoacetic acid

$$CH_3CH\begin{matrix}COOH\\COOH\end{matrix} + Br_2 \longrightarrow CH_3\,CBr\begin{matrix}COOH\\COOH\end{matrix}$$

Methyl malonic acid — α-bromomethyl-malonic acid

$$\downarrow -CO_2,\ \Delta$$

$$CH_3\,CHBrCOOH$$

α-Bromopropionic acid

4. From α-hydroxy acids. α-hydroxy acids can be converted into α-halo acids on treatment with halogen acids.

$$CH_3CH(OH)COOH + HBr \longrightarrow CH_3CHBrCOOH + H_2O$$

Lactic acid — α-bromopropionic acid

5. From α, β unsaturated carboxylic acids. Such acids on addition of halogens and halogen acids produce halo acids as given below.

(*i*) $CH_3CH = CHCOOH + Br_2 \longrightarrow CH_3CHBr\,CHBrCOOH$

Crotonic acid — α, β-dibromobutyric acid

(*ii*) $CH_3CH = CHCOOH + HBr \longrightarrow CH_3CHBr\,CH_2COOH$

Crotonic acid — β-bromobutyric acid

This affords us a method to prepare β haloacids. It may be mentioned that in the above reaction (*ii*), the addition of HBr to the double bond takes place contrary to the Markownikoff's rule. This is because of the electron withdrawing (– I) inductive effect of the carboxylic group. Consequently β-carbon becomes slightly positive and α-carbon becomes slightly negative. Therefore Br^- and H^+ (of HBr) are attached to α and β positions respectively.

$$CH_3 - \overset{\delta^+}{C}H = \overset{\delta^-}{C}H - COOH \xrightarrow{\overset{\delta^+}{H} - \overset{\delta^-}{Br}} CH_3\overset{+}{C}HCH_2COOH$$

$$\downarrow Br^-$$

$$CH_3CHBrCH_2COOH$$

6. From α, β unsaturated aldehydes. Addition of halogen acid to α, β unsaturated aldehydes followed by oxidation yields β-halogen acids

$$\overset{\delta^+}{C}H_2 = \overset{\delta^-}{C}H - CHO \xrightarrow{\overset{\delta^+}{H} - \overset{\delta^-}{Cl}} CH_2ClCH_2CHO$$

Acrolein — β-chloropropion aldehyde

$$\downarrow [O]$$

$$CH_2ClCH_2COOH$$

β-chloropropionic acid

Q. 3. Give the properties of halo acids.

Ans. Halo acids give all the properties of the parent carboxylic acid. In addition, they give the properties due to the halogen group. Important properties of halo acids are as follows.

1. Acidic strength. They show acidic properties due to the presence of carboxylic group in the molecule. In fact, the halo acids are more acidic than the normal acids because of the inductive effect of halogen. Halogen withdraws the electron towards itself, thereby helping in the release of proton.

$$R-\underset{\underset{X}{\downarrow}}{\overset{H}{\overset{|}{C}}}\leftarrow\overset{O}{\overset{\|}{C}}\leftarrow O\leftarrow H \longleftrightarrow R-\underset{\underset{X}{|}}{\overset{H}{\overset{|}{C}}}-\overset{O}{\overset{\|}{C}}-O^- + H^+$$

Following observations are made in this connection.

(*i*) Among various halo acids, chloro acids are stronger in acidic nature than bromo acids which are in turn, stronger than iodoacids. This is due to greater inductive effect of chlorine compared to bromine and greater effect of bromine compared to iodine.

(*ii*) For halo acids, having the same halogen atom, the acid strength decreases as the position of halogen goes away from the carboxy group. Thus 3-chloropropionic acid will be a weaker acid than 2-chloropropionic acid.

2. Reactions due to the carboxylic group. The halogen does not interfere in the reactions of carboxylic group and these properties are given in the normal manner. Halo acids decompose carbonates and bicarbonates, react with alcohols to form esters and with phosphorus pentachloride to form acids chlorides.

$$\underset{\text{Chloroacetic acid}}{ClCH_2COOH} + NaHCO_3 \longrightarrow \underset{\text{Sod. chloro-acetate}}{ClCH_2COONa} + CO_2 + H_2O$$

$$2ClCH_2COOH + Na_2CO_3 \longrightarrow 2ClCH_2COONa + CO_2 + H_2O$$

$$ClCH_2COOH + C_2H_5OH \xrightarrow{H^+} \underset{\text{Ethyl-chloroacetate}}{ClCH_2COOC_2H_5} + H_2O$$

$$\underset{\text{Chloroacetic acid}}{ClCH_2COOH} + PCl_5 \longrightarrow \underset{\text{Chloroacetyl chloride}}{ClCH_2COCl} + POCl_3 + HCl$$

3. Reactions due to the halogen atom. (*i*) **Action of alkalis.** Different products are obtained when halo acids are heated with aqueous solution of alkali, depending upon the position of halogen atom with respect to the carboxylic group.

α-halogen substituted acids give α-hydroxy acids on hydrolysis

$$\underset{\alpha\text{-halo acid}}{\underset{\underset{Cl}{|}}{RCH}-COOH} \xrightarrow{NaOH} \underset{\alpha\text{-hydroxy acid}}{R-\underset{\underset{OH}{|}}{CH}-COOH}$$

$$\underset{\alpha\text{-chloropropionic}}{CH_3CHClCOOH} \xrightarrow{NaOH} \underset{\text{Lactic acid}}{CH_3CH(OH)COOH}$$

β-haloacids on treatment with alkali yield a mixture of β-hydroxy acid and α, β unsaturated acid. This is because β-hydroxy acid is unstable and eliminates a water molecule to give more stable α, β unsaturated acid.

$$CH_2ClCH_2COOH \xrightarrow{NaOH} CH_2(OH)CH_2COOH$$

β-chloropropionic acid → β-hydroxypropionic acid

$$\downarrow -H_2O$$

$$CH_2 = CHCOOH$$

Acrylic acid

(*ii*) **Nucleophilic substitution reactions.** Halogen of the halo acid can be substituted by nucleophiles like CN^-, NH_3, $-OC_2H_5$.

$$\underset{\substack{|\\Cl}}{CH_3CH} - COOH + KCN \longrightarrow \underset{\substack{|\\CN}}{CH_3CH} - COOH + HCl$$

α-chloropropionic acid → α-cyanopropionic acid

$$\underset{\substack{|\\Cl}}{CH_3CH} - COOH + NH_3 \longrightarrow \underset{\substack{|\\NH_2}}{CH_3CH} - COOH + HCl$$

α-chloropropionic acid → α-aminopropionic acid

$$\underset{\substack{|\\Cl}}{CH_3CH} - COOH + \underset{\text{Sod. ethoxide}}{NaOC_2H_5} \longrightarrow CH_3 - \underset{\substack{|\\OC_2H_5}}{CH} - COOH$$

α-ethoxypropionic acid

Q. 4. Explain the following:

(*a*) Chloroacetic acid is stronger than acetic acid.

(*b*) Dichloroacetic acid is stronger than chloroacetic acid.

(*c*) Trichloroacetic is strong like a mineral acid.

(*d*) Monochloroacetic acid is stronger than monobromoacetic acid.

Ans. (*a*) Because of – I inductive effect of – Cl, chloroacetic acid is stronger than acetic acid.

$$Cl \leftarrow CH_2 \leftarrow COOH \quad > \quad CH_3 - COOH$$

Chloroacetic acid > Acetic acid

(*b*)

$$\begin{matrix} Cl \\ Cl \end{matrix} \!\!\gt\! CH - COOH \quad > \quad Cl \leftarrow CH_2COOH$$

Dichloroacetic acid > Chloroacetic acid

There are two chloro groups in dichloroacetic acid and one chloro group in chloroacetic acid. Hence in the former compound, there is twice the electron withdrawing inductive effect and hence greater acid strength.

(*c*)

$$Cl \leftarrow \underset{\substack{\downarrow\\Cl}}{\overset{\substack{Cl\\\uparrow}}{C}} - COOH \quad \text{Trichloroacetic acid}$$

There is the electron-withdrawing effect of three chloro groups. The effect is so strong that it produces a great acid strength equivalent to that of a mineral acid.

(*d*)

$$Cl \leftarrow CH_2 - COOH \quad > \quad Br \leftarrow CH_2COOH$$

Chloroacetic acid > Bromoacetic acid

Both chloro and bromo have – I inductive effect but out of the two, chloro has greater inductive effect than bromo. Therefore chloroacetic acid is stronger than bromoacetic acid.

Q. 5. What are hydroxy acids? Give their nomenclature.

Ans. A carboxylic acid with a hydroxy group in its hydrocarbon chain is called hydroxy acid. It can be obtained by substituting a hydrogen atom in its hydrocarbon chain by hydroxy group. Hydroxy group could be present in α, β, γ, δ positions with respect to the carboxylic group (according to common system) or 1, 2, 3, 4 positions (according to IUPAC system). The nomenclature of hydroxy acids is given below:

Hydroxy acid	*Common name*	*IUPAC name*
$HOCH_2COOH$	Hydroxyacetic acid (Glycollic acid)	Hydroxyethanoic acid
$HOCH_2CH_2COOH$	β-hydroxypropionic acid	3-hydroxypropanoic acid
$CH_3CHOHCOOH$	α-hydroxypropionic acid (lactic acid)	2-hydroxypropanoic acid

Q. 6. Give the methods of preparation of hydroxy acids. *(MS Baroda, 2004)*

Ans. [A] Preparation of α-hydroxy acids.

1. Hydrolysis of α-haloacids. Hydrolysis of α-haloacids using alkali solution or moist silver oxide yields α-hydroxy acids.

$$R-\underset{\underset{Cl}{|}}{CH}-COOH + NaOH \longrightarrow R-\underset{\underset{OH}{|}}{CH}-COOH + NaCl$$

$$R-\underset{\underset{Cl}{|}}{CH}-COOH + AgOH \longrightarrow R-\underset{\underset{OH}{|}}{CH}-COOH + AgOH$$

$$\underset{\text{Chloroacetic acid}}{CH_2ClCOOH} + AgOH \longrightarrow \underset{\text{Glycollic acid}}{CH_2OHCOOH} + AgCl$$

2. Action of Nitrous acids. α-amino acid reacts with nitrous acid to produce α-hydroxy acids. A mixture of sodium nitrite and hydrochloric acid yields the nitrous acid.

$$\underset{\alpha\text{-aminopropionic acid}}{CH_3-\underset{\underset{NH_2}{|}}{CH}-COOH} + HNO_2 \xrightarrow[HCl]{NaNO_2} \underset{\text{Lactic acid}}{CH_3-\underset{\underset{OH}{|}}{CH}-COOH} + N_2 + H_2O$$

3. Hydrolysis of Cyanohydrins. Cyanohydrins are obtained by the action of HCN on aldehydes or ketones. These are then hydrolysed in the presence of mineral acid to product hydroxy acids.

$$\underset{\text{Acetaldehyde}}{CH_3\overset{\overset{O}{||}}{C}-H} + HCN \longrightarrow \underset{\text{Acetaldehyde cyanohydrin}}{CH_3-\overset{\overset{OH}{|}}{\underset{\underset{H}{|}}{C}}-CN} \xrightarrow[H^+]{H_2O} \underset{\text{Lactic acid}}{CH_3-\overset{\overset{OH}{|}}{\underset{\underset{H}{|}}{C}}-COOH}$$

$$\underset{\text{Acetone}}{CH_3-\overset{\overset{O}{||}}{\underset{\underset{CH_3}{|}}{C}}} + HCN \longrightarrow CH_3-\overset{\overset{OH}{|}}{\underset{\underset{H}{|}}{C}}-CN \xrightarrow[H^+]{H_2O} \underset{\alpha\text{-hydroxy }\alpha\text{-methyl-propionic acid}}{CH_3-\overset{\overset{OH}{|}}{\underset{\underset{CH_3}{|}}{C}}-COOH}$$

4. Reduction of ketoacids. Reduction of keto acid with sodium amalgam yields hydroxy acid

$$\underset{\substack{\alpha\text{-ketopropionic acid}\\ \text{pyruvic acid}}}{CH_3COCOOH} + 2[H] \xrightarrow{Na/Hg} \underset{\text{Lactic acid}}{CH_3CHOHCOOH}$$

5. Oxidation of 1, 2 dihydroxy compounds. 1, 2 dihydroxy compounds of the following type upon oxidation with suitable oxidising agents give hydroxy acids.

$$\underset{\text{Glycol}}{OHCH_2CH_2OH} + 2[O] \xrightarrow{\text{dil. }HNO_3} \underset{\text{Glycollic acid}}{OHCH_2COOH} + H_2O$$

$$\underset{\text{1,2 Propylene glycol}}{CH_3CHOHCH_2OH} + 2[O] \xrightarrow[HNO_3]{\text{dil.}} \underset{\text{Lactic acid}}{CH_3CHOHCOOH} + H_2O$$

[B] Preparation of β-hydroxy acids

1. Action of HNO_2. β-amino acid reacts with nitrous acid to produce β-hydroxy acid

$$\underset{\beta\text{-aminopropionic acid}}{\underset{\substack{|\\ NH_2}}{CH_2}CH_2COOH} + HNO_2 \longrightarrow \underset{\beta\text{-hydroxypropionic acid}}{\underset{\substack{|\\ OH}}{CH_2}CH_2COOH} + N_2 + H_2O$$

2. From chlorohydrin. A chlorohydrin is treated with potassium cyanide and then hydrolysed with water.

$$\underset{\text{Ethylene chlorohydrin}}{HOCH_2CH_2Cl} + KCN \xrightarrow[-KCl]{} OHCH_2CH_2CN \xrightarrow{H_2O} \underset{\beta\text{-hydroxypropionic acid}}{OHCH_2CH_2COOH}$$

3. Oxidation of 1, 3 dihydroxy compounds

$$\underset{\text{1,3 Propyleneglycol}}{HOCH_2CH_2CH_2OH} + 2[O] \xrightarrow[HNO_3]{\text{dil.}} \underset{\beta\text{-hydroxypropionic acid}}{HOCH_2CH_2COOH}$$

4. Reformatsky reaction. An aldehyde or ketone is treated with an α-bromoester and zinc metal in ethereal solution. The addition product initially formed is hydrolysed by dilute mineral acid to yield β-hydroxy acid.

$$\underset{\text{Acetaldehyde}}{CH_3\overset{\substack{O\\ \|}}{C}H} + BrCH_2COOC_2H_5 + Zn \xrightarrow{\text{Ether}} CH_3-\overset{\substack{OZnBr\\ |}}{\underset{\substack{|\\ H}}{C}}-CH_2COOC_2H_5 \xrightarrow[H^+]{H_2O} \underset{\beta\text{-hydroxybutyric acid}}{CH_3CH(OH)CH_2COOH}$$

Q. 7. (*a*) Describe the properties of hydroxy acids.

(*b*) What is the action of heat on α- and β-hydroxy acids.

(*Mysore, 2003; Dibrugarh, 2003; Garwal 2004, MS Baroda, 2004; Purvanchal, 2007*)

Ans. (*a*) **1. Solubility.** Hydroxy acids are soluble in water. This is because of the presence of hydroxy and carboxylic groups both of which form hydrogen bonds with water.

2. Reaction with acetyl chloride. Acetyl derivatives of hydroxy acids are obtained on treatment with acetyl chloride or acetic anhydride.

$$\underset{\text{Lactic acid}}{CH_3\underset{\substack{|\\ OH}}{C}HCOOH} + CH_3COCl \longrightarrow \underset{\text{Acetyl deri. of lactic acid}}{CH_3\underset{\substack{|\\ OCOCH_3}}{C}H - COOH} + HCl$$

$$\underset{\beta\text{-hydroxy propionic acid}}{HOCH_2CH_2COOH} + CH_3COCl \longrightarrow \underset{\text{Acetyl derivative}}{CH_3COOCH_2CH_2COOH} + HCl$$

3. Oxidation. Hydroxyl group of the hydroxy acid is oxidised to aldehydic or ketonic group (as the case may be)

$$\underset{\text{Glycollic acid}}{HOCH_2COOH} \xrightarrow[\text{dil. } HNO_3]{[O]} \underset{\text{Glyoxalic acid}}{OHCCOOH} \xrightarrow{[O]} \underset{\text{Oxalic acid}}{HOOCCOOH}$$

$$\underset{\text{Lactic acid}}{CH_3CH(OH)COOH} \xrightarrow[\text{dil. } HNO_3]{[O]} \underset{\text{Pyruvic acid}}{CH_3\overset{\substack{O\\ \|}}{C}COOH}$$

4. Reduction. On heating with hydroiodic acid, hydroxy acids are reduced to the parent acids.

$$\underset{\text{Lactic acid}}{CH_3\underset{\substack{|\\ OH}}{C}HCOOH} + 2HI \xrightarrow{P} \underset{\text{Propionic acid}}{CH_3CH_2COOH} + H_2O + I_2$$

5. Formation of esters. On treatment with alcohols, hydroxy acid form esters

$$\underset{\beta\text{-hydroxypropionic acid}}{\underset{\substack{|\\ OH}}{C}H_2CH_2COOH} + C_2H_5OH \xrightarrow{H^+} \underset{\text{Ethyl } \beta\text{-hydroxy propionate}}{\underset{\substack{|\\ OH}}{C}H_2CH_2COOC_2H_5} + H_2O$$

$$\underset{\text{Lactic acid}}{CH_3\underset{\substack{|\\ OH}}{C}HCOOH} + C_2H_5OH \xrightarrow{H^+} \underset{\text{Ethyl lactate}}{CH_3\underset{\substack{|\\ OH}}{C}HCOOC_2H_5} + H_2O$$

6. Formation of salts. Salts are obtained when hydroxy acids are treated with alkali.

$$\underset{\text{Lactic acid}}{CH_3\underset{\substack{|\\ OH}}{C}HCOOH} + NaOH \longrightarrow \underset{\text{Sod. lactate}}{CH_3\underset{\substack{|\\ OH}}{C}HCOONa} + H_2O$$

7. Reaction with sodium metal. Both the hydroxy and carboxy groups are reacted when treated with sodium metal.

$$\underset{\text{Lactic acid}}{CH_3\underset{\substack{|\\ OH}}{C}HCOOH} + Na \longrightarrow \underset{\text{Disodium salt}}{CH_3\underset{\substack{|\\ ONa}}{C}HCOONa} + H_2$$

8. Action of PCl_5. Hydroxy acids react with PCl_5, when both the hydroxy and carboxy groups are chlorinated.

$$\underset{\text{Lactic acid}}{\underset{\displaystyle\overset{|}{OH}}{CH_3CHCOOH}} + PCl_5 \longrightarrow \underset{\text{2-chloropropanoyl chloride}}{\underset{\displaystyle\overset{|}{Cl}}{CH_3CHCOCl}} + POCl_3 + H_2O$$

(b) Action of heat. Hydroxy acids on heating yield different products depending upon the position of the hydroxy group relative to the carboxylic group.

(*i*) **α-hydroxy acids.** Two molecules of α-hydroxy acid on heating form cyclic diesters known as lactides by the interaction of – OH from one molecule and – COOH from the other molecule

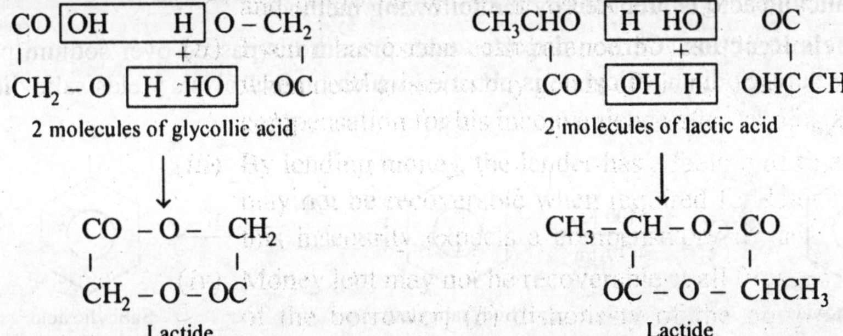

(*ii*) **β-hydroxy acid.** β-hydroxy acids on heating liberate water molecule and give α, β unsaturated carboxylic acid.

$$\underset{\beta\text{-hydroxy acid}}{\underset{\displaystyle\overset{|}{OH}}{R-CH}-CH_2COOH} \xrightarrow[-H_2O]{\Delta} \underset{\alpha,\ \beta \text{ unsaturated acid}}{R-CH=CHCOOH}$$

$$\underset{\beta\text{-hydroxy propionic acid}}{\underset{\displaystyle\overset{|}{OH}}{CH_2}-CH_2COOH} \xrightarrow[-H_2O]{\Delta} \underset{\text{Acrylic acid}}{CH_2=CHCOOH}$$

This reaction is catalysed by acids. The mechanism of reaction is as follows:

$$\underset{\displaystyle\overset{|}{OH}}{R-CH}-CH_2COOH \xrightarrow{H^+} R-\underset{\displaystyle\overset{|}{\overset{+}{O}H_2}}{CH}-\overset{\displaystyle H}{\overset{|}{CH}}-COOH \xrightarrow{-H_2O,\ H^+} \underset{\alpha,\ \beta\text{ unsaturated acid}}{R-CH=CH-COOH}$$

Q. 8. Of the following pairs of hydroxy acids, which are more acidic? Explain.

(*a*) (*i*) $HOCH_2COOH$ (*ii*) $HOCH_2CH_2COOH$

(*b*) (*i*) $CH_3CHOHCOOH$ (*ii*) $OHCH_2CH_2COOH$

(*c*) (*i*) $OHCH_2COOH$ (*ii*) $CH_3CHOHCOOH$

Ans. (*a*) (*i*) is more acidic than (*ii*)

(*b*) (*i*) is more acidic than (*ii*)

(*c*) (*i*) is more acidic than (*ii*)

Hydroxy group is an electron withdrawing group. Thus this is expected to increase the acidity of the acid. However, electron withdrawing effect of the hydroxy group depends upon its position. Nearer its position from the carboxy group, greater the electron withdrawing effect and hence stronger the acid. Thus we would expect a hydroxy group at α position to cause greater increase in the acid strength than at β position. Based on above principle, compound (*i*) is stronger than (*ii*) in pairs (*a*) and (*b*). In pair (*c*) compound (*i*) is stronger acid than (*ii*) although in both of the compounds, the hydroxy group is at α position. This is so because in compound (*ii*), the methyl group has electron donating effect which diminishes the electron withdrawing effect of the hydroxy group.

Q. 9. Give the methods of preparation of salicylic acid giving mechanism of reactions wherever applicable. *(Meerut, 2000)*

Ans. Salicylic acid is prepared by the following methods.

1. Kolbe's Reaction. Carbon dioxide under pressure is passed over sodium phenoxide at 400 K temperature. Sodium salicylate is obtained which on hydrolysis yields salicylic acid.

$$C_6H_5ONa + CO_2 \xrightarrow[4\text{-}7\ \text{Atm}]{400\ \text{K}} C_6H_4(OH)COONa \xrightarrow{H^+} C_6H_4(OH)COOH$$

Sod. phenoxide Sod. salicylate Salicylic acid

Mechanism. Mechanism of the reaction is as follows:

1. Carbon dioxide attacks the phenoxide ion by nucleophilic addition mechanism. A proton is removed from the position where CO_2 is attached. Salicylic ion is produced.

2. Hydrolysis leads to the final product viz. salicylic acid. These steps are illustrated as under.

$$C_6H_5O^{\ominus} + O{=}C{=}O \longrightarrow \text{(cyclohexadienone with } \underset{\substack{\|\\ O}}{\overset{H}{C}}{-}O^{\ominus}\text{)} \longrightarrow C_6H_4(O^{\ominus})\underset{\substack{\|\\ O}}{C}{-}O^{\ominus} \xrightarrow{H^{\oplus}} C_6H_4(OH)\underset{\substack{\|\\ O}}{C}{-}OH$$

2. Reimer-Tiemann's Reaction. Sodium phenoxide on heating with carbon tetrachloride in alkaline medium gives sodium salicylate which upon hydrolysis gives salicylic acid.

$$C_6H_5ONa + CCl_4 \xrightarrow{NaOH} C_6H_4(OH)CCl_3 \xrightarrow{NaOH} C_6H_4(ONa)COONa \xrightarrow{H^+} C_6H_4(OH)COOH$$

Sodium phenoxide Salicylic acid

Mechanism of the reaction is given below

$$\text{C}_6\text{H}_5\text{O}^{\ominus} + \text{CCl}_3 \xrightarrow{-\text{Cl}^-} \text{cyclohexadienone–CH(CCl}_3) \rightarrow \text{o-HOC}_6\text{H}_4\text{CCl}_3 \xrightarrow[^{\ominus}\text{OH}]{-\text{Cl}^{\ominus}} \text{o-HOC}_6\text{H}_4\text{C(Cl)}_2\text{OH} \xrightarrow[^{\ominus}\text{OH}]{-\text{Cl}^{\ominus}} \text{o-HOC}_6\text{H}_4\text{C(=O)Cl}$$

$$\rightarrow \text{o-HOC}_6\text{H}_4\text{C(OH)(Cl)O}^{\ominus} \xrightarrow{-\text{Cl}^{\ominus}} \text{o-HOC}_6\text{H}_4\text{COCl} \xrightarrow{^{\ominus}\text{OH}} \text{o-}^{\ominus}\text{OC}_6\text{H}_4\text{COO}^{\ominus} \xrightarrow{\text{H}^{\oplus}} \text{o-HOC}_6\text{H}_4\text{COOH}$$

3. Action of nitrous acid on anthranilic acid

$$\underset{\text{Anthranilic acid}}{\text{o-H}_2\text{NC}_6\text{H}_4\text{COOH}} \xrightarrow[273\text{ K}]{\text{HNO}_2/\text{HCl}} \text{o-ClN}_2\text{C}_6\text{H}_4\text{COOH} \xrightarrow[\text{Warm}]{\text{Water}} \text{o-HOC}_6\text{H}_4\text{COOH} + \text{N}_2 + \text{HCl}$$

Q. 10. Describe the properties of salicylic acid. *(Meerut, 2000; Nagpur, 2002)*

Ans. 1. Reaction with alkalis and carbonates

Alkalis react with both the phenolic and carboxylic groups whereas carbonates react only with carboxylic group

$$\underset{\text{Salicylic acid}}{\text{o-HOC}_6\text{H}_4\text{COOH}} + 2\text{NaOH} \longrightarrow \underset{\text{Disodium salicylate}}{\text{o-NaOC}_6\text{H}_4\text{COONa}} + 2\text{H}_2\text{O}$$

$$2\,C_6H_4(OH)COOH + Na_2CO_3 \longrightarrow 2\,C_6H_4(OH)COONa + CO_2 + H_2O$$

Salicylic acid → Monosodium salicylate

2. Reaction with alcohols and phenols. Formation of esters takes place with both alcohols and phenols.

$$C_6H_4(OH)COOH + C_2H_5OH \xrightarrow{\text{HCl gas}} C_6H_4(OH)COOC_2H_5 + H_2O$$

Salicylic acid, Ethanol → Ethyl salicylate

$$C_6H_4(OH)COOH + C_6H_5OH \xrightarrow{POCl_3} C_6H_4(OH)COOC_6H_5 + H_2O$$

Phenyl salicylate (salol)

3. Reaction with phosphorus pentachloride

$$C_6H_4(OH)COOH + 2PCl_5 \longrightarrow C_6H_4(Cl)COCl + 2POCl_3 + 2HCl$$

Salicylic acid → *o*-chloro benzoyl chloride

4. Reaction with Acetyl chloride and Acetic anhydride

$$C_6H_4(OH)COOH + CH_3COCl \longrightarrow C_6H_4(O.COCH_3)COOH + HCl$$

Salicylic acid, Acetyl chloride → Acetyl salicylic acid (Aspirin)

$$C_6H_4(OH)COOH + (CH_3CO)_2O \longrightarrow C_6H_4(O.COCH_3)COOH + CH_3COOH$$

Salicylic acid, Acetic anhydride → Acetyl salicylic acid

5. Reaction with soda lime. Carboxylic group is eliminated on heating with soda lime leaving the phenolic group.

$$C_6H_4(OH)COOH + CaO \longrightarrow C_6H_5OH + CaCO_3$$

Phenol

6. Reaction with Zn dust. On heating salicylic acid with Zn dust, phenolic group is eliminated leaving behind the carboxy group.

$$\text{Salicylic acid } (o\text{-}HOC_6H_4COOH) + Zn \xrightarrow{\text{Distil.}} \text{Benzoic acid } (C_6H_5COOH) + ZnO$$

7. Electrophilic substitution reactions. Halogenation, nitration and sulphonation can be performed on salicylic acid.

Sulphonation. Sulphonic group takes the para position with respect to phenolic group.

$$\text{Salicylic acid} + H_2SO_4 \longrightarrow \text{Salicylic acid sulphonate } (HO_3S\text{–}C_6H_3(OH)COOH) + H_2O$$

Bromination. During bromination, carboxy group is decomposed and a tribromo product is formed.

$$\text{Salicylic acid} + Br_2 \text{ (Aq. solution)} \longrightarrow \text{2, 4, 6-Tribromophenol} + 3HBr + CO_2$$

Nitration. As in case of bromination, nitration takes place with the simultaneous decomposition of carboxylic group.

$$\text{Salicylic acid} + 3HNO_3 \xrightarrow{H_2SO_4} \text{2, 4, 6-Trinitrophenol} + 3H_2O + CO_2$$

Mechanism of decomposition of carboxylic group during the course of halogenation and nitration is illustrated below.

$$o\text{-}HOC_6H_4COO^- + Br\text{–}Br \xrightarrow{-Br^-} \text{intermediate (}\overset{+}{O}H\text{, C(Br)–COO}^-\text{)} \xrightarrow{-CO_2} o\text{-Bromophenol}$$

Q. 11. *o*-hydroxy benzoic acid (Salicylic acid) melts at 432 K, *m*-hydroxy benzoic acid at 474 K and *p*-hydroxy benzoic acid at 487 K. Explain the difference in melting points.

Ans. Intramolecular (within the same molecule) hydrogen bonding takes place in *o*-hydroxy benzoic acid due to the close proximity of the phenolic and carboxylic groups.

O—H
C = O
OH

This results in shrinking of size of the molecule. Consequently the melting point is lowered.

On the other hand *m*- & *p*-isomers exhibit intermolecular (between different molecules) hydrogen bonding resulting in increase in the size of molecules.

OH OH OH
··· HO — C=O ··· H — O — C=O ··· HO — C=O ···

Consequently there is an increase in the melting point. In the *p*-isomer, there is more effective H-bonding because there is no steric hindrance. *m*-isomer faces some steric hindrance.

Q. 12. What are α, β unsaturated acids? Give their nomenclature.

Ans. α, β unsaturated acids are organic carboxylic acids having a double bond between α and β positions *i.e.* at positions next and still next to the carboxylic group.

Nomenclature of a few important α, β unsaturated acids is given below

α, β unsaturated acid	*Common name*	*IUPAC name*
$CH_2=CH-COOH$	Acrylic acid	Propenoic acid
$CH_3-CH=CH-COOH$	Crotonic acid	But-2-enoic acid
$CH_2=C(CH_3)-COOH$	Methyl acrylic acid	2-methylpropenoic acid
$C_6H_5-CH=CHCOOH$	Cinnamic acid	3-phenylpropenoic acid

Q. 13. Describe the methods of preparation of α, β-unsaturated acids giving mechanisms wherever necessary. (*Nagpur, 2005*)

Ans. Such acids can be prepared by the following general methods

1. Oxidation of α, β unsaturated aldehyde.

$$\underset{\text{Acrolein}}{CH_2=CH-CHO}+O \xrightarrow[\text{reagent}]{\text{Tollen's}} \underset{\text{Acrylic acid}}{CH_2=CH-COOH}$$

2. Dehydrohalogenation of β-halo acids. α, β unsaturated acids can be obtained by the dehydrohalogenation of β-halo acid using alcoholic potassium hydroxide

$$\underset{\text{3-chlorobutanoic acid}}{CH_3-\underset{\underset{Cl}{|}}{CH}-CH_2-COOH} \xrightarrow[KOH]{Alc.} \underset{\text{But-2-enoic acid}}{CH_3-CH=CH-COOH} + HCl$$

The reaction is supposed to take place by the following mechanism

$$R-\underset{\underset{Cl}{|}}{CH}-\overset{\overset{H}{|}}{CH}-COOH \xrightarrow{OH^-} R-CH=CHCOO^- + H_2O + Cl^-$$

3. From alkynes. Acetylene is treated with HCN to give an addition product which is subjected to hydrolysis.

$$\underset{\text{Acetylene}}{CH\equiv CH} + HCN \longrightarrow \underset{\text{Vinyl cyanide}}{CH_2=CHCN} \xrightarrow{H_2O} \underset{\text{Acrylic acid}}{CH_2=CHCOOH}$$

4. Dehydration of β-hydroxy acid

$$\underset{\beta\text{-hydroxy acid}}{R-\underset{\underset{OH}{|}}{CH}-CH_2-COOH} \xrightarrow{\Delta} \underset{\alpha,\ \beta\text{-unsaturated acid}}{RCH=CHCOOH}$$

$$\underset{\beta\text{-hydroxypropanoic acid}}{CH_2OH-CH_2-COOH} \xrightarrow{\Delta} \underset{\text{Acrylic acid}}{CH_2=CH-COOH}$$

5. From Grignard reagent

$$\underset{\text{Vinyl magnesium halide}}{CH_2=CHMgX} + CO_2 \xrightarrow[\text{furan (THF)}]{\text{Tetrahydro}} \underset{\text{(Addition compound)}}{CH_2=CH-\overset{\overset{O}{||}}{C}-OMgX}$$

$$\xrightarrow{H_2O/H^+} \underset{\text{Acrylic acid}}{CH_2=CH-COOH} + Mg\begin{matrix} OH \\ X \end{matrix}$$

6. Perkin's reaction

$$\underset{\text{Benzaldehyde}}{C_6H_5-CHO} + \underset{\text{Acetic anhydride}}{(CH_3CO)_2O} \xrightarrow[\text{acetate}]{\text{Sod.}} \underset{\text{Cinnamic acid}}{C_6H_5-CH=CH-COOH}$$

7. Knoevenagel reaction

$$C_6H_5CHO + CH_2(COOC_2H_5)_2 \xrightarrow{\text{Pyridine}} C_6H_5CH = C(COOC_2H_5)_2 \xrightarrow{H_2O} C_6H_5CH = C(COOH)_2 \xrightarrow[-CO_2]{\text{Heat}} C_6H_5CH = CHCOOH$$

Cinnamic acid

8. By isomerisation. When unsaturated acids other than α, β unsaturated acids are boiled with strong alkali, the double bond shifts to the α, β position. On acidification, α, β unsaturated acid is obtained.

$$\underset{\beta,\ \gamma \text{ unsaturated acid}}{CH_3CH = CH - CH_2 - COOH} \xrightarrow[\text{Boil}]{\text{NaOH}} \underset{\alpha,\ \beta \text{ unsaturated acid}}{CH_3CH_2CH = CHCOOH}$$

Mechanism

$$CH_3 - CH = CH - CH(H) - \overset{O}{\overset{\|}{C}} - O - H + OH^-$$

$$\downarrow$$

$$CH_3 - CH = CH - \overset{\ominus}{CH} - \overset{O}{\overset{\|}{C}} - O - H$$

$$\downarrow$$

$$CH_3 - \overset{\ominus}{CH} - CH = CH - \overset{O}{\overset{\|}{C}} - O^-$$

$$\downarrow H^+$$

$$CH_3 - CH_2 - CH = CH - \overset{O}{\overset{\|}{C}} - O^- \xrightarrow{H^+} CH_3CH_2CH = CHCOOH$$

Q. 14. Describe with mechanism, wherever required, properties of unsaturated acids.

(*Nagpur, 2005*)

Ans. 1. Electrophilic addition reactions. α, β unsaturated acids undergo addition reactions characteristics of carbon-carbon double bond. Addition of halogen, acids and water takes place on the ehthylenic double bond but the reaction is slow because of the conjugation of the double bond with carboxylic acid.

It may be noted that during hydration and hydrohalogenation, hydroxyl and halogen always add on β-carbon, which is contrary to the Markownikoff's rule.

(*i*) $$\underset{\text{Acrylic acid}}{CH_2 = CH - COOH} + Cl_2 \longrightarrow \underset{\alpha,\ \beta \text{ dichloropropionic acid}}{\underset{Cl}{\underset{|}{CH_2}} - \underset{Cl}{\underset{|}{CH}} - COOH}$$

(*ii*) $CH_2 = CH - COOH + HBr \longrightarrow \underset{\underset{Br}{|}}{CH_2} - CH_2 - COOH$

β-bromopropionic acid

(*iii*) $CH_2 = CH - COOH + H_2O \longrightarrow \underset{\underset{OH}{|}}{CH_2} - CH_2 - COOH$

β-hydroxypropionic acid

Mechanism of addition

1. A proton gets attached to carbonyl oxygen of the carboxyl group giving resonance stabilised positive ion.

2. The nucleophile (X^- or OH^-) attacks the carbonium ion at β position to give enol form of the acids.

3. The enol form tautomerises to more stable keto form.

These steps are mechanistically shown below.

$$CH_2 = CH - \overset{\overset{O}{||}}{C} - OH \underset{}{\overset{H^{\oplus}}{\rightleftharpoons}} \left[CH_2 = CH - \overset{\overset{\overset{\oplus}{O}H}{||}}{C} - \ddot{O}H \right]$$

$$\updownarrow$$

$$CH_2 = CH - \underset{\oplus}{\overset{\overset{OH}{|}}{C}} - OH \longleftrightarrow CH_2 = CH - \overset{\overset{OH}{|}}{C} = \underset{\oplus}{\ddot{O}H}$$

$$\updownarrow$$

$$\overset{\oplus}{C}H_2 - CH = \overset{\overset{OH}{|}}{C} - OH \equiv \left[CH_2 - CH - \overset{\overset{OH}{\vdots}}{C} \mathbin{\text{⋯}} OH \right]^{\oplus}$$

Resonance stabilised cation

$$\downarrow Br^-$$

$$\underset{\underset{Br}{|}}{CH_2} - CH_2 - \overset{\overset{O}{||}}{C} - OH \xleftarrow{\text{Tautomerism}} \underset{\underset{Br}{|}}{CH_2} - CH = \overset{\overset{OH}{|}}{C} - OH$$

Keto form — Enol form

β-Bromopropionic acid

2. Reaction of the carboxylic group

(*i*) **Formation of salt**

$$\underset{\text{Crotonic acid}}{CH_3 - CH = CH - COOH} + NaOH \longrightarrow \underset{\text{Sod. crotonate}}{CH_3 - CH = CHCOONa} + H_2O$$

$$\underset{\text{Acrylic acid}}{2CH_2 = CHCOOH} + Na_2CO_3 \longrightarrow \underset{\text{Sod. acrylate}}{2CH_2 = CHCOONa} + CO_2 + H_2O$$

(*ii*) **Esterification**

$$\underset{\text{Acrylic acid}}{CH_2 = CHCOOH} + C_2H_5OH \xrightarrow{H_2SO_4} \underset{\text{Ethyl acrylate}}{CH_2 = CHCOOC_2H_5} + H_2O$$

(iii) **Action of PCl_5**

$$\underset{\text{Crotonic acid}}{CH_3CH = CHCOOH} + PCl_5 \longrightarrow \underset{\text{Crotonoyl chloride}}{CH_3CH = CHCOCl} + POCl_3 + HCl$$

3. Reduction. Double bond of the unsaturated acid gets saturated on catalytic hydrogenation or treatment with nascent hydrogen.

$$\underset{\text{Acrylic acid}}{CH_2 = CHCOOH} + 2[H] \xrightarrow{Na / C_2H_5OH} \underset{\text{Propanoic acid}}{CH_3CH_2COOH}$$

4. Reaction with alkali. Unsaturated acid (α, β or others) on fusion with alkali break to give acetic acid and one more acid with two carbons less.

$$\underset{\beta,\ \gamma \text{ unsaturated acid}}{R - CH = CH - CH_2 - COOH} \xrightarrow{NaOH} \underset{\alpha,\ \beta \text{ unsaturated acid}}{R - CH_2 - CH = CHCOOH}$$

$$\downarrow \text{NaOH fuse}$$

$$R - CH_2COOH + CH_3COOH$$

Q. 15. What are keto acids? Give their nomenclature.

Ans. Organic acids having a ketonic group (> C = O) in the molecule are known as keto acids. Nomenclature of the keto acids will be clear from the following table.

Keto acid	*Common name*	*IUPAC name*
$CH_3\overset{O}{\overset{\parallel}{C}} - COOH$	Pyruvic acid	2-oxopropanoic acid
$CH_3 - \overset{O}{\overset{\parallel}{C}} - CH_2COOH$	Acetoacetic acid	3-oxobutanoic acid
$CH_3\overset{O}{\overset{\parallel}{C}}CH_2CH_2COOH$	Laevulinic acid	4-oxopentanoic acid

Some of the keto acids are of biological and synthetic importance.

Q. 16. Give the methods of preparation of pyruvic acid.

Ans. Important methods of preparation of pyruvic acid, which is the simplest keto acid, are as follows.

1. By oxidation of lactic acid

$$\underset{\text{Lactic acid}}{CH_3CH(OH)COOH} \xrightarrow{[O]} \underset{\text{Pyruvic acid}}{CH_3COCOOH} + H_2O$$

2. By the hydrolysis of acetyl cyanide

$$\underset{\text{Acetyl cyanide}}{CH_3COCN} \xrightarrow{H_2O} \underset{\text{Pyruvic acid}}{CH_3COCOOH}$$

3. By the hydrolysis of dibromopropionic acid

$$\underset{\substack{\alpha,\,\alpha\text{ dibromo-}\\\text{propionicacid}}}{CH_3CBr_2COOH} \xrightarrow{H_2O} \underset{\text{(Unstable)}}{CH_3C(OH)_2-COOH} \xrightarrow{-H_2O} \underset{\text{Pyruvic acid}}{CH_3COCOOH}$$

4. By heating tartaric acid. On dry distillation or heating in the presence of $KHSO_4$, tartaric acid yields pyruvic acid

$$\begin{matrix} CH(OH)COOH \\ | \\ CH(OH)COOH \end{matrix} \xrightarrow[KHSO_4]{\text{Heat}} \underset{\text{Pyruvic acid}}{CH_3-CO-COOH}$$

The mechanism of this transformation is believed to be is follows.

$$HOOC-CH(\ddot{O}H)-CH(OH)-COOH \xrightarrow{H^+} HOOC-CH(\overset{+}{O}H_2)-CH(OH)-COOH \xrightarrow{-H_2O,\,-H^+} HOOC-CH=C(OH)-COOH \rightleftharpoons$$

$$H_2C-C(=O)-O-H\ \text{(cyclic, with HOOC on C)} \xrightarrow{-CO_2} \underset{\text{Enol form of pyruvic acid}}{CH_2=C(OH)-COOH} \rightleftharpoons \underset{\text{Keto form of pyruvic acid}}{CH_3-C(=O)-COOH}$$

Q. 17. Describe with mechanism, wherever applicable, the properties of pyruvic acid.

Ans. Pyruvic acid contains a keto group and a carboxylic group. Hence it exhibits properties due to both of these groups. Important properties of pyruvic acid are given below

1. Properties due to the ketonic group

(*i*) $$\underset{\text{Pyruvic acid}}{CH_3COCOOH} + HCN \longrightarrow \underset{\text{Pyruvic acid cyanohydrin}}{CH_3C(OH)(CN)-COOH}$$

(*ii*) $$\underset{\text{Pyruvic acid}}{CH_3C(=O)COOH} + \underset{\text{Hydroxyl amine}}{H_2NOH} \longrightarrow \underset{\text{Pyruvic acid oxime}}{CH_3C(=N-OH)COOH} + H_2O$$

(*iii*) $$CH_3C(=C)COOH + \underset{\text{Hydrazine}}{H_2NNH_2} \longrightarrow \underset{\text{Pyruvic acid hydrazone}}{CH_3C(=N-NH_2)COOH} + H_2O$$

2. Reactions due to the carboxyl group

$$\underset{\text{Pyruvic acid}}{CH_3COCOOH} + NaOH \longrightarrow \underset{\text{Sod. pyruvate}}{CH_3COCOONa} + H_2O$$

$$\underset{\text{Pyruvic acid}}{2CH_3COCOOH} + Na_2CO_3 \longrightarrow \underset{\text{Sod. pyruvate}}{2CH_3COCOONa} + CO_2 + H_2O$$

$$\underset{\text{Pyruvic acid}}{CH_3COCOOH} + C_2H_5OH \xrightarrow{H^+} \underset{\text{Ethyl pyruvate}}{CH_3COCOOC_2H_5} + H_2O$$

$$\underset{\text{Pyruvic acid}}{CH_3COCOOH} + PCl_5 \longrightarrow \underset{\text{Pyruvyl chloride}}{CH_3COCOCl} + POCl_3 + HCl$$

3. Reduction. Pyruvic acid changes into lactic acid an reduction or hydrogenation.

$$CH_3COCOOH + 2[H] \xrightarrow[H_2O]{Na-Hg} CH_3CH(OH)COOH$$

4. Oxidation. On oxidation with pot. permanganate, pyruvic acid is changed into acetic acid

$$CH_3COCOOH + [O] \xrightarrow{H_2O_2} CH_3COOH + CO_2$$

5. Dicarboxylation. Pyruvic acid gets decarboxylated into acetaldehyde, on heating with dil. H_2SO_4.

$$\underset{\text{Pyruvic acid}}{CH_3\overset{\overset{O}{\|}}{C}-COOH} \xrightarrow{\text{dil. } H_2SO_4} CH_3CHO + CO_2$$

Mechanism of the reaction

$$\underset{\text{Pyruvic acid}}{CH_3-\overset{\overset{O}{\|}}{C}-\overset{\overset{O}{\|}}{C}-O-H} \overset{H^+}{\rightleftharpoons} CH_3-\overset{\overset{\overset{+}{O}H}{\|}}{C}-\overset{\overset{O}{\|}}{C}-\ddot{O}-H \longrightarrow CH_3-\overset{\overset{\overset{+}{O}H}{\|}}{C}: + C=\overset{+}{O}-H$$

$$\xrightarrow{+H^+} CH_3-\overset{\overset{\overset{+}{O}-H}{\|}}{C}-H \; CO_2 + H^+$$

$$CH_3-\overset{\overset{\overset{+}{O}-H}{\|}}{C}-H \longrightarrow H^+ + CH_3-\overset{\overset{O}{\|}}{C}-H$$

6. Action of hot conc. sulphuric acid. When heated with conc. sulphuric acid, CO is eliminated and acetic acid is formed.

$$CH_3-\overset{\overset{O}{\|}}{C}-\overset{\overset{O}{\|}}{C}-OH \xrightarrow[\text{heat}]{\text{Conc. } H_2SO_4} CH_3COOH + CO$$

Q. 18. Give the methods of preparation and properties of acetoacetic acid (CH_3COCH_2COOH).

Ans. Methods of Preparation

1. It is prepared by the hydrolysis of ethyl acetoacetate with sodium hydroxide in the cold when sodium salt of the acetoacetic acid is formed, the solution is then treated with a mineral acid to obtain acetoacetic acid.

$$\underset{\text{Ethyl acetoacetate}}{CH_3COCH_2COOC_2H_5} \xrightarrow[H^+]{NaOH} \underset{\text{Acetoacetic acid}}{CH_3COCH_2COOH}$$

$$\underset{\beta\text{-hydroxybutyric acid}}{CH_3CH(OH)CH_2COOH} \xrightarrow{[O]} \underset{\text{Acetoacetic acid}}{CH_3COCH_2COOH}$$

2. Properties of acetoacetic acid. Some important reactions of acetoacetic acid are given below.

(*i*) **Haloform reaction.** A compound containing CH_3CO- gives the halogen reaction. Such compounds on heating with iodine and sodium hydroxy give a yellow ppt. of iodoform.

(*ii*) **Keto-enol Tautomerism.** Acetoacetic acid exhibits keto-enol tautomerism.

$$\underset{\text{Keto form}}{CH_3\overset{\overset{O}{\|}}{C}-CH_2COOH} \longrightarrow \underset{\text{Enol form}}{CH_3\overset{\overset{OH}{|}}{C}=CHCOOH}$$

(*iii*) **Decarboxylation.** Acetoacetic acid undergoes decarboxylation on warming to give acetone

$$\underset{\text{Acetoacetic acid}}{CH_3-\overset{\overset{O}{\|}}{C}-CH_2COOH} \xrightarrow[\text{Acetone}]{\text{Warm}} CH_3-\overset{\overset{O}{\|}}{C}-CH_3+CO_2$$

Decarboxylation involves both the carboxylic acid molecule and the carboxylate ion. Mechanism involving both species is given below.

$$\underset{\text{Acetoacetate ion}}{CH_3-\underset{\underset{O}{\|}}{C}-CH_2-\underset{\underset{O}{\|}}{C}-\overset{\ominus}{O}} \xrightarrow{-CO_2} \left[CH_3-\underset{\underset{O}{\|}}{C}-\overset{\ominus}{C}H_2 \longleftrightarrow CH_3-\underset{\underset{O^{\ominus}}{|}}{C}=CH_2\right]$$

$$CH_3-\underset{\underset{O}{\|}}{C}-CH_3 \xleftarrow{H_2O} \underset{\text{Resonance stabilised carbanion}}{CH_3-\underset{\underset{O}{\|}}{C}=CH_2}$$

$$\underset{\text{Acetoacetic acid}}{CH_3-\underset{\underset{O}{\|}}{C}\overset{CH_2}{\frown}\underset{\underset{O-H}{|}}{C=O}} \xrightarrow[\text{slow}]{-CO_2} \underset{\text{Enol}}{CH_3-\underset{\underset{OH}{|}}{C}=CH_2} \underset{\text{Acetone}}{\rightleftharpoons} \underset{\text{Ketone}}{CH_3-\underset{\underset{O}{\|}}{C}-CH_3}$$

Q. 19. Describe the method of formation of ethyl acetoacetate from ethyl alcohol and sodium ethoxide. (*Gulberga 2002; Saugar 2003; Delhi, 2005*)

Or

Give the methods of preparation of ethyl acetoacetate.

(*Awadh, 2000; Garhwal, 2000; Kurukshetra, 2001; Nagpur, 2003*)

Or

What is Claisen condensation? *(Delhi 2000; Goa 2003, Kalyani 2004, Andhra 2010)*

Ans. Claisen condensation. Two molecules of ethyl acetate undergo self condensation in the presence of a strong base such as sodium ethoxide, giving ethyl acetoacetate

$$2CH_3COOC_2H_5 \xrightarrow{Na\,/\,Alcohol} CH_3COCH_2COOC_2H_5 + C_2H_5OH$$

Mechanism

1st step. A proton is removed by ethoxide ion from α-carbon of ethyl acetate to form resonance stabilised carbanion.

$$\underset{\text{Ethyl acetate}}{CH_2(H)-\overset{\overset{O}{\|}}{C}-OC_2H_5} + C_2H_5O^{\ominus} \underset{}{\overset{-C_2H_5OH}{\rightleftharpoons}} CH_2=\overset{\overset{O^{\ominus}}{|}}{C}-OC_2H_5 \longleftrightarrow \underset{\text{Resonance stabilised carbanion}}{:\overset{\ominus}{C}H_2-\overset{\overset{O}{\|}}{C}-OC_2H_5}$$

2nd step. The carbanion attacks the carbonyl carbon of the second molecule of ethyl acetate and displaces the ethoxide ion to give ethyl acetoacetate.

$$\underset{\substack{\text{Ethyl acetate}\\\text{(Second molecule)}}}{CH_3-\overset{\overset{O}{\|}}{\underset{\underset{OC_2H_5}{|}}{C}}} + \underset{\text{Carbanion}}{:\overset{\ominus}{C}H_2-COOC_2H_5} \rightleftharpoons \left[CH_3-\overset{\overset{O^{\ominus}}{|}}{\underset{\underset{OC_2H_5}{|}}{C}}-CH_2COOC_2H_5\right] \rightleftharpoons C_2H_5O^- + CH_3-\overset{\overset{O}{\|}}{C}-CH_2COOC_2H_5$$

3rd step. Ethoxide ion reacts with ethyl acetoacetate to give ethanol and anion of acetic ester which on acidification with acetic acid yields ethyl acetoacetate in the final step.

$$CH_3-\overset{\overset{O}{\|}}{C}-\underset{\underset{H}{|}}{CH}-COOC_2H_5 + C_2H_5O^{\ominus} \overset{-C_2H_5OH}{\rightleftharpoons} \underset{\substack{\text{Anion of ethyl acetoacetate}\\\text{as sodium salt}}}{CH_3-\overset{\overset{O^{\ominus}}{|}}{C}=CH-COOC_2H_5} \xrightarrow{H^+ \;|\; CH_3COOH}$$

$$CH_3-\overset{O}{\overset{\|}{C}}-CH_2COOC_2H_5 \rightleftharpoons CH_3-\overset{OH}{\overset{|}{C}}=CH-COOC_2H_5$$

Keto form — Enolic form

2. From ketene. Commercially, acetoacetic ester is prepared by dimerisation of ketene in acetone solution followed by treatment with alcohol.

$$\underset{\text{Ketene}}{2CH_2=C=O} \longrightarrow \begin{array}{c} CH_2=C-O \\ | \quad\; | \\ H_2C-C=O \end{array} \xrightarrow{C_2H_5OH} \underset{\text{Acetoacetic ester}}{CH_3COCH_2COOC_2H_5}$$

Q. 20. Describe giving mechanism the following properties of ethyl acetoacetate.

(*i*) Acidic nature of methylene

(*ii*) Ketonic hydrolysis

(*iii*) Acidic hydrolysis.

Ans. 1. Acidic nature of methylene group. Methylene group in acetoacetic ester is flanked by two electron withdrawing groups viz., an acetyl group and an ester group. The hydrogens of this methylene group are ionisable because of the electron withdrawing effect of the surrounding groups. Also the negative ion obtained after losing the proton gets stabilised due to resonance as shown below.

$$\underset{\text{Active methylene group}}{CH_3-\overset{O}{\overset{\|}{C}}-CH_2-\overset{O}{\overset{\|}{C}}-OC_2H_5} \xrightleftharpoons{-H^+} \left[CH_3-\overset{O}{\overset{\|}{C}}-\overset{\ominus}{C}H-\overset{O}{\overset{\|}{C}}-OC_2H_5 \leftrightarrow CH_3-\overset{O^{\ominus}}{\overset{|}{C}}=CH-\overset{O}{\overset{\|}{C}}-OC_2H_5 \leftrightarrow CH_3-\overset{O}{\overset{\|}{C}}-CH=\overset{O^{\ominus}}{\overset{|}{C}}-OC_2H_5 \right]$$

$$\underset{\text{Active methylene group}}{CH_3-\overset{O}{\overset{\|}{C}}=CH=\overset{O}{\overset{\|}{C}}-OC_2H_5} \quad (\ominus \text{ delocalised}) \quad \text{Equivalent to}$$

As a consequence of above property, acetoacetic ester reacts with sodium ethoxide to form sodium salt of acetoacetic ester

$$\underset{\text{Acetoacetic acid}}{CH_3COCH_2COOC_2H_5} \xrightarrow{C_2H_5ONa} \underset{\text{Sod. salt}}{[CH_3COCHCOOC_2H_5]^- Na^+}$$

This sodium salt on treatment with an alkyl or acyl halide gives a monoalkyl or acyl derivative of acetoacetic ester

$$[CH_3COCHCOOC_2H_5]^- Na^+ + RX \longrightarrow CH_3CO\underset{R}{\underset{|}{C}H}COOC_2H_5 + NaX$$

Another such treatment can produce dialkyl derivatives.

2. Ketonic hydrolysis. Acetoacetic ester or its alkyl derivatives on hydrolysis with dil. aqueous alkali form the corresponding acids and then undergo decarboxylation to yield ketones. This process is known as *ketonic hydrolysis*.

$$\underset{\text{Acetoacetic ester}}{CH_3COCH_2COOC_2H_5} \xrightarrow[\text{NaOH}]{\text{dil. aqeous}} \underset{\text{Sodium salt of acetoacetic acid}}{CH_3COCH_2COONa}$$

$$\xrightarrow{H^+} \underset{\text{Acetoacetic acid}}{CH_3COCH_2COOH} \xrightarrow{-CO_2} \underset{\text{Acetone}}{CH_3COCH_3}$$

3. Acid hydrolysis. If hydrolysis of acetoacetic acid or its alkyl derivative is carried out in the presence of conc. alkali, two molecules of carboxylic acid are obtained.

$$CH_3-\overset{\overset{O}{\|}}{C}-CH_2-\overset{\overset{O}{\|}}{C}-OC_2H_5 \xrightarrow{2NaOH} 2CH_3COONa + C_2H_5OH$$

Synthetic Uses of Acetoacetic ester (*Jadavpur 2003, Baroda, 2004*)

Based on above three characteristics of the ester, a large variety of compounds can be obtained.

(*i*) **Synthesis of methyl ketones.** As explained above under the heading *ketonic hydrolysis*, a desired ketone may be prepared from the appropriate alkyl derivative of acetoacetic ester. For example, if it is desired to prepare 2-pentanone by ketonic hydrolysis, then the alkyl derivative of acetoacetic ester that needs to be taken is

$$CH_3CO\underset{\underset{CH_3}{|}}{\underset{CH_2}{\underset{|}{C}}}HCOOC_2H_5$$

On ketonic hydrolysis, it will produce 2-pentanone

$$CH_3CO\underset{\underset{CH_3}{|}}{\underset{CH_2}{\underset{|}{C}}}HCOOC_2H_5 \xrightarrow[\text{NaOH} \; -C_2H_5OH]{\text{dil. aq.}} CH_3CO\underset{\underset{CH_3}{|}}{\underset{CH_2}{\underset{|}{C}}}HCOOH$$

$$\xrightarrow{-CO_2} \underset{\text{2-pentanone}}{CH_3COCH_2CH_2CH_3}$$

As a general rule, if it is desired to prepare CH_3COCH_2R, where R is any alkyl group, linear or branched, then the starting compound will be

$$CH_3CO\underset{\underset{R}{|}}{C}HCOOC_2H_5$$

(*ii*) **Synthesis of diketone.** Acyl derivative of acetoacetic ester on ketonic hydrolysis yields diketones. Acetylacetone, for example, may be prepared as under:

$$CH_3COCH_2COOC_2H_5 \xrightarrow[\text{ethoxide}]{\text{Sod.}} [CH_3COCHCOOC_2H_5]^- Na^+$$

Acetoacetic ester

$$\xrightarrow{CH_3COCl} CH_3COCH(COCH_3)COOC_2H_5 \xrightarrow[-CO_2]{\text{dil. alkali}} CH_3COCH_2COCH_3$$

Acetyl acetone

(*iii*) **Synthesis of mono and dialkyl acetic acids.** Acidic hydrolysis of mono alkyl derivative of acetoacetic ester produces monoalkyl acetic acid. Similarly dialkyl derivatives would produce dialkyl acetic acid.

$$\underset{\text{Acetoacetic ester}}{CH_3COCH_2COOC_2H_5} \xrightarrow[C_3H_7Br]{\text{Sod. ethoxide}} \underset{\text{Propyl derivative}}{CH_3COCH(C_3H_7)COOC_2H_5}$$

$$\downarrow \text{Conc. NaOH}$$

$$C_2H_5OH + \underset{\text{Acetic acid}}{CH_3COOH} + \underset{n\text{-valeric acid}}{C_3H_7CH_2COOH}$$

In general to produce RCH_2COOH, the alkyl derivatives required will be

$$CH_3COCH(R)COOC_2H_5$$

It is also possible to synthesise dialkyl acetic acid *e.g.*

$$\underset{\text{Diethyl derivative}}{CH_3COC(CH_2CH_3)_2 \cdot COOC_2H_5} \xrightarrow[-C_2H_5OH]{\text{Conc. NaOH}} CH_3COOH + \underset{\text{Diethyl acetic acid}}{(CH_3CH_2)_2CHCOOH}$$

(*iv*) **Synthesis of dicarboxylic acids.** Reaction of sod. salt of acetoacetic ester with halogen derivative of an ester followed by acidic hydrolysis, produces dicarboxylic acid.

$$\underset{\text{Ethyl acetoacetate}}{CH_3COCH_2COOC_2H_5} \xrightarrow[\text{Ethoxide}]{\text{Sod.}} \underset{\text{Sod. salt}}{[CH_3COCHCOOC_2H_5]^- Na^+}$$

$$\xrightarrow{ClCH_2COOC_2H_5} CH_3COCH(CH_2COOC_2H_5)COOC_2H_5 \xrightarrow[\text{NaOH}]{\text{Conc.}} CH_3COOH + \underset{\text{Succinic acid}}{CH_2COOH\text{-}CH_2COOH} + 2C_2H_5OH$$

(*v*) **Synthesis of α, β unsaturated acids.** Acetoacetic ester condenses with aldehydes and ketones in the presence of pyridine (Knoevenagal Reaction)

$$CH_3CHO + CH_3COCH_2COOC_2H_5 \xrightarrow{\text{Pyridine}} \underset{\underset{CHCH_3}{\|}}{CH_3COCCOOC_2H_5}$$

$$\xrightarrow{\text{Hydrolysis}} CH_3COOH + \underset{\text{Crotonic acid}}{CH_3CH = CHCOOH} + C_2H_5OH$$

(*vi*) **Condensation with urea.** Acetoacetic ester in enol form condenses with urea to form 4-methyl uracil

$$\begin{matrix} CH_3C = CHCOOC_2H_5 \\ | \\ OH \end{matrix} + \begin{matrix} H \quad\quad H \\ | \quad\quad | \\ NH - CO - NH \end{matrix} \longrightarrow \begin{matrix} CH_3C = CH - CO \\ | \quad\quad\quad | \\ NH - CO - NH \end{matrix} + C_2H_5OH + H_2O$$

(*vii*) **Condensation with hydroxylamine**

$$\begin{matrix} CH_3 - C = O \\ | \\ CH_2 - CO - OC_2H_5 \end{matrix} + \begin{matrix} H_2N \\ | \\ HO \end{matrix} \longrightarrow \underset{\text{Methyl iso-oxazolone}}{\begin{matrix} CH_3 - C = N \\ | \\ CH_2 - CO \end{matrix} > O} + C_2H_5OH + H_2O$$

(*viii*) **Synthesis of pyrrole derivatives.** Reaction of sod. salt of acetoacetic ester with iodine followed by treatment with ammonia gives pyrrole derivatives

$$2[CH_3COCHCOOC_2H_5]^- Na^+ + I_2 \longrightarrow \begin{matrix} CH_3COCHCOOC_2H_5 \\ | \\ CH_3COCHCOOC_2H_5 \end{matrix} + 2NaI$$

$$\xrightarrow{NH_3} \underset{\text{(Pyrrole derivative)}}{\begin{matrix} CH_3COCHCO \\ | \\ CH_3COCHCO \end{matrix} > NH}$$

Q. 21. Give a detailed account of keto-enol tautomerism.

(*Delhi, 2002; Nagpur, 2008; Lucknow, 2010*)

Ans. 1. An isomerism in which two forms of the compound exist simultaneously as an equilibrium mixture is called *tautomerism*. This is different from the ordinary isomerism phenomenon in which there are separately different compounds comforming to the same molecular formula.

2. Keto-enol tautomerism is a particular case of tautomerism in which two co-existing substances have the ketonic group and ethylenic double bond.

3. A typical example of a compound exhibiting keto-enol tautomerism is acetoacetic ester. It exists as an equilibrium mixture of two substances one of which has a ketonic group and the other ethylenic bond

$$\underset{\text{Keton form}}{CH_3 - \overset{\overset{O}{\|}}{C} - CH_2COOC_2H_5} \longrightarrow \underset{\text{Enol form}}{CH_3 - \overset{\overset{OH}{|}}{C} = CHCOOC_2H_5}$$

Enol is a combination of *ene* + *ol i.e.* a double bond and an alcoholic group.

4. This conclusion is based on the observation that acetoacetic ester gives the properties expected of a ketone and those expected of an ethylenic compound containing an alcoholic group. For example, Acetoacetic ester gives reaction with HCN to form cyanohydrin, with hydroxy amine to form oxime and with phenyl hydrazine to form yellow crystalline hydrazone. All these tests point to the presence of a ketonic group.

At the same time, acetoacetic ester liberates hydrogen with metallic sodium, it forms acetyl derivative on treatment with acetyl chloride. These are the characteristic reactions expected of alcohols.

It decolourises bromine solution, which indicates the presence of a double bond. Thus it supports our belief that acetoacetic ester exists as a mixture of two forms one of which is a keto compound and the other an enol *i.e.* ethylenic compound along with an alcoholic group.

5. The two tautomers differ in respect of point of attachment of hydrogen. In the *enol* form, the hydrogen is attached to oxygen whereas it is attached to carbon in the *ketonic* form. The two forms interconvert into each other as follows.

$$\underset{\text{Keto form}}{CH_3-\overset{\overset{O}{\parallel}}{C}-CH_2-\overset{\overset{O}{\parallel}}{C}-OC_2H_5} \rightleftharpoons \underset{\text{Carbanion}}{CH_3-\overset{\overset{O}{\parallel}}{C}=CH=\overset{\overset{O}{\parallel}}{C}-OC_2H_5} + H^+$$

$$\Updownarrow$$

$$\underset{\text{Carbanion}}{CH_3-\overset{\overset{OH}{|}}{C}=CH-\overset{\overset{O}{\parallel}}{C}-OC_2H_5}$$

6. Keto form of the ester has a percentage of 93 and enol form has the percentage of 7.

7. Normally it is not possible to isolate either pure keto form or enol form. This is because, as we try to remove keto form from the mixture, some of the enol form in the resulting mixture converts into keto form to maintain 93 : 7 ratio and *vice versa*.

8. However, under conditions of low temp. and using specific solvents, it is possible to affect separation. If a solution of acetoacetic ester in ether and hexane is cooled to 195 K, pure keto form of the ester is obtained. Enol form of the ester can be obtained by treating a suspension of sodium salt of the ester in petroleum ether with dry HCl at 195 K.

Q. 22. Starting from ethyl acetoacetate, give a reaction scheme to prepare each of the following:

(A) Methyl ethyl ketone (B) Acetyl acetone (C) 2, 3 dimethyl butanoic acid (D) succinic acic F) Crotonic acid. (*Mddurai ,2004; Magadh 2004; Panjab, 2005, Vidyasagar 2007; Nagpur 2008*)

Ans. (A) $\underset{\text{Acetoacetic ester}}{CH_3COCH_2COOC_2H_5} \xrightarrow{C_2H_5ONa} \underset{\text{Sod. salt}}{[CH_3CC\cdot CHCOOC_2H_5]^- Na^+}$

$$\xrightarrow{CH_3Br} \underset{\alpha\text{-methyl acetoacetate}}{CH_3CO\underset{\underset{CH_3}{|}}{C}HCOOC_2H_5} \xrightarrow[NaOH]{dil.} \underset{\text{Methyl ethyl ketone}}{CH_3COCH_2CH_3}$$

(B) $CH_3COCH_2COOC_2H_5 \xrightarrow{C_2H_5ONa} [CH_3COCHCOOC_2H_5]^- Na^+$

$\xrightarrow{CH_3COCl} \underset{\displaystyle COCH_3}{CH_3COCHCOOC_2H_5} \xrightarrow[NaOH]{dil.} \underset{\text{Acetylacetone}}{CH_3COCH_2COCH_3}$

(C) $CH_3COCH_2COOC_2H_5 \xrightarrow[(ii)\ CH_3Br]{(i)\ C_2H_5ONa} CH_3CO\overset{\displaystyle CH_3}{C}HCOOC_2H_5$

$\xrightarrow[(ii)\ \text{Isopropyl bromide}]{(i)\ C_2H_5ONa} CH_3CO-\overset{\displaystyle CH_3}{C}(-CH(CH_3)-CH_3)-COOC_2H_5 \xrightarrow[NaOH]{Conc.}$

$CH_3COOH + CH_3-\overset{\displaystyle CH_3}{CH}-\overset{\displaystyle CH_3}{CH}-COOH + C_2H_5OH$

2, 3 dimethyl butyric acid

(D) $CH_3COCH_2COOC_2H_5 \xrightarrow{C_2H_5ONa} [CH_3COCHCOOC_2H_5]^- Na^+$

$\xrightarrow{ClCH_2COOC_2H_5} \underset{\displaystyle CH_2COOC_2H_5}{CH_3COCHCOOC_2H_5} \xrightarrow[NaOH]{Conc.}$

$CH_3COOH + \underset{\displaystyle CH_2COOH}{CH_2COOH} + 2C_2H_5OH$

(E) $\underset{\text{Acetoacetic ester}}{CH_3COOCH_2COOC_2H_5} + \underset{\text{Acetaldehyde}}{CH_3CHO} \xrightarrow[-H_2O]{Pyridine} \underset{\displaystyle \| \atop CHCH_3}{CH_3COCCOOC_2H_5}$

$\downarrow$ Acid hydrolysis Conc. NaOH

$CH_3COOH + \underset{\text{Crotonic acid}}{CH_3CH = CHCOOH} + C_2H_5OH$

Q. 23. How will you obtain the following from acetoacetic ester

(a) Isobutyric acid

(b) Methyl propyl ketone *(Lucknow 2000)*

(c) 4-Methyl uracil from acetoacetic ester

Ans. (a) Isobutyric acid

$\underset{\text{Acetoacetic ester}}{CH_3COCH_2COOC_2H_5} \xrightarrow[CH_3Br]{C_2H_5ONa} \underset{\text{Methyl derivative}}{\underset{\displaystyle CH_3}{CH_3COCHCOOC_2H_5}}$

$$\xrightarrow[CH_3Br]{C_2H_5ONa} \underset{\text{Dialkyl derivative}}{CH_3COC(CH_3)(CH_3) - COOC_2H_5} \xrightarrow[H^+]{\text{Conc. NaOH}} CH_3COOH$$

$$+ \underset{\text{Isobutyric acid}}{(CH_3)_2CHCOOH} + C_2H_5OH$$

(b) Methylpropyl ketone

$$\underset{\text{Acetoacetic ester}}{CH_3COCH_2COOC_2H_5} \xrightarrow[C_2H_5Br]{C_2H_5ONa} \underset{\text{Ethyl derivative}}{CH_3COCH(C_2H_5)COOC_2H_5}$$

$$\xrightarrow{\text{dil. NaOH}} \underset{\text{Sod. salt of ethyl acetoacetic acid}}{CH_3COCH(C_2H_5)COONa} \xrightarrow[\Delta]{H^+} \underset{\text{Methylpropyl ether}}{CH_3COC_3H_7} + CO_2 \uparrow$$

(c) 4-Methyl uracil. Acetoacetic ester in *enol* form condenses with urea to form 4-methyl uracil.

$$CH_3C(OH) = CHCOOC_2H_5 + \underset{}{H-NH - CO - NH-H} \longrightarrow \underset{\text{4-Methyl uracil}}{CH_3C = CH - CO,\ NH - CO - NH} + C_2H_5OH + H_2O$$

Q. 24. Maleic acid forms an anhydride on heating, but fumaric acid does not. Why?

(Kerala, 2000)

Ans. In maleic acid, the two carboxy groups are on the same side of the double bond and are nearer. For anhydride formation, it is essential that two carboxy groups should be close enough to eliminate a water molecule.

$$\underset{\text{Maleic acid}}{\begin{matrix} CHCOO\ \boxed{H} \\ \| \\ CHCO\ \boxed{OH} \end{matrix}} \xrightarrow[-H_2O]{\Delta} \underset{\text{Maleic anhydride}}{\begin{matrix} CHCO \\ \| \\ CHCO \end{matrix}\!\!>\!O}$$

$$\underset{\text{Fumaric acid}}{\begin{matrix} CHCOOH \\ \| \\ HOOCC \end{matrix}} \xrightarrow{\Delta} \text{No reaction}$$

This condition is satisfied in maleic acid. It forms maleic anhydride on heating.

In fumaric acid, the two carboxy groups are at a distance. Hence removal of a water molecule from two carboxy groups is not possible. Therefore, no anhydride is formed.

MISCELLANEOUS QUESTIONS

Q. 25. Sketch the transformation of ethyl acetoacetate into the following:

(*i*) *n*-valaric acid (*ii*) Crotonic acid (*iii*) 4-methyl uracil.

Ans. See Q. 20.

Q. 26. Describe the method of preparation of the following compounds using ethyl acetoacetate

(*i*) Methyl ethyl ketone (*ii*) Butanoic acid (*iii*) Succinic acid

Ans. See Q. 20.

Q. 27. How would you convert ethyl acetoacetate into

(*i*) 2-ethyl butanoic acid (*ii*) Crotonic acid (*iii*) Laevulinic acid

Ans. For (*i*) and (*ii*), see Q. 20.

(*iii*) **Leavulinic acid**

$$CH_3COCH_2COOC_2H_5 \xrightarrow{C_2H_5O^-Na^+} [CH_3COCHCOOC_2H_5]^- Na^+$$

$$\xrightarrow{ClCH_2COOC_2H_5} \underset{\displaystyle CH_2COOC_2H_5}{CH_3COCH - COOC_2H_5} \xrightarrow{NaOH} \underset{\displaystyle CH_2COONa}{CH_3COCHCOONa}$$

$$\xrightarrow{H^+} \underset{\text{Laevulinic acid}}{CH_3COCH_2CH_2COOH}$$

Q. 28. How will you prepare cinnamic acid, applying Knoevenagal reaction? (*Kerala, 2001*)

Ans. See Q. 13 (7)

Q. 29. How many isomers of salicylic acid are known? How these isomers compare with benzoic acid as regards to acid strength? (*Meerut, 2000*)

Ans. Isomers of salicylic acid :

ortho: $C_6H_4(OH)COOH$ (OH and COOH on adjacent carbons) — *ortho*

meta: $C_6H_4(OH)COOH$ (OH and COOH in 1,3-positions) — *meta*

para: $C_6H_4(OH)COOH$ (OH and COOH in 1,4-positions) — *para*

Acid strength –See Q.8

Q. 30. How are following prepared from acetoacetic ester ?

(*i*) Butanone (*Lucknow, 2000, Kerala, 2001*)

(*ii*) Succinic acid

(*Lucknow, 2000, Garhwal 2000, M. Dayanand 2000, Awadh 2000, Kerala 2000*)

(*iii*) Methyl phenylprazolone (*Awadh, 2000*)

(*iv*) Acetyl acetone (*Awadh, 2000; Garhwal, 2000*)

(*v*) Antipyrine (*Kerala, 2000; Garhwal, 2000*)

(*vi*) Methyl uracil (*Garhwal, 2000*)

***(vii)* Butanoic acid** (*Kanpur 2001*)

Ans. (*i*) See Q. 20 (3) (*i*). In place of take $CH_3COCH(CH_2CH_3)COOC_2H_5$, take $CH_3COCH(CH_3)COOC_2H_5$

(*ii*) See Q. 20 (3) (*iv*), (*iii*) See Q. 20 (3) (*viii*), (*iv*) See Q. 20 (3) (*ii*)

(*v*)...............(*vi*) See Q. 20 (3) (*vi*), See Q. 20 (3) (*iii*)

Q. 31. Write a note on the chemistry of cinnamic acid (*Kerala, 2001*)

Ans. See Q. 13 and Q. 14

Q. 32. How will you get optically active lactic acid from pyruvic acid?

(*M. Dayanand, 2000*)

Ans. See Q. 17 (3)

Q. 33. Complete the following reaction : (*Calcutta, 2000*)

$$\text{Salicylic acid} + Br_2 \longrightarrow$$

Ans. See Q. 10 (7)

Q. 34. How will you synthesize the following from acetoacetic ester ?

(*i*) acetonylacetate **(*ii*) Crotonic acid**

(*iii*) *n*-butane **(*iv*) Methyluracil** (*Purvanchal 2003*)

Ans. See Q. 20

Q. 35. An aromatic hydrocarbon A, C_6H_6 on treatment with conc. H_2SO_4 gives a compound B ($C_6H_6SO_3$). Compound B on fusion with NaCN yields C (C_7H_5N). C on treatment with Sn Cl_2/ HCl followed by steam distillation gives D (C_7H_6O). D reacts with sodium acetate and acetic anhydride to give an unsaturated acid E ($C_9H_8O_2$). Identify the compounds and name the reactions involved. (*Delhi 2003*)

Ans.

$$\underset{A}{C_6H_6} \xrightarrow[\text{Sulphonation}]{\text{Conc. } H_2SO_4} \underset{B}{C_6H_5SO_3H} \xrightarrow[\text{NaCN}]{\text{Fusion}} \underset{C}{C_6H_5CN}$$

$$\xrightarrow[\text{HCl, Stephen's reaction}]{SnCl_2} \underset{D}{C_6H_5CHO} \xrightarrow[(CH_3CO)_2O]{CH_3COONa} \underset{\text{E (cinnamic acid)}}{C_6H_5CH=CHCOOH}$$

Q.36. How will you synthesise lactic acid from acetaldehyde?

(*Patna 2003, Cochin, 2003; Panjab 2005*)

Ans. Following steps are involved

$$CH_3-\overset{\overset{O}{\|}}{C}-H \xrightarrow{HCN} CH_3-\overset{\overset{OH}{|}}{CH}-CN \xrightarrow{H^+/H_2O} \underset{\text{Lactic acid}}{CH_3-\overset{\overset{OH}{|}}{CH}-COOH}$$

23

POLYCARBOXYLIC ACIDS AND THEIR DERIVATIVES

CONTENTS

Q. 1. What are polycarboxylic acids? Give their nomenclature.

Ans. Polycarboxylic acids are those organic compounds which contain more than one carboxylic group in the molecule. Those containing two such groups are called dicarboxylic acids, those containing three such groups are called tricarboxylic acids and so on.

Nomenclature of some prominent polycarboxylic acids are given below:

Compound	*Common name*	*IUPAC name*
COOH | COOH	Oxalic acid	Ethanedioic acid
$CH_2<$ COOH, COOH	Malonic acid	Propane-1, 3-dioic acid
CH_2COOH | CH_2COOH	Succinic acid	Butane-1, 4-dioic acid
CH_2COOH | CH(OH)COOH	Malic acid	Butan-2-ol-1, 4-dioic acid
CH(OH)COOH | CH(OH)COOH	Tartaric acid	Butan-2, 3 diol-1, 4-dioic acid
CHCOOH ‖ CHCOOH	Maleic acid	But-2-ene-1, 4-dioic acid
CH_2COOH | CH(OH)COOH | CH_2COOH	Citric acid	3-hydroxy-3-carboxypentane-1, 5-dioic acid

Q. 2. Give the methods of preparation and properties of malonic acid.

Ans. Methods of Preparation

1. **By oxidation of malic acid**

$$\underset{\text{Malic acid}}{\begin{array}{c}COOH\\|\\CH(OH)\\|\\CH_2\\|\\COOH\end{array}} \xrightarrow[K_2Cr_2O_7]{[O]} \underset{\text{Oxalacetic acid}}{\begin{array}{c}COOH\\|\\CO\\|\\CH_2\\|\\COOH\end{array}} \xrightarrow[K_2Cr_2O_7]{[O]} \underset{\text{Malonic acid}}{\begin{array}{c}COOH\\|\\CH_2\\|\\COOH\end{array}}$$

2. **From potassium chloroacetate**

$$\underset{\text{Potassium chloroacetate}}{\begin{array}{c}CH_2Cl\\|\\COOK\end{array}} + KCN \xrightarrow{-KCl} \underset{\text{Potassium cyanoacetate}}{\begin{array}{c}CH_2CN\\|\\COOK\end{array}} \xrightarrow[H^+]{H_2O} \underset{\text{Malonic acid}}{\begin{array}{c}CH_2COOH\\|\\COOH\end{array}}$$

Properties

1. **Decarboxylation (Action of heat).** On heating to 410–420 K malonic acid decomposes to give acetic acid

$$CH_2\begin{cases}COOH\\COOH\end{cases} \xrightarrow[410\text{–}420\text{ K}]{\Delta} CH_3COOH + CO_2$$

Mechanism of decarboxylation

$$\begin{array}{c}HO\quad CH_2\quad O\\ \backslash \;\;/\;\;\backslash\;\;//\\ C \qquad C\\ \| \qquad |\\ O \qquad O\\ \quad\; |\\ \quad\; H\end{array} \xrightarrow{-CO_2} \underset{\text{Enolic form of acetic acid}}{\begin{array}{c}HO \quad CH_2\\ \backslash\;//\\ C\\ |\\ OH\end{array}} \rightleftharpoons \underset{\text{Acetic aicd}}{CH_3COOH}$$

2. **Dehydration.** Malonic acid loses two water molecules on heating in the presence of P_2O_5

$$\begin{array}{c}\boxed{H \quad OH}\\ O=C—C—C=O\\ \boxed{OH \quad H}\end{array} \xrightarrow{-2H_2O} \underset{\text{Carbon suboxide}}{O=C=C=C=O}$$

3. **Action of nitrous acid.** On treatment with nitrous acid followed by hydrolysis, malonic acid gives mesoxalic acid

$$\underset{\text{Malonic acid}}{CH_2\begin{cases}COOH\\COOH\end{cases}} \xrightarrow[-H_2O]{HNO_3} HON=C\begin{cases}COOH\\COOH\end{cases} \xrightarrow{H_2O} \underset{\text{Mesoxalic acid}}{O=C\begin{cases}COOH\\COOH\end{cases}}$$

4. **Action of bromine**

$$\underset{\text{Malonic acid}}{CH_2\begin{cases}COOH\\COOH\end{cases}} + Br_2 \longrightarrow \underset{\text{Monobromo-malonic acid}}{BrCH\begin{cases}COOH\\COOH\end{cases}} + HBr$$

5. Action with aldehyde

$$CH_2\begin{cases}COOH\\COOH\end{cases} + RCHO \longrightarrow RCH=C\begin{cases}COOH\\COOH\end{cases} \xrightarrow{\text{Heat}} RCH=CHCOOH + CO_2$$

Malonic acid; α, β unsaturated acid

Q. 3. Describe the methods of preparation and properties of maleic acid.

Ans. Methods of Preparation.

1. From malic acid (Note the spelling difference between malic and maleic acids)

$$\begin{matrix}CH_2COOH\\|\\CH(OH)COOH\end{matrix} \xrightarrow[-2H_2O]{523\text{ K}} \begin{matrix}CHCO\\||\\CHCO\end{matrix}\!\!>O \xrightarrow[\text{Boil}]{H_2O} \begin{matrix}CHCOOH\\||\\CHCOOH\end{matrix}$$

Malic acid; Maleic anhydride; Maleic acid

2. From benzene (Manufacture)

$$\text{Benzene} + 9O_2 \xrightarrow[-4CO_2,\ -2H_2O]{V_2O_5,\ 673\text{–}700\text{ K}} \begin{matrix}CHCO\\||\\CHCO\end{matrix}\!\!>O \xrightarrow{2H_2O} \begin{matrix}CHCOOH\\||\\CHCOOH\end{matrix}$$

Benzene; Maleic anhydride; Maleic acid

Properties

1. Formation of salts. It reacts with alkalis to form salts. Two series of salts viz., mono and dialkali metal maleates can be obtained.

$$\begin{matrix}CHCOOH\\||\\CHCOOH\end{matrix} + NaOH \xrightarrow[-H_2O]{} \begin{matrix}CHCOONa\\||\\CHCOOH\end{matrix} \xrightarrow[-H_2O]{NaOH} \begin{matrix}CHCOONa\\||\\CHCOONa\end{matrix}$$

Maleic acid; Mono sodium maleate; Disodium maleate

2. Formation of esters. Two series of esters viz., mono and dialkyl maleates are formed with alcohols.

$$\begin{matrix}CHCOOH\\||\\CHCOOH\end{matrix} + C_2H_5OH \xrightarrow[-H_2O]{\text{Conc.}H_2SO_4} \begin{matrix}CHCOOC_2H_5\\||\\CHCOOH\end{matrix} \xrightarrow[-H_2O]{C_2H_5OH,\ H^+} \begin{matrix}CHCOOC_2H_5\\||\\CHCOOC_2H_5\end{matrix}$$

Mono ethyl maleate; Diethyl maleate

3. Action of heat. Maleic acid on heating forms anhydride.

$$\begin{matrix}CHCOOH\\||\\CHCOOH\end{matrix} \xrightarrow[\Delta\ (403\text{ K})]{-H_2O} \begin{matrix}CH-CO\\||\\CH-CO\end{matrix}\!\!>O$$

Maleic anhydride

The ease of formation of anhydride suggests that the two carboxylic groups are on the same side of the double bond *i.e.* maleic acid is a *cis* compound. Two carboxylic groups in *trans* positions would make it difficult for dehydration to take place.

4. Isomerisation. On heating continuously for sufficient time at 423 K or exposure to UV light, it changes into fumaric acid, its *trans* isomers, which is more stable.

$$\underset{\text{Maleic acid}}{\begin{matrix}H-C-COOH\\ \| \\ H-C-COOH\end{matrix}} \xrightarrow[\Delta]{423\ K} \underset{\text{Fumaric acid}}{\begin{matrix}H-C-COOH\\ \| \\ HOOC-C-H\end{matrix}}$$

5. Addition of hydrogen. On reduction with sod-amalgam and water or catalytic hydrogenation in presence of Ni, maleic acid converts into succinic acid.

$$\underset{\text{Maleic acid}}{\begin{matrix}CHCOOH\\ \| \\ CHCOOH\end{matrix}} + 2[H] \text{ or } H_2 \xrightarrow[\text{or } H_2, Ni]{Na/Hg-H_2O} \underset{\text{Succinic acid}}{\begin{matrix}CH_2COOH\\ | \\ CH_2COOH\end{matrix}}$$

6. Addition of bromine

$$\underset{\text{Maleic acid}}{\begin{matrix}CHCOOH\\ \| \\ CHCOOH\end{matrix}} + Br_2 \longrightarrow \underset{\text{Dibromo-succinic acid } (d \text{ and } l)}{\begin{matrix}CHBrCOOH\\ | \\ CHBrCOOH\end{matrix}}$$

7. Addition of HBr

$$\underset{\text{Maleic acid}}{\begin{matrix}CHCOOH\\ \| \\ CHCOOH\end{matrix}} + HBr \longrightarrow \underset{\text{Bromosuccinic acid}}{\begin{matrix}CH_2COOH\\ | \\ CHBrCOOH\end{matrix}}$$

8. Oxidation. On oxidation with alkaline potassium permanganate, maleic acid changes into mesotartaric acid.

$$\underset{\text{Maleic acid}}{\begin{matrix}CHCOOH\\ \| \\ CHCOOH\end{matrix}} + H_2O + [O] \xrightarrow[KMnO_4]{Alk.} \underset{\text{Mesotartaric acid}}{\begin{matrix}CH(OH)COOH\\ | \\ CH(OH)COOH\end{matrix}}$$

Q. 4. Describe the methods of preparation and properties of fumaric acid.

Ans. Methods of preparation

1. From maleic acid. On heating for a long time at 423 K maleic acid converts into more stable isomer fumaric acid because of less steric hindrance.

$$\underset{\text{Maleic acid}}{\begin{matrix}H-C-COOH\\ \| \\ H-C-COOH\end{matrix}} \xrightarrow[423\ K]{\Delta} \underset{\text{Fumaric acid}}{\begin{matrix}H-C-COOH\\ \| \\ HOOC-C-H\end{matrix}}$$

2. Action of alcoholic potash on bromosuccinic acid

$$\underset{\text{Bromosuccinic acid}}{\begin{matrix}CHBrCOOH\\ | \\ CH_2COOH\end{matrix}} \xrightarrow[KOH]{Alc.} \underset{\text{Fumaric acid}}{\begin{matrix}H-C-COOH\\ \| \\ HOOC-C-H\end{matrix}} + KBr + H_2O$$

3. From malonic acid

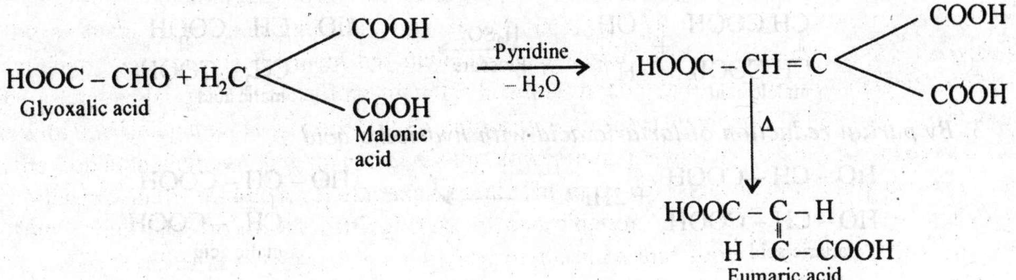

4. Industrial preparation. Fermentation of carbohydrate using *Rhizopus nigricans* produces fumaric acid

$$\text{Carbohydrate} \xrightarrow[\text{nigricans}]{\text{Rhizopus}} \text{Fumaric acid}$$

Properties

1. Dehydration. Compared to maleic acid, fumaric acid gets dehydrated with greater difficulty at 540 K to give maleic anhydride.

$$\underset{\text{Fumaric acid}}{\begin{matrix} H-C-COOH \\ \| \\ HOOC-C-H \end{matrix}} \xrightarrow[-H_2O]{540\ K} \underset{\text{Maleic anhydrides}}{\begin{matrix} CHCO \\ \| \\ CHCO \end{matrix} > O}$$

2. Reduction

$$\underset{\text{Fumaric acid}}{\begin{matrix} H-C-COOH \\ \| \\ HOOC-C-H \end{matrix}} + 2[H] \longrightarrow \underset{\text{Succinic acid}}{\begin{matrix} CH_2COOH \\ | \\ CH_2COOH \end{matrix}}$$

3. Addition of Br_2

$$\underset{\text{Fumaric acid}}{\begin{matrix} H-C-COOH \\ \| \\ HOOC-C-H \end{matrix}} + Br_2 \longrightarrow \underset{\substack{d \text{ and } l \\ \text{Dibromosuccinic acid}}}{\begin{matrix} CHBrCOOH \\ | \\ CHBrCOOH \end{matrix}}$$

4. Addition of HBr

$$\underset{\text{Maleic acid}}{\begin{matrix} CHCOOH \\ \| \\ CHCOOH \end{matrix}} + HBr \longrightarrow \underset{\text{Bromosuccinic acid}}{\begin{matrix} CH_2COOH \\ | \\ CH_2BrCOOH \end{matrix}}$$

5. Oxidation. When oxidised with alkaline $KMnO_4$, fumaric acid gives a racemic mixture of *d* and *l* tartaric acids.

$$\underset{\text{Fumaric acid}}{\begin{matrix} H-C-COOH \\ \| \\ HOOC-C-H \end{matrix}} + H_2O + O \xrightarrow[KMnO_4]{Alk.} \underset{\substack{d \text{ and } l \\ \text{Tartaric acid}}}{\begin{matrix} CH(OH)COOH \\ | \\ CH(OH)COOH \end{matrix}}$$

Q. 5. Give the methods of preparation and properties of malic acid HOOC – CHOH CH_2 – COOH

Ans. Methods of Preparation

1. It is best prepared from green mountain-ash berries. The extract from the fruit is treated with lime when malic acid gets precipitated as insoluble calcium malate. Calcium malate is filtered off and treated with dilute sulphuric acid to obtain malic acid.

2. *By heating maleic acid with dilute sulphuric acid under pressure*

$$\underset{\text{maleic acid}}{\begin{matrix} CH\,COOH \\ \| \\ CH\,COOH \end{matrix}} + \begin{matrix} OH \\ | \\ H \end{matrix} \xrightarrow[\text{Pressure}]{H_2SO_4} \underset{\text{malic acid}}{\begin{matrix} HO-CH-COOH \\ | \\ CH_2-COOH \end{matrix}}$$

3. *By partial reduction of tartaric acid with hydriodic acid*

$$\underset{\text{tartaric acid}}{\begin{matrix} HO-CH-COOH \\ | \\ HO-CH-COOH \end{matrix}} + 2HI \longrightarrow \underset{\text{malic acid}}{\begin{matrix} HO-CH-COOH \\ | \\ CH_2-COOH \end{matrix}}$$

4. *Laboratory preparation.* By hydrolysis of bromosuccinic acid with moist silver oxide

$$\underset{\text{bromosuccinic acid}}{\begin{array}{l}Br-CH-COOH\\ \quad\ \ |\\ \quad CH_2-COOH\end{array}} + Ag\,OH \longrightarrow \underset{\text{malic acid}}{\begin{array}{l}HO-CH-COOH\\ \qquad\ |\\ \qquad CH_2-COOH\end{array}} + Ag\,Br$$

Properties

1. *Reduction with HI.* With HI, succinic acid is obtained

$$\underset{\text{malic acid}}{\begin{array}{l}HO-CH-COOH\\ \qquad\ |\\ \qquad CH_2-COOH\end{array}} + 2\,HI \longrightarrow \underset{\text{succinic acid}}{\begin{array}{l}CH_2-COOH\\ \ \ |\\ CH_2-COOH\end{array}} + H_2O + I_2$$

2. *Action of Heat.* On heating it forms fumaric acid, maleic acid and maleic anhydride

$$\begin{array}{l}HO-CH-COOH\\ \qquad\ |\\ \quad\ CH_2-COOH\end{array} \xrightarrow[-H_2O]{\Delta} \begin{array}{l}HOOC-C-H\\ \qquad\quad\ \|\\ \qquad H-C-COOH\end{array} \xrightarrow{\Delta} \begin{array}{l}H-C-COOH\\ \quad\ \ \|\\ H-C-COOH\end{array} \xrightarrow{\Delta} \begin{array}{l}H-C-CO\searrow\\ \quad\ \ \|\qquad\quad\ O\\ H-C-CO\nearrow\end{array}$$

3. *Oxidation.* On oxidation, it is converted into oxal-acetic acid which exhibits keto-enol tautomerism.

$$\begin{array}{l}HO-CH-COOH\\ \qquad\ |\\ \quad\ CH_2-COOH\end{array} + [O] \xrightarrow{-H_2O} \underset{\text{(keto from)}}{\underset{\text{oxal acetic acid}}{\begin{array}{l}O=C-COOH\\ \qquad |\\ \quad\ CH_2-COOH\end{array}}} \rightleftharpoons \underset{\text{(enol form)}}{\begin{array}{l}HO-C-COOH\\ \qquad\ \ \|\\ \qquad CH-COOH\end{array}}$$

Q. 6. Give the methods of preparation and properties of tartaric acid.

Ans. Methods of preparation

1. From grape juice. Tartaric acid occurs as potassium hydrogen tartarate in the grape juice. On fermentation, it separates out as a brown mass known as *argol*, which is crystallised to obtained colourless crystals of pure substance, known as *cream of tartar*. This cream of tartar is treated with lime and subsequently with calcium chloride. Calcium tartarate obtained during the process is separated by filtration. This is neutralised with calculated quantities of sulphuric acid. Calcium sulphate produced during the reaction is removed by filtration. Filtrate is concentrated to obtained crystals of dextro –tartaric acid.

$$\underset{\substack{\text{Pot.hydrogen}\\ \text{tartarate}}}{2KHC_4H_4O_6} + \underset{\substack{\text{Cal.}\\ \text{hydroxide}}}{Ca\,(OH)_2} \longrightarrow \underset{\text{Pot. Tartarate}}{K_2C_4H_4O_6} + \underset{\text{Cal. Tartarte}}{CaC_4H_4O_6} + 2H_2O$$

$$\underset{\text{Pot.tartarate}}{K_2C_4H_4O_6} + CaCl_2 \longrightarrow \underset{\text{Cal. tartarate}}{CaC_4H_4O_6} + 2KCl$$

$$\underset{\substack{\text{Calcium}\\ \text{tartarate}}}{CaC_4H_4O_6} + H_2SO_4 \longrightarrow \underset{\text{(+) Tartaric acid}}{H_2C_4H_4O_4} + CaSO_4\downarrow$$

2. By hydroxylation of maleic and fumaric acids. Maleic acid and fumaric acids on hydroxylation with alkaline pot. permanganate give meso tartaric acid and racemic tartaric acid respectively.

$$\underset{\text{Maleic acid}}{\begin{array}{l}HC-COOH\\ \ \|\\ HC-COOH\end{array}} + H_2O + O \xrightarrow{KMnO_4} \begin{array}{c}COOH\\ |\\ H-C-OH\\ |\\ H-C-OH\\ |\\ COOH\end{array} \quad \text{Mesotartaric acid}$$

$$\underset{\text{Fumaric acid}}{\begin{array}{l}\quad\ H-C-COOH\\ \qquad\quad \|\\ HOOC-C-H\end{array}} + H_2O + O \xrightarrow{KMnO_4} \underset{\text{(+) Tartaric acid}}{\begin{array}{c}COOH\\ |\\ H-C-OH\\ |\\ HO-C-OH\\ |\\ COOH\end{array}} + \underset{\text{(-) Tartaric acid}}{\begin{array}{c}COOH\\ |\\ HO-C-H\\ |\\ H-C-OH\\ |\\ COOH\end{array}}$$

3. Synthesis from acetylene

$$\underset{\text{Acetylene}}{\begin{array}{c}CH\\ |||\\ CH\end{array}} \xrightarrow[BaSO_4]{H_2/Pd} \underset{\text{Ethylene}}{\begin{array}{c}CH_2\\ ||\\ CH_2\end{array}} + Br_2 \longrightarrow \underset{\text{Dibromo ethane}}{\begin{array}{c}CH_2Br\\ |\\ CH_2Br\end{array}} \xrightarrow[KCN]{Alc.} \underset{\text{Ethylene cyanide}}{\begin{array}{c}CH_2CN\\ |\\ CH_2CN\end{array}}$$

$$\xrightarrow{H_2O/H^+} \begin{array}{c}CH_2COOH\\ |\\ CH_2COOH\end{array} \xrightarrow{H_2O/H^+} \underset{\text{Dibromo-succinic acid}}{\begin{array}{c}CHBrCOOH\\ |\\ CHBrCOOH\end{array}} \xrightarrow{AgOH} \underset{\text{Tartaric acid}}{\begin{array}{c}CH(OH)COOH\\ |\\ CH(OH)COOH\end{array}}$$

Properties

1. Optical isomerism. Due to the presence of two chiral carbon atoms in the molecule of tartaric acid, it exists in two optically active forms and one optically inactive meso form.

2. Formation of salts and esters. As this compound contains two carboxylic groups, treatment with alkalis and alcohols gives salts and esters respectively. Two series of salts and esters are obtained depending upon whether one or two hydrogens of tartaric acid are replaced.

$$\underset{\text{Tartaric acid}}{\begin{array}{c}CH(OH)COOH\\ |\\ CH(OH)COOH\end{array}} + NaOH \xrightarrow{-H_2O} \underset{\text{Monosodium tartarate}}{\begin{array}{c}CH(OH)COONa\\ |\\ CH(OH)COOH\end{array}} \xrightarrow[-H_2O]{NaOH} \underset{\text{Disodium tartarate}}{\begin{array}{c}CH(OH)COONa\\ |\\ CH(OH)COONa\end{array}}$$

$$\underset{\text{Tartaric acid}}{\begin{array}{c}CH(OH)COOH\\ |\\ CH(OH)COOH\end{array}} + C_2H_5OH \xrightarrow[-H_2O]{H^+} \underset{\text{Monoethyl tartarate}}{\begin{array}{c}CH(OH)COOC_2H_5\\ |\\ CH(OH)COOH\end{array}} \xrightarrow[-H_2O]{C_2H_5OH} \underset{\text{Diethyl tartarate}}{\begin{array}{c}CH(OH)COOC_2H_5\\ |\\ CH(OH)COOC_2H_5\end{array}}$$

It also forms mixed salts as given below

$$\underset{\text{Sod. pot. tartarate (Rochelle salts)}}{\begin{array}{c}CH(OH)COONa\\ |\\ CH(OH)COOK\end{array}} \qquad \underset{\text{Pot. antimonyl tartarate (Tartar emetic)}}{\begin{array}{c}CH(OH)COO.SbO\\ |\\ CH(OH)COOK\end{array}}$$

3. Dehydrogenation. On treatment with hydrogen peroxide in the presence of ferrous salts, tartaric acid is dehydrogenated to dihydroxy fumaric acid.

$$\underset{\text{Tartaric acid}}{\begin{array}{c}CH(OH)COOH\\ |\\ CH(OH)COOH\end{array}} \xrightarrow[Fe^{2+}]{H_2O_2} \underset{\text{Dihydroxy fumaric acid}}{\begin{array}{c}HO-C-COOH\\ ||\\ HOOC-C-OH\end{array}}$$

4. Action of heat

$$\begin{array}{c}CH(OH)COOH\\ |\\ CH(OH)COOH\end{array} \xrightarrow{\Delta} \underset{\text{Pyruvic acid}}{CH_3-CO-COOH} + CO_2 + H_2O$$

On prolonged heating at 423 K, cyclic anhydride is obtained.

$$\underset{\text{Tartaric acid}}{\begin{array}{c}CH(OH)COOH\\ |\\ CH(OH)COOH\end{array}} \xrightarrow[\text{Prolonged heating}]{\Delta} \begin{array}{l}CH(OH)-C{=}O\\ |\qquad\qquad\quad >O\\ CH(OH)-C{=}O\end{array}$$

5. Reduction. Hydrogen iodide reduces tartaric acid first to malic acid and then to succinic acid

$$\underset{\text{Tartaric acid}}{\begin{array}{l}CH(OH)COOH\\ |\\ CH(OH)COOH\end{array}} \xrightarrow{HI} \underset{\text{Malic acid}}{\begin{array}{l}CH_2COOH\\ |\\ CH(OH)COOH\end{array}} \xrightarrow{HI} \underset{\text{Succinic acid}}{\begin{array}{l}CH_2COOH\\ |\\ CH_2COOH\end{array}}$$

6. Oxidation

$$\underset{\text{Tartaric acid}}{\begin{array}{l}CH(OH)COOH\\ |\\ CH(OH)COOH\end{array}} \xrightarrow{[O]} \underset{\text{Tartronic acid}}{\begin{array}{l}CH(OH)COOH\\ |\\ COOH\end{array}} \xrightarrow{[O]} \underset{\text{Oxalic acid}}{\begin{array}{l}COOH\\ |\\ COOH\end{array}}$$

Q. 7. Give the methods of preparation and properties of citric acid.

(A.N. Bahuguna 2000; Kumaon 2000; Awadh 2000)

Ans. Methods of Preparation

1. From Lemon Juice. Lemon juice is boiled to coagulate proteinous matter which is filtered off. The hot filtrate is neutralised with lime. Citric acid in the filtrate is converted into calcium citrate which is filtered off. Calcium citrate is decomposed to citric acid by treating with a calculated amount of sulphuric acid. Solution containing citric acid is concentrated to crystallise out pure citric acid.

2. From Molasses. Fermentation of molasses with the micro-organism *Aspergillus niger* in the presence of some inorganic salts yields citric acid. The fermentation takes place at a pH of 1-2 and takes about 10 days at a temperature 301-303 K.

3. Synthesis of citric acid from glycerol

$$\underset{\text{Glycerol}}{\begin{array}{l}CH_2OH\\ |\\ CHOH\\ |\\ CH_2OH\end{array}} \xrightarrow{Cl_2} \underset{\text{1, 3-dichloro-propanol-2}}{\begin{array}{l}CH_2Cl\\ |\\ CHOH\\ |\\ CH_2Cl\end{array}} \xrightarrow[HNO_3]{[O]} \underset{\alpha,\alpha'\text{ Dichloro-acetone}}{\begin{array}{l}CH_2Cl\\ |\\ C=O\\ |\\ CH_2Cl\end{array}} \xrightarrow[HNO_3]{HCN} \underset{\alpha,\alpha'\text{ Dichloroacetone cyanohydrin}}{\begin{array}{l}CH_2Cl\\ |\\ C\langle^{OH}_{CN}\\ |\\ CH_2Cl\end{array}}$$

$$\xrightarrow{KCN} \begin{array}{l}CH_2CN\\ |\\ C\langle^{OH}_{CN}\\ |\\ CH_2CN\end{array} \xrightarrow{H_2O/H^+} \underset{\text{Citric acid}}{\begin{array}{l}CH_2COOH\\ |\\ C(OH)COOH\\ |\\ CH_2COOH\end{array}}$$

Properties

1. Formation of salts and esters. As it is a tricarboxylic compound, there is a possibility to form three series of salts and esters depending upon whether one, two or three carboxy hydrogens are substituted

$$\begin{array}{l}CH_2COOH\\ |\\ C(OH)COOH\\ |\\ CH_2COOH\end{array} \xrightarrow{NaOH} \underset{\text{Mono sodium citrate}}{\begin{array}{l}CH_2COONa\\ |\\ C(OH)COOH\\ |\\ CH_2COOH\end{array}} \xrightarrow{NaOH} \underset{\text{Di. sodium citrate}}{\begin{array}{l}CH_2COONa\\ |\\ C(OH)COOH\\ |\\ CH_2COONa\end{array}}$$

$$\xrightarrow{NaOH} \underset{\text{Tri. sodium citrate}}{\begin{array}{l}CH_2COONa\\ |\\ C(OH)COONa\\ |\\ CH_2COONa\end{array}}$$

Similarly formation of mono, di and tri alkyl citrate takes place on reaction with an alcohol.

2. Acetylation. Acetyl derivatives are obtained with acetic anhydride or acetyl chloride.

$$\begin{array}{l} CH_2COOH \\ | \\ C(OH)COOH \\ | \\ CH_2COOH \end{array} + CH_3COCl \longrightarrow \begin{array}{l} CH_2COOH \\ | \\ C(OCOCH_3)\,COOH \\ | \\ CH_2COOH \end{array} + HCl$$

Citric acid → Monoacetyl citric acid

3. Reduction. On reduction with HI, citric acid gives tricarballylic acid.

$$\begin{array}{l} CH_2COOH \\ | \\ C(OH)COOH \\ | \\ CH_2COOH \end{array} + 2HI \longrightarrow \begin{array}{l} CH_2COOH \\ | \\ CHCOOH \\ | \\ CH_2COOH \end{array} + H_2O + I_2$$

Citric acid → Tricarballylic acid

4. Action of fuming sulphuric acid

$$\begin{array}{l} CH_2COOH \\ | \\ C(OH)COOH \\ | \\ CH_2COOH \end{array} \xrightarrow[H_2SO_4]{\text{Fuming}} \begin{array}{l} CH_2COOH \\ | \\ CO \\ | \\ CH_2COOH \end{array} + H_2O + CO$$

Acetone dicarboxylic acid

5. Action of heat. When heated to 423 K, citric acid loses a water molecule forming aconitic acid. This reaction is characteristic of B-hydroxy acids.

$$\begin{array}{l} CH_2COOH \\ | \\ C(OH)COOH \\ | \\ CH_2COOH \end{array} \xrightarrow[-H_2O]{\Delta} \begin{array}{l} CH_2COOH \\ \| \\ C-COOH \\ | \\ CH_2COOH \end{array}$$

Aconitic acid

On heating to higher temperatures, aconitic acid further loses a molecule of carbon dioxide to form a mixture of itaconic acid and citraconic acid and mesaconic acid which change into their anhydrides as shown below.

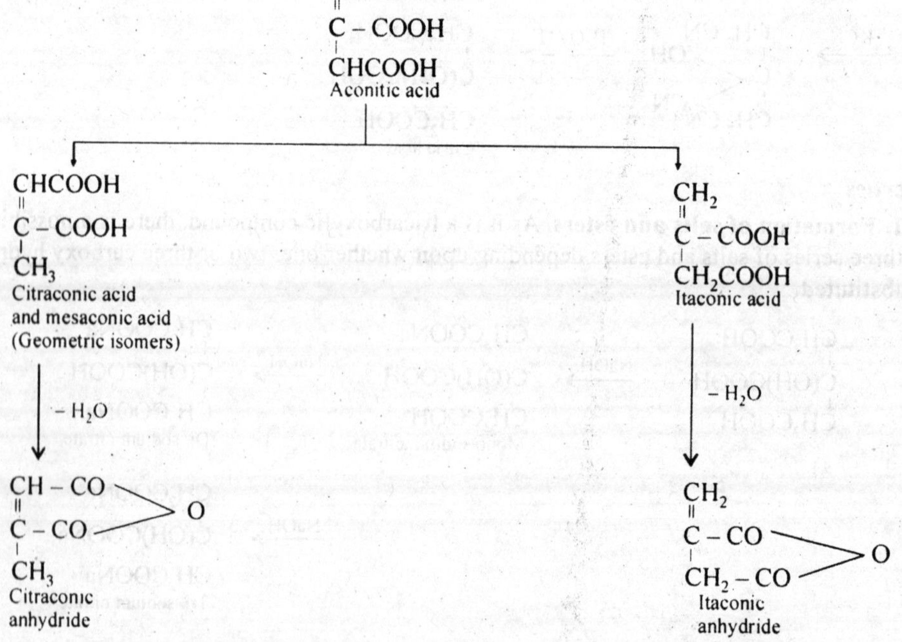

6. Complex Formation. Citric acid prevents the precipitation of heavy metal hydroxides by forming soluble complexes. This property of citric acid is used in the making of *Benedict solution* which contains a mixture of copper sulphate, sodium carbonate and sodium citrate.

Q. 8. Describe a reaction sequence depicting synthesis of diethyl malonate.

(*Sri Venkateswara, 2004; Vidyasagar, 2007; Nagpur, 2008; Patna, 2010*

Ans. Synthesis of diethyl malonate

1. Malonic ester is prepared by passing dry HCl gas through a solution of pot. cyano acetate in ethanol

$$\underset{\text{Pot. cyanoacetate}}{\begin{matrix} CH_2COOK \\ | \\ CN \end{matrix}} + 2\,C_2H_5OH + 2HCl \longrightarrow \underset{\text{Diethyl malonate}}{\begin{matrix} CH_2COOC_2H_5 \\ | \\ COOC_2H_5 \end{matrix}} + NH_4Cl + KCl$$

Cyanoacetate itself, is prepared from acetic acid as under:

$$\underset{\text{Acetic acid}}{CH_3COOH} \xrightarrow{Cl_2,\ P} \underset{\text{Chloroacetic acid}}{CH_2\begin{matrix} Cl \\ COOH \end{matrix}} \xrightarrow{K_2CO_3} \underset{\text{Pot. chloroacetate}}{CH_2\begin{matrix} Cl \\ COOK \end{matrix}} \xrightarrow{KCN} CH_2\begin{matrix} CN \\ COOK \end{matrix}$$

2. **From malonic acid.** Malonic acid is esterified with an alcohol in the presence of conc. H_2SO_4 to yield malonic ester

$$CH_2\begin{matrix} COOH \\ COOH \end{matrix} + 2\,C_2H_5OH \xrightarrow{H^+} CH_2\begin{matrix} COOC_2H_5 \\ COOC_2H_5 \end{matrix} + 2H_2O$$

Q. 9. Describe properties of malonic ester.

(*Annamalai, 2003; Bangalore, 2004; Amravati, 2004; Patna, 2010*

Ans. Acidic nature of methylene hydrogens

The methylene group in the molecule of malonic ester contains hydrogens which are ionisable and are removed in the form of hydrogen ions. This is because the anion left after releasing the proton is stabilized by resonance as shown below.

$$\underset{\text{Active methylene group}}{C_2H_5O-\overset{\overset{O}{\|}}{C}-CH_2-\overset{\overset{O}{\|}}{C}-OC_2H_5} \underset{}{\overset{-H^+}{\rightleftharpoons}} \left[\begin{matrix} C_2H_5O-\overset{\overset{O}{\|}}{C}-\overset{\ominus}{C}H-\overset{\overset{O}{\|}}{C}-OC_2H_5 \\ \updownarrow \\ C_2H_5O-\overset{\overset{O^{\ominus}}{|}}{C}=CH-\overset{\overset{O}{\|}}{C}-OC_2H_5 \\ \updownarrow \\ C_2H_5O-\overset{\overset{O}{\|}}{C}-CH=\overset{\overset{O^{\ominus}}{|}}{C}-OC_2H_5 \end{matrix} \right]$$

$$C_2H_5O-\overset{\overset{O}{\|}}{C}=CH=\overset{\overset{O}{\|}}{C}-OC_2H_5 \ (\ominus \text{ delocalised over both O}) \overset{\text{Equivalent to}}{\rightleftharpoons}$$

This imparts synthetic importance to malonic ester. When treated with sodium ethoxide, it forms the sodium salt. Sodium salt of malonic ester is capable of reacting with alkyl or acyl halides to give substituted products.

$$CH_2(COOC_2H_5)_2 \xrightarrow{C_2H_5ONa} [CH(COOC_2H_5]^- Na^+$$

RCOCl ↙ ↘ RX

$$\underset{\substack{|\\R}}{CH(COOC_2H_5)_2} \qquad \underset{\substack{|\\CO\\|\\R}}{CH(COOC_2H_5)_2}$$

Mono alkyl malonic ester — Mono acyl malonic ester

There is a possibility of further substitution as still one more active hydrogen is left on the molecule after monosubstitution. Therefore the monosubstituted product may be subjected to the above treatment once more to obtain disubstituted products.

$$\underset{\substack{|\\R}}{CH(COOC_2H_5)_2} \xrightarrow[RX]{C_2H_5CNa} \overset{\substack{R'\\|}}{\underset{\substack{|\\R}}{C(COOC_2H_5)_2}}$$

Disubstitutd product

Action of heat on substituted malonic acid

Another property of malonic ester and substituted malonic ester is that out of two carboxylic groups, one can be decarboxylated easily. This property is utilized in a number of synthetic reactions.

$$\underset{\text{Alkyl malonic ester}}{RCH(COOC_2H_5)_2} \xrightarrow{NaOH} RCH\,(COONa)_2$$

$$\xrightarrow{NaOH} RCH(COOH)_2 \xrightarrow{\Delta} RCH_2COOH + CO_2$$

Based on the above two properties or principles, following synthetic reactions can be performed.

1. Synthesis of monoalkylacetic acid. Malonic ester is treated with an appropriate alkyl halide after converting the ester into sodium salt. Alkylated malonic ester is then hydrolysed and heated. Heating brings about decarboxylation of one of the carboxyl group. Thus in order to prepare RCH_2COOH, the alkyl halide to be used is RX.

$$\underset{\text{Malonic ester}}{CH_2(COOC_2H_5)_2} \xrightarrow{C_2H_5ONa} \underset{\text{Sod. salt}}{[CH(COOC_2H_5)_2]^- Na^+}$$

$$\xrightarrow{RX} \underset{\substack{|\\R\\\text{Alkyl malonic ester}}}{CH\,(COOC_2H_5)_2} \xrightarrow{NaOH} \underset{\substack{|\\R\\\text{Sod. salt of}\\\text{alkyl malonic acid}}}{CH\,(COONa)_2}$$

$$\xrightarrow{HCl} \underset{\substack{|\\R\\\text{Alkyl malonate}}}{CH\,(COOH)_2} \xrightarrow{\Delta} \underset{\text{Monocarboxy acid}}{RCH_2\,COOH + CO_2}$$

Butyric acid can be prepared by treating malonic ester with CH_3CH_2I followed by hydrolysis & decarboxylation

$$CH_2(COOC_2H_5)_2 \xrightarrow[CH_3CH_2I]{C_2H_5ONa} \underset{\underset{CH_3}{|}}{\underset{CH_2}{|}}{CH(COOC_2H_5)_2}$$

$$\xrightarrow{\text{Hydrolysis}} \underset{\underset{CH_3}{|}}{\underset{CH_2}{|}}{CH(COOH)_2} \xrightarrow{\Delta} CH_3CH_2CH_2COOH + CO_2$$

Isovaleric acid $CH_3 - \underset{\underset{CH_3}{|}}{CH} - CH_2\,COOH$ can be prepared by treating malonic ester with isopropyl bromide followed by hydrolysis and decarboxylation.

2. Synthesis of dialkyl acetic acid. Malonic ester is first converted into dialkyl malonic ester. For this sodium ethoxide and alkyl halide are taken in double the quantities of malonic ester (in molar quantities). Dialkyl malonic ester is then hydrolysed and decarboxylated.

$$\underset{\text{Malonic ester}}{CH_2(COOC_2H_5)_2} \xrightarrow[CH_3I]{C_2H_5ONa} \underset{\text{Methyl malonic ester}}{\underset{\underset{CH_3}{|}}{CH(COOC_2H_5)_2}}$$

$$\xrightarrow[CH_3CH_2I]{C_2H_5ONa} \underset{\text{Ethyl methyl malonic ester}}{\overset{CH_3-CH_3}{\overset{|}{\underset{\underset{CH_3}{|}}{C(COOC_2H_5)_2}}}} \xrightarrow{\text{Hydrolysic}} \underset{\text{Ethyl methyl malonic acid}}{\overset{CH_2-CH_3}{\overset{|}{\underset{\underset{CH_3}{|}}{C(COOH)_2}}}}$$

$$\xrightarrow{\Delta} \underset{\text{2-Methyl butanoic acid}}{CH_3 - CH_2 - \underset{\underset{CH_3}{|}}{CHCOOH}}$$

3. Synthesis of-keto acids $RCOCH_2COOH$. Sodium salt of malonic ester is treated with acyl chloride followed by hydrolysis and decarboxylation.

$$CH_2(COOC_2H_5)_2 \xrightarrow{C_2H_5ONa} [CH(COOC_2H_5)_2]^- Na^+$$

$$\xrightarrow{CH_3COCl} \underset{\underset{CH_3}{|}}{\underset{CO}{|}}{CH(COOC_2H_5)_2} \xrightarrow{\text{Hydrolysic}} \underset{\underset{CH_3}{|}}{\underset{CO}{|}}{CH(COOH)_2}$$

$$\xrightarrow{-CO_2,\ \Delta} \underset{\text{Acetyl acetic acid}}{CH_3COCH_2COOH}$$

4. Synthesis of dicarboxylic acids.

(*a*) Using α-haloester

$$\underset{\text{Malonic ester}}{CH_2(COOC_2H_5)_2} \xrightarrow{C_2H_5ONa} \underset{\text{Sod. salt of malonic ester}}{[CH(COOC_2H_5)_2]^- Na^+}$$

$$\xrightarrow[\text{(a-bromoethyl acetate)}]{BrCH_2COOC_2H_5} \begin{array}{l} CH(COOC_2H_5)_2 \\ | \\ CH_2COOC_2H_5 \end{array} \xrightarrow{\text{Hydrolysis}} \begin{array}{l} CH_2(COOH)_2 \\ | \\ CH_2COOH \end{array}$$

$$\xrightarrow{\Delta} \underset{\text{Succinic acid}}{\begin{array}{l} CH_2COOH \\ | \\ CH_2COOH \end{array}} + CO_2$$

(*b*) Using I_2

$$\underset{\text{Malonic ester}}{CH_2(COOC_2H_5)_2} \xrightarrow{C_2H_5ONa} \underset{\text{Sod. salt}}{[CH(COOC_2H_5)_2]^- Na^+}$$

$$\xrightarrow[-2NaI]{I_2} \begin{array}{l} CH(COOC_2H_5)_2 \\ | \\ CH(COOC_2H_5)_2 \end{array} \xrightarrow{\text{Hydrolysis}} \begin{array}{l} CH(COOH)_2 \\ | \\ CH(COOH)_2 \end{array}$$

$$\xrightarrow{\Delta} \underset{\text{Succinic acid}}{\begin{array}{l} CH_2COOH \\ | \\ CH_2COOH \end{array}}$$

(*c*) Using α, ω dihalide

$$\underset{\text{Ethylene bromide}}{\begin{array}{l} CH_2Br \\ | \\ CH_2Br \end{array}} + \underset{\text{Sod. salt of malonic ester}}{2[CH(COOC_2H_5)_2]^- Na^+} \longrightarrow \begin{array}{l} CH_2CH(COOC_2H_5)_2 \\ | \\ CH_2CH(COOC_2H_5)_2 \end{array}$$

$$\xrightarrow{\text{Hydrolysis}} \begin{array}{l} CH_2CH(COOH)_2 \\ | \\ CH_2CH(COOH)_2 \end{array} \xrightarrow{\Delta} \underset{\text{Adipic acid}}{\begin{array}{l} CH_2CH_2COOH \\ | \\ CH_2CH_2COOH \end{array}}$$

5. Synthesis of cycloalkane carboxylic acid

$$\underset{\text{1, 3 dibromo-propane}}{\begin{array}{l} CH_2Br \\ | \\ CHCH_2Br \end{array}} + \underset{\text{Sod. salt of malonic ester}}{2[CH(COOC_2H_5)_2]^- Na^+} \longrightarrow \begin{array}{l} CH_2-CH(COOC_2H_5)_2 \\ | \\ CH_2CH_2Br \end{array}$$

$$\xrightarrow{C_2H_5ONa} \begin{array}{ll} CH_2 - & C\,(COOC_2H_5)_2 \\ | & | \\ CH_2 - & CH_2 \end{array} \xrightarrow{\text{Hydrolysis}} \begin{array}{ll} CH_2 - & C\,(COOH)_2 \\ | & | \\ CH_2 - & CH_2 \end{array}$$

$$\xrightarrow[-CO_2]{} \underset{\text{Cyclobutane carboxylic acid}}{\begin{array}{ll} CH_2 - & CHCOOH \\ | & | \\ CH_2 - & CH_2 \end{array}}$$

6. Synthesis of α, β unsaturated acids. Malonic ester and its derivatives condense with aldehydes or ketones. Resulting unsaturated esters on hydrolysis and decarboxylation give α, β unsaturated acids.

$$C_6H_5-CHO \xrightarrow[\text{Pyridine}]{CH_2(COOC_2H_5)_2} C_6H_5-CH=C(COOC_2H_5)_2$$

$$\xrightarrow{\text{Hydrolysis}} C_6H_5-CH=C(COOH)_2 \xrightarrow{\Delta} \underset{\text{Cinnamic acid}}{C_6H_5-CH=CHCOOH}$$

7. Synthesis of amino acids. Malonic ester is treated with nitrous acid. α-oximino malonic ester formed is reduced to give aminomalonic ester. This on hydrolysis and decarboxylation gives amino acetic acid.

$$\underset{\text{Nitrous acid}}{HO-N=O} + \underset{\text{Malonic ester}}{H_2C(COOC_2H_5)_2} \longrightarrow \underset{\text{Oximino malonic ester}}{HO-N=C(COOC_2H_5)_2}$$

$$\xrightarrow{\text{Zn / Acetic acid}} \underset{\text{Amino malonic ester}}{H_2N-CH(COOC_2H_5)_2} \xrightarrow[-HCl]{CH_3COCl}$$

$$\underset{\text{N-acetyl amino malonic ester}}{CH_3CONH-CH(COOC_2H_5)_2} \xrightarrow{\text{Hydrolyis}} H_2N-CH(COOH)_2 \xrightarrow[\text{Heat}]{-CO_2} \underset{\text{Amino acetic acid}}{H_2N-CH_2COOH}$$

8. Synthesis of heterocyclic compounds. Malonic ester undergoes condensation with urea in the presence of sodium ethoxide to form malonyl urea commonly known as barbituric acid

$$\underset{\text{Urea}}{O=C\begin{matrix} NH_2 \\ NH_2 \end{matrix}} + \begin{matrix} H_5C_2O-\overset{O}{\overset{\|}{C}} \\ H_5C_2O-\underset{O}{\underset{\|}{C}} \end{matrix} CH_2 \longrightarrow \underset{\text{Malonyl urea (Barbituric acid)}}{O=C\begin{matrix} NH-\overset{O}{\overset{\|}{C}} \\ NH-\underset{O}{\underset{\|}{C}} \end{matrix} CH_2}$$

Substituted malonic esters react with urea to give substituted barbituric acids known as barbiturates. Such compounds are used medicinally as sleep-inducing drugs (hypnotics).

Q. 10. How will you convert

(*a*) Acetylene ⟶ Tartaric acid

(*b*) Glycerol ⟶ Citric acid

(*c*) Diethyl malonate ⟶ Succinic acid

(*d*) Tartaric acid ⟶ Maleic acid.

Ans. (*a*) Acetylene ⟶ Tartaric acid

$$\underset{\text{Acetylene}}{\begin{matrix} CH \\ ||| \\ CH \end{matrix}} \xrightarrow[BaSO_4]{H_2/Pd} \underset{\text{Ethylene}}{\begin{matrix} CH_2 \\ || \\ CH_2 \end{matrix}} + Br_2 \longrightarrow \underset{\text{Dibromoethane}}{\begin{matrix} CH_2Br \\ | \\ CH_2Br \end{matrix}} \xrightarrow[KCN]{Alc.} \underset{\text{Ethylene cyanide}}{\begin{matrix} CH_2CN \\ | \\ CH_2CN \end{matrix}}$$

$$\xrightarrow{H_2O/H^+} \underset{\text{Succinic acid}}{\begin{array}{l} CH_2COOH \\ | \\ CH_2COOH \end{array}} \xrightarrow{Br_2/P} \underset{\text{Dibromo-succinic acid}}{\begin{array}{l} CH_2BrCOOH \\ | \\ CH_2BrCOOH \end{array}} \xrightarrow{AgOH} \underset{\text{Tartaric acid}}{\begin{array}{l} CH(OH)COOH \\ | \\ CH(OH)COOH \end{array}}$$

(b) Glycerol ⟶ Citric acid

$$\underset{\text{Glycerol}}{\begin{array}{l} CH_2OH \\ | \\ CHOH \\ | \\ CH_2OH \end{array}} \xrightarrow{HCl} \underset{\text{1,3 Dichloro-propanol-2}}{\begin{array}{l} CH_2Cl \\ | \\ CHOH \\ | \\ CH_2Cl \end{array}} \xrightarrow[HNO_3]{[O]} \underset{\alpha\ \alpha'\text{ dichloro-acetone}}{\begin{array}{l} CH_2Cl \\ | \\ C=O \\ | \\ CH_2Cl \end{array}} \xrightarrow{HCN} \underset{\alpha\ \alpha'\text{ Dichloro-acetone cyanohydrin}}{\begin{array}{l} CH_2Cl \\ | \\ C \langle^{OH}_{CN} \\ | \\ CH_2Cl \end{array}}$$

$$\xrightarrow{KCN} \begin{array}{l} CH_2CN \\ | \\ C \langle^{OH}_{CN} \\ | \\ CH_2CN \end{array} \xrightarrow{H_2O/H} \underset{\text{Citric acid}}{\begin{array}{l} CH_2COOH \\ | \\ C(OH)COOH \\ | \\ CH_2COOH \end{array}}$$

(c) Diethyl malonate ⟶ Succinic acid

$$\underset{\text{Malonic ester}}{CH_2(COOC_2H_5)_2} \xrightarrow{C_2H_5ONa} \underset{\text{Sod. salt}}{[CH(COOC_2H_5)_2]Na}$$

$$\xrightarrow{BrCH_2COOC_2H_5} \begin{array}{l} CH(COOC_2H_5)_2 \\ | \\ CH\ COOC_2H_5 \end{array} \xrightarrow{\text{Hydrolysis}} \begin{array}{l} CH(COOH)_2 \\ | \\ CH_2COOH \end{array}$$

$$\xrightarrow{\Delta} \underset{\text{Succinic acid}}{\begin{array}{l} CH_2COOH \\ | \\ CH_2COOH \end{array}} + CO_2$$

(d) Tartaric acid ⟶ Maleic acid.

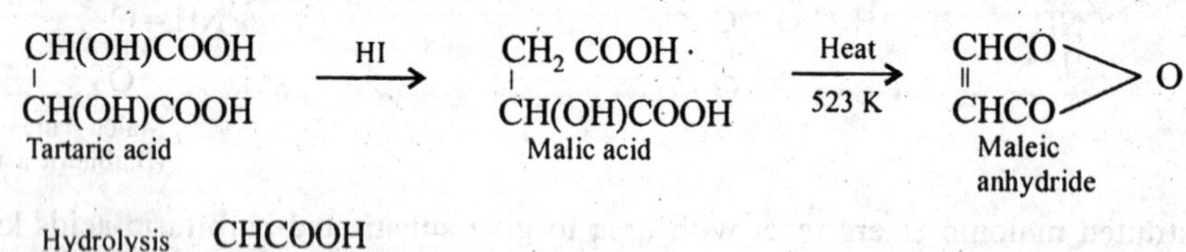

$$\xrightarrow{\text{Hydrolysis}} \underset{\text{Maleic acid}}{\begin{array}{l} CHCOOH \\ \| \\ CHCOOH \end{array}}$$

Q. 11. Sketch the transformation of diethyl malonate into the following:

(i) Succinic acid **(ii) Adipic acid** **(iii) Barbituric acid**

(*M. Dayanand, 2000; Kumaon, 2000; Kurukshetra, 2000; Jabalpur 2003, Utkal 2003*)

Ans. See Q. 9.

Q. 12. Convert diethyl malonate into

(i) Glutaric acid **(ii) Malonyl urea** **(iii) Cinnamic acid**

(*Kurukshetra 2000; M. Dayanand, 2000*)

Ans. See Q. 9.

Q. 13. Prepare the following from diethyl malonate

(*i*) Adipic acid **(*ii*) Cinnamic acid**

Ans. See Q. 9.

Q. 14. Write a note on stereisomerism in tartaric acid.

Ans. Stereoisomers (optical isomers) of tartaric acid are:

COOH	COOH	COOH
H – C – OH	HO – C – H	H – C – OH
HO – C – H	H – C – OH	H – C – OH
COOH	COOH	COOH
d-tartaric acid	*l*-tartaric acid	*meso*-tartaric acic

Q. 15. Name a tricarboxylic acid and write its structure.

Ans. Citric acid is a tricarboxylic acid. For structure, see Q. 6.

Q. 16. What is the action of heat on

(*i*) glutaric acid

(*ii*) adipic acid

Ans. (*i*) $CH_2<\begin{matrix}CH_2COOH\\CH_2COOH\end{matrix} \xrightarrow{\Delta} CH_3CH_2CH_2COOH + CO_2\uparrow$

Glutaric acid → Butyric acid

(*ii*) $\begin{matrix}CH_2-CH_2COOH\\|\\CH_2-CH_2-COOH\end{matrix} \xrightarrow{\Delta} CH_3CH_2CH_2CH_2COOH + CO_2\uparrow$

Valeric acid

Q. 17. Prove that citric acid is a monohydroxy tribasic acid and has three –COOH groups on different carbon atoms.

Ans. 1. Citric acid forms three series of salts and esters corresponding to successive removal of one, two or three carboxy hydrogens.

2. On treatment with HI, citric acid produces tricarballylic acid which is known to be a tricarboxylic acid with three carboxy groups on different carbon atoms. Action with HI, reduces one –OH group to –H.

The above reactions prove that citric acid is a monohydroxy tribasic acid.

Q. 18. How will you synthesize following compounds from diethyl malonate?

(*i*) Butyric acid (*ii*) Succinic acid (*iii*) Barbituric acid (*iv*) crotonic acid (*v*) Amino acetic acid. (*Kurukshetra, 2000; Meerut, 2000; Agra 2004; Shivaji 2004*)

Ans. (*i*) See Q. 9 (*i*) (*ii*) See Q. 11 (*i*) (*iii*) See Q. 11 (*iv*) See Q. 9 (6), take acetaldehyde in place of benzaldehyde (*v*) See Q. 9 (7).

Q. 19. Starting from diethyl malonate, how will you prepare :

(*i*) An unsaturated acid (*ii*) a dicarboxylic acid (*iii*) Malonyl urea (*iv*) a fatty acid.

(*Kerala, 2000*)

Ans. (*i*) See Q. 9 (6) (*ii*) See Q. 10 (c) (*iii*) See Q. 12 (*ii*) (*iv*) See Q. 13 (*i*)

Q. 20. How will you do the following conversions ? (*Meerut 2000*)

(*i*) Acetic acid ⟶ Malonic acid

(*ii*) Glycerol ⟶ Citric acid

Ans.

(*i*) $CH_3COOH \xrightarrow{Cl_2,\ P} CH_2\begin{cases}Cl \\ COOH\end{cases} \xrightarrow{KCN} CH_2\begin{cases}CN \\ COOH\end{cases} \xrightarrow{KMnO_4} CH_2\begin{cases}COOH \\ COOH\end{cases}$

Acetic acid — Chloroacetic acid — Malonic acid

(*ii*) See Q. 7(3)

Q.21. Write the names and structures of three isomeric dicarboxylic acid obtained when three isomeric xylenes are oxidized. Indicate, which diacarboxylic acid gives anhydride on dehydration ? (*Lucknow, 2000*)

Ans. 1. o-Xylene ($C_6H_4(CH_3)_2$, 1,2-) $\xrightarrow{O}$ Phthalic acid ($C_6H_4(COOH)_2$, 1,2-) $\xrightarrow{\text{dehydration}}$ Phthalic anhydride ($C_6H_4(CO)_2O$)

o-Xylene — Phthalic acid — Phthalic anhydride

2. m-Xylene ($C_6H_4(CH_3)_2$, 1,3-) $\longrightarrow$ Isophthalic acid ($C_6H_4(COOH)_2$, 1,3-)

m-Xylene — Isophthalic acid

3. p-Xylene ($C_6H_4(CH_3)_2$, 1,4-) $\longrightarrow$ Terephthalic acid ($C_6H_4(COOH)_2$, 1,4-)

p-Xylene — Terephthalic acid

Q. 22. How many optically active stereoisomers of tartaric acid are possible? Draw their configurations. (*Delhi, 2002; Nagpur, 2002*)

Ans.

I (*l*-form)	II (*d*-form)	III (meso)
COOH	COOH	COOH
HO – C – OH	H – C – OH	H – C – OH
H – C – OH	HO – C – OH	H – C – OH
COOH	COOH	COOH

I and II are optically active forms while III is optically non active.

Q. 23. How will you distinguish between maleic acid and fumaric acid ? (*Delhi 2005*)

Ans. Maleic acid on heating gives maleic anhydride (m.p. 250°C). No anhydride is formed when furmaric acid is heated.

$$\begin{matrix}CHCOOH \\ \| \\ CHCOOH\end{matrix} \xrightarrow{\text{Heat}} \begin{matrix}CHCO \\ \| \\ CHCO\end{matrix}\!\!>O + H_2O$$

Maleic acid — Maleic anhydride

24

FATS, OILS, SOAPS, DETERGENTS AND WAXES

Q. 1. Describe the occurrence and composition of fats and oils.

OR

Write briefly about triglycerides. (*Panjab 2003*)

Ans. Occurrence

Fats and oils are found in abundance in plants and animals. Seeds store the fats and oils in plants whereas in animals, they are located under the skin and in muscles.

Composition

Fats and oils are triesters of glycerol and their general structure can be written as follows:

$$\begin{array}{l} CH_2OOCR \\ | \\ CHOOCR' \\ | \\ CH_2OOCR'' \end{array}$$

where R, R′, R″ are parts of the higher fatty acids forming ester with glycerol. The three acids could be same, two of them same, or all different from one another. Accordingly, they are called simple or mixed glycerides.

$$\begin{array}{l} CH_2OCOC_{15}H_{31} \\ | \\ CHOCOC_{15}H_{31} \\ | \\ CH_2OCOC_{15}H_{31} \end{array} \qquad \begin{array}{l} CH_2OCOC_{15}H_{31} \\ | \\ CHOCOC_{17}H_{35} \\ | \\ CH_2OCOC_{15}H_{31} \end{array}$$

Glyceryl tripalmitate (Tripalmitin) Simple glyceride — Glyceryl stearato dipalmitate (Mixed glyceride)

Natural fats and oils contain a mixture of mixed and simple glycerides.

Q. 2. What is the difference between a fat and an oil?

(*Purvanchal, 2003; Sambalpur, 2004; Garhwal, 2010*)

Ans. 1. Oils are liquid and fats are solid at room temperature (293 K). However, we cannot rely on this definition of oils and fats. Coconut, for example, would look to be an oil in summer and fat in winter, although this difference in physical state is due to wide temperature variation in summer and winter.

2. It is found that fats contain a high percentage of saturated acids in glycerides whereas oils contain a high percentage of unsaturated acids in glycerides. Thus fats are saturated and oils are unsaturated in nature. The important saturated acids in fats are palmitic, stearic and lauric acids. Oleic, linoleic and linolenic acids are important unsaturated acids in oils.

Q. 3. Name some substances which look like or behave like fats and oils but are not accepted as such.

Ans. Following substances are not accepted in the category of oils and fats:

Mineral oils. Kerosene oil, diesel oil etc. have mineral origin. They are mixtures of hydrocarbons. Hence they are not acceptable as fats and oils.

Essential oils. Pleasant-smelling liquids found in plants, clove oil, lemon oil, turpentine oil, although resembling fats and oils in physical properties, are again not accepted in this category as such.

Q. 4. Describe the extraction of fats and oils from animals and plants.

Ans. Following steps are employed for the extraction of fats and oils from animal or plant sources:

1. Chopping or rendering. This method is employed for extraction of fats and oils from animals. Animal tissues richer in fats are chopped off and subjected to heat-treatment either as such or in combination with water till the fat melts and form a separate layer.

2. Crushing. This method is employed for extraction of fats or oils from plant sources. Oil-seeds like cotton-seed, castor seeds, mustard-seeds and linseed are crushed between steel rollers and crushed seeds are pressed in a hydraulic press. The oil is extracted and the residue called **deoiled-cake** is left.

3. Solvent extraction. In another process, the crushed animal tissues and seeds are extracted with suitable solvents like benzene, petroleum, ether etc. to take out the maximum amount of fats and oils. The solvent is distilled off leaving behind the fat or oil. The same solvents is recycled. This method gives better yield of fats and oils than conventional crushing and passing through rollers.

4. Refining. The fat or oil obtained above may not be in pure state. It is purified as follows:

(*a*) **Neutralisation.** Some free fatty acid might be produced as a result of heating or crushing or hydrolysis. This is neutralised with small amounts of alkali.

(*b*) The fat or oil is decolourised by warming with animal charcoal.

(*c*) Odour is removed from the oil or fat by passing superheated steam through it and then separating the aqueous layer from the oily layer.

Q. 5. Describe chemical properties of fats and oils.

Ans. Following are some of the important chemical properties of oils and fats:

1. Hydrolysis. Fats and oils are hydrolysed by the action of acids, alkalis or superheated steam.

$$\begin{array}{l} CH_2OCOR \\ | \\ CHOCOR' + 3H_2O \\ | \\ CH_2OCOR'' \end{array} \longrightarrow \begin{array}{l} CH_2OH \\ | \\ CHOH \\ | \\ CH_2OH \\ \text{Glycerol} \end{array} \begin{array}{r} + RCOOH + R'COOH \\ \\ + R''COOH \\ \text{Mixture of fatty acids} \end{array}$$

It hydrolysis is carried out by alkali, a mixture of alkali salts of fatty acids and glycerol is obtained. Salts of fatty acids are used as soaps and therefore, this hydrolysis using alkali is also called **saponification.**

$$\begin{array}{l} CH_2OCOR \\ | \\ CHOCOR' + 3NaOH \\ | \\ CH_2OCOR'' \end{array} \longrightarrow \begin{array}{l} CH_2OH \\ | \\ CHOH \\ | \\ CH_2OH \\ \text{Glycerol} \end{array} \begin{array}{r} RCOO\,Na + R'COONa \\ + R''COONa \\ \text{(Soap)} \end{array}$$

This reaction forms the basis of soap industry. Luxury soaps, beauty soaps and shaving soaps are prepared by using potassium hydroxides instead of sodium hydroxide.

2. Hydrogenation. Oils contain unsaturated hydrocarbon chain as part of the constituent fatty acid of the glyceride. If hydrogen gas is passed through such oils in the presence of finely divided nickel, the double bonds in the hydrocarbon chain is saturated and we obtain hydrogenated oil which is a solid (fat).

$$\begin{array}{l} CH_2OCOC_{17}H_{33} \\ | \\ CHOCOC_{17}H_{33} \\ | \\ CH_2OCOC_{17}H_{33} \\ \text{Triolein (Oil)} \end{array} + 3H_2 \xrightarrow[450\,K]{Ni} \begin{array}{l} CH_2OCOC_{17}H_{35} \\ | \\ CHOCOC_{17}H_{35} \\ | \\ CH_2OCOC_{17}H_{35} \\ \text{Tristearin (Oil)} \end{array}$$

The hydrogenated fat has a longer shelf life and does not become rancid easily.

3. Hydrogenolysis. If hydrogen gas is passed in excess through a fat or oil at high temperature and under high pressure, glycerol and long chain aliphatic alcohols are obtained.

$$\begin{array}{l} CH_2OCOC_{15}H_{31} \\ | \\ CHOCOC_{15}H_{31} \\ | \\ CH_2OCOC_{15}H_{31} \\ \text{Tripalmitin} \end{array} + 6H_2 \xrightarrow[\text{Chromite}]{\text{Copper}} \begin{array}{l} CH_2OH \\ | \\ CHOH \\ | \\ CH_2OH \\ \text{Glycerol} \end{array} + \begin{array}{c} 3C_{15}H_{31}CH_2OH \\ \text{Hexadecyl} \\ \text{alcohol} \end{array}$$

4. Trans-esterification. Simple esters can be prepared by treating oils or fats with lower alcohols in the presence of sod. alkoxides

$$\begin{array}{l} CH_2OCOC_{17}H_{35} \\ | \\ CHOCOC_{17}H_{35} \\ | \\ CH_2OCOC_{17}H_{35} \\ \text{Stearin} \end{array} + 3CH_3OH \xrightarrow[\text{Chromite}]{\text{Copper}} \begin{array}{l} CH_2OH \\ | \\ CHOH \\ | \\ CH_2OH \end{array} + \begin{array}{c} 3C_{17}H_{35}COOCH_3 \\ \text{Methyl} \\ \text{stearate} \end{array}$$

Q. 6. Write a note on rancidification. *(Lucknow 2010)*

Ans. Fats and oils, on exposure to air and light, start giving a foul smell and taste. We say that the fat or oil has become **rancid.** This phenomenon is called rancidification. This change takes place due to the formation of fatty acids and carbonyl compounds, which are foul smelling. The low hydrolysis of the fats or oils under the effect of atmospheric moisture produces these fatty acids. Similarly attack of oxygen at the double bonds in oil produces foul-smelling carbonyl compounds. These reasons contribute together towards the phenomenon of rancidification.

Q. 7. Write a note on the manufacture of vanaspati (hydrogenation of oil).

Ans. Hydrogenation is commercially used in the manufacture of vegetable ghee (vanaspati). Groundnut oil, cotton seed oil, coconut oil etc. are used as raw materials for the preparation of vanaspati. The hydrogenator which is used for this purpose is shown in Fig. 23.1.

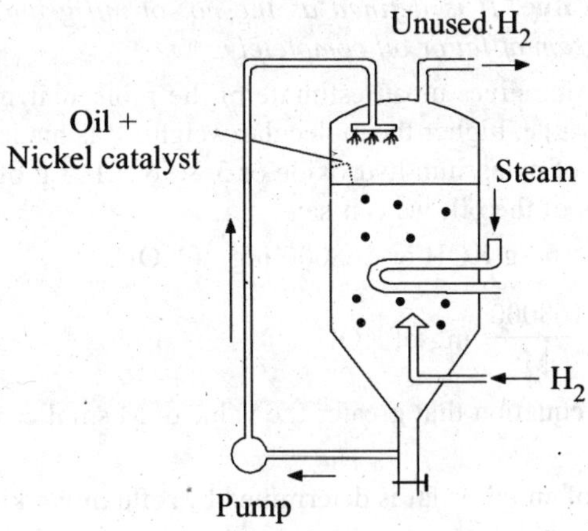

Fig. 23.1. Manufacture of vegetable ghee.

The oil is filtered and refined and taken in the hydrogenator fitted with steam coils for heating. Finely divided nickel is added to the oil. It is heated to 450 K and a current of hydrogen gas is passed through the oil. The oil is cycled again and again into the hydrogenator to ensure complete saturation of double bonds in the oil molecules. After the complete hydrogenation has taken place, the molten fat is taken out and filtered to remove nickel catalyst.

Q. 8. How do you isolate carboxy acids and alcohols from fats and oils?

Ans. Isolation of carboxylic acids.

Fats or oils are subjected to acidic hydrolysis *i.e.* hydrolysis in the presence of a mineral acid like HCl. Ester linkage is broken and glycerol and fatty acids are produced. Fractional distillation of these acids is carried out to separate them.

In an alternative process, alkaline hydrolysis of the ester is carried out. We obtain glycerol and a mixture of sodium salts of the fatty acids. The sodium salts are treated with a mineral acid to liberate a mixture of carboxylic acids which are separated by fractional distillation. In still another alternative method, trans-esterification is carried out with methyl alcohol. A mixture of methyl esters is obtained. It is separated into individual esters by fractional distillation. Esters are hydrolysed separately to obtain carboxylic acids.

Isolation of Alcohols

The fats or oils are subjected to hydrogenation at high temperature and under high pressure in the presence of copper chromite. A mixture of long-chain alcohols is obtained which is separated into individual alcohols by fractional distillation.

Q. 9. Write notes on

(*a*) Acid value

(*b*) Saponification value (*Nagpur, 2008; Lucknow, 2010*)

(*c*) Iodine value of an oil or fat (*Berhampur, 2004; Nagpur, 2008; Lucknow, 2010*)

(*d*) Reichert-Meissl value.

Ans. (*a*) **Acid value.** *It is defined as the no. of milligrams of potassium hydroxide required to neutralise one gram of oil or fat.*

It indicates the amount of free fatty acid present in an oil or fat. A high acid value indicates a stale oil.

The acid value is determined by titrating a solution of oil or fat in pure alcohol against standard potassium hydroxide solution.

(*b*) **Saponification value.** *It is defined as the no. of miligram of potassium hydroxide required to saponify one gram of fat or oil completely.*

The saponification value gives us an estimate of the molecular mass of the fat or oil, the smaller the saponification value, higher the molecular weight. It is made clear like this. One mole of the oil requires 3 moles of potassium hydroxide or $3 \times 56 = 168$ g of KOH for saponification. If M is the molecular mass of the oil, we can say

M g of oil requires = 168 g KOH or 168000 mg of KOH.

$$1 \text{ g of oil requires} = \frac{168000}{M} \text{ mg of KOH.}$$

It is evident from the equation that greater the value of M smaller will be the saponification value.

Saponification value of an oil or fat is determined by refluxing a known amount of a sample ...ss of standard alcoholic potassium hydroxide solution and titrating the unused alkali ...dard acid solution.

(*c*) **Iodine value.** *It is the no. of grams of iodine that combine with 100 g of oil or fat.*

It gives a measure of unsaturation in an oil or fat. Iodine value is determined by adding a known excess of Wij's solution, which is iodine monochloride in glacial acetic acid, to a solution of known weight of the oil or fat in carbon tetrachloride. Unused iodine is estimated by titration with standard hypo solution. We use Wij's solution because iodine as such does not react with the unsaturated oils.

(*d*) **Reichert-Meissl Value (R.M. Value).** *It is the number of millilitre of N/10 potassium hydroxide solution required to neutralise the distillate of 5 g of hydrolysed fat or oil.*

It is a measure of steam volatile fatty acids present as esters in fats or oils. It is used for checking the purity of butter or ghee.

To determine Reichert-Meissl value, 5 g of the sample is hydrolysed with NaOH and the mixture is acidified with dil. H_2SO_4 and then steam distilled. The distillate containing acids upto C_{10} is cooled, filtered and titrated against N/10 alkali.

Q. 10. Calculate saponification value of the following ester:

$$\begin{array}{l} CH_2OCOC_{15}H_{31} \\ | \\ CHOCOC_{15}H_{31} \\ | \\ CH_2OCOC_{15}H_{31} \end{array}$$

Ans. Saponification value of an oil or fat is the no. of milligram of potassium hydroxide required to saponify one gram of fat or oil completely. The above ester (fat) requires three moles of KOH per mole of fat for saponification.

$$\underset{\substack{\text{Tripalmitin} \\ (806\text{ g})}}{\begin{array}{l} CH_2OCOC_{15}H_{31} \\ | \\ CHOCOC_{15}H_{31} \\ | \\ CH_2OCOC_{15}H_{31} \end{array}} + \underset{(168\text{ g})}{3KOH} \longrightarrow \begin{array}{l} CH_2OH \\ | \\ CHOH \\ | \\ CH_2OH \end{array} + \underset{\text{Pot. palmitate}}{3C_{15}H_{31}COOK}$$

806 g of ester requires for saponification = 168000 mg KOH

$$1 \text{ g of ester requires for saponification} = \frac{168000}{806} \text{ mg}$$

$$= 208.4 \text{ mg}$$

Hence saponification value of the ester = 208.4.

Q. 11. What is drying or hardening of oil? Classify oils according to their behaviour on exposure to air.

Ans. Some oils which contain glycerides of unsaturated acids, like linoleic acid or linolenic acid, become thick and harden to resin-like solid, when they are exposed to air. *This phenomenon of hardening of oil on exposure to air and light is called drying.* It is believed that this hardening process takes place due to the oxidation of double bonds in the carboxylic acid component of the glyceride (oil). There is also some possibility of polymerisation taking place. This drying process is quickened or catalysed by certain manganese or lead salts.

Based on behaviour on exposure to air, the oils can be divided into three categories:

(*a*) **Drying oils.** These oils show strong tendency towards drying and within 4-5 hours of exposure to air, harden to a resin-like solid. It is found that such oils are constituted of linoleic and linolenic acids. Examples of oils of this category are linseed oil and hemp seed oil. Such materials are used in paints and varnish industries. They leave a shining surface in the coating of paint or varnish. Iodine value of such oils lies above 130.

(*b*) **Semi-drying oils.** These oils do have some tendency towards drying although they take a much longer time. This is because the proportion of linoleic and linolenic acid component is much less in the oils. Examples of oils of this category are sunflower and cotton seed oils. They absorb oxygen slowly and gradually convert into a solid mass. However, they are not commercially exploited for use in paints and varnishes because of poor results. Their iodine value ranges between 100 and 130.

(*c*) **Non-drying oils.** These oils remain stable to air. They don't dry on exposure for any long period. Examples of this type of oils are olive oil and almond oil. Their iodine value is below 100

Q. 12. What is the difference between soap and detergent? How is soap manufactured?

(*Bangalore, 2002; Meerut, 2003; Nagpur, 2008; Lucknow, 2010*

Ans. Soaps. Soaps are sodium salts of higher fatty acids. A typical example of soap is sodium stearate $C_{17}H_{35}\,COONa$. Soaps are obtained by alkaline hydrolysis of oils and fats which are glyceryl esters of higher fatty acids. Upon hydrolysis, sodium salt of the higher fatty acid *i.e.* soap and glycerol are obtained.

$$\underset{\text{Oil or fat}}{\begin{array}{l} CH_2OCOC_{17}H_{35} \\ | \\ CHOCOC_{17}H_{35} \\ | \\ CH_2OCOC_{17}H_{35} \end{array}} + 3\,NaOH \longrightarrow \underset{\text{Glycerol}}{\begin{array}{l} CH_2OH \\ | \\ CHOH \\ | \\ CH_2OH \end{array}} + \underset{\text{Soap}}{3C_{17}H_{35}COONa}$$

This process is called *saponification.*

There are two methods for soap manufacture : kettle process and hydrolyser process.

The *kettle* process is an old process and is now used in small factories or for production on a small scale.

The *hydrolyser process* is the modern continuous process which is far more economical and gives pure stuff in better yield.

Soaps are not effective cleansing effects with water containing dissolved impurities of calcium, magnesium and iron salt. This impurities are generally present in hard water. In such cases, insoluble stearates are formed which hinder the process of cleansing.

$$2C_{17}H_{35}COONa + CaCl_2 \longrightarrow \underset{\text{Cal. stearate (insoluble)}}{(C_{17}H_{35}COO)_2\,Ca} + 2NaCl$$

Detergents. Detergents are used for effective cleansing in cases where poor results are obtained with soaps. Detergents are substances like soap, having long alkyl chains attached to polar groups *e.g.* sulphonic acids. A typical detergent is sodium lauryl sulphate $CH_3(CH_2)_{10}$ $CH_2OSO_3^-\,Na^+$.

Also there are detergents from the aromatic series. An aromatic detergent from the sulphonic acid group may be represented by the general formula

$$R-C_6H_4-SO_3Na^+$$

where R is an alkyl group with 9-31 carbon atoms. The sodium salts of such sulphonic acids having alkyl groups are oil soluble.

These detergents are better cleansing agents than ordinary soaps. There is no problem of precipitation of these compounds as insoluble salts with calcium ions. Therefore detergents can be used for cleaning even with hard water.

Q. 13. Explain the washing (cleansing) action of soaps and detergents.

(*Kerala, 2001; Guwahati 2002, Purvanchal 2003 ; Goa 2003; Gulberga 2004*)

Ans. Soaps are composed of molecules that contain:

1. Large hydrocarbon groups called lipophilic groups (water repelling or fat loving).
2. Polar group which is hydrophilic (water loving).

The lipophilic group dissolves in greases and oils and the polar group is soluble in water. Soap molecules when dissolved in water aggregate to form a micelle. A micelle is a spherical assembly of large molecules having polar and non-polar groups.

The washing or cleansing action of soaps or detergents can be visualised as under.

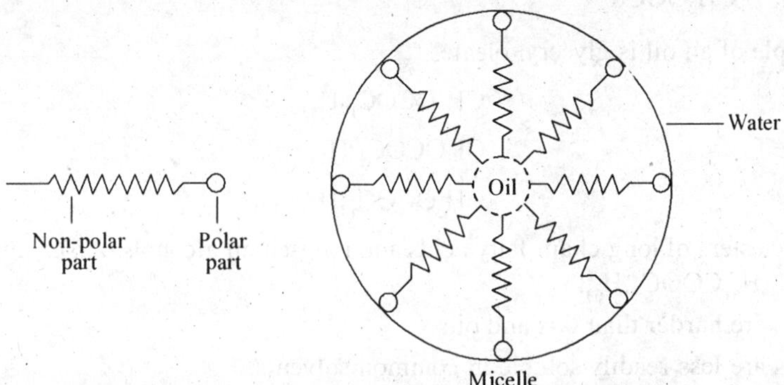

Fig. 23.2.

The non-polar lipophilic or hydrophobic portions of a number of molecules are directed towards grease or oil particles and the polar ends of these molecules are surrounded by water molecules.

Thus grease or oil particles are arrested by these soap or detergent molecules in the form of micelles. With free use of water, these micelles containing dirt particles in grease are washed away.

Q. 14. What are waxes? Describe some common waxes. (*Kuvempu 2005*)

Ans. Waxes are the esters of long chain fatty acids and long-chain alcohols. Both the acid and the alcohol contain even number of carbon atoms. The general formula of waxes is RCOOR' where R' and are long chain alkyl groups. The acids present in waxes are:

Palmitic acid	$C_{15}H_{31}COOH$
Cerotic acid	$C_{25}H_{51}COOH$
Melissic acid	$C_{30}H_{61}COOH$

The alcohols present in waxes are:

Cetyl alcohol	$C_{16}H_{33}OH$
Ceryl alcohol	$C_{26}H_{53}OH$
Milissyl or Myricyl alcohol	$C_{31}H_{63}OH$

Some common waxes are described as under:

(*a*) **Carnuba wax.** This wax is extracted from the leaves of Brazilian palm tree and consists of mericyl cerotate $C_{25}H_{41}COOC_{31}H_{63}$.

This wax is used for floor and automobile polishing.

(*b*) **Bee's wax.** This wax is obtained from bee's honey comb. It consists of myricyl palmitate ($C_{15}H_{31}COOC_{31}H_{63}$) and ceryl myristate ($C_{13}H_{27}COOC_{26}H_{53}$).

This wax is used for preparing candles, shoe-polishes and water proof coatings.

(*c*) **Spermaceti.** This wax is obtained from sperm whale. It consists of ceryl palmitate $C_{13}H_{27}COOC_{26}H_{53}$.

This wax is white odourless and tasteless and is used as a base for cosmetics and pharmaceuticals.

Q. 15. Giving one example in each case, explain the difference between waxes and oils.

Ans. (*a*) Oils are glyceryl esters of long-chain fatty acids having the general formula.

$$\begin{array}{l} CH_2OOCR \\ | \\ CHOOCR' \\ | \\ CH_2OOCR'' \end{array}$$

where R, R′, R″, are alkyl parts of long chain fatty acid, R, R′, R″ could be same or different form one another

An example of an oil is glyceryl oleate.

$$\begin{array}{l} CH_2OCOC_{17}H_{33} \\ | \\ CHOCOC_{17}H_{33} \\ | \\ CH_2OCOC_{17}H_{33} \end{array}$$

Waxes are esters of long chain fatty acids and long-chain alcohols. An example of a wax is carnuba wax $C_{25}H_{51}COOC_{31}H_{63}$.

(*b*) Waxes are harder than fats and oils.

(*c*) Waxes are less readily soluble in common solvents.

(*d*) Waxes are saponified with difficulty compared to oils which are easily saponified.

(*e*) Waxes cannot be used for making soaps because the alcohol produced is insoluble in water, unlike glycerol which is water-soluble.

25

SULPHONIC ACIDS

Q. 1. What are sulphonic acids? Give the structure and nomenclature of sulphonic acids.

Ans. Sulphonic acids are organic compounds having the general formula R–SO_2OH or R–SO_3H where R is an alkyl or aryl group. It is mainly the aromatic compounds which are prominent and useful and hence well shall confine ourselves to the study of aromatic sulphonic acids in this chapter.

Sulphonic acids may be regarded as derivatives of sulphuric acid $HOSO_2OH$ in which – OH group is replaced by an aryl group.

Sulphonic acid has the following structure.

$$\begin{array}{c} O \\ | \\ Ar - S - OH \\ | \\ O \end{array}$$

Studies indicate that there is some double bond character in S–O bonds.

Nomenclature. Aromatic sulphonic acids are named by adding the words sulphonic acid to the parent aromatic compound. For example,

Compound	*Name*
C_6H_5–SO_3H (benzene ring bearing SO_3H)	Benzenesulphonic acid
benzene ring with SO_3H at top and CH_3 at bottom	*p*-Toluenesulphonic acid
benzene ring with Br at top and SO_3H at meta position	*m*-Bromobenzenesulphonic acid

SO_3H / SO_3H

m-Benzenedisulphonic acid

Q. 2. Give the methods of preparation of sulphonic acid. How will you isolate sulphonic acids?

(*Nagpur 2008*)

Ans. Sulphonic acids are prepared by sulphonation. Various sulphonating agents are used depending upon the substance to be sulphonated and the number of sulphonic acid groups to be introduced. Different sulphonating agents are as follows.

1. Concentrated H_2SO_4.
2. Chlorosulphonic acid ($ClSO_3H$).
3. Fuming H_2SO_4 (Oleum) containing varying amount of SO_3 dissolved in conc. H_2SO_4.

Also different temperatures are employed for different substances.

(*a*) Benzene sulphonic acid is prepared by the sulphonation of benzene using oleum as the sulphonating agent at 303–323 K.

$$\text{Benzene} + H_2SO_4\ \text{(Oleum)} \xrightarrow{303\text{-}323\ K} \text{Benzene sulphonic acid } (C_6H_5SO_3H)$$

(*b*) If the benzene ring contains an activating group like phenolic or alkyl group, conc. H_2SO_4 is used as the sulphonating agent at a temperature of 303 K.

$$\text{Phenol} + H_2SO_4 \xrightarrow{303\ K} \text{o-hydroxy benzene sulphonic acid} + H_2O$$

Sulphonic group takes the *o*- and *p*-position with respect to the activating group already present.

(*c*) If the benzene ring contains a deactivating group like $-NO_2$ or $-COOH$ or $-SO_3H$ itself, the sulphonating agent used is oleum at 473–513 K. It requires heating for several hours. Sulphonic group enters the meta position with respect to the group present.

$$\text{Nitrobenzene} \xrightarrow[473\text{-}513\ K]{\text{Oleum}} \text{m-nitrobenzene sulphonic acid}$$

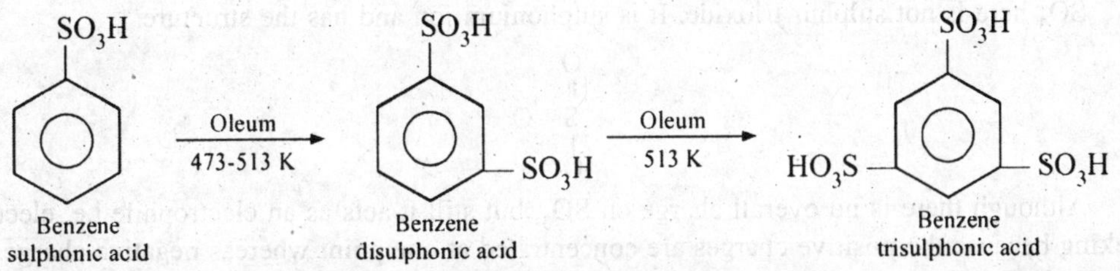

SO_3H (Benzene sulphonic acid) $\xrightarrow[473\text{-}513\ K]{Oleum}$ Benzene disulphonic acid (1,3-SO_3H) $\xrightarrow[513\ K]{Oleum}$ Benzene trisulphonic acid (HO_3S, SO_3H, SO_3H)

Benzene sulphonic acid | Benzene disulphonic acid | Benzene trisulphonic acid

(*d*) There is a marked influence of temperature on the relative proportion of isomers obtained. In the case of an activating group already present in the benzene ring, we obtain mixture of *o-p* sulphonic acids. It is experienced that a lower temperature favours the formation of ortho sulphonic acid, whereas at higher temperature, para isomer is obtained predominantly.

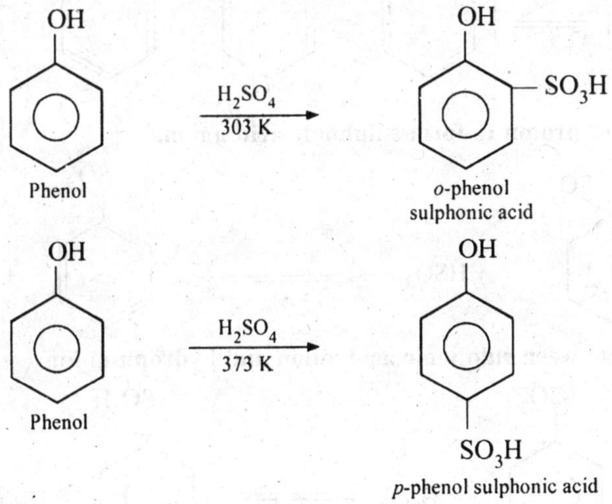

Phenol (OH) $\xrightarrow[303\ K]{H_2SO_4}$ *o*-phenol sulphonic acid (OH, SO_3H)

Phenol (OH) $\xrightarrow[373\ K]{H_2SO_4}$ *p*-phenol sulphonic acid (OH, SO_3H)

Isolation of sulphonic acids

The reaction mixture is diluted with cold water. Calcium or magnesium carbonate or hydroxide is added to it. Calcium or magnesium sulphonate which is formed as a result of reaction with sulphonic acid remains in the solution whereas calcium or magnesium sulphate formed by reaction with unused sulphuric acid is thrown out and removed by filtration. The filtrate is then treated with calculated quantity of sodium carbonate when calcium or magnesium carbonate is precipitated and filtered out.

Calcium sulphonate + Sodium carbonate

$\longrightarrow$ Sod. sulphonate + Calcium carbonate

Magnesium sulphonate + Sodium carbonate

$\longrightarrow$ Sod. sulphonate + Magnesium carbonate

The filtrate containing sod. sulphonate is evaporated to obtain crystals of sod. sulphonate which is used as such in the reactions.

Q. 3. Give the mechanism of sulphonation of benzene. (*Panjab 2005*)

Ans. Sulphonation of benzene is an electrophilic reaction in which hydrogen of the benzene ring is substituted by sulphonic acid group. The electrophile here is SO_3 sulphonium ion which is produced from sulphuric acid. The mechanism involves the following steps.

(*i*) **Generation of electrophile**

$$2H_2SO_4 \rightleftharpoons SO_3 + H_3O^+ + HSO_4^-$$

SO_3 here is not sulphur trioxide. It is sulphonium ion and has the structure.

$$\overset{\cdot\cdot}{S}(=O)(-O)(-O^-)$$

Although there is no overall charge on SO_3, but still it acts as an electrophile i.e. electron-seeking because the positive charges are concentrated at one point whereas negative charges are scattered. Thus it acts as an electrophile.

(*ii*) **Attachment of electrophile to the benzene nucleus**

$$SO_3 + C_6H_6 \rightleftharpoons \left[\text{resonance structures of } C_6H_6(H)(SO_3^-)^+ \right] \text{ Or } C_6H_5(H)(SO_3^-)^+$$

(*iii*) **Removal of proton to form sulphonic acid anion.**

$$C_6H_5(H)(SO_3^-)^+ + HSO_4^- \rightleftharpoons C_6H_5SO_3^- + H_2SO_4$$

(*iv*) **Reaction between sulphonic acid anion and hydronium ion**

$$C_6H_5SO_3^- + H_3O^+ \rightleftharpoons C_6H_5SO_3H + H_2O$$

Q. 4. Why are sulphonic acids stronger than carboxylic acids? (*Garhwal, 2000*)

Ans. Sulphonic acids are strong acids and are completely ionised in water to form sulphonate ion and hydronium ion.

$$C_6H_5SO_3H + H_2O \longrightarrow C_6H_5SO_3^- + H_3O^+$$

The acidity of sulphonic acids is due to the ease of release of protons. This becomes possible due to presence of two strongly electronegative oxygens and formation of a stable sulphonate ion in which the negative charge is dispersed over three oxygen atoms.

$$\left\{ C_6H_5-S(O)(O)-O \right\}^-$$ Sulponate ion stabilized by dispersal of negative charge over three oxygens.

Sulphonic acid is stronger than a carboxy acid because in the carboxylate ion, the negative charge is dispersed over two oxygens only

$$\left\{ R-C(\cdots O)(\cdots O) \right\}^-$$

Q. 5. Describe the chemical properties of benzene sulphonic acid. (*Udaipur, 2004*)

Ans. 1. Salt formation. As sulphonic acids are strongly acidic, they form salts with hydroxides, carbonates and bicarbonates

$$C_6H_5-SO_3H + NaOH \longrightarrow C_6H_5-SO_3^-\overset{+}{Na} + H_2O$$

Benzene sulphonic acid — Sod. benzene sulphonate

$$2\,C_6H_5-SO_3H + Na_2CO_3 \longrightarrow 2\,C_6H_5-SO_3^-\overset{+}{Na} + H_2O + CO_2$$

2. Formation of sulphonyl chloride. Sulphonic acid forms sulphonyl chloride with thionyl chloride or phosphorus pentachloride.

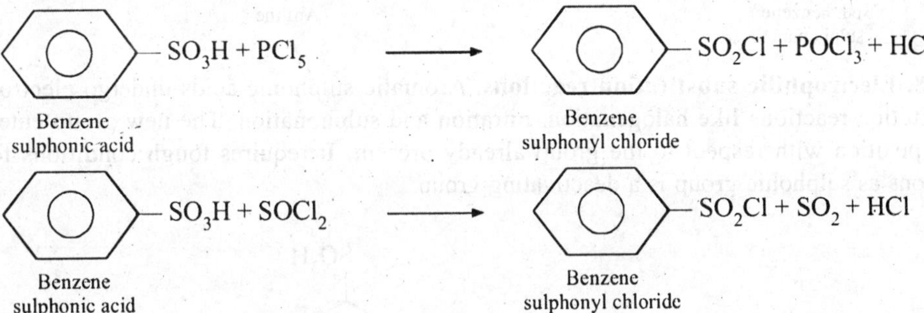

3. Formation of sulphonic esters and sulphonamides. Sulphonyl chloride reacts with alcohols and ammonia to give sulphonic esters and sulphonamide respectively.

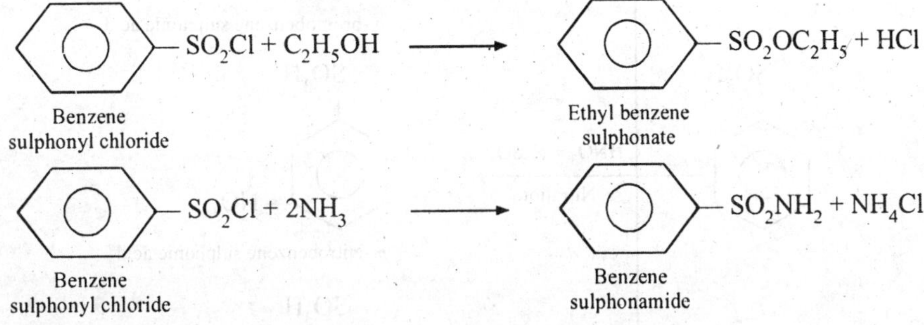

4. Replacement of sulphonic group by phenolic group

$$C_6H_5-SO_3Na + NaOH \xrightarrow{\text{Fuse}} C_6H_5-OH + Na_2SO_3$$

Sod. benzene sulphonate — Phenol

5. Replacement of – SO_3H by – CN. On fusing sod. benzene sulphonate with sodium cyanide, benzonitrile is obtained.

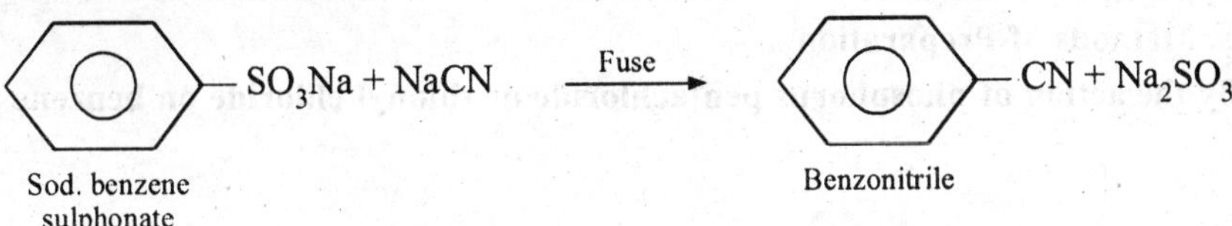

6. Replacement of – SO_3H by – H. When sulphonic acids are heated with mineral acids at 425 K, sulphonic acid group is replaced by hydrogen to give aromatic hydrocarbon.

$$C_6H_5\text{–}SO_3H + H_2O \xrightarrow{H^+,\ 425\ K} C_6H_6 + H_2SO_4$$

Benzene sulphonic acid → Benzene

7. Replacement of – SO_3H by – NH_2. The fusion of sodium salt of sulphonic acid with sodamide leads to the formation of amines.

$$C_6H_5\text{–}SO_3Na + NaNH_2 \xrightarrow{\text{Fuse}} C_6H_5\text{–}NH_2 + Na_2SO_3$$

Sod. benzene sulphonate → Aniline

8. Electrophilic substitution reactions. Aromatic sulphonic acids undergo electrophilic substitution reactions like halogenation, nitration and sulphonation. The new group enters the meta position with respect to the group already present. It requires tough conditions for the reactions as sulphonic group is a deactivating group.

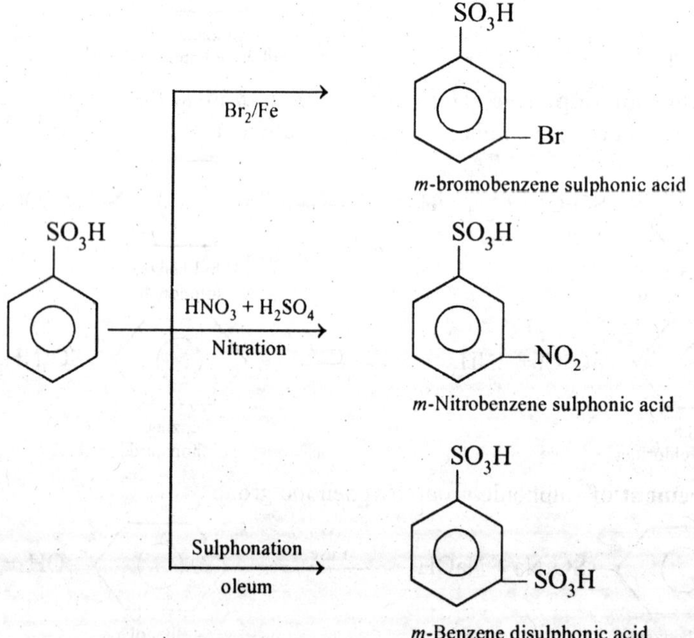

Q. 6. Describe the preparation and properties of benzene sulphonyl chloride ($C_6H_5SO_2Cl$).

Ans. Methods of Preparation

1. By the action of phosphorus pentachloride or thionyl chloride on benzene sulphonic acid

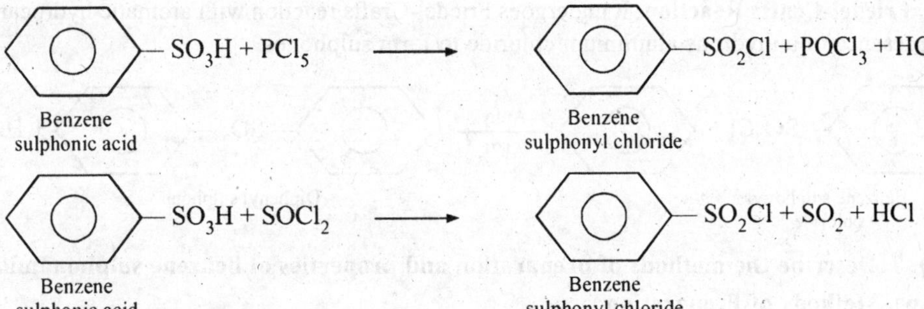

2. By the action of chlorosulphonic acid on benzene

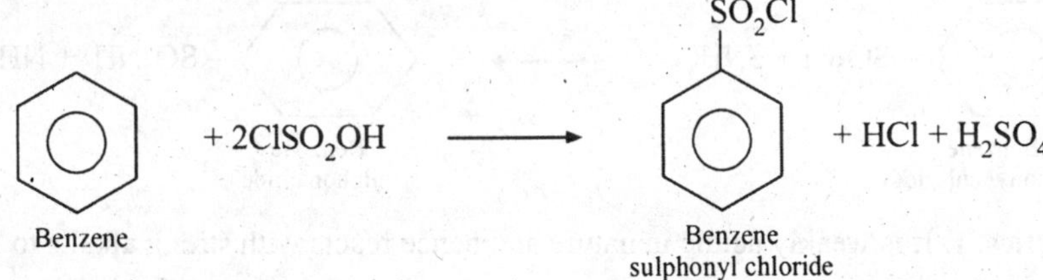

Properties. 1. Hydrolysis. On hydrolysis in the presence of alkali, it hydrolyses into benzene sulphonic acid.

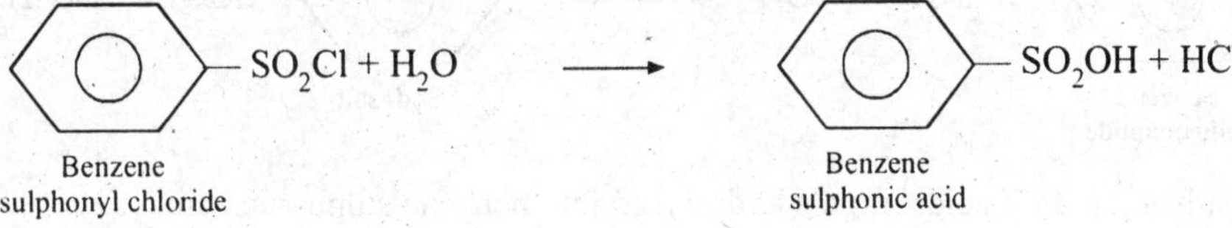

2. Formation of esters. It reacts with alcohols or phenols to form sulphonic esters.

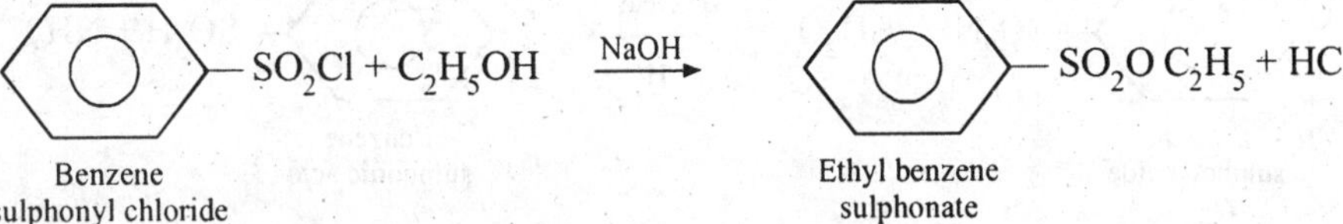

3. Reaction with ammonia and amines.

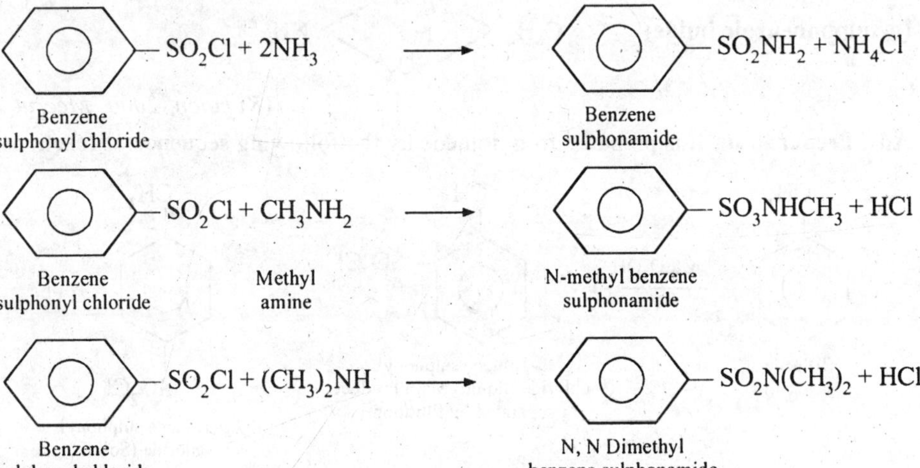

4. Friedel-Crafts Reaction. It undergoes Friedel-Crafts reaction with aromatic hydrocarbons in the presence of anhydrous aluminium chloride to form sulphones.

$$C_6H_5-SO_2Cl + C_6H_6 \xrightarrow[AlCl_3]{Anhy.} C_6H_5-SO_2-C_6H_5 + HCl$$

Benzene sulphonyl chloride; Diphenyl sulphone

Q. 7. Describe the methods of preparation and properties of benzene sulphonamide.

Ans. Methods of Preparation

Benzene sulphonamide is prepared by the action of ammonia on benzene sulphonyl chloride.

$$C_6H_5-SO_2Cl + 2NH_3 \longrightarrow C_6H_5-SO_2NH_2 + NH_4Cl$$

Benzene sulphonyl chloride; Benzene sulphonamide

Properties. 1. It is weakly acidic in nature and hence reacts with strong alkalis to form water soluble salts

$$C_6H_5-SO_2NH_2 + NaOH \longrightarrow C_6H_5-SO_2\overset{-}{N}H\overset{+}{N}a + H_2O$$

Benzene sulphonamide; Sod. salt

2. On heating with acids, it gets hydrolysed into benzene sulphonic acid.

$$C_6H_5-SO_2NH_2 + H_2O \xrightarrow[H^+]{Heat} C_6H_5-SO_3H + NH_3$$

Benzene sulphonamide; Benzene sulphonic acid

Q. 8. Give the preparation and properties of saccharin.

(o-sulphobenzoic imide) $C_6H_4 \begin{matrix} \diagup CO \diagdown \\ \diagdown SO_2 \diagup \end{matrix} NH$

(Kumaon, 2000; Meerut 2000)

Ans. Preparation. It is prepared from toluene by the following sequence of reactions.

$$C_6H_5CH_3 \xrightarrow{ClSO_2OH} o\text{-}CH_3C_6H_4SO_2Cl + p\text{-}CH_3C_6H_4SO_2Cl$$

Toluene; *o*-Toluene sulphonyl chloride (liquid) Main Product separated by filtration; *p*-Toluene sulphonyl chloride (Solid) o-Toluene sulphonyl chloride $\xrightarrow{NH_3}$ o-Toluene sulphonamide

$\xrightarrow{KMnO_4}$ o-Sulphonamide benzoic acid $\xrightarrow[-HO]{Heat}$ Saccharin

Properties. 1. It is a white solid about 500 times as sweet as sugar.

2. It forms a salt with alkali.

$$C_6H_4(CO)(SO_2)NH + NaOH \longrightarrow C_6H_4(CO)(SO_2)\overset{-}{N}\overset{+}{Na} + H_2O$$

Sod. salt of Saccharin (water soluble)

Q. 9. Which one is the strongest acid?

(*i*) C_6H_5COOH (*ii*) $OCH_3C_6H_6SO_3H$ (*iii*) $NO_2C_6H_4SO_3H$

(*iv*) $C_6H_5SO_3H$.

Ans. Benzene sulphonic acid is definitely stronger than benzoic acid because in benzene sulphonate ions the negative charge is dispersed over three oxygens whereas in carboxylate ion the negative charge is dispersed over two oxygens.

An electron withdrawing group at *o*- or *p*-position in the benzene ring of benzene sulphonic acid increases the acid strength whereas an electron-donating group decreases the acid strengths. Hence the order of acid strength in the compounds is as follows:

o-nitrobenzene sulphonic acid > benzene sulphonic acid > *o*-methoxybenzene sulphonic acid > benzoic acid.

Q. 10. Predict the product of monosulphonation of the following

(*a*) Toluene at 373 K (*b*) *p*-nitrophenol (*c*) nitrobenzene (*d*) *m*-dimethylbenzene.

Ans. (*a*) *p*-Toluene sulphonic acid (CH_3, SO_3H)

(*b*) 2-Hydroxy-5-nitrobenzene sulphonic acid (OH, SO_3H, NO_2)

(*c*) *m*-Nitrobenzene sulphonic acid (NO_2, SO_3H)

(*d*)

CH_3, CH_3, SO_3H

2, 4, Dimethyl benzene sulphonic acid

Q. 11. How will you prepare *o*-bromotoluene from toluene without using direct bromination, which gives a large amount of unwanted *p*-isomer?

Ans.

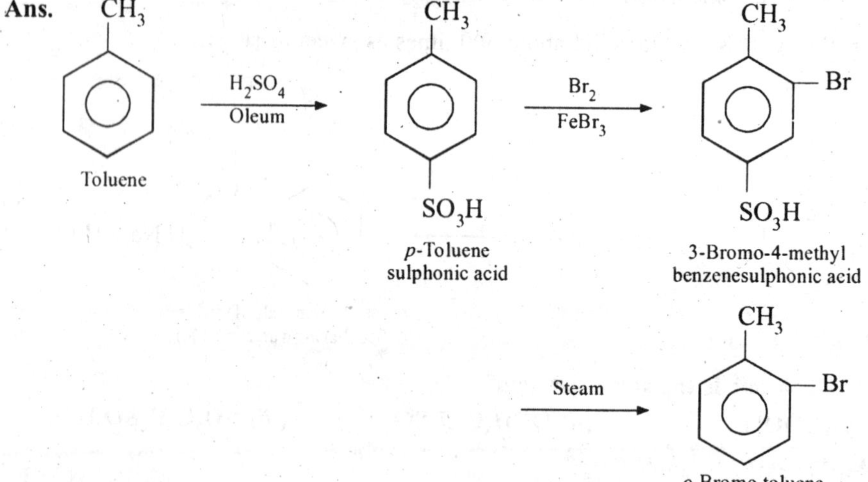

Toluene

p-Toluene sulphonic acid

3-Bromo-4-methyl benzenesulphonic acid

o-Bromo toluene

Q. 12. Identify the compound $C_9H_{11}O_2SCl$ (A) and the compound (B), (C) and (D) in the following reactions.

$$\textbf{AgCl} \xleftarrow[\text{Rapid}]{AgNO_3} C_9H_{11}O_2SCl\ \textbf{(A)} \xrightarrow[H_3O^+]{NaOH} \textbf{Water soluble (B)}$$

$$\textbf{(B)} \xrightarrow[\text{heat}]{H_3O^+} C_9H_{12}\ \textbf{(C)} \xrightarrow[Fe]{Br_2} \textbf{One monobromo derivative (D)}$$

Ans. Compound (A) contains sulphur and gives a precipitate with $AgNO_3$. It indicates an active chlorine and the probable group is – SO_2Cl. Sulphonyl chloride group on treatment with sodium hydroxide gives a sodium salt – SO_2Na and on hydrolysis with an acid is converted into – SO_2OH. On heating, the sulphonic acid group in removed. The following sequence explains the above reactions.

SO_2Cl, H_3C, CH_3, CH_3 $\xrightarrow[\text{Oleum, } H_3O^+]{NaOH}$ SO_3H, H_3C, CH_3, CH_3 $\xrightarrow[\text{Heat}]{H_3O^+}$ CH_3, CH_3, CH_3

2, 4, 6 Trimethyl benzene sulphonyl chloride (A)

2, 4, 6 Trimethyl benzene sulphonic acid (B)

Mesitylene (1, 3, 5 trimethyl benzene) (C)

$$\xrightarrow[Fe]{Br} \text{Mesitylene bromide (D)}$$

Br
CH_3 — CH_3
CH_3
Mesitylene bromide
(D)

Q. 13. Write commercial name and structural formula of ortho sulphobenzoic imide and give its important uses. (*Purvanchal 2003*)

Ans. Commercial name of orthosulphobenzoic imide is saccharin and its structural formula is:

CO
NH
SO_2

Uses of Saccharin

1. It is used as an artificial sweetening agent.
2. It is used by patients of diabetes as a substitute for sugar.

Q. 14. How does saccharin react with sodium hydroxide? Give chemical equation.

Ans. See Q. 8.

Q. 15. Write the name of the following compound. (*Awadh, 2000*)

O
||
C
NH
SO_2

Ans. Saccharin

Q. 16. What happens when sodium benzene sulphonate is fused with solid sodium hydroxide? (*Kerala, 2000*)

Ans. $C_6H_5SO_3Na + NaOH \longrightarrow C_6H_5OH + Na_2SO_3$

Sodium benzene sulphonate (solid) — phenol

Q. 17. What is sulphonation ? How is benzene sulphonic acid prepared in laboratory and on large scale ? Give its important properties and uses. (*Meerut, 2000*)

Ans. See Q. 2 and Q. 5

Q. 18. What happens when benzene sulphonic acid reacts with

(*i*) NaOH Sol (*ii*) Conc HNO_3

Ans. $C_6H_5-SO_3H \xrightarrow[-H_2O]{NaOH} C_6H_5-SO_3Na$

$C_6H_5-SO_3H \xrightarrow[HNO_3]{Conc} C_6H_4(NO_2)-SO_3Na$

Q. 19. How is chloramine – T prepared ? What are its uses ?

Ans. It is prepared as follows :

$$CH_3C_6H_5 \xrightarrow{ClSO_3H} CH_3C_6H_4SO_2Cl \xrightarrow{NH_3} \underset{\text{p-Toluene sulphonamide}}{CH_3C_6H_4SO_2NH_2} \xrightarrow[NaOH]{NaOCl} \underset{\text{Chloramine-T}}{CH_3C_6H_4SO_2N\overset{-}{Cl}\overset{+}{Na}} + H_2O$$

Chloramine – T reacts with water to liberate hypochlorous acid.

It is an effective antiseptic for wounds.

Q. 20. How is sulphanilic acid prepared? What are its uses ?

(*Mysore 2003: Bhopal 2004*)

Ans. It is prepared as under :

$$C_6H_5NH_2 + H_2SO_4 \xrightarrow[-H_2O]{180°C} C_6H_5NHSO_3H \xrightarrow{\text{Rearranges}} \underset{\text{Sulphanilic acid}}{NH_2C_6H_4SO_3H}$$

Uses. It is an important dye intermediate. Its substituted amides form sulpha drugs.

26

NITRO COMPOUNDS

Q. 1. What are nitro compounds? Give the structure of nitro group. How does it differ from nitrite group?

Ans. Nitro compounds are the compounds which contain at least one nitro group ($-NO_2$) in the molecule *e.g.*, nitromethane (CH_3NO_2), nitrobenzene ($C_6H_5NO_2$) etc.

Structure of Nitro group. Nitro group has the structure

$$-N\begin{matrix} \nearrow\!\!=O \\ \searrow O \end{matrix} \qquad (I)$$

Thus there is one nitrogen-oxygen double bond and one nitrogen-oxygen co-ordinate bond. It is also represented as

$$-\overset{+}{N}\begin{matrix} \diagup\!\!\!=O \\ \diagdown O^- \end{matrix} \qquad (II)$$

Structure II contributes to a greater extent to the actual structure of nitro group, as it is resonance stabilized

$$-\overset{+}{N}\begin{matrix} \diagup\!\!\!=O \\ \diagdown O^- \end{matrix} \longleftrightarrow -\overset{+}{N}\begin{matrix} \diagup O^- \\ \diagdown\!\!\!=O \end{matrix}$$

The above two resonating structures have equal energies. This imparts extra stability to the structure of nitro group. Single and double bonds are exchanging their positions, hence single structure of $-NO_2$ group could be written as

$$-\overset{\oplus}{N}\begin{matrix} \diagup\!\!\!\!\cdots O \\ \diagdown\!\!\!\!\cdots O \end{matrix}\Bigg\}\ominus$$

Orbital diagram of nitro group is represented as

Difference between nitro and nitrite groups

Consider the following two compounds

$$R-\overset{+}{N}\begin{matrix}\nearrow O \\ \searrow O^-\end{matrix} \qquad\qquad R-O-N=O$$

(*a*) (*b*)

Out of the two, compound (*a*) is a nitro compound whereas compound (*b*) is a nitrite. The difference between the two is obvious.

In nitro compound, the alkyl (or aryl) group is directly linked to nitrogen atom, whereas in the nitrite, alkyl group is linked to oxygen.

Some points of distinction between nitro compounds and nitrites are given in the table below.

Distinction between Nitroalkanes and Alkyl nitrites

No.	Property	Nitroalkanes	Alkyl Nitrite
1.	*Boiling point*	High	Low
2.	*Reduction (Sn/HCl)*	A primary amine is formed. $RNO_2 + 6H \xrightarrow{Sn/HCl} RNH_2 + 2H_2O$	Alcohol is formed. $RNO_2 + 4H \longrightarrow ROH + NH_2OH$
3.	*Action of alkalis*	Salt formation takes places. $RCH_2NO_2 + NaOH \longrightarrow RCHNaNO_2 + H_2O$	Hydrolysis takes place with the formation of alcohol. $RNO_2 + NaOH \longrightarrow ROH + NaNO_2$
4.	*Action of HNO_2*	Nitrolic acid is formed. $RCH_2NO_2 + HNO_2$ $R-\underset{\underset{NOH}{\parallel}}{C}-NO_2 + H_2O$	No action.

Q. 2. Give the nomenclature of nitro compounds.

Ans. Nitro compounds are named by writing the word nitro before the name of parent compound. Nitro compounds may be divided into two categories *viz.*, aliphatic nitro compounds and aromatic nitro compounds. The nomenclature of compounds from both categories is given below.

Aliphatic nitro compounds

Compound	*Name*
CH_3NO_2	Nitromethane
$CH_3CH_2NO_2$	Nitroethane
$CH_3CH_2CH_2NO_2$	1-nitropropane
$CH_3CH_2\underset{\underset{NO_2}{\mid}}{CH}-CH_3$	2-nitrobutane

Aromatic nitro compounds

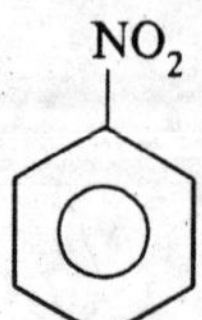

Nitrobenzene

NO_2 / benzene ring / NO_2 (meta) — *m*-dinitrobenzene

CH_3 / benzene ring / NO_2 (para) — *p*-nitrotolune

NH_2 / benzene ring / NO_2 (ortho) — *o*-nitroaniline

$NHCOCH_3$ / benzene ring / NO_2 (para) — *p*-nitroacetanilide

Q. 3. Give the methods of preparation of nitroalkanes or aliphatic nitro compounds.

(Nagpur 2005)

Ans. Methods of preparation

1. Nitration of alkanes. It is carried out by heating the alkane with nitric acid in vapour phase at 700 K. One hydrogen of the alkane is replaced by nitro group.

$$\underset{\text{Alkane}}{R-H} + HNO_3 \longrightarrow \underset{\text{Nitroalkane}}{R\,NO_2} + H_2O$$

However this reaction does not give one single product in pure state. At the high temp. of 700 K, hydrocarbons are broken into smaller fragments and consequently we obtain a mixture of different nitroalkanes.

$$\underset{\text{Propane}}{CH_3-CH_2-CH_3} \xrightarrow[700\ K]{HNO_3} \underset{\text{1-Nitropropane}}{CH_3-CH_2-CH_2\,NO_2} + \underset{\text{2-Nitropropane}}{CH_3\,\underset{}{\overset{NO_2}{\overset{|}{C}H}}-CH_3}$$

$$+ \underset{\text{Nitroethane}}{CH_3\,CH_2\,NO_2} + \underset{\text{Nitromethane}}{CH_3\,NO_2}$$

2. Action of nitrous acid on α-halogen acid. On boiling an aqueous solution of sodium nitrite with an α-halogen carboxylic acid, nitroalkane is obtained

$$\underset{\text{Chloroacetic acid}}{\underset{Cl}{\underset{|}{CH_2}}COOH} + NaNO_2 \longrightarrow \underset{\text{Nitroacetic acid}}{\underset{NO_2}{\underset{|}{CH_2}}COOH} \xrightarrow{\Delta} \underset{\text{Nitro-methane}}{CH_3\,NO_2} + CO_2$$

3. Action of silver nitrite on alkyl halide. Alkyl halides undergo nucleophilic substitution with silver nitrite to yield nitroalkanes.

$$\underset{\text{Ethyl iodide}}{C_2H_5I} + AgNO_2 \longrightarrow \underset{\text{Nitroethane}}{C_2H_5NO_2} + AgI$$

4. Oxidation of amines. An amino group attached to a tertiary carbon atom is oxidised with potassium permanganate to give nitroalkane.

$$\underset{\text{Ter.butylamine}}{H_3C-\overset{\overset{CH_3}{|}}{\underset{\underset{CH_3}{|}}{C}}-NH_2} + 3[O] \longrightarrow \underset{\text{Tert. nitrobutane}}{CH_3-\overset{\overset{CH_3}{|}}{\underset{\underset{CH_3}{|}}{C}}-NO_2} + H_2O$$

5. Hydrolysis of α-nitroalkenes. An α-nitroalkene on hydrolysis in the presence of an acid or alkali produces nitroalkane.

$$\underset{\text{2-Methyl-1 nitro-propene-1}}{CH_3-\overset{\overset{CH_3}{|}}{C}=CHNO_2} + H_2O \longrightarrow \underset{\text{Acetone}}{CH_3-\overset{\overset{CH_3}{|}}{C}=O} + \underset{\text{Nitromethane}}{CH_3NO_2}$$

6. Oxidation of oximes. Aldoximes and ketoximes on oxidation with trifluoroperoxy acetic acid yield primary and secondary nitroalkanes respectively.

$$\underset{\text{Aldoxime}}{RCH=NOH} \xrightarrow{[O]} RCH=\overset{\overset{O^-}{|}}{N^+}-OH \rightleftharpoons \underset{\text{1° Nitroalkane}}{RCH_2NO_2}$$

$$\underset{\text{Ketoxime}}{R_2C=NOH} \xrightarrow{[O]} R_2C=\overset{\overset{O^-}{|}}{N^+}-OH \rightleftharpoons \underset{\text{2° Nitroalkane}}{R_2CHNO_2}$$

Q. 4. Describe the properties of nitroalkanes.

Ans. 1. Acidic nature of nitroalkanes. (Tautomerism) Niroalkanes containing α-hydrogen atom exhibit acidic behaviour and form salts with strong alkalis.

$$RCH_2NO_2 + NaOH \longrightarrow [RCHNO_2]^- Na^+ + H_2O$$

Acidic nature of α-hydrogen containing nitroalkanes is due to the electron-withdrawing inductive effect of $-NO_2$ group. Also the anion obtained after releasing one proton gets resonance stabilized.

$$R-\overset{\overset{H}{|}}{\underset{\underset{H}{|}}{C}}-\overset{+}{N}\begin{smallmatrix}O\\O^-\end{smallmatrix} \xrightarrow{OH} \left[R-\overset{-}{C}H-\overset{+}{N}\begin{smallmatrix}O\\O^-\end{smallmatrix} \longleftrightarrow R-CH=\overset{+}{N}\begin{smallmatrix}O^-\\O^-\end{smallmatrix}\right] + H_2O$$

Primary and secondary nitroalkanes exhibit tautomerism, the tautomeric forms being nitro form and aci-nitro form

$$\underset{\text{Nitro form}}{R-CH_2-\overset{+}{N}\begin{smallmatrix}O\\O^-\end{smallmatrix}} \rightleftharpoons \underset{\text{Acintro form}}{R-CH=\overset{+}{N}\begin{smallmatrix}OH\\O^-\end{smallmatrix}}$$

2. Reduction. Nitroalkanes on reduction under different conditions give different products (*a*) *In acidic medium*

$$\underset{\text{Nitroalkane}}{RNO_2 + 6[H]} \xrightarrow{\text{Metal / Acid}} \underset{\text{Primary amine}}{RNH_2} + 2H_2O$$

(*b*) *Catalytic reduction*

$$\underset{\text{Nitro alkane}}{RNO_2 + 3H_2} \xrightarrow{\text{Ni}} \underset{\text{Primary amine}}{RNH_2} + 2H_2O$$

(*c*) *In neutral reducing medium*

$$RNO_2 + 4[H] \xrightarrow{\text{Zn / NH}_4\text{Cl}} \underset{\text{N-alkyl hydroxylamine}}{RNHOH} + H_2O$$

(*d*) *With $SnCl_2$ and HCl*

$$RCH_2\,NO_2 \xrightarrow[\text{HCl}]{\text{SnCl}_2} \underset{\text{N-alkyl hydroxylamine}}{RCH_2\,NHOH} + \underset{\text{Aldoxime}}{RCH = NOH}$$

3. Hydrolysis. (*a*) Primary nitro compounds on boiling with HCl are converted into a mixture of carboxylic acid and hydroxylamine

$$\underset{1^\circ\text{ Nitroalkane}}{RCH_2NO_2 + H_2O} \xrightarrow{\text{HCl}} \underset{\text{Carboxylic acid}}{RCOOH} + \underset{\text{Hydroxyl amine}}{NH_2OH}$$

(*b*) Secondary nitroalkanes are hydrolysed to ketones on boiling with HCl

$$\underset{2^\circ\text{ Nitroalkane}}{2R_2CHNO_2} \xrightarrow{\text{HCl}} \underset{\text{Ketone}}{2R_2CO} + \underset{\text{Nitrous oxide}}{N_2O} + H_2O$$

4. Reaction with nitrous acid. (*a*) Primary nitroalkanes react with nitrous acid to form nitrolic acid which dissolves in alkali to give red coloured solution.

$$\underset{\text{Primary nitro compound}}{R-\underset{\underset{NO_2}{|}}{CH_2}} + O = NOH \longrightarrow \underset{\text{Nitrolic acid}}{R-\underset{\underset{NO_2}{|}}{C} = NOH} + H_2O$$

(*b*) Secondary nitroalkanes react with nitrous acid to produce pseudo nitroles which give blue colour with alkali.

$$\underset{2^\circ\text{ Nitro compound}}{R_2C\begin{smallmatrix} H \\ NO_2 \end{smallmatrix}} + \underset{\text{Nitrous acid}}{HO-NO} \longrightarrow R_2C\begin{smallmatrix} NO \\ NO_3 \end{smallmatrix} \quad \text{Pseudo nitrole}$$

5. Halogenation. Primary and secondary nitro-compounds are halogenated in α position in alkaline solution.

(*a*)
$$\underset{\text{Primary nitro alkane}}{RCH_2NO_2} \longrightarrow [RCH = N^-O_2]Na^+ \xrightarrow[\text{NaOH}]{\text{Br}_2} \underset{\alpha\text{-bromo nitroalkane}}{RCH - \underset{\underset{Br}{|}}{NO_2}} + NaBr$$

$$\xrightarrow{\text{NaOH}} [RCBr = NO_2^-]\,Na^+ \xrightarrow{\text{Br}_2/\text{NaOH}} \underset{\alpha,\,\alpha\text{dibromo-nitroalkane}}{RCBr_2NO_2} + NaBr$$

(*b*)
$$\underset{\text{Secondary nitroalkane}}{R_2CHNO_2} \xrightarrow{\text{NaOH}} [R_2C = NO_2^-]Na^+ \xrightarrow[\text{NaOH}]{\text{Br}_2} \underset{\text{2-bromo product}}{R_2CBrNO_2} + NaBr$$

Nitromethane has three α-hydrogens and gives trihalogen product

$$\underset{\text{Nitro-methane}}{CH_3NO_2} + 3Cl_2 + NaOH \longrightarrow \underset{\text{Chloro-picrin}}{Cl_3\,CNO_2} + 3NaCl + 3H_2O$$

6. Reaction with Grignard Reagent. The aci-form of nitroalkane reacts with Grignard reagent as follows.

$$\underset{\text{Nitroalkane}}{RCH=N{<}^{OH}_{O}} + CH_3MgI \longrightarrow \underset{\text{Methane}}{CH_4} + RCH=N{<}^{OMgI}_{O}$$

7. Action of heat. Nitroalkanes on heating to 573 K produce alkenes

$$\underset{\text{Nitroalkane}}{RCH_2CH_2NO_2} \xrightarrow{573\text{ K}} \underset{\text{Alkene}}{RCH=CH_2} + HNO_2$$

8. Condensation with aldehydes. Primary and secondary nitroalkanes undergo aldol type reaction with aldehydes and ketones in the presence of a base to form nitroalcohols

$$\underset{\text{Acetaldehyde}}{CH_3-\underset{H}{C}=O} + \underset{\text{Nitroethane}}{\overset{CH_3}{CH_2}-NO_2} \longrightarrow \underset{\text{3-nitro-2-butanol}}{CH_3-\underset{OH}{CH}-\overset{CH_3}{CH}-NO_2}$$

$$CH_3-\overset{H}{\underset{H}{C}}-\overset{+}{N}{<}^{O}_{O^-} + OH^- \longrightarrow \left[CH_3-\underset{H}{\overset{-}{C}}-\overset{+}{N}{<}^{O}_{O^-} \longrightarrow CH_3-\underset{H}{C}=\overset{+}{N}{<}^{O^-}_{O^-}\right] + H_2O$$

$$CH_3-\overset{H}{C}=O + \left[\overset{CH_3}{\underset{H}{\overset{-}{C}}}-\overset{+}{N}{<}^{O}_{O^-} \longleftrightarrow \overset{CH_3}{\underset{H}{C}}=\overset{+}{N}{<}^{O^-}_{O^-}\right] \longrightarrow CH_3-\overset{H}{\underset{O^-}{C}}-\overset{CH_3}{CH}-\overset{+}{N}{<}^{O}_{O^-}$$

$$\xrightarrow{H-O-H} CH_3-\overset{H}{\underset{OH}{C}}-\overset{CH_3}{CH}-NO_2 + OH^-$$

9. Mannich reaction. Nitro compounds having α-hydrogen atom condense with formaldehyde and ammonia as follows.

$$R_2CHNO_2 + HCHO + NH_4Cl \longrightarrow \underset{\text{Nitroamine hydrochloride}}{R_2\overset{NO_2}{C}-CH_2NH_2HCl} + H_2O$$

Q. 5. (*a*) Give the methods of preparation of aromatic nitro compounds.(*Nagpur, 2005*)
(*b*) Give the mechanism of nitration of benzene.

(*M. Dayanand, 2000; Panjab 2003, Baroda 2004*)

Ans. (*a*) Aromatic nitro compounds are prepared by the nitration of benzene or its derivatives. Following nitrating agents are generally employed.

(*i*) In the case of aromatic compounds containing electron releasing agents such as $-NH_2$, $-OH$ and $-CH_3$, the reaction takes place with conc. HNO_3 as the nitrating agent. In some cases even dil. HNO_3 can carry out nitration at room temp.

Phenol + HNO_3 (dil.) $\xrightarrow{-H_2O}$ *o*-nitrophenol + *p*-nitrophenol

As the electron releasing groups are activating groups, further substitution in the benzene ring is easier and the nitro group takes either the *ortho* or *para* position relative to the group already present.

(*ii*) For aromatic compounds containing electron-withdrawing group in the ring, or when it is desired to introduce more than one nitro group, a strong nitrating agent is required and the reaction requires a higher temperature. Here a mixture of conc. HNO_3 and conc. H_2SO_4 is used as the nitrating agent

Benzene $\xrightarrow[\text{Conc. } H_2SO_4,\ 328\text{ K}]{\text{Conc. } HNO_3}$ Nitrobenzene $\xrightarrow[\text{Conc. } H_2SO_4,\ 373\text{ K}]{\text{Conc. } HNO_3}$ *m*-dinitrobenzene

(*b*) **Mechanism of Nitration**

Nitration of aromatic compounds is an electrophilic substitution reaction, in which the hydrogen of the ring is substituted by $-NO_2$ group. The electrophile involved is NO_2^+ nitronium ion which is released from the mixture of HNO_3 and H_2SO_4.

Different steps involved in the process of nitration are as follows:

(*i*) **Formation of electrophile**

$$HNO_3 + H_2SO_4 \longrightarrow \overset{+}{N}O_2 + H_3\overset{+}{O} + 2HSO_4^-$$

Nitronium ion

(*ii*) **Attachment of NO_2^+ to the ring.**

$\overset{\oplus}{N}O_2$ + benzene $\longrightarrow$ [resonance structures] Or Intermediate product

Intermediate product

(*iii*) **Proton transfer to yield the final product.**

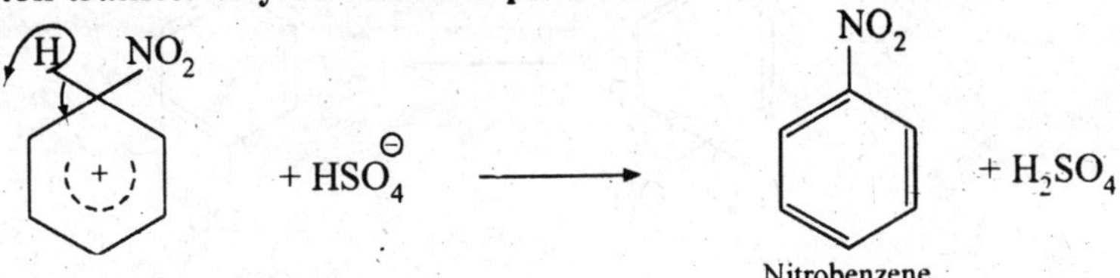

Nitrobenzene

Q. 6. Describe the properties of aromatic nitro compounds. (*Nagpur, 2008*)

Ans. 1. Melting and boiling points of nitrobenzenes gradually increase with increase in the number of nitro groups. This is evident from the table below:

Compound	**M.P. (K)**	**B.P. (K)**
Ntrobenzene	287.7	483.0
m-Dinitrobenzene	363.8	576.0
Sym-Trinitrobenzene	395.0	–

2. Electrophilic Substitution. Nitro group is a deactivating group. So further substitution does not take place easily in nitro compounds. To convert nitro benzene into *m*-dinitrobenzene, fuming HNO_3 is required.

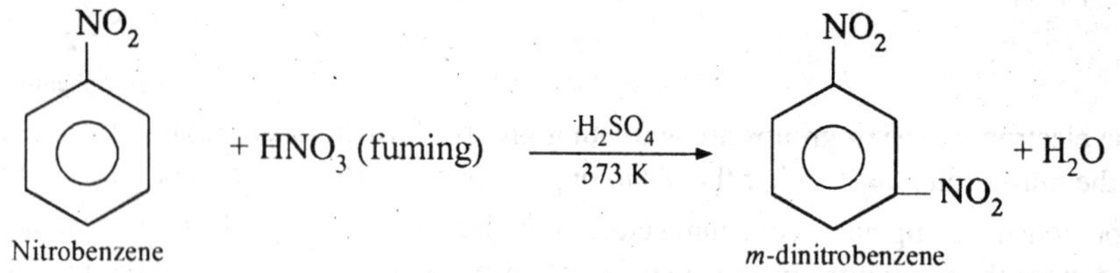

$$\text{Nitrobenzene} + HNO_3\ (\text{fuming}) \xrightarrow[373\ K]{H_2SO_4} m\text{-dinitrobenzene} + H_2O$$

3. Nucleophilic substitution. Nitro compounds easily undergo nucleophilic substitution reaction because of positive charge on the ring as a result of electron withdrawing effect of nitro group.

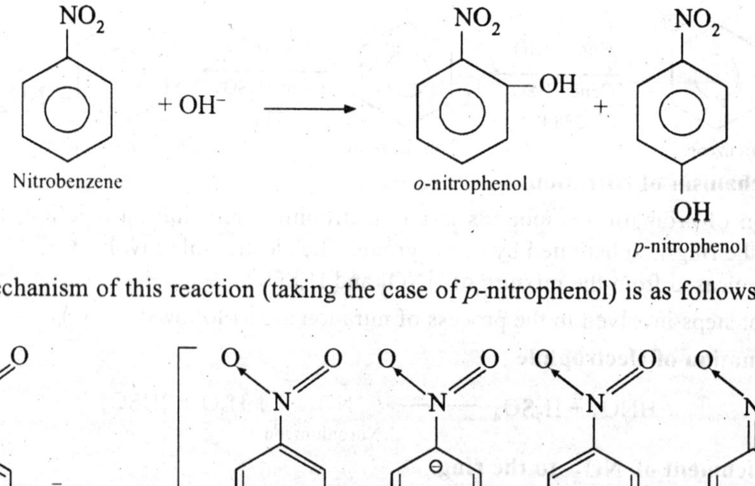

$$\text{Nitrobenzene} + OH^- \longrightarrow o\text{-nitrophenol} + p\text{-nitrophenol}$$

The mechanism of this reaction (taking the case of *p*-nitrophenol) is as follows:

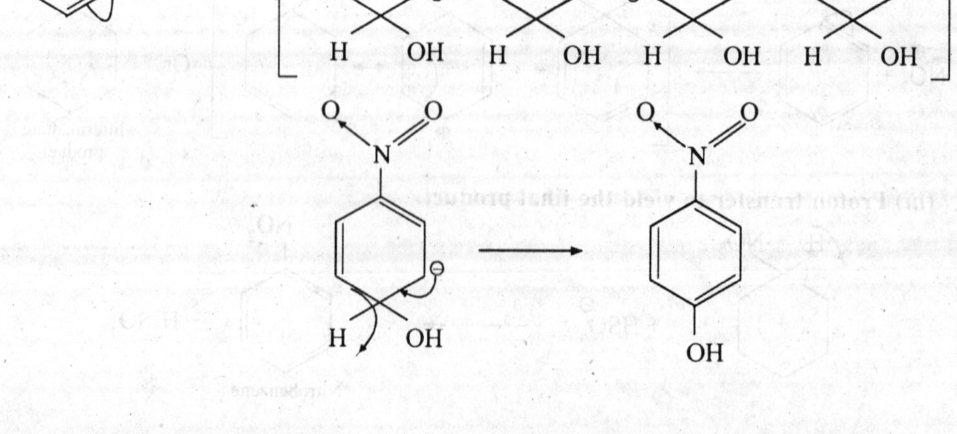

$$\text{Nitrobenzene} + \bar{O}H \longleftrightarrow \left[\text{resonance structures of the intermediate (H, OH at C-4)}\right] \longrightarrow p\text{-nitrophenol}$$

4. Substitution of nitro group in polynitro compounds. On or more nitro groups can be replaced by other nucleophilic groups

p-Dinitrobenzene $\xrightarrow{2NH_3}$ p-nitroaniline $+ N_2 + 2H_2O$

p-Dinitrobenzene $\xrightarrow{KOH}$ p-nitrophenol $+ KNO_2$

p-Dinitrobenzene $\xrightarrow{C_2H_5ONa}$ p-nitrophenetole $+ NaNO_2$

5. Reduction. Nitro compounds undergo reduction to give different products with different reducing agents. The reduction proceeds in stages involving the following intermediate compounds.

Nitrobenzene $\xrightarrow{[H]}$ Nitrosobenzene $\xrightarrow{[H]}$ Phenyl hydroxylamine $\xrightarrow{[H]}$ Aniline

Actual product obtained depends upon the reducing agent used

(*i*) **Reduction in acidic medium.**

Nitrobenzene $\xrightarrow[\text{or Sn/HCl}]{\text{Zn/HCl}}$ Aniline $+ 2H_2O$

(*ii*) **Reduction in neutral medium**

$$C_6H_5NO_2 \xrightarrow{Zn/NH_4Cl} C_6H_5NHOH$$

Nitrobenzene → Phenylhydroxyl amine

(*iii*) **Reduction in alkaline medium**

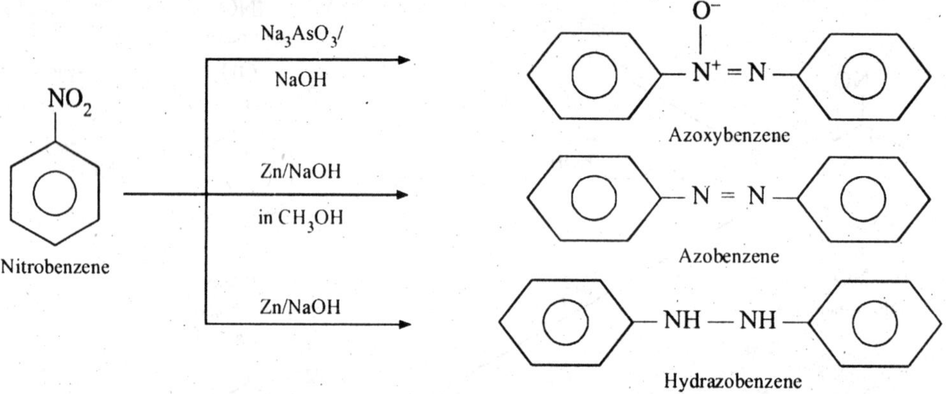

(*iv*) **Electrolytic Reduction.** On electrolytic reduction in strongly acidic solution, nitrobenzene yields *p*-aminophenol after the rearrangement of phenylhydroxylamine, first formed.

$$C_6H_5NO_2 \xrightarrow{\text{Reduction}} C_6H_5NHOH \xrightarrow{\text{Rearrange-ment}} HO-C_6H_4-NH_2$$

Nitrobenzene → Phenyl-hydroxylamine → *p*-aminophenol

(*v*) **Catalytic Reduction**

$$C_6H_5NO_2 + 3H_2 \xrightarrow{Ni} C_6H_5NH_2 + 2H_2O$$

Nitrobenzene → Aniline

(*vi*) **Selective reduction.** It is possible to reduce one nitro group out of two in a dinitro compound by taking a mild reducing agent such as amm. hydrosulphide NH_4SH or sodium polysulphide Na_2S_x.

$$C_6H_4(NO_2)_2 + 3NH_4SH \longrightarrow NO_2C_6H_4NH_2 + 3S + 2H_2O + 3NH_3$$

m-Dinitrobenzene → *m*-Nitroaniline

Q. 7. What are different dinitrobenzenes? Give their methods of preparation and properties.

Ans. There are three dinitrobenzenes. They differ in the relative position of nitro groups.

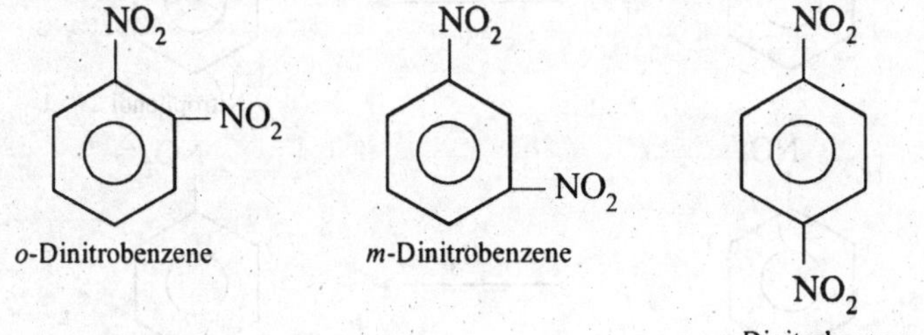

o-Dinitrobenzene *m*-Dinitrobenzene *p*-Dinitrobenzene

Preparation of *o*- and *p*-dinitrobenzene

These are obtained from corresponding nitroanilines, by subjecting them to diazotisation followed by treatment with sodium nitrite in the presence of copper.

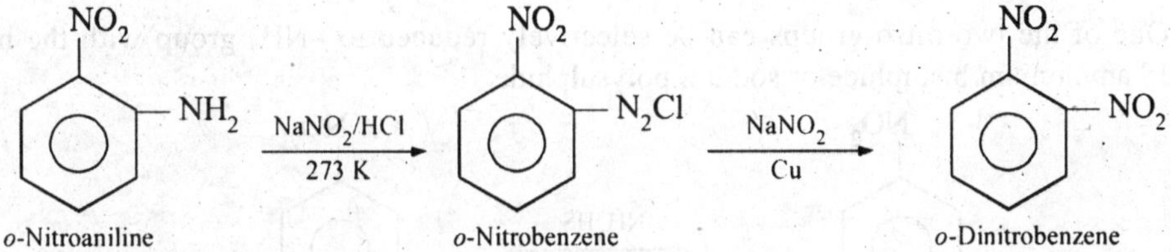

$$\text{o-Nitroaniline} \xrightarrow[273\ K]{NaNO_2/HCl} \text{o-Nitrobenzene diazonium chloride } (N_2Cl) \xrightarrow[Cu]{NaNO_2} \text{o-Dinitrobenzene}$$

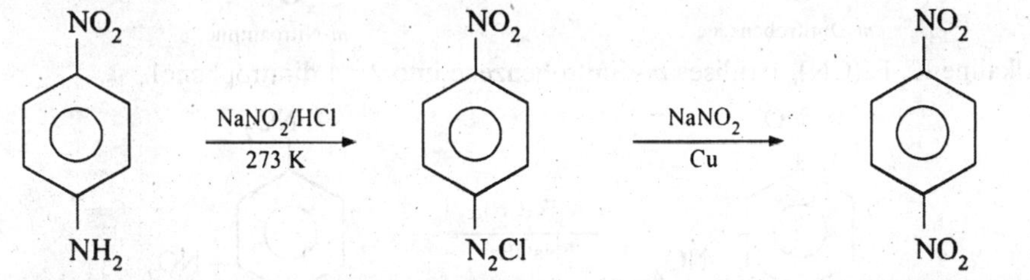

$$\text{p-Nitroaniline} \xrightarrow[273\ K]{NaNO_2/HCl} \text{p-Nitrobenzene diazonium chloride } (N_2Cl) \xrightarrow[Cu]{NaNO_2} \text{p-Dinitrobenzene}$$

Preparation of *m*-dinitrobenzene

m-dinitrobenzene is prepared by heating nitrobenzene with a mixture of conc. HNO_3 and conc. H_2SO_4 at 373 K.

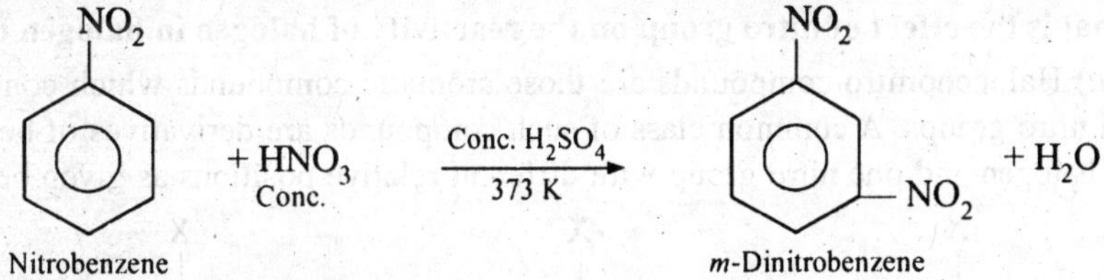

$$\text{Nitrobenzene} + \underset{\text{Conc.}}{HNO_3} \xrightarrow[373\ K]{\text{Conc. } H_2SO_4} \text{m-Dinitrobenzene} + H_2O$$

Properties of *o*- and *p*-dinitrobenzenes

o- and *p*-dinitrobenzenes allow one of the nitro groups to be replaced by groups like $-NH_2$ and $-OH$ in nucleophilic substitution reactions.

NO$_2$ NO$_2$ $\xrightarrow[\text{aq.}]{\text{NaOH}}$ NO$_2$ OH

o-Nitrophenol

NO$_2$ NO$_2$ $\xrightarrow{\text{NH}_3}$ NO$_2$ NH$_2$

p-Dinitrobenzene *p*-Nitroaniline

Properties of *m*-dinitrobenzenes

One of the two nitro groups can be selectively reduced to $-NH_2$ group with the help of NH_4HS ammonium bisulphide or sodium polysulphide.

NO$_2$ NO$_2$ $\xrightarrow{\text{NH}_4\text{HS}}$ NO$_2$ NH$_2$

m-Dinitrobenzene *m*-Nitroaniline

Alkaline K_3 $Fe(CN)_6$ oxidises *m*-dinitrobenzene into 2 : 4 dinitrophenol.

NO$_2$ NO$_2$ $\xrightarrow[\text{NaOH}]{\text{K}_3\text{Fe(CN)}_6}$ NO$_2$ NO$_2$ OH

m-Dinitrobenzene 2, 4 dinitrophenol

Q. 8. (*a*) What are halogenonitro compounds? Give their methods of preparation and properties.

(*b*) What is the effect of nitro group on the reactivity of halogen in halogen compounds?

Ans. (*a*) Halogenonitro compounds are those aromatic compounds which contain both the halogen and nitro groups. A common class of such compounds are derivatives of benzene which contain one halogen and one nitro group with different relative positions as given below:

X NO$_2$ X NO$_2$ X NO$_2$

o-halogennitrobenzene *m*-halogenonitrobenzene *p*-halogenonitrobenzene

Preparation of *o*- and *p*-halogenonitrobenzenes

Halogen compound will be the starting materials. It is subjected to nitration with a mixture of nitric and sulphuric acids. As the halogen group is *o-p* directing group, we obtain a mixture of *o*- and *p*-halogenonitrobenzene. The mixture can be subjected to fractional distillation to separate them into pure substances.

$$\text{Chlorobenzene} + HNO_3 \xrightarrow[H_2SO_4]{\text{Conc.}} o\text{-chloronitrobenzene} + p\text{-chloronitrobenzene}$$

Preparation of *m*-halogenonitrobenzene

Here the starting material is nitrobenzene. It is subjected to halogenation in the presence of iron. As the nitro group is meta directing, we shall obtain *m*-halogenonitrobenzene.

$$\text{Nitrobenzene} \xrightarrow[Fe]{Cl_2} m\text{-chloronitrobenzene} + HCl$$

Properties of halogenonitrobenzenes

Nucleophilic substitution reactions. *o*- and *p*-halogenonitrobenzenes give nucleophilic substitution reactions easily. Thus the halogen group is easily replaced by –OH, $-NH_2$, –OR groups. There is one more observation that greater the member of nitro groups in *ortho* or *para* position easier is the nucleophilic substitution. We compare the case of substitution of chloro group in 4-chloronitrobenzene, 2, 4-Dinitrochlorobenzene and 2, 4, 6-trinitrochlorobenzene. In the first case substitution reaction takes place with 15% NaOH solution, in the second case with a much weaker base, sodium carbonate and in the third case, warming with water is enough to perform the reaction.

$$\text{4-Chloronitrobenzene} \xrightarrow[423\text{ K}]{15\%\text{ Aq. NaOH}} \text{4-Nitrophenol} + NaCl$$

$$\text{2, 4-Dinitrochlorobenzene} \xrightarrow[403\text{ K}]{\text{Aq. } Na_2CO_3} \text{2, 4-Dinitrophenol}$$

Cl, O_2N, NO_2, NO_2 —Warm with water→ OH, O_2N, NO_2, NO_2

2, 4, 6 Trinitrochlorobenzene → 2, 4, 6 Trinitrophenol

(*b*) Influence of nitro groups on the reactivity of halogen towards nucleophilic substitution. Compare the reactivity of two compounds *viz.*, chlorobenzene and *o*-nitrochlorobenzene towards nucleophilic substitution reaction.

Chlorobenzene + NaOH —High Temp., High Pressure→ Phenol + NaCl

o-Nitrochlorobenzene + NaOH (15% Aq. sol) ⟶ *o*-Nitrophenol + NaCl

We find that in the second case, much easier conditions are required to perform the reaction. In a nucleophilic substitution reaction, the following steps are involved.

(I) Ar – X (Halogen compound) + Z^- (Nuclephile) ⟶ $\left[Ar{<}^{X}_{Z}\right]^-$ (Intermediate compound)

(2) $\left[Ar{<}^{X}_{Z}\right]^-$ (Intermediate compound) ⟶ $ArZ + X^-$ (New Product)

Thus for the reaction to take place efficiently, the intermediate compound needs to be stabilized. The intermediate compound, a carbonion, is resonance-stabilized in the presence of nitro group. But stabilisation is more if the nitro group is present is ortho or para position with respect to halogen group. This is shown below.

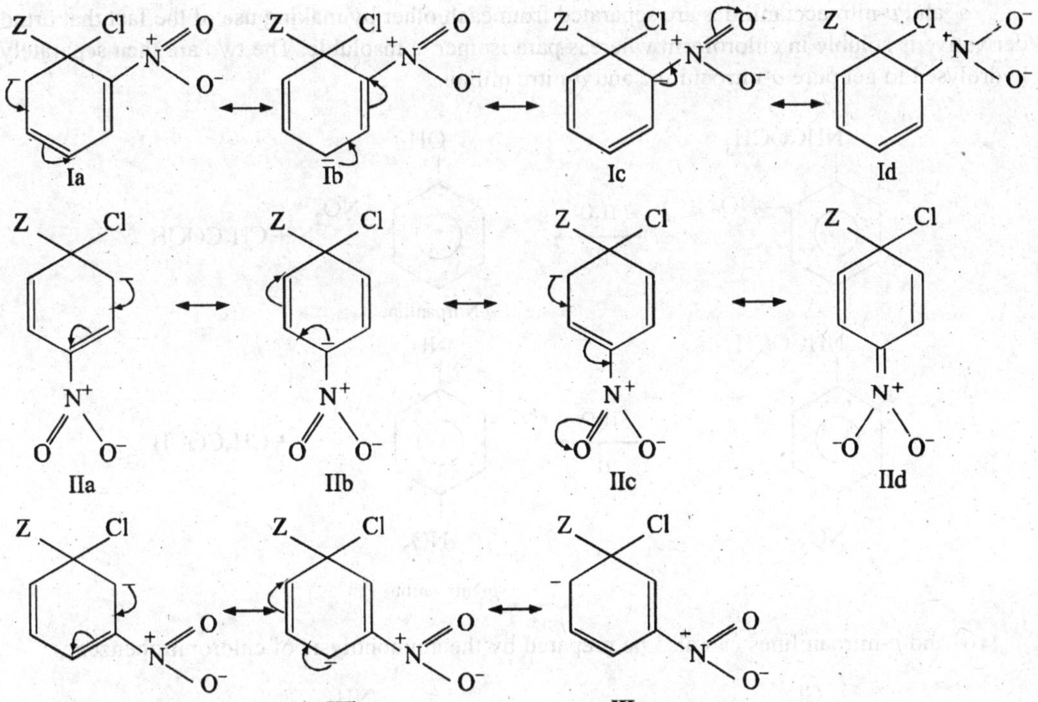

Ia ⟷ Ib ⟷ Ic ⟷ Id

IIa ⟷ IIb ⟷ IIc ⟷ IId

IIIa ⟷ IIIb ⟷ IIIc

It may be noted that in case of *ortho* or *para* nitrochlorobenzene, the intermediate product has 4 resonating structures each Ia, Ib, Ic and Id (for *ortho*) and IIa, IIb, IIc and IId (for *para*) whereas there are only three resonating structures in case of meta IIIa, IIIb and IIIC. Greater the number of resonating structures for a compound, greater is the stabilization. Hence intermediate compound is more stabilized in case of *ortho* and *para* nitrochlorobenzenes compared to *m*-nitrochlorobenzene.

Moreover structures Ic and IIb are particularly stable as there is a negative charge directly to neutralize the +ve charge on nitro group. No such stable structure is present in the case of meta isomer. Thus it is proved that nitro group if present in ortho or para position with respect to chloro group, enhances nucleophilic substitution for the chloro group.

Again, every nitro group has its effect to increase the nucleophilic substitution ability of chloro group. Hence greater the number of nitro groups in the ring, greater is the ease of nucleophilic substitution reaction.

Q. 9. Give the methods of preparation and properties of *o*-, *m*- and *p*-nitroanilines.

Ans. Preparation of o- and p-nitroanilines

These are prepared by the nitration of aniline. But before nitration, it is necessary to protect the $-NH_2$ group by acetylation.

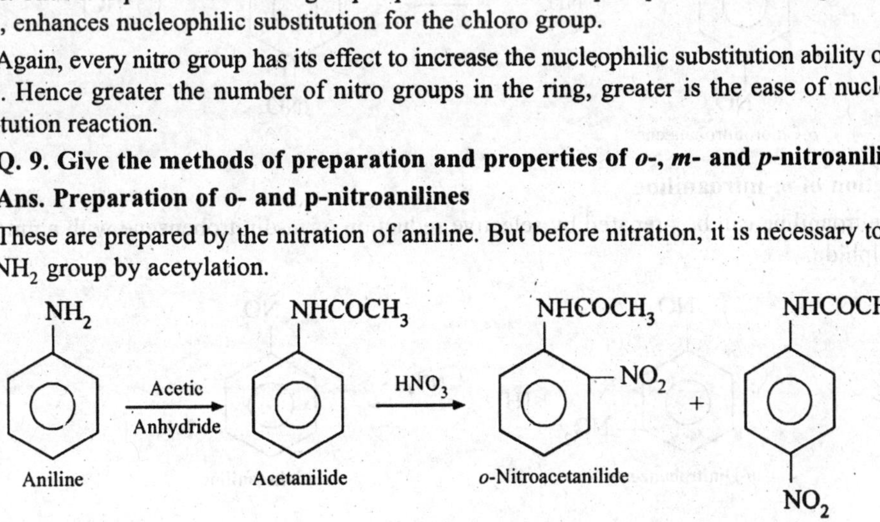

NH_2 (Aniline) $\xrightarrow[\text{Anhydride}]{\text{Acetic}}$ $NHCOCH_3$ (Acetanilide) $\xrightarrow{HNO_3}$ $NHCOCH_3$, NO_2 (*o*-Nitroacetanilide) + $NHCOCH_3$, NO_2 (*p*-Nitroacetanilide)

o- and *p*-nitroacetanilides are separated from each other by making use of the fact that ortho derivative is soluble in chloroform whereas para isomer is insoluble. The two are then separately hydrolysed to get pure *o*-nitroaniline and *p*-nitroaniline.

$$o\text{-}NO_2C_6H_4NHCOCH_3 \xrightarrow[H^+]{H_2O} o\text{-}NO_2C_6H_4OH + CH_3COOH$$

o-Nitroaniline

$$p\text{-}NO_2C_6H_4NHCOCH_3 \xrightarrow[H^+]{H_2O} p\text{-}NO_2C_6H_4NH_2 + CH_3COOH$$

p-Nitroaniline

o- and *p*-nitroanilines can also be prepared by the ammonolysis of chloronitrobenzenes.

$$o\text{-}ClC_6H_4NO_2 + NH_3 \longrightarrow o\text{-}NH_2C_6H_4NO_2 + HCl$$

o-Chloronitrobenzene

$$p\text{-}ClC_6H_4NO_2 + NH_3 \longrightarrow p\text{-}NH_2C_6H_4NO_2 + HCl$$

p-Chloronitrobenzene

Preparation of *m*-nitroaniline

m-nitroaniline can be prepared by selective reduction of m-dinitrobenzene with ammonium hydrosulphide.

$$m\text{-}C_6H_4(NO_2)_2 + [H] \xrightarrow{NH_4HS} m\text{-}NO_2C_6H_4NH_2$$

m-Dinitrobenzene *m*-Nitroaniline

Preperties of nitroanilines

(*i*) *o*-, *m*- and *p*-nitroanilines on reduction with metal and acid, yield the corresponding phenylene diamines.

$$o\text{-Nitroaniline} \xrightarrow{Sn/HCl} o\text{-Phenylene diamine}$$

$$m\text{-Nitroaniline} \xrightarrow{Sn/HCl} m\text{-Phenylene diamine}$$

$$p\text{-Nitroaniline} \xrightarrow{Sn/HCl} p\text{-Phenylene diamine}$$

(*ii*) On boiling with sodium hydroxide solution, *o*-and *p*-nitroanilines change into corresponding nitrophenols.

$$o\text{-Nitroaniline} \xrightarrow{NaOH} o\text{-Nitrophenol}$$

$$p\text{-Nitroaniline} \xrightarrow{NaOH} p\text{-Nitrophenol}$$

Q. 10. Give the methods of preparation and properties of nitrophenols.

Ans. There are three nitrophenols as given below:

o-Nitrophenol *m*-Nitrophenol *p*-Nitrophenol

Methods of preparation of *o*-and *p*-nitrophenols

(*i*) **By nitration of phenols.** Phenolic group is *o-p* directing group. Hence it will direct the nitro group in *o*- and *p*-positions on nitration of phenols

$$\text{Phenol} \xrightarrow[\text{HNO}_3]{\text{Dil.}} o\text{-Nitrophenol} + p\text{-Nitrophenol} + H_2O$$

The two are separated by steam distillation. *Ortho* isomer distils over leaving behind the *para* isomer.

(*ii*) **By hydrolysis of *o*- and *p*-chloronitrobenzenes**

$$o\text{-Chloronitrobenzene} + NaOH \longrightarrow o\text{-Nitrophenol} + NaCl$$

$$p\text{-Chloronitrobenzene} + NaOH \longrightarrow p\text{-Nitrophenol} + NaCl$$

Preparation of *m*-nitrophenol. It is prepared from *m*-dinitrobenzene as follows:

$$m\text{-Dinitrobenzene} \xrightarrow{NH_4HS} m\text{-Nitroaniline} \xrightarrow[273\text{ K}]{NaNO_2/HCl} m\text{-Nitrobenzene diazonium chloride} \xrightarrow[\text{warm}]{H_2O} m\text{-Nitrophenol}$$

Properties. 1. *Ortho* nitrophenol behaves differently as regards solubility in water. Out of the three isomers, *ortho* nitrophenol is least soluble in water. It can however be steam distilled out of the three isomers, *ortho* has the lowest m.p. followed by *meta* and *para* isomers. The different physical properties of *ortho* isomer from the other two can be explained in terms of *intramolecular* hydrogen bonding *i.e.* hydrogen bonding within the same molecule. The nitro and phenolic groups are so close to each other that a hydrogen bond, as shown below is formed.

O — H····O=N→O (o-nitrophenol, intramolecular hydrogen bond)

This eliminates the chances of hydrogen bond between the molecules of *o*-nitrophenol and water and hence accounts for its insolubility in water. The other isomers form intermolecular hydrogen bond with water as shown below and are water soluble.

······O—H······O—H······O—H······ (p-nitrophenol, H_2O, p-nitrophenol; NO_2, H, NO_2)

······O—H······O—H······O—H······O—H······ (m-nitrophenol, H_2O, m-nitrophenol, H_2O; NO_2, H, NO_2, H)

Intramolecular hydrogen bonding also accounts for lowest m.p. for *ortho* nitrophenol. As the molecular size shrinks due to hydrogen bonding, it melts and boils at a lower temp. compared to *m*- and *p*-isomers.

2. Nitrophenols are acidic compounds and acidity is enhanced due to electron withdrawing $-NO_2$ group. They turn blue litmus into red and react with sodium carbonate to give salts.

$$2\ C_6H_4(OH)(NO_2) + Na_2CO_3 \longrightarrow 2\ C_6H_4(ONa)(NO_2) + CO_2 + H_2O$$

p-nitro sod. phenoxide

3. Reduction of nitrophenols gives aminophenols.

$$C_6H_4(OH)(NO_2) + 6\,[H] \longrightarrow C_6H_4(ONa)(NH_2) + 2H_2O$$

o-nitrophenol → *o*-aminophenol

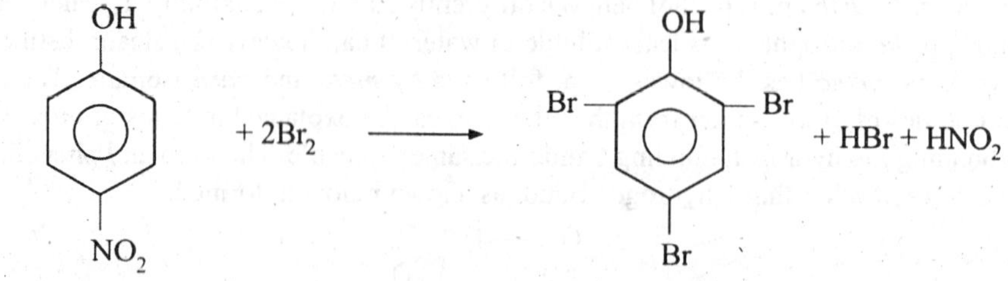

$$p\text{-nitrophenol} + 2Br_2 \longrightarrow \text{2, 4, 6-Tribromophenol} + HBr + HNO_2$$

p-nitrophenol

2, 4, 6-Tribromophenol

Q. 11. Write a note on effect of nitro group on the reactivity of phenolic group.

Ans. Nitro group is an electron-withdrawing group. Hence it helps in the release of protons. The phenoxide ion gets resonance stabilized by the nitro group. This is the reason why nitrophenols are more acidic then phenols.

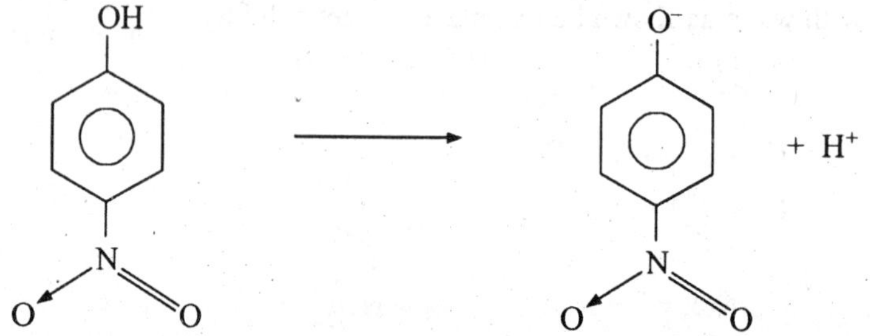

Phenoxide ion

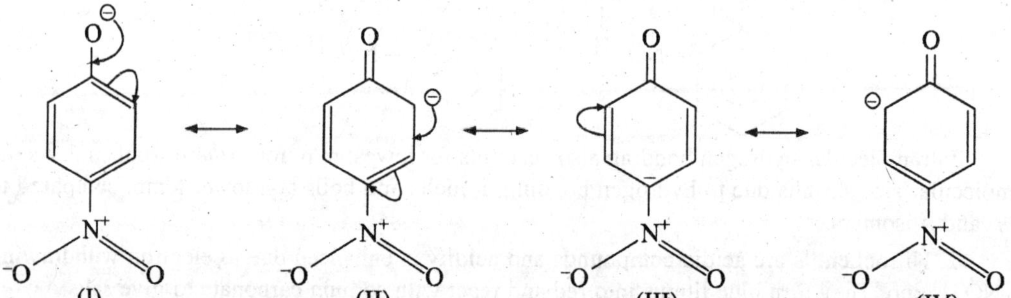

(I) (II) (III) (IV)

Out of resonating structures I, II, III & IV, structure III is particularly stable as it has a negative charge to directly neutralise the positive charge on nitro nitrogen. But, it may be noted that position of nitro group relative to the phenolic group is also important. We observe that nitro group in *ortho* and *para* positions and not *meta* position, enhance the acidity of nitrophenols.

Q. 12. Write the methods of preparation and properties of 2, 4, 6-trinitrophenol (picric acid).

Ans. Methods of preparation

1. From chlorobenzene

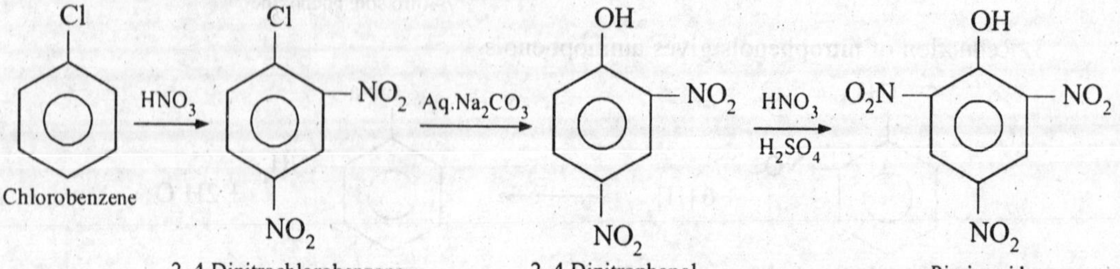

$$\text{Chlorobenzene} \xrightarrow{HNO_3} \text{2, 4 Dinitrochlorobenzene} \xrightarrow{Aq.Na_2CO_3} \text{2, 4 Dinitrophenol} \xrightarrow[H_2SO_4]{HNO_3} \text{Picric acid}$$

Chlorobenzene

2, 4 Dinitrochlorobenzene

2, 4 Dinitrophenol

Picric acid

2. From phenol

$$C_6H_5OH \xrightarrow{H_2SO_4} C_6H_3(OH)(SO_3H)_2 \xrightarrow{HNO_3} C_6H_2(OH)(NO_2)_3$$

Picric acid

Properties. 1. Due to the presence of three NO_2 groups, it is much stronger acid than phenol. It can decompose bicarbonates, a property normally given by carboxylic compounds and not phenolic compounds.

$$C_6H_2(OH)(NO_2)_3 + NaHCO_3 \longrightarrow C_6H_2(ONa)(NO_2)_3 + CO_2 + H_2O$$

Picric acid — Sod. picrate

2. With phosphorus pentachloride, it gives picryl chloride

$$C_6H_2(OH)(NO_2)_3 \xrightarrow{PCl_5} C_6H_2(Cl)(NO_2)_3$$

3. On selective reduction with sodium sulphide, picric acid gives picramic acid.

$$C_6H_2(OH)(NO_2)_3 \xrightarrow[\text{Reduction}]{Na_2S} C_6H_2(OH)(NO_2)_2(NH_2)$$

Picric acid — Picramic acid

Q. 13. A compound $C_7H_5O_6N_3$ (A) undergoes oxidation with acidified potassium dichromate to give a monocarboxylic acid $C_7H_3O_8N_3$ (B). When (B) is heated in acetic acid solution, $C_6H_3O_6N_3$ (C) is formed. What are (A), (B) and (C)? Give the reactions involved.

Ans. Compound A contains nitrogen, it could be a nitro or amino compound. Upon oxidation, (A) gives a carboxylic acid (B). It is the alkyl groups in benzene rings that are oxidised to –COOH. On heating (B), the compound (C) is formed which contains two oxygens less than (B). This step appears to be decarboxylation reaction. Based on these observations, the compounds, (A), (B) and (C) are:

CH_3, O_2N, NO_2, NO_2 (A) COOH, O_2N, NO_2, NO_2 (B) O_2N, NO_2, NO_2 (C)

$$\text{Trinitrotoluene (A)} \xrightarrow{K_2Cr_2O_7} \text{2, 4, 6 Trinitrobenzoic acid (B)} \xrightarrow[\text{CH}_3\text{COOH solution}]{\text{Heat}} \text{1, 3, 5 Trinitrobenzene (C)}$$

Q. 14. Identify the unknown compounds (A), (B) and (C) in the following:

(*i*) $$\text{Benzene} \xrightarrow[H_2SO_4]{HNO_3} \textbf{(A)} \xrightarrow[H_2SO_4]{HNO_3} \textbf{(B)} \xrightarrow{(NH_4)_2S} \textbf{(C)}$$

(*ii*) $$\text{Benzene} \xrightarrow[H_2SO_4]{HNO_3} \textbf{(A)} \xrightarrow[FeBr_3]{Br_2} \textbf{(B)} \xrightarrow{Sn/HCl} \textbf{(C)}$$

Ans. (*i*) $$\text{Benzene} \xrightarrow[H_2SO_4]{HNO_3} \text{Nitrobenzene (A)} \xrightarrow[H_2SO_4]{HNO_3} m\text{-Dinitrobenzene (B)} \xrightarrow{(NH_4)_2S} m\text{-Nitroaniline (C)}$$

(*ii*) $$\text{Benzene} \xrightarrow[H_2SO_4]{HNO_3} \text{Nitrobenzene (A)} \xrightarrow[FeBr_3]{Br_3} m\text{-Bromonitrobenzene (B)} \xrightarrow{Sa/HCl} m\text{-Bromoaniline}$$

Q. 15. How will you synthesise the following:

(*a*) 2, 4, 6 Trinitrotoluene from benzene. (*Arunachal, 2004*)

(*b*) Nitrosobenzene from nitrobenzene.

Ans. (*a*) **2, 4, 6 Trinitrotoluene from benzene**

$$\text{Benzene} \xrightarrow[AlCl_3]{CH_3Cl} \text{Toluene} \xrightarrow[H_2SO_4,\ 500\ K]{HNO_3} \text{Trinitrotoluene}$$

(*b*) **Nitrosobenzene from nitrobenzene**

$$C_6H_5NO_2 \xrightarrow[NH_4Cl]{Zn} C_6H_5NHOH \xrightarrow[H_4SO_4]{Na_2Cr_2O_7} C_6H_5NO$$

Nitrobenzene — Phenyl hydroxylamine — Nitrosobenzene

Q. 16. Explain the following:

(*i*) Nitration of toluene is easier than that of benzene.

(*ii*) Nitration of nitrobenzene is more difficult than that of benzene.

(*iii*) Nitrobenzene is often used as a solvent in Friedel-Crafts reaction.

(*iv*) s-Trinitrobenzene is prepared from TNT rather than by direct nitration of benzene.

(*v*) p-nitrophenol decomposes sodium carbonate whereas phenol does not.

Ans. (*i*) Nitration of aromatic hydrocarbons is an electrophilic substitution reaction. Toluene contains a methyl group which exerts an electron repelling inductive effect. It makes the benzene ring richer in electron cloud thereby facilitating the nitration process.

(*ii*) Nitrobenzene is difficult to nitrate than benzene. Nitro group is an electron withdrawing group. It deactivates the benzene ring for further electrophilic substitution because it creates a positive centre on the ring due to the electron withdrawing effect of nitro group. Hence further nitration becomes difficult.

(*iii*) Nitrobenzene is often used as a solvent in Friedel-Crafts reaction because nitrobenzene itself does not participate in Friedel-Crafts reaction i.e. it does not get alkylated or acylated itself. This is due to the deactivating effect of $-NO_2$ on further electrophilic substitution.

(*iv*) *s*-Trinitrobenzene is prepared from trinitrotoluene rather than by direct nitration of benzene. This is so because of the deactivating effect of $-NO_2$. It is quite difficult to prepare *m*-dinitrobenzene from nitrobenzene and still more difficult to prepare *s*-trinitrobenzene from *m*-dinitrobenzene. Hence toluene is trinitrated to get trinitrotoluene which is oxidised to trinitrobenzoic acid and finally the carboxylic group is decarboxylated to obtain *s*-trinitrobenzene. This is achieved easily because $-CH_3$ in toluene is an activating group.

$$\text{Toluene} \xrightarrow[H_2SO_4]{HNO_3} \text{T.N.T.} \xrightarrow{[O]} \text{2,4,6-trinitrobenzoic acid (COOH)} \xrightarrow[\text{heat}]{\text{Sodalime}} s\text{-Trinitrobenzene}$$

s-Trinitrobenzene (*s* stands for symmetrical or 1, 3, 5 substituted)

(*v*) Phenol is not strong enough an acid to decompose sodium carbonate. When a nitro group is inducted in *p*-position, the acid strength of phenol is increased due to the electron-withdrawing effect of $-NO_2$. It is now able to decompose sodium carbonate.

Q. 17. Arrange the following compounds in order of their decreasing activity towards the hydroxide ion.

Benzene, chlorobenzene, *p*-nitrochlorobenzene, *m*-nitrochlorobenzene, 2, 4 dinitrochlorobenzene and 2, 4, 6-trinitrochlorobenzene.

Ans. Nitro group in the *ortho* and *para* position has the effect of increasing reactivity towards nucleophilic substitution in aromatic compounds. Greater the number of such nitro groups, more pronounced will be the effect. Hence the order of decreasing activity towards hydroxide ion will be as follows.

2, 4, 6-trinitrochlorobenzene > 2, 4-dinitrochlorobenzene > *p*-nitrochlorobenzene > *m*-nitrochlorobenzene (due to –I effect) > chlorobenzene > benzene.

Q. 18. Which reagent is used for the reduction of one of nitro groups of *m*-dinitrobenzene?

Or

How will you synthesise *m*-nitroaniline from nitrobenzene. (*Delhi 2003*)

Ans. Ammonium hydrogen sulphide is used for partial reduction of *m*-dinitrobenzene.

NO_2 NO_2 NO_2

$\xrightarrow[H_2SO_4]{HNO_3}$ $-NO_2$ $\xrightarrow{NH_4HS}$ $-NH_2$

nitrobenzene *m*-dinitrobenzene *m*-nitroaniline

Q. 19. Explain why nitroethane dissolves in alkali.

Ans. See Q. 4. Acidic nature of nitroalkanes.

Q. 20. Explain why nitrobenzene undergoes electrophilic as well as nucleophilic substitution reactions.

Ans. See Q. 6.

Q. 21. (*a*) *o*-nitrophenol has lower m.p. and b.p. as compared to *m*- and *p*-isomers. Explain.

(*b*) Nitrophenols are more acidic than phenols. Explain.

Ans. (*a*) *o*-nitrophenol has lower m.p. and b.p. as compared to *m*- and *p*-isomers because hydrogen bonding takes place in *o*-isomer as depicted below.

O — H
O
N
O

This results in shrinking of size of the molecule resulting in lower m.p. and b.p.

(*b*) Nitrophenols are more acidic than phenols because nitro group is electron attracting. This helps in the release of protons.

Also nitrophenoxide ion obtained after the release of proton stabilizes itself by resonance

Q. 22. *o*-nitrophenol is steam volatile but *p*-nitrophenol is not. Explain.

Ans. *o*-nitrophenol exhibits intramolecular hydrogen bonding which results in lowering of boiling point. Hence it can be steam distilled. *p*-nitrophenol shows no such phenomenon, as such it has a higher boiling point. It is not steam volatile.

Q. 23. What products are obtained when nitrobenzene is reduced under different conditions. Give necessary reactions.

Ans. See Q. 6.

Q. 24. Why *p*-nitrochlorobenzene can be more easily hydrolysed by KOH than chlorobenzene?

Ans. See Q. 8.

Q. 25. Account for the following conversions giving only equation with reaction conditions. *(Lucknow, 2000)*

(*i*) $H_2N-C_6H_4-C_6H_4-NH_2 \longrightarrow F-C_6H_4-C_6H_4-F$

(*ii*) p-$CH_3C_6H_4NH_2 \longrightarrow$ p-$CH_3C_6H_4NO_2$

(*iii*) p-$NO_2C_6H_4NH_2 \longrightarrow$ p-$NO_2C_6H_4NO_2$

Ans.

(*i*) $H_2N-C_6H_4-C_6H_4-NH_2 \longrightarrow ClN=N-C_6H_4-C_6H_4-N=NCl \xrightarrow[\Delta]{HBF_4} F-C_6H_4-C_6H_4-F$

(*ii*) $CH_3-C_6H_4-NH_2 \xrightarrow[HBF_4,\ NaNO_2,\ \Delta]{NaNO_2/HCl} CH_3-C_6H_4-NO_2$

(*iii*) $NO_2-C_6H_4-NH_2 \xrightarrow[HBF_4,\ NaNO_2,\ \Delta]{NaNO_2/HCl} NO_2-C_6H_4-NO_2$

Q. 26. Trinitrochlorobenzene reacts simply with water to form picric acid, while chlorobezene requires vigorous conditions to change into phenol. (*Garhwal, 2000*)

Ans. See Q. 8 (*b*)

Q. 27. Explain the facts that chlorination of nitrobenzene gives *m*-chloronitrobenzene, while nitration of chlorobenzene gives *o*-and *p*-chloronitrobenzene when both the chloro and nitro groups are deactivating. (*Garhwal, 2000; Punjab, 2003*)

Ans. In nitrobenzene, nitro group is *meta* directing, therefore, we shall obtain *m*-chloronitrobenzene on chlorination. While in chlorobenzene, chloro group is *ortho* and *para* directing, therefore, we shall obtain *o*-and *p*-chloronitrobenzene on nitration.

Q. 28. Explain the reduction of nitrobenzene in neutral medium. (*Garhwal, 2000*)

Ans. See Q. 4, 2 (*c*)

Q. 29. How will you do the following conversion : (*Garhwal, 2000*)

Nitrobenzene → *m*-bromochlorobenzene

Ans.

$C_6H_5NO_2 \xrightarrow[Fe,\ 413\ K]{Br_2} m\text{-}Br-C_6H_4-NO_2 \xrightarrow[H_2]{Ni} m\text{-}Br-C_6H_4-NH_2 \xrightarrow[HBF_4,\ NaNO_2,\ \Delta]{NaNO_2/HCl} m\text{-}Br-C_6H_4-NO_2$

nitrobenzene — *m*-bromochlorobenzene

Q. 30. What happens when nitrobenzene is reduced in an alkaline solution ? (*Kerala, 2000*)

Ans. See Q. 6(5), (*iii*)

Q. 31. Complete the following reaction : (*M. Dayanand, 2000*)

NO_2, NO_2 $+ NH_4SH \longrightarrow$

Ans. See Q. 6 (5) (*vi*).

Q. 32. Explain the reduction of nitrobenzene under different conditions.

(*Kumaon, 2000*)

Ans. See Q. 6 (5).

Q. 33. Starting from nitrobenzene, how will you get the following compounds:

(*Meerut, 2000*)

(*i*) **Phenol**
(*ii*) **benzoic acid**
(*iii*) **benzene**
(*iv*) **bromobenzene**
(*v*) **benzylamine**

Ans. (*i*) NO_2 (nitrobenzene) $\xrightarrow{Sn/HCl}$ NH_2 $\xrightarrow[273\ K]{NaNO_2\ /\ HCl}$ N_2Cl $\xrightarrow{H_2O}$ OH (phenol)

(*ii*) NO_2 (nitrobenzene) $\xrightarrow{Sn/HCl}$ NH_2 $\xrightarrow[273\ K]{NaNO_2\ /\ HCl}$ N_2Cl $\xrightarrow{CuCN}$ CN $\xrightarrow{H_2O}$ COOH (benzoic acid)

(*iii*) & (*iv*) NO_2 (nitrobenzene) $\xrightarrow{Sn/HCl}$ NH_2 $\xrightarrow[273\ K]{NaNO_2\ /\ HCl}$ N_2Cl $\xrightarrow[H_2O]{H_3PO_2}$ benzene $\xrightarrow[Br_2]{Fe}$ Br (Bromobenze)

Q. 34. An aromatic compound A ($C_6H_5O_2N$) when reduced with Fe/HCl gave compound B(C_6H_7N), which on diazotisation gave compound C which on boiling with acidified water gave compound D (C_6H_6O). D in alkaline medium reacts with carbon dioxide to form E ($C_7H_6O_3$), E on heating with soda lime gives back D. With explanations and equations, identify compounds A to E. (*Delhi 20002*)

Ans. $A = C_6H_5NO_2$ $B = C_6H_5NH_2$ $C = C_6H_5N_2Cl$

$D = C_6H_5OH$ $E = C_6H_5(OH)\ COOH$

Equations

$$\underset{A}{C_6H_5NO_2} \xrightarrow{Fe/HCl} \underset{B}{C_6H_5NH_2} \xrightarrow[HCl]{NaNO_2} \underset{C}{C_6H_5N_2Cl}$$

$$\underset{C}{C_6H_5N_2Cl} \xrightarrow[H_2O]{H^+} \underset{D}{C_6H_5OH}$$

$$\underset{D}{C_6H_5OH} \xrightarrow[NaOH]{CO_2} \underset{\text{E, Salicylic acid}}{C_6H_4(OH)COOH} \longrightarrow \underset{\text{Phenol, D}}{C_6H_5OH}$$

27

AMINES

Q. 1. What are amines? What are the different types of amines? Give their nomenclature.

Ans. Amines are organic compounds containing nitrogen. They may be considered as derivatives of ammonia (NH_3), in which one or more hydrogen atoms have been replaced by alkyl or aryl groups. They may be classified as under:

Primary amines. These are the compounds in which one hydrogen atom in ammonia is replaced by an alkyl or aryl group. For example,

$CH_3 - NH_2$	$CH_3 - CH_2 - NH_2$	$C_6H_5NH_2$
Methyl amine	Ethyl amine	Phenyl amine

Secondary amines. These are the compounds in which two hydrogen atoms are replaced by two, same or different, alkyl or aryl groups.

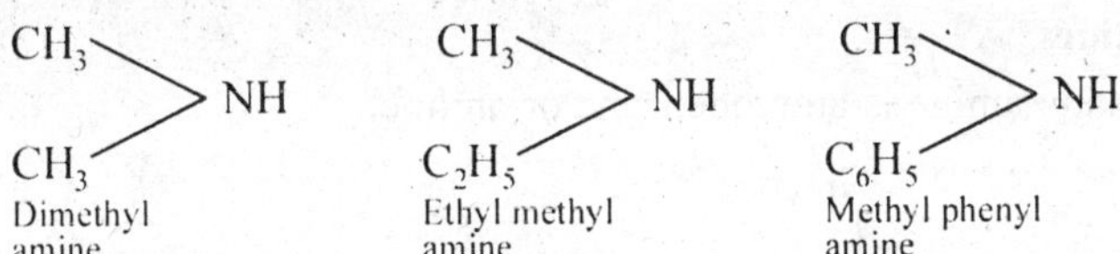

Tertiary amines. These are the compounds in which all the three hydrogens of ammonia are replaced by three alkyl or aryl groups.

$(CH_3)(CH_3)(CH_3) > N$	$(CH_3)(CH_3)(C_2H_5) > N$	$(CH_3)(C_2H_5)(C_6H_5) > N$
Trimethyl amine	Ethyl dimethyl amine	Ethyl methyl Phenyl amine

There is another classification of amines, too.

Alkyl amines. The amines which contain only alkyl groups in the molecule are known as alkyl amines.

Aryl amines. Amines containing only aryl groups are known as aryl amines.

Alkylaryl amines. Amines containing both alkyl and aryl groups in the molecule are known as alkylarly amines.

Secondary and tertiary amines are classified as under:

Simple amines. Amines containing same alkyl or aryl groups attached to nitrogen are called single amines.

Mixed amines. If the alkyl or aryl groups attached to nitrogen are different, the amine is called mixed amine.

NOMENCLATURE OF AMINES

Compound	*Common Name*	*Name IUPAC*
Primary Amines		
CH_3-NH_2	Methyl amine	Aminomethane
$CH_3-CH_2-NH_2$	Ethyl amine	Aminoethane
$CH_3-CH_2-CH_2-NH_2$	Propyl amine	Aminopropane
$CH_3-\underset{\underset{CH_3}{\vert}}{CH}-NH_2$	Isopropyl amine	2-methylaminoethane
Secondary Amines		
$CH_3-NH-CH_3$	Dimethyl amine	N-methylaminomethane
$CH_3-NH-C_2H_5$	Ethyl methyl amine	N-methylaminoethane
$CH_3-NH-C_3H_7$	methyl propyl amine	N-methylaminopropane
Tertiary Amines		
$CH_3-\underset{\underset{CH_3}{\vert}}{N}-CH_3$	Trimethyl amine	N, N-dimethylaminomethane
$CH_3-\underset{\underset{C_2H_5}{\vert}}{N}-C_2H_5$	Diethyl methyl amine	N-ethly-N-methylaminoethane

It may be noted that when different alkyl groups are attached to nitrogen, the biggest alkyl group forms part of aminoalkane. The notation N-methyl means that methyl group is attached to nitrogen.

Aromatic amines

Simplest aromatic amine is aminobenzene or aniline.

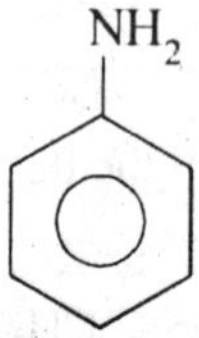

Aminobenzene or aniline

IUPAC name of aniline is benzenamine.

There are amines in which other positions in the benzene ring are occupied by other groups. Their position, relative to $-NH_2$ group is identified. For this purpose, we give number 1 to the position where $-NH_2$ group is attached. Then the number increases as we move clockwise.

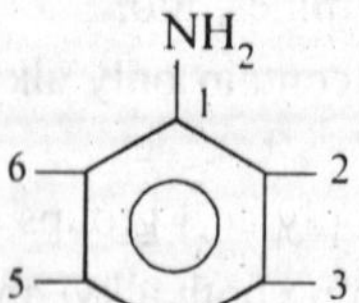

By another convention, the position next to the amine group is ortho (*o*), with gap of one position, meta (*m*) and that vertically opposite to $-NH_2$ group is para (*p*).

NH$_2$ with positions: ortho, ortho, meta, meta, para

The compounds are named accordingly. Sometimes there are special names, too.

Compound	Name	Special name
$C_6H_5NH_2$	Aminobenzene (IUPAC name is benzenamine)	Aniline
NH$_2$, CH$_3$ (ortho)	2-methylaniline (IUPAC name is 2-methylbenzenamine)	*o*-Toluidine
NH$_2$, CH$_3$ (para)	4-methylaniline	*p*-Toluidine
NH$_2$, NH$_2$ (para)	*p*-diaminobenzene	*p*-Phenylene diamine
NH$_2$, OCH$_3$ (para)	*p*-methoxy aniline	Anisidine

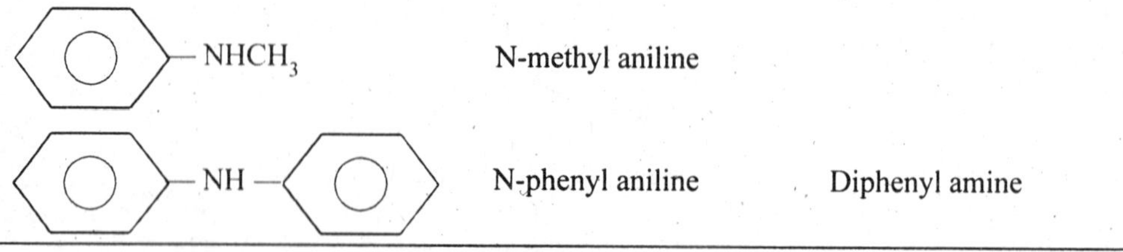

Q. 2. Describe the shape of amine molecules.

Ans. As amines are derivatives of ammonia, the shape of the former is basically the same as that of the latter. We know that ammonia uses sp^3 hybridisation. Out of the four, three hybridised orbitals are oriented towards the corners of a tetrahedron. The fourth orbital contains a lone pair of electrons. Thus the shape of NH_3 could be represented as:

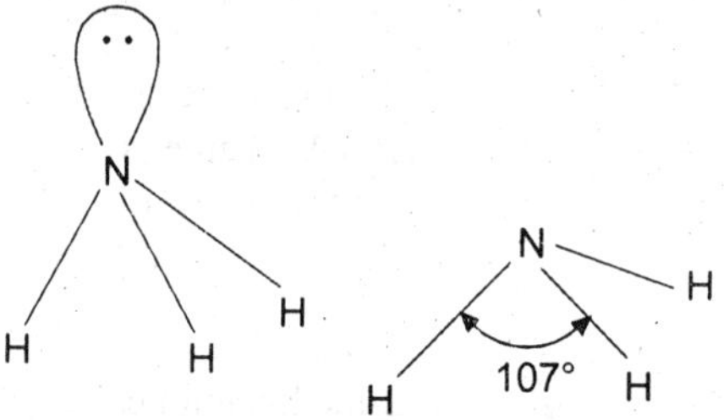

As amines molecules are obtained by replacing hydrogen by alkyl or aryl groups, their shape could be represented as:

N — R, R or H, R or H

As the angle ∠HNH in ammonia is known to be 107°, we expect an equal angle in case of amines also. It has been observed that trimethyl amine has an angle ∠CNC equal to 108°.

Q. 3. Describe general methods of preparation of amines.

(*Bangalore, 2004; Nagpur 2005*)

Ans. Common methods for preparing primary, secondary and tertiary amines.

1. Ammonolysis (Hoffmann's method). This reaction involves treating an alkyl halide or an aryl halide of the type $C_6H_5CH_2$ X with ammonia. Ammonia is a nucleophile. It attaches itself with the alkyl group and the halogen is removed as halide to give a molecule of primary amine or 1° amine. But the reaction does not stop here. The primary amine formed acts as a nucleophile and attaches itself to the alkyl halide molecule giving rise to secondary amine or 2° amines. Again secondary amine acts as nucleophile and attaches itself to a molecule of alkyl halide giving rise to tertiary amine or 3° amine. Finally tertiary amine attaches itself to the alkyl halide molecule to form the quaternary ammonium salt (4°). These reactions are explained as under.

$$H_3\ddot{N} + R - X \longrightarrow H_3\overset{\oplus}{N} - R + \overset{\ominus}{X}$$

$$\big\downarrow -H^{\oplus}$$

$$R - \ddot{N}H_2$$

1° Amine

$$R - \ddot{N}H_2 + R - X \longrightarrow R - \overset{\oplus}{N}H_2 - R + \overset{\ominus}{X}$$

$$\xrightarrow{-H^{\oplus}} \underset{(2^\circ \text{ Amine})}{R - NH - R}$$

$$R_2\ddot{N} + R - X \xrightarrow{-X^{\ominus}} R_3\overset{\oplus}{N}H \xrightarrow{-H^{\oplus}} \underset{(3^\circ \text{ Amine})}{R_3N}$$

$$R_3\ddot{N} + R - X \longrightarrow \overset{\oplus\ominus}{R_4NX}$$

Quaternary ammonium salt (4°)

For example, $CH_3 - Cl + NH_3 \xrightarrow{-HCl} \underset{\text{Methyl amine}}{CH_3NH_2} \xrightarrow[-HCl]{CH_3Cl} \underset{\text{Dimethyl amine}}{(CH_3)_2NH}$

$$\xrightarrow[-HCl]{CH_3Cl} \underset{\text{Trimethyl amine}}{(CH_3)_3N} \xrightarrow{CH_3Cl} \underset{\text{Tetramethyl ammonium chloride}}{(CH_3)_4N^+Cl^-}$$

Thus, we generally obtain a mixture of the above substance on the ammonolysis of methyl chloride.

However, reaction can be made to stop at the first step *i.e.* at the stage of primary amine by taking large excess of NH_3 so that alkyl halide is not available to the primary amine once it is formed.

For preparing aryl amines, aryl halides are treated with ammonia, but drastic conditions are used.

$$\underset{\text{Chlorobenzene}}{C_6H_5Cl} + NH_3 \xrightarrow[60\text{ Atm.}]{473\text{ K}} \underset{\text{Aniline}}{C_6H_5NH_2} + HCl$$

But it is easier to ammonolise aromatic compounds with a side chain.

$$\underset{\text{Benzyl chloride}}{C_6H_5CH_2Cl} + NH_3 \longrightarrow \underset{\text{Benzyl amine}}{C_6H_5CH_2NH_2} + HCl$$

2. By the action of ammonia on alcohols or phenols. On passing a mixture of alcohol and ammonia over heated alumina at 633 K, a mixture of primary, secondary and tertiary amines is obtained

$$C_2H_5OH + NH_3 \xrightarrow[633\ K,\ -H_2O]{Al_2O_3} C_2H_5NH_2 \xrightarrow[Al_2O_3\ (633\ K),\ -H_2O]{C_2H_5OH}$$

$$\underset{\text{Diethyl amine}}{(C_2H_5)_2NH} \xrightarrow[Al_2O_3\ (633\ K),\ -H_2O]{C_2H_5OH} \underset{\text{Triethyl amine}}{(C_2H_5)_3N}$$

$$\underset{\text{Phenol}}{C_6H_5OH} + NH_3 \xrightarrow{\text{Anhy. AlCl}_3} \underset{\text{Aniline}}{C_6H_5NH_2} + H_2O$$

3. Reductive amination of aldehydes and ketones. Aldehydes and ketones on reduction with hydrogen and ammonia change into primary amines at 373 K and 150 atm. pressure and in the presence of nickel as catalyst.

$$R-\overset{H}{\overset{|}{C}}=O + NH_3 \xrightarrow{-H_2O} \underset{\text{Aldimine}}{R-\overset{H}{\overset{|}{C}}=NH} \xrightarrow{H_2,\ Ni} \underset{\text{Primary amine}}{RCH_2NH_2}$$

$$R-\overset{R'}{\overset{|}{C}}=O + NH_3 \xrightarrow{-H_2O} \underset{\text{Ketimine}}{R-\overset{R'}{\overset{|}{C}}=NH} \xrightarrow{H_2,\ Ni} \underset{\text{Primary amine}}{R-\overset{R'}{\overset{|}{C}}HNH_2}$$

$$\underset{\text{Propanal}}{CH_3CH_2CHO} \xrightarrow[373\ K,\ 150\ Atm.]{NH_3,\ H_2,\ Ni} \underset{\text{Propyl amine}}{CH_3CH_2CH_2NH_2}$$

$$(CH_3)_2C=O \xrightarrow[373\ K,\ 150\ Atm.]{NH_3,\ H_2,\ Ni} \underset{\text{Isopropyl amine}}{(CH_3)_2CH-NH_2}$$

Mechanism

Reductive amination involves the formation of amine as an intermediate and proceeds through the following mechanism.

$$>C=O + :NH_3 \longrightarrow >C(O^-)(\overset{+}{N}H_3) \longrightarrow \left[>C(OH)(N-H)\underset{|}{} H\right] \xrightarrow{-H_2O} >C=NH \xrightarrow{H_2,\ Ni} -\overset{|}{C}H-NH_2$$

4. Reduction of nitro compounds. Nitro compounds on reduction or hydrogenation give rise to amines. Because of ready availability of nitrobenzene, this method is used to prepare aniline particularly.

(*i*) **Catalytic hydrogenation.** Nitro compound is treated with hydrogen gas in the presence of finely divided nickel or platinum.

$$C_6H_5-NO_2 + 3H_2 \xrightarrow{\text{Ni or Pt}} C_6H_5-NH_2 + 2H_2O$$

Nitrobenzene Aniline

We can't use this method to hydrogenate the nitro group in a compound containing ethylenic double bond, as the double bond would also get saturated.

(*ii*) **Reduction.** Here the nitro compound is dissolved in hydrochloric acid and treated with tin metal.

$$C_6H_5-NO_2 \xrightarrow[\text{6[H]}]{\text{Sn/HCl}} C_6H_5-NH_2 + 2H_2O$$

(*iii*) Lithium aluminium hydride could also be used as a reducing agent.

(*iv*) **Partial Reduction.** One of the two nitro groups in a dinitro compound can be reduced keeping the second intact, using ammonium hydrogen sulphide as the reducing agent.

$$m\text{-}C_6H_4(NO_2)_2 \xrightarrow{NH_4HS} m\text{-}NO_2C_6H_4NH_2$$

m-dinitro benzene *m*-nitro aniline

5. Reduction of nitriles or cyanides. Aliphatic and aromatic nitriles (of the type $C_6H_5CH_2CN$), can be reduced by catalytic hydrogenation or chemical reduction to produce amines. When the reducing agent is sodium and alcohol, the reduction is known as **Mendius Reaction or Reduction.**

$$R-C\equiv N + 2H_2 \xrightarrow[\text{433 K}]{\text{Ni}} RCH_2NH_2$$

Nitrile Primary amine

$$CH_3CN + 2H_2 \xrightarrow[\text{433 K}]{\text{Ni}} CH_3CH_2NH_2$$

Ethane nitrile Ethylamine

$$C_6H_5-CH_2CN + 2H_2 \xrightarrow[\text{433 K}]{\text{Ni}} C_6H_5-CH_2CH_2NH_2$$

Phenyl methyl cynide Phenyl ethylamine

6. Reduction of oximes and amides. Hydrogenation or chemical reduction of oximes (which are obtained from aldehydes and ketones) and amides yields amines.

$$CH_3CH = NOH + 4[H] \xrightarrow[\text{Alcohol}]{\text{Na}} CH_3CH_2NH_2 + H_2O$$

Acetaldoxime

$$\begin{matrix} CH_3 \\ CH_3 \end{matrix}\!\!>\!C = NOH + 4H \xrightarrow[\text{Alcohol}]{\text{Na}} \begin{matrix} CH_3 \\ CH_3 \end{matrix}\!\!>\!CHNH_2 + H_2O$$

Acetone oxime — Isopropyl amine

$$CH_3CONH_2 + 4[H] \longrightarrow CH_3CH_2NH_2 + H_2O$$

7. Hofmann's degradation of amides. On treatment with bromine and potassium hydroxide, amides are converted into amines.

$$\underset{\text{Amide}}{R-\overset{O}{\overset{\|}{C}}NH_2} + Br_2 + 4KOH \longrightarrow \underset{\text{Amine}}{RNH_2} + K_2CO_3 + 2KBr + 2H_2O$$

$$\underset{\text{Acetamide}}{CH_3CONH_2} + Br_2 + 4KOH \longrightarrow \underset{\text{Methyl amine}}{CH_3NH_2} + K_2CO_3 + 2KBr + 2H_2O$$

Mechanism

They hypobromite ion OBr^- produced by alkaline solution of Br_2 attacks the amide molecule. Various steps which are involved are illustrated hereunder.

(1) $R-C(=O)NH_2 + OBr^- \longrightarrow R-C(=O)-N(Br)-H + OH^-$

(2) $R-C(=O)-N(Br)-H + OH^- \longrightarrow R-C(=O)-N^-(Br) + H_2O$

(3) $R-C(=O)-N^-(Br) \longrightarrow R-C(=O)-\ddot{N}: + Br^-$

(4) $R-C(=O)-\ddot{N}: \longrightarrow R-N=C=O$

$$R-N=C=O + 2OH^- \longrightarrow R-NH_2 + CO_3^{2-}$$

There is rearrangement in step 4 where the alkyl group leaves carbon and attaches itself to nitrogen to form isocyanate which is hydrolysed to amine.

8. Gabriel Synthesis. Phthalimide is converted into potassium salt on treatment with KOH. Pot. phthalimide on treatment with an alkyl or arylalkyl halide forms N-substituted phthalimide, which is hydrolysed to produce phthalic acid and a primary amine. Hydrolysis is performed with 20% HCl under pressure.

$$\text{Phthalimide } C_6H_4(CO)_2NH \xrightarrow{KOH} \text{Pot. salt of phthalimide } C_6H_4(CO)_2NK \xrightarrow[-KX]{RX}$$

$$\text{N-alkyl Phthalimide } C_6H_4(CO)_2NR \xrightarrow{H^+,\ H_2O} \text{Phthalic acid } C_6H_4(COOH)_2 + \underset{\text{Primary amine}}{RNH_2}$$

9. From Grignard Reagent. Grignard regents react with chloramine in ether to produce amines.

$$CH_3MgBr + ClNH_2 \longrightarrow \underset{\text{Methyl amine}}{CH_3NH_2} + \underset{\text{Magnesium chloro bromide}}{Mg(Cl)Br}$$

Additional Methods for Preparing Secondary Amines

(*i*) **Reduction of isonitriles or isocyanides**

$$\underset{\text{Alkyl isocyanide}}{R-NC} + 2H_2 \xrightarrow{Ni} \underset{\text{Methyl alkyl amine}}{RNHCH_3}$$

(*ii*) **Reduction of N-alkyl amides**

$$\underset{\text{N-alkyl amide}}{R-\overset{O}{\overset{\|}{C}}-NHR'} \xrightarrow{LiAlH_4} \underset{\text{Sec. amine}}{RCH_2NHR'}$$

$$\underset{\text{N-methyl acetamide}}{CH_3CONHCH_3} + 2H_2 \xrightarrow[-H_2O]{Ni} \underset{\text{Ethyl methyl amine}}{CH_3CH_2NHCH_3}$$

(*iii*) From Grignard reagent. Substituted chloramine reacts with Grignard reagent to produce a secondary amine.

$$\underset{\text{Methyl mag. chloride}}{CH_3MgCl} + \underset{\text{Methyl chloramine}}{ClNHCH_3} \longrightarrow \underset{\text{Dimethyl amine}}{CH_3NHCH_3} + MgCl_2$$

Additional Methods for Preparing Tertiary Amines

(*i*) **Reduction of N, N Disubstituted amides**

$$\underset{\text{N, N Disubstituted amide}}{R\overset{O}{\overset{||}{C}}-N\begin{cases}R'\\R''\end{cases}} + 2H_2 \longrightarrow \underset{\text{Tertiary amine}}{RCH_2N\begin{cases}R'\\R''\end{cases}} + H_2O$$

(*ii*) **From Grignard reagent.** Disubstituted chloramine reacts with Grignard reagent to produce tertiary amine.

$$RMgCl + Cl\,N\begin{cases}R'\\R''\end{cases} \longrightarrow \underset{\text{Tertiary amine}}{R-N\begin{cases}R'\\R''\end{cases}} + MgCl_2$$

(*iii*) Decomposition of quaternary ammonium hydroxide. Quaternary ammonium salts are hydrolysed with moist silver oxide to produce quaternary ammonium hydroxides which are decomposed on heating to give tertiary amine.

$$[R_4N]^+X^- + AgOH \longrightarrow [R_4N]^+OH^- + AgX$$

$$[R_4N]^+OH^- \xrightarrow{\Delta} \underset{\text{Tert amine}}{R_3N} + ROH$$

Q. 4. (*a*) Describe Gabriel phthalimide synthesis. (*Garhwal, 2000; Coimbatore, 2000; A.N. Bhauguna, 2000; Agra, 2003; Purvanchal, 2007; Nagpur, 2008*)

(*b*) **How will you prepare ethylamine and benzylamine?**

(*c*) **Can this method be used to prepare aniline? Give reasons for your answer.**

Ans. (*a*) For description of Gabriel phthalimide synthesis see Q. No. 3 (Methods of preparation of amines).

(*b*) **Preparation of ethylamine and benzylamine.**

$$C_6H_4(CO)_2NH \xrightarrow{KOH} C_6H_4(CO)_2N^-K^+$$

$C_6H_5CH_2Cl$ ↓ $-KCl$ | $-KCl$ ↓ C_2H_5Cl

N-Benzyl phthalimide ($C_6H_4(CO)_2N-CH_2C_6H_5$) | N-Ethyl phthalimide ($C_6H_4(CO)_2NC_2H_5$)

↓ H^+ H_2O | ↓ H^+ H_2O

$C_6H_4(COOH)_2$ (Phthalic acid) $+ C_6H_5CH_2NH_2$ (Benzylamine) | $C_6H_4(COOH)_2$ (Phthalic acid) $+ C_2H_5NH_2$ (Ethylamine)

(*c*) Aniline cannot be prepared by this method. This is because chlorobenzene has no tendency to combine with potassium phthalimide to give N-phenyl phthalimide. This step is nucleophilic substitution and nuclear aromatic halogen compounds have no reactivity towards nucleophilic substitution reactions.

Q. 5. What do you understand by ammonolysis of halides? How will you prepare benzyl amine and ethylene diamine using this method?

Ans. For ammonolysis, see Q. No. 3 (Methods of preparation of amines)

Preparation of benzylamine

$$\underset{\text{Benzyl chloride}}{C_6H_5CH_2Cl} + NH_3 \longrightarrow C_6H_5CH_2NH_2 + HCl$$

Preparation of ethylene diamine.

$$\underset{\text{Ethylene dichloride}}{ClCH_2-CH_2-Cl} + 2NH_3 \longrightarrow \underset{\text{Ethylene diamine}}{NH_2CH_2CH_2NH_2} + 2HCl$$

Q. 6. Write a note on Hofmann degradation of amides.

(*Manipur2004; Meerut 2004; Nagpur 2005*)

Or

Write the mechanism and reagent used in the preparation of aniline from benzamide.

Ans. See Q. No. 3. Write C_6H_5- in place of R –

Q. 7. Describe methods for separating a mixture of primary, secondary and tertiary amines. (*Madurai 2002, Arunachal 2003; Burdwan 2004; Purvanchal 2007; Calcutta 2007; Nagpur 2008*)

Ans. Separation of Primary, Secondary and Tertiary Amines

Certain methods of preparation of amines lead to the formation of a mixture of primary, secondary and tertiary amines and quaternary ammonium salts. When a mixture of amines (or salts or amines) and quaternary ammonium salts is to be separated, it is first distilled with potassium hydroxide. The quaternary ammonium salt, being non-volatile, is left behind while the mixture of amines distils over. This mixture is then separated by the following methods:

(1) Hofmann's method. The mixture of amines is treated with diethyl oxalate when the primary amine forms a substituted oxamide (a solid), the secondary amine forms dialkyl oxamic ester (**a liquid**) and the tertiary amine does not react.

$$\begin{array}{l} CO\,OC_2H_5 \quad H\,NHR \\ | \\ CO\,OC_2H_5 \quad H\,NHR \end{array} \longrightarrow \begin{array}{l} CONHR \\ | \\ CONHR \end{array} + 2C_2H_5OH$$

Diethyl oxalate; Primary amine; Substituted oxamide (*Solid*)

$$\begin{array}{l} COOC_2H_5 \\ | \\ COOC_2H_5 \end{array} + HNR_2 \longrightarrow \begin{array}{l} CONR_2 \\ | \\ COOC_2H_5 \end{array} + C_2H_5OH$$

Diethyl oxalate; Secondary amine; Dialkyl oxamic ester (*Liquid*)

On distilling the reaction mixture, the tertiary amine distils over first and is collected as the distillate. The remaining mixture is cooled and filtered when oxamic ester passes as the filtrate and oxamide is left as the residue.

Oxamide and oxamic ester are now separately boiled with alkali to get back the original amines.

$$\begin{array}{l} CONHR \\ | \\ CONHR \end{array} + 2KOH \longrightarrow \begin{array}{l} COOK \\ | \\ COOK \end{array} + 2RNH_2$$

Oxamide; Pot. oxalate; Primary amine

$$\begin{array}{l} CONR_2 \\ | \\ COOC_2H_5 \end{array} + 2KOH \longrightarrow \begin{array}{l} COOK \\ | \\ COOK \end{array} + C_2H_5OH + R_2NH$$

Oxamic ester; Pot. oxalate; Sec. amine

(2) Hinsberg's method. The mixture of amines is treated with *p*-toluenesulphonyl chloride, also known as tosyl chloride (TSCl) or Hinsberg's reagent (in the earlier days benzene sulphonyl chloride was used as Hinsberg's reagent) in the presence of sodium hydroxide. The primary amine forms mono-alkyl sulphonamide which dissolves in sodium hydroxide with the formation of sodium salt.

$$p-CH_3-C_6H_4-SO_2Cl + H_2NR \longrightarrow p\text{-}CH_3C_6H_4-SO_2NHR$$

p-Toluene sulphonyl chloride; Primary amine; Alkyl sulphonamide

$$\downarrow NaOH$$

$$p-CH_3-C_6H_4-SO_2NNaR$$

Sod. salt of alkyl sulphonamide (*Soluble*)

The secondary amine forms dialkyl sulphonamide which is insoluble in alkali.

$$p-CH_3-C_6H_4-SO_2Cl + HNR_2 \longrightarrow p-CH_3-C_6H_4-SO_2NR_2 + HCl$$

Secondary amine — Dialkyl sulphonamide (*Insoluble in alkali*)

The tertiary amine does to react at all.

The alkaline mixture obtained above is extracted with ether when the unreacted tertiary amine and sulphonamide of secondary amine pass into ether layer while the sodium salt of sulphonamide of primary amine remains in aqueous layer. From the ethereal layer, ether is removed by evaporation and the remaining mixture is distilled. As a result tertiary amine (being volatile) distils off while sulphonamide of secondary amide is left behind. This is hydrolysed by hydrochloric acid when the hydrochloride of secondary amine is obtained. This is distilled over sodium hydroxide to get the secondary amine.

$$p-CH_3-C_6H_4-SO_2NR_2 + H_2O + HCl \longrightarrow p-CH_3C_6H_4-SO_2NHR + R_2NH.HCl$$

Dialkyl sulphonamide — *p*-Toluene sulphonic acid — Sec. amine hydrochloride

$$R_2NH.HCl + NaOH \longrightarrow R_2NH + NaCl + H_2O$$

Sec. amine

The aqueous layer containing the sodium salt of sulphonamide of primary amine is acidified with hydrochloric acid when the sodium salt changes to free sulphonamide which gets hydrolysed to yield the hydrochloride of primary amine. This is followed by distillation over sodium hydroxide to get the free amine.

$$p-CH_3-C_6H_4-SO_2NNaR + HCl \longrightarrow p-CH_3-C_6H_4-SO_2NHR + NaCl$$

Sod. salt of sulphonamide

$$\downarrow H.OH + HCl$$

$$p-CH_3-C_6H_4SO_2OH + RNH_2.HCl$$

Primary amine hydrochloride

$$RNH_2.HCl + NaOH \longrightarrow RNH_2 + NaCl + H_2O$$

Primary amine

Q. 8. Discuss physical properties and stereochemistry of amines.

Ans. (*i*) Lower amines are colourless gases, higher ones are liquids and still higher are solids.

(*ii*) Lower aliphatic amines have ammoniacal odour which vanishes as we move to higher one.

(*iii*) Amines are polar compounds and form intermolecular hydrogen bonding as shown below

$$\begin{array}{c} \quad R \qquad\qquad R \qquad\qquad R \\ \text{-----}\underset{H}{\overset{|}{N}}\text{—H-----}\underset{H}{\overset{|}{N}}\text{—H-----}\underset{H}{\overset{|}{N}}\text{—N} \end{array}$$

Primary amine

$$\begin{array}{c} \quad R \qquad\qquad R \qquad\qquad R \\ \text{-----}\underset{R}{\overset{|}{N}}\text{—H-----}\underset{R}{\overset{|}{N}}\text{—H-----}\underset{R}{\overset{|}{N}}\longrightarrow N \end{array}$$

Secondary amine

Tertiary amines do not form hydrogen bonding as there is no hydrogen attached to nitrogen.

As a result of hydrogen bonding, amines have higher boiling points as compared to non-polar compounds of comparable molecular masses. However, the strength of hydrogen bond in amines is lower as compared to that in alcohols and carboxylic acids. Therefore, amines boil at lower temperatures as compared to alcohols and carboxylic acids of comparable molecular masses.

(*iv*) Lower amines are soluble in water and solubility decreases with the increase in size of the alkyl group. This solubility is because of hydrogen bond formation between the molecules of amine and water

$$\cdots\overset{\overset{R}{|}}{\underset{\underset{H}{|}}{N}}-H\cdots\overset{\overset{H}{|}}{O}-H\cdots\overset{\overset{R}{|}}{\underset{\underset{R}{|}}{N}}-H\cdots\overset{\overset{H}{|}}{O}-H$$

Stereochemistry of amines

Amines in which nitrogen carries three different alkyl groups can exist in two forms, which are non-superimposable mirror images of each other. Such an amine is dissymmetric and should exhibit optical activity. In actual practice, amines have not been reported to exhibit optical activity.

$$(CH_3)(C_2H_5)(C_3H_7)N \xrightleftharpoons{\text{Rapid inversion}} N(CH_3)(C_2H_5)(C_3H_7)$$

This is because the two forms undergo rapid inversion from one form to the other and hence the two forms become indistinguishable.

Q. 9. Describe with mechanism chemical properties of amines. (*Nagpur, 2005*)

Ans. 1. Basic character. Due to the presence of a lone pair of electrons on nitrogen, amines act as bases and give the following reactions:

(*a*) **Formation of salts.** They react with acids to form salts

$$RNH_2 + HCl \longrightarrow R\overset{+}{N}H_3\overset{-}{Cl} \quad \text{Alkyl ammonium chloride}$$

$$R_2NH + HCl \longrightarrow R_2\overset{+}{N}H_2Cl^- \quad \text{Dialkyl ammonium chloride}$$

$$R_3N + HCl \longrightarrow R_3\overset{+}{N}HCl^- \quad \text{Trialkyl ammonium chloride}$$

(*b*) **Formation of hydroxide.** Water soluble amines form substituted ammonium hydroxides which ionise to give hydroxyl ions.

$$RNH_2 + H_2O \rightleftharpoons RNH_3OH \rightleftharpoons RNH_3^+ + OH^-$$

$$R_2NH + H_2O \rightleftharpoons R_2NH_2OH \rightleftharpoons R_2NH_2^+ + OH^-$$

$$R_3N + H_2O \rightleftharpoons R_3NHOH \rightleftharpoons R_3NH^+ + OH^-$$

Thus the aqueous solution of amines are basic in nature.

2. Alkylation. All the three types of amines, (primary, secondary and tertiary) undergo alkylation with alkyl halides. The ultimate compound is quaternary ammonium halide.

$$\underset{\text{Primary amine}}{RNH_2} \xrightarrow[-HX]{RX} \underset{\text{Secondary amine}}{R_2NH} \xrightarrow[-HX]{RX} \underset{\text{Tertiary amine}}{R_3NH} \xrightarrow[-HX]{RX} \underset{\text{Quatenary amm. salt.}}{[R_4N]^+X^-}$$

3. Acylation. Primary and secondary amines on treatment with acetyl chloride or acetic anhydride undergo acylation to give N-substituted amides.

$$RCOCl + R'NH_2 \longrightarrow \underset{\text{N-alkyl amide}}{RCONHR'} + HCl$$

$$RCOCl + R'_2NH \longrightarrow \underset{\text{N, N dialkyl amide}}{RCONR'_2} + HCl$$

Tertiary amine has no hydrogen and as such it does not participate in acylation.

In case the acid chloride or anhydride is aromatic, the reaction is known as Schotten-Baumann's reaction as given below:

$$\underset{\text{Benzoyl chloride}}{C_6H_5COCl} + \underset{\text{Aniline}}{H_2NC_6H_5} \longrightarrow \underset{\text{N-Phenyl benzamide}}{C_6H_5CONHC_6H_5}$$

The mechanism of the reaction is as follows:

$$R-\underset{N}{\overset{R'}{N}}: + R-\overset{O}{\overset{\|}{C}}-Cl \longrightarrow R-\underset{H}{\overset{R'}{\overset{\oplus}{N}}}-\overset{O}{\overset{\|}{C}}-R + Cl^{\ominus}$$

(R = ALKYL or ARYL GROUP)
(R′ = H or ALKYL or ARYL GROUP)

$$\xrightarrow{R-\ddot{N}H-R'} R-\overset{R'}{N}-\overset{O}{\overset{\|}{C}}-R + R-\overset{\oplus}{N}H_2-R'$$

Since tertiary amines cannot lose the proton after attachment to carbon, they do not react.

4. Reaction with Hinsberg's reagent (sulphonyl chloride). Primary and secondary amines react with sulphonyl chlorides such as benzene or toluene sulphonyl chloride to form substituted sulphonamides.

$$\underset{\substack{p\text{-Toluene} \\ \text{sulphonyl chloride}}}{H_3C-C_6H_4-SO_2Cl} + CH_3CH_2NH_2 + NaOH \xrightarrow{\text{Ethyl amine}} \underset{\substack{\text{N-Ethyl } p\text{-Toluene} \\ \text{sulphonamide}}}{C_6H_5-SO_2NHCH_2CH_3} + NaCl + H_2O$$

This reaction is used for the separation of three classes of amines.

5. Reaction with nitrous acid. (*i*) **Primary aromatic amines** react with nitrous acid (or a mixture of $NaNO_2$ and HCl) in cold to form diazonium salts.

$$C_6H_5-NH_2 + NaNO_2 + HCl \xrightarrow{273\text{ K}} \underset{\substack{\text{Benzene} \\ \text{diazonium chloride}}}{C_6H_5-N_2Cl} + NaCl + 2H_2O$$

(*ii*) **Secondary amines,** both aliphatic and aromatic, react with nitrous acid to give N-nitroso-amines which are yellow oily compounds insoluble in dilute mineral acids.

$$R_2NH + NaNO_2 + HCl \longrightarrow \underset{\text{N-nitrosodialkyl. amine}}{R_2N-N=O} + NaCl + H_2O$$

$$C_6H_5-\underset{CH_3}{NH} + NaNO_2 + HCl \longrightarrow \underset{\text{N-nitroso N-methyl aniline}}{C_6H_5-\underset{CH_3}{N}-N=O} + NaCl + H_2O$$

(*iii*) **Tertiary aromatic amines** react with nitrous acid to give substituted product *p*- nitrosoamine which is green in colour.

$$(CH_3)_2N-C_6H_5 + NaNO_2 + HCl \xrightarrow{273\text{ K}} \underset{\text{p-Nitroso, n-Dimethyl aniline}}{(CH_3)_2N-C_6H_4-N=O} + NaCl + H_2O$$

Tertiary aliphatic amines react with nitrous acid to form water soluble amine salts, trialkyl ammonium nitrite. There is no change in colour.

$$R_3N + NaNO_2 + HCl \longrightarrow \underset{\text{Trialkyl ammonium nitrite}}{[R_3NH]^+NO_2^-} + NaCl$$

6. Carbylamine Reaction. Alkyl isocyanide is formed when a primary amine is heated with chloroform and alcoholic potash. Alkyl isocyanide gives a very disagreeable odour.

$$\underset{\text{Primary amine}}{RNH_2} + CHCl_3 + KOH\ (alc.) \longrightarrow \underset{\text{Carbyl-amine}}{RNC} + 3KCl + 3H_2O$$

This reaction is not given by secondary and tertiary amine and hence this reaction can be used as a test to distinguish primary amines from secondary and tertiary amines.

Mechanism of the reaction

The reaction is believed to involve the formation of a **nitrogen ylid** by the addition of dichlorocarbene initially formed. This is followed by proton transfer and elimination of hydrogen chloride to give the product.

$$HO^- + H-CCl_3 \xrightarrow[-H_2O]{} :\bar{C}Cl_3 \longrightarrow \underset{\text{Dichloro carbene}}{:CCl_2} + Cl^-$$

$$R-\underset{H}{\overset{H}{N}}: + :CCl_2 \longrightarrow R-\underset{H}{\overset{H}{N^+}}-\bar{\ddot{C}}Cl_2$$

$$\xrightarrow{-H^+,\ +H^+} R-\underset{H}{\overset{}{N}}-\underset{Cl}{\overset{H}{C}}-Cl \longrightarrow R-\underset{H}{\overset{+}{N}}=C\begin{smallmatrix}H\\Cl\end{smallmatrix}$$

$$\xrightarrow{-H^+} R-\ddot{N}=C\begin{smallmatrix}H\\Cl\end{smallmatrix} \longrightarrow R-\overset{+}{N}\equiv C-H \xrightarrow{OH^-} R-N\equiv C^-$$

7. Formation of Schiff's base. Schiff's base is obtained when primary amine reacts with aldehyde

$$\underset{\text{Aldehyde}}{RCHO} + \underset{\text{Primary amine}}{H_2NR'} \longrightarrow \underset{\text{Schiff's base}}{RCH = NR'} + H_2O$$

8. Action of carbon disulphide. On warming with carbon disulphide in the presence of mercuric chloride, primary amines yield isothiocyanate which has the smell of mustard oil.

$$RNH_2 + S = C = S \xrightarrow{HgCl_2} \underset{\text{Isothiocyanate}}{R - N = C = S} + H_2S$$

9. Acidic nature. Although predominantly basic in nature, primary and secondary amines also exhibit some acidic properties when treated with alkali metals or Grignard reagents.

$$\underset{1^\circ\text{ Amine}}{2RNH_2} + 2Na \longrightarrow R\,NHNa^+ + H_2$$

$$R_2NH + R'Mg\,X \longrightarrow R_2NMgX + R'H$$

10. Ring substitution in aromatic amines. Halogen, nitro or sulphonic acid groups can be introduced in the benzene ring by electrophilic substitution mechanism. Amino group is an activating group, it activates the *ortho* and *para* positions for substitution. This is because, there is a lone pair of electrons on nitrogen of the amines group. This lone pair of electrons is donated to the benzene ring by resonance at *ortho* and *para* positions which become negative and attract the electrophiles like X^+, NO_2^+ and SO_3. The products obtained are therefore, *ortho* and *para* substituted amines.

Halogenation

$$\underset{\text{Aniline}}{C_6H_5NH_2} + 3Br_2 \longrightarrow \underset{\text{2, 4, 6 Tribromo aniline}}{C_6H_2Br_3NH_2} + 3HBr$$

If a monohalogen derivatives is desired, amino group is first acetylated. Acetylation slows down the rate of electrophilic substitution as well as protects the amino group. Thus

$$\underset{\text{Aniline}}{C_6H_5NH_2} \xrightarrow{CH_3COCl} \underset{\text{Acetanilide}}{C_6H_5NHCOCH_3} \xrightarrow{Br_2} \underset{p\text{-bromo acetanilide}}{Br{-}C_6H_4{-}NHCOCH_3} \xrightarrow{H_2O} \underset{p\text{-bromo aniline}}{Br{-}C_6H_4{-}NH_2}$$

Nitration. Nitric acid is a strong oxidising agent. Hence there is a danger that amines will get decomposed by nitric acid. However, if aniline is first protected by acetylation and then nitrated, we obtain *p*-nitro aniline.

NH_2 (Aniline) $\xrightarrow{CH_3COCl}$ $NHCOCH_3$ (Acetanilide) $\xrightarrow[H_2SO_4]{HNO_3}$ $NHCOCH_3$ / NO_2 $\xrightarrow[H^+]{H_2O}$ NH_2 / NO_2

Aniline Acetanilide *p*-nitroaniline

Sulphonation

NH_2 $\xrightarrow{H_2SO_4}$ NH_2 / SO_3H

Sulphanilic acid

Sulphanilic acid contains both an acidic and basic group. These groups react to form intramolecular salt. This type of internal salts are known as dipolar or Zwitter ions as shown below:

$\overset{+}{N}H_3$ / SO_3^-

Q. 10. Write the products of the reaction of the following with HNO_2 (*i*) Aniline (*ii*) Diethylamine (*iii*) N, N-dimethylaniline.

Ans.

(*i*) $C_6H_5-NH_2 + HNO_2 \xrightarrow{273\ K} C_6H_5-N_2Cl + 2H_2O$

Aniline Benzene diazonium chloride

(*ii*) $(C_2H_5)_2NH + HNO_2 \longrightarrow C_2H_5 - \underset{\substack{| \\ C_2H_5}}{N} - N = O + H_2O$

N-nitroso diethyl amine

(*iii*) $(CH_3)_2N-C_6H_4 + HNO_2 \longrightarrow (CH_3)_2N-C_6H_4-N = O + H_2O$

N-nitroso-N, N-dimethyl aniline

Q. 11. How will you distinguish between primary, secondary and tertiary amines?

(*Kuruksheta, 2001; Kerala 2001; Garhwal, 2000; HImachal 2000*)

Ans. 1. Nitrous acid method

Aliphatic primary amines react with HNO_2 (or a mixture of $NaNO_2$ and HCl) to give an effervescence of nitrogen gas.

$$RNH_2 + NaNO_2 + HCl \longrightarrow [RN_2Cl] \xrightarrow{H_2O} ROH + HCl + N_2$$

Aromatic primary amines react with HNO_2 to form diazonium salt.

$$ArNH_2 + NaNO_2 + HCl \xrightarrow{273\text{ K}} ArN^+{}_2Cl^- + NaCl + H_2O$$

Secondary amines react with HNO_2 to give yellow oily N-nitrosoamines

$$R_2NH + NaNO_2 + HCl \longrightarrow R_2N - N = O + NaCl + H_2O$$

N-nitroso dialkyl amine

Tertiary aromatic amines react with nitrous acid to yield green-coloured *p*-nitrosoamine.

$$(CH_3)_2N-C_6H_5 + NaNO_2 + HCl$$

$$\xrightarrow{273\text{ K}} (CH_3)_2N-C_6H_4-N = O + NaCl + H_2O$$

p-nitroso N, N-Dimethyl aniline

Tertiary aliphatic amines react with HNO_2 to form water soluble amine salts trialkyl ammonium nitrites with no change in colour.

$$R_3N + NaNO_2 + HCl \longrightarrow [R_3NH]^+ NO_2^- + NaCl$$

Trialkyl amm. nitrite

2. Hinsberg's method. It is a very reliable test for distinguishing between amines and consists of following steps:

(*i*) Amine is shaken with toluene sulphonyl chloride in the presence of an alkali.

(*ii*) Primary and secondary amines form mono and disubstituted sulphonamides while tertiary amines do not react.

(*iii*) The reaction mixture is treated with excess of sodium hydroxide. Monosubstituted sulphonamide forms a soluble sodium salt which on acidification regenerates the insoluble amide.

(*iv*) The disubstituted sulphonamide from secondary amine produces an insoluble compound unaffected by the action of acids.

(v) If the unreacted amine remains insoluble in alkali but dissolves on acidification, it indicates a tertiary amines.

For reactions, see Question 4 in this Chapter.

3. Carbylamine test. This test can be used to distinguish primary amines from secondary and tertiary amines. The amine is heated with chloroform and alcoholic potassium hydroxide. If a strong offensive smell (or RNC) is noticed, it is a primary amine.

$$\underset{\text{Primary amine}}{R\,NH_2} + CHCl_3 + 3KOH \longrightarrow \underset{\text{Carbyl amine}}{R - NC} + 3KCl + 3H_2O$$

Q. 12. Discuss basicity of amines. What is meant by pK_b of an amine? Discuss relative basicities of amines.

Ans. An aqueous solution of an amine exists in the following equilibrium state.

$$RNH_2 + H_2O \rightleftharpoons R\overset{+}{N}H_3 + OH^-$$

For a reaction in equilibrium, we can have the following equation:

$$K_b = \frac{[R\overset{+}{N}H_3][OH^-]}{[RNH_2]}$$

Concentration of water has been omitted from the above equation as it remains constant.

K_b is called the basicity constant of the base. It reflects the extent of basic character of a base. Higher the value of K_b, stronger is the base. It reflects the extent to which amine can accept the proton from water.

***p*K$_b$.** This is another term which is used to express the basic strength of a base. It is related to K_b as follows:

$$pK_b = -\log K_b$$

Thus, it is negative logarithm of the basicity constant K_b. It will be observed that higher the K_b value, lower is the pK_b value. For example, for methylamine the values of K_b and pK_b are as follows:

$$K_b = 4.4 \times 10^{-4}$$
$$pK_b = 3.36$$

So for two bases X and Y, if X has a higher value of K_b, it will have a lower value of pK_b. pK_b values of a few important amines are given in the table below:

pK_b values of some amines

(pK_b of ammonia = 4.75)

Compound	*Formula*	*pK_b*
Methylamine	CH_3NH_2	3.36
Dimethylamine	$(CH_3)_2NH$	3.29
Trimethylamine	$(CH_3)_3N$	4.2
Ethylamine	$CH_3CH_2NH_2$	3.34
Diethylamine	$(CH_3CH_2)_2NH$	3.02
Triethylamine	$(CH_3CH_2)_3N$	3.26
n-Propylamine	$CH_3CH_2CH_2NH_2$	3.42

Benzylamine	$—CH_2NH_2$	4.64
Aniline	$—NH_2$	9.38
N-Methylaniline	$—N(CH_3)_2$	9.15
N, N-Dimethylaniline	$—NHCH_3$	8.95
Diphenylaniline	—NH—	13.15

Following observations may be made on perusal of the above table.

(*i*) Aliphatic amines are stronger bases compared to ammonia.

(*ii*) Aromatic amines are weaker than ammonia and aliphatic amines.

(*iii*) Mixed aliphatic-aromatic secondary and tertiary amines are slightly stronger than pure aromatic amines.

Explanation of relative strength of basic characeter

(*a*) According to modern acid-base theory, strength of a base depends upon the availability of electrons on nitrogen in the amine and also upon the stability of the protonated amine formed. Any factor that can increase electron concentration on nitrogen will bring about increase in the basic strength. Similarly, if the protonated amine is more stable than the amine, we can expect greater basic strength from the amine.

(*b*) Why aliphatic amines are more basic than ammonia can be explained on the above logic.

$$\underset{\text{Ammonia}}{H-\overset{\displaystyle H}{\overset{|}{N}}-H} \qquad \underset{\text{Primary amine}}{R\rightarrow\overset{\displaystyle H}{\overset{|}{N}}-H} \qquad \underset{\text{Secondary amine}}{R\rightarrow\overset{\displaystyle R}{\overset{\downarrow}{N}}-H}$$

In aliphatic amines, there are alkyl groups attached to nitrogen. Alkyl groups are known as electron repelling groups. Thus there will be greater electron density on nitrogen in aliphatic amines compared to ammonia. Consequently aliphatic amines are stronger bases than ammonia. Primary amine is stronger than ammonia and secondary amine is stronger than primary amine because of obvious double electron repelling by two alkyl group. But this trend stops here and does not continue to tertiary amines. This is because of additional steric and solution factors. Although there will be maximum electron density of nitrogen on account of three alkyl groups, but there is difficulty in stabilising the protonated amine because of steric hindrance between the bulky groups and hydrogen.

$$\underset{CH_3\quad CH_3}{\overset{\displaystyle CH_3}{\overset{|}{\overset{+}{N}}}-H} \qquad \text{Sterically unstable}$$

Moreover tertiary amine is not easily solvated by water.

(*c*) Aromatic amines are weaker bases than ammonia and aliphatic amines. As explained earlier, smaller the electron concentration on nitrogen, weaker will be the amine in basic strength. In aniline, electron pair on nitrogen enters into resonance with benzene ring giving rise to a number of resonance structure which impart greater stability to the resonance structure. On the other hand the protonated aniline or aniline ion has just two resonance structures.

$:NH_2$ $:NH_2$ $+NH_2$ $+NH_2$ $+NH_2$

Ia Ib Ic Id Ie

$+NH_3$ $+NH_3$

IIa IIb

As a result of resonance structures of aniline, the electron density on nitrogen is decreased thereby weakening the basic strength. Moreover anilinium ion is less stable (with only two resonating structures) than aniline (with five resonating structure).

Q. 13. Discus the effect of substituents on the basicity of aromatic amines.

Ans. Aniline has the pK_b value of 9.387. If some group is introduced in the benzene ring of aniline molecule, we find that there is appreciable change in the pK_b value. It means substituents change the basic strength of aniline. The pK_b values of the some substituted anilines are given in the table below:

pK_b values of ring substituted anilines

(pK_b of aniline = 9.387)

Substituent	pK_b		
	Para isomer	Meta isomer	Ortho isomer
$-CH_3$	8.93	9.31	9.6
$-OCH_3$	8.83	9.7	9.53
$-NH_2$	7.83	9.13	9.5
$-Cl$	9.82	10.53	11.31
$-NO_2$	13	11.5	13.46

Some generalisations about the substituents and their effect on pK_b are made as under:

(*a*) Electron-releasing groups such as $-CH_3$, $-OCH_3$ and $-NH_2$ have the effect of increasing the basic strength of the amine. It can be understood in terms of increased electron density on nitrogen.

(*b*) Electron-withdrawing groups such as $-Cl$ and $-NO_2$ have the effect of decreasing basic strength of amine. It can be explained in terms of decreased electron density on nitrogen.

(*c*) Electron withdrawing substituents manifest the maximum decreasing effect on basic strength when present in ortho position.

(*d*) Even electron-releasing groups decrease the basic strength when present in ortho position. The phenomenon of decreasing of basic strength with electron-releasing as well as electron withdrawing groups in the ortho position is known as **ortho effect.** There is no satisfactory explanation of this observation.

Q. 14. Explain why aniline is less basic than cyclohexylamine? (*Delhi 2006*)

Ans.

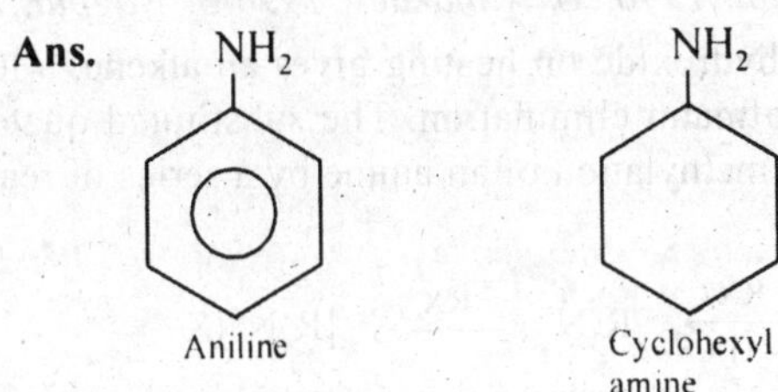

Smaller basicity of aniline is due to delocalisation of non-bonding electrons on the amino group in aniline

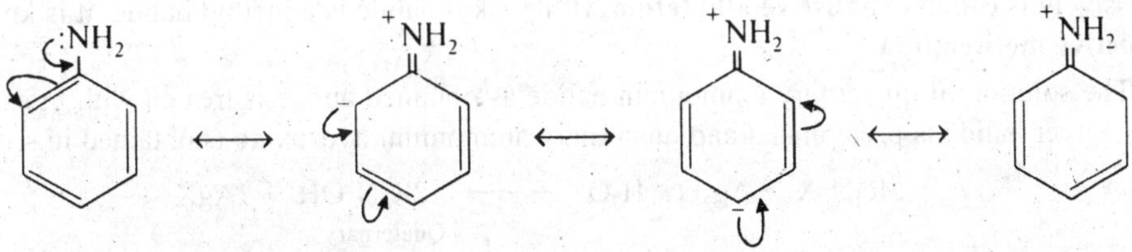

No such delocalisation of electrons takes place in cyclohexyl amine.

Q. 15. Arrange the following compounds in increasing order of their basic strength NH_3, CH_3NH_2, $(CH_3)_2NH$, $(CH_3)_3N$ in

(*a*) Polar solvent like ethanol (or water)

(*b*) In non-polar solvent or in gas phase (*Delhi 2006*)

Ans. (*a*) In polar solvent. It is expected that basicity will increase in the order:

$$NH_3 < CH_3NH_2 < (CH_3)_2NH < (CH_3)_3N$$

This is because of the greater and greater inductive effect of methyl groups as we move from left to right. Because of increasing inductive effect (methyl is an electron repelling group), electron density on nitrogen and hence basicity of the compound should increase. Here, the solvation effect also plays the role.

Due to the combined effect of inductive effect and solvation effect, the basicity increases in the order:

$$NH_3 < (CH_3)_3N < CH_3NH_2 < (CH_3)_2NH$$

This is explained by saying that the proton of alcohol (or water) finds it difficult to attach to nitrogen in tertiary amine because of steric hindrance as there are already three bulky methyl groups attached. Moreover tertiary amine is not easily solvated by alcohol or water

(*b*) In non-polar solvent or in gas phase

As expected, the basicity will increase in the order

$$NH_3 < CH_3NH_2 < (CH_3)_2NH < (CH_3)_3N$$

Then is no attachment of protons from the solvent in this case. In the gas phase, no solvent is involved.

Q. 16. What is exhaustive methylation of amines and Hofmann's elimination.

(*Guwahati, 2002; Panjab 2003; Calcutta 2007*)

Or

Discuss Hofmann elimination of quaternary ammonium hydroxides.

(*Punjabi, 1996; G. Nanakdev, 1996 S; Nagpur, 2005*)

Ans. A substituted quaternary ammonium hydroxide on heating gives an alkene, a tertiary amine and water. This reaction is known as Hofmann elimination. The substituted quaternary ammonium hydroxide is prepared by exhaustive methylation of an amine by a series of reactions given below:

$$\underset{1^\circ \text{ Amine}}{RNH_2} \xrightarrow[-HX]{RX} \underset{2^\circ \text{ Amine}}{R_2NH} \xrightarrow[-HX]{RX} \underset{3^\circ \text{ Amine}}{R_3N} \xrightarrow{RX} \underset{\text{Quaternary amm. halide}}{[R_4N^+]X^-}$$

Conversion of an amine into quaternary ammonium salt by treatment with alkyl halide successively is called **exhaustive alkylation.** If the alkyl halide is a methyl halide, it is known as **exhaustive methylation.**

The solution of quaternary ammonium halide as obtained above is treated with moist silver oxide. Silver halide is precipitated and quaternary ammonium hydroxide is obtained in solution.

$$2R_4\overset{+}{N}\bar{X} + Ag_2O + H_2O \longrightarrow \underset{\text{Quaternary hydroxide}}{2R_4\overset{+}{N}\overline{O}H} + 2AgX$$

Quaternary ammonium hydroxide on heating strongly at 400 K undergoes an elimination reaction as under.

$$\left[\begin{array}{c} R \\ | \\ R-\overset{+}{N}-R \\ | \\ R \end{array}\right] OH^- \xrightarrow{\text{Heat}} \begin{array}{c} R \\ | \\ R-N \\ | \\ R \end{array} + \text{Alkene} + \text{Water}$$

Taking a particular example

$$\underset{\text{Trimethyl-ethyl ammonium hydroxide}}{\left[\begin{array}{c} CH_3 \\ | \\ CH_3-\overset{+}{N}-CH_2CH_3 \\ | \\ CH_3 \end{array}\right] OH^-} \xrightarrow{\text{Heat}} \underset{\text{Trimethyl amine}}{\begin{array}{c} CH_3 \\ | \\ CH_3-N \\ | \\ CH_3 \end{array}} + \underset{\text{Ethylene}}{CH_2=CH_2} + H_2O$$

This reaction is known as Hofmann's elimination.

Utility of Hofmann's elimination. This reaction is of great importance in determining the structure of unknown amines. Consider for example that we obtain a mixture of trimethylamine, 1-butene and water in Hofmann elimination *i.e.* after subjecting the amine to exhaustive methylation, treatment with moist silver oxide, followed by heating. By tracing the steps backward in the above reaction, we can see that the amine in the present case is $CH_3CH_2CH_2NH_2$ or *n*-butyl amine.

$$CH_3CH_2CH_2CH_2NH_2 \xrightarrow[\text{Methylation}]{\text{Exhaustive}} CH_3CH_2CH_2CH_2N(CH_3)_3X$$

$$\xrightarrow[H_2O]{Ag_2O} [CH_3CH_2CH_2CH_2N^+(CH_3)_3]OH^-$$

$$\xrightarrow{\text{Heating}} \underset{\text{1-Butene}}{CH_3CH_2CH = CH_2} + \underset{\text{Trimethyl amine}}{(CH_3)_3N} + H_2O$$

Mechanism of Hofmann's elimination

The reaction involves the mechanism given below:

The hydroxide ion removes the hydrogen from the carbon as shown hereunder, creating a double bond and separating the tertiary amine from rest of the molecule.

$$(CH_3)_3\overset{+}{N} - \overset{|}{\underset{|}{C}} - \overset{|}{\underset{H}{C}} - \longrightarrow \rangle C = C \langle + (CH_3)_3 N + H_2O$$
$$OH^-$$

Q. 17. Explain Cope elimination with the help of a suitable example.

(*Himachal, 2000; Kurukshetra, 2001*)

Ans. Cope elimination is a reaction which is performed to identify the unknown tertiary amine. It involves treatment of tertiary amine with hydrogen peroxide to obtain amine oxide which on heating gives N, N dialkylhydroxyl amine and an alkene.

$$\underset{\text{Methyl ethyl propyl amine}}{(CH_3)(C_2H_5)N{-}C_3H_7} \xrightarrow{H_2O_2} \underset{\text{Amine oxide}}{(CH_3)(C_2H_5)(C_3H_7)N^+{-}O^-} \xrightarrow{\text{Heat}}$$

$$\underset{\text{Methyl ethyl hydroxyl amine}}{(CH_3)(C_2H_5)N - OH} + \underset{\text{Propylene}}{CH_3CH = CH_2}$$

The dialkyl hydroxylamine can be again converted into tertiary amine on treatment with CH_3I and Ag_2O. Again Cope elimination reaction is performed provided there is still β-hydrogen atom in the molecule.

$$(CH_3)(CH_3CH_2)N - OH \xrightarrow[Ag_2O]{CH_3I} \underset{\text{Dimethyl ethyl amine}}{(CH_3)(CH_3CH_2)N - CH_3} \xrightarrow{H_2O_2}$$

$$(CH_3)(\overset{\beta\ \alpha}{CH_3CH_2})(CH_3)\overset{+}{N}{-}O^- \xrightarrow{H_2O_2} (CH_3)_2N - OH + \underset{\text{Ethene}}{CH_2 = CH_2}$$

It may be noted that in the first Cope elimination process propylene was obtained and in the second Cope elimination process ethene was obtained. This lead us to say that one propyl and one ethyl groups were present in the original tertiary amine. Of course, the third alkyl group was methyl because further repetition of Cope elimination will not give us an alkene. In fact, no β-hydrogen is left in the substituted dimethyl hydroxylamine that we obtain in the second Cope reaction and hence there is no possibility of Cope elimination taking place any more.

The following mechanism has been proposed for Cope elimination reaction:

$$R_2\overset{+}{N}(\ddot{O}:^-)-C-C-H \longrightarrow R_2N-OH + >C=C<$$

Q. 18. Arrange the following in decreasing order of basicity:

(*i*) Aniline (*ii*) *p*-Nitroaniline (*iii*) *m*-nitroaniline (*iv*) Diphenylamine

Ans. (*i*) Nitro group has electron-withdrawing effect. Hence nitro anilines are expected to be weaker bases than aniline. Out of *p* and *m* nitroanilines, *p*-nitroaniline is weaker because nitro group has greater electron-withdrawing effect at *p*-position.

(*ii*) Diphenylamine is expected to be weaker than aniline because there are two benzene rings which draw the electrons by resonance. In fact, of the four amines, this is the weakest amine. Hence order of decreasing basicity of the amines is as under:

Aniline > *m*-nitroaniline > *p*-nitroaniline > diphenylamine

Q. 19. K_b values of *p*-nitroaniline and m-nitroaniline are respectively 0.001×10^{-10} and 0.032×10^{-10}. Which is a weaker base and why?

Ans. Higher the K_b value of a base, greater is the strength of the base. In this problem *m*-nitroaniline has a greater value of K_b, hence it is a stronger base than *p*-nitroaniline. It may be noted that a reverse sequence is followed when we use pK_b values.

Q. 20. Arrange the following compounds in increasing order of basicity:

Aniline, *p*-nitroaniline, *p*-methoxyaniline, ammonia, ethylamine, diethylamine, benzylamine and *o*-toluidine.

Ans. Based upon the rules discussed in this chapter, the compounds are arranged in increasing order of basic strength as given below:

p-nitro aniline (NH_2–C_6H_4–NO_2) < *o*-Toluidine (NH_2–C_6H_4–CH_3) < *p*-methoxy aniline (NH_2–C_6H_4–OCH_3) < Aniline ($C_6H_5NH_2$) < NH_3 Ammonia

< Benzyl amine ($C_6H_5CH_2NH_2$) < $C_2H_5NH_2$ Ethylamine < $(C_2H_5)_2NH$ Diethylamine

Q. 21. Explain the following:

(*a*) Aniline is less basic than N-methyl aniline.

(*M. Dayanand, 2000; Devi Dhilya 2000; Awadh 2000*)

(*b*) Diphenylamine is a much weaker base than aniline.

(c) Triphenylamine and N, N-dimethyl aniline are both tertiary amines. Triphenylamine is insoluble in HCl but N, N-dimethylaniline readily dissolves in HCl.

(d) 2, 4, 6-trinitroaniline is termed picramide even though it contains no amide linkage.

Ans. (*a*) Aniline is less basic than N-methylaniline

NH_2 — Aniline

$HN \leftarrow CH_3$ — N-Methylaniline

In N-methylamine, there is electron donating methyl group which increases electron density on nitrogen thereby increasing the basicity of the compound. Hence, N-methylaniline is a stronger base than aniline.

(*b*) Diphenylamine is a much weaker base than aniline.

NH_2 — Aniline

NH — Diphenylamine

Phenyl groups are electron-attracting groups. They attract the electron pair on nitrogen, towards themselves, thereby reducing the basic strengths. Hence diphenylamine is a weaker base than aniline.

(*c*)

N — Triphenylamine

$N(CH_3)_2$ — N, N-Dimethyl aniline

Triphenylamine is a very weak base because of three electron-withdrawing phenyl groups. Hence it does not dissolve in HCl.

N, N, Dimethylaniline is comparatively a strong base because of electron donating effect of two methyl groups. Hence it dissolves in HCl.

(*d*)

NH_2, O_2N, NO_2, NO_2

2, 4, 6-Trinitro aniline

The combined electron-withdrawing effect of three nitro groups makes it almost a neutral compound comparable to an amide.

Q. 22. Explain the following:

(*a*) 3°Amines have greater pK_b values (less basic) than 1° and 2° amines.

(*b*) Aromatic amines are weaker bases than aliphatic amines.

(*Meerut, 2000; Delhi, 2002; Nagpur 2002*)

(*c*) Aniline is weaker base than ammonia.

(*d*) Aniline is weaker base than ethylamine. (*Himachal, 2000; Delhi, 2007*)

Ans. (*a*) Basicity of a compound depends upon the availability of electrons (on nitrogen in case of amines).

Alkyl groups, by virtue of their inductive effect, increase the electron density on nitrogen in the amine.

$$R - \ddot{N}H_2$$

Therefore, primary amine is more basic than ammonia, secondary amine is more basic than primary amine because of the inductive effect of two alkyl groups. But this trend does not extend further. Tert. amine is not more basic than secondary amine. In fact, it is weaker than both primary and secondary (1° and 2°) amines. Therefore, it has a smaller value of K_b and greater value of pK_b. This is because of steric hindrance. Protonated tert. amine is unstable because of steric hindrance.

$$R_3N: + H^+ \longrightarrow R_3N^+H$$

Unstable

(*b*) Aromatic amines are weaker bases than aliphatic amines. This is because the electron pair on nitrogen in aromatic compounds takes part in resonance with benzene ring. This decreases the availability of electrons on nitrogen making it a weaker base.

$$C_6H_5\ddot{N}H_2 \longrightarrow {}^{+}NH_2{=}C_6H_5^{-}$$ and so on

(*c*) Aniline is weaker base than ammonia. This is again because of the same reasons as in part (*b*) of this question.

(*d*) Aniline is a weaker base than ethylamine.

$$C_6H_5\ddot{N}H_2 \qquad CH_3CH_2\ddot{N}H_2$$

Electron pair on nitrogen in aniline is comparatively less available because of resonance with the benzene ring, thus making it less basic.

Q. 23. Given below are pK_a values of a few aromatic amines with substituent

Z in Z—C_6H_4—NH_2

Z =	**pK_a values**
CH_3	**5.12**
NO_2	**1.02**
Br	**3.91**
MeO	**5.29**
NH_2	**6.08**

Arrange them in order to decreasing basic strength.

Ans. Greater the strength of a base, lower the pK_a value.

Note: Here the term pK_a has been used in place of pK_b. The increasing order of pK_a values is

1.02, 3.91, 5.12, 5.29, 6.08

Hence the order of decreasing basic strength will be

NO_2—C_6H_4—NH_2 > Br—C_6H_4—NH_2 > CH_3—C_6H_4—NH_2 >

MeO—C_6H_4—NH_2 > NH_2—C_6H_4—NH_2

Q. 24. Arrange the following pairs of substances in order of increasing strength. Show your reasoning.

(*i*) $(CH_3)_3$ N and $(F_3C)_3$ N

(ii) C_6H_5—NH_2 and $C_2H_5NH_2$

Ans. (*i*) $(F_3C)_3N$ (I) < $(CH_3)_3N$ (II)

In compound II, there are three electron-donating methyl groups, attached to nitrogen. They increase electron density on nitrogen. In compound I, there are three electro-withdrawing CF_3-groups attached to nitrogen (fluorine is highly electronegative). They decrease the electron density on nitrogen. Hence, compound II is more basic than compound I.

(*ii*) C_6H_5—NH_2 (I) < $C_2H_5 \rightarrow NH_2$ (II)

In compound I, the electron pair on nitrogen is drawn by the ring due to resonance, as shown below:

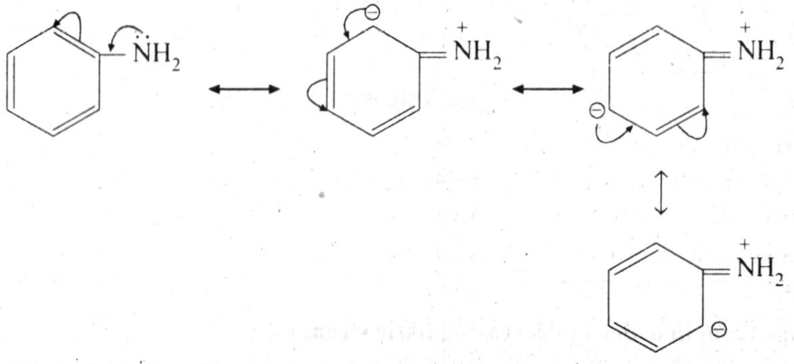

This results in decrease of basic strength.

In ethylamine II, the electron-releasing effect of ethyl group increases the electron density and basic strength. Hence, ethylamine is more basic than aniline.

Q. 25. How will you convert

(*i*) Methylamine to ethylamine

(*ii*) Ethylamine to methylamine

(*iii*) Aniline to benzoic acid?

Ans. (*i*) Methylamine to ethylamine.

$$\underset{\text{Methylamine}}{CH_3NH_2} \xrightarrow{HNO_2} CH_3OH \xrightarrow{PCl_5} CH_3Cl$$

$$\xrightarrow{KCN} CH_3CN \xrightarrow[\text{[H]}]{\text{Na/Alcohol}} \underset{\text{Ethylamine}}{CH_3CH_2NH_2}$$

(*ii*) Ethylamine to methylamine.

$$\underset{\text{Ethylamine}}{CH_3CH_2NH_2} \xrightarrow{HNO_2} CH_3CH_2OH \xrightarrow{[O]} CH_3CHO$$

$$\xrightarrow{[O]} CH_3COOH \xrightarrow[\Delta]{NH_3} CH_3CONH_2 \xrightarrow[KOH]{Br_2} \underset{\text{Methylamine}}{CH_3NH_2}$$

(*iii*) Aniline to benzoic acid.

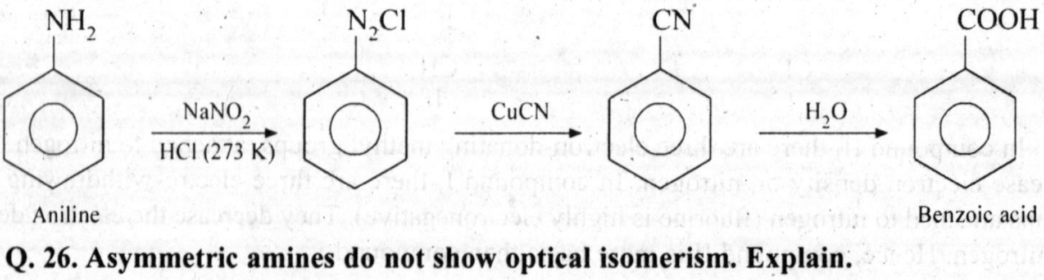

$$\underset{\text{Aniline}}{C_6H_5NH_2} \xrightarrow[\text{HCl (273 K)}]{NaNO_2} C_6H_5N_2Cl \xrightarrow{CuCN} C_6H_5CN \xrightarrow{H_2O} \underset{\text{Benzoic acid}}{C_6H_5COOH}$$

Q. 26. Asymmetric amines do not show optical isomerism. Explain.

Ans. This is because the mirror images of the compound undergo rapid inversion or flipping into each other.

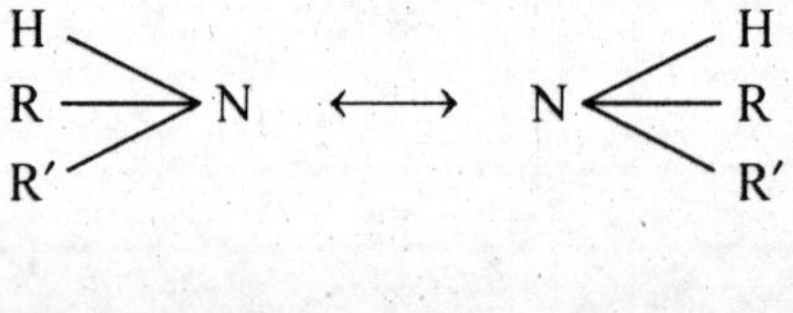

$$(H)(R)(R')N \longleftrightarrow N(H)(R)(R')$$

Therefore, it is not possible to distinguish between the two different forms.

Q. 27. Which is more basic amine in the following pairs and why?

(*i*) $(C_2H_5)_3$ N and $(C_2H_5)_2$ NH

(*ii*) Aniline and benzylamine

(*iii*) Diphenylamine and methylphenylamine

(*iv*) Triethylamine and N, N-dimethylaniline.

Ans. (*i*) $(C_2H_5)_2$ NH is more basic than $(C_2H_5)_3$ N contrary to expectation. This is because triethylamine after getting protonated is difficult to stabilize due to steric hindrance.

(*ii*) Benzylamine is more basic than aniline. This is because in aniline, there is the electron-withdrawing phenyl group attached directly to $-$ NH_2 but in benzylamine, no such electron-withdrawing group is present with the result that electrons are not delocalised and therefore it is a stronger base.

(*iii*) Methylphenylamine is a stronger base than diphenylamine because in the former there is only one electron-withdrawing benzene ring whereas in diphenylamine there are two benzene rings.

(*iv*) Triethylamine is a stronger base than N, N-dimethylaniline. This is because the latter contains an electron withdrawing group. Steric factors are common to both.

Q. 28. How will you convert ethylamine to acetic acid?

Ans. The following steps are involved in the conversion.

$$C_2H_5NH_2 \xrightarrow{HNO_3} CH_3CH_2OH \xrightarrow{[O]} CH_3CHO \xrightarrow{[O]} CH_3COOH$$

Q. 29. How will you separate a mixture of CH_3NH_2 , $(CH_3)_2$ NH and $(CH_3)_3$ N using benzene sulphonyl chloride and aqueous KOH?

Ans. See Q. 11 under Hinsberg method.

Q. 30. Give the mechanism of Hofmann bromamide reaction.

(*Punjab 2003; Calcutta 2007*)

Ans. See Q. 3, Reaction 7

Q. 31. Why are amides weaker bases than amines? (*Annamalai 2002*)

Ans. This is because carbonyl group is electron-attracting. It attracts the lone pair of electrons on nitrogen and thus reduces its basicity

$$R-\overset{\overset{O}{\|}}{C}-\ddot{N}H_2 \longleftrightarrow R-\overset{\overset{O^-}{|}}{C}=\overset{+}{N}H_2$$

Q. 32. Give the name, products and mechanism of the following reaction.

$$C_6H_5-CO-NH_2 + Br_2 + KOH \longrightarrow$$ (*Coimbatore, 2000*)

Ans. See Q. 3 (7).

Q. 33. Complete the following reactions :

(*i*) C_6H_5Br (bromobenzene) $\xrightarrow[\text{liq. } NH_3]{NaNH_2}$ (*Lucknow, 2000*)

(*ii*) $C_6H_5NH_2$ (aniline) + ClCOC_6H_5 $\xrightarrow{NaOH}$ (*Awadh, 2000*)

Ans. (*i*) See Q. 3 (1).

(*ii*) See Q. 9 (3).

Q. 34. Compare the basicities of methylamine, dimethylamine and trimethyl amine.

(*Kerala, 2000*)

Ans. See Q. 12.

Q. 35. Write the reactions of primary, secondary and tertiary amines with nitrous acid.

(*Garhwal, 2000*)

Ans. See Q. 11 (1).

Q. 36. Arrange the following compounds in order of their increasing basic character.

(*Garhwal, 2000*)

(*i*) aniline ($C_6H_5NH_2$) (*ii*) p-toluidine (NH_2, CH_3 para) (*iii*) p-phenylenediamine (NH_2, NH_2 para) (*iv*) m-nitroaniline (NH_2, NO_2 meta)

Ans. m-nitroaniline (NH_2, NO_2) < aniline (NH_2) < p-toluidine (NH_2, CH_3) < p-phenylenediamine (NH_2, NH_2)

Q. 37. How will you do the following conversions :

(*i*) Aniline ⟶ Azobenzene (*Garhwal, 2000*)

(*ii*) o–Chlorotoluene ⟶ *o*–Toluic acid (*Lucknow, 2000*)

Ans. (*i*)

$$\underset{\text{aniline}}{C_6H_5NH_2} \xrightarrow{NaNO_2/HCl} C_6H_5N_2Cl \xrightarrow{C_6H_5OH} C_6H_5-N=N-C_6H_4-OH \xrightarrow[\Delta]{Zn} \underset{\text{azobenzene}}{C_6H_5-N=N-C_6H_5}$$

(*ii*)

$$\underset{o\text{-chlorotoluene}}{CH_3C_6H_4Cl} \xrightarrow{KCN} CH_3C_6H_4CN \xrightarrow{H_2O} \underset{o\text{-toluic acid}}{CH_3C_6H_4COOH}$$

Q. 38. How will you distinguish between primary, secondary and tertiary nitroalkanes?

(*Meerut, 2000*)

Ans. Firstly, convert primary, secondary and tertiary nitroalkanes into amines with Sn/HCl reagent, After that see Q. 11.

Q. 39. Why can not benzylamine be diazotised? (*Meerut, 2000*)

Ans. Benzylamine cannot be diazotised because – NH_2 group is not directly attached to benzene ring.

Q. 40. What is the action of aniline with (*i*) chloroform and alc. KOH and (*ii*) benzaldehyde? (*Kerala, 2001*)

Ans. (*i*) See Q. 9 (6)

(*ii*) $C_6H_5-NH_2 + OHC-C_6H_5 \longrightarrow C_6H_5-N=CH-C_6H_5$

Q. 41. Explan, why,

(*i*) the basic character of acetamide is less than ethylamine ?

(*ii*) all aliphatic amines are more basic than ammonia ?

Ans. (*i*) It is due to electron attracting effect of $>C=O$ group, thereby reducing electron density on nitrogen.

(*ii*) See Q. 12.

Q. 42. Complete the following reactions :

(*i*) Primary amine + CS_2 $\xrightarrow{HgCl_2}$ (*Awadh, 2000*)

(*ii*) aniline + acetic anhydride $\longrightarrow$ (*M. Dayanand, 2000*)

Ans. (*i*) See Q. 9 (8)

(*ii*) $C_6H_5NH_2 + (CH_3CO)_2O \longrightarrow C_6H_5NHCOCH_3 + CH_3COOH$

Acetanilide

Q. 43. Suggest a method for conversion of aniline into nitrobenzene. (*Himachal, 2000*)

Ans.

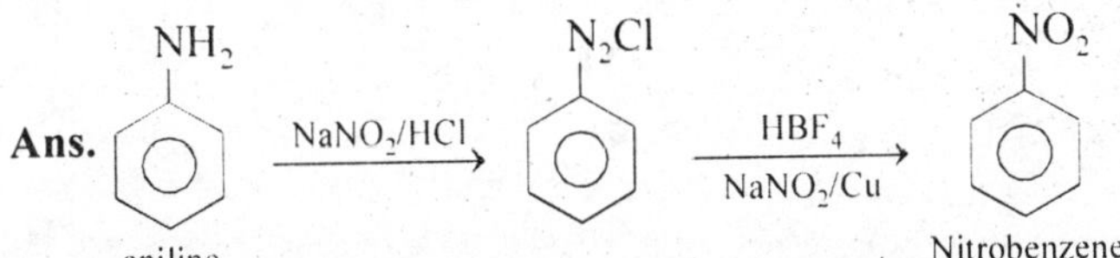

Q. 44. Which of the following pairs are more basic, and why ? (*Kurukshetra, 2001*)

(*i*) $C_6H_5\ CH_2NH_2$, $C_6H_5NH_2$

(*ii*) $p-ClC_6H_4NH_2$, $o-NO_2C_6H_4NH_2$

Ans. (*i*) See Q. 27 (*ii*).

(*ii*) See Q. 13.

Q. 45. Explain why aniline is a poorer base than cyclohexyl amine. What happens when

(*i*) Aniline is heated with a mixture of chloroform and sodium hydroxide (*W. Bengal 2002*)

(*ii*) Aniline is heated with CS_2 in presence of a base

(*iii*) Aniline is treated with sodium nitrite and HCl and the product is boiled with ethanol (*Guwahati 2002*)

Ans.

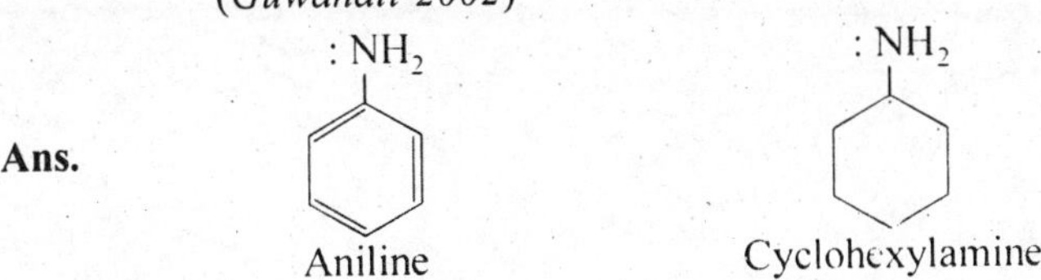

The electron pair on nitrogen in aniline is delocalised due to resonance with benzene ring. This is not so in the case of cyclohexyl amine.

Thus, aniline is less basic.

For reactions of aniline, see Q. 9.

Q. 46. Explain why dimethylamine is a stronger base than trimethylamine.(*Delhi, 2005*)

Ans. See Q. 12.

Q. 47. How will you distinguish between N-methylaniline and N, N-dimethylaniline?

(*Punjab 2005*)

Hint. Use Hinsberg's test. Q.11.

28

DIAZONIUM SALTS

Q. 1. What are diazonium salts? Give their nomenclature.

Ans. Diazonium salts are aromatic organic compounds containing $-N_2X$ group where X is a halogen. It could also be a $-NO_2$, $-HSO_4$ or $-BF_4$ group. The name **diazo** stands for two nitrogens. The word **onium** is obtained from ammonium as these compounds resemble ammonium salts in some respects.

Nomenclature

Diazonium salts are named by adding **diazonium** to the aromatic compound to which it is attached, followed by the name of the anion. Names of some diazonium salts are given below:

Compound	*Name*
$C_6H_5-N_2Cl$	Benzenediazoniumchloride
$C_6H_5-N_2HSO_4$	Benzenediazoniumhydrogensulphate
$CH_3-C_6H_4-N_2Br$	*p*-Toluenediazoniumbromide
$O_2N-C_6H_4-N_2BF_4$	*p*-Nitrobenzenediazoniumfluoroborate

Q. 2. Discuss the structure of benzene diazonium chloride. (*Awadh, 2000; Purvanchal 2007*)

Ans. (*i*) It has the molecular formula $C_6H_5N_2Cl$ as confirmed by qualitative and quantitative analysis.

(*ii*) It is a colourless crystalline solid soluble in water but insoluble in alcohol and ether.

(*iii*) It undergoes substitution reaction in which $-N_2Cl$ group is replaced by a monovalent group like $-Cl$, $-Br$, $-CN$ or $-NO_2$.

(*iv*) Benzene diazonium chloride exhibits salt-like character. This is indicated by conductivity measurement.

(*v*) It gives coupling reaction with phenols and amines to form azocompounds of the type

$$C_6H_5-N=N-C_6H_4-NH_2$$

(*vi*) On reduction with stannous chloride and hydrochloric acid, it gives phenylhydrazine.

(*vii*) It forms salts called diazotate, in alkaline solution, which exist as geometrical isomers.

Based on the above observations, different scientists gave their structure for benzene diazonium chloride, which are represented below.

Griess Formula. Griess was the discoverer of benzene diazonium chloride. He proposed the following structure for this compound:

However this formula was not consistent with the properties of the compound. It did not explain well the following properties:

(*a*) Its salt-like character.

(*b*) Replacement of $-N_2Cl$ by other monovalent groups.

(*c*) Reduction to phenylhydrazine in which only one nitrogen is attached to the benzene ring.

(*d*) Formation of azocompounds in coupling reactions.

Kekule Structure

Kekule gave his structure of benzene diazonium chloride as under:

$$C_6H_5-N=N-Cl$$

This formula explained satisfactorily replacement of $-N_2Cl$ by monovalent groups, coupling reactions and reduction to phenylhydrazine. But it didn't account for the salt-like or ionic nature of the compound.

Blomstrand formula

Blomstrand gave the following structure for benzene diazonium chloride.

$$C_6H_5-\underset{\underset{Cl}{|}}{N}\equiv N$$

But this has been modified to

$$\left[C_6H_5-\overset{+}{N}\equiv N:\right]Cl^-$$

This modified structure resembles that a ammonium chloride, which can be written as

$$[H-\overset{+}{N}\equiv H_3]\,Cl^-$$

This structure very well explain the following:

(*a*) Salt-like or ionic structure for the compound as indicated by conductivity measurement.

(*b*) Solubility in water and isolubility in organic solvents. It is a common observation that polar compounds are soluble in polar solvents.

(*c*) Presence of nitrogen-nitrogen triple bond as indicated by infrared spectrum.

(*d*) Existence of geometrical isomers of diazoates

$$\underset{\text{Benzene diazonium chloride}}{C_6H_5N^+\equiv N-Cl} \overset{NaOH}{\rightleftharpoons} \underset{\text{Benzene diazonium hydroxide}}{C_6H_5N^+\equiv NOH^-}$$

$$\updownarrow$$

$$\underset{\text{Sod. diazotate}}{C_6H_5-N=NO^-Na^+} \longleftarrow \underset{\text{Diazohydroxide}}{C_6H_5N=N-OH}$$

Due to the presence of double bond between N–N, there exists a possibility of occurrence of geometrical isomers as given below:

$$\begin{matrix} C_6H_5N \\ \| \\ Na^+O^- - N \end{matrix} \qquad \begin{matrix} C_6H_5 - N \\ \| \\ N - O^- Na^+ \end{matrix}$$

(Cis) diazotate (Trans) diazotate

Q. 3. What is diazotisation? How is benzene diazonium chloride prepared in the laboratory? Give its mechanism. *(Kurukshetra, 2001; Mysore, 2002; Annamalai, 2003; Saurashtra, 2004; Bangalore 2004; Calcutta 2007)*

Ans. Diazonium salts are prepared by reaction between aromatic primary amine and nitrous acid obtained from sodium nitrite and dil. hydrochloric acid in *situ*. For preparing benzene diazonium chloride, the following method is adopted.

Aniline is dissolved in dil. hydrochloric acid and the solution is cooled to 273-278 K. Now a cooled solution of sodium nitrite is added to the first solution taking care that the temperature does not rise beyond 278 K. A few drops of the mixture are added to starch-potassium iodide solution. As soon as a blue colour is obtained, further addition of sodium nitrite is stopped. Thus the excess of nitrous acid is avoided. This is the completion of diazotisation.

$$C_6H_5NH_2 + NaNO_2 + HCl \xrightarrow{273-278\ K} C_6H_5N_2^+Cl^- + NaCl + 2H_2O$$

Benzene diazonium chloride obtained above is used as such in the solution. It is not extracted or crystallised out. This is because dry benzene diazonium chloride is explosive.

Mechanism of diazotisation. It is believed that the amine molecule undergoes nucleophilic reaction by attaching itself to nitrous anhydride obtained from nitrous acid. The complete mechanism is given below:

$$H-\ddot{O}-\ddot{N}=O + \overset{\oplus}{H} \rightleftharpoons H-\overset{H}{\underset{\oplus}{O}}-N=O \rightleftharpoons \overset{\oplus}{N}=O + H_2O$$

Mineral acid (Nitrosoniumion)

$$Ar-\ddot{N}H_2 + \overset{\oplus}{N}=O \longrightarrow Ar-\overset{H}{\underset{H}{\overset{\oplus}{N}}}-N=O \xrightarrow{-H^{\oplus}} Ar-\ddot{N}H-\ddot{N}=\ddot{O}:$$

$$\xrightarrow{H^+} Ar-\underset{H}{\ddot{N}}-\ddot{N}=\overset{\oplus}{O}H \xrightarrow{-H^{\oplus}} Ar-N=N-\ddot{O}H \xrightarrow{H^{\oplus}} Ar-\ddot{N}=\ddot{N}-\overset{\oplus}{O}H_2$$

$$\xrightarrow{-H_2O} \left[Ar-\overset{\oplus}{N}\equiv N \longleftrightarrow Ar-\ddot{N}=\ddot{N}^{\oplus} \right]$$

Aromatic amines carrying electron withdrawing groups in the benzene ring are difficult to diazotise. This is because mechanistically, it is a nucleophilic reaction. Electron withdrawing groups decrease the electron density on amino nitrogen, thereby reducing the chances of attack on nitrous acid and nitrous anhydride.

$$O=N(O^-)-C_6H_4-\ddot{N}H_2 \longleftrightarrow (O^-)(O^-)N=C_6H_4=\overset{\oplus}{N}H_2$$

Q. 4. (*a*) Give the synthesis and mechanism of diazotization.

(*b*) Which of the following gives most stable diazo compound on treatment with HNO_2 and why?

$C_2H_5NH_2$, Aniline, *o*-nitroaniline (*Kurukshetra, 2001*)

Ans. (*a*) For synthesis and mechanism, see Q. 3.

(*b*) Aniline gives the most stable diazo compound viz. benzene diazonium chloride on treatment with HNO_2.

Aliphatic amines don't give diazo compound, alcohol is formed on treatment with HNO_2. Extent of diazotisation is decreased if an electron-withdrawing group like $-NO_2$ group is attached to the benzene ring See Q. 3 for details.

Q. 5. Describe, with mechanism wherever applicable, chemical properties of benzene diazonium chloride. (*Kerala, 2001*)

Ans. Properties of benzene diazonium chloride can be discussed under two heads.

(A) Reaction in which N_2 is evolved and the $-N_2Cl$ group is replaced by some other monovalent group.

(B) Coupling reaction of reduction, in which nitrogen atoms are retained.

These reactions are being given separately as under:

Reactions in which N_2Cl is replaced by some other group.

1. Replacement by – Cl, – Br or – CN. This is a reaction in which group is replaced by one of above mentioned groups. The reaction is carried out by treating a solution of diazonium salt with cuprous chloride, bromide or cyanide. The reactions are given as under:

$$C_6H_5\overset{+}{N}_2Cl^- \xrightarrow{CuCl} C_6H_5N_2Cl + N_2\uparrow$$

$$C_6H_5\overset{+}{N}_2\overset{-}{Cl} + CuBr \longrightarrow C_6H_5Br + N_2\uparrow + CuCl$$

$$C_6H_5\overset{+}{N}_2\overset{-}{Cl} + CuCN \longrightarrow C_6H_5CN + N_2\uparrow + CuCl$$

This reaction is known as **Sandmeyer's reaction.**

In a slighly different reaction, known as **Gattermann reaction,** replacement by – Cl, – Br or – CN is done by taking copper powder and halogen acid.

$$C_6H_5\overset{+}{N}_2Cl^- \xrightarrow{Cu/HCl} C_6H_5Cl + N_2$$

$$C_6H_5\overset{+}{N}_2Cl^- \xrightarrow{Cu/HBr} C_6H_5Br + N_2$$

$$C_6H_5\overset{+}{N}_2Cl^- \xrightarrow{Cu/KCN} C_6H_5CN + N_2$$

2. Replacement by iodine. On warming diazonium salt solution with potassium iodide solution, $-N_2Cl$ group in replaced by – I group.

$$C_6H_5\overset{+}{N}_2\overset{-}{Cl} + KI \longrightarrow C_6H_5I + N_2 + KCl$$

3. Replacement by fluorine. Replacement by fluorine can be brought about by the following two-step reaction.

$$\underset{}{C_6H_5\overset{+}{N}_2\,Cl^-} + \underset{\text{Fluoro boric acid}}{HBF_4} \longrightarrow \underset{\text{Diazonium fluoborate (insoluble)}}{C_6H_5\overset{+}{N}_2\,BF_4^-} + HCl$$

$$C_6H_5\overset{+}{N}_2BF_4^- \xrightarrow{\text{Heat}} \underset{\text{Fluorobenzene}}{C_6H_5F} + N_2 + BF_3$$

4. Replacement by – OH. On warming the solution of diazonium salt, $-N_2Cl$ group is replaced by – OH group to yield phenol.

$$\underset{\text{Benzene diazonium chloride}}{C_6H_5N_2Cl} + H_2O \longrightarrow \underset{\text{Phenol}}{C_6H_5OH} + N_2 + HCl$$

5. Replacement by $-NO_2$. Substitution by nitro group is achieved by decomposing diazonium fluoborate with sodium nitrite solution in the presence of copper powder.

$$C_6H_5\overset{+}{N}_2\,Cl^- + HBF_4 \longrightarrow C_6H\overset{+}{N}_2\,BF_4^- \xrightarrow[\text{Cu}]{NaNO_2} C_6H_5-NO_2 + N_2 + NaBF_4$$

6. Replacement by aryl group. Diazonium salts react with other aromatic compounds in the presence of alkali to give diphenyl.

$$C_6H_5-\overset{+}{N}_2\overset{-}{Cl} + C_6H_6 \xrightarrow{\text{NaOH}} C_6H_5-C_6H_5 + N_2 + HCl$$

7. Replacement by hydrogen. If diazonium salt solution is allowed to stand in contact with hypophosphorus acid in the presence of cuprous ions, $-N_2Cl$ group is replaced by – H.

$$C_6H_5N_2Cl + H_3PO_2 + H_2O \longrightarrow C_6H_6 + H_3PO_3 + N_2 + HCl$$

Mechanism of the reaction is as follows:

$$Ar\overset{+}{N}\equiv N + \underset{\text{(Cuprous)}}{Cu^+} \xrightarrow{-Cu^{2+}} Ar-N\equiv N + \underset{\text{(Cupric)}}{Cu^{2+}}$$

$$\downarrow$$

$$\dot{Ar} + N_2$$

$$\dot{Ar} + H-\overset{\overset{O}{\|}}{\underset{\underset{H}{|}}{P}}-OH \longrightarrow ArH + H-\overset{\overset{O}{\|}}{\underset{\bullet}{P}}-OH$$

$$ArN_2^+ + H-\overset{\overset{O}{\|}}{\underset{\bullet}{P}}-OH \longrightarrow \dot{Ar} + N_2 + H-\overset{\overset{O}{\|}}{\underset{+}{P}}-OH \xrightarrow{H_2O} H-\overset{\overset{O}{\|}}{\underset{\underset{OH}{|}}{P}}-OH + H^+$$

Reactions in which nitrogen atoms are retained in the product:

8. Reduction. On treatment with stannous chloride and HCl, diazonium salts are reduced to hydrazines.

$$C_6H_5\overset{+}{N}_2Cl^- + 4[H] \xrightarrow[\text{or } Na_2SO_3]{SnCl_2 / HCl} C_6H_5NHNH_2HCl$$

Q. 6. What is Coupling reaction? Give the mechanism of coupling reaction with

(*a*) Phenol

(*b*) Tertiary amine. (*Himachal 2000; Kerala, 2001; Vikram 2002; Panjab, 2003*)

Ans. Coupling reaction. It is an important reaction of diazonium salts and involves reaction of diazonium ions which act as electrophilic agents, with aromatic compounds containing strong electron-releasing groups such as – OH, – NHR and – NHR_2. The reaction is known as coupling and leads to the formation of azo compounds having the general formula

$$C_6H_5 - N = N - C_6H_4 - G$$

where G is one of above mentioned electron releasing groups. The coupling reaction in general may be written as

$$C_6H_5 - N_2^+ + H - C_6H_4 - G \longrightarrow C_6H_5 - N = N - C_6H_4 - G + H^+$$

(*a*) Coupling with phenol. Benzene diazonium salt reacts with phenol in weakly alkaline solution to form hydroxy azo compound.

$$C_6H_5 - \overset{+}{N}_2\overset{-}{Cl} + C_6H_5 - OH \xrightarrow{(pH\ 9-10)} C_6H_5 - N = N - C_6H_4 - OH + HCl$$

p-hydroxyazobenzene

Mechanism of reaction

It involves the electrophilic attack of diazonium ion on phenoxide ion as shown below:

$$C_6H_5 - \overset{+}{N} \equiv N + C_6H_5\ddot{O}:^- \longrightarrow C_6H_5 - N = N - C_6H_5(H)=\ddot{O}:$$

Benzene diazonium ion Phenoxide ion

$$\downarrow H^+$$

$$C_6H_5 - N = N - C_6H_4 - \ddot{O}$$

(*b*) Coupling with tertiary amine. This coupling takes place in acidic solution. Diazonium salt reacts with tertiary amine to form dialkylamino azo compound.

$$C_6H_5 - \overset{+}{N}_2 - \overset{-}{Cl} + C_6H_5 - N(CH_3)_2 \xrightarrow{(pH\ 4-5)} C_6H_5 - N = N - C_6H_4 - N(CH_3)_2$$

Benzene diazonium chloride; N, N-Dimethyl aniline; *p*-Dimethyl amino azobenzene

The mechanism of the reaction is as follows:

$$C_6H_5-\overset{+}{N}\equiv N + C_6H_5N(CH_3)_2 \longrightarrow C_6H_5-N=N-\text{(H)}C_6H_4=\overset{+}{N}(CH_3)_2$$

Benzenediazonium ion N, N-Dimethyl aniline

$$\xrightarrow{-H^+} C_6H_5-N=N-C_6H_4-N(CH_3)_2$$

Q. 7. How are diazo salts prepared? Compare the stability of aliphatic and aromatic diazo compounds and give their synthetic applications.

(*Shivaji 2002; Panjab 2003; Annamalai 2004*)

Ans. For preparation of diazo salts see Q. 3. For synthetic applications, see Q. 5 and Q. 6.

Q. 8. How does benzene diazonium chloride react with aniline in mildly acidic medium? Give the mechanism of this coupling reaction. (*Panjabi 2003*)

Ans. Primary and secondary amines react with diazonium salts in weakly alkaline medium to form colourless diazoamino compounds. However, these compounds undergo rearrangement when heated in acidic medium to form coloured azo compounds.

$$C_6H_5-\overset{+}{N_2}\overset{-}{Cl} + H_2N-C_6H_5 \xrightarrow[-HCl]{CH_3COONa} C_6H_5-N=N-NH-C_6H_5$$

Diazoaminobenzene

$$\xrightarrow{\text{Heat} \quad H^+} C_6H_5-N=N-C_6H_4- \quad + NH_2$$

p-aminoazobenzene

Mechanism of reaction

$$C_6H_5-\overset{+}{N}\equiv N + C_6H_5-\ddot{N}H_2 \longrightarrow C_6H_5-N=N-\text{(H)}C_6H_4=\overset{+}{N}H_2$$

$$\xrightarrow{-H^+} C_6H_5-N=N-C_6H_4-NH_2$$

p-Aminoazobenzene

Q. 9. How will you synthesise:

(*a*) *o*-Toluic acid from toluene

(*b*) *m*-bromophenol from nitrobenzene

(*c*) 1, 3, 5-Tribromobenzene from aniline (*Kalyani 2003, Nagpur 2004*)

(*d*) *o*- and *p*-dinitrobenzne from aniline

Ans. (*a*) *o*-Toluic acid from toluene

$$\underset{\text{Toluene}}{C_6H_5CH_3} \xrightarrow[H_2SO_4]{HNO_3} \underset{o\text{-Nitrotoluene}}{CH_3C_6H_4NO_2} \xrightarrow[SnCl_2/HCl]{[H]} \underset{o\text{-Toluidine}}{CH_3C_6H_4NH_2}$$

$$\xrightarrow[HCl]{NaNO_2} \underset{o\text{-Toluene diazonium chloride}}{CH_3C_6H_4N_2Cl} \xrightarrow{CuCN} CH_3C_6H_4CN \xrightarrow{H_2O} \underset{o\text{-Toluic acid}}{CH_3C_6H_4COOH}$$

(*b*) *m*-bromophenol from nitrobenzene

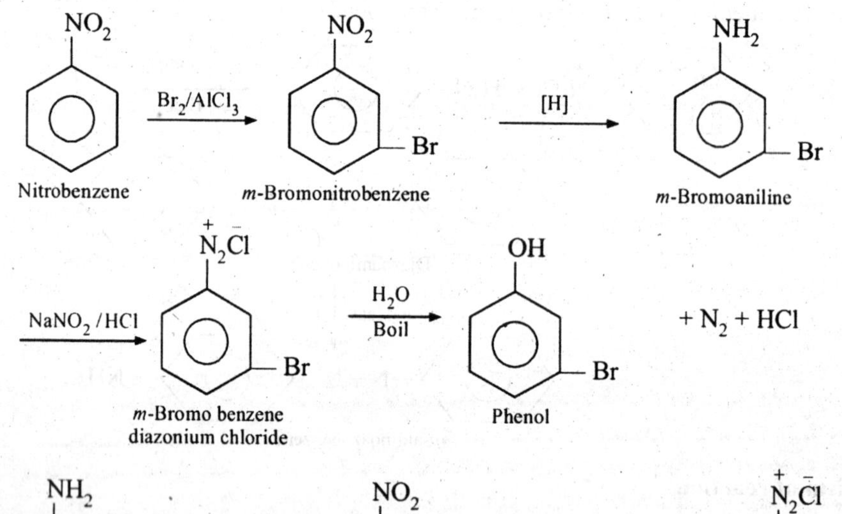

(*c*)

$$\underset{\text{Aniline}}{C_6H_5NH_2} \xrightarrow[\text{Water}]{Br_2} \underset{\text{2, 4, 6-Tribromoaniline}}{C_6H_2Br_3NH_2} \xrightarrow[HCl]{NaNO_2} \underset{\text{2, 4, 6-Tribromobenzene diazonium chloride}}{C_6H_2Br_3\overset{+}{N_2}\overset{-}{Cl}}$$

$$\xrightarrow{H_3PO_2} \underset{\text{1, 3, 5-Tribromobenzene}}{C_6H_3Br_3}$$

(*d*) *o*- and *p*-dinitrobenzene from aniline

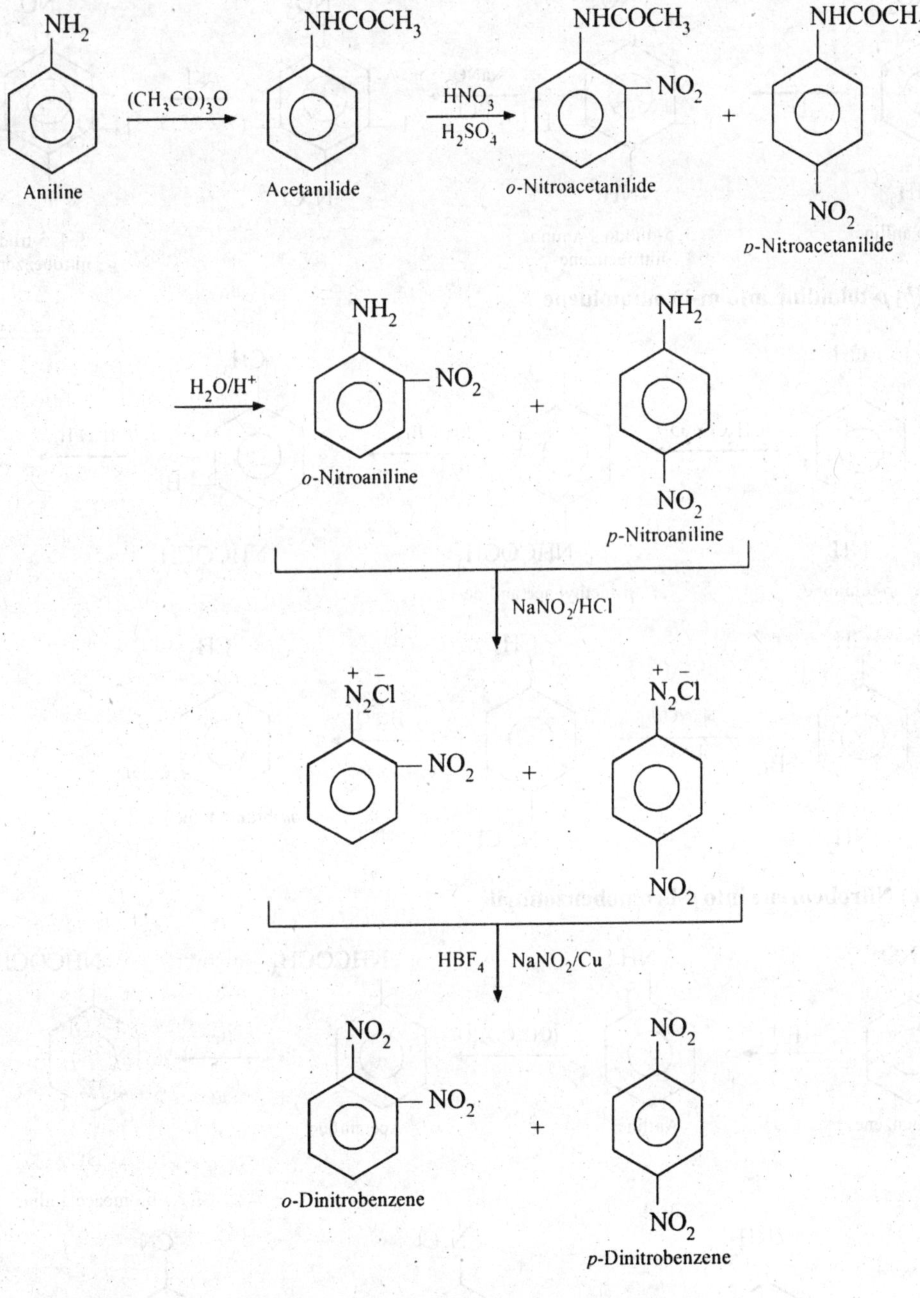

Aniline $\xrightarrow{(CH_3CO)_2O}$ Acetanilide $\xrightarrow[H_2SO_4]{HNO_3}$ *o*-Nitroacetanilide + *p*-Nitroacetanilide

$\xrightarrow{H_2O/H^+}$ *o*-Nitroaniline + *p*-Nitroaniline

$\xrightarrow{NaNO_2/HCl}$ $\overset{+}{N_2}\overset{-}{Cl}$ (*o*-nitro) + $\overset{+}{N_2}\overset{-}{Cl}$ (*p*-nitro)

$\xrightarrow{HBF_4 \mid NaNO_2/Cu}$ *o*-Dinitrobenzene + *p*-Dinitrobenzene

Q. 10. How will you bring about the following conversions?

(*a*) *p*-nitroaniline into 3, 4, 5-triiodo nitrobenzene.

(*b*) *p*-toluidine into *m*-bromotoluene. (*c*) Nitrobenzene into *p*-bromobenzonitrile.

(*d*) *m*-dinitrobenzene into *m*-nitrochlorobenzene.

Ans. (*a*) *p*-nitroaniline into 3, 4, 5-triiodo nitrobenzene:

p-nitro aniline (NO$_2$, NH$_2$) $\xrightarrow[\text{Fe}]{I_2}$ 3, 5-diiodo 4-amino nitrobenzene $\xrightarrow[\text{HCl}]{\text{NaNO}_2}$ (N$_2$Cl) $\xrightarrow{\text{KI}}$ 3, 4, 5-triido nitrobenzene

(*b*) *p*-toluidine into m-bromotoluene

p-toluidine (CH$_3$, NH$_2$) $\xrightarrow{(CH_2CO)_2O}$ *p*-methyl acetanilide (CH$_3$, NHCOCH$_3$) $\xrightarrow{Br_2}$ (CH$_3$, Br, NHCOCH$_3$) $\xrightarrow{H_2O/H^+}$

(CH$_3$, Br, NH$_2$) $\xrightarrow[\text{HCl}]{\text{NaNO}_2}$ (CH$_3$, Br, $N_2^+Cl^-$) $\xrightarrow{H_3PO_2}$ *m*-bromo toluene

(*c*) Nitrobenzene into *p*-bromobenzonitrile

Nitrobenzene (NO$_2$) $\xrightarrow{[H]}$ Aniline (NH$_2$) $\xrightarrow{(CH_3CO)_3O}$ Acetanilide (NHCOCH$_3$) $\xrightarrow{Br_2}$ *p*-bromoacetanilide (NHCOCH$_3$, Br)

$\xrightarrow[H^+]{H_2O}$ *p*-bromoaniline (NH$_2$, Br) $\xrightarrow[\text{HCl}]{\text{NaNO}_2}$ *p*-bromobenzene diazonium chloride ($\overset{+}{N_2}\overset{-}{Cl}$, Br) $\xrightarrow{\text{CuCN}}$ *p*-bromobenzonitrile (CN, Br)

(*d*) *m*-dinitrobenzene into *m*-nitrochlorobenzene

$$m\text{-dinitrobenzene} \xrightarrow{NH_4HS} m\text{-nitroaniline} \xrightarrow{NaNO_2} m\text{-nitrobenzene diazonium chloride} (\overset{+}{N_2}\overset{-}{Cl}) \xrightarrow{CuCl} m\text{-nitrochlorobenzene}$$

Q. 11. Explain the following:

(*i*) During diazotisation of arylamines, excess of mineral acid is used.

(*ii*) A weakly basic solution favours coupling with phenol.

(*iii*) 2, 6 Dimethyl-N, N-dimethylaniline has a free *p*-position, but it does not undergo coupling with benzene diazonium chloride.

Ans. (*i*) To avoid coupling reaction due to formation of $Ar\overset{+}{N}H_3$, excess mineral acid is used.

(*ii*) High acidity suppresses the formation of the more reactive $C_6H_5O^-$. In the presence of a weak base, $C_6H_5O^-$ is formed but $C_6H_5N = \overset{+}{N}OH$ is not formed.

(*iii*) Due to steric hindrance of methyl groups the *p*-position of 2, 6 dimethyl -N, N-dimethylaniline is not sufficiently activated for coupling reaction.

$N(CH_3)_2$; H_3C, CH_3 (2,6-dimethyl-N,N-dimethylaniline)

Q. 12. Starting with benzene outline synthesis of the following compounds using diazonium salts as intermediates:

(*i*) *p*-hydroxyazobenzene

(*ii*) phenylhydrazine

(*iii*) phenol

(*iv*) bromobenzene

Ans. First Step. Convension of benzene into benzene diazonium chloride.

$$\text{Benzene} + HNO_3 \xrightarrow{H_2SO_4} \text{Nitrobenzene } (NO_2) \xrightarrow{[H]} \text{Aniline } (NH_2) \xrightarrow[HCl]{NaNO_2} \text{Benzene diazonium chloride } (N_2Cl)$$

$$C_6H_5N_2Cl \xrightarrow[\text{(pH 9-10)}]{C_6H_5OH} C_6H_5-N=N-C_6H_4-OH + HCl$$

p-hydroxy azobenzene

$$C_6H_5N_2Cl \xrightarrow{SnCl_2/HCl} C_6H_5-NHNH_2$$

Phenyl hydrazine

$$C_6H_5N_2Cl \xrightarrow{H_2O} C_6H_5-OH + N_2 + HCl$$

Phenol

$$C_6H_5N_2Cl \xrightarrow{Cu/HBr} C_6H_5-Br + N_2\uparrow$$

Bromobenzene

Benzene diazonium chloride

DIAZOMETHANE

Q. 13. Give the structure, methods of preparation and synthetic uses of diazonethane

(*Merut, 2000; Guwahati, 2007*)

Ans. Structure

It exists as a resonance hybrid of two structures

$$H_2C=\overset{+}{N}=\overset{-}{\ddot{N}}: \longleftrightarrow H_2\overset{-}{\ddot{C}}-\overset{+}{N}\equiv N$$

Methods of Preparation 1. It can be obtained from N-nitroso-N-methyl compounds as shown below:

$$\underset{\text{N-nitroso-N-methylurethane}}{CH_3-N(CO-OC_2H_5)-N=O} + KOH \xrightarrow[\text{boil}]{\text{Ether}} \underset{\text{Diazomethane}}{CH_2N_2} + C_2H_5OH + KHCO_3$$

$$\underset{\text{N-nitroso-N-methyl urea}}{CH_3-N(CO-NH_2)-N=O} + KOH \xrightarrow{\text{Ether}} \underset{\text{Diazomethane}}{CH_2N_2} + KCN + 2H_2O$$

2. By passing nitrous oxide (N_2O) through an ethereal solution of methylithium (CH_3Li)

$$N_2O + CH_3Li \longrightarrow CH_3-N=N-OLi \xrightarrow[-\,LiOH]{\Delta} CH_2N_2$$

Properties of diazomethane

1. Action of heat. When heated or exposed to light, it decomposes to form methylene

$$\overset{-}{C}H_2-\overset{+}{N}\equiv N \xrightarrow[\text{or } h\nu]{\Delta} \underset{\text{Methylene}}{CH_2:} + N_2$$

Methylene is very reactive and adds to alkanes to form higher homologues

$$\underset{\text{Ethane}}{CH_3\,CH_3} + CH_2: \longrightarrow \underset{\text{Propane}}{CH_3CH_2CH_3}$$

2. Reduction. On treatment with Na / Hg, it gives methyl hydrazine

$$CH_2N_2 + 4[H] \xrightarrow{\text{Na / Hg}} \underset{\text{Methylhydrazine}}{CH_3\,NHNH_2}$$

3. Reaction with minereal acids.

$$CH_2N_2 + HCl \longrightarrow \underset{\text{Intermediate compound}}{[CH_3N_2Cl]} \longrightarrow \underset{\text{Methyl chloride}}{CH_3Cl + N_2\uparrow}$$

4. Reaction with carboxylic acids. Diazomethane reacts with acids to form methyl esters

$$\underset{\text{Acetic acid}}{CH_3 - \overset{\overset{\displaystyle O}{||}}{C} - OH} + CH_2N_2 \xrightarrow{\text{Ether}} \underset{\text{Methyl acetate}}{CH_3 - \overset{\overset{\displaystyle O}{||}}{C} - OCH_3} + N_2$$

5. Reaction with phenols. On treatment with diazomethane, phenols get methylated

$$C_6H_5OH + CH_2N_2 \xrightarrow{\text{Ether}} \underset{\text{Anisole}}{C_6H_5OCH_3} + N_2$$

6. Reaction with alcohols and amines. Hydrogen of the alcohol or amine is replaced by methyl group in the presence of BF_3.

$$C_2H_5OH + CH_2N_2 \xrightarrow{BF_3} \underset{\text{Ethyl methyl ether}}{C_2H_5OCH_3} + N_2$$

$$C_2H_5NH_2 + CH_2N_2 \xrightarrow{BF_3} \underset{\text{N-methyl ethanamine}}{C_2H_5NHCH_3} + N_2$$

7. Reaction with carbonyl compounds. Diazomethane converts aldehydes into ketones, while ketones are converted into higher homologues

$$\underset{\text{Acetaldehyde}}{CH_3 - \overset{\overset{\displaystyle O}{||}}{C} - H} + CH_2N_2 \longrightarrow \underset{\text{Acetone}}{CH_3 - \overset{\overset{\displaystyle O}{||}}{C} - CH_3} + N_2$$

$$CH_3 - \overset{\overset{\displaystyle O}{||}}{C} - CH_3 + CH_2N_2 \longrightarrow \underset{\text{Methyl ethyl ketone}}{CH_3 - \overset{\overset{\displaystyle O}{||}}{C} - CH_2CH_3} + N_2$$

8. Ring expansion. Cyclohexanone reacts with diazomethane to produce cycloheptanone.

$$\underset{\text{Cyclohexanone}}{C_6H_{10}O} + CH_2N_2 \longrightarrow \underset{\text{Cycloheptanone}}{C_7H_{12}O}$$

9. Arndt-Eistert Synthesis. This method is used to produce the higher homologue of a carboxylic acid by the following sequence of reactions:

$$\underset{\text{Acid}}{R-COOH} \xrightarrow{SO_2Cl_2} RCOCl \xrightarrow{CH_2N_2} RCOCHNH_2 \xrightarrow{Ag_2O} \underset{\text{Higher acid}}{RCH_2COOH}$$

Q. 14. Complete the following reaction :

(*i*) $C_6H_5-N=NCl \xrightarrow{Cu_2Br_2}$

(*ii*) $C_6H_5N_2Cl \xrightarrow{SnCl_2/HCl}$

Ans. (*i*) See Q. 5 (1)

(*ii*) See Q. 5 (8).

Q. 15. Starting from benzene diazonium chloride, how you get the following compounds? (*Kumaon, 2000*)

(*i*) C_2H_5OH (*ii*) $C_6H_5NO_2$ (*iii*) $C_6H_5-N=N-NHO-C_6H_5$

Ans. See Q. 5(4) (*ii*) See Q. 5(5) (*iii*) See Q. 8

Q. 16. Explain why this reaction is carried out in weakly alkaline medium (*Kurukshetra, 2001*)

$$C_6H_5N_2Cl + C_6H_5OH \xrightarrow[\text{medium}]{\text{weakly alkaline}}$$

Ans. Weakly alkaline medium helps in removal of the proton in the intermediate step of the reaction.

Q. 17. What is Sandmeyers reaction? (*Punjab 2003*)

Ans. Reactions in which $-N_2Cl$ group of benzene diazonium chloride is replaced by $-Cl$, $-Br$ or $-CN$ group is called Sandmeyer's reaction. For details, see Q. 5.

Q.18. How will you synthesise benzylamine from aniline. (*Delhi, 2005*)

Ans. The synthesis involves the following steps :

$$\underset{\text{Aniline}}{C_6H_5NH_2} \xrightarrow[HCl,\ 0-5°C]{NaNO_2} C_6H_5N_2^+Cl^- \xrightarrow{CuCN} C_6H_5CN \xrightarrow{LiAlH_4} \underset{\text{Benzylamine}}{C_6H_5CH_2NH_2}$$

29

SPECTROSCOPY AND STRUCTURE (U.V., I.R., N.M.R. and Mass Spectroscopy)

Q. 1. What is meant by spectrum of light? What is electromagnetic spectrum?

Ans. When a beam of sunlight is passed through a prism, it splits into seven different colours. This set of colours obtained by splitting the white light is called *spectrum*. Different colours are associated with different energies and wavelengths. Red light has the smallest energy and the longest wavelength. Energy goes on increasing and wavelength goes on decreasing as we move from red colour to violet colour, thus the violet light has the maximum energy and minimum wavelength.

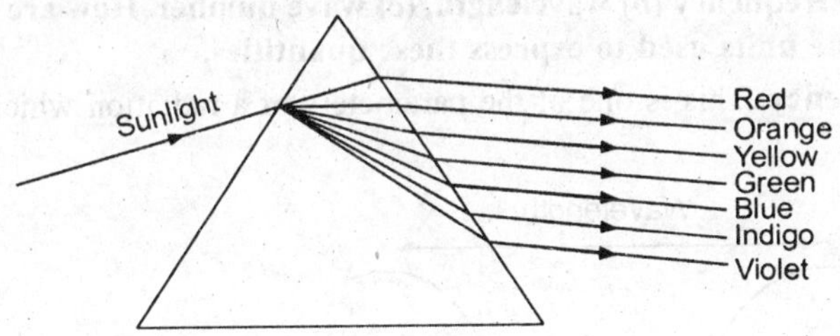

Fig. 29.1.

Electromagnetic spectrum. In addition to the white light with its seven constituent radiations, there are many more different types of radiations. Some of them are more energetic and some less energetic than the visible white light. The electromagnetic spectrum comprises the following radiations.

Cosmic rays. These rays come from sun and are known to be the radiations of highest energy. They have wavelength less than 10^{-3} nm (nanometre).

γ-rays and X-rays. γ-rays and X-rays are less energetic than cosmic rays. Their wavelength ranges between $10^{-3} - 10^{-1}$ nm.

Utraviolet light (U.V. light). These are the rays which are less energetic than X-rays but more energetic than visible light. The range of wavelength of U.V. light is 10–400 nm.

Visible light. It is ordinary sunlight and is comprised of seven different radiations (from red to violet). It has the wavelength range 400 nm–860 nm.

Infrared light are radiations weaker in energy but larger in wavelength than visible light. They have their wavelength in the range $8 \times 10^2 - 3 \times 10^5$ of nm.

Microwaves are radiations of still smaller energy but larger wavelengths lying in the range of $3 \times 10^5 - 1 \times 10^7$ nm.

Radiowaves have the longest wavelength, more than 10^7 nm.

All the above radiations are part of the electromagnetic spectrum. Their wavelengths are shown as follows:

Cosmic Rays	γ-rays	X-rays	UV	Visible	IR	Micro waves	Radio waves
$<10^3$		$10^{-3} - 10^{-1}$	10 – 400	400 – 860	$8 \times 10^2 - 3 \times 10^5$	$3 \times 10^5 - 1 \times 10^7$	$>10^7$

Wavelength in nanometre (nm)

The wavelengths of different constituents of visible light are given below:

Colour	*Wavelength, (nm)*
Violet, indigo	400–435
Blue	435–480
Green-blue	480–490
Blue-green	490–500
Green	500–560
Yellow-green	560–580
Yellow	580–595
Orange	595–610
Red	610–750

Q. 2. Define (*a*) frequency (*b*) wavelength, (*c*) wave number. How are they related to one another? Mention the units used to express these quantities.

Ans. (*a*) **Frequency.** This is one of the parameters of a radiation which has the form of a wave as shown below:

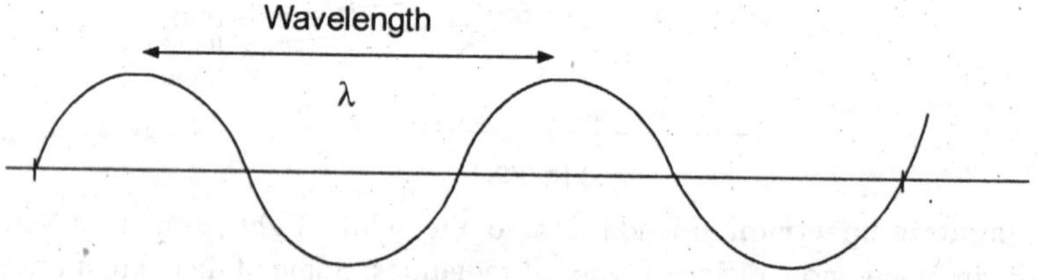

Fig. 29.2.

Frequency is the no. of times a wave crosses a particular point in one second. It is denoted by the Greek letter ν (nu). It is expressed as cycles per second or Hertz.

$$1 \text{ Hz} = 1 \text{ cycle sec}^{-1}$$

Bigger units of frequency are kilocycle and megacycle per second.

$$1 \text{ kilocycle} = 10^3 \text{ Hz}$$

$$1 \text{ megacycle} = 10^6 \text{ Hz}$$

(*b*) **Wavelength.** This is another parameter of a wave. It is the distance between two successive crests of the wave. It is denoted by the Greek letter λ (lambda). Different regions of the electromagnetic spectrum possess different wavelengths. Different units for expressing wavelength have been used in the past. National Bureau of Standards 1963 has recommended nano metre (nm) as the unit of wavelength.

$$1 \text{ nm} = 10^{-9} \text{ m}$$

Sometimes, one comes across other units of wavelength, too. These are given here for the benefit of readers.

The other units that are used are, centimetre (cm), angstrom (Å), micron (μ), millimicron (mμ). Their relations between one another are reproduced below:

(*i*) 1 cm = 10^{-2} m

(*ii*) 1 Å = 10^{-8} cm = 10^{-10} m

(*iii*) 1 μ = 10^{-4} cm = 10^{-6} m

(*iv*) 1 mμ = 10^{-7} cm = 10^{-9} m

(*v*) 1 nm = 10 Å= 1 mμ

(*c*) **Wave number $\bar{\nu}$ (nu bar).** Frequency of a wave is more conveniently used in the form of wave number. It is defined on the number of waves per centimetre. Its relation with wavelength is

$$\bar{\nu} = \frac{1}{\lambda}$$

No. of waves in λ cm = 1

No. of waves in 1 cm = $\frac{1}{\lambda}$

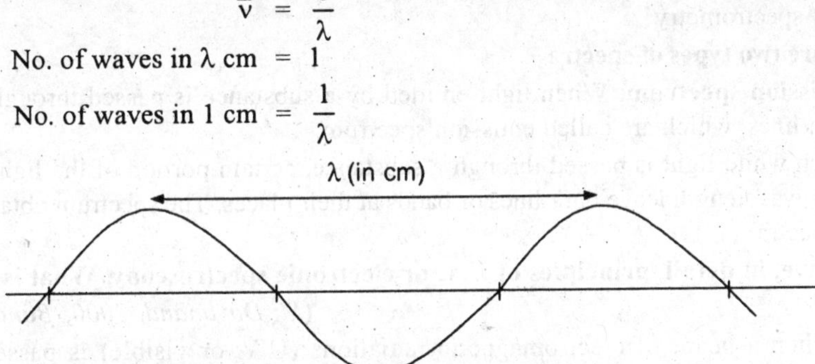

Wave no. has the units of cm^{-1}

Fig. 28.3.

Relations between wavelength, frequency, wave number.

Wavelength and frequency are related by the equation

$$\lambda \times \nu = c$$

where λ and ν stand for wavelength and frequency of the wave, *c* is the velocity of light.

$$\bar{\nu} = \frac{1}{\lambda} \qquad ...(2)$$

But $$\lambda = \frac{c}{\nu} \text{ or } \frac{1}{\lambda} = \frac{\nu}{c} \qquad ...(3)$$

From (2) and (3)

$$\bar{\nu} = \frac{\nu}{c}$$

A wave having frequency ν has energy given by the relation

E = *h*ν where *h* is Planck's constant

Q. 3. What are the advantages of spectroscopic methods of analysis over the conventional methods? (*Himachal, 2000*)

Ans. 1. These methods require microquantities of the sample for investigation compared to a few grams required in classical methods.

2. These methods are quick and very sensitive.

3. These methods generally give reliable and reproducible results and provide a permanent record of observations in the form of spectra.

4. The material can be retrieved unchanged after the investigation and can be reused for other studies, except in case of mass spectrometry, where the sample is destroyed.

5. Continuous monitoring of changes taking place is possible in spectroscopic methods which can give information on kinetics of changes.

Q. 4. Name different spectroscopic techniques which are employed for structure elucidation of organic compounds. How do you classify different spectra?

Ans. Different spectroscopic studies are as follows:

1. Ultraviolet spectroscopy
2. Visible spectroscopy
3. Infrared spectroscopy
4. Nuclear magnetic resonance
5. Proton magnetic resonance
6. Mass spectrometry

There are two types of spectra

(*a*) **Emission spectrum.** When light emitted by a substance is passed through a prism we obtain certain lines, which are called emission spectrum.

(*b*) When white light is passed through a substance, certain portion of the light is absorbed by it. Missing wavelength leave dark lines or bands at their places. The spectrum obtained is called absorption spectra.

Q. 5. Give, in detail, principles of U.V. or electronic spectroscopy. What is its range?

(*M. Dayanand, 2000; Bangalore 2002*)

Ans. When a beam of electromagnetic radiations (U.V. or visible) is passed through a compound, certain radiations (or wavelengths) are absorbed by the substance. There is some energy associated with every radiation given by the relation

$$E = h\nu$$

where h is the Planck's constant and ν is the frequency of radiation.

This energy absorbed by the substance produces some changes within the molecule. If the energy absorbed is high (in UV or visible range), it causes electronic excitation *i.e.*, electrons are excited to higher levels.

It is important to note the molecule does not absorb just any radiation. It absorbs only that radiation which possess the appropriate energy to bring about permitted transition to higher electronic, vibrational or rotational levels. This all depends upon the structure of the molecule. The radiation absorbed is something which is characteristic of a substance. The absorption of radiation is observed with the help of a spectrophotometer by passing radiations of continuously changing wavelength through a substance and analysing the intensity of transmitted radiations. The graph between the amount of radiation absorbed by the sample and the wavelength of radiation is called *absorption spectrum.* The absorption spectrum consists of bands with peaks of maximum intensity.

The range of electronic spectroscopy is 200–800 nm.

Q. 6. Describe the working of a UV Spectrophotometer?

Ans. UV Spectrophotometers record the spectra of compounds in the range 200–800 nm. This range consists of two parts, 200–400 nm is the range of ultraviolet radiations whereas 400–800 nm is the range of visible light, spectra cannot be taken below 200 nm because oxygen present in the spectrophotometer absorbs this radiation. To record spectra of substances below 200 nm, special vacuum techniques are needed.

The figure below gives the schematic representation of spectrophotometer.

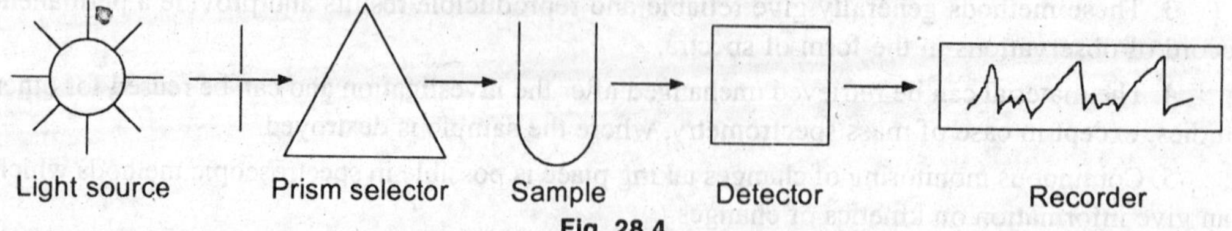

Fig. 28.4

For recording the UV spectrum, the given sample is dissolved in a suitable solvent which itself does not absorb light in that range. Commonly employed solvents are 95% ethanol, hexane and water. The positions of the absorption peaks are slightly shifted with the change of solvent. A quartz cell of 1 cm path length is used as a container for the sample solution. The solution is exposed to UV/visible light by the prism selector. Prism selector is rotating continuously to emit lights of varying wave lengths. Hydrogen discharge lamp and tungsten lamps and used for emitting UV and visible light respectively.

The instruments provides a running graph between wavelength of radiation absorbed and the intensity of absorption. There are a number of peaks or humps in the spectrum. The wavelengths corresponding to tops of the humps correspond to the absorption wavelengths and are denoted as λ_{max}. Intensity of absorption corresponding to that wavelength is called molar extinction coefficient expressed as $\in$ (epsilon).

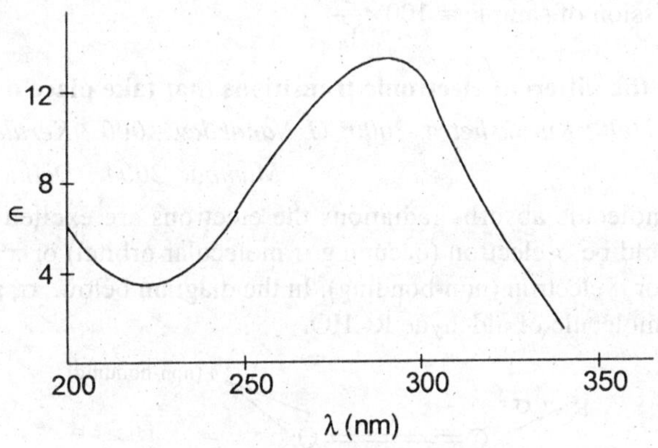

Fig. 28.5

A specimen UV spectrum is given in the graph above. It indicates λ_{max} at 290 nm.

Q. 7. State Beer-Lamert's Law and give mathematical expression for the same.

(Guwahati, 2002)

Ans. Spectroscopy works on the laws governing absorption of light by a substance. There are two laws which govern the absorption of light.

1. Beer's Law. When a beam of monochromatic light is passed through a substance dissolved in a non-absorbing medium, the absorption of light is directly proportional to the molar concentration of the substance. Mathematically.

$$\log_{10} \frac{I_0}{I} \propto c$$

2. Lamberts's Law. When a beam of light is passed through substance, the absorption of light is proportional to the path length of the substance. Mathematically,

$$\log_{10} \frac{I_0}{I} \propto l$$

The two laws are combined to obtain the absorption of light by a substance.

Mathematically, Beer Lambert's Law can be expressed as:

$$\log_{10} \frac{I_0}{I} \propto c.l.$$

or $$\log_{10} \frac{I_0}{I} = \in c.l. \quad ...(1)$$

where I_0 = Intensity of incident light
I = Intensity of transmitted light
c = Conc. of substance absorbing light in *moles per litre.*
l = Path length of the substance in *cm*
$\in$ = Proportionality constant known as molar extinction constant.

The expression $\log_{10} I_0/I$ is termed as optical density or **absorbance** of the substance and denoted as A. So equation (1) above can be written as:

$$A = \log_{10} \frac{I_0}{I} = \in .c.l.$$

$$\text{Percent transmission of sample} = 100 \times \frac{I}{I_0}$$

Q. 8. What are the different electronic transitions that take place on absorption of light?

(*Himachal, 2000; Kurukshetra, 2000; G. Nanakdev, 2000 ; Kerala, 2001;Panjab 2003; Mumbai, 2004 ; Delhi 2005 ; Nagpur 2005*)

Ans. When a molecule absorbs radiations the electrons are excited to higher levels. The electron involved could be σ electron (occuping σ molecular orbital) or π electron (occupying a π molecular orbital) or *n* electron (non-bonding). In the diagram below, σ, π and *n* electrons have been indicated, in a molecule of aldehyde RCHO.

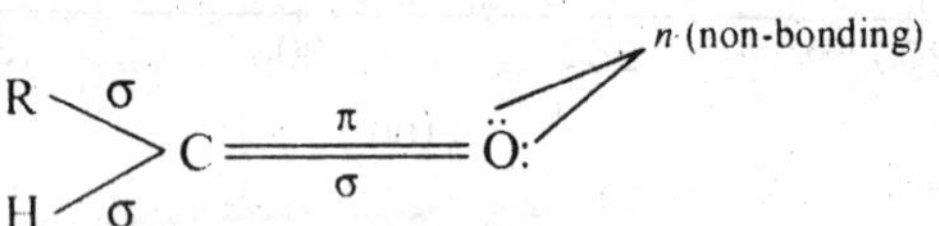

The following electronic transitions are possible

(*a*) σ ⟶ σ* (σ antibonding)
(*b*) *n* ⟶ σ* (σ antibonding)
(*c*) *n* ⟶ π* (π antibonding)
(*d*) π ⟶ π* (π antibonding)

A brief description of each is given below:

(*a*) σ ⟶ σ* transition. Transition in which a σ electron is excited to **σ*** orbital is called σ ⟶ σ* transition. A high amount of energy is required for this transition, this energy is supplied by UV radiation of wavelength less than 200 nm. Such a transition does not take place in ordinary UV range > 200 nm.

(*b*) *n* ⟶ σ* transitions. This transition corresponding to the excitation of a non-bonding electron to σ* molecular orbital. Compounds having lone-pair of electrons, not participating in bonding, show this transition. These transitions are associated with much lower energies than in σ ⟶ σ* transitions.

(*c*) *n* ⟶ π* transitions. These are the transitions in which non-bonding electrons are excited to π* molecular orbitals. Compounds having double or triple bonds, such as C = O, C = S and N = O display these transitions and weak absorption bands are obtained in the UV spectrum. Small amounts of energies are required for these transitions which take place in the ordinary UV range.

(*d*) π ⟶ π* transitions. These are the transitions in which π electrons are excited to π* molecular orbitals. These transitions requires a large amount of energy but the intensity of absorption is very high.

However conjugated double bonds require a small amount of energy for $\pi - \pi^*$ transition. Molecules containing group such as $> C = C - C = C <$ and $> C = C - C = O$ show $\pi - \pi^*$ absorption bands in the ultraviolet range. Whereas a simple alkene or ketone gives the absorption peak at 170 nm, a conjugated diene like butadiene $CH_2 = CH - CH = CH_2$ shows λ_{max} at 217 nm ($\in$ = 20,900). As the no. of conjugated bonds increases, the absorption takes place at longer wavelength. If there are enough conjugated bonds in the molecule, λ_{max} will move into visible region and the compound will look coloured. β-carotene, which contains 11 double bonds in conjugation, is coloured because it absorbs radiation of wavelengths 451 nm.

In the study of organic compounds, we are particularly interested in

$$n \longrightarrow \pi^* \text{ transitions}$$

and

$$\pi \longrightarrow \pi^* \text{ transitions}$$

Q. 9. Write notes on (*a*) allowed transitions (*b*) forbidden transitions.

Ans. (*a*) Allowed transitions. Molecules don't absorb just any radiation. They absorb only such radiations which have appropriate energy to excite the electrons to allowed higher levels. These are the transitions which have $\varepsilon = 10^4$ or more. $\pi - \pi^*$ transitions conform to this requirement. 1–3 butadiene which shows absorption at 217 nm and has $\varepsilon = 21000$ undergoes an **allowed transition.**

(*b*) Forbidden transitions. These are the transitions having ε value less than 10^4, $n - \pi^*$ transitions generally belong to this category. A saturated aldehyde or ketone shows an absorption band at 290 nm at $\varepsilon < 100$. It involves a $n \longrightarrow \pi^*$ forbidden transition.

Q. 10. Write notes on (*a*) chromophore (*b*) auxochrome.

(*Kerala 2000; Mumbai, 2004; Nagpur 2008*)

Ans. (*a*) Chromophore. Originally, chromophore was considered as functional group which has a capability or property to impart a colour to a compound. In a broader sense, a chromophore is defined as a group which absorbs electromagnetic radiation in the visible or UV range. Imparting a colour to the compound is no more a criteria. Some prominant chromophores along with their λ_{max} and ε_{max} are listed in the following table:

Chromophore	*Transition*	λ_{max} *(nm)*	ε_{max}	*Solvent*
$> C = C <$	$\pi - \pi^*$	175	15000	Vapour
$- C \equiv C -$	$\pi - \pi^*$	175	10000	Hexane
$> C = O$	$\pi - \pi^*$	180	10000	
	$n - \pi^*$	160	18000	Hexane
	$n - \pi^*$	285	15	
$-\overset{OH}{\overset{\mid}{C}} = O$	$n - \pi^*$	205	60	Methanol
$- N = N -$	$n - \pi^*$	338	5	Ethanol
$- NO_2$	$n - \pi^*$	274	15	Methanol

(*b*) Auxochromes. An auxochrome is a group which when attached to chromophore shifts the absorption maximum towards longer wavelengths along with an increase in the intensity of absorption. By itself, it does not show any absorption above 200 nm. We may compare anxochromes with promoters of catalysts. Promoters by themselves have no catalytic ability but they improve upon the activity of the catalyst. Some prominent auxochromic groups are –OH, $-NH_2$, –OR, –NHR and $-NR_2$. When auxochrome $-NH_2$ is attached to benzene ring, the absorption changes from 255 nm (ε_{max} = 203) to 280 nm (ε_{max} = 1430).

The secret behind the capability of auxochromes, to shift absorption maximum towards longer wavelength, is that it helps in extending conjugation by providing lone pair of electrons as shown below:

$$\underset{\text{Chromophore}}{CH_2 = CH} — \underset{\text{Auxochrome}}{\ddot{N}R_2} \longleftrightarrow :\bar{C}H_2 - CH = \overset{+}{N}R_2$$

Q. 11. Write notes on (*a*) Bathochromic or red shift (*b*) Hypsochromic or blue shift (*c*) Hyperchromic effect (*d*) Hypochromic effect.

(*M. Dayanand, 2000; Panjab, 2003; Mumbai, 2004; Nagpur 2005*)

Ans. (*a*) Bathochromic or red shift. Auxochromes bring about an increase in the value of λ_{max}. The absorption is shifted towards higher wavelength or towards red portion of the spectrum. This shift of absorption of light towards higher wavelength is known as **bathochromic or red shift.** This effect may be produced by change of solvent. It is also produced if two or more chromophores are present in conjugation. Ethylene, for example, shows $\pi - \pi^*$ transition at 170 nm whereas 1, 3 butadiene shows absorption at 217 nm.

(*b*) Hypsochromic or blue shift. If we remove conjugation from a system, the absorption maximum is shifted towards lower wavelength or towards blue portion of the spectrum. This shifting of absorption maximum towards shorter wavelength by removing conjugation from a system is called **hypsochromic or blue shift.** For example, protonation of aniline causes a blue shift from 280 nm to 203 nm because the aniline ion has no electrons to participate in conjugation with benzene ring.

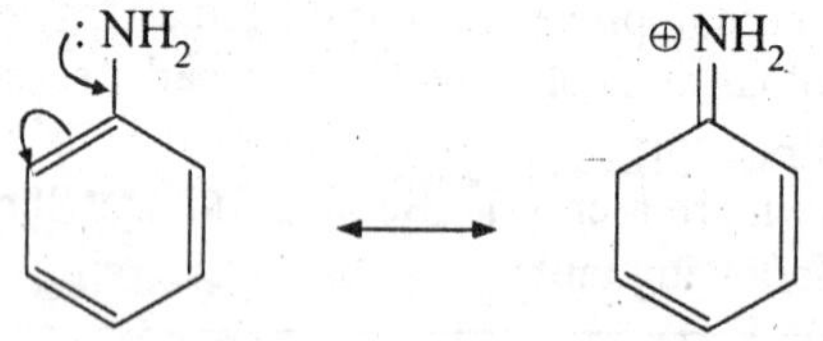

$:NH_2$ (Aniline) $\longleftrightarrow$ $\oplus NH_2$ $\qquad \lambda_{max} = 280$ nm

Aniline

$\oplus NH_3$

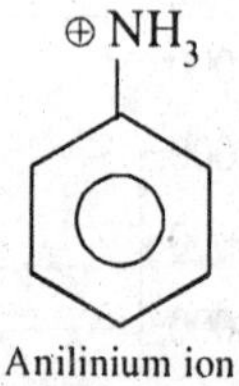

Anilinium ion $\qquad$ No conjugation $\qquad \lambda_{max} = 203$ nm

(*c*) Hyperchromic effect. Introduction of an auxochrome into a system brings about an increase in the intensity of absorption. This effect is known as **hyperchromic effect.** A methyl group at no. 2 position in pyridine increases $\in_{max}$ from 2750 to 3560 for $\pi - \pi^*$ transition.

(*d*) Hypochromic effect. Groups which distort the geometry of molecule bring about a decrease in intensity of absorption ($\in_{max}$). If we introduce a methyl group at position no. 2 in biphenyl, distortion of the geometry of the molecule causes the hydrochromic effect.

Q. 12. Give the UV absorption bands displayed by some prominent compounds, indicating the chromophores, λ_{max} and $\in_{max}$.

Ans. Absorption bands displayed by some prominent compounds are given in the table below:

Table: Some typical ultraviolet absorption bands

Group	*Example of Compound*	λmax (*in mμ or nm*)	Emax	*Solvent*
>C = C<	$CH_2 = CH_2$	171	15.530	Vapour
– C ≡ C –	$CH_3 - C \equiv CH$	187	450	Cyclohexane
>C = O	$CH_3(H)>C = O$	160 180 290	20,000 10,000 17	Vapour Hexane
	$(CH_3)_2>C = O$	166 189 279	16,000 900 17	Vapour Hexane
– C(=O)OH	$CH_3 - C(=O)OH$	208	32	Ethanol
>C = C – C = C<	$CH_2 = CH - CH = CH_2$	217	20,900	Hexane
>C = C – C ≡ C –	$CH_2 = CH - C \equiv CH$	219 228	7600 7800	Ethanol
>C = C – C = O	$CH_3 - CH = CH - CH = O$	218 320	18,000 30	Hexane
	$CH_3 - CH = CH - C(CH_3) = O$	224 314	9750 38	Ethanol
Benzenoid ring	Benzene	203.5 254	7400 204	Water
	$C_6H_5OCH_3$	217 269	600	Water
	$C_6H_5 - C(=O) - CH_3$	245.5	9800	Water

Q. 13. Discuss the applications of ultraviolet spectroscopy.

(*Kerala, 2001; Kanpur, 2001; Nagpur, 2002*)

Ans. The important applications of the study of UV spectra of compounds are as given below:

***(i)* Detection of conjugation.** With the help of UV spectrum, we can establish the presence of conjugation in a compound. Conjugation can be

$$-C = C - C = \quad \text{or} \quad -C \equiv C - C \equiv \quad \text{or} \quad C = C - C = O \quad \text{or} \quad C_6H_5-C = C$$

By observing the λ_{max} values, we can also predict the location of substituents.

(ii) **Detection of functional groups.** It is possible to detect certain functional groups with the help of UV spectrum. The negative test is also of value. Absence of absorption above 200 nm is a sure indication of the absence of conjugation, carbonyl group and benzene rings in the compound.

(iii) **Detection of geometrical isomers.** When a compound exhibits geometrical isomerism, the trans isomer shows absorption at higher wavelength with larger values of extinction coefficients ϵ_{max}) compared to cis isomers. Out of two stilbenes ($C_6H_5CH = CH\ C_6H_5$), the trans isomer shows absorption at 294 nm, ($\epsilon_{max} = 24000$) while the cis isomer absorbs at 278 nm ($\epsilon_{max} = 9350$).

SOME STRUCTURAL PROBLEMS

Q. 14. The following dienes have λ_{max} at 176 nm, 211 nm and 215 nm. Find out which is which? *(Himachal 2000)*

(a) $CH_2 = CH(H)C = C(CH_3)(H)$ — $CH_2=CH$ and H on one carbon; CH_3 and H on the other (CH₂=CH and CH₃ on the same side)

(b) $CH_2 = CH(H)C = C(H)(CH_3)$ — $CH_2=CH$ and H on one carbon; H and CH_3 on the other (CH₂=CH and CH₃ on opposite sides)

$CH_2 = CH - CH_2 - CH = CH_2$
(c)

Ans. *(i)* $\lambda_{max} = 176$ nm indicates that there is no conjugation in the molecule. Compound *(c)* has no conjugation. Hence the compound showing λ_{max} at 176 nm is *(c)* *i.e.,* $CH_2 = CH - CH_2 - CH = CH_2$ or 1, 4 Pentadiene.

(ii) Structure *(a)* and *(b)* are geometrical isomers of the compound 1, 3, pentadiene, *(a)* is cis isomer and *(b)* is *trans* isomer. As a rule the trans isomer shows absorption at higher wavelength. hence the compound showing λ_{max} at 215 is *(b)* or *trans*-1, 3 pentadiene.

(iii) Compound showing λ_{max} at 211 nm is *(a)* *i.e., cis*-1, 3 pentadiene.

Q. 15. Which of the following compounds, if any, can give a UV spectrum showing λ_{max} at *(i)* 218 ($\varepsilon = 18000$) *(ii)* 245.5 ($\varepsilon = 9800$) *(iii)* 186.5 ($\varepsilon = 450$) and 415 ($\varepsilon = 63000$).

(a) $CH_3CH_2CH_2CH_3$ **(b)** $CH_3 - C \equiv CH$

(c) $CH_3 - CH = CH - CHO$ **(d)** $C_6H_5 - \overset{O}{\overset{\|}{C}} - CH_3$

Ans. *(a)* $CH_3CH_2CH_2CH_3$ does not have any conjugation. It does not have even one double or triple bond. Therefore none of the λ_{max} values correspond to this compound.

(b) $CH_3 - C \equiv CH$ has a triple bond but the electrons are not much mobile. This compound absorbs at $\lambda_{max} = 186.5$ nm.

(c) $CH_3 - CH = CH - \overset{\|}{\underset{O}{C}}H$

This compound has conjugation with possibility of resonance. The compound absorbs at 218 nm ($\lambda_{max} = 218$).

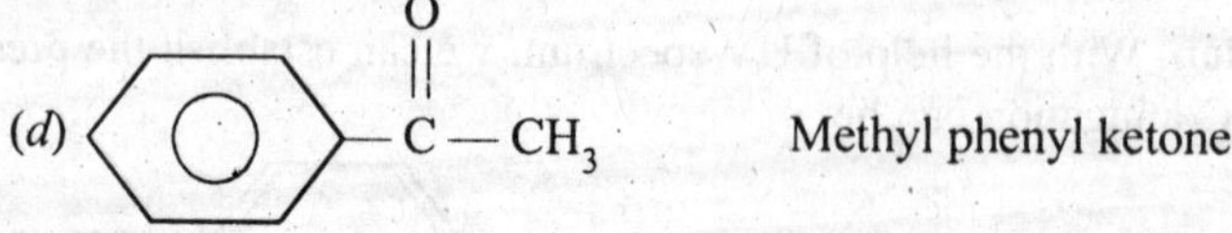

(d) $C_6H_5 - \overset{O}{\overset{\|}{C}} - CH_3$ Methyl phenyl ketone

In this compound, three double bonds of the ring and double bond in C = O are in conjugation. The no. of double bonds in conjugation is much more than in compound *(c)*. Thus compound *(d)* absorbs at 245.5 nm.

Q. 16. Indicate the increasing order of wavelength of λ_{max} of the following compounds in the UV region of the spectrum.

(*a*) CH = CH – CH = CH

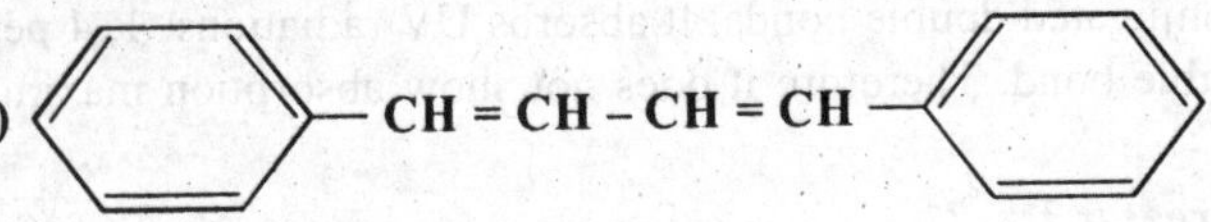

(*b*) $CH_2 = CH - CH = CH_2$

(*c*) $CH_2 = CH - CH_2 - CH = CH_2$

Ans. Compound (*c*) has no conjugation in the molecule. Therefore it will have the smallest value of λ_{max}.

Compound (*a*) has 8 double bonds in conjugation (three each from two benzene rings and two outside the ring). Therefore it will have the maximum value of λ_{max}).

Compound (*b*) having two double bonds in conjugation will have intermediate value of λ_{max} between the compounds (*a*) and (*c*).

Hence increasing order of wavelength of λ_{max} for the compounds (*a*), (*b*) and (*c*) is (*c*) < (*b*) < (*a*).

Q. 17. The UV spectrum of acetone shows two important peaks at λ_{max} 279 nm (ε_{max} 15) and λ_{max} 198 nm (ε_{max} 900). Identify the electronic transition for each. *(Calcutta, 2000)*

Ans. (*i*) λ_{max} 279 nm ($\in_{max}$ 15) corresponds to $n - \pi^*$ transition.

(*ii*) λ_{max} 198 nm ($\in_{max}$ 900) corresponds to $\pi - \pi^*$ interactions.

Q. 18. What type of compounds absorb UV radiations? Select the compounds in support of your answer from the following list.

(*i*) 1, 3-butadiene (*ii*) 1-hexene (*iii*) cyclobutane (*iv*) 1, 3-cyclohexadiene (*v*) chlorobenzene (*vi*) nitrobenzene and (*vii*) chlorocyclohexane

Ans. Compounds having conjugated double bonds absorbs U.V. radiations. Out of the list given, the following compounds will absorb U.V. radiations.

1, 3-butadiene, 1, 3-cyclohexadiene, chlorobenzene and nitrobenzene.

Q. 19. Arrange the following in increasing order of U.V. absorption maxima.

(*i*) anthracene (*ii*) ethylene (*iii*) naphathalene (*iv*) butadiene

Ans. UV absorption maxima depends upon the no. of conjugated double bonds in the compounds. Greater the no. of such double bonds, greater will be absorption maxima. Out of the list given above, anthracene has 7 conjugated double bonds, naphthalene has 5 conjugated double bonds, butadiene has two conjugated double bonds and ethylene has no conjugated double bond. Therefore increasing order of U.V. absorption maximum of the compounds will be as under:

Ethylene < butadiene < naphthalene < anthracene

Q. 20. Can UV spectral data be useful to distinguish between the following pairs? Give reasons?

(*i*) Ethylbenzene and Styrene

(*ii*) $CH_2 = CH - CH_2 - CH = CH_2$ and $CH_3 - CH = CH - CH = CH_2$

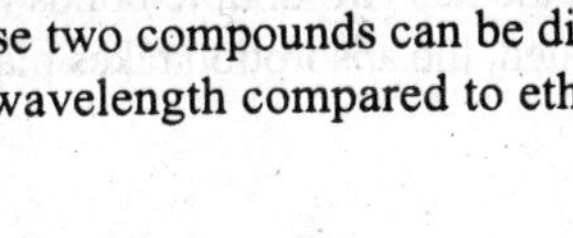

CH_2CH_3

Ethylbenzene

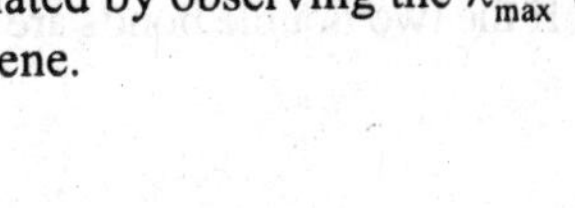

$CH{=}\ CH_2$

Styrene

Ans. (*i*) Ethylbenzene has 3 conjugated double bonds and styrene has four conjugated double bonds. Hence these two compounds can be differentiated by observing the λ_{max} value. Styrene will absorb at higher wavelength compared to ethylbenzene.

(*ii*) $CH_2=CH-CH-CH=CH_2$ (1, 4 pentadiene) $CH_3-CH=CH-CH=CH_2$ (1, 3 pentadiene)

1, 3 pentadiene has two conjugated double bonds. It absorbs UV radiations. 1, 4 pentadiene does not have a conjugated double bond. Therefore it does not show absorption maxima in UV range.

Q. 21. Which will have greater λ_{max}?

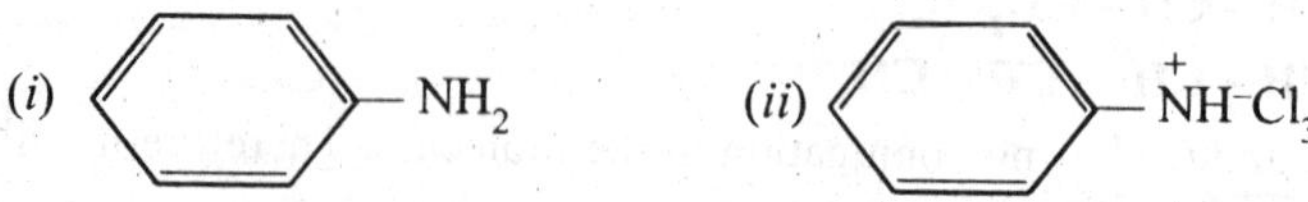

Ans. Lone pair of electrons on nitrogen is available for conjugation with the double bonds of benzene in structure (*i*). In structure (*ii*), there are no electrons on nitrogen for conjugation with the benzene ring. Hence structure (*i*) shows greater λ_{max} compared to structure (*ii*).

Q. 22. Mesityl oxide, a condensation product of acetone, is known to consist of two isomers (I) and (II) shown below. One isomer exhibits a maximum at 235 nm with ε = 12000 while the other shows no high intensity absorption above 220 nm. Which of the two isomers absorbs at 235 nm?

$$CH_2=\underset{\underset{CH_3}{|}}{C}-CH_2-COCH_3 \quad \text{(I)} \qquad CH_3-\underset{\underset{CH_3}{|}}{CH}=C-COCH_3 \quad \text{(II)}$$

Ans. Isomers (II) absorbs at 235 nm because its double bond system is conjugated while (I) has unconjugated double bonds and would be expected to absorb at shorter wavelengths.

Q. 23. How will you differentiate between the following compounds using UV spectroscopy?

CH_3COCH_3 and $CH_3COC_6H_5$ (*Kurukshetra, 2001*)

Ans. In $CH_3COC_6H_5$, there is a sort of conjugation within the ring as well as outside the ring.

$$C_6H_5-\overset{\overset{O}{||}}{C}-CH_3$$

whereas CH_3COCH_3 gives absorption maximum of 166 nm (∈ = 16000), methylphenyl ketone will show absorption maximum at a much higher wave length due to conjugated double bonds in the ring. Also C = O bond is in conjugation with the double bonds of benzene ring.

Q. 24. Which of the following absorb and which don't absorb U.V. radiations?

(*i*) Ethyl alcohol (*ii*) Benzene (*iii*) Water (*iv*) Acetone

Ans. Benzene and acetone absorb U.V. radiations.

Ethyl alcohol and water don't absorb U.V. radications.

Q. 25. Compounds A, B and C have the formula C_5H_8 and on hydrogenation yield *n*-pentane. Their ultraviolet spectra show the following values of λ_{max}:

A : 176 mμ B : 211 mμ

C : 215 mμ

1-pentene has λ_{max} = 178 mμ

What are the likely structures for A, B and C?

Ans. The formula C_5H_8 suggests that the molecule has two double bonds (C_nH_{2n-2}) in all the three compounds. If the two double bonds are conjugated, the absorption takes place around 220 mμ.

Accordingly

(*i*) Compound A is 1, 4-Pentadiene

(*ii*) Compounds B and C are both 1, 3-Pentadiene. Out of the two, B is cis isomer and C is trans isomer. This is because trans isomer absorbs at a higher wavelength. Thus

A $CH_2 = CH - CH_2 - CH = CH_2$

B $\begin{array}{l} CH_2 = CH \\ \quad\quad\quad\quad\quad > C = C < \\ H \end{array} \begin{array}{l} CH_3 \\ \\ H \end{array}$

C $\begin{array}{l} CH_2 = CH \\ \quad\quad\quad\quad\quad > C = C < \\ H \end{array} \begin{array}{l} H \\ \\ CH_3 \end{array}$

INFRARED SPECTROSCOPY

Q. 26. What are the principles of infrared spectroscopy?

(*M. Dayanand, 2000; Madras, 2003*)

Or

Explain bending and stretching vibrations.

(*Guwahati, 2002; Mumbai, 2004; Bhopal, 2004*)

Ans. The atoms in a molecule are in a state of constant vibratory and rotatory motion. The different ways in which the vibrations can take place are as follows:

For diatomic molecules

Let us say we have a molecule XY. It can stretch the X-Y bond inwards or outwards in the following ways:

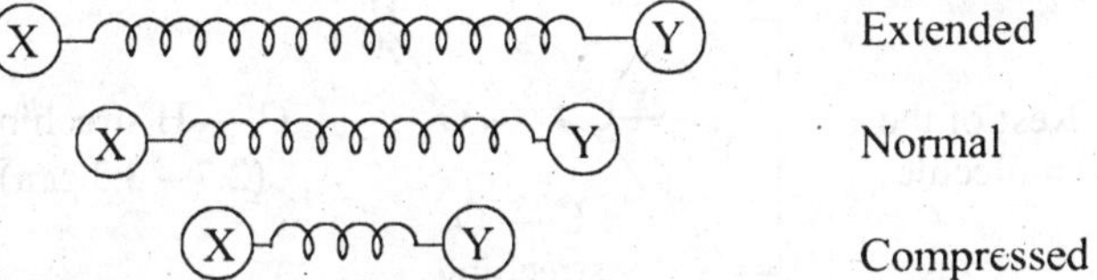

Thus the compression and extension of the bond takes place along the X–Y bond (see Fig. above).

For triatomic molecule

Let there be a molecule XYZ with a bond angle ∠ XYZ. Here we have two types of movements.

Stretching. As in the case of XY diatomic molecule, there will be stretching in the XYZ molecule, too. Here two bonds are involved viz, XY and YZ bonds. Let us assume that the molecule is linear. The different stretching forces that are manifested are as follows:

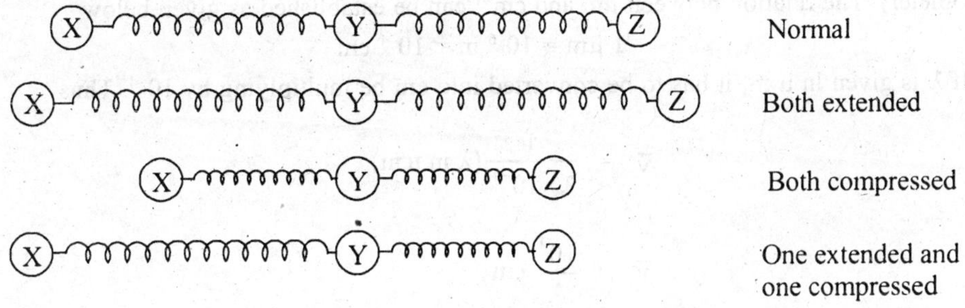

Bending. If the molecule is angular, there will still be stretching of XY and YZ bonds. Besides there will be some bending forces between XY and YZ bonds, as a result of which the ∠ XYZ will change, it will increase and decrease continuously as shown below:

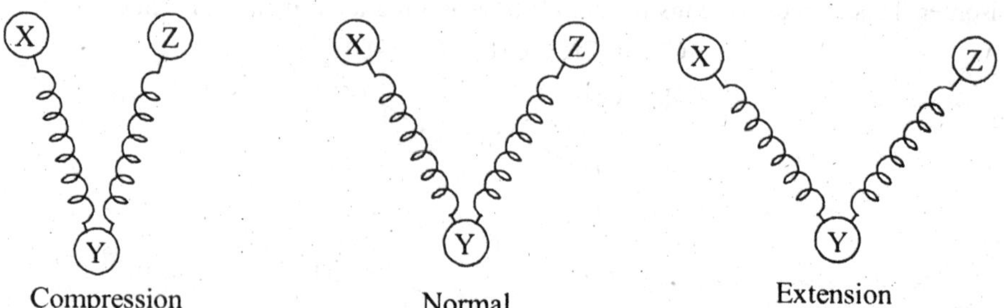

Thus there are a number of vibrational and rotational changes taking place in the molecule. On absorption of light, the molecules are excited from lower vibrational and rotational levels to higher ones. For affecting such transitions, frequencies in the infrared range are good enough. Molecules will absorb such frequencies as are needed to excite molecules from lower vibrational levels to permitted higher levels. Absorption of energy by a bond increases the amplitude of vibrations. Every bond and every functional group has a specific absorption frequency which is calculated to excite the molecule to higher vibrational or rotational level. A graph between the amount of absorption (or emission) of infrared light against the frequency (or wavelength) of this light is called **Infrared (or IR) spectrum.** The following diagram illustrates various frequencies in the IR range associated with various bonds in an alcohol molecule.

C — O — H
(6.7 – 8.3 μm)

Rest of the molecule — C ⟷ O — H

O — H streching (2.7 – 3.3 μm)

C — O streching (8.3 – 10 μm)

Q. 27. What is the range of infrared radiations? What are the units used in infrared spectroscopy?

Ans. IR region has the range of 0.8 to 20 μm or 800 to 20000 nm. The region between 0.8 to 2.5 μm is called the near infrared region and the one between 15 to 20 μm is called far infrared region. The region between 2.5 to 15 μm is the proper infrared region. Generally wave number ($\bar{\nu}$) which is the no. of vibration of the radiation per centimeter, is used for describing IR spectra. The unit of wave number is cm^{-1}. The other unit which is used in connection with IR spectra is μm (micrometer). The relation between μm and cm^{-1} can be established as given below:

$$1\ \mu m = 10^{-6}\ m = 10^{-4}\ cm$$

If λ is given in μ m, it has to be converted into cm be multiplying by 10^{-4}. Thus

$$\bar{\nu} = \frac{1}{\lambda \times 10^{-4}} (\lambda \text{ in } \mu m)$$

or

$$\bar{\nu} = \frac{10^4}{\lambda}\ cm^{-1}$$

Q. 28. Give the IR absorption bands of some important bonds and functional groups.

Ans. The table below gives the bond type of compound, the absorption wavelength and frequency (in cm^{-1}) and intensity of absorption.

In the column of intensity, *s*, *m*, *w* and *v* stand for strong, medium, weak and variable respectively.

Table: Some characteristic infrared absorption frequencies.

Bond	*Type of compound*	*Wavelength (in μm)*	*Frequency (in cm⁻¹)*	*Intensity*
$-\overset{\vert}{\underset{\vert}{C}}-H$	Alkanes	3.28–3.51	2850–2960	*m–s*
$=\overset{\vert}{C}-H$	Alkenes	3.22–3.32	3010–3100	*m*
$\equiv C-H$	Alkynes	3.03	3300	*s*
$>–C-H$	Aromatic rings	3.22–3.33	3000–3100	*m*
$-\overset{\Vert}{C}-H$	Aldehydes	3.47–3.77	2650–2880	*w* (often two bands)
$-\overset{\vert}{\underset{\vert}{C}}-\overset{\vert}{\underset{\vert}{C}}-$	Alkanes	6.66–16.66	600–1500	*w* (seldom useful for identification)
$>C=C<$	Alkenes	5.95 –6.2	1620–1680	*v*
$-C\equiv C-$	Alkynes	4.42–4.76	2100–2600	*v*
$C=C$	Aromatic rings	6.25–6.66	1500–1600	*v*
$-\overset{\vert}{\underset{\vert}{C}}-O-$	Alcohols, ethers, carboxylic acids, esters	7.7–10	1000–1300	*s*
$>C=O$	Aldehydes, ketones, carboxylic acids, esters	5.68–5.95	1680–1760	*s*
$-O-H$	Monomeric alcohols and phenols	2.74–2.78	3600–3650	*v*
$-O-H$	Hydrogen bonded alcohols and phenols Hydrogen bonded acids	2.94–3.11 3.33–4	3200–3400 2500–300	*s (broad)* *v (broad)*
$-\overset{\vert}{\underset{\vert}{N}}-H$	Amines	2.86–3.03	3300–3500	*m*
$-\overset{\vert}{\underset{\vert}{C}}-N$	Amines	7.36–8.49	1180–1360	*v*
$-C\equiv N$	Nitriles	4.42–4.52	2210–2260	*v*
$-NO_2$	Nitro compounds	6.5–7.5	1330–1540	Pair of strong bands

Q. 29. Write notes on (*i*) Functional group region (*ii*) Fingerprint region.

(Mumbai, 2004)

Ans. (*i*) Functional group region. The region of 2.5 to 7.4 μm (4000 to 1430 cm^{-1}) is related with vibrational changes in the functional groups in the molecule. By observing absorption in this range, we can draw some conclusions regarding the functional groups present in the molecule. Hence this region is called the functional group region.

(*ii*) **Fingerprint region.** It is found that the region 7–11 μm (1430–910 cm^{-1}) gives a good deal of information regarding structure of the compound and various other groups besides the functional groups in the molecule and their relative positions. This region is called the fingerprint region. Frequencies in the fingerprint region can lead us to identify a given compound.

Q. 30. How does the absorption frequency of a group change from compound to compound? (*Kerala 2001*)

Ans. In the table of IR absorption frequencies, it is the range and not a particular frequency, which is given. This is because the absorption frequency depends not only upon the group but also upon the environments of that group *i.e.* the other bonds and groups which are associated with that particular group in the molecule. We find that there is a slight variation in the absorption frequency if the environment are changed. This is illustrated in the table below which gives the absorption frequencies due to the carbonyl group in different compounds.

Compound	*Frequency*	*Compound*	*Frequency*
$CH_3-\overset{O}{\overset{\parallel}{C}}-CH_3$ Acetone	1700 cm^{-1}	Cyclobutanone	1780 cm^{-1}
$CH_3-\overset{O}{\overset{\parallel}{C}}-CH_2CH_3$ Methyl ethyl ketone	1710 cm^{-1}	Cyclopentanone	1750 cm^{-1}
$CH_3-\overset{O}{\overset{\parallel}{C}}-CH=CH_2$ Methyl vinyl ketone	1680 cm^{-1}	Cyclohexanone	1710 cm^{-1}

Important

Negative evidence is very reliable in interpreting the IR spectrum of a compound. If an absorption frequency corresponding to a particular group is missing from the spectrum, it is almost sure that the group is absent in the molecule.

Q. 31. How do you record the IR spectra of a compound?

Ans. IR spectrophotometer consists of the following components.

(*a*) IR radiation source. A Nernst glower, which is a rod made of an alloy of zirconium, yttrium and erbium oxides is used as IR radiation source. It is electrically heated to 1750 K.

(b) Rock-salt cells and prisms. Glass and quartz absorb IR radiations, hence these materials cannot be used for making cells for the sample and for prism which are used in the optics of the IR spectrophotometers. Rock-salt or some similar material is used for this purpose.

Sample preparation. Generally the sample in the form of solid is avoided because it scatters the radiations too much. For obtaining better spectra the sample is taken in the form of potassium bromide disc (a mixture of the sample and KBr). Second technique is to grind the sample with nujol which is a highly purified form of petroleum to obtain nujol mull which is subjected to investigation. Third technique is to dissolve the sample in pure $CHCl_3$ or CCl_4 . Solvents like water and alcohol are avoided. In special cases the substance, if gaseous, may be taken as such.

Recording of spectra. The sample to be investigated is placed in the path of IR rays. The intensity of transmitted light is measured automatically and a graph between the intensity of transmitted light and the wavelength is obtained.

Q. 32. How is the frequency of vibration of a diatomic molecule related to reduced mass?

Or

What is Hooke's Law? *(Allahabad, 1997)*

Ans. The vibrations of a diatomic molecule may be compared to those of a simple harmonic oscillator in which the force tending to restore an atom to its original position is proportional to the displacement of the atom from that position. This is called *Hooke's Law*. In other words, a covalent bond may be considered as a weightless spring. The restoring force F acting upon it when it is stretching to a distance (x) is given by Hooke's Law:

$$F = -kx$$

where k is the force constant of the spring (or bond).

A molecule may, therefore, be considered as a collection of balls and springs, where the balls are atoms and the springs are the chemical bonds. As a direct consequence, a vibrating diatomic molecule can function as a simple harmonic oscillator.

The **frequency of its vibration** (ν) is given by

$$\nu = \frac{1}{2\pi c}\sqrt{\frac{k}{R}}$$

where R is termed the **reduced mass** of the diatomic system. It is calculated from the expression, $R = m_1 m_2/(m_1 + m_2)$; where m_1 and m_2 and are the masses of the two balls, c is the velocity of light.

Such a simple mechanical analogy explains the position (frequency or wavelength) of various absorptions in the infrared spectrum from the nature of the atoms and the strengths of the bonds between them as illustrated below.

(*i*) *The* **smaller** *the consituent atoms of a diatomic molecule or group, smaller is the reduced mass (R) and, therefore,* **higher** *its vibrational frequency.*

The bonds between hydrogen and other atoms have the highest known fundamental frequencies, but covalent compounds of halogens of most metals show infrared absorptions of very low frequencies which are too low to be recorded on a routine spectrophotometer.

(*ii*) The force constant (k) is a measure of the strength of the bond and, therefore, it increases with the increase in bond order. As such vibrational frequency also increases with the increase in bond strength or bond order. For example the stretching frequencies increase as we move from C–C to C = C to C ≡ C (see Table in Q. 28).

Q. 33. What is the number of fundamental vibrations in case of diatomic and polyatomic molecules. *(Nagpur 2002)*

Ans. The number of fundamental absorption bands exhibited by a molecule is related to the fundamental ways of vibrating or **vibrational modes** available to a molecule. In general:

$$\frac{\textit{Number of fundamental}}{\textit{absorption band}} = \frac{\textit{Number of fundametal}}{\textit{vibrational modes}}$$

Since a diatomic molecule has only one fundamental vibrational mode (*i.e.* stretching vibrations), it gives rise to one fundamental absorption band. However, polyatomic molecules can have more than one kind of vibrational movement, they exhibit more than one fundamental absorption bands as given below.

(a) Non-linear polyatomic molecules. It has been found that a non-linear polyatomic molecule can have $3n - 6$ modes of fundamental vibrations where n is the number of atoms present in the molecule. For example, water ($n = 3$) has three vibrational modes (*i.e.* $3 \times 3 - 6$), each of which can give an absorption band.

(b) Linear polyatomic molecules. A linear polyatomic molecule having n atoms can have $3n - 5$ fundamental vibrations. For example, carbon dioxide ($n = 3$) has four vibrational modes (*i.e.*, $3 \times 3 - 5$).

In case of larger molecules, the number of fundamental vibrations increases very rapidly with the increase in the value of n. But it has been observed that *the number of fundamental bands actually obtained is usually less than that expected from the theoretical number of fundamental vibrations*. This may be due to the following reasons:

(*i*) Some of the fundamental vibrations may be very weak and therefore, may not be recorded as bands.

(*ii*) Some of the fundamental vibrations may be very close and as a result overlapping of bands may take place.

Q. 34. What are Selection Rules?

Ans. Selections Rules

Most of the molecules possess dipole moment due to unequal sharing of electrons within the bonds linking the constituents atoms. When a polar bond undergoes stretching vibration along the internuclear axis, the electron distribution changes. As a result dipole moment also undergoes a change. In other words, a vibration produces a fluctuating dipole moment. Interaction between the oscillating dipole moment and the infrared radiation leads to the absorption of energy from the radiation. This interaction can occur only if the dipole moment at one extreme of the vibration is different from the dipole moment at the other extreme of the vibration in a molecule.

There are certain rules, known a **selection rules,** which determine whether a vibration would be effective for a particular type of spectrum or not. For infrared spectrum, the rule is that *only those vibrations are effective in causing absorption which are not* **centro-symmetric** (*i.e.* the vibrations are not symmetrical about the centre of the molecule). Since most of the functional groups in organic chemistry are not centro-symmetric, they respond very well to infrared spectroscopy.

Q. 35. Give the applications of Infrared spectroscopy? (*Kerala, 2000; Panjab, 2000 Bangalore, 2001; Bombay, 1998; Bhopal, 2004*)

Ans. Some of the important applications of IR spectroscopy are given below:

(*i*) **Identification of the functional group.** Every functional group absorbs IR radiation in a particular range irrespective of the compound in which such group is present. For example, a carbonyl group gives IR absorption band in the range of 1650–1750 cm^{-1}. Hence by observing the absorption frequency, we can tell something about the functional group in the compound.

(*ii*) **Determination of structure.** Investigation of the IR spectra in the range 2.5–7 μ m gives us information regarding functional groups present in the molecule. Further, the IR spectra in the fingerprint region gives us an insight into various other groups and their relative positions. The two studies together help us to arrive at the exact compound.

(*iii*) **Testing purity of a compound.** Generally, a pure compound has well defined and sharp absorption frequencies whereas the spectra of an impure compound has diffused and blurred lines due to additional impurity materials present in the sample. IR spectra of many pure compounds are available in literature. We can compare the IR spectra of the unknown sample with that of the pure sample. If the two tally exactly, the sample is pure, otherwise not.

(*iv*) **Following progress of a reaction.** During the course of a reaction, many intermediate compounds are formed which change into the products. These intermediate compounds can be identified by IR spectroscopy. IR spectrophotometer records the spectra continuously and by interpreting them, we can follow the reaction. The mechanism of the reaction can be proposed.

STRUCTURAL PROBLEMS

Q. 36. IR spectrum of acetone gives two maxima due to C–H vibrations at 1360 cm^{-1} and 3000 cm^{-1}. Identify the stretching and bending mode. Express the maxima in μm.

Ans. Bending vibrations require less energy then stretching vibrations. Hence absorption of IR in case of bending vibration will occur at longer wavelength or lower frequency than in case of stretching vibrations. Hence

(*i*) The absorption frequency of 1360 cm^{-1} is due to C–H bending vibration.

(*ii*) The absorption frequency of 3000 cm^{-1} is in due to C–H stretching vibrations.

Maxima in μm

Using the relation $\bar{\nu} = \dfrac{10^4}{\lambda}$

or $\lambda = \dfrac{10^4}{\nu}$

(*i*) Substituting $\bar{\nu} = 1360, \quad \lambda = \dfrac{10^4}{1360} = 7.35\ \mu m$

(*ii*) Substituting $\bar{\nu} = 3000, \quad \lambda = \dfrac{10^4}{3000} = 3.33\ \mu m$

Q. 37. Indicate the expected absorption regions in the IR spectra of the following compounds: isopropyl alcohol, dimethyl ether and toluene.

Ans. Isopropyl alcohol

The absorption bands anticipated in the IR spectra of isopropyl alcohol $(CH_3)_2$ CHOH are as follows:

(*i*) A strong and broad band due to hydrogen bonded O–H stretching	3200–3400 cm^{-1}
(*ii*) Due to C–H stretching	2850–2950 cm^{-1}
(*iii*) Due to C–O stretching	1000–1300 cm^{-1}

```
  H  H
  |  |
H-C--C--O-H      <- 3200–3400 cm⁻¹ (O–H)
  |  |           <- 1000–1300 cm⁻¹ (C–O)
  H  |
   H-C-H         <- 2850–250 cm⁻¹ (C–H)
     |
     H
```

Dimethyl ether

(*i*) Due to C–O stretching	1060–1150 cm^{-1}
(*ii*) Due to C–H stretching of methyl groups	2850–2950 cm^{-1}

Toluene 

(i) Due to C–H stretching of methyl groups 2850–2950 cm^{-1}
(ii) Due to C–H stretching of aromatic ring 3000–3100 cm^{-1}
(iii) Due to C–C stretching of aromatic ring 1500–1600 cm^{-1}

Q. 38. Give the approximate IR bonds in the following compounds

(a) CH_3CH_2OH **(b)** $CH_3-\underset{\underset{O}{||}}{C}-CH_3$ **(c)** $C_6H_5-NO_2$

Ans. IR absorption bonds in the compounds are obtained as given below:

CH_3CH_2OH

(i) Due to hydrogen bonded O–H stretching 3200–3400 cm^{-1}
(ii) Due to C–H stretching 2850–2950 cm^{-1}
(iii) Due to C–O stretching 1000–1300 cm^{-1}

$CH_3-\underset{\underset{O}{||}}{C}-CH_3$

(i) Due to C–H stretching 2850–2950 cm^{-1}
(ii) Due to C = O stretching 1700 cm^{-1}

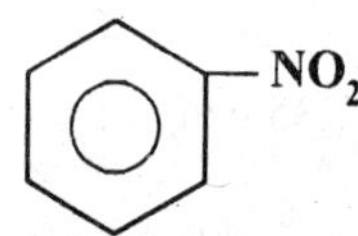

(i) Due to C – H stretching of aromatic ring 3000–3100 cm^{-1}
(ii) Due to C – C stretching of aromatic ring 1500–1600 cm^{-1}
(iii) Due to $-NO_2$ group stretching 1330–1540 cm^{-1}

Q. 39. Give the approximate position of characteristic absorption bands of carbonyl group in the IR spectra of the following compounds.

(i) CH_3CHO **(ii)** C_6H_5CHO **(iii)** $C_6H_5COCH_3$

(iv) HO–C_6H_4–$COCH_3$ **(v)** C_6H_4(OH)–$COCH_3$

Ans. The absorption bands of the compounds are as given below:

(i) CH_3CHO ~ 1740 cm^{-1}
(ii) C_6H_5CHO ~ 1700 cm^{-1}
(iii) C_6H_5CHO ~ 1690 cm^{-1}
(iv) HO–C_6H_4–$COCH_3$ ~ 1690 cm^{-1}
(v) C_6H_4(OH)–$COCH_3$ ~ 1690 cm^{-1}
(Due to chelation)

Q. 40. How will you distinguish between the following pairs of compounds using IR spectra?

(a) $CH_3CH_2COCH_3$ and $CH_3OCH = CH_2$

(b) CH_3COOH and CH_3COCH_3 *(Kurukshetra, 2000)*

(c) o-OHC_6H_4COOH and m-$OH.C_6H_4COOH$.

Ans. It is possible to distinguish between the compounds as given below :

(a) $CH_3CH_2COCH_3$ and $CH_3OCH = CH_2$ (methyl ethyl ketone and methyl vinyl ether)

(*i*) A strong absorption band at about 1715 cm^{-1} due to C–O stretching in methyl ethyl ketone will be obtained. No such band will be seen in the spectrum of methyl vinyl ether.

(*ii*) An absorption band at about 3050 cm^{-1} due to C–H stretching of $CH_2 = CH$ – groups and another band at about 1100 cm^{-1} due to C–O stretching will be observed in the IR spectrum of methyl vinyl ether. Such bands will not be formed in the spectrum of methyl ethyl ketone.

(b) CH_3COOH and CH_3COCH_3 (acetic acid and acetone)

(*i*) Both the compound will show an absorption band at about 1700cm^{-1} due to C = O stretching.

(*ii*) Spectrum of acetic acid will show an absorption band between 2500 – 3000 cm^{-1} due to hydrogen bonded carboxy group which will not be observed in the spectrum of acetone.

(c) o-hydroxy benzoic acid and m-hydroxybenzoic acid

In *o*-hydroxy benzoic acid, there is a possibility of hydrogen bonding between the two groups as they are quite near to each other. Hence, we shall observe a band in the range 2500–3000 cm^{-1} while in *m*-hydroxy benzoic acid (no hydrogen bonding) we shall see an absorption band in the range 3200–3400 cm^{-1}.

Q. 41. Give the approximate positions of characteristic absorption bands of the carbonyl group in the IR spectrum of the following compounds.

CH_3CHO, $CH_3COCH_2CH_3$, $CH_3COCH = CH_2$

[cyclopentanone structure] =O **and** [benzene ring]–**CHO**

Ans. The approximate positions of the absorptions bands in the above compounds are given below:

(*i*) CH_3CHO 1740 cm^{-1}

(ii) $CH_3COCH_2CH_3$ 1715 cm^{-1}

(iii) CH_3 $COCH = CH_2$ 1675 cm^{-1}

(iv) [cyclopentanone structure] =O 1740 cm^{-1}

(*v*) [benzene ring]–CHO 1700 cm^{-1}

Q. 42. Deduce the structure of saturated open chain compound C_3H_8O, which has an infrared absorption band at 2950 cm^{-1} but none near 3300 cm^{-1} and 1725 cm^{-1}.

Ans. (*i*) A saturated open chain compound having the formula C_3H_8O can be either an alcohol or ether. Thus the possible structures are $CH_3OCH_2CH_3$, $CH_3CH_2CH_2$ OH and $CH_3CHOHCH_3$.

(*ii*) There is no absorption band at 3300 cm^{-1} (due to hydrogen bonded alcohol). Hence the possibility of $CH_3CH_2CH_2OH$ and $CH_3CHOHCH_3$ is ruled out.

(*iii*) Absence of a band at 1725 cm^{-1} rules out the possibility of a carbonyl group (aldehyde or ketone).

(*iv*) The only possibility is $CH_3OCH_2CH_3$. The band at 2950 cm^{-1} is due to C–H stretching of alkyl groups.

Q. 43. Which of the following compounds could give rise to infrared spectrum in the figure below?

Acetone, propionic acid, *n*-propyl alcohol or ehtyl formate?

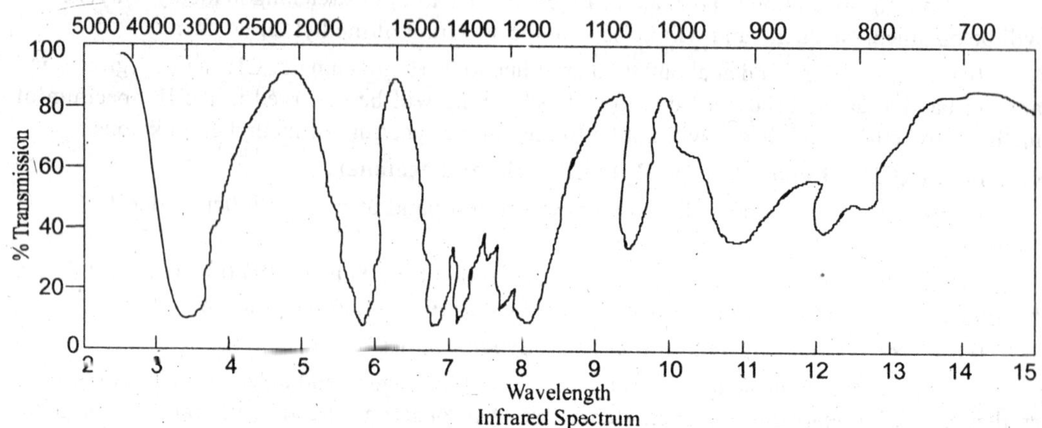

Fig. 28.6

Ans. (*i*) We observe a band at 3000 cm^{-1}. This can be due to hydrogen bonded alcoholic or carboxylic group.

(*ii*) Since acetone and ethyl formate are not expected to give this band, the possibility of the compound being acetone or ethyl formate is ruled out.

(*iii*) The spectrum shows a band at 1700 cm^{-1} which is due to C = O of the propionic acid. No such group is present in propyl alcohol.

(*iv*) The band near 1300 cm^{-1} confirms the presence of propionic acid. Hence the spectrum relates to propionic acid.

Q. 44. IR spectrum of ethyl acetate shows three important bands at 3002 cm^{-1}, 1742 cm^{-1} and 1240 cm^{-1}. Attribute these peaks to the following features of the molecule: CH_3 or CH_2 –CH_3, C = O, C–O.

Ans. The features responsible for each of the bands are as follows:

3002 cm^{-1}	C–H stretching of CH_3 or CH_2–CH_3
1742 cm^{-1}	C = O stretching
1240 cm^{-1}	C–O stretching

Q. 45. Two isomers (A) and (B) having the molecular formula C_3H_6O exhibit the following peaks in the IR spectrum. (A) 1710 cm^{-1} (B) 3300 cm^{-1}, 1640 cm^{-1}. Write the structures of A and B on this basis.

Ans. (*i*) The compound with molecular formula C_3H_6O could be an aldehyde, ketone or an unsaturated alcohol, A and B are isomers of each other.

(*ii*) In the compound (A), we do not have an absorption band at 2600–2800 which is due to aldehydic $-\overset{\|}{C}-H$ stretching. Hence the possibility of the compound being an aldehyde is ruled out. A peak at 1710 cm^{-1} shows that it is CH_3COCH_3 *i.e.* acetone.

(*iii*) In the compound (B), the peaks are obtained at 3300 cm^{-1} and 1640 cm^{-1}. There is no peak at 1700 cm^{-1} thus ruling out the possibility of carboxyl group. The band at 3300 cm^{-1} is due to hydrogen bonded alcoholic group and at 1640 cm^{-1} is due to C = C stretching. Thus simultaneous presence of an alcoholic group and a C = C is indicated. The probable structure is $CH_2 = CH - CH_2OH$ *i.e.* allyl alcohol.

Q. 46. The infrared spectrum of methyl salicylate

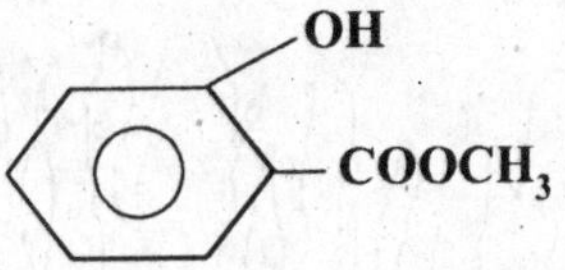

OH

— $COOCH_3$ shows the following peaks: 3300, 1700, 3050, 1540, 1590 and 2950 cm^{-1}. Which of these represent which of the following structures?

(*i*) CH_3 (*ii*) C = O (*iii*) OH group on the ring (*iv*) aromatic ring.

Ans. Various peaks correspond to the following structures:

3300 cm^{-1}	Hydrogen bonded O – H group
1700 cm^{-1}	C = O stretching
3050 cm^{-1}	C – H stretching in benzene ring
1540, 1590 cm^{-1}	C – C stretching in benzene ring
2950 cm^{-1}	C – H stretching in methyl group

Q. 47. Which peaks in the IR spectrum could be used to distinguish between the two compounds in each of the given pairs?

(*i*) CH_3COOH and $CH_3COOC_2H_5$ (*ii*) $(CH_3)_3$ N and $(CH_3)_2$ $CHNH_2$

Ans. CH_3COOH and $CH_3COOC_2H_5$

A peak around 2500–3000 cm^{-1} can distinguish between acetic acid and ethyl acetate because such a peak will be given by acetic acid only. This peak corresponds to hydrogen bonded –OH stretching.

$(CH_3)_3$ N and $(CH_3)_2$ $CHNH_2$

A peak around 3300–3500 cm^{-1} can distinguish between the two compounds. The peak corresponds to N–H stretching and will be given by $(CH_3)_2$ $CHNH_2$ only.

Q. 48. A compound having molecular formula C_8H_8O shows strong peaks at 1685 cm^{-1}. Which of the following is the likely structure of the compound (*i*) C_6H_5 $COCH_3$ (*ii*) C_6H_5 CH_2CHO (*iii*) $C_6H_5OCH = CH_2$?

Ans. (*i*) A band at 1685 cm^{-1} shows the presence of a carbonyl group. Hence the structure $C_6H_5OCH = CH_2$ is ruled out.

(*ii*) There are two possibilities, phenyl methyl ketone and phenyl acetaldehyde. Aliphatic aldehydes and phenyl substituted aliphatic aldehydes give a peak at about 1740 cm^{-1}. Hence the possibility of $C_6H_5CH_2$ CHO which is phenyl substituted acetaldehyde is ruled out. The only possibility left out is $C_6H_5COCH_3$.

Q. 49. An aromatic compound having molecular formula C_8H_8O shows an IR absorption band at 1690 cm^{-1} but no band at 3300 cm^{-1}. Deduce the structure of the compound.

Ans. An absorption band at 1690 cm^{-1} shows that a carbonyl group (> C = O) is present in the molecule. Absence of a band at 3300 cm^{-1} shows that the molecule does not contain an alcoholic group. Hence the possible structure of the compound is

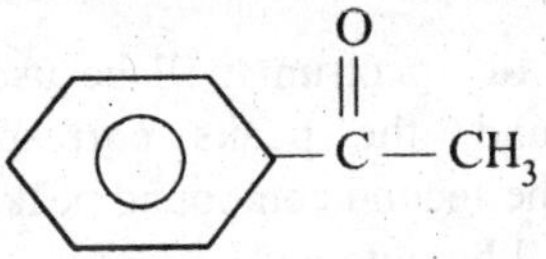

O
||
— C — CH_3

methyl phenyl ketone

Q. 50. Draw conclusion about the structure of the compound from the following IR spectrum. The molecular formula of the compound is C_7H_8O.

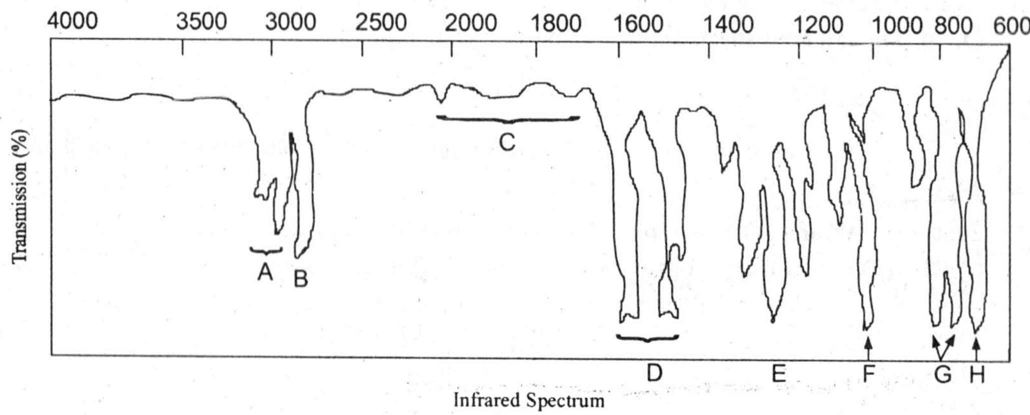

Fig. 28.7

Ans. The absorption frequencies as observed in the spectrum are related to the groups as follows:

A	Aromatic C–H stretching	3060, 3030 & 3000 cm^{-1}
B	Methyl C–H stretching	2950, 2835 cm^{-1}
C	Overtone	2000–1650 cm^{-1}
D	C ········ C ring stretching	1590, 1480 cm^{-1}
E & F	C–O–C stretching	1245, 1030 cm^{-1}
G	C–H bending	800, 740 cm^{-1}
H	C ········ C bending	680 cm^{-1}

Based in the above observations the structure of the compound is

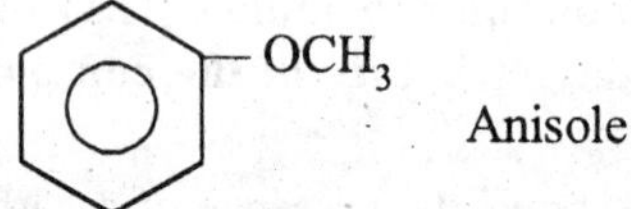

Q. 51. Indicate whether you will use IR or UV spectroscopy for distinguishing between the following pairs of compounds. Explain on what basis distinction can be made.

(*i*) $CH_2 = CH - CH_2 - OCH_3$ and $CH_3CH_2COCH_3$

(*ii*) $CH_3 - O - CH_3$ and $CH_3 - CH_2 - OH$

(*iii*) $CH_2 = CH\overset{O}{\overset{\|}{C}}(CH_2)_2 - CH_3$ and $CH_2 = CH - (CH_2)_2 - \overset{O}{\overset{\|}{C}} - CH_3$

Ans. (*i*) We shall use IR spectroscopy for the first pair because in the first compound, we shall obtain characteristic absorption peaks corresponding to the C = C, C–O–C and C–H stretching. In the second compound, the absorption peaks will correspond to the C = O and C–H stretching vibrations.

(*ii*) In the second pair, again IR spectrum will be used for distinguishing between the compounds. In the first compound, the peaks corresponding to C–O–C and methyl C–H stretching will be obtained. In the second compound peaks corresponding to methyl C–H and hydrogen bonded –OH stretching will be obtained.

(*iii*) Here we shall record the UV spectra of the two compound. We shall make use of the fact that only conjugated double bond gives an absorption peak above 200 nm. The first compound gives the absorption maximum above 200 nm. The second compound does not.

Q. 52. A compound has the molecular formula $C_7H_6O_2$. It gives an IR absorption band at 1771 cm^{-1}. On treatment with $LiAlH_4$, it gets converted into compound B which shows characteristic IR absorption bands at 3330 cm^{-1} and 1050 cm^{-1}. Assign structures to compounds A and B.

Ans. IR absorption band at 1771 cm^{-1} corresponds to C = O stretching in the compound $C_7H_6O_2$.

In compound B, band at 3330 cm^{-1} corresponds to hydrogen bonded alcoholic group. Band at 1050 cm^{-1} is also characteristic of alcoholic group.

With the formula, $C_7H_6O_2$, there are two possible compounds which conform to the given data.

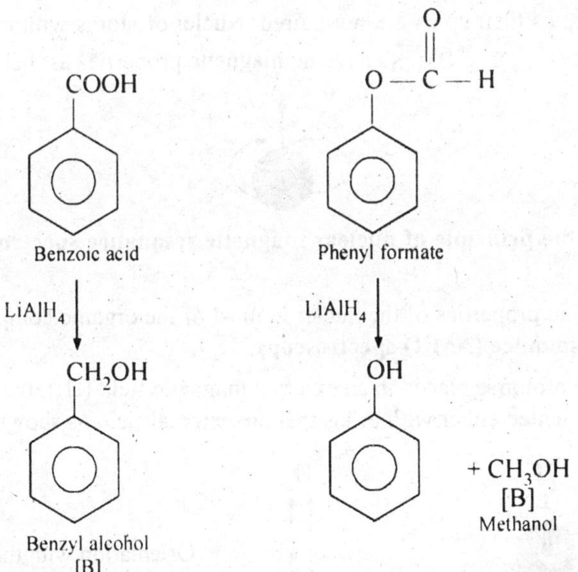

Q. 53. Distinguish between $CH_2 = CH - CH_2$ OH and CH_3 CH_2 CHO with IR spectroscopy. (*Nagpur 2008*)

Ans. (*i*) The compound that absorbs around 1700 cm^{-1} is. CH_3CH_2CHO

(*ii*) The compound that absorbs around 3400 cm^{-1} (due to O–H stretching) and around 1620 cm^{-1} (due to C = C stretching) is $CH_2 = CH - CH_2$ OH.

Q. 54. What absorption in I.R. spectrum could be used to distinguish between acetone and ethanol?

Ans. Acetone gives a band at 1700 cm^{-1} due to C = O stretching whereas ethanol gives a band at 3400 cm^{-1} due to O–H stretching (hydrogen bonded).

Q. 55. Oxidation of 2-propanol to propanone is being carried out. How would you study the progress of the reaction with I.R. spectrum?

Or

2-propanal on oxidation gives a ketone and on dehydration gives an olefin. Explain the use of IR spectroscopy in the identification of the starting material and the products. (*Kerala, 2000*)

Ans. 2-propanol gives a strong band due to hydrogen bonded O–H stretching at 3400 cm^{-1}. On being converted into propanone on oxidation, it gives an absorption band at 1700 cm^{-1} due to C = O stretching.

NUCLEAR MAGNETIC RESONANCE (NMR)

Q. 56. What is nuclear magnetic resonance?

Ans. Like electrons, the protons and neutrons also spin. If the particles in a nucleus don't have their spins paired, there is a net spin. A charged particle like proton, if it is spinning, will produce a magnetic field and magnetic moment along the axis of spin. Such a proton (or nucleus) acts like a tiny magnet. Thus a nucleus spinning in the anticlockwise direction will be associated with a magnetic field with moment acting upwards. If the nucleus is spinning clockwise, the magnetic moment will act downwards. Nuclei of atoms having odd numbered masses such as ^{1}H, ^{17}F, ^{31}P as or those having even numbered masses but odd atomic numbers such as $^{2}_{1}H$, $^{10}_{5}B$, $^{14}_{7}N$ are magnetic in nature as their spins are not paired. Nuclei of atoms with even masses and even atomic numbers, such as, $^{12}_{6}C$, $^{16}_{8}B$, $^{32}_{16}S$ have no magnetic properties as there is no resultant spin.

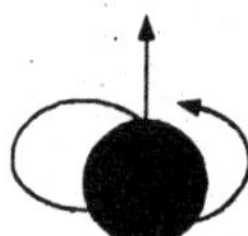

Q. 57. What is the principle of nuclear magnetic resonance spectroscopy?

(Banglalore, 2001)

Ans. The magnetic properties of the nuclei in most of the organic compounds is the basis of **nuclear magnetic resonance (NMR) spectroscopy.**

If a nucleus like proton is placed in an external magnetic field (H_0), the magnetic moment of proton (H^+) will be oriented either with or against the external field as shown in the figure below:

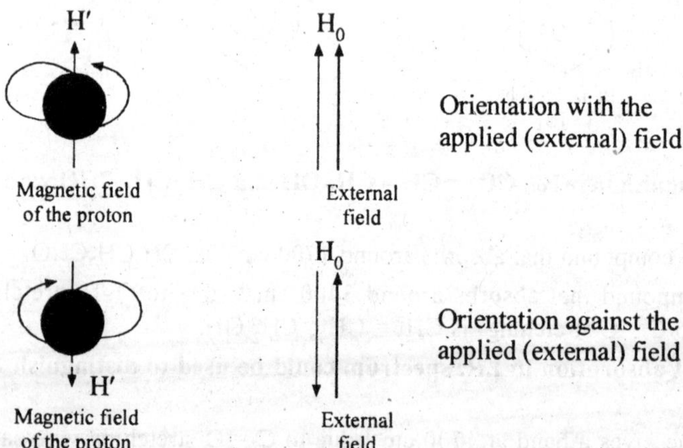

Out of the two orientations the one along the applied field is more stable (or associated with smaller energy) than the one against the applied field. The difference in the energies of the two orientations is designated as ΔE. Its magnitude varies with the magnitude of the applied field. The difference is, however, small and lies within the range of radio frequency region of electromagnetic spectrum. Thus if we desire to flip (shift) the nucleus from lower energy state to higher energy state, an amount of energy ΔE will have to the absorbed by the nucleus.

In organic chemistry, we are more interested in the protons as the nuclei, as hydrogen is a constituent of almost every organic compound. The particular branch of NMR spectroscopy, where the nucleus is a proton, is called **proton magnetic resonance (PMR).** However principally, NMR and PMR spectroscopies are the same.

The relationship between the ΔE *i.e.* energy difference between the higher and lower energy states of the proton (or nucleus), frequency (ν) of the radiation and the strength (H_0) of the magnetic field is given by the following equations:

$$\Delta E = h\nu$$

and
$$h\nu = \frac{\gamma . h . H_0}{2\pi}$$

or
$$\nu = \frac{\gamma . H_0}{2\pi} \qquad \text{... (i)}$$

where γ = nuclear constant known as gyromagnetic ratio

h = Placks's constant

For a proton γ = 26750

From eq. (*i*), it is obvious that higher the value of H_0 (applied magnetic field), higher will be the frequency of radiation required to flip the proton from lower to higher energy state.

A PMR spectrum can be recorded by placing the substance containing hydrogen nuclei in a magnetic field of constant strength and passing radiations of changing frequency through the substance and noting the frequency at which the absorption of energy corresponding to flipping of proton, from lower to higher energy state, takes place. However for ease of operation, the frequency of radiation is kept constant (at 40, 60 or 100 megacycles/sec) and the strength of the magnetic field is changed. We obtain a graph of absorption of energy versus magnetic field strength. A signal is obtained which signifies the absorption of energy at a particular field strength.

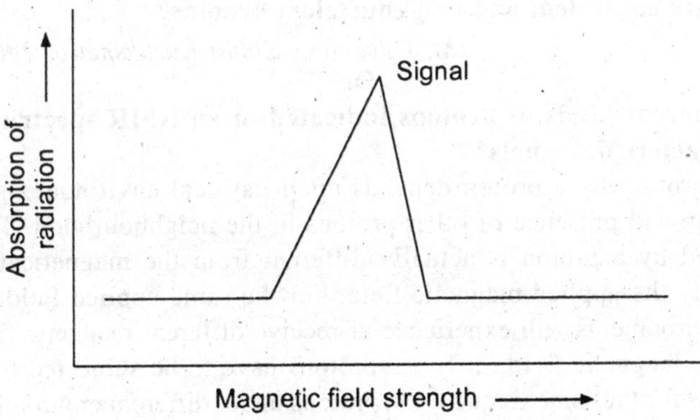

Fig. 28.8. A typical PMR signal

Q. 58. Give the schematic diagram of a PMR spectrophotmeter.

Ans. The main constituents of a PMR spectrophotometer are shown in the figure ahead.

Absorption of radiofrequency energy (resonance) occurs at a particular field strength when a compound containing protons is placed below the magnetic pole gap and irradiated with radiofrequency radiations. The sample is subjected to rapid rotation to ensure uniform exposure to the radiations. The solution of the compound, if solid, is prepared in a solvent like deuterio-chloroform ($CDCl_3$), deuterioacetone (CD_3COCD_3) or deuterium oxide (D_2O). If liquid, the sample is taken as such.

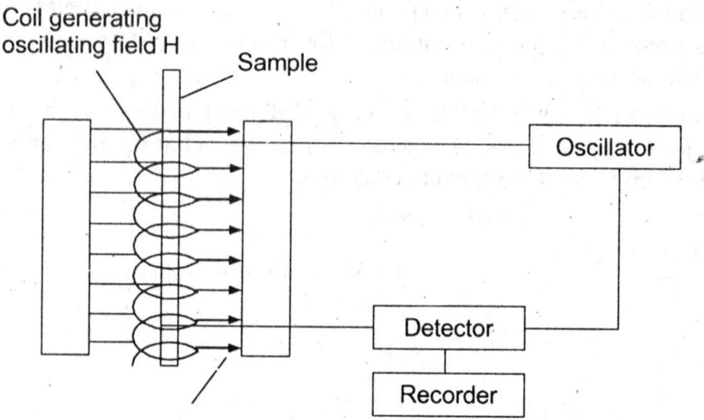

Fig. 28.9

Q. 59. Enumerate various applications of NMR spectroscopy.

Ans. Various applications of NMR spectroscopic technique are given as under:

1. Identification of functional groups. Every functional group gives a characteristic signal in the NMR spectrum. By studying the chemical shift of compound, it becomes possible to establish what kind of functional group is present in the compound.

2. Structure of an unknown compound. It is possible to elucidate the structure of an unknown compound from the NMR studies. This is because protons under different environments give different chemical shifts. By observing doublets, triplets and multiplets, it is possible to place hydrogens at appropriate place in the formula and hence to establish the structure.

3. Comparison of two compounds. NMR spectrum is like fingerprint of a compound. Two compound showing same NMR spectrum must be structurally identical.

Q. 60. What are equivalent and non-equivalent protons?

(*M. Dayanand, 2000; Kurukshetra, 2001; Nagpur 2008*)

Or

What are different kinds of protons indicated in an NMR spectrum? How do they produce their characteristic signals?

Ans. The absorption by a proton depends upon its local environments such as electron density at the proton and presence of other protons in the neighbourhood. The magnetic field strength experienced by a proton is actually different from the magnetic field applied. The environments modify the applied magnetic field. For the same applied field, different protons having different environments will experience or receive different magnetic fields. Therefore to experience the same magnetic field, different protons have to be subjected to different applied fields. In other words for the same frequencies, protons under different environments absorb at the same effective magnetic field but at different applied fields.

Equivalent protons. Protons in a molecule having the same environments absorb at the same magnetic field strength, such protons are called equivalent protons.

Non-equivalent protons. Protons which have different environments absorb at different magnetic fields, such protons are called non-equivalent protons.

All equivalent protons give rise to one signal in the NMR spectrum. From the no. of signals, we can tell how many different types of protons are there in the molecule.

Magnetically equivalent protons are also chemically equivalent and *vice-versa*. We can judge whether the two (or more) protons are chemically equivalent or not, by the isomer number method. Let us take the case of ethyl chloride $CH_3 - CH_2 - Cl$. We have three methyl proton and two methylene protons in the molecule. The methyl protons are evidently different from the methylene protons. This is because, if we substitute a methyl proton and a methylene proton separately by another group, we obtain two products as follows:

$$\underset{\text{I}}{CH_2ZCH_2Cl} \qquad \underset{\text{II}}{CH_3CHZCL}$$

Compounds I and II are clearly different compounds therefore, we say that methyl protons are different from methylene protons or in other words, the two types of protons are non-equivalent.

Now let us see whether the three methyl protons amongst themselves are equivalent or not. If we substitute any of the three hydrogens by a group Z, we obtain the same compound, CH_2ZCH_2Cl. Therefore, three methyl protons are chemically equivalent. They will provide just one signal in the spectrum because they will absorb at the same field strength.

Again let us see whether the two methylene protons are equivalent or not. If we replace, two hydrogens separately by a group Z, we obtain structure III and IV which are enantiomers of each other: Such protons are also considered chemically equivalent.

$$H-\underset{Cl}{\overset{CH_3}{C}}-H \xrightarrow[+Z]{-H} H-\underset{Cl}{\overset{CH_3}{C}}-Z \qquad Z-\underset{Cl}{\overset{CH_3}{C}}-H$$

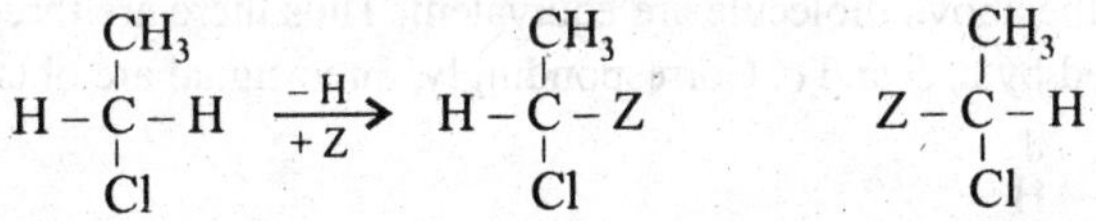

Mirror

The spectrum does not distinguish between such enantiomeric protons. We get one signal corresponding to these two protons. In all, ethyl chloride will give rise to two signals in NMR spectrum.

Q. 61. Identify different types of protons in the following compounds.
$CH_3-CBr_2-CH_3$, $CH_3-CHCl-CH_3$ and $CH_3-CH_2-CH_2Cl$

Ans. (*i*) $\overset{a}{C}H_3-CBr_2-\overset{a}{C}H_3$ **(2, 2 dibromopropane)**

Six methyl protons on the two extremes are equivalent. This has been indicated by putting a letter "a" on the methyl protons. All the protons being equivalent, only one signal will be obtained

(*ii*) $\overset{a}{C}H_3-\overset{b}{C}HCl-\overset{a}{C}H_3$ **(Isopropyl chloride)**

Three protons on the extreme left are equivalent. Three protons on the extreme right are again equivalent and equivalent to extreme left protons. The middle proton is of different type. Thus there are three types of protons. This has been indicated by putting small letters *a* and *b* on them.

(*iii*) $\overset{a}{C}H_3-\overset{b}{C}H_2-\overset{c}{C}H_2-Cl$ **(*n*-propyl chloride)**

There are three types of protons in the above molecule as indicated by the letters *a*, *b*, *c*.

Q. 62. Identify different protons in the following compounds.

(*i*) $(CH_3)(Br)C=C(H)(H)$ (*ii*) $(CH_3)(CH_3)C=C(H)(H)$

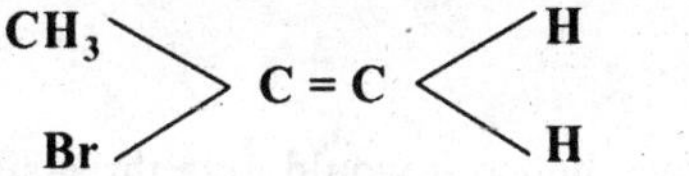

(*iii*) $(H)(Cl)C=C(H)(H)$ (*iv*) **1, 2 dichloropropene**

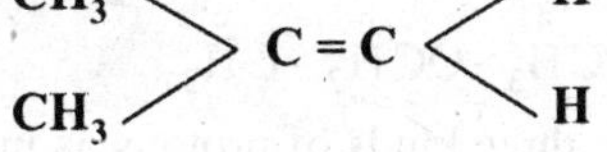

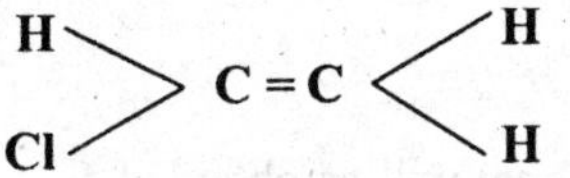

Ans. (*i*) $(\overset{a}{C}H_3)(Br)C=C(\overset{b}{H})(\overset{c}{H})$ 2-bromopropene

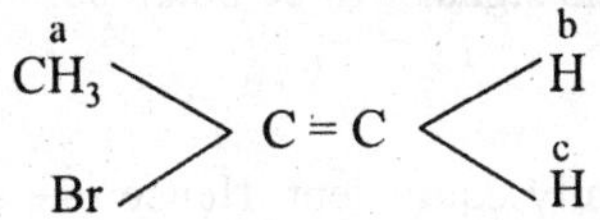

A close look at the molecule reveals that there are three different types of protons, indicated by the letters *a, b, c*. We may follow the rules explained earlier to dicide whether any two protons

in a molecule are equivalent or not. Replace the two protons separately by another group Z. If the two new products obtained are the same or enantiomers of each other, the protons are equivalent, otherwise not.

(ii) $\overset{a}{CH_3}$ and $\underset{a}{CH_3}$ on one carbon, $>C=C<$, $\overset{b}{H}$ and $\underset{b}{H}$ on the other carbon.

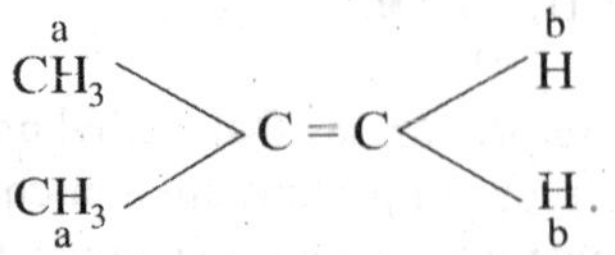

There are two types of protons. Six methyl protons on the L.H.S. are equivalent. Two protons on R.H.S. are again equivalent.

(iii) $\overset{a}{H}$ and Cl on one carbon, $>C=C<$, $\overset{b}{H}$ and $\underset{c}{H}$ on the other carbon.

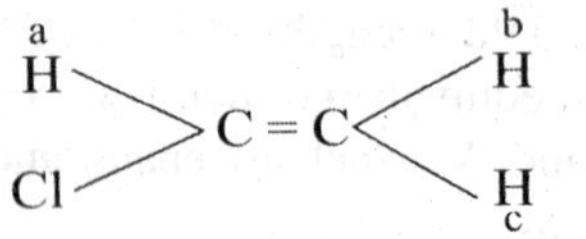

No two proton in the above molecule are equivalent. Thus there are three types of protons in the above molecule indicated by *a, b* and *c*. Correspondingly, three signal are obtained in NMR spectrum.

(iv) $\underset{a}{CH_3}-\underset{b}{CHCl}-\overset{\overset{d}{H}}{\underset{\underset{c}{H}}{C}}-Cl$ (1, 2, dichloropropane)

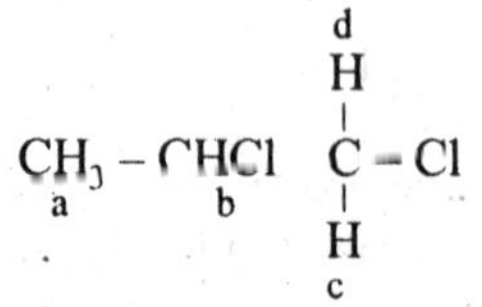

Protons marked *c* and *d*, appear to be equivalent, but they are not actually so. It becomes clear when we have a look, at its stereochemical structure.

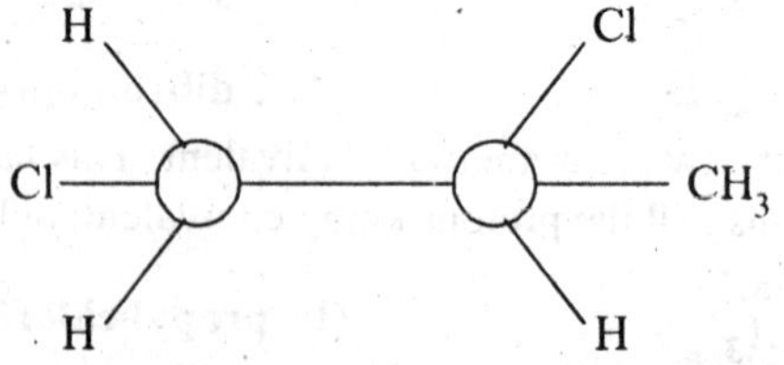

(Newman projection: front carbon bearing H, Cl, H; rear carbon bearing Cl, CH_3, H)

Thus it has four types of protons and hence it would provide four signals.

Q. 63. How many NMR signals would be obtained in the case of following compounds?

(*i*) $CH_3OCH_2CH_3$ (*ii*) CH_2ClCH_2Cl (*iii*) CH_3OCH_3

(*iv*) Cis and trans 1, 2 dibromocyclopropane (*v*) Cyclopentane

(*vi*) Cyclohexane (*vii*) CH_3CH_2CHO (*viii*) $C_6H_5-CH_3$

(*ix*) $(CH_3)_2CH-CH_2-CH_3$ (*x*) $CH_3-CH=CH_2$

(*xi*) $CH_3-CH_2-CHCl-CH_3$

Ans. (*i*) $\overset{a}{C}H_3-O\overset{b}{C}H_2-\overset{c}{C}H_3$

There are three kinds of protons as indicated above, hence it would give three signals.

(*ii*) $Cl-\overset{a}{C}H_2-\overset{a}{C}H_2-Cl$

All the four protons are equivalent. Hence only one signal will be obtained.

(*iii*) $\overset{a}{C}H_3-O-\overset{a}{C}H_3$

All the six protons are magnetically and chemically equivalent. Hence one signal will be obtained.

(*iv*)

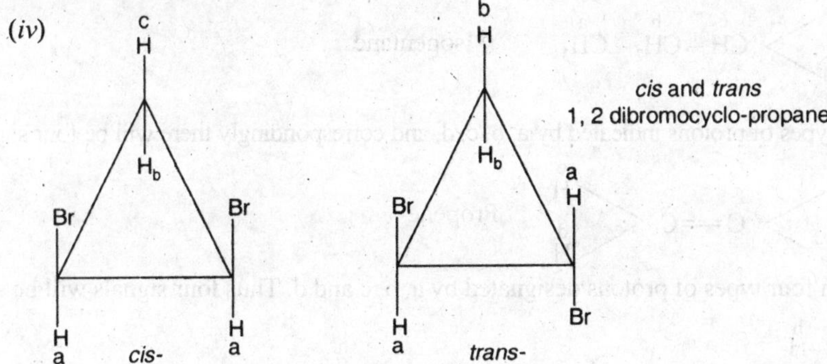

Protons labelled "a" in the cis isomer are equivalent and will give one signal. Protons labelled b and c are non-equivalent giving separate signals. In all, there will be three signals from the cis-isomer.

Trans-isomer has two protons indicating by "*a*" which are equivalent. Similarly protons labelled '*b*' are also equivalent. Thus two signals will be obtained in this case.

(*v*) Cyclopentane: five $\overset{a}{CH_2}$ groups (CH_2, H_2C, CH_2, $\underset{a}{CH_2}$, $\underset{a}{CH_2}$) joined in a ring.

Here all the protons are equivalent. Hence only one signal will be obtained.

(*vi*) Cyclohexane: six $\overset{a}{CH_2}$ groups (CH_2, ^aH_2C, CH_2, ^aH_2C, $CH_2{}^a$, $\underset{a}{CH_2}$) joined in a ring.

Here, again, all the protons are equivalent giving rise to only one PMR signal.

(*vii*) $\overset{a}{CH_3}-\overset{\overset{b}{H}}{C}(\underset{b}{H})-\overset{c}{CHO}$ Propionaldehyde

There are three kinds of protons giving rise to three signals.

(*viii*) $C_6H_5-CH_3$ Toluene

There are two types of protons, benzene ring protons and the methyl protons. Two signals will be observed.

(*ix*) $\overset{d}{CH_3}$ and $\overset{d}{CH_3}$ $>$ $\overset{c}{CH} - \overset{b}{CH_2} - \overset{a}{CH_3}$ Isopentane

There are four types of protons indicated by a, b, c, d, and correspondingly there will be four signals.

(*x*) $\overset{a}{CH_3}$ and $\overset{b}{H}$ $>$ C=C $<$ $\overset{c}{H}$ and $\overset{d}{H}$ Propene

There are again four types of protons designated by a, b, c and d. Thus four signals will be seen.

(xi) $\overset{a}{CH_3} - \underset{\underset{c}{H}}{\overset{\overset{b}{H}}{C}} - \overset{d}{CHCl} - \overset{e}{CH_3}$

It offers an interesting case. Methylenic protons are not equivalent. These protons are distereomeric protons. Thus we have a total of 5 different types of protons displaying five signals in PMR spectrum.

Q. 64. What is meant by shielding and unshielding of protons?

(*Panjab, 2003; Nagpur, 2008*)

Ans. In the NMR spectrum of a compound, the electrons around the protons also play their role. When a compound is placed in a magnetic field, the electrons around the protons also generate a magnetic field called *induced magnetic field.* The induced magnetic field may reinforce (support) or oppose the applied field. Thus two cases arise.

(*i*) If the induced field opposes the applied field, the effective field strength experienced by the protons decreases. Thus a greater applied field is required for the excitation of protons to higher level. It is expressed by saying that the proton absorbs **upfield**. The proton is said to be **shielded** by electrons in this case.

(*ii*) If the induced field reinforces or supports the applied field, an enhanced field strength will be experienced by the proton. Proton is said to be **deshielded** in this case. A deshielded proton absorbs **downfield** as a smaller field strength will be sufficient for absorption of energy to give a signal in NMR spectra. Shielding and deshielding of protons is a result of factors, like the inductive and electromeric effects of groups and hydrogen bonding.

As a result of shielding and deshielding of protons, there is a **shift** in the position of the signal.

Q. 65. What is chemical shift? What are the factors influencing chemical shift?

(*Bangalore, 2002; Nagpur, 2002; Bhopal, 2004; Lucknow, 2010*)

Ans. The shift in the positions of PMR signal, compared with a standard substance, as a result of shielding and deshielding by electrons, is referred to as **chemical shift.**

Factors influencing chemical shift.

(*i*) Nature of groups. Atoms and groups which are electron withdrawing, deshield the protons. The effective magnetic field experienced by the protons, in such a case, is more than normal. Hence the protons will absorb downfield. Atoms like halogens, oxygen and nitrogen are electron-withdrawing. If such atoms are linked with protons, they deshield the protons resulting in downfield absorption. For example the proton in CH_3OCH_3 shows a downfield abosorption compared to that in $CH_3 - CH_3$. An electron releasing group like methyl increases the electron density around the proton and thus shields it.

Greater the electronegativity of the atom, greater will be the extent of deshielding and hence lower will be the absorption energy. Proton in $CH_3 - F$ are deshielded to a greater extent than in $CH_3 - Cl$. Consider the case of $\overset{b}{CH_3}\overset{a}{CH_2}Cl$. It has two types of protons. The protons labelled a are more deshielded than protons labelled b because of different distance of Cl from the two types of protons.

Electron withdrawing groups reduce the electron density around the proton or deshield the proton, thereby resulting in downfield absorption. Electron releasing groups increase the electron density around protons and result in upfield absorption.

(*ii*) **Hydrogen bonding.** If the proton is linked to some electronegative atom in the form of a hydrogen bond, it is deshielded. As such, the absorption will occur downfield.

The effect of various atoms, bonds or groups is summarised as under:

(*a*) Protons on a primary carbon atom are shielded to the maximum followed by secondary and tertiary protons.

$$-\overset{|}{\underset{|}{C}}-H \;<\; -\overset{H}{\underset{|}{\overset{|}{C}}}-H \;<\; -\overset{H}{\underset{H}{\overset{|}{\underset{|}{C}}}}-H$$

(*b*) Electronegative atoms such as halogens, oxygen and nitrogen deshield the protons resulting in donwfield absorption.

(*c*) π electrons of the carbonyl group, alkene and benzene ring have the deshielding effect on protons. The order of deshielding is

$$-\overset{O}{\overset{\|}{C}}-H \;>\; C_6H_5-H \;>\; \overset{|}{\underset{|}{C}}=\overset{|}{C}-H$$

(*d*) Carbonyl group has the effect of deshielding protons on the neighbouring carbon.

$$O=C-\overset{|}{\underset{|}{C}}-H$$

(*e*) Hydrogen bonding deshields the proton.

(*f*) Strongly electropositive atoms such as silicon increase the electron density on the proton and thus shield the proton.

Q. 66. Why do we choose tetramethylsilane (TMS) $(CH_3)_4$ Si as a standard substance for recording chemical shift? (*Nagpur, 2008; Lucknow, 2010*)

Ans. In order to measure or record chemical shifts, we need some standard substance. The choice falls on TMS for the reasons given below.

(*i*) It has 12 equivalent protons and thus gives a sharp signal.

(*ii*) Electronegativity of silicon (1.8) is lower than that of carbon. Consequently the shielding of equivalent protons in TMS is more than that of almost all organic compounds. Therefore the signals of almost all organic compounds appear in the downfield direction with respect to TMS.

(*iii*) TMS is chemically inert and possesses a low b.p. of 300 K. It can be easily evaporated after the experiment to recover the compound.

Q. 67. What are the units for expressing chemical shifts?

Ans. Chemical shifts are expressed in frequency units (cycles per second). Frequency and magnetic field are related to each other by the following equation.

$$\nu = \frac{\gamma H_0}{2\pi}$$

Chemical shift expressed in cps (cycles per second) is directly proportional to the strength of applied field which in turn is equivalent to radiofrequency used. Since different PMR spectrophotometer use different radiofrequencies of 40, 60 or 100 magacycle/sec (Mcps) it is more appropriate to use units in which the radiofrequency of the instrument has been cancelled out. This is done by dividing the shift in cps by the radiofrequency employed and multiplying by a factor of 10^6 so that shifts are obtain in **parts per million (ppm).**

If a signal is obtained at 120 cps downfield with reference to TMS using 60 Mcps (or 60×10^6 cps), then the chemical shifts is given by

$$\text{Chemical shift } (\delta) = \frac{120\times10^6}{60\times10^6} = 2 \text{ ppm}.$$

There are two scales for expressing the chemical shift. These are the δ (delta) scale and τ (tau) scale. Both of them are in **parts per million.**

$$\delta = \frac{\Delta v(\text{in cps})\times10^6}{\text{Radio frequency used in cps}}$$

and $\tau = 10 - \delta$

Taking the position of TMS signal as zero ppm, most chemical shifts have values between 0–10. A small value of δ indicates a small downfield shift whereas a large δ value indicates a large downfield shift.

On the τ scale, the positions of the TMS signal is taken as 10.0 ppm and most compounds have chemical shift values between 10–0. This is explained with the help of the following diagram.

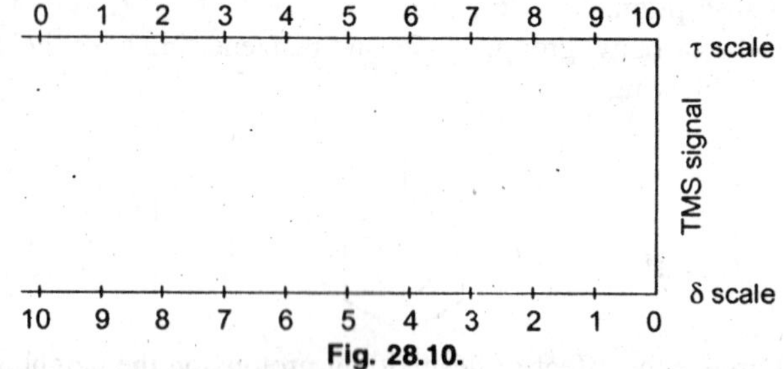

Fig. 28.10.

Q. 68. Give the characteristic proton chemical shifts of various types of protons.

Ans. Chemical shifts for various types of protons are given in the table below:

Table: Typical proton chemical shifts

	Types of proton	*Chemical sift ppm*	
		δ	τ
Primary	$R-CH_3$	0.9	9.1
Secondary	R_2CH_2	1.3	8.7
Tertiary	R_3CH	1.5	8.5
Vinylic	$-\overset{\vert}{C}=\overset{\vert}{C}-H$	4.6 – 5.9	4.1 – 5.4
Acetylene	$-C\equiv C-H$	2 – 3	7 – 8
Aromatic	Ar – H	6 – 8.5	1.5 – 4
Benzylic	$Ar-\underset{\vert}{\overset{\vert}{C}}-H$	2.2 – 3	7 – 7.8
Allylic	$-\overset{\vert}{C}=\overset{\vert}{C}-\underset{\vert}{\overset{\vert}{C}}-H$	1.7	8.3
Alkyl fluoride	$-\overset{\vert}{C}H-F$	4 – 4.5	5.5 – 6
Alkyl chloride	$-\overset{\vert}{C}H-Cl$	3 – 4	6 – 7
Alkyl bromide	$-\overset{\vert}{C}H-Br$	2.5 – 4	6 – 7.5
Alkyl iodide	$-\overset{\vert}{C}H-I$	2 – 4	6 – 8
Alcohol	$-\overset{\vert}{C}H-OH$	3.4 - 4	6 – 6.6
Ethers	$-\overset{\vert}{C}H-O-R$	1.3 – 4	6 – 6.7

Carbonyl compounds	$-\overset{\vert}{C}H-\overset{O}{\overset{\Vert}{C}}-$	2 – 2.7	7.3 – 8
Acids	$-\overset{\vert}{C}H-COOH$	2 2.6	7.4 – 8
Hydroxylic	R – O – H	1 – 5.5	4.5 – 9
Phenolic	Ar – O – H	4 – 12	(–2) – 6
Aldehydic	$R-\overset{O}{\overset{\Vert}{C}}-H$	9 – 10	0 – 1
Carboxylic	R – COOH	10.5 – 12	(–2) – (–0.5)

Q. 69. Explain the various signals in the PMR spectrum of benzyl alcohol.

(Kurukshetra, 2001)

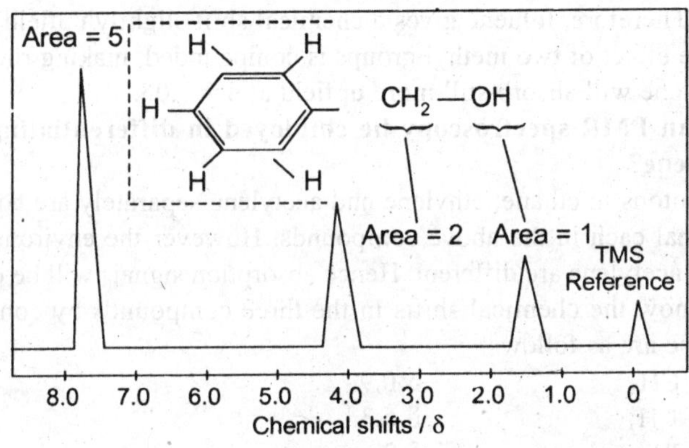

Fig. 28.11

Ans. The no. of signals in a spectrum tells us how many different types of protons are present and the position of the signals (chemical shift) tells us about the nature of different protons (or their environments).

In the spectrum of benzyl alcohol shown above, we observe four signals which are due to the protons as detailed below:

(*i*) The small but sharp signal at δ = 0 is the reference signal of TMS.

(ii) Signal at δ = 7.3 is due to five equivalent ring protons.

(*iii*) The signal at δ = 4.6 is due to two chemically equivalent methylene protons.

(*iv*) The peak at δ = 2.4 is due to hydroxyl proton.

Q. 70. What is the significance of peak area? *(Kurukshetra, 2001)*

Ans. Peaks obtained in the PMR spectrum make different areas with the base line. It is observed that area under a PMR signal is directly proportional to the number of equivalent protons giving rise to that signal. By comparing the areas subtended by different signals, we can calculate the relative proportion of different types of protons. For example, in the case of peaks of benzyl alcohol. The areas under the peaks are in the ratio of 1 : 2 : 5 (see figure in Q. 69 above) indicating that the three types of protons are in the ratio 1 : 2 : 5. This is actually so. There are one hydroxyl proton, two methylene protons and five ring protons.

The peak areas of different signals are measured by an automatic electronic integrator. Also heights of peaks are proportional to areas.

Q. 71. How can you explain the difference in chemical shifts of aromatic protons in the following compounds? Benzene δ = 7.37, Toluene δ = 7.17, *p*-xylene δ = 7.05.

Ans.

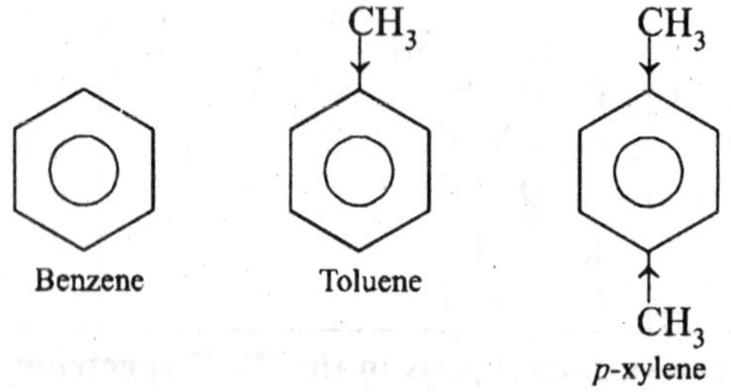

Benzene | Toluene (CH_3) | *p*-xylene (CH_3, CH_3)

Methyl group is an electron releasing group. Hence in toluene, the effect of methyl group will be to increase electron density around the ring protons. Thus the protons will be shielded and they will absorb upfield. Therefore, toluene gives a chemical shift slightly upfield at δ = 7.17.

In *p*-xylene, the effect of two methyl groups is compounded, making ring protons still more shielded. Thus *P*-xylene will absorb still more upfield at δ = 7.05.

Q. 72. How can PMR spectroscopy be employed in differentiating between ethane, ethylene and acetylene? (*Kerala, 2000*)

Ans. All the protons in ethane, ethylene and acetylene separately are equivalent. Hence we expect only one signal each in the above compounds. However the environments of protons in ethane, ethylene and acetylene are different. Hence absorption signals will be obtained at different positions. We can know the chemical shifts in the three compounds by consulting the table of chemical shifts. These are as follows:

$CH_3 - CH_3$ δ 0.9

$CH_2 = CH_2$ δ 5.3

$CH \equiv CH$ δ 2

Position of the signal can help us differentiate one compound from the rest.

Q. 73. How many signals do you expect to have in the NMR spectrum of $CH_3 - CH_2 - Cl$? What would be the approximate absorption positions of the signals?

Ans. $CH_3 - CH_2 - Cl$ (Ethyl chloride) has two types of protons, three equivalent methyl protons and two equivalent methylene protons. Hence we expect two signals. Methylene protons are attached to carbon which also has an electron withdrawing chlorine attached to it. Thus the electron withdrawing or shielding effect of chlorine will be more on methylene protons than on the methyl protons. So methylene protons will absorb downfield compared to methyl protons.

The signals are obtained as detailed below:

Methyl protons δ 1– 1.5

Methylene protons δ 3.5

Q. 74. Account for the splitting of PMR signals. (*Guwahati 2002*)

Or

Explain spin-spin coupling.

(*Himachal, 2000; M.Dayanand, 2000; Bangalore 2001; Panjab, 2003*)

Ans. We assume that one type of protons give rise to one signal or peak in PMR spectrum. In actual practice it is not so. Consider, for example, the following compounds.

$BrCH_2 - CHBr_2$	$CH_3 - CHBr_2$	$CH_3 - CH_2Br$
1, 1, 2 Tribromo ethane	1, 1 dibromo ethane	Ethyl bromide

(*a*) An examination of the above compounds reveals that each one of them contains two types of protons. We expect to observe, therefore, two signals in each case but this is actually not the case.

(*b*) We observe that 1, 1, 2 tribromoethane gives five peaks (two peaks split into five), 1, libromc ethane gives six peaks and ethyl bromide gives seven peaks. This phenomenon o splitting of a peak into several peaks is called **splitting of signals.**

(*c*) Splitting of signals takes place when there are non-equivalent protons on adjacent carbon atoms. Thus in $CH_3 - CH_3$, all the six protons are equivalent, hence no splitting is observed.

(*d*) Splitting of signals is explained on the basis of **spin-spin coupling** of the absorbing and neighbouring protons. Let us consider a case of two adjacent carbon atoms carrying non-equivalent protons, one having two secondary protons and the other a tertiary protons.

$$-\overset{t}{\underset{|}{C}H}-\overset{s}{C}H_2-$$

Letters *t* and *s* stand for tertiary and secondary respectively. Consider the absorption by one of the secondary protons. Magnetic field produced by neighbouring proton (*i.e.* tertiary proton) may have two possible orientations with respect to applied magnetic field.

(*i*) If the magnetic field produced by the proton is oriented along the applied field, the secondary protons experience some increased magnetic field strength. Consequently, the secondary proton will absorb at lower applied field (downfield) than it would have done in the absence of *t*-protons.

(*ii*) If the magnetic field produced by the proton is oriented against the applied field, the secondary protons experience some decreased magnetic field strength. Consequently, the secondary proton will absorb at higher applied magnetic field (upfield) than it would have done in the absence of *t*-protons.

(*e*) Considering that there is an equal probability of the *t*-proton magnetic field being aligned along or against the applied magnetic field, half of the secondary protons will absorb downfield and the other half will absorb upfield. Thus one peak will be split into two peaks **(doublet)** with equal peak intensities as shown below:

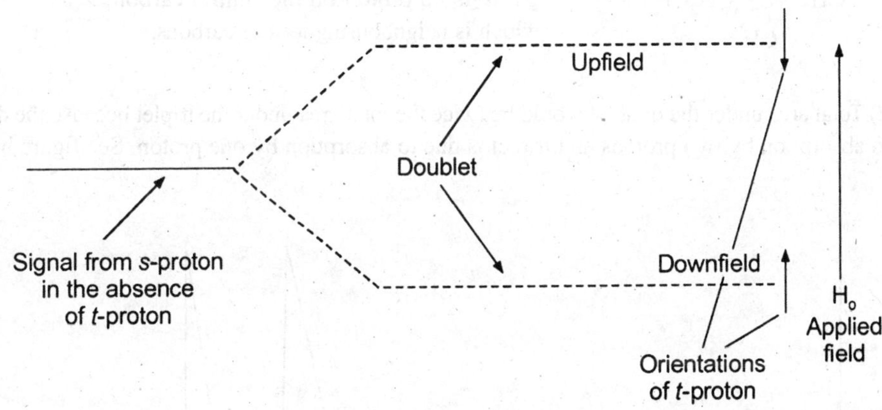

Fig. 28.12

(*f*) Now consider the absorption by *t*-protons. In this case, the two neighbouring s-protons may have three probable combinations of spin alignments as given below:

Case (*i*) : The two protons are aligned with the applied magnetic field.

Case (*ii*) : The two protons are aligned against the applied magnetic field.

Case (*iii*): One proton is aligned along the field and the other proton is aligned against the field.

In the first case, the tertiary proton will absorb downfield.

In the second case, the tertiary proton will absorb upfield.

In the third case, the position of the signal will not change.

As a result of above modification, the peak duc to tertiary proton will be split into three peaks **(triplet)**. The relative intensities are in the ratio 1 : 2 : 1. The central peak is of double intensity because it belongs to the third case (as given above) in which one proton is along and the other against the field. If we designate the secondary proton as no. 1 & 2, then proton 1 is along and proton 2 is against the field or proton 2 is along and proton 1 is against the field. The middle peak corresponds to case (*iii*) which takes into account both these possibilities. Hence the middle peak of double intensity. This is diagrammatically shown below.

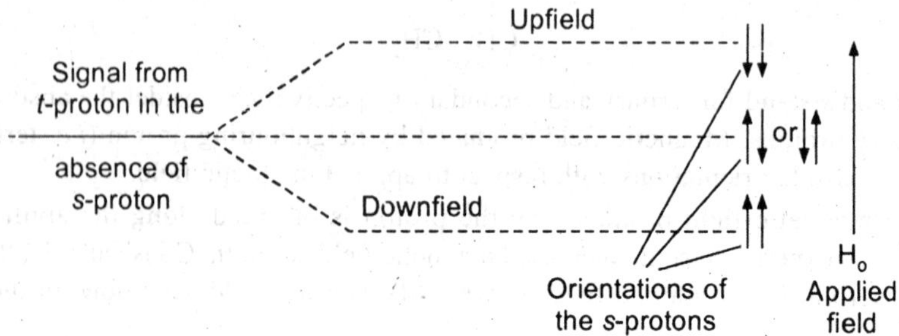

Fig. 28.13

Q. 75. What are the rules governing splitting of signals?

Ans. (*i*) Only the neighbouring or vicinal protons cause the splitting of a signal provided these vicinal protons are non-equivalent to the absorbing protons. Thus we don't expect splitting in the following compounds.

(*a*) CH_2Cl-CH_2Cl Adjacent carbon has equivalent protons

(*b*) $CH_3-\underset{CH_3}{\overset{CH_3}{\underset{|}{\overset{|}{C}}}}-CH_2Cl$ There is no proton on the central carbon which is neighbouring to four carbons.

(*ii*) Total area under the doublet would be twice the total area under the triplet because the doublet is due to absorption by two protons and triplet is due to absorption by one proton. See figure below.

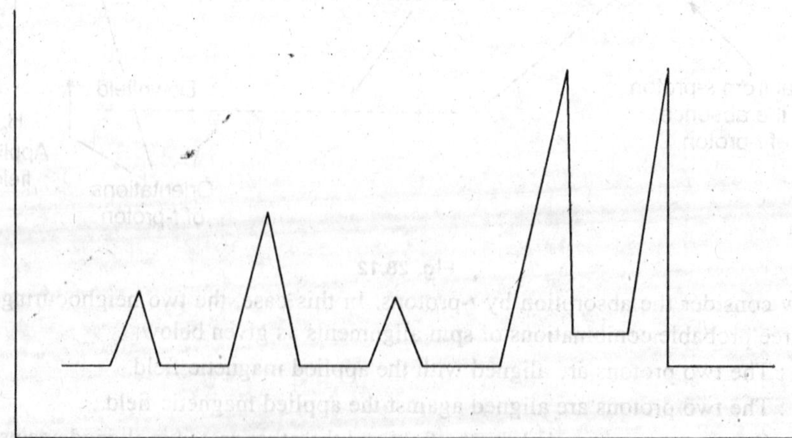

PMR spectrum of 1, 1, 2, tribromomethane

Fig. 28.14

(*iii*) The no. of peaks obtained after splitting is one more than the no. of non-equivalent protons on the adjacent carbon atoms. This is illustrated below for different types of protons.

Y–C(H)–C(H)–X	–C(H)(H)–C(H)	–C(H)(H)(H)–C(H)(H)–H
Doublet (1:1), Doublet (1:1)	Triplet (1:2:1), Doublet (1:1)	Quartet (1:3:3:1), Doublet (1:1)

Q. 76. How many signals would you expect from the following? Also indicate the multiplicity of various signals.

(*i*) H_3C–C_6H_4–$C(CH_3)_3$ (*ii*) C_2H_5OH (*Calcutta, 2000*)

(*iii*) $CH_3 - O - CH_3$ (*iv*) $CH_3OCH_2CH_3$

Ans. (*i*) There are three types of protons viz. methyl protons on the L.H.S. of the ring, methyl hydrogens on the R.H.S. of the ring and the ring protons. Hence three signals will be obtained.

All the peaks will be singlet because, if we consider any absorbing hydrogen, there are no non-equivalent protons in the molecules. So no splitting will take place.

(*ii*) $CH_3 - CH_2 - OH$

There are three types of protons, hence three signals. It will be a triplet for methyl protons, a quartet for methylene protons and a singlet for OH.

(*iii*) $CH_3 - O - CH_3$

There are only one type of protons, hence one signal will be obtained and it will be a singlet.

(*iv*) $CH_3 - O - CH_2 - CH_3$

There are three types of protons, hence three signals will be obtained. It will be a singlet for methoxy protons, a quartet for methylene protons and a triplet for methyl protons.

Q. 77. Give the structure consistent with the following set of NMR data. Molecular formula C_9H_{12}, singlet τ 3.22, 3H, singlet τ 7.75, 9H.

Ans. (*i*) The formula C_9H_{12}, corresponds to the general formula of aromatic hydrocarbon C_nH_{2n-6}. The compound seems to be a substituted benzene.

(*ii*) The compound gives two signals. This shows the presence of two types of protons.

The above conditions are satisfied by 1, 3, 5, trimethyl benzene.

CH_3 / H_3C — C_6H_3 — CH_3

This structure explains the existence of singlet τ 3.22 due to three ring protons and singlet τ 7.75 due to nine methyl protons.

Q. 78. How many NMR signals would you expect from the following? Toluene, 1, 1-dichloroethane and allyl alcohol.

Ans. (*i*)

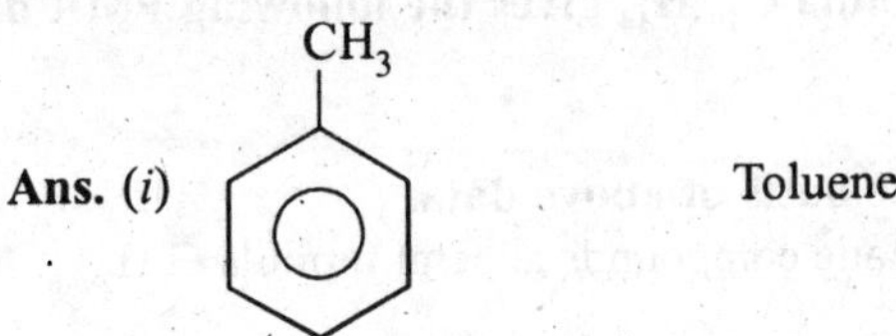

It contains two types of protons, the ring protons and methyl protons. Hence two signals will be observed.

(ii) $\overset{b}{C}H\,Cl_2-\overset{a}{C}H_3$ 1, 1 dichloroethane

There are two kinds of protons indicated by a & b. Hence two signals will be obtained.

$$\text{(iii)}\quad \underset{e}{CH_2}=\underset{d}{CH}-\underset{\underset{\underset{c}{H}}{|}}{\overset{\overset{\overset{b}{H}}{|}}{C}}-\underset{a}{OH}$$

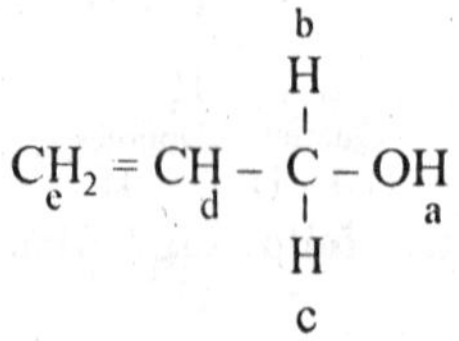

There are five kinds of protons designated as a, b, c, d, e. So five signals will be obtained.

Q. 79. Of the following molecules, which will exhibit spin-spin coupling? What will be the multiplicity of each kind of proton in such molecules?

(*i*) $BrCH_2CH_2\,Br$ **(*ii*) $CH_3-CH\,Br_2$** **(*iii*) $ClCH_2CH_2Br$**

$$\text{(iv)}\quad CH_3-\underset{\underset{CH_3}{|}}{\overset{\overset{CH_3}{|}}{C}}-Br \qquad \text{(v)}\quad \begin{matrix}H\\Cl\end{matrix}\!>C=C<\!\begin{matrix}H\\Cl\end{matrix}$$

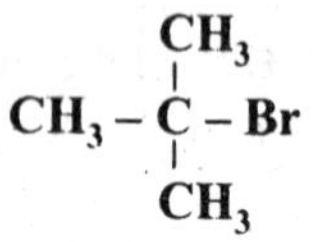

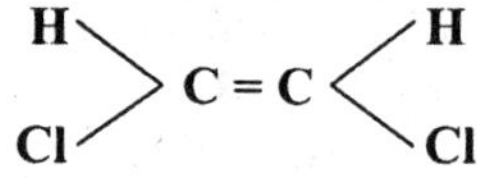

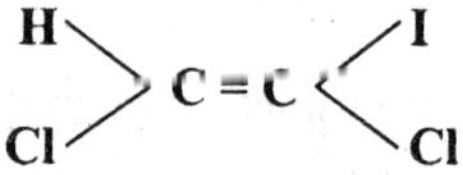

$$\text{(vi)}\quad \begin{matrix}H\\Cl\end{matrix}\!>C=C<\!\begin{matrix}I\\Cl\end{matrix} \qquad \text{(vii)}\quad \begin{matrix}I\\Cl\end{matrix}\!>C=C<\!\begin{matrix}H\\H\end{matrix}$$

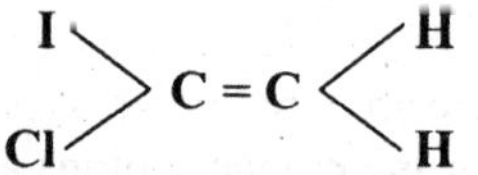

Ans. (*i*) All the protons are of the same type, hence no spin-spin coupling takes place.

(*ii*) The proton of $-CHBr_2$ is split into quartet and the protons of methyl are split into a doublet.

(*iii*) Protons of each methylene group are split into triplets.

(*iv*) No spin-spin coupling takes place because there is no hydrogen on the bromine-bearing carbon. Consequently, there are no non-equivalent protons on the neighbouring carbon with reference to any methyl group.

(*v*) No spin-spin coupling takes place because both the protons in the molecule are equivalent.

(*vi*) Each proton is split into doublet.

(*vii*) As the two protons on one carbon are not equivalent, each is split into doublet.

Q. 80. For the compound

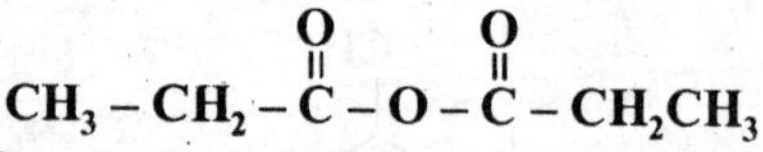

$$CH_3-CH_2-\overset{\overset{O}{||}}{C}-O-\overset{\overset{O}{||}}{C}-CH_2CH_3$$

predict the number of signals, their positions, relative intensities and splitting.

Ans. (*i*) As there are two types of protons, the methyl and methylene protons, two signals are expected.

(*ii*) Signals for methyl protons are obtained at δ 1 – 1.5 and for methyl protons at δ 2 – 2.5 (consult table).

(*iii*) Methyl protons are split into triplet and methylene protons into quartet. So the overall spectrum can be expressed as; Triplet δ 1 – 1.5 (6H), quartet δ 2 – 2.5 (4H).

Q. 81. A compound having molecular formula $C_{10}\,H_{14}$ gives the following PMR data.

(a) Singlet τ 9.12 (δ 0.88), 9H

(b) Singlet τ 2.72 (δ 7.28), 5H

Assign a structure to the compound on the basis of above data.

Ans. (*i*) The formula corresponds to the aromatic compounds general formula C_nH_{2n-6}. Hence it appears to be a substituted benzene.

(*ii*) There are nine protons of one kind and 5 protons of another kind. It suggests that there is mono-substitution in the benzene ring.

(*iii*) The two kinds of protons are not spin-spin coupled as we are getting the singlets.

(*iv*) The singlet τ 2.72 (δ 7.28) can be due to C_6H_5 *i.e.* phenyl group, confirming monosubstitution.

(*v*) The remaining part of the molecule *i.e.* C_4H_9 can have nine equivalent protons only if it is in the form of a *t*-butyl group.

$$C_6H_5-C(CH_3)_3$$

The singlet at τ 9.12 (δ 0.8) is due to the methyl protons.

Q. 82. On the basis of the following data, assign the structure to the compound having molecular formula $C_{10}H_{14}$.

NMR (*i*) **A singlet (2.7 τ, 5H)**
(*ii*) **A doublet (7.5 τ, 2H)**
(*iii*) **A multiplet (8.0 τ, 1H)**
(*iv*) **A doublet (9.0 τ, 6H)**

Ans. (*i*) Singlet at 2.7 τ corresponds to hydrogen atoms of benzene ring.

(*ii*) Doublet at 7.5 τ corresponds to $-CH_2$.

(*iii*) A multiplet at 8.0 τ shows that there is branching in the carbon side chain.

(*iv*) A doublet at 9.0 τ corresponds to hydrogens of terminal methyl groups.

On the basis of above, the compound has the following structure:

$$C_6H_5-CH_2-CH(CH_3)_2$$

2.7 τ (ring H); 7.5 τ (CH_2); 8.0 τ (CH); 9.0 τ (CH_3)

(Isobutyl benzene)

Q. 83. A chloropropane gives a PMR spectrum as follows:

(*a*) Triplet δ – 0.9.

(*b*) Triplet more downfield as compared with above but an intensity of 2/3 of above.

(c) Multiplet in between the above triplets.

Is the compound 1-chloropropane or 2-chloropropane?

Ans. Let us consider both possibilities

$$CH_3-CH_2-CH_2-Cl \qquad CH_3-CH(Cl)-CH_3$$

1-Chloropropane 2-Chloropropane

(*i*) 2-chloropropane has only two types of protons and hence would give only two signals whereas we are getting three signals. Hence the possibility of the compound being 2-chloropropane is ruled out and it appears to be 1-chloropropane.

(*ii*) Now 1-chloropropane contains three types of protons and hence three signals are expected.

(*iii*) Triplet δ ~ 0.9 is due to methyl protons (consult table). Triplet is obtained because of methylene protons.

(*iv*) A triplet more downfield as compared to the methyl protons is because of the protons of $> C - Cl$. This is due to deshielding effect of chlorine. The triplet, which is again due to methylene protons will appears at δ 3–4.

(*v*) A signal due to methylene protons will be obtained at about δ 1.3 (consult table). It will however be a sixlet (multiplet) due to 5 protons on the two sides.

The data given conforms to the structures of 1 chloropropane.

Q. 84. Suggest a structure on the basis of PMR data of the following compounds.

(*i*)	**C_3H_7Br**	**δ 1.7 (*d*, 6H),**	**δ 4.3 (septet, 1H)**
(*ii*)	**$C_4H_{10}O$**	**δ 1.28 (*s*, 9H),**	**δ 1.35 (*s*, 1H)**
(*iii*)	**$C_5H_9Cl_3$**	**δ 1.0 (*d*, 6H),**	**1.5 (multiplet, 1H)**
		δ 3.3 (*s*, 2H)	

Ans. (*i*) The structure is $CH_3 - CHBr - CH_3$

There are two types of protons, hence two signals. Six protons of one type and one proton of another type as given in the data conforms to this structure. The middle proton will give a septet because of six surrounding protons and doublet will be obtained in respect of methyl protons due to the middle proton.

(*ii*) $C_4H_{10}O$ is the molecular formula of butyl alcohol. A no. of butyl alcohols are possible with this formula as given below

$$\underset{\text{I}}{CH_3CH_2CH_2CH_2\,OH} \qquad CH_3-\underset{\underset{\text{II}}{CH_3}}{\underset{|}{CH}}-CH_2OH \qquad CH_3CH_2-\underset{\underset{\text{III}}{CH_3}}{\underset{|}{CHOH}}$$

$$\text{and } CH_3-\overset{\overset{CH_3}{|}}{\underset{\underset{\text{IV}}{CH_3}}{\underset{|}{C}}}-OH$$

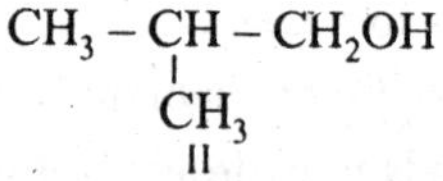

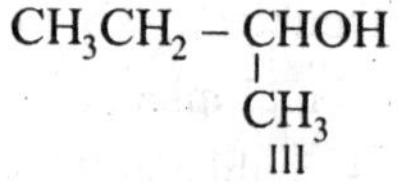

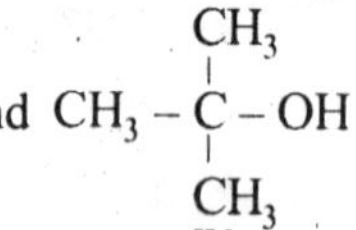

The data given above conforms to structure IV.

Structure IV has only two types of protons and gives rise to two signals. Structures I, II & III would have given more than two signals and are therefore, discarded.

There are nine protons of one type (all methyl protons) and one proton of one type. This is supported by the data. Singlet δ 1.28 is because of methyl protons and singlet δ 1.35 is due to the hydroxy proton.

(*iii*) Out of the various structures possible, the structure which is in conformity with the data is

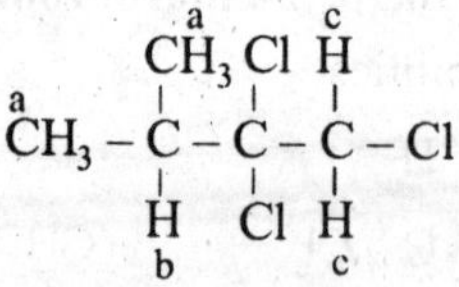

$$\overset{a}{CH_3}-\overset{\overset{\overset{a}{CH_3}}{|}}{\underset{\underset{\underset{b}{H}}{|}}{C}}-\overset{\overset{Cl}{|}}{\underset{\underset{Cl}{|}}{C}}-\overset{\overset{\overset{c}{H}}{|}}{\underset{\underset{\underset{c}{H}}{|}}{C}}-Cl$$

There are three types of protons and hence three signals are obtained. Protons designated "a"will give a doublet δ 1.0, protons designated "b" will give a multiple at δ 1.5 and protons designated "c" will give singlet. All these properties are satisfied with the structure given above.

Q. 85. A compound having a molecular formula C_4H_9Br gave the following PMR data:

(*a*) doublet τ 8.96 (δ 1.04), 6H

(*b*) multiplet τ 8.05 (δ 1.95), 1H

(*c*) doublet τ 6.67 (δ 3.33), 2H

What structure can the compound have?

Ans. There are four structures possible with this molecular formula

$CH_3 - CH_2 - CH_2 - CH_2Br$ — 1-Bromobutane (I)

$CH_3 - CH_2 - CH_2Br - CH_3$ — 2-Bromobutane (II)

$$CH_3 - \underset{\textstyle}{\overset{\displaystyle CH_3 \atop |}{CH}} - CH_2\,Br$$

1-Bromo-2-methylpropane (III)

$$CH_3 - \underset{\displaystyle | \atop CH_3}{\overset{\displaystyle CH_3 \atop |}{C}} - Br$$

2-Bromo-2 methylpropane (IV)

(*a*) Structure I and II have four types of protons each. Structure IV has only one type of protons. The data shows that the compound contains three types of protons. Thus structures I, II & IV are ruled out. Structure III satisfies this condition and is the likely structure.

(*b*) Structure III has six protons of one type, one protons of second type and two protons of third type which tallies with the data.

(*c*) Six methyl protons give a doublet due to tertiary proton. The tertiary proton gives a multiplet due to six methyl protons and protons in $-CH_2Br$.

Finally protons in $-CH_2Br$ give a doublet due to *t*-proton.

Thus III is the structure which is consistent with the data.

Q. 86. Give a structure consistent with the following set of PMR data. Molecular formula of the compound is $C_9H_{11}Br$.

(*i*) Multiplet τ 7.85 (2H) **(*ii*) Triplet τ 7.25 (2H)**

(*iii*) Triplet τ 6.62, (2H) **(*iii*) Singlet τ 2.78, (5H)**

Ans. The data reveals that

(*a*) There are four types of protons.

(*b*) There is a benzene ring in the compound as shown by singlet τ 2.78 (5H).

(*c*) The side chain in the benzene ring consists of $- C_3\ H_6$ Br and the six protons in the side chain form three pairs of different types of protons. The following structure satisfies the data completely.

$$C_6H_5-\overset{3}{C}H_2-\overset{2}{C}H_2-\overset{1}{C}H_2Br$$

The signals given by the side-chain are explained as under:

(*a*) The two protons on C_3 give a signal τ 7.25. It splits into three signals due to two protons on C_2.

(*b*) The two protons on C_2 give a signal τ 7.85 and it splits into 5 peaks (multiplet) due to 4 neighbouring protons on C_3 and C_1. Slight downfield shift is due to the unshielding effect of bromine.

(*c*) The two protons on C give a downfield signal τ 6.62 due to deshielding effect of bromine. The signal is split into a triplet due to the neighbouring protons on C_2. Thus the structure assigned above is consistent with the given data.

Q. 87. A compound having molecular formula $C_9H_{11}Br$ furnished the following set of PMR data:

1. Singlet δ 7.25, 5H **2. Doublet δ 7.25, 2H**

3. Multiplet δ 3.40, 1H **4. Doublet δ 1.45, 3H**

Assign a structure to this compound showing your reasoning.

Ans. (*a*) The data reveals that there are four types of protons.

(*b*) There is a benzene ring in the compound as shown by δ 7.25 (5H)

(*c*) The side chain in the benzene ring consists of

$$-CH_2-\underset{\underset{Br}{|}}{\overset{\overset{H}{|}}{C}}\ CH_3$$

and the six protons in the side chain form three different sets of protons. The following structure for the compound satisfies the data completely.

$$C_6H_5-\overset{3}{C}H_2-\underset{\underset{Br}{|}}{\overset{\overset{H}{|}}{\overset{2}{C}}}-\overset{1}{C}H_3$$

The signals given by the side-chain are explained as under:

(*a*) Three protons on C-1 give the signal δ 1.45. It splits into doublet due to the presence of one proton on C-2.

(*b*) One proton on C-2 gives the signal δ 3.40. It splits into multiplet due to the presence of five protons on neighbouring carbons.

(*c*) Two protons on C-3 give the signal δ 2.75. It splits into doublet due to the presence of one proton on neighbouring C-2.

Q. 88. An organic compound with molecular formula $C_3H_3Cl_5$ gave the following NMR data

(*i*) A triplet 4.49 τ (4.52 δ) 1H

(ii) A doublet 3.93 τ (6.07 δ) 2H

Assign a structural formula to the compound consistent with the above data

Ans. The molecular formula $C_3H_3Cl_5$ suggests that it is a saturated molecules with no multiple bonds. Since we are getting two peaks, there are two kinds of protons. Hence protons are attached to two carbon atoms.

As a triplet and a doublet are obtained, it means there are protons on the neighbouring carbon atoms. If the protons were present on terminal carbon atoms, only singlet would be obtained. In the light of above, the structure is

$$Cl-\underset{\underset{H^b}{|}}{\overset{\overset{H^b}{|}}{C}}-\underset{\underset{Cl}{|}}{\overset{\overset{H^a}{|}}{C}}-\underset{\underset{Cl}{|}}{\overset{\overset{Cl}{|}}{C}}-Cl$$

Q. 89. A compound with molecular formula $C_{10}H_{12}O$ shows a strong absorption at 1705 cm^{-1} in its IR spectrum and NMR spectrum of the compound shows the following peaks:

δ 7.22 (Singlet 5H)

δ 3.59 (Singlet 2H)

δ 2.77 (Quartet 2H)

δ 0.97 (Triplet, 3H)

Giving reasons assign a structure to the compound.

Ans. (*a*) IR absorption at 1705 cm^{-1} shows the presence of a carbonyl group (–CHO or > C = O) (*b*) The molecular formula suggests the presence of a benzene ring.

(*c*) A peak at δ 7.22 singlet, 5H suggests that it is a monosubstituted aromatic compound i.e. there is only one chain in the ring, thereby retaining five equivalent protons in the ring.

(*d*) The possibility of an aldehyde is ruled out because in that case, there will be two protons each of three different types. This is not in accordance with the data.

$$C_6H_5-\overset{4}{C}-\overset{3}{C}-\overset{2}{C}-\overset{1}{C}$$

(with C=O at position 2, 3 or 4)

(*e*) The only possibility left is a ketone, with the ketonic position at 2 or 3 or 4

$$\underset{4}{C_6H_5-CH_2}-\underset{3}{\overset{O}{\overset{\|}{C}}}-\underset{2}{CH_2}-\underset{1}{CH_3}$$

(*f*) A ketonic group at C-3 satisfies the observation.

(*g*) Three protons at C-1 will give a peak, it will be a triplet due to two protons at C-2. Two protons at C-2 will give a quartet due to three protons on C-1. Two protons at C-4 will give a singlet because there is no proton on the neighbouring carbon atoms.

Q. 90. A compound C_3H_6O contains a carbonyl group. How could NMR establish whether this compound is an aldehyde or ketone?

Ans. The aldehvde with the above formula will have the structure

$$\overset{c}{C}H_3-\overset{b}{C}H_2-\overset{a}{C}HO$$

It will give three peaks corresponding to three types of protons as follows :

triplet, 1H

pentet, 2H

triplet, 3H

The ketone with the above formula will have the structure

$$\overset{a}{C}H_3-\overset{O}{\overset{\|}{C}}-\overset{a}{C}H_3$$

Here all the six protons are of one type and will give one peak which will be singlet.

This is how the identity of the compound can be established.

Q. 91. Propose structural formulae for the compounds with the following molecular formulae which give only one NMR signal:

(*i*) C_5H_{12}

(*ii*) C_2H_6O

(*iii*) $C_2H_4Br_2$

(*iv*) C_8H_{18}

Ans. Since only one NMR signal is obtained, the compounds must possess only one kind of protons. The structural formulae of the compounds accordingly are:

(*i*) $CH_3-\underset{CH_3}{\overset{CH_3}{\overset{|}{\underset{|}{C}}}}-CH_3$ (*ii*) CH_3-O-CH_3
Methoxymethane

(*iii*) $Br-CH_2-CH_2Br$
1, 2-Dibromoethane

(*iv*) $CH_3-C(CH_3)_2-C(CH_3)_2-CH_3$
2, 2, 3, 3-Tetramethylbutane

Q. 92. How many NMR signals would you expect from the following?

(*i*) Toluene **(*ii*) 1, 1-Dichloroethane** **(*iii*) Allyl alcohol**

Ans. (*i*)

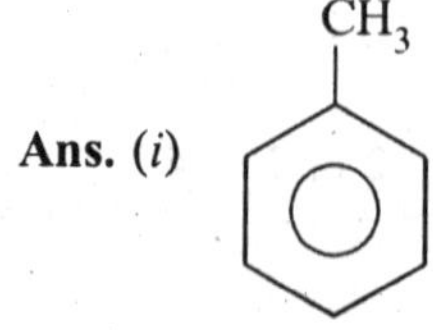

There are two types of protons the ring protons and the methylprotons. Hence two signals will be obtained.

(*ii*) $\overset{b}{C}H_3-\overset{a}{C}HCl_2$ Two types of protons marked a and b. Hence two signals

(*iii*) $\overset{d}{C}H_3-\overset{c}{C}H=\overset{b}{C}H_2-\overset{a}{O}H$ Four types of protons marked a, b, c & d. Hence four signals will be obtained.

Q. 93. What is coupling constant? What are the factors that govern coupling constant?

(*Kerala, 2001; Nagpur 2002, Panjab, 2003*)

Ans. The distance between adjacent peaks is called *coupling constant* and denoted by J. Coupling constant gives a measure of the splitting effect. The values of J are independent of applied field strength and depend upon the molecular structure only. The unit of J is Hertz (Hz). Mutually coupled protons show the same value of J.

It is observed that J depends upon the number and type of bonds in between and the spatial relations between the protons. Following examples will illustrate.

(*i*) For the protons attached to same carbon *i.e.* **geminal protons,** J varies from 0-20 Hz depending upon the dihedral angle and the structure of the molecule.

$>C(H)(H)$ J = 0 – 20Hz

(*ii*) For protons attached to adjacent carbon atoms *i.e.* **vicinal protons,** J varies from 2-18 Hz depending upon the spatial position of protons and the structure of the molecule. Protons with anti-conformation have the value of J between 5 – 12 Hz and those with gauche conformation have J = 2 – 4 Hz.

H, H (anti) J = 5 – 12 Hz

H, H (gauche) J = 2 – 4 Hz

In case of vinylic protons, where there is a restricted rotation because of the double bond, the *cis* protons have J = 6 – 14 Hz whereas *trans* protons have J = 11 – 18 Hz.

H, H > C = C (Cis, $J = 6 - 14$ Hz) H, H > C = C (Trans, $J = 11 - 18$ Hz)

Q. 94. Discuss the NMR spectra of the following molecules: Ethyl bromide, 1, 1 dibromoethane and 1, 1, 2 tribromoethane. *(Kerala, 2001)*

Ans. (1) *Ethyl bromide* CH_3CH_2Br. The NMR spectra of ethyl bromide is represented as follows:

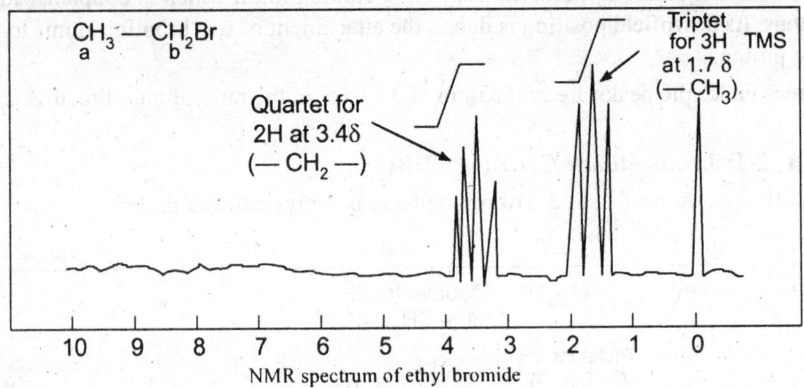

NMR spectrum of ethyl bromide

Fig. 28.15

The following peaks can be identified in the spectrum

(*a*) Triplet, δ 1.7, 3H

(*b*) Quartet, δ 3.4, 2H

The triplet at δ 1.7 is given by the three methyl protons which are magnetically equivalent and are coupled with the two methylene protons to give an upfield triplet.

The quartet at δ 3.4 is from the two equivalent methylene protons which are coupled with the three methyl protons to produce a downfield quartet as a result of deshielding influence of bromine.

The relative areas under the respective signals are in the ratio of the number of protons involved *i.e.* 3 : 2.

(2) 1, 1 Dibromoethane, CH_3–$CHBr_2$. NMR spectrum of 1, 1 Dibromoethane is represented as under.

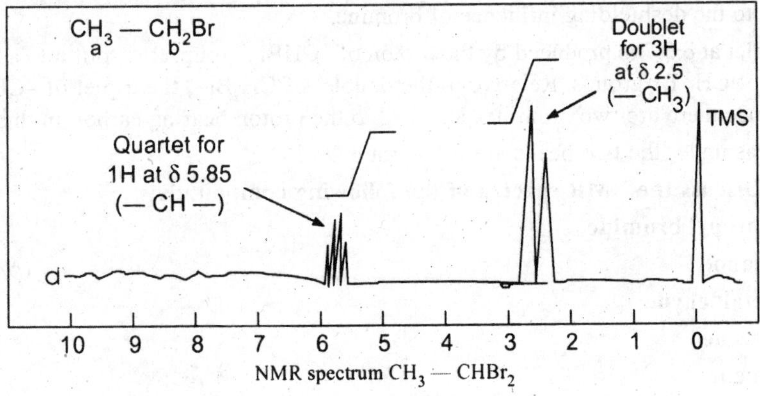

NMR spectrum CH_3 — $CHBr_2$

Fig. 28.16

The following peaks can be identified in the spectrum.

(*a*) Doublet, δ 2.5, 3H

(*b*) Quartet, δ 5.85, 1H

The doublet at δ 2.5 is due to the three equivalent methyl protons which are coupled with the methine (CH) protons. As compared with the signal from the methyl protons of $CH_3 - CH_2$ Br, the signal here is somewhat downfield because of the presence of two bromine atoms on adjacent carbons.

The downfield quartet at δ 5.85 is from the methine proton which is coupled with the three methyl protons. Its downfield position is due to the attachment of two bromine atoms to the carbon carrying the protons.

The areas under the peaks are in the ratio of 3 : 1 *i.e.* in the ratio of no. of methyl and methine protons.

(3) 1, 1, 2-Tribromoethane $CH_2Br - CHBr_2$

The NMR spectrum of 1, 1, 2-Tribromoethane is represented as under:

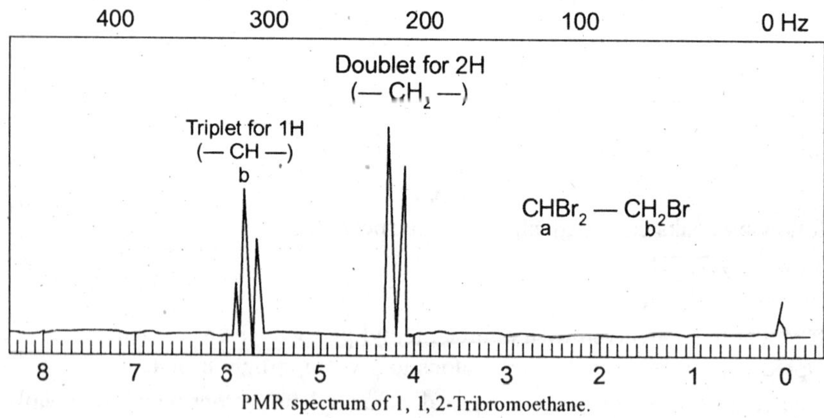

PMR spectrum of 1, 1, 2-Tribromoethane.

Fig. 28.17

The following peaks can be identified in the spectrum:

(*a*) Doublet, δ 4.2, 2H

(*b*) Triplet, δ 5.7, 1H

The doublet at δ 4.2 is due to two equivalent protons of CH_2 Br– group which is split into a doublet by the neighbouring proton of $-CHBr_2$ group. The downfield position of this doublet can be attributed to the deshielding influence of bromine.

The triplet at δ 5.7 is produced by the proton of $-CHBr_2$ group. It is split into a triplet by the two protons of CH_2 Br groups. Relative to the doublet of CH_2Br–, the triplet of $-CHBr_2$ is more downfield since there are two bromines attached to the proton bearing carbon in this case.

The areas under the two peaks are in the ratio 2 : 1.

Q. 95. Discuss the NMR spectra of the following compounds:

(*i*) Isopropyl bromide

(*ii*) Ethanol (*Nagpur 2002*)

(*iii*) Acetaldehyde

(*iv*) Benzene

(*v*) Toluene

(*vi*) *p*-nitrotoluene

Ans. (*i*) Isopropyl bromide, $CH_3 - CHBr - CH_3$

Spectrum of isopropyl bromide is represented as under.

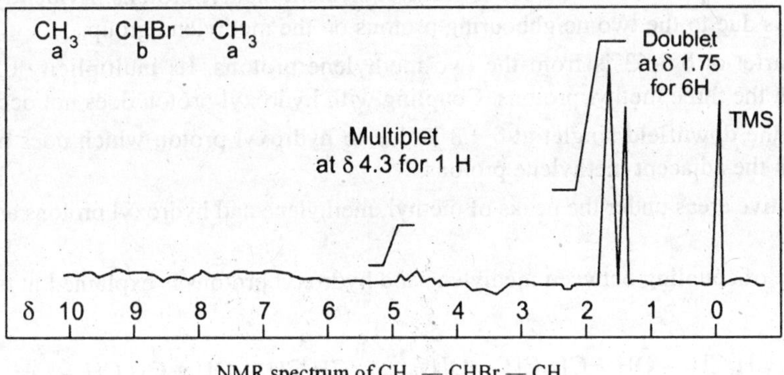

NMR spectrum of $CH_3 — CHBr — CH_3$

Fig. 28.18

Following peaks can be identified in the spectrum:

(*a*) Doublet, δ 1.75, 6H

(*b*) Multiplet, δ 4.3, 1H

It can be seen that the six protons on the two methyl groups are all equivalent and different from the proton of –CHBr– group. These six protons give rise to an upfield signal at δ 1.75 which is split into a doublet due to coupling with the lone proton –CHBr– group.

The proton of –CHBr– gives a downfield signal at 4.3 which is split into a multiplet by the six protons, on adjacent carbons.

The two peak areas are in the ratio 6 : 1.

(*ii*) Ethanol, CH_3–CH_2OH

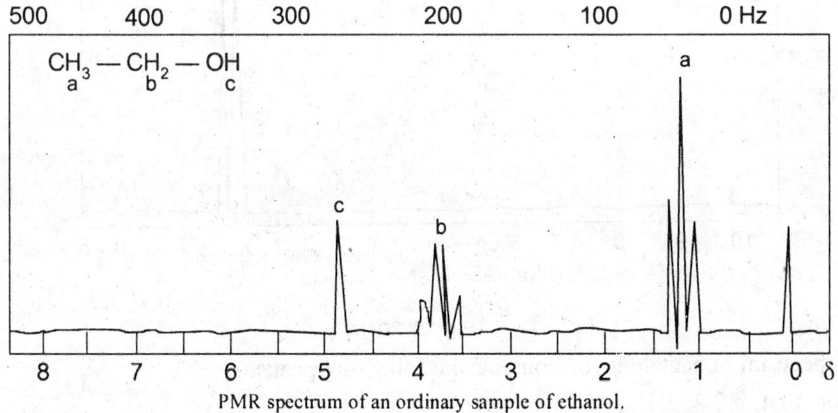

PMR spectrum of an ordinary sample of ethanol.

Fig. 28.19

Following figure shows the NMR spectrum of ordinary ethanol.

The spectrum of an ordinary sample of ethanol may be described as follows:

(*a*) Triplet, δ 1.2, 3H

(*b*) Quartet, δ 3.63, 2H

(*c*) Singlet, δ 4.8, 1H

Ethanol contains three kinds of protons and consequently exhibits three signals as shown above.

The upfield triplet of δ 1.2 is due to three equivalent methyl protons. Evidently its splitting into a triplet is due to the two neighbouring protons on the methylene group.

The quartet at δ 3.63 is from the two methylene protons. Its multiplicity of 4 is due to coupling with the three methyl protons. Coupling with hydroxyl proton does not occur.

Finally the downfield singlet at δ 4.8 is due to hydroxyl proton which does not show any coupling with the adjacent methylene protons.

The relative areas under the peaks of methyl, methylene and hydroxyl protons are in the ratio 3 : 2 : 1.

Absence of coupling between methylene and hydroxyl protons is explained in terms of rapid exchange.

$$\underset{\text{Molecule 1}}{CH_3CH_2 - \overset{*}{O}H} + \underset{\text{Molecule 2}}{CH_3CH_2 - OH} \qquad \underset{\text{Molecule 1}}{CH_3CH_2 - OH} + \underset{\text{Molecule 2}}{CH_3CH_2 - \overset{*}{O}H}$$

The hydroxyl protons undergoes exchange with another molecule in which the alignment of methylene protons is different from that in the first molecule and so on. In other words the hydroxyl proton does not stay in the same environment long enough for its coupling with the methylene protons to be recorded.

NMR spectrum of pure ethanol shows the signal from hydroxyl proton split into triplet and signal from methylene protons split into octet. This is because coupling between hydroxyl and methylene protons can be recorded now as the exchange process slows down.

(iii) Acetaldehyde, CH_3–CHO. NMR spectrum of acetaldehyde is reproduced as under.

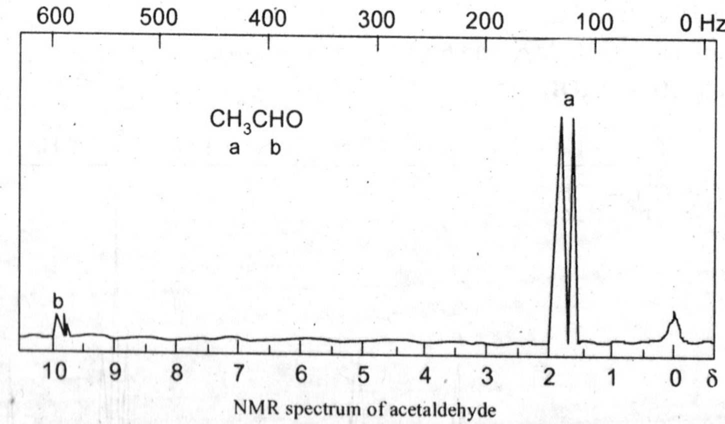

NMR spectrum of acetaldehyde

Fig. 28.20

The spectrum of acetaldehyde contains the following peaks:

(*a*) Doublet, δ 2.2, 3H

(*b*) Quartet, δ 9.8, 1H

The doublet at δ 2.2 is due to three equivalent protons of methyl group coupled with a single proton of aldehyde group.

The aldehydic proton absorbs far downfield at δ 9.8. The signal appears in the form of a quartet due to coupling by the three protons of the neighbouring methyl groups.

Peaks areas are in the ratio 3 : 1.

***(iv)* Benzene.** NMR spectrum of benzene is reproduced as under.

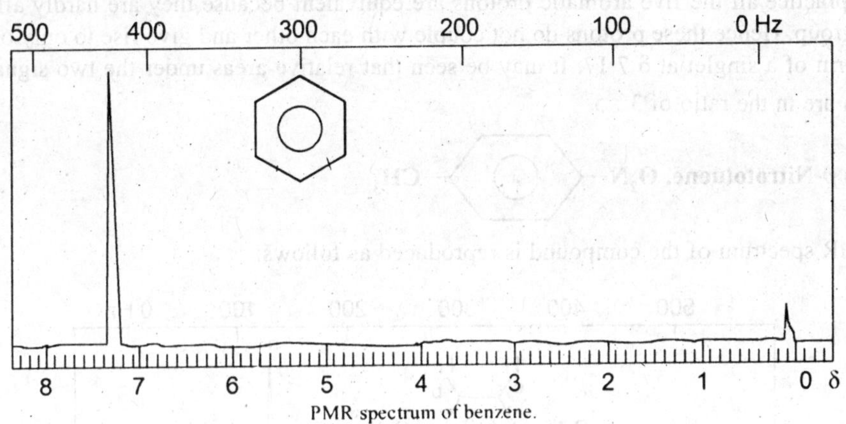

PMR spectrum of benzene.

Fig. 28.21

The NMR spectrum of benzene exhibits only a sharp singlet at δ 7.37 from the six equivalent protons.

Such a downfield signal can be explained in terms of π electron system of benzene which is considerably polarisable. Therefore, the application of a magnetic field induces a flow of these electrons around the ring. This ring current in turn generates an induced magnetic field which reinforces the applied field. As a result the benzene protons get deshielded and the absorption moves downfield.

***(v)* Toluene.** C_6H_5—CH_3

NMR spectrum of toluene is reproduced as under.

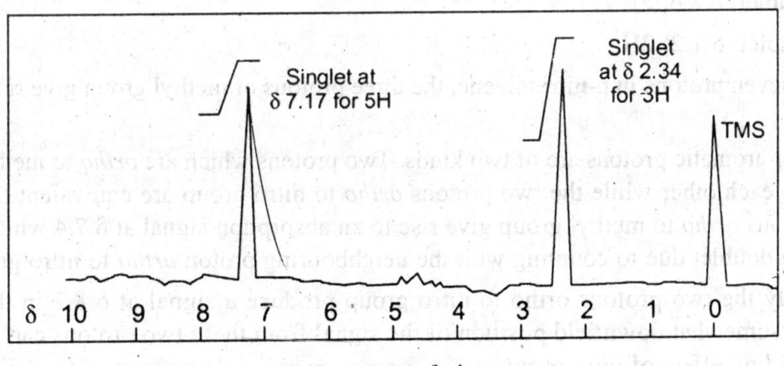

NMR spectrum of toluene.

Fig. 28. 22

The following peaks are observed in the NMR spectrum of toluene:

(*a*) Singlet, δ 2.34, 3H

(*b*) Singlet, δ 7.17, 5H

Toluene has eight protons, five of which are aromatic and remaining three from methyl group.

The signal for three protons of methyl group which is joined to an aromatic ring appears as a singlet at δ 2.34.

In practice all the five aromatic protons are equivalent because they are hardly affected by methyl group. Hence these protons do not couple with each other and give rise to only one signal in the form of a singlet at δ 7.17. It may be seen that relative areas under the two signals in the spectrum are in the ratio of 3 : 5.

(*vi*) *p*-Nitrotoluene, $O_2N-C_6H_4-CH_3$

NMR spectrum of the compound is reproduced as follows:

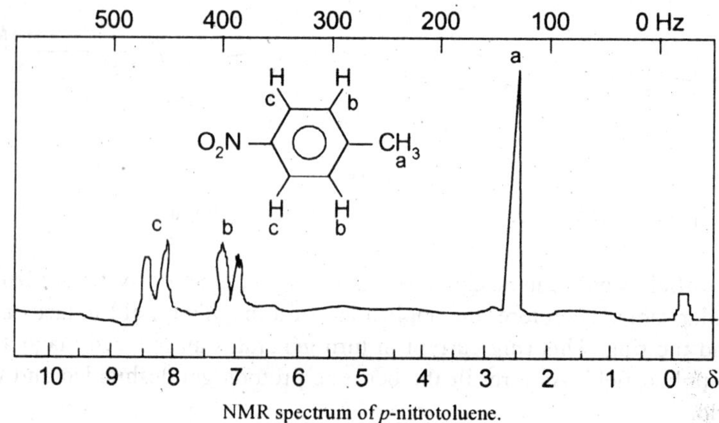

NMR spectrum of *p*-nitrotoluene.

Fig. 28.23

The spectrum of *p*-nitrotoluene has the following peaks:

(*a*) Singlet, δ 2.4, 3H

(*b*) Doublet, δ 8.4, 2H

(*c*) Doublet, δ 1.2, 2H

Of the seven protons in p-nitrotoluene, the three protons of methyl group give rise to a singlet at δ 2.4.

The four aromatic protons are of two kinds. Two protons which are *ortho* to methyl group are equivalent to each other while the two protons *ortho* to nitro group are equivaient to each other. The two protons *ortho* to methyl group give rise to an absorption signal at δ 7.4 which appears in the form of a doublet due to coupling with the neighbouring proton *ortho* to nitro group.

Similarly the two protons ortho to nitro group produce a signal at δ 8.2 in the form of a doublet. The somewhat downfield position of the signal from these two protons can be attributed to the deshielding effect of nitro group.

The peak areas underneath the various signals are, as expected, in the ratio of 3 : 2 : 2.

Q. 96. Discuss the NMR spectrum of benzaldehyde.

Ans. The NMR spectrum of benzaldehyde is shown as under.

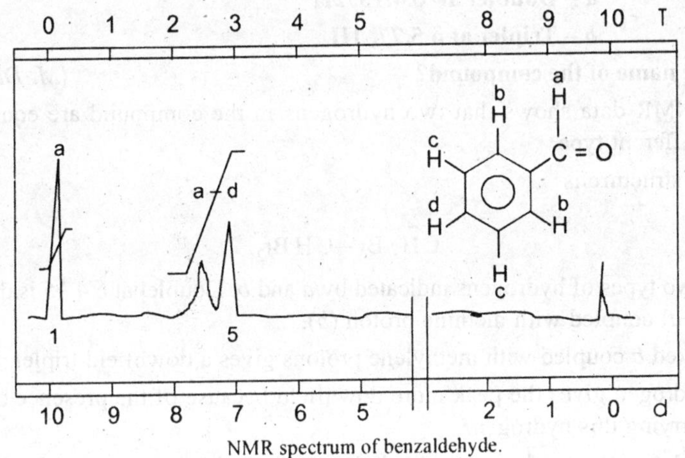

NMR spectrum of benzaldehyde.

Fig. 28.24

In this spectrum, there are two peaks.

(*i*) A singlet due to aldehydic proton is obtained near δ 10 because of lack of coupling.

(*ii*) A doublet at δ 7.5 is obtained due to five hydrogens of benzene ring. Splitting takes place due to aldehydic proton.

MISCELLANEOUS QUESTIONS

Q. 97. A compound has the molecular formula $C_7H_6O_2$. It gives an infrared absorption bond at 1771 cm^{-1}. On treatment with $LiAlH_4$, it gets converted into compound B which shows characteristic infrared absorption bands at 3330 cm^{-1} and 1050 cm^{-1}. Assign structures to compound A and B.

Ans. $LiAlH_4$ is a reagent which reduces aldehydes, ketones and carboxy acids into alcohols. The formula $C_7 H_6 O_2$ rules out the possibility of an aldehyde or ketone which contain one oxygen. The compound appears to be benzoic acid with molecular formula $C_7H_6O_2$.

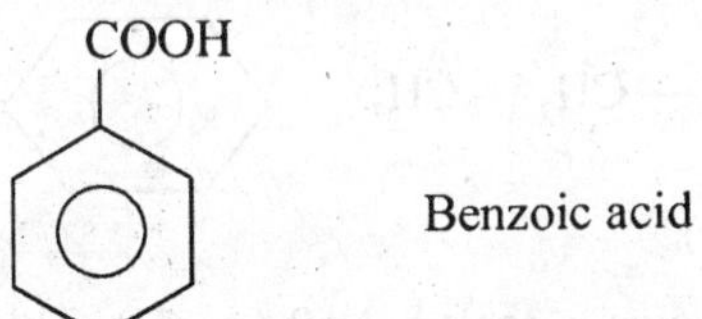

Benzoic acid

An absorption band at 1771 cm^{-1} support this assumption. On treatment with $LiAH_4$, benzoic acid is reduced to benzyl alcohol.

$$C_6H_5COOH \xrightarrow[{[H]}]{LIAlH_4} C_6H_5CH_2OH$$

Absorption bands at 3330 cm^{-1} is due to aromatic ring and at 1050 cm^{-1} due to alcoholic group.

Q. 98. Give a structure consistent with the following data:

Molecular formula – $C_2H_3Br_2$

a – Doublet at δ 4.15, 2H

b – Triplet at δ 5.77, 1H

What is the name of the compound? (*M. Dayanand, 2000*)

Ans. The NMR data shows that two hydrogens in the compound are equivalent and one hydrogen is of different type.

A probable structure is

$$\overset{a}{C}H_2Br-\overset{b}{C}HBr_2$$

There are two types of hydrogens indicated by a and b. Doublet at δ 4.15 is due to methylene protons (marked a) coupled with methine proton (b).

Proton marked b coupled with methylene protons gives a downfield triplet.

Methine hydrogen gives the peak more downfield because of the presence of two bromines on the carbon carrying this hydrogen.

Q. 99. PMR spectrum of a compound shows the following peaks δ 7.22 (*s*, 5H), δ 3.59 (*s*, 2H), δ 2.77 (*q*, 2H), δ 0.97 (*t*, 3H).

In the IR spectrum there is a strong absorption at 1705 cm^{-1}. Giving reasons, find out which of the following structures is in keeping with the above data.

(*a*) $CH_3-C_6H_4-CH_2OCH_3$ (*b*) $C_6H_5-CH_2-\overset{O}{\overset{\|}{C}}-CH_2CH_3$

(*c*) $C_6H_5-CH_2-CH_2-CH_2OH$ (*d*) $C_6H_5-\overset{O}{\overset{\|}{C}}-CH_2CH_2CH_3$

Ans. (*i*) IR absorption band is obtained at 1705 cm^{-1}. This is characteristic of a carbonyl group. Hence structures (*a*) and (*c*) are ruled out as they don't contain the carbonyl group.

The probable structure is out of (*b*) or (*d*)

$$C_6H_5-\overset{4}{C}H_2-\overset{O}{\overset{\|}{\overset{3}{C}}}-\overset{2}{C}H_2-\overset{1}{C}H_3 \qquad C_6H_5-\overset{O}{\overset{\|}{C}}-\overset{3}{C}H_2\overset{2}{C}H_2-\overset{1}{C}H_3$$

(*b*) (*d*)

(*ii*) The PMR spectrum shows a peak δ 3.59 (*s*, 2H). This is a singlet in respect of two hydrogens. We don't expect to observe it in structure (*d*) because protons on carbon 1 will give a quartet, protons on carbon 2 will give a sixlet and protons on carbon 3 will give a triplet. Hence structure (*d*) is ruled out.

(*iii*) The only structure left is (*b*). Protons on carbon 2 will give a singlet. There will be no splitting of peak as there is no hydrogen on the neighbouring carbons. Hence structure (*b*) is confirmed.

Q. 100. An organic compound having the molecular formula C_4H_8O gives a characteristics band at 275 nm (ϵ_{max} 17) in its UV spectrum. Its infrared spectrum exhibits two important peaks at 2940 – 2855 cm^{-1} and 1715 cm^{-1}. PMR spectrum of the compound is as follows:

δ 2.5 (*q*, 2H), δ 2.12 (*s*, 3H) and δ 1.07 (*t*, 3H). Assign a structural formula to the compound. (*Shivaji, 2000*)

Ans. (*a*) The compound shows a UV absorption at 275 nm. This absorption is characteristic of carbonyl group.

(*b*) IR absorption peaks at 2940 – 2855 cm^{-1} and 1715 cm^{-1} are due to C–H stretching of methyl groups and a saturated ketonic group respectively.

(*c*) With the formula C_4H_8O, there are two possibilities.

$$\underset{\text{I}}{CH_3CH_2CH_2CHO} \qquad\qquad \underset{\text{II}}{CH_3COCH_2CH_3}$$

(*d*) Formula I is ruled out as it contains four types of protons and would give four signals which is not in agreement with the data.

(*e*) The structure left is $\overset{4}{C}H_3\,\overset{3}{C}O\overset{2}{C}H_2\,\overset{1}{C}H_3$ There are two protons of one type, three of second and three of third type and will give rise to three signals. C_2 protons will give a quartet δ 2.5, C_4 protons will give singlet, δ 2.12 and C_1 positions will give a triplet. Observations made with structure II agree with the data.

MASS SPECTROSCOPY

Q. 101. What is mass spectroscopy? What is the principle of mass spectroscopy?

(*Guwahati 2002; Mumbai, 2004*)

Ans. Mass spectroscopy involves the bombardment of molecules of a substance by energetic electrons and analyses the charged particles and fragments. The instrument used for this purpose is called *mass spectrometer.*

Principle of mass spectroscopy

In mass spectroscopy, the molecules of a substance are bombarded with medium energy electrons. The energetic electrons knock out generally one most loosely bound electron in the outermost orbit. In rare cases, two electrons may be removed. This process produces cations or molecular ions. The molecular ion being energetic is fragmented further in steps to produce smaller positive ions called **daughter ions.** Each of the daughter ions has certain *m/e* ratio *i.e.* ratio of mass to charge of the ions. The sequence of changes that take place in the case of ammonia molecule, is shown as under:

$$NH_3 + e^- \xrightarrow{-2e} \underset{\substack{m/e\,=\,17\\ \text{molecular}\\ \text{ion}}}{[NH_3]^+} \xrightarrow{-H} \underset{16}{[NH_2]^+} \xrightarrow{-H} \underset{15}{[NH]^+} \xrightarrow{-H} \underset{14}{[N]^+}$$

Daughter ions

These molecular and daughter ions having different *m/e* values are passed through magnetic and electric fields and are separated from one another. A signal is obtained corresponding to each *m/e* value. The intensity of a signal is proportional to the number of ions giving rise to that signal. A graph between the intensity and *m/e* values of the ions is called *mass spectrum.*

The most intense peak is referred to as **base peak** and its intensity is taken as 100. The intensities of other ions are measured with respect to the base peak. In the mass spectrum analysis of NH_3 , the following results are obtained, which are expressed in graphical and tabular form as given below:

In most cases the base peak corresponds to molecular ion (molecule minus one electron). But in some cases the base **peak** may correspond to other cations. The main job is to locate and identify the base peak. The small peak at *m/e* 18 is due to the isotope ^{15}N in the molecule of ammonia $^{15}NH_3$.

m/e	Intensity (% of base peak)
14	2.2
15	7.5
16	80.0
17	100.0 (Base peak)
18	0.4

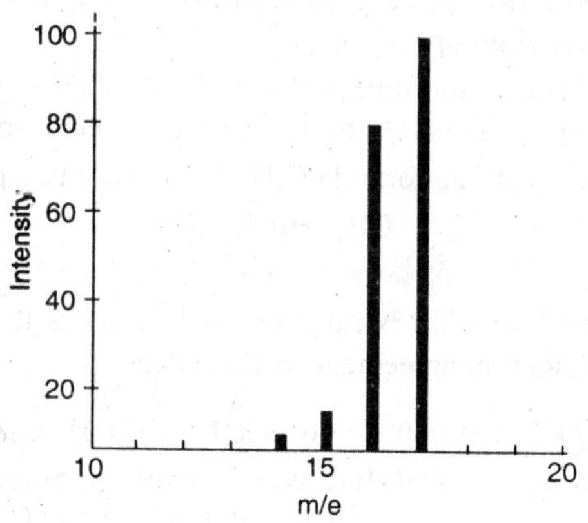

Q. 102. Describe the applications of mass-spectroscopy.

Ans. 1. Identification of substances. Mass spectrum is highly characteristic of a substance. The mass spectra of two identical samples superimpose exactly on each other. By comparing the spectra of an unknown substance with those of known samples, it is possible to identify the unknown samples.

2. Determination of molecular mass and formula. Molecular mass of a compound is given *m/e* value of the molecular ion in the mass spectrum

$$M + e^- \longrightarrow \underset{\text{Molecular ion}}{M^+} + 2e^-$$

All that has to be done in interpreting the mass spectrum is to identify the M^+ peak (In most cases, it is the base peak). The molecular mass obtained in this way is the mass corresponding to common isotopes of the elements.

Once the molecular mass has been obtained, it is not difficult to arrive at the molecular formula.

3. Determination of molecular structure. Mass spectrum gives us various values of *m/e* which correspond to various fragments of the molecular ion. By establishing the nature of different fragments, we can known about the structure of the compound by joining those fragments together.

Q. 103. Write down the different fragments with their *m/e* values obtained in the mass spectrum of ethanol.

Ans.

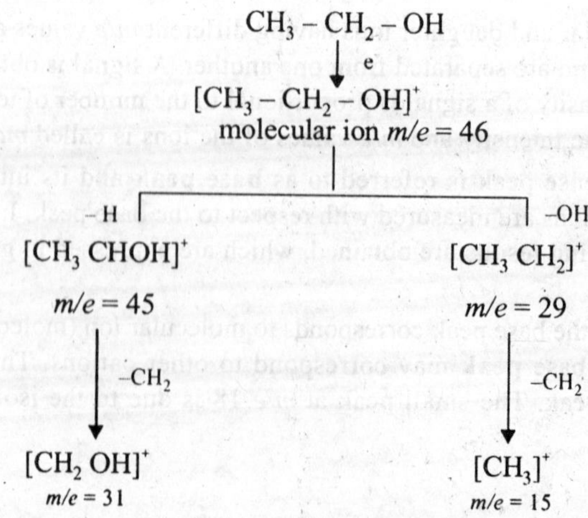

Q. 104. Suggest molecular formulae for the fragment ions obtained from isobutane *m/e* 43, 28, 27.

Ans. The formula of isobutane is

$$\underset{\displaystyle CH_3}{CH_3 - \overset{}{CH} - CH_3} \quad \text{Mol. wt.} = 58$$

(*i*) Difference between the mol. weights of isobutane and the first fragment is 58–43 = 15. It is equivalent to $-CH_3$. Hence

$$\underset{m/e = 58}{\left[\begin{array}{c} CH_3 - CH - CH_3 \\ | \\ CH_3 \end{array}\right]^+} \xrightarrow{-CH_3} \underset{m/e = 43}{[CH_3 - CH - CH_3]^+}$$

(*ii*) Difference between the molecular weights of the first and second fragments is 43–28 = 15. Again, it appears that $-CH_3$ is lost.

$$\underset{m/e = 43}{[CH_3 - CH - CH_3]^+} \xrightarrow{-CH_3} \underset{m/e = 28}{[CH_3 - CH]^+}$$

(*iii*) The difference between the masses of the second & third fragments is 28–27 = 1. Thus one hydrogen is lost.

$$\underset{m/e = 28}{[CH_3 - CH]^+} \xrightarrow{-H} \underset{m/e = 27}{[CH_2 - CH]^+}$$

Q. 105. Discuss the fragmentation pattern of 2-pentene.

Ans. Mass spectrum of 2-pentene shows *m/e* peaks at 70, 55, 41, 39, 29 and 27. The formation of signals is explained as under:

$$\underset{\text{2-Pentene}}{CH_3 - CH_2 - CH = CH - CH_3} \xrightarrow{+e} \underset{\overset{+}{M}\ (m/e\ 70)}{[CH_3 - CH_2 - CH = CH - CH_3]^+}$$

$$[CH_3 - CH_2 - CH = CH - CH_3]^+ \longrightarrow$$

$$\xrightarrow{-CH_3} \underset{(m/e\ 55)}{[CH_2CH = CH - CH_3]}$$

$$\xrightarrow{-[CH_3 - CH = CH]} \underset{(m/e = 29)}{[CH_3 - CH_2]^+} \xrightarrow{+e} \underset{(m/e = 27)}{[CH_2 = CH]^+}$$

$$\longrightarrow \underset{(m/e = 41)}{[CH_3 - CH = CH]^+}$$

Q. 106. The mass spectrum of hydrocarbon shows as abundant molecular ion at *m/e* = 120. IR spectrum of the compound shows that it is aromatic in nature. PMR spectrum exhibits three signals at δ 1.2 (*d*, 6H), δ 2.8 (*m*, 1H) and δ 7.2 (*s*, 5H). Work out a structure for the compound.

Ans. (*a*) The molecular ion has the value *m/e* = 120, it means the molecular mass of the compound is 120.

(*b*) The molecule contains a benzene ring. The mass of the side-chain/s can be calculated by substracting 77 (mass of phenyl ring) from 120. It comes to 43. The side chain thus could be $-C_3H_7$. The possible structures are:

$C_6H_5-CH_2-CH_2-CH_3$
I

$C_6H_5-CH(CH_3)-CH_3$
II

Also there is a possibility of the compound being disubstituted benzene. The possible structures are

$CH_3-C_6H_4-CH_2CH_3$
o, m or *para*
(III)

(*c*) Structures (I) and (III) are ruled out because, each one of them contains four types of protons and thus would give four signals which is inconsistent with the data.

(*d*) The only structure left is (II)

$C_6H_5-\overset{b}{CH}(\overset{a}{CH_3})-\overset{a}{CH_3}$

This structure has 6 protons of one type (methyl protons), 1 proton designated *b* of second type and 5 ring protons of third type.

(*e*) Six methyl protons give a doublet δ 1.2 due to splitting by protons *b*, one proton designated *b* gives a multiplet δ 2.8 due to splitting by methyl protons and 5 ring protons give a singlet δ 7.2.

All the observations made with structure (II) agree with the data. Hence this is the correct structure.

Q. 107. A compound containing C, H and O exhibits molecular ion at *m/e* = 7. Its IR spectrum shows a strong band at 1715 cm^{-1}. PMR spectrum shows a triplet, a singlet and a quartet (at increasing values of δ) in the ratio 3 : 3 : 2. What is the structure of the compound?

Ans. (*a*) Since the mass spectrum of the compound exhibits molecular ion at *m/e* = 72, the compound has a molecular weight = 72.

(*b*) The IR spectrum shows an absorption peak at 1715 cm^{-1}. This is characteristic of a carbonyl compound. It could be an aldehyde or ketone. Possible structures with molecular mass 72 are as follows:

$CH_3CH_2CH_2\,CHO$ (I) $\quad (CH_3)_2CHCHO$ (II) $\quad CH_3COCH_2CH_3$ (III)

(*c*) Structure I contains four types of protons. It would therefore give four PMR signals which is inconsistent with the data. Hence structure I is ruled out.

(*d*) We would expect a doublet, a multiplet and again a doublet from structure (II). Hence this structure is also ruled out.

(*e*) Structure III is the only structure left

$$\overset{1}{C}H_3\overset{2}{C}O\overset{3}{C}H_2\overset{4}{C}H_3$$

Three protons on carbon 4 will give a triplet due to splitting by protons of carbon 3. Singlet will be given by three protons on carbon 1. There will be no splitting here. Again two hydrogens on carbon 3 will give quartet due to interaction with the protons of carbon 4. Thus the ratio of various peak intensities is 3 : 3 : 2 *i.e.* in the ratio of hydrogen atoms forming the peaks. If we consult the table of PMR chemical shift, we find the values of δ in increasing order. Thus these observations are in conformity with the data. Therefore the correct structure is (III).

We obtain three peaks.

Q. 108. A hydrocarbon show a mass parent peak at *m/e* 102. It exhibits NMR signals δ 7.4 (5H, singlet) and δ 3.08 (1H, singlet). What is the probable structure of the hydrocarbon?

Ans. (*i*) Mass parent peak at *m/e* = 102 shows that the molecular mass of the compound is 102.

(*ii*) NMR signals show that there are two types of protons, 5 protons of one type and 1 proton of another type. Five protons are possibly the protons of monosubstituted benzene ring.

(*iii*) The size of the side-chain can be obtained as follows :

102	–	77	=	25
(Mol mass of compound)		(Mass of $-C_6H_5$)		Mol. mass of side-chain

A value of 25 for mass of side-chain suggests it to be $-C\equiv CH$

This is verified from NMR spectrum which tells that the second singlet is due to one hydrogen only. Thus, the structure of the hydrocarbon is

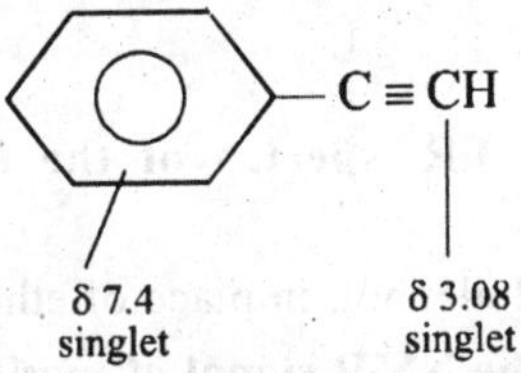

$C_6H_5-C\equiv CH$ (Phenyl acetylene)

δ 7.4 singlet (ring H); δ 3.08 singlet (≡CH)

Q. 109. An aromatic hydrocarbon C_8H_{10} shows two singlets in its PMR at δ 2.3 and 7.05 with proton counting in the ratio of 3 : 2. Giving reasons assign structure to it.

Ans. (*a*) There are two possible hydrocarbons with the formula C_8H_{10} as given below:

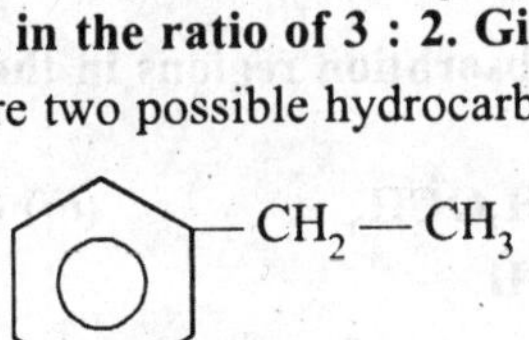

I: $C_6H_5-CH_2-CH_3$

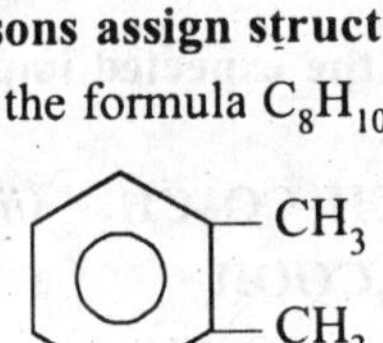

II: $C_6H_4(CH_3)_2$ (two adjacent CH_3 groups on the ring)

Structure I would give three peaks as there are three kinds of protons in it. Structure II would give two signals as there are two kinds of protons. It may be noted that six protons of two methyl groups are equivalent. As the compound under examination gives two peaks, it can be only II.

(*b*) The peak at δ 2.3 corresponds to methyl protons and that at δ 7.05 corresponds to ring hydrogens.

(*c*) There are six methyl protons and four ring protons. Thus, they are in the ratio 6 : 4 or 3 : 2. This ratio agrees with the given value.

Thus, the structure of the hydrocarbon is

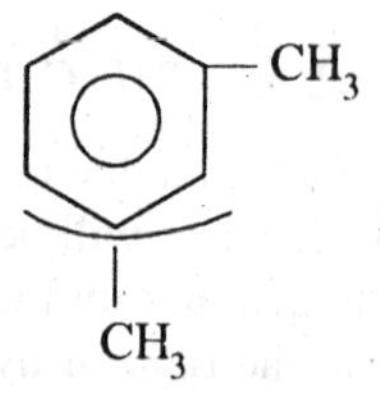

o-, *m*- or *p*-xylene

Q. 110. How would you differentiate between CH_3COCH_3 and $CH_3\ COCH = CH_2$ using IR and UV spectra?

Ans. Using IR.

Both the compounds would give an absorption band at around 1700 cm^{-1} due to the carbonyl group. $CH_3COCH = CH_2$ would give an additional band between 3000–3100 cm^{-1} due to the double bond.

Using UV

The compound $CH_3 - \overset{O}{\overset{||}{C}} - CH_3$ has no conjugated double bond, hence, it gives an absorption band below 200 nm.

The compound $CH_3 - \overset{O}{\overset{||}{C}} - CH = CH_2$ has conjugated double bonds in the molecule. Such compounds give an absorption band above 200 nm.

This is how the two compounds can be differentiated.

Q. 111. UV spectroscopy is suitable for the identification of conjugated system of double bonds.

Ans. See. Q. 8.

Q. 112. What is the difference in the I.R. spectra of the following pairs of the compounds : methyl alcohol and acetic acid. (*Himachal , 2000*)

Ans. See Q. 40 (b) and Q. 54. (take methyl alcohol, in place of ethanol).

Q. 113. Would there be any splitting in the NMR signal of mesitylene ?

Ans. See Q. 77

Q. 114. Indicate the expected important absorbtion regions in the IR spectrum of the following compounds.

(*i*) $CH_2 = CH–CO–CH_3$ (*ii*) $CH_3\ CH_2OCH_3$ (*iii*) $CH_3\ COOC_2H_5$

(*iv*) CH_3CH_2COOH (*v*) $C_6H_5NH_2$

Ans. (*i*) $CH_2 = CH–CO–CH_3$ ———— 1675 cm^{-1}

(*ii*) $CH_3–CH_2.O.CH_3$ ———— 2950 cm^{-1}

(*iii*) $CH_3COOC_2H_5$ ———— 1680–1760 cm^{-1}

(*iv*) CH_3CH_2COOH ———— 1300 cm^{-1}

(*v*) $C_6H_5NH_2$ ———— 3000–31000 cm^{-1}, 3300–3500 cm^{-1}

Q. 115. Calculate λ_{max} for the following compounds (*Sihvaji, 2000*)

(*i*) (*ii*)

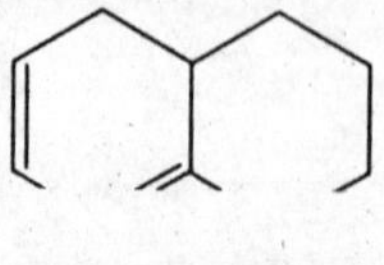

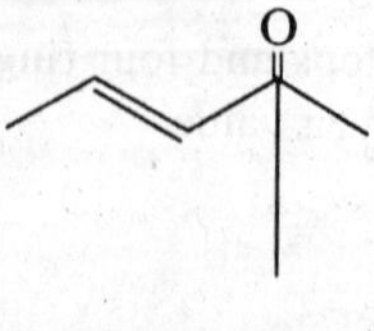

Ans. (*i*) λ_{max} = 217nm and 203.5 nm (*ii*) λ_{max} = 224 nm and 314 nm.

Q. 116. Will there be any spin-spin splitting in the spectrum of molecules?

(*M. Dayanand, 2000*)

(*i*) $ClCH_2 - CH_2Cl$ (*ii*) $CH_3 - CCl_2 - CH_2Cl$

Ans. See Q. 75, in place of $CH_3 - C(CH_3)_2 - CH_2Cl$, take $CH_3 - CCl_2 - CH_2Cl$

Q. 117. A compound was believed to be either diphenyl ether or diphenyl methane could PMR spectrum be used to distinguish between these two compounds How?

(*M. Dayanand, 2000*)

Ans. Diphenyl ether does not show spin-spin splitting while diphenyl methane shows.

diphenyl ether diphenyl methane

Q. 118. Ethylene falis to absorb at ~1600 cm^{-1} in its IR spectrum. Explain.

Ans. It is because, it contains a terminal hydrogen atom linked to carbn. (*Calcutta, 2000*)

Q. 119. How will you distinguish between the three dibromobenzens by their NMR spectra.

(*Visva Bharti, 2003*)

Ans. *o*-Dibromobenzene will show *two* peaks : *m*-dibromobenzene will show three peaks: *p*-dibromobenzene will show only *one* peak.

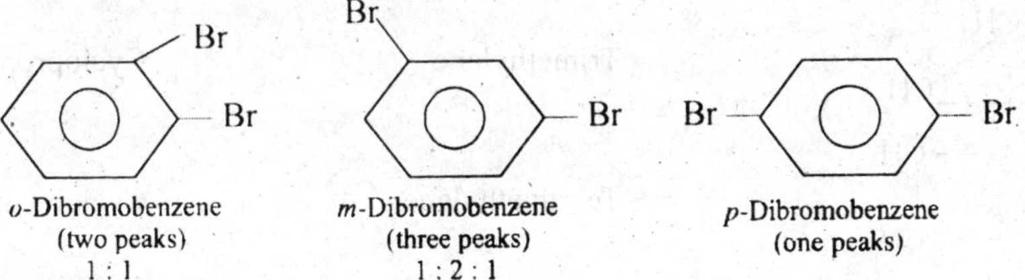

o-Dibromobenzene (two peaks) 1 : 1 *m*-Dibromobenzene (three peaks) 1 : 2 : 1 *p*-Dibromobenzene (one peaks)

Q. 120. Suggest a structure consistent with the following NMR data : Molecular formula = C_9H_{12}

(*a*) Singlet at δ 6.78, 3H **(*b*) singlet at δ 2.25, 9H** (*Mysore, 2004*)

Ans. The compound is

Mesitylene

Q. 121. How would you distinguish between the following pairs of compounds by NMR spectroscopy.

(*a*) $CH_3 - \overset{O}{\overset{\|}{C}} - CH_3$ and $CH_3CH_2 - \overset{O}{\overset{\|}{C}} - H$

(*b*) $CH_3 - \overset{O}{\overset{\|}{C}} - CH_3$ and $CH_3 - \overset{O}{\overset{\|}{C}} - OCH_3$ (*Punjab, 2005*)

Ans. (*a*) CH_3COCH_3 will give only *one* signal while CH_3CH_2CHO will give *three* signals.

(*b*) CH_3COCH_3 will give only *one* signal while CH_3COOCH will give *two* signals.

30

CYCLOALKANES

Q. 1. What are cycloalkanes? Give their nomenclature? (*Bhavnagar, 2004*)

Ans. Cycloalkanes are cyclic compounds containing closed hydrocarbon chains. They are also called *alicyclic* compounds and resemble alkanes in many respects. Examples of cycloalkanes are cyclopropane, cyclobutane etc. They can be expressed by a general formula $(CH_2)_n$ where n is 3, 4, 5...

Nomenclature. Common and IUPAC systems of nomenclature of cycloalkanes are given below.

Compound	Common name	IUPAC name
CH_2 CH_2—CH_2	Trimethylene	Cyclopropane
CH_2—CH_2 | | CH_2—CH_2	Tetramethylene	Cyclobutane
CH_2 CH_2 CH_2 | | CH_2—CH_2	Pentamethylene	Cyclopentane
CH_2 CH_2 CH_2 | | CH_2 CH_2 CH_2	Hexamethylene	Cyclohexane

For the sake of simplicity, cyclopropane, cyclobutane, cyclopentane and cyclohexane are represented by the following geometrical figures.

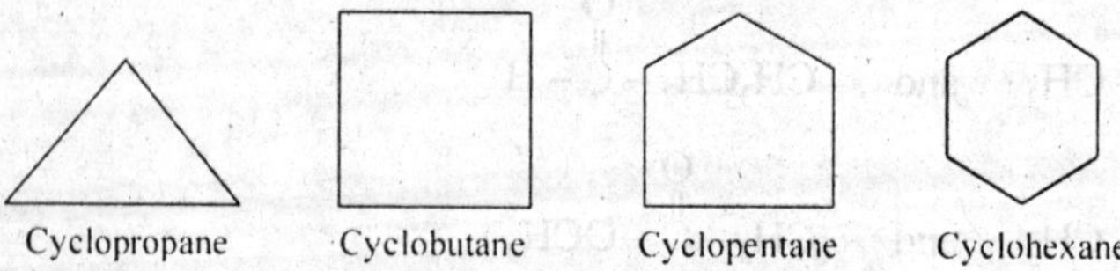

Q. 2. Give two methods of preparation of cycloalkanes. (*Kerala, 2000; Guwahati 2006*)

Ans. 1. By dehalogenation of ω-dihalogen derivatives of alkanes (Freund's method).

A dihalogen compound containing halogens on terminal positions, on treatment with Zn, in the presence of NaI catalyst gets dehalogenated. The terminal halogen atoms are removed thereby closing the ring. The starting material to be taken depends upon the desired cycloalkane, as explained below.

$$H_2C\langle^{CH_2Br}_{CH_2Br} + Zn \xrightarrow[\text{Catalyst}]{\text{NaI}} H_2C\langle^{CH_2}_{CH_2} + ZnBr_2$$

1, 3-dibromopropane → Cyclopropane

$$\begin{matrix} CH_2-CH_2Br \\ | \\ CH_2-CH_2Br \end{matrix} + Zn \xrightarrow[\text{Catalyst}]{\text{NaI}} \begin{matrix} CH_2-CH_2 \\ | \quad\quad | \\ CH_2-CH_2 \end{matrix} + ZnBr_2$$

1, 4-dibromobutane → Cyclobutane

$$CH_2\langle^{CH_2-CH_2Br}_{CH_2-CH_2}\rangle CH_2Br \xrightarrow{\text{Zn}} CH_2\langle^{CH_2-CH_2}_{CH_2-CH_2}\rangle CH_2 + Br_2$$

1, 6-Dibromohexane → Cyclohexane

2. Distillation of calcium or barium salts of dicarboxylic acids followed by reduction (Wislicenus method). A cyclic ketone results when a calcium or barium salt of a dicarboxylic acid is distilled. Ketone obtained is subjected to Clemmenson reduction, giving the cycloalkane as illustrated by the following reaction.

$$\begin{matrix} CH_2-CH_2COO \\ | \\ CH_2-CH_2COO \end{matrix}\rangle Ca \xrightarrow[-\,CaCO_3]{\text{Distil}} \begin{matrix} CH_2-CH_2 \\ | \\ CH_2-CH_2 \end{matrix}\rangle CO \xrightarrow{\text{Zn / Hg} \quad \text{HCl}} \begin{matrix} CH_2-CH_2 \\ | \\ CH_2-CH_2 \end{matrix}\rangle CH_2$$

Calcium adipate → Cyclopentanone → Cyclopentane

3. Reduction of aromatic compounds. Cyclohexane and its derivatives can be prepared by passing hydrogen gas through benzene or its derivatives, in the presence of Ni at 423-473 K temperature.

$$\text{Benzene} + 3H_2 \xrightarrow[\text{423-473 K}]{\text{Ni}} \text{Cyclohexane}$$

4. Diels-Alder addition. Two unsaturated compounds can be combined to produce a cycloalkane. When a mixture of 1, 3, butadiene and ethylene is heated to 475 K, cyclohexene is obtain. It can be reduced with hydrogen to yield cyclohexane.

$$CH_2=CH-CH=CH_2 + CH_2=CH_2 \xrightarrow[\Delta]{475\ K} \text{Cyclohexene} \xrightarrow[Ni]{H_2} \text{Cyclohexane}$$

1, 3-butadiene Cyclohexene Cyclohexane

5. Addition of carbenes to alkenes. Cyclopropane derivatives can be prepared by treating an alkene with diazomethane. Methylene group in the form of carbene : CH_2 is attached across the double bond in the alkene molecule.

$$\underset{\text{Butene - 2}}{CH_3-CH=CH-CH_3} + CH_2N_2 \xrightarrow{\text{Light}} CH_3-\underset{\diagdown\ CH_2\ \diagup}{CH-CH}-CH_3$$

1, 2 Dimethylcyclopropane

$$\underset{\text{Butene}}{CH_3-CH_2-CH=CH_2} + CH_2N_2 \xrightarrow{\text{Light}} CH_3-CH_2-\underset{\diagdown\ CH_2\ \diagup}{CH-CH_2}$$

Ethylcyclopropane

6. Dieckmann reaction. Cyclopentane can be obtained when an ester of adipic acid is treated with sodium or sodium ethoxide. Intramolecular change produces cyclic ketone. Ketone is reduced to obtain cyclopentane.

$$\underset{\text{Deithyl adipate}}{\begin{matrix} CH_2-CH_2COOC_2H_5 \\ | \\ CH_2-CH_2COOC_2H_5 \end{matrix}} \xrightarrow[C_2H_5OH]{Na} \underset{\text{2-Carbethoxycyclopentanone}}{\begin{matrix} CH_2-CH_2 \\ | \\ CH_2-CH \\ | \\ COOC_2H_5 \end{matrix} \!\!> CO} \xrightarrow[-C_2H_5OH]{H_2O/H^+}$$

$$\underset{\text{2-Carboxycyclopentanone}}{\begin{matrix} CH_2-CH_2 \\ | \\ CH_2-CH \\ | \\ COOH \end{matrix} \!\!> CO} \xrightarrow{\Delta} \underset{\text{Cyclopentanone}}{\begin{matrix} CH_2-CH_2 \\ | \\ CH_2-CH_2 \end{matrix} \!\!> CO} \xrightarrow[HCl]{Zn / Hg} \underset{\text{Cyclopantane}}{\begin{matrix} CH_2-CH_2 \\ | \\ CH_2-CH_2 \end{matrix} \!\!> CH_2}$$

7. By [2 + 2] Photochemical Cycloaddition reaction. Cyclobutane derivatives can be obtained by addition of two alkenes under photochemical conditions. [2 + 2] means combination of two molecules using two electrons each.

$$\begin{matrix} CH_2 \\ \| \\ CH_2 \end{matrix} + \begin{matrix} CH_2 \\ \| \\ CH_2 \end{matrix} \xrightarrow{hv} \underset{\text{Cyclobutane}}{\begin{matrix} CH_2 - CH_2 \\ | \quad\quad | \\ CH_2 - CH_2 \end{matrix}}$$

$$R_2C{=}CR_2 + R_2C{=}CR_2 \xrightarrow{h\nu} \text{(cyclobutane ring bearing eight R groups)}$$

8. Tharpe-Ziegler reaction. A dinitrile is treated with $LiN(C_2H_5)_2$ in a large volume of benzene or toluene as solvent. A cyclic imino compound is obtained which on hydrolysis yields a cycloketone.

$$\begin{matrix} CH_2-CH_2CN \\ | \\ CH_2-CH_2CN \end{matrix} \xrightarrow{LiN(C_2H_5)_2} \begin{matrix} CH_2-CH_2 \\ | \\ CH_2-CH(CN) \end{matrix}\!\!>C{=}NH$$

1, 6-Hexanedinitrile

$$\begin{matrix} CH_2-CH_2 \\ | \\ CH_2-CH_2 \end{matrix}\!\!>CH_2 \xleftarrow[HCl]{Zn/Hg} \begin{matrix} CH_2-CH_2 \\ | \\ CH_2-CH_2 \end{matrix}\!\!>C{=}O \xleftarrow{-CO_2} \begin{matrix} CH_2-CH_2 \\ | \\ CH_2-CH(COOH) \end{matrix}\!\!>C{=}O$$

Cyclopentanone

9. From Molonic ester and Acetoacetic ester.

$$CH_2\!<\!\begin{matrix} CH(COOEt)_2 \\ CH(COOEt)_2 \end{matrix} \xrightarrow{Na} CH_2\!<\!\begin{matrix} \overset{\ominus}{C}(COOEt)_2 \\ \underset{\ominus}{C}(COOEt)_2 \end{matrix} \xrightarrow{CH_2I_2} \text{cyclobutane-1,1,3,3-tetracarboxylic ester (EtOOC)}_2C(CH_2)_2C(COOEt)_2$$

$$\xrightarrow[\Delta]{HCl} \text{cyclobutane-1,3-dicarboxylic acid (COOH-CH)(CH}_2)_2\text{(CH-COOH)} \xrightarrow{\Delta} \text{Cyclobutane } (CH_2)_4$$

Cyclobutane

10. By Demjanov rearrangement. This is a convenient method of expanding and contracting alicylic ring systems through the formation of carbocation intermediates. This is achieved by deamination of alicyclic amines with nitrous acid. The carbocations may undergo reactions such as addition of nucleophiles, elimination of a proton and rearrangement to a more stable carbocation. The following examples will illustrate.

Ring expansion. When cyclobutyl methyl amine is treated with nitrous acid, the product is a mixture of cyclopentanol and cyclopentene. Both these products are formed by ring expansion. In addition to this, cyclobutyl carbinol and methylene cyclobutane are also formed where there is no ring expansion. The series of reactions may be shown as follows:

$$\underset{\text{Cyclobutylmethyl amine}}{\begin{matrix} H_2C - CH - CH_2NH_2 \\ | \quad\;\; | \quad\quad\quad \\ H_2C - CH_2 \quad\quad\quad \end{matrix}} \xrightarrow{HNO_2} \begin{matrix} H \quad\; \\ | \quad\; \\ H_2C - \overset{\oplus}{C} - CH_2 \\ | \quad\;\; | \quad\quad \\ H_2C - CH_2 \quad\quad \end{matrix} \xrightarrow{-H^{\oplus}} \underset{\substack{\text{Methylene Cyclobutane} \\ \text{(No ring expansion)}}}{\begin{matrix} H_2C - C = CH_2 \\ | \quad\;\; | \quad\quad \\ H_2C - CH_2 \quad\quad \end{matrix}}$$

$$\downarrow (i)\ H_2O;\ (ii) - H^+$$

$$\underset{\substack{\text{Cyclobutyl carbinol} \\ \text{(No ring expansion)}}}{\begin{matrix} H_2C - CH - CH_2OH \\ | \quad\;\; | \quad\quad\quad \\ H_2C - CH_2 \quad\quad\quad \end{matrix}}$$

In the above changes, rearrangement of carbocation has not taken place.

If the carbocation rearranges itself to a more stable secondary carbocation, it is accompanied by ring expansion as shown below. This is called *Demjanov rearrangement.*

$$\begin{matrix} H_2C - CH - \overset{\oplus}{C}H_2 \\ | \quad\;\; | \quad\quad \\ H_2C - CH_2 \quad\quad \end{matrix} \xrightarrow[\text{Rearrangement}]{\text{Demjanov}} \begin{matrix} H_2C - \overset{\oplus}{C}H \\ | \quad\quad \\ H_2C - CH_2 \end{matrix} \!\!> CH_2 \xrightarrow{-H^+} \underset{\substack{\text{Cyclopentene} \\ \text{(Ring expansion)}}}{\begin{matrix} H_2C - CH \\ | \quad\quad \\ H_2C - CH_2 \end{matrix} \!\!> CH}$$

$$\downarrow (i)\ H_2O;\ (ii) - H^+$$

$$\underset{\substack{\text{Cyclopentanol} \\ \text{(Ring expansion)}}}{\begin{matrix} OH \\ | \\ H_2C - CH \\ | \quad\quad \\ H_2C - CH_2 \end{matrix} \!\!> CH_2}$$

Ring contraction. The ring contraction of the cyclobutyl amine gives cyclopropyl carbinol by Demjanov rearrangement. In addition to this, cyclobutanol is also formed without any ring contraction. The reactions involved are shown below:

$$\underset{\text{Cyclobutyl amine}}{\begin{matrix} H_2C - CH - NH_2 \\ | \quad\;\; | \quad\quad \\ H_2C - CH_2 \quad\quad \end{matrix}} \xrightarrow{HNO_2} \begin{matrix} H_2C - \overset{\oplus}{C}H \\ | \quad\;\; | \\ H_2C - CH_2 \end{matrix} \xrightarrow[(ii) - H^+]{(i)\ H_2O} \underset{\substack{\text{Cyclobutanol} \\ \text{(No ring contraction)}}}{\begin{matrix} H_2C - CH - OH \\ | \quad\;\; | \quad\quad \\ H_2C - CH_2 \quad\quad \end{matrix}}$$

$$\begin{matrix} H_2C - \overset{\oplus}{C}H \\ | \quad\;\; | \\ H_2C - CH_2 \end{matrix} \xrightarrow[\text{Rearrangement}]{\text{Demjanov}} \begin{matrix} H_2C - CH - \overset{\oplus}{C}H_2 \\ | \;/ \quad\quad\quad \\ H_2C \quad\quad\quad\quad \end{matrix} \xrightarrow[(ii) - H^+]{(i)\ H_2O} \underset{\text{Cyclopropyl carbinol}}{\begin{matrix} H_2C - CH - CH_2 - OH \\ | \;/ \quad\quad\quad\quad \\ H_2C \quad\quad\quad\quad\quad \end{matrix}}$$

It can thus be generalised that in the Demjanov rearrangement, ring contraction takes place when the carbocation intermediate has a positive charge on the alicyclic carbon atom. Similarly, ring expansion takes place when the carbocation has a positive charge on the carbon atom α-to the alicyclic ring.

Q. 3. What is Blanc's Rule? What is its utility?

Ans. Blanc's Rule states that:

Pyrolysis of 1, 4- and 1-5 linear dicarboxylic acids yields cyclic anhydrides while 1, 6- and 1, 7-dicarboxylic acids give cyclic ketones, 1-8 and higher dicarboxylic *acids remain unaffected.*

Thus

$$\begin{matrix} CH_2COOH \\ | \\ CH_2COOH \end{matrix} \xrightarrow[-H_2O]{\Delta} \begin{matrix} CH_2CO \\ | \\ CH_2CO \end{matrix} \!\!>\! O$$

Succinic acid (1, 4 Dicarboxylic acid) → Succinic anhydride

$$\begin{matrix} CH_2CH_2COOH \\ \\ CH_2CH_2COOH \end{matrix} \xrightarrow[-CO_2,\ -H_2O]{\Delta} \begin{matrix} CH_2CH_2 \\ | \\ CH_2CH_2 \end{matrix} \!\!>\! CO$$

Adipic acid (1, 6 Dicarboxylic acid) → Cyclopentanone

$$\begin{matrix} CH_2CH_2CH_2COOH \\ | \\ CH_2CH_2CH_2COOH \end{matrix} \xrightarrow{\Delta} \text{No Reaction}$$

1,8 dicarboxylic acid

Utility of the Blanc's Rule

This rule finds use in the determination of size of the ring and chain length of the carboxylic acid. The number of carbons in the ring is reduced by one when the cyclization of the dicarboxylic compound takes place (Refer to cyclization of adipic acid into cyclopentanone).

In an unknown ring compound, whose size is desired to be determined, a double bond is created by a known method. The compound is then subjected to oxidation with $KMnO_4$ when the fission of double bond takes place and a dicarboxylic compound is obtained. This dicarboxylic compound is pyrolysed to obtain a cyclic compound and this can provide information about the size of the original ring.

Q. 4. How can you obtain

(*i*) Ethylcyclopentane

(*ii*) Methylcyclopentane?

Ans. (*i*) Ethylcyclopentane

$$\begin{matrix} CH_2CH_2COO \\ | \\ CH_2CH_2COO \end{matrix} \!\!>\! Ca \xrightarrow[-CaCO_3]{\Delta} \begin{matrix} CH_2CH_2 \\ | \\ CH_2CH_2 \end{matrix} \!\!>\! CO \xrightarrow{C_2H_5MgBr}$$

Calcium adipate → Cyclopentanone

$$\begin{matrix} CH_2CH_2 \\ | \\ CH_2CH_2 \end{matrix} \!\!>\! C \begin{matrix} -OMgBr \\ \diagdown C_2H_5 \end{matrix} \xrightarrow[-Mg(OH)Br]{H_2O} \begin{matrix} CH_2CH_2 \\ | \\ CH_2CH_2 \end{matrix} \!\!>\! C \begin{matrix} -OH \\ \diagdown C_2H_5 \end{matrix} \xrightarrow[distil]{Zn}$$

Addition product

$$\begin{matrix} CH_2CH_2 \\ | \\ CH_2CH_2 \end{matrix} \!\!>\! CH-C_2H_5$$

Ethyl cyclopentane

It can altermatively be prepared as under

$$\begin{matrix} CH_2-CH_2Br \\ | \\ CH_2-CH_2Br \end{matrix} + CH_2 \begin{matrix} \diagup COCH_3 \\ \diagdown COC_2H_5 \end{matrix} \xrightarrow{C_2H_5ONa}$$

1, 4 Dibromo butane — Acetoacetic ester

$$\begin{matrix} CH_2-CH_2 \\ | \\ CH_2-CH_2 \end{matrix} > C \begin{matrix} \diagup COCH_3 \\ \diagdown COOC_2H_5 \end{matrix} + 2C_2H_5OH + 2NaBr$$

$$\Big\downarrow \text{Dil KOH} \mid H^+ / H_2O$$

$$\begin{matrix} CH_2-CH_2 \\ | \\ CH_2-CH_2 \end{matrix} > C \begin{matrix} \diagup COCH_3 \\ \diagdown COOC_2H_5 \end{matrix} \xrightarrow[-CO_2]{Heat} \begin{matrix} CH_2-CH_2 \\ | \\ CH_2-CH_2 \end{matrix} > CHCOCH_3$$

Cyclopentyl methyl ketone

$$\Big\downarrow \text{Zn / Hg} \mid HCl$$

$$\begin{matrix} CH_2-CH_2 \\ | \\ CH_2-CH_2 \end{matrix} > CHCH_2CH_3$$

Ethylcyclopentane

(*ii*) **Methylcyclopentane**

$$\begin{matrix} CH_2CH_2COO \\ | \\ CH_2CH_2COO \end{matrix} > Ca \xrightarrow[-CaCO_3]{\Delta} \begin{matrix} CH_2CH_2 \\ | \\ CH_2CH_2 \end{matrix} > CO \xrightarrow{C_2H_5MgBr}$$

Calcium adipate — Cyclopentanone

$$\begin{matrix} CH_2CH_2 \\ | \\ CH_2CH_2 \end{matrix} > C \begin{matrix} -OMgBr \\ \diagdown CH_3 \end{matrix} \xrightarrow[-Mg(OH)Br]{H_2O} \begin{matrix} CH_2CH_2 \\ | \\ CH_2CH_2 \end{matrix} > C \begin{matrix} -OH \\ \diagdown CH_3 \end{matrix} \xrightarrow[distil]{Zn}$$

Addition product

$$\begin{matrix} CH_2CH_2 \\ | \\ CH_2CH_2 \end{matrix} > CH-CH_3$$

Methylcyclopentane

Q. 5. Describe chemical properties of cycloalkanes. (*Kerala, 2000; Guwahati 2006*)

Ans. Cycloalkanes are saturated compounds like alkanes and therefore, exhibit substitution reactions. But cycloalkanes having ring of 3- or 4-carbon atoms are unstable and tend to form open chain aliphatic compounds by the addition of the reagent. Thus, they exhibit the chemical properties of both alkanes and alkenes.

(1) Reaction with halogens. Generally cycloalkanes undergo free radical substitution with halogens at high temperature or in the presence of light, for example,

$$\text{Cyclohexane } (C_6H_{12}) + Br_2 \longrightarrow \text{Bromocyclohexane } (C_6H_{11}Br) + HBr$$

Cyclohexane   Bromocyclohexane

whereas in case of cyclopropane, bromine and chlorine in the presence of break the ring system and give open chain addition products. For example,

$$\text{Cyclopropane } (C_3H_6) + Br_2 \longrightarrow CH_2Br-CH_2-CH_2Br$$

Cyclopropane   1-3-Dibromopropane

(2) Reaction with hydrogen. When heated with hydrogen in the presence of nickel, cyclopropane and cyclobutane give addition products whereas higher members do not give this reaction.

$$\text{Cyclopropane } (C_3H_6) + H_2 \longrightarrow CH_3-CH_2-CH_3$$

Cyclopropane   *n*-Propane

$$\text{Cyclohexane } (C_6H_{12}) + H_2 \longrightarrow \text{No reaction}$$

Cyclohexane

(3) Reaction with halogen acids. Generally cycloalkanes do not react with halogen acids. However cyclopropane and cyclobutane add on halogen acids to give open chain alkyl halides.

$$\text{Cycloproapane } (C_3H_6) + HI \longrightarrow CH_3-CH_2-CHI$$

Cycloproapane   *n*-Propyl iodide

$$\text{Cyclohexane } (C_6H_{12}) + HI \longrightarrow \text{No reaction}$$

Cyclohexane

Hence it may be pointed that cyclopropane is the most reactive cycloalkane because it exhibits many addition reactions accompanied by ring cleavage. Then comes cyclobutane which shows some addition reactions. The rest of the cycloalkanes do not form addition products.

Q. 6. Explain Baeyer Strain theory. Why is it not applicable to cyclohexane?

(*Venkateswara, 2005; Purvanchal, 2007; Nagpur, 2008; Pune, 2010 Garhwal, 2010; Andhra, 2010*)

Ans. According to Le Bel and Vant Hoff, the four valencies of a carbon atom are directed towards the corners of a regular tetrahedron and hence the angle between any two valencies (any two bonds) is 109°28′ . Baeyer held that any departure from this position leads to a strain in the molecule resulting in the decrease of the stability of the molecule.

In the light of this concept Baeyer gave a theory which is known as Baeyer's Strain Theory. The main points of the theory are as given below:

(*i*) *The carbon atoms constituting the rings lie in the same plane. Hence bond angles between the adjacent carbon atoms of the ring no longer remain 109°28′. Different rings have different values of this angle.* For example, the cyclopropane ring is a triangle having C–C–C angle of 60° only.

(*ii*) *Any deviation, positive or negative from the normal tetrahedral bond angle of 109°28′ during the formation of a ring creates a strain in the molecule which makes the molecule unstable.*

(*iii*) *The larger (more) the deviation from the normal angle, the greater is the strain and thus lesser is the stability. However, it should be noted that the sign of deviation does not make any difference.*

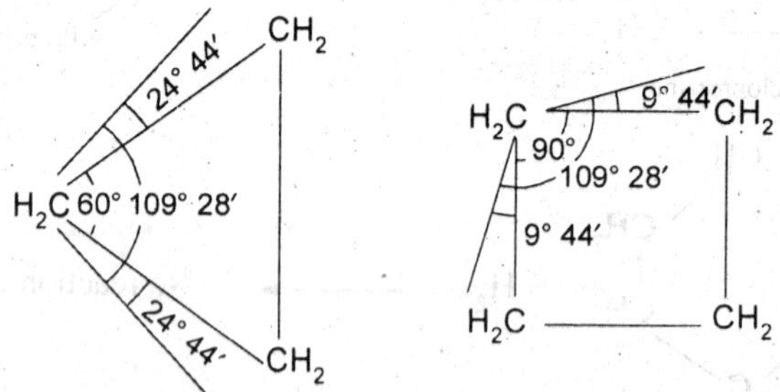

Baeyer worked out the deviation from normal angle or what can be called angle strain involved in the formation of cycloalkanes of various ring sizes. For instance the C–C–C angle of a cyclopropane ring is 60° because each carbon atom occupies a corner of an equilateral triangle. This shows that the angle strain in cyclopropane would be 1/2 (109°28′ – 60°) = 24° – 44′. In the same way C–C–C angle in cyclobutane would be 90° because each carbon atom is situated at the corner of the square. The angle strain in cyclobutane is 1/2 (109°28′ – 90°) = 9° – 44′ . The angle strain for cycloalkanes calculated for various ring sizes are given in following table.

Table: Angle Strains in Various Cycloalkanes

Compound	C–C–C *Angle*	*Angle strain*
Cyclopropane, C_3H_6	60°	1/2 (109°28′ – 60°) = + 24°44′
Cyclobutane, C_4H_8	90°	1/2 (109°28′ – 90°) = + 9°44′
Cyclopentane, C_5H_{10}	108°	1/2 (109°28′ – 108°) = + 0°44′
Cyclohexane, C_6H_{12}	120°	1/2 (109°28′ – 120°) = – 5°16′
Cycloheptane, C_7H_{14}	128°35′	1/2 (109°28′ – 128°34°) = – 9°33′
Cyclooctane, C_8H_{16}	134°	1/2 (109°28′ – 135°) = – 14°46′

It is clear from the above data that angle strain decreases as we move from cyclopropane to cyclopentane. This predicts ease of formation and stability of the compound from cyclopropane to cyclopentane. This is actually found to be so.

However the stability of cyclohexane and higher members cannot be explained by the angle strain values. According to angle strains, cyclohexane and higher members should be quite unstable as there is a lot of strain in the molecule. But this is not true. Actually cyclohexane and higher members are not planar molecules. Different carbon atoms in such molecules lie in different planes, thereby relieving the strain.

Q. 7. Giving examples of first three cycloalkanes, explain the relationship, of relative stabilities with the heat of combustion. *(G. Nanakdev, 2000; Kerala 2000)*

Ans. Relationship of relative stability with heat of combustion.

Size of the ring	*Heat of combustion per – CH_2 group kJ mol^{-1}*	*Size of ring*	*Heat of combustion per – CH_2 group kJ mol^{-1}*
3	697	6	658.5
4	686	7	662.3
5	664	8	663.6

Heat of combustion per CH_2 group of open chain alkanes = 658.0 kJ mol^{-1}.

Heat of combustion may be defined as the quantity of heat evolved when one mole of the compound is burnt in excess of air.

Heat of combustion per CH_2 group gives a fair idea of the relative stabilities of cycloalkanes. If the heat of combustion is large, the compound contains more energy and is therefore less stable.

With this principle, let us examine the stabilities of cyclopropane, cyclobutane and cyclopentane. Heat of combustion of cyclopropane is 697 kJ mol^{-1}. It is 39 kJ more than that of open chain alkane. Thus it is less stable than the open chain alkane by this amount.

In the case of cyclobutane, the heat of combustion is more by an amount 28 kJ as compared to open-chain alkane. Thus cyclobutane is less stable than open chain alkanes but it is more stable than cyclopropane. Cyclopentane having heat of combustion 664 kJ is more stable than cyclobutane. The relative stability of first three members of cycloalkanes in decreasing order is given an under.

Cyclopentane > Cyclobutane > Cyclopropane

On having a look at the heat of combustion of cyclohexane and higher members, we observe that the values do not differ much. Thus higher members are expected to display almost uniform stability which is confirmed by their reactions.

Q. 8. Describe Sachse Mohr Theory of strainless rings.

(Kurukshetra, 2001; Nagpur, 2003; Venkateswara, 2005, Garhwal, 2010; Patna, 2010; Andhra, 2010)

Ans. To explain the greater stability of cyclohexane as compared with cyclopentane and stability of large rings, Sachse in 1890 advanced the Theory of Strainless Rings. *According to this theory large rings due to their bigger size get twisted in different planes with the result that the carbon atoms do not lie in the same plane as is assumed by Baeyer.* Thus, because of carbon atoms being in the different planes, their valency bonds are not distorted from their original angle and consequently no strain is set up in the molecules. In other words, the rings with six or more carbon atoms get puckered and thus become strainless.

Thus according to Sachse cyclohexane exists in the following strain-free forms called boat and chair forms.

Mohr however suggested that the two forms cyclohexane are continuously undergoing transformation from one form to other.

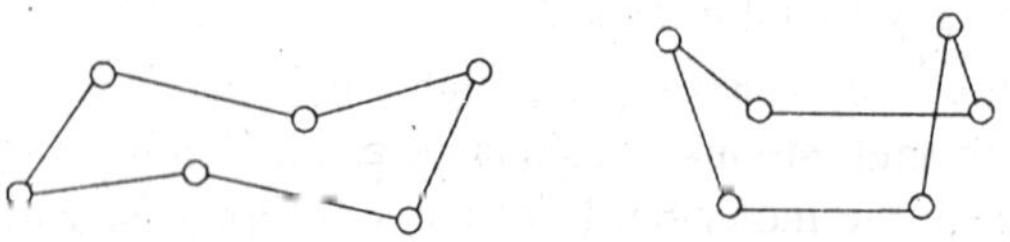

Chair Form Boat Form

Q. 9. Discuss the difficulties encountered in the synthesis of large-membered rings. How are these difficulties overcome?

Ans. It has been observed that heat of combustion per – CH_2 group in cycloalkanes from cyclopentane onwards remains nearly constant and is comparable to that of alkanes. Thus these cycloalkanes (5-membered and higher) should be as stable as corresponding alkanes. But still it is not easy to prepare such higher cyclohexane. To obtain a cycloalkane from a linear compound, it is necessary that the terminal groups of the linear compound should come sufficiently close to each other so that ring closure can take place.

In normal circumstances, the probability of end-groups coming close to provide cyclization is not very bright. This, then, results in intermolecular reaction leading to polymeristation. Baeyer advanced the views that because of tremendous strain, higher cycloalkanes are not formed easily. But this is not the case. As postulated by Sachse-Mohr, such higher cycloalkanes are strain-free because the constituent atoms are in different planes and they retain the normal valency angle of 109° – 28′. It is rather improbability of the end groups of the linear molecules coming closer that prevents the formation of ring compounds.

How is the difficulty overcome?

High dilution technique is resorted to in the preparation of higher cycloalkanes. The substance is taken in low concentration which minimises the collision between two molecules preventing polymerisation. Due to flopping, the two ends of the molecule come closer enough to react, giving rise to cyclization.

It may be remarked that during the synthesis of cyclic compounds, two processes are competing with each other. Process I is the ring closure by the coming together of the end groups and Process II is chain lengthening (polymerization) by the interaction of end group of one molecule with that of the other molecules as illustrated below. By reducing concentration of the reactant, the chances of Process II are minimised

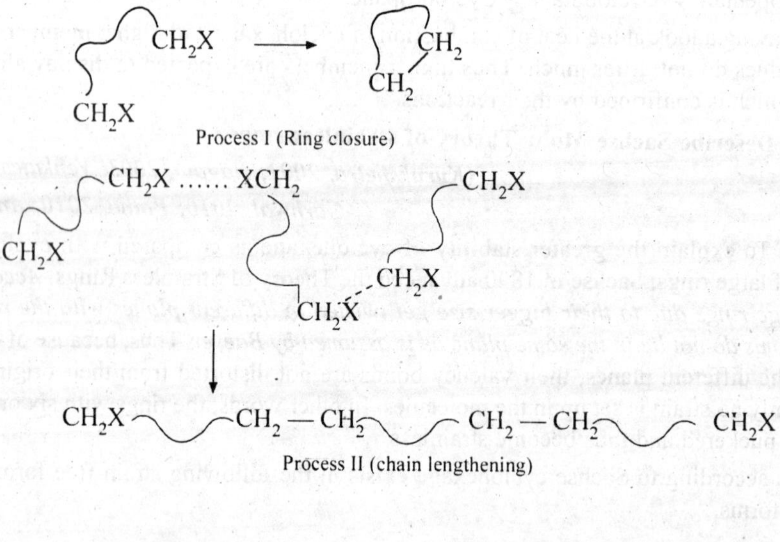

CH_2X ⟶ CH_2
CH_2X CH_2

Process I (Ring closure)

CH_2X ...... XCH_2 CH_2X
CH_2X CH_2X
CH_2X

↓

CH_2X ∼∼ $CH_2 — CH_2$ ∼∼ $CH_2 — CH_2$ ∼∼ CH_2X

Process II (chain lengthening)

It is amazing that, once formed, these higher cyclic compounds are as stable as their straight chain counterparts.

Q. 10. Discuss the orbital picture of angle strain in cycloalkanes.

Ans. In normal alkanes, there is perfect overlapping between the hybrid orbitals of one carbon with the sp^3 hybrid orbitals of other carbon atoms. This provides stability to the structure.

The sp^3 hybrid orbitals of carbon are oriented towards the corners of a regular tetrahedron *i.e.* their orbitals make an angle of 109° – 28′ to one another. If this angle is maintained during overlapping of orbitals, it will provide a complete overlapping of the orbitals. In cyclopentane this angle is, more or less, maintained. There is no strain in the molecule and it offers a case of perfect overlapping of orbitals.

However, the situation is different in cyclopropane. Firstly, the orbitals have been twisted inwards from an angle of 109°28' to 60°. Secondly, the overlap between the orbitals is not perfect. That is, the overlapping orbitals are not symmetrically oriented around the internuclear axis. Such an overlapping is not going to give strength to the bond. It will be akin to bonding which is weaker than bonding. Hence such compounds are definite to be unstable and reactive.

Q. 11. What are cycloalkanes? Write the formulae, common and IUPAC names of any three cycloalkanes. Do they exhibit isomerism with any class of open chain compounds? If so give examples.

Ans. See question 1 for answer to: What are cycloalkanes? Write the formulae, common and IUPAC names of any three cycloalkanes.

Isomerism. Cycloalkanes, which can be represented by the general formula $(CH_2)_n$ are isomeric with alkenes (open-chain compounds) having the general formula C_nH_{2n}.

Thus cyclopropane is iosmeric with propene.

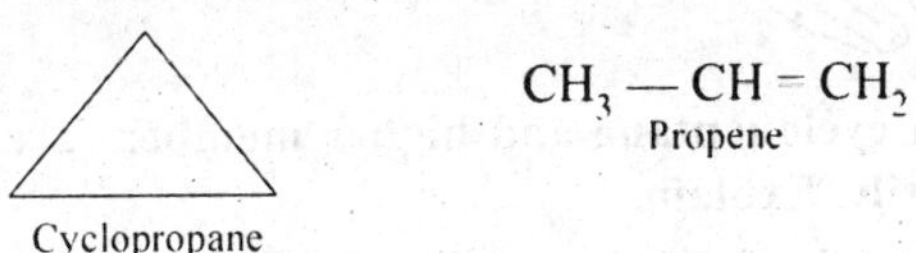

And cyclobutane is isomeric with butene-1 and butene-2

$CH_3 — CH = CH — CH_3$ (Butene-2) $\quad CH_3 — CH_2 — CH = CH_2$ (Butene-1)

Cyclobutane

Q. 12. Cyclopropane is the least stable member of cycloalkanes. How do you justify this is terms of orbital picture of 3-membered rings? *(Panjab 2005)*

OR

There are three banana bonds in cyclopropane. Explain.

Ans. In cyclopropane, the carbon atoms are hybridised. The hybrid orbitals are inclined at an angle of 109°28′ to one another. In order to form a single bond between two carbon atoms, the orbitals of the two carbon atoms must overlap to a maximum extent. Therefore when a carbon atom forms single bonds with two adjacent carbon atoms, the angle C–C–C should be 109°28′ as shown in the diagram below. This is not actually so in cyclopropane molecule. The angle C–C–C in this case is only 60°. This orientation will not allow perfect overlapping of the orbitals and therefore the bonds formed will be weaker.

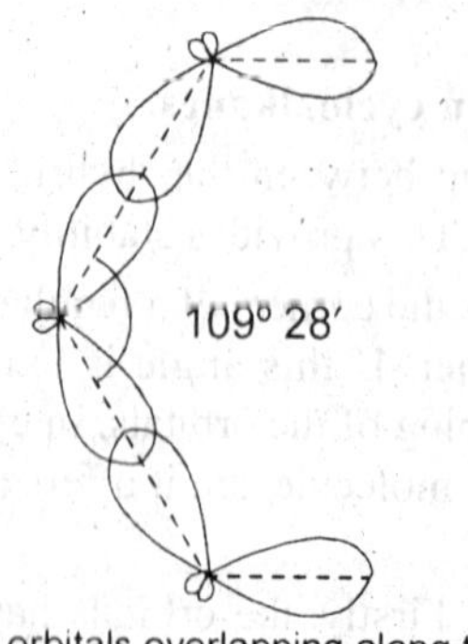

sp^3 orbitals overlapping along their axis; Maximum overlap, strong bond.

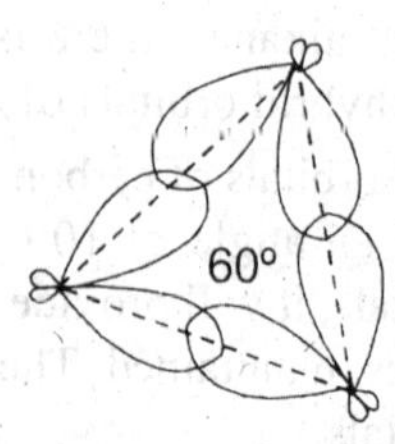

Overlap of sp^3 orbitals in cyclopropane; Poor overlap, weak bond.

This is the reason why cyclopropane molecule is unstable. Calculations show (on geometrical considerations) that the maximum deviation from the normal tetrahedral angle of 109°28' is in the case of cyclopropane. Hence this is the least stable member of cycloalkanes.

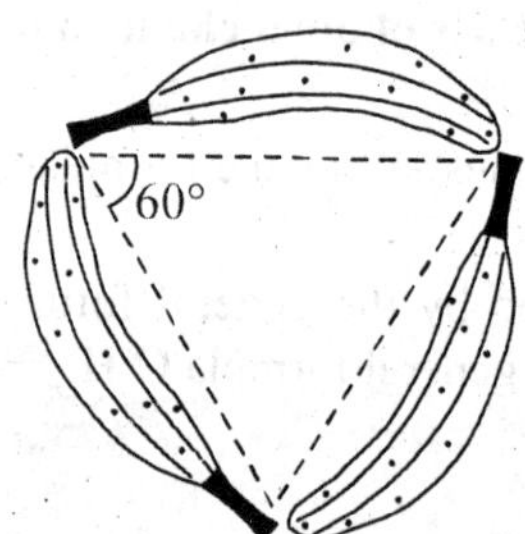

The C-C bonds in cyclopropane are weaker than the C-C bonds in propane. They are called Banana bonds or bent bonds.

Q. 13. Even though cycloheptane and higher members are free of angle strain, they cannot be synthesised easily. Explain. (*Panjab 2003*)

Ans. It is necessary for the formation of a cyclic compound that its terminal carbon atoms should come closer. In the case of a large hydrocarbon chain, the possibility that end carbons will come sufficiently closer and form the closed ring is very small. That is why cycloheptane and higher members are difficult to synthesise.

Q. 14. Give the methods of preparation of cyclopropene, cyclobutene, cyclopentane and cyclohexene

Ans.

1. Cyclopropene 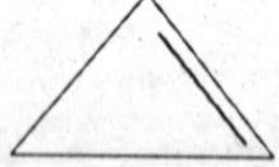

Cyclopene may be prepared by the following sequence of reactions

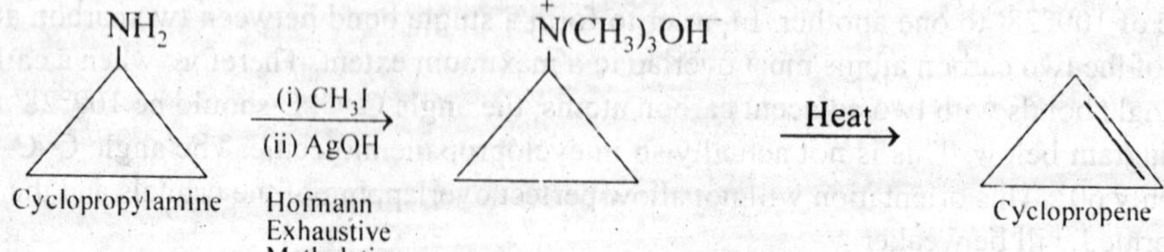

2. Cyclobutene,

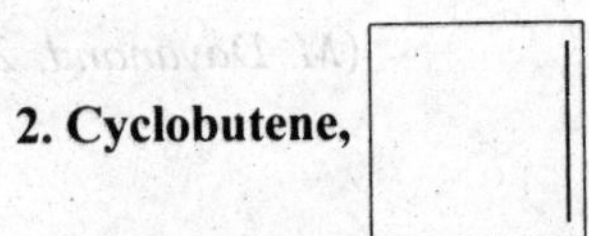

Cyclobutene may be prepared by Hofmann exhaustive methylation of cyclobutylamine in a similar process as in cyclopropene

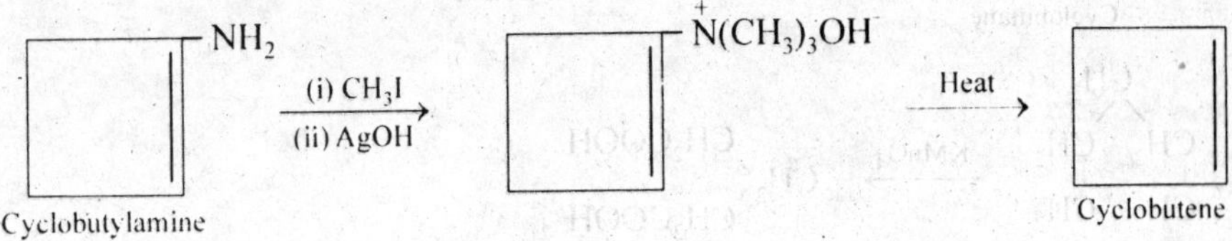

3. Cyclopentene,

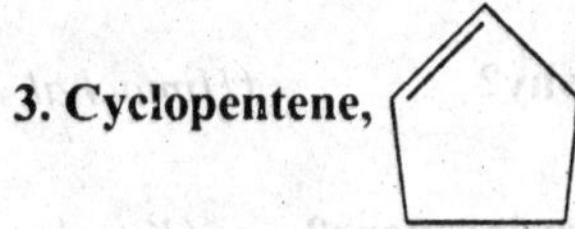

(*i*) It may be prepared by dehydrating cyclopentanol

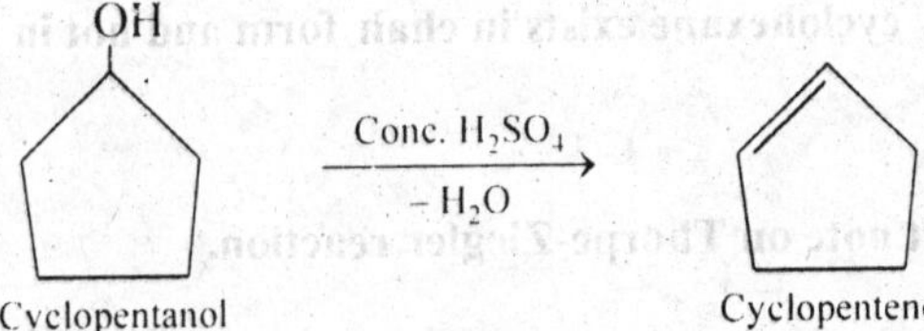

(*ii*) It may also be prepared by heating cyclopentyl bromide with ethanolic potassium hydroxide

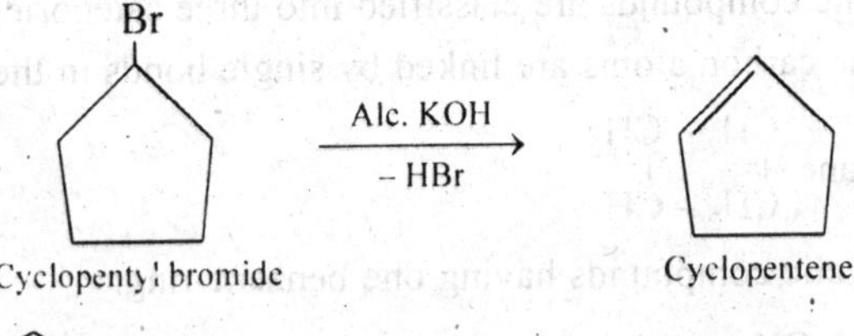

4. Cyclohexene

It may be prepared by dehydrating cyclohexanol with sulphuric acid

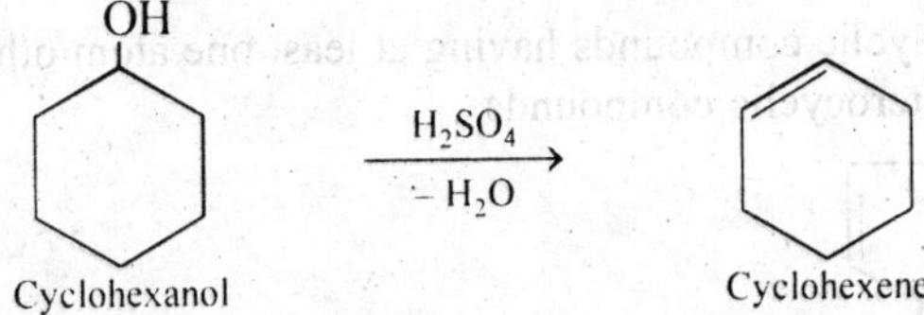

Q. 15. Discuss the preparation of cycloalkanes by pyrolysis of barium or calcium salt of dicarboxylic acid in the light of Blanc's rule. (*M. Dayanand, 2000*)

Ans. See Q. 2 (2) and Q. 3

Q. 16. Complete the following reactions : (*M. Dayanand, 2000*)

(*i*) Cyclobutane + H_2/Ni ⟶

(*ii*) Cyclopentane + $KMnO_4$

Ans. (*i*)

$$\begin{matrix} CH_2 - CH_2 \\ | \quad\quad | \\ CH_2 - CH_2 \end{matrix} + H_2 \xrightarrow[373\ K]{Ni} CH_3CH_2CH_2CH_3$$

Cyclobutane $\qquad$ *n*-Butane

(*ii*)

$$\begin{matrix} CH_2 \\ / \quad \backslash \\ CH_2 \quad CH_2 \\ | \quad\quad | \\ CH_2 - CH_2 \end{matrix} \xrightarrow{KMnO_4} CH_2 \begin{matrix} \diagup CH_2COOH \\ \diagdown CH_2COOH \end{matrix}$$

Cyclopentane $\qquad$ Glutaric acid

Q. 17. Cyclopropane is more reactive than cyclohexane, why? (*Himachal, 2000*)

Ans. See Q. 12

Q. 18. How will you prepare cyclopropane from 1, 3-dichloropropane? (*Kerala, 2001*)

Ans. See Q. 2 (1)

Q. 19. Explain, why cyclohexane exists in chair form and not in a plane form? (*Kerala, 2001*)

Ans. See Q. 6

Q. 20. Write a short note on Thorpe-Ziegler reaction.

Ans. See Q. 2 (8)

Q. 21. How are cyclic organic compounds classified? Give one example for each class. (*Kurukshetra, 2001*)

Ans. Cyclic organic compounds are classified into three categories:

(i) Alicyclic – The carbon atoms are linked by single bonds in the form of a closed ring.

e.g., Cyclobutane

$$\begin{matrix} CH_2 - CH_2 \\ | \quad\quad | \\ CH_2 - CH_2 \end{matrix}$$

(ii) Aromatic – Cyclic compounds having one benzene ring.

e.g., phenol

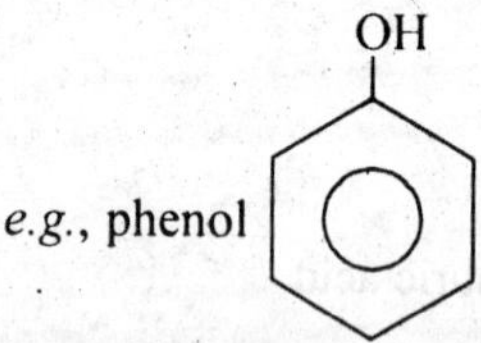

(iii) Heterocyclic – Cyclic compounds having at least one atom other than carbon in the ring are termed as heterocyclic compounds.

e.g., Furan (five-membered ring with O)

Q. 22. How will you get the following compounds? (*Kerala, 2000*)

(*i*) Cyclobutane from ethylene bromide

(*ii*) Cyclopentane from diethyl adipate

Ans. (*i*)

$$\begin{matrix} CH_2-Br \\ | \\ CH_2-Br \end{matrix} + 2\,Zn + \begin{matrix} Br-CH_2 \\ | \\ Br-CH_2 \end{matrix} \xrightarrow[-2\,ZnBr]{\Delta} \begin{matrix} CH_2-CH_2 \\ | \quad\quad | \\ CH_2-CH_2 \end{matrix}$$

Ethylene bromide Cyclobutane

(*ii*) See Q. 2 (6)

Q. 23. Whether the following transformation can be effected under Δ or *hν* condition?

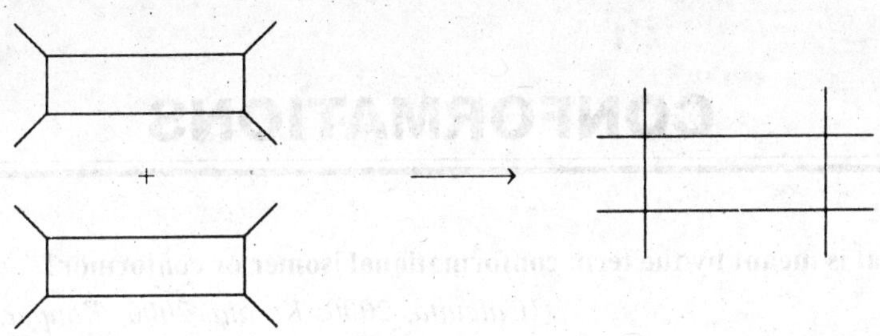

Ans. See Q. 2 (7)

Q. 24. Chair form of cyclohexane is more stable than boat form. Explain.

(Punjab, 2003; Mumbai, 2010

Ans. All the angles in boat form are tetrahedral. But hydrogens on four carbons are eclipsed. As a result there is some torsional strain. Also the flagship hydrogens are very close to each other causing van der Waal strain. Due to these factors, chair form of cyclohexane is more stable.

31

CONFORMATIONS

Q. 1. What is meant by the term conformational isomer or conformor?

(*Calcutta, 2000; Kerala, 2000; Kanpur, 2001; Nagpur 2004; Kuvempur 2005*)

Ans. Carbon-carbon single bond in alkanes is a sigma bond formed by the overlapping of sp^3 hybrid atomic orbitals along the inter-nuclear axis. The electron distribution in such a bond is symmetrical around internuclear axis; so that free rotation of one carbon against the other is possible without breaking sigma or single covalent bond. Consequently such compound can have different arrangements of atoms in space, which can be converted into one another simply by rotation around single bond, without breaking it. These different arrangements are known as *Conformational Isomers or Rotational Isomers or Conformers.* Since the potential energy barrier for their inter-coversion is very low, it is not possible to isolate them at room temperature. At least 60-85 kJ/mole must be the energy difference between two conformers to make them isolatable at room temperature.

Hence, conformations can be defined as different arrangements of the atoms which can be converted into one another by rotation around single bonds.

Q. 2. Discuss Newman and Sawhorse representations for the conformations of ethane.

(*A.N. Bhauguna, 2000; Devi Ahilya, 2000; Nagpur, 2003*)

Ans. Alkanes can have an infinite number of conformations by rotation around carbon-carbon single bonds. In ethane two carbon atoms are linked by a single bond and each carbon atom is further linked with three hydrogen atoms. If one of the carbon atoms is allowed to rotate about carbon-carbon single bond keeping the other carbon stationary, an infinite number of arrangements of the hydrogens of one carbon, with respect to those of the other, are obtained. All these arrangements are called conformations (Bond angles and bond lengths remain the same).

Newman representation

This can be easily understood with the help of Newman Projection formulae. The molecule is viewed from front to back in the direction of carbon-carbon single bond. The carbon nearer to the eye is represented by a point and three hydrogens attached to it are shown by three lines at an angle of 120° to one another. The carbon atom away from the eye is represented by a circle and three hydrogens attached to it are shown by shorter lines at an angle of 120° to each other.

Out of infinite number of conformations, Newman Projection formulae for two extreme cases are as shown below:

Staggered Eclipsed

Newman Projection Formulae for conformations of ethane.

The conformation in which the H atoms of two carbons are as far apart as possible, is known as *Staggered conformation* and the conformation in which the H atoms of back carbon are just behind those of the front carbon is known as **Eclipsed conformation.** These are converted into one another by rotation of one carbon against the other through 60°. The other conformations, in between these two, are known as *skew conformations.*

Sawhorse representation

In this representation, the molecule is visualised slightly from above and from the right and then projected on the paper. The bond between two carbons is drawn diagonally and is drawn a bit longer for the sake of clarity. The lower left hand carbon is taken as front carbon and the upper right hand carbon is taken as back carbon. The Sawhorse representation of staggered and eclipsed conformations of ethane are given below:

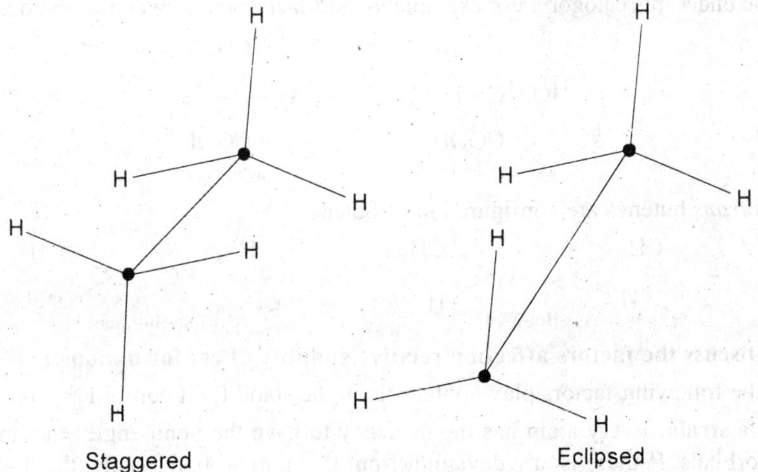

Sawhorse representation for conformations of ethane.

Pitzer in 1936 found that the rotation is not completely free. Rather there exists a potential energy barrier which restricts the free rotations. It means that the molecule spends most of its time in the most stable staggered conformation and it spends least time in the least stable eclipsed conformation; the energy difference being 12 kJ/mole in the case of ethane.

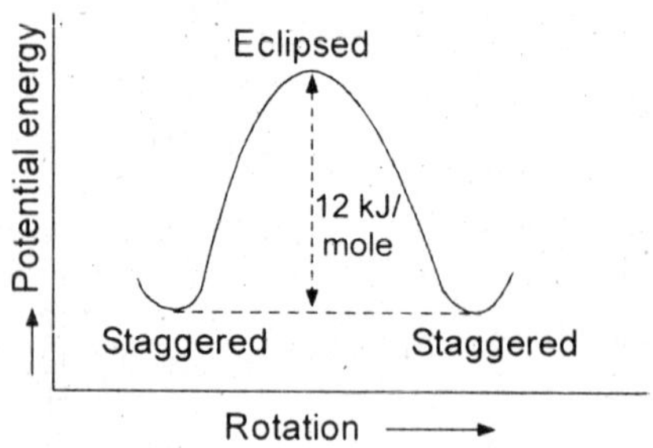

Fig. 30.3 Rotational or torsional energy of ethane.

The energy required to rotate the molecule about carbon-carbon bond is called rotational or torsional energy.

Q. 3. Bring out clearly the difference between conformation and configuration by giving examples. (*Lucknow, 2000; Nagpur 2005; Delhi, 2007*)

Ans. *Conformation.* Structures containing different arrangement of atoms of a molecule in space which can arise by rotation about a single bond are called conformers. For example ethane exists in different conformations called staggered, eclipsed and skew conformations. The energy difference between different conformers is rather small. This phenomenon is called conformation. The phenomenon of conformation is also exhibited by *n*-butane, cyclohexane, stilbene dichloride etc.

Configuration

Structures of a compound differencing in the arrangement of atoms or groups around a particular atom in space are called *configurations.* Enantiomers, distereomers and geometrical isomers come under this catogory. For example *d*-and *l*-lactic acids are configurations of lactic acid

$$\begin{array}{c} CH_3 \\ | \\ HO-C-H \\ | \\ COOH \end{array} \qquad \qquad \begin{array}{c} CH_3 \\ | \\ H-C-OH \\ | \\ COOH \end{array}$$

d-lactic acid *l*-lactic acid

cis and *trans* butenes are configuration of butene

$$\begin{array}{l} CH_3 \\ H \end{array}\!\!>C=C<\!\!\begin{array}{l} CH_3 \\ H \end{array} \qquad \qquad \begin{array}{l} H \\ CH_3 \end{array}\!\!>C=C<\!\!\begin{array}{l} CH_3 \\ H \end{array}$$

cis-But-2-ene *trans*-But-2-ene

Q. 4. Discuss the factors affecting relative stability of conformations.

Ans. The following factors play a vital role in the stability of conformations.

1. Angle strain. Every atom has the tendency to have the bond angles that match those of its bonding orbitals. If there is any deviation from this normal bond angle, the molecule suffers from *angle strain.* Conformations suffering from angle strain are found to be less stable.

2. Torsional strain. There is a tendency on the part of two carbons linked to each other to have their bonds staggered. That is why the staggered form of any molecule like ethane, *n*-butane is most stable. As the bonds of two connected carbons move towards eclipsed state, a torsional strain is set up in the molecule thus raising its energy. Thus the staggered conformations have the least and eclipsed the highest torsional strain. The energy required to rotate the molecule around the carbon-carbon bond is called *torsional energy.*

3. Steric strain (van der Waals Strain). Groups attached to two linked carbons can interact in different ways depending upon their size and polarity. These interactions can be attractive or repulsive. If the distance between the groups or atoms is just equal to the sum of their van der Waals radii, there will be attractive interactions between them. And if these atoms or groups are brought closer than this distance, there will be repulsions leading to van der *Waals strain or steric strain* in the molecule.

4. Dipole-dipole Interactions. Atoms or groups attached to bonded carbons orient or position themselves to have favourable dipole-dipole interactions. It will be their tendency to have maximum dipole-dipole attractions. Hydrogen bond is a particular case of powerful dipole-dipole attractions.

The stability of a conformer is determined by the net effect of all the above factors.

Q. 5. Discuss the conformations and change in dipole-moment of 1, 2-dibromoethane with temperature. (*Kuvempu 2005*)

Ans. The conformations of 1, 2 dibromoethane have been extensively studied by dipole-moment measurement. The conformations are depicted as under:

In the liquid state, the percentage of *anti* forms is 65 corresponding to conformational free energy of 3.5 kJ mol^{-1} in favour of *anti*.

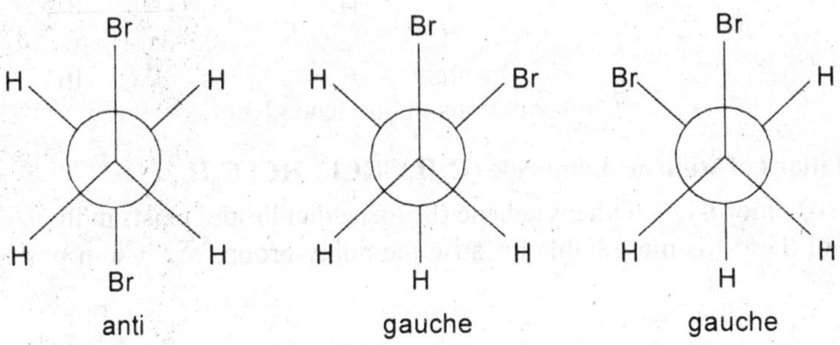

In the gaseous state at 22°C, it contains 85% of *anti* conformer. Thus there is an increase in the proportion of anti form with increase of temperature. The greater stability of *anti* form is due to combined effect of steric factor and dipole-dipole interactions. It goes in favour of formation of *anti* conformer. Dipole moment increases with increase of temperature.

Q. 6. Discuss the conformations of the following:

(*a*) Chlorohydrin

(*b*) Ethylene glycol

(*c*) Stilbene dichloride (*meso* & *dl* forms)

Ans. Conformations of chlorohydrin and ethylene glycol.

A lot of stability (amounting to 8 – 20 kJ mol^{-1}) is imparted by intramolecular hydrogen bonding between two neighbouring groups to a conformer. The donor and acceptor groups must be sufficiently close to each other for an effective hydrogen bonding. It could take place in an *eclipsed* or *gauche* form, there being no possibility of such hydrogen bonding in *anti* form because of remoteness in the groups.

Even in eclipsed form, van der Waals repulsive forces come into play making conformation unstable. Gauche conformation is thus most suited for hydrogen bonding. It is therefore the preferred conformation in case of chlorohydrin and erthylene glycol. Structures II and III are preferred conformations in both cases.

Conformations of chlorohydrin.

Conformations of ethylene glycol.

Conformations of stilbene dichloride ($C_6H_5CHCl\ CHCl\ C_6H_5$)

Meso-1, 2-Dichloro-1, 2-diphenylethane (Stilbene dichloride) exists in three conformations I, II and III. Out of these I is most stable because the bulky groups are in anti position

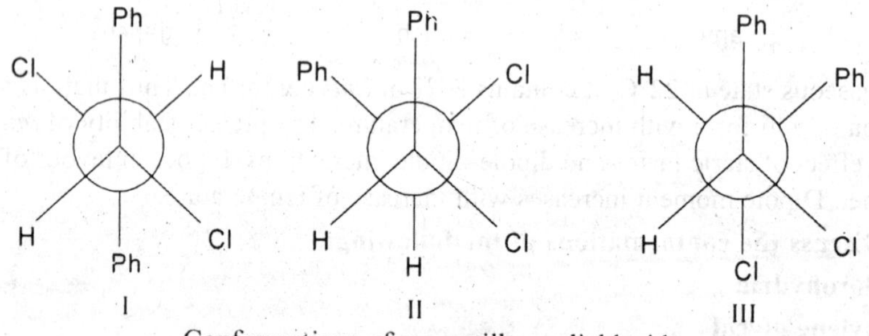

Conformations of mesostilbene dichloride.

(1S, 2S) enantiomer of the optically active form of stlbene dichloride also exists in three forms I, II and III as shown below.

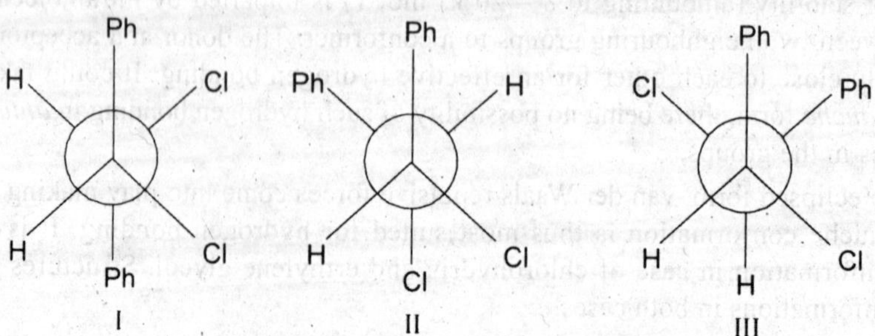

Fig. 30.4. Conformors of optically active (1S, 2S) stilbene dichloride.

Here probably conformor II is more stable.

Q. 7. Explain the conformations of butane. *(Kerala, 2001; Kurukshetra, 2001. Devi Ahilya 2001; Madras 2003; H.S. Gaur 2004; Delhi, 2006*

Ans. *n*-butane is an alkane with four carbon atoms, which can be considered to be derived from ethane by replacing one hydrogen on each carbon with a methyl group. If we consider the rotation about the central carbon-carbon bond ($C_2 - C_3$), the situation is somewhat similar to ethane; but *n*-butane has more than one staggered and eclipsed conformations (unlike ethane which has only one staggered and one eclipsed conformation). Newman Projection formulae for various staggered and eclipsed conformations of *n*-butane are as given below:

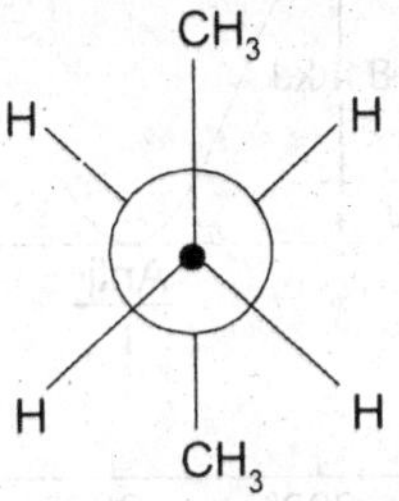

Completely staggered or anti form I

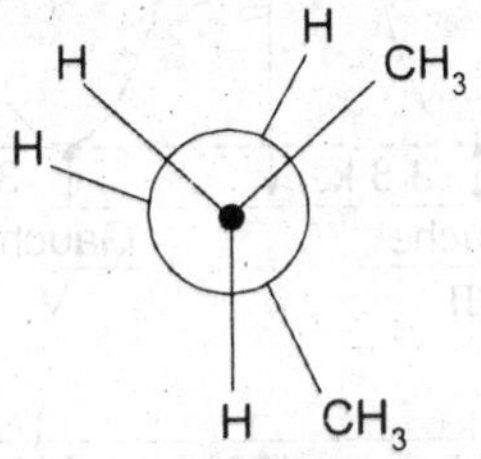

Partially eclipsed form II

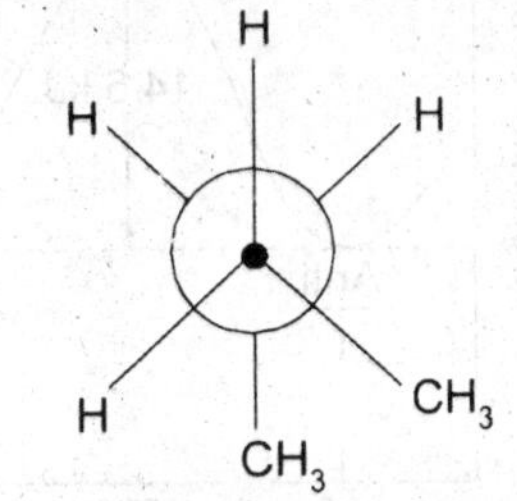

Partially staggered or gauche form or skew form III

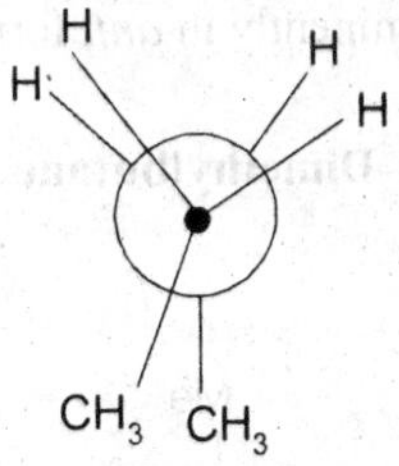

Completely eclipsed form IV

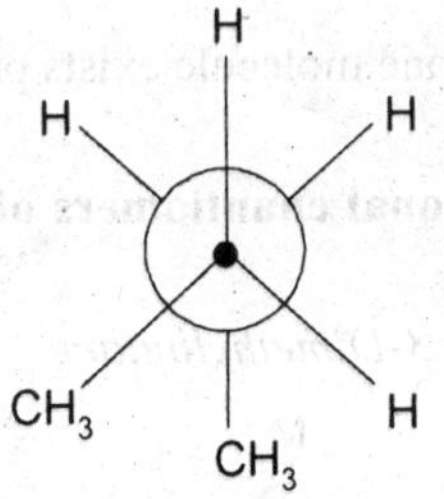

Partially staggered or gauche V

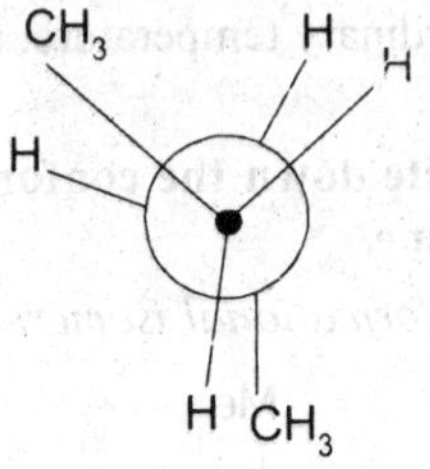

Partially eclipsed from VI

The completely staggered conformation, (I) also known as anti form, is having the methyl groups as far apart as possible. Let us see how these forms have been obtained. Let us start from structure I. Holding the back carbon (represented by circle) fixed along with its groups, – H, – H and – CH_3, rotate the front carbon (shown by a dot) in clockwise direction by an angle of 60°. Groups attached to it will also move. *Partially eclipsed* form (II) is obtained. In this conformation (II), methyl group of one carbon is at the back of hydrogen of the other carbon. Further rotation of 60° leads to a *partially staggered* conformation (III), also known as gauche form, in which the two methyl groups are at an angle of 60°. Rotation by another 60° gives rise to a *fully eclipsed* form (IV) having two methyl groups at the back of each other. Further rotation of 60°, again leads to *partially staggered or gauche* form (V) having methyl groups at an angle of 60° (as in III). Still further rotation of 60° leads to *partially eclipsed* form (VI), having methyl group of one carbon at the back of hydrogen of the other (as in II). If a further rotation of 60° is operated (completing the rotation of 360°), again form I is obtained. Of course, there will be an infinite number of other conformations in between these six conformations (I to VI). (*Gauche* form is also known as *skew* form).

Out of these six conformations, the *completely staggered or anti conformation* (I) is most stable and *partially staggered or gauche* conformation (III or V) is slightly less stable: the energy difference being only 3.8 kJ/mole. On the other hand the *completely eclipsed* conformation (IV) is least stable and *partially eclipsed* conformation (II or VI) is slightly more stable, again the energy difference being 3.8 kJ/mole. (This is due to presence of steric strain between two methyl groups). The energy difference between most stable conformation (I) and least stable conformation (IV) is about 18.4 kJ/mole while that between I and II (or VI) is about 14.6 kJ/mole.

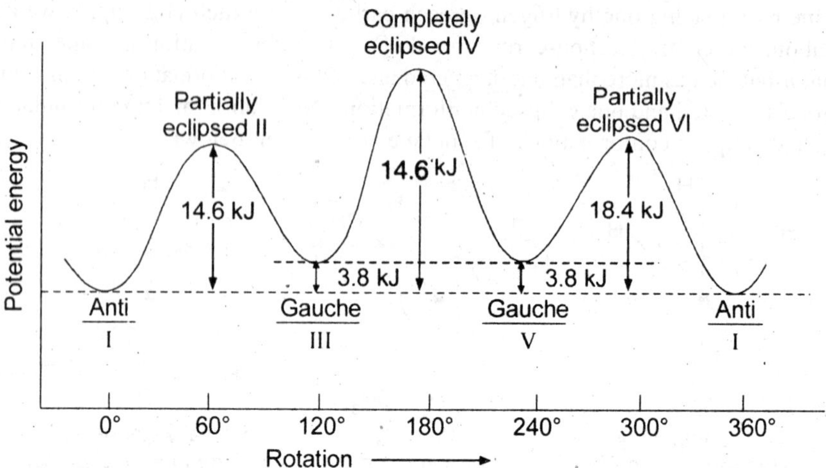

Rotational or torsional energy of *n*-butane.

Thus at ordinary temperature, n-butane molecule exists predominently in *anti* form with some gauche forms.

Q. 8. Write down the conformational enantiomers of 2, 3-Dimethylbutane and 2, 2, 3-Trimethylbutane.

Ans. *Conformational isomers of 2, 3-Dimethylbutane.*

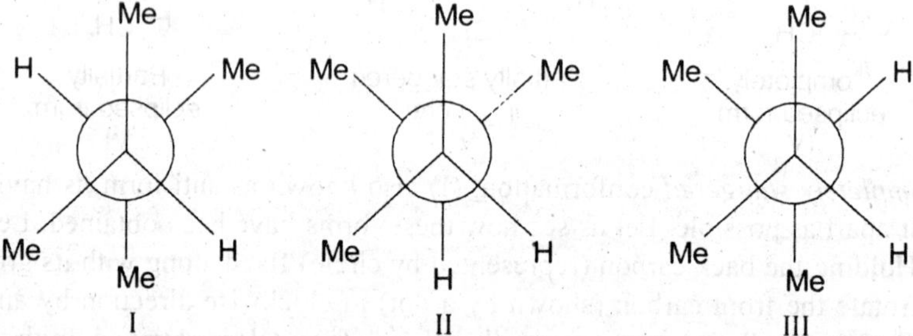

Structure I is supposed to have the maximum stability out of these three isomers, II and III are conformational enantiomers (mirror-images) also. This can be visualized after rotation of III as a whole.

Conformational isomers of 2, 2, 3-Trimethylbutane :

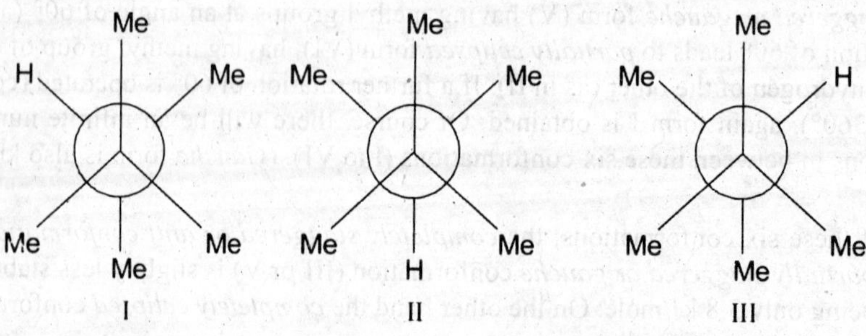

All these structures have the same stability and the compound may be assumed to occur in one conformation only.

Q. 9. Discuss the relative stabilities of chair and boat conformations of cyclohexane.

(*Purvanchal 2003; Bhopal, 2004; Mumbai, 2004; Guwahati 2005; Nagpur 2005*)

Or

Draw Newman projection formulae for the chair and boat form of cyclohexane. Which form is more stable and why? (*Meerut, 2000; Lucknow, 2000; M. Dayanand, 2000; G. Nanakdev, 2000; Kanpur, 2001; Nagpur 2003*)

Ans. Chair conformation.

Chair conformation of cyclohexane is represented below:

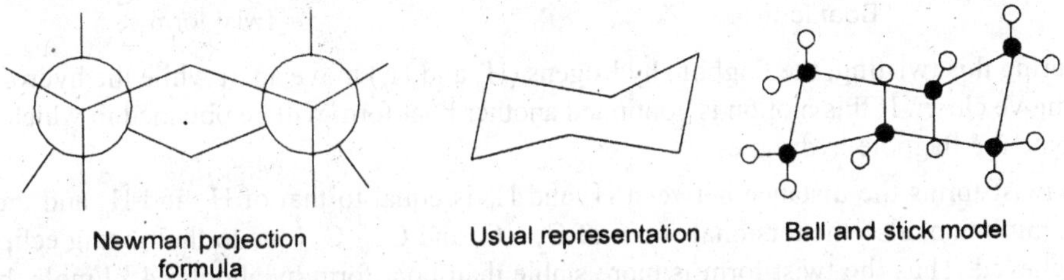

Fig. 30.5. Chair conformations of cyclohexane

This is the most stable conformation of cyclohexane. In this conformation, all the bond angles are tetrahedral and the C–H bonds on adjacent carbons are in a staggered position. This conformation has no strain and has minimum energy.

Boat conformation

There is no angle strain in the molecule as all the angles are tetrahedral. But hydrogens on four carbon atoms (C_2 and C_3, C_5 and C_6) are eclipsed. As a result, there is considerable torsional strain. Also, two hydrogens pointing towards each other at C_1 and C_4 (called flagpole hydrogens) are very close to each other. This brings in van der Waals strain in the molecule. Due to these reasons boat conformation is less stable than chair conformation by an amount 28.8 kJ/mole.

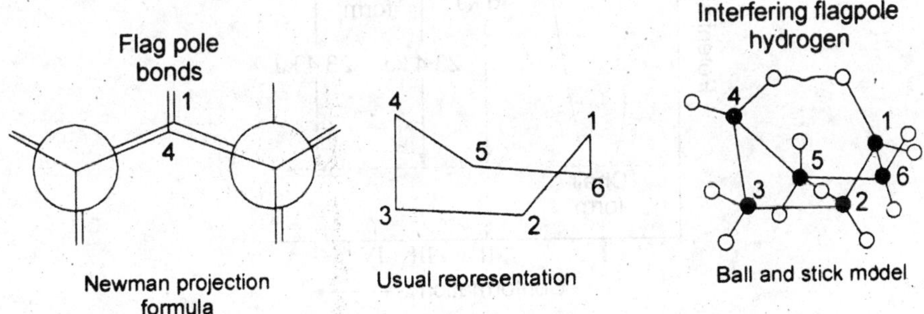

Fig. 30.7. Boat conformation of cyclohexane

Q. 10. What is twist conformation of cyclohexane? Give the sequence of changes in going from chair form to boat form. (*Punjabi, 1992*)

Ans. Besides chair and boat conformations, cyclohexane can have several other possible conformations. Consider model of boat conformation of cyclohexane. Hold C_2–C_3 bond in one hand and C_5–C_6 in the other and twist the model so that C_2 and C_5 come upwards and C_3 and C_6 go downwards. We will get another conformation known as *twist form or skew boat* form.

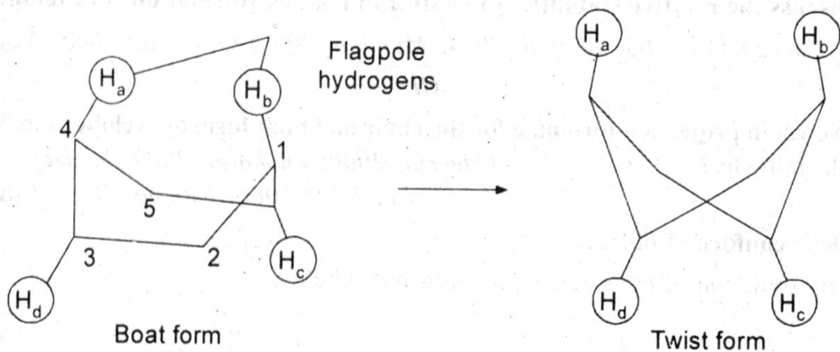

During this twisting, the flagpole hydrogens (H_a and H_b) move apart while the hydrogens H_c and H_d move closer. If this motion is continued another boat form will be obtained in which H_c and H_d become the flagpole hydrognes.

In twist forms the distance between H_a and H_b is equal to that of H_c and H_d and the steric strain is minimum; also the torsional strain of $C_2 - C_3$ and $C_2 - C_6$ (due to their being eclipsed) is partly relieved. Thus the twist form is more stable than boat form by about 5.4 kJ/mole, but it is much less stable than chair form by 23.4 kJ/mole.

If we want to convert the chair form into boat form it will have to pass through a half chair form having considerable angle strain and torsional strain. The energy difference between chair form and half chair form being about 11 kcal/mole, half chair form is quite unstable.

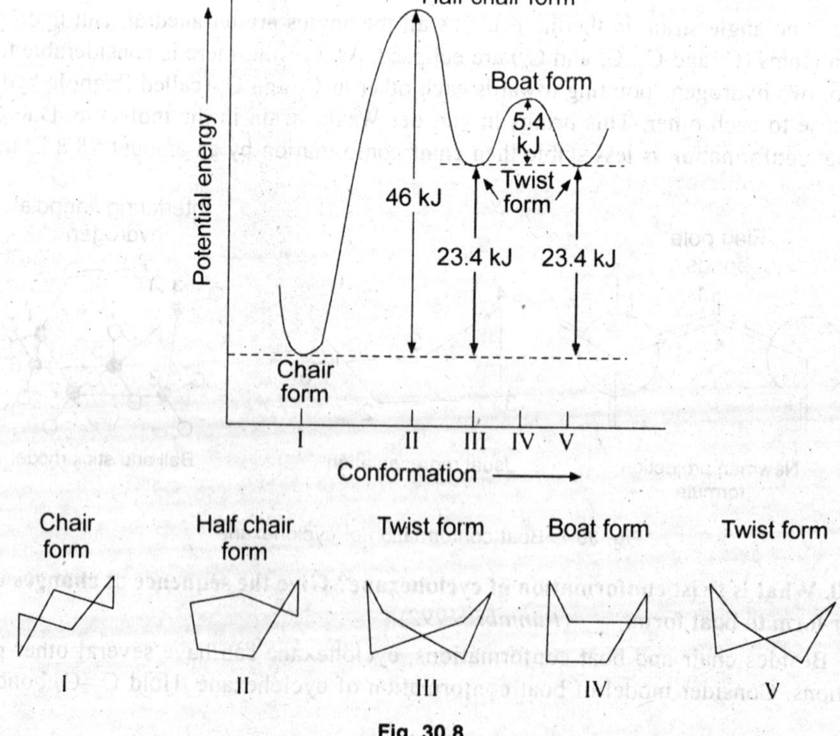

Fig. 30.8.

Q. 11. Write notes on equatorial and axial bonds in cyclohexane.

(Nagpur, 2003; Bhopal, 2003; Delhi, 2006)

Ans. Consider the structure of chair form of cyclohexane as given below:

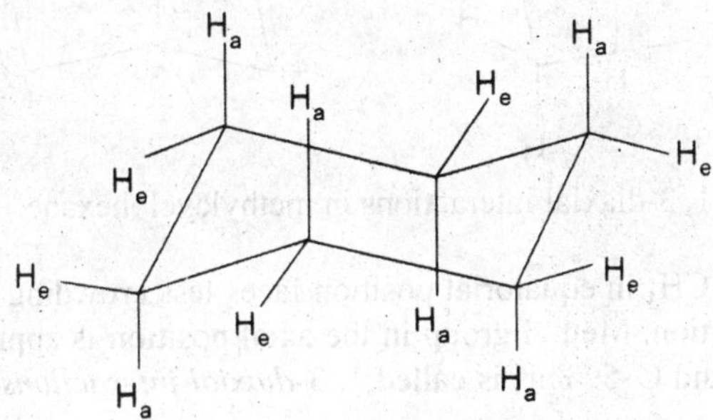

Fig. 30.9. Equatorial and axial bonds in cyclohexane.

Although, the cyclohexane ring is not planar completely, but for approximation, we can take it as planar. Consider the position of various hydrogens in the chair conformation. There are two distinct kinds of hydrogens. Six of the hydrogens which are marked H_e are almost oriented within the plane of cyclohexane ring. These are called Equatorial hydrogen atoms. The bonds by which these are held to the ring are called Equatorial bonds.

We again observe that six hydrogen atoms which are shown as H_a in the figure above are oriented perpendicular to the cyclohexane ring. These are called Axial hydrogen atoms and the bonds by which they are held to the ring are called Axial bonds.

It may be noted that there is one axial and one equatorial hydrogen on each carbon in the chair conformation of cyclohexane.

Q. 12. What is meant by 1, 3-diaxial interaction? *(Madras, 2004)*

Ans. Consider the chair model of cyclohexane. Looking at the molecule, we find that six hydrogens lie in the plane while six lie above or below the plane. Six bonds holding hydrogens in the plane are called *equatorial bonds* and six bonds holding hydrogens above or below the plane are called *axial bonds*. By and large, there is no stress in the molecule and it is as stable as staggered ethane.

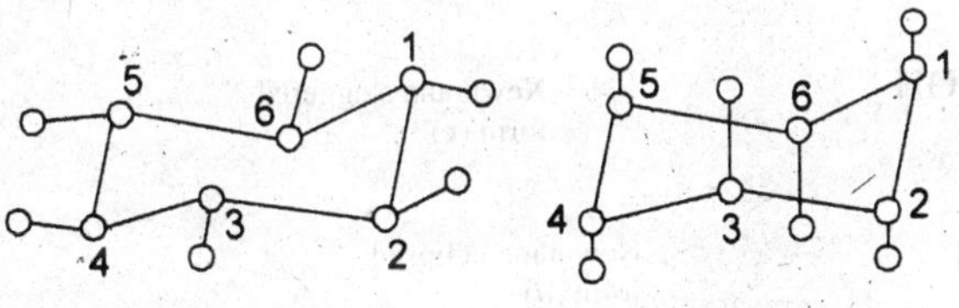

Equatorial bonds Axial bonds

If a hydrogen is replaced by a larger group or atom, than crowding occurs. Atoms linked by axial bonds on the same side face severe crowding, This interaction is called 1, 3-diaxial interaction. Generally speaking, atoms or groups have more free space in equatorial position than in axial position.

There are two possible chair conformations of methylcyclohexane one with $-CH_3$ in equatorial position and the other with $-CH_3$ in axial position.

H CH$_3$ H H H
H 5 H CH$_3$ 1 4 3 2

1, 3-diaxial interactions in methylcyclohexane.

It is observed that – CH_3 in equatorial position faces less crowding by hydrogens compared to – CH_3 in the axial position. Methyl group in the axial position is approached more closely by axial hydrogens on C–3 and C–5. This is called 1, 3-*diaxial interactions.*

Q. 13. Out of the following two conformational forms of 1, 2 - dibromoethane, which will have higher dipole moment ? (*Delhi,2006*)

(*a*)

Br H H H H Br

Br H Br H H H

Ans. In structure (*b*) two Br groups are close to each other. This will result in repulsion between them leading to separation of charge and thus creation of higher dipole moment.

Q. 14. Convert the Fischer's projection of D(–) erythrose sugar (*a*) to the following:

(*a*) CHO
H — OH
H — OH
CH_2OH
D (–) Erythrose

→ Sawhorse eclipsed form (*b*)

→ Newmann staggered form (*e*)

→ Newmann eclipsed form (*d*)

→ Sawhorse staggered form (*c*)

(*Delhi 2006*)

Ans.

CHO; H — OH; H — OH; CH_2OH

D (–) Erythrose

Newmann staggered form (*e*): CH_2OH, OH, H, H, OH, CHO

Sawhorse eclipsed form (*b*): CHO, CH_2OH, H, OH, H, OH

Newmann staggered form (*d*): CH_2OH, CHO, H, H, OH, OH

Sawhorse staggered form (*c*): OH, H, CH_2OH, CHO, H, OH

Q. 15. Draw the Newmann projections (eclipsed and staggered) of meso dichlorobutane. Which is more stable? (*Delhi 2007*)

Ans. Meso dichlorobutane may be written as:

$$CH_3-\underset{Cl}{\overset{H}{C}}-\underset{Cl}{\overset{H}{C}}-CH_3$$

Eclipsed and staggered Newmann projections may be written as

(staggered): CH_3, H, Cl, H, Cl, CH_3

(Eclipsed): CH_3, CH_3, H, H, Cl, Cl

Staggered form is more stable because the groups attached to the middle carbon atoms are as apart as possible and there is least interaction between the attached groups.

Q. 16. Discuss conformations of substituted cyclohexanes

Ans. A methyl group is bulkier than a hydrogen atom. When the methyl group in methylcyclohexane is in the axial position, the methyl group and the axial hydrogens of the ring repel each other. These interacions are called **Axial-Axial Interactions.** When the methyl group is in the equatorial position, the repulsions are minimum. Thus, the energy of the conformation with equatiorial methyl group is lower. At room temperature, about 95% methylcyclohexane molecules are in the conformation in which the methyl group is equatorial.

The bulkier the group, the greater is the energy difference between equatorial and axial conformations. In other words, a cyclohexane ring with a bulky substituent (*e.g.*, *t*-Butyl group) is more likely to have that group in the equatorial position.

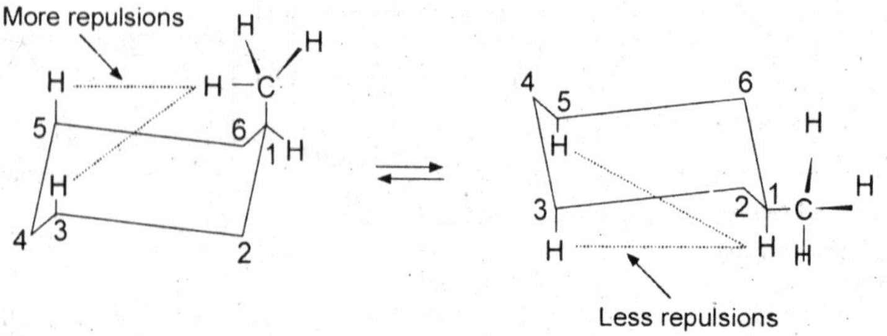

More repulsions

Less repulsions

Q. 17. Show with the help of a diagram how the axial bonds become equatorial bonds and *vice versa* taking the example of cyclohexane.

Ans. Each of the six carbon atoms of cyclohexane has one equatorial and one axial hydrogen atom. Thus, there are six equatorial hydrogens and six axial hydrogens. In the flipping and reflipping between conformations, axial hydrogens become equatorial hydrogen while equatorial hydrogens become axial. This is represented in Fig. below :

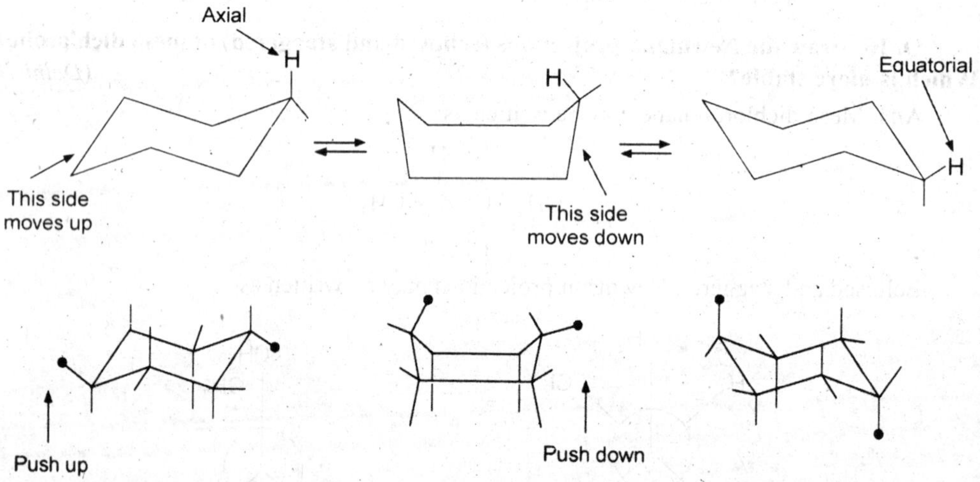

Axial

Equatorial

This side moves up

This side moves down

Push up

Push down

Q. 18.The gauche conformation of ethylene glycol is preferred as compared to its anti conformation. explain. (*Kurukshetra, 2001*)

Ans. See Q. 6

Q. 19.Draw structural formulas of 3 bromo-2-hydroxy butane according to Sawhorse, Fisher's project and Newman projection. (*Kurukshetra 2001*)

Ans. proceed as in Q. 2

Q. 20. Write a short note on relative configuration. (*Lucknow 2000*)

Ans. See Q. 3

32

CARBOHYDRATES

Q. 1. What are carbohydrates?

(*Panjab, 2000; Kurukshetra, 2001; Bangalore, 2001; Bhopal, 2004; Patna 2005*)

Ans. Carbohydrates are important organic compounds widely distributed in nature. These include sugars such as glucose, fructose and sucrose as well as non-sugars such as starch and cellulose. Sugars are crystalline substances with a sweet taste and are soluble in water while non-sugars are non-crystalline substances which are not sweet and are insoluble or less soluble in water.

In the green plants, carbohydrates are produced by a process called **photosynthesis.** This process involves the conversion of simple compounds carbon dioxide and water into glucose and is catalysed by green colouring matter chlorophyll present in the leaves of plants. The energy required to effect the conversion is supplied by the sun in the form of sunlight.

$$6CO_2 + 6H_2O \xrightarrow[\text{Chlorophyll}]{\text{Sunlight}} C_6H_{12}O_6 + 6O_2$$

Thousands of glucose molecules can then be combined to form much larger molecules of cellulose which constitutes the supporting framework of the plants.

Q. 2. What is the usefulness of carbohydrates to human beings?

Ans. Carbohydrates are very useful for human beings. They provide us all the three basic necessities of life *i.e., food* (starch containing grain), *clothes* (cellulose in the form of cotton, linen and rayon) and *shelter* (cellulose in the form of wood used for making our houses and furniture etc.). Not only this, our present civilisation depends on cellulose to a surprising degree, particularly in the form of paper. Carbohydrates are also important to the economy of many nations. For example, sugar is one of the most important commercial commodities. Many things of daily use, such as paper, photographic films, plastics etc. are derived from carbohydrates.

Q. 3. What is the general formula for carbohydrates?

Ans. Carbohydrates are so called, because the general formula for most of them could be written as $C_x(H_2O)_y$ and thus they may be regarded as hydrates of carbon. However, this definition was not found to be correct *e.g.*, rhamnose, a carbohydrate, is having the formula $C_6H_{12}O_5$ while acetic acid having formula $C_2H_4O_2$ is not a carbohydrate.

Chemically, carbohydrates contain mainly two functional groups, carbonyl group (aldehyde or ketone) and a number of hydroxyl groups. Accordingly carbohydrates are now defined as *polyhydroxy aldehydes or polyhydroxy ketones or the compound that can be hydrolyzed to either of them.*

The carbonyl group, however does not occur as such but is combined with hydroxyl groups to form intramolecular hemiacetal or acetal linkages. All the carbohydrates contain more than one asymmetric carbon atoms and are optically active.

Q. 4. Give the classification of carbohydrates.

(*Bhopal, 2004; Patna, 2005; Pune, 2010; Mumbai, 2010*

Ans. Carbohydrates are broadly classified into the following types:

(*a*) **Monosaccharides.** These are the simplest carbohydrates which cannot be hydrolyzed further into smaller units. They contain four, five or six carbon atoms and are known as *tetroses, pentoses* or *hexoses* respectively. Depending upon whether they contain an *aldehyde* or *keto* groups, they may be called aldoses or ketoses. For example, a five carbon monosaccharide having aldehyde group is called aldopentose and a six-carbon monosaccharide containing a keto group is called keto-hexose. A few examples of monosaccharides are given below:

Aldotetroses. Erythrose and Threose; $CH_2OH(CHOH)_2$ CHO.

Ketotetroses. Erythrulose, $CH_2OHCOCHOHCH_2OH$.

Aldopentoses. Ribose, Arabinose, Xylose and Lyxose.

$$CH_2OH(CHOH)_3CHO$$

All have a common molecular formula but different structures.

Ketopentoses. Ribulose and xylulose;

$$CH_2OHCO(CHOH)_2CH_2OH$$

Aldohexoses. Glucose, Mannose, Galactose etc;

$$CH_2OH(CHOH)_4CHO$$

Ketohexoses. Fructose, Sorbose etc.

$$CH_2OHCO(CHOH)_3CH_2OH$$

(*b*) **Oligosaccharides.** These are the carbohydrates which can be hydrolysed into a definite number of monosaccharide molecules. Depending upon the number of monosaccharides that are obtained from them on hydrolysis, they may be called *di-, tri- or tetra*-saccharides. For example:

(*i*) **Disaccharides.** These give two molecules of monosaccharides on hydrolysis. For example:

$$\underset{\substack{\text{Sucrose}\\ \text{(A disaccharide)}}}{C_{12}H_{22}O_{11}} + H_2O \longrightarrow \underset{\text{Glucose}}{C_6H_{12}O_6} + \underset{\text{Fructose}}{C_6H_{12}O_6}$$

Other examples of disaccharides are Maltose and Lactose having molecular formula, $C_{12}H_{22}O_{11}$.

(*ii*) **Trisaccharides.** These give three molecules of monosacharides on hydrolysis. For example:

$$\underset{\substack{\text{Raffinose}\\ \text{(A trisaccharide)}}}{C_{18}H_{32}O_{16}} + 2H_2O \longrightarrow \underset{\text{Galactose}}{C_6H_{12}O_6} + \underset{\text{Glucose}}{C_6H_{12}O_6} + \underset{\text{Fructose}}{C_6H_{12}O_6}$$

(c) Polysaccharides. These are high molecular weight carbohydrates yielding a large number of monosaccharide molecules (hundreds or even thousands) on hydrolysis, *e.g.*, Starch and cellulose have molecular formula $(C_6H_{10}O_5)_n$, where *n* is very large number.

$$\underset{\substack{\text{Starch}\\ \text{(A polysaccharide)}}}{(C_6H_{10}O_5)_n} + nH_2O \longrightarrow \underset{\text{Glucose}}{nC_6H_{12}O_6}$$

Q. 5. Discuss the occurrence and physical properties of glucose.

Ans. Glucose is the most important and abundant sugar and occurs in honey, sweet fruits (ripe grapes and mangoes), blood and urine of animals. It is a normal constituent of blood (0.1%, also known as blood sugar) and is present in the urine (8–10%) of a diabetic person. In the combined form, it is a constituent of many disaccharides such as sucrose and polysaccharides such as starch, cellulose etc.

Physical Properties

1. It is a colourless sweet crystalline compound having m.p. 419 K.

2. It is readily soluble in water, sparingly soluble in alcohol and insoluble in ether.

3. It forms a monohydrate having m.p. 391 K.

4. It is optically active and its solution is dextrorotatory (hence the name *dextrose*). The specific rotation of fresh solution is + 112°C.

5. It is about three-fourth as sweet as cane-sugar *i.e.*, sucrose.

Q. 6. Describe the chemical properties of glucose. *(Kerala, 2000; Mumbai, 2010)*

Ans. (A) Reactions of the aldehydic group

1. Oxidation. (Glucose gets oxidised to gluconic acid with mild oxidising agents like bromine water

$$\underset{\text{Glucose}}{CH_2OH\,(CHOH)_4CHO} \xrightarrow[Br_2]{[O]} \underset{\text{Gluconic acid}}{CH_2OH(CHOH)_4COOH}$$

only –CHO group is affected.

(*b*) A strong oxidising agent like nitric acid oxidises both the terminal groups viz. $-CH_2OH$ and –CHO groups and saccharic acid or glucaric acid is obtained.

$$\underset{\text{Glucose}}{CH_2OH(CHOH)_4CHO} \xrightarrow[{[O]}]{HNO_3} \underset{\text{Saccharic or glucaricacid}}{COOH(CHOH)_4\,COOH}$$

(*c*) Glucose gets oxidised to gluconic acid with ammoniacal silver nitrate (*Tollen's reagent*) and alkaline copper sulphate (*Fehling solution*). Tollen's reagent is reduced to metallic silver and Fehling solution to cuprous oxide which is a red precipitate.

(*i*) **With Tollen's reagent**

$$AgNO_3 + NH_4OH \longrightarrow AgOH + NH_4NO_3$$

$$2AgOH \longrightarrow Ag_2O + H_2O$$

$$\underset{\text{Glucose}}{CH_2OH(CHOH)_4CHO} + Ag_2O \longrightarrow \underset{\text{Gluconic acid}}{CH_2OH(CHOH)_4COOH} + \underset{\text{Silver mirror}}{2Ag\downarrow}$$

(*ii*) **With Fehling solution**

$$CuSO_4 + 2NaOH \longrightarrow Cu(OH)_2 + Na_2SO_4$$

$$Cu(OH)_2 \longrightarrow CuO + H_2O$$

$$\underset{\text{Glucose}}{CH_2OH(CHOH)_4CHO} + 2CuO \longrightarrow \underset{\text{Gluconic acid}}{CH_2OH(CHOH_4COOH} + \underset{\text{Red ppt}}{Cu_2O}$$

2. Reduction. (*a*) Glucose is reduced to sorbitol on treatment with sodium amalgam and water.

$$\underset{\text{Glucose}}{CH_2OH(CHOH)_4CHO} + 2[H] \xrightarrow[\text{Water}]{Na/Hg} \underset{\text{Sorbitol}}{CH_2OH(CHOH)_4CH_2OH}$$

(*b*) On reduction with conc. HI and red phosphorus at 373 K glucose gives a mixture of *n*-hexane and 2-iodohexane.

$$\underset{\text{Glucose}}{CH_2OH(CHOH)_4CHO} \xrightarrow{HI/P} \underset{n\text{–hexane}}{CH_3(CH_2)_4CH_3} + \underset{\text{2–Iodohexane}}{CH_3(CH_2)_3CHICH_3}$$

3. Reaction with HCN. An addition compound glucose cyanohydrin is obtained which on hydrolysis gives a hydroxy acid. On treatment with HI, hydroxy acid gives heptanoic acid.

$$CH_2OH(CHOH)_4CHO + HCN \longrightarrow CH_2OH(CHOH)_4CH\begin{matrix} OH \\ CN \end{matrix}$$

Glucose cyanohydrin

$\downarrow$ Hydrolysis

$$CH_3(CH_2)_5COOH \xleftarrow{HI} CH_2OH(CHOH)_4CH(OH)-COOH$$

Heptanoic acid — Hydroxy acid

4. Reaction with hydroxyl amine. Glucose forms glucoseoxime

$$\underset{\text{Glucose}}{CH_2OH(CHOH)_4CH}\boxed{O + H_2}\underset{\text{Hydroxyl amine}}{NOH} \longrightarrow \underset{\text{Glucose oxime}}{CH_2OH(CHOH)_4CH = NOH} + H_2O$$

5. Reaction with phenyl hydrazine. Glucose reacts with phenyl hydrazine in stages as shown below:

(*i*) One molecule of phenyl hydrazine reacts with one molecule of glucose to form glucose phenyl hydrazone.

$$\underset{\text{Glucose}}{\begin{matrix} CH\boxed{O + H_2}NNHC_6H_5 \\ | \\ CHOH \\ | \\ (CHOH)_3 \\ | \\ CH_2OH \end{matrix}} \longrightarrow \underset{\text{Glucose phenyl hydrazone}}{\begin{matrix} CH = NNHC_6H_5 \\ | \\ CHOH \\ | \\ (CHOH)_3 \\ | \\ CH_2OH \end{matrix}}$$

(*ii*) The product obtained above reacts further with a molecule of phenyl hydrazine as under:

$$\begin{matrix} CH = NNHC_6H_5 \\ | \\ CHOH \\ | \\ (CHOH)_3 \\ | \\ CH_2OH \end{matrix} + C_6H_5NHNH_2 \longrightarrow \underset{\text{Glucosone phenyl hydrazone}}{\begin{matrix} CH = NNHC_6H_5 \\ | \\ CO \\ | \\ (CHOH)_3 \\ | \\ CH_2OH \end{matrix}} + NH_3 + \underset{\text{Aniline}}{C_6H_5NH_2}$$

(*iii*) The product in step (*ii*) reacts further with a molecule of phenyl hydrazine to give glucosazone, a yellow solid having a sharp melting point.

$$\begin{matrix} CH = NNHC_6H_5 \\ | \\ C = \boxed{O + H_2}\underset{\text{Phenyl hydrazine}}{NNHC_6H_5} \\ | \\ (CHOH)_3 \\ | \\ CH_2OH \end{matrix} \longrightarrow \underset{\text{Glucosazone}}{\begin{matrix} CH = NNHC_6H_5 \\ | \\ C = NNHC_6H_5 \\ | \\ (CHOH)_3 \\ | \\ CH_2OH \end{matrix}} + H_2O$$

The above steps to explain the formation of glucosazone were given by Fischer.

Lately, however, the above mechanism of glucosazone formation has been superseded by the mechanism of Weygand. It is now explained in terms of *Amadori rearrangement* explained below.

$$\begin{array}{l} CHO \\ | \\ CHOH \\ | \\ (CHOH)_3 \\ | \\ CH_2OH \end{array} \xrightarrow{C_6H_5NHNH_2} \begin{array}{l} CH=NNHC_6H_5 \\ | \quad H \\ C \\ | \quad OH \\ (CHOH)_3 \\ | \\ CH_2OH \end{array} \rightleftharpoons \begin{array}{l} \quad NH \\ CH \quad NHC_6H_5 \\ \| \\ C \quad H \\ | \quad O \\ (CHOH)_3 \\ | \\ CH_2OH \end{array}$$

Glucose phenyl hydrazone

$$\xrightarrow{-C_6H_5NH_2} \begin{array}{l} CH=NH \\ | \\ C=O \\ | \\ (CHOH)_3 \\ | \\ CH_2OH \end{array} \xrightarrow[(-H_2O)]{C_6H_5NHNH_2} \begin{array}{l} CH=NH \\ | \\ C=NNHC_6H_5 \\ | \\ (CHOH)_3 \\ | \\ CH_2OH \end{array}$$

Iminoketone

$$\xrightarrow{C_6H_5NHNH_2} \begin{array}{l} CH=NNHC_6H_5 \\ | \\ C=NNHC_6H_5 + NH_3 \\ | \\ (CHOH)_3 \\ | \\ CH_2OH \end{array}$$

Glucosazone

(B) Reactions of the hydroxyl groups

1. Reaction with acetic anhydride or acetyl chloride. Glucose forms penta-acetate with acetic anhydride

$$\underset{\text{Glucose}}{\begin{array}{l} CHO \\ | \\ (CHOH)_4 \\ | \\ CH_2OH \end{array}} + \underset{\text{Acetic anhydride}}{5\,(CH_3CO)_2O} \longrightarrow \underset{\text{Glucose penta-acetate}}{\begin{array}{l} CHO \\ | \\ (CHOCOCH_3)_4 \\ | \\ CH_2OCOCH_3 \end{array}} + 5CH_3COOH$$

2. Reaction with methyl alcohol. Glucose reacts with methyl alcohol in the presence of dry HCl gas to form methyl glucoside.

$$\underset{\text{Glucose}}{C_6H_{11}O_5}\ \boxed{OH + H}\ \underset{\text{Methyl alcohol}}{OCH_3} \xrightarrow[\text{Gas}]{\text{HCl}} \underset{\text{Methyl glycoside}}{C_6H_{11}O_5OCH_3} + H_2O$$

3. Reaction with metallic hydroxides. Glucose reacts with calcium hydroxide to form calcium glucosate which is water-soluble.

$$\underset{\text{Glucose}}{C_6H_{11}O_5}\ \boxed{OH + H}\ \underset{\text{Cal.hydroxide}}{OCaOH} \longrightarrow \underset{\text{Cal.glucosate}}{C_6H_{11}O_5OCaOH} + H_2O$$

(C) Miscellaneous Reactions

1. Action of acids. On being heated with conc. HCl, glucose forms 5-hydroxy methyl furfural, which on further reaction gives laevulinic acid.

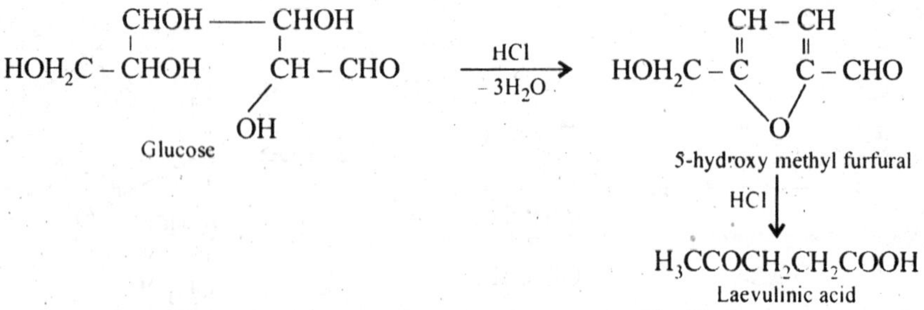

$$\underset{\text{Glucose}}{\text{HOH}_2\text{C}-\underset{|}{\overset{\text{CHOH}}{}}\text{CHOH}-\overset{\text{CHOH}}{}\text{CH(OH)}-\text{CHO}} \xrightarrow[-3H_2O]{HCl} \underset{\text{5-hydroxy methyl furfural}}{\text{HOH}_2\text{C}-\text{C}=\text{CH}-\text{CH}=\text{C}(-\text{O}-)-\text{CHO}}$$

$$\xrightarrow{HCl} \underset{\text{Laevulinic acid}}{H_3CCOCH_2CH_2COOH}$$

2. Fermentation. Glucose undergoes fermentation into ethyl alcohol in the presence of the enzyme *zymase.*

$$\underset{\text{Glucose}}{C_6H_{12}O_6} \xrightarrow{\text{Zymase}} \underset{\text{Ethyl alcohol}}{2C_2H_5OH + 2CO_2}$$

3. Reaction with alkalis. (*i*) Glucose is turned into brown resins in the presence of conc. alkalis.

(*ii*) In the presence of dilute solutions of alkalis, *Lobry de Bruyn-Van Ekenstein rearrangement* takes place which produces a mixture of (+) glucose, (+) mannose and (–) fructose. The reaction proceeds *via* 1, 2 enolisation.

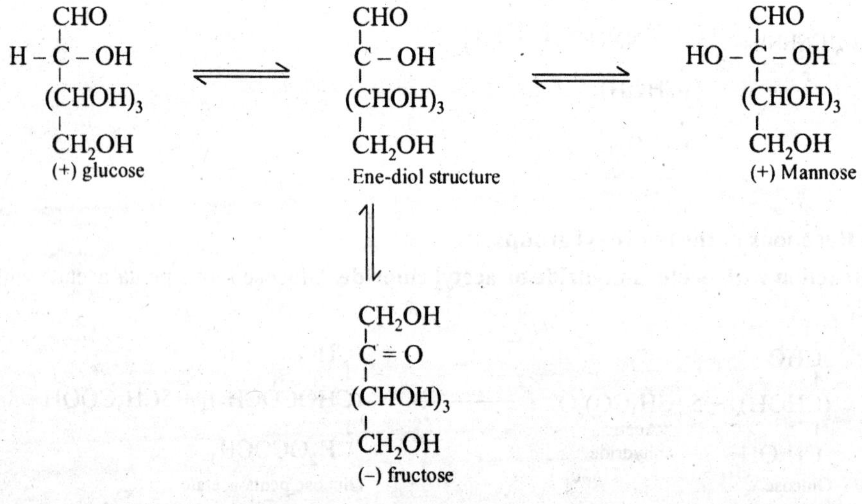

$$\underset{\text{(+) glucose}}{\begin{array}{c}CHO\\|\\H-C-OH\\|\\(CHOH)_3\\|\\CH_2OH\end{array}} \rightleftharpoons \underset{\text{Ene-diol structure}}{\begin{array}{c}CHO\\|\\C-OH\\|\\(CHOH)_3\\|\\CH_2OH\end{array}} \rightleftharpoons \underset{\text{(+) Mannose}}{\begin{array}{c}CHO\\|\\HO-C-OH\\|\\(CHOH)_3\\|\\CH_2OH\end{array}}$$

$$\Big\Updownarrow$$

$$\underset{\text{(–) fructose}}{\begin{array}{c}CH_2OH\\|\\C=O\\|\\(CHOH)_3\\|\\CH_2OH\end{array}}$$

Q. 7. Establish the open-chain structure of glucose.

(*Banaras 2002; Osmania 2003; Nagaland 2004; Vidyasagar 2007*)

Ans. 1. On the basis of elemental analysis and molecular weight determination the molecular formula of glucose is $C_6 H_{12} O_6$.

2. The reduction of glucose with red phosphorus and HI gives *n*-hexane.

$$\underset{\text{Glucose}}{C_6H_{12}O_6} \xrightarrow{HI} \underset{n\text{-hexane}}{C_6H_{14}}$$

Therefore, the six carbon atoms of glucoses form a straight chain.

3. It forms a penta acetate on treatment with acetic anhydride which indicates the presence of five hydroxyl groups in the molecule.

4. Glucose reacts with hydroxyl amine to form an oxime and with hydrogen cyanide to form a cyanohydrin. It indicates the presence of a carbonyl group. It also forms phenylhydrazone on treatment with phenylhydrazine.

5. The mild oxidation of glucose with bromine water or sodium hypobromite yields a monocarboxylic acid (gluconic acid) containing same number of carbon atoms as in glucose, *i.e.*, six. This indicates that the carbonyl group must be an aldehyde group. This is further confirmed by the fact that hydrolysis of cyanohydrin of glucose followed by reduction with hydroidic acid yields *n*-heptanoic acid.

6. The catalytic reduction of glucose yields a hexahydric alcohol (glucitol or sorbitol) which gives hexaacetate on treatment with acetic anhydride. The sixth hydroxy group must be obtained by the reduction of aldehyde group, thus further confirming the presence of an aldehyde group and five hydroxyl groups in glucose.

7. Oxidation of gluconic acid with nitric acid yields a dicarboxylic acid (glucaric acid or saccharic acid) with the same number (six) of carbon atoms as in glucose or gluconic acid. Thus besides aldehyde (–CHO) group, glucose must contain a primary alcoholic group (–CH_2 OH) also, which generates the second carboxylic group on oxidation. The two groups being monovalent must be present on either end of straight chain of six carbon atoms.

8. Glucose is a stable compound and does not undergo dehydration easily, indicating that not more than one hydroxyl group is bonded to a single carbon atom. Thus all the hydroxyl groups are attached to different carbon atoms.

On the basis of the above evidence, the structure of glucose may be written as:

$$CH_2OH - CHOH - CHOH - CHOH - CHOH - CHO$$

This structure explains all the reactions discussed above.

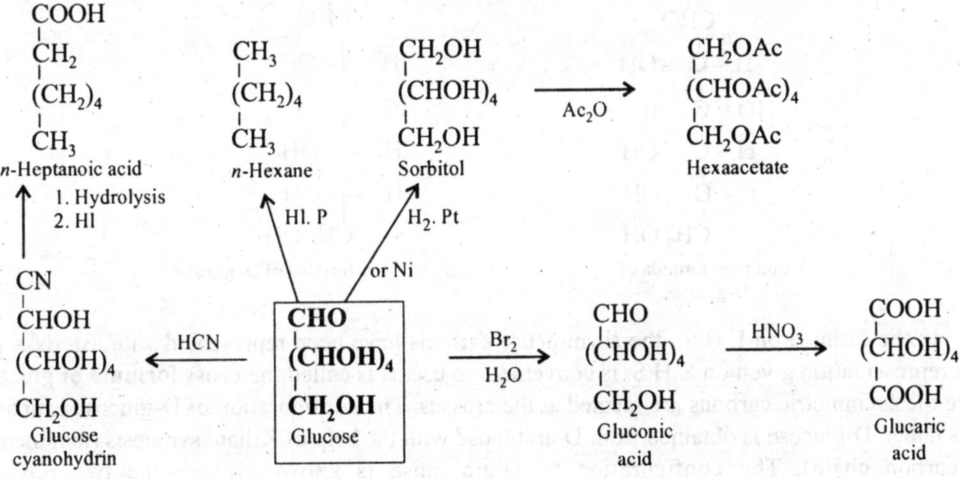

9. The above structure of glucose is also confirmed by the cleavage reaction of glucose with periodic acid. Five moles of periodic acid are consumed by one mole of glucose giving five moles of formic acid and one mole of formaldehyde.

$$\underset{\text{Glucose}}{C_6H_{12}O_6} + \underset{\text{Periodic acid}}{5HIO_4} \longrightarrow HCHO + 5HCOOH + 5HIO_3$$

Stepwise cleavage of glucose molecule is given below:

$$\begin{array}{c}CHO\\|\\CHOH\\|\\CHOH\\|\\CHOH\\|\\CHOH\\--|--\\CH_2OH\end{array}\xrightarrow{HIO_4}\begin{array}{c}CHO\\|\\CHOH\\|\\CHOH\\|\\CHOH\\|\\CHO\\+HCOOH\end{array}\xrightarrow{HIO_4}\begin{array}{c}CHO\\|\\CHOH\\|\\CHOH\\--|--\\CHO\\+HCOOH\end{array}\xrightarrow{HIO_4}$$

$$\begin{array}{c}CHO\\|\\CHOH\\--|--\\CHO\\+HCOOH\end{array}\xrightarrow{HIO_4}\begin{array}{c}CHO\\|\\CHO\\+HCOOH\end{array}\xrightarrow{HIO_4}\begin{array}{c}HCOOH\\+HCOOH\end{array}$$

Configuration of (+) glucose

10. *Configuration of a compound means the arrangement of various groups in the molecule.* In a molecule of glucose, there are four asymmetric carbon atoms, i.e., carbon atoms linked to four different groups. This gives rise to a possibility of 2^4, *i.e.*, 16 different configurations for glucose in which the different arrangements of –H and –OH groups exist. All these sixteen compounds having different names have been prepared. Emil Fischer and his coworkers gave the following configuration for D-glucose:

$$\begin{array}{c}CHO\\|\\H-\overset{*}{C}-OH\\|\\HO-\overset{*}{C}-H\\|\\H-\overset{*}{C}-OH\\|\\H-\overset{*}{C}-OH\\|\\CH_2OH\end{array}\qquad\qquad\begin{array}{c}CHO\\H\text{---}|\text{---}OH\\HO\text{---}|\text{---}H\\H\text{---}|\text{---}OH\\H\text{---}|\text{---}OH\\CH_2OH\end{array}$$

Common formula of D-glucose

Cross formula of D–glucose

In the formula on L.H.S., the asymmetric carbons have been represented with asterisks (*). The representation given on R.H.S. is convenient to use. It is called the **cross formula** of glucose. Here the asymmetric carbons are located at the crosses. The configuration of D-glucose is arrived at as under. D-glucose is obtained from D-arabinose with the help of Kiliani synthesis (lengthening of carbon chain). The configuration of D-arabinose is known. Hence the two possible configurations of D-glucose are given below:

$$\underset{\text{D–arabinose}}{\begin{array}{c}CHO\\|\\HO-C-H\\|\\H-C-OH\\|\\H-C-OH\\|\\CH_2OH\end{array}}\xrightarrow[\text{Synthesis}]{\text{Kiliani's}}\underset{\text{I}}{\begin{array}{c}CHO\\|\\H-C-OH\\|\\HO-C-H\\|\\H-C-OH\\|\\H-C-OH\\|\\CH_2OH\end{array}}\ \text{or}\ \underset{\text{II}}{\begin{array}{c}CHO\\|\\HO-C-H\\|\\HO-C-H\\|\\H-C-OH\\|\\H-C-OH\\|\\CH_2OH\end{array}}$$

11. Two different compounds glucose and gulose on oxidation give saccharic acid in which the end groups –CHO and – CH_2 OH are converted into – COOH groups. This means that the configuration around the asymmetric carbons is the same in glucose and gulose. Only the end groups have been interchanged. Structure I is found to be the true configuration for glucose as it can give rise to a different compound gulose in which the end groups have been interchanged. Structure II will not give a different compound after interchanging the end groups as explained below:

$$\begin{array}{c} CHO \\ | \\ H-C-OH \\ | \\ HO-C-H \\ | \\ H-C-OH \\ | \\ H-C-OH \\ | \\ CH_2OH \\ \text{I} \end{array} \xrightarrow[\text{end groups}]{\text{Interchanged}} \begin{array}{c} CH_2OH \\ | \\ H-C-OH \\ | \\ HO-C-H \\ | \\ H-C-OH \\ | \\ H-C-OH \\ | \\ CHO \\ \text{III Different compound (gulose)} \end{array}$$

$$\begin{array}{c} CHO \\ | \\ HO-C-H \\ | \\ HO-C-H \\ | \\ H-C-OH \\ | \\ H-C-OH \\ | \\ CH_2OH \\ \text{II} \end{array} \xrightarrow[\text{end groups}]{\text{Interchanged}} \begin{array}{c} CH_2OH \\ | \\ HO-C-H \\ | \\ HO-C-H \\ | \\ H-C-OH \\ | \\ H-C-OH \\ | \\ CHO \\ \text{IV} \end{array}$$

In fact there is no difference between structures II and IV. If structure IV is rotated through an angle of 180°, structure II is obtained. This leads us to the conclusion that D-glucose has the configuration I. The configuration II is for D-mannose.

Q. 8. Discuss the evidence leading to cyclic structure of D (+) glucose.

(Guwahati, 2002; Delhi, 2003; Kuvempu, 2005 Patna, 2005; Nagpur, 2008; Pune, 2010)

Or

Point out the limitations in the open chain structure of glucose. How these defects have been removed in the cyclic structure?

Ans. The open chain structure for glucose fails to explain some observations given below:

(*i*) Glucose does not restore the pink colour of Schiff's reagent, a characteristic property of aldehydic compounds.

(*ii*) Glucose does not react with ammonia or sodium bisulphite to form addition compounds.

It means the aldehydic group in glucose is not free to give these reactions.

(*iii*) D-glucose exists in two isomeric forms which undergo mutarotation. *The phenomenon of change of specific rotation of an optically active compound with time to an equilibrium value of specific rotation is called mutarotation.*

Glucose is an optically active compound because of the presence of asymmetric carbon atoms. It rotates the plane of polarised light. It is observed that when D-glucose having m.p. 419 K is dissolved in water, its specific rotations gradually drops from an initial value of + 112° to a value + 52.7°. On the other hand, when D-glucose having m.p. 423 K is dissolved in water, the specific rotation gradually increases from + 19° to 52.7°. The form of D-glucose having m.p. 419 K and having a higher specific rotation is called α-D-glucose and the other one is called β-D-glucose.

(*iv*) These two forms of glucose form methyl α-D-glucoside and methyl β-D-glucoside on treatment with methyl alcohol in the presence of dry HCl. These glucosides do not exhibit mutarotation and fail to give the Tollen's reagent or Fehling solution tests.

To account for the above, ring structure for glucose was given by Fischer, Tanret and Haworth, as shown below :

H – C – OH
H – C – OH
HO – C – H   O
H – C – OH
H – C
CH_2OH

α-D (+) glucose
m.p. 419 K, [α] = +112°

HO – C – H
H – C – OH
HO – C – H   O
H – C – OH
H – C
CH_2OH

β-D (+) glucose
m.p. 419 K, [α] = +19°

Ring structure explains the facts which are not explained by open-chain structure as follows:

(*a*) The existence of two forms of glucose, α and β, is explained. The two forms differ in configuration at C–1. Such a carbon atom is called *anomeric carbon atom and the two forms are called anomers.*

(*b*) Mutarotation is explained by the interconversion of cyclic forms into each other till 36% of α-forms and 64% of β-form are present in the equilibrium mixture. This conversion takes place through the open-chain structure. Thus whether we take the α-form or β form, on dissolution in water, we shall obtain an equilibrium mixture with the percentage given above.

H – C – OH
H – C – OH
HO – C – H   O
H – C – OH
H – C
CH_2OH

α-D (+) glucose

⇌

CHO
H – C – OH
HO – C – H
H – C – OH
H – C – OH
CH_2OH

Open-chain form

⇌

HO – C – H
H – C – OH
HO – C – H   O
H – C – OH
H – C
CH_2OH

β-D (+) glucose

(*c*) α and β methyl glucosides do not exhibit mutarotation, they are not readily hydrolysed to the open-chain form. The structures of methyl glucosides are given below. They are acetals.

H – C – OCH_3
H – C – OH
HO – C – H   O
H – C – OH
H – C
CH_2OH

Methyl α-D-glucoside

H_3CO – C – H
H – C – OH
HO – C – H   O
H – C – OH
H – C
CH_2OH

Methyl β-D-glucoside

(*ii*) Reaction with Tollen's reagent and Fehling solution and osazone formation is explained by the presence of the open-chain form in traces.

Size of the Ring

Heworth and coworkers have shown that D-glucose and glucosides have a six-membered ring and not five-membered ring as proposed earlier. They made these inferences based on the following reactions:

(*i*) Methyl β-D-glucoside (A) on treatment with methyl sulphate and sodium hydroxide is converted to compound (B) as under:

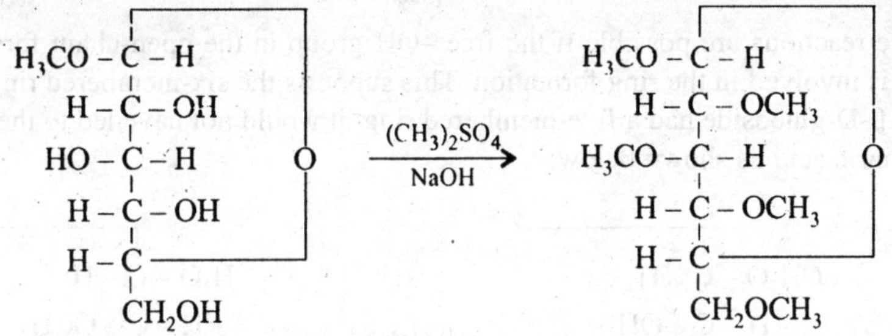

$$\begin{array}{c} H_3CO-C-H \\ H-C-OH \\ HO-C-H \\ H-C-OH \\ H-C \\ CH_2OH \end{array}\ O \quad \xrightarrow[NaOH]{(CH_3)_2SO_4} \quad \begin{array}{c} H_3CO-C-H \\ H-C-OCH_3 \\ H_3CO-C-H \\ H-C-OCH_3 \\ H-C \\ CH_2OCH_3 \end{array}\ O$$

Methyl βD-glucose (A) — Methyl β2, 3, 4, 6 tetra O-methyl D-glucoside (B)

The compound (B) is hydrolysed by dil. HCl to compound C which is a cyclic hemi-acetal and is in equilibrium with open-chain structure. Only the $-OCH_3$ at C-1 in the compound B is hydrolysed because acetal is more easily hydrolysed than an ether, ($-OCH_3$ are ether groups at C-2, C-3, C-4 and C-6).

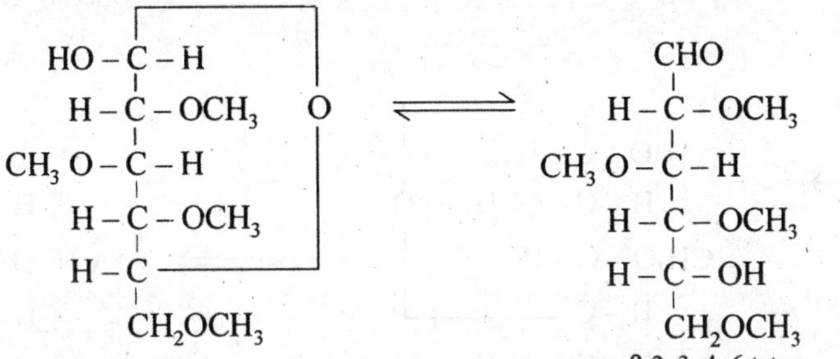

$$\begin{array}{c} HO-C-H \\ H-C-OCH_3 \\ CH_3O-C-H \\ H-C-OCH_3 \\ H-C \\ CH_2OCH_3 \end{array}\ O \quad \rightleftharpoons \quad \begin{array}{c} CHO \\ H-C-OCH_3 \\ CH_3O-C-H \\ H-C-OCH_3 \\ H-C-OH \\ CH_2OCH_3 \end{array}$$

β-2, 3, 4, 6 tetra O-methyl glucose (C) — β-2, 3, 4, 6 tetra O-methyl glucose (Open-chain form)

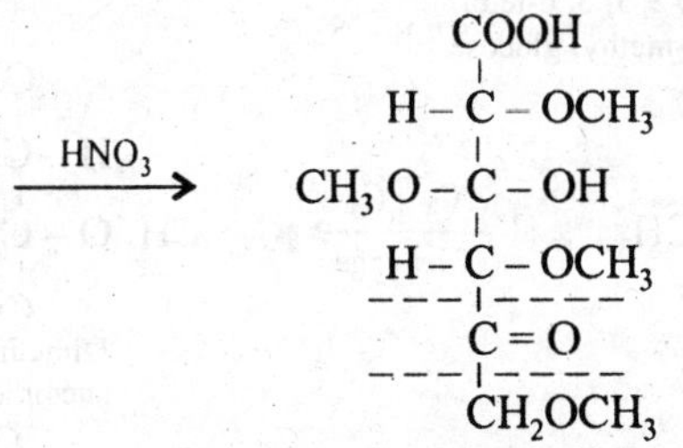

$$\xrightarrow{HNO_3} \quad \begin{array}{c} COOH \\ H-C-OCH_3 \\ CH_3O-C-OH \\ H-C-OCH_3 \\ C=O \\ CH_2OCH_3 \end{array}$$

(keto acid) (D)

Compound (C) on oxidation with HNO_3 gives a mixture of trimethoxy glutaric acid and dimethoxy succinic acid. These may be considered to have been formed via an intermediate keto acid (D). The two acids may be supposed to have been formed by cleavage between $C_5 - C_6$ and $C_4 - C_5$ respectively.

Compound D above $\longrightarrow$

$$\begin{array}{c} COOH \\ | \\ H-C-OCH_3 \\ | \\ CH_3O-C-H \\ | \\ H-C-OCH_3 \\ | \\ COOH \end{array} \quad + \quad \begin{array}{c} COOH \\ | \\ H-C-OCH_3 \\ | \\ CH_3O-C-H \\ | \\ COOH \end{array}$$

Trimethoxy gluatric acid   Dimethoxy succinic acid

The above reactions are possible if the free –OH group in the open-chain form above is at C-5. Thus C-5 is involved in the ring formation. This supports the six-membered ring for glucose.

If methyl β-D-glucoside had a five-membered ring, it would not have led to the formation of trimethoxy glutaric acid as shown below:

$$\begin{array}{c} CH_3O-C-H \\ | \\ H-C-OH \\ | \\ HO-C-H \\ | \\ H-C \\ | \\ H-C-OH \\ | \\ CH_2OH \end{array} \text{(ring O between C-1 and C-4)} \xrightarrow[NaOH]{(CH_3)_2SO_4} \begin{array}{c} CH_3O-C-H \\ | \\ H-C-OCH_3 \\ | \\ CH_3O-C-H \\ | \\ H-C \\ | \\ H-C-OCH_3 \\ | \\ CH_2OCH_3 \end{array} \text{(ring O between C-1 and C-4)}$$

Methyl β-D-glucoside (A)   Methyl β-2, 3, 5, 6 tetra O-methyl glucoside

$$\xrightarrow[HCl]{dil.} \begin{array}{c} HO-C-H \\ | \\ H-C-OCH_3 \\ | \\ CH_3O-C-H \\ | \\ H-C \\ | \\ H-C-OCH_3 \\ | \\ CH_2OCH_3 \end{array} \text{(ring O)} \rightleftharpoons \begin{array}{c} CHO \\ | \\ H-C-OCH_3 \\ | \\ CH_3O-C-H \\ | \\ H-C-OH \\ | \\ H-C-OCH_3 \\ | \\ CH_2OCH_3 \end{array}$$

β-2, 3, 5, 6-tetra O-methyl glucose   Open-chain form

$$\longrightarrow \begin{array}{c} COOH \\ | \\ H-C-OCH_3 \\ | \\ CH_3O-C-H \\ -\,-\,-|-\,-\,- \\ C=O \\ -\,-\,-|-\,-\,- \\ H-C-OCH_3 \\ | \\ CH_2OCH_3 \end{array} \quad \begin{array}{l} \xrightarrow[\text{Cleavage}]{C_4-C_5} \begin{array}{c} CHO \\ | \\ H-C-OCH_3 \\ | \\ CH_3O-C-H \\ | \\ COOH \end{array} \text{ Dimethoxy succinic acid} \\ \\ \xrightarrow[\text{Cleavage}]{C_3-C_5} \begin{array}{c} COOH \\ | \\ H-C-OCH_3 \\ | \\ COOH \end{array} \text{ Methoxy malonic acid} \end{array}$$

Instead, dimethoxy succinic acid and methoxy malonic acid would be formed, which is not the case. Hence glucose has a six-membered ring.

Q. 9. Draw Haworth projection formulae of α-and β-D-glucose.

(*Lucknow, 2010; Mumbai, 2010*)

Howorth Pyranose Structure (Projection Formula)

Haworth suggested that D-glucose can be represented in the form of a cyclic structure shown below:

CH_2OH, 5, O, H, H, 4, 1, OH, H, OH, OH, 3, 2, H, OH

α - D-Glucose

CH_2OH, O, H, H, OH, 4, 1, OH, OH, H, H, 3, 2, H, OH

β - D-Glucose

The ring is in the same plane. It consists of five carbons (not shown) C_1 to C_5. One position of the ring is occupied by oxygen. At each carbon C_1 to C_4, two groups –H and –OH are attached. These groups are perpendicular to the plane of the ring. The bonds going up are above the plane and going down are below the plane. At C-5, $-CH_2$ OH is above the plane and –H below the plane. α-D-glucose and β-D-glucose differ in the respect that at C-1, –OH is below the plane in α-isomer whereas it is above the plane in β-isomer.

Q. 10. What is fructose (fruit sugar)? Mention its physical properties.

Ans. Fructose is another commonly known monosaccharide having the same molecular formula as glucose. It is laevorotatory because it rotates the plane of polarised light towards the left. It is present abundantly in fruits. That is why it is called *fruit-sugar* also.

Physical Properties

1. It is the sweetest of all known sugars.
2. It is readily soluble in water, sparingly soluble in alcohol and insoluble in ether.
3. It is a white crystalline solid with m.p. 375 K.
4. Fresh solution of fructose has a specific rotation –133° which changes to –92°.at equilibrium due to mutarotation.

Q. 11. Describe the chemical properties of fructose.

[A] Reactions due to ketonic group

1. Reaction with hydrogen cyanide. Fructose reacts with HCN to form cyanohydrin.

$$\begin{matrix} CH_2OH \\ | \\ CO \\ | \\ (CHOH)_3 \\ | \\ CH_2OH \end{matrix} + HCN \longrightarrow \begin{matrix} CH_2OH \\ | \\ C \begin{smallmatrix} OH \\ CN \end{smallmatrix} \\ | \\ (CHOH)_3 \\ | \\ CH_2OH \end{matrix}$$

Fructose   Fructose cyanohydrin

2. Reaction with hydroxylamine. Fructose reacts with hydroxylamine to form an oxime

$$\begin{matrix} CH_2OH \\ | \\ C \;\boxed{O + H_2}\; NOH \\ | \\ (CHOH)_3 \\ | \\ CH_2OH \end{matrix} \longrightarrow \begin{matrix} CH_2OH \\ | \\ C = NOH \\ | \\ (CHOH)_3 \\ | \\ CH_2OH \end{matrix}$$

Fructose   Fructose oxime

3. Reaction with phenylhydrazine

$$\begin{array}{l} CH_2OH \\ | \\ C \;\boxed{O + H_2}\; NNHC_6H_5 \\ | \\ (CHOH)_3 \\ | \\ CH_2OH \\ \text{Fructose} \end{array} \quad \begin{array}{c} \\ \\ \text{Phenyl} \\ \text{hydrazine} \end{array} \longrightarrow \begin{array}{l} CH_2OH \\ | \\ C = NNHC_6H_5 + H_2O \\ | \\ (CHOH)_3 \\ | \\ CH_2OH \\ \text{Fructose phenyl} \\ \text{hydrazone} \end{array}$$

With another molecule of phenyl hydrazine, the reaction is as under:

$$\begin{array}{l} CH_2OH \\ | \\ C = NNHC_6H_5 + C_6H_5NHNH_2 \\ | \\ (CHOH)_3 \\ | \\ CH_2OH \end{array} \longrightarrow \begin{array}{l} CHO \\ | \\ C = NNHC_6H_5 + \underset{\text{Aniline}}{C_6H_5NH_2} + NH_3 \\ | \\ (CHOH)_3 \\ | \\ CH_2OH \\ \text{(Intermediate} \\ \text{product)} \end{array}$$

The intermediate product containing the –CHO group finally reacts with the third molecule of phenyl hydrazine to give fructosazone.

$$\begin{array}{l} CH \;\boxed{O + H_2}\; NNHC_6H_5 \\ | \\ C = NNHC_6H_5 \\ | \\ (CHOH)_3 \\ | \\ CH_2OH \end{array} \longrightarrow \begin{array}{l} CH = NNHC_6H_5 \\ | \\ C = NNHC_6H_5 + H_2O \\ | \\ (CHOH)_3 \\ | \\ CH_2OH \\ \text{Frutosazone} \end{array}$$

It has been found that fructosazone has the same structure as glucosazone. The m.p. is 478 K. A modern mechanism due to Weygand explaining the formation of fructosazone is given below:

$$\begin{array}{l} CH_2OH \\ | \\ C = O \\ | \\ (CHOH)_3 \\ | \\ CH_2OH \end{array} \xrightarrow{C_6H_5NHNH_2} \begin{array}{l} CH - OH \\ |\swarrow H \\ C = NNHC_6H_5 \\ | \\ (CHOH)_3 \\ | \\ CH_2OH \end{array} \rightleftharpoons \begin{array}{l} CH - O - H \\ \| \\ C - NH - NHC_6H_5 \\ | \\ (CHOH)_3 \\ | \\ CH_2OH \end{array}$$

$$\xrightarrow{-C_6H_5NH_2} \begin{array}{l} CHO \\ | \\ C = NH \\ | \\ (CHOH)_3 \\ | \\ CH_2OH \\ \text{Iminoaldehyde} \end{array} \xrightarrow[-H_2O]{C_6H_5NHNH_2} \begin{array}{l} CH = NNHC_6H_5 \\ | \\ C = NH \\ | \\ (CHOH)_3 \\ | \\ CH_2OH \end{array}$$

$$\xrightarrow{C_6H_5NHNH_2} \begin{array}{l} CH = NNHC_6H_5 \\ | \\ C = NNHC_6H_5 + NH_3 \\ | \\ (CHOH)_3 \\ | \\ CH_2OH \\ \text{Fructosazone} \end{array}$$

4. Reduction. Fructose gives a mixture of sorbitol and mannitol on reduction with Na–Hg and water or catalytic hydrogenation.

$$\underset{\text{Fructose}}{\begin{matrix} CH_2OH \\ | \\ CO \\ | \\ (CHOH)_3 \\ | \\ CH_2OH \end{matrix}} + 2[H] \longrightarrow \underset{\text{Mannitol}}{\begin{matrix} CH_2OH \\ | \\ HO-C-H \\ | \\ (CHOH)_3 \\ | \\ CH_2OH \end{matrix}} + \underset{\text{Sorbitol}}{\begin{matrix} CH_2OH \\ | \\ H-C-OH \\ | \\ (CHOH)_3 \\ | \\ CH_2OH \end{matrix}}$$

5. Oxidation. (*i*) There is no action of a mild oxidising agent like bromine water on fructose.

(*ii*) Strong oxidising agents like nitric acid oxidise fructose into a mixture of trihydroxy glutaric, glycollic and tartaric acids.

$$\underset{\text{Fructose}}{\begin{matrix} CH_2OH \\ | \\ CO \\ | \\ (CHOH)_3 \\ | \\ CH_2OH \end{matrix}} \xrightarrow{[O]} \underset{\text{Trihydroxy glutaric acid}}{\begin{matrix} COOH \\ | \\ (CHOH)_3 \\ | \\ COOH \end{matrix}} + \underset{\text{Glycollic acid}}{\begin{matrix} CH_2OH \\ | \\ COOH \end{matrix}} + \underset{\text{Tartaric acid}}{\begin{matrix} COOH \\ | \\ (CHOH)_2 \\ | \\ COOH \end{matrix}}$$

(*iii*) Unlike other ketones, it reduces Tollen's reagent (ammoniacal silver nitrate) and Fehling solution (alkaline copper sulphate). This is due to the presence of traces of glucose in alkaline medium.

[B] Reactions of the alcoholic group

1. Acetylation. With acetic anhydride or acetyl chloride, fructose forms penta-acetate.

$$\underset{\text{Fructose}}{\begin{matrix} CH_2OH \\ | \\ CO \\ | \\ (CHOH)_3 \\ | \\ CH_2OH \end{matrix}} + \underset{\text{Acetic anhydride}}{5\,(CH_3CO)_2O} \longrightarrow \underset{\text{Fructose penta-acetate}}{\begin{matrix} CH_2OCOCH_3 \\ | \\ CO \\ | \\ (CHOCOCH_3)_3 \\ | \\ CH_2OCOCH_3 \end{matrix}} + 5\,CH_3COOH$$

2. Reaction with methyl alcohol (glucoside formation). Fructose reacts with methyl alcohol in the presence of dry HCl gas forming methyl fructoside.

$$\underset{\text{Fructose}}{C_6H_{11}O_5}\ \boxed{OH + H}\ OCH_3 \xrightarrow[\text{gas}]{\text{HCl}} \underset{\text{Methyl fructoside}}{C_6H_{11}O_5OCH_3} + H_2O$$

3. Reaction with metallic hydroxides (fructosate formation)

$$\underset{\text{Fructose}}{C_6H_{11}O_5OH} + \underset{\text{Calcium hydroxide}}{HOCaOH} \longrightarrow \underset{\text{Cal. fructosate}}{C_6H_{11}O_5OCaOH} + H_2O$$

Q. 12. Describe various steps leading to the open chain structure of fructose.

Or

Give evidence to show that fructose is a ketohexose.

Ans. 1. Elemental analysis and molecular weight determination of fructose show that it has the molecular formula $C_6H_{12}O_6$.

2. Fructose on reduction gives sorbitol which on reduction with HI and red phosphorus gives a mixture of *n*-hexane and 2-iodohexane. This reaction indicates that the six carbon atoms in fructose are in a straight chain.

$$C_6H_{12}O_6 \xrightarrow[\text{Na / Hg / water}]{\text{Reduction}} \underset{\text{Sorbitol}}{\begin{array}{c}CH_2OH\\|\\(CHOH)_4\\|\\CH_2OH\end{array}} \xrightarrow{\text{HI / P}} \underset{\text{2-iodohexane}}{\begin{array}{c}CH_3\\|\\CHI\\|\\(CH_2)_3\\|\\CH_3\end{array}} + \underset{n\text{-Hexane}}{\begin{array}{c}CH_3\\|\\(CH_2)_4\\|\\CH_3\end{array}}$$

3. Fructose reacts with hydroxylamine, hydrogen cyanide and phenyl hydrazine. It shows the presence of –CHO or > C = O group in the molecule of fructose.

4. On treatment with bromine water, no reaction takes place. This rules out the possibility of presence of –CHO group.

5. On oxidation with nitric acid, it gives glycollic acid and tartaric acids which contain smaller number of carbon atoms than fructose. This shows that a ketonic group is present at position 2. It is at this point that the molecule is broken.

$$\begin{array}{l}1\ CH_2OH\\ \quad |\\2\ CO\\ --|---\\3\ CHOH\\ \quad |\\4\ CHOH\\ \quad |\\5\ CHOH\\ \quad |\\6\ CH_2OH\end{array} \xrightarrow{[O]} \underset{\text{Glycollic acid}}{\begin{array}{c}CH_2OH\\|\\COOH\end{array}} + \underset{\text{Tartaric acid}}{\begin{array}{c}COOH\\|\\CHOH\\|\\CHOH\\|\\COOH\end{array}}$$

It also shows the presence of – CH_2OH at position 1 and 6.

6. The presence of five –OH groups is also indicated by the formation of pentaacetate with acetic anhydride. Moreover, the five –OH groups must be linked to separate carbon atoms, because the presence of two –OH groups to one carbon atom makes the compound to lose water on heating which is not the case with fructose. Based on the above observations, the straight chain structure of fructose is as:

$$\overset{1}{C}H_2OH\overset{2}{C}O\overset{3}{C}HOH\overset{4}{C}HOH\overset{5}{C}HOH\overset{6}{C}H_2OH$$

Facts in support of straight-chain structure

(*i*) Fructose cyanohydrin on hydrolysis gives a mono carboxylic acid which on reduction with HI gives α-methyl caproic acid.

$$\underset{\text{Cyanohydrin}}{\begin{array}{c}CH_2OH\\|\\C(OH)CN\\|\\(CHOH)_3\\|\\CH_2OH\end{array}} \xrightarrow{H_2O} \underset{\text{Hydroxy acid}}{\begin{array}{c}CH_2OH\\|\\C(OH)COOH\\|\\(CHOH)_3\\|\\CH_2OH\end{array}} \xrightarrow{H_2O} \underset{\alpha\text{-Methyl caproic acid}}{\begin{array}{c}CH_3\\|\\CHCOOH\\|\\(CH_2)_3\\|\\CH_3\end{array}}$$

(*ii*) Fructose gives on reduction, a mixture of mannitol and sorbitol, which confirm the presence of a ketonic group. On reduction the ketonic group is transformed into –CHOH. There are two possible arrangements of –H and –OH groups giving rise to two isomers.

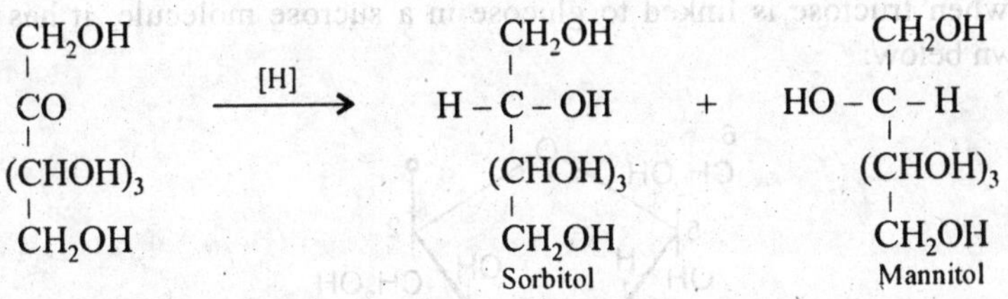

$$\begin{array}{c} CH_2OH \\ | \\ CO \\ | \\ (CHOH)_3 \\ | \\ CH_2OH \end{array} \xrightarrow{[H]} \underset{\text{Sorbitol}}{\begin{array}{c} CH_2OH \\ | \\ H-C-OH \\ | \\ (CHOH)_3 \\ | \\ CH_2OH \end{array}} + \underset{\text{Mannitol}}{\begin{array}{c} CH_2OH \\ | \\ HO-C-H \\ | \\ (CHOH)_3 \\ | \\ CH_2OH \end{array}}$$

Configuration of D(–) fructose

Glucose and fructose form the same osazone. And in this reaction, only C-1 and C-2 are involved. This means the configuration of rest of the molecule (arrangement of –H and –OH) should be the same as in glucose. In fructose C-3, C-4 and C-5 are chiral carbon atoms, whereas in glucose, C-2, C-3, C-4 and C-5 are chiral atoms. In view of the above, the configuration of fructose is as given below:

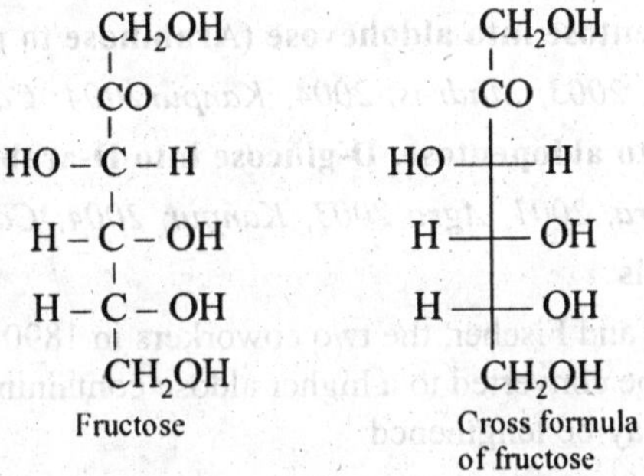

$$\underset{\text{Fructose}}{\begin{array}{c} CH_2OH \\ | \\ CO \\ | \\ HO-C-H \\ | \\ H-C-OH \\ | \\ H-C-OH \\ | \\ CH_2OH \end{array}} \qquad \underset{\text{Cross formula of fructose}}{\begin{array}{c} CH_2OH \\ | \\ CO \\ HO-\!\!\!+\!\!\!-H \\ H-\!\!\!+\!\!\!-OH \\ H-\!\!\!+\!\!\!-OH \\ CH_2OH \end{array}}$$

Q. 13. Establish the ring structure of fructose. *(Mumbai, 2010)*

Ans. Fructose shows the property of mutarotation. This means that it exists in two forms α-fructose and β-fructose which are cyclic in structure and change into each other via the open chain structure. The cyclic and pyranose structures of α-D-fructose and β-D-fructose are represented below:

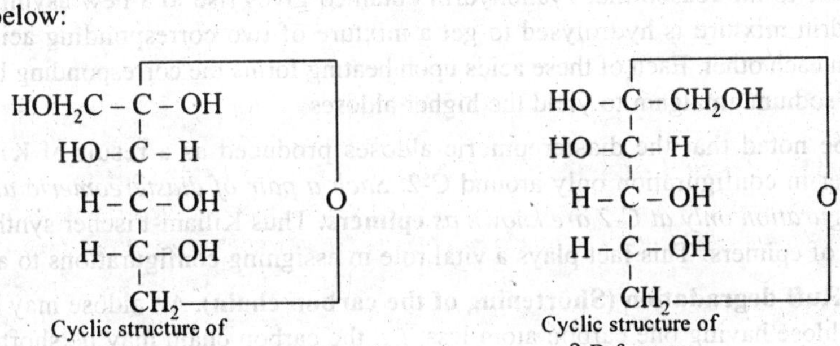

$$\underset{\text{Cyclic structure of α-D-fructose}}{\begin{array}{c} HOH_2C-C-OH \\ | \\ HO-C-H \\ | \\ H-C-OH \\ | \\ H-C-OH \\ | \\ CH_2 \end{array} \; O} \qquad \underset{\text{Cyclic structure of β-D-fructose}}{\begin{array}{c} HO-C-CH_2OH \\ | \\ HO-C-H \\ | \\ H-C-OH \\ | \\ H-C-OH \\ | \\ CH_2 \end{array} \; O}$$

Haworth Pyranose structure

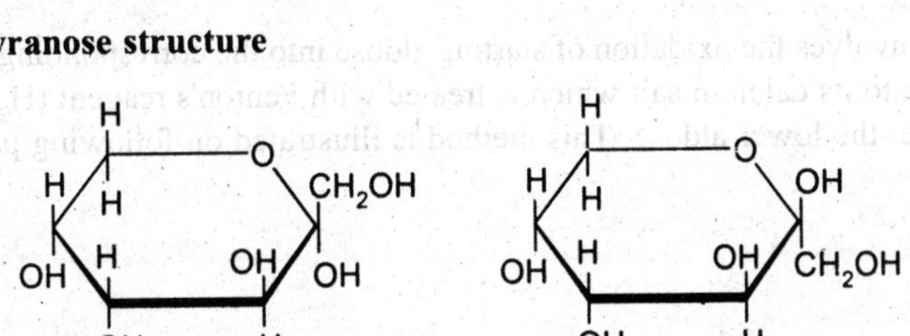

Pyranose structure of α-D-fructose

Pyranose structure of β-D-fructose

However, when fructose is linked to glucose in a sucrose molecule, it has the furanose structure as shown below:

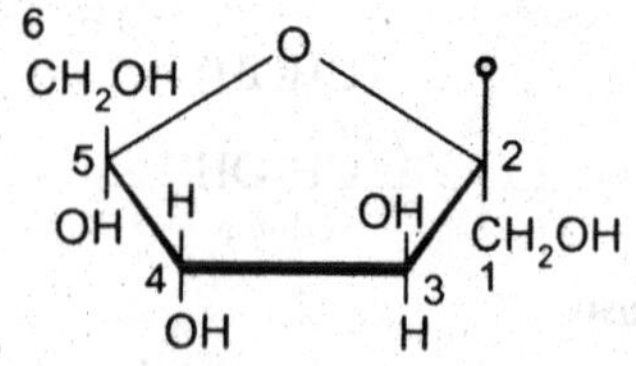

β-D-fructofuranose

Q. 14. Write notes on

(*a*) Kiliani synthesis (*Kerala, 2001; Nehru 2004*)

Or

Describe the conversion of an aldopentose into aldohexose (Arabinose to glucose).

(*Delhi, 2003; Madras, 2004; Kanpur2004; Calcutta 2007*)

(*b*) Ruff degradation (Aldohexose into aldopentose, D-glucose into D-arabinose)

(*Kurukshetra, 2001; Agra 2003; Kanpur, 2004; Calcutta 2007*)

Ans. (*a*) The Kiliani-Fischer Synthesis

Lengthening of carbon chain. Kiliani and Fischer, the two coworkers in 1890, developed a prominent method by which an aldose may be converted to a higher aldose containing one carbon atom more *i.e.* the carbon chain of aldose may be lengthened.

The complete process followed by them has been illustrated on next page by taking into consideration the conversion of an aldopentose into a mixture of two aldohexoses.

The process involves the reaction of given aldose with HCN, when a mixture of two diastereomeric cyanohydrins is formed.

This is due to the reason that cyanohydrin obtained gives rise to a new asymmetric centre. The cyanohydrin mixture is hydrolysed to get a mixture of two corresponding acids which are separated from each other. Each of these acids upon heating forms the corresponding lactone which is reduced by sodium amalgam to yield the higher aldoses.

It may be noted that the diastereomeric aldoses produced as a result of Kiliani-Fischer synthesis differ in configuration only around C-2. *Such a pair of diastereomeric aldoses which differ in configuration only at C-2 are known as* **epimers.** Thus Kiliani-Fischer synthesis leads to the formation of epimers. This fact plays a vital role in assigning configurations to aldoses.

(*b*) The Ruff degradation (Shortening of the carbon chain). An aldose may be converted into a lower aldose having one carbon atom less, *i.e.* the carbon chain may be shortened by *Ruff degradation.*

The method involves the oxidation of starting aldose into the corresponding aldonic acid. The acid is converted into its calcium salt which is treated with Fenton's reagent (H_2O in the presence of Fe^{3+} ions) to get the lower aldose. This method is illustrated on following pages:

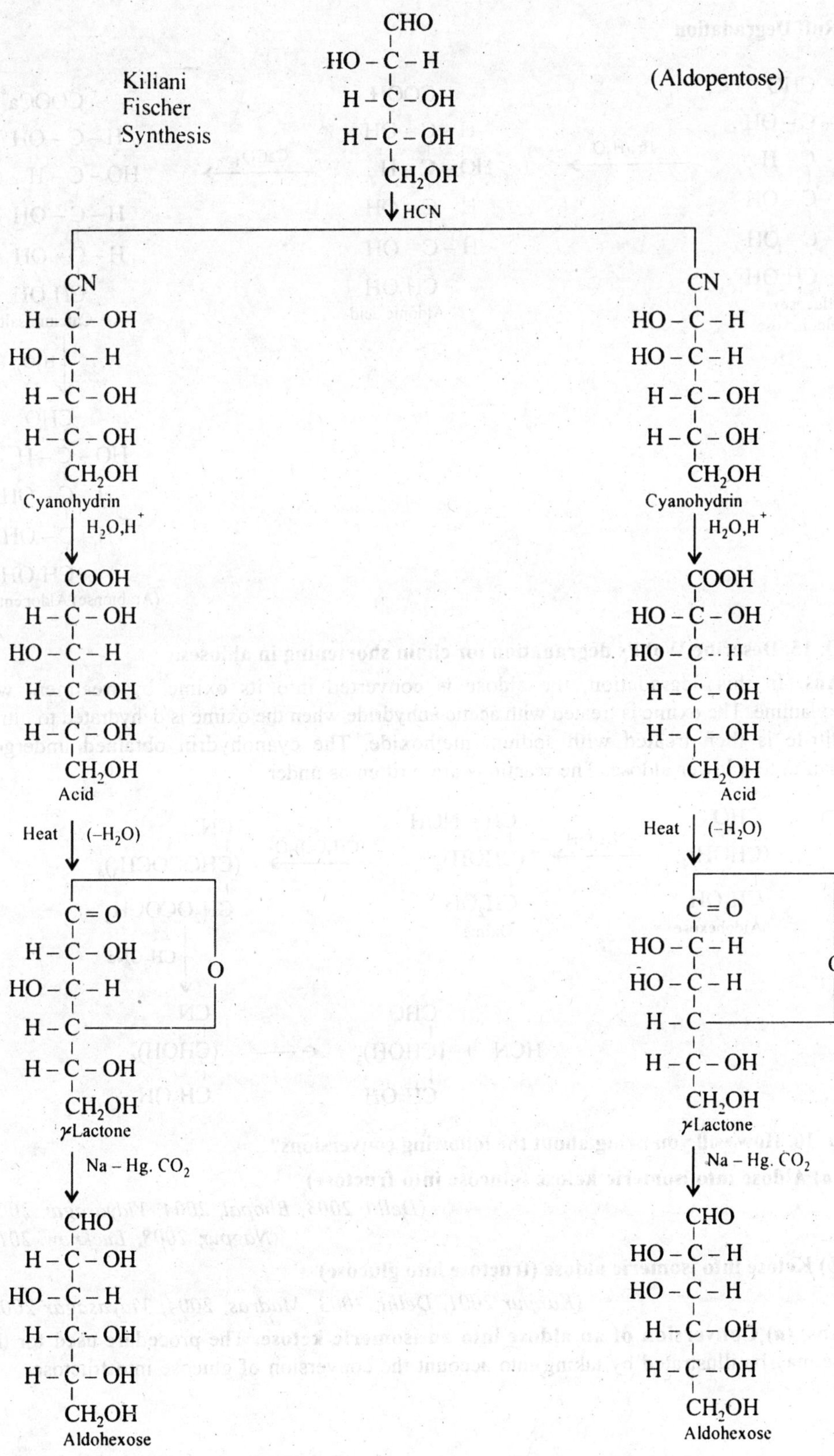
Kiliani
Fischer
Synthesis
(Aldopentose)
HCN
Cyanohydrin
Cyanohydrin
H_2O,H^+
Acid
Acid
Heat
$(-H_2O)$
γ-Lactone
γ-Lactone
Na – Hg. CO_2
Aldohexose
Aldohexose

Ruff Degradation

$$\begin{array}{c} CHO \\ | \\ H-C-OH \\ | \\ HO-C-H \\ | \\ H-C-OH \\ | \\ H-C-OH \\ | \\ CH_2OH \end{array} \xrightarrow{Br_2/H_2O} \begin{array}{c} COOH \\ | \\ H-C-OH \\ | \\ HO-C-H \\ | \\ H-C-OH \\ | \\ H-C-OH \\ | \\ CH_2OH \end{array} \xrightarrow{CaCO_3} \begin{array}{c} COOCa^{2+}/2 \\ | \\ H-C-OH \\ | \\ HO-C-H \\ | \\ H-C-OH \\ | \\ H-C-OH \\ | \\ CH_2OH \end{array}$$

(Glucose) An aldohexose — Aldonic acid — Calcium salt

$$\xrightarrow{H_2O_2,\ Fe^{3+}} \begin{array}{c} CHO \\ | \\ HO-C-H \\ | \\ H-C-OH \\ | \\ H-C-OH \\ | \\ CH_2OH \end{array}$$

(Arabionse) Aldopentose

Q. 15. Describe Wohl's degradation for chain shortening in aldoses.

Ans. In this degradation, the aldose is converted into its oxime by treatment with hydroxylamine. The oxime is treated with acetic anhydride when the oxime is dehydrated to nitrile. The nitrile is then treated with sodium methoxide. The cyanohydrin obtained undergoes degradation to a lower aldose. The reactions are written as under

$$\underset{\text{Aldohexose}}{\begin{array}{c} CHO \\ | \\ (CHOH)_4 \\ | \\ CH_2OH \end{array}} \xrightarrow{NH_2OH} \underset{\text{Oxime}}{\begin{array}{c} CH=NOH \\ | \\ (CHOH)_4 \\ | \\ CH_2OH \end{array}} \xrightarrow{(CH_3CO)_2O} \begin{array}{c} CN \\ | \\ (CHOCOCH_3)_4 \\ | \\ CH_2OCOCH_3 \end{array}$$

$$\xrightarrow{CH_3ONa} \begin{array}{c} CN \\ | \\ (CHOH)_4 \\ | \\ CH_2OH \end{array} \longrightarrow HCN + \begin{array}{c} CHO \\ | \\ (CHOH)_3 \\ | \\ CH_2OH \end{array}$$

Q. 16. How will you bring about the following conversions?

***a*) Aldose into isomeric ketose (glucose into fructose)**

(*Delhi, 2003; Bhopal, 2004; Vidyasagar, 2007; Nagpur, 2008; Lucknow, 2010*)

(*b*) Ketose into isomeric aldose (fructose into glucose)

(*Kanpur 2001, Delhi, 2003; Madras, 2004; Vidyasagar 2007*)

Ans. (*a*) Conversion of an aldose into an isomeric ketose. The procedure used for this purpose may be illustrated by taking into account the conversion of glucose into fructose.

$$\begin{array}{c}\text{CHO}\\ |\\ \text{CHOH}\\ |\\ (\text{CHOH})_3\\ |\\ \text{CH}_2\text{OH}\\ \textbf{Glucose}\end{array} \xrightarrow[\text{(2 moles)}]{C_6H_5NHNH_2} \begin{array}{c}\text{CH}=\text{NNHC}_6\text{H}_5\\ |\\ \text{C}=\text{NNHC}_6\text{H}_5\\ |\\ (\text{CHOH})_3\\ |\\ \text{CH}_2\text{OH}\\ \textbf{Glucosazone}\end{array} \xrightarrow{H_2O,\ H^-} \begin{array}{c}\text{CHO}\\ |\\ \text{C}=\text{O}\\ |\\ (\text{CHOH})_3\\ |\\ \text{CH}_2\text{OH}\\ \textbf{Glucosone}\end{array}$$

$$\xrightarrow[\text{Zn/CH}_3\text{COOH}]{\text{Reduction}} \begin{array}{c}\text{CH}_2\text{OH}\\ |\\ \text{C}=\text{O}\\ |\\ (\text{CHOH})_3\\ |\\ \text{CH}_2\text{OH}\\ \textbf{Fructose}\end{array}$$

(b) Conversion of a ketose into an isomeric aldose. The procedure used here may be illustrated by taking into account the conversion of fructose into a mixture of epimeric aldoses, *viz.*, glucose and mannose.

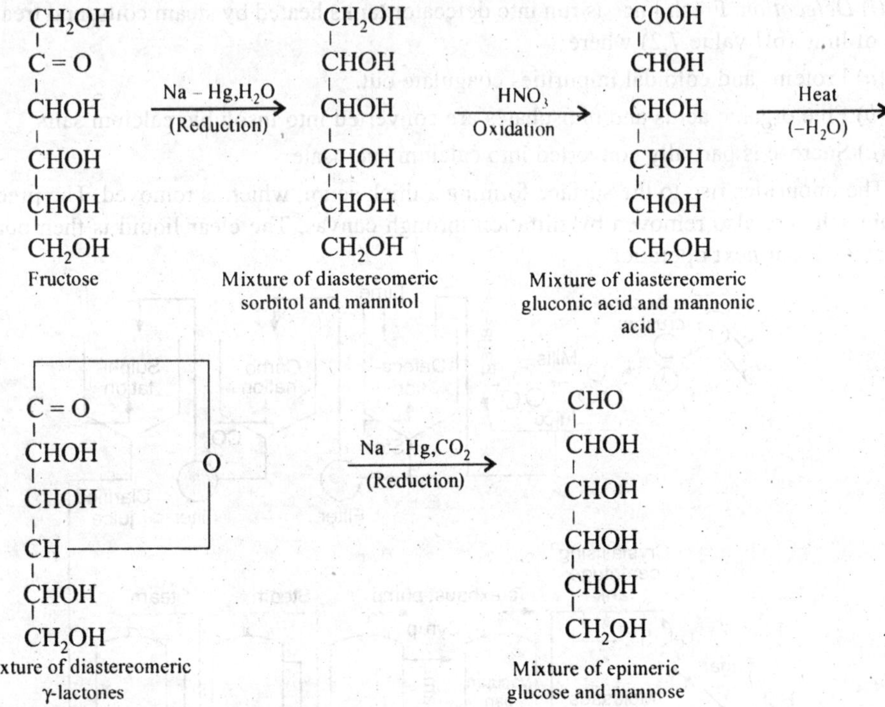

Q. 17. Write the products of hydrolysis of different disaccharides. What are the enzymes used for carrying out hydrolysis?

Ans. The disaccharides yield on hydrolysis two monosaccharides. Those disaccharides which yield two hexoses on hydrolysis have a general formula $C_{12}H_{22}O_{11}$. The hexoses obtained on hydrolysis may be same or different *e.g.*,

$$\underset{\text{Sucrose}}{C_{12}H_{22}O_{11}} \longrightarrow \underset{\text{Glucose}}{C_6H_{12}O_6} + \underset{\text{Fructose}}{C_6H_{12}O_6}$$

$$\text{Lactose} \xrightarrow{+H_2O} \text{Glucose + Galactose}$$

$$\text{Maltose} \xrightarrow{+H_2O} \text{Glucose + Glucose.}$$

The disaccharides may be hydrolysed by dil. acids or enzymes. The enzymes which bring about the hydrolysis of sucrose, lactose and maltose are *invertase*, *lactase* and *maltase* respectively.

Q. 18. Describe the manufacture of cane sugar from sugar cane. Also mention the by-products of sugar industry.

Ans. It consists of the following steps:

1. Extraction of the juice. Sugar cane is cleaned and cut into small pieces. Fresh pieces are passed through a series of crushers to squeeze out the juice. Hot or cold water is sprayed over fibrous material and it is again passed through rollers to ensure complete extraction. About 90–95% of the juice is usually extracted. The residual cellulosic material is known as **bagasse**.

2. Purification of the juice. The cane juice so obtained is a dark brown opaque liquid containing 15 to 20% of sucrose and some glucose, fructose, organic acids (oxalic and citric), vegetable proteins, mineral salts, colouring matter, gums and fine particles of bagasse suspended in it in colloidal form. The juice is passed through screens to remove suspended impurities and then subjected to purification as described below:

(*i*) *Defecation.* Fresh juice is run into defecator tanks heated by steam coils, and treated with 2-3% of lime (pH value 7.2) where

(*a*) Proteins and coloidal impurities coagulate out.

(*b*) Free organic acids and phosphates are converted into insoluble calcium salts.

(*c*) Sucrose is partially converted into calcium sucrosate.

The impurities rise to the surface forming a thick scum, which is removed. The precipitated calcium salts are also removed by filtration through canvas. The clear liquid is then pumped to conical tanks for next operation.

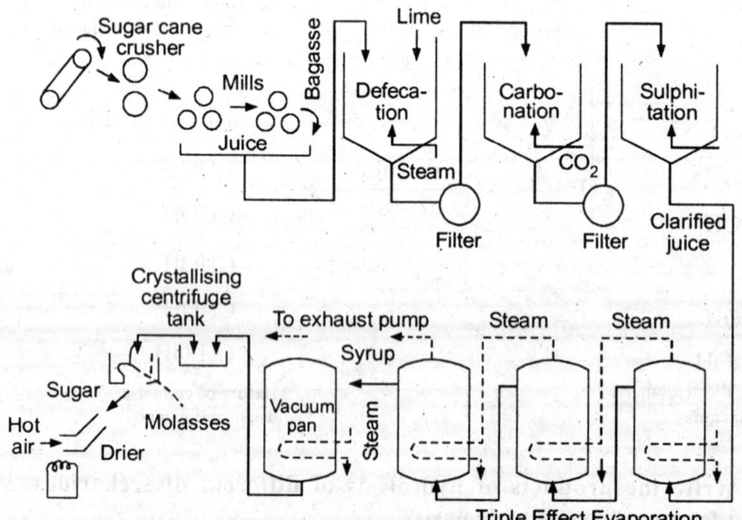

Fig. 31.1. Plant for the manufacture of cane sugar

(*ii*) *Carbonation and Sulphitation.* The juice contains excess of lime and calcium sucrosate. CO_2 is then passed through this juice (*carbonation*) which removes excess of lime as insoluble $CaCO_3$ and decomposes calcium sucrosate to give back sugar.

$$C_{12}H_{22}O_{11}.3CaO + 3CO_2 \rightarrow \underset{\text{Cane sugar}}{C_{12}H_{22}O_{11}} + \underset{\text{ppt.}}{3CaCO_3}$$
Cal. sucrosate

If SO_2 is used in this operation (*sulphitation*), it being a bleaching agent produces a juice with much lighter colour. At times, carbonation is followed by sulphitation in order to ensure complete neutralisation of lime and decomposition of calcium sucrosate.

$$\underset{\text{Cal.sucrosate}}{C_{12}H_{22}O_{11}.3CaO} + 3SO_2 \rightarrow \underset{\text{Cane sugar}}{C_{12}H_{22}O_{11}} + \underset{\text{ppt.}}{3CaSO_3}$$

3. Concentration and Crystallisation. The clear juice is evaporated to a syrup under reduced pressure in '*multiple effect* vacuum pans.

In this arrangement steam from boilers passes through steam coils in the first pan only. The steam produced in first pan is used to boil the juice in second pan at a lower pressure and steam from this pan is sent to the third pan at a still lower temperature and so on. This process of concentration saves a lot of fuel.

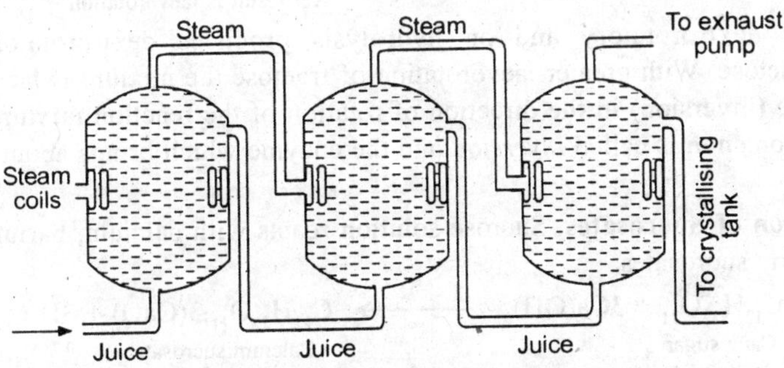

Fig. 32.2 The Multiple effect evaporator

The concentrated juice is then sent to vacuum pans where evaporation further reduces water content to 6–8%. On cooling here, partial crystallisation takes place. The contents of this pan are sent to another tank fitted with centrifugal machines.

4. Separation and drying of crystals. Centrifugal machines separate out sugar crystals from mother liquor known as `molasses'. Sugar crystals are then washed with a little water to remove any impurities sticking to their surface. The crystals are then dried by a current of hot air. In case the sugar crystals so obtained are brown in colour and have slight odour, these are further purified. The crystals are dissolved in water and filtered through animal charcoal or *norit* (activated coconut charcoal). The colourless filtrate is concentrated in vacuum pans and the crystals obtained by centrifuging the syrup.

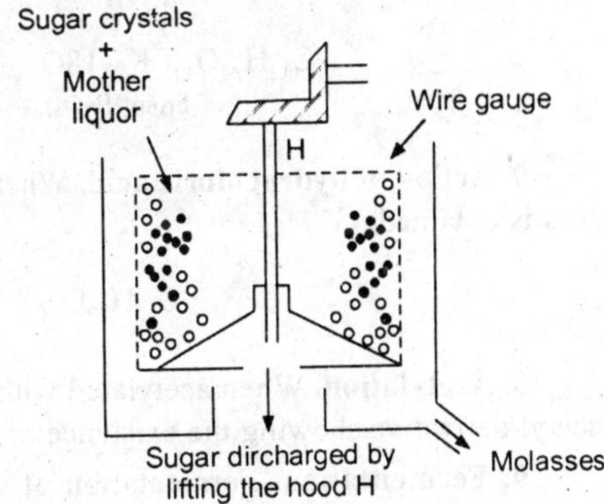

Fig. 32.3

By-products of Sugar Industry

(*i*) *Bagasse*. It is the cellulosic material left after the extraction of juice from sugar cane. It is used as a fuel and fodder in our country but in U.S.A., it is used in the manufacture of building and insulating material known as *Celotex*. It is also used in the manufacture of low grade paper.

(*ii*) *Molasses*. It is the mother liquor left after the separation of sugar crystals. It contains nearly 60% fermentable sugar. It is largely used now in the manufacture of ethyl alcohol and also CO_2. It is also used to some extent as a manure and road binder.

Q. 19. Describe the properties of cane sugar $C_{12}H_{22}O_{11}$.

Ans. 1. It is a colourless, crystalline substance, sweet in taste. It is very soluble in water and the solution is dextrorotatory $[(\alpha)_D = +66.5]$.

Chemical Properties

2. Effect of heat. Sucrose on heating slowly and carefully melts and then if allowed to cool, it solidifies to pale-yellow glassy mass called '*Barley sugar*'.

When heated to 473 K, it loses water to form a brown amorphous mass called *caramel.* On strong heating it chars to almost pure carbon giving characteristic smell of burnt sugar.

3. Hydrolysis. Sucrose when boiled with dilute acids or hydrolysed by enzyme *invertase*, yields an equimolecular mixture of glucose and fructose.

$$\underset{\substack{\text{Cane sugar} \\ [\alpha]_D = 66.5}}{C_{12}H_{22}O_{11}} + H_2O \longrightarrow \underset{\substack{\text{Glucose} \\ [\alpha]_D = +52.7}}{C_6H_{12}O_6} + \underset{\substack{\text{Fructose} \\ [\alpha]_D = -92.4}}{C_6H_{12}O_6}$$

Net result is laevorotation

Sucrose is dextrorotatory and on hydrolysis produces dextrorotatory glucose and laevorotatory fructose. With greater laevorotation of fructose the mixture is laevorotatory. Thus, there is a change (inversion) in the direction of rotation of the reaction mixture from dextro to laevo. This phenomenon is called *inversion* and the enzyme which brings about this inversion is called *invertase.*

4. Formation of Sucrosates. Sucrose solution reacts with calcium, barium and strontium hydroxides to form sucrosates.

$$\underset{\text{Cane sugar}}{C_{12}H_{22}O_{11}} + 3Ca(OH)_2 \longrightarrow \underset{\text{Calcium sucrosate}}{C_{12}H_{22}O_{11}.3(CaO)} + 3H_2O$$

The sucrosate decomposes when carbon dioxide is passed in the solution.

5. Action of sulphuric acid. Concentrated sulphuric acid abstracts water and charring takes place. Carbon produced gets oxidised to carbon dioxide and sulphur dioxide is produced due to reduction of the acid.

$$C_{12}H_{22}O_{11} + H_2SO_4 \longrightarrow 12C + [11H_2O + H_2SO_4]$$

$$C + 2H_2SO_4 \longrightarrow CO_2 + SO_2 + 2H_2O$$

6. Action of nitric acid. Concentrated nitric acid oxidises cane sugar to oxalic acid.

$$C_{12}H_{22}O_{11} + \underset{(\text{From } HNO_3)}{18O} \longrightarrow 6\ \underset{\text{Oxalic acid}}{\begin{matrix} COOH \\ | \\ COOH \end{matrix}} + 5H_2O$$

7. Action of hydrochloric acid. When boiled with concentrated hydrochloric acid, laevulinic acid is obtained.

$$C_{12}H_{22}O_{11} \xrightarrow{\text{Conc. HCl}} \underset{\text{Laevulinic acid}}{CH_3.CO.CH_2.CH_2COOH}$$

8. Acetylation. When acetylated with acetic anhydride and sodium acetate, it gives an octa-acetyl derivative showing the existence of eight –OH groups in sucrose molecule.

9. Fermentation. Fermentation of sucrose is brought about by yeast when the enzymes *invertase* hydrolyses sucrose to glucose and fructose and *zymase* converts them to ethyl alcohol.

$$C_{12}H_{22}O_{11} + H_2O \xrightarrow{\text{Invertase}} \underset{\text{Glucose}}{C_6H_{12}O_6} + \underset{\text{Fructose}}{C_6H_{12}O_6}$$

$$C_6H_{12}O_6 \xrightarrow{\text{Zymase}} 2C_2H_5OH + 2CO_2$$

10. Methylation. When treated with dimethyl sulphate in presence of alkali it gives octamethyl derivative.

Note. *Sucrose does not give positive test with Fehling's solution and Tollen's reagent nor does it react with reagents like HCN, NH_2OH and phenyl hydrazine. This shows the absence of monosaccharide character and absence of aldehydic and ketonic groups. It is stable to alkali and does not show mutarotation.*

Q. 20. How will you analyse cane sugar in a sample?

Ans. Tests. (1) When heated in a test tube it gives a characteristic odour which is named as 'smell of burnt sugar'.

(2) It gives positive Molische's test. Add alcoholic solution of α-naphthol to cane sugar solution followed by conc. H_2SO_4 along the side of the test tube. A violet ring is obtained.

(3) When it is kept in contact with concentrated sulphuric acid, charring takes place in cold. When warmed a mixture of carbon dioxide and sulphur dioxide are evolved.

(4) Upon adding Fehling's solution to a sucrose solution which has been boiled with dilute hydrochloric acid, a red precipitate of cuprous oxide is obtained. *It is to be noted that no red precipitate is obtained if sucrose solution has not been hydrolysed into glucose and fructose by dilute hydrochloric acid.*

(*Distinction from glucose and fructose*):

It does not turn yellow or brown when heated with caustic soda solution nor gives osazone when treated with phenyl hydrazine.

Q. 21. Describe various steps for establishing the structure of sucrose (cane sugar).

(*Coimbatore, 2000; M. Dayanand, 2000; Madras, 2004; Kuvempu, 2005*)

Ans. (1) It has a molecular formula $C_{12}H_{22}O_{11}$.

(2) It does not reduce Fehling's solution or Tollen's reagent. It also does not react with carbonyl group reagents like hydroxyl amine and phenyl hydrazine. Sucrose does not show mutarotation and forms no methyl glycosides like monosaccharides.

All these negative tests show that sucrose has no free aldehydic or ketonic group.

(3) If hydrolysed, sucrose yields equimolecular mixture of D(+) glucose and D(–) fructose. Since sucrose does not give any indication of the presence of free aldehydic or ketonic group, it is obvious that the union between glucose and fructose must have taken place through C-1 carbon of glucose (carrying aldehydic group) and C-2 carbon of fructose (carrying ketonic group).

(4) It gives octa-acetyl derivative upon acetylation.

(5) Upon complete methylation, octamethyl sucrose is obtained. Formation of octa-acetyl and octamethyl derivatives shows *the presence of eight –OH groups in the molecule.*

(6) The hydrolysis of octa-methyl sucrose gives 2, 3, 4, 6-tetra-O-methyl derivative of glucose and 1, 3, 4, 6-tetra-O-methyl fructose. Formation of 2, 3, 4, 6, tetra-O-methyl derivative of glucose indicates a ring between C-1 and C-5 of glucose (as shown in structure of glucose) and 1, 3, 4, 6-tetra-O-methyl fructose indicates a ring between C_2 and C_5 of 'fructose'. Hence glucose is pyranoside and fructose is furanoside.

(7) Study of optical rotation and the behaviour of sucrose towards various enzymes have verified that D-glucose unit has α-configuration and D-fructose unit has β-configuration.

With all these facts in view Haworth (1927) suggested the following structure for sucrose:

Sucrose

Hudson has designated sucrose as 1-α-D-glucopyranoside-2-β-D-fructofuranoside.

α-D-glucopyranose unit β-D-fructofuranose unit

Sucrose

Its conformational structure is written as under:

Q. 22. Discuss the structure of maltose. *(Nagpur 2008)*

Ans. Maltose is obtained by hydrolysis of starch using β-amylase.

$$\text{starch} \xrightarrow{\beta\text{-amylase}} \text{Maltose}$$

1. Molecular formula of maltose as determined by usual analytical techniques comes out to be $C_{12}H_{22}O_{11}$.

2. Maltose on heating with dil HCl yields two equivalents of D(+) glucose. It is thus made up of two units of glucose.

$$\underset{\text{Maltose}}{C_{12}H_{22}O_{11}} + H_2O \xrightarrow{H^+} \underset{\text{Glucose}}{2C_6H_{12}O_6}$$

3. Maltose reduces Tollen's reagent and Fehling solution. This means it is a reducing sugar. It also forms an osazone and shows the phenomenon of mutarotation.

$$\alpha\text{-form} \rightleftharpoons \text{Equilibrium Mixture} \rightleftharpoons \beta\text{-form}$$

4. Maltose on treatment with bromine water gives a mono-carboxylic acid. This shows that maltose contains a carbonyl group present in a reactive hemiacetal form as in case of glucose.

5. The following structure has been proposed for maltose. This structure explains the observations described above

Reducing half

Non-reducing half

The structure can also be represented as Haworth projection formula

Maltose (α-form)

Maltose (β-form)

Conformational formula of α-form of maltose may be represented as

Q. 23. Discuss the structure of lactose. (*Kurukshetra, 2001;Bangalore, 2002*)

Ans. Lactose is found in the milk of mammal and in *whey*, a bye-product in the manufacture of cheese.

1. Its molecular formula is determined to be $C_{12}H_{22}O_{11}$.

2. It reduces Fehling solution, forms an osazone. It means its is a reducing sugar. It also exhibits mutarotation.

$$\alpha\text{-form} \rightleftharpoons \text{Equilibrium mixture} \rightleftharpoons \beta\text{-form}$$

3. Lactose on hydrolysis either with an acid or with an enzyme yields equivalent amounts of D(+) glucose and D(+) galactose.

$$\text{Lactose} \xrightarrow{\text{Hydrolysis}} \text{Glucose + Galactose}$$

It is therefore, made up of a molecule of glucose and a molecule of galactose linked together. The linkage between the two units is shown as under:

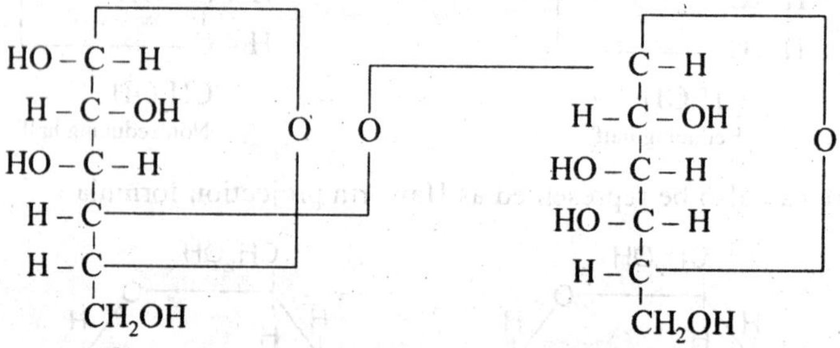

Haworth projection formula of lactose can be represented as

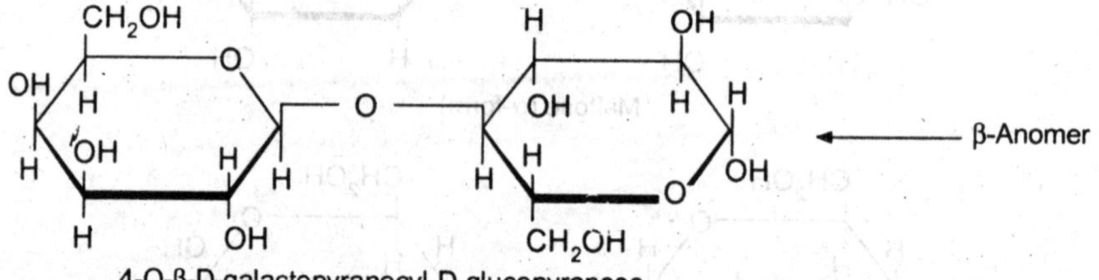

β-Anomer

4-O-β-D-galactopyranocyl-D-glucopyranose

Its conformational formula may be written as

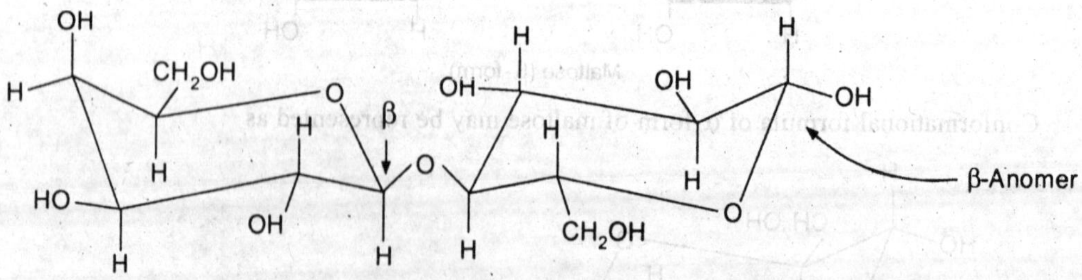

β-Anomer

Q. 24. Describe the occurrence and properties of starch amylum ($C_6H_{10}O_5$),

Ans. Starch is most widely distributed in vegetable kingdom. In nature, it is transformed into complex polysaccharides like gum and cellulose and into simpler mono and disaccharides by enzymes working in vegetable kingdom. Its rich sources are potatoes, wheat, maize, rice, barley and arrow root. It is interesting to note that no two sources give identical starch.

Physical Properties. It is a white, amorphous, substance with no taste or smell. It is insoluble in water but when starch is added to boiling water the granules swell and burst forming colloidal, translucent suspension.

Chemical Properties

(*i*) When heated to a temperature between 200 to 250°C it changes into dextrin. At higher temperature charring takes place.

(*ii*) Starch, when boiled with dilute acid, yields ultimately glucose.

$$\underset{\text{Starch}}{(C_6H_{10}O_5)_n} \longrightarrow \underset{\text{Dextrin}}{(C_6H_{10}O_5)_{n1}} \longrightarrow \underset{\text{Maltose}}{C_{12}H_{22}O_{11}} \longrightarrow \underset{\text{Glucose}}{C_6H_{12}O_6}$$

When hydrolysed with enzyme diastase, maltose is obtained.

$$\underset{\text{Starch}}{2(C_6H_{10}O_5)_n} + nH_2O \xrightarrow{\text{Diastase}} \underset{\text{Maltose}}{nC_{12}H_{22}O_{11}}$$

(*iii*) Starch solution gives a blue colour with a drop of iodine solution. The blue colour disappears on heating and reappears on cooling. *In fact it is the amylose that gives a blue colour with iodine; the amylopectin gives a red brown colour with iodine.*

(*iv*) When heated with a mixture of concentrated sulphuric and nitric acids it gives nitro-starch.

Q. 25. Discuss the structure of starch.

Ans. The exact chemical nature of starch varies from source to source. Even the starch obtained from same source consists of two fractions (i) *amylose*, and (ii) *amylopectin*, present in 1 : 3 or 1 : 4 ratio.

Molecular weight of amylose from ultracentrifuge method has been shown to lie between 15,000 to 225,000 whereas osmotic pressure measurement suggests that it may have value between 1,00,000 to 20,00,000. Molecular weight of amylopectin from osmotic pressure data has been shown to be between 10,00,000 to 60,00,000 and from light-scattering measurement between 10,00,000 and much higher value. Both of them on hydrolysis yield α-D-glucose. When hydrolysed by enzyme β-*maltase* it yields maltose –a disaccharide of known structure having 1, 4-α-glycosidic linkage.

Thus, it has been shown that amylose is a linear polymer containing glucopyranose units joined by 1, 4-α-glycosidic linkages.

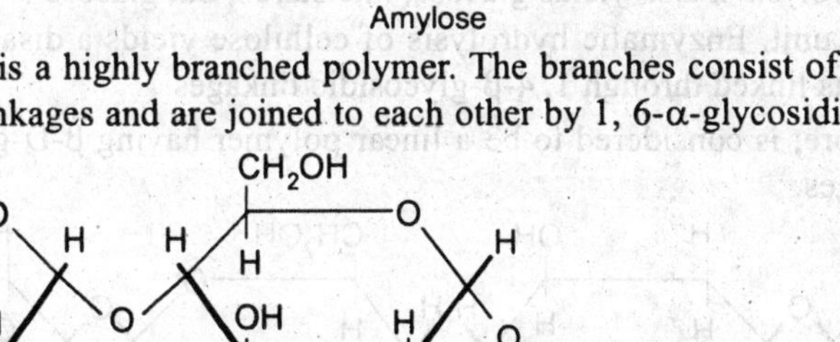

Amylose

Amylopectin is a highly branched polymer. The branches consist of 20 to 25 glucose units joined by 1, 4-α-linkages and are joined to each other by 1, 6-α-glycosidic linkages.

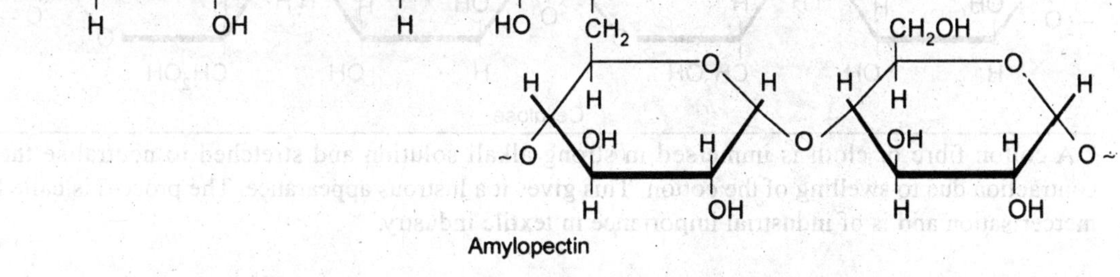

Amylopectin

Q. 26. What is cellulose? Describe the preparation and properties of cellulose.

Ans. Cellulose $(C_6H_{10}O_5)_n$ is widely distributed in plant kingdom and plants maintain their structure due to the support of fibrous material which is cellulose. Cotton is almost pure cellulose and jute, hemp, wood, paper etc. are all different forms of cellulose.

Preparation. (1) *From Cotton.* Cotton is treated with organic solvents to remove fats and waxes. It is then treated with hydrofluoric acid to remove mineral matter. The crude cellulose thus obtained is bleached by alkali and sodium hypochlorite to get pure cellulose as white amorphous mass.

(2) *From Wood.* Cellulose can also obtained by taking wood shavings or wood chips and boiling them with sodium bisulphite or caustic soda to remove lignin and resins. It is then treated successively with dilute acid, water and organic solvents to remove other substances. Lastly it is bleached with sodium or calcium hypochlorite to get white amorphous cellullose.

Physical Properties. Cellulose is a colourless, amorphous substance with organised fibrous structure which is characteristic of the source. It is insoluble in water but dissolves in ammoniacal solution of cupric hydroxide (Schweitzer's reagent). Cellulose also dissolves in a solution of zinc chloride in hydrochloric acid.

Chemical Properties. (1) When cellulose is treated with concentrated sulphuric acid in cold it slowly passes into solution. The solution when diluted with water, precipitates a starch like substances *amyloid.*

(2) When boiled with dilute sulphuric acid it is completely hydrolysed into D-glucose.

$$\underset{\text{Cellulose}}{(C_6H_{10}O_5)_n} + nH_2O \longrightarrow \underset{\text{Glucose}}{nC_6H_{12}O_6}$$

(3) It forms *hydrocellulose* with dilute acids and water on mechanical communication.

(4) When treated with 20% caustic soda solution its appearance becomes smooth and lustrous. This property was noted by John Mercer* in 1884.

(5) Cellulose has D-glucose unit in its molecule which has 3 hydroxyl groups free for esterification. Thus when treated with a mixture of concentrated nitric and sulphuric acids it forms mono, di and tri nitrates of cellulose. Cellulose nitrates are used in preparation of explosives.

Similarly when treated with a mixture of glacial acetic acid and acetic anhydride, cellulose yields a mixture of di and tri acetates. Cellulose acetate is used in the manufacture of synthetic fibres and paints

Q. 27. Discuss the structure of cellulose.

Ans. Like starch the molecular weight of cellulose also varies with the source. Its molecular weight as determined by ultracentrifugal method lies between 1–2 million.

On complete hydrolysis it also yields glucose, like starch, but glucose is present in cellulose as β-D-glucopyranose unit. Enzymatic hydrolysis of cellulose yields a disaccharide cellobiose having the glucose units linked through 1, 4-β-glycosidic linkages.

Cellulose, therefore, is considered to be a linear polymer having β-D-glucopyranose units joined by 1, 4-β-linkages.

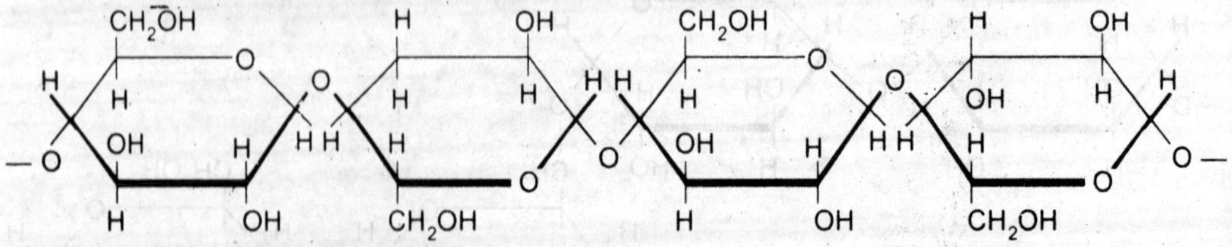

Cellulose

* A cotton fibre or cloth is immersed in strong alkali solution and stretched to neutralise the contraction due to swelling of the cotton. This gives it a lustrous appearance. The process is called mercerisation and is of industrial importance in textile industry.

Q. 28. Give an account of principal uses of cellulose. *(Kerala, 2001)*

Ans. (1) It is used as such in the manufacture of paper and cloth.

(2) Cellulose nitrates are used in the manufacture of explosives, medicines, paints and lacquers.

(3) Cellulose nitrate with camphor yields celluloid which is used in manufacturing toys, decorative articles and photographic films.

(4) Cellulose acetate is used in rayon manufacture and in plastics.

Q. 29. What are reducing and non-reducing sugars? *(Himachal, 2000; Panjab, 2003)*

Or

While glucose is a reducing sugar, sucrose is not. Explain.

Ans. Reducing sugars are those carbohydrates which reduce the solution of alkaline copper sulphate (Fehling solution) into a red precipitate of Cu_2O and reduce Tollen's reagent into silver mirror. Non-reducing sugars are those which don't respond to these tests. In reducing sugars, there is an aldehydic group or an alcoholic group adjacent to a ketonic group. Such groups are easily oxidised and hence compounds containing such group act as reducing agents.

(1) Fehling solution test

$$CuSO_4 + 2NaOH \longrightarrow Cu(OH)_2 + Na_2SO_4$$

$$Cu(OH)_2 \longrightarrow CuO + H_2O$$

$$\underset{\text{(from sugar)}}{-CHO} + 2CuO \longrightarrow -COOH + \underset{\text{Red ppt.}}{Cu_2O}$$

(2) Tollen's reagent test

$$AgNO_3 + NH_4OH \longrightarrow AgOH + NH_4NO_3$$

$$2AgOH \longrightarrow Ag_2O + H_2O$$

$$\underset{\text{(from carbohydrate)}}{-CHO} + Ag_2O \longrightarrow -COOH + \underset{\text{Silver mirror}}{2Ag\downarrow}$$

As non-reducing sugars do not contain these groups in the free state, they do not respond to these tests.

Glucose & fructose are reducing sugars because they contains either an aldehydic group or an alcoholic group adjacent to a ketonic group.

Take the case of glucose. Although there is no free aldehydic group in the cyclic structure, but as the cyclic form is in equilibrium with straight-chain structure, there is always some concentration of aldehydic species and hence it gives the Fehling solution and Tollen's reagent tests.

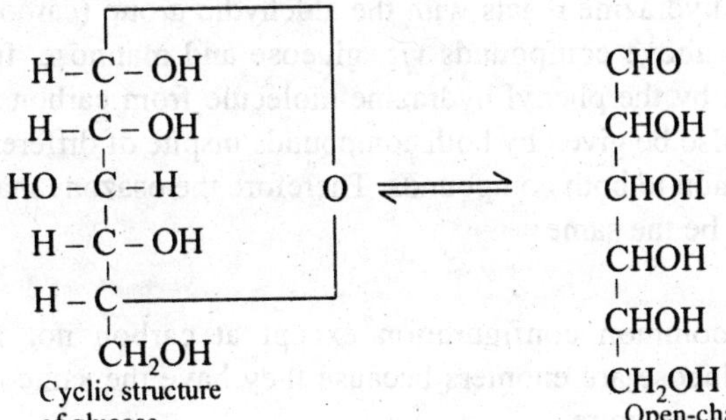

```
  H – C – OH ─────┐                 CHO
      |           |                  |
  H – C – OH      |                 CHOH
      |           |                  |
 HO – C – H       O  ⇌              CHOH
      |           |                  |
  H – C – OH      |                 CHOH
      |           |                  |
  H – C ──────────┘                 CHOH
      |                              |
    CH₂OH                           CH₂OH
```

Cyclic structure of glucose α-D-Glucose

Open-chain structure of glucose

Similarly, cyclic structure of fructose has no reducing group, but it is in equilibrium with straight chain structure which has alcoholic groups adjacent to a ketonic group, because of which it gives reducing properties.

$$\begin{array}{l}
HOH_2C-C-OH \\
HO-C-H \qquad\qquad O \\
H-C-OH \\
H-C-OH \\
H-C-OH \\
CH_2
\end{array}
\qquad\qquad
\begin{array}{c}
CH_2OH \\
| \\
C=O \\
| \\
CHOH \\
| \\
CHOH \\
| \\
CHOH \\
| \\
CH_2OH
\end{array}$$

Cyclic structure of glucose α-D-Glucose

Straigth-chain structure of fructose

Sucrose has no such groups in free state. Hence it does not respond to Fehling solution and Tollen's reagent tests. Therefore, sucrose is a non-reducing sugar

Q. 30. Differentiate between epimers and anomers. *(Delhi, 2003 ; Kuvempu, 2005)*

Ans. Epimers. Sugars having a common configuration (arrangement of –H and –OH around carbon atoms), except at α-carbon or carbon no. 2 are called epimers. Examples are glucose and mannose. They are epimers of each other.

$$\begin{array}{c}
\overset{1}{CHO} \\
| \\
H-\overset{2}{C}-OH \\
| \\
HO-\overset{3}{C}-H \\
| \\
H-\overset{4}{C}-OH \\
| \\
H-\overset{5}{C}-OH \\
| \\
\overset{6}{C}H_2OH
\end{array}
\qquad\qquad
\begin{array}{c}
\overset{1}{CHO} \\
| \\
HO-\overset{2}{C}-H \\
| \\
HO-\overset{3}{C}-H \\
| \\
H-\overset{4}{C}-OH \\
| \\
H-\overset{5}{C}-OH \\
| \\
\overset{6}{C}H_2OH
\end{array}$$

Glucose Mannose

We observe that glucose and mannose differ in their structures only at carbon no. 2.

Both the compounds give the same osazone with phenyl hydrazine. This reactions is completed in 3 steps. (Refer to the chemical properties of glucose in this chapter). In this first step, one molecule of phenyl hydrazine reacts with the aldehydic group (carbon no. 1). This step will be common to both the above compounds *viz.* glucose and mannose. In the second step, two hydrogens are taken out by the phenyl hydrazine molecule from carbon no. 2 to convert it into $>C=O$. This step will also be given by both compounds inspite of different configuration at C-2. Third step is again common to both compounds. Therefore the osazone that will be obtained from the two compounds will be the same.

Anomers

Sugars having a common configuration except at carbon no. 1 are called anomers. α-D-glucose and β-D-glucose are enomers because they have the same configuration except at carbon no. 1 in the cyclic structure.

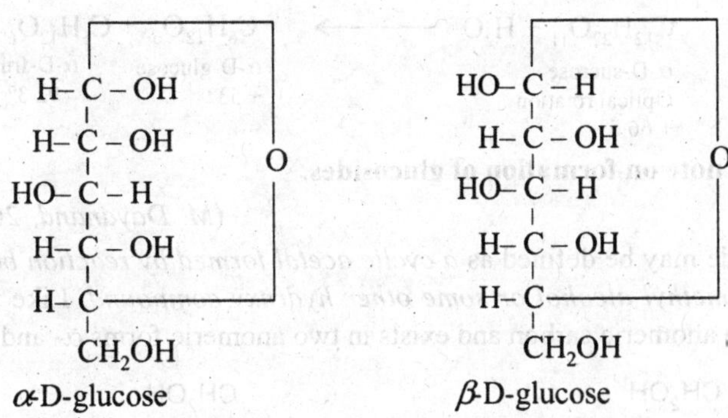

Q. 31. What is mutarotation ? Why does D-glucose show pnenomenon of mutarotatior

(Punjab, 2003; Mumbai, 2004; Calcutta, 2007; Lucknow, 2010; Pune, 2010

Ans. Mutarotation. Glucose contains asymmetric carbon atoms, therefore it exhibits optica' properties like optical rotation i.e. rotation of plane polarised light when it passes through glucose solution. It is observed that when D-glucose having m.p. 419 K is dissolved in water, it shows an optical rotation of + 112°. Gradually it drops to + 52.7°.

On the other hand, when D-glucose having m.p. 423 K is dissolved in water, it shows a value of + 19° for optical rotation. Gradually the value rises to + 52.7°. This happens because they attain a state of equilibrium having a fixed composition after sometime thereby giving the same value of optical rotation *i.e.* 52.7°. The compound having m.p. 419 K is α-D-glucose with optical rotation value as + 112° and the compound having m.p. equal to 423 K is β-D-glucose and has optical rotation + 19°. On dissolving either α or β-D-glucose in water, we shall obtain, after some time, an equilibrium mixture of the two forms having 36% of α-form and 64% of β-form having optical rotation value 52.7°. This can be calculated mathematically also.

$$\text{Optical rotation of mixture} = \frac{\begin{array}{c}\text{Percentage of } \alpha\text{ - form} \times \text{opt. rotation of } \alpha\text{ - form} \\ + \text{Percentage of } \beta\text{ - form} \times \text{opt. rotation of } \beta\text{ - form}\end{array}}{100}$$

$$= \frac{36 \times 112 + 64 \times 19}{100} = 52.48$$

The slight difference is due to approximation that has been made in the percentage of the two isomers.

Thus mutarotation may be defined as the phenomenon of change in specific rotation of an optically active compound with time to an eqiulibrium value of specific rotation. (For reason of mutarotation, see question on ring structure of glucose.)

Q. 32. What is meant by inversion of sugar?

Ans. A change in the sign of specific rotation of a sugar solution, after hydrolysis, is called inversion of sugar.

Sucrose in an optically active compound having sp. rotation of + 66.5°. When sucrose solution is subjected to hydrolysis in the presence of an acid or an enzyme, we observe that the sp. rotation has reversed its sign from positive to negative or from dextrorotatory to laevorotatory. This is because sucrose gets hydrolysed to glucose and fructose. Out of the two substances, glucose and fructose, the latter has a negative sp. rotation and to a larger magnitude. As a consequence, the resulting mixture is laevorotatory.

$$\underset{\substack{\alpha\text{-D-sucrose} \\ \text{Optical rotation} \\ +66.5^\circ}}{C_{12}H_{22}O_{11}} + H_2O \longrightarrow \underset{\substack{\alpha\text{-D-glucose} \\ +53^\circ}}{C_6H_{12}O_6} + \underset{\substack{\alpha\text{-D-fructose} \\ -92.3^\circ}}{C_6H_{12}O_6}$$

Q. 33. Write a note on formation of glucosides.

(*M. Dayanand, 2000; Panjab 2003*)

Ans. A glycoside may be defined as *a cyclic acetal formed by reaction between glucose and an alcohol such as methyl alcohol or some other hydroxy compound.* Like the parent glucose, glycoside also has an anomeric carbon and exists in two anomeric forms α- and β- as shown below.

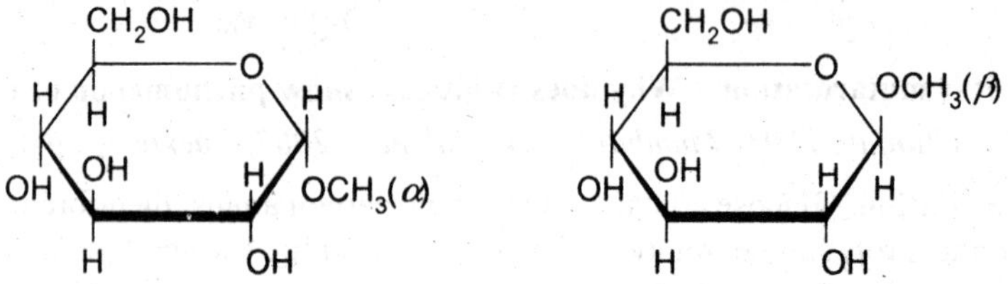

Methyl-α-D-glucoside

Methyl-β-D-glucoside

Unlike acetals, glucosides do not reduce Fehling solution or Tollen's reagent. They also do not exhibit mutarotation. That shows when either α-or β-form is dissolved in water, it stays in its form and does not give an equilibrium mixture of two forms. This is due to the fact that the mobile hydrogen of the hydroxy group at carbon no. 1 has been substituted by alkyl group. Thus the configuration around the anomeric carbon gets fixed.

MISCELLANEOUS QUESTIONS

Q. 34. What happens when (+) glucose is treated with a dilute alkali solution?

(*Awadh, 2000*)

Or

What is Lobry-de-Bruyn Van Ekenstein rearrangement.

Ans. On treatment with dilute alkali solution, Lobry-de-Bruyn Van Ekenstein rearrangement takes place which leads to a mixture of glucose, mannose and fructose.

For complete reaction, see the question on properties of (+) glucose (Miscellaneous reactions).

Q. 35. Glucose and fructose have the same formula. Glucose has an aldehydic group, while fructose has a ketonic group but both reduce Fehling solution. Explain.

(*Kurukshetra, 2001; Guwahati; 2002*)

Ans. This is because Fehling solution is alkaline in nature. In alkaline medium, fructose undergoes Lobry-de-Bruyn Van Ekenstein rearrangement when an equilibrium mixture of glucose, mannose and fructose is obtained (for complete reaction, see question on properties of glucose, miscellaneous reactions).

It is because of the glucose fraction in the solution that fructose reduces Fehling solution.

Q. 36. Why does D-glucose not give Schiff's test? (*Garhwal, 2000*)

Ans. D-glucose does not give Schiff's test because the aldehydic group in glucose is not completely free. It is involved in acetal formation with the hydroxyl group at carbon no. 5 to give the cyclic structure.

Q. 37. Draw the chair conformations of α-D(+) glucose and β-D(+) glucose. Which is more stable and why? (*Calcutta, 2000*)

Ans. The chair conformations of the two forms of glucose are given below:

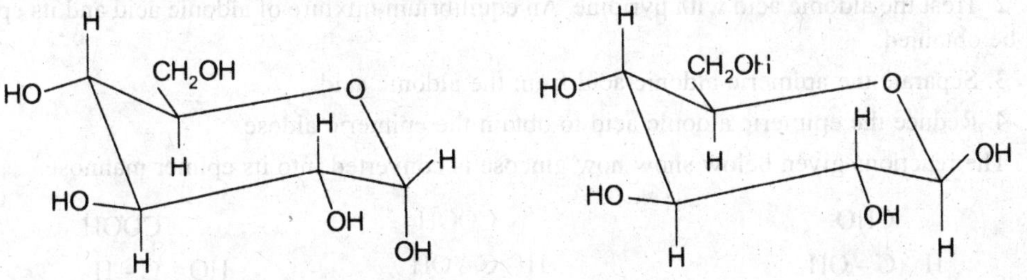

Out of α- and β-forms of glucose, β-form is more stable. This is because all bulky groups like –OH and – CH_2 OH are in equitorial positions in β-form while the bulky group –OH at carbon no.1 in α-form is in axial position. In general, there is less crowding and hence greater stability if the groups are oriented in equatorial positions. Hence, β-form is more stable than α-form.

Q. 38. Explain why unlike glucose, neither α-nor β-methyl glucoside reduce Tollen's reagent or Fehling solution.

Ans. Tollen's reagent and Fehling solution tests are given by glucose because its cyclic structure is converted into linear structure which contains –CHO group. Aldehydic group gives Tollen's and Fehling solution tests. But methyl glucosides are not hydrolysed readily to obtain cyclic and open-chain structure of glucose. In the absence of –CHO groups, these tests are not given.

Q. 39. What is amylose? Give its structure.

Ans. Amylose is a fraction of starch. It has the structure as given below

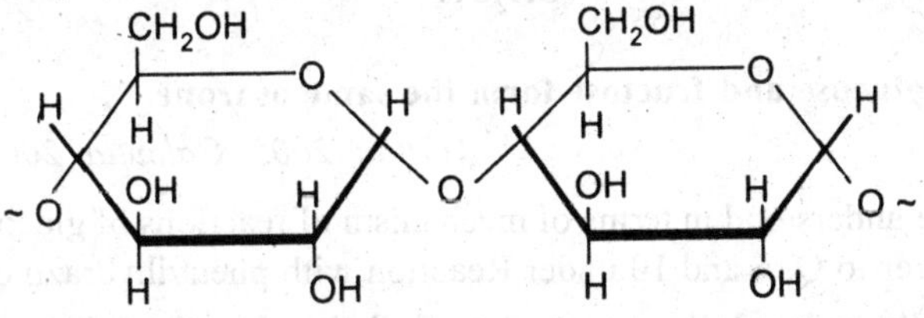

Q. 40. Why sucrose is known as invert sugar? How do you account for the fact that sucrose does not reduce Fehling solution or form oxime? Write down the structure of sucrose molecule.

Ans. Sucrose is an optically active compound. It is dextrorotatory. But on hydrolysis, it produces glucose and fructose in equivalent amounts. Glucose is dextrorotatory and fructose is laevorotatory. But the laevorotation of fructose is much more in magnitude than the dextrorotation of glucose, with the result, that solution of sucrose exhibits laevorotation. As inversion in optical rotation takes place, it is called invert sugar.

Sucrose does not reduce Fehling solution nor does it form oxime. These reactions are give by a compound containing a carbonyl group. Although the constituents of sugar *viz.*, glucose and fructose both contain carbonyl group but sucrose does not. This is because the two units are linked to each other through their carbonyl group. Thus carbonyl groups are not free in the molecule of sugar. For structure of sucrose, see Q. No. 20.

Q. 41. Write a note on epimerization.

(*Coimbatore, 2000; Kerala, 2001; Meerut, 2000; M. Dayanand, 2000*)

Ans. Aldoses which differ in their configuration only at C-2 are called epimers and the interconversion of such compounds is called epimerisation. An aldose can be converted into its epimer as follows:

1. Oxidise the aldose into aldonic acid by treatment with bromine water.

2. Treat the aldonic acid with pyridine. An equilibrium mixture of aldonic acid and its epimer will be obtained.

3. Separate the apimeric aldonic acid from the aldonic acid.

4. Reduce the epimeric aldonic acid to obtain the epimeric aldose.

The reactions given below show how glucose is converted into its epimer mannose.

$$\begin{array}{c}CHO\\ |\\ H-C-OH\\ |\\ HO-C-H\\ |\\ H-C-OH\\ |\\ H-C-OH\\ |\\ CH_2OH\\ \text{Glucose}\end{array} \xrightarrow{Br_2/H_2O} \begin{array}{c}COOH\\ |\\ H-C-OH\\ |\\ HO-C-H\\ |\\ H-C-OH\\ |\\ H-C-OH\\ |\\ CH_2OH\\ \text{Gluconic acid}\end{array} + \begin{array}{c}COOH\\ |\\ HO-C-H\\ |\\ HO-C-H\\ |\\ H-C-OH\\ |\\ H-C-OH\\ |\\ CH_2OH\\ \text{Mannonic acid}\end{array}$$

$$\text{Mannonic acid} \xrightarrow[-H_2O]{\Delta} \begin{array}{c}CO\\ |\\ HO-C-H\\ |\\ HO-C-H\\ |\\ H-C\\ |\\ H-C-OH\\ |\\ CH_2OH\end{array}\text{(ring through O between CO and C-4)} \xrightarrow{Na/Hg} \begin{array}{c}CHO\\ |\\ HO-C-H\\ |\\ HO-C-H\\ |\\ H-C-OH\\ |\\ H-C-OH\\ |\\ CH_2OH\\ \text{D-Mannose}\end{array}$$

Q. 42. Why do glucose and fructose form the same osazone.

(*Patna, 2005; Calcutta, 2007; Garhwal, 2010*)

Ans. This can be understood in terms of mechanism of reactions of glucose and fructose with phenyl hydrazine. Refer to Q. 6 and 10 under Reaction with phenylhydrazine.

Q. 43. Explain the methylation method to find the ring size of glucose and fructose in sucrose molecule?

Ans. See Q. 21.

Q. 44. Fructose is laevorotatory whereas it is written as D(–) fructose. Explain.

(*Kurukshetra, 2001; M. Dayanand, 2000*)

Ans. Terms leavorotatory or dextrorotatory denote the sign of rotation of plane-polarised light when passed through an optically active substance. For *leavo*, the sign (–) is used. Capital letters D and L stand for absolute configuration of the substance *i.e.* relative arrangement of –H and –OH groups around different carbon atoms. Fructose has the absolute D-configuration and rotates the plane of polarised light in the anticlockwise (or left) direction. Hence it is written as D(–) fructose.

Q. 45. How does the treatment with HIO_4 confirm that glucose is an aldohexose?

Ans. See. Q. 7.

Q. 46. How will you distinguish between glucose and fructose by a chemical test?

(*Kanpur, 2001; Meerut, 2000*)

Ans. Glucose gives a silver mirror with ammoniacal solution of silver nitrate (Tollen's test) and a red precipitate with alkaline solution of copper sulphate (*Fehling solution test*). Fructose does not give these tests.

Q. 47. C_1 and C_2 atoms of glucose and fructose are only involved in osazone formatsion. Explain. (*Awadh, 2000*)

Ans. See Q.6

Q. 48. Identify compound A and B in the following reactions : (*Lucknow, 2000*)

(*a*) **D-glucose** $\xrightarrow{HI/P}$ **B**

(*b*) **D-glucose** $\xrightarrow[\text{oxdn}]{Br_2}$ **A** $\xrightarrow[\text{oxdn}]{HNO_3}$ **B**

(*c*) **D-glucose** $\xrightarrow{NH_2OH}$ **A** $\xrightarrow[\text{warm or AcONa}]{Ac_2O}$ **B**

(*d*) **D-frucose** $\xrightarrow{HNO_3}$ **A + B**

Ans. (*a*) See Q. 6, (2) (*b*)

(*b*)$CH_2OH(CHOH)_4CHO \xrightarrow[\text{oxdn}]{Br_2} CH_2OH(CHOH)_4CHO \xrightarrow[\text{oxdn}]{HNO_3} COOH(CHOH)_4COOH$

D-glucose gluconic acid glucaric acid

(*c*) See Q.15

(*d*) See Q. 11, (A), (5)

Q. 49. Account for the following conversions: (*Lucknow, 2000*)

```
      CHO                          CHO
 HO ──┼── H                  HO ──┼── H
 HO ──┼── H      ──────→      H ──┼── OH
  H ──┼── OH                  H ──┼── OH
  H ──┼── OH                      CH2OH
      CH2OH
```

Ans. See Q. 14 (*b*)

Q. 50. How is glucose obtained from starch? (*Panjab, 2000*)

Ans. See Q. 24 (*ii*)

Q. 51. Give the products fo reaction of D(+) glucose with
(*i*) phenyl hydrazine (*ii*) bromine water. (*Panjab, 2000; Kerala, 2001*)

Ans. (*i*) See Q. 6 (5) (*ii*) See Q. 6 (i)

Q. 52. Glucose does not react with sodium bisulphite even thought it has an aldehyde group. Explain. (*M. Dayanand, 2000; Kurukshetra 2001*)

Ans. This is because the aldehydic group is not completely free. It is involved in acetal formation with hydroxyl group at C–5.

Q. 53. Write a short note on osozone (*Meerut, 2000*)

Ans. See q. 6 (5)

Q. 54. Complete the following reactions: (*Kerala, 2001*)

(*i*) fructose + conc. HNO_3

(*ii*) fructose + phenyl hydrazine

Ans. (*i*) See Q. 11, (A) (5) (*ii*)

(*ii*) See Q. 11 (A), (3)

Q. 55. Write the structures of the following: (*Panjab, 2000*)
(*i*) B–D–glucopyranose (*ii*) methyl–D–glucoside (*iii*) maltose

Ans. (*i*)

CH$_2$OH, O, OH, H, H, 4, 1, OH, OH, H, H, 3, 2, H, OH

B–D–glucopyronose

(*ii*) See Q. 33 (*iii*) See Q. 22

Q. 56. Discuss the structure of glucose on the basis of following reaction :
(*i*) HCN (*ii*) PCl_5 (*iii*) Br_2/H_2O (*iv*) HI/red P (*v*) CH_3OH/HCl (*Awadh, 2000*)
Ans. See Q. 7

Q. 57. Determine the size of glucose ring using periodic acid. (*Delhi 2003*)
Ans. See Q. 7

Q. 58. How would you prove that the carbonyl group in fructose is a ketonic group (*Nagpur 2002*)
Ans. See Q. 12

Q. 59. Explain the high solubility of sugars in water (*Delhi 2007*)
Ans. There is extensive hydrogen bonding in water because of which it is highly soluble in water. This is explained by taking the example of glucose.

$$\underset{\text{Glucose}}{\ldots\,\overset{\overset{CHO}{|}\,(CHOH)_4\,|\,CH_2}{|}\!\!O-H} \;\ldots\; \underset{\text{Water}}{\overset{H}{\overset{|}{O}}-H} \;\ldots\; \underset{\text{Glucose}}{\overset{\overset{CHO}{|}\,(CHOH)_4\,|\,CH_2}{|}\!\!O-H} \;\ldots\; \underset{\text{Water}}{\overset{H}{\overset{|}{O}}-H\ldots}$$

Q. 60. How will you convert D-glucose into D-mannose?

Or

Write a note on epimerisation.
Ans.

$$\underset{\text{D–glucose}}{\begin{array}{c}CHO\\|\\H-C-OH\\|\\HO-C-H\\|\\H-C-OH\\|\\H-C-OH\\|\\CH_2OH\end{array}} \xrightarrow{Br_2/H_2O} \underset{\text{D–gluconic acid}}{\begin{array}{c}COOH\\|\\H-C-OH\\|\\HO-C-H\\|\\H-C-H\\|\\H-C-OH\\|\\CH_2OH\end{array}} + \underset{\text{D–mannonic acid}}{\begin{array}{c}COOH\\|\\HO-C-H\\|\\HO-C-H\\|\\H-C-OH\\|\\H-C-OH\\|\\CH_2OH\end{array}}$$

$$\xrightarrow{\text{Pyridine}} \underset{\gamma\text{-D-Gluconolactone}}{\begin{array}{c}C=O\\|\alpha\\H-C-OH\\|\beta\\HO-C-OH\\|\gamma\\H-C\\|\\H-C-OH\\|\\CH_2OH\end{array}} O \rightleftharpoons \underset{\gamma\text{-D-Mannonolactone}}{\begin{array}{c}{}^1C=O\\{}^2|\\HO-C-H\\{}^3|\\HO-C-H\\{}^4|\\H-C\\{}^5|\\H-C-OH\\{}^6|\\CH_2OH\end{array}} O \xrightarrow[H^+]{Na-Hg} \underset{\text{D–mannose}}{\begin{array}{c}CHO\\|\\HO-C-H\\|\\HO-C-H\\|\\H-C-OH\\|\\H-C-OH\\|\\CH_2OH\end{array}}$$

D-Glucose and D-mannose differ in the configuration at C – 2. Such isomers (having the same formula and groups) differing in configuration at C–2 or C–4 are called epimers.

Conversion of a compound into its epimer, *e.g.*, D–glucose into D-mannose is called epimerisation.

Q. 61. Give the structures of ribose and deoxyribose

Ans. There are the sugars that have been isolated from the hydrolysis products of nucleic acids. Nucleic acids are named according to the sugar produced in hydrolysis. Nucleic acid which contains ribose is called ribonucleic acid (RNA). Nucleic acid which contains deoxyribose as a component is called deoxyribonucleic acid (DNA). Structures of ribose and deoxyribose are given below:

$$\begin{array}{c} CHO \\ | \\ (CHOH)_3 \\ | \\ CH_2OH \end{array} \qquad \begin{array}{c} CHO \\ | \\ CH_2 \\ | \\ (CHOH)_2 \end{array}$$

Ribose — 2-Deoxyribose

33

POLYNUCLEAR HYDROCARBONS

Q. 1. What are polynuclear hydrocarbons? What are the different types?

Ans. Polynuclear hydrocarbons are those compounds which possess more than one aromatic ring in their molecules. They may be divided into two types.

1. Those in which the aromatic rings are linked directly by single bonds or through other carbon atoms. For example:

Diphenyl
I

$-CH_2-$

Diphenyl methane
III

Terphenyl
II

$-CH_2-CH_2-$

Dibenzyl
IV

2. Those in which the rings are fused by sharing pairs of ortho carbon atoms. For example:

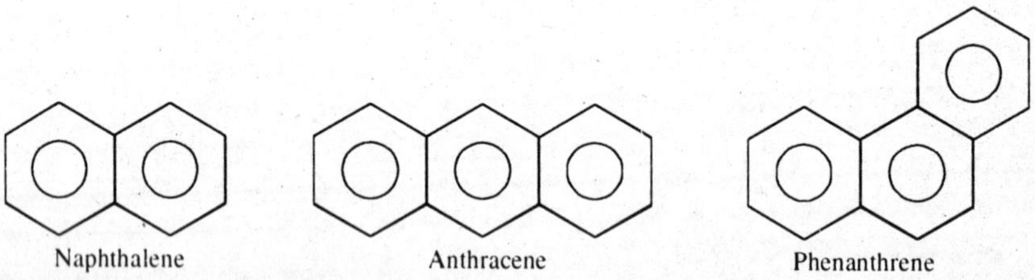

Naphthalene Anthracene Phenanthrene

This type of compounds are generally known as *fused* or *condensed* polynuclear hydrocarbons. Actually they are the true polynuclear compounds and are much more important than the first type compounds. We would discuss here naphthalene, phenanthrene and anthracene.

Q. 2. Discuss nomenclature and isomerisation of naphthalene derivatives.

Ans.

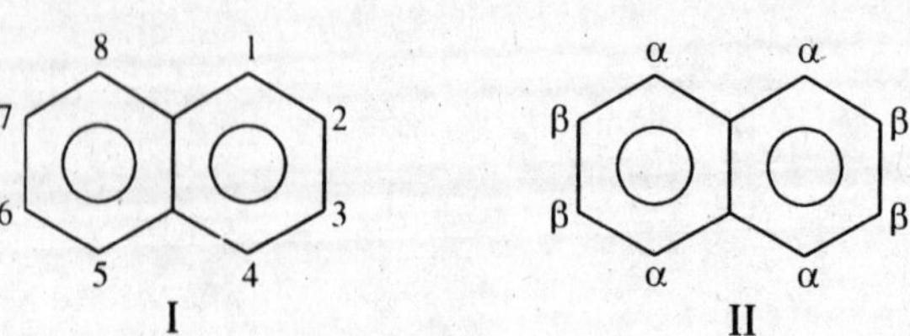

I II

Naphthalene molecule contains ten carbon atoms which are numbered or designated as shown above. It may be noted that positions in naphthalene ring system are indicated either by numbering the various positions or by designating them with Greek letters α and β. Positions 1, 4, 5 and 8 are equivalent and are designated as α-positions. Similarly 2, 3, 6 and 7 positions are termed as β-positions.

In naming monosubstituted derivatives of naphthalene, the prefixes 1- and 2- or α- and β- are used. But in di and more highly substituted compound, the positions of substituents are indicated by numbers. For example:

NO_2 SO_3H NO_2

1-Nitronaphthalene or α-Nitronaphthalene

2-Naphthalene sulphonic acid or β-Naphthalene sulphonic acid

NO_2

1, 5-Dinitronaphthalene

Derivatives of naphthalene show position isomerism. Thus it is clear that in a mono-substituted derivatives, the substituents occupies either α or β position. This conveys that mono-substituted derivatives can exist in two isomeric forms.

The disubstitution products in which both the substituents are identical can exist in ten isomeric forms. However, if the two substituents are different then fourteen isomers are possible.

Q. 3. How do you isolate naphthalene from coal-tar? *(Andhra, 2010)*

Ans. Isolation from coal-tar. Naphthalene, $C_{10}H_8$ is the single largest constituent of coal-tar. It constitutes about 6-10% of coal-tar and is present in the middle oil fraction of coal-tar distillation. On allowing the middle oil fraction to cool, most of the naphthalene crystallises out and is separated by centrifugation or by pressing out the oil in a hydraulic press. The resultant solid mass is washed with hot water and aqueous alkali to remove traces of oil and phenols. It is then washed with a little conc. sulphuric acid which removes the basic impurities. Crude naphthalene is purified by sublimation. For further purification it may be recrystallised by using petroleum ether.

Manufacture. Naphthalene these day is also made synthetically from petroleum. When petroleum fractions are passed over copper catalyst at 953 K, naphthalene and higher aromatics are formed. Naphthalene can be separated by usual methods from the complex mixture.

Q. 4. How do you establish the structure of naphthalene?

(Gulberga 2003 ; Delhi, 2003; Purvanchal 2003; Venkateswara, 2005)

Or

Give chemical reactions to show that naphthalene contains two equivalent benzene rings fused together. *(Bangalore, 2004; Patna, 2010; Pune, 2010)*

Ans. 1. Molecular formula. On the basis of analytical data, its molecular formula is found to be $C_{10}H_8$.

2. Chemical behaviour. (*i*) Although it appears from its formula, that naphthalene is a highly unsaturated molecule, still it is *resistant* (though less than benzene) to addition reactions, characteristic of unsaturated compounds.

(*ii*) Like benzene, it undergoes electrophilic *aromatic substitution reactions* like halogenation, nitration, etc., in which hydrogen is displaced as hydrogen ion.

(*ii*) Like benzene, it undergoes electrophilic *aromatic substitution reactions* like halogenation, nitration, etc., in which hydrogen is displaced as hydrogen ion.

(*iii*) Its nuclear substituted hydroxy compounds are *phenolic* and amino derivatives can be *diazotised*. Further these compounds undergo the usual *coupling reaction* also. It resembles benzene in this respect.

(*iv*) Like benzene, naphthalene is unusually stable. Its heat of combustion is 255 kJ less than that calculated for an open-chain structure corresponding to the formula $C_{10}H_8$.

The above observations suggest that naphthalene should have a ring structure like benzene.

3. Oxidation. On oxidation with air in the presence of vanadium pentoxide, naphthalene yields phthalic acid (*o*-benzene dicarboxylic acid) showing thereby the presence of at *least one aromatic ring and two side chains in ortho position to each other*.

$$\underset{\text{Naphthalene}}{C_{10}H_8} \xrightarrow[V_2O_5]{[O]} \underset{\text{Phthalic acid}}{C_6H_4(COOH)_2}$$

Therefore the formula of naphthalene may be written as:

$$C_6H_4 \langle \; \rangle \; C_4H_4 \quad \text{or} \quad C_6H_4 \; \rangle \; C_4H_4$$

4. Possible side chain structure. The two side-chains, on the basis of valency requirement, must be unsaturated. One such structure of naphthalene could be

$$C_6H_4\begin{matrix} -C \equiv CH \\ -CH = CH_2 \end{matrix}$$

However, such a structure is in contrast to known properties of naphthalene and is, therefore, ruled out.

5. Erlenmeyers fused ring structure. Naphthalene on nitration yields nitronaphthalene. This on oxidation with potassium permanganate gives 3–nitrophthalic acid. However, if nitronaphthalene is first reduced to aminonaphthalene and then oxidised, it gives phthalic acid. To account for above reactions, Erlenmeyers suggested that a naphthalene molecule is a fusion of two benzene rings (A & B) together. The reactions are represented as under

$$\underset{\text{Naphthalene}}{\text{A B}} \xrightarrow{\text{Nitration}} \underset{\text{Nitronaphthalene}}{\text{A B} \; (NO_2 \text{ on A})} \xrightarrow{\text{Oxidation}} \underset{\text{Nitronaphthalic acid}}{\text{A} \; (NO_2, COOH, COOH)}$$

$$\text{Nitronaphthalene} \xrightarrow{\text{Reduction}} \underset{\text{Aminonaphthalene}}{\text{A B} \; (NH_2 \text{ on A})} \xrightarrow{\text{Oxidation}} \text{HOOC, HOOC} - \text{B}$$

The reason for the difference in behaviour on oxidation is that the nitro group makes an aromatic ring harder to oxidise than an unsubstituted benzene ring, whereas the amino group increases the ease of oxidation of the ring to which it is attached.

6. Evidence in support of Erlenmeyers structure. The Erlenmeyers structure is supported by the following observations:

(*i*) It shows the presence of five double bonds in the molecule and naphthalene is known to add to a maximum of five molecules of hydrogen to form decalin.

Naphthalene $+ 5H_2 \longrightarrow$ Decalin

(*ii*) This formula suggests the formation of two mono substituted (α and β-isomer) and 10 disubstituted derivatives (in case the two substituents are same), which is in agreement with known facts.

(*iii*) **Synthesis of naphthalene.** Many syntheses of naphthalene support Erlenmeyer formula. Some are described hereunder.

(*a*) If 4-phenyl butene–1 is passed over red hot calcium oxide, naphthalene is formed.

$C_6H_5-CH_2-CH_2-CH=CH_2 \xrightarrow[D]{CaO}$ Naphthalene $+ 2H_2$

4-Phenyl but-1-ene

(*b*) **Fittig's Synthesis.** Cyclization of 4-phenyl-3-butenoic acid in the presence of concentrated sulphuric acid gives α-naphthol which on distillation with zinc dust yields naphthalene.

$C_6H_5-CH=CH-CH_2-C(=O)-OH \xrightarrow[-H_2O]{Conc.\ H_2SO_4}$ (cyclic ketone) $\xrightarrow{Isomerisation}$ α-Naphthol (OH) $\xrightarrow{Zinc,\ Heat}$ Naphthalene

4-Phenyl-3-butenoic acid

(*c*) **Haworth's Synthesis (1932).** Friedel-Crafts reaction of succinic anhydride with benzene in the presence of anhydrous aluminium chloride yields a ketonic acid (I) which is reduced to II.

β-Benzoyl propionic acid (I)

Zn/Hg HCl

γ-Phenyl butyric acid (II)

Conc. H_2SO_4 — H_2O

α-Tetralone (III)

Zn/Hg HCl

Tetralin (IV)

Se or palladised charcoal/Δ — $2H_2O$

Naphthalene

which on cyclization gives ketone III. This on Clemmenson reduction followed by selenium dehydrogenation yields naphthalene.

Note: Birch *et al* in 1946 have shown that heating α-tetralone with 1 : 1 mixture of NaOH — KOH at 493 K gives naphthalene (58%). The proposed mechanism is as given below:

α-tetralone ⇌ ⇌ ⇌ Naphthalene

7. Resonance Concept. In terms of resonance theory naphthalene is considered to be a resonance hybrid of mainly three resonating structures. (I), (II) and (III) as shown below:

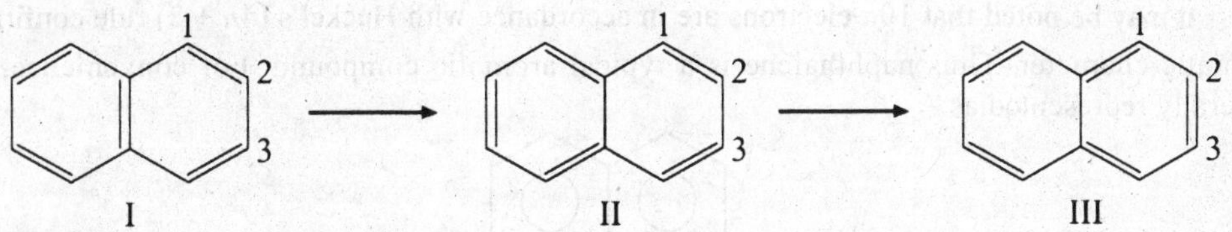

I $\longrightarrow$ II $\longrightarrow$ III

The resonance hybrid structure of naphthalene is supported by various observations as given below:

(*i*) Naphthalene is a planar molecule.

(*ii*) X-ray studies have shown that all *carbon-carbon bonds* in *naphthalene* are not equivalent. The C_1 - C_2 bond is having relatively greater double bond character and is 1.361 Å as compared to C_2 - C_3 bond having greater single bond character and length 1.421 Å. This is obvious, if we look at the contributing structures where C_1 - C_2 bond is a double bond in structure I & II and a single bond in structure III. Thus it has a *two-third* double bond character. On the other hand C_2 - C_3 bond is a single bond in structures I & II and a double bond in structure III.

(*iii*) The resonance energy of naphthalene is about 255 kJ/mole. This value is less than twice the amount for a single benzene ring (150 kJ mole). This means that the two rings of naphthalene are slightly less stabilised or in other words slightly more reactive than benzene. This is actually so and is shown by the fact that naphthalene undergoes addition reactions relatively more easily than benzene.

8. Orbital Concept. In the light of molecular orbital theory, all the ten carbon atoms of naphthalene (lying at the corner of two fused hexagons) are in sp^2 hybridised state. Each carbon atom is linked to three other atoms by σ bond resulting from the overlapping of sp^2 orbitals of one carbon with similar orbitals of other carbons or *s* orbitals of hydrogens. As a result, all carbon and hydrogen atoms lie in a single plane, *i.e.*, *naphthalene* is a flat molecule. The unhybridized *p*-orbital at each carbon overlaps with the *p*-orbitals on its sides forming a π electron cloud above and below the plane of the ring containing all the carbon and hydrogen atoms. This π electron bond has a shape of '8' and consists of two partially overlapping sextets—thus imparting aromatic character to naphthalene. However, since a pair of π electrons is common to both the rings, it possesses less aromatic character than benzene. Its orbital diagram is given below:

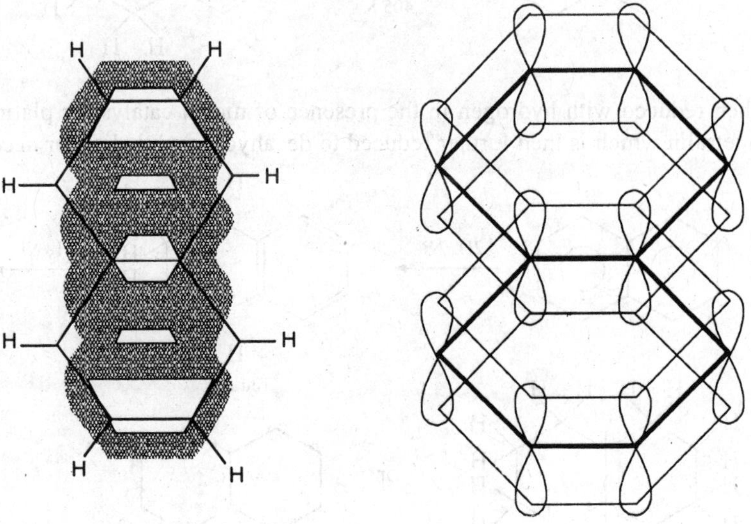

It may be noted that 10π electrons are in accordance with Huckel's $(4n + 2)$ rule confirming aromatic character. Thus naphthalene is a typical aromatic compound. For convenience, it is generally represented as:

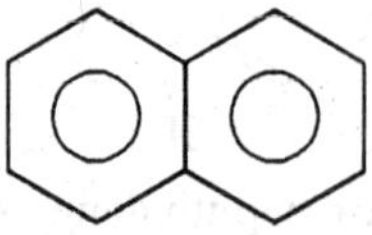

Q. 5. Describe the general reactions of naphthalene. (*Kerala, 2004 ; Bangalore, 2006*)

Ans. It resembles benzene in its chemical behaviour. However, it is more reactive than benzene and forms the addition and substitution products much more readily. It is also more susceptible to oxidation and reductions. Consequently, it is somewhat less aromatic than benzene. Important reactions of naphthalene are:

[A] Addition Reactions.

1. Addition of hydrogen. Unlike benzene, naphthalene can be reduced to a number of products depending upon the reducing agent used.

(*i*) When reduced with sodium and ethanol, it gives 1, 4-dihydronaphthalene (1, 4-*dialin*), m.p. 298 K which readily isomerises to 1, 2-dialin.

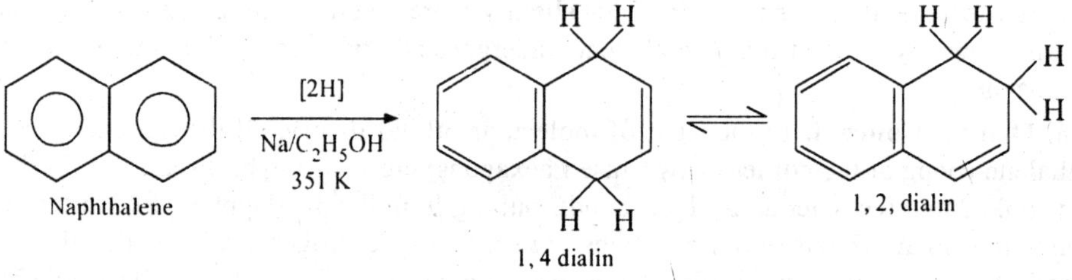

Naphthalene →[2H], Na/C_2H_5OH, 351 K→ 1, 4 dialin ⇌ 1, 2, dialin

(*ii*) On reduction with sodium and *iso* amyl alcohol it forms 1, 2, 3, 4-tetrahydronaphthalene (*tetralin*, b.p. 479 481 K).

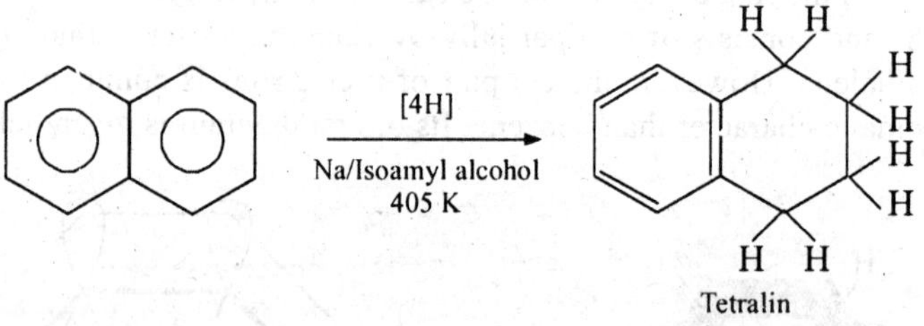

→[4H], Na/Isoamyl alcohol, 405 K→ Tetralin

(*iii*) When reduced with hydrogen in the presence of nickel catalyst or platinum, it is first converted to tetralin which is then further reduced to decahydronaphthalene or *decalin*.

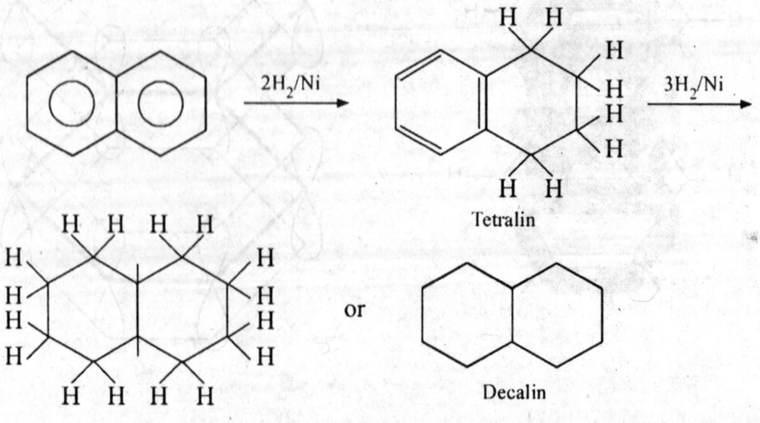

→$2H_2/Ni$→ Tetralin →$3H_2/Ni$→ or Decalin

Note. Tetralin and decalin are used as solvents for varnishes, lacquers etc.

2. Addition of halogens. Naphthalene adds chlorine or bromine to yield naphthalene *di-* and *tetra* chlorides or *di-* and *tetra* bromides with the halogens in the same ring.

$$\text{Naphthalene} \xrightarrow[CCl_4]{Cl_2} \text{Naphthalene dichloride} \xrightarrow{Cl_2} \text{Naphthalene tetrachloride}$$

On heating, these halides undergo dehydrohalogenation to give substituted halonaphthalenes. On oxidation they are converted to phthalic acid.

$$\text{1, Chloronaphthalene} \xleftarrow[-\,HCl]{313\text{ K}} \text{Naphthalene dichloride} \xrightarrow{\text{Oxidation}} \text{Phthalic acid}$$

$$\text{1, 3--Dichloronaphthalene + other isomers} \xleftarrow[\Delta]{NaOH} \text{Naphthalene tetrachloride} \xrightarrow{\text{Oxidation}} \text{Phthalic acid}$$

3. Ozonolysis. When treated with ozone, it forms a diozonide and this on reaction with water gives phthalaldehyde.

$$\text{Naphthalene} \xrightarrow{O_3} \text{Ozonide} \xrightarrow{H_2O/Zn} \text{Phthaladehyde}$$

[B] Substitution Reactions

As in case of benzene, the important reactions of naphthalene are electrophilic substitution reactions which can be carried out by using same reagents and catalysts. However in case of mono substitution it may form either an α- or β-derivatives unlike benzene. In general the substituent enters the α- or (1–) position with some exceptions.

1. Halogenation. Naphthalene undergoes halogenation much more readily as compared to benzene. Chlorination with chlorine in the presence of Fe yields 95% of α-chloronaphthalene and about 5% β-isomers.

Cl

S_2Cl_2 298 K ← Naphthalene → Cl_2/Fe

Cl Cl

α-chloro-naphthalene (95%) + β-chloro-naphthalene (5%)

However chlorination with chlorine in the presence of iodine at higher temperatures gives a mixture of α- and β-derivatives in 1 : 1 ratio.

Cl Cl

Cl_2/I_2 573 K → +

Br_2/CCl_4 ↓

Br

α-Bromonaphthalene 75%

Bromination with bromine in boiling CCl_4 gives mainly the α-isomer. Further halogenation of monohalogen derivatives yields mainly 1, 4-dihalonaphthalene. The halogen atom in naphthalene *i.e.*, nuclear halogen acts similarly as in benzene.

Cl Cl

Cl_2 Fe →

Cl

1, 4-Dichloronaphthalene

2. Nitration. Nitration of naphthalene with a mixture of concentrated nitric acid and sulphuric acid yields (at 323–333 K) almost exclusively α-nitronaphthalene (95%) with minor amounts (5%) of β-nitronaphthalene.

$\xrightarrow{HNO_3/H_2SO_4}$

Naphthalene α-Nitronaphthalene (95%) + β-Nitronaphthalene (5%)

However at higher temperatures, a mixture of 1, 5- and 1, 8-dinitronaphthalene is obtained. Reduction of nitronaphthalene followed by diazotisation yields diazonium salts which find use in the preparation of many other derivatives of naphthalene.

3. Sulphonation. Sulphonation of naphthalene with conc. H_2SO_4 at 355 K gives a product containing 96% α-naphthalene sulphonic acid whereas at 425 K or above the mixture of sulphonic acids obtained contains 85% β-naphthalene sulphonic acid, as the main product.

SO_3H

β-Naphthalene sulphonic acid $\xleftarrow[425\ K]{Conc.\ H_2SO_4}$ Naphthalene $\xrightarrow[355\ K]{Conc.\ H_2SO_4}$

SO_3H

α-Nitronaphthalene sulphonic acid (95%)

4. Friedel-Crafts reaction. Naphthalene is attacked by aluminium chloride when vigorous conditions are employed (binaphthyls and compound, with one of the naphthalene rings opened, are formed).

Thus to carry out the *Friedel-Crafts reaction* successfully *mild* conditions (low temperatures) should be used, and even then the maximum yield is about 60%.

(*i*) With methyl iodide, 1- and 2-methyl naphthalenes are formed.

(*ii*) With ethyl bromide only 2-ethyl naphthalene; and with *n*-propyl bromide, 2-isopropyl naphthalene are formed.

(*iii*) With alcohols and aluminium chloride, 2, 5-dialkyl naphthalene, and with alcohols and boron trifluoride, 1, 4-dialkyl naphthalene are obtained.

Friedel-Crafts acylation with accetyl chloride in the presence of aluminium chloride gives a mixture of α- and β-isomers. The solvent used as reaction medium determines their relative ratios.

Thus in CS_2 medium 45% α- and 15% β-isomers are obtained whereas in nitrobenzene medium, the β-isomer is obtained predominantly.

Reactions

CH_3 CH_3

(*i*) Naphthalene $\xrightarrow{CH_3I/AlCl_3}$ 1-Methyl naphthalene + 2-Methyl naphthalene

(ii) Naphthalene $\xrightarrow{C_2H_5Br/AlCl_3}$ 2-Ethyl naphthalene

Naphthalene $\xrightarrow{n\text{-}C_3H_7I/AlCl_3}$ 2-Isopropyl naphthalene ($CH(CH_3)_2$)

(iii) Naphthalene $\xrightarrow[CS_2 \text{ (as solvent)}]{CH_3COCl/AlCl_3}$ Methyl α-naphthyl ketone (Main Product) ($COCH_3$)

Naphthalene $\xrightarrow[C_6H_5NO_2 \text{ (as solvent)}]{CH_3COCl/AlCl_3}$ Methyl β-naphthyl ketone (Main Product) ($COCH_3$)

Note : The effect of nitrobenzene as solvent has been attributed to its forming a complex with acid chloride and aluminium chloride, which because of its bulkiness, attacks naphthalene at the less crowded β-position. (*The effect of nitrobenzene in directing the orientation to β-position has been explained on steric grounds*).

[C] Other Reactions

Oxidation. Naphthalene upon oxidation gives a number of products as described below:

(*i*) When oxidised with air in the presence of V_2O_5 at 733–753 K or with concentrated H_2SO_4 and mercuric sulphate at 473 K, naphthalene gives phthalic anhydride.

$$2\ \text{Naphthalene} + 9O_2 \xrightarrow{V_2O_5/733\ K} 2\ \text{Phthalic anhydride} + 4CO_2 + 4H_2O$$

Phthalic anhydride can be easily hydrolysed to phthalic acid.

(*ii*) Acidic $KMnO_4$ oxidises naphthalene to phathalic acid directly.

Naphthalene —Acidic $KMnO_4$→ Phthalic acid (COOH, COOH)

(*iii*) Chromic acid oxidises naphthalene to 1, 4-naphthaquinone.

Naphthalene —CrO_3, CH_3COOH, 298 K→ 1, 4-Naphthaquinone

Q. 6. Give the mechanism of electrophilic substitution in naphthalene.

(*Annamalai, 2002; Panjab, 2005*)

Ans. Naphthalene undergoes a number of substitution reactions in most of which the attack occurs mainly on the α-carbon atom.

Considering mechanistic point of view, these reactions are typical electrophilic aromatic substitution reactions as described below:

Y : Z ⟶ Y^+ + : Z^- (where Y^+ is the electrophile)

I + Y^+ ⟶ [II ⟷ III ⟷ IV etc.]

+ Z^- ⟶ (1-Y-naphthalene) + H : Z

It may be observed that the *carbonium ion* intermediate in electrophilic substitution in naphthalene has more contributing structures than in case of benzene.

As such, this carbonium ion is more resonance stabilised and therefore, more readily formed. Due to all this, naphthalene is more reactive than benzene towards electrophilic substitution.

Q. 7. Discuss orientation of electrophilic substitution in naphthalene.

(*Guwahati, 2002; Purvanchal, 2003*)

Or

Electrophilic substitution in naphthalene proceeds faster at α-position compared to that at β-position. Explain. (*Garhwal, 2000; Lucknow, 2000; Himachal, 2000; Delhi, 2003*)

Ans. In many substitution reactions of naphthalene, the attack occurs on the α-carbon atoms. In order to have a clear understanding of greater reactivity of α-carbon atom in these reactions, let us examine the contributing structures of the intermediate carbonium ion obtained from the attack at the α and β positions. As seen above in previous question the intermediate carbonium ion resulting from the α-attack is a resonance hybrid of structures II and III (in which the positive charge remains on the ring under attack) and structure like IV (where the positive charge resides on the other ring).

In the same fashion, the carbonium ion intermediate in case of β-attack would be a resonance hybrid of structures V and VI (where the positive charge is carried by the ring under attack) and structure like VII (where the positive charge is held by the other ring) as depicted below:

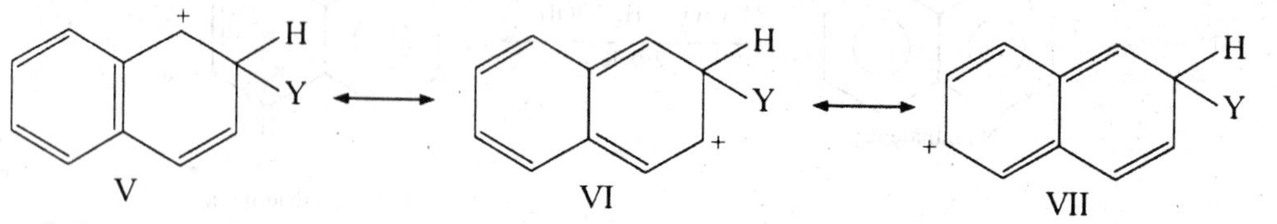

V ⟷ VI ⟷ VII

It is noteworthy that unlike other contributing structures, structures II and III (in case of α-attack) and V (in case of β-attack) *still possess one benzene ring (resonance stabilisation 150 kJ per mole) completely intact.*

Hence these three structures *i.e.* II, III and V are more stable than the others. Another point to be noted is that two of these (*i.e.* II and III) more stable structures are involved during α-attack, while only one *i.e.* V is involved in β-attack. It is therefore clear that the carbonium ion formed during α-attack and also the transition state yielding this ion, would be much more stable than the carbonium ion as well as the corresponding transition state obtained during β-attack. Thus we can conclude that α-substitution would be more favourable than β-substitution.

Q. 8. Describe the orientation of sulphonation of naphthalene.

Ans. Sulphonation of naphthalene at 355 K gives α-naphthalene sulphonic acid and at 425 K gives β-naphthalene sulphonic acid.

Sulphonation takes place more readily at α-position because the intermediate α-substituted carbonium ion is more stable than its β-counterpart. But α-naphthalene sulphonic acid is less stable due to steric hindrance. Moreover sulphonation is a reversible reaction, desulphonation also takes place rapidly. β-naphthalene sulphonic acid although less readily formed, is quite stable because of no steric hindrance.

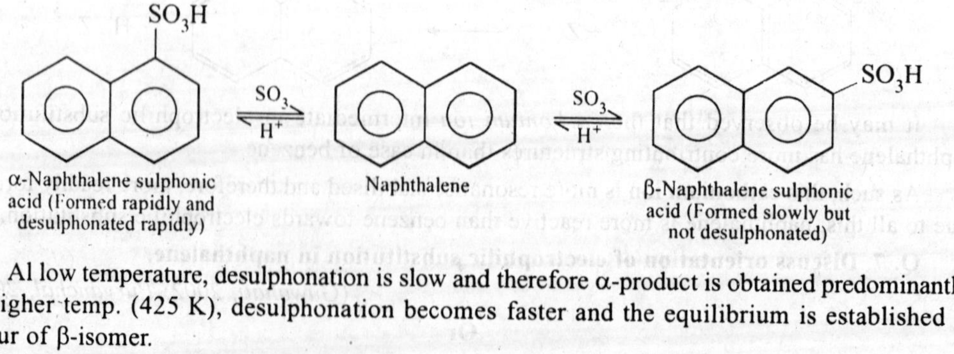

α-Naphthalene sulphonic acid (Formed rapidly and desulphonated rapidly) $\underset{H^+}{\overset{SO_3}{\rightleftharpoons}}$ Naphthalene $\underset{H^+}{\overset{SO_3}{\rightleftharpoons}}$ β-Naphthalene sulphonic acid (Formed slowly but not desulphonated)

At low temperature, desulphonation is slow and therefore α-product is obtained predominantly. At higher temp. (425 K), desulphonation becomes faster and the equilibrium is established in favour of β-isomer.

In other words, it is the kinetics which is the controlling factor at low temperature, whereas position of equilibrium is the controlling factor at high temperature.

Q. 9. Discuss the orientation of Friedel-Crafts acylation and alkylation of naphthalene.

Ans. Substitution of alkyl and acyl groups in naphthalene can be explained in terms of steric factors. It is observed that the alkyl group has a tendency to go to less hindered β-position in Friedel-Crafts reaction, more so, if the alkyl group involved is bigger. Thus with methyl iodide, a mixture of α- and β-methyl napthalenes is obtained. With ethyl and propyl bromides, it is the β-alkylated naphthalene, which is formed in excess.

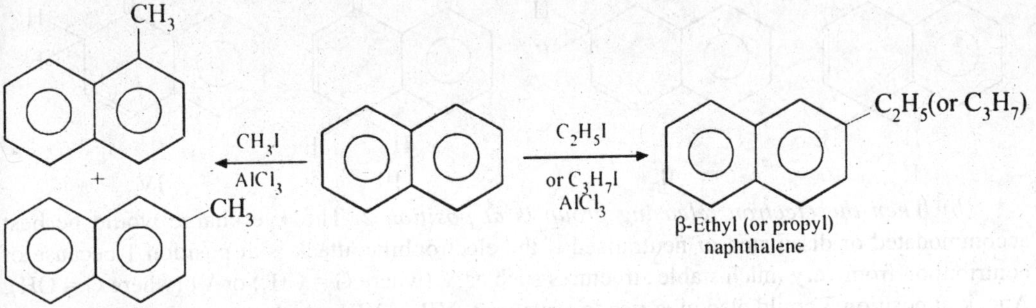

CH_3I, $AlCl_3$ / C_2H_5I or C_3H_7I, $AlCl_3$; CH_3; CH_3; C_2H_5(or C_3H_7)

Mix. of α- and β-Methyl naphthalenes

β-Ethyl (or propyl) naphthalene

Again it is observed that if we perform Friedel-Crafts acylation of naphthalene with nitrobenzene as the solvent, we obtain β-acetyl naphthalene in place of α-isomer. It is believed that nitrobenzene form a complex with acetyl chloride and aluminium chloride (which are the reagents for Friedel-Crafts acylation) and this complex attacks the less hindered β-position.

Q. 10. Discuss orientation of substitution in monosubstituted naphthalenes.

Ans. Orientation in disubstitution in naphthalene follows essentially the same rules as for benzene, but is somewhat more complicated. In case of naphthalene seven different positions are available for the second substituent to enter. The second substituent can enter the ring already having a substituent giving rise to *homonuclear* substitution or the other ring, leading to *heteronuclear* substitution. However the following rules are generally observed.

(I) *Activating groups* (electrons releasing and usually *o*- and *p*-directing groups like –OH, –OR, $-NH_2$, –NHR, –R etc.), when present, favour further substitution in the same ring.

When such a group is present at position 1, the second substituent goes to position 4 and to some extent to position 2. In case the group already present occupies position 2, the second substituent goes to position 1.

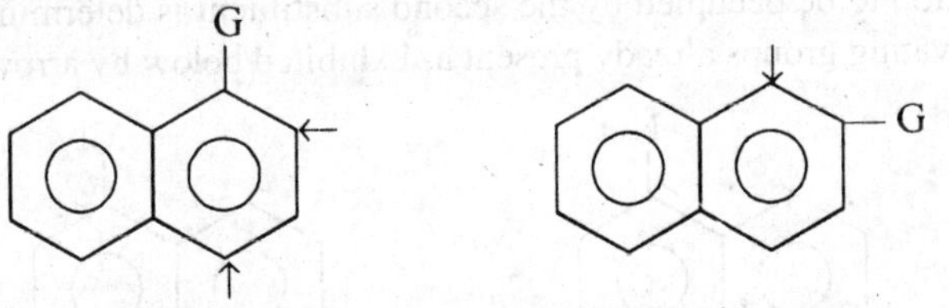

G G

Electron releasing group at position 1, second substituent goes to position 4 or 2.

Electron releasing group at position 2, second substituent goes to position 1.

Remember that the electron releasing group already present can be rightly assumed to exert the main effect at the ring to which it is directly attached. This way the electrophilic attack (Say by $\overset{+}{E}$) would be favoured to occur at this very ring because it can best accommodate or dissipate the positive charge

(a) *When the electron releasing group is at position 1, the positive charge* would be best accommodated if the electrophilic attack occurs at position 4 or 2. This is so because the intermediate carbonium ion would have maximum resonance stabilization in these positions due to contribution from stable structures such as I, II in particular (where G = CH_3) or III, IV (where G = OH).

Similar structure are not obtained in case of electrophilic attack occurring at position 3.

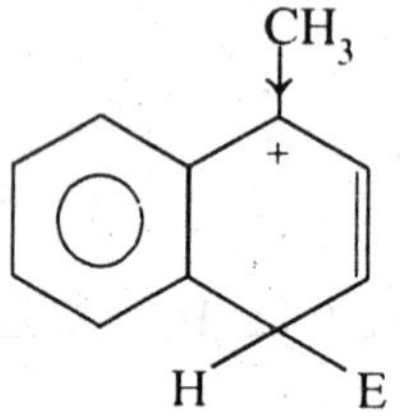

I 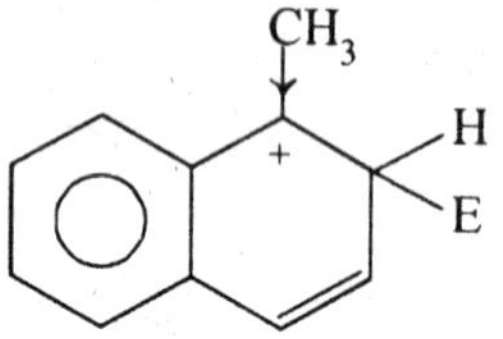 II 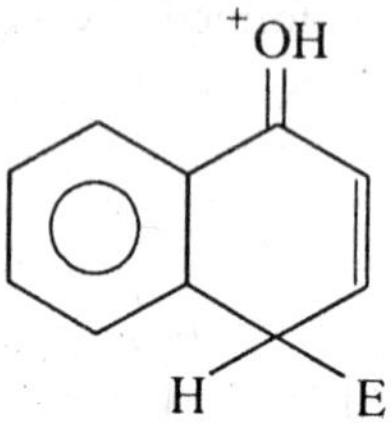 III 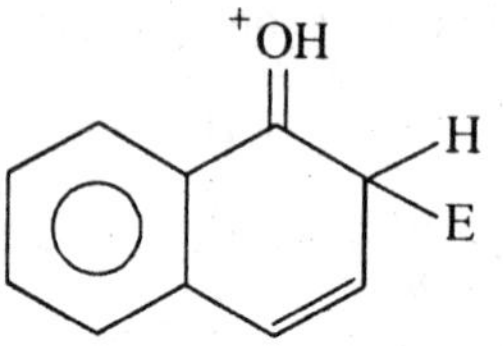 IV

(b) *When the electron releasing group is at position 2*. The +ve charge would be best accommodated or dissipated or neutralised if the electrophilic attack is at position 1 because of contribution from very much stable structures such as V (where G = CH_3) or VI (where G = OH). Attack at position 3 could also give rise to structures, VII or VIII which can easily accommodate +ve charge, but such a structure is comparatively much less stable since in this case even the second ring loses its aromatic character.

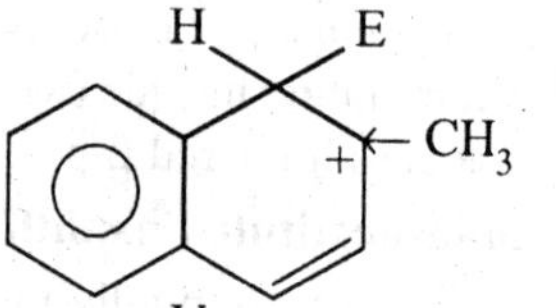

V 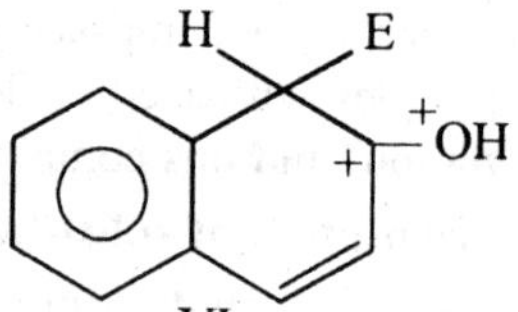 VI

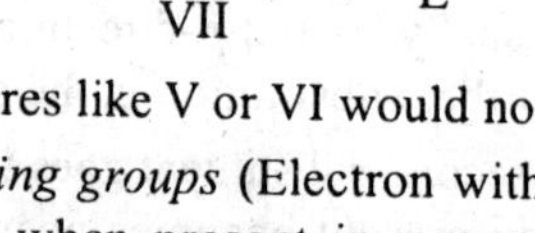

VII 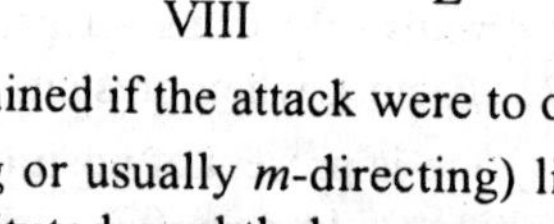 VIII

Stable structures like V or VI would not be obtained if the attack were to occur at position 4.

(c) *Deactivating groups* (Electron withdrawing or usually *m*-directing) like $-NO_2$, $-SO_3H$, -COOH, –Cl etc. when present in a mono substituted naphthalene generally direct further substitution into the other ring.

The true position to be occupied by the second substituent is determined by the position and nature of the deactivating groups already present as exhibited below by arrow marks in some cases.

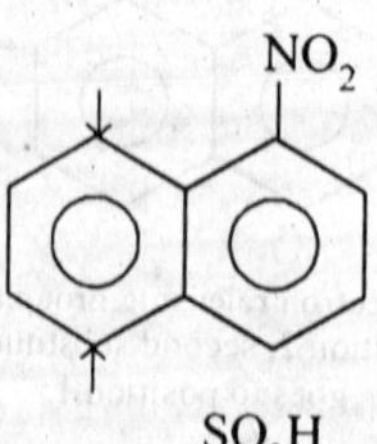

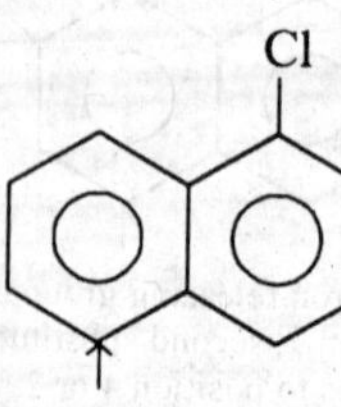

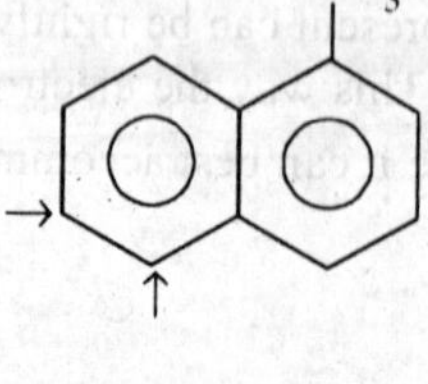

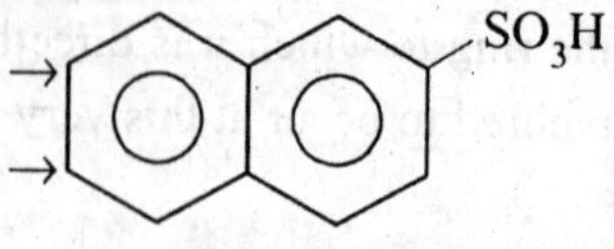

It is very much clear that an electron withdrawing group would not easily permit further substitution in the same ring, so the second substituent would go only to the other ring. The actual orientation in the second ring is, however dictated by various factors including the stabilisation of the intermediate carbonium ions.

ANTHRACENE

Q. 11. How would you obtain anthracene ($C_{14}H_{10}$) from anthracene oil?

Ans. Anthracene is a tricyclic fused ring aromatic hydrocarbon, which occurs in the coal tar distillation fraction known as anthracene oil (Greek anthrax = coal) or green oil because of its dark green fluorescence. This fraction is collected between 543–633 K and contains phenanthrene and carbazole besides 1% anthracene.

The anthracene oil is cooled and allowed to crystallise over a period of one week. It is then filtered off, the solid cake is known as '*anthracene cake*', it is pressed free from liquid. The crude anthracene contains phenanthrene and carbazole $(C_6H_4)_2$ NH.

The anthracene cake is powdered and washed with '*solvent naphtha*' which dissolves phenanthrene; and the remaining solid is then washed with pyridine which dissolves the carbazole leaving behind anthracene. Anthracene is purified by sublimation. Alternatively, after removal of phenanthrene, the remaining solid is fused with KOH, whereby potassio-carbazole is formed; unreacted anthracene is sublimed out of the melt and recovered.

In a recent technique, after removal of phenanthrene, the remaining mixture is oxidised catalytically by air in the presence of V_2O_5 at 573–773 K. Under these conditions carbazole is oxidised to CO_2 etc. and the anthracene to anthraquinone.

Q. 12. Discuss nomenclature and isomerism of anthracene derivatives.

Ans. Anthracene molecule contains fourteen carbon atoms which are indicated by numbers or Greek letters as shown below. It may be noted that positions 1, 4, 5, 8 are identical. Similarly positions 2, 3, 6, 7 and position 9, 10 are identical.

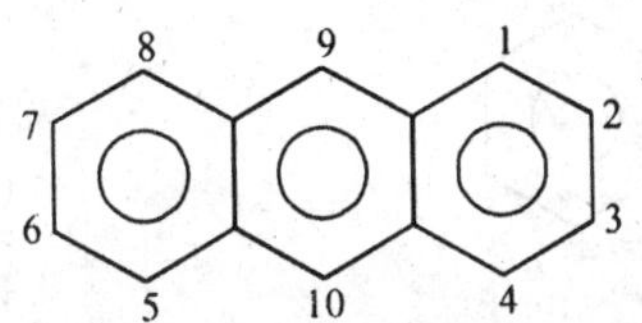

α γ α
β β
β β
α γ α

It is clear from these formulae that in anthracene three different position are available for a substituent and hence it can give rise to three isomeric monosubstitution products *viz.*, α- or 1-; β- or 2- and γ- or 9-(or meso).

In disubstitution, fifteen isomers are possible when the two substituents are identical. When, however, the two substituents are different, the number of isomers is still larger.

Orientation in anthracene derivatives follows roughly the same course as in naphthalene derivatives, though it is much more complex.

Q. 13. Discuss the structure of anthracene. (*Aligarh, 2003; Karnataka, 2004; Andhra, 2010*)

Or

How do you show that anthracene contains three benzene rings fused together at adjacent carbon atoms?

Ans. 1. Molecular formula. The molecular formula of anthracene as calculated from its analytical data is $C_{14}H_{10}$.

2. Chemical behaviour. (*i*) The formula suggests that anthracene may be related to benzene and naphthalene

$$\underset{\text{Benzene}}{C_6H_6} \xrightarrow{+4C;\,+2H} \underset{\text{Naphthalene}}{C_{10}H_8} \xrightarrow{+4C;\,+2H} \underset{\text{Anthracene}}{C_{14}H_{10}}$$

(*ii*) Like benzene and naphthalene it gives electrophilic substitution reactions, *e.g.*, halogenation, nitration and sulphonation and also it undergoes addition reactions under suitable conditions. This supports the view that anthracene may be related to benzene and naphthalene.

3. Fused ring structure. (*i*) Bromination of anthracene yields bromoanthracene, $C_{14}H_9Br$, which on fusion with KOH gives hydroxy anthracene $C_{14}H_9OH$. Strong oxidation of hydroxyanthracene gives phthalic acid and small amounts of *o*-benzoyl benzoic acid.

Reaction

$$\underset{\text{Anthracene}}{C_{14}H_{10}} \xrightarrow{Br_2} \underset{\text{Bromo-anthracene}}{C_{14}H_9Br} \xrightarrow{KOH} \underset{\text{Hydroxy anthracene}}{C_{14}H_9OH}$$

↓ Oxidation

COOH, CO — *o*-Benzyl benzoic acid + COOH, COOH — Phthalic acid

These reactions suggest the presence of at least two benzene rings in anthracene and that its skeleton (I) is similar to that of *o*-benzyl benzoic acid

C, C

(I)

(*ii*) In the presence of sodium dichromate and H_2SO_4, anthracene forms anthraquinone, which on fusion with KOH at 523 K gives two molecules of benzoic acid.

$$\underset{\text{Anthracene}}{C_{14}H_{10}} \xrightarrow{Na_2Cr_2O_7,\ H_2SO_4} \underset{\text{Anthraquinone}}{C_{14}H_8O_2} \xrightarrow{KOH,\ 523\ K} 2\ \underset{\text{Benzoic acid}}{C_6H_5COOH}$$

This confirms the presence of at least two benzene rings in anthracene and conveys the existence of an anthraquinone like skeleton.

The structure of anthracene has been further established on the basis of the following synthesis of anthraquinone:

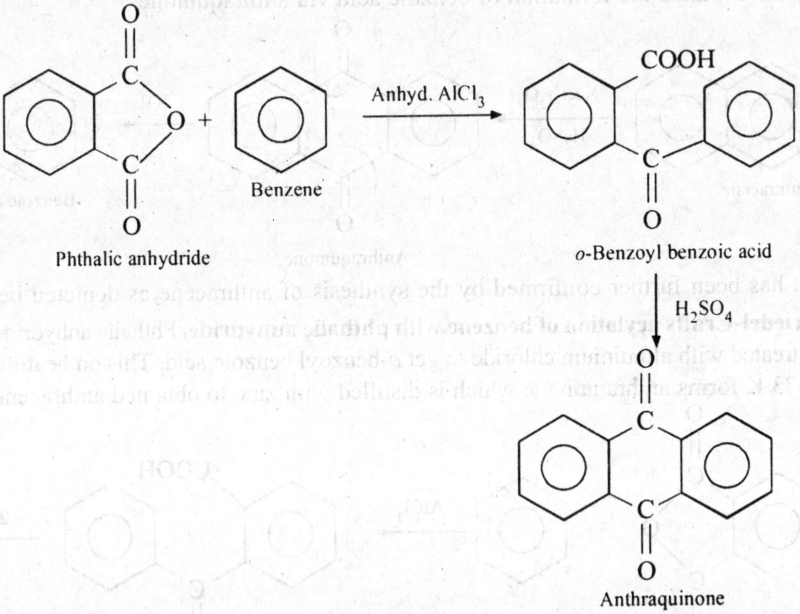

Phthalic anhydride + Benzene $\xrightarrow{\text{Anhyd. } AlCl_3}$ *o*-Benzoyl benzoic acid $\xrightarrow{H_2SO_4}$ Anthraquinone

In the light of above arguments and the tetracovalency of carbon, the following tricyclic structure was suggested for anthracene.

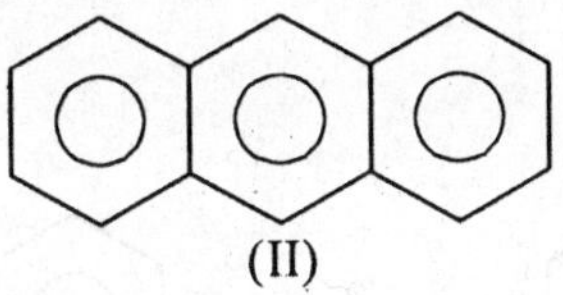

(II)

4. Evidence in favour of above structure. (*i*) It accounts for the formation of 3 mono substituted and 15 disubstituted isomers.

(*ii*) It shows the presence of 7 double bonds which is supported by the fact that anthracene on drastic hydrogenation adds 7 molecules of hydrogen to form perhydroanthracene.

(*iii*) This structure explains the bromination of naphthalene, followed by fusion with KOH and oxidation to give a mixture of phthalic acid and *o*-benzoyl benzoic acid

Anthracene $\xrightarrow{Br_2}$ 9-Bromo anthracene $\xrightarrow{KOH}$ 9-Hydroxy anthracene

$\xrightarrow{[O]}$ *o*-Benzoyl benzoic acid + Phthalic acid

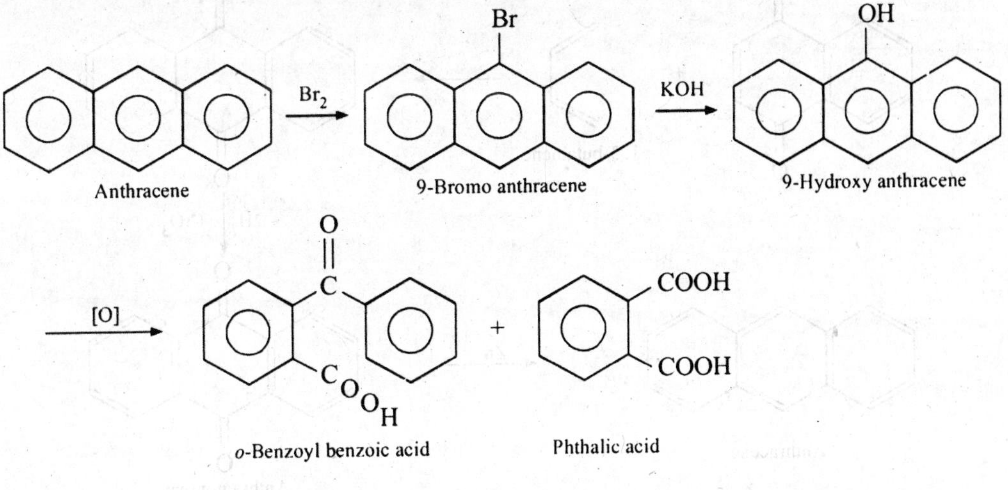

This also explains the formation of benzoic acid via anthraquinone

Anthracene $\xrightarrow[H_2SO_4]{Na_2Cr_2O_7}$ Anthraquinone $\xrightarrow{[O]}$ 2 Benzoic acid (COOH)

(*iv*) It has been further confirmed by the synthesis of anthracene as depicted below:

(*a*) **Friedel-Crafts acylation of benzene with phthalic anhydride.** Phthalic anhydride in benzene solution is treated with aluminium chloride to get *o*-benzoyl benzoic acid. This on heating with conc. H_2SO_4 at 373 K forms anthraquinone which is distilled with zinc to obtained anthracene.

Phthalic anhydride + Benzene $\xrightarrow{AlCl_3}$ *o*-Benzoyl benzoic acid (COOH) $\xrightarrow{-2\,(H)}$

o-Benzoyl benzoic acid $\xrightarrow{H_2SO_4}$ Anthraquinone $\xrightarrow{Zn}$ Anthracene

(*b*) **Diels-Alder reaction between 1, 3-butadiene and 1, 4-naphthaquinone**

1,4-Naphthaquinone + 1, 3-butadiene $\longrightarrow$ adduct $\xrightarrow[CrO_3]{-2H_2}$ Anthraquinone $\xrightarrow{Zn}$ Anthracene

(*c*) On heating *o*-bromobenzyl bromide with sodium, dihydroanthracene is formed, which on mild oxidation yields anthracene.

CH_2Br + 4Na + Br, BrH_2C —Δ, − 4NaBr→ (H_2C, CH_2)

o-Bromobenzyl bromide

9, 10-Dihydroanthracene

[O]

Anthracene

(*d*) **Friedel-Crafts condensation.** Two molecules of benzyl chloride give dihydroanthracene which under the conditions of the reaction readily eliminates two hydrogen atoms to yield anthracene

CH_2Cl + $ClCH_2$ —$AlCl_3$→ [Dihydroanthracene] —−2[H]→

Dihydroanthracene

Anthracene

5. **Resonance concept.** X-ray studies reveal that all carbon and hydrogen atoms in anthracene lie in the same plane. Like benzene and naphthalene, anthracene is considered to be a resonance hybrid of the following contributing structures.

I ⟷ II (1, 2, 3) ⟷ III ⟷ IV

The resonance energy of anthracene is 351.5 kJ mol^{-1} which is only 51.5 kJ more than the resonance energy of 2 benzene rings (150 kJ each). Thus anthracene is highly reactive.

The greater reactivity of 9, 10 positions—giving a product with two benzene rings intact is also explained properly on this basis, as the reaction at any other position will leave a naphthalene system with lesser amount of resonance energy. The above contributing structures also convey that

$C_1 — C_2$ bond distances have 3/4 double bond character while $C_2 — C_3$ bond distances possess 1/4 double bond character. Out of the resonating structures, structures II and III contribute more towards the actual structure because both of them contain two benzenoid rings and one quininoid ring. Benzenoid rings have greater resonance energy and hence more stability. Structures I and IV contribute to a smaller extent because they contain one benzenoid ring and two quininoid rings each. Because of greater contribution of structures II and III, $C_1 — C_2$ bond length has 3/4 double bond and $C_2 — C_3$ has 1/4 double bond characters. If all the structures had contributed equally, the double bond characters in $C_1 — C_2$ and $C_2 — C_3$ would have been 1/2. Total bond characters in all the bonds would then be $1^1/_2$.

It could also be explained by saying that the no. of structures in which there is a double bond between $C_1 — C_2$ is 3 (I, II, III) and no. of structures having double bond between $C_2 — C_3$ is only 1 (IV).

Molecular orbital concept

(*a*) In the light of molecular orbital theory, all the fourteen carbon atoms of anthracene are in sp^2 hybridisation state and are linked to three other atoms (one hydrogen and two carbon atoms or three carbon atoms) by the σ bonds formed as a result of overlapping of trigonal sp^2 orbitals. The molecule thus gets a state formed by the fusion of three hexagons. Hence all carbon and hydrogen atoms lie in a single plane or in other words, anthracene is a flat molecule.

(*b*) The unhybridised *p*-orbitals at carbon atoms give rise to a delocalised π-electron cloud on either side of the plane of the ring of carbon atoms. This π-electron cloud consists of three partially overlapping sextets possessing two pairs of electrons in common.

The number of total π electrons in anthracene is 14 which is in keeping with Huckel's $4n + 2$ rule thus confirming aromatic character.

For convenience sake, anthracene is generally represented as:

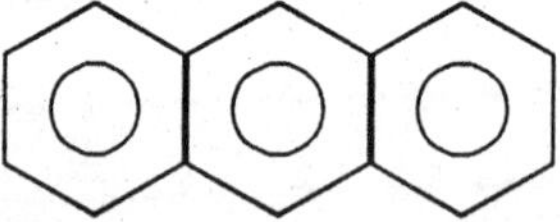

Molecular orbital diagram of anthracene is shown as under.

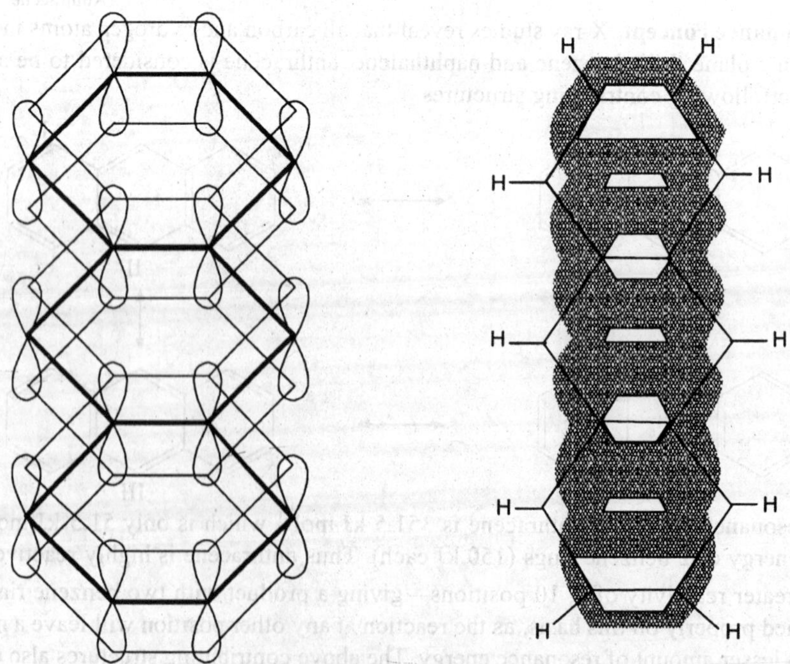

Molecular Orbital Picture of Anthracene.

Q. 14. Describe chemical properties of anthracene. (*Osmania 2003*)

Ans. Chemically, it resembles the aromatic hydrocarbons like benzene and naphthalene in many respects. The positions 9 and 10 are very reactive. The attack at these positions leaves two benzene rings in tact. Thus there is a sacrifice of only 51.5 kJ of resonance energy (351.5 – 2 × 150). Its important chemical reactions are:

1. Halogenation. Anthracene when treated with chlorine or bromine yields both addition and substitution products depending upon the reaction conditions. However, the addition products originally formed also get converted into substitution products at higher temperature. For example, the reaction with chlorine proceeds as depicted below under different sets of conditions.

Cl_2 in CS_2, Room temp. → Anthracene dichloride —Δ, – HCl→ 9-Chloroanthracene

Cl_2, 373 K → 9-Chloroanthracene (Main product) + 9-10-Dichloroanthracene

Bromine reacts in the similar fashion.

2. Nitration. Anthracene, when treated with conc. HNO_3, yields mono and dinitro products.

Anthracene —HNO_3 in $(CH_3CO)_2O$, 290 K→ 9-Nitroanthracene

9-10 Dinitroanthracene

3. Sulphonation. Anthracene, when treated with dil H_2SO_4, gives a mixture of two isomeric mono sulphonic acids.

Anthracene + H_2SO_4 ⟶ Anthracene-1-sulphonic acid + Anthracene-2-sulphonic acid

4. Reduction. Anthracene upon reduction with sodium and *iso*-amyl alcohol gives 9, 10-dihydroanthracene.

Upon catalytic reduction anthracene may give tetra, hexa and octahydroanthracenes depending upon the amounts of hydrogen used. Hence, we have:

Anthracene —(Na, iso $C_5H_{11}OH$)→ 9, 10-Dihydroanthracene

Anthracene —(H_2, Ni, 523 K)→ Perhydroanthracene

5. Oxidation. (*i*) Anthracene is readily oxidised by sodium dichromate and sulphuric acid to give anthraquinone.

Anthracene —($Na_2Cr_2O_7/H_2SO_4$, [O])→ 9, 10-Anthraquinone

(*ii*) It adds on a molecule of *oxygen* in the presence of light to form a colourless peroxide, which is believed to have the structure.

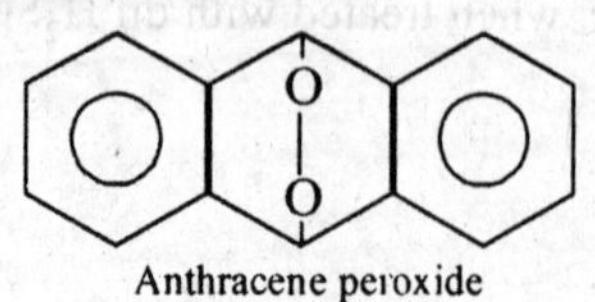

Anthracene peroxide

6. Dimerisations. When a saturated solution of anthracene and xylene is exposed to sunlight, it forms crystals of dimer-*dianthracene* or *paranthracene*.

Light

Anthracene

Dianthracene

7. Picrate formation. Anthracene on treatment with picric acid gives a red crystalline addition (m.p. 411 K).

8. Diels Alder reaction.

CHCO, CHCO, O

Maleic anhydride

Diels Alder Adduct

Q. 15. Discuss mechanism of substitution and addition reactions of anthracene.

Ans. The substitution reactions of anthracene (such as halogenation, sulphonation and nitration), as well as addition reaction with halogen proceed by electrophilic mechanism as outlined below.

$$Y-Z \longrightarrow Y^+ + :Z^-$$

$+ Y^+ \longrightarrow$

Intermediate carbonium ion
(Most stable contributing structure)

$-H^+$ $+ :Z^-$

Substitution

Addition

PHENANTHRENE

Q. 16. Describe the nomenclature and isomerism of phenanthrene derivatives.

Ans. Phenanthrene can be represented by either of the following structures given below. The numbering system is also shown in the structure.

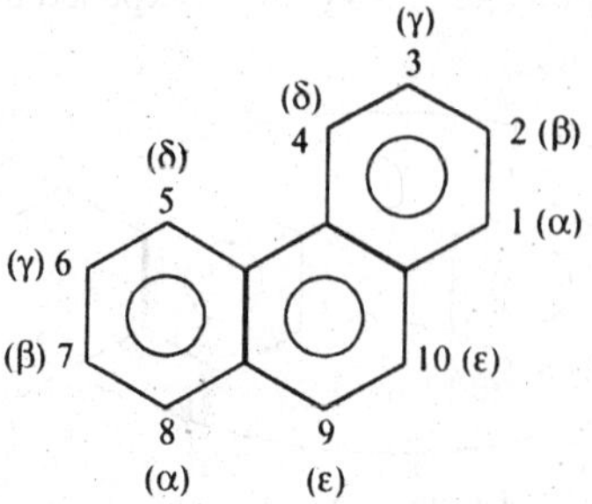

(γ) 3, (δ) 4, 2 (β), (δ) 5, 1 (α), (γ) 6, (β) 7, 10 (ε), 8 (α), 9 (ε)

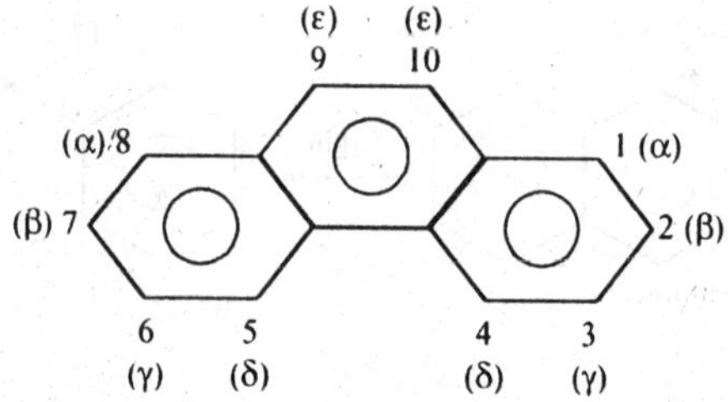

(ε) 9, (ε) 10, (α) 8, 1 (α), (β) 7, 2 (β), 6 (γ), 5 (δ), 4 (δ), 3 (γ)

There are five different positions α, β, γ, δ and ε. Thus phenanthrene forms five different mono-substituted compounds. In disubstitution, if the groups are the same, 25 different isomers are possible.

In naming a phenanthrene derivative, the positions of substituents are indicated by numbers.

Q. 17. Elucidate the structure of phenanthrene. Confirm it by Haworth and Pschorr's synthesis.

Ans. 1. Elemental analysis and mol. weight determination of the compound show that it has the molecular formula $C_{14}H_{10}$ which is the same as that of anthracene.

2. *On oxidation* it yields phenanthraquinone which upon further oxidation gives *diphenic acid.* This on distillation with soda lime forms *diphenyl.* The formulae of diphenic acid and diphenyl are known and are given below:

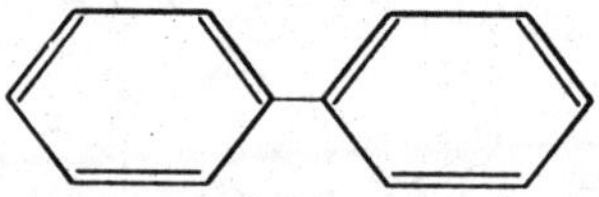

COOH COOH

Diphenyl 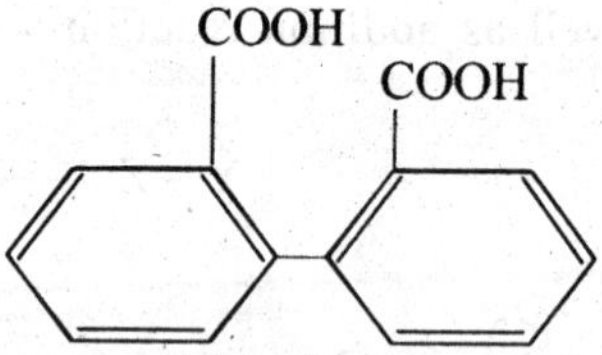 Diphenic acid

3. From above it is clear that phenanthrene must possess the skeleton (A).

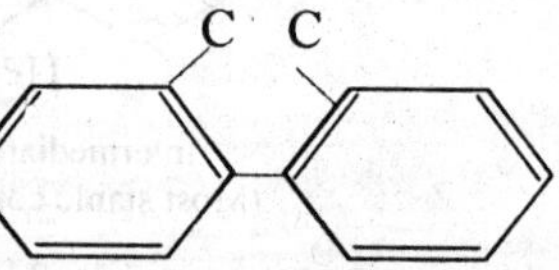

C C

(A)

The above skeleton has two hydrogens less than phenanthrene formulae $C_{14}H_{10}$. Hence closing the middle ring, the possible structure for phenanthrene could be as given below:

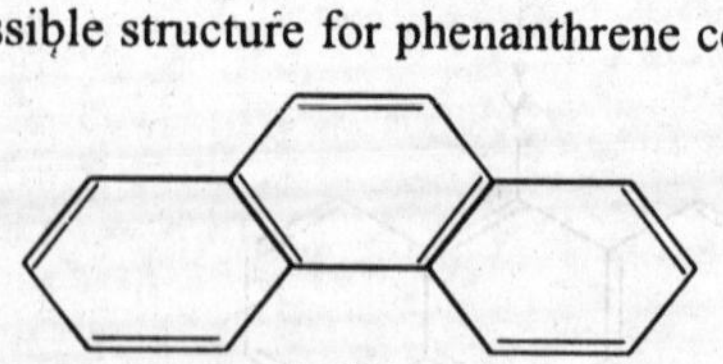

(B)

(*c*) The structure (B) is further confirmed by its synthesis.

(*i*) Phenanthrene can be obtained by passing dibenzyl through a red hot tube.

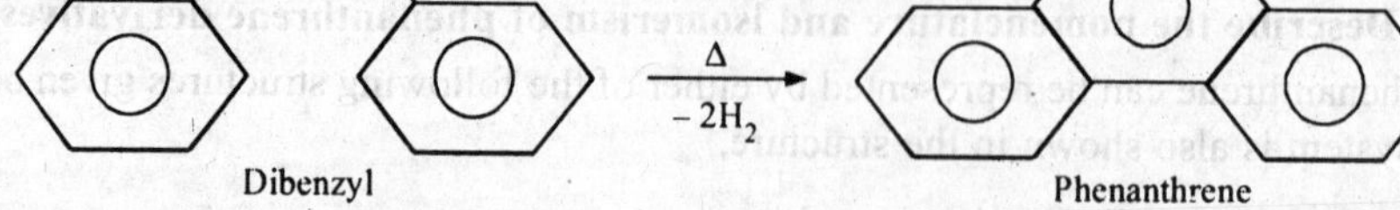

CH_2-CH_2

$\xrightarrow[-2H_2]{\Delta}$

Dibenzyl Phenanthrene

(*ii*) It can also be obtained by cyclodehydrogenation of 2, 2′ dimethyl-diphenyl.

CH_3 CH_3

$\xrightarrow{-2H_2}$

2, 2′ dimethyl diphenyl

Phenanthrene

(*iii*) **Haworth Synthesis.** It is quite a useful method for preparing phenanthrene or its substituted derivatives using naphthalene (or substituted naphthalene) and succinic anhydride.

Naphthalene + Succinic anhydride $\xrightarrow{AlCl_3}$ β-Naphthoyl propionic acid

$\xrightarrow{Zn/Hg/HCl}$ γ-(1-Naphthyl) butyric acid $\xrightarrow{H_2SO_4}$ 1-Keto-1, 2, 3, 4-tetrahydro-phenanthrene

$\xrightarrow{Zn/Hg\ HCl}$ 1, 2, 3, 4-tetra hydrophenanthrene $\xrightarrow{Pd/C/\Delta}$ Phenanthrene

(*iv*) **Pschorr's Synthesis**

CHO, NO_2 + CH_2 — COONa $\xrightarrow{\text{Acetic Anhydride}}$ COOH, NO_2

$\downarrow$ Sn/HCl; $NaNO_2$, H_2SO_4

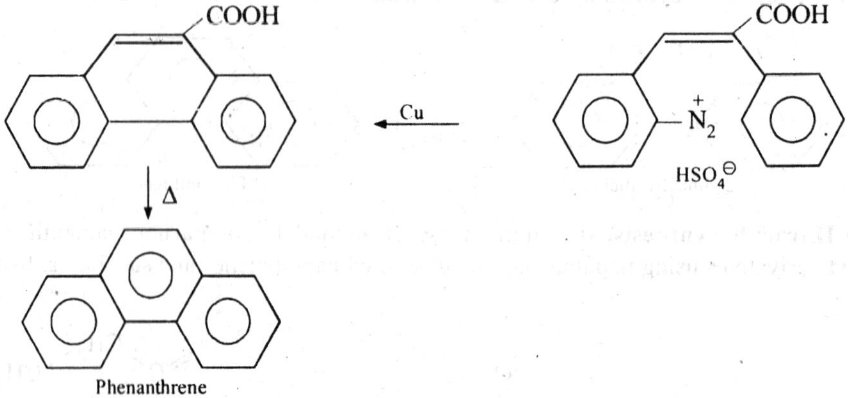

COOH

Cu

$\overset{+}{N}_2$

$HSO_4^{\ominus}$

Δ

Phenanthrene

(*d*) Similar to other aromatic hydrocarbons, it is also a *resonance hybrid* of contributing structures as shown below:

Its, resonance energy is 387.0 kJ mol^{-1}) and it is more resonance stabilised than anthracene.

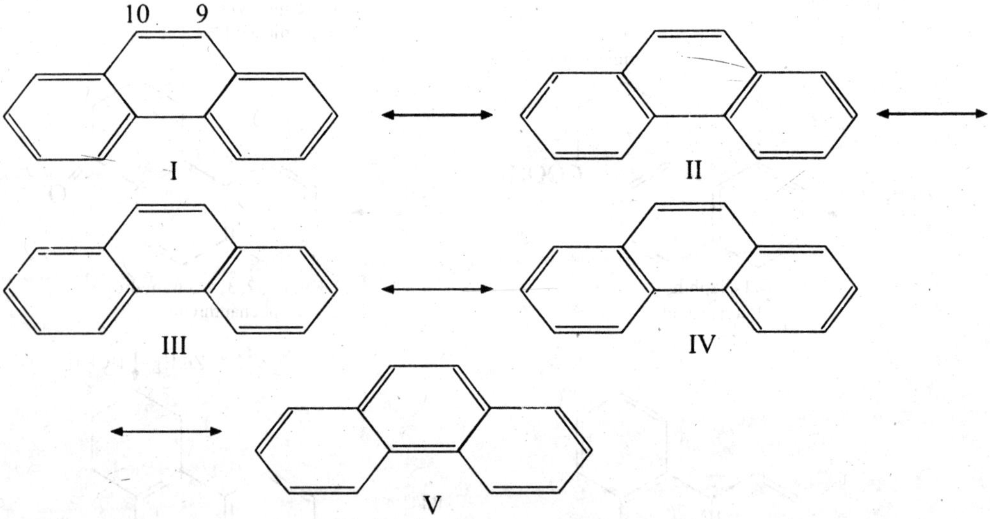

10 9

I

II

III

IV

V

The double bond character of the various bonds is shown in figure below:

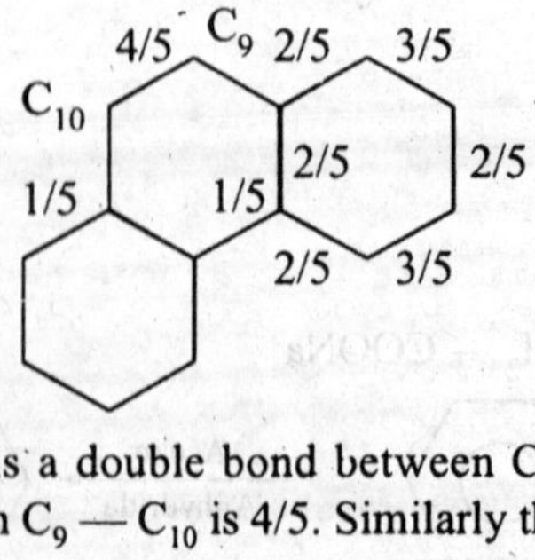

4/5 C_9 2/5 3/5

C_{10}

2/5 2/5

1/5 1/5

2/5 3/5

It may be noted that there is a double bond between C_9 — C_{10} in 4 out of 5 structures. Therefore double bond character in C_9 — C_{10} is 4/5. Similarly the fraction of double bond between other carbons can be calculated.

Resonance energy of phenanthrene shows that it is more stable than anthracene This is due to the fact that the resonating structures of phenanthracene are more stable. Structures II has all the benzenoid rings. Similarly 3 other structures have two benzenoid rings.

Because of higher resonance stabilization energy (387 kJ mol^{-1}) of phenanthrene, it shows smaller reactivity compared to anthracene.

Orbital Concept

Phenanthrene is a planar molecule in which all the fourteen carbon atoms are sp^2 hybridised. Each carbon is linked to three other atoms by σ bonds resulting from overlapping of sp^2 orbitals of one carbon with similar orbitals of other carbons or *s* orbital of hydrogens. The unhybridised *p*-orbitals of each carbon overlap laterally with unhybridised orbitals of other carbons. There is an electron cloud of 14 π electrons above and below the ring structure. Presence of 14 π electrons satisfies $4n + 2$ Huckel's rule.

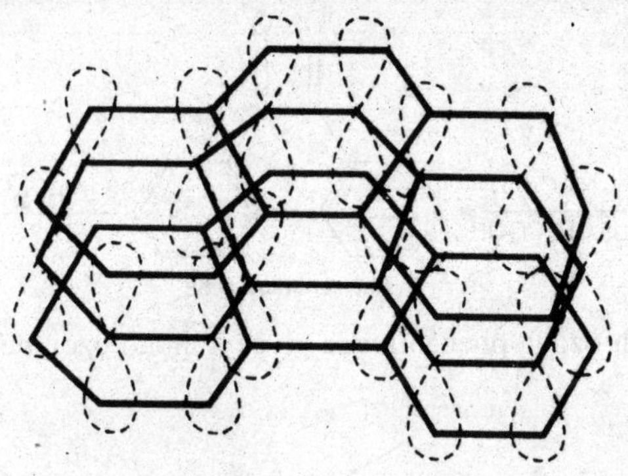

Orbital diagram of phenanthrene

Q. 18. Describe chemical properties of phenanthrene. *(Bangalore 2006)*

Ans. 1. Reduction. On catalytic reduction with H_2 (using $CuO + Cr_2O_3$ as catalyst), it gives 9, 10 dihydrophenanthrene.

Phenanthrene $+ H_2$ $\xrightarrow{CuO/Cr_2O_3}$ 9, 10-Dihydrophenanthrene

2. Reaction with bromine. It adds bromine to yield 9, 10 phenanthrene dibromide as readily as ethylenic compound. When treated with bromine in presence of iron (halogen carrier) it gives 9-bromophenanthrene.

Phenanthrene $\xrightarrow{Br_2}$ Phenanthrene dibromide (Br H Br H)

Phenanthrene $\xrightarrow[Fe]{Br_2}$ 9-Bromophenanthrene (Br)

9-Bromophenanthrene is employed in the preparation of other derivatives *e.g.*

$$\text{9-bromo derivative} \xrightarrow[\Delta]{\text{CuCN}} \text{9-Cyano derivative} \xrightarrow{H_2O} \text{9-Carboxylic acid}$$

3. Oxidation. Phenanthrene on oxidation with dichromate in glacial acetic acid gives phenanthraquinone which on further oxidation with dichromate and H_2SO_4 yields diphenic acid.

$$\text{Phenanthrene} \xrightarrow[\text{in } CH_3COOH]{Na_2Cr_2O_7} \text{Phenanthraquinone} \xrightarrow[H_2SO_4]{Na_2Cr_2O_7} \text{Diphenic acid}$$

4. With ozone. With ozone phenanthrene gives a mono ozonide which on oxidation forms diphenic acid.

$$\text{Phenanthrene} \xrightarrow{O_3} \text{Ozonide} \xrightarrow[KMnO_4]{(O)} \text{Diphenic acid}$$

5. Nitration. On nitration, it gives a mixture of three mono nitro derivatives, the 3-isomer dominates.

$$\text{Phenanthrene} \xrightarrow{HNO_3} \text{3-Nitrophenanthrene}$$

6. Sulphonation. Phenanthrene on sulphonation yields a mixture of 1-, 2-, 3- and 9- phenanthrene sulphonic acid.

7. Friedel-Crafts reaction. Phenanthrene undergoes Friedel Crafts reaction in 3-position mainly and in 2-position only to a smaller extent.

Q. 19. Which are reactive positions of phenanthrene towards electrophiles?

Or

Account for the following: substitution and addition reactions on phenanthrene take place at positions 9 and 10.

Ans. Phenanthrene undergoes electrophilic attack mainly at C–9. This is so because the intermediate carbonium ion formed with this position is more stable compared to intermediate carbonium ions obtained when the electrophile attacks other positions. Particular stability of the intermediate ion with attack at C–9 lies in the fact that there are two benzene rings intact which represents a stable system (because of high resonance energy).

Similarly reduction and oxidation of phenanthrene also prefer this position because two benzene rings remain intact.

$$Y:Z \longrightarrow \underset{\text{Electrophile}}{Y^+} + :Z^-$$

Phenanthrene → Intermediate ion $\xrightarrow{:Z^-}$ 9-Substitution product + H : Z

Intermediate ion $\xrightarrow{:Z^-}$ 9, 10-Addition product

It may be noted that intermediate ion is the same whether it is substitution or addition. From the stage of intermediate product, the reaction could proceed to substitution or addition, depending upon the reaction conditions.

Q. 20. Explain the following:

(*i*) Benzene is more aromatic than naphthalene. *(Panjab 2000)*

(*ii*) Methylbenzene on oxidation gives benzoic acid while 2-methylnaphthalene on oxidation gives 2-methyl 1, 4 naphthaquinone.

(*iii*) Nitration and halogenation of naphthalene gives mainly the α-isomer.

(*iv*) Sulphonation of naphthalene gives α-isomer at low temperature and β-isomer at high temperature.

(*v*) Friedel-Craft's acylation of naphthalene in nitrobenzene mainly gives β-aceto-naphthalene.

(*vi*) Electrophilic substitution of anthracene and phenanthrene mainly takes place at 9 and 10 positions.

Ans. (*i*) Benzene is more aromatic than naphthalene. Aromaticity of a compound can be judged by resonance stabilization energy. Benzene has a resonance energy 150 kJ mol^{-1}. Naphthalene to be equally aromatic should have resonance energy equal to 300 kJ mole^{-1} *i.e.* 150 kJ mole^{-1} per ring. But we find that naphthalene has resonance energy 255 kJ mole^{-1} or less than 300 kJ. Hence benzene is more aromatic than naphthalene.

(ii) Methylbenzene on oxidation gives benzoic acid.

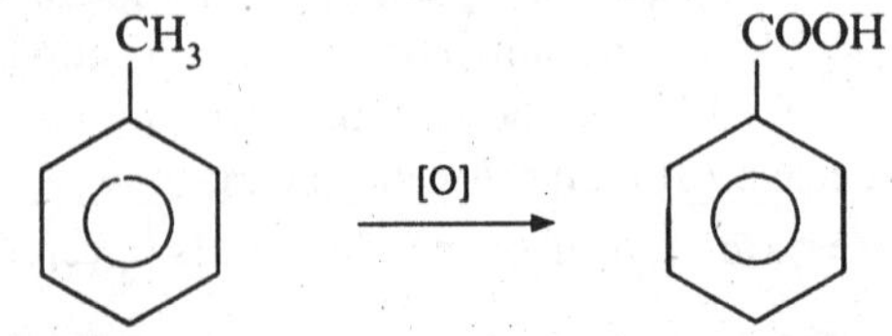

CH_3 [O] → COOH

Methylbenzene Benzoic acid

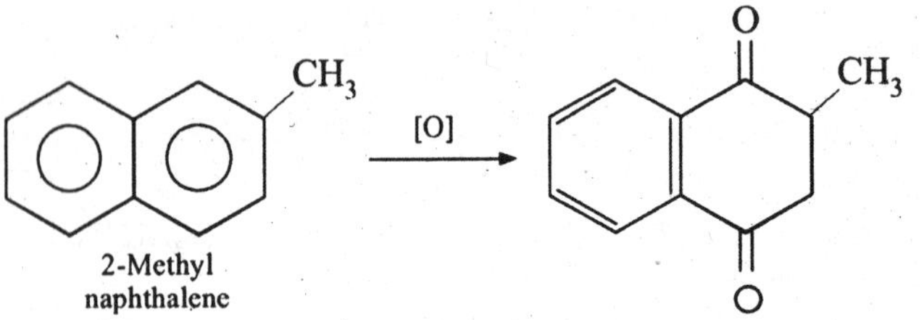

CH_3 [O] → O CH_3 O

2-Methyl naphthalene

2, Methyl-1, 4 naphthaquinone

This is so because reactions of naphthalene take place by electrophilic attack. Methyl group is an activating group. It activates the α-position for oxidation. Hence 2-methyl-1, 4 naphthaquinone is formed.

(iii) Nitration and halogenation of naphthalene give mainly the α-isomer. Nitration and halogenation of naphthalene are electrophilic substitution reactions. There are two possibilities of attack by the electrophile, the α-position and the β-position. That position will be favoured for attack which can provide more stable intermediate carbocations. It is found that the intermediate species are more stable in case of α-attack. Hence nitration and halogenation gives mainly α-isomers.

(iv) Sulphonation of naphthalene gives α-isomer at low temperature and β-isomer at high temperature. As a general rule, substitution is preferred at α-position because of greater stability of α-substituted intermediate ion. So at low temperature, we obtain α-isomer. At higher temperature, desulphonation takes place (At lower temp., desulphonation is very slow) and during sulphonation—desulphonation process, α-isomer changes into β-isomer because there is some steric hindrance at α-position. β-position is free from steric hindrance.

(v) Friedel-Crafts acylation of naphthalene in nitrobenzene mainly gives β-aceto-naphthalene. Nitrobenzene forms a complex with acetyl chloride and aluminium chloride. As this complex is large in size, it suffers steric hindrance at α-position and therefore proceeds to β-position during electrophilic attack.

(vi) Electrophilic substitution of anthracene and phenanthrene mainly takes place at 9 and 10 positions. This is so because the intermediate ion formed during electrophilic attack at 9 or 10 position is more resonance stabilized.

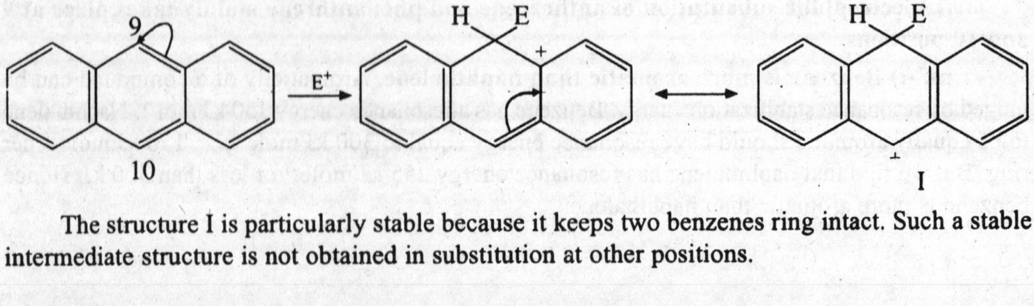

9 E^+ → H E + ⟷ H E

10 I

The structure I is particularly stable because it keeps two benzenes ring intact. Such a stable intermediate structure is not obtained in substitution at other positions.

Q. 21. Arrange benzene, naphthalene, anthracene and phenanthrene in increasing order of reactivity giving reasons.

Or

Explain that anthracene and phenanthracene are more reactive than benzene.

Ans. The stability or reactivity of a compound can be gauzed by the resonance stabilization energy of the compound. The resonance energies of the compounds and resonance energy per ring are tabulated below.

Compound	*Resonance energy kJ/mole*	*Resonance energy kJ/mole/ring*
Benzene	150	150.0
Naphthalene	225	127.5
Anthracene	351.5	117.2
Phenanthrene	387.0	129.0

The table shows that benzene having the maximum resonance energy is most stable followed by phenanthrene, naphthalene and anthracene. The order of reactivity will be the reverse of above order. Greater the stability, smaller is the reactivity. Hence the increasing order of reactivity will be

Benzene < phenanthrene < naphthalene < anthracene

Q. 22. A neutral solid (A) has the formula $C_{10}H_7N$. Vigorous oxidation of (A) gives a dibasic acid (B), $C_8H_5O_6N$. (A) can be reduced to a compound (C) $C_{10}H_9N$ which on oxidation gives a dibasic acid (D), $C_8H_6O_4$. The compound (A) can be obtained by direct nitration of a particular aromatic hydrocarbon. Identify compounds (A) to (D).

Ans.

$$(A) \xrightarrow{[O]} C_6H_3O_2N\left]\begin{matrix}COOH\\COOH\end{matrix}\right. \quad (B)$$

$$\downarrow [H]$$

$$C_{10}H_9N \ (C) \xrightarrow{[O]} C_6H_4\left]\begin{matrix}COOH\\COOH\end{matrix}\right. \quad (D)$$

In all probability (B) appears to be nitrophthalic acid and (D) appears to be phthalic acid. Retracing the reaction, we can say that (A) is nitronaphthalene and since (A) on reduction gives (B), (B) should be aminonaphthalene. The reactions are reproduced as under.

NO_2 (naphthalene ring) $\xrightarrow{[O]}$ HOOC, HOOC, NO_2 (benzene ring)

α-Nitronaphthalene (A) — α-Nitrophthalic acid (B)

$\downarrow$ [H]

NH_2 (naphthalene ring) $\xrightarrow{[O]}$ (benzene ring) COOH, COOH

α-Amino naphthalene (C) — Phthalic acid (D)

Q. 23. What is aromaticity? Describe Huckel's "4*n* + 2" rule.

Or

What is aromatic character? What are the conditions for a compound to show aromatic character? Will cycloheptatrienyl cation show aromatic character? Give reasons.

Ans. Benzene and its derivatives were, to start with, described as aromatic compounds because they possessed some special properties which were not given by straight chain unsaturated compounds and alicyclic compounds. There properties are:

(*i*) In spite of high degree of unsaturation in the molecule as indicated by the molecular formula, such compounds do not give addition reactions readily but give substitution reactions instead.

(*ii*) They are cyclic compounds but are relatively more stable than the corresponding open-chain compounds.

(*iii*) These molecules are planar.

Such compounds were termed as aromatic. *The property which renders such compounds stable and accounts for their substitution reactions rather than addition reactions is called* **aromaticity**.

Theoretical Requirement for Aromaticity (Huckel's Rule)

Huckel in 1931 presented theoretical basis for aromaticity by studying molecular orbital pictures of various aromatic compounds. He came to certain conclusions regarding aromaticity and aromatic compounds, which are presented below:

(*a*) In order to exhibit aromaticity, a compound should have a cyclic cloud of delocalised π electrons above and below the coplaner ring of the molecule.

(*b*) The number of electrons in π electron cloud should be $4n + 2$ where *n* is a positive whole number like 1, 2, 3... etc.

According to Huckel's rule an aromatic compound will have 6, 10, 14... electrons in the electron cloud. Examples of aromatic compounds with the no. of π electrons are given ahead:

Monocyclic Systems

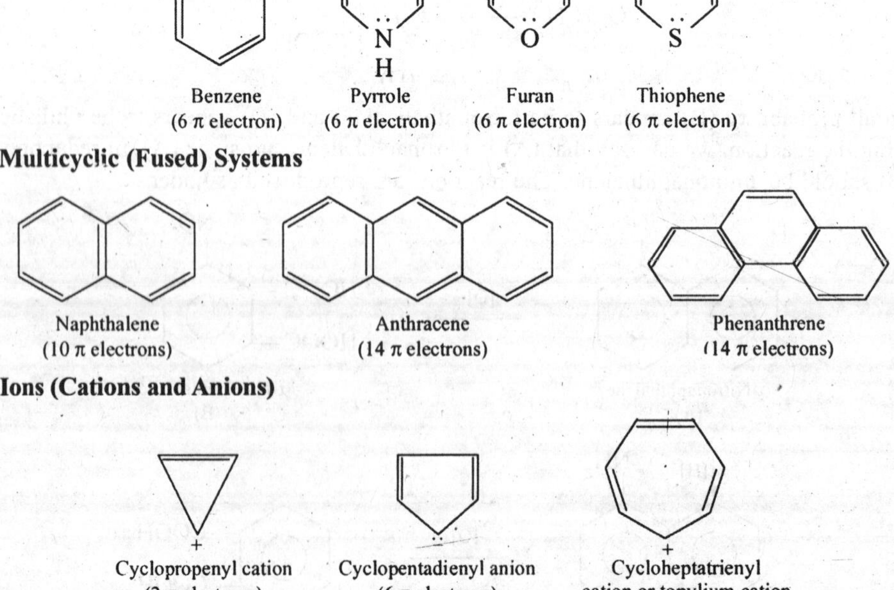

Benzene (6 π electron) Pyrrole (6 π electron) Furan (6 π electron) Thiophene (6 π electron)

Multicyclic (Fused) Systems

Naphthalene (10 π electrons) Anthracene (14 π electrons) Phenanthrene (14 π electrons)

Ions (Cations and Anions)

Cyclopropenyl cation (2 π electrons) Cyclopentadienyl anion (6 π electrons) Cycloheptatrienyl cation or topylium cation (6 π electrons)

In all the above cases, we find that Huckel's $4n + 2$ rule is followed. Hence they are aromatic compounds showing extra stabily and giving substitution reactions.

Q 24. What is anti-aromaticity? *(Kerala, 2000)*

Ans. There are certain cyclic compounds which are much less stable than their acylic counterparts. Such compounds are called anti-aromatic and this phenomenon is called anti-aromaticity. A very well known example of anti-aromaticity is cyclobutadiene.

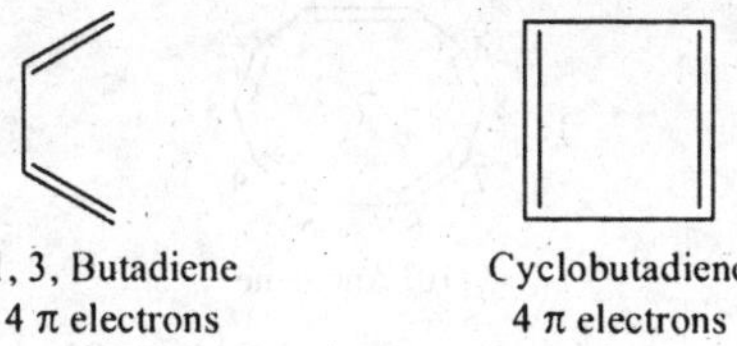

1, 3, Butadiene 4 π electrons — Cyclobutadiene 4 π electrons

Cyclobutadiene is much less stable than open-chain 1, 3 butadiene. The former could be prepared only recently after careful attempts.

It has been revealed that anti-aromaticity is exhibited by those species which have 4 π electrons in their cyclic electron clouds. A few more example of anti-aromatic compounds are given below.

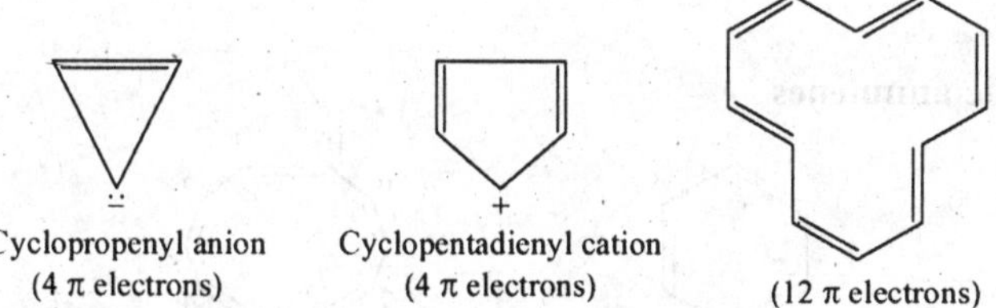

Cyclopropenyl anion (4 π electrons) — Cyclopentadienyl cation (4 π electrons) — (12 π electrons)

The destabilisation caused by $4n$ π electrons is attributed to molecular orbitals electrons configuration.

Q. 25. What is homoaromaticity?

Ans. There are certain species which contain one or more sp^3 hybridised carbon atoms in a conjugated cyclic system having $4n + 2$ π electrons. Such species are called **homoaromatic** species and this phenomenon is known as **homoaromaticity**.

Homopropylium ion formed by the addition of a proton to cyclooctatetraene is an example of homoaromatic species.

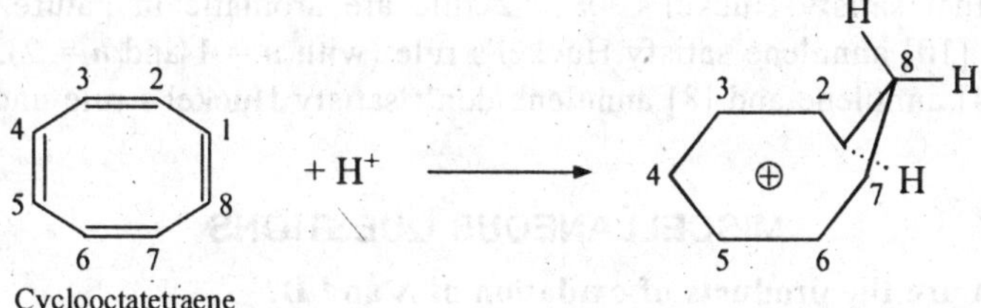

Cyclooctatetraene

It is noteworthy that atom with sp^2 hybridised orbitals are in a plane almost vertically perpendicular to the rest of the ring formed. This facilitates overlap of π orbitals in the newly formed aromatic ring.

Q. 26. What are annulenes ? *(Himachal, 2000; G. Nanakdev, 2000; S; Punjab 2003)*

Ans. Annulenes are monocyclic compounds having a system of single and double bonds in alternate positions. They may be called cyclic polyenes and can be represented by a general formula $(-CH = CH-)_n$ where n is a positive integer. They are named by writing the ring size followed by the word "annulene". Some examples of annulenes are given below:

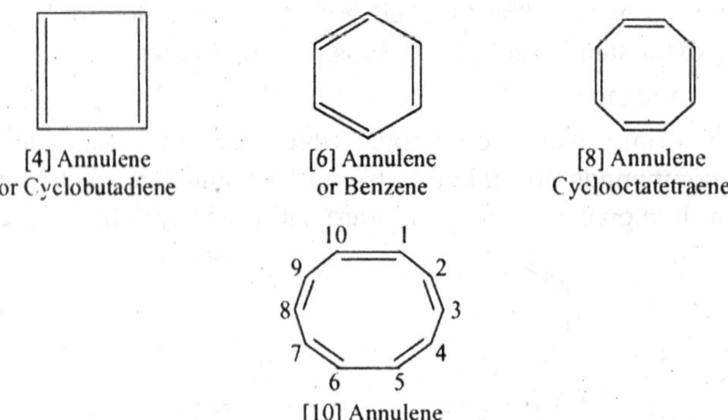

According to Huckel's rule, annulenes containing $(4n + 2)$ π electrons and having a coplanar ring should be aromatic in nature. The synthesis of a no. of annulenes has confirmed this. Thus [6] annulene or benzene is aromatic. [4] annulene and [8] annulene are not aromatic because they don't conform to Huckel's rule. They are rather anti-aromatic as discussed earlier. [10] annulene although expected to be aromatic as per Huckel's rule is not so because the ring is not co-planar.

Q. 27. Write the formula of two aromatic and two non-aromatic annulenes.

(Punjab, 2003)

Ans. Aromatic annulenes

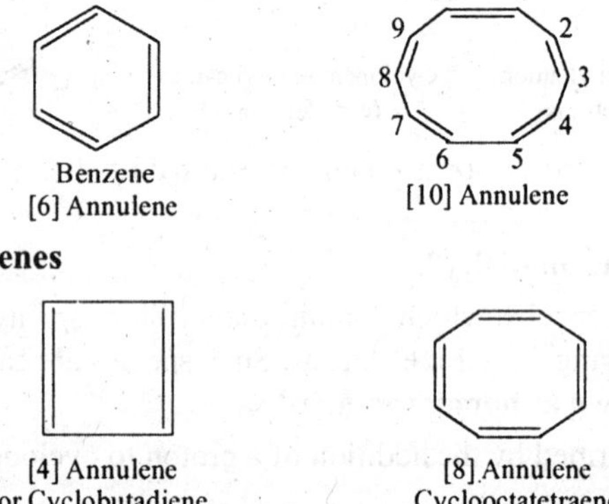

Annulenes that satisfy Huckel's $4n + 2$ rule are aromatic in nature, others are not. [6] annulene and [10] annulene satisfy Huckel's rule (with $n = 1$ and $n = 2$), hence, they are aromatic while [4] annulene and [8] annulene don't satisfy Huckel's rule and hence are non-aromatic.

MISCELLANEOUS QUESTIONS

Q. 28. What are the products of oxidation of A and B?

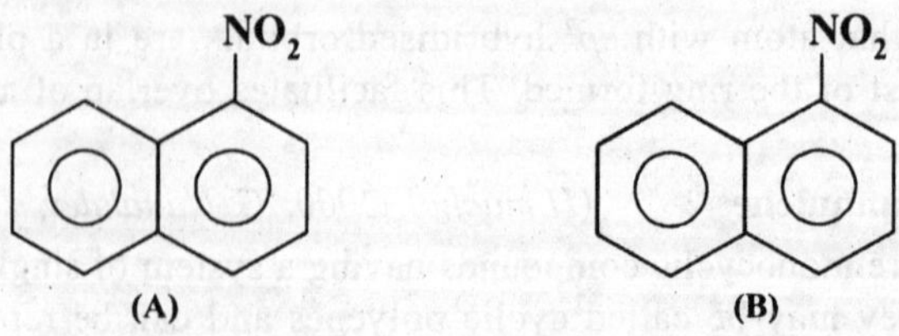

What information do these oxidation products convey regarding the structure of naphthalene?

Ans.

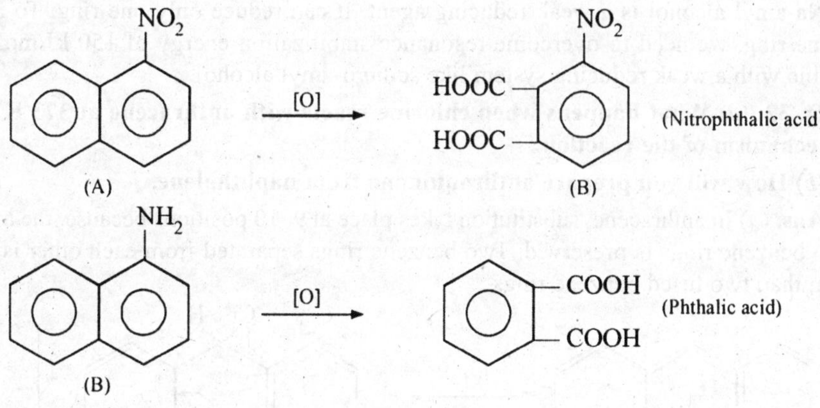

NO$_2$ (A) $\xrightarrow{[O]}$ HOOC, HOOC, NO$_2$ (B) (Nitrophthalic acid)

NH$_2$ (B) $\xrightarrow{[O]}$ COOH, COOH (Phthalic acid)

The above reactions show that naphthalene molecule consists of two benzene rings fused together.

The difference in the behaviour of (A) and (B) towards oxidation can be explained as under.

Nitro group makes a ring harder to oxidise than an unsubstituted benzene ring. That is why in compound A, the ring containing the nitro group remains intact and the other ring is oxidised giving two carboxy groups. On the other hand, an amino group increases the ease of oxidation of the ring to which it is attached. That is why, in compound B, the ring containing amino group is oxidised giving two carboxy groups.

Q. 29. Describe Haworth's synthesis of naphthalene. (*Kerala, 2004; Goa, 2004*)

Ans. See question on structure of naphthalene.

Q. 30. Give two methods of preparation of phenanthrene.

Ans. Methods of preparation of phenanthrene are described as under:

1. By passing dibenzyl through a red hot tube.

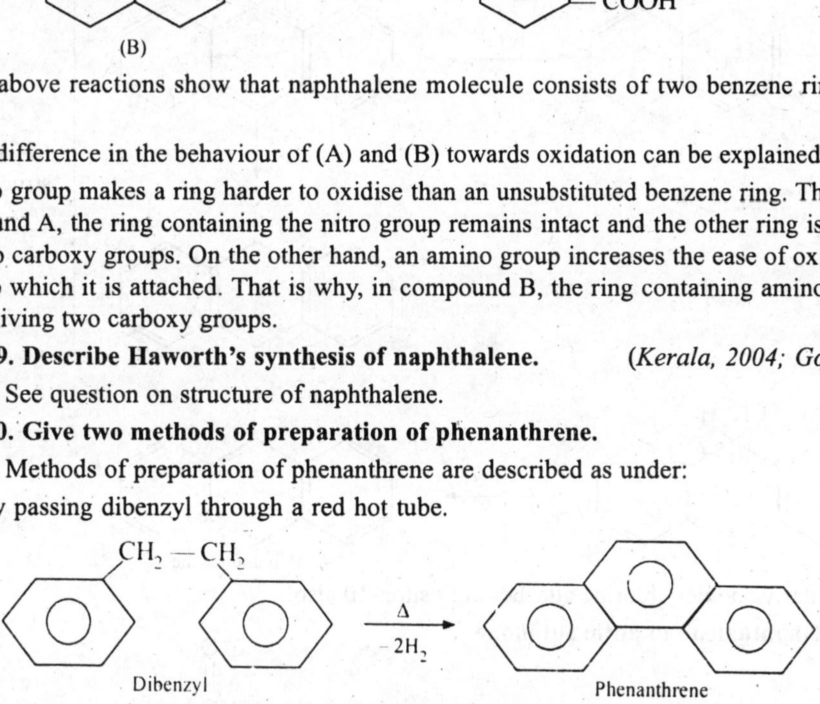

CH$_2$ — CH$_2$ Dibenzyl $\xrightarrow[-2H_2]{\Delta}$ Phenanthrene

2. By cyclodehydrogenation of 2, 2′ dimethyl diphenyl.

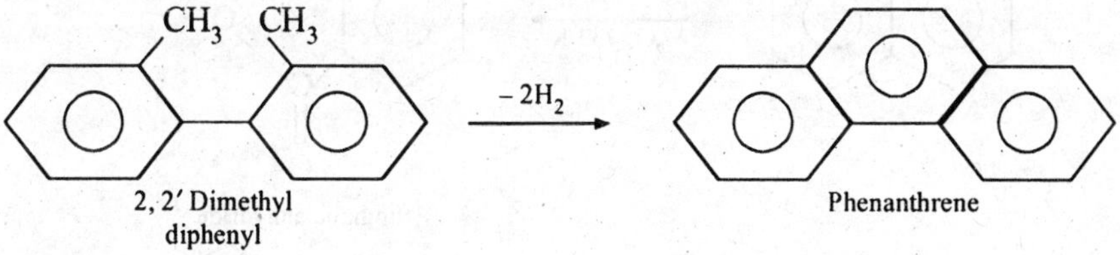

CH$_3$ CH$_3$ 2, 2′ Dimethyl diphenyl $\xrightarrow{-2H_2}$ Phenanthrene

Q. 31. Only one benzene ring is reduced when naphthalene is treated with sodium-amyl alcohol. Why?

Ans.

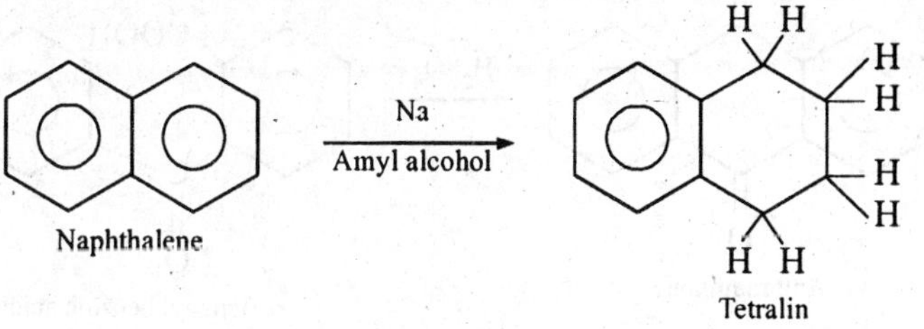

Naphthalene $\xrightarrow[\text{Amyl alcohol}]{\text{Na}}$ Tetralin (H H, H, H, H, H, H H)

Na-amyl alcohol is a weak reducing agent. It can reduce only one ring. To reduce second benzene ring, we need to overcome resonance stabilization energy of 150 kJ/mole. This is not available with a weak reducing system like sodium-amyl alcohol.

Q. 32. (*a*) What happens when chlorine reacts with anthracene at 373 K? Also sketch the mechanism of the reaction.

(*b*) How will you prepare anthraquinone from naphthalene. (*Delhi, 2003*)

Ans. (*a*) In anthracene, substitution takes place at 9, 10 positions because, the aromatic nature of two benzene rings is preserved. Two benzene rings separated from each other is a more stable system than two fused benzene rings

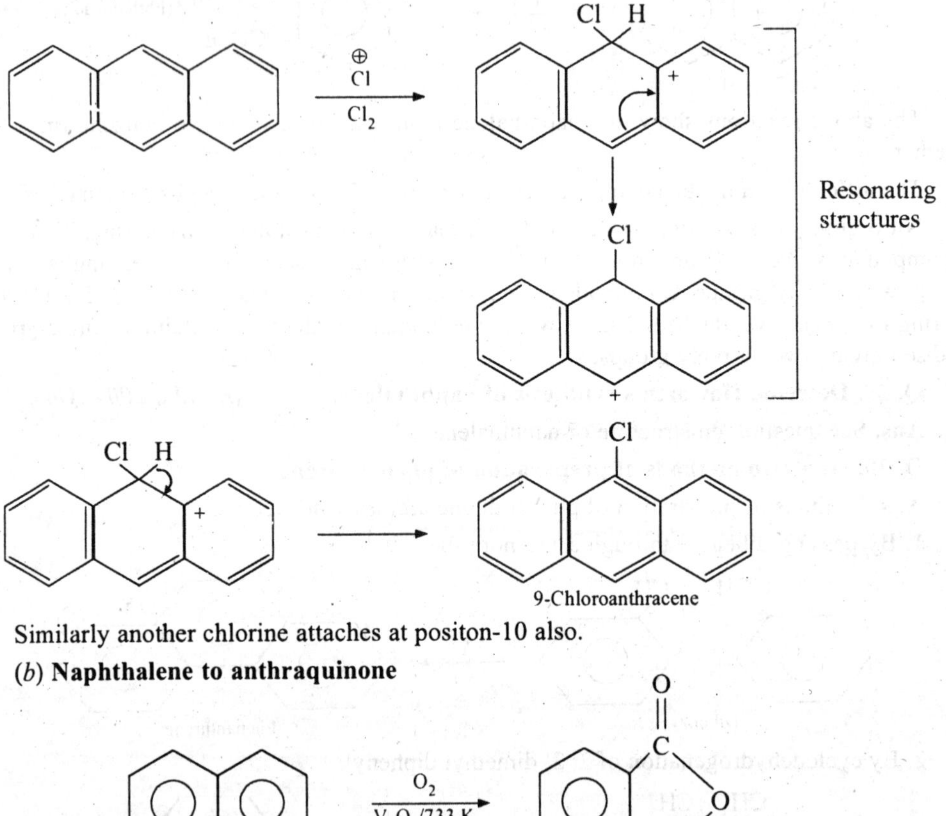

$\overset{\oplus}{Cl}$ / Cl_2

Cl H

Cl

Resonating structures

Cl H

Cl

9-Chloroanthracene

Similarly another chlorine attaches at positon-10 also.

(*b*) **Naphthalene to anthraquinone**

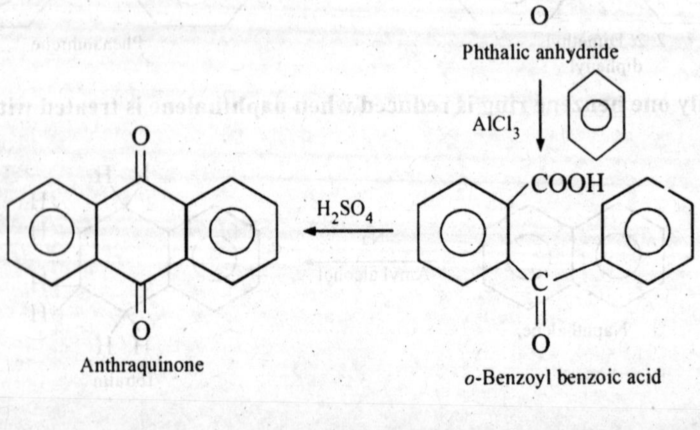

O_2 / V_2O_5/733 K

Phthalic anhydride

$AlCl_3$

COOH

H_2SO_4

Anthraquinone

o-Benzoyl benzoic acid

Q. 33. How do you account for the following? C_1 — C_2 bond of naphthalene has greater double bond character than C_2 — C_3.

Ans. This can be established as under:

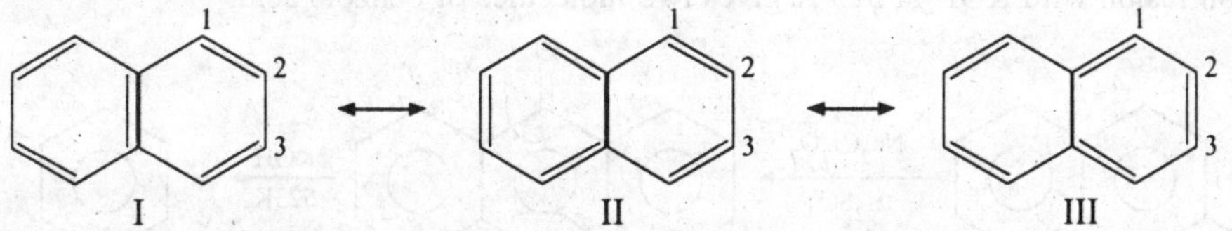

I ⟷ II ⟷ III

There are three resonating structures of naphthalene I, II and III.

Out of three structures, structure I and II have double bonds between C_1 and C_2. Structure III has a single bond between C_1 and C_2. Thus it has two-third of double bond character. Similarly bond between C_2 and C_3 has one-third of the double bond character.

Q. 34. Describe one method for synthesis of anthracene.

(*Kera'ı, 2001; Coimbatore, 2000; Rajasthan, 1997; Allahabad, 1996*)

Ans. See Q. 13.

Q. 35. Why only one benzene ring is reduced when naphthalene is reduced with sodium and amyl alcohol? (*M. Dayanand, 1999*)

Ans. This is because sodium and amyl alcohol form a weak reducing agent. Moreover, extra stability in the form of resonance energy is associated with the benzene rings. The activation energy that is available with sodium and amyl alcohol system can reduce only one of the two benzene rings.

Q. 36. Complete the following reactions: (*Lucknow, 2000*)

(*i*) 1-Naphthol (OH) + HNO_3 ⟶

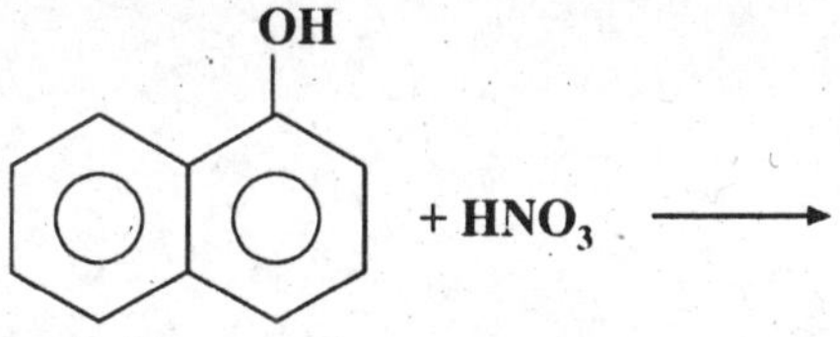

(*ii*) Anthracene $\xrightarrow[\Delta]{Na_2Cr_2O_7/H_2SO_4}$

Ans. (*i*) 1-Naphthol (OH) + HNO_3 ⟶ 2-nitro-1-naphthol (OH, NO_2) + 4-nitro-1-naphthol (OH, NO_2)

(*ii*) Anthracene $\xrightarrow[H_2SO_4, \Delta]{Na_2Cr_2O_7}$ Anthraquinone (O, O)

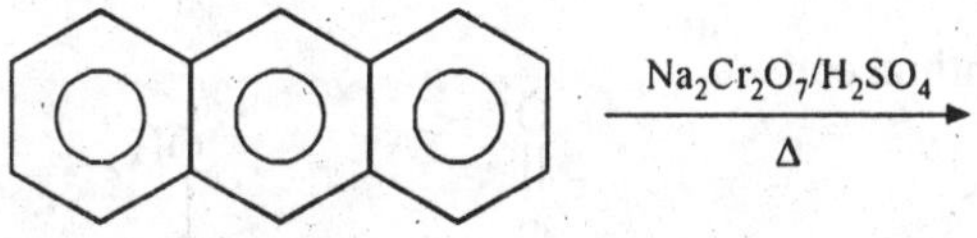

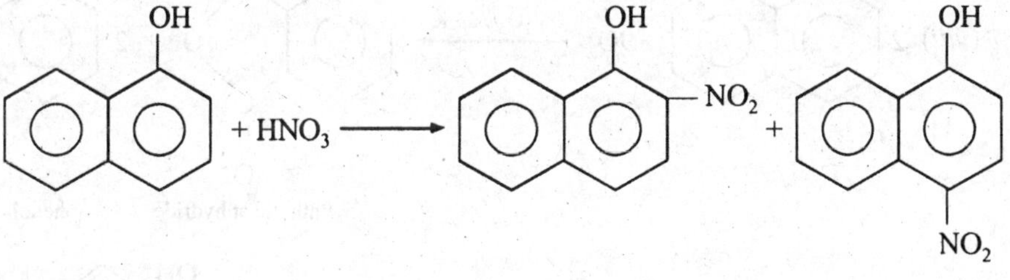

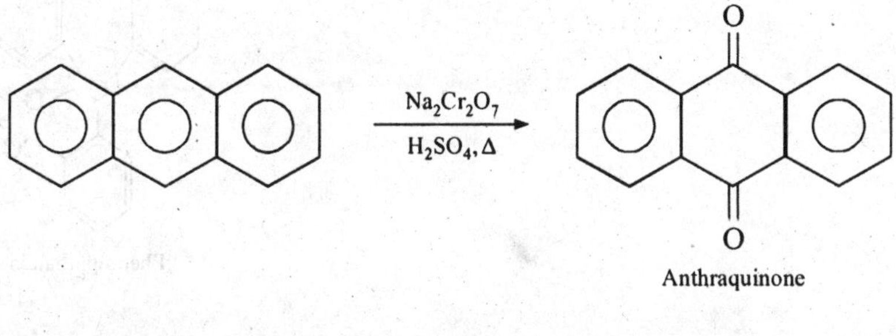

Anthraquinone

Q. 37. How will your differentiate between naphthalene and anthracene?

(*Lucknow, 2000*)

Ans. In the presence of sodium dichromate and H_2SO_4, anthracene forms anthraquinone, which on fusion with KOH at 523 K gives two molecules of benzoic acid.

$$\text{anthracene} \xrightarrow[H_2SO_4]{Na_2Cr_2O_7} \text{anthraquinone} \xrightarrow[523K]{KOH} 2\ C_6H_5COOH \text{ (benzoic acid)}$$

anthracene

anthraquinone

benzoic acid

while naphthalene on oxidation in the presence of V_2O_5 gives phthalic anhydride.

Q. 38. How will you synthesize?

(*i*) Naphthalene from benzene (*Lucknow, 2000*)

(*ii*) Anthracene from phthalic anhydride (*Lucknow, 2000*)

(*iii*) Anthraquinone from Anthracene (*Panjab, 2000*)

(*iv*) Diphenic acid from phenanthrene (*Panjab, 2000*)

(*v*) Phenanthrene from naphthalene (*Kurukshetra, 2001*)

(*vi*) 1, 4-dihydronaphthalene from naphthalene (*Calcutta, 2000*)

(*vii*) Anthracene from naphthalene (*Calcutta, 2000*)

(*viii*) Phenol phthalein from naphthalene (*Kanpur, 2001*)

Ans. (*i*) See Q. 4, (6), (*iii*), (*c*)

(*ii*) See Q. 13 (4) (*iv*) (*a*)

(*iii*) See Q. 13, (4), (*iv*) (*a*)

(*iv*) See Q. 18. (4)

(*v*) See Q. 17, (3), (*iii*)

(*vi*) See Q. 5, (A), (*i*), (*i*)

(*vii*) 32 (*b*) reduce anthraquinone to anthracene

(*viii*) $$2\ C_{10}H_8 + 9O_2 \xrightarrow{V_2O_5/733\ K} \text{Phthalic anhydride} + 2\ \text{phenol} \longrightarrow \text{Phenolphthalein}$$

Phthalic anhydride

phenol

Phenolphthalein

Q. 39. Complete the following reactions: *(Himachal, 2000)*

(*i*) Naphthalene $\xrightarrow{\text{nitration}}$

(*ii*) Naphthalene $\xrightarrow[\text{750 K}]{\text{air, } V_2O_5}$

(*iii*) Anthracene $\xrightarrow{\text{oxidation}}$

(*iv*) Anthracene $\xrightarrow[\text{heat}]{\text{Na / } C_2H_5OH}$

Ans. (*i*) See Q. 5 (B), (2)

(*ii*) See Q. 5 (C), (1)

(*iii*) See Q. 14, (5)

(*iv*) See Q. 14, (4)

Q. 40. Give Diels Alder reaction of anthracene with maleic anhydride. *(Himachal, 2000)*

Ans. See Q. 14. (8)

Q. 41. From naphthalene, how will you get the following compound:

(*i*) 2-naphthylamine

(*ii*) 2-ethylnaphthalene

Ans. (*i*) See Q. 5, (B) (2), then perform reduction

(*ii*) See Q. 5, (B), (4)

Q. 42. Write the reactions of Haworth's synthesis of phenanthrene.

(Garhwal, 2000; Kanpur, 2001)

Ans. See Q. 17, (3) (*iii*)

Q. 43. Identify compound A and B in the following reaction:

anthracene (1 mole) + Br_2 **(1 mole) A** $\xrightarrow{\Delta}$ **B** *(Calcutta, 2000)*

Ans. See Q. 14. (*i*)

Q. 44. Give Haworth synthesis of anthracene *(Delhi, 2003; Nehru, 2004)*

Ans. $2C_6H_6 + 2CH_2Br_2 \xrightarrow[\text{F.C. Reaction}]{AlCl_3}$ 9,10-dihydro anthracene

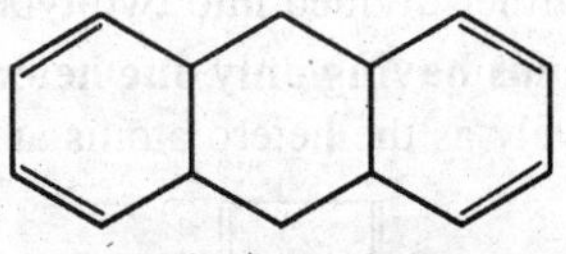

$\xrightarrow[-2H]{\text{Se}}$ Anthracene

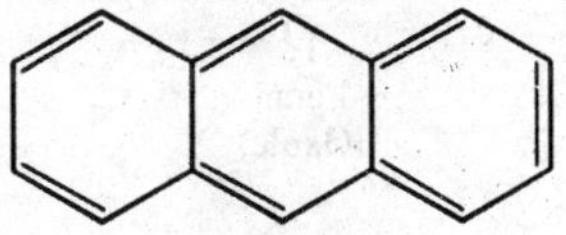

34

HETEROCYCLICS

Q. 1. What are heterocyclic aromatic compounds? (*Awadh, 2000; Panjab, 2000*)

Ans. *Cyclic compounds having at least one atom other than carbon in the ring are termed as heterocyclic compounds.* Although any atom capable of forming two or more covalent bond can be the ring element yet the most commonly found heterocyclic compounds contain only nitrogen, oxygen and sulphur as *hetero atoms*.

Substances such as cyclic anhydrides, alkene oxides and lactones, are heterocyclic in the strict sense of the term but are not generally considered as heterocyclic, because they are highly unstable, also they change into open-chain compounds upon hydrolysis.

Hetero atom in the ring, in place of carbon does not alter the strain relationship significantly. The most stable heterocyclic compounds are those having five or six membered rings.

Thus the heterocyclic compounds may also be defined as *five* or *six membered ring compounds with at least one hetero atom as the ring member. They are relatively stable and exhibit aromatic character.*

Heterocyclic compounds are abundantly available in plant and animal products and form one half of the natural organic compounds. Alkaloids, dyes, drugs, proteins, enzymes, etc. are the important representatives of this class of compounds.

Q. 2. Give the classification of heterocyclic compound.

Ans. Heterocyclic compounds may be divided into three categories.

1. Five-membered heterocyclic compounds. These can be considered to be derived from benzene by replacement of a (C = C) by a hetero-atom with an unshared electron pair.

These are further divided into two types.

(*i*) **Compounds having only one hetero atom.** Furan, thiophene, and pyrrole, containing O, S and N respectively as the hetero atoms are examples of this category.

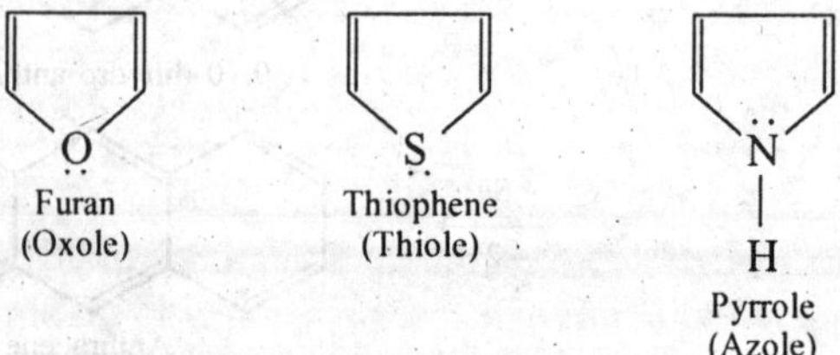

(*ii*) **Compounds having more than one hetero-atom.** For example,

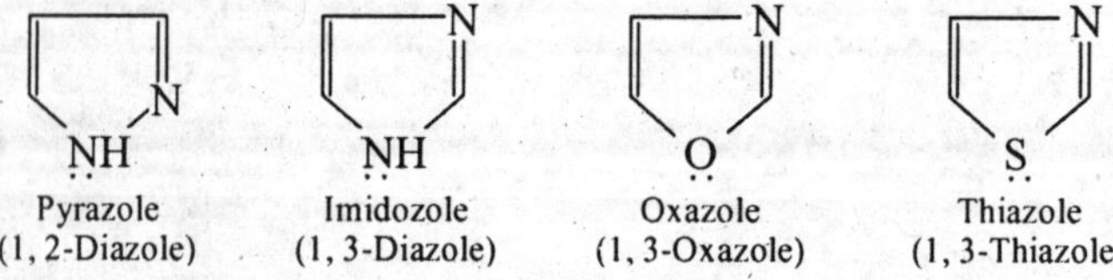

1, 2, 3-Triazole Tetrazole

2. Six-membered heterocyclic compounds. These are obtained by the replacement of a carbon of benzene by an isoelectronic hetero-atom. These are further classified into two types.

(*i*) **Compounds having only one hetero atom.** Pyridine, pyran and thiopyran are some of the examples.

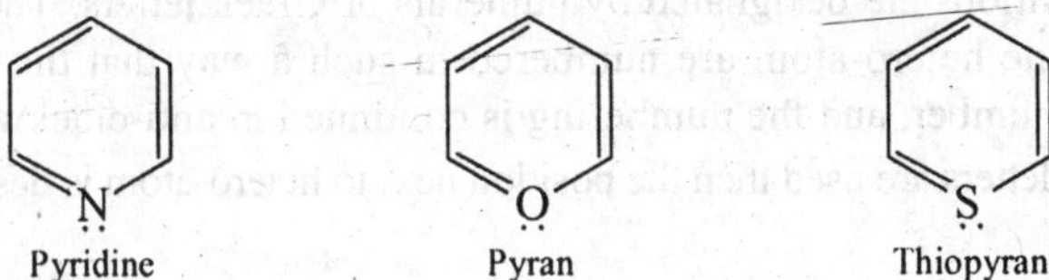

Pyridine Pyran Thiopyran

(*ii*) **Compounds having more than one hetero-atom.** Pryrimidine, pyrazine, pyridazine are some of the examples.

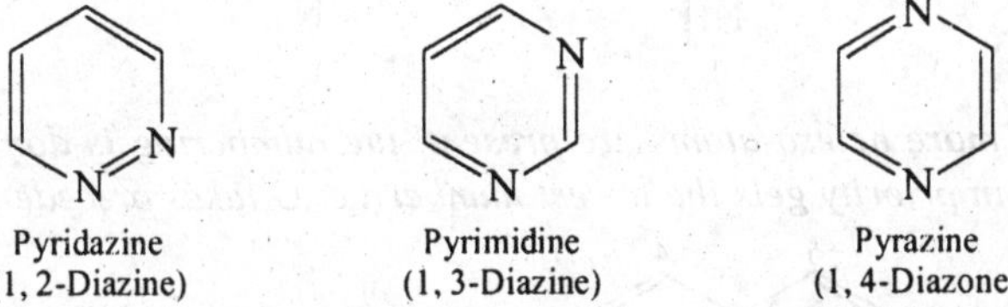

Pyridazine (1, 2-Diazine) Pyrimidine (1, 3-Diazine) Pyrazine (1, 4-Diazone)

3. Condensed Heterocyclic compounds. They consist of two or more fused rings which can be partly carbocyclic, and partly heterocyclic, *e.g.*, indole, quinoline, carbazole etc. or may be fully heterocyclic, *e.g.*, purine, pteridine.

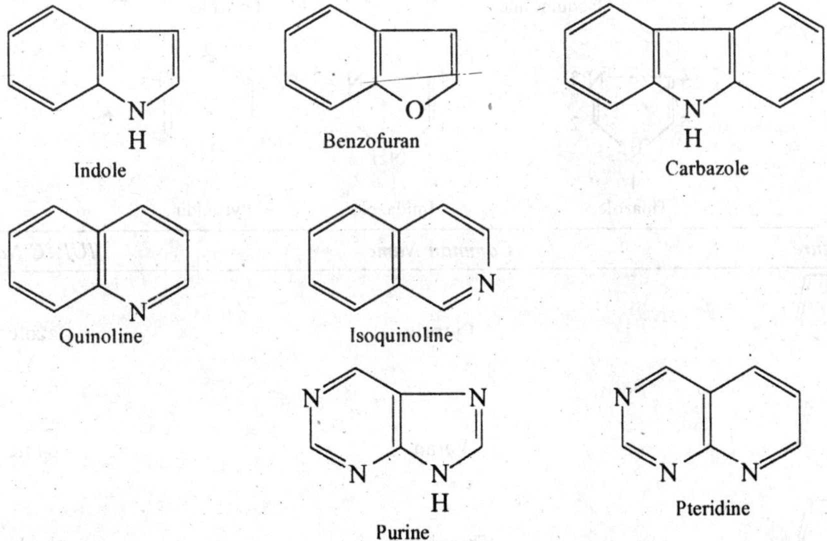

Q. 3. Describe the nomenclature of heterocyclic compounds. (*Panjab, 2000*)

Ans. Heterocyclic compounds are popularly known by their common names. Even the IUPAC system has accepted some of these trivial name as such. IUPAC name of heterocyclic compound is obtained by combining prefixes and suffixcs as given below:

Order of Priority	*Prefixes*	*Suffixes*
1.	*oxa*—for oxygen	*–ole* 5-membered ring
2.	thia—for sulphur	
3.	*aza*—for nitrogen	*–ine* 6-membered ring
4.	*phospha*—for phosphorus	*–epine* 7-membered ring

(*a*) The terminal '*a*' of the prefixes is generally removed when combining prefixes and suffixes. In case more than one hetero-atoms are involved, the prefixes are placed in the order of priority *i.e.*, oxa-first, than thio and so on.

(*b*) The ring positions are designated by numerals or Greek letters. The simple heterocyclic compounds having one hetero-atom are numbered in such a way that the hetero-atom gets the number 1 (or lowest number, and the numbering is continued in anti-clockwise direction).

(*c*) When Greek letters are used then the position next to hetero-atom is designated as α-followed by β- and so on.

In case two or more hetero-atoms are present, the numbering is done in such a way that hetero-atom highest in priority gets the lowest number i.e. O takes precedence over S and so on.

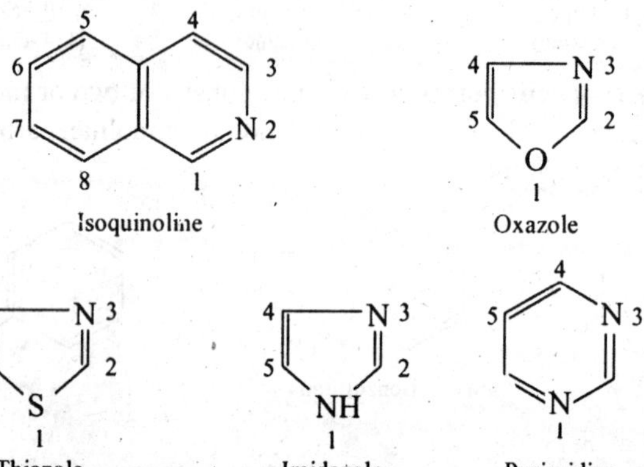

Structure	*Common Name*	*IUPAC Name*
	Pyrrole	azole
	Furan	oxole
	Thiophene	thiole
	Pyridine	azine

Q. 4. Give the nomenclature of furan derivatives.

Ans. The derivatives of furan are named, giving the substituents and the positions to which they are attached. The numbering system starts with the oxygen in the ring.

Furan (positions: O 1, α 2, β 3, β′ 4, α′ 5) — 2, 3-dimethylfuran

The 2, 5 positions are frequently referred to α, α′ and 3, 4 positions as β, β′ positions respectively.

Q. 5. Describe the methods of preparation of Furan. *(Mumbai, 2010)*

Ans. (*i*) **From mucic acid.** *Dry distillation of mucic acid gives furan.*

$$\underset{\text{Mucic acid}}{HOOC-[CHOH]_4-COOH} \xrightarrow[-CO_2,\ -CH_2O]{\text{Dry Distil.}} \underset{\text{Furoic acid}}{C_4H_3O-COOH} \xrightarrow[-CO_2]{\text{Heat}} \underset{\text{Furan}}{C_4H_4O}$$

(*ii*) **From furfural.** Furfural and steam are passed over a catalyst consisting of a mixture of zinc and manganese chromites at 673 K, CO is eliminated with formation of furan.

$$\underset{\text{Furfural}}{C_4H_3O-CHO} \xrightarrow[\text{Zn and Mn chromites}]{\text{Steam/673 K}} \underset{\text{Furan}}{C_4H_4O} + CO$$

(*iii*) **Paal-Knorr synthesis.** (*Dehydration of 1, 4-dicarbonyl compounds*)

$$\underset{\text{Acetonyl acetone}}{H_3C-CO-CH_2-CH_2-CO-CH_3} \xrightarrow{P_2O_5,\ \text{Heat}} \underset{\alpha,\ \alpha'\ \text{dimethyl furan}}{H_3C-C_4H_2O-CH_3} + H_2O$$

(*iv*) **Fiest-Benary synthesis**

$$\underset{\beta\text{-ketoester}}{EtOOC-CH_2-CO-CH_3} + \underset{\beta\text{-haloketone}}{CH_3-CO-CH_2-Cl} \xrightarrow{\text{Pyridine}} \text{(furan with EtOOC, } CH_3 \text{ and } H_3C \text{ substituents)}$$

Q. 6. Describe the chemical properties of furan. *(Pune, 2010)*

Ans. 1. Electrophilic substitution reactions. Electrophilic substitutions of furan occurs preferentially at the 2- or 5-position. The intermediate carbonium ion in the case of 2- or 5-substitution is resonance stabilised to a greater extent than in the case of 3 or 4-substitution. Substitution at 3- or 4-position, however, may take place in case 2- and 5-positions of furan are already occupied.

Resonance stabilisation of the intermediate products in substitution at positions 2 and 3 is illustrated below. $Y^{\oplus}$ represents the electrophile.

Attack at Position 3 — Hybrid of two resonating structures

Furan — $Y^{\oplus}$

Attack at Position 2 — Hybrid of three resonating structures

(*a*) **Nitration.** When furan is treated with acetyl nitrate, 2-nitrofuran is obtained.

$$\text{Furan} \xrightarrow{CH_3COONO_2} \text{2 Nitrofuran}$$

(*b*) **Sulphonation.** Furan upon sulphonation with pyridine—sulphur trioxide forms Furan-2 sulphonic acid.

$$\text{Furan} \xrightarrow{SO_3/\text{Pyridine}} \text{Furan-2-sulphonic acid}$$

(*c*) **Halogenation.** Furan readily reacts with halogens leading to destruction of furan ring and forming halogen acids. However halogen derivatives of furan may be obtained indirectly. For example, furoic acid on bromination gives 5-bromo furoic acid which upon decarboxylation yields 2-bromofuran.

$$\text{Furoic acid} \xrightarrow{Br_2} \text{5-Bromo-furoic acid} \xrightarrow[-CO_2]{\text{Heat}} \text{2-Bromofuran}$$

(*d*) **Friedel-Craft's reaction.** Acylation of furan is done with boron trifluoride in ether.

$$\text{Furan} \xrightarrow[BF_3 \text{ in ether}]{(CH_3CO)_2O} \text{2-Acetyl Furan}$$

Alkylation reactions with furan results in polymerisation and are not, therefore, possible.

(*e*) **Gattermann Koch Synthesis.** Furan when treated with a mixture of HCN and HCl in the presence of $AlCl_3$ followed by decomposition yields furfural.

$$HCl + HCN \xrightarrow{AlCl_3} HN = CHCl$$

$$\text{Furan} + HN = CHCl \xrightarrow{AlCl_3} HCl + \text{(furan-2-)}CH = NH$$

$$\xrightarrow{H_2O} NH_3 + \text{(furan-2-)}CHO$$

Furfural

(*f*) **Gomberg reaction.** Furan on treatment with diazonium salt in alkaline solution gives aryl furan.

$$\text{Furan} + Ar\overset{+}{N_2}\overset{-}{Cl} \xrightarrow{NaOH} \text{Aryl furan} + N_2$$

(*g*) **Reaction with *n*-butyl-lithium.** Furan gives, 2-lithium furan when treated with *n* butyl-lithium.

$$\text{Furan} + C_4H_9Li \longrightarrow C_4H_{10} + \text{2-Lithium furan}$$

(*h*) **Formation of 2-chloromercurifuran.** Treatment with a mixture of mercuric chloride and sod. acetate gives 2-chloromercurifuran which is a useful synthetic intermediate as the mercuric group is replaceable.

$$\text{Furan} \xrightarrow[\text{Sod. acetate}]{HgCl_2} \text{2-Chloromercuric furan} \xrightarrow{I_2} \text{2-Iodofuran}$$

2. Reduction. Furan upon catalytic reduction using Raney Ni forms tetrahydrofuran [THF] which breaks upon treatment with HCl and ultimately the formation of adipic acid and 1, 6 diamino hexane, which are starting materials in the formation of nylon, takes place.

$$\text{Furan} \xrightarrow[\text{100 atm. 298 K}]{H_2/Ni} \text{THF} \xrightarrow{HCl} CH_2Cl{-}CH_2{-}CH_2{-}CH_2OH \xrightarrow[-H_2O]{HCl} CH_2Cl{-}CH_2{-}CH_2{-}CH_2Cl$$

$$CH_2Cl{-}CH_2{-}CH_2{-}CH_2Cl \xrightarrow{NaCN} \underset{(x)}{NC-[CH_2]_4-CN} \xrightarrow[H_2SO_4]{H_2O} \underset{\text{Adipic acid}}{HOOC-[CH_2]_4-COOH}$$

$$\underset{(x)}{NC-[CH_2]_4-CN} \xrightarrow[N_i]{H_2} \underset{\text{1,6-diaminohexane}}{H_2N-[CH_2]_6NH_2}$$

3. Diel's Alder Reaction. Furan is the only heterocyclic compound which undergoes Diels Alder reactions.

$$\text{Furan} + \underset{\text{Maleic anhydride}}{CHCO{-}O{-}COCH} \longrightarrow \text{Adduct}$$

Q. 7. Establish the structure of furan.

Ans. 1. Analytical studies and molecular weight determination prove the molecular formula of furan as C_4H_4O.

2. Keeping in view the tetravalency of carbon and divalency of oxygen a possible structure of furan is given below:

3. The properties of furan are not in keeping with the above structure which has a pair of conjugated double bonds and an ether group. Rather it gives substitution reactions. This is characteristic of aromatic compounds. Its heat of combustion indicates that it is resonance stabilised by an amount of 65 kJ mol^{-1}. Resonance stabilization on account of conjugated double bonds would have provided 12.5 kJ mol^{-1} resonance energy. Therefore the diene structure of furan is ruled out.

4. Resonance structures. To account for extra stability and high value of resonance energy, furan is considered to be resonance hybrid of the following resonating structures.

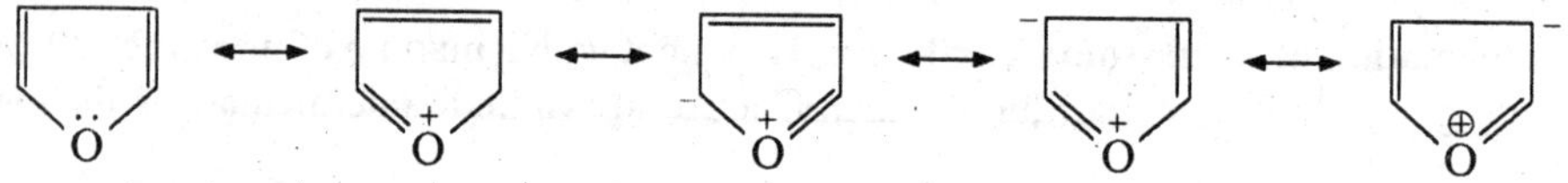

To be aromatic in nature, a monocyclic molecule must be planar and should have $(4n + 2)$ π electrons.

Out of the six electrons ($n = 1$) needed for aromaticity, two electrons are provided by oxygen in the form of its lone pair. The resonance hybrid structure of furan has been confirmed by X-ray diffraction studies which shows that C — O bond length in furan is 1.37 Å which is shorter than normal C — O single bond length of 1.43 Å. It establishes that this bond has some double bond character as required by the resonance phenomenon.

5. Orbital structure. Furan has been established to have a flat pentagonal ring involving sp^2 hybridised carbon and oxygen atoms with cyclic π electron cloud containing six electrons each lying above and below the plane of the ring. Oxygen forms two sigma bonds with two adjacent carbon atoms. Of the hybrid orbitals of oxygen, two have got one electron each while the third has a lone pair of electrons.

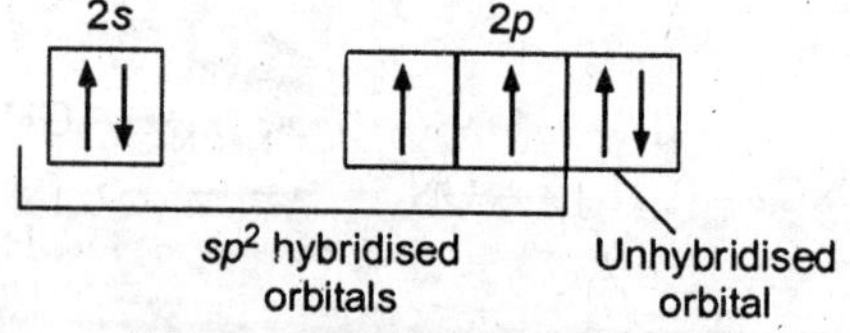

Electronic configuration of oxygen

The unhybridised p-orbital also has a pair of electron. The two hybridised orbitals having single electrons each are involved in the formation of σ-bonds with the two adjacent carbon atoms while the unhybridised p-orbital having two electrons from a π bond by overlapping sideways with the p-orbitals of carbon. The pair of electrons in the third hybrid orbital of oxygen is left unshared. The orbital picture of furan is shown as under:

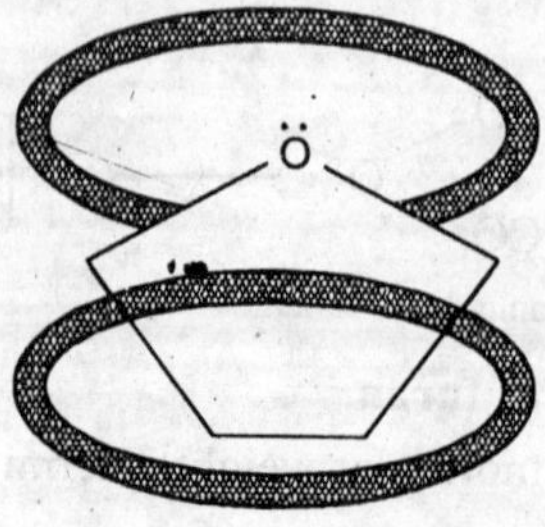

Orbital structure of furan

Q. 8. Describe the methods of preparation of thiophene. *(Coimbatore, 2000)*

Ans. Thiophene is prepared by the following methods.

1. Laboratory Method. On heating sodium succinate with phosphorus sulphide, thiophene is obtained.

$$\underset{\text{Sod. succinate}}{\begin{matrix} CH_2COONa \\ | \\ CH_2COONa \end{matrix}} \xrightarrow{P_2S_3} \underset{\text{Thiophene}}{C_4H_4S}$$

2. From *n*-butane. On heating *n*-butane with sulphur at 923 K, thiophene is formed.

$$\underset{n\text{-Butane}}{\begin{matrix} CH_2 - CH_2 \\ | \quad\quad | \\ CH_3 \quad CH_3 \end{matrix}} + 4S \xrightarrow{923\ K} \underset{\text{Thiophene}}{C_4H_4S} + 3\,H_2S$$

3. Paal-Knorr synthesis of thiophene derivatives. It involves the action of heat on enolisable 1, 4 diketone in the presence of pentasulphide.

$$\underset{\text{Acetonyl acetone}}{\begin{matrix} CH_2 - CH_2 \\ | \quad\quad | \\ H_3C - C \quad\quad C - CH_3 \\ \| \quad\quad \| \\ O \quad\quad O \end{matrix}} \xrightarrow[\text{Heat}]{P_2S_5} \underset{\text{2, 5-Dimethyl thiophene}}{H_3C - C_4H_2S - CH_3} + H_2O$$

4. Manufacture. Thiophene is manufactured by passing a mixture of acetylene and hydrogen sulphide through a tube containing Al_2O_3 at 673 K.

$$2CH \equiv CH + H_2S \longrightarrow C_4H_4S + H_2$$

Q. 9. Describe the properties of thiophene.

Ans. Properties of Thiophene

1. Electrophilic Substitution Reactions. Thiophene undergoes electrophilic substitution reactions. On the basis of charge distribution and stability of carbonium ions, the electrophilic substitution would be expected to take place at position 2 [or 5] rather than at position 3 [or 4].

(*a*) Nitration. Thiophene when nitrated with fuming HNO_3 in acetic anhydride gives 2-nitrothiophene.

$$C_4H_4S \xrightarrow[\text{in acetic anhydride}]{\text{Fuming } HNO_3} \underset{\text{2-Nitrothiophene}}{C_4H_3S - NO_2}$$

(*b*) Sulphonation. Sulphonation of thiophene with cold conc. H_2SO_4 gives 2-thiophene sulphonic acid.

$$C_4H_4S \xrightarrow[\text{Conc. } H_2SO_4]{\text{Cold}} \underset{\text{2-Thiophene sulphonic acid}}{C_4H_3S - SO_3H}$$

(*c*) Bromination. When thiophene is treated with Br_2 in the absence of any halogen carrier, the formation of 2, 5-dibromothiophene takes place.

Thiophene $+ Br_2 \longrightarrow$ 2, 5-Dibromothiophene (Br at 2 and 5 positions)

2, 5-Dibromothiophene

(*d*) **Iodination.** When thiophene is treated with I_2 in the presence of yellow mercuric oxide, 2-Iodothiophene is obtained.

$$\text{Thiophene} + I_2 \xrightarrow{HgO} \text{2-Iodothiophene}$$

2 Iodothiophene

(*e*) **Friedel-Crafts Acylation.** Thiophene is acetylated with acetic anhydride in the presence of phosphoric acid.

$$\text{Thiophene} + (CH_3CO)_2O \xrightarrow{H_3PO_4} \text{Thiophene}-COCH_3$$

Methyl thiophenyl ketone

(*f*) **Chloromethylation.** Thiophene may be chloromethylated with the mixture of HCHO and HCl.

$$\text{Thiophene} \xrightarrow{HCHO + HCl} \text{Thiophene}-CH_2Cl$$

2-Chloromethyl thiophene

(*g*) **Mercuration.** On mercuration with mercuric chloride in the presence of sod. acetate it gives 2-mercuri-chloride.

$$\text{Thiophene} \xrightarrow[CH_3COONa]{HgCl_2} \text{Thiophene}-HgCl$$

2-Mercuri-chloride thiophene

2. Reduction. (*a*) Catalytic hydrogenation of thiophene using large amounts of catalysts yields tetrahydrothiophene [Thiophan].

$$\underset{\text{Thiophene}}{\text{Thiophene}} \xrightarrow{Pd/H_2} \underset{\text{[Thiophane]}}{CH_2-CH_2,\ CH_2\ CH_2,\ S}$$

Thiophene

Tetrahydrothiophene

[Thiophane]

(*b*) When reduced with sodium in liquid NH_3, thiophene yields, 2, 3 and 2, 5 dihydrothiophene (Birch reaction).

$$\text{Thiophene} \xrightarrow{Na/NH_3} \text{2 : 3 Dihydrothiophene} + \text{2 : 5 Dihydrothiophene}$$

Thiophene; 2 : 3 Dihydro-thiophene; 2 : 5 Dihydro-thiophene

(*c*) Catalytic reduction of thiophene with Raney Ni as catalyst gives *n*-butane as the main product.

3. Formation of lithium derivative. On treatment with *n*-butyl lithium in ether, thiophene gives 2-lithium thiophene. It is useful in the synthesis of various 2-substituted thiophenes.

$$\text{Thiophene} \xrightarrow[\text{Ether}]{C_4H_9Li} \text{2-Lithium thiophene}$$

$$\text{2-Lithium thiophene} \xrightarrow[(ii)\ H^+]{(i)\ CO_2} \text{Thiophene-2 carboxylic acid (–COOH)}$$

Q. 10. Deduce the structure of thiophene.

Ans. 1. Analytical studies and molecular weight determination reveal the formula of thiophene as C_4H_4S.

2. An unsaturated structure is indicated by the tormula. A possible structure is shown below:

This structure has a pair of conjugated double bonds. But this structure does not fully explain the observed behaviour of thiophene.

3. Thiophene undergoes electrophilic substitution reactions like nitration, sulphonation, halogenation and Friedel-Crafts reactions, quite unexpected from a compound with the above-written unsaturated compound. It is observed that it is stabilised by 125 kJ mol^{-1} whereas we expect stabilisation to the extent of 12.5 kJ mol^{-1} on account of conjugated double bonds. This is revealed by heat of combustion studies.

4. In the light of 125 kJ mol^{-1} as stabilisation energy, thiophene was considered as a resonance hybrid of the fcllowing resonating structures.

I ⟷ II ⟷ III ⟷ IV ⟷ V

The resonance structure has been confirmed from measurement of bond lengths and dipole moments.

5. Orbital structure. Although the stabilization energy is explained with the help of resonance phenomenon, electrophilic substitution remains unexplained. This is explained by the orbital structure of thiophene. Thiophene has a flat pentagonal ring of sp^2 hybridised carbon and sulphur atoms. Structure of thiophene is similar to that of furan with the difference that oxygen has been replaced by sulphur.

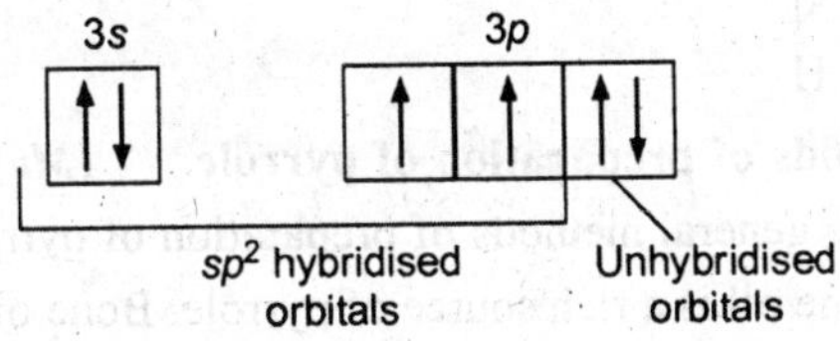

Hybridisation of the sulphur atom orbitals

Two of the hybrid orbitals of sulphur have got one electron each while the third has got a pair of electrons. The two hybridised orbitals containing single electron overlap with the sp^2 hybrid orbitals of carbons on either side and form the σ bonds. Unhybridised orbital of sulphur containing two electrons form part of the delocalised electron cloud by lateral (sideways) overlapping with the unhybridised orbitals of carbon atoms. Electron pair in one of the hybrid orbitals remains unshared. Orbital structure of thiophene may be given as:

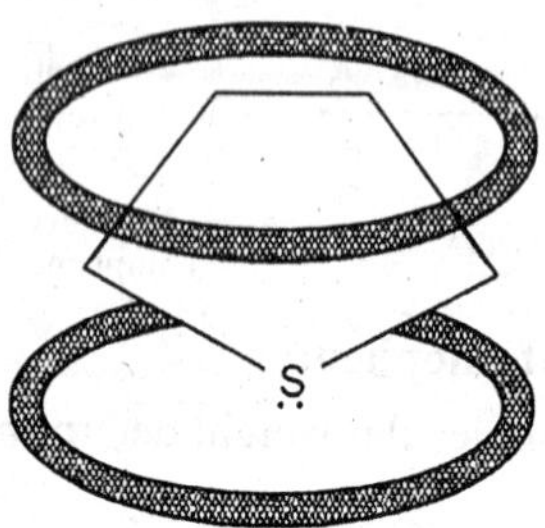

Structure of thiophene

Q. 11. How will you show that thiophene is aromatic in nature, giving its molecular orbital diagram?

Ans. For molecular orbital diagram, see. Q. 10. The no. of π-electrons in thiophene is found to be six *i.e.* four from two carbon-carbon double bonds and two from the hybridised orbitals of nitrogen. Thus the Huckels rule for aromaticity is satisfied.

Q. 12. What is pyrrole? What is the importance of this compound? Give the nomenclature of pyrrole and its derivatives.

Ans. Pyrrole is the nitrogen analogue of furan. It is of interest as it is related to many naturally occurring materials. Pyrrole nucleus is found in many alkaloids and bile pigments, in the chlorophyll of plants and in haemoglobin. It is also found in morphine. It derives its name from the fact that its vapours give a bright red colour on coming in contact with a pure splint moistened with conc. HCl.

The numbering of positions in pyrrole ring is done as under:

(β′)4 3(β)

(α′)5 2(α)

N 1

H

Thus the following compounds will be named as given against each.

$-CH_3$ N H — 2-Methylpyrrole

C_2H_5- $-CH_3$ N H — 2-Methyl-5-ethylpyrrole

Q. 13. Give the methods of preparation of pyrrole. (*Mumbai, 2004; Garhwal, 2010*)

Ans. Following are the general methods of preparation of pyrrole.

(*i*) From Bone oil. Bone oil is a rich source of pyrrole. Bone oil is thoroughly treated with a solution of dilute alkali to remove acidic impurities and then it is treated with a solution of dilute acid to remove basic impurities. It is then subjected to fractional distillation. The fraction appearing between 373 K and 423 K is collected. It is further purified by fusing with KOH when solid potassiopyrrole is formed. This on steam distillation gives pure pyrrole.

$$C_4H_4NK + H_2O \longrightarrow C_4H_4NH + KOH$$

(ii) By distilling succinimide with Zn dust

$$\begin{matrix} CH_2-CO \\ | \\ CH_2-CO \end{matrix} \Big> NH \xrightarrow{Zn} \begin{matrix} CH=CH \\ | \\ CH=CH \end{matrix} \Big> NH$$

Pyrrole.

(iii) From ammonium mucate. When ammonium mucate is distilled in the presence of glycerol at (473 K), pyrrole is obtained.

$$\begin{matrix} (CHOH)_2COONH_4 \\ | \\ (CHOH)_2COONH_4 \end{matrix} \xrightarrow[(473\ K)]{Glycerol} C_4H_5N + 2CO_2 + NH_3 + 4H_2O$$

Amm. Mucate

(iv) Synthesis. By passing a mixture of acetylene and NH_3 through red hot tube.

$$2C_2H_2 + NH_3 \longrightarrow C_4H_5N + H_2$$

(v) By the action of acetylene on formaldehyde and then heating with NH_3 under pressure.

$$CH \equiv CH + 2HCHO \xrightarrow{Cu_2C_2} HOH_2C \cdot C \equiv CCH_2OH$$

$$\downarrow NH_3, \text{ Pressure}$$

Pyrrole

(vi) Paal-Knorr synthesis. In this synthesis, a 1, 4-diketone is heated with ammonia or a primary amine. Pyrrole derivatives are obtained according to the following sequence:

$$CH_3-\underset{\underset{O}{\|}}{C}(CH_2-CH_2)\underset{\underset{O}{\|}}{C}-CH_3 \xrightarrow{NH_3} CH_3-\underset{\underset{NH}{\|}}{C}(H-CH-CH_2)\underset{\underset{O}{\|}}{C}-CH_3$$

$$\downarrow$$

$$H_3C-(\text{pyrrole ring, N-H})-CH_3 \xleftarrow{-H_2O} \text{cyclic intermediate: } CH-CH(H),\ C,\ C(CH_3)(OH),\ N-H$$

***(vii)* Manufacture.** Pyrrole can be manufactured by passing mixture of furan, ammonia and steam over heated alumina at catalyst.

$$\text{Furan} + NH_3 \xrightarrow[(723\ K)]{Al_2O_3} \text{Pyrrole}$$

Q. 14. Describe the general properties of pyrrole.

Ans. Chemically it shows the reactions of aromatic compounds. It is less aromatic than thiophene but more aromatic than furan. Some important reactions of pyrrole are as under:

1. Electrophilic Substitution Reactions. *(a)* Nitration. With HNO_3 in acetic anhydride, pyrrole gives 2-nitropyrrole.

$$\text{Pyrrole} \xrightarrow[\text{at 263 K}]{HNO_3 \text{ in } (CH_3CO)_2O} \text{2-Nitropyrrole}$$

***(b)* Sulphonation.** With pyridine – SO_3 mixture in ethylene chloride, pyrrole is sulphonated to give 2-pyrrole sulphonic acid.

$$\text{Pyrrole} \xrightarrow[\text{in ethylene chloride}]{\text{Pyridine} - SO_3} \text{2-pyrrole-}SO_3H$$

***(c)* Halogenation.** With iodine solution, it gives tetraiodopyrrole (Iodole) which is used as a substitute for iodoform.

$$\text{Pyrrole} + 4I_2 \longrightarrow \text{Iodole} + 4HI$$

Iodole

***(d)* Friedel-Crafts reaction.** Pyrrole upon acetylation with acetic anhydride in the presence of $SnCl_4$ forms 2-acetylpyrrole.

$$\text{Pyrrole} \xrightarrow[SnCl_4]{(CH_3CO)_2O} \text{2-Acetylpyrrole}$$

***(e)* Addition of dichlorocarbene.** In the presence of a strong base and chloroform, pyrrole undergoes Reimar Tiemann's reaction. The attacking species is dichlorocarbene (: CCl_2). The reaction takes place as under:

Pyrrole $\xrightarrow[KOH]{CHCl_3}$ intermediate (H, Cl, C, Cl, H; $OH^{\ominus}$) $\xrightarrow{H_2O}$ 2-($CHCl_2$)pyrrole $\xrightarrow{2H_2O}$ Pyrrole-2-aldehyde (CHO)

Pyrrole-2-aldehyde

2. Reduction. (*i*) Pyrrole gives pyrroline (2 : 5 dihydropyrrole) on reduction with Zn and acetic acid.

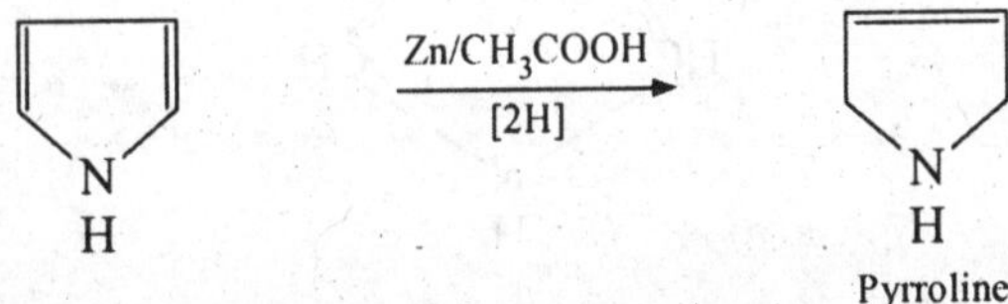

Pyrroline

(*ii*) Pyrrole gives pyrrolidine (Tetrahydropyrrole) on hydrogenation in the presence of Ni.

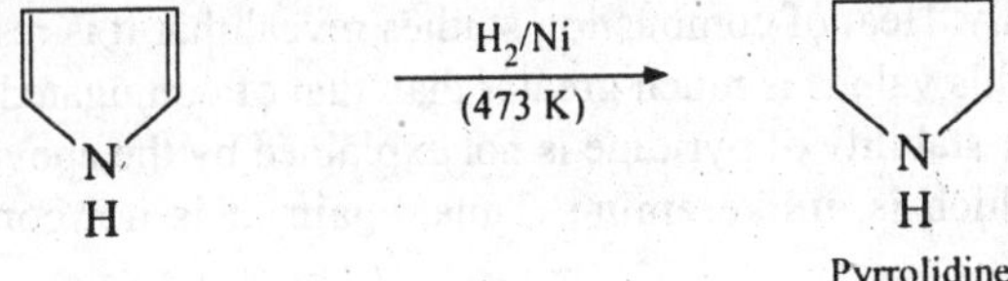

Pyrrolidine

3. Oxidation. When oxidised with chromium trioxide in H_2SO_4, it gives malecimide

Pyrrole $\xrightarrow[CrO_3 + H_2SO_4]{(O)}$ CH = CH, O = C C = O, N–H

Malecimide

4. Ring expansion. When pyrrole is treated with sodium methoxide and methylene iodide, it gives pyridine (a six membered ring).

Pyrrole $+ 2CH_3ONa + CH_2I_2 \longrightarrow$ Pyridine $+ 2CH_3OH + 2NaI$

Pyridine

5. Acidic character. Pyrrole unlike furan and thiophene is weakly acidic and forms alkali metal salts.

Pyrrole (N–H) $+ KOH \longrightarrow$ Pyrrole (N–K) $+ H_2O$

6. Ring fission. When pyrrole is treated with hydroxylamine, the ring is ruptured and succinaldehyde dioxime is formed.

$$\text{Pyrrole (N–H)} + 2NH_2OH \longrightarrow \begin{matrix} H_2C - CH = NOH \\ | \\ H_2C - CH = NOH \end{matrix}$$

Succinaldehyde dioxime

Q. 15. Deduce the structure of pyrrole.

(Madhya Pradesh, 2004; Punjab, 2005; Mumbai, 2010)

Ans. 1. Analytical studies and molecular weight determination establish its molecular formula as C_4H_5H.

2. In terms of tetravalency of carbon and trivalency of nitrogen, a possible structure is

CH — CH, HC, CH, N, H

3. The above structure does not explain the electrophilic substitution reactions given readily by pyrrole. We would rather expect pyrrole to give addition reactions. Pyrrole does give addition reactions but not so readily. Heat of combustion studies reveal that it is resonance stabilized to the extent of 100 kJ mol^{-1}. This values is much greater than that of conjugated double bonds. Thus the aromatic nature and extra-stability of pyridine is not explained by the above structure. Also pyrrole lacks basic character which is unlike amine. Thus, again, it is not consistent with the above structure.

4. Pyrrole has been found to be a resonance hybrid of the following resonating structures.

I ⟷ II ⟷ III ⟷ IV ⟷ V

5. Orbital structure. Although the resonance hybrid structure explains the resonance stabilization energy of pyrrole, it does not explain the aromatic nature and electrophilic substitution reactions of pyrrole. The orbitals of carbon and nitrogen in pyrrole are *sp*^{2} hybridised. The four carbons and a nitrogen atom form a flat pentagon.

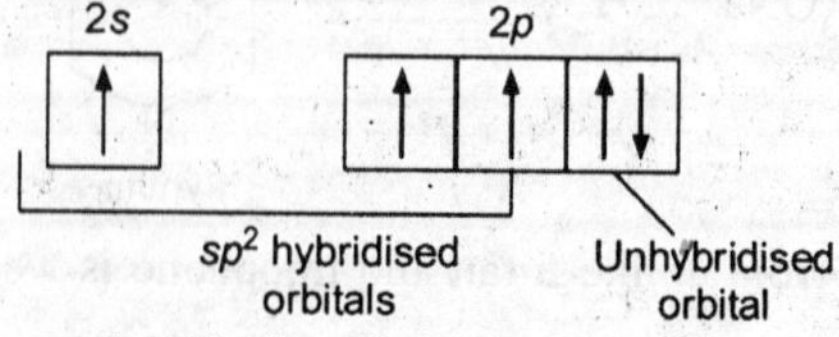

Hybridisation of N atom orbitals

The three *sp*^{2} hybrid orbitals of nitrogen contain single electrons. The unhybridised orbital contains two electrons. The two hybrid orbitals of nitrogen containing single electrons overlap with the hybrid orbitals of two carbon atoms on either side forming σ bonds. The third hybrid orbital overlaps with the orbital of hydrogen and forms N — H bond. Similarly each carbon forms two —

σ bonds with the neighbouring carbon atoms and one σ bond with hydrogen. Now each carbon has one unhybridised orbitals. These unhybridised orbitals of carbon along with the unhybridised orbital of nitrogen form a delocalised electron cloud above and below the pentagonal ring of pyrrole as shown in the figure below.

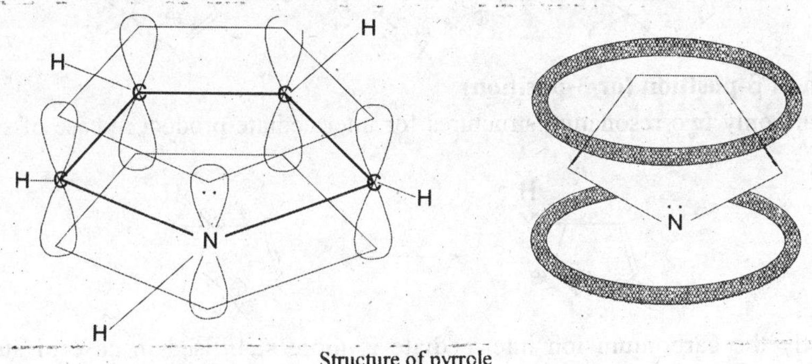

Structure of pyrrole

This accounts for the aromatic nature of pyrrole.

Q. 16. Discuss the mechanism of orientation and substitution in furan, thiophene and pyrrole.

Ans. The mechanism of electrophilic substitution in furan, thiophene and pyrrole which occurs preferentially at α-position is represented as under.

The three compounds may be represented as under

$\ddot{X}$

where X stands for O, S or NH is case of furan, thiophene or pyrrole respectively.

(*i*) Y : Z ⟶ Y^+ + Z^- :
Electrophile

(*ii*) $\ddot{X}$ + Y^+ ⟶ [$\ddot{X}$ (+) H Y ⟷ ⊕ $\ddot{X}$ H Y ⟷ ⊕X H Y]

(*iii*) $\ddot{X}$ (+) H Y + $:Z^-$ ⟶ X —Y + H : Z

Q. 17. Explain why the electrophilic substitution takes place preferably at α-position in furan, thiophene and pyrrole.

Or

Explain clearly what you know about the reactive site for electrophilic substitution of five-membered heterocyclic compounds.

Or

Why position-2 is more reactive than position-3 in pyrrole? (*Mumbai, 2004; Lucknow, 2010*)

Ans. Substitution takes place preferably at that position where the intermediate products is more stabilised. In the case of furan, thiophene and pyrrole, there are two positions 2 or 3 (α or β) where the substitution could take place. Let us see what are the intermediate products in the two cases.

Substitution at α-position (or 2-position)

The intermediate products in case of substitution at α-position has three resonating structure are given below :

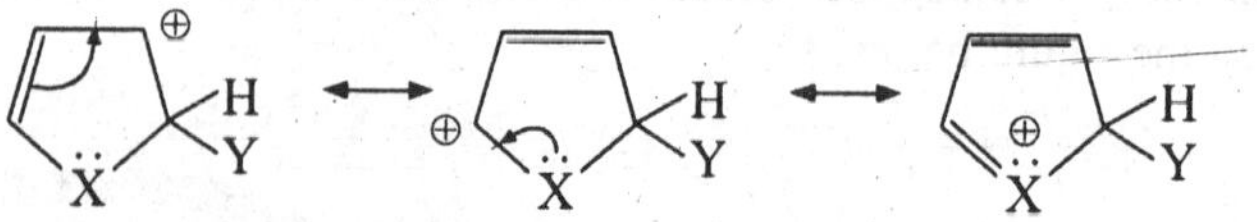

Substitution at β-position (or 3-position)

There are only two resonating structures for intermediate product in case of substitution at β-position.

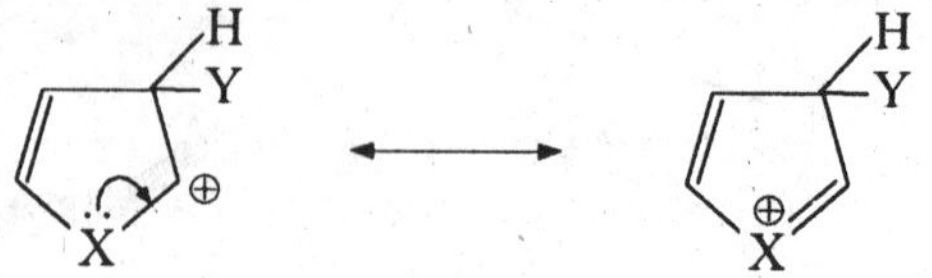

Evidently the carbonium ion intermediate is more stabilised in case of substitution at α-position. Hence substitution is preferred at α-position.

Q. 18. Pyrrole is acidic in character like phenol. Explain.

(Awadh, 2000; Coimbatore, 2000; Nagpur 2002)

Ans. Both the compounds release the protons and the resulting anions are stabilised by resonance as shown below:

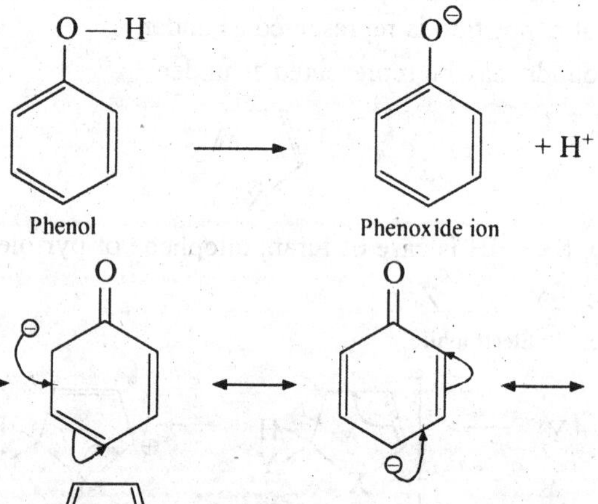

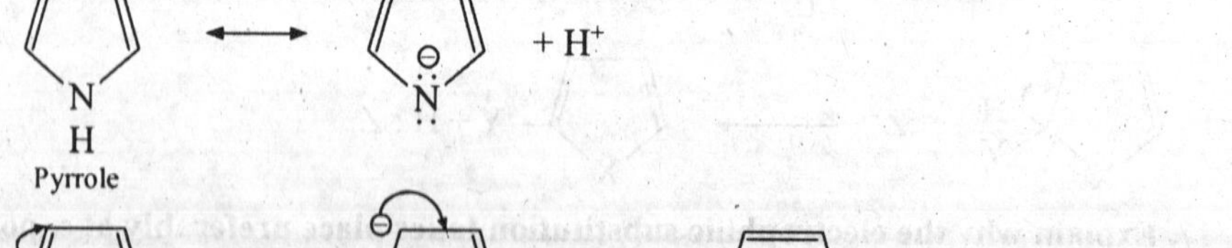

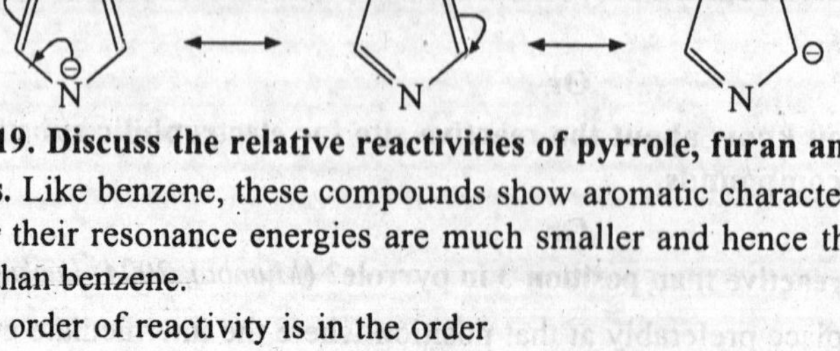

$+ H^+$

Phenol

Phenoxide ion

$+ H^+$

Pyrrole

Q. 19. Discuss the relative reactivities of pyrrole, furan and thiophene.

Ans. Like benzene, these compounds show aromatic character and are resonance stabilised. However their resonance energies are much smaller and hence they are less aromatic or more reactive than benzene.

The order of reactivity is in the order

Pyrrole > Furan > Thiophene > Benzene

The increasing order of aromatic character in pyrrole, furan and thiophene is as follows:

Pyrrole < Furan < Thiophene

(*i*) Thiophene is least reactive or most stable of the three because it can use its *d*-electrons to form a larger no. of resonating structures and hence gets stabilised to a greater extent.

(*ii*) Furan is less reactive than pyrrole because oxygen accommodates a positive charge less readily than nitrogen.

Q. 20. Pyrrole, furan and thiophene are more reactive than benzene to electrophilic attack. Explain. (*Panjab, 2003*)

Ans. The above heterocyclics are more reactive than benzene in giving electrophilic reactions because of the formation of intermediate carbonium ions which is specially stabilised by the following resonating structure.

: Y^+ is the electrophile and X is the heterc atom in heterocyclic compound.

This structure is particularly stable because all the atoms have their complete octet.

Q. 21. Describe the nomenclature and isomerism in pyridine derivatives.

Ans. According to IUPAC nomenclature, name for pyridine is azine but it is rarely used. The ring atoms of pyridine are denoted by numerals or Greek letters and are numerically counted anti-clockwise with nitrogen occupying position 1.

4(γ), (β)5, 3(β), (α)6, 2(α), N 1

Pyridine

α-Picoline (2-Picoline)

β-Picoline (3-Picoline)

γ-Picoline (4-Picoline)

2, 4-Lutidine

2, 4, 6-Collidine

Pyridine gives *three isomers on mono-substitution*. The nomenclature of pyridine derivative needs special attention because the common names are in general used for methyl pyridine and pyridine carboxylic acids. So mono methyl pyridines are known as *picolines*, the dimethyl pyridines as *lutidines* and trimethyl pyridines as *collidines*. The three isomeric α-, β- and γ- pyridine carboxylic acids are called *picolinic*, *nicotinic* and *isonicotinic* acids respectively.

Picolinic acid (Pyridine-2-carboxylic acid)

Nicotinic acid (Pyridine-3-carboxylic acid)

Isonicotinic acid (Pyridine-4-carboxylic acid)

Q. 22. Give the methods of preparation of pyridine.

Ans. Following are the important methods of preparation of pyridine.

1. Isolation from coal-tar. The light oil fraction of coal-tar contains aromatic hydrocarbons and phenols besides pyridine and alkyl pyridines. The light oil fraction is thus heated with dil.

H_2SO_4 which dissolves pyridine and other basic substances. This aqueous acid layer is neutralised using sodium hydroxide, when bases are liberated as a dark brown oily liquid. This oily layer is separated and pyridine is obtained by fractional distillation.

2. It is prepared by passing a mixture of acetylene and hydrogen cyanide through a red hot tube.

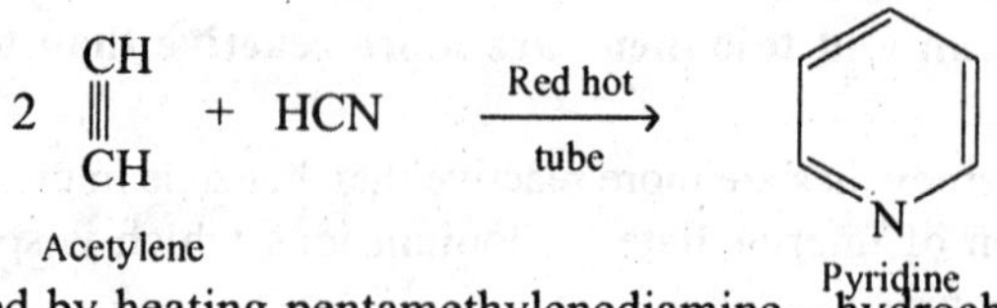

$$2 \begin{array}{c} CH \\ ||| \\ CH \end{array} + HCN \xrightarrow[\text{tube}]{\text{Red hot}} \text{Pyridine}$$

Acetylene — Pyridine

3. It can be prepared by heating pentamethylenediamine—hydrochloride and oxidising the product piperidine with concentrated sulphuric acid at 573 K.

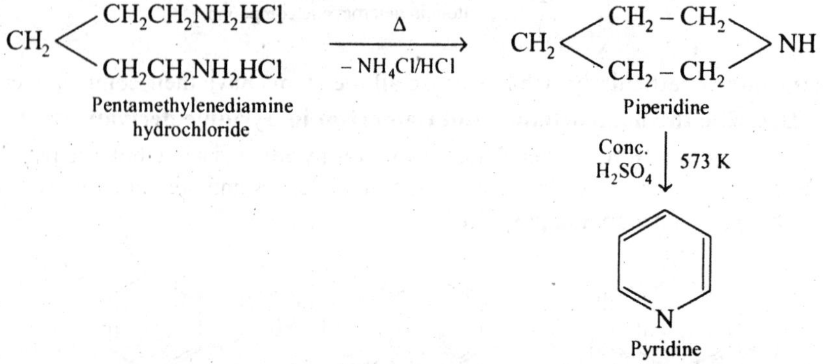

$$CH_2 \begin{cases} CH_2CH_2NH_2HCl \\ CH_2CH_2NH_2HCl \end{cases} \xrightarrow[-NH_4Cl/HCl]{\Delta} CH_2 \begin{cases} CH_2-CH_2 \\ CH_2-CH_2 \end{cases} NH$$

Pentamethylenediamine hydrochloride — Piperidine

$$\xrightarrow[\text{573 K}]{\text{Conc. } H_2SO_4} \text{Pyridine}$$

Pyridine

4. Hantzsch Synthesis (1882). It involves the condensation of a β-dicarbonyl compound (2 mole), an aldehyde (1 mole) and ammonia (1 mole). The dihydropyridine derivative formed on oxidation with HNO_3 yields pyridine derivative.

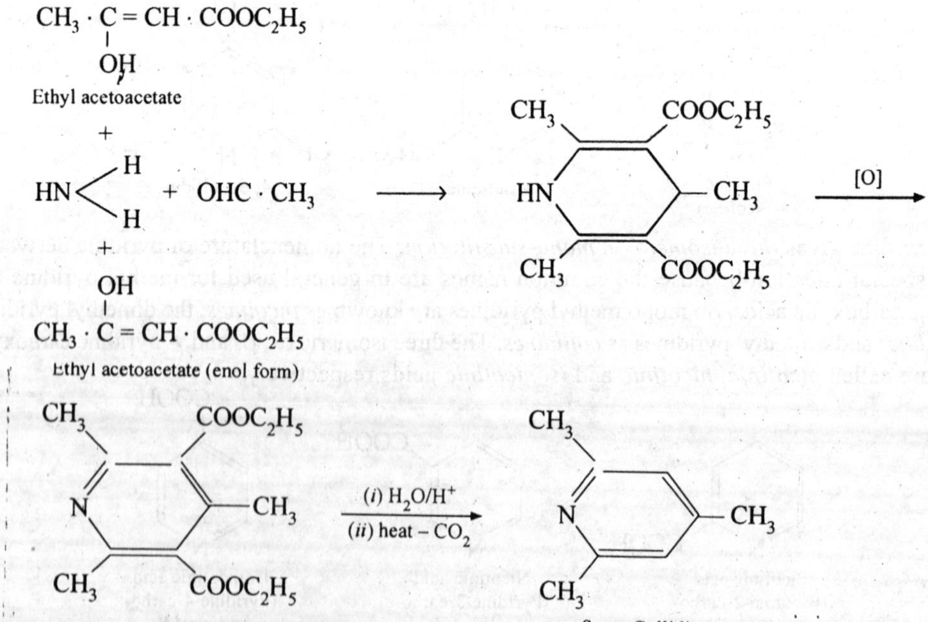

$$CH_3 \cdot \underset{\displaystyle OH}{\underset{|}{C}} = CH \cdot COOC_2H_5$$

Ethyl acetoacetate

+

$$HN\langle^{H}_{H} + OHC \cdot CH_3 \longrightarrow$$

+

$$CH_3 \cdot \overset{\displaystyle OH}{\overset{|}{C}} = CH \cdot COOC_2H_5$$

Ethyl acetoacetate (enol form)

(dihydropyridine: 2,6-dimethyl ring with CH_3, $COOC_2H_5$, HN, CH_3, $COOC_2H_5$, CH_3) $\xrightarrow{[O]}$

(pyridine ring with CH_3, $COOC_2H_5$, N, CH_3, CH_3, $COOC_2H_5$) $\xrightarrow[(ii)\ \text{heat} - CO_2]{(i)\ H_2O/H^+}$ (pyridine ring with CH_3, N, CH_3, CH_3)

Sym–Collidine (2, 4, 6-Trimethyl pyridine)

5. From Pyrrole. When heated with methylene dichloride in presence of sod. ethoxide, pyrrole gives pyridine.

$$\text{Pyrrole} + CH_2Cl_2 + 2C_2H_5ONa \xrightarrow{\text{Heat}} \text{Pyridine} + 2NaCl + 2C_2H_5OH$$

6. Pyridine of very high purity needed for the synthesis of medicinal compounds is obtained on a large scale by passing acetylene, formaldehyde hemi methyl and ammonia over alummia-silica catalyst at 773 K.

$$2\ \begin{matrix} CH \\ ||| \\ CH \end{matrix} + 2CH_2(OH)OCH_3 + NH_3 \xrightarrow[773\ K]{Al_2O_3/SiO_2} \underset{\text{Pyridine}}{C_5H_5N} + H_2O + 3CH_3OH$$

Note. In this synthesis acetylene can be replaced by acetaldehyde.

7. A new **industrial method** for the preparation of pyridine is to heat tetrahydro furfuryl alcohol, obtained by the catalytic reduction of furfuryl alcohol, with ammonia at 773 K.

$$\text{Tetrahydro furfuryl alcohol} + NH_3 \xrightarrow[773\ K]{Al_2O_3} \text{Pyridine} + 2H_2O + 2H_2$$

Q. 23. Describe chemical properties of pyridine. *(Bhopal, 2004; Garhwal, 2010*

1. Basic nature. Pyridine is a base with pK_b value 8.8 comparable in strength to aniline (pK_b = 9.4). It is much stronger base than pyrrole, (pK_b = 13.6) but is much weaker than aliphatic tertiary amines (pK_b = 4). Basic nature of pyridine is due to the presence of the lone pair of electrons on nitrogen atom. It behaves as a mono acid tertiary base and reacts with alkyl halides to give quaternary salts.

$$\underset{\text{Pyridine}}{C_5H_5N:} + CH_3I \longrightarrow \underset{\text{N-methyl pyridinium iodide}}{[C_5H_5\overset{+}{N}CH_3]I^-}$$

The fact that pyridine is much weaker base than aliphatic tertiary amines can also be explained in terms of the state of hybridisation of the orbitals having the lone pair of electrons. In case of aliphatic tertiary amines, the unshared pair of electrons is in sp^3 hybrid orbital while in case of pyridine unshared pair is in sp^2 hybrid orbital. Electrons are held more tightly by the nucleus in an sp^2 orbital than in sp^3 orbital. Hence they are less available for sharing with acids. This accounts for low basicity of pyridine.

2. Reduction. Pyridine undergoes catalytic hydrogenation to form hexahydropyridine, called *piperidine*. Reduction with Na/C_2H_5OH also produces piperidine.

$$\text{Pyridine} + 3H_2 \xrightarrow[\text{or } Na/C_2H_5OH]{\text{Ni or Pt.}} \text{Piperidine}$$

3. Addition of halogen. It adds halogen to yield a dihalide at room temperature and in the absence of catalyst.

$$C_5H_5N + Br_2 \longrightarrow [C_5H_5\overset{+}{N}Br]\,Br^-$$

Dibromopyridine

4. Electrophilic substitution reactions. Pyridine behaves as a highly deactivated aromatic nucleus towards electrophilic substitution reactions, and vigorous reaction condition should be used for these reactions to take place. The low reactivity of pyridine towards electrophilic substitutions is due to the following two reasons.

(*i*) *Greater electronegativity of nitrogen atom decreases electron density* (–I effect) *of the ring thereby deactivating it.*

(*ii*) *In acidic medium it forms a pyridinium cation with a positive charge at nitrogen atom and this decreases the electron density on nitrogen very much*, thus deactivating the ring. However the position – 3 is least affected and is comparatively the position of highest electron density in pyridine.

It undergoes electrophilic substitution like halogenation, nitration and sulphonation only under drastic conditions and does not undergo Friedel-Crafts reaction at all. The substitution occurs preferentially at β- or 3-position.

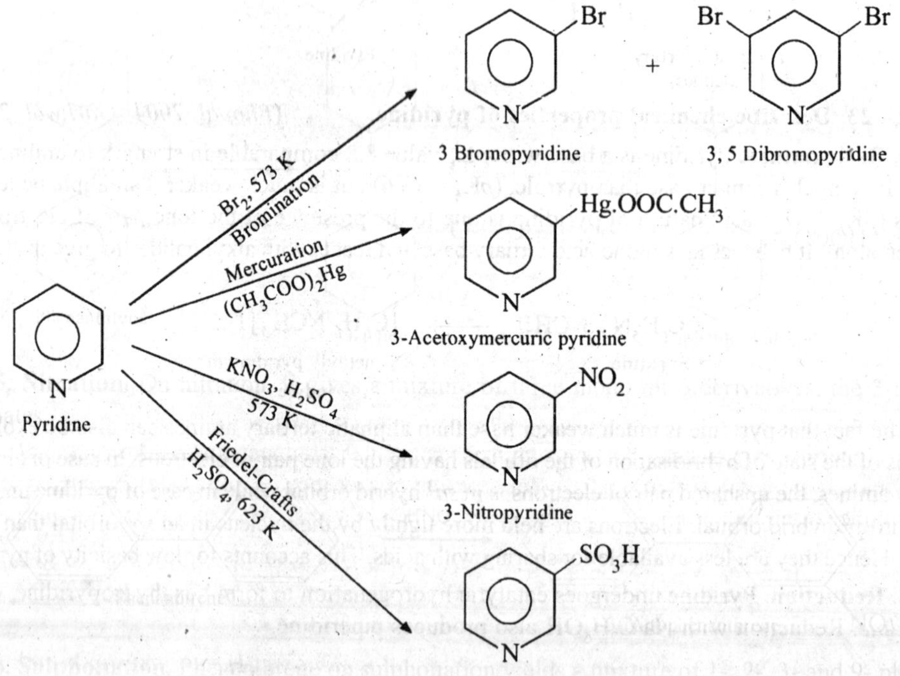

5. Nucleophilic substitution reactions. Pyridine is a highly deactivated system. Due to decrease of electron density from ring carbon atoms it becomes susceptible to nucleophilic attack. Since the positions 2- and 4- are very much electron deficient than position-3, hence nucleophilic substitution occurs readily at positions (2- and 4-), the 2-position being more preferred. The pyridinium ion is still more reactive than pyridine towards nucleophilic substitutions, because of the presence of full positive charge. Some of the important nucleophilic substitution reactions are given as under:

Pyridine → (KOH 573 K) → 2-Pyridinol ⇌ 2-Pyridone

Pyridine → (n-C_4H_9Li 373 K) → 2-Butyl pyridine + LiH

Pyridine → ($C_6H_5^-Li^+$ 383 K) → 2-Phenyl pyridine

Pyridine → ($Na^+NH_2^-$ 383 K) → 2-Amino pyridine

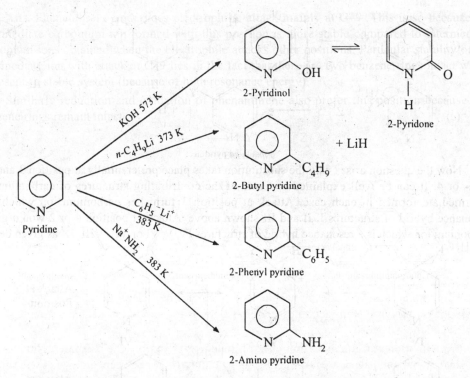

Reaction with sodamide to give 2-aminopyridine is also called as *Chichibabin Reaction.*

6. Oxidation. Like other tertiary amines, pyridine is oxidised by peroxy benzoic acid to yield pyridine N-oxide.

$$\text{Pyridine} \xrightarrow{C_6H_5-\overset{O}{\overset{\|}{C}}OOH} \text{Pyridine N-oxide } (N^{\oplus}-O^-)$$

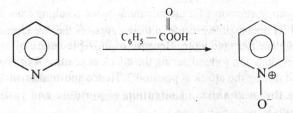

The N-oxide is more reactive than pyridine and finds use is preparing many pyridine derivatives.

Q. 24. Give the mechanism and orientation of electrophilic substitution in pyridine.

Or

Electrophilic attack in pyridine occurs in position-3. Explain.

(Punjab, 2003; Mumbai, 2004; Patna, 2010; Lucknow, 2010

Or

Why 3-position in pyridine is more reactive?

Ans. The mechanism of electrophilic substitution in pyridine which occurs preferentially at position 3 can be described in the following three steps.

(*a*) $$Y : Z \longrightarrow \underset{\text{Electrophile}}{Y^{\oplus}} + :Z^-$$

(b) [pyridine + $\overset{\oplus}{Y}$ → resonance structures I, II, III]

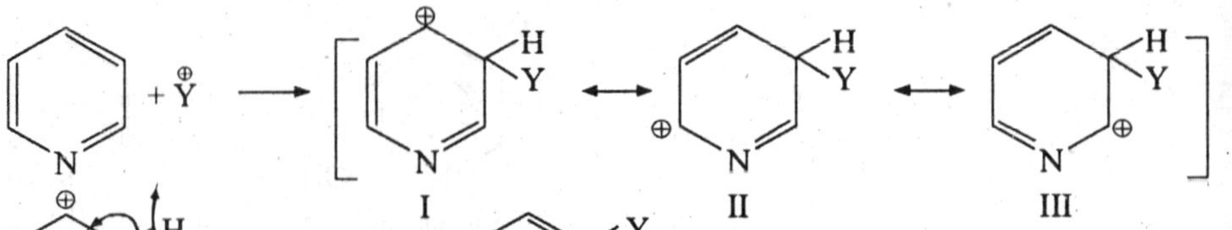

(c) [intermediate + $Z^{\ominus}$ → substituted pyridine + H : Z]

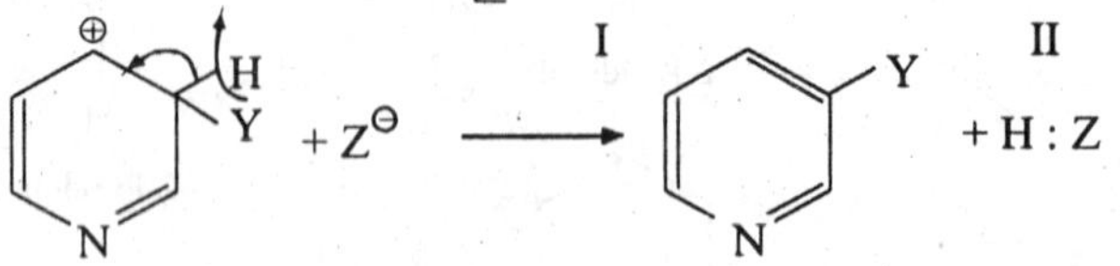

Substituted pyridine

Now the question arises why the substitution takes place preferentially at position 3 and not at 2- or 4. It can be well explained in terms of the contributing structures of carbonium ion intermediate formed in each case. Attack at position 3 furnishes carbonium ion which is a resonance hybrid of structures I, II and III shown above. Attack at position 4 or 2 would give a carbonium ion which is a resonance hybrid of structures IV, V, VI or VII, VIII, IX as given below:

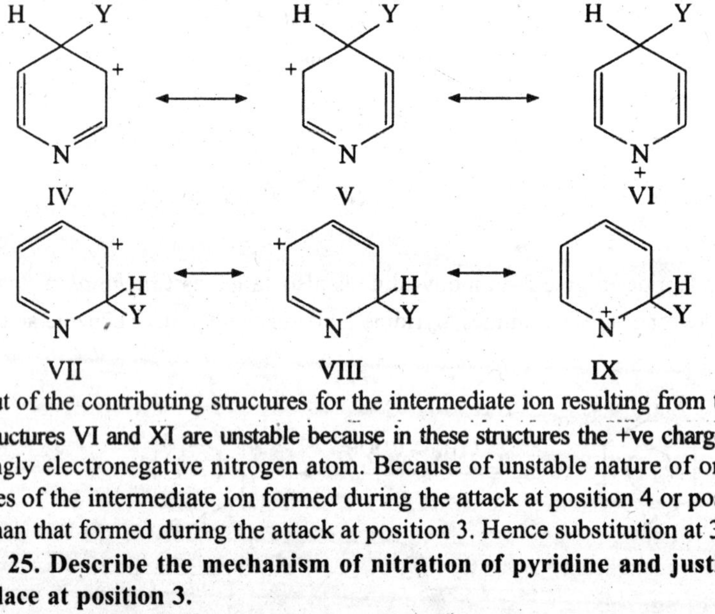

IV ⟷ V ⟷ VI — Attack at Position 4

VII ⟷ VIII ⟷ IX — Attack at Position 2

Out of the contributing structures for the intermediate ion resulting from the attack at position 4 or 2 structures VI and XI are unstable because in these structures the +ve charge is particularly carriec by strongly electronegative nitrogen atom. Because of unstable nature of one of the contributing structures of the intermediate ion formed during the attack at position 4 or position 2 this ion is less stable than that formed during the attack at position 3. Hence substitution at 3-position is favoured.

Q. 25. Describe the mechanism of nitration of pyridine and justify that substitution takes place at position 3.

Ans. Proceed as in Q. 24, Replace Y^+ by NO_2^+.

Q. 26. Give the mechanism and orientation of nucleophilic substitution in pyridine.

(*Guwahati 2002*)

Ans. The mechanism of nucleophilic substitution in pyridine which usually occurs at position 2- or 4- is outlined as follows:

Reaction with sodamide has been considered.

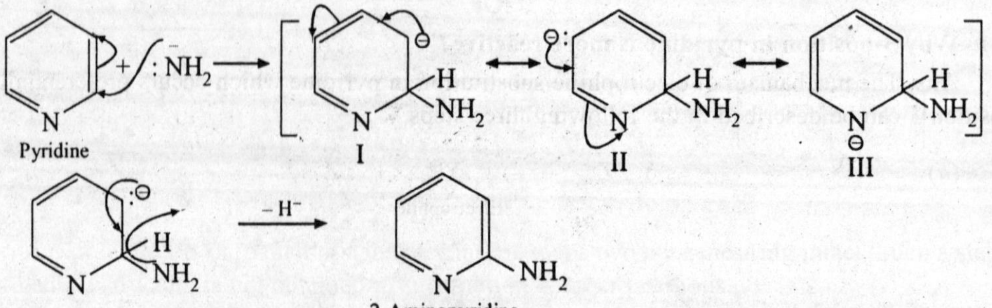

Pyridine + $:\overset{-}{N}H_2$ → [I ⟷ II ⟷ III]

→ (− H^-) → 2-Aminopyridine

We can decide about the position of substitution by assessing the stability of intermediate carbanion obtained in various positions.

Attack at position 2 furnishes a carbanion which in a resonance hybrid of structures I, II, III as shown above. Attack at position 4 or 3 would furnish a carbanions which are resonance hybrids of structures IV, V, VI or VII, VIII, IX as shown below:

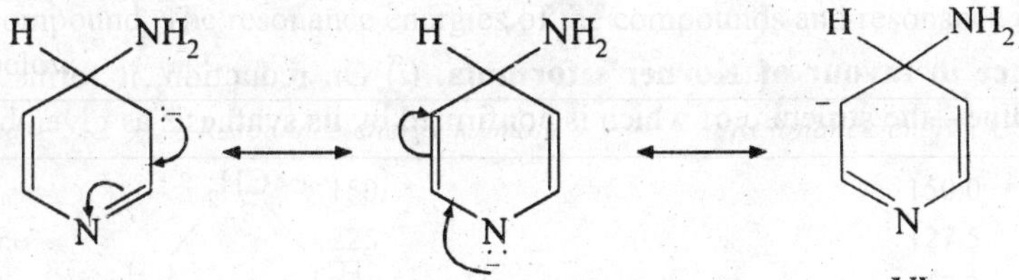

IV ⟷ V ⟷ VI

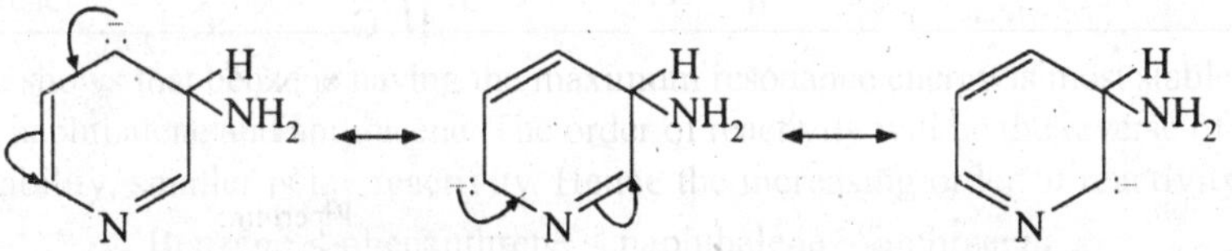

VII ⟷ VIII ⟷ IX

It may be noted that when the attack occurs at position 2, the contributing structure III of the *intermediate carbanion is particularly stable, because the negative charge is present on the electronegative nitrogen atom which can be easily accommodated by it. Similarly contributing structure V during attack at position 4 is also quite stable.*

There is no such stable contributing structure of intermediate ion resulting from attack at position 3. This means that the intermediate carbanion resulting from attack at positions 2 or 4 would be more stable, and hence more readily formed than that involved in attack at position 3. That is the reason why nucleophilic substitution is favoured at positions 2 or 4.

Q. 27. How will you establish the structure of pyridine? (*Meerut 2000; Dibrugarh 2003, Mysore 2004; Annamalai, 2004; Nagpur 2008*)

Ans. 1. Molecular formula. On the basis of analytical data and molecular weight determination, the molecular formula of pyridine comes out to be C_5H_5N.

2. Chemical behaviour. (*i*) Pyridine is basic in nature and forms salt with acids

$$\underset{\text{Pyridine}}{C_5H_5N} + HCl \longrightarrow \underset{\text{Pyridine hydrochloride}}{C_5H_5N \cdot HCl}$$

(*ii*) It does not react with acetyl chloride or nitrous acid indicating the absence of primary or secondary amino group. *The above reactions reveal that pyridine is a mono acid tertiary base.*

(*iii*) It reacts with *equimolar* amount of methyl iodide to form a quaternary ammonium salt.

$$\underset{\text{Pyridine}}{C_5H_5N} + CH_3I \longrightarrow \underset{\text{N-Methyl pyridinium iodide}}{[C_5H_5N^+(CH_3)]\, I^-}$$

(*iv*) It molecular formula indicates that it is a highly unsaturated compound, yet pyridine resists addition reactions and is stable towards usual oxidising agents.

(*v*) Pyridine exhibits aromatic character like benzene and gives electrophilic substitution reactions such as halogenation, nitration and sulphonation.

The last two reactions indicate that pyridine has aromatic character.

3. Korner's formula. On the basis of above evidence Korner assigned the following cyclic structure to pyridine which is quite analogous to Kekule structure of benzene.

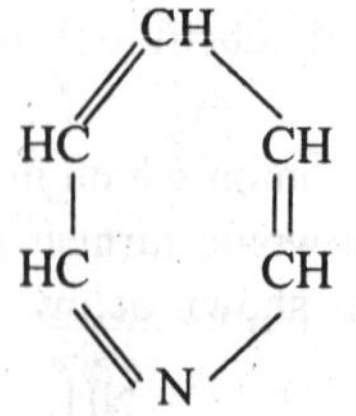

4. Evidence in favour of Korner's formula. (*i*) On reduction, it forms piperidine—a hexahydropyridine—the structure of which is confirmed by its synthesis as given below:

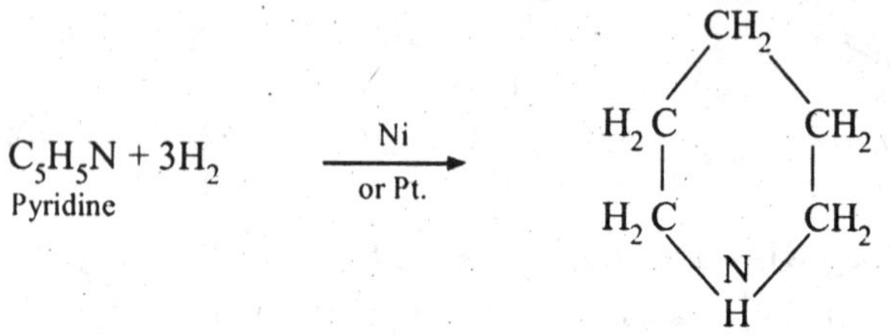

$$\underset{\text{Pyridine}}{C_5H_5N} + 3H_2 \xrightarrow[\text{or Pt.}]{\text{Ni}} \text{Piperidine}$$

Piperidine

The synthesis of piperidine from penta-methylene-1, 5-diamine hydrochloride and subsequent conversion to pyridine by oxidation with conc. H_2SO_4 further confirms the cyclic structure of piperidine and pyridine both.

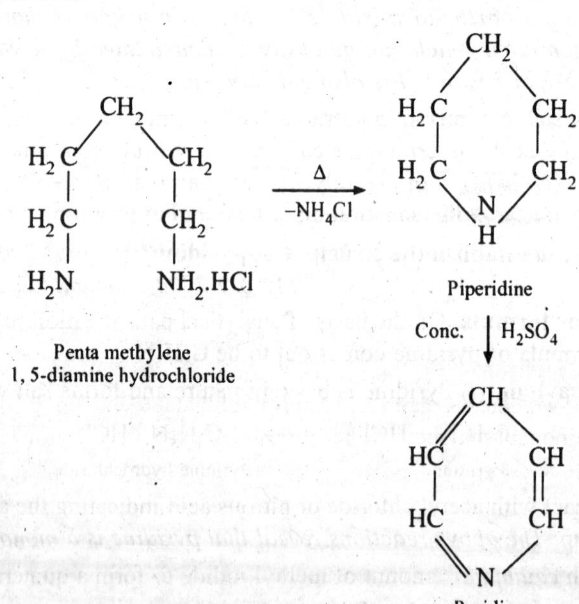

$$H_2N-CH_2-CH_2-CH_2-CH_2-CH_2-NH_2.HCl \xrightarrow[-NH_4Cl]{\Delta} \text{Piperidine} \xrightarrow{\text{Conc. } H_2SO_4} \text{Pyridine}$$

Penta methylene 1, 5-diamine hydrochloride

Piperidine

Pyridine

(*ii*) Pyridine gives three mono substituted products and six disubstitution products (in case the two substituents are identical) quite expected from the Korner's formula. Thus we have:

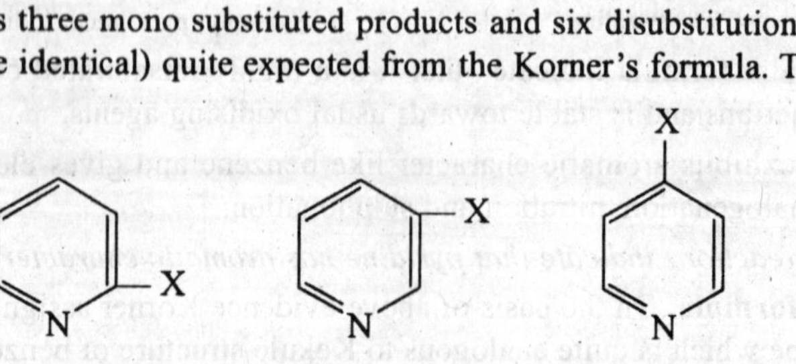

Three mono substitution products of pyridine, are shown above.

5. Objections of Korner formula. It fails to the explain following:

(*i*) its aromatic character.

(*ii*) its resistance to addition reactions.

(*iii*) its electrophilic and nucleophilic substitution reactions.

6. Alternative formulae. Following alternative formulae were suggested for pyridine, but were soon rejected as none was consistent with the behaviour and properties of pyridine.

Riedels formula

Bamberger and Vonpechmann's formula

7. Latest position. (I) Resonance Concept

Pyridine is considered to be resonance hybrid of the following structures:

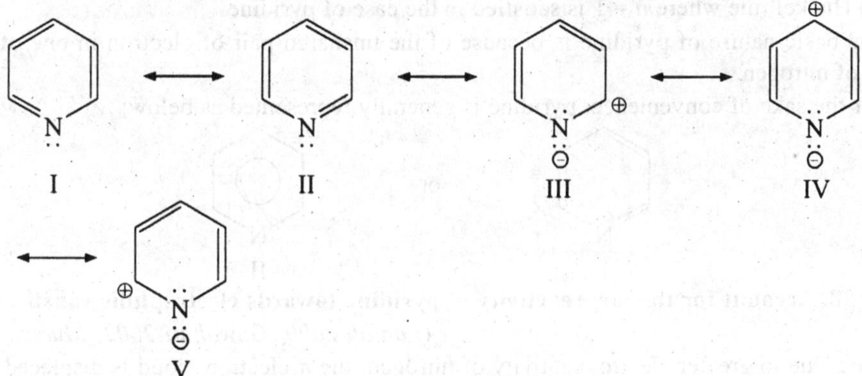

Resonance concept is supported by the following points:

(*i*) All the carbon, nitrogen and hydrogen atoms lie in the same plane.

(*ii*) All the carbon-carbon bonds in pyridine are of equal length (1.39 Å) which is intermediate between C — C single bond (1.54 Å) and C = C double bond (1.34 Å) length.

(*iii*) The two carbon-nitrogen bonds are also of equal length (1.37 Å) which is again intermediate between C — N single bond (1.47 Å) and C = N double bond (1.28 Å) length.

(*iv*) It resists addition due to the absence of true double bonds as due to resonance all bond distances are intermediate between single and double bonds.

(*v*) Since positive charge is present at 2- and 4-positions in the contributing structures it behaves as deactivated system in electrophilic substitutions—which take place at position 3.

(*vi*) Its resonance energy is 96 kJ mol^{-1}.

II. Molecular Orbital Concept

(*i*) The nitrogen and each of the five carbon atoms in the pyridine are in sp^2 *hybridization* state. These carbon and nitrogen atoms combine with each other to form a ring making use of two of their sp^2 hybrid trigonal orbitals for forming σ (sigma) bonds.

(*ii*) The third sp^2 orbital of each the five carbon atoms overlaps with '*s*' orbital of hydrogen to give rise to σ bonds whereas *third* sp^2 *orbital of nitrogen possesses the lone pair of electrons.* The nitrogen lone pair electrons in sp^2 hybrid orbital do not interact with π molecular orbital.

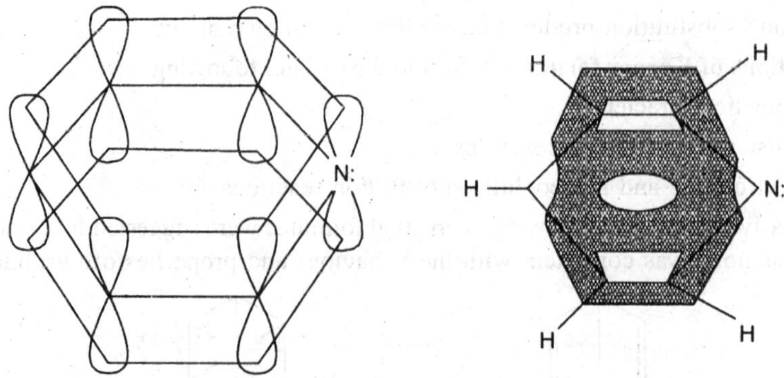

M.O. picture of pyridine.

(*iii*) The unhybridised '*p*' orbitals each carbon atom and nitrogen atom which are perpendicular to the plane of ring atoms, overlap with each other sideway to form a π electron cloud below and above the plane of ring as in the case of benzene.

(*iv*) These delocalised orbitals (having 6 π electrons) which are in the form of π electron clouds above and below the plane of ring are responsible for the stability and aromatic character of pyridine. $(4n + 2)$ Huckel rule where $n = 1$ is satisfied in the case of pyridine.

The basic nature of pyridine is because of the unshared pair of electron in one of the sp^2 orbitals of nitrogen.

For the sake of convenience, pyridine is generally represented as below:

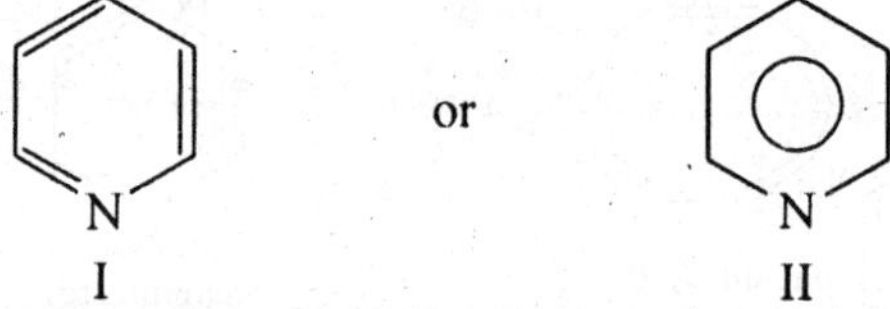

Q. 28. Account for the low reactivity of pyridine towards electrophilic substitutions.

(*Panjab 2000; Guwahati 2002; Mumbai, 2004*)

Ans. Due to greater electronegativity of nitrogen, the π electron cloud is displaced slightly towards nitrogen and this causes the carbons of pyridine ring to have smaller electron density (than the carbons of benzene) and consequently deactivates pyridine towards electrophilic substitution reactions. This is further deactivated, because the basic nitrogen atom of pyridine is readily attacked by the protons present in the reaction mixture (*e.g.* nitration and sulphonation, or the reacting electrophile itself (say Y^+) to yield *pyridinium ion* having a positive charge on nitrogen.

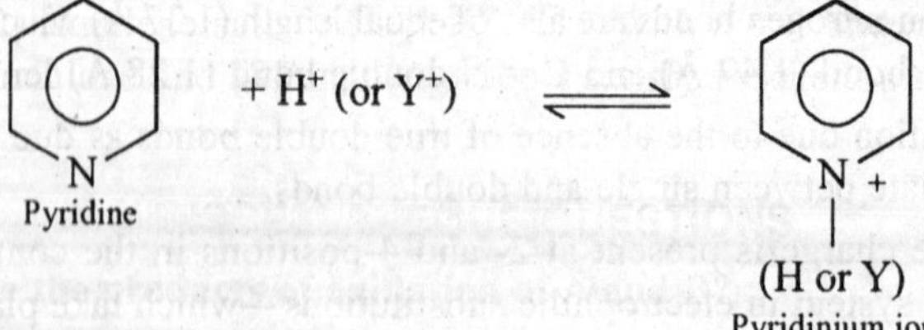

Hence, nitrogen withdraws electrons from the ring quite strongly and makes it very difficult for the electrophilic reagent to attack the ring.

Q. 29. While benzene undergoes exclusively electrophilic substitution reactions, pyridine can undergo electrophilic and nucleophilic substitution reactions. Explain.

(*M. Dayanand, 2000*)

Ans. Pyridine undergoes nucleophilic addition reactions because the intermediate carbanion gets stabilised due to special features. For example in the reaction of pyridine with sodamide, which is an example of nucleophilic substitution reaction, one of the resonating structures of the intermediate carbanion, when the attack is at number 2 position, is:

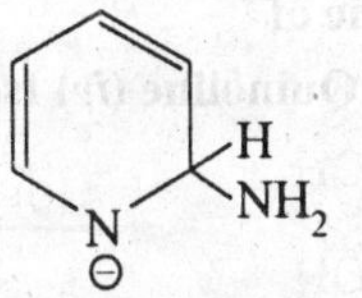

H

NH_2

N

⊖

This is a particularly stable structure because nitrogen an electronegative element can easily absorb the negative charge.

No such stable intermediate carbanions are obtained with benzene. Hence benzene does not exhibit nucleophilic substitution reactions.

Q. 30. Explain why pyridine is more basic than pyrrole? *(M. Dayanand, 2000; Kerala, 2001; Delhi, 2007; Nagpur. 2008)*

Ans. (*i*) In case of pyridine, two sp^2 orbitals of nitrogen overlap with the sp^2 orbitals of two carbons on either side. The unhybridised orbital forms part of delocalised electron cloud. The third sp^2 orbital of nitrogen is unshared and is available to the acids.

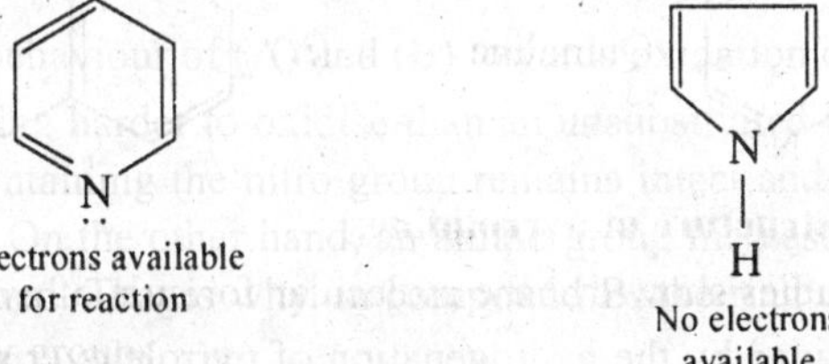

(*ii*) In case of pyrrole, all the electrons are shared. No electrons is free and available to react with acids. Hence pyridine is more basic than pyrrole.

Q. 31. Pyrrole is more reactive than pyridine towards electrophilic substitution reactions. Explain.

Ans. Pyrrole is more reactive than pyridine towards electrophilic substitution because the intermediate carbocation formed with pyrrole is more stable as every atom has a complete octet in one of the intermediate structures as shown below:

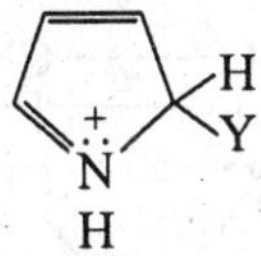

H

Y

$\overset{+}{\ddot{N}}$

H

Here Y^+ is the electrophile.

In pyridine, the intermediate carbocation, after attachment of the electrophile has nitrogen with six electrons and hence less stable.

Q. 32. How will you synthesise the following compounds:

(*i*) Piperidine (*ii*) 3- Nitropyridine?

Ans.

(*i*) Pyridine $+ 3H_2$ $\xrightarrow[\text{or Sod./Alcohol}]{\text{Ni or Pt}}$ Piperidine

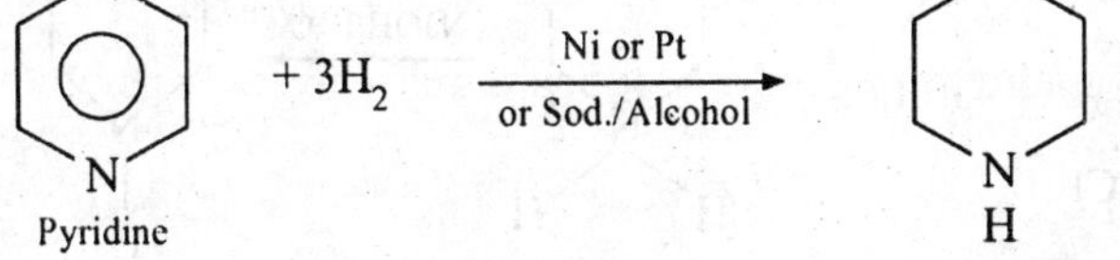

(*ii*) Pyridine $\xrightarrow[\text{573 K}]{KNO_3,\ H_2SO_4}$ 3-Nitropyridine

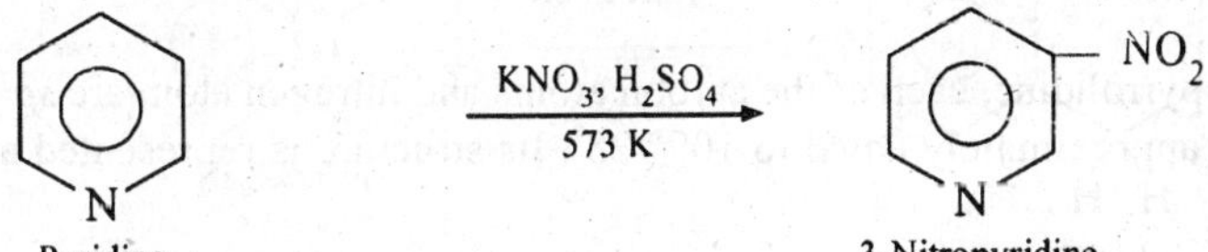

Q. 33. Give the structural formulae of
(*i*) Pyrrolidine (*ii*) Piperidine (*iii*) Quinoline (*iv*) Isoquinoline. (*Bangalore, 2004*)

Ans. (*i*) 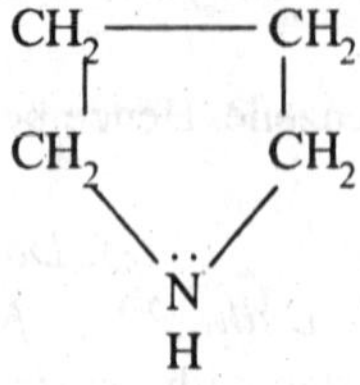 Pyrrolidine (*ii*) 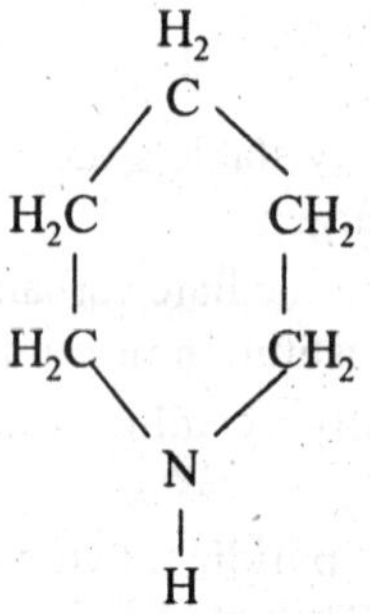 Piperidine

(*iii*) 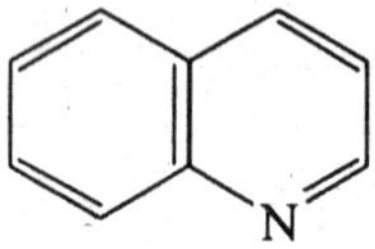 Quinoline (*iv*) 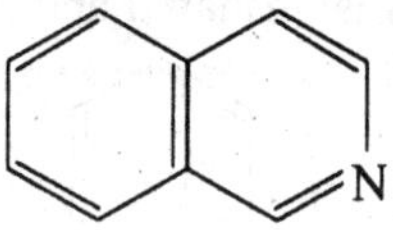 Isoquinoline

Q. 34. Establish the structure of pyrrolidine.

Ans. 1. Analytical studies show that the molecular formula of pyrrolidine is C_4H_9N.

2. Pyrrolidine is obtained by the hydrogenation of pyrrole over nickel at about 473 K.

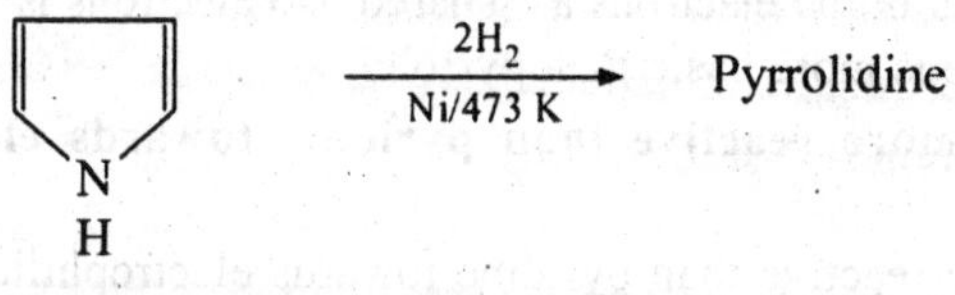

$$\text{Pyrrole} \xrightarrow[\text{Ni/473 K}]{2H_2} \text{Pyrrolidine}$$

Structure of pyrrole is known. It contains two double bonds. From this the structure of pyrrolidine can be derived as follows:

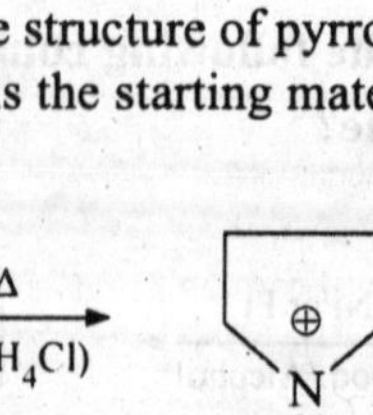

3. Confirmation by synthesis. The structure of pyrrolidine has been confirmed by synthesis. Tetramethylene diamine hydrochloride is the starting material.

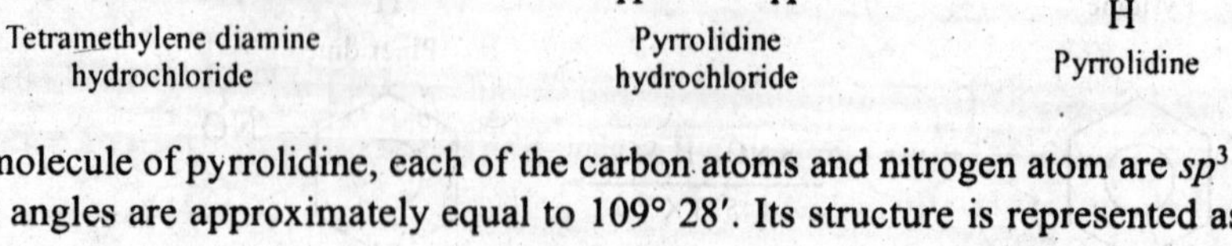

$$\text{Tetramethylene diamine hydrochloride} \xrightarrow[-(NH_4Cl)]{\Delta} \text{Pyrrolidine hydrochloride} \xrightarrow{NaOH/H_2O} \text{Pyrrolidine}$$

In the molecule of pyrrolidine, each of the carbon atoms and nitrogen atom are sp^3 hybridised and all bond angles are approximately equal to 109° 28′. Its structure is represented as:

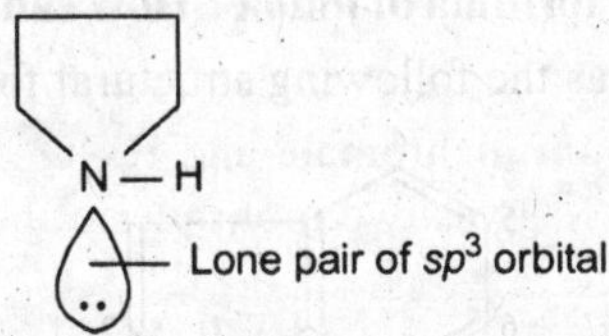

Q. 35. Establish the structural formula of piperidine.

Ans. 1. The molecular formula of piperidine, from molecular wt. determination and elemental analysis, is found to be $C_5H_{11}N$.

2. It is obtained by the hydrogenation of pyridine over nickel at about 473 K.

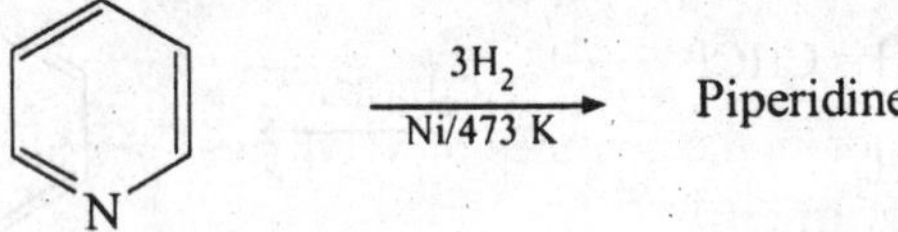

The structure of pyridine is known. It contains three double bonds.

From this, the structure of piperidine can be derived as follows:

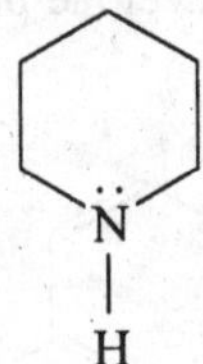

3. The structure of piperidine has been confirmed by synthesis starting from pentamethylene diamine hydrochloride.

CH_2, H_2C, CH_2, H_2C, CH_2, $\overset{+}{Cl}NH_3$, $\overset{+}{N}H_3Cl^-$ $\xrightarrow[-(NH_4Cl)]{\Delta}$ Piperidine hydrochloride ($\overset{\oplus}{N}H_2$ Cl^-) $\xrightarrow{NaOH/H_2O}$ Piperidine (N—H)

Pentamethylene diamine hydrochloride — Piperidine hydrochloride — Piperidine

In the structure of piperidine, each of the carbon atoms and nitrogen atom are sp^3 hybridised and all bond angles are approximately equal to 109° 28′. It adopts a conformation similar to the chair form of cyclohexane.

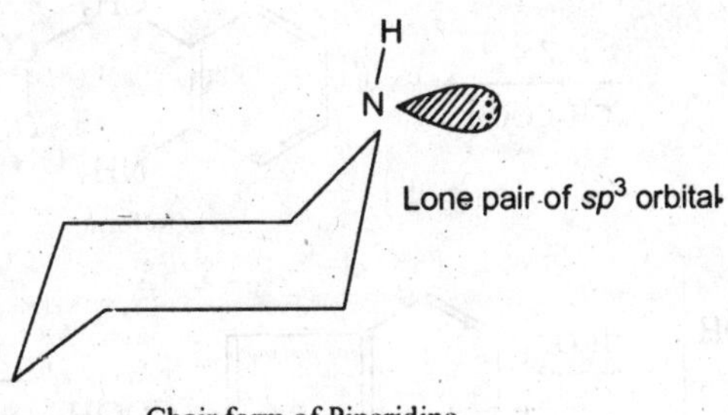

Chair form of Piperidine

Q. 36. What is the structural formula of indole? How can it be synthesised? (*Nagpur 2008*)

Ans. Indole (benzopyrrole) has the following structural formula

4, 5, 6, 7, 3, 2, 1 NH

Different positions on the molecule of indole are numbered as shown above.

It can be synthesised in a number of ways.

1. Lipp's synthesis. It is carried out by heating *o*-amino-ω-chlorostyrene with sodium ethoxide.

$CH = CHCl$, NH_2 $\xrightarrow{C_2H_5ONa}$ NH

2. Fischer's indole synthesis. It is carried out by heating phenylhydrazone or substituted phenylhydrazone of an appropriate aldehyde or ketone with zinc chloride as catalyst. Synthesis of 2-methyl indole can be achieved by taking acetone phenylhydrazone as shown below:

CH_3, $C - CH_3$, N, NH $\rightleftharpoons$ CH_2, $C - CH_3$, N, NH

Phenyl hydrazone of acetone (Tautomer form)

$\longrightarrow$ CH, $C - CH_3$, NH_2, NH_2 $\xrightarrow{-NH_3}$ NH, CH_3

2-methyl indole

3. Reissert synthesis. This is carried out with *o*-nitrotoluene (or its derivatives) and ethyl oxalate as follows:

CH_3, NO_2 + $COOC_2H_5$, $COOC_2H_5$ $\xrightarrow{C_2H_5ONa}$ $CH_2COCOOC_2H_5$, NO_2 $\xrightarrow{HCl}$

o-nitrotoluene Diethyl oxalate

$CH_2COCOOH$, NO_2 $\xrightarrow[CH_3COOH]{Zn}$ CH_2, CO, COOH, NH_2 $\rightleftharpoons$

keto form

[CH, C – OH, COOH, NH_2] $\xrightarrow{-H_2O}$ NH, COOH $\xrightarrow[-CO_2]{Heat}$ NH

enol form Indole

Q. 37. Describe the structure of indole.

Ans. All the ring atoms in indole are sp^2 hybridised. sp^2 hybrid orbitals overlap with each other and with s orbitals of hydrogen to form C – C, C – N, C – H and N – H σ bonds. Each ring atom also possesses a p-orbital. These are perpendicular to the plane of the ring. Lateral overlap of these p orbitals produces a π molecular orbital containing 10 electrons. Indole is aromatic because it satisfies $4n + 2$ Huckel's rule. Here $n = 2$.

Indole is a hybrid of several resonance structures as shown below:

Q. 38. Describe important chemical properties of indole.

Ans. Important chemical properties of indole are given as under:

1. Nitration

Indole + $C_2H_5O\overset{+}{N}O_2$ $\xrightarrow{5°C}$ 3-nitroindole

All the electrophilic substitution reactions take place at position – 3 because the intermediate product is stabilized more at this position than at position – 2.

2. Sulphonation

Sulphonation is found to occur at position -2

Indole + SO_3 $\xrightarrow[\Delta]{\text{pyridine}}$ Indole-3-sulphonic acid (SO_3H ← Unusual orientation)

3. Bromination

Indole + Br_3 $\xrightarrow[\Delta]{\text{dioxane}}$ 3-bromoindole

4. Friedel crafts alkylation

Indole + CH_3I $\xrightarrow[\text{Dimethyl sulphoxide}]{\text{DMF, }\Delta}$ 3-methylindole

5. Friedel crafts acylation

Indole + CH_3COCl $\xrightarrow{SnCl_4}$ 3-acetylindole ($COCH_3$ at position 3)

3-acetylindole

6. Diazocoupling

Indole + $C_6H_5N_2Cl$ (Benzene diazonium chloride) $\longrightarrow$ 3-phenylazoindole ($-N=N-C_6H_5$) + HCl

7. Reimer Tiemann formylation

Indole + $CHCl_3$ $\xrightarrow{NaOH}$ Indole-3-aldehyde (CHO) + 3-chloroquinoline (Cl)

8. Oxidation. Indole is oxidised by ozone in formamide to give 2-formamido benzaldehyde

Indole + O_3 $\xrightarrow{H-\overset{O}{\overset{\|}{C}}-NH_2}$ 2-Formamido benzaldehyde (CHO, NH–CHO)

2-Formamido benzaldehyde

9. Reduction. Reduction under different conditions give different products as shown below:

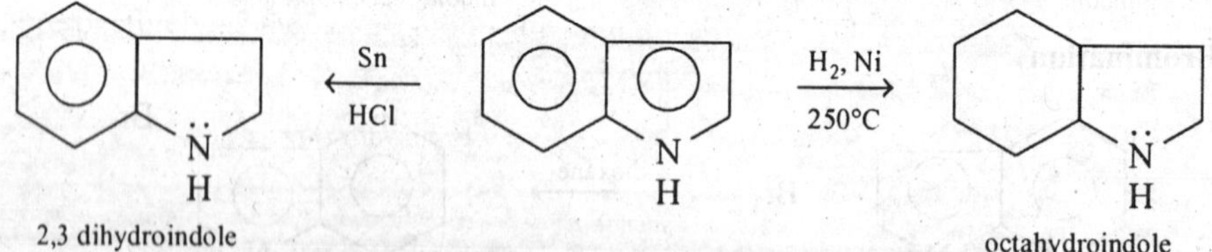

Q. 39. Establish the structure of quinoline.

(*Garhwal, 2000; Kerala 2001*)

Ans. 1. Molecular formula. From the elemental analysis and molecular mass determination, the molecular formula of quinoline is found to be C_9H_7N.

2. Electrophilic substitution reactions. It undergoes electrophilic substitution reactions such as nitration, sulphonation etc. like benzene. This property shows the presence of one or more benzene rings in the molecule.

3. Action of alkyl halide. When treated with alkyl halide, it forms a quaternary salt. This shows the presence of tertiary nitrogen atom in it.

4. Oxidation. On oxidation with alkaline potassium permanganate, it forms quinolinic acid and oxalic acid.

C_9H_7N (Quinoline) $\xrightarrow{\text{Alk. } KMnO_4}$ Quinolinic acid + Oxalic acid (COOH–COOH)

It appears that in quinoline, one pyridine ring is fused with a benzene ring which is broken during oxidation. Thus, the structural formula of quinoline can be written as:

Quinoline

Confirmation by Synthesis. The structure of quinoline has been confirmed by synthesis.

Friedlander's Synthesis. Quinoline is obtained by condensing *o*-amino benzaldehyde with acetaldehyde in sodium hydroxide solution.

o-Amino Benzaldehyde + Acetaldehyde $\xrightarrow[-H_2O]{NaOH}$ $\xrightarrow[-H_2O]{\Delta}$ Quinoline

Skraup Synthesis. A mixture of aniline and glycerol is heated in the presence of sulphuric acid and a mild oxidising agent, usually nitrobenzene.

Aniline + Glycerol (CH_2OH–$CHOH$–CH_2OH) $\xrightarrow[C_6H_5NO_2,\ \Delta]{\text{Conc. } H_2SO_4}$ Quinoline

All ring atoms in quinoline are sp^2 hybridised. Like pyridine, the electron pair on nitrogen atom resides in an sp^2 orbital and is not involved in the formation of the delocalised π-molecular orbital. Due to this electron pair, quinoline can readily form an N-oxide and undergoes quaternisation. It satisfies the Huckel's rule as it has 10 delocalised π electrons.

Quinoline exists in the following resonating structures:

(1) (2) (3)

(4) (5) (6) (7)

Q. 40. Establish the structure of isoquinoline. (*Bhopal, 2004*)

Ans. 1. Molecular formula. From the elemental analysis and molecular mass determination, the molecular formula of isoquinoline is found to be C_9H_7N.

2. Electrophilic substitution reactions. It undergoes electrophilic substitution reactions in the same way as benzene. This points to the presence of one or more aromatic rings in the molecule.

3. Action with alkyl halide. It reacts with alkyl halide to form a quaternary salt. This shows the presence of a tertiary nitrogen atom in it.

4. Oxidation. When oxidised with potassium permanganate, it gives a mixture of phthalic acid and cinchomeronic acid.

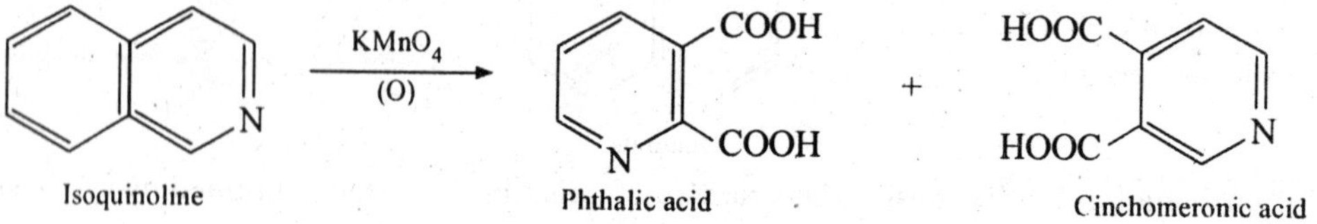

Isoquinoline $\xrightarrow[(O)]{KMnO_4}$ Phthalic acid (COOH, COOH, N) + Cinchomeronic acid (HOOC, HOOC, N)

Obviously isoquinoline molecule consists of one benzene ring fused with a pyridine ring in β-γ position.

In cinchomeronic acid, the positions of two —COOH groups show the positions where benzene ring is fused. From all these observation the structural formula of isoquinoline can be written as:

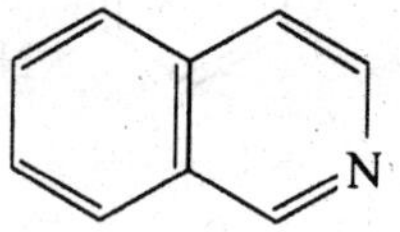

Isoquinoline

It may be noticed that

(*i*) Phthalic acid is produced when pyridine ring is ruptured and

(*ii*) Cinchomeronic acid is obtained when benzene ring is ruptured.

Confirmation by Synthesis. The structure of isoquinoline can be confirmed by synthesis.

(*a*) From Cinnamaldehyde. It involves the condensation of cinnamaldehyde with hydroxylamine to form the corresponding oxime. The oxime on heating with P_2O_5 yields isoquinoline.

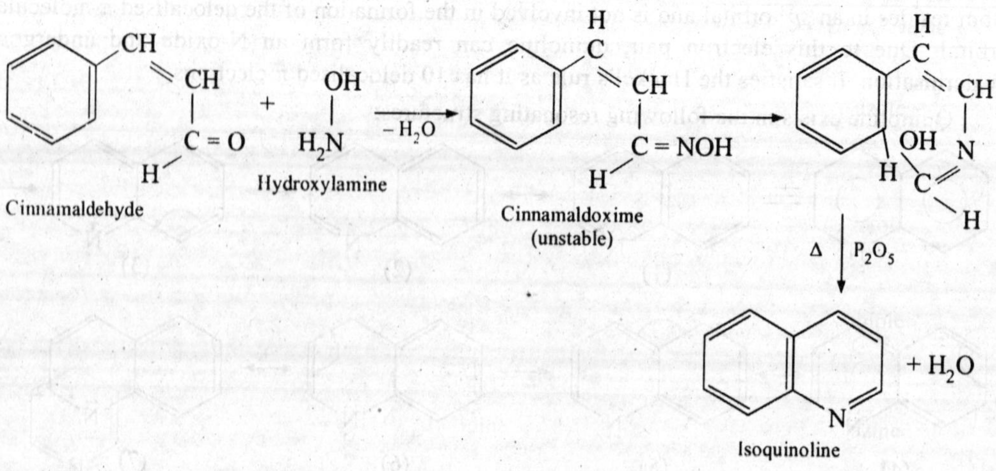

Cinnamaldehyde ($C_6H_5CH{=}CH{-}CHO$) + Hydroxylamine ($H_2N{-}OH$) $\xrightarrow{-H_2O}$ Cinnamaldoxime (unstable) ($C_6H_5CH{=}CH{-}CH{=}NOH$) ⟶ $\xrightarrow[\Delta]{P_2O_5}$ Isoquinoline $+ H_2O$

(b) Bischler-Napieralski Synthesis. The starting material for the synthesis of isoquinoline is 2-Phenylethylamine.

2-Phenylethylamine + Formyl chloride $\xrightarrow[(-\,HCl)]{\bar{O}H}$ N-Formyl-2-Phenylethylamine $\rightleftharpoons$ (iminol form) $\xrightarrow[-H_2O]{P_2O_5}$ 3, 4-Dihydroisoquinoline $\xrightarrow[-H_2]{Se/\Delta}$ Isoquinoline

All the ring atoms in isoquinoline are sp^2 hybridised. The electron pair on nitrogen atom resides in an sp^2 orbital and is not involved in the formation of the delocalised π-molecular orbital. Isoquinoline constitutes delocalised 10 π-electron system and satisfies the Huckel's rule. Isoquinoline can be written in the following resonating structures:

MISCELLANEOUS QUESTIONS

Q. 41. Electrophilic substitution in pyrrole takes place at 2-position whereas in pyridine at 3 position. Why?

Ans. This can be explain in terms of stability of intermediate carbonium ions. For details see Qs. 17 and 24.

Q. 42. Electrophilic substitution in 5-membered heterocyclic ring is easier than in pyridine. Why?

Ans. Nitrogen being more electronegative than carbon draws the electrons towards itself in the molecule of pyridine. This is possible because there is a double bonds attached to nitrogen in pyridine. This reduces the availability of π-electrons on the ring which is so essential for electrophilic substitution. However in 5-member heterocyclics (say pyrrole), this is not effectively possible because, there is no double bond (or π-electrons) in contact with it.

Q. 43. In acidic solutions, pyridine undergoes electrophilic substitution with great difficulty. Explain.

Ans. Nitrogen atom in pyridine gets protonated due to its electronegative nature as shown below:

Protonated nitrogen attracts the π-electrons from the ring. This reduces the electron density on the ring, thereby reducing chances of electrophilic attack.

N + H^+ ⟶ $\overset{+}{N}$–H

Q. 44. Explain the aromatic character of pyrrole on the basis of its resonance and molecular orbital structures. *(Meerut, 2000)*

Ans. There are two carbon-carbon bonds in the molecule. This constitutes four π-electrons. There is a lone pair of electrons on nitrogen. This lone pair joins the four π-electrons, making a total of 6 π-electrons. Huckel's (4*n* + 2) rule for aromaticity is satisfied.

In terms of molecular orbital structure for pyrrole, there is a cycle of electron clouds, above and below the pyrrole ring, giving it aromatic nature.

Q. 45. Explain the reactions of pyridine with:

(*i*) KNO_3 and conc. H_2SO_4

(*ii*) Sodamide

(*iii*) Phenyllithium *(Garhwal, 2000; Lucknow, 2000)*

Ans. (*i*) This is electrophilic substitution reaction

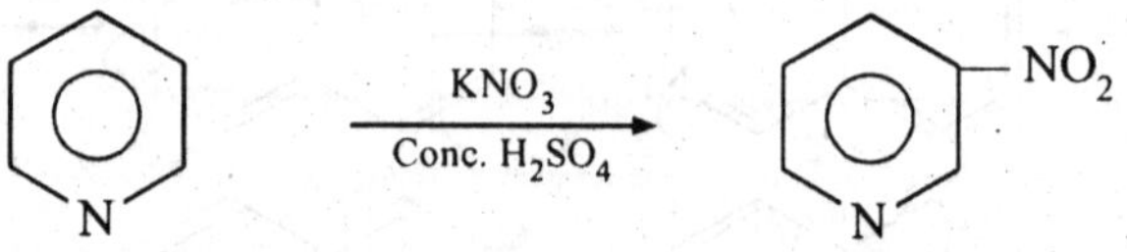

3-Nitropyridine

(*ii*) This is nucleophilic substitution reaction

N —$NaNH_2$⟶ N, NH_2

2-Aminopyridine

This is also nucleophilic substitution reaction.

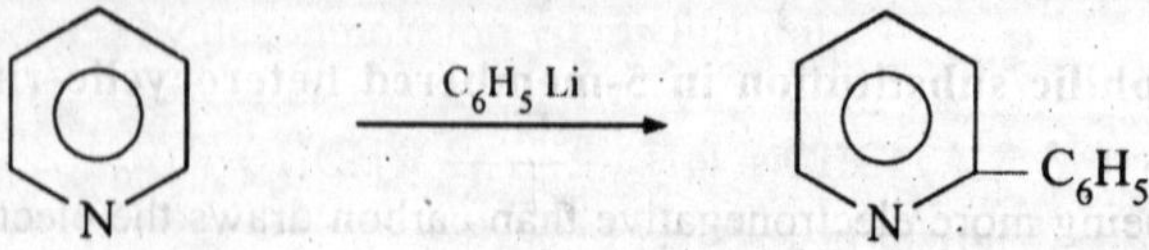

2-Phenylpyridine

Q. 46. How will you obtain

(*i*) Pyrrole from furan

(*ii*) Pyridine from pyrrole?

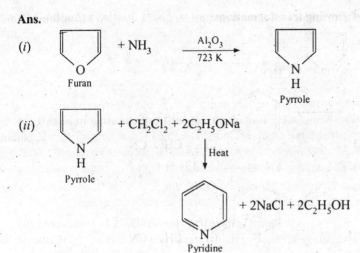

Ans.

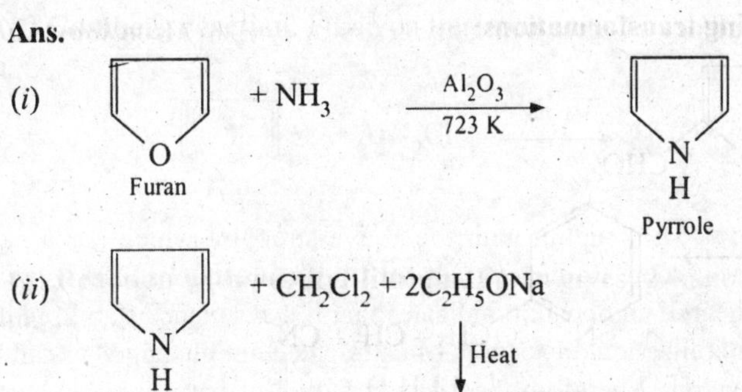

+ $2NaCl + 2C_2H_5OH$

Pyridine

Q. 47. Why pyrrole, furan and thiophene are classified as aromatics?

(Kurukshetra, 2001; Mumbai, 2010)

Ans. A compound is called aromatic if it satisfies the $(4n + 2)$ Huckel's rule *i.e.* the no. of π electrons in the molecule is $4n + 2$ where n is an integer including zero. In pyrrole, furan and thiophene molecules, there are two carbon-carbon double bonds *i.e.* 4 π electrons. The lone pair of electrons on nitrogen, oxygen and sulphur also joins the four π electrons to make it $2 + 4 = 6$. Thus Huckel's rule is satisfied.

Q. 48. How do you account for the formation of 3-nitropyridine when pyridine is treated with KNO_3/H_2SO_4 at 573 K and 2-aminopyridine when pyridine is treated with $NaNH_2$ at 373 K.
(Kerala, 2000; Calcutta, 2000)

Ans. (*i*) Reaction of pyridine with KNO_3/H_2SO_4 at 573 K is an electrophilic substitution reaction.

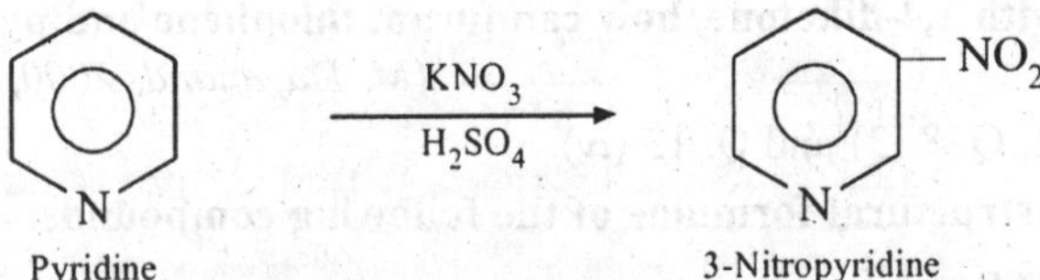

In such reactions, the electrophile preferably attaches itself at position-3. This can be explained in terms of stability of intermediate products.

(*ii*) Reaction of pyridine with $NaNH_2$ at 373 K is a nucleophilic substitution reaction.

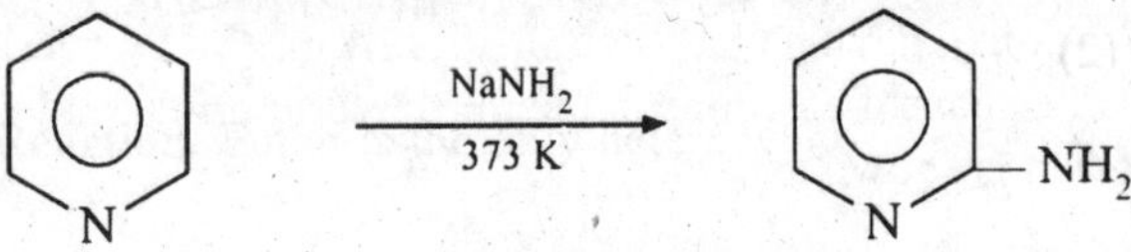

2-Aminopyridine

Again this can be explained in terms of stability of intermediate ion.

Q. 49. Quinoline undergoes electrophilic attack in the benzene ring while the nucleophilic attack takes place in the pyridine ring. Explain.

Ans. Quinoline molecule consists of two rings, the benzene ring and the pyridine ring Pyridine ring gives electrophilic reactions less readily (see Q. 28). Pyridine ring, on the other hand. gives nucleophilic reactions readily because of the formation of a stable intermediate structure (see Q. 29).

Q. 50. Account for the following transformations: (*Lucknow, 2000*)

(*i*) Furan $\longrightarrow$ Furan-2-CHO

(*ii*) 2-vinylpyridine ($CH=CH_2$) $\longrightarrow$ 2-pyridyl CH_2-CH_2-CN

Ans. (*i*) See Q. 6 (1), (*e*).

(*ii*) 2-vinylpyridine ($CH=CH_2$) $\xrightarrow{HCN}$ 2-pyridyl CH_2-CH_2-CN

Q. 51. Discuss the aromaticity of five membered heterocyclic compounds. (*Lucknow, 2000*)

Ans. See Q. 47.

Q. 52. Draw resonating structure of pyrrole, thiophene and furan. (*Kerala, 2001; Lucknow, 2000*)

Ans. See Q. 15, Q. 10 and Q. 7.

Q. 53. Thiophene is more aromatic in nature than furan. Explain. (*M. Dayanand, 2000*)

Or

Explain the following:

Furan < Pyrrole < Thiophene. (*Garhwal, 2000*)

Ans. See Q. 19.

Q. 54. Starting with 1,4-diketone, how can furan, thiophene and pyrrole derivatives be synthesised? (*M. Dayanand, 2000; Kurukshetra, 2001*)

Ans. See Q. 5 (*ii*), Q. 8 (3) and Q. 13 (*iv*).

Q. 55. Write the structural formulae of the following compounds: (*Punjab, 2000*)

(*i*) tetrahydrofuran

(*ii*) α-pyridone

(*iii*) isoquinoline

(*iv*) 2-acetyl furan

Ans. (*i*) See Q. 6 (2)

(*ii*) See Q. 23 (5)

(*iii*) See Q. 33 (*iv*)

(*iv*) See Q. 6 (1) (*d*)

Q. 56. Pyrrole is very weakly basic as well as weakly acidic in nature. Explain. (*Punjab, 2000*)

Ans. See Q. 18.

Q. 57. Electrophilic substitution in pyrrole takes place at 2-position, whereas in pyridine at 3-position. Explain.

Ans. See Q. 17 and Q. 24.

Q. 58. Write down the structures and names of the products of the following reactions:

(*i*) **Thiophene** $\xrightarrow{CH_3COCl}$

(*ii*) **Furan** $\xrightarrow{H_2 / Ni}$

(*iii*) **Pyrrole** $\xrightarrow{CHCl_3 / KOH}$

(*iv*) **Pyridine** $\xrightarrow{Br_2 / 575\ K}$

Ans. (*i*) See Q. 9 (1) (*e*).

(*ii*) See Q. 6 (2).

(*iii*) See Q. 14 (1) (*e*).

(*iv*) See Q. 23 (4).

Q. 59. Identify compounds A and B in the following reactions:

(*i*) **Pyrrole + $CHCl_3$ + Sodium ethoxide** $\longrightarrow$ **A** (*Kerala, 2001*)

(*ii*) **Furan + maleic anhydride** $\longrightarrow$ **A** $\xrightarrow[\Delta]{HBr}$ **B** (*Kerala, 2001*)

(*iii*) **Pyridine + acetyl chloride** $\xrightarrow{anhy.\ AlCl_3}$ **A** (*Garhwal, 2001*)

(*iv*) **Furan** $\xrightarrow[Ni]{H_2}$ **A** (*Kerala, 2001*)

(*v*) **Quinoline** $\xrightarrow[KMnO_4]{alkaline}$ **A** (*Kerala, 2001*)

(*iv*) **Pyridine** $\xrightarrow[anhy.\ AlCl_3]{CH_3I}$ (*Awadh, 2000*)

Ans. (*i*) See Q. 14 (1) (e).

(*ii*) See Q. 6 (3).

(*iii*)
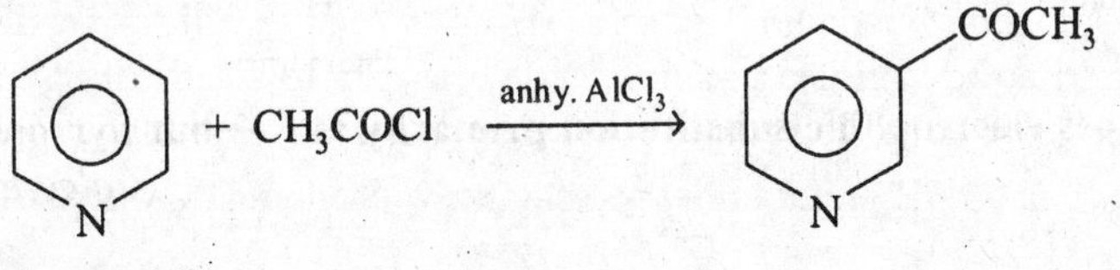

(*iv*) See Q. 6 (2).

(*v*) See Q. 39 (4).

(*vi*)
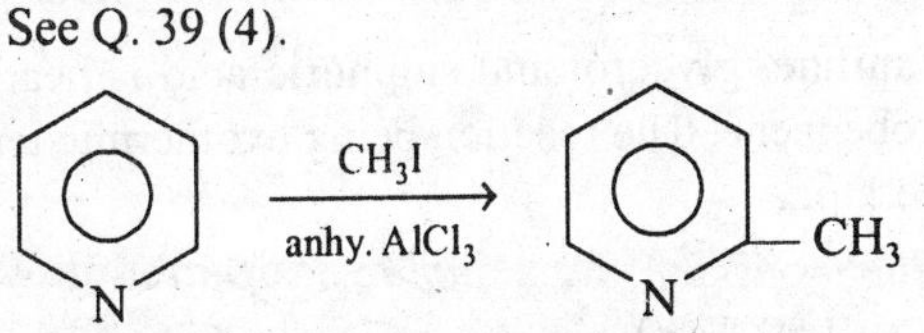

Q. 60. Pyrrole is an aromatic compound and resembles phenol in properties. Explain. (*Kanpur, 2001*)

Ans. See Q. 44.

Q. 61. Explain, why pyridine is more basic than aniline. (*Kanpur, 2001*)

Ans. As in Q. 30.

Q. 62. How does pyridine reacts with: (*Coimbatore, 2000*)

(*i*) Sod amide

(*ii*) methyl iodide

(*iii*) sodium and ethanol

Ans. (*i*) See Q. 23 (5).

(*ii*) See Q. 23 (1).

(*iii*) See Q. 23 (2).

Q. 63. Write a note on Bischler-Napieralski synthesis. (*Coimbatore, 2000*)

Ans. See Q. 40, 4 (*b*).

Q. 64. How will you do the following conversions?

(*i*) Pyrrole ⟶ an open chain compound (*Meerut, 2000*)

(*ii*) Pyrrole ⟶ a pyridine derivative (*Meerut, 2000*)

(*iii*) Pyridine ⟶ α-picoline (*Awadh, 2000*)

Ans. (*i*) See Q. 14 (6).

(*ii*) See Q. 14 (4).

(*iii*) Pyridine $\xrightarrow[\text{anhy. } AlCl_3]{CH_3I}$ 2-methylpyridine (C_5H_4N–CH_3) (α-Picoline)

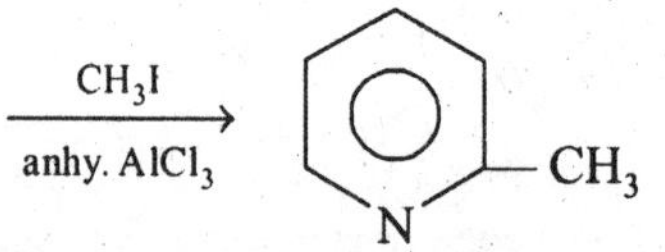

Q. 65. Explain, why furan is an enol ether? (*Calcutta, 2000*)

Ans. Furan is considered to be enol ether of the following structures:

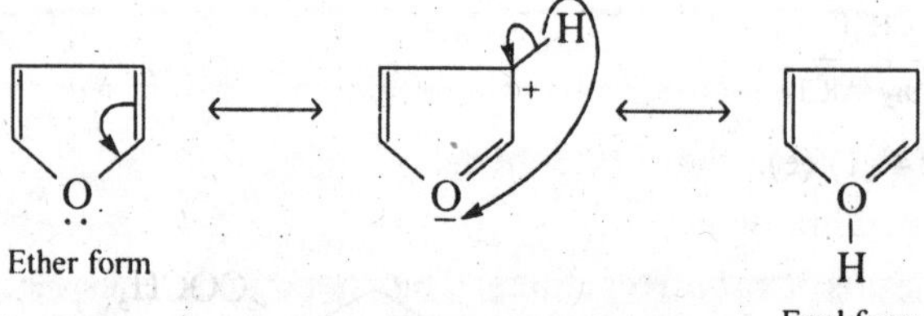

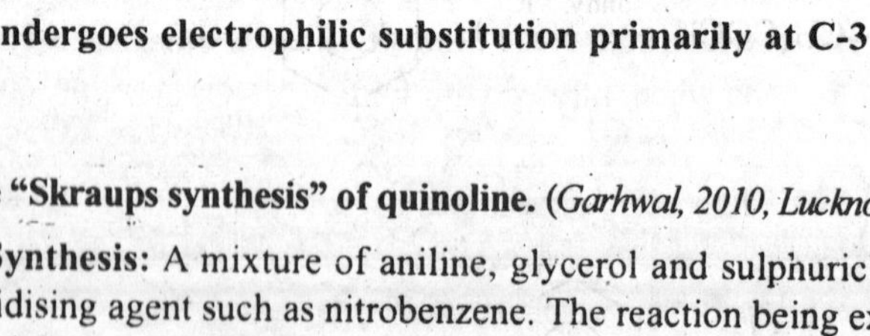

Q. 66. Indole undergoes electrophilic substitution primarily at C-3 but pyrrole at C-2. Explain, why? (*Calcutta, 2000*)

Ans. See Q. 17.

Q. 67. Describe "Skraups synthesis" of quinoline. (*Garhwal, 2010, Lucknow, 2010; Nagpur 2008*)

Ans. Skraups Synthesis: A mixture of aniline, glycerol and sulphuric acid is heated in the presence of a mild oxidising agent such as nitrobenzene. The reaction being exothermic tends to be violent and ferrous sulphate is added as moderator.

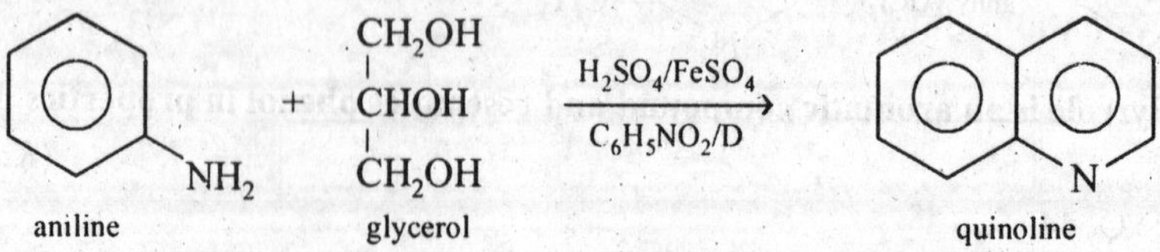

$$\underset{\text{aniline}}{C_6H_5NH_2} + \underset{\text{glycerol}}{CH_2OH\text{–}CHOH\text{–}CH_2OH} \xrightarrow[C_6H_5NO_2/D]{H_2SO_4/FeSO_4} \underset{\text{quinoline}}{C_9H_7N}$$

Q. 68. Explain the formation of 3-chloropyridine during the Reimer Tiemann reaction of pyrrole. (*Delhi, 2003*)

Ans.

Pyrrole (N–H) $\xrightarrow{:CCl_2}$ dichlorocyclopropane adduct (N–H, Cl, Cl) $\xrightarrow[-Cl^-]{OH^-}$ 3-chloropyridine (N, Cl)

3-chloropyridine

Q. 69. Pyrrole is much more acidic than *sec*-alkylamine, suggest a reason. (*Delhi, 2003*)

Ans. Pyrrole anion can stabilize itself while *sec*-alkyl amine cannot.

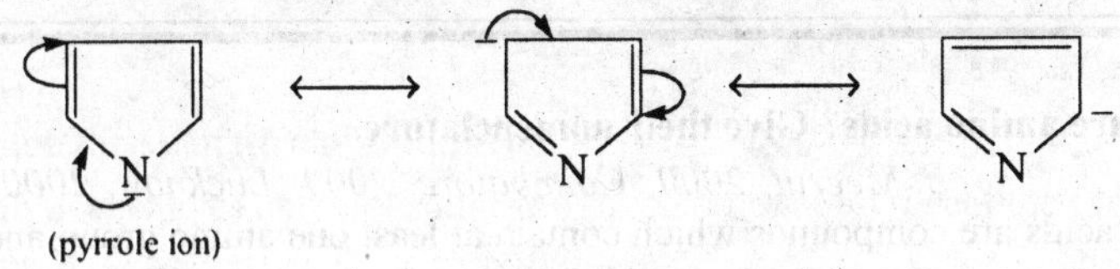

$CH_3 - \bar{N} - CH_3$

Sec-amine anion

Sec-amine anion rather gets destabilised due to the electron repelling effect of — CH_3 groups.

35

AMINO ACIDS AND PROTEINS

Q. 1. What are amino acids? Give their nomenclature.

(Meerut, 2000; Coimbatore, 2000; Lucknow, 2000; Nagpur, 2008)

Ans. Amino acids are compounds which contain at least one amino group and one carboxylic group. The amino group could be linked to carbon just next to carboxylic group or with a gap of one or two carbons. They are called α, β or γ acids respectively. Some examples of amino acids are:

$$\underset{\displaystyle NH_2}{\underset{|}{CH_2}} — COOH$$

α - amino acetic acid (or 2-aminoethanoic acid)

$$\underset{\displaystyle NH_2}{\underset{|}{CH_2}} — CH_2 — COOH$$

β - amino propionic acid (or 3-aminopropanoic acid)

$$\underset{\displaystyle NH_2}{\underset{|}{CH_2}} — CH_2 — CH_2 — COOH$$

γ - amino butyric acid (or 4-aminobutanoic acid)

Common system of nomenclature is still more popular in naming amino acids and proteins. It is mostly α-amino acids, which are involved in the building up of protein molecules. The simplest α-amino acid is α-amino acetic acid also called *glycine*. In general, an α-amino acid can be represented as

$$R — \underset{\displaystyle NH_2}{\underset{|}{CH}} — COOH$$

where R is an alkyl or aryl group. It could also represent highly branched or unsaturated carbon chain or heterocyclic ring.

Q. 2. Give evidence supporting dipolar nature of amino acids.

Or

Explain the term Zwitter ion. *(Coimbatore, 2000; Kerala, 2000; Himachal 2000; 2001; Kurukshetra, 2001; Delhi, 2003; Panjab 2003)*

Or

Why are amino acids called amphoteric compounds?

Ans. It has been found that an amino acid molecule appears as a dipole, one part of it carrying positive charge and the second negative charge. The dipolar ionic structure of amino acids can be represented as:

$$R — \underset{\displaystyle +NH_3}{\underset{|}{CH}} — COO^-$$

This is also called a *Zwitter ion* or *Internal salt*. There is no free amino or carboxylic group present in the molecule. This is supported by the following evidence:

1. Amino acids are insoluble in non-polar solvents and soluble in polar solvents like water. This behaviour can be expected of the polar substances.
2. Amino acids are non-volatile crystalline solids, which melt at high temperature. This is quite like ionic substances which have high melting points and unlike amines and carboxyiic acids which have low melting points.
3. They have high dipole moments indicating polar nature of the molecule.
4. Dissociation constant K_a and K_b give us an idea about the acid and base strengths. Amino acids have very low values of K_a and K_b indicating that the molecule does not possess these groups in tiie normal forms.

Q. 3. Give explanation of low values of K_a and K_b of amino acids.

Ans. Because of the dipolar nature of amino acids, it is the substituted ammonium ion, and not the carboxylic group, which acts as a proton donor and it is the acidic centre. K_a, therefore, refers to the acidity of substituted ammonium ion, as shown below:

$$\overset{+}{N}H_3-\underset{\underset{R}{|}}{CH}-COO^- + H_2O \rightleftharpoons NH_2-\underset{\underset{R}{|}}{CH}-COO^- + H_3O^+$$

$$K_b = \frac{\left[NH_2-\underset{\underset{R}{|}}{CH}-COO^-\right][H_3O^+]}{\left[\overset{+}{N}H_3-\underset{\underset{R}{|}}{CH}-COO\right]}$$

The concentration of H_2O remains constant and hence it is not taken up in the denominator.

Similarly it is the carboxylate ion, and not the amino group, which acts as the proton acceptor and it is the basic centre. K_b, therefore, refers to the basicity of carboxylate ion as given below:

$$\overset{+}{N}H_3-\underset{\underset{R}{|}}{CH}-COO^- + H_2O \rightleftharpoons \overset{+}{N}H_3-\underset{\underset{R}{|}}{CH}-COOH + OH^-$$

or

$$K_b = \frac{\left[\overset{+}{N}H_3-\underset{\underset{R}{|}}{CH}-COOH\right][OH^-]}{\left[\overset{+}{N}H_3-\underset{\underset{R}{|}}{CH}-COO^-\right]}$$

The measured values of K_a and K_b of amino acids, of say, glycine (α-amino acetic acid), can be justified, in terms of dipolar structure, by calculating the values for conjugate bases or conjugate acids and comparing them with observed values.

Let us take into consideration K_a and K_b values for glycine $H_3\overset{+}{N}CH_2COO^-$. It shows K_a values equal to 1.6×10^{-10}. K_b of its conjugate base $NH_2-CH_2-COO^-$ can be calculated by using the generalisation that the product of K_a and K_b in an aqueous solution is 1×10^{-14} Thus,

$$K_a \times K_b = 1.0 \times 10^{-14}$$

$$1.6 \times 10^{-10} \times K_b = 1.0 \times 10^{-14}$$

or
$$K_b = \frac{1.0 \times 10^{-14}}{1.6 \times 10^{-10}}$$
$$= 6.3 \times 10^{-3}$$

This value is quite reasonable for an aliphatic amine. Basicity constant K_b of glycine is measured as 2.5×10^{-12}. Acidity constant K_a of its conjugate acid $\overset{+}{N}H_3 CH_2COOH$ can be calculated as per the above criteria.

$$K_a \times K_b = 1.0 \times 10^{-14}$$
$$K_a \times 2.5 \times 10^{-12} = 1.0 \times 10^{-14}$$

or
$$K_a = \frac{1.0 \times 10^{-14}}{2.5 \times 10^{-12}}$$
$$= 4 \times 10^{-3}$$

This value is quite reasonable for an aliphatic acid with an electron-withdrawing group.

Q. 4. How does the dipolar amino acid change its equilibrium in acidic or alkaline medium?

Or

What is the effect of pH on the structure of amino acids?

Ans. 1. When the solution of an amino acid is made acidic, the dipolar ion (A) gets converted nto a cation as shown below. This is because the stronger acid H_3O^+ releases a proton to the carboxylate ion and displaces a weaker acid.

$$\underset{(A)}{\overset{+}{N}H_3-\underset{\underset{R}{|}}{CH}-CO\bar{O}} + \underset{\text{Stronger acid}}{H_3O^+} \rightleftharpoons \underset{\underset{(I)}{\text{(Weaker acid)}}}{\overset{+}{N}H_3-\underset{\underset{R}{|}}{CH}-COOH} + H_2O$$

2. Similarly when the solution of amino acid is made alkaline with NaOH, the dipolar ion is converted into an anion as shown below :

$$\underset{(A)}{\overset{+}{N}H_3-\underset{\underset{R}{|}}{CH}-CO\bar{O}} + \underset{\text{Stronger base}}{OH^-} \rightleftharpoons \underset{(II)}{NH_2-\underset{\underset{R}{|}}{CH}-CO\bar{O}} + \underset{\text{Weaker base}}{H_2O}$$

The stronger base OH^- abstracts a proton from the amino acid to give a weaker base.

It is to be borne in mind that the ions I and II shown above are in equilibrium with the dipolar ion (A) such that amino acids can still exhibit properties of amines as well as carboxylic acid.

$$\underset{(I)}{\overset{+}{N}H_3-\underset{\underset{R}{|}}{CH}-COOH} \underset{H^+}{\overset{OH^-}{\rightleftharpoons}} \underset{(A)}{\overset{+}{N}H_3-\underset{\underset{R}{|}}{CH}-COO^-} \underset{H^+}{\overset{OH^-}{\rightleftharpoons}} \underset{(II)}{NH_2-\underset{\underset{R}{|}}{CH}-CO\bar{O}}$$

Q. 5. Give two examples each of neutral amino acids, basic amino acids and acidic amino acids.

Ans. Neutral Amino Acids

Amino acids having one amino and one carboxylic groups are called neutral amino acids. Examples are glycine and alanine.

Basic Amino Acids

Amino acids containing two amino (or imino) and one carboxylic groups are called basic amino acids. Examples: lysine and arginine.

Acidic Amino Acids

Amino acids having one amino (or imino) and two carboxylic groups are called acidic amino acids. Examples are aspartic acid and glutamic acid.

Q. 6. Explain isoelectric point with reference to amino acid.

(Garhwal, 2000; Himachal 2000; A.N. Bhauguna, 2000; Coimbatore, 2000; Panjab, 2000; Delhi, 2005)

Ans. As amino acids are polar in nature, they show electrical properties. On applying electrical field to the solution of amino acids, they migrate to one or the other electrode depending upon the following factors:

(*a*) If the solution is acidic, then the equilibrium lies towards positively charged amino acid ($NH_3^+CHR - COOH$). Hence, on passing an electric current through the solution of the amino acid, it moves towards the cathode.

(*b*) If the solution is alkaline, then the equilibrium is predominantly lying towards the negatively charged amino acid ($NH_2CHR - COO^-$). Hence on passing electricity, amino acid molecule which is in the form of anion, moves towards the anode.

(*c*) At a certain pH of the solution, the anionic and cationic structures will be in equal concentrations. On passing electricity we shall observe that there is no movement of the amino acid.

The pH at which a particular amino acid does not migrate under the influence of the electrical field is called isoelectric point. Every amino acid has a characteristic isoelectric point. Glycine has an isoelectric point at pH 6.1. It may be noted that amino acids have the minimum solubility at the isoelectric point. This is because at isoelectric point, there is maximum concentration of dipolar ions which are relatively less soluble.

Q. 7. Give a list of natural amino acids. *(Delhi, 2003)*

Ans. Table below gives a list of 26 amino acids which are essential for the synthesis of proteins in living system. Some of them have been synthesised.

S.No.	*Name*	*Structure*
	Neutral	
1.	Glycine	$\underset{^+NH_3}{CH_2}\ COO^-$
2.	(+) Alanine	$CH_3\ \underset{^+NH_3}{CH}\ COO^-$
3.	(+) Valine	$(CH_3)_2\ CH\underset{^+NH_3}{CH}COO^-$

S.No.	*Name*	*Structure*
4.	(–) Leucine	$(CH_3)_2CHCH_2CH(\overset{+}{N}H_3)COO^-$
5.	(+) Isoleucine	$CH_3CH_2CH(CH_3)-CH(\overset{+}{N}H_3)-COO^-$
6.	(–) Phenylalanine	$C_6H_5-CH_2CH(\overset{+}{N}H_3)COO^-$
7.	(–) Serine	$HOCH_2CH(\overset{+}{N}H_3)COO^-$
8.	(–) Threonine	$CH_3CHOHCH(\overset{+}{N}H_3)COO^-$
9.	(–) Cysteine	$HSCH_2CH(\overset{+}{N}H_3)COO^-$
10.	(–) Cystine	$^-OOCCH(\overset{+}{N}H_3)CH_2S-SCH_2CH(\overset{+}{N}H_3)COO^-$
11.	(–) Methionine	$CH_3SCH_2CH_2CH(\overset{+}{N}H_3)COO^-$
12.	(–) Tyrosine	$HO-C_6H_4-CH_2CH(\overset{+}{N}H_3)COO^-$
13.	(+) 3, 5-Dibromotyrosine	$HO-C_6H_2Br_2-CH_2CH(\overset{+}{N}H_3)COO^-$
14.	(+) 3, 5-Diiodotyrosine	$HO-C_6H_2I_2-CH_2CH(\overset{+}{N}H_3)COO^-$

S.No.	*Name*	*Structure*
15.	(+) Thyroxine	$HO-C_6H_2I_2-O-C_6H_2I_2-CH_2CH(\overset{+}{N}H_3)COO^-$
16.	(–) Proline	pyrrolidine ring $-COO^-$, ring $\overset{+}{N}H_2$
17.	(–) Hydroxyproline	HO-substituted ring $-COO^-$, ring $\overset{+}{N}H_2$
18.	(–) Tryptophan	indole (N–H) $-CH_2CH(\overset{+}{N}H_3)COO^-$
	Basic	
19.	(+) Lysine	$H_3N^+ CH_2 CH_2 CH_2 CH_2 CH(NH_2) COO^-$
20.	(–) Hydroxylysine	$H_3N^+ CH_2 CHOHCH_2 CH_2 CH(NH_3) COO^-$
21.	(+) Arginine	$H_2NC(\overset{+}{N}H_3)H - NH - CH_2 CH_2 CH_2 CH(NH_2)COO^-$
22.	(–) Histidine	imidazole ring (N, N–H) $-CH_2CH(\overset{+}{N}H_3)COO^-$
	Acidic	
23.	(+) Aspartic acid	$HOOCCH_2 CH(\overset{+}{N}H_3)COO^-$
24.	(–) Asparagine	$H_2NCOCH_2 CH(\overset{+}{N}H_3)COO^-$
25.	(+) Glutamic Acid	$HOOCCH_2CH_2 CH(\overset{+}{N}H_3)COO^-$
26.	(+) Glutamine	$H_2NCOCH_2CH_2 CH(\overset{+}{N}H_3)COO^-$

Q. 8. Describe some characteristics of natural amino acids.

Ans. 1. Some of the amino acids are acidic, some neutral and some basic. It may be noted that neutral amino acids are those which have one amino and one carboxy group thus neutralizing each other. Cystine has two amino and carboxy groups each. Compounds containing more amino groups than carboxy groups are expected to show basic behaviour. Similarly compounds having more carboxy groups (or acidic groups) than amino groups are supposed to exhibit acidic properties.

2. Valine, leucine, isoleucine, phenylalanine, threonine, methionine, tryptophan, lysine, histidine, asparagine cannot be synthesised by animals. And these are very important for the growth of animals. Hence these must be supplied in the diet as such.

3. All these acids except glycine have four different groups attached to the α-carbon atom. Thus the α-carbon is asymmetric and hence there is a possibility of optical isomerism. However they belong to the same stereochemical series. They have the same configuration as that of L (–) glyceraldehyde which is shown below:

$$\begin{array}{c} CHO \\ | \\ HO - C - H \\ | \\ CH_2OH \end{array} \qquad\qquad \begin{array}{c} COO^- \\ | \\ H_3N^+ - C - H \\ | \\ R \end{array}$$

L (–) Glyceraldehyde L - Amino acid

They are optically active compounds and rotate the plane of polarised light. It may be mentioned that the sign L refers to the *configuration* whereas the (+) or (–) sign refers to the *direction* of rotation of plane-polarised light.

4. The amino acids isolated from the proteins rotate the plane of polarised light i.e. they are *optically active* whereas the same amino acids if synthesised are *optically inactive.* It is because during the synthesis, both dextro and laevo forms are obtained in equal forms which counterbalance the rotation of each form. Such mixtures are called *racemic mixtures.*

5. Another interesting observation may be made. All the proteins which are resynthesised by human body have L-configuration. This happens so in all humans. That is why all human beings have almost similar characteristics. Had different persons produced different types (L and D) of proteins, then perhaps there would have been vast differences in the characteristics of different persons. This would have resulted into two different genetic breeds of persons (L type and D type).

Q. 9. Describe two methods of preparation of amino acids.

(Kerala, 2001;North Eastern Hill, 2004; Shivaji, 2004; Calcutta 2007)

Ans. Methods commonly used for preparation of amino acids are described below:

1. From α-halogeno acids. Treatment of α-halogeno acids with ammonia or ammonium hydroxide gives α-amino acids.

$$\underset{\text{α-Chloroacetic acid}}{\begin{array}{l} CH_2COOH \\ | \\ Cl \end{array}} + 2NH_4OH \longrightarrow \underset{\text{Glycine}}{\begin{array}{l} CH_2 - COO^- \\ | \\ {}^+NH_3 \end{array}} + NH_4Cl + H_2O$$

A conc. solution of ammonia is needed to obtain a good yield of the amino acid. The α-halogen used as a starting substance is obtained by Hell-Vohlard-Zelinsky method as follows:

$$\underset{\text{Acetic acid}}{CH_3COOH} + Cl_2 \xrightarrow[P]{Red} \underset{\text{Chloroacetic acid}}{\begin{array}{l} CH_2COOH \\ | \\ Cl \end{array}} + HCl$$

2. From sodium salt of malonic ester

$$\underset{\text{Monosodium diethyl malonate}}{Na^+ CH(COOC_2H_5)_2} + \underset{\text{Isobutyl bromide}}{(CH_3)_2CHCH_2Br}$$

$$\xrightarrow{-NaBr} (CH_3)_2CHCH_2CH(COOC_2H_5)_2 \xrightarrow{\text{Hydrolysis}} (CH_3)_2CHCH_2CH(COOH)_2$$

$$\xrightarrow{Br_2} (CH_3)_2CHCH_2CBr(COOH)_2 \xrightarrow[-CO_2]{\text{Heat}} (CH_3)_2CHCH_2\underset{\displaystyle Br}{\underset{|}{C}}HCOOH$$

$$\xrightarrow{NH_4OH} \underset{\text{Leucine}}{(CH_3)_2CHCH_2\underset{\displaystyle {}^+NH_3}{\underset{|}{C}}HCOO^-}$$

Additional Method

3. Phthalimido malonic ester synthesis. It is a modification of the Gabriel phthalimide synthesis.

$$\underset{\text{Potassium phthalimide}}{C_6H_4(CO)_2N^-K^+} + \underset{\text{Bromo ethyl malonate}}{BrCH(COOC_2H_5)_2}$$

$$\longrightarrow C_6H_4(CO)_2NCH(COOC_2H_5)_2$$

$$\xrightarrow{Na} C_6H_4(CO)_2\bar{N}CNa^+(COOC_2H_5)_2$$

$$\xrightarrow[\text{Chloroethyl acetate}]{\text{Heat}} C_6H_4(CO)_2NC(COOC_2H_5)_2\underset{|}{}CH_2COOC_2H_5$$

$$\xrightarrow[\text{2 KOH}]{\text{1. HCl, Heat}} \underset{\text{Aspartic acid}}{H_3N^+\underset{\displaystyle CH_2COOH}{\underset{|}{C}}HCOO^-}$$

Q. 10. Give Gabriel Phthalimide synthesis of α-amino acids.

(Bhopal 2004; Mumbai 2004)

Ans. Gabriel Phthalimide Synthesis. In this method, potassium salt of phthalimide is made to react with an α-haloester and the product is hydrolysed to get the amino acid.

$$C_6H_4(CO)_2\overset{-}{N}K^+ + ClCH_2COOC_2H_5$$

Potassium phthalimide; Chloroethyl acetate

$$\xrightarrow{-KCl} C_6H_4(CO)_2NCH_2COOC_2H_5$$

$$\xrightarrow{H_2O} C_6H_4(COOH)_2 + H_3\overset{+}{N}CH_2COO^- + C_2H_5OH$$

Phthalic acid; Glycine; Ethanol

Q. 11. Give Strecker synthesis of α-amino acid.

Ans. Strecker's synthesis. Here we treat an aldehyde with ammonia and hydrogen cyanide followed by hydrolysis. Thus

$$C_6H_5CH_2CHO + NH_3 \rightleftharpoons C_6H_5CH_2-CH=NH + H_2O$$

Phenyl acetaldehyde; Aldimine

$$\xrightarrow{HCN} C_6H_5CH_2CH(NH_2)-CN \xrightarrow[H^+]{\text{Hydrolysis}} C_6H_5CH_2CH(\overset{+}{N}H_3)COO^-$$

α-Amino nitrile; Phenyl alanine

Q. 12. Give Erlenmeyer azlactone synthesis of α-amino acids.

(Kurukshetra, 2001; Bhopal, 2004; Mumbai, 2004)

Ans. This method is used for the preparation of aromatic amino acid. The sequence of reactions given below give the synthesis of phenylalanine from glycine.

$$\underset{\text{Glycine}}{CH_2(NH_2)-COOH} \xrightarrow[NaHCO_3]{C_6H_5COCl} \underset{\text{Benzoyl glycine}}{CH_2(NHCOC_6H_5)-COOH}$$

$$\xrightarrow[\Delta]{C_6H_5CHO,\ AC_2O/ACONa} \underset{\text{Azlactone}}{C_6H_5-HC=C-C=O \ (\text{ring: } N=C(C_6H_5)-O)}$$

$$\xrightarrow[\text{Warm}]{\text{dil. NaOH}} C_6H_5-CH=C(NHCOC_6H_5)-COOH$$

$$\xrightarrow[\text{Na/Hg}]{\text{Reduction}}$$

$$C_6H_5-CH_2-\underset{\displaystyle |\atop NHCOC_6H_5}{CH}-COOH \xrightarrow[\Delta]{HCl} C_6H_5-CH_2-\underset{\displaystyle |\atop NH_2}{CH}-COOH + C_6H_5COOH$$

Phenylalanine

Q. 13. Describe chemical properties of α-amino acids. *(Kerala, 2001)*

Ans. Some important properties of α-amino acids are given below:

1. With acids. The dipolar ionic structure of amino acid exists in equilibrium with structure X in acidic medium as follows :

$$R-\underset{\displaystyle |\atop {}^{+}NH_3}{CHCOO^-} \overset{H^+}{\rightleftharpoons} \underset{\displaystyle |\atop {}^{+}NH_3}{RCHCOOH}$$

X

There is free carboxy group in structure X. Hence it gives the reactions of carboxy group.

2. With alkalis. The dipolar ionic structure Y exists in alkaline medium as follows:

$$R-\underset{\displaystyle |\atop {}^{+}NH_3}{CHCOO^-} \overset{OH^-}{\rightleftharpoons} R-\underset{\displaystyle |\atop NH_2}{CHCOO^-} + H_2O$$

Y

There is a free amino group in structure Y. Hence an amino acid in alkaline medium gives the reactions of amines.

3. Alkylation. In basic medium, the amino group of the acid reacts with an alkyl halide as follows.

$$\underset{\displaystyle |\atop NH_3}{RCHCOO'} + R'X \xrightarrow{Base} \underset{\displaystyle |\atop NHR'}{RCHCOO^-} + HX$$

N-Alkyl amino acid

$$\underset{\displaystyle |\atop NH_2}{CH_2COO^-} + CH_3Cl \xrightarrow{Base} \underset{\displaystyle |\atop NHCH_3}{CH_2COO^-} + HCl$$

N-Methyl glycine

4. Acetylation. In basic medium, the amino group of the acid reacts with acid chloride or acid anhydride to form acetyl derivatives.

$$\underset{\displaystyle |\atop NH_2}{RCHCOO^-} + CH_3COCl \xrightarrow{Base} \underset{\displaystyle |\atop NHCOCH_3}{RCHCOO^-} + HCl$$

$$\underset{\displaystyle |\atop NH_2}{RCHCOO^-} + (CH_3CO)_2O \xrightarrow{Base} \underset{\displaystyle |\atop NHCOCH_3}{RCHCOO^-} + CH_3COOH$$

5. Esterification. In acidic medium, the amino acids, give the reactions of carboxy group like formation of esters, acid chlorides and acid anhydrides.

$$\underset{\overset{|}{{}^{+}NH_3}}{RCHCOOH} + C_2H_5OH \xrightarrow[HCl]{Anhy.} \underset{\underset{\text{Ester}}{\overset{|}{{}^{+}NH_3}}}{RCHCOOC_2H_5} + H_2O$$

$$R-\underset{\overset{|}{{}^{+}NH_3}}{CH}-COOH + PCl_5 \longrightarrow R-\underset{\underset{\text{Acid chloride}}{\overset{|}{{}^{+}NH_3}}}{CH}-COCl + POCl_3 + HCl$$

6. Reaction with nitrous acid. Nitrous acid reacts with amino acids to liberate nitrogen gas. This method is used to analyse amino acids and is known as **Van Slyke's method**.

$$\underset{\overset{|}{{}^{+}NH_3}}{RCHCOO^-} + HNO_2 \longrightarrow N_2 + RCH(OH)COOH + H_2O$$

7. Reaction with formaldehyde. Amino acids with formaldehyde to form N-methylene amino acids.

$$HCHO + H_3\overset{+}{N}-\underset{\underset{R}{|}}{CH}-COO^- \longrightarrow CH_2=N-\underset{\underset{\underset{\text{N-methylene amino acid}}{R}}{|}}{CHCOOH} + H_2O$$

As a result of this change, the amino group of the amino acid gets blocked and the resulting product is acidic in nature. It can be titrated with alkali and forms the basis of **Sorenson formol titration** method for the estimation of proteins.

8. Reduction. Amino acids are converted into amino alcohols on reduction with lithium aluminium hydride.

$$\underset{\underset{R}{|}}{H_3\overset{+}{N}CHCOO^-} \xrightarrow[\text{[H]}]{LiAlH_4} \underset{\underset{\underset{\text{Amino alcohol}}{R}}{|}}{H_2NCHCH_2OH}$$

9. Decarboxylation. Amino acids undergo decarboxylation in the presence of acids or bases or enzymes.

$$\underset{\underset{R}{|}}{H_3\overset{+}{N}CHCOO^-} + Ba(OH)_2 \xrightarrow{\Delta} \underset{\text{Primary amine}}{RCH_2NH_2} + BaCO_3 + H_2O$$

10. Action of heat. (*a*) α-amino acids. On heating, α-amino acids undergo dehydration by interaction between two amino acid molecules.

$$\begin{matrix} R & & O \\ & CH-C & \\ H_2N & & OH \\ HO & & NH_2 \\ & C-C & \\ O & & R \end{matrix} \xrightarrow[-2H_2O]{Heat} \begin{matrix} R & & O \\ & CH-C & \\ HN & & NH \\ & C-CH & \\ O & & R \end{matrix}$$

Diketopiperazine

(*b*) **β-amino acids.** β-amino acids, on heating, lose ammonia to form α, β unsaturated acids.

$$\underset{\beta\text{-amino acid}}{R-\underset{\underset{NH_2}{|}}{CH}-CH_2-COOH} \xrightarrow{\text{Heat}} \underset{\alpha,\beta\text{- unsaturated acid}}{R-CH=CHCOOH} + NH_3$$

(*c*) **γ and δ-amino acids.** γ and δ-amino acids, on heating undergo intramolecular dehydration to form cyclic amides called *lactums.*

$$\underset{\gamma\text{-amino butyric acid}}{CH_2(C(=O)OH)-CH_2-CH_2NH_2} \xrightarrow[-H_2O]{\text{Heat}} \underset{\gamma\text{-butyrolactum}}{\text{cyclic } CH_2-C(=O)-NH-CH_2-CH_2}$$

11. Reaction with metallic ions. Amino acids react with heavy metal ions in aqueous solution to form deep coloured complexes.

$$\underset{\text{Glycine}}{2H_2N-CH_2-\overset{\overset{O}{||}}{C}-OH} + Cu^{2+} \longrightarrow \underset{\text{(Deep blue colour)}}{\underset{\text{Complex}}{(O=C-O,\ H_2C-NH_2)\,Cu\,(NH_2-CH_2,\ O-C=O)}}$$

Q. 14. What are peptides?

(Himachal, 2000; Panjab, 2000; Mumbai, 2004; Patna, 2005; Nagpur, 2008)

Ans. Peptides are amides obtained by interaction between the amino and carboxylic groups of two or more amino acid molecules. Two molecules of glycine, for example, combine to form amide substance known as glycyl glycine.

$$\underset{\text{2 molecules of glycine}}{\overset{+}{N}H_3CH_2COO^- + \overset{+}{N}H_3CH_2COO^-} \xrightarrow{-H_2O} \underset{\text{Glycyl glycine}}{H_3\overset{+}{N}CH_2CONHCH_2COO^-}$$

The amide group —CO—NH— in the peptides is called peptide linkage

Q. 15. Give the classification of peptides.

Ans. Peptides are classified as under:

Dipeptide. A peptide obtained by the condensation of two amino acid molecules is called dipeptide.

Tripeptide. A peptide obtained by the condensation of three amino acid molecules is called tripeptide.

Tetrapeptide. A peptide obtained by the condensation of four amino acid molecules is called tetrapeptide.

Polypeptide. A peptide obtained by the condensation of more than four amino acid molecules is called polypeptide. A polypeptide may be represented as:

$$H_3\overset{+}{N}-\underset{\underset{R}{|}}{CH}-CO(NH\underset{\underset{R'}{|}}{CH}CO)_n-NH\underset{\underset{R''}{|}}{CH}COO^-$$

Peptides of molecular weight upto 10000 are known as **polypeptides** whereas peptides of higher molecular weight are the **proteins**. Conventionally, N-terminal amino acid residue is written at the left end while the C-terminal amino acid residue is written at the right end.

Q. 16. Explain the geometry of peptide linkage.

Ans. X-ray diffraction studies reveal that the peptide linkage is flat *i.e.* carbonyl carbon, nitrogen and atoms attached to them lie in the same plane. The C—N bond distance comes out to be 1.32Å compared to usual C—N single bond distance of 1.47 Å, indicating that C—N bond has 50% double bond character. Further, measurement of bond angles shows that bonds to nitrogen are similar to those about trigonal carbon atom as shown below:

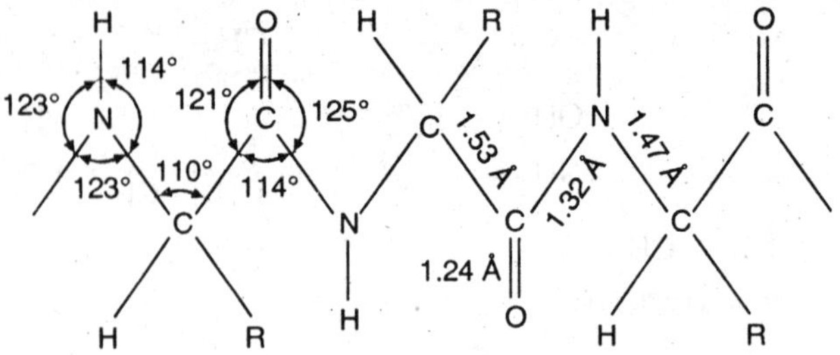

Q. 17. How will you synthesise peptides from amino acids? What are the difficulties encountered in their synthesis?

Or

Using carbobenzoxychloride as a N-protecting agent, sketch the synthesis of glycylalanine

(Madras 2004; Nagpur 2008)

Ans. Synthesis of a peptide from amino acids involves the following steps:

(*a*) Protection of the —NH_2 group of the amino acid with chlorobenzoxy group.

(*b*) Conversion of the carboxyl group into acid chloride.

(*c*) Formation of peptide linkage.

(*d*) Removal of chlorobenzoxy group.

It is observed that condensation of the amino acid molecules does not take place rapidly. The carboxy group has to be made reactive in order that the reaction takes place with appropriate kinetics. This can be done by converting amino acid into acid chloride. But before this is done, the amino group of the amino acid is protected with a suitable reagent. Ordinary acetylation or benzylation is found to be untenable. Protection is done with carbobenzoxy chloride, which is prepared as follows:

$$\underset{\text{Benzyl chloride}}{C_6H_5CH_2Cl} + \underset{\text{Carbonyl chloride}}{Cl-\overset{\overset{O}{\parallel}}{C}-Cl} \longrightarrow \underset{\text{Carbobenzoxy chloride}}{C_6H_5CH_2-O-\overset{\overset{O}{\parallel}}{C}-Cl}$$

This group is easy to remove at the completion of peptide synthesis. Stepwise synthesis of glycylalanine (dipeptide) is illustrated as under:

(*i*) Protection of amino group

$$\underset{\text{Glycine}}{H_3\overset{+}{N}CH_2COO^-} + \underset{\text{Carboxybenzoxy chloride}}{C_6H_5CH_2OCOCl} \xrightarrow{-HCl} \underset{\text{Carbobenzoxy glycine}}{C_6H_5CH_2OCONHCH_2COOH}$$

(ii) Formation of acid chloride

$$C_6H_5 - CH_2OCONHCH_2COOH + SOCl_2 \xrightarrow[-HCl]{-SO_2} \underset{\text{Acid chloride of carbobenzoxy glycine}}{C_6H_5CH_2OCONHCH_2COCl}$$

(iii) Formation of peptide linkage

$$C_6H_5CH_2OCONHCH_2COCl + \underset{\text{Alanine}}{H_3\overset{+}{N}CH(CH_3)COO^-} \xrightarrow{-HCl} \underset{\text{Carbobenzoxy glycylalanine}}{C_6H_5CH_2OCONHCH_2CONHCH(CH_3)COOH}$$

(iv) Removal of protecting group

$$C_6H_5CH_2OCONHCH_2CONHCH(CH_3)COOH \xrightarrow{Hg-Pd} \underset{\text{Glycylalanine}}{H_3\overset{+}{N}CH_2CONHCH(CH_3)COO^-} + \underset{\text{Toluene}}{C_6H_5CH_3} + CO_2$$

Repeating the above steps could yield tri, tetra and polypeptides.

Q. 18. What are proteins? *(Lucknow, 2000; Shivaji, 2000; Meerut, 2000; Panjab, 2003)*

Ans. Proteins are complex organic compounds essential for growth and maintenance of life. These are nitrogenous compounds obtained from α-amino acids. They are the constituents of all living organisms and found in every part of plants and animals. They are present in muscles, skin, hair, nails, blood, tendons and arteries.

Proteins present in different plants and animals differ from one another in composition and biological action. Proteins perform diverse functions in life processes. Some proteins are responsible only for structural shapes of parts of the body *e.g.*, keratin in hair, some are responsible for regulating metabolic processes *e.g.*, insulin in blood sugar level whereas some act as catalysts for biological reactions *e.g.*, enzymes.

Plants synthesise proteins from carbon dioxide, water and other nitrogenous materials. Animals consume proteins from plants. Inside animal body, proteins are hydrolysed to a mixture of amino acids from which a number of different or same proteins are resynthesised. Besides carbon, hydrogen oxygen, nitrogen, sulphur is also present in some proteins. Phosphorus, iron and magnesium may be present in select proteins. Molecular weights of proteins are abnormally high as a protein molecule is constituted of several thousands of amino acid molecules. In all there are 26 different natural amino acids which are the building blocks of different proteins.

Q. 19. How would you classify proteins on the basis of structure?

(Delhi, 2003; Pune, 2010)

Ans. Proteins are classified into two types on the basis of structure:

(a) Fibrous proteins. These proteins consist of thin linear molecules which lie side by side to form fibres. Intramolecular hydrogen bonding holds the peptide chain together. Fibrous proteins are insoluble in water.

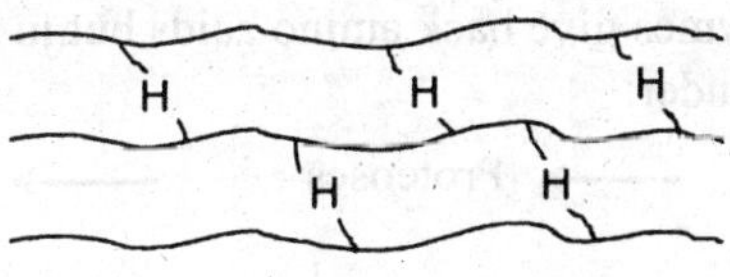

Structure of fibrous proteins

Fibrous proteins are the main structural materials of tissues. Examples of important fibrous proteins are *keratin* in skin and hair, *callagen* in tendon, *fibroin* in silk and *myosin* in muscles.

(*b*) Globular proteins. In these proteins, polypeptides are folded into compact spheroidal shapes. Intramolecular hydrogen bonding holds the peptide chain in shape in such proteins. These are soluble in water or aqueous solutions of acids and bases.

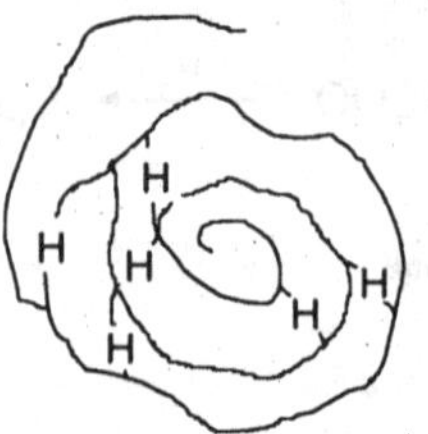

Structure of globular proteins

These proteins have the role to regulate and maintain life processes. Examples of this class of proteins are enzymes, hormones, haemoglobin and albumin.

Q. 20. How do you classify proteins according to hydrolysis products? *(Delhi, 2003)*

Ans. Proteins are classified according to hydrolysis products as follows:

(*a*) Simple proteins. Proteins which on hydrolysis yield only amino acids are called simple proteins. Examples of this class of proteins are albumins (such as egg albumin, serum albumin), globulins (such as tissue globulin) and glutelins (such as wheat gluteline).

(*b*) Conjugated proteins. Proteins which are a combination of two parts, a proteinous part and non-proteinous part, are called conjugate proteins. The non-proteinous part is called prosthetic group. Prosthetic group plays its part in biological function of the protein.

Conjugated proteins are further classified as nucleoproteins, glycoproteins and chromoproteins. The prosthetic groups in such proteins are nucleic acid, carbohydrate and haemoglobin (or chlorophyll) respectively.

Q. 21. Describe general characteristics of proteins.

Ans. 1. Composition. The elements generally present in proteins are carbon, hydrogen, oxygen and nitrogen.

2. High molecular weights. As protein molecules are obtained from hundreds and thousands of amino acid molecules, their molecular masses run into several thousands and sometimes into several lakhs.

3. Physical state. Generally speaking, proteins are colourless, tasteless, amorphous solids having no sharp melting points. They are colloidal size particles. This property is used to separate proteins from crystalline salts by the process of dialysis.

4. Optical activity. Because of the presence of asymmetric carbon atoms, proteins show optical activity. However, all naturally occurring proteins have the same configuration *viz*. the L-glyceraldehyde configuration.

5. Amphoteric nature. Protein molecules exist as dipolar ions. Hence they react with both acids and alkalis and are thus amphoteric.

6. Hydrolysis. As amino acids are the constituents of proteins, the latter on hydrolysis in the presence of acids, bases or enzymes give back amino acids but in steps. The steps involved during hydrolysis are represented as under:

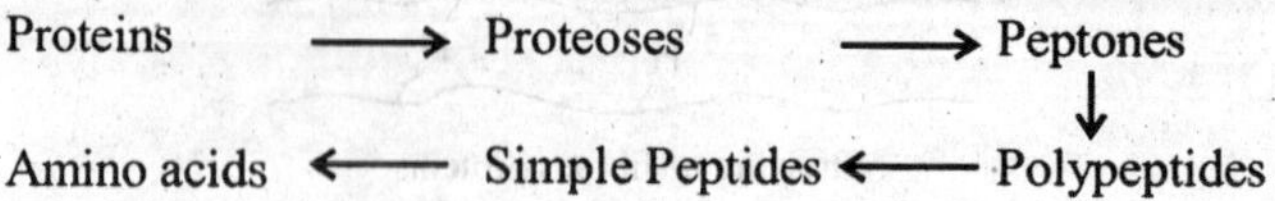

Q. 22. Explain the term denaturation of protein.

(Bangalore, 2001; Kerala, 2000; Gulberga 2003; Bhopal 2004, Kuvempu 2005)

Ans. Denaturation. Proteins are very tender and delicate substances. When subjected to heat or action of acids or alkalis, they lose their biological activity. They are said to be **denatured**. This phenomenon in which the proteins lose their biological activity and other characteristics under the effect of temperature, is called **denaturation**. The denatured proteins can be brought back to its original state by cooling the protein solution very slowly. This process is called **renaturation.**

During denaturation, there is a rearrangement in the secondary and tertiary structure of the protein but the primary structure remains unchanged. Coagulation of egg-white by the action of heat is an example of irreversible denaturation of proteins.

Q. 23. How will you show by tests that the given substance is a protein?

(Coimbatore, 2000; Shivaji, 2000)

Ans. Following tests are performed to identify proteins:

1. Biuret test. A drop of copper sulphate is added to an alkaline solution of protein. A bluish colour develops. This test is also given by proteoses and peptones which are the hydrolytic products of proteins.

2. Ninhydrin test. Ninhydrin is triketo hydrindene hydrate. A blue to red-violet is obtained when a protein is treated with ninhydrin.

3. Xanthoproteic test. This tests is given by proteins containing tyrosine or tryptophane. When warmed with conc. HNO_3, such proteins give a yellow colour.

4. Millon's test. Millon's reagent is a mixture of mercurous and mercuric nitrates. Proteins containing tyrosine give a white precipitate turning red when treated with Millon's reagent.

5. Heller's test. This test is commonly employed for detecting albumin in urine. When conc. HNO_3 is poured along the side of a test tube containing protein solution, a white precipitate is obtained.

Q. 24. Explain biological importance of proteins.

Or

What functions do proteins perform in human body? *(Himachal, 2000; Panjab, 2003)*

Ans. Proteins are vital to animal life. They perform a number of biological processes and play a pivotal role in the running and maintenance of body. They are hydrolysed inside the body into its constituent amino acids before performing various functions.

In the stomach, the proteins are hydrolysed, in the presence of hydrochloric acid and enzyme pepsin, into lower molecular weight polypeptides.

In the intestine, proteins are hydrolysed in the presence of enzymes trypsin and pepsin. The simple amino acids are then assimilated in the blood stream and transported to various cells of the body. It is interesting to note that some of the amino acids are reconverted into some specific proteins which are needed by the body for specific requirements and other amino acids are oxidised to produce energy. Thus there is a continuous cycle of decomposition and synthesis of proteins in the body.

Q. 25. Write notes on (*a*) enzymes (*b*) antibodies (*c*) haemoglobins (*d*) structural proteins.

Ans. (*a*) Enzymes. These are the proteins which catalyse biological reactions in the body. Every enzyme is very specific in its action. They are able to perform reactions without requiring any change in temperature, pressure or pH. They carry out degradation and synthetic reaction in the body with much greater speed and efficiency than would be achieved in the laboratory. For example, enzyme carbonic anhydrase present in the blood catalyses the decomposition of over 30 million molecules of carbonic acid into carbon dioxide in just a minute, thereby maintaining carbon dioxide level in the body fluids.

Enzymes pepsin and trypsin catalyse the hydrolysis of peptide linkage in protein molecules in digestive tract.

The enzyme ptyalum, which is present in saliva catalyses the conversion of starch into maltose during chewing of food.

The enzymes maltase, lactase, sucrase catalyse the decomposition of disaccharides into monosaccharides.

(*b*) Antibodies. These proteins defend the body against attack from foreign organism. These proteins are produced by the body during attack by foreign infectious species called *antigens*. The newly produced proteins combine with these antigens and thus protect the body against the destructive action of antigens. Gamma globulins present in blood are examples of antibodies.

(*c*) Haemoglobin. This is an example of transporting proteins. Haemoglobin present in blood transports oxygen from lungs to all parts of the body. Haemoglobin consists of two parts. The protein part is *globin* and the non-protein part is *haemo*. It is haemo which is responsible for transporting oxygen and is also responsible for the red colour of blood.

(*d*) Structural proteins. Structural proteins are responsible for providing distinct structure to various parts of the body. Keratin is a structural protein present in skin, hair, nails, horn and features.

Myosin is a structural protein present in muscles and *collagen* in tendons.

Q. 26. Describe the primary, secondary and tertiary structure of proteins.

(Bangalore, 2001; Punjab, 2003; Lucknow, 2010)

Ans. Studying structure of a protein is a very complex exercise. If we divide the work into smaller steps, the task becomes easier. For structure elucidation of protein, the work is divided under four headings as listed below:

(*a*) Primary structure (*b*) Secondary structure
(*c*) Tertiary structure (*d*) Quaternary structure

PRIMARY STRUCTURE

It refers to the number, nature and sequence of the amino acids in polypeptide chains. Primary structure determination of protein involves the following steps:

(*i*) Determination of amino acid composition. The protein under investigation is hydrolysed by means of acid, alkali or enzyme to its constituent amino acids. The amino acids are separated and identified by means of ion-exchange chromatography. The weights of amino acids produced are noted. Thus the number of moles of each amino acid and the relative number of various amino acid residues in the protein are estimated. Knowing the molecular weights of the protein, which is determined by a suitable method, the actual number of amino acid residues in the molecules is calculated.

(*ii*) Sequence of arrangement of amino acids. The next job and a difficult job is to arrive at the sequence of arrangement of amino acids. For this a controlled hydrolysis of protein is carried out to produce small polypeptides. The sequence of arrangement of amino acids is determined by terminal and residue analysis.

Terminal residue analysis. The amino acid residues at the two extremes of a peptide chain are different from all other amino acid residue and also from each other. Amino acid present at one end of the chain has free — NH_2 group. It is called as *N-terminal amino acid*. The amino acid present on the other end has a free — COOH group. It is called as *C-terminal amino acid.*

$$\underset{\text{N-terminal residue}}{NH_2-\overset{\displaystyle R}{\overset{|}{C}}HCO}\left[NH\overset{\displaystyle R'}{\overset{|}{C}}HCO\right]_n\underset{\text{C-terminal reside}}{NH\overset{\displaystyle R''}{\overset{|}{C}}HCOOH}$$

The two terminal amino acid residues are identified as follows:

N-terminal residue analysis

(*i*) Sanger's method. The polypeptide is treated with 2, 4-dinitrofluorobenzene (DNFB) commonly known as Sanger's reagent, in presence of sod. bicarbonate. The reagent reacts with free amino group of N-terminal amino acid residue to form a 2, 4-dinitrophenyl (DNP) derivative of the polypeptide. The product obtained is hydrolysed by an acid to form dinitrophenyl (DNP) derivative of N-terminal amino acid and a smaller polypeptide molecule. The DNP derivative of the amino acid is isolated and analysed chromatographically or by thin layer chromatography. The reactions are given as under:

$$O_2N-C_6H_3(NO_2)-F + H_2N\underset{}{\overset{R}{C}}HCONH\overset{R'}{C}HCO\ldots$$

Dinitrofluorobenzene

$$\longrightarrow O_2N-C_6H_3(NO_2)-NH\overset{R}{C}HCONH\overset{R'}{C}HCO\ldots$$

$$\xrightarrow{H^+} O_2N-C_6H_3(NO_2)-NH\overset{R}{C}HCOOH + H_2N\overset{R'}{C}HCO\ldots$$

DNP derivative of amino acid — Smaller polypeptide

***(ii)* Enzymatic method.** The enzyme amino peptidase attacks the protein (or peptides) at the terminal which has free amino group. When the terminal amino acid is liberated, as new smaller peptide having amino group is produced, which is again attacked by the enzymes. Different amino acids liberated thus are analysed and identified.

C-terminal residue analysis

***(i)* Hydrazinolysis method.** In this method, the peptide is heated with anhydrous hydrazine at 373 K. All the amino acid residues of the peptide chain except the C-terminal one, are converted into hydrazides. The C-terminal residue is isolated as free amino acid and is identified.

$$\ldots HN\overset{R}{C}HCONH\overset{R'}{C}HCONH\overset{R''}{C}HCOOH$$

$$\Big\downarrow \text{Heat} \mid NH_2NH_2$$

$$H_2N\overset{R}{C}HCONHNH_2 + H_2N\overset{R'}{C}HCONHNH_2 + H_2N\overset{R''}{C}HCOOH$$

Hydrazides — Free C-terminal amino acid

***(ii)* Enzymatic method.** The enzyme carboxypeptidase specifically hydrolyses the peptide chain adjacent to the free carboxy group. The C-terminal amino acid is thus set free and identified. The process is repeated on smaller peptide chains to set free amino acids one by one from the side of the free carboxy group.

SECONDARY STRUCTURE

Secondary structure tells us about the shape and conformation of the peptide chains in protein molecule. X-ray studies show that proteins have two different conformations viz. α-helix and β-structure or pleated sheet structure.

The β-structure. Peptide chains in silk fibroin are fully extended to form flat zig-zags. These chains of polypeptides lie side by side and are held to one another by hydrogen bonds as shown ahead:

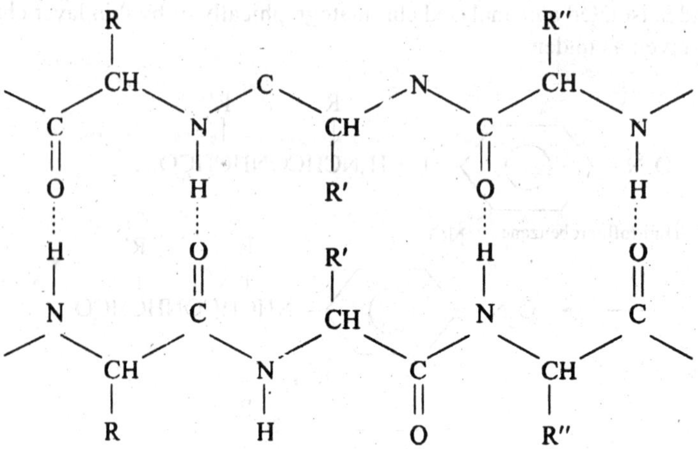

β-Flat sheet structure of a protein

Due to steric hindrance between R, R', R" side chains, the flat sheet structure contracts and adopts pleated sheet structure (β-structure). The exact contraction depends upon the size of the side chains.

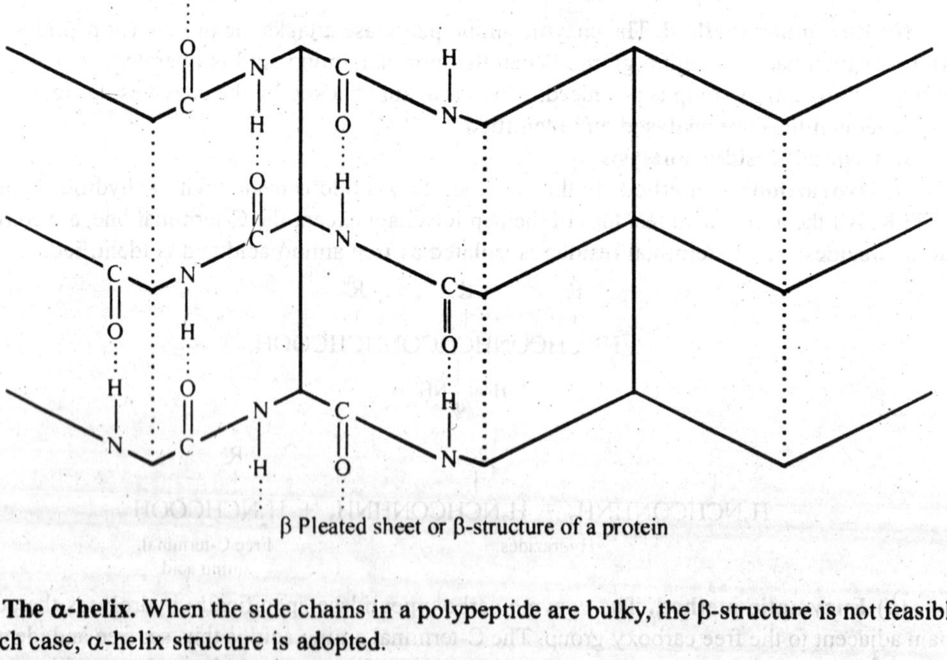

β Pleated sheet or β-structure of a protein

The α-helix. When the side chains in a polypeptide are bulky, the β-structure is not feasible. In such case, α-helix structure is adopted.

Pauling proposed α-helix structure for α-keratin, which is the constituent structural protein in hair, nails etc. In α-helix structure, each peptide chain is coiled to form helix as shown in the figure below. Each turn of the helix has about 3-7 amino acid residues and the distance between two helices is 5.4 Å. Two adjacent turns are linked by means of hydrogen bonds which involve NH— group of one amino group and the carbonyl oxygen of the fourth residue in the chain. Hydrogen bonding keeps the structure tight and prevents free rotation keeping the helix intact.

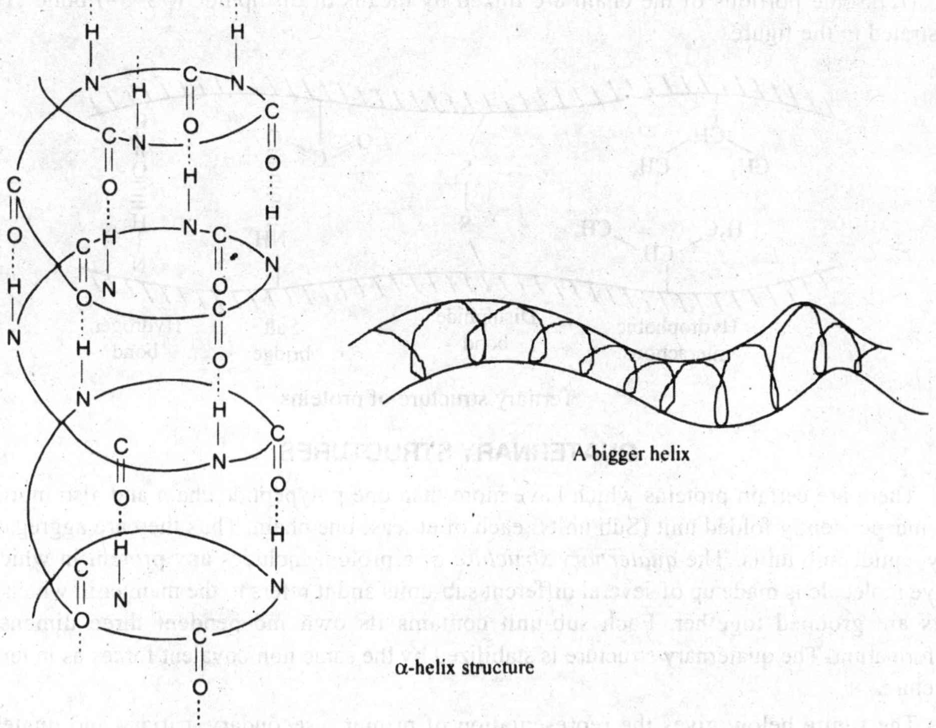

α-helix structure

α-helix may be left or right-handed. But it is observed that the right-handed helix is more stable. Diameter of the helix is about 10 Å.

TERTIARY STRUCTURE

Tertiary structure of a protein refers to the three-dimensional structure i.e. the folding and bending of the long peptide chains. Types of bonds that are responsible for the tertiary structure of a protein. are:

(*i*) Hydrogen bond

(*ii*) Salt bridge (Ionic bond)

(*iii*) Hydrophobic bond

(*iv*) Disulphide bond

(*i*) Hydrogen bond. Hydrogen bond may arise due to free —OH groups or free $—NH_2$ groups in the peptide chain or due to peptide backbone.

(*ii*) Ionic bond. Whenever there are positively or negatively charged groups present on the side chain e.g. $—COO^-$ and $—\overset{+}{N}H_3$, these bonds are formed. These bonds have a tendency to be on the exterior of the molecule.

(*iii*) Hydrophobic bond. Such bonds are formed between the two methyl or two phenyl groups in the side chain. As they are hydrophobic groups, they prefer to remain in the interior where there is less water.

(*iv*) Some portions of the chain are linked by means of disulphide (–S–S–) bond. This is illustrated in the figure.

Hydrophobic interactions   Disulphide bond   Salt bridge   Hydrogen bond

Tertiary structure of proteins

QUATERNARY STRUCTURES

There are certain proteins which have more than one polypeptide chain and also more than one independently folded unit (Sub units) each of at least one chain. Thus they are aggregates of polypeptide sub units. The *quaternary structure* of a protein includes any protein in which the native molecule is made up of several different sub-units and it refers to the manner in which these units are grouped together. Each sub-unit contains its own independent three dimensional conformation. The quaternary structure is stabilized by the same non-covalent forces as in tertiary structure.

The figure below gives the representation of primary, secondary, tritiary and quaternary structures of proteins.

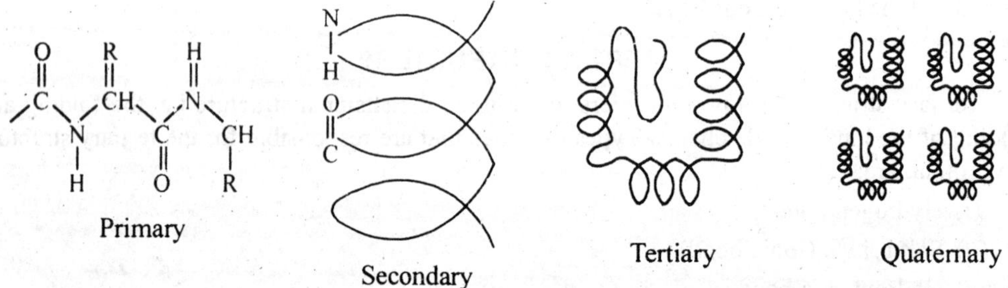

Diagrammatic representation of primary, secondary, tertiary and quaternary structure of protein

Q. 27. How can lysine and glycine be separated from each other? The isoelectric points are pH 9.6 for lysine and pH 5.97 for glycine.

Ans. An aqueous solution of the mixture is placed between two electrodes. The pH is adjusted to 9.6 and an electric current is passed. Lysine does not migrate at this pH. Glycine can be collected at the anode. Now the pH is adjusted to 5.97. At this pH, glycine does not migrate. Lysine can be collected at the cathode.

Q. 28. How many different peptides can be synthesised from (*a*) glycine and alaine and (*b*) gly. ala$_2$?

Ans. (*a*) Four products can be obtained:

(*i*) Gly. Gly. (*ii*) Ala. Gly. (*iii*) Ala. Ala. (*iv*) Gly. Ala.

(*b*) Three products are possible:

(*i*) Gly. Ala. Ala. (*ii*) Ala. Gly. Ala. (*iii*) Ala. Ala. Gly.

Q. 29. A tripeptide is partially hydrolysed to two dipeptides viz. Gly. Leu. and Asp. Gly. Assign a possible structure to the tripeptide.

Ans. As Gly is present in both the dipeptides, it must be present in the middle of tripeptide. The free $\overset{+}{N}H_3$— is in aspartic acid and free –COOH is in leucine. Therefore the tripeptide is Asp. Gly. Leu.

Q. 30. Explain how the following reagents denature proteins (*a*) Ag^+ and Pb^{2+} (*b*) ethanol (*c*) urea and (*d*) heat.

Ans. (*a*) Heavy metal cations Ag^+ and Pb^{2+} form insoluble salts with the —COO^- group.

(*b*) Ethanol interferes with the hydrogen bonding by offering its own competing hydrogens.

(*c*) Urea can form a good hydrogen bond and thus interferes with the original hydrogen bonding in the protein.

(*d*) Heat produces more random conformations in the protein and thus hampers the formation of helix.

Due to these reasons, the denaturation of proteins takes place.

Q. 31. Write the structure of dipeptide Leu-Ala.

Ans. Dipeptides are compounds obtained by the condensation of two amino acid (same or different) molecules. Hydroxy from the carboxy group of one molecule and hydrogen from amino group of second molecule are removed as water, resulting in the formation of peptide linkage as given below:

$$\underset{\text{Amino acid}}{NH_2—\underset{\substack{|\\R}}{C}HCO}\ \boxed{OH + H}\ \underset{\text{Amino acid}}{NH—\underset{\substack{|\\R'}}{C}HCOOH}$$

$$\downarrow$$

$$\underset{\text{Dipeptide}}{NH_2—\underset{\substack{|\\R}}{C}HCONH\underset{\substack{|\\R'}}{C}HCOOH}$$

Dipeptide Leu-Ala will be obtained from the combination of leucine and alanine.

$$\underset{\text{Leucine}}{NH_2\underset{\substack{|\\CH_2\\|\\CH(CH_3)_2}}{C}HCO}\ \boxed{OH + H}\ \underset{\text{Alanine}}{NH\underset{\substack{|\\CH_3}}{C}HCOOH}$$

$$\downarrow -H_2O$$

$$NH_2\underset{\substack{|\\CH_2\\|\\CH(CH_3)_2}}{C}HCONH\underset{\substack{|\\CH_3}}{C}HCOOH$$

Since peptides exist as dipolar compounds, the above compound Leu-Ala dipeptide may be correctly written as

$$\overset{+}{N}H_3\underset{\substack{|\\CH_2\\|\\CH(CH_3)_2}}{C}H—CO—NH—\underset{\substack{|\\CH_3}}{C}HCO\overset{-}{O}$$

NUCLEIC ACIDS, RNA AND DNA

Q. 32. What are nucleic acids? Explain their structure. *(Kavempur, 2005; Garhwal, 2010,*

Ans. Nucleic acids form an important class of compounds, having high molecular weights. They play an important role in the development and reproduction of all forms of life. Living cells contains nucleic acids in the form of nucleoproteins. Nucleoproteins consists of a protein and a nucleic acid.

There are two types of nucleic acids

(*i*) Deoxyribonucleic acid or DNA

(*ii*) Ribonucleic acid or RNA

These names for nucleic acids are derived from their hydrolysis products.

A nucleic acid on hydrolysis gives a nucleotide which on further hydrolysis gives a nucleoside and phosphoric acid. A nucleoside on hydrolysis in the presence of inorganic acid gives a mixture of sugar, purines and pyrimidines.

$$\text{Nucleic acid} \xrightarrow{Ba(OH)_2} \text{Nucleotide} \xrightarrow[NH_3]{\text{Aqueous}} \text{Nucleoside} \xrightarrow{\text{Inorganic acid}} \text{Sugar + Purines + Pyrimidines}$$

Only two sugars have been isolated from the hydrolysis products of nucleic acids. They are D(–) ribose and 2-deoxy-D (–) ribose.

CHO
|
$(CHOH)_3$
|
CH_2OH
Ribose

CHO
|
CH_2
|
$(CHOH)_2$
|
CH_2OH
2-Deoxyribose

Nucleic acids are named according to the sugar produced by its hydrolysis. Nucleic acid which contains ribose is called *ribonucleic* acid and is abbreviated as RNA. Nucleic acid which contains deoxyribose as a component is called *deoxyribonucleic* acid and is abbreviated as DNA. Correspondingly, there are ribonucleoproteins and deoxyribonucleo proteins. Nucleic acids are derived from sugars, phosphoric acid and heterocyclic bases (purines and pyrimidines). The linkage between them is shown as under:

```
       base        O       base        O
        |          ||       |          ||
...... Sugar — O — P — O — Sugar — O — P — O ......
                   |                   |
                   OH                  OH
```

Q. 33. Describe the structure of DNA and RNA. *(Pune, 2010)*

Ans. Structure of DNA. DNA is a huge molecule with mol. mass ranging between 10^6 and 10^9. On hydrolysis of DNA, we obtain 2-deoxyribose, phosphoric acid and the bases adenine, guanine, cytosine, 2-methyl cytosine and thymine. Different DNA molecules contain the bases in different order. It is this different sequence of bases in the DNA molecule that is responsible for genetic variations. A general sequence of DNA molecule is shown :

OH$_2$C O Base 1
H H H H
O H
HO — P = O
OH$_2$C O Base 2
H H
O
HO — P = O
OH$_2$C O Base 3
H H H H
O H

Watson and Crick in 1953 gave the secondary structure of DNA. They described DNA molecule as two identical polynucleotide chains twisted about each other to form double helix with a diameter of 18 Å in which the heads of two chains are in opposite directions. Both helices are right handed and have ten nucleotide resides per turn. It is shown schematically below:

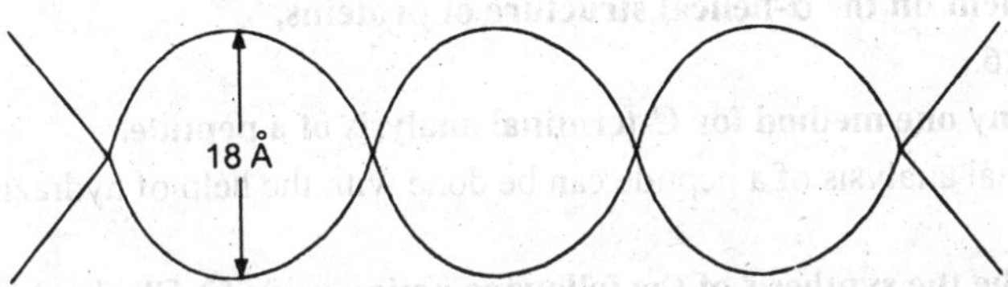

Structure of RNA

The difference between DNA and RNA is that the former contains the sugar 2-deoxyribose whereas the latter contains sugar ribose. There is also some difference between the base content of the two nucleic acids. RNA contains the bases adenine, guanine, cytosine and uracil. DNA has 2-methyl cytosine and thymine in place of uracil. RNA is lighter than DNA in weight by 33%.

In general, the polynucleotide chain of RNA may be represented in the same way as that of DNA except that the 2-deoxyribose sugar unit is replaced by ribose unit.

Q. 34. Give the main importance of RNA and DNA. *(Panjab, 2000; Bangalore, 2001)*

Ans. Importance (Functions) of RNA

1. There are three types of RNA in the living cell and they perform different roles as given below:

(*i*) Messenger RNA (m-RNA) carries the message of DNA for performing a particular synthesis.

(*ii*) Ribosomal RNA (*r*-RNA) makes provision for a site for performing synthesis of protein.

(*iii*) Transfer RNA (*t*-RNA) directs the amino acids to the site for the purpose of synthesis.

2. Processes of learning and memory storage are believed to be controlled by RNA.

Importance (Functions) of DNA

1. It has a property of self replication and is responsible for maintaining hereditary traits from one generation to the next.

2. It controls the synthesis of RNA which guides the synthesis of proteins.

3. The base sequence of DNA gets altered by the action of UV light, X-rays and certain chemicals. This process is called *mutation*. These mutations are responsible for changing hereditary traits of animals and their off-springs.

Q. 35. Explain why amino acids are weaker acids than carboxylic acids?

Ans.
$$R-\overset{\overset{\displaystyle O}{\|}}{C}-O-H$$

$$R-\underset{\underset{\displaystyle H_2\ddot{N}}{|}}{CH}-\overset{\overset{\displaystyle O}{\|}}{C}-O-H$$

This is because the proton after leaving the carboxy group attaches itself to the amino group instead of going into the solution.

Q. 36. What is the action of heat on α, β and γ amino acids separately. (*Kanpur, 2001*)

Ans. See Q. 13 under Action of heat.

Q. 37. Explain the steps involved in the synthesis of a dipeptide taking the example of glycylalanine. (*Kerala, 2000*)

Ans. See Q. 17.

Q. 38. Comment on the α-helical structure of proteins.

Ans. See Q. 26.

Q. 39. Give any one method for C-terminal analysis of a peptide.

Ans. C-terminal analysis of a peptide can be done with the help of hydrazinolysis. For details, see Q. 26.

Q. 40. Describe the synthesis of the following amino acids (*i*) Glycine by Strecker synthesis (*ii*) Alanine by Gabriel phthalimide synthesis.

Ans. See Qs. 10 and 11. Amend suitably.

Q. 41. What is meant by primary structure of proteins? Explain with the help of suitable examples.

Ans. See Q. 26.

Q. 42. Write the structure of phenylalanine. How will you establish it by DNFB method?

Ans. Refer to Sanger's method in Q. 26.

Q. 43. How are amino acids classified based on the polarity of their side chains at neutral pH? Give example. (*Bangalore, 2001*)

Ans. See Q. 5.

Q. 44. What is a nucleotide? Write the structure of nucleotide containing uracil. (*Bangalore, 2001*)

Ans. See Q. 32 and Q. 33 (structure of RNA).

Q. 45. Explain the different specificities exhibited by enzymes. Give example.

(*Bangalore, 2001*)

Ans. See Q. 25 (*a*).

Q. 46. Write a short note on primary and secondary structure of proteins.

(*Kerala, 2000, Kurukshetra, 2001*)

Ans. See Q. 26 (*a*) and (*b*).

Q. 47. What is "biuret reaction"? (*Kerala, 2000, A.N. Bhauguna, 2000*)

Ans. See Q. 23 (1)

Q. 48. What are the characteristics of amino acids present in neutral protein?

(*Kerala, 2001*)

Ans. See Q.2.

Q. 49. Discuss any two methods that are used for N-terminal amino acid determination.

(*Kerala, 2001*)

Ans. See Q. 26

Q. 50. What is the action of heat on alanine? (*Coimbatore, 2000*)

Ans. See Q. 13 (10).

Q. 51. Give the structure and IUPAC name of glycine. How will you obtain glycine from chloroacetic acid (*Nagpur 2003*)

Ans. $CH_2—COOH$ with NH_2 on the CH_2

$$\underset{\text{Chloroacetic acid}}{ClCH_2—COOH} \xrightarrow[HCl]{NH_3} \underset{\text{2-Aminoethanoic acid}}{\overset{+}{N}H_3—CH_2—COO^-}$$

Q. 52. Describe Edman's method for the identification of N-terminal amino acid

(*Delhi 2003*)

Ans. This method is based on the reaction between one amino group and phenyl isocyanate to form substituted thiourea. Hydrolysis with HCl selectively removes the N-terminal residue as shown below:

$$C_6H_5NCS + H_2N\underset{R}{\underset{|}{C}H}—\overset{O}{\overset{\|}{C}}—NH—\underset{R}{\underset{|}{C}H}—\overset{O}{\overset{\|}{C}}\text{-----} \xrightarrow{\text{Alk. Medium}}$$

$$C_6H_5\overset{H}{\overset{|}{N}}—\underset{S}{\underset{\|}{C}}—NH—\underset{R}{\underset{|}{C}H}—\overset{O}{\overset{\|}{C}}—NH—\underset{R}{\underset{|}{C}H}—\overset{O}{\overset{\|}{C}}\text{----} \xrightarrow{H_2O/HCl}$$

$$\text{(ring: } C_6H_5—N\text{, } C{=}S\text{, NH, } CH(R)\text{, } C{=}O\text{)} + H_2N\,CH—\overset{O}{\overset{\|}{C}}\text{-----}$$

36

SYNTHETIC DRUGS, INSECTICIDES AND PESTICIDES

DRUGS

Q. 1. What are (a) antiseptics (b) antipyretics (c) analgesics (d) tranquilisers and hypnotics (e) antimalarials (f) antibiotics (g) non-antibiotic antimicrobial drugs.

(*Meerut, 2000*)

Ans. **(*a*) Antiseptics.** Substances, which are applied externally to the infected skin to stop micro-organism growth, are called antiseptics. Phenol, cresol, xylenol, chloramine-T, potassium permanganate and boric acid in dilute solutions are used as antiseptics for dressing, mouth wash and gargles etc.

(*b*) Antipyretics. Substances which lower down body temperature are called antipyretics. Patients suffering from high fever are administered a dose of antipyretic substance. Commonly used antipyretics are paracetamol, aspirin and phenacetin.

(*c*) Analgesics. Substances which relieve the pain in body are called analgesics. Such substances depress the central nervous system thereby relieving the pain. It is found that substances which lower down the temperature also act as pain-relieving agents. Commonly used analgesics are aspirin, codeine and morphine.

(*d*) Tranquilisers and hypnotics. Substances which induce sleep by reducing anxiety are called tranquilisers and hypnotics. Their effect is by way of action on nerve centres. Derivatives of barbituric acid are commonly used as tranquilisers.

(*e*) Anti-malarials. Medicines used in the treatment of malaria are called anti-malarials. Quinine, chloroquin, plasmoquin and proguanil are some of the commonly used anti-malarials.

(*f*) Antibiotics. Chemical substances produced by some specific micro-organisms like bacteria, fungi or moulds and used to kill some other organisms are called antibiotics. The first antibiotic substance penicillin was discovered by Fleming in 1929. We have a long list of antibiotics including streptomycin, gentamycin, erythromycin, tetracyclin, chloroamphenicol, ampicillin, amoxcillin, which have been discovered since then. Every antibiotic substance besides having a general effect, has a specific action, too.

(*g*) Non-antibiotic antimicrobial drugs. These drugs are not produced by micro-organism but have the capacity to fight against certain organisms. Sulphanilamide, sulphadiazine and sulpha guanidine belong to this category.

Q. 2. Give a brief description of the synthesis and uses of aspirin (acetyl salicylic acid).

(*Awadh, 2000; Garhwal 2000; Kerala, 2001; Nagpur, 2002*)

Ans. It is synthesised in two steps:

(*i*) Conversion of sod. phenoxide into salicylic acid. Phenol is treated with sod. metal or a conc. solution of sod. hydroxide to obtain sodium phenoxide. Carbon dioxide is then passed through it at 400 K under pressure to obtain sod. salicylate. It is hydrolysed with an acid to produce salicylic acid.

Phenol —(Na metal)→ Sod. phenate —(CO_2, 400 K, Pressure)→ Sod. salicylate —(H^+)→ Salicylic acid

(*ii*) **Conversion of salicylic acid into aspirin.** Salicylic acid is subjected to acetylation with acetic anhydride in the presence of conc. H_2SO_4 to give aspirin.

Salicylic acid —($(CH_3CO)_2O$, H^+)→ Acetyl salicylic acid (aspirin)

Uses. 1. It is widely used as an analgesic and antipyretic.

2. It has been long noticed that salicylic acid produced as a result of hydrolysis in the stomach is dangerous and can cause bleeding from the stomach wall, when aspirin is consumed freely.

Q. 3. Give a brief description and uses of phenacetin.

Ans. *p*-aminophenol is subjected to acetylation with the help of acetic anhydride to obtain *p*-hydroxy acetanilide. It is then treated with sod. ethoxide and ethyl iodide to give phenacetin.

p-Amino phenol —($(CH_3CO)_2O$)→ *p*-Hydroxy acetanilide —(C_2H_5ONa)→ Phenacetin

Uses. (*i*) It is used as a general antipyretic and analgesic.

(*ii*) It has long been used in APC tablets which contain aspirin, phenacetin, and caffeine for curing common coughs and colds.

Q. 4. Briefly describe the synthesis and uses of paracetamol.

(*M. Dayanand, 2000; Garhwal, 2000; Kurukshetra, 2001; Nagpur, 2003; Delhi, 2003*)

Ans. Synthesis. The starting material for obtaining paracetamol is *p*-aminophenol. *p*-Aminophenol is acetylated with acetic anhydride to give *p*-hydroxy acetanilide or paracetamol.

$$\underset{p\text{-Amino phenol}}{HO-C_6H_4-NH_2} + (CH_3CO)_2O \longrightarrow \underset{\text{Paracetamol}}{HO-C_6H_4-NHCOCH_3} + CH_3COOH$$

Uses. As a safe antipyretic for curing fevers.

Q. 5. Describe the synthesis, physiological action and uses of sulphanilamide (p-amino-benzene sulphonamide). (*Kerala, 2000; Garhwal, 2000; Guwahati 2002, Nagpur 2002*)

Ans. Synthesis. Acetanilide is treated with chlorosulphonic acid to produce *p*-acetamido benzene sulphonyl chloride which is treated with NH_3 to produce *p*-acetamido benzene sulphonamide. The latter on hydrolysis in the presence of an acid yields sulphanilamide.

$$\underset{\text{Acetanilide}}{C_6H_5NHCOCH_3} \xrightarrow{ClSO_3H} \underset{p\text{-Acetamido benzene sulphonyl chloride}}{CH_3COHN-C_6H_4-SO_2Cl} \xrightarrow{NH_3} \underset{p\text{-Acetamido benzene sulphonamide}}{CH_3COHN-C_6H_4-SO_2NH_2}$$

$$\xrightarrow[H^+]{H_2O} \underset{\text{Sulphanilamide}}{H_2N-C_6H_4-SO_2NH_2}$$

It has got antibacterial properties. The antibacterial activity of sulphanilamide is associated with the group.

$$H_2N-C_6H_4-SO_2-N<$$

p-aminobenzoic acid is an essential growth factor for most bacteria susceptible to sulphonamide. The theory of action is, that due to similarity in structure, bacteria absorb sulphonamide by mistake and the bacteria cease to grow in number. Thus sulphonamides are bactericidal as well as baçteriastatic.

Uses. 1. It is used as antibacterial agent.

2. It is used in medicine to cure cocci-infections, streptococci, gonococci and pneumococci.

Q. 6. Describe the synthesis and uses of sulphaguanidine.

(*Awadh, 2000; Kerala, 2000; Kurukshetra, 2000; M. Dayanand, 2000; Nagpur, 2008*

Ans. The starting material for sulphaguanidine is the same as for sulphanilamide. *p*-acetamido benzene sulphonyl chloride obtained in the first step is treated with guanidine to obtain sulphaguanidine.

Acetanilide $\xrightarrow{Cl\text{-}SO_3H}$ *p*-acetamido benzene sulphonyl chloride $\xrightarrow[\text{(Guanidine)}]{NH_2-C(=NH)-NH_2}$ $CH_3CONH-C_6H_4-SO_2-NH-C(=NH)-NH_2$ $\xrightarrow{H_2O/H^+}$ $NH_2-C_6H_4-SO_2-NH-C(=NH)-NH_2$

Sulphaguanidine

Uses. It is used in the treatment of bacillary dysentry.

Q. 7. Describe the synthesis and uses of chloramphenicol (chloromycetin).

(*Bangalore, 2002; Delhi, 2003*)

Ans. It is a laevorotatory compound which is produced by **streptomyces venezuelase.**

$O_2N-C_6H_4-CO-CH_3 \xrightarrow{Br_2} O_2N-C_6H_4-CO-CH_2-Br$

$\xrightarrow[(ii)\ HCl-C_2H_5OH]{(i)\ (CH_2)_4N_4} O_2N-C_6H_4-CO-CH_2-NH_2 \xrightarrow{(CH_3CO)_2O}$

$O_2N-C_6H_4-CO-CH_2-NH-CO-CH_3$

$\xrightarrow[Na_2CO_3\ (aq)]{HCHO} O_2N-C_6H_4-CO-CH(NH-CO-CH_3)(CH_2OH) \xrightarrow[\text{Aluminium isopropoxide}]{[(CH_3)_2CHO]_3Al]}$

$$O_2N-C_6H_4-CH(OH)-CH(NH-CO-CH_3)(CH_2OH) \xrightarrow{HCl} O_2N-C_6H_4-CH(OH)-CH(NH_2)(CH_2OH)$$

$$\xrightarrow[(ii)\ CHCl_2-COOCH_3]{(i)\ \text{Resolved}} O_2N-C_6H_4-CH(OH)-CH(NHCOCHCl_2)(CH_2OH)$$

(–) Chloramphenicol

Uses. 1. It is very effective in the treatment of typhoid fever.

2. It is used for curing diarrhoea, pneumonia and whooping cough.

Q. 8. Give the synthesis and uses of chloroquine. (*Kurukshetra, 2001; Kerala, 2001*)

Ans. The structural formula of chloroquine is given below:

$$\text{(7-chloroquinolin-4-yl)}-NH-CH(CH_3)-CH_2-CH_2-CH_2-N(C_2H_5)_2$$

Synthesis of chloroquine involves three stages as given below:

(*i*) *Synthesis of 4, 7 dichloroquinoline*

$$\underset{\text{2-amino-4-chloro-benzoyl chloride}}{Cl-C_6H_3(NH_2)-COCl} + \underset{\text{Acetaldehyde}}{CH_3CHO} \xrightarrow[-2H_2O]{NaOH} \underset{\text{4, 7-Dichloroquinoline}}{C_9H_5NCl_2}$$

(*ii*) *Synthesis of 4-amino-1-diethylaminopentane*

$$\underset{\text{Ethylene chlorohydrin}}{CH_2OH-CH_2Cl} \xrightarrow[-HCl]{HN(C_2H_5)_2} CH_2OH-CH_2N(C_2H_5)_2 \xrightarrow[-HCl,\ -H_2O]{SOCl_2} CH_2Cl-CH_2N(C_2H_5)_2$$

$$\begin{matrix} CH_2Cl \\ | \\ CH_2N(C_2H_5)_2 \end{matrix} + \underset{\text{Sod. salt of aceto-acetic ester}}{[CH_3COCHCOOC_2H_5]^- Na^+} \xrightarrow[-NaCl]{ClCH_2-CH_2N(C_2H_5)_2}$$

$$\begin{matrix} CH_3COCHCOOC_2H_5 \\ | \\ CH_2CH_2N(C_2H_5)_2 \end{matrix} \xrightarrow[\text{Hydrolysis}]{\text{Ketonic}} CH_3COCH_2CH_2CH_2N(C_2H_5)_2$$

$$\xrightarrow[\text{Amination}]{\text{Reductive}} \begin{matrix} CH_3-CH-CH_2CH_2CH_2N(C_2H_5)_2 \\ | \\ NH_2 \end{matrix}$$

NH_3, Ni, H_2

4-amino-1-diethylaminopentane

(*iii*) *Condensation of 4, 7 dichloroquinoline and 4-amino -1, 1-diethyl aminopentane.*

$$\text{4, 7-Dichloroquinoline} + \begin{matrix} H_2N-CH-(CH_2)_3-N(C_2H_5)_2 \\ | \\ CH_3 \end{matrix}$$

4-amino-1, 1-diethyl aminopentane

$\downarrow$ (– HCl) C_2H_5OH, reflux

$$\begin{matrix} NH-CH-(CH_2)_3-N(C_2H_5)_2 \\ | \\ CH_3 \end{matrix}$$

Chloroquine

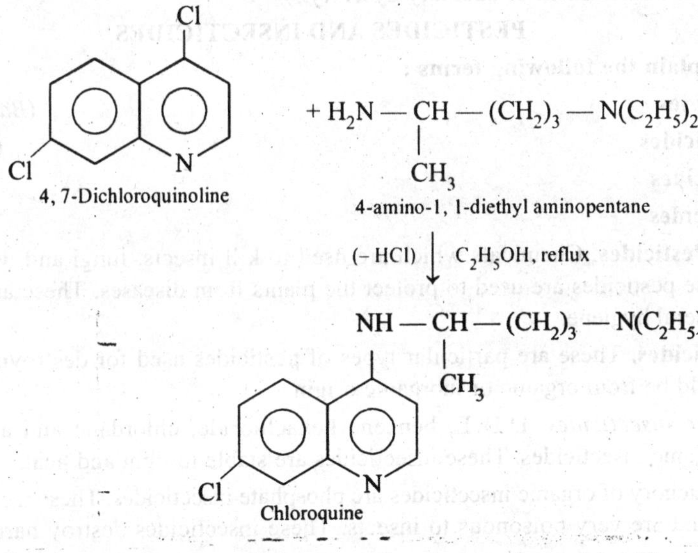

Uses. It is used as a common medicine to cure malaria without side effects.

MISCELLANEOUS QUESTIONS

Q. 9. What are drugs? (*Shivaji, 2000; Panjab 2000*)

Ans. Drug is a broad term used for chemical substances, obtained from natural sources or synthesised in the laboratory, used to cure ailments and diseases.

Antiseptics used to stop growth of microorganism on a wound, antipyretics used to lower down body temperature, analgesics used to relieve pain, antimalarials used to combat malaria and antibiotics used to kill organism, are all examples of drugs.

Q. 10. Give name structure and one method of preparation for an

(*i*) Antimalarial (*ii*) Antipyretic

(*iii*) Antibacterial drug.

Ans. An example of an antimalarial drug is quinine, that of an antipyretic is paracetmol. An example, of anti-bacterial drug is chloramphenicol.

For structure and method of preparation, see questions 8, 4 and 7 respectively in this chapter.

Q. 11. Name a drug which is used in treatment of typhoid. How it can be synthesised? (*M. Dayanand, 2000; Bangalore, 2001*)

Ans. Chloramphenicol is used in the treatment of typhoid. It is an anti-bacterial drug.

For the synthesis of chloramphenicol see Q. 7.

Q. 12. Write the structural formula of sulphaguanidine? It is used for the treatment of which disease?

Ans. Sulphaguanidine has the structure

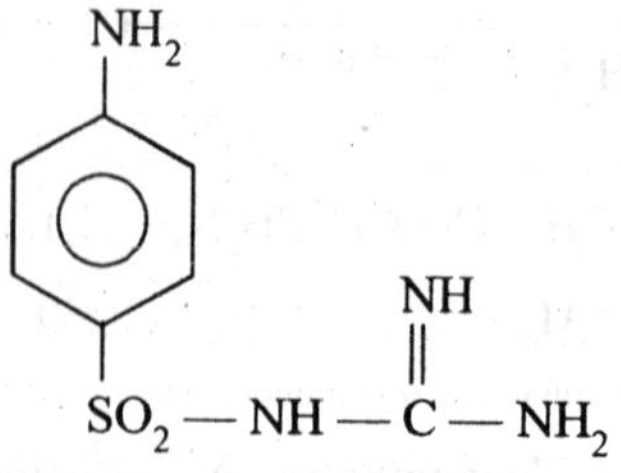

$$H_2N-C_6H_4-SO_2-NH-\underset{\Large\|}{\overset{NH}{C}}-NH_2$$

It is used in the treatment of bacillary dysentry.

PESTICIDES AND INSECTICIDES

Q. 13. Explain the following terms :

(*a*) **Pesticides** (*Bangalore, 2002*)

(*b*) **Insecticides** (*Kanpur, 2001*)

(*c*) **Herbicides**

(*d*) **Fungicides**

Ans. (*a*) **Pesticides.** Chemicals which are used to kill insects, fungi and weeds are called pesticides. These pesticides are used to protect the plants from diseases. These are also used for maintaining general hygiene.

(*b*) **Insecticides.** These are particular types of pesticides used for destroying insects. The insecticides could be from organic or inorganic origin.

(*i*) *Organic insecticides.* D.D.T., benzene hexachloride, chlordane and aldrin are some examples of organic insecticides. These insecticides are stable to light and heat.

Another category of organic insecticides are phosphate insecticides. These include malathion, parathion etc. and are very poisonous to insects. These insecticides destroy harmful as well as useful insects.

(*ii*) *Inorganic insecticides.* Some common inorganic insecticides are given below.

Fluorine compounds in the form of sodium, calcium and barium fluosilicates.

($Na_2\ SiF_6$, $CaSiF_6$ and $BaSiF_6$)

Arsenic compounds like arsenates of calcium and lead $Ca_3(AsO_4)_2$, $Pb_3(AsO_4)_2$.

Mercuric chloride and calomel (mixture of mercury and mercury chloride) also act as insectides.

(*c*) **Herbicides.** Pesticides which are used to destroy unwanted weeds in the crops are called herbicides. Some examples of herbicides are simazine, 2, 4-D and 2, 4, 5-T.

(*d*) **Fungicides.** These are many fungi which are responsible for producing plant diseases. Chemicals which destroy such fungi are called fungicides. Some examples of fungicides are:

Bordeaux, which is a mixture of lime, $CuSO_4$ and water.

Organo-metallic compounds of mercury and tin.

Dithiocarbamates and pentachlorophenol.

Q. 14. What are the merits and demerits of using pesticides? (*Bangalore, 2002*)

Ans. There is no doubt that pesticides protect the plants from insects, fungi and weeds. Lot of food which was spoilt earlier by insects, weeds and fungi is saved now with the discovery and use of more and more pesticides. Thus it has proved to be a boon to the mankind. But there is a darker side of pesticides also. Some of them are very toxic and leave a permanent effect. Such pesticides are called **hard pesticides.** D.D.T. which is an example of hard pesticide is not easily degraded and destroyed by environment processes. Such pesticides continue to remain in soil, water and plants. They also make their presence felt in the tissues of animals and is thus a health hazard.

Use of pesticides is, therefore, a boon and a curse for the mankind.

Q. 15. Write down the preparation of D.D.T. What are its limitations in use?

(Meerut, 2000; Kurukshetra, 2000; Devi Ahilya, 2001; M. Dayanand, 2002; Bangalore, 2002)

Ans.

$$Cl-C_6H_4-CH(C_6H_4Cl)-CCl_3$$

2, 2, di(*p*-chlorophenyl) 1, 1, 1-trichloroethane (D.D.T.)

Preparation. It is obtained by condensation between chlorobenzene and chloral in the presence of an acid.

$$\underset{\text{Chlorobenzene}}{2Cl-C_6H_5} + \underset{\text{Chloral}}{H-\overset{O}{\overset{||}{C}}-CCl_3} \xrightarrow{H_2SO_4} \underset{\text{(D.D.T.)}}{Cl-C_6H_4-CH(C_6H_4Cl)-CCl_3}$$

Uses. It destroys insects carrying diseases of malaria and typhus. However being a hard pesticide, its continuous use poses an environment problem and hence its use is being discontinued slowly.

Q. 16. Write the preparation and uses of BHC (benzene hexachloride).

(Delhi, 2003; Purvanchal, 2007; Garhwal, 2011)

Ans.

$$C_6H_6Cl_6$$

Benzene hexachloride

Preparation. It is prepared by the action of chlorine on benzene. Action takes place in the presence of sunlight by free-radical mechanism. A mixture of products containing isomers of 1, 2, 3, 4, 5, 6-hexachlorocyclohexane (or benzene hexachloride) is obtained. This mixture is called BHC or 666.

$$\text{Benzene} + 3Cl_2 \xrightarrow{\text{Sunlight}} \text{Benzene hexachloride } (C_6H_6Cl_6)$$

There is one hydrogen atom linked at every corner which has not been shown.

There are 9 possible stereoisomers of BHC, out of which 8 have been identified. The toxic properties of BHC are attributed to *r*-isomer which constitutes 10–18% of the mixture. It is known as gammaxene or lindane. It has the following chair-form structure:

Structure formula of lindane

Uses. It is used as a potent insecticide.

Q. 17. Give the preparation and uses of malathion. (*Bangalore, 2001; Kurukshetra 2001*)

Ans.

$$(CH_3O)_2-\overset{\overset{S}{\|}}{P}-S-\underset{\underset{CH_2COOC_2H_5}{|}}{CHCOOC_2H_5}$$

Preparation. This important member of the class of organophosphates is prepared by reacting methyl alcohol (or ethyl alcohol) with phosphorus pentasulphide to get dimethyl dithiophosphate. The latter on treatment with diethyl maleate undergoes Michael addition to give malathion.

$$4CH_3OH + P_2O_5 \longrightarrow 2(CH_3O)_2-\overset{\overset{S}{\|}}{P}-SH + H_2S$$

$$(CH_3O)_2-\overset{\overset{S}{\|}}{P}-SH + \underset{\text{Dimethyl maleate}}{\underset{\underset{CHCOOC_2H_5}{\|}}{CHCOOC_2H_5}} \longrightarrow \underset{\text{Malathion}}{(CH_3O)_2-\overset{\overset{S}{\|}}{P}-S-\underset{\underset{CH_2COOC_2H_5}{|}}{CHCOOC_2H_5}}$$

Uses. 1. Malathion is an effective insecticide. It has the merit of being less toxic to mammals. 2. It is not a hard insecticide. It can be easily degraded by environmental conditions.

ADDITIONAL QUESTIONS

Q. 18. What is gammaxene? How can BHC be prepared?

Ans. Gammaxene is one of the conformational isomers of hexachlorocyclohexane. Arrangement of hydrogen and chlorine groups on the cyclohexane ring in the chair-form is shown in the figure in Q. 16.

It has three chlorines in axial and three in equitorial positions. Gammaxene is also known as lindane. This is a very toxic compound and constitutes 10–18% of BHC which is a mixture of different isomers of hexachlorochclohexane.

For preparation of BHC see Q. 16.

Q. 19. Give one method of preparation of malathion, its importance to farmers and show how is it better than DDT.

Ans. For method of preparation of malathion, see Q. 17.

Importance of malathion.

Malathion is a potent and effective pesticide against insects, fungi and weed. It is a real boon to the farmer to protect the crops from pests and insects.

Malathion has the advantage that it is a soft insecticide. This means that it can be degraded by ordinary environmental conditions. Thus it leaves no ill effects. On the contrary D.D.T. is a hard insecticide. It is not easily degraded and leaves its toxic effects long after its use.

Q. 20. Give two examples of organophosphorus insectides. What is their mode of action? *(Delhi, 2003)*

Ans. One of the organophosphorus insecticides is malathion (see Q. 17). The other is tetratethyl pyrophosphate.

$$(C_2H_5O)_2P(=O)-O-P(=O)(OC_2H_5)_2$$

The insects are lured by smell and come to poison bait.

37

PHOTOCHEMISTRY

Q. 1. What is photochemistry? List its important applications. (*Kerala, 2001*)

Ans. Photochemistry is the branch of chemistry which deals with reactions initiated by radiations. The radiations generally considered lie in the range of 1000–10000 Å.

Applications

1. The process of *photosynthesis* which the plants carry out takes place with the help of sunlight radiations. Carbon dioxide, water, oxygen, ammonia and methane react under the influence of sunlight and in the presence of chlorophyll to produce carbohydrates, proteins etc., which is so important for humans and animals.

2. Photochemistry displays its importance in the synthesis of some important compounds which was not possible with dark reactions (without radiations). Some examples of industrially important substances produced by photochemical methods are : Vitamin D from ergosterol, caprolactum which is a manomer for nylon-6, insecticides and halogenated aromatics.

3. Fluorescent tubes and X-rays work on the phenomenon of fluorescence and phosphorescence. These are all photo-induced processes.

4. Photography and litho printing work on the principle of photopolymerisation.

5. Photochromic materials, like spiropyrans are used in sunglasses. These materials are photo-sensitive. They change their colour when some radiation is incident on them and revert to the original colour when the radiation is cut-off.

6. Their latest and very important application is in Laser technology. Lasers are intense monochromatic and coherent beams of radiations. These laser beams represent electronically excited molecular systems. Because of their ability to concentrate on a tiny point, lasers find use in cutting and boring diamonds and in eye surgery to remove a defect.

7. Techniques like flash photolysis and pulsed laser photolysis, used to study high energy systems work on the broad phenomenon of photochemistry.

8. Solar energy conversion and storage systems are being worked out based on the principles of photochemistry.

Q. 2. What are the main points of difference between thermochemical and photochemical reactions?

Or

What rules govern thermal and photochemical reactions?

Ans. Main points of difference between thermal and photochemical reactions are summarized below:

1. The reactants in **thermal reactions** are in their ground electronic states whose vibrational, rotational and translational energies are in the upper range of the distribution as per the Maxwell-Boltzmann law. Reaction can occur, between any of the molecules which have energies above the minimum amount (Threshold energy) necessary for the reaction. In thermal reactions, it is not possible to know as to how many molecules are taking part in the reaction. Therefore, the exact excitation energy for the reaction is not known.

In case of **photochemical reactions,** the exact degree of excitation can be controlled with the help of radiations of desired wavelength provided the molecules are capable of a absorbing them. For a usual photochemical reaction, each molecule that takes part in the reaction absorbs one quantum of radiation. The reactant molecules in the photochemical reactions are not in the ground electronic state.

2. A high temperature is generally used during a *thermally excited reaction.* The reaction intermediates free radical or molecular fragments are formed. They have short life times and less concentration. Due to this, the nature of the reactant intermediates cannot be studied.

In a **photochemical reaction,** specific bond cleavage and other forms of chemical changes can be brought about at any desired initial temperature of the reactant molecules.

3. A photochemical reaction follows an entirely different potential energy path from the one followed by the thermally excited molecules even when the reaction is carried at the temperature equivalent to the energy of absorbed light. The reason is that in case of thermal reactions, there are many alternative paths of low energy that the molecules can make use of. Configuration which is reached in photochemical reactions is never attained in case of thermal reactions. Hence we say that the products of photochemical reactions may be unique, thermodynamically highly unstable and structurally strained.

4. As a rule only those **thermochemical reactions** are feasible which are accompanied by the decrease in free energy (ΔG is negative).

On the contrary many **photochemical reactions** are known which accompany an increase in free energy. Some examples are (*i*) conversion of oxygen into ozone (*ii*) Photosynthesis of plants etc. The energy of the radiations get converted into the free energy of the products.

5. Thermally excited reactions are affected by the change in temperature of the reactants. The reason is that with the rise in temperature, the number of effective collisions increases. Effect of temperature is so dominating that the rate of reaction becomes double for every 10° rise in temperature. In certain cases, where the activation energy is very high, the reaction occurs at high temperature. For example, the reduction of acidified potassium permanganate with sodium oxalate occurs at high temperature.

The rate of **photochemical reaction** is independent of the temperature of the reactants but depends upon the intensity of the light absorbed.

Q. 3. Explain the process of electronic excitation in photochemical reactions.

Ans. When a molecule absorbs light in the visible and ultraviolet region, electrons are promoted from bonding to non-bonding orbitals. This is shown in the diagram below:

Fig. 36.1

σ and π represent bonding orbitals, σ* and π* are corresponding antibonding orbitals. *n* represent a non-bonding electron. There is nothing like *n** as this electron does not participate in the bonding process. These orbitals σ, π, π* and σ* have their wave function Ψ_1, Ψ_2, Ψ_3 and Ψ_4 respectively. On absorption of radiation, the electrons may be excited from σ to σ*, σ to π*, π to π*, *n* to π* or *n* to σ*. This depends upon the frequency or wavelength of the radiation involved. The energy of visible light varies from 160 kJ/mole to 290 kJ/mole whereas U.V. light has the energy 590 kJ corresponding to wavelength of 200 nm. The actual transition involved depends upon the energy of the radiation.

Consider the case of butadiene $CH_2 = CH - CH = CH_2$.

In this case π – π* transition takes place as illustrated below:

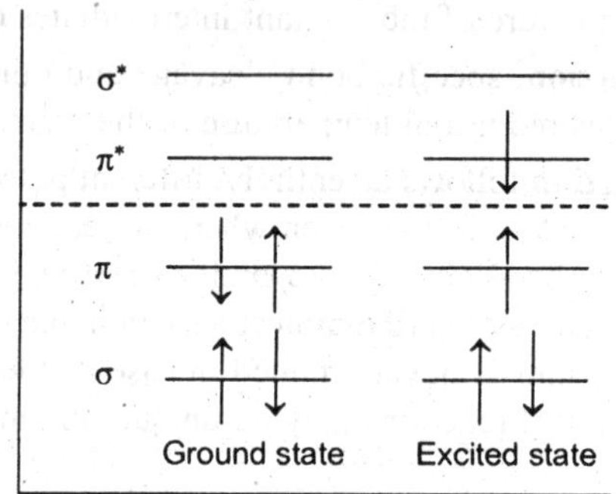

Fig. 36.2

In the ground state, there are 2 electrons in the π (bonding) orbital. On absorption of radiation, one of these is excited to π* orbital.

Q. 4. Explain the terms (*i*) singlet state (*ii*) triplet state. (*Panjab, 2003; Bhopal, 2004*)

Ans. (*i*) **Singlet State.** A molecule is said to be in singlet state if all the electrons in it are paired. It is represented by the symbol S.

(*ii*) **Triplet State.** A molecule is said to be in triplet state if all the electrons in the molecule are not paired. It is represented by the symbol T.

According to Hund's rule, the lower energy state will be one in which the electrons have parallel spins *i.e.* they are unpaired. Thus according to this rule, triplet state is associated with smaller energy, as there are lesser inter-electronic repulsions. Let us consider a molecule having two paired electrons in the ground state. This state is represented by S_0. On absorption of radiation, in changes to the excited state S_1 (singlet), when one electron is lifted from bonding to antibonding orbital. The two electrons are still paired. Sometimes, it so happen that this electronic excitation is followed by spin-inversion, giving rise to the triplet state (shown as T_1). Direct conversion from S_0 to T_1 is not possible.

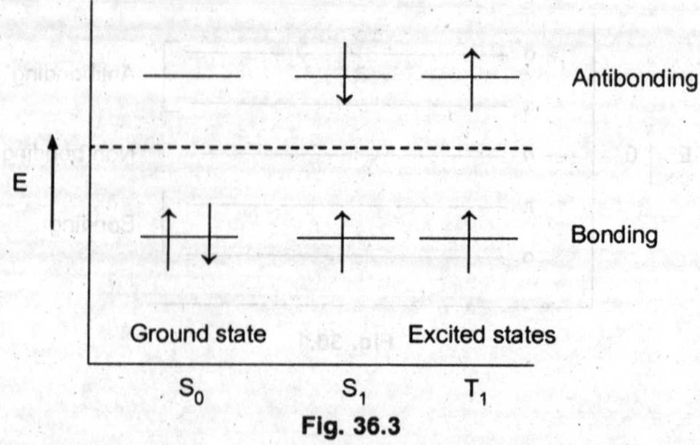

Fig. 36.3

Q. 5. Explain $n - \pi^*$ and $\pi - \pi^*$ excitation with the example of benzophenone. *(Kerala, 2001)*

Ans. The excitations (transitions) of interest to a chemist are $n - \pi^*$ and $\pi - \pi^*$ as these are associated with wavelength with the range of usual instruments. Benzophenone absorbs both these radiations, one around 285 nm and the other at 180 nm. Absorption of 285 nm is due to excitation of non-bonding electron to lowest unoccupied antibonding π orbital (π^*), whereas, the one at 180 nm is due to the promotion of electron from π (bonding) orbital to π^* orbital. Evidently, the absorption at 285 nm (the higher wavelength) is associated with lower amount of energy than at 180 nm. The figure below shows two absorption maxima corresponding to absorption of radiation of wavelength.

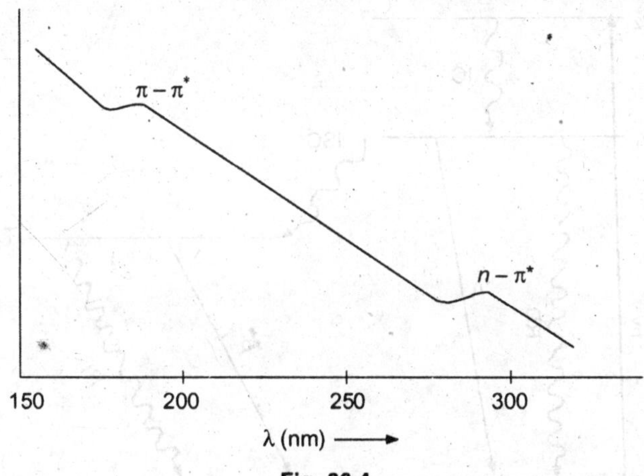

Fig. 36.4

Q. 6. Give Jablonski's diagram describing dissipation (decay) of energy from higher to lower levels. *(Kerala, 2000)*

Or

Describe photophysical processes in electronically excited molecules. *(Punjab, 2003)*

Or

Explain fluorescence and phosphorescence by means of a labelled diagram. *(Mumbai, 2010)*

Ans. 1. A molecule on absorption of radiation is excited from the ground state S_0 to S_1 where S_1 is the excited singlet state.

2. From S_1, there may be further excitation to S_2 where S_2 represents the higher vibrational level of S_1.

3. Now when it loses the energy, S_2 state changes rapidly to S_1 because the lifetime of S_2 is less than 10^{-11} sec. This conversion from S_2 to S_1 is called **internal conversion.** It is also called **vibrational cascade.**

4. S_1 which has lifetime ~ 10^{-8} s is longest living of all excited singlet states. It undergoes energy losing process, although with a slow pace, in one of the following ways.

(*i*) It may emit energy as a photon and revert to S_0. This process is called **fluorescence.**

(*ii*) It may drop to S_0 by non-radiative process in which the excess energy of the excited state is shuffled into vibrational mode.

(*iii*) S_1 may undergo chemical reaction.

(*iv*) It may drop to T_1 (triplet) state. Although T_1 state is associated with smaller energy, this transition is a slow process in conformity with the spectroscopic rule that *change of multiplicity is a forbidden process.* The transition $S_1 \rightarrow T_1$ is called intersystem crossing and is abbreviated as ISC. Intersystem crossing is an important phenomenon because of long lifetime. In cases where there is a smaller energy gap between S_1 and T_1, this transition takes place efficiently. These transition are shown in the modified Jablonski's diagram below.

5. T_1 state may undergo the following energy-degradation processes.

(*i*) It may drop to S_0 by emission of photon. This step is called *phosphorescence*, and has a lifetime of $10^{-3} - 10$ sec.

(*ii*) It may return to S_0 by radiationless decay.

(*iii*) It may start a reaction.

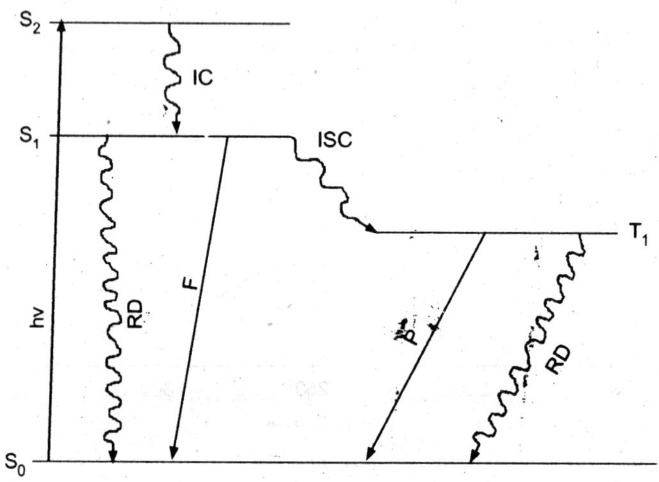

Fig. 36.5. Modified Jablonski diagram

S_0 = Ground state	IC = Internal conversion
S_1 = Singlet excited state	ISC = Intersystem crossing
T_1 = Triplet excited state	F = Fluorescence
S_2 = Higher vibrational level of S_1	P = Phosphorescence
	RD = Radiationless decay

Q. 7. Describe the mode of energy transfer from one molecule to another.

Ans. As T_1 has a greater lifetime compared to S_1, there is a greater probability of energy transfer through T_1. There are many substances which have a wide energy difference in the S_1 and T_1 states. For such substances, conversion from S_1 to T_1 is not feasible *i.e.* it is not possible to create T_1-state for such molecule. However it can be achieved indirectly, by transfer of energy of triplet state of some other substances. The mechanism is like this:

Donor molecule is excited from S_0 to S_1 and subsequently converted into T_1. Collision of T_1 donor with the ground state acceptor molecule produces T_1 acceptor and ground state donor molecule.

$$D \xrightarrow{h\nu} D^1 \qquad D = \text{Donor}$$

$$D^1 \xrightarrow{ISC} D^3 \qquad A = \text{Acceptor}$$

$$A + D^3 \xrightarrow[\text{Transfer}]{\text{Energy}} D + A^3 \qquad D^1 = \text{Donor (excited singlet)}$$

$A^3 \longrightarrow$ Products D^3 = Donor (excited triplet)

A^3 = Acceptor (excited triplet)

Q. 8. What is meant by quantum yield of a photochemical reaction? Describe the mechanism of photoreduction of benzophenone to benzopinacol.

(Panjab, 2003; Bhopal, 2004)

Ans. Usually, absorption of one photon of a suitable wavelength produces one molecule of the product. But in many cases, this is not so because of the decay of some radiation by fluorescence and phosphorescence.

Quantum yield or quantum efficiency of a photochemical reaction, ϕ, may be defined as,

$$\phi = \frac{\text{No. of molecules reacting or formed in unit time}}{\text{No. of quantas of radiation absorbed in unit time}}$$

Photoreduction

Benzophenone is one of the substances whose photoreduction has been studied in detail. Photoreduction of this substance is carried out by taking a solution of benzophenone in isopropyl alcohol and irradiating it with a radiation of 345 nm. Isopropyl alcohol does not absorb this radiation but $n - \pi^*$ transition takes place in benzophenone. S_1 state of benzophenone drops to T_1 state. The excited triplet of benzophenone abstracts a hydrogen atom from isopropyl alcohol as shown below:

$$\underset{\text{Benzophenone}}{(C_6H_5)_2CO} \xrightarrow{h\nu} \underset{\text{Excited Singlet } S_1}{[(C_6H_5)_2CO]^1} \longrightarrow \underset{\text{Excited Triplet } T_1}{[(C_6H_5)_2CO]^3}$$

$$[(C_6H_5)_2CO]^3 + CH_3CHOHCH_3 \longrightarrow \underset{\text{Benzophenone ketyl radical}}{(C_6H_5)_2\dot{C}-OH} + \underset{\text{Acetone ketyl}}{CH_3\overset{\bullet}{\underset{\underset{OH}{|}}{C}}CH_3}$$

$$(C_6H_5)_2CO + CH_3-\overset{\bullet}{\underset{\underset{OH}{|}}{C}}-CH_3 \longrightarrow (C_6H_5)_2\dot{C}OH + CH_3-\underset{\underset{O}{||}}{C}-CH_3$$

$$2(C_6H_5)_2\dot{C}OH \longrightarrow \underset{\text{Benzopinacol}}{(C_6H_5)_2-\overset{\overset{OH}{|}}{C}-\overset{\overset{OH}{|}}{C}-(C_6H_5)_2}$$

Acetone ketyl transfers its hydrogen to another molecule of benzophenone. Two benzophenone ketyl radicals unite to give benzopinacol.

The evidence in favour of this mechanism is:

(*i*) The quantum efficiency of the reactions is one. This means one molecule of the product is obtained by the absorption of one photon of radiations.

(*ii*) It may be noted from the above reactions that one molecule of the pinacol is obtained from two benzophenone ketyl radicals and the two benzophenone ketyl radicals are obtained by the absorption of one photon only (the 1st step) and not two photons.

Q. 9. What are Norrish type I or α-cleavage photochemical reactions?

Or

Write a note on photochemistry of carbonyl compounds.

Ans. 1. This reaction concerns the irradiation of a carbonyl compound. It is observed that bond dissociation energy of C – C bond adjacent to the carbonyl group is comparatively small.

This bond undergoes homolytic fission. Acetone undergoes homolytic fission at the C – C bond adjacent to the carbonyl group. Methyl and acetyl free radicals are formed.

$$CH_3-\overset{\overset{\large O}{\|}}{C}-CH_3 \xrightarrow{h\nu} \dot{C}H_3 + CH_3-\overset{\overset{\large O}{\|}}{C}\bullet$$

(*i*) At room temperature, two acetyl free radicals combine to form diacetonyl

$$2CH_3-\overset{\overset{\large O}{\|}}{C}\bullet \longrightarrow CH_3-\overset{\overset{\large O}{\|}}{C}-\overset{\overset{\large O}{\|}}{C}-CH_3$$

(*ii*) Above 373 K, secondary photochemical process takes place, giving rise to the following products.

$$CH_3-\overset{\overset{\large O}{\|}}{C}\bullet \longrightarrow \dot{C}H_3 + CO$$

$$2\dot{C}H_3 \longrightarrow CH_3-CH_3$$

2. When unsymmetrical ketones are involved, the free radicals produced are the ones, which are more stable. Consider the following ketones.

(*a*)

$$CH_3COCH_2CH_3 \xrightarrow{h\nu} CH_3C\dot{O} + CH_3\dot{C}H_2$$

$$2CH_3C\dot{O} \longrightarrow CH_3COCOCH_3$$

$$2CH_3\dot{C}H_2 \longrightarrow \underset{\text{n-Butane}}{CH_3CH_2CH_2CH_3}$$

(*b*)

$$CH_3COCH(CH_3)_2 \xrightarrow{h\nu} CH_3C\dot{O} + (CH_3)_2\dot{C}H$$

$$2CH_3C\dot{O} \longrightarrow CH_3COCOCH_3$$

$$2(CH_3)_2\dot{C}H \longrightarrow (CH_3)_2CHCH(CH_3)_2$$

3. Consider the photolytic decarboxylation of the cyclic ketone

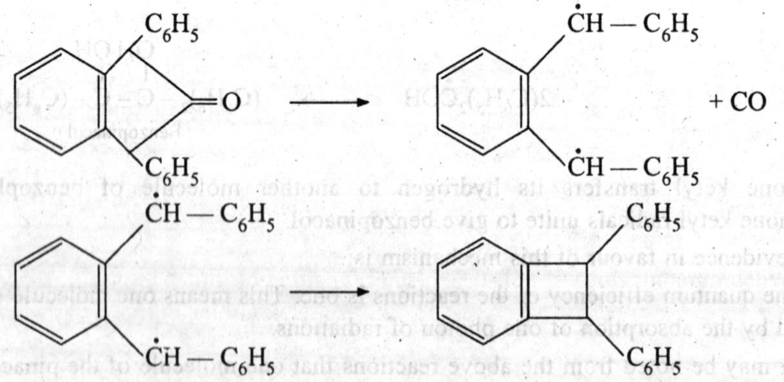

[Structure: 1,3-diphenylindan-2-one (C_6H_5 groups on the two carbons flanking C=O, fused to benzene ring) $\longrightarrow$ o-xylylene diradical (benzene ring bearing two $\dot{C}H-C_6H_5$ groups) + CO]

[Structure: 2 × (benzene ring bearing two $\dot{C}H-C_6H_5$ groups) $\longrightarrow$ benzocyclobutene bearing two C_6H_5 groups]

Q. 10. Explain the term photodissociation with the help of an example.

Ans. See Q. 9.

Q. 11. What are Norrish type II reactions? Explain with examples.

Ans. These reactions relate to the photochemical changes taking place in ketones possessing γ-hydrogen atom. In this process an olefin and enol of a smaller ketone is formed.

$$(i)\quad R-\overset{O}{\overset{\|}{C}}\overset{\alpha}{C}H_2\overset{\beta}{C}H_2\overset{\gamma}{C}H_2CH_3 \xrightarrow{h\nu} R-\overset{O\cdots H}{\overset{\|}{C}}\ \ CH_2\text{—}CH_2\text{—}CH\text{—}CH_3$$

$$\downarrow$$

$$R-\overset{OH}{\overset{|}{C}}=CH_2 + CH_3CH=CH_2 \longleftarrow R-\overset{OH}{\overset{|}{\dot{C}}}\ \ CH_2-CH_2-\dot{C}HCH_3 \quad \text{Biradical}$$

$$\rightleftharpoons$$

$$R-\overset{O}{\overset{\|}{C}}-CH_3$$

The reaction takes place by initial γ-hydrogen abstraction by oxygen, thereby providing a 1, 4 biradical. The biradical is broken into an olefin and an enol.

(*ii*) α-hydrogen abstraction is sometimes followed by ring closure of the biradical produced above.

$$R-\overset{OH}{\overset{|}{\dot{C}}}-CH_2-CH_2-\dot{C}HCH_3 \longrightarrow R-\overset{OH}{\overset{|}{C}}\text{—}CH-CH_3 \quad (\text{ring: } CH_2-CH_2)$$

(*iii*) It is sometimes observed that the products obtained correspond to Norrish type I and type II both. Irradiation of 2-pentanone gives the following products:

$$CH_3COCH_2CH_2CH_3 \xrightarrow{h\nu}$$

Norrish type I:

$$\begin{bmatrix} \dot{C}H_3 + CH_3CH_2CH_2\dot{C}O \\ CH_3\dot{C}O + CH_3CH_2\dot{C}H_2 \end{bmatrix}$$

$$\downarrow$$

$CH_3COCOCH_3$
$+ CH_3COCH_3$
$+ (CH_3CH_2CH)_2$

Norrish type II:

$CH_3COCH_3 + CH_2=CH_2$

$+ H_3C-$ (1-hydroxycyclobutyl, OH on the cyclobutane ring)

Q. 12. Describe Paterno-Buchi reaction. (*Guwahati, 2002*)

Ans. This reactions concerns irradiation of carbonyl compounds in the presence of olefins to produce oxetanes.

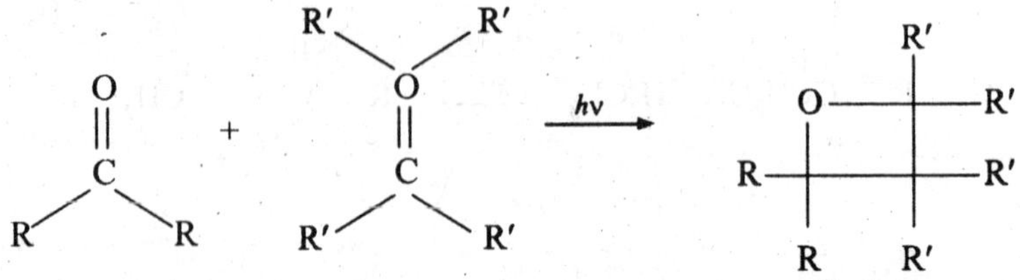

$$R_2C{=}O + R'_2C{=}CR'_2 \xrightarrow{h\nu} \text{Oxetanes}$$

Oxetanes

In this reaction, the radiation to be used is such as can be absorbed only by the carbonyl compound and not by the olefin.

Mechanism of the reaction is as follows:

$$(C_6H_5)_2CO \xrightarrow{h\nu} [(C_6H_5)_2CO]^1$$

Excited singlet

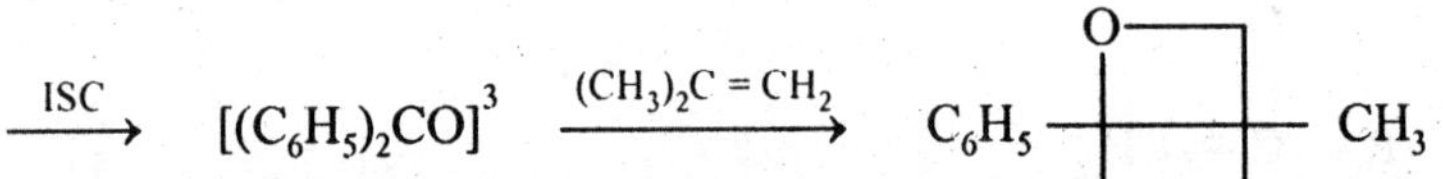

$$\xrightarrow{ISC} [(C_6H_5)_2CO]^3 \xrightarrow{(CH_3)_2C=CH_2} \text{oxetane (}C_6H_5, C_6H_5, CH_3, CH_3\text{)}$$

Excited triplet

Q. 13. Explain photocycloaddition reactions. (*Andhra, 2010*)

Ans. If the carbonyl group is conjugated with a carbon-carbon double bond, as in the case of α, β unsaturated ketone, the energy gap between π and π^* orbitals is lowered and therefore a radiation of longer wavelength can bring about $\pi - \pi^*$ transition. There will be a simultaneous lowering of energy gap between non-bonding orbital and π^* orbital. Thus $n - \pi^*$ transitions will also take place conveniently. Important reactions involving this principle are olefin addition across double bond and photocyclodimerisation.

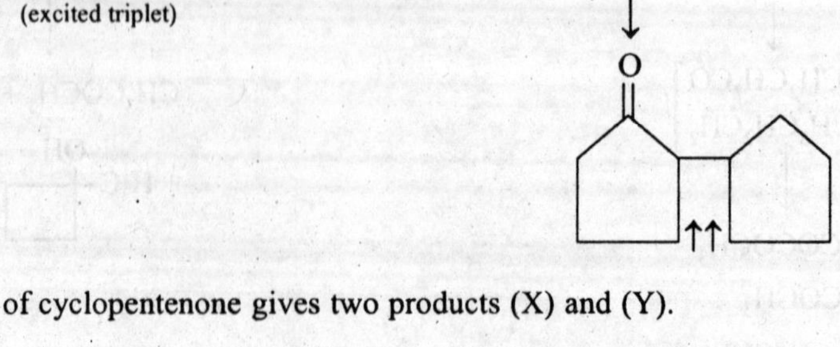

Cyclopentenone (excited triplet)

Irradiation of cyclopentenone gives two products (X) and (Y).

(X) (Y)

Q. 14. What is photochemical isomerisation?

Ans. 1. Olefins absorb U.V. light and undergo geometrical isomerisation. This transformation can take place through singlet or triplet excited species. Interconversion of fumaric and maleic acid is illustrated below:

$$\text{HOOC(H)C=C(H)COOH} \xrightarrow{h\nu} \text{HOOC(H)C=C(H)COOH}$$

Fumaric acid (*Trans*) E isomer → Maleic acid (*Cis*) Z isomer

2. A common reaction among photoisomerisations is cis-trans isomerisation of stilbene.

$$C_6H_5(H)C=C(H)C_6H_5 \xrightarrow{h\nu} C_6H_5(H)C=C(H)C_6H_5$$

cis-stilbene → *trans*-stilbene

Many of the photoisomerisation reactions of alkenes are carried out in the presence of triplet photosensitisers. The mechanism of photoisomerisation of stilbene in the presence of benzophenone as sensitiser is given below:

$$(C_6H_5)_2CO \xrightarrow{h\nu} [(C_6H_5)CO]^1 \longrightarrow [(C_6H_5)_2CO]^3$$

Excited singlet; Excited triplet

$$[(C_6H_5)_2CO]^3 + (C_6H_5)_2\ C_2H_2 \longrightarrow (C_6H_5)_2CO + \text{H(C}_6\text{H}_5)\dot{\text{C}}-\dot{\text{C}}(\text{H})\text{C}_6\text{H}_5$$

Triplet; cis-stilbene; Triplet

$$[(C_6H_5)_2CO]^3 + (C_6H_5)_2\ C_2H_2 \longrightarrow (C_6H_5)_2CO + \text{C}_6\text{H}_5(\text{H})\dot{\text{C}}-\dot{\text{C}}(\text{H})\text{C}_6\text{H}_5$$

Triplet; trans-stilbene; Triplet

$$\text{H(C}_6\text{H}_5)\dot{\text{C}}-\dot{\text{C}}(\text{H})\text{C}_6\text{H}_5 \xrightarrow{\text{Spin inversion}} \text{H(C}_6\text{H}_5)\text{C}=\text{C(H)C}_6\text{H}_5$$

$$\text{C}_6\text{H}_5(\text{H})\dot{\text{C}}-\dot{\text{C}}(\text{H})\text{C}_6\text{H}_5 \xrightarrow{\text{Spin inversion}} \text{C}_6\text{H}_5(\text{H})\text{C}=\text{C(H)C}_6\text{H}_5$$

3. Photochemical isomerisation can also be brought about in the presence of bromine or iodine. The mechanism is given below:

$CH_3(H)C{=}C(H)CH_3 \xrightarrow{Br^{\bullet}} CH_3(H)\dot{C}{-}C(H)(Br)CH_3 \rightleftharpoons H(CH_3)\dot{C}{-}C(H)(Br)CH_3$

Cis-2-butene I II

I $\xrightarrow{-Br^{\bullet}}$ $CH_3(H)C{=}C(H)CH_3$ (Cis-2-butene)

II $\xrightarrow{-Br^{\bullet}}$ $H(CH_3)C{=}C(H)CH_3$ (Trans-2-butene)

It may be mentioned that mechanism assumes rotation around the carbon-carbon single bond in species I and II to change the configuration from cis to trans and vice-versa.

4. Oximes or azocompounds can also be isomerised likewise.

$C_6H_5{-}N{=}N{-}C_6H_5 \overset{h\nu}{\rightleftharpoons} C_6H_5{-}N{=}N{-}C_6H_5$

E-azobenzene Z-azobenzene

Q. 15. Give photochemical dimerisation reactions of olefins.

Ans. 1. This reaction involves the generation of excited triplet molecule which reacts with a ground state olefin molecule. Acetone sensitised photodimerisation of norbornene is explained as under.

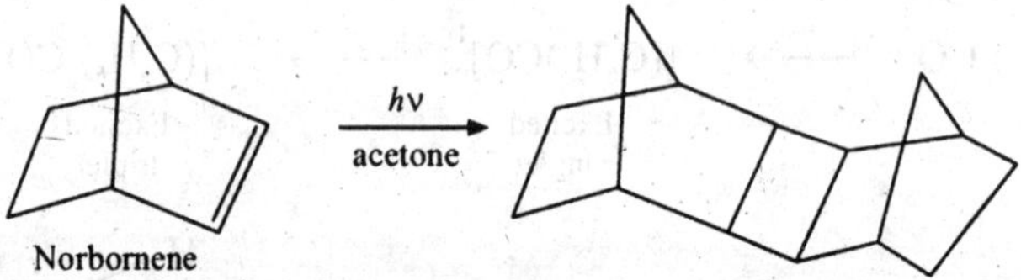

2. Photoisomerisation of norbornadiene to quadricyclene on irradiation is an example of intramolecular change, as shown below:

Nibornadiene $\xrightarrow{h\nu}$ [norbornadiene][1] Excited singlet $\longrightarrow$ [norbornadiene][3] Excited triplet $\longrightarrow$ Quadricyclene

Q. 16. What are the products you expect from the irradiation of following compound?

$$\begin{matrix} CH_3 \\ H \end{matrix} \Big\rangle H = C \Big\langle \begin{matrix} CH_2C_6H_5 \\ H \end{matrix}$$

Ans. Two products that are expected from the irradiation of the above compound are:

(1) $$\begin{matrix} CH_3 \\ H \end{matrix} \Big\rangle C = C \Big\langle \begin{matrix} H \\ CH_2C_6H_5 \end{matrix}$$ (Isomerisation)

(2) $$\begin{matrix} CH_3 & & CH_2C_6H_5 \\ & C - C & \\ & | \quad\; | & \\ & C - C & \\ CH_3 & & CH_2C_6H_5 \end{matrix}$$ (Dimerisation)

Q. 17. Compounds A and B undergo photodecarbonylation resulting in the formation of I and II (from A) and III and IV (from B). Give the structures of I, II, III and IV and mechanism of their formation.

(*i*) [A] $\xrightarrow{h\nu}$ **CO + I + II**

(*ii*) [B] $\xrightarrow{h\nu}$ **CO + III + IV**

Ans. (*i*) Norrish type I reaction takes place in (*i*)

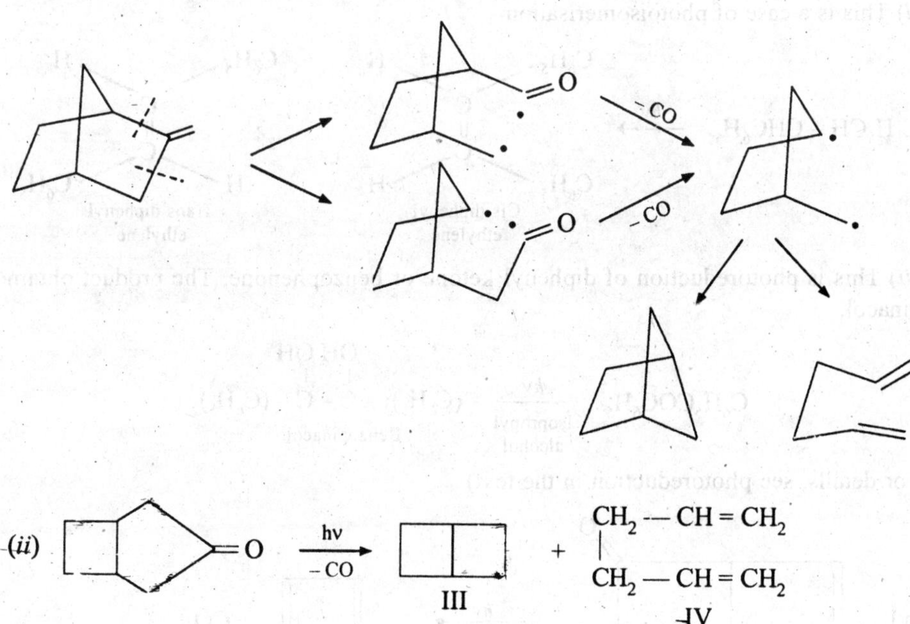

Q. 18. What are the likely products in the following photochemical reactions?

(*i*) $CH_2=CH-CH=CH_2 \xrightarrow{h\nu}$; $\xrightarrow[\text{Benzophenone}]{h\nu}$

(*ii*) $C_6H_5CH=CHC_6H_5 \xrightarrow{h\nu}$

(*iii*) $C_6H_5COC_6H_5 \xrightarrow[\text{Isopropyl alcohol}]{h\nu}$

(*iv*) (bicyclic β-lactone with cyclobutene ring, C=O and ring O) $\xrightarrow[77\ K]{h\nu}$

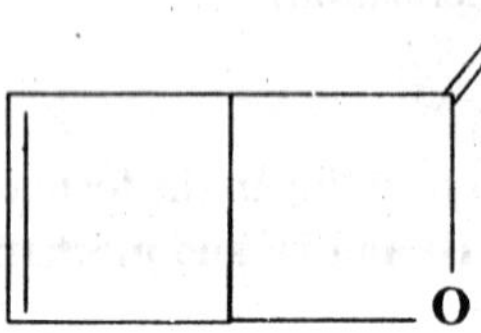

Ans. (*i*) $CH_2=CH—CH=CH_2 \xrightarrow{S_1}$ (I) + (II) ; $\xrightarrow{T_1}$ (III) + (IV) + (V)

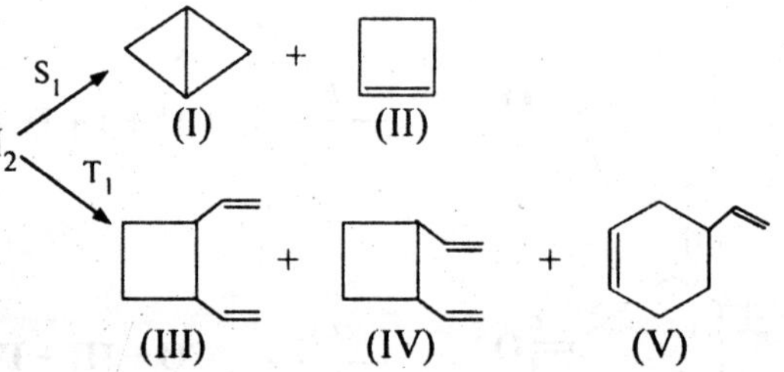

On irradiation, butadiene forms S_1 (excited singlet) and gives the products I and II.

On irradiation in the presence of benzophenone, triplet formation will give rise to products III, IV and V.

(*ii*) This is a case of photoisomerisation

$C_6H_5CH=CHC_6H_5 \xrightarrow{h\nu}$ $(C_6H_5)HC=CH(C_6H_5)$ (Cis-diphenyl ethylene) + $(C_6H_5)HC=CH(C_6H_5)$ (Trans-diphenyl ethylene)

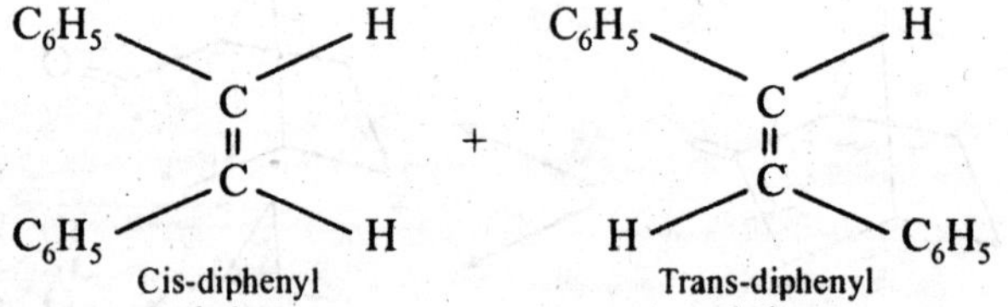

(*iii*) This is photoreduction of diphenyl ketone or benzophenone. The product obtained is benzopinacol.

$$C_6H_5COC_6H_5 \xrightarrow[\text{Isopropyl alcohol}]{h\nu} (C_6H_5)_2-\underset{|}{\overset{OH}{C}}-\underset{|}{\overset{OH}{C}}-(C_6H_5)_2$$

Benzopinacol

(For details, see photoreduction in the text)

(*iv*) (bicyclic β-lactone) $\xrightarrow[77\ K]{h\nu}$ (cyclobutadiene) $+ CO_2$

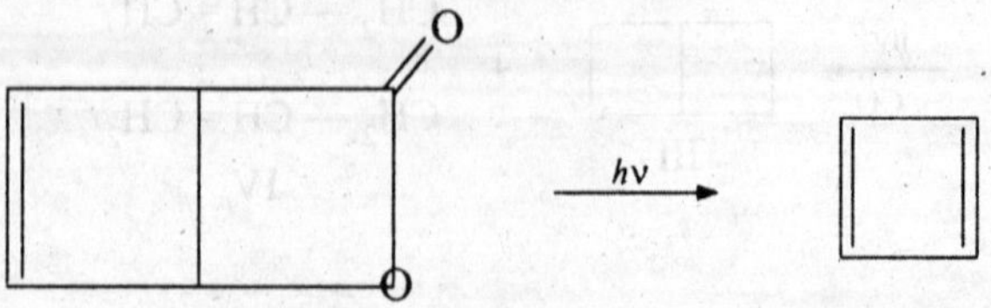

Q. 19. Complete the following reactions:

(*i*) $CH_3-CH=CH-CH_3 \xrightarrow[h\nu]{CH_2N_2}$

trans

(*ii*) $Ph_2CO \xrightarrow[h\nu]{\text{Toluene}}$

Ans. (*i*) This is photoisomerisation of the olefin

$$\underset{\text{Trans-2-butene}}{CH_3-CH=CH-CH_3} \xrightarrow{h\nu} \underset{\text{Cis-2-butene}}{\overset{CH_3}{\underset{H}{>}}C=C\overset{CH_3}{\underset{H}{<}}}$$

(*ii*) This is photoreduction of benzophenone. Benzophenone triplet abstracts hydrogen from toluene to give benzopinacol as the final product.

$$C_6H_5COC_6H_5 \xrightarrow[\text{Toluene}]{h\nu} \underset{\text{Benzopinacol}}{(C_6H_5)_2-\overset{OH}{\overset{|}{C}}-\overset{OH}{\overset{|}{C}}-(C_6H_5)_2}$$

Q. 20. What products are expected in the following photochemical reactions?

(*Kerala, 2000*)

(*i*) 2-hexanone $\xrightarrow[\text{Vapour phase}]{h\nu}$

(*ii*) Cyclopentanone (ring)$=O \xrightarrow[\text{V.P.}]{h\nu}$

(*iii*) $CH_3CHO \xrightarrow{h\nu}$

(*iv*) $CH_3CH_2CHO \xrightarrow{h\nu}$

(*v*) $CH_3CH=CHCHO \xrightarrow{h\nu}$

Ans. (*i*)

$$CH_3-CH(H)-CH_2-CH_2-\overset{O}{\overset{||}{C}}-CH_3 \xrightarrow{h\nu} \underset{\text{Acetone}}{CH_3COCH_3} + \underset{\text{Propene}}{CH_3-CH=CH_2}$$

This is Norrish type II reaction.

(*ii*)

$$\text{Cyclopentanone} \xrightarrow{h\nu} \left[\cdot CH_2CH_2CH_2CH_2\dot{C}=O\right] \longrightarrow CO + \cdot CH_2CH_2CH_2CH_2\cdot$$

$$\cdot CH_2CH_2CH_2CH_2\cdot \longrightarrow \underset{\text{Ethene}}{2CH_2=CH_2} \quad \text{and} \quad \longrightarrow \underset{\text{Cyclobutane}}{\square}$$

(*iii*)

$$CH_3CHO \xrightarrow{h\nu} \dot{C}H_3 + H\dot{C}O$$

$$H\dot{C}O \longrightarrow \dot{H} + CO$$

$$\dot{C}H_3 + \dot{H} \longrightarrow CH_4 \text{ (Methane)}$$

(*iv*)

$$\underset{\text{Propanal}}{CH_3CH_2CHO} \xrightarrow{h\nu} \dot{C}H_3CH_2 + \dot{C}HO$$

$$\dot{C}HO \longrightarrow \dot{H} + CO$$

$$2CH_3\dot{C}H_2 \longrightarrow CH_3CH_2CH_2CH_3$$

$$CH_3\dot{C}H_2 + \dot{H} \longrightarrow CH_3CH_3$$

(*v*)

$$CH_3 - CH = CHCHO \xrightarrow{h\nu} CH_3CH = \dot{C}H + \dot{C}HO$$

$$\dot{C}HO \longrightarrow \dot{H} + CO$$

$$CH_3CH = \dot{C}H + \dot{H} \longrightarrow CH_3CH = CH_2$$

38

REACTIVE INTERMEDIATES

CONTENTS

CARBOCATIONS

Q. 1. What are carbocations? How are they formed? Discuss the stability of carbocations.

(*G. Nanakdev, 2000; Panjab, 2003; Guwahati, 2006*)

Ans. Carbocations. These are defined as the species in which the positive charge is carried by the carbon atom with six electrons in its valence shell. These are formed by the heterolytic fission in which an atom or group along with its pair of electrons leaves the carbon. In heterolytic fission, the shared pair of electrons between two atoms goes to one atom only.

$$R-CH_2-X \xrightarrow[\text{Fission}]{\text{Heterolytic}} \overset{\oplus}{R}CH_2 + \overset{\ominus}{\ddot{X}}$$

Stability

1. The order of stability among simple alkyl carbocations is: tertiary > secondary > primary. In most of the reactions, primary and secondary carbocations get rearranged to tertiary carbocations. Both *n*-propyl fluoride and isopropyl fluoride form the same isopropyl cation (2° carbocation). Similarly all the four butyl fluorides viz. *n*-, iso-, sec. and tertiary butyl fluorides form the same tert-butyl cation. There are two factors which determine the stability:

(*i*) *Hyperconjugation or resonance.*

(*ii*) *Field effect or inductive effect of groups.*

Hyperconjugation

A large number of canonical forms can be written for tertiary carbocation compared to those for primary carbocation. Consider hyperconjugation in primary and tert-carbocation.

$$R-\underset{H}{\overset{H}{C}}-\underset{H}{\overset{H}{C^{\oplus}}} \longleftrightarrow R-\underset{H}{\overset{H^{\oplus}}{C}}=\underset{H}{\overset{H}{C}} \longleftrightarrow R-\underset{\underset{\oplus}{H}}{\overset{H}{C}}=\underset{H}{\overset{H}{C}}$$

Hyperconjugation in primary carbocation

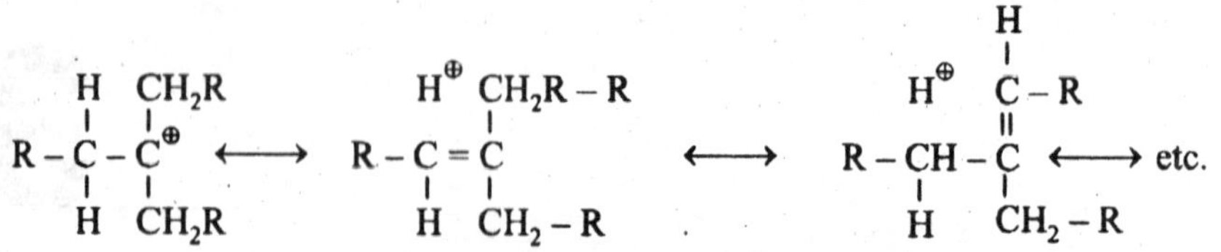

$$R-\underset{H}{\overset{H}{C}}-\underset{CH_2R}{\overset{CH_2R}{C^{\oplus}}} \longleftrightarrow R-\underset{H}{C}=\underset{CH_2-R}{\overset{H^{\oplus}\ CH_2R-R}{C}} \longleftrightarrow R-\underset{H}{CH}-\underset{CH_2-R}{\overset{H^{\oplus}\ \overset{H}{C}-R}{C}} \longleftrightarrow \text{etc.}$$

Hyperconjugation in tertiary cation

Greater the number of canonical forms, greater is the stability. Clearly, the number of canonical forms is greater in tertiary cation than in primary or secondary cation.

Field effect

Electron donating effect of alkyl groups increases the electron density around the +vely charged carbon. This results in reducing the magnitude of positive charge on it and thus the charge is delocalised on α-carbon. Dispersal of positive charge increases the stability of carbocation. Of all the simple cation, tert-butyl cation being most stable. Even tert-pentyl and tert-hexyl cations produce tert-butyl cation at high temperatures. Lower alkanes like methane, ethane and propane when treated with super acid also yield tert-butyl cation as the main product. Salts of tert-butyl cation, like $(CH_3)_3\ C^+\ SbF^-_3$ have been prepared from super acid solutions.

2. The stability of carbocation containing a conjugated double bond is usually greater due to increased delocalisation by resonance. In such carbocations, the positive charge is dispersed on at least two carbon atoms. Consider the following carbocation.

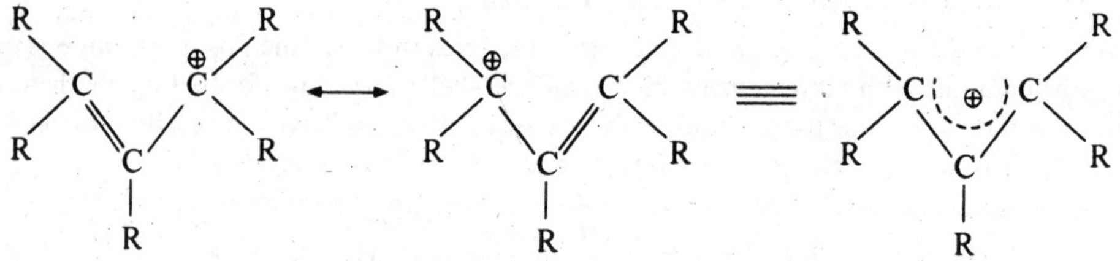

$$R_2C=C(R)-C^{\oplus}R_2 \longleftrightarrow R_2C^{\oplus}-C(R)=CR_2 \equiv R_2C\overset{\oplus}{\cdots}C(R)\cdots CR_2$$

3. Allylic type carbocations have been prepared from the solution of conjugated diene in conc. sulphuric acid and are found to be most stable.

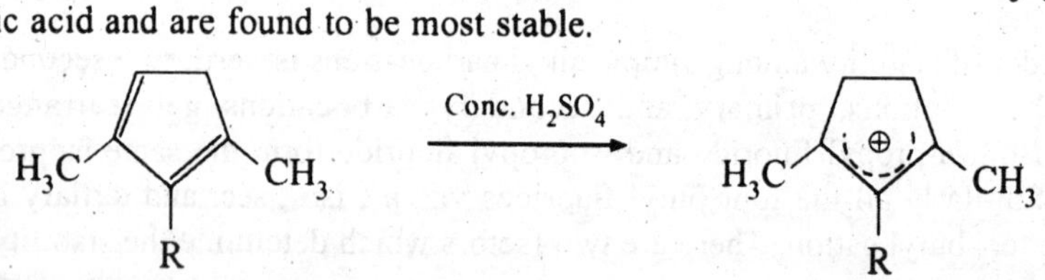

$$\text{1,3-dimethyl-2-R-cyclopentadiene (}H_3C,\ CH_3,\ R\text{)} \xrightarrow{\text{Conc. } H_2SO_4} \text{allylic cyclopentenyl cation (}H_3C,\ \oplus,\ CH_3,\ R\text{)}$$

4. Carbocations can be stabilised through aromatisation also. 1-Bromocyclohepta-2, 4, 6-triene tropylium bromide is a crystalline solid. It is highly soluble in water and form bromide ions in solution. Clearly, the compound is not covalent in nature. The reason for such a behaviour is the stability of tropylium cation which follows Huckel's rule (6 π electrons) for aromaticity.

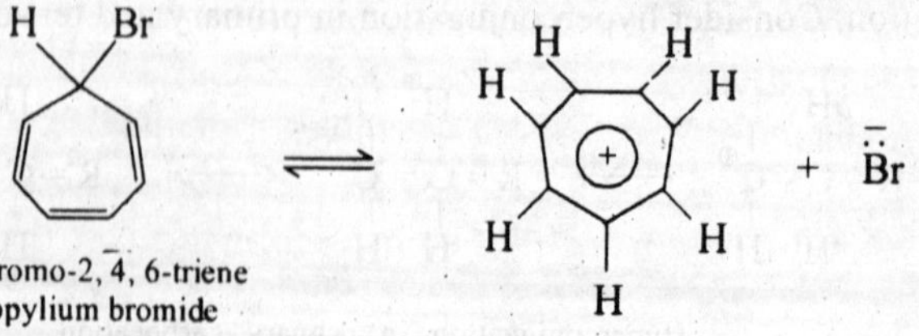

$$C_7H_7Br \rightleftharpoons C_7H_7^{+} + \overset{-}{Br}$$

1, Bromo-2, 4, 6-triene tropylium bromide

Tropylium cation

Its stability is also explained in terms of canonical structures which stabilise the tropylium cation.

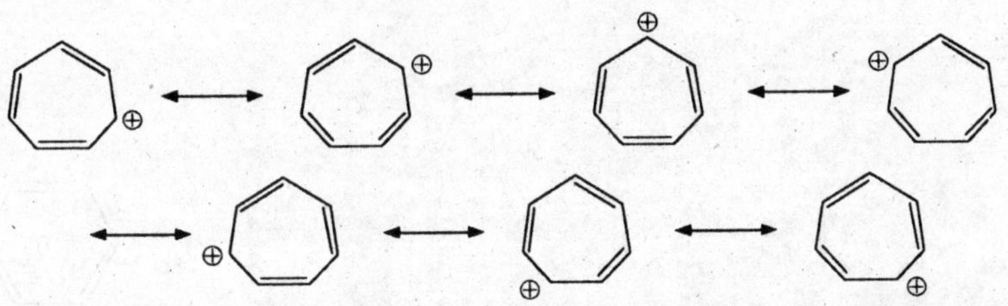

5. Substituted cyclopropenyl cation possesses even more aromatic stabilisation ($n = 0$).

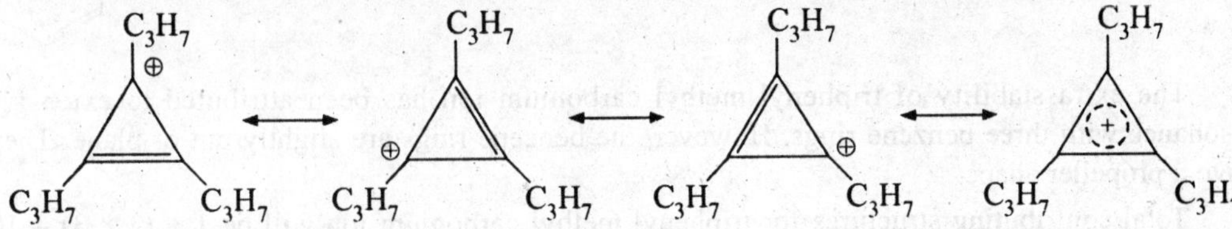

In the above tripropyl substituted cation, all the carbons are sp^2 hybridised. Thus all the p-orbitals including the vacant p-orbital of the positively charged carbon overlap forming a delocalised molecular orbital, leading to stabilization.

6. Benzyl cation can be written in the following canonical forms:

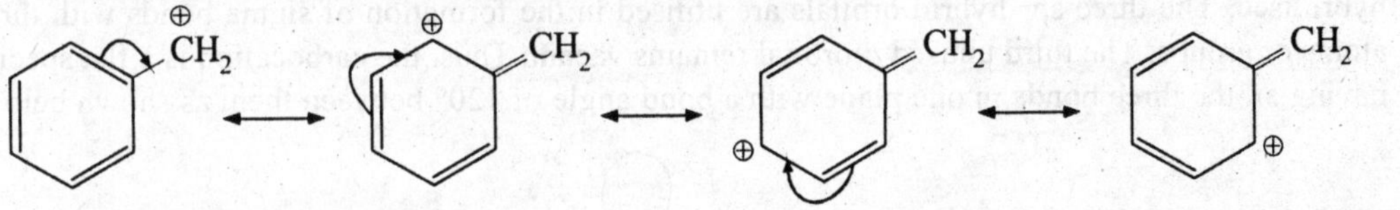

Due to larger number of canonical forms and greater dispersal of positive change, benzyl carbocation is still more stable.

7. Triphenyl chloromethane ionises in SO_2 as

$$(C_6H_5)_3C \cdot Cl \rightleftharpoons (C_6H_5)_3C^{\oplus} + Cl^-$$

Triphenyl methyl cation has been isolated as solid salt such as $(C_6H_5)_3\ C^+ \cdot BF_4^-$. The stability of triarylmethyl cation is further increased if the rings are substituted at ortho and para positions by electron donating groups.

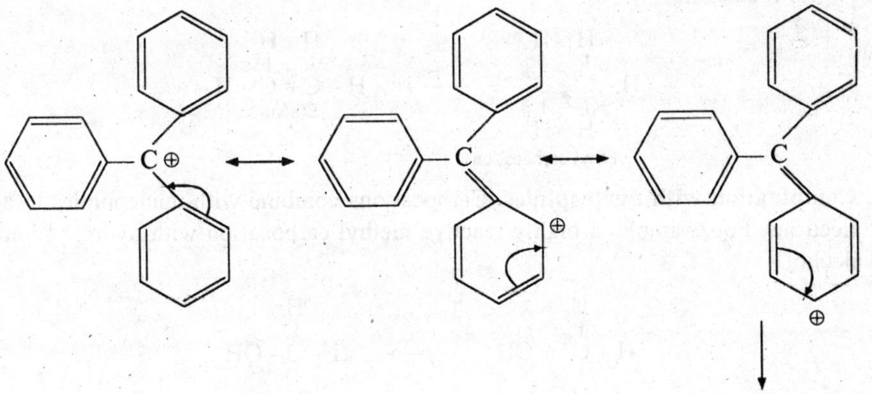

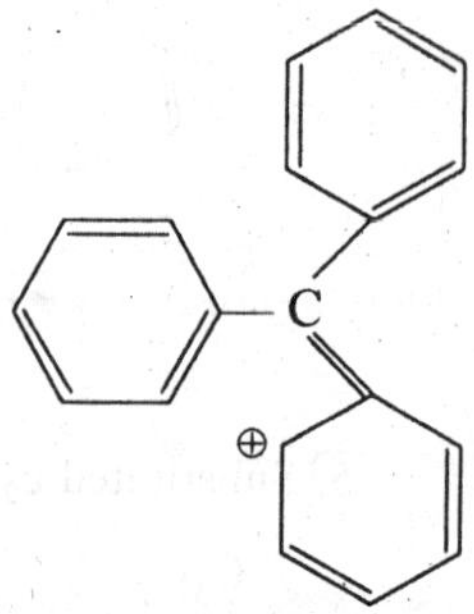

The extra stability of triphenyl methyl carbonium ion has been attributed to extensive resonance with three benzene rings. However, the benzene rings are slightly out of plane. They have a propeller shape.

Total contributing structures for triphenyl methyl carbonium ion will be 1 + (3 × 3) = 10 because 3 contributing structures result from one ring, one structure being the original one.

Q. 2. Write down the structure and reactions of carbocations.

(*G. Nanakdev, 2000; M. Dayanand, 2000; Panjab 2002*)

Ans. Structure. The carbon atom bearing the positive charge in a carbocation is sp^2 hybridised. The three sp^2 hybrid orbitals are utilised in the formation of sigma bonds with three atoms or groups. The third unused *p*-orbital remains vacant. Thus, the carbocation is a flat species having all the three bonds in one plane with a bond angle of 120° between them as shown below:

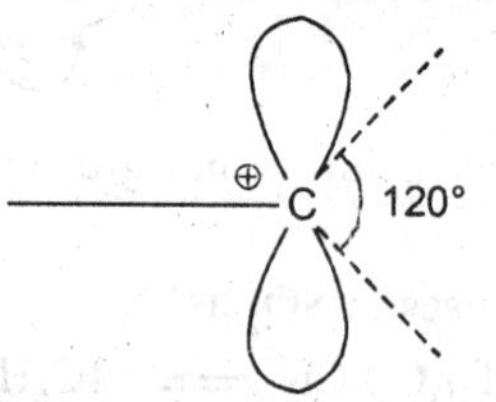

120°

Reactions

Carbocations, which are shortlived species and very reactive give the following reactions:

1. Proton loss. Carbocation may lose a proton to form an alkene. An ethyl carbocation loses a proton to form ethene.

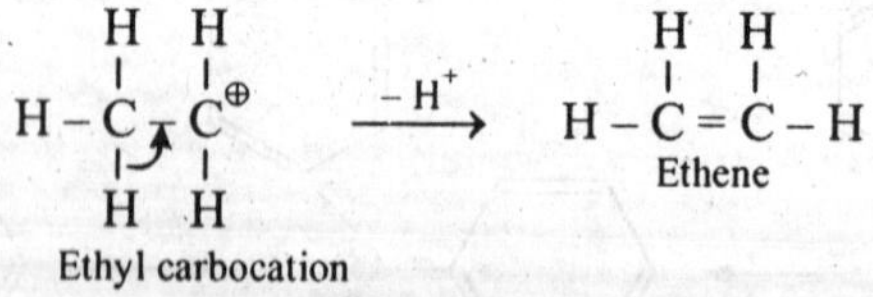

$$H_3C-CH_2^{\oplus} \xrightarrow{-H^+} H_2C=CH_2$$

Ethyl carbocation → Ethene

2. Combination with nucleophiles. Carbocations combine with nucleophiles to acquire a pair of electrons. For example, a highly reactive methyl carbocation with hydroxyl ion to form methyl alcohol.

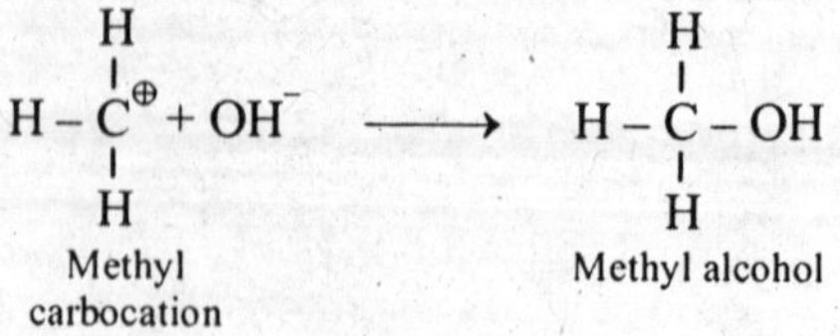

$$CH_3^{\oplus} + OH^- \longrightarrow CH_3-OH$$

Methyl carbocation → Methyl alcohol

3. Addition to alkene. A carbocation may add to an alkene to produce another carbocation possessing higher molecular weight.

$$\overset{\oplus}{C}H_3 + H_2C{=}CH_2 \longrightarrow H{-}CH_2{-}CH_2{-}\overset{+}{C}H_2$$

Methyl carbocation + Ethylene → *n*-Propyl carbocation

4. Rearrangement. The migration of alkyl or aryl or hydrogen along with an electron pair to the positive centre takes place which results in the formation of more stable carbocation. Some examples are given below:

$$\underset{(1°)}{CH_3-\overset{H}{\overset{|}{C}H}-\overset{\oplus}{C}H_2} \xrightarrow{\text{Hydride shift}} \underset{(2°)}{CH_3-\overset{\oplus}{C}H-CH_3}$$

$$C_6H_5-CH_2-\overset{OH}{\overset{|}{C}}(CH_3)_2 \xrightarrow{FSO_3H} C_6H_5-\overset{H}{\overset{|}{C}H}-\overset{\oplus}{C}(CH_3)_2 \xrightarrow[\text{Shift}]{\text{Hydride}} C_6H_5-\overset{\oplus}{C}H-\overset{H}{\overset{|}{C}}(CH_3)_2$$

(More stable)

Such a rearrangement takes place because a tertiary carbocation is more stable than secondary carbocation, which in turn is more stable then primary carbocation. Similarly, a carbocation attached to a phenyl group (benzene ring) is more stable than others.

Q. 3. Arrange the following in order of decreasing stability and give suitable explanation for it.

I: cyclopropyl–$\overset{+}{C}H_2$ ($(CH_2)_2CH-\overset{+}{C}H_2$)

II: $\overset{+}{C}$ bearing three cyclopropyl groups ($(CH_2)_2CH-\overset{+}{C}(-CH(CH_2)_2)-CH(CH_2)_2$)

III: $(CH_2)_2CH-\overset{+}{C}H-CH(CH_2)_2$

Ans. The stability of a carbocation depends upon two factors:

(*i*) Hyperconjugation

(*ii*) Inductive effect of groups attached.

In this problem, it is hyperconjugation or resonance which is important. Structure II can form maximum number of resonating structures and hence this is most stable. It can form resonating structures with the help of hydrogen from three sides (from three cyclopropyl groups) whereas carbocations III and I can form resonating structures from two and one side respectively. Thus the order of stability in decreasing order is:

(II) > (III) > (I)

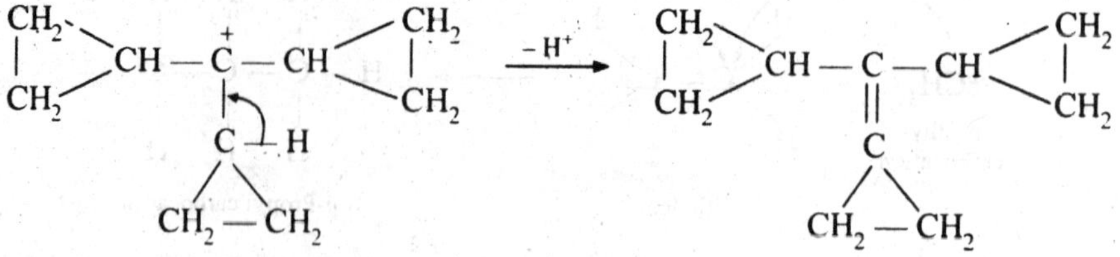

$$\begin{matrix} CH_2 \\ | \\ CH_2 \end{matrix}\!\!>\!CH-\overset{+}{C}-CH\!<\!\!\begin{matrix} CH_2 \\ | \\ CH_2 \end{matrix} \quad (\text{with } \underset{CH_2-CH_2}{\overset{|}{C}-H} \text{ on the central carbon; II}) \xrightarrow{-H^+} \begin{matrix} CH_2 \\ | \\ CH_2 \end{matrix}\!\!>\!CH-\underset{\underset{CH_2-CH_2}{C}}{\overset{}{C}}\!=\!-CH\!<\!\!\begin{matrix} CH_2 \\ | \\ CH_2 \end{matrix}$$

and so on.

Q. 4. Arrange the following in order of increasing stability and explain the order on the basis of hyperconjugation.

$$CH_3-\overset{+}{C}H_2, \quad CH_3-\underset{CH_3}{\underset{|}{\overset{+}{C}}}-CH_3, \quad \overset{+}{C}H_3, \quad CH_3-\overset{+}{C}H-CH_3$$

(I) (II) (III) (IV)

Ans. Greater the number of resonating structures, greater is the stability attached to a carbocation. Carbocation (II) which is a tertiary carbocation will give the maximum number of resonating structures, involving hydride shift, like

$$\overset{+}{C}H_2-\underset{CH_3}{\underset{|}{CH}}-CH_3 \qquad CH_3-\underset{CH_3}{\underset{|}{CH}}-\overset{+}{C}H_2 \qquad CH_3-\underset{{}^+CH_2}{\underset{|}{CH}}-CH_3$$

Greater the number of hydrogens on carbon atoms in the immediate neighbourhood of carbocation, greater the number of resonating structures and hence greater the stability.

This is followed by secondary carbocation IV, primary carbocation I and methyl carbocation (III). Hence the order of stability in increasing order is:

$$\overset{+}{C}H_3 < CH_3-\overset{+}{C}H_2 < CH_3-\overset{+}{C}H-CH_3 < CH_3-\underset{CH_3}{\underset{|}{\overset{+}{C}}}-CH_3$$

Q. 5. Order of stability of carbocation is as under. Explain.

$$(C_6H_5)_3\overset{+}{C} > (C_6H_5)_2\overset{+}{C}H > C_6H_5\overset{+}{C}H_2$$

(I) (II) (III)

Ans. Carbocation (I) gives the maximum number of resonating structures, as this is linked to three benzene rings, hence this will have maximum stability.

For details, see Q. 1.

The order of stability in decreasing order is:

I > II > III

Q. 6. Explain why benzyl carbocation is more stable than cyclohexyl carbocation.

(Delhi, 2007)

Ans. Benzyl carbocation is able to stabilise itself through resonance as shown below:

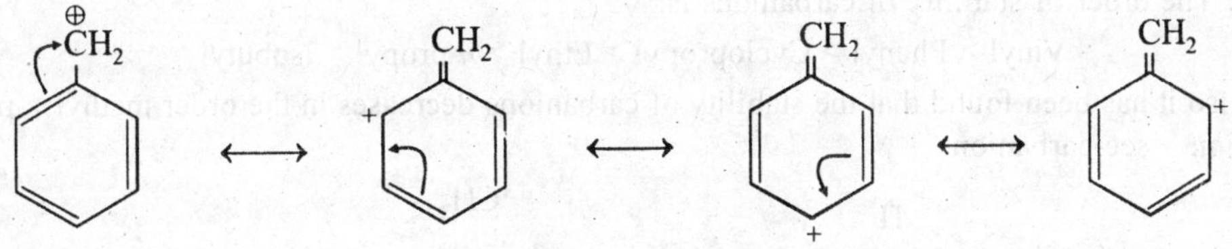

There is no such possibility in cyclohexyl carbocation

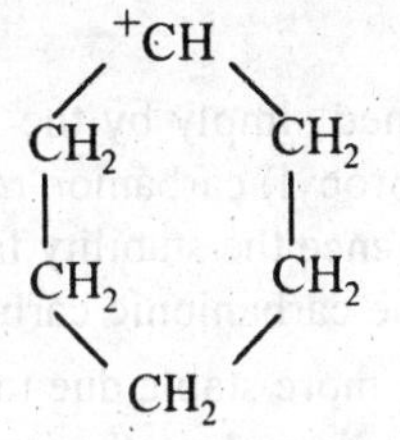

Cyclohexyl carbocation

Q. 7. Why tropyllium bromide gives ppt. with $AgNO_3$ and CH_3Br does not?

Ans. This is because tropyllium bromide behaves like an ionic compound whereas methyl bromide does not. Ionic nature of tropyllium ion is because of extra stability of tropyllium ion. For details refer to Question 1.

CARBANIONS

Q. 8. What are carbanions? Discuss their stability. (*G. Nanakdev, 2000; Kerala, 2001*)

Ans. Carbanions

These are defined as the species with an unshared pair of electrons and a negative charge on the central carbon atom. A carbanion may be formed in one of the following ways:

(*i*) An atom or group leaves carbon without the electron pair (Heterolytic fission).

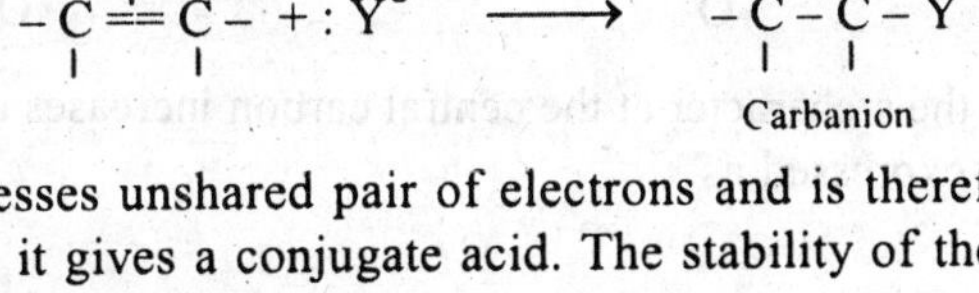

$$R-CH_2-Y \xrightarrow[\text{Fission}]{\text{Heterolytic}} R-\overset{\oplus}{C}H_2: + Y^+$$

Carbanion

(*ii*) An anion adds to a carbon-carbon double or triple bond

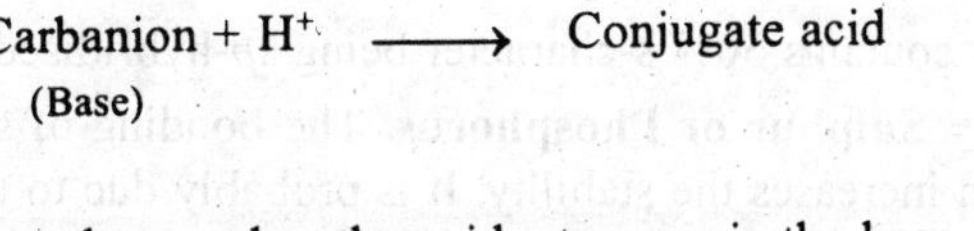

$$-C{=}C- + :Y^{\ominus} \longrightarrow -\overset{\ominus}{C}-C-Y$$

Carbanion

Every carbanion possesses unshared pair of electrons and is therefore a base. When the carbanion accepts a proton, it gives a conjugate acid. The stability of the carbanion is directly related with the strength of the conjugate acid.

$$\underset{\text{(Base)}}{\text{Carbanion}} + H^+ \longrightarrow \text{Conjugate acid}$$

Stability

1. It is important to note that weaker the acid, stronger is the base and hence lower is the stability of the carbanion. Clearly, the order of stability of carbanion can be determined from the order of the strength of conjugate acids. Carbanions are highly unstable in solution as compared to carbocations.

2. The order of stability of carbanions is:

Vinyl > Phenyl > Cyclopropyl > Ethyl > *n*-propyl > Isobutyl

Also it has been found that the stability of carbanions decreases in the order methyl > prim-carbanion > sec-carbanion.

$$H-\overset{\ominus}{\underset{H}{\overset{H}{C}}} > CH_3 \longrightarrow \overset{\ominus}{C}H_2 > (CH_3 \rightarrow)_2\overset{\ominus}{C}H$$

This stability order can be explained simply by the field effects. The presence of electron donating alkyl groups in secondary (Isopropyl) carbanion results in greater localisation of negative charge on the central carbon atom and hence the stability falls. Cyclopropyl carbanion has greater stability due to greater *s*-character on the carbanionic carbon.

3. Vinyl and phenyl carbanions are more stable due to resonance. In cases where a double or triple bond is located at α to the carbanionic carbon, the ion is stabilised by resonance in which the unshared pair overlaps with π-electrons of the double bond. The stability of allylic and benzylic carbanions can be illustrated as under.

$$R-CH=CH-\overset{\ominus}{C}H_2 \longleftrightarrow R-\overset{\ominus}{C}H-CH=CH_2$$

$$C_6H_5-\overset{\ominus}{C}H_2 \longleftrightarrow \text{(ortho carbanion)}=CH_2 \longleftrightarrow \text{(para carbanion)}=CH_2 \longleftrightarrow \text{(ortho carbanion)}=CH_2$$

4. Diphenyl methyl and triphenyl methyl carbanions are more stable then even benzylic and can be kept in solution for a longer time if water is completely excluded.

5. When the carbanionic carbon is conjugated with a carbon-oxygen or carbon-nitrogen multiple bond, then the stability of carbanion is increased. The reason is that the presence of electronegative atoms helps in the dispersal of negative charge and thus such carbanions are better capable of bearing the negative charge. Thus,

$$R-\overset{\ominus}{C}H-\underset{\underset{O}{\|}}{C}-R \longleftrightarrow R-CH=\underset{\underset{:\ddot{O}:^{\ominus}}{|}}{C}-R$$

6. The increase in the s-character at the central carbon increases the stability of carbon. Thus, the order of stability is expressed as:

$$R-C\equiv\overset{\ominus}{\ddot{C}} > R_2C=\overset{\ominus}{\ddot{C}}H > \overset{\ominus}{Ar} > R_3C-\overset{\ominus}{\ddot{C}}H_2$$

Acetylenic carbon contains 50% s-character being *sp*-hybridised.

7. Stabilisation by Sulphur or Phosphorus. The bonding of sulphur or phosphorus atom with carbanionic carbon increases the stability. It is probably due to the overlap of unshared pair with an empty *d*-orbital. Consider a carbanion containing sulphur atom:

$$R-\underset{\underset{O}{\|}}{\overset{\overset{O}{\|}}{S}}-\underset{\underset{R}{|}}{\overset{\overset{R}{|}}{C}}{}^{\oplus}: \longleftrightarrow R-\underset{\underset{:\overline{O}:}{|}}{\overset{\overset{\ominus O}{\|}}{S}}=\underset{\underset{R}{|}}{\overset{\overset{R}{|}}{C}} \longleftrightarrow \text{etc.}$$

8. Aromatic nature. Some carbanions are stable as they are aromatic. *e.g.* Cyclopentadienyl anion is a stable carbanion.

Huckel's $(4n + 2)$ rule is satisfied here with 6π electrons in the ring.

Q. 9. Describe the structure and reactions of the carbanions.

(Panjab, 2003; Guwahati, 2007)

Ans. Structure. The carbanions, being unstable have not been isolated. Therefore, their structure is not known with certainty. Most likely, the central carbon atom with an unshared pair of electrons (and negative charge) is sp^3 hybridised. The three sp^3 hybrid orbitals are used in forming three sigma bonds with the other atoms. The fourth sp^3-orbital is occupied by a lone pair of electron. In fact, the structure of carbanion is quite similar to that of ammonia as shown in the figure below.

If different alkyl groups are attached to the central carbon, the carbanion will be chiral. Reactions in which such a carbanion is formed, give a product with retention of configuration. Carbanions which get stabilized by resonance involving the lone pair of electrons and the π electrons of multiple bonds. should be planar as this is the necessary condition for resonance to take place

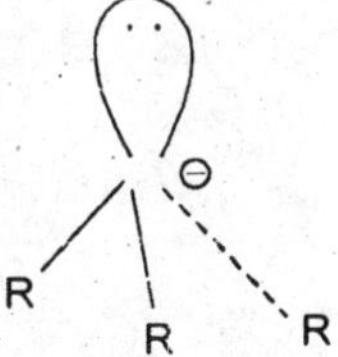

Reactions

Some important reactions of carbanions are described below:

1. Reaction with positive species or electrophiles. A carbanion reacts with a proton or with another cation.

$$R-\underset{R}{\overset{R}{|}}{C}:^{\ominus} + \overset{\oplus}{Y} \longrightarrow R-\underset{R}{\overset{R}{|}}{C}-Y$$

2. Reactions involving the displacement of an atom or a group. Such reactions are nucleophilic substitution (SN^2) reactions observed in alkyl halides.

$$\overset{\ominus}{\ddot{R}} + \rangle C - X \longrightarrow R - C\langle + \overset{\ominus}{\ddot{X}}$$

3. Addition to carbonyl compounds. Carbanions attack an electron deficient carbon in the carbonyl compounds. Such reactions are called nucleophilic addition reactions.

$$\overset{\ominus}{\ddot{R}} + -\underset{|}{C}=\ddot{O} \longrightarrow -\underset{|}{\overset{R}{\overset{|}{C}}}-\ddot{O}:^{\ominus}$$

4. Rearrangement. In some cases, carbanions may rearrange to form more stable species. Consider the rearrangement in triphenyl methyl carbanion.

$$(C_6H_5)_3\,C-\overset{\ominus}{\ddot{C}}H_2 \longrightarrow (C_6H_5)_2\,\overset{\ominus}{\ddot{C}}-CH_2-C_6H$$

Here the two phenyl groups directly attached to the central carbon help in dispersing the negative charge.

FREE RADICALS

Q. 10. What are free radical? How are they produced? Discuss their stability.

(*Meerut, 2000; Kerala, 2001; Panjab, 2003*)

Ans. Free radicals are produced by the homolytic fission of a covalent bond. These are odd electron neutral species which are formed by the homolytic fission of a covalent bond. Free radicals are paramagnetic due to the presence of unpaired electron. Formation of free radicals is favoured by the presence of UV light, heat and organic peroxides. Reactions involving radicals widely occur in the gas phase. Such reactions also occur in solutions, particularly if carried in non-polar solvents. An important characteristic of free radical reactions is that, once initiated, they proceed very fast. The free radicals can be detected by magnetic susceptibility measurements."

$$R-R \xrightarrow[\text{Fission}]{\text{Homolytic}} \underset{\text{Free radical}}{2R\bullet}$$

$$Cl-Cl \xrightarrow{h\nu} \underset{\text{Free radical}}{2Cl\bullet}$$

$$\underset{\text{Acyl peroxide}}{RCO-O-OCOR} \xrightarrow{\text{Thermal}} \underset{\text{Free radical}}{2RCO\dot{O}}$$

Stability

1. Simple alkyl free radicals are highly reactive like carbocations and carbanions. Their life time is extremely short in solution. The relative stability of simple alkyl radicals has the order:

$$\underset{3^\circ}{R_3\dot{C}} > \underset{2^\circ}{R_2\dot{C}H} > \underset{1^\circ}{R\dot{C}H_2} > \underset{\text{Methyl}}{\dot{C}H_3}$$

It can be explained on the basic of hyperconjugation similar to that in carbocations.

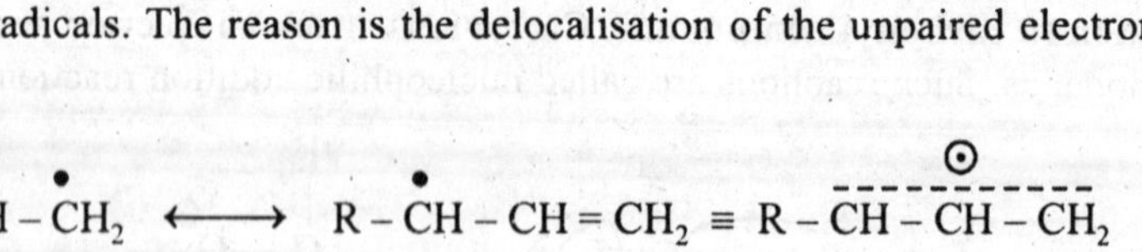

$$R-CH_2-\dot{C}H_2 \longleftrightarrow R-CH_2=CH_2 \;(\text{with } \dot{H}) \longleftrightarrow R-CH=CH-H \;(\text{with } H\bullet)$$

Greater the number of the resonating structures, greater is the stability of the free radical.

2. Allylic and benzylic type of free radicals are more stable and comparatively less reactive than simple alkyl radicals. The reason is the delocalisation of the unpaired electron over the π-orbital system.

$$\underset{\text{Allylic radical}}{R-CH=CH-\dot{C}H_2} \longleftrightarrow R-\dot{C}H-CH=CH_2 \equiv R-\overset{\odot}{\overline{CH-CH-CH_2}}$$

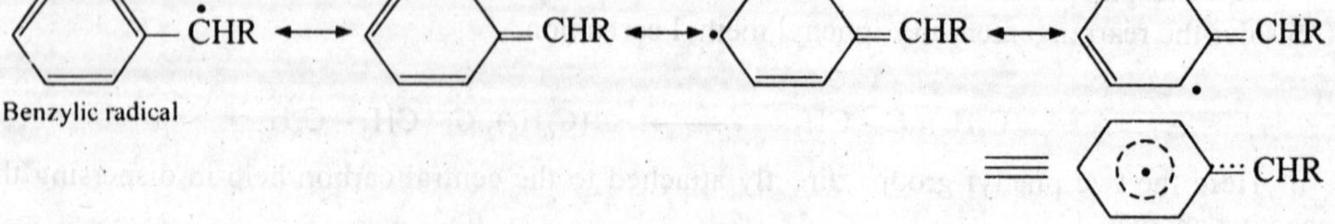

$$\underset{\text{Benzylic radical}}{C_6H_5-\dot{C}HR} \longleftrightarrow \text{(ortho radical)}=CHR \longleftrightarrow \text{(para radical)}=CHR \longleftrightarrow \text{(ortho radical)}=CHR \equiv \text{(delocalised ring)}\cdots CHR$$

3. Triphenyl methyl and triarylmethyl radicals are much more stable in solution at room temperature. The stability of such radicals is due to resonance.

$$(C_6H_5)_3\dot{C} \longleftrightarrow \text{(cyclohexadienyl radical)}=C(C_6H_5)_2 \longleftrightarrow \cdot\text{(cyclohexadiene)}=C(C_6H_5)_2 \longleftrightarrow \text{etc.}$$

Steric hindrance to dimerisation is probably the major cause of their stability. If each aromatic nucleus in the radical has a bulky *p*-substituent, then irrespective of any substitution at the *o*-positions, dimerisation will be greatly inhibited and hence radical stability increases.

Q. 11. Give the structure and reactions of free radicals.

(*G. Nanakdev, 2000; Lucknow, 1998; Panjab, 1995*)

Ans. Structure. The state of hybridisation of carbon atom having the unpaired electron is not clearly established. However, it is believed to be either pyramidal (like ammonia) or planar. But experimental evidence suggests strongly that the alkyl radicals such as methyl radical are actually planar with sp^2 hybridisation. The three coplanar hybrid orbitals are used in the formation of three sigma bonds with other atoms. The unhybridised orbital which lies in a plane at right angles to the plane of the hybrid orbitals, carries the unpaired electron. ESR spectra of $\dot{C}H_3$ and other simple alkyl radicals indicate that these radicals have planar structures. This is also in accordance with the fact that optical activity is lost when a free radical is generated at an asymmetric carbon. As a general rule we can say that simple alkyl free radicals prefer a planar or near planar shape. However, the free radicals in which the carbon is connected to atoms of high electronegativity prefer a pyramidal shape. The increase in electronegativity causes the deviation from the planar geometry.

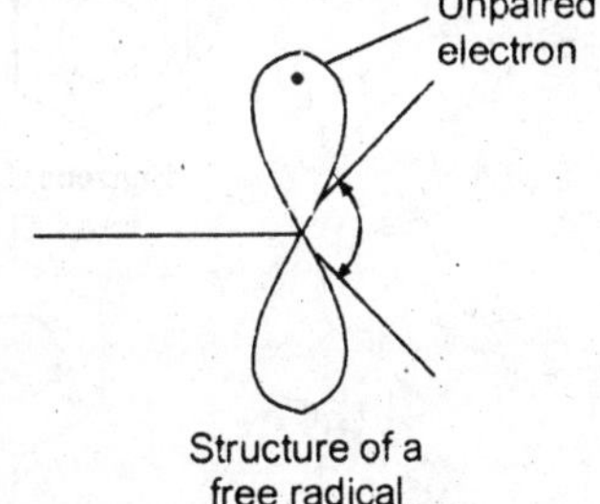

Structure of a free radical

Reactions

Some important reactions of free radicals are described below:

1. Halogenation of aliphatic hydrocarbons

$$Cl-Cl \xrightarrow{h\nu} 2Cl\bullet$$
$$CH_4 + Cl\bullet \longrightarrow CH_3\bullet + HCl$$
$$CH^{\bullet}_3 + Cl_2 \longrightarrow CH_3Cl + Cl\bullet$$

In the presence of sunlight, chlorination of methane gives chloromethane, dichloromethane, chloroform and carbon tetrachloride.

2. Addition. Halogen addition to alkenes takes place with free radical mechanism. The addition of chlorine to tetrachloroethene is photochemically catalysed. A molecule of chlorine undergoes homolytic fission giving two chlorine radicals. Each radical is capable of initiating a reaction chain.

$$Cl-Cl \underset{}{\overset{h\nu}{\rightleftharpoons}} Cl\bullet + Cl\bullet$$

$$Cl_2C=CCl_2 + Cl\bullet \longrightarrow Cl_2\dot{C}-CCl_3$$

$$\Big\downarrow \; Cl-Cl$$

$$\dot{C}l + Cl_3C-CCl_3$$

$\bullet$
Cl radical further adds to tetrachloroethene. This continues till the whole of tetrachloroethene has been converted into hexachloroethane. Chain termination occurs through radical-radical collision. The radical ractions are inhibited by the presence of oxygen. The reason is that the molecule of oxygen has two unpaired electrons and behaves as a biradical. This biradical combines with highly reactive radical intermediate and converts it to less reactive peroxy radical which is unable to propagate the chain reaction.

3. Consider the addition of chlorine to benzene in the presence of light. The reaction proceeds by radical intermediates and gives benzene hexachloride. But the attack of chlorine radical on toluene results in preferential hydrogen abstraction giving substitution in CH_2 group. Clearly, it is because of the greater stability of benzyl radical due to delocalisation.

$$\text{Benzene} + 3Cl_2 \xrightarrow[h\nu]{\text{Sunlight}} \text{Benzene hexachloride } (C_6H_6Cl_6)$$

Benzene

Benzene hexachloride

$$\text{Toluene} \xrightarrow{\dot{C}l} [\text{Benzyl radical } C_6H_5\dot{C}H_2] \xrightarrow{Cl-Cl} C_6H_5CH_2Cl$$

More stable

Benzyl chloride (Major)

$$\text{Toluene} \xrightarrow{\dot{C}l} \text{(Cl, H adduct radical of ring)}$$

Toluene

Less stable, no longer aromatic

4. The addition of HBr to propene in presence of peroxides yields 1-bromopropane. The eaction involves the formation of radical intermediates giving anti-Markownikoff addition. In this eaction, Br radical attacks propene to form a more stable secondary free radical.

$$ROOR \longrightarrow 2R\dot{O}$$

$$R\dot{O} + H-Br \longrightarrow RO-H + \dot{B}r$$

$$CH_3CH=CH_2 + \dot{B}r \longrightarrow CH_3\dot{C}H\ CH_2Br$$

$$\big\downarrow HBr$$

$$\dot{B}r + CH_3-CH_2-CH_2Br$$

1-Bromopropane

5. Vinyl polymerisation. Radical reactions also produce polymers of great importance. Like other radical reactions, polymerisation reactions also constitute three step viz.; intiation, propagation and termination. In initiation step, a free radical is formed under the influence of peroxide. The radical formed then propagates the chain reaction.

$$\dot{R} + CH_2 = CH_2 \longrightarrow RCH_2 - \dot{C}H_2 \xrightarrow{CH_2 = CH_2} RCH_2 - CH_2 - CH_2 - \dot{C}H_2 \text{ and so on.}$$

In the final step, the collisions between the radicals terminate the reaction.

$$R(CH_2)_n^{\bullet} + \dot{R} \longrightarrow R(CH_2)_n R$$

$$R(CH_2^{\bullet})_n + R(\dot{C}H_2)_n \longrightarrow R(CH_2)_n R$$

Polythene, PVC, teflon etc. are the polymers which are formed from the respective monomers by radical pathways.

CARBENES

Q. 12. What are carbenes? How are they formed? Discuss their stability.

(*M. Dayanand, 2000; Meerut, 2000; Kerala, 2001; Delhi 2006*)

Ans. These are defined as the neutral organic species containing a divalent carbon atom having a sextet of electrons but no charge on it. For example

(*a*) $:CH_2$ — Methyl carbene (*ii*) $:CCl_2$ — Dichlorocarbene

Formation of carbenes

They can be generated by the following methods:

(*b*) **By the action of UV light on diazomethane**

$$CH_2N_2 \longrightarrow \underset{\substack{\text{Methylene} \\ \text{(Carbene)}}}{:CH_2} + N_2$$

(*ii*) **By the action of sodium ethoxide on chloroform**

$$\underset{\text{Chloroform}}{CHCl_3} + C_2H_5ONa \longrightarrow \underset{\text{Dichlorocarbene}}{:CCl_2} + C_2H_5OH + NaCl$$

These are highly reactive species.

Stability

1. The two unbonded electrons of a carbene may be either paired or unpaired. When the two electrons on the carbon atom are paired, then it is called *singlet* carbene. In the *triplet* state, it is called a biradical as the two electrons on carbon atom are unpaired.

$$\underset{\text{Singlet methylene}}{H - \ddot{C} - H} \qquad \underset{\text{Triplet methylene}}{H - \dot{\underset{\bullet}{C}} - H}$$

In the triplet state, carbene is relatively more stable as it has lower energy content.

2. The singlet and the triplet species can be distinguished by the common addition reaction of carbene to double bond to form cyclopropane derivatives. The addition of singlet species to cis-Butene forms a cis product. The reason is that the movement of two pairs of electrons occur either simultaneously or with one rapidly succeeding the other.

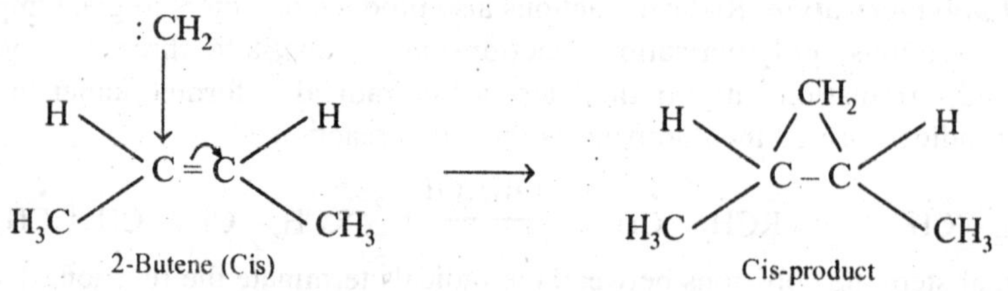

2-Butene (Cis) → Cis-product

3. In case of triplet carbene, the two unpaired electrons cannot form a new covalent bond simultaneously as they have parallel spins. Thus, one of the unpaired electrons will form a bond with the electron from the double bond that has opposite spin. Now, two unpaired electrons with the same spin are left and thus no bond is formed. The bond formation is possible by collision process when one of the electrons can reverse its spin. This is illustrated in the diagram below:

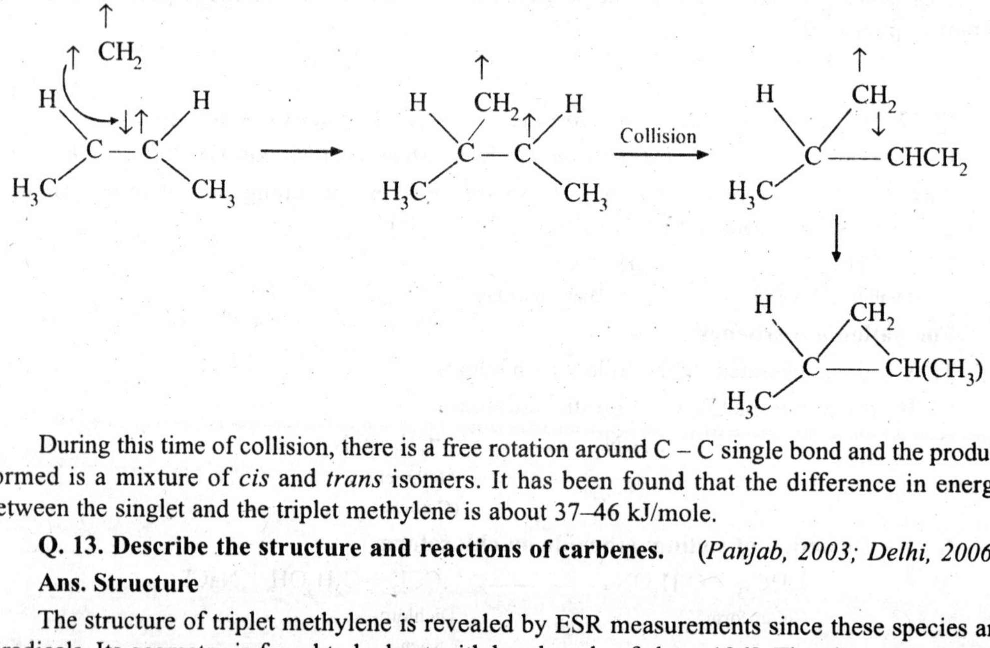

Collision

During this time of collision, there is a free rotation around C – C single bond and the product formed is a mixture of *cis* and *trans* isomers. It has been found that the difference in energy between the singlet and the triplet methylene is about 37–46 kJ/mole.

Q. 13. Describe the structure and reactions of carbenes. (*Panjab, 2003; Delhi, 2006*)

Ans. Structure

The structure of triplet methylene is revealed by ESR measurements since these species are biradicals. Its geometry is found to be bent with bond angle of about 136°. The electronic spectra of singlet methylene formed in flash photolysis of diazomethane tell that it is also bent and the bond angle is about 103°.

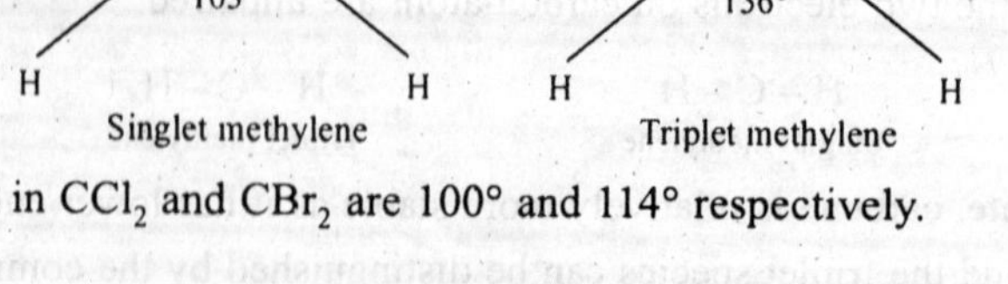

Singlet methylene (103°) Triplet methylene (136°)

The bond angles in CCl_2 and CBr_2 are 100° and 114° respectively.

Reactions

Some important types of reactions of carbenes are described below:

1. Addition to carbon-carbon double bonds. The addition reactions of singlet and triplet methylene to 2-Butene has been discussed above. Carbenes also add to aromatic systems and the intermediate products rearrange with ring enlargement reaction. Carbenes also show addition reactions to C = N bonds.

$$\text{Benzene} + :CH_2 \longrightarrow \text{Cycloheptatriene}$$

2. **Addition reactions to alkanes.** Methylene ($:CH_2$) reacts with ethane to form propane. Propane adds methylene to yield *n*-butane and isobutane. This shows the greater reactivity of carbene.

$$\underset{\text{Propane}}{CH_3-CH_2-CH_3} \xrightarrow{:CH_2} \underset{n\text{-Butane}}{CH_3-CH_2-CH_2-CH_3} + \underset{\text{Isobutane}}{CH_3-\underset{|}{\overset{}{CH}}(CH_3)-CH_3}$$

The reaction of carbenes with higher alkanes yield a number of possible products. It shows that the addition of carbenes is not at all selective. Dichlorocarbenes do not give reactions involving insertion.

3. An important reaction of carbenes is dimerisation

$$R_2\ddot{C} + R'_2\ddot{C} \rightleftharpoons R_2C=CR'_2$$

As the carbenes are highly reactive, the dimer formed carries so much energy that it dissociates again.

4. We also come aross rearrangement reactions in carbenes with the migration of alkyl group or hydrogen atom. These rearrangements are much more rapid. Thus, the addition and insertion reactions of carbenes are seldom encountered with alkyl or dialkyl carbenes. Many rearrangements of carbenes directly give stable molecules. Some rearrangements involved in carbenes are:

$$\underset{\text{Carbene}}{CH_3-CH_2-CH(H)-\ddot{C}H} \longleftrightarrow \underset{\text{1-Butane}}{CH_3-CH_2-CH=CH_2}$$

$$\underset{\text{Acyl Carbene}}{R-C(=O)-\ddot{C}H} \longrightarrow \underset{\text{Ketene}}{O=C=CHR} \text{ (Wolff rearrangement)}$$

$$\text{Cyclopropyl}-\ddot{C}H \longrightarrow \text{Cyclobutene}$$

NITRENES

Q. 14. What are nitrenes? How are they produced? Discuss their stability.

(Purvanchal, 2007; Nagpur 2008)

Ans. Organic species having general formula as R – N: are called nitrenes. These are analogous to carbenes and are sometimes referred to as azocarbenes. The structure R – N: shows that nitrogen is bonded to one atom or group only and there are two non-bonded electron pairs. Thus, there is only a sextet of electrons around nitrogen. This explains why nitrenes are electron deficient and reactive species.

The most common method of nitrene generation involves photolytic or thermolytic decomposition of azides, acyl azides and isocyanates

$$HN_3 \xrightarrow[\text{or }\Delta]{h\nu} H-\ddot{N}: + N_2\uparrow$$

$$CH_3-N_3 \xrightarrow{h\nu} CH_3-\ddot{N}: + N_2\uparrow$$

$$C_6H_5-N_3 \xrightarrow{h\nu} C_6H_5-\ddot{N}: + N_2\uparrow$$

$$RCON_3 \xrightarrow{h\nu} N_2 + RCO\ddot{\underset{..}{N}}$$

Acyl nitrene

$$R-N=C=O \xrightarrow{h\nu} CO + R-\ddot{\underset{..}{N}}$$

Stability

Nitrenes are highly reactive species and cannot be isolated under ordinary conditions. Alkyl nitrenes have been isolated by trapping in matrices at 4 K while aryl nitrenes (less reactive) can be trapped at 77 K. Nitrenes can be generated both in the singlet and the triplet states but its ground state is probably the triplet state.

$R-\ddot{\underset{..}{N}}$ Singlet $\quad$ $R-\dot{\underset{..}{N}}$ Triplet

Q. 15. Give the reactions of nitrenes. (*Purvanchal 2007*)

Ans. Reactions

Nitrenes show the following reaction:

1. Addition to C = C bonds. Nitrenes add to alkenes to form cyclic product. Such reactions are common for acyl nitrenes.

$$R-\ddot{\underset{..}{N}} + R_2C=CR_2 \longrightarrow R_2C-CR_2 \text{ (ring bridged by N–R)}$$

Cycloaddition of nitrenes to alkenes produces acridines.

$$H-\ddot{\underset{..}{N}} + CH_2=CH_2 \longrightarrow \text{(three-membered ring with N–H)}$$

Ethylenimine

$$\underset{\text{2-Butene}}{CH_3(H)C=C(H)CH_3} + :N-H \longrightarrow H_3C-HC-CH-CH_3 \text{ (ring bridged by N–H)}$$

Acridine

2. Insertion. Nitrenes, especially acyl and sulphonyl nitrenes can insert into C–H and certain other bonds.

$$R'-\overset{O}{\overset{\|}{C}}-\ddot{N}: + R_3C-H \longrightarrow R'\overset{O}{\overset{\|}{C}}-\overset{H}{\overset{|}{N}}-CR_3$$

3. Abstraction. Nitrenes are also capable of extracting hydrogen atom from an alkane to form an alkyl radical.

$$R-\ddot{N} + R-H \longrightarrow R\dot{N}-H + \dot{R}$$

4. Rearrangements. Nitrenes undergo rearrangements also due to the migration of a group from adjacent atom to the electron deficient nitrogen. Insertion and addition to C = C bond reactions do not proceed in alkyl nitrenes as the rearrangement is much more faster.

$$R-\underset{H}{\underset{|}{CH}}-\ddot{N} \longrightarrow RCH{=}NH$$

5. Dimerisation. Nitrenes also dimerise to form a di-imide. Aryl nitrenes yield azobenzenes.

$$2C_6H_5-\ddot{N} \longrightarrow C_6H_5-N=N-C_6H_5$$

6. Ring enlargement. Azepines are prepared by the ring enlargement of phenyl azide.

$$C_6H_5-N=N^+=\bar{N} + (C_2H_5)_2NH \xrightarrow{h\nu} \text{azepine ring}-N(C_2H_5)_2$$

MISCELLANEOUS QUESTIONS

Q. 16. Write down possible products in the following reactions:

1. $CH_2(COOEt)_2 \xrightarrow[C_2H_5I]{NaOMe}$

2. $CH_3-CH_2-CH_2^+ \xrightarrow{OH^-}$

Ans.

(1) $CH_2(COOEt)_2 \xrightarrow[-MeOH]{NaOMe} \bar{C}H(COOEt)_2$ (Carbanion)

$$\xrightarrow{C_2H_5I} \underset{C_2H_5}{\underset{|}{CH(COOEt)_2}} \xrightarrow{Hydrolysis} C_2H_5CH_2COOH$$

Butanoic acid

(2) $CH_3-CH_2-\overset{+}{C}H_2$ (*n*-Propyl carbocation)

$\downarrow$

$CH_3-\overset{+}{C}H-CH_3$ (more stable carbocation)

$\downarrow OH^-$

$CH_3-CH(OH)CH_3$

Propanol-2

Q. 17. Write down the possible products of the following reaction:

1. $\mathbf{CHCl_3 + Sod.\ ethoxide \longrightarrow}$

2. $\mathbf{CH_3 - CH_2 - \overset{+}{C}H_2 \xrightarrow{OH^-}}$

Ans. (1) $CHCl_3 + C_2H_5ONa \longrightarrow \underset{\text{Dichloro carbene}}{:CCl_2} + C_2H_5OH + NaCl$

(2) See Q. 16.

Q. 18. How do you explain that

(*i*) Benzyl carbocation is much more stable than phenylethyl carbocation?

(*ii*) Triphenyl methyl carbocation is a very stable carbocation?

Ans.

(*i*) $C_6H_5-\overset{\oplus}{C}H_2$ $C_6H_5-CH_2-\overset{\oplus}{C}H_2$

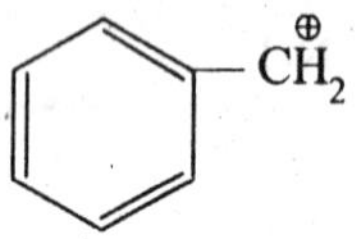

Benzyl carbocation 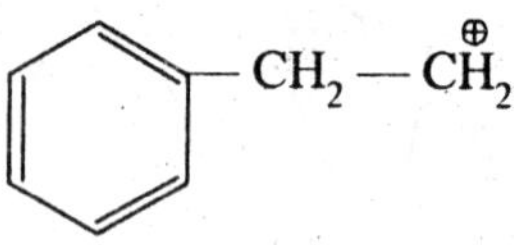   Phenylethyl carbocation

Benzyl carbocation exists in a number of canonical forms and hence is more stable than phenyl ethyl carbocation in which there is no possibility of resonance.

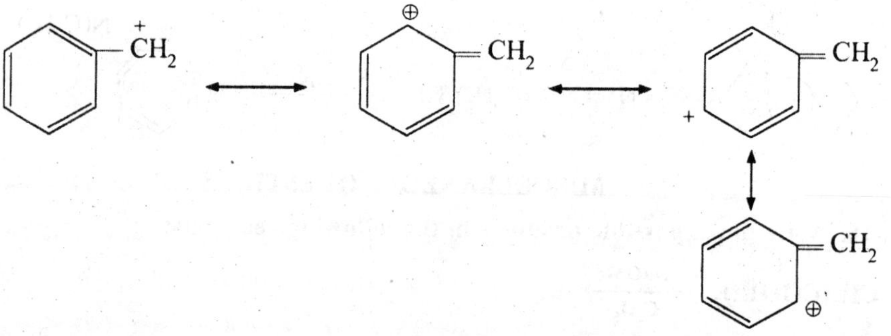

$C_6H_5-\overset{+}{C}H_2 \longleftrightarrow$ (ring with $\oplus$, $=CH_2$) $\longleftrightarrow$ (ring with $+$, $=CH_2$) $\updownarrow$ (ring with $\oplus$, $=CH_2$)

(*ii*) Triphenylmethyl carbocation is a very stable carbocation. See Q. 1.

Q. 19. What happens when 2, 3-Dimethyl butan-2, 3-diol is treated with dil. $\mathbf{H_2SO_4}$?

(*G Nanakdev, 1996*)

Ans.

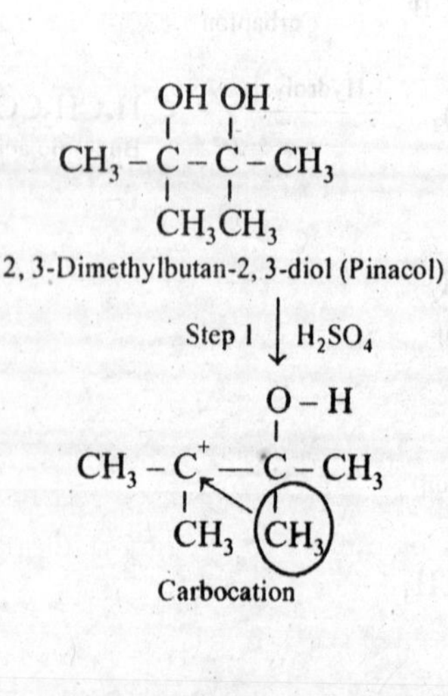

$$CH_3-\underset{CH_3}{\overset{OH}{C}}-\underset{CH_3}{\overset{OH}{C}}-CH_3$$

2, 3-Dimethylbutan-2, 3-diol (Pinacol)

Step I $\Big\downarrow H_2SO_4$

$$CH_3-\underset{CH_3}{\overset{+}{C}}-\underset{CH_3}{\overset{O-H}{C}}-CH_3$$

Carbocation

$$\downarrow \text{Step II}$$

$$\begin{array}{c} \quad CH_3\ :\ddot{O}-H \\ \quad | \qquad | \\ CH_3-C-\overset{+}{C}-CH_3 \\ | \\ CH_3 \end{array} \rightleftharpoons \begin{array}{c} \quad CH_3\ \overset{+}{O}H \\ \quad | \qquad \| \\ CH_3-C-C-CH_3 \\ | \\ CH_3 \end{array}$$

$$\downarrow \text{Step III}$$

$$\begin{array}{c} CH_3 \\ | \\ CH_3-C-C-CH_3 \\ | \qquad \| \\ CH_3 \quad O \end{array}$$

Pinacolone

Migration of methyl group in step II is facilitated by the resonance possible in the product.

Q. 20. Mention the methods by which you can infer that a particular reaction involves a free radical intermediate or not.

Ans. Methods :

1. Formation of free radicals is favoured by the presence of u.v. ligh, heat and organic peroxides.
2. Free radicals are paramagnetic due to the presence of unpaired electrons.
3. Reactions involving free radicals widely occur in the gas phase.

Q. 21. Give the orbital structure of $\bar{C}H_3$, Mention its hybridisation and its shape.

(Delhi, 2007)

Ans. The central carbon atom in $\bar{C}H_3$ with a pair of unshared electrons is sp^3 hybridised. The three sp^3 hybrid orbitals are used in forming three sigma bonds with hydrogen. The fourth sp^3 hybrid orbital is occupied by a lone pair of electrons. The structure is similar to that of ammonia in which one hybrid orbital of nitrogen is occupied by a lone pair of electrons. The structure may be depicted as :

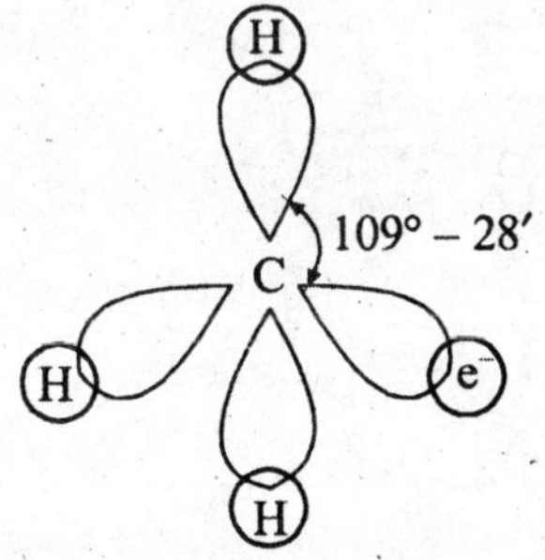

Thus it has a tetrahedral geometry. Three hydrogens are directed towards three corners of the tetrahedron. The fourth position is occupied by unshared pair of electrons.

Q. 22. Arrnage the following in increasing order of stability with explanation.

$$(C_6H_5)_3\overset{+}{C},\ (C_6H_5)_2\overset{+}{C}H,\ C_6H_5\overset{+}{C}H_2,\ (CH_3)_3\overset{+}{C},$$

$$(CH_3)_2\overset{+}{C}H,\ CH_3\overset{+}{C}H_2,\ \overset{+}{C}H_3,\ CH_2=CH-\overset{+}{C}H_2$$

Ans. The increasing order of stability is:

$$\overset{+}{C}H_3 < CH_3\overset{+}{C}H_2 < (CH_3)_2\overset{+}{C}H < (CH_3)_3\overset{+}{C} < CH_2=CH-\overset{+}{C}H_2$$

$$< C_6H_5\overset{+}{C}H_2 < (C_6H_5)_2\overset{+}{C}H < (C_6H_5)_3\overset{+}{C}$$

Aromatic carbocations $(C_6H_5)_3\overset{+}{C}$, $(C_6H_5)_2\overset{+}{C}H$ and $C_6H_5\overset{+}{C}H_2$ are more stable than aliphatic carbocations $(CH_3)_3\overset{+}{C}$, $(CH_3)_2\overset{+}{C}H$, $CH_3\overset{+}{C}H_2$ and $\overset{+}{C}H_3$. This is because in aromatic carbocations, stability is attained through resonance effect while in aliphatic carbocations, stability is attained through inductive effect. Resonance effect is observed to be stronger than inductive effect.

Within the aromatic carbocations, greater the number of benzene rings attached to the carbon carrying positive charge, greater is the number of resonance structures. Hence, greater is the stability.

Similarly, within aliphatic carbocations, greater the number of alkyl groups attached to the positive carbon, greater is the inductive effect and hence greater is the stability.

Allyl carbocation is stabilised through resonance effect, hence it occupies the central position between aromatic and aliphatic carbocations. It exists as a resonance hybrid of only two structures of equal energy.

$$\underset{\text{I}}{CH_2=CH-\overset{+}{C}H_2} \longleftrightarrow \underset{\text{II}}{\overset{+}{C}H_2-CH=CH_2}$$

39

REARRANGEMENTS

CONTENTS

Q. 1. What is meant by a rearrangement reaction? Explain the following terms–nucleophilic rearrangement, electrophilic rearrangement, free-radical rearrangement.

Ans. A rearrangement reaction is one which involves the shuffling of the sequence of atoms in a molecule to form a new product.

Nucleophilic rearrangement. These are the rearrangement reactions in which the migrating group (nucleophile) gets migrated to an electron deficient centre. Such a rearrangement takes place in order to give more stable carbocations and involves the migration of hydride ion, alkyl or aryl groups. Migration of such groups could take place from carbon to carbon, carbon to oxygen or carbon to nitrogen. When the migration of a group takes place to a neighbouring carbon, nitrogen or oxygen, it is called **1, 2 shift.**

Electrophilic rearrangement. In these reactions, an electrophile migrates from one position to an electron rich atom. These rearrangements are initiated by such basic reagents which remove a positively charged species like H^+.

Free-radical rearrangement. Free-radical rearrangement is one in which there is a change in the position of free radical in the structure. Generally there is a tendency on the part of a primary free radical to change into secondary free-radical and so on. But this tendency is much less than that observed in the case of carbocations.

$$-\overset{|}{\underset{|}{C}}(R)-C^{\oplus}- \longrightarrow -\overset{+}{\underset{|}{C}}-\overset{R}{\underset{|}{C}}-$$ migration to C

$$-\overset{|}{\underset{|}{C}}(R)-\ddot{N}: \longrightarrow -\underset{|}{C}=\ddot{N}-R$$ migration to N

$$-\overset{|}{\underset{|}{C}}(R)-\ddot{O}: \longrightarrow -\underset{|}{C}=O-R$$ migration to O

Q. 2. Describe Benzidine rearrangement giving examples. Explain the mechanism of the reaction. (*Kerala, 2000; Nagpur 2002*)

Ans. Benzidine rearrangement is the acid-catalysed rearrangement of aromatic hydrazo compounds. The name of the rearrangement has been derived from the major product, benzidine, which is formed when hydrazobenzene (1, 2-diphenylhydrazine) is warmed with dilute HCl or H_2SO_4. The other product of the rearrangement is the isomeric diphenylene.

$$C_6H_5-NH-NH-C_6H_5 \xrightarrow[H_2SO_4]{\text{Dil. HCl or}}$$

Hydrazobenzene

$$H_2N-C_6H_4-C_6H_4-NH_2$$

Benzidine (70%)

$$\overset{+}{N}H_2\text{ (ortho)}-C_6H_4-C_6H_4-NH_2$$

Diphenylene (30%)

In addition to these two products, less frequently, three other products, namely *o*-benzidine, *o*-semidine and *p*-semidine are also obtained in smaller amounts. For example, when solid hydrazobenzene is treated with dry hydrogen chloride, all the possible products, as shown below, are obtained.

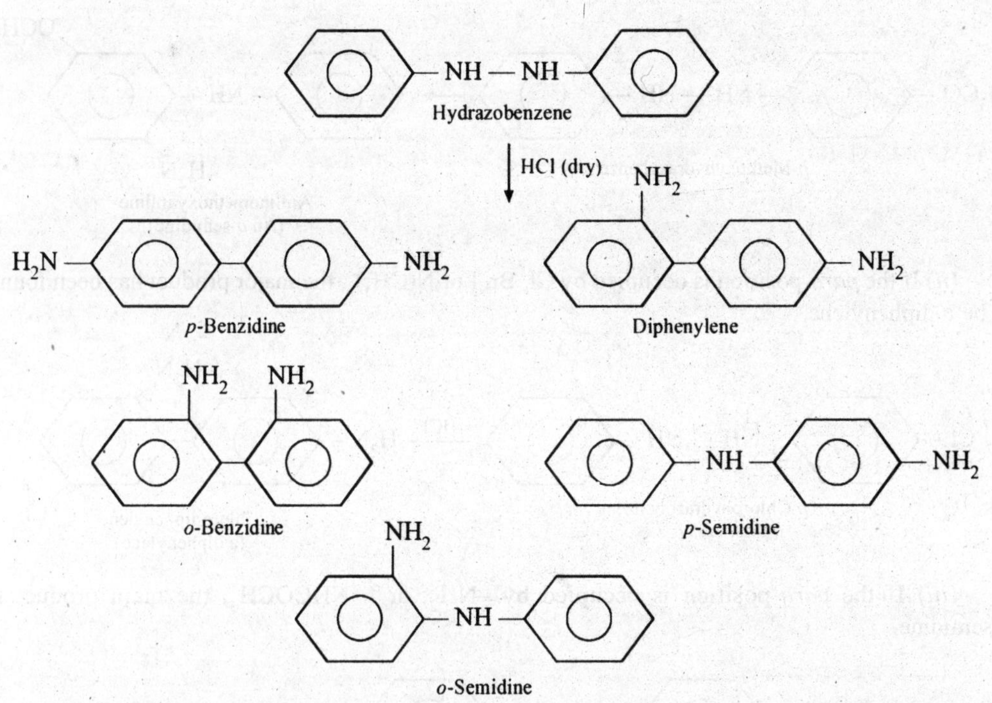

When 1-hydrazonaphthalene is treated with $HClO_4$, it undergoes benzidine rearrangement forming *p*-benzidine type compound as well as *o*-benzidine type compound (major product), the latter of which may cyclise to form pyrrole derivative.

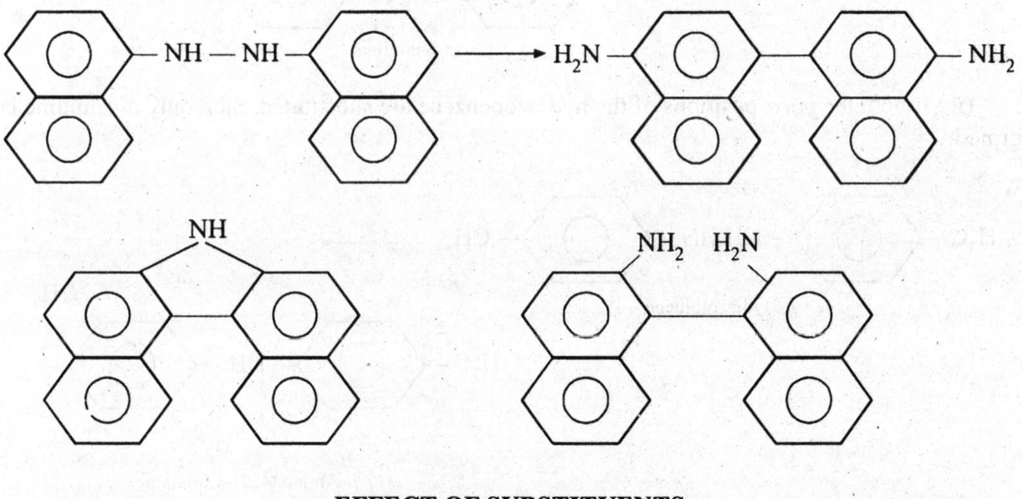

EFFECT OF SUBSTITUENTS

Products formed during benzidine rearrangement are determined by the nature of the substituent, if any, especially in the *p*-position in the benzene nucleus of hydrazobenzene. This is illustrated with the help of the following examples:

(*i*) If the *para*-position is occupied by a methyl, methoxy or an ethoxy group, the main product has been found to the *o*-semidine.

OCH_3

H_3CO — NH — NH — → NH —

p-Methoxyhydrazobenzene

H_2N

Anilinomethoxyaniline
(An *o*-semidine)

(*ii*) If the *para*-position is occupied by Cl, Br, I or $N(CH_3)_2$, the major product has been found to be a diphenylene.

H_2N

Cl — NH — NH — $\xrightarrow{HCl}$ H_2N —

p-Chlorohydrazobenzene

Chlorodiphenylene Cl
(a diphenylene)

(*iii*) If the *para*-position is occupied by $-NH_2$, or $-NHCOCH_3$, the main product is *p*-semidine.

AcHN — NH.NH — →

AcHN — NH — NH_2

(*p*-Semidine)

(*iv*) If both the *para*-positions of the hydrazobenzene are substituted, then only *o*-semidine is formed.

H_3C — NH.HN — CH_3 →

p-Hydrazotoluene

CH_3

H_3C — NH —

H_2N

p-Amino (4, 3)-ditolamine

(*v*) Some groups especially carboxyl and sulphonic acid get decomposed during the benzidine rearrangement.

HOOC–C₆H₄–NH.NH–C₆H₅ $\xrightarrow{-CO_2}$

Hydrazobenzene-4-carboxylic acid

H_2N–C₆H₄–C₆H₄–NH_2

Benzidine

Mechanism. Although the mechanism of benzidine rearrangement is still uncertain, involvement of an intramolecular reaction has been established by a number of experiments.

(*i*) When a mixture of *o, o'*-dimethoxyhydrazobenzene and *o, o'*-diethoxybenzidine is subjected to benzidine rearrangement, they rearrange to form *m, m'*-diethoxybenzidine. No cross product, ethoxymethoxybenzidine, is detected.

o, o'-Dimethoxy hydrazobenzene (OCH_3, OCH_3; NH.NH) + *o, o'*-Diethoxy hydrazobenzene (OC_2H_5, OC_2H_5; NH.NH) ⟶ *m, m'*-Dimethoxy benzidine (H_3CO, OCH_3; H_2N, NH_2) + *m, m'*-Diethoxy benzidine (H_5C_2O, OC_2H_5; H_2N, NH_2)

(*ii*) When a mixture of two hydrazobenzenes, one having labelled methyl group ^{14}C as a substituent, is subjected to benzidine rearrangement, it is found that the whole of the tracer atom ^{14}C gets incorporated into the product derived from the labelled hydrazobenzene.

C₆H₅–NH — NH–C₆H₄($^{14}CH_3$) + (CH_3)C₆H₄–NH.NH–C₆H₄(CH_3) ⟶ H_2N–C₆H₄–C₆H₃($^{14}CH_3$)–NH_2 + NH_2–C₆H₃(H_3C)–C₆H₃(CH_3)–NH_2

(*iii*) By comparing the rates of rearrangement in H_2O and D_2O, it was proved that the protonation at one or both of the nitrogens is not a slow and determining step. If the protonation were rate determining step of the rearrangement, there should have occurred a retardation in the rate of reaction in D_2O. In actual practice, there occurred a two-fold increase in the rate of reaction in D_2O. This is expected because D_3O^+ in D_2O is a stronger acid than H_3O^+ in H_2O.

(*iv*) As the replacement of the ring protons in the hydrazo compounds by deuterium or tritium could not slow down the rate of their rearrangement, it means that the loss of two ring protons required for the formation of benzidine and diphenylene is also not a slow and rate-determining step.

A number of theories have been postulated to explain the mechanism of the benzidine rearrangement. Some of these theories are described as follows:

1. π-Complex mechanism. As the benzidine rearrangement is kinetically of first order in acid, Dewar postulated that the monoprotonated hydrazo compound undergoes a rate-determining heterolytic fission to form a π-complex. This complex does not represent transition state but represents a definite compound in which the covalent bond between the nitrogen atoms has been replaced by a delocalised π-bond between the aromatic rings.

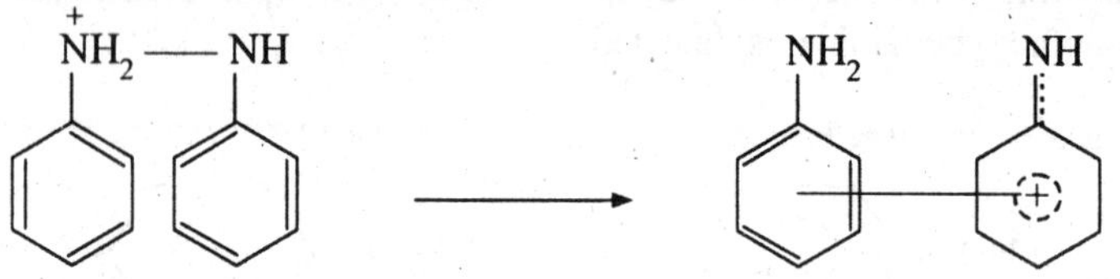

$\overset{+}{N}H_2$—NH → NH_2 NH

(I)

The π-bonds in (I) can hold the aromatic rings in parallel planes. These rings can rotate and collapse to form the various products.

When it was found that the rearrangement is kinetically of second order in acid, the original theory was modified and it was postulated that the attack of second proton on (I) was rate determining step.

2. Polar transition state theory. This theory was proposed by Bandhorpe, Hughes and Ingold. According to this theory, it is assumed that a polar transition state is developed in the rate-determining step of the rearrangement. We shall illustrate this theory by considering the arrangement of hydrazobenzene (II) *via.* a two-proton mechanism. As the diprotonated molecule of (II) has two adjacent positively charged nitrogen atoms, the repulsion between them facilitates the cleavage of the N-N bond. As a result of heterolysis two loosely bound aromatic rings are formed. One of these rings carries two units of positive charge and thus behave like a dication. However, the other ring does not have charge but has the character of an aryl amine. This possesses fractional dipolar charge which is distributed in the same manner as in an aniline molecule.

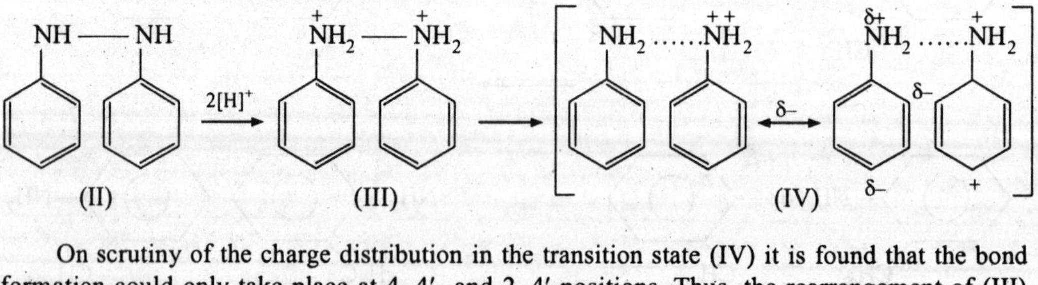

NH—NH $\xrightarrow{2[H]^+}$ $\overset{+}{N}H_2$—$\overset{+}{N}H_2$ → [NH_2......$\overset{++}{N}H_2$ ↔ $\overset{\delta+}{N}H_2$......$\overset{+}{N}H_2$]

(II) (III) (IV)

On scrutiny of the charge distribution in the transition state (IV) it is found that the bond formation could only take place at 4, 4′- and 2, 4′-positions. Thus, the rearrangement of (III) involves a transition state in which the heterolysis of the N-N bond is converted with the formation of new bonds at the most favourable positions in the transition state. On the basis of these assumptions, a two-proton mechanism for the formation, of (V) and (VI) from the rearrangement of (II) can be postulated as follows:

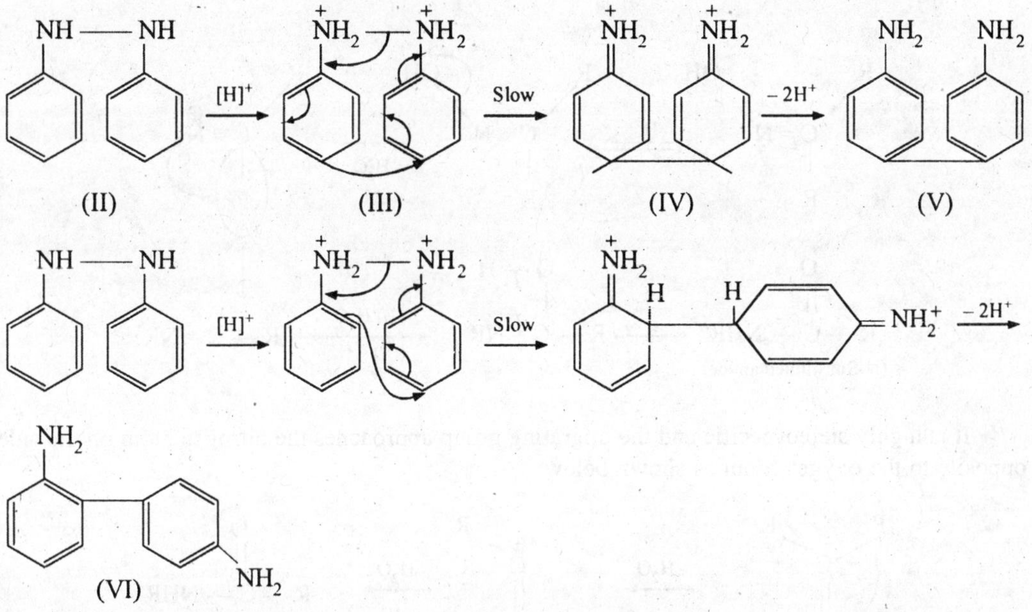

(II) (III) (IV) (V)

(VI)

Applications. As the alkaline reduction of the nitrocompounds yields hydrazobenzenes very readily benzidine rearrangement provides an important route for the preparation of 4, 4′-disubstituted biphenyls.

The 4, 4′-diaminodiphenyl (benzidine) and its derivatives are especially used in the synthesis of dyes. An example is benzopurpurin 4B which is an important red dye and is obtained from *o*-toluidine as a starting material.

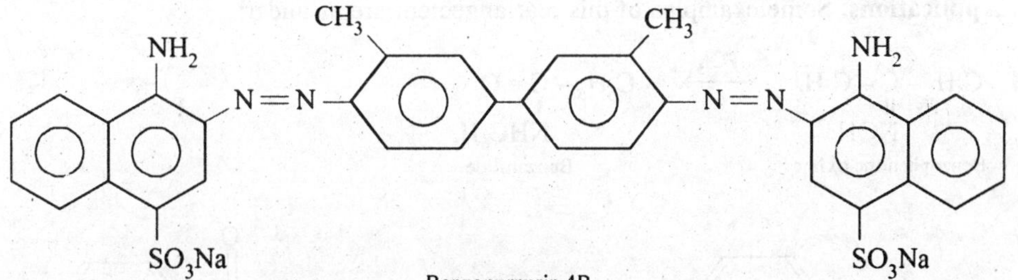

Benzopurpurin 4B

Q. 3. Describe the mechanism and applications of Beckmann rearrangement.

(Punjab, 2003; Mumbai, 2010)

Ans. Principle. This rearrangement relates to the conversion of ketoximes into N-substituted acid amides. The rearrangement is catalysed by acidic reagents such as thionyl chloride, phosphorus pentachloride, sulphuric acid and phosphorus pentoxide.

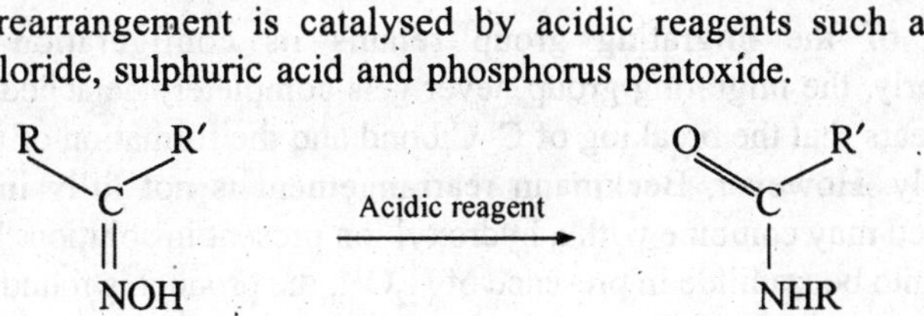

$$R(R')C{=}NOH \xrightarrow{\text{Acidic reagent}} R'C(=O)NHR$$

Ketoxime → Substituted amide

Mechanism. The following steps are involved in the mechanism of the reaction.

$$\underset{R'}{\overset{R}{\ }}C{=}\ddot{N}{-}OH \underset{}{\overset{H^+}{\rightleftharpoons}} \underset{R'}{\overset{R}{\ }}C{=}N{-}\overset{+}{O}H_2 \xrightarrow[-H_2O]{} \underset{R'}{\overset{R}{\ }}C{=}\overset{\oplus}{N} \longrightarrow R{-}\overset{\oplus}{C}{=}NR' \xrightarrow[(ii)\ -H^+]{(i)\ H_2O} R{-}\overset{O-H}{C}{=}NR' \longrightarrow R{-}\overset{O}{\overset{\|}{C}}{-}NHR'$$

(N-Substituted amide)

It is highly stereospecific and the migrating group approaches the nitrogen atom on the side opposite to the oxygen atom as shown below:

$$\overset{R'\quad R}{C}{=}N{-}\overset{\oplus}{O}H_2 \xrightarrow{-H_2O} R{-}\overset{\oplus}{C}{=}N{-}R' \xrightarrow[-H^+]{H_2O} R{-}\overset{O}{\overset{\|}{C}}{-}NHR'$$

From the above mechanism, it appears that either the breakage of the N–O bond and the group migration are spontaneous or the two steps follow each other quickly. The migration of the anti-group is so definite in the Beckmann rearrangement that it is possible to establish the configuration of a particular oxime by identifying its rearrangement products.

Applications. Some examples of this rearrangement are as under:

1. $C_6H_5-\underset{NOH}{\underset{\|}{C}}-C_6H_5 \xrightarrow{PCl_5} C_6H_5-\underset{NHC_6H_5}{\underset{|}{C}}=O$

Benzophenone oxime → Benzanilide

2. $C_6H_5-\underset{NOH}{\underset{\|}{C}}-C_6H_4-OCH_3 \xrightarrow{PCl_5} C_6H_5-NH-\overset{O}{\overset{\|}{C}}-C_6H_4-OCH_3$

p-Methoxy benzophenone oxime → *p*-Methoxy benzanilide

Carbon atom of the migrating group retains its configuration in the Beckmann rearrangement. Clearly, the migrating group never gets completely detached from the rest of the molecule and it appears that the breaking of C–C bond and the formation of new C–N bond takes place simultaneously. However, Beckmann rearrangement is not fully intramolecular as the carbonium ion formed may combine with a hydroxyl ion present in solution. When benzophenone oxime is converted into benzanilide in presence of H_2O^{18}, the product is found to contain the heavy isotope.

$$C_6H_5(C_6H_5)C=N-OH \xrightarrow[(ii)\ H_2O^{18}]{(i)\ PCl_5} C_6H_5-\overset{O^{18}}{\overset{\|}{C}}-NHC_6H_5$$

3. Enlargement of rings is possible when cyclic ketoximes are subjected to this rearrangement. Ring expansion takes place and a cyclic amide is formed. An important example is the conversion of cyclohexanone oxime into caprolactum.

Cyclohexanone $\xrightarrow{NH_2OH}$ Cyclohexanone oxime $\xrightarrow[\text{rearrangement}]{\text{Beckmann}}$ Caprolactum (2-Keto hexamethylene amine)

Cyclohexanone

Cyclohexanone oxime

Caprolactum

(2-Keto hexamethylene amine)

$$C_6H_{10}{=}NOH \xrightarrow{PCl_5} C_6H_{10}{=}N-Cl \xrightarrow{-\ddot{C}l^{\ominus}} \text{N}\equiv\text{C}^{+} \xrightarrow{HOH} \text{N}{=}\text{C}-\overset{+}{O}H_2 \xrightarrow{-H^+} \text{N}{=}\text{C}-OH \rightleftharpoons \text{N}-H,\ \text{C}{=}O$$

Caprolactum

4. The Beckmann rearrangement is applicable in the condensation of the oxime of cinnamic aldehyde in the presence of phosphorus pentoxide. The expected product is quinoline but the product actually obtained is isoquinoline. It can be explained on the basis of Beckmann rearrangement.

Cinnamaldehyde $\xrightarrow{NH_2OH}$ Cinnamaldehyde oxime $\xrightarrow{H^+}$ [A]

[A] → intermediate $\xrightarrow{H_2O}$ $\underset{-H^+}{\rightleftharpoons}$ $\xrightarrow[-H_2O]{P_2O_5}$ Isoquinoline

5. It was established by Kenyon and Young that there is a complete retention of configuration in this rearrangement. They also correlated Beckmann and Hofmann rearrangement. An optically active acid was converted to its amide and the amide was subjected to Hofmann rearrangement to form an amine. In another set of experiment, the acid was converted to ketone which in turn is changed to oxime. The oxime was then subjected to Beckmann rearrangement. The product in this case was also found to be optically active amine without any racemic product.

6. Beckmann rearrangement is not fully intramolecular in nature. It is partly intermolecular. This can be proved when we consider the rearrangement of benzophenone oxime with the help of PCl_5 carried with water containing heavy oxygen (H_2O^{18}). The final product formed was found to contain the labelled oxygen.

$$\underset{\text{Benzophenone oxime}}{(C_6H_5)_2C=N-OH} \xrightarrow[H_2O_{18}]{PCl_5} (C_6H_5)_2C=N-OPCl_4 \xrightarrow{-(OPCl_4)} C_6H_5-\overset{\oplus}{C}=NC_6H_5$$

$$C_6H_5-\overset{\oplus}{C}=NC_6H_5 \xrightarrow{H_2O^{18}} C_6H_5-C(-\overset{\oplus}{O}^{18}H_2)=NC_6H_5 \underset{}{\overset{-H^+}{\rightleftharpoons}} C_6H_5-C(-{}^{18}O-H)=NC_6H_5 \longrightarrow \underset{\text{Benzanilide}}{C_6H_5-\overset{^{18}O}{\overset{\|}{C}}-NHC_6H_5}$$

Q. 4. Describe the mechanism and applications of Schmidt reaction.

(Punjab, 2003; Mumbai, 2010

Ans. Schmidt reaction is a modification of Curtius reaction.

In this reaction, carboxylic acids react with hydrazoic acid in the presence of strong acid to form primary amines. In this reaction, an acid (aliphatic or aromatic) is first dissolved in concentrated sulphuric acid and then heated with sodium azide slowly. Consider the following general reaction:

$$RCOOH + HN_3 \xrightarrow{H_2SO_4} RNH_2 + CO_2 + N_2$$

Mechanism. In the first step, the acid gets protonated in sulphuric acid. The protonated acid then adds a molecule of hydrazoic acid (nucleophilic addition).

$$R-\overset{O}{\overset{\|}{C}}-O-H \underset{H_2SO_4}{\overset{H^+}{\rightleftharpoons}} R-\overset{OH}{\overset{|}{\underset{\oplus}{C}}}-OH \overset{H-\bar{N}-\overset{+}{N}\equiv N}{\rightleftharpoons} R-\overset{OH}{\overset{|}{C}}(-OH)-\underset{H}{N}-\overset{+}{N}\equiv N$$

$$\overset{-H_2O}{\rightleftharpoons} R-\overset{OH}{\overset{|}{C}}=N-\overset{+}{N}\equiv N \xrightarrow{-N_2} \overset{OH^-}{\overset{|}{\underset{\oplus}{C}}}=N-R \xrightarrow[-H_2O]{OH^-} R-N=C=O \xrightarrow{2OH^-} \underset{\text{1°-Amine}}{RNH_2} + CO_3^{2-}$$

The rearrangement involves the migration of an alkyl or aryl groups from the carboxyl carbon to the adjacent nitrogen. This rearrangement is intermolecular.

Some Examples

1. Hexanoic acid on dissolving in concentrated sulphuric acid is heated with sodium azide.

$$\underset{\text{Hexanoic acid}}{CH_3(CH_2)_4COOH} \xrightarrow{H_2SO_4/NaN_3} \underset{n\text{-Pentyl amine}}{CH_3(CH_2)_4NH_2}$$

2. Benzoic acid is converted to aniline

$$\underset{\text{Benzoic acid}}{C_6H_5-COOH} \xrightarrow{\text{Conc. } H_2SO_4/NaN_3} \underset{\text{Aniline}}{C_6H_5-NH_2}$$

3.
$$CH_3CHO + NH_3 \xrightarrow{H_2SO_4} \underset{\text{Methyl cyanide}}{CH_3CN} + \underset{\text{Formyl derivative of amine}}{CH_3NH.CHO}$$

4. The reaction between a ketone and hydrazoic acid is a method for insertion of NH between the carboxyl group and a R group. The ketone is thus converted into a N– substituted amide or an anilide.

$$R-\overset{O}{\overset{\|}{C}}-R' + HN_3 \xrightarrow{H^+} R-\overset{O}{\overset{\|}{C}}-NHR'$$

One or both the R-groups may be aryl. It has been found that dialkyl ketones and cyclic ketones react more rapidly than alkyl aryl ketones which, in turn, react more rapidly than diaryl ketones. With alkyl aryl ketones, it is the aryl group that generally migrates to the nitrogen atom, except when the alkyl group is bulky. The mechanism with ketones is described as under:

$$\underset{\text{Ketone}}{R-\overset{O}{\overset{\|}{C}}-R} \xrightarrow{H^+} R-\overset{\oplus}{C}(OH)-R' \xrightarrow{HN_3} R-C(OH)(-\ddot{N}H-\overset{\oplus}{N}\equiv N)-R' \xrightarrow{-H_2O} R-C(=N-\overset{+}{N}\equiv N)-R'$$

$$\downarrow$$

$$\underset{\text{Anilide}}{R'-\underset{O}{\underset{\|}{C}}-NHR} \longleftarrow R'-C(O-H)=\ddot{N}R \xleftarrow{-H^+} R'-C(\overset{\oplus}{O}H_2)=\ddot{N}-R \xleftarrow{H_2O} R'-\overset{\oplus}{C}=NR$$

5. Cyclic ketones under these conditions undergo ring expansion and form lactams.

$$\underset{\text{Cyclohexanone}}{C_6H_{10}=O} \xrightarrow[H^+]{HN_3} \underset{\text{Caprolactum}}{\text{(ring NH, C=O)}}$$

Q. 5. Describe completely Wagner-Meerwein rearrangement.

(Panjab, 2003; Mumbai, 2004)

Ans. These arrangements take place in tertiary amyl and neopentyl type of alcohols.

Such alcohols on treatment with acids generally produce substitution and elimination products. Alcohols in which two or more alkyl or aryl groups are in the β-carbon, the products get rearranged. These rearrangements are called Wagner-Meerwein rearrangements.

1. If the hydrolysis of 1-Bromo-2, 2-dimethyl propane (neopentyl bromide) is carried under the conditions of SN^1 mode, the product formed is 2-methylbutan-2-ol instead of the expected product *i.e.* neopentyl alcohol. In this reaction, neopentyl rearrangement takes place.

$$\underset{\text{Neo-pentyl bromide}}{CH_3-\overset{\overset{CH_3}{|}}{\underset{\underset{CH_3}{|}}{C}}-CH_2Br} \xrightarrow{SN^1} CH_3-\overset{\overset{\boxed{CH_3}}{|}}{\underset{\underset{CH_3}{|}}{C}}-\overset{\oplus}{C}H_2 \xrightarrow{H_2O} \underset{\text{Neo-pentyl alcohol}}{CH_3-\overset{\overset{CH_3}{|}}{\underset{\underset{CH_3}{|}}{C}}-CH_2OH}$$

$$\downarrow \text{1,2-alkyl shift}$$

$$\underset{\text{2-Methyl-2-Butene}}{(H_3C)_2C{=}C(CH_3)H} \xleftarrow{-H^+} \underset{\text{(3°) (More stable carbocation)}}{CH_3-\overset{\oplus}{\underset{\underset{CH_3}{|}}{C}}-CH_2-CH_3} \xrightarrow{H_2O} \underset{\text{2-Methyl-2-butanol}}{CH_3-\overset{\overset{CH_3}{|}}{\underset{\underset{OH}{|}}{C}}-CH_2-CH_3}$$

The encircled methyl group along with the pair of shared electrons, migrates to the carbocation.

The greater stability of tertiary carbocation provides the driving force for the migration of methyl group. Such changes in carbon skeleton involving carbocations are collectively known as Wagner-Meerwein rearrangements. The formation of tertiary carbocation is further confirmed by the formation of 2-methyl-2-butene by the loss of proton (Saytzev product).

2. It is important to note that neo-pentyl bromide undergoes rearrangement during SN^1 hydrolysis but no such rearrangement occurs with its phenyl analogue.

$$CH_3-\overset{\overset{CH_3}{|}}{\underset{\underset{CH_3}{|}}{C}}-\underset{\underset{CH_3}{|}}{CH}-Br \xrightarrow{SN^1} CH_3-\overset{\overset{CH_3}{|}}{\underset{\underset{CH_3}{|}}{C}}-\underset{\underset{CH_3}{|}}{\overset{\oplus}{C}H} \xrightarrow[\text{Shift}]{\text{1, 2-alkyl}} \underset{\text{I}}{CH_3-\overset{\oplus}{\underset{\underset{CH_3}{|}}{C}}-CH(CH_3)}$$

$$\text{I} \longrightarrow \text{Substitution and elimination products}$$

$$CH_3-\overset{\overset{CH_3}{|}}{\underset{\underset{CH_3}{|}}{C}}-\underset{\underset{C_6H_5}{|}}{CH}-Br \xrightarrow{SN^1} CH_3-\overset{\overset{CH_3}{|}}{\underset{\underset{CH_3}{|}}{C}}-\overset{\oplus}{C}HC_6H_5 \not\longrightarrow \underset{\text{Not formed}}{CH_3-\overset{\oplus}{\underset{\underset{CH_3}{|}}{C}}-\underset{\underset{CH_3}{|}}{CH}-C_6H_5}$$

$$\downarrow$$

Substitution and elimination products

3. In pentyl bromide, the 1, 2-alkyl shifts result in the formation of more stable carbocation. In certain reactions, a, 1, 2-hydride shift also takes place. For example, when *n*-propyl bromide is heated with anhydrous aluminium bromide, isopropyl bromide results.

$$\underset{\text{1-Bromopropane}}{CH_3-CH_2-CH_2-Br} \xrightarrow[\Delta]{AlBr_3} \underset{\text{2-Bromopropane}}{CH_3-\underset{\displaystyle Br}{\underset{|}{CH}}-CH_3}$$

4. In cyclic compounds, a carbonium ion is first formed with positive charge on one of the ring carbon atoms. Then it gets rearranged by using σ electrons of C–C bond of the ring. The rearranged product then combines with the nucleophile or loses a proton, forming the final product. Pinene hydrochloride on heating gets rearranged into bornyl chloride. Bornyl chloride on dehydrohalogenation gives camphene.

(*i*) *Rearrangement of α-pinene hydrochloride*

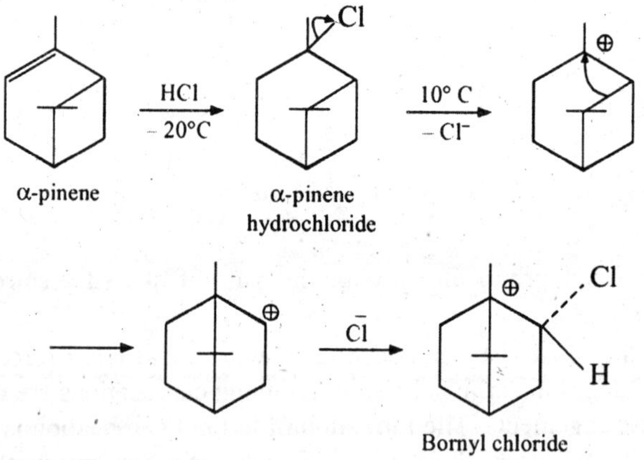

HCl / 20°C; 10° C / $-Cl^-$; $\bar{Cl}$

α-pinene

α-pinene hydrochloride

Bornyl chloride

(*ii*) *Dehydrohalogenation of bornyl chloride*

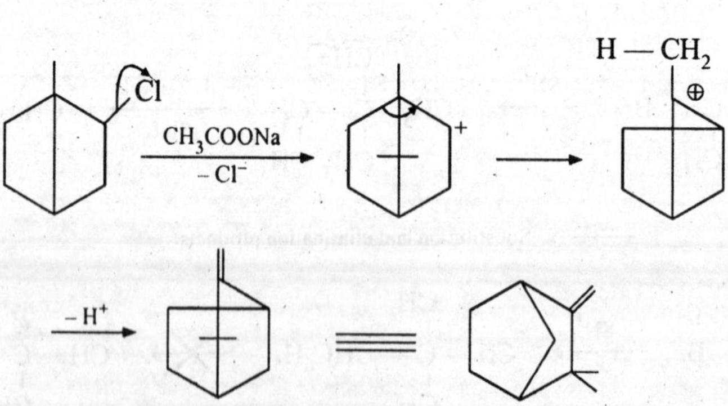

CH_3COONa / $-Cl^-$; H — CH_2; $-H^+$

Camphene

Q. 6. Discuss Cope rearrangement.

Ans. 1. This is an example of thermal rearrangement.

The Cope rearrangement. The thermal conversion of 3, 4-dimethyl-1, 1, 5-hexadiene to 2, 6-octadiene is a typical example of Cope rearrangement. The Cope rearrangement is a reversible process and the equilibrium mixture has a greater proportion of the thermodynamically more stable isomer. It follows first order kinetics. It is an intramolecular process because rearrangement of a mixture of two different 1, 5-hexadienes does not lead to the formation of cross products. The reaction proceeds through a six membered cyclic transition state.

H_3C, H_3C —(473 K ⇌)→ H_3C, H_3C ⇌ H_3C, H_3C

3, 4 Dimethyl Hexa 1, 5-diene — Transition state Six membered — 2, 6-octadiene

2. When the diene is symmetrical about 3, 4-bond, we have the unusual situation where a reaction gives a product identical with the starting material.

3. In 1, 5-dienes, substitution by phenyl or carbethoxy group at 3 or 4-positions facilitates the reaction. The reaction is highly stereospecific.

H_5C_6 —(Δ)→ $C_6H_5—CH$, CH, CH_2, H_2C, CH_2, CH

CH_3, C, $(H_5C_2OOC)_2C$, CH_2, H_2C, $CH—CH_3$, CH —(Δ)→ CH_3, C, $(H_5C_2OOC)_2C$, CH, H_2C, $CH—CH_3$, HC

4. In thermal rearrangements, a six membered cyclic transition state in the chair conformation is preferred. It is shown by the fact that a 1, 5-diene (meso form) yields almost exclusively the *cis-trans* form out of three possible geometrical isomerides, *i.e. cis-cis*, *cis-trans* and *trans-trans*.

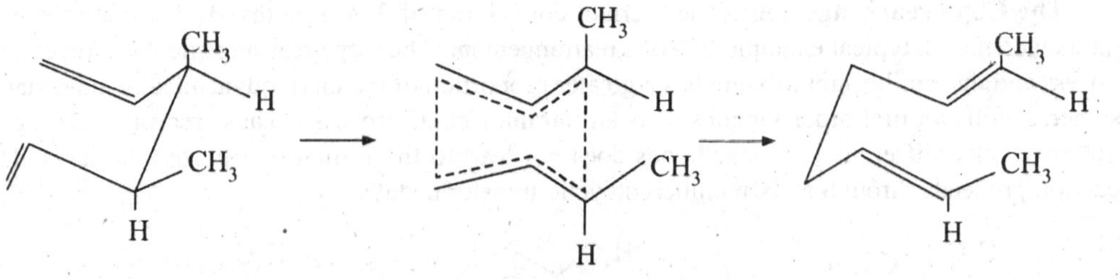

(meso) Transition state (Six membered) (cis-trans)

5. It may be noted that if a diene is not symmetrical about 3, 4-bond, Cope rearrangement occurs. Thus, when 3-methyl-1, 5-hexadiene is heated to 673 K, we get 1, 5-heptadiene. Also the reaction takes place even at a low temperature when there is a double bond at 3- or 4-carbon atom with which the new double bond can conjugate. Clearly, the reaction is reversible and produces an equilibrium mixture to two 1, 5-dienes which is rich in thermodynamically more stable isomer.

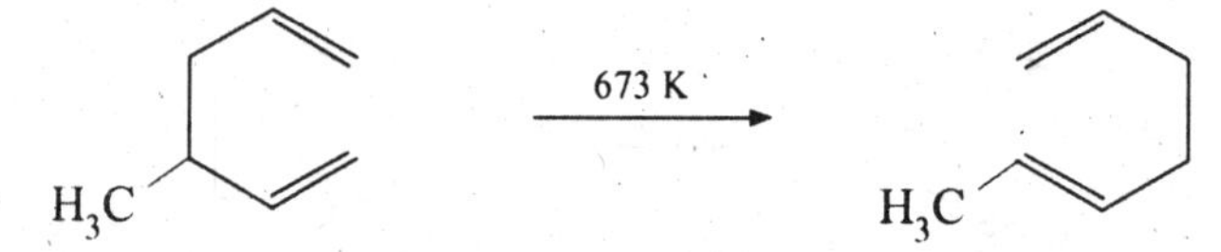

3 Methyl-1, 5-Hexadiene $\xrightarrow{673\ K}$ 1, 5-Heptadiene

6. In case of 3-hydroxy-1, 5-diene, the product tautomerises to an aldehyde or a ketone.

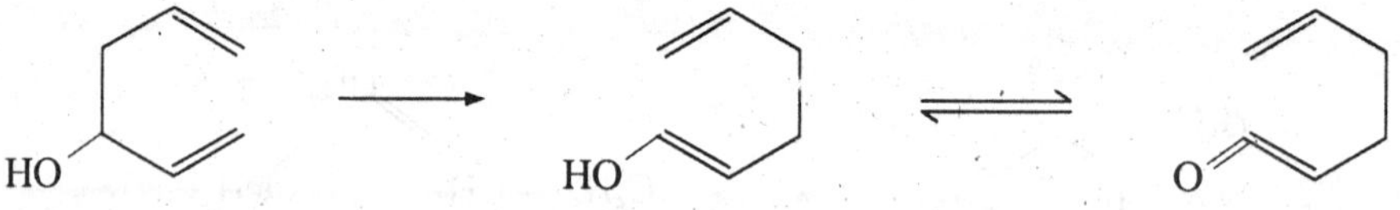

3-Hydroxy-1, 5-Diene A ketone

This rearrangement is called Oxy-Cope-rearrangement and is greatly accelerated in presence of alkoxide ions. When alkoxide is used, the product is the enolate ion, which is hydrolysed to a ketone:

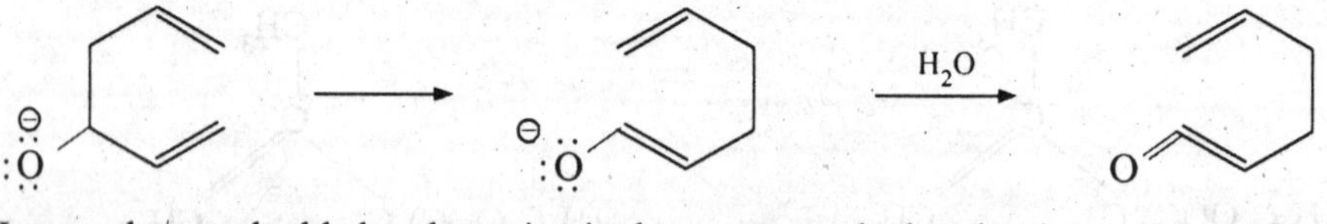

7. In case the two double bonds are in vinyl groups, attached to the ring positions, the product is a ring four carbons larger. Consider Cope rearrangement as applied to divinyl cyclopropanes and divinylcyclobutanes given below:

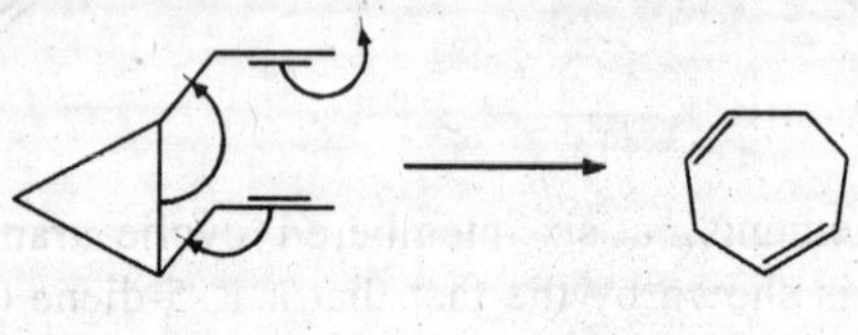

Similarly,

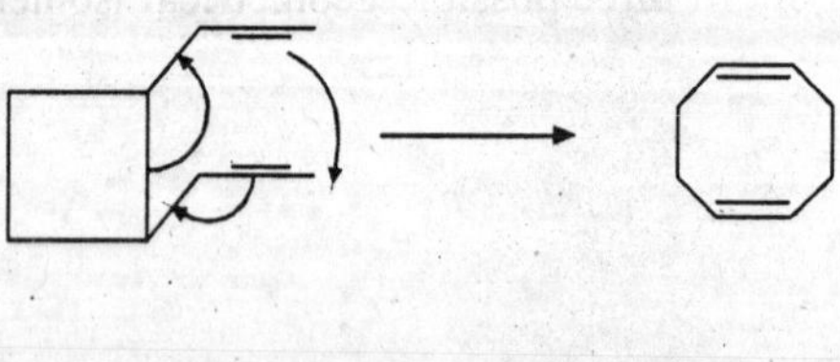

8. Thermally, 1, 5-diynes can also be converted to 3, 4-dimethylenecyclobutenes.

$$\begin{array}{l} H_2C-C\equiv CH \\ \quad | \\ H_2C-C\equiv CH \end{array} \xrightarrow{\Delta} \begin{array}{l} H_2C=C=CH \\ \qquad\quad | \\ H_2C=C=CH \end{array} \longrightarrow \begin{array}{l} H_2C=C-CH \\ \qquad\quad |\quad \| \\ H_2C=C-CH \end{array}$$

But-1, 5-diyne — 3, 4-Dimethylene cyclobutene

Conformation

The mechanism of the uncatalysed Cope rearrangement involves a simple six membered transition state. Now let us study whether the six membered transition state is in the boat form or the chair form. As already indicated in case of 3, 4-dimethyl-1, 5 hexadiene, the transition state formed is in the chair form preferably. The meso isomer is found to yield cis-trans product while the racemic mixture yields, *trans-trans* diene. Consider the transition state from the meso isomer. If the transition state is in the chair form, then one methyl must be axial and the other must be equitorial and hence the product must be a *cis-trans* olefin. But if the transition state is in the boat form, then the meso form yields the *trans-trans* and the *cis-cis* products.

(CHAIR) → (CIS) (TRANS)

(BOAT) → (TRANS) (TRANS)

The nature of the products actually obtained reveals that the transition state in the chair form and not in the boat form.

9. Cope rearrangements do not proceed via six membered cyclic transition state. Some reactions are also found to proceed via diradical mechanism. Consider Cope rearrangement on divinyl cyclobutane.

10. In Cope rearrangement, valence tautomerism is shown by 1,2,3-tri-tert-butylcyclobutadiene.

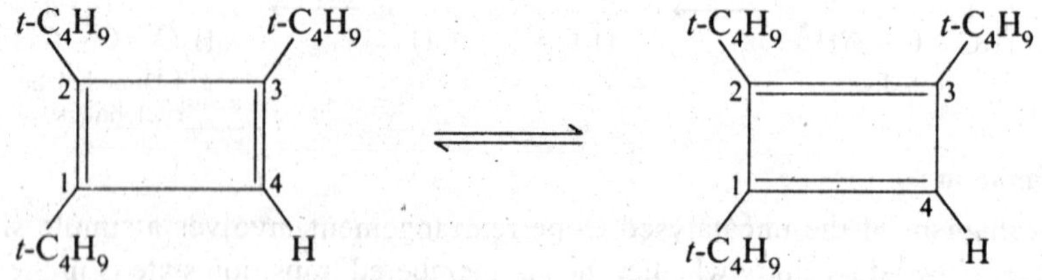

t-C_4H_9 t-C_4H_9 2 3 1 4 t-C_4H_9 H $\rightleftharpoons$ t-C_4H_9 t-C_4H_9 2 3 1 4 t-C_4H_9 H

From the NMR studies, these two isomers are found to exist in dynamic equilibrium even at a very low temperature.

Q. 7. Discuss Claisen rearrangement. *(Kerala, 2001; Purvanchal 2007)*

Ans. When a simple or substituted allyl aryl ether is heated at 473 K, an isomerisation takes place in which the allyl group migrates from the ether oxygen atom to a carbon atom of the aromatic ring. The allyl group goes to the ortho position preferentially.

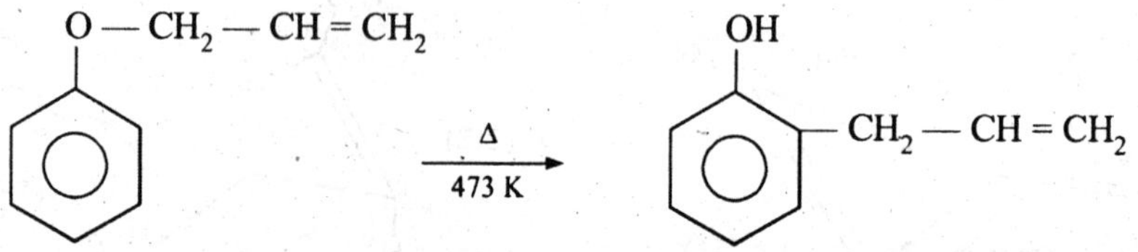

$$C_6H_5-O-CH_2-CH=CH_2 \xrightarrow[473\ K]{\Delta} o\text{-}HO-C_6H_4-CH_2-CH=CH_2$$

If both the ortho positions are occupied, migration takes place to the para position.

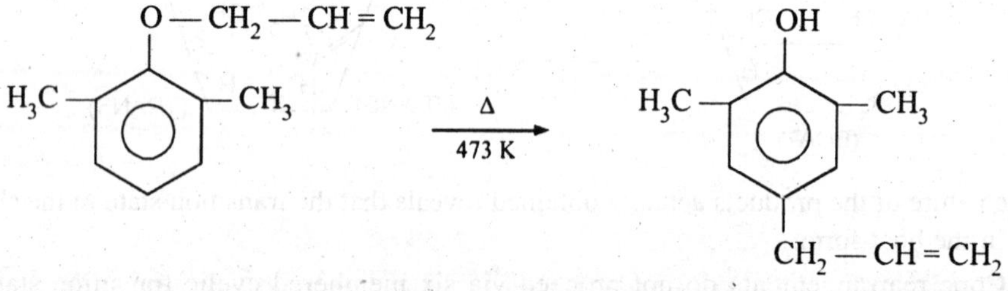

$$2,6\text{-}(H_3C)_2C_6H_3-O-CH_2-CH=CH_2 \xrightarrow[473\ K]{\Delta} 2,6\text{-}(H_3C)_2\text{-}4\text{-}(CH_2-CH=CH_2)C_6H_2-OH$$

Migration to meta position is not observed. In the ortho migration, the allyl group always undergoes an allylic shift. In the para migration, there is never an allylic shift and the allyl group is found exactly as it was in the original ether. Mechanism for the ortho rearrangement is given below:

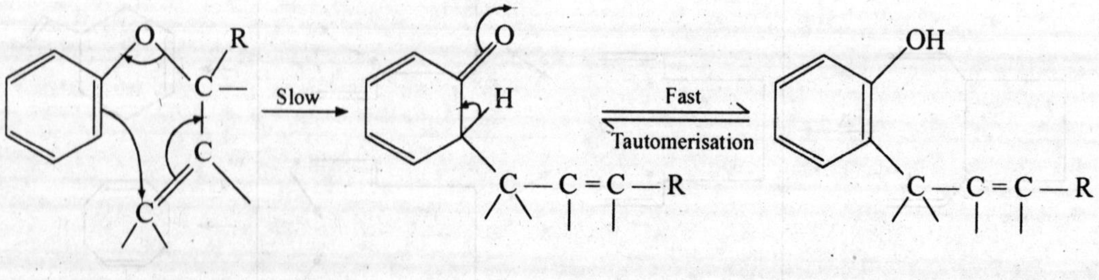

O R C— C C — Slow → O H C—C=C—R — Fast ⇌ Tautomerisation — OH C—C=C—R

The mechanism of Claisen rearrangement when both the ortho positions are blocked is given below:

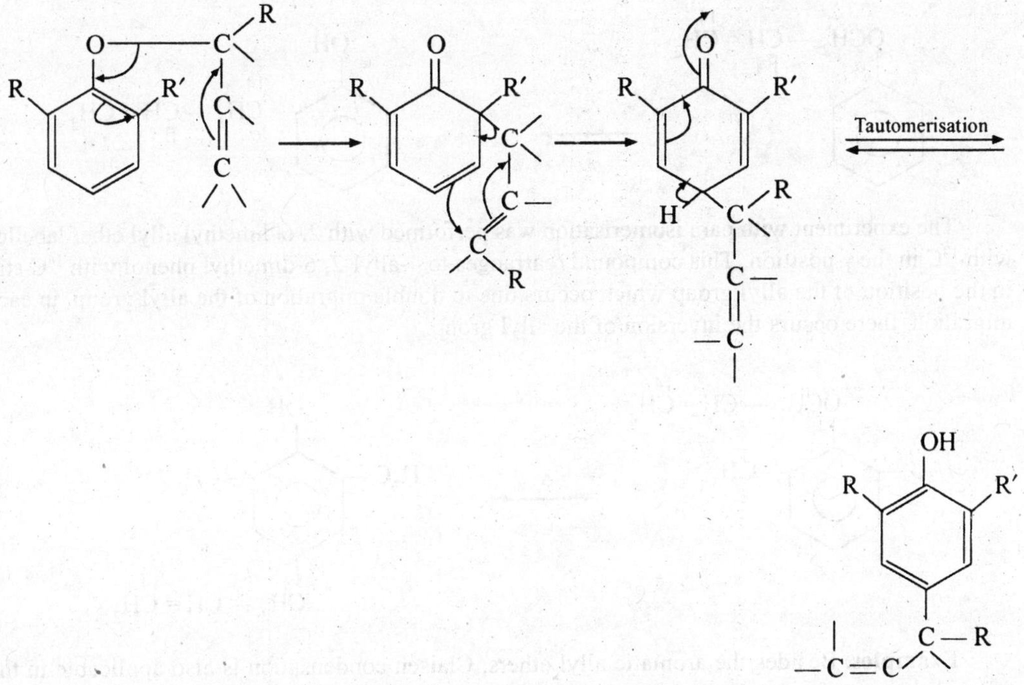

$$\ce{OCH2-CH=CH-CH3} \xrightarrow{\Delta}$$

As Claisen condensation involves a cyclic mechanism, it is not influenced by the substituents on the ring. Ortho isomerisation is accompanied by inversion of the position of substituent with respect to that of the starting material.

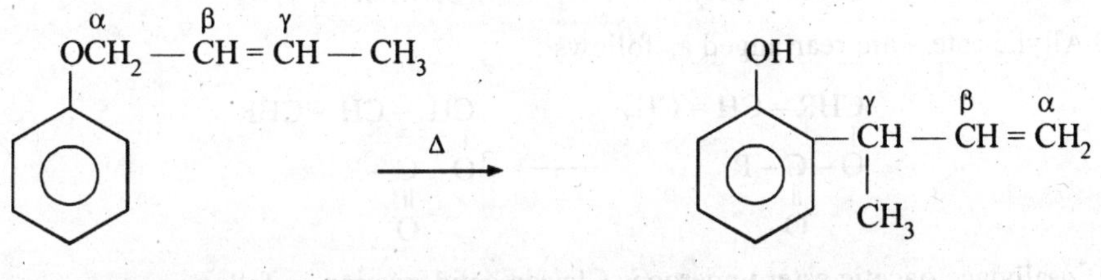

$$\overset{\alpha}{OCH_2}-\overset{\beta}{CH}=\overset{\gamma}{CH}-CH_3 \xrightarrow{\Delta} \text{o-}C_6H_4(OH)-\overset{\gamma}{CH}(CH_3)-\overset{\beta}{CH}=\overset{\alpha}{CH_2}$$

But no such inversion takes place when para isomerisation is involved.

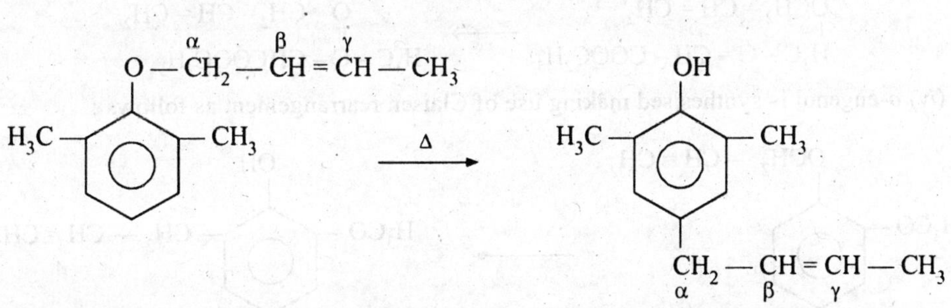

$$O-\overset{\alpha}{CH_2}-\overset{\beta}{CH}=\overset{\gamma}{CH}-CH_3 \text{ (2,6-dimethylphenyl ether)} \xrightarrow{\Delta} \text{4-}(\underset{\alpha}{CH_2}-\underset{\beta}{CH}=\underset{\gamma}{CH}-CH_3)\text{-2,6-}(CH_3)_2C_6H_2OH$$

To prove that ortho isomerisation takes place with inversion of position, Schmidt carried out the rearrangement of phenyl allyl ether labelled with ^{14}C in the γ-position and obtained *o*-allyl phenol having ^{14}C in the α-position.

$$\underset{\text{Allyl phenyl ether}}{C_6H_5-OCH_2(\alpha)-CH(\beta)=\overset{14}{C}H_2(\gamma)} \longrightarrow o\text{-}HO-C_6H_4-\overset{14}{C}H_2(\alpha)-CH(\beta)=CH_2(\gamma)$$

The experiment with para isomerisation was performed with 2, 6 dimethyl allyl ether labelled with ^{14}C in the γ-position. This compound rearranges to 4-allyl-2, 6-dimethyl phenol with ^{14}C still in the position of the allyl group which occurs due to double migration of the allyl group, in each migration, there occurs the inversion of the allyl group.

$$2,6\text{-}(H_3C)_2C_6H_3-OCH_2-CH=\overset{14}{C}H_2 \xrightarrow{\Delta} 2,6\text{-}(H_3C)_2\text{-}4\text{-}(CH_2-CH=\overset{14}{C}H_2)C_6H_2-OH$$

Examples. Besides the aromatic allyl ethers, Claisen condensation is also applicable in the following types of compounds.

(*i*) Allyl vinyl ethers undergo this rearrangement.

$$\begin{matrix} CH_2-CH=CH_2 \\ | \\ OCH=CH_2 \end{matrix} \longrightarrow \begin{matrix} CH_2-CH=CH_2 \\ | \\ CH_2CHO \end{matrix}$$

(*ii*) Allylic esters are rearranged as follows :

$$\begin{matrix} CHR-CH=CH_2 \\ | \\ O-\underset{\underset{O}{\|}}{C}-R \end{matrix} \longrightarrow \begin{matrix} CH_2-CH=CHR \\ | \\ O-\underset{\underset{O}{\|}}{C}-R \end{matrix}$$

(*iii*) *o*-allylacetoacetic ester undergoes Claisen condensation as follows :

$$\begin{matrix} OCH_2-CH=CH_2 \\ | \\ H_3C-C=CH-COOC_2H_5 \end{matrix} \longrightarrow H_3C-\overset{\overset{O}{\|}}{C}-\overset{\overset{CH_2-CH=CH_2}{|}}{C}HCOOC_2H_5$$

(*iv*) *o*-eugenol is synthesised making use of Claisen rearrangement as follows :

$$\underset{\text{Guaiacol allyl ether}}{2\text{-}H_3CO\text{-}C_6H_4-OCH_2-CH=CH_2} \longrightarrow \underset{o\text{-Eugenol}}{2\text{-}H_3CO\text{-}6\text{-}(CH_2-CH=CH_2)C_6H_3-OH}$$

Claisen rearrangement does not involve the formation of free-radicals or ions as the presence or absence of substituents does not affect the rearrangement.

Q. 8. Describe pinacole-pinacolone rearrangement. (*Madras, 2004; Purvanchal, 2007*)

or

Name a rearrangement involving a carbonium ion and discuss its mechanism.

(*Coimbatore 2000*)

Ans. Conversion of pinacols (substituted 1, 2 diols) to aldehydes or ketones by means of mineral acids, acid chlorides, zinc chloride or other electrophilic reagents, is known as **pinacole-pinacolone** rearrangement.

$$\begin{array}{c} CH_3\ CH_3 \\ |\quad\ | \\ CH_3-C-C-CH_3 \\ |\quad\ | \\ OH\ \ OH \\ \text{Pinacole} \end{array} \xrightarrow[\Delta]{\text{Dil. } H_2SO_4} \begin{array}{c} CH_3 \\ | \\ CH_3-C-C-CH_3 \\ |\quad \| \\ CH_3\ O \\ \text{Pinacolone} \end{array}$$

Mechanism. The mechanism of the reaction involves the following four steps:

(*i*) The basic OH group accepts a proton from sulphuric acid to form protonated glycol.

(*ii*) The protonated glycol then loses a molecule of water to form a carbonium ion.

(*iii*) The carbonium ion then undergoes 1, 2-alkyl shift or hydride shift to yield a more stable carbonium ion.

(*iv*) Finally, the more stable carbonium ion loses a proton to form pinacolone.

$$\begin{array}{c} CH_3\ CH_3 \\ |\quad\ | \\ CH_3-C-C-CH_3+H^+ \\ |\quad\ | \\ OH\ \ OH \\ \text{Pinacole} \end{array} \xrightarrow{H_2SO_4} \begin{array}{c} CH_3\ CH_3 \\ |\quad\ | \\ CH_3-C-C-CH_3 \\ |\quad\ | \\ OH\ \ \overset{\oplus}{O}H_2 \end{array}$$

$$\downarrow -H_2O$$

$$\begin{array}{c} CH_3 \\ \oplus\quad | \\ CH_3-C-C-CH_3 \\ |\quad\ | \\ :\ddot{O}H\ \ CH_3 \end{array} \xleftarrow[\text{Shift}]{1,2\text{-alkyl}} \begin{array}{c} CH_3\ CH_3 \\ |\quad\ | \\ CH_3-C-\overset{\oplus}{C}-CH_3 \\ | \\ OH \end{array}$$

$$\longleftrightarrow \begin{array}{c} CH_3 \\ \oplus\quad | \\ CH_3-C-C-CH_3 \\ \|\quad\ | \\ O\quad CH_3 \\ | \\ H \end{array} \xrightarrow{-H^+} \begin{array}{c} CH_3 \\ | \\ CH_3-C-C-CH_3 \\ \|\quad | \\ O\ \ CH_3 \\ \text{Pinacolone} \end{array}$$

The intermediate carbonium ion formed in this rearrangement can also be demonstrated. When the pinacole is rearranged in H_2O^{18} and the product is recovered before the completion of the reaction, pinacolone is found to contain O^{18} isotope. This shows the reversible formation of carbonium ion.

$$\underset{\text{Pinacole}}{CH_3-\overset{CH_3}{\underset{OH}{C}}-\overset{CH_3}{\underset{\oplus}{C}}-CH_3} + H_2O^{18} \rightleftharpoons CH_3-\overset{CH_3}{\underset{OH}{C}}-\overset{CH_3}{\underset{^{18}OH_2^{\oplus}}{CH_3}} \rightleftharpoons CH_3-\overset{CH_3}{\underset{OH}{C}}-\overset{CH_3}{\underset{\underset{18}{OH}}{C}}-CH_3 + H^+$$

Examples

$$\text{1. } \underset{\text{2-Methyl propane 1, 2 diol}}{CH_3-\overset{CH_3}{\underset{OH}{C}}-\underset{OH}{CH_2}} \xrightarrow{H_2SO_4} \underset{\text{Isobutyraldehyde}}{CH_3-\overset{CH_3}{\underset{H}{C}}-\underset{O}{\overset{\|}{CH}}}$$

$$\underset{\text{3-Phenylpropane 2, 3-diol}}{CH_3-\underset{OH}{CH}-\underset{OH}{CH}-C_6H_5} \xrightarrow{H_2SO_4} \underset{\text{Phenyl acetone}}{CH_3-\underset{O}{\overset{\|}{C}}-CH_2-C_6H_5}$$

2. In case of unsymmetrical glycols, the positive charge resides on that carbon atom which forms a more stable carbonium ion. A phenylated carbonium ion is more stable due to resonance stabilisation.

$$C_6H_5-\overset{C_6H_5}{\underset{OH}{C}}-\overset{CH_3}{\underset{OH}{C}}-CH_3 \xrightarrow[(-H_2O)]{H^+} \underset{\text{Less stable carbonium ion}}{C_6H_5-\overset{C_6H_5}{\underset{OH}{C}}-\overset{CH_3}{\underset{\oplus}{C}}-CH_3}$$

$$C_6H_5-\overset{C_6H_5}{\underset{OH}{C}}-\overset{CH_3}{\underset{OH}{C}}-CH_3 \xrightarrow[(-H_2O)]{H^+} \underset{\text{More stable carbonium ion}}{C_6H_5-\overset{C_6H_5}{\underset{\oplus}{C}}-\overset{CH_3}{\underset{OH}{C}}-CH_3}$$

$$\downarrow$$

$$\underset{\text{3, 3 diphenyl-2-butanone}}{C_6H_5-\overset{CH_3}{\underset{C_6H_5}{C}}-\underset{O}{\overset{\|}{C}}-CH_3}$$

In this case, phenylated carbonium ion (+ve carbon carrying two phenyl groups) is more stable.

3. In the rearrangements of α, β diol esters, we can know the relative stabilities of carbonium ions. The partial positive charge on the carbonyl carbon of the ester group does not permit the formation of carbonium ion at α-carbon due to mutual repulsion between two similar positive charges.

$$H_2C(OH)-CH(OH)-COOC_2H_5 \xrightarrow{H^+} \overset{+}{C}H_2-C(H)(OH)-COOC_2H_5 \xrightarrow{-H^+} CH_3-\underset{\underset{O}{\|}}{C}-COOC_2H_5$$

$$CH_3-C(H_3C)(OH)-C(C_6H_5)(OH)-COOC_2H_5 \xrightarrow[-H_2O]{H^+} CH_3-\overset{\oplus}{C}(CH_3)-C(C_6H_5)(O-H)-COOC_2H_5 \longrightarrow CH_3-C(C_6H_5)(CH_3)-\underset{\underset{O}{\|}}{C}-COOC_2H_5$$

4. If the two groups present on the non-ionic carbon are different, then the more powerful electron donor towards carbon, gets migrated preferentially. In the example given below, *p*-Me OC_6H_4 migrates preferentially to phenyl, due to electron donating effect.

$$C_6H_5-C(OH)(C_6H_4OCH_3)-C(OH)(C_6H_4OCH_3)-C_6H_5 \xrightarrow{H^+} C_6H_5-\underset{\underset{O}{\|}}{C}-C(C_6H_4OCH_3)(C_6H_5)-C_6H_4-OCH_3$$

5. A study on the large number of compounds for this rearrangement reveals that a more stable carbonium ion is always formed. The stability of the carbonium ion is in the order.

Trityl > benhydryl > benzyl > allyl > tert. carbonium > sec. carbonium > primary carbonium

6. Other compounds in which a positive charge can be placed on a carbon α - to one leaving as OH group can also give this rearrangement. This is true for β-amino alcohols which rearrange on treatment with nitrous acid. The rearrangement is called "Semipinacol rearrangement". A similar rearrangement is given by epoxides when treated with acidic reagents like BF_3 in ether.

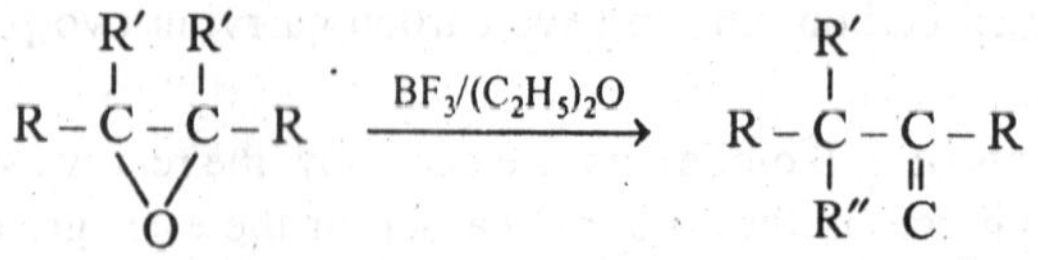

$$\underset{\underset{O}{\diagdown\diagup}}{R-\overset{R'}{\overset{|}{C}}-\overset{R'}{\overset{|}{C}}-R} \xrightarrow{BF_3/(C_2H_5)_2O} R-\overset{R'}{\overset{|}{\underset{R''}{\underset{|}{C}}}}-\underset{C}{\underset{\|}{C}}-R$$

where R = alkyl, aryl or hydrogen

It has been found that epoxides are intermediates in the pinacol rearrangements of certain glycols.

Epoxides can also rearrange to aldehydes or ketones on treatment with certain transition metal catalysts. For example, when α, β-epoxy ketone in toluene is heated at 80–140°C with small amounts of $[(C_6H_5)_3P]_4$ Pd, a β-diketone results.

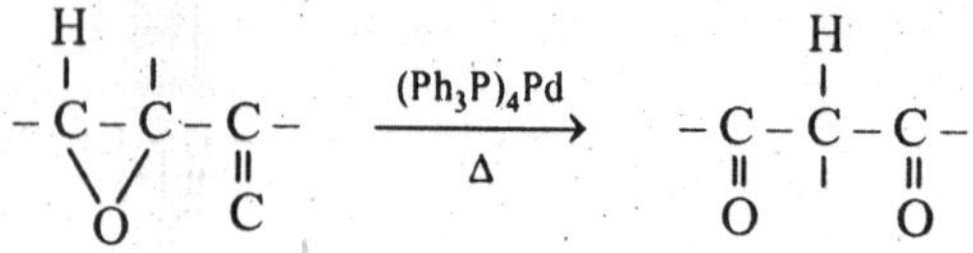

$$\underset{\underset{O}{\diagdown\diagup}}{-\overset{H}{\overset{|}{C}}-\overset{|}{C}}-\underset{C}{\underset{\|}{C}}- \xrightarrow[\Delta]{(Ph_3P)_4Pd} -\underset{O}{\underset{\|}{C}}-\overset{H}{\overset{|}{\underset{|}{C}}}-\underset{O}{\underset{\|}{C}}-$$

It involves 1, 2-hydride shift.

7. It may be noted that in this rearrangement, the migrating group does not become free. It is supported by the fact that *cis*-1, 2-dimethylcyclopentane 1, 2-diol undergoes this rearrangement (Pinacol-Pinacolone) while the trans isomer does not undergo this rearrangement.

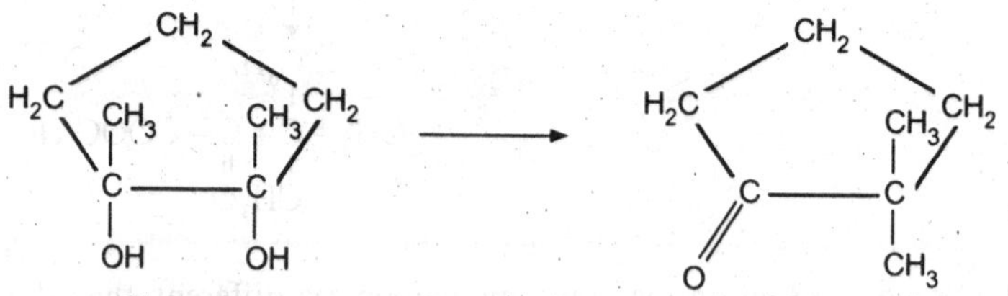

[Cis isomer: cyclopentane ring (CH_2, H_2C, CH_2) with both C bearing CH_3 above and OH below] → [cyclopentanone with C=O and adjacent C bearing two CH_3 groups]

Cis isomer

[Trans isomer: one C bearing CH_3 above and OH below; other C bearing OH above and CH_3 below] → No rearrangement

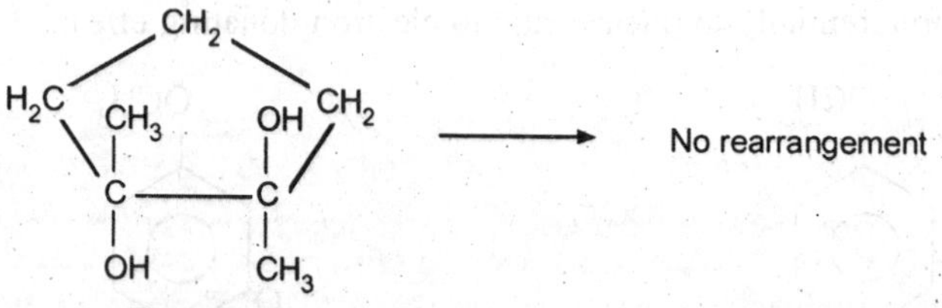

Trans isomer

It is difficult to explain why the methyl group does not migrate in the *trans* form if it is assumed that the methyl group migrates in the free state in *cis* form. If the migrating group does not become free, it will not have any access to the neighbouring carbon atom in the *trans* form because the carbon atom of the methyl group is also tetrahedral in nature. Hence, we say that the alkyl or aryl group is not completely detached to form a carbonium ion intermediate.

8. Migrating aptitude of Group. The important question in Pinacole Pinacolone rearrangement is: which of the two groups migrates preferentially? The migrating aptitude greatly depends upon the inducting influence of the group. The group with greater –I effect as compared to the other will have greater tendency to migrate. The migrating aptitude of few groups is in the following order:

Aryl > methyl > ethyl > *n*-propyl > iso-propyl > tert-butyl

If the migratory aptitude of phenyl is assigned the value 1, then the values for the migratory aptitude of substituted phenyl groups are as follows:

o-Cl-C_6H_4– : 0.0 o-OCH_3-C_6H_4– : 0.3 p-Cl-C_6H_4– : 0.7 C_6H_5– : 1

p-C_6H_5-C_6H_4– : 12 p-CH_3-C_6H_4– : 16 p-CH_3O-C_6H_4– : 500

This is illustrated as under

$$\text{Ph}-\underset{\text{OH}}{\overset{\text{p-Tolyl}}{\text{C}}}-\underset{\text{OH}}{\overset{\text{p-Tolyl}}{\text{C}}}-\text{Ph} \longrightarrow \text{Ph}-\underset{\text{O}}{\overset{}{\text{C}}}-\underset{\text{p-Tolyl}}{\overset{\text{p-Tolyl}}{\text{C}}}-\text{Ph} \quad 94\%$$

$$+\ \text{p-Tolyl}-\underset{\text{O}}{\text{C}}-\underset{\text{Ph}}{\overset{\text{p-Tolyl}}{\text{C}}}-\text{Ph} \quad 6\%$$

In a number of cases, it has been found that the expected group does not migrate. The important condition for migration is that the migrating group must lie in the same plane in which the molecule is present.

Q. 9. Discuss acid catalysed rearrangements of aldehydes and ketones.

Ans. In this type of rearrangement, certain aldehydes are converted into ketones. Under drastic conditions, the conversion of some ketones into other ketones has also been reported. But a ketone cannot be converted into aldehyde. One possible mechanism is described as:

$$R_2-\underset{R_3}{\overset{R_1}{C}}-\underset{O}{\overset{}{C}}-H \xrightarrow{H^+} R_2-\underset{R_3}{\overset{R_1}{C}}-\underset{OH}{\overset{\oplus}{C}}-H \longrightarrow R_2-\overset{R_3}{\underset{\oplus}{C}}-\underset{H}{\overset{R_1}{C}}-OH \longrightarrow R_2-\underset{H}{\overset{R_3}{C}}-\overset{R_1}{\underset{\oplus}{C}}-O-H$$

$$\xrightarrow{-H^+} R_2-\underset{H}{\overset{R_3}{C}}-\underset{R_1}{C}=O \quad \text{Ketone}$$

where R_1, R_2, R_3 are alkyl groups.

In an alternative mechanism for the above reaction, there appears to be a possibility of the formation of an epoxide.

$$\underset{\text{(An aldehyde)}}{R_2-\overset{\overset{R_1}{|}}{\underset{\underset{R_3}{|}}{C}}-\underset{\underset{O}{||}}{C}-H} \xrightarrow{H^+} R_2-\overset{\overset{R_1}{|}}{\underset{\underset{R_3}{|}}{C}}-\overset{\oplus}{\underset{\underset{OH}{|}}{C}}-H \longrightarrow R_2-\overset{R_3}{C}-\overset{R_1}{C}-H\ (\text{epoxide},\ \overset{\oplus}{O}-H) \longrightarrow R_2-\overset{\oplus}{\underset{\underset{O-H}{|}}{C}}-\overset{\overset{R_1}{|}}{\underset{\underset{H}{|}}{C}}-R_3$$

$$\xrightarrow{-H^+} \underset{\text{A ketone}}{R_2-\underset{\underset{O}{||}}{C}-\overset{\overset{R_1}{|}}{\underset{\underset{H}{|}}{C}}-R_3}$$

If the acid catalysed rearrangement is carried with a ketone (labelled ^{14}C in C = O group), then by the first mechanism the product (ketone) contains ^{14}C in C = O group.

In the case of α-hydroxy aldehydes and ketones, the process may stop only after one migration.

$$R_2-\overset{\overset{R_1}{|}}{\underset{\underset{OH}{|}}{C}}-\underset{\underset{O}{||}}{C}-R_3 \xrightarrow{H^+} R_2-\underset{\underset{O}{||}}{C}-\overset{\overset{R_1}{|}}{\underset{\underset{OH}{|}}{C}}-R_3$$

It is called α-ketol rearrangement. It can also be brought about by base catalysis only when the alcohol is tertiary. In case the alcohol is primary ($R_1 = R_2 = H$), enolisation is favoured than the rearrangement.

$$R_1-\overset{\overset{R_2'}{|}}{\underset{\underset{OH}{|}}{C}}-\underset{\underset{O}{||}}{C}-R_3 \xrightarrow{\ddot{B}^{\ominus}} R_2-\overset{\overset{R_1}{|}}{\underset{\underset{\ominus:O:}{|}}{C}}-\underset{\underset{O}{||}}{C}-R_3 \xrightarrow{BH} R_2-\underset{\underset{O}{||}}{C}-\overset{\overset{R_1}{|}}{\underset{\underset{OH}{|}}{C}}-R_3$$

Thermal Rearrangement of Amino ketones. Ketones containing an α-secondary amino group may undergo rearrangement involving 1, 2-alkyl shift (carbon to carbon) of alkyl group. An alkyl or aryl group shifts from one carbon atom to the other. An example of this thermal rearrangement is given below:

$$R_1-\underset{\underset{O}{||}}{C}-\overset{\overset{R_2}{|}}{\underset{\underset{NHR_4}{|}}{C}}-R_3 \xrightarrow{\text{Heat}} R_2-\underset{\underset{O}{||}}{C}-\overset{\overset{R_1}{|}}{\underset{\underset{NHR_4}{|}}{C}}-R_3 \qquad (R = \text{alkyl or aryl})$$

The mechanism of this reaction can be described as given below:

$$R_1-\underset{\underset{O}{||}}{C}-\overset{\overset{R_2}{|}}{\underset{\underset{NR_4}{|}}{C}}-R_3\ (\text{N}-H) \longrightarrow R_2-\overset{\overset{R_1}{|}}{\underset{\underset{O}{|}}{C}}-\underset{\underset{NR_4}{||}}{C}-R_3\ (\text{O}-H) \longrightarrow R_2-\underset{\underset{O}{||}}{C}-\overset{\overset{R_1}{|}}{\underset{\underset{NHR_4}{|}}{C}}-R_3$$

Q. 10. Discuss Benzilic acid Rearrangement. *(Coimbatore, 2000; Kerala, 2001)*

Ans. It is a base-catalysed reaction whereby 1, 2 diketones are converted into α-hydroxy acids. When benzil is heated with potassium hydroxide solution, it undergoes molecular rearrangement with the formation of the potassium salt of benzilic acid. It is base catalysed. In this rearrangement, one of the alkyl or aryl groups migrates and one of the ketonic groups is reduced to an alcohol. The other ketonic group is oxidised to carboxylate group ($-COO^-$ group).

Consider the action of potassium hydroxide on benzil.

$$\underset{\text{Benzil}}{C_6H_5-\overset{O}{\overset{\|}{C}}-\overset{O}{\overset{\|}{C}}-C_6H_5} \xrightarrow[\text{Ethanol}]{\text{KOH}} \underset{\text{Benzilic acid}}{C_6H_5-\underset{C_6H_5}{\underset{|}{\overset{OH}{\overset{|}{C}}}}-\overset{O}{\overset{\|}{C}}-OH}$$

Another example of benzilic acid rearrangement is the conversion of furil into furilic acid.

$$\underset{\text{Furil}}{C_4H_3O-\overset{O}{\overset{\|}{C}}-\overset{O}{\overset{\|}{C}}-C_4H_3O} \xrightarrow[\text{Ethanol}]{\text{KOH}} \underset{\text{Furilic acid}}{C_4H_3O-\underset{OH}{\underset{|}{\overset{COOH}{\overset{|}{C}}}}-C_4H_3O}$$

Mechanism. The mechanism of this reaction is described as under :

$$\underset{\text{Benzil}}{C_6H_5-\overset{O}{\overset{\|}{C}}-\overset{O}{\overset{\|}{C}}-C_6H_5} + \overset{\ominus}{O}H \xrightarrow{\text{KOH}} C_6H_5-\overset{O}{\overset{\|}{C}}-\underset{C_6H_5}{\underset{|}{\overset{:\ddot{O}:^{\ominus}}{\overset{|}{C}}}}-O-H$$

$$\downarrow$$

$$\underset{\text{Benzilic acid}}{C_6H_5-\underset{C_6H_5}{\underset{|}{\overset{OH}{\overset{|}{C}}}}-\overset{OH}{\overset{|}{C}}=O} \xleftarrow{H^+} \left[C_6H_5-\underset{C_6H_5}{\underset{|}{\overset{OH}{\overset{|}{C}}}}-\overset{:\ddot{O}:^{\ominus}}{\overset{|}{C}}=O \rightleftharpoons C_6H_5-\underset{C_6H_5}{\underset{|}{\overset{:\ddot{O}:^{\ominus}}{\overset{|}{C}}}}-\overset{O}{\overset{\|}{C}}-O-H \right]$$

The rearrangement is of the first order with respect to both benzil and hydroxyl ion. It has been shown that when benzil is heated with a methanolic sodium hydroxide solution in water containing O^{18}, benzil recovered contained O^{18}. It is explained by assuming that the first step is rapid reversible addition of hydroxyl ion to benzil.

$$C_6H_5-\overset{O}{\overset{\|}{C}}-\overset{O}{\overset{\|}{C}}-C_6H_5 + {}^{18}OH^- \rightleftharpoons C_6H_5-\overset{O}{\overset{\|}{C}}-\underset{{}^{18}OH}{\underset{|}{\overset{:\ddot{O}^{\ominus}}{\overset{|}{C}}}}-C_6H_5 \rightleftharpoons$$

$$C_6H_5-\overset{\overset{O}{\|}}{C}-\underset{\underset{^{18}O^{\ominus}}{|}}{\overset{\overset{OH}{|}}{C}}-C_6H_5 \rightleftharpoons C_6H_5-\overset{\overset{O}{\|}}{C}-\overset{\overset{O^{18}}{\|}}{C}-C_6H_5+OH^-$$

The rearrangement involves 1, 2-shift. It may be noted that when the two aryl groups are different, the one which is more electron releasing will tend to neutralise the positive charge on the carbon atom to which it is attached when the carbonyl group (C = O) is polarised. Clearly, it will be the other C = O group that will link with the hydroxyl ion and hence it will be aryl group attached to this carbonyl group which migrates preferentially.

The formation of the esters is a strong evidence that the first step in the rearrangement is the addition of nucleophilic reagent.

$$\underset{\text{Benzil}}{C_6H_5-\overset{\overset{O}{\|}}{C}-\overset{\overset{O}{\|}}{C}-C_6H_5}+\overset{\ominus}{O}C_2H_5 \longrightarrow C_6H_5-\overset{\overset{O}{\|}}{C}-\underset{\underset{C_6H_5}{|}}{\overset{\overset{:\ddot{O}:^{\ominus}}{|}}{C}}-OC_2H_5$$

$$\overset{\ominus}{HO}+C_6H_5-\underset{\underset{C_6H_5}{|}}{\overset{\overset{OH}{|}}{C}}-\overset{\overset{O}{\|}}{C}-OC_2H_5 \xleftarrow{H_2O} C_6H_5-\underset{\underset{C_6H_5}{|}}{\overset{\overset{^{\ominus}:\ddot{O}:}{|}}{C}}-\overset{\overset{O}{\|}}{C}-OC_2H_5$$

Other Examples

1. Cyclohexane 1, 2-dione forms 1-hydroxy cyclopentane carboxylic acid.

$$\text{Cyclohexane-1,2-dione} + KOH \longrightarrow \text{(ring)}\overset{OH}{\underset{COOK}{C}}\text{ (with C=O)}$$

2. Although this rearrangement is generally applicable to aromatic diketone, it can also be applied to aliphatic diketones and α-ketone acids. Further, barium hydroxide and thorium hydroxide can also be used in place of caustic alkali. It may be noted that hydroxyl ions are not specific in this rearrangement and sodium methoxide, sodium ethoxide and potassium tert-butoxide can also be employed and in such cases, the product formed is α-hydroxy ester.

$$C_6H_5-\overset{\overset{O}{\|}}{C}-\overset{\overset{O}{\|}}{C}-C_6H_5 \xrightarrow{CH_3ONa} \underset{\text{(Methyl ester)}}{C_6H_5-\underset{\underset{OH}{|}}{\overset{\overset{C_6H_5}{|}}{C}}-COOCH_3}$$

Phenoxide or aroxide ions cannot be employed as these are not strong enough to bring about the rearrangement. The change appears to be similar to Cannizzaro's reaction in which one out the two carbonyl groups is reduced to alcoholic group while the other is oxidised to carboxyl group.

Strictly speaking it resembles more with internal Cannizzaro's reaction in which oxidation and reduction take place in the same molecule.

$$\begin{matrix} O \\ \| \\ C-H \\ | \\ C-H \\ \| \\ O \end{matrix} \xrightarrow{OH^-} H-\underset{\underset{H}{|}}{\overset{\overset{OH}{|}}{C}}-COO^-$$

3. Migrating aptitude. If the two aryl groups attached to the carbonyl carbons in benzil happen to the different, the one which is more electron releasing due to the presence of some suitable substituent will have a lesser tendency to migrate than the aryl group which is less electron releasing. Consider the following compound:

$$C_6H_5-\overset{\overset{O}{\|}}{C}-\overset{\overset{O}{\|}}{C}-\underbrace{C_6H_4\,(p\text{-}OCH_3)}_{p\text{-anisyl group}}$$

In this example, the migration of *p*-anisyl group is about 32% and rest is the migration of phenyl group. The reason is that the *p*-anisyl group will tend to increase the electron density of the carbonyl carbon atom as due to this, the attack of the nucleophile (the base) will be comparatively less at this carbon. The base will tend to attack the other carbonyl carbon atom which carries the phenyl group with lesser electron releasing tendency.

4. Action of potassium hydroxide on phenanthraquinone.

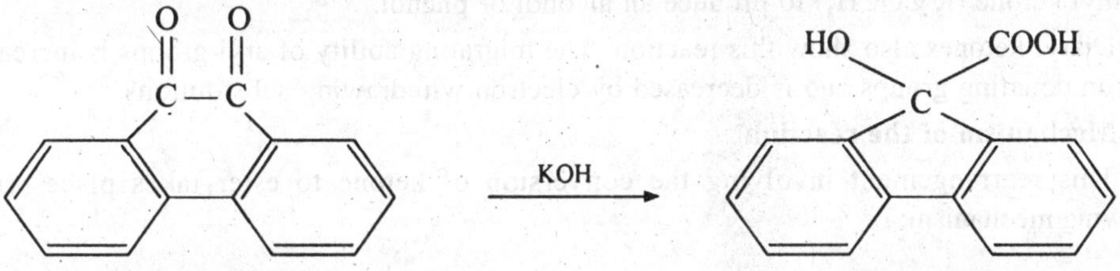

Phenanthraquinone $\xrightarrow{KOH}$ 9-Hydroxyl fluorene 9-Carboxylic acid

5. Action of potassium hydroxide on furil.

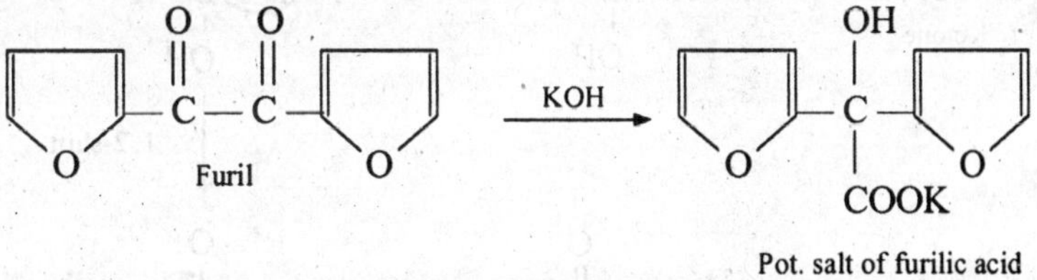

Furil $\xrightarrow{KOH}$ Pot. salt of furilic acid

6. Some aliphatic α-hydroxy ketones when treated with 20% KOH solution bubbled with air are oxidised to diketone followed by benzilic acid rearrangement.

$$(CH_3)_2\,CH-\overset{\overset{O}{\|}}{C}-\underset{\underset{OH}{|}}{CH}-CH(CH_3)_2 \xrightarrow{20\%\ KOH}$$

$$(CH_3)_2CH-\overset{\overset{O}{||}}{C}-\overset{\overset{O}{||}}{C}-CH(CH_3)_2 \xrightarrow{\text{Rearrangement}} (CH_3)_2CH-\underset{\underset{CH(CH_3)_2}{|}}{\overset{\overset{OH}{|}}{C}}-COOK$$

Q. 11. Describe Baeyer-Villiger rearrangement.

Ans. This rearrangement involves the migration of an alkyl or aryl group from carbon to oxygen. A ketone on treatment with a peracid such as peracetic acid or perbenzoic acid in the presence of acid catalyst forms an ester.

The best oxidising agent is peroxy trifluoroacetic acid.

$$\underset{\text{Ketone}}{R-\overset{\overset{O}{||}}{C}-R'} + C_6H_5-\overset{\overset{O}{||}}{C}-OOH \longrightarrow \underset{\text{Ester}}{R-\overset{\overset{O}{||}}{C}-OR'}$$

For acyclic compounds, R′ must usually be secondary, tertiary or vinylic. For unsymmetrical ketones, the order of migration is:

tert. alkyl > *sec.* alkyl > *pri.* alkyl > methyl.

As the methyl group has a low migrating ability, the reaction provides a means of cleaving a methyl ketone ($R'COCH_3$) to produce an alcohol or phenol.

Diaryl ketones also show this reaction. The migrating ability of aryl groups is increased by electron donating groups and is decreased by electron withdrawing substitutions.

Mechanism of the reaction

This rearrangement involving the conversion of ketone to ester takes place with the following mechanism:

$$\underset{\text{Ketone}}{R'-\overset{\overset{O}{||}}{C}-R} \xrightarrow{H^+} R'-\underset{\underset{OH}{|}}{\overset{\oplus}{C}}-R \xrightarrow{R''CO.OOH} R'-\underset{\underset{OH}{|}}{\overset{\overset{:\ddot{O}-O-\overset{\overset{O}{||}}{C}-R''}{|}}{C}}-R$$

$$\xrightarrow{\text{1, 2-shift}} \overset{\oplus}{C}(-\ddot{O}-R')(-R)(-O-H) \xrightarrow{-H^+} \underset{\text{Ester}}{R'-O-\overset{\overset{O}{||}}{C}-R}$$

C-14 isotope effect studies on acetophenones have shown that migration of aryl groups takes place in the rate determining step. Also the labelled carbonyl oxygen atom of ketone becomes the carbonyl oxygen atom of the ester.

$$C_6H_5-\overset{\overset{O^{18}}{\parallel}}{C}-C_6H_5 \xrightarrow{C_6H_5COOOH} C_6H_5-\overset{\overset{O^{18}}{\parallel}}{C}-OC_6H_5$$

The reaction is often applied to cyclic ketones to give lactones. If this rearrangement is conducted with aldehydes, the migration of hydrogen atom gives carboxylic acids. Migration of some alkyl group might give a formate but this seldom happens.

Examples

1.

$$\text{Cyclopentanone} \xrightarrow{CF_3COOOH} \text{lactone}$$

$$\text{Cyclohexanone} \xrightarrow{RCO_3H} \in\text{-Caprolactone}$$

Cyclohexanone ∈-Caprolactone

Thus cyclic ketones give lactones.

2. Anhydrides are obtained when α-diketones are treated with peracids.

$$C_6H_5-\overset{\overset{O}{\parallel}}{C}-\overset{\overset{O}{\parallel}}{C}-C_6H_5 \xrightarrow{C_6H_5COOOH} C_6H_5\overset{\overset{O}{\parallel}}{C}-O-\overset{\overset{O}{\parallel}}{C}-C_6H_5$$

Benzoic anhydride

$$\text{1, 2-Naphthadiquinone} \xrightarrow{RCO_3H} C_6H_4(\overset{\overset{O}{\parallel}}{C}-O-C{=}O)(CH{=}CH)$$

1, 2-Naphthadiquinone

Q. 12. Discuss the rearrangement reactions of hydroperoxides.

Ans. Hydroperoxides can be cleaved by proton or Lewis acids in a reaction whose principal step is rearrangement.

$$R-\underset{\underset{R}{|}}{\overset{\overset{R}{|}}{C}}-O-OH \xrightarrow{H^+} R-\overset{\overset{O}{\parallel}}{C}-R+ROH$$

Mechanism. The rearrangement involving 1, 2-shift of alkyl group from carbon to oxygen atom can be described by the following mechanism:

$$R-\overset{R}{\underset{R}{C}}-O-OH \underset{}{\overset{H^+}{\rightleftharpoons}} R-\overset{(R)}{\underset{R}{C}}-\ddot{O}-\overset{\oplus}{O}H_2 \longrightarrow R-\overset{\oplus}{\underset{R}{C}}-\ddot{O}-R$$

Hydroperoxide 1, 2-Shift

$$\xrightarrow{H_2O} R-\overset{\oplus OH_2}{\underset{R}{C}}-OR \xrightarrow{-H^+} R-\overset{O-H}{\underset{R}{C}}-OR \longrightarrow R-\underset{R}{C}=O + ROH$$

When aryl or alkyl groups are both present, migration of alkyl group takes place preferentially. Among the alkyl groups, the migratory order is *tert* R > *sec* R > $C_3H_7 \approx H > C_2H_5 > CH_3$. For this rearrangement the preparation and isolation of hydroperoxides is not necessary. The rearrangement takes place when an alcohol is treated with hydrogen peroxide and acids. The migration of an alkyl group of a primary hydroperoxide provides a means of converting an alcohol to its next lower homologue. For example, consider the following conversion.

$$\underset{\text{Ethyl hydroperoxide}}{CH_3-CH_2-OOH} \longrightarrow CH_3OH + CH_2=O$$

Examples

1. An important example is the decomposition of hydroperoxide obtained by the air oxidation of cumene. It is often employed for the large scale preparation of phenol and acetone.

$$C_6H_5-\overset{CH_3}{\underset{O-OH}{C}}-CH_3 \overset{H^+}{\rightleftharpoons} C_6H_5-\overset{CH_3}{\underset{O-\overset{\oplus}{O}H_2}{C}}-CH_3 \xrightarrow{-H_2O} \oplus\overset{CH_3}{\underset{OC_6H_5}{C}}-CH_3 \xrightarrow{(i)\ H_2O,\ (ii)\ -H^+} HO-\overset{CH_3}{\underset{OC_6H_5}{C}}-CH_3$$

$$HO-\overset{CH_3}{\underset{OC_6H_5}{C}}-CH_3 \xrightarrow{H^+/H_2O} \underset{\text{Phenol}}{C_6H_5OH} + \underset{\text{Acetone}}{CH_3-\overset{CH_3}{C}=O}$$

In this mechanism, the loss of water and migration of C_6H_5 group to the resulting electron deficient oxygen atom are almost certainly concerted. Addition of water to the carbonium ion forms hemiketal. Hemiketal undergoes hydrolysis under the reaction conditions to form phenol and acetone. It is observed that phenyl group migrates in preference to methyl group. It may be noted that preferred migratory aptitude of phenyl group results from its migration via a bridged transition state.

$$\overset{CH_3}{\overset{|}{\delta+C}}-CH_3 \quad \overset{\delta+}{C_6H_5} \quad \underset{\delta+}{O}\ldots.\overset{\delta+}{O}H_2$$

In these examples, we considered the essentially heterolytic fission of peroxide linkages

$-O:O- \rightarrow -\overset{\oplus}{O}:O^{\ominus}-$ in polar solvents.

Cumene hydroperoxide required for the preparation of phenol and acetone is obtained as follows:

$$\text{Benzene} + CH_3-CH=CH_2 \,(\text{Propane}) \xrightarrow[\text{Friedel-Crafts reaction}]{H_3PO_4} C_6H_5CH(CH_3)_2 \,(\text{Cumene})$$

$$\xrightarrow[\text{Auto-oxidation}]{\text{Air blown in}} C_6H_5C(OOH)(CH_3)_2 \,(\text{Cumene hydroperoxide})$$

Q. 13. Discuss Hofmann rearrangement. *(Kerala, 2001)*

Ans. It is a rearrangement which involves the migration of a group from carbon to nitrogen.

When an amide (aliphatic or aromatic) is treated with bromine and potassium hydroxide, a primary amine having one carbon atom less than the starting amide results. It is also called Hofmann degradation reaction.

$$\underset{\text{Amide}}{R-\overset{O}{\overset{\|}{C}}-NH_2} + Br_2 + 4KOH \xrightarrow{\Delta} \underset{1^\circ\text{-amine}}{RNH_2} + K_2CO_3 + 2KBr + 2H_2O$$

$$\underset{\text{Acetamide}}{CH_3-CONH_2} + Br_2 + 4KOH \xrightarrow{\Delta} \underset{\text{Methyl amine}}{CH_3NH_2} + K_2CO_3 + 2KBr + 2H_2O$$

$$\underset{\text{Benzamide}}{C_6H_5-\overset{O}{\overset{\|}{C}}-NH_2} + Br_2 + 4KOH \xrightarrow{\Delta} \underset{\text{Aniline}}{C_6H_5NH_2} + K_2CO_3 + 2KBr + 2H_2O$$

Mechanism. The mechanism of Hofmann rearrangement involves the following steps.

(*i*) $2NaOH + Br_2 \longrightarrow NaBr + NaBrO + H_2O$

(*ii*) $CH_3-\overset{O}{\overset{\|}{C}}-NH_2 + \bar{O}Br \longrightarrow \underset{\text{N-Bromoamide}}{R-\overset{O}{\overset{\|}{C}}-\ddot{N}HBr}$

(*iii*) $\underset{\text{N-Bromoamide}}{R-\overset{O}{\overset{\|}{C}}-NHBr} + \bar{O}H \longrightarrow \underset{\text{Acyl nitrene ion}}{R-\overset{O}{\overset{\|}{C}}-\overset{\ominus}{\ddot{N}}Br}$

(iv) $R-\overset{O}{\overset{\|}{C}}-\overset{\ominus}{\ddot{N}}-Br \xrightarrow[\text{Carbon to nitrogen}]{\text{1, 2-Shift}} O=C=NR + \bar{B}r$

Alkyl isocyanate

(v) $R-N=C=O + \bar{O}H \longrightarrow R-\bar{N}-\underset{\underset{O-H}{|}}{C}-\overset{\ominus}{O} \xrightarrow{\text{Rearrangement}} R-NH-\underset{\underset{O}{\|}}{C}-\overset{\ominus}{\ddot{O}}:$

$\downarrow H_2O$

$CO_2 + \underset{1°\text{-amine}}{RNH_2} \longleftarrow \underset{\text{Carbamic acid}}{R-\ddot{N}H-\underset{\underset{O}{\|}}{C}-O-H} + \bar{O}H$

In this mechanism, a few of the intermediates have been isolated which lend support to the above mechanism. It is intramolecular rearrangement and involves 1, 2-shift from carbon to nitrogen atom. Unlike 1, 2-hydride or alkyl shift in carbonium ion, the migration of alkyl anion to electron deficient nitrogen leads to the displacement of halide anion.

Stereochemistry. + (α)-Phenyl propionamide undergoes Hofmann rearrangement to form (–) α-Phenyl ethyl amine. We observe a complete retention of configuration about the asymmetric carbon atom.

CH3, H, C(=O)NH2, C6H5 — Hofmann Degradation → CH2, H, NH2, C6H5

(+) α-Phenylpropionamide (–) α-Phenylethylamine

It is observed that

(i) $-NH_2$ group occupies the same relative position on the asymmetric carbon as was originally occupied by $-CONH_2$ group.

(ii) The asymmetric carbon atom does not separate from carbonyl carbon of amide until it has started attaching itself to nitrogen. Thus, the migrating group never becomes free otherwise a carbonium ion would have formed and there would have been a loss of configuration. The transition stat which is actually formed during migration is shown below:

H, H_5C_6, CH_3, O=C–N: → [H, H_5C_6, CH_3, O=C.........N:] → H, H_5C_6, CH_3, O=C=N:

(Nitrene) Transition state An isocyanate

This is how the configuration of the amide is preserved.

Effect of substituents on the rate of reaction

The rate of Hofmann rearrangement is more when benzamide contains electron releasing groups (such as $-CH_3$, $-OCH_3$ etc.). On the other hand, a decrease in the rate is observed if benzamide contains electron withdrawing groups such as –Cl, $-NO_2$ etc. Clearly, the rearrangement step (1, 2-shift) is facilitated by +I groups and retarded by the presence of –I groups. More closely, the migration of aryl group (in benzamide) is an example of electrophilic aromatic substitution in which the electron deficient nitrogen is the attacking reagent. Such reactions, like other electrophilic substitution reactions are facilitated by electron releasing substituents. Hence, Hofmann rearrangement of benzamide is a kind of electrophilic substitution reaction.

The rate of Hofmann bromamide is more in benzamides containing electron releasing substituents. The decreasing rate other of the various substituents is:

$$OCH_3 > CH_3 > H > Cl > NO_2$$

More Examples

1. Anthranilic acid can be prepared by the action of bromine and aqueous KOH solution on phthalimide.

Phthalimide $\xrightarrow[(ii)\ KOH]{(i)\ Br_2}$ Anthranilic acid (COOH, NH_2)

2. β-amino pyridine can be prepared from nicotinamide (available from natural sources) because it cannot be prepared in good yield by the nitration of pyridine.

Nicotinamide ($CONH_2$) $\xrightarrow[KOH]{Br_2}$ 3-aminopyridine (NH_2)

3. Consider the Hofmann rearrangement of an amide which is optically active due to restricted rotation around single bond joining the phenyl and the naphthyl groups. The restricted rotation is due to the presence of bulky groups ($-NO_2$ and $-CONH_2$) at the meta positions. The product formed as a result of rearrangement is also optically active.

(NO_2, O_2N, $CONH_2$ substituted phenyl–naphthyl) $\xrightarrow[\text{rearrangement}]{\text{Hofmann}}$ (NO_2, O_2N, NH_2 substituted phenyl–naphthyl) (Optically active)

From this reaction, it is clear that this rearrangement is intramolecular and the migratory group remains partially bonded to the two centres in the transition.

Q. 14. Describe Curtius rearrangement. *(Coimbatore, 2000)*

Ans. This is another rearrangement in which the migration of a group takes place from carbon to nitrogen.

This rearrangement involves the conversion of an acid chloride into acid amide by treatment with sodium azide. The azide formed gives off nitrogen gas when warmed in alcohol solution. Also an isocyanate results which gets converted into urethane in presence of alcohol. Urethane is hydrolysed to free amine by alkali. The reaction can be shown as:

$$\underset{\text{Acid chloride}}{R-\overset{O}{\overset{\|}{C}}-Cl} \xrightarrow{NaN_3} \underset{\text{Acid azide}}{R-\overset{O}{\overset{\|}{C}}-N_3} \xrightarrow[-N_2]{C_2H_5OH/\Delta} \underset{\text{Alkyl isocyanate}}{R-N=C=O}$$

$$\underset{\text{Alkyl isocyanate}}{R-N=C=O} \xrightarrow{C_2H_5OH} \underset{\text{Urethane}}{R-NH-\overset{H}{\overset{\|}{C}}-C_2H_5} \xrightarrow{OH} \underset{1^\circ\text{-amine}}{RNH_2} + C_2H_5OH + CO_3^{2-}$$

Mechanism. The acid azide decomposes with the evolution of nitrogen forming nitrene which rearranges to an isocyanate with the migration of alkyl group by 1, 2-shift.

$$\underset{\text{An Azide}}{R-\overset{O}{\overset{\|}{C}}-\overset{-}{\ddot{N}}-\overset{+}{N}\equiv N} \xrightarrow[-N_2]{} \underset{\text{Nitrene}}{R-\overset{O}{\overset{\|}{C}}-\ddot{N}:} \longrightarrow \underset{\text{Alkyl isocyanate}}{\overset{O}{\overset{\|}{C}}=N-R}$$

From alkyl isocyanate to the final product, the steps involved are as follows:

$$R-N=C=O \xrightarrow{H_2O} \left[R-NH-C\begin{matrix}\nearrow O\\ \searrow OH\end{matrix}\right] \xrightarrow{-CO_2} \underset{\text{Amine}}{R-NH_2}$$

$$R-N=C=O \xrightarrow{C_2H_5OH} \underset{\text{Urethane}}{R-NH-COOC_2H_5} \xrightarrow{\text{Hydrolyse}} \left[R-NH-C\begin{matrix}\nearrow O\\ \searrow OH\end{matrix}\right]$$

Examples

1. Acetyl chloride can be converted into methyl amine by the rearrangement.

$$\underset{\text{Acetyl chloride}}{CH_3COCl} \xrightarrow[\text{rearrangement}]{\text{Curtius}} \underset{\text{Methyl amine}}{CH_3NH_2}$$

2. Preparation of aldehydes. Curtius reaction can be used for converting α-hydroxy acids into aldehydes.

$CH_3CH(OH)COOH \rightarrow CH_3CH(OH)CON_3 \rightarrow CH_3CH(OH)NH_2 \rightarrow CH_3CHO + NH_3$

3. Cycloalkyl azides give product involving ring expansion.

$$\text{1-R-1-}N_3\text{-cyclohexane} \xrightarrow{H^+} \underset{80\%}{\text{2-R-azacycloheptene (ring-expanded imine)}} + \underset{20\%}{\text{cyclohexylidene}=NR}$$

4. Aryl azides are also found to show ring expansion on heating.

$$C_6H_5N_3 \xrightarrow[\Delta]{C_6H_5NH_2} \text{(azepine)}-NHC_6H_5$$

Q. 15. Describe Lossen rearrangement. *(Mumbai 2010)*

Ans. This rearrangement involves the decomposition of derivatives of hydroxamic acid into primary amines.

When hydroxamic acid is treated with acid chloride or anhydride, we get *o*-acyl hydroxamic acid. This compound when heated in an alkaline solution produces primary amine. This reaction is called Lossen rearrangement. It is similar to Hofmann and Curtius rearrangement. Consider the action of acid halide on hydroxamic acid.

$$\underset{\text{Hydroxamic acid}}{R-\overset{O}{\overset{\|}{C}}=NH-OH} + R'-\overset{O}{\overset{\|}{C}}-X \xrightarrow[-HX]{\text{Base}} \underset{o\text{-Acyl hydroxamic acid}}{R-\overset{O}{\overset{\|}{C}}-NH-O-\overset{O}{\overset{\|}{C}}-R'}$$

$$\downarrow \bar{O}H$$

$$\underset{\text{Nitrene}}{R-\overset{O}{\overset{\|}{C}}-\ddot{N}} \xleftarrow[R'COO^-]{\Delta} R-\overset{O}{\overset{\|}{C}}-\bar{N}-O-\overset{O}{\overset{\|}{C}}-R'$$

$$\downarrow \text{Rearrangement}$$

$$O=C=N-R \xrightarrow{2OH^-} \underset{1°\text{-Amine}}{RNH_2} + CO_3^{2-}$$

It is possible to convert C_6H_5COOH to $C_6H_5NH_2$ (aniline) in one step by heating the acid with nitromethane in polyphosphoric acid. The hydroxamic acid is an intermediate and it is actually a Lossen rearrangement.

The rearrangement gets facilitated if some electron-withdrawing group is present in meta or para position in the acid chloride which is combined with hydroxamic acid.

MISCELLANEOUS QUESTIONS

Q. 16. Give the products of the following transformations and name the reaction involved.

(*i*) 2,6-dimethylphenyl allyl ether ($OCH_2-CH=CH_2$ on ring flanked by H_3C and CH_3) $\xrightarrow{\Delta}$?

(*ii*) $CH_2=CH-CH(CH_3)-CH(CH_3)-CH=CH_2$ $\xrightarrow{\Delta}$?

Ans. (*i*) This is an example of Claisen rearrangement.

$$\text{2,6-dimethylphenyl allyl ether: } H_3C\text{–}C_6H_3(OCH_2-CH=CH_2)\text{–}CH_3 \xrightarrow{\Delta} H_3C\text{–}C_6H_2(OH)(CH_2-CH=CH_2)\text{–}CH_3$$

(*ii*) This is an example of Cope rearrangement.

$$\underset{\text{3, 4-Dimethylhexa-1, 5-diene}}{H_3C-CH(CH=CH_2)-CH(CH=CH_2)-CH_3} \xrightarrow{\Delta} \underset{\text{2, 6-Octadiene}}{H_3C-CH=CH-CH_2-CH_2-CH=CH-CH_3}$$

Q. 17. Describe the reaction of ketoximes which involves enlargement of rings giving a suitable example.

Ans. See Q. 3.

Q. 18. What happens when trans, cis, trans-2, 4, 6,-octatriene is heated?

Ans. $CH_3-CH=CH-CH=CH-CH=CH-CH_3$

2, 4, 6, Octatriene

The reaction takes place as follows :

$$\underset{\text{2, 4, 6-Octatriene}}{CH_3-CH=CH-CH=CH-CH=CH-CH_3} \xrightarrow{\Delta} \text{ring: } CH(CH_3)-CH=CH-CH=CH-CH(CH_3)\text{ (cyclised, } CH-CH \text{ bond formed)}$$

This is an example of Cope rearrangement.

Q. 19. Describe the formation of isoquinoline from cinnamaldehyde by means of Beckmann's rearrangement.

Ans. See Q. 3.

Q. 20. Give the product and mechanism for the following raction :

$$CH_3-\underset{CH_3}{\overset{CH_3}{\underset{|}{\overset{|}{C}}}}-CH_2Br \xrightarrow{OH^-/H_2O} ?$$

Ans. This is an example of Wagner-Meerwein rearrangement.

$$CH_3-\underset{CH_3}{\overset{CH_3}{\underset{|}{\overset{|}{C}}}}-CH_2Br \xrightarrow[\dot{B}r]{SN^1} CH_3-\underset{CH_3}{\overset{CH_3}{\underset{|}{\overset{|}{C}}}}-\overset{+}{C}H_2$$

$$CH_3-\underset{CH_3}{\overset{CH_3}{\underset{|}{\overset{|}{C}}}}-\overset{+}{C}H_2 \xrightarrow[\text{Shift}]{\text{1, 2 Alkyl}} CH_3-\underset{CH_3}{\underset{|}{\overset{+}{C}}}-CH_2-CH_3$$

Elimination $-H^+$ → $CH_3-\underset{CH_3}{\underset{|}{C}}=CH-CH_3$

Substitution OH^- → $CH_3-\underset{CH_3}{\overset{OH}{\underset{|}{\overset{|}{C}}}}-CH_2-CH_3$

Q. 21. Name thermal rearrangements. What do you know about Claisen rearrangement? Give the principles and mechanism with suitable examples.

Ans. Cope and Claisen rearrangements are examples of thermal rearrangement. For mechanism of Claisen rearrangement. See Q. 7.

Q. 22. Describe the isomerism exhibited by ketoximes. Indicate the role of Beckmann rearrangement in determining the isomerism exhibited by ketoximes. *(Kerala 2000)*

Ans. See Q. 3.

40

OXIDATION AND REDUCTION (Some Synthetic Reagents)

CONTENTS

1. Selenium
2. Selenium dioxide
3. Periodic acid
4. Lead tetraacetate
5. Ozone
6. Chromic acid
7. Hydrogen peroxide
8. Osmium tetroxide
9. Lithium aluminium hydride
10. Sodium boro-hydride
11. Raney nickel
12. Platinum and palladium
13. Sodium/liquid ammonia

Q. 1. Describe the uses of selenium (Se) for dehydrogenation of hydrocarbons.

Ans. Selenium was first used by Diels in 1927. Selenium is prepared by the treatment of sodium selenite with sulphur dioxide as given below:

$$Na_2SeO_3 + 2SO_2 + H_2O \longrightarrow Na_2SO_4 + H_2SO_4 + Se$$

Its utility as dehydrogenating agent.

The compound to be dehydrogenated is heated with calculated amounts of selenium at 520–560 K when hydrogen is removed as hydrogen selenide. Some examples of reaction in which Se is used for dehydrogenation are given below:

1. Dehydrogenation of hydrocarbons

$$\text{Tetrahydronaphthalene} \xrightarrow[-2H_2]{Se} \text{Naphthalene}$$

$$\text{Tetrahydrophenanthrene} \xrightarrow[-2H_2]{Se} \text{Phenanthrene}$$

Tetrahydrophenanthrene

Phenanthrene

2. Dehydrogenation of ketones. Cyclic ketones are dehydrogenated by Se to phenols whereas side-chain ketonic groups are not affected.

$$\text{Cyclohexanone} \xrightarrow{Se} \text{Phenol}$$

Cyclohexanone

Phenol

3. Migration of methyl group. During selenium dehydrogenation, a methyl group may migrate to another position.

$$\text{5, 6, 7, 8 Tetrahydro 1, 5 dimethylphenanthrene} \xrightarrow{Se} \text{1, 8 Dimethylphenanthrene}$$

5, 6, 7, 8 Tetrahydro
1, 5 dimethylphenanthrene

1, 8 Dimethylphenanthrene

4. Ring contraction. On prolonged heating a high temperature, ring contraction might take place.

$$\text{Cycloheptane} \xrightarrow{Se} \text{Toluene}$$

Cycloheptane

Toluene

We find that a seven-membered ring above has been contracted to a six-membered ring.

5. Cyclisation

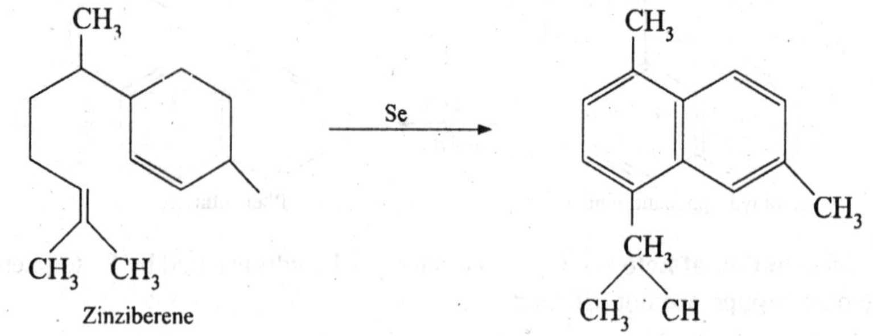

Zinziberene $\xrightarrow{Se}$ Cadelene

Q. 2. Describe the action of selenium dioxide in dehydrogenation and oxidation of organic compounds.

Ans. 1. Selenium dioxide can be used for dehydrogenating hydrocarbons in either acidic or alkaline medium. The reagent is particularly used for introducing unsaturation at α, β-position with respect to the ketonic group.

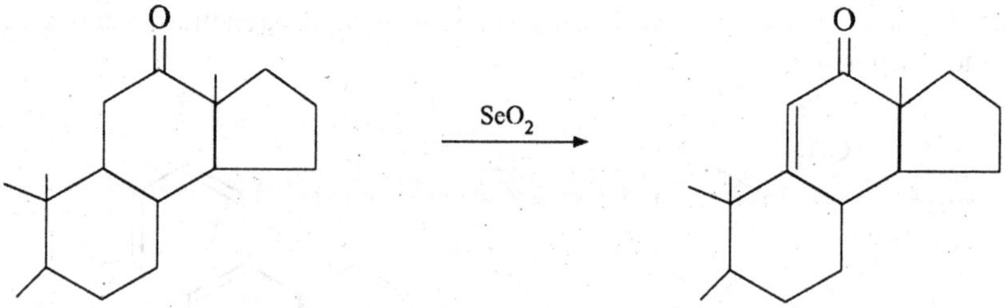

$\xrightarrow{SeO_2}$

The mechanism of the reaction is as follows:

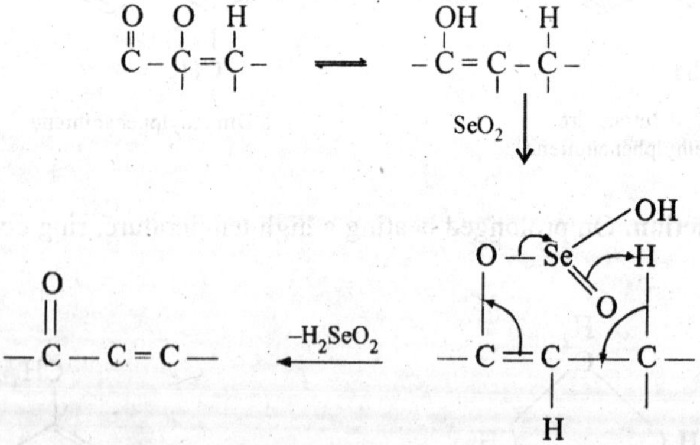

$-C(=O)-CH-CH- \rightleftharpoons -C(OH)=C-CH-$

$\downarrow SeO_2$

$-C(=O)-C=C- \xleftarrow{-H_2SeO_3}$

2. SeO_2 is employed to dehydrogenate 1, 4 diketones and 1, 2 diarylalkanes.

$$RCOCH_2CH_2COR \xrightarrow{SeO_2} RC(=O)-CH=CH-C(=O)-R + H_2$$

$$Ar-CH_2-CH_2-Ar \xrightarrow{SeO_2} ArCH=CH-Ar + H_2$$

3. SeO_2 dissolved in a little hot water can oxidise $-CH_2CO-$ group to $-COCO-$ group. Acetophenone is oxidised to phenyl glyoxal.

$$SeO_2 + H_2O \longrightarrow H_2SeO_3$$

$$\underset{\text{Acetophenone}}{C_6H_5COCH_3} + H_2SeO_3 \longrightarrow \underset{\text{Phenyl glyoxal}}{C_6H_5COCHO} + Se + H_2O$$

4. Desoxybenzoin is oxidised to benzil.

$$\underset{\text{Desoxy benzoin}}{C_6H_5CH_2COC_6H_5} \xrightarrow{H_2SeO_3} \underset{\text{Benzil}}{C_6H_5\overset{O}{\overset{\|}{C}}-\overset{O}{\overset{\|}{C}}-C_6H_5}$$

5. Methyl or methylene group alpha to a carbonyl group can be oxidised with selenium dioxide to form α-keto aldehyde and α-diketones respectively.

$$R-\overset{O}{\overset{\|}{C}}-CH_2-R' \xrightarrow{SeO_2} \underset{\text{Diketone}}{R-\overset{O}{\overset{\|}{C}}-\overset{O}{\overset{\|}{C}}-R'}$$

Q. 3. Discribe the utility of mercuric acetate in the dehydrogenation of amines.

Ans. Mercuric acetate $(CH_3COO)_2Hg$ plays the specific role of dehydrogenating tertiary amines, the solvent used being acetic acid. This reagent is generally used for dehydrogenating cyclic amines. The following reaction illustrates the dehydrogenating property of mercuric acetate.

H

N̈ $\xrightarrow[\text{CH}_3\text{COOH, 373 K}]{(CH_3COO)_2Hg}$ N

Enamine

The action of the reagent takes place by the following mechanism:

N $\xrightarrow[(CH_3COO)_2Hg]{CH_3COOH}$ H, N⊕, Hg — O — $\overset{O}{\overset{\|}{C}}$ — CH_3 $\xrightarrow{CH_3COO^-}$

H

N ⊕ + CH_3COOH

↓ CH_3COO^-

N + CH_3COOH

Enamine

Two hydrogens are removed from the molecule in two steps leading to a double bond in the molecule. It is a very specific action brought about by acetate ions in the presence of mixture of mercuric acetate and acetic acid. Mercuric acetate breaks into positive and negative species as given below:

$$(CH_3COO)_2Hg \longrightarrow [CH_3COOHg]^+ + CH_3COO^-$$

Q. 4. Write what you know about periodic acid as an oxidant. Describe the mechanism involved with a suitable example.

Or

Describe the oxidation reactions of periodic acid.

Ans. Periodic acid may be prepared by one of the following methods:

(*a*) By action of iodine on an aqueous solution of perchloric acid

$$2HClO_4 + I_2 \longrightarrow 2HIO_4 + Cl_2$$

(b) Electrolytic method. It may be prepared by electrolysing a conc. solution of iodic acid.

Applications of periodic acid

Periodic acid is used as an oxidising agent. It can oxidise compounds having

(*i*) two or more –OH groups attached to adjacent carbon atoms.

(*ii*) an aldehyde or ketonic group adjacent to an alcoholic group.

(*iii*) adjacent carbonyl group.

The above generations can be illustrated with the help of reactions given below:

(*a*) $CH_3-\underset{OH}{CH}-\underset{OH}{CH}-CH_2-CH_3 \xrightarrow{HIO_4} \underset{\text{Ethanal}}{CH_3CHO} + \underset{\text{Propanal}}{CH_3CH_2CHO}$

2, 3-Pentanediol

(*b*) $CH_3-\underset{OH}{CH}-CH_2-\underset{OH}{CH}-CH_3 \xrightarrow{HIO_4} \text{No action}$

2, 4-Pentanediol

(*c*) $CH_3-\underset{OH}{\overset{CH_3}{C}}-\underset{OH}{CH}-CH_3 \xrightarrow{HIO_4} \underset{\text{Propanone}}{CH_3-\overset{CH_3}{C}=O} + \underset{\text{Ethanal}}{CH_3CHO}$

2-methylbutan-2, 3-diol

(*d*) $CH_3-\underset{OH}{CH}-\underset{O}{\overset{}{C}}-CH_2CH_3 \xrightarrow{HIO_4} \underset{\text{Ethanal}}{CH_3CHO} + \underset{\text{Propanoic acid}}{CH_3CH_2COOH}$

2-hydroxypentanone -3

(*e*) $CH_3\underset{O}{C}-\underset{O}{C}-CH_3 \xrightarrow{HIO_4} CH_3COOH + CH_3COOH$

Butan-2, 3-dione Ethanoic acid

(*f*) $CH_3\underset{OH}{\overset{3}{CH}}-\underset{OH}{\overset{2}{CH}}-\underset{OH}{\overset{1}{CH_2}} \xrightarrow{HIO_4} \underset{\text{Ethanal}}{CH_3\overset{3}{C}HO} + \underset{\text{Methanoic acid}}{H\overset{3}{C}OOH} + \underset{\text{Methanal}}{H\overset{1}{C}HO}$

Butan-1, 2, 3-triol

(*g*) $\begin{matrix} CHO \\ | \\ H-C-OH \\ | \\ H-C-OH \\ | \\ H \end{matrix} \xrightarrow{HIO_4} \underset{\text{Formic acid}}{H-\overset{O}{\overset{\|}{C}}-OH} + \underset{\text{Formic acid}}{H-\overset{O}{\overset{\|}{C}}-OH} + \underset{\text{Formaldehyde}}{H-\overset{H}{\overset{|}{C}}=O}$

Glyceraldehyde

(*h*) $\begin{array}{l} CH_2OH \\ | \\ CO \\ | \\ CH_2OH \end{array} \xrightarrow{HIO_4} \underset{\text{Formaldehyde}}{HCHO} + \underset{\text{Carbon dioxide}}{O=C=O} + \underset{\text{Formaldehyde}}{HCHO}$

Dihydroxy acetone

(*i*) It is used to cleave epoxides to aldehydes.

$$\text{Cyclohexene oxide} \xrightarrow{HIO_4} \underset{\text{Hexan-1, 6-dial}}{H-\overset{\overset{\large O}{\|}}{C}-(CH_2)_4-CHO}$$

Mechanism. Mechanism of oxidation of a diol proceeds through a cyclic transition state as shown below:

$$\begin{array}{l} -C-OH \\ -C-OH \end{array} + IO_4^- \longrightarrow \text{cyclic periodate ester (} -C-O-I(=O)(=O)(O^-)-O-C- \text{)}$$

$$\longrightarrow -\overset{|}{\underset{|}{C}}=O + -\overset{|}{\underset{|}{C}}=O + IO_3^-$$

Q. 5. How oxidation with periodic acid helps in the elucidation of structure of carbohydrates?

Ans. Oxidation of a carbohydrate with HIO_4 can help to arrive at a structure of the compound. For example, one mole of glucose, a carbohydrate consumes five moles of periodic acid to form five moles of formic acid and one mole of formaldehyde as shown below:

$$\begin{array}{l} CHO \\ | \\ CHOH \\ | \\ CHOH \\ | \\ CHOH \\ | \\ CHOH \\ | \\ CH_2OH \end{array} \xrightarrow{HIO_4} \begin{array}{l} CHO \\ | \\ CHOH \\ | \\ CHOH \\ | \\ CHOH \\ | \\ CHO \\ + HCHO \end{array} \xrightarrow{HIO_4} \begin{array}{l} CHO \\ | \\ CHOH \\ | \\ CHOH \\ | \\ CHO \\ + HCOOH \end{array} \xrightarrow{HIO_4} \begin{array}{l} CHO \\ | \\ CHOH \\ | \\ CHO \\ +HCOOH \end{array}$$

$$\xrightarrow{HIO_4} \begin{array}{l} CHO \\ | \\ CHO \\ +HCOOH \end{array} \xrightarrow{HIO_4} \begin{array}{l} HCOOH \\ + HCOOH \end{array}$$

Thus, knowing the amounts of formaldehyde and formic acid produced, we can estimate the carbon chain in the compound.

Q. 6. Describe the oxidation reactions of lead tetraacetate.

(*Kanpur, 2001; Devi Ahilya, 2001; Delhi, 2003*)

Ans. Lead tetraacetate may be prepared by the gradual addition of red lead to a mixture of acetic acid and acetic anhydride at 350 K.

$$Pb_3O_4 + 8CH_3COOH \longrightarrow Pb(CH_3COO)_4 + 2Pb(CH_3COO)_2 + 4H_2O$$

Applications. (*a*) Oxidation of vicinal diols. This reagent is used for breaking vicinal diols to aldehydes, ketones or both according to the structure of diol.

$$R'-\underset{\underset{R''-\underset{\underset{R'''}{|}}{C}-OH}{|}}{\overset{\overset{R}{|}}{C}}-OH \xrightarrow{Pb(OAc)_4} \underset{\text{Ketone}}{R'-\overset{\overset{R}{|}}{C}=O} + \underset{\text{Ketone}}{R''-\underset{\underset{R'''}{|}}{C}=O}$$

Vicinal diol

$$\underset{\text{Pinacol}}{(CH_3)_2\underset{OH}{C}-\underset{OH}{C}(CH_3)_2} \xrightarrow{Pb(OAc)_4} 2\ \underset{\text{Acetone}}{(CH_3)_2C=O}$$

(*b*) Oxidation of alcohols

$$\underset{\text{Propanol-1}}{CH_3CH_2CH_2OH} \xrightarrow{Pb(OAc)_4} \underset{\text{Propanal}}{CH_3CH_2CHO}$$

$$\underset{\text{Hexan-2, 5-diol}}{CH_3CHOHCH_2CH_2CHOHCH_3} \xrightarrow{Pb(OAc)_4} \underset{\text{Hexan-2, 5-dione}}{CH_3COCH_2CH_2COCH_3}$$

$$C_6H_5CH=CHCH_2OH \xrightarrow{Pb(OAc)_4} \underset{\text{Cinnamic aldehyde}}{C_6H_5CH=CHCHO}$$

(*c*)

$$\underset{\text{Cyclohexane 1, 2-diol}}{C_6H_{10}(OH)_2} \xrightarrow{Pb(OAc)_4} \underset{\text{Adipaldehyde}}{OHC(CH_2)_4CHO}$$

(*d*) Oxidation of hydrøquinol

$$\underset{\text{Quinol}}{HO-C_6H_4-OH} \xrightarrow{Pb(OAc)_4} \underset{\text{Quinone}}{O=C_6H_4=O}$$

(*e*) Oxidation of cis-cyclopentane-1, 2-diol

$$\text{cis-}C_5H_8(OH)_2 \xrightarrow{(CH_3COO)_4Pb} \underset{\text{Pentane-1, 5-dione}}{H-\overset{\overset{O}{\|}}{C}-CH_2-CH_2-CH_2-\overset{\overset{O}{\|}}{C}-H}$$

It may be mentioned that oxidation of *trans* isomer is very slow because the cyclic intermediate is not easily formed.

(f) Decarboxylation of carboxylic acids

$$\underset{\text{Butanoic acid}}{CH_3CH_2CH_2COOH} \xrightarrow{(CH_3COO)_4Pb} \underset{\text{Propene}}{CH_3CH=CH_2}$$

(g) Oxidation of furan

$$\text{Furan} \xrightarrow{(CH_3COO)_4Pb} \text{2,5-diacetoxy-2,5-dihydrofuran } (H-C=C-H,\ H-C\ \ C-H \text{ bridged by O, each C bearing } O-C(=O)-CH_3)$$

$$\xrightarrow[\text{Hydrolysis}]{\text{Heat}} OHC-CH=CH-CHO \longrightarrow \underset{\text{Butane dial}}{\begin{matrix} CH_2-CH_2 \\ | \quad\quad | \\ CHO\ \ CHO \end{matrix}}$$

Q. 7. Discuss the mechanism of a reaction in which lead tetraacetate is used as an oxidising agent.

Ans. Mechanism for the oxidation of 1, 2 glycol with lead tetraacetate.

$$\begin{matrix} -C-OH \\ -C-OH \end{matrix} + (CH_3COO)_4Pb \rightleftharpoons \begin{matrix} -C-O-Pb(OCOCH_3) \\ -C-OH \end{matrix} + CH_3COOH$$

$$\begin{matrix} -C-O-Pb(OCOCH_3)_3 \\ -C-OH\ (H \end{matrix} \xrightarrow{\text{Slow}} \begin{matrix} -C-O \\ -C-O \end{matrix}\!\!>Pb(OCOCH_3)_2 + CH_3COOH$$

$$\begin{matrix} -C-O \\ -C-O \end{matrix}\!\!>Pb(OCOCH_3)_2 \longrightarrow 2-\overset{|}{C}=O + \underset{\text{Lead acetate}}{(CH_3COO)_2Pb}$$

Q. 8. Ozonolysis helps greatly in locating double bonds in alkenes. Comment on the statement taking a few examples. Give the mechanism of ozonolysis.

Ans. Ozonolysis of a compound means treating the compound with ozone followed by hydrolysis of the intermediate compound called ozonide.

$$\text{Compound} \xrightarrow{O_3} \text{Ozonide} \xrightarrow{\text{Hydrolysis}} \text{Products}$$

Ozone forms a bridge like structure with compounds containing a double bond or triple bond.

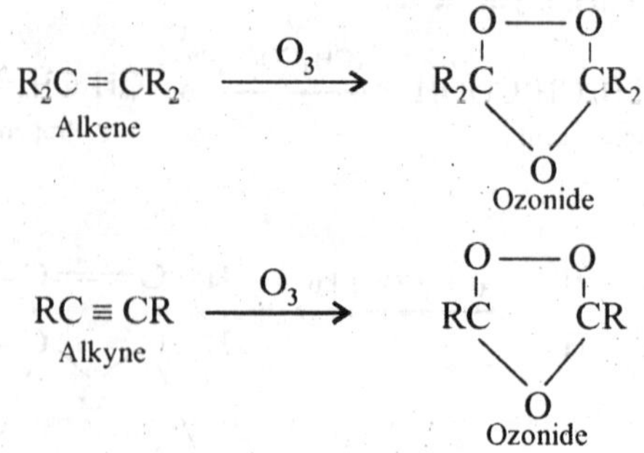

$$R_2C = CR_2 \xrightarrow{O_3} R_2C(-O-O-)(-O-)CR_2$$

Alkene → Ozonide

$$RC \equiv CR \xrightarrow{O_3} RC(-O-O-)(-O-)CR$$

Alkyne → Ozonide

These ozonides are subsequently hydrolysed to give aldehydes and ketones depending upon the structure of the compound.

To locate the position of double bond in a compound

Let us say we have a sample of butene. We don't know whether it is 1-butene or 2-butene and it is required to establish that. This can be done with the help of ozonolysis. We shall see what are the ozonolysis products with 1-butene and 2-butene.

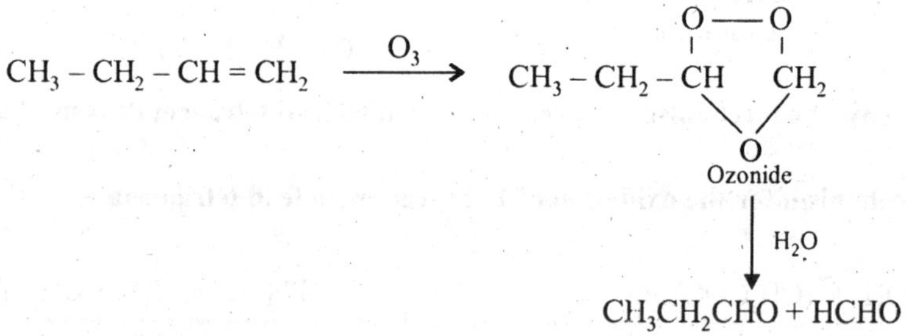

$$CH_3-CH_2-CH=CH_2 \xrightarrow{O_3} CH_3-CH_2-CH(-O-O-)(-O-)CH_2 \text{ (Ozonide)} \xrightarrow{H_2O} CH_3CH_2CHO + HCHO$$

Thus, we obtain a mixture of acetaldehyde and formaldehyde if we start with 1-butene.

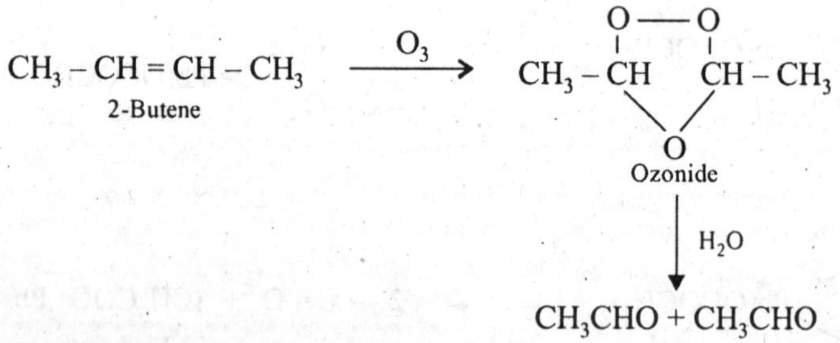

$$\underset{\text{2-Butene}}{CH_3-CH=CH-CH_3} \xrightarrow{O_3} CH_3-CH(-O-O-)(-O-)CH-CH_3 \text{ (Ozonide)} \xrightarrow{H_2O} CH_3CHO + CH_3CHO$$

We get only acetaldehyde if we start from 2-butene.

Thus after identifying the ozonolysis product, we can tell what the initial compound was. In other words, the position of the double bond can be located.

Mechanism of ozonisation (formation of ozonide)

Ozone molecule may be represented as a resonance hybrid of the following resonating structure:

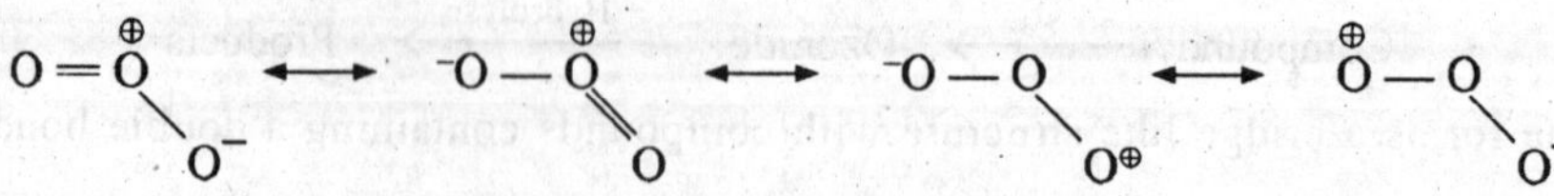

$$O=\overset{\oplus}{O}-O^- \longleftrightarrow {}^-O-\overset{\oplus}{O}=O \longleftrightarrow {}^-O-O-O^{\oplus} \longleftrightarrow \overset{\oplus}{O}-O-O^-$$

Ozonisation involves electrophilic addition of ozone molecule to the alkene as depicted below:

Ozonide

The ozonide is then subjected to hydrolysis (ozonolysis) to obtain the final products as shown below:

$$\xrightarrow{H_2O} \;\; >C=O + >C=O + H_2O_2$$

Hydrolysis is preferably carried out in a reducing atmosphere to prevent hydrogen peroxide from oxidising further the products of ozonolysis. Aldehydes and ketones are the products obtained after ozonolysis.

Q. 9. (*a*) How can we establish the structure of an unknown compound using ozonolysis?

(*b*) What is obtained on ozonolysis of alkynes?

Ans. (*a*) Establishing the structure of unknown compounds. Ozonolysis has been used to establish the structure of unknown compounds. Kekule's oscillation formula of benzene has been confirmed by the ozonolysis of *o*-xylene because the three carbonyl compounds could not be possibly obtained from a single structure of benzene.

o-xylene (*i*) $\xrightarrow{3O_3}$ Ozonide $\xrightarrow{3H_2O}$ 2CHO–CHO (Glyoxal I) + CH_3CO–CH_3CO (Dimethyl glyoxal II)

o-xylene (*ii*) $\xrightarrow{3O_3}$ Ozonide $\xrightarrow{3H_2O}$ $2CH_3CO$–CHO (Monomethyl glyoxal III) + CHO–CHO (Glyoxal)

As we obtained compounds I, II and III, the resonating structures of benzene are confirmed.

Structure of terpenoids. Structure of a no. of terpenoids have been confirmed making use of ozonolysis, we reproduce here the ozonolysis products obtained from terpenoids myrecene and ocimene.

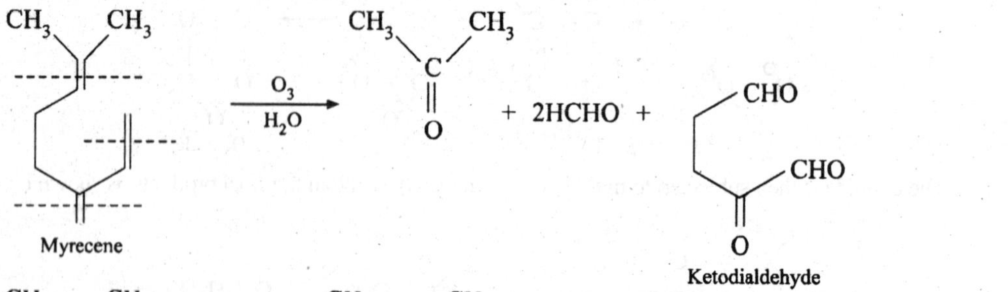

$$\text{Myrecene} \xrightarrow[H_2O]{O_3} (CH_3)_2C{=}O + 2HCHO + OHC{-}CH_2{-}CH_2{-}\underset{\underset{O}{\|}}{C}{-}CHO$$

Myrecene

Ketodialdehyde

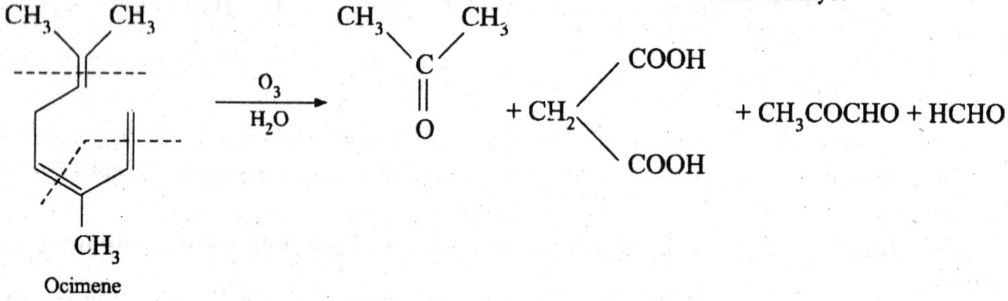

$$\text{Ocimene} \xrightarrow[H_2O]{O_3} (CH_3)_2C{=}O + CH_2(COOH)_2 + CH_3COCHO + HCHO$$

Ocimene

(*b*) **Ozonolysis of alkynes.** Alkynes on ozonolysis give acids instead of carboxyl compound. Consider the case of ozonolysis of acetylene.

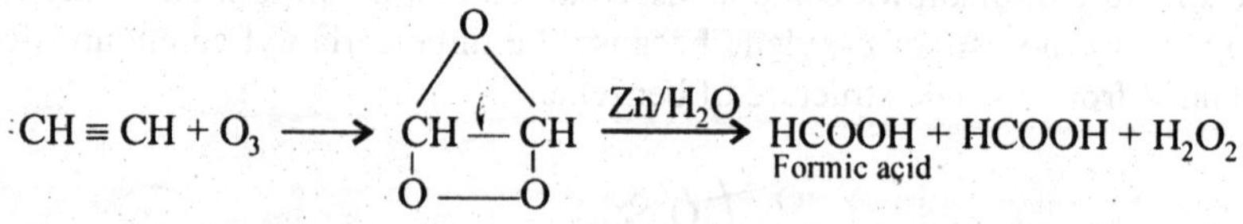

$$CH \equiv CH + O_3 \longrightarrow \text{CH}\overset{O}{—}\text{CH (ozonide with O—O)} \xrightarrow{Zn/H_2O} \underset{\text{Formic acid}}{HCOOH} + HCOOH + H_2O_2$$

Q. 10. Describe the use of chromic acid in the oxidation of alcohols.

Ans. Chromic acid (H_2CrO_4) is a very suitable reagent for the oxidation of primary alcohols to aldehydes and secondary alcohols to ketone. An additional advantage of this reagent is that it does not affect double bond present in the alcohol.

Chromic acid reagent is used in the form of a mixture of potassium dichromatic and sulphuric acid. This mixture mixed with water is known as **Jones reagent.**

Applications of chromic acid

1. Oxidation of primary alcohols.

$$RCH_2OH \xrightarrow{H_2CrO_4} RCHO$$

Mechanism of the reaction is as follows:

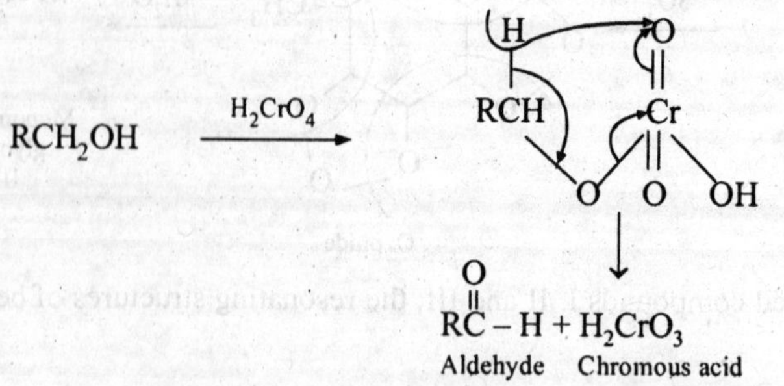

$$RCH_2OH \xrightarrow{H_2CrO_4} \text{RCH(H)–O–Cr(=O)(=O)–OH} \longrightarrow \underset{\text{Aldehyde}}{R\overset{O}{\overset{\|}{C}}-H} + \underset{\text{Chromous acid}}{H_2CrO_3}$$

2. Oxidation of secondary alcohols

$$\underset{\text{Sec. alcohols}}{R_2CHOH} \xrightarrow{H_2CrO_4} \underset{\text{Ketone}}{RCOR}$$

Mechanism of the reaction is as follows:

$$\underset{\displaystyle OH}{R_2\overset{}{C}} - H + HCrO_4^- + H^+ \rightleftharpoons \underset{\displaystyle OCrO_3H}{R_2C} - H$$

$$\begin{matrix} R_2C - H \\ O - CrO_3H \end{matrix} \xrightarrow{\text{Base}} R_2CO + HCrO_3^- + \text{Base} - H$$

3. Oxidation of ketones

$$(i)\ CH_3\overset{O}{\overset{\|}{C}} - CH_2CH_3 \xrightarrow[\Delta]{H_2CrO_4} CH_3\overset{O}{\overset{\|}{C}} - OH + CH_3 - \overset{O}{\overset{\|}{C}} - OH$$

Ethanoic acid

It has been noticed that the ketone first gets converted into enolic form and then cleavage takes place at the double bond.

$$CH_3 - \overset{O}{\overset{\|}{C}} - CH_2 - CH_3 \rightleftharpoons CH_3 - \overset{OH}{\overset{\|}{C}} = CH - CH_3$$

$$\downarrow \Delta\ H_2CrO_4$$

$$\underset{\text{Acetic acid}}{CH_3COOH + CH_3COOH}$$

$$(ii)\ CH_3CH_2 - \overset{O}{\overset{\|}{C}} - CH_2 - CH_2 - CH_3 \xrightarrow[\Delta]{H_2CrO_4} \underset{\text{Propanoic acid}}{CH_3CH_2COOH} + \underset{\text{Ethanoic acid}}{CH_3COOH}$$

Q. 11. Describe the oxidation reactions of hydrogen peroxide.

Ans. Hydrogen peroxide is an important oxidising reagent in organic chemistry. Its applications are illustrated with the help of following examples.

1. Hydroxylation of alkenes

$$\underset{\text{Ethylene}}{\begin{matrix} CH_2 \\ CH_2 \end{matrix}} + H_2O + O \xrightarrow{H_2O} \underset{\text{Ethylene glycol}}{\begin{matrix} CH_2OH \\ | \\ CH_2OH \end{matrix}}$$

2. Oxidation of carboxylic acids. Carboxy acids on treatment with hydrogen peroxide are converted into per acids.

$$\underset{\text{Carboxy acid}}{R - \overset{O}{\overset{\|}{C}} - OH} \xrightarrow{H_2O} \underset{\text{Per acid}}{R - \overset{O}{\overset{\|}{C}} - O - OH}$$

Some mineral acid is added as a catalyst.

Mechanism of the reaction is as follows:

$$\underset{\text{Acid}}{R-\overset{O}{\overset{\|}{C}}-OH} \xrightarrow{H^+} R-\underset{\oplus}{\overset{O-H}{\overset{|}{C}}}-OH \xrightarrow{H-\ddot{O}-\ddot{O}-H} R-\underset{H-O-O-H}{\overset{OH}{C}}-OH$$

$$\downarrow$$

$$\underset{\text{Per acid}}{R-\overset{O}{\overset{\|}{C}}-O-OH} \xleftarrow{H^+} R-\underset{\oplus}{\overset{O-H}{\overset{|}{C}}}-O-O-H \xleftarrow{-H_2O} R-\underset{O-O-H}{\overset{O-H}{C}}-\overset{\oplus}{O}H_2$$

3. Action on aromatic carbonyl compound. An aromatic aldehyde or ketone containing a hydroxy or amino group in *ortho* or *para* position is converted into a phenol.

$$H_2O_2 + OH^- \longrightarrow H_2O + \bar{O}OH$$

$$HO-C_6H_4-\overset{O}{\overset{\|}{C}}-CH_3 + \bar{O}OH \xrightarrow{H_2O_2/OH^-} HO-C_6H_4-\underset{O-O-H}{\overset{CH_3}{C}}-\ddot{O}:$$

$$\downarrow$$

$$\underset{\text{Quinol}}{HO-C_6H_4-HO} \xleftarrow{\text{Hydrolysis}} HO-C_6H_4-O-\overset{CH_3}{C}=O$$

4. Action on α-keto acid

$$R-\overset{O}{\overset{\|}{C}}-COOH \xrightarrow{H_2O_2} R-\overset{O}{\overset{\|}{C}}-OH$$

Q. 12. Give in detail the role of OsO_4 as an oxidant in the hydroxylation reactions of alkenes. Give the mechanism of reaction involved.

Or

Give synthetic applications of osmium tetroxide OsO_4.

Ans. Osmium tetroxide can be prepared by heating metallic osmium to red hot in air.

$$Os + 2O_2 \xrightarrow[\text{hot}]{\text{Red}} OsO_4$$

Applications. Osmium tetroxide is used for hydroxylation of olefinic compounds. To carry out the reaction, OsO_4 is dissolved in dry ether and this solution is added to the olefin in the presence of pyridine. A cyclic osmic ester is formed which is hydrolysed to produce dihydroxy compound. Mechanism of the reaction is as given below:

$$>C=C< + OsO_4 \xrightarrow{\text{Pyridine}} >\underset{O}{\underset{|}{C}}-\underset{O}{\underset{|}{C}}< \;(\text{O, O bridged by } OsO_2) \xrightarrow{H_2O} >\underset{OH}{\underset{|}{C}}-\underset{OH}{\underset{|}{C}}< + OsO_3^-$$

Osmium tetroxide is a highly specific reagent for *cis*-hydroxylation on the less hindered side of the alkene.

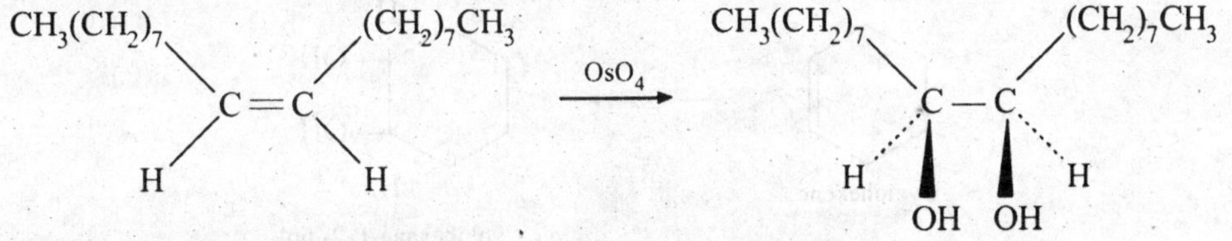

$$CH_3(CH_2)_7(H)C{=}C(H)(CH_2)_7CH_3 \xrightarrow{OsO_4} CH_3(CH_2)_7(H)(OH)C{-}C(OH)(H)(CH_2)_7CH_3$$

(*i*) Preparation of cis-1, 2-diols

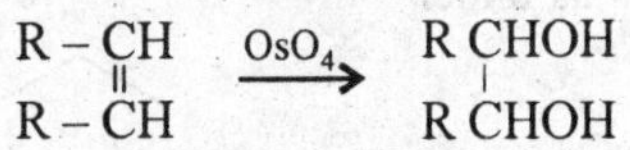

$$\begin{matrix} R-CH \\ \| \\ R-CH \end{matrix} \xrightarrow{OsO_4} \begin{matrix} R\,CHOH \\ | \\ R\,CHOH \end{matrix}$$

(*ii*) Maleic and fumeric acids yield *dl*-tartaric acid and *meso* tartaric acid respectively.

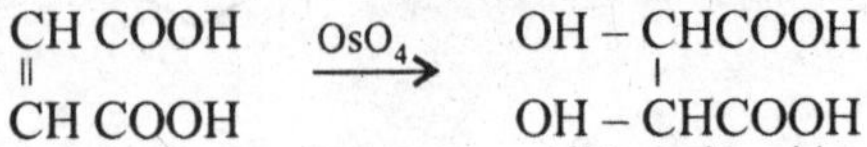

$$\begin{matrix} CH\,COOH \\ \| \\ CH\,COOH \end{matrix} \xrightarrow{OsO_4} \begin{matrix} OH-CHCOOH \\ | \\ OH-CHCOOH \end{matrix}$$

dl-tartaric acid

(*iii*) Phenanthrene derivatives are hydroxylated as under:

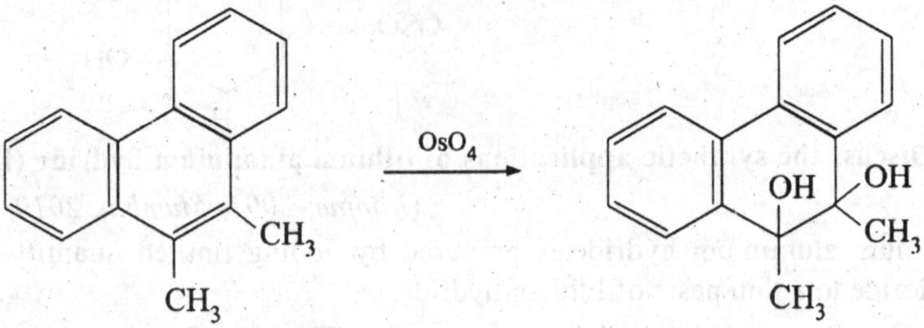

9,10-dimethylphenanthrene $\xrightarrow{OsO_4}$ 9,10-dihydroxy-9,10-dimethyl-9,10-dihydrophenanthrene (OH, OH, CH_3, CH_3)

(*iv*) **Hydroxylation of cyclopentene**

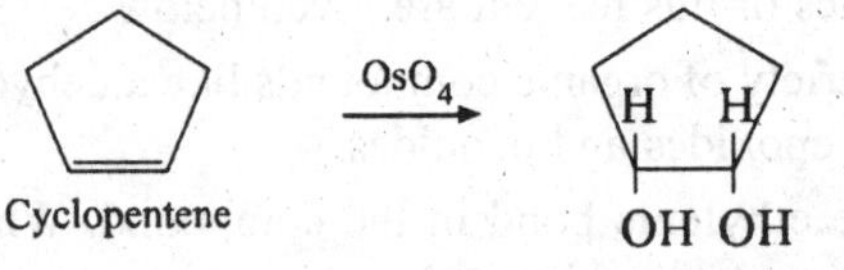

Cyclopentene $\xrightarrow{OsO_4}$ Cyclopentane-1, 2-diol (H H, OH OH)

(*v*) **Hydroxylation of 2-butene**

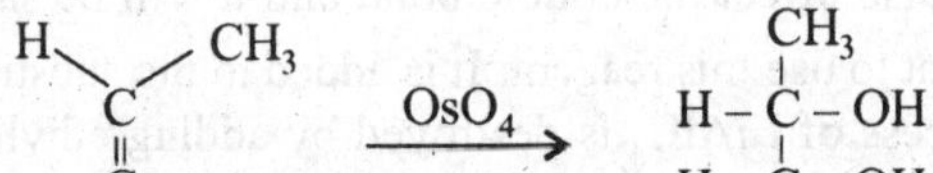

$$\begin{matrix} H\diagdown\ \ \diagup CH_3 \\ C \\ \| \\ C \\ H\diagup\ \ \diagdown CH_3 \end{matrix} \xrightarrow{OsO_4} \begin{matrix} CH_3 \\ | \\ H-C-OH \\ | \\ H-C-OH \\ | \\ CH_3 \end{matrix}$$

Cis-2-butene — *Meso,* 2, 3-butane diol

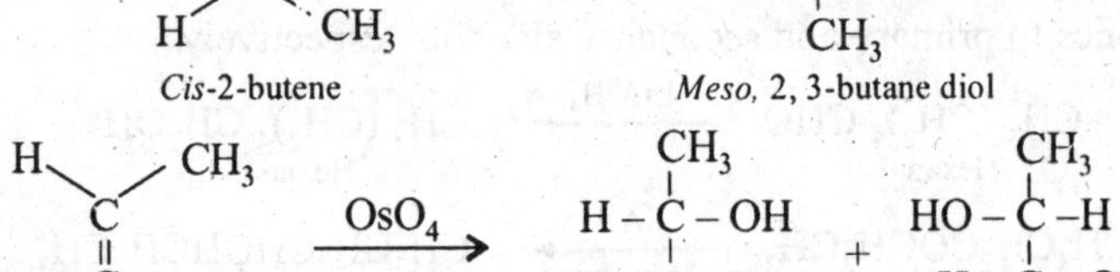

$$\begin{matrix} H\diagdown\ \ \diagup CH_3 \\ C \\ \| \\ C \\ CH_3\diagup\ \ \diagdown H \end{matrix} \xrightarrow{OsO_4} \begin{matrix} CH_3 \\ | \\ H-C-OH \\ | \\ HO-C-H \\ | \\ CH_3 \end{matrix} + \begin{matrix} CH_3 \\ | \\ HO-C-H \\ | \\ H-C-OH \\ | \\ CH_3 \end{matrix}$$

Rans-s-butene — ± 2, 3-butane diol

Q. 13. Predict the product of the following reaction and give the mechanism.

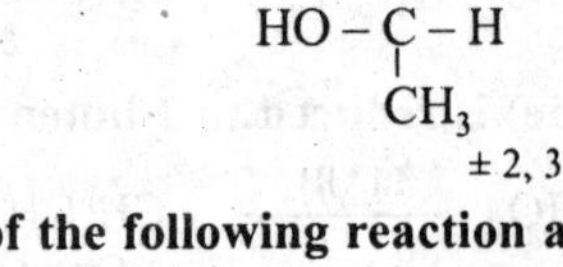

Cyclohexene $\xrightarrow{OSO_4}$

Ans. Cyclohexene reacts with osmium oxide to give cis-cyclohexane-1, 2-diol as shown below:

$$\text{Cyclohexene} \longrightarrow \textit{cis}\text{-Cyclohexane-1, 2-diol}$$

Mechanism of the reaction is as under :

$$\text{Cyclohexene} + OsO_4 \longrightarrow \text{cyclic osmate ester} \xrightarrow{H_2O} OSO_3^- + \textit{cis}\text{-cyclohexane-1,2-diol}$$

Q. 14. Discuss the synthetic applications of lithium aluminium hydride ($LiAlH_4$).

(Kanpur, 2001; Mumbai, 2010; Patna, 2010

Ans. Lithium aluminium hydride is prepared by adding limited quantity of anhydrous aluminium chloride to a thin paste of lithium hydride.

$$4LiH + AlCl_3 \longrightarrow LiAlH_4 + 3LiCl$$

Some of the characteristics of this reagent are given below :

(*i*) It can reduce a vast variety of organic compounds like aldehydes, ketones, acids and their derivatives, nitro compounds, epoxides and ozonides.

(*ii*) It does not touch the ethylenic bond in the compound. When it is required to reduce carbonyl groups without affecting the double bond in the compound, it is perhaps the best reagent. However if a phenyl group is attached at β-position in an α, β unsaturated carbonyl compound, treatment with $LiAlH_4$ will also affect the double bond and it will be saturated.

(*iii*) It is very convenient to use this reagent. It is added to the substrate solution. The mixture is stirred for some time. Excess of $LiAlH_4$ is destroyed by adding ethyl acetate.

Applications. (*a*) Reduction of carbonyl compounds. Lithium aluminium hydride reduces aldehydes and ketones to primary and secondary alcohols respectively.

$$\underset{\text{Hexanal}}{CH_3(CH_2)_4CHO} \xrightarrow{LiAlH_4} \underset{\text{Hexanol-1}}{CH_3(CH_2)_4CH_2OH}$$

$$\underset{\text{Diethyl ketone}}{CH_3CH_2COCH_2CH_3} \xrightarrow{LiAlH_4} \underset{\text{3-Pentanol}}{CH_3CH_2CHOHCH_2CH_3}$$

(*b*) 2-Butenal (Crotonaldehyde) is reduced to 2-butenol

$$\underset{\text{Crotonaldehyde}}{CH_3CH = CHCHO} \xrightarrow{LiAlH_4} \underset{\text{Crotyl alcohol (2-butenol)}}{CH_3CH = CHCH_2OH}$$

$$\underset{\text{Cinnamaldehyde}}{C_6H_5CH = CHCHO} \xrightarrow{LiAlH_4} \underset{\text{3-Phenyl propanol-1}}{C_6H_5CH_2CH_2CH_2OH}$$

Here double bond is also reduced.

(*c*) Cyclic ketones are reduced as under:

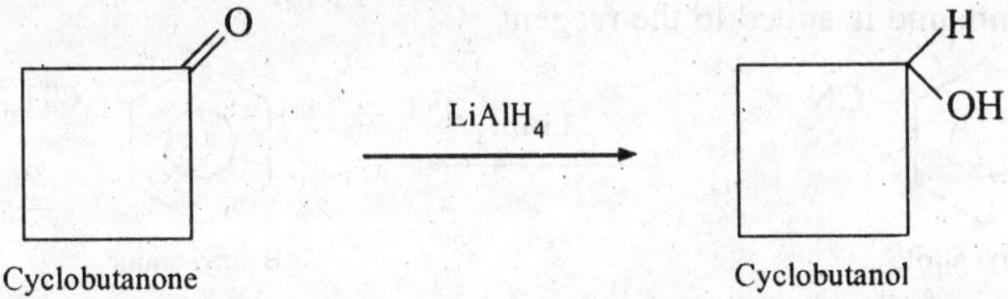

Cyclobutanone → Cyclobutanol

(*d*) Reduction of acids

$$CH_3COOH \xrightarrow{LiAlH_4} CH_3CH_2OH$$
Acetic acid → Ethyl alcohol

$$(CH_3)_2CHCOOH \xrightarrow{LiAlH_4} (CH_3)_2CHCH_2OH$$
Isobutyric acid → Isobutyl alcohol

$$C_6H_5CH = CHCOOH \xrightarrow{LiAlH_4} C_6H_5CH_2CH_2CH_2OH$$
Cinnamic acid → 3-phenylpropanol-1

(*e*) Reduction of acid derivatives

$$CH_3COCl \xrightarrow{LiAlH_4} CH_3CH_2OH$$
Acetyl chloride → Ethanol

$$CHCl_2COCl \xrightarrow{LiAlH_4} CHCl_2CH_2OH$$
2, 2-Dichloroethanol

$$CH_3COOC_2H_5 \xrightarrow{LiAlH_4} 2CH_3CH_2OH$$
Ethyl acetate

$$C_6H_5COOC_2H_5 \xrightarrow{LiAlH_4} C_6H_5CH_2OH + C_2H_5OH$$
Ethyl benzoate → Phenyl ethanol + Ethanol

$$CH_3CH = CHCOOC_2H_5 \xrightarrow{LiAlH_4} CH_3CH = CHCH_2OH$$
Ethyl crotonic acid → Crotonyl alcohol

$$CH_3CONH_2 \xrightarrow{LiAlH_4} CH_3CH_2NH_2$$
Acetamide → Ethyl amine

$$CH_3CONHCH_3 \xrightarrow{LiAlH_4} CH_3CH_2NHCH_3$$
N-Methyl acetamide → Ethyl methyl amine

(*f*) Reduction of epoxide

$$\underset{\text{(epoxide, O bridging CH and CH}_2)}{CH_3CH - CH_2} \xrightarrow{LiAlH_4} CH_3CH_2CH_2OH + CH_3\underset{\underset{OH}{|}}{CH} - CH_3$$
Propylene oxide

(*g*) Reduction of nitro compounds

$$CH_3CH(NO_2)CH_2CH_3 \xrightarrow{LiAlH_4} CH_3CH(NH_2)CH_2CH_3$$
2-nitrobutane → 2-amino butane

$$C_6H_5NO_2 \xrightarrow{LiAlH_4} C_6H_5NO + C_6H_5NHOH$$
Nitrobenzene → Nitroso benzene + Phenyl hydroxylamine

(*h*) Reduction of nitrile.

(*i*) If the compound is added to the reagent

$$C_6H_5-CN \xrightarrow{LiAlH_4} C_6H_5-CH_2NH_2$$

Phenyl nitrile → Benzyl amine

(*ii*) If the reagent is added to the compound

$$C_6H_5-CN \xrightarrow{LiAlH_4} C_6H_5-CHO$$

Benzaldehyde

(*i*) Reduction of oximes

$$(CH_3)_2C=NOH \xrightarrow{LiAlH_4} (CH_3)_2CHNH_2$$

1-methyl-1-ethyl amine

Q. 15. Discuss the synthetic applications of sodium borohydride ($NaBH_4$). (*Kanpur, 2001*)

Ans. The reducing properties of this substance were first used by Schlesinger and Brown in 1943. This reagent can be prepared by the reaction of sodium hydride with methyl borate at 520 K.

$$4NaH + B(OCH_3)_3 \xrightarrow{520\ K} NaBH_4 + 3NaOCH_3$$

It is a mild reducing agent. Its advantages over $LiAlH_4$ as a reducing agent are:

(*i*) It is more selective than lithium aluminium hydride. Whereas $LiAlH_4$ can reduce groups like esters, carboxy acids, lactones, epoxides, nitro and nitrile besides the carboxyl group, sodium borohydride reduces only the carbonyl group. It reduces aldehydes into primary and ketones into secondary alcohols.

(*ii*) Like lithium aluminium hydride, it does not affect the ethylenic bond in the molecule.

(*iii*) It is much milder and softer reducing agent than lithium aluminium hydride.

(*iv*) This reagent, unlike $LiAlH_4$, can be used in aqueous medium also. Therefore it is widely used as a reducing agent in carbohydrate chemistry.

(*v*) It reduces lactones and acid chlorides but not esters.

Applications. (*a*) Reduction of aldehydes and ketones

$$CH_3COCH_2COCH_3 \xrightarrow{NaBH_4} CH_3CH(OH)CH_2CH(OH)CH_3$$

Pentane-2,4-dione → Pentane-2, 4-diol

$$CH_3CH=CHCHO \xrightarrow{NaBH_4} CH_3CH=CHCH_2OH$$

Crotonaldehyde → Crotonyl alcohol

$$CH_2OH(CHOH)_4CHO \xrightarrow{NaBH_4} CH_2OH(CHOH)_4CH_2OH$$

Glucose → Sorbitol

(*b*) Selective reduction. If the compound contains some group besides the carbonyl group, then only the latter is reduced.

$$NO_2CH_2CH_2CHO \xrightarrow{NaBH_4} NO_2CH_2CH_2CH_2OH$$

3-nitropropionaldehyde → 3-nitropropanol-1

$$C_6H_5COCH_2CH_2COOH \xrightarrow{NaBH_4} C_6H_5CH(OH)CH_2CH_2COOH$$

β-benzoyl propionic acid → γ-hydroxyphenyl butyric acid

$$\text{m-nitrobenzaldehyde } (C_6H_4(NO_2)CHO) \xrightarrow{NaBH_4} \text{m-nitrobenzyl alcohol } (C_6H_4(NO)CH_2OH)$$

m-nitrobenzaldehyde — *m*-nitrobenzyl alcohol

(*c*) **Reduction of ozonide.** It reduces ozonide into alcohol.

$$\text{Cyclohexene} \xrightarrow{O_3} \text{ozonide} \xrightarrow{NaBH_4} OH(CH_2)_6OH$$

Cyclohexene — Hexane-1, 6-diol

(*d*) **Exceptional case.** With methanol as the solvent, reducing action of sodium borohydride becomes violent.

$$CH_3CH = CHCOOC_2H_5 \xrightarrow{NaBH_4} CH_3CH_2CH_2CH_2OH$$

It reduces even the ester group and the double bond.

Q. 16. How does $NaBH_4$ differ from $LiAlH_4$ as a reducing agent? Explain with example. Name the products formed by the reduction of crotonaldehyde with these reagents.

Ans. See Q. 15.

Q. 17. Describe synthetic uses of Raney nickel.

Ans. The active form of nickel was discovered by Murray Raney in 1927.

It is prepared by digesting an alloy, containing 50% nickel and 50% aluminium, in sodium hydroxide. Aluminium is dissolved in sodium hydroxide as sodium aluminate and the residue is finely divided nickel. The surface of nickel is covered with hydrogen (Ni–H) during the preparation.

$$Al - Ni + NaOH + H_2O \longrightarrow NaAlO_2 + \underset{\text{Raney nickel}}{Ni - H} + H_2$$

Activity of nickel depends upon the adsorbed hydrogen. More the adsorbed hydrogen, more is the activity of catalyst. Raney nickel is also sometimes expressed as $Ni–H_2$. Raney nickel is more active than supported nickel catalysts.

(*i*) Raney nickel can perform hydrogenation at lower temperature.

(*ii*) It is capable of reducing all unsaturated groups, ethylenic and acetylenic bonds, aromatic rings and performing hydrogenolysis of sulphur compounds.

Applications. (*a*) Reduction of alkenes

$$\underset{\text{Propene}}{CH_3CH = CH_2} + H_2 \xrightarrow[\text{Ni}]{\text{Raney}} \underset{\text{Propane}}{CH_3CH_2CH_3}$$

Compounds having conjugated double bonds first undergo 1, 2 and 1, 4 addition and finally reduced completely.

$$CH_2 = CH - CH = CH_2 \xrightarrow[H_2]{\text{Raney Ni}} \underset{\text{Butene-1}}{CH_3 - CH_2CH = CH_2} + \underset{\text{Butene-2}}{CH_3 - CH = CH - CH_3}$$

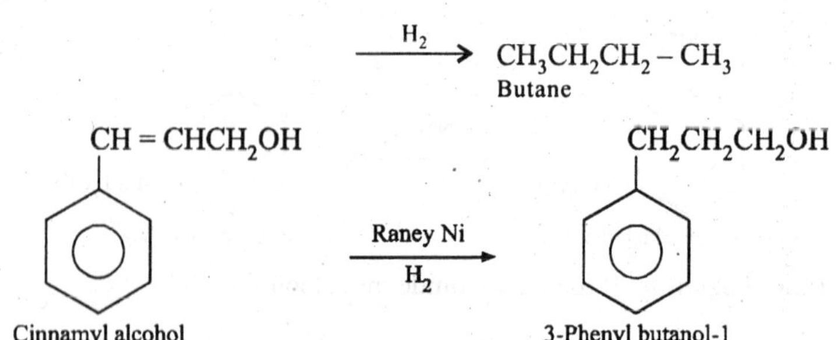

(b) Reduction of aromatic compounds

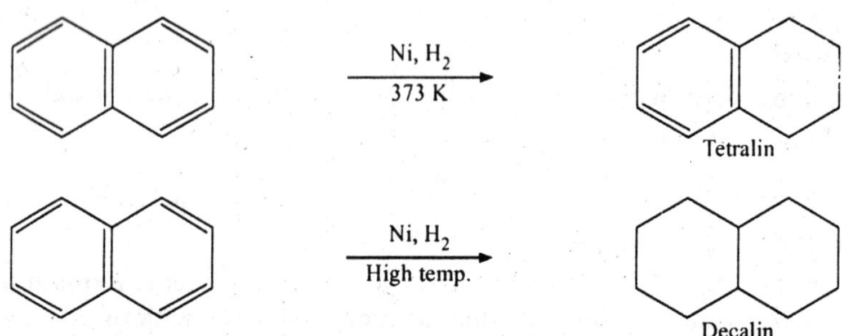

(c) Reduction of phenols

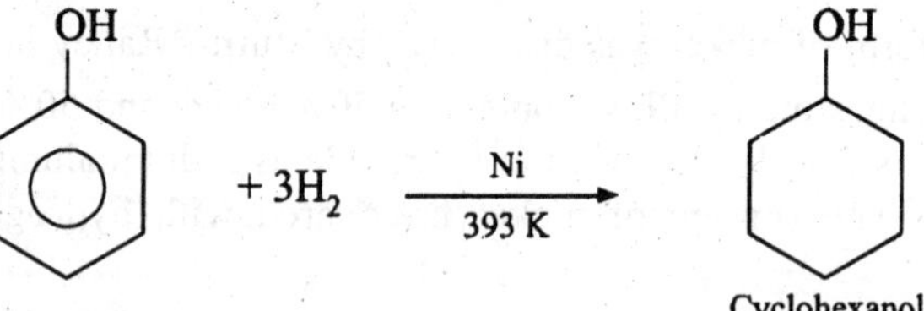

(d) Reduction of nitro compounds

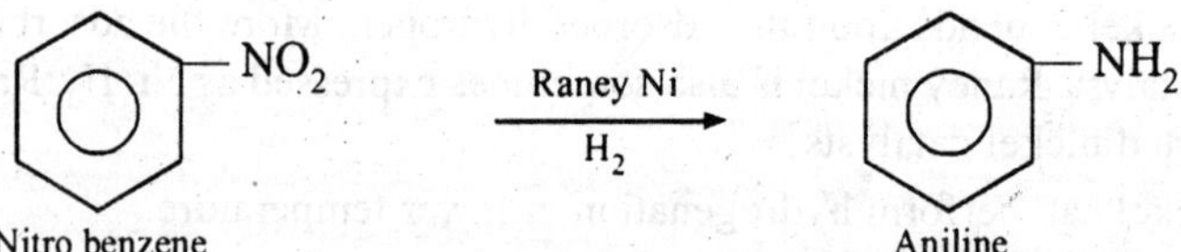

(e) Reduction of nitriles to amines. When nitriles are reduced under pressure in the presence of Raney Ni, they yield the corresponding primary amines.

$$RCN \xrightarrow[Ni]{H_2} RCH = NH \xrightarrow[Ni]{H_2} \underset{\text{Primary amine}}{RCH_2NH_2}$$

(f) Reduction of oximes

$$\underset{\text{Oxime}}{R - CH = N - OH} + 2H_2 \xrightarrow[Ni]{H_2} \underset{\text{Amine}}{RCH_2NH_2} + H_2O$$

(g) Reduction of carbonyl group. Aldehydes and ketones are reduced to primary and secondary alcohols.

$$\underset{\text{Propanal}}{CH_3CH_2CHO} \xrightarrow[H_2]{\text{Raney Ni}} \underset{\text{Propanol-1}}{CH_3CH_2CH_2OH}$$

$$\underset{\text{Acetone}}{CH_3COCH_3} \xrightarrow[H_2]{\text{Raney Ni}} \underset{\text{Propanol-2}}{CH_3CHOHCH_3}$$

$$\underset{\text{Cinnamaldehyde}}{C_6H_5-CH=CH-CHO} \xrightarrow[H_2]{\text{Raney Ni}} \underset{\text{3-Phenylpropanol}}{C_6H_5-CH_2CH_2CH_2OH}$$

(*h*) Desulphurisation

$$\underset{\text{Tetrahydro-thiophene}}{C_4H_8S} \xrightarrow[H_2]{Ni} \underset{\text{Butane}}{CH_3CH_2CH_2CH_3} + NiS$$

Q. 18. Describe the use of platinum and palladium as hydrogenation catalysts.

Ans. Platinum and palladium are used as catalysts in two forms viz. colloidal and amorphous.

Colloidal form. Colloidal form of these metals is prepared by reducing their salt solution with hydrazine hydrate. On complete reduction, the salts are removed by dialysis and the solutions are concentrated to obtain colloidal metals.

Amorphous form. Amorphous form of the metals is prepared by reducing their salts with sodium borohydride. It may also be prepared by treating oxides of these metals with sodium nitrate in the presence of support. Supports like activated carbon, asbestos, barium sulphate, calcium carbonate have been found extremely useful during hydrogenation.

Both platinum and palladium are active catalysts for hydrogenation and dehydrogenation reactions.

It is possible to increase the catalytic activity of Pt or Pd catalysts by adding small quantities of promoters usually salts of Pt or Pd.

Applications. (*a*) These metals can reduce most of olefin compounds below 373 K and at atmospheric pressure. Out of Pt and Pd, the former is used for bringing about exhaustive reduction.

$$\underset{\text{Butene-2}}{CH_3CH=CHCH_3} \xrightarrow[H_2]{\text{Pt or Pd}} \underset{\text{Butane}}{CH_3CH_2CH_2CH_3}$$

$$\underset{\text{Propyne}}{CH_3C\equiv CH} \xrightarrow[H_2]{\text{Pt or Pd}} \underset{\text{Propane}}{CH_3CH_2CH_3}$$

$$\underset{\text{2-Benzyl cyclopentanol}}{\text{(OH)}C_5H_8CH_2C_6H_5} \xleftarrow[Pt]{H_2} \underset{\text{2-Benzylidene cyclopentanone}}{\text{(O=)}C_5H_6=CHC_6H_5} \xrightarrow{H_2/Pd} \underset{\text{2-Benzyl cyclopentanone}}{\text{(O=)}C_5H_7CH_2C_6H_5}$$

(*b*) Partial hydrogenation. Partial hydrogenation of acetylenes to ethylenes can be brought about by Pd–$CaCO_3$ catalyst. The theory for partial hydrogenation of acetylenes to olefins is that more electrophilic acetylenic compounds are preferentially absorbed in comparison to olefins, on the catalyst surface hindering further hydrogenation.

$$\underset{\text{Propyne}}{CH_3C \equiv CH} \xrightarrow[H_2]{Pd - CaCO_3} \underset{\text{Propene}}{CH_3CH = CH_2}$$

(*c*) Hydrogenation of nitriles, oximes and nitro compounds

$$C_6H_5CH = CHNO_2 \xrightarrow[\text{Room temp.}]{H_2/Pt - C} \underset{\text{2-Phenylethylamine}}{C_6H_5CH_2CH_2NH_2}$$

$$\underset{\text{Ethyl cyanide}}{CH_3CH_2CN} \xrightarrow[H_2]{\text{Pt or Pd}} \underset{\text{Propylamine}}{CH_3CH_2CH_2NH_2}$$

$$\underset{\text{Acetaldoxime}}{CH_3CH = NOH} \xrightarrow[H_2]{\text{Pt or Pd}} \underset{\text{Ethylamine}}{CH_3CH_2NH_2}$$

(*d*) Reduction of acid chloride

$$\underset{\text{Acetyl chloride}}{CH_3COCl} + H_2 \xrightarrow{Pd/BaSO_4} \underset{\text{Acetaldehyde}}{CH_3CHO} + HCl$$

$$\underset{\text{Cinnamoyl chloride}}{C_6H_5CH = CHCOCl} \xrightarrow[Pd/BaSO_4]{H_2} \underset{\text{Cinnamaldehyde}}{C_6H_5CH = CHCHO}$$

$$\underset{p\text{-Nitrobenzoyl chloride}}{p\text{-}NO_2C_6H_4COCl} \xrightarrow[Pd/BaSO_4]{H_2} \underset{p\text{-Nitrobenzaldehyde}}{p\text{-}NO_2C_6H_4CHO}$$

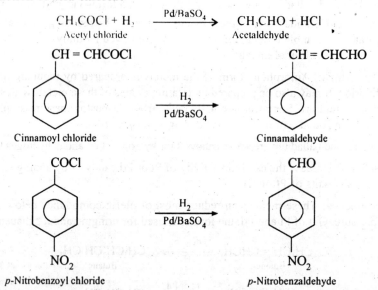

(*e*) As dehydrogenation catalyst. Both Pt and Pd are also used as dehydrogenation catalysts.

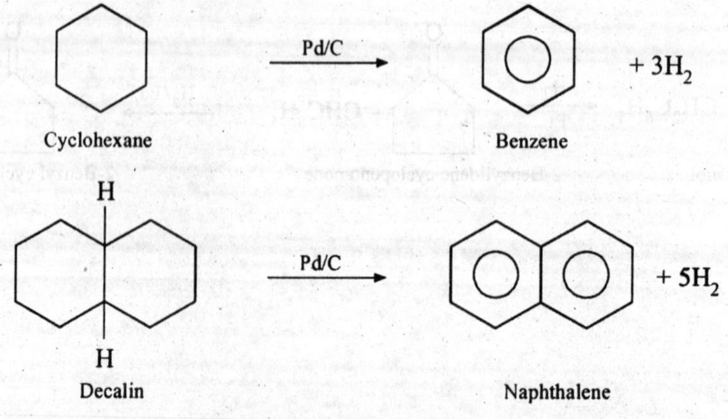

Q. 19. Describe in detail alkali metal-ammonia reductions.

Ans. Alkali metals like Li, Na or K in combination with liquid ammonia as a solvent act as efficient reducing agents. Some of their applications are given below:

(*a*) Reduction of aromatic ring (Birch reduction). I involves partial reduction of aromatic ring.

CH_2CH_2OH (Phenyl ethanol) $\xrightarrow[\text{Ethanol}]{\text{Li-NH}_3}$ CH_2CH_2OH

Naphthalene $\xrightarrow[\text{Ethanol}]{\text{Li-NH}_3}$ Dihydronaphthalene (α, β, γ, δ)

In the Birch reduction involving ammonia as a solvent, hydrogen atoms are added at positions that are α, δ to each other. One double bond that is left in dihydronaphthalene is difficult to reduce.

If, however, ethylamine is used as the solvent, the α, δ dihydro derivative gets rearranged to α, β dihydro derivative.

Naphthalene $\xrightarrow[C_2H_5NH_2]{\text{Li-NH}_3}$ (α, β)

(*b*) Reduction of olefins. Reduction of aliphatic compounds takes place if the double bond is conjugated. This is because the resonance stabilizes the intermediate product.

$$\underset{\text{Isoprene}}{CH_2 = \overset{\overset{CH_3}{|}}{C} - CH = CH_2} \xrightarrow[\text{Liq.NH}_3]{\text{Na}} \underset{\text{Intermediate}}{CH_2 \cdots\cdots \overset{\overset{CH_3}{|}}{C} \cdots\cdots CH \cdots\cdots CH_2}$$

$$\longrightarrow \underset{\text{2-methyl-2-butene}}{CH_3 - \overset{\overset{CH_3}{|}}{C} = CH - CH_3}$$

(*c*) Reductions of terminal double bonds. Terminal double bond is reduced with Na–NH_3 system.

$$\underset{\text{Propene}}{CH_3CH = CH_2} \xrightarrow{\text{Na} - \text{NH}_3} \underset{\text{Proppane}}{CH_3CH_2CH_3}$$

(*d*) Reduction of acetylenes. Although reduction of alkenes (except in terminal position) is difficult, alkynes can be comparatively easily reduced to alkene giving trans isomers.

$$\underset{\text{Butyne-2}}{CH_3C \equiv C - CH_3} \xrightarrow{\text{Na} - \text{NH}_3} \underset{\text{Trans 2-butene}}{\overset{CH_3}{\underset{H}{>}}C = C\overset{H}{\underset{CH_3}{<}}}$$

(e) Reduction of cyclononyne

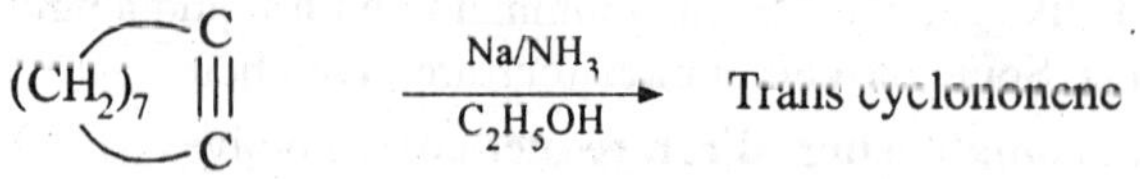

$$(CH_2)_7 \begin{matrix} C \\ ||| \\ C \end{matrix} \xrightarrow[C_2H_5OH]{Na/NH_3} \text{Trans cyclononene}$$

Mechanism of reduction. Mechanism of reduction by means of Na/NH_3 in the presence of ROH is explained below. Taking the example of benzene.

$$Na + (x + y)\ NH_3 \longrightarrow Na^+ (NH_3)_x + \bar{e}\ (NH_3)_y$$

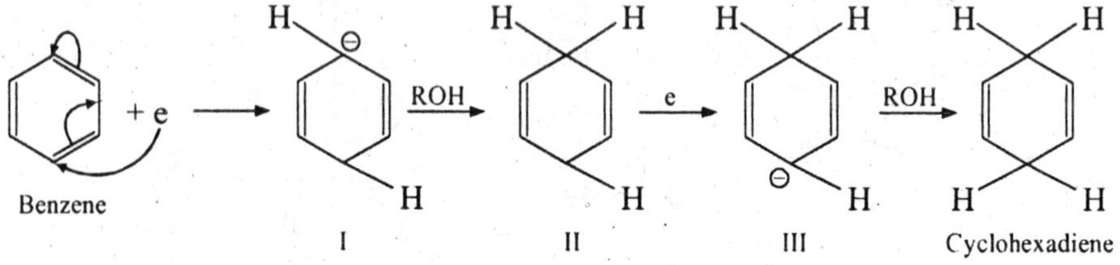

$$\text{Benzene} + e \longrightarrow \text{I} \xrightarrow{ROH} \text{II} \xrightarrow{e} \text{III} \xrightarrow{ROH} \text{Cyclohexadiene}$$

Benzene I II III Cyclohexadiene

Q. 20. State whether the following transformations involve oxidation or reduction.

(i) $RCOCH_3 \longrightarrow RCONH_2$

(ii) $RH \longrightarrow RCl$

(iii) $RCOOH \longrightarrow RCH_2OH$

(iv) $-\overset{|}{C}-\overset{|}{C}- \longrightarrow -C=C-$

(v) $R-NH_2 \longrightarrow R-H$

Ans. Oxidation is a process which involves :

Addition of oxygen or electronegative radical or atom

Or

Removal of hydrogen or electropositive radical or atom.

Similarly,

Reduction is a process which involves:

Addition of hydrogen or electropositive radical or atom.

Or

Removal of oxygen or electronegative radical or atom.

In the light of above generalisation, we can decide whether the above reactions involve oxidation or reduction.

(*i*) $RCOCH_3 \longrightarrow RCONH_2$

It involves replacement of carbon by more electronegative nitrogen, therefore, it is an oxidation process.

(*ii*) $RH \longrightarrow RCl$

Again, it involves replacement of hydrogen by electronegative atom Cl, therefore, it is also an oxidation process.

(*iii*) $R-COOH \longrightarrow RCH_2OH$

One oxygen of R–COOH has been replaced by two hydrogens, therefore, it is a reduction process.

(*iv*) $-\overset{|}{C}-\overset{|}{C}- \longrightarrow -C=C-$

This is actually dehydrogenation process. The reactant is an alkane. It has been converted into alkene by removal of hydrogen. Therefore, it is an oxidation process (*v*) $R - NH_2 \longrightarrow R - H$

This involves removal of electronegative element nitrogen, therefore, it is a reduction process.

Q. 21. Describe the mechanism of reduction of a ketone to an alcohol using $LiAlH_4$. Can you use $LiAlH_4$ to reduce cinnamic acid to cinnamyl alcohol? Give reason for your answer.

Ans. (*i*) Mechanism of reduction of a ketone to an alcohol is given below:

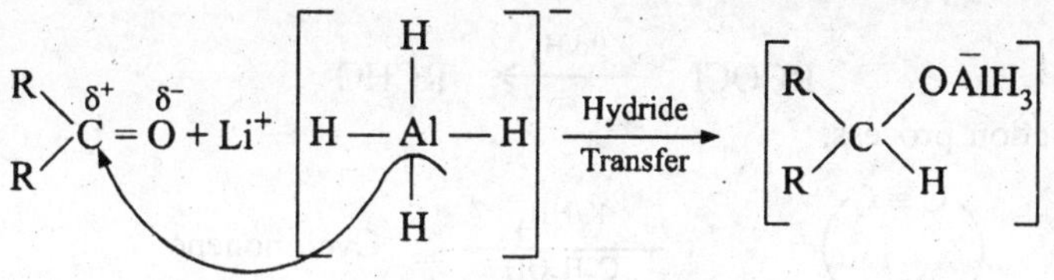

$$\underset{R}{\overset{R}{>}}\overset{\delta^+}{C}=\overset{\delta^-}{O} + Li^+ \left[H-\underset{H}{\overset{H}{Al}}-H \right]^- \xrightarrow[\text{Transfer}]{\text{Hydride}} \left[\underset{R}{\overset{R}{>}}C\underset{H}{\overset{O\bar{A}lH_3}{<}} \right] Li^+$$

$$R_2CHO\bar{A}lH_3Li + 3R_2CO \longrightarrow (R_2CHO)_4\overset{-}{Al}\overset{+}{Li} \xrightarrow{4H_2O} \underset{\text{Sec.alcohol}}{4R_2CHOH} + Al(OH)_3 + LiOH$$

(*ii*) Cinnamic acid cannot be reduced to cinnamyl alcohol with the help of $LiAlH_4$ because this reagent reduces the double bond also if present in α-position relative to the carbonyl or carboxyl group. The compound obtained on reduction will be

$$\underset{\text{Cinnamic acid}}{C_6H_5CH = CHCOOH} \xrightarrow{LiAlH_4} \underset{\text{3-Phenylpropanol}}{C_6H_5CH_2CH_2CH_2OH}$$

Q. 22. Complete the following, specifying the transformation as oxidation or reduction.

(*a*) [tricyclic diol: two cyclohexane rings fused to a cyclobutane, OH OH on the ring-junction carbons] $\xrightarrow{?}$ 2 [cyclohexanone]

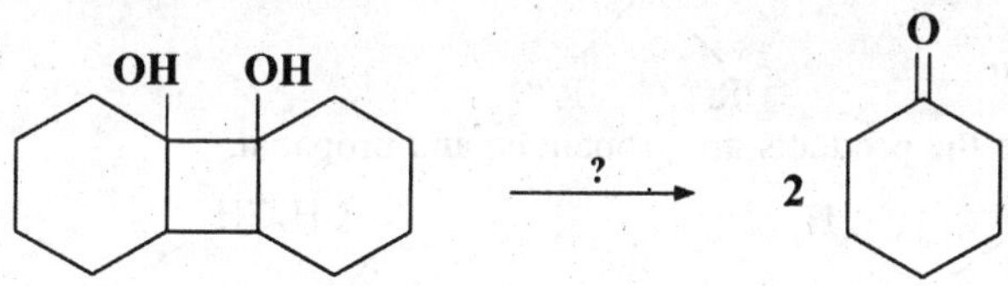

(*b*) $Ph - CH = CH - COOEt \xrightarrow{AlH_3}$

(*c*) [benzene] $\xrightarrow{?}$ [1,4-cyclohexadiene]

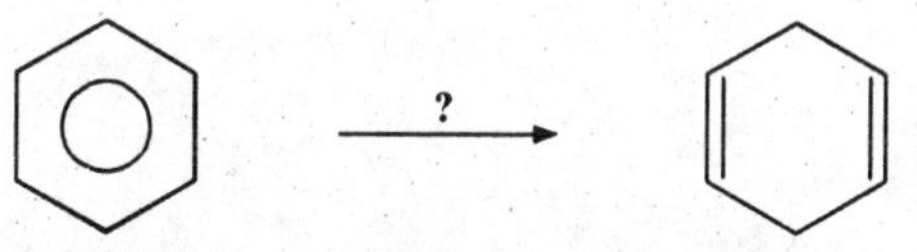

(*d*) $? \xrightarrow{Pd/H_2} RCHO$

(*e*) [cyclodecyne: $C \equiv C$ bridged by $(CH_2)_8$] $\xrightarrow[C_2H_5OH]{K/NH_3}$

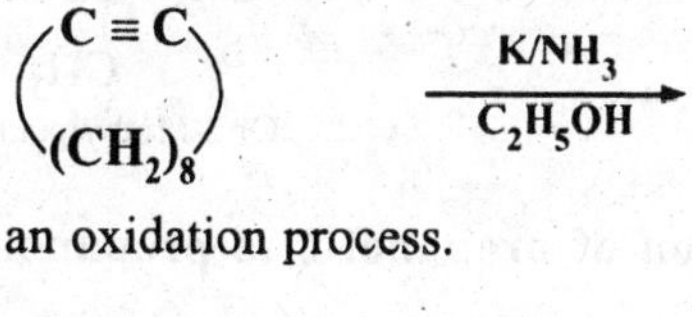

Ans. (*a*) It is an oxidation process.

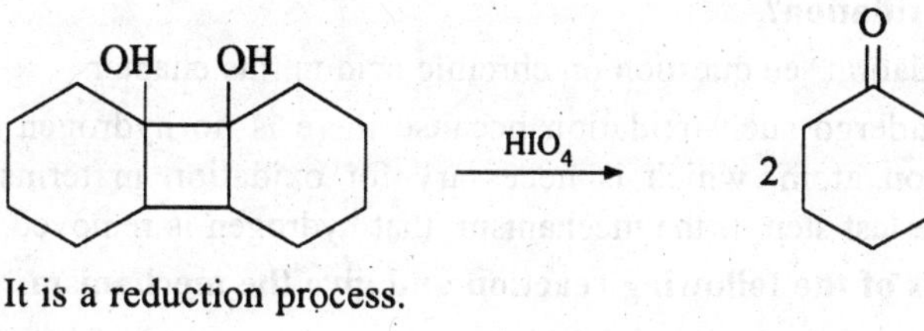

[tricyclic diol with OH OH] $\xrightarrow{HIO_4}$ 2 [cyclohexanone]

(*b*) It is a reduction process.

$$Ph - CH = CH - COOEt \xrightarrow{AlH_3} PhCH_2CH_2CH_2OH$$

(*c*) It is a reduction process.

$$\text{Benzene} \xrightarrow{K/NH_3} \text{Dihydro benzene}$$

(*d*) It is a reduction process.

$$RCOCl \xrightarrow{Pd/H_2} RCHO$$

(*e*) It is a reduction process.

$$\left(\begin{matrix} C \equiv C \\ (CH_2)_8 \end{matrix}\right) \xrightarrow[C_2H_5OH]{K/NH_3} \text{Cyclononene}$$

Q. 23. Describe ozonolysis of alkene. The products of ozonolysis of an alkene are propanone and propanal. Write the structural formula of alkene and give its IUPAC name.

Ans. An alkene on treatment with ozone forms an ozonide. The ozonide on hydrolysis gives a mixture of carbonyl compounds. By identifying the products, we can identify the alkene.

$$RR'C{=}CR''R''' + O_3 \longrightarrow \text{ozonide} \xrightarrow{H_2O} RR'C{=}O + R''COR'''$$

In the present case the products are propanone and propanal.

$$CH_3COCH_3 \quad \text{and} \quad CH_3CH_2CHO$$

Therefore, the alkene must be

$$(CH_3)_2C{=}O \;\; O{=}CH(CH_2CH_3) \quad \text{or} \quad CH_3-\underset{\displaystyle CH_3}{\underset{|}{C}}=CH-CH_2CH_3$$

Or 2-Methylpentene-2

Q. 24. Describe chromic acid oxidation of propanol-2 to propanone. Why tertiary alcohols do not undergo such oxidation?

Ans. For mechanism of oxidation see question on chromic acid in this chapter.

Tertiary alcohols do not undergo such oxidation because there is no hydrogen directly attached to –OH carrying carbon atom, which is necessary for oxidation in terms of the mechanism of the reaction. In the last step of the mechanism, that hydrogen is removed.

Q. 25. Predict the products of the following reaction and give the mechanism.

$$CH_3-CH=CH-CHO \xrightarrow{NaBH_4}$$

Ans. Sodium borohydride ($NaBH_4$) is quite selective in its reducing action. It does not affect the double bond and only reduces the carbonyl group to alcoholic group.

$$\underset{\text{Crotonaldehyde}}{CH_3-CH=CH-CHO} \xrightarrow{NaBH_4} \underset{\text{Crotonyl alcohol}}{CH_3-CH=CH-CH_2OH}$$

The mechanism of reduction of a carbonyl group with sodium borohydride is illustrated as under

$$\underset{\text{Carbonyl compound}}{>C=O} + Na^+[H-BH_2-H]^- \longrightarrow \left[>C{<}^{OBH_3}_{H}\right]^- Na^+$$

$$\downarrow >C=O$$

$$\underset{\text{Alcohol}}{>CHOH} \xleftarrow{H^+/H_2O} [(>CHO)_4\bar{B}]\overset{+}{Na}$$

Q. 25. Give two methods for dehydrogenation of hydrocarbons by taking suitable examples.

Ans. Se and SeO_2 are two reagents used for dehydrogenation of hydrocarbons. Examples given below illustate.

$$\text{Tetrahydro naphthalene} \xrightarrow[-2H_2]{Se} \text{Naphthalene}$$

$$\text{Tetrahydro phenanthrene} \xrightarrow[-2H_2]{SeO_2} \text{Phenanthrene}$$

Q. 26. Why is sodium borohydride preferred over lithium aluminium hydride for the reduction of aldehydes and ketones ? Illustrate with example.

Ans. Sodium borohydride selectively reduces the carbonyl group without affecting multiple bonds or groups like $-NO_2$, $-CN$ in the molecule.

$$C_6H_5CH=CHCHO \xrightarrow{NaBH_4} C_6H_5CH=CH-CH_2OH$$

$$C_6H_5CH=CHCHO \xrightarrow{LiAlH_4} C_6H_5CH_2-CH_2-CH_2OH$$

41

NATURAL PRODUCTS

CONTENTS

Terpenes — **Citral, comphor and carvone**

Steroids — **Estrone and cholesterol**

Alkaloids — **Quinine, Nicotine, piperine and atropine**

TERPENES

Q. 1. What are terpenes and terpenoids? (*Panjab, 2000*)

Ans. Terpenes are a group of compounds having the general formula $(C_5H_8)_n$ where n is an integer. These compounds occur in plant kingdom. Compounds having the value of n as 2 and 3 viz. $C_{10}H_{16}$ and $C_{15}H_{24}$ are the chief constituents of essential oils used in perfumery. These essential oils are obtained from gums and resins of plants.

The term 'terpenoids' is used these days to call the compounds having the general formula $(C_5H_8)_n$ and their oxygen derivatives.

Q. 2. Describe the classification of terpenoids. (*Panjab, 2000 ; Sambalpur 2004*)

Ans. Terpenoids are classified on the basis of the number of isoprene units which they contain. The formula of isoprene (C_5H_8) is 2-methyl-1, 3-butadiene

$$CH_2 = \overset{\overset{\displaystyle CH_3}{|}}{C} - CH = CH_2$$

On thermal decomposition, almost all the terpenoids yields isoprene as one of the products.

$$\text{Terpenoid} \xrightarrow{\Delta} CH_2 = \overset{\overset{\displaystyle CH_3}{|}}{C} - CH = CH_2 \text{ (isoprene)}$$

Some of the classes of terpenoids along with the isoprene units and molecular formulae are given below:

Classification of Terpenoid Hydrocarbons

S.No.	Class	No. of isoprene units (C_6H_8)	Molecular formulae
1.	Monoterpenes	2	$C_{10}H_{16}$
2.	Sesquiterpenes	3	$C_{15}H_{24}$
3.	Diterpenes	4	$C_{20}H_{32}$
4.	Triterpenes	6	$C_{30}H_{48}$
5.	Tetraterpenes or Carotenoids	8	$C_{40}H_{64}$
6.	Polyterpenes or Rubber	n	$(C_5H_8)_n$

In addition to these terpenoid hydrocarbons, there are oxygenated derivatives of each class. These are mainly alcohols, aldehydes or ketones.

Terpenoids are classified on the basis of the number of rings present in the molecule, as given below:

(*a*) *Acyclic*. They have an open chain structure.

(*b*) *Monocyclic*. The terpenes which have one ring in their structure.

(c) *Bicyclic*. Those having two rings in their structure.

(d) *Tricyclic*. Those having three rings in their structure.

Volatile oils present in different parts of plants (roots, leaves, fruits, etc.) are separated by steam distillation. These are called *essential oils* and possess strong and *pleasant odour*. Essential oils are mixtures of terpenoid hydrocarbons and their oxygenated derivatives.

Q. 3. What is isoprene rule? (*Avadh, 2003; Bernampur, 2003; Delhi, 2005 Kuvempu, 2005; Garhwal, 2010*)

Ans. Nearly all the terpenoids are composed of isoprene which has the formula as given below:

$$CH_2=\overset{\overset{\displaystyle CH_3}{|}}{C}-CH=CH_2 \quad \text{or} \quad CH_2{=}\overset{\overset{\displaystyle CH_3}{|}}{C}-CH{=}CH_2$$

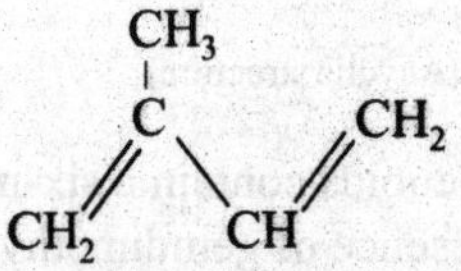

Thus terpenes can be represented by the general formula $(C_5H_8)_n$ where n is an integer.

This inference was based on the following observations:

(*i*) Empirical formula of most terpenoids is C_5H_8 which is the molecular formula of isoprene.

(*ii*) Thermal decomposition of terpenoids yields isoprene as one of the products.

Isoprene Rule which was given by Wallach can be stated as:

Skeleton structures of all naturally occurring terpenoids are built up of isoprene units.

Ingold in 1925 observed that isoprene units in terpenoids are usually joined in *head to tail* or 1, 4 linkage. This is known as *special isoprene rule*. The branched end of isoprene is called *head* and is linked to the *tall* of the other molecule as shown below:

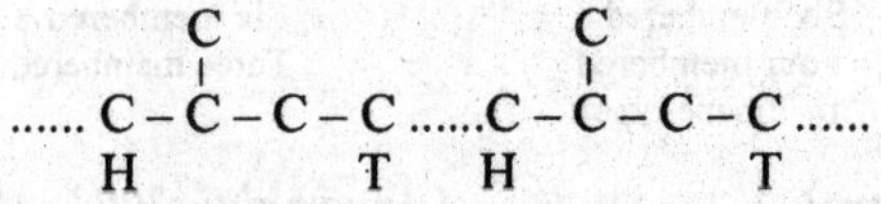

$$\ldots\ldots \underset{H}{C}-\overset{\overset{\displaystyle C}{|}}{C}-C-\underset{T}{C}\ldots\ldots\underset{H}{C}-\overset{\overset{\displaystyle C}{|}}{C}-C-\underset{T}{C}\ldots\ldots$$

H means head
T means tail

or

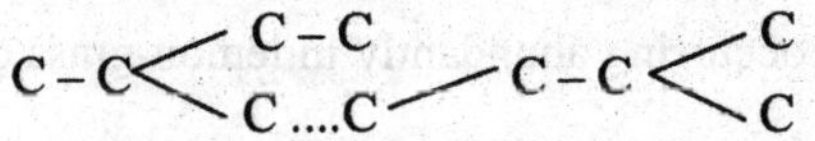

$$C-C\!\begin{smallmatrix}C-C\\C\ldots C\end{smallmatrix}\!-C-C\!\begin{smallmatrix}C\\C\end{smallmatrix}$$

It may be noted that there are several exceptions to this rule. Thus, this rule can only be used as a guiding principle and not as an absolute rule. Monoterpenoid molecule $(C_{10}H_{16})$ is short of six hydrogen atoms than the number required in corresponding alkane. Each double bond in molecule reduces the number of hydrogen atoms by two. The shortage of hydrogen atoms can be accounted for by the presence of :

(*a*) Three double bonds if the structure is open chain or an acyclic monoterpenoid.

(*b*) One monocyclic ring and two double bonds.

(*c*) Two bicyclic rings and one double bond.

Monocyclic monoterpenoids contain a six-membered ring. According to Ingold, the presence of gem-dialkyl group makes the cyclohexane ring more stable. Thus, the *gem-dialkyl rule*, limits the number of possible structures obtained by closing the open chain to a cyclohexane ring. Clearly the monoterpenoid open chain can give rise to only monocyclic monoterpenoid. It is found that most of the naturally occurring monocyclic monoterpenoids are the derivatives of *p*-cymene.

(Acyclic structure) (*p*-Cymene structure)

Bicyclic monoterpenoids contain a six-membered ring along with another, three, four or five membered ring. The presence of gemdimethyl group in these cyclopropane and cyclobutane rings provides them the stabilising effect.

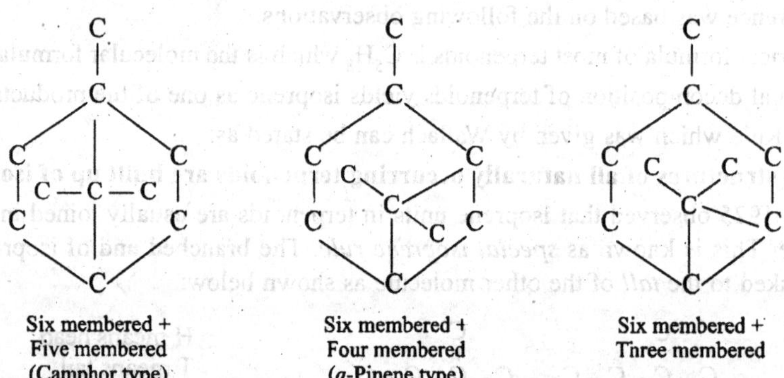

Six membered + Five membered (Camphor type) Six membered + Four membered (*a*-Pinene type) Six membered + Three membered

Q. 4. Establish the structure of citral. (*Annamalai, 2002; Mumbai, 2010*)

Ans. Citral is a monoterpene aldehyde occurring abundantly in lemon-grass oil.

Strûcture

1. Molecular formula, $C_{10}H_{16}O$, has been arrived at by the usual analytical studies.

2. Citral gives the following addition reactions.

$$\underset{\text{Citral dibromide}}{C_{10}H_{16}OBr_2} \xleftarrow{Br_2} C_{10}H_{16}O \xrightarrow{2H_2(Pd)} \underset{\text{Tetrahydrocitral}}{C_{10}H_{20}O}$$

This shows the presence of two carbon-carbon double bonds.

3. The oxygen present in citral is in the form of a carbonyl group since it forms a bisulphite derivative, an oxime, a semicarbozone and a red 2, 4-dinitrophenylhydrazone. That the carbonyl group is aldehydic is proved by the fact that it reduces Fehling solution and is oxidized to geranic acid with silver oxide and reduced to geraniol, a primary alcohol, with sodium amalgam. Both geranic acid and geranoil have the same number of carbon atoms as citral.

$$\underset{\text{Geraniol}}{C_{10}H_{18}O} \xleftarrow{Na(Hg)} \underset{\text{Citral}}{C_{10}H_{16}O} \xrightarrow{Ag_2O} \underset{\text{Geranic acid}}{C_{10}H_{18}O_2}$$

4. Complete reduction to the corresponding saturated hydrocarbon gives $C_{10}H_{22}$ which corresponds to an acyclic structure. Thus citral is acyclic.

5. One of the two double bonds is present as an isopropylidene moiety as is evidenced by the appearance of an absorption bond at ~800 cm^{-1}.

6. Citral gives both *mono* and *di*-bisulphite addition compounds which indicates that one of the double bond is conjugated with the carbonyl group. This is confirmed by UV-spectrum (λ_{max} 238 nm, ε 13500).

$$>C=CH-CHO \xrightarrow{NaHSO_3} \underset{\text{Monobisulphite}}{>C=CH-CH(SO_3Na)-OH} \quad \underset{\text{Dibisulphite}}{>C(SO_3Na)-CH_2-CH(SO_3Na)-OH}$$

Further the α, β-nature of one of the double bonds is proved by prolonged heating with aqueous potassium carbonate when citral undergoes hydration and *retro-aldol reaction* to give 6-methylhept-5-ene-2-one which on ozonolysis yields acetone and levulinic acid. Such a reaction is given by α, β-unsaturated carbonyl compounds.

Citral $\xrightarrow{H_2O,\ K_2CO_3}$ (hydrated intermediate: O—H, $\bar{O}H$, CH_2—CHO) ⟶ 6-methylhept-5-ene-2-one + $\bar{C}H_2CHO$ $\xrightarrow{H_2O}$ CH_3CHO

6-methylhept-5-ene-2-one $\xrightarrow{O_3}$ Levulinic acid (CO_2H) + $CH_3—CO—CH_3$

The structure of 6-methylhept-5-ene-2-one is proved by the following synthesis.

Acetone $\xrightarrow{HC\equiv CNa}$ HO—C(CH$_3$)$_2$—C≡CH $\xrightarrow{Zn\text{-}Cu/H_2O}$ HO—C(CH$_3$)$_2$—CH=CH$_2$ $\xrightarrow{PBr_3}$ Br—CH$_2$CH=C(CH$_3$)$_2$ $\xrightarrow[NaOEt]{CH_3CO.CH_2CO_2Et}$ EtO$_2$C—CH(COCH$_3$)—CH$_2$CH=C(CH$_3$)$_2$ $\xrightarrow{\text{dil. NaOH}}$ 6-methylhept 5-ene-2-one

7. On heating with $KHSO_4$ citral forms *p*-cymene, a compound of known structure. Thus the relative position of the methyl group and the isopropylidene moiety are decided in citral. The conversion of citral to *p*-cymene can be visualized as under :

CHO
Citral
$H^+(KHSO_4)$
CH = $\overset{+}{O}H$
CH
$\overset{+}{C}H$ — OH
Cyclization
Hydride shift
H
OH
H
OH
$\overset{\oplus}{-H}$
H
H
OH
H
$-H_2O$
p-Cymene

8. The structure arrived at from the above experiments is supported by oxidation of citral with alkaline permanganate followed by chromic acid, when acetone, oxalic acid and levulic acid are obtained.

CHO
IV
$KMnO_4$
OH
OH
CHO
OH
OH
CrO_3
O
CO_2H
+
O
+
CO_2H
|
CO_2H

9. *Synthesis.* The above structure of citral is confirmed by its synthesis from 6-methyl hept-5-ene-2-one through two different routes.

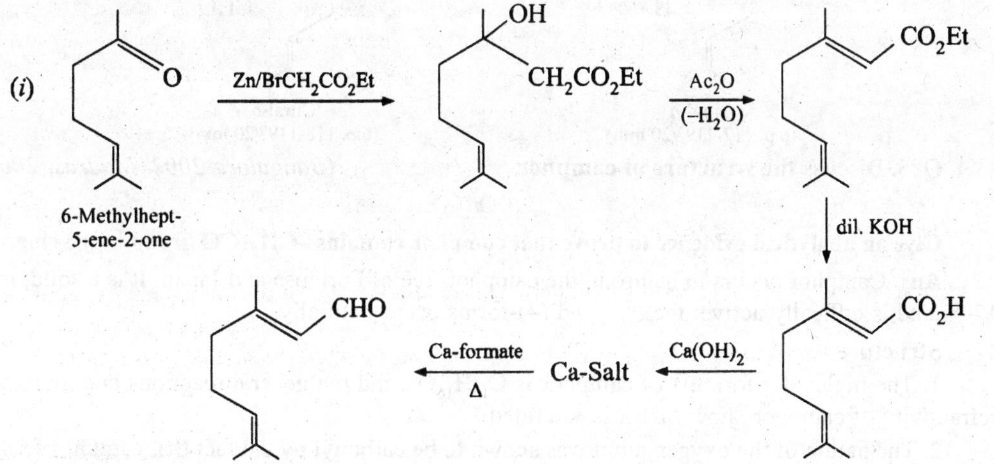

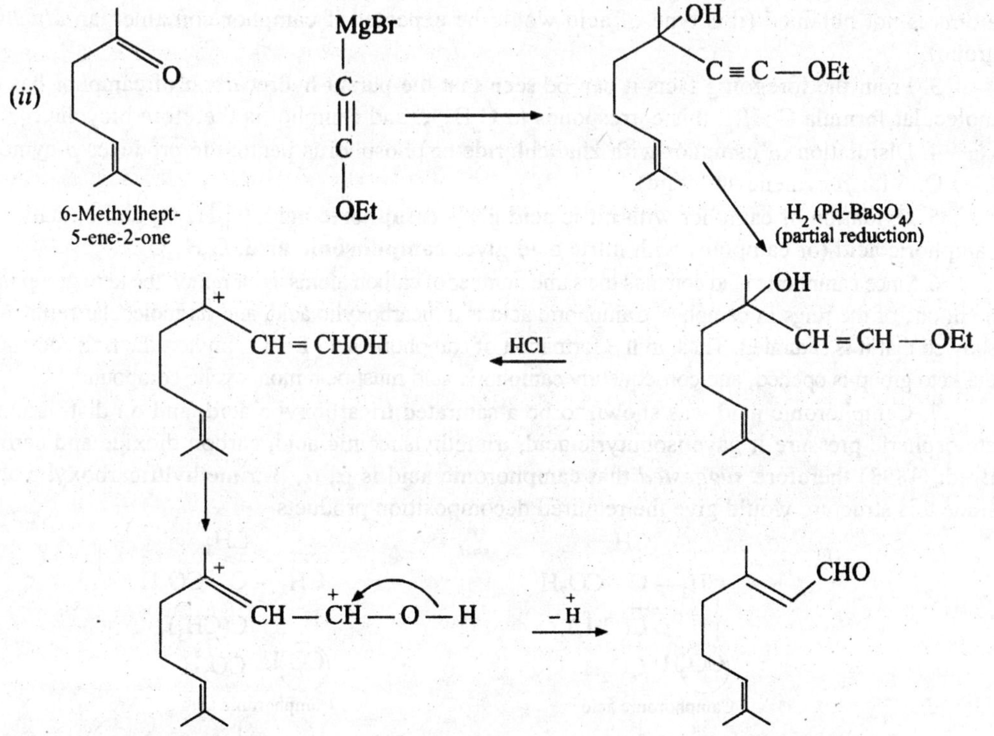

(*i*) 6-Methylhept-5-ene-2-one $\xrightarrow{Zn/BrCH_2CO_2Et}$ (OH, CH_2CO_2Et) $\xrightarrow[(-H_2O)]{Ac_2O}$ (CO_2Et) $\xrightarrow{\text{dil. KOH}}$ (CO_2H) $\xrightarrow{Ca(OH)_2}$ Ca-Salt $\xrightarrow[\Delta]{\text{Ca-formate}}$ (CHO)

(*ii*) 6-Methylhept-5-ene-2-one + MgBr—C≡C—OEt $\longrightarrow$ (OH, C ≡ C — OEt) $\xrightarrow[\text{(partial reduction)}]{H_2\ (Pd\text{-}BaSO_4)}$ (OH, CH = CH — OEt) $\xrightarrow{HCl}$ (+, CH = CHOH) $\longrightarrow$ (+, CH — $\overset{+}{C}$H — O — H) $\xrightarrow{-\overset{+}{H}}$ (CHO)

10. *Cis-trans isomerism in citral.* Since one of the double bonds has different substituents at each end, citral shows *cis-trans* isomerism. One of the isomers has the aldehydic group and the main chain on the same side and is known as citral-*b* and the other in which they are on opposite sides is called citral-*a*. Both occur together in nature.

Citral-*a*
(b.p. 117-119°/20 mm)

Citral-*b*
(b.p. 118-119°/20 mm)

Q. 5. Discuss the structure of camphor. (*Bangalore 2002, Madras, 2004*)

Or

Give an analytical evidence to prove that comphor contains –CH_2–CO group in the ring.

Ans. Camphor occurs in nature in the camphor tree of Formosa and Japan. It is a solid, m.p. 179°, and is optically active; the (+)- and (–)-forms occur naturally.

Structure

1. The molecular formula of camphor is $C_{10}H_{16}O$, and the general reactions and molecular refractivity of camphor show that it is saturated.

2. The nature of the oxygen atom was shown to be carbonyl by the fact that camphor formed an oxime and that it was a keto group was decided from the fact that oxidation of camphor gives a *dicarboxylic* acid containing 10 carbon atoms; a *monocarboxylic* acid containing 10 carbon atoms is not obtained (this type of acid would be expected if camphor contained an *aldehyde* group).

3. From the foregoing facts it can be seen that the parent hydrocarbon of camphor has the molecular formula $C_{10}H_{18}$; this corresponds to C_nH_{2n-2}, and camphor is therefore bicyclic.

4. Distillation of camphor with zinc chloride or phosphorus pentoxide produces *p*-cymene. (See Q. 6 for *p*-cymene structure).

5. Oxidation of camphor with nitric acid gives **camphoric acid**, $C_{10}H_{16}O_4$ and oxidation of camphoric acid (or camphor) with nitric acid gives **camphoronic acid**, $C_9H_{14}O_6$.

6. Since camphoric acid contains the same number of carbon atoms as camphor, the keto group must be in one of the rings in camphor. Camphoric acid is a dicarboxylic acid, and its molecular refractivity showed that it is saturated. Thus, in the formation of camphoric acid from camphor, the ring containing the keto group is opened, and consequently camphoric acid must be a monocyclic compound.

7. Camphoronic acid was shown to be a saturated tricarboxylic acid, and on distillation at atmospheric pressure it gave isobutyric acid, trimethylsuccinic acid, carbon dioxide and carbon. Bredt (1893) therefore *suggested* that camphoronic acid is α, α, β-trimethyltricarboxylic acid, since this structure would give the required decomposition products.

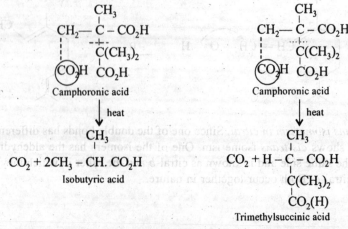

$$\begin{array}{c} CH_3 \\ | \\ CH_2-C-CO_2H \\ | \quad\; | \\ | \quad C(CH_3)_2 \\ | \quad\; | \\ CO_2H \;\; CO_2H \end{array}$$

Camphoronic acid

↓ heat

$$\begin{array}{c} CH_3 \\ | \\ CO_2 + 2CH_3-CH.\,CO_2H \end{array}$$

Isobutyric acid

$$\begin{array}{c} CH_3 \\ | \\ CH_2-C-CO_2H \\ | \quad\; | \\ | \quad C(CH_3)_2 \\ | \quad\; | \\ CO_2H \;\; CO_2H \end{array}$$

Camphoronic acid

↓ heat

$$\begin{array}{c} CH_3 \\ | \\ CO_2 + H-C-CO_2H \\ | \\ C(CH_3)_2 \\ | \\ CO_2(H) \end{array}$$

Trimethylsuccinic acid

8. Hence, if camphoronic acid has above structure, then camphoric acid (and camphor) must contain *three methyl groups*. On this basis, the formula of camphoric acid, $C_{10}H_{16}O_4$, can be written as $(CH_3)_3C_5H_5(CO_2H)_2$. The parent (saturated) hydrocarbon of this is C_5H_{10}, which corresponds to $C_n H_{2n}$, *i.e.*, camphoric acid is a *cyclo*-pentane derivative (this agrees which the previous evidence that camphoric acid is monocyclic). Thus the oxidation of camphoric acid to camphoronic acid may be written:

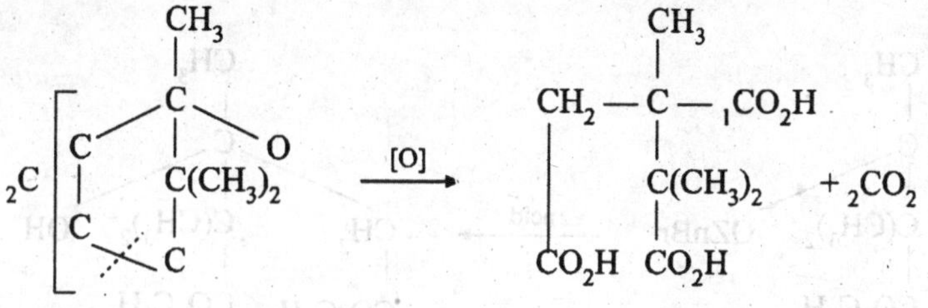

9. This skeleton, plus one carbon atom, arranged with two carboxyl groups, will therefore be the structure of camphoric acid. Now camphoric anhydride forms only one monobromo derivative (bromine and phosphorus); therefore there is only *one* α-hydrogen atom in camphoric acid. Thus the carbon atom of one carboxyl group must be $_1C$ (this is the only carbon atom joined to a tertiary carbon atom). Furthermore, $_1C$ must be the carbon of the keto or methylene groups in camphor, since it is these two groups which produce the two carboxyl groups in camphoric acid. The problem is now to find the position of the other carboxyl group in camphoric acid. Its position must be such that when the *cyclo* pentane ring is opened to give camphoronic acid, one carbon atom is readily lost. Using this as a working hypothesis, there are only two reasonable structures for camphoric acid, I and II, I may be rewritten as Ia, and since the two carboxyl groups are produced from the $-CH_2.CHO-$ group in camphor, the precursor of Ia (*i.e.*, camphor) will contain

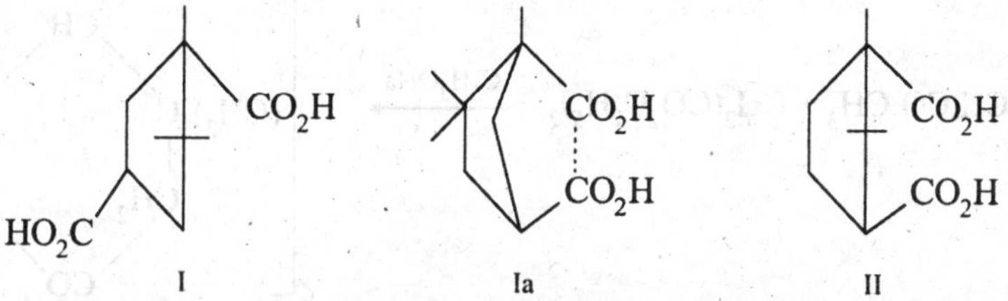

I Ia II

a six-membered ring with a *gem*-dimethyl group. This structure cannot account for the conversion of camphor into *p*-cymene. One the other hand, II accounts for all the facts given in the foregoing discussion. Bredt therefore assumed that II was the structure of camphoric acid, and that VI was the structure of camphor.

10. The relationships between camphor, camphoric acid and camphoronic acid, are depicted as under:

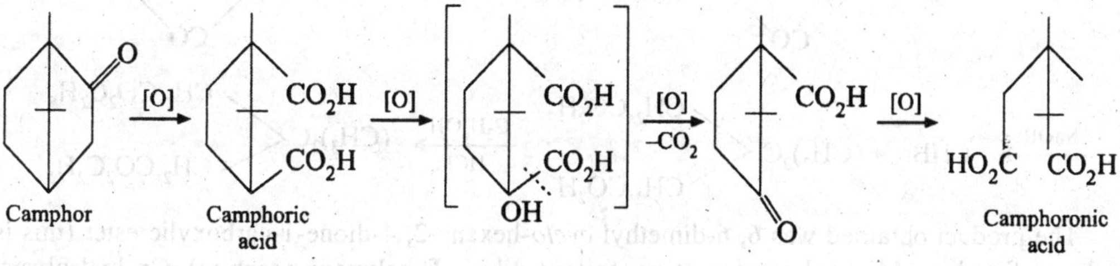

Camphor Camphoric acid Camphoronic acid

11. The structure of camphor was finally confirmed by the synthesis of camphoronic acid, camphoric acid and camphor.

Synthesis of (±)-camphoronic acid.

$$CH_3COCH_2CO_2C_2H_5 \xrightarrow[(ii)\ CH_3I]{(i)\ C_2H_5ONa} CH_3COCH(CH_3)CO_2C_2H_5 \xrightarrow[(ii)\ CH_3I]{(i)\ C_2H_5ONa} CH_3COC(CH_3)_2CO_2C_2H_5 \xrightarrow[\text{(Reformatsky reaction)}]{Zn + CH_2Br.CO_2C_2H_5}$$

$$C_2H_5O_2C.CH_2-C(CH_3)(OZnBr)-C(CH_3)_2.CO_2C_2H_5 \xrightarrow{\text{acid}} C_2H_5O_2C.CH_2-C(CH_3)(OH)-C(CH_3)_2.CO_2C_2H_5 \xrightarrow[(ii)\ KCN]{(i)\ PCl_5}$$

$$C_2H_5O_2C.CH_2-C(CH_3)(CN)-C(CH_3)_2.CO_2C_2H_5 \xrightarrow{HCl} HO_2C.CH_2-C(CH_3)(CO_2H)-C(CH_3)_2.CO_2H$$

Synthesis of (±)-camphoric acid. Komppa (1899) first synthesised β, β-dimethylglutaric ester as follows, starting with mesityl oxide and ethyl malonate.

$$(CH_3)_2C=CH.CO.CH_3 + CH_2(CO_2C_2H_5)_2 \xrightarrow{C_2H_5ONa} \left[(CH_3)_2C(CH_2COCH_3)-CH(CO_2C_2H_5)_2\right]$$

$$\xrightarrow{C_2H_5ONa} \text{(ring: } (CH_3)_2C-CH(CO_2C_2H_5)-CO-CH_2-CO-CH_2\text{)} \xrightarrow[(ii)\ HCl]{(i)\ Ba(OH)_2} \text{(ring: } (CH_3)_2C-CH_2-CO-CH_2-CO-CH_2\text{)}$$

$$\xrightarrow{NaOBr} CHBr_3 + (CH_3)_2C{<}^{CH_2.CO_2H}_{CH_2.CO_2H} \xrightarrow[HCl]{C_2H_5OH} (CH_3)_2C{<}^{CH_2.CO_2C_2H_5}_{CH_2.CO_2C_2H_5}$$

The product obtained was 6, 6-dimethyl *cyclo*-hexane-2, 4-dione-1-carboxylic ester (this is produced first by a Michael condensation, followed by a Dieckmann reaction). On hydrolysis, followed by oxidation with sodium hypobromite, β, β-dimethylglutaric acid was obtained.

Komppa (1903) then prepared comphoric acid as follows:

$CO_2C_2H_5$ | $CO_2C_2H_5$ + $CH_2.CO_2C_2H_5$ | $C(CH_3)_2$ | $CH_2.CO_2C_2H_5$ $\xrightarrow{C_2H_5ONa}$ Diketoapocamphoric ester

$\xrightarrow[(ii)\ CH_3I]{(i)\ Na}$ Diketocamphoric ester $\xrightarrow[NaOH]{Na\text{-}Hg}$ $\xrightarrow{HI}$ $\xrightarrow{HBr}$ $\xrightarrow[CH_3CO_2H]{Zn}$

Synthesis of camphor (Haller, 1896). Haller started with camphoric acid prepared by the oxidation of camphor, but since the acid was synthesised later by Komppa, we now have a total synthesis of camphor.

Camphoric acid $\xrightarrow{CH_3COCl}$ Camphoric anhydride $\xrightarrow{Na\text{-}Hg}$ α-campholide $\xrightarrow{KCN}$

$\xrightarrow{hydrolysis}$ Homocamphoric acid $\xrightarrow[heat]{Ca\ Salt}$ Camphor

Q. 6. Write down boat structures for camphor, borneol and isoborneol.

Ans. The boat structures for camphor, borneol and isoborneol are given as under.

Camphor Borneol Isoborneol

Q. 7. Give the method of commercial preparation of camphor.

Ans. Synthetic camphor is usually obtained as the racemic modification. The starting material is α-pinene, and the formation of camphor involves the Wagner-Meerwein rearrangements. Two different routes, as given below are possible.

(*i*) α-Pinene $\xrightarrow[10°C]{HCl\ gas}$ Bornyl chloride $\xrightarrow[-HCl]{CH_3.CO_2Na}$ Camphene $\xrightarrow[H_2SO_4]{CH_3.CO_2H}$ *iso* Bornyl acetate $\xrightarrow{NaOH}$ *iso* Borneol $\xrightarrow{C_6H_5NO_2}$ Camphor.

α-Pinene $\xrightarrow[10°C]{HCl\ gas}$ Bornyl chloride $\xrightarrow[-HCl]{CH_3.CO_2Na}$ Camphene $\xrightarrow{H.CO_2H}$ *iso* Bornyl formate $\xrightarrow{NaOH}$ *iso* Borneol $\xrightarrow[Ni,\ 200°C]{O_2}$ Camphor.

Q. 8. Describe the synthesis of carvone.

Ans. Carvone is a terpenoid having the structure

O

Its synthesis involves the following three steps:

1. Synthesis of δ-keto-hexahydrobenzoic acid. β-iodopropionic ester is treated with the di-sodium derivative of cynoacetic ester which is very reactive at ordinary temperature to form γ-cyanopentane α, μ-tricarboxylic ester in presence of sodium ethoxide in alcohol.

$$\begin{matrix} H_5C_2-COOCH_2-CH_2-I \\ H_5C_2-COOCH_2-CH_2-I \end{matrix} + Na_2C\begin{matrix} CN \\ COOC_2H_5 \end{matrix}$$

β-iodopropionic ester

$$\downarrow$$

$$\begin{matrix} H_5C_2-COOCH_2-CH_2 \\ H_5C_2-COOCH_2-CH_2 \end{matrix} > C < \begin{matrix} CN \\ COOC_2H_5 \end{matrix}$$

$$\downarrow \text{Conc. HCl}$$

$$\begin{matrix} HOOC-CH_2-CH_2 \\ HOOC-CH_2-CH_2 \end{matrix} > C < \begin{matrix} COOH \\ COOH \end{matrix}$$

$$\downarrow \text{(i) Boil with acetic anhydride (ii) Distil}$$

$$O=C < \begin{matrix} CH_2-CH_2 \\ CH_2-CH_2 \end{matrix} > CH-COOH + CO_2 + H_2O$$

δ-keto hexahydrobenzoic acid

2. Conversion of δ-keto hexahydrobenzoic acid into α-terpineol. The ester of δ-ketohexahydrobenzoic acid is made to react with methyl magnesium iodide (CH_3MgI). The keto group is attacked and δ-hydroxy hexahydrotoluic ester is formed.

$$H_5C_2OOC-CH < \begin{matrix} CH_2-CH_2 \\ CH_2-CH_2 \end{matrix} > C=O + CH_3MgI$$

$$\longrightarrow H_5C_2OOC-CH < \begin{matrix} CH_2-CH_2 \\ CH_2-CH_2 \end{matrix} > C < \begin{matrix} CH_3 \\ OH \end{matrix}$$

δ-hydroxy hexahydrotoluic ester

The hydroxyl group of this ester is replaced by bromine by the action of fuming HBr and subsequently a molecule of hydrobromic acid is eliminated by means of pyridine to give an ester. The ester so formed is treated with excess of methyl magnesium iodide. The ester group is attacked and the product formed on hydrolysis yields α-Terpineol.

$CH_3-C(OH)<(CH_2-CH_2)(CH_2-CH_2)>CH-COOC_2H_5 \xrightarrow{HBr/\Delta} CH_3-C(Br)<(CH_2-CH_2)(CH_2-CH_2)>CH-COOC_2H_5$

Pyridine ↓

$CH_3-C<(CH_2-CH_2)(=CH-CH_2)>CH-COOC_2H_5$ (Ester) $\xrightarrow{CH_3MgI} CH_3-C<(CH_2-CH_2)(=CH-CH_2)>CH-C(=O)-CH_3$

Excess of CH_3MgI ↓

$CH_3-C<(CH_2-CH_2)(=CH-CH_2)>CH-C(OH)(CH_3)_2$ or

α-Terpineol

Alternative method to obtain α-terpineol

Meldrum and Fischer obtained α-terpineol starting from p-toluic acid as given below:

COOH, CH₃ (*p*-toulic acid) $\xrightarrow{H_2SO_4}$ COOH, SO_3H, CH_3 $\xrightarrow{KOH}$ COOH, OH, CH_3 $\xrightarrow[\text{Alcohol}]{Na}$ COOH, OH, CH_3

$\xrightarrow{HBr}$ COOH, Br, CH_3 $\xrightarrow[\text{Pyridine (–HBr)}]{\text{Heat in}}$ COOH, CH_3 $\xrightarrow[HCl]{C_2H_5OH}$ $COOC_2H_5$

$\xrightarrow{2CH_3MgI}$ OMgI $\xrightarrow{\text{Acid}}$ OH

(±) α-terpineol

3. Conversion of α-terpineol into carvone

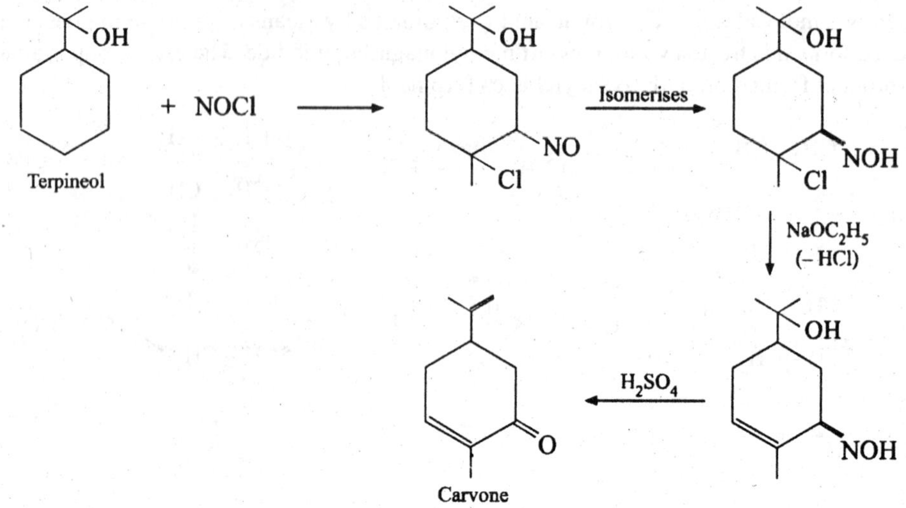

Terpineol

+ NOCl

OH

NO

Cl

Isomerises

NOH

$NaOC_2H_5$
(– HCl)

H_2SO_4

O

Carvone

STEROIDS

Q. 9. What are steroids?

Ans. Steroids are structurally related compounds which are widely distributed in plants and animals. Sterols, bile acids, sex hormones, vitamin D, adrenal cortex hormones and some carcinogenic hydrocarbons are included in the category of steroids. The structure of steroids is based on 1, 2-cyclopentenophenanthrene skeleton. All steroids on selenium dehydrogenation form Diel's hydrocarbon. **A steroid can be defined as a compound which on selenium dehydrogenation at 360°C forms Diel's hydrocarbon.** At higher temperatures, i.e. at 420°C, steroids on dehydrogenation form chrysene. The studies on Diel's hydrocarbon reveal that it is 3′–methyl-1, 2-cyclopentenophenanthrene. The structure of Diel's hydrocarbon is further confirmed by synthesis as follows:

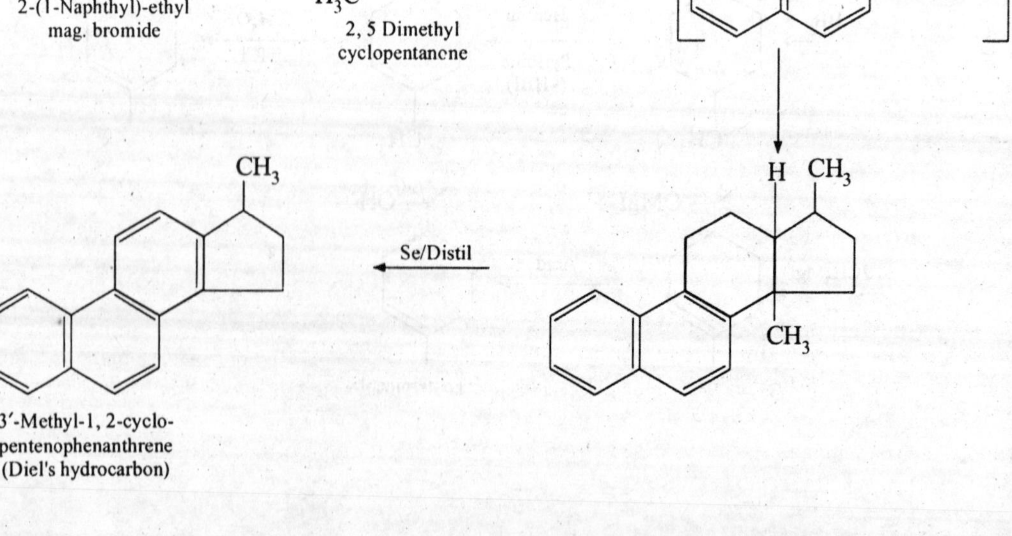

$CH_2 — CH_2MgBr$

2-(1-Naphthyl)-ethyl mag. bromide

+

CH_3

H_3C

2, 5 Dimethyl cyclopentanone

P_2O_5/140°C

Reduced pressure

CH_3

CH_3

H CH_3

CH_3

Se/Distil

CH_3

3′-Methyl-1, 2-cyclopentenophenanthrene (Diel's hydrocarbon)

Q. 10. Write short notes on

(*a*) Cholesterol **(*b*)Ergosterol**

Ans. (*a*) Cholesterol : Cholesterol, $C_{27}H_{46}$ O is one of the most widely distributed sterols which is found in abundance in the brain, spinal cord and gallstones and in small amounts in almost all animal tissues.

Sterols are solid alcohols having a hydroxyl group at position 3, a double bond between C_5 and C_6, a side chain at C_{17}, and methyl groups linked to C_{10} and C_{13}.

Deposition of cholesterol in the arteries block the flow of blood. It results into high blood pressure and some form of cardiovascular disease.

Structure of Cholesterol

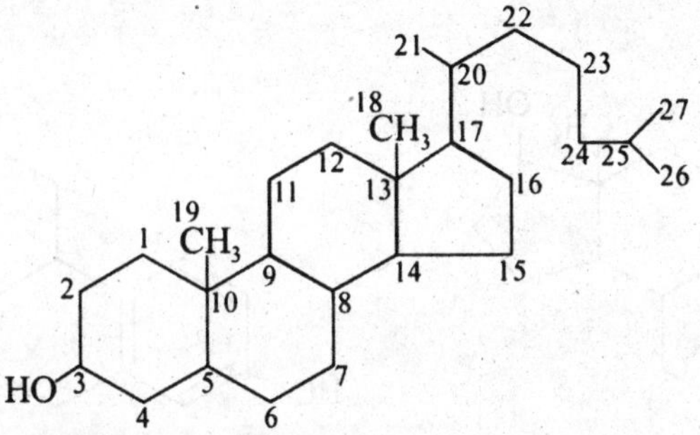

22 21 20 23 18 CH_3 17 27 24 25 26 12 16 11 13 19 1 CH_3 9 14 15 2 10 8 HO 3 5 7 4 6

(*b*) Ergosterol

A large number of compounds resembling cholesterol occur in plants. These compounds are collectively known as phytosterols. A member of this class of compounds is *ergosterol* (C_{28} $H_{44}O$). It is produced by yeast. On irradiation, ergosterol yields calciferol or vitamin D_2. The structure of ergosterol is given below.

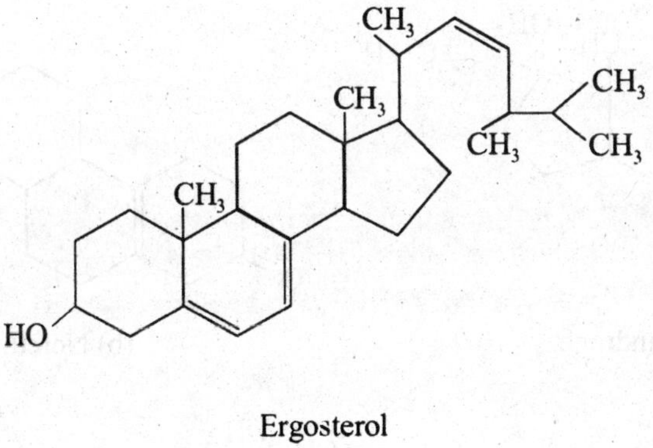

CH_3 CH_3 CH_3 CH_3 CH_3 CH_3 HO

Ergosterol

Q.11. Write a note on sex hormones. (*Mumbai, 2010*)

Ans. Male and female sex hormones are steroidal compounds. They account for the development of sex characteristics and sexual processes in animal. Sex hormones are produced in the ovaries and testes as a natural process. The female sex hormones control menstrual cycle, the changes in the uterus and pregnancy related matters. The structure of principal sex hormones are given below.

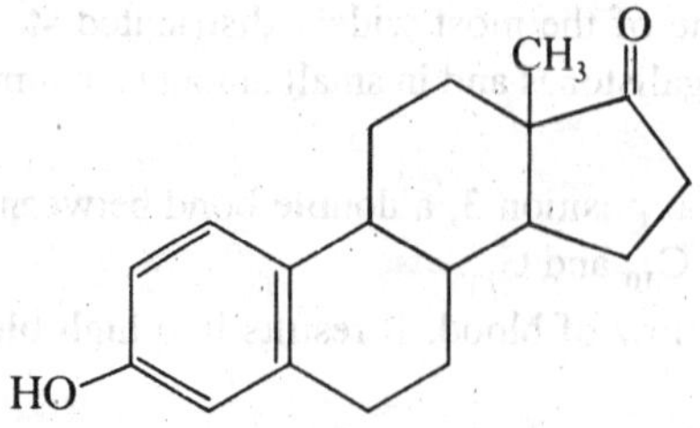

Estrone

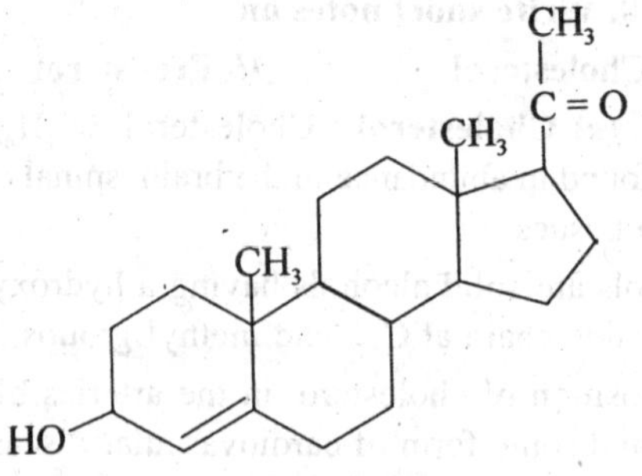

Progesterone

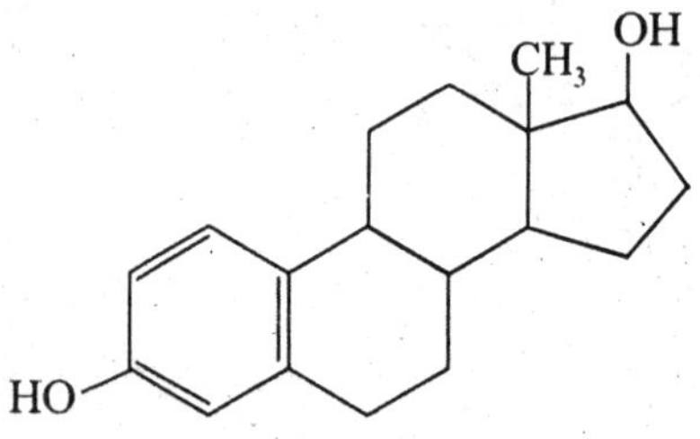

Estradiol

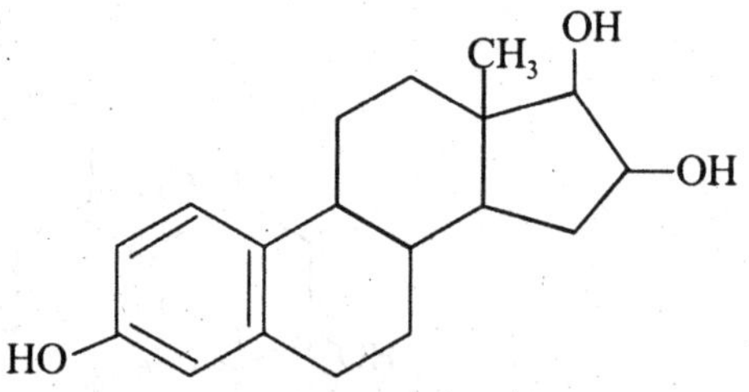

Estriol

Oral contraceptives are synthetic compounds resembling progesterone and estriol in structure. These compounds assimilate easily into the bloodstream and suppress ovulation and thus prevent pregnancy.

The structures of two synthetic progesterone (a) and (b) and synthetic estrogen (c) are given below.

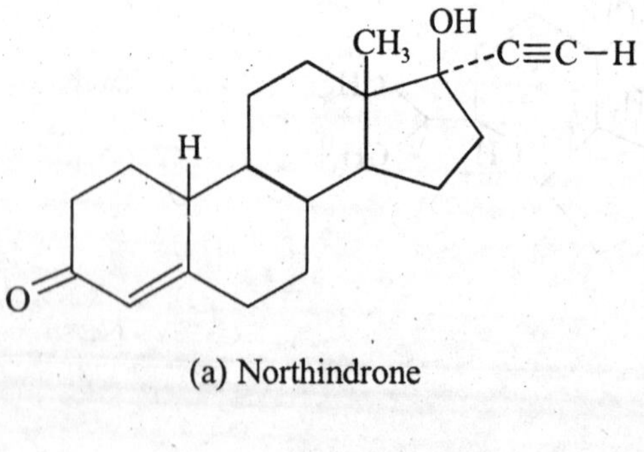

(a) Northindrone

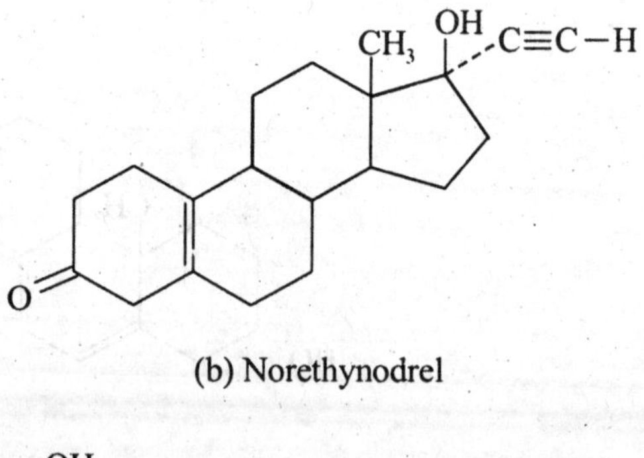

(b) Norethynodrel

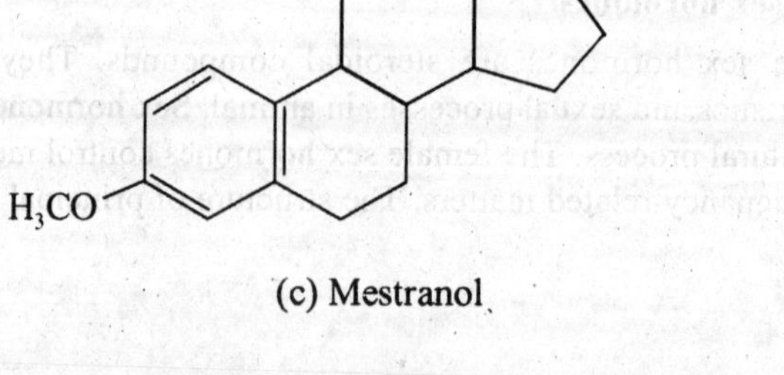

(c) Mestranol

It is either the combination of (a) + (c), called orthonovum or (b) + (c), called enovide, which is used commercially as oral contraceptive to prevent pregnancy in women.

The male sex hormones, known as androgens are similar in structure to female sex hormones, the difference being that male hormones do not contain a benzene ring. Total number of rings in the two categories is, however, the same. The male sex hormones are responsible for the developemnt of male genital tract and male characteristics including voice, beard, etc.

The structure of male sex hormones are given below.

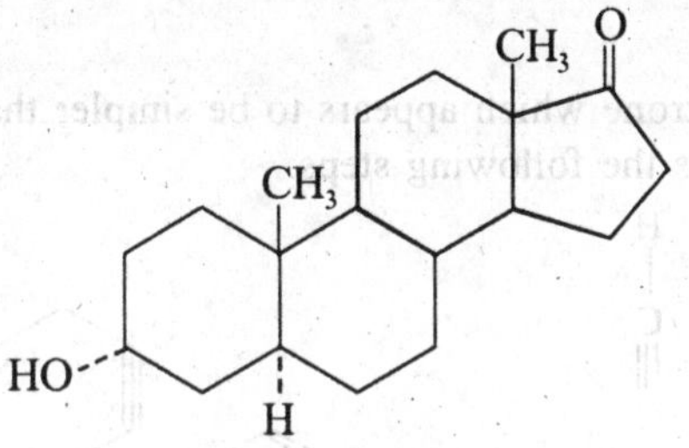

Androsterone

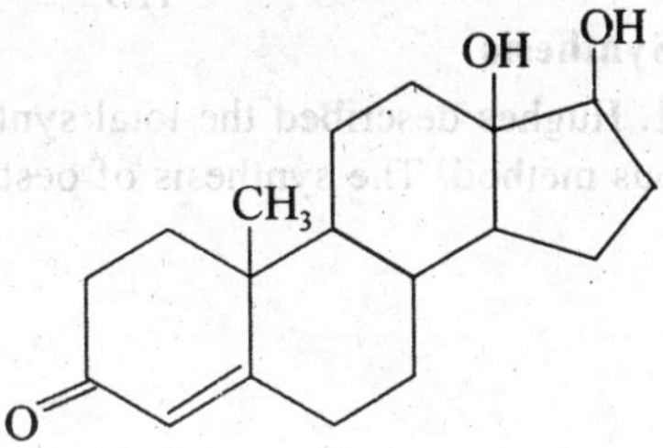

Testosterone

Q. 12. Write a note on adrenal steroids.

Ans. Steroids prodcued by adrenal cortex are called adrenal steroids. Cortisone and its derivative 17-hydroxycorticosterone are the two steroids in this category. These compounds have been found effective in the treatment of inflammatory and allergic problems. Their structures are given below:

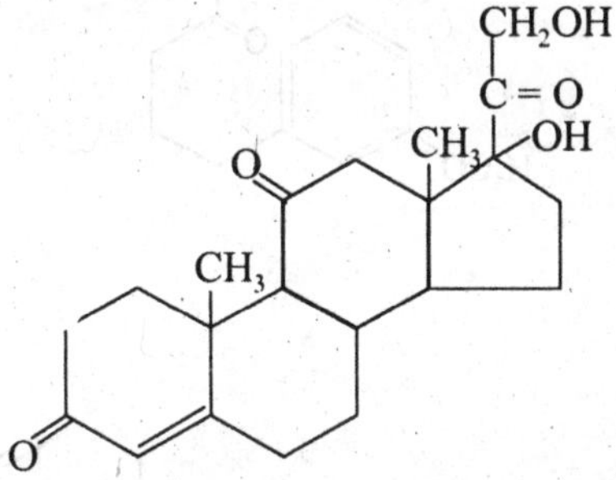

Cortisone

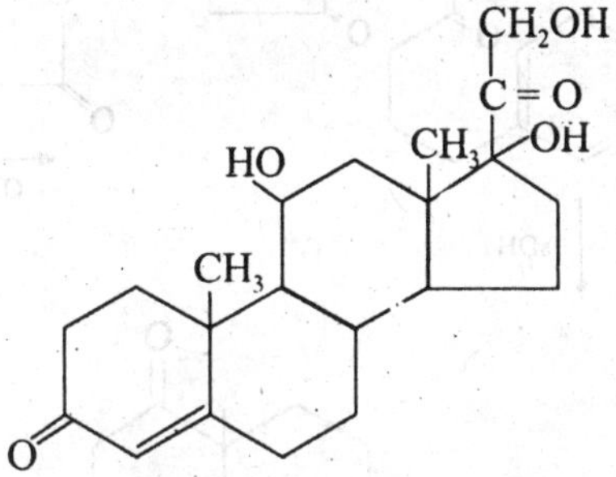

17-Hydroxycorticosterone

Q. 13. What are anabolic steroids?

Ans. It is a group of compounds structurally and chemically resembling male sex hormones. These synthetic compounds were developed in 1930's to supplement natural hormones in men whose bodies produced insufficient amounts of such hormones. Masculine feautres occuring at puberty, such as change of voice and growth of hair are observed only if the body produces enough male sex hormones.

These anabolic steroids were meant to treat men who for lack of sufficient of male sex hormones were unable to develop male characteristics, etc.

But, during the past many years, it has been observed that these steroids cause serious side effects demaging different parts of the body and causing several complications. Hence, their use has been on the decline now.

Q. 14. Describe synthesis of estrone (Oestrone). *(Panjab, 2003)*

Ans. The molecular formula of estrone is $C_{18}H_{22}O_2$. It is a female sex hormone and controls the uterine cycle. It has the structural formula as given below:

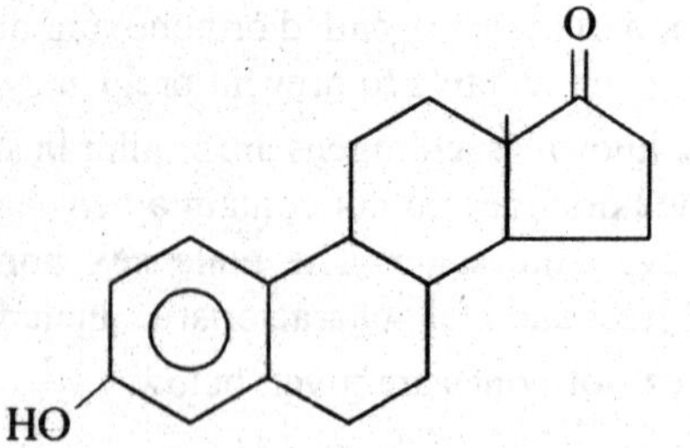

HO

Synthesis

1. Hughes described the total synthesis of oestrone which appears to be simpler than any previous method. The synthesis of oestrone involves the following steps:

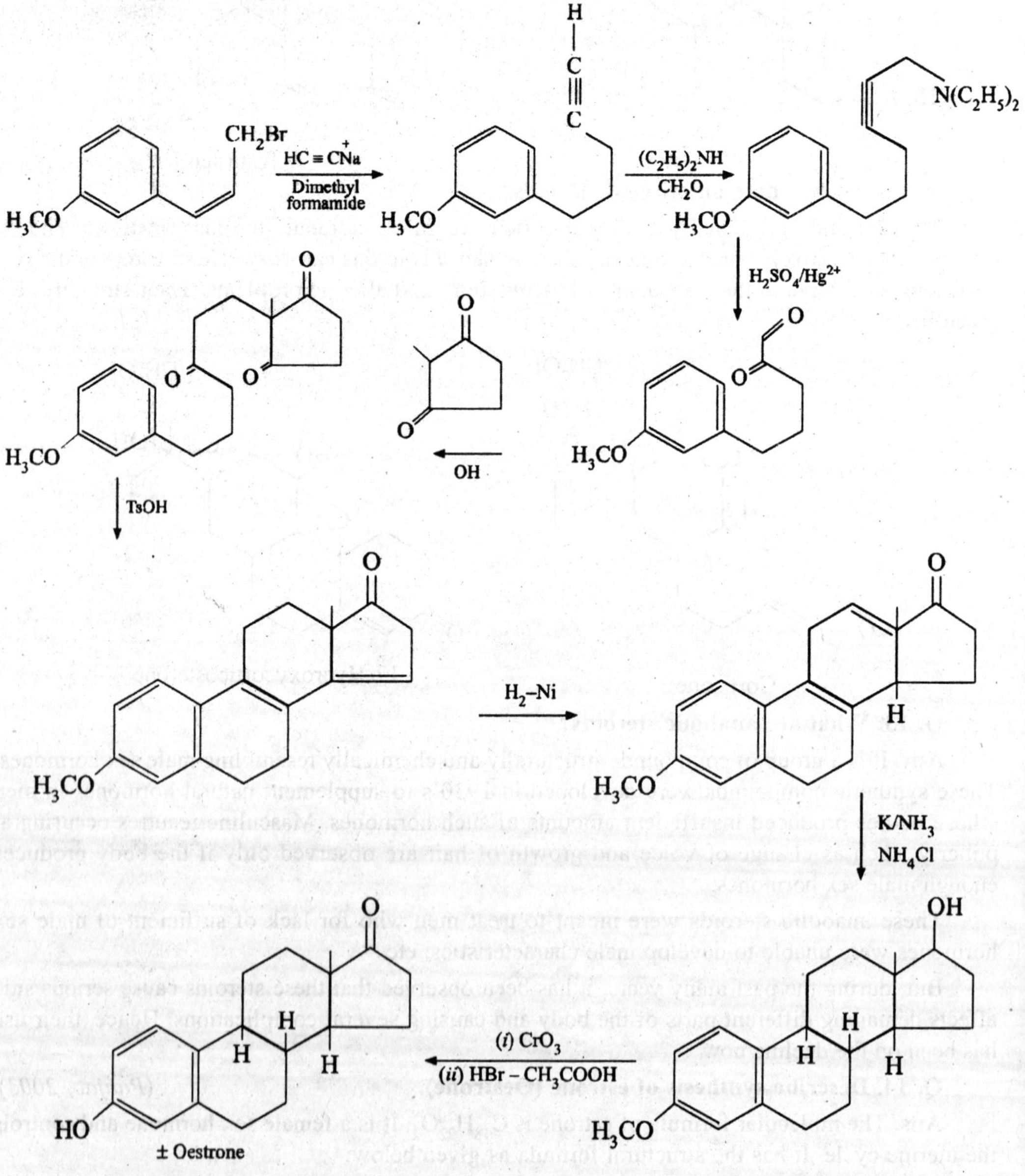

CH_2Br

$HC \equiv C\overset{-}{N}\overset{+}{a}$

Dimethyl formamide

H_3CO

H

C

C

H_3CO

$(C_2H_5)_2NH$

CH_2O

$N(C_2H_5)_2$

H_3CO

H_2SO_4/Hg^{2+}

O

O

H_3CO

OH

O

O

O

O

O

H_3CO

TsOH

O

H_3CO

H_2–Ni

O

H

H_3CO

K/NH$_3$

NH_4Cl

OH

H

H

H

H_3CO

(i) CrO_3

(ii) HBr – CH_3COOH

O

H

H

H

HO

± Oestrone

2. Torgov *et at* (1960-62) gave the synthesis of oestrone which is described as under.

$CH_2 = CH.MgBr$

H_3CO OH

TsOH

H_2–Ni

(*i*) K/NH_3
(*ii*) CrO_3

$C_5H_5N.HCl$
Pyridine hydrochloride

HO

(±) Oestrone

Q. 15. Establish the structure of cholesterol.

Ans. This is the sterol of higher animals occurring free or as fatty ester in all animal cells, particularly in brain and spinal cord. The main sources of cholesterol are fish-lever oils and the brains and spinal cord of cattle.

Cholesterol is a white crystalline solid and is optically active. The structure of cholesterol was elucidated after a tremendous work done by Wieland, Windaus and their co-workers (1903-32). The numbering of the carbon atoms in the established structure of cholesterol is described below:

The molecule consists of (*i*) a nucleus which is composed of four rings (*ii*) a side chain at C_{17}-position (*iii*) a hydroxyl group at C_3 (*iv*) a double bond at C_5–C_6 and (*v*) angular methyl groups at C_{10} and C_{13} positions.

Structure educidation

1. Molecular formula. The study of the analytical data reveals that the molecular formula of cholesterol is $C_{27}H_{46}O$.

2. Presence of double bond and hydroxyl group. The usual tests for functional groups show that cholesterol contains one double bond and one hydroxyl group.

(*i*) The conversion of cholesterol into cholestanol shows the presence of double bond.

$$\underset{\text{Cholesterol}}{C_{27}H_{46}O} \xrightarrow{H_2-Pt} \underset{\text{Cholestanol}}{C_{27}H_{48}O}$$

(*ii*) The oxidation of cholestanol with CrO_3 into cholestanone shows that cholesterol has a secondary alcoholic group.

$$\underset{\text{Cholestanol}}{C_{27}H_{48}O} \xrightarrow{CrO_3} \underset{\text{Cholestanone}}{C_{27}H_{46}O} \quad \text{(a ketone)}$$

(*iii*) Further reduction of cholestanone with Zn-Hg/conc. HCl (Clemmenson's reduction) yields cholestane which is a saturated hydrocarbon.

$$\underset{\text{Cholestanone}}{C_{27}H_{46}O} \xrightarrow{Zn/Hg-HCl} \underset{\text{Cholestane}}{C_{27}H_{48}}$$

The molecular formula of cholestane corresponds to the general formula C_nH_{2n-6}. Thus, cholestane and hence cholesterol is tetracyclic in nature.

A. Structure of ring system. When cholesterol is distilled with selenium at 360°C, Diel's hydrocarbon is obtained.

$$\underset{\text{Cholesterol}}{C_{27}H_{46}O} \xrightarrow{Se/360°C} \text{3'-Methyl-1, 2-cyclopentenophenanthrene (Diel's hydrocarbon)} + \text{Chrysene}$$

This shows the presence of cyclopentenophenanthrene nucleus in cholesterol and thus cholesterol is a steroid. The various rings in the steroid nucleus were opened to form discarboxylic acid. The relative positions of the two carboxylic groups with respect to each other were determined by the application of **Blanc's rule** which states : *On heating with acetic anhydride, 1, 5-dicarboxylic acids form cyclic anhydrides and 1, 6-dicarboxylic acids form cyclopentanones with the elimination of carbon dioxide.*

Ring A. Cholesterol and cholic acid* were converted into the dicarboxylic acid (A) which on heating with acetic anhydride gave a cyclopentanone. This shows that **the ring A in cholesterol is six membered.**

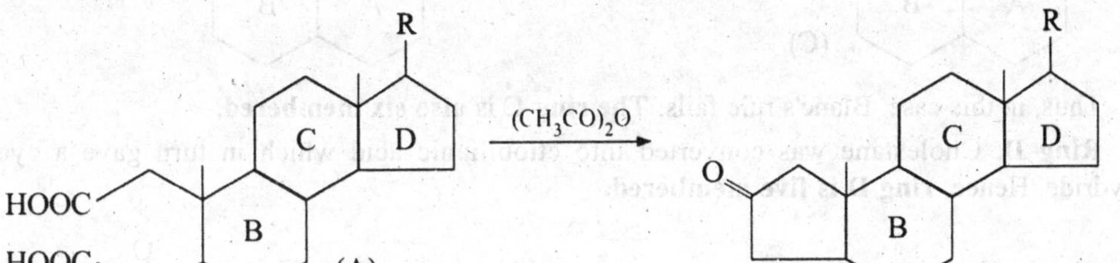

R C D HOOC HOOC B (A) $\xrightarrow{(CH_3CO)_2O}$ R C D O B

Ring B. Now cholesterol was converted into tricarboxylic acid (B) which on heating with acetic anhydride gave cyclopentanone derivative as shown below. This reaction shows that the **ring B is six membered.**

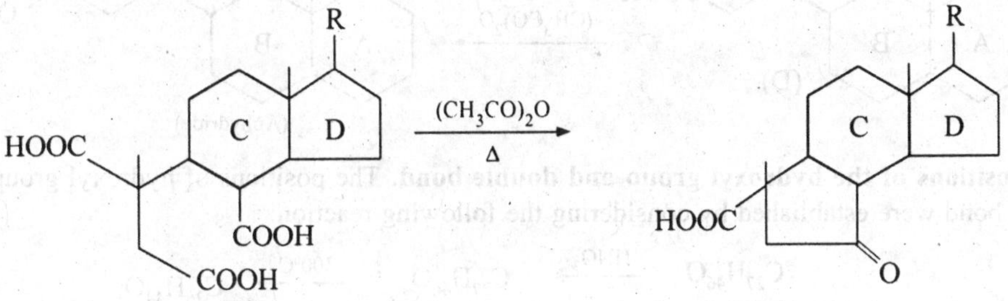

R C D HOOC COOH COOH (B) Tricarboxylic acid $\xrightarrow[\Delta]{(CH_3CO)_2O}$ R C D HOOC O

Ring C. Deoxycholic acid* was converted into a dicarboxylic acid which gave a cyclic anhydride. Thus, the ring C was assumed to be five membered. On the basis of this, Windaus and Wieland proposed the following formula for cholesterol.

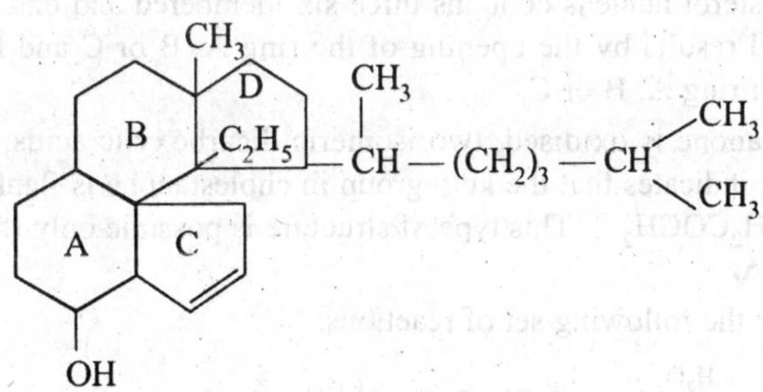

CH_3 D B C_2H_5 CH_3 $CH-(CH_2)_3-CH(CH_3)_2$ A C OH

The uncertain point was the nature of the two extra carbon atoms. They were thought to be present as ethyl group at C_{10} position. But Wieland *et al* (1930) proved that there was no ethyl group at this position. These two `homeless' carbon atoms were not placed until it was proposed that steroids contain chrysene nucleus and cyclopentenophenanthrene structure was suggested. If we use the correct structure of cholesterol, the cyclisation reaction results in the formation of seven membered cyclic anhydride.

*Cholesterol, Cholic acid and deoxycholic acid are structurally similar compounds.

$$\text{(C)} \xrightarrow{(CH_3CO)_2O} \text{anhydride}$$

Thus, in this case, Blanc's rule fails. **The ring C is also six membered.**

Ring D. Cholestane was converted into etiobilianic acid which in turn gave a cyclic anhydride. Hence, **ring D is five membered.**

$$\text{(D)} \xrightarrow{(CH_3CO)_2O} \text{(Anhydride)}$$

Positions of the hydroxyl group and double bond. The positions of hydroxyl group and double bond were established by considering the following reaction:

(*i*) $\underset{\substack{\text{Cholestanone}\\(a)}}{C_{27}H_{46}O} \xrightarrow{HNO_3} \underset{\substack{\text{Dicarboxylic acid}\\(b)}}{C_{27}H_{46}O_4} \xrightarrow{300°C} \underset{\substack{\text{Ketone}\\(c)}}{C_{26}H_{44}O}$

As dicarboxylic acid contains the same number of carbon atom as ketone, we say that keto group must be in the ring. Further pyrolysis of dicarboxylic acid produces a ketone with the loss of one carbon atom. Thus, from Blanc's rule, (*b*) may be 1, 6 or 7-dicarboxylic acid. We have already seen that cholesterol nucleus contains three six membered and one five membered ring. Thus, dicarboxylic acid results by the opening of the ring A, B or C and hence –OH group in cholesterol should be in ring A, B or C.

(*ii*) When cholestanone is oxidised, two isomeric dicarboxylic acids are actually formed. Formation of two acids indicates that the keto group in cholestanone is flanked on either side by a methylene group ($-CH_2COCH_2-$). This type of structure is possible only if the **hydroxyl group is present in the ring A.**

(*iii*) Now consider the following set of reactions:

$$\underset{\substack{\text{Cholesterol}\\ \text{I}}}{C_{27}H_{46}O} \xrightarrow[CH_3COOH]{H_2O_2} \underset{\substack{\text{Cholestanetriol}\\ \text{II}}}{C_{27}H_{48}O_3} \xrightarrow{CrO_3} \underset{\substack{\text{Hydroxycholestanedione}\\ \text{III}}}{C_{27}H_{44}O_3}$$

$$\text{III} \xrightarrow{(i)\ -H_2O;\ (ii)\ Zn-CH_3COOH} \underset{\substack{\text{Cholestane dione}\\ \text{IV}}}{C_{27}H_{44}O_2} \xrightarrow{CrO_3} \underset{\substack{\text{Tetracarboxylic acid}\\ \text{V}}}{C_{27}H_{44}O_8}$$

In the above reactions, we see that cholestane triol (II) is converted to dione (III). This shows that two –OH groups (secondary) are oxidised to ketonic groups while the third –OH group, being resistant to oxidation must be tertiary. Dehydration of III followed by reduction of double bond and oxidation yields tetracarboxylic acid without loss of carbon atoms. This shows that the two keto groups must be in different rings. If the keto groups were in the same ring, then there would have been a loss of carbon atom. Hence, we say that **hydroxyl group** and double bond must be in different rings.

(*iv*) Compound IV above (cholestanedione) forms a pyridazine derivative with hydrazine. Thus, diketone must be γ-diketone. As we have already placed hydroxyl group in ring A, the above reactions can be readily explained if we place –OH group at C_3 and double bond between C_5–C_6. (Only two rings A and B are shown).

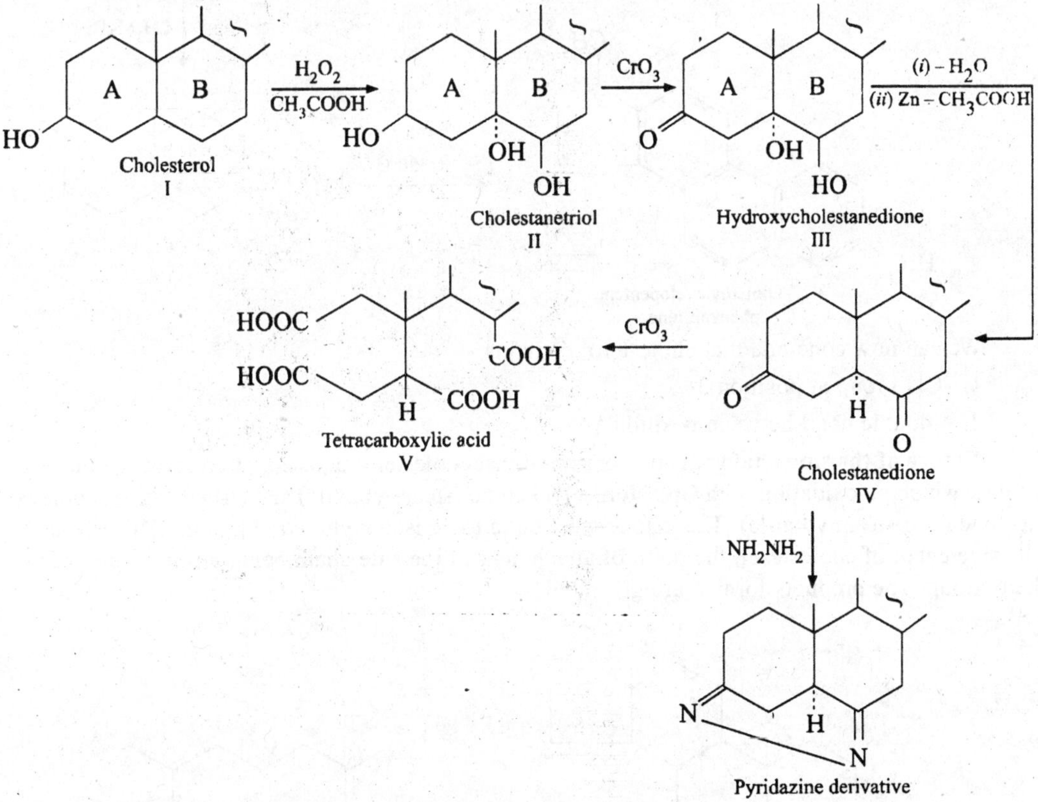

A B HO Cholesterol I $\xrightarrow[CH_3COOH]{H_2O_2}$ A B HO OH OH Cholestanetriol II $\xrightarrow{CrO_3}$ A B O OH HO Hydroxycholestanedione III $\xrightarrow[(ii)\ Zn - CH_3COOH]{(i) - H_2O}$

O H O Cholestanedione IV $\xrightarrow{CrO_3}$ HOOC HOOC COOH H COOH Tetracarboxylic acid V

IV $\xrightarrow{NH_2NH_2}$ N H N Pyridazine derivative

(*v*) The above explanation is supported by the fact that when cholesterol is heated with copper oxide at 290°C, cholestenone is formed which on oxidation with permanganate forms keto acid with a loss of carbon atom. It is explained as under:

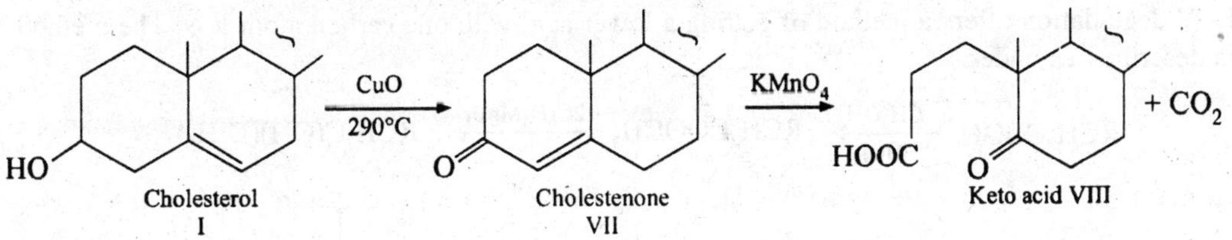

HO Cholesterol I $\xrightarrow[290°C]{CuO}$ O Cholestenone VII $\xrightarrow{KMnO_4}$ HOOC O Keto acid VIII $+ CO_2$

The formation of keto acid shows that the keto group and the double bond in cholestenone are in the same ring. These results can only be explained if we assume that the double bond in cholesterol migrates in the formation of cholestenone.

(*vi*) The position of hydroxyl group at position 3 is conclusively proved by Kon *et al.* The formation of 3′, 7-dimethylcyclopentenophenanthrene from cholesterol by the following steps is possible only if –OH group is considered at position C_3.

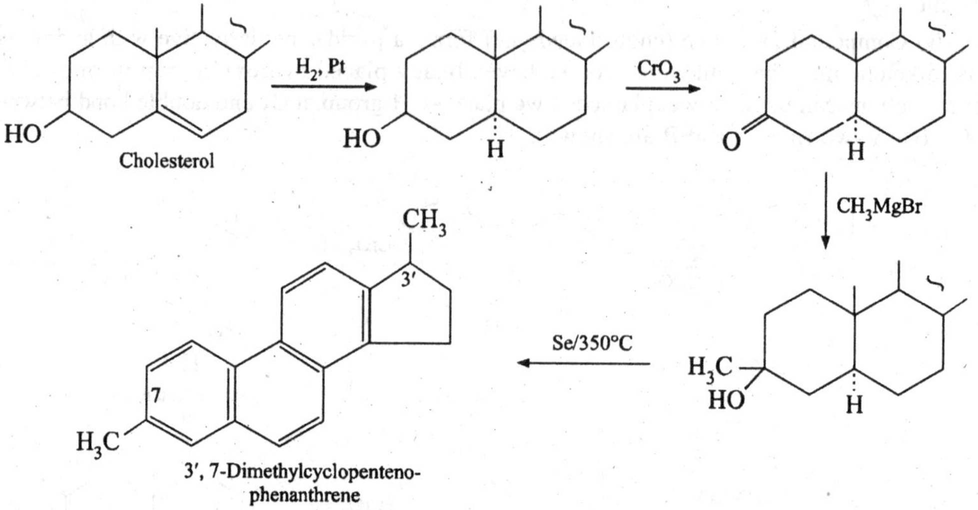

HO Cholesterol $\xrightarrow{H_2,\ Pt}$ HO H $\xrightarrow{CrO_3}$ O H

$\downarrow CH_3MgBr$

3′, 7-Dimethylcyclopenteno-phenanthrene $\xleftarrow{Se/350°C}$ H_3C HO H

We can now conclude that cholesterol contains

1. –OH group at position 3
2. A double bond between position 5–6.

Nature of the side chain : (*i*) It is observed that cholesterol on acetylation forms cholesteryl acetate which on oxidation with CrO_3 forms (*i*) ketone (steam volatile) and (*ii*) acetate of hydroxy acetone (non-steam volatile). The ketone was found to be isohexyl methyl ketone. This ketone is the side chain of cholesterol, the point of attachment of the side chain being at the carbon of the keto group. The products formed are given below:

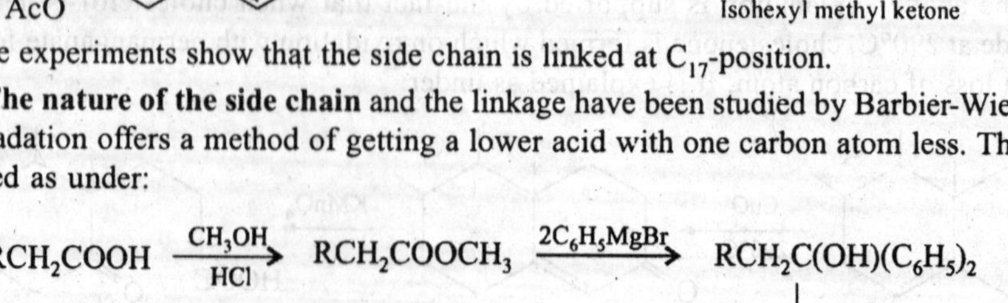

AcO + Isohexyl methyl ketone

These experiments show that the side chain is linked at C_{17}-position.

(*ii*) **The nature of the side chain** and the linkage have been studied by Barbier-Wieland. The B-W degradation offers a method of getting a lower acid with one carbon atom less. The method is described as under:

$$RCH_2COOH \xrightarrow[HCl]{CH_3OH} RCH_2COOCH_3 \xrightarrow{2C_6H_5MgBr} RCH_2C(OH)(C_6H_5)_2$$

$$\downarrow -H_2O$$

$$(C_6H_5)_2C = O + RCOOH \xleftarrow{CrO_3} RCH = C(C_6H_5)_2$$

Cholesterol was first converted into 5β-cholestane.* If the nucleus of 5β-cholestane is represented as Ar and side chain as C_n, then degradation can be expressed as under :

$$\underset{Ar—C_n}{5\beta\text{Cholestane (or coprostane)}} \xrightarrow{CrO_3} CH_3COCH_3 + \underset{Ar—C_{n-3}}{\text{Cholanic acid}}$$

$$\xrightarrow{B—W} (C_6H_5)_2CO + \underset{Ar—C_{n-4}}{\text{Norcholanic acid}} \xrightarrow{B.W.}$$

$$(C_6H_5)_2CO + \underset{Ar—C_{n-5}}{\text{Bisnorcholanic acid}} \xrightarrow{B.W.}$$

$$(C_6H_5)_2CO + \underset{Ar—C_{n-6}}{\text{etiocholylmethyl ketone}} \xrightarrow{CrO_3} \underset{Ar—C_{n-7}}{\text{Etianic acid}}$$

B.W. stands for Barbier–Wieland degradation

The formation of acetone from 5β-cholestane shows that side chain terminates in an **isopropyl group.** The conversion of bisnorcholanic acid into a ketone means that there is an alkyl group on the α-carbon in bisnorcholanic acid. As etiocholyl methyl ketone is oxidised to etianic acid with a loss of one carbon atom, we say that ketone must be methyl ketone. So, alkyl group on α-carbon atom in bisnorcholanic acid is a methyl group.

(*iii*) When etianic acid is subjected to one more B–W degradations, a ketone (etiocholanone) is obtained which on oxidation with nitric acid yields dicarboxylic acid (etiobilianic acid) without loss of any carbon atoms. Clearly, etiocholanone must be a cyclic ketone. It reveals that there are eight carbon atoms in the side chain. The above degradation can only be explained if the side chain has the following structure.

$$\underset{\text{Nucleus}}{Ar} - \overset{CH_3}{\overset{|}{CH}} - CH_2 - CH_2 - CH_2 - CH\begin{matrix} \diagup CH_3 \\ \diagdown CH_3 \end{matrix}$$

Side chain

Position of side chain : The dicarboxylic acid (etiobilianic acid) forms an anhydride when heated with acetic anhydride. According to the Blanc's rule etiocholanone may be a five membered ketone and therefore the side chain is attached to the five membered ring D. The formation of Diel's hydrocarbon from cholesterol suggests that the side chain is at position 17 as dehydrogenation degrades a side chain to methyl group. Further we know that 5β-cholanic acid may be obtained by the oxidation of 5β cholestane. 5β-cholanic acid may also be obtained by the oxidation of deoxycholic acid followed by Clemmenson reduction. Thus, the side chains in cholesterol and deoxycholic acid are in the same position. Now as the nature and the position (C_{17-}) of the side chain are known, conversion of 5β cholestane into etiobilianic acid can be expressed as under.

*5β-cholestane is a stereoisomer of cholestane.

5β-Cholestane $\xrightarrow{CrO_3}$ $CH_3-\overset{O}{\overset{\|}{C}}-CH_3$ + 5β-Cholanic acid $\xrightarrow{B-W}$ Nor-5β Cholanic acid $\xrightarrow{B-W}$ Bis-nor-5β-cholanic acid $\xrightarrow{B-W}$ Etiocholyl methyl ketone $\xrightarrow{CrO_3}$ Etianic acid $\xrightarrow{B-W}$ Etiocholanone $\xrightarrow{HNO_3}$ Etiobilianic acid

It has been found that when anhydride of etiobilianic acids is distilled with selenium, 1, 2-dimethylphenanthrene is obtained. It shows the presence of phenanthrene nucleus in cholesterol and also an angular methyl at position 13.

Positions of the two angular methyl groups

(*i*) The cyclopentenophenanthrene nucleus of cholesterol accounts for seventeen carbon atoms and the side chain for eight carbons. This accounts for 25 carbon atoms and the remaining two are assumed to be angular methyl groups.

When the position of the hydroxyl group and double bond are determined, one of the compounds formed was keto acid (VIII). When this compound is put to Clemmensen reduction followed by two Barbier-Wieland degradations, we get an acid which is difficult to esterify and gives CO on warming with conc. sulphuric acid. Clearly, –COOH group must be linked to tertiary carbon atom and side chain must be of the following type:

$$C-\underset{C}{\overset{C}{\overset{|}{\underset{|}{C}}}}-C-C-COOH \xrightarrow{2(B-W)} C-\underset{C}{\overset{C}{\overset{|}{\underset{|}{C}}}}-COOH$$

The reactions are shown below:

Keto acid (VIII) $\xrightarrow{Zn/Hg-HCl}$ HOOC… $\xrightarrow{2\,(B-W)}$ Acid group with 3° carbon atom

This shows an alkyl group at C_{10}–position.

(*ii*) To determine the position of second methyl group, consider the selenium dehydrogenation of cholesterol. Chrysene and Diel's hydrocarbons are obtained. The formation of chrysene can be explained by saying that an angular methyl group is present at position 13 and on selenium dehydrogenation, this methyl group enters the five membered ring to give a six membered ring. The positions of two angular methyls at positions 10 and 13 are supported by the following reactions:

OH HO Deoxycholic acid $\xrightarrow{CrO_3}$ O O Dehydrodeoxy cholic acid COOH $\xrightarrow{HNO_3}$ O COOH HOOC HOOC Deoxybilianic acid

$\downarrow \Delta$

O COOH HOOC O Diketo-dicarboxylic acid $\xleftarrow{KMnO_4}$ O COOH O

$\downarrow HNO_3$

COOH HOOC COOH (A) + HOOC — CH_2 HOOC COOH HC COOH (B) $\xrightarrow{Heat}$ O COOH H COOH (C)

$\downarrow HNO_3$

COOH COOH COOH H (D)

The compound (A) was found to be butane-2, 2, 4-tricarboxylic acid. This shows an angular methyl group at position 10. Compound (B) is a tetracarboxylic acid containing a cyclopentane ring with a side chain $-\underset{}{\overset{CH_3}{\overset{|}{C}}}H-CH_2-CH_2-COOH$. Also (D) is a tricarboxylic acid containing a five membered ring. In compound (D), a –COOH group is found to be linked with tertiary carbon atom and we say that methyl group is either at position 13 or 14. The position of angular methyl at **position 13** is confirmed by the fact that etiobilianic acid on heating forms an anhydride which on selenium dehydrogenation gives 1, 2-dimethylphenanthrene.

COOH COOH Heat → O O O Se → CH_3 CH_3

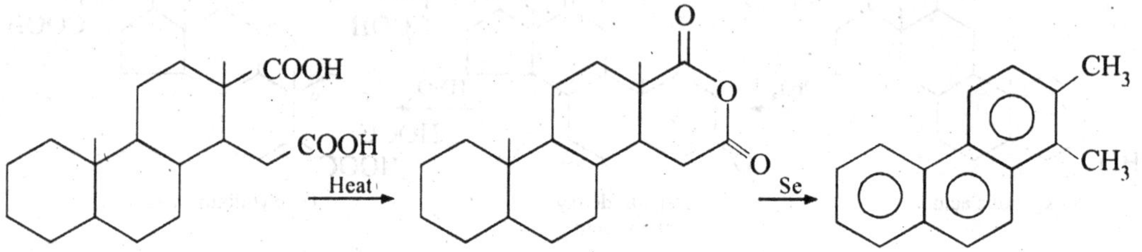

Etiobilianic acid Anhydride 1, 2-Dimethyl-phenanthrene

Had the angular methyl group been at position 14, then 1-methyl phenanthrene would have been formed.

ALKALOIDS

Q. 16. What are alkaloids? How are they obtained? *(Bangalore, 2002; Utkal, 2003; Mumbai, 2010)*

Ans. Alkaloids are defined as the basic nitrogenous compounds of vegetable origin. They have marked physiological action and they often contain pyridime, pyrrole, quinoline or similar nitrogenous nuclei. Alkaloids act as medicines in small dose but in large doses, they are poisons.

They are generally extracted from plants in the form of their salts with organic acids. These are generally present in the seeds and also in the bark of trees.

Extraction

The plant material is powdered and treated with water. It is then acidified with hydrochloric acid. The alkaloids form salts with acid and get dissolved in water. Thus, the water extract contains the hydrochlorides of alkaloids along with dyestuffs, carbohydrates and other products from plant tissue. It is then treated with alkali. The alkaloids get precipitated. In case the alkaloid is volatile, the acidic solution may be subjected to steam distillation. The crude product so obtained may be purified by special methods.

Q. 17. Describe the general properties of alkaloids. *(Shivaji, 2000, Kerala, 2000; Himachal, 2000; Panjab, 2000; Utkal 2003)*

Ans. The general properties of alkaloids are given below:

1. Physical state. Alkaloids are generally crystalline solids, only a very few alkaloids are liquids, e.g., nicotine.

2. Solubility. Mostly alkaloids are insoluble in water and some are sparingly soluble. Nicotine and caffein are exceptions. These are readily soluble in alcohol.

3. Optical activity. These are optically active compounds. Most of the alkaloids are laevo-rotatory.

4. Basic character. Alkaloids are basic in nature and hence form salts with acids. Their salts such as hydrochloride, sulphate, oxalate, etc., crystallise well.

5. Precipitation of alkaloids. The alkaloids, being basic, dissolve in acids. From their acidic solution, they can be precipitated by reagents such as picric acid, tannic acid, perchloric acid, etc. Precipitation is often done for the purification or the isolation of an alkaloid.

6. Physiological action. Alkaloids in small doses possess curative properties and are of great medicinal use.

Q. 18. Write notes on

(*i*) Hofmann exhaustive methylation (*Garhwal, 2010*)

(*ii*) Emde degradation

(*iii*) Reductive degradation

(*iv*) Oxidative degradation (*Punjab, 2005*)

Ans. Degradation is an exercise which is carried out to elucidate the structure of complex compounds. The compound is degraded (broken) into simpler and smaller compounds which are easy to identify. From the structure of simpler parts, structure of the original compound is arrived at. Different degradations are discussed separately as under.

(*i*) Hoffmann exhaustive methylation. It is a method by which heterocyclic rings are opened, then nitrogen is eliminated and finally aliphatic compounds results. Quaternary ammonium hydroxide is obtained by the complete methylation of the amine and its subsequent treatment with moist silver oxide.

$$CH_3-CH_2-CH_2-N(CH_3)_2 \xrightarrow[(ii)Ag_2O]{(i)CH_3I} CH_3CH_2CH_2-\overset{+}{N}(CH_3)_3OH$$

Quaternary amm.hydroxide

$$\downarrow \Delta$$

$$CH_3CH=CH_2 + (CH_3)_3N + H_2O$$

Propene

It may be noted that the method is also applicable to the reduced ring compounds. If nitrogen is present in a cyclic structure, then the ring is first reduced and then subjected to exhaustive methylation followed by hydrolysis and heating. For example, consider pyridine. Pyridine is reduced to piperidine by catalytic hydrogenation. Then it is heated with excess of methyl iodide to give quaternary ammonium iodide. Then the product is treated with moist silver oxide to give quaternary ammonium hydroxide which on heating lose a molecule of water with the cleavage of C–N bond from the side from which β-hydrogen atom is eliminated. The process of (*i*) methylation (*ii*) treatment with moist silver oxide and (*iii*) heating is repeated till nitrogen is eliminated as trimethylamine along with the formation of water and an unsaturated compound.

Pyridine $\xrightarrow{H_2/Ni}$ Piperidine $\xrightarrow{2CH_3I}$ Dimethyl piperidonium iodide ($N^+(CH_3)_2\ I^-$) $\xrightarrow{Ag_2O}$ Dimethyl piperidonium hydroxide ($N^+(CH_3)_2\ OH^-$)

$\xrightarrow[-H_2O]{Heat}$ 4-Pentenyl dimethylamine ($N(CH_3)_2$) $\xrightarrow[(ii)\ Ag_2O]{(i)\ CH_3I}$ $\overset{+}{N}(CH_3)_3\ OH^-$ $\xrightarrow[-H_2O]{\Delta}$ diene $+ (CH_3)_3N$

$\downarrow$ Isomerises

Piperylene

It may be noted that in the elimination of a molecule of water from quaternary ammonium hydroxide, hydrogen atom always eliminates from the β-position. Otherwise the reaction does not proceed.

Isoquinoline $\xrightarrow{Na/C_2H_5OH}$ NH $\xrightarrow[(ii)\ Ag_2O]{(i)\ 2CH_3I}$ $\overset{+}{N}(CH_3)_2\overset{-}{O}H$ $\xrightarrow{\Delta/-H_2O}$ $CH_2N(CH_3)_2$ $\xrightarrow[(ii)\ AgOH]{(i)\ CH_3I}$ $CH_2\overset{+}{N}(CH_3)_2\overset{-}{O}H$

In this case, there is no β-hydrogen atom with respect to nitrogen, hence it does not undergo Hofmann degradation.

(*ii*) **Emde's degradation.** The compounds, for which Hofmann degradation is not applicable, can be successfully degraded by Emde method. In this method, aqueous or alcoholic solution of quaternary ammonium halide is reduced with sodium amalgam in ethanol.

$CH_2-\overset{+}{N}(CH_3)_3I^-$ $\xrightarrow{Na-Hg/H_2O-C_2H_5OH}$ CH_3 $+ (CH_3)_3N$

with no hydrogen in β-position; *o*-Methylstyrene

Tetrahydroisoquinoline can be directly degraded by Emde degradation.

$\overset{+}{N}I^-$ (H_3C, CH_3) $\xrightarrow{Hofmann}$ $N(CH_3)_2$ $\xrightarrow[(ii)\ Na/Hg]{(i)\ CH_3I}$ CH_3 $+ (CH_3)_3N$

$\overset{+}{N}I^-$ (H_3C, CH_3) $\xrightarrow{Na/NH_3}$ $N(CH_3)_2$, CH_3 $\xrightarrow{Hofmann}$ CH_3 $+ (CH_3)_3N$

(*iii*) **Reductive degradation.** This is a simple method whereby pyridine or piperidine nuclei in some cases may be estimated as ammonia and *n*-pentane by heating with hydriodic acid at 573 K.

N OR N–H $\xrightarrow[573\ K]{Hi}$ H_3C CH_3 $+ NH_3$

n-Pentane

(*iv*) **Oxidative degradation.** We can obtain important information about the fundamental structure of alkaloids and the position and the nature of some of the functional groups or side chain such as $>C=C<$, –CHOH etc., by oxidative degradation.

For example, oxidation of coniine to picolinic acid suggests that coniine is α-substituted pyridine.

$$\text{Connine} \xrightarrow{(O)} \text{Picolinic acid (2-pyridinecarboxylic acid, N, COOH)}$$

Type of products obtained depends upon whether strong, moderate or mild oxidising agent has been used.

(*i*) Mild oxidation is usually done by : (*a*) hydrogen peroxide; (*b*) iodine in ethanolic solution; or (*c*) alkaline potassium ferricyanide.

(*ii*) Moderate oxidation is carried by means of alkaline potassium permanganate or chromium oxide in acetic acid.

(*iii*) Strong oxidation is carried by using $K_2Cr_2O_7$–H_2SO_4, conc. HNO_3 and MnO_2–H_2SO_4.

Q. 19. Establish the structure of quinine. (*Kerala, 2001*)

Ans. Various steps involved in the elucidation of structure of quinine, an important antimalarial, are given below.

1. Molecular formula. The molecular composition of quinine is found to be $C_{20}H_{24}N_2O_2$ by elemental analysis.

2. It reacts with two moles of methyl iodide to form two series of quaternary salts. It shows the presence of two tertiary nitrogen atoms of different types.

3. It can be acetylated to form a monoacetate and benzoylated to form a benzoate. It can be converted into a chlorocompound also. It shows the presence of –OH group. Quinine on oxidation yields a ketone which shows that it contains a secondary alcoholic group.

4. On treatment with HI, it yields one mole of CH_3I which shows the presence of methoxy group in the molecule.

$$—OCH_3 \xrightarrow[\Delta]{HI} —OH + CH_3I$$

5. Quinine decolourises bromine water and consumes one mole of hydrogen per mole of quinine, in the presence of a catalyst. These properties show unsaturation to the extent of one double bond in the molecule.

6. When oxidised with mild pot. permanganate, it forms a monocarboxylic acid along with formic acid.

$$\underset{\text{Quinine}}{\sim CH=CH_2} \xrightarrow[KMnO_4]{[O]} \sim COOH + \underset{\text{Formic acid}}{HCOOH}$$

This property shows that quinine contains a vinyl group.

7. Fusion with KOH. When fused with KOH, two major products formed are (*i*) 6-methoxy-quinoline and (*ii*) lepidine. These two products show that quinine contains quinoline nucleus.

$$\underset{\text{Quinine}}{C_{20}H_{24}N_2O_2} \xrightarrow[\text{Fuse}]{KOH} \underset{\text{6-Methoxy quinoline}}{\text{(quinoline with } H_3CO \text{ at C-6)}} + \underset{\text{Lepidine}}{\text{(quinoline with } CH_3 \text{ at C-4)}}$$

These two products are also formed from cinchonine under similar conditions. This shows that quinine is methoxy-cinchonine.

8. On oxidation with chromic acid, quinine yields two products viz. quininic acid and meroquinene.

$$\underset{\text{Quinine}}{C_{20}H_{24}N_2O_2} \xrightarrow{CrO_3} \underset{\text{Quininic acid}}{C_{11}H_9NO_3} + \underset{\text{Meroquinene}}{C_9H_{15}NO_2}$$

Thus, to arrive at the structure of quinine, it is necessary to have an insight into the structures of quininic acid and meroquinene.

9. Structure of Quininic acid ($C_{11}H_9NO_3$). (*a*) Action with sodalime. On heating with sodalime, quininic acid forms 6-methoxy quinoline. It shows that quininic acid is 6-methoxy quinoline derivative. This derivative on treatment with HCl gives 6-hydroxyquinoline showing methoxy group at position 6.

$C_{11}H_9NO_3$ (Quinine) —Sodalime→ 6-Methoxyquinoline (H_3CO, N) —HCl→ 6-Hydroxyquinoline (HO, N)

(*b*) Oxidation with Chromic acid. On oxidation with chromic acid, quinine forms pyridine 2, 3, 4-tricarboxylic acid. It shows that (*i*) benzene ring containing methoxyl group is oxidised (*ii*) –COOH group is present at position 4 in quininic acid. The reaction is explained as under:

Quininic acid (COOH, H_3CO, N) —CrO_3→ Pyridine-2, 3, 4-tricarboxylic acid (COOH, HOOC, HOOC, N)

(*c*) Confirmation of the structure of quininic acid by synthesis. The structure of quininic acid is confirmed by synthesis by Rube *et al.*

o-Aminoanisol (H_3CO, NH_2) —$CH_3COCH_2COOC_2H_5$ (Aceto acetic ester)→ (Ketonic) (H_3CO, NH–C=O–CH_2–$COCH_3$) ⇌ (H_3CO, N=C–OH, CH, HO–C–CH_3) —H_2SO_4 (Cyclisation)→ (CH_3, H_3CO, N, O) —$POCl_3$→ (CH_3, H_3CO, N, Cl) —Al/CH_3COOH, – HCl→ (CH_3, H_3CO, N) —$ZnCl_2$ | C_6H_5CHO→

$$\text{6-Methoxy-4-styrylquinoline } (H_3CO, CH=CHC_6H_5) \xrightarrow[\text{[O]}]{KMnO_4} \text{Quininic acid } (H_3CO, COOH)$$

10. Structure of Meroquinene. (*a*) The molecular formula of meroquinene comes out to be $C_9H_{15}NO_2$ from analytical data and molecular weight determination. It decolourises bromine water, showing the presence of a double bond in the molecule.

Hinsberg test and reaction with nitrous acid reveal that it contains a secondary amino group.

Reaction with sod. bicarbonate shows that it contains a carboxy group.

(*b*) **Oxidation with cold acidic pot. permanganate.** Meroquinene forms formic acid and cincholoiponic acid (dicarboxylic acid). Cincholoiponic acid on further oxidation yields loiponic acid (another dicarboxylic acid).

$$\underset{\text{Meroquinene}}{C_9H_{15}NO_2} \xrightarrow[\text{[O]}]{KMnO_4} HCOOH + \underset{\text{Cincholoiponic acid}}{C_8H_{13}NO_4}$$

$$\underset{\text{Cincholoiponic acid}}{C_8H_{13}NO_4} \xrightarrow[\text{[O]}]{KMnO_4} \underset{\text{Loiponic acid}}{C_7H_{15}NO_4}$$

The formation of formic acid shows the presence of vinyl group in meroquinene unit. The molecular formulae reveal that cincholoiponic acid is a higher homologue of loiponic acid with a difference of $-CH_2$.

(*c*) Loiponic acid is less stable and isomerises to more stable, hexahydrocinchomeronic acid $C_7H_{11}NO_4$ (Piperidine-3, 4-dicarboxylic acid) on treatment with KOH at about 200°C. Clearly, loiponic acid should also be piperidine-3, 4-dicarboxylic acid. Thus, loiponic acid is:

COOH (C-4), COOH (C-3), N–H piperidine ring

(Loiponic acid)

As cincholoiponic acid has one more $-CH_2$ group than loiponic acid, it can have one of the following two structures.

I: CH_2COOH (C-4), COOH (C-3), N–H piperidine ring

OR

II: COOH (C-4), CH_2COOH (C-3), N–H piperidine ring

Cincholoiponic acid is found to be I since on heating with conc. H_2SO_4, it forms γ-picoline.

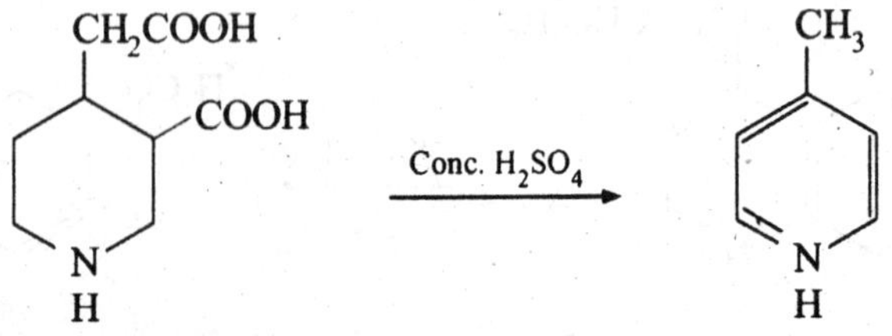

Cincholoipionic acid $\xrightarrow{\text{Conc. } H_2SO_4}$ γ-Picoline

The structure of cincholoipionic acid is confirmed by synthesis.

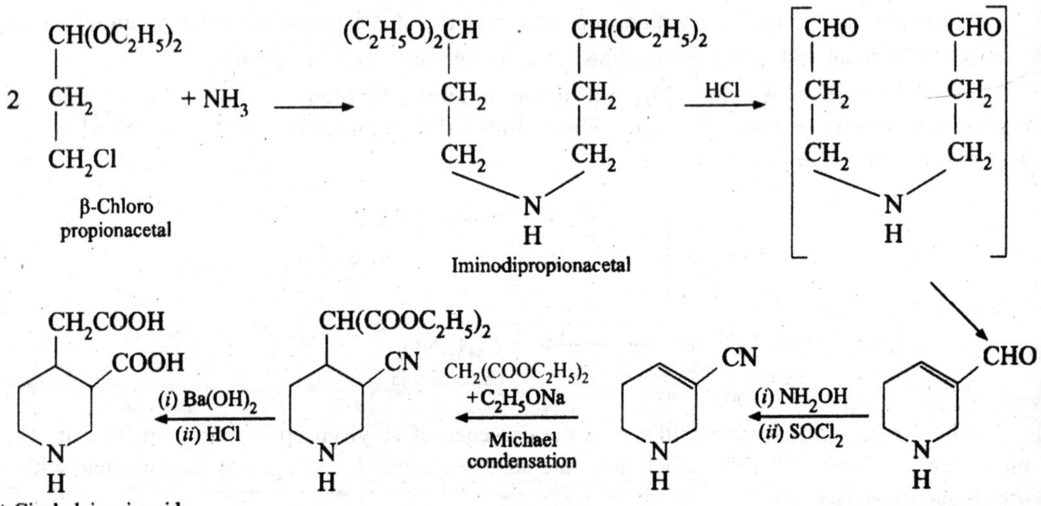

2 $CH(OC_2H_5)_2-CH_2-CH_2Cl$ (β-Chloro propionacetal) + NH_3 ⟶ Iminodipropionacetal $\xrightarrow{HCl}$ [dialdehyde] ⟶ aldehyde $\xrightarrow{(i)\ NH_2OH,\ (ii)\ SOCl_2}$ nitrile $\xrightarrow[\text{Michael condensation}]{CH_2(COOC_2H_5)_2 + C_2H_5ONa}$ $\xrightarrow{(i)\ Ba(OH)_2,\ (ii)\ HCl}$ ± Cincholoipionic acid

(*d*) As cincholoipionic acid is obtained along with formic acid (HCOOH) by oxidation of meroquinene, it appears that meroquinene should have grouping of the type ($-CH = CH_2$). The meroquinene should have one of the following two structures.

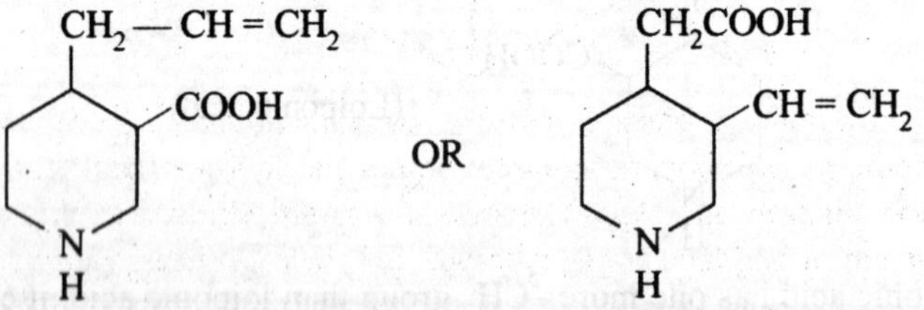

III OR IV

Structure IV seems to be more probable due to the following observation:

On heating with hydrochloric acid at 240°C, meroquinene yields 3-ethyl-4-methylpyridine.

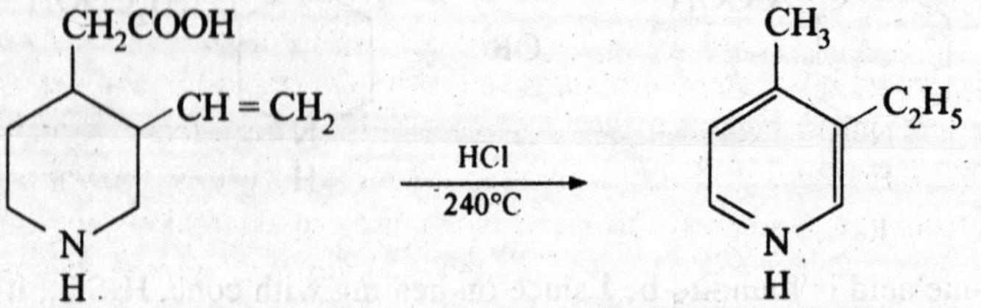

Meroquinene $\xrightarrow[240°C]{HCl}$ 3-Ethyl-4-methylpyridine

This product cannot be obtained from structure III above.

Oxidation of meroquinene and its degraded products are shown below:

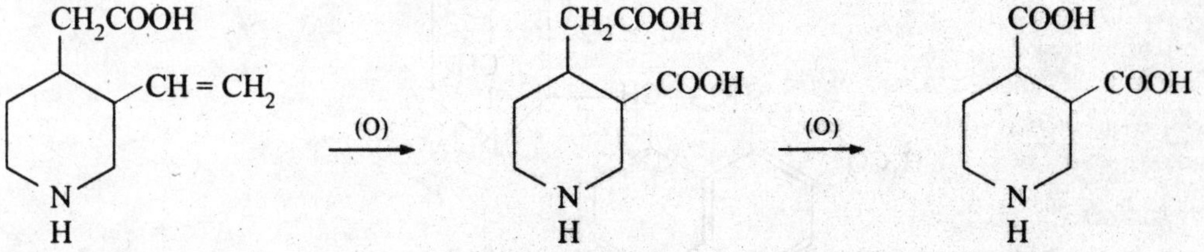

Meroquinene (4-CH_2COOH, 3-$CH=CH_2$ piperidine, N–H) $\xrightarrow{(O)}$ Cincholoiponic acid (4-CH_2COOH, 3-COOH piperidine, N–H) $\xrightarrow{(O)}$ Loiponic acid (4-COOH, 3-COOH piperidine, N–H)

11. Linkage between quininic acid and meroquinene. We know that quinine has no carboxylic group. But its oxidation product, viz., quininic acid and meroquinene contain free carboxylic groups. This shows that the two units are linked to each other by carboxylic carbon atoms. Also we know that quinine is a di-tertiary base while meroquinene has a secondary nitrogen atom. It indicates that during the oxidation of quinine, the second half unit is oxidised in such a way so that the tertiary nitrogen atom is converted into secondary with the introduction of carboxyl group. It can be explained by saying that the precursor of meroquinene has the following structure.

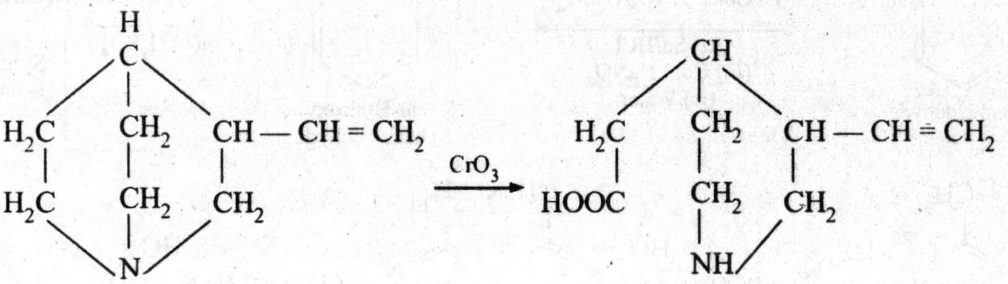

2-Vinylquinuclidine (H–C, H_2C, H_2C, CH_2, CH_2, $CH—CH=CH_2$, CH_2, N) $\xrightarrow{CrO_3}$ Meroquinene (CH, H_2C, HOOC, CH_2, CH_2, $CH—CH=CH_2$, CH_2, NH)

OR

(CH_2COOH, $CH=CH_2$ piperidine, N–H)

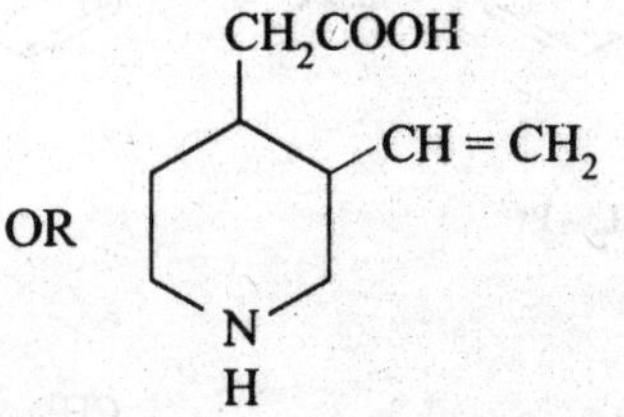

The existence of 2-vinyl quinuclidine in the above form is possible as 3-ethyl quinuclidine can be synthesised. Hence, we say that quinoline unit is joined at position-4 to the quinuclidine unit at position 8.

12. Position of the alcoholic group. The last thing to be decided is to know the position of secondary alcoholic group. The gentle oxidation of quinine with CrO_3 forms quininone. Also quininone on treatment with amyl nitrite and hydrogen chloride gives quininic acid and an oxime. It is a characteristic reaction of –CO–CH– group. The reaction can be explained as follows:

$$-CO-CH< \xrightarrow{HONO} \underset{\text{Acid}}{-COOH} + \underset{\text{NO}}{CH}< \xrightarrow{\text{Isomerises}} \underset{\underset{\text{(Oxime)}}{\text{NOH}}}{C}<$$

When the oxime formed above is hydrolysed we get hydroxyl amine and meroquinene. Hence, it is clear that quinoline and quinuclidine units are linked through –CHOH–, group. Hence, the structure of quinine can be written as:

Quinine

Q. 20. Describe the synthesis of quinine.

Ans. The structure of quinine is confirmed by its synthesis by R.B. Woodward and W.E. Doering. The starting material for its synthesis is *m*-hydroxy benzaldehyde. Benzaldehyde is converted into *m*-hydroxy benzaldehyde in a series of steps:

Benzaldehyde —(*i*) Conc. HNO_3/H_2SO_4 (*ii*) Sn/HCl (*iii*) Cold $NaNO_2$ (*iv*) Water→ *m*-Hydroxy benzaldehyde —(*i*) $NH_2CH_2CH(OC_2H_5)_2$ (*ii*) H_2SO_4, Synthesis of Isoquinoline→ 7-hydroxyisoquinoline —$CH_2O + C_5H_{11}N$, Mannich reaction→ (a) ($CH_2NC_5H_{10}$) —CH_3ONa, 220°C→ (b) (CH_3)

(b) —$H_2 - Pt$→ —(*i*) $(CH_3CO)_2O$ (*ii*) $H_2 - Ni$→ N.$COCH_3$ —(*i*) CrO_3/CH_3COOH (*ii*) Separate the isomers→ N.$COCH_3$ ketone —$C_2H_5NO_2$, C_2H_5ONa→ HO—N=C(CH_3), H_5C_2OOC, N.$COCH_3$ —$H_2 - Pt$→ H_2N, CH_3, HC, H_5C_2OOC, N.$COCH_3$ —(*i*) CH_3I (*ii*) 60% KOH (*iii*) Heat; Hofmann exhaustive methylation→

$HOOC$ ... NH (with $CH_2=HC$ side chain) $\xrightarrow[(ii)\ C_6H_5COCl]{(i)\ C_2H_5OH - H^+}$ H_5C_2OOC ... $N.COC_6H_5$

$\xleftarrow[\text{(Crossed claisen condensation)}]{C_2H_5ONa}$ H_3CO (quinoline with $COOC_2H_5$) + ($CH=CH_2$, H_2C, CH_2, H_2C, CH_2, H_5C_2OOC, N, COC_6H_5)

(O, C—CH, $CH=CH_2$, CH_2, CH_2, H_3CO, N, C, O, OC_2H_5, COC_6H_5) $\xrightarrow{HCl}$

$O=C$—CH_2 ($CH=CH_2$, CH_2, CH_2, H_3CO, N, NH)

+ Quintoxine

(i) Resolve and take (+) Isomer
(ii) NaOBr

($O=C$—CH, C, $CH=CH_2$, CH_2, CH_2, H_3CO, N, N)

(+) Quininone

Al – C_2H_5ONa | C_2H_5OH

(H, C, $CH=CH_2$, H, CH_2, HO — C — C — H, CH_2, H_3CO, N, N)

+ Quinine

Q. 21. Discuss the structure (constitution) of nicotine. (*Panjab, 2000; Shivaji, 2000; Rohilkhand, 2004; Kuvempu, 2005*)

Ans. Nicotine which is an alkaloid occurs abundantly in tobacco. Its structure was established as follows:

1. Elemental analysis and molecular weight of nicotine are consistent with the molecular formula, $C_{10}H_{14}N_2$.

2. On oxidation with chromic acid, nicotine is degraded to smaller fragments. One of these fragments possessing a carboxyl group has the composition $C_6H_5O_2N$ and has been shown to be identical with pyridine-3 carboxylic acid, or nicotinic acid which on decarboxylation yields pyridine.

$$\underset{\text{Nicotine}}{C_{10}H_{14}N_2} \longrightarrow \underset{\text{Nicotinic acid}}{\text{(pyridine ring with COOH at 3-position)}} \longrightarrow \underset{\text{Pyridine}}{\text{(pyridine ring)}}$$

It can be inferred that nicotine contains a pyridine ring substituted at the 3 position by a group $C_5H_{10}N$.

3. It was further shown that the formation of nicotinic acid by oxidation of nicotine with $KMnO_4$ is accompanied by the formation of methylamine, and that nicotine behaves towards ethyl iodide like a ditertiary base. Exact composition of the $C_5H_{10}N$ residue was established by bromination studies. Treatment of an acetic acid solution of nicotine with bromine gives an oily perbromide which is converted to a free base by the action of sulphurous acid. This base, dibromocotinine $C_{10}H_{10}ON_2Br_2$, forms a methiodide, $C_{10}H_{10}ON_2Br_2 \cdot CH_3I$, but reacts neither with bezoyl chloride nor hydroxylamine. On heating with a mixture of sulphurous and sulphuric acids at 140°, dibromocotinine is decomposed into 3-acetylpyridine, methylamine and oxalic acid.

$$\underset{\text{Dibromocotinine}}{C_{10}H_{10}ON_2Br_2} \xrightarrow[H_2SO_4]{H_2SO_3} \underset{\text{3-Acetylpyridine}}{\text{(pyridine ring with }COCH_3\text{ at 3-position)}} + CH_3NH_2 + \begin{matrix} COOH \\ | \\ COOH \end{matrix}$$

4. Another product which has been isolated from the reaction mixture obtained by the treatment of nicotine with bromine has the molecular formula $C_{10}H_{10}O_2N_2Br_2$ (dibromoticonine). On heating this compound in a sealed tube at 100° with barium hydroxide, nicotinic acid, methylamine and malonic acid are obtained.

$$\underset{\text{Dibromoticonine}}{C_{10}H_8O_2N_2Br_2} \xrightarrow{Ba(OH)_2} \underset{\text{Nicotinic acid}}{\text{(pyridine ring with }COCH_3\text{ at 3-position)}} + CH_3NH_2 + \underset{\text{Malonic acid}}{\begin{matrix} COOH \\ | \\ CH_2 \\ | \\ COOH \end{matrix}}$$

5. All these reactions can be explained by assuming structure A for nicotine.

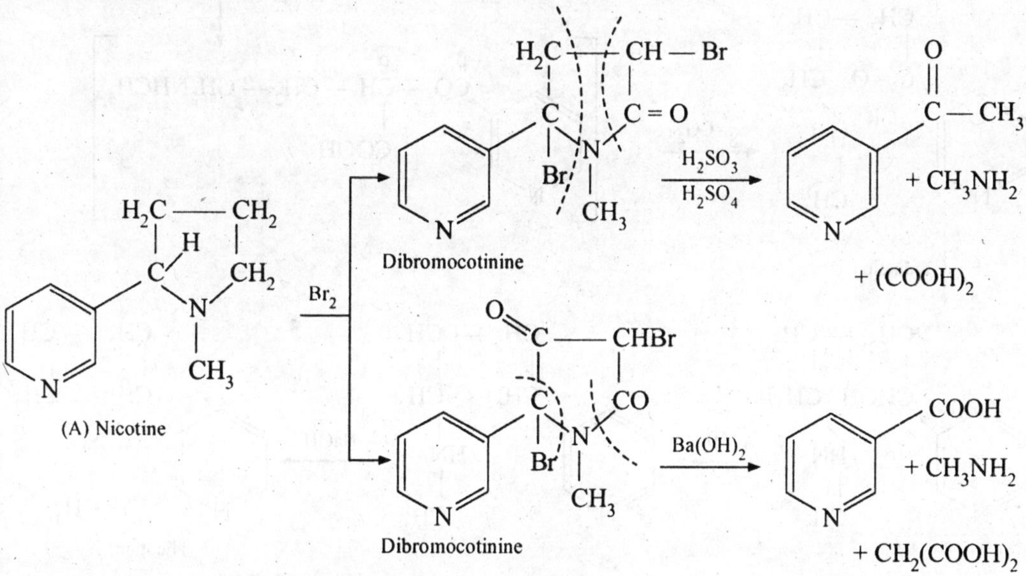

(A) Nicotine

Br_2

Dibromocotinine

H_2SO_3 / H_2SO_4 → 3-acetylpyridine + CH_3NH_2 + $(COOH)_2$

Dibromocotinine

$Ba(OH)_2$ → nicotinic acid (COOH) + CH_3NH_2 + $CH_2(COOH)_2$

Nicotine is thus N-methyl-(3′-pyridyl)-2-pyrrolidine.

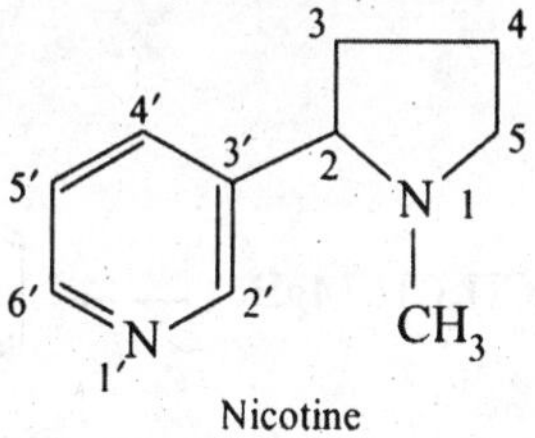

Nicotine

6. Final confirmation of structure of nicotine is provided by its synthesis.

1. Pictet Synthesis

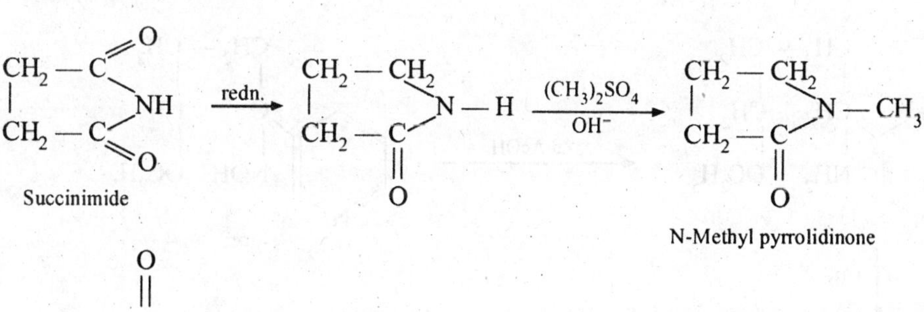

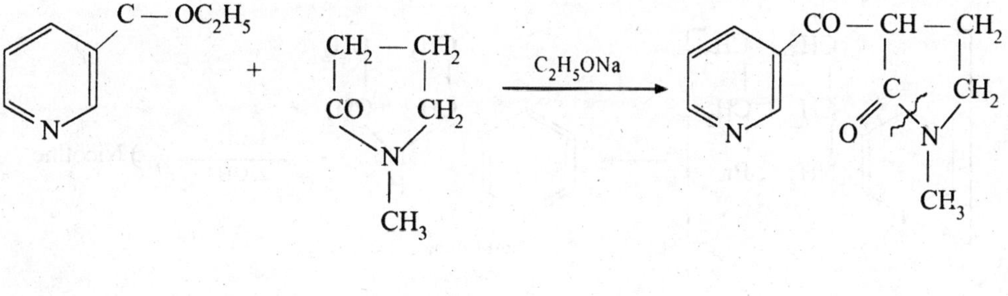

Succinimide —redn.→ pyrrolidinone —$(CH_3)_2SO_4$, OH^-→ N-Methyl pyrrolidinone

Ethyl nicotinate ($C—OC_2H_5$) + N-methyl pyrrolidinone —C_2H_5ONa→ product

$-CO_2$

reduction

HI

NaOH

Nicotine

2. Craig Synthesis

Nicotinonitrile + $C_2H_5OCH_2CH_2CH_2MgBr$

γ-Ethoxypropylmagnesium bromide

NH_2OH

Zn/AcOH

HBr

Nornicotine

1. CH_3I
2. OH^-

(±) Nicotine

Q. 22. Give the preparation and establish the structure of piperine. *(Madras, 2004 Garhwal, 2010)*

Ans. Piperine occurs naturally in black pepper.

Extraction. the powdered black peppers are heated with milk of lime to dryness and the piperine is recovered from the residue by extraction with ether.

Preparation. Piperine is prepared by heating piperoyl chloride with piperidine in benzene solution.

$$\underset{\text{Peperoyl chloride}}{C_{11}H_9O_2COCl} + \underset{\text{Piperidine}}{C_5H_{10}NH} \longrightarrow \underset{\text{Piperine}}{C_{11}H_9O_2CONC_5H_{10}} + HCl$$

Properties. Piperine is a colourless crystalline solid melting at 128–129°C. It is insoluble in water but soluble in common organic solvents.

On hydrolysis with base, piperine gives piperic acid and piperidine.

$$C_{17}H_{19}O_2N + H_2O \xrightarrow{KOH} \underset{\text{Piperic acid}}{C_{11}H_9O_2COOH} + \underset{\text{Piperidine}}{C_5H_{10}NH}$$

Structure. 1. Elementary analysis and molecular weight determination show the molecular formula of piperine is $C_{17}H_{19}O_3N$.

2. When boiled with alcoholic potash, piperine gets hydrolysed into piperic acid and piperidine.

$$\underset{\text{Piperine}}{C_{17}H_{19}O_3N} + H_2O \xrightarrow{KOH} \underset{\text{Piperic acid}}{C_{11}H_9O_2COOH} + \underset{\text{Piperidine}}{C_5H_{10}NH}$$

Thus piperine is the piperidine amide of piperic acid. Since piperidine is hexahydropyridine, the structural formula of piperine can be written if we decide that of piperic acid.

3. Structure of piperic acid. (*i*) The molecular formula of piperic acid is $C_{12}H_{10}O_4$.

(*ii*) The addition reactions and action on bicarbonates indicate the presence of two ethylenic bonds and one carboxyl group in the molecule of piperic acid. For example,

(*a*) Piperic acid reacts with bromine to form tetrabromo derivative, indicating the presence of two ethylenic double bonds.

$$C_{12}H_{12}O_4 + 2\,Br_2 \xrightarrow{CCl_4} C_{12}H_{10}O_4Br_4$$

(*b*) Piperic acid decomposes $NaHCO_3$ to liberate CO_2, indicating the presence of carboxyl group in the molecule of piperic acid.

$$C_{11}H_9O_2(COOH) + NaHCO_3 \longrightarrow C_{11}H_9O_2(COONa) + CO_2 + H_2O$$

(*iii*) On oxidation with potassium permanganate, piperic acid produces first piperonal and then piperonylic acid. Piperonylic acid ($C_7H_5O_2COOH$) contains 4 carbon atoms and 4 hydrogen atoms less than piperic acid ($C_{11}H_9O_2COOH$).

The structure of piperonylic acid has been confirmed by its synthesis from protocatechuic acid and methylene iodide.

$$\underset{\text{Protocatechuic acid}}{(HO)_2C_6H_3COOH} + CH_2I_2 \xrightarrow[\text{Heat}]{NaOH} \underset{\text{Piperonylic acid}}{CH_2\!<\!\!\begin{smallmatrix}O\\O\end{smallmatrix}\!\!>\!C_6H_3COOH}$$

Piperonlyic acid produces protocatechuic acid on heating with hydrochloric at 200°C under pressure.

$$C_8H_6O_5 \text{ or } C_7H_5O_2COOH + H_2O \xrightarrow{HCl}$$

piperonylic acid

$$(HO)_2C_6H_3COOH + HCHO$$

Protocatechuic acid

Hence, piperonylic acid is the methylene ether of protocatechuic acid, named as 3, 4-methylene dioxybenzoic acid.

Since piperonal (an aldehyde) forms piperonlyic acid on oxidation, piperonal is 3, 4-methylene dioxylbenzaldehyde.

$$CH_2O_2C_6H_3CHO \xrightarrow[KMnO_4]{[O]} CH_2O_2C_6H_3COOH$$

Piperonal — Piperonylic acid

(*iv*) The formation of piperonal and piperonylic acid on oxidation of piperic acid leads to the opinion that piperic acid is a benzene derivative containing only one side chain. It is this side chain which gives addition reaction with bromine forming tetrabromo derivative. Thus, there are two double bonds in the side chain.

The side chain should contain four carbon and four hydrogen atoms (piperic acid $C_{12}H_{10}O_4$– piperonlyic acid, $C_8H_6O_4$; $C_{12}H_{10}O_4 - C_8H_6O_4 = C_4H_4$). Therefore piperic acid may be assigned the following structure.

$$H_2CO_2C_6H_3-CH=CH-CH=CH-COOH$$

(*v*) The structure of piperic acid is confirmed by its synthesis is given below

$$(HO)_2C_6H_4 + CHCl_3 \xrightarrow{NaOH} (HO)_2C_6H_3CHO \xrightarrow[NaOH]{CH_2I_2} CH_2O_2C_6H_3CHO$$

Catechol — Piperonal

$$\xrightarrow[NaOH]{CH_3CHO} CH_2O_2C_6H_3-CH=CHCHO \xrightarrow[CH_3COONa]{(CH_3CO)_2O} CH_2O_2C_6H_3-CH=CH \cdot CH=CH \cdot COCl$$

Cinnamaldehyde derivative — Piperic acid

4. Since piperine is the piperidine amide of piperic acid it should have following structural formula.

$$CH_2\langle O_2 \rangle C_6H_3-CH=CH-CH=CH-CO-N\langle (CH_2-CH_2)_2 \rangle CH_2$$

5. The structure of piperine, is confirmed by its synthsis from piperyl chloride and piperidine.

$$\underset{\text{Piperoyl chloride}}{CH_2\langle O_2 \rangle C_6H_3-CH=CHCH=CHCOCl} + \underset{\text{Piperidine}}{HN\langle (CH_2-CH_2)_2 \rangle CH_2} \xrightarrow[\text{heat}]{-HCl} \underset{\text{Piperine}}{CH_2\langle O_2 \rangle C_6H_3-CH=CHCH=CHCO-N\langle (CH_2-CH_2)_2 \rangle CH_2}$$

Q. 23. Discuss the occurrence, properties and structure of atropine.

Ans. Occurrence. Atropine occurs in the stems and roots of the plant Atropa belladonna, together with hyoscyamine. Hyoscyamine is optically active and readily recemises to atropine when it is warmed in an ethanolic alkaline solution. Thus, atropine may be regarded as (±) hyoscyamine.

Isolation. Atropine is obtained by treating the plant juice with alkali followed by extraction with ether. The crude alkaloid isolated in this way is purified by oxalate formation, fractional recrystallisation and finally the treatment of the salt with alkali.

Properties. Atropine is a white crystalline solid, melting at 116–117°C. It is bitter in taste and basic in nature. When warmed with barium hydroxide solution, it undergoes hydrolysis to give (±)–tropic acid and tropine, an alcohol). Thus, atropine is the tropine ester of tropic acid.

Atropine dilates the eye pupil and paralyses the accommodation muscles of the eye temporarily. Hence, it is used in dilute solutions in ophthalmic examination.

Structure. (1) Elemental analysis and molecular mass determination show that atropine has the molecular formula $C_{17}H_{23}NO_3$.

(2) When atropine is hydrolysed with HCl at 130°C or with alkali at 60°C, it gives (±) tropic acid and tropine an alcohol.

$$\underset{\text{Atropine}}{C_{17}H_{23}NO_3} + H_2O \longrightarrow \underset{(\pm)\text{ Tropic acid}}{C_9H_{10}O_3} + \underset{\text{Tropine}}{C_8H_{15}NO}$$

This reaction leads to the opinion that atropine is an ester of tropic acid and tropine and the structural formula for atropine can be arrived at by elucidating the structure of tropic acid and tropine.

(3) **Structure of tropic acid.**

(*i*) The molecular formula of tropic acid is $C_9H_{10}O_3$.

(*ii*) Tropic acid does not give addition reactions. This indicates that it is a saturated compound.

(*iii*) The usual tests of tropic acid show that its molecule contains one carboxyl group and one alcoholic group.

(*iv*) On heating strongly, tropic acid loses a molecule of water to yield atropic acid ($C_9H_8O_2$) which on oxidation forms benzoic acid (C_6H_5COOH).

$$\underset{\text{Tropic acid}}{C_6H_5(C_2H_3OH)COOH} \xrightarrow[-H_2O]{\text{heat}} \underset{\underset{(a)}{\text{Atropic acid}}}{C_6H_5-CH=CH-COOH} \text{ or } \underset{(b)}{C_6H_5-\overset{\overset{\displaystyle CH_2}{\|}}{C}-COOH}$$

$$\xrightarrow{\text{oxid.}} \underset{\text{Benzoic acid}}{C_6H_5COOH}$$

These reactions suggest that tropic acid and atropic acid contain benzene ring with one side chain. Further, it follows that atropic acid has two possible structures (*a*) and (*b*). Since the structure (*a*) is known to be cinnamic acid, the structure (*b*) must therefore be atropic acid. The structure (*b*) for atropic acid is supported by the fact that the oxidation of atropic acid with permanganate produces phenylglyoxylic acid ($C_6H_5CO\ COOH$).

$$C_6H_5-\overset{\overset{\displaystyle CH_2}{\|}}{C}-COOH \xrightarrow{[O]} \underset{\text{Phenyl glyoxalic acid}}{C_6H_5\overset{\overset{\displaystyle O}{\|}}{C}-COOH}$$

The formula of tropic acid can be obtained from the structure (b) by adding a molecule of water to the unit $-\overset{\overset{\displaystyle CH_2}{\|}}{C}-$. This can be done in two ways and consequently, tropic acid must be either (*c*) or (*d*).

$$\underset{(c)}{C_6H_5-\overset{\overset{\displaystyle CH_3}{|}}{\underset{\underset{\displaystyle OH}{|}}{C}}-COOH} \qquad \underset{(d)}{C_6H_5-\overset{\overset{\displaystyle CH_2OH}{|}}{\underset{\underset{\displaystyle H}{|}}{C}}-COOH}$$

Tropic acid has been shown to have structure (*d*) by its synthesis as given below

$$\begin{matrix}C_6H_5\\CH_3\end{matrix}\!\!>C=O \xrightarrow{HCN} \begin{matrix}C_6H_5\\CH_3\end{matrix}\!\!>C<\!\!\begin{matrix}OH\\CN\end{matrix} \xrightarrow[H_2O]{H^+} \underset{\underset{\text{Atrolactic acid}}{(c)}}{\begin{matrix}C_6H_5\\CH_3\end{matrix}\!\!>C<\!\!\begin{matrix}OH\\COOH\end{matrix}}$$

$$\xrightarrow[\text{reduced pressure}]{\text{Heat}} \underset{\text{Atropic acid}}{C_6H_5-\overset{\overset{\displaystyle CH_3}{|}}{C}-COOH} \xrightarrow[\text{ether}]{HCl} C_6H_5-\overset{\overset{\displaystyle CH_2Cl}{|}}{\underset{\underset{\displaystyle H}{|}}{C}}-COOH \xrightarrow{K_2CO_3} \underset{\text{Tropic acid}}{C_6H_5-\overset{\overset{\displaystyle CH_2OH}{|}}{\underset{\underset{\displaystyle H}{|}}{C}}-COOH}$$

(4) **Structure of Tropine.**

(*i*) the molecular formula of tropine from elemental analysis and mol. mass determination comes out to be $C_8H_{15}NO$.

(*ii*) Usual tests show that tropine is a saturated compound and contains an alcoholic group.

(*iii*) On oxidation with chromic acid, tropine gives rise to the formation of a ketone called tropinone.

This oxidation reaction indicates the presence of secondary alcoholic group in the molecule of tropine.

$$\underset{\text{Tropine}}{\left.C_7H_{13}N\right]}>CHOH \xrightarrow{\text{oxid.}} \underset{\text{Tropinone}}{\left.C_7H_{13}N\right]}>CO$$

(*iv*) Tropine takes up one molecule of methyl iodide forming methiodide.

$$\underset{\text{Tropine}}{C_8H_{15}ON} + CH_3I \longrightarrow \underset{\text{Methiodide}}{C_8H_{15}ON \cdot CH_3I}$$

This reaction shows the presence of a tertiary nitrogen atom in tropine.

(*v*) it is observed that tropinone (an oxidation product of tropine) reacts with benzaldehyde to form a dibenzylidene derivative. This suggested that in tropinone, the keto group is flanked by two methylene groups, $-CH_2-CO-CH_2-$. In view of this, tropinone may be written as II.

$$\underset{\text{I (tropine)}}{C_5H_9N\left[\begin{array}{l}-CH_2\\ \;\;|\\ \;\;CHOH\\ \;\;|\\ -CH_2\end{array}\right.} \xrightarrow{\text{oxid.}} \underset{\text{II (tropinone)}}{C_5H_9N\left[\begin{array}{l}-CH_2\\ \;\;|\\ \;\;CO\\ \;\;|\\ -CH_2\end{array}\right.} \xrightarrow{C_6H_5CHO} \underset{\text{III (dibenzylidene derivative)}}{C_5H_9N\left[\begin{array}{l}-C=CHC_6H_5\\ \;\;|\\ \;\;CO\\ \;\;|\\ -C=CHC_6H_5\end{array}\right.}$$

(*vi*) On oxidation, tropinone yields (±) tropinic acid ($C_8H_{13}NO_4$) and this on further oxidation yields N-methylsuccinimide.

$$\underset{\text{Tropinone}}{C_5H_9N\left[\begin{array}{l}-CH_2\\ \;\;|\\ \;\;CO\\ \;\;|\\ -CH_2\end{array}\right.} \xrightarrow[CrO_3]{\text{oxid.}} \underset{\text{(±) tropinic acid}}{C_5H_9N\left[\begin{array}{l}-COOH\\ \;\;COOH\\ \;\;|\\ -CH_2\end{array}\right.} \xrightarrow[H_2SO_4]{CrO_3} \underset{\text{N-methyl succinimide}}{\begin{array}{l}CH_2-C(=O)\\ |\qquad\qquad \diagdown\\ |\qquad\qquad\;\; N-CH_3\\ |\qquad\qquad \diagup\\ CH_2-C(=O)\end{array}}$$

These reactions lead to the conclusion that:

(*a*) Tropinic acid contains the same number of carbon atoms as tropinone and that it is a cyclic ketone.

(*b*) Tropinone contains N-methyl pyrrolidine ring system.

(*vii*) Exhaustive methylation of tropine produces tropilidene (cycloheptatriene), C_7H_8 and the exhaustive methylation of tropinic acid produces an unsaturated di-carboxylic acid which on reduction forms pimelic acid.

$$\text{Tropine} \xrightarrow[\text{methylation}]{\text{Exhaustive}} \underset{\text{Tropilidene}}{\begin{array}{l}CH_2-CH=CH\\ |\qquad\qquad\quad\;\; |\\ |\qquad\qquad\quad\;\; CH\\ |\qquad\qquad\quad\;\; \|\\ CH=CH-CH\end{array}}$$

$$\text{Tropinic acid} \xrightarrow[\text{methylation}]{\text{Exhaustive}} \text{unsaturated dicarboxylic acid}$$

$$\xrightarrow[\text{Catalyst}]{H_2} \underset{\text{Pimelic acid}}{\begin{array}{l}CH_2-CH_2-COOH\\ |\\ CH_2\\ |\\ CH_2-CH_2-COOH\end{array}}$$

These reactions suggest that tropine as well as tropinone must have a seven-carbon ring and tropinic acid a straight chain of seven carbon atoms.

(*viii*) In view of the above discussion, tropine, tropinone and tropinic acid may be represented as under:

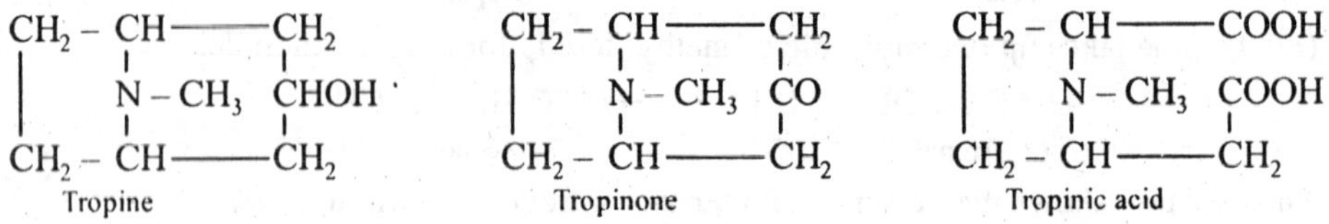

$$\begin{array}{l} CH_2-CH——CH_2 \\ |\qquad\quad |\qquad\quad | \\ |\quad N-CH_3\ \ CHOH \\ |\qquad\quad |\qquad\quad | \\ CH_2-CH——CH_2 \end{array} \qquad \begin{array}{l} CH_2-CH——CH_2 \\ |\qquad\quad |\qquad\quad | \\ |\quad N-CH_3\ \ CO \\ |\qquad\quad |\qquad\quad | \\ CH_2-CH——CH_2 \end{array} \qquad \begin{array}{l} CH_2-CH——COOH \\ |\qquad\quad | \\ |\quad N-CH_3\ \ COOH \\ |\qquad\quad |\qquad\quad | \\ CH_2-CH——CH_2 \end{array}$$

Tropine — Tropinone — Tropinic acid

As is evident, tropine contains a pyridine and pyrrole nucleus with the nitrogen atom common to both.

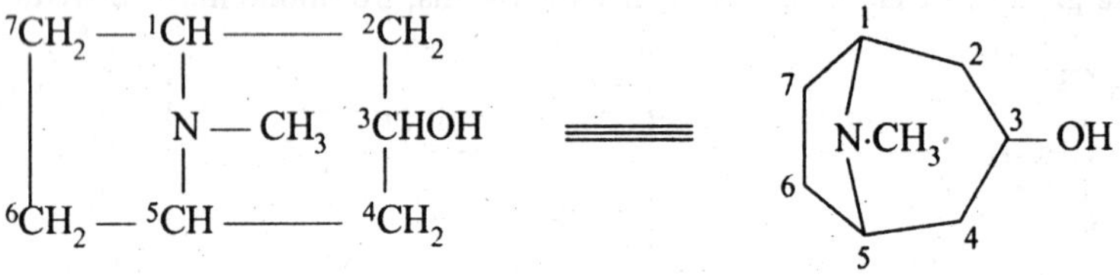

$$\begin{array}{l} {}^7CH_2-{}^1CH——{}^2CH_2 \\ |\qquad\quad |\qquad\quad | \\ |\quad N-CH_3\ \ {}^3CHOH \\ |\qquad\quad |\qquad\quad | \\ {}^6CH_2-{}^5CH——{}^4CH_2 \end{array} \equiv \text{(bicyclic ring with positions 1–7, N·CH}_3\text{ bridge, 3–OH)}$$

(*ix*) The structure of tropine was confirmed by its synthesis. Robinson synthesised tropine from succindialdehyde, methylamine and acetone as follows:

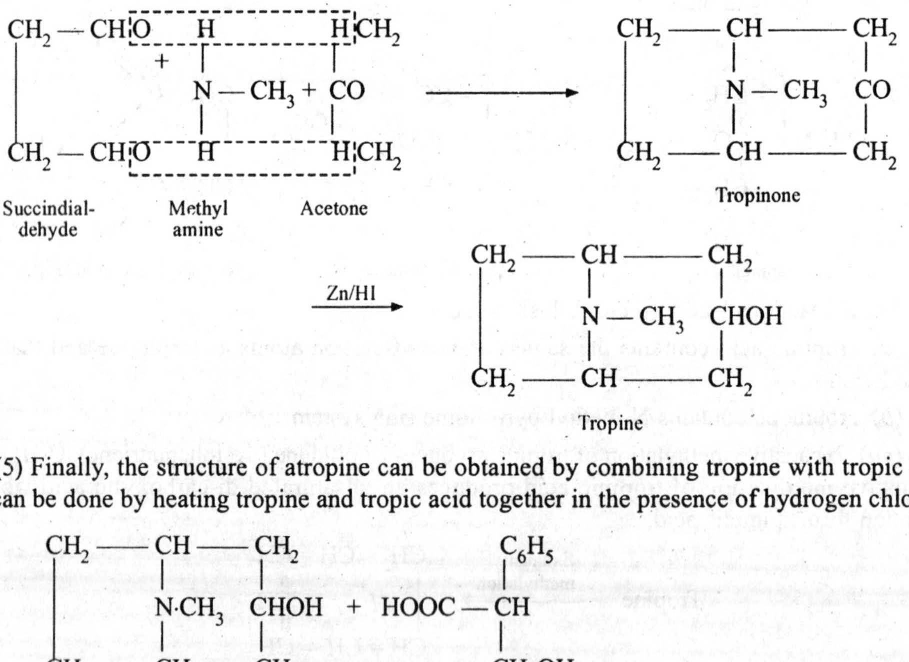

$$\begin{array}{l} CH_2-CHO\quad H\qquad HCH_2 \\ |\qquad\qquad +\ |\qquad\quad | \\ |\qquad\quad N-CH_3+CO \\ |\qquad\qquad\quad |\qquad\quad | \\ CH_2-CHO\quad H\qquad HCH_2 \end{array} \longrightarrow \begin{array}{l} CH_2——CH——CH_2 \\ |\qquad\quad |\qquad\quad | \\ |\quad N-CH_3\ \ CO \\ |\qquad\quad |\qquad\quad | \\ CH_2——CH——CH_2 \end{array}$$

Succindial-dehyde — Methyl amine — Acetone — Tropinone

$$\xrightarrow{Zn/HI} \begin{array}{l} CH_2——CH——CH_2 \\ |\qquad\quad |\qquad\quad | \\ |\quad N-CH_3\ \ CHOH \\ |\qquad\quad |\qquad\quad | \\ CH_2——CH——CH_2 \end{array}$$

Tropine

(5) Finally, the structure of atropine can be obtained by combining tropine with tropic acid. This can be done by heating tropine and tropic acid together in the presence of hydrogen chloride.

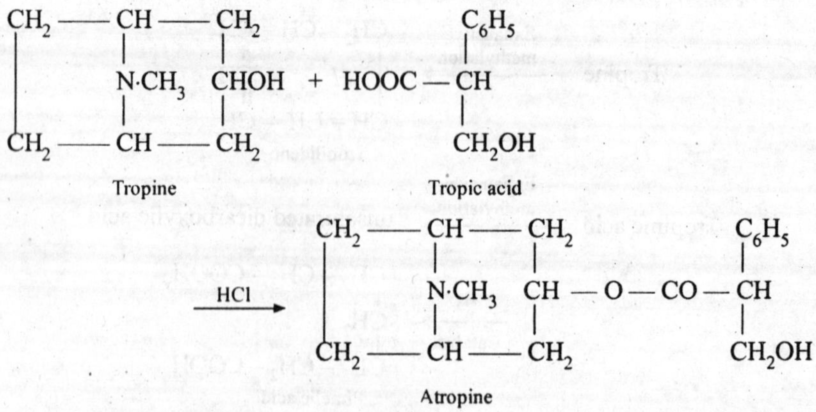

$$\begin{array}{l} CH_2——CH——CH_2 \\ |\qquad\quad |\qquad\quad | \\ |\quad N{\cdot}CH_3\ \ CHOH \\ |\qquad\quad |\qquad\quad | \\ CH_2——CH——CH_2 \end{array} + \begin{array}{c} C_6H_5 \\ | \\ HOOC-CH \\ | \\ CH_2OH \end{array}$$

Tropine — Tropic acid

$$\xrightarrow{HCl} \begin{array}{l} CH_2——CH——CH_2\qquad\qquad\quad C_6H_5 \\ |\qquad\quad |\qquad\quad |\qquad\qquad\qquad\quad | \\ |\quad N{\cdot}CH_3\ \ CH-O-CO-CH \\ |\qquad\quad |\qquad\quad |\qquad\qquad\qquad\quad | \\ CH_2——CH——CH_2\qquad\qquad\quad CH_2OH \end{array}$$

Atropine

i.e., (tropane ring with $N \cdot CH_3$)$-O-CO-CH(C_6H_5)(CH_2OH)$

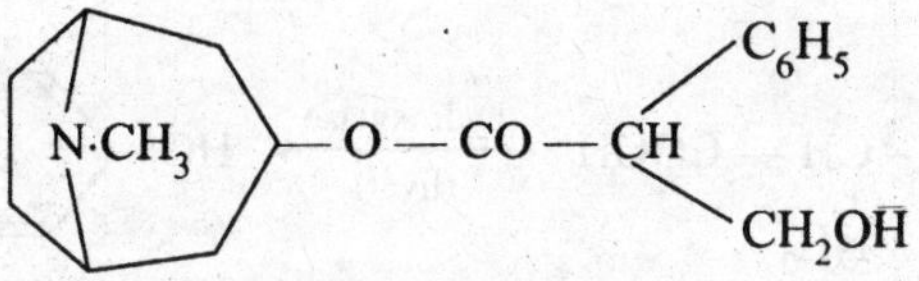

Q. 24. Give the synthesis of nicotine from nicotinonitrite.

Ans. See Q. 21.

Q. 25. Discuss analytical evidence to establish the structure of camphoric acid.

Ans. See Q. 5.

Q. 26. Give analytical evidence to prove the presence of pyridine and N-methyl pyrrolidine rings and their points of linkage in nicotine.

Ans. See Q. 21.

Q. 27. Give the synthesis of camphor.

Ans. See Q. 5.

Q. 28. Complete the following reaction : *(Kerala, 2001)*

Ans. Tropine $\xrightarrow[<150°C]{HI}$ **A** $\xrightarrow{HI}$ **B** $\xrightarrow[\text{hydrochloride}]{\text{distil}}$ **C** $\xrightarrow[\text{Zn dust}]{\text{distil}}$ **D**

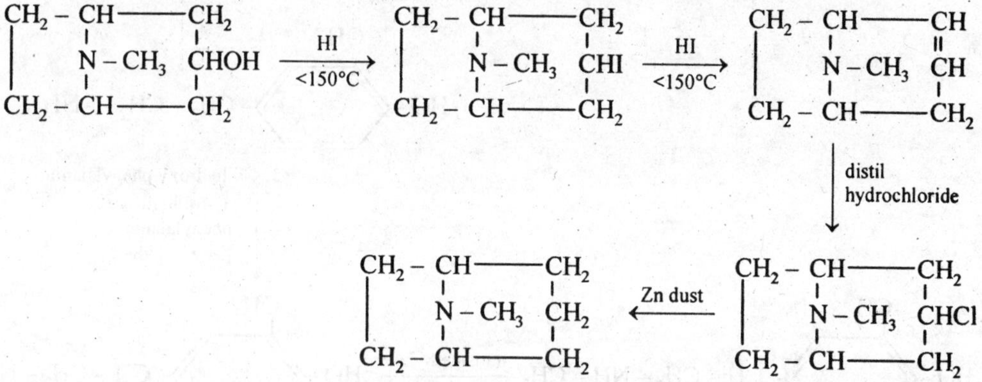

$$\begin{array}{l} CH_2-CH-CH_2 \\ \quad\; N-CH_3 \;\; CHOH \\ CH_2-CH-CH_2 \end{array} \xrightarrow[<150°C]{HI} \begin{array}{l} CH_2-CH-CH_2 \\ \quad\; N-CH_3 \;\; CHI \\ CH_2-CH-CH_2 \end{array} \xrightarrow[<150°C]{HI} \begin{array}{l} CH_2-CH-CH \\ \quad\; N-CH_3 \;\; \overset{\|}{CH} \\ CH_2-CH-CH_2 \end{array}$$

$$\xrightarrow[\text{hydrochloride}]{\text{distil}} \begin{array}{l} CH_2-CH-CH_2 \\ \quad\; N-CH_3 \;\; CHCl \\ CH_2-CH-CH_2 \end{array} \xrightarrow{\text{Zn dust}} \begin{array}{l} CH_2-CH-CH_2 \\ \quad\; N-CH_3 \;\; CH_2 \\ CH_2-CH-CH_2 \end{array}$$

Q. 29. How will you effect the following transformation and account ? *(Kerala, 2001)*

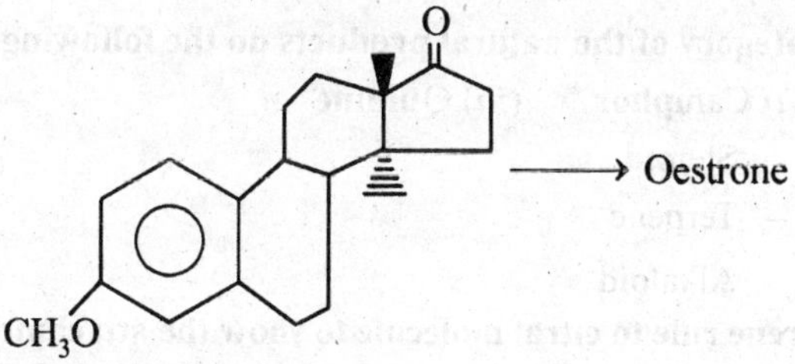

(Steroid structure with CH_3O on aromatic ring A and ketone O on ring D) $\longrightarrow$ Oestrone

Ans. See Q. 11 (2).

Q. 30. How will you arrive at the structure of adrenaline ? Confirm it by synthesis. *(Shivaji, 2000)*

Ans. Synthesis : Following reactions are involved in the synthesis of adrenaline.

$C_6H_5-CH_2-CH(NH_2)-COOH \xrightarrow[\text{(liver)}]{\text{hydroxylase}} HO-C_6H_4-CH_2-CH(NH_2)-COOH$

Phenylalanine

↓ tyrosine hydroxylase

$(HO)_2C_6H_3-CH_2-CH(NH_2)-COOH$

tyrosine

3, 4-dihydroxy phenylalanine (DOPA)

↓ DOPA-decarboxylase (CO)

$(HO)_2C_6H_3-CH_2-CH_2-NH_2$

3, 4-dihydroxy phenylalanine

↓ 3, 4-dihydroxy phenylalanine β-hydroxylase

$(HO)_2C_6H_3-CH(OH)-CH_2-NH_2 \xrightarrow[CH_3]{\text{transmethylase}} (HO)_2C_6H_3-CH(OH)-CH_2-NH-CH_3$

Adrenaline

Q.31. To which category of the natural products do the following belong ?

(*i*) Cholesterol (*ii*) Camphor (*iii*) Quinine (*Panjab 2003*)

Ans. Cholesterol – Steroid

Camphor – Terpene

Quinine – Alkaloid

Q. 32. Apply isoprene rule to citral molecule to show the attachment of isoprene unit. What happens when citral is boiled with an aqueous solution of alkali ? How can you convert the following ketone to citral ? (*Guwahati 2002*)

$\xrightarrow{?}$ Citral

Ans. See Q. 3 and 4

Q. 33. Write the structure of cholesterol and indicate the location number of functional groups in it. How many chiral atoms are present in the molecule ? (*Guwahati 2002*)

Ans. See Q. 15

Q. 34. How is nicotine isolated from tobacco leaves ? (*Goa 2003*)

Ans. Tobacco leaves are dried and finely powdered. Dil. hydrochloric acid is added to the powdered leaves to extract nicotine. Acid extract is treated with NaOH solution and steam distillation is carried out. Crude nicotine is obtained in the form of oily layer. It is purified by fractional distillation in vacuum.

Q. 35. What is conine ? Give its structure. How is it extracted ?

Ans. It is an alkaloid belonging to piperidine group. Its structure can be represented as :

N (ring) — $CH_2-CH_2-CH_3$

|

H

Conine occurs in seeds of the hemlock herb. Socrates, the great Greek philosopher was executed in 400 BC by drinking a cup of hemlock. It is a deadly poisonous substance.

Isolation. The powdered hemlock seeds are distilled with NaOH solution. The distillate is extracted with ether. The ethereal extracts is evaporated and the oily extract is fractionated to obtain conine (b.p. 167°C).

42

BIOSYNTHESIS OF NATURAL PRODUCTS

Q. 1. What is meant by biosynthesis? What types of reactions take place in living organism?

Ans. Description of the actual pathways by which the living organism produce the natural products, is called biosynthesis.

Following types of reactions take place in biosynthesis :

1. Oxidation
2. Hydrogenation
3. Hydrolysis
4. Carboxylation
5. Dehydration
6. Dehydrogenation
7. Decarboxylation
8. O—, N—and C— methylation
9. Acetylation

Q. 2. What are the main points of difference between biosynthesis and chemical synthesis?

Ans.

Biosynthesis	*Chemical Synthesis*
1. Biosynthesis involves the use of enzymes.	1. Chemical synthesis involves the use of catalysts.
2. Most of the biosynthetic reactions like carboxylation, decarboxylation, hydrogenation etc. take place at ordinary temperature and pressure.	2. Reactions taking place in chemical synthesis usually require hard conditions of temperature and pressure.
3. While describing biosynthesis pathways, we usually divide the natural product into several units and then, we imagine the synthesis of individual units.	3. Synthesis of the natural product considered straight in the chemical synthesis.

Q. 3. Describe biosynthetic pathways for terpenes.

Ans. According to the isoprene rule, terpenes are made up of isoprene units joined head to tail. The question is how this unit (isoprene) is formed and how these units link to form various types of terpenoids. It is assumed that the fundamental units used in the cell in synthesis are water, carbon dioxide, formic acid and acetic acid. The active compounds are acyl derivative of coenzyme A. This coenzyme is a complex thiol derivative and is written as CoA—SH. The biosynthesis of cholesterol from acetic acid labelled with ^{14}C in the methyl group (C_m) and in the carboxyl group (C_c) has led to the suggestion that the carbon atoms in the isoprene units are as shown below:

$$(C_m)(C_m)>C_c-C_m-C_c$$

This distribution is confirmed by the formation of senecioic acid (3-methylbut-2-enoic acid). Formation of this carbon skeleton is supported by the fact that labelled isovaleric acid forms cholesterol in which the isopropyl group and the carboxyl group have been incorporated."

$$(CH_3)(CH_3)>CH-CH_2-COOH \quad \text{Isovaleic acid}$$

The **biosynthesis** of terpenoids can be subdivided into three steps :

(*i*) The conversion of acetate into mevalonic acid takes place as given below:

$$2CH_3COOH + 2CoA-SH \longrightarrow 2H_2O + 2CH_3Co-SCoA$$

CoA—SH; S—CoA; + CoA—SH; CH_3Co — SCoA; + CoA — SH; OH; HOOC; O; SCoA; R(HMG) +; NADPH; NADP; OH; HOOC; CHO; Mevaldic acid; NADPH; NADP; OH; HOOC; CH_2OH; di-Mevalonic acid (di-MVA)

NADPH stands for nicotinamide-adenine dinucleotide phosphate and HMG stands for hydroxy methyl glutarate.

(*ii*) Conversion to Isopentyl pyrophosphate. Mevalonic acid contains six carbon atoms. It must lose one carbon to give isopentane unit. If we start with labelled MVA, it is found that it is the carboxyl group in MVA which is lost. The various steps are shown below:

OH; HOOC; CH_2OH; MVA; ATP; ADP; OH; HOOC; CH_2OP; ATP; ADP; OH; C; CH_2OP; H—O; O; $-CO_2$; $-H_2O$; CH_2OPP

Pyrophosphate
3-Methyl-but-3 enyl
(Isopentyl)

ADP and ATP stand for adenosine diphosphate and adenosine triphosphate respectively.

On the basis of biosynthetic studies, another isoprene rule has been formulated. It is called *biogenetic isoprene rule.* This rule states that members of the isopentane group can be derived from simple hypothetical precursors such as geraniol, farnesol and squalene. The rule also includes those compounds which originate from regular isoprenoid precursors. The biogenetic isoprene unit is 3-methylbut-3 enyl pyrophosphate. It exists in two resonating structures in equilibrium in the presence of suitable enzymes.

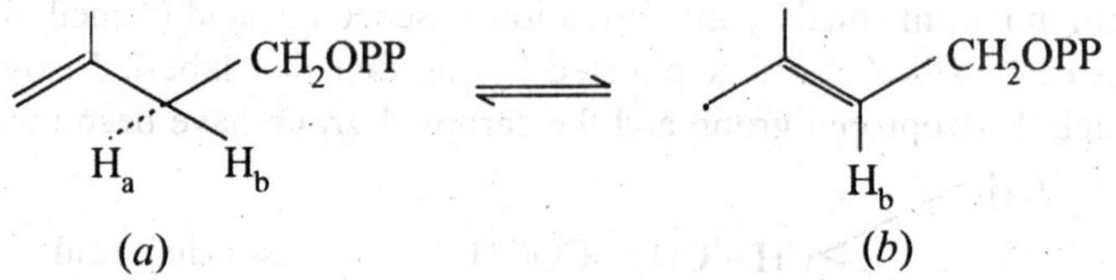

CH_2OPP H_a H_b ⇌ CH_2OPP H_b

(*a*) (*b*)

(*iii*) Convension into terpenoids. It has been experimentally found that the units (*a*) and (*b*) combine to form geranyl pyrophosphate (trans). In this synthesis (*a*) acts as a nucleophile reagent and (*b*) acts as an electrophile and they react to give head to tail union. The reaction is shown as:

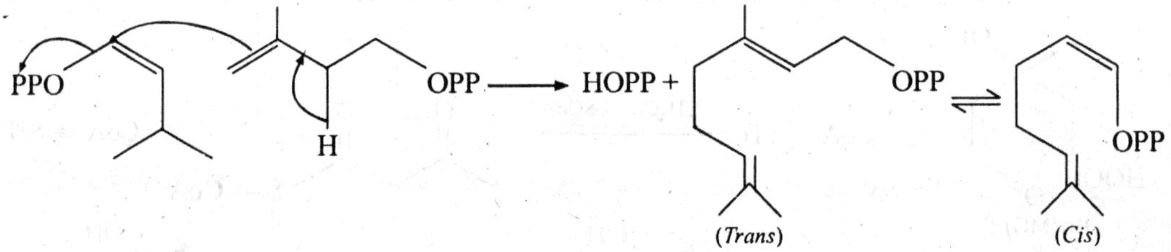

PPO OPP H ⟶ HOPP + OPP ⇌ OPP

(*Trans*) (*Cis*)

Now geranyl pyrophosphate acts as a precursor for the monocyclic monoterpenoids via the cis isomer (nerol). Acid catalysed cyclisation of geraniol into α-Terpineol is shown below:

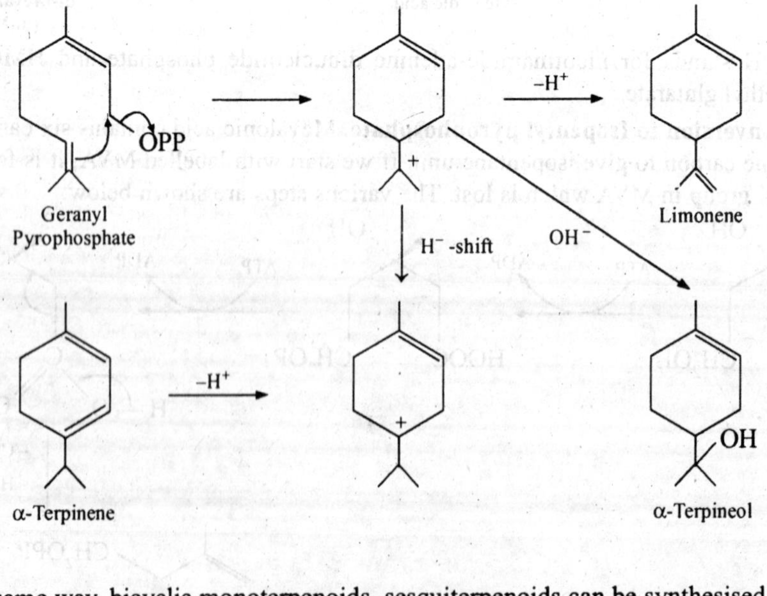

OPP — Geranyl Pyrophosphate ⟶ + —$-H^+$⟶ Limonene

H^--shift; OH^-

α-Terpinene ←$-H^+$— + ; OH — α-Terpineol

In the same way, bicyclic monoterpenoids, sesquiterpenoids can be synthesised.

Q. 4. Describe biosynthetic pathways for steroids.

Ans. Cholesterol is a content of fats produced by animals. The biosynthesis of cholesterol was carried from acetic acid labelled in methyl or carboxyl or in both the groups ($^{14}CH_3$ $^{14}COOH$). Tracer studies established the distribution of carbon atoms shown below:

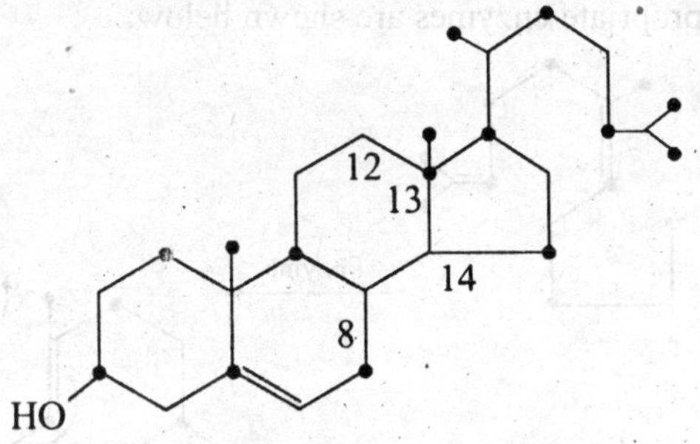

In this structure, the carbon atoms derived from methyl group of acetic acid are shown by dots. Clearly, acetic acid molecule is regarded as the fundamental unit. It has been shown experimentally that mevalonic acid (MVA) can be completely converted into cholesterol by rat liver. The conversion of acetic acid into mevalonic acid has already been considered. How mevalonic acid is converted into cholesterol is discussed here.

Heilbron *et al* suggested that squalene is a precursor of cholesterol. Biosynthetic experiments reveal that squalene is produced by the linkage of two farnesyl bromide molecules joined head to tail by Wurtz reaction.

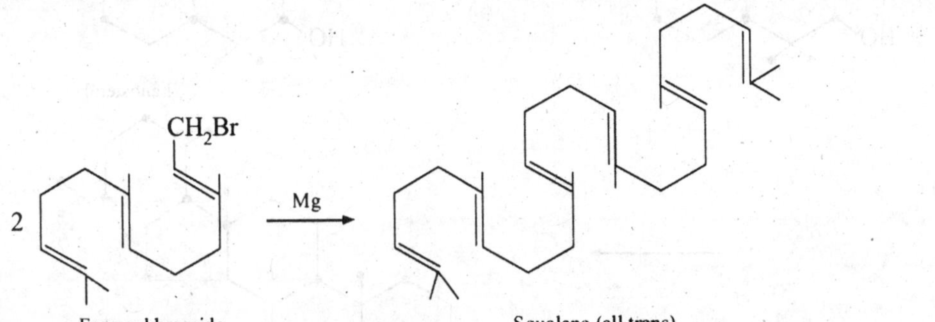

Famesyl bromide Squalene (all trans)

Robinson *et al* proposed a scheme for the cyclisation of squalene molecule with the loss of three methyl groups. The broken lines show the loss of methyl groups.

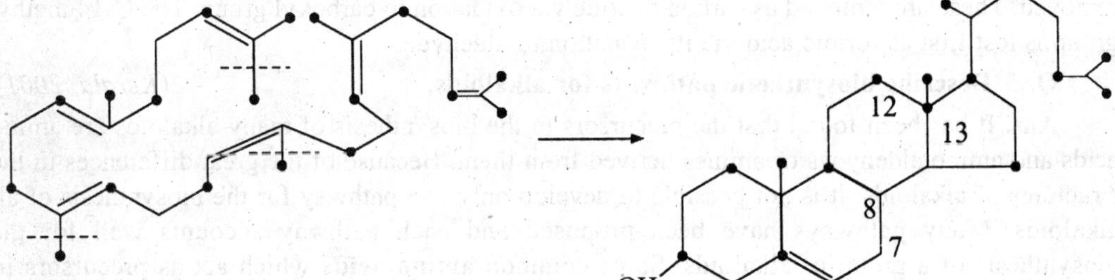

According to Woodward *et at* (1955), squalene is first cyclised to lanosterol and then it loses three methyl groups to form cholesterol. It is also found that lanosterol is converted into cholesterol in rats. Block *et al* also carried out the synthesis of lanosterol from labelled acetate. Woodward suggested that squalene ring closes to form lanosterol. Block *et al* then proposed a 1, 2-shift of the methyl group from C_{14} to C_{13} and another 1, 2-shift from C_8 to C_{14}. The various steps under the influence of appropriate enzymes are shown below:

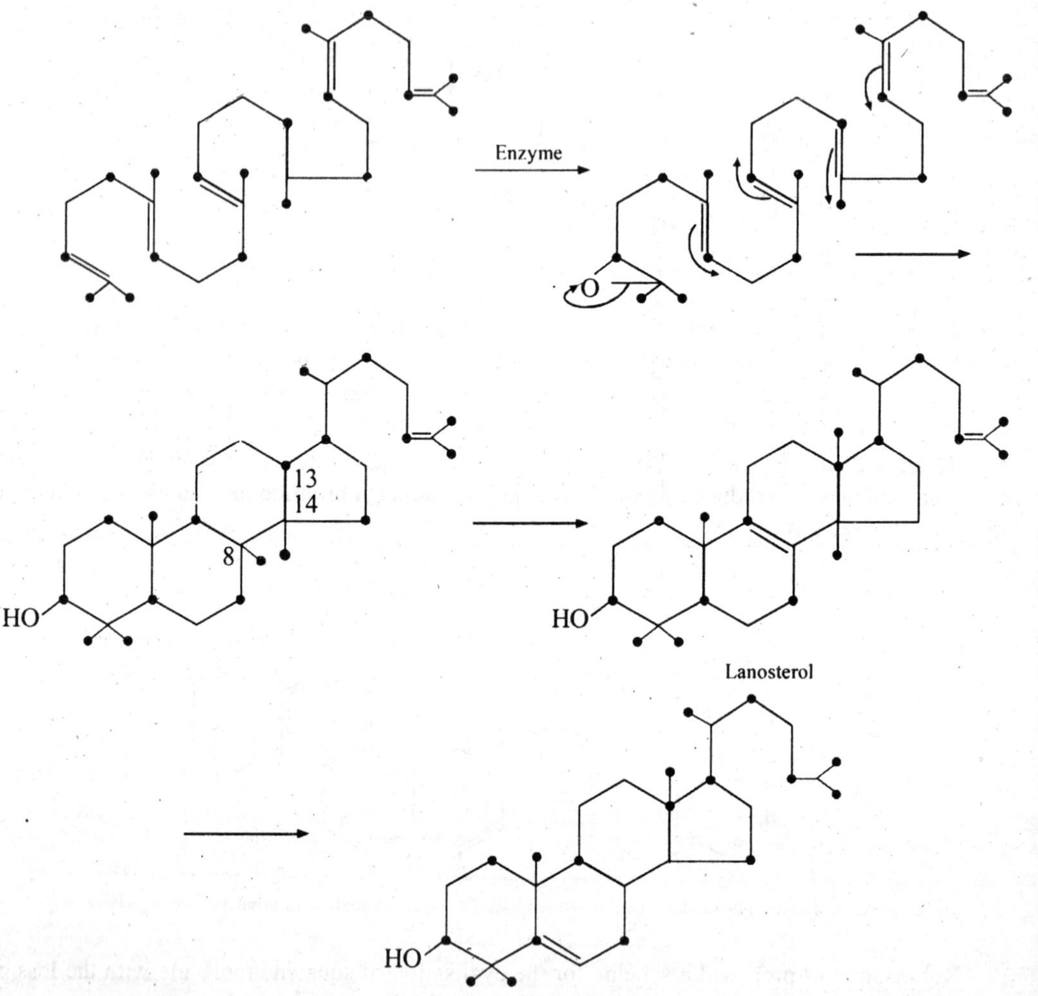

Enzyme

13
14
8
HO

HO

Lanosterol

HO

Cholesterol

In the conversion of lanosterol into cholesterol, the methyl groups at C-4, 4′ and 14 are removed. These are removed as carbon dioxide via oxidation to carboxyl group. The C-14 methyl group is lost first as formic acid via its oxidation to aldehyde.

Q. 5. Describe biosynthetic pathways for alkaloids. (*Kerala, 2001*)

Ans. It has been found that the precursors in the biosynthesis of many alkaloids are amino acids and amino aldehydes or amines derived from them. Because of the great differences in the structures of alkaloids, it is not possible to develop only one pathway for the biosynthesis of all alkaloids. Many pathways have been proposed and each pathway accounts well for the biosynthesis of a group of alkaloids. Some common **amino acids** which act as precursors in alkaloid biosynthesis are:

$H_2N-(CH_2)_3-CH(NH_2)COOH$
Ornithine

$H_2N-(CH_2)_4CH(NH_2)COOH$
Lysine

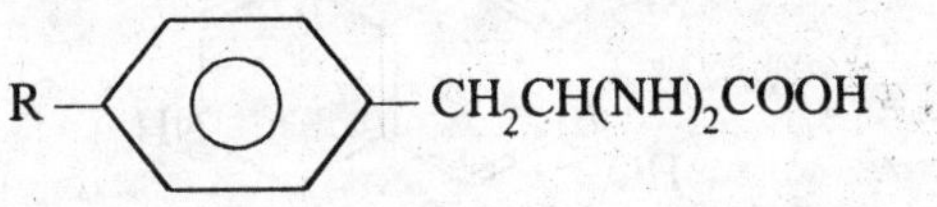

$R-C_6H_4-CH_2CH(NH)_2COOH$
Phenylalanine (when R = H)

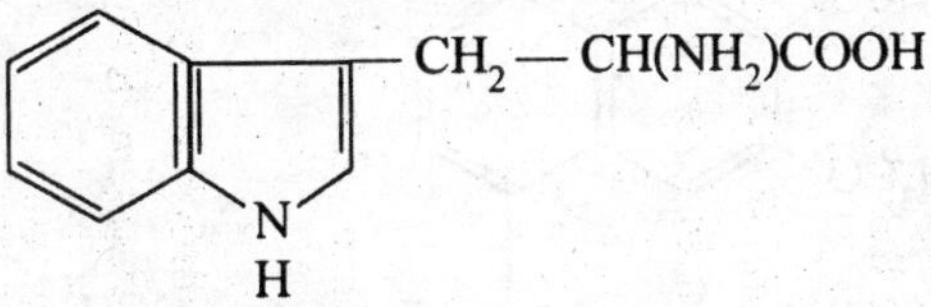

$CH_2-CH(NH_2)COOH$
Tryptophan

Some important reactions of amino acids under the influence of enzymes are:

(*i*) **Decarboxylation.** This results in the formation of an amine

$$RCH(NH_2)COOH \xrightarrow{\text{Enzyme}} RCH_2NH_2 + CO_2$$

(*ii*) **Oxidative deamination.** It can occur in two different ways:

$$RCH(NH_2)COOH \xrightarrow{\text{Enzyme}} RCO.COOH \longrightarrow RCHO$$

$$RCH(NH_2)COOH \xrightarrow{\text{Enzyme}} RCH_2NH_2 \longrightarrow RCHO$$

(*iii*) **Mannich reaction.** This is the reaction between a molecule with an active hydrogen atom, an aldehyde and an amine.

$$HCHO + (C_2H_5)_2NH \longrightarrow (C_2H_5)_2N = CH_2 + H_2O$$

Oxidation, reduction, dehydration, alkylation, acylation, etc. are other reactions involved in the biosynthesis of alkaloids.

For elucidating biosynthetic pathways, some postulated routes are tested by means of labelled precursors administered to plants. After a suitable interval of time, the alkaloid is isolated from the plants and the isotopic content in it is examined.

Q. 6. Describe biosynthesis of papaverine and laudanosine belonging to isoquinoline group.

Ans. Papaverine together with nearly 24 other alkaloids, some of the important ones being laudanosine, narcotine, morphine, codein and thebaine occur in the opium *poppy*. All these are known as *opium alkaloids*.

It has been shown by tracer techniques that papaverine is derived from tyrosine by the enzymatic action. Tyrosine can be obtained from phenyl alanine by the cation of some enzymes. Tyrosine produces dopamine and 3, 4-dihydroxyphenyl acetaldehyde and these two products in turn undergo condensation.

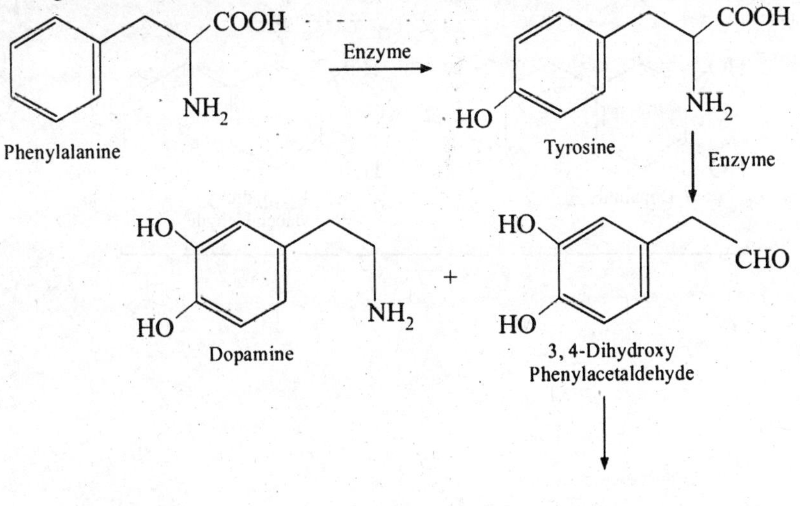

Phenylalanine (COOH, NH_2) $\xrightarrow{\text{Enzyme}}$ Tyrosine (HO, COOH, NH_2) $\xrightarrow{\text{Enzyme}}$

Dopamine (HO, HO, NH_2) + 3, 4-Dihydroxy Phenylacetaldehyde (HO, HO, CHO)

↓

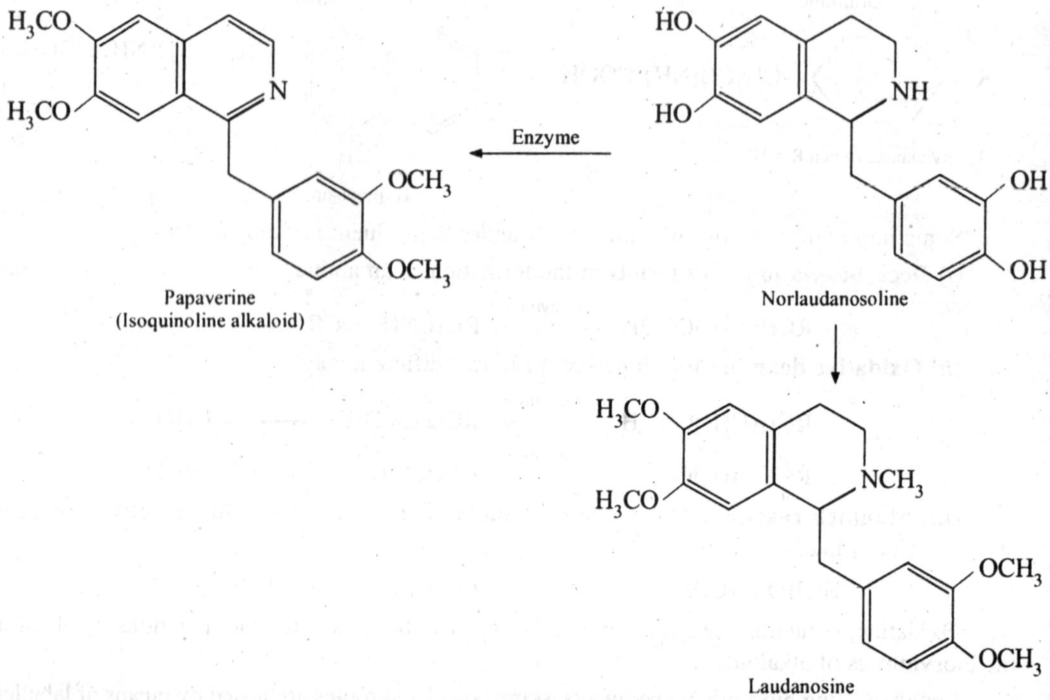

Norlaudanosoline —Enzyme→ Papaverine (Isoquinoline alkaloid)

Norlaudanosoline → Laudanosine

Q. 7. Describe biosynthesis of morphine alkaloids.

Ans. It has been observed that opium alkaloids (*i.e.* morphine, codein, thebaine) are biosynthesised from the alkaloid, reticuline. Reticuline can be derived from norlaudanosoline which in turn can be biosynthesised from phenylalanine. The various steps involved are illustrated as under:

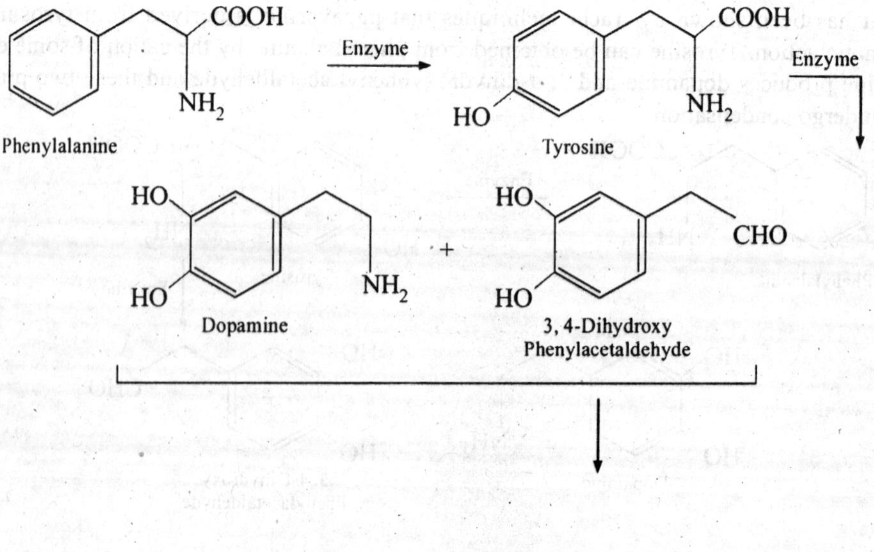

Phenylalanine —Enzyme→ Tyrosine —Enzyme→ Dopamine + 3, 4-Dihydroxy Phenylacetaldehyde →

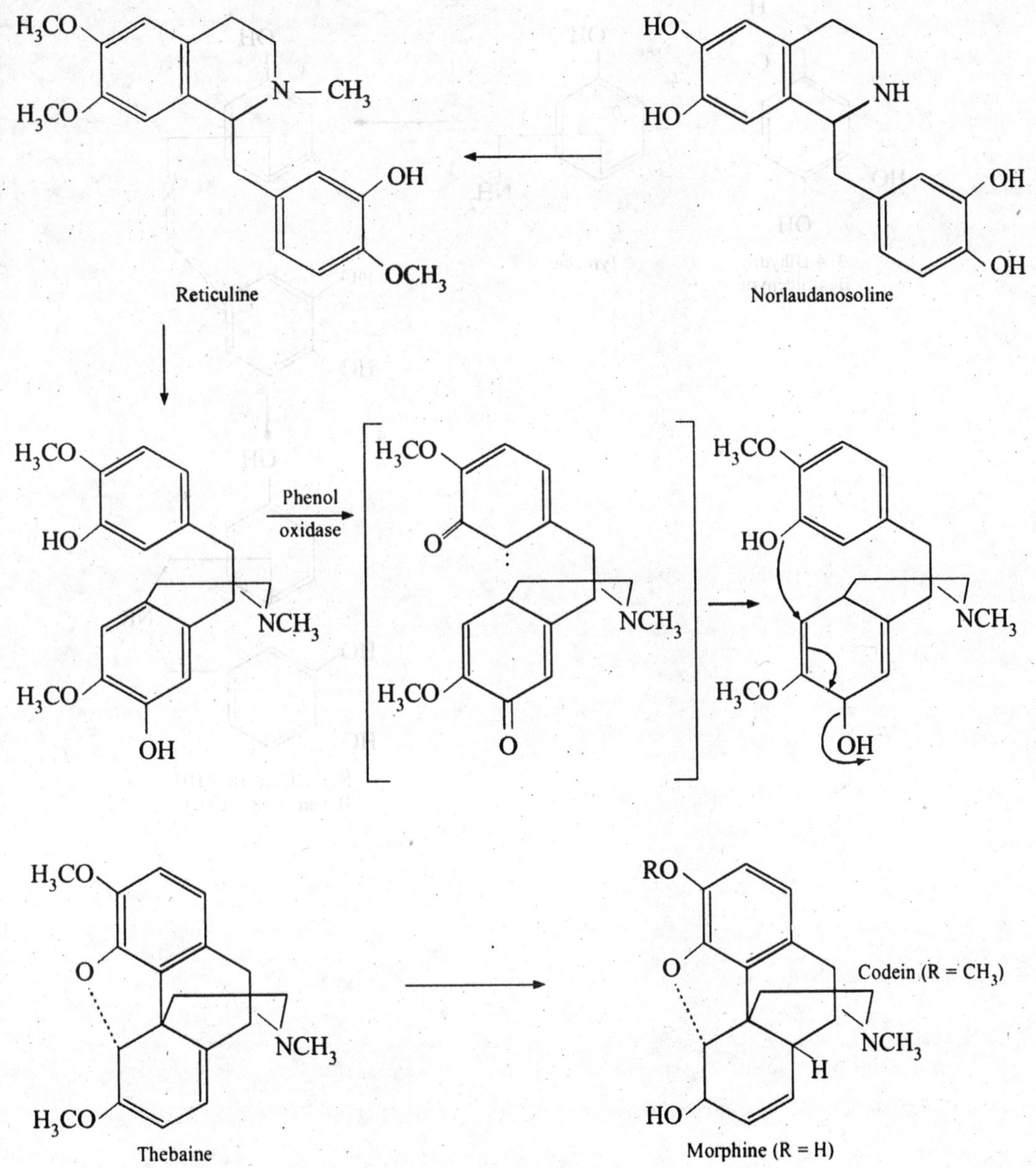

H$_3$CO, H$_3$CO, N—CH$_3$, OH, OCH$_3$
Reticuline

HO, HO, NH, OH, OH
Norlaudanosoline

H$_3$CO, HO, NCH$_3$, H$_3$CO, OH

Phenol oxidase

H$_3$CO, O, NCH$_3$, H$_3$CO, O

H$_3$CO, HO, NCH$_3$, H$_3$CO, OH

H$_3$CO, O, NCH$_3$, H$_3$CO
Thebaine

RO, O, Codein (R = CH$_3$), NCH$_3$, H, HO
Morphine (R = H)

From this representation, the rotation of one part of the molecule around the dotted line, the relationship of this alkaloid to morphine and similar alkaloids becomes clear.

Q. 8. Describe biosynthesis of amaryllidacea alkaloids.

Ans. The alkaloids occurring in the amaryllidaceae represent a biogenetically uniform group. The compounds originate from two precursors both of which are derived from phenyl propane bodies (**Protocatechuic aldehyde**), which originate from caffec acid. The second precursor is **tyramine.** Both these compounds first form a Schiff's base from which the protoalkaloids **norbelladine** and **belladine** originate. The various steps involved in the biosynthesis of norbelladine and belladine are given below:

3, 4-Dihydro Benzaldehyde + Tyrosine →

Norbellidine (R = H)
Belladine (R = CH_3)

43

DYES
(With Theories of Colour and Constitution)

Q. 1. What is a dye? What are the characteristics of a dye?

(Madras, 2004; Delhi 2004; Baroda 2004)

Ans. *A dye may be defined as a coloured substance which when applied to the fabrics imparts a permanent colour and the colour is not removed by washing with water, soap or on exposure to light.* All coloured organic substances are not necessarily dyes. For instance, both picric acid and trinitrotoluene are yellow in occur, but only the former can fix to a cloth and is a dye, while, the latter does not fix to a cloth and is not a dye. Characteristics of a dye are:

1. A dye should be a coloured substance
2. It should be able to fix itself to the material from solution or be capable of being fixed on it.
3. It should resist the action of water, soap and light.

Q. 2. Describe in brief various dyeing processes.

Ans. The following four methods are usually employed for dyeing, depending upon the nature of the cloth (fabric) and dye itself.

(*a*) *Substantive dyeing method,*

(*b*) *Adjective dyeing (mordant) method,*

(*c*) *Developed dyeing (ingrain) method,*

(*d*) *Vat dyeing method.*

(*a*) **Substantive dyeing method.** This is also called *direct dyeing* method. In this method, a cloth is dyed directly by immersing it in a solution of dye. It is largely used for dyeing of animal fabrics (wool and silk), but not for vegetable fabrics (cotton and artificial silk).

(*b*) **Adjective dyeing method.** This is also known as *mordant dyeing* method. In this method, the fabric is first impregnated with certain substances referred to as mordants which combine with the dye. The mordants used are basic substances for acidic dyes and acidic substances for basic dyes. The impregnated fabric is then immersed in the solution of a dye to form insoluble *lakes* on the cloth. Dyeing with alizarin is done by this method.

(*c*) **Developed dyeing method.** This is also known as *ingrain dyeing method.* It consists in preparing and applying the dye on the cloth itself. For example for azo-dyes the cloth to be dyed is first impregnated with a phenol solution and then treated with a solution of diazotised amine to produce azo-dye on the cloth.

(*d*) **Vat dyeing method.** The water insoluble coloured compound which are used in their reduced state (leuco-compounds) for dyeing purpose are called **Vat dyes.**

In this method, the cloth to be dyed is soaked in the solution of reduced dye and then spread in the air so that the leuco-compound is oxidised back to generate the original dye.

Q. 3. Give the classification of dyes based on chemical constitution.

(*Mumbai, 2004; Delhi 2004; Baroda 2004*)

Ans. The chemical classification is based on the common parent structure of the dye. The number of dyes based on this classification is fairly large. We shall give here an account of important classes of dyes.

(*i*) **Azo-Dyes.** These dyes contain one or more azo groups (–N = N–) which form bridges between two or more aromatic rings. The azo-dyes form the largest and the most important group of dyes. Azo group is the chromophore. Aniline yellow methyl orange congo red and bismark brown are some examples of this class.

(*ii*) **Nitroso dyes.** These dyes contain a nitroso group (–NO) as the chromophore and hydroxyl group (–OH) as the auxo chrome. Fast green O (Dinitrosoresorcinol) and Gambine Y (α-nitroso-β-naphthol) are some examples of this class.

(*iii*) **Nitro dyes.** These dyes contain a nitro group ($-NO_2$) as the chromphore and hydroxyl group (–OH) as the auxo-chrome. They are generally nitro-derivatives of phenols containing at least one nitro group in ortho or para position to the hydroxyl group. Pictric acid, (2, 4, 6-trinitrophenol), martius yellow (2, 4-dinitro-l-naphthol) belong to this category.

(*iv*) **Phthaleins.** These dyes are obtained by the condensation of phthalic anhydride with phenols in the presence of some dehydrating agent like conc. H_2SO_4, anhydrous $ZnCl_2$. An important example of this class is phenolphthalein.

(*v*) **Xanthene dyes.** These dyes are usually derived from Xanthen. These dyes have a common structural feature and are obtained by the condensation of phenols with phthalic anhydride in the presence of a dehydrating agent like anhydrous $ZnCl_2$, H_2SO_4, etc. Examples of this class are fluorescein and eosin.

(*vi*) **Triphenyl methane dyes.** The nucleus of this group is triphenyl methane. These dyes are usually obtained by introducing groups like NH_2, NR_2 or OH into the rings of this nucleus which contains three phenyl radicals. The typical instances of triphenyl-methane dyes are rosaniline, malachite green and crystal violet.

Q. 4. Give a classification of dyes based on application to fibre. (*Mumbai, 2004*)

Ans. Dyes can be classified as follows based on application to fibre:

(*i*) **Acid dyes.** These dyes are essentially the sodium salts of acids which may contain sulphonic acid or phenolic acid group. The colour of an acid dye is in its negative ion. These dyes give very bright colour and have a wide range of fastness. These dyes are also known as *anionic dyes.*

The acid dyes are always used in acidic solution. The fabric is stirred in the hot solution of the dye in the presence of either an acid or salt till it is smoothly dyed. It is then removed and dried. Some typical instances of acid dyes are picric acid, orange-II, naphthol yellow, etc.

(*ii*) **Basic dyes.** These are also called *Cationic dyes*. The basic dyes are those which contain a basic amino group and it is protonated under the acid conditions of fibres by formation of salt linkages with anionic or acidic group in the fibres. Some examples of basic dyes are crystal violet, methylene blue and methyl violet.

(*iii*) **Direct dyes.** These dyes are also known as *substantive dyes*.

They dye cotton as well as wool and silk. The dye is applied to the fabric by immersing it in its hot boiling solution, removing and then drying the fabric. Some typical members of direct dyes are congo red, naphthol yellow S and martius yellow.

(*iv*) **Developed dyes.** These dyes are also called *azotic* or *ingrain dyes*. These are the dyes which are produced within the cloth itself as a result of chemical action between the two reactants producing the dye. For example, the cloth is first dipped in an alkaline solution of phenol, resorcinol or β-naphthol and then immersed in an alkaline solution of diazo compound. The coupling reaction takes place between the phenols and diazo compound within the textile fibres giving rise to the formation of a dye.

(*v*) **Mordant dyes.** These dyes are also called *adjective dyes.* They cannot directly dye cotton, silk or wool but require the help of a mordant. A mordant is a substance which is taken up by the fibres and which in turn takes up the dye. There are basic and acidic mordants. If the dye is acidic, the mordant must be basic, *e.g.*, salts of Cr, Al, Sn and Fe. On the other hand, if the dye is basic, the mordant must be acidic, *e.g.*, tannin or tannic acid containing some amount of tartar emetic.

The mordanted cloth is dipped into the solution of the dye. The dye is absorbed by the mordant forming an insoluble mordant dye compound which gets firmly fixed within the fibres. Certain dyes give different colours with different mordants. Alizarin gives red, violet or purple colour with Al, Fe or Ba salts, respectively.

(*vi*) **Vat dyes.** These dyes are insoluble in water but their reduced forms are soluble. On reduction with alkaline sodium bisulphite, the vat dyes are converted into water soluble compounds called *leuco-compounds.* They dye both vegetable and animal fibres directly. The vat dyes are mostly used to dye cotton. Some typical examples of vat dyes are indigo and anthraquinone dyes.

Q. 5. Give Witt's theory of colour and constitution of dyes. (*Kerala, 2001; Bangalore, 2002; Delhi 2003; Purvanchal, 2007; Nagpur 2008*)

Or

Explain (1) Chromophores (2) Auxochromes.

Or

What are (*i*) bathochromic groups (*ii*) hyprochromic groups?

Ans. According to Witt's theory of colour and chemical constitution of dyes, a coloured substance or a dye is essentially composed of two parts namely, **chromophores and auxochromes.**

(1) Chromophores. *The colour in an organic compound is due to the presence of certain groups with multiple bonds.* Witt designated these groups with multiple bonds as **chromophores.** The chromophores are the colour bearing groups and their mere presence produces a colour in the molecule of an organic compound.

The important chromophoric groups are:

1. Nitro, $-N\begin{smallmatrix}\nearrow O\\ \searrow O\end{smallmatrix}$ 2. Nitroso, $-N=O$

3. Azo, $-N=N-$ 4. Azoxy, $-N=\underset{\underset{O}{\downarrow}}{N}N-$

5. Carbonyl, $>C=O$

6. *p*-quinonoid, $=\langle\quad\rangle=$

7. *o*-quinonoid,

The organic compound containing a chromophoric group in its molecule is referred to as **chromogen.** The presence of any one of the above stated chromophoric groups in the molecule is sufficient to impart colour to an organic compound. For instance,

	Compound	Colour
1.	Benzene, C_6H_6	Colourless
2.	Nitrobenzene, $C_6H_5.NO_2$	Pale-yellow
3.	Azobenzene, $C_6H_5 - N = N - C_6H_5$	Red
4.	*p*-Benzoquinone $O = C_6H_4 = O$	Yellow
5.	*o*-Benzoquinone $O = C_6H_4 = O$	Orange-red

It has been observed that the chromogen containing only one chromophoric group is usually coloured (yellow). The intensity of colour generally increases with number of chromophoric groups. A single $C = C$ group as in ethene $CH_2 = CH_2$ does not produce any colour, but if a number of these groups are present in conjugation, the colour may develop. For instance, $CH_3 -(CH = CH)_6 - CH_3$ is yellow in colour.

In case of weaker chromophoric groups, more than one group is needed to develop a visible colour.

(2) **Auxochromes.** *Certain groups (which are not chromophores) when present in the chromogen tend to intensify its colour.* Such groups are referred to as **auxochromes.** The auxochromes may be acidic or basic in character.

The most effective auxochromes are as under:

1. Hydroxyl	$- OH$	5. Sulphonic acid	$- SO_3H$
2. Alkoxy	$- OR$	6. Carboxyl	$- COOH$
3. Amino	$- NH_2$	7. Phenolic	$- OH$

4. Alkylated amino $-$ NHR or $- NR_2$

Auxochromes are salt-forming groups and perform two functions:

(*i*) *Auxochromes deepen the colour of chromogen.*

(*ii*) *The presence of auxochromes is essential to make the chromogen a dye.*

Further, the groups which deepen the colour are referred to as **bathochromic groups.** The groups which bring about the opposite effect are known as **hypsochromic groups.** The replacement of H in NH_2 group by R or Ar has bathochromic effect while replacement of H in OH group by acetyl group has a hypsochromic effect. Often a colourless chromogen becomes coloured when an auxochrome is introduced in the molecules. For example, benzophenone (colourless) becomes yellow when an amino ($- NH_2$) group is introduced into it. Nitroaniline is deeper in colour than nitrobenzene.

Q. 6. Describe quinonoid theory of colour and chemical constitution. (*Agra 2003*)

Ans. According to this theory, all colouring substances may be represented by quinonoid structures (*o* – or – *p*). If a particular substance can be formulated in a quinonoid form, it is coloured otherwise it is colourless.

(*i*) According to the quinonoid theory, benzene is colourless while benzoquinone is coloured.

Benzene (Colourless)

Benzoquinone (Yellow)

(*ii*) On the basis of this theory, the dye like phenolphthalein is coloured when present in *p*-quinonoid structure but is colourless when *p*-quinonoid structure is absent.

Phenolphthalein (Colourless)

Phenolphthalein (Red coloured)

It has been observed that quinonoid theory fails to explain the colouring characteristics of all the compounds, *e.g.*, iminoquinone and di-iminoquinone have a quinonoid structure but they are colourless.

Iminoquinone

Di-iminoquinone

Similarly, many coloured compounds like diacetyl, and azobenzene are coloured but they cannot be represented by quinonoid structures.

Q. 7. Describe valence bond theory of colour and constitution.

(Shivaji 2000; Nagpur 2002, Kuvempu 2005)

Ans. This theory is also called **resonance theory.** The various important postulates of valence bond theory are as under;

(*a*) Chromophores are the groups of atoms in which the π-electrons may get transferred from ground state to excited state by the absorption of radiations. This absorption gives the colour.

(*b*) Auxochromes are the groups which tend to increase resonance by interaction of the unshared pair of electrons on nitrogen or oxygen atoms of the auxochromes with the π-electrons of the aromatic rings. This increase in resonance results in an increase of the intensity of absorption of light and shifts the absorption band to longer wavelength. As a result, there occurs deepening of the colour. It has been observed that increase in resonance always deepens the colour of a compound.

(*c*) The dipole moment changes as a consequence of oscillation of electron pairs. The ease of excitation of different groups follows the order:

$$N=O > C=S > N=N > C=O > C=C$$

(*d*) Valence bond theory explains the relation between colour and the symmetry of the molecule or transition dipole of the molecule. As the number of charged canonical structures increases, the colour of the compound deepens.

Now we shall consider some cases to illustrate the resonance theory.

(*i*) Benzene is colourless, nitrobenzene is pale-yellow and nitroaniline is orange red.

Benzene is a resonance hybrid of two Kekule structures I and II. In addition, a small number of charged canonical structures of the type III contributes relatively little to the resonance hybrid of benzene molecule.

I ⟷ II ⟷ III ⟷ ------

Thus benzene absorbs light only in the ultraviolet region and the absorption is weak due to the symmetry of the benzene molecule. Therefore, benzene is colourless. Unsymmetrical molecules absorb strongly and appear coloured.

On the other hand, in nitrobenzene the charged structures contribute much more than in case of benzene. Consequently nitrobenzene absorbs light of longer wavelength producing a pale yellow colour. Further, the intensity of absorption is increased in nitrobenzene owing to the loss of symmetry of the molecule.

In a similar manner, in *p*-nitroaniline, the contribution of the charged structure is still larger and hence the light of longer wavelength is absorbed and further, the deepening of the colour to orange red is produced.

$:NH_2$ $\oplus NH_2$

(*ii*) Valence bond or resonance theory also explains why pure *p*-nitrophenol is colourless but has yellow colour in alkaline solution. This is explained on the basis that in alkaline solution, *p*-nitrophenol exists as nitrophenoxide ion in which only the charged structures are contributing to the resonance hybrid and hence the compound absorbs light of higher wavelength.

(*iii*) The resonance theory also explains why *p*-aminobenzene is yellow but in acidic medium, it becomes violet. The deepening of colour in acidic solution is due to the fact that in the yellow form it is a resonance hybrid of the two structures out of which only one is charged while in the acidic solution (violet form) both the contributing structures are charged. The absorption band is shifted to longer wavelength and the colour gets deepened in the acidic solution.

(*iv*) As the conjugated system of double bonds provides a long path for resonance, it plays an extremely important role in producing deep colour. The longer the conjugation in a molecule, the deeper will be the colour.

Q. 8. Describe molecular orbital theory of colour and constitution. (*Delhi, 2002*)

Ans. According to molecular orbital theory, the excitation of an atom or a molecule involves the transference of one electron from an orbital of lower energy (a bonding orbital) to that of higher energy (an antibonding orbital). The electrons involved may be σ (sigma), π (pi) or *n* (non-bonding) electrons. The higher energy states are usually known as *antibonding orbitals* while the lower energy states are called *bonding orbitals*. The antibonding orbitals associated with σ and π bond are represented by σ* and π* orbitals, respectively. There are, however, no antibonding orbitals associated with *n* (non-bonding) electrons as they do not form bonds. Energy levels of various orbitals are shown in the figure below:

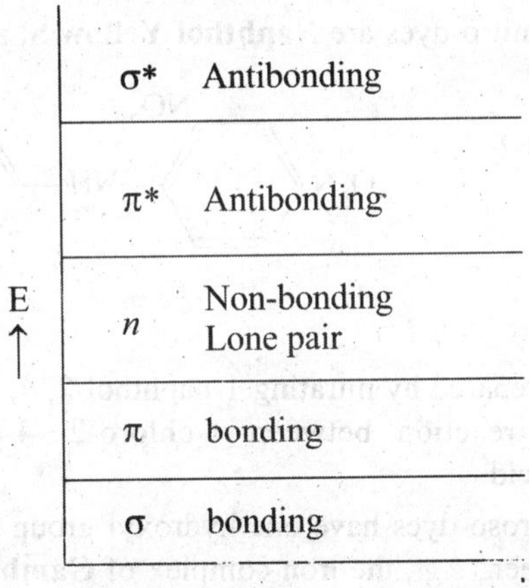

σ*	Antibonding
π*	Antibonding
n	Non-bonding Lone pair
π	bonding
σ	bonding

E ↑

Fig. 42.1. Pattern showing energy levels (molecular orbitals)

The electronic transitions can occur by the absorption of energy. Several electronic transitions are possible but permissible or allowed transitions are given below.

(*i*) $\sigma \rightarrow \sigma^*$. In this electronic transition, a bonding σ-electron is excited to an antibonding σ-orbital.

(*ii*) $\pi \rightarrow \pi^*$. In this electronic transition, a bonding π electron is raised to an antibonding π-orbital.

(*iii*) $n \rightarrow \sigma^*$ In this transition, one electron of a lone pair (*i.e.* a non-bonding pair) of electrons is excited to antibonding σ-orbital.

(*iv*) $n \rightarrow \pi^*$. In this electronic transition, a non-bonding electron is raised to antibonding π-orbital.

Energy changes occurring in various transitions follow the sequence

$$\sigma \to \overset{*}{\sigma} > n \to \overset{*}{\sigma} > \pi \to \overset{*}{\pi} > n \to \overset{*}{\pi}$$

We shall consider some examples of electronic transitions in organic compounds as under:

1. Alkanes contain C–H and C – C bonds and can show only $\sigma - \overset{*}{\sigma}$ transitions. The energy required to bring about this transition is very high and is available in the ultraviolet or far ultraviolet region. For example, ethane has a absorption maximum λ_{max} at 135 nm (C – C) which lies in the far ultraviolet region and hence appears colourless.

2. In ethylene, $\left\{ \begin{matrix} H \\ H \end{matrix} \right\rangle C = C \left\langle \begin{matrix} H \\ H \end{matrix} \right\}$, two types of electronic transitions $\sigma \to \overset{*}{\sigma}$ and $\pi \to \overset{*}{\pi}$ are possible. Ethylene has λ_{max} 175 nm and this absorption band is due to a $\pi \to \overset{*}{\pi}$ transition in the far ultraviolet region. For this reason, ethylene is colourless. Comparatively the absorption band in ethylene is in the region of longer wavelength due to the presence of a double bond in it.

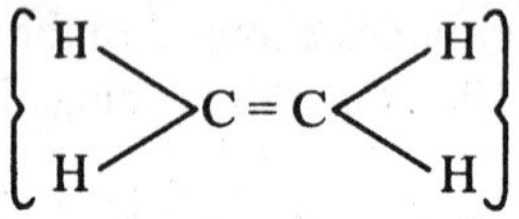

3. Butadiene shows λ_{max} 217 nm for a $\pi \to \overset{*}{\pi}$ transition. The conjugation in butadiene shifts the absorption band to a longer wavelength (217 nm) although still not into the visible region. This compound also appears colourless.

Q. 9. Give examples of nitro and nitroso-dyes. How are they prepared?

Ans. Two examples of nitro-dyes are **Naphthol Yellow S**, and **Amido Yellow E.**

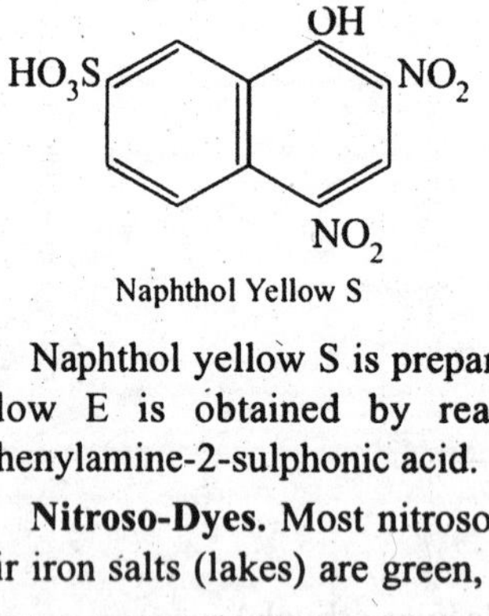

OH, HO_3S, NO_2, NO_2

Naphthol Yellow S

NO_2, SO_3H, O_2N, NH, NH

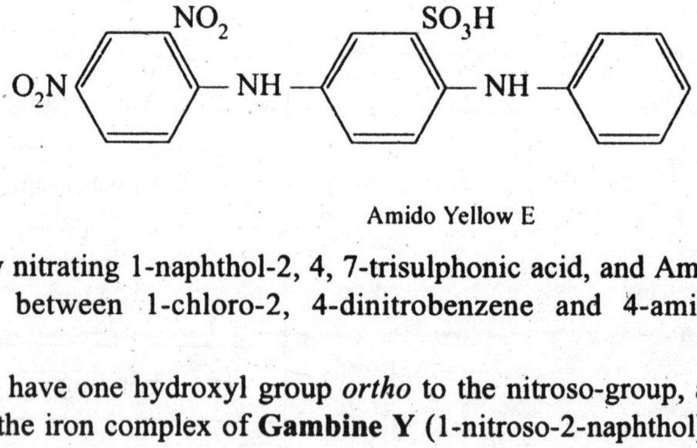

Amido Yellow E

Naphthol yellow S is prepared by nitrating 1-naphthol-2, 4, 7-trisulphonic acid, and Amido yellow E is obtained by reaction between 1-chloro-2, 4-dinitrobenzene and 4-amino-diphenylamine-2-sulphonic acid.

Nitroso-Dyes. Most nitroso-dyes have one hydroxyl group *ortho* to the nitroso-group, and their iron salts (lakes) are green, *e.g.*, the iron complex of **Gambine Y** (1-nitroso-2-naphthol).

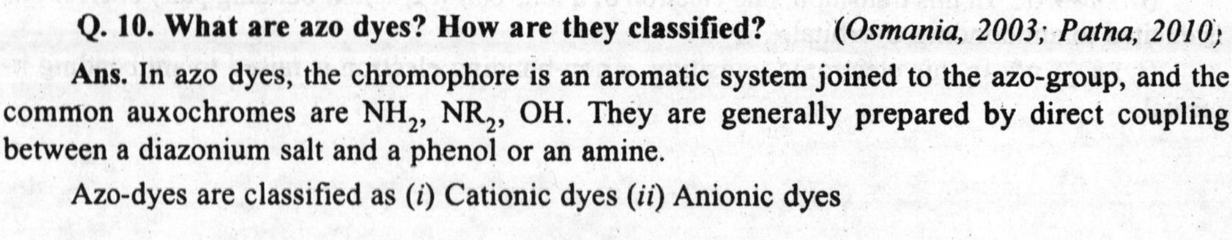

NO, OH → NOH, O $\xrightarrow{Fe^{3+}}$ (Fe, O, N, O)$_3$

Gambine Y

Q. 10. What are azo dyes? How are they classified? (*Osmania, 2003; Patna, 2010*)

Ans. In azo dyes, the chromophore is an aromatic system joined to the azo-group, and the common auxochromes are NH_2, NR_2, OH. They are generally prepared by direct coupling between a diazonium salt and a phenol or an amine.

Azo-dyes are classified as (*i*) Cationic dyes (*ii*) Anionic dyes

Cationic dyes

Chrysoidine is a cationic dye and is prepared by coupling benzenediazonium chloride with *m*-phenylenediamine. It is used for dyeing paper, leather, and jute.

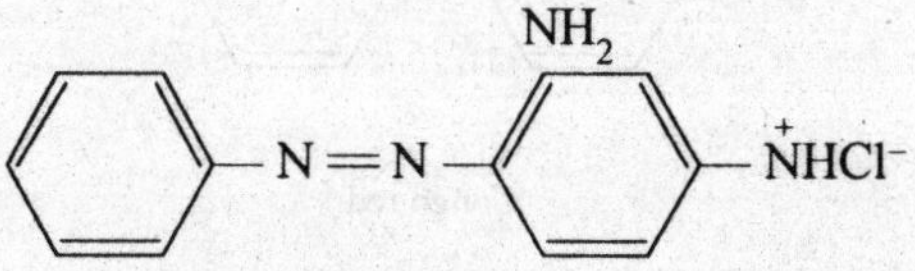

Chrysoidine

Anionic dyes

These dyes contain a sulphonic or a carboxylic group. They belong to the class of *soluble dyes*. **Methyl Orange** (*Helianthin*), an anionic dye is prepared by coupling diazotised sulphanilic acid with dimethylaniline. It is used as an indicator, being orange in alkaline solution and red in acid solution.

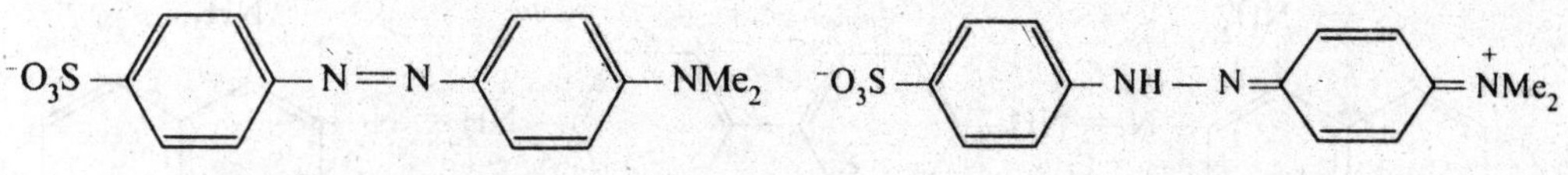

Orange (alkaline medium) Methyl orange Red (acid solution)

Methyl orange is not being used as a dye.

Q. 11. What are mordant azo-dyes? How is mordanting carried out? Give an example of mordant dye.

Ans. Mordant dyes are the substances with a hydroxyl group *ortho* to the azo-group. The most important mordanting metal is chromium. The fibre is mordanted by boiling with potassium dichromate solution along with a reducing agent such as formic or oxalic acid when the dichromate is converted into chromium hydroxide. Metal salts (lakes) are the formed by reaction with various azo-dyes. **Mordant Brown** formed by coupling diazotised *p*-aminophenol with pyrogallol is an example of such dyes.

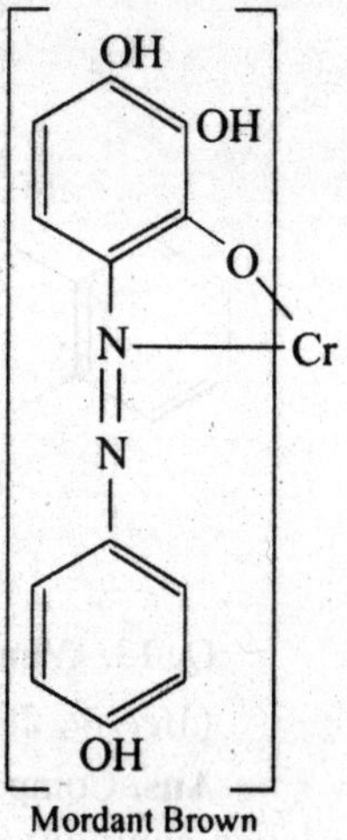

These complexes are anionic and are capable of dyeing wool in neutral solution. These dyes are fast to light and washing.

Mordant Brown

Q. 12. What is Congo red? How is it prepared? (*Kerala, 2001; Delhi 2003; Nagpur 2008*)

Ans. Congo red is a member of diazo dyes. It is prepared by the coupling reaction between tetrazotised benzidine with two molecules of naphthionic acid.

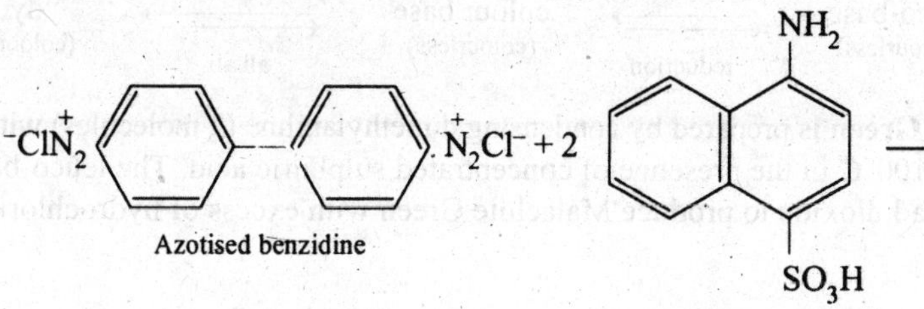

$^{-}Cl\overset{+}{N_2}$–C_6H_4–C_6H_4–$\overset{+}{N_2}Cl^{-}$ + 2 (naphthionic acid: NH_2, SO_3H) ⟶

Azotised benzidine Naphthionic acid

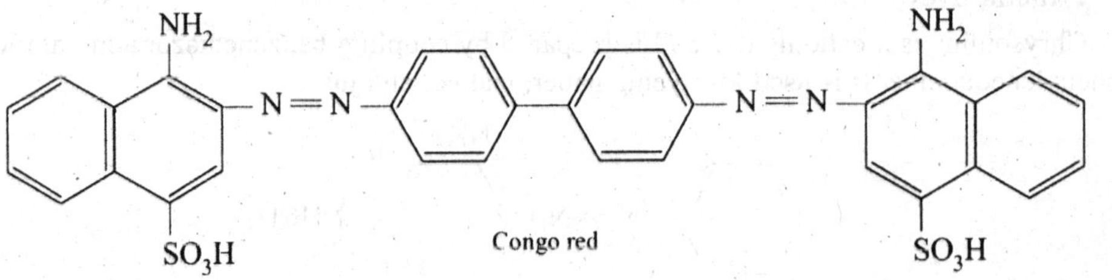

Congo red

This is red in alkaline solution, and its sodium salt dyes cotton perfectly. Congo red was the first synthetic dye produced to dye cotton directly. This dye is sensitive to acids and changes the colour from red to blue in the presence of inorganic acids. The blue colour, may be attributed to resonance among charged resonating structures shown below:

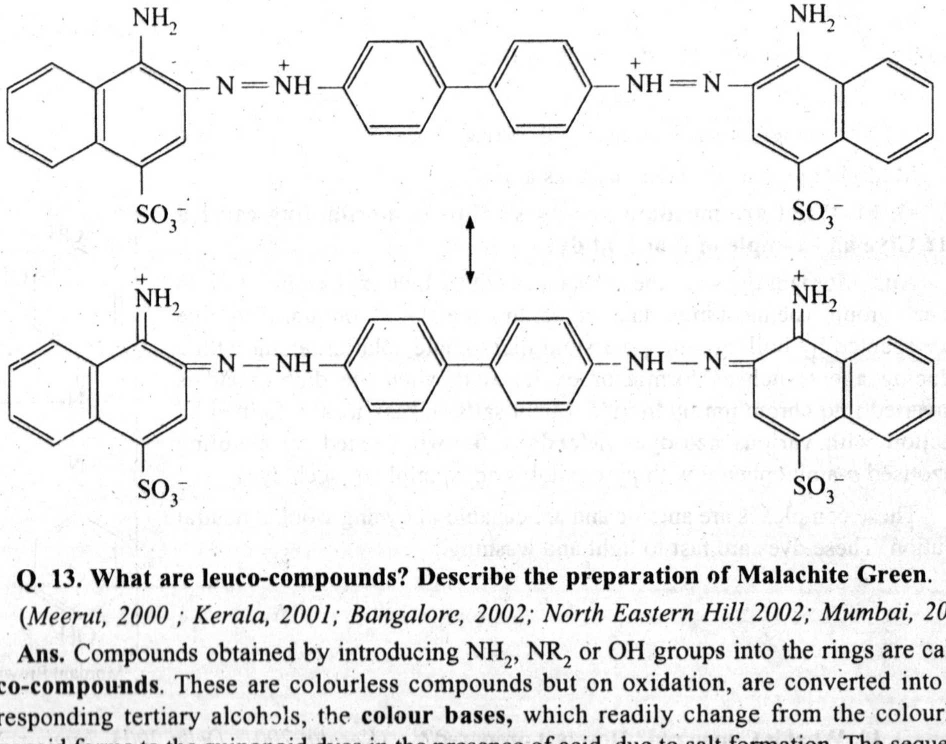

Q. 13. What are leuco-compounds? Describe the preparation of Malachite Green.

(*Meerut, 2000 ; Kerala, 2001; Bangalore, 2002; North Eastern Hill 2002; Mumbai, 2004*)

Ans. Compounds obtained by introducing NH_2, NR_2 or OH groups into the rings are called **leuco-compounds**. These are colourless compounds but on oxidation, are converted into the corresponding tertiary alcohols, the **colour bases,** which readily change from the colourless benzenoid forms to the quinonoid dyes in the presence of acid, due to salt formation. The sequence of changes is given below:

$$\underset{\text{(colourless)}}{\text{leuco-base}} \underset{\text{reduction}}{\overset{\text{oxidation}}{\rightleftharpoons}} \underset{\text{(colourless)}}{\text{colour base}} \underset{\text{alkali}}{\overset{\text{acid}}{\rightleftharpoons}} \underset{\text{(coloured)}}{\text{dye}}$$

Malachite Green is prepared by condensing dimethylaniline (2 molecules) with benzaldehyde (1 molecule) at 100°C in the presence of concentrated sulphuric acid. The leuco-base produced is oxidised with lead dioxide to produce Malachite Green with excess of hydrochloric acid:

C_6H_5–CHO + 2 C_6H_5–NMe_2 $\xrightarrow{H_2SO_4}$ C_6H_5–CH($C_6H_4NMe_2$)$_2$ $\xrightarrow[PbO_2]{[O]}$

Benzaldehyde Dimethylaniline

C_6H_5–C(OH)($C_6H_4NMe_2$)$_2$ $\xrightarrow{HCl}$ C_6H_5–C($C_6H_4NMe_2$)=C_6H_4=$\overset{+}{N}Me_2Cl^-$

Malachite Green

Malachite Green dyes wool and silk directly, but for cotton, mordanting with tannin is required.

Q. 14. Give the preparation and properties of Rosaniline or Magenta or Fuschine.

Ans. Rosaniline or Magenta or Fuchsine is produced by oxidising an equimolecular mixture of aniline, *o*- and *p*-toluidines with nitrobenzene in the presence of iron filings. The product is a mixture of rosaniline and pararosaniline (no methyl group present), in which the former predominates:

NH_2–C_6H_4–CH_3 + 2[O] + *o*-Toluidine + Aniline ⟶ NH_2–C_6H_4–CH($C_6H_3(Me)NH_2$)($C_6H_4NH_2$) + $2H_2O$ $\xrightarrow{[O]}$

p-Toluidine

NH_2–C_6H_4–C(OH)($C_6H_3(Me)NH_2$)($C_6H_4NH_2$) $\xrightarrow{HCl}$ $\overset{-}{Cl}\overset{+}{N}H_2$=$C_6H_4$=C(OH)($C_6H_3(Me)NH_2$)($C_6H_4NH_2$)

Crystals of rosaniline show a green metallic lustre, and dissolve in water to form a deep-red solution. This solution is decolourised by sulphur dioxide and is then known as Schiff's reagent, which is used as a test for aldehydes.

Rosaniline (and pararosaniline) dyes wool and silk directly and produce a violet-red colour. Pre-mordanting with tannin is required for dyeing cotton.

Q. 15. Give the preparation and properties of crystal violet. (*Nagpur, 2008*)

Ans. Crystal violet may be prepared by heating *Michler's ketone* with dimethylaniline in the presence of phosphoryl chloride or carbonyl chloride.

$$COCl_2 + 2\ C_6H_5-NMe_2 \longrightarrow CO(C_6H_4-NMe_2)_2$$

Carbonyl chloride; Dimethyl aniline

$$CO(C_6H_4-NMe_2)_2 + C_6H_5-NMe_2 + COCl_2 \longrightarrow Me_2N-C_6H_4-C(C_6H_4-NMe_2)=C_6H_4=\overset{+}{N}Me_2Cl^- + HCl + CO_2$$

Michler's ketone; Dimethylaniline; Crystal violet

A weakly acid solution of Crystal violet gives a purple colour. The colour changes with the acid strength of the solution. Thus the colour changes to green in strongly acidic medium, whereas a yellow colour is obtained in still stronger acidic medium.

Q. 16. Describe the preparation and properties of (*i*) Phenolphthalein (*ii*) Fluorescein (*Kerala, 2001; Punjab, 2005; Garhwal, 2010*)

Ans. (*i*) **Phenolphthalein** is prepared by heating phthalic anhydride (1 molecule) with phenol (2 molecule) in the presence of concentrated sulphuric acid:

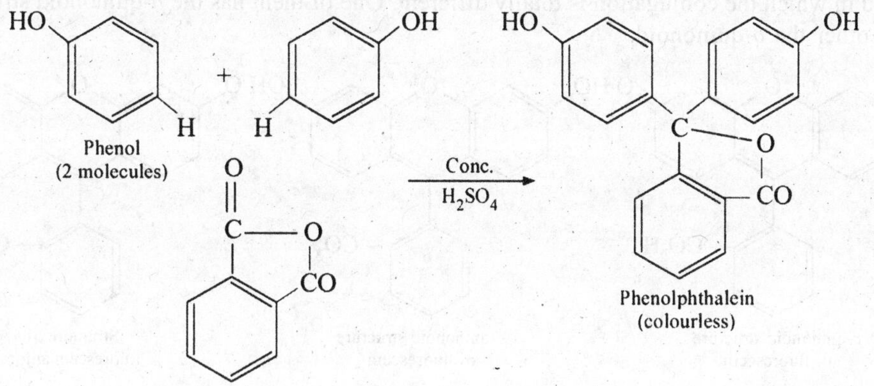

HO OH HO OH

H H

Phenol (2 molecules) + Phthalic anhydride $\xrightarrow[H_2SO_4]{Conc.}$ Phenolphthalein (colourless)

It is a white crystalline solid, insoluble in water, but soluble in alkalis to form deep red solutions:

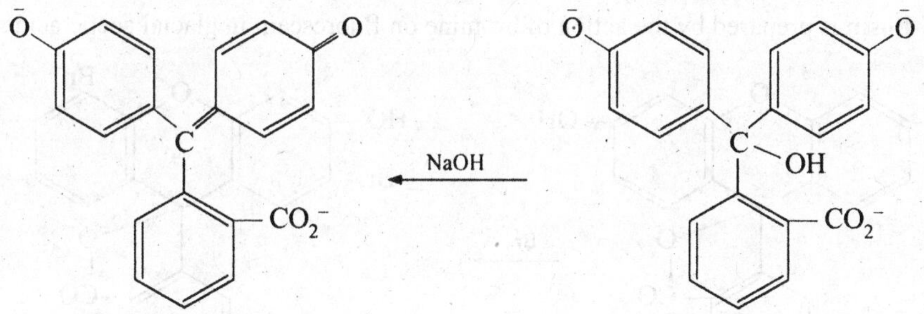

Phenolphthalein (deep red) $\xleftarrow{NaOH}$ Phenolphthalein (colourless)

In the presence of *excess* of strong alkali, the solution of phenolphthalein becomes colourless again due to the loss of the quinonoid structure and resonance.

(*ii*) **Fluorescein** is prepared by heating phthalic anhydride (1 molecule) with resorcinol (2 molecules) at 473 K, or at 373-383 K with anhydrous oxalic acid:

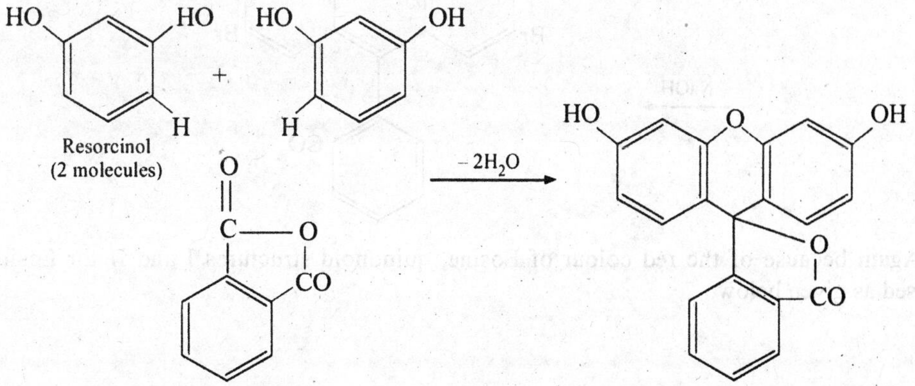

Resorcinol (2 molecules) + Phthalic anhydride $\xrightarrow{-2H_2O}$ Fluorescein

Fluorescein is a red powder insoluble in water. Since it is a coloured compound the non-quinonoid uncharged structure has been considered improper. Two quinonoid structures are proposed in which the conjugation is totally different. One of them has the *p*-quinonoid structure, and the other the *o*-quinonoid.

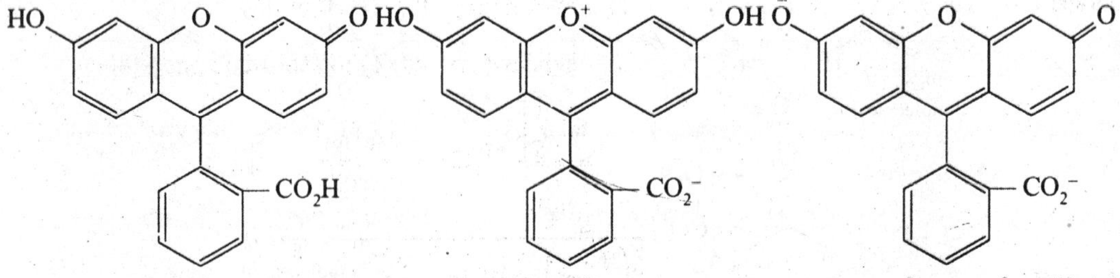

p-quinonoid structure of fluorescein

o-quinonoid structure of fluorescein

Structure of fluorescein anion

Fluorescein dissolves in alkalis to give a reddish-brown solution which gives a strong yellowish-green fluorescence on dilution. The structure of the fluorescein anion is given above. The sodium salt of fluorescein is known as **Uranine**, and is used to dye wool and silk yellow from an acid bath.

Q. 17. Give the preparation, structure and uses of Eosin.

Ans. Eosin is prepared by the action of bromine on fluorescein in glacial acetic acid solution:

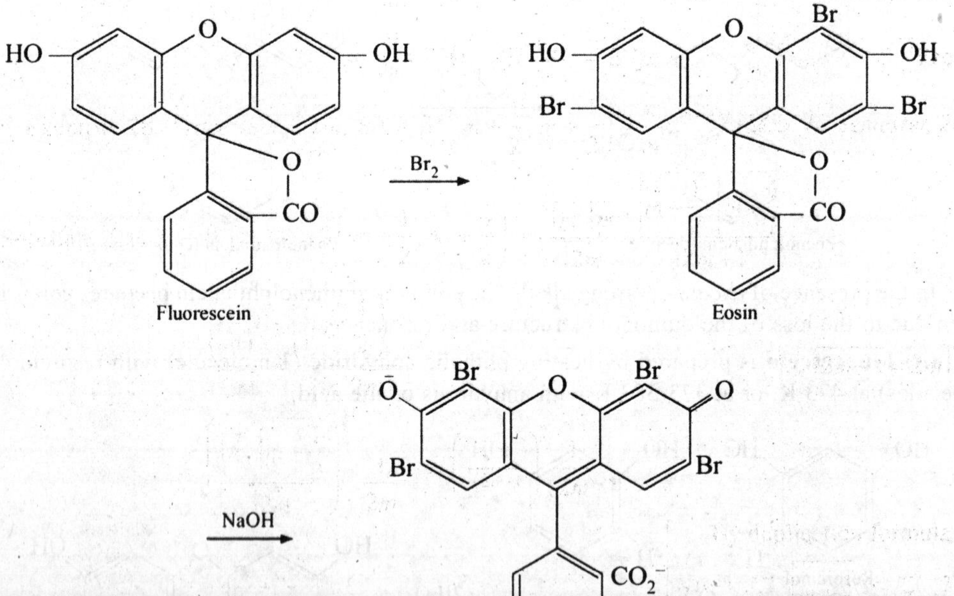

Fluorescein $\xrightarrow{Br_2}$ Eosin

$\xrightarrow{NaOH}$

Again because of the red colour of Eosine, quinonoid structures I and II for Eosin are proposed as given below:

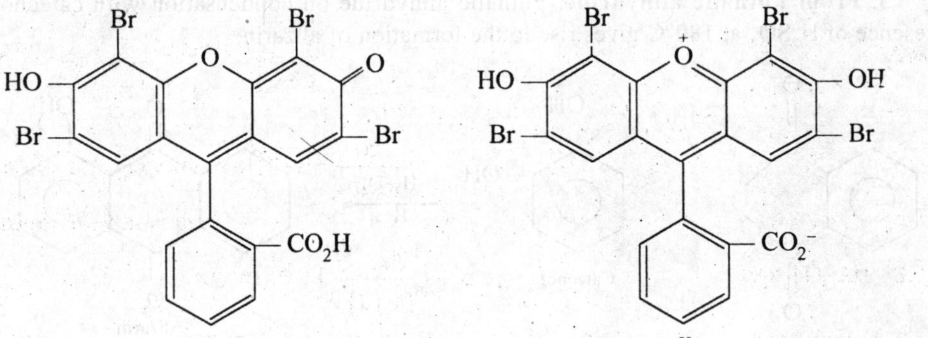

I

II

Eosin dyes wool and silk a pure red. A yellow fluorescence is obtained in the case of silk. Most red inks are dilute solutions of eosin in water-alcohol mixture.

Q. 18. Describe the preparation of Alizarin, 1, 2-Dihydroxyquinone, $C_{14}H_6O_2(OH)_2$.

(Delhi, 2003; North Gujarat 2004)

Ans. Alizarin is one of the most important anthraquinoid dyes. It is a dye of splendid colour. Alizarin is the chief constituent of the madder root wherein it is present as the glucoside, *ruberythric acid.* It can be prepared by the following methods:

1. **From madder root.** Alizarin is extracted from the root of the madder plant. It contains alizarin as glucoside, ruberythric acid, which forms alizarin on hydrolysis in the presence of enzyme or dilute acid.

$$\underset{\text{Ruberythric acid}}{C_{26}H_{28}O_{14}} + 2H_2O \xrightarrow[\text{(or Enzyme)}]{H^+} \underset{\text{Glucose}}{2C_6H_{12}O_6} + \underset{\text{Alizarin}}{C_{14}H_6O_2(OH)_2}$$

2. **From anthraquinone.** Anthraquinone on sulphonation with fuming H_2SO_4 at 140°C yields anthraquinone-2-sulphonic acid which is converted into its sodium salt by treatment with NaOH. This sodium salt on fusion with NaOH in the presence of $KClO_3$ at 200°C under pressure produces alizarin (1, 2, dihydroxyquinone).

Anthraquinone $\xrightarrow[140° C]{\text{Fum. } H_2SO_4}$ Anthraquinone-2-sulphonic acid (SO_3H) $\xrightarrow{NaOH}$

Sodium salt (SO_3Na) $\xrightarrow[\text{fused}]{NaOH,\ KClO_3}$ Alizarin

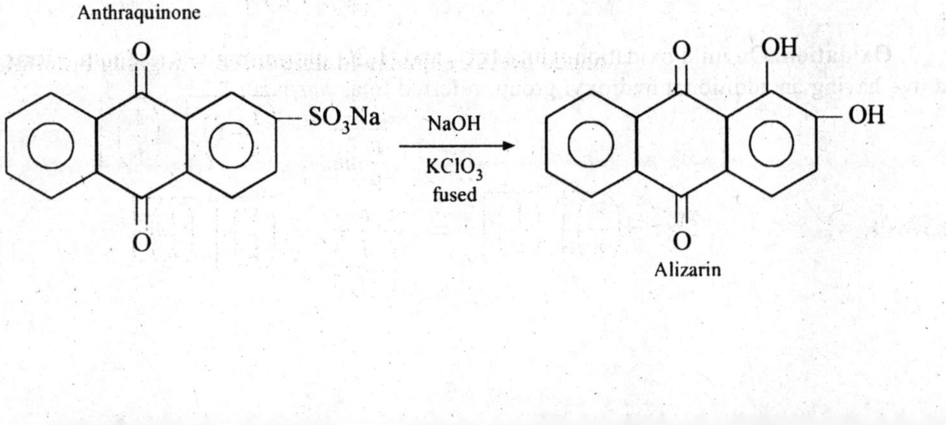

3. **From Phthalic anhydride.** Phthalic anhydride on condensation with catechol in the presence of H_2SO_4 at 180°C gives rise to the formation of alizarin.

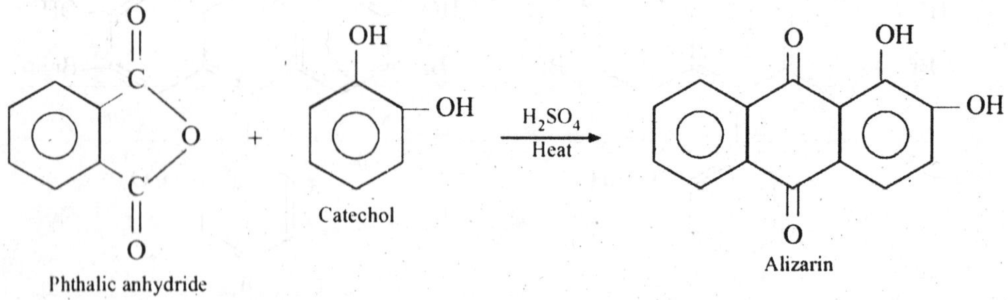

Phthalic anhydride + Catechol $\xrightarrow[\text{Heat}]{H_2SO_4}$ Alizarin

Q. 19. Describe properties and uses of alizarin (mordant dye). (*Madras, 2004*)

Ans. Physical properties

1. It melts at 290°C.
2. It sublimes on heating.
3. Alizarin forms ruby red crystals.
4. Alizarin dissolves in caustic alkalis to give a purple coloured solution.
5. It is insoluble in water and sparingly soluble in ethanol.

Chemical Properties

Alizarin is a dihydroxyquinone. It is a mordant dye and forms characteristic lake depending upon the nature of metal used. The important chemical reactions of alizarin are as under:

1. **As a valuable dye**. Alizarin is one of the most important anthraquinoid dyes. It is a typical mordant dye yielding lakes of different colours depending upon the nature of metal used.

With chromium alizarin forms a brown violet lake. With barium a blue coloured lake is formed.

2. **Reduction.** Reduction with zinc dust and ammonia, produces a valuable medicine which is referred to as anthrarobin (dihydroxy anthranol) for curing skin diseases

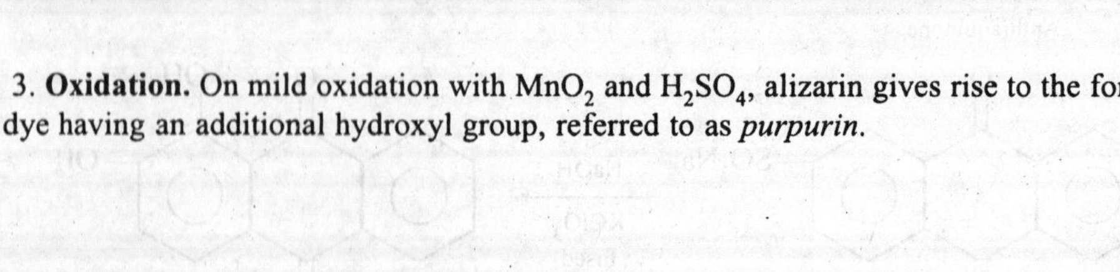

Alizarin $\xrightarrow{Zn/NH_3}$ Anthrarobin

3. **Oxidation.** On mild oxidation with MnO_2 and H_2SO_4, alizarin gives rise to the formation of a dye having an additional hydroxyl group, referred to as *purpurin*.

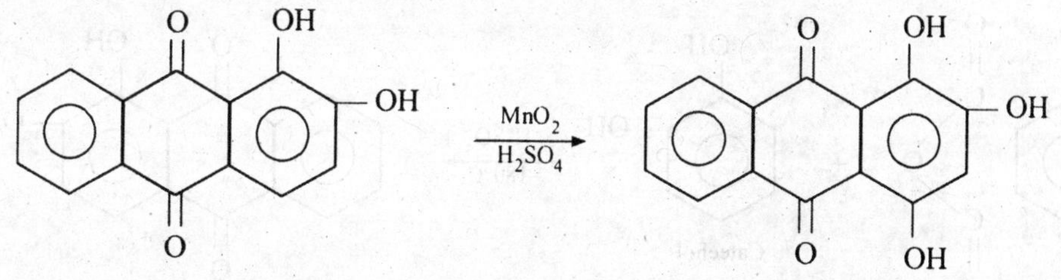

$$\xrightarrow[H_2SO_4]{MnO_2}$$

Purpurin

Vigorous oxidation of alization yields phthalic acid.

4. Action of alkalis. Alizarin dissolves in caustic alkalis forming purple solution of alizarate.

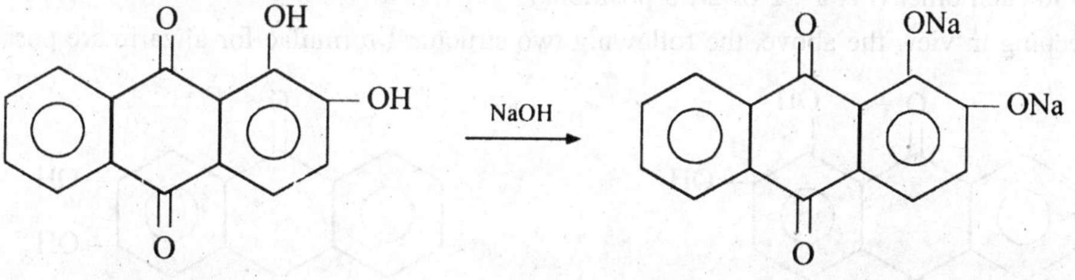

$$\xrightarrow{NaOH}$$

Alizarin Sodium alizarate

Uses. Alizarin is used:

1. As a mordant dye.
2. In the preparation of anthrarobin.
3. In the manufacture of printing inks.
4. As a purgative in medicine.

Q. 20. Establish the structure (constitution) of alizarin.

(Panjab, 2000; Kerala, 2001; Bangalore, 2001)

Ans. 1. Molecular formula of alizarin. The elemental analysis and molecular weight determination show that the molecular formula of alizarin is $C_{14}H_6O_4$.

2. Carbon skeleton of alizarin. On distillation with zinc dust, alizarin is converted into anthracene. This indicates that alizarin has the same carbon skeleton as that of anthracene. Thus, it has been derived from anthracene.

3. Presence of two –OH groups. (*a*) Alizarin on treatment with acetic anhydride forms a diacetate showing the presence of two –OH groups in the molecule of alizarin.

(*b*) Anthraquinone on heating with a calculated amount of brominc yields dibromoanthraquinone which on fusion with KOH gives rise to the formation of *alizarin*. This indicates that alizarin is dihydroxy-anthraquinone.

4. Possible position of two –OH groups. (*a*) Alizarin on vigorous oxidation gives rise to the formation of phthalic acid. This indicates that the two OH groups are in the same ring – the ring which is destroyed during the vigorous oxidation, otherwise either hydroxyphthalic acid would have been formed or the entire molecule would have been broken down completely.

5. Possible formulae of alizarin. Alizarin is obtained by condensing phthalic anhydride and catechol in the presence of H_2SO_4 at 180°C in the same way as anthraquinone is formed by heating phthalic anhydride with benzene.

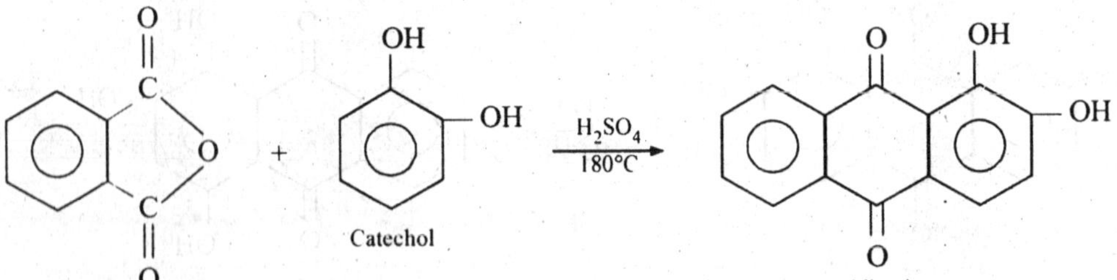

Phthalic anhydride + Catechol $\xrightarrow[180°C]{H_2SO_4}$ Alizarin (1, 2 dihydroxyquinone)

This shows that the two –OH groups in alizarin must be in the same ring and in the ortho-position to each other (*i.e.* 1 : 2 or 2 : 3 positions).

Keeping in view the above, the following two structural formulae for alizarin are possible.

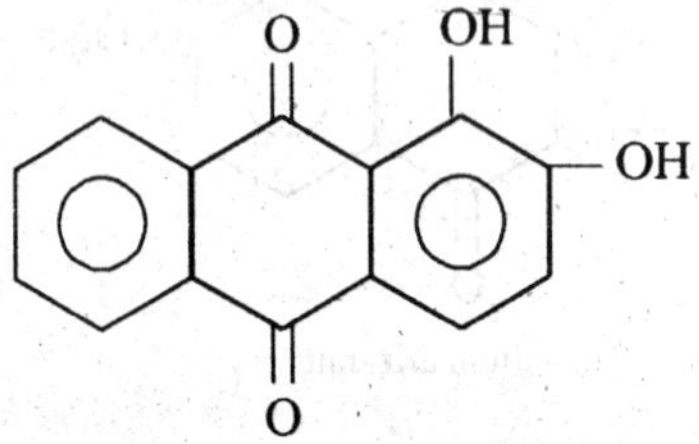

I

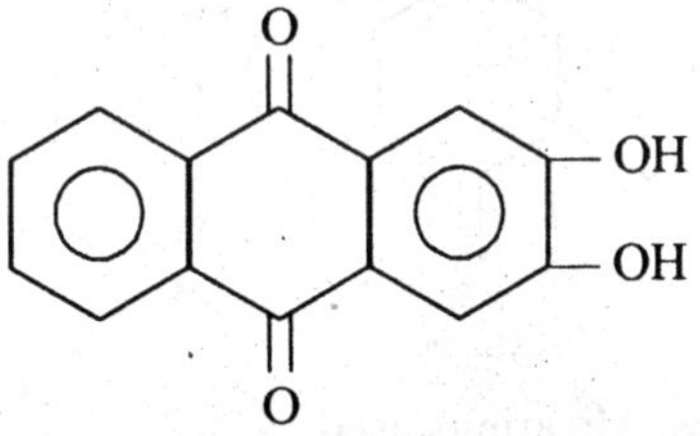

II

6. Confirmation of structural formula. (*a*) On nitration, alizarin gives rise to the formation of two isomeric mononitro derivatives having –OH and –NO_2 groups in the same ring. Further, both these derivatives on oxidation yield phthalic acid. This indicates that in each of these derivatives, the –NO_2 group is inducted on the benzene ring containing –OH groups.

The possible mononitro derivatives from structure I are as under:

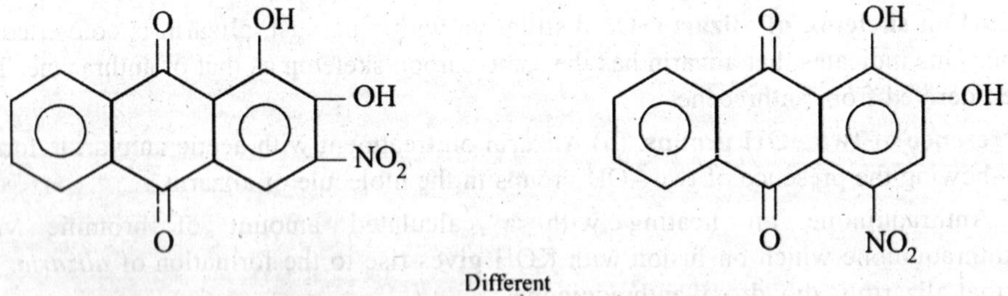

Different

The possible mono-nitro derivatives from structure II are as under:

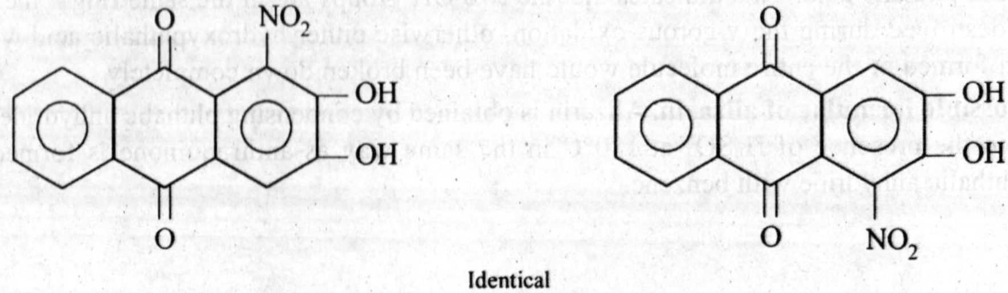

Identical

Only structure I permits the formation of two isomeric mononitro-derivatives and hence, structure I gives the true representation of alizarin.

(*b*) Final confirmation of structure is provided by the synthesis of alizarin from 1 : 2 dibromoanthraquinone. Alizarin is formed by fusing 1 : 2 dibromoanthraquinone with NaOH.

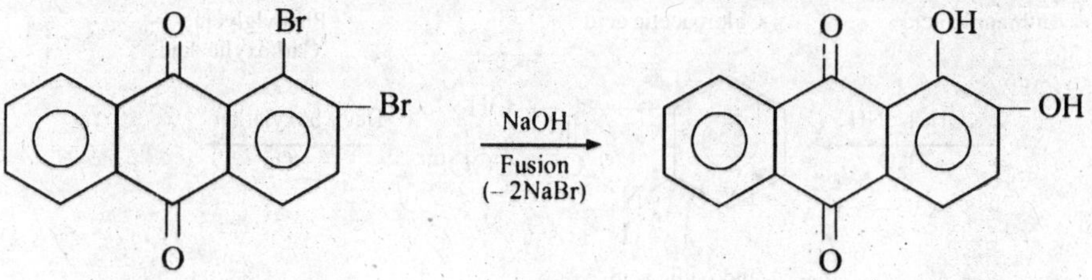

NaOH, Fusion ($-$2NaBr)

(1, 2, Dihydroxyanthraquinone)

Alizarin
(1, 2, Dihydroxyanthraquinone)

Q. 21. Give the methods of preparation, properties and uses of indigo.

(*Madras, 2003; Lucknow, 2010*)

Ans. Indigo is prepared by the following methods:

1. From isatin chloride. Isatin chloride on reduction with zinc dust in glacial acetic acid yields indoxyl which on oxidation in air forms indigo.

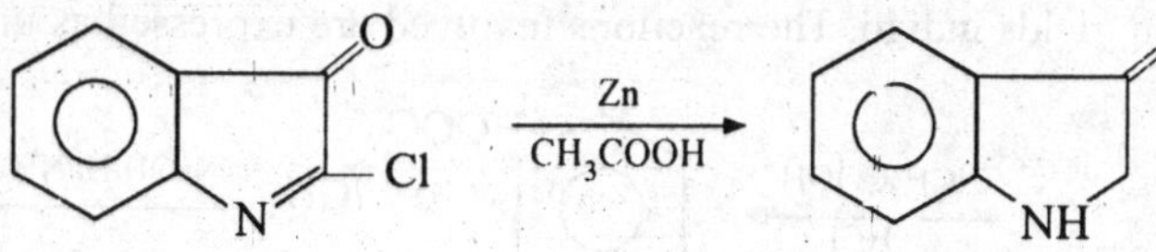

Zn, CH_3COOH

Isatin Chloride

Indoxyl

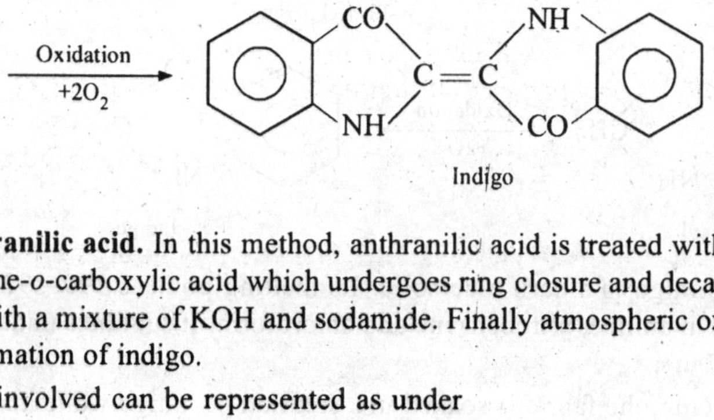

Oxidation, $+2O_2$

Indigo

2. From anthranilic acid. In this method, anthranilic acid is treated with chloroacetic acid to give phenyl glycine-*o*-carboxylic acid which undergoes ring closure and decarboxylation to give indoxyl on fusion with a mixture of KOH and sodamide. Finally atmospheric oxidation of indoxyl gives rise to the formation of indigo.

The reactions involved can be represented as under

$$\text{Anthranilic acid (}C_6H_4(COOH)NH_2\text{)} + \underset{\text{Chloroacetic acid}}{ClCH_2COOH} \longrightarrow \underset{\text{Phenylglycine-}o\text{-Carboxylic acid}}{C_6H_4(COOH)NHCH_2COOH}$$

$$\xrightarrow[-H_2O]{KOH,\ NaNH_2} \underset{\text{Indoxylic acid}}{C_6H_4\text{–}C(OH){=}C(COOH)\text{–}NH} \xrightarrow[-CO_2]{\text{Decarboxylation}}$$

$$\underset{\text{Indoxyl}}{C_6H_4\text{–}CO\text{–}CH_2\text{–}NH} \xrightarrow[\text{Oxidation } +2O_2]{\text{Atmospheric}} \underset{\text{Indigo}}{C_6H_4(NH)(CO)C{=}C(CO)(NH)C_6H_4}$$

3. From aniline. In this method, aniline is heated with chloroacetic acid to form N-phenylglycine which on fusion with caustic soda and sodamide at 300–350°C gives indoxyl. This product on oxidation yields indigo. The reactions involved are expressed as under:

$$\underset{\text{Aniline}}{C_6H_5NH_2} \xrightarrow[-HCl]{ClCH_2COOH} \underset{\text{N-Phenyl glycine}}{C_6H_5NHCH_2COOH} \xrightarrow{NaOH/NaNH_2}$$

$$\underset{\text{Indoxyl}}{C_6H_4\text{–}CO\text{–}CH_2\text{–}NH} \xrightarrow[+2O_2]{\text{Oxidation}} \underset{\text{Indigo}}{C_6H_4(CO)(NH)C{=}C(NH)(CO)C_6H_4}$$

Properties. Indigo is a dark blue solid. Its melting point is 390–392°C. It is insoluble in water. It dissolves in aniline, nitrobenzene and chloroform. It sublimes under reduced pressure to give deep red vapours.

Uses. For dyeing, the fabric is soaked in a solution of indigo white and then exposed to air. On exposure, the fabric gets a dark blue colour which is extremely fast to washing.

Q. 22. Establish the structure of indigo or indigotin. *(Panjab 2000)*

Ans. 1. Empirical and molecular formulae. From the elemental analysis, the empirical formula of indigo is found to be C_8H_5ON. Vapour density determination reveals that its molecular formula is $C_{16}H_{10}O_2N_2$.

2. Degradation of indigo. (*i*) Vigorous oxidation of indigo with HNO_3 forms two molecules of isatin, showing the presence of two similar units joined by a double bond (point of attack). Each unit on oxidation yields one molecule of isatin.

$$\underset{\text{Indigo}}{C_{16}H_{10}O_2N_2} \xrightarrow{2(O)} 2C_8H_5O_2N$$

(*ii*) On distillation with zinc dust at high temperature, indigo gives a new product known as indole C_8H_7N.

$$\underset{\text{Indigo}}{C_{16}H_{10}O_2N_2} \xrightarrow[\text{High Temp.}]{\text{Zn dust}} \underset{\text{Indole}}{C_8H_7N}$$

From the above reactions, it follows that there exists a close structural similarity among indigo, isatin and indole. Further indigo and indoxyl are structurally related to each other. Indoxyl on oxidation yields indigo. Indoxyl has been shown to be 3-hydroxy indole and is isomeric with oxindole.

3. Structure of isatin. (*i*) **Molecular formula of isatin.** The molecular formula of isatin has been found to be $C_8H_5O_2N$.

(*ii*) **Action of phosphorus pentachloride.** On treatment with PCl_5, it forms isatin chloride indicating the presence of a hydroxyl group in isatin.

$$\underset{\text{Isatin}}{C_8H_4ON(OH)} \xrightarrow[-HCl,\ -POCl_3]{PCl_5} \underset{\text{Isatin chloride}}{C_8H_4ONCl}$$

(*iii*) **Action of hydroxylamine.** On treatment with hydroxylamine, it forms an oxime showing the presence of a carbonyl group in isatin.

(*iv*) **Action of alkali.** On boiling with an alkali like NaOH, it forms *o*-aminobenzoyl formic acid indicating that Isatin has structure I. The reaction involved is represented as under:

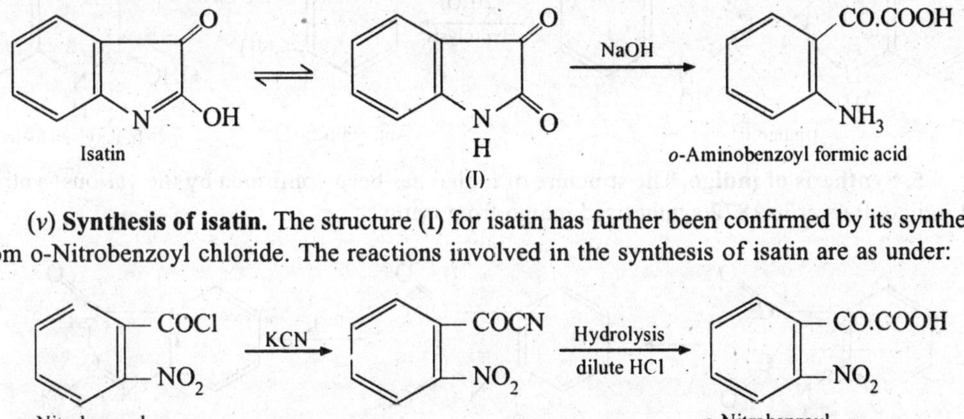

O, N, OH — Isatin ⇌ O, N H, O — (I) $\xrightarrow{\text{NaOH}}$ CO.COOH, NH_2 — *o*-Aminobenzoyl formic acid

(*v*) **Synthesis of isatin.** The structure (I) for isatin has further been confirmed by its synthesis from o-Nitrobenzoyl chloride. The reactions involved in the synthesis of isatin are as under:

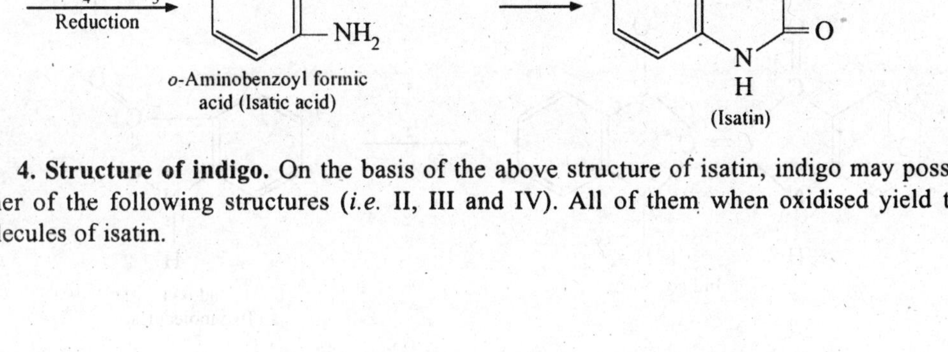

COCl, NO_2 — *o*-Nitrobenzoyl chloride $\xrightarrow{\text{KCN}}$ COCN, NO_2 $\xrightarrow[\text{dilute HCl}]{\text{Hydrolysis}}$ CO.COOH, NO_2 — *o*-Nitrobenzoyl formic acid

$\xrightarrow[\text{Reduction}]{FeSO_4 - NH_3}$ CO.COOH, NH_2 — *o*-Aminobenzoyl formic acid (Isatic acid) $\xrightarrow{\text{Heat}}$ O, O, N H — (Isatin)

4. Structure of indigo. On the basis of the above structure of isatin, indigo may possess either of the following structures (*i.e.* II, III and IV). All of them when oxidised yield two molecules of isatin.

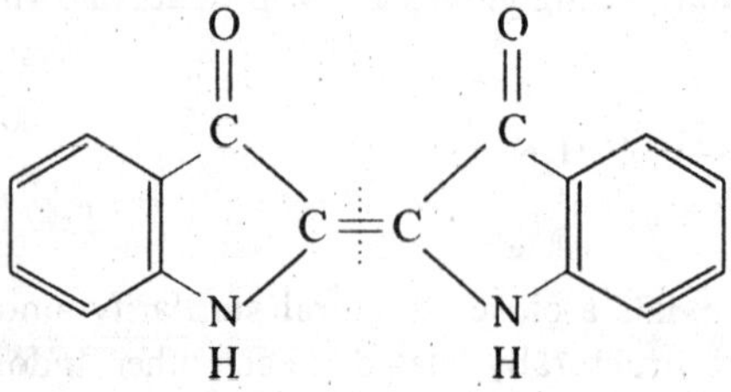

(II) 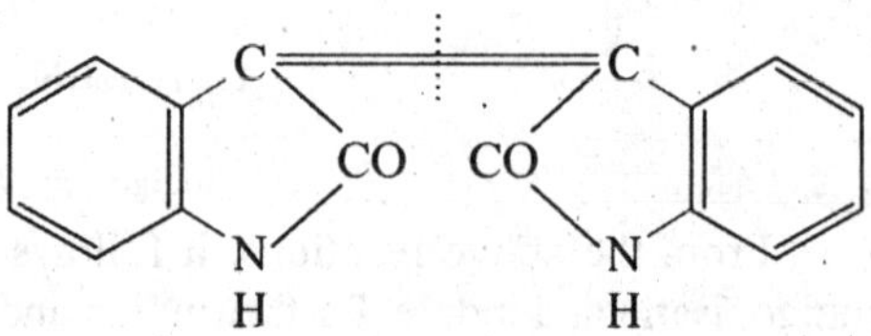 (III)

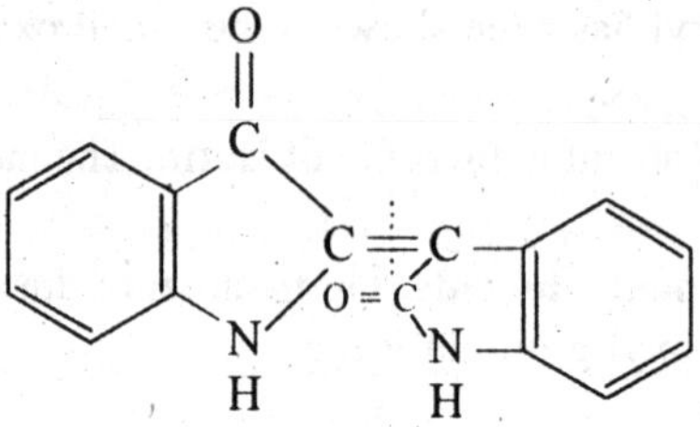

(IV)

Out of these structures, the structure (II) is found to be correct because indigo when hydrolysed with dilute alkali like NaOH yields anthranilic acid and indoxyl-2-aldehyde.

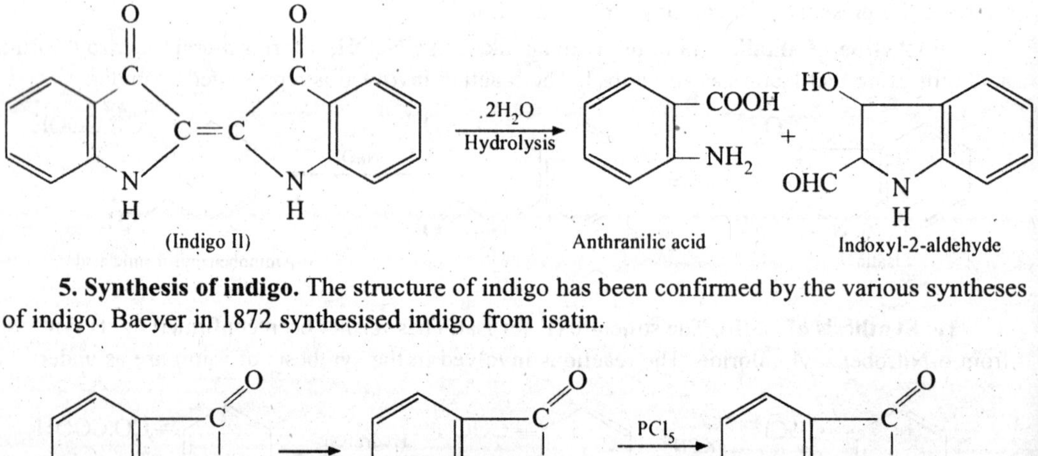

(Indigo II) $\xrightarrow[\text{Hydrolysis}]{2H_2O}$ Anthranilic acid + Indoxyl-2-aldehyde

5. Synthesis of indigo. The structure of indigo has been confirmed by the various syntheses of indigo. Baeyer in 1872 synthesised indigo from isatin.

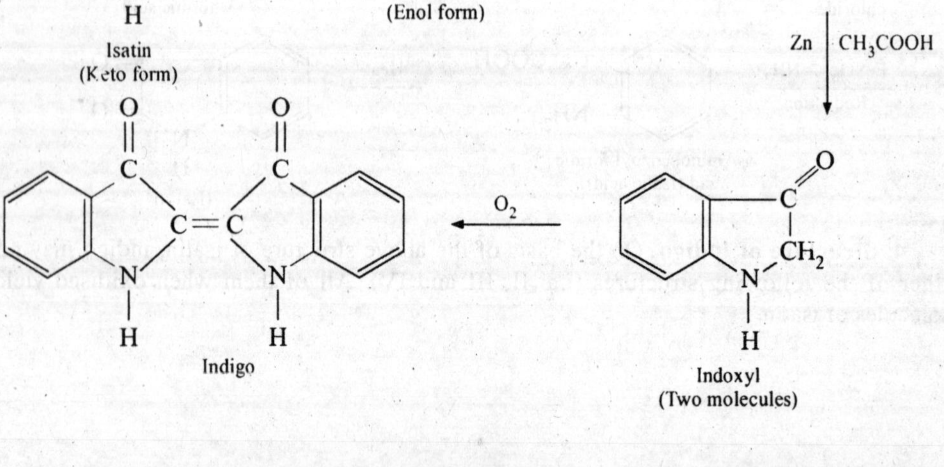

Isatin (Keto form) → Isatin (Enol form) $\xrightarrow{PCl_5}$ Isatin chloride $\xrightarrow{Zn \;\; CH_3COOH}$ Indoxyl (Two molecules) $\xrightarrow{O_2}$ Indigo

44

VITAMINS

Q. 1. Establish the structure of vitamin A. (*Kuvempu, 2005*)

Ans. The structure of vitamin A (also known as vitamin A_1) is established as under:

1. Elemental analysis and molecular weight determination of vitamin A shows its molecular formula to be $C_{20}H_{30}O$.

2. On catalytic hydrogenation, vitamin A_1 is converted into perhydro vitamin A, $C_{20}H_{40}O$. It shows the presence of five double bonds in vitamin A.

$$\underset{\text{Vitamin A}}{C_{20}H_{30}O} \xrightarrow{5H_2} \underset{\text{Perhydro vitamin A}}{C_{20}H_{40}O}$$

3. Vitamin A forms an ester with *p*-nitrobenzoic acid. It means vitamin A contains a hydroxyl group. It also follows that parent hydrocarbon of vitamin A is $C_{20}H_{40}$.

4. Ozonolysis of vitamin A produces one molecule of geronic acid (a substance of known structure) per molecule of vitamin A. This indicates the presence of one β-ionone nucleus.

5. Oxidation of vitamin A with $KMnO_4$ produces acetic acid. This shows the presence of some $-C(CH_3)=$ groups in the chain.

6. We know from common knowledge that β-carotene is converted into vitamin A in the intestinal mucosa. This suggests that vitamin A is half the β-carotene structure.

7. On heating vitamin A_1 in an ethanolic solution of hydrogen chloride, compound I is obtained which on dehydrogenation with selenium forms 1 : 6-dimethylnaphthalene. The following structure of vitamin A explains these reactions

CH, CH, C.CH_3, CH, CH_3, CH = CH.C = CH.CH_2OH

Vitamin A

$\xrightarrow{HCl - C_2H_5OH}$

H, CH_3, CH_3, CH_3, CH = CH.C = CH.CH_2OH

I

$\xrightarrow{Se}$ CH_3, CH_3

1, 6-dimethylnaphthalene

Perhydro vitamin A_1 (step 2 above) has been synthesised from β-ionone which is identical with that obtained by reducing vitamin A. This lends support to the above structure of vitamin A.

Final confirmation of the structure is provided by synthesis. Two synthesis are given below:

I. This synthesis is due to van Dorp *et al* (1946) who prepared vitamin A_1 acid which was then reduced by means of lithium aluminium hydride to vitamin A_1 by Tishler (1949); β-ionone and methyl γ-bromocrotonate are the starting materials.

CH_3

$-CH=CHCO + CH_2BrCH=CH-COOCH_3$

β-ionone

↓ γ-bromocrotonate Zn (Reformatsky)

β-ionone

CH_3

$-CH=CHCCH_2CH=CHCOOCH_3$

OH

↓ (*i*) $(COOH)_2$ $[-H_2O]$ (*ii*) KOH

CH_3

$-CH=CH{\cdot}C=CH{\cdot}CH=CH-COOH$

↓ (*i*) $SOCl_2$ (*ii*) CH_3Li

CH_3 CH_3

$-CH=CH\ C=CH-CH=CH-CO$

↓ $CH_2Br{\cdot}CO_2C_2H_5/Zn$ (Reformatsky)

CH_3 CH_3

$-CH=CH-C=CH-CH=CH-C-CH_2-COOC_2H_5$

OH

↓ (*i*) $-H_2O$ (*ii*) KOH

CH_3 CH_3

$-CH=CH-C=CH-CH=CH-C=CH-COOH$

Vitamin A_1 acid

↓ $LiAlH_4$

$$CH=CH-\overset{CH_3}{\overset{|}{C}}=CH-CH=CH-\overset{CH_3}{\overset{|}{C}}=CH-CH_2OH$$

Vitamin A

II. Attenburrow *et al* (1952) have also synthesised vitamin A starting from 2-methylcyclohexanone

$$\xrightarrow[CH_3I]{NaNH_2} \quad \xrightarrow[Na/NH_3]{CH\equiv CH} \quad C\equiv CH,\ OH$$

$$\xrightarrow[(ii)\ CH_3\cdot CO\cdot CH=CH\cdot CH=C(CH_3)\cdot CH=CH_2]{(i)\ Et\,MgBr} \quad C\equiv C-\overset{CH_3}{\overset{|}{C}}(OH)-CH=CH-CH=CH-\overset{CH_3}{\overset{|}{CH}}-CH=CH_2,\ OH$$

$$\xrightarrow[(rearr.)]{acid} \quad C\equiv C\cdot\overset{CH_3}{\overset{|}{C}}=CH\cdot CH=CH\cdot\overset{CH_3}{\overset{|}{C}}=CH\cdot CH_2OH,\ OH$$

$$\xrightarrow[(ii)\ (CH_3\cdot CO)_2O]{(i)\ LiAlH_4} \quad CH=CH-\overset{CH_3}{\overset{|}{C}}=CH-CH=CH-\overset{CH_3}{\overset{|}{C}}=CH-CH_2O-CO-CH_3,\ OH$$

$$\xrightarrow[(-H_2O)]{pMe\cdot C_6H_4\cdot SO_3H} \quad CH=CH-\overset{CH_3}{\overset{|}{C}}=CH-CH=CH-\overset{CH_3}{\overset{|}{C}}=CH-CH_2OH$$

Q. 2. Establish the structure of vitamin B_1, thiamine (aneurin). (*Mumbai, 2004*)

Ans. The structure of thiamine is established as under:

1. Elemental analysis and molecular mass determination of thiamine hydrochloride (the crystalline salt of thiamine) shows the molecular formula as $C_{12}H_{18}ON_4Cl_2S$.

2. On treatment with sodium sulphite at room temperature, thiamine is decomposed quantitatively into two compounds A and B as shown below:

$$C_{12}H_{18}ON_4Cl_2S + Na_2SO_3 \longrightarrow C_6H_9ONS + C_6H_9O_3N_3S + 2NaCl$$

Thiamine chloride hydrochloride — A — B

Thus we shall first discuss structures of A and B separately.

Compound A (C_6H_9ONS)

1. This compound shows basic nature and does not react with HNO_2. It can be held that nitrogen is present as tertiary amino group.

2. Compound A reacts with HCl to produce a chloro group. This shows that oxygen is present in the molecule in the form of hydroxyl group.

3. The absorption spectrum of the chloro compound obtained by treating thiamine with HCl is the same as that of thiamine, proving, therefore, that the alcoholic group is in the side-chain.

4. Compound A does not give reactions of a mercapto group or of a sulphide and the sulphur in the compound is unreactive. This indicates that sulphur is present in a heterocyclic ring. We observe that spectrum of compound A has similarity with that of thiazole.

5. Based upon the above observations the following structure for compound A is proposed:

```
N——C-CH3
‖    ‖
CH   C-CH2-CH2OH
  \ /
   S
```

Compound A

6. The alcoholic group in compound B is primary and not secondary. This is borne out by the fact that this compound does not give iodoform test.

The structure of compound A is confirmed by its synthesis.

Synthesis of compound A

(*i*)

```
CH3
|
CO
|   -  +                                CH3  COOC2H5              SO2Cl2
CH Na     + BrCH2·CH2OC2H5  ——→         |    |                 ——————→
|                                       CO–CH–CH2–CH2OC2H5
COOC2H5
```

```
CH3 COOC2H5             "ketonic           CH3
|   |                  ——————→             |
CO–CCl–CH2–CH2OC2H5     hydrolysis"        CO–CHCl–CH2–CH2OC2H5
```

(*ii*)

```
   NH       HO  C-CH3                                      N——C-CH3
  //            ‖                                          ‖   ‖
CH      +       C-CH2-CH2OC2H5  ——→  H2O + HCl +           CH  C-CH2-CH2OH
  \                                                          \ /
   SH       Cl                                                S
thioformamide
```

```
 HCl    N——C-CH3                 Boil with    N——C-CH3
——→     ‖   ‖                    ———————→     ‖   ‖
        CH  C-CH2-CH2Cl            H2O        CH  C-CH2-CH2OH
          \ /                                   \ /
           S                                     S
```

Londergan synthesis

2-methylfuran $\xrightarrow[HCl]{H_2O/Pd\text{-}C}$ $CH_2(CH_2OH)\text{—}CH_2(CO\text{—}CH_3)$ $\xrightarrow{-H_2O}$ (cyclic: $CH_2\text{—}CH{=}C(CH_3)\text{—}O\text{—}CH_2$)

$\xrightarrow{Cl_2}$ (cyclic: $CH_2\text{—}CHCl\text{—}C(Cl)CH_3\text{—}O\text{—}CH_2$) $\xrightarrow[\text{in } HCO_2H]{H-CS-NH_2}$ (thiazole: $N\text{—}C(CH_3){=}C(CH_2\text{—}CH_2OH)\text{—}S$)

Compound B ($C_6H_9O_3N_3S$)

1. Compound B on heating with steam at 200°C gives sulphuric acid. Thus it is a sulphonic acid.

2. On treatment with HNO_2, it evolves nitrogen. Therefore, it must contain one or more amino groups. Since the treatment with HNO_2 produces hydroxyl groups, the quantitative analysis showed that compound B contains only one $-NH_2$ group

$$-NH_2 + HNO_2 \longrightarrow -OH + N_2 + H_2O$$

3. Evolution of nitrogen as mentioned above and the reaction of compound B with benzoyl chloride are slow. This suggests that B contains an amidine structure.

4. When compound B is heated with HCl at 150°C under pressure, a compound C along with ammonia are obtained. It indicates that an amino group is replaced by hydroxyl group. This type of reactions is characteristic of 2- and 6-aminopyrimidines. Also the spectrum of compound C is similar to that of synthetically prepared 6-hydroxypyrimidine. Thus compound B is probably 6-amino pyrimidine.

Thus B has the following structure:

(pyrimidine ring: NH_2 at C-6, $CH_2{\cdot}SO_3H$ at C-5, CH_3 at C-2)

B

This structure is confirmed by synthesis

Acetamidine ($CH_3.C(NH_2){=}NH$) + Ethoxymethylenemalononitrile ($(CN)_2C{=}C(H)OC_2H_5$) $\xrightarrow{C_2H_5ONa}$ 6-amino-5-cyano 2-methylpyrimidine $\xrightarrow[(ii)\ H_2\text{—}Pd\text{—}C]{(i)\ CH_3.CO_2H + HCl \text{ gas}}$

6-amino-5-aminomethyl-2-methylpyrimidine ($CH_2{\cdot}NH_2$ at C-5) $\xrightarrow[(ii)\ HBr]{(i)\ HNO_2}$ (CH_2Br at C-5) $\xrightarrow[SO_2]{NaHSO_3}$ ($CH_2{\cdot}SO_3H$ at C-5)

The compounds A and B are linked together as follows to give thiamine chloride hydrochloride

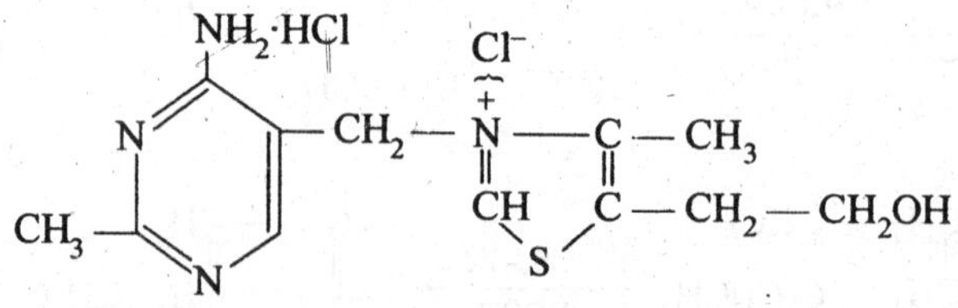

CH_3–(pyrimidine, 4-$NH_2 \cdot HCl$)–CH_2–$\overset{+}{N}$(Cl^-) thiazolium ring with C–CH_3 and C–CH_2–CH_2OH, CH, S

thiamine chloride hydrochloride

This structure has been confirmed by synthesis.

Williams Synthesis

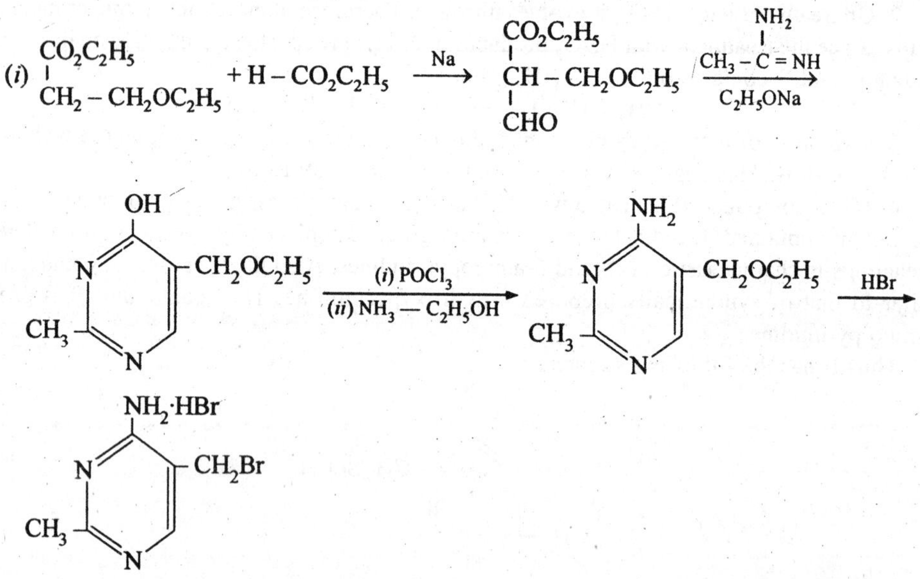

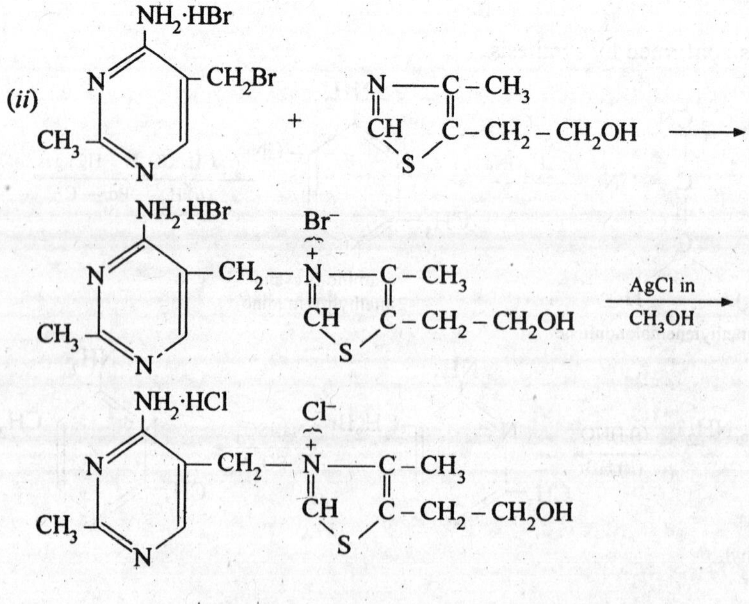

(*i*) $CO_2C_2H_5$–CH_2–$CH_2OC_2H_5$ + $H-CO_2C_2H_5$ $\xrightarrow{Na}$ $CO_2C_2H_5$–$CH(CHO)$–$CH_2OC_2H_5$ $\xrightarrow[C_2H_5ONa]{CH_3-C(NH_2)=NH}$

4-OH, 2-CH_3, 5-$CH_2OC_2H_5$ pyrimidine $\xrightarrow[(ii)\ NH_3 - C_2H_5OH]{(i)\ POCl_3}$ 4-NH_2, 2-CH_3, 5-$CH_2OC_2H_5$ pyrimidine $\xrightarrow{HBr}$

4-$NH_2 \cdot HBr$, 2-CH_3, 5-CH_2Br pyrimidine

(*ii*) 4-$NH_2 \cdot HBr$, 2-CH_3, 5-CH_2Br pyrimidine + thiazole (N—C–CH_3, CH, C–CH_2–CH_2OH, S) $\longrightarrow$

4-$NH_2 \cdot HBr$ pyrimidine–CH_2–$\overset{+}{N}$(Br^-) thiazolium (C–CH_3, CH, C–CH_2–CH_2OH, S) $\xrightarrow[CH_3OH]{AgCl\ in}$

4-$NH_2 \cdot HCl$ pyrimidine–CH_2–$\overset{+}{N}$(Cl^-) thiazolium (C–CH_3, CH, C–CH_2–CH_2OH, S)

Q. 3. Elucidate the structure of Vitamin B_2, riboflavin (lactoflavin), $C_{17}H_{20}O_6H_4$.

Ans. The structure of lactoflavin is arrived at as under:

1. When exposed to light, lactoflavin in sodium hydroxide solution forms mainly lumi-lactoflavin, $C_{13}H_{12}O_2N_4$. Lumi-lactoflavin, on boiling with barium hydroxide solution, is hydrolysed to one molecule of urea and one molecule of the barium salt of a β-ketocarboxylic acid, I, $C_{12}H_{12}O_3N_2$. The nature of this acid is shown by the fact that, on acidification of the barium salt, the free acid immediately eliminates carbon dioxide to form the compound, II, $C_{11}H_{12}ON_2$. This compound showed the properties of a lactam, and on vigorous hydrolysis by boiling with sodium hydroxide solution, it forms one molecule of glyoxylic acid and one molecule of the compound $C_9H_{14}N_2$ (III).

$$\underset{\text{Lactoflavin}}{C_{17}H_{20}O_6N_4} \xrightarrow[\text{Light}]{\text{NaOH}} \underset{\text{Lumi-lactoflavin}}{C_{13}H_{12}O_2N_4} \xrightarrow{Ba(OH)_2}$$

$$CO(NH_2)_2 + \underset{\text{I}}{[C_{12}H_{12}O_3N_2]} \xrightarrow[-CO_2]{\text{acid}}$$

$$\underset{\text{II}}{C_{11}H_{12}ON_2} \xrightarrow{\text{NaOH}} CHO{\cdot}COOH + \underset{\text{III}}{C_9H_{14}N_2}$$

2. The structure of III was decided as follows. Preliminary tests showed that III was an aromatic diamino compound. Then it was found that it gave a blue precipitate with ferric chloride, and since this reaction is characteristic of monomethyl-*o*-phenylenediamine, it suggests that III contains the following nucleus, IV. The molecular formula of IV is $C_7H_{10}N_2$, and since III is $C_9H_{14}N_2$, two carbon and four hydrogen atoms must be accounted for. This can be done by assuming the presence of an ethyl group or of two methyl groups in the benzene ring.

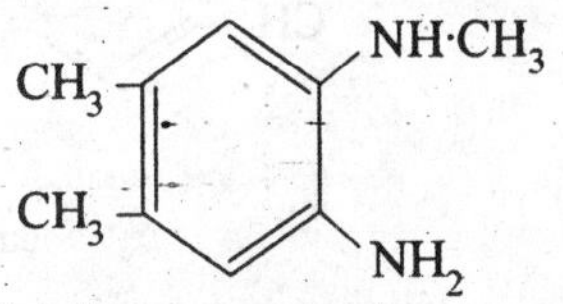

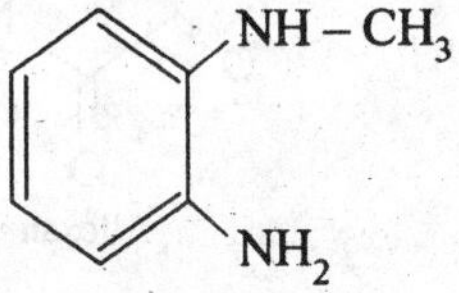

3. Kuhn *et al.* carried out a series of synthetic experiments and showed that III has the structure given, *N*-methyl-4, 5-diamino-o-xylene. Kuhn then proposed II as the structure of the precursor of III, as explained below:

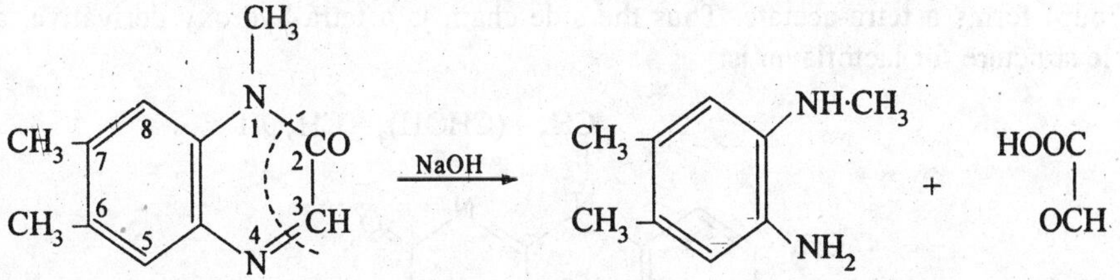

4. Now the precursor of II can be proposed. II could have been produced from the β-ketocarboxylic acid I.

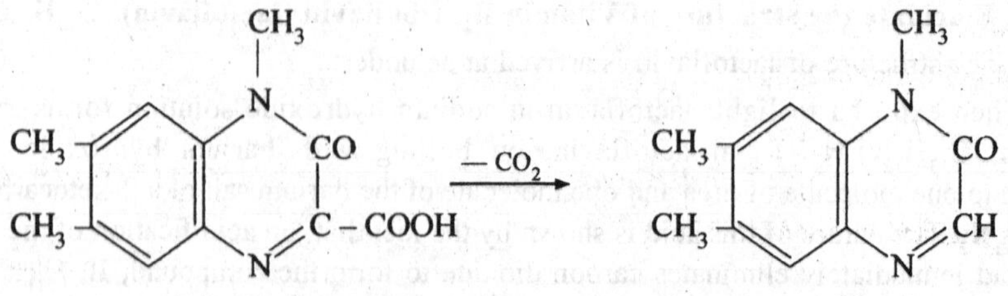

5. Since I and a molecule of urea are obtained from lumi-lactoflavin, the latter could be 6, 7, 9-trimethyl *iso* alloxazine (6, 7, 9-trimethylflavin).

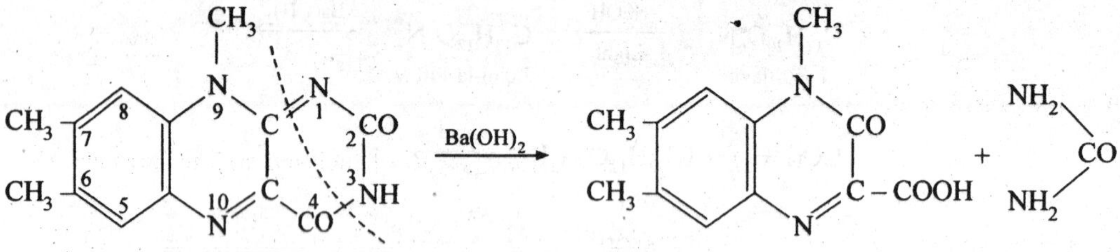

Lumi-lactoflavin

This structure for lumi-lactoflavin has been confirmed by synthesis as given below. *N*-Methyl-4, 5-diamino-*o*-xylene is condensed with alloxan hydrate in aqueous solution at 50-60°C.

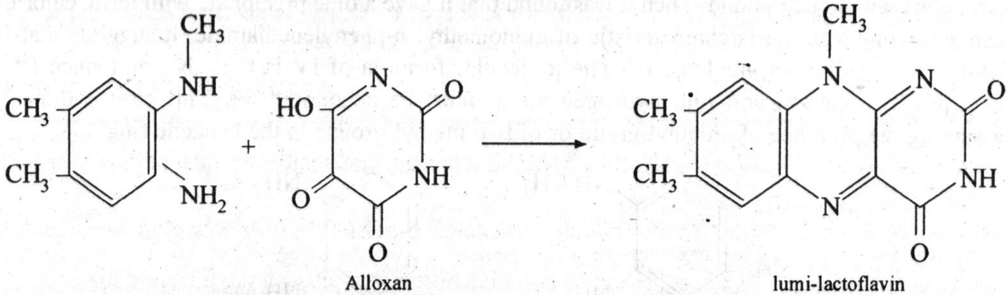

Alloxan

lumi-lactoflavin

6. *Side-chain in lactoflavin*. Experiments suggest that lactoflavin contains a side-chain (of five carbon atoms) attached to N. The Zerewitnoff procedure shows that lactoflavin contains five active hydrogen atoms; thus the molecule contains four hydroxyl groups (one active hydrogen atom is the hydrogen of the NH group at position 3). The presence of these four hydroxyl groups is supported by the fact that the silver salt of lactoflavin (the silver atom replaces the hydrogen of the NH group) forms a tetra-acetate. Thus the side-chain is a tetra-hydroxy derivative, and so a possible structure for lactoflavin is:

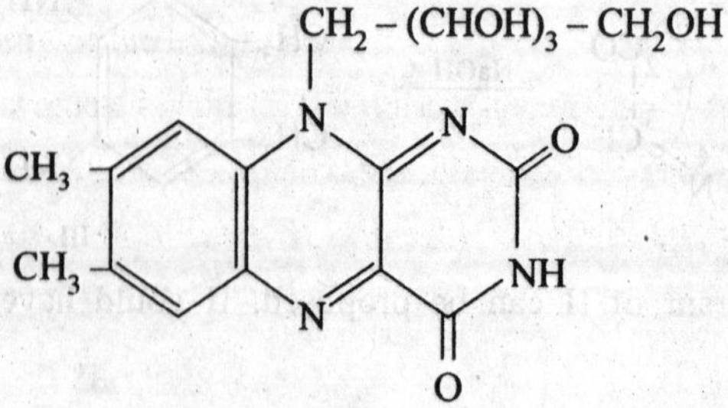

$CH_2-(CHOH)_3-CH_2OH$

Lactoflavin

This structure is confirmed by the following synthesis.

Karrer synthesis

$$\text{3,4-}(CH_3)_2C_6H_3NH_2 + CHO-CH(OH)-CH(OH)-CH(OH)-CH_2OH \xrightarrow[(ii)\ H_2-Pd]{(i)\ \text{condensation}}$$

$$\text{3,4-}(CH_3)_2C_6H_3-NH-CH_2-CH(OH)-CH(OH)-CH(OH)-CH_2OH \xrightarrow{NO_2C_6H_4N_2Cl}$$

$$CH_2-(CHOH)_3-CH_2OH \text{ (on NH)}, \ (CH_3)_2C_6H_2(NH-)(N=N-C_6H_4-NO_2) \xrightarrow[\text{pressure}]{H_2-Ni} (CH_3)_2C_6H_2(NH-CH_2-(CHOH)_3-CH_2OH)(NH_2)$$

$$\xrightarrow[H_3BO_3]{\text{alloxan}}$$ riboflavin: 7,8-dimethyl-10-$[CH_2-(CHOH)_3-CH_2OH]$-isoalloxazine

$$\text{3,4-}(CH_3)_2C_6H_3NH_2 + Cl-COOC_2H_5 \longrightarrow (CH_3)_2C_6H_3-NH-COOC_2H_5 \xrightarrow{HNO_3}$$

$$(CH_3)_2C_6H_2(NO_2)-NH-COOC_2H_5 \xrightarrow{H_2-Pd} (CH_3)_2C_6H_2(NH_2)-NH-COOC_2H_5 \xrightarrow{D(-)\text{-ribose}}$$

$$\left[\text{(CH}_3)_2\text{C}_6\text{H}_2(\text{N=CH(CHOH)}_3\text{CH}_2\text{OH})(\text{NHCOOC}_2\text{H}_5)\right] \xrightarrow{H_2 - Ni} (\text{CH}_3)_2\text{C}_6\text{H}_2(\text{NH–CH}_2\text{(CHOH)}_3\text{CH}_2\text{OH})(\text{NHCOOC}_2\text{H}_5) \xrightarrow{NaOH}$$

$$(\text{CH}_3)_2\text{C}_6\text{H}_2(\text{NH–CH}_2\text{(CHOH)}_3\text{CH}_2\text{OH})(\text{NH}_2) \xrightarrow[H_3BO_3]{\text{Alloxan}} \text{Lactoflavin}$$

Lactoflavin

Q. 4. Establish the structure of pantothenic acid, $C_9H_{17}O_5N$.

Ans. The structure of pantothenic acid is established as under:

1. Pantothenic acid shows the reactions of a monocarboxylic acid, *e.g.*, it can be esterified to form monoesters.

2. The experiment for determining active hydrogen atoms shows that pantothenic acid contains two hydroxyl groups, and since the acid condenses with benzaldehyde (to form a benzylidene derivative) and with acetone (to form an *iso* propylidene derivative), this suggests that the two hydroxy groups are in either the 1 : 2- or 1 : 3-position.

3. When warmed with dilute hydrochloric acid, pantothenic acid is hydrolysed into compounds I and II. Compound I has been identified as β-alanine

$$C_9H_{17}O_5N \xrightarrow{HCl} \underset{I}{C_3H_7O_2N} + \underset{II}{C_6H_{10}O_3}$$

4. When hydrolysed with alkali, pantothenic acid forms β-alanine (I) and the salt of an acid which, on acidification, spontaneously forms the lactone II. Thus the free acid of II is probably a γ- or δ-hydroxycarboxylic acid. Since the rate of lactonisation is fast, II is more likely a γ-lactone than a δ-lactone.

5. As pointed out above (step 2) pantothenic acid contains two hydroxyl groups. One of these has now been accounted for, and so the problem is to find the position of the second one. This was shown to be α- by the fact that the sodium salt of the acid of the lactone II gives a canary-yellow colour with ferric chloride (a test characteristic of α-hydroxy acids), and also by the fact that II, on warming with concentrated sulphuric acid, liberates carbon monoxide (a test also characteristic of α-hydroxy acids). Thus II is most probably the γ-lactone of an α-hydroxy acid.

6. II was shown to contain one active hydrogen atom, and the application of the Kuhn-Roth methyl side-chain determination showed the presence of a *gem*-dimethyl group; the presence of this group is confirmed by the formation of acetone when the lactone II is oxidised with barium permanganate. Thus a possible structure for II is α-hydroxy-β, β-dimethyl-γ-butyrolactone:

$$\underbrace{CH_2 - C(CH_3)_2 - CHOH - CO}_{O} \equiv C_6H_{10}O_3$$

II

This has been confirmed as follows. Treatment of the lactone with methylmagnesium iodide, followed by hydrolysis, gives a trihydric alcohol which, on oxidation with lead tetraacetate, gives acetone and an aldehyde. This aldehyde, on oxidation with silver oxide, gave a compound III, which was shown to be β-hydroxy-α, α-dimethylpropionic acid. The foregoing reactions may be formulated as follows:

$$\underbrace{CH_2 - C(CH_3)_2 - CHOH - CO}_{O} \xrightarrow[(ii)\ H_2O]{(i)\ CH_3MgI}$$

II

$$CH_2OH - C(CH_3)_2 - CHOH - C(OH)(CH_3)_2 \xrightarrow{(CH_3-CO_2)_4Pb}$$

$$CH_3 - CO - CH_3 + CH_2OH - C(CH_3)_2 - CHO \xrightarrow{Ag_2O}$$

$$CH_2OH - C(CH_3)_2 - COOH$$

III

7. In pantothenic acid, the nitrogen atom is not basic. Also, since hydrolysis of pantothenic acid produces a free amino group (in β-alanine), this suggests that the group — CONH — is present, i.e., pantothenic acid is an amide. Thus the hydrolysis may be formulated:

$$CH_2OH - C(CH_3)_2 - CHOH - CO - NH - CH_2 - CH_2 - COOH$$

Pantothenic acid

$$\downarrow HCl$$

$$[CH_2OH - C(CH_3)_2 - CHOH - COOH] + NH_2 - CH_2 - CH_2 - COOH$$

$$\downarrow \text{Lactonises}$$

$$\underbrace{CH_2 - C(CH_3)_2 - CHOH - CO}_{O}$$

Synthesis

This interpretation of the results has been proved by the synthesis of pantothenic acid. Stiller *et al.* (1940) warmed pantolactone with the ethyl ester of β-alanine, and removed the ester group by hydrolysis with a cold solution of barium hydroxide.

$$\underbrace{CH_2 - C(CH_3)_2 - CHOH - CO}_{O} + NH_2 - CH_2 - CH_2COOC_2H_5$$

Pantolactone — Ethyl ester of β-alanine

$$\downarrow \text{Warm}$$

$$CH_2OH - C(CH_3)_2 - CHOH - CO - NH - CH_2 - CH_2 - COOC_2H_5$$

$$\downarrow Ba(OH)_2$$

$$CH_2OH - C(CH_3)_2 - CHOH - CO - NH - CH_2 - CH_2 - COOH$$

Q. 5. Establish the structure of vitamin B_6 (pyridoxin, adermin).

Ans. The structure is established as under:

1. Elemental analysis and molecular mass determination show the formula of pyridoxin as $C_8H_{11}O_3N$.

2. Pyridoxin behaves as a weak base, and the usual tests showed the absence of methoxyl and methylamino groups.

3. Experiments showed the presence of three active hydrogen atoms. When treated with diazomethane, pyridoxin formed active hydrogen atoms. When treated with diazomethane, pyridoxin formed a monomethyl ether which, on acetylation, gave a diacetyl derivative. Therefore, three oxygen atoms in pyridoxin are present as hydroxyl groups, and since one is readily methylated, this one is probably phenolic. This conclusion is supported by the fact that pyridoxin gives the ferric chloride colour reaction of phenols. Hence the other two hydroxyl groups are present as alcoholic.

4. Examination of the ultraviolet absorption spectrum of pyridoxin showed that it is similar to that of 3-hydroxypyridine. It was, therefore, inferred that pyridoxin is a pyridine derivative with the phenolic group in position 3.

5. Since lead tetraacetate has no action on the monomethyl ether of pyridoxin, the two alcoholic groups are not on adjacent carbon atoms in a side-chain.

6. When this methyl ether is *very carefully* oxidised with alkaline potassium permanganate, the product is a methoxypyridinetricarboxylic acid, $C_9H_7O_7N$. This acid gave a blood-red colour with ferrous sulphate, a reaction which is characteristic of pyridine-2-carboxylic acid; thus one of the three carboxyl groups is in the 2-position.

7. When the methyl ether of pyridoxin was oxidised with alkaline permanganate under the usual conditions, the products were carbon dioxide and the anhydride of a dicarboxylic acid, $C_8H_5O_4N$; thus these two carboxyl groups are in the *ortho*-position. Furthermore, since this anhydride, on hydrolysis to its corresponding acid, did not give a red colour with ferrous sulphate, there is no carboxyl group in the 2-position. It, therefore, follows that, on decarboxylation, the tricarboxylic acid eliminates the 2-carboxyl group to form the anhydride.

The above reactions can be explained on the basis of following structure of pyridoxin.

CH_2OH / HOH_2C / OH / CH_3 / N

Pyridoxin

$\xrightarrow{CH_2N_2}$ (CH_2OH, HOH_2C, OCH_3, CH_3, N) $\xrightarrow{(CH_3CO)_2O}$ (CH_2OAc, $AcOH_2C$, OCH_3, CH_3, N)

$\xrightarrow{KMnO_4 \text{ (carefully)}}$

COOH / HOOC / OCH_3 / COOH (2) / N (1)

Methoxy pyridine tricarboxylic acid

$\xrightarrow[-CO_2,\ -H_2O]{KMnO_4^-}$

O / CO / CO / OCH_3 / N

Anhydride of dicarboxylic acid

(No red colour with $FeSO_4$)

This structure has been confirmed by synthesis.

Ethoxyacetylacetone ($CH_2OC_2H_5$–CO–CH–CH_3·CO) + Cyanoacetamide (CH_2ON, CO, H_2N) $\xrightarrow[C_2H_5OH]{\text{Piperidine}}$ (pyridine with $CH_2OC_2H_5$, CN, OH, CH_3) $\xrightarrow[(CH_3 \cdot CO)_2O]{HNO_3}$

(with O_2N) $\xrightarrow{PCl_5}$ (OH → Cl) $\xrightarrow{H_2 - Pt}$ (H_2N, CN, Cl) $\xrightarrow{H_2 - Pt + Pd - C}$

(H_2N, CH_2NH_2, CH_3, $CH_2OC_2H_5$) $\xrightarrow[175°C]{HCl}$ (CH_2OH, H_2N, CH_2NH_2, CH_3) $\xrightarrow{HNO_2}$ (CH_2OH, HO, CH_2OH, CH_3) Pyridoxin

Q. 6. Discuss the structure of vitamin B_{12}, cyanocobalamin.

Ans. The structure of cyanocobalamin is discussed as under:

1. The metal cobalt has been found to be attached to a cyano groups.

2. The hydrolysis of vitamin B_{12} with hydrochloric acid under different conditions produces ammonia, 1-aminopropan-2-ol (I), 5, 6-dimethylbenzimidazole (II), 5, 6-dimethylbenzimidazole-1-α-D-ribofuranoside (III) and the 3′-phosphate of III.

3. Chromic acid oxidation of hydrolysed vitamin B_{12} produces compound IV besides I, II and III.

$CH_3CHOHCH_2NH_2$
1-aminopropane-2-ol
(I)

5, 6-dimethylbenzimidazole
(II)

(III)

Me(Me)C(CO–NH–CO)–CH – CH_2 – CH_2 – COOH
(IV)

4. It has been shown that six amido groups are present in the molecule. Also, alkaline hydrolysis of vitamin B_{12} gives a mixture consisting mainly of a penta- and a hexacarboxylic acid, in both of which the nucleotide fragment is absent.

5. As the result of a detailed X-ray analysis of the hexacarboxylic acid, vitamin B_{12} has been assigned the following structure.

$NH_2-CO-CH_2-CH_2$ Me Me CH_2-CONH_2
$NH_2-CO-CH_2$
Me CN $-CH_2CH_2-CONH_2$
N + N
Me Co
N N
$NH_2-CO-CH_2$ Me
Me
$NH-CO-CH_2-CH_2$ Me Me $CH_2-CH_2-CONH_2$
CH_2
MeCH N Me
O $\bar{O}$ Me
P N
O O OH
H
H H
$HOCH_2$ O

Vitamin B_{12}

Q. 7. Establish the structure of vitamin C (ascorbic acid).

Ans. Structure of vitamin C is established as under:

1. Elemental analysis and molecular weight determination of vitamin C show that it has molecular formula $C_6H_8O_6$.

2. Vitamin C gives the reactions of an unsaturated compound and gives the reaction of a strong reducing agent.

3. Vitamin C forms a phenylhydrazone (property of a carbonyl group) and gives a violet colour with ferric chloride (property of alcoholic group). This suggests a keto-enol system in the compound

$$-CO-\overset{|}{C}H- \rightleftharpoons C(OH)=\overset{|}{C}-$$

4. Vitamin C does not give Schiff's reaction. Hence possibility of an aldehydic group is ruled out, retaining the possibility of a ketonic group.

5. When boiled with hydrochloric acid, ascorbic acid gives furfuraldehyde quantitatively

$$C_6H_8O_6 \xrightarrow{HCl} \text{(furan ring: } CH=CH-CH=C(CHO)-O\text{)} + CO_2 + 2H_2O$$

This reaction shows the presence of at least five carbon atoms in a straight chain. Presence of a number of hydroxyl groups is also indicated.

6. Vitamin C (ascorbic acid) is oxidised by iodine solution to dehydroascorbic acid. In this reaction, two atoms of iodine are used and two molecules of hydrogen iodide are removed. Thus there is a net removal of two hydrogen atoms from ascorbic acid.

```
     CO ─┐                      ┌  CO ──────┐  ┐                   CO ─┐
     |   |                      |  |        |  |                   |   |
HO ─ C   |                      |  C (OH)2  |  |                   CO  |
     ‖   O     I2               |  |        O  |     -2H2O         |   O
HO ─ C   |   ──────→            |  C (OH)2  |  |    ──────→        CO  |
     |   |    2H2O              |  |        |  |                   |   |
 H ─ C ──┘                      |H─C ───────┘  |               H ─ C ──┘
     |                          |  |           |                   |
HO ─ C ─ H                      HO─C─H         |              HO ─ C ─ H
     |                          |  |           |                   |
    CH2OH                       └ CH2OH        ┘                  CH2OH
Ascorbic acid                                               Dehydroascorbic acid
```

7. Dehydroascorbic acid is a neutral substance and behaves as a lactone (internal ester) of a monobasic hydroxy acid. It is reconverted into ascorbic acid on treatment with H_2S.

8. Ascorbic acid is capable of forming salts with alkalis. This can be explained in terms of the *enol* group. Thus the molecule contains an α-hydroxy ketone grouping in ascorbic acid. This group explains the observations as under:

```
    |              |                            |
H ─ C ─ OH         C ─ OH      I2 + 2H2O        C (OH)2
    |              ‖          ──────────→       |          + 2HI
    C = O          C ─ OH       Iodine          C (OH)2
    |              |            solution        |
                                                |
Formation of     Unsaturation                   |  -2H2O
hydrazone        (colour with FeCl3)            ↓
                                                |
                                                C = O
                                                |
                                                C = O
                                                |
                                         Molecule smaller by
                                        2 hydrogen atoms than
                                            ascorbic acid
```

9. Dehydroascorbic acid on oxidation with sodium hypoiodite produces oxalic and L-threonic acids in quantitative yield. This reaction can be explained as under:

```
      CO ─┐                                         CO2H
      |   |                                         |
      CO  O     NaOI          COOH              H ─ C ─ OH
 - - -|- -|   ──────→         |           +         |
      CO  |                   COOH             HO ─ C ─ H
      |   |                   Oxalic acid           |
  H ─ C ──┘                                         CH2OH
      |                                        L-Threonic acid
 HO ─ C ─ H
      |
      CH2OH
```

L-Threonic acid was identified by methylation and conversion into a crystalline amide.

10. The ring in ascorbic acid has been assumed to be five- and not six-membered because the lactone (*i.e.*, ascorbic acid) is stable towards alkali.

11. We finally confirm the structure of vitamin C (ascorbic acid) by synthesis. Two syntheses are given below. The first one is by Haworth and Hirst (1933) starting from L-Lyxose. The second is the commercial one followed these days.

I. Haworth Synthesis

CHO
HO – C – H
H – C – OH
HO – C – H
CH_2OH
L-Lyxose

$\xrightarrow[\text{HCl}]{\text{Phenyl hydrazine}}$

CHO
CO
H – C – OH
HO – C – H
CH_2OH
L(–)Xylosone

$\xrightarrow{\text{KCN}}$

[CN
CHOH
CO
H – C – OH
HO – C – H
CH_2OH]
β-ketocyanide

$\xrightarrow{H_2O}$

CO_2H
CHOH
CO
H – C – OH
HO – C – H
CH_2OH
Pseudo-L-ascorbic acid

⇌

[COOH
C – OH
‖
C – OH
H – C – OH
HO – C – H
CH_2OH]

$\xrightarrow{-H_2O}$

CO
HO – C
‖ O
HO – C
H – C
HO – C – H
CH_2OH

L(+) ascorbic acid

II. Modern synthesis

CH_2OH
HO – C – H
HO – C – H
HO – C – H
HO – C – H
CHO
D-glucose

$\xrightarrow[\text{Cu–Cr}]{H_2}$

CH_2OH
HO – C – H
HO – C – H
HO – C – H
HO – C – H
CH_2OH
(+)-Sorbitol

$\xrightarrow[\text{Suboxydans}]{\text{Acetobacter}}$

CH_2OH
CO
HO – C – H
HO – C – H
HO – C – H
CH_2OH

⇌

H O OH
H HO
$HOCH_2$ CH_2OH
OH H
(–)-sorbose

$\xrightarrow[H_2SO_4]{2Me_2CO}$

Me_2
C
H O O
H O
CH_2 CH_2OH
O O H
C
Me_2
Diacetone(–)-sorbose

$\xrightarrow[NaOH]{KMnO_4}$ (diacetone derivative, Me_2C, O, H, CH_2, COONa, H, Me_2) $\xrightarrow{H_2SO_4}$

COOH
|
CO
|
HO – C – H
|
H – C – OH
|
HO – C – H
|
CH_2OH

2-ketogulonicacid

$\xrightarrow{CHCl_3\ soln.}$

CO
|
HO – C
||
HO – C (O ring)
|
H – C
|
HO – C – H
|
CH_2OH

L-ascorbic acid

Q. 8. Establish the structure of Vitamin D_2 (ergocalciferol, calciferol).

Ans. The structure of vitamin D_2 is established as under:

1. Elemental analysis and molecular mass determination show the molecular formula of vitamin D_2 as $C_{28}H_{44}O$.

2. Vitamin D_2 forms an ester with an acid. That shows that oxygen in the compound is present as hydroxyl group.

3. Vitamin D_2 (ergocalciferol) on oxidation forms a ketone (which is identified by usual tests). That shows that the hydroxyl group is a secondary alcoholic group.

4. On ozonolysis of vitamin D_2, methyl *iso*-propylacetaldehyde is obtained among other products. That shows that the side chain in ergocalciferol is the same as that in ergisterol (a compound with known structure).

5. Catalytic hydrogenation of ergocalciferol gives a saturated compound octahydro ergocalciferol ($C_{28}H_{52}O$). This supports the presence of four double bonds in the molecule. Since one of the double bonds is in the side chain, the other three must be in the nucleus. Thus the parent hydrocarbon of vitamin D_2 is $C_{27}H_{52}$. As this conforms to the general formula C_nH_{2n-2}, a tricyclic molecule is indicated.

6. Ergocalciferol does not give Diel's hydrocarbon on distillation with Se, hence presence of a four-ring system in ergocalciferol is ruled out.

7. If we assume the following structure for ergocalciferol (I), the reactions given below can be properly explained:

(*i*) Oxidation of ergocalciferol with chromium oxide gives compounds II and III. The absorption spectrum of $C_{21}H_{34}O$ (Compound II) is characteristic of α, β-unsaturated aldehydes. The absence of hydroxyl group and carbon content indicates the absence of ring A. It is also suggested that in ergocalciferol, *ring B* is open between C_9 and C_{10} and that compound II is obtained by rupture of the molecule at a double bond in 5, 6 positions. It can be an α, β–unsaturated aldehyde only if there is a double bond at 7, 8 positions.

(*ii*) Isolation of a ketone (compound III) confirms the presence of a double bond at 7, 8 positions.

(*iii*) Ozonolysis of ergocalciferol gives compounds IV, V and VI. Formation of formaldehyde (compound IV) indicates the presence of an exocyclic methylene group at position 10. This is in accordance with the opening of *ring B* at 9, 10 positions.

(*iv*) Formation of compound V, a keto acid suggests that ring B is open at 9, 10 position and that there are two double bonds at 7, 8 and 22, 23 positions.

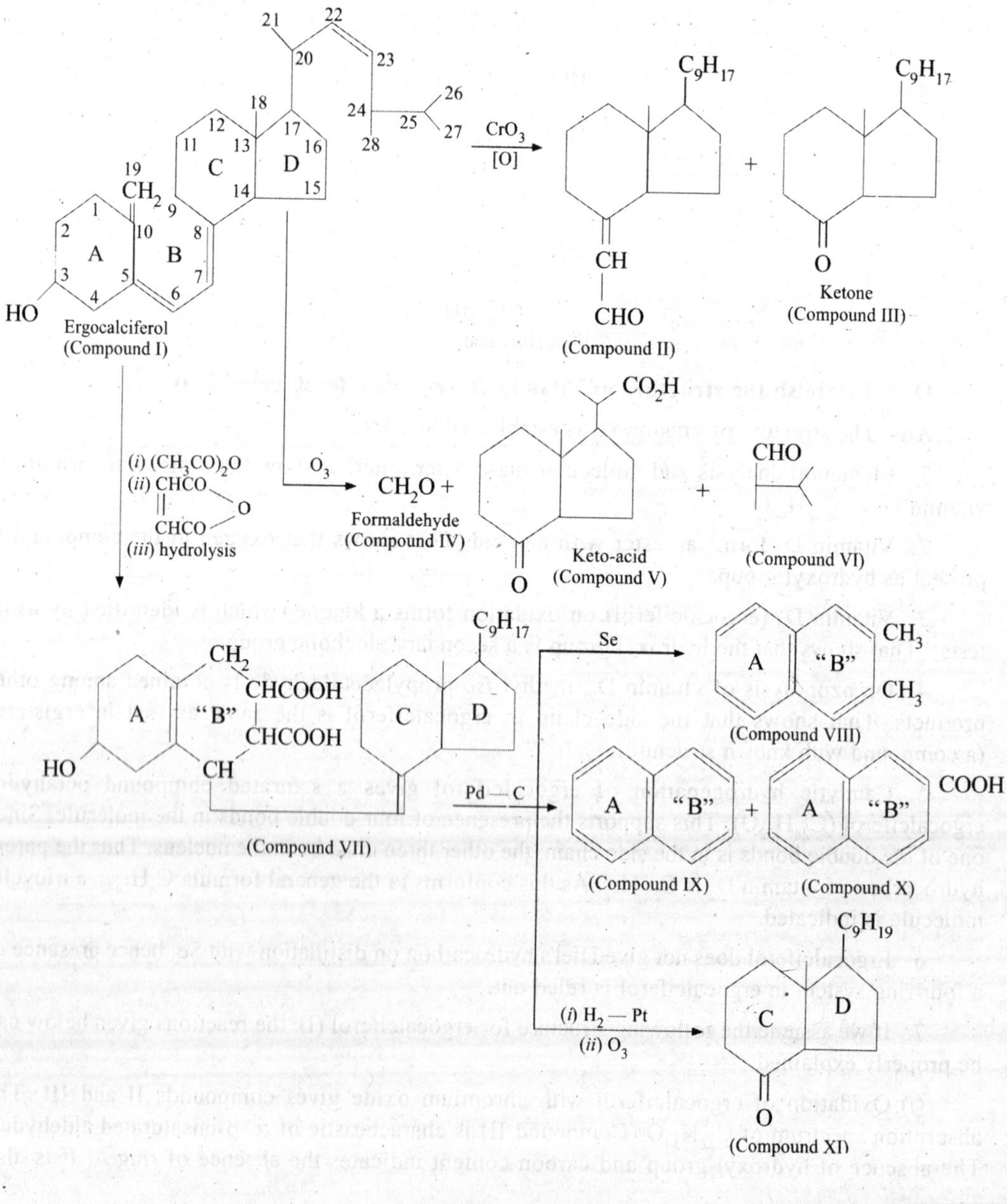

8. Structure I for ergocalciferol is also supported by the formation of compound VII on treatment with acetic anhydride and maleic anhydride followed by hydrolysis.

(*i*) Compound VII on selenium treatment gives compound VIII, 2, 3, dimethyl naphthalene. On selenium dehydrogenation, carboxyl groups sometimes give methyl groups. Formation of compounds IX and X point to the presence of rings A and B in compound VII.

(*ii*) Catalytic reduction of compound VII (which reduce side-chain double bonds only) followed by ozonolysis produce the compound XI. This again supports the structure I for ergocalciferol.

Q. 9. Establish the structure of α-tocopherol (vitamin E group).

Ans. The following points establish the structure:

1. Molecular formula of α-tocopherol, based on elemental analysis and molecular mass determination comes out to be $C_{29}H_{50}O_2$.

2. Oxidation of α-tocopherol with chromic acid forms dimethylmaleic anhydride and a compound $C_{21}H_{40}O_2$.

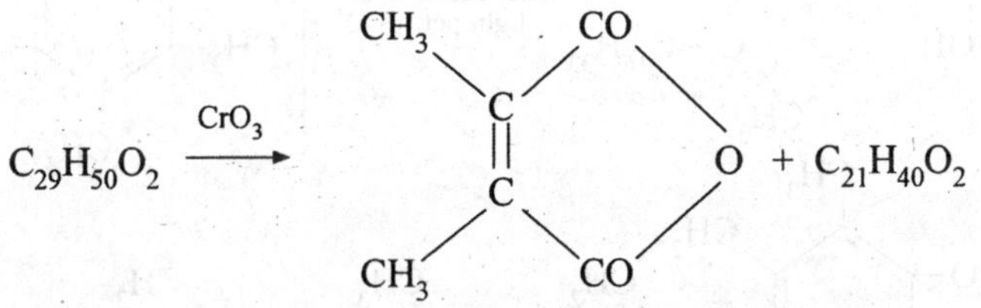

$$C_{29}H_{50}O_2 \xrightarrow{CrO_3} \underset{\text{(CH}_3\text{C(CO)=C(CO)CH}_3\text{, cyclic anhydride -CO-O-CO-)}}{} + C_{21}H_{40}O_2$$

This latter compound was shown to be an optically active saturated lactone. This lactone was then shown to be derived from a γ-hydroxy acid in which the hydroxyl group is tertiary, *e.g.*, the acid lactonised immediately its salt was acidified, and also could not be oxidised to a keto acid. Thus the structure of this lactone may be written (R + R′ = 17C):

$$R-\overset{R'}{\overset{|}{C}}CH_2CH_2CO \quad (\text{C and CO linked through } -O-)$$

3. Now α-tocopherol acetate, on oxidation with chromic acid, forms an acid, $C_{16}H_{32}O_2$, I and a ketone, $C_{18}H_{36}O$, II. Both of these compounds must be produced by the oxidation of the lactone at different points in the chain. Fernholz, therefore, suggested that if in the lactone R = $C_{16}H_{33}$ and R′ = CH_3, then the products I and II can be accounted for; thus:

(*i*) $$C_{16}H_{33}-\overset{CH_3}{\overset{|}{C}}CH_2CH_2CO \;(\text{lactone, }-O-) \xrightarrow{CrO_3} \underset{\text{I}}{C_{16}H_{32}O_2}$$

(*ii*) $$C_{16}H_{33}-\overset{CH_3}{\overset{|}{C}}-CH_2CH_2CO \;(\text{lactone, }-O-) \xrightarrow{CrO_3} \underset{\text{II}}{C_{16}H_{33}COCH_3}$$

Fernolz then showed that the acid (I) contained methyl groups and was led to propose to structure based on the isoprene unit, viz.,

$$CH_3\overset{CH_3}{\overset{|}{C}H}(CH_2)_3\overset{CH_3}{\overset{|}{C}H}(CH_2)_3\overset{CH_3}{\overset{|}{C}H}(CH_2)_2COOH$$

The evidence obtained so far points to the presence of a substituted benzene ring and a long side-chain in α-tocopherol.

The following structure for α-tocopherol explains all the above reactions

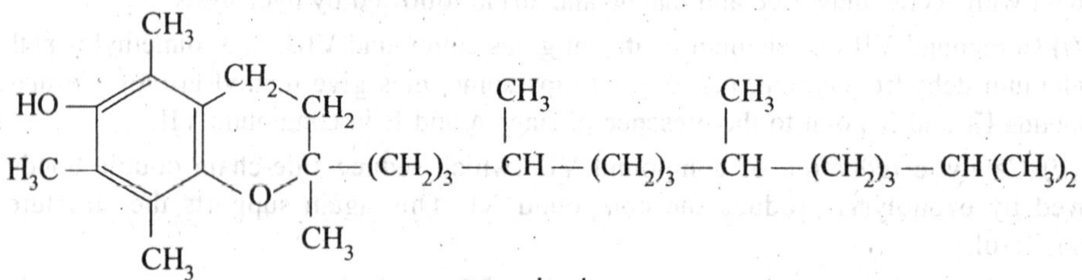

$$C-(CH_2)_3-CH(CH_3)-(CH_2)_3-CH(CH_3)-(CH_2)_3-CH(CH_3)_2$$

α-tocopherol

Karrer *et al.* have synthesised (±)-α-tocopherol by condensing trimethylquinol with phytyl bromide.

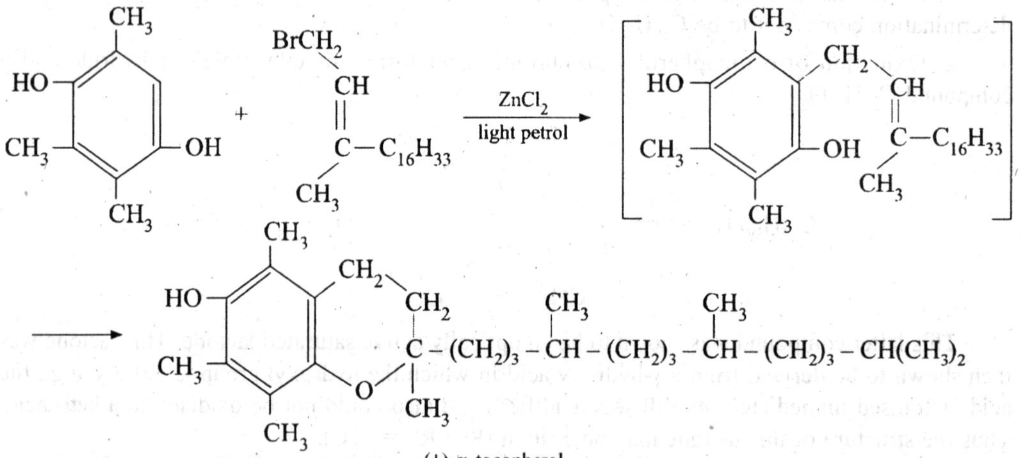

$$\xrightarrow[\text{light petrol}]{ZnCl_2}$$

(±)-α-tocopherol

Q. 10. Study the structures of β-tocopherol and γ-tocopherol.

Ans. β-Tocopherol, $C_{28}H_{48}O_2$.

1.This formula differs from that of α-tocopherol by CH_2. Thermal decomposition of β-tocopherol gives trimethylquinol, I, and heating with hydriodic acid *p*-xylenol, II.

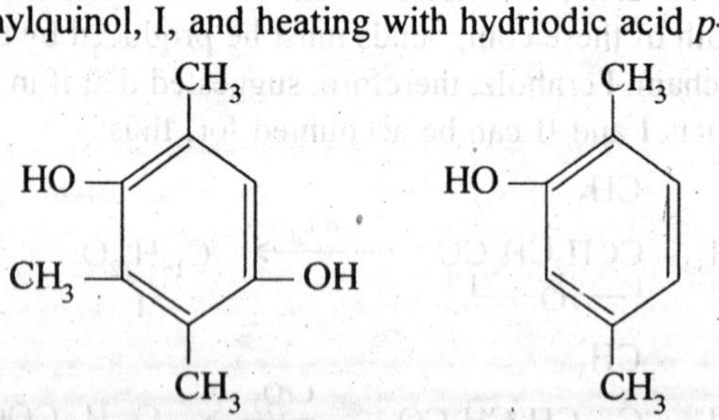

Trimethylquinol
I

p-xylenol
II

2.When oxidised with chromic acid, β-tocopherol gives the same lactone ($C_{21}H_{40}O_2$) as that obtained from α-tocopherol. Thus the only difference between the two tocopherols is that the α-compound has one more methyl group in the benzene ring than the β-; hence the latter is

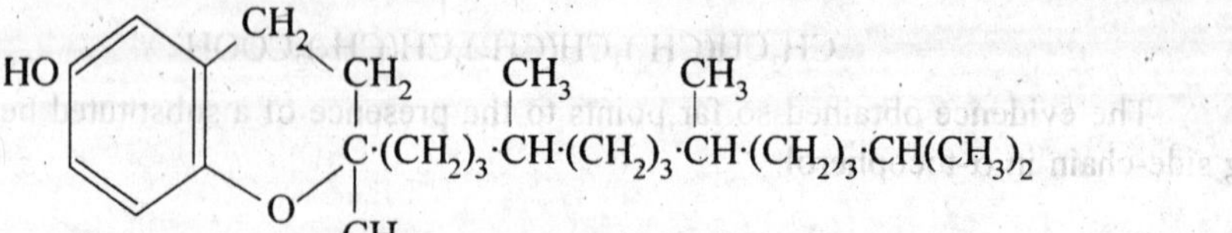

$$C\cdot(CH_2)_3\cdot CH(CH_3)\cdot(CH_2)_3\cdot CH(CH_3)\cdot(CH_2)_3\cdot CH(CH_3)_2$$

β-tocopherol

3. This has been confirmed by synthesis, starting from the monoacetate of *p*-xyloquinol and phytyl bromide.

γ-Tocopherol, $C_{28}H_{48}O_2$.

This is isomeric with β-tocopherol; the only difference is in the positions of the two methyl groups in the benzene ring, *e.g.*, when heated with hydriodic acid, γ-tocopherol gives *o*-xyloquinol. Thus γ-tocopherol is

γ-tocopherol

This structure has been confirmed by synthesis, starting from the monoacetate of *o*-xyloquinol and phytyl bromide.

Q. 11. Establish the structure of β-Biotin (vitamin H).

Ans. 1. Qualitative and quantitative analysis of β-Biotin show that its molecular formula is $C_{10}H_{16}O_3N_2S$.

2. β-Biotin does not give the usual tests of ethylenic double bond. Thus, it behaves as a saturated compound.

3. β-Biotin forms a monomethyl ester $C_{11}H_{18}O_3N_2S$ which, on hydrolysis, gives an acid the titration curve of which corresponds to a monocarboxylic acid; thus the formula of β-biotin may be written $C_9H_{15}ON_2S - COOH$.

4. When heated with barium hydroxide solution at 140°, β-biotin is hydrolysed to carbon dioxide and diaminocarboxylic acid $C_9H_{18}O_2N_2S$ which, by the action of carbonyl chloride, is reconverted into β-biotin. These reactions suggest that β-biotin contains a cyclic ureide structure. Since the diaminocarboxylic acid condenses with phenanthraquinone to form a quinoxaline derivative, it follows that the two amino-groups are in the 1 : 2-positions, and thus the cyclic ureide is five-membered. Hence, we may write the above reactions as follows:

```
      CO
     /  \
   NH    NH        Ba(OH)2          NH2    NH2
   |      |       ========>          |      |
   C ---- C        COCl2             C ---- C
  / \    / \                        / \    / \

   β-biotin                        diamino-compound
```

5. When this diaminocarboxylic acid is oxidised with alkaline permanganate, adipic acid is produced. One of the carboxyl groups in adipic acid was shown to be that originally present in β-biotin as follows. When the carbomethoxyl group of the methyl ester of β-biotin was replaced by an amino-group by means of the Curtius reaction (ester → hydrazide → azide → urethan → NH_2), and the product hydrolysed with barium hydroxide solution, a triamine was obtained which did not give adipic acid on oxidation with alkaline permanganate. It was therefore inferred that β-biotin contains a – $(CH_2)_4$ – COOH side-chain (*n*-valeric acid side-chain).

6. The nature of the sulphur atom in β-biotin was shown to be of the thio-ether type (*i.e.*, C – S – C) since:

(*i*) Oxidation of β-biotin with hydrogen peroxide produced a sulphone.

(*ii*) When the methyl ester of β-biotin was treated with methyl iodide, a sulphonium iodide was formed.

7. As reported above, β-biotin does not contain a double bond; hence, from its molecular formula, it was deduced that β-biotin contains two rings. As molecular formula of β-biotin is $C_{10}H_{16}O_3N_2S$, the carboxyl group may be regarded as a substituent group, and so the parent compound will be $C_9H_{16}ON_2S$. Also, since two NH groups are present, these may be replaced by CH_2 groups; thus the parent compound is $C_{11}H_{18}OS$. The CO group may be replaced by a CH_2 group and the sulphide atom also by a CH_2 group. This gives a compound of formula $C_{12}H_{22}$ which has the same "structure" as β-biotin. Now, the formula $C_{12}H_{22}$ corresponds to the general formula C_nH_{2n-2}, and this, for a *saturated* compound, corresponds to a system containing two rings.

8. When heated with Raney nickel, β-biotin formed *dethiobiotin* by elimination of the sulphur atom. Dethiobiotin, on hydrolysis with hydrochloric acid, gave a diaminocarboxylic acid which, on oxidation with periodic acid, gave pimelic acid. These results can be explained by assuming that the sulphur atom is in a five-membered ring and the *n*-valeric acid side-chain is in the position shown.

```
       CO                                              CO
      /  \                                            /  \
    NH    NH                     Raney              NH    NH
    |      |                    ------>             |      |
    CH --- CH                     Ni                CH --- CH
   /         \                                      |      |
  CH2        CH-(CH2)4-COOH                         CH3    CH-(CH2)4-COOH
    \       /
       S                                                 dethiobiotin
  β-biotin

          NH2    NH2
  HCl     |      |              HIO4        COOH
------>   CH --- CH            ------>      |
          |      |                          CH2-(CH2)4-COOH
          CH3    CH2-(CH2)4-COOH
                                            pimelic acid
```

9. Further evidence for this structure is given by the fact that the exhaustive methylation of the diaminocarboxylic acid (produced from β-biotin), followed by hydrolysis, gave δ-(2-thienyl)-valeric acid (du Vigneaud *et al.*, 1942); the structure of this compound was confirmed by synthesis.

Thiophen + Glutaric anhydride [$(CH_2)_3(CO)_2O$] $\xrightarrow{AlCl_3}$ 2-thienyl–$CO-(CH_2)_3-COOH$ $\xrightarrow[HCl]{Zn-Hg}$

NH_2–CH–CH(NH_2)–CH$-(CH_2)_4-COOH$ with CH_2 and S in a ring (diaminocarboxylic acid) $\xrightarrow[(ii)\ Conc.\ HCl]{(i)\ (CH_3)_2SO_4\cdot NaOH}$ 2-thienyl–$(CH_2)_3-COOH$

δ-(2-thienyl)-valeric acid

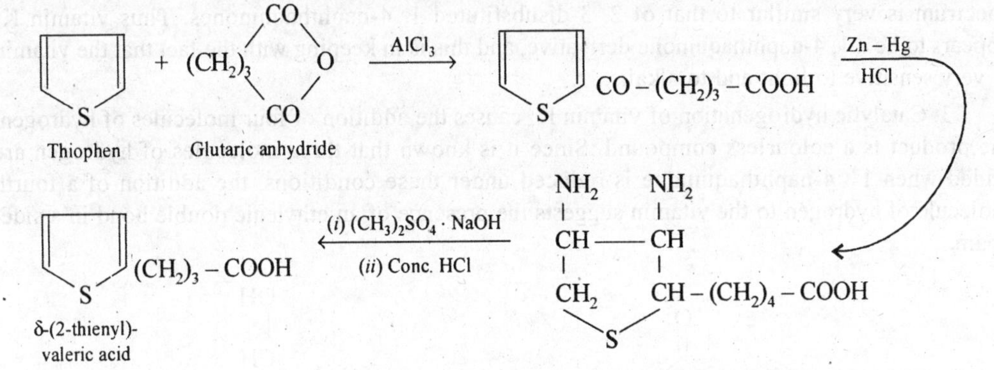

10. The above structure for β-biotin has been confirmed by synthesis (Harris *et al.*).

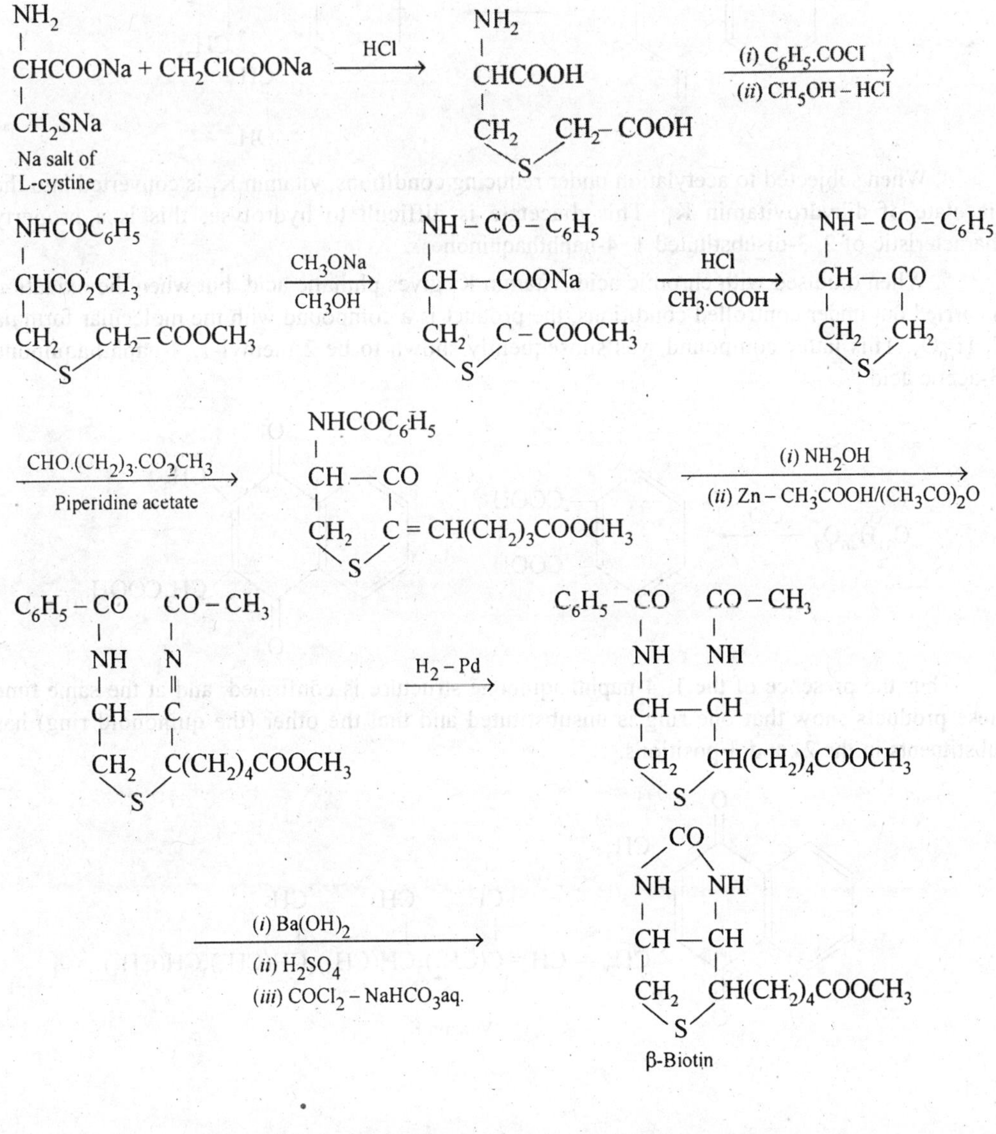

NH_2–$CHCOONa$–CH_2SNa (Na salt of L-cystine) + $CH_2ClCOONa$ $\xrightarrow{HCl}$ NH_2–$CHCOOH$–CH_2–S–CH_2-COOH $\xrightarrow[(ii)\ CH_5OH-HCl]{(i)\ C_6H_5.COCl}$

$NHCOC_6H_5$–$CHCO_2CH_3$–CH_2–S–$CH_2-COOCH_3$ $\xrightarrow[CH_3OH]{CH_3ONa}$ $NH-CO-C_6H_5$–CH–COONa, CH_2–S–$C-COOCH_3$ (ring) $\xrightarrow[CH_3.COOH]{HCl}$ $NH-CO-C_6H_5$–CH–CO, CH_2–S–CH_2 (ring)

$\xrightarrow[Piperidine\ acetate]{CHO.(CH_2)_3.CO_2CH_3}$ $NHCOC_6H_5$–CH–CO, CH_2–S–$C=CH(CH_2)_3COOCH_3$ (ring) $\xrightarrow[(ii)\ Zn-CH_3COOH/(CH_3CO)_2O]{(i)\ NH_2OH}$

C_6H_5-CO–NH–CH–CH_2–S–$C(CH_2)_4COOCH_3$, with $CO-CH_3$–N= attached to C (ring) $\xrightarrow{H_2-Pd}$ C_6H_5-CO–NH–CH–CH–NH–$CO-CH_3$, CH_2–S–$CH(CH_2)_4COOCH_3$ (ring)

$\xrightarrow[(ii)\ H_2SO_4\ (iii)\ COCl_2-NaHCO_3aq.]{(i)\ Ba(OH)_2}$ CO(NH)(NH)–CH–CH, CH_2–S–$CH(CH_2)_4COOCH_3$ (ring)

β-Biotin

Q. 12. Establish the structure of vitamin K_1(α-phylloquinone).

Ans. 1. It is found to have a molecular formule $C_{31}H_{46}O_2$.

2. The redox potential of vitamin K_1 is very similar to that of 1, 4-quinones and its absorption spectrum is very similar to that of 2, 3-disubstituted 1, 4-naphthaquinones. Thus vitamin K_1 appears to be a 1, 4-naphthaquinone derivative, and this is in keeping with the fact that the vitamin is very sensitive to light and to alkalis.

3. Catalytic hydrogenation of vitamin K_1 causes the addition of four molecules of hydrogen; the product is a colourless compound. Since it is known that three molecules of hydrogen are added when 1 : 4-naphthaquinone is reduced under these conditions, the addition of a fourth molecule of hydrogen to the vitamin suggests the presence of an ethylenic double bond in a side-chain.

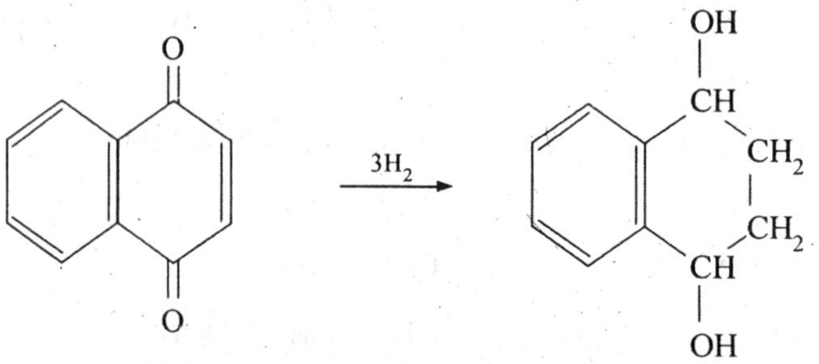

O, O, $3H_2$, OH, CH, CH_2, CH_2, CH, OH

4. When subjected to acetylation under reducing conditions, vitamin K_1 is converted into the diacetate of dihydrovitamin K_1. This diacetate is difficult to hydrolyse; this is a property characteristic of 2, 3-disubstituted 1, 4-naphthaquinones.

5. When oxidised with chromic acid, vitamin K_1 gives phthalic acid, but when the oxidation is carried out under controlled conditions, the product is a compound with the molecular formula $C_{13}H_{10}O_4$. This latter compound was subsequently shown to be 2-methyl-1, 4-naphthaquinone -3-acetic acid.

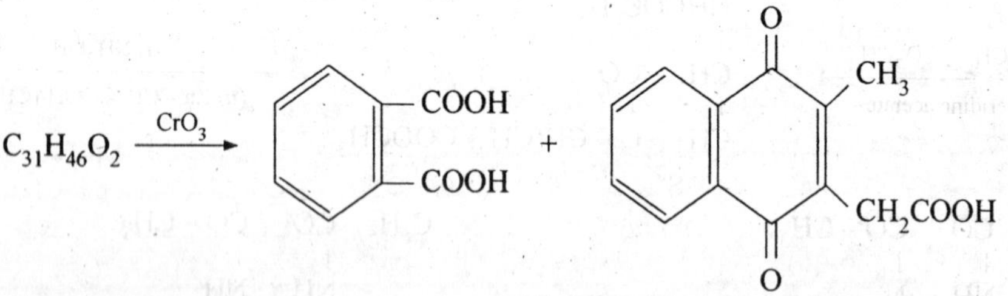

$C_{31}H_{46}O_2$ $\xrightarrow{CrO_3}$ COOH, COOH + O, CH_3, CH_2COOH, O

Thus the presence of the 1, 4-naphthaquinone structure is confirmed, and at the same time these products show that one ring is unsubstituted and that the other (the quinonoid ring) has substituents in the 2- and 3-positions.

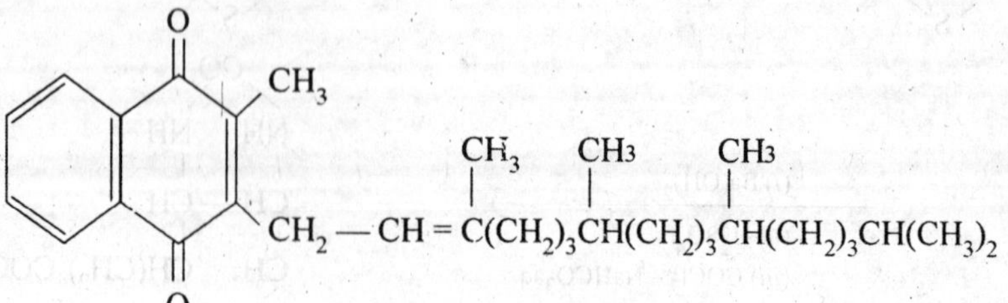

O, CH_3, CH_3, CH3, CH3, $CH_2-CH=C(CH_2)_3CH(CH_2)_3CH(CH_2)_3CH(CH_3)_2$, O

Vitamin K_1

Hence, on the evidence obtained above, vitamin K_1 is 2-methyl-3-phytyl-1, 4-naphthaquinone. This structure has been confirmed by synthesis:

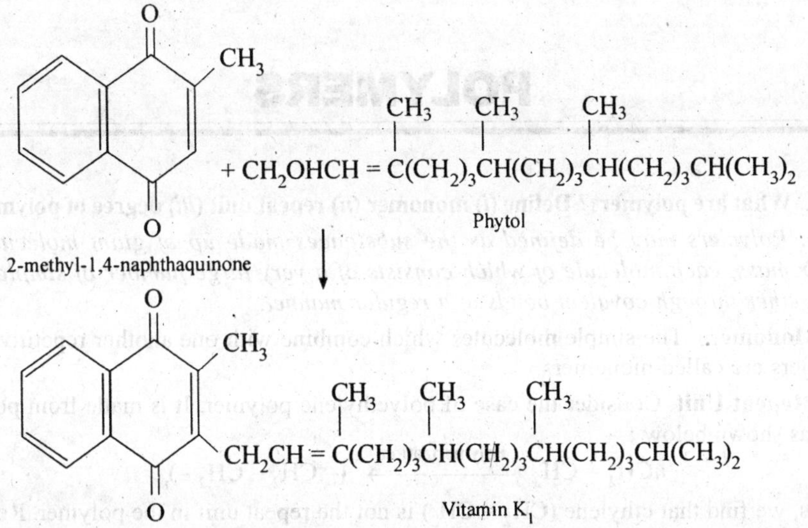

O

CH_3

O

2-methyl-1,4-naphthaquinone

$$+\ CH_2OHCH = \underset{\displaystyle CH_3}{C}(CH_2)_3\underset{\displaystyle CH_3}{CH}(CH_2)_3\underset{\displaystyle CH_3}{CH}(CH_2)_3CH(CH_3)_2$$

Phytol

↓

O

CH_3

$$CH_2CH = \underset{\displaystyle CH_3}{C}(CH_2)_3\underset{\displaystyle CH_3}{CH}(CH_2)_3\underset{\displaystyle CH_3}{CH}(CH_2)_3CH(CH_3)_2$$

O

Vitamin K_1

Q.13. Vitamin C contains a $-CH_2-CO-$ group in the ring system. Explain.

(*Shivaji 2000*)

Ans. See Q. 7

45

POLYMERS

Q. 1. What are polymers? Define (*i*) monomer (*ii*) repeat unit (*iii*) degree of polymerisation.

Ans. *Polymers may be defined as the substances made up of giant molecules of high molecular mass, each molecule of which consists of a very large number of simple molecules joined together through covalent bonds in a regular manner.*

(*i*) **Monomers.** The simple molecules which combine with one another repetitively to form the polymers are called monomers.

(*ii*) **Repeat Unit.** Consider the case of polyethylene polymer. It is made from polymerizing ethylene as shown below :

$$n\text{CH}_2 = \text{CH}_2 \xrightarrow{\text{Polymerization}} (-\text{CH}_2 - \text{CH}_2 -)_n$$

Thus, we find that ethylene ($CH_2 = CH_2$) is not the repeat unit in the polymer. Repeat unit is $-CH_2-CH_2-$ which is repeated over and over again to produce the polymer. *Thus, repeat unit is the smallest chain which is repeated to provide the polymer.*

(*iii*) **Degree of Polymerisation.** It is the number of times a repeat unit is contained in the polymer molecule. In the above example of polyethylene, n denotes the degree of polymerisation. Usually, n is very large, of the order of a few thousands or tens of thousands.

Q. 2. What are homopolymers and copolymers? Explain giving examples.

Ans. Homopolymers. *Polymers whose repeat-unit is derived from only one type of monomer are called* **homopolymers.** *For example, in case of polyethylene, polymer which is obtained by polymerization of ethylene molecules, the repeat unit, i.e.,* $-CH_2-CH_2-$ is derived from only one type of monomer, *i.e.*, ethylene.

$$\underset{\substack{\text{Ethylene} \\ \text{(Monomer)}}}{n\text{CH}_2 = \text{CH}_2} \xrightarrow{\text{Polymerization}} \underset{\substack{\text{Polyethylene} \\ \text{(Polymer)}}}{(-\text{CH}_2 - \text{CH}_2 -)_n}$$

Polypropylene, polyvinyl chloride (PVC), polyisoprene, neoprene (polychloroprene), polyacrylonitrile (PAN), nylon-6, polybutadiene, teflon (polytetrafluoroethylene), etc., are other examples of homopolymers.

Copolymers. Polymers whose repeat units are derived from two or more types of monomers are called **copolymers.** For example, in case of nylon-66, the repeat-unit, *i.e.*,

$-NH-(CH_2)_6-NH-CO-(CH_2)_4-CO-$ is derived from two monomer, *i.e.*, hexamethylenediamine and adipic acid.

$$\underset{\substack{\text{Hexamethylenidiamine} \\ \text{(Monomer)}}}{n\text{H}_2\text{N} - (\text{CH}_2)_6 - \text{NH}_2} + \underset{\substack{\text{Adipic acid} \\ \text{(Monomer)}}}{n\text{HOOC} - (\text{CH}_2)_4 - \text{COOH}} \xrightarrow{\text{Polymerization}}$$

$$\underset{\substack{\text{Nylon-66} \\ \text{(Polymer)}}}{[-\text{NH} - (\text{CH}_2)_6 - \text{NH} - \text{CO} - (\text{CH}_2)_4 - \text{CO} -]_n} + n\text{H}_2\text{O}$$

Buna-S, polyesters, bakelite, melamine-formaldehyde polymer, etc., are other examples of compolymers.

Q. 3. Classify polymers based upon the source from which they are obtained.

Depending upon the source from which they are obtained, polymers are broadly divided into two classes

(1) *Natural polymers* **and** **(2) *Synthetic polymers.***

Ans. 1. Natural Polymers. Polymers which are obtained from animals and plants are called **natural polymers**. Starch cellulose, proteins, nucleic acids and natural rubber are some examples of this type of polymers. A brief description of these natural polymers is given below :

Starch. It is a polymer of α-glucose. It is the chief food reserve of the plants and is made up of two fractions — amylose and amylopectin. Amylose is a *linear polymer* of α-glucose and amylopectin is a *branched polymer* of α-glucose.

Cellulose. It is a polymer of β-glucose. It is the chief structural material of the plants and is obtained from wood and cotton. About 50% of wood is cellulose. Cotton contains about 90–95% cellulose.

Both starch and cellulose are made by plants from glucose produced during *photosynthesis.*

Proteins. Proteins are polypeptides or polyamides. These polymers contain a large number of α-amino acids joined together through peptide (NH–CO) bonds in a particular sequence. These are either long chain or crossed linked polymers.

Natural rubber. It is prepared from latex which, in turn, is obtained from rubber trees. It is a polymer of the hydrocarbon isoprene (2-methyl-1, 3-butadiene). Polymerization is illustrated as under :

$$\underset{\substack{\text{Isoprene}\\ \text{(2-Methyl-1, 3-butadiene)}}}{nCH_2=\overset{\overset{\displaystyle CH_3}{|}}{C}-CH=CH_2} \xrightarrow{\text{Polymerization}} \underset{\substack{\text{Polyisoprene}\\ \text{(Natural rubber)}}}{(-CH_2-\overset{\overset{\displaystyle CH_3}{|}}{C}=CH-CH_2-)_n}$$

2. Sythetic polymers. Polymers which are made in the laboratory or industry are called synthetic polymers. Natural rubber swells and loses elasticity after prolonged exposure to petrol and motor oil. Stability and melting points of many natural polymers are such that they cannot be melted and cast into desired shapes. Further, natural fibres such as cotton, wool, silk, etc., cannot meet the increasing demands. All these needs led man to synthesize polymers in the laboratory and industry. *These man-made polymers are called* **synthetic polymers.** Some important synthetic polymers are : polyethylene, polystyrene, polyvinyl chloride (PVC), bakelite, nylon and dacron.

Q. 4. Classify polymers based upon structure.

Ans. On the basis of structure, polymers are classified into three types :

1. Linear Polymers. In such polymers, the monomers are joined together to form long straight chains of polymer molecules. The various polymeric chains are then stacked over one another to give a well packed structure (Fig. 45.1a). *Linear polymers have high melting points, high densities and high tensile strength because of close packing of chains.* Nylon and polyesters are examples of linear polymers.

2. Branched Chain Polymers. In such polymers, the monomer units not only combine to form the linear chain but also form branches along the main chain (Fig. 45.1b).

These polymer molecules do not pack well because of branches. As a result, branched chain polymers have lower melting points, densities and tensile strength compared to linear polymers. An important example of a branched chain polymer is low density polythene.

3. Three-dimensional Network Polymers. In such polymers, the linear polymer chains formed initially are joined together to form a three-dimensional network structure (Fig. 45.1c). Because of the presence of crosslinks, these polymers are also called **cross-linked polymers.** These polymers are hard, rigid and brittle. Bakelite, urea-formaldehyde polymer and melamine-formaldehyde polymer are some examples of this class.

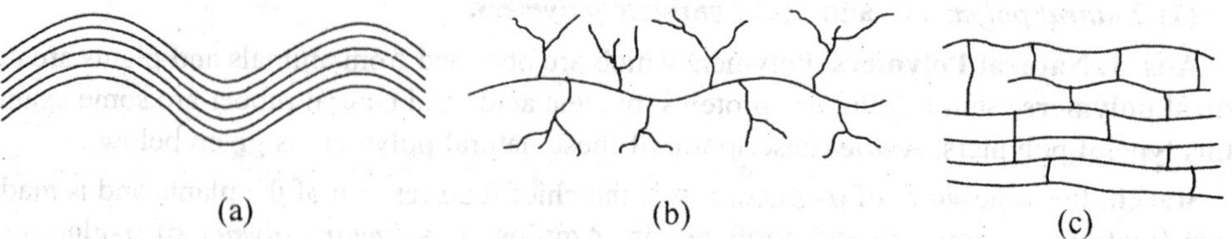

Fig. 45.1. Different structures of polymers (a) linear structure. (b) branched chain structure and (c) three-dimensional network structure.

Q. 5. Classify polymers based on synthesis. Explain giving examples. *(Nagpur 2008)*

Ans. Polymerization mainly occurs by two modes :

1. *Addition polymerization* *and* 2. *Condensation polymerization*

Corresponding to these two modes of synthesis, polymers have been classified into two types:

1. *Addition polymers* *and* 2. *Condensation polymers.*

1. Addition Polymerization. In this type of polymerization, the molecules of the same or different monomers simply add to one another in repetition leading to the formation of a polymer in which the molecular formula of the repeat unit is the same as that of the monomer. The polymers, thus, formed are called **addition polymers.**

Addition polymerization generally occurs among molecules containing double bonds.

$$\underset{\text{Ethylene}}{nCH_2 = CH_2} \longrightarrow \underset{\text{Polyethylene}}{(-CH_2-CH_2-)_n}$$

$$\underset{\text{Propylene}}{nCH_3-CH=CH_2} \longrightarrow \underset{\text{Polypropylene}}{(-CH_2-\overset{\displaystyle CH_3}{\overset{|}{C}H}-)_n}$$

2. Condensation Polymerization. In this type of polymerization, a large number of monomer molecules combine together usually with the loss of simple molecules like water, alcohol, ammonia, carbon dioxide, hydrogen chloride, etc., to form a polymer in which the molecular formula of the repeat unit is not the same as that of the monomer. The polymers, thus, formed are called **condensation polymers.** Condensation polymerization generally occurs between bifunctional compounds.

Nylon-66, obtained by the condensation between two monomers, *viz.*, hexamethylenediamine and adipic acid, each containing two functional groups with the loss of water molecules, is an example of condensation polymer.

$$\underset{\text{Hexamethylenediamine}}{nH_2N-(CH_2)_6-NH_2} + \underset{\text{Adipic acid}}{nHOOC-(CH_2)_4-COOH} \xrightarrow{525\text{ K}}$$

$$\underset{\text{Nylon-66}}{[-NH-(CH_2)_6-NH-\overset{\overset{\displaystyle O}{\|}}{C}-(CH_2)_4-\overset{\overset{\displaystyle O}{\|}}{C}-]_n} + nH_2O$$

Terylene and bakelite are other examples of condensation polymers.

Q. 6. What is meant by

***(i)* Chain growth polymer**

***(ii)* Step-growth polymer?**

Ans. A more rational method of classification based upon the *mode of addition of the monomer units to the growing chain has been proposed.* According to this method of classification, there are two types of polymers

(*i*) *Chain-growth polymers* and (*ii*) *Step-growth polymers.*

(*i*) **Chain-growth polymers** (*also called as addition polymers*). These polymers are formed by the successive addition of monomer units to the growing chain carrying a reactive intermediate such as a free-radical, a carbocation or a carbanion.

Monomer	Polymer
Ethylene	Polyethylene
Propylene	Polypropylene
Butadiene	Polybutadiene
Tetrafluoroethylene	Polytetrafluoroethylene (PTFE) or Teflon
Vinyl chloride	Polyvinyl chloride (PVC)
Isoprene	*cis*-Polyisoprene (natural rubber)

Thus, chain-growth polymers are formed by a process which involves chain reactions.

The most commonly used radical initiator is benzoyl peroxide.

Examples of chain growth polymers are given in the table above :

All the examples given in the table are those of homopolymers. An important example of a chain-growth copolymer is Buna-S involving 1, 3-butadiene and styrene as monomers.

(*ii*) **Step-growth polymers** (*also called as condensation polymers*). These polymers are formed through a series of reactions. Each such reaction involves the condensation (bond formation) between two difunctional monomer molecules to produce dimers which, in turn, produce tetramers and so on with the loss of simple molecules like H_2O, NH_3, HCl, etc. Since in this process, *the polymer is formed in a step-wise manner, it is called* **step-growth polymer** *and the process is called* **step growth polymerization.**

In contrast to chain growth polymers, the formation of step-growth polymers does not occur through chain reactions involving free radicals, carbanions or carbocations as reactive chemical species. Some typical examples of step-growth polymers are given in the table below :

S.No.	Monomers	Polymer
(*i*)	Hexamethylenediamine and adipic acid.	Nylon-66
(*ii*)	Phenol and formaldehyde.	Bakelite
(*iii*)	Terephthalic acid or its methyl ester and ethylene glycol.	Polyester (Terylene)

Q. 7. Exaplain the following

1. Elastomers **2. Fibres**

3. Thermoplastics. **4. Thermosetting polymers** (*Garhwal, 2010*)

Ans. (1) Elastomers. Elastomers are those polymers in which the intermolecular forces of attraction between the polymer chains are the weakest.

Elastomers are amorphous polymers having high degree of elasticity. They have the ability to stretch out many times their normal length and return to original position when the force is withdrawn. These polymers consist of randomly coiled molecular chains of irregular shape having a few cross links (Fig. 45.2). When the force applied, these coiled chains open out and the polymer is stretched. Since the *van der Waals'* forces of attraction between the polymer chains are very weak, these cannot maintain this stretched form. Therefore, as soon as the force is withdrawn, the polymer chains return to its original coiled state. Thus, we observe that weak *van der Waals' forces of attraction allow the polymer chains to be stretched on applying the force but the cross links bring the polymer back to the original position when the force is withdrawn.* The most important example of an elastomer is natural rubber.

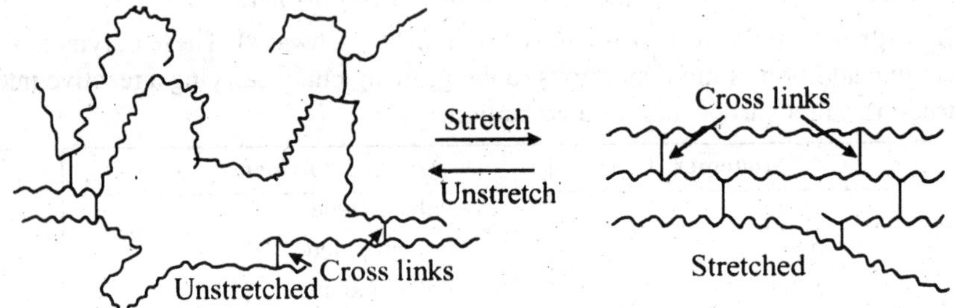

Fig. 45.2. Unstretched and stretched forms of an elastomer.

(2) Fibres. Polymers having the strongest intermolecular forces of attraction are called **fibres.** These forces are either due to H-bonding or dipole-dipole interactions. In case of nylons, the intermolecular forces are due to H-bonding (Fig. 45.3) while in polyesters and polyacrylonitrile they are due to powerful dipole-dipole interactions between the polar carbonyl (C = O) groups and between carbonyl and cyano (–C ≡ N) groups respectively.

Fibres show high tensile strength and minimum elasticity due to strong intermolecular forces.

The molecules of these polymers are long, thin and thread-like and, hence, polymer chains can be easily packed over one another. As a result, they have high melting points and low solubility.

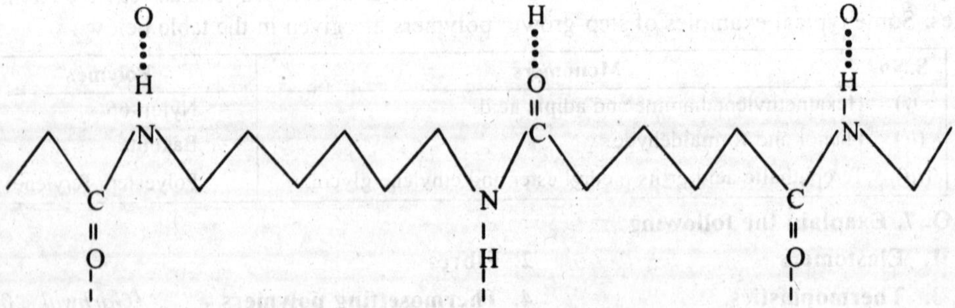

Fig. 45.3. Hydrogen bonding in nylon-66.

(3) Thermoplastics. Thermoplastics are those polymers in which the intermolecular forces of attraction are in between those of elastomers and fibres. These polymers which are linear in shape and hard at room temperature, become soft and viscous on heating and again become rigid on cooling. The process of softening and cooling can be repeated as many times as desired without any change in properties of the plastic. As a result, these plastics can be moulded into toys, buckets, telephone and television cases, etc. Thermoplastics have little or no cross linking and, hence, the individual polymer chains can slip past one another on heating. Some common examples of thermoplastics are polythene, polystyrene, polyvinyl chloride, teflon, polyvinyl acetate and polyacrylonitrile.

Plasticizers. Those plastics which do not soften easily on heating can be made soft by the addition of certain organic compound called *plasticizers.* Dialkylphthalates or cresyl phosphates are the commonly used plasticizers.

4. Thermosetting polymers. Thermosetting polymers are semifluid substances with low molecular weights which on heating in a mould, undergo change in chemical composition to give a hard, infusible and insoluble mass. This hardening on heating takes place due to extensive cross-linking between different polymer chains to give a three-dimensional network solid (Fig. 45.4).

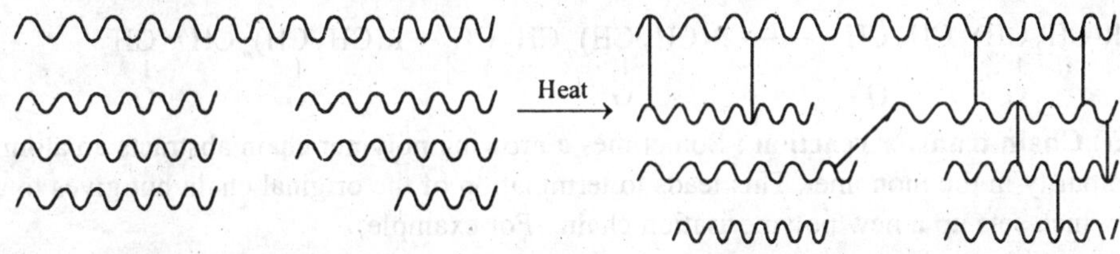

Fig. 45.4. Conversion of uncross-linked polymer into highly cross-linked thermosetting polymer.

Thus, a **thermoplastic polymer** *can be melted again and again without any change, while a* **thermosetting polymer** *can be heated only once when it permanently sets into a solid which cannot be remelted and reworked.* Some examples of thermosetting polymers are phenol-formaldehyde (bakelite), urea-formaldehyde and melamine-formaldehyde.

Q. 8. Explain the mechanism of free-radical polymerisation. Give examples

Ans. This type of polymerisation involves chain reactions and makes use of catalysts which are known to generate free radicals. The most commonly used catalysts are the organic and inorganic peroxides and the salts of the peracids *e.g.*, benzoyl peroxide, hydrogen peroxide, potassium perborate, etc.

The monomers used in this process are generally monosubstituted alkenes such as :

vinyl chloride ($CH_2 = CHCl$), styrene ($CH_2 = CH -C_6H_5$) and propylene ($CH_2 = CHCH_3$).

The mechanism of free radical addition polymerisation may be described as follows by considering the polymerisation of an ethylenic compound, $CH_2 = CHG$, in the presence of an organic peroxide. Here G stands for some group like $-Cl$, $-C_6H_5$ or $-CH_3$, etc.

(1) Chain initiation step : $\underset{\text{Peroxide}}{(RCOO)_2} \longrightarrow 2RCOO^{\bullet} \longrightarrow 2R^{\bullet} + 2CO_2$

$$R^{\bullet} + CH_2 = \underset{\substack{|\\G}}{CH} \longrightarrow RCH_2 - \underset{\substack{|\\G}}{CH^{\bullet}}$$

(2) Chain propagation step :

$$RCH_2 - \underset{\substack{|\\G}}{CH^{\bullet}} + CH_2 = \underset{\substack{|\\G}}{CH} \longrightarrow RCH_2 - \underset{\substack{|\\G}}{CH} - CH_2 - \underset{\substack{|\\G}}{CH^{\bullet}}$$

and repetition of the steps similar to (2) a number of times.

(3) Chain termination step : Chain termination could take place in one of the following ways:

(i) Coupling : Collision between growing chains or a growing chain and a catalyst radical to form a deactivated molecule could take place as follows :

$$2R(CH_2\underset{G}{CH})_nCH_2-\underset{G}{CH^{\bullet}} \longrightarrow R(CH_2\underset{G}{CH})_n\,CH_2\underset{G}{CH}-\underset{G}{CH}-CH_2(\underset{G}{CH}-CH_2)_n\,R$$

$$R(CH_2\underset{G}{CH})_n\,CH_2\underset{G}{CH^{\bullet}}+R^{\bullet} \longrightarrow R(CH_2\underset{G}{CH})_n\,CH_2\underset{G}{CHR}$$

(ii) Disproportionation : One free radical acquiring hydrogen from the other leading to their deactivation could take place.

$$2R(CH_2\underset{G}{CH})_n\,CH_2\underset{G}{CH^{\bullet}} \longrightarrow R(CH_2\underset{G}{CH})_n\,CH_2\underset{G}{CH_2}+R(CH_2\underset{G}{CH})_n\,CH=\underset{G}{CH}$$

(iii) Chain transfer reaction : Sometimes a growing polymer chain abstracts an atom from some impurity in the monomer. This leads to termination of the original chain but gives to a new radical which sets up a new polymerisation chain. For example,

$$R(CH_2\underset{G}{CH})_n\,CH_2\underset{G}{CH^{\bullet}}+ASH \longrightarrow R(CH_2\underset{G}{CH})_n\,CH_2\underset{G}{CH_2}+AS^{\bullet}$$

ASH stands for the impurity in the monomer.

Thus, the free radical addition polymerisation is generally initiated by the decomposition fo the catalyst to generate a suitable free radical which adds to a multiple bond to generate a new free radical. The new free radical, in turn, adds to a second alkene molecule to form a bigger new free radical. This process of addition of each newly formed free radical to a molecule of alkene continues to form a large polymeric molecule.

Some examples of free radical addition polymerisation :

(i)
$$\underset{\text{Acrylonitrile}}{CH_2=\underset{CN}{CH}} \xrightarrow{\text{Peroxide}} \underset{\text{Polyacrylonitrile or Orlon}}{-CH_2-\underset{CN}{CH}-CH_2-\underset{CN}{CH}-CH_2-\underset{CN}{CH}-}$$

(ii)
$$\underset{\text{Chloroprene}}{CH_2=\underset{Cl}{C}-CH=CH_2} \xrightarrow[\text{persulphate}]{\text{Potassium}} \underset{\text{Polychloroprene}}{-CH_2-\underset{Cl}{C}=CH-CH_2}$$

(iii)
$$\underset{\text{Styrene}}{CH_2=\underset{C_6H_5}{CH}} \xrightarrow{\text{Peroxide}} \underset{\text{Polystyrene}}{-CH_2-\underset{C_6H_5}{CH}-CH_2-\underset{C_6H_5}{CH}-CH_2-\underset{C_6H_5}{CH}-}$$

(iv)
$$\underset{\text{Styrene}}{CH_2=\underset{C_6H_5}{CH}}+\underset{\text{Butadiene}}{CH_2=CH-CH=CH_2} \xrightarrow[\text{persulphate}]{\text{Potassium}}$$

$$\underset{\text{Styrene butadiene rubber (A copolymer)}}{-CH_2-\underset{C_6H_5}{CH}-CH_2-CH=CH-CH_2-}$$

Q. 9. Explain the mechanism of cationic addition polymerisation. (*Mumbai, 2010*)

Ans. This type of polymerisatioin is initiated by acids. The commonly used acid catalysts being H_2SO_4, HF and Lewis acids such as $AlCl_3$, $SnCl_2$ or BF_3.

Mechanism of cationic addition polymerisation may be explained as follows by considering the polymerisation of a substituted alkene $CH_2 = CHG$, in the presence of an acid A. Here G stands for an alkyl or aryl or halogen atom.

(1) Chain initiation step. The process is initiated by the reaction of an acid with the monomer to form a carbocation.

$$\underset{\text{Acid}}{\overset{+}{A}} + CH_2 = \underset{\underset{G}{|}}{CH} \longrightarrow \underset{\text{Carbocation}}{A - CH_2 - \underset{\underset{G}{|}}{\overset{+}{CH}}}$$

(2) Chain propagation step. The carbocation formed above being electrophilic adds to another molecule of the monomer to yield a second carbocation. This process continues till a large polymer molecule is formed.

$$A - CH_2 - \underset{\underset{G}{|}}{\overset{+}{CH}} + CH_2 = \underset{\underset{G}{|}}{CH} \longrightarrow A - CH_2 - \underset{\underset{G}{|}}{CH} - CH_2 - \underset{\underset{G}{|}}{\overset{+}{CH}}$$

(3) Chain terminating step. The chain reaction is usually terminated by the loss of a hydrogen ion from the growing carbocation.

$$A - CH_2 - \underset{\underset{G}{|}}{CH} - \left[CH_2 - \underset{\underset{G}{|}}{CH} \right]_n - CH_2 - \underset{\underset{G}{|}}{\overset{+}{CH}} \xrightarrow{-H^+} A - CH_2 - \underset{\underset{G}{|}}{CH} - \left[CH_2 - \underset{\underset{G}{|}}{CH} \right]_n - CH = \underset{\underset{G}{|}}{CH}$$

Polymerisation of isobutylene in the presence of BF_3 and a trace of water to form butyl rubber as shown below, is an example of cationic addition polymerisation.

$$BF_3 + H_2O \rightleftharpoons H^+ + BF_3OH$$

$$\overset{+}{H} + H_2C = \underset{\underset{CH_3}{|}}{\overset{\overset{CH_3}{|}}{C}} \longrightarrow H_3C - \underset{\underset{CH_3}{|}}{\overset{\overset{CH_3}{|}}{C^+}} \xrightarrow{CH_2 = \overset{\overset{CH_3}{|}}{C} - CH_3} H_3C - \underset{\underset{CH_3}{|}}{\overset{\overset{CH_3}{|}}{C}} - CH_2 - \underset{\underset{CH_3}{|}}{\overset{\overset{CH_3}{|}}{C^+}}$$

and so on to yield ultimately a large molecule of butyl rubber, *i.e.*, $\left[- CH_2 - \underset{\underset{CH_3}{|}}{\overset{\overset{CH_3}{|}}{C}} - \right]_n$

Q. 10. Explain the mechanism of anionic addition polymerisation. (*Mumbai, 201(*

Ans. This type of polymerisation is initiated by bases. The catalysts usually employed for thi. purpose includes the alkali metals, alkali metal alkyls, alkali metal amides and Grignard reagents. The monomers which generally undergo this type of polymerisation are alkenes carrying electron withdrawing substituents.

The mechanism of anionic addition polymerisation may be explained as follows by considering polymerisation of a substituted alkene $\underset{\displaystyle G}{CH_2 = \underset{|}{CH}}$ in the presence of a base, B:, G is an electron withdrawing group like $-C_6H_5$.

(1) Chain initiation step. The reaction is initiated by some base with the formation of carbanion intermediate.

$$\underset{\text{Base}}{B:} + CH_2 = \underset{\underset{\displaystyle G}{|}}{CH} \longrightarrow \underset{\text{Base}}{B} - CH_2 - \underset{\underset{\displaystyle G}{|}}{\bar{C}H}$$

(2) Chain propagation step. The carbanion formed above reacts with another monomer molecule and this step is repeated unitl a polymeric molecule is formed.

$$B - CH_2 - \underset{\underset{\displaystyle G}{|}}{\bar{C}H} + CH_2 = \underset{\underset{\displaystyle G}{|}}{CH} \longrightarrow B - CH_2 - \underset{\underset{\displaystyle G}{|}}{CH} - CH_2 - \underset{\underset{\displaystyle G}{|}}{\bar{C}H}$$

(3) Chain termination step. The chain reaction gets terminated by combination with H^+ ion or some Lewis acid present in the reaction mixture:

$$B - CH_2 - \underset{\underset{\displaystyle G}{|}}{CH} \left[- CH_2 - \underset{\underset{\displaystyle G}{|}}{CH} \right]_n - CH_2 - \underset{\underset{\displaystyle G}{|}}{\bar{C}H} + \overset{+}{H} \longrightarrow$$

$$B - CH_2 - \underset{\underset{\displaystyle G}{|}}{CH} \left[- CH_2 - \underset{\underset{\displaystyle G}{|}}{CH} \right]_n - CH_2 - \underset{\underset{\displaystyle G}{|}}{CH_2}$$

It is noteworthy that anionic addition polymerisation takes place with the formation of carbanion intermediates. That is why this type of polymerisation occurs only if the carbanions are stabilised by electron withdrawing substituents.

Another feature of the mechanism is that chain termination can take place only if there is some source of H^+ ions or some Lewis acid in the reaction mixture. Otherwise, the reaction goes on till all monomer molecule are used up and a **living** polymer which is capable of reacting further is obtained.

The polymerisation of styrene in the presence of the base $K^+NH_2^-$ to form polystyrene is an example of anionic addition polymerisation.

$$\bar{N}H_2 + CH_2 = \underset{\underset{\displaystyle C_6H_5}{|}}{CH} \longrightarrow NH_2 - CH_2 - \underset{\underset{\displaystyle C_6H_5}{|}}{\bar{C}H} \xrightarrow{CH_2 = \underset{\underset{\displaystyle C_6H_5}{|}}{CH}} NH_2 - CH_2 - \underset{\underset{\displaystyle C_6H_5}{|}}{CH} - CH_2 - \underset{\underset{\displaystyle C_6H_5}{|}}{\bar{C}H}$$

and so on till polymeric molecule of polystyrene, $(- CH_2 - \underset{\underset{\displaystyle C_6H_5}{|}}{CH} -)_n$ is formed.

Q. 11. What is natural rubber?

Ans. Natural rubber is polyisoprene. Polymerisation of isoprene can be represented as under

$$\underset{\text{Isoprene}}{H_2C = \overset{\overset{\displaystyle CH_3}{|}}{C} - CH = CH_2} \xrightarrow{\text{Polymerisation}} \underset{\substack{\text{Natural rubber} \\ \text{Polyisoprene}}}{\left[CH_2 - \overset{\overset{\displaystyle CH_3}{|}}{C} = CH - CH_2 \right]_n}$$

Raw rubber is obtained from latex which is tapped from rubber tree particularly *Hevea brasiliensis.*

An isoprene polymer can have all *cis*, all *trans* or a random distribution of *cis* and *trans* configurations about the double bond. Natural rubber obtained from rubber tree is *cis*-polyisoprene and can be represented as under

$$-CH_2 \quad CH_2-CH_2 \quad CH_2-CH_2 \quad CH_2-$$
$$C=C \quad\quad C=C \quad\quad C=C$$
$$CH_3 \quad H \quad CH_3 \quad H \quad CH_3 \quad H$$

All *cis* configuration of natural rubber (*cis*-polyisoprene)

There is another variant of natural rubber called gutta-percha. This is also obtained from plants (other than *Hevea brasiliensis*). This form is all *trans* configuration and can be represented as under :

$$-CH_2 \quad H \quad CH_3 \quad CH_2-CH_2 \quad H$$
$$C=C \quad\quad C=C \quad\quad C=C$$
$$H_3C \quad CH_2-CH_2 \quad H \quad CH_3 \quad CH_2-$$

All *trans* configuration of gutta-percha (*trans*-polyisoprene)

While natural rubber (all *cis*-form) is soft and elastic, gutta percha (all *trans*-form) is tough and nonelastic.

Q. 12. What ase Ziegler-Natta catalysts? What are they used for?

Ans. Ziegler-Natta catalysts are a mixture of alkyl aluminium and titanium halides [for example $(C_2H_5)_3Al/TiCl_4$].

Free-radical polymerisation of isoprene produces non-stereospecific rubbers, which don't possess the desired qualities of natural rubber since they contain both *cis* and *trans* isomers.

To overcome this problem, we carry out polymerisation by ionic mechanism producing stereospecific polymers. This is done by using Ziegler-Natta catalysts, which enables us to obtain polyisoprene with all *cis* configuration. We obtain the product with properties comparable to those of natural rubber.

Q. 13. Explain the term vulcanisation.

Ans. Properties like tensile strength, elasticity and resistance to abrasion of natural rubber can be improved by a process called vulcanization. It consists of heating rubber with 3-5% sulphur. During vulcanization, sulphur cross-links between polymer chains are introduced (Fig. 45.5). The process of vulcanization was discovered Charles Goodyear in 1839. Goodyear is now a famous brand of automobile tyres and tubes. Fig. 45.5 below demonstrates how sulphur cross-links between the polymer chains are formed.

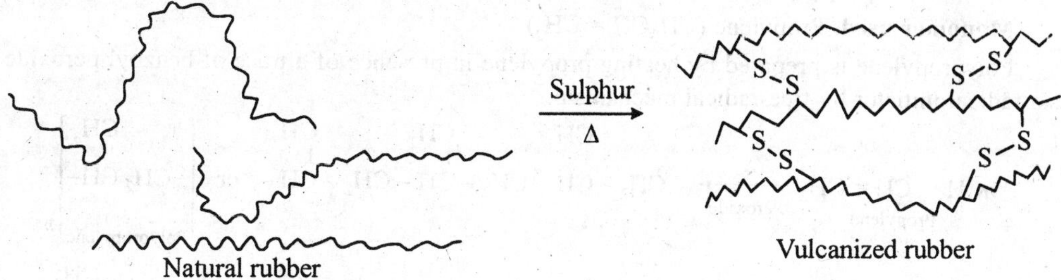

Fig. 45.5. In vulcanized rubber, the polymer chains are held together by polysulphide bridges or cross links.

Q. 14. What is meant by

1. Low-density polythene

2. High density polythene

Give their properties and uses

Ans. 1. Low density polythene. In this polymer, the repeat unit is $-CH_2-CH_2-$. It is manufactured by heating ethylene to 473 K under a pressure of 1500 atmospheres and in the presence of a trace of oxygen. This polymerization occurs by a free radical mechanism initiated by oxygen.

$$nCH_2=CH_2 \xrightarrow[\text{Traces of oxygen}]{473\text{ K}, 1500\text{ atm}} \underset{\text{Polythene}}{(-CH_2-CH_2-)_n}$$

The polythene, thus, produced has a molecular mass of about 20,000 and *has a branched structure.* There branched polythene molecules do not pack well and, hence, this type of polythene has a low density (0.92 g/cm^3) and a low melting point (384 K). That is why polythene prepared by free radical polymerization is called low density polythene.

Properties and uses. Low density polythene is a *transparent polymer* of moderate tensile strength and high toughness. It is chemically inert, and is a poor conductor of electricity.

It is widely used as a packaging material (in the form of thin plastic films, bags etc.) and as insulation for electrical wires and cables.

2. High density polythene. It is prepared by *co-ordinaion polymerization* of ethylene. In this process, enthylene in a solvent is heated to 333-343 K under a pressure of 6-7 atmospheres in the presence of a catalyst consisting of triethylaluminium and titanium tetrachloride called *Ziegler-Natta catalyst.*

$$\underset{\text{Ethylene}}{n(CH_2=CH_2)} \xrightarrow[\text{Ziegler- Natta catalyst}]{333-343\text{ K}, 6-7\text{ atm.}} \underset{\text{Polythene}}{(-CH_2-CH_2-)_n}$$

This polythene consists of linear chains of polymer molecules. These polymer molecules pack well and hence, this polythene has higher density (0.97 g/cm^3) and higher melting point (403 K) than the polymer produced by *free-radical* polymerisation. That is why polythene prepared by coordination polymerisation is called high density polythene.

Properties and Uses. High density polythene is a *translucent polymer.* It is also chemically inert but has greater toughness, hardness and tensile strength compared to low density polythene. It is used in the manufacture of containers, housewares, pipes and bottles.

Q. 15. Write notes on the following

1. Polypropylene or polypropene *(Mumbai, 2010)*

2. Polystyrene or Styron *(Mumbai, 2010)*

Ans. 1. Polypropylene

Monomer used. Propylene ($CH_3CH=CH_2$)

Polypropylene is prepared by heating propylene in presence of a trace of benzoyl peroxide as a radical initiator by free-radical mechanism.

$$\underset{\text{Propylene}}{nCH_3-CH=CH_2} \xrightarrow[\text{Peroxide}]{\Delta} -CH_2-\overset{\overset{CH_3}{|}}{CH}-CH_2-\overset{\overset{CH_3}{|}}{CH}-CH_2-\overset{\overset{CH_3}{|}}{CH}- \text{ or } \underset{\text{Polypropylene}}{\left[-CH_2\overset{\overset{CH_3}{|}}{CH}-\right]_n}$$

The monomer of polypropylene is $CH_3-CH=CH_2$ and the repeat unit is $-CH_2\overset{\overset{CH_3}{|}}{CH}-$

Uses. It is harder and stronger than polythene. It is used :

(*i*) in the manufacture of stronger pipes and bottles.

(*ii*) for packing of textiles and foods.

(*iii*) for making liners for bags and heat shrinkable wraps for records.

(*iv*) for making automotive mouldings, seat covers, carpet fibres and ropes.

2. Polystyrene or Styron

Monomer used. Styrene ($C_6H_5CH = CH_2$).

In the presence of peroxides, styrene polymerises to form polystyrene.

$$\underset{\text{Styrene}}{n\,CH_2 = \underset{\substack{|\\ C_6H_5}}{CH}} \xrightarrow{\text{Peroxides}} \left[-CH_2 - \underset{\substack{|\\ C_6H_5}}{CH} - \right]_n$$

Uses. It is a transparent polymer and is used for making plastic toys, household wares, radio and television bodies.

Q. 16. Explain the preparation and use of

1. Neoprene

2. Buna S (*Lucknow, 2010*)

Ans. 1. Neoprene. It is a polymer of chloroprene and is a synthetic rubber.

Monomer used. Chloroprene ($CH_2 = CH - \overset{\substack{Cl\\ |}}{C} = CH_2$)

Chloroprene is prepared by the Markownikov's addition of HCl to vinylacetylene at the triple bond.

$$\underset{\text{Vinylacetylene}}{CH_2 = CH - C \equiv CH} + HCl \longrightarrow \underset{\text{Chloroprene}}{CH_2 = CH - \overset{\substack{Cl\\ |}}{C} = CH_2}$$

Vinylacetylene needed for the purpose is prepared by dimerization of acetylene as given below :

$$\underset{\text{Acetylene (2 molecules)}}{CH \equiv CH + HC \equiv CH} \xrightarrow[343\ K]{NH_4Cl,\ CuCl} \underset{\text{Vinylacetylene}}{CH_2 = CH - C \equiv CH}$$

Chloroprene polymerises very fast. No specific catalysts are needed but the polymerization is slower in absence of oxygen. The reaction occurs by 1, 4-addition of one chloroprene molecule to the other as shown below :

$$\underset{\text{Chloroprene}}{\overset{1}{CH_2} = \overset{2}{\underset{\substack{|\\ Cl}}{C}} - \overset{3}{CH} = \overset{4}{CH_2}} + \underset{\text{Chloroprene}}{\overset{1}{CH_2} = \overset{2}{\underset{\substack{|\\ Cl}}{C}} - \overset{3}{CH} = \overset{4}{CH_2}} + \ldots$$

$$\Big\downarrow \ O_2 \text{ or Peroxide}$$

$$\underset{\text{Polychloroprene or Neoprene}}{-CH_2 - \underset{\substack{|\\ Cl}}{C} = CH - CH_2 - CH_2 - \underset{\substack{|\\ Cl}}{C} = CH - CH_2 -}$$

The polymer may be represented as $(-CH_2-\underset{\underset{Cl}{|}}{C}=CH-CH_2-)_n$
Neoprene

It has excellent rubber like properties.

Uses. Neoprene is inferior to natural rubber in some properties but is quite stable to aerial oxidation and resistant to oils, gasoline and other solvents. It is, therefore, used in the manufacture of hoses, shoe heels, stoppers, etc.

2. Buna-S. It is a synthetic rubber and is a copolymer of 1, 3-butadiene and styrene. **Bu** stands for 1, 3-butadiene, **na** for sodium which is used as the polymerizing agent and **S** stands for styrene. It is also called SBR (Styrene, Butadiene, Rubber).

Materials used. 1. 3-Butadiene ($CH_2 = CH - CH = CH_2$) and styrene ($C_6H_5CH = CH_2$).

It is obtained by copolymerization of 1, 3-butadiene and styrene in the ratio 3:1 in the presence of sodium.

$$\underset{\text{1, 3-Butadiene}}{nCH_2=CH-CH=CH_2} + \underset{\text{Styrene}}{n\underset{\underset{C_6H_5}{|}}{C}H=CH_2} \xrightarrow{\text{Na, heat}} \underset{\text{Buna-S or SBR}}{\left[-CH_2-CH=CH-CH_2-\underset{\underset{C_6H_5}{|}}{C}H-CH_2-\right]_n}$$

Uses. It is used in the manufacture of tyres, rubber soles, water-proof shoes, etc.

Q. 17. Give the methods of preparation and uses of the following polymers:

1. Poly(methyl methacrylate), PMMA

2. Poly(ethyl acrylate)

3. Poly(acrylonitrile), PAN

4. Poly(vinyl chloride), PVC (*Mumbai, 2010*)

5. Poly(tetrafluoroethylene), PTFE, Teflon

Ans. 1. Poly(methyl methacrylate), PMMA.

Monomer used. Methyl methacrylate ($CH_2 = \overset{\overset{CH_3}{|}}{C}H - COOCH_3$)

The monomer methyl methacrylate is obtained by treating acetone cyanohydrin with $CH_3OH - H_2SO_4$ which brings about simultaneous hydrolysis, dehydration and esterification. This is polymerized in the presence of a radical initiator to give poly (methyl methacrylate).

$$\underset{\text{Acetone}}{CH_3-\overset{\overset{CH_3}{|}}{C}=O} \xrightarrow{\text{HCN}} \underset{\text{Acetone cyanohydrin}}{CH_3-\overset{\overset{CH_3}{|}}{\underset{\underset{OH}{|}}{C}}-CN} \xrightarrow{CH_3OH-H_2SO_4} \underset{\text{Methyl methacrylate}}{CH_2=\overset{\overset{CH_3}{|}}{C}-COOCH_3}$$

$$\underset{\text{Methyl methacrylate}}{nCH_2=\overset{\overset{CH_3}{|}}{C}-COOCH_3} \xrightarrow{\text{Peroxides}} \underset{\text{Poly (methyl methacrylate)}}{\left[-CH_2-\overset{\overset{CH_3}{|}}{\underset{\underset{COOCH_3}{|}}{C}}-\right]_n}$$

Poly (methyl methacrylate) is a hard and transparent substance

The most important property of poly (polymethyl methacrylate) is its clearness and excellent light transmission even better than glass.

Uses. It is used in the manufacture of lenses, light covers, light shades, transparent domes and skylights, aircraft windows, dentures and plastic jewellery.

2. Poly (ethyl acrylate)

Monomer used. Ethyl acrylate ($CH_2 = CH - COOC_2H_5$).

Ethyl acrylate on polymerization in the presence of perc..ides gives poly (ethyl acrylate).

$$\underset{\text{Ethyl acrylate}}{n\underset{\displaystyle COOC_2H_5}{CH_2 = \underset{|}{CH}}} \xrightarrow{\text{Peroxides}} \underset{\text{Poly (ethylacrylate)}}{\left[-CH_2 - \underset{\displaystyle COOC_2H_5}{\underset{|}{CH}} - \right]_n}$$

Poly (ethyl acrylate) is tough but with somewhat rubber-like properties.

3. Polyacrylonitrile (PAN)

Monomer used. Acrylonitrile ($CH_2 = CH - CN$).

Acrylonitrile polymerizes in presence of peroxides to give polyacrylonitrile.

$$\underset{\text{Acrylonitrile}}{nCH_2 = \underset{\displaystyle CN}{\underset{|}{CH}}} \xrightarrow{\text{Peroxides}} \underset{\text{Polyacrylonitrile}}{\left[-CH_2 - \underset{\displaystyle CN}{\underset{|}{CH}} - \right]_n}$$

The monomer acrylonitrile is manufactured by either of the following reactions :

$$\underset{\text{Acetylene}}{CH \equiv CH} + HCN \xrightarrow{\text{CuCl– HCl}} \underset{\text{Acrylonitrile}}{CH_2 = CH - CN}$$

$$\underset{\text{Propylene}}{2CH_3CH = CH_2} + \underset{\text{(From air)}}{3O_2} + 2NH_3 \xrightarrow[\text{Mo, Co and Al, 723 K}]{\text{Oxides of}} \underset{\text{Acrylonitrile}}{2CH_2 = CH - CN} + 6H_2O$$

Uses. Polyacrylonitrile is a hard and high melting material.

(*i*) It is used in the manufacture of *Orlon* and *Acrilan* fibres used for making clothes, carpets and blankets.

(*ii*) It is blended with other polymers to improve their qualities.

4. Polyvinyl chloride (PVC)

Monomer used. Vinyl chloride ($CH_2 = CH - Cl$).

Vinyl chloride polymerises in the presence of peroxides to form polyvinyl chloride.

$$\underset{\text{Vinyl chloride}}{nCH_2 = \underset{\displaystyle Cl}{\underset{|}{CH}}} \xrightarrow{\text{Peroxides}} \underset{\text{Polyvinyl chloride}}{(-CH_2 - \underset{\displaystyle Cl}{\underset{|}{CH}} -)_n}$$

The monomer vinyl chloride is itself manufactured by the addition of HCl to acetylene in the presence of mercury salts as catalyst or by dehydrochlorination of ethylene dichloride.

$$\underset{\text{Acetylene}}{CH \equiv CH} + HCl \xrightarrow{Hg^{2+}} \underset{\text{Vinylchloride}}{CH_2 = CH - Cl}$$

$$CH_2Cl-CH_2Cl \xrightarrow{873-923\ K} \underset{\text{Vinyl chloride}}{CH_2=CH-Cl} + HCl$$

Uses. (*i*) It is a good electrical insulator and hence is used for coating wires and cables.

(*ii*) It is also used in making gramophone records and pipes.

(*iii*) It is used for making raincoats, hand bags, plastic dolls, upholstery, shoe soles and vinyl flooring.

5. Polytetrafluoroethylene (PTFE) or Teflon

Monomer used. Tetrafluoroethylene ($F_2C = CF_2$).

Tetrafluoroethylene polymerises in the presence of oxygen to give polytetrafluoroethylene popularly called **Teflon.**

$$\underset{\text{Tetrafluoroethylene}}{nCF_2=CF_2} \xrightarrow{O_2} \underset{\text{Polytetrafluoroethylene or Teflon}}{(-CF_2-CF_2-)_n}$$

Uses. Teflon is unaffected by solvents, boiling acids and aqua regia upto 598 K.

(*i*) Because of its great chemical inertness and high thermal stability, teflon is used for making non-sticks utensils.

(*ii*) It is also used for making gaskets, pump packings, valves, seals, non-lubricated bearings, etc.

Q. 18. Explain the method of formation and uses of the following polymers

1. Terylene or dacron

2. Glyptal or alkyd resin

3. Nylon 6, 6

4. Nylon 6, 10

5. Nylon 6 or perlon

Ans. 1. Terylene or Dacron. It is prepared by condensation polymerization of ethylene glycol and terephthalic acid with elimination of water at 425-475 K.

$$\underset{\text{Ethylene glycol}}{n(HO-CH_2-CH_2-OH)} + \underset{\text{Terephthalic acid}}{n(HO-\overset{O}{\overset{\|}{C}}-C_6H_4-\overset{O}{\overset{\|}{C}}-OH)} \xrightarrow{425-475\ K}$$

$$\underset{\text{Terylene or Dacron}}{\left[-O-CH_2-CH_2-O-\overset{O}{\overset{\|}{C}}-C_6H_4-\overset{O}{\overset{\|}{C}}-\right]_n} + nH_2O$$

Uses. (*i*) The fibre of terylene is highly crease-resistant, durable and has low moisture content. It is, therefore used for the manufacture of wash and wear fabrics, tyre cords, seat belts and sails. It is also blended with cotton and wool to increase their durability.

(*ii*) The Mylar film made from dacron is flexible and resistant to ultraviolet degradation. It is, therefore, used for making magnetic recording tapes.

2. Glyptal or Alkyd resin. Glyptal, *i.e.*, poly (ethylene phthalate) is formed by the condensation of ethylene glycol and phthalic acid.

$$n(HO-CH_2-CH_2-OH) + n\ C_6H_4(COOH)_2 \xrightarrow[-nH_2O]{\Delta} \left[-O-CH_2-CH_2-O-\overset{O}{\overset{\|}{C}}-C_6H_4-\overset{O}{\overset{\|}{C}}- \right]_n$$

Ethylene glycol — Phthalic acid — Poly Ethylene Phthalate

Uses. Poly (ethylene phthalate) is a thermoplastic. Its solution on evaporation leaves a tough and inflexible film. It is, used in the manufacture of paints and lacquers.

3. Nylon-6, 6. * It is manufactured by the condensation polymerization of adipic acid and hexamethylenediamine at about 525 K when water is lost as steam and the nylon is produced in the molten state. It is then cast into a sheet or fibres by passing through a spinneret.

$$nHO-\overset{O}{\overset{\|}{C}}-(CH_2)_4-\overset{O}{\overset{\|}{C}}-OH + nH_2N-(CH_2)_6-NH_2 \xrightarrow[\text{Polymerization}]{525\text{ K}} \left[-\overset{O}{\overset{\|}{C}}-(CH_2)_4-\overset{O}{\overset{\|}{C}}-NH-(CH_2)_6-NH- \right]_n + n(H_2O)$$

Adipic acid — Hexamethylenediamine — Nylon-6, 6

It is called nylon-6, 6 (read as nylon six, six) since both adipic acid and hexamethylenediamine contain six carbon atoms each.

4. Nylon-6, 10. Another commonly used nylon is nylon-6, 10 (read as nylon six, ten) which is obtained by the condensation of hexamethylenediamine (containing six carbon atoms) and sebacic acid [$HOOC(CH_2)_8COOH$], dibasic acid containing ten carbon atoms.

$$nH_2N-(CH_2)_6-NH_2 + nHOOC(CH_2)_8COOH \xrightarrow{-n\,H_2O} \left[-HN-(CH_2)_6-NH-\overset{O}{\overset{\|}{C}}-(CH_2)_8-\overset{O}{\overset{\|}{C}}- \right]_n$$

Hexamethylenediamine — Sebacic acid — Nylon-6, 10

Nylon fibres possess high tensile strength and are abrasion resistant. They also possess some elasticity.

Uses. (*i*) These are used in the manufacture of carpets, textile fibres and bristles for brushes.

(*ii*) Crinkled nylon fibres are used for making elastic hosiery.

(*iii*) Being tough nylon is used as a substitute for metals in bearings and gears.

---

* Nylon was originally prepared in New York and London simultaneously. (*Ny* – New York, *Lon* – London). This is how the polymer got its name.

5. Nylon-6 or Perlon. Monomer *caprolactam* on polymerization gives nylon-6.

Caprolactam needed for the purpose is manufactured from cyclohexane as described below :

Cyclohexane $\xrightarrow{\text{Oxidation}}$ Cyclohexane (C=O) $\xrightarrow[-H_2O]{NH_2OH}$ Cyclohexanone oxime (C=NOH) $\xrightarrow[\text{Beckmann rearrangement}]{\text{conc. } H_2SO_4}$ Caprolactum

Caprolactam is heated with a trace of water, it hydrolyses to ε-aminocaproic acid which upon continued heating undergoes polymerization to give nylon-6. It is called as nylon-6 (read as nylon six) since the monomer (caprolactam) contains six carbon atoms.

$$\text{Caprolactam} \xrightarrow{H_2O,\ \Delta} \underset{\epsilon\text{-Aminocaproic acid}}{[H_3\overset{+}{N}-(CH_2)_5-COO^-]} \xrightarrow{\Delta,\ \text{Polymerises}} \underset{\text{Nylon-6}}{[-NH-(CH_2)_5-\overset{O}{\overset{\|}{C}}-]_n}$$

The fibres of nylon-6 are obtained when molten polymer is forced through a spinneret.

Uses. It is used for the manufacture of tyre cords, fabrics and mountaineering ropes.

Q. 19. Explain method of formation and uses of phenol-formaldehyde resin (bakelite)

(Nagpur, 2008; Mumbai, 2010; Lucknow, 2010)

Ans. Phenol formaldehyde resin (bakelite). Phenol on treatment with formaldehyde in the presence of a basic catalyst undergoes condensation polymerization to form either a linear or a cross-linked polymer called **phenol-formaldehyde resin or bakelite.** Methylene bridges are formed either at *ortho* or *para*-position or both at *ortho* and *para*-positions with respect to the phenolic group.

$$\underset{\text{Phenol}}{C_6H_5OH} + \underset{\text{Formaldehyde}}{H_2C=O} \xrightarrow{OH^-} \underset{o\text{-Hydroxymethyl phenol}}{o\text{-}HOC_6H_4CH_2OH} + \underset{p\text{-Hydroxymethyl phenol}}{p\text{-}HOC_6H_4CH_2OH}$$

$$n\ \underset{o\text{-Hydroxymethyl phenol}}{o\text{-}HOC_6H_4CH_2OH} \xrightarrow[\text{Polymerization}]{-(n-1)\,H_2O} \underset{\text{Linear polymer}}{\left[-C_6H_3(OH)-CH_2-C_6H_3(OH)-CH_2-\right]_n}$$

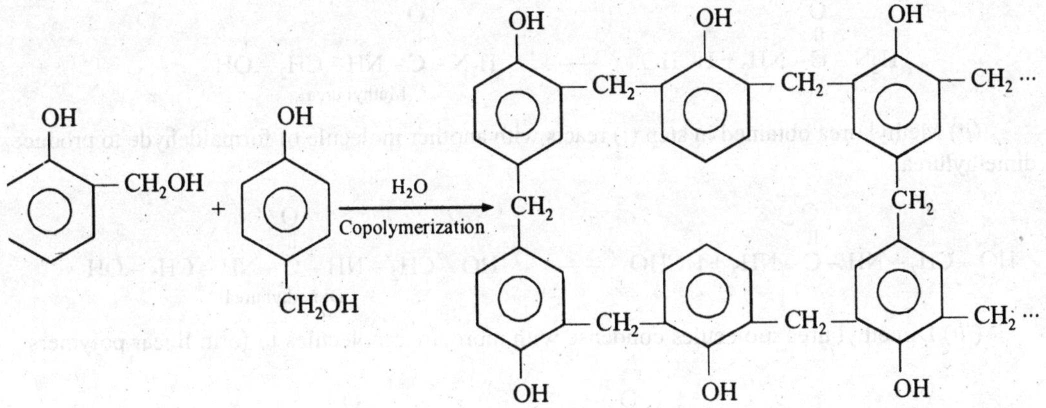

Uses. (*i*) Soft bakelites are used as binding glue for laminated wooden planks, in varnishes and lacquers.

(*ii*) Hard bakelite which is highly cross-linked is used as a *thermosetting polymer.*

(*iii*) It is used for the manufacture of combs, formica table-tops, electrical switches and gramophone records, etc.

Q. 20. Explain the method of formation and uses of melamine-formaldehyde resin.

Ans. Melamine-formaldehyde resin. Melamine and formaldehyde on heating undergo copolymerization to for melamine-formaldehyde resin.

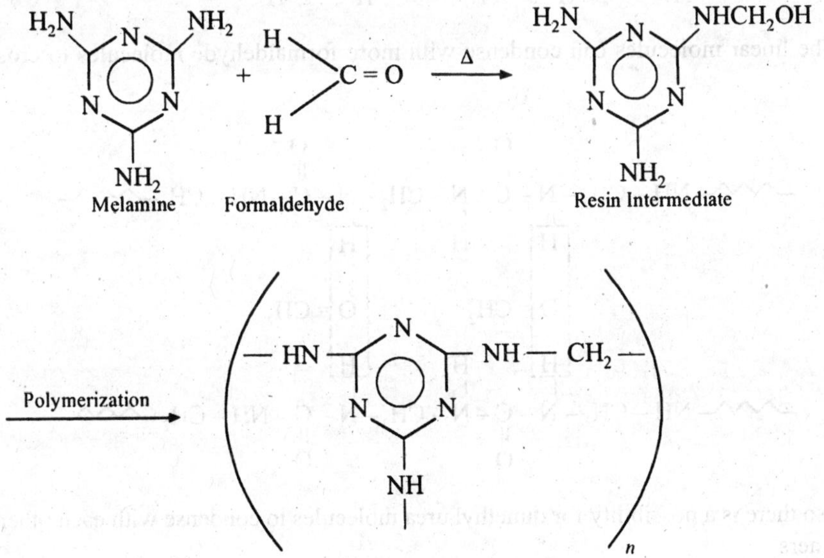

Melamine + Formaldehyde $\xrightarrow{\Delta}$ Resin Intermediate $\xrightarrow{\text{Polymerization}}$

MELAMINE-FORMALDEHYDE POLYMER

Use. It is used for making unbreakable crockery.

Q. 21. How is urea-formaldehyde resin obtained? Give its uses

Ans. preparation. (*i*) This polymer is obtained by a reaction between urea and formaldehyde in the presence of a base or an acid

$$H_2N-\overset{\overset{O}{||}}{C}-NH_2+HCHO \longrightarrow \underset{\text{Methyl urea}}{H_2N-\overset{\overset{O}{||}}{C}-NH-CH_2-OH}$$

(*ii*) Methyl urea obtained in step (*i*) reacts with another molecule of formaldehyde to produce dimethylurea

$$HO-CH_2-NH-\overset{\overset{O}{||}}{C}-NH_2+HCHO \longrightarrow \underset{\text{Dimethyl urea}}{HO-CH_2-NH-\overset{\overset{O}{||}}{C}-NH-CH_2-OH}$$

(*iii*) Dimethyl urea molecules condense with more urea molecules to form linear polymers

$$\sim\!\!\sim\underset{\text{Urea}}{\underset{H}{\underset{|}{N}}}-\boxed{H+HO}-\underset{\text{Dimethyl urea}}{CH_2-NH-\overset{\overset{O}{||}}{C}-NH}-\boxed{OH+H}-NH-\overset{\overset{O}{||}}{C}-\underset{\text{Urea}}{\underset{H}{\underset{|}{N}}}-H+\underset{\text{Dimethyl urea}}{HO-CH_2}\sim\!\!\sim$$

$$\Big\downarrow -nH_2O$$

$$\sim\!\!\sim\underset{H}{\underset{|}{N}}-CH_2-\underset{H}{\underset{|}{N}}-\overset{\overset{O}{||}}{C}-\underset{H}{\underset{|}{N}}-CH_2-\underset{H}{\underset{|}{N}}-\overset{\overset{O}{||}}{C}-\underset{H}{\underset{|}{N}}-CH_2\sim\!\!\sim$$

(*iv*) The linear molecules can condense with more formaldehyde molecules to cross-linked structures

$$\sim\!\!\sim NH-CH_2-\underset{\boxed{H}}{\underset{|}{N}}-\overset{\overset{O}{||}}{C}-\underset{H}{\underset{|}{N}}-CH_2-\underset{\boxed{H}}{\underset{|}{N}}-\overset{\overset{O}{||}}{C}-NH-CH_2\sim\!\!\sim$$

$$\boxed{O}=CH_2 \qquad\qquad \boxed{O}=CH_2$$

$$\sim\!\!\sim NH-CH_2-\overset{\boxed{H}}{\overset{|}{N}}-\underset{\underset{O}{||}}{C}-\overset{H}{\overset{|}{N}}-CH_2-\overset{\boxed{H}}{\overset{|}{N}}-\underset{\underset{O}{||}}{C}-NH-CH_2\sim\!\!\sim$$

(*v*) Also there is a possibility for dimethyl urea molecules to condense with each other to form linear polymers

$$\begin{array}{ccccc} H-N-CH_2-\boxed{OH\quad H}-N-CH_2-\boxed{OH\quad H}-N-CH_2-OH \\ | \qquad\qquad | \qquad\qquad | \\ C=O \qquad + \qquad C=O \qquad + \qquad C=O \\ | \qquad\qquad | \qquad\qquad | \\ H-N-CH_2-\boxed{OH\quad H}-N-CH_2-\boxed{OH\quad H}-N-CH_2-OH \end{array}$$

Depending upon the reaction conditions, resins of different types are obtained

Uses. 1. These polymers tind use as ingradients of varnishes and lacquers

2. These are used in the manufacture of electrical fittings, radio and TV cabinets, telephones, table tops and toys

3. These are used for gluing and impregnating timber

Q. 22. Write notes on

(*i*) epoxy resins (*ii*) polyurethanes (*Mumbai, 2010*)

Ans. (*i*) epoxy resins

Preparation. Epoxy resins are prepared by the condensation of excess of chloroepoxy alkanes with dihydric phenols in the presence of NaOH catalyst at 50-60°C

$$nCH_2 - \overset{O}{\widehat{\quad}} CH - CH_2Cl + nHO - C_6H_4 - \underset{CH_3}{\overset{CH_3}{\underset{|}{\overset{|}{C}}}} - C_6H_4 - OH$$

$$\longrightarrow \left[O - CH_2 - \overset{OH}{\overset{|}{CH}} CH_2O - C_6H_4 - \underset{CH_3}{\overset{CH_3}{\underset{|}{\overset{|}{C}}}} - C_6H_4 - \right]_n$$

Cross-linking of chains is achieved by heating the above product with amines, when terminal epoxy groups of the chain react with amine.

Uses

1. Epoxy resins are known for their bonding properties and as such are used in adhesives such as araldite. Epoxy resins are used to bind glass, porcelain, metal and wood.
2. Epoxy resins possess the properties of inertness, handness and flexibility and as such are used as protective coating. Fibreglass parts of boats and automobiles have a metal frame coated with a layer of a mix of spun glass and epoxy resin

(*ii*) Polyurethanes

Preparations. Polyurethanes are obtained by the reaction between diisocyanates and diols in the presence of stannic chloride at 50-100°C, as shown below:

$$n\,OH(CH_2)_4OH + n\,OCN - C_6H_4 - CH_2 - C_6H_4 - NCO$$

$$\longrightarrow \left[CONH - C_6H_4 - CH_2 - C_6H_4 - NHCOO(CH_2)_4O \right]_n$$

Cross linking between chains is possible if triols and di- or triisocyanates are used. Cross linking helps to obtain a thermosetting polymer.

If polyurethanes are prepared in the presence of a low boiling liquid, it will evaporate as the reaction proceeds and a semi-rigid polyurethane is obtained

Uses. Polyurethane foams are used in packing, construction and interior decoration of buildings.

DIRECTORY OF NAME REACTIONS

A directory of name reactions is given below for the convenience of readers. With the help of this directory, it is possible to locate the place where a particular name reaction is described. The reactions are arranged in alphabetic order

	Name Reaction	Chapter
1.	Aldol condensation	Aldehydes and ketones
2.	Balz-Schiemann's reaction	Diazonium salts
3.	Beauveault Balanc reaction	Carboxy Acids & Acid Der.
4.	Benzoin condensation	Aldehydes & ketones"
5.	Cannizzaro reaction	Aldehydes & ketones
6.	Carbylamine reaction	Amines
7.	Claisen Schmidt condensation	Aldehydes & ketones
8.	Crossed Aldol condensation	Aldehydes & ketones
9.	Clemmenson reduction	Aldehydes & ketones
10.	Cope elimination	Amines
11.	Coupling reaction	Diazonium salts
12.	Curtius rearrangement	Amines
13.	Etard's reaction	Aldehydes & ketones
14.	Fischer's esterification	Carboxylic Acid Der.
15.	Friedel-Crafts acylation	Ald. & ketones, Acid Der.
16.	Gattermann's reaction	Ald.-ketones, Diaz. salts
17.	Gattermann Koch reaction	Aldehydes & ketones
18.	Haloform reaction	Aldehydes & ketones
19.	Hell-Volhard-Zelinsky reaction	Aldehydes & ketones
20.	Hinsberg methods	Amines
21.	Hoffmann reaction	Amines
22.	Hoffmann's bromamide or degradation	Acid Der.
23.	Hunsdiecker reaction	Carboxylic acids
24.	Knoevenagal reaction	Aldehydes & ketones
25.	Kolbe's reaction	Acid Der.
26.	Kolbe's electrolytic method	Carboxylic acids
27.	Libermann's nitroso method	Amines
28.	Mendius reduction	Amines
29.	Mannich reaction	Ald. ketones & Nitro compound
30.	Meerwein-Ponndorf-Verley reduction	Aldehydes & ketones
31.	Oppenauer oxidation	Aldehydes & ketones
32.	Perkin's reaction	Aldehydes & ketones
33.	Reformatsky reaction	Aldehydes & ketones
34.	Reimer-Tiemann reaction	Aldehydes
35.	Rosenmund's reduction	Aldehydes & ketones
36.	Sandmeyer's reaction	Diazonium salts
37.	Schmidt rearrangement	Amines
38.	Schotten Baumann reaction	Amines
39.	Stephen's reduction	Aldehydes & ketones
40.	Trans-esterification	Acid Der.
41.	Wittig reaction	Aldehydes & ketones
42.	Wolf-Kishner reduction	Aldehydes & ketones

ASCENT AND DESCENT OF SERIES

In the conversion of organic compounds, it is sometimes required to increase or decrease the number of carbon atoms. This can be achieved as under :

ASCENT OF SERIES

The conversion of an organic compound into its higher homologue is called *ascent of series.* The length of the carbon chain is increased by one or more carbon atoms at a time.

Following methods are used for this purpose :

(*i*) By Wurtz reaction (*ii*) Through Grignard reagents (*iii*) Through introducing cyanide group

(*i*) By Wurtz reaction Conversion of a lower alkane, R–H to a higher alkane, R–R through Wurtz reaction.

$$\underset{\text{Alkane}}{\text{R–H}} \xrightarrow{X_2/h\nu} \text{R–X} \xrightarrow[\text{alkyl halide}]{\text{Na/Dry ether}} \text{R–R}$$

Where R stands for an alky! group and *X* stand for a halogen.

(*ii*) Through Grignard reagent

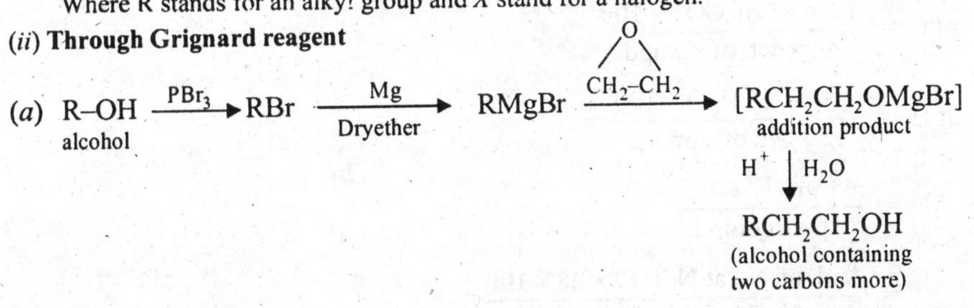

(*b*) $$\underset{\text{alcohol}}{\text{ROH}} \xrightarrow{PBr_3} \text{R–Br} \xrightarrow[\text{in dry ether}]{\text{Mg}} \underset{\text{Grignard reagent}}{\text{R–MgBr}} \xrightarrow{\text{HCHO}} \underset{\text{addition product}}{[RCH_2OMgBr]} \xrightarrow{H^+,\ H_2O} RCH_2OH$$

(*iii*) Through introducing cyanide group

$$\underset{\text{alcohol}}{\text{ROH}} \xrightarrow{P+I_2} \underset{\text{alkyl iodide}}{\text{R–I}} \xrightarrow[\text{Heat}]{\text{Alc.KCN}} \underset{\text{alkyl cyanide}}{\text{R–CN}} \xrightarrow[\text{or Na/alcohol reduction}]{LiAlH_4} RCH_2NH_2$$

DESCENT OF SERIES The conversion of an organic compound into lower homologue is called descent of series. The length of the carbon chain is decreased by one carbon atom at a time. We can use the following methods for this purpose.

(*i*) By heating sodium salt of a fatty acid with soda lime (NaOH/CaO) (decarboxylation).

(*ii*) By Hofmann bromamide reaction.

Examples

(*i*) By heating sodium salt of the fatty acid with soda lime (NaOH/CaO) (decarboxylation)

$$\underset{\text{alcohol}}{CH_3CH_3} \xrightarrow[-HCl]{Cl_2/hv} \underset{\text{Ethyl chloride}}{CH_3CH_2Cl} \xrightarrow{\text{Aq.KOH}} \underset{\text{Ethanol}}{CH_3CH_2OH} \xrightarrow{K_2Cr_2O_7} \underset{\text{Acetaldehyde}}{CH_3CHO} \xrightarrow{K_2Cr_2O_7 + H_2SO_4} \underset{\text{Acetic acid}}{CH_3COOH}$$

$$\underset{\text{Acetic acid}}{CH_3COOH} \xrightarrow[\text{NaOH}]{-H_2O} CH_3COONa \xrightarrow[\Delta]{\text{CaO + NaOH}} \underset{\text{Methane}}{CH_4}$$

(*ii*) By Hofmann bromamide reaction.

$$\underset{\text{Ethanol}}{CH_3CH_2OH} \xrightarrow[H_2SO_4]{K_2Cr_2O_4} \underset{\text{Acetaldehyde}}{CH_3CHO} \xrightarrow[H_2SO_4]{K_2Cr_2O_4} \underset{\text{Acetic acid}}{CH_3COOH} \xrightarrow{NH_3}$$

$$\underset{\text{Ammonium acetate}}{CH_3COONH_4} \xrightarrow[-H_2O]{\text{Heat}} \underset{\text{Acetamide}}{CH_3CONH_2} \xrightarrow[\text{Hofmann bromamide reaction}]{Br_2/KOH} \underset{\text{Methanamine}}{CH_3NH_2} \xrightarrow{\text{HONO}} \underset{\text{Methanol}}{CH_3OH}$$

Appendix 3

SYSTEMATIC APPROACH TO SOLVING A STRUCTURAL PROBLEM

Solving a structural problem is an interesting job. What is required to accomplish this work is the thorough knowledge of organic reactions. Before that, it is required to determine the molecular formula of the compound by making use of percentage of elements and molecular mass data. Given below is the summary of various relations used and sequence of investigation made.

(*a*) **Estimation of the elements.**

$$\text{Percentage of C} = \frac{12 \times \text{wt. of } CO_2 \times 100}{44 \times \text{wt. of compd.}}$$

$$\text{Percentage of H} = \frac{2 \times \text{wt. of } H_2O \times 100}{18 \times \text{wt. of compd.}}$$

$$\text{Percentage of N} = \frac{\text{wt. of } N_2 \times 100}{\text{wt. of compd.}}$$

$$= \frac{\text{Vol. of } N_2 \text{ at N.T.P.} \times 28 \times 100}{22400 \times \text{wt. of compd.}}$$

(*Duma's Method*)

$$= \frac{1.4 \times \text{Normality of acid} \times \text{Vol. of acid used}}{\text{wt. of compound}} \times 100$$

(*Kjeldahl's Method*)

$$\text{Percentage of Cl} = \frac{35.5 \times \text{wt. of AgCl} \times 100}{143.5 \times \text{wt. of compd.}}$$

(*Carius Method*)

(*b*) **Calculation of empirical formula** *i.e.* the simplest ratio between the atoms of various elements.

(*c*) **Determination of molecular formula** which is *n* times the empirical formula where

$$n = \frac{\text{Molecular weight}}{\text{Empirical Formula weight}}$$

(*i*) *Mol. wt. of acid* = Eq. wt. of acid × basicity.

Eq. wt. of acid is generally determined by silver salt method.

$$\text{Eq. wt acid} = \frac{\text{wt. of Ag salt} \times 108}{\text{wt. of Ag}} - 107$$

(*ii*) *Mol. wt. of Base* = Eq. wt. of base × Acidity

Eq. wt of base is determined by platinic chloride method.

$$\frac{\text{wt. of chloroplatinate}}{\text{wt. of Pt.}} = \frac{2B + 410}{195}$$

(where B is the eq. wt. of the base)

Molecular weight by cryoscopic method

This method involves the measurement of depression of freezing point by dissolving a known amount of an organic substance in a definite quantity of solvent. The depression of freezing point is related to its molal depression constant, K_f, as follows :

$$\text{Molecular weight} = \frac{1000 \times K_f \times \text{weight of the sample}}{\Delta T_f \times \text{weight of solvent}}$$

where ΔT_f = depression of freezing point

and K_f = molal depression constant.

(*d*) Determination of structural formulae

REMEMBER THESE POINTS FOR ALIPHATIC COMPOUNDS

(i) An unsaturation in a compound is indicated by addition of bromine (*i.e.*, bromine water decolourized) and also by oxidation with 1% alk. $KMnO_4$ soln. (decolourised).

$$\underset{\text{Propylene}}{CH_3—CH = CH_2} + Br_2 \longrightarrow \underset{\text{Propylene dibromide}}{CH_3CHBr.CH_2Br}$$

$$CH_3—CH = CH_2 + HO_2 + O \longrightarrow \underset{\text{Propylene glycol}}{CH_3—\underset{OH}{\underset{|}{C}H}—\underset{OH}{\underset{|}{C}H_2}}$$

***(ii)* A compound containing C, H and O, if it**

(*a*) gives violet colour with Schiff's reagent or

(*b*) gives silver mirror with Tollen's reagent and red ppt. with Fehling's solution, it contains — CHO group (*Aldehyde*).

(*c*) forms oxime, phenyl hydrazone, but does not reduce Fehling's solution, it contains ketonic >C = O group (*Ketones*).

(*d*) gives an aldehyde on oxidation, it contains primary alcoholic —CH_2OH group and if yields ketone, it contains secondary alcoholic >CHOH group (*Alcoholic*).

(*e*) gives effervescence with $NaHCO_3$, it contains —COOH group (*Acids*).

***(iii)* If a hydroxy acid** on heating loses water, it contains —OH group on β-carbon atom.

$$\underset{\beta\text{-Hydroxy propionic acid}}{\overset{OH}{\overset{|}{C}H_2}.CH_2COOH} \longrightarrow \underset{\text{Acrylic acid}}{CH_2 = CHCOOH} + H_2O$$

***(iv)* A dibasic acid on heating**

(*a*) If evolves CO_2, the two —COOH groups are present on same C-atom.

$$\underset{\text{Malonic acid}}{CH_2 \begin{matrix} COOH \\ COOH \end{matrix}} \longrightarrow CH_3COOH + CO_2$$

(*b*) If loses water, the two —COOH groups are present on adjacent C-atom (*i.e.* succinic acid).

$$\underset{\text{Succinic acid}}{\begin{matrix} CH_2COOH \\ | \\ CH_2COOH \end{matrix}} \longrightarrow \underset{\text{Succinic anhydride}}{\begin{matrix} CH_2CO \\ | \\ CH_2CO \end{matrix} > O} + H_2O$$

***(v)* A compound containing C, H, N and O, if it**

(*a*) loses NH_3 on heating, it may be an amide, ammonium salt of an acid or urea.

$$NH_2CONH \boxed{H + H_2N} CONH_2 \longrightarrow NH_2CONHCONH_2 + NH_3$$

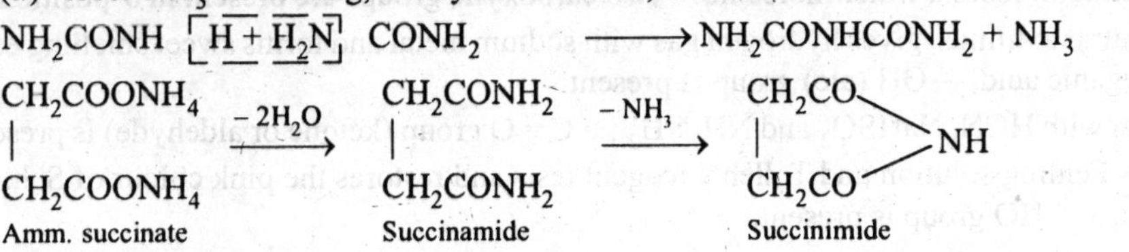

(*b*) Gives primary amine with Br_2 and KOH, it contains—$CONH_2$ group or (*Amide*).

(*c*) On heating with NaOH gives off NH_3, it contains amide group or is ammonium salt.

$$\underset{\text{Acetamide}}{CH_3CONH_2} + NaOH \longrightarrow \underset{\text{Sod. acetate}}{CH_3COONa} + NH_3$$

(*d*) Gives an amine on reduction, it is nitro compound and if it gives alcohol and NH_3, it is alkyl nitrite.

$$\underset{\text{Nitroethane}}{C_2H_5NO_2} + 6H \longrightarrow \underset{\text{Ethylamine}}{C_2H_5NH_2} + 2H_2O$$

$$\underset{\text{Ethyl nitrite}}{C_2H_5ONO} + 6H \longrightarrow \underset{\text{Ethyl alcohol}}{C_2H_5OH} + NH_3 + H_2O$$

(*vi*) **A compound containing C, H and N only** may be either amine (prim., sec., tert.), cyanide or isocyanide.

(*a*) If it is soluble in dil. HCl, it may be amine.

(*b*) If on hydrolysis with dil. HCl yields acid and NH_3, it is cyanide but if gives amine and formic acid, it is isocyanide.

$$\underset{\text{Methyl cyanide}}{CH_3CN} + 2H_2O \longrightarrow \underset{\text{Acetic acid}}{CH_3COOH} + NH_3$$

$$\underset{\text{Methyl isocyanide}}{CH_3NC} + 2H_2O \longrightarrow \underset{\text{Methylamine}}{CH_2NH_2} + HCOOH$$

(*c*) If on reduction yields primary amine, it is cyanide and if yields secondary amine, it is iso-cyanide.

$$\underset{\text{Methyl cyanide}}{CH_3\ CN} + 4H \longrightarrow \underset{\text{Ethyl amine}}{CH_3CH_2NH_2}$$

$$\underset{\text{Methyl isocyanide}}{CH_3NC} + 4H \longrightarrow \underset{\text{Dimethyl amine}}{CH_3\ NHCH_3}$$

(*vii*) **A compound containing asymmetric C atom is optically active.**

REMEMBER THESE POINTS FOR AROMATIC COMPOUNDS

(*i*) **Compounds containing C and H**

(*a*) On oxidation, if it gives a carboxylic acid, it contains side chain.

(*b*) On oxidation, if it gives dicarboxylic acid, it contains two side chains.

(*c*) If the compound decolourises alk. $KMnO_4$ solution and gives precipitate with decolourisation with Br_2 water, it contains unsaturation in the side-chain.

(*ii*) **Compounds containing C, H and O**

(*a*) If it turns blue litmus red, it could be a carboxy or phenolic compound.

(*b*) It turns blue litmus red but does not give effervescence with $NaHCO_3$, phenolic group is present.

(*c*) It gives a red, blue, green or violet colour with $FeCl_3$—phenolic group is present.

(*d*) It undergoes Reimer-Tiemann, Kolbe's-Schmidt reaction—phenolic group is present.

(*e*) It decomposes $NaHCO_3$—Carboxy group is present.

(*f*) Its sodium salt distilled with soda-lime, undergoes decarboxylation—carboxy group present.

(*g*) On heating, it loses a water molecule—Two carboxylic groups are present in *o*-position.

(*h*) It is neutral to litmus, gives hydrogen gas with sodium metal and forms sweet smelling ester with organic acid, —OH (alc) group is present.

(*i*) It reacts with HCN, $NaHSO_3$ and NH_2NH_2 ; $>C=O$ group (ketone or aldehyde) is present.

(*j*) It gives Fehling solution and Tollen's reagent tests and restores the pink colour of Schiff's reagent, —CHO group is present.

(*k*) Give Cannizzaro reaction, —CHO group is present.

(*l*) It gives reaction with acetic anhydride and sodium acetate (Perkin's reaction) to give unsaturated acid, —CHO group is present.

(*m*) Compound with pleasant smell, hydrolysed to a mixture of carboxylic acid and alcohol — ester is present.

(*n*) Burning sugar smell on heating, gives Molisch's test—carbohydrate is present.

(*iii*) Compounds containing C, H, O and N

(*a*) Neutral compound, when heated with NaOH it gives carboxy acid and NH_3, —$CONH_2$ group indicated.

(*b*) It gives primary amine with one carbon less with Br_2 + KOH (Hoffmann Bromamide reaction), —$CONH_2$ group indicated.

(*c*) It gives amine and carboxylic acid on hydrolysis, —NHCOR group is present.

(*d*) On reduction with metal and mineral acid, it gives amine, —NO_2 group is present.

(*iv*) Compounds containing C, H and N

(*a*) It turns red litmus blue, soluble in HCl, —NH_2 group is present.

(*b*) It gives unpleasant smell with $CHCl_3$ and NaOH (carbylamine test)—NH_2 group is present.

(*c*) It gives diazotisation with $NaNO_2$ and HCl followed by coupling reaction with phenols, —NH_2 group is present.

(*d*) With HNO_2, N_2 is evolved with the formation of alcohol,—NH_2 in the side-chain.

(*e*) It produces ammonia and an acid on hydrolysis and on reduction it gives a primary amine, —CN group.

(*f*) It gives a sec. amine on reduction, —NC group is present.

(*v*) Compounds containing C, H and Cl

(*i*) It gives easily substitution reactions with NH_3, NaOH and NaCN, —Cl group in the side-chain.

(*ii*) Oxidation of compound gives an acid not containing chlorine, —Cl present in the side-chain.

(*iii*) Oxidation of compound gives an acid containing chlorine,—Cl attached to benzene nucleus.

(*vi*) Compounds containing C, H, O, S

(*i*) It is acidic to litmus, reacts with PCl_5 and $SOCl_2$, —SO_3H group is present.

(*ii*) On fusion with NaOH, it forms phenol, —SO_3H group is present.

(*iii*) When heated with dil. mineral acid in the presence of steam, aromatic hydrocarbon is produced, —SO_3H group is present.

ALIPHATIC COMPOUNDS WITH MOL. FORMULAE

(*i*) **C_2H_6O** $CH_3.CH_2.OH$ (Ethyl alcohol); $CH_3.O.CH_3$ (Dimethyl ether)

(*ii*) **C_3H_6O** $CH_3CH_2.C(H)=O$ (Propionaldehyde); $CH_3.C(=O)—CH_3$ (Acetone); $CH_2=CH—CH_2OH$ (Allyl alcohol)

(*iii*) **$C_3H_6O_2$** C_2H_5COOH (Propionic acid); $H.COOC_2H_5$ (Ethyl formate); $CH_3.COOCH_3$ (Methyl acetate)

(*iv*) **$C_3H_6O_3$** $(HCHO)_3$ (Metaformaldehyde); $CH_3.CHOH—COOH$ (α-Hydroxy propionic acid); $CH_2OH.CH_2.COOH$ (β-Hydroxy propionic acid) (*v*) C_3H_8O $CH_3.CH_2.CH_2OH$ — *n*-Propyl alcohol; $CH_3CHOHCH_3$ — iso-Propyl alcohol

$CH_3—O—CH_2.CH_3$ — Methyl ethyl ether

(*vi*) C_4H_8O $\begin{matrix} C_2H_5 \\ CH_3 \end{matrix}\!\!>C=O$ — Methyl ethyl ketone; C_3H_7CHO — Butyraldehyde

(*vii*) $C_4H_8O_2$ $CH_3.CH_2.CH_2COOH$; $\begin{matrix} CH_3 \\ CH_3 \end{matrix}\!\!>CH.\ COOH$ — Iso-butyric acid

Esters $HCOOC_3H_7$ — Propyl formate; $CH_3COOC_2H_5$ — Ethyl acetate; $C_2H_5COOCH_3$ — Methyl propionate

Aldol $CH_3.CHOH.CH_2CHO$ (β-hydroxy butryic aldehyde)

(*viii*) $C_4H_6O_4$ $H—C—COOH \parallel H—C—COOH$ — Maleic acid; $H—C—COOH \parallel HOOC—C—H$ — Fumaric acid

$CH_3—CH\!<\!\begin{matrix} COOH \\ COOH \end{matrix}$ — Iso-succinic acid (Methylmalonic acid); $\begin{matrix} CH_2COOH \\ | \\ CH_2COOH \end{matrix}$ — Succinic acid

(*ix*) $C_4H_{10}O$ $C_2H_5—O—C_2H_5$ — Diethyl ether; $CH_3—O—C_3H_7$ — Methyl propyl ether

(*x*) C_2H_7N $CH_3—NH—CH_3$ — Dimethylamine; $C_2H_5NH_2$ — Ethylamine

(*xi*) C_3H_5N $C_2H_5—C \equiv N$ — Ethyl cyanide; $C_2H_5 \equiv NC$ — Ethyl isocyanide

AROMATIC COMPOUNDS WITH MOL. FORMULAE

1. Hydrocarbons

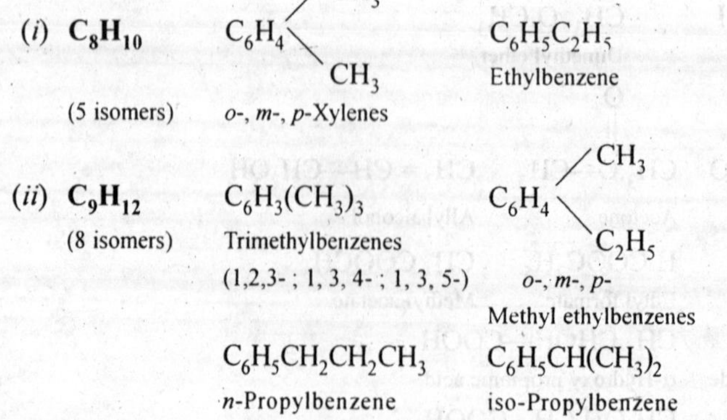

(*i*) C_8H_{10} (5 isomers) $C_6H_4\!<\!\begin{matrix} CH_3 \\ CH_3 \end{matrix}$ — *o*-, *m*-, *p*-Xylenes; $C_6H_5C_2H_5$ — Ethylbenzene

(*ii*) C_9H_{12} (8 isomers) $C_6H_3(CH_3)_3$ — Trimethylbenzenes (1,2,3- ; 1, 3, 4- ; 1, 3, 5-); $C_6H_4\!<\!\begin{matrix} CH_3 \\ C_2H_5 \end{matrix}$ — *o*-, *m*-, *p*- Methyl ethylbenzenes

$C_6H_5CH_2CH_2CH_3$ — *n*-Propylbenzene; $C_6H_5CH(CH_3)_2$ — iso-Propylbenzene

2. Halogenated hydrocarbons, *e.g.* C_7H_7Cl (4 isomers).

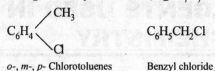

o-, m-, p- Chlorotoluenes — Benzyl chloride

3. Phenols, e.g. C_7H_8O (5 isomers).

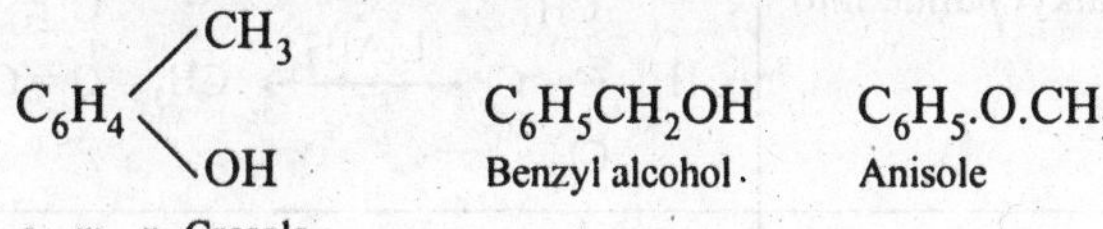

o-, m-, p- Cresols

4. Carboxylic acids

(*i*) $\mathbf{C_8H_8O}$ (4 isomers)

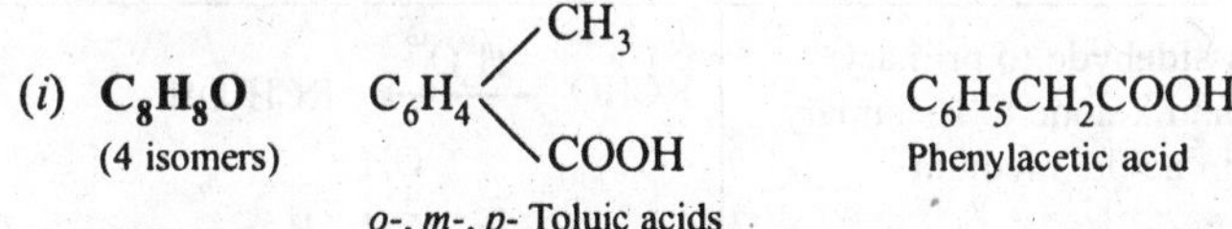

o-, m-, p- Toluic acids

5. Aldehydes

(*i*) $\mathbf{C_7H_6O_2}$ (4 isomers)

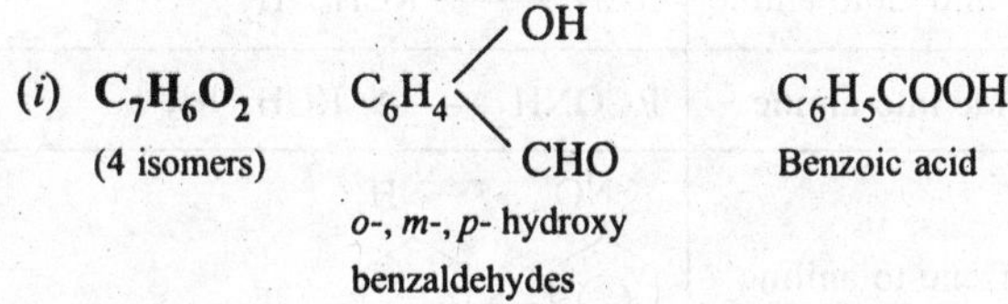

o-, m-, p- hydroxy benzaldehydes

(*ii*) $\mathbf{C_7H_5ClO}$ (4 isomers)

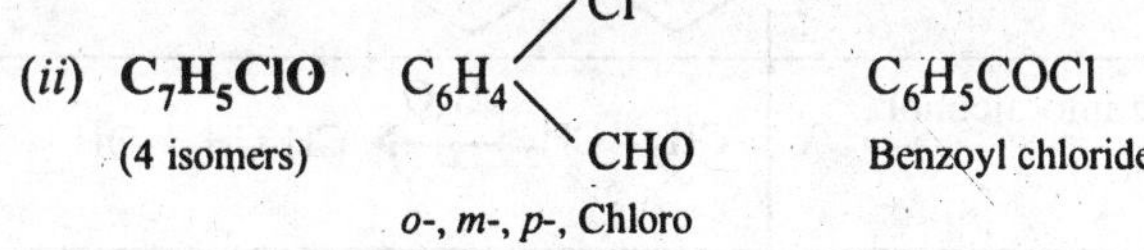

o-, m-, p-, Chloro benzaldehydes

6. Amines

(*i*) $\mathbf{C_7H_9N}$ (5 isomers)

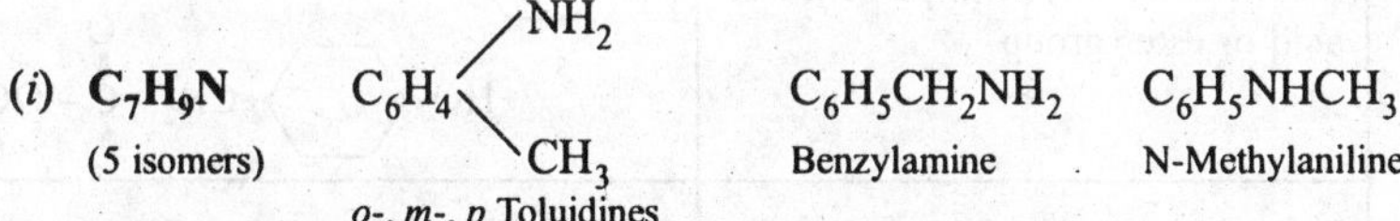

o-, m-, p Toluidines

7. Nitro compounds

(*i*) $\mathbf{C_7H_7O_2}$ (8 isomers)

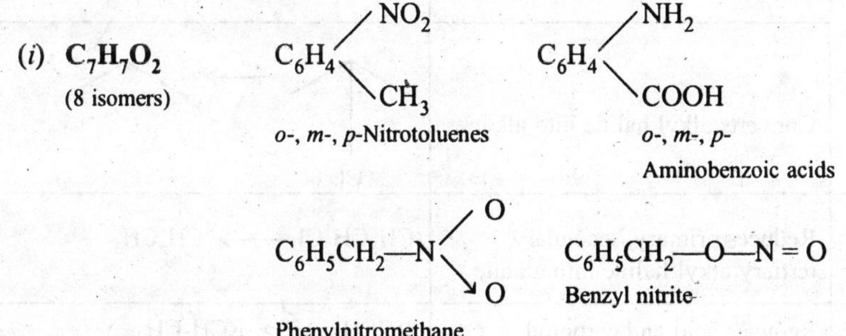

8. Nitriles and Isonitriles

$\mathbf{C_7H_5N}$ (2 isomers)

$C_6H_5\ C \equiv N$ — Phenyl cyanide

$C_6H_5—NC$ — Phenyl isocyanide

IMPORTANT REAGENTS USED IN ORGANIC CHEMISTRY

	REaGENT	USES	EXAMPLE
1.	$LiAlH_4$ Lithium aluminium hydride	Converts tertiary alkyl halide into alkenes	$CH_3-C(CH_3)_2-Cl \xrightarrow{LiAlH_4} CH_3-C(CH_3)=CH_2$
		Converts 1-chloro and 2-chloro propane to propane	$CH_3CH_2CH_2Cl \longrightarrow CH_3CH_2CH_3$
		Reduces aldehyde to primary alcohol and ketone to secondary alcohol in acidic medium	$RCHO \xrightarrow{H_3O^{\oplus}} RCH_2OH$ $RC(=O)-R \longrightarrow RCH(OH)-R$
		Converts alkyl cyanide into amine	$RCN \longrightarrow RCH_2NH_2$
		Converts acid amide into amine	$RCONH_2 \longrightarrow RCH_2NH_2$
		Converts nitrobenzene to aniline	$C_6H_5NO_2 \rightarrow C_6H_5NH_2$
		Reduces epoxide into alcohol in acidic medium	CH_2-CH_2 (epoxide, bridged by O) $\xrightarrow{H_3O^+} CH_3CH_2-OH$
2	$NaBH_4$ Sodium borohydride	For reducing a simple ketone or aldehyde in presence of acid or ester group	$O=C_6H_{10}-CH_2-C(=O)-OCH_3 \longrightarrow HO-C_6H_{10}-CH_2-C(=O)-OCH_3$
		Reduces double bond but does not affect carboxyl group or ester	$RCH=CH_2 \longrightarrow RCH_2CH_3$
		Converts alkyl halide into alkanes	(skeletal: alkyl chloride with Cl) $\longrightarrow$ (skeletal alkane)
3	TPH (Ph_3 SnH) Triphenyl tin hydride	Reduces primary, secondary, tertiary alkyl halide into alkane	$CH_3CH_2Cl \longrightarrow CH_3CH_3$
4	Red phosphorus and HI	Reduces acid and carbonyl compound into alkane	$RC(=O)CH_3 \xrightarrow{\Delta} RCH_2CH_3$

REAGENT	USES	EXAMPLE
5 Red phosphorus and halogen	Replaces a hydrogen atom in carboxylic acids by halogen	$CH_3COOH \longrightarrow X-CH_2COOH$
6 Zn(Hg) + Conc.HCl	Reduces carbonyl compound into alkane (Clemmensen reduction)	$RCOCH_3 \longrightarrow RCH_2CH_3$
7 Hydrazine NH_2-NH_2/OH^-	Reduces carbonyl compound into alkane (Wolf-Kishner reaction)	$R-\underset{\underset{O}{\parallel}}{C}-CH_3 \rightarrow R-\underset{\underset{NNH_2}{\parallel}}{C}-CH_3 \xrightarrow[450°C]{C_2H_5ONa} RCH_2CH_3$
8 ***Grignard reagent*** $R-Mg-X$	Grignard reagent on treatment with water or dilute acid or alcohol is decomposed to alkane	$RMgX + H_2O \longrightarrow R-H + MgXOH$
	Reacts with α, β unsaturated ketone predominantly by 1, 4 addition	$\overset{4}{R-CH}=\overset{3}{CH}-\overset{2}{CR'}=\overset{1}{O} \xrightarrow{R''MgX}$ $RR''CH-CH=CR'-OMgX \xrightarrow{H_3O^+}$ $RR''CH-\underset{Enol}{CH=CR'OH}] \rightarrow \underset{Keto}{RR''CH-CH_2CR'=O}$
9 Phosphorus oxychloride ($POCl_3$)	Converts alcohol to alkene	$\begin{matrix}R\\R\end{matrix}>\overset{H}{C}-\overset{OH}{C}<\begin{matrix}R\\R\end{matrix} \rightarrow \begin{matrix}R\\R\end{matrix}>C=C<\begin{matrix}R\\R\end{matrix}$
10 Orgonolithium reagent R – Li	Reacts with carbonyl group to form alcohol	$RLi + >C=O \rightarrow R-\overset{\mid}{\underset{\mid}{C}}-OLi \xrightarrow{H_3O^+} R-\overset{\mid}{\underset{\mid}{C}}-OH$
11 NBS N–bromosuccinimide	Used for bromination at allylic and benzylic position	$C_6H_5-CH_3 \rightarrow C_6H_5-CH_2Br$
12 *Gilman reagent* R_2CuLi	Reacts with acid chloride or ester to form ketone	$R_2CuLi + R'\underset{\underset{O}{\parallel}}{C}-Cl \rightarrow R\underset{\underset{O}{\parallel}}{C}R' + RCu + LiCl$
13 Alcoholic KOH	Dehydrohalogenation of compound	$CH_3CH_2CH_2Cl \longrightarrow CH_3CH=CH_2$
14 ***Lucas reagent*** Mixture of conc. HCl and anhyd. zinc chloride	Converts alcohol to alkyl halides (Order of reactivity tertiary > secondary > primary)	$CH_3\underset{\underset{OH}{\mid}}{CH}-CH_3 \rightarrow CH_3-\underset{\underset{Cl}{\mid}}{CH}-CH_3$
15 ***Jones reagent*** Chromic acid in aq. acetone solution	Oxidises primary alcohol to aldehyde and secondary alcohol to ketone without affecting C = C bond	cyclohex-2-en-1-ol (OH) → cyclohex-2-en-1-one (O)
16 Manganese dioxide MnO_2	Selectively oxidises the –OH group of allylic and benzylic primary and secondary alcohol to give aldehydes and ketones respectively	cyclohex-2-en-1-ol (OH) → cyclohexanone (O)

SUMMARY OF SOME IMPORTANT NAME REACTIONS

Reaction	Starting Compound	Products Formed
1. Aldol Condensation	$CH_3CHO + HCH_2.CHO \xrightarrow[NaOH]{dil.} CH_3.CHOH.CH_2CHO \xrightarrow{\Delta}$ Crotonaldehyde $(CH_3)_2CO + HCH_2.COCH_3 \xrightarrow{Ba(OH)_2} (CH_3)_2COH.CH_2CO.CH_3$ Acetone (2moles) — Diacetone alc. (Keto) $\xrightarrow{\Delta}$ Mesityl oxide	
2. Benzoin Condensation	$2C_6H_5CHO \xrightarrow[H_2O, C_2H_5OH]{KCN, heat} C_6H_5CHOH.CO.C_6H_5$ Benzaldehyde (2 moles) — Benzoin	
3. Carbylamine Reaction	$CHCl_3 + C_2H_5NH_2 + 3KOH \longrightarrow C_2H_5NC + 3KCl + 3H_2O$ Chloroform, Ethylamine — Ethyl isocyanide $CHCl_3 + C_6H_5NH_2 + 3KOH \longrightarrow C_6H_5NC + 3KCl + 3H_2O$ Aniline — Phenyl isocyanide	
4. Cannizzaro's Reaction	$2HCHO + NaOH \longrightarrow HCOONa + CH_3OH$ Formaldehyde — Sod. formate, Methanol $2C_6H_5CHO + NaOH \longrightarrow C_6H_5COONa + C_6H_5CH_2OH$ Benzaldehyde — Sod. benzoate, Benzyl alcohol	
5. Clemmensen Reduction	$>C=O \xrightarrow[NH_2-NH_2, NaOH]{Zn-Hg, HCl} >CH_2$ Carbonyl group Ald. or Ketone — Methylene group	
6. Claisen Schmidt reaction	$C_6H_5CHO + CH_3CHO \xrightarrow{OH^-} C_6H_5CHOHCH_2CHO$ Acetaldehyde $\xrightarrow[-H_2O]{} C_6H_5CH=CHCHO$ Cinnamaldehyde	
7. Coupling Reaction	$C_6H_5-N=NCl + C_6H_5NH_2 \xrightarrow[40^oC]{H^+} C_6H_5-N=N-C_6H_4-NH_2$ Benz. diazonium chloride, Aniline — *p*-Amino azobenzene(Azo dye)	
8. Crossed Cannizzaro's Reaction	$C_6H_5CHO + HCHO \xrightarrow[heat]{NaOH} C_6H_5CH_2OH + HCOONa$ Benzaldehyde — Benzyl alc., Sod.formate	
9. Diazotisation	$C_6H_5NH_2 \xrightarrow[HCl]{NaNO_2} C_6H_5-\overset{+}{N}\equiv NCl^-$ Aniline — Benz. diazonium.chloride	
10. Friedel-Crafts' Alkylation	$C_6H_6 + ClCH_3 \xrightarrow[AlCl_3]{Anhy} C_6H_5CH_3 + HCl$ Benzene — Toluene	
11. Friedel-Crafts' Acylation	$C_6H_6 + ClCOCH_3 \xrightarrow[AlCl_3]{Anhy} C_6H_5COCH_3 + HCl$ Acetyl chloride — Acetophenone	

12. Gattermann's Reaction

$$\underset{\text{Benz. diazonium chloride}}{C_6H_5N_2Cl} \xrightarrow{Cu+HCl} \underset{\text{Chlorobenzene}}{C_6H_5Cl} + N_2$$

13. Gatterman Aldehyde Synthesis

$$C_6H_6 + \underset{\text{or (HCl + HC}\equiv\text{N)}}{ClCH=NH} \xrightarrow{AlCl_3} C_6H_5-CH=NH$$

$$\xrightarrow{H_2O} \underset{\text{Benzaldehyde}}{C_6H_5CHO} + NH_3$$

14. Haloform Reaction

$$\underset{\text{Methyl ketone}}{R-\overset{O}{\overset{\|}{C}}-CH_3} + 3Br_2 + 4NaOH \longrightarrow R-\overset{O}{\overset{\|}{C}}-ONa + \underset{\text{Bromoform}}{CHBr_3} + 3NaBr + 3H_2O$$

15. Hofmann Bromamide Reaction

$$\underset{\text{Benzamide}}{C_6H_5CONH_2} + Br_2 + 4KOH \longrightarrow \underset{\text{Anline}}{C_6H_5NH_2} + 2KBr + K_2CO_3 + 2H_2O$$

16. Kolbe-Schmidt Reaction

Sod. phenoxide (ONa) $+ CO_2 \xrightarrow[6\text{ atm.}]{398K}$ Sod. salicylate (OH, COONa) $\xrightarrow{H^+}$ Salicylic acid (OH, COOH)

17. Kolbe Electrolysis

$$\underset{\text{Sod. acetate}}{2CH_3COONa} + 2H_2O \xrightarrow{\text{Electrolysis}} \underset{\text{At Anode}}{CH_3-CH_3 + 2CO_2} + \underset{\text{At Cathode}}{NaOH + H_2}$$

18. Reimer–Tiemann's Reaction

Phenol (OH) $+ \underset{\text{(or }CCl_4)}{CHCl_3} + 3NaOH \longrightarrow$ (OH, CHO) $+ 3NaCl + 2HO$

Salicyldehyde (or Salicylic acid)

19. Rosenmund Reduction

$$\underset{\text{Acid chloride}}{RCOCl} \xrightarrow[Pd/BaSO_4]{H_2} \underset{\text{Aldehyde}}{RCHO}$$

20. Sandmeyer's Reaction

$$\underset{\text{Benz. diazonium chloride}}{C_6H_5N_2Cl} + CuCN \longrightarrow \underset{\text{Phenyl benzoate}}{C_6H_5CN} + N_2 + CuCl$$

21. Schotten Baumann Reaction

$$\underset{\text{Phenol}}{C_6H_5OH} + \underset{\text{Benoyl chloride}}{C_6H_5COCl} \xrightarrow{NaOH} \underset{\text{Phenyl benzoate}}{C_6H_5OCOC_6H_5} + HCl$$

22. Stephen Reduction

$$\underset{\text{Methyl cyanide}}{CH_3C\equiv N} + 2H \xrightarrow[HCl]{SnCl_2} \underset{\text{Aldimine}}{CH_3CH=NH} \xrightarrow{H_2O} CH_3CHO + NH_3$$

23. Tischenko Reaction

$$\underset{\text{Benzaldehyde}}{2C_6H_5CHO} \xrightarrow{Al(OC_2H_5)_3} \underset{\text{Benzyl benzoate}}{C_6H_5CH_2OOC.C_6H_5}$$

24. Ulmann Reaction

$$\underset{\text{Iodobenzene}}{C_6H_5I} + 2Cu + IC_6H_5 \xrightarrow{\text{Heat}} \underset{\text{Diphenyl}}{C_6H_5-C_6H_5} + 2CuI$$

25. Williamson Synthesis

$$\underset{\text{Sod. alkoxide}}{RONa} + \underset{\text{Alkyl halide}}{XR} \xrightarrow{\text{Heat}} \underset{\text{Ether}}{R'-O-R} + NaX$$

26. Wurtz-Fittig Reaction

$$\underset{\text{Bromobenzene}}{C_6H_5Br} + 2Na + \underset{\text{Methyl bromide}}{BrCH_3} \xrightarrow[\text{ether}]{\text{Dry}} \underset{\text{Toluene}}{C_6H_5CH_3} + 2NaBr$$